Commonly Encountered Structural Types of Organic Chemistry

Functional Group	Example	IUPAC Name	Common Name (if applicable)	Suffix	Prefix
Principal Groups[a]					
Alkane	CH_3-CH_3	Ethane	—	-ane	—
Alkyne	$H-C\equiv C-H$	Ethyne	Acetylene	-yne	—
Alkene	$H_2C=CH_2$	Ethene	Ethylene	-ene	—
Amines					
Primary amine	$CH_3CH_2-NH_2$	Ethanamine	Ethylamine	-amine	amino-
Secondary amine	$CH_3CH_2N-CH_3$ (H)	N-Methylethanamine	Ethylmethylamine	-amine	amino-
Tertiary amine	$CH_3CH_2N(CH_3)CH_3$	N,N-Dimethylethanamine	Ethyldimethylamine	-amine	amino-
Quaternary amine	$(CH_3CH_2)_4N^+$	N,N,N-Triethylethanammonium	Tetraethylammonium	-ammonium	—
Imine	(NH)	1-Propanimine	Propanal imine	-imine	imino-
Thiol	CH_3CH_2-SH	Ethanethiol	Ethyl mercaptan	-thiol	mercapto-
Alcohol	CH_3CH_2-OH	Ethanol	Ethyl alcohol	-ol	hydroxy-
Borane	(Triethylborane structure, B)	Triethylborane	—	-borane	boro-
Ketone	$H_3C-C(=O)-CH_3$	Propanone	Acetone	-one	oxo-
Aldehyde	$H_3C-C(=O)-H$	Ethanal	Acetaldehyde	-al	oxo-
Nitrile	CH_3CH_2-CN	Propanenitrile	Ethyl cyanide	-nitrile	cyano-
Amides					
Primary amide	$H_3C-C(=O)-NH_2$	Ethanamide	Acetamide	-amide	amido-
Secondary amide	$H_3C-C(=O)-N(H)-CH_3$	N-Methylethanamide	—	-amide	amido-
Tertiary amide	$H-C(=O)-N(CH_3)-CH_3$	N,N-Dimethyl-methanamide	Dimethylformamide (DMF)	-amide	amido-

Commonly Encountered Structural Types of Organic Chemistry (*continued*)

Functional Group	Example	IUPAC Name	Common Name (if applicable)	Suffix	Prefix
Acid chloride	$H_3C-\overset{\overset{O}{\|\|}}{C}-Cl$	Ethanoyl chloride	Acetyl chloride	-oyl chloride	chlorocarbonyl-
Ester	$H_3C-\overset{\overset{O}{\|\|}}{C}-O-CH_2CH_3$	Ethyl ethanoate	Ethyl acetate	-oate	alkyoxycarbonyl-
Anhydride	$H_3C-\overset{\overset{O}{\|\|}}{C}-O-\overset{\overset{O}{\|\|}}{C}-CH_3$	Ethanoic anhydride	Acetic anhydride	-oic anhydride	acyloxycarbonyl-
Sulfonic acid	$CH_3CH_2-\overset{\overset{O}{\|\|}}{\underset{\underset{O}{\|\|}}{S}}-OH$	Ethanesulfonic acid	—	-sulfonic acid	sulfo-
Carboxylic acid	$H_3C-\overset{\overset{O}{\|\|}}{C}-OH$	Ethanoic acid	Acetic acid	-oic acid	carboxy-

Subordinate Groups[a]

Functional Group	Example	IUPAC Name	Common Name (if applicable)	Suffix	Prefix
Alkyl	$CH_3CH_2\overset{\overset{CH_3}{\|}}{C}HCH_2CH_3$	3-Methylpentane	—	—	alkyl-
Aryl	⬡—CH_2OH	Phenylmethanol	Benzyl alcohol	—	phenyl-
Halides					
Fluoride	CH_3-F	Fluoromethane	Methyl fluoride	—	fluoro-
Chloride	CH_3CH_2-Cl	Chloroethane	Ethyl chloride	—	chloro-
Bromide	$CH_2{=}CH-Br$	Bromoethene	Vinyl bromide	—	bromo-
Iodide	⬡—I	Iodobenzene	Phenyl iodide	—	iodo-
Ether	$CH_3CH_2OCH_2CH_3$	Ethoxyethane	Diethyl ether	—	alkoxy-
Sulfide	$CH_3CH_2SCH_2CH_3$	Ethylthioethane	Diethyl sulfide	—	alkylthio-
Sulfoxide	$H_3C-\overset{\overset{O}{\|\|}}{S}-CH_3$	Methylsulfinylmethane	Dimethyl sulfoxide (DMSO)	—	sulfinyl-
Sulfone	$H_3C-\overset{\overset{O}{\|\|}}{\underset{\underset{O}{\|\|}}{S}}-CH_3$	Methylsulfonylmethane	Dimethyl sulfone	—	sulfonyl-
Nitro	$H_3C-\overset{+}{N}\overset{\diagup O}{\diagdown}_{O^-}$	Nitromethane	—	—	nitro-
Nitroso	$H_3C-N{=}O$	Nitrosomethane	—	—	nitroso-
Azide	$H_3C-N{=}\overset{+}{N}{=}\overset{-}{N}$	Azidomethane	Methyl azide	—	azido-

[a] The principal groups are listed in increasing order of priority. The subordinate groups have no established priority and can only be referenced as Prefixes.

PERIODIC TABLE OF THE ELEMENTS

Legend:
```
  1 —— Atomic number
  H —— Symbol
Hydrogen —— Name
1.00794 —— Average atomic mass
```

Metals
Metalloids
Nonmetals

1 / 1A	2 / 2A	3 / 3B	4 / 4B	5 / 5B	6 / 6B	7 / 7B	8 / 8B	9 / 8B	10	11 / 1B	12 / 2B	13 / 3A	14 / 4A	15 / 5A	16 / 6A	17 / 7A	18 / 8A
1 **H** Hydrogen 1.00794																	2 **He** Helium 4.002602
3 **Li** Lithium 6.941	4 **Be** Beryllium 9.012182											5 **B** Boron 10.811	6 **C** Carbon 12.0107	7 **N** Nitrogen 14.0067	8 **O** Oxygen 15.9994	9 **F** Fluorine 18.9984032	10 **Ne** Neon 20.1797
11 **Na** Sodium 22.98976928	12 **Mg** Magnesium 24.3050											13 **Al** Aluminum 26.9815386	14 **Si** Silicon 28.0855	15 **P** Phosphorus 30.973762	16 **S** Sulfur 32.065	17 **Cl** Chlorine 35.453	18 **Ar** Argon 39.948
19 **K** Potassium 39.0983	20 **Ca** Calcium 40.078	21 **Sc** Scandium 44.955912	22 **Ti** Titanium 47.867	23 **V** Vanadium 50.9415	24 **Cr** Chromium 51.9961	25 **Mn** Manganese 54.938045	26 **Fe** Iron 55.845	27 **Co** Cobalt 58.933195	28 **Ni** Nickel 58.6934	29 **Cu** Copper 63.546	30 **Zn** Zinc 65.409	31 **Ga** Gallium 69.723	32 **Ge** Germanium 72.64	33 **As** Arsenic 74.92160	34 **Se** Selenium 78.96	35 **Br** Bromine 79.904	36 **Kr** Krypton 83.798
37 **Rb** Rubidium 85.4678	38 **Sr** Strontium 87.62	39 **Y** Yttrium 88.90585	40 **Zr** Zirconium 91.224	41 **Nb** Niobium 92.90638	42 **Mo** Molybdenum 95.94	43 **Tc** Technetium [98]	44 **Ru** Ruthenium 101.07	45 **Rh** Rhodium 102.90550	46 **Pd** Palladium 106.42	47 **Ag** Silver 107.8682	48 **Cd** Cadmium 112.411	49 **In** Indium 114.818	50 **Sn** Tin 118.710	51 **Sb** Antimony 121.760	52 **Te** Tellurium 127.60	53 **I** Iodine 126.90447	54 **Xe** Xenon 131.293
55 **Cs** Cesium 132.9054519	56 **Ba** Barium 137.327	57 **La** Lanthanum 138.90547	72 **Hf** Hafnium 178.49	73 **Ta** Tantalum 180.94788	74 **W** Tungsten 183.84	75 **Re** Rhenium 186.207	76 **Os** Osmium 190.23	77 **Ir** Iridium 192.217	78 **Pt** Platinum 195.084	79 **Au** Gold 196.966569	80 **Hg** Mercury 200.59	81 **Tl** Thallium 204.3833	82 **Pb** Lead 207.2	83 **Bi** Bismuth 208.98040	84 **Po** Polonium [209]	85 **At** Astatine [210]	86 **Rn** Radon [222]
87 **Fr** Francium [223]	88 **Ra** Radium [226]	89 **Ac** Actinium [227]	104 **Rf** Rutherfordium [261]	105 **Db** Dubnium [262]	106 **Sg** Seaborgium [266]	107 **Bh** Bohrium [264]	108 **Hs** Hassium [277]	109 **Mt** Meitnerium [268]	110 **Ds** Darmstadtium [271]	111 **Rg** Roentgenium [272]							

6 Lanthanides

58 **Ce** Cerium 140.116	59 **Pr** Praseodymium 140.90765	60 **Nd** Neodymium 144.242	61 **Pm** Promethium [145]	62 **Sm** Samarium 150.36	63 **Eu** Europium 151.964	64 **Gd** Gadolinium 157.25	65 **Tb** Terbium 158.92535	66 **Dy** Dysprosium 162.500	67 **Ho** Holmium 164.93032	68 **Er** Erbium 167.259	69 **Tm** Thulium 168.93421	70 **Yb** Ytterbium 173.04	71 **Lu** Lutetium 174.967

7 Actinides

90 **Th** Thorium 232.03806	91 **Pa** Protactinium 231.03588	92 **U** Uranium 238.02891	93 **Np** Neptunium [237]	94 **Pu** Plutonium [244]	95 **Am** Americium [243]	96 **Cm** Curium [247]	97 **Bk** Berkelium [247]	98 **Cf** Californium [251]	99 **Es** Einsteinium [252]	100 **Fm** Fermium [257]	101 **Md** Mendelevium [258]	102 **No** Nobelium [259]	103 **Lr** Lawrencium [262]

We have used the United States system as well as the system recommended by the International Union of Pure and Applied Chemistry (IUPAC) to label the groups in this periodic table. The system used in the United States includes a letter and a number (1A, 2A, 3B, 4B, etc.), which is close to the system developed by Mendeleev. The IUPAC system uses numbers 1–18 and has been recommended by the American Chemical Society (ACS). Elements with higher atomic numbers have been reported but not yet fully authenticated.

Organic Chemistry

Organic Chemistry

FOURTH EDITION

Maitland Jones, Jr.

NEW YORK UNIVERSITY

Steven A. Fleming

TEMPLE UNIVERSITY

W. W. Norton & Company
New York • London

Drawing structures and reactions is critical to understanding them. The front and back covers show chair cyclohexanes, which are introduced in Chapter 5 and appear over and over again in the natural world. These structures were hand-drawn and have ever-so-small imperfections. Evaluate these chairs, and after you finish Chapter 5, we suggest you take a very close look to see how you might improve them.

W.W. Norton & Company has been independent since its founding in 1923, when William Warder Norton and Mary D. Herter Norton first published lectures delivered at the People's Institute, the adult education division of New York City's Cooper Union. The Nortons soon expanded their program beyond the Institute, publishing books by celebrated academics from America and abroad. By mid-century, the two major pillars of Norton's publishing program—trade books and college texts—were firmly established. In the 1950s, the Norton family transferred control of the company to its employees, and today—with a staff of four hundred and a comparable number of trade, college, and professional titles published each year—W.W. Norton & Company stands as the largest and oldest publishing house owned wholly by its employees.

Editor: Erik Fahlgren

Project editor: Carla L. Talmadge

Marketing manager: Kelsey Volker

Editorial assistant: Mary Lynch

Production manager: Chris Granville

Managing editor, College: Marian Johnson

Ancillaries editor: Matthew A. Freeman

Media editor: Rob Bellinger

Developmental editor: Irene Nunes

Design director: Rubina Yeh

Photo editor: Junenoire Mitchell

Cover design: John Hamilton

Composition: Prepare, Inc.

Illustration studio: Penumbra Design, Inc.

Manufacturing: Transcontinental Interglobe

Library of Congress Cataloging-in-Publication Data

Jones, Maitland, 1937–
    Organic chemistry / Maitland Jones, Jr.—4th ed. / Steven A. Fleming.
    p. cm.
    Includes bibliographical references and index.
    ISBN 978-0-393-93149-5 (hardcover)
    1. Chemistry, Organic—Textbooks. I. Fleming, Steven A. II. Title.
QD253.2.J66 2010
547—dc22   2009033807

W. W. Norton & Company, Inc., 500 Fifth Avenue, New York, N.Y. 10110
www.wwnorton.com

W. W. Norton & Company Ltd., Castle House, 75/76 Wells Street, London W1T 3QT

1 2 3 4 5 6 7 8 9 0

Brief Contents

1 Atoms and Molecules; Orbitals and Bonding 1

2 Alkanes 50

3 Alkenes and Alkynes 97

4 Stereochemistry 147

5 Rings 185

6 Alkyl Halides, Alcohols, Amines, Ethers, and Their Sulfur-Containing Relatives 223

7 Substitution and Elimination Reactions: The S_N2, S_N1, E1, and E2 Reactions 261

8 Equilibria 331

9 Additions to Alkenes 1 363

10 Additions to Alkenes 2 and Additions to Alkynes 409

11 Radical Reactions 467

12 Dienes and the Allyl System: $2p$ Orbitals in Conjugation 511

13 Conjugation and Aromaticity 571

14 Substitution Reactions of Aromatic Compounds 623

15 Analytical Chemistry: Spectroscopy 694

16 Carbonyl Chemistry 1: Addition Reactions 762

17 Carboxylic Acids 828

18 Derivatives of Carboxylic Acids: Acyl Compounds 876

19 Carbonyl Chemistry 2: Reactions at the α Position 931

20 Special Topic: Reactions Controlled by Orbital Symmetry 1030

21 Special Topic: Intramolecular Reactions and Neighboring Group Participation 1080

22 Special Topic: Carbohydrates 1124

23 Special Topic: Amino Acids and Polyamino Acids (Peptides and Proteins) 1173

Contents

Selected Applications   xix

Organic Reaction Animations   xxi

Preface to the Fourth Edition   xxiii

Introduction   xxxiii

1   Atoms and Molecules; Orbitals and Bonding   1

1.1   Preview   2

1.2   Atoms and Atomic Orbitals   4

1.3   Covalent Bonds and Lewis Structures   13

1.4   Resonance Forms   22

1.5   Hydrogen (H_2): Molecular Orbitals   30

1.6   Bond Strength   36

1.7   An Introduction to Reactivity: Acids and Bases   41

1.8   Special Topic: Quantum Mechanics and Babies   42

1.9   Summary   43

1.10   Additional Problems   45

2   Alkanes   50

2.1   Preview   51

2.2   Hybrid Orbitals: Making a Model for Methane   52

2.3   The Methyl Group (CH_3) and Methyl Compounds (CH_3X)   60

2.4   The Methyl Cation ($^+CH_3$), Anion ($^-:CH_3$), and Radical ($\cdot CH_3$)   62

2.5   Ethane (C_2H_6), Ethyl Compounds (C_2H_5X),
        and Newman Projections   64

2.6   Structure Drawings   70

2.7   Propane (C_3H_8) and Propyl Compounds (C_3H_7X)   71

2.8   Butanes (C_4H_{10}), Butyl Compounds (C_4H_9X),
        and Conformational Analysis   73

2.9   Pentanes (C_5H_{12}) and Pentyl Compounds ($C_5H_{11}X$)   76

2.10   The Naming Conventions for Alkanes   78

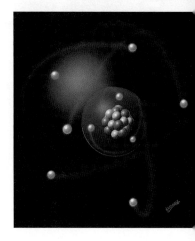

2.11   Drawing Isomers   82

2.12   Rings   83

2.13   Physical Properties of Alkanes and Cycloalkanes   86

2.14   Nuclear Magnetic Resonance Spectroscopy   88

2.15   Acids and Bases Revisited: More Chemical Reactions   90

2.16   Special Topic: Alkanes as Biomolecules   92

2.17   Summary   93

2.18   Additional Problems   94

3   **Alkenes and Alkynes   97**

3.1   Preview   98

3.2   Alkenes: Structure and Bonding   99

3.3   Derivatives and Isomers of Alkenes   107

3.4   Nomenclature   110

3.5   The Cahn–Ingold–Prelog Priority System   112

3.6   Relative Stability of Alkenes: Heats of Formation   115

3.7   Double Bonds in Rings   118

3.8   Physical Properties of Alkenes   123

3.9   Alkynes: Structure and Bonding   123

3.10   Relative Stability of Alkynes: Heats of Formation   126

3.11   Derivatives and Isomers of Alkynes   126

3.12   Triple Bonds in Rings   128

3.13   Physical Properties of Alkynes   129

3.14   Acidity of Alkynes   129

3.15   Molecular Formulas and Degrees of Unsaturation   130

3.16   An Introduction to Addition Reactions of Alkenes and Alkynes   131

3.17   Mechanism of the Addition of Hydrogen Halides to Alkenes   132

3.18   The Regiochemistry of the Addition Reaction   137

3.19   A Catalyzed Addition to Alkenes: Hydration   139

3.20   Synthesis: A Beginning   141

3.21   Special Topic: Alkenes and Biology   142

3.22   Summary   143

3.23   Additional Problems   144

4   Stereochemistry   147

4.1   Preview   148

4.2   Chirality   149

4.3   The (*R/S*) Convention   152

4.4   Properties of Enantiomers: Physical Differences   155

4.5   The Physical Basis of Optical Activity   157

4.6   Properties of Enantiomers: Chemical Differences   159

4.7   Interconversion of Enantiomers by Rotation about a Single Bond:
       gauche-Butane   163

4.8   Diastereomers and Molecules Containing More than One Stereogenic
       Atom   164

4.9   Physical Properties of Diastereomers: Resolution, a Method of Separating
       Enantiomers from Each Other   169

4.10   Determination of Absolute Configuration (*R* or *S*)   172

4.11   Stereochemical Analysis of Ring Compounds (a Beginning)   173

4.12   Summary of Isomerism   176

4.13   Special Topic: Chirality without "Four Different Groups Attached to One
       Carbon"   177

4.14   Special Topic: Stereochemistry in the Real World: Thalidomide, the
       Consequences of Being Wrong-Handed   180

4.15   Summary   181

4.16   Additional Problems   182

5   Rings   185

5.1   Preview   186

5.2   Rings and Strain   187

5.3   Quantitative Evaluation of Strain Energy   193

5.4   Stereochemistry of Cyclohexane: Conformational Analysis   197

5.5   Monosubstituted Cyclohexanes   199

5.6   Disubstituted Ring Compounds   204

5.7   Bicyclic Compounds   211

5.8   Special Topic: Polycyclic Systems   216

5.9   Special Topic: Adamantanes in Materials and Biology   217

5.10   Summary   219

5.11   Additional Problems   220

6   Alkyl Halides, Alcohols, Amines, Ethers, and Their Sulfur-Containing Relatives   223

6.1   Preview   224

6.2   Alkyl Halides: Nomenclature and Structure   225

6.3   Alkyl Halides as Sources of Organometallic Reagents: A Synthesis of Hydrocarbons   227

6.4   Alcohols   230

6.5   Solvents in Organic Chemistry   238

6.6   Diols (Glycols)   240

6.7   Amines   240

6.8   Ethers   249

6.9   Special Topic: Thiols (Mercaptans) and Thioethers (Sulfides)   251

6.10   Special Topic: Crown Ethers   254

6.11   Special Topic: Complex Nitrogen-Containing Biomolecules—Alkaloids   255

6.12   Summary   256

6.13   Additional Problems   258

7   Substitution and Elimination Reactions: The S$_N$2, S$_N$1, E1, and E2 Reactions   261

7.1   Preview   262

7.2   Review of Lewis Acids and Bases   263

7.3   Reactions of Alkyl Halides: The Substitution Reaction   267

7.4   Substitution, Nucleophilic, Bimolecular: The S$_N$2 Reaction   268

7.5   The S$_N$2 Reaction in Biochemistry   288

7.6   Substitution, Nucleophilic, Unimolecular: The S$_N$1 Reaction   289

7.7   Summary and Overview of the S$_N$2 and S$_N$1 Reactions   296

7.8   The Unimolecular Elimination Reaction: E1   298

7.9   The Bimolecular Elimination Reaction: E2   301

7.10   What Can We Do with These Reactions? How to Do Organic Synthesis   312

7.11   Summary   320

7.12   Additional Problems   323

8   Equilibria   331

8.1   Preview   332

8.2   Equilibrium   334

8.3   Entropy in Organic Reactions   337

8.4   Rates of Chemical Reactions   339

8.5   Rate Constant   341

8.6   Energy Barriers in Chemical Reactions:
       The Transition State and Activation Energy   342

8.7   Reaction Mechanism   349

8.8   The Hammond Postulate: Thermodynamics versus Kinetics   351

8.9   Special Topic: Enzymes and Reaction Rates   357

8.10   Summary   358

8.11   Additional Problems   360

9   Additions to Alkenes 1   363

9.1   Preview   364

9.2   Mechanism of the Addition of Hydrogen Halides to Alkenes   365

9.3   Effects of Resonance on Regiochemistry   366

9.4   Brief Review of Resonance   372

9.5   Resonance and the Stability of Carbocations   374

9.6   Inductive Effects on Addition Reactions   378

9.7   HX Addition Reactions: Hydration   380

9.8   Dimerization and Polymerization of Alkenes   384

9.9   Rearrangements during HX Addition to Alkenes   386

9.10   Hydroboration   390

9.11   Hydroboration in Synthesis: Alcohol Formation   398

9.12   Special Topic: Rearrangements in Biological Processes   401

9.13   Summary   402

9.14   Additional Problems   404

10   Additions to Alkenes 2 and Additions to Alkynes   409

10.1   Preview   410

10.2   Addition of H_2 and X_2 Reagents   410

10.3   Hydration through Mercury Compounds: Oxymercuration   421

10.4   Other Addition Reactions Involving Three-Membered Rings: Oxiranes
       and Cyclopropanes   423

10.5   Dipolar Addition Reactions: Ozonolysis and the Synthesis
       of Carbonyl (R_2C=O) Compounds   436

10.6   Addition Reactions of Alkynes: HX Addition   444

10.7   Addition of X_2 Reagents to Alkynes   447

10.8   Hydration of Alkynes   448

10.9   Hydroboration of Alkynes   450

10.10   Hydrogenation of Alkynes: Alkene Synthesis
        through syn Hydrogenation   452

10.11   Reduction by Sodium in Ammonia: Alkene Synthesis
        through anti Hydrogenation   452

10.12   Special Topic: Three-Membered Rings in Biochemistry   455

10.13   Summary   456

10.14   Additional Problems   460

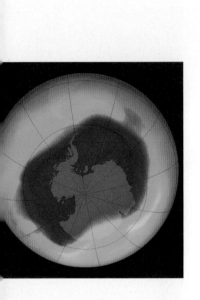

11   Radical Reactions   467

11.1   Preview   468

11.2   Formation and Simple Reactions of Radicals   469

11.3   Structure and Stability of Radicals   477

11.4   Radical Addition to Alkenes   481

11.5   Other Radical Addition Reactions   487

11.6   Radical-Initiated Addition of HBr to Alkynes   489

11.7   Photohalogenation   490

11.8   Allylic Halogenation: Synthetically Useful Reactions   497

11.9   Special Topic: Rearrangements (and Nonrearrangements)
        of Radicals   501

11.10   Special Topic: Radicals in Our Bodies; Do Free Radicals Age Us?   504

11.11   Summary   505

11.12   Additional Problems   507

12   Dienes and the Allyl System: 2p Orbitals in Conjugation   511

12.1   Preview   512

12.2   Allenes   513

12.3   Related Systems: Ketenes and Cumulenes   515

12.4   Allenes as Intermediates in the Isomerization of Alkynes   516

12.5   Conjugated Dienes   519

12.6   The Physical Consequences of Conjugation   521

12.7   Molecular Orbitals and Ultraviolet Spectroscopy   525

12.8   Polyenes and Vision   533

12.9   The Chemical Consequences of Conjugation:
        Addition Reactions of Conjugated Dienes   534

12.10   Thermodynamic and Kinetic Control of Addition Reactions   537

12.11 The Allyl System: Three Overlapping $2p$ Orbitals   541

12.12 The Diels–Alder Reaction of Conjugated Dienes   544

12.13 Special Topic: Biosynthesis of Terpenes   554

12.14 Special Topic: Steroid Biosynthesis   559

12.15 Summary   563

12.16 Additional Problems   564

13   Conjugation and Aromaticity   571

13.1   Preview   572

13.2   The Structure of Benzene   573

13.3   A Resonance Picture of Benzene   575

13.4   The Molecular Orbital Picture of Benzene   578

13.5   Quantitative Evaluations of Resonance Stabilization in Benzene   580

13.6   A Generalization of Aromaticity: Hückel's $4n + 2$ Rule   582

13.7   Substituted Benzenes   595

13.8   Physical Properties of Substituted Benzenes   598

13.9   Heterobenzenes and Other Heterocyclic Aromatic Compounds   598

13.10 Polynuclear Aromatic Compounds   602

13.11 Introduction to the Chemistry of Benzene   606

13.12 The Benzyl Group and Its Reactivity   610

13.13 Special Topic: The Bio-Downside, the Mechanism of Carcinogenesis
       by Polycyclic Aromatic Compounds   614

13.14 Summary   617

13.15 Additional Problems   619

14   Substitution Reactions of Aromatic Compounds   623

14.1   Preview   624

14.2   Hydrogenation of Aromatic Compounds   626

14.3   Diels–Alder Reactions   628

14.4   Substitution Reactions of Aromatic Compounds   631

14.5   Carbon–Carbon Bond Formation: Friedel–Crafts Alkylation   639

14.6   Friedel–Crafts Acylation   643

14.7   Synthetic Reactions We Can Do So Far   646

14.8   Electrophilic Aromatic Substitution of Heteroaromatic Compounds   652

14.9   Disubstituted Benzenes: Ortho, Meta, and Para Substitution   655

14.10 Inductive Effects in Aromatic Substitution   666

14.11   Synthesis of Polysubstituted Benzenes   668

14.12   Nucleophilic Aromatic Substitution   674

14.13   Special Topic: Stable Carbocations in "Superacid"   679

14.14   Special Topic: Benzyne   680

14.15   Special Topic: Biological Synthesis of Aromatic Rings;
        Phenylalanine   682

14.16   Summary   685

14.17   Additional Problems   688

15   Analytical Chemistry: Spectroscopy   694

15.1   Preview   695

15.2   Chromatography   697

15.3   Mass Spectrometry (MS)   699

15.4   Infrared Spectroscopy (IR)   707

15.5 ^{1}H Nuclear Magnetic Resonance Spectroscopy (NMR)   713

15.6   NMR Measurements   717

15.7 ^{13}C NMR Spectroscopy   740

15.8   Problem Solving: How to Use Spectroscopy to Determine Structure   742

15.9   Special Topic: Dynamic NMR   746

15.10   Summary   750

15.11   Additional Problems   751

16   Carbonyl Chemistry 1: Addition Reactions   762

16.1   Preview   763

16.2   Structure of the Carbon–Oxygen Double Bond   764

16.3   Nomenclature of Carbonyl Compounds   767

16.4   Physical Properties of Carbonyl Compounds   770

16.5   Spectroscopy of Carbonyl Compounds   770

16.6   Reactions of Carbonyl Compounds: Simple Reversible Additions   773

16.7   Equilibrium in Addition Reactions   777

16.8   Other Addition Reactions: Additions of Cyanide and Bisulfite   781

16.9   Addition Reactions Followed by Water Loss: Acetal Formation   783

16.10   Protecting Groups in Synthesis   788

16.11   Addition Reactions of Nitrogen Bases:
        Imine and Enamine Formation   790

16.12   Organometallic Reagents   797

16.13   Irreversible Addition Reactions: A General Synthesis of Alcohols   799

16.14   Oxidation of Alcohols to Carbonyl Compounds   802

16.15   Retrosynthetic Alcohol Synthesis   807

16.16   Oxidation of Thiols and Other Sulfur Compounds   809

16.17   The Wittig Reaction   811

16.18   Special Topic: Biological Oxidation   813

16.19   Summary   816

16.20   Additional Problems   821

17   **Carboxylic Acids   828**

17.1   Preview   829

17.2   Nomenclature and Properties of Carboxylic Acids   830

17.3   Structure of Carboxylic Acids   832

17.4   Infrared and Nuclear Magnetic Resonance
         Spectra of Carboxylic Acids   833

17.5   Acidity and Basicity of Carboxylic Acids   834

17.6   Synthesis of Carboxylic Acids   839

17.7   Reactions of Carboxylic Acids   841

17.8   Special Topic: Fatty Acids   862

17.9   Summary   867

17.10   Additional Problems   870

18   **Derivatives of Carboxylic Acids: Acyl Compounds   876**

18.1   Preview   877

18.2   Nomenclature   879

18.3   Physical Properties and Structures of Acyl Compounds   884

18.4   Acidity and Basicity of Acyl Compounds   885

18.5   Spectral Characteristics   887

18.6   Reactions of Acid Chlorides: Synthesis of Acyl Compounds   889

18.7   Reactions of Anhydrides   894

18.8   Reactions of Esters   895

18.9   Reactions of Amides   901

18.10   Reactions of Nitriles   904

18.11   Reactions of Ketenes   907

18.12   Special Topic: Other Synthetic Routes to Acid Derivatives   907

18.13   Special Topic: Thermal Elimination Reactions of Esters   912

18.14   Special Topic: A Family of Concerted Rearrangements
        of Acyl Compounds   914

18.15   Summary   921

18.16   Additional Problems   925

19   Carbonyl Chemistry 2: Reactions at the α Position   931

19.1   Preview   932

19.2   Many Carbonyl Compounds Are Weak Brønsted Acids   933

19.3   Racemization of Enols and Enolates   944

19.4   Halogenation in the α Position   946

19.5   Alkylation in the α Position   954

19.6   Addition of Carbonyl Compounds to the α Position:
        The Aldol Condensation   965

19.7   Reactions Related to the Aldol Condensation   980

19.8   Addition of Acid Derivatives to the α Position:
        The Claisen Condensation   985

19.9   Variations on the Claisen Condensation   992

19.10   Special Topic: Forward and Reverse Claisen
        Condensations in Biology   996

19.11   Condensation Reactions in Combination   998

19.12   Special Topic: Alkylation of Dithianes   1001

19.13   Special Topic: Amines in Condensation Reactions,
        the Mannich Reaction   1003

19.14   Special Topic: Carbonyl Compounds without α Hydrogens   1004

19.15   Special Topic: The Aldol Condensation in the Real World, an
        Introduction to Modern Synthesis   1007

19.16   Summary   1010

19.17   Additional Problems   1017

20   Special Topic: Reactions Controlled by Orbital Symmetry   1030

20.1   Preview   1031

20.2   Concerted Reactions   1032

20.3   Electrocyclic Reactions   1034

20.4   Cycloaddition Reactions   1043

20.5   Sigmatropic Shift Reactions   1048

20.6   The Cope Rearrangement   1059

20.7   A Molecule with a Fluxional Structure   1063

20.8   How to Work Orbital Symmetry Problems   1071

20.9   Summary   1073

20.10   Additional Problems   1074

21   Special Topic: Intramolecular Reactions and Neighboring Group Participation   1080

21.1   Preview   1081

21.2   Heteroatoms as Neighboring Groups   1083

21.3   Neighboring π Systems   1096

21.4   Single Bonds as Neighboring Groups   1108

21.5   Coates' Cation   1117

21.6   Summary   1118

21.7   Additional Problems   1118

22   Special Topic: Carbohydrates   1124

22.1   Preview   1125

22.2   Nomenclature and Structure of Carbohydrates   1126

22.3   Formation of Carbohydrates   1138

22.4   Reactions of Carbohydrates   1140

22.5   The Fischer Determination of the Structure of D-Glucose (and the 15 Other Aldohexoses)   1153

22.6   Special Topic: An Introduction to Di- and Polysaccharides   1161

22.7   Summary   1168

22.8   Additional Problems   1170

23   Special Topic: Amino Acids and Polyamino Acids (Peptides and Proteins)   1173

23.1   Preview   1174

23.2   Amino Acids   1175

23.3   Reactions of Amino Acids   1186

23.4   Peptide Chemistry   1189

23.5   Nucleosides, Nucleotides, and Nucleic Acids   1210

23.6   Summary   1215

23.7   Additional Problems   1218

Glossary   G-1

Credits   C-1

Index   I-1

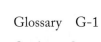

Selected Applications

Helium   5

Methane   60

Ethylene: A Plant Hormone   106

Strychnine   172

Thalidomide   180

Carboranes: Weird Bonding   217

Synthetic diamonds   219

Ajoene   252

Malic Acid   274

Cholesterol Formation   358

Ethyl Alcohol   400

Pyrethrins   456

Radical inhibitors as food preservatives   484

Vitamin E   505

Vitamin A and vision   533

Bombykol   534

The birth-control pill   562

Vanillin   582

Carcinogens   605

Carcinogenesis   614

Aniline   648

Magnetic resonance imaging (MRI)   713

Maitotoxin   745

Civetone   769

Salicylic Acid   838

Nylon and polyesters   852

Fats, oils, soaps, and detergents   862

Velcro   898

Eat Your Broccoli!   921

Anticancer drugs   980

Palytoxin   1009

Chorismate to Prephenate:
         A Biological Cope Rearrangement   1063

Mustard Gas   1089

Sugar Substitutes   1164

Cellulose and starch   1168

Canavanine: An Unusual Amino Acid   1177

DNA and RNA   1211

Organic Reaction Animations

Bimolecular nucleophilic substitution: S_N2 267

Halide formation   285

Unimolecular nucleophilic substitution: S_N1 289

Unimolecular elimination: E1   298

Bimolecular elimination: E2   301

Hofmann elimination   308

Acetylide addition   312

Intramolecular S_N2 316

S_N2 with cyanide   344

Alkene hydrohalogenation   365

Alkene hydration   381

Alkene polymerization   385

Carbocation rearrangement E1   389

Alkene hydroboration   390

Alkene halogenation   414

Stabilized alkene halogenation   417

Halohydrin formation   418

Alkene epoxidation   424

Basic epoxide ring opening   426

Acidic epoxide ring opening   427

Dihydroxylation of alkenes   443

Radical alkene hydrohalogenation   485

Alkane halogenation   494

1,2-Hydrohalogenation of dienes   534

1,4-Hydrohalogenation of dienes   534

Diels–Alder reaction   544

Benzylic oxidation   613

Friedel–Crafts acylation   644

Arene halogenation   670

Nucleophilic aromatic substitution   674

Benzyne formation   681

Carbonyl hydration   775

Acetal formation   786

Imine formation   792

Grignard reaction   801

Carbonyl reduction   802

Alcohol oxidation   804

Cleavage of vicinal diols   807

Wittig reaction   812

Fischer esterification   841

Acid chloride formation   854

Acid chloride aminolysis   854

Acid chloride aminolysis   889

Ester hydrolysis   895

Nitrile hydrolysis   904

Baeyer–Villiger reaction   907

Enol halogenation   947

Decarboxylation   960

Malonate alkylation   962

Aldol condensation   965

Michael addition   976

Mixed aldol condensation   983

Claisen condensation   987

Cope rearrangement   1063

Intramolecular S_N2 1081

Preface to the Fourth Edition

Most students in our organic chemistry courses are not chemistry majors. We wrote this book for anyone who wants a broad yet modern introduction to the subject. We stress general principles because it is impossible to memorize all the details of this vast subject. We want students to learn to make connections and to apply a set of broad organizing principles in order to make the material more manageable and understandable.

The Fourth Edition

Quite a bit has changed from the third to the fourth edition. New coauthor Steven Fleming brought his experience with this book and with his students to this revision. Developmental editor Irene Nunes read the third edition from the student perspective and made extensive comments. Irene helped us identify places where we may have assumed knowledge some students would not have or where we could further clarify a point or figure. We made substantial changes that will benefit students using this book.

The voice of the book remains the same. It is personal, and talks directly to the student not only about the material at hand, but also about the "how and why" of organic chemistry. We think it is much easier to enjoy, and learn, organic chemistry if a strong focus on "Where are we and why are we here?" and "What is the best way to do this?" is maintained. On occasion, we try to help students through a tough part of the subject by pointing out that it *is* tough and then suggesting ways to deal with it. Everyone who has taught this subject knows there are such places—when we talk to students ourselves, we try to use our experience to help students succeed, even when we know the going is likely to be difficult for some, and the book tries to do the same thing.

Every chapter begins with a **Preview** section in which the coming chapter is outlined. At the end of the Preview, we describe the **Essential Skills and Details** students will need for the chapter. At exam time, students can use these sections as guides for study and review.

Organic chemistry is a highly visual subject. Organic chemists think by constructing mental pictures of molecules and communicate with each other by drawing pictures. This book favors series of figures over long discussions in the text. The text serves to point out the changes in successive figures. Color is used to highlight

---

ESSENTIAL SKILLS AND DETAILS

1. Sidedness and Handedness. You have the broad outlines of structure under control now—acyclic alkanes, alkenes, and alkynes have appeared, as have rings. Now we come to the details, to stereochemistry. "Sidedness"—cis/trans isomerism—is augmented by questions of chirality—"handedness." Learning to see one level deeper into three-dimensionality is the next critical skill.

2. Difference. The topic of difference and how difference is determined arises in this chapter. The details may be nuts and bolts, and, indeed, any way that you work out to do the job will be just fine, but there is no avoiding the seriousness of the question. When are two atoms the same (in exactly the same environment) or different (not in the same environment)? This question gets to the heart of structure and is much tougher to answer than it seems. By all means, concentrate on this point. This chapter will help you out by introducing a method—an algorithm for determining whether or not two atoms are in different environments. It is well worth knowing how it works.

3. There is no way out—the (R) and (S) priority system must be learned.

4. Words—Jargon. This chapter is filled with jargon: Be certain that you learn the difference between enantiomers and diastereomers. Learn also what "stereogenic," "chiral," "racemic," and "meso" mean.

change and often to track the fates of atoms and groups in reactions. We have used text bubbles to show important steps in a process or to note an important point.

One factor that can make organic chemistry difficult is that new language must be learned. Organic chemists talk to each other using many different conventions and at least some of that language must be learned or communication is impossible. In addition to general treatments of nomenclature at the beginning of many chapters, we have incorporated numerous **Convention Alerts** in which aspects of the language that chemists use are highlighted.

Throughout the book, reference is made to the connection between organic chemistry and the world of biology. Almost every chapter has a section devoted to the biological relevance of new reactions discussed. We have also added **Applications Boxes** to illustrate the relevance of the subject to students' lives.

CHOLESTEROL FORMATION

Sometimes, our well-being depends on interfering with an enzyme-mediated rate acceleration. Cholesterol (p. 121) is formed in the body by a lengthy process in which the thioester **A** is reduced to mevalonic acid, which is then converted in a series of reactions into cholesterol.

Controlling cholesterol levels in the body is an important part of healthy living. Recent advances in medicinal and organic chemistry have allowed the development of the "statin" drugs (i.e., atorvastatin, fluvastatin, lovastatin, pravastatin, simvastatin, and rosuvastatin), which act to reduce the level of "unhealthy" low-density lipoprotein (LDL) cholesterol. Examples of a healthy artery and a partially clogged

artery, which can result from high levels of LDL cholesterol, are shown below. The benefit of taking a statin drug is a significant reduction in the risk of heart attacks and strokes.

The statins function by inhibiting the enzyme HMG-CoA reductase, which is involved in the rate-limiting or slowest step in the formation of mevalonic acid, and hence cholesterol. Without the enzyme, the reduction to mevalonic acid in the body is much too slow. We have learned about hydride reagents (p. 315), and we do have biological hydride sources. You will meet one, NADH, in Chapter 16. But simply mixing this "natural" hydride reagent with the HMG-CoA thioester **A** results in no reaction. However, HMG-CoA reductase can bring the hydride source and the thioester **A** together in a fashion that allows the reaction to occur. The energy barrier for the reaction is lowered, and mevalonic acid is formed and goes on to make cholesterol. The statins interfere with the reduction by inhibiting the enzyme necessary—no reduction, no cholesterol!

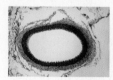

3-Hydroxy-3-methylglutaryl
(or HMG)   **CoA**

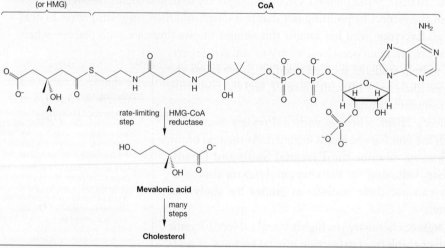

A

rate-limiting step | HMG-CoA reductase

Mevalonic acid

many steps

Cholesterol

It can be a great help to be shown profitable (and unprofitable) approaches to problem solving, and the number of **Problem Solving** sections has greatly increased in the fourth edition. Although there is no substitute for a thorough understanding of the material, our subject is an experience-intensive one, and by definition students are short on that commodity.

There are many moments in organic chemistry when it is important to take stock of where we are. **Summary** sections have been incorporated into every

> **PROBLEM SOLVING**
>
> Whenever you see the word "rate" in a problem, or when you see words such as "faster" that talk about rates, you are very likely to have to answer the question by drawing the transition state for the reaction. Remember that the rate of a reaction is determined by $\Delta G^{\ddagger}$, the energy difference between starting material and the transition state, and not by the energy difference between starting material and product. You need a pull-down menu that says, "Think about the transition state" when the word "rate" appears in a problem.

chapter. Here the narrative is broken and the reader is brought up to date on the important points of the previous topic. These summaries serve as excellent "reminder and review" sections when a student is studying for an exam.

Each chapter ends with a summary of **New Concepts, Key Terms,** new **Reactions, Mechanisms, and Tools,** new **Syntheses,** and **Common Errors.** These sections recapitulate and reinforce the material of the chapter, and serve as excellent study tools.

We incorporate unsolved problems in two ways. There are many such problems scattered throughout the text and more problems, of all degrees of difficulty, are found at the end of each chapter. They range from drill exercises and simple examples, designed to emphasize important skills and illustrate techniques, to sophisticated, challenging problems. In those last cases, we are careful to provide hints and references to material useful for the solution. All these problems are solved in the *Study Guide*, which does much more than provide a bare-bones answer. It, along with the new Problem Solving sections in the fourth edition, tries to show students problem-solving techniques that will help them solve future problems. There are also many solved problems in the text, each designed to reinforce a point just made.

The fourth edition contains many new problems including ones that require the use of the Organic Reactions Animations software.

Highlights of the Content and Organizational Changes in the Fourth Edition

The fourth edition has:

- A much more complete discussion of resonance in Chapter 1. This change is in response to a request by reviewers who felt this central theme should be presented as early as possible. We agree. Understanding resonance structures is critical, and early coverage provides a strong foundation. The student will have a chance to review the topic in Chapter 9.

- A reorganization of Chapter 15. The new outline emphasizes important topics in NMR spectroscopy. Also, the chapter was modified so that it is even more movable than it was in the third edition, in case the instructor wants to present spectroscopy early in the first semester. We believe spectroscopy is a topic that should be taught early. Students can use the tools (UV, IR, NMR, MS) to help understand symmetry, resonance, electronegativity, aromaticity, acidity, and reactivity of molecules. Spectroscopy is an important tool to link structure and activity.

- Enolates discussed as a single topic in Chapter 19. Earlier editions of this text had the enolate topic covered in two chapters separated by a chapter on carboxylic acids. Because enolates of aldehydes, ketones, esters, and carboxylic acids all share similar reactions, we have combined them into one chapter. It is the chemistry of carban-ions formed by deprotonating the carbon alpha to a carbonyl that sews this chapter together. Students will benefit from this unified approach to enolate chemistry.

• New structures and topics in the carbohydrate chapter, Chapter 22. This chapter was rewritten in an attempt to provide more current carbohydrate material. Many instructors do not cover carbohydrates in their organic chemistry course, but those who do will find both the historical Fischer proof and modern synthetic methods presented in this chapter. We have drawn most of the carbohydrates as chair struc-tures so that students can recognize and remember why glucose is the most com-mon carbohydrate.

• Special Topics in most chapters (except Chapters 7, 20, 21, and 23). This material deals with logical extensions and applications of related subjects. Instructors who choose to move through the chapters more quickly may opt to leave these sections out of lecture material. We hope students will read the Special Topic sections. The material is provided because we believe it is important, but we recognize that it is not critical.

• A substantially increased use of the terms *electrophile* and *nucleophile* throughout the text. There is a unifying theme that the interaction between Lewis acids (elec-trophiles) and Lewis bases (nucleophiles) in organic chemistry is stabilizing. In orbital terms, that statement simply means that the interaction between a filled and empty orbital is stabilizing. The text has a focus on this principle and the frequent use of these terms will help direct the student to see their significance.

• More end-of-chapter problems. Over 100 new problems have been added in this edition. The added problems are not the mind-numbing sort; these are practice problems that will give the student more experience and confidence with the sub-ject. Organic Reaction Animations (ORA) problems have also been added. These questions will help the student connect the ORA visualizations with the principles being discussed in the chapter.

Overall Organization

To understand atoms and molecules one must first think sensibly about electrons, and for that we need to explore a bit of what quantum mechanics tells us. That does *not* mean we will all have to become mathematicians. Far from it. Our discussion here will be purely qualitative, since we need only grasp qualitatively what the math-ematicians have to say to us. Qualitative molecular orbital theory is not too compli-cated a subject for students, and requires no mathematics. Yet, this simple theory is amazingly powerful in its ability to rationalize and, especially, to predict structure and reactivity. The abbreviated tutorial in Chapter 1 on qualitative applications of molecular orbital theory is likely to be new to students. This material is important, as it enables us to emphasize explanations throughout the rest of the book. Not only are traditional subjects such as aromaticity and conjugation (Chapters 12–14) more accessible with the background of Chapter 1, but explanations for the essential, build-ing-block reactions of organic chemistry (Chapters 7 and 9, for example) become possible. There is, after all, no essential difference between the classic statement "Lewis acids react with Lewis bases" and the idea that the interaction of a filled and empty orbital is stabilizing. The latter formulation allows all sorts of seemingly dis-parate reactions to be gathered together—unified—in a very useful way. (For exam-ple, a hydride shift and the S_N1 reaction become partners in a unified theory rather than two wildly different reactions that must be memorized in all their detail.)

The language that makes both the macro scale and the micro scale accessible to us is mathematics. Although we need not do the mathematical operations ourselves, we do need to appreciate some of the things that quantum mechanics has to say to us. Chapter 1 also focuses strongly on Lewis structures—pictorial representations of atoms and ions. The ability to write good Lewis structures easily and to determine the locations of charges in molecules with ease is an essential skill. This skill is part of the language of chemistry and will be as important in Chapter 23 as it is in Chapter 1.

After the introductory chapter comes a sequence of four chapters devoted largely to aspects of structure (Chapters 2–5). Here the details of the archetypal structures of organic chemistry are introduced. Hybridization appears, and the wonderful three-dimensionality of the subject begins to grow in. Some functional groups are introduced, and stereochemistry is dealt with in depth. A particularly vexing and fundamental question concerns what makes two atoms or molecules the same or different. The fourth edition includes a preview of NMR spectroscopy in Chapter 2. This section is not detailed—it is only an introduction—but it allows a real discussion of that elusive question of "difference." It also allows reinforcement through a series of new problems introduced throughout the first half of the book.

In Chapter 3, the addition of HX molecules to alkenes allows an introduction to synthesis, as well as a discussion of selectivity, catalysis, and reaction mechanism in general.

After the series of "structure" chapters comes Chapter 6 on alcohols, amines, halides, and the properties of solvents. This chapter functions as a lead in to a discussion of several building-block reactions, the S_N2, S_N1, E2, and E1 reactions.

Chapter 7 is one of the key chapters, and the reactions discussed here—substitution and elimination—serve as reference points throughout the book; they are fundamental reactions to which we return over and over in later chapters in order to make analogies.

Once we have these basic reference reactions under control, a general discussion of the role of energetics—kinetics and thermodynamics—becomes appropriate (Chapter 8). Chapters 9 and 10 introduce other building-block reactions, and other functional groups, in the context of an expansion of the earlier discussion of addition reactions in Chapter 3. Even at these early stages we introduce the biological applications of organic chemistry. For example, in the chapters devoted to the structure of alkanes, and, especially, alkenes, biorelevant examples appear. These do not obscure the essential information of the chapters, however. They are kept as examples, potential extensions, and applications of what we have learned at this point. Later on in the book their role is expanded, with whole chapters (Chapters 22 and 23) devoted to biological topics.

The basic reactions of the middle chapters (Chapters 7–14) provide a foundation for the chemistry of carbonyl compounds, the subject of a series of chapters in the second half of the book (Chapters 16–19).

Chapter 15, set in the midst of this run of chapters, is devoted to spectroscopy. Sections of earlier chapters have already dealt with parts of this subject in an introductory way (Chapter 2 for NMR, Chapter 12 for UV/vis). So, in order to allow flexible usage of this chapter, we have tried to make this chapter freestanding.

To make an analogy to the study of a language, in the first sequence of reaction chapters (Chapters 7, 9, and 10) we write sentences constructed from the vocabulary and grammar developed in the early, structural chapters. We will go on in later chapters in the book to more complicated mechanisms and molecules, and to write whole paragraphs and even short essays in organic chemistry. Some of those essays are contained in the Special Topics chapters toward the end of the book (Chapters 20–23) in which biological chemistry and physical-organic chemistry are further explored with the material of the early chapters as a foundation.

Flexibility. There is no consensus on the precise order in which to take up many subjects in organic chemistry. This book makes different decisions possible. For example, as already pointed out, the spectroscopy chapter is largely freestanding. Traditionally, it comes where it is here, roughly at the midpoint of the book. But cogent arguments can be made that spectroscopy should be introduced earlier, and that is possible, if one is willing to pay the price of delaying the serious discussion of chemical reactions one more chapter.

The last few chapters explicitly constitute a series of Special Topics. No one really hopes to finish everything in an organic textbook in one year, and this book provides a number of choices. One might emphasize biological aspects of our science, for example, and Chapters 22 and 23 provide an opportunity to do this. Alternatively, a more physical approach would see the exciting chemistry of Chapters 20 and 21 as more appropriate.

Instructor Resources

- **PowerPoint slides, available at wwnorton.com/nrl.** Both lecture slides and slides containing textbook art are available for download from the instructor site. Lecture slides include questions for classroom response systems (also known as clickers).
- **Transparency set,** with approximately 150 key figures from the text.
- **Test Bank** (Tim Minger, Mesa Community College). New to the fourth edition, the Test Bank contains 1,150 questions from which to choose. Questions are organized by chapter section, and each question is ranked by difficulty and type. The Test Bank is available as a printed book, in Word RTF, in PDF, and in ExamView Assessment Suite.

Student Resources

- **Study Guide/Solutions Manual** (Maitland Jones, Jr., New York University; Henry L. Gingrich, Princeton University; Steven A. Fleming, Temple University). Written by the textbook authors, this guide provides students with fully worked solutions to all unworked problems that appear in the text. In addition to the solutions presented for each specific problem, the authors present good problem-solving strategies for solving organic chemistry problems in general.

- **StudySpace, available at wwnorton.com/studyspace.** This free and open Web site is available to all students. StudySpace includes more than 350 interactive, 3-D molecules from the text (formerly hosted on Norton's Orgo3D Web site). These structures were made in Chem3D and can be manipulated in space and viewed in several ways (ball-and-stick, space-filling, etc.). In addition, there is a short writeup and usually a few questions (and answers!) for most of the molecules.

  StudySpace will house two review features from the text: Essential Skills and Details and Convention Alerts. The site will also provide links to the ebook and SmartWork.

- **SmartWork: an online tutorial and homework program for organic chemistry, available at wwnorton.com/smartwork.** SmartWork is the most intuitive online tutorial and homework-management system available for organic chemistry. Powerful quizzing engines support an unparalleled range of questions from both the book and a supplementary problem set, focusing on the students' ability to understand and draw molecules. Answer-specific feedback and hints coach students through solving problems. Integration of ebook and multimedia content completes this learning system. The guiding principle at the heart of this system is that, given enough time and effort, every student should be able to earn an "A" on every assignment. Assignments are scored automatically. SmartWork includes equally sophisticated and flexible tools for managing class data and determining how assignments are scored.

Assigning, editing, and administering homework within SmartWork is easy. WYSIWYG (What You See Is What You Get) authoring tools allow instructors to modify existing problems or develop new content.

- **Organic Reaction Animations.** Since its second year, this book has been graced by its association with the Organic Reaction Animations software created by Steven Fleming, Paul Savage, and Greg Hart. There are now over 50 different reactions in this splendid collection, which are fully integrated into the text, with icons identifying each reaction and its place in the book. All versions of ORA 2.3 also include tutorials on the reactions themselves. ORA problems are found at the end of most chapters in the fourth edition.

Acknowledgments from Mait Jones

Books don't get written by setting an author on his or her way and then waiting for the manuscript to appear. There is a great deal more work to be done than that. In general, it is an editor's job to make it possible for the author to do the best of which he is capable. Don Fusting, Joe Wisnovsky, Vanessa Drake-Johnson, and for this fourth edition, Erik Fahlgren at W. W. Norton were exemplary in their execution of that role. My special thanks go to Erik for keeping the big picture in mind, and for keeping two authors more or less on track. Jeannette Stiefel was copyeditor for the first three editions; Philippa Solomon and Connie Parks copyedited this edition. Kate Barry and Christopher Granville were early project editors at Norton. Carla Talmadge succeeded them for the fourth edition, and was exceptionally helpful and creative in her dealings with too many author-produced problems.

This book also profited vastly from the comments and advice of an army of reviewers, and I am very much in their debt. Their names and affiliations follow this preface. Two special reviewers, Henry L. Gingrich of Princeton and Ronald M. Magid of the University of Tennessee, read the work line by line, word by word, comma by missing comma. Their comments, pungent at times but helpful always, were all too accurate in uncovering both the gross errors and lurking oversimplifications in the early versions of this work.

Finally, MJ gives special thanks to Steven Fleming for joining up and adding so much more than his chemical expertise to this project. It has been a great pleasure to work with him!

Cape North, June 2009

Acknowledgments from Steven Fleming

I am honored to be involved with the Jones text. I thank Mait for letting me join forces to produce the fourth edition. My involvement wouldn't have happened without the indefatigable Erik Fahlgren. It has been a pleasure to work with Mait, Erik, and the W. W. Norton team. My interest in understanding and teaching "why" things happen led me to this text. I can't remember when I developed such an interest in cause and effect, but it must have been my genetics or environment. My brother, Ron Fleming, is also an organic chemist and he has been my chemistry hero. I have appreciated the opportunities to discuss this topic with him for the past 40 years. It has also been my good fortune to have Paul Savage as a close colleague. He has taught me much. There are over 100 students (mostly undergraduates) who have attended my weekly group meetings since 1986. I have enjoyed having them along for the ride, and it has been a joy to work with them and learn from them. I look forward to the next 100 unsuspecting souls and the topics we will learn together in the next 23 years. This is "for my children" Melissa, Nathan, Amy, and Erin.

Philadelphia, June 2009

Despite all the efforts of editors and reviewers, errors will persist. These are our fault only. When you find them, let us know.

Fourth Edition Reviewers

Margaret Asirvatham, University of Colorado, Boulder

France-Isabelle Auzanneau, University of Guelph

K. Darrell Berlin, Oklahoma State University

Brian M. Bocknack, University of Texas, Austin

Peter Buist, Carleton University

Arthur Cammers, University of Kentucky

Paul Carlier, Virginia Tech

Dana Chatellier, University of Delaware

Tim Clark, Western Washington University

Barry A. Codens, Northwestern University

Gregory Dake, University of British Columbia

Bonnie Dixon, University of Maryland

Tom Eberlein, Penn State, Harrisburg

Amy Gottfried, University of Michigan

Eric J. Kantorowski, California Polytechnic State University

Rizalia Klausmeyer, Baylor University

Masato Koreeda, University of Michigan

Brian Kyte, Saint Michael's College

Tim Minger, Mesa Community College

Susan J. Morante, Mount Royal College

Jonathan Parquette, Ohio State University

Chris Pigge, University of Iowa

John Pollard, University of Arizona

T. Andrew Taton, University of Minnesota

Alexander Wurthmann, University of Vermont

Previous Editions' Reviewers

Mark Arant, University of Louisiana at Monroe

Arthur Ashe, University of Michigan

William F. Bailey, University of Connecticut

John Barbaro, University of Georgia

Ronald J. Baumgarten, University of Illinois at Chicago

Michael Biewer, University of Texas at Dallas

David Birney, Texas Tech University

John I. Brauman, Stanford University

Peter Buist, Carleton University

Jeffrey Charonnat, California State University, Northridge

Marc d'Arlacao, Tufts University

Donald B. Denney, Rutgers University

Robert Flowers, Lehigh University

David C. Forbes, University of South Alabama

B. Lawrence Fox, University of Dayton

John C. Gilbert, University of Texas at Austin

Henry L. Gingrich, Princeton University

David Goldsmith, Emory University

Nancy S. Goroff, State University of New York, Stony Brook

David N. Harpp, McGill University

Richard K. Hill, University of Georgia

Ian Hunt, University of Calgary

A. William Johnson, University of Massachusetts

Guilford Jones II, Boston University

Richard Keil, University of Washington

S. Bruce King, Wake Forest University

Grant Krow, Temple University

Joseph B. Lambert, Northwestern University

Philip Le Quesne, Northeastern University

Steven V. Ley, Imperial College of Science, Technology and Medicine

Robert Loeschen, California State University, Long Beach

Carl Lovely, University of Texas at Arlington

Ronald M. Magid, University of Tennessee–Knoxville

Eugene A. Mash, Jr., University of Arizona

John McClusky, University of Texas at San Antonio

Lydia McKinstry, University of Nevada, Las Vegas

Robert J. McMahon, University of Wisconsin–Madison

Keith Mead, Mississippi State University

Andrew F. Montana, California State University, Fullerton

Kathleen Morgan, Xavier University of Louisiana

Roger K. Murray, Jr., University of Delaware

Thomas W. Nalli, State University of New York at Purchase

R. M. Paton, University of Edinburgh

Patrick Perlmutter, Monash University

Matthew S. Platz, Ohio State University

Lawrence M. Principe, Johns Hopkins University

Kathleen S. Richardson, Capital University

Christian Rojas, Barnard College

Alan M. Rosan, Drew University

Charles B. Rose, University of Nevada–Reno

Carl H. Schiesser, Deakin University

Martin A. Schwartz, Florida State University

John F. Sebastian, Miami University

Jonathan L. Sessler, University of Texas at Austin

Valerie V. Sheares, Iowa State University

Robert S. Sheridan, University of Nevada–Reno

Philip B. Shevlin, Auburn University

Matthew Sigman, University of Utah

William Tam, University of Guelph

Edward Turos, University of South Florida

Harry H. Wasserman, Yale University

David Wiedenfeld, New Mexico Highlands University

Craig Wilcox, University of Pittsburgh

David R. Williams, Indiana University

Introduction

These days, a knowledge of science must be part of the intellectual equipment of any educated person. Of course, that statement may always have been true, but we think there can be no arguing that an ability to confront the problems of concern to scientists is especially important today. Our world is increasingly technological, and many of our problems, and the answers to those problems, have a scientific or technological basis. Anyone who hopes to understand the world we live in, to evaluate many of the pressing questions of the present and the future—and to vote sensibly on them—must be scientifically literate.

The study of chemistry is an ideal way to acquire at least part of that literacy. Chemistry is a central science in the sense that it bridges such disparate areas as physics and biology, and connects those long-established sciences to the emerging disciplines of molecular biology and materials science. Similarly, as this book shows, organic chemistry sits at the center of chemistry, where it acts as a kind of intellectual glue, providing connections between all areas of chemistry. One does not have to be a chemist, or even a scientist, to profit from the study of organic chemistry.

The power of organic chemistry comes from its ability to give insight into so many parts of our lives. How does penicillin work? Why is Teflon nonstick? Why does drinking a cup of coffee help me stay awake? How do plants defend themselves against herbivores? Why is ethyl alcohol a depressant? All these questions have answers based in organic chemistry. And the future will be filled with more organic chemistry—and more questions. What's a buckyball or a nanotube, and how might it be important to my life? How might an organic superconductor be constructed? Why is something called the Michael reaction important in a potential cancer therapy? Read on, because this book will help you to deal with questions such as these, and many more we can't even think up yet.

What Is Organic Chemistry? Organic chemistry is traditionally described as the chemistry of carbon-containing compounds. Until the nineteenth century, it was thought that organic molecules were related in an immutable way to living things, hence the term "organic." The idea that organic compounds could be made only from molecules derived from living things was widespread, and gave rise to the notion of a vital force being present in carbon-containing molecules. In 1828 Friedrich Wöhler (1800–1882) synthesized urea, a certified organic substance, from heating ammonium cyanate, a compound considered to be inorganic.[1] Wöhler's experiment

[1]Wöhler's urea is an end product of the metabolism of proteins in mammals, and is a major component of human urine. An adult human excretes about 25 g (6–8 level teaspoons) of urea each day. The formation of urea is our way of getting rid of the detritus of protein breakdown through a series of enzymatic reactions. If you are missing one of the enzymes necessary to produce urea, it's very bad news indeed, as coma and rapid death result.

really did not speak to the question of vital force, and he knew this. The problem was that at the time there were no sources of ammonium cyanate that did not involve such savory starting materials as horns and blood—surely "vital" materials. The real coup de grâce for vitalism came some years later when Adolph Wilhelm Hermann Kolbe (1818–1884) synthesized acetic acid from elemental carbon and inorganic materials in 1843–1844 (see structures below).

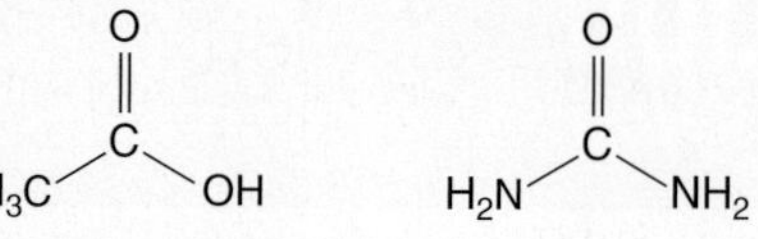

Acetic Acid   **Urea**

Despite the demise of the vital-force idea, carbon-containing molecules certainly do have a strong connection to living things, including ourselves. Indeed, carbon provides the backbone for all the molecules that make up the soft tissues of our bodies. Our ability to function as living, sentient creatures depends on the properties of carbon-containing organic molecules, and we are about to embark on a study of their structures and transformations.

Organic chemistry has come far from the days when chemists were simply collectors of observations. In the beginning, chemistry was largely empirical, and the questions raised were, more or less, along the lines of "What's going to happen if I mix this stuff with that stuff?" or "I wonder how many different things I can isolate from the sap of this tree?" Later, it became possible to collate knowledge and to begin to rationalize the large numbers of collected observations. Questions now could be expanded to deal with finding similarities in different reactions, and chemists began to have the ability to make predictions. Chemists began the transformation from the hunter–gatherer stage to modern times, in which we routinely seek to use what we know to generate new knowledge.

Many advances have been critical to that transformation; chief among them is our increased analytical ability. Nowadays the structure of a new compound, be it isolated from tree sap or produced in a laboratory, cannot remain a mystery for long. Today, the former work of years can often be accomplished in hours. This expertise has enabled chemists to peer more closely at the *why* questions, to think more deeply about reactivity of molecules. This point is important because the emergence of unifying principles has allowed us to teach organic chemistry in a different way, to teach in a fashion that largely frees students from the necessity to memorize organic chemistry. That is what this book tries to do: to teach concepts and tools, not vast compendia of facts. The aim of this book is to provide frameworks for generalizations, and the discussions of topics are all designed with this aim in mind.

We will see organic molecules of all types in this book. Organic compounds range in size from hydrogen (H_2)—a kind of honorary organic molecule even though it doesn't contain carbon—to the enormously complex biomolecules, which typically contain thousands of atoms and have molecular weights in the hundreds of thousands. Despite this diversity, and the apparent differences between small and big molecules, the study of all molecular properties always begins the same way, with structure. Structure determines reactivity, which provides a vehicle for navigating from the reactions of one kind of molecule to another and back again. So, early on, this book deals extensively with structure.

What Do Organic Chemists Do? Structure determination has traditionally been one of the things that practicing organic chemists do with their lives. In the early days, such activity took the form of uncovering the gross connectivity of the atoms in the molecule in question: What was attached to what? Exactly what are those molecules isolated from the Borneo tree or made in a reaction in the lab? Such questions are quickly answered by application of today's powerful spectroscopic techniques, or, in the case of solids, by X-ray diffraction crystallography. And small details of structure lead to enormous differences in properties: morphine, a pain-killing agent in wide current use, and heroin, a powerfully addictive narcotic, differ only by the presence of two acetyl groups (CH_3CO units), a tiny difference in their large and complex structures.

Today, much more subtle questions are being asked about molecular structure. How long can a bond between atoms be stretched before it goes, "Boing," in its quiet, molecular voice, and the atoms are no longer attached? How much can a bond be squeezed? How much can a bond be twisted? These are structural questions, and reveal much about the properties of atoms and molecules—in other words, about the constituents of us and the world around us.

Many chemists are more concerned with how reactions take place, with the study of "reaction mechanisms." Of course, these people depend on those who study structure; one can hardly think about how reactions occur if one doesn't know the detailed structures—connectivity of atoms, three-dimensional shape—of the molecules involved. In a sense, every chemist must be a structural chemist. The study of reaction mechanisms is an enormously broad subject. It includes people who look at the energy changes involved when two atoms form a molecule or, conversely, when a molecule is forced to come apart to its constituent atoms, as well as those who study the reactions of the huge biomolecules of our bodies—proteins and polynucleotides. How much energy is required to make a certain reaction happen? Or, how much energy is given off when it happens? You are familiar with both kinds of processes. For example, burning is clearly a process in which energy is given off as both heat and light.

Chemists also want to know the details of how molecules come together to make other molecules. Must they approach each other in a certain direction? Are there catalysts—molecules not changed by the reaction—that are necessary? There are many such questions. A full analysis of reaction mechanism requires a knowledge of the structures and energies of all molecules involved in the process, including species called intermediates, molecules of fleeting existence that cannot usually be isolated because they go on quickly to other species. One also must have an idea of the structure and energy of the highest energy point in a reaction, called the transition state. Such species cannot be isolated—they are energy maxima, not energy minima—but they can be studied nonetheless. We will see how.

Still other chemists focus on synthesis. The goal in such work is the construction of a target molecule from smaller, available molecules. In earlier times the reason for such work was sometimes structure determination. One set out to make a molecule one suspected of being the product of some reaction of interest. Now, determination of structure is not usually the goal. And it must be admitted that Nature is still a *much* better synthetic chemist than any human. There is simply no contest; evolution has generated systems exquisitely designed to make breathtakingly complicated molecules with spectacular efficiency. We cannot hope to compete. Why, then, even try? The reason is that there is a cost to the evolutionary development of synthesis, and that is specificity. Nature can make a certain molecule in an extraordinarily competent way, but Nature can't make changes on request. The much less efficient syntheses devised by humans are far more flexible than Nature's, and one reason for the chemist's interest in synthesis is the possibility of generating molecules of Nature in systematically modified forms. We hope to make small changes in the structures and to study the influence on biological properties induced by those changes. In that way it could be possible to find therapeutic agents of greatly increased efficiency, for example, or to stay ahead of microbes that become resistant to certain drugs. Nature can't quickly change the machinery for making an antibiotic molecule to which the microbes have become resistant, but humans can.

What's Happening Now? It's Not All Done. In every age, some people have felt that there is little left to be done. All the really great stuff is behind us, and all we can hope for is to mop up some details; we won't be able to break really new ground. And every age has been dead wrong in this notion. By contrast, the slope of scientific discovery continues to increase. We learn more every year, and not just details. Right now the frontiers of molecular biology—a kind of organic chemistry of giant molecules, we would claim—are the most visibly expanding areas, but there is much more going on.

In structure determination, completely new kinds of molecules are appearing. For example, just a few years ago a new form of carbon, the soccer ball–shaped C_{60}, was synthesized in bulk by the simple method of vaporizing a carbon rod and collecting the products on a cold surface. Even more recently it has been possible to capture atoms of helium and argon inside the soccer ball. These are the first neutral compounds of helium ever made. Molecules connected as linked chains or as knotted structures are now known. What properties these new kinds of molecules will have no one knows. Some will certainly turn out to be mere curiosities, but others will influence our lives in new and unexpected ways.

The field of organic reaction mechanisms continues to expand as we become better able to look at detail. For example, events on a molecular time scale are becoming visible to us as our spectrometers become able to look at ever smaller time periods. Molecules that exist for what seems a spectacularly short time—microseconds or nanoseconds—are quite long-lived if one can examine them on the femtosecond time scale. Indeed, the Nobel Prize in Chemistry in 1999 was given to Ahmed H. Zewail (b. 1946) of Caltech for just such work. Nowadays we are moving ever further into the strange realm of the attosecond time regime. We are sure to learn much more about the details of the early stages of chemical reactions in the next few years.

At the moment, we are still defining the coarse picture of chemical reactions. Our resolution is increasing, and we will soon see micro details we cannot even imagine at the moment. It is a very exciting time. What can we do with such knowledge? We can't answer that question yet, but chemists are confident that with more detailed knowledge will come an ability to take finer control of the reactions of molecules. At the other end of the spectrum, we are learning how macromolecules react, how they coil and uncoil, arranging themselves in space so as to bring two reactive molecules to just the proper orientation for reaction. Here we are seeing the bigger picture of how much of Nature's architecture is designed to facilitate positioning and transportation of molecules to reactive positions. We are learning how to co-opt Nature's methods by modifying the molecular machinery so as to bring about new results.

We can't match Nature's ability to be specific and efficient. Over evolutionary time, Nature has just had too long to develop methods of doing exactly the right thing. But we are learning how to make changes in Nature's machinery—biomolecules—that lead to changes in the compounds synthesized. It is likely that we will be able to co-opt Nature's methods, deliberately modified in specific ways, to retain the specificity but change the resulting products. This is one frontier of synthetic chemistry.

The social consequences of this work are surely enormous. We are soon going to be able to tinker in a controlled way with much of Nature's machinery. How does humankind control itself? How does it avoid doing bad things with this power? Those questions are not easy, but there is no hiding from them. We are soon going to be faced with the most difficult social questions of human history, and how we deal with them will determine the quality of the lives of us and our children. That's one big reason that education in science is so important today. It is not that we will need more scientists; rather it is that we must have a scientifically educated population in order to deal sensibly with the knowledge and powers that are to come. So, this book is not specifically aimed at the dedicated chemist-to-be. That person can use this book, but so can anyone who will need to have an appreciation of organic chemistry in his or her future—and that's nearly everyone these days.

How to Study Organic Chemistry

Work with a Pencil. We were taught very early that "Organic chemistry must be read with a pencil." Truer words were never spoken. You can't read this book, or any chemistry book, in the way you can read books in other subjects. You *must* write things as you go along. There is a real connection between the hand and the brain in this business, it seems. When you come to the description of a reaction, especially where the text tells you that it is an important reaction, by all means take the time to draw out the steps yourself. It is not enough to read the text and look at the drawings; it is not sufficient to highlight. Neither of these procedures is reading with a pencil. Highlighting does not reinforce the way working out the steps of the synthesis or chemical reaction at hand does. You might even make a collection of file cards labeled "Reaction descriptions" on which you force yourself to write out the steps of the reaction. Another set of file cards should be used to keep track of the various ways to make molecules. At first, these cards will be few in number, and sparsely filled, but as we reach the middle of the course there will be an explosion in the number of synthetic methods available. This subject can sneak up on you, and keeping a catalog will help you to stay on top of this part of the subject. We will try to help you to work in this interactive way by interrupting the text with problems, with solutions that follow immediately when we think it is time to stop, take stock, and reinforce a point before going on. These problems are important. You can read right by them of course, or read the answer without stopping to do the problem, but to do so will be to cheat yourself and make it harder to learn the subject. Doing these in-chapter problems is a part of reading with a pencil and should be very helpful in getting the material under control. There is no more important point to be made than this one. Ignore it at your peril!

Don't Memorize. In the old days, courses in organic chemistry rewarded people who could memorize. Indeed, the notorious dependence of medical school admission committees on the grade in organic chemistry may have stemmed from the need to memorize in medical school. If you could show that you could do it in organic, you could be relied upon to be able to memorize that the shin bone was connected to the foot bone, or whatever. Nowadays, memorization is the road to disaster; there is just too much material. Those who teach this subject have come to see an all too familiar pattern. There is a group of people who do very well early and then crash sometime around the middle of the first semester. These folks didn't suddenly become stupid or lazy; they were relying on memorization and simply ran out of memory. Success these days requires generalization, understanding of principles that unify seemingly disparate reactions or collections of data. Medical schools still regard the grade in organic as important, but it is no longer because they look for people who can memorize. Medicine, too, has outgrown the old days. Now medical schools seek people who have shown that they can understand a complex subject, people who can generalize.

Work in Groups. Many studies have shown that an effective way to learn is to work in small groups. Form a group of your roommates or friends, and solve problems for each other. Assign each person one or two problems to be solved for the group. Afterward, work through the solution found in the chapter or *Study Guide*. You will find that the exercise of explaining the problem to others will be enormously useful. You will learn much more from "your" problems than from the problems solved by others. When Mait teaches organic chemistry at Princeton, and now at NYU, he increasingly replaces lecture with small-group problem solving.

Work the Problems. As noted above, becoming good at organic chemistry is an interactive process; you can't just read the material and hope to become expert. Expertise in organic chemistry requires experience, a commodity that by definition you are very low on at the start of your study. Doing the problems is vital to gaining the necessary experience. Resist the temptation to look at the answer before you have tried to do the problem. Disaster awaits you if you succumb to this temptation, for you cannot learn effectively that way and there will be no answers available on the examinations until it is too late. That is not to say that you must be able to solve all the problems straight away. There are problems of all difficulty levels in each chapter, and some of them are very challenging indeed. Even though the problem is hard or very hard, give it a try. When you are truly stuck, that is the time to gather a group to work on it. Only as a last resort should you take a peek at the *Study Guide*. There you will find not just a bare bones answer, but, often, advice on how to do the problem as well. Giving hard problems is risky, because there is the potential for discouraging people. Please don't worry if some problems, especially hard ones, do not come easily or do not come at all. Each of us in this business has favorite problems that we still can't solve. Some of these form the basis of our research efforts, and may not yield, even to determined efforts, for years. A lot of the pleasure in organic chemistry is working challenging problems, and it would not be fair to deprive you of such fun.

Use All the Resources Available to You. You are not alone. Moreover, everyone will have difficulty at one time or another. The important thing is to get help when you need it. Of course the details will differ at each college or university but there are very likely to be extensive systems set up to help you. Professors have office hours, there are probably teaching assistants with office hours, and there will likely be help, review, or question sessions at various times. Professors are there to help you, and they will not be upset if you show enough interest to ask questions about a subject they love. "Dumb questions" do not exist! You are not expected to be an instant genius in this subject, and many students are too shy to ask perfectly reasonable questions. Don't be one of those people!

If you feel uncertain about a concept or problem in the book—or lecture—get help soon! This subject is highly cumulative, and ignored difficulties will come back to haunt you. We know that many teachers tell you that it is impossible to skip material and survive, but this time it is true. What happens in December or April depends on September, and you can't wait and wait, only to "turn it on" at the end of the semester or year. Almost no one can cram organic chemistry. Careful, attentive, daily work is the route to success, and getting help with a difficult concept or a vexing problem is best done immediately. Over the life of the early editions of this book, Mait interacted with many of you by e-mail, much to his pleasure. Of course, we can't begin to replace local sources of help, and we can't be relied upon in an emergency, as we might be out of touch with e-mail, but we can usually be reached at mj55@nyu.edu or sfleming@temple.edu. We look forward to your comments and questions.

Atoms and Molecules; Orbitals and Bonding

1.1 Preview
1.2 Atoms and Atomic Orbitals
1.3 Covalent Bonds and Lewis Structures
1.4 Resonance Forms
1.5 Hydrogen (H_2): Molecular Orbitals
1.6 Bond Strength
1.7 An Introduction to Reactivity: Acids and Bases
1.8 Special Topic: Quantum Mechanics and Babies
1.9 Summary
1.10 Additional Problems

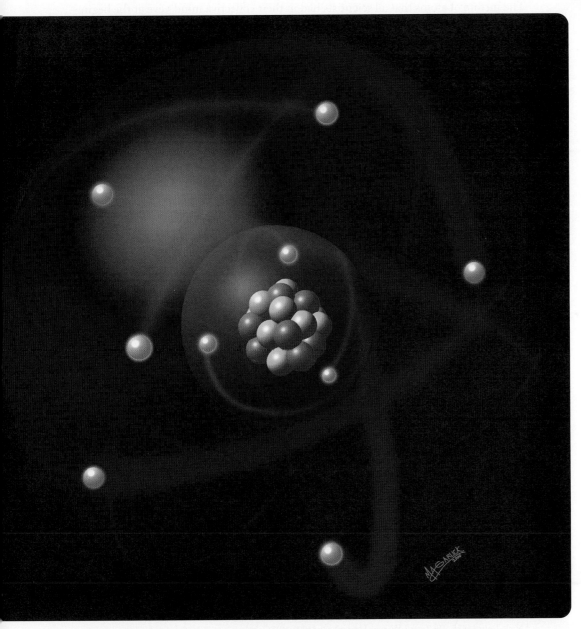

RUTHERFORD'S ATOM This photo represents Rutherford's view of electrons orbiting the nucleus. We know now that electrons are not traveling in circular paths around the nucleus, but this was the early model.

When it comes to atoms, language can be used only as in poetry. The poet too, is not nearly so concerned with describing facts as with creating images.

—NIELS BOHR TO WERNER HEISENBERG[1]

1.1 Preview

The picture of atoms that all scientists had in their collective mind's eye at the beginning of the last century was little different from that of the ancient Greek philosopher Democritus (~460–370 B.C.), who envisioned small, indivisible particles as the constituents of matter. These particles were called **atoms**, from the Greek word for indivisible. The British chemist John Dalton (1766–1844) had the idea that different atoms might have different characteristic masses, but he did not abandon the notion of a solid, uniform atom. That picture did not begin to change dramatically until 1897, when the English physicist J. J. Thomson (1856–1940) discovered the negatively charged elementary particle called the **electron**. Thomson postulated a spongelike atom with the negatively charged electrons embedded within a positively charged material, rather like raisins in a pudding. Ernest Rutherford's (1871–1937) discovery, in 1909, that an atom was mostly empty space demolished the pudding picture and led to his celebrated planetary model of the atom, in which electrons were seen as orbiting a compact, positively charged **nucleus**, the core of positively charged protons and neutral neutrons at the center of the atom.

It was Niels Bohr who made perhaps the most important modification of the planetary model. He made the brilliant and largely intuitive[2] suggestion that electrons were required to occupy only certain orbits. Because an electron's energy depends on the distance of its orbit from the positively charged nucleus, Bohr's suggestion amounted to saying that electrons in atoms can have only certain energies. An electron might have energy x or energy $2x$, but nothing in between. Whenever a property is restricted to certain values in this way, we say the property is *quantized*. Although this notion may seem strange, there are similar phenomena in the everyday world. One cannot create just any tone by blowing across the mouth of a bottle, for example. Say you start by blowing gently and creating a given tone. Of course the tone you hear depends on the size and shape of the particular bottle, but if you gradually increase how hard you blow, which gradually increases the energy you are supplying, the tone does not change smoothly. Instead you hear the first tone unchanged over a certain range of energy input, then a *sudden* change in tone when just the right "quantum" of energy has been provided.

In the 1920s and 1930s, a number of mathematical descriptions emerged from the need to understand Bohr's quantum model of the atom. It became clear that one must take a probabilistic view of the subatomic world. Werner Heisenberg discovered that it was not possible to determine simultaneously both the position and momentum (mass times speed) of an electron.[3] Thus, one can determine where an

---

[1] Niels Bohr (1885–1962) and Werner Heisenberg (1901–1976) were pioneers in the development of quantum theory, the foundation of our current understanding of chemical bonding.

[2] Some people's intuitions are better able than others' to cope with the unknown!

[3] This idea is extraordinarily profound—and troubling. The **Heisenberg uncertainty principle** (which states that the product of the uncertainty in position times the uncertainty in momentum is a constant) seems to limit fundamentally our access to knowledge. For an exquisite exposition of the human consequences of the uncertainty principle, see Jacob Bronowski, *The Ascent of Man*, Chapter 11 (Little Brown, New York, 1973).

electron is at any given time only in terms of probability. One *can* say, for example, that there is a 90% probability of finding the electron in a certain volume of space, but one *cannot* say that at a given instant the electron is at a particular point in space.

The further elaboration of this picture of the atom has given us the conceptual basis for all modern chemistry: the idea of the orbital. Loosely speaking, an **orbital** describes the region of space surrounding an atomic nucleus that may be occupied by either an electron or a pair of electrons of a certain energy. Both the combining of atoms to form molecules and the diverse chemical reactions these molecules undergo involve, at a fundamental level, the interactions of electrons in orbitals. This notion will appear throughout this book; it is the most important unifying principle of organic chemistry. In atoms, we deal with **atomic orbitals**, and in molecules, we deal with **molecular orbitals**.

Various graphic conventions are used in this book to represent atoms and molecules—letters for atoms, dots for electrons not involved in bonding, and lines for electrons in bonds—but it is important to keep in mind from the outset that the model that most closely approximates our current understanding of reality at the atomic and molecular level is the cloudy, indeterminate—one might even say poetic—image of the orbital.[4]

There is great conceptual overlap between the concept of an orbital and the notion you probably encountered in general chemistry of shells of electrons surrounding the atomic nucleus. For example, you are accustomed to thinking of the noble gas elements as having filled shells of electrons, two electrons for helium in the first shell, two electrons in the first shell and eight in the second shell for neon, and so on. In the noble gases, the outermost, or valence shells are filled. We will speak of those valence shells as valence orbitals. We shall say much more about orbitals in a moment, especially about their shapes, but the point to "get" here is the move from the old word *shell* to the new word *orbital*.

ESSENTIAL SKILLS AND DETAILS

The following list of Essential Skills and Details, a version of which will appear in every chapter, is designed to alert you to the important parts of the chapter and, especially, to aid you in reviewing. After you finish the chapter, or before an examination, it is a good idea to return to this list and make sure you are clear on all the Essential Skills and Details.

1. Writing correct Lewis dot structures for atoms, ions (charged atoms and molecules), and neutral molecules is an absolutely critical skill that will be essential throughout this book.

2. Take charge! It is necessary to be able to determine the formal charge of an atom, especially an atom in a molecule.

3. You have to be able to write the resonance forms (different electronic structures) that, taken together, give a more accurate picture of molecules than does any single structure.

4. Learn how to use the curved arrow formalism to "push" pairs of electrons in writing resonance forms and in sketching electron flow in chemical reactions.

5. Remember the sign convention for exothermic ($\Delta H°$ is negative) and endothermic ($\Delta H°$ is positive) reactions.

---

[4] In the wonderful quote that opens this chapter, Niels Bohr points out that once we transcend the visible world, all that is possible is modeling or image-making. To us, what is even more marvelous about the quote is the simple word *too*: It was obvious to Bohr that scientists speak in images, and he was pointing out to Heisenberg that there was another group of people out there in the world who did the same thing—poets.

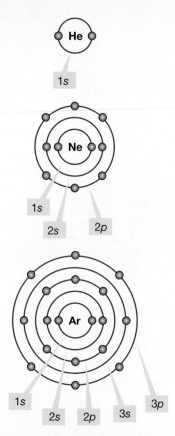

FIGURE 1.1 Highly schematic representations of He, Ne, and Ar.

1.2 Atoms and Atomic Orbitals

In a neutral atom, the nucleus, or core of positively charged protons and neutral neutrons, is surrounded by a number of negatively charged electrons equal to the number of protons. If the number of electrons and protons is not equal, the atom must be charged and is called an **ion**. A negatively charged atom or molecule is called an **anion**; a positively charged species is a **cation**. (One of the tests of whether or not you know how to "talk organic chemistry" is the pronunciation of the word *cation*. It is *kat*-eye-on, not *kay*-shun.)

The energy required to remove an electron from an atom to form a cation is called the **ionization potential**. In general, the farther away an electron is from the nucleus the easier it is to remove the electron and the lower the ionization potential.

Much of the chemistry of atoms is dominated by gaining or losing electrons in order to achieve the electronic configuration of one of the noble gases (He, Ne, Ar, Kr, Xe, and Rn). The noble gases have especially stable filled shells of electrons: 2 for He, 10 for Ne (2 + 8), 18 for Ar (2 + 8 + 8), and so on. The idea that filling certain shells creates especially stable configurations is known as the **octet rule**. With the exception of the first shell, called the 1*s* orbital, which can hold only two electrons, all shells fill with eight electrons; thus this rule specifies an octet. The second shell can hold eight electrons, two in the subshell called the 2*s* orbital and six in the subshell 2*p* orbitals. We will explain these numbers and names shortly, but we need the labels first. Figure 1.1 shows highly schematic pictures of three noble gases and uses this terminology. These pictures do not give good three-dimensional representations of these species, but they do show orbital occupancy. Better pictures are forthcoming.

The two electrons surrounding the helium (He) nucleus completely fill the first shell, and for this reason it is most difficult to remove an electron from He. Helium has an especially high ionization potential, 24.6 eV/atom = 566 kcal/mol.[5] Likewise, the chemical inertness of the other noble gases, which also have high ionization potentials, is the result of the stability of their filled valence shells.

Electrons can be added to atoms as well as removed. The energy that is released by adding an electron to an atom to form an anion is called the atom's **electron affinity**, measured in electron volts. The noble gases have very low electron affinities. Conversely, atoms to which adding an electron would complete a noble gas configuration have high electron affinities. The classic example is fluorine: the addition of a single electron yields a fluoride ion, F^-, which has the electronic configuration of Ne. Both F^- and Ne are 10-electron species, as Figure 1.2 shows.

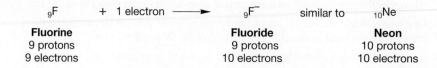

$_9$F	+ 1 electron ⟶	$_9$F$^-$	similar to	$_{10}$Ne
Fluorine		**Fluoride**		**Neon**
9 protons		9 protons		10 protons
9 electrons		10 electrons		10 electrons

FIGURE 1.2 Addition of an electron to a fluorine atom gives a fluoride ion.

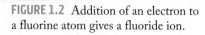

[5] There are several units of energy in use. Organic chemists commonly use kilocalories per mole (kcal/mol); physicists use the electron volt (eV). One electron volt/molecule translates into about 23 kcal/mol. Recently, the International Committee on Weights and Measures suggested that still another energy unit, the kilojoule (kJ), be substituted for kilocalorie. So far, organic chemists in some countries, including the United States, seem to have resisted this suggestion (1 kcal is equal to 4.184 kJ).

HELIUM

Helium (He) is the only substance that remains liquid under its own pressure at the lowest temperature recorded. There are only about five parts per million of helium in the Earth's atmosphere, but it reaches substantially higher concentrations in natural gas, from which it is obtained. Helium is formed from the radioactive decay of heavy elements. For example, a kilogram of uranium gives 865 L of helium after complete decay. There's not much helium on Earth, but there is *a lot* in the universe. About 23% of the known mass of the universe is helium, mostly produced by thermonuclear fusion reactions between hydrogen nuclei in stars. So, an outside observer of our universe (whatever that means!) would probably conclude that helium is some of the most important stuff around.

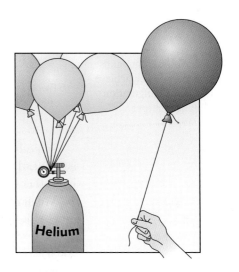

Single atoms, such as the fluorine atom shown on the left in Figure 1.2, are often written in the form $^{W}_{Z}\text{C}$, where W is the mass number (number of protons and neutrons in nucleus), Z is the atomic number (number of protons in nucleus), and C is the element symbol, here C for carbon. Thus, this notation for fluorine is $^{19}_{9}\text{F}$. In this book, the superscript W is omitted, which leads to the $_9\text{F}$ and $_{10}\text{Ne}$ you see in Figure 1.2.

Table 1.1 shows ionization potentials and electron affinities of some elements arranged as in the periodic table. Notice that with the exception of hydrogen, atoms with low ionization potentials, which are atoms that have easily removed electrons, cluster on the left side of the periodic table and atoms with high electron affinities, which are atoms that accept electrons easily, are on the right side (excluding the noble gases).

CONVENTION ALERT

TABLE 1.1   Some Ionization Potentials (black) and Electron Affinities (red) in eV

H 13.60 0.75									He 24.59 ~0
Li 5.39 0.62	Be 9.32 ~0			B 8.30 0.24	C 11.26 1.27	N 14.53 ~0	O 13.62 1.47	F 17.42 3.34	Ne 21.56 ~0
Na 5.14 0.55	Mg 7.65 ~0			Al 5.99 0.46	Si 8.15 1.24	P 10.49 0.77	S 10.36 2.08	Cl 12.97 3.61	Ar 15.75 ~0

Atoms having low ionization potentials often transfer an electron to atoms having high electron affinities, forming **ionic bonds**. In an ionically bonded species, such as sodium fluoride (Na^+F^-), the atoms are held together by the electrostatic attraction of the opposing charges. In sodium fluoride, both Na^+ and F^- have filled second shells, and each has achieved the stable electronic configuration of the noble gas Ne. In potassium chloride (K^+Cl^-), both ions have the electronic configuration of Ar, another noble gas (Fig. 1.3).

FIGURE 1.3 In the ionic compounds NaF and KCl, each atom can achieve a noble gas electronic configuration with a filled octet of electrons.

$_{11}Na^+$ and $_9F^-$ Resemble $_{10}Ne$ $_{19}K^+$ and $_{17}Cl^-$ Resemble $_{18}Ar$

10 Electrons   10 Electrons   18 Electrons   18 Electrons
each   each

Ionically bonded compounds are traditionally the province of inorganic chemistry. Nearly all of the compounds of organic chemistry are bound not by ionic bonds but rather by **covalent bonds**, which are bonds formed by the sharing of electrons.

We have just developed pictures of some atoms and ions. It is time now to elaborate a bit to provide a fuller picture of atomic orbitals. Your reward for bearing with an increase in complexity will be a much-increased ability to think about structure and reactivity. The most useful models for explaining and predicting chemical behavior focus on the qualitative aspects of atomic (and molecular) orbitals, so it is time to learn more about them.

The electrons in atoms do not occupy simple circular orbits. To describe an electron in the vicinity of a nucleus, Erwin Schrödinger (1887–1961) developed a formula called a *wave equation*. Schrödinger recognized that electrons have properties of both particles and waves. The solutions to Schrödinger's wave equation, called **wave functions** and written ψ (pronounced "sigh"), have many of the properties of waves. They can be positive in one region, negative in another, or zero in between.

An orbital is mathematically described by a wave function, ψ, and the square of the wave function, ψ^2, is proportional to the probability of finding an electron in a given volume. There are regions or points in space where ψ and ψ^2 are both 0 (zero probability of finding an electron in these regions) and such regions or points are called **nodes**. However, ψ^2 does not vanish at a large distance from the nucleus, but maintains a finite value, even if inconsequentially small.

Each electron is described by a set of four **quantum numbers**. Quantum numbers are represented by the symbols n, l, m_l, and s. The first two quantum numbers correspond to the orbital of the electron. The first one, called the principal quantum number, is represented by n and may have the integral values $n = 1, 2, 3, 4$, and so on. It is related to the distance of the electron from the nucleus and hence to the energy of the electron. It describes the atomic shell the electron occupies. The higher the value of n, the greater the average distance of the electron from the nucleus and the greater the electron's energy. The principal quantum number of the highest-energy electron of an atom also determines the row occupied by the atom in the periodic table. For example, the electron in hydrogen and the two electrons in helium are all $n = 1$, as Table 1.2 shows, and so these two atoms are in the first row of the table. The principal quantum number in Li, Be, B, C, N, O, F, and Ne is $n = 2$, which tells us that electrons can be in the second shell for these atoms and places the atoms in the second row. The elements Na, Mg, Al, Si, P, S, Cl, and Ar are in the third row, and orbitals for the electrons in these atoms correspond to $n = 1, 2$, or 3.

The second quantum number, l, is related to the shape of the orbital and depends on the value of n. It may have only the integral values $l = 0, 1, 2, 3, \ldots, (n - 1)$. So, for an orbital for which $n = 1$, l must be 0; for $n = 2$, l can be 0 or 1; and for $n = 3$, the three possible values of l are 0, 1, and 2.

TABLE 1.2 Principal Quantum Number (n) of the Highest Energy Electron

Atom	n
H, He	1
Li, Be, B, C, N, O, F, Ne	2
Na, Mg, Al, Si, P, S, Cl, Ar	3

Each value of l signifies a different orbital shape. We shall learn about these shapes in a moment, but for now just remember that each shape is represented by a letter. The orbital for which $l = 0$ is spherical, and the letter s is used to designate all spherical orbitals. For higher values of l, we do not have the convenience of easily remembered letters the way we do with "s for spherical." Instead, you just have to remember that p is used for orbitals for which $l = 1$, d is used for those for which $l = 2$, and f is used for $l = 3$. These letters associated with the various values of l lead to the common orbital designations shown in Table 1.3.

The third quantum number, m_l, depends on l. It may have the integral values $-l \ldots 0 \ldots +l$, and is related to the orientation of the orbital in space. Table 1.4 presents the possible values of n, l, and m_l for $n = 1, 2$, and 3. Orbitals of the same shell (n) and the same shape (l) are at the same energy regardless of the m_l value.

TABLE 1.3 Relationship between n and l

n	l	Orbital Designation
1	0	1s
2	0	2s
2	1	2p
3	0	3s
3	1	3p
3	2	3d

TABLE 1.4 Relationship between n, l, and m_l

n	l	m_l	Orbital Designation
1	0	0	1s
2	0	0	2s
2	1	−1	2p
2	1	0	2p
2	1	+1	2p
3	0	0	3s
3	1	−1	3p
3	1	0	3p
3	1	+1	3p
3	2	−2	3d
3	2	−1	3d
3	2	0	3d
3	2	+1	3d
3	2	+2	3d

Finally, there is s, the spin quantum number, which may have only the two values $\pm 1/2$.

Table 1.5 lists all the possible combinations of quantum numbers through $n = 3$.

TABLE 1.5 Possible Combinations of Quantum Numbers for $n = 1, 2$, and 3

n	l	m_l	s	Orbital Designation
1	0	0	$\pm\frac{1}{2}$	1s
2	0	0	$\pm\frac{1}{2}$	2s
2	1	−1, 0, +1	$\pm\frac{1}{2}$ each value of m_l	2p
3	0	0	$\pm\frac{1}{2}$	3s
3	1	−1, 0, +1	$\pm\frac{1}{2}$ each value of m_l	3p
3	2	−2, −1, 0, +1, +2	$\pm\frac{1}{2}$ each value of m_l	3d

Be careful. The s that designates the spin quantum number is not the same as the s in a 1s or 2s orbital. What is electron spin anyway? The word *spin* tries to make an analogy with the macroscopic world—in many ways, the electron behaves like a spinning top that can spin either clockwise or counterclockwise.

As shown in Tables 1.3–1.5, orbitals are designated with a number and a letter—1s, 2s, 2p, and so on. The number in the designation tells us the n value for a given

CONVENTION ALERT

orbital, and the letter tells us the l value. Thus, the notation $1s$ means the orbital for which $n = 1$ and $l = 0$. Because l values can be only 0 to $n - 1$, $1s$ is the only orbital possible for $n = 1$. The $2s$ orbital has $n = 2$ and $l = 0$, but now l can also be 1 ($n - 1 = 2 - 1 = 1$), and so we also have $2p$ orbitals, for which $n = 2, l = 1$. For the $2p$ orbitals, m_l, which runs from 0 to $\pm l$, may take the values $-1, 0, +1$. Thus there are three $2p$ orbitals, one for each value of m_l. These equi-energetic orbitals are differentiated by arbitrarily designating them as $2p_x$, $2p_y$, or $2p_z$. A little later we will see that the x, y, z notation indicates the relative orientation of the $2p$ orbitals in space.

For $n = 3$, we have orbitals $3s$ ($n = 3, l = 0$) and $3p$ ($n = 3, l = 1$). Just as for the $2p$ orbitals, m_l may now take the values $-1, 0, +1$. The three $3p$ orbitals are designated as $3p_x$, $3p_y$, and $3p_z$. With $n = 3$, l can also be 2 ($n - 1 = 2$), so we have the $3d$ ($n = 3, l = 2$) orbitals, and m_l may now take the values $-2, -1, 0, +1$, and $+2$. Thus there are five $3d$ orbitals. These turn out to have the complicated designations $3d_{x^2-y^2}$, $3d_{z^2}$, $3d_{xy}$, $3d_{yz}$, and $3d_{xz}$. Mercifully, in organic chemistry we only very rarely have to deal with $3d$ orbitals and need not consider the f orbitals, for which $n = 4$ and are even more complicated.

For all of the orbitals in Tables 1.4 and 1.5, the spin quantum number s may be either $+1/2$ or $-1/2$. We may now designate an electron occupying the lowest-energy orbital, $1s$, as either $1s$ with $s = +1/2$ or $1s$ with $s = -1/2$, *but nothing else*. Similarly, there are only two possibilities for electrons in the $2p_x$ orbital: $2p_x$ with $s = +1/2$ or $2p_x$ with $s = -1/2$. The same is true for all orbitals—$3p_z$, $4d_{xy}$, or whatever. Only two values are possible for the spin quantum number. That is why it is impossible for more than two electrons to occupy any orbital!

The convention used to designate electron spin shows the electrons as up ($\uparrow$) and down ($\downarrow$) pointing arrows. Two electrons having opposite spins are denoted $\uparrow\downarrow$ and are said to have **paired spins**. Two electrons having the same spin are denoted $\uparrow\uparrow$ and are said to have parallel, or **unpaired spins**.

How many electrons occupy a given orbital in an atom is indicated with a superscript. When we write $1s^2$ we mean that the $1s$ orbital is occupied by two electrons, and these electrons must have opposite (paired, with $s = +1/2$ and $s = -1/2$) spin quantum numbers. The designation $1s^3$ is meaningless because there is no way to put a *different* third electron in any orbital. No two electrons may have the same values of the four quantum numbers. This rule is called the **Pauli principle**, after Wolfgang Pauli (1900–1958), who first articulated it in 1925. These ideas are summarized in Figure 1.4.

CONVENTION ALERT

TABLE 1.6 Electronic Descriptions of Some Atoms

Atom	Electronic Configuration
$_1$H	$1s$
$_2$He	$1s^2$
$_3$Li	$1s^2 2s$
$_4$Be	$1s^2 2s^2$
$_5$B	$1s^2 2s^2 2p_x$

FIGURE 1.4 Two electrons in the same orbital must have opposite (paired) spins.

$1s^2$ means the 1s orbital contains two electrons:

	Electron No.1	Electron No. 2
	$n = 1, l = 0, m_l = 0, s = +\frac{1}{2}$	$n = 1, l = 0, m_l = 0, s = -\frac{1}{2}$

CONVENTION ALERT

In Table 1.6, we see the entry $1s^2$, which means there are two electrons in the $1s$ orbital. It would seem that $1s^1$ would be appropriate notation for a $1s$ orbital occupied by one electron. That notation is rarely used, however, and the superscript "1" is almost always understood.

We can now use the quantum numbers n, l, m_l, and s to write electronic descriptions, called configurations, for atoms using what is known as the **aufbau principle** (*aufbau* is German for building up or construction). This principle simply makes the reasonable assumption that we should fill the available orbitals in order of their energies, starting with the lowest-energy orbital. To form these descriptions of neutral atoms, we add electrons until they are equal to the number of protons in the nucleus. Table 1.6 gives the electronic configurations of H, He, Li, Be, and B.

For carbon, the atom after boron in the periodic table, there is a choice to be made in adding the last electron. The first five electrons are placed as in boron, but where does the sixth electron go? One possibility would be to put it in the same orbital as the fifth electron to produce the electronic configuration $_6C = 1s^2 2s^2 2p_x^2$ (Fig. 1.5a). In such an atom, the spins of the two electrons in the $2p_x$ orbital must be paired (opposite spins).

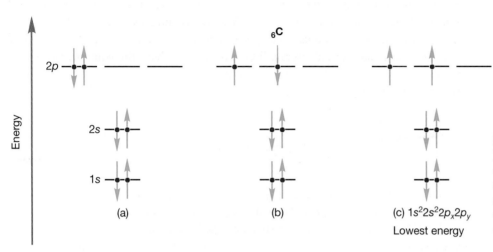

FIGURE 1.5 An application of Hund's rule. The electronic configuration with the largest number of parallel (same direction) spins is lowest in energy. Note the use of the arrow convention to show electron spin.

WORKED PROBLEM 1.1 Explain why the two electrons in the $2p_x$ orbital of carbon $(_6C = 1s^2 2s^2 2p_x^2)$ must have paired spins ($+1/2$ and $-1/2$).

ANSWER[6] Because they are in the same orbital ($2p_x$), both electrons have the same values for the three quantum numbers n, l, and m_l ($n = 2$, $l = 1$, $m_l = +1$, 0, or -1). If the spin quantum numbers were not opposite, one $+1/2$, the other $-1/2$, the two electrons would not be different from each other! The Pauli principle (no two electrons may have the same values of the four quantum numbers) ensures that two electrons in the same orbital must have different spin quantum numbers.

Alternatively, the sixth electron in the carbon atom could be placed in another $2p$ orbital to produce $_6C = 1s^2 2s^2 2p_x 2p_y$ (Fig. 1.5b). The only difference is the presence of two electrons in a *single* $2p$ orbital ($_6C = 1s^2 2s^2 2p_x^2$) in Figure 1.5a, versus one electron in each of two *different* $2p$ orbitals ($_6C = 1s^2 2s^2 2p_x 2p_y$) in Figure 1.5b. The three $2p$ orbitals are of equal energy (because each has $n = 2$, $l = 1$), and so how is one to make this choice? Electron–electron repulsion would seem to make the second arrangement the better one, but there is still another consideration. When two electrons occupy different but equi-energetic orbitals, their spins can either be paired ($\uparrow\downarrow$) or unpaired ($\uparrow\uparrow$). **Hund's rule** (Friedrich Hund, 1896–1997) states that for a given electron configuration, the state with the greatest number of unpaired (parallel) spins has the lowest energy. So the third possibility for carbon's sixth electron (Fig. 1.5c) is actually the best one in this case. Essentially, giving the two

---

[6] Worked Problems are answered in whole or in part in the text. When only part of a problem is worked, that part will have an asterisk. Complete answers can be found in the *Study Guide*.

electrons the same spin (↑↑) ensures that they cannot occupy the same orbital and tends to keep them apart. So, Hund's rule tells us that the configuration $_6C = 1s^2 2s^2 2p_x 2p_y$ with **parallel spins** for the two electrons in the $2p$ orbitals is the lowest-energy state for a carbon atom.

> **PROBLEM 1.2** Explain why the fifth and sixth electrons in a carbon atom may not occupy the same orbital as long as they have parallel spins.

Now we can write electronic configurations for the rest of the atoms in the second row of the periodic table (Table 1.7). Notice that for nitrogen we face the same kind of choice just described for carbon. Does the seventh electron go into an already occupied $2p$ orbital or into the empty $2p_z$ orbital? The configuration in which the $2p_x$, $2p_y$, and $2p_z$ orbitals are all singly occupied by electrons that have the same spin is lower in energy than any configuration in which a $2p$ orbital is doubly occupied. Again, Hund rules!

> **PROBLEM 1.3** Write the electronic configurations for the atoms in the third row in the periodic table, $_{11}Na$ through $_{18}Ar$.

TABLE 1.7 Electronic Descriptions of Some Atoms in the Second Row

Atom	Electronic Configuration
$_6C$	$1s^2 2s^2 2p_x 2p_y$
$_7N$	$1s^2 2s^2 2p_x 2p_y 2p_z$
$_8O$	$1s^2 2s^2 2p_x^2 2p_y 2p_z$
$_9F$	$1s^2 2s^2 2p_x^2 2p_y^2 2p_z$
$_{10}Ne$	$1s^2 2s^2 2p_x^2 2p_y^2 2p_z^2$

In order to understand the properties and reactions of the molecules you will encounter in organic chemistry, it will be necessary to know the shapes of the molecules. It turns out that molecular shape can be best understood by knowing where the electrons are. Recall that ψ^2 is related to the electron density around a nucleus. Therefore a graph that plots ψ^2 as a function of r, the distance from the nucleus, can describe the shape of the orbital. Figure 1.6 plots ψ^2 as a function of r for the lowest-energy solution to the Schrödinger equation, which describes the $1s$ orbital. The graph shows that the probability of finding an electron falls off sharply in all directions (x, y, and z) as we move out from the nucleus. Because there is no directionality to r, we find the $1s$ orbital is symmetrical in all directions—that is, it is spherical. As noted earlier, it is the quantum number l that is associated with orbital shape, and all s orbitals, for which l always equals zero, are spherically symmetric. Note also that the electron density never goes to zero, that there is a finite (though very, very small) probability of finding an electron at a distance of several angstroms from the nucleus.

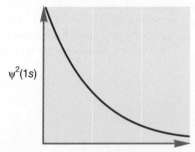

$\psi^2(1s)$

Distance from nucleus r

FIGURE 1.6 A plot of $\psi^2(1s)$ versus distance r for hydrogen. Note that the value of ψ^2 does not go to zero, even at very large r.

FIGURE 1.7 A three-dimensional picture of the $1s$ orbital. The surface of the sphere denotes an arbitrary cutoff.

Figure 1.7 translates the two-dimensional graph of Figure 1.6 into a three-dimensional picture of the $1s$ orbital, a spherical cloud of electron density that has its maximum near the nucleus. Although this cloud of electron density is shown as bound by the spherical surface at some arbitrary distance from the nucleus, it does not really terminate sharply there. This spherical boundary surface simply indicates the volume within which we have a high probability of finding the electron. We can choose to put the boundary at any percentage we like—the 95% confidence level, the 99% confidence level, or any other value. It is this picture that we have been approaching throughout these pages. The representation of the electron density in Figure 1.7 is the one you will probably remember best and the one that will be most useful in our study of chemical reactions. This sphere centered on the nucleus is the region of space occupied by an electron in the $1s$ orbital.

Figure 1.6 correctly gives the probability of finding an electron at a particular point a distance r from the nucleus, but it doesn't recognize that as r increases, a given change in r (denoted by the symbol Δr) produces a greater volume of space

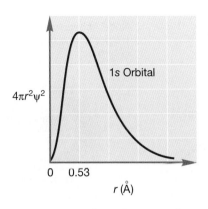

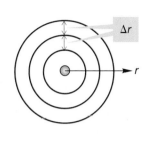

FIGURE 1.8 The graph on the left shows a slice through the three-dimensional $1s$ orbital. At relatively low values of r (short distance from the nucleus) Δr will contain a smaller volume than at higher values of r (larger distance from the nucleus). The outer spherical shell contains a greater volume than the middle shell even though they are both Δr wide. This plot is weighted for the increasing influence of Δr as r gets larger.

and thus a greater number of points where the electron might be. To see why this is so, look at Figure 1.8. Even though the increase, Δr, for each concentric circle is the same, the volume is much greater for the outer spherical shell. The change in volume is not linear with the change in radius.

We account for this increasing influence of Δr with the graph shown in Figure 1.8. Here the horizontal axis is again r, but the vertical axis $(4\pi r^2 \psi^2)$ accounts for the volume of the sphere. The resulting graph gives a picture of the probability of finding an electron at *all* points at a distance r from the nucleus. This new picture takes account of the increasing volume of spherical shells of thickness Δr as the distance from the nucleus, r, increases.

Summary

1. An electron is described by four quantum numbers (n, l, m_l, and s).
2. Electrons are filled into the available orbitals starting with the lowest-energy orbital (the aufbau principle).
3. An orbital can contain only two electrons, which must have paired spin quantum numbers, $s = +1/2$ and $s = -1/2$.
4. The quantum number l specifies the shape of the orbital, and all s orbitals have spherical symmetry.
5. We can do no better than to say that within some degree of certainty that an electron is within a certain volume of space. ψ^2 gives us the shape of that volume.
6. The cutoff point for our degree of certainty is arbitrary because, as Figures 1.6 and 1.8 show, the chance of an electron being far from the nucleus is never zero.

As noted earlier, the mathematical function ψ that describes an atomic orbital has many of the properties of waves. There are, for example, the hills and valleys evident in waves in liquids and in a vibrating string. Hills correspond to regions for which the mathematical sign of ψ is positive, and valleys correspond to regions for which the mathematical sign of ψ is negative. In a vibrating string, when we pass from a length of string in which the amplitude is positive to one where it is negative, we find a stationary point called a node. Nodes also appear in wave functions at the points at which the sign of ψ changes, where $\psi = 0$. Of course, if $\psi = 0$, then $\psi^2 = 0$, and so the probability of finding an electron at a node is also zero. In an atomic orbital, therefore, a node is a region of space in which the electron density is zero. The number of nodes in an orbital is always one fewer than n, the principal quantum number. Thus, the $n = 1$ orbital ($1s$) has no nodes, all $n = 2$ orbitals have a single node, all $n = 3$ orbitals have two nodes, and so on. As we turn our attention from the $1s$ orbital to higher-energy orbitals, we will have to pay attention to the presence of nodes.

FIGURE 1.9 (a) A plot of $\psi^2(r)$ versus r for a 2s orbital and (b) a cross section through the orbital.

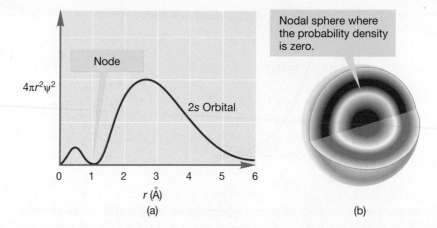

Nodal sphere where the probability density is zero.

(a)   (b)

The next higher energy orbital is the *2s* ($n = 2$, $l = 0$). Because $n = 2$, there must be a single node in this orbital, and we expect spherical symmetry, as with all *s* orbitals. Figure 1.9a shows a plot of $4\pi r^2\psi^2$ versus r. Figure 1.9b is a three-dimensional representation of the *2s* orbital, which shows a spherical node, representing the spherical region at which ψ and ψ^2 are zero.

When $n = 2$, l can equal 1 as well as zero. As described on page 7, the quantum number combination of $n = 2$, $l = 1$ gives the $2p_x$, $2p_y$, and $2p_z$ orbitals, each of which must possess a single node. Because l is not zero, the *2p* orbitals are not spherically symmetrical. As you may have guessed from the *x*, *y*, *z* designations, the *2p* orbitals are directed along the *x*, *y*, and *z* axes of a Cartesian coordinate system (Fig. 1.10). Each *2p* orbital is made up of two lobes that are slightly flattened spheres, and the orbital as a whole is shaped roughly like a dumbbell. The node is the plane separating the two halves of the dumbbell. In one lobe of the orbital, the sign of the wave function is positive; in the other the sign is negative. To avoid confusion with electrical charge these signs are usually indicated by a change of color rather than + and −. The three *2p* orbitals together are shown in Figure 1.11. As you can see,

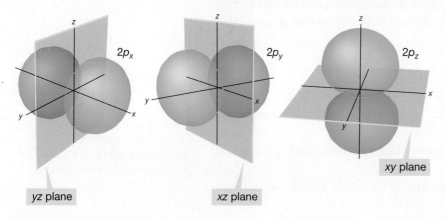

FIGURE 1.10 An accurate three-dimensional representation of three *2p* orbitals. Note the nodal planes separating the lobes where the sign of ψ differs. Color is used to emphasize the opposite signs of the lobes.

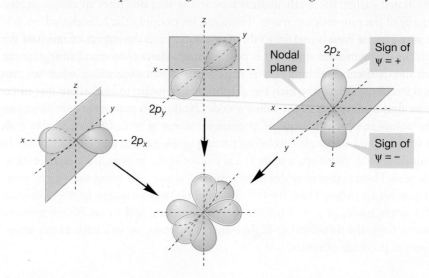

FIGURE 1.11 A schematic three-dimensional representation of three *2p* orbitals, shown separately and in combination.

the stylized dumbbells used to represent $2p$ orbitals are easier to draw than the more accurate picture shown in Figure 1.10.

For the first time, we begin to get hints of the causes of the complicated three-dimensional structures of molecules: Electrons are the "glue" that holds the atoms of molecules together, and electrons are confined to regions of space that are by no means always spherically symmetrical.

As Figure 1.12 shows, we can now make quite detailed pictures of the electrons in atoms. The $1s$ orbitals have been omitted in the atoms past He. The other electrons are shown as dots.

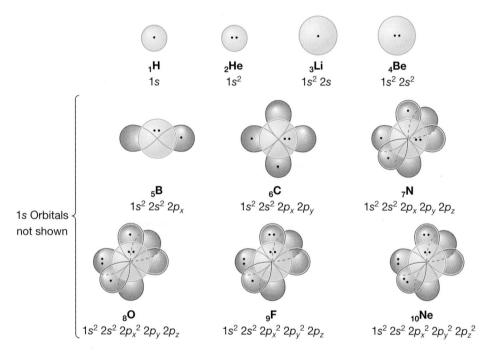

$_1$H
$1s$

$_2$He
$1s^2$

$_3$Li
$1s^2\,2s$

$_4$Be
$1s^2\,2s^2$

1s Orbitals not shown

$_5$B
$1s^2\,2s^2\,2p_x$

$_6$C
$1s^2\,2s^2\,2p_x\,2p_y$

$_7$N
$1s^2\,2s^2\,2p_x\,2p_y\,2p_z$

$_8$O
$1s^2\,2s^2\,2p_x^{\,2}\,2p_y\,2p_z$

$_9$F
$1s^2\,2s^2\,2p_x^{\,2}\,2p_y^{\,2}\,2p_z$

$_{10}$Ne
$1s^2\,2s^2\,2p_x^{\,2}\,2p_y^{\,2}\,2p_z^{\,2}$

FIGURE 1.12 Schematic pictures of the first 10 atoms in the periodic table. Each dot represents an electron, although the $1s$ electrons for the atoms past He are not shown.

Summary

Electrons in atoms are confined to certain volumes of space, and these volumes are not all alike in shape. A dumbbell-shaped $2p$ orbital is quite different from a spherical $2s$ orbital, for example.

1.3 Covalent Bonds and Lewis Structures

The formation of molecules through covalent bonding is the subject of much of the remainder of this chapter. Because a covalent bond is formed by the sharing of a pair of electrons between two atoms, the bond is shown either as a pair of dots between two atoms or as a line joining the two (Fig. 1.13). These two-electron bonds will be prominent in almost all of the molecules shown in this book. For many simple molecules, even the line between the atoms is left out, and the compound is written simply as A_2 or AB. You are left to supply mentally the missing two electrons.

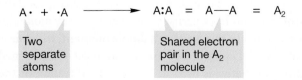

A· + ·A ⟶ A:A = A—A = A_2

Two separate atoms

Shared electron pair in the A_2 molecule

FIGURE 1.13 Bonding through the sharing of electrons is not ionic, but covalent. Covalent bonding can also result in stable electronic configurations.

The idea of covalent bonding was largely the creation of the American chemist Gilbert Newton Lewis (1875–1946), and the molecules formed by covalent bonds are usually written as what are called **Lewis structures** or Lewis dot structures. In almost all covalently bonded molecules, every atom can achieve a noble gas electronic configuration by sharing electrons. In the H_2 molecule, each hydrogen atom shares two electrons and thus resembles the noble gas helium (Fig. 1.14). In F_2, each fluorine has a pair of electrons in the lowest-energy $1s$ orbital and eight other electrons in $n = 2$ orbitals, six unshared (called either **nonbonding electrons** or **lone-pair electrons**) and two shared in the fluorine–fluorine bond as shown in Figure 1.14. Thus, each fluorine in F_2 has a filled octet and, by virtue of sharing electrons, resembles the noble gas neon. The Lewis dot structure of HF in Figure 1.14 shows that the hydrogen is helium-like and the fluorine is neon-like.

FIGURE 1.14 Lewis dot structures for H_2, F_2, and HF. Every atom has a noble gas electronic configuration.

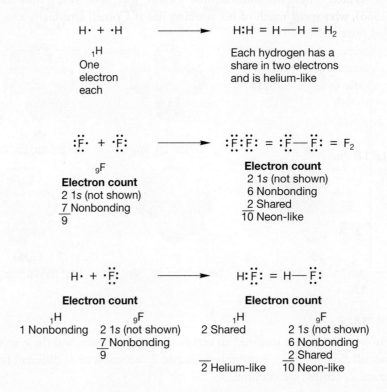

With the exception of H, electrons in $1s$ orbitals are not involved in bonding and therefore are not shown in Lewis structures. Anytime you need to count electrons in one of these structures, you have to remember to account for the unshown $1s$ electrons.

Hydrogen fluoride is different from any of the molecules mentioned so far in that the two electrons in the bond are shared unequally between the two atoms. In any diatomic molecule made from two different atoms, there cannot be equal sharing of the electrons because one atom must attract the electrons more strongly than the other atom does. Covalent bonds in which the two electrons are shared unequally are **polar covalent bonds**. Because electrons carry a negative charge, the atom bearing the larger share of electrons carries a partial negative charge, and the atom bearing the lesser share of electrons has a partial positive charge. Any molecule in which electrical charge is separated in this way has a **dipole moment**, which is a measurement of the polarity of a bond. Dipole literally means two poles. Often a small arrow pointing from positive to negative is used to indicate the direction of the dipole, or δ^+

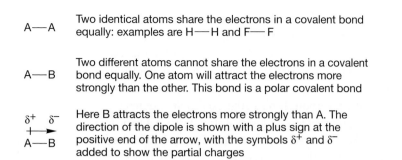

FIGURE 1.15 Comparison of covalent and polar covalent bonds.

A—A   Two identical atoms share the electrons in a covalent bond equally: examples are H——H and F——F

A—B   Two different atoms cannot share the electrons in a covalent bond equally. One atom will attract the electrons more strongly than the other. This bond is a polar covalent bond

δ^+ δ^-

A—B   Here B attracts the electrons more strongly than A. The direction of the dipole is shown with a plus sign at the positive end of the arrow, with the symbols δ^+ and δ^- added to show the partial charges

(partial positive charge) and δ^- (partial negative charge) signs are placed on the appropriate atoms (Fig. 1.15). The dipole moment is measured in *debye* units named after the Dutch physicist and chemist Petrus Josephus Wilhelmus Debye (1884–1966), who spent much of his working life at Cornell University and won the Nobel prize in Chemistry in 1936.

The tendency of an atom to attract electrons is called **electronegativity**. Atoms with high electron affinities are the most *electronegative* and are found at the right of the periodic table (Table 1.8). Atoms with low electronegativities and low ionization potentials are called *electropositive* and lie at the left of the table. The extreme case for unequal electron sharing is the ionic bonding seen in Na^+F^- and K^+Cl^-.

TABLE 1.8 Some Electronegativities

1 (IA)	2 (IIA)					3 (IIIA)	4 (IVA)	5 (VA)	6 (VIA)	7 (VIIA)
H 2.3										
Li 0.9	Be 1.6					B 2.1	C 2.5	N 3.1	O 3.6	F 4.2
Na 0.9	Mg 1.3					Al 1.6	Si 1.9	P 2.3	S 2.6	Cl 2.9
K 0.7										Br 2.7
Rb 0.7										I 2.4

PROBLEM 1.4 Show the direction of the dipole in the indicated bonds of the following molecules:

$$H—Cl, \; H—F, \; Li—F, \; H_3C—Cl, \; HO—Br, \; Br—Cl$$

Not all molecules containing polar covalent bonds have dipole moments. Many do, including all those shown in Problem 1.4; but consider carbon dioxide (CO_2), a linear molecule of the structure O=C=O (the double lines represent double bonds in which the carbon atom shares two pairs of electrons with each oxygen atom). Because oxygen is more electronegative than carbon (Table 1.8), both bonds

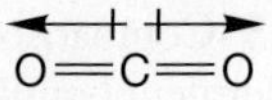

$$\overset{\longleftarrow\;+\;\longrightarrow}{O = C = O}$$

Carbon dioxide

FIGURE 1.16 Carbon dioxide has no dipole moment.

CONVENTION ALERT

in CO_2 are polar, as shown in Figure 1.16. Even though carbon dioxide contains these two polar bonds, however, there is no net dipole moment in the molecule because the bond dipoles cancel. The dipole moment of a molecule is the vector sum of all dipoles in the molecule. In $O{=}C{=}O$, the two dipoles are equal in magnitude and pointed in exactly opposite directions; they cancel. There are two polar carbon–oxygen bonds in carbon dioxide, but no dipole moment.

In representing three-dimensional structures on a two-dimensional surface such as a chalkboard or book page, it is necessary to devise some scheme for showing bonds directed away from the surface, either toward the front (up and out of the board or page) or toward the rear (down and into the board or page). In the figure for Problem 1.5, and in many, many future figures, the solid wedges (▬) represent covalent bonds that are coming out toward you and the dashed wedges (┅) represent covalent bonds that are retreating into the page.

PROBLEM 1.5 Which of these two molecules has a dipole moment and which does not? Look carefully at the shapes of the two tetrahedral molecules. The blue dashed lines show the outlines of a tetrahedron. If visualizing the three-dimensional aspects of these molecules is hard for you at first, by all means use molecular models.

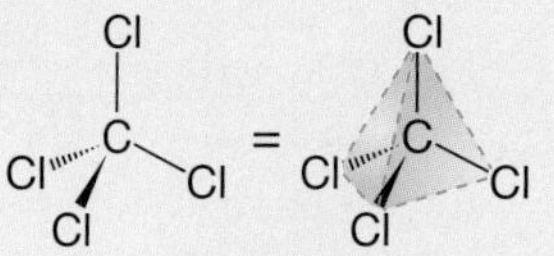

WEB 3D [7]

Carbon tetrachloride
(CCl_4)

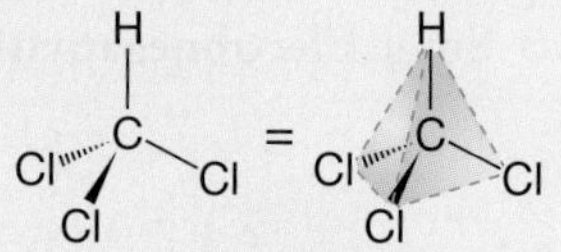

Chloroform
($CHCl_3$)

WEB 3D

In order to write Lewis dot structures for molecules containing only hydrogen and second row atoms (most organic molecules), we need to know the number of electrons in the atoms. For atoms in the second row of the periodic table, we can tell immediately how many electrons are available for bonding by knowing the atomic number of the atom or by knowing the column the atom is in. The number of available electrons is equal to the total number of electrons in the atom, which is the same as the atomic number, less the two $1s$ electrons. One can also use the column number, which corresponds to the number of electrons in the outer shell (see Table 1.8). The electrons in the outermost shell, which are the least tightly held electrons, are called the **valence electrons**.

For atoms in the third row of the periodic table, neither the two $1s$ electrons nor the eight electrons in the second shell ($2s^2 2p^6$) are shown in the Lewis structure. Instead, only the valence electrons in the third shell appear.

In complete Lewis structures, every valence electron is written as a dot. In another form of a Lewis structure, pairs of electrons forming covalent bonds are shown as lines and only the nonbonding valence electrons appear as dots. We will use this kind of Lewis structure throughout this text. Some examples of Lewis structures are

---

[7] This icon means that you can find the calculated three-dimensional structure of the molecule on the Web at www.wwnorton.com/studyspace. That structure can be manipulated so that you can view it from any direction, and changed so that you can see different versions of it (space filling, ball-and-stick, and so on). The structures can also be interrogated; they will tell you bond lengths and interatomic angles, if asked. You will also find comments, problems, and answers related to the molecule on the Web site.

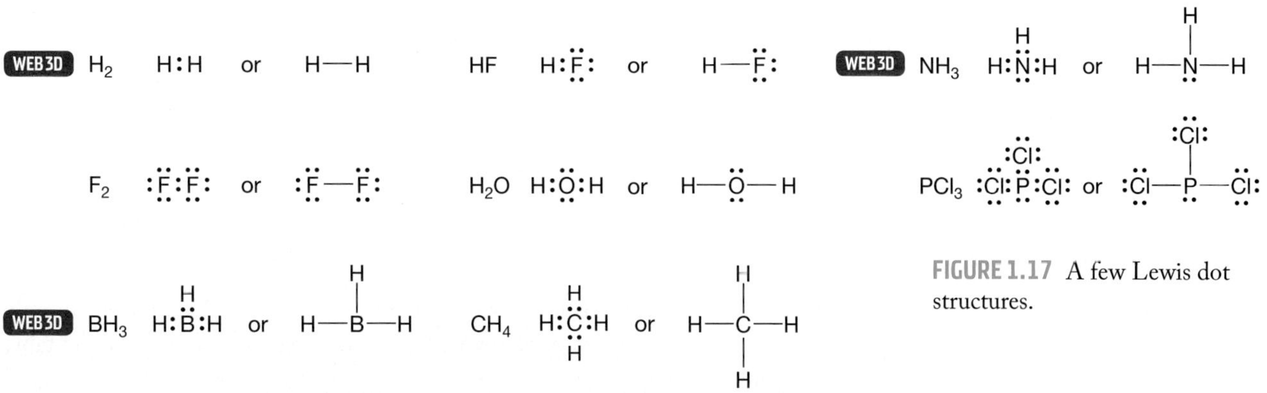

WEB 3D H_2 H:H   or   H—H   HF   H:$\ddot{\text{F}}$:   or   H—$\ddot{\text{F}}$:   **WEB 3D** NH_3 H:$\ddot{\text{N}}$:H   or   H—$\underset{\text{H}}{\overset{\text{H}}{\text{N}}}$—H

F_2 :$\ddot{\text{F}}$:$\ddot{\text{F}}$:   or   :$\ddot{\text{F}}$—$\ddot{\text{F}}$: H_2O H:$\ddot{\text{O}}$:H   or   H—$\ddot{\text{O}}$—H PCl_3 :$\ddot{\text{Cl}}$:$\ddot{\text{P}}$:$\ddot{\text{Cl}}$:   or   :$\ddot{\text{Cl}}$—$\underset{}{\overset{:\ddot{\text{Cl}}:}{\text{P}}}$—$\ddot{\text{Cl}}$:

FIGURE 1.17 A few Lewis dot structures.

WEB 3D BH_3 H:$\dot{\text{B}}$:H   or   H—$\overset{\text{H}}{\text{B}}$—H CH_4 H:$\overset{\text{H}}{\underset{\text{H}}{\text{C}}}$:H   or   H—$\overset{\text{H}}{\underset{\text{H}}{\text{C}}}$—H

shown in Figure 1.17 Another variation of the Lewis structure uses a line to represent the nonbonding electrons. We will not use this representation because the line can be easily confused with a negative charge.

Let's draw Lewis structures for methane and ammonia (Fig. 1.18). To construct the Lewis dot structure of methane (CH_4), we first determine that carbon ($_6$C) has four valence electrons for bonding. Carbon has a total of six electrons, but two are in the unused $1s$ orbital. Note that carbon is in column 4 of the periodic table. Each hydrogen contributes one electron. Thus, four two-electron covalent bonds are formed, with no electrons left over (Fig. 1.18a). For ammonia (:NH_3), nitrogen ($_7$N) contributes five electrons (seven minus the $1s$ pair; nitrogen is in column 5), and the three hydrogens contribute one electron each. Therefore, three two-electron nitrogen–hydrogen bonds are formed, each containing one electron from N and one from H, and there is a nonbonding pair left over (Fig. 1.18b).

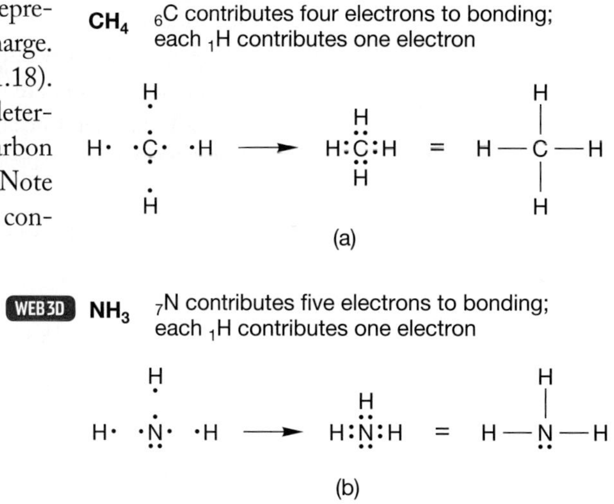

CH_4 $_6$C contributes four electrons to bonding; each $_1$H contributes one electron

H· ·$\dot{\text{C}}$· ·H   ⟶   H:$\overset{\text{H}}{\underset{\text{H}}{\text{C}}}$:H   =   H—$\overset{\text{H}}{\underset{\text{H}}{\text{C}}}$—H

(a)

WEB 3D **NH_3** $_7$N contributes five electrons to bonding; each $_1$H contributes one electron

H· ·$\dot{\text{N}}$· ·H   ⟶   H:$\ddot{\text{N}}$:H   =   H—$\overset{\text{H}}{\ddot{\text{N}}}$—H

(b)

FIGURE 1.18 Construction of a Lewis structure for (a) methane, CH_4, and (b) ammonia, NH_3.

WORKED PROBLEM 1.6 Construct Lewis structures for the following neutral molecules:

*(a) BF_3 (b) H_2Be (c) BH_3 (d) $ClCH_3$ (e) $HOCH_3$ (f) H_2N—NH_2

ANSWER (a) The first task is to determine the gross structure of BF_3. Is it B—F—F—F, or some other structure containing fluorine–fluorine bonds, or a structure containing only boron–fluorine bonds? We start by working out the number of electrons available for bonding. Fluorine ($_9$F) has seven valence electrons available for bonding ($9 - 2$ $1s$, or fluorine is in column 7) and thus has a single unpaired, or "odd" electron. Boron ($_5$B) has three valence electrons ($5 - 2$ $1s$, or boron is in column 3).

$_9$F $_5$B
$1s^2\,2s^2\,2p_x^2\,2p_y^2\,2p_z$ $1s^2\,2s^2\,2p$

There are seven electrons available for bonding   There are three electrons available for bonding

:$\dot{\ddot{\text{F}}}$·   ·$\dot{\text{B}}$·

Notice that boron has three electrons available for bonding and a structure with three boron–fluorine bonds can be nicely accommodated. There is no easy way to form a molecule containing both boron–fluorine and fluorine–fluorine bonds.

(continued)

A structure with covalent bonds in which the seventh electron on each fluorine is shared with one of boron's available three electrons gives the correct answer:

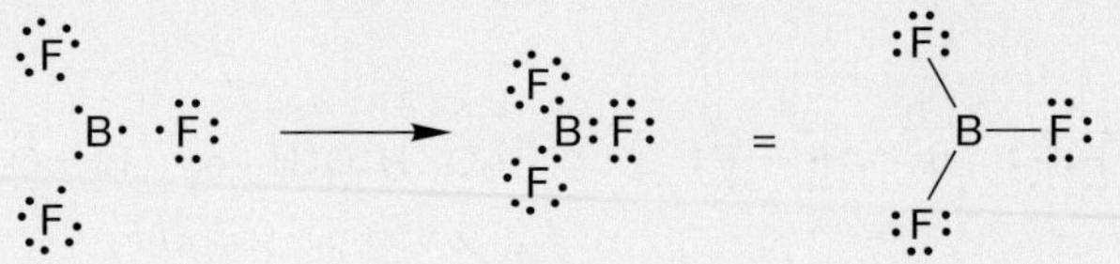

Single bonds such as the ones shown in Figures 1.17 and 1.18 are not the only kind of covalent bonds. Atoms of elements below the first row of the periodic table can form double and triple bonds, as we saw earlier when we looked at the covalent bonds in carbon dioxide. Usually, the structural formula will give you a clue when multiple bonding is necessary. For example, ethane (H_3CCH_3) requires no multiple bonds; indeed it permits none, as can be seen from construction of the Lewis structure (Fig 1.19). Each carbon is attached to four other atoms—all four of the valence electrons in each carbon are shared in single bonds to the three hydrogen atoms and the other carbon. Therefore all electrons can be accounted for in single bonds between the atoms.

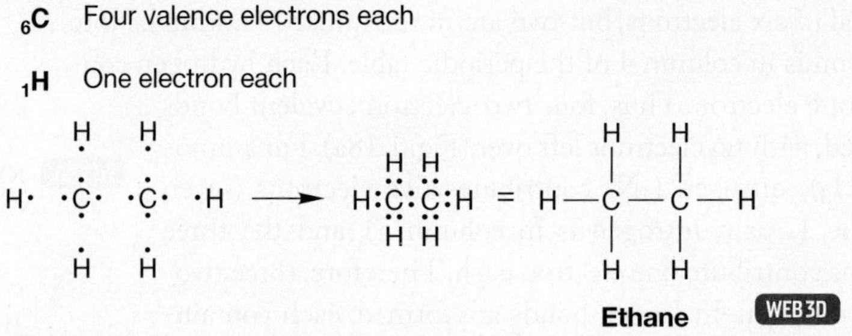

FIGURE 1.19 Construction of a Lewis structure for ethane.

In the molecule known as ethene or ethylene (H_2CCH_2), only three of carbon's four bonding electrons are used up in forming single covalent bonds to two hydrogens and the other carbon (Fig. 1.20). There is an unbonded electron remaining on each carbon, and these two electrons are shared in a second carbon–carbon bond. Thus, ethylene contains a carbon–carbon double bond.

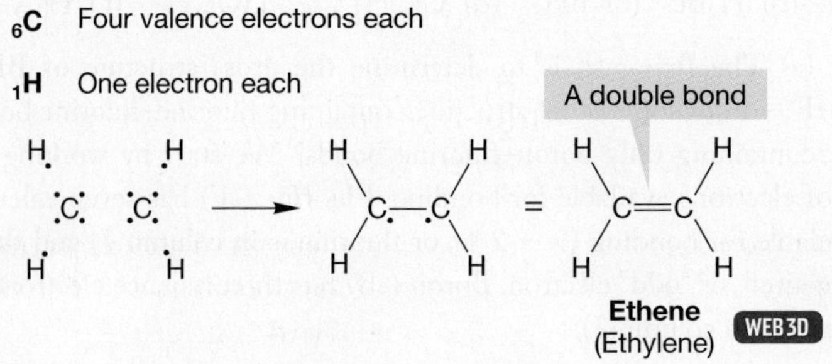

FIGURE 1.20 Construction of a Lewis structure for ethene (ethylene).

In ethyne or acetylene (HCCH), shown in Figure 1.21, two of the four bonding electrons in each carbon are used in forming single bonds to one hydrogen and the other carbon. Two electrons are left over on each carbon, and these are shared to create a triple bond between the carbons.

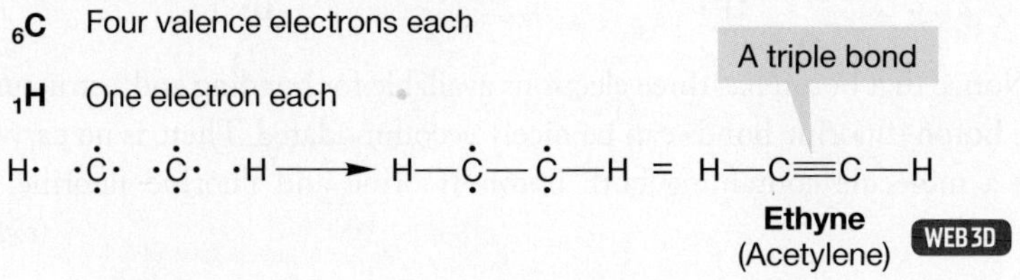

FIGURE 1.21 Construction of a Lewis structure for ethyne (acetylene).

WORKED PROBLEM 1.7 Draw Lewis structures for the following neutral species. Use lines to indicate electrons in bonds, and dots to indicate nonbonding electrons.

*(a) CH_3 *(b) CH_2 (c) Br   (d) OH   (e) NH_2 (f) H_3C—N

ANSWER The first step is to determine the number of bonding electrons in each atom. Then make the possible bonds and see how many electrons are left over. Because each compound is neutral, the number of bonding electrons is equal to the atomic number less the $1s$ electrons. For neutral hydrogen, there is always only the single $1s$ electron.

(a) Carbon has four bonding electrons and forms covalent bonds with three hydrogens, each of which contributes a single electron. There is one electron left over on carbon, which is shown as a dot in the Lewis structure:

H
:
·C:H   =   ·C—H
:
H

(b) In CH_2, carbon can form only two covalent bonds because only two hydrogen atoms are available. Two nonbonding electrons are left over:

H.
:C: =   C:
H

WORKED PROBLEM 1.8 In each of the following compounds, there is at least one multiple bond. Draw a Lewis structure for each molecule. Use lines to indicate electrons in bonds, and dots to indicate nonbonding electrons.

(a) F_2CCF_2 *(b) H_3CCN (c) H_2CO
(d) H_2CCO (e) $H_2CCHCHCH_2$ (f) H_3CNO

$$\overset{\displaystyle O}{\underset{\displaystyle \parallel}{}}$$

(g) H_3COCOH

ANSWER (b) If you did Problems 1.6 and 1.7, the methyl (CH_3) group should be familiar by now. Its three covalent bonds are constructed by sharing three of carbon's four valence electrons with the single electrons of the three hydrogens:

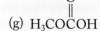

H
:
H:C·   =   H—C·
:
H

The second carbon and the nitrogen bring four and five electrons, respectively. Carbon–carbon and carbon–nitrogen covalent bonds can be formed, which leaves two unshared electrons on carbon and four on nitrogen:

H   H
| |
H—C· ·C· ·N: =   H—C—C—N:
| |
H   H

Formation of a triple bond between carbon and nitrogen completes the picture, and nitrogen is left with a nonbonding pair of electrons:

H   H
| |
H—C—C—N: =   H—C—C≡N:
| |
H   H

As we examine the nature of the bonding in the various molecules you will study in organic chemistry, it is vital that you be able to draw Lewis structures quickly and easily. It is also extremely important to be able to determine the charge on atoms, sometimes referred to as a **formal charge**, in a given structure. For single atoms in the second row of the periodic table, this process is easy as long as you remember to count the two $1s$ electrons, which are never shown. Figure 1.22 gives Lewis structures and charge determinations for six species. Be sure you understand how the charge is determined in each case.

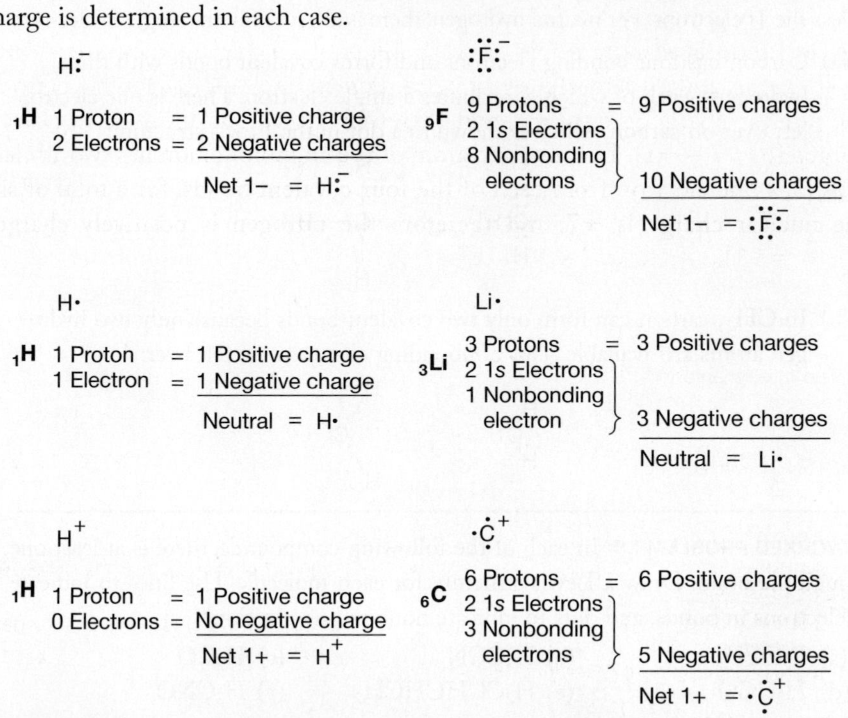

FIGURE 1.22 A few examples of charge determination.

H:⁻

$_1$H   1 Proton   = 1 Positive charge
        2 Electrons = 2 Negative charges
                      Net 1– = H:⁻

:F:⁻

$_9$F   9 Protons   = 9 Positive charges
        2 1s Electrons ⎫
        8 Nonbonding   ⎬
        electrons   ⎭ 10 Negative charges
                         Net 1– = :F:⁻

H·

$_1$H   1 Proton   = 1 Positive charge
        1 Electron   = 1 Negative charge
                      Neutral = H·

Li·

$_3$Li 3 Protons   = 3 Positive charges
        2 1s Electrons ⎫
        1 Nonbonding   ⎬
        electron   ⎭ 3 Negative charges
                         Neutral = Li·

H⁺

$_1$H   1 Proton   = 1 Positive charge
        0 Electrons = No negative charge
                      Net 1+ = H⁺

·C⁺

$_6$C   6 Protons   = 6 Positive charges
        2 1s Electrons ⎫
        3 Nonbonding   ⎬
        electrons   ⎭ 5 Negative charges
                         Net 1+ = ·C⁺

In molecules, the determination of charge is a bit harder because you must take care to account for shared electrons in the right way. There are two steps to this process. First, we need to count the electrons that the atom "owns." Remember that every second-row atom has a pair of unshown $1s$ electrons. Count all nonbonding valence electrons as owned by the atom and count one electron for each bond to the atom. The second step is to consider the atomic number (positive charge of the nucleus) for that particular atom and subtract the number of electrons that the atom owns (negative charges). The resulting value is the formal charge on the atom. In H_2, for example, each hydrogen has only a share in the pair of electrons binding the two nuclei, and therefore each hydrogen is neutral (Fig. 1.23).

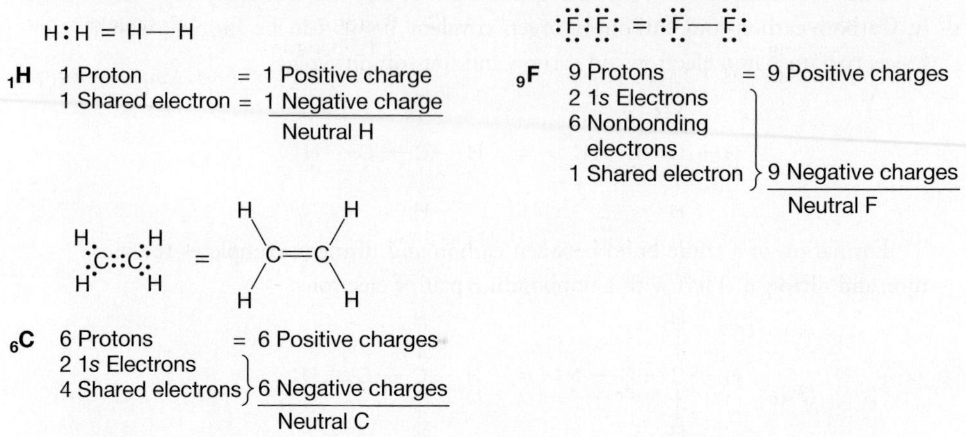

H:H = H—H

$_1$H   1 Proton   = 1 Positive charge
        1 Shared electron = 1 Negative charge
                           Neutral H

:F:F: = :F—F:

$_9$F   9 Protons   = 9 Positive charges
        2 1s Electrons   ⎫
        6 Nonbonding   ⎬
        electrons   ⎬
        1 Shared electron⎭ 9 Negative charges
                           Neutral F

H₂C::CH₂ = C=C structure

$_6$C   6 Protons   = 6 Positive charges
        2 1s Electrons   ⎫
        4 Shared electrons⎬ 6 Negative charges
                           Neutral C

FIGURE 1.23 Electron counting in some simple molecules.

In F_2 each fluorine has two $1s$ electrons, six nonbonding electrons, and a share in the single covalent bond between the two fluorine atoms. This count gives a total of nine electrons, exactly balancing the nine positive nuclear charges ($9 - 9 = 0$). In $H_2C=CH_2$, each carbon has a pair of $1s$ electrons and a share in four covalent bonds for a total of six. Because each carbon has a nuclear charge of $+6$, both are neutral.

Let's see why the carbon of the methyl anion ($^-:CH_3$) is negatively charged and why the nitrogen of the ammonium ion ($^+NH_4$) is positively charged (Fig. 1.24). In the methyl anion, carbon has a pair of $1s$ electrons, two nonbonding electrons, and a share in three covalent bonds, for a total of seven electrons. The nuclear charge is only $+6$, and therefore the carbon must be negatively charged ($6 - 7 = -1$). The nitrogen atom of the ammonium ion has two $1s$ electrons plus one electron from each of the four covalent bonds, for a total of six. The nuclear charge is $+7$, and therefore the nitrogen is positively charged ($7 - 6 = +1$).

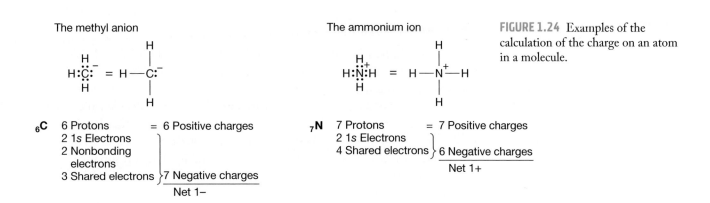

The methyl anion

<table>
<tr><td>$_6C$</td><td>6 Protons</td><td>= 6 Positive charges</td></tr>
<tr><td></td><td>2 1s Electrons
2 Nonbonding
electrons
3 Shared electrons</td><td>} 7 Negative charges</td></tr>
<tr><td></td><td colspan="2" style="text-align:center">Net 1−</td></tr>
</table>

The ammonium ion

<table>
<tr><td>$_7N$</td><td>7 Protons</td><td>= 7 Positive charges</td></tr>
<tr><td></td><td>2 1s Electrons
4 Shared electrons</td><td>} 6 Negative charges</td></tr>
<tr><td></td><td colspan="2" style="text-align:center">Net 1+</td></tr>
</table>

FIGURE 1.24 Examples of the calculation of the charge on an atom in a molecule.

Figure 1.25 shows two more ions and works through the calculations of charge.

<table>
<tr><td>$_5B$</td><td>5 Protons</td><td>= 5 Positive charges</td></tr>
<tr><td></td><td>2 1s Electrons
4 Shared electrons</td><td>} 6 Negative charges</td></tr>
<tr><td></td><td colspan="2" style="text-align:center">Net 1−</td></tr>
</table>

<table>
<tr><td>$_6C$</td><td>6 Protons</td><td>= 6 Positive charges</td></tr>
<tr><td></td><td>2 1s Electrons
3 Shared electrons</td><td>} 5 Negative charges</td></tr>
<tr><td></td><td colspan="2" style="text-align:center">Net 1+</td></tr>
</table>

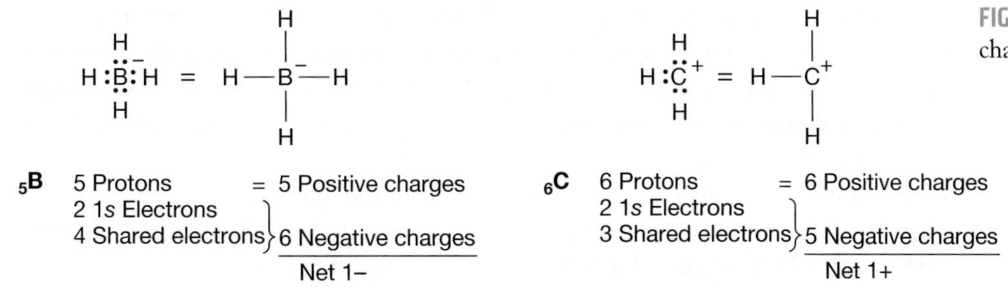

FIGURE 1.25 More calculations of charge in molecules.

PROBLEM 1.9 Draw Lewis structures for the following charged species. In each case, the charge is shown closest to the charged atom.

(a) ^-OH (b) $^-BH_4$ (c) $^+NH_4$ (d) ^-Cl (e) $^+CH_3$ (f) $^+OH_3$

(g) $^+NO_2$ (*Hint*: Nitrogen is the central atom.)

PROBLEM 1.10 Add charges to the following compounds wherever necessary:

(a) :CH$_2$

(b) ·CH$_3$

(c) :ĊH

(d) :ÖH

(e) :OH$_3$

(f) H$_2$C═Ö:

(g) :ṄH$_2$

(h) HṄ═C═ṄH

PROBLEM 1.11 Add electrons to complete the following Lewis structures. In each case, the charge is placed as close as possible to the charged atom.

(a) $^+$CH$_2$

(b) $^-$CH$_2$CH$_3$

(c) HC̄═CH$_2$

(d) $^+$OH$_3$

(e) $^-$OH

(f) $^+$NH$_2$

(g) $^-$NH$_2$

(h) CH$_3$—C≡N⁺—H

1.4 Resonance Forms

Often, there will be more than one possible Lewis structure for a molecule. This statement is especially true for charged molecules, but it applies to many neutral species as well. (You may already have encountered this phenomenon in comparing your answers to Problems 1.9 and 1.10 with ours.) How do we decide which Lewis structure is the correct one? The answer almost always is that neither Lewis structure is complete all by itself. Instead, the molecule is best described as a combination of all *reasonable* Lewis structures, that is, as a combination of **resonance forms**, which are different *electronic* representations of the same molecule. Whether or not a Lewis structure is reasonable, whether or not it is an important contributor, will be addressed below. The word *electronic* is emphasized to remind you that the only differences allowed in a set of resonance forms are differences in electron distribution. There can be no change in the position of the *atoms* in resonance forms.

Let's use nitromethane, H$_3$CNO$_2$, as an example. The nitro group, NO$_2$, is a **functional group**, a collection of atoms that behaves more or less the same way wherever it appears in a molecule. As we work through the various structural types of molecules in organic chemistry, you will become familiar with many functional groups, but right now, at this early stage, almost everyone needs to look up a structure now and then. The problem here is that the condensed formula NO$_2$ doesn't contain structural information about how the atoms are connected to one another. For that matter, neither does CH$_3$. Over time, you will become familiar with many of the functional groups and will be able to write the appropriate structures without even thinking about it. For now, you can consult the collection of functional groups and their structures on the inside front cover of this book.

Even though nitromethane is a neutral molecule, there is no good way to draw it without separated charges. Figure 1.26 shows a Lewis structure for nitromethane that has the nitrogen positive and one of the oxygens negative. But this is not the only Lewis structure possible! We can draw the molecule so the other oxygen bears the negative charge. The two renderings of nitromethane in Figure 1.27 are resonance forms of the molecule. Note that the arrangement of the atoms is identical in the two forms. Resonance forms are different electronic representations of the same molecule, not pictures of different molecules. The real molecule is the combination of all its resonance forms and is often called a resonance hybrid.

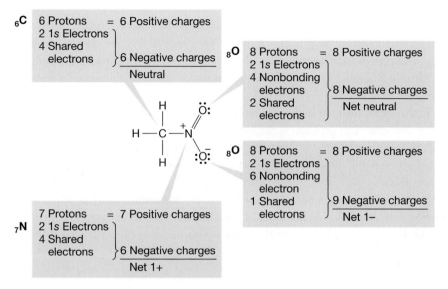

FIGURE 1.26 One electronic representation of nitromethane.

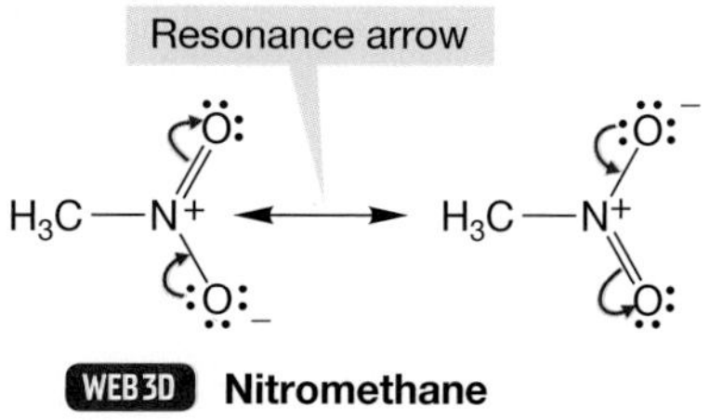

Resonance arrow

WEB 3D Nitromethane

FIGURE 1.27 Two equivalent resonance forms for nitromethane.

PROBLEM 1.12 Draw a Lewis structure for nitric acid ($HO—NO_2$), and verify that the nitrogen is positive and one of the oxygens is negative (see Fig. 1.26).

Neither Lewis structure in Figure 1.27 is completely accurate because in the real nitromethane molecule, the two oxygens share the negative charge equally. Thus, the two resonance forms in Figure 1.27 are *equivalent* electronic representations for nitromethane. Notice the special, double-headed arrow between the two forms. This symbol is a **resonance arrow**, and in the notation of chemistry is used exclusively to indicate resonance forms. Also notice the red, curved arrows in Figure 1.27. These arrows are a notation used to show how we move a pair of electrons from one point to another. In this case, such movement converts one resonance form into the other. Using curved arrows to show how electrons move from one point to another is called the **curved arrow formalism**. Other terms you might encounter that are used to describe this type of analysis are **arrow formalism** or **electron pushing**.

PROBLEM 1.13 Use the arrows to convert your Lewis structure for nitric acid ($HO—NO_2$, Problem 1.12) into a resonance form.

WORKED PROBLEM 1.14 Draw another structure for nitromethane in which every atom is neutral. *Hint*: There are only single bonds in this structure.

ANSWER You can arrive at the answer by using arrows to push electron pairs in the following way:

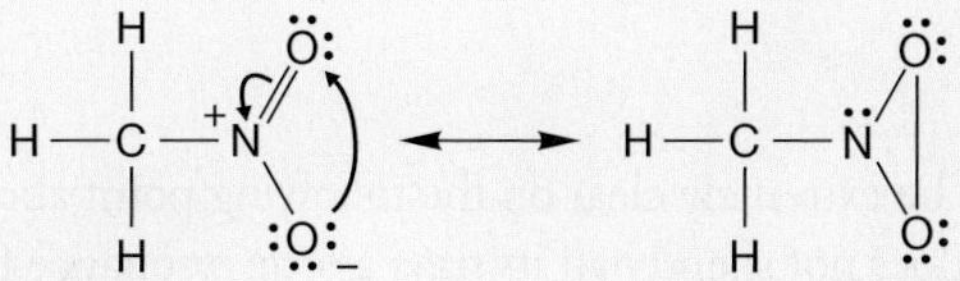

(continued)

Note that, as a new bond is formed between the two oxygen atoms, the two charges originally on oxygen and nitrogen are canceled. Notice also the long oxygen–oxygen bond. *Remember*: In drawing resonance forms, you move only electrons, not atoms. The oxygen–oxygen distance must be the same in each resonance form.

Is this cyclic form an important resonance form? Is it a good representation of the molecule? The problem is the long bond between the oxygens. This bond must be weak just because it is so long. In the language of organic chemistry, we would say that the cyclic resonance form contributes little to the structure of nitromethane.

Is the cyclic structure below a resonance form? Resonance involves different electronic structures that do not differ in the positions of atoms. If atoms have been moved, as in the figure below in which there is a normal oxygen–oxygen bond, the two structures are in equilibrium and are not resonance forms.

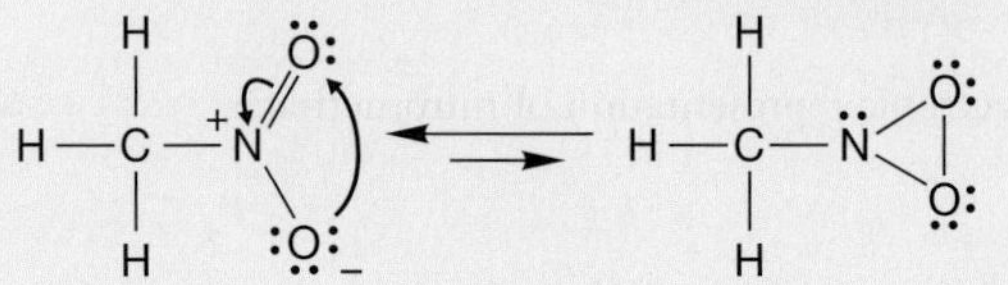

CONVENTION ALERT

Two things must be stressed here. First, although the curved arrows in Figure 1.27 have no physical significance, they do constitute an extraordinarily important bookkeeping device. They map out the motions of electrons. But be careful—these arrows, useful as they are, do not represent more than a bookkeeping process. We will use curved arrows throughout this book, and all organic chemists use them to keep track of electron movement, not only in drawing resonance forms but in writing chemical reactions as well. Being able to draw resonance forms quickly and accurately is an essential skill for anyone wishing to master organic chemistry.

The second point to be stressed is that the bookkeeping represented by curved arrows in resonance forms is accomplished by moving or "pushing" *pairs* of electrons. Be careful when doing this sort of thing not to violate the rules of valence, not to make more bonds than are possible, for this mistake is easy to make. Figure 1.28 gives some examples of this kind of electron-pair pushing in molecules best represented as combinations of resonance forms.

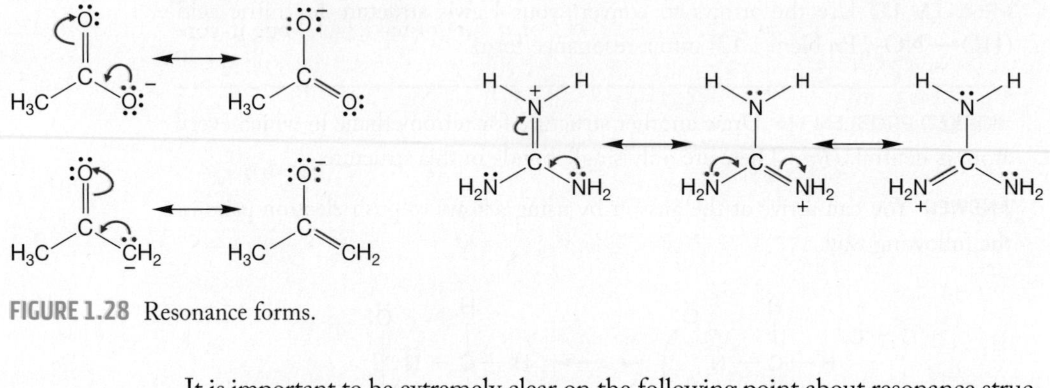

FIGURE 1.28 Resonance forms.

It is important to be extremely clear on the following point about resonance structures. Nitromethane does not spend half its time as one resonance form and half as the other. *There is no equilibration between the resonance forms.* Nitromethane is best described as the combination of the two electronic structures shown in Figure 1.27. In nitromethane, the two resonance forms are equivalent because there is no difference between having the negative charge on one oxygen or the other, and thus nitromethane can be reasonably described as a 50:50 combination (average) of the two forms.

Here's an analogy that might help: Frankenstein's monster was always a monster. One might describe that poor constructed creature as part monster and part human, but he was always that combination—he did not oscillate between the two. In chemical terms, one would say that Frankenstein was a resonance hybrid of monster and human. On the other hand, Dr. Jekyll and Mr. Hyde were in equilibrium. When the good Dr. Jekyll drank the potion, he became the monstrous Hyde. Later, when the potion wore off, he reverted to Jekyll. Part of the time he was Jekyll, and part of the time he was Hyde. The two were in equilibrium, not resonance.

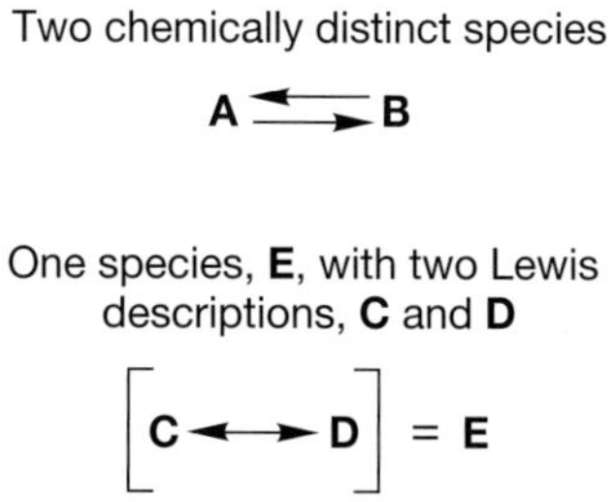

The special double-headed arrow ($\longleftrightarrow$) used in Figures 1.27 and 1.28 is reserved for resonance phenomena and is never used for anything but resonance. A pair of arrows ($\rightleftarrows$) indicates equilibrium, the interconversion of two chemically distinct species, and is never used for resonance (Fig. 1.29). This point is most important in learning the language of organic chemistry. It is difficult because it is arbitrary. There is no way to reason out the use of the different kinds of arrows; they simply must be learned.

The carbon–oxygen double bond in formaldehyde ($H_2C{=}O$, Fig. 1.30), gives us another opportunity to write resonance forms. In the resonance form on the left in Figure 1.30, carbon has a pair of $1s$ electrons and shares in four covalent bonds; therefore it is neutral ($6 - 6 = 0$). Oxygen has a pair of $1s$ electrons, four nonbonding electrons, and a share in two covalent bonds, for a total of eight electrons ($8 - 8 = 0$). Oxygen is also neutral. However, we can push electrons to generate the resonance form shown on the right in Figure 1.30, in which the carbon is positive and the oxygen negative. The real formaldehyde molecule is a combination, *not a mixture*, of these two resonance forms, a resonance hybrid. Because charge separation is energetically unfavorable, these two resonance forms do not contribute equally to the structure of the formaldehyde molecule. Still, neither by itself is a perfect representation of the molecule, and in order to represent formaldehyde well, both electronic descriptions must be considered. Formaldehyde is the weighted average of the two resonance forms in Figure 1.30. How we determine a weighted average is what we look at next.

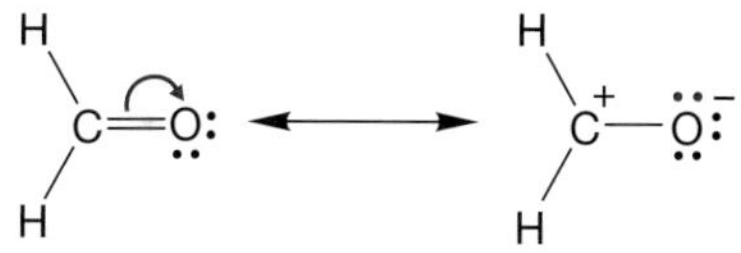

Two chemically distinct species

$$A \rightleftarrows B$$

One species, **E**, with two Lewis descriptions, **C** and **D**

$$\left[C \longleftrightarrow D \right] = E$$

FIGURE 1.29 The difference between equilibrium (two different species, **A** and **B**; two arrows) and resonance (different electronic representations, **C** and **D**, for the same molecule, **E**; double-headed arrow).

CONVENTION ALERT

WEB 3D   Resonance forms for formaldehyde

FIGURE 1.30   Formaldehyde.

PROBLEM 1.15 There is a third resonance form for formaldehyde, but it contributes very little to the structure and is usually ignored. Can you find it and explain why it is relatively unimportant?

PROBLEM 1.16 Use the arrow formalism to convert each of the following Lewis structures into another resonance form. Notice that part (e) of this question asks you to do something new—to move electrons one at a time in writing Lewis forms.

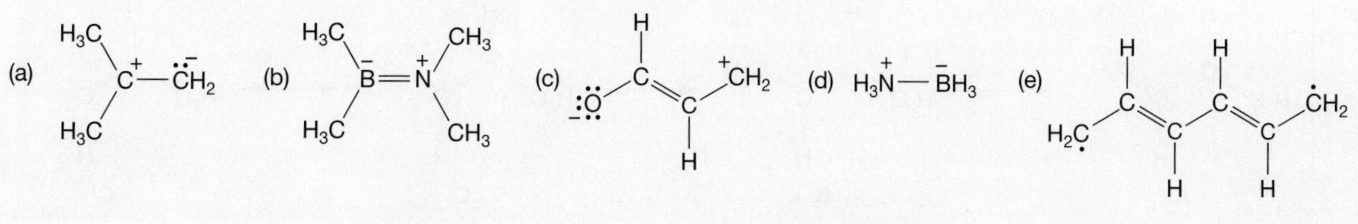

WORKED PROBLEM 1.17 Use the arrow formalism to write resonance forms that contribute to the structures of the following molecules:

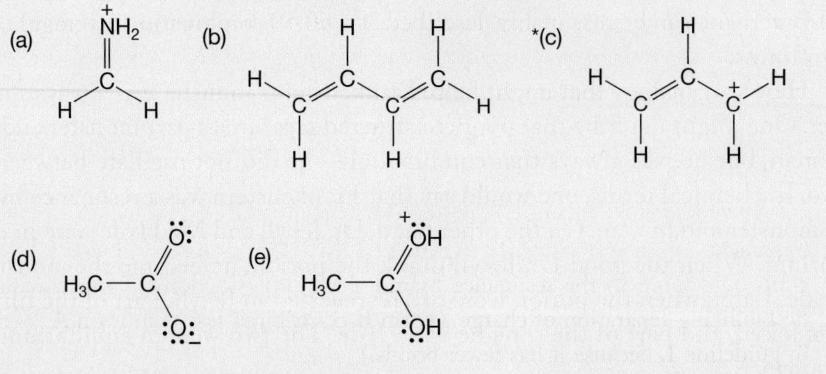

ANSWER (c) As we have seen, it is not only electron pairs (nonbonding electrons) that can be redistributed (pushed) in writing resonance forms, but bonding electrons as well. In this case, a pair of electrons in a carbon–carbon double bond is moved.

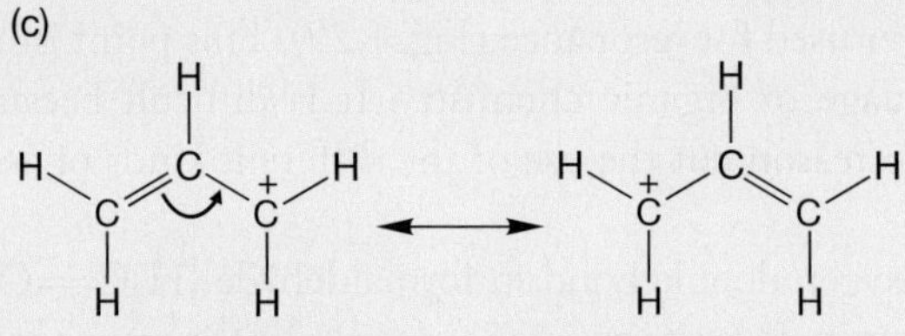

The positive charge on the right-hand carbon disappears, but reappears on the left-hand carbon. In the real structure, the two carbons share the positive charge equally, each bearing one-half the charge. The structure is sometimes written with dashed bonds to show this sharing:

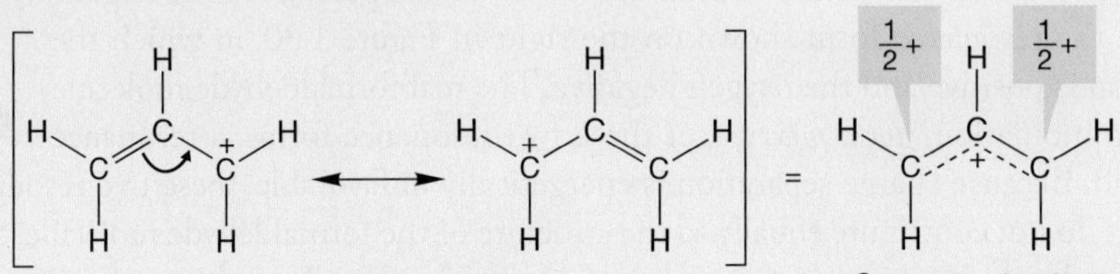

Summary structure

It is important to be able to estimate the relative importance of resonance forms in order to get an idea of the best way to represent a molecule. To do so, we can assign a **weighting factor**, c, to each resonance form. The weighting factor is a number that indicates the percent contribution of each resonance form to the resonance hybrid. Some guidelines for assigning weighting factors are listed below. Fortunately, this is an area in which common sense does rather well.

1. The more bonds in a resonance form, the more important a contributor the form is. For example, 1,3-butadiene can be written as a resonance hybrid of four structures (Fig. 1.31).

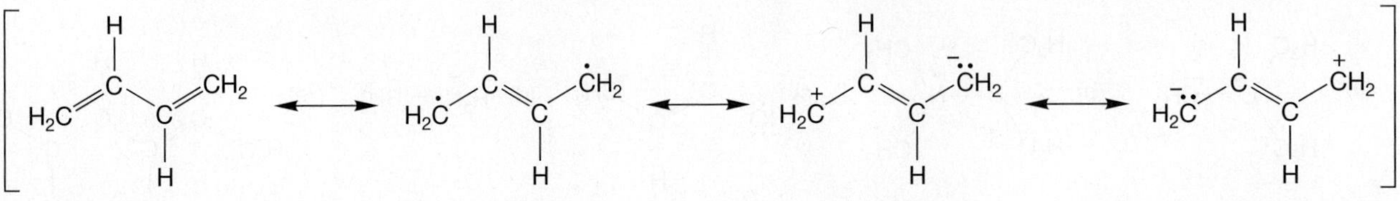

FIGURE 1.31 Four resonance forms contributing to 1,3-butadiene. Form **A** is by far the best.

It is reasonably well described by form **A**, which has a total of 11 bonds. Forms **B**, **C**, and **C′** each contain 10 bonds, and will contribute far less than **A** to the actual 1,3-butadiene structure. Forms **C** and **C′** are equivalent (this is why they are labeled **C** and **C′** rather than **C** and **D**) and contribute equally to the hybrid. Even though **A** is clearly the major contributor, this does not mean that **B**, **C**, and **C′** do not contribute at all! It simply means that in our equation

$$\text{Butadiene} = c_1(\mathbf{A}) + c_2(\mathbf{B}) + c_3(\mathbf{C}) + c_3(\mathbf{C'})$$

the weighting factor c_1 is much larger than the other coefficients. Therefore, 1,3-butadiene looks much more like **A** than **B**, **C**, or **C′**.

2. Separation of charge is bad. In Figure 1.31, the neutral resonance forms **A** and **B** contribute more to the resonance hybrid than **C** or **C′**, both of which require a destabilizing separation of charge. (Form **B** contributes less than form **A**, as noted in guideline 1, because it has fewer bonds.)

3. In ions, **delocalization** of electrons (distributing them over as many atoms as possible) is especially important. Delocalization of electrons is a process that allows more than one atom to share electrons and it is almost always stabilizing. Electrons are like water, if allowed to spread out, they will. The allyl cation is a good example—the two end carbons each bear one-half of the positive charge (Fig. 1.32).

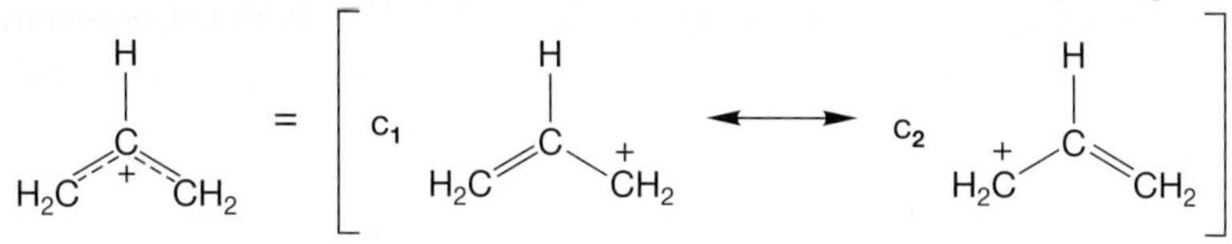

FIGURE 1.32 The allyl cation is an equal combination of **A** and **A′**.

The electrons of the double bond have spread into the empty p orbital. The ion shown in Figure 1.33 is another good example. Here the charge is shared by the carbon and the chlorine, as the two resonance forms show.

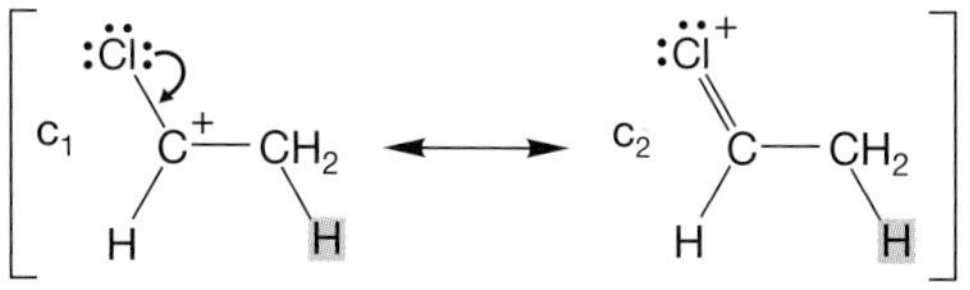

FIGURE 1.33 In this carbocation, the charge is shared by carbon and chlorine.

4. Electronegativity is important. In the allyl anion (Fig. 1.34a), for example, there are two equivalent resonance forms. As in the allyl cation, the charge resides

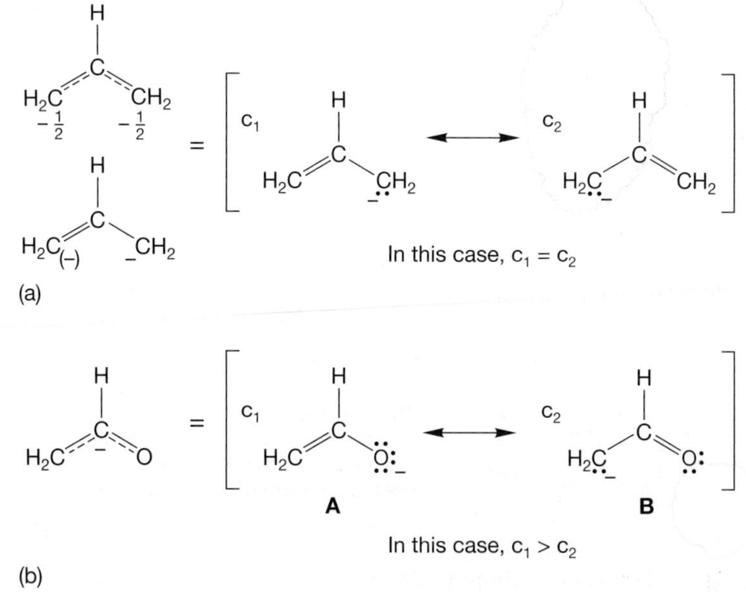

(a)

In this case, $c_1 = c_2$

(b)

In this case, $c_1 > c_2$

FIGURE 1.34 (a) The allyl anion. (b) The enolate anion.

equally on the two end carbons. Figure 1.34a shows two summary structures for the allyl anion. In one, each of the two end carbons is shown with half a negative charge; in the other, one resonance form is drawn, and the positions sharing the charge are indicated by drawing the charge in parentheses ($-$). The related *enolate anion*, in which one CH_2 of allyl is replaced with an oxygen (Fig. 1.34b), also has two resonance forms, but they do not contribute equally because they are not equivalent. Each has the same number of bonds, and so we cannot choose the better representation that way. However, form **A** has the charge on the relatively electronegative oxygen while form **B** has it on carbon, which is considerably less electronegative than oxygen. Form **A** is the better representation for the enolate anion, although both forms contribute. Mathematically, we would say that the weighting factor for **A**, c_1, is larger than that for **B**, c_2.

5. Resonance forms that are equivalent contribute equally to a resonance hybrid. The two forms of nitromethane in Figure 1.27 and the two forms of the allyl cation (**A** and **A'**) in Figure 1.32 are good examples. In both cases the resonance forms are completely equal to each other (indistinguishable, same number of bonds, same number of charges, charges on same atoms).

   In cases such as these, the weighting factors for the two forms, c_1 and c_2, must be equal. An equation to represent the mathematics involved in our allyl cation example is

$$\text{Allyl} = c_1(\text{form } \mathbf{A}) + c_2(\text{form } \mathbf{A}')$$

where c_1 is the weighting factor for resonance form **A**, c_2 is the weighting factor for form **A'**, and $c_1 = c_2$.

6. All resonance forms for a given species must have the same number of paired and unpaired electrons. As we saw in Figure 1.5, two electrons occupying the same orbital must have opposite spin quantum numbers. Most organic molecules, including our example molecule 1,3-butadiene, have all paired electrons. Therefore, resonance forms for 1,3-butadiene must also have all paired electrons. Consider form **B** in Figure 1.31. It is often the convention to emphasize that all electrons are paired in this molecule by appending electron-spin arrows alongside each single electron in this form (Fig. 1.35). Opposed arrows ($\downarrow$ and $\uparrow$) indicate paired spins, arrows in the same direction ($\uparrow$ and $\uparrow$ or $\downarrow$ and $\downarrow$) indicate parallel (same-direction) spins.

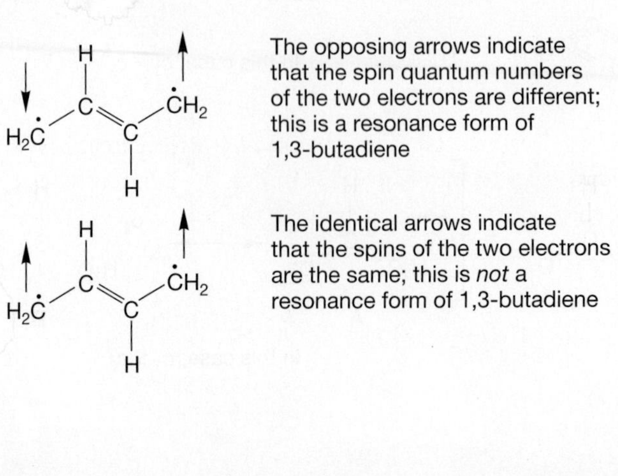

The opposing arrows indicate that the spin quantum numbers of the two electrons are different; this is a resonance form of 1,3-butadiene

The identical arrows indicate that the spins of the two electrons are the same; this is *not* a resonance form of 1,3-butadiene

FIGURE 1.35 In 1,3-butadiene, all electrons are paired, and all resonance forms contributing to the structure must also have all electrons paired.

WORKED PROBLEM 1.18 Add dots for the electron pairs and write resonance forms for the following structures:

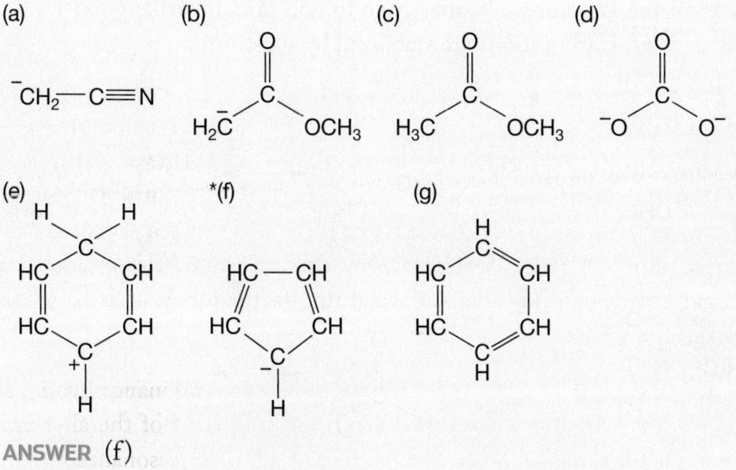

(a)   (b)   (c)   (d)

(e)   *(f)   (g)

ANSWER (f)

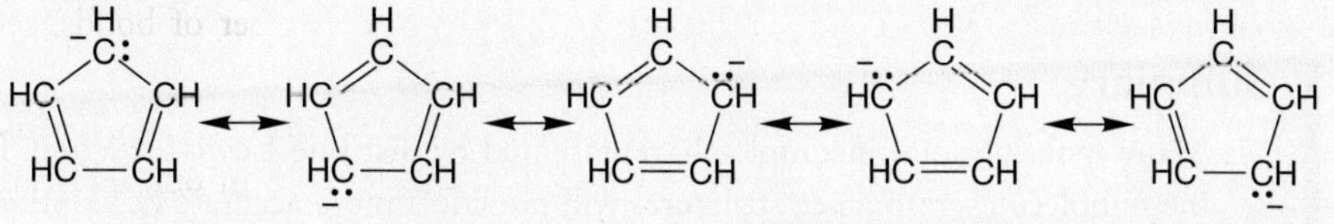

PROBLEM 1.19 Write Lewis structures and resonance forms for the following compounds. If you have problems visualizing the structure of some of these molecules, see the inside front cover of this book.

(a) $HONO_2$ (b) $^-OSO_2OH$ (c) CH_3COO^-

WORKED PROBLEM 1.20 Which of the following pairs of structures are not resonance forms of each other? Why not? You may have to add dots to make good Lewis structures first.

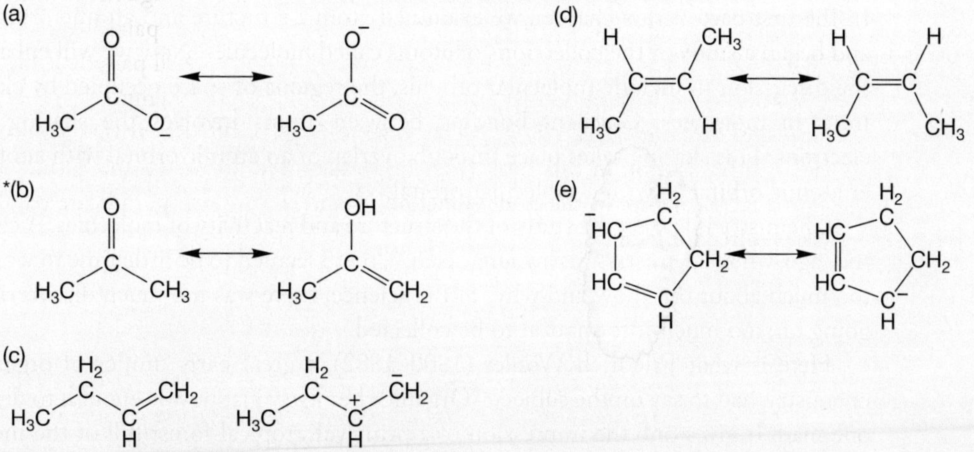

(a)   (d)

*(b)   (e)

(c)

ANSWER (b) These two are not resonance forms because atoms have been moved, not just electrons. These two molecules are in equilibrium. Molecules related through the change of position of a single hydrogen are called **tautomers**.

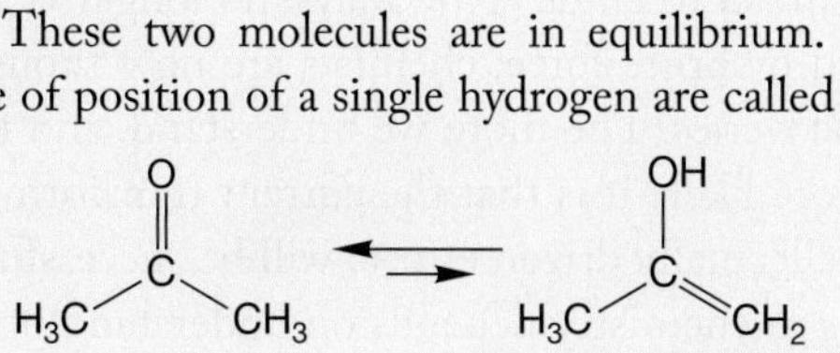

PROBLEM 1.21 In the following pairs of resonance forms, assign weighting factors. In each case, indicate which form you think is more important, and therefore contributes more to the structure. You may have to add dots to make good Lewis structures first. *Hint*: Carbocations are stabilized by substitution.

(a)

(b)

(c)

(d)

Summary

1. Many molecules are incompletely represented by just one Lewis structure. For these molecules, resonance structures will provide a more accurate description.

2. Resonance forms are different electronic representations of the same molecule.

3. Resonance forms are not species in equilibrium—molecules do not oscillate back and forth between resonance forms. The more accurate structure of the molecule—called the resonance hybrid—is the weighted average of all forms.

4. Some resonance forms contribute more than others to the resonance hybrid.

WEB3D 1.5 Hydrogen (H_2): Molecular Orbitals

In the first part of this chapter, we examined atomic structure and atomic orbitals and began a study of the collections of atoms called molecules. Now we will enlarge the discussion to include molecular orbitals, the regions of space occupied by electrons in molecules. Covalent bonding between atoms involves the sharing of electrons. This sharing takes place through overlap of an atomic orbital with another atomic orbital or with a molecular orbital.

Chemistry is largely the study of the structure and reactivity of molecules. A century ago, when chemistry was a young science, there seemed to be little time to worry too much about the "how and why" of the science; there was too much discovering going on, too much information to be collected.

Here is what Friedrich Wöhler (1800–1882), a great early unifier of organic chemistry, had to say on the subject: "Organic chemistry just now is enough to drive one mad. It gives one the impression of a primeval, tropical forest full of the most remarkable things, a monstrous and boundless thicket, with no way of escape, into which one may well dread to enter." Nowadays, no longer is some metaphoric vast jungle being explored by brute force; chemists are now thoughtfully aiming more and more at new discoveries. The more we understand, and the better our models of Nature are, the more likely it is that the current transformation from trial-and-error methods to intellectually driven efforts will be successful.

Directed progress in chemistry depends on understanding how molecules react and on the application of that understanding in creative ways. These days, the idea that both the structures of molecules and the reactions they undergo can be understood by examining the shapes and interactions of atomic and molecular orbitals has gained widespread acceptance in the chemical community, and an important branch of theoretical chemistry embraces these ideas. Although the study of molecular orbitals can be approached in a highly mathematical way, we will not follow that path. One of the great benefits of our nonmathematical approach is that it can be appreciated quite readily by nonmathematicians! We'll find that even a very "low level," highly qualitative molecular orbital theory can provide striking insights into structure and reactivity.

In Section 1.3, we constructed Lewis dot structures for molecules. Our next task is to elaborate on this theme to produce better pictures of the bonds that hold atoms together in molecules. We'll need to consider how electrons act to bind nuclei together, and we'll take as our initial example hydrogen (H_2), the second simplest molecule.

> **PROBLEM 1.22** What is the simplest molecule? If H_2 is the second simplest molecule, the answer to this question must be "H_2 minus something." What might the "something" be? The answer will appear farther along in the text, so think about this question for a while now.

We start with two hydrogen atoms, each consisting of a single proton, the nucleus, surrounded by an electron in the spherically symmetrical 1s orbital. In principle, the situation is this simple only as long as the two hydrogen atoms are infinitely far apart. As soon as they come closer than this infinite distance, they begin to "feel" each other. *Remember*: Wave functions do not vanish as the distance from the nucleus increases (Fig. 1.6). In practice, we can ignore the influence of one hydrogen atom on the other until they come quite close together, but as soon as they do, the energy of the system changes greatly. As one hydrogen atom approaches the other, the energy of the two-atom system decreases until the two hydrogen atoms are 0.74 angstrom (Å, where 1 Å = 10^{-8}cm) apart (Fig. 1.36). From this point the energy of the system rises sharply, asymptotically approaching infinity as the distance between the atoms approaches zero.

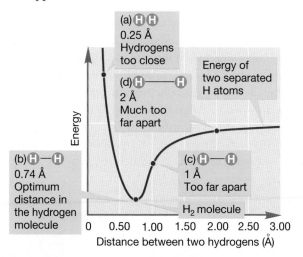

FIGURE 1.36 A plot of the energy for the hydrogen molecule (H_2) as a function of the distance between the two hydrogen nuclei. The point of minimum energy corresponds to the equilibrium internuclear separation, the bond distance.

In the hydrogen molecule, two 1s electrons serve to bind the two nuclei together. The system H—H is *more stable than two separated hydrogen atoms*. In the molecule, the two negatively charged electrons attract both nuclei and hold them together. When the atoms approach too closely, the positively charged nuclei begin to repel each other, and the energy goes up sharply.

We begin our mathematical description of bonding by combining the two $1s$ atomic orbitals (symbol ψ) of the two hydrogen atoms to produce two new molecular orbitals. The first molecular orbital we will consider is called the **bonding molecular orbital**. Its symbol is $\Phi_{bonding}$ (this Greek letter is pronounced "fy"). We will use Φ_B as a shorter notation for the bonding molecular orbital in this discussion. Wave function Φ_B mathematically represents the bonding molecular orbital and results from a simple addition of the two atomic orbitals. It can be written as $\psi(H_a, 1s) + \psi(H_b, 1s) = \Phi_{bonding} = \Phi_B$. This bonding molecular orbital is drawn in Figure 1.37, which shows that Φ_B looks much like what one would expect from a simple addition of two spherical $1s$ orbitals.

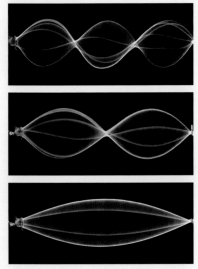

$\psi(H_a) = \psi(H_a, 1s)$ $\psi(H_b) = \psi(H_b, 1s)$ $\Phi_B (H_2) = \psi(H_a) + \psi(H_b)$

FIGURE 1.37 The bonding molecular orbital (Φ_B) of H_2.

CONVENTION ALERT

Many people use the Greek ψ for both atomic orbitals and molecular orbitals, but we will use Φ for molecular orbitals. Be careful in your other reading, though, because you will encounter situations in which ψ is used for all orbitals and even cases where Φ is used for atomic orbitals and ψ for molecular orbitals!

Because the bonding orbital Φ_B is concentrated between the two nuclei (Fig. 1.37), two electrons in it can interact strongly with both nuclei. This attraction explains why the molecular orbital Φ_B, in which the electron density is highest between the nuclei, is strongly bonding.

Quantum mechanics tells us an important principle: *The number of wave functions (orbitals) resulting from the mixing process must equal the number of wave functions (orbitals) going into the calculation.* Because we started by mixing two hydrogen atomic orbitals, there must be another molecular orbital that is formed. It is called an **antibonding molecular orbital**, $\Phi_{antibonding}$ or Φ_A (Fig. 1.38), and results from a subtraction of one atomic orbital from the other to give $\psi(H_a, 1s) - \psi(H_b, 1s) = \Phi_{antibonding} = \Phi_A$.

Electrons can occupy both bonding and antibonding molecular orbitals. An electron in a bonding orbital acts to hold the two nuclei together. As the name "antibonding" suggests, an electron in an antibonding orbital does just the opposite—it contributes to the dissociation of the two nuclei.

There is a third kind of molecular orbital; one in which an electron is neutral in its effect on the two nuclei. Such an orbital is called a **nonbonding orbital**. We will see some nonbonding orbitals when we come to molecules larger than H_2.

By comparing Figures 1.37 and 1.38 you can see that in the antibonding molecular orbital Φ_A there is a nodal plane between the two hydrogen nuclei. Recall that a node is a region in which the sign of the wave function is zero. As we pass from a region where the sign of the wave function is positive to a region in which it is negative, the sign of Φ must go through zero. As noted earlier, the square of the wave function corresponds to the electron density, and if Φ is equal to zero, then

Molecular orbitals increase in energy as the number of nodes increase. A good analogy is the energy required to increase the number of nodes in a jump rope. This photo shows a rope with zero nodes in the bottom picture. A rope with one node requires more energy to generate, and a rope with two nodes even more.

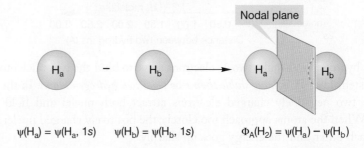

Nodal plane

$\psi(H_a) = \psi(H_a, 1s)$ $\psi(H_b) = \psi(H_b, 1s)$ $\Phi_A(H_2) = \psi(H_a) - \psi(H_b)$

FIGURE 1.38 The antibonding molecular orbital of H—H (Φ_A).

Φ^2 is also zero. Therefore the probability of finding an electron at a node must be zero as well. Accordingly, when a pair of electrons occupies the antibonding molecular orbital of Figure 1.38, the two nuclei are poorly shielded from each other, and electrostatic repulsion forces the two positively charged nuclei apart.

In summary, the combination of two hydrogen 1s atomic orbitals yields two new, molecular orbitals, Φ_B and Φ_A. This idea seems intuitively reasonable. There are, after all, only two ways in which two 1s atomic orbitals can be combined to construct two new molecular orbitals. The signs of the wave functions can be either the same, as in the bonding orbital Φ_B, or opposite, as in the antibonding orbital Φ_A. One common analogy for this phenomenon is the jump rope, which has wavelike properties. If we start with a jump rope that has an amplitude of 3 meters and we add an in-phase wave of the same amplitude, then the result will be a jump rope that has an amplitude of 6 meters. If, however, we start with the original amplitude and we subtract a wave (or add an out-of-phase wave) of the same amplitude, then we get a zero amplitude wave. Subtracting the waves cancels them out.

Figure 1.39 shows a very simple and most useful graphic device for summing up the energy comparison between atomic orbitals and the molecular orbitals formed from mixing the orbitals. The graph is called an **orbital interaction diagram**, or an interaction diagram for short. The atomic orbitals (ψ) going into the calculation are shown at the left and right sides of the figure. For molecular hydrogen, these orbitals are the two equivalent hydrogen 1s atomic orbitals, which must be of the same energy. They combine in a constructive, bonding way ($1s + 1s$) to give the lower-energy Φ_B and in a destructive, antibonding way ($1s - 1s$) to give the higher-energy Φ_A. Constructive combination for wave functions means that they are in-phase with each other. Destructive combination means that they are out-of-phase.

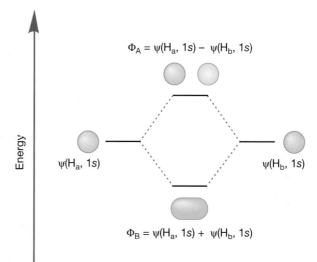

$\Phi_A = \psi(H_a, 1s) - \psi(H_b, 1s)$

$\psi(H_a, 1s)$ $\psi(H_b, 1s)$

$\Phi_B = \psi(H_a, 1s) + \psi(H_b, 1s)$

FIGURE 1.39 An orbital interaction diagram—a graphical representation of the combination of two atomic 1s orbitals to form a new bonding and antibonding pair of molecular orbitals, Φ_B and Φ_A.

PROBLEM 1.23 Sketch the molecular orbitals produced through the interaction of two carbon 2s atomic orbitals.

WORKED PROBLEM 1.24 Sketch the orbitals produced through the interaction of a carbon 2s atomic orbital overlapping end-on with a carbon 2p atomic orbital.

ANSWER The two atomic orbitals can interact in a bonding way ($2s + 2p$) or in an antibonding way ($2s - 2p$):

2s − 2p   Antibonding orbital—note
the new node (bar)

2s + 2p   Bonding orbital

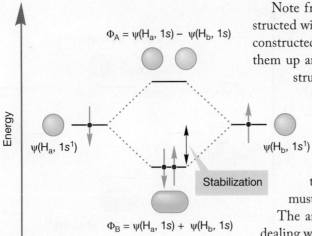

$\Phi_A = \psi(H_a, 1s) - \psi(H_b, 1s)$

$\psi(H_a, 1s^1)$

Energy

$\psi(H_b, 1s^1)$

Stabilization

$\Phi_B = \psi(H_a, 1s) + \psi(H_b, 1s)$

FIGURE 1.40 The electronic occupancy for the molecule H_2.

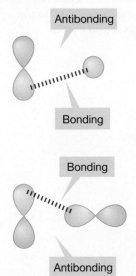

Antibonding

Bonding

Bonding

Antibonding

FIGURE 1.41 Two examples of noninteracting (orthogonal) orbitals.

Note from Figure 1.39 that the orbital interaction diagram is constructed without reference to electrons. Only after the diagram has been constructed do we have to worry about electrons. But now let's count them up and put them in the available molecular orbitals. In the construction of H_2, each hydrogen atom brings one electron. In Figure 1.40, these electrons are placed in the appropriate $1s$ orbitals $\psi(H_a)$ and $\psi(H_b)$. The spin direction of the electrons is shown as paired, but we could also show them as parallel. In the H_2 molecule, we put the two electrons into the lower-energy Φ_B. The electrons must be paired, because they are in the same orbital and their spin quantum numbers must be different. The Pauli principle made this point earlier (p. 8).

The antibonding molecular orbital (Φ_A) is empty because we are dealing with only two electrons and they are both accommodated in the bonding molecular orbital.[8]

So far we have dealt only with mixing one atomic orbital with another atomic orbital. But it is also possible to mix a molecular orbital with another molecular orbital or an atomic orbital with a molecular orbital. Not all combinations of orbitals are productive. If two orbitals approach each other in such a way that the new bonding interactions are exactly balanced by antibonding interactions, there is no net interaction between the two and no bond would form. Such orbitals are called **orthogonal orbitals**, orbitals that do not mix. Therefore, the way in which orbitals approach each other in space—how the lobes overlap—is critically important. Figure 1.41 shows two cases of atomic orbitals that are orthogonal to each other. As the orbitals approach each other, the number of bonding overlaps (shown with red dashed line) and antibonding overlaps are the same. The result is no net bonding. In each case, as the orbitals are brought together the bonding interactions (blue–blue starting to overlap) are exactly canceled by the antibonding interactions (blue–green starting to overlap).

Here are some "rules" for orbital construction:

1. The number of orbitals produced must equal the number of orbitals you begin with. If you start with n orbitals, you must produce n new orbitals. Here is a way to check your work as you proceed.

2. Keep the process as simple as you can. Use what you know already, and combine orbitals in as symmetrical a fashion as you can.

3. The closer in energy two orbitals are, the more strongly they interact. At this primitive (but useful) level of theory, you need mix only the pairs of orbitals closest in energy to each other.

4. When two orbitals interact in a bonding way (wave functions for the two orbitals have same sign), the energy of the resulting orbital is lowered; when they interact in an antibonding way (wave functions have different signs), the energy of the resulting orbital is raised.

5. When two orbitals interact, the only options for mixing are adding (in-phase mixing) or subtracting (out-of-phase mixing). When three orbitals interact we will have

---

[8] There can be no denying that the concept of an empty antibonding orbital is slippery! It makes physicists very uneasy, for example. Chemists see the empty orbital of Figures 1.39 and 1.40 as "the place the next electron would go."

three ways to mix the orbitals. With four orbitals there are four combinations possible, and so on.

6. To put new orbitals in order according to their energy, count the nodes. For a given molecule, the more nodes in an orbital, the higher it is in energy.

There is a connection between stability and energy. The lower the energy of an orbital, the greater the stability of an electron in it. A consequence of this stability is that the strongest bonds in molecules are formed by electrons occupying the lowest-energy molecular orbitals. Throughout this chapter we will be taking note of what factors lead to low-energy molecular orbitals and thus to strong bonding between atoms.

WORKED PROBLEM 1.25 Contrast the interactions between two $2p$ orbitals approaching in the two different ways shown below.

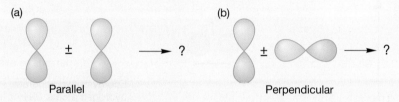

ANSWER (a) A pair of $2p$ atomic orbitals aligned parallel to each other interacting side by side produces two new molecular orbitals, one bonding ($2p + 2p$) and one antibonding ($2p - 2p$). (b) The orbitals aligned perpendicular to each other produce no molecular orbitals because there is no net bonding or antibonding. The two exactly cancel, producing no net interaction. In this case, the two orbitals are orthogonal.

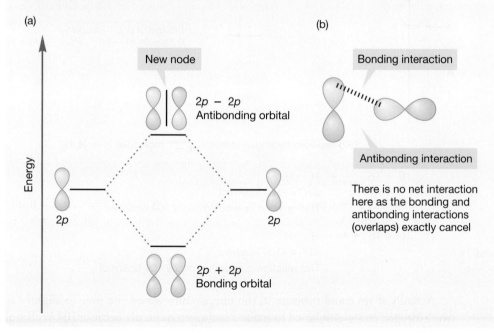

Summary

Overlap of two atomic orbitals produces two new molecular orbitals, one bonding and the other antibonding. Two electrons can be accommodated in the lower-energy bonding molecular orbital. In this way, two atoms (or groups of atoms) can be bound through the sharing of electrons in a covalent bond.

1.6 Bond Strength

In Figure 1.40, the energy of the electrons in the two atomic orbitals is greater than their energy in the bonding molecular orbital Φ_B. As noted earlier, lower energy means greater stability, but just how much energy is the stabilization apparent in Figure 1.40 worth? The difference between the energies of the two electrons in Φ_B and the energies of the two separate electrons in atomic $1s$ orbitals is 104 kcal/mol. This amount of energy is very large, and consequently the H_2 molecule is held together (bound) by a substantial amount of energy; this amount of energy is released into the environment if H_2 molecules are formed from separated hydrogen atoms. It would require the application of exactly this amount of energy to generate two hydrogen atoms from a molecule of H_2. The 104 kcal/mol represents the stabilization of two electrons in Φ_B (Fig. 1.42).

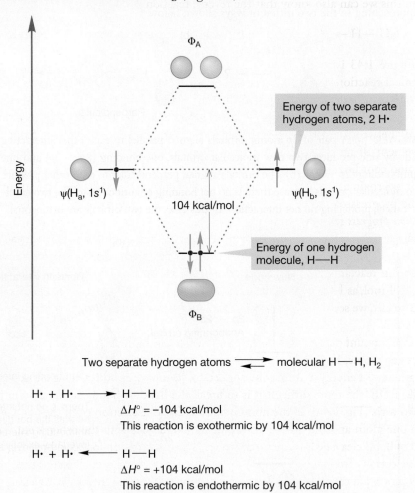

$$H\cdot + H\cdot \longrightarrow H\!-\!H$$
$$\Delta H° = -104 \text{ kcal/mol}$$
This reaction is exothermic by 104 kcal/mol

$$H\cdot + H\cdot \longleftarrow H\!-\!H$$
$$\Delta H° = +104 \text{ kcal/mol}$$
This reaction is endothermic by 104 kcal/mol

FIGURE 1.42 Molecular hydrogen (H_2) is more stable than two isolated hydrogen atoms by 104 kcal/mol.

Actually, if we could measure it, the temperature would rise ever so slightly as two hydrogen atoms combined to make a hydrogen molecule because 104 kcal/mol would be released as heat energy into the vessel. The environment in the vessel would warm up as the process proceeded. A reaction that liberates heat, one where the products are more stable than the starting materials, is called an **exothermic reaction**. The opposite situation, in which the products are less stable than the starting material, requires the application of heat and is called an **endothermic reaction**.

The **enthalpy change ($\Delta H°$)** of any chemical reaction is estimated as the difference between the total bond energy of the products and the total bond energy of the reactants. The value of $\Delta H°$ for a reaction between hydrogen atoms to give H_2 is -104 kcal/mol.

The sometimes troublesome sign convention shows $\Delta H°$ for an exothermic reaction as a negative quantity. For an endothermic reaction, $\Delta H°$ is positive. For example,

If A + B → C $\Delta H° = -x$ kcal/mol, then the reaction is exothermic

If A + B → C $\Delta H° = +x$ kcal/mol, then the reaction is endothermic

In Figure 1.42

H· + H· → H—H $\Delta H° = -104$ kcal/mol, is an exothermic reaction

From this we can also know that the reverse reaction

H—H → H· + H· $\Delta H° = +104$ kcal/mol, is an endothermic reaction

Figure 1.43 is a graph that plots the energy of reactants and products in a chemical reaction against a variable known either as the *reaction progress* or the *reaction coordinate*. A reaction coordinate monitors a change or changes taking place as the reaction proceeds. These changes may be distance between atoms, angles between bonds, or a combination of relationships that result from the movement of atoms. For example, in the reaction H· + H· → H_2, a suitable reaction coordinate might be the distance between the two hydrogen atoms. It is a fair approximation at this point to take this "reaction coordinate" as indicative of the progress of the overall reaction, and for that reason we will use the term *reaction progress* from now on.

A reaction progress graph can be read either left to right or right to left. Reading Figure 1.43 left to right tells us that the energy of the product is lower than the energy of the reactants, and so the graph shows the release of energy, in this case, 104 kcal/mol, as H_2 is formed in an exothermic reaction. If we read the graph from right to left, we see the endothermic formation of two hydrogen atoms from the H_2 molecule.

This amount of energy that must be added to break the H_2 bond to produce two hydrogen atoms, 104 kcal/mol, is called the **bond dissociation energy (BDE)**. It is the amount of energy that must be applied for **homolytic bond cleavage**. The higher the BDE, the more difficult it is to break the bond. A bond can break in either of two ways (Fig. 1.44). In the homolytic cleavage, one electron from the bond goes with one atom and the other electron of the bond goes with the other atom. This kind of bond cleavage gives neutral species. The other option for bond breaking has both electrons from the bond go to the same atom creating an anion–cation pair. This process is called **heterolytic bond cleavage**. Normally, heterolytic cleavage is not the path followed because it usually costs more energy to develop separated charges than to make neutral species.

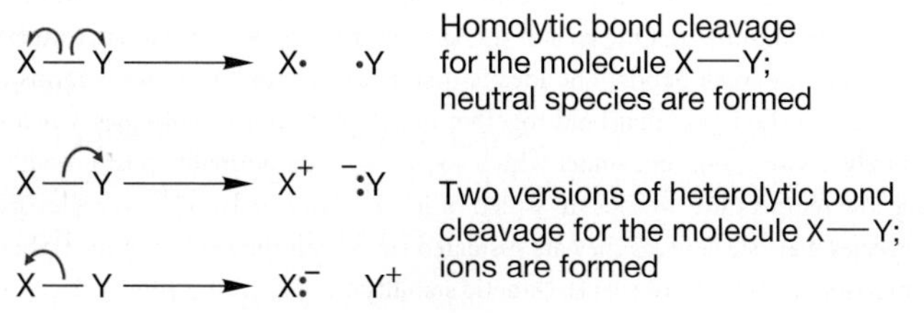

CONVENTION ALERT

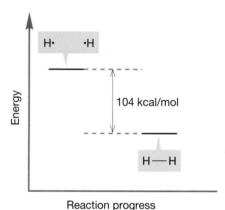

FIGURE 1.43 Another schematic picture of the formation of molecular hydrogen from two hydrogen atoms.

Homolytic bond cleavage for the molecule X—Y; neutral species are formed

Two versions of heterolytic bond cleavage for the molecule X—Y; ions are formed

FIGURE 1.44 In one kind of bond cleavage, the pair of binding electrons is split evenly between the atoms, giving a pair of neutral species. This lower-energy process is called homolytic cleavage. Heterolytic cleavage produces a pair of ions.

CONVENTION ALERT | Notice the difference in the red curved arrows in the two parts of Figure 1.44. The curved arrow showing electron movement in the heterolytic cleavage has the standard *double-barbed* arrows, representing the movement of *two* electrons to the same atom. The homolytic cleavage pathway uses *single-barbed* or "fishhook" arrows representing the movement of *one* electron to each atom.

WORKED PROBLEM 1.26 Sketch the profile of an endothermic reaction. See Figure 1.43 for the sketch of an exothermic reaction.

ANSWER Aha! A trick question. If you have written an exothermic reaction, you have already written an endothermic reaction as well. You need only read your answer backward. We are psychological prisoners of our tendency to read from left to right. Nature has no such hangups! The formation of molecular hydrogen from two hydrogen atoms (Fig. 1.43, left to right) is exothermic (104 kcal/mol of energy is given off as heat), and the formation of two hydrogen atoms from a single hydrogen molecule is endothermic by the same amount (104 kcal/mol of heat energy must be applied).

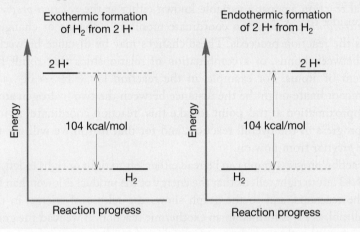

In practice, a reaction that requires about 15–20 kcal/mol of thermal energy proceeds quite reasonably at room temperature, about 25 °C. It is important to begin to develop a feeling for which bonds are strong and which are weak, to start to build up a knowledge of approximate bond strengths.[9] The bond in the hydrogen molecule is a strong one. As we continue our discussion of structure, we'll make a point of noting bond energies as we go along. Unfortunately, there is no way to acquire this knowledge except by learning some bond strengths. Fortunately, there are not too many numbers to remember. Table 1.9 gives a few important bond dissociation

---

[9] To describe a bond as strong or weak is an arbitrary function of human experience. A bond that requires 100 kcal/mol to break is "strong" in a world where room temperature supplies much less thermal energy. It would not be strong on the sunlit side of Mercury (where it's *hot*: the average temperature is about 377 °C, or 710 °F). Similarly, if you were a life form that evolved on Pluto (where it's *cold*: the average temperature is about −220 °C, or −361 °F), you would regard as strong all sorts of "weak" (on Earth) interactions, and your study of chemistry would be very different indeed. One area of research in organic chemistry focuses on extremely unstable molecules held together by weak bonds. The idea is that by understanding extreme forms of weak bonding we can learn more about the forces that hold together more conventional molecules. Chemists who work in this area deliberately devise conditions under which species that are normally most unstable can be isolated. They create very low temperature "worlds" in which other reactive (predatory) molecules are absent. In such a world, exotic species may be stable, as they are insulated from both the ravages of heat (thermodynamic stability) and the predations of other molecules (kinetic stability).

TABLE 1.9 Some Average Bond Dissociation Energies

Bond	BDE (kcal/mol)	Bond	BDE (kcal/mol)
C—H	96–105	C—I	55–57
N—H	93–107	C=C	~175
O—H	110–119	C=N	~143
S—H	82–87	C=O	173–181
C—C	83–90	C≡C	~230
C—O	85–96	C≡N	~204
C—N	69–75	H—I	71
C—F	105–115	H—Br	88
C—Cl	83–85	H—Cl	103
C—Br	72–74	H—F	136

energies. These numbers do not have to be known precisely, but it is important to have a rough idea of the bond strengths of common covalent bonds.

Table 1.9 gives averages over a range of compounds. A few compounds may lie outside these values. In later chapters more precise specific values will appear. Let's consider the simplest molecule, H_2^+, which is H_2 with only one electron and a positive charge. Even this molecule is bound quite strongly.[10] The small amount of information already in hand—a molecular orbital picture of H_2 and the bond strength of the H—H bond (104 kcal/mol)—enables us to construct a picture of this exotic molecule and to estimate its bond strength. We can imagine making H_2^+ by allowing a hydrogen atom, H·, to combine with H^+, a bare proton. All the work necessary to create an orbital interaction diagram for this reaction was done in the construction of Figure 1.39. We are still looking at the combination of a pair of hydrogen 1s atomic orbitals, so the building of a diagram for H_2^+ produces precisely the same orbital diagram, reproduced in Figure 1.45. The only difference comes when we put in the electrons. Instead of having two electrons, as does H_2, with one coming from each hydrogen atom, H_2^+ has only one electron. Naturally, it goes into the lower-energy, bonding molecular orbital, Φ_B. The electron spin is shown down, but it could equally well be shown up. There is no difference in energy between the two spins and we have no way of knowing which way a single spin is oriented.

What might we guess about the bond energy of this molecule H_2^+? If two electrons in Φ_B result in a bond energy of 104 kcal/mol, it seems reasonable to guess first that stabilization of a single electron would be worth half the amount, or about 52 kcal/mol. That is, when an H· atom and an H^+ ion combine to form the species H_2^+, we are estimating that 52 kcal/mol of heat is given off. And we are amazingly close to being correct! The H_2^+ molecule is bound by 64 kcal/mol (Fig. 1.45), which means that the system H_2^+ is 64 kcal/mol more stable than the system of a separated H· and H^+.

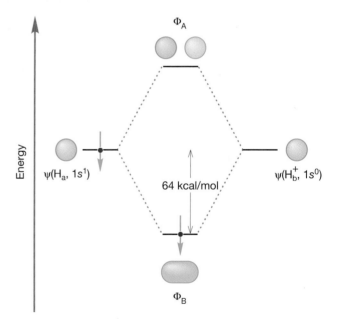

FIGURE 1.45 An orbital interaction diagram for H_2^+.

In the figure: Φ_A ; $\psi(H_a, 1s^1)$; $\psi(H_b^+, 1s^0)$; 64 kcal/mol ; Φ_B ; Energy

---

[10] Here is the answer to Problem 1.22. The simplest molecule is hydrogen (H_2) minus one electron. Another electron cannot be removed to give something even simpler because H_2^{2+} is not a molecule—there are no electrons to bind the two nuclei.

PROBLEM 1.27 Although our guess 52 kcal/mol is quite close to the actual value, it is a bit low. In other words, the H_2^+ molecule is more stable (lower in energy) than we thought. Why is our estimate of bond strength a little low? To ask the same question another way, why might the stabilization of two electrons in an orbital be less than twice the stabilization of one electron in the orbital? *Hint*: Consider electron–electron repulsion.

It may seem somewhat counterintuitive that a low-energy (strong) bond is associated with a large number for the BDE. The lower the energy of a bond, the higher the number representing bond strength! The low-energy, strong bond in H_2 has a BDE of 104 kcal/mol, for example, whereas the higher-energy, weaker bond in H_2^+ has a BDE of only 64 kcal/mol. A strong bond means a low-energy, stable species. The diagram in Figure 1.42 should help keep this point straight. A strong bond means a more stable bond and it means a lower position on an energy diagram.

Remember the sign convention: $2\,H\cdot \rightarrow H{-}H$, $\Delta H° = -104\ \text{kcal/mol}$. The minus sign tells us that this reaction is exothermic as read from left to right. An exothermic reaction will have products that are more stable than the starting materials. The products will be lower than the reactants on the energy diagram.

The antibonding molecular orbital (Φ_A) has been empty in all of the examples we have considered. Why the emphasis on Φ_A? What is an empty orbital, anyway? The easiest, nonmathematical way to think of Φ_A in H_2 is as the place the next electron would go. If the bonding molecular orbital Φ_B is filled, as it is in H_2, for example (Fig. 1.42), another electron cannot occupy Φ_B and must go instead into the antibonding orbital, Φ_A. This would create H_2^- ($H_2 + e^- \rightarrow H_2^-$).

Dihelium, He_2, a strictly hypothetical species, is an example of a molecule in which electrons must occupy antibonding orbitals. Each He, like each H in H_2, brings only a $1s$ orbital to the mixing. Just as in the molecules H_2 and H_2^+, the molecular orbitals for He_2 are created by the combination of two $1s$ atomic orbitals (Fig. 1.46). Each helium atom brings two electrons to the molecule, though, so the electronic occupancy of the orbitals will be different from that for H_2 or H_2^+. The bonding orbital (Φ_B) can hold only two electrons (Pauli principle, p. 8) so the next two must occupy Φ_A. The two electrons in this antibonding orbital are destabilizing to the molecule, just as the two electrons in Φ_B are stabilizing. As a result, there

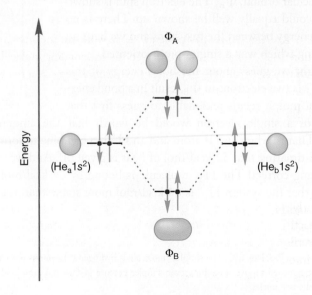

FIGURE 1.46 An orbital interaction diagram for He_2. There is no net bonding because the stabilization owing to the pair of electrons in the bonding molecular orbital is offset by the destabilization from the two electrons in the antibonding orbital.

is no net bonding in the hypothetical molecule He_2. The stabilization afforded by the two electrons in the bonding molecular orbital is cancelled by the destabilization resulting from the two electrons in the antibonding molecular orbital. Molecular helium He_2 is, in fact, unknown.

Summary

The energy of an electron in a bonding molecular orbital generated through mixing of two orbitals depends primarily on the extent of mixing between the two orbitals that combine to produce it. Extensive interaction leads to a low-energy bonding molecular orbital (and a high-energy antibonding molecular orbital). Thus, one or two electrons in such an orbital will be well stabilized. The bond joining the two atoms is strong. Orbitals that are orthogonal do not mix.

PROBLEM 1.28 Draw the orbital interaction diagram for $He_2{}^+$.

WORKED PROBLEM 1.29 Estimate the bond strength for $He_2{}^+$. Show how you arrived at your estimate.

ANSWER The molecular orbital diagram for this molecule can be easily derived from Figure 1.46 by removing one electron. The $He_2{}^+$ molecule can be constructed from He and He^+. This molecule will have only three electrons. Assuming that an electron in the antibonding orbital Φ_A destabilizes the molecule about as much as one in the bonding orbital Φ_B stabilizes, there is one net bonding electron in this molecule (2 bonding electrons – 1 antibonding electron = 1 net bonding electron). The $He_2{}^+$ should have a bond strength about the same amount as $H_2{}^+$, another molecule with a single electron in the bonding molecular orbital. This estimate turns out to be correct. The bond energy for $He_2{}^+$ has been experimentally measured to be about 60 kcal/mol.

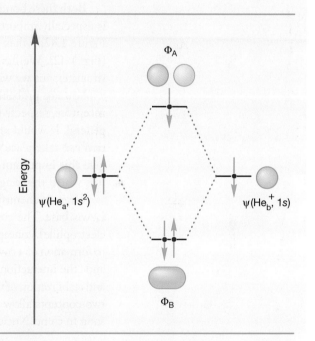

PROBLEM 1.30 Use the answer to Problem 1.27 to work out the answer to a more subtle question. In He_2, both the bonding and the antibonding molecular orbitals are filled with two electrons. Consider electron–electron repulsion to explain why the stabilization of the two electrons in Φ_B is less than the destabilization of the electrons in Φ_A.

1.7   An Introduction to Reactivity: Acids and Bases

Even at this early point we can begin to consider what many see as the real business of chemistry—reactivity. The orbital interaction diagrams we have just learned to draw lead us to powerful unifying generalizations. Remember that we can fit no more than two electrons into any atomic or molecular orbital (p. 8). There are two ways to provide the electrons that fill the bonding molecular orbital so as to stabilize two electrons.

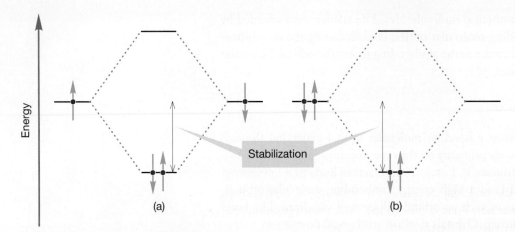

FIGURE 1.47 Two interactions that lead to stabilization of a pair of electrons.

Figure 1.47a shows a pair of electrons, each in a singly occupied orbital, combining to produce a bond in which the two electrons are stabilized (both electrons move down to a lower energy value). Alternatively, we could imagine the situation outlined in Figure 1.47b, in which a filled orbital mixes with an empty orbital to produce a similar stabilization. An atom or molecule that is a two-electron donor is called a **Lewis base**. An atom or molecule that is a two-electron acceptor is called a **Lewis acid**. These labels were introduced by G. N. Lewis (p. 14).

Both these bond-forming scenarios are common, but the scenario in Figure 1.47b is especially important. We have already seen an example of the process shown in Figure 1.47a, when we considered a pair of hydrogen atoms combining to form H_2 (Fig. 1.42). We have not yet seen any real examples of the Lewis acid–Lewis base situation, but we will encounter hundreds as we work through this book.

You are probably familiar with Brønsted acids and bases (proton donors and acceptors, respectively), but the situation described by Figure 1.47b is much more general. It would seem that a stabilizing (energy-lowering) bond-forming interaction can take place whenever a filled orbital overlaps with an empty orbital. We will give this important idea the careful treatment it deserves in Chapters 3 and 7, but it is worth thinking about it a bit right now.

Organic chemists use the terms **electrophile** for Lewis acid and **nucleophile** for Lewis base. The combination of a Lewis base and a Lewis acid (or nucleophile + electrophile), conceptualized in the orbital interaction diagram of Figure 1.47b, leads to formation of a covalent bond. The concepts that "Lewis bases react with Lewis acids" and "the interaction of a filled orbital with an empty orbital leads to bond formation and stabilization of two electrons" run throughout organic (and other) chemistry. These two concepts allow us to generalize and not to be swamped by all the prolific detail soon to come. Nucleophile + electrophile is the unifying theme of organic chemistry.

PROBLEM 1.31 Identify the Lewis base (nucleophile) and Lewis acid (electrophile) in each of these reactions. Write the molecule formed in each case.

(a) H^+ $H{:}^- \rightarrow$?   (b) H^+ $^-{:}\overset{..}{\underset{..}{O}}H \rightarrow$?

(c) H_3C^+ $^-{:}CH_3 \rightarrow$?   (d) H_3C^+ $:NH_3 \rightarrow$?

1.8 Special Topic: Quantum Mechanics and Babies

We might imagine that molecular orbitals and quantum mechanics, so critical in the microscopic world, would have little real influence on our day-to-day lives because we really do live in a world of baseballs, not electrons. Or so it would seem. Yet it has been argued[11] that "a new-born baby becomes conscious of . . . the consequences

---

[11] By Walter Kauzmann (1916–2009) in an early book, *Quantum Chemistry*, Academic Press, New York, 1957.

[of the Pauli principle] long before he finds it necessary to take account of the consequences of Newton's laws of motion." How so? Ernest Rutherford (p. 2) told us long ago that an atom is essentially empty space because the nucleus is very small compared to the size of a typical atom, and electrons are even tinier than the nucleus. Why is it then, that when "solid" objects come together they do not smoothly pass through one another, as they sometimes do in 3 A.M. science fiction movies?

The answer is that the apparent solidity of matter is the result of Pauli forces. When your hand encounters the table top, the electrons in the atoms of your hand and the electrons in the atoms of the table top have the same spins and nearly the same energies. The Pauli principle ensures that these electrons cannot occupy the same regions of space, that the table top feels solid to your touch, even though Rutherford showed us that it is not. Were it not for the Pauli principle our apprehension of the world would be vastly different.

1.9 Summary

New Concepts

Atomic orbitals (mathematically described by wave functions) are defined by four quantum numbers, n, l, m_l, and s. Electrons may have only certain energies determined by these quantum numbers. Orbitals have different, well-defined shapes: s orbitals are spherically symmetric, p orbitals are roughly dumbbell shaped, d and f orbitals are more complicated. An orbital may contain a maximum of two electrons.

Some molecules cannot be well described by a single Lewis structure but are better represented by two or more different electronic descriptions. This phenomenon is called *resonance*. Be sure you are clear on the difference between an equilibrium between two different molecules and the description of a single molecule using a number of different resonance forms.

When two orbitals overlap, two new orbitals are formed: one lower in energy than the starting orbitals, the other higher in energy than the starting orbitals. Electrons placed in the new, lower-energy orbital are stabilized because their energy is *lowered*. The overlap of atomic orbitals to form molecular orbitals and the attendant stabilization of electrons as they move from the atomic orbitals to the lower-energy molecular orbitals are the basis of covalent bonding.

This orbital-forming process can be generalized: when you mix n orbitals into the calculation, the mathematical solution will produce n new orbitals.

Key Terms

anion (p. 4)
antibonding molecular orbital (p. 32)
arrow formalism, curved arrow formalism, or electron pushing (p. 23)
atom (p. 2)
atomic orbital (p. 3)
aufbau principle (p. 8)
bond dissociation energy (BDE) (p. 37)
bonding molecular orbital (p. 32)
cation (p. 4)
covalent bond (p. 6)
delocalization (p. 27)
dipole moment (p. 14)
electron (p. 2)
electron affinity (p. 4)
electronegativity (p. 15)
electrophile (p. 42)
endothermic reaction (p. 36)
enthalpy change ($\Delta H°$) (p. 36)

exothermic reaction (p. 36)
formal charge (p. 20)
functional group (p. 22)
Heisenberg uncertainty principle (p. 2)
heterolytic bond cleavage (p. 37)
homolytic bond cleavage (p. 37)
Hund's rule (p. 9)
ion (p. 4)
ionic bond (p. 5)
ionization potential (p. 4)
Lewis acid (p. 42)
Lewis base (p. 42)
Lewis structure (p. 14)
lone-pair electrons (p. 14)
molecular orbital (p. 3)
node (p. 6)
nonbonding electrons (p. 14)
nonbonding orbital (p. 32)
nucleophile (p. 42)

nucleus (p. 2)
octet rule (p. 4)
orbital (p. 3)
orbital interaction diagram (p. 33)
orthogonal orbitals (p. 34)
paired spin (p. 8)
parallel spin (p. 10)
Pauli principle (p. 8)
polar covalent bond (p. 14)
quantum numbers (p. 6)
resonance arrow (p. 23)
resonance forms (p. 22)
tautomers (p. 29)
unpaired spin (p. 8)
valence electrons (p. 16)
wave function (ψ) (p. 6)
weighting factor (p. 26)

Reactions, Mechanisms, and Tools

There are no reactions to speak of yet, but we have developed a number of tools in this first chapter.

Lewis structures are drawn by representing each valence electron by a dot. These dots are transformed into vertical arrows if it is necessary to show electron spin. Exceptions are the 1s electrons, which are held too tightly to be importantly involved in bonding except in hydrogen. These electrons are not shown in Lewis structures except in H and He. Somewhat more abstract representations of molecules are made by showing electron pairs in bonding orbitals as lines—the familiar bonds between atoms.

The double-headed resonance arrow is introduced to cope with those molecules that cannot be adequately represented by a single Lewis structure. Different *electronic* representations, called *resonance forms*, are written for the molecule. The molecule is better represented by a combination of all the resonance forms. Be very careful to distinguish the resonance phenomenon from chemical equilibrium. Resonance forms give multiple descriptions of a single species. Equilibrium describes two (or more) different molecules.

The curved arrow formalism is introduced to show the flow of electrons. This critically important device is used both in drawing resonance forms and in sketching electron flow in reactions throughout this book.

The formation of a bonding molecular orbital (lower in energy) and an antibonding molecular orbital (higher in energy)

from the overlap of two atomic orbitals can be shown in an orbital interaction diagram (Fig. 1.48). Electrons are represented by vertical arrows which also show electron spin ($\uparrow$ or $\downarrow$). Remember that only two electrons can be stabilized in the bonding orbital and that two electrons in the same orbital must have opposite spins.

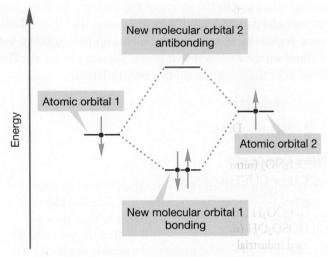

FIGURE 1.48 The overlap of two atomic orbitals produces two new molecular orbitals.

Common Errors

Here, and in similar sections throughout the book, we will take stock of some typical errors made by those who attempt to come to grips with organic chemistry.

Electrons are not baseballs. Nothing is harder for most students to grasp than the consequences of this observation. Electrons behave in ways that the moving objects in our ordinary lives do not. No one who has ever kicked a soccer ball or caught a fly ball can doubt that on a practical level it is possible to determine both the position and speed of such an object at the same time. However, Heisenberg demonstrated that this is *not* true for an electron. We cannot know both the position and speed of an electron at the same time.

Baseballs move at a variety of speeds (energies), and a baseball's energy depends only on how hard we throw it. Electrons are restricted to certain energies (orbitals) determined by the values of quantum numbers. Electrons behave in other strange and counterintuitive ways. For example, we have seen that a node is a region of space of zero electron density. Yet, an electron occupies an entire $2p$ orbital in which the two halves are separated by a nodal plane. A favorite question, How does the electron move from one lobe of the orbital to the other? simply has no meaning. The electron is *not* restricted to one lobe or the other but occupies the *whole orbital* (Fig. 1.49). Mathematics makes these properties seem inevitable; intuition, derived from our experience in the macroscopic world, makes them very strange.

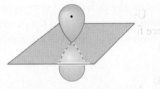

FIGURE 1.49 An electron occupies an entire $2p$ (or other) orbital—it is not restricted to only one lobe.

It is easy to confuse resonance with equilibrium. On a mundane, but nonetheless important level, this confusion appears as a misuse of the arrow convention. Two arrows separate two entirely different molecules (**A** and **B**), each of which might be described by several resonance forms. The amount of **A** and **B** present at equilibrium depends on the equilibrium constant. The double-headed resonance arrow separates two different electronic descriptions (**C** and **D**) of the same species, **E** (Fig. 1.29).

Constructing molecular orbitals through combinations of atomic orbitals (or other molecular orbitals) can be daunting, at least in the beginning. Remember these hints:

1. The number of orbitals produced at the end must equal the number at the beginning. If you start with n orbitals, you must produce n new orbitals. Here is a way to check your work as it proceeds.

2. Keep the process as simple as you can. Use what you know already, and combine orbitals in as symmetrical a fashion as you can.
3. The closer in energy two orbitals are, the more strongly they interact. At this level of theory, you need only interact the pairs of orbitals closest in energy to each other.
4. When orbitals interact in a bonding way (same sign of the wave function), the energy of the resulting orbital is lowered; when they interact in an antibonding way (different signs of the wave function), the energy of the resulting orbital is raised.
5. To order the product orbitals in energy, count the nodes. For a given molecule, the more nodes in an orbital, the higher it is in energy.

These methods will become much easier than they seem at first, and they will provide a remarkable number of insights into structure and reactivity not easily available in other ways.

The sign convention for exothermic and endothermic reactions seems to be a perennial problem. In an exothermic reaction, the products are more stable than the starting materials. Heat is given off in the reaction and $\Delta H°$ is given a minus sign. Thus, for the reaction $A + B \rightarrow C$, $\Delta H° = -x$ kcal/mol. An endothermic reaction is just the opposite. Energy must be applied to form the less stable products from the more stable starting material. The sign convention gives $\Delta H°$ a plus sign. Thus, $A + B \rightarrow C$, $\Delta H° = +x$ kcal/mol.

1.10 Additional Problems

PROBLEM 1.32   Draw Lewis dot structures for the following compounds:

(a) CH_3NO_2 (nitromethane, used for fuel in stock car racing)
(b) $CH_2{=}CHCl$ (vinyl chloride used to make polyvinyl chloride)
(c) CH_3CO_2H (acetic acid, the acid in vinegar)
(d) $HOSO_2OH$ (sulfuric acid, H_2SO_4, the world's most widely used industrial chemical)

See the inside front cover for structures, if you don't know them.

PROBLEM 1.33   Draw two resonance structures for each of the compounds in the previous problem. Show the arrow pushing for interconverting the resonance forms for each compound.

PROBLEM 1.34   Use the arrow formalism to write structures for the resonance forms contributing to the structures of the following ions:

(a)

:O:
‖
C
⁻:O: :O:⁻

Carbonate ion

(b)

:O:
‖
⁻:O—S—O:⁻
‖
:O:

Sulfate ion

(c)

:O:
‖
⁻:O—N⁺—O:⁻

Nitrate ion

(d)

⁺NH₂
‖
C
H₂N   NH₂

Guanidinium ion

(e)

H
\
C
/   \
H₂C   ⁺N(CH₃)₃

A vinyl ammonium ion

PROBLEM 1.35   Use the arrow formalism to write resonance forms contributing to the structures for the following molecules:

(a)

H₂C̈⁻—N̈=N⁺:

(b)

H₃C—N̈⁻—N̈=N⁺:

(c)

H₃C—C⁺=N̈—N̈⁻—CH₃

(d)

H₃C
\
C—N̈=C⁺—CH₃
/
H₃C

(e)

H₃C   N̈⁻—CH₃
\   /
C⁺—N̈:
/   \
H₃C   CH₃

(f)

⁻:Ö—N̈=C⁺—CH₃

PROBLEM 1.36   Draw three resonance structures for each of the following:

(a) $^{-}CH_2NO_2$
(c) $^{-}CH_2CO_2{}^{-}$
(b) $CH_3CO_2CH_3$
(d) $HOSO_2O^{-}$

PROBLEM 1.37   Draw resonance forms for the following cyclic molecules:

(a)

H   H
\   /
C=C
:C⁻
|
H

(b)

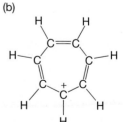

(c)

H   H
\   /
C
H—C   C—H
‖   ‖
C   C
H   ⁺C   H
|
H

PROBLEM 1.38 Draw resonance forms for the following acyclic molecules:

(a)

$H_2C=C(H)-C(H)=CH_2$ with C anion

(b)

$H_2C=C(H)-C(H)=CH_2$ with anion

PROBLEM 1.39 Ozone (O_3) resembles the molecules in Problem 1.35. These days it has a rather bad press, as it is present in too small an amount in the stratosphere and too great an amount in cities. Write a Lewis "dot" structure for ozone and sketch out contributing resonance forms. Write one neutral resonance form. Be careful with this last part; the answer is tricky.

O—O—O

PROBLEM 1.40 Draw two resonance structures for each of the compounds shown below.

(a)

$H_3C-C(H)=C(H)-C(O)-CH_3$

(b)

$H_3C-N=C=O$

PROBLEM 1.41 Which "resonance structure" does not contribute to the molecule CH_3NOCH_2. Why doesn't it contribute?

(a) $H_3C-N^+(=\ddot{O}:)-CH_2^-$

(b) $H_3C-N(-\ddot{O}:)=CH_2$

(c) $H_3C-N^+(-:\ddot{O}:^-)=CH_2$

PROBLEM 1.42 Add charges to the following molecules where necessary:

(a) :O: over H_2C-CH_2

(b) :N: over H_2C-CH_2

(c) :Br: over H_2C-CH_2

(d) :S: over H_2C-CH_2

PROBLEM 1.43 Add charges to the following molecules where necessary:

(a) $H_3C-\ddot{O}-H$

(b) $H_3C-\ddot{O}:$

(c) $H_3C-\ddot{O}(-H)-H$

(d) $H_3C-\ddot{S}-H$

(e) $H_3C-\ddot{S}:$

(f) $H_3C-\ddot{S}(-H)-H$

(g) $H_3C-\ddot{N}(-H)$ (N with lone pair, H above, H to right)

(h) $H_3C-N(-H)(-H)(-H)$ (H above and below)

(i) $H_3C-\ddot{N}(-H)$ (H above)

(j) $H_3C-\ddot{P}-H$

(k) $H_3C-P(-H)(-H)-H$ (H above and below)

(l) $H_3C-\ddot{P}-H$ (H above)

PROBLEM 1.44 Determine the formal charge, if there is one, for each of the nitrogens in the following molecules:

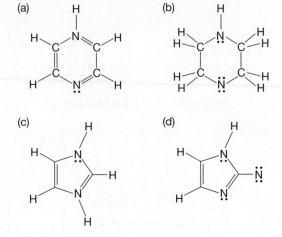

PROBLEM 1.45   Determine the formal charges, if any, for the molecules shown below.

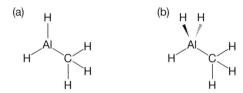

(a)

(b)

(c)

(d)

PROBLEM 1.46   Write Lewis dot structures for the neutral diatomic molecules F_2 and N_2. In F_2, there is a single bond between the two atoms, but in N_2 there is a triple bond between the two atoms.

PROBLEM 1.47   Atomic carbon can exist in several electronic states, one of which is (of course) lowest in energy and is called the "ground state." Write the electronic description for the ground state and at least two higher energy, "excited states."

PROBLEM 1.48   Write the electronic configurations for the following ions:

(a) Na^+ (b) F^- (c) Ca^{2+}

PROBLEM 1.49   Write the electronic configurations for the atoms in the fourth row of the periodic table, $_{19}K$ through $_{36}Kr$. *Hint*: The energy of the $3d$ orbitals falls between that of the $4s$ and $4p$ orbitals. Don't worry about the m_l designations of the $3d$ orbitals.

PROBLEM 1.50   Write electronic configurations for $_{14}Si$, $_{15}P$, and $_{16}S$. Indicate the spins of the electrons in the $3p$ orbitals with a small up or down arrow.

PROBLEM 1.51   There is an instrument, called an electron spin resonance (ESR) spectrometer, that can detect "unpaired spin." In which of the following species would the ESR machine find unpaired spin? Explain.

(a) O   (b) O^+ (c) O^{2-} (d) Ne^+ (e) F^-

PROBLEM 1.52   For the Lewis structure of carbon monoxide shown below, first verify that both the carbon and the oxygen atoms are neutral.

Second, indicate the direction of the dipole moment in this Lewis structure:

$$:C{=}\ddot{O}:$$

As you have just shown, based on this Lewis structure, carbon monoxide should have a substantial dipole moment. In fact, the experimentally determined dipole moment is very small, 0.11 D. Draw a second resonance structure for carbon monoxide, verify the presence of any charges, and indicate the direction of any dipole in this second resonance form. Finally, rationalize the observation of only a very small dipole moment in carbon monoxide.

PROBLEM 1.53   Would you expect formaldehyde, shown below, to have a greater dipole moment than carbon monoxide (see Problem 1.52)? Why or why not?

PROBLEM 1.54   Consider three possible structures for methylene fluoride (CH_2F_2), one tetrahedral (structure **A**), the others flat (structures **B** and **C**). Does the observation of a dipole moment in CH_2F_2 allow you to decide between structures **A** and **B**? What about structures **A** and **C**?

A   **B**   **C**

PROBLEM 1.55   Draw the electron pushing for the homolytic cleavage of Br_2. Draw the electron pushing for heterolytic cleavage of Br_2.

PROBLEM 1.56   While wandering in an alternative universe you find yourself in a chemistry class and, quite naturally, you glance at the periodic table on the wall. It looks (in part) like this:

$_1H$	$_2He$								
$_3Li$	$_4Be$								
$_5B$	$_6C$	$_7N$	$_8O$	$_9F$	$_{10}Ne$	$_{11}Na$	$_{12}Mg$		
$_{13}Al$	$_{14}Si$	$_{15}P$	$_{16}S$	$_{17}Cl$	$_{18}Ar$	$_{19}K$	$_{20}Ca$		

Deduce the allowed values for the four quantum numbers in the alternative universe.

PROBLEM 1.57   In a different alternative universe, the following restrictions on quantum numbers apply:

$$n = 1, 2, 3, \ldots$$
$$l = n - 1, n - 2, n - 3, \ldots, 0$$
$$m_l = l + 1, l, \ldots, 0, \ldots, -l - 1$$
$$s = \pm\tfrac{1}{2}$$

Call the elements in this universe **1, 2, 3**, . . . and assume that Hund's rule and the Pauli principle still apply.

(a) For $n = 1, 2$, and 3 show what orbital subshells are available and label them as s, p, d, and so on.

(b) How many electrons can be accommodated in the first three shells ($n = 1, 2$, and 3) in this universe?

(c) Provide electronic descriptions for elements 1 through 14 in this universe. (In *our* universe Li would be written $1s^2 2s$.)

(d) Construct a periodic table for elements 1 through 23 in this universe.

PROBLEM 1.58   In Problem 1.25, you looked at two possible interactions of a pair of $2p$ orbitals: side-by-side and perpendicular, or end-on. Now let a pair of $2p$ orbitals interact end to end, and draw the two new molecular orbitals produced.

PROBLEM 1.59   Indicate whether the following reactions are exothermic or endothermic. Estimate by how much. Use Table 1.9 (p. 39) and 66 kcal/mol for the "double" part of C═C.

(a) $H_3C\cdot$ + $\cdot\ddot{B}r:$ ⇌ H_3C—$\ddot{B}r:$

(b) H_3C—$\ddot{C}l:$ ⇌ $H_3C\cdot$ + $\cdot\ddot{C}l:$

(c) H_2C═CH_2 + H—H ⇌ $\underset{\underset{H}{|}}{H_2C}$—$\underset{\underset{H}{|}}{CH_2}$

(d) H_2C═CH_2 + H—$\ddot{C}l:$ ⇌ $\underset{\underset{H}{|}}{H_2C}$—$\underset{\underset{:\ddot{C}l:}{|}}{CH_2}$

PROBLEM 1.60   Let's extend our discussion of H—H a little bit to make the orbitals for the molecule linear HHH. Use the molecular orbitals for H_2 and the $1s$ atomic orbital of H. Place the new H in between the two hydrogen atoms of H—H. Watch out for net-zero (orthogonal) interactions!

H—H ± H ⟶ H—H—H

H_2, Φ_B, Φ_A H, $1s$

H_2 Molecular ± H Atomic ⟶ Molecular orbitals
   orbitals   orbital   for HHH

(a) How many orbitals will there be in linear HHH?
(b) Use the molecular orbitals of H_2 and the $1s$ orbital of hydrogen to produce the molecular orbitals of linear HHH. Sketch the new orbitals.
(c) Order the new orbitals in energy (count the nodes).

The following molecular orbital problems are more challenging than the earlier ones, and may be fairly regarded as "special topics." Nonetheless, they do provide some remarkable insights, and for those who like orbital manipulation, they will actually be fun. Just remember the "rules" outlined on p. 34.

PROBLEM 1.61   Make a set of molecular orbitals very much like the ones you made for linear HHH in Problem 1.60, but this time use $2p$ orbitals, not $1s$ orbitals. Place one $2p$ orbital between the other two. Use the molecular orbitals $2p + 2p$ and $2p - 2p$ that you constructed in Problem 1.25. Order the new orbitals in energy.

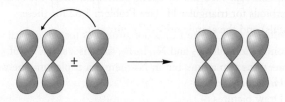

PROBLEM 1.62   One can generate the molecular orbitals for triangular H_3 simply by bending the orbitals generated in Problem 1.60 to transform the linear molecule into the bent one.

(a) Bend the three molecular orbitals for HHH to make the orbitals for the triangle. Make careful drawings. Note that as the old H(1) and H(3) come closer together in the triangle they will create a new bonding or antibonding interaction.
(b) Order the new molecular orbitals in energy by counting the nodes.

PROBLEM 1.63   Which will be lower in energy, linear or triangular H_3^+? What a sophisticated question! Yet, given the answer to Problems 1.60 and 1.62, it is easy.

PROBLEM 1.64   In the chapter, we made the two molecular orbitals for H_2 (p. 34), and in Problem 1.60 we used the two molecular orbitals of H_2 and a $1s$ orbital to make HHH. This time, generate the molecular orbitals for HHHH, linear H_4, from the molecular orbitals of two H_2 molecules placed end to end. *Remember*: At this level of "theory," you need only interact orbitals closest in energy.

H—H ± H—H ⟶ H—H—H—H

Order the new molecular orbitals in energy by counting nodes, and place the proper number of electrons in the orbitals.

PROBLEM 1.65 Make the molecular orbitals for square H_4 by allowing the molecular orbitals of H_2 to interact as shown below:

Order the new molecular orbitals by counting nodes, and add the proper number of electrons. You might check your answer by deriving the same orbitals. Do this by bending the molecular orbitals for linear HHHH developed in Problem 1.64.

PROBLEM 1.66 Generate the molecular orbitals for planar ammonia, NH_3. Do this by taking combinations of the molecular orbitals for triangular H_3 (see Problem 1.62 for these orbitals) and the atomic orbitals of nitrogen.

(a) Show clear pictures of the molecular and atomic orbitals you are using to make the molecular orbitals of planar ammonia.
(b) How many molecular orbitals will ammonia have?
(c) Draw pictures of the molecular orbitals for planar ammonia. Show clearly how these are generated.
(d) Order the bonding and nonbonding molecular orbitals in terms of energy. Place them on a scale relative to the energy of a lone, nonbonding $2p$ orbital. You do not have to order the antibonding orbitals.
(e) Place the appropriate number of electrons in the orbitals. Be careful to indicate the spin quantum number for each electron (use an up or down arrow to show spin).

PROBLEM 1.67 Generate the molecular orbitals for linear methylene, H—C—H, by combining the atomic orbitals of carbon with the molecular orbitals of hydrogen, H_2.

(a) Show clear pictures of the molecular and atomic orbitals you are using.
(b) How many molecular orbitals will linear methylene have?
(c) Draw pictures of the molecular orbitals for linear methylene.

(d) Order these molecular orbitals in terms of energy. Place them on a scale relative to the energy of a lone, nonbonding carbon $2p$ orbital.
(e) Place the appropriate number of electrons in the orbitals, being careful to indicate the spin quantum number for each electron (use an up or down arrow to show spin).

Use Organic Reaction Animations (ORA) to answer the following questions:

PROBLEM 1.68 Choose the reaction titled "Unimolecular nucleophilic substitution" and click on the Play button. Do you suppose the first step of this reaction is a homolytic or a heterolytic cleavage? Observe the Highest Occupied Molecular Orbital (HOMO) track by clicking on the HOMO button. The location (orbital) of the most available electrons will be shown throughout the reaction. Notice that the electron density goes with the bromine as it comes off. That should help you answer this question.

PROBLEM 1.69 Choose the "Introduction" on the bottom left of the Table of Contents page. Read this short document. Under the "Technical Issues" heading there is a discussion of solvent effects. After reading this section, how do you think using a polar solvent in the "Unimolecular nucleophilic substitution" reaction might affect the answer to the previous question? That is, would a polar solvent have more impact on a homolytic or a heterolytic cleavage?

PROBLEM 1.70 Choose the reaction "Alkene hydrohalogenation" and observe the molecule that initially comes to the screen. It has a carbon–carbon double bond. Click on the HOMO button. Observe the calculated area for the π bond that is shown to answer the following questions. Is the electron density of a π bond constrained to the space between the carbons? Do you suppose the π bond electrons are held more or less tightly than σ bond electrons?

2 Alkanes

2.1 Preview

2.2 Hybrid Orbitals: Making a Model for Methane

2.3 The Methyl Group (CH_3) and Methyl Compounds (CH_3X)

2.4 The Methyl Cation ($^+CH_3$), Anion ($^-:CH_3$), and Radical ($\cdot CH_3$)

2.5 Ethane (C_2H_6), Ethyl Compounds (C_2H_5X), and Newman Projections

2.6 Structure Drawings

2.7 Propane (C_3H_8) and Propyl Compounds (C_3H_7X)

2.8 Butanes (C_4H_{10}), Butyl Compounds (C_4H_9X), and Conformational Analysis

2.9 Pentanes (C_5H_{12}) and Pentyl Compounds ($C_5H_{11}X$)

2.10 The Naming Conventions for Alkanes

2.11 Drawing Isomers

2.12 Rings

2.13 Physical Properties of Alkanes and Cycloalkanes

2.14 Nuclear Magnetic Resonance Spectroscopy

2.15 Acids and Bases Revisited: More Chemical Reactions

2.16 Special Topic: Alkanes as Biomolecules

2.17 Summary

2.18 Additional Problems

A CELESTIAL VIEW Does it include molecular orbitals?

God never saw an orbital.

—WALTER KAUZMANN, FEBRUARY, 1964[1]

2.1 Preview

The chemical reactions a given compound undergoes are a function of its structure. No compound reacts exactly like another, no matter how similar their structures. Yet there are gross generalizations that can be made, and some order can be carved out of the chaos that would result if compounds of similar structure did not react in somewhat the same fashion. To a first approximation, the reactivity of a molecule depends on the functional groups (p. 22) it contains.

The inside front cover of this book shows some of the functional groups that are important in organic chemistry. This material is not to be memorized, and it is certainly not complete. You will want to work through it, however, to be certain you can write correct Lewis structures for each group and to begin to become familiar with the variety of functional groups important in organic chemistry. These different kinds of molecules will all be discussed in detail in later chapters. We start in this chapter with **alkanes**, the simplest members of the family of organic molecules called **hydrocarbons**—compounds composed entirely of carbon and hydrogen.

Alkanes all have the molecular formula C_nH_{2n+2}. The simplest alkane is **methane**, CH_4. Vast numbers of related molecules can be constructed from methane by replacing one or more of its hydrogens with other carbons and their attendant hydrogens. Linear arrays can be made, as well as branched structures and even rings (ring compounds called **cycloalkanes**, have a slightly different formula: C_nH_{2n}). Figure 2.1 shows a few schematic representations for these molecules.

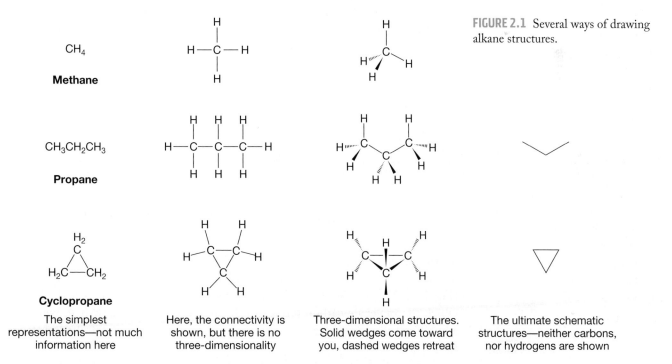

FIGURE 2.1 Several ways of drawing alkane structures.

The simplest representations—not much information here

Here, the connectivity is shown, but there is no three-dimensionality

Three-dimensional structures. Solid wedges come toward you, dashed wedges retreat

The ultimate schematic structures—neither carbons, nor hydrogens are shown

---

[1] Walter Kauzmann (p. 42) was chairman of the Princeton University chemistry department in 1964 when MJ interviewed for a job. He made this comment in response to a typically convoluted molecular orbital answer to some simple question.

Before we can go further, we need to make sense of these schematics, and as usual, we must start with structure. The first task is to describe the shape of methane. Once we have done this, we can go on to investigate other members of the alkane family.

PROBLEM 2.1 Draw Lewis structures for the linear alkanes butane (C_4H_{10}) and pentane (C_5H_{12}).

ESSENTIAL SKILLS AND DETAILS

1. Hybridization. Above all, it is important to master the structural model introduced in this chapter, hybridization. The hybridization model does a good job of allowing us to predict the general structures of compounds and it is nicely suited for following electron flow in chemical reactions. Thus, it fits in well with the curved arrow formalism introduced in Chapter 1 (p. 23).

2. Structures. It is important to be able to use the various structural formulas, which range from richly detailed three-dimensional representations to the ambiguous condensed formulas that give no hint of the three-dimensional complexity often present in a molecule.

3. Difference. The concept of difference is essential in organic chemistry. When are two atoms the same (in exactly the same environment) and when are they different (not in the same environment)? This question gets to the heart of structure and is much tougher to answer than it seems. In this chapter, we will introduce this subject and we will return to it in Chapter 4.

4. Names. In truth, it is not necessary to know every nuanced detail of the naming convention for alkanes, but you do need to know some nomenclature.

5. The cis/trans convention. Cyclic molecules (rings) have sides—above the ring and below the ring—and attached groups can be on the same or opposite sides of the ring.

6. The Newman projection. Drawing and "seeing" Newman projections is a critical skill in organic chemistry.

7. Bond strengths. You surely do not have to memorize all possible bond strengths, but by the end of this chapter you should have a good idea of some of the important ones, such as C—C, C—H, and H—H.

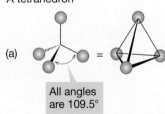

A tetrahedron

(a)

All angles are 109.5°

Methane, a tetrahedral molecule

(b)

H
1.09 Å
H•••C•••H
H 109.5° WEB3D

FIGURE 2.2 (a) A tetrahedron. (b) A tetrahedral molecule, methane, showing the arrangement of the four hydrogens about the central carbon.

2.2 Hybrid Orbitals: Making a Model for Methane

2.2a Hybridization Our task is to devise a bonding model for the structure of the simplest alkane, methane, CH_4. Physical chemists tell us the structure: methane is a tetrahedron, with all carbon–hydrogen bonds the same length (Fig. 2.2). We had an earlier look at a tetrahedron in Problem 1.5 (p. 16), in which the molecule carbon tetrachloride appeared. Our job now is to work out a bonding scheme that leads to the experimentally determined tetrahedral structure of methane.

The method we use to generate our model of methane goes by the name **hybridization**. Remember that we are working out a *model*, which means that no matter how useful it is, what we produce here is bound to be flawed in some respects. It is not the "ultimate truth" by any means! Our strategy will be to combine the four atomic orbitals of a carbon atom ($2s$, $2p_x$, $2p_y$, $2p_z$) to produce four new orbitals for the atom, called **hybrid orbitals**, that can overlap with the $1s$ orbitals of four hydrogens to produce methane. Despite the fancy name *hybridization*, we are really doing something quite similar to the combining of atomic orbitals we saw often in Chapter 1. In Chapter 1 we combined orbitals from different atoms to make molecular orbitals and here we will combine orbitals from the same atom to make hybrid atomic orbitals. Remember that we already know that a combination of *four* atomic orbitals must produce *four* new orbitals (p. 34).

It will greatly help our understanding of the hybridization process if we back up just a bit and look at two simpler hybridization models before we return to methane.

2.2b *sp* Hybridization   Let's first combine the $2s$ and $2p_x$ atomic orbitals from some arbitrary atom, as shown in Figure 2.3. The two can be combined in a constructive ($2p + 2s$) or a destructive ($2p - 2s$) way to create a pair of hybrid orbitals.

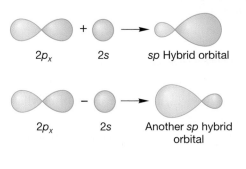

2p_x 2s *sp* Hybrid orbital

2p_x 2s Another *sp* hybrid orbital

$2s \pm 2p_x \longrightarrow$ Two *sp* hybrid orbitals

One *sp* orbital in schematic form...

...and more correctly:

FIGURE 2.3 *sp* Hybridization.

Note in Figure 2.3 that the hybrid orbitals no longer have equal-sized lobes, as in a $2p$ orbital. In the $2p + 2s$ combination, we get an expansion of one of the original lobes (the one with the same sign as the $2s$ orbital) and a shrinking of the other (the one with the opposite sign). The second combination, ($2p - 2s$), is also lopsided, but in the opposite sense. These two new atomic orbitals are called ***sp* hybrid** orbitals. The designation *sp* means that the orbitals are composed of 50% *s* orbital (they have 50% *s* character) and 50% *p* orbital (they have 50% *p* character).

Figure 2.3 also shows the real shape of an *sp* hybrid orbital. It is traditional to use the schematic form in drawings, rather than attempt to reproduce the detailed structure of the orbital.

Overlap between an *sp* orbital of one atom and an orbital of some other atom will be especially good if the fat lobe of the *sp* orbital is used (Fig. 2.4a). Contrast this overlap with the overlap when a $1s$ orbital overlaps with one lobe of a $2p$ orbital of another atom (Fig. 2.4b). The better the overlap, the greater the stabilization and the stronger the bond. In a sense, overlap of a $1s$ orbital with an unhybridized atomic $2p$ orbital "wastes" the back lobe of the $2p$ orbital. Hybridization improves overlap!

Hybridization also minimizes electron–electron repulsion. The two new *sp* hybrid atomic orbitals formed from $2p + 2s$ and $2p - 2s$ (Fig. 2.3) are aimed 180° from each other. The electrons in the bonds are as remote from each other as possible.

(a)

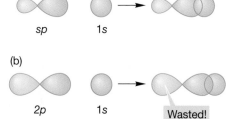

sp   1*s*

(b)

2*p*   1*s*   Wasted!

FIGURE 2.4 A comparison between overlap of a hydrogen $1s$ orbital with (a) an *sp* hybrid orbital and (b) an unhybridized $2p$ orbital. With the *sp* hybrid orbital, overlap is maximized because the non-overlapping back lobe is small. With the unhybridized $2p$ orbital, all of the non-overlapping back lobe is "wasted."

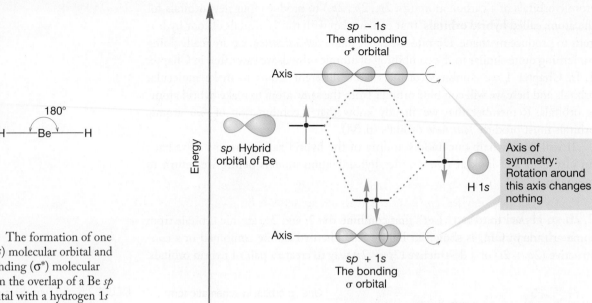

H —— Be —— H   180°

FIGURE 2.5 The formation of one bonding (σ) molecular orbital and one antibonding (σ*) molecular orbital from the overlap of a Be *sp* hybrid orbital with a hydrogen 1*s* atomic orbital. Note that the interaction diagram shows the 1*s* atomic orbital is at a lower energy than the *sp* hybrid orbital. The figure shows formation of only one of the two beryllium–hydrogen bonds in BeH₂. The other Be—H bond is made in identical fashion through overlap of the second Be *sp* hybrid orbital with the 1*s* atomic orbital of a second H atom.

CONVENTION ALERT

There are many compounds in which the bonding can be nicely described by *sp* hybridization. The classic example is H—Be—H, a linear hydride of beryllium (Fig. 2.5). Beryllium ($_4$Be) has the electronic configuration $1s^2 2s^2$ and thus brings only two electrons to the bonding scheme (*Remember*: The filled shell 1*s* electrons are not used for bonding). We construct our picture of beryllium hydride (BeH₂) by making two *sp* hybrid atomic orbitals from the Be 2*s* atomic orbital and one of the three equivalent Be 2*p* orbitals, arbitrarily taking the $2p_x$ orbital (which exists even though the configuration $1s^2 2s^2$ tells us there are no electrons in this atomic orbital). The overlap of the $2p_x + 2s$ hybrid orbital with a singly occupied hydrogen 1*s* orbital and of the $2p_x - 2s$ hybrid orbital with a second hydrogen 1*s* orbital produces two Be—H bonds directed at an angle of 180° with respect to each other. As a result of the linear shape of this compound we say that the beryllium is *sp-hybridized*.

Bonds of this type, with cylindrical symmetry, are called **sigma bonds** (σ bonds). The orbital is unchanged through rotation around the axis between the Be and H. The combination of a Be *sp* orbital and a H 1*s* orbital produces the sigma bonding orbital (*sp* + 1*s*) and must also yield an antibonding orbital (*sp* − 1*s*), which in this case is empty, as shown in Figure 2.5. This antibonding orbital also has cylindrical symmetry and is therefore properly called a sigma antibonding orbital.

In Chapter 1, we used subscripts A and B to indicate antibonding and bonding orbitals (Φ_A, Φ_B), but from now on we can use the simpler notation of an asterisk (*) to indicate an antibonding orbital. Thus, in Figure 2.5, σ is the bonding molecular orbital *sp* + 1*s* and σ* is the antibonding molecular orbital *sp* − 1*s*.

It is no accident that we use two of beryllium's atomic orbitals (2*s* and $2p_x$) in our description of the bonding in BeH₂. *Two* hybrid orbitals are needed for bonding to *two* hydrogens, which means we must combine *two* unhybridized atomic orbitals to get them.[2]

---

[2] Nothing in this discussion so far is specific to the element Be. Even though we are using this element as our example, we have been constructing hybrid orbitals for any second row linear HXH molecule.

PROBLEM 2.2 What happens to the $2p_y$ and $2p_z$ atomic orbitals of Be in BeH_2? Elaborate the picture in Figure 2.5 to show the unused, empty $2p$ atomic orbitals on Be.[3]

WORKED PROBLEM 2.3 Use the hybridization model to build a picture of the bonding in the linear molecule H—C—H. (Recall that a molecular orbital picture of this linear molecule was constructed in Problem 1.67, p. 49). If you didn't do that problem, you might go back, try it now, and contrast your molecular orbital answer with your hybridization answer.

ANSWER As with BeH_2, the central atom, here carbon, is hybridized sp because the CH_2 carbon only has two bonds. The $2s$ and $2p_x$ atomic orbitals of carbon are combined to form two sp hybrid orbitals ($2s + 2p_x$ and $2s - 2p_x$). One of the bonds in CH_2 is made through overlap of the $2s + 2p_x$ sp hybrid orbital and the $1s$ atomic orbital of one hydrogen. The other bond is made through overlap of the $2s - 2p_x$ hybrid orbital and the $1s$ atomic orbital of the other hydrogen.

The $2p_y$ and $2p_z$ atomic orbitals in the carbon remain unchanged. The atom uses two of its four bonding electrons in the carbon–hydrogen bonds, leaving two for the remaining $2p$ orbitals. Each contains a single electron.

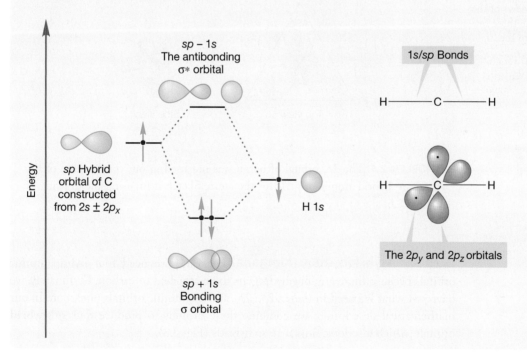

2.2c sp^2 Hybridization

In BH_3, we must make three boron–hydrogen bonds to form the molecule. The electronic configuration of boron ($_5B$) is $1s^2 2s^2 2p_x$, and thus there are three electrons available to form bonds with the three hydrogen $1s$

---

[3] This kind of thing generally drives physical chemists crazy. As an orbital is nothing more than the electron probability region, how can there be an "empty orbital"? The problem is at least partially avoided if the empty orbital is thought of as "the place the next electron would go" as mentioned in Chapter 1.

atomic orbitals. This time we combine the boron $2s$, $2p_x$, and $2p_y$ orbitals to produce the three orbitals we need in our model. These are called, naturally, **sp^2 hybrid** orbitals (constructed from one s orbital and two p orbitals) and have 33% s character and 67% p character. They resemble the sp hybrid orbitals, shown in Figure 2.3. As with the sp orbital, the fat lobe of the sp^2 orbital allows for efficient overlap with a hydrogen $1s$ or with any other orbital.

$$2s + 2p_x + 2p_y \rightarrow \text{Three } sp^2 \text{ hybrid orbitals}$$

PROBLEM 2.4 Boron also forms a compound BF_3. Create a bonding scheme and draw a Lewis structure for this molecule.

How are the three sp^2 hybrid orbitals needed to form BH_3 arranged in space? Your intuition may well give you the answer. They are directed so that electrons in them will be as far as possible from one another. The angle between them is 120°, and they are aimed toward the corners of a triangle (Fig. 2.6). An sp^2 carbon and the three atoms surrounding it lie in the same plane.

FIGURE 2.6 Top and side views of the arrangement in space of the three sp^2 hybrid orbitals formed from the $2s$, $2p_x$, and $2p_y$ atomic orbitals of a boron atom.

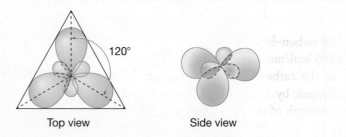

120°

Top view   Side view

PROBLEM 2.5 The $2p_z$ orbital of boron was not used in our construction of the three sp^2 hybrid atomic orbitals for the atom. Sketch it in, using the drawing of Figure 2.6.

2.2d sp^3 Hybridization: Methane

For CH_4 we need four hybrid atomic orbitals because there are four hydrogens to be attached to carbon. Fortunately, we have just what we need in the $2s$, $2p_x$, $2p_y$, and $2p_z$ atomic orbitals of carbon. In our mathematical description, we combine these orbitals to produce four **sp^3 hybrid** orbitals, which also look similar to sp hybrids (Fig. 2.3).

$$2s + 2p_x + 2p_y + 2p_z \rightarrow \text{Four } sp^3 \text{ hybrid orbitals}$$

The familiar lopsided shape appears again, and thus we can expect strong bonding to the four hydrogen $1s$ atomic orbitals. By now you will surely anticipate that these orbitals will be aimed in space so as to keep them, and any electrons in them, as far apart as possible. How are four objects arranged in space so as to maximize the distance between them? The answer is, at the corners of a tetrahedron (Fig. 2.2).

The carbon in methane is hybridized sp^3 (25% s character; 75% p character) and the four hybrid atomic orbitals are directed toward the corners of a tetrahedron. The $1s$ atomic orbital in each of four hydrogen atoms overlaps with one of the four

sp^3 orbitals to form a bonding molecular orbital and an antibonding molecular orbital (Fig. 2.7). The resulting H—C—H angles are each 109.5°, and the length of a carbon–hydrogen bond in methane is 1.09 Å (Fig. 2.2).

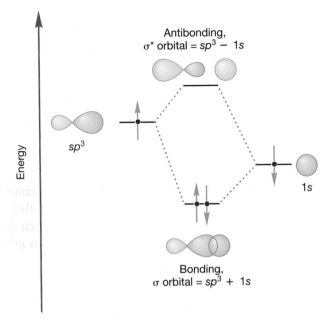

FIGURE 2.7 The formation of a carbon–hydrogen bond through the overlap of an sp^3 hybrid orbital with a hydrogen $1s$ atomic orbital. Note the formation of both the bonding (σ) molecular orbital and the antibonding (σ*) molecular orbital.

The carbon–hydrogen bond in methane is very strong; its bond dissociation energy is 105 kcal/mol. This means that 105 kcal/mol must be applied in order to break one of the carbon–hydrogen bonds in methane homolytically. This reaction is endothermic by 105 kcal/mol (Fig. 2.8). Compare this value to the 104 kcal/mol bond strength of the very strong H—H bond in H_2 (p. 36). The C—H bond in methane and the H—H bond in H_2 have very similar strengths.

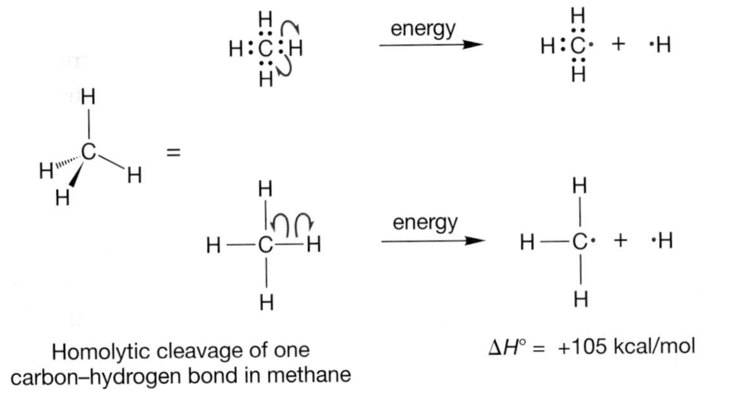

Homolytic cleavage of one
carbon–hydrogen bond in methane

$\Delta H° = +105$ kcal/mol

FIGURE 2.8 Two ways of showing homolytic cleavage of one carbon–hydrogen bond in methane. Each single-barbed arrow represents movement of *one* electron.

2.2e Why Hybridization?   Why do we need to develop this new structural model called hybridization? If we want to form methane, why don't we just overlap the $1s$ atomic orbitals of four hydrogens with the $2s$ and $2p$ atomic orbitals of carbon? Well, we certainly could have done just that, but the results are not very satisfying—this approach doesn't give us a decent approximation of the tetrahedral structure of methane. However, let's follow this flawed procedure. We can learn a lot from examining what is wrong with models that are too simple. Further, what we do here approximates what really happens in science in that we make a series of successive structural approximations in which we come increasingly close to the real methane. In Section 2.2d, we went right to a satisfactory model for methane, thus shortcutting the intellectual process.

FIGURE 2.9 A structure for methane, CH_4. Bonds are formed by the overlap of four hydrogen $1s$ orbitals with the $2p$ and $2s$ atomic orbitals of one carbon atom. This model requires some 90° bond angles and requires the bonds to be of different lengths. Carbon's $1s$ electrons are shown in parentheses because they are not used in bonding.

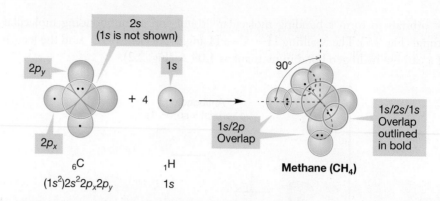

Figure 2.9 shows the overlap of the occupied carbon $2s$, $2p_x$, and $2p_y$ atomic orbitals with four hydrogen $1s$ orbitals. Two predictions are clear: First, we should form one C—H bond from the overlap of the carbon $2p_x$ orbital and one hydrogen $1s$ orbital and one C—H bond from the overlap of the carbon $2p_y$ orbital and another hydrogen $1s$ orbital; second, these bonds must be at 90° to each other because of the 90° angle between the carbon $2p_x$ and $2p_y$ orbitals. The third and fourth bonds are difficult to locate exactly, as they are formed from the overlap of a carbon $2s$ orbital with two hydrogen $1s$ orbitals. Both the $2s$ and $1s$ orbitals are spherically symmetric, and thus no directionality can be induced by the shape of the orbitals. We might guess that these two new carbon–hydrogen bonds would be aimed so as to keep the electrons in the bonds as far from each other as possible, thus minimizing repulsions.

PROBLEM 2.6 In a quick analysis there appear to be too many electrons in the region of Figure 2.9 where the carbon $2s$ orbital and the two hydrogen $1s$ orbitals overlap. You might think that an atomic orbital with two electrons can't make a bond with two other atoms, each bringing an electron. It is true that no orbital can have more than two electrons (the Pauli principle). You are used to thinking of two-electron bonds, and our new system is clearly more complicated than that. But there is no violation of the Pauli principle in the unhybridized model illustrated in Figure 2.9. Explain. *Hint*: Remember that overlap of *n* orbitals produces *n* new orbitals. See Problem 1.60.

Using unhybridized carbon atomic orbitals to form methane, a model that yields the structure shown in Figure 2.9, has a fatal flaw. Modern spectroscopic methods show that there are no 90° H—C—H angle in methane and that all carbon–hydrogen bonds in the molecule are of equal length! Knowing that the Figure 2.9 structure is wrong, we can begin to construct a model that does a better job of describing Nature. Don't be offended by this—science works this way all the time. The goal is to use what we learn from our wrong guesses to come closer next time. We can already anticipate one problem in this model—the electrons in some of the bonds are only 90° apart. They could be farther apart, and thus electron–electron repulsion could be reduced.

PROBLEM 2.7 Consider the ammonium ion ($^+NH_4$). Predict the hybridization of the nitrogen and the shape of the molecule.

A second problem with the model used in Figure 2.9 might be labeled "inefficient overlap" or "wasted orbitals." Recall from Figure 2.4 that a $2p$ atomic orbital is poorly designed for efficient end-on overlap because overlap can take place only with one lobe. The back lobes of $2p$ orbitals are unused and thus wasted (Fig. 2.10). Our unhybridized model for methane has two such overlapping pairs of orbitals.

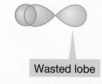

Wasted lobe

$1s/2p$ End-on overlap; notice the wasted lobe not involved in bonding

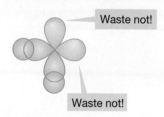

Waste not!

Waste not!

This partial picture of methane contains two such $1s/2p$ end-on overlaps; two rear lobes are wasted

FIGURE 2.10 End-on overlap of a $2p$ atomic orbital with a $1s$ atomic orbital.

PROBLEM 2.8 Why not allow a $2p$ orbital to overlap sideways with an s orbital? Why would this not solve the problem of waste of the rear lobe (see below and Fig. 2.10)?

Summary

There are two severe problems with the too-simple model for methane shown in Figure 2.9. The hybridization model solves them both!

1. The electrons in the bonds are far from being as remote from each other as possible. The hybridization model yields four hybrid orbitals directed toward the corners of a tetrahedron, the best possible arrangement.

2. Bonds are formed through overlap between the carbon $2p$ atomic orbitals and hydrogen $1s$ orbitals, thus wasting the back lobes of the $2p$ orbitals. In the hybridization model, overlap is much improved in the bonds formed using the fat lobes of sp^3 orbitals and hydrogen $1s$ orbitals. Table 2.1 reviews the properties of sp, sp^2, and sp^3 hybrid orbitals.

TABLE 2.1 **Properties of Hybrid Orbitals**

Hybridization	Constructed from	Angle between Bonds (°)	Examples
sp	50% s, 50% p	180	BeH_2, linear HCH
sp^2	33.3% s, 66.7% p	120	BH_3, BF_3
sp^3	25% s, 75% p	109.5	CH_4, $^+NH_4$

So why didn't we just tell you that methane is tetrahedral and be done with it rather than taking several pages to explore that experimental observation? First of all, learning science is not about gathering a collection of facts. Rather it is learning a way of thinking—a way of solving problems—of reaching an approximate idea of how Nature works. In science, we look at experimental data (methane has the formula CH_4, for example) and then postulate a model to explain our observations. In the beginning, that model is almost certain to be either flawed or at the very least severely incomplete. So we test it against new data—against what can be learned from new experiments, and modify our hypothesis. In this case, that process looks like this: We find out that methane has no 90° angles—how can we change our model to get rid of those errors and come to a better model of Nature? Ultimately, we come to the tetrahedral structure. We have learned much from looking at why our early, too-simple models were inadequate and from our process of investigation. You will forget most of the facts in this book, eventually. No matter, you can always look them up here or elsewhere. What we hope you will not forget is the investigative method used in organic chemistry and all other sciences—and many, many other disciplines as well.

But it is a utilitarian world these days, and we can hear cries of, "just gimme the facts, buster; I'm not in this course to learn about science, I just want to get on with my career goals." No! No! No! We would argue with you forever that if you feel this way, you are making a profound, lifetime error, but even if such a goal is granted, there is a fatal flaw in just giving the facts. Put simply, there is too much material to be learned this year to memorize your way through it. You can survive at first, but somewhere about the middle of the first semester you will run out of memory, and there is no way (yet) to install any more. If you try to get through this course by memorizing a set of facts, you will almost certainly not succeed. Success in this course—and in much of life—depends on learning how to think, how to reason sensibly from new data. So that's what we will try to do throughout this book, and we have begun right here.

METHANE

There is a lot of methane in very surprising places. Some anaerobic bacteria degrade organic matter to produce methane. When that process occurs deep in the ocean, water can crystallize in a cubic arrangement, encapsulating methane molecules to form hydrates called clathrates. And those clathrates can hold *a lot* of methane. One cubic meter of this hydrate could contain as much as 170 m³ of methane! Nor are those clathrates merely rare curiosities. Estimates vary, but the Arctic might hold as much as 400 gigatons, and worldwide estimates (guesses, really) run to as much as 10,000 gigatons. That's a good news–bad news story. If we ever solve the daunting economic problems involved in finding these clathrates and getting the methane out, our hydrocarbon scarcity problems vanish for a long time. On the other hand, if that methane is ever released in an uncontrolled fashion (as it regularly is in science fiction disaster stories), watch out!

Lest you think that last scenario fanciful, as you read this, there are Siberian and Alaskan lakes formed from melting permafrost that bubble with escaping methane. The University of Alaska's Katey Walter, shown here setting one of those lakes afire, is investigating those burbling lakes. She and her associates suggest that earlier atmospheric methane spikes were at least partially the result of the escape of large amounts of this greenhouse gas.

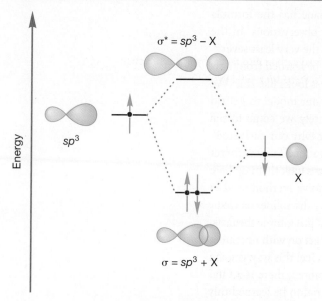

FIGURE 2.11 Formation of a C—X bond from overlap of an sp^3 hybrid orbital of C with an s atomic orbital of X.

$\sigma^* = sp^3 - X$

sp^3

X

$\sigma = sp^3 + X$

Energy

2.3 The Methyl Group (CH₃) and Methyl Compounds (CH₃X)

As soon as one of the four hydrogens surrounding the central carbon in methane is replaced with another atom, pure tetrahedral symmetry is lost. We might well anticipate that one of the four sp^3 hybrid orbitals of carbon could overlap in a stabilizing way with any atom X offering an electron in any atomic orbital (Fig. 2.11). For maximum stabilization, we want to fill the new, bonding molecular orbital created from this overlap, which requires two, and only two, electrons.

All manner of derivatives of methane, called **methyl compounds** (CH_3—X), are possible. The atom or group replacing H is called a **substituent**, and one option for naming these substituted compounds is to drop the "ane" from the name of the parent molecule, methane, and append "yl."

The *methyl* is followed by the name of the X group as a separate word. Table 2.2 shows a number of methyl derivatives with their melting points, boiling points, and physical appearances.

TABLE 2.2 Some Simple Derivatives of Methane, Otherwise Known as Methyl Compounds

H₃C—X	Common Name	mp (°C)	bp (°C)	Physical Properties
H₃C—H	Methane	−182.5	−164	Colorless gas
H₃C—OH	Methyl alcohol or methanol	−93.9	65.0	Colorless liquid
H₃C—NH₂	Methylamine	−93.5	−6.3	Colorless gas
H₃C—Br	Methyl bromide	−93.6	3.6	Colorless gas/liquid
H₃C—Cl	Methyl chloride	−97.7	−24.2	Colorless gas
H₃C—CN	Methyl cyanide or acetonitrile	−45.7	81.6	Colorless liquid
H₃C—F	Methyl fluoride	−141.8	−78.4	Colorless gas
H₃C—I	Methyl iodide	−66.5	42.4	Colorless liquid
H₃C—SH	Methyl mercaptan or methanethiol	−123	6.2	Colorless gas/liquid

The bonding in methyl compounds closely resembles that in methane, and it is conventional to speak of the carbon atom in any H₃C—X as being sp^3 hybridized, just as it is in the more symmetrical CH_4. Strictly speaking, this is wrong because the bond from C to X is not the same length as that from C to H and the H—C—H bond angle cannot be exactly the same as the X—C—H angle. Because sp^3 hybridization yields four exactly equivalent bonds directed toward the corners of a tetrahedron, the bonds to H and X in H₃C—X cannot be *exactly sp^3* hybrids, only *approximately sp^3* ($sp^{2.8}$, say). This point is very often troubling to students, but it is really quite simple and, once one has seen it, even obvious. Consider converting an sp^2 hybridized carbon into an sp^3 hybridized carbon, as in Figure 2.12.

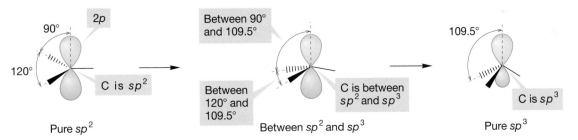

FIGURE 2.12 A thought experiment. The conversion of an sp^2 hybridized carbon into an sp^3 hybridized carbon. What is the hybridization at a point intermediate between the starting point (sp^2) and end point (sp^3)?

We start with the carbon having three sp^2 hybrid orbitals and a pure, unhybridized $2p$ orbital (H—C—H angle = 120°). As we bend the sp^2 hybrid orbitals, we will come to a point at which the H—C—H angle has contracted to 109.5°, which is sp^3 hybridization. Our orbitals, originally one-third s and two-thirds p (sp^2: 33.3% s, 66.7% p character), have become one-fourth s and three-fourths p (sp^3: 25% s, 75% p character). *They have gained p character in the transformation from sp^2 to sp^3.* Now look at the $2p$ atomic orbital in Figure 2.12. As we bend, this orbital goes from pure p to one-fourth s, three-fourths p. *It has gained s character in the transformation.* There is a smooth transformation as we pyramidalize the orbitals from sp^2 to sp^3, and between these two extremes, the hybridization of carbon is intermediate between sp^2 and sp^3, something like $sp^{2.8}$. The hybridizations sp^2 and sp^3 are *limiting* cases, applicable only in completely symmetrical situations. The hybridization of a carbon (or other) atom is intimately related to the bond angles! If you know one, you know the other.

In general, methyl compounds (H_3C—X) will closely resemble methane in their approximately tetrahedral geometry; sp^3 is a good approximation, but methyl compounds cannot be perfect tetrahedra.

In addition to CH_3X, other substituted molecules are possible in which more than one hydrogen is replaced by another atom or group of atoms. Thus, even for one-carbon molecules, there are many possible substituted structures.

PROBLEM 2.9 Use the halogens (X = F, Cl, Br, or I) to draw all possible molecules CH_2X_2. For example, CH_2BrCl is one answer.

PROBLEM 2.10 Draw all possible molecules of the formula CH_2X_2, CHX_3, and CX_4 when X is F or Cl.

2.4 The Methyl Cation ($^+CH_3$), Anion ($^-:CH_3$), and Radical ($\cdot CH_3$)

Table 2.2 could be augmented to include $^+CH_3$, $^-:CH_3$, and $\cdot CH_3$, which are members of a class of compounds called **reactive intermediates**. This name implies that these molecules are too unstable to be isolable under normal conditions and must usually be studied by indirect means, often by looking at what they did during their brief lifetimes, or sometimes by isolating them at very low temperature. These three species are especially important because they are *prototypes*—the simplest examples of whole classes of molecules to be encountered later when the study of chemical reactions takes over our attention.

We call $^+CH_3$ a **methyl cation**. To make this molecule, one could imagine removing a **hydride** ($^-:H$) from methane to produce $^+CH_3$. It is the simplest example of a **carbocation**, a molecule containing a positively charged carbon[4] (Fig. 2.13a).

Now imagine forming the **methyl anion** ($^-:CH_3$) by removing a proton (H^+) from methane (Fig. 2.13b), leaving behind a pair of nonbonding or lone-pair electrons in the negatively charged methyl anion. The resulting $^-:CH_3$ is a simple example of a carbon-based anion, or **carbanion**.

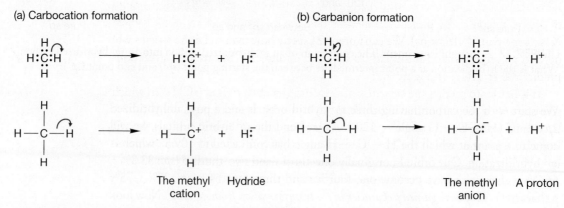

FIGURE 2.13 The formation of (a) the methyl cation ($^+CH_3$) and (b) the methyl anion ($^-:CH_3$) by two different heterolytic cleavages of a carbon–hydrogen bond in CH_4.

---

[4] The name for carbocations was the focus of a long and too intense argument in the chemical community. A carbocation is sometimes called a carbonium ion by traditionalists or a carbenium ion by others. The compromise carbocation is both aptly descriptive and avoids the emotional reactions of the staunch defenders of the other terms.

Both of these ways of producing $^+CH_3$ and $^-:CH_3$ involve the concept of breaking a carbon–hydrogen bond in unsymmetrical fashion, a process known as heterolytic bond cleavage (p. 37 and Fig. 2.13). Remember the curved arrow formalism—the red arrows of Figure 2.13 move the *pair* of electrons in the carbon–hydrogen bond to the hydrogen or to the carbon.

Recall from p. 37 that there is another way of breaking a two-electron bond, and that is to allow one electron to go with each atom involved in the breaking bond (Fig. 2.14). This homolytic bond cleavage in methane gives a hydrogen atom (H·) and leaves behind the neutral **methyl radical** ($\cdot CH_3$). Note the single-barbed "fish-hook" curved arrow convention is used to represent movement of one electron.

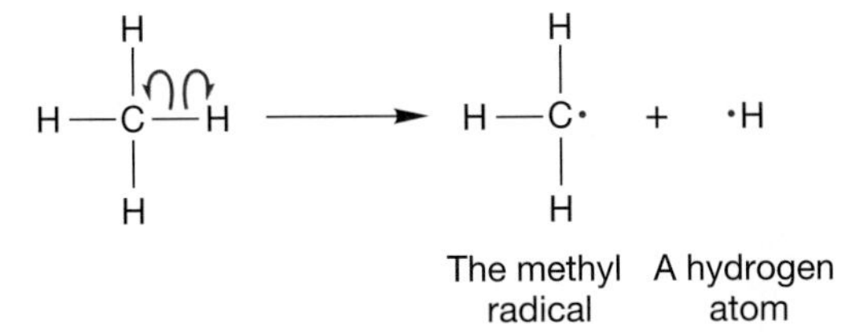

The methyl   A hydrogen
radical   atom

FIGURE 2.14 The homolytic cleavage of a carbon–hydrogen bond in methane to give a hydrogen atom and the methyl radical.

The methyl cation, anion, and radical have all been observed, although each is extremely reactive, and thus, short-lived. They exist, though, and we can make some predictions of structure for at least two of them. In the methyl cation ($^+CH_3$), carbon is attached to three hydrogens, suggesting the need for three hybrid atomic orbitals (recall BH_3, p. 55), and therefore sp^2 hybridization (Fig. 2.15).

Unlike the methyl cation, the carbon in the methyl anion is not only attached to three hydrogens but also has a pair of nonbonding electrons. The cation has an empty pure p orbital (zero s character) and therefore the species is as flat as a pancake (H—C—H angle = 120°). The methyl anion has two more electrons than the cation and we have to consider them in arriving at a prediction of the anion's shape. Recall from Figure 1.7 that s orbitals have density at the nucleus. Because the nucleus is positively charged and electrons are negatively charged, it is reasonable to assume that an electron is more stable (lower in energy) in an orbital with a lot of s character. A pyramidal structure seems appropriate for the methyl anion, although of course it cannot be a perfect tetrahedron because this is a CH_3X molecule (where X is a lone pair of electrons). We can't predict exactly how pyramidal the species will be, and an anion's shape is difficult to measure in any case, but recent calculations predict the structure in Figure 2.16.

It is harder to predict the structure of the neutral methyl radical ($\cdot CH_3$), in which there is only a single nonbonding electron. At present, it is not possible to choose between a planar species and a rapidly inverting and very shallow pyramid, although it is clear that the methyl radical is close to planar (Fig. 2.17). Do not be disturbed by this! Chemists still do not know many seemingly simple things (such as the shape and hybridization of the methyl radical). There is still lots to do!

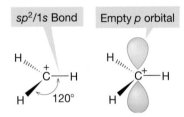

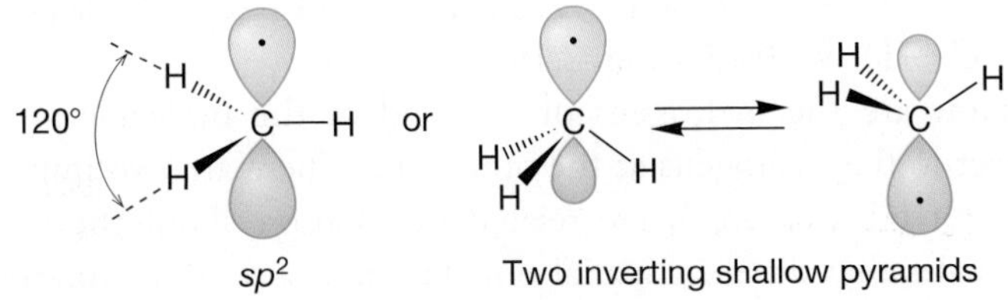

FIGURE 2.15 The sp^2 hybridized methyl cation, $^+CH_3$. The three bonds shown are the result of overlap between carbon's sp^2 hybrid orbitals and the 1s atomic orbital of each hydrogen. The four atoms all lie in the same plane, which is perpendicular to the plane of the page.

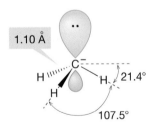

FIGURE 2.16 The structure of the methyl anion, $^-:CH_3$. The hybridization of the carbon in this carbanion is approximately sp^3. The molecular shape is pyramidal.

120°

sp^2 Two inverting shallow pyramids

FIGURE 2.17 The methyl radical ($\cdot CH_3$) is either planar or a rapidly inverting shallow pyramid. The carbon is close to sp^2 hybridized.

PROBLEM 2.11 Draw a structure for the methyl radical at the halfway point for the inversion shown in Figure 2.17. What is the hybridization of the carbon atom in the structure you drew?

Summary

Methane can be substituted in many ways through replacement of one or more hydrogens with another atom or groups of atoms. In principle, removal of a hydrogen from methane can lead to the methyl anion ($^-$:CH_3), the methyl radical (·CH_3), or the methyl cation ($^+CH_3$) depending on the nature of the hydrogen removed (^+H, ·H or $^-$:H). In this chapter, we have discussed only the shapes of these intermediates—reactions are coming later. It will be important to remember that carbocations are flat and sp^2 hybridized and that simple carbanions are pyramidal and approximately sp^3 hybridized.

2.5 Ethane (C_2H_6), Ethyl Compounds (C_2H_5X), and Newman Projections

There is a special substituent X that produces a most interesting H_3C—X. In this case, we let X = CH_3, another methyl group. Now H_3C—X becomes H_3C—CH_3 which can also be written as C_2H_6. This molecule is ethane (Fig. 2.18), the second member of the alkane family. You might try to anticipate the following discussion by building a model of ethane and examining its structure now.

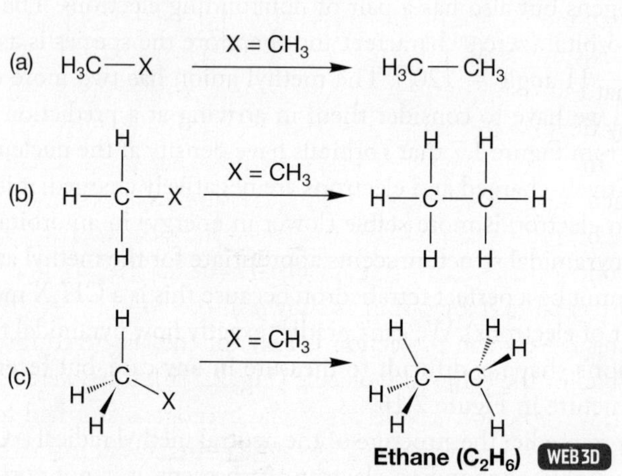

FIGURE 2.18 The construction of ethane (H_3C—CH_3) by the replacement of X in H_3C—X with a methyl group.

Ethane (C_2H_6)   WEB 3D

Even in a molecule as simple as ethane there are a number of interesting structural questions. From the point of view of one carbon, the attached methyl group takes up more room than the much smaller hydrogens. So we expect the H_3C—C—H angle to be slightly larger than the H—C—H angle. And so it is: H—C—CH_3 = 111.0° and H—C—H = 109.3° (Fig. 2.19).

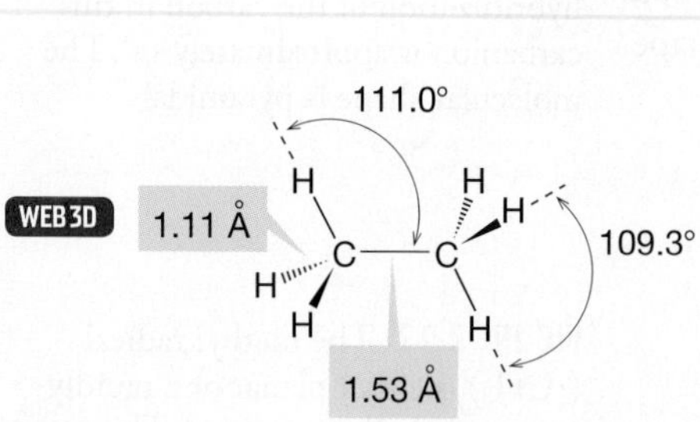

How are the hydrogens of one end of the molecule arranged with respect to the hydrogens at the other end? There are two important structures for ethane, which are related by rotation about the C—C bond. Molecules that differ in spatial orientation as a result of rotation around a

FIGURE 2.19 The detailed structure of ethane.

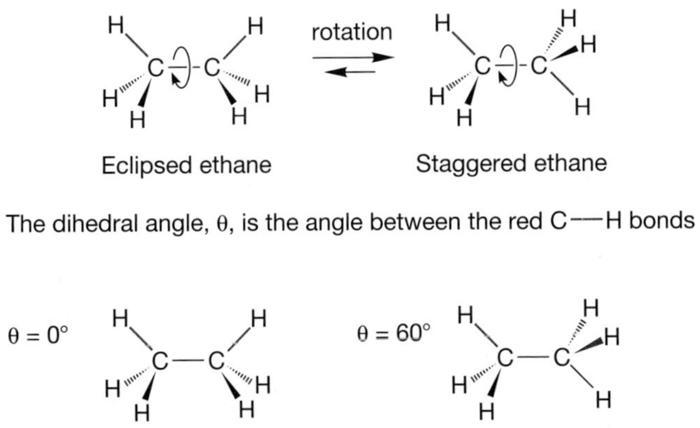

FIGURE 2.20 Two conformations of ethane: the eclipsed and staggered forms.

Eclipsed ethane   Staggered ethane

The dihedral angle, θ, is the angle between the red C—H bonds

$\theta = 0°$ $\theta = 60°$

sigma bond are called **conformations**. The two important conformations for ethane are **eclipsed ethane** and **staggered ethane**, each shown in a side view in Figure 2.20. If one looks down the C—C bond (Fig. 2.21) one can see that in the eclipsed conformation each of the three hydrogens on the front methyl eclipses a hydrogen on the back methyl. In the staggered conformation, the view down the C—C shows the hydrogens in front arranged so that each hydrogen in back can be seen. When we look down the C—C bond, the angle between a hydrogen in front and a hydrogen in back is called the **dihedral angle**. The dihedral angle is measured in degrees θ (pronounced thā-ta). The dihedral angle between the hydrogens in the eclipsed ethane is 0°. The dihedral angle in the staggered conformation is 60°. Ethane can exist in an infinite number of conformations, depending on the value of the dihedral angle. We focus here on the two extreme structures, eclipsed and staggered ethane. Be sure you see how these two forms are interconverted by rotation about the carbon–carbon bond. Use a model!

A particularly effective way of viewing different conformations of a molecule is called a **Newman projection**, after its inventor, Melvin S. Newman (1908–1993) of Ohio State University. Like all devices, the Newman projection contains arbitrary conventions that can only be learned, not reasoned out. Its utility will repay you for the effort many times over, however, so do it! We first imagine looking down a particular bond, in this case the carbon–carbon bond in ethane. The three carbon–hydrogen bonds attached to the front carbon are now drawn in as shown in the middle row of Figure 2.21. Notice that from this end-on view the H—C—H angle on the front carbon is 120°. The rear carbon is next represented as a circle, and the H atoms attached to it are put in as shown in the bottom row of Figure 2.21.

Constructing a Newman projection is easy for the staggered form (Fig. 2.21a) but not so simple for the eclipsed molecule (Fig. 2.21b). If we were strictly accurate in the drawing, we wouldn't be able to see the hydrogens attached to the rear carbon of the eclipsed conformation because each is directly behind an H on the front carbon (the dihedral angle is 0°). So we cheat a little and offset these eclipsed hydrogens just enough so we can see them.

Newman projections are extraordinarily useful in making three-dimensional structures clear on a two-dimensional surface. For example, Figure 2.21a makes it obvious that all six hydrogens in staggered ethane are the same—they are *equivalent* in the language of organic chemistry.

CONVENTION ALERT

PROBLEM 2.12 Write Newman projections for the staggered conformations of ethyl chloride (CH$_3$—CH$_2$—Cl) and 1,2-dichloroethane (Cl—CH$_2$—CH$_2$—Cl). In the second case, there are two staggered conformations of different energy. Can you estimate which is more stable?

FIGURE 2.21 (a) A Newman projection for staggered ethane. The dihedral angle (θ) between two carbon–hydrogen bonds is 60°. (b) A Newman projection for eclipsed ethane. The dihedral angle (θ) between two carbon–hydrogen bonds is 0°.

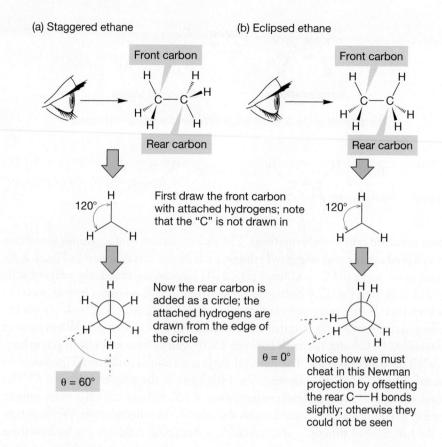

(a) Staggered ethane

(b) Eclipsed ethane

First draw the front carbon with attached hydrogens; note that the "C" is not drawn in

Now the rear carbon is added as a circle; the attached hydrogens are drawn from the edge of the circle

Notice how we must cheat in this Newman projection by offsetting the rear C—H bonds slightly; otherwise they could not be seen

Which of the two limiting conformations of ethane represents the more stable molecule? We would surely guess that it would be the staggered conformation, and that guess would be exactly right. But the reason has little to do with the "obvious" spatial requirement for the hydrogens. The hydrogens do not compete for the same space in the eclipsed conformation. There may be some electrostatic repulsion between the electrons in the bonds in the eclipsed bonds, but the eclipsed conformation is mostly destabilized by the repulsive interaction of two *filled* orbitals in each of the three pairs of eclipsed carbon–hydrogen bonds (Fig. 2.21b).

PROBLEM 2.13 Use an orbital interaction diagram like the one for He_2 in Figure 1.46, p. 40, to show the destabilization in eclipsed ethane. How many eclipsing filled orbital–filled orbital interactions are present?

PROBLEM 2.14 There is a related orbital effect that *stabilizes* staggered ethane. Use an orbital interaction diagram like the one for H_2 in Figure 1.47, p. 42, to show the stabilization in staggered ethane. This problem is much harder than Problem 2.13, so here is some help in the form of a set of tasks. The explanation that this problem leads to was first pointed out to MJ by an undergraduate just like you about 25 years ago.

(a) Draw staggered ethane in a Newman projection.
(b) Draw the antibonding orbital for one of the C—H bonds of the front carbon.
(c) Consider the C—H bond that is directly behind the antibonding orbital you have drawn. Can you see the overlap between the filled bonding orbital in back and the empty antibonding orbital you drew?

(continued)

(d) What factor stabilizes the staggered conformation of ethane over the eclipsed conformation? How many interactions in the staggered conformation of ethane have overlap of bonding and antibonding orbitals?

If you can fight your way through this problem, you are in excellent shape in terms of manipulating orbitals!

We can now make a plot of dihedral angle as a function of energy (Fig. 2.22). Calculations show that the eclipsed form of ethane is an energy maximum, the top of the energy barrier separating two staggered forms. Such a maximum-energy point is called a **transition state (TS)**.

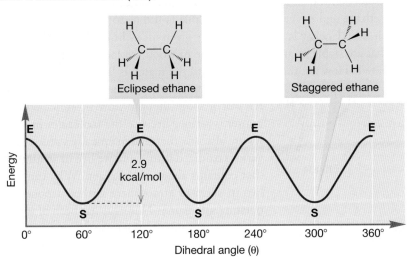

FIGURE 2.22 A plot of the energy of ethane as rotation around the carbon–carbon bond occurs.

PROBLEM 2.15 You recently saw a transition state, although the term wasn't used. Where? *Hint*: Where have we described two forms of a molecule that interconvert by passing through another species? It is close by.

How big is the energy difference between the staggered and eclipsed conformations of ethane? This question asks how high the energy barrier is between two of the minimum-energy staggered forms in Figure 2.22. The barrier turns out to be a small number, 2.9 kcal/mol, an amount of energy easily available at room temperature. At normal ambient temperatures, ethane is said to be freely rotating because there is ample thermal energy to traverse the barrier separating any two staggered forms. On some very cold planet, though, where there would be much less thermal energy available than on Earth, alien organic chemists would have to worry more about this rotational barrier. Thus, their lives would be even more complicated than ours.

PROBLEM 2.16 Draw the low-energy Newman projection for the structure depicted below by looking down the carbon–carbon bond.

$$H_3C - CH_2OH$$

Eye

PROBLEM 2.17 Imagine 1,2-dideuterioethane (DCH$_2$—CH$_2$D), where D = deuterium, on some cold planet such as Pluto. If there were not enough energy available to overcome the barrier to rotation, how many 1,2-dideuterioethanes would there be?

The chemical and physical properties of ethane closely resemble those of methane. Both are gases at room temperature, and both are quite unreactive under most chemical conditions. But not all conditions! One need only light a match in a room containing either ethane or methane and air to find that out. This reaction is actually very interesting. If we examine the debris after the explosion, we find two new molecules, water and carbon dioxide:

$$2\ H_3C\!-\!CH_3 \xrightarrow[\text{match}]{7\ O_2} 4\ CO_2 \ + \ 6\ H_2O \ + \ \text{Heat and Light}$$
Ethane

A thermochemical analysis indicates that water and carbon dioxide are more stable than ethane and oxygen. Presumably, this is the reason ethane explodes when the match is lit. Energy is all too obviously given off as heat and light. But why do objects that are stable in air, such as ethane or wood, explode or burn continuously when ignited? How are they protected until the match is lit?

We'll approach these questions in Chapter 8, but it's worth some thought now in anticipation. This matter is serious because the molecules in our bodies are also less thermodynamically stable than their various oxidized forms, and we live in an atmosphere that contains about 20% oxygen. Why can humans live in such an atmosphere without spontaneously bursting into flame? This issue has been of some concern, and not only for scientists.[5]

We can imagine making ethane by allowing two methyl radicals to come together (Fig. 2.23). The two sp^3 hybrid orbitals, one on each carbon, overlap to form a new carbon–carbon bond. This orbital overlap produces a bonding molecular orbital, which is a σ orbital because of its cylindrical symmetry. Of course, the process simultaneously creates an antibonding molecular orbital (σ*). The bonding orbital is filled by the two electrons originally in the two methyl radicals and the antibonding orbital is empty. We might guess that the overlap of the two equal-energy sp^3 hybrid orbitals would be very favorable, and it is. The carbon–carbon bond in ethane has a strength of about 90 kcal/mol. This value is the amount of energy by which ethane is stabilized relative to a pair of separated methyl radicals and is therefore the amount of energy required to break this strong carbon–carbon bond. The formation of ethane from two methyl radicals is exothermic by 90 kcal/mol:

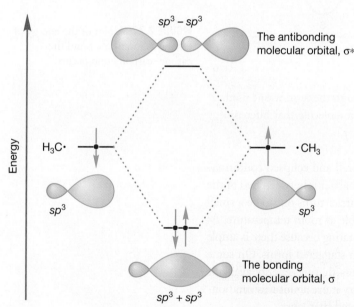

FIGURE 2.23 An orbital interaction diagram for the formation of ethane through the combination of a pair of methyl radicals.

$$H_3C\!\cdot\ +\ \cdot CH_3 \rightarrow H_3C\!-\!CH_3 \qquad \Delta H° = -90 \text{ kcal/mol}$$

[5]Eliot . . . claimed to be deeply touched by the idea of an inhabited planet with an atmosphere that was eager to combine violently with almost everything the inhabitants held dear . . . "When you think of it, boys," he said brokenly, "that's what holds us together more than anything else, except maybe gravity. We few, we happy few, we band of brothers—joined in the serious business of keeping our food, shelter, clothing and loved ones from combining with oxygen."

—Kurt Vonnegut,
God Bless You, Mr. Rosewater

and the dissociation of ethane into two methyl radicals is endothermic by 90 kcal/mol:

$$H_3C—CH_3 \rightarrow H_3C\cdot + \cdot CH_3 \qquad \Delta H° = +90 \text{ kcal/mol}$$

Just as we imagined replacing one of the hydrogens in methane with an X group to generate the series of methyl compounds, so we can produce **ethyl compounds** ($CH_3CH_2—X$) from ethane. Indeed, we can now generalize: Substituted alkanes are called **alkyl compounds**, often abbreviated R—X, where R stands for a generic alkyl group. Table 2.3 shows some common ethyl compounds and a few of their physical properties.

TABLE 2.3 Some Simple Derivatives of Ethane: Ethyl Compounds

$CH_3CH_2—X$	Common Name	mp (°C)	bp (°C)	Physical Properties
$CH_3CH_2—H$	Ethane	−183.3	−88.6	Colorless gas
$CH_3CH_2—OH$	Ethyl alcohol or ethanol	−117.3	78.5	Colorless liquid
$CH_3CH_2—NH_2$	Ethylamine	−81	16.6	Colorless gas/liquid
$CH_3CH_2—Br$	Ethyl bromide	−118.6	38.4	Colorless liquid
$CH_3CH_2—Cl$	Ethyl chloride	−136.4	12.3	Colorless gas/liquid
$CH_3CH_2—CN$	Ethyl cyanide or propionitrile	−92.9	97.4	Colorless liquid
$CH_3CH_2—F$	Ethyl fluoride	−143.2	−37.7	Colorless gas
$CH_3CH_2—I$	Ethyl iodide	−108	72.3	Colorless liquid
$CH_3CH_2—SH$	Ethanethiol or ethyl mercaptan	−144	35	Colorless liquid
$^+CH_2CH_3$	Ethyl cation (reactive intermediate)			
$^-:CH_2CH_3$	Ethyl anion (reactive intermediate)			
$\cdot CH_2CH_3$	Ethyl radical (reactive intermediate)			

How many different $CH_3CH_2—X$ compounds are there? From a two-dimensional representation of ethane (Fig. 2.24a), there appear to be two ethyl compounds with the same formula but different structures. Compounds of this type, which have the same molecular formula but are not the same structure, are called

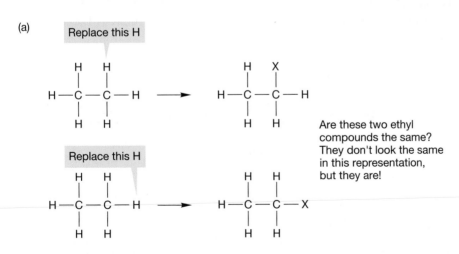

FIGURE 2.24 (a) A two-dimensional representation of ethane suggests that there are two different kinds of ethyl compound, $CH_3CH_2—X$. (b) We must look at ethane in three dimensions to see why there is only one $CH_3CH_2—X$.

isomers (from the Greek words meaning "the same parts"). If the two ethyl compounds of Figure 2.24a existed, they would be isomers of each other. However, the two different positions shown for X (at the "top" of the molecule and on the "side") do not in fact exist. They are artifacts of our attempt to represent a three-dimensional molecule in two dimensions. This common misconception for beginners demonstrates how important it is to know the three-dimensional structures of compounds. As we saw in Figure 2.21a, all six hydrogens in ethane are the same (equivalent) and therefore there is only one kind of substituted ethyl compound (Fig. 2.24b). Of course there are different *conformations* of a substituted ethyl compound. We know about the staggered and eclipsed conformations and we can call these **conformational isomers**, because the two conformations have the same molecular formula but different structures. However, even though most conformational isomers have relatively low barriers to rotation, we usually focus on the most stable representation. In the ethyl series, we will work with the staggered structure.

Summary

Even a molecule as simple as ethane, the two-carbon alkane, has a complex structure! For example, staggered and eclipsed conformational isomers exist, separated by a low barrier to rotation (~3 kcal/mol). Because this barrier is so low, there is rapid rotation around the carbon–carbon bond. All six hydrogens in ethane are equivalent; there is only one ethane and only one kind of ethyl derivative. You may need a ball-and-stick model set to convince yourself.

2.6 Structure Drawings

CONVENTION ALERT

In the preceding sections, there are many different representations of the very simple molecule ethane. Figure 2.25 recapitulates them and adds some new ones. Methyl groups are sometimes written as Me and ethyl groups as Et, especially in colloquial use.

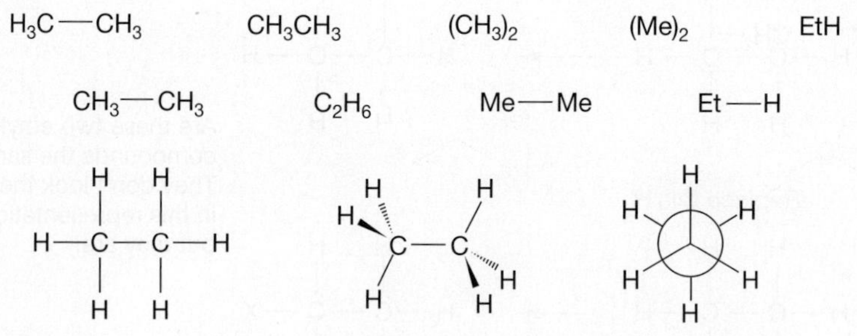

FIGURE 2.25 Twelve different representations of ethane.

The difficulty of representing the real, dynamic, three-dimensional structure of ethane should be more apparent to you now. In the real world, there is not often the time to draw out carefully a good representation of even as simple a molecule as ethane. The solid and dashed wedges of Figure 2.25 are the traditional attempts at adding a three-dimensional feel to the two-dimensional drawing. More complicated molecules can raise horrendous problems. Adequate codes are needed, and you must learn to see past the coded structures to the real molecules both easily and quickly. It's worth considering here, at this very early point, some of the pitfalls of the various schemes.

First of all, it is not simple to represent even ethane effectively in a linear fashion. Clearly, neither C_2H_6 nor Et—H is as descriptive as H_3C—CH_3, as those formulas don't show the connectivity of the atoms in a proper way. The representation H_3C—CH_3 is better, but even this picture lacks three dimensionality. We used this representation in this chapter, but you will sometimes see the variant CH_3—CH_3. The two formulations are equivalent, even though the CH_3—CH_3 seems to indicate that the bond to the right-hand carbon comes from one of the hydrogens on the left-hand carbon. Of course, it does not; it is the two carbons that are bonded, as H_3C—CH_3 shows. Yet it is easier to write CH_3—CH_3, and you will see this form often.

PROBLEM 2.18 Draw a three-dimensional structure for $(Me)_2CH_2$, $(CH_3)_4C$, $(CH_3)_3CH$, $EtCH_3$, $(Et)_2$, $CH_3CH_2CH_3$, EtMe, MeEt.

PROBLEM 2.19 Start with the two "different" structures in Figure 2.24a and replace the X group with CH_3 in each. Make three-dimensional drawings of the "two" molecules you've created and convince yourself that both your three-dimensional drawings represent the same molecule; there is only one $CH_3CH_2CH_3$. By all means, use your models.

2.7 Propane (C₃H₈) and Propyl Compounds (C₃H₇X)

Propane, the third member of the alkane series, contains three carbon atoms and has many of the chemical and physical properties of ethane and methane. Figure 2.26 shows several representations for propane, including two new ones. In Figure 2.26d, only the carbon frame is retained, and all eight hydrogen atoms are implied.

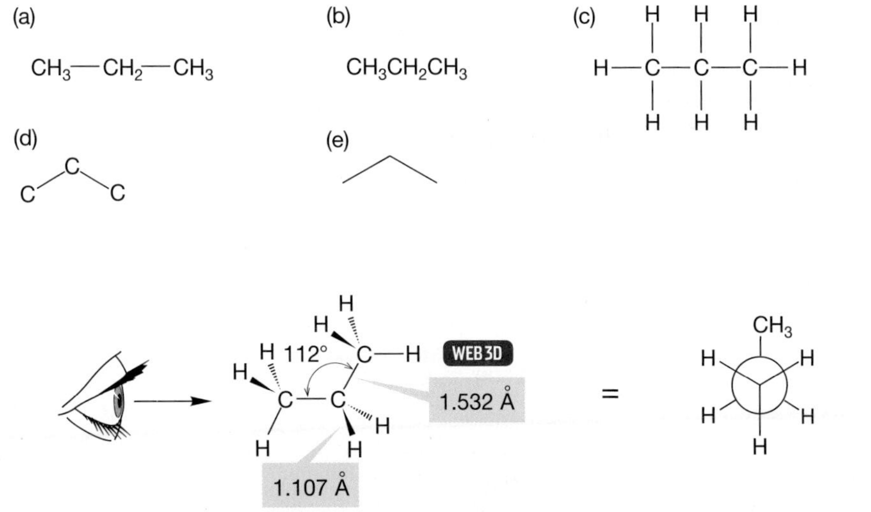

FIGURE 2.26 Different representations of propane.

Not even the carbons are shown in Figure 2.26e, and the reader is left to fill in mentally all the atoms. This most abstract representation—called a *line drawing* of the molecule—is the representation of choice for organic chemists, and we will gradually slip into this way of drawing molecules in subsequent chapters. In a line drawing of a hydrocarbon, every angle vertex and terminus is a carbon, never a hydrogen, and you have to add hydrogens to every carbon so that carbon's valence of four bonds is satisfied.

Notice that some, but not all of the schematic representations for propane show that it is not a strictly linear species. All carbons are hybridized approximately sp^3, and the C—C—C angle is close to 109° (in fact, it is 112°).

The Newman projection for propane (Figs. 2.26 and 2.27), constructed by looking down one of the two equivalent carbon–carbon bonds, shows the structure of propane particularly well. A plot of energy versus dihedral angle for propane, shown in Figure 2.27, is similar to that for ethane, except that the rotational barrier of 3.4 kcal/mol is slightly higher. In propane, a large methyl group is eclipsed with a hydrogen at the top of the barrier (the transition state), and this spatial interaction accounts for the higher energy.

FIGURE 2.27 A graph of energy versus dihedral angle (θ) for propane.

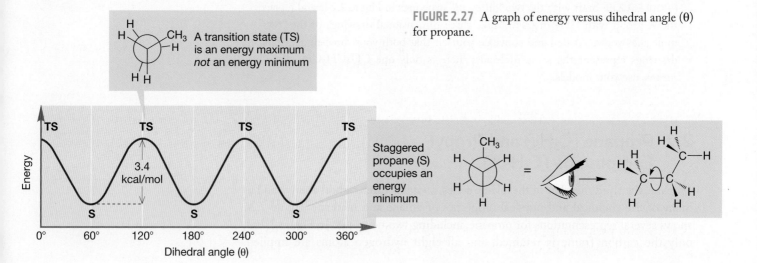

Replacement of a hydrogen in propane with an X group yields propyl derivatives, but the situation is more complicated than in methane or ethane. In both methane and ethane, all hydrogens are equivalent, and there can be only one methyl or one ethyl derivative. In propane, there are two different kinds of hydrogen. There is a central CH_2 group, called a **methylene group**, and two equivalent methyl groups at the two ends of the molecule (Fig. 2.28). The carbon and hydrogens in a methylene group (CH_2) are not the same as the carbons and hydrogens in methyl groups (CH_3).

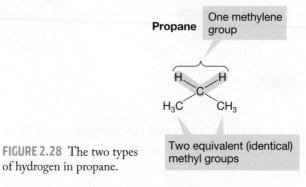

FIGURE 2.28 The two types of hydrogen in propane.

Either of these two kinds of hydrogen can be replaced by an X to give a propyl compound (Fig. 2.29). The linear compound $CH_3—CH_2—CH_2—X$, in which X replaces a methyl hydrogen, is called **propyl** X (in the old days, *n*-propyl X). The branched compound $CH_3—CHX—CH_3$, in which X replaces a hydrogen on the methylene carbon, is called **isopropyl** X (and sometimes *i*-propyl X). Propyl compounds and isopropyl compounds are **structural isomers** of each other; they are compounds with the same formula but different structures.

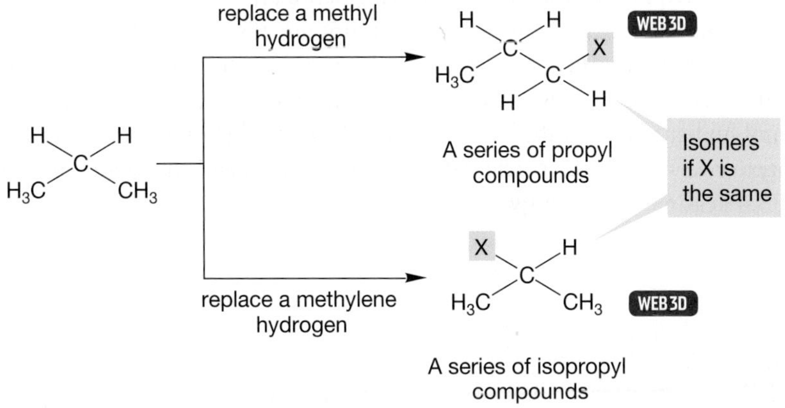

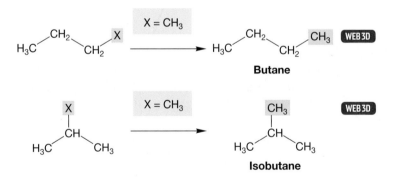

FIGURE 2.29 Propyl and isopropyl compounds.

PROBLEM 2.20 Make a three-dimensional drawing of propyl alcohol ($CH_3—CH_2—CH_2—OH$) and one of the related isopropyl alcohol ($CH_3—CHOH—CH_3$).

2.8 Butanes (C₄H₁₀), Butyl Compounds (C₄H₉X), and Conformational Analysis

If, in the two propyl isomers on the right in Figure 2.29, we let X be methyl, we generate the isomers butane and isobutane (Fig. 2.30). The structure of butane

FIGURE 2.30 If X = CH₃, we produce the isomers butane and isobutane.

(Fig. 2.31) provides a nice example of **conformational analysis**, the study of the relative energies of conformational isomers. Let's begin this analysis by constructing a Newman projection by looking down the C(2)—C(3) bond of butane

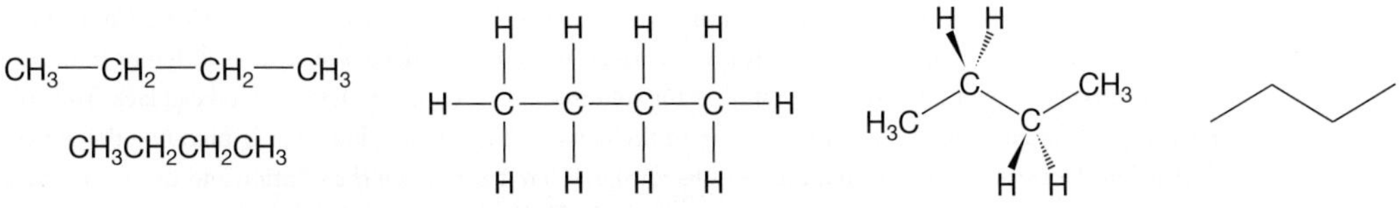

FIGURE 2.31 Different representations of butane.

(Fig. 2.32a). The notation C(2), C(3), and so on will be used to refer to the carbon atom number. If we make the reasonable assumption that the relatively large methyl groups attached to these two carbons are best kept as far from each other as possible, we have the most stable staggered conformation **A** in Figure 2.32a, which is called the anti form. As in ethane (see Fig. 2.22), a 120° clockwise rotation of the C(2)—C(3) bond keeping C(2) fixed, passes over a transition state (an energy maximum); this is the first transition state in Figure 2.32b. As the rotation continues, the molecule acquires a new staggered conformation, **B**. Conformation **B** is less stable than **A** because in **B** the two methyl groups are closer to each other than they are in **A**. Conformation **B** is called *gauche*-butane, and this kind of methyl–methyl interaction is a gauche interaction. The gauche form of butane (**B**) is about 0.6 kcal/mol higher in energy than the anti form (**A**).[6] The eclipsed transition state for the interconversion of staggered conformation **A** and staggered conformation **B** lies about 3.4 kcal/mol higher than **A**.

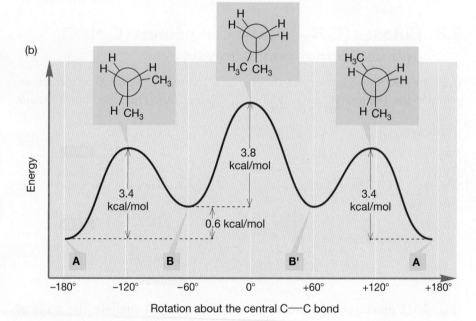

FIGURE 2.32 (a) Newman projections for butane. The staggered conformations are shown. (b) A graph of dihedral angle (θ) versus energy for butane as the molecule changes through rotation about the C(2)—C(3) bond. All three eclipsed conformations are shown.

---

[6]This number, 0.6 kcal/mol, is somewhat controversial. Most books give this value as 0.9 kcal/mol. That 0.9 figure is correct for the gas phase, but is not correct in solution. Indeed, the value is different in different solvents. It seemed best to give a number for solution, in which most chemistry takes place. You will find slight differences in almost every version of the curve in Figure 2.32, but the big picture—the general shape of the curve—will not vary. The reasons for the difference between the solution and the vapor phase are complex, but if you would like to read more, see E. E. Eliel and S. H. Wilen, *Stereochemistry*, Wiley, 1994, pp. 600 ff.

Another 120° clockwise rotation brings us to a second *gauche*-butane, **B′**. The transition state for the conversion of **B** to **B′**, the second transition state in Figure 2.32b, involves an eclipsing of two C—CH₃ bonds and lies 3.8 kcal/mol above **B** and **B′**, a total of about 4.4 kcal/mol above **A**. A final 120° clockwise rotation returns us to **A**.

PROBLEM 2.21 Recall from Figure 2.22 that the staggered ethane conformation is 2.9 kcal/mol more stable than its eclipsed conformation. Given that value and the information that the transition state for the interconversion of staggered conformations **A** and **B** in butane is 3.4 kcal/mol higher in energy than **A** (Fig. 2.32b), calculate how much energy each eclipsed C—CH₃/C—H interaction in the transition state for the interconversion of **A** and **B** is worth. How much energy is the eclipsed C—CH₃/C—CH₃ interaction in the transition state for the conversion of **B** to **B′** worth?

PROBLEM 2.22 Draw Newman projections constructed by looking down the indicated carbon–carbon bond in the following molecules. For the molecule on the right, make a graph of energy versus dihedral angle.

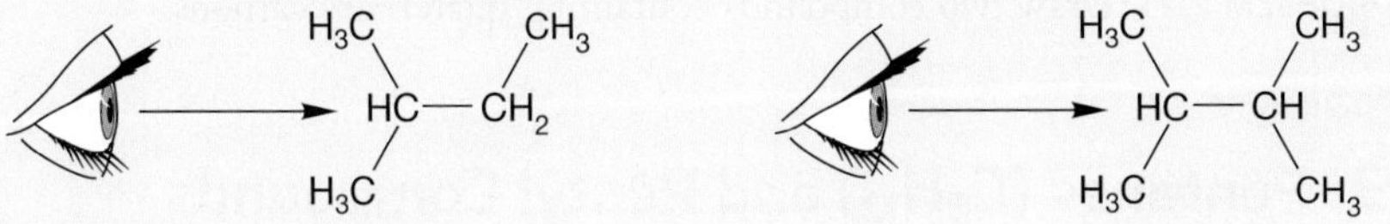

We start our analysis of branched compounds with the butyl compounds, C₄H₉—X. Butane contains two kinds of hydrogen atoms in its pairs of equivalent methylene (CH₂) groups and methyl (CH₃) groups. Replacement of hydrogen a or b produces two types of butyl compounds, shown in Figure 2.33a and b. Similarly, isobutane contains a single **methine group** (CH) and three equivalent methyl groups and also yields two types of butyl compounds (Fig. 2.33c and d).

The four different types of butyl compounds are called **butyl, isobutyl, *sec*-butyl** (short for *secondary*-butyl), and ***tert*-butyl** (or sometimes *t*-butyl, short for *tertiary*-butyl).

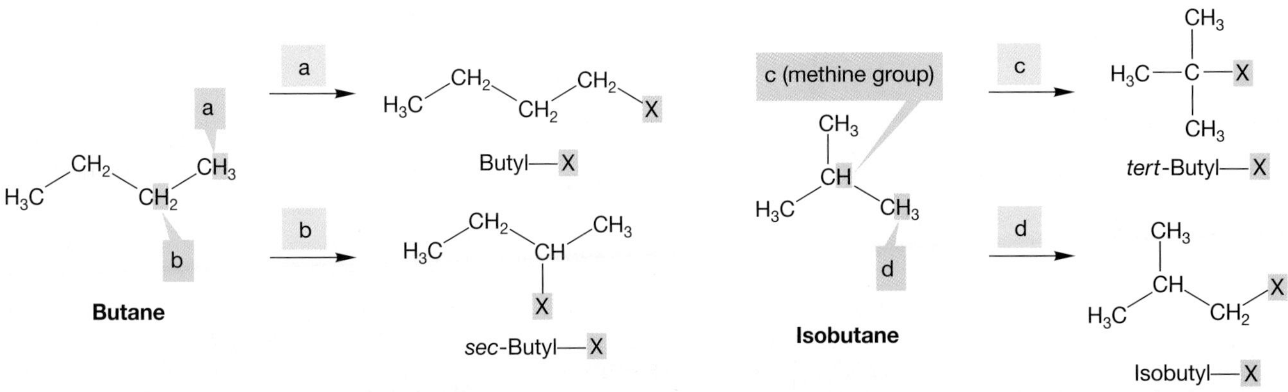

FIGURE 2.33 (a) and (b) Replacement of one of the hydrogens of butane with X yields two kinds of butyl derivatives, butyl—X and *sec*-butyl—X. (c) and (d) Replacement of hydrogen with X in isobutane yields two more butyl compounds, *tert*-butyl—X and isobutyl—X.

To see what the prefixes *sec-* and *tert-* mean, we must introduce new and important terminology (Fig. 2.34). A **primary carbon** is a carbon that is attached to only one other carbon atom. A **secondary carbon** is a carbon attached to two other carbons, and a **tertiary carbon** is a carbon attached to three other carbons. A **quaternary** (*not* qua~r~ternary) **carbon** (not shown in Figure 2.34, but we'll see one in a moment) is a carbon that is attached to four other carbons. Thus, the names *sec-*butyl and *tert-*butyl tell you something about the structure.

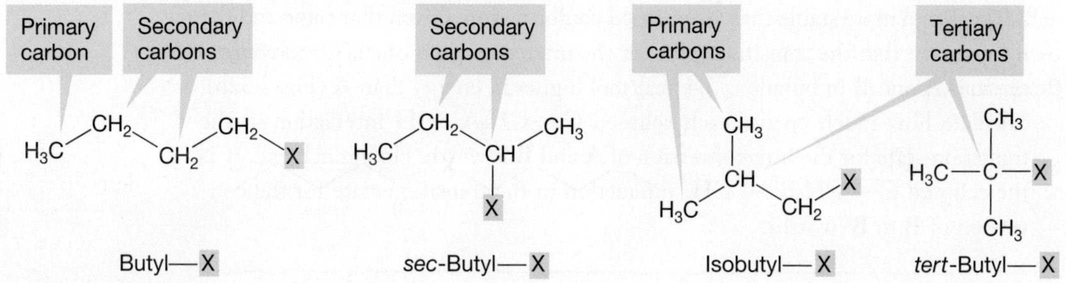

FIGURE 2.34 The four kinds of butyl compounds used to illustrate the bonding in primary, secondary, and tertiary carbons (X is not carbon or hydrogen).

PROBLEM 2.23 Draw two compounds containing quaternary carbons.

2.9 Pentanes (C₅H₁₂) and Pentyl Compounds (C₅H₁₁X)

We might now anticipate that four pentane compounds would emerge when we transform our four butyl—X compounds into five-carbon compounds by letting X = CH₃ (Fig. 2.35). Yet only three different pentanes exist! Somehow the

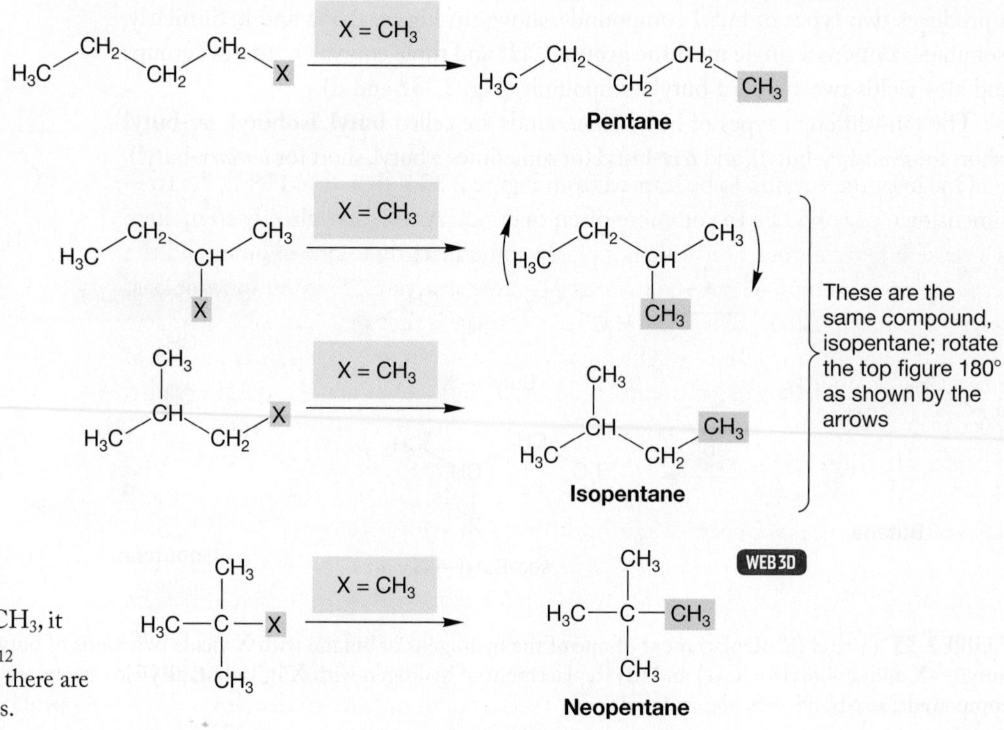

FIGURE 2.35 If we let X = CH₃, it appears as though four C₅H₁₂ compounds should exist. Yet there are only three C₅H₁₂ compounds.

technique of generating larger alkanes by letting a general X become a methyl group has failed us. The problem lies again in the difficulty of representing three-dimensional structures on a two-dimensional surface. Two of the four structures in Figure 2.35 represent the same compound, even though in the two dimensional representation they look different. The common names for the branched pentanes are isopentane and neopentane. We will not usually use these common names. Section 2.10 will describe a more systematic procedure for naming alkanes, including these two.

Even though there are only three pentane compounds—pentane, isopentane, and neopentane—Figure 2.36 shows that replacing one H in a pentane leads to no fewer than eight pentyl derivatives (C$_5$H$_{11}$—X)! In the next section, we'll see how to name these derivatives.

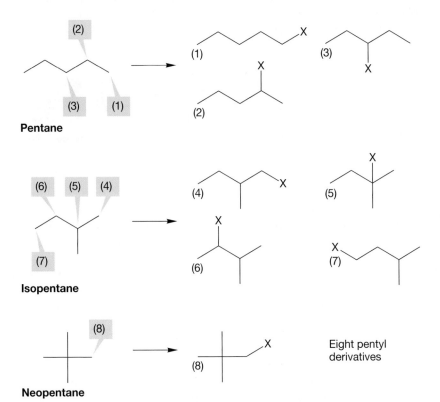

FIGURE 2.36 The eight pentyl derivatives formed by replacing hydrogens at the three different positions of pentane, the four different positions in isopentane, and the lone position of neopentane.

One important lesson to be learned from Figure 2.35 is that, if you let it, the two-dimensional page will lie to you more often than not. As we have already seen, there is a tension between our need to talk quickly and efficiently to one another and the accurate representation of these complicated organic structures. This tension is at least partly resolved by using codes, but the price of this use is that you must *always* be able to see past the code to the real, three-dimensional structure if you expect to be able to think effectively about organic chemistry. Because these codes become quite abstract, it is worth going through the process of developing the abstraction one more time.

Consider the representations for isopentane in Figure 2.37. The first drawing (Fig. 2.37a) only vaguely represents the real molecule. One then goes to the more schematic structure of Figure 2.37b, and then to a structure in which the hydrogens are left out (Fig. 2.37c). Finally, one removes even the carbon labels, as in Figure 2.37d, to give the ultimate abbreviation for the structure of isopentane. Here nothing survives from the original picture except a representation of the backbone. You must be able to translate these highly schematic structures into three dimensions and to see the way these skeletal pictures transform into molecules.

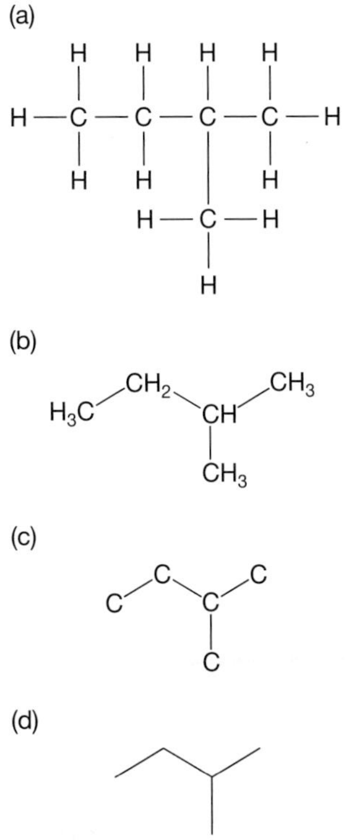

FIGURE 2.37 Four increasingly abstract representations of isopentane.

PROBLEM 2.24 Draw all the isomers of pentane, first in the format used in Figure 2.37b and then in the format used in Figure 2.37d.

Actually, there is even more to be done in our escape from the two dimensions of a book page. Line drawings can be approximated with models in which balls and sticks are used to represent atoms and the bonds between atoms. Other models concentrate on the bonds and leave it to you to imagine the atoms. Regardless of what kind of models your class uses, they probably do a relatively poor job of showing the **steric requirements** of the real molecules. When we refer to steric requirements, we mean the volumes of space that atoms occupy. Chemists have developed space-filling models, which attempt to show the volumes of space carved out by the atoms. Figure 2.38 shows some representations of isopentane.

FIGURE 2.38 Views of isopentane.

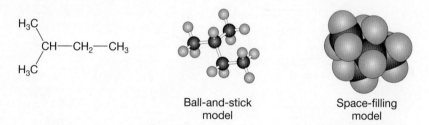

Ball-and-stick model   Space-filling model

Summary

There are several ways to represent molecules as summarized in Figures 2.37 and 2.38. The line drawings are the preferred choice for most cases. The convention of dashes and wedges gives more detail about the three-dimensional structures. Propane (C_3H_8) has staggered and eclipsed conformations and the barrier between them is relatively low (3.4 kcal/mol). Two isomers are possible for substituted propane (C_3H_7—X), with the common names propyl and isopropyl. Conformational analysis of butane (C_4H_{10}) introduces several new concepts. Butane has anti and gauche forms. The energy diagram for rotation around the central bond of butane (Fig. 2.32b) shows the effects of the steric interactions of eclipsing methyl groups. There are four isomers for the substituted butanes (C_4H_9—X), with the common names of butyl—X, *sec*-butyl—X, *tert*-butyl—X, and isobutyl—X. Pentane has two structural isomers isopentane and neopentane. There are eight isomers of the substituted pentanes (C_5H_{11}—X). A helpful tool for distinguishing between the different substituted compounds is the very important system of classifying carbons as primary, secondary, and tertiary.

2.10 The Naming Conventions for Alkanes

Clearly, things are rapidly getting out of hand. Although it is not too difficult to remember the four types of butyl compounds (Fig. 2.34), remembering the eight types of pentyl compounds (Fig. 2.36) constitutes a tougher task—and the problem rapidly gets worse as the number of carbons increases. A system is absolutely necessary. In practice, the old common, nonsystematic names are retained through the butanes, and for a few other old favorite molecules. Once we reach five carbons, a systematic naming protocol developed by the International Union of Pure and Applied Chemistry (IUPAC) takes over. The IUPAC system is designed to handle any organic structure. There are IUPAC names for all the common names we have learned.

In general, the stem of the name tells you the number of carbon atoms (meth = 1, eth = 2, prop = 3, but = 4, and so on), and the suffix "ane" tells you that the molecule is an alkane. The alkane name is sometimes called the root word. The names of some straight-chain alkanes are collected in Table 2.4.

TABLE 2.4 **Some Straight-Chain Alkanes**

Name	Formula	mp (°C)	bp (°C)
Methane	CH_4	−182.5	−164
Ethane	CH_3CH_3	−183.3	−88.6
Propane	$CH_3CH_2CH_3$	−189.7	−43.1
Butane	$CH_3(CH_2)_2CH_3$	−138.4	−0.5
Pentane	$CH_3(CH_2)_3CH_3$	−129.7	36.1
Hexane	$CH_3(CH_2)_4CH_3$	−95	69
Heptane	$CH_3(CH_2)_5CH_3$	−90.6	98.4
Octane	$CH_3(CH_2)_6CH_3$	−56.8	125.7
Nonane	$CH_3(CH_2)_7CH_3$	−51	150.8
Decane	$CH_3(CH_2)_8CH_3$	−29.7	174.1
Undecane	$CH_3(CH_2)_9CH_3$	−25.6	195.9
Dodecane	$CH_3(CH_2)_{10}CH_3$	−9.6	216.3
Eicosane	$CH_3(CH_2)_{18}CH_3$	36.8	343.0
Triacontane	$CH_3(CH_2)_{28}CH_3$	66	449.7
Pentacontane	$CH_3(CH_2)_{48}CH_3$	92	

It will be important to know how to interpret the IUPAC names, because this skill is part of learning the language of organic chemistry. This ability also will help you understand the contents listed on your cereal boxes and shampoo bottles. Here are the most important rules of the IUPAC naming convention for alkanes and substituted alkanes:

1. Find the longest chain of continuously connected carbons. Choose the root word that matches that chain length (see Table 2.4). The compound is named as a derivative of this parent hydrocarbon. Finding the longest chain can be a bit tricky at first because the angled, two-dimensional line drawings are not always drawn in a convenient horizontal line. The horizontal part of the drawing may not be the longest chain. In the examples in Figure 2.39 it is perfectly reasonable to count in any possible way in order to find the longest chain.

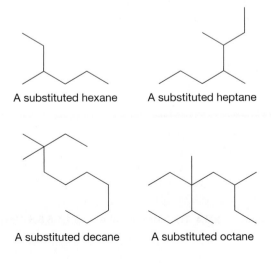

A substituted hexane   A substituted heptane

A substituted decane   A substituted octane

FIGURE 2.39 Finding the longest straight chain of carbons in alkanes.

2. In a substituted alkane, the substituent is given a number based on its position in the parent hydrocarbon. The longest chain is numbered so as to make the number of the substituent position as low as possible. The substituent is listed as a prefix to the root word, with the number indicating its location on the chain. Some examples are shown in Figure 2.40. In a polysubstituted alkane, all the substituents receive numbers. The longest chain is numbered so that the lowest possible numbers are used. A useful trick is to number the chain from the end closest to the first substituent or branch point.

FIGURE 2.40 Number the chain so as to produce the smallest number for a substituent.

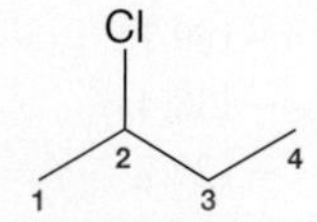

2-Chlorobutane

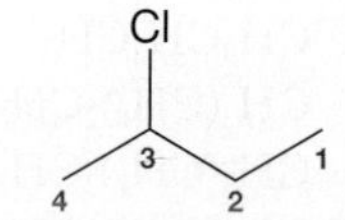

not **3-Chlorobutane**

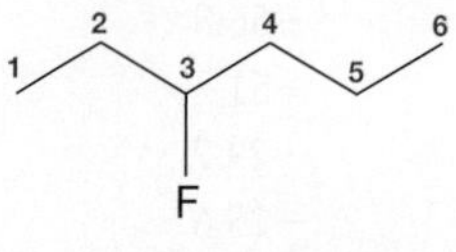

3-Fluorohexane

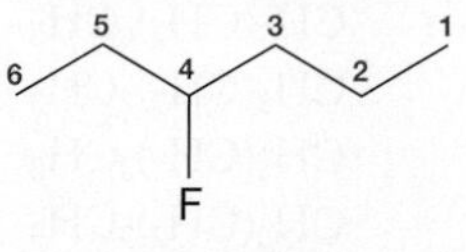

not **4-Fluorohexane**

3. When there are multiple substituents, they are always ordered alphabetically in the prefix (Fig. 2.41). If the multiple substituents are identical, then di-, tri-, tetra-, and

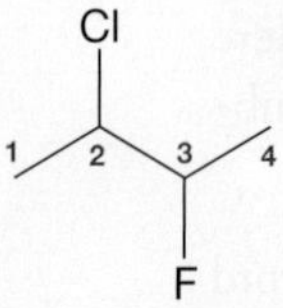

2-Chloro-3-fluorobutane
not 3-chloro-2-fluorobutane
not 2-fluoro-3-chlorobutane

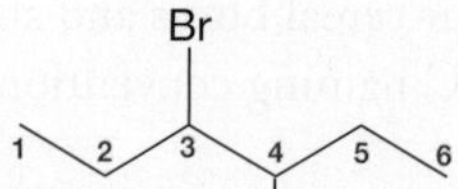

3-Bromo-4-methylhexane
not 4-bromo-3-methylhexane
not 3-methyl-4-bromohexane

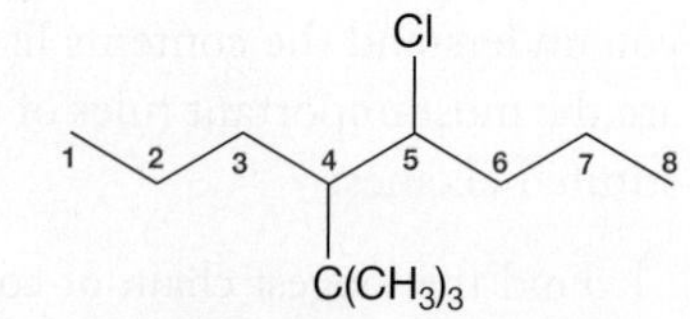

4-*tert*-Butyl-5-chlorooctane
not 4-chloro-5-*tert*-butyloctane

FIGURE 2.41 Nonidentical substituents are incorporated into the name alphabetically.

so on are used as prefixes to the substituent name (Fig. 2.42). The modifiers di-, tri-, tetra-, and so on are ignored in determining alphabetical order. Use of the common names for substituents is not strictly proper, but you might encounter it. In such cases the *sec-* and *tert-* are ignored in determining alphabetical order, but the *iso-* and *neo-* are used. Thus, *tert*-butyl would appear in the name earlier than chloro, and isopropyl would come before methyl.

FIGURE 2.42 When the substituents in a polysubstituted alkane are the same, prefixes di-, tri-, and so on tell you the number of the multiple substituents.

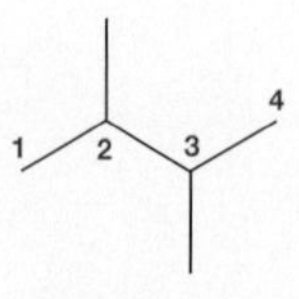

2,3-Dimethylbutane 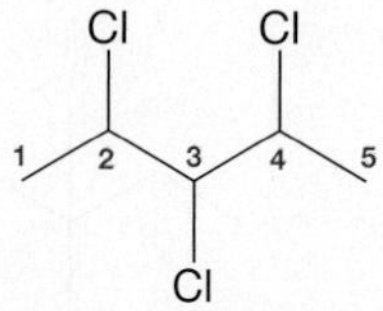   **2,3,4-Trichloropentane** 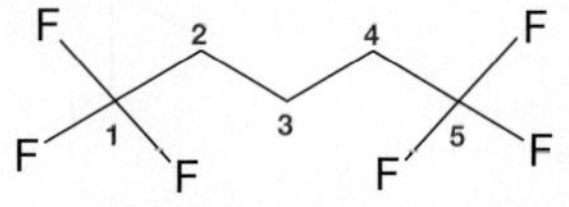   **1,1,1,5,5,5-Hexafluoropentane**

4. When rules 1 through 3 do not resolve the issue, the name that *starts* with the lower number is used, as demonstrated in Figure 2.43. This rule is sometimes referred to as the alphabetical preference rule. If both numbering choices on the alkane use the same numbers, then the preference is given to the alphabet.

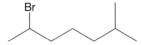

2-Bromo-6-methylheptane
not 6-bromo-2-methylheptane

FIGURE 2.43 In unresolvable cases, start with the lowest possible number. The carbon chain is numbered from the end that gives the first-named substituent the lower number.

PROBLEM 2.25 Write systematic names for the following compounds:

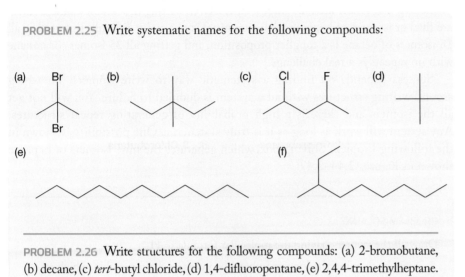

PROBLEM 2.26 Write structures for the following compounds: (a) 2-bromobutane, (b) decane, (c) *tert*-butyl chloride, (d) 1,4-difluoropentane, (e) 2,4,4-trimethylheptane.

Although the instruction to name substituents on alkanes in alphabetical order looks simple, the process is a bit complex because some modifiers count in the alphabet and others do not. Table 2.5 contains a brief summary.

TABLE 2.5 Some Common Prefixes

Prefix	Use	Counts in Alphabetical Order?
di-	Any two identical groups	No
tri-	Any three identical groups	No
tetra-	Any four identical groups	No
iso	Isopropyl = $(CH_3)_2CH-$	Yes
neo	Neopentyl = $(CH_3)_3CCH_2-$	Yes
sec-	sec-Butyl = $CH_3CH_2CH(CH_3)-$	No
tert-	tert-Butyl = $(CH_3)_3C-$	No

PROBLEM 2.27 Use the information in Table 2.5 to name the following two compounds:

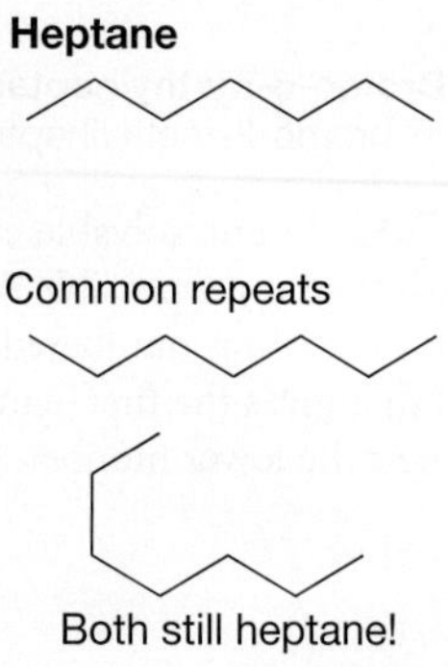

Heptane

Common repeats

Both still heptane!

FIGURE 2.44 Heptane.

Hexanes—
one C atom must
be added as a branch

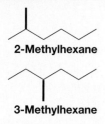

2-Methylhexane

3-Methylhexane

Common repeats

Still 3-methylhexane

Still 3-methylhexane

FIGURE 2.45 Methylhexanes.

2.11 Drawing Isomers

A common question on organic chemistry tests is, Write all the isomers of pentane (or hexane, or heptane, etc.). Such problems force us to cope with the translation of two-dimensional representations into three-dimensional reality. Figuring out the pentanes is easy; there are only three isomers. Even writing the hexane isomers (there are five) or isomers of heptane (there are nine) is not really difficult. But getting all 18 isomers of octane is a tougher proposition, and getting all 35 isomers of nonane with no repeats is a real challenge.

Success depends on finding a systematic way to write isomers. Thrashing around writing structures without a system is doomed to failure. You will not get all the isomers and there is a high probability of generating repeat structures. Any system will work as long as it is truly systematic. One possibility is shown in the following Problem Solving box, which generates the nine isomers of heptane shown in Figures 2.44–2.47.

PROBLEM SOLVING

Draw all the isomers having the molecular formula C_7H_{16}.

1. Start by drawing the longest unbranched carbon chain possible, which is heptane in this case (Fig. 2.44).
2. Shorten the chain by one carbon, and add the "extra" carbon as a methyl group at all possible positions, starting at the left and moving to the right (Fig. 2.45). Note that you can't add this carbon to the end of the chain because doing that would just regenerate heptane. You have to start the addition process one carbon in from the end. This step generates all the methylhexanes. You must check each isomer you make to be sure it is not a repeat. One way to be absolutely certain is to name each molecule as you generate it. If you repeat a name, you have repeated an isomer.
3. Shorten the chain by one more carbon. The longest straight chain is now five carbons. Two carbons are to be added as two methyl groups or as a single ethyl group (Fig. 2.46). Two methyl groups can be placed either on the same carbon or on different carbons. Again, start at the carbon one in from the left end of the chain and move to the right, checking each isomer you create by naming it to be sure it is not a repeat.
4. Shorten the chain by one carbon again, and try to fit the three "extra" carbons on a butane chain. There is only one way to do this (Fig. 2.47).

Pentanes—
two C atoms must be added as branches

2,2-Dimethylpentane   **2,3-Dimethylpentane**   **2,4-Dimethylpentane**   **3,3-Dimethylpentane**   **3-Ethylpentane**

Common repeat   Common repeat

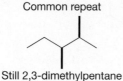   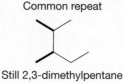

Still 2,3-dimethylpentane   Still 2,3-dimethylpentane

FIGURE 2.46 3-Ethylpentane and the dimethylpentanes.

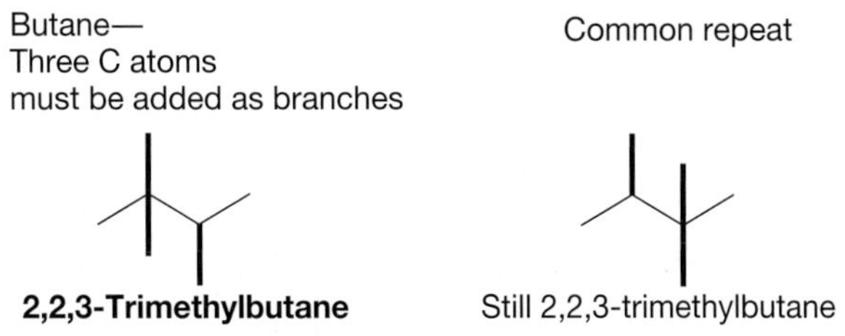

Butane—
Three C atoms
must be added as branches

Common repeat

2,2,3-Trimethylbutane

Still 2,2,3-trimethylbutane

FIGURE 2.47 Trimethylbutane.

PROBLEM 2.28 Write and name all the isomers of octane, C_8H_{18}.

2.12 Rings

All the hydrocarbons we have met so far have the molecular formula C_nH_{2n+2}. Because these molecules are linear or branched chains, we refer to them as either *noncyclic* or *acyclic* alkanes. Molecules of this formula are also called **saturated hydrocarbons**, which means that the carbon–carbon bonds in the molecule are all single bonds. There is a class of closely related molecules that shares most chemical properties with the noncyclic alkanes but not the general formula. These molecules have the composition C_nH_{2n} and are the cycloalkane ring compounds mentioned briefly in Section 2.1. Cycloalkanes with this formula are also saturated. Molecules that have the formula C_nH_{2n} but have no rings are called **unsaturated hydrocarbons**. Unsaturated compounds have carbon–carbon double (or triple) bonds and will be discussed in Chapter 3.

What might the bonding in cyclic molecules be? It probably will not deviate much from what we already know because the chemical properties of cycloalkanes resemble very closely those of the noncyclic alkanes. There is no essential difference between the process used to make ethane from two methyl radicals (Fig. 2.48a) and this construction of a ring compound (Fig. 2.48b). Nor would we expect to see big differences in chemical properties, because the bonding in pentane is not significantly different from that in cyclopentane.

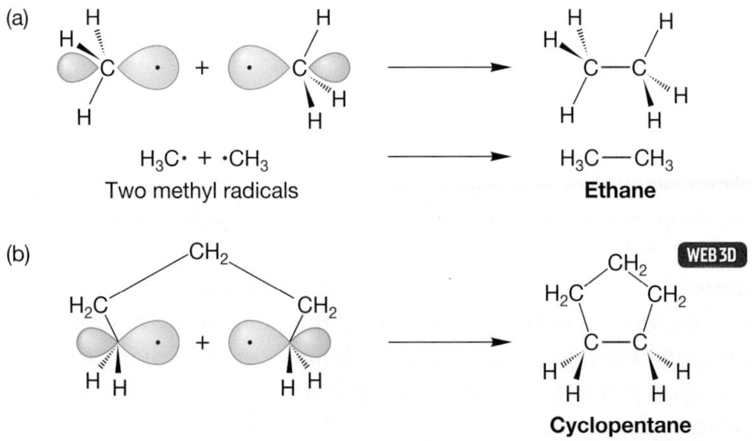

(a)

$H_3C\cdot + \cdot CH_3$
Two methyl radicals

$H_3C—CH_3$
Ethane

(b)

Cyclopentane

FIGURE 2.48 (a) The formation of ethane through the overlap of two singly occupied sp^3 hybrid orbitals. (b) The closely related formation of a ring compound, cyclopentane, through the overlap of two singly occupied sp^3 hybrid orbitals.

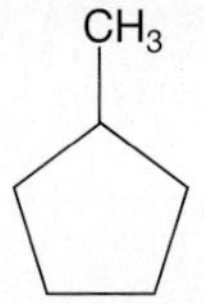

CH3

Methylcyclopentane

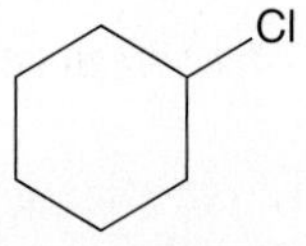

Cl

Chlorocyclohexane

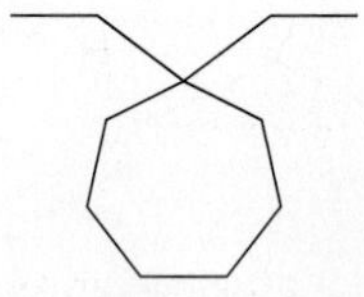

1,1-Diethylcycloheptane

F

I

1-Fluoro-4-iodocyclooctane
not 4-iodo-1-fluorocyclooctane
or 4-fluoro-1-iodocyclooctane

FIGURE 2.49 Cycloalkanes are named and numbered in a fashion similar to the system for acyclic alkanes.

Cycloalkanes are named by attaching the prefix cyclo- to the name of the parent hydrocarbon (Table 2.6).

TABLE 2.6 Some Cyclic Alkanes (Cycloalkanes)

Name	Formula	mp (°C)	bp (°C, 760 mmHg)
Cyclopropane	$(CH_2)_3$	−127.6	−32.7
Cyclobutane	$(CH_2)_4$	−90	12
Cyclopentane	$(CH_2)_5$	−93.9	49.2
Cyclohexane	$(CH_2)_6$	6.5	80.7
Cycloheptane	$(CH_2)_7$	−12	118.5
Cyclooctane	$(CH_2)_8$	14.3	148–149 (749 mmHg)

Monosubstituted cycloalkanes require no numbers for naming. In polysubstituted ring compounds, substituents are assigned numbered positions, and the ring atoms are numbered so that the lowest possible numbers are used. Multiple substituents are named in alphabetical order in the prefix (Fig. 2.49). Prefix modifiers (di-, tri-, etc.) are dealt with as shown in Table 2.5.

There is one important new structural feature that appears in cycloalkanes. There is only a single methylcyclopropane, which is why there is no number used in naming it. But there are two isomers of 1,2-dimethylcyclopropane. As shown in Figure 2.50, there is no "sidedness" to methylcyclopropane. The molecule with the methyl group "up" can be transformed into the molecule with the methyl group "down" by rotating the molecule 180° about the axis shown in Figure 2.50, in other words, by simply turning the molecule over.

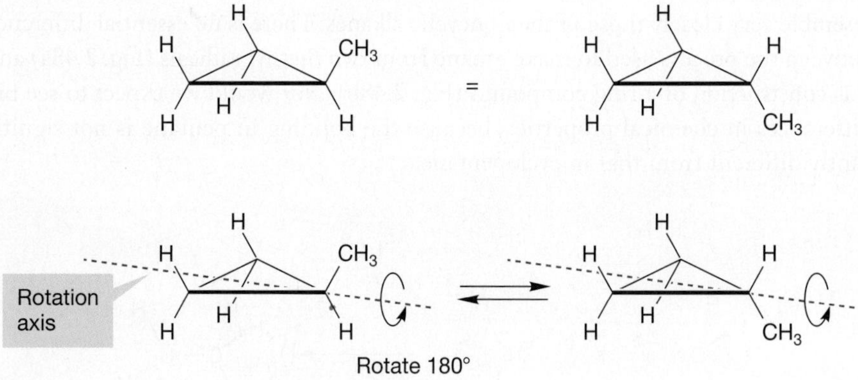

FIGURE 2.50 There is only one isomer of methylcyclopropane. The thicker line in these drawings is used to show the side of the cyclopropane nearest to you.

H
H   CH3
H'
H   H   =   H   H
H   H
H'   CH3

H
H   CH3
Rotation   H'
axis   H   H
⇌   H   H
H'
H   CH3
Rotate 180°

However, no number of translational or rotational operations can change the 1,2-dimethylcyclopropane with both methyl groups on the same side of the ring, called **cis**, into the 1,2-dimethylcyclopropane with the methyl groups on opposite sides, called **trans** (Fig. 2.51). The two molecules are certainly different from each other. The use of models is absolutely mandatory at this point. Make models of these two compounds and convince yourself that nothing short of breaking carbon–carbon bonds will allow you to turn *cis*-1,2-dimethylcyclopropane into the isomeric *trans*-1,2-dimethylcyclopropane.

PROBLEM 2.29 1,2-Dimethylcyclopropane is even more complicated than just described, as you will see in Chapter 4, because there are two isomers of *trans*-1,2-dimethylcyclopropane! Use your models to construct the mirror image of the *trans*-1,2-dimethylcyclopropane you just made and see if it is identical to your first molecule.

Cyclic compounds are extremely common. Both small and large varieties are found in Nature, and many kinds of exotic cyclic molecules not yet found outside the laboratory have been made by chemists. Moreover, ring molecules can be combined in a number of ways to form polycyclic molecules. Here is an opportunity for you to think ahead. How might two rings be attached to each other? Some ways are obvious, but others require some thought. We will work through this topic as a series of four problems, two worked in the chapter, two not. A number of different structural types can be created from two rings. These problems lead you through them.

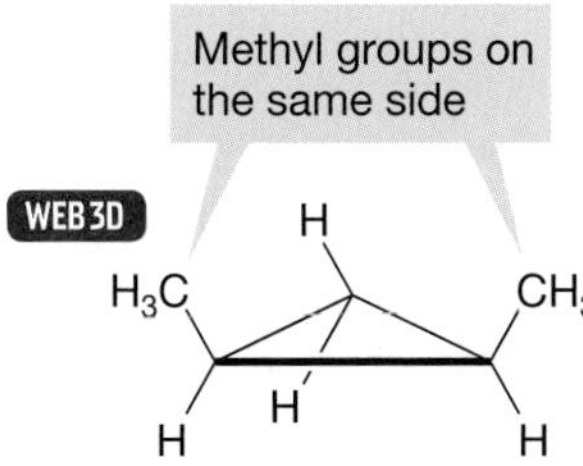

WORKED PROBLEM 2.30 Draw a molecule that has the formula $C_{10}H_{18}$ and is composed of two five-membered rings.

ANSWER Two five-membered rings can be joined in a very simple way to make bicyclopentyl. There is no real difference between this process and the formation of ethane from two methyl radicals (p. 68).

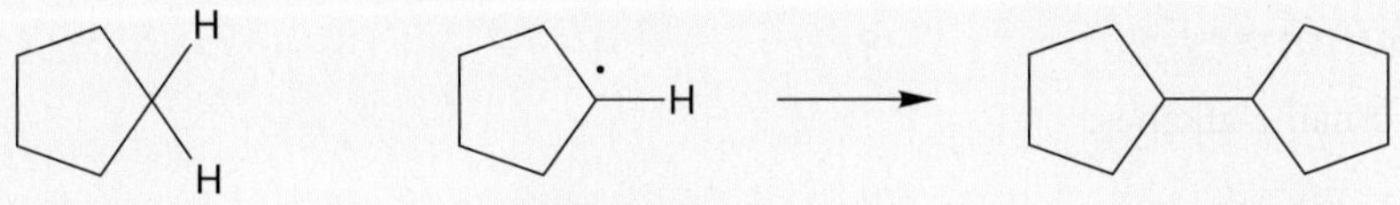

H
Cyclopentane
(C_5H_{10})

H
—H
Cyclopentyl
(C_5H_9)

⟶

Bicyclopentyl
$(C_{10}H_{18})$

WORKED PROBLEM 2.31 Another molecule containing two five-membered rings has the formula C_9H_{16}. Clearly, we are not dealing with a simple combination of two cyclopentanes here because we are short one carbon—it's C_9, not C_{10}, as in bicyclopentyl. The two rings must share one carbon somehow. Draw this compound.

ANSWER The way to have two rings sharing a carbon is to let one carbon be part of both rings:

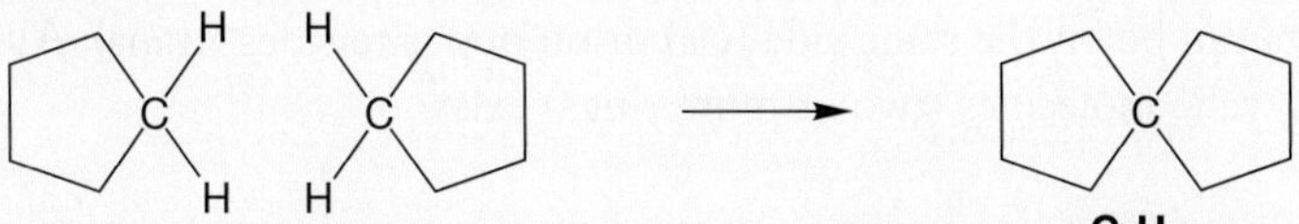

H H
C
H H
C

⟶

C
C_9H_{16}
The two rings share one carbon

PROBLEM 2.32 Two five-membered rings can share more than one carbon. Two similar molecules that are structured this way both have the formula C_8H_{14}. Draw them. *Hint*: Focus on the two shared carbons and the hydrogens attached to them. (Make a model.) That's all the help you get here. Use Problem 2.31 as a guide.

PROBLEM 2.33 Finally, and most difficult, there is a molecule, still constructed from five-membered rings, that has the formula C_7H_{12}. In this molecule, the two five-membered rings must share three carbons. Draw this molecule.

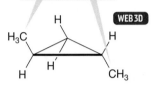

Methyl groups on the same side

WEB3D

H
H_3C CH_3
H'
H   H

***cis*-1,2-Dimethylcyclopropane**

Methyl groups on opposite sides

WEB3D

H
H_3C H
H'
H CH_3

***trans*-1,2-Dimethylcyclopropane**

FIGURE 2.51 Two isomers of 1,2-dimethylcyclopropane.

Polycyclic compounds can be exceedingly complex. Indeed, much of the fascination that organic chemistry holds for some people is captured nicely by the beautifully architectural structures of these compounds. Figure 2.52 shows three examples of cyclic molecules. Aflatoxin B_1 and progesterone are found in Nature and we will refer to such compounds as **natural products**. The compound [1.1.1]propellane is not (yet) found outside the laboratory. Aflatoxin B_1 is a highly toxic fungal metabolite. Progesterone is one of a class of molecules called steroids; it has an antiovulatory effect if taken during the middle days of the menstrual cycle.

FIGURE 2.52 Some polycyclic molecules.

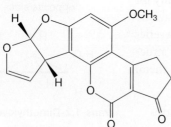

Aflatoxin B₁
made in 1966 by G. Büchi
and his research group at MIT

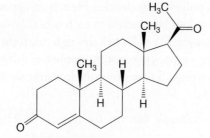

Progesterone
made in 1967 by G. Stork and his
group at Columbia University

[1.1.1]Propellane
made in 1982 by K. B. Wiberg
and his group at Yale; this
molecule was also made
slightly later in a particularly
simple way by the group of
G. Szeimies at Munich

Summary

To name alkanes:

1. Find the longest chain and use the appropriate root word.
2. Number the chain to give the lowest possible numbers for the substituents.
3. Arrange the substituents in the prefix alphabetically, and include the carbon number for each to describe its location.

Drawing isomers requires finding a system that allows you to consider all the possible perturbations. One approach is to start with the longest possible chain and then reduce the chain length one carbon at a time, considering the possible locations of the displaced methyl group with each reduction.

Cycloalkanes are bonded in the same way as noncyclic alkane molecules—by overlap of sp^3 hybrid orbitals. Ring compounds have sides, which means that substituents can be on the same side (cis) or on opposite sides (trans). All manner of polycyclic molecules (two or more rings) exist.

2.13 Physical Properties of Alkanes and Cycloalkanes

At room temperature and atmospheric pressure, simple saturated hydrocarbons and cycloalkanes are colorless gases, clear liquids, or white solids, depending on their molecular weight. To many people, they smell bad, although some of us think that these molecules have been the victims of a bad press and don't smell bad at all. Cooking gas, which is mostly saturated hydrocarbons, has an odor that comes from a mercaptan (RSH) (*Remember*: R stands for a general alkyl group, p. 69), put in specifically so that escaping gas can be detected by smell.

Tables 2.4 and 2.6 show some physical properties of straight-chain and cyclic alkanes. Why do the boiling points increase as the number of carbons in the molecule increases? The boiling point is a measure of the ease of breaking up intermolecular attractive forces. There is a factor that stabilizes the liquid phases of hydrocarbons called **van der Waals forces** (Johannes Diderik van der Waals, 1837–1923). When two clouds of electrons approach each other, dipoles (molecules with two poles, p. 14) are induced as the clouds polarize in such a fashion as to stabilize each other by opposing plus and minus charges (Fig. 2.53).

Separated molecules
(gas phase)

Aggregated molecules
(solution) held together
by opposite charges

$\delta-$
$\delta+$

$\delta-$
$\delta+$

Attraction between
induced opposite
charges in
liquid phase

$\delta-$
$\delta+$

FIGURE 2.53 The stabilization of molecules through van der Waals forces.

Of course, many alkanes have small dipoles to begin with, but they are very small and do not serve to hold the molecules together strongly. So alkanes have relatively low boiling points. Other molecules are much more polar, and this polarity makes a big difference in boiling point. Polar molecules can associate quite strongly with each other by aligning opposite charges. This association has the effect of increasing the boiling point.

The more extended a molecule is, the stronger its induced dipole can be. More compact, more spherical molecules have smaller induced dipoles and therefore lower boiling points. A classic example is the difference between pentane and neopentane (Fig. 2.54).

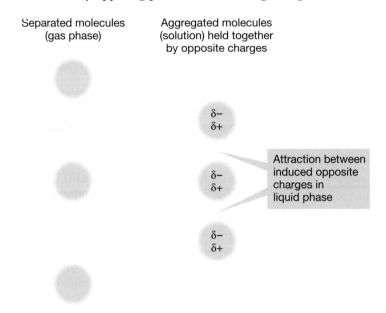

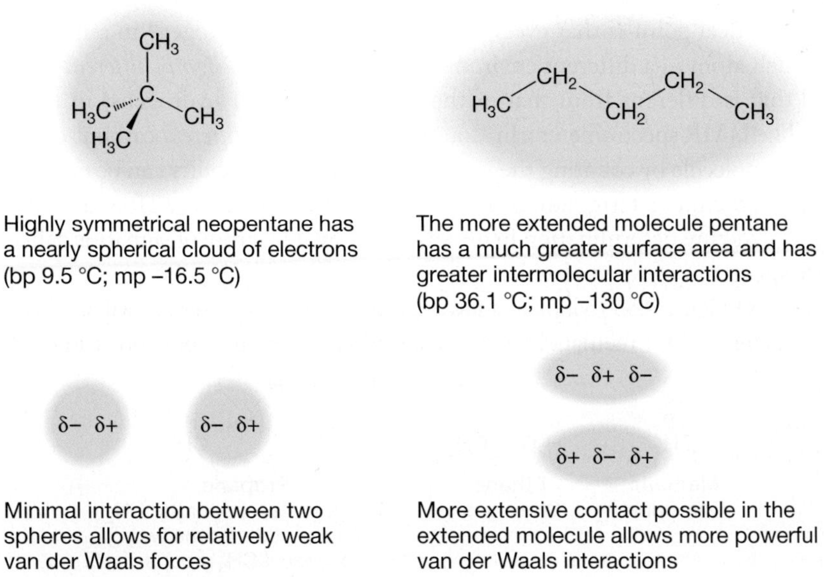

Highly symmetrical neopentane has a nearly spherical cloud of electrons (bp 9.5 °C; mp –16.5 °C)

The more extended molecule pentane has a much greater surface area and has greater intermolecular interactions (bp 36.1 °C; mp –130 °C)

$\delta-$ $\delta+$ $\delta-$ $\delta+$

$\delta-$ $\delta+$ $\delta-$

$\delta+$ $\delta-$ $\delta+$

Minimal interaction between two spheres allows for relatively weak van der Waals forces

More extensive contact possible in the extended molecule allows more powerful van der Waals interactions

FIGURE 2.54 The more extended pentane boils at a higher temperature than the more compact neopentane does.

The more spherical neopentane boils about 25 °C lower than the straight-chain isomer. Isopentane is less extended than pentane but more extended than neopentane, and its boiling point is right between the two, 30 °C.

Symmetry is especially important in determining melting point because highly symmetric molecules pack well into crystal lattices. (Think of the computer game Tetris and how easy packing would be if every shape were a highly symmetrical square.) The better the packing of the lattice, the more energy it takes to break it up. So neopentane, for example, melts 113 °C higher than pentane does.

2.14 Nuclear Magnetic Resonance Spectroscopy

Spectroscopy is the study of molecules through the investigation of their interaction with electromagnetic radiation. There are many kinds of spectroscopy (as we shall see in Chapter 15). One version is called **nuclear magnetic resonance (NMR) spectroscopy** and is particularly valuable, both in chemistry as a device for determining molecular structure and in medicine as an imaging tool. You have heard of this form of spectroscopy before if you have ever read an article about magnetic resonance imaging (MRI). NMR and MRI are the same process, but the dreaded word "nuclear" must be hidden from public view.

Although we won't go into much detail yet, this early introduction to nuclear magnetic resonance does allow us to address the critical question of *difference*. When are two atoms the same and when are they different?

Like electrons, nuclei of many atoms have a property called spin. A nonzero nuclear spin is necessary for a nucleus to be NMR active and thus detectable by an NMR spectrometer. The ^{13}C and ^{1}H nuclei each have spins of $\pm 1/2$, just like the electron. Although ^{13}C is present in only 1.1% abundance in ordinary carbon, which is mostly ^{12}C, that small amount can be detected.

Like the electron, the ^{13}C and ^{1}H nuclei can be thought of as spinning in one of two directions. In the presence of a strong magnetic field, those two spin states differ in energy, but by only a tiny amount. Nonetheless, transitions between the two states can be detected by NMR spectrometers tuned to the proper frequency. We'll have more to say about those spectrometers and those transitions in Chapter 15, but there is really not much more to it than that. So we can see a signal whenever a transition between the lower and the higher energy nuclear spin states is induced. So what? It would seem that we have simply built a (very expensive) machine to detect carbon or hydrogen in a molecule, and it would be hardly surprising to find such atoms in organic molecules!

The critical point is that every different carbon (or hydrogen) in a molecule— every such atom in a different environment, *no matter how slightly different*—gives a signal that is different from that of the other carbons (or hydrogens) in the molecule. The NMR spectrometer can "count" the number of *different* carbons or hydrogens in a molecule by counting the number of signals. That ability can be enormously useful in structure determination, and NMR spectroscopy is very often used by "the pros" of structure determination in exactly that way. The array of signals is called the **NMR spectrum** of the molecule.

Let's use Figure 2.55 to look at a few examples. How many signals will a ^{13}C NMR spectrometer "see" for methane? One, of course, because there is only one carbon. How

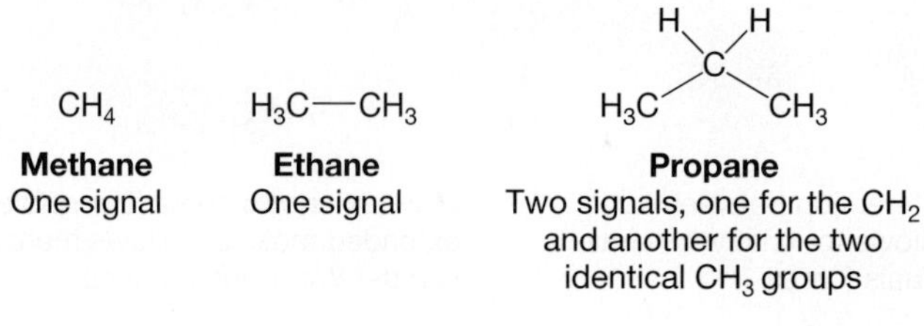

FIGURE 2.55 Carbon-13 NMR signals for three alkanes.

CH_4

Methane
One signal

$H_3C{-}CH_3$

Ethane
One signal

$H_3C{\diagdown}\overset{\overset{H\quad H}{\diagdown\,\diagup}}{\underset{}{C}}{\diagup}CH_3$

Propane
Two signals, one for the CH_2
and another for the two
identical CH_3 groups

about ethane? There are two carbons here, but they are in identical environments and thus equivalent: one signal again. How about propane? Three carbons are here, two methyl groups and a single methylene. Surely, we must observe a different signal for that CH_2 than for the CH_3 groups. And we do—there are two signals in the ^{13}C NMR spectrum for propane, one for the CH_2 and another for the two equivalent CH_3 groups.

Counting the number of signals is trivial if (and only if) we can discern whether the molecular environments of the carbons are the same or not. Here is a slightly harder example. The molecule tetrahydrofuran produces two ^{13}C signals (Fig. 2.56). It is tempting to say that because there are only methylene (CH_2) groups in this molecule, there should only be one signal. But those methylenes are not all the same. Two of them are adjacent to the oxygen, and two are not. So there are two signals, one for each set of two different methylenes.

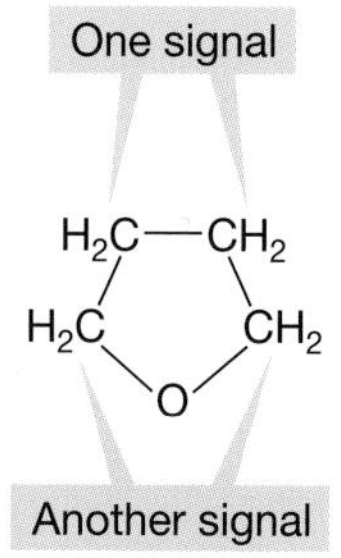

We will pick these points up again as we discuss the various structural types in the next few chapters, but for the moment, work through the following Problem Solving box and then try a few problems on your own.

One signal

Another signal

FIGURE 2.56 Tetrahydrofuran has two different methylene groups.

PROBLEM SOLVING

Nearly everyone has initial problems with seeing differences in the atoms making up any molecule. Why, for example, are there three different methylene (CH_2) groups in heptane? Maybe it is intuitively obvious (maybe!) that the central methylene group is different from the others, but almost everyone worries about the other methylene groups.

The best technique to use when you are in doubt is to list all the groups to which the carbons in question are attached. And be *very* detailed in making that list—do not just look at nearest neighbors. Here is an example using heptane:

$$CH_3-CH_2-CH_2-CH_2-CH_2-CH_2-CH_3$$
$$\quad 1 \qquad 2 \qquad 3 \qquad 4 \qquad 5 \qquad 6 \qquad 7$$

Carbon 1 methyl: Attached to C—C—C—C—C—C, same as carbon 7 methyl

Carbon 2 methylene: Attached to C on one side and to C—C—C—C—C on the other, same as carbon 6 methylene

Carbon 3 methylene: Attached to C—C and to C—C—C—C, same as carbon 5 methylene

Carbon 4 methylene: Attached to C—C—C and to C—C—C

Carbon 5 methylene: Attached to C—C—C—C and to C—C, same as carbon 3 methylene

Carbon 6 methylene: Attached to C—C—C—C—C and to C, same as carbon 2 methylene

Carbon 7 methyl: Attached to C—C—C—C—C—C, same as carbon 1 methyl

Therefore, heptane will have four signals in the ^{13}C NMR. One signal for the equivalent methyl groups (carbons 1 and 7), one signal for the methylenes of carbons 2 and 6, one signal for the methylenes of carbons 3 and 5, and one signal for the carbon 4 methylene.

PROBLEM 2.34 How many ^{13}C signals will we see in an NMR spectrum of the molecules in Figures 2.45–2.47?

PROBLEM 2.35 How many ^{13}C signals will an NMR spectrometer detect in the molecules of Figure 2.49?

PROBLEM 2.36 How many ^{13}C signals will an NMR spectrometer detect in the molecules of Figure 2.52?

Hydrogen NMR (^{1}H NMR) spectroscopy is similar to ^{13}C spectroscopy. The areas of the signals we observe in ^{1}H NMR spectroscopy are in the ratio of the numbers of different hydrogens giving rise to those signals. For example, the signal recorded for six hydrogens of one kind will give a signal three times as big as a signal for two hydrogens and so on. These ratios from the signals are more reliably determined in ^{1}H NMR than they are in ^{13}C NMR. There are other very useful complications introduced by the abundance of the NMR active isotope (^{1}H) in organic molecules, but we can leave them for Chapter 15. We will use some of the molecules introduced in this chapter to work through a bit of hydrogen NMR. We'll find a few more subtleties, but the overall picture will not be very different from what we have seen for ^{13}C.

PROBLEM 2.37 What will be the ratios of the signals in the ^{1}H NMR spectra for propane (Fig. 2.55), and THF (Fig. 2.56)?

PROBLEM 2.38 How many signals would be seen in the ^{1}H NMR spectrum for heptane (Fig. 2.44)? What will the relative sizes of these signals be?

WEB3D

PROBLEM 2.39 How many signals would be seen in the ^{1}H NMR spectrum for *cis*-1,2-dimethylcyclopropane (Fig. 2.51)? *Caution*: This question is a bit tricky—look carefully at a model, or use the website to see *cis*-1,2-dimethylcyclopropane in three dimensions.

2.15 Acids and Bases Revisited: More Chemical Reactions

In Section 1.7 (p. 41), we introduced acids and bases. Now we know quite a bit more about structure and can return to the important subject of acids and bases in greater depth. In particular, we know about carbocations and carbanions, which play an important role in acid–base chemistry in organic chemistry. The Lewis definition of acids and bases is far more inclusive than the Brønsted[7] definition, which focuses solely on proton donation (**Brønsted acid**) and acceptance (**Brønsted base**). The archetypal Brønsted acid–base reaction is the reaction between KOH and HCl to transfer a proton from HCl to HO$^-$. This reaction is a competition between the hydroxide and the chloride for a proton. In this case, the stronger base hydroxide wins easily (Fig. 2.57).

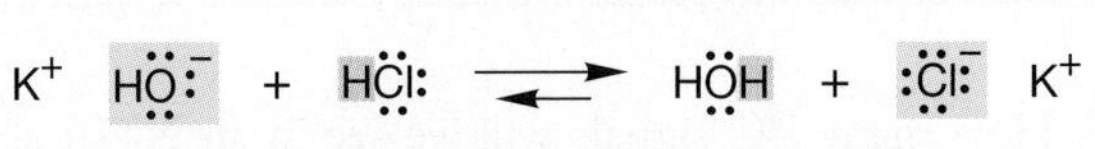

FIGURE 2.57 The two Brønsted bases, HO$^-$ and Cl$^-$, compete for the proton, H$^+$.

$$K^+ \quad H\ddot{O}\colon^- \;+\; H\ddot{C}l\colon \;\rightleftharpoons\; H\ddot{O}H \;+\; \colon\ddot{C}l\colon^- \;\; K^+$$

---

[7]These acids and bases were named after Johannes Nicolaus Brønsted (1879–1947).

In the Lewis description of acids and bases, any reaction between an empty orbital and a reactive pair of electrons in a filled orbital is an acid–base reaction. Another way of putting this is to point out that the interaction between an empty orbital (Lewis acid) and a filled orbital (Lewis base) is stabilizing (Fig. 2.58).

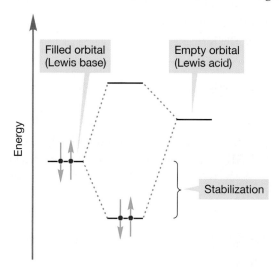

FIGURE 2.58 The interaction of a filled orbital (Lewis base) with an empty orbital (Lewis acid) is stabilizing.

Consider the methyl cation (Fig. 2.59). That empty $2p$ orbital makes $^+CH_3$ a powerful Lewis acid, very reactive toward all manner of Lewis bases with their electron pairs ready to react. Figure 2.59 gives three examples of such Lewis acid–Lewis base reactions, each of which fits the orbital interaction diagram of Figure 2.58.

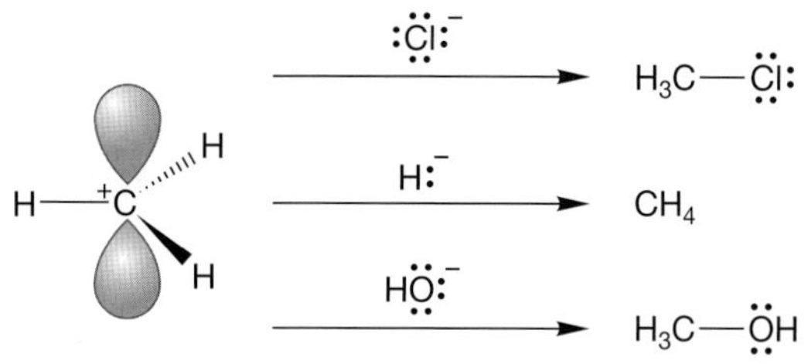

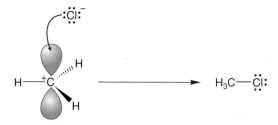

FIGURE 2.59 Three Lewis bases reacting with the methyl cation, a Lewis acid.

In Figure 2.60, the curved arrow formalism shows the electron flow as the new bond is formed in this illustrated nucleophile–electrophile (Lewis base–Lewis acid) reaction.

FIGURE 2.60 The curved arrow formalism for the formation of methyl chloride from the methyl cation (Lewis acid) and a chloride ion (Lewis base).

WORKED PROBLEM 2.40 Write arrow formalisms and products for the following reactions. Identify the participants as acids and bases. Sketch in the empty orbital for each Lewis acid, and fill in the complete Lewis dot structures for the Lewis bases. Remember that charge must be conserved in each of these reactions.

(a) $(CH_3)_3C^+$ Br^- $\rightarrow$

(b) $(CH_3)_3C^+$ NH_3 $\rightarrow$

(c) H^+ $^-NH_2$ $\rightarrow$

*(d) H_3B HO^- $\rightarrow$

(continued)

ANSWER (d) Hydroxide must be the Lewis base, but where is the empty orbital? Where is the Lewis acid? Recall the structure of the sp^2 hybridized borane (p. 56). Boron is neutral, but there is an empty $2p$ orbital on boron, and borane is very definitely a Lewis acid. Reaction with the Lewis base hydroxide looks just like the reaction of the methyl cation with hydroxide (Fig. 2.59) except that the product is negatively charged.

Lewis base–Lewis acid reactions, even the very simple ones, form the basis for understanding much of organic chemistry. Essentially all polar reactions can be looked at in exactly these terms—the ultimately stabilizing reaction of a filled orbital with an empty orbital. Figure 2.58 really does describe much of the next 1000 pages or so! But, of course, the details will change and there are surely devils in those details. Those all-too-prolific details will be easier to manage if you keep Figure 2.58 in mind at all times—search for the Lewis base and the Lewis acid in every reaction, and you will be able to generalize.

2.16 Special Topic: Alkanes as Biomolecules

In truth, there really isn't very much biochemistry of alkanes. The carbon–carbon and carbon–hydrogen bonds of alkanes are just too strong to enter easily into chemical reactions. These molecules are not totally inert, as we shall see shortly, but saturated alkane chains serve more as frameworks in our bodies, holding less strongly bonded atoms in the proper place for reactions to occur, than as reactive species themselves. However, as we have seen already (p. 68), in the presence of oxygen and activating energy (a spark or a lighted match, for example), combustion occurs and heat is given off. We use that energy, that exothermicity of reaction, to warm our homes and propel our automobiles. Gasoline is largely a mixture of saturated hydrocarbons. Combustion is a very biological process. Life depends on it.

In the movie *Mad Max: Beyond Thunderdome*, the trading village Bartertown is powered by the methane produced by herds of pigs kept in the depths of Undertown. How did those piggies produce that methane? Bacteria in the stomachs of both pigs and cattle are capable of producing methane from plant material, and the belching and flatulence of those animals is a major source of planetary methane. Contrary to conventional wisdom, it is mostly belching that produces the methane.

Methane may also have been essential to the formation of life. The early atmosphere on Earth was relatively rich in methane and ammonia (as are the current atmospheres of most of the outer planets). Given an energy source such as ultraviolet radiation or lightning, methane and ammonia react to form hydrogen cyanide (HCN), a molecule that polymerizes to give adenine, a common component of ribonucleic acid (RNA) and other molecules of biological importance. In the presence of water and an energy source, methane and ammonia react to give amino acids, the building blocks of proteins.

2.17 Summary

New Concepts

In this chapter, a bonding scheme for the alkanes is developed. We continue, as in Chapter 1, to form bonds through the overlap of atomic orbitals. We describe a model in which the atomic orbitals of carbon are combined to form new, hybrid atomic orbitals. The four orbitals resulting from a combination of three $2p$ orbitals and one $2s$ orbital are called sp^3 hybrid atomic orbitals, reflecting the 75% p character and 25% s character in the hybrid. These hybrid orbitals solve the problems encountered in forming bonds between pure atomic orbitals. The sp^3 hybrids are asymmetric, and have a fat and a thin lobe. Overlap between a hydrogen $1s$ orbital and the fat lobe provides a stronger bond than that between hydrogen $1s$ and carbon $2p$ orbitals. In addition, these hybrid orbitals are directed toward the corners of a tetrahedron, which keeps the electrons in the bonds as far apart as possible, thus minimizing destabilizing interactions. Other hybridization schemes are sp^2, in which the central atom is bonded to three other atoms, and sp, in which the central atom is bonded to two other atoms. Simple molecules in which a central carbon is hybridized sp^2 are planar, with the three attached groups at the corners of an equilateral triangle. Molecules with sp hybridized carbons are linear.

Even simple molecules have complicated structures. Methane is a perfect tetrahedron, but one need only substitute a methyl group for one hydrogen of methane for complexity to arise. In ethane, for example, we must consider the consequences of rotation about the carbon–carbon bond. The minimum energy conformation for ethane is the arrangement in which all carbon–hydrogen bonds are staggered. About 3 kcal/mol above this energy minimum form is the eclipsed form, the transition state, the high-energy point (not an energy minimum, but an energy maximum) separating two staggered forms of ethane. The 3 kcal/mol constitutes the rotational barrier between the two staggered forms. This barrier is small compared to the available thermal energy at room temperature, and rotation in ethane is fast.

This chapter covers again the concept of Lewis acids and bases. The familiar Brønsted bases compete for a proton, but Lewis bases are far more versatile. A Lewis base is defined as anything with a reactive pair of electrons, and a Lewis acid is anything that reacts with a Lewis base. We are paid back for these very general definitions with an ability to see as similar all manner of seemingly different reactions. These concepts will run through the entire book.

Key Terms

alkanes (p. 51)
alkyl compounds (p. 69)
Brønsted acid (p. 90)
Brønsted base (p. 90)
butyl group (p. 75)
sec-butyl group (p. 75)
tert-butyl group (p. 75)
carbanion (p. 62)
carbocation (p. 62)
cis (p. 84)
conformation (p. 65)
conformational analysis (p. 73)
conformational isomers (p. 70)
cycloalkanes (p. 51)
dihedral angle (p. 65)
eclipsed ethane (p. 65)
ethyl compounds (p. 69)
hybridization (p. 52)
hybrid orbitals (p. 53)

hydride (p. 62)
hydrocarbon (p. 51)
isobutyl group (p. 75)
isomers (p. 70)
isopropyl group (p. 73)
methane (p. 51)
methine group (p. 75)
methyl anion (p. 62)
methyl cation (p. 62)
methyl compounds (p. 60)
methylene group (p. 72)
methyl radical (p. 63)
natural product (p. 86)
Newman projection (p. 65)
NMR spectrum (p. 88)
nuclear magnetic resonance (NMR)
    spectroscopy (p. 88)
primary carbon (p. 76)
propyl group (p. 73)

quaternary carbon (p. 76)
reactive intermediates (p. 62)
saturated hydrocarbons (p. 83)
secondary carbon (p. 76)
sigma bond (p. 54)
spectroscopy (p. 88)
sp hybrid (p. 53)
sp^2 hybrid (p. 56)
sp^3 hybrid (p. 56)
staggered ethane (p. 65)
steric requirements (p. 78)
structural isomers (p. 73)
substituent (p. 60)
tertiary carbon (p. 76)
trans (p. 84)
transition state (TS) (p. 67)
unsaturated hydrocarbons (p. 83)
van der Waals forces (p. 87)

Reactions, Mechanisms, and Tools

In this chapter, new molecules are built up by first constructing a generic substituted molecule such as methyl—X (CH$_3$—X). The new hydrocarbon is then generated by letting X = CH$_3$. In principle, each different carbon–hydrogen bond in a molecule could yield a new hydrocarbon when X = CH$_3$. In practice, this procedure is not so simple. The problem lies in seeing which hydrogens are really different. The two-dimensional drawings are deceptive. One really must see the molecule in three dimensions before the different carbon–hydrogen bonds can be identified with certainty.

The coding or drawing procedures can get quite abstract. It is vital to be able to keep in mind the real, three-dimensional structures of molecules even as you write them in two-dimensional code. The Newman projection, an enormously useful device for representing molecules three-dimensionally, is described in this chapter.

The naming convention for alkanes is introduced. There are several common or trivial names that are often used and which, therefore, must be learned.

Both ^{1}H and ^{13}C NMR spectroscopy are introduced in this chapter. At this point, the NMR spectrometer functions basically as a machine to determine the numbers of different hydrogens or carbons in a molecule. That function is important, however, both in these early stages of your study of chemistry and in the "real world" of structure determination.

Common Errors

It is easy to become confused by the abstractions used to represent molecules on paper or blackboards. Do not trust the flat surface! It is easy to be fooled by a "carbon" that does not really exist, or to be bamboozled by the complexity introduced by ring structures. One good way to minimize such problems (none of us ever really becomes free of the necessity to think about structures presented on the flat surface) is to work lots of isomer problems, and play with models!

2.18 Additional Problems

PROBLEM 2.41   Provide the IUPAC name for the following compounds:

(a)

(b)

(c)

(d)

(e)

(f)

(g)

(h)

(i)

PROBLEM 2.42   Write Lewis structures for ethane, ethylene, acetylene, ethanol, ethanethiol, tetraethylammonium ion, diethylphosphine, the imine of diethyl ketone, diethylborane, tetraethylborate ion, diethyl ether, diethyl sulfide, acetaldehyde, acetone, acetic acid, ethyl acetate, acetamide, acetyl chloride, propanenitrile, ethyl fluoride, and ethyl chloride. Show nonbonding electrons as dots and electrons in bonds as lines.

PROBLEM 2.43   Draw all the isomers of "bromopentane," C$_5$H$_{11}$Br. Give systematic names to all of them.

PROBLEM 2.44   How many signals would a ^{13}C NMR spectrometer "see" for each of the molecules in your answer to Problem 2.43?

PROBLEM 2.45   Draw all the isomers of "chlorohexane," C$_6$H$_{13}$Cl. Give proper systematic names to all of them. *Hint*: There are 17 isomers.

PROBLEM 2.46 Use your model set to look down the C(2)—C(3) bond of 2-chlorobutane. Draw the Newman projection for all possible staggered conformations. Circle the one you think is the most stable and explain why you've chosen that one.

PROBLEM 2.47 Use your model set to look down the C(2)—C(3) bond of 2-bromo-3-methylbutane. Draw the Newman projection for all possible staggered conformations. Circle the one you think is the most stable and explain why you've chosen that one. Determine the number of gauche interactions in each projection.

PROBLEM 2.48 Draw the Newman projections for the different eclipsed and staggered conformations of 2,3-dichlorobutane. Label each projection as either eclipsed or staggered. In each staggered projection, determine the number and type of gauche interactions.

PROBLEM 2.49 Draw Newman projections of the eclipsed and staggered conformations of 1,2-dichloropropane by looking down the C(1)—C(2) bond.

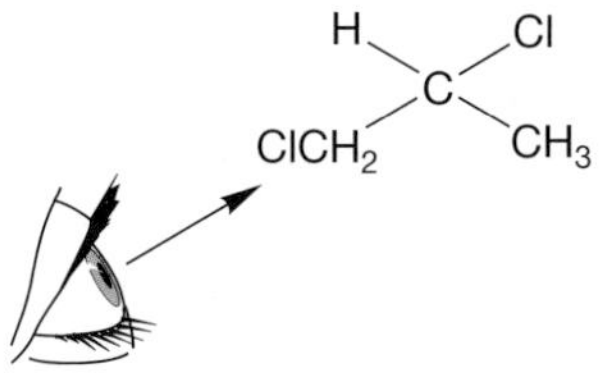

PROBLEM 2.50 Draw Newman projections constructed by looking down the C(1)—C(2) bond of 2-methylpentane. Repeat this process looking down the C(2)—C(3) bond. In each case, indicate which conformations will be the most stable.

PROBLEM 2.51 Although a chlorine is about the same size as a methyl group, the conformation of 1,2-dichloroethane in which the two chlorines are eclipsed is higher in energy than the conformation of butane in which the two methyl groups are eclipsed. Explain. *Hint*: Size isn't everything!

PROBLEM 2.52 What is the approximate hybridization of the indicated carbon in the following compounds?

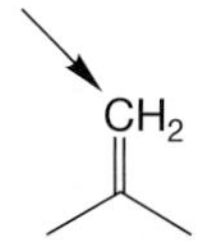

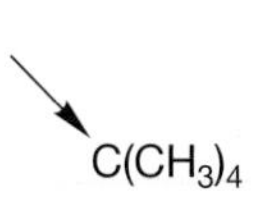

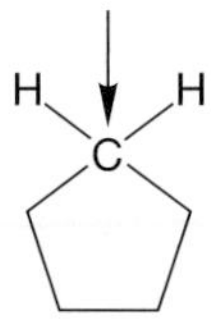

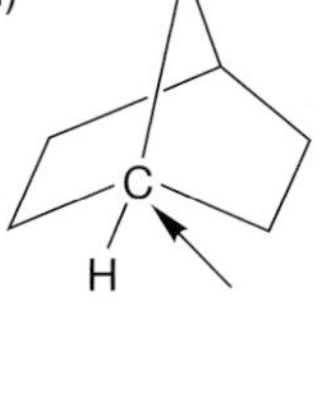

PROBLEM 2.53 Indicate the hybridization for each carbon, nitrogen, and oxygen in the molecules shown below. Put a circle around the sp^3-hybridized atoms, a triangle around sp^2-hybridized atoms and a box around the sp-hybridized atoms.

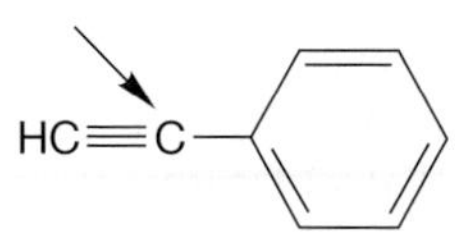

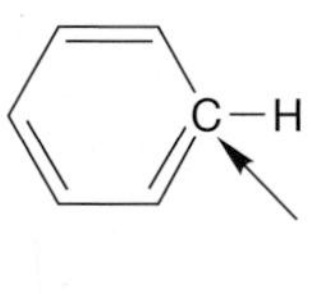

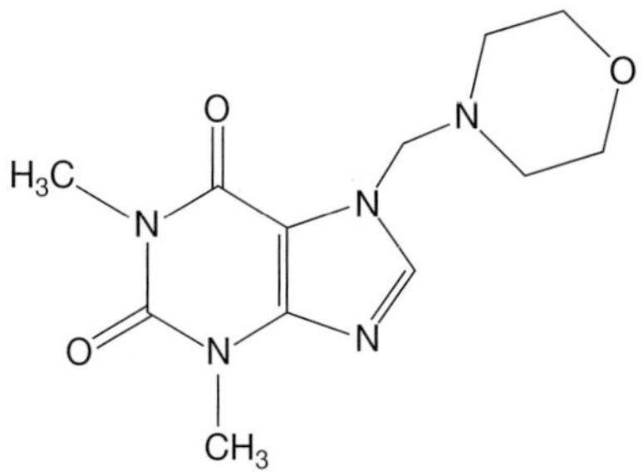

(a) **Xanturil** is a diuretic

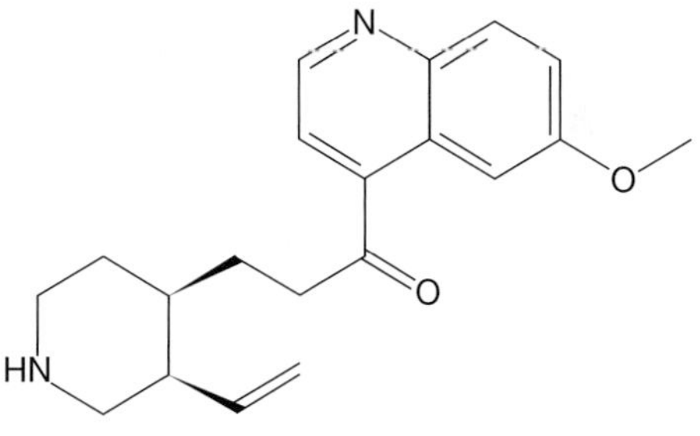

(b) **Viquidil** is a vasodilator and antiarrhythmic found in cinchona bark

PROBLEM 2.54 Use the boiling point data in Table 2.4 to estimate the boiling point of pentadecane, $C_{15}H_{32}$.

PROBLEM 2.55 Rationalize the differences in the melting points for the isomeric pentanes shown below.

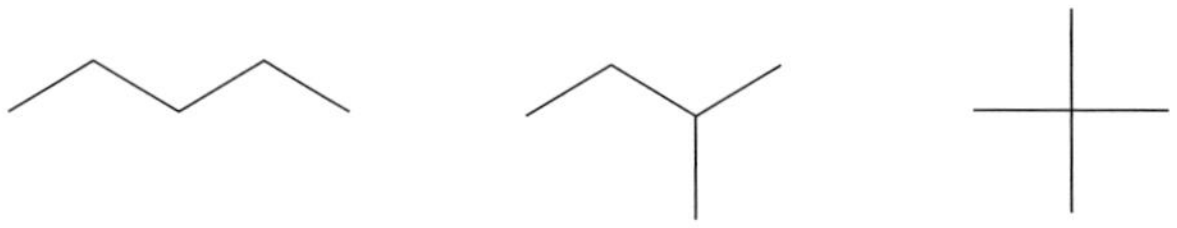

Pentane mp −129.7 °C   **Isopentane** mp −159.9 °C   **Neopentane** mp −16.5 °C

PROBLEM 2.56 Imagine replacing two hydrogens of a molecule with some group X. For example, methane (CH_4) would become CH_2X_2. (a) What compounds would be produced by replacing two hydrogens of ethane? (b) What compounds would be produced by replacing two hydrogens of propane? (c) What compounds would be produced by replacing two hydrogens of butane?

PROBLEM 2.57 Write systematic names for all the pentanes and hexanes.

PROBLEM 2.58 How many signals would a ^{13}C NMR spectrometer "see" for each of the molecules in your answer to Problem 2.57?

PROBLEM 2.59   Draw and write systematic names for all the isomers of nonane (C$_9$H$_{20}$). (*Hint*: There is a rumor that there are 35 isomers!) This problem is hard to get completely right.

You may want to continue isomer drawing and naming on your own. But be careful, this is no trivial task. For example, there are 75 structural isomers of decane and over 6500 for tetradecane.

PROBLEM 2.60   Write systematic names for the following compounds:

(a)

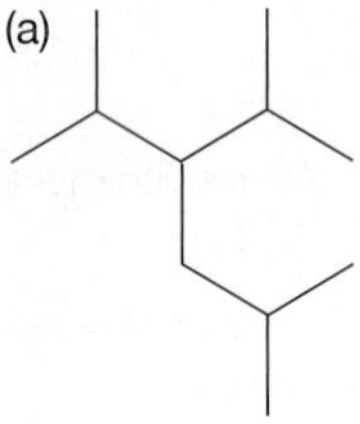

(b)

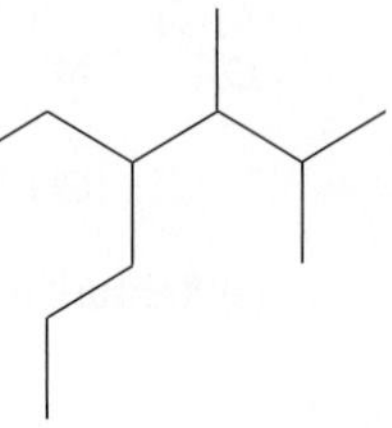

(c)

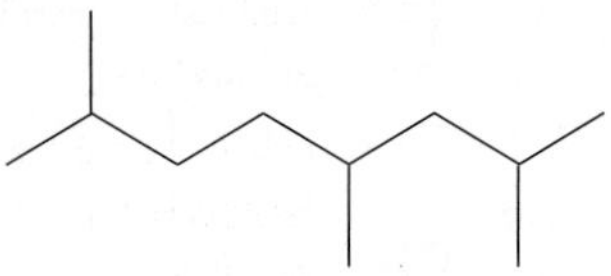

(d)

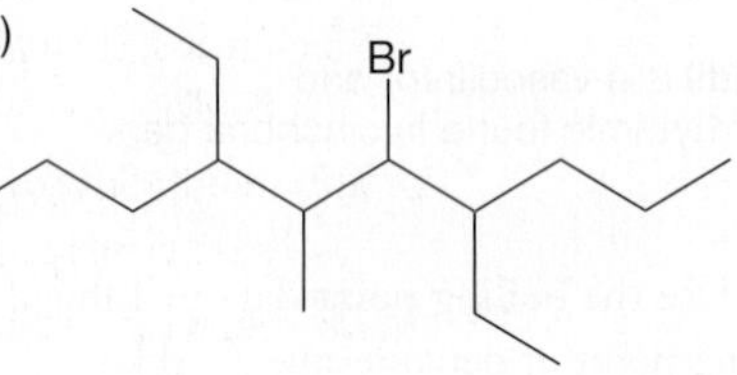

PROBLEM 2.61   Draw structures for the following compounds:

(a) 4-ethyl-4-fluoro-2-methylheptane
(b) 2,7-dibromo-4-isopropyloctane
(c) 3,7-diethyl-2,2-dimethyl-4-propylnonane
(d) 2-chloro-7-iodo-5-isobutyldecane

PROBLEM 2.62   Although you should be able to draw structures for the following compounds, each is named incorrectly below. Give the correct systematic name for each of the following compounds:

(a) 2-methyl-2-ethylpropane
(b) 2,6-diethylheptane
(c) 1-bromo-3-propylpentane
(d) 5-fluoro-8-methyl-3-*tert*-butylnonane

PROBLEM 2.63   One of the octane isomers of Problem 2.28, 3-ethyl-2-methylpentane, could have been reasonably named otherwise. Draw this compound and find the other reasonable name. The answer will tell you why the name given is preferred.

PROBLEM 2.64   Name the following compound:

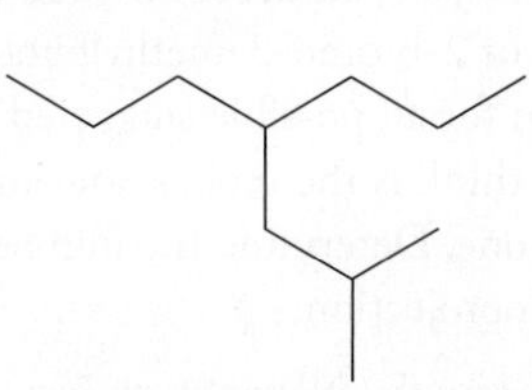

PROBLEM 2.65   Show the curved arrow formalism (electron pushing or arrow pushing) for the protonation of water in sulfuric acid (HOSO$_2$OH = H$_2$SO$_4$).

PROBLEM 2.66   Show the curved arrow formalism (electron pushing or arrow pushing) for the protonation of ammonia (NH$_3$) in hydrochloric acid.

PROBLEM 2.67   Show the curved arrow formalism for the protonation of an alcohol (ROH) in sulfuric acid.

Use Organic Reaction Animations (ORA) to answer the following questions:

PROBLEM 2.68   Choose the reaction titled "S$_N$2 with cyanide" and click on the Play button. As you watch the reaction occur, notice that the methyl group on the left rotates during the process. Explain why there is rotation.

PROBLEM 2.69   Choose the reaction "Alkene hydrohalogenation" and click on the Play button. Notice that there are two steps to this reaction. Identify the electrophile and the nucleophile in the first step. Identify the electrophile and the nucleophile in the second step.

PROBLEM 2.70   Choose the reaction "Bimolecular nucleophilic substitution: S$_N$2" and click on the Play button. Notice that the S$_N$2 is a one-step reaction. What is the nucleophile in this reaction? You can click on the highest occupied molecular orbital (HOMO) button to observe the electron density of the nucleophile. What is the electrophile? The electrophile is shown in the LUMO movie.

Alkenes and Alkynes

3.1   Preview

3.2   Alkenes: Structure and Bonding

3.3   Derivatives and Isomers of Alkenes

3.4   Nomenclature

3.5   The Cahn–Ingold–Prelog Priority System

3.6   Relative Stability of Alkenes: Heats of Formation

3.7   Double Bonds in Rings

3.8   Physical Properties of Alkenes

3.9   Alkynes: Structure and Bonding

3.10   Relative Stability of Alkynes: Heats of Formation

3.11   Derivatives and Isomers of Alkynes

3.12   Triple Bonds in Rings

3.13   Physical Properties of Alkynes

3.14   Acidity of Alkynes

3.15   Molecular Formulas and Degrees of Unsaturation

3.16   An Introduction to Addition Reactions of Alkenes and Alkynes

3.17   Mechanism of the Addition of Hydrogen Halides to Alkenes

3.18   The Regiochemistry of the Addition Reaction

3.19   A Catalyzed Addition to Alkenes: Hydration

3.20   Synthesis: A Beginning

3.21   Special Topic: Alkenes and Biology

3.22   Summary

3.23   Additional Problems

HOT CHEMISTRY The smallest alkyne is acetylene ($HC\equiv CH$). It is a gas and its most common use is for welding. An acetylene/oxygen mixture burns at the very high temperature of 3200 °C (5800 °F).

On the atomic and subatomic levels, weird electrical forces are crackling
and flaring, and amorphous particles . . . are spinning simultaneously
forward, backward, sideways, and forever at speeds so uncalculable that
expressions such as "arrival," "departure," "duration," and "have a nice
day" become meaningless. It is on such levels that magic occurs.

—TOM ROBBINS,[1] *SKINNY LEGS AND ALL*

3.1 Preview

Not all hydrocarbons have the formula C_nH_{2n+2}. Indeed, we have already seen some
that do not: the ring compounds, C_nH_{2n} (Fig. 3.1). We noted in Chapter 2 (p. 83)
that the chemical properties of these ring compounds closely resemble those of the

FIGURE 3.1 The chemical properties
of saturated alkanes, C_nH_{2n+2}, are
very similar to those of the
cycloalkanes, C_nH_{2n}.

$$CH_3CH_2CH_2CH_3$$

Butane
(C_4H_{10})
a saturated alkane

$$H_2C—CH_2$$
$$|\quad\quad|$$
$$H_2C—CH_2$$

Cyclobutane
(C_4H_8)
a cycloalkane

acyclic saturated hydrocarbons. There is another family of hydrocarbons that also
has the formula C_nH_{2n}, many of whose chemical properties are sharply different from
those of the ring compounds and saturated chains we have seen so far. These compounds are different from other hydrocarbons in that they contain carbon–carbon
double bonds. One bond is a sigma bond similar to the sigma bond of alkanes, the
other is a weaker bond formed by overlapping pi orbitals (see Section 3.2). They are
called **alkenes** to distinguish them from the saturated alkanes.

There are also hydrocarbons of the formula C_nH_{2n-2}, whose chemical properties resemble those of the alkenes but not those of the alkanes or cycloalkanes. These
compounds are the **acetylenes**, or **alkynes**.

Alkenes and alkynes contain fewer hydrogen atoms than alkanes with the same
number of carbons. Because they are not "saturated" with hydrogen, they are called
unsaturated hydrocarbons. The structures and some of the properties of the fami-
lies of alkenes and alkynes are discussed in this chapter.

Our study of the structures of alkenes and alkynes allows us to begin to examine chemical reactivity in a serious way.

ESSENTIAL SKILLS AND DETAILS

1. The hybridization model for sp^2 and sp bonding. It is critical that you be able to use the
   hybridization model to derive structures for the alkenes (double bonds) and the alkynes
   (triple bonds).

2. Reactivity—the addition reaction. In this chapter, we encounter reactivity, as addition
   reactions to alkenes appear. It is most important to see these reactions not as the first of
   a near-endless series of different processes, but as the initial exemplars of classes of
   reactions. Be sure you can relate what you learn about these addition reactions to the
   discussion of acids and bases in Chapters 1 and 2.

3. Priorities! The Cahn–Ingold–Prelog priority system is essential for determining E and
   Z, the arrangement of groups around a double bond.

4. Character counts. It is important to see the relationship between the amount of s
   character of an orbital and the stability of an electron in that orbital.

---

[1] Thomas Eugene Robbins is an American author born in 1936 in Blowing Rock, NC.

3.2 Alkenes: Structure and Bonding

The simplest alkene is properly called **ethene** but is almost universally known by its common name, **ethylene**.[2]

   Several spectroscopic measurements and chemical reactions show that ethylene has the formula C_2H_4 and is a symmetrical compound composed of a pair of methylene groups: H_2CCH_2. Earlier, when we encountered methane (CH_4), we constructed a bonding scheme designed to reproduce the way four hydrogens were attached to the central carbon. In ethylene, we have a slightly different arrangement. Each carbon atom in ethylene is attached not to four other atoms, as is the carbon in methane or ethane, but to three. It seems we will need a different bonding rationale with which to describe Nature now. Our strategy will be to develop a bonding scheme for the simplest trivalent compound of carbon, methyl (CH_3), and then extend it to ethylene, in which each carbon is attached not to three hydrogens as in methyl, but to two hydrogens and the other methylene (CH_2) group (Fig. 3.2).

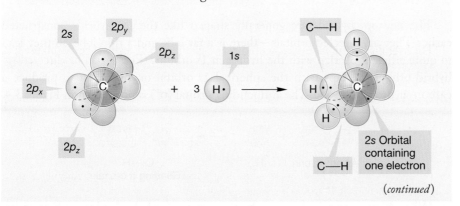

Three bonds are needed to make attachments to the three hydrogens in CH_3. Using the atomic orbitals of carbon to overlap with the three hydrogen $1s$ orbitals leads to problems, just as it did in our earlier construction of methane (see Section 2.2), as Worked Problem 3.1 shows.

FIGURE 3.2 Replacement of one hydrogen in methyl (CH_3), with a methylene (CH_2) group leads to the framework of ethylene (H_2CCH_2). Note that each carbon so far has only three bonds. In this drawing the full bonding scheme for ethylene is not yet in place.

WORKED PROBLEM 3.1 Produce a bonding model for neutral methyl ($\cdot CH_3$) using the unhybridized atomic orbitals of carbon and the hydrogen $1s$ orbitals. Critically discuss the shortcomings of this model. What's wrong with it?

ANSWER There is more than one way to construct such a model. Start by determining how many electrons are available. Carbon ($_6C$) supplies four (six less the two low-energy $1s$ electrons), and each hydrogen supplies one, for a total of seven. We might form three carbon–hydrogen bonds by overlap of the carbon $2p_x$, $2p_y$, and $2p_z$ orbitals with three hydrogen $1s$ orbitals, for example. This process uses six electrons (two in each of the three carbon–hydrogen bonds), and leaves the carbon $2s$ orbital to hold the remaining electron.

(continued)

---

[2]We like the old names and will use many of them. The newer, systematic names are fine when complexity develops—we obviously could not have a common, or "trivial" name for every compound. Yet some of the flavor of organic chemistry is lost when the system is applied too universally. The trivial names connect to history, to the quite correct image of the bearded geezer slaving over the boiling retort, and we like that.

However, this model gives a "methyl" with carbon–hydrogen bonds 90° apart. It is possible for a central carbon to be surrounded by three hydrogens 120° apart, and so we might suspect that this unhybridized model with its 90° angles will be destabilized by electron–electron repulsion. Moreover, overlap of a 1*s* orbital with a 2*p* orbital is inefficient because the rear lobes of the 2*p* orbitals are wasted. Stronger bonds can be constructed from directed, hybrid orbitals.

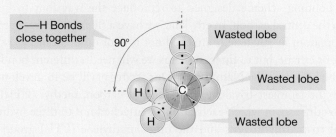

Other possible answers include making two carbon–hydrogen bonds from 2*p*/1*s* overlap and the third from 2*s*/1*s* overlap. This model leaves the last electron in a 2*p* orbital and consequently has the same problems as noted previously.

As in our earlier discussion of sp^2 hybridization (p. 55), we will create three new hybrid orbitals that will do a better job of attaching the carbon atom to the three hydrogens than do the pure atomic orbitals. Three hybrid orbitals are needed, and our mathematical operations will therefore involve combining three wave functions (atomic orbitals) to produce the three new hybrid orbitals. Recall that in such quantum mechanical calculations the number of orbitals created always equals the number of orbitals combined in the calculation, here three. So, let's combine the carbon $2s$, $2p_x$, and $2p_y$ orbitals to produce three new, sp^2 hybrids. The $2p_z$ orbital unused in our calculation remains, unhybridized and waiting to be incorporated in our picture (Fig. 3.3). Remember that the choice of 2*p* orbitals to be combined is arbitrary. Any pair of *p* orbitals can be used, leaving the third left over.

FIGURE 3.3 The combination of three atomic orbitals of carbon.

C (1s²) 2s² 2p_x 2p_y

Combine 2s, 2p_x, 2p_y $\longrightarrow$ three sp^2 hybrid orbitals; $2p_z$ is not used

The new sp^2 hybrids are generally shaped like the sp^3 hybrids constructed earlier. They are directed orbitals—there is a fat lobe and a thin lobe, so they lead to quite efficient overlap with the hydrogen 1*s* orbitals (Fig. 3.4). The directed sp^2 hybrid orbital overlaps with the spherical 1*s* orbital on hydrogen to produce a carbon–hydrogen σ bond. The antibonding orbital (σ*) is not shown in Figure 3.4.

FIGURE 3.4 The bonding C—H σ orbital.

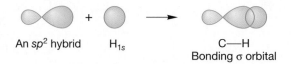

What do you guess is the angle between the three hybrid orbitals? Remember that one of the problems with using unhybridized carbon atomic orbitals is that this process produces bonds that are too close together, and repulsions between filled orbitals are not minimized. Hybrid orbitals produce strong bonds (good overlap) that are directed so as to minimize repulsions. As Figure 3.5 shows, the best way to arrange three things in space (the hydrogens in this case) surrounding a central object (here, the carbon) so they are as far from one another as possible, is to put them at the corners of an equilateral triangle. And indeed, the angles between the three sp^2 hybrids are exactly 120°. Recall the discussion of BH_3 in Section 2.2c.

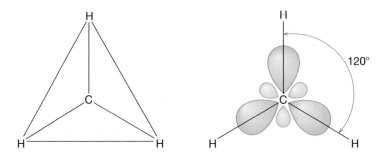

FIGURE 3.5 In planar CH_3, the angle between the carbon–hydrogen bonds is 120°. Repulsive overlap between the filled carbon–hydrogen bonds is minimized in this arrangement.

Each of the three $sp^2/1s$ overlapping systems produces a bonding ($\sigma = sp^2 + 1s$) and an antibonding ($\sigma^* = sp^2 - 1s$) molecular orbital. Stabilization is maximized if we form two-electron bonds, as the stabilized bonding molecular orbital will be filled and the antibonding, high-energy molecular orbital is empty (Fig. 3.6).

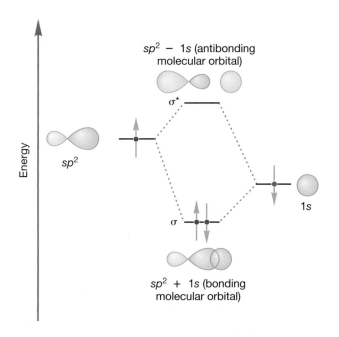

$sp^2 - 1s$ (antibonding molecular orbital)

σ^*

Energy

sp^2

$1s$

σ

$sp^2 + 1s$ (bonding molecular orbital)

FIGURE 3.6 The overlap of sp^2 and $1s$ orbitals can occur in a bonding, stabilizing way ($sp^2 + 1s$), or an antibonding, destabilizing way ($sp^2 - 1s$).

Each hydrogen supplies a single electron. Therefore, in order to form three two-electron bonds we need a single electron from carbon for each of the three sp^2 hybrids (Fig. 3.7).

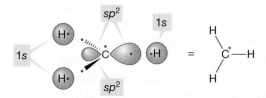

$1s$ sp^2 $1s$

sp^2

FIGURE 3.7 Three two-electron carbon–hydrogen bonds can be formed using one electron from each hydrogen and three electrons from carbon (all electrons are shown in red). This leaves one electron on carbon (shown in red) left over. In this figure, two of the sp^2 orbitals are shown schematically as wedges.

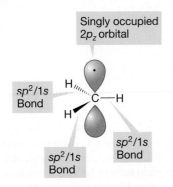

FIGURE 3.8 A view of sp^2 hybridized CH_3.

Remember that we did not use the carbon $2p_z$ orbital in creating the sp^2 hybrids. That's where the last electron of carbon goes. So our sp^2 bonding scheme leads to the picture in Figure 3.8: a carbon atom surrounded by a planar array of three hydrogen atoms, with the unhybridized $2p_z$ orbital extending above and below the plane of the four atoms. The carbon–hydrogen bonds are familiar two-electron bonds (don't forget the empty antibonding orbitals, though) and there is a single electron in the $2p_z$ orbital. Don't be confused by the electron shown in only one lobe of the $2p$ orbital. The two lobes are not separate, and the electron occupies the whole orbital, not just one lobe. We have seen this molecule, CH_3, before! It is nothing more than a methyl radical.

In the text so far, we have built up a picture that emphasizes localized bonds. In methyl, for example, electrons occupy three bonding orbitals made up of overlapping sp^2 and $1s$ orbitals. In Problem 3.2, we develop a picture of methyl in which the delocalization of electrons throughout the molecule is emphasized. Delocalizing electrons, spreading them out over several atoms rather than localizing them between two atoms, is almost always energetically favorable. But there are advantages to both schemes. In one sense the delocalized picture is probably more "real," because electrons are not limited to the regions of space the hybridization scheme suggests. However, we do not make horrible energetic mistakes if we ignore the delocalization that the orbitals you develop in Problem 3.2 show so clearly, and the hybridization scheme is excellent for "bookkeeping" purposes. It helps us to keep track of electrons in the chemical reactions that follow in later chapters, for example.

PROBLEM 3.2 Construct the bonding molecular orbitals for planar methyl (CH_3). Use the molecular orbitals for cyclic H_3 (**A**, **B**, and **C**) given below. The molecular orbitals for H_3 are shown in the drawing, but if they are unfamilar, take a look now at Problems 1.62 and 1.63. Allow these orbitals to interact with the appropriate atomic orbitals of a carbon atom placed at the center of the triangle of hydrogens. Next, place the bonding molecular orbitals in order of energy, lowest first. The dot in **B** shows the position of the third hydrogen atom.

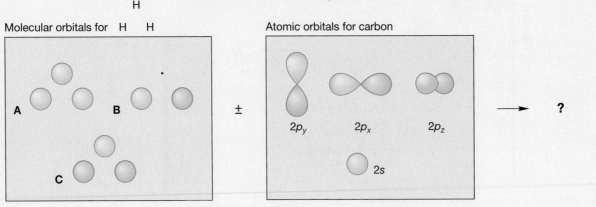

It is important that we do not view the molecular orbital and hybridization schemes as being in conflict or as giving substantially different pictures of the bonding in CH_3. Note, for example, that the geometry is exactly the same in the two schemes. We are humans, stuck with our inability to apprehend the properties of electrons easily, and needing approximations in order to represent Nature. Different approximate bonding schemes have been developed that emphasize different properties of the molecules. The molecular orbital picture does an excellent job of showing the distribution of electrons throughout the molecule. The hybridization picture sacrifices an ability to show this delocalization for the advantages of clarity, and ease of following the course of chemical reactions. We need to keep both representations in mind as we study chemical reactions.

We could easily imagine exchanging one of the three hydrogens of CH_3 for another atom or group. In constructing ethylene (H_2CCH_2), this other group would be another methylene (CH_2; Fig. 3.9). This transformation gives us our first orbital picture of

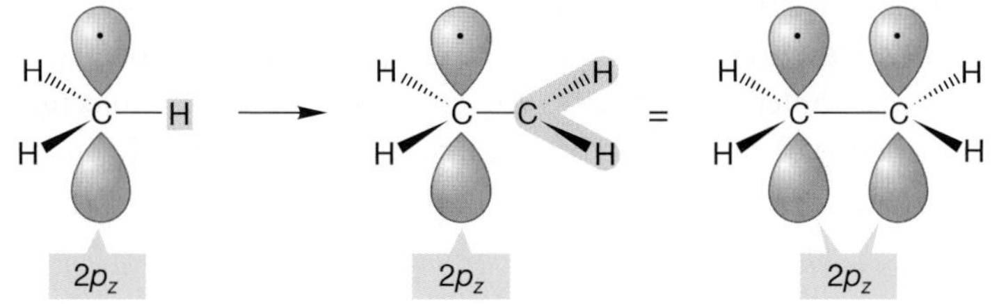

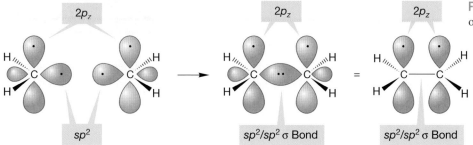

FIGURE 3.9 Our first orbital picture of ethylene.

ethylene ($H_2C{=}CH_2$), although the carbon–carbon double bond is not yet complete-ly drawn in Figure 3.9. So our first picture of ethylene is derived from joining a pair of methylene groups. The two sp^2 hybrids not involved in the carbon–hydrogen bonds over-lap to make an sp^2/sp^2 σ bond joining the two carbons of ethylene. The two $2p_z$ orbitals remain, and extend above and below the plane defined by the six atoms (Fig. 3.10).

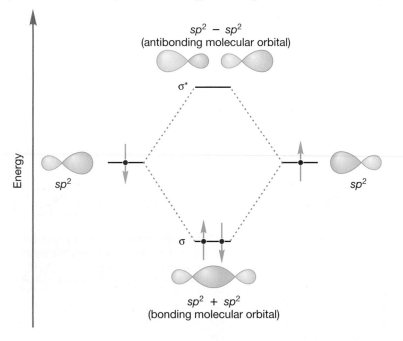

FIGURE 3.10 Another orbital picture of ethylene.

The structure of ethylene shown in Figure 3.10 is quite analogous to that pro-duced in the construction of ethane from a pair of sp^3 hybridized carbons (Section 2.5). In ethane, two sp^3 hybrids overlapped to form a bonding σ orbital and an anti-bonding σ* orbital (p. 68). Construction of the carbon–carbon sigma bond system in ethylene begins with the similar overlap of a pair of sp^2 hybrids. Once more, a bonding σ orbital and antibonding σ* orbital are produced, this time by the con-structive and destructive overlap of a pair of sp^2 hybrids (Fig. 3.11). Because there are only two electrons, only the low-energy, bonding molecular orbital is filled.

FIGURE 3.11 Overlap of the two sp^2 hybrids to form a bonding and antibonding pair of molecular orbitals.

So far, we have ignored the $2p_z$ orbital on each carbon. It is time to take them into account. These two carbon $2p_z$ orbitals are only the distance of a single bond apart and will interact quite strongly. Moreover, they are of equal energy and that too contributes to a strong interaction. Orbitals of equal energy interact most strongly (p. 34). We know what happens when atomic orbitals overlap—bonding and antibonding molecular orbitals are created. The new molecular orbitals formed from overlap of a pair of $2p$ orbitals are shown schematically in Figure 3.12.

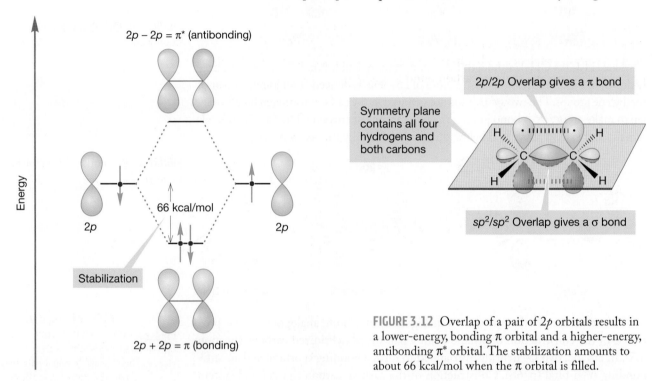

FIGURE 3.12 Overlap of a pair of $2p$ orbitals results in a lower-energy, bonding π orbital and a higher-energy, antibonding π^* orbital. The stabilization amounts to about 66 kcal/mol when the π orbital is filled.

These new orbitals are not σ orbitals because they do not have cylindrical symmetry. There is a plane of symmetry instead, and this makes them **pi (π) orbitals**, here called π (bonding) and π^* (antibonding). There are only two electrons to be put into the new system of two molecular orbitals, one from each carbon $2p_z$ orbital. These are accommodated very nicely in the highly stabilized bonding π molecular orbital.

So the two carbons of ethylene are held together by two bonds. One is made up of sp^2/sp^2 σ overlap and the other of $2p/2p$ π overlap. There is a double bond between the two carbons made up of the σ and π bonds. The convention is to simply draw two lines between the atoms, making no distinction between the two bonds, although, as you have just seen, this approximation is far from justified (Fig. 3.13).

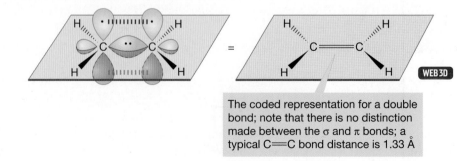

FIGURE 3.13 The π and σ bonds of a carbon–carbon double bond are not differentiated in the coded representation of an alkene.

The coded representation for a double bond; note that there is no distinction made between the σ and π bonds; a typical C=C bond distance is 1.33 Å

The overall bonding scheme for a carbon–carbon double bond includes both σ and π bonds (and their empty antibonding counterparts) and is also shown in Figure 3.13. Carbon–carbon double bonds are quite short, with a typical bond distance being 1.33 Å.

In ethane, there is nearly free rotation about the carbon–carbon bond (~3 kcal/mol, p. 65). Is the same true for ethylene? Actually, the argument is a bit more subtle. There isn't really free rotation about the carbon–carbon bond in ethane. There is a 3-kcal/mol barrier, produced by the need to pass through a structure in which the carbon–hydrogen bonds are eclipsed. Our real question should be, How high is the barrier to rotation about the carbon–carbon double bond in ethylene? To begin the answer we construct Newman projections by sighting along the carbon–carbon bond (Fig. 3.14).

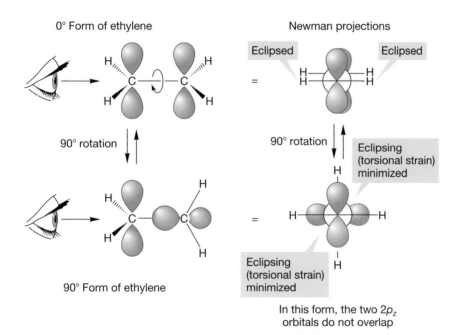

FIGURE 3.14 Although rotation about the carbon–carbon bond relieves torsional strain, it also destroys the overlap between the two $2p_z$ orbitals on the adjacent carbons. This loss of orbital overlap leads to a high barrier to rotation. When one $2p$ orbital is rotated 90°, orbital overlap is completely lost.

Note that in the 0° conformation of ethylene there is torsional strain induced by the eclipsing of the carbon–hydrogen bonds. This strain is removed in the 90° rotated arrangement because now the electrons are as far apart as possible. In ethane, each pair of eclipsed hydrogens induces approximately 1 kcal/mol of destabilization. Therefore our initial guess might be that the 0° form of ethylene would be about 2 kcal/mol higher in energy than the 90° form. That estimate would be approximately right *if torsional strain were the whole story.* But it is not, as we have ignored the pair of $2p_z$ orbitals, and their interaction completely overwhelms the small amount of torsional strain.

In the 0° form, the two *p* orbitals overlap; in the 90° form they do not (Fig. 3.14). When rotation occurs, the $2p/2p$ overlap declines and so does the stabilization derived from occupying the new bonding molecular orbital with two electrons. There is an enormous stabilizing effect in the 0° arrangement that overcomes the relatively minor effects of torsional strain.

PROBLEM 3.3 Show that in the 90° rotated form of an alkene there is zero over-lap between the pair of carbon $2p_z$ orbitals.

Figure 3.15 plots the energy change as rotation occurs. The barrier to rota-tion here is no paltry 3 kcal/mol as it is in ethane but 65.9 kcal/mol, an amount far too high to be available under normal conditions. Alkenes are "locked" in a planar conformation by this amount of energy. We will soon see some conse-quences of this locking.

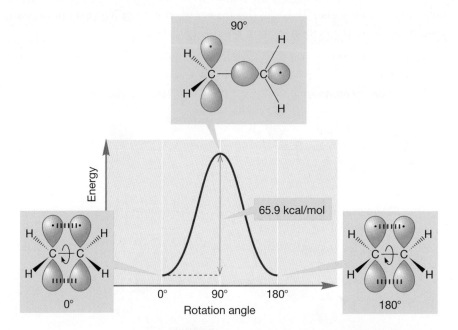

FIGURE 3.15 A plot of potential energy versus rotation angle in an alkene.

ETHYLENE: A PLANT HORMONE

It's amazing, but the simplest of all alkenes, ethylene, is an important plant hormone. Among other functions, ethyl-ene acts to promote ripening of fruit. Moreover, produc-tion of ethylene is autocatalytic; that is, a little ethylene induces the formation of more from the amino acid methionine, and its effects are magnified. Accordingly, fruits such as tomatoes and bananas are now typically shipped green in well-ventilated containers so they will arrive unspoiled. Ripening can then be started by exposure to ethylene.

Ethylene also induces abscission; the falling of leaves or flowers, by promoting the formation of the enzyme cellulase that weakens cell walls by destroying the cellulose from which they are made. A weakened abscission layer forms at the base of the leaf or flower, and wind or rain then breaks the stem. Indeed, it was this phenomenon that indirectly led

to the discovery of ethylene as a plant hormone. Trees sur-rounding gas lamps were often defoliated when gas leaks occurred. A principal component of the illuminating gas was ethylene, and a little investigation showed that this chemical was the cause of the defoliation.

Summary

Two carbons joined by a double bond are held together by a σ bond made up of overlapping sp^2 orbitals and a π bond made from $2p$ orbitals overlapping side to side. There is no free rotation about a carbon–carbon double bond, because such rotation diminishes overlap between the $2p$ orbitals making up the π bond and costs energy. A full 90° rotation requires about 66 kcal/mol of energy. Therefore, alkenes are planar.

In the hybridization model, we have constructed alkenes with both relatively strong σ bonds and weaker π bonds. The weaker, higher-energy π bonds will be more reactive than the stronger, lower-energy σ bonds. This point is critical and is often a source of confusion. A bond with lower energy is more stable and less reactive than a higher-energy bond. The weaker π bonds are likely to be the locus of reactivity of alkenes. In fact, alkenes and alkynes undergo all sorts of reactions that are unknown for alkanes with their exclusive set of strong, unreactive σ bonds. As soon as we work a bit more on nuts and bolts (nomenclature, isomers, and some thermochemistry), we'll get to what is usually the most fun in organic chemistry—reactions—and they will be triggered by the relatively weak π bonds in alkenes.

3.3   Derivatives and Isomers of Alkenes

After we developed a structure for ethane (p. 68), we examined substituted alkanes. Here we do the same thing for ethene, and discover a new kind of isomerism. There is but one ethylene (or ethene, if you insist). We can make derivatives of this molecule by mentally replacing a hydrogen with some "X" group. Figure 3.16 shows a few such compounds. These are not called "ethenyl" or even "ethylenyl" compounds but instead bear the delightfully trivial name, so far untampered with, of **vinyl**.

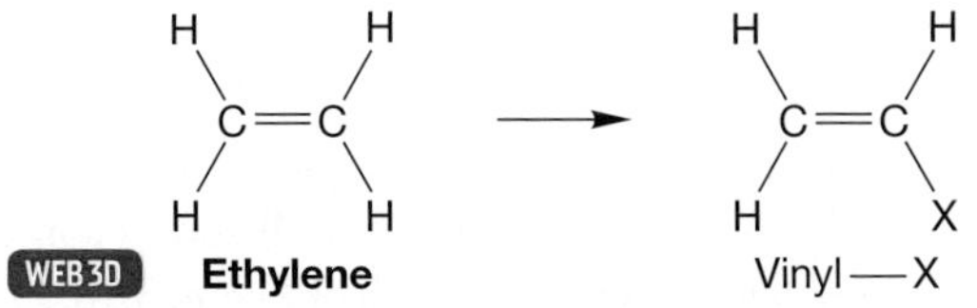

X =	Br	Vinyl bromide	OH	Vinyl alcohol
	Cl	Vinyl chloride	CN	Vinyl cyanide
	F	Vinyl fluoride	SiH$_3$	Vinyl silane
	I	Vinyl iodide	NH$_2$	Vinylamine

FIGURE 3.16 Some substituted alkenes called vinyl compounds.

If we now let X = CH$_3$ (methyl), we get a compound, C$_3$H$_6$ (propene or, commonly, propylene, Fig. 3.17). Notice how the root prefix *prop*- has been retained for this three-carbon species, and only the vowel changed to note the difference between saturated prop**a**ne and unsaturated prop**e**ne.

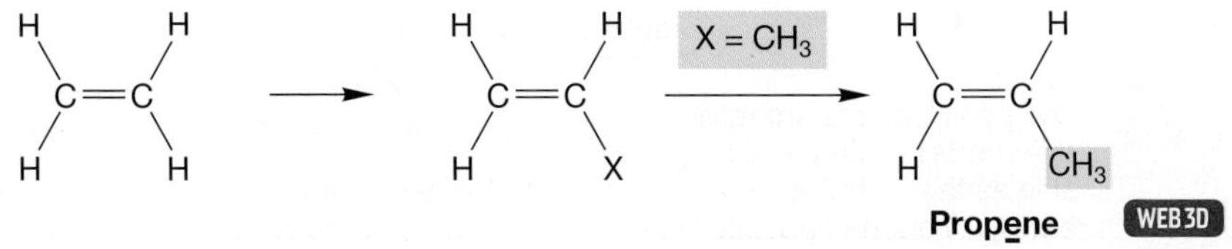

FIGURE 3.17 When X = CH$_3$ (methyl), we get propene (C$_3$H$_6$). Note the change of vowel from propane to propene.

We can now imagine replacing a hydrogen in propene with an "X" atom or group, but the situation is not as simple as it was with ethylene. Although there is only one kind of hydrogen in ethylene, there are several different hydrogens in propene that might be replaced. A quick count of the different hydrogens in propene is misleading. Care must be taken from now on to draw out the three-dimensional structures of the alkenes on which we are doing our mental replacements. In practice, this task is quite simple, because alkenes are planar, and most structures are easily drawn. If we just write a shorthand structure, we predict only three new kinds of compounds when a hydrogen is replaced with an X atom or group. As shown in Figure 3.18, these

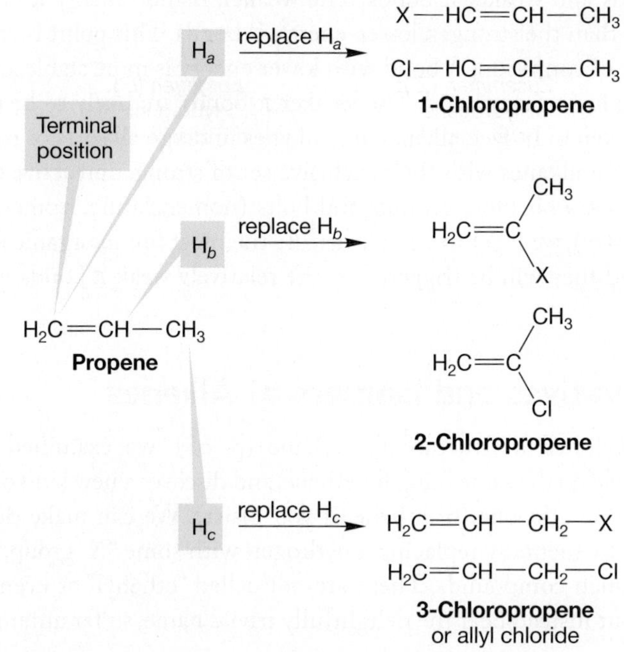

H₂C=CH—CH₂
The allyl group

H₂C=CH—CH₂—Cl
Allyl chloride

H₂C=CH—CH₂—Br
Allyl bromide

H₂C=CH—CH₂—NH₂
Allylamine

WEB 3D H₂C=CH—CH₂—OH
Allyl alcohol

H₂C=CH—CH₂—CN
Allyl cyanide

FIGURE 3.19 Some allyl compounds.

FIGURE 3.18 Substitution of three different hydrogens in propene appears to give three different propenyl—X compounds.

compounds are all known. Note the use of the frequently encountered common name **allyl** for the H₂C=CH—CH₂ group. Figure 3.19 shows some allyl compounds.

There is one severe problem with the approach of Figure 3.18. In reality, there are *not* three kinds of substituted propenes, but four. The more detailed drawing in Figure 3.20 shows why. The schematic structure for propene used in Figure 3.18 is inadequate

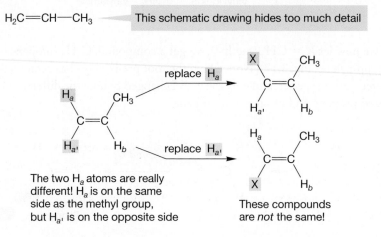

FIGURE 3.20 A detailed drawing of propene reveals four different potential substitution sites.

to show that the two hydrogens on the end (called terminal) methylene group are *not* the same. One, H_a, is on the same side of the double bond as the methyl group, whereas the other, $H_{a'}$, is on the opposite side from the methyl. Replacing hydrogens H_a and $H_{a'}$ with X gives two different compounds.

The two molecules on the right of Figure 3.20 are interconvertible only by rotation about the double bond (Fig. 3.21). We know the rotation shown in Figure 3.21

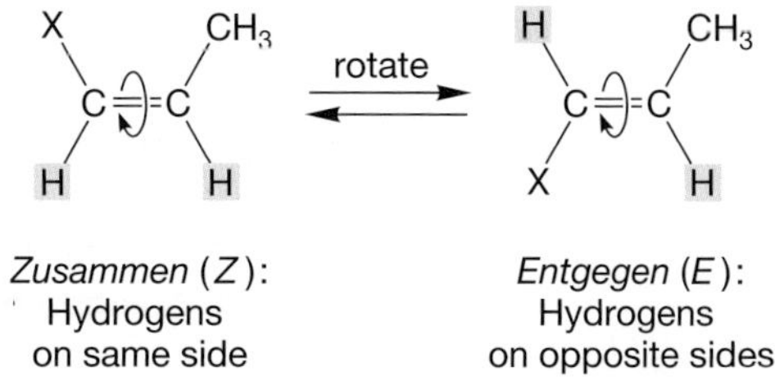

Zusammen (Z):
Hydrogens
on same side

Entgegen (E):
Hydrogens
on opposite sides

FIGURE 3.21 The (hypothetical) interconversion of two isomers of $CH_3CH{=}CH{-}X$ by rotation about the carbon–carbon double bond.

requires about 66 kcal/mol, too high an amount of energy for interconversion to be common. The molecule with the two hydrogens on the same side of the double bond is designated as cis or (Z) (for *zusammen*, German for "together"). The compound with the hydrogens on the opposite sides is called trans or (E) (for *entgegen*, German for "opposite"). Recall the use of cis and trans in our discussion of ring compounds (p. 84).

If we let X = methyl we get the butenes (C_4H_8). The four isomeric butenes are 1-butene, *cis*-2-butene, *trans*-2-butene, and 2-methylpropene (or isobutene, but also sometimes called isobutylene). These are shown in Figure 3.22. The numbers in the

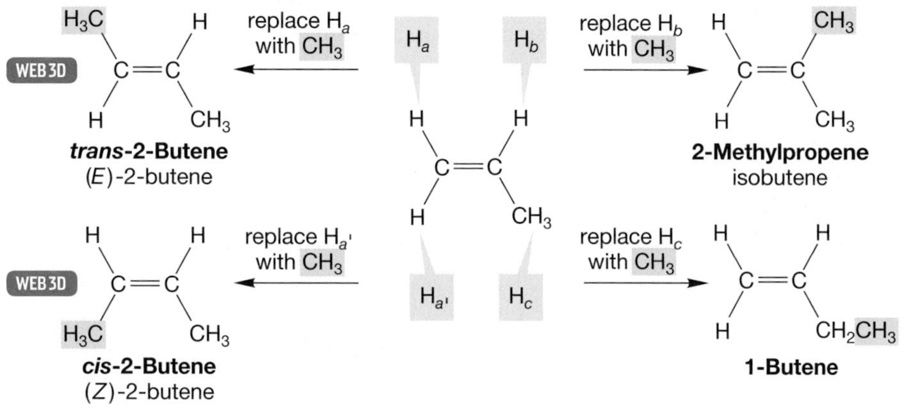

CONVENTION ALERT

trans-2-Butene
(E)-2-butene

cis-2-Butene
(Z)-2-butene

2-Methylpropene
isobutene

1-Butene

FIGURE 3.22 Replacement of the four different hydrogens in propene by a methyl group leads to the four butene isomers.

names 1-butene, *cis*-2-butene, and *trans*-2-butene are used to locate the position of the double bond. We will shortly discuss the naming convention for alkenes, but you might try to figure it out here from the names and structures of the butenes.

Not every alkene is capable of cis/trans (Z/E) isomerism! Ethylene and propene are not, and of the butenes only 2-butene has such isomers. Dealing with cis and trans isomers is an essential skill that can be "cemented" forever right now. Future difficulties can be avoided if you try Problem 3.4. It is simple but worthwhile.

Once we get to alkenes containing more than four carbons, the systematic naming protocol takes over. Five-carbon compounds (C_5) are called pentenes, six-carbon compounds (C_6) are called hexenes, and so on. Be alert for isomerism of the cis/trans (Z/E) kind. It takes some practice to find it.

PROBLEM SOLVING

Alkenes are *flat*! In the minimum energy form of an alkene, the two carbon atoms making up the double bond and the four atoms directly attached to it (atoms 1, 2, 3, and 4) are coplanar. It is very easy to forget that fact, or ignore it. Don't, because it leads immediately to the idea that alkenes have discernable sides. Thus, cis and trans isomers become possible for many alkenes.

$$\begin{matrix} 1 \\ 2 \end{matrix} \diagdown C = C \diagup \begin{matrix} 3 \\ 4 \end{matrix}$$

PROBLEM 3.4 Which of the following alkenes is capable of cis/trans (*Z/E*) isomerism?

PROBLEM 3.5 Write all the isomers of the pentenes (C_5H_{10}) and hexenes (C_6H_{12}). Be alert for isomerism of the cis/trans (*Z/E*) kind in these molecules.

3.4 Nomenclature

Most of the rules for naming alkenes are similar to those used for alkanes, with a number attached to indicate the position of the double bond. Chains are numbered so as to give the double bond the smallest possible number. One important rule that is different in the alkenes is that the name is based on the longest chain containing a double bond whether or not it is also the longest chain in the molecule. When molecules contain more than one double bond, they are called **polyenes**, and are named as dienes, trienes, and so on. In these cases, the longest straight chain is defined as the one with the greatest number of double bonds. Cyclic molecules are named as **cycloalkenes**. These rules are illustrated with the molecules in Figure 3.23.

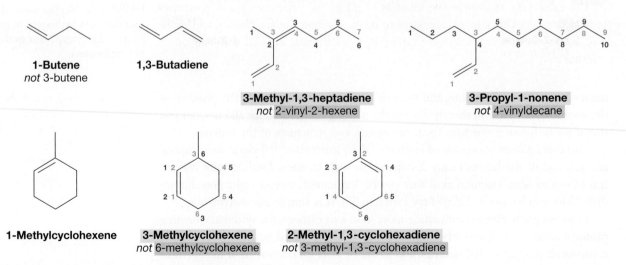

1-Butene
not 3-butene

1,3-Butadiene

3-Methyl-1,3-heptadiene
not 2-vinyl-2-hexene

3-Propyl-1-nonene
not 4-vinyldecane

1-Methylcyclohexene

3-Methylcyclohexene
not 6-methylcyclohexene

2-Methyl-1,3-cyclohexadiene
not 3-methyl-1,3-cyclohexadiene

FIGURE 3.23 Examples of the naming protocols for alkene nomenclature.

PROBLEM 3.6 Name the pentenes and hexenes of Problem 3.5.

PROBLEM 3.7 Recall p. 88 in Chapter 2 where the ^{13}C NMR spectrometer was described. That machine will find a signal for each different carbon in a molecule. How many signals will each of the following molecules show in its ^{13}C NMR spectrum?

(a) ethylene
(b) propene
(c) 1-butene
(d) *cis*-2-butene
(e) *trans*-2-butene
(f) isobutene (2-methylpropene)

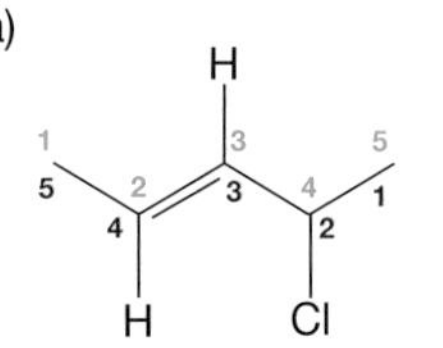

PROBLEM 3.8 How many signals will each of the molecules in Problem 3.4 show in its ^{13}C NMR spectrum?

Finally, there are examples in which the numbering of a substituent becomes important. Usually, it is the double bond numbering that must be considered first. The double bond is usually given the lowest possible number (Fig. 3.24a). An important exception is the hydroxyl (OH) group, which has priority over the double bond, and is given the lower number (Fig. 3.24b). Only if there are two possible names in which the double bond has the same number do you have to consider the position of the substituent. In such a case, give preference to the name with the lower number for the substituent (Fig. 3.24c).

Usually the cis/trans naming system is adequate to distinguish pairs of isomers, but not always. Problem 3.10 and Figure 3.25 give examples in which cis/trans naming is inadequate. In the molecule 1-bromo-1-chloropropene there are clearly two isomers, but it is not obvious which is cis and which is trans. It was to solve such problems that the old cis/trans system was elaborated into the (*Z/E*) nomenclature by three European chemists, R. S. Cahn (1899–1981), C. K. Ingold (1893–1970), and V. Prelog (1906–1998). The **Cahn–Ingold–Prelog priority system** ranks the groups attached to the double bonds, with the higher priority being 1 and the next 2. The compound with the higher priority groups on the same side of the double bond is (*Z*), and the other is (*E*). This priority system has other, very important uses to be encountered in Chapter 4, and so we will spend some time here elaborating it, even though we will see it again in another context.

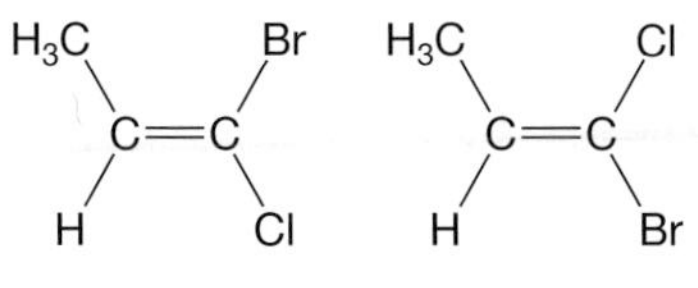

(a)

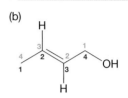

(*E*)-4-Chloro-2-pentene
trans-4-Chloro-2-pentene
not *trans*-2-chloro-3-pentene

(b)

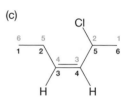

(*E*)-2-Buten-1-ol
trans-2-Buten-1-ol
not *trans*-2-buten-4-ol

(c)

(*Z*)-2-Chloro-3-hexene
cis-2-Chloro-3-hexene
not *cis*-5-chloro-3-hexene

FIGURE 3.24 The double bond takes precedence over most substituents, but when two names put the double bond at the same numbered position, we use the name in which the substituent has the lower number.

FIGURE 3.25 Sometimes the cis/trans convention is inadequate to distinguish two isomers. Which of these two isomers of 1-bromo-1-chloropropene would you call cis and which trans?

PROBLEM 3.9 Draw the following molecules: *cis*-2-pentene, 2-chloro-1-pentene, *trans*-3-penten-2-ol, 4-bromocyclohexene, 1,3,6-cyclooctatriene.

WORKED PROBLEM 3.10 In Figure 3.23, the structure of 3-methyl-1,3-heptadiene is not completely specified by that name. Even if the cis/trans nomenclature is applied, problems remain. What's the difficulty? Can you devise a system for resolving the ambiguity?

ANSWER In describing fully the structure of 3-methyl-1,3-heptadiene, we need to specify how groups are oriented in space (the stereochemistry), in order to distinguish the two possible forms. In one of the structures, the two saturated alkyl groups (methyl and propyl) are cis while in the other they are trans.

trans CH_3 and $CH_2CH_2CH_3$

cis CH_3 and $CH_2CH_2CH_3$

Often, cis/trans will do the job, but here these terms fail. There is but one H, so you cannot find two hydrogens "on the same side" (cis), and two hydrogens "on opposite sides" (trans). That is the problem; it is up to you to devise a protocol for resolving it. For more about the device used by organic chemists, the Z/E system, read on in the text.

CONVENTION ALERT | ## 3.5 The Cahn–Ingold–Prelog Priority System

1. The first step in using the priority system is to distinguish atoms on the basis of atomic number. The atom of higher atomic number has the higher priority (Fig. 3.26). Thus, a methyl group, attached to the double bond through a carbon atom (atomic number 6), has a higher priority than the hydrogen attached to the same carbon (atomic number 1). Similarly, in the second compound, oxygen (atomic number 8) has a higher priority than the boron attached to the same carbon (atomic number 5).

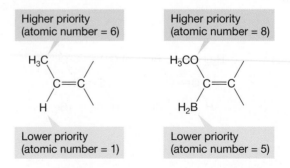

FIGURE 3.26 The substituent with the higher atomic number gets the higher priority.

How is this first rule used in practice? Consider the two isomers in Figure 3.25. Bromine has a higher atomic number than chlorine, so the isomer in which the higher-priority methyl group and the higher-priority bromine are on the same side is (*Z*), and the isomer in which the lower-priority hydrogen is on the same side as the higher-priority bromine is (*E*) (Fig. 3.27).

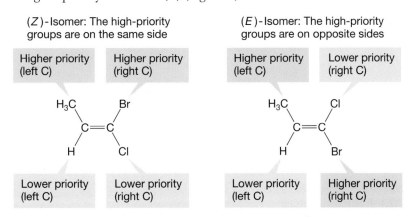

(*Z*)-Isomer: The high-priority groups are on the same side

Higher priority (left C)

Higher priority (right C)

Lower priority (left C)

Lower priority (right C)

(*E*)-Isomer: The high-priority groups are on opposite sides

Higher priority (left C)

Lower priority (right C)

Lower priority (left C)

Higher priority (right C)

FIGURE 3.27 Use of the Cahn–Ingold–Prelog priority system in naming alkenes.

2. For isotopes, atomic mass is used to break the tie in atomic number. Thus deuterium (atomic number 1, atomic mass 2) has a higher priority than hydrogen (atomic number 1, atomic mass 1) (Fig. 3.28).

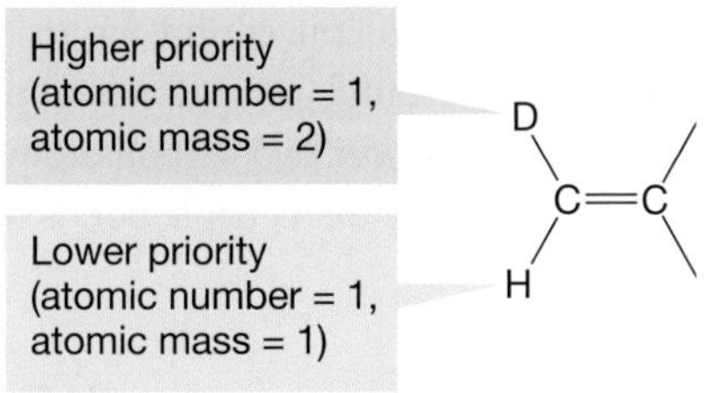

Higher priority (atomic number = 1, atomic mass = 2)

Lower priority (atomic number = 1, atomic mass = 1)

FIGURE 3.28 When the atomic numbers are the same, the heavier isotope gets the higher priority.

3. Nonisotopic ties are broken by looking at the groups attached to the tied atoms. For example, 1-chloro-2-methyl-1-butene has a double bond to which a methyl and an ethyl group are attached. Both these groups are connected to the double bond through carbons, so Rule 1 does not differentiate them (Fig. 3.29). To break the tie, we look at the groups attached to these carbons. The carbon of the methyl group is attached to three hydrogens, whereas the "first" carbon of the ethyl group is attached to two hydrogens and a carbon. Therefore the ethyl group gets the higher priority (Fig. 3.30). If the groups are still tied after this procedure, one simply looks farther out along the chain to break the tie, as in Problem 3.11.

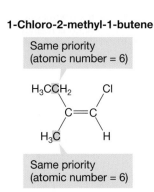

1-Chloro-2-methyl-1-butene

Same priority (atomic number = 6)

Same priority (atomic number = 6)

FIGURE 3.29 In this molecule, the two C atoms attached to the double bond of course have the same atomic number and, therefore, the same priority according to Cahn–Ingold–Prelog Rule 1. This isomer is *Z*.

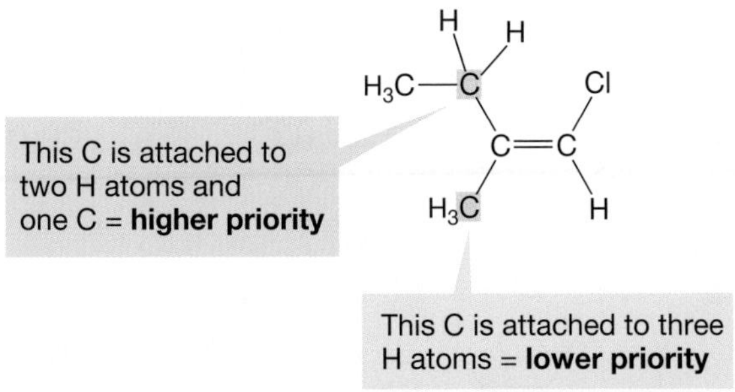

This C is attached to two H atoms and one C = **higher priority**

This C is attached to three H atoms = **lower priority**

FIGURE 3.30 The red methyl C is attached to three H atoms. The red ethyl C is attached to C, H, and H. The ethyl C gets the higher priority. Because the two higher priority groups, ethyl and Cl, are on the same side of the double bond, this isomer is *Z*.

PROBLEM 3.11 Determine which of the following molecules is (Z) and which is (E).

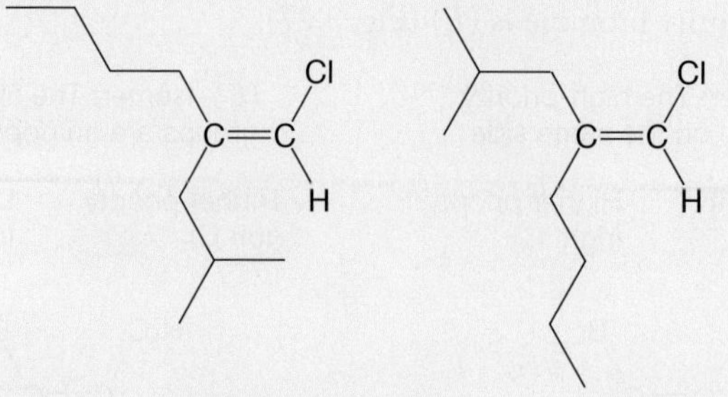

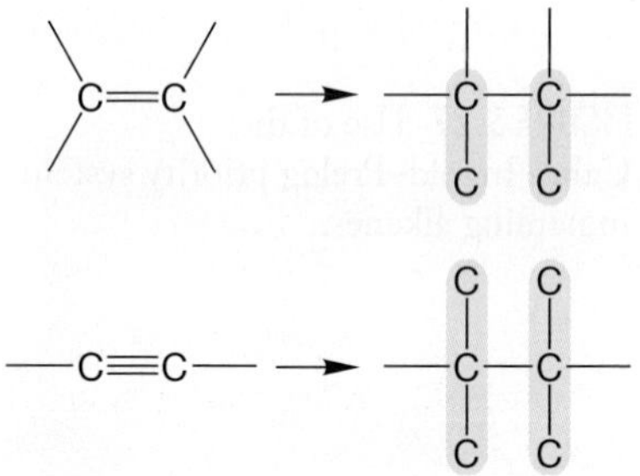

PROBLEM 3.12 Make drawings of the following molecules:
(a) (E)-3-fluoro-3-hexene
(b) (E)-4-ethyl-3-heptene
(c) (Z)-1-bromo-2-chloro-2-fluoro-1-iodoethylene

FIGURE 3.31 The priority of doubly bonded carbons is determined by adding two single carbon bonds as shown. A triple bond is treated similarly.

4. Multiple bonds attached to alkenes are treated as multiplied single bonds. A double bond to carbon is considered to be two single bonds to the carbons in the double bond, as shown in Figure 3.31. This convention results in an isopropenyl group being of higher priority than a *tert*-butyl group, for example (Fig. 3.32). As we will shortly see, it is also possible to connect two carbon atoms through a **triple bond** (see Section 3.9). The priority system treats triple bonds in a similar fashion.

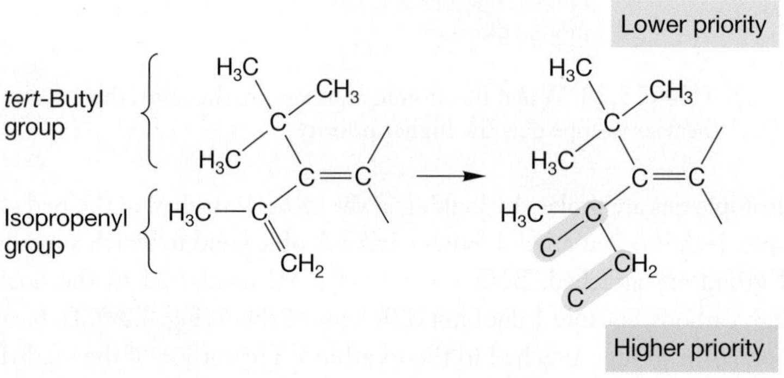

FIGURE 3.32 An isopropenyl group has a higher priority than a *tert*-butyl group.

PROBLEM 3.13 In each of the following pairs of isomers, which is (Z), and which is (E)?

(a)

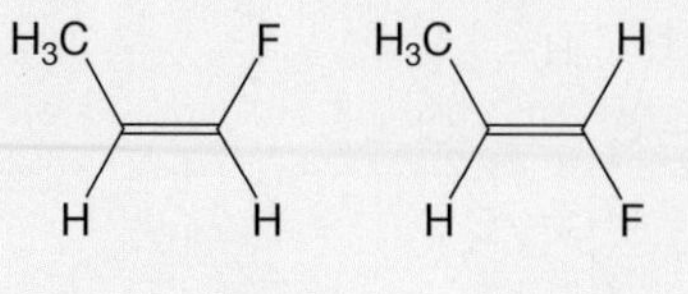

(b)

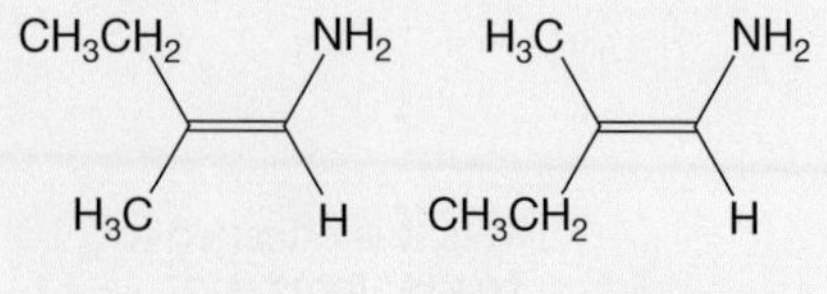

(c)

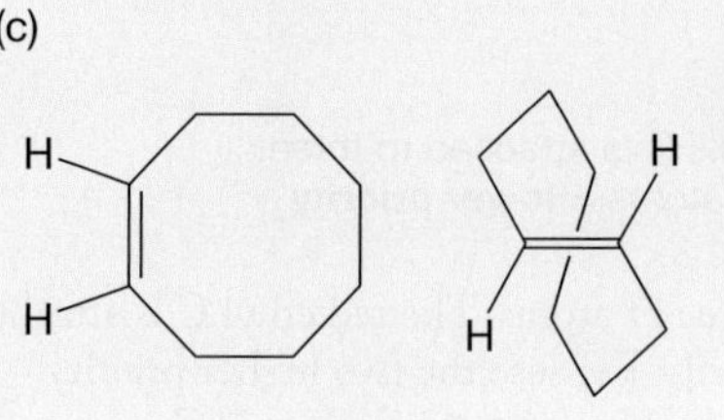

(d)

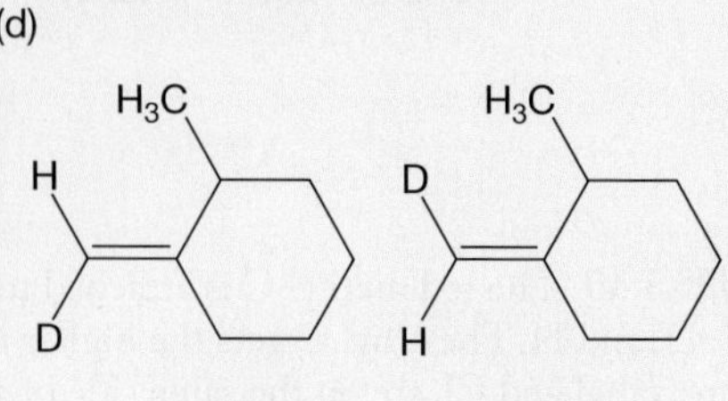

Summary

The technique of assigning (Z) and (E) is to determine the priority (high or low) at each carbon of the double bond. The isomer with the two high-priority groups on the same side of the double bond is (Z). The isomer with the two high-priority groups on opposite sides is (E). Two examples are given in Figure 3.33.

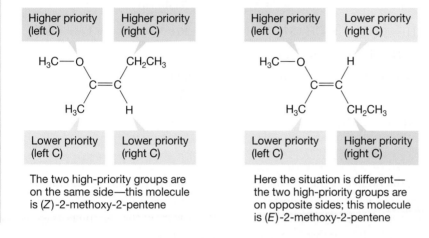

Higher priority (left C)   Higher priority (right C)

Lower priority (left C)   Lower priority (right C)

The two high-priority groups are on the same side—this molecule is (Z)-2-methoxy-2-pentene

Higher priority (left C)   Lower priority (right C)

Lower priority (left C)   Higher priority (right C)

Here the situation is different— the two high-priority groups are on opposite sides; this molecule is (E)-2-methoxy-2-pentene

FIGURE 3.33 Two examples of the Cahn–Ingold–Prelog priority system at work.

3.6 Relative Stability of Alkenes: Heats of Formation

The **heat of formation (ΔH_f°)** of a compound is the enthalpy of formation from its constituent elements in their standard states. The standard state of an element is the most stable form of the element at 25 °C and 1 atm pressure. For an element in its standard state, ΔH_f° is taken as zero. For carbon, the standard state is graphite. Thus, for graphite, as well as simple gases (e.g., H_2, O_2, and N_2), ΔH_f° is 0 kcal/mol. The more negative—or less positive—a compound's ΔH_f° is, the more stable the compound is. A negative ΔH_f° for a compound means that its formation from its constituent elements is *exothermic*—heat is liberated in the reaction. In contrast, a positive ΔH_f° means that the constituent elements are more stable than the compound and its formation is *endothermic*—energy must be applied.

Remember: Bonding is an energy-releasing process. We expect the formation of a molecule from its constituent atoms to be an exothermic process. For example, consider the simple formation of methane from carbon and hydrogen. Carbon in its standard state (graphite) and gaseous H_2 have $\Delta H_f^\circ = 0$ kcal/mol, whereas for methane $\Delta H_f^\circ = -17.8$ kcal/mol. The formation of methane from graphite and hydrogen releases 17.8 kcal/mol. In other words, this reaction is exothermic by 17.8 kcal/mol (Fig. 3.34).

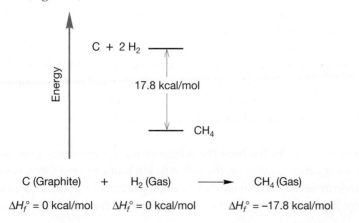

FIGURE 3.34 The formation of methane from graphite and gaseous hydrogen is exothermic.

Table 3.1 gives the heats of formation of a number of hexenes. First look at the pair of disubstituted alkenes, the isomers *cis*- and *trans*-3-hexene. "Substituted" simply means that an atom other than hydrogen is attached to the double bond.

TABLE 3.1 Heats of Formation for Some Hexenes

Isomer	ΔH_f° (kcal/mol)	
$CH_2{=}CHCH_2CH_2CH_2CH_3$	−10.0	Least stable
cis $CH_3CH_2CH{=}CHCH_2CH_3$	−11.2	
trans $CH_3CH_2CH{=}CHCH_2CH_3$	−12.1	
$(CH_3)_2C{=}CHCH_2CH_3$	−16.0	
$(CH_3)_2C{=}C(CH_3)_2$	−16.6	Most stable

Thus, the double bond in 1-hexene is monosubstituted and the double bond in 2-hexene or 3-hexene is disubstituted. The trans isomer is the more stable compound (more negative ΔH_f°). Figure 3.35 shows the reason; in the cis isomer there is a cis ethyl–ethyl eclipsing that is absent in the trans compound. In the trans isomer, both alkyl groups are eclipsed by hydrogen, and the destabilizing ethyl–ethyl repulsion is missing.

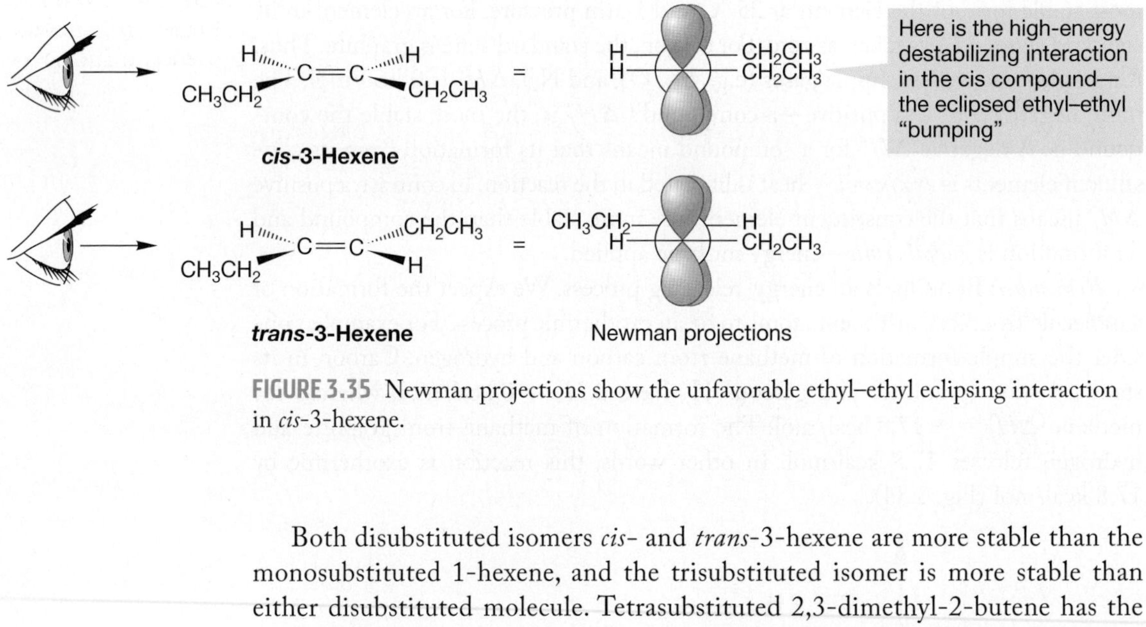

FIGURE 3.35 Newman projections show the unfavorable ethyl–ethyl eclipsing interaction in *cis*-3-hexene.

Both disubstituted isomers *cis*- and *trans*-3-hexene are more stable than the monosubstituted 1-hexene, and the trisubstituted isomer is more stable than either disubstituted molecule. Tetrasubstituted 2,3-dimethyl-2-butene has the lowest energy of all (lowest heat of formation, most stable). From these observations, it would appear that the degree of substitution (the number of alkyl groups attached to the double bond) is important in stabilizing the molecule. In general, the more substituted the double bond, the more stable it is (Table 3.1).

The reason for this has been the subject of some controversy. One good way to look at the question focuses on the different kinds of carbon–carbon bonds present in isomeric molecules of different substitution patterns. In all the molecules in Table 3.1, three kinds of σ carbon–carbon linkages are present: sp^2/sp^2, sp^2/sp^3, and sp^3/sp^3

bonds (Fig. 3.36). An electron in a *2s* orbital is at lower energy than an electron in a *2p* orbital. The more *s* character in an orbital, the more an electron in it is stabilized. In these isomeric hexenes, the more substituted the carbon–carbon double bond is, the more relatively low-energy (strong) sp^2/sp^3 bonds are present.

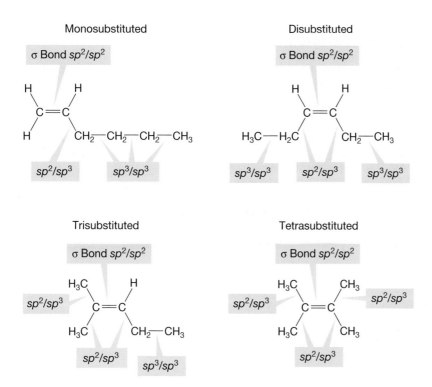

FIGURE 3.36 Three kinds of carbon–carbon σ bonds in hexene isomers: sp^2/sp^2 overlap, sp^2/sp^3 overlap, and sp^3/sp^3 overlap.

By contrast, less substituted alkenes have more relatively high-energy (weaker) sp^3/sp^3 bonds. The more strong bonds present, the more stable is the molecule. Figure 3.37 shows the number of sp^3/sp^3, sp^2/sp^3, and sp^2/sp^2 bonds in the isomeric hexenes of Table 3.1. 2,3-Dimethyl-2-butene, the molecule with the strongest bonds, is the most stable isomer of the set, and 1-hexene, with the weakest set of bonds, is the least stable.[3]

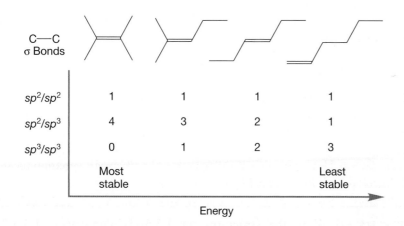

FIGURE 3.37 The isomeric hexenes have different numbers of three types of carbon–carbon σ bonds.

---

[3]This analysis is somewhat primitive because it ignores the carbon–hydrogen bonds, which are also different in differently substituted alkenes. However, the changes in carbon–carbon bonds dominate. You might try to do a full comparison using detailed bond energies.

PROBLEM 3.14 Carry out an analysis similar to that of Figure 3.37 for the series of molecules: 1-pentene, 2-methyl-2-butene, (*Z*)-2-pentene.

3.7 Double Bonds in Rings

Figure 3.38 shows a number of cycloalkenes, ring compounds containing one or more double bonds. Note that even small rings can contain double bonds.

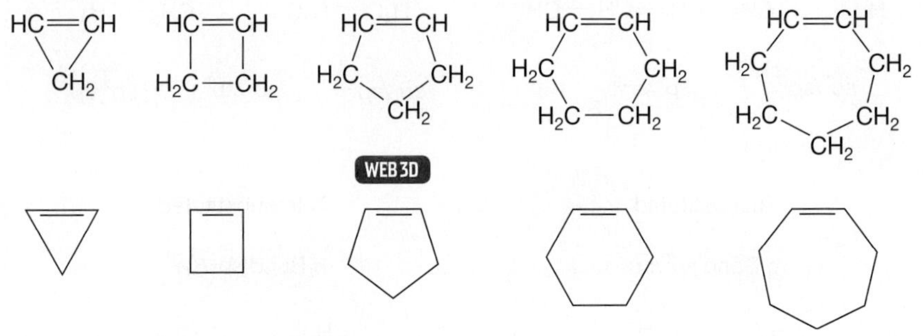

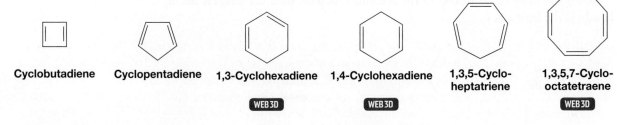

FIGURE 3.38 Some cyclic alkenes (cycloalkenes).

PROBLEM 3.15 Name all the compounds in Figure 3.38.

Once the ring becomes larger than cyclopropene, it becomes possible for it to contain more than one double bond. Several such molecules are shown in Figure 3.39.

Cyclobutadiene   Cyclopentadiene   1,3-Cyclohexadiene   1,4-Cyclohexadiene   1,3,5-Cyclo-heptatriene   1,3,5,7-Cyclo-octatetraene

FIGURE 3.39 Some cyclic polyenes, ring compounds containing more than one double bond.

PROBLEM 3.16 How many signals will appear in the ^{13}C NMR spectrum (p. 88) of the compounds in Figure 3.39?

PROBLEM 3.17 Write the structures for 1,3,5-cyclohexatriene, 1,3,5,7-cyclooctatetraene, 3-methyl-1,4-cyclohexadiene, 2-fluoro-1,3-cyclohexadiene, and 2-bromo-1,4-cycloheptadiene.

WORKED PROBLEM 3.18 Draw (*E*)- and (*Z*)-1-methylcycloheptene. Which isomer would you expect to be more stable? Explain.

ANSWER In (*Z*)-1-methylcycloheptene, the two higher priority groups are on the same side of the double bond, whereas in (*E*)-1-methylcycloheptene they are on opposite sides. On carbon 2, hydrogen is lower priority than CH_2, and on carbon 1, CH_3 is lower priority than CH_2—C. The compound with the higher priority groups on the same side is (*Z*), and the compound with the higher priority groups on opposite sides is (*E*).

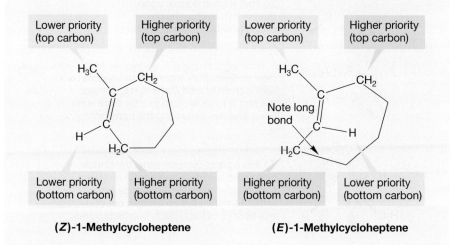

| Lower priority (top carbon) | Higher priority (top carbon) | Lower priority (top carbon) | Higher priority (top carbon) |

(Z)-1-Methylcycloheptene   **(E)-1-Methylcycloheptene**

| Lower priority (bottom carbon) | Higher priority (bottom carbon) | Higher priority (bottom carbon) | Lower priority (bottom carbon) |

In this example, even the drawings give a clue to the relative stabilities. It is trivial to draw the (*Z*) form, but the (*E*) form requires you to stretch bonds (arrow) in order to make the necessary connections. If you made a model, you saw that the (*E*) isomer contains a badly twisted double bond, with poor overlap between the $2p$ orbitals making up the π bond. There are no such problems in the (*Z*) isomer, which is therefore much more stable.

All the double bonds in Figures 3.38 and 3.39 are cis. For some reason this is often a sticky point for students. Make sure you understand why all the double bonds shown in the figures are appropriately called cis. To make the point absolutely clear we will use cyclopentene as an example. Note that the two hydrogens on the double bond are on the same side. The two alkyl groups (part of the ring) are on the other side (Fig. 3.40). The molecule is properly called cis, or (*Z*).

It is extremely difficult to put a trans double bond in a small ring, and it's worth our time to see why. As we saw in Section 3.2, double bonds depend on overlap of p orbitals for their stability. There is no great problem in a ring containing a cis double bond (Fig. 3.40), but if the cycloalkene is trans, the double bond becomes severely twisted and highly destabilized (see Worked Problem 3.18). Remember that π bonds derive their stability from overlap of $2p$ orbitals, and that a fully twisted (90°) π bond, with its $2p$ orbitals oriented at 90° to one another, is about 66 kcal/mol higher in energy than the 0° form (p. 106). For a small or medium-sized ring, it is not possible to bridge opposite sides of the double bond without very

cis-**Cyclopentene**

The two hydrogens are on the same side of the double bond—the double bond in this molecule is cis, or (*Z*)

FIGURE 3.40 The double bond in cyclopentene is cis.

severe strain. Let's use cyclopentene as an example again. A single CH$_2$ group cannot span the two carbons attached to opposite sides of a trans double bond (Fig. 3.41). This difficulty is easy to see in models. Use your models to try to make *trans*-cyclopentene. Hold the double bond planar, and then try to introduce the three bridging methylene groups. They won't reach. Next, allow the π bond to relax as you connect the methylenes. The methylenes will reach this time, but now the 2p orbitals making up the π bond no longer overlap efficiently. Make sure you see this point.

FIGURE 3.41 It is difficult to incorporate a trans double bond in a small ring. Either the bridge of methylene groups is too small to span the trans positions or the 2p orbitals making up the π bond must be twisted out of overlap.

To start a trans double bond, put the two hydrogens on opposite sides of the π bond

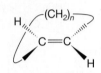

Now connect the carbons of the double bond with a chain of methylene (CH$_2$)$_n$ groups; if the ring is large enough ($n \geq 6$), there is no great problem in forming the trans isomer

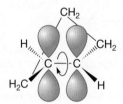

To make *trans*-cyclopentene, the double bond carbons must be bridged by only three methylenes; this cannot be done without twisting the 2p orbitals out of overlap; this twisting is very costly in energy terms

Of course, the larger the ring, the less problem there is because more atoms are available to span the trans positions of the double bond. In practice, the smallest trans cycloalkene stable at room temperature is *trans*-cyclooctene. Even here, the trans isomer is 11.4 kcal/mol less stable than its cis counterpart (Fig. 3.42).

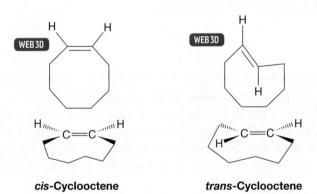

cis-Cyclooctene   **trans-Cyclooctene**

FIGURE 3.42 *cis-* and *trans-*Cyclooctene.

PROBLEM 3.19 Calculate the equilibrium distribution of *cis-* and *trans-*cyclooctene at 25 °C given an energy difference of 11.4 kcal/mol between the two isomers (of course, this question assumes that equilibrium can be reached—something not generally true for alkenes). The relationship between the energy difference ($\Delta G°$) and the equilibrium constant (K) is $\Delta G° = -RT \ln K$, or $\Delta G° = -2.3RT \log K$, where R is the gas constant (1.986 cal/deg · mol) and T is the absolute temperature.

Many cyclic and polycyclic (more than one ring) compounds containing double bonds are known, including a great many natural products. Figure 3.43 shows some common structures incorporating cycloalkenes.

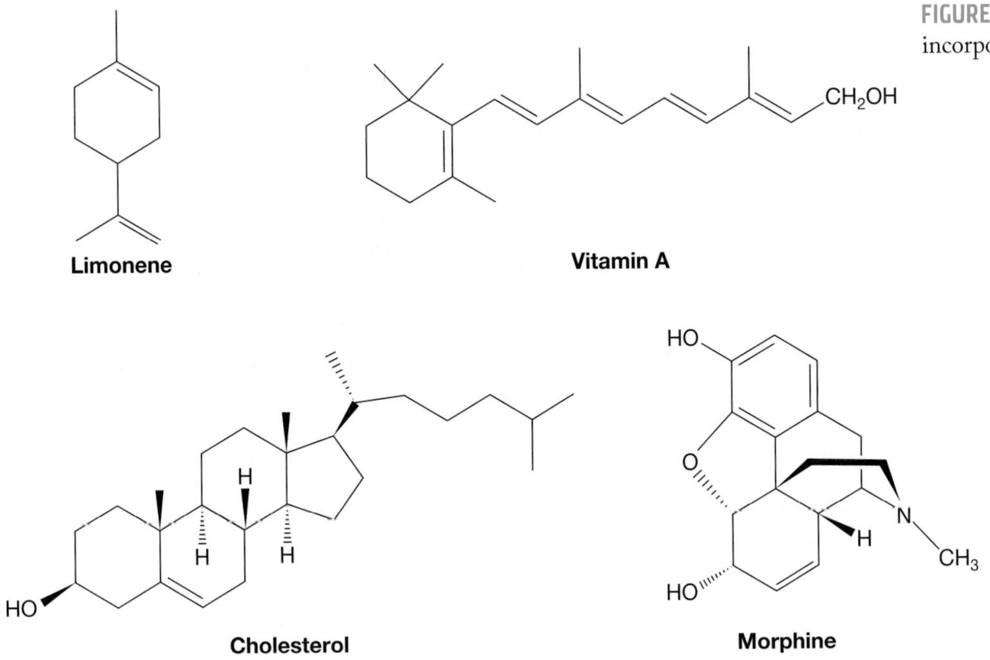

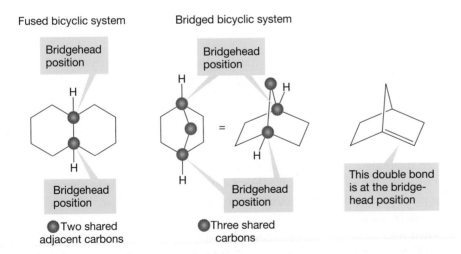

FIGURE 3.43 Some natural products incorporating cycloalkenes.

It was noticed in 1924 by the German chemist Julius Bredt (1855–1937) that one structural type was conspicuously absent among the myriad compounds isolated from sources in Nature. There seemed to be no double bonds attached to the **bridgehead position** in what are called "bridged bicyclic molecules." The bridgehead position is the point at which the bridges meet (Fig. 3.44). Apparently, great

FIGURE 3.44 Fused and bridged bicyclic molecules.

instability is caused by the incorporation of a double bond at the bridgehead position. You have met such structures before in Chapter 2 (p. 85), and they will reappear in some detail in Chapter 5, but you might make a model now of a simple bridged bicyclic structure. Figure 3.44 shows two kinds of bicyclic molecules, "bridged" and "fused." In fused compounds, two rings share a pair of adjacent carbons. In bridged bicyclic compounds, more than two carbons are shared.

Bredt was unable to explain this absence of double bonds at the bridgehead position, but was alert enough to note the phenomenon—the absence of molecules with double bonds at the bridgehead. That there could not be such compounds has become deservedly known as **Bredt's rule**. What is the reason behind this rule? First of all, the rigid cage structure with its *pyramidal* bridgehead carbons (Fig. 3.44) requires that the π bond, the "double" part of the alkene, be formed not from $2p/2p$ overlap, but by $2p$/hybrid orbital overlap. Overlap is not as good as in a normal alkene π system (Fig. 3.45). There is even more to this question, however, and a

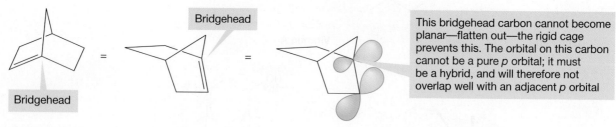

This bridgehead carbon cannot become planar—flatten out—the rigid cage prevents this. The orbital on this carbon cannot be a pure *p* orbital; it must be a hybrid, and will therefore not overlap well with an adjacent *p* orbital

FIGURE 3.45 Three views of a bridged bicyclic molecule containing a double bond at the bridgehead.

careful three-dimensional drawing of the molecule best reveals the story. As is so often the case, a Newman projection is the most informative way to look at the molecule (Fig. 3.46). Note that the orbitals making up the (hypothetical) double bond

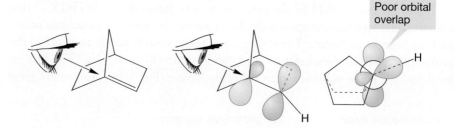

Poor orbital overlap

FIGURE 3.46 A Newman projection shows that in these bridgehead alkenes the orbitals making up the "π bond" cannot overlap well.

do not overlap at all well. Just as trans cycloalkenes are severely twisted, so are bridgehead alkenes. Moreover, the rigid structure of these compounds (make a model!) allows for no relief. This is a good example of how the two dimensions of the paper can fool you. There is no difficulty in drawing the lines making up the double bond on the paper, and unless you can see the structure, you will almost certainly be fooled.

As the bridges in bicyclic molecules get longer, flexibility returns and bridgehead "anti-Bredt" alkenes become stable (Fig. 3.47). Clever syntheses have been devised so that even quite unstable compounds can be made and studied, and old Bredt's rule violated. In simple bridgehead alkenes, the limits of room temperature stability are reached with bicyclo[3.3.1]non-1-ene (do not worry yet about the naming system; we will deal with it in Chapter 5), which contains a trans double bond in an eight-membered ring. As with the simple trans cycloalkenes, it is the eight-carbon compound that is the first molecule stable at room temperature. It doesn't seem unreasonable that there should be a rough correspondence between the trans cycloalkenes and the bridgehead alkenes, which also contain a trans double bond in a ring.

FIGURE 3.47 The smallest bridgehead alkene stable under normal conditions is bicyclo[3.3.1]non-1-ene. This molecule contains a *trans*-cyclooctene, shown in color.

PROBLEM 3.20 Is the bicyclo[3.3.1]non-1-ene shown in Figure 3.47 (*Z*) or (*E*)?

PROBLEM 3.21 Draw the other form of bicyclo[3.3.1]non-1-ene (*Z* or *E*). Why is it less stable than the one in Figure 3.47? *Hint*: Focus on the rings in which the double bonds are contained.

3.8 Physical Properties of Alkenes

There is little difference in physical properties between the alkenes and their saturated relatives, the alkanes. The odors of the alkenes are a bit more pungent and perhaps justify being called "evil-smelling." In fact, the old trivial name for alkenes, **olefins**, readily evokes the sense of smell. Tables 3.2 and 3.3 list some data for alkenes and cycloalkenes.

TABLE 3.2 Some Simple Alkenes

Name	Formula	mp (°C)	bp (°C)
Ethene (ethylene)	$H_2C{=}CH_2$	−169	−103.7
Propene (propylene)	$H_2C{=}CH{-}CH_3$	−185.2	−47.4
1-Butene	$H_2C{=}CH{-}CH_2{-}CH_3$	−185	−6.3
cis-2-Butene	$CH_3{-}CH{=}CH{-}CH_3$	−138.9	3.7
trans-2-Butene	$CH_3{-}CH{=}CH{-}CH_3$	−105.5	0.9
2-Methylpropene (isobutene)	$(CH_3)_2CH{=}CH_2$	−40.7	−6.6
1-Pentene	$H_2C{=}CH{-}(CH_2)_2{-}CH_3$	−165	30.1
1-Hexene	$H_2C{=}CH{-}(CH_2)_3{-}CH_3$	−139.8	63.3
1-Heptene	$H_2C{=}CH{-}(CH_2)_4{-}CH_3$	−119	93.6
1-Octene	$H_2C{=}CH{-}(CH_2)_5{-}CH_3$	−101.7	121.3
1-Nonene	$H_2C{=}CH{-}(CH_2)_6{-}CH_3$	−83	146
1-Decene	$H_2C{=}CH{-}(CH_2)_7{-}CH_3$	−66.3	170.5

TABLE 3.3 Some Simple Cycloalkenes

Name	mp (°C)	bp (°C)
Cyclobutene		2
Cyclopentene	−135	44.2
Cyclohexene	−103.5	83
Cycloheptene	−56	115
cis-Cyclooctene	−12	138
trans-Cyclooctene	−59	143
cis-Cyclononene		167–169
trans-Cyclononene		94–96 (at 30 mmHg)

3.9 Alkynes: Structure and Bonding

Like the simplest alkene, ethylene, the simplest alkyne is generally known by its trivial name, acetylene, not the systematic name, ethyne. Recall that more complicated, substituted alkynes are often named as derivatives of acetylene. Acetylene itself is a symmetrical compound of the formula HCCH in which the two carbon atoms are attached to each other by a triple bond. Each carbon is attached to two other atoms: one hydrogen and the other carbon. Accordingly, a reasonable bonding scheme must yield only two hybrid orbitals, one to be used in bonding to the hydrogen, and one to be used in bonding to the carbon. As in the discussion of alkanes and alkenes (p. 64; p. 99), the atomic orbitals of carbon are combined to yield hybrid orbitals, which do a better job of bonding than the unchanged atomic orbitals. Because only two bonds are needed, one for the bond to hydrogen, and another for the bond to carbon, we need to combine only two of carbon's atomic orbitals to give us our hybrids. A combination of the 2*s* and 2*p_x* orbitals will yield a pair of *sp* hybrids.

We can anticipate that these new *sp* hybrids will be directed so as to keep the bonds, and the electrons in them, as far apart as possible, which produces 180° angles (Fig. 3.48). (Recall our discussion of BeH_2, p. 53.)

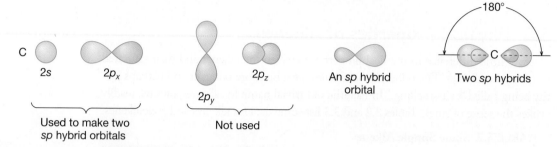

FIGURE 3.48 To make an *sp* hybrid we combine the wave functions for the carbon 2*s* and 2p_x orbitals. The 2p_y and 2p_z orbitals are not used in the hybridization scheme.

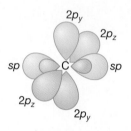

FIGURE 3.49 An *sp*-hybridized carbon atom. Note the leftover, unhybridized 2p_y and 2p_z orbitals.

These new *sp* hybrid orbitals generally resemble the sp^2 and sp^3 hybrids we made before (compare Figure 3.48 with Figures 2.6 and 2.23). Because we used only two of carbon's four available atomic orbitals, the 2p_y and 2p_z orbitals are left over. Figure 3.49 shows an *sp*-hybridized carbon atom. (From now on, we will not show the small lobes.)

Each carbon–hydrogen bond in acetylene is formed by overlap of one *sp* hybrid with the hydrogen 1*s* orbital, and the carbon–carbon bond is formed by the overlap of two *sp* hybrids. This process forms the σ bonds of the molecule (Fig. 3.50).

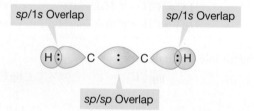

FIGURE 3.50 The σ bonding system of acetylene: two C—H bonds and one C—C bond.

Figure 3.50 shows the bonds formed by *sp/sp* and *sp/1s* overlap, but don't forget that these overlapping hybrid and atomic orbitals create both bonding and antibonding molecular orbitals. The empty antibonding orbitals are not shown in Figure 3.50, but are shown in the schematic construction of Figure 3.51.

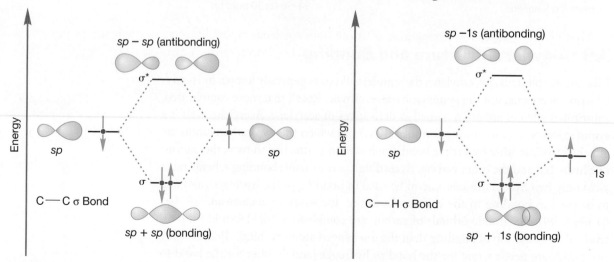

FIGURE 3.51 Orbital interaction diagrams showing the bonding and antibonding orbitals formed by *sp/sp* and *sp/1s* overlap.

As in the alkenes, the remaining unhybridized 2*p* orbitals in the alkynes overlap to form π bonds. This time there are two *p* orbitals remaining on each carbon of acetylene, and we can form a pair of π bonds that are directed, as are the 2*p* orbitals making up the π bonds, at 90° to each other (Fig. 3.52).

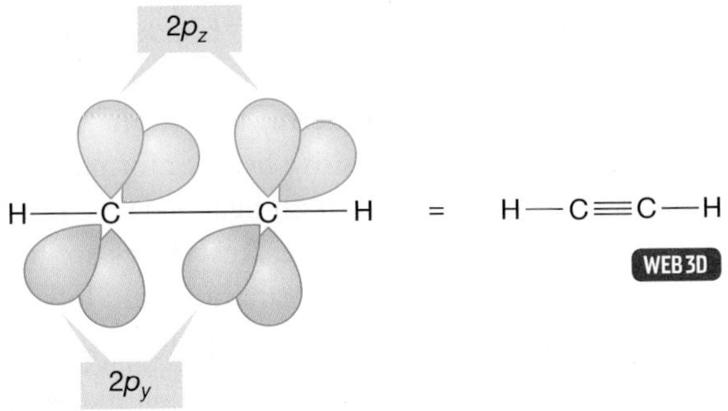

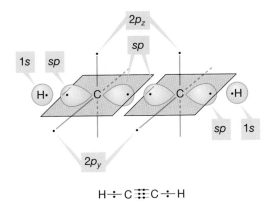

FIGURE 3.52 Overlap of two 2*p*$_y$ and two 2*p*$_z$ orbitals forms a pair of π bonds in alkynes.

Both carbons in acetylene have four valence electrons (*Remember*: All carbons have four bonding electrons because we ignore the two very low energy 1*s* electrons) and each can participate in four two-electron bonds. In acetylene, these are the σ bond to hydrogen and the carbon–carbon triple bond composed of one carbon–carbon σ bond and two carbon–carbon π bonds (Fig. 3.53).

FIGURE 3.53 A highly schematic construction of the π and σ bonds in acetylene.

PROBLEM 3.22 Draw the orbital interaction diagrams for construction of the π bonds of acetylene from the 2*p* orbitals.

One of the structural consequences of triple bonding is an especially short carbon–carbon bond distance (~1.2 Å), considerably shorter than either carbon–carbon single or double bonds (Fig. 3.54).

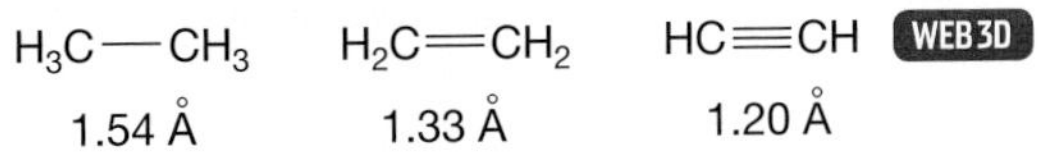

$$H_3C—CH_3 \qquad H_2C=CH_2 \qquad HC≡CH \quad \boxed{WEB\,3D}$$
$$1.54\ \text{Å} \qquad\quad 1.33\ \text{Å} \qquad\quad 1.20\ \text{Å}$$

FIGURE 3.54 Bond lengths in angstroms (Å) of simple two-carbon hydrocarbons.

Summary

The σ bonding schemes we used to construct alkanes and cycloalkanes are not the only way that atoms can be held together to form molecules. In alkenes and alkynes, there is π bonding as well as σ bonding. In the construction of a π bond, 2*p* orbitals overlap side-to-side to form a bonding and antibonding pair of π molecular orbitals. Alkenes and cycloalkenes contain one π bond, whereas alkynes have two π bonds. Ring compounds can also contain double (and sometimes triple) bonds.

3.10 Relative Stability of Alkynes: Heats of Formation

Heats of formation show that alkynes are very much less stable than their constituent elements. Table 3.4 lists heats of formation for some alkynes, alkenes, and alkanes. The virtues of using hybrid orbitals for σ bonding rather than p orbitals for π bonding are apparent. The more π bonds in a molecule, the more positive the heat of formation, and the more endothermic the formation of the compound from its constituent elements. The very positive heats of formation for alkynes have their practical consequences. When a very high heat is desired, as for many welding applications, acetylene is the fuel of choice. Acetylenes are high-energy compounds (with correspondingly high heats of formation) and lots of energy is given off when they react with oxygen to form the much more stable molecules water and carbon dioxide.

Because of their triple bonds, alkynes can be only mono- or disubstituted. Disubstituted alkynes are more stable than their monosubstituted isomers. For example, we can see from Table 3.4 that the heat of formation of 1-butyne is about 5 kcal/mol more positive than that for 2-butyne. This phenomenon echoes the increasing stability of alkenes with increasing substitution (Section 3.6).

TABLE 3.4 Heats of Formation for Some Small Hydrocarbons

Alkanes	ΔH_f° (kcal/mol)	Alkenes	ΔH_f° (kcal/mol)	Alkynes	ΔH_f° (kcal/mol)
Ethane	−20.1	Ethylene	12.5	Acetylene	54.5
Propane	−25.0	Propylene	4.8	Propyne	44.6
Butane	−30.2	1-Butene	−0.1	1-Butyne	39.5
		2-Butene	−1.9 (cis)	2-Butyne	34.7
			−2.9 (trans)		

3.11 Derivatives and Isomers of Alkynes

We can imagine replacing one of the hydrogens of acetylene with an atom or group X to give substituted alkynes. Figure 3.55 gives an example of an "ethynyl" compound. In practice, two naming systems are used here, and the figure shows them. Chloroethyne can also be called chloroacetylene, and both naming systems are commonly used.

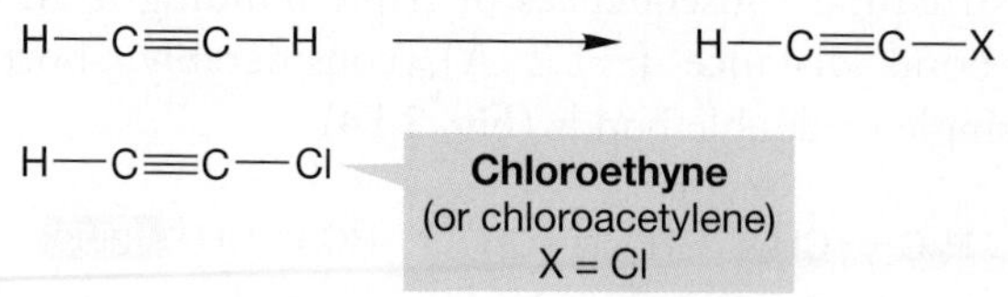

FIGURE 3.55 An ethynyl compound, chloroethyne, a monosubstituted alkyne.

We can start constructing the family of alkynes by replacing X in Figure 3.55 with a methyl group, CH_3. This process produces propyne (or methylacetylene), the lone three-carbon alkyne (Fig. 3.56).

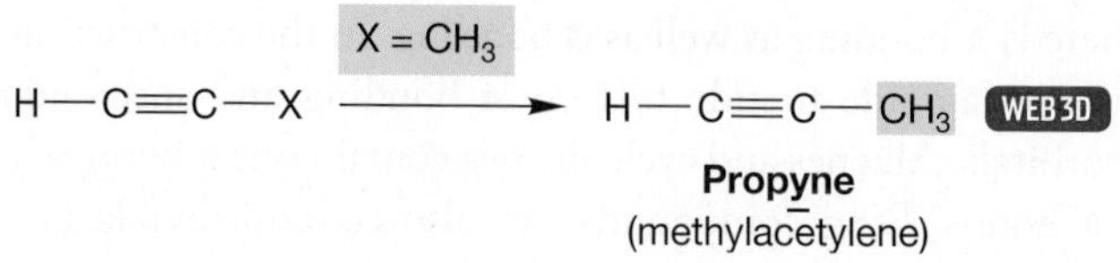

FIGURE 3.56 Replacement of X with a methyl (CH_3) group gives propyne.

There are two different hydrogens in propyne, and therefore two different ways to produce a substituted compound (Fig. 3.57). When the acetylenic hydrogen is replaced, the compounds can be named either as acetylenes or as 1-propynyl compounds. 3-Propynyl compounds result from replacement of a methyl hydrogen with X.

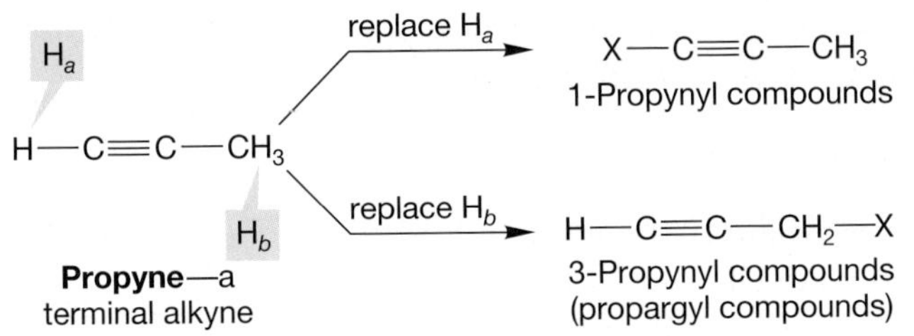

FIGURE 3.57 Examples of the two kinds of hydrogen in propyne, each replaced with X to give derivatives.

The common name **propargyl** is reserved for the $HC\equiv C-CH_2$ group and is often seen. Note the relationship between the allyl (Fig. 3.19) and propargyl groups.

When $X = CH_3$ we produce the two four-carbon alkynes, which must be butynes (Fig. 3.58).

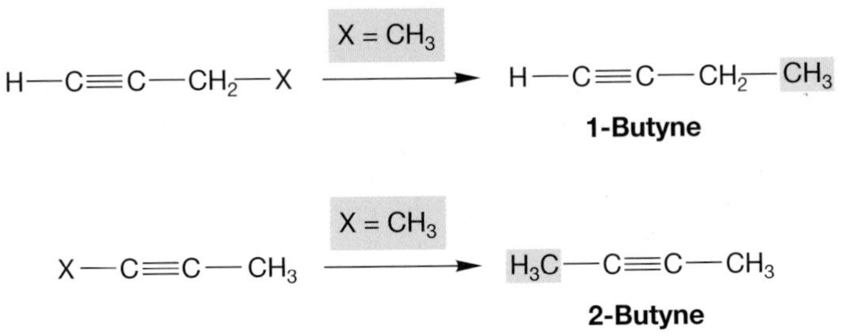

FIGURE 3.58 If $X = CH_3$, we get the two butynes.

The naming protocol for alkynes is similar to that used for alkenes. As with the double bond of alkenes, the triple bond position is designated by assigning it a number, which is kept as low as possible. When both double and triple bonds are present, the compounds are named as enynes, and the numbers designating the positions of the multiple bonds kept as low as possible. If the numbering scheme could produce two names in which the lower number could go to either the "-ene" or the "-yne," the "-ene" gets it (Fig. 3.59).

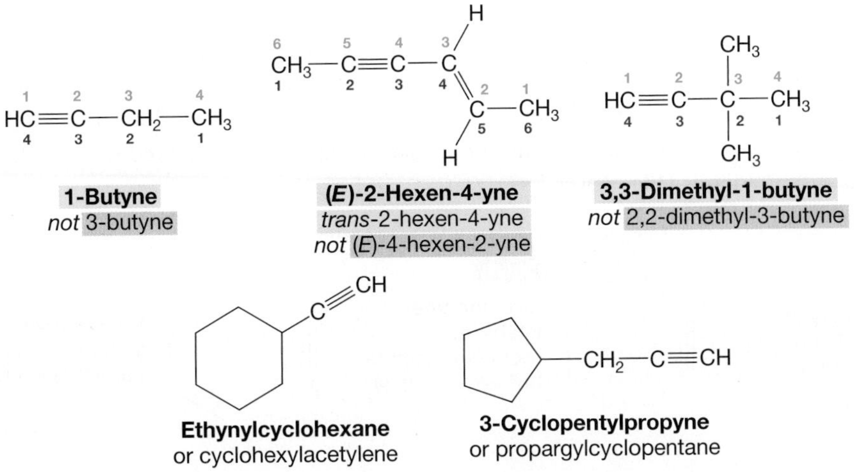

1-Butyne
not 3-butyne

(E)-2-Hexen-4-yne
trans-2-hexen-4-yne
not (E)-4-hexen-2-yne

3,3-Dimethyl-1-butyne
not 2,2-dimethyl-3-butyne

Ethynylcyclohexane
or cyclohexylacetylene

3-Cyclopentylpropyne
or propargylcyclopentane

FIGURE 3.59 Some examples of the naming convention for alkynes.

PROBLEM 3.23 Draw all the pentynes, hexynes, and heptynes.

PROBLEM 3.24 Name all the isomers you drew in Problem 3.23.

PROBLEM 3.25 How many signals will appear in the ^{13}C NMR spectra of the compounds in Figure 3.59 (see Section 2.14)?

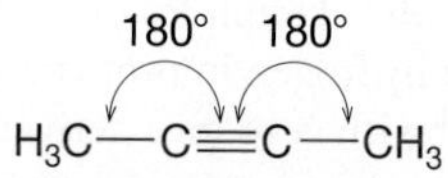

180° 180°
H3C—C≡C—CH3

FIGURE 3.60 There are 180° angles in 2-butyne.

Can alkynes exhibit cis/trans isomerism? It takes only a quick look to see that there can be no such isomerism in a linear compound (Fig. 3.60). The *sp* hybridization requires 180° angles, and therefore there can be no cis/trans isomerism in alkynes. The alkynes are simpler to analyze structurally than the alkenes.

3.12 Triple Bonds in Rings

Like double bonds, triple bonds can occur in rings, although very small ring alkynes are not known. The difficulty comes from the angle strain induced by the preference for linear, 180° bond angles in an acetylene. When a triple bond is incorporated in a ring, it becomes difficult to accommodate these 180° angles. Deviation from 180° reduces the overlap between the *p* orbitals making up one of the π bonds of the acetylene and that raises the energy of the compound (Fig. 3.61). *Remember*: "R" is shorthand for a general group.

Second π bond
in the acetylene
R
C
C
R
bend
R
C
C
R
Reduced
overlap
A cyclic acetylene

FIGURE 3.61 Incorporation of a triple bond in a ring is difficult because orbital overlap between the 2*p* orbitals of one of the π bonds is reduced by the bending required by the ring.

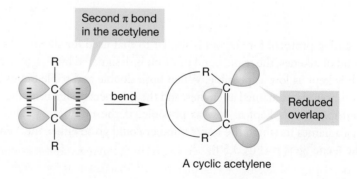

A similar problem with angles exists for the cycloalkenes, but it is less severe. In practice, the smallest ring in which an alkyne is stable under normal conditions is cyclooctyne. Thus, it and larger cycloalkynes (cyclononyne, etc.) are known, but the smaller cycloalkynes are either unknown or have been observed only as fleeting intermediates (Fig. 3.62).

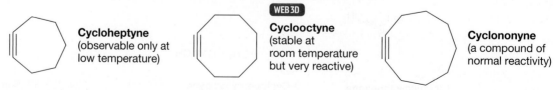

Cycloheptyne
(observable only at
low temperature)
WEB 3D
Cyclooctyne
(stable at
room temperature
but very reactive)
Cyclononyne
(a compound of
normal reactivity)

FIGURE 3.62 Some cycloalkynes.

If you got this problem right, feel really good. If it was a struggle, that is alright as well. Not all problems are easy! If they were, Nature would be trivial to understand, and life would be boring indeed. In any case, try to learn from these in-chapter answered problems. There is a "take-home lesson" intended in each of them.

Now let's try to estimate qualitatively the overall change in energy as this reaction proceeds. We will plot energy on the vertical axis against something we might call "reaction progress" on the horizontal axis. Roughly speaking, we are asking how the energy changes as the reaction takes place. Notice in Figure 3.74 that two bonds

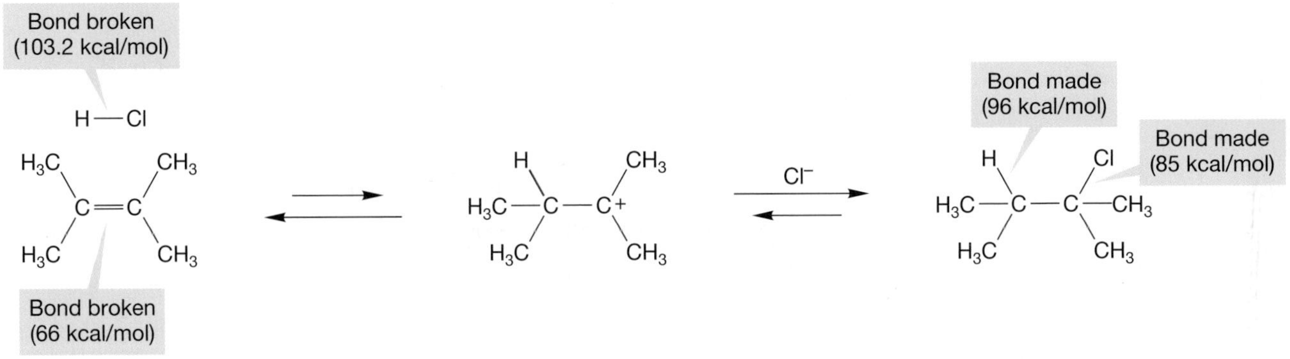

Overall: Bonds broken (103 + 66) = 169 kcal/mol
Bonds made (96 + 85) = 181 kcal/mol

Net: 181 − 169 = about 12 kcal/mol exothermic

FIGURE 3.74   An estimation of the energetics of the addition of HCl to 2,3-dimethyl-2-butene.

must be broken in the first step of the reaction. These are the carbon–carbon π bond (66 kcal/mol), and the bond in HCl (103.2 kcal/mol). In the final product, two bonds have been made, the new carbon–hydrogen bond (∼96 kcal/mol), and the new carbon–chlorine bond (∼85 kcal/mol). It requires almost 170 kcal/mol to break the bonds in this reaction, but that quantity is more than fully compensated by the roughly 181 kcal/mol released by the new bonds formed (Fig. 3.74).

As we have seen, in the chemistry trade, this kind of reaction, in which the products are more stable than the starting materials, is called an exothermic reaction. But this reaction actually takes place in two steps—we do not just suddenly arrive at a product from the starting material. There is an intermediate carbocation in the reaction, a transient charged species relatively high in energy. Not only is the carbocation charged—a bad deal in terms of energy—but the carbon does not have an octet of electrons, which is another factor contributing to its high energy. There are two factors that make this first step difficult. First, the bonds to be broken, the carbon–carbon π bond and the σ bond between hydrogen and chlorine, are collectively stronger than the bond made, the new carbon–hydrogen bond shown in red in Figure 3.74. Second, charged species such as carbocations are high in energy, and are not usually easily formed. Thus, the first step in this reaction is surely "uphill" in energy, and thus is an endothermic reaction.

By contrast, the second step of this reaction, capture of the carbocation by chloride, is sure to be very energy-releasing—an exothermic reaction. Charges are annihilated and a new carbon–chlorine bond is made, at no bond-breaking cost. If we plot these estimated energy changes, we get Figure 3.75, the first of many such energy diagrams to appear in this book as we estimate the energetics of other reactions. The overall reaction is exothermic—the products are more stable than the starting materials,

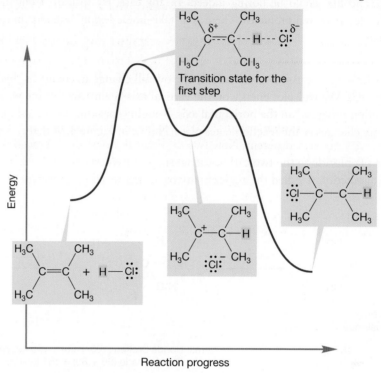

FIGURE 3.75 In this simple addition of hydrogen chloride to an alkene, the first step, the endothermic formation of a carbocation, is the slow step. The developing partial charges are shown as δ^+ and δ^-.

but an initial endothermic step, the formation of the intermediate carbocation, must occur before product can be formed. There is an energy hill—a barrier—to be passed.

The tops of the hills in Figure 3.75 are transition states—the high points in energy separating starting material and product. We encountered transition states earlier when we discussed the barrier to rotation in ethane and other alkanes (p. 67). Although these energy maxima cannot be isolated, we can describe their structures. In Figure 3.75, the making and breaking bonds are shown as dashed partial bonds. The developing partial positive and negative charges appear as δ^+ and δ^-.

With just a little thought we have a very clear picture of the addition reaction. An unstable carbocationic intermediate is formed and then destroyed as the chloride ion adds to give the final product. Notice two important things about this reaction. First, it cannot proceed unless enough energy is supplied so that the first transition state can be reached. This amount of energy—that required to pass over the highest-energy point, the highest transition state—is called the **activation energy ($\Delta G^{\ddagger}$)** for the reaction.

Second, we get a bonus from this analysis. If we understand the pathway from left to right, we also understand the reverse reaction—the one that runs from right to left. It follows exactly the same energy curve, only in the opposite direction.

PROBLEM 3.29 Generalize from the discussion of HCl addition to predict the products of the following reactions:

3.18 The Regiochemistry of the Addition Reaction

In the reaction introduced in Figure 3.68, the addition of HCl to 2,3-dimethyl-2-butene, the two ends of the alkene are the same, which means that there is no choice to be made in the protonation step. Addition to either end of the alkene gives exactly the same carbocation. Let's extend our discussion just a bit and examine an alkene in which the two ends are different. We'll do more of this sort of thing in Chapter 9, when we take a long look at many kinds of addition reactions, but there is much to be learned from even a small extension of the symmetrical addition reaction.

So, let's allow HCl to react with 2-methylpropene, an alkene in which the two ends are certainly very different. Now two different protonations are possible, and addition of chloride to the two carbocations will give two products (Fig. 3.76). This kind of distinction is called the **regiochemistry** of the reaction. In fact, only one of the two possible products is formed. Let's see if we can guess why.

Each potential reaction pathway in Figure 3.76 involves a different carbocationic intermediate. Perhaps formation of a tertiary carbocation requires less energy than

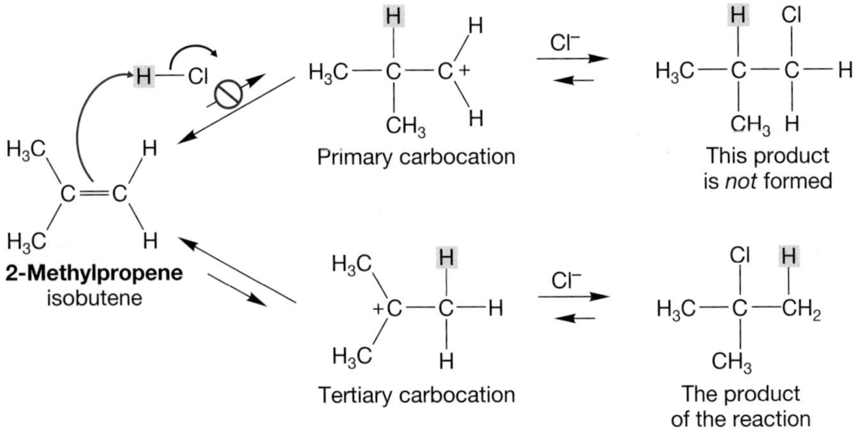

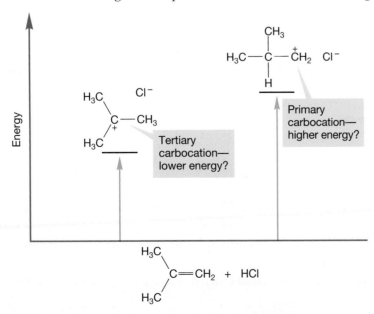

FIGURE 3.76 Two possible additions of HCl to 2-methylpropene.

formation of a primary carbocation. If that idea were correct, the energy hill to climb to give the observed product (formed from the tertiary carbocation) would be lower than that leading to the product that is not observed (Fig. 3.77).

FIGURE 3.77 Relative stability of the *tert*-butyl cation (tertiary) and the isobutyl cation (primary).

This idea leads to a prediction. If our surmise that the stability of carbocations increases with substitution is correct, that is, if a tertiary carbocation is more stable than a primary carbocation, presumably a tertiary carbocation will also be more stable than a secondary carbocation, and a secondary carbocation will be more stable than a primary carbocation. The complete order of cationic stability would be tertiary > secondary > primary > methyl (Fig. 3.78).

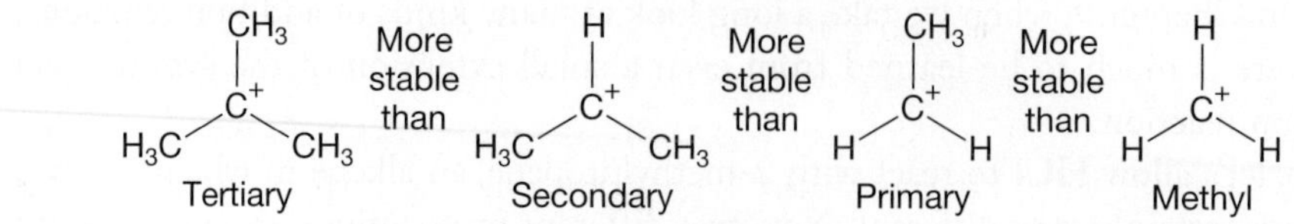

FIGURE 3.78 Predicted stability order for carbocations.

If the order of carbocation stability shown in Figure 3.78 is correct, we should always get the product resulting from formation of the more stable carbocation (Fig. 3.79).[5] We can test this idea. Just allow HCl to add to 2-methyl-2-butene or propene, each an unsymmetrical alkene. If we are right in our ideas about carbocation stability, the products will be 2-chloro-2-methylbutane and isopropyl chloride, respectively. When these experiments are run, those are exactly the products observed (Fig. 3.79).

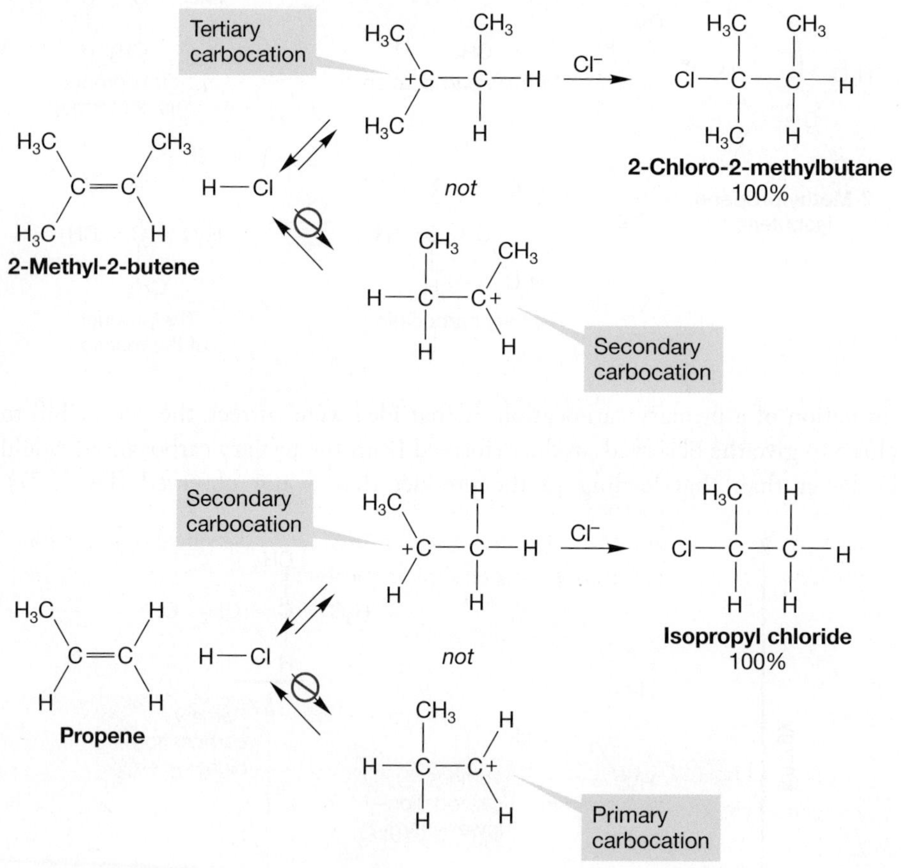

FIGURE 3.79 Addition of HCl to two unsymmetrical alkenes.

The results shown in Figure 3.79 are in accord with our mechanistic ideas; our proposed reaction mechanism is supported by the new data. Do we now know how this addition of HCl to alkenes works? Is our mechanism proved? Sadly, the answer

[5]We hope you are wondering why this order exists. We will get to that question in Chapter 9.

is "no," and in fact, we can never know for certain that we are absolutely right. A mechanistic hypothesis survives only pending the next clever test. Someone—you perhaps—may develop a new test of the ideas outlined in this section that reveals an error, or at least an area that needs further thinking and working out of details. It's sad but true: It is easy to disprove a scientific hypothesis—just find some contradictory data. It is impossible ever to prove a mechanistic hypothesis in chemistry beyond a shadow of a doubt.

We all have to get used to this ambiguity—it is part of the human condition. We can be quite certain we are right in our mechanistic ideas, and the more time that goes by, and the more tests our ideas pass, the more confident we can be that we are basically right. However, there have been some big surprises in science, as conventional wisdom is reversed every once in a while, and there will be more. To be honest, we sort of like it when that happens.

PROBLEM 3.30 Predict the products of the following reactions:

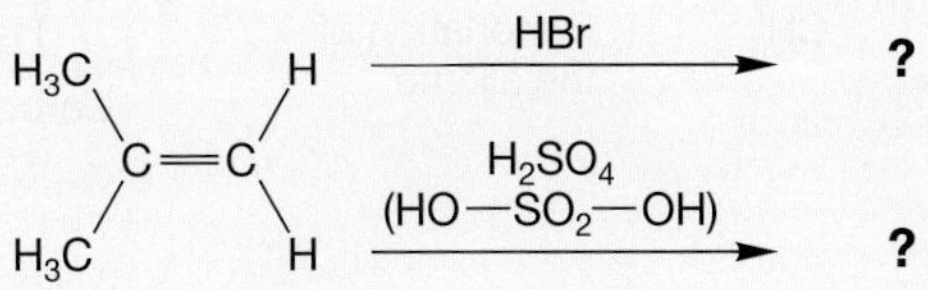

PROBLEM SOLVING

GO

All "which direction does HX add" problems can be approached—and solved—in the same way. Always draw out the two possible carbocations formed by initial protonation of the double bond. For an unsymmetrical (ends different) double bond, there are always two and only two possibilities. Next evaluate those two carbocations—ask yourself which of them is more stable? It is that one that will lead to the major product. The answer to Problem 3.30 is typical. Use it as a prototype, a template for answering all similar questions.

3.19 A Catalyzed Addition to Alkenes: Hydration

Not every HX is a strong enough acid to protonate an alkene. Water, HOH (here X = OH), is an example of such a weak acid. Yet it would be very useful to be able to add the elements of water to an alkene, to "hydrate" an alkene as in Figure 3.80,

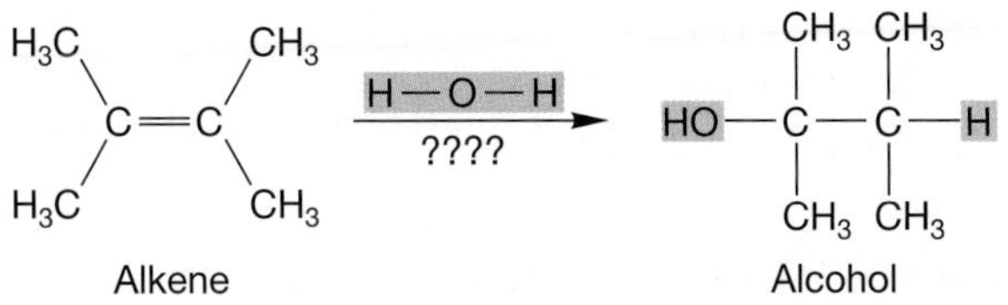

FIGURE 3.80 The highly desirable addition of water to an alkene.

because the products (R—OH, called **alcohols**) are useful in themselves and are also versatile starting materials for the synthesis of other molecules. The solution to this problem is to use a small amount of an acid that is strong enough to protonate the alkene. The added acid acts as a **catalyst** for the addition. A catalyst is a material that increases the rate of a reaction without being consumed.

Let's work through the mechanism for the hydration of alkenes. The first step is protonation of the alkene by the catalyst, H_3O^+. As in the reaction of HCl, this addition leads to a carbocation intermediate (Fig. 3.81). Now there is no halide

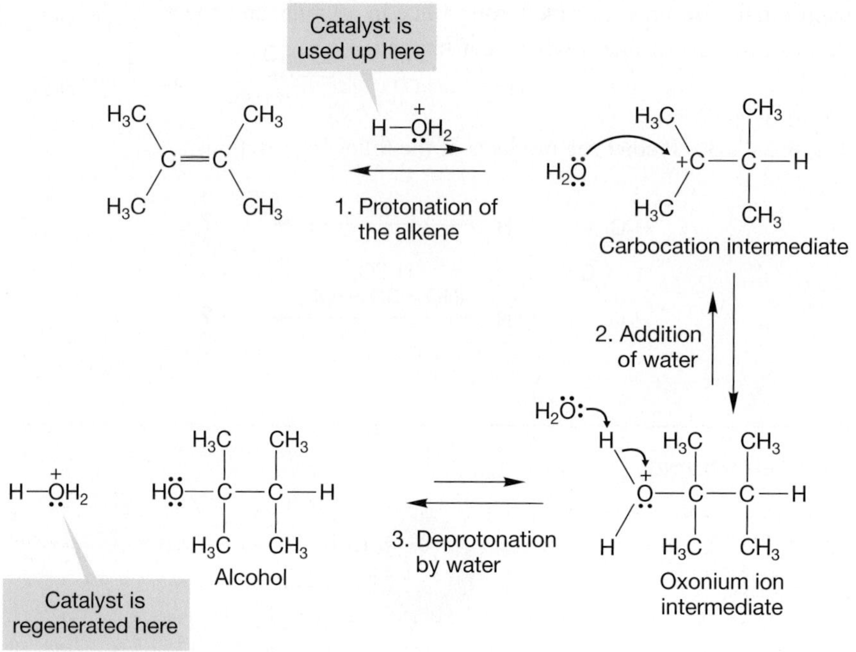

FIGURE 3.81 The acid-catalyzed addition of water to 2,3-dimethyl-2-butene—a hydration reaction.

ion (I^-, Br^-, Cl^-, F^-) to capture the carbocation, but there are lots of water molecules available. Capture by water leads not to the final product, the alcohol, but to an intermediate **oxonium ion**, in which a trivalent oxygen bears a positive charge. Water then deprotonates the oxonium ion to give the alcohol, and regenerate the catalyst, H_3O^+.

You can see from this last step why only a catalytic amount of acid is necessary. Although it is consumed in the first protonation step, it is regenerated in the last step and can recycle to start the hydration reaction again (Fig. 3.81).

WORKED PROBLEM 3.31 Only one alcohol is produced in the following reaction. What is it, and why is only one alcohol, not two, produced?

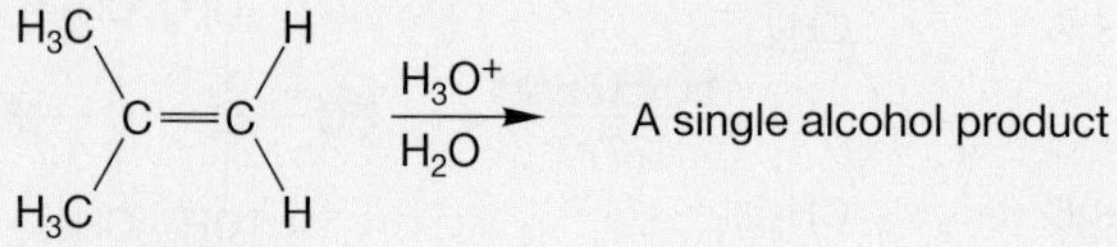

(continued)

ANSWER Formation of the more stable tertiary carbocation rather than the much less stable primary carbocation inevitably leads to the single product observed, *tert*-butyl alcohol.

More stable

H_3C ... ^+C—C—H ... $\xrightarrow{H_2O}$ ... HO—C—CH_3

tert-**Butyl alcohol**

H_3C—C=C—H ... $\xrightarrow{H_3O^+}$

H—C—C^+ ... $\xrightarrow{H_2O}$ ... H—C—CH_2OH

Much less stable ... Not formed

PROBLEM SOLVING

The order of carbocation stability is: tertiary ($3°$, R_3C^+) more stable than secondary ($2°$, R_2HC^+), more stable than primary ($1°$, RH_2C^+), more stable than methyl (H_3C^+). We'll explain this stability order later, but you should remember that simple primary and methyl carbocations are too high in energy to be formed in common organic reactions. A primary or methyl carbocation is a mechanistic stop sign. Nothing you write after it can be correct. Don't go through the stop sign!

What does "simple" mean? It means that no special factors such a resonance stabilization can be present—a "simple" carbocation is one that is not significantly delocalized by resonance.

3.20 Synthesis: A Beginning

Organic chemistry is not only understanding how reactions occur. A very important part of chemistry is the use of those reactions to make molecules. The construction of target molecules from smaller pieces is called synthesis, and we are now able to do a surprising amount of this kind of thing. Let's just take stock of what kinds of transformations we can do—of what kinds of molecules we can make out of other kinds of molecules.

The addition of hydrogen halides to alkenes gives us a way to make alkyl bromides, chlorides, and iodides (fluorides don't work very well) as long as we have the appropriate starting alkene. We will have to be careful when using unsymmetrical alkenes, because there is a choice to be made here. As you saw in Section 3.18, the addition of HX will only lead to the more substituted alkyl halide, as the regiochemistry of the addition is controlled by the stability of the intermediate carbocation.

We can also make alcohols through the acid-catalyzed addition of water to alkenes (Section 3.19).

In each chapter, the synthetic methods developed and discussed are collected on the Summary page. In this chapter, these reactions appear on pages 143 and 144.

> **PROBLEM 3.32** Devise syntheses of the following three compounds, starting in each case with any alkene that contains four carbons or fewer. You do not have to write mechanisms, although at this point it may be very helpful for you to do so.
>
> $$(CH_3)_3COH \qquad (CH_3)_2CHOH \qquad CH_3CH_2CHOHCH_3$$

3.21 Special Topic: Alkenes and Biology

In sharp contrast to the alkanes, the unsaturated alkenes are highly active biologically. As we saw in Section 3.2, alkenes contain π bonds that are substantially weaker than almost all σ bonds of alkanes. Accordingly, their reactivity is higher, and Nature "uses" this reactivity to accomplish things. One thing that can be accomplished is the formation of carbon–carbon single bonds. It is extraordinarily fortunate for us that Nature is able to do that reaction easily, because our tissues are composed mainly of carbon–carbon single bonds. We'll just sketch one example of alkene-mediated carbon–carbon single bond formation here. We will return to more details of this and related transformations in later chapters.

For its own purposes, the tuberculosis-causing organism *Mycobacterium tuberculosis* needs a molecule called 10-methylstearate. It makes this compound from a modified amino acid, *S*-adenosylmethionine, and oleate, a salt of oleic acid (Fig. 3.82).

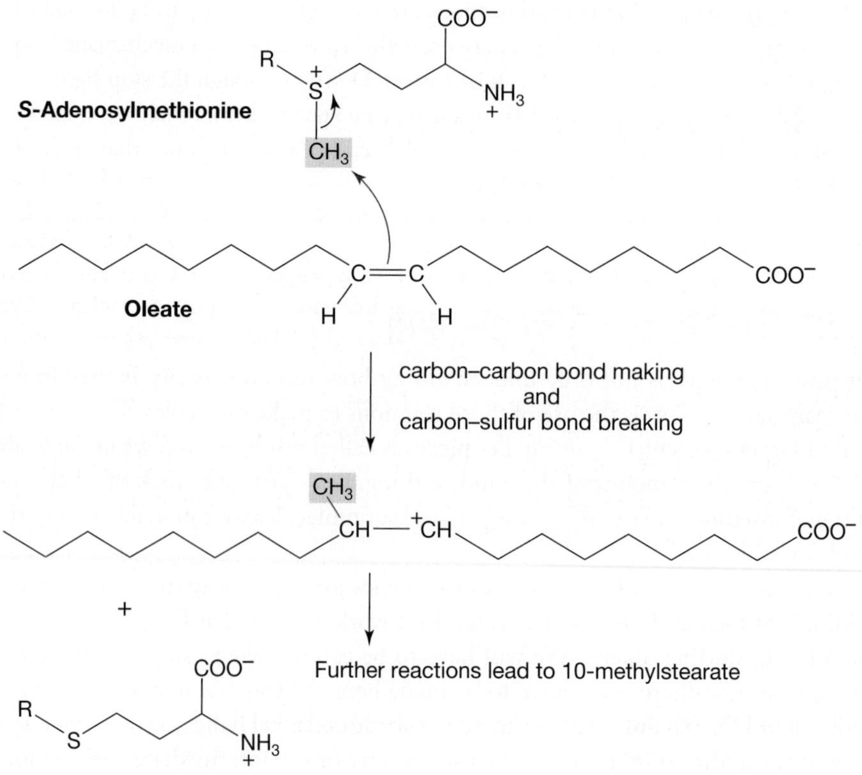

FIGURE 3.82 The reaction of a nucleophilic alkene (Lewis base) to produce a carbon–carbon single bond, shown in red.

The first interesting step in what turns out to be, ultimately, a reaction chock full of interesting steps is the transfer of the methyl group to the alkene in oleate. The double bond acts as the trigger for the whole sequence. Note also that here the π bond acts as a Lewis base, much as it does in its reaction with HCl (p. 132), where it acts as a Brønsted base.

3.22 Summary

New Concepts

This chapter deals mainly with the structural consequences of sp^2 and sp hybridizations. The doubly bonded carbons in alkenes are hybridized sp^2 and the triply bonded carbons in alkynes are hybridized sp. In these molecules, atoms are bound not only by the σ bonds we saw in Chapter 2 but by π bonds as well.

These π bonds, composed of overlapping $2p$ orbitals, have substantial impact upon the shape (stereochemistry) of the molecule. Alkenes can exist in cis (Z) or trans (E) forms—they have sides. Unlike the alkanes, which contain only σ bonds which have very low barriers to rotation, there is a substantial barrier

(~66 kcal/mol) to rotation about a π bond. The linear alkynes do not have cis/trans isomers and the question of rotation does not arise.

Double and triple bonds can be contained in rings if the ring size is large enough to accommodate the strain incurred by the required bond angles. The smallest trans cycloalkene stable at room temperature is *trans*-cyclooctene. The smallest stable cycloalkyne is cyclooctyne.

More substituted alkenes and alkynes are more stable than their less substituted relatives.

Key Terms

acetylenes (p. 98)
acetylide (p. 129)
activation energy ($\Delta G^{\ddagger}$) (p. 136)
alcohol (p. 140)
alkenes (p. 98)
alkynes (p. 98)
allyl (p. 108)
Bredt's rule (p. 122)
bridgehead position (p. 121)

Cahn–Ingold–Prelog priority system (p. 111)
catalyst (p. 140)
cycloalkenes (p. 110)
degree of unsaturation (Ω) (p. 131)
double bond (p. 98)
ethene (p. 99)
ethylene (p. 99)
heat of formation (ΔH_f°) (p. 115)
HOMO (p. 133)

LUMO (p. 133)
olefins (p. 123)
oxonium ion (p. 140)
pi (π) orbitals (p. 104)
polyenes (p. 110)
propargyl (p. 127)
regiochemistry (p. 137)
triple bond (p. 114)
vinyl (p. 107)

Reactions, Mechanisms, and Tools

The Cahn–Ingold–Prelog priority system is introduced. It is used in the (Z/E) naming system. We will encounter it again very soon in Chapter 4.

In this chapter, we continue the idea of constructing larger hydrocarbons out of smaller ones by replacing the different available hydrogens with an X group. When X is methyl (CH_3), a larger hydrocarbon is produced from a smaller one.

Addition reactions between alkenes and acids, H—X, are introduced. Many acids (H—Br, H—Cl, H—I, $HOSO_2OH$, generally, H—X) add directly to alkenes. The first step is addition of a proton to give the more stable carbocation. In the second step of the reaction, the negative end of the original H—X

dipole usually adds to complete the addition process. The regiochemistry of the addition is determined by the formation of the more stable carbocation in the original addition.

Other H—X molecules are not strong enough acids to protonate alkenes. Water (HOH) is an excellent example. However, such molecules will add if the reaction is acid-catalyzed. Enough acid catalyst is added to give the protonated alkene, which is attacked by water. The catalyst is regenerated in the last step and recycles to carry the reaction further.

The removal by a base of a hydrogen from the terminal position of acetylenes to give acetylides is mentioned. Terminal alkynes are moderately acidic molecules.

Syntheses

1. Acetylides

$$CH_3—C\equiv C—H + B:^- \longrightarrow CH_3—C\equiv C:^- + B—H$$
Base

2. Alkyl Halides

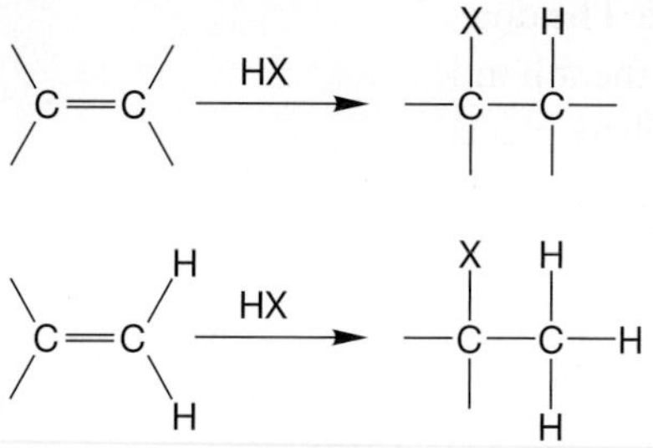

X = Cl, Br, I, sometimes F

Addition gives the compound
with the halide, X, on the
more substituted carbon

3. Alcohols

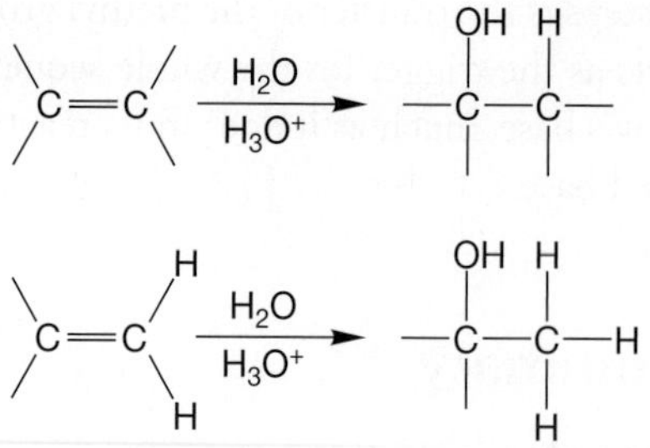

The addition reaction of water
is catalyzed by acid

Common Errors

The concept of cis/trans (Z/E) isomerism is a continuing problem for students. Errors of omission (the failure to find a possible isomer, for example) and of commission (failure to recognize that a certain double bond, so easily drawn on paper, really cannot exist) are all too common. Unlike σ bonds, π bonds have substantial barriers to rotation (~66 kcal/mol). The barrier arises because of the shape of the p orbitals making up the π bond. Rotation decreases overlap and raises energy. Accordingly, double bonds have sides, which are by no means easily interchanged. Substituents cannot switch sides, and are locked in position by the high barrier to rotation. It is vital to see why there are two isomers of 2-butene, *cis*- and *trans*-2-butene, but only one isomer of 2-methyl-2-butene (Fig. 3.83).

This relatively easy idea has more complicated implications. For example, the need to maintain planarity (p/p overlap) in a π bond leads to the difficulty of accommodating a trans double bond in a small ring, or at the bridgehead position of bridged bicyclic molecules. In each case, the geometry of the molecule results in a twisted (poorly overlapping) pair of p orbitals.

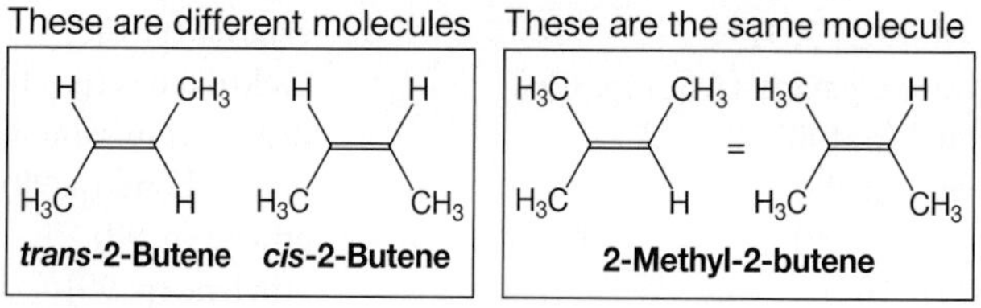

FIGURE 3.83

3.23 Additional Problems

In this chapter, we encounter the determination of molecular formulas. The following three review problems (problems 3.33–3.35) deal with this subject.

PROBLEM 3.33   Alkanes combine with oxygen to produce carbon dioxide and water according to the following scheme:

$$C_nH_{2n+2} + (3n + 1)/2\,O_2 \rightarrow (n + 1)H_2O + nCO_2$$

This process is generally referred to as combustion. An important use of this reaction is the quantitative determination of elemental composition (elemental analysis). Typically, a small sample of the compound is completely burned and the water and carbon dioxide produced are collected and weighed. From the weight of water the amount of hydrogen in the compound can be determined. Similarly, the amount of carbon dioxide formed allows us to determine the amount of carbon in the original compound. Oxygen, if present, is usually determined by difference. The determination of the relative molar proportions of

carbon and hydrogen in a compound is the first step in deriving its molecular formula.

If combustion of 5.00 mg of a hydrocarbon gives 16.90 mg of carbon dioxide and 3.46 mg of water, what are the weight percents of carbon and hydrogen in the sample?

PROBLEM 3.34   Calculate the weight percents for each element in the following compounds: (a) C_5H_{10} and (b) C_9H_6ClNO.

PROBLEM 3.35   A compound containing only carbon, hydrogen, and oxygen was found to contain 70.58% carbon and 5.92% hydrogen by weight. Calculate the empirical formula for this compound. If the compound has a molecular weight of approximately 135 g/mol, what is the molecular formula?

PROBLEM 3.36   How many degrees of unsaturation are there in compounds of the formula C_5H_8? Write at least eight isomers including ones with no π bonds and ones with no rings.

PROBLEM 3.37 Determine the degrees of unsaturation for the following compounds.

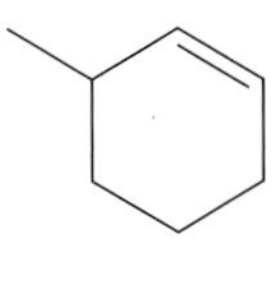

(a) 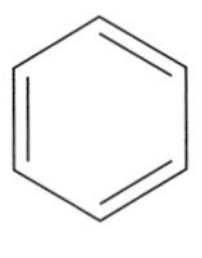   (b) 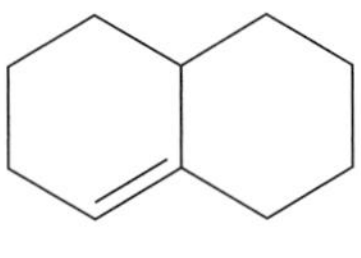   (c) 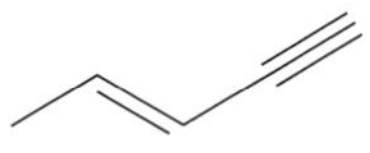   (d)

PROBLEM 3.38 Provide the IUPAC name for the following compounds:

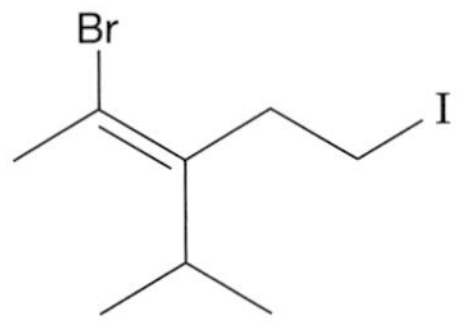

(a) 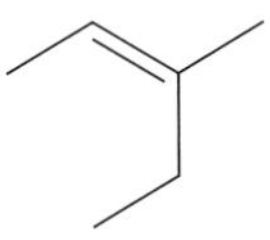   (b) 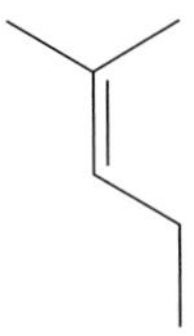   (c)

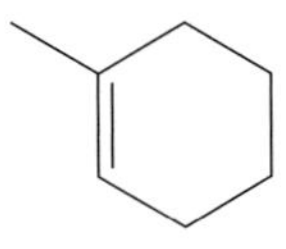

(d) 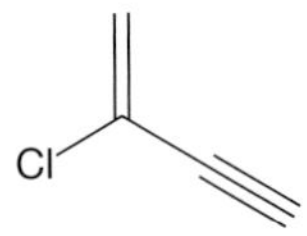   (e)

PROBLEM 3.39 Are the two structures below the same molecule? Draw the Newman projection for each looking down the C(2)—C(3) bond. Which form do you think is more stable? Why?

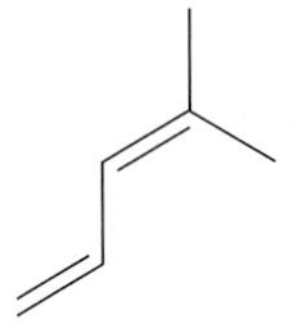

(a) 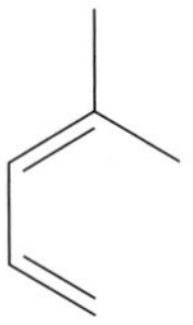   (b)

PROBLEM 3.40 Draw and write the systematic names for all of the acyclic isomers of 1-heptene (C_7H_{14}). There are 36 isomers, including 1-heptene.

PROBLEM 3.41 Draw and write the systematic names for all of the acyclic isomers of 1-octyne (C_8H_{14}) containing a triple bond. There are 32 isomers, including 1-octyne.

PROBLEM 3.42 Draw and write the systematic names for all of the isomers of dichlorobutene ($C_4H_6Cl_2$). There are 27 isomers, according to usually reliable sources.

PROBLEM 3.43 Write systematic names for the following compounds:

(a)

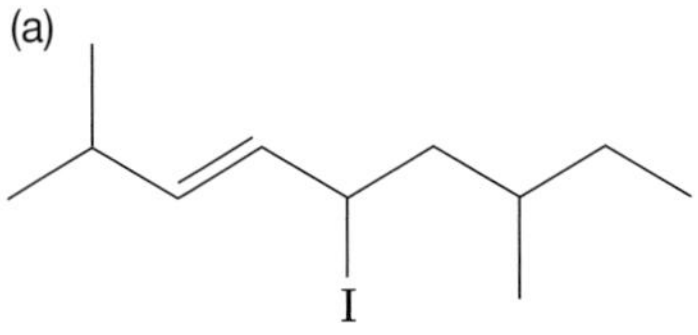

(b)

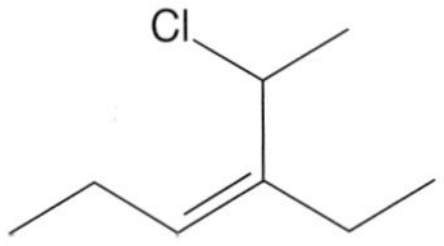

(c)

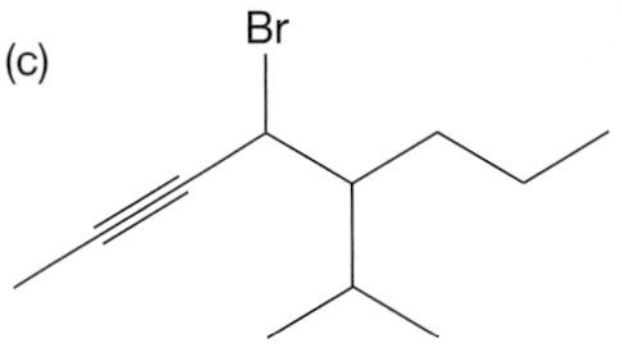

(d)

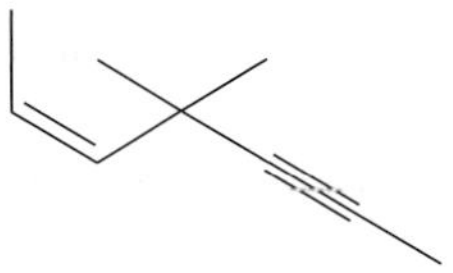

PROBLEM 3.44 Draw structures for the following compounds:
(a) (Z)-3-fluoro-2-methyl-3-hexene
(b) 2,6-diethyl-1,7-octadien-4-yne
(c) *trans*-3-bromo-7-isopropyl-5-decene
(d) 5-chloro-4-iodo-6-methyl-1-heptyne

PROBLEM 3.45 Although you should be able to draw structures for the following compounds, they are named incorrectly. Give the correct systematic name for each compound:
(a) 5-chloro-2-methyl-6-vinyldecane
(b) 3-isopropyl-1-pentyne
(c) 4-methyl-6-hepten-1-yne
(d) (E)-3-propyl-2,5-hexadiene

PROBLEM 3.46 Find all the compounds of the formula $C_4H_6Br_2$ in which the dipole moment is zero. It is easiest to divide this long problem into sections. First, work out the number of degrees of unsaturation so that you can see what kinds of molecules to look for. Then find the acyclic molecules. Next work out the compounds containing a ring.

PROBLEM 3.47 Find all the compounds of the formula C_4H_6 in which there are four different carbon atoms. There are only nine compounds possible, but two of them may seem quite strange at this point. *Hint*: One of these molecules contains a carbon that is part of two double bonds, and the other is a compound containing only three-membered rings.

PROBLEM 3.48 Find all the isomers of the formula C_4H_5Cl. As in Problem 3.46, it is easiest to divide this problem into sections. First, find the degrees of unsaturation so that you know what types of molecule to examine. There are 23 isomers to find. *Hint*: See the hint for Problem 3.47.

PROBLEM 3.49 Find the compound of the formula $C_{10}H_{16}$, composed of two six-membered rings, that has only three different carbons.

PROBLEM 3.50 Develop a bonding scheme for the compound H_2CO in which both carbon and oxygen are hybridized sp^2.

PROBLEM 3.51 (a) Develop a bonding scheme for a ring made up of six CH groups, $(CH)_6$. Each carbon is hybridized sp^2. (b) This part is harder. The molecule you built in (a) exists, but all the carbon–carbon bond lengths are equal. Is this consistent with the structure you wrote in (a), which almost certainly requires two different carbon–carbon bond lengths? How would you resolve the problem?

PROBLEM 3.52 In this chapter, we developed a picture of the methyl cation ($^+CH_3$), a planar species in which carbon is hybridized sp^2. In Chapter 2, we briefly saw compounds formed from several rings, bicyclic compounds. Explain why it is possible to form carbocation (a) in the bicyclic compound shown below, but very difficult to form carbocation (b). It may be helpful to use models to see the geometries of these molecules.

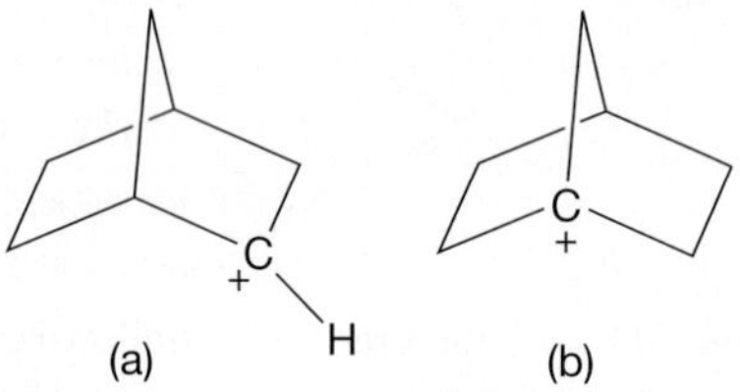

(a)   (b)

PROBLEM 3.53 Show the curved arrow formalism (electron pushing or arrow pushing) for the following reactions. Which side of the equilibrium is favored in each reaction? Use the pK_a table in the endpaper of the book to answer this part of the problem.

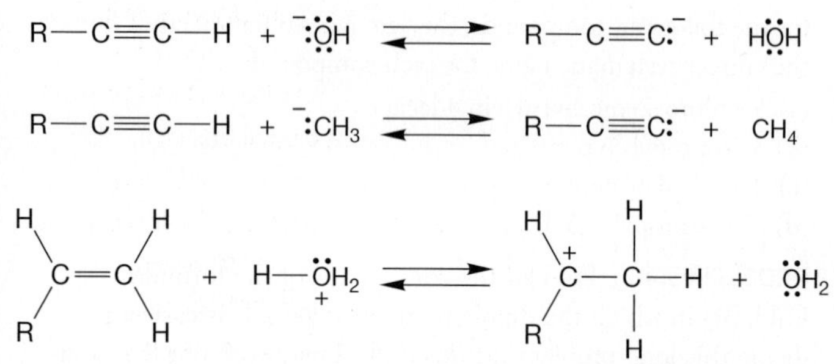

PROBLEM 3.54 Predict the products of the following reactions:

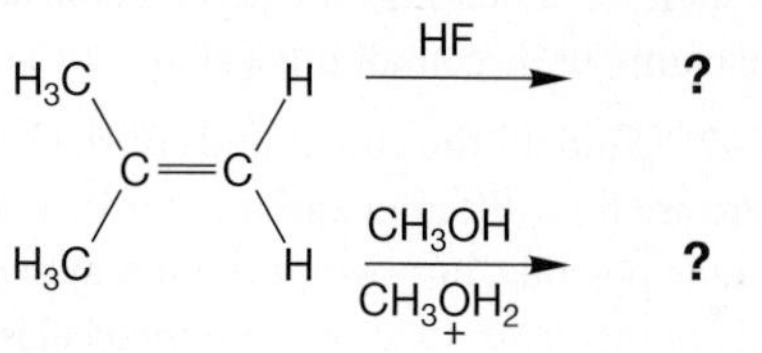

PROBLEM 3.55 Explain the formation of two products in the following reaction:

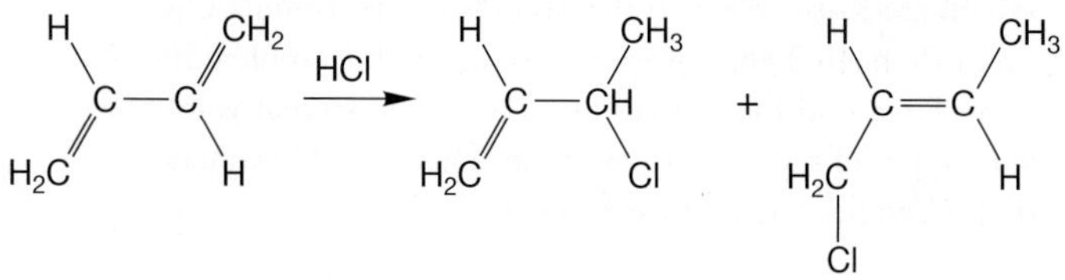

PROBLEM 3.56 Explain the formation of the two products in the reaction shown.

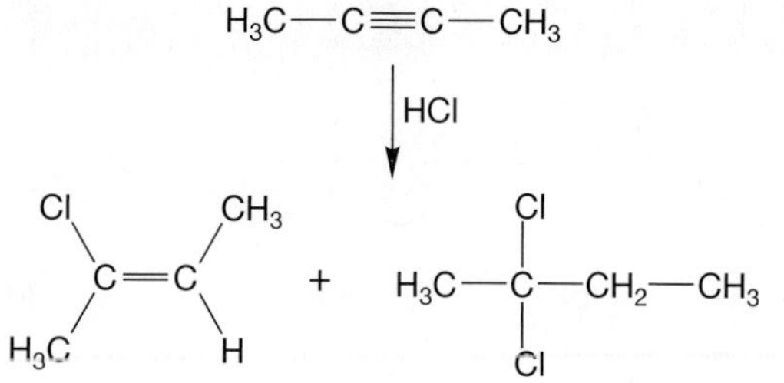

PROBLEM 3.57 In principle, there is a second "double addition" possible in Problem 3.56. Explain why only one compound is formed. *Hint:* Think "resonance."

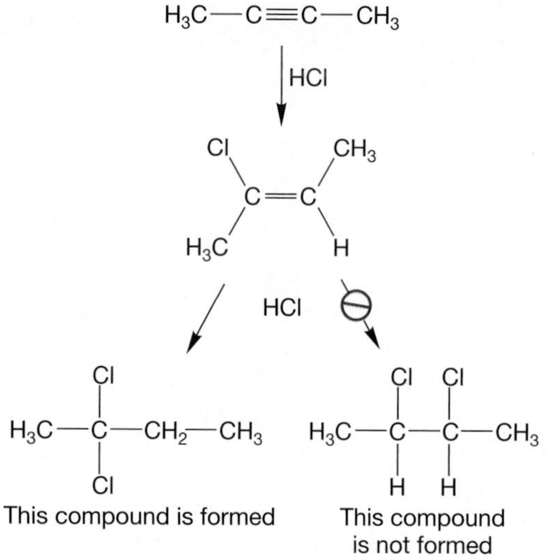

This compound is formed   This compound is not formed

Use Organic Reaction Animations (ORA) to answer the following questions:

PROBLEM 3.58 Choose the reaction titled "Alkene hydration" and click on the HOMO button. Observe the calculated highest occupied molecular orbital for 2-methylpropene. One hydrogen on each methyl group is not involved in the HOMO. Why is that the case?

PROBLEM 3.59 In the "Alkene hydration" reaction, click on Play and observe the last step of the reaction, which is removal of an acidic hydrogen. It is shown in the following figure:

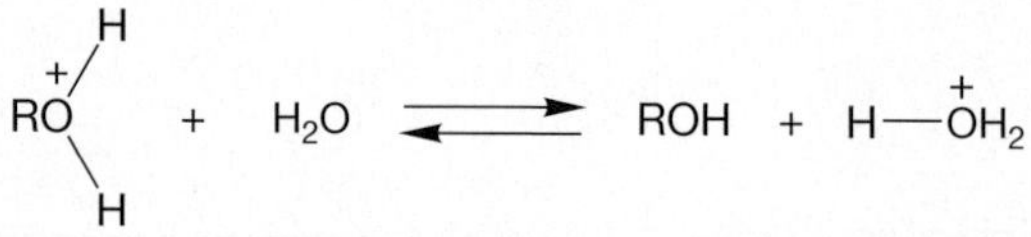

Show the electron pushing for the reaction going from left to right. Show the electron pushing for the reaction going from right to left. Which direction is favored? How can the neutral alcohol be obtained from this reaction?

PROBLEM 3.60 Choose the reaction "Acetylide addition" and click on the HOMO button and then play the reaction. Stop the reaction before the acetylide reacts with the bromoethane. Observe the calculated highest occupied molecular orbital of the propyne anion. Are the π orbitals involved in the anion? Why or why not?

Stereochemistry

4.1 Preview
4.2 Chirality
4.3 The (*R/S*) Convention
4.4 Properties of Enantiomers: Physical Differences
4.5 The Physical Basis of Optical Activity
4.6 Properties of Enantiomers: Chemical Differences
4.7 Interconversion of Enantiomers by Rotation about a Single Bond: *gauche*-Butane
4.8 Diastereomers and Molecules Containing More than One Stereogenic Atom
4.9 Physical Properties of Diastereomers: Resolution, a Method of Separating Enantiomers from Each Other
4.10 Determination of Absolute Configuration (*R* or *S*)
4.11 Stereochemical Analysis of Ring Compounds (a Beginning)
4.12 Summary of Isomerism
4.13 Special Topic: Chirality without "Four Different Groups Attached to One Carbon"
4.14 Special Topic: Stereochemistry in the Real World: Thalidomide, the Consequences of Being Wrong-Handed
4.15 Summary
4.16 Additional Problems

MIRROR IMAGES This photo shows a woman and her mirror image. In this chapter we will see that some molecules exist in mirror image forms.

> I held them in every light. I turned them in every attitude. I surveyed their characteristics. I dwelt upon their peculiarities. I pondered upon their conformation.
>
> —EDGAR ALLAN POE,[1] *BERENICE*

4.1 Preview

If you worked Problem 2.29 (p. 85), you saw that there are two isomers of *trans*-1,2-dimethylcyclopropane. We will see in this chapter why two *trans*-1,2-dimethylcyclopropanes exist and continue with a detailed discussion of **stereochemistry**, the structural and chemical consequences of the arrangement of atoms in space. We have already seen several examples of stereoisomeric molecules, compounds that differ only in the spatial arrangement of their constituent parts. The compounds *cis*- and *trans*-1,2-dimethylcyclopropane are stereoisomers (as are all cis/trans pairs). Substituents on both ring compounds and alkenes can be attached in two ways: cis and trans (Fig. 4.1). We will soon see that the two *trans*-1,2-dimethylcyclopropanes are stereoisomers of a different kind.

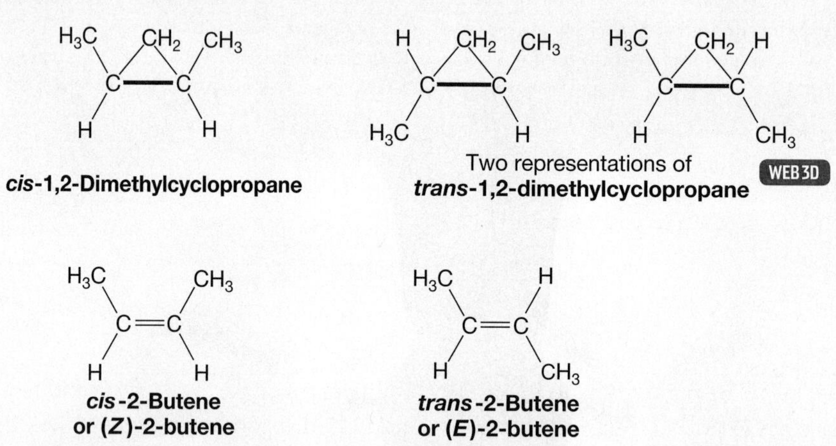

FIGURE 4.1 *cis*- and *trans*-1,2-Dimethylcyclopropane and *cis*- and *trans*-2-butene are pairs of stereoisomeric molecules. In the two cyclopropanes, the thicker bond indicates the side of the ring closest to you.

cis-1,2-Dimethylcyclopropane

Two representations of *trans*-1,2-dimethylcyclopropane   `WEB 3D`

cis-2-Butene or (*Z*)-2-butene

trans-2-Butene or (*E*)-2-butene

In this chapter, we will examine subtle questions of stereoisomerism. Most molecules of Nature are **chiral**. The word "chiral" is derived from the Greek and means "handed." Chiral molecules are related to their mirror images in the same way that your left hand is related to your right. Your hands are not identical—they cannot be superimposed. They are mirror images of one another. Many molecules of Nature (for example, the amino acids) are handed in the same way: One might say that our amino acids are all left-handed. The source of the ubiquity of left-handedness of the amino acids is one of the great remaining questions in chemistry. Was it chance? Did an accidental left-handed start determine all that followed? When we encounter extraterrestrial civilizations will their amino acids (assuming they have any) also be left-handed or will we find right-handed amino acids? As it now appears we will find no cohabitants within our own solar system, and as physical contact with our galactic neighbors is unlikely, you might consider how you would convey to an invisible resident of the planet Altair 4 our concepts of left and right.

Chirality, or handedness, is a complicated but important subject that is absolutely essential to organic and biological chemistry. When we return to our examination of reaction mechanisms—of the way molecules come together to

---

[1]Edgar Allan Poe (1809–1849) was an American poet and writer probably most famous for his macabre stories and poems. As the quotation shows, he was also an early devotee of stereochemistry.

make other compounds—we will often use chiral molecules to provide the level of detail necessary to work out how reactions occur. Moreover, because most of the molecules of Nature are chiral, an understanding of chirality is essential to virtually all of biochemistry and molecular biology.

You are very strongly urged to work through this chapter with models. It is a rare person who can see easily in three dimensions without a great deal of practice.

ESSENTIAL SKILLS AND DETAILS

1. Sidedness and Handedness. You have the broad outlines of structure under control now—acyclic alkanes, alkenes, and alkynes have appeared, as have rings. Now we come to the details, to stereochemistry. "Sidedness"—cis/trans isomerism—is augmented by questions of chirality—"handedness." Learning to see one level deeper into three-dimensionality is the next critical skill.

2. Difference. The topic of difference and how difference is determined arises in this chapter. The details may be nuts and bolts, and, indeed, any way that you work out to do the job will be just fine, but there is no avoiding the seriousness of the question. When are two atoms the same (in exactly the same environment) or different (not in the same environment)? This question gets to the heart of structure and is much tougher to answer than it seems. By all means, concentrate on this point. This chapter will help you out by introducing a method—an algorithm for determining whether or not two atoms are in different environments. It is well worth knowing how it works.

3. There is no way out—the (R) and (S) priority system must be learned.

4. Words—Jargon. This chapter is filled with jargon: Be certain that you learn the difference between enantiomers and diastereomers. Learn also what "stereogenic," "chiral," "racemic," and "meso" mean.

4.2 Chirality

The phenomenon of handedness, or chirality, actually surfaces long before we encounter a molecule as complicated as *trans*-1,2-dimethylcyclopropane. We went right past it when we wrote out the heptane isomers in Figure 2.45 (p. 82). One of these heptanes, 3-methylhexane, exists in two forms. To see this, draw out the molecule in tetrahedral form using C(3) as the center of the tetrahedron (Fig. 4.2a).

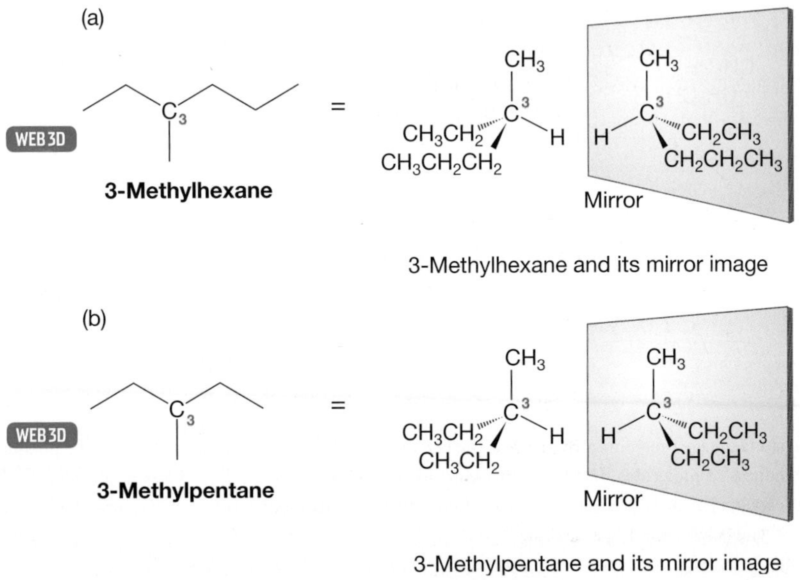

(a)

3-Methylhexane

3-Methylhexane and its mirror image

(b)

3-Methylpentane

3-Methylpentane and its mirror image

FIGURE 4.2 3-Methylhexane (a) and 3-methylpentane (b) reflected to show their mirror images.

Next, draw the mirror image of this molecule. Compare these structures with 3-methylpentane treated the same way, with C(3) still the center of the tetrahedron (Fig. 4.2b).

Now see if the reflections—the mirror images—are the same as the originals. For 3-methylpentane, one need only rotate the molecule 180° around the C(3)—CH$_3$ bond, and slide it over to the left to superimpose it on the original molecule (Fig. 4.3a). The molecule and its mirror image are clearly identical, and hence, 3-methylpentane is **achiral** (not chiral).[2] However, Nature seems to lay a trap for us, as there are many motions of the mirror image that will not directly generate a molecule superimposable on the original. In such cases, we may be led to think we have uncovered chirality. For example, try rotating 3-methylpentane as shown in Figure 4.3b. The original and the newly rotated molecule are not superimposable as drawn. But chiral? No, for as long as there is one motion (Fig. 4.3a, top) that does generate the mirror image, the molecule is achiral.

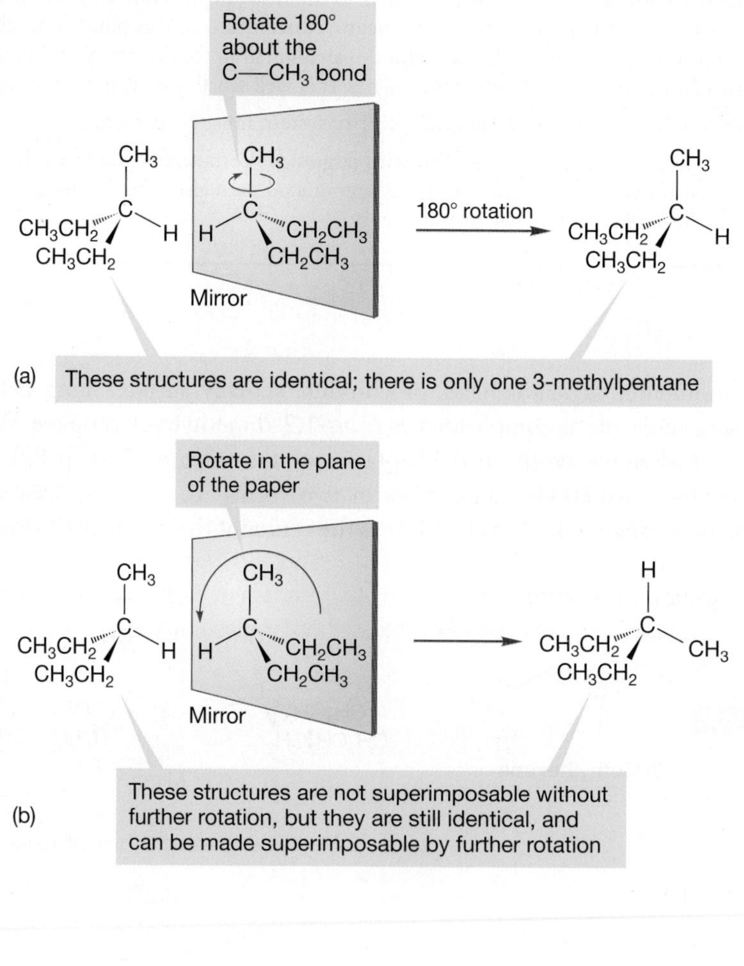

FIGURE 4.3 (a) 3-Methylpentane and its mirror image are identical. Rotate the mirror image at the top of the figure 180° about the carbon–methyl bond to see this point. (b) Although motion that produces a nonsuperimposable mirror image tells us nothing about the chirality of the molecule, a motion that produces a superimposable mirror image tells us that the molecule is not chiral (achiral).

Rotate 180° about the C—CH$_3$ bond

180° rotation

Mirror

(a)   These structures are identical; there is only one 3-methylpentane

Rotate in the plane of the paper

Mirror

(b)   These structures are not superimposable without further rotation, but they are still identical, and can be made superimposable by further rotation

---

[2] There is a language problem introduced by the choice of the Greek word *chiral* to describe the phenomenon of handedness of molecules. In Greek, the word for *not chiral* is *achiral*. This poses no problem in Greek, but, at least in spoken English it does. It is often difficult to distinguish between "Compound X is a chiral molecule" and "Compound X is achiral."

Now consider Figure 4.4. Here we apply to 3-methylhexane the same rotation that we applied before to 3-methylpentane. In this case, we cannot superimpose the original molecule onto the newly rotated molecule no matter how many ways we try. The propyl group in one form winds up where the ethyl group is in the other. Indeed, no amount of twisting and turning can do the job: The two are irrevocably different. 3-Methylhexane is chiral.

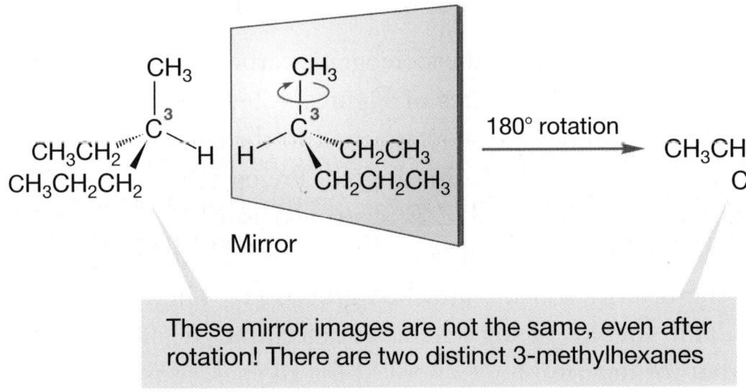

These mirror images are not the same, even after rotation! There are two distinct 3-methylhexanes

FIGURE 4.4 3-Methylhexane and its mirror image are not identical.

Models are essential now, as you must be absolutely certain of the difference between 3-methylpentane and 3-methylhexane. The more symmetric 3-methylpentane is achiral, whereas 3-methylhexane and your hand are chiral. The two stereoisomers of 3-methylhexane are related in exactly the same way your two hands are. A chiral compound can be defined as a molecule for which the mirror image is not superimposable on the original. The two nonsuperimposable mirror images of 3-methylhexane are examples of **enantiomers**.

Our next tasks are to learn how to differentiate enantiomers in words (we do have to be able to talk to each other!) and to see how these stereoisomers differ physically. Which physical properties do enantiomers share and which are different? This topic leads us to a brief discussion of how the physical differences arise. Next, we explore how enantiomers differ chemically. Which chemical properties are shared by enantiomers and which are different? Next, we need to explore the circumstances under which chirality will appear. What structural features will suffice to ensure chirality? Will, for example, the phenomenon we see in Figure 4.5 of four different groups surrounding a carbon be a sufficient condition to ensure chirality? Will it be a necessary condition? This chapter discusses such questions.

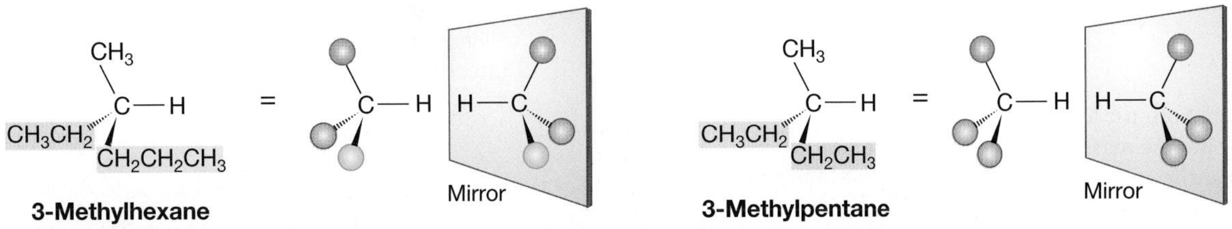

FIGURE 4.5 3-Methylhexane, 3-methylpentane, and their mirror images.

PROBLEM 4.1 Are the following molecules chiral? (a) bromochlorofluoroiodomethane, (b) dibromochlorofluoromethane, (c) 3-methylheptane, (d) 4-methylheptane

4.3 The (R/S) Convention

In order to specify the **absolute configuration** of a molecule, the three-dimensional arrangement of its atoms, one first identifies the stereogenic center, very often a carbon atom.[3] A **stereogenic atom** can be simply defined as follows: "An atom (usually carbon) of such nature and bearing groups of such nature that it can have two nonequivalent configurations," in other words, having a stereogenic atom means enantiomers can exist.

In our example of 3-methylhexane, the stereogenic carbon, C(3), is the one upon which we based the tetrahedral structures of Figures 4.2–4.5. Next, one identifies the four groups (atoms or groups of atoms) attached to the stereogenic carbon and gives them priorities according to the Cahn–Ingold–Prelog scheme we first met in Chapter 3 (p. 111) when we discussed Z/E (cis/trans) isomerism. The system is applied here so that any stereogenic atom can be designated as either (R) or (S) from a consideration of the priorities of the attached groups (1 = high, 4 = low). The application of the Cahn–Ingold–Prelog scheme to determine (R) or (S) is a bit more complicated than that for determining (Z) or (E) for alkenes, so the system will be summarized again here.

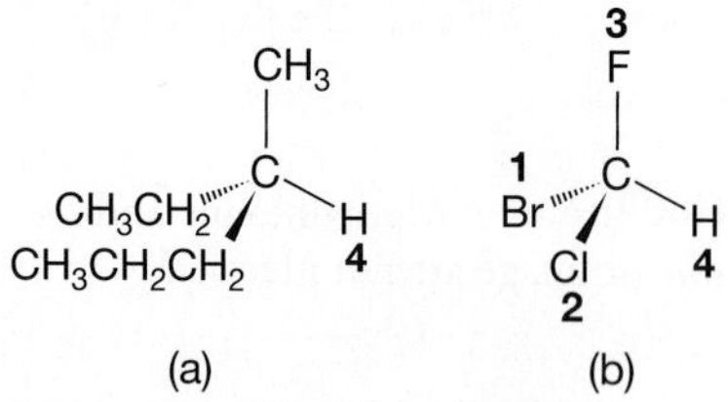

The atom of lowest atomic number is given the lowest priority number, 4. In 3-methylhexane, this atom is hydrogen (Fig. 4.6a). In some molecules, one can give priority numbers to all the atoms by simply ordering them by atomic number. Thus, bromochlorofluoromethane is easy: H is 4, F is 3, Cl is 2, and Br is 1 (Fig. 4.6b). The priorities are assigned in order of increasing atomic number so that the highest priority, 1, is given to the atom of highest atomic number, in this case bromine (Fig. 4.6b).

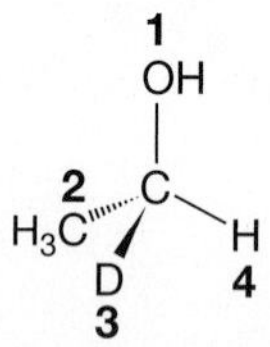

A subrule is illustrated by another easy example, 1-deuterioethyl alcohol (Fig. 4.7). We break the tie between the isotopes H and D by assigning the lower priority number to the atom of lower mass, H. So H is 4; D is 3; C is 2; and O, with the highest atomic number, is priority 1.

There can be more difficult ties to break, however. The molecule *sec*-butyl alcohol (2-hydroxybutane) illustrates this point. The lowest priority, 4, is H and the highest priority, 1, is O. But how do we choose between the two carbons shown in red in Figure 4.8? The tie is broken by working outward from the tied atoms until a difference is found. In this example, it's easy. The methyl carbon is attached to three hydrogens and the methylene carbon is attached to two hydrogens and a carbon.

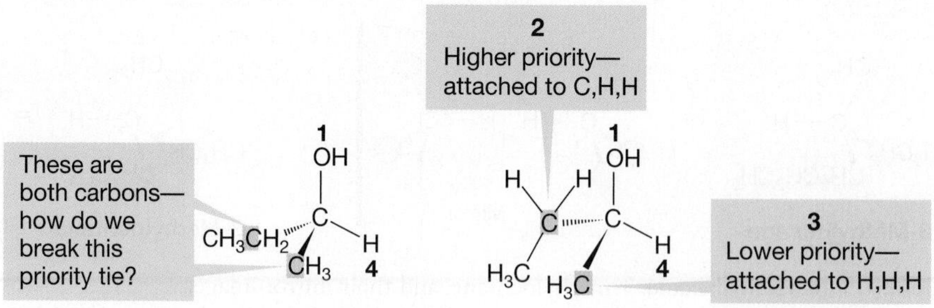

FIGURE 4.6 The atom of lowest atomic number, often hydrogen, is given the lowest priority, 4. The remaining priorities are assigned in order of increasing atomic number.

FIGURE 4.7 When the atomic numbers are equal, the atomic masses are used to break the tie.

FIGURE 4.8 When both the atomic numbers and the atomic masses are equal, one looks at the atoms to which the tied atoms are attached and determines their atomic numbers.

---

[3]The word "stereogenic," though officially suggested in 1953, came into general use only recently when it was urged by Princeton professor Kurt M. Mislow (b. 1923) as a replacement for a plethora of unsatisfactory terms such as "chiral carbon," "asymmetric carbon (or atom)," and many others. These old terms are slowly disappearing, but will persist in the literature (and textbooks) for many more years.

The methyl carbon gets the lower priority number 3 by virtue of the lower atomic numbers of the atoms attached to it (H,H,H = 1,1,1) and the ethyl carbon (attached to C,H,H = 6,1,1) gets the higher priority number, 2 (Fig. 4.8).

Sometimes there will be even more difficult ties to break, and another look at 3-methylhexane illustrates this problem well (Fig. 4.9). It is easy to identify H as the atom with the lowest atomic number, but the other three atoms are all carbons. How do we break this tie? We work outward from the identical atoms until a difference appears. In 3-methylhexane, the methyl group is attached to three hydrogens, but each methylene group is attached to two hydrogens and one carbon. Because H has a lower atomic number than C, the methyl group gets priority number 3 (Fig. 4.9).

Now we are faced with yet another tie, and must find a way to distinguish the pair of methylene groups. Again we work outward. Look at the carbons attached to the two tied methylenes (Fig. 4.10). The red one is a methyl carbon and is attached to three hydrogens. The green carbon is a methylene carbon and is attached to two hydrogens and a carbon. The methylene attached to the methyl gets priority 2 and the other methylene is left with the highest priority, 1. In practice, this detailed bookkeeping is often unnecessary. In this example, it is really quite easy to see that the smaller ethyl group is of lower priority than the larger propyl group, and thus its carbon will get the lower priority number. More difficult examples can be devised, however!

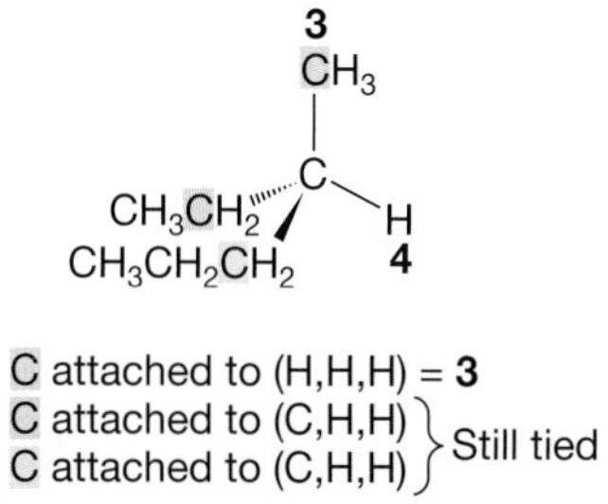

FIGURE 4.9 An example of a complicated tie-breaking procedure. In this case, we start by looking at the atoms attached to the three tied carbons. This procedure does not completely resolve the tie.

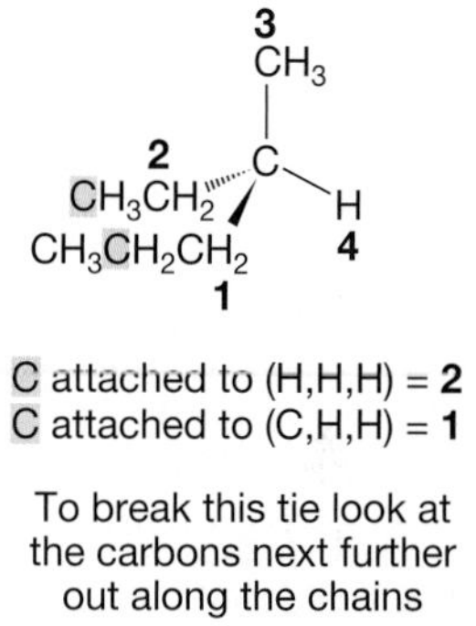

FIGURE 4.10 The final assignment of priority numbers in 3-methylhexane.

PROBLEM 4.2 Identify the priorities of the groups attached to the stereogenic carbons in the compounds in the following figure:

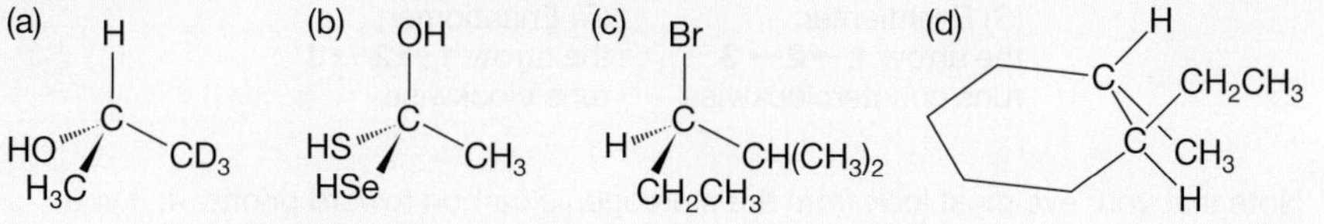

Note that in (d) there are two stereogenic carbon atoms

PROBLEM 4.3 Draw the mirror images of the chiral compounds (a–d) in Problem 4.2.

How do we deal with multiple bonds in the Cahn–Ingold–Prelog scheme? Carbon–carbon double bonds, carbon–carbon triple bonds, and carbon–oxygen double bonds are treated as shown in Figure 4.11. Hydrogen obviously gets the lowest priority, 4, and the bromine gets the highest priority, 1. But both the other

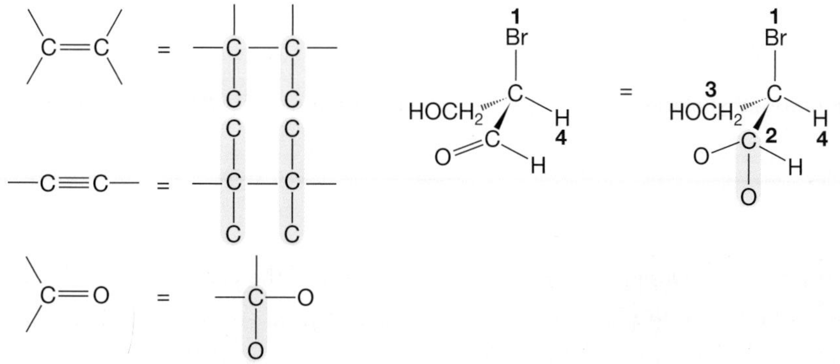

FIGURE 4.11 Carbon–carbon double bonds and carbon–carbon triple bonds are elaborated by adding new carbon bonds as shown. The carbon–oxygen double bond is treated by adding a new bond from carbon to oxygen. Thus, the carbon is treated as being attached to one hydrogen and two oxygens and gets a higher priority number than the carbon that is attached to two hydrogens and one oxygen.

substituents on the stereogenic carbon are carbons. The tie is resolved by treating the carbon atom in the carbon–oxygen double bond as if it were attached to two oxygens as shown in Figure 4.11. Therefore it gets the higher priority, 2, and the carbon attached to only one oxygen gets the remaining priority, 3.

Once priority numbers have been established, the next step in determining absolute configuration (R or S) is to imagine looking down the bond from the stereogenic atom (usually carbon) toward the atom of lowest priority (4, often H), as in Figure 4.12. The other three substituents (1, 2, 3) will be facing you. Connect these three with an arrow running from highest to lowest priority number (1 → 2 → 3). If this arrow runs clockwise from your perspective, the enantiomer is called (R) (Latin: rectus, "right"); if it runs counterclockwise, the enantiomer is called (S) (Latin: sinister, "left") (Fig. 4.12).

FIGURE 4.12 The (R) and (S) naming convention. It often helps to put the priority 4 in the back.

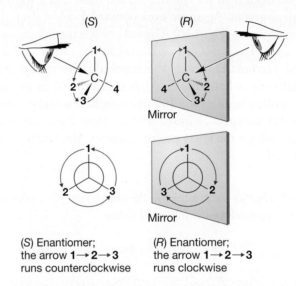

(S) Enantiomer; the arrow 1→2→3 runs counterclockwise

(R) Enantiomer; the arrow 1→2→3 runs clockwise

Note that your eye must look from the stereogenic carbon toward priority 4; if you look in the other direction, or if you draw the arrow 3→2→1, you will get the convention backward! For clarity, we have completed the circle by drawing the arrow 3→1

The label (R) or (S) is added to the name as an italic prefix in parentheses. Figure 4.13 shows the (R) and (S) forms of some examples from the previous paragraphs.

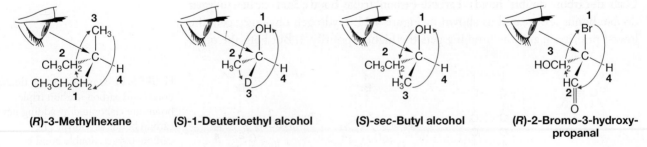

(R)-3-Methylhexane   (S)-1-Deuterioethyl alcohol   (S)-sec-Butyl alcohol   (R)-2-Bromo-3-hydroxy-propanal

FIGURE 4.13 Some (R) and (S) molecules.

PROBLEM 4.4 Designate the stereogenic carbons of the compounds in Problem 4.2 as either (R) or (S).

PROBLEM SOLVING

The (*R/S*) convention is difficult only because it is utterly arbitrary. A bunch of folks in Europe thought up some conventions and we all use them. You can't "think" your way through an (*R/S*) assignment. You have to know the rules. For example, you have to look from the stereogenic atom (almost always a carbon) toward the priority 4 atom (often H). But what if the drawing makes this view difficult, as in the molecule below? It is easy to assign priorities and draw the arrow, but the molecule has the H (priority 4) pointing right at us, and it is not easy to put your eye in the proper position. The "fix" is to look the other (wrong) way, from priority 4 to the stereogenic C, and to reverse your answer. In this case, the arrow, seen from the anti-conventional "wrong" direction is counterclockwise, so the molecule is *R*. This technique is a bit tricky, because you have to remember that you are violating the convention, but it is often easier, and safer, than redrawing the molecule in a different orientation.

Arrow is counterclockwise, but we are looking "backward!"— molecule is (*R*)

PROBLEM 4.5 Draw the enantiomers of the molecules in Figure 4.13.

The (*R/S*) convention looks a bit complicated. It is easier than it appears right now, but it just must be learned, and cannot be reasoned out. It's quite arbitrary at a number of points. For the price of learning this convention we get the return of being able to specify great detail most economically. Time and again we will use the subtle differences between enantiomers to test ideas about reaction mechanisms—of how chemical reactions occur. The study of the effects of stereochemistry on reaction mechanisms has been, and remains, one of the most powerful tools used by chemists to unravel how Nature works. Practice using the (*R/S*) convention and it will become easy. It's well worth it—indeed, it's essential.

4.4 Properties of Enantiomers: Physical Differences

Almost all physical properties of enantiomers are identical. Their melting points, boiling points, densities, and many other physical properties do not serve to distinguish two enantiomers. But they do differ physically in one somewhat obscure way: They rotate the plane of plane-polarized light to an equal degree but in opposite directions. What is **plane-polarized light**? Light consists of electric (and magnetic) fields that oscillate in all planes. Certain filters are able to isolate light in which the electric field oscillates in only one plane—hence, plane-polarized light. Rotation of the plane of polarization is called **optical activity**, and molecules causing such a rotation are said to be optically active. The enantiomer that rotates the plane clockwise (as one faces a beam of light passing through the sample) is called **dextrorotatory**. The enantiomer that rotates the plane counterclockwise is called **levorotatory**. The direction of rotation is indicated by placing a (+) for dextrorotatory, and a (−) for levorotatory, before the name of the enantiomer (Fig. 4.14). *Note*: There is no connection between (*R*) and (*S*) and the (+) and (−) used to denote the direction of rotation of the plane of plane-polarized light.

FIGURE 4.14 The (+)-enantiomer rotates the plane of plane-polarized light clockwise. The (−)-enantiomer rotates the light an equal amount counterclockwise.

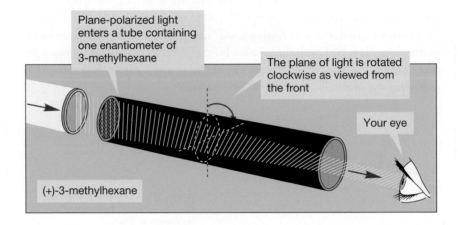

Plane-polarized light enters a tube containing one enantiometer of 3-methylhexane

The plane of light is rotated clockwise as viewed from the front

Your eye

(+)-3-methylhexane

Very often, equal amounts of the dextrorotatory and levorotatory enantiomers are present as a mixture. In such a case, no rotation can be observed as there are equal numbers of molecules rotating the plane in clockwise and counterclockwise directions, and the effects cancel. Such a mixture is called either a **racemate**, or a **racemic mixture**. The presence of a racemic mixture of enantiomers is often indicated by placing (±) before the name as in Figure 4.15.

FIGURE 4.15 There is no net rotation by a racemic mixture of enantiomers. One isomer rotates the plane of plane-polarized light to the right and the other rotates it an equal amount to the left.

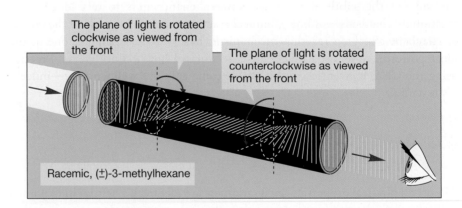

The plane of light is rotated clockwise as viewed from the front

The plane of light is rotated counterclockwise as viewed from the front

Racemic, (±)-3-methylhexane

When we are talking about a chiral molecule, how do we know whether the subject of our discussion is one enantiomer or the racemic mixture? Often the text will tell you by indicating (+), (−), or (±). If no indication is made, we are almost certainly talking about the racemic mixture. Drawings present other problems. If the subject is one enantiomer, the drawing will certainly point this out. Sometimes an asterisk (*) is used to indicate optical activity.

There are times when a three-dimensional representation is important, but the subject is a racemic mixture of the two enantiomers rather than one enantiomer. Unless both enantiomers are drawn, the figure appears to focus only on a single enantiomer. Our molecule, 3-methylhexane, illustrates this point in Figure 4.16. If we are talking about a racemic mixture of 3-methylhexane but want to show the three-dimensional structure of this molecule, we should, strictly speaking, draw both of the enantiomers. In practice this is rarely done, and you must be alert for the problem. Unless optical activity is specifically indicated, it is usually the racemate that is meant.

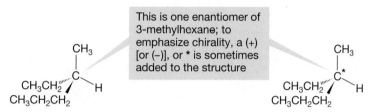

This is one enantiomer of 3-methylhexane; to emphasize chirality, a (+) [or (−)], or * is sometimes added to the structure

When it is the racemic mixture that is meant, sometimes (±) is added. However, this is not always done, and you must be alert for such times

Indicates the racemic mixture of the two enantiomers

FIGURE 4.16 A variety of conventions is used to indicate the presence of a single enantiomer.

4.5 The Physical Basis of Optical Activity

We have covered the way in which enantiomers differ from each other physically: They rotate the plane of plane-polarized light in opposite directions. Before we go on to talk about how enantiomers differ in their chemical reactivity, let's take a little time to say something about the phenomenon of optical activity. How does a collection of atoms (a molecule) interact with a wave phenomenon (light), and how does the property of chirality (handedness) induce the observed rotation?

Light consists of electromagnetic waves vibrating in all possible directions perpendicular to the direction of propagation (Fig. 4.17).

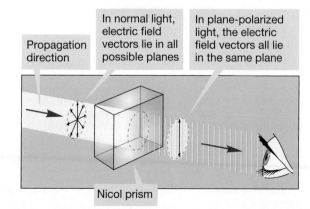

Propagation direction

In normal light, electric field vectors lie in all possible planes

In plane-polarized light, the electric field vectors all lie in the same plane

Nicol prism

FIGURE 4.17 A Nicol prism transmits only plane-polarized light.

Passing ordinary light through a Nicol prism (or a piece of Polaroid film) eliminates all waves except those whose electric field vectors are vibrating in a single plane. The classic experiment to demonstrate this uses two pairs of Polaroid sunglasses. If we observe light through one pair, we can still see, although the light has been plane polarized.

The second pair of glasses transmits light only when oriented in the same way as the first pair. If the second pair is turned 90° to the first pair, none of the plane-polarized light can pass, and darkness results (Fig. 4.18).

FIGURE 4.18 A classic experiment using Polaroid sunglasses: The first pair of sunglasses acts as a Nicol prism and transmits only plane-polarized light. If the second pair is held in the same orientation, light will still be transmitted. If it is held at 90° to the first pair, no light can be transmitted.

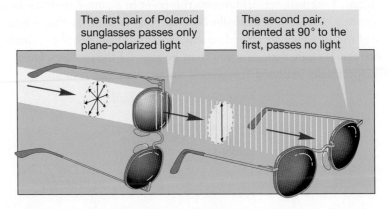

Now we are ready to see why some molecules are optically active and some are not. The electrons in a molecule, moving charged particles, generate their own electric fields, which interact with the electric field of the light. As we have already seen, many molecules are not highly symmetric, and the interactions of their unsymmetric electric fields with that of the plane-polarized light change the electric field and rotate the plane of polarization slightly. Achiral molecules, such as 3-methylpentane, do not give a net rotation of plane-polarized light because equivalent molecules exist in mirror-image positions, and thus the net rotation must be 0° (Fig. 4.19).

FIGURE 4.19 The net rotation of an achiral molecule must be 0°.

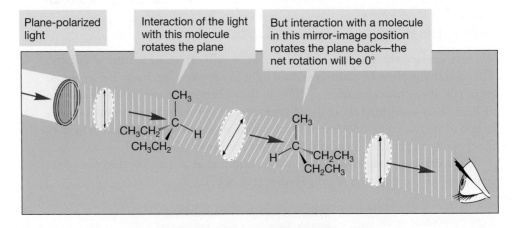

The result is no net rotation of the plane. However, when the interaction is with a chiral molecule, there are no compensating mirror-image molecules (Fig. 4.20).

FIGURE 4.20 Optically active molecules, such as (*R*)-3-methylhexane, will rotate the plane of plane-polarized light. There are no compensating mirror-image molecules for this kind of molecule.

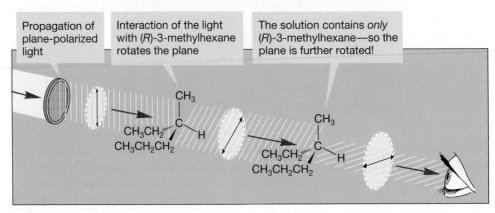

There must be a net change in the plane of polarization, and this rotation is generally large enough to be observed by a device called a **polarimeter** (Fig. 4.21). Because the amount of rotation varies with the wavelength of light, monochromatic light, light of only one wavelength, must be used to produce results easily reproduced in other laboratories. Generally, the sodium D line is used, which is light that

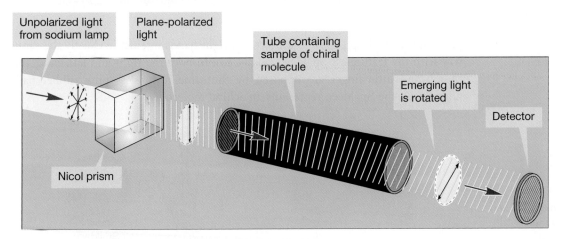

FIGURE 4.21 A highly schematic drawing of a polarimeter.

has a wavelength of 589 nm (5890 Å). The light is passed through a Nicol prism to polarize it, and then through a tube containing a solution of the molecule in known concentration. The amount of rotation is proportional to the number of molecules in the path of the light, and so the length of the tube and the concentration of the solution are important. For this reason, rotation of a sample is reported as the **specific rotation**, $[\alpha]$, which is the rotation induced by a solution of concentration 1 g/mL in a tube 10 cm long. The specific rotation $[\alpha]$ is related to the observed rotation α in the following way:

$$[\alpha] = \alpha/cl, \text{ where } c = \text{concentration in grams per milliliter (g/mL)},$$
$$\text{and } l = \text{length of the tube in decimeters (dm)}$$

Subscripts and superscripts are used to denote the wavelength and temperature of the measurement. Thus, $[\alpha]_D^{20}$ means the sodium D line was used and the measurement was made at 20 °C.

> **PROBLEM 4.6** Create a scheme like those in Figures 4.19 and 4.20 to show what happens when plane-polarized light is passed through a solution of racemic 3-methylhexane.

4.6 Properties of Enantiomers: Chemical Differences

Now let's look at how the two enantiomers of a chiral molecule differ from each other chemically. We still can't discuss specific chemical reactions, but we can make some important general distinctions. Note that there is no difference in how the two enantiomers of 3-methylhexane interact with a "Spaldeen," an old-fashioned smooth

FIGURE 4.22 (R)- and (S)-3-methylhexane interact identically with the achiral Spaldeen.

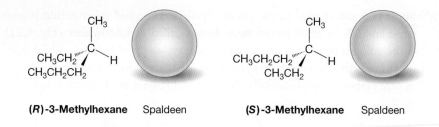

(R)-3-Methylhexane   Spaldeen   (S)-3-Methylhexane   Spaldeen

pink rubber ball (Fig. 4.22).[4] Your left and right hands interact in exactly the same way with the featureless Spaldeen (Fig. 4.23). In both the right and left hand, the thumb and little finger are opposed by the same featureless pink rubber ball. Chemically, the Spaldeen represents an achiral molecule.

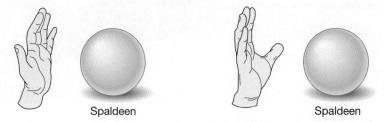

FIGURE 4.23 Mirror image left and right hands also interact identically with the achiral Spaldeen.

Spaldeen   Spaldeen

WORKED PROBLEM 4.7 Actually, the Spaldeen is achiral only if the name has been worn off by too much stoopball playing. Show that the two enantiomers interact differently with a Spaldeen with the name "Spalding" still visible.

ANSWER In the approach shown, (R)-3-methylhexane interacts with the new Spaldeen to place the ethyl group near the "S" and the propyl group near the "G." These interactions are opposite in (S)-3-methylhexane; the propyl group approaches the "S" and the ethyl group approaches the "G."

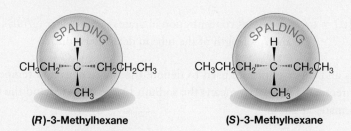

(R)-3-Methylhexane   (S)-3-Methylhexane

You should be certain that this result is not caused by the particular approach chosen for the two objects. Try interacting 3-methylhexane by bringing it toward the Spaldeen "H and methyl first" or simply turn it over (H down, methyl up) in the figure above.

---

[4]The inclusion of the term "Spaldeen" was the source of one of the few minor skirmishes between author and editor of this book. MJ maintained that a Spaldeen was part of our cultural heritage, and that all educated people would immediately know what one was. Perhaps he was overly optimistic. Here's what the editors of the *New Yorker* magazine replied when queried as to the meaning of the phrase "a kid bouncing a Spaldeen off a wall of the Boston Museum of Fine Arts": "Spaldeens are the small rubber balls beloved by generations of New York stoopball and stickball players. In Boston, those balls are often known as 'pinkies,' but what would you have thought if we had written 'bouncing a pinkie off a wall'?"

Figure 4.24 elaborates this subject—the interactions of chiral and achiral objects—by showing the two enantiomers of 3-methylhexane approaching an achiral molecule, propane.

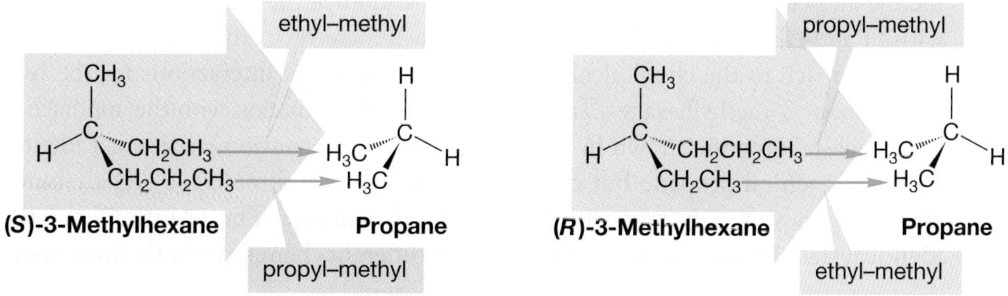

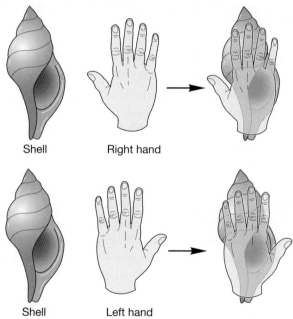

FIGURE 4.24   The two enantiomers of the chiral molecule 3-methylhexane interact in exactly the same way with the achiral molecule propane.

In each instance, the 3-methyl group of 3-methylhexane approaches the hydrogen of propane and both the ethyl and propyl groups are opposite methyl groups of propane. Because of the symmetry of propane, one can always find exactly equivalent interactions for the two enantiomers and propane, no matter what approach you pick. Try it with models.

Now imagine your hands, or the two enantiomers of 3-methylhexane, interacting with a chiral object. The Spaldeen with the label will do, but let's use a more common object, such as the shell in Figure 4.25.

FIGURE 4.25   All sorts of common objects are chiral. A shell is one example. The interactions between a chiral object and each of a pair of enantiomers—here represented by a shell and two hands—are different. The enantiomeric right and left hands interact differently with the chiral shell.

PROBLEM 4.8   Convince yourself that the shell in Figure 4.25 is chiral.

As your right hand approaches the shell (Fig. 4.25), it is the little finger that opposes the opening in the shell, and the thumb that is directed away from it. For the left hand the situation is reversed: It is the thumb that opposes the opening and the little finger that does not. The interactions of the two hands with the shell are distinctly different.

PROBLEM 4.9   Convince yourself that turning the shell upside down doesn't change the situation.

For the molecular counterpart of what Figure 4.25 shows, look at the approach of the two enantiomers of 3-methylhexane to another chiral molecule, (*R*)-*sec*-butyl alcohol (Fig. 4.26). As the (*S*) enantiomer approaches the alcohol, propyl is opposite methyl and ethyl is opposite H. For the (*R*) enantiomer, the interactions are propyl with H and ethyl with methyl. The two approaches are very different. Indeed, there is no approach to the chiral alcohol that can give identical interactions for the two enantiomeric 3-methylhexanes. Try some. Contrast this situation with the approaches to the achiral propane shown in Figure 4.24. The two enantiomers interact identically with achiral propane but differently with the chiral molecule (*R*)-*sec*-butyl alcohol. This situation is general for any chiral molecule: The enantiomers have identical chemistries with achiral reagents but different chemistries with chiral ones.

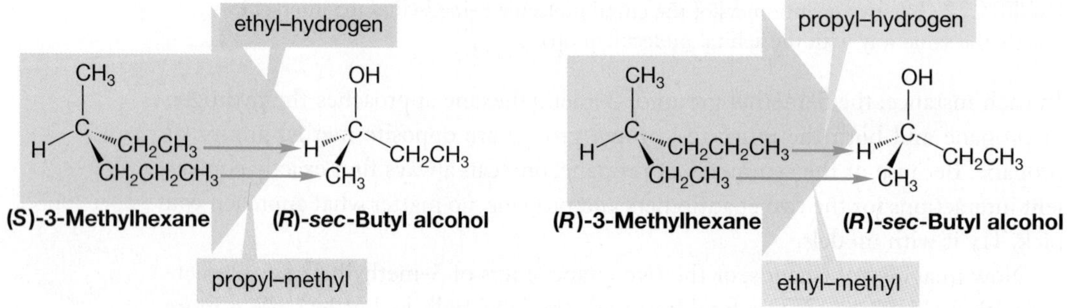

FIGURE 4.26 The approach of (*S*)- and (*R*)-3-methylhexane to (*R*)-*sec*-butyl alcohol. The two approaches are very different and in sharp contrast to the interaction shown in Figure 4.24 with the achiral molecule propane.

WORKED PROBLEM 4.10 A reader of an early version of this chapter suggested that students would ask, quite correctly, Why can't I turn the picture of (*R*)-3-methylhexane upside down, and then the ethyl–hydrogen and propyl–methyl interactions with (*R*)-*sec*-butyl alcohol would be just like the interactions with (*S*)-3-methylhexane, propyl–methyl and ethyl–hydrogen? Explain carefully why turning the molecule upside down doesn't produce a situation identical to that in Figure 4.24.

ANSWER It is true that turning the molecule upside down will make two of the interactions the same. Propyl will be opposite methyl and ethyl will be opposite hydrogen. However, everything else is wrong! No longer are methyl and OH being opposed, for example (red bonds). Simply turning (*R*)-3-methylhexane over does not produce the same interactions—or even interactions similar to those between (*S*)-3-methylhexane and (*R*)-*sec*-butyl alcohol.

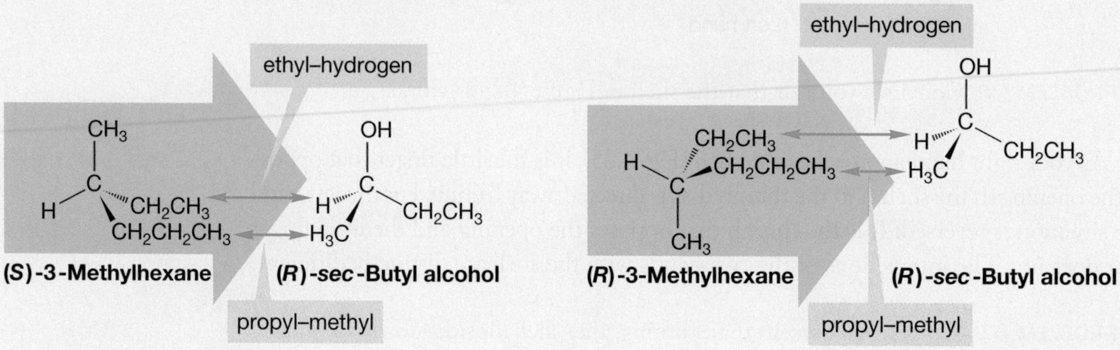

A practical example of the consequences of chiral interactions is the difference in smells of enantiomeric compounds. One smells a molecule by binding it in a decidedly chiral receptor in the nose. There are many examples in which one enantiomer binds either differently from the other or not at all. For example, (*R*)-carvone smells like spearmint and (*S*)-carvone like caraway (Fig. 4.27).

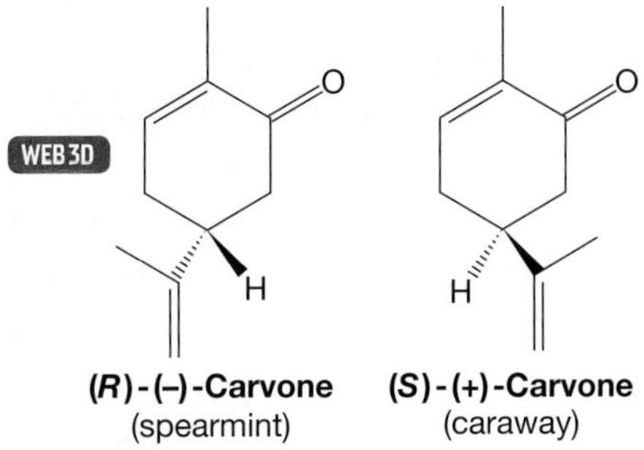

FIGURE 4.27 (*R*)- and (*S*)-Carvone.

WEB 3D

(*R*)-(–)-Carvone   **(*S*)-(+)-Carvone**
(spearmint)   (caraway)

PROBLEM 4.11 Convince yourself that the assignment of (*R*) and (*S*) in Figure 4.27 is correct.

It is not at all easy to show (*R*)- and (*S*)-carvone fitting into different (and not yet exactly known) nasal receptors, but it is easy to show in principle how fit and non-fit works. Imagine a chiral binding site as the blue tetrahedron in Figure 4.28, with the smaller red and green tetrahedra representing two enantiomeric molecules. The red enantiomer fits perfectly inside, with the "proper" attachments, shown as 1–1, 2–2, 3–3, and 4–4. If we try the green tetrahedron, the enantiomer of the red one, there is no possible way for the proper attachments to be achieved. The green tetrahedron cannot bind to the blue one with the proper 1–1, 2–2, 3–3, and 4–4 interactions.

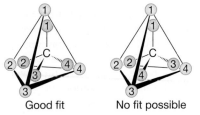

Good fit   No fit possible

FIGURE 4.28 The red tetrahedron fits inside the blue one with the proper interactions (1–1, 2–2, 3–3, and 4–4), but the green one does not.

4.7   Interconversion of Enantiomers by Rotation about a Single Bond: *gauche*-Butane

In Chapter 2 (p. 74), we analyzed the conformations of butane, finding an anti form lower in energy than either of the two enantiomeric gauche forms. Figure 4.29 shows the anti and gauche forms of butane as both three-dimensional drawings and Newman projections.

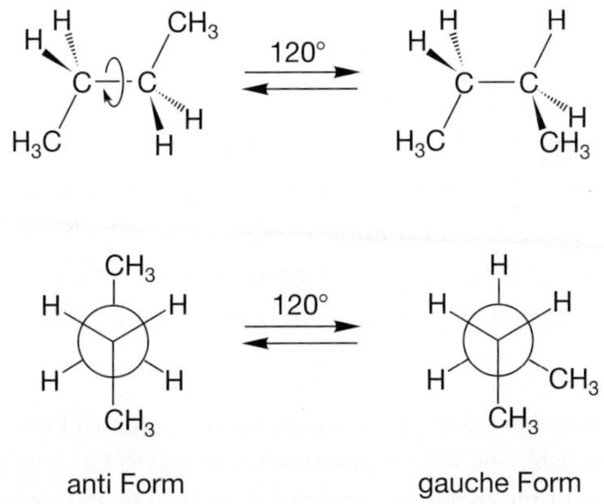

anti Form   gauche Form

FIGURE 4.29 *anti*- and *gauche*-Butane.

The Newman projections clearly reveal the high-energy, destabilizing gauche methyl–methyl interaction. Moreover, they show clearly that the gauche form of butane is chiral. It cannot be superimposed on its mirror image (Fig. 4.30).

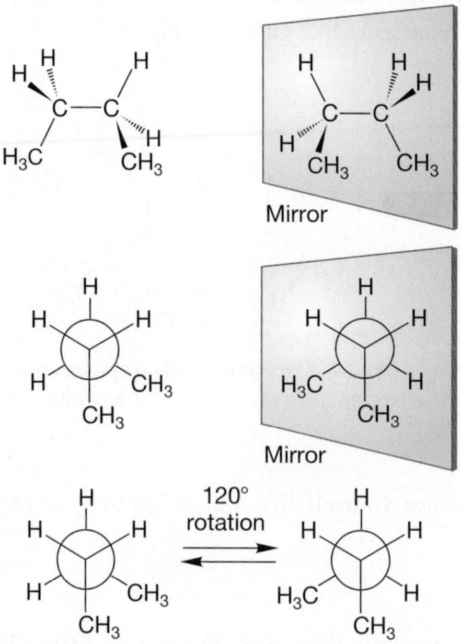

Mirror

Mirror

120° rotation

FIGURE 4.30 The gauche form of butane is chiral! The mirror images are *not* superimposable. However, rapid rotation around carbon–carbon single bonds interconverts these two "conformational enantiomers."

Why then is butane achiral? No matter how hard we try, no optical rotation can be observed for butane. The nearly free rotation about the carbon–carbon bond in butane rapidly interconverts the two enantiomers of the gauche form. If we were able to work at the extraordinarily low temperatures required to "freeze out" the equilibrium between the two gauche forms we could separate butane into enantiomers (Fig. 4.30). The two gauche forms are **conformational enantiomers**, enantiomers that are interconverted through rotation about carbon-carbon single bonds. We will encounter this phenomenon again when we study the stereochemistry of cyclic compounds.

4.8 Diastereomers and Molecules Containing More than One Stereogenic Atom

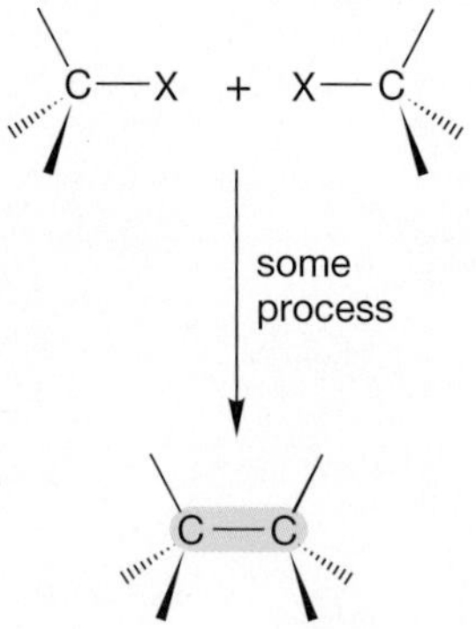

We have just looked at the different interactions between a single enantiomer of a chiral molecule and an enantiomeric pair of molecules (Fig. 4.26). Let's take this interaction business to its limit and see what happens if we actually form bonds to produce new molecules. We can't worry yet about real chemical reactions—we are still in the tool-building stages—but that doesn't matter. We can imagine some process that replaces the two X substituents in Figure 4.31 with a chemical bond, shown in all succeeding figures with a blue screen.[5]

If we follow this hypothetical process (which mimics some very real chemical reactions), we generate a new molecule containing two potentially stereogenic carbons. First, let's look again at what happens when a single chiral molecule reacts with each isomer of a racemic (50:50) pair of enantiomers. It's worth taking a moment to see if we can figure out what's going to happen in general before writing real mol-

some process

FIGURE 4.31 A process that removes the two X substituents and joins the carbons. The new connection (bond) is shown in blue.

---

[5]In this and many succeeding figures, the molecules are drawn in eclipsed forms for clarity. As you know, these arrangements are not stable and will be converted into staggered forms by rotation about the central carbon–carbon bond. None of the arguments presented is changed by this rotation, and in the staggered arrangements it is more difficult to visualize the relationships involved.

ecules. Let's assume we are using the (*R*) form of our single enantiomer. It will react with the (*R*) form of the enantiomeric pair to generate a molecule we might call (*R–R*), and with the (*S*) form to generate (*R–S*) (Fig. 4.32).

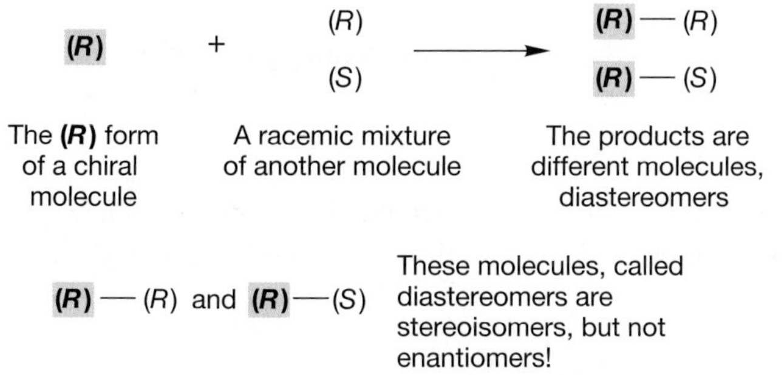

(R)   +   (R)   ⟶   **(R)** — (R)
                        (S)   **(R)** — (S)

The **(R)** form   A racemic mixture   The products are
of a chiral   of another molecule   different molecules,
molecule   diastereomers

(R) — (R) and **(R)** — (S)   These molecules, called
diastereomers are
stereoisomers, but not
enantiomers!

FIGURE 4.32 A schematic analysis of the reaction of one enantiomer with a racemic mixture of another molecule. The products of this reaction are stereoisomers but not enantiomers. They are called diastereomers.

PROBLEM 4.12 What is the mirror image of (*R–S*)? This question sounds insultingly simple, but most people get it wrong!

We can tell a lot even from the schematic coded picture of Figure 4.32. First of all, (*R–R*) and (*R–S*) are clearly neither identical nor enantiomers (nonsuperimposable mirror images), because the mirror image of (*R–R*) is (*S–S*), not (*R–S*). Yet these molecules are stereoisomers—they differ only in the arrangement of their constituent parts in space. There is a new word for such molecules—they are **diastereomers**, "stereoisomers that are not mirror images of each other—in other words, stereoisomers but not enantiomers."

To sum up, there are two kinds of stereoisomers: enantiomers (nonsuperimposable mirror images) and diastereomers.

WORKED PROBLEM 4.13 If the two gauche forms of butane shown in Figure 4.30 are enantiomers, what is the relationship between one gauche form and the anti form?

ANSWER The two gauche forms are mirror images and, therefore, enantiomers. Each gauche form is a stereoisomer of the anti form, but not its mirror image. Stereoisomers that are not mirror images are diastereomers.

PROBLEM 4.14 We have seen diastereomers twice before. Show two different kinds of diastereomeric molecules. *Hint*: Start with a good definition of diastereomer.

Now let's make all this theory a little more real, and look at some molecules. Imagine a process (Figure 4.31, X = Cl) in which the (*S*) enantiomer of 2-chlorobutane (*sec*-butyl chloride) reacts with the two enantiomers of 2-chloro-1,1,1-trifluoropropane (Fig. 4.33).

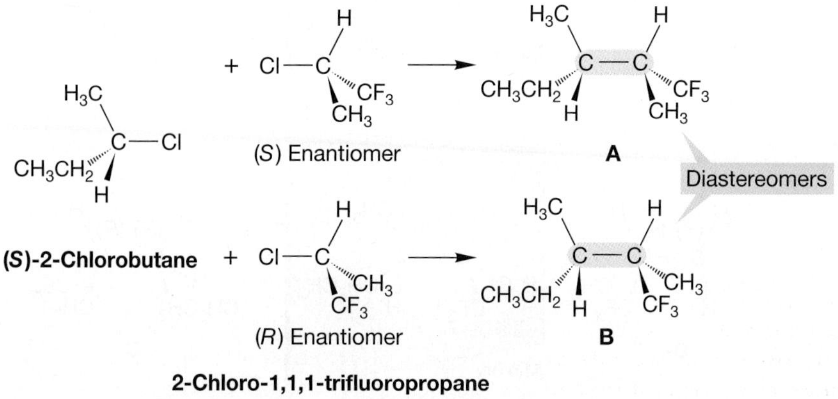

(S)-2-Chlorobutane   +   2-Chloro-1,1,1-trifluoropropane

FIGURE 4.33 The hypothetical reaction of (*S*)-2-chlorobutane with the two enantiomers of 2-chloro-1,1,1-trifluorochloropropane. Two stereoisomeric products, **A** and **B**, are formed. They are not mirror images (enantiomers) so they must be diastereomers.

Look carefully at the two new products, **A** and **B**, produced from our reaction. Are they identical? Certainly not. Are they enantiomers? No, again. They are stereoisomers, though, and so they must be diastereomers.

> **PROBLEM 4.15** Verify the claims of the preceding paragraph. Are molecules **A** and **B** diastereomers?

The (*R*) form of 2-chlorobutane would also produce two new products, **C** and **D** (Fig. 4.34) that are neither identical nor enantiomeric and thus must also be diastereomers.

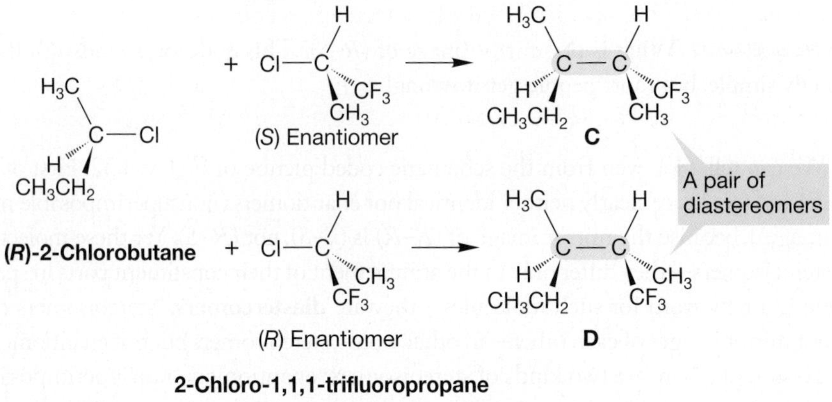

FIGURE 4.34 The hypothetical reaction of (*R*)-2-chlorobutane with the two enantiomers of 2-chloro-1,1,1-trifluoropropanes gives another pair of diastereomers, **C** and **D**.

Now the question is, What are the relationships among molecules **A**, **B**, **C**, and **D**? As Figure 4.35 shows, molecules **A** and **D** are a pair of enantiomers and **B** and **C** are another pair of enantiomers. So all possible combinations of 2-chlorobutane with the two enantiomers of 2-chloro-1,1,1-trifluoropropanes have produced two pairs of enantiomers, or four stereoisomers in all. The (*R*) and (*S*) relationships are also shown in Figure 4.35.

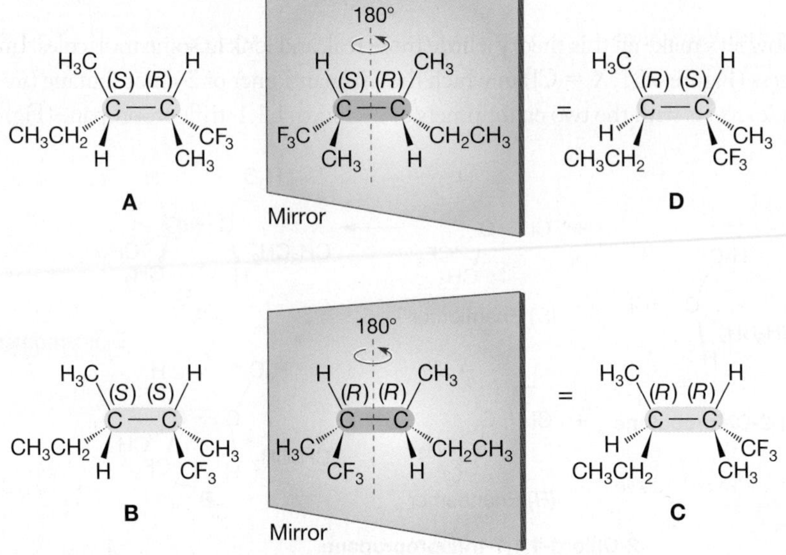

FIGURE 4.35 Two pairs of enantiomers (**A** and **D**, **B** and **C**) are formed from this hypothetical reaction. Combinations **A** and **B**, **A** and **C**, **B** and **D**, and **C** and **D** are diastereomeric pairs. Each of these pairs is composed of two stereoisomers that are not mirror images.

As a simple theoretical analysis would predict (Fig. 4.36), we have formed all possible isomers in this combination [(S–R), **A**; (S–S), **B**; (R–S), **D**; and (R–R), **C**].

> **PROBLEM 4.16** (S)-2-Chlorobutane and (S)-2-chloro-1,1,1-trifluoropropane (Fig. 4.33) are used to produce isomer **A**, which is properly designated as the (S,R) isomer (Fig. 4.35). Check that there is no mistake in this analysis, and then explain how the two (S) compounds can produce the (S,R) isomer, **A**.

We can generalize even at this early stage. A molecule with one stereogenic center has two possible stereoisomers, a single pair of (R) and (S) enantiomers. A molecule containing two stereogenic atoms has two pairs of enantiomers, or four stereoisomers in all. It seems that for n stereogenic carbons, we get 2^n stereoisomers. We can immediately predict that a molecule with $n = 3$ stereogenic atoms has 2^3 or 8 stereoisomers. This surmise is easy to check—what stereoisomers are possible? The easiest way to tell is not to examine some complicated molecule and thus incur the task of drawing a large number of stereoisomers, but to analyze the problem in a simpler fashion. A stereogenic carbon can be only (R) or (S). Figure 4.37 shows the possibilities for any molecule containing three stereogenic atoms.

The eight possible isomers appear easily in this coded form. Note that there is nothing in this analysis that requires the stereogenic atoms to be adjacent—they need only be in the same molecule.

This procedure reliably gives you the *maximum* number of possible stereoisomers. There are some interesting (and important) special cases, however, and this very general method won't spot them. Let's imagine using our chemical coupling reaction again, but this time let's combine two molecules of 2-chlorobutane. This procedure is exactly the same as the one we used before to generate the four stereoisomers of Figures 4.33 and 4.34, **A**, **B**, **C**, and **D**. We allow our (R) and (S) starting materials to combine in all possible ways (Fig. 4.38).

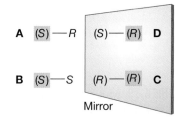

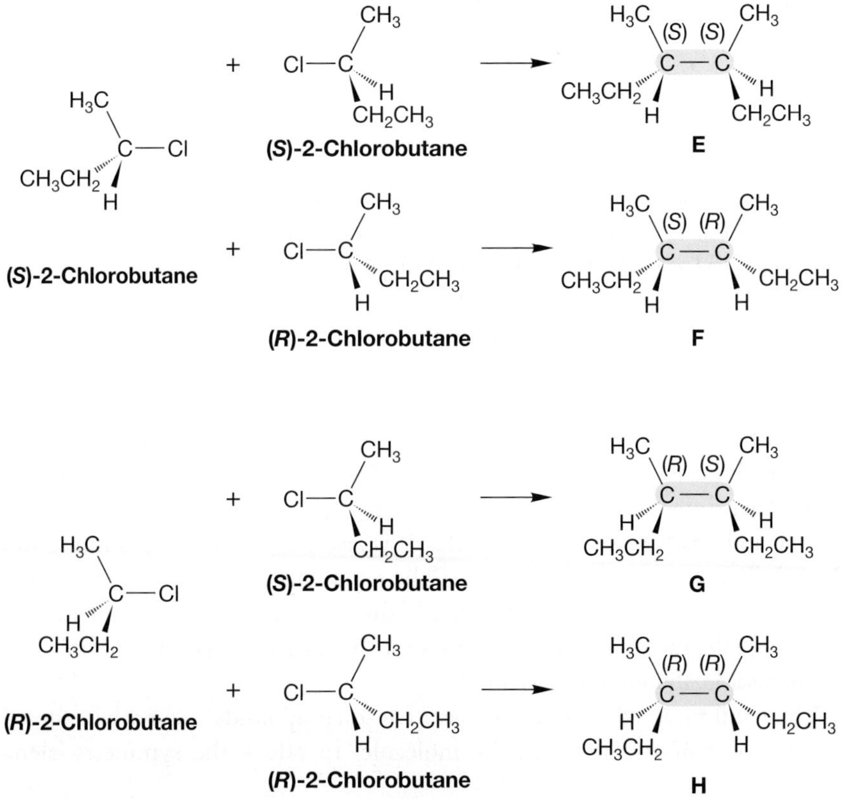

FIGURE 4.36 (R)- and (S)-2-chlorobutane reacting with (R)- and (S)-2-chloro-1,1,1-trifluoropropane should give four stereoisomers: two pairs of enantiomers.

(RRR)   All (R)

(RRS) (RSR) (SRR)   Two (R), one (S)

(SSR) (SRS) (RSS)   One (R), two (S)

(SSS)   All (S)

FIGURE 4.37 The eight possible stereoisomers for a molecule containing three stereogenic atoms.

FIGURE 4.38 All possible combinations of (R)- and (S)-2-chlorobutane (sec-butyl chloride) produce four isomers, **E**, **F**, **G**, and **H**.

In the previous example in Figure 4.35, we found that we had made two pairs of enantiomers, (*S–R*) and (*R–S*), **A** and **D**, and (*S–S*) and (*R–R*), **B** and **C**. When we combine two molecules of 2-chlorobutane, one pair of enantiomers, **E** and **H**, does appear (Fig. 4.39).

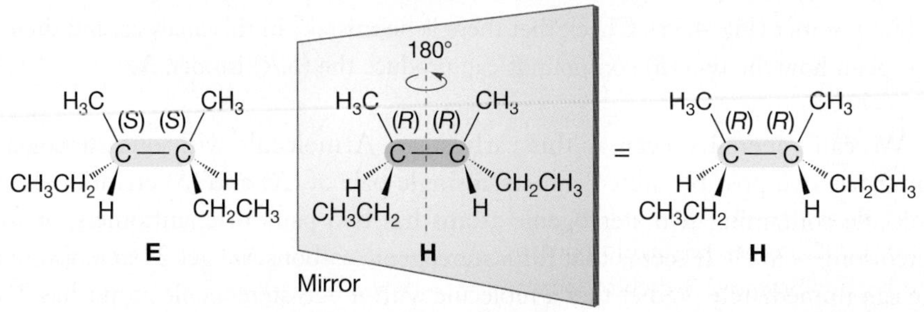

FIGURE 4.39 Molecules **E** and **H** in Figure 4.38 are enantiomers.

However, the second potentially enantiomeric pair of molecules, **F** and **G**, are in fact identical species (Fig. 4.40)! So in this example of a molecule containing two stereogenic atoms, we get only three stereoisomers, not the maximum number of $2^2 = 4$. Clearly, we are one stereoisomer short because of the identity of **F** and **G**. This molecule, **F** = **G**, contains stereogenic carbons but is not chiral. Such a molecule is called a **meso compound**.

Twice before we have seen molecules that for different reasons give no observable rotation of plane-polarized light. Neither achiral molecules nor racemic mixtures of enantiomers can induce optical rotation. A racemic mixture of chiral

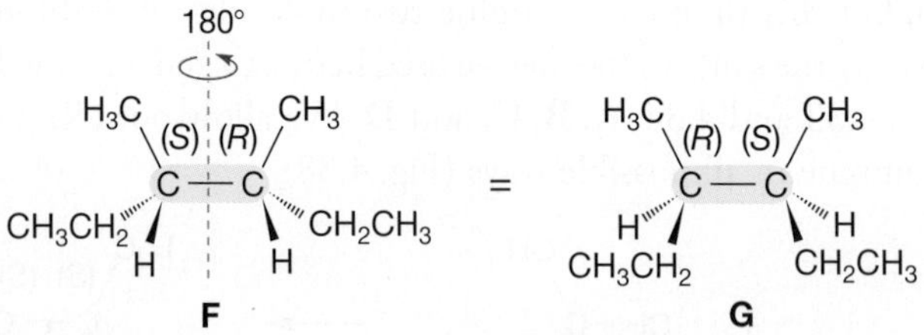

FIGURE 4.40 The **F** and **G** molecules are identical. You can see this identity if you rotate **F** 180° around the axis shown. This compound **F** = **G** is meso.

molecules contains equal numbers of right- and left-handed compounds, which rotate the light equally in clockwise and counterclockwise directions, respectively, for a net zero rotation. Our third example, a meso compound, does contain stereogenic atoms, but it is achiral and can induce no optical rotation.

On page 151 we asked if finding four different groups attached to a stereogenic carbon was a sufficient condition for chirality. We just answered this question! No, "four different groups attached to a stereogenic carbon" is not a sufficient condition for optical activity. Our meso compound **F** = **G** certainly has carbons to which four different groups are bonded, but it is equally clearly not chiral. Even though searching for carbons attached to four different groups is definitely a good way to begin a hunt for chirality, you must be careful. Finding such a carbon does not guarantee that you have found a chiral molecule, unless you are certain that the molecule contains one and only one such carbon. Alas, it is also the case that not finding such a carbon does not guarantee that there will be no chirality. The ultimate test remains the superimposability of mirror images. If the molecule and its mirror image are superimposable, the molecule is not chiral; if the mirror images are different, nonsuperimposable, the molecule is chiral.

When will meso compounds appear? Meso compounds occur when there is a plane or point of symmetry in the molecule. In effect, the symmetry element divides the molecule into halves that contribute equally and oppositely to the rotation of plane-polarized light. Compound **F** = **G** provides an example of a molecule with a plane of symmetry in the eclipsed forms we have been using, and a point of symmetry if we draw the molecule in its energy minimum, staggered arrangement (Fig. 4.41).

PROBLEM 4.17 Draw all the stereoisomers of the following molecules, in which there is free rotation about carbon–carbon single bonds. Some of the molecules may be achiral:

(a) 2,2,3,3-tetrabromobutane
(b) 2,2-dibromo-3,3-dichlorobutane
(c) 2,3-dibromo-2,3-dichlorobutane
(d) 2,3-dibromo-2-chloro-3-fluorobutane

Hint: It is easier to analyze these molecules in their eclipsed forms.

PROBLEM 4.18 How does your answer to Problem 4.17d change if there is no free rotation about carbon–carbon single bonds for 2,3-dibromo-2-chloro-3-fluorobutane?

PROBLEM 4.19 Identify each stereogenic carbon in the compound of Problem 4.17c as either (*R*) or (*S*).

Plane of symmetry
in eclipsed form

Point of symmetry
in staggered form

FIGURE 4.41 Two views of the meso compound of Figure 4.40, **F** = **G**.

4.9   Physical Properties of Diastereomers: Resolution, a Method of Separating Enantiomers from Each Other

The formation of diastereomers allows the separation of enantiomers from each other. Separation of enantiomers, called **resolution**, is a serious experimental difficulty. So far we have ignored it. Enantiomers have identical physical properties except for the ability to rotate the plane of plane-polarized light, and one might legitimately wonder how in the world we are ever going to get them apart. At several points, we used a single enantiomer without giving any hint of how a pair of enantiomers might be separated. The key to this puzzle is that diastereomers, unlike enantiomers, have different physical properties—melting point, boiling point, and so on.

Dextro   Levo

Crystals of (+) tartaric acid and (−) tartaric acid are shown here. These crystals are non-superimposable mirror images. Although tedious, a careful visual analysis would allow separation of the two crystal forms. This would be a "visual resolution."

One general procedure for separating enantiomers is to allow them to react with a naturally occurring chiral molecule to form a pair of diastereomers. These can then be separated from each other by taking advantage of one of their different physical properties. For example, we can often separate such a pair by crystallization because the members of the diastereomeric pair have different solubilities in a given solvent. Then, if the original chemical reaction can be reversed, we have the pair of enantiomers separated. Figure 4.42 outlines the general scheme and begins with a schematic recapitulation of Figure 4.33, which first described the reaction of a single enantiomer with a racemic mixture to give a pair of diastereomers. Be sure to compare the two figures.

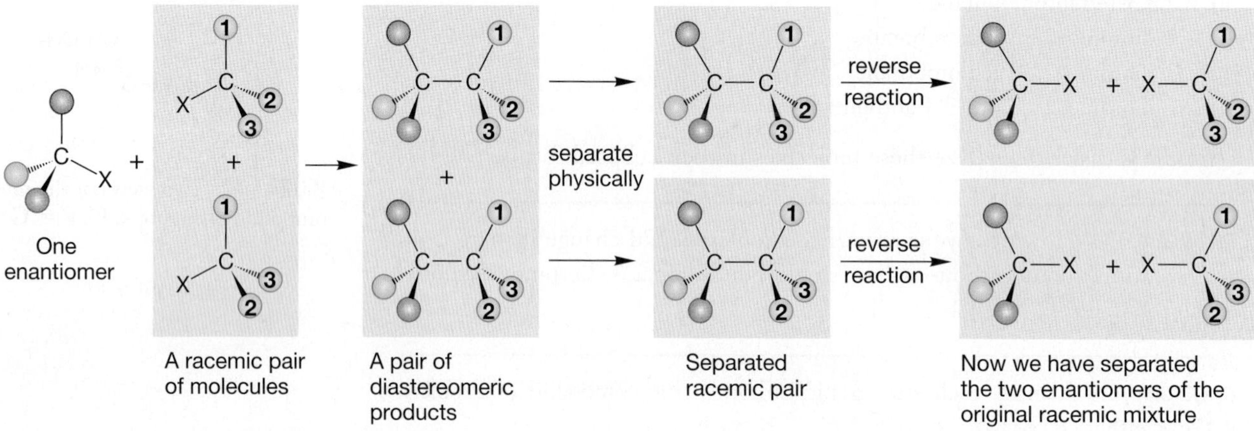

| One enantiomer | A racemic pair of molecules | A pair of diastereomeric products | Separated racemic pair | Now we have separated the two enantiomers of the original racemic mixture |

FIGURE 4.42 Resolution is a general method for separating the constituent enantiomers of a racemic mixture. A single enantiomer of a chiral molecule is used to form a pair of diastereomers, which can be separated physically. If the original chemical reaction can be reversed, the enantiomers can be isolated.

It is not even necessary to form covalent bonds in order to separate enantiomers. For example, in the traditional method for separating enantiomers of organic acids, optically active nitrogen-containing molecules, called alkaloids (see Section 6.11), are used to form a pair of diastereomeric salts, which can then be separated by crystallization.

The term **alkaloid** refers to any nitrogen-containing compound extracted from plants, although the word is used loosely and some compounds of nonplant origin are also commonly known as alkaloids. These biologically active compounds are basic like all amines, and it is this basicity that led to the name. Presumably, basicity was also important in the relative ease of extraction of alkaloids from the myriad compounds present in an organism as complex as a plant. Extraction of plant mass with acid yields any amines present as water-soluble ammonium ion salts, from which the amines can easily be regenerated. Figure 4.43 shows this process for a simple alkaloid known as coniine. Coniine is the active ingredient in hemlock, and it is the physiological activity of coniine that led to the demise of Socrates.

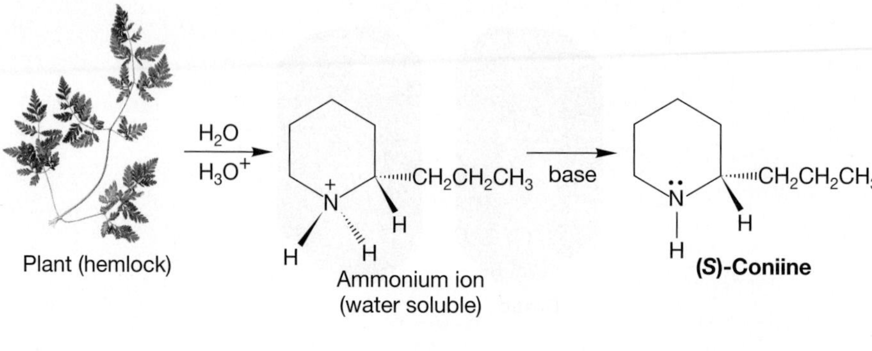

FIGURE 4.43 A scheme for isolation of alkaloids from plant materials.

Other alkaloids have wondrously complex structures. Two further examples, brucine and strychnine, are shown in Figure 4.44 along with the general procedure for resolution.

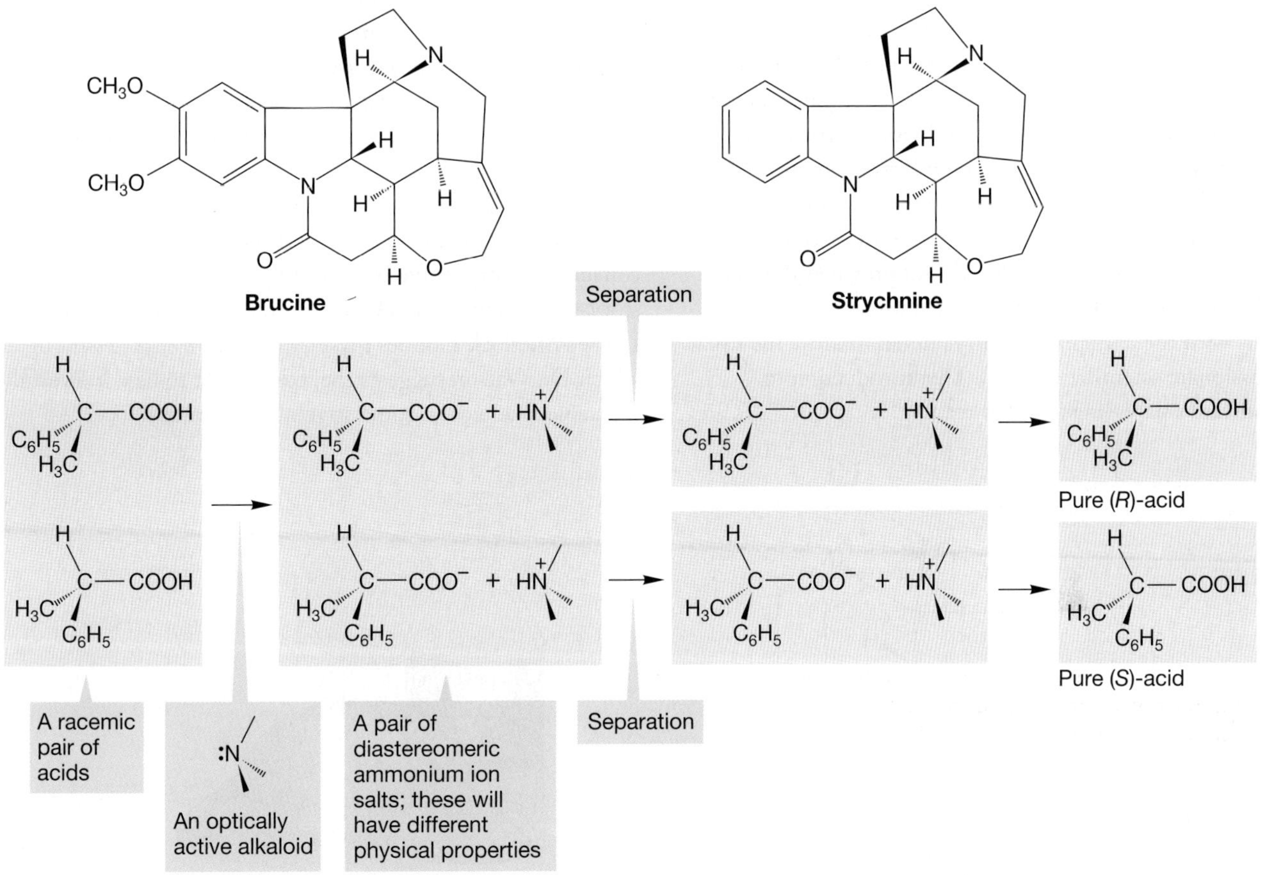

FIGURE 4.44 Two alkaloids, brucine and strychnine, are commonly used to separate the enantiomers of chiral organic acids. Diastereomeric salts are first formed, separated by crystallization, and the individual enantiomeric acids are regenerated.

WORKED PROBLEM 4.20 Identify with an asterisk (*) all the stereogenic carbons in brucine.

ANSWER

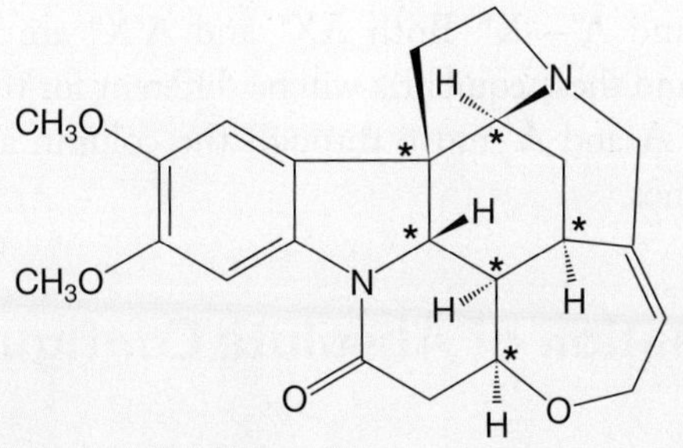

PROBLEM 4.21 Identify each stereogenic carbon in brucine as (R) or (S).

STRYCHNINE

The notorious poison strychnine was first isolated from the beans of *strychnos ignati Berg* by Pelletier and Caventou in 1818. It constitutes about one-half of the alkaloids present in the beans and makes up 5–6% of their weight. Its structure is obviously complicated and was only determined correctly by Sir Robert Robinson in 1946. It was a mere two years more before a physical determination of the structure by X-ray diffraction was reported by Bijvoet, confirming Robinson's deductions, and presaging the demise of chemical, as opposed to physical, structure determination as a viable enterprise. Robert B. Woodward provided the first synthesis of strychnine in 1954 in a landmark paper that begins with the exclamation "Strychnine!" (hardly the usual dispassionate scientific writing). The introduction to Woodward's paper makes good reading, and it provides an interesting defense of the art of synthesis in the face of critics who thought that the profession would surely be rendered obsolete by the increasingly powerful physical methods. The ensuing years have shown Woodward's defense to be correct. (See *Tetrahedron*, **1963**, *19*, 247; your chemistry library probably has it.)

Of course, much interest in strychnine centers on its pharmacological properties. It is a powerful convulsant, lethal to an adult human in a dose as small as 30 mg. Death comes from central respiratory failure and is preceded by violent convulsions. Strychnine is the deadly agent in many a murder story, real and imagined. One example is Sir Arthur Conan Doyle's Sherlock Holmes mystery, "*The Sign of the Four*," in which Dr. Watson suggests the lethal agent to be a "powerful vegetable alkaloid . . . some strychnine-like substance."

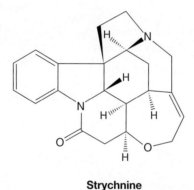

Strychnine

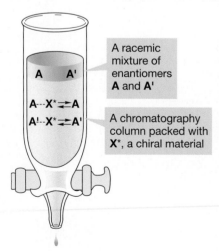

FIGURE 4.45 Separation of enantiomers through chromatography.

A racemic mixture of enantiomers **A** and **A'**

A chromatography column packed with **X***, a chiral material

These days, the general procedure outlined in Figure 4.42 has been extended so that all manner of enantiomeric pairs can be separated by chromatography. In chromatography, as in the salt formation in Figure 4.44, covalent chemical bonds are not formed. Rather, advantage is taken of the formation of complexes with partial bonds as the pair of enantiomers passes over an optically active substrate (Fig. 4.45). The optically active substrate (X^*) forms a complex with the enantiomers **A** and **A'**. These complexes are diastereomers and so have different physical properties, including bond strengths of **A**---**X*** and **A'**---**X***. Both **AX*** and **A'X*** are in equilibrium with the free enantiomers, and these equilibria will be different for the two diastereomeric complexes. Therefore, **A** and **A'** move through the column at different rates and emerge at different times.

4.10 Determination of Absolute Configuration (*R* or *S*)

Now that we have achieved the separation of our racemic mixture of enantiomers into a pair of optically active stereoisomers, we face the difficult task of finding out which enantiomer is (*R*) and which is (*S*). This problem is not trivial! Indeed, in

Chapter 23, when we deal with sugars, we'll find that until the mid 1900s, there was no way to be certain, and one just had to guess (correctly in the case of the sugars, it turns out). One would like to peer directly at the structures, of course, and under some circumstances this kind of inspection is possible.

X-ray crystallography can determine the relative positions of atoms in a crystal, and a special kind of X-ray diffraction called anomalous dispersion can tell the absolute configuration of the molecule. But this technique is not generally applicable— to do X-ray analysis one needs a crystalline compound, for example. It does serve to give us some benchmarks, though. If we know the absolute configuration (R or S) of some compounds, we may be able to determine the absolute configurations of other molecules by relating them to those compounds of known absolute configuration. We must be very careful, however. The chemical reactions that interconvert the molecules of known and unknown absolute configuration must not alter the stereochemical arrangement at the stereogenic atoms, or if they do, it must be in a known fashion. How do we know whether a given chemical reaction will or will not change the stereochemistry? We need to know the reaction mechanism—to know how the chemical changes occur—in order to answer this question. This reason is just one of many for the study of reaction mechanisms, and we'll take great pains to examine carefully the stereochemical consequences of the various chemical reactions we encounter. Figure 4.46 shows the general case, as well as examples of two chemical reactions that do not alter stereochemistry at the stereogenic carbon.

FIGURE 4.46 Two reactions that do not change the configuration at an adjacent carbon (*).

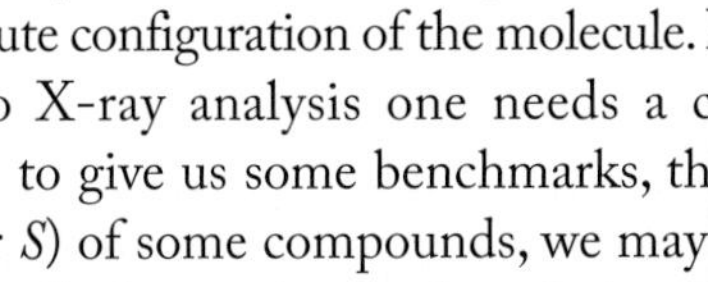

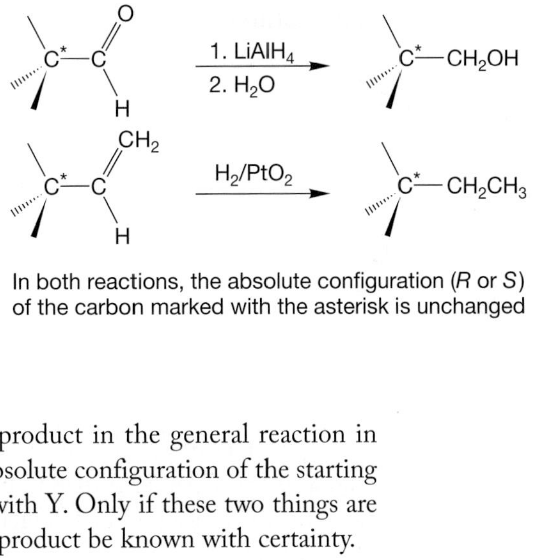

If we know the absolute configuration (R or S) of the starting material and the mechanism of the reaction, we can know the absolute configuration of the product

In both reactions, the absolute configuration (R or S) of the carbon marked with the asterisk is unchanged

Summary

To know the absolute configuration of the product in the general reaction in Figure 4.46 you must know two things: the absolute configuration of the starting material and the mechanism of the reaction with Y. Only if these two things are known will the absolute configuration of the product be known with certainty.

4.11 Stereochemical Analysis of Ring Compounds (a Beginning)

Most of the stereochemical principles discussed in the previous sections apply quite directly to ring compounds. We'll start with a simple question. How many isomers of chlorocyclopropane exist? That looks like an easy question, and it is. There is only one (Fig. 4.47).

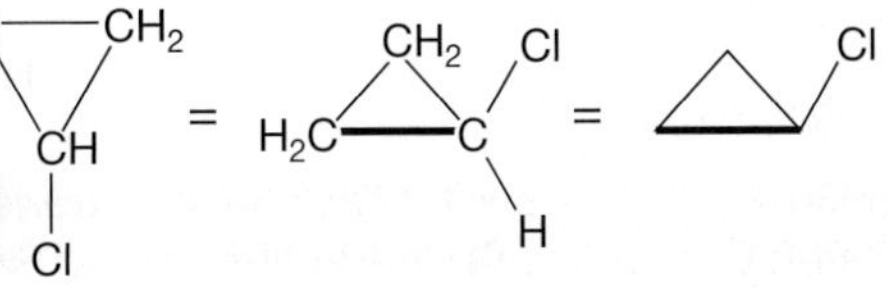

FIGURE 4.47 Different representations of the single isomer of chlorocyclopropane. The bold (thick) bond is toward you.

To be certain there is only one chlorocyclopropane, however, we should examine this compound for chirality. Again this process is simple. As Figure 4.48 shows, the mirror image is easily superimposed on the original, through a 120° rotation. Chlorocyclopropane is an achiral molecule.

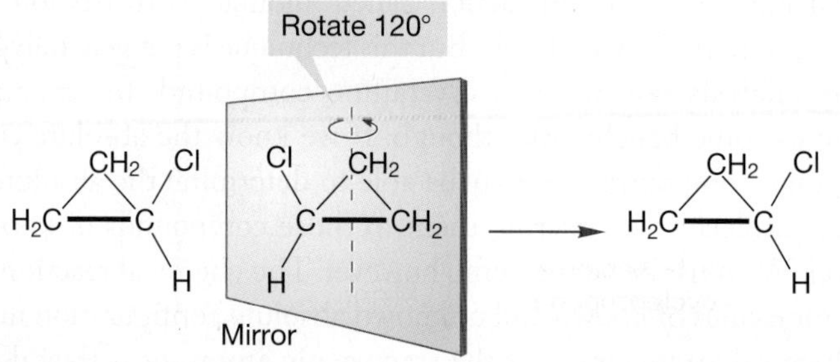

FIGURE 4.48 Chlorocyclopropane is superimposable on its mirror image and is, therefore, an achiral molecule. A rotation of 120° about the dotted axis passing through the center of the ring easily shows that the mirror image and the original are identical.

Rotate 120°

Mirror

How many isomers of dichlorocyclopropane exist? This question is much tougher than the last one. There are four isomers of dichlorocyclopropane. Most people don't get them all at this point, so it's worth going carefully through an analysis. Where can the second chlorine be placed in the single monochlorocyclopropane? As Figure 4.49 shows, there are only two places, either on the same carbon as the first chlorine, or on one of the two equivalent adjacent carbons. This analysis yields two compounds called structural isomers, molecules with the same formula, but differing atom-to-atom connectivity. Structural isomers are sometimes called **constitutional isomers**.

FIGURE 4.49 Replacement of one hydrogen of chlorocyclopropane with another chlorine can occur in only two places, on the carbon to which the first chlorine is attached or on one of the other two equivalent carbons. This procedure generates the structural isomers 1,1-dichlorocyclopropane and 1,2-dichlorocyclopropane.

Structural isomers

replace a hydrogen with a chlorine

1,1-Dichloro-cyclopropane   **1,2-Dichloro-cyclopropane**

Figure 4.49 brings us only two isomers; we're still two short of reality. As first mentioned in Chapter 2 (p. 84), however, rings have sides. We can find one more isomer if we notice that in 1,2-dichlorocyclopropane the second chlorine can be on the same side (the cis isomer) or on the opposite side (the trans isomer) of the ring (Fig. 4.50). These compounds are stereoisomers, more precisely, diastereomers (stereoisomers that are not mirror images). Now we are just one isomer short.

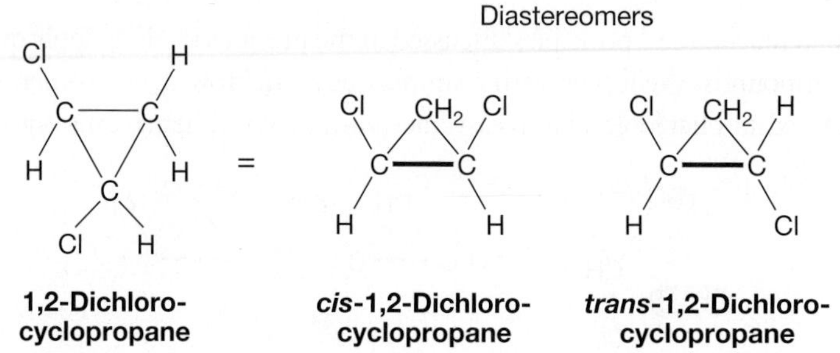

FIGURE 4.50 1,2-Dichlorocyclopropane can exist in cis and trans forms. In the figure, all the ring hydrogens have been drawn in to help you see the difference between these two molecules. Be absolutely certain you see why these two molecules cannot be interconverted, no matter how they are rotated or translated in space. Use models.

Diastereomers

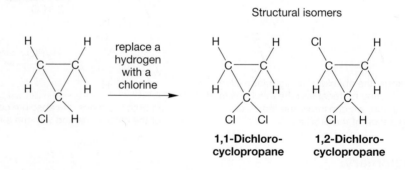

1,2-Dichloro-cyclopropane   *cis*-**1,2-Dichloro-cyclopropane**   *trans*-**1,2-Dichloro-cyclopropane**

Once we are sure we have found all the different positions on which to put the two chlorines, it's time to look for chirality. Figure 4.51 shows that neither 1,1-dichlorocyclopropane nor *cis*-1,2-dichlorocyclopropane is chiral. The mirror images are easily superimposable on the originals. Note that *cis*-1,2-dichlorocyclopropane is a meso compound. It contains stereogenic carbons but is superimposable on its mirror image, and therefore is achiral.

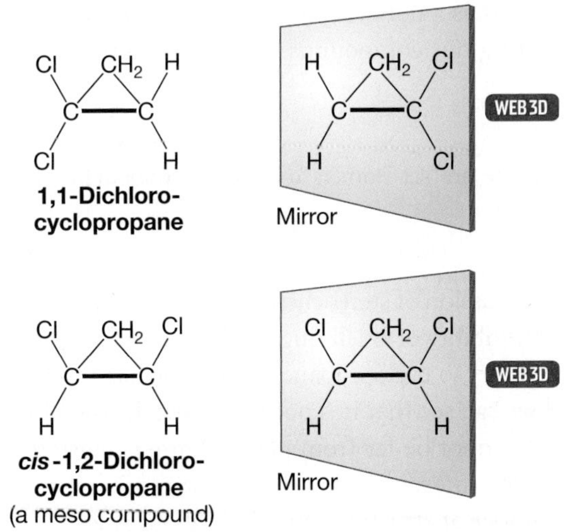

FIGURE 4.51   Both 1,1-dichlorocyclopropane and *cis*-1,2-dichlorocyclopropane are achiral.

trans-1,2-Dichlorocyclopropane is different from the cis molecule we just examined for chirality. In the trans diastereomer, the mirror image is not superimposable on the original and this stereoisomer is chiral (Fig. 4.52).

FIGURE 4.52   *trans*-1,2-Dichlorocyclopropane is *not* superimposable on its mirror image. It is a chiral molecule and therefore two enantiomers exist. This stereochemical relationship is probably the most difficult you've encountered so far, so be sure you can see why no manipulation of one enantiomer will make it superimposable on the other. This is a point at which models are very useful.

Now we have the four isomers: 1,1-dichlorocyclopropane, *cis*-1,2-dichlorocyclopropane, *trans*-(1*S*,2*S*)-dichlorocyclopropane, and *trans*-(1*R*,2*R*)-dichlorocyclopropane (Fig. 4.53). Notice that we have to designate each stereogenic carbon—two in each trans stereoisomer—as either *R* or *S*.

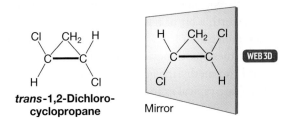

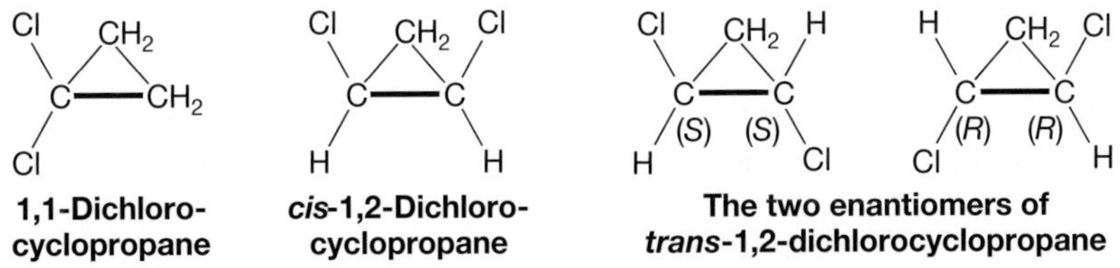

1,1-Dichloro-cyclopropane   **cis-1,2-Dichloro-cyclopropane**   **The two enantiomers of trans-1,2-dichlorocyclopropane**

FIGURE 4.53   The four isomers of dichlorocyclopropane.

PROBLEM 4.22   Verify the assignment of absolute configuration in the two enantiomers of *trans*-1,2-dichlorocyclopropane in Figure 4.53.

PROBLEM 4.23   Identify the stereochemical relationships among the four isomers of Figure 4.53. In other words, find all the pairs of enantiomers and diastereomers.

This section illustrates one procedure for working out an isomer problem, but any rational technique will work. The important thing is to devise some systematic way of searching. We happen to like the one described above in which one first finds all the structural isomers (in this case 1,1- and 1,2-dichlorocyclopropane), then searches for stereoisomerism of the cis/trans type, and finally examines each isomer for chirality. But it doesn't matter what procedure you use as long as you are systematic. What will not work is a nonrational scheme. So don't start a "find the isomers" problem by just writing out compounds without thinking first. No one can do it that way.

PROBLEM 4.24 There are six isomers of dichlorocyclobutane. Find them all and name them carefully.

We started this discussion of stereochemistry in ring compounds with small rings for a reason—the inflexibility of small rings makes them flat, or very nearly so. Three points determine a plane, so cyclopropane must be planar. Cyclobutane need not be absolutely flat, and we can see that it is not flat if we take the time to make a model. At the same time, it cannot be far from planar. Larger rings are more complicated though, and we'll defer an examination of their stereochemistries until Chapter 5, which looks at a number of structural questions about rings.

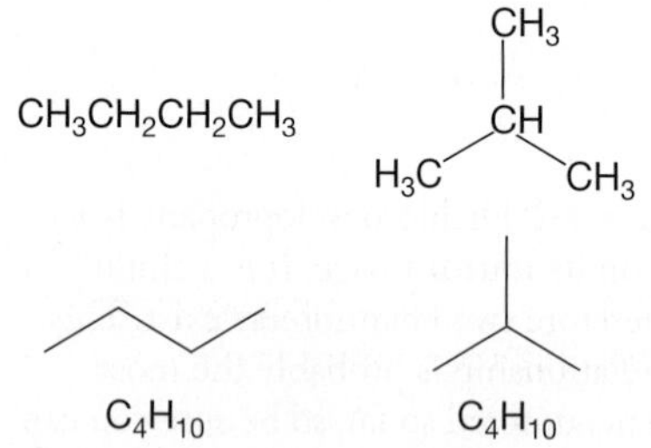

FIGURE 4.54 Two structural isomers.

4.12 Summary of Isomerism

The words isomer and stereoisomer are often tossed around quite loosely, and the subject of isomerism is worth a bit of review. Structural isomers (often called constitutional isomers) are molecules of the same formula, but differing atom-to-atom connectivity. Their constituent parts may well be different (but need not be). Butane (two methyl groups and two methylene groups) and isobutane (three methyl groups and one methine group) are typical examples of structural isomers (Fig. 4.54).

PROBLEM 4.25 Find a pair of structural isomers whose constituent parts are not different from one another.

Stereoisomers have the same connectivity, but differ in the arrangement of their parts in space. Two kinds of stereoisomers exist: enantiomers and diastereomers. Enantiomeric molecules are nonsuperimposable mirror images of each other. Simple examples are (R)- and (S)-3-methylhexane. Slightly more complicated are (2S,3S)-dichlorobutane and its enantiomer, (2R,3R)-dichlorobutane. These pairs are not structural isomers because they have the same connectivity (Fig. 4.55).

FIGURE 4.55 Two pairs of enantiomers.

Diastereomers are stereoisomers that are not mirror images; *cis-* and *trans-*2-butene are examples, as are *meso-*2,3-dichlorobutane and either the (*R*,*R*) or the (*S*,*S*) isomer of 2,3-dichlorobutane in Figure 4.55 (Fig. 4.56).

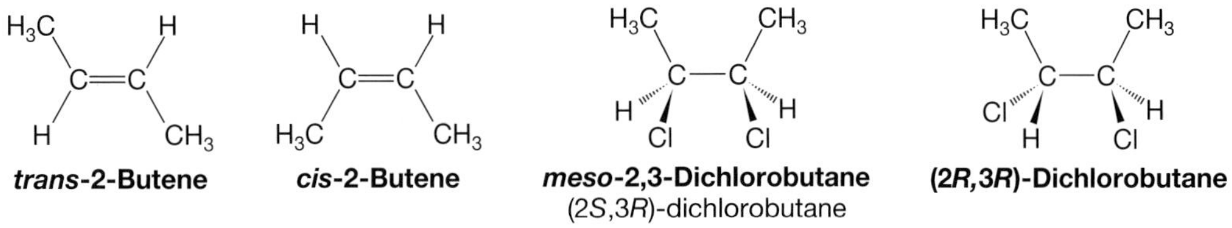

trans-2-Butene **cis-2-Butene** **meso-2,3-Dichlorobutane**
(2S,3R)-dichlorobutane

(2R,3R)-Dichlorobutane

FIGURE 4.56 Two pairs of diastereomers.

Stereoisomers also include conformational isomers, in which different isomers are generated through rotations about bonds. Conformational isomers are often called "conformers." Eclipsed and staggered ethane are typical examples. Note that a conformational isomer need not be an energy minimum—the eclipsed conformation of ethane is an energy maximum, for example. Conformational isomers can be either enantiomeric or diastereomeric. The two gauche forms of butane are conformational enantiomers, but the gauche and the anti forms of butane are conformational diastereomers (Fig. 4.57).

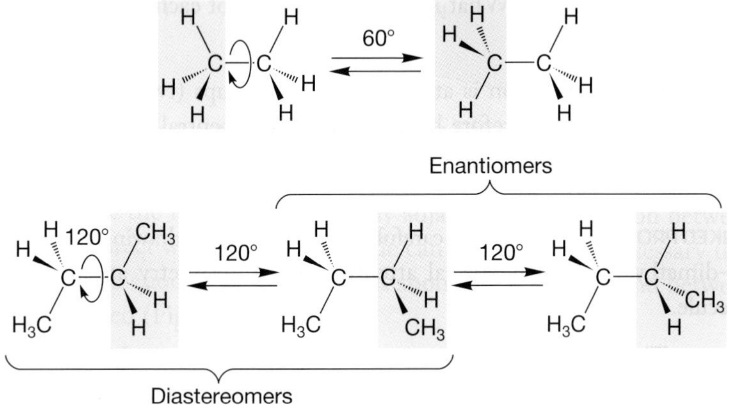

Enantiomers

Diastereomers

FIGURE 4.57 Examples of conformational isomers.

WORKED PROBLEM 4.26 Show that the two conformational isomers labeled "enantiomers" in Figure 4.57 really are mirror images.

ANSWER The trick is to find a way to make an easy comparison. In this case, it takes only a simple 180° rotation to put the molecules into the proper orientation. Now it is easy to see that the two molecules from Figure 4.57 (labeled **A** and **B** below) are indeed mirror images.

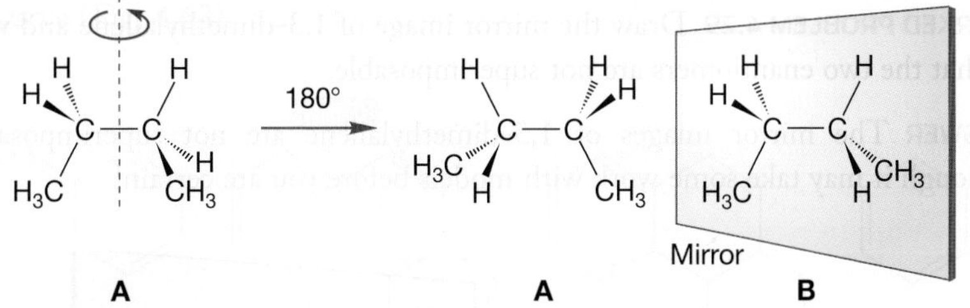

A **A** **B**

Mirror

4.13 Special Topic: Chirality without "Four Different Groups Attached to One Carbon"

We already know from Section 4.8 that finding four different groups attached to one carbon is not a sufficient condition for chirality. There are achiral meso compounds that satisfy this condition. But is the presence of a carbon atom attached

4.16 Additional Problems

PROBLEM 4.32   Find all the chiral isomers of the heptanes (pp. 82–83). Designate the stereogenic atoms with an asterisk (*).

PROBLEM 4.33   Draw the (S) enantiomers of the chiral heptanes.

PROBLEM 4.34   Find all the chiral isomers of the octanes (see Problem 2.28).

PROBLEM 4.35   Draw the (R) enantiomers of all of the chiral isomers of octane whose name must end in "pentane."

PROBLEM 4.36   Now find all the chiral isomers of the nonanes (see Problem 2.59). Indicate those with more than one stereogenic carbon.

PROBLEM 4.37   Draw three-dimensional pictures of all the stereoisomers of 3,4-dimethylheptane and 3,5-dimethylheptane. It is probably easiest to draw them in the eclipsed arrangement, even though this is not a low-energy conformation. Determine the absolute configuration (R or S) of each stereogenic carbon. Designate the stereogenic carbons with an asterisk.

PROBLEM 4.38   You are given a bottle labeled 3,4-dimethylheptane. Presumably, all of the 3,4-dimethylheptane isomers you found in your answer to Problem 4.37 are in that bottle. How many signals will appear in the ^{13}C NMR spectrum of the contents of the bottle?

PROBLEM 4.39   You are given a bottle labeled 3,5-dimethylheptane. Presumably, all of the 3,5-dimethylheptane isomers you found in your answer to Problem 4.37 are in that bottle. How many signals will appear in the ^{13}C NMR spectrum of the contents of the bottle?

PROBLEM 4.40   Find all the chiral isomers of the hexenes (see Problem 3.5) and heptenes (see Problem 3.40).

PROBLEM 4.41   Find all the chiral isomers of the hexynes, heptynes (see Problem 3.23), and octynes (see Problem 3.41).

PROBLEM 4.42   For each of the following compounds determine the number of stereogenic carbons and draw all possible stereoisomers. You can treat the cyclohexanes as if they were flat.

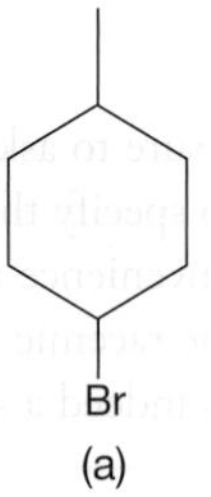

(a) 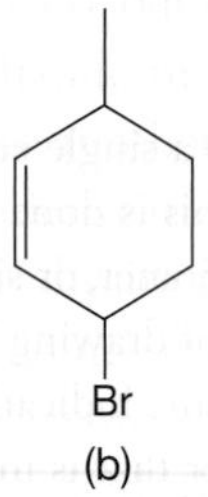   (b) 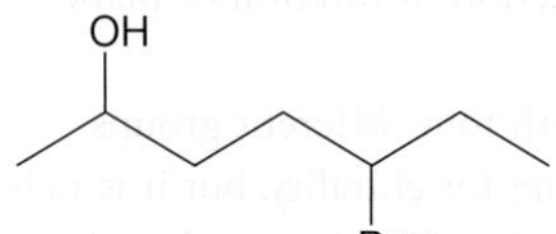   (c) 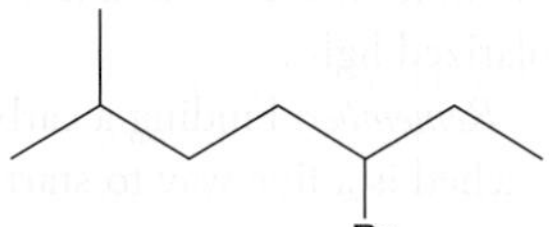   (d)

PROBLEM 4.43   Which of the following compounds are chiral? Determine the configuration (R or S) for all stereogenic atoms.

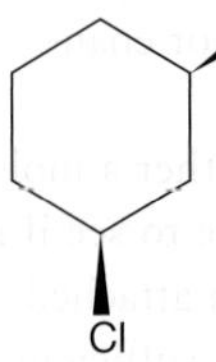

(a) 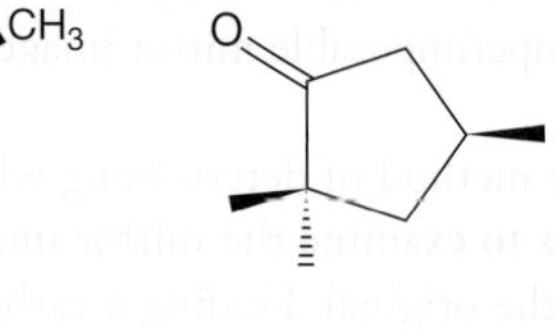   (b) 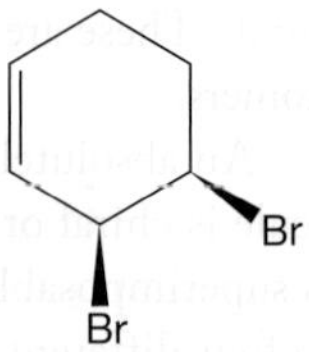   (c)

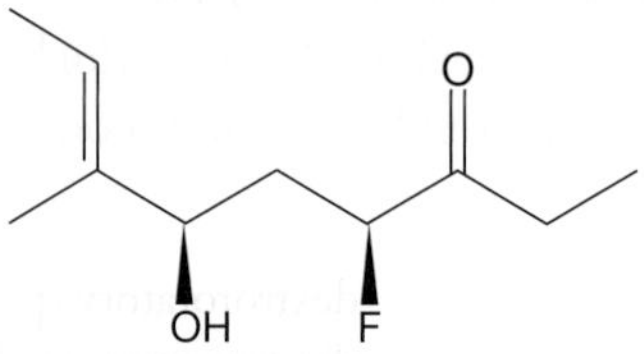

(d) 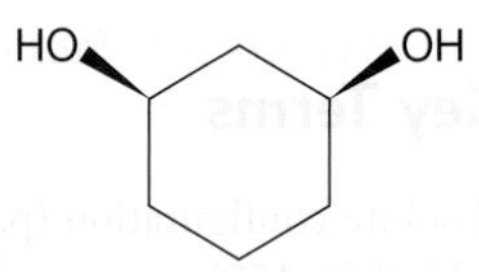   (e)

PROBLEM 4.44   Estradiol (shown below) is the active ingredient in one birth control method. Indicate the stereogenic carbons in the molecule and determine the configuration for each.

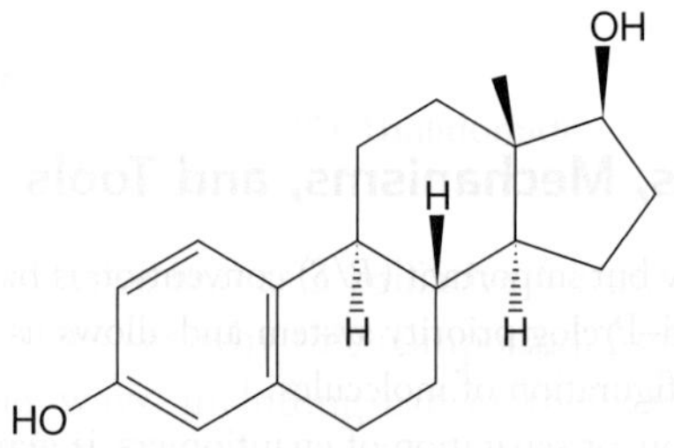

Estradiol

PROBLEM 4.45   Narbomycin is an antibiotic substance produced by *Streptomyces narbonensis*. It is shown below. Put an asterisk by each stereogenic carbon.

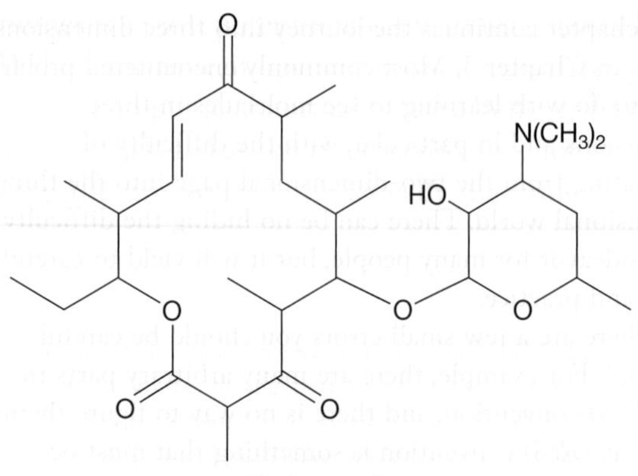

Narbomycin

PROBLEM 4.46 Which of the following molecules are chiral? Specify each stereogenic atom as (R) or (S).

PROBLEM 4.47 Taxol is a natural product with antitumor activity. It is a component of the Yew tree. Put an asterisk by each stereogenic atom.

Taxol

PROBLEM 4.48 In Problem 2.45, you drew all the isomers of "chlorohexane" $C_6H_{13}Cl$. Which of these isomers contain stereogenic carbons? Be sure to be alert for molecules containing more than one stereogenic carbon. Designate the stereogenic carbons with an asterisk.

PROBLEM 4.49 Draw three-dimensional pictures of the stereoisomers of the molecules in your answer to Problem 4.48 that contain more than one stereogenic carbon.

PROBLEM 4.50 Which of the following molecules can exist as cis/trans (Z/E) isomers? Draw the (E) and (Z) forms. Which molecules possess stereogenic carbons? Designate the stereogenic carbons with an asterisk.

(a) $(CH_3)_2C=CHCH_2CH_2CH(CH_3)CH_2CH_2OH$

(b) $(CH_3)_2C=CHCH_2CH_2C(CH_3)=CHCH_2OH$

(c) (d)

PROBLEM 4.51 Draw the (R) enantiomer for each molecule containing a stereogenic carbon in Problem 4.50.

PROBLEM 4.52 Draw all the stereoisomers of the following molecule. Identify all pairs of enantiomers and diastereomers.

PROBLEM 4.53 In Problem 3.48, you worked out all the isomers of the formula C_4H_5Cl. Find all the chiral isomers of the ring compounds with this formula. Indicate the stereogenic carbons with an asterisk.

PROBLEM 4.54 Draw the four chiral isomers found in Problem 4.53, show the mirror images, and indicate the absolute configuration (R or S) of the stereogenic carbons.

PROBLEM 4.55 In Problem 3.46, you found 10 cyclopropanes of the formula $C_4H_6Br_2$. Find the chiral ones. Indicate the stereogenic carbons with an asterisk.

PROBLEM 4.56 One of the isomers in the answer to Problem 4.55 has a pair of bromines on one carbon. Draw the stereoisomers of this molecule, and use the (R/S) convention to show the absolute configuration of the stereogenic carbons.

PROBLEM 4.57 Write the IUPAC name for each of the following compounds. Don't forget the (*R/S*) or (*E/Z*) designation if relevant.

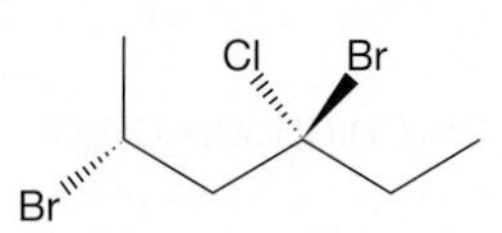

(a) 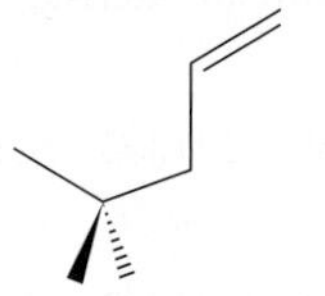   (b)

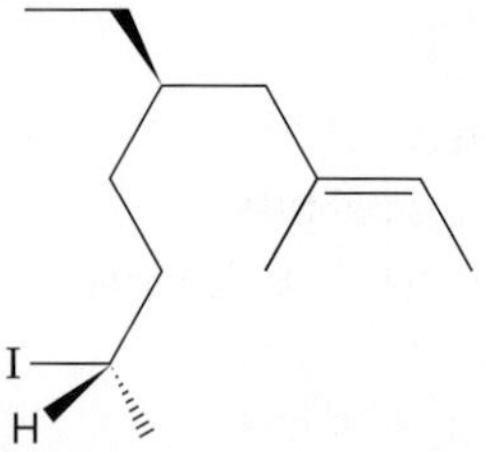

(c) 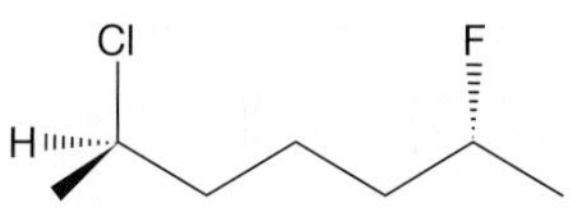   (d)

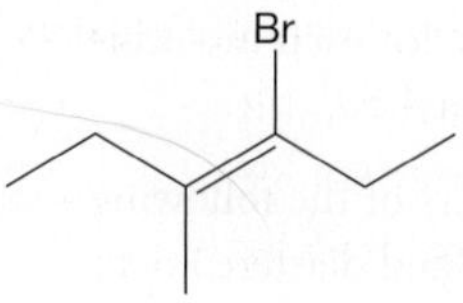

(e) 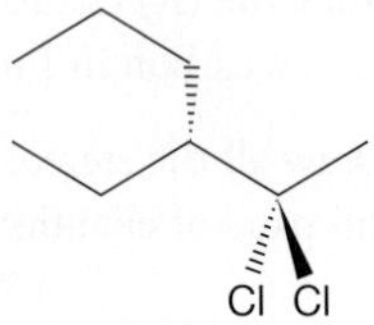   (f) 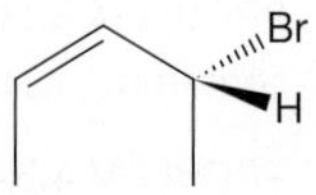   (g)

PROBLEM 4.58 In Dorothy Sayers' and Robert Eustace's mystery novel, *The Documents in the Case* (Harper Paperbacks, 1995), mushroom expert George Harrison is murdered by Harwood Latham. Latham doses Harrison's stew of *Amanita rubescens* with muscarine, the toxin of the related, but deadly, *Amanita muscaria*. Lathom hopes to make it look like an acci-

dental poisoning through a mistaken identification by Harrison. See if you can guess how Latham is caught. Here is a clue: Latham obtains the poison muscarine not from the mushroom itself, but from a laboratory in which it has been synthesized by an unwitting accomplice from simple starting materials. Thanks go to Ms. Dana Guyer, a former student of organic chemistry at Princeton University, for finding this book.

Use Organic Reaction Animations (ORA) to answer the following questions:

PROBLEM 4.59 Observe the reactions titled "Acetylide addition" and "S_N2 with cyanide." What stereochemical event happens at the carbon being attacked in each of these reactions?

PROBLEM 4.60 Observe the reactions "Unimolecular nucleophilic substitution" and "Unimolecular elimination." The first intermediate in each reaction has a carbon that is planar. What is the hybridization of that carbon in each case? What would be the outcome if a stereogenic carbon were involved in the reaction? Would chirality be retained? Why or why not?

PROBLEM 4.61 Observe the "Bimolecular elimination" reaction. When the base reacts with the hydrogen in the first step of the reaction, what is the spatial relationship between that hydrogen and the bromide leaving group? Do you suppose this relationship is important? Why or why not?

PROBLEM 4.62 Is the starting material shown in the reaction "Carbocation rearrangement" chiral? If yes, what is the configuration of the molecule shown in the animation? Is the product chiral? Is this a case of: (a) a chiral molecule becoming achiral, (b) a chiral molecule retaining its chirality, (c) an achiral molecule staying achiral, or (d) an achiral molecule becoming chiral?

Rings

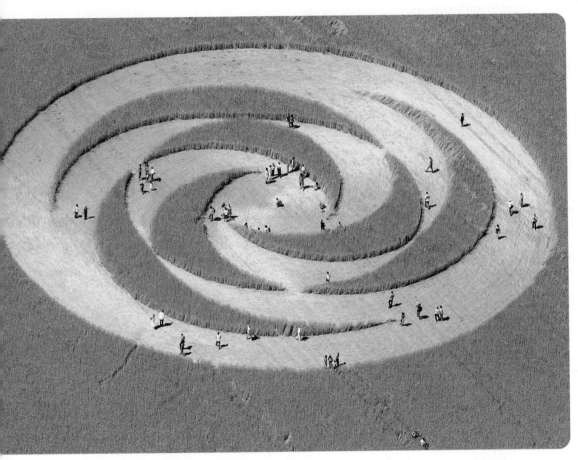

5.1 Preview

5.2 Rings and Strain

5.3 Quantitative Evaluation of Strain Energy

5.4 Stereochemistry of Cyclohexane: Conformational Analysis

5.5 Monosubstituted Cyclohexanes

5.6 Disubstituted Ring Compounds

5.7 Bicyclic Compounds

5.8 Special Topic: Polycyclic Systems

5.9 Special Topic: Adamantanes in Materials and Biology

5.10 Summary

5.11 Additional Problems

RING FORMATION This chapter is about rings, from the smallest to the largest. We'll learn about their properties here, but their formation will remain a mystery for later chapters to solve.

Clearly the ring had an unwholesome power . . .

—J. R. R. TOLKIEN,[1] *THE LORD OF THE RINGS*

5.1 Preview

In these early chapters, we have used our powers of imagination to picture the three dimensionality of organic molecules. But nowhere is thinking in three dimensions more important than in depicting the myriad structures formed from chains of atoms tied into rings. It's not always easy to see these structures clearly, so do not be reluctant to work with models, especially at the beginning. All organic chemists use them. Polycyclic compounds are especially complicated. No one can see all the subtlety of these compounds without models. Much of chemistry involves interactions of groups in proximity, and the two-dimensional page is often quite ineffective in showing which atoms are close to others. As you go on in this subject, your ability to use the two-dimensional surface to depict the three-dimensional molecular world will increase, but none of us will ever outgrow the need for models. Rings were first encountered in Section 2.12, and Figure 5.1 recalls some simple and complex ring compounds.

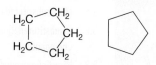

Two schematic views
of cyclopentane

A complicated, polycyclic compound,
bicyclo[2.2.1]heptane

A really complicated, polycyclic
compound that nobody tries to name
systematically—it's called
"dodecahedrane"

FIGURE 5.1 Some simple and complex ring structures.

ESSENTIAL SKILLS AND DETAILS

1. Craftsmanship. This chapter is entirely about the structures of rings. Drawing those rings accurately is an essential skill for any chemist. Ring compounds abound in Nature, and the positions of groups attached to rings can be known quite well. If you know for certain the shape of the ring compound, you also know where the groups attached to it are located. The distance between groups and the orientation of groups can be controlled in this way if you know the ring structure. It is really more than a detail, but drawing the chair form of a cyclohexane perfectly (well, at least very accurately) is a necessary skill. Fortunately, this skill may look hard but, in fact, it is easy. Just follow the directions in Section 5.2.

2. Flexibility. Many rings are highly mobile. One example of this phenomenon is the ring "flip" of cyclohexane. You must be able to "flip" cyclohexanes and draw the two interconverting chair forms well.

3. Strain. You have to be able to identify two sources of instability in rings. Eclipsing (torsional) strain and angle strain both act to make rings less stable and thus more reactive.

4. Intermediate versus transition state. The distinction between an intermediate and a transition state reappears in this chapter. An intermediate is a potentially isolable compound in an energy well. However, often it is high in energy and thus not easy to isolate. On the other hand, a transition state is the top of an energy hill—the top of the barrier separating two compounds in energy wells—and cannot be isolated.

---

[1] John Ronald Reuel Tolkien (1892–1973) was Merton Professor of English language at Oxford University.

5.2 Rings and Strain

Much can be learned from the simplest of all rings, cyclopropane, $(CH_2)_3$. Indeed, even at this early stage we can make some preliminary judgments about this compound. In doing so, we will be recapitulating the thoughts of Adolf von Baeyer (1835–1917), who in 1885 appreciated the importance of the deviation of the internal angles in most cycloalkanes from the ideal tetrahedral angle, 109.5° (Fig. 5.2). He

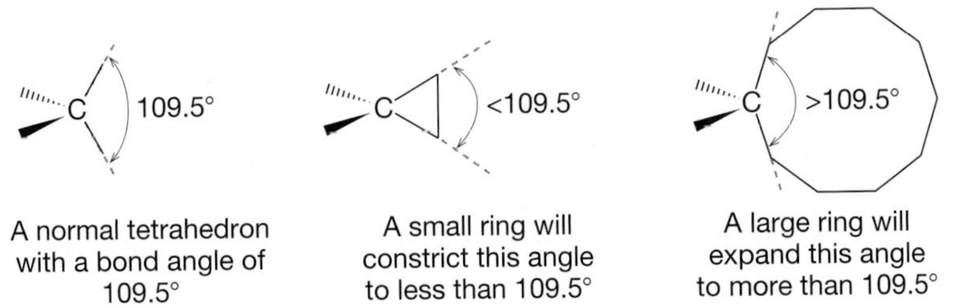

A normal tetrahedron with a bond angle of 109.5°

A small ring will constrict this angle to less than 109.5°

A large ring will expand this angle to more than 109.5°

FIGURE 5.2 Ring compounds have angle strain if angles are forced to be significantly smaller or larger than the ideal 109.5°.

suggested that the instabilities of ring compounds should parallel the deviations of their internal angles from the ideal; the further the angle from tetrahedral, the more **angle strain** the molecule has, and the less stable (higher energy) it is. The idea that angle strain is an important factor in ring stability has continued to this day. Cyclopropane represents an extreme example of the effects of angle strain. The strain induced by the reduction of the ideal tetrahedral angle of 109.5° to 60° is so great as to make other bonding arrangements possible. One way of looking at cyclopropane views the bonds between the carbons as "bent." The *internuclear* angle in an equilateral triangle is, of course, 60° and cannot be otherwise. However, the *interorbital* angle is apparently quite a bit larger (Fig. 5.3) and is estimated at 104°. Thus, the internuclear angle (60°) and the interorbital angle (~104°) are different for cyclopropane.

WEB 3D

Cyclopropane
(internuclear angle is 60°)

Cyclopropane
(interorbital angle is 104°)

FIGURE 5.3 Bent bonds in cyclopropane.

Of course, the carbon skeleton of cyclopropane is flat—three points determine a plane—and planarity causes still other problems for cyclopropane. Figure 5.4 shows a Newman projection looking down one of the three equivalent carbon–carbon bonds of cyclopropane. In cyclopropane, all carbon–hydrogen bonds are eclipsed. *Remember*: The solid wedges are coming toward you, and the dashed wedges are retreating from you.

We have seen eclipsing before in the eclipsed form of ethane (and other alkanes). Ethane has a simple way to avoid this destabilizing opposition of atoms and electrons; it adopts the staggered conformation and minimizes the problem. There is no way for cyclopropane to do likewise. Its hydrogens are stuck in the eclipsed arrangement, and there is no possible release. We can even make an estimate of how much damage this **torsional strain** (eclipsing strain) might cause in cyclopropane, of how

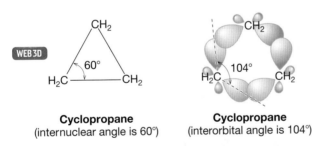

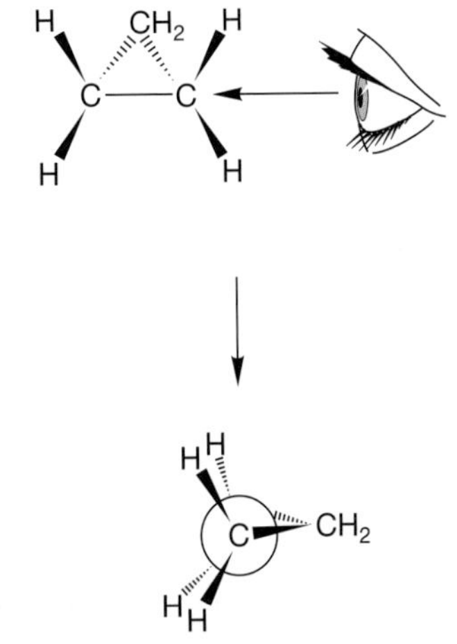

This Newman projection attempts to show the eclipsed C—H bonds; in cyclopropane all the C—H bonds are eclipsed

FIGURE 5.4 Eclipsing of C—H bonds in cyclopropane.

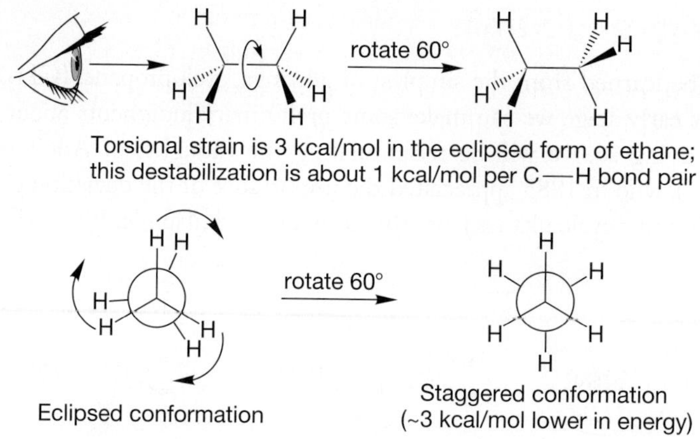

FIGURE 5.5 The destabilization introduced by a pair of eclipsed carbon–hydrogen bonds is about 1 kcal/mol.

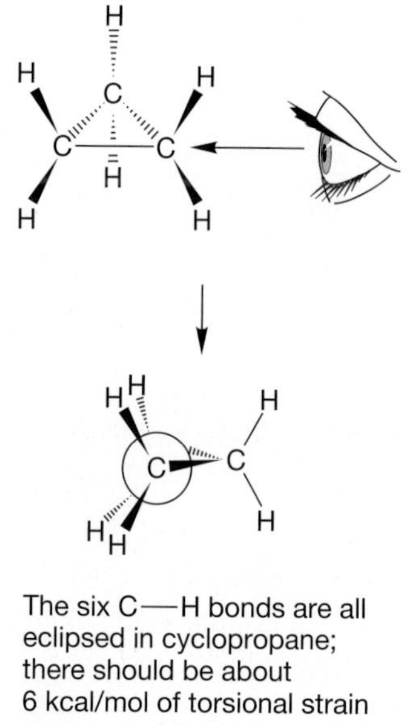

Torsional strain is 3 kcal/mol in the eclipsed form of ethane; this destabilization is about 1 kcal/mol per C—H bond pair

rotate 60°

Eclipsed conformation

Staggered conformation
(~3 kcal/mol lower in energy)

The six C—H bonds are all eclipsed in cyclopropane; there should be about 6 kcal/mol of torsional strain

FIGURE 5.6 There are six pairs of eclipsed C—H bonds in cyclopropanes.

much torsional strain raises the energy of cyclopropane. The rotational barrier in ethane is the result of three pairs of eclipsed hydrogens (Chapter 2, p. 67), and amounts to 3 kcal/mol (Fig. 5.5). Thus, a first guess would put the energy cost of each pair of eclipsed hydrogens at about 1 kcal/mol. That estimate is a maximum value, however, because the carbon–hydrogen bonds in cyclopropane point away from each other more than do the ones in ethane. There is some debate about the amount of eclipsing strain in cyclopropane, but the consensus now is that it is substantially less than that 6 kcal/mol maximum. The total strain of cyclopropane is composed of torsional strain plus angle strain, which is even more difficult to estimate without the help of calculations or experiments (Fig. 5.6).

The effects of high strain in cyclopropane show up in a number of ways, including the unusual reactivity of the carbon–carbon bonds. Breaking the central carbon–carbon bond in butane requires 88 kcal/mol, but the carbon–carbon bond in cyclopropane needs only 65 kcal/mol (Fig. 5.7). This difference allows a first quantitative estimate of the strain of cyclopropane (on p. 193 we will see two more quantitative estimates). Strain destabilizes cyclopropane by about (88 − 65) kcal/mol = 23 kcal/mol.

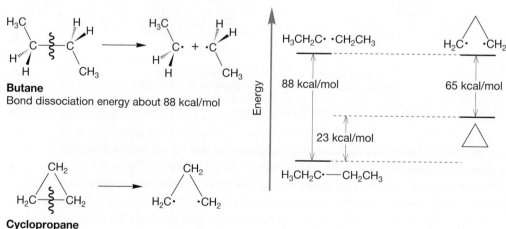

Butane
Bond dissociation energy about 88 kcal/mol

Cyclopropane
Bond dissociation energy about 65 kcal/mol

FIGURE 5.7 The bond dissociation energy of a carbon–carbon bond in cyclopropane is only 65 kcal/mol. Cyclopropane is strained by about 23 kcal/mol.

At first glance, the other cycloalkanes would seem to have similar difficulties. In planar cyclobutane, for example, angle strain is less of a problem because cyclobutane's 90° C—C—C bond angle is a smaller deviation from the ideal 109.5° than is cyclopropane's 60°. Along with this relief, however, comes increased torsional strain, as there are eight pairs of hydrogens eclipsed as opposed to cyclopropane's six pairs (Fig. 5.8). One might estimate the maximum torsional strain in planar cyclobutane at 8 kcal/mol (1 kcal/mol strain for each pair of eclipsed carbon–hydrogen bonds).

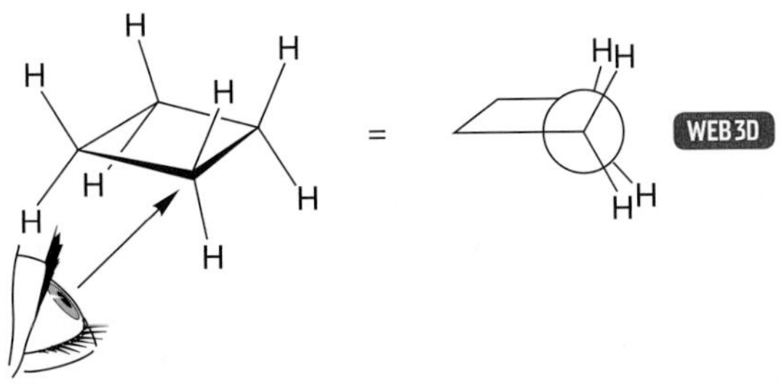

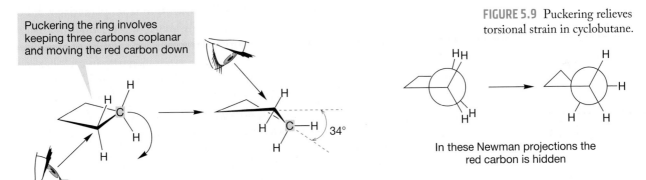

FIGURE 5.8 There are eight pairs of eclipsed carbon–hydrogen bonds in cyclobutane and therefore about 8 kcal/mol of torsional strain.

Unlike cyclopropane, cyclobutane has a way to balance torsional and angle strain. The four-membered ring need not be planar but can distort, or "pucker" somewhat if this will result in an energy lowering. Let's look at the consequences of puckering the ring by moving one methylene group out of the plane of the other three (Fig. 5.9).

FIGURE 5.9 Puckering relieves torsional strain in cyclobutane.

In these Newman projections the red carbon is hidden

As the ring puckers, torsional strain is reduced, but angle strain is increased. A balance between the two effects is struck in which the ring puckers about 34° and the C—C—C angle closes to about 88°. This form is not static, however; the ring is in motion through rotation about the carbon–carbon bonds, much as are the acyclic alkanes. The planar form of cyclobutane is like the eclipsed form of ethane and lies not in an energy well, but at the top of an energy barrier separating a pair of puckered cyclobutanes. This barrier is very small, about 1.4 kcal/mol (Fig. 5.10). Cyclobutane is a mobile, not static, molecule in which different nonplanar forms rapidly interconvert. In Figure 5.10, the bending is exaggerated—the flap is really only 34° out of plane.

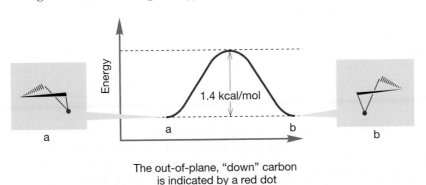

The out-of-plane, "down" carbon is indicated by a red dot

FIGURE 5.10 Mobile cyclobutane.

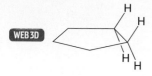

Planar cyclopentane
showing two pairs of
eclipsed hydrogens

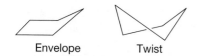

Envelope   Twist

The two forms of
nonplanar cyclopentane

FIGURE 5.11 Two distorted
cyclopentanes.

Let's now look at cyclopentane, $(CH_2)_5$, the next cycloalkane, and see how it distorts in a similar way. Were cyclopentane planar, it would suffer the torsional strain induced by ten pairs of eclipsed hydrogens. The internal angle in a planar pentagon is 108°, however, so angle strain would be quite small. As in cyclobutane, the ring distorts from planarity, relieving some eclipsing at the cost of increased angle strain. For nonplanar cyclopentane, there are two forms of comparable energy, the "envelope" form and the "twist" form (Fig. 5.11).

> **PROBLEM 5.1** Draw a Newman projection looking down one of the carbon–carbon bonds in planar and envelope cyclopentane. Use models!

Neither the puckered envelope form nor the twist form of cyclopentane is static. In the envelope form the "flap" moves around the ring, generating the five possible puckered isomers. This motion requires only a series of rotations around carbon–carbon bonds and closely resembles the motions in cyclobutane shown in Figure 5.10. A model will help you to visualize this motion. Hold two adjacent carbons, sight down the bond that connects them, and convert one form of the envelope into another.

By now you must be expecting that a similar distortion from planarity will be present in cyclohexane, $(CH_2)_6$, but you are probably prepared for neither the magnitude of the effect nor its consequences for the structure and stability of all cyclohexanes. In contrast to the smaller rings, distortion from planarity in cyclohexane relieves both the angle and torsional strain of the planar structure. The internal angle in a planar hexagon is 120°, larger, not smaller, than the ideal sp^3 angle of 109.5°. Deviation from planarity will *decrease* this angle and thus *decrease* angle strain. Torsional strain from the twelve pairs of eclipsed carbon–hydrogen bonds in planar cyclohexane can be expected to contribute about 12 kcal/mol of strain, which is also decreased in a nonplanar form. So in cyclohexane, *both* angle and torsional strain will be relieved by relaxing from a planar structure. Remarkably, this relaxation produces a molecule in which essentially all of the torsional and angle strain is gone. The formation of the energy-minimum cyclohexane, called the "chair" form, is shown in Figure 5.12. The internal angle in real cyclohexane is 111.5°, close to the ideal C—C—C angle in a simple straight-chain alkane (112° in propane), and the carbon–hydrogen bonds are nicely staggered. Newman projections looking down the carbon–carbon bonds show this, as will (even better) a look at a model.

FIGURE 5.12 Planar and chair
cyclohexane.

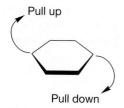

Pull up

Pull down

Planar cyclohexane:
120° C—C—C bond angles and
about 12 kcal/mol of torsional strain
from C—H bond eclipsing; there is
also angle strain from the too-wide
120° C—C—C internal angles

Chair cyclohexane:
111.5° C—C—C bond angles;
about 0 kcal/mol of torsional strain

A double Newman projection of
the chair form of cyclohexane;
note the perfect staggering of
the C—H bonds

The details of cyclohexane stereochemistry are important enough to warrant a lengthy discussion (Section 5.4, p. 197), but we can do a few things here in preparation. First of all, it is necessary to learn to draw a decent cyclohexane. No person can truly be described as educated unless he or she can do this, and anyone can, regardless of artistic ability. So learn how to do it and the next time your roommate mentions some obscure European writer, impressing you with his or her erudition and calling into question your sophistication, confound your tormenter with a perfectly drawn cyclohexane!

There are a few tricks that make the construction of a perfect drawing easy. First of all, within the ring, opposite bonds are parallel to each other (Fig. 5.13). With a little practice, keeping the proper bonds parallel should let you easily draw the carbon framework of cyclohexane.

It is a bit more difficult to get the hydrogens right. It helps *greatly* to tip the ring a little so that the "middle" bonds are not parallel to the top and bottom of the paper (Fig. 5.14). Now we see the molecule as it would rest on a flat surface. Six of the carbon–hydrogen bonds are easy to draw; three of them point straight up and three straight down. The "up" and "down" hydrogens alternate. The hydrogens in these six bonds are called **axial hydrogens**, and the bonds to them are called the axial carbon–hydrogen bonds.

FIGURE 5.13 In chair cyclohexane, there are three pairs of parallel carbon–carbon bonds.

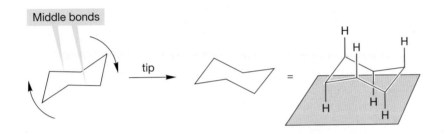

FIGURE 5.14 The set of six "straight up and down" or "axial" carbon–hydrogen bonds of chair cyclohexane are shown in red. All axial carbon–hydrogen bonds are parallel.

So far, so good—but it is the positioning of the last six hydrogens, the **equatorial hydrogens**, that gives people the most trouble. To get them right, take advantage of the parallel carbon–carbon bonds shown in Figure 5.13. Each member of this set of six equatorial carbon–hydrogen bonds is parallel to the two carbon–carbon bonds one bond away. Like the six axial carbon–hydrogen bonds, the equatorial bonds also alternate up and down, but this set points only slightly up or slightly down (Fig. 5.15).

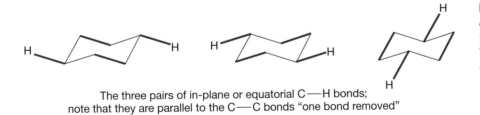

The three pairs of in-plane or equatorial C—H bonds; note that they are parallel to the C—C bonds "one bond removed"

FIGURE 5.15 The equatorial carbon–hydrogen bonds are parallel to the ring carbon–carbon bonds one bond away. The drawing of a perfect chair cyclohexane is now complete.

The six equatorial hydrogens of chair cyclohexane

All 12 C—H bonds of chair cyclohexane; axial hydrogens are shown in red, equatorial hydrogens in blue

Up hydrogens in green, down hydrogens in gray

WEB 3D

Summary

You have learned to draw a perfect cyclohexane ring in its energy-minimum chair form. This exercise in drawing lets us see one important thing immediately. Note that there are two *different* kinds of hydrogen. There is the "straight up and down" set of six, which are called the axial hydrogens, and the other set of six, which lie roughly along the equator of the molecule and are called the equatorial hydrogens (Fig. 5.16).

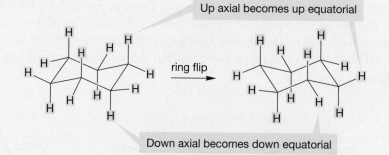

The six axial hydrogens   The six equatorial hydrogens
of chair cyclohexane   of chair cyclohexane

FIGURE 5.16 Axial and equatorial hydrogens.

WORKED PROBLEM 5.2 It is possible to convert one chair cyclohexane into another. Get out your models, grasp the two "end" carbons (at the 1 and 4 positions). Move the carbon that is "up" downward, and move the one that is "down" upward to generate another chair. You have just "flipped" the ring. We'll have much more to say about this process in Section 5.4. What happens to the set of axial hydrogens when you convert one chair into another? This problem is important and simple.

ANSWER The set of axial hydrogens becomes the set of equatorial hydrogens! Whether axial or equatorial, an *up* hydrogen always remains *up*, however. Similarly, a *down* hydrogen is always *down*, whether axial or equatorial.

Up axial becomes up equatorial

ring flip

Down axial becomes down equatorial

PROBLEM 5.3 How many carbons will the ^{13}C NMR spectrum of cyclohexane reveal?

PROBLEM 5.4 How many hydrogens will the ^{1}H NMR spectrum of cyclohexane reveal? *Caution:* This problem is much harder than Problem 5.3. Try to find two answers to this question, one correct at high temperature, the other correct at low temperature.

Larger rings than cyclohexane will also be nonplanar; but no more spectacular surprises such as cyclohexane's energy-minimum form remain to be found. Planar heptagons and octagons have internal angles of 129° and 135°, respectively, and thus have substantial angle strain. Both seven- and eight-membered rings would have severe torsional strain were they to remain planar. There is relaxation to nonplanar forms, but some amount of strain always remains in these medium-sized rings. It is a complicated business to analyze medium-sized rings completely, and we will not embark upon it. Rings with minimal angle strain will have severe interactions—often, but not always, eclipsings of carbon–hydrogen bonds, and thus substantial torsional strain. Relieving those eclipsings usually induces increased angle strain. As in cyclopentane, there is a compromise to be made between minimizing angle and torsional strain, and Nature does the best it can, finding the lowest energy structural compromise. Unlike cyclohexane, however, in the medium-sized rings there is no obvious low-energy solution. Often there are several somewhat different minima, close to one another in energy, and separated by quite low barriers.

Fragments of chair cyclohexanes are sometimes seen in the structures of medium-sized rings. Cyclodecane is an example, as Figure 5.17 shows. Several destabilizing interactions between carbon–hydrogen bonds should be apparent, especially if you make a model.

As rings get even bigger, the strain energy decreases. The limit is an infinitely large ring, which resembles an endless chain of methylene groups. In such a species, it is possible to stagger all carbon–hydrogen bonds, and, of course, in an infinite ring there is no angle strain. So after cyclohexane, the next strain-free species is reached when the ring size has grown large enough to approximate an infinitely large ring.

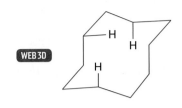

Cyclodecane

FIGURE 5.17 A medium-sized ring.

5.3 Quantitative Evaluation of Strain Energy

Strain is an important factor in chemical reactivity. The more strained a compound, the higher its energy. The higher in energy a compound, the more likely it is to react. We have just seen how a combination of torsional and angle strains can act to change the energy of a ring compound, cyclopropane. Now we move on to some quantitative measures of the energies of ring compounds. We introduce some general techniques and look ahead to our detailed consideration of energy in Chapter 8.

5.3a Heats of Formation To recapitulate material from Chapter 3 (p. 115), the heat of formation (ΔH_f°) of a compound is the enthalpy of its formation by the reaction of its constituent elements in their standard states. The standard state of an element is the most stable state at 25 °C and 1 atm pressure. For an element in its standard state, ΔH_f° is taken as zero. The more negative (or less positive) a compound's ΔH_f°, the more stable it is. A negative ΔH_f° for a compound means that its formation from its constituent elements would be *exothermic*—heat would be liberated. By contrast, a positive ΔH_f° means that the constituent elements are more stable than the compound and its formation would be *endothermic*—energy would have to be applied (Fig. 5.18).

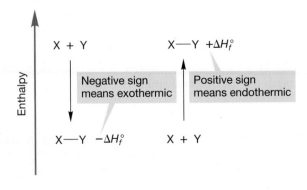

FIGURE 5.18 For a compound more stable than its constituent elements, ΔH_f° is negative. For a compound less stable than its constituent elements, ΔH_f° is positive.

Columns 1 and 2 of Table 5.1 collect the heats of formation for some cycloalkanes and calculate the ΔH_f° *per methylene group* for these hydrocarbons.

TABLE 5.1 Strain Energies for Some Cycloalkanes

Molecule	1 Measured ΔH_f° (kcal/mol)	2 ΔH_f° per CH_2 (kcal/mol)	3 Calcd ΔH_f° for Strain-free Molecule (kcal/mol)	4 Strain E (kcal/mol)	5 Strain E per CH_2 (kcal/mol)
Cyclopropane	+12.7	+4.2	−14.7	27.4	9.1
Cyclobutane	+6.8	+1.7	−19.6	26.4	6.6
Cyclopentane	−18.7	−3.7	−24.5	5.8	1.2
Cyclohexane	−29.5	−4.9	−29.4	0.1	0
Cycloheptane	−28.3	−4.0	−34.3	6.0	0.9
Cyclooctane	−29.7	−3.7	−39.2	9.5	1.2
Cyclodecane	−36.9	−3.7	−49.0	12.1	1.2
Cyclododecane	−55.0	−4.6	−58.8	3.8	0.3
$(CH_2)_\infty$		−4.9			

> **PROBLEM 5.5** How would one get the value of ΔH_f° for a CH_2 group in an infinite chain, $(CH_2)_\infty$? Use the following data to estimate the value of ΔH_f° of a strain-free methylene group in kilocalories per mole: ΔH_f° hexane = −39.9, heptane = −44.8, octane = −49.8, nonane = −54.5, decane = −59.6 kcal/mol.

Note from the final line of Table 5.1 that the ΔH_f° for a methylene (CH_2) group in a strain-free straight-chain alkane, $(CH_2)_\infty$, is −4.9 kcal/mol. This value is exactly the ΔH_f° we calculate for a methylene group in cyclohexane. Cyclohexane really is strain-free. We can now use these data to calculate strain energies for the cycloalkanes. First, calculate (Table 5.1, column 3) the ΔH_f° for the ring constructed from strain-free methylene groups. The difference between the calculated, strain-free ΔH_f° (column 3) and the real, measured ΔH_f° (column 1) is the strain energy (strain E, column 4). For example, cyclopropane has a measured ΔH_f° of +12.7 kcal/mol. A strain-free cyclopropane ring would have a ΔH_f° of 3 × −4.9 kcal/mol = −14.7 kcal/mol. The difference between these two values (27.4 kcal/mol) is the strain energy (column 4). The strain energy per CH_2 (column 5) is (27.4)/3 = 9.1 kcal/mol.

Now look at the other rings. Cyclopropane and cyclobutane are about equally strained, although the strain per methylene is significantly higher for cyclopropane. Cyclopentane and cycloheptane are slightly strained, and the strain gets worse in the medium-sized rings until we reach a 12-membered ring, in which strain decreases markedly. Strain continues to decline in the larger rings until (with some aberrations) we reach the hypothetical strain-free infinite ring.

> **PROBLEM 5.6** Use models to look for the sources of strain in medium-sized rings.

5.3b Strain Analyzed by Heats of Combustion

Strain energies can also be determined from an analysis of heats of combustion (ΔH_c°), the energy released—or consumed—when a compound reacts with oxygen. This method is attractive for analyzing strain because the measurements can be made and compared directly.

$$C_3H_8 + 5\,O_2 \longrightarrow 3\,CO_2 + 4\,H_2O + Energy$$

Combustion of a hydrocarbon is an exothermic reaction. The products, H_2O and CO_2, are more stable than the starting hydrocarbon and oxygen. This energy difference, the exothermicity of the reaction, appears as heat and light (Fig. 5.19), which is easily apprehended by looking at the light and feeling the heat evolved by a propane stove.

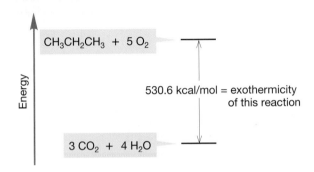

FIGURE 5.19 The combustion of propane to give CO_2, H_2O, and energy.

CONVENTION ALERT

The conventions are confusing here. When writing energy as the product of a reaction it is conventional to show it as "+ energy." The enthalpy of this exothermic reaction, ΔH, is negative, however! So, when writing an equation for the reaction, we say:

$$C_6H_{12} + 9\,O_2 \longrightarrow 6\,CO_2 + 6\,H_2O + 937 \text{ kcal/mol}$$

but, when indicating the enthalpy for the reaction, we say:

$$C_6H_{12} + 9\,O_2 \longrightarrow 6\,CO_2 + 6\,H_2O \qquad \Delta H = -937 \text{ kcal/mol}$$

Combustion of hydrocarbons can often be used to establish energy differences between molecules. Compare, for example, the heats of combustion for octane and an isomer, 2,2,3,3-tetramethylbutane (Fig. 5.20).[2]

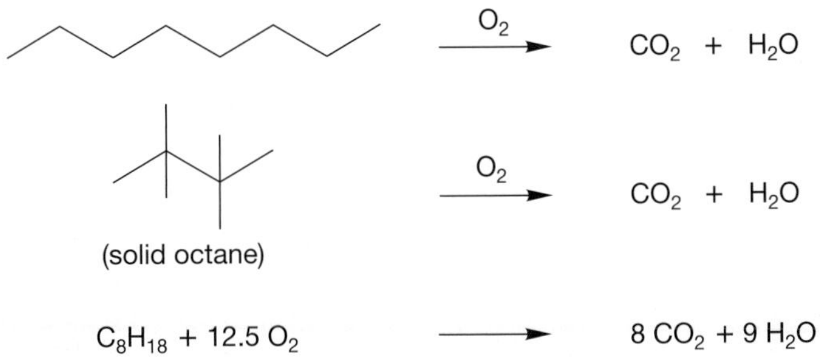

FIGURE 5.20 Combustion of two isomeric octanes.

---

[2] A reviewer of this book, RMM, who sat with me (MJ) in chemistry classes at Yale, many, many years ago, reminds me that we were once told that this molecule, the only isomer of octane with a melting point above room temperature (mp 104 °C), was known as "solid octane." We were then told that this was likely to be the *only* fact we retained from our organic chemistry course! Now you are stuck with this knowledge.

The measured heats of combustion of octane and 2,2,3,3-tetramethylbutane are 1307.60 kcal/mol and 1303.04 kcal/mol, respectively. Figure 5.21 graphically shows the relationship between these values and the energies of the molecules. Because these isomeric molecules produce exactly the same products on burning, the difference in their heats of combustion is the difference in energy between them. Of course, this analysis also tells you which isomer is more stable, and by how much.

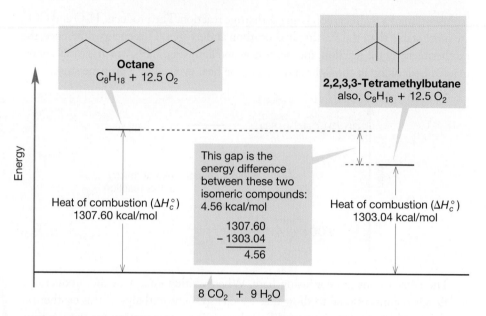

FIGURE 5.21 A quantitative picture of the combustion of two octanes.

PROBLEM 5.7 The heats of combustion of heptane, 3-methylhexane, and 3,3-dimethylpentane, respectively, are 1149.9, 1148.9, and 1147.9 kcal/mol. Carefully draw a diagram showing the relative stabilities of these molecules.

We can also use heats of combustion to measure the relative strain energies of the cycloalkanes. The higher the ΔH_c°, the less stable the compound (Fig. 5.22).

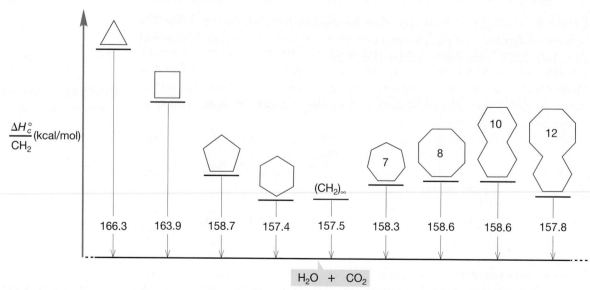

FIGURE 5.22 The heats of combustion per CH_2 for a series of cycloalkanes and a strain-free sequence of methylene (CH_2) groups. The higher the heat of combustion, the less stable the compound.

Once more we need a baseline, and we use the ΔH_c° of a strain-free methylene group in an infinite chain of methylenes, 157.5 kcal/mol. This value is determined in much the same way as was used to get the ΔH_f° for a strain-free methylene (p. 194). Then we measure the ΔH_c° values for the series of cycloalkanes, finding the ΔH_c° per methylene for each ring. Subtraction of 157.5 kcal/mol, the ΔH_c° of a strain-free methylene group, gives the strain energy per CH_2 and, therefore, the strain energy of the ring compound. As Table 5.2 shows, there is a good correspondence between the strain energies measured from ΔH_f° and ΔH_c°. Actually this correspondence is no surprise, because ΔH_f° values are often derived from ΔH_c°.

TABLE 5.2 Strain Energies for Some Cycloalkanes from ΔH_c° and ΔH_f°

Molecule	Measured ΔH_c° per CH_2 (kcal/mol)	Strain E per CH_2 (ΔH_c°−157.5) (kcal/mol)	Strain E from ΔH_c° (kcal/mol)	Strain E from ΔH_f° (kcal/mol)
Cyclopropane	166.3	8.8	26.4	27.4
Cyclobutane	163.9	6.4	25.6	26.4
Cyclopentane	158.7	1.2	6.0	5.8
Cyclohexane	157.4	−0.1	−0.6	0.1
Cycloheptane	158.3	0.8	5.6	6.0
Cyclooctane	158.6	1.1	8.8	9.5
Cyclodecane	158.6	1.1	11.0	12.1
Cyclododecane	157.8	0.3	3.6	3.8
$(CH_2)_\infty$	157.5			

5.4 Stereochemistry of Cyclohexane: Conformational Analysis

Earlier, when we were considering the ways in which ring compounds distort so as to minimize strain, we saw that cyclohexane adopted a strain-free chair conformation. Now it is time to look at cyclohexane in detail. Six-membered rings are most common in organic chemistry, and a great deal of effort has been made over the years at understanding their structure and reactivity. One chair form of cyclohexane can easily be converted into another. We are now going to look in detail at this transformation and estimate the energies of the various intermediates and transition states along the path from one chair to another. This process is an example of conformational analysis.

A little manipulation of a model will show the overall conversion, but it's not easy to come to the actual pathway for the process. If we move carbons 1, 2, 3, and 4 in Figure 5.23 into one plane, with carbon 6 above the plane and carbon 5 below the plane, we come to a "half-chair" structure.

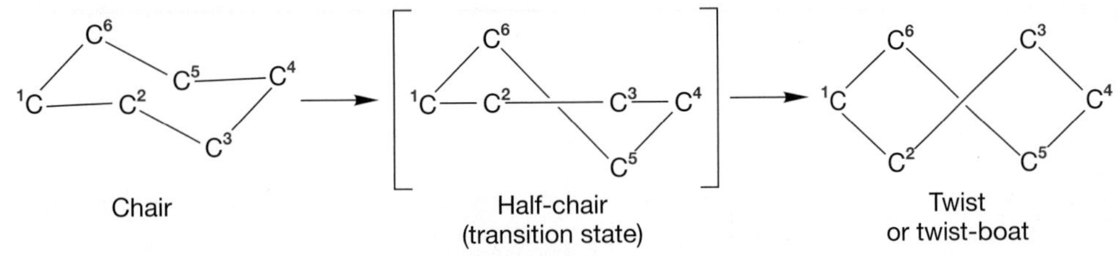

FIGURE 5.23 Conversion of the energy-minimum chair cyclohexane into the half-chair and then the twist form.

Chair   Half-chair (transition state)   Twist or twist-boat

The half-chair conformation does not represent a stable molecule, but is instead a picture of the top of the energy "mountain pass" (a transition state) leading to a lower energy molecular "valley" called the twist conformation or, sometimes the twist-boat conformation (Fig. 5.23). The half-chair contains many eclipsed carbon–hydrogen bonds, which become staggered somewhat in the twist conformation, and angle strain is partially relieved as well. The twist conformation can pass through a second half-chair to give another chair cyclohexane. Kinetic measurements allow an evaluation of the energies involved (Fig. 5.24). The half-chair and the twist lie 10.8 and 5.5 kcal/mol above the chair, respectively.

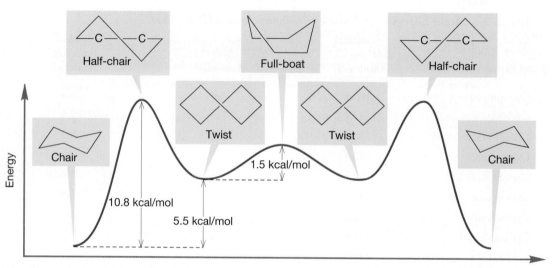

FIGURE 5.24 The interconversion of two chair cyclohexanes. The two chairs and the two twist forms are intermediates (energy minima), and the two half-chairs and the full-boat are transition states (energy maxima).

Do not confuse the twist with the full-boat shown at the top of a "mountain" in Figure 5.24. The full-boat is not an energy minimum but is, like the half-chair, an energy maximum. It lies at the top of the energy barrier (like the half-chair, it, too, is a transition state) separating two twist forms. The full-boat is only 1.5 kcal/mol higher than the twist (7.0 kcal/mol higher than the chair), and can never be isolated because it is not an energy minimum.

Both the half-chair and the full-boat suffer from the kinds of strain we have seen before. In the half-chair, much of the ring is planar, and there is both angle and torsional strain. The full-boat also has angle and torsional strain, but there is another hydrogen–hydrogen interaction that is destabilizing. This new interaction is between the "prow" and "stern" carbons and between the two "inside" hydrogens at the prow and stern of the boat. This new kind of strain is induced when two atoms come too close to each other and is called **van der Waals strain** (Fig. 5.25).

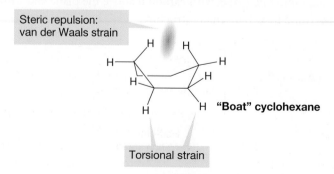

FIGURE 5.25 van der Waals strain in the full-boat form of cyclohexane. Don't confuse this strain with attractive van der Waals forces (p. 87).

WORKED PROBLEM 5.8 Draw a Newman projection looking down the "side" $H_2C—CH_2$ bond of the full-boat cyclohexane.

ANSWER This task is relatively easy. Once again, set your eye slightly to one side so as to see the carbon–hydrogen bonds on the rear carbon.

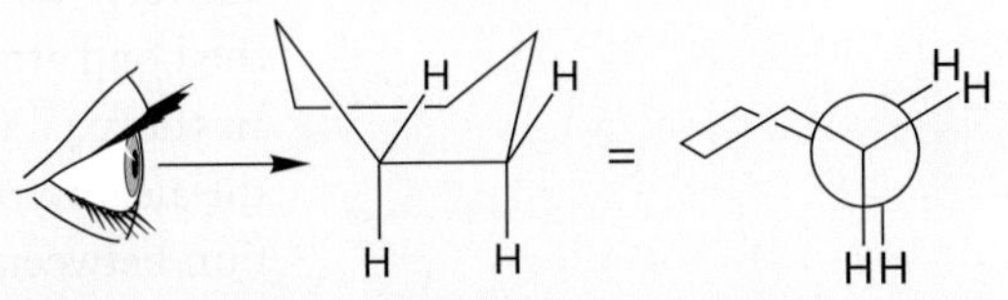

One chair cyclohexane will equilibrate with another as long as the environment supplies the requisite 10.8 kcal/mol to traverse the energy barrier. Under normal conditions this process is easy, but at very low temperature, we can freeze out one chair form by lowering the temperature to the point at which there isn't enough energy available to cross the energy barrier to the other chair.

How much of the relatively high-energy twist form is there at 25 °C? We will see many such calculations in Chapter 8, but the chair–twist equilibrium can be treated like any equilibrium process. A calculation shows that a relatively small energy difference of $\Delta G = 5.5$ kcal/mol results in an enormous preference for the more stable isomer of the pair, about 10^4:1. The "take-home lesson" here is that small energy differences between two molecules in equilibrium result in a very large excess of the more stable isomer.

5.5 Monosubstituted Cyclohexanes

The danger of relying too strongly on two-dimensional representations of molecules is shown nicely by methylcyclohexane. Two-dimensional structures give no hint of the richness of the complicated, three-dimensional structure of this molecule; they even hide the presence of two conformational isomers of methylcyclohexane (Fig. 5.26).

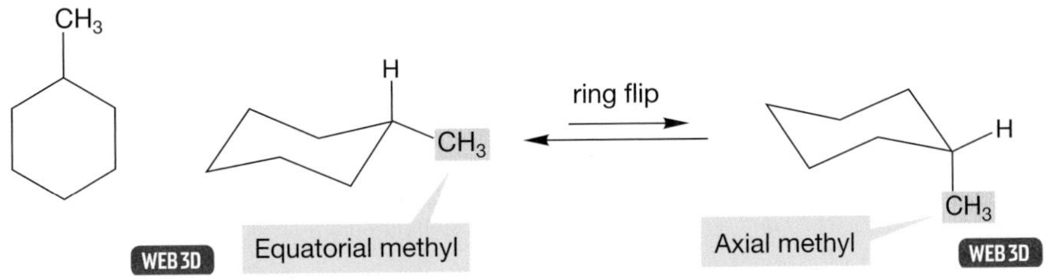

Equatorial methyl   Axial methyl

FIGURE 5.26 The uninformative two-dimensional structure hides the existence of two conformational isomers of methylcyclohexane. In one isomer, the methyl group is equatorial, and in the other it is axial.

Recall (Problems 5.2–5.4, p. 192) that flipping the cyclohexane chair forms interconverts the set of six axial hydrogens and the set of six equatorial hydrogens (Fig. 5.27).

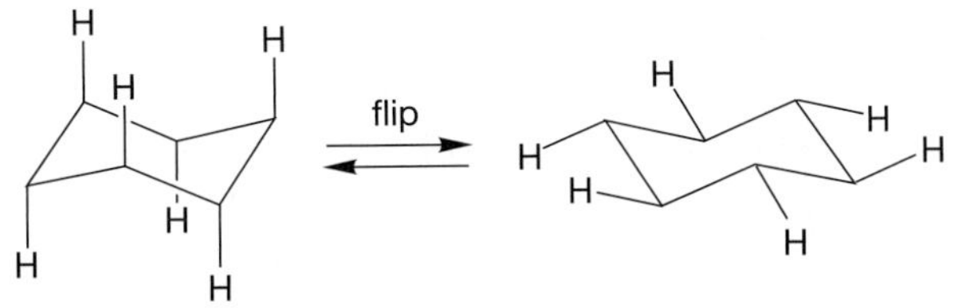

FIGURE 5.27 In cyclohexane, a ring flip interconverts axial and equatorial hydrogens.

Therefore, a methyl group, or any substituent, can be in either an axial or equatorial position. These two molecules are diastereomers. (*Remember*: Diastereomers are stereoisomers that are not mirror images—see Chapter 4, p. 165.)

Given that two isomers of methylcyclohexane exist, our next job is to determine which is more stable. As it turns out, we can even make a reasonable guess at the magnitude of the difference in energy between the two diastereomers. First of all, notice that axial and equatorial methylcyclohexane are interconverted by a chair flip (Fig. 5.26). What turns out to be the crucial factor in creating the energy difference between the two forms is shown by the Newman projections made by sighting down the bond attaching a ring methylene group to the carbon bearing the methyl group. In the axial form, there are two gauche interactions between the methyl group and the nearby ring CH_2 group (Fig. 5.28). Note the destabilizing 1,3-diaxial interaction between the axial hydrogen shown in the figure and the axial methyl group.

Although it is easy to see one of the gauche interactions, the perspective of the drawing hides the other unless you are careful. No amount of words can help you here as much as working out this stereochemical problem yourself. It is very important to try Problem 5.9. The answer follows immediately.

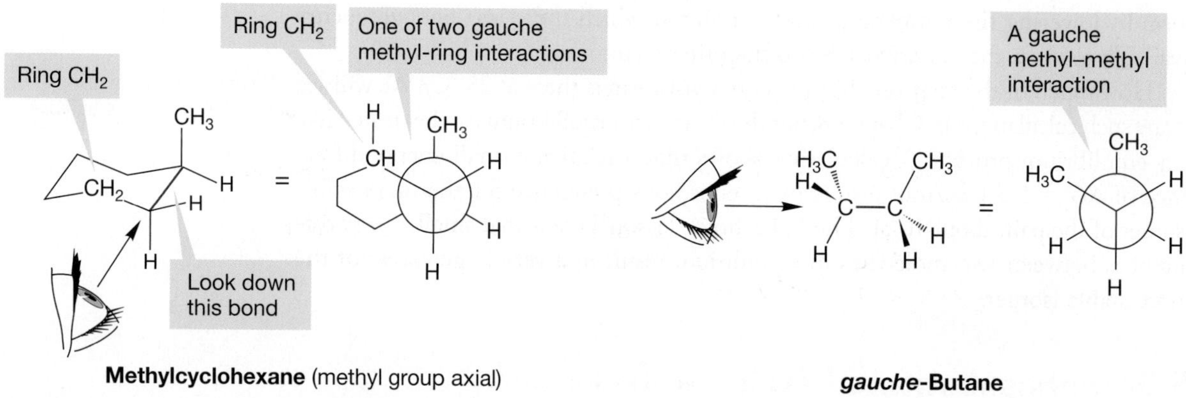

FIGURE 5.28 Methylcyclohexane with the methyl group axial. The Newman projection shows the gauche interaction with one of the two equivalent ring methylene groups. Each of the two gauche interactions in this compound resembles a gauche interaction in *gauche*-butane.

WORKED PROBLEM 5.9 Draw the Newman projection of methylcyclohexane looking from the other adjacent methylene group toward the methyl-bearing carbon. Be sure that you see the second gauche methyl–ring interaction. Next, compare this gauche interaction to that in *gauche*-butane. Are the two exactly the same?

ANSWER It is difficult to place your "eye" correctly for this view, but if you look along the red carbon–carbon bond, you can construct the appropriate Newman projection.

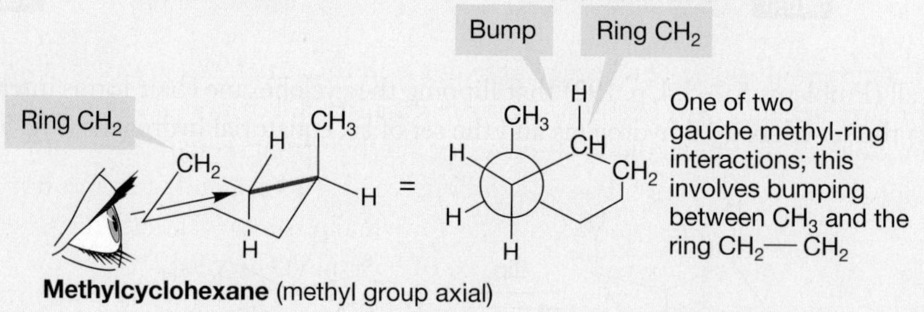

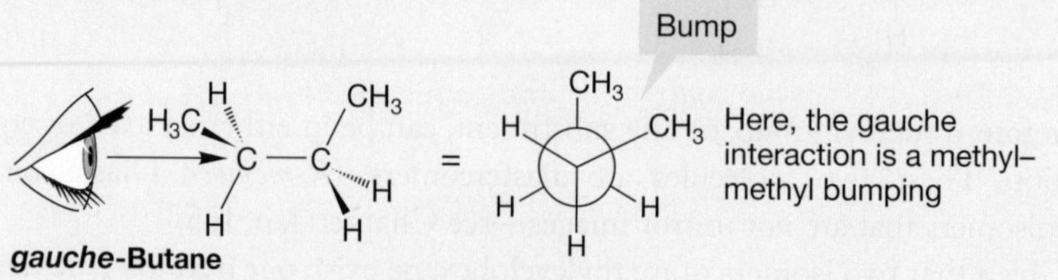

The two interactions are very similar, but not exactly the same. In butane two methyl groups bump, whereas in axial methylcyclohexane a methyl group bumps with a CH_2—CH_2.

A Newman projection of the conformation of methylcyclohexane with the equatorial methyl group made from the same perspective reveals none of these gauche interactions—this arrangement resembles the anti form of butane (Fig. 5.29).

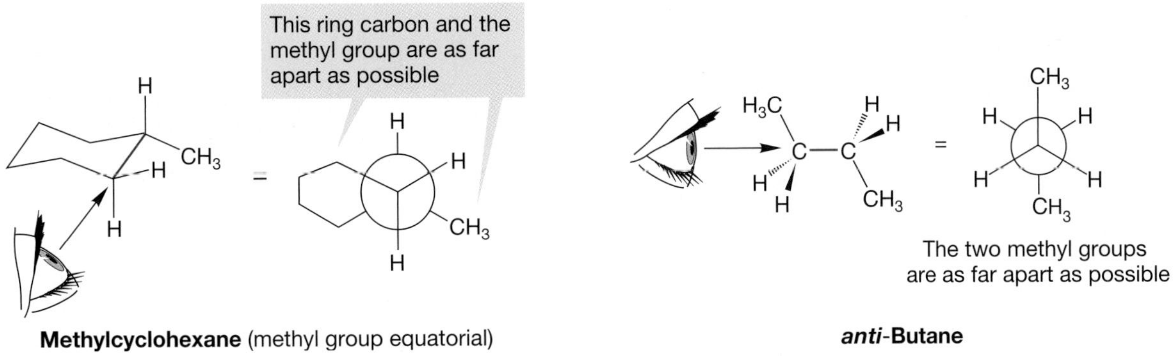

FIGURE 5.29 Methylcyclohexane with an equatorial methyl group. There are no gauche methyl–ring interactions. This isomer is 1.74 kcal/mol more stable than the molecule with the methyl group axial. Note the resemblance to the anti form of butane.

Recall our conformational analysis of butane in Chapter 2 (p. 73). The gauche form of butane was about 0.6 kcal/mol less stable than the anti form, in which the terminal methyl groups of butane were as far apart as possible. Accordingly, we might guess that the isomer with the methyl group equatorial would be more stable than the axial isomer, and that the energy difference between the two would be approximately twice the difference between *gauche-* and *anti*-butane, $2 \times 0.6 = 1.2$ kcal/mol. We'd be close. The real difference is 1.74 kcal/mol.

PROBLEM 5.10 Why isn't the difference exactly 1.2 kcal/mol? Point out some of the differences between the gauche interaction in butane and the gauche interaction in axial methylcyclohexane.

Calculations of the percentage of isomers in the equatorial and axial forms of methylcyclohexane will become clearer in Chapter 8. For now it's fair to say that about 95% of the molecules will have the methyl group equatorial at any given instant. Once more, a small energy difference (1.74 kcal/mol) has resulted in a large excess of the more stable isomer. Don't make the mistake of thinking that methylcyclohexane consists of a mixture of 95% molecules locked forever in the conformation with the methyl group equatorial, and 5% of similarly frozen molecules with their methyl groups axial. These molecules are in rapid equilibrium. At any given moment 95% will have their methyl groups equatorial, but all molecules are continuously equilibrating between equatorial and axial methyl isomers.

PROBLEM 5.11 The formula for doing equilibrium calculations is $\Delta G = -RT \ln K$, where R is about 2 cal/deg·mol, T is the absolute temperature, and K the equilibrium constant sought. Calculate the amounts at equilibrium at 25 °C of two compounds differing in energy by $\Delta G = 2.8$ kcal/mol.

A group larger than methyl would result in an even greater dominance of the equatorial conformation over the less stable axial conformation. The 1,3-diaxial bumpings responsible for the destabilizing gauche interactions will be magnified, and reflected in the equilibrium constant (Fig. 5.30; Table 5.3).

FIGURE 5.30 The larger R, the more favored will be the partner in the equilibrium with the R group equatorial.

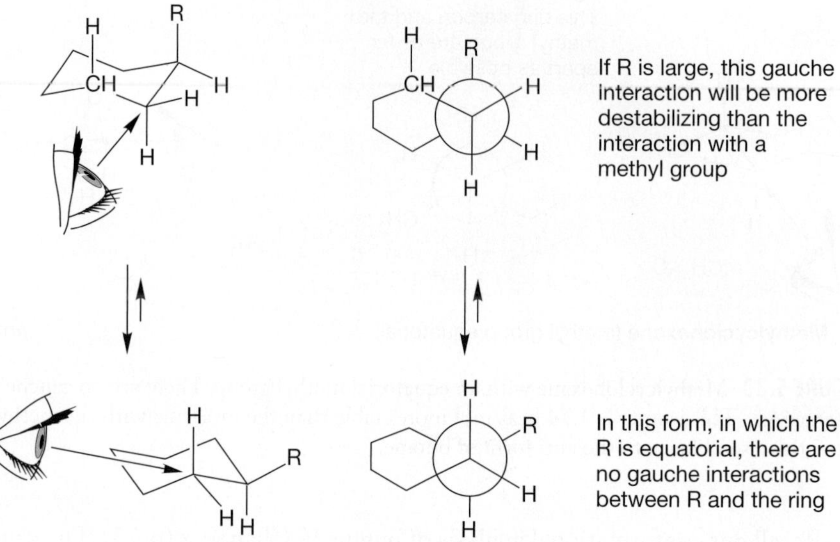

If R is large, this gauche interaction will be more destabilizing than the interaction with a methyl group

In this form, in which the R is equatorial, there are no gauche interactions between R and the ring

TABLE 5.3 Axial–Equatorial Energy Differences for Some Alkylcyclohexanes at 25 °C

Compound	ΔG (ax/eq) (kcal/mol)	K
Methylcyclohexane	1.74	19.5
Ethylcyclohexane	1.79	21.2
Propylcyclohexane	2.21	43.4
Isopropylcyclohexane	2.61	86.0
tert-Butylcyclohexane	5.5	11,916

Note the large axial–equatorial (ax/eq) energy difference for the *tert*-butyl group. This approximately 5–5.5 kcal/mol translates into an equilibrium constant of almost 12,000 at 25 °C. The situation is complicated for the *tert*-butyl group, because the conformation with an axial *tert*-butyl group is so greatly destabilized that a twist conformation, with an energy about 5 kcal/mol higher than the conformation with an equatorial *tert*-butyl group, may be lower in energy than the pure chair with an axial *tert*-butyl group (Fig. 5.31). Regardless, the conformation of *tert*-butylcyclohexane with the *tert*-butyl group equatorial is favored enormously.

FIGURE 5.31 The large *tert*-butyl group distorts the equilibrium far toward the much more stable equatorial form. The geometrical relationships in the molecule can be estimated with confidence because this form dominates the equilibrum mixture so strongly.

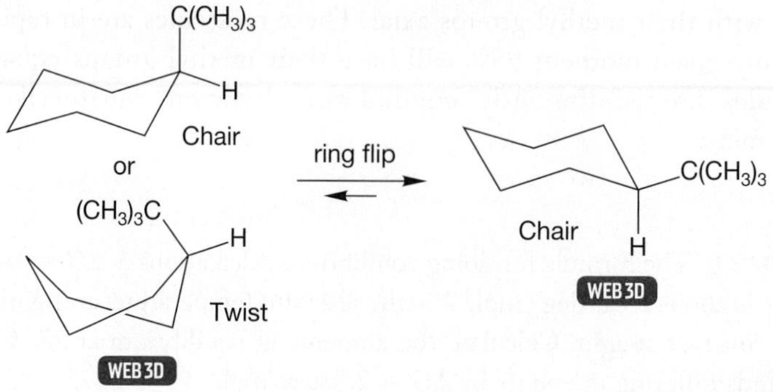

WORKED PROBLEM 5.12 It is often said that the *tert*-butyl group locks *tert*-butylcyclohexane into the form with the *tert*-butyl group equatorial. Is this a good way to put it? Why not?

ANSWER No! The molecule is not locked at all. The forms with the large *tert*-butyl group axial and equatorial are in rapid equilibrium, with the equatorial form greatly predominating. Very few molecules are in the axial form at any one time, but the equilibrium is mobile. This molecule is not locked into one form and immobilized forever.

PROBLEM 5.13 How many carbons will the ^{13}C NMR spectrum of *tert*-butylcyclohexane reveal at 25 °C?

PROBLEM 5.14 How many carbons will the ^{13}C NMR spectra of *trans*-1,4-di-*tert*-butylcyclohexane and all-*cis*-1,3,5-tri-*tert*-butylcyclohexane reveal?

PROBLEM SOLVING

GO

In many problems there are clues to the answer that are exceptionally important to recognize. Here is a perfect example. Whenever you see a *tert*-butyl group attached to a cyclohexane, you can be certain that it is there in order to fix the cyclohexane in one of the possible chairs. *tert*-Butyl groups are *always* equatorial! That observation allows you to fix the positions of any other groups of the ring with certainty. If the problem involves a *tert*-butyl-substituted cyclohexane, you can be sure that the solution will require you to draw the ring in three dimensions, and that you must put the *tert*-butyl group in an equatorial position.

In studies of reaction mechanisms it is often very helpful to know with some precision the spatial relationships among groups. Ring compounds are quite helpful in this regard. For example, the rigid frame of cyclopropane allows quite accurate estimations of angles and distances to be made. But cyclopropanes are so strained as to be sometimes too reactive for mechanistic work. In fact, the ring is often opened in chemical reactions. One wouldn't want one's rigid framework disappearing as the reaction occurred, but this is all too likely for cyclopropanes. Other rings are flexible and don't allow us the firm predictions of angles and distances we want.

However, a large group such as *tert*-butyl enables us to prejudice the mobile equilibrium between cyclohexane rings so strongly in favor of the form with the *tert*-butyl group equatorial that we can determine with confidence the positions of other atoms in the molecule (Fig. 5.31). We will see later that this technique is used frequently.

Summary

All cycloalkanes except cyclopropane distort from planarity so as to minimize strain. The amount of strain can be measured in a variety of ways, including measurements of heats of combustion or heats of formation. Cyclohexane adopts an energy-minimum chair form in which there are six hydrogens in the axial position and six in the equatorial position. Cyclohexane is a mobile system, as chair forms interconvert. This interconversion exchanges substituents in the axial and equatorial positions.

5.6 Disubstituted Ring Compounds

We have already mentioned that cis and trans diastereomers exist for dichlorocyclopropanes (Chapter 4, pp. 173–176) and that the trans form is chiral (Fig. 5.32).

FIGURE 5.32 *cis*- and *trans*-1,2-Dichlorocyclopropane.

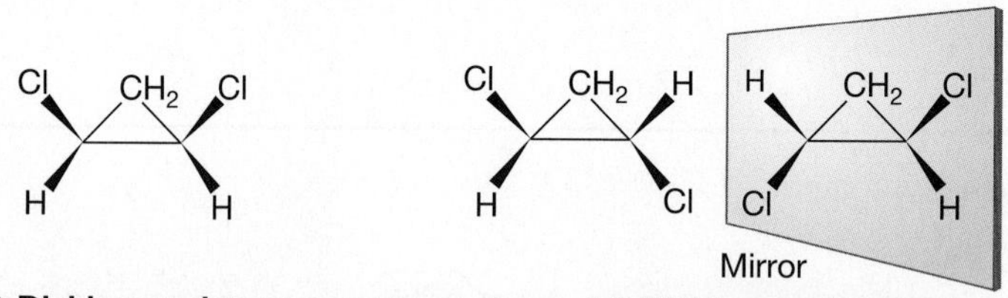

cis-1,2-Dichlorocyclopropane
(an achiral molecule, a meso compound)

trans-1,2-Dichlorocyclopropane
(chiral—there is a pair of enantiomers)

When the two substituents on the ring are different, as in 1-bromo-2-chlorocyclopropane, two pairs of enantiomers are possible, and both the cis and trans forms of this molecule are chiral (Fig. 5.33).

FIGURE 5.33 *cis*- and *trans*-1-Bromo-2-chlorocyclopropane.

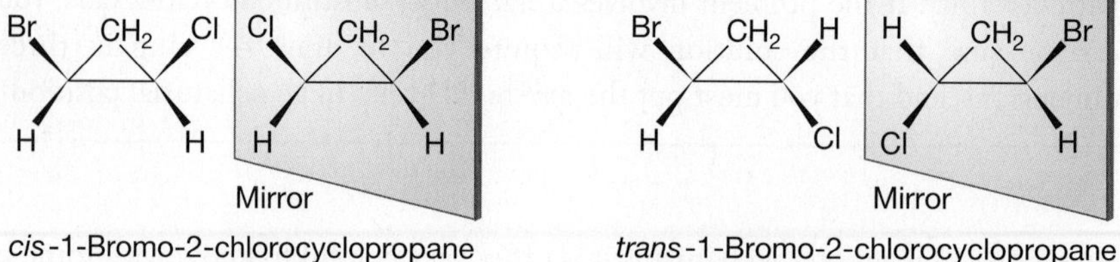

cis-1-Bromo-2-chlorocyclopropane
is a chiral molecule—there is
a pair of enantiomers

trans-1-Bromo-2-chlorocyclopropane
is also chiral, and, of course, there is
a pair of enantiomers

Stereochemical relationships become harder to see in cyclohexanes, where the molecules are not planar, but they don't really change. We will first take a look at 1,1-dialkylcyclohexanes and then at the somewhat more complicated 1,2-dialkylcyclohexanes.

5.6a 1,1-Disubstituted Cyclohexanes 1,1-Dimethylcyclohexane is a simple, achiral molecule in which the axial–equatorial equilibration induced by flipping one chair to the other interconverts equivalent molecules, each of which has one methyl group axial and one equatorial. The ring flip converts the axial methyl group into an equatorial methyl group and the equatorial methyl group into an axial methyl group. There is no net change and, as for cyclohexane itself, the equilibrium constant for the equilibration must be 1 (Fig. 5.34).

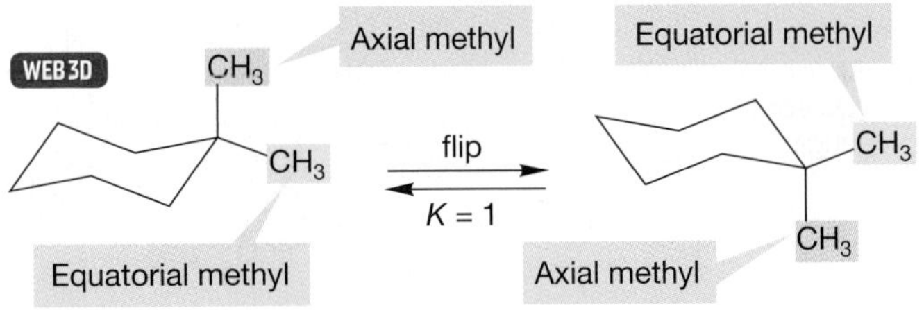

FIGURE 5.34 1,1-Dimethylcyclohexane.

1-Isopropyl-1-methylcyclohexane is only slightly more complicated. There are two isomers, nearly equal in energy, neither of which is chiral (Fig. 5.35).

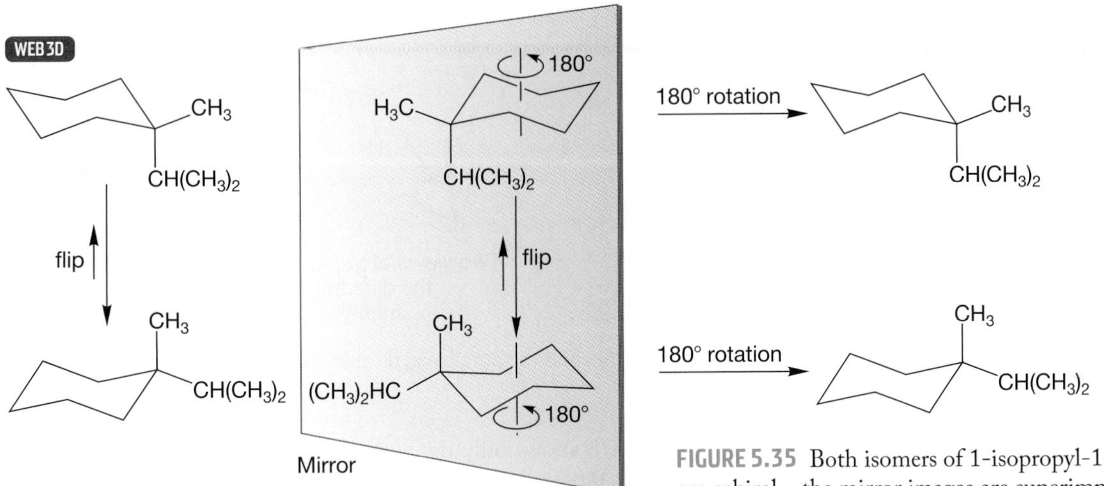

FIGURE 5.35 Both isomers of 1-isopropyl-1-methylcyclohexane are achiral—the mirror images are superimposable.

PROBLEM 5.15 Use the data in Table 5.3 to calculate the energy difference between the possible isomers of 1-isopropyl-1-methylcyclohexane.

5.6b 1,2-Disubstituted Cyclohexanes Just like 1,2-dimethylcyclopropane, 1,2-dimethylcyclohexane can exist in cis and trans forms (as can any 1,2-disubstituted ring compound), but the chair form of cyclohexane makes the compounds look somewhat different from the more familiar rigid isomers of cis and trans disubstituted cyclopropanes (Fig. 5.36).

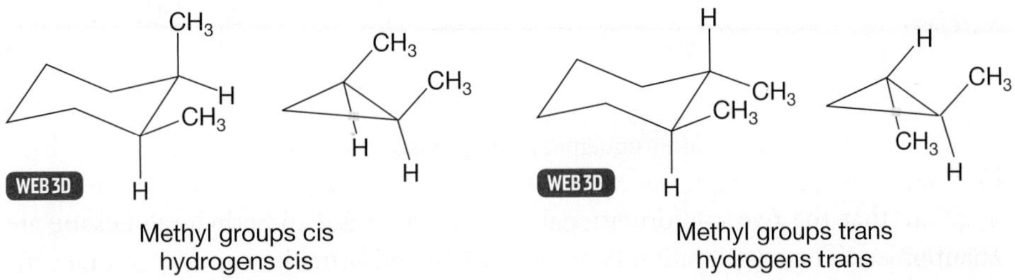

Methyl groups cis
hydrogens cis

Methyl groups trans
hydrogens trans

FIGURE 5.36 Three-dimensional representations of *cis*- and *trans*-1,2-dimethylcyclohexane and *cis*- and *trans*-1,2-dimethylcyclopropane.

The words cis and trans refer to the "sidedness" of a molecule and do not depend strictly on the angles between the groups. In a cis disubstituted cyclopropane, the dihedral angle between the cis groups is 0°, whereas it is 60° in the disubstituted cyclohexane (Fig. 5.37a). The methyl groups are referred to as cis in either case. In *trans*-1,2-dimethylcyclopropane, the dihedral angle between the methyl groups is 120°, whereas in the six-membered ring it is 60° (Fig. 5.37b).

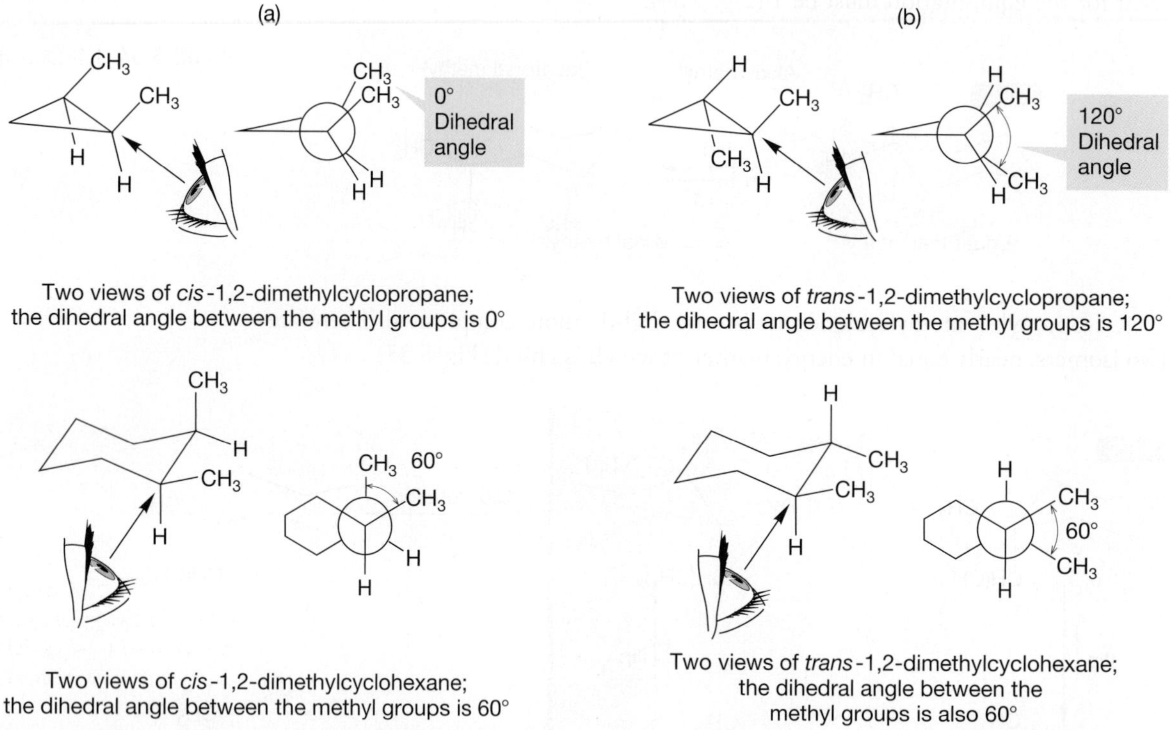

(a)   (b)

Two views of *cis*-1,2-dimethylcyclopropane; the dihedral angle between the methyl groups is 0°

Two views of *trans*-1,2-dimethylcyclopropane; the dihedral angle between the methyl groups is 120°

Two views of *cis*-1,2-dimethylcyclohexane; the dihedral angle between the methyl groups is 60°

Two views of *trans*-1,2-dimethylcyclohexane; the dihedral angle between the methyl groups is also 60°

FIGURE 5.37 Isomerism in the planar, inflexible cyclopropanes and the nonplanar, flexible cyclohexanes: (a) cis, (b) trans.

PROBLEM 5.16 Figure 5.37b shows only the equatorial–equatorial form of *trans*-1,2-dimethylcyclohexane. What is the dihedral angle between the methyl groups in the much less stable axial–axial *trans*-1,2-dimethylcyclopropane isomer?

In *cis*-1,2-dimethylcyclohexane, one methyl group is axial and the other is equatorial (Fig. 5.38). The conformational ring flip does not alter the structure—the axial methyl becomes equatorial, and the equatorial methyl becomes axial. The equilibrium constant between these two equivalent compounds must be 1.

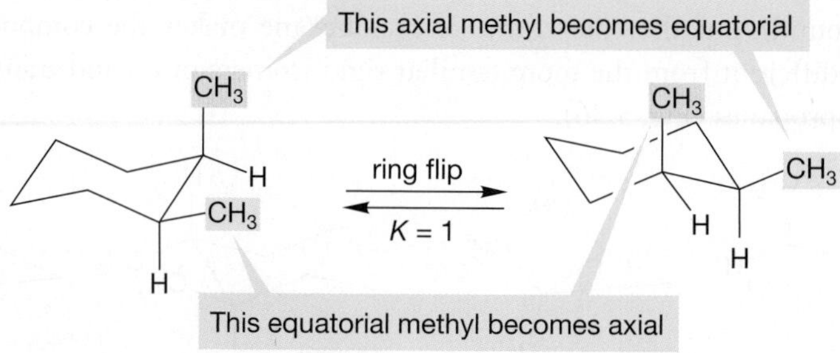

This axial methyl becomes equatorial

ring flip
$K = 1$

This equatorial methyl becomes axial

FIGURE 5.38 *cis*-1,2-Dimethylcyclohexane.

Note that the two conformational isomers of *cis*-1,2-dimethylcyclohexane are enantiomers. The equilibration between the two conformations produces a racemic mixture of the two enantiomers. Have we seen a similar situation before? Indeed we have; just recall our discussion of the gauche forms of butane (Chapter 4, p. 163). The equilibration of the two *cis*-1,2-dimethylcyclohexanes is very similar to the equilibration of the two gauche forms of butane (Fig. 5.39).

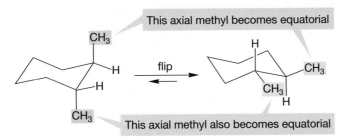

FIGURE 5.39 The ring flip in *cis*-1,2-dimethylcyclohexane converts one enantiomer into the other. A very similar process interconverts the two enantiomers of *gauche*-butane.

The trans form of 1,2-dimethylcyclopropane presents a different picture. *trans*-1,2-Dimethylcyclohexane can either have both methyl groups axial or both equatorial (Fig. 5.40). The ring flip converts the diaxial form into the diequatorial form. Let's apply conformational analysis to predict which of these two diastereomers is more stable.

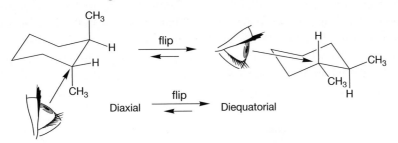

FIGURE 5.40 The ring flip in *trans*-1,2-dimethylcyclohexane converts a molecule with two axial methyl groups into one with two equatorial methyl groups.

As usual, Newman projections are vital, and in Figure 5.41 we sight along the carbon–carbon bond attaching the two methyl-substituted carbons. In the diequatorial isomer, there is one gauche methyl–methyl interaction costing 0.6 kcal/mol.

FIGURE 5.41 Diequatorial *trans*-1,2-dimethylcyclohexane is more stable than the diaxial isomer.

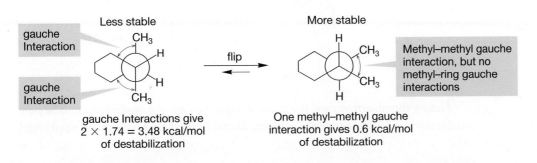

In the diaxial isomer, the two methyl groups occupy anti positions, and therefore their interaction with one another is not destabilizing. However, for each axial methyl group we still have the two gauche interactions with the ring methylene groups (see p. 199), and this will cost the molecule 2×1.74 kcal/mol = 3.48 kcal/mol. We can predict that the diequatorial isomer should be $(3.48 - 0.6)$ kcal/mol = 2.88 kcal/mol more stable than the diaxial isomer. In turn, this calculation predicts that at 25 °C there will be more than 99% of the diequatorial form present.

> **PROBLEM 5.17** Show that the ring flip of the diequatorial form to the diaxial form will not racemize optically active *trans*-1,2-dimethylcyclohexane. What is the relation between the equatorial–equatorial and axial–axial forms of *trans*-1,2-dimethylcyclohexane?

In principle, both the diaxial and diequatorial forms should be separable into a pair of enantiomers, resolvable (p. 169)—they are both chiral molecules. In practice, we cannot isolate *trans*-diaxial-1,2-dimethylcyclohexane; it simply flips to the much more stable diequatorial form. The diequatorial form can be isolated and resolved.

Summary

Neither *cis*-1,2-dimethylcyclopropane nor *cis*-1,2-dimethylcyclohexane can be resolved, because the cyclopropane is a meso compound (Chapter 4, p. 168), and the cyclohexane flips into its mirror image. However, both *trans*-1,2-dimethyl-cyclopropane and *trans*-1,2-dimethylcyclohexane *can* be resolved; they are not superimposable on their mirror images. The point is that in a practical sense the planar cyclopropanes and nonplanar cyclohexanes behave in the same way. This finding has important consequences. In deciding questions of stereochemistry, we can treat the decidedly nonplanar 1,2-dimethylcyclohexanes *as if* they were planar. Indeed, all cyclohexanes can be treated as planar for the purposes of stereo-chemical analysis, because the planar forms represent the average positions of ring atoms in the rapid chair–chair interconversions.

Now let's look at some slightly more complicated molecules in which the two substituents on the ring are different. In 1-isopropyl-2-methylcyclohexanes cis and trans isomers exist, and once more we will have to worry about the effects of flip-ping from one chair form to the other. In the cis compound, there are two differ-ent conformational isomers (Fig. 5.42). In one isomer, the isopropyl group is axial and the methyl group is equatorial. When the ring flips, the axial isopropyl group becomes equatorial and the equatorial methyl group becomes axial.

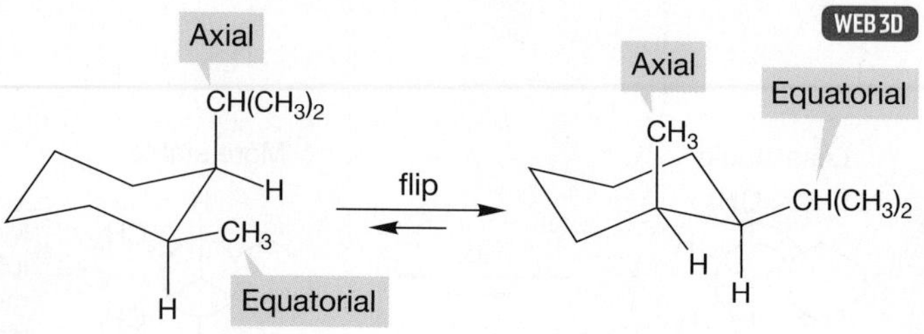

FIGURE 5.42 *cis*-1-Isopropyl-2-methycyclohexane flips from the conformer with an axial isopropyl group and an equatorial methyl, to the conformer with an equatorial isopropyl group and an axial methyl.

Unlike the dimethyl case, these two compounds are not enantiomeric. They are conformational diastereomers—conformational isomers that have different physical and chemical properties. In principle, each could be isolated and resolved. So there is a total of four possible cis isomers, the two shown in Figure 5.42 and their mirror images (Fig. 5.43).

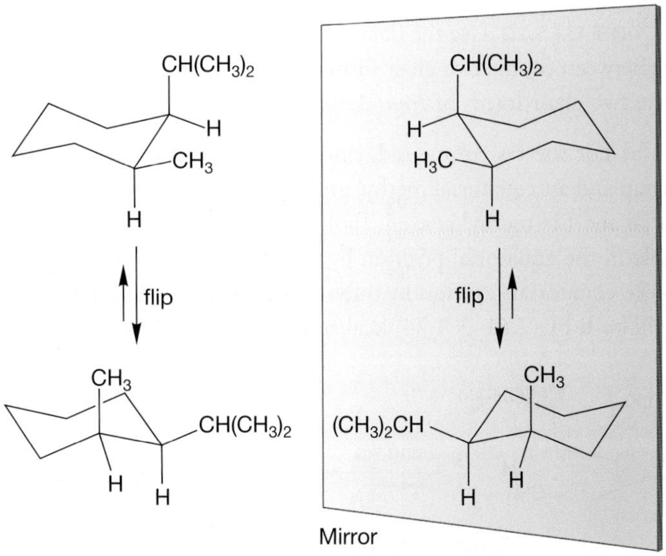

FIGURE 5.43 The four stereoisomers of *cis*-1-isopropyl-2-methyl-cyclohexane. Two pairs of enantiomers exist.

The rigid molecule *cis*-1-isopropyl-2-methylcyclopropane cannot undergo any ring flip, and there is only a single pair of enantiomers. The lack of a possible ring flip reduces the total number of possible stereoisomers (Fig. 5.44).

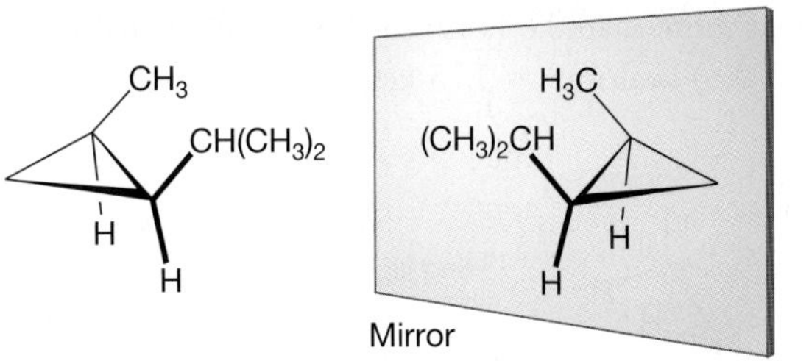

FIGURE 5.44 The two enantiomers of *cis*-1-isopropyl-2-methylcyclopropane. The rigid ring prevents any ring flip.

PROBLEM 5.18 Label the stereogenic carbons of Figures 5.43 and 5.44 as (*R*) or (*S*).

The situation is similar in the case of the trans form. Although the inflexible cyclopropane ring has but a single pair of enantiomers (Fig. 5.45), there are two pairs of enantiomers in the flexible cyclohexane (Fig. 5.46).

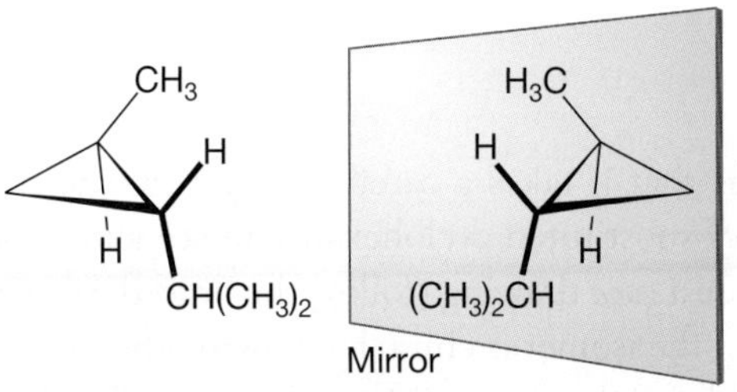

FIGURE 5.45 The single pair of enantiomers of *trans*-1-isopropyl-2-methylcyclopropane. Again, the rigid cyclopropane ring prevents any ring flip.

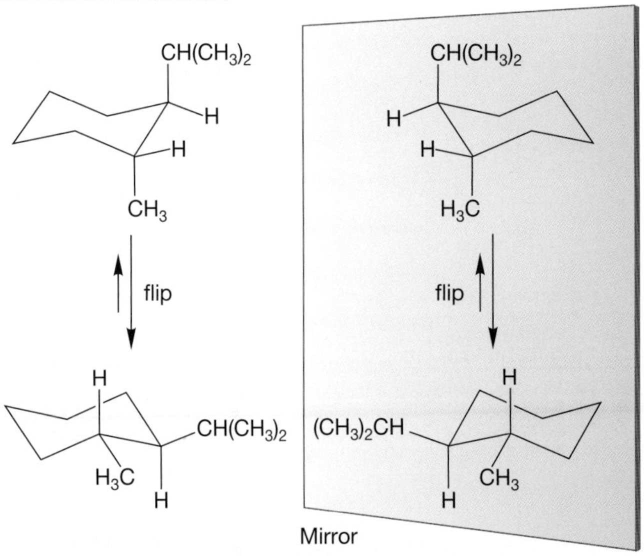

FIGURE 5.46 The four stereoisomers of *trans*-1-isopropyl-2-methylcyclohexane. Two pairs of enantiomers exist.

PROBLEM 5.19 Point out the pairs of enantiomers and diastereomers in Figure 5.46. Assign all the stereogenic carbons as (*R*) or (*S*).

WORKED PROBLEM 5.20 Use the data of Table 5.3 (p. 202) to estimate the energy difference between (a) the two chair forms of *cis*-1-isopropyl-2-methylcyclohexane and (b) the two chair forms of *trans*-1-isopropyl-2-methylcyclopropane.

ANSWER (a) For the cis compound, ring flipping converts **A** with an axial isopropyl group and an equatorial methyl group into **B** in which the isopropyl group is equatorial and the methyl is axial. Table 5.3 tells you that an isopropyl group is more stable in the equatorial position by 2.61 kcal/mol. A methyl group is more stable in the equatorial position by only 1.74 kcal/mol. The more stable conformation will be **B** by (2.61 − 1.74) kcal/mol = 0.87 kcal/mol.

(b) The trans compound has both groups axial (**C**) or both groups equatorial (**D**). Conformation **D** will be preferred by (2.61 + 1.74) kcal/mol = 4.35 kcal/mol. However, **D** suffers a methyl–isopropyl gauche interaction that will be destabilizing by somewhat more than 0.6 kcal/mol. So, our final answer would be approximately (4.35 − 0.6) kcal/mol = 3.75 kcal/mol.

PROBLEM 5.21 Estimate the energy difference between the two conformational isomers of *cis*-1,3-dimethylcyclohexane.

PROBLEM 5.22 Does ring flip racemize *trans*-1,3-dimethylcyclohexane, as it does *cis*-1,2-dimethylcyclohexane (Fig. 5.39, p. 207)?

Summary

The "take-home lesson" here is that it takes a careful analysis of the stereoisomers formed by ring flipping of substituted cyclohexanes to see all the possibilities. First, be sure you have made a *good* drawing (follow the procedure on p. 190). Next, determine whether or not the isomer is chiral by drawing the mirror image and seeing if it is superimposable on the original. Now do a ring flip of the original isomer. Is the ring-flipped isomer the same as the original? Is it the mirror image? Is it completely different? Differently substituted cyclohexanes give all three possibilities.

5.7 Bicyclic Compounds

Now that we have dealt with both simple and complicated cycloalkanes (molecules containing a single ring), it is time to look briefly at molecules containing more than one ring. We have met such structures before, in both Chapter 2 (p. 85) and Chapter 3 (p. 121).

Of course, one can imagine all sorts of molecules containing two separated rings. There is nothing special about such molecules. However, there are other molecules in which two rings share a carbon or carbons, and these compounds are more interesting. There are three general ways in which two rings can be connected by sharing carbons. They can share a single carbon, two carbons, or more than two carbons. These modes of attachment are called **spiro** (one carbon shared), **fused** bicyclic (two carbons shared), and **bridged** bicyclic (more than two carbons shared). Fused is really a special case of bridged, with $n = 0$ (Fig. 5.47).

Spiro

Spiro substitution
(one carbon is shared)

Fused bicyclic

Fused substitution
(two carbons are
shared)

Bridged bicyclic

Bridged substitution
(more than two carbons
are shared; if $n = 0$,
the molecule is fused)

FIGURE 5.47 Rings can share a single carbon (spiro), two carbons (fused bicyclic), or more than two carbons (bridged bicyclic).

Spiro substitution is common in both natural products and in molecules so far encountered only in the laboratory. Note in Figure 5.48 how poorly a two-dimensional picture represents the real shape of such molecules. Even the simplest spirocyclic compound, spiropentane, is badly served by the flat page. The central carbon is approximately tetrahedral, and we really must use wedges to give a three-dimensional feel to this kind of molecule.

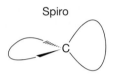

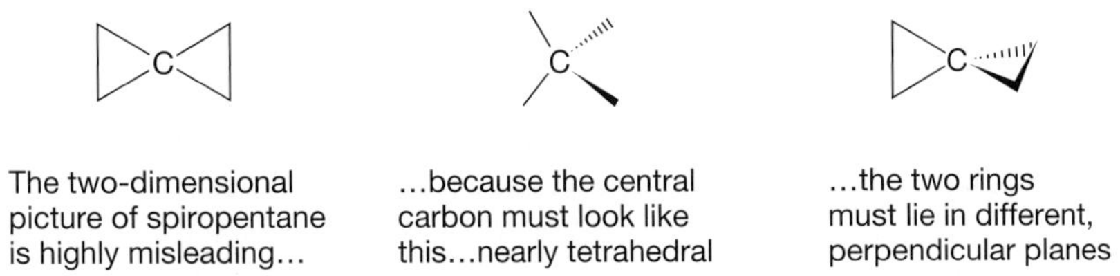

The two-dimensional
picture of spiropentane
is highly misleading...

...because the central
carbon must look like
this...nearly tetrahedral

...the two rings
must lie in different,
perpendicular planes

FIGURE 5.48 A two-dimensional picture of spiropentane is misleading. The two rings lie in perpendicular planes.

Figure 5.49 shows some "natural" (the first two compounds) and "unnatural" (the second two compounds) spiro compounds. Of course, the distinction is purely formal—the chemical laboratory is part of Nature.

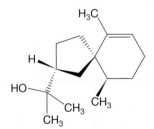

WEB3D Agarospirol

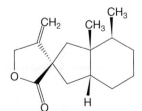

Bakkenolide-A

A [5]triangulane

Tricyclo[4.1.0.0^{4,6}]heptane

FIGURE 5.49 Some spiro compounds. Don't worry about the exotic names.

Bicyclic molecules in which two rings share two or more carbons are even more important than the spirocyclic compounds. The simplest way in which two rings can share more than one carbon is for two adjacent carbons to be shared in a fused structure. The "fusion" positions, or "bridgeheads" (p. 121), are shown in red (Fig. 5.50).

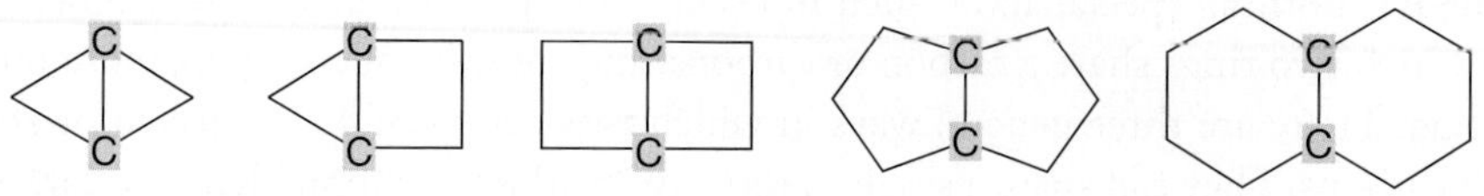

FIGURE 5.50 Some schematic, two-dimensional structures for fused bicyclic compounds in which two rings share a pair of adjacent carbon atoms.

The hydrogens attached at the ring junction, or fusion positions, can be either on the same side (cis) or on opposite sides (trans). In practice, both stereochemistries are possible for larger rings, but for the small rings only the cis form is stable (see Problem 5.23). Figure 5.51 shows two compounds that contain fused rings.

FIGURE 5.51 Two fused polycyclic molecules. Note the cis and trans ring fusions.

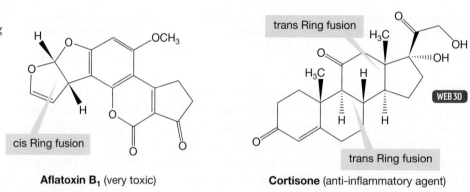

trans Ring fusion

cis Ring fusion

Aflatoxin B₁ (very toxic)

trans Ring fusion

Cortisone (anti-inflammatory agent)

PROBLEM 5.23 Use your models to convince yourself that trans stereochemistry is not possible for the three molecules on the left of Figure 5.50.

Particularly important are *cis*- and *trans*-decalin, the compounds formed by the fusion of two cyclohexanes. Although Figure 5.52 shows these compounds in schematic form, by now we can certainly guess that the real shape will be much more intricate. It's easy enough to draw them though, if one goes about it carefully.

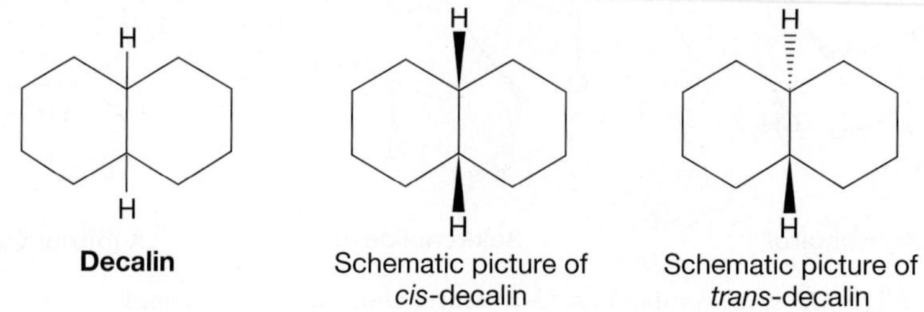

FIGURE 5.52 In decalin, the two hydrogens at the bridgeheads can be either on the same side of the rings (cis), or on opposite sides (trans).

Decalin

Schematic picture of *cis*-decalin

Schematic picture of *trans*-decalin

For *trans*-decalin first draw a single, perfect chair cyclohexane, and put in the axial and equatorial bonds at two adjacent carbons (Fig 5.53a). As for any trans di-substituted cyclohexane (p. 207), the two rings of *trans*-decalin will be attached either through two equatorial bonds or through two axial bonds. As shown in Figure 5.53b, attachment through two axial bonds is impossible—the distance to be bridged is too great (try it with models). So we are left with only the two equatorial bonds for trans

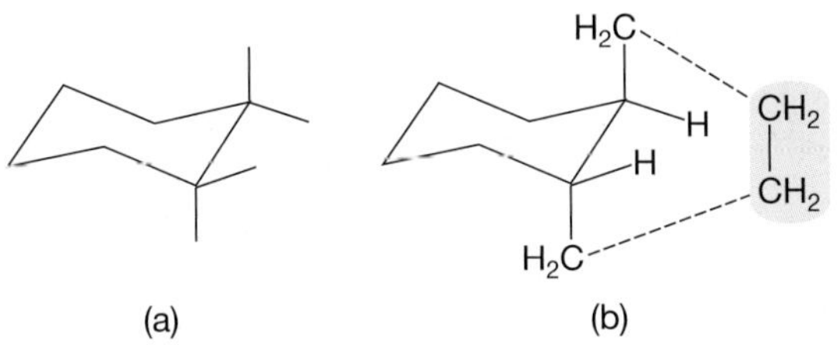

(a)   (b)

FIGURE 5.53 A futile attempt to construct *trans*-decalin by connecting two axial carbons with a pair of methylene groups. It is not possible to span the two axial positions with only four methylene groups; two methylene groups occupy axial positions leaving only two others (blue) to complete the ring.

attachment of a second ring, and this time the new chair fits in easily. *trans*-Decalin can be made by connecting two adjacent equatorial positions through a chain of four methylene groups (Fig. 5.54).

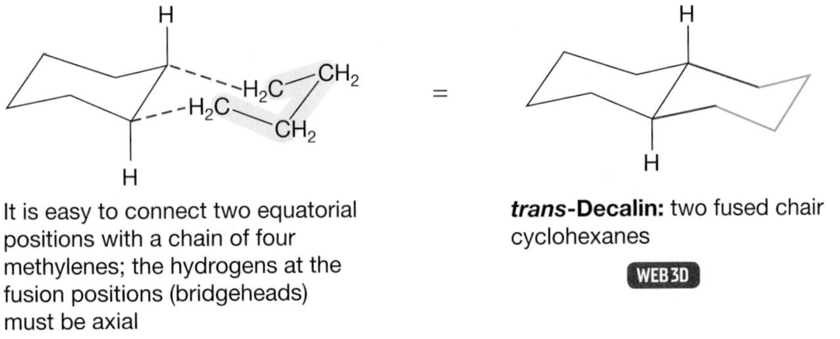

FIGURE 5.54 *trans*-Decalin.

It is easy to connect two equatorial positions with a chain of four methylenes; the hydrogens at the fusion positions (bridgeheads) must be axial

trans-**Decalin:** two fused chair cyclohexanes

cis-Decalin is quite different because cis substitution must involve one axial and one equatorial bond in the attachments to the second ring (Fig. 5.55). If we pay atten-tion to the rules for drawing perfect chairs, it's easy to produce a good drawing for this molecule. Remember that every ring bond in a cyclohexane is parallel to the bond directly across the ring (p. 191).

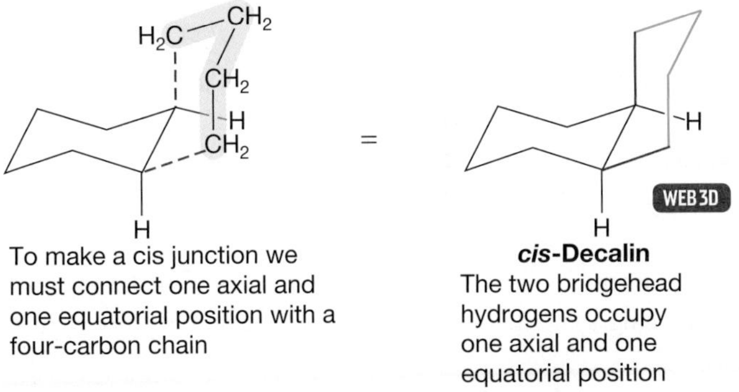

FIGURE 5.55 *cis*-Decalin.

To make a cis junction we must connect one axial and one equatorial position with a four-carbon chain

cis-**Decalin**
The two bridgehead hydrogens occupy one axial and one equatorial position

PROBLEM 5.24 Use models to convince yourself that *cis*-decalin is a mobile mol-ecule, undergoing easy double chair–double chair interconversions, whereas *trans*-decalin is rigidly locked.

Now imagine two rings sharing three carbons in a bridged bicyclic molecule. Two cyclopentanes, for example, can share two or three carbons. We have seen the first case before (Fig. 5.50) and the second is drawn in Figure 5.56. In each case, the carbons at the fusion points are called the bridgehead positions (p. 121).

FIGURE 5.56 Two five-membered rings sharing three carbons. The shared carbons are shown in red.

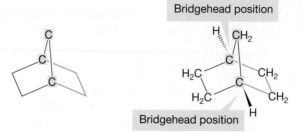

The shared carbons are shown in red in Figure 5.56; the carbon–carbon bonds in blue complete one cyclopentane. A more complete structure for this bicyclic hydrocarbon is also shown. Such bridged molecules are extremely common. Three compounds found in Nature are shown in Figure 5.57.

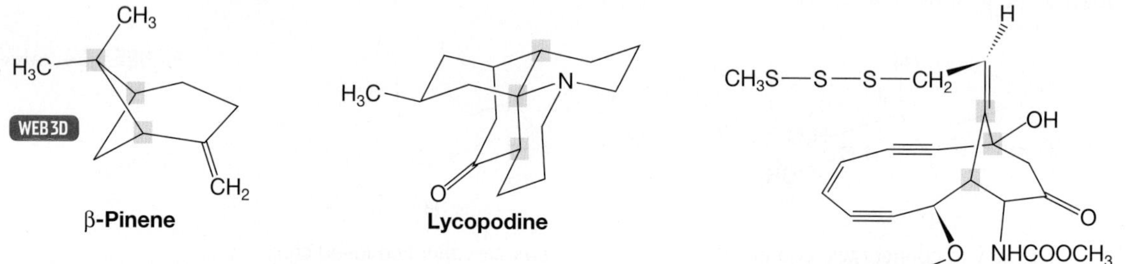

β-Pinene

Lycopodine

Calicheamicin

Several sugars

FIGURE 5.57 Three naturally occurring bicyclic molecules. The shared carbons are highlighted in red.

WORKED PROBLEM 5.25 Design a molecule in which two rings share four carbons.

ANSWER One sure-fire way to do a problem such as this is to start by drawing the carbons to be shared (four, in this example). Then add the remainders of the other rings.

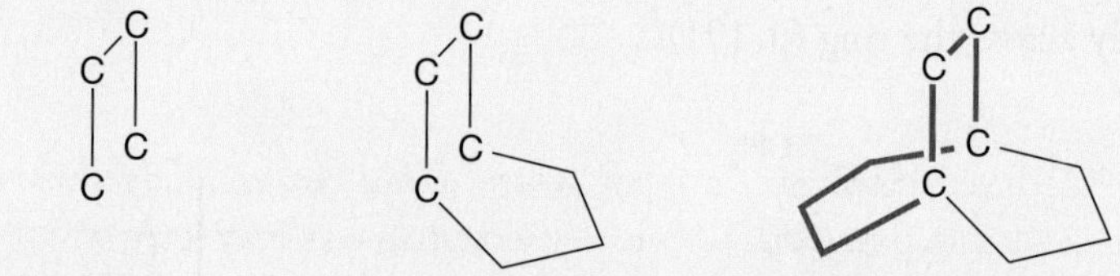

The four carbons to be shared

Here are those four carbons incorporated into one ring

Here they are incorporated into another ring

CONVENTION ALERT

Bicyclic compounds are named in the following way: One first counts the number of carbons in the ring system. The molecule in Figure 5.58 has eight ring carbons.

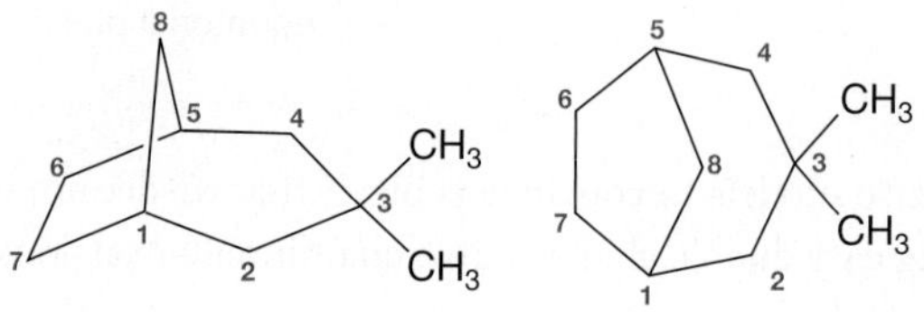

FIGURE 5.58 Part of the naming protocol for a typical bicyclic compound.

Thus the base name is "octane." The molecule is numbered by counting from the bridgehead carbon [carbon number 1, C(1)] around the longest bridge first, proceeding to the other bridgehead. One then continues counting around the second-longest bridge, and finally numbers the shortest bridge. Any substituents can now be assigned a number. So the compound in Figure 5.58 is a 3,3-dimethylbicyclooctane.

The bridges are counted from the bridgeheads and assigned numbers equal to the number of atoms in the bridges, *not counting the bridgehead atoms*. These numbers are enclosed in brackets, largest number first, between the designation "bicyclo" and the rest of the name (Fig. 5.59). So this compound is called 3,3-dimethylbicyclo[3.2.1]octane.

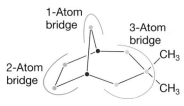

3,3-Dimethylbicyclo[3.2.1]octane

FIGURE 5.59 More of the naming protocol for bridged bicyclic compounds. This molecule is 3,3-dimethylbicyclo[3.2.1]octane. The red dots show the bridgehead atoms. These are not counted in sizing the bridges, but are counted in numbering the compound.

PROBLEM 5.26 Make drawings of (a) *cis*-bicyclo[3.3.0]octane, (b) 1-fluorobicyclo[2.2.2]octane, (c) *trans*-9,9-dimethylbicyclo[6.1.0]nonane.

PROBLEM 5.27 Name the following compounds:

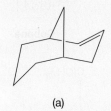

(a)

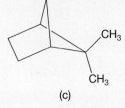

CH₃

CH₃

(b)

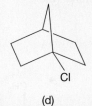

CH₃

CH₃

(c)

Cl

(d)

(e)

PROBLEM 5.28 How many signals would appear in the ¹³C NMR spectrum for each of the molecules in Problem 5.27?

As with 1,2-fused systems, we must worry about the stereochemistry at the bridgehead positions in bridged molecules. At first, this task may seem trivial—where could these hydrogens be but where they are shown in the figures? The answer is, "inside the cage!" In the compounds we have already examined, it is not easy for hydrogens to occupy the inside position. As shown in Figure 5.60, all four bonds of one or both of the bridgehead carbons of bicyclo[3.2.1]octane would be pointing in the same direction if one or both bridgehead hydrogens were inside the cage.

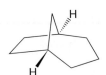

"Out, out" bicyclo[3.2.1]octane

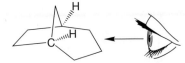

"In, out" bicyclo[3.2.1]octane

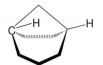

Side view of "in, out" bicyclo[3.2.1]octane as seen by the eye

"In, in" bicyclo[3.2.1]octane

FIGURE 5.60 Three stereoisomers of bicyclo[3.2.1]octane. Only the "out, out" isomer is known.

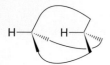

FIGURE 5.61 If the bridges are long enough, there is no angle strain in an "in, out" isomer.

But now imagine increasing the size of the bridges. In principle, it should be possible to make the bridge chains long enough so that a normal, nearly tetrahedral, arrangement can be achieved with one or both bridgehead hydrogens inside the cage (Fig. 5.61). And so it is. The molecules shown in Figure 5.62 are both known, as are several other examples.

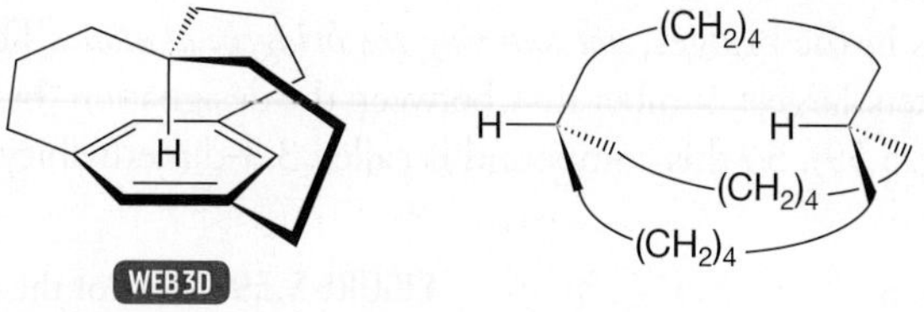

WEB 3D

FIGURE 5.62 Two known "in" or "in, out" compounds.

5.8 Special Topic: Polycyclic Systems

We needn't stop here. More rings can be attached in fused or bridged fashion to produce wondrously complex structures. The naming protocols are complicated, if ultimately logical, and won't be covered here. Many of the compounds have common names that are meant to be evocative of their shapes. Prismane and cubane are examples. The molecular versions of the Platonic solids, tetrahedrane, cubane, and dodecahedrane are all known, although the parent, unsubstituted tetrahedrane still evades synthesis (Fig. 5.63).

FIGURE 5.63 Polycyclic natural (cholesterol) and "unnatural" (the rest) products.

Prismane   **Cubane**

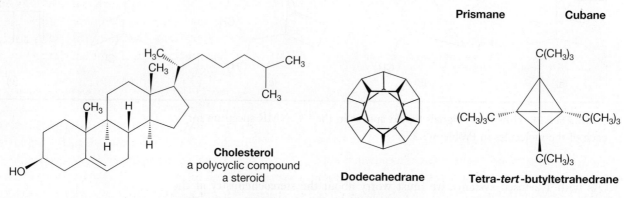

Cholesterol
a polycyclic compound
a steroid

Dodecahedrane   **Tetra-*tert*-butyltetrahedrane**

Spectacular new molecules containing rings are always appearing. In 1984, Professor Kenneth B. Wiberg (b. 1927) of Yale University reported the construction of the polycyclic molecule, tricyclo[1.1.1.0^{1,3}]pentane (also known as [1.1.1]propellane, Fig. 5.64), a marvelously exciting molecule. You are now only five chapters into your study of organic chemistry, yet it is possible for you to appreciate why the chemical world was knocked out by this compound.

FIGURE 5.64 Two polycyclic small ring compounds.

Tricyclo[1.1.1.0^{1,3}]pentane
([1.1.1]propellane)   **Bicyclo[1.1.1]pentane**

PROBLEM 5.29 Why is tricyclo[1.1.1.0^{1,3}]pentane unusual (Fig. 5.64)? Would you expect it to be especially stable or unstable with respect to its cousin bicyclo[1.1.1]pentane? Why?

Wiberg's molecule was synthesized in a chemistry laboratory, and probably (?) does not occur in Nature. Nature is by no means played out as a source of fascinating polycyclic molecules, however. For example, in 2003 Sanae Furuya and Shiro Terashima reported the synthesis of optically active tricycloillinone, a molecule isolated from the wood of *Illicium tashiori* (Fig. 5.65). This molecule enhances the activity of choline acetyltransferase, an agent that catalyzes the synthesis of acetylcholine. Why should we care about tricycloillinone? A form of senile dementia (Alzheimer's disease) is associated with reduced levels of acetylcholine, and a molecule that might be useful in increasing levels of acetylcholine is of obvious importance to all of us.

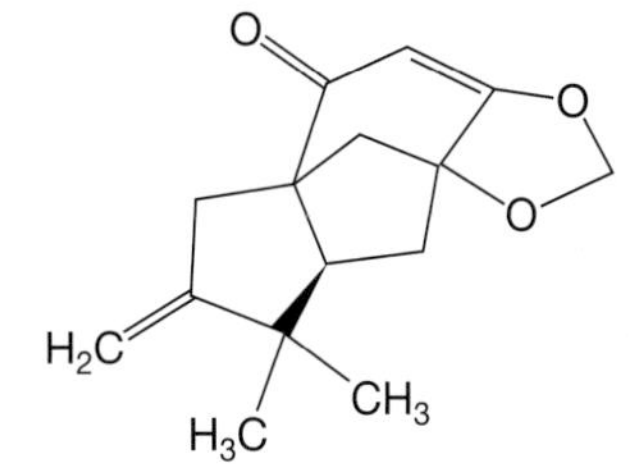

FIGURE 5.65 Tricycloillinone.

CARBORANES: WEIRD BONDING

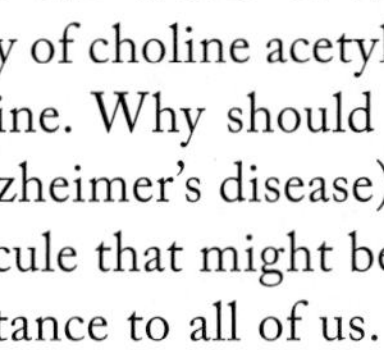

o-Carborane
The dots are carbons; every other vertex is a boron. There is a hydrogen atom at every vertex

As you've seen in this chapter, ring compounds can be straightforward (cyclopentane, p. 190, is a nice example), moderately complex (the mobile cyclohexanes, p. 197), or exotic ([1.1.1]propellane, p. 216). Here is a compound that surely qualifies as exotic, if not downright weird. It is composed of two carbons (the dots) and ten borons (the other 10 vertices), and contains no fewer than 20 three-membered rings of carbons and borons. Why "weird?" Count the bonds to carbon. There are six bonds emanating from the carbon! Six bonds? How can that be? If you draw a carbon with five

or six bonds, the dreaded red "X" is sure to follow. How does Nature get away with it? If you did Problems 1.62 and 1.63, you encountered triangular H_3, and H_3^+, molecules related to the carboranes in that they, too, contain "too many" bonds, in this case, two bonds to hydrogen. The answer to this seeming impossibility is that those bonds are not the simple two-electron bonds we are becoming used to, but partial bonds containing fewer than two electrons.

Far from being "weird" and thus presumably exotic in properties, the carboranes are almost unbelievably stable compounds, sitting in bottles seemingly forever, and showing a rich history and chemistry. Professor William Lipscomb (b. 1919) won the Nobel Prize for chemistry in 1976 for explaining the bonding in carboranes. They are now being used in both medicinal chemistry and materials science. In Japan, for example, carboranes are used in treating certain brain tumors in "Boron-neutron capture therapy." However, it must be admitted that the practical development of these compounds was slow to happen. Why? Perhaps we chemists were wary of all those potential red X's, sure to arrive if we drew too many bonds to carbon!

5.9 Special Topic: Adamantanes in Materials and Biology

Consider constructing a polycyclic molecule by expanding a chair cyclohexane. First, connect three of the axial bonds to a cap consisting of three methylene (CH$_2$) groups all connected to a single methine (CH) group. This process produces the beautiful molecule tricyclo[3.3.1.1^{3,7}]decane, better known as "adamantane," $C_{10}H_{16}$ (Fig. 5.66).

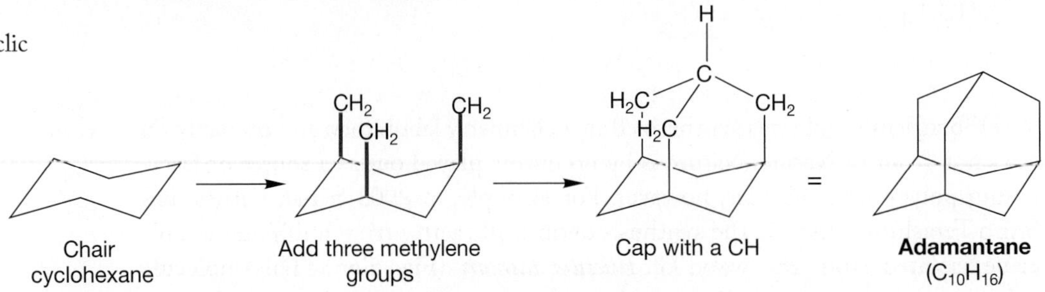

FIGURE 5.66 A schematic construction of the polycyclic molecule adamantane.

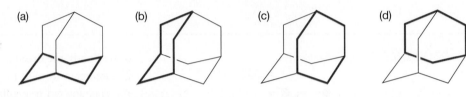

Chair cyclohexane → Add three methylene groups → Cap with a CH = Adamantane ($C_{10}H_{16}$)

As Figure 5.67 shows, adamantane is composed entirely of perfect chair six-membered rings. As you would expect, this molecule is nearly strain-free and constitutes the thermodynamic minimum for all the $C_{10}H_{16}$ isomers.

FIGURE 5.67 Adamantane is composed entirely of chair cyclohexanes. Those in (a) and (b) are easy to see as chairs, but those in (c) and (d) may require the use of models.

(a)   (b)   (c)   (d)

Adamantane was first found in the 1930s in trace amounts in petroleum residues by the Czech chemist Stanislav Landa (1898–1981), but it can now be made easily in quantity by a simple process discovered by the American chemist Paul von R. Schleyer (b. 1930).

Consider what happens when we continue the capping process begun in our transformation of chair cyclohexane into adamantane in Figure 5.66. Adamantane itself is composed of only chair cyclohexanes, so we have a number of possible places to start the process (Fig. 5.68). Addition of one more four-carbon cap

FIGURE 5.68 The continuation of the capping process leads to polyadamantanes and, ultimately, to diamond.

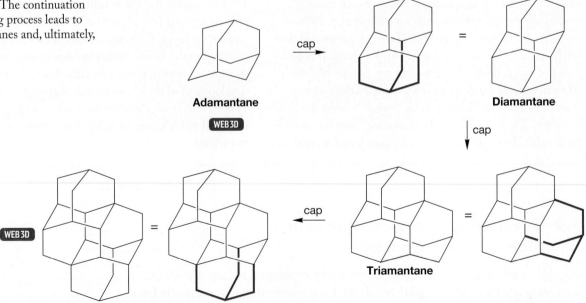

Adamantane
WEB 3D

cap →

= Diamantane

↓ cap

← cap   Triamantane   =

WEB 3D

=

One isomer of tetramantane

produces "diamantane," two caps gives us "triamantane," three caps produces "tetramantane," and so on. The ultimate result of the capping process is a molecule composed of a network of adamantanes, and this is the structure of diamond.

Adamantane chemistry remains full of astonishing surprises. For years it has been known that synthetic diamonds could be made from another polymeric form of carbon, graphite, by treatment at high temperature (~2300 K) and very high pressure (7×10^4 kg/cm^2). However, in the 1980s in the Soviet Union and Japan, it was discovered that diamond films could be grown at low temperature and pressure by passing a stream of hydrogen gas containing a few percent methane through an electric discharge (Fig. 5.69). The mechanism of this reaction remains obscure. Any ideas?

The diamond in this impressive bling is composed of a multi-ring carbon framework of adamantanes.

$$H_2 \; + \; CH_4 \xrightarrow[\text{discharge}]{\text{electrical}} \text{Diamond}$$

FIGURE 5.69 A simple (and astonishing) synthesis of diamond.

PROBLEM 5.30 What is the empirical formula of diamond? For a large polymeric molecule such as diamond, the edges of the molecule are insignificant in figuring out the formula.

PROBLEM 5.31 Can you draw other isomers of tetramantane?

Adamantane remains a favorite of chemists because of its intrinsic symmetry and beauty, and probably because of the difficulty of working out the mechanism of its formation. It does not seem of much practical interest. However, some of its simple derivatives have remarkable properties and some are quite active biologically. For example, 1-aminoadamantane, a compound easily made from adamantane itself, is one of the few antiviral agents known. This remarkable property was discovered during routine empirical screening at duPont in the 1960s, and 1-aminoadamantane has since been marketed, mostly as an agent against influenza A and C. It apparently works by migrating through the cell membrane to attack the virus within. It is the adamantane cage, acting as a molecular ball of grease, that helps in the membrane penetration. It has been speculated that other, large, symmetrical hydrocarbon "blobs" might act in the same way. Unfortunately, they are not so easy to make.

5.10 Summary

New Concepts

This chapter deals exclusively with the structural properties of ring compounds. Two kinds of destabilizing effects on rings are discussed. Torsional strain, the destabilizing effect of eclipsed carbon–hydrogen bonds, was mentioned in Chapter 2 when the acyclic hydrocarbons were discussed. The planar forms of ring compounds are particularly subject to torsional strain. In addition, rings contain varying amounts of angle strain, depending on the ring size. Any deviation from the ideal tetrahedral angle of 109.5° will introduce angle strain. Rings adopt nonplanar forms in order to minimize the combination of torsional and angle strain.

One nonplanar ring deserves special mention. Cyclohexane avoids torsional and angle strain by adopting a chair conformation in which a nearly ideal tetrahedral angle is achieved and all carbon–hydrogen bonds are perfectly staggered. In cyclohexane, there is a set of six axial hydrogens that is converted into the set of six equatorial hydrogens through rotations about carbon–carbon bonds called a ring flip.

There are many possible quantitative evaluations of strain. This chapter uses related heats of formation (ΔH_f°) and heats of combustion (ΔH_c°) to arrive at values for the strain energies of various rings. As rings increase in size, strain initially decreases, reaching a minimum at the strain-free cyclohexane. Strain then increases until large ring sizes are reached.

By combining rings in spiro, fused, and bridged fashion, very complex polycyclic molecules can be constructed.

Key Terms

angle strain (p. 187)
axial hydrogens (p. 191)
bridged (p. 211)

equatorial hydrogens (p. 191)
fused (p. 211)
spiro (p. 211)

torsional strain (p. 187)
van der Waals strain (p. 198)

Reactions, Mechanisms, and Tools

The few reactions encountered here are not really new. There is, however, a somewhat more detailed treatment of the formation of carbon dioxide and water from hydrocarbons and oxygen

(combustion) than we saw in Chapter 2 (p. 68). There is also a discussion of the effects of strain on the bond energy of the carbon–carbon bond.

Common Errors

You are presented in this chapter with the most difficult challenge so far in visualizing molecules. Moreover, the difficulty in translating from two dimensions to three is compounded by the mobile equilibria present in many ring compounds. In analyzing complicated phenomena, such as the equilibrations of disubstituted cyclohexanes, be sure to use models, at least at first.

Two small points seem to give students trouble. First, one hears constantly, I'm no artist, I can't draw a good chair cyclohexane! Nonsense. It does not take an artist, merely an artisan. Go slowly, do not scribble, and follow the simple procedure outlined in this chapter (p. 191ff). Drawing a perfect chair cyclohexane is one thing that everyone can do.

Determining which groups are cis and trans in ring compounds causes problems. *Remember*: cis means "on the same side,"

and trans, "on opposite sides." Bonds to cis groups do not have to be parallel, merely both up or both down. Figure 5.70 gives two common examples, one straightforward, the other not so obvious.

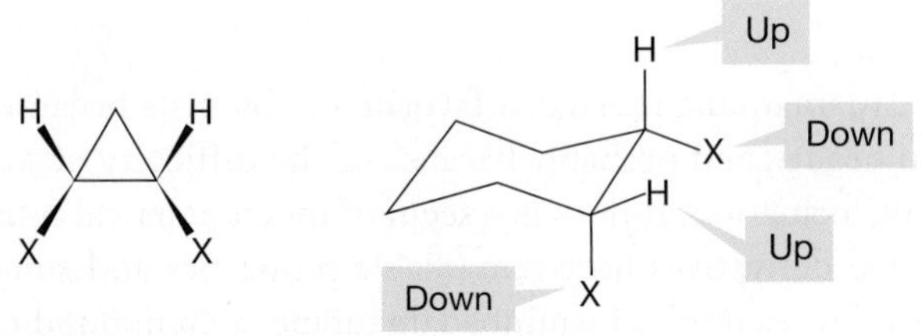

FIGURE 5.70 In both these compounds the two hydrogens shown are cis, as are the two X groups. This relationship is easy to see for the flat cyclopropane, but harder in the nonplanar cyclohexane.

5.11 Additional Problems

PROBLEM 5.32   By now you have encountered most structural types and have some experience with rings. Write and name all the isomers of the formula C_5H_8. Stay alert for isomerism of the cis/trans (*Z/E*) and (*R/S*) types. Indicate chiral molecules with an asterisk. This problem is very hard at this point.[3] There are all sorts of odd structural types among these isomers. It should not be hard to get most of them, but getting them all is really tough. Think in three dimensions and *organize*. *Hint*: The total number of isomers, not including enantiomers, is 28.

PROBLEM 5.33   Write out the pairs of enantiomers for the chiral isomers in Problem 5.32.

PROBLEM 5.34   Assign (*R*) and (*S*) configurations for the versions of the chiral cyclic molecules shown on the left in the answers to Problem 5.33 given in the Study Guide.

PROBLEM 5.35   Draw the most stable chair conformation of *tert*-butylcyclohexane. Put a circle around each of the axial hydrogens on the cyclohexane. Put a square around each of the equatorial hydrogens attached to the ring.

---

[3]Indeed, this problem constituted the entire first hour examination in MJ's first course in organic chemistry.

PROBLEM 5.36   Draw the two possible chair forms of *cis*- and *trans*-1,4-dimethylcyclohexane. Are the two forms identical, enantiomeric, or diastereomeric? In each case, indicate which chair form will be more stable and explain why. Is either of these molecules chiral?

PROBLEM 5.37   Draw the two possible chair forms of *cis*- and *trans*-1-isopropyl-4-methylcyclohexane. Are the two forms identical, enantiomeric, or diastereomeric? In each case, indicate which chair form will be more stable and explain why. Is either of these molecules chiral?

PROBLEM 5.38   Draw the two possible chair forms of *cis*- and *trans*-1,3-dimethylcyclohexane. Are the two forms identical, enantiomeric, or diastereomeric? In each case, indicate which chair form will be more stable and explain why. Is either of these molecules chiral?

PROBLEM 5.39   Draw the possible chair forms of *cis*- and *trans*-1-isopropyl-3-methylcyclohexane. Are the two forms identical, enantiomeric, or diastereomeric? In each case, indicate which chair form will be more stable and explain why. Is either of these molecules chiral?

PROBLEM 5.40   Draw the double Newman projection as shown in Figure 5.12 for the following compounds:

(a) 1,1-dimethylcyclohexane looking down the C(2)—C(1) and C(4)—C(5) bonds.
(b) *cis*-1,2-dimethylcyclohexane looking down the C(1)—C(2) and C(5)—C(4) bonds.
(c) the more stable conformation of *trans*-1,2-dimethylcyclohexane looking down the C(1)—C(2) and C(5)—C(4) bonds.
(d) *trans*-1,3-dimethylcyclohexane looking down the C(1)—C(2) and C(5)—C(4) bonds.

PROBLEM 5.41   In Section 5.6b (p. 205), you saw that for purposes of stereochemical analysis you could treat the decidedly nonplanar *cis*- and *trans*-1,2-dimethylcyclohexanes as if they were planar. The planar forms represent the average positions of ring atoms in the rapid chair–chair interconversions. Use planar representations of the following cyclohexanes to determine which molecules are chiral.

(a) *cis*-1,3-dimethylcyclohexane
(b) *trans*-1,3-dimethylcyclohexane
(c) *cis*-1-isopropyl-3-methylcyclohexane
(d) *trans*-1-isopropyl-3-methylcyclohexane
(e) *cis*-1,4-dimethylcyclohexane
(f) *trans*-1,4-dimethylcyclohexane
(g) *cis*-1-isopropyl-4-methylcyclohexane
(h) *trans*-1-isopropyl-4-methylcyclohexane

PROBLEM 5.42   Draw the planar structure for each of the following compounds. Next, draw the most stable conformation for each in three dimensions.

(a) *trans*-1,2-dibromocyclohexane

(b) *cis*-1,3-dichlorocyclohexane
(c) *trans*-1-chloro-4-fluorocyclohexane

PROBLEM 5.43   Which of the molecules in Problem 5.42 are chiral and which are achiral?

PROBLEM 5.44   Draw a chair conformation of *cis*-1-bromo-4-fluorocyclohexane. Draw the ring-flipped structure. Which structure do you expect to be favored? Why?

PROBLEM 5.45   Draw a chair conformation of *cis*-1-bromo-3-chlorocyclohexane. Draw the ring-flipped structure. Which is favored? Why? Is the molecule chiral?

PROBLEM 5.46   Use the data in Table 5.3 (p. 202) to calculate the energy difference between the possible isomers of *cis*- and *trans*-1,4-dimethylcyclohexane.

PROBLEM 5.47   Use the data in Table 5.3 (p. 202) to calculate the energy difference between the possible isomers of *cis*- and *trans*-1-isopropyl-4-methylcyclohexane.

PROBLEM 5.48   Cholestanol (shown below) is a natural product found in gallstones and eggs. How many stereogenic atoms are there in this molecule? Draw the compound with the rings in their most stable conformation. *Hint*: The molecule is relatively flat when viewed in three dimensions.

Cholestanol

PROBLEM 5.49   How many signals will appear in the ^{13}C NMR spectrum of 1,1-dimethylcyclohexane?

(a) At low temperature.
(b) At high temperature.

Hint: Start by figuring out what effect the increase in temperature will have.

PROBLEM 5.50   How many signals will appear for *cis*-1,3-dimethylcyclohexane in its ^{13}C NMR spectrum? How many signals will appear for *trans*-1,3-dimethylcyclohexane in its ^{13}C NMR spectrum?

(a) At low temperature.
(b) At high temperature.

PROBLEM 5.51   How many signals will appear in the ^{13}C NMR spectrum of adamantane (p. 218)?

PROBLEM 5.52 Draw both chair forms for the following cyclo-hexanes. Indicate the more stable chair form in each case. Not all examples will be obvious; some choices may be too close to call.

(a)

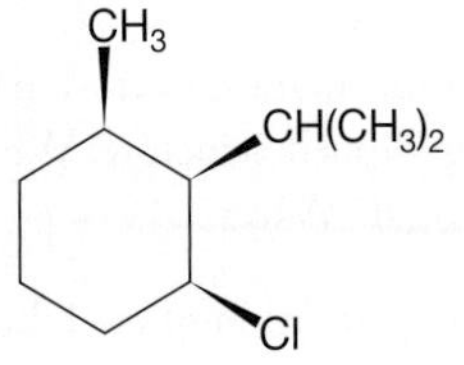

(b)

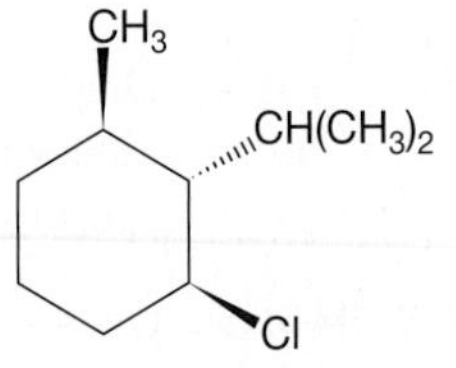

(c)

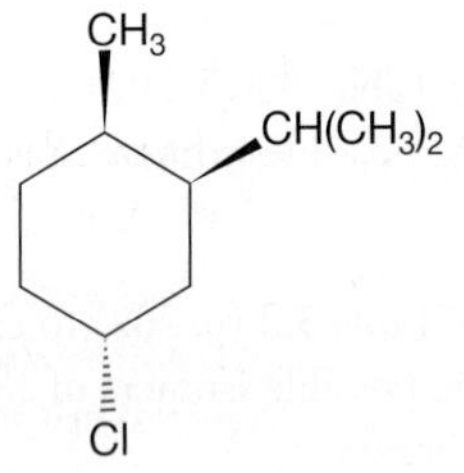

(d)

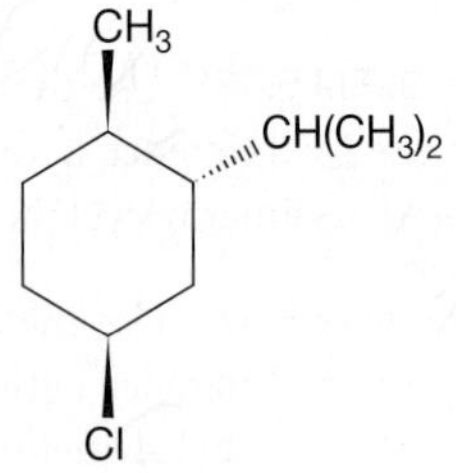

(e)

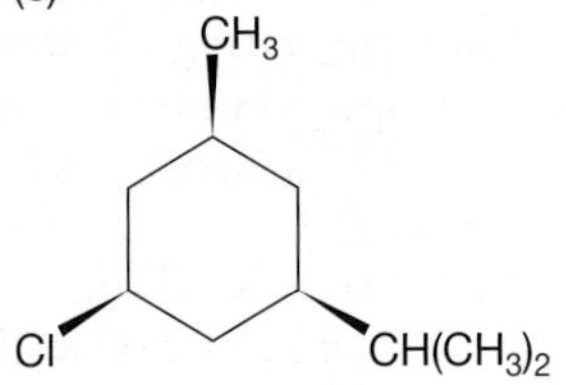

(f)

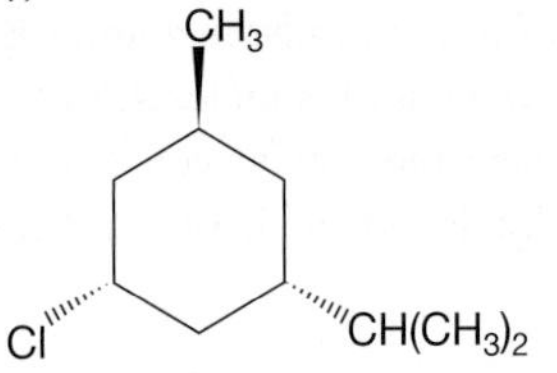

PROBLEM 5.53 Name the following compounds:

(a)

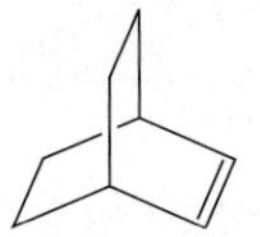

(b)

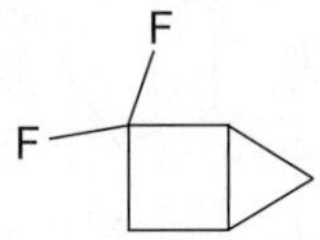

(c)

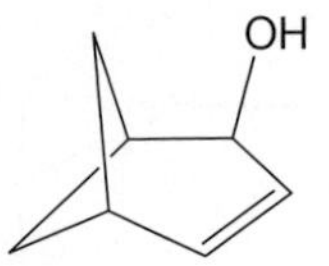

PROBLEM 5.54 Write structures for the following compounds:

(a) 2,2-dibromo-3,3-dichloro-5,5-difluoro-6,6-diiodobicy-clo[2.2.1]heptane
(b) bicyclo[1.1.1]pentan-1,3-diol
(c) hexamethylbicyclo[2.2.0]hexa-2,5-diene

PROBLEM 5.55 The sugar glucopyranose is a six-membered ring containing an oxygen atom. There are groups attached at each of the five ring carbons as shown. The red "squiggly" bond merely means that the hydroxyl group at the 1-position may be either up or down; that is, there are two structures for gluco-pyranose. Draw these two molecules in three dimensions.

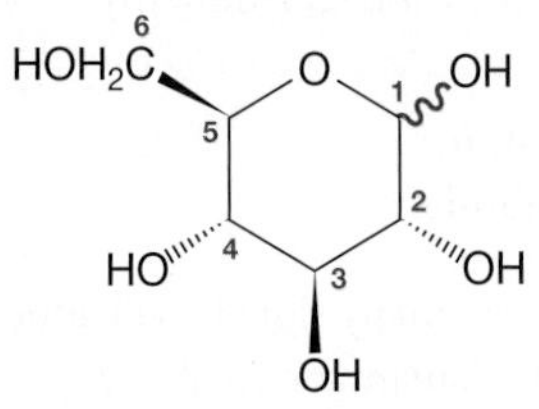

PROBLEM 5.56 One of the isomers of methylbicyclo[2.2.1]hep-tane can exist in two diastereomeric forms (we are not count-ing mirror images, which increases the total number of stereoisomers to four). Write the possible isomers of methylbi-cyclo[2.2.1]heptane and explain how one of them can exist in two forms.

PROBLEM 5.57 (a) Can one of the isomers of methylbicy-clo[2.2.2]octane also exist in two diastereomeric forms? That is, does it behave just like the methylbicyclo[2.2.1]heptane men-tioned in Problem 5.56? (b) Can one of the isomers of methyl-bicyclo[2.2.2]octane exist in two enantiomeric forms?

PROBLEM 5.58 Draw the favored envelope conformation of the cyclopentane shown below.

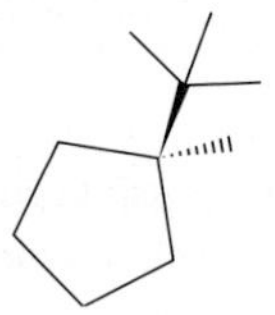

Use Organic Reaction Animations (ORA) to answer the fol-lowing questions. In these problems you will encounter new reactions. Don't be psyched out or intimidated—follow the ani-mation carefully and you should be able to get these problems.

PROBLEM 5.59 Observe the reaction titled "Bimolecular nucleophilic substitution." Notice that a bromide is being dis-placed from a carbon atom, in this case C(2) of 2-bromo-propane. We can imagine a closely related reaction in which displacement takes place in bromocyclopropane. All we have (mentally) done is to connect the two end carbons of 2-bromo-propane with a carbon–carbon bond. Would the reaction with bromocyclopropane be faster or slower (it can't be the same!) as the reaction with 2-bromopropane? *Hint*: Think about the tran-sition state for this reaction—what is the hybridization of car-bon in the transition state?

PROBLEM 5.60 Now consider the same substitution reaction with axial bromocyclohexane. What would be the conformation of the product? Would this reaction be faster or slower than the reaction with 2-bromopropane?

PROBLEM 5.61 Would the rate of reaction with equatorial bromocyclohexane be faster, slower, or the same as that with axial bromocyclohexane? Would the reaction with equatorial bromocyclohexane be faster or slower than the reaction with 2-bromopropane?

PROBLEM 5.62 Observe the reaction titled "Bimolecular elim-ination." This reaction also starts with 2-bromopropane. Look at the orientation of the breaking carbon–hydrogen and carbon–bromine bonds in this reaction. Now imagine the same bimolecular elimination occurring in axial and equatorial bro-mocyclohexane. Which stereoisomer would undergo this reac-tion faster, the axial or equatorial bromide?

Alkyl Halides, Alcohols, Amines, Ethers, and Their Sulfur-Containing Relatives

6

6.1 Preview

6.2 Alkyl Halides: Nomenclature and Structure

6.3 Alkyl Halides as Sources of Organometallic Reagents: A Synthesis of Hydrocarbons

6.4 Alcohols

6.5 Solvents in Organic Chemistry

6.6 Diols (Glycols)

6.7 Amines

6.8 Ethers

6.9 Special Topic: Thiols (Mercaptans) and Thioethers (Sulfides)

6.10 Special Topic: Crown Ethers

6.11 Special Topic: Complex Nitrogen-Containing Biomolecules—Alkaloids

6.12 Summary

6.13 Additional Problems

CANOLA Canola oil is obtained from the rapeseed plant. Rapeseed also pumps a significant amount of methyl bromide into the atmosphere.

So, if Sunday you're free,
Why don't you come with me,
And we'll poison the pigeons in the park.

My pulse will be quickenin',
With each drop of strych-i-nin',
We feed to a pigeon,
(It just takes a smidgin),
To poison a pigeon in the park.

—TOM LEHRER[1]

6.1 Preview

In this chapter, we will see some old friends, alkyl halides and alcohols, as well as some new molecules rather closely related to them: amines, ethers, thiols, and thioethers. Both alkyl halides and alcohols can be made through the addition reactions encountered in Chapter 3 (p. 131). Amines somewhat resemble alcohols, and understanding the chemistry of amines is a necessary prelude to working through Chapter 23, which discusses the chemistry of some of the many biologically important molecules containing nitrogen. A nitrogen atom provides one of the crucial linking agents in the formation of polyamino acids, called **peptides** or **proteins** (Fig. 6.1). In this chapter, we examine the qualities that make these molecules similar and prepare for Chapter 7, where we will take a long, detailed look at four prototypal new reactions in which alkyl halides and alcohols are especially important.

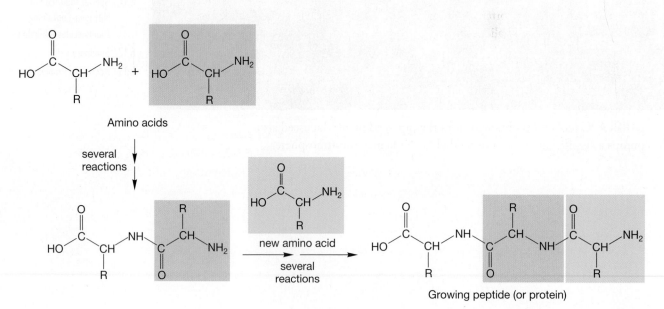

FIGURE 6.1 A nitrogen-to-carbon link is vital in the polymeric molecules known as proteins or peptides.

---

[1]Thomas Lehrer (b. 1928) is an American mathematician and songwriter, most of whose work was done in the 1950s and 1960s. Despite the unfortunate fact that there has been little new material for many years, his songs, often sardonic and always very funny, retain their appeal. (Lyrics © 1953 Tom Lehrer. Reprinted with permission.)

ESSENTIAL SKILLS AND DETAILS

1. So much of chemistry, including organic chemistry, involves acids and bases. In this chapter, we review what we know already about this subject, and focus on a measure of acidity, pK_a. It is important to have a reasonable idea of the pK_a values of the general classes of compounds. There is a table of pK_a values on the inside of the back cover of this book.

2. A low pK_a means that the molecule in question is a strong acid, a good proton donor. Conversely, a high pK_a means that the compound is a poor acid.

3. We also encounter solvation in this chapter. Being able to choose an appropriate solvent for a reaction, and to understand how solvents operate, is an important skill both in the practical world and in this course.

4. The synthetic sequence "alkene → halide → Grignard reagent" is one you will use many times in solving synthetic problems.

6.2 Alkyl Halides: Nomenclature and Structure

Alkyl iodides, bromides, chlorides, and fluorides are collected under the name of **alkyl halides**. Alkyl halides have the formula $C_nH_{2n+1}X$, where X = F, Cl, Br, or I. Quite appropriately, alkyl halides are named as fluorides, chlorides, bromides, and iodides. Both common and systematic nomenclature names are used, although as the complexity of the molecule increases, the system naturally takes over. The important rules to remember for naming saturated halides are (1) minimize the number given to the substituent, (2) name multiple substituents in alphabetical order, and (3) base the name on the longest straight chain that contains the halide. If a carbon–carbon double bond is present, the double bond takes precedence over the halogen when positional numbers are assigned. Table 6.1 gives some examples with both common and systematic names given. There are some important common names that are used very often. For example, haloforms are CHX_3, vinyl halides are $H_2C=CHX$, and allyl halides are $H_2C=CH—CH_2X$.

TABLE 6.1 Some Alkyl Halides, Names, and Known Properties

	Compound	Common Name	Systematic Name	bp (°C)	mp (°C)
WEB3D	$CH_3—F$	Methyl fluoride	Fluoromethane	−78.4	−141.8
WEB3D	$CH_3CH_2—Br$	Ethyl bromide	Bromoethane	38.4	−118.6
WEB3D	$(CH_3)_2CH—I$	Isopropyl iodide	2-Iodopropane	89.4	−90.1
WEB3D	$(CH_3)_3C—Cl$	*tert*-Butyl chloride	2-Chloro-2-methylpropane	51.0	−25.4
WEB3D	CH_2Br_2	Methylene bromide	Dibromomethane	97	−52.7
WEB3D	$CHCl_3$	Chloroform	Trichloromethane	61.5	−63.5
WEB3D	$H_2C=CHBr$	Vinyl bromide	Bromoethene	15.8	−139.5
WEB3D	$H_2C=CH—CH_2Cl$	Allyl chloride	3-Chloropropene	45	−134.5
			2-Bromo-1-chlorobutane	147	
			3-Chloro-1-pentene	94	
			Bromocyclopentane	138	

Halides are classified as methyl, primary, secondary, tertiary, and vinyl (Fig. 6.2). Chapter 2 (p. 76) introduced this nomenclature. *Remember*: A primary carbon is attached to only one other carbon, a secondary carbon to two others, and a tertiary carbon to three others. In a primary halide, the halogen atom is attached to a primary carbon, and so on. In vinyl halides, there is always a carbon–carbon double bond to which the halide is directly attached.

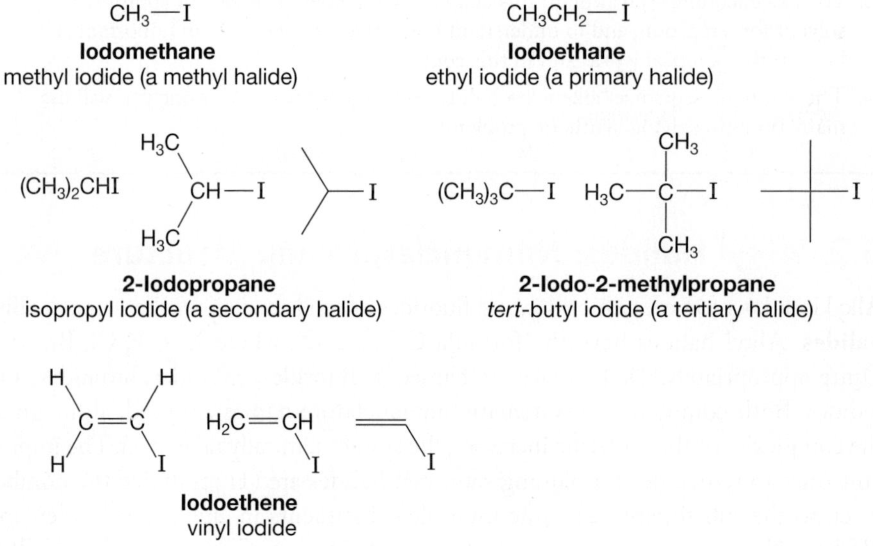

FIGURE 6.2 Some alkyl and vinyl halides at various levels of abstraction.

$CH_3{-}I$
Iodomethane
methyl iodide (a methyl halide)

$CH_3CH_2{-}I$
Iodoethane
ethyl iodide (a primary halide)

$(CH_3)_2CHI$

2-Iodopropane
isopropyl iodide (a secondary halide)

$(CH_3)_3C{-}I$

2-Iodo-2-methylpropane
tert-butyl iodide (a tertiary halide)

Iodoethene
vinyl iodide

Substituted alkanes are polar molecules. The halogens are relatively electronegative atoms and strongly attract the electrons in the carbon–halogen bond. Alkyl halides have dipole moments of approximately 2 debye, abbreviated D (Fig. 6.3; Table 6.2).

TABLE 6.2
Dipole Moments of Some Simple Alkyl Halides

Alkyl Halide	Dipole Moment (D)
Methyl fluoride	1.85
Ethyl fluoride	1.94
Methyl chloride	1.87
Ethyl chloride	2.05
Methyl bromide	1.81
Ethyl bromide	2.03
Methyl iodide	1.62
Ethyl iodide	1.91

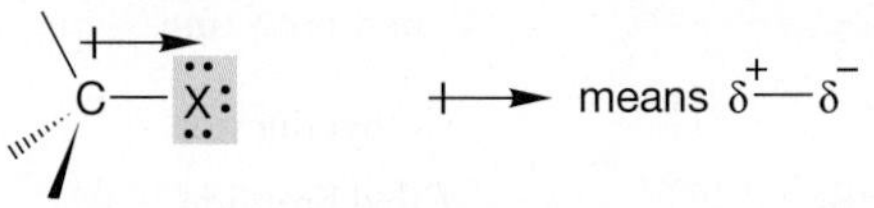

means $\delta^+{-}\delta^-$

FIGURE 6.3 A substantial dipole moment exists in the polar carbon–halogen bond, X = F, Cl, Br, or I.

Simple alkyl halides are nearly tetrahedral, and therefore the carbon to which the halogen is attached is approximately sp^3 hybridized. The bond to halogen involves an approximately sp^3 carbon orbital overlapping with an orbital on the halogen for which the principal quantum number varies from $n = 2$ for fluorine to $n = 5$ for iodine. The C—X bond lengths increase and the bond strengths decrease as we read down the periodic table. Some typical values for methyl halides are shown in Figure 6.4, which gives both C—X bond lengths (in angstrom units) and C—X bond strengths (in kilocalories per mole). *Remember*: These values refer to homolytic bond breaking (formation of two neutral radicals, Fig. 1.44).

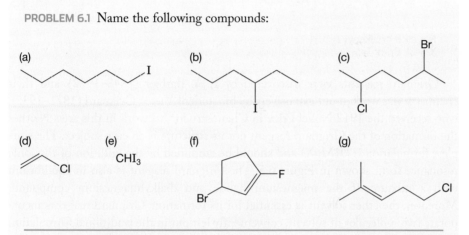

FIGURE 6.4 The C—X bond of an alkyl halide is formed by the overlap of a singly occupied carbon sp^3 orbital with a singly occupied halogen orbital.

PROBLEM 6.1 Name the following compounds:

(a)

(b)

(c)

(d)

(e) CHI$_3$

(f)

(g)

PROBLEM 6.2 Draw structures for the following compounds:

(a) 3,3-dichlorohexane

(b) 1-bromocyclohexene

(c) 2-bromopropene

(d) *meso*-3,4-dichlorohexane (three dimensions!)

(e) (1*R*, 2*R*)-dichlorocyclopropane

6.3 Alkyl Halides as Sources of Organometallic Reagents: A Synthesis of Hydrocarbons

Halides have all sorts of uses in the "real" practical world. They appear as anesthetics, insecticides, herbicides, the building blocks of polymers such as polyvinyl chloride (PVC), and myriad other products. As you read this paragraph you are probably thinking, "Sure, but haven't there been serious problems with insecticides such as DDT and herbicides like Agent Orange, and aren't chlorofluorocarbons eating up the ozone layer?" The answer is a resounding, Yes! There seems to be no "free lunch" as unintended consequences often appear when we try to solve problems with technology. The issues are complex, but it is clear that the chemists of today and the future need to supply the technical expertise to avoid these problems while retaining the benefits.

Halides are important in synthetic chemistry because they allow for the formation of carbon–hydrogen and carbon–carbon bonds. Central to these uses of alkyl and other halides is their easy conversion into **organometallic reagents**, molecules containing carbon–metal bonds. We look at only one limited use of these synthetically important compounds in this chapter, but many other uses will appear later.

When an alkyl halide (fluorides are generally exceptions) is added to a cold mixture of magnesium or lithium metal and an ether solvent, the metal begins to disappear in an exothermic reaction. In general, the solvent used for this reaction, an ether, has the structure R—O—R. Diethyl ether or tetrahydrofuran (THF) is most often used. The end result of the reaction is either a **Grignard reagent** or an **organolithium reagent** (Fig. 6.5).

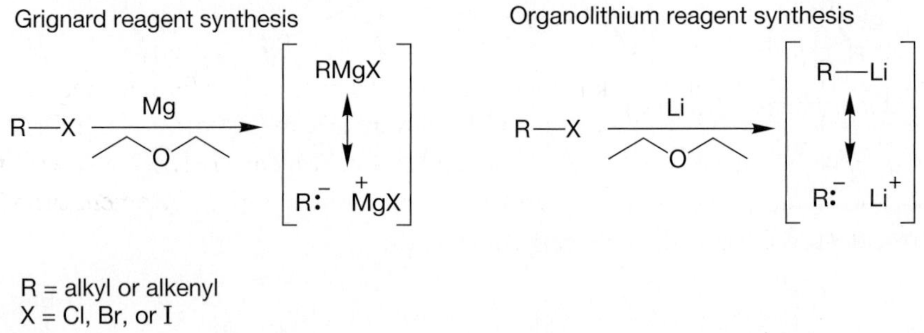

FIGURE 6.5 Formation of organometallic reagents.

Grignard reagent synthesis

Organolithium reagent synthesis

R = alkyl or alkenyl
X = Cl, Br, or I

Grignard reagents were discovered by P. A. Barbier (1848–1922), and their chemistry was worked out extensively by his student Victor Grignard (1871–1935), who received the 1912 Nobel Prize in Chemistry for his work in this area. Neither the formation of the Grignard reagent nor its structure is an easy subject. The simplest formulation is "RMgX" and should be amplified by the addition of the ionic resonance form shown in Figure 6.5. The Grignard reagent is also in equilibrium with a mixture of the magnesium halide and dialkylmagnesium compound. Moreover, the ether solvent is essential for its formation. Grignard reagents incorporate two molecules of solvent, conveniently left out in the traditional formulation as RMgX (Fig. 6.6). Whatever the detailed structure, the result is a highly polar reagent that is a very strong Lewis base.

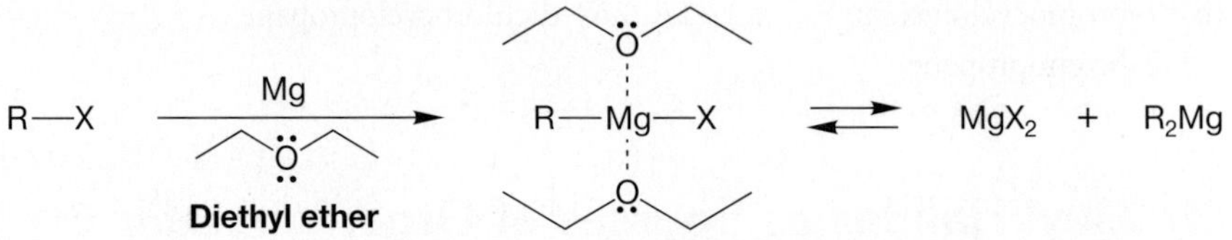

FIGURE 6.6 The Grignard reagent has a complex structure.

The mechanism of formation of the Grignard reagent involves radical-transfer reactions in which a transient alkyl radical is formed in the presence of a magnesium-centered radical (Fig. 6.7).

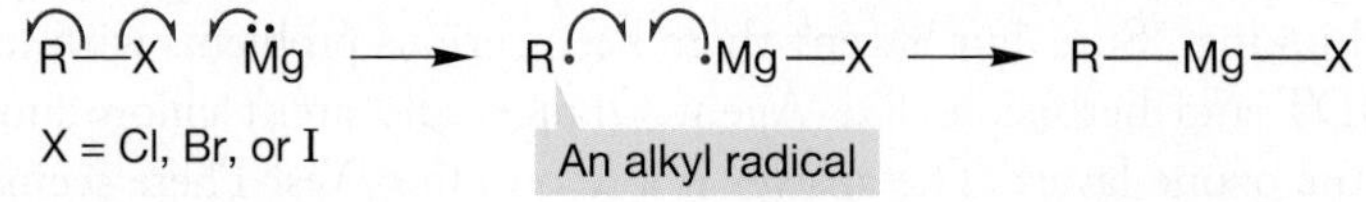

FIGURE 6.7 Formation of a Grignard reagent.

X = Cl, Br, or I

An alkyl radical

CONVENTION ALERT

We met the methyl radical in Chapter 2 (p. 62). Radicals, sometimes called **free radicals**, are neutral species that have a single nonbonding electron and can undergo many reactions, as we will see in Chapter 11. Typical and important reactions of radicals include combining with other radicals to form bonds. The combination of two carbon-based radicals is especially important because it results in the formation of a carbon–carbon bond. The arrow formalism for radical reactions uses single-barbed arrows.

PROBLEM 6.3 Draw an interaction diagram to show the stabilization when two methyl radicals combine. How great is that stabilization? For an example of an interaction diagram see Figure 1.39, p. 33.

Coupling of R· and ·MgX generates RMgX, and further radical reactions can equilibrate this species with the dialkylmagnesium. Most halides (except fluorides) can be made into Grignard reagents.

Organolithium reagents are even more complex. Here the ether solvent is not essential, but the nature of the organolithium reagent does depend on the solvent. Organolithium reagents are known not to be monomeric; they form aggregates the size of which depends on the nature of the solvent and the structure of the R group. As in RMgX, the representation of organolithium reagents as RLi, even with the addition of the polar resonance form shown in Figure 6.5, is a convenient simplification (Fig. 6.8).

$$R\!-\!X \ + \ Li \ \longrightarrow \ (R\!-\!Li)_n$$

FIGURE 6.8 Organolithium reagents are not monomeric, but oligomeric. The exact value of n depends on the solvent and the structure of R.

Both of these organometallic reagents are extraordinarily strong bases, and they must be carefully protected from moisture and oxygen. Chemists form them under a strictly dry and inert atmosphere. Although these reagents do not contain free carbanions, *they act as if they did*. They are sources of R:⁻. The very polar carbon–metal bond attacks all manner of Lewis and Brønsted acids. Water is more than strong enough to protonate a Grignard or organolithium reagent (Fig. 6.9).

$$R\!-\!Mg\!-\!X \ + \ H\!-\!\ddot{O}H \ \longrightarrow \ R\!-\!H \ + \ H\ddot{O}\!-\!Mg\!-\!X$$

$$R\!-\!Li \ + \ H\!-\!\ddot{O}H \ \longrightarrow \ R\!-\!H \ + \ Li\ddot{O}H$$

FIGURE 6.9 The powerfully basic organometallic reagents are easily protonated by water to give hydrocarbons and metal hydroxides.

One of the end products of the reactions in Figure 6.9 is a hydrocarbon. This sequence constitutes a synthesis of hydrocarbons from most halides. This reaction can often be put to good use in the construction of isotopically labeled reagents, because D_2O can be used in place of H_2O to produce a specifically deuteriolabeled hydrocarbon (Fig. 6.10).

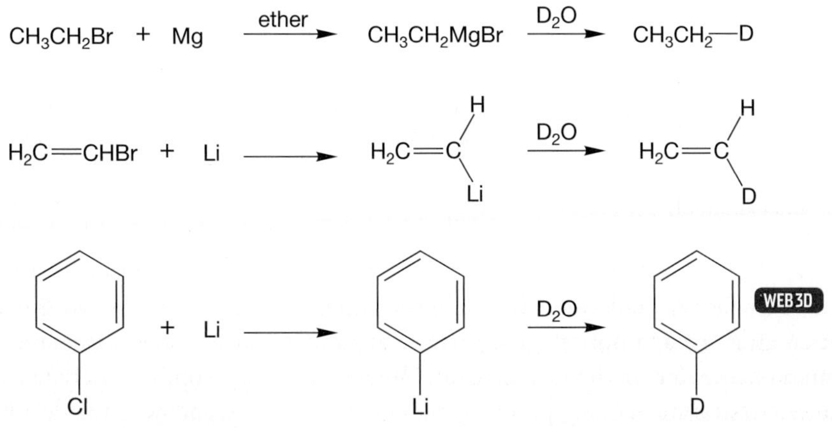

FIGURE 6.10 Treatment of an organometallic reagent with D_2O gives deuteriolabeled molecules.

PROBLEM 6.4 Given inorganic reagents of your choice (including D_2O), devise syntheses of the following molecules from the indicated starting materials:

6.4 Alcohols

6.4a Nomenclature There are two ways to name alcohols: the IUPAC systematic way, and the way chemists often do it, by using the splendidly vulgar common names. As usual, these common names are used only for the smaller members of the class, and the system takes over as complexity increases.

Alcohols are named systematically by dropping the final "e" of the parent hydrocarbon and adding the suffix "ol." The functional group, OH, is given the lowest number possible, taking precedence over other groups in the molecule including the related thiol group, SH. The longest carbon chain is identified and the substituents numbered appropriately (Fig. 6.11).

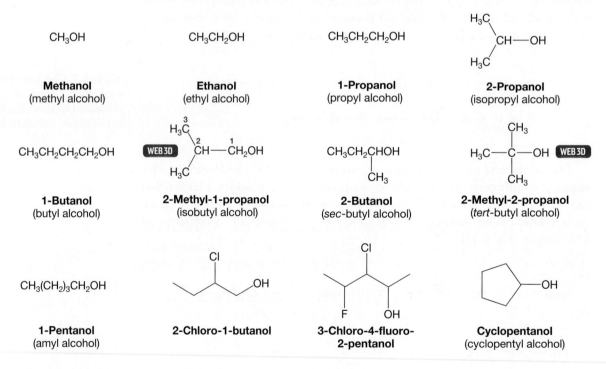

FIGURE 6.11 Some systematic and commonly used names for alcohols.

The smaller alcohols, however, are given common names with the word *alcohol* written after the appropriate group name. Figure 6.11 gives some systematic and common names for the smaller alcohols. When we get past approximately five carbons, the systematic naming protocol takes over completely, and even the delightful name amyl for the five-carbon alcohols is disappearing.[2] However, some common names seem to be quite hardy. For example, the correct IUPAC name for hydroxy-benzene is *benzenol*, but the common name *phenol* is still accepted in the IUPAC system and probably will survive for a long time (Fig. 6.12).

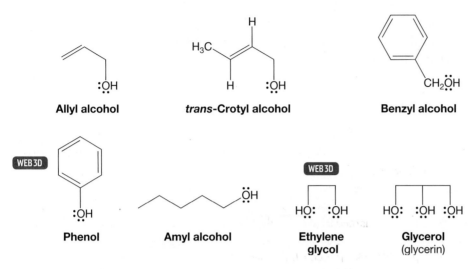

FIGURE 6.12 Some widely used common names for alcohols.

PROBLEM 6.5 Provide systematic names for the following alcohols:

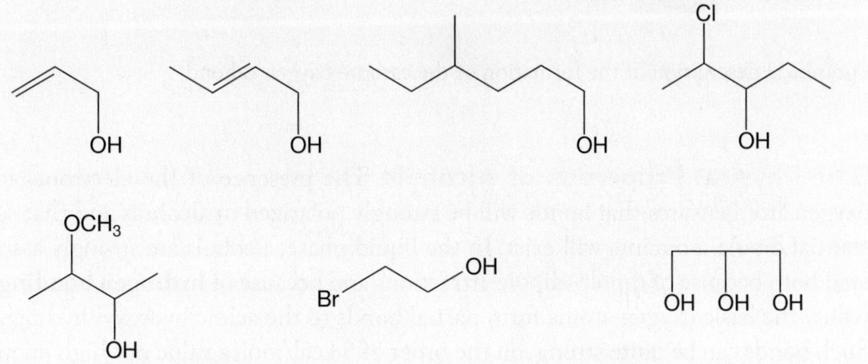

6.4b Structure of Alcohols

Alcohols are derivatives of water, and it will be no surprise to see that they rather closely resemble water structurally. The bond angle is expanded a bit from 104.5° in H—O—H to approximately 109° in simple alcohols R—O—H.

The oxygen–hydrogen bond length is very little changed from that of water, but the carbon–oxygen bond is a bit shorter and stronger than the carbon–carbon bond in ethane (Fig. 6.13).

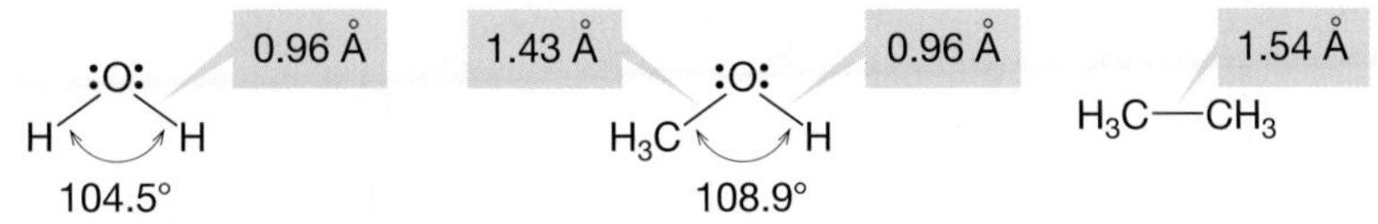

FIGURE 6.13 Bond lengths and angles for water, methyl alcohol, and ethane.

---

[2] *Amyl* derives from the Latin word for starch, *amylum*. The first amyl alcohol was isolated from the fermentation of potatoes.

PROBLEM 6.6 What is the approximate hybridization of the oxygen atom in methyl alcohol (H_3C—O—H angle = 108.9°). How do you know?

The orbital interaction diagram of Figure 6.14 shows the construction of the carbon–oxygen bond, as well as typical bond energies. Notice that an electron in an orbital near the very electronegative oxygen is more stable than it would be in an orbital near carbon.

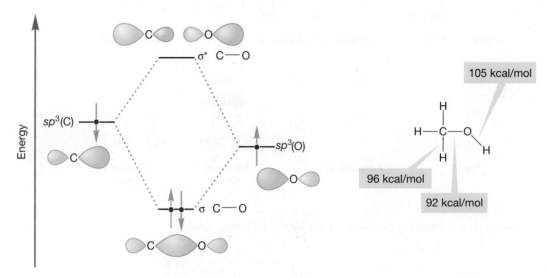

FIGURE 6.14 A graphical description of the formation of the carbon–oxygen σ bond.

6.4c Physical Properties of Alcohols The presence of the electronegative oxygen atom ensures that bonds will be strongly polarized in alcohols and that substantial dipole moments will exist. In the liquid phase, alcohols are strongly associated both because of dipole–dipole attractions and because of **hydrogen bonding** in which the basic oxygen atoms form partial bonds to the acidic hydroxyl hydrogens. Such bonds can be quite strong, on the order of 5 kcal/mol, a value not high enough to make permanent dimeric or polymeric structures but quite sufficient to raise the boiling points of alcohols far above those of the much less strongly associated alkanes. Figure 6.15 shows an ordered, hydrogen-bonded structure for an alcohol. In the absence of hydrogen bonding, water with a molecular weight of only 18, and the low molecular weight alcohols, would surely be gases. The polarity of alcohols

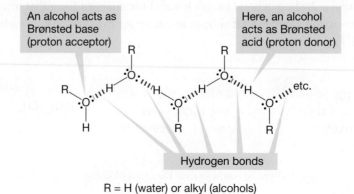

FIGURE 6.15 Water and alcohols are both Brønsted acids and Brønsted bases. This figure shows hydrogen bonding between molecules of water (R = H) and alcohols.

An alcohol acts as Brønsted base (proton acceptor)

Here, an alcohol acts as Brønsted acid (proton donor)

Hydrogen bonds

R = H (water) or alkyl (alcohols)

makes them quite water soluble, and most of the smaller molecules are miscible (soluble in all proportions) with, or at least highly soluble in, water. Table 6.3 gives some physical properties of alcohols. For comparisons with the parent alkanes see Table 2.4 (p. 79).

TABLE 6.3 Some Physical Properties of Alcohols

Compound	bp (°C)	mp (°C)	Density (g/cm³)	Dipole Moment (D)
CH_3OH Methyl alcohol	65.15	−93.9	0.79	1.7
CH_3CH_2OH Ethyl alcohol	78.5	−117.3	0.79	1.69
$(CH_3)_2CHOH$ Isopropyl alcohol	82.4	−89.5	0.80	1.68
$CH_3CH_2CH_2CH_2OH$ Butyl alcohol	117.2	−89.5	0.81	1.66
$(CH_3)_2CHCH_2OH$ Isobutyl alcohol	108	−108	0.80	1.64
$CH_3CH_2CH(CH_3)OH$ sec-Butyl alcohol	99.5	−115	0.81	
$(CH_3)_3COH$ tert-Butyl alcohol	82.3	25.5	0.79	
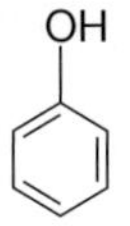 OH Phenol	181.7	43	1.06	1.45

Even though these intramolecular hydrogen bonds are relatively weak (~5 kcal/mol) they are critically important. For example, life as we know it is clearly not possible without water and there would be no liquid water without hydrogen bonds. Moreover, hydrogen bonds are used to maintain the proper structures in proteins and nucleic acids, polymeric structures essential to our existence that we will meet later in this book.

6.4d Acid and Base Properties of Alcohols We have just seen water and alcohols acting as Brønsted acids and bases. Hydrogen bonding is a perfect example of such behavior. In Chapter 7, we are going to look closely at a number of reactions of alcohols. These processes depend on the ability of alcohols to act as both acids and bases. Accordingly, we first make sure that we can see these roles clearly.

A Brønsted acid is any compound that can donate a proton. A Brønsted base is any compound that can accept a proton. Let's look first at a very simple process, the reaction of solid potassium hydroxide (KOH) with gaseous hydrochloric acid (HCl). This reaction is nothing but a competition of two Brønsted bases (Cl⁻ and HO⁻) for a proton, H⁺. The stronger Brønsted base (HO⁻) wins the competition (Fig. 6.16).

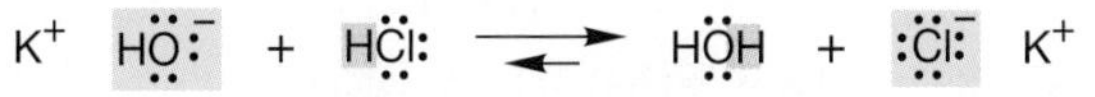

$$K^+ \ \ H\ddot{O}\!:^- \ + \ H\ddot{C}l\!: \ \rightleftharpoons \ H\ddot{O}H \ + \ :\ddot{C}l\!:^- \ K^+$$

FIGURE 6.16 Two Brønsted bases (blue) competing for a proton.

Hydrochloric acid (HCl) is called the **conjugate acid** of Cl⁻, and Cl⁻ is the **conjugate base** of hydrochloric acid. Conjugate bases and acids are related to each other through the gain and loss of a proton (Fig. 6.17).

FIGURE 6.17 Conjugate acids and bases.

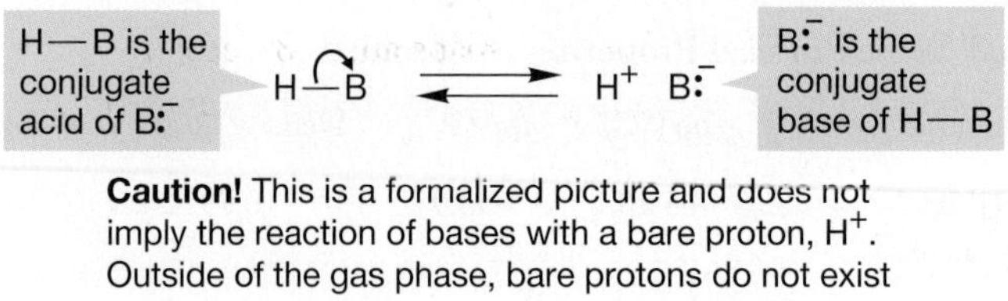

H—B is the conjugate acid of B:⁻   H⌒B ⇌ H⁺ B:⁻   B:⁻ is the conjugate base of H—B

Caution! This is a formalized picture and does not imply the reaction of bases with a bare proton, H⁺. Outside of the gas phase, bare protons do not exist

PROBLEM 6.7 What are the conjugate acids of the following molecules?
(a) H_2O (b) ⁻OH (c) NH_3 (d) CH_3OH (e) $H_2C{=}O$ (f) ⁻CH_3

PROBLEM 6.8 What are the conjugate bases of the following molecules?
(a) H_2O (b) ⁻OH (c) NH_3 (d) CH_3OH (e) CH_4 (f) $HOSO_2OH$ (g) ⁻OSO_2OH

As we saw in Figure 6.15, a molecule may be *both* a Brønsted acid and a Brønsted base. Consider water, for example. Water can both donate a proton (act as a Brønsted acid) and accept a proton (act as a Brønsted base). Figure 6.18 shows two reactions, (a) the protonation of water by hydrogen chloride, in which water acts as a base, and (b) the protonation of hydroxide ion by water, in which water is acting as the acid.

FIGURE 6.18 Water is both a Brønsted acid and base.

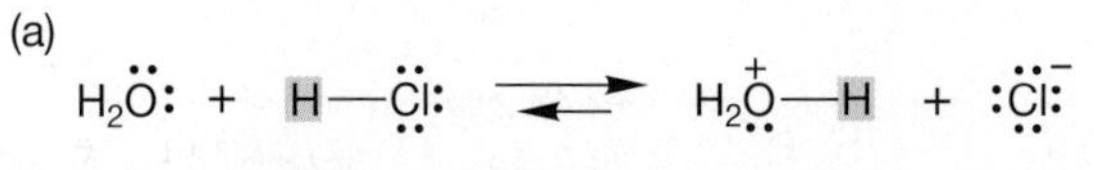

(a)
$H_2\ddot{O}{:}$ + H—$\ddot{\underset{..}{C}}l{:}$ ⇌ $H_2\overset{+}{\underset{..}{O}}$—H + ${:}\ddot{\underset{..}{C}}l{:}^-$

(b)
$H\ddot{\underset{..}{O}}{:}^-$ + H—$\ddot{O}H$ ⇌ $H\ddot{\underset{..}{O}}$—H + $H\ddot{\underset{..}{O}}{:}^-$

Many reactions begin with a protonation step. Because strong acids are better able to donate a proton than weaker acids, it will be important for us to know which molecules are strong acids and which are weak acids. The dissociation of an acid HA in water can be described by the equation

$$HA + H_2O \rightleftharpoons H_3O^+ + A^-$$

This equation leads to an expression for the equilibrium constant, K.

$$K = \frac{[H_3O^+]\,[A^-]}{[HA]\,[H_2O]}$$

Because it is present as solvent, in vast excess, the concentration of water remains constant in the ionization reaction. This equation is usually rewritten to give the acidity constant, K_a,

$$K_a = \frac{[H_3O^+]\,[A^-]}{[HA]}$$

where the square brackets indicate concentration.

By analogy with pH, we can define a quantity **pK_a**.

$$pK_a = -\log K_a$$

The stronger the acid, the lower is its pK_a. Table 6.4 gives pK_a values for some representative compounds. This short list spans about 80 powers of 10 (pK_a is a log function). In practice, any compound with a pK_a lower than about +5 is regarded as a reasonably strong acid; those with pK_a values below 0 are very strong. As we discuss new kinds of molecules, more pK_a values will appear. Should you memorize this list? We think not, but you will need to have a reasonable idea of the approximate acidity of different kinds of molecules.

TABLE 6.4 Some pK_a Values for Assorted Molecules[a]

Compound	pK_a	Compound	pK_a
HI	−10	H_4N^+	9.2
H_2SO_4	−3	$CH_3O\underline{H}$	15.5
HBr	−9	H_2O	15.7
HCl	−8	$RO\underline{H}$	16–18
$R\overset{+}{O}H_2$	−2	$HC\equiv CH$	24
H_3O^+	−1.7	NH_3	38
HNO_3	−1.3	$H_2C=CH_2$	50
HF	3.2	$(CH_2)_3$ (cyclopropane)	46
$RCOO\underline{H}$	4–5	CH_4	50–60
H_2S	7.0	$(CH_3)_3C\underline{H}$	50–70

[a]When there is a choice, it is the underlined $\underline{H}$ that is lost. A longer list is provided on the back inside cover.

In alcohols and water, Brønsted basicity is centered on the nonbonding electrons of the oxygen atom, which form bonds with a variety of acids. Protonation converts an alcohol, ROH, into an oxonium ion (RO^+H_2). This conversion will have profound chemical consequences, which we can preview here. Breaking the R—OH bond to form ions is very difficult, because both positive (R^+) and negative (HO^-) ions must be formed. Hydroxide, HO^-, is very difficult to form— one says that hydroxide is a poor leaving group—and alcohols do not ionize easily (Fig. 6.19). After protonation, however, the "leaving group" is no longer hydroxide, but water. Formation of the neutral molecule water is a much easier process (Fig. 6.19).

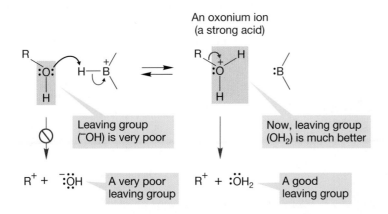

FIGURE 6.19 Protonation of an alcohol leads to an oxonium ion. In this process, the very poor leaving group OH is transformed into the good leaving group OH_2.

TABLE 6.5 Some pK_a Values for Simple Oxonium Ions

Compound	pK_a
$H_3\overset{+}{O}$	−1.74
$CH_3\overset{+}{O}H_2$	−2.2
$CH_3CH_2\overset{+}{O}H_2$	−1.9
$CH_3CH_2CH_2CH_2\overset{+}{O}H_2$	−2.3
$CH_3CH_2CHO\overset{+}{H}CH_3$	−2.2
$(CH_3)_3C\overset{+}{O}H_2$	−2.6
$RO\overset{+}{H}R'$	About −3.5

Protonated alcohols are very strong Brønsted acids. Therefore, in order to protonate an alcohol, it takes a very strong acid indeed. The conjugate base (B:⁻) of the protonating acid (HB) must be a weaker base than the alcohol. The base must be an ineffective competitor in the equilibrium of Figure 6.19.

Oxonium ions are much better proton donors than are the alcohols themselves. Table 6.5 gives the pK_a values for a few protonated alcohols and water. Oxonium ions formed from alcohols have pK_a values in the −2 range, and those from ethers (R—O—R′) are even more acidic.

Brønsted acidity, of course, is centered on the hydroxyl hydrogen. Loss of this hydrogen to an acceptor, a Brønsted base, leads to the conjugate base of the alcohol, the **alkoxide ion** (RO⁻). The proton is not just lost as H⁺ but removed by a base (Fig. 6.20).

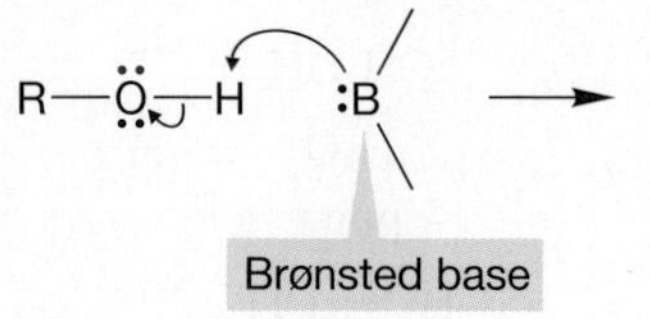

FIGURE 6.20 Loss of a proton to a general Brønsted base leads to an alkoxide ion. Alkoxides are named very simply as oxides of the parent alcohol. Thus, methyl alcohol, CH_3OH, is deprotonated to give methoxide, CH_3O^-, ethyl alcohol, CH_3CH_2OH, is deprotonated to give ethoxide, $CH_3CH_2O^-$, and so on.

Alcohols are approximately as acidic as water. Table 6.6 gives the pK_a values for some simple alcohols in aqueous solution as well as water itself.

TABLE 6.6 Some pK_a Values for Simple Alcohols in Water

Compound		pK_a (loss of underlined H)
Water	H_2O	15.7 Strongest acid
Methyl alcohol	$CH_3O\underline{H}$	15.5
Ethyl alcohol	$CH_3CH_2O\underline{H}$	15.9
Isopropyl alcohol	$(CH_3)_2CHO\underline{H}$	16.5
tert-Butyl alcohol	$(CH_3)_3CO\underline{H}$	17.0 Weakest acid

Note that the acidity order in water is $CH_3OH > CH_3CH_2OH > (CH_3)_2CHOH > (CH_3)_3COH$. Apparently, the more alkyl groups on the alcohol, the weaker an acid it is. For many years, this pK_a order was explained by assuming that alkyl groups are intrinsically electron donating. If this were true, the product alkoxides would be destabilized by alkyl groups and the acidity of the corresponding alcohol would be reduced (Fig. 6.21).

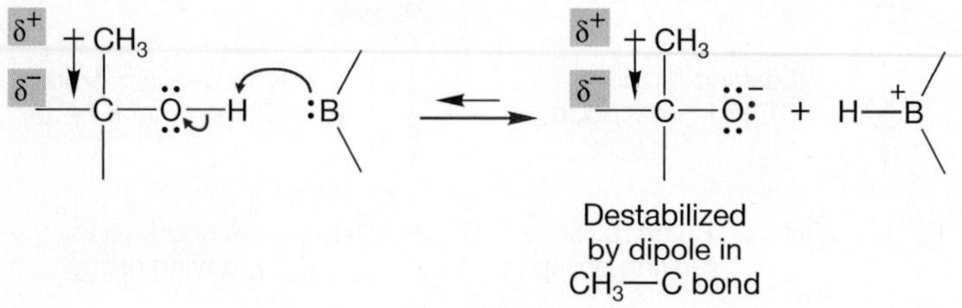

FIGURE 6.21 If alkyl groups were electron-releasing, formation of an alkoxide from an alcohol would be retarded by alkyl groups.

Inductive effects, the result of polarized sigma bonds, *are* important. For example, the pK_a of 2,2,2-trifluoroethanol ($CF_3CH_2O\underline{H}$) is 12.5, whereas the pK_a of ethyl alcohol is 15.9. The fluorinated alcohol is more than a thousand times as acidic as ethyl alcohol (remember the pK_a scale is logarithmic). Fluorinated alcohols are always stronger acids than their hydrogen-substituted counterparts.

WORKED PROBLEM 6.9 Explain in detail why 2,2,2-trifluoroethanol is a stronger acid than ethyl alcohol.

ANSWER As fluorine is very electronegative, the fluorines are strongly electron-withdrawing and will help stabilize the alkoxide ion. They will also stabilize the transition state leading to the alkoxide because partial negative charge has begun to develop on the oxygen atom. So the fluorines will have an effect on both the kinetic and thermodynamic properties of the alcohol.

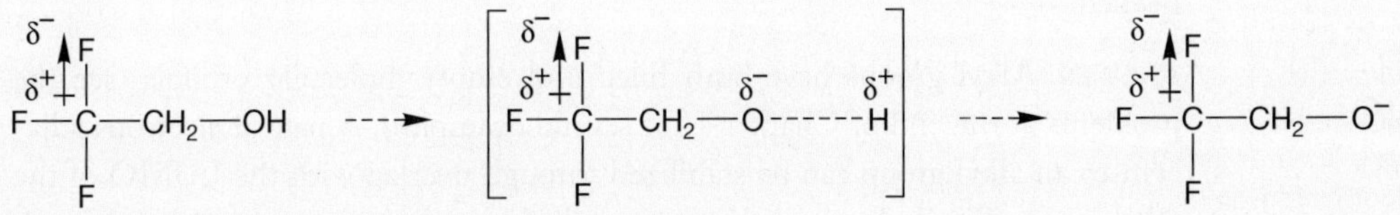

Yet this traditional, and simple, explanation that treats alkyl groups as intrinsically electron-donating is not correct. In 1970, Professor John Brauman (b. 1937) and his co-workers at Stanford University showed that in the gas phase the *opposite* acidity order obtained. The intrinsic acidity of the four alcohols of Table 6.6 is exactly opposite to that found in solution. The acidity order measured in solution reflects a powerful effect of the solvent, not the natural acidities of the alcohols themselves. Organic ions are almost all unstable species, and the formation of the alkoxide anions depends critically on how easy it is to stabilize them through interaction with solvent molecules, a process called **solvation**. *tert*-Butyl alcohol is a weaker acid in solution than methyl alcohol because the large *tert*-butyl alkoxide ion is difficult to solvate. The more alkyl groups, the more difficult it is for the stabilizing solvent molecules to approach (Fig. 6.22). Of course, in the gas phase where solvation is impossible, the natural acidity order is observed.

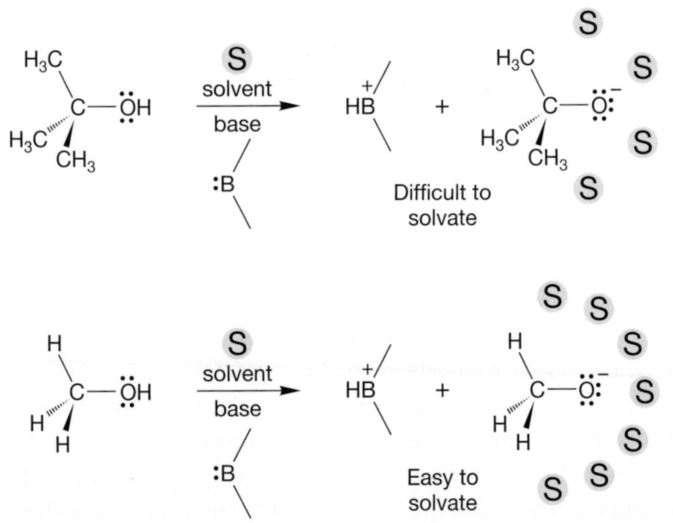

FIGURE 6.22 The smaller the alkoxide ion, the easier it is for solvent molecules to approach and stabilize it.

In the gas phase, the alkyl groups are actually operating so as to *stabilize* the charged alkoxide ions, presumably by withdrawing electrons, exactly the opposite from what had been thought (Fig. 6.23).

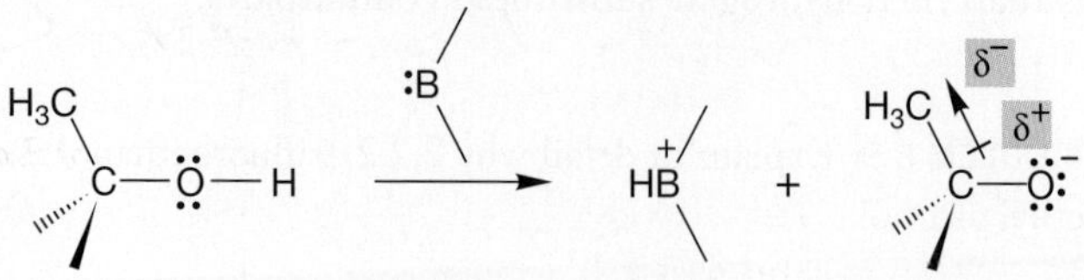

FIGURE 6.23 Alkyl groups can be electron-withdrawing.

WORKED PROBLEM 6.10 Can you describe this phenomenon—the stabilization of a pair of electrons by an adjacent alkyl group—in orbital terms? No orbital construction or complicated argument is necessary. A simple statement is all that is needed.

ANSWER Alkyl groups have both filled and empty molecular orbitals (see the problems at the end of Chapter 1 for several examples). A pair of electrons adjacent to an alkyl group can be stabilized through overlap with the LUMO of the alkyl group. (Similarly, an alkyl group stabilizes an adjacent empty orbital through overlap with the alkyl HOMO.)

What lessons are to be drawn from this discussion? First, solvation is important and not completely understood. We are certain to find other phenomena best explained in terms of solvation in the future. Second, the gas phase is the ideal medium for revealing the intrinsic properties of molecules. (Some would go further and say that calculation is the best way.) However, the practical world of solvated reactions is certainly real, and for us to understand reactivity we can no longer afford to ignore the solvent.

6.5 Solvents in Organic Chemistry

We have just seen an example of a solvent playing a crucial role in determining the acidities of alcohols. The greater acidity of methyl alcohol over *tert*-butyl alcohol in a polar solvent is a result of the greater ease of solvating the conjugate base of methyl alcohol, the methoxide ion, over the *tert*-butoxide ion (Fig. 6.22). Let's take a quick look at the properties of solvents and the process of solvation.

6.5a Polar Solvents We have seen that one kind of stabilization by solvent—solvation—is a consequence of hydrogen bonding. For hydrogen bonding to be possible, there clearly must be both a proton donor (usually a hydroxy or amino group, OH or NH_2) and a proton acceptor available (a pair of electrons). Solvents that can donate a proton are called **protic solvents**. Protic solvents are usually also quite polar; that is, they have relatively high dielectric constants, ε, a measure of the ability of a solvent to separate charges. Solvents without available protons—in other words, solvents that are not proton donors—are **aprotic solvents** and can be either polar or nonpolar. It is distinctly possible for a solvent to be quite polar, to have a high ε, but not be a proton donor. Of course, many molecules are both nonpolar and aprotic; hydrocarbons are obvious examples you know. Table 6.7 collects a few common polar solvents and their descriptions.

TABLE 6.7 Some Common Polar Solvents

Compound	Name	Dielectric Constant (ε)	Description
H_2O	Water	78	Protic, polar
HCOOH	Formic acid	59	Protic, polar
CH_3OH	Methyl alcohol	33	Protic, polar
CH_3CH_2OH	Ethyl alcohol	25	Protic, polar
$(CH_3)_2SO$	Dimethyl sulfoxide (DMSO)	47	Aprotic, polar
CH_3CN	Acetonitrile	38	Aprotic, polar
$(CH_3)_2NCHO$	Dimethylformamide (DMF)	37	Aprotic, polar
CH_3NO_2	Nitromethane	36	Aprotic, polar

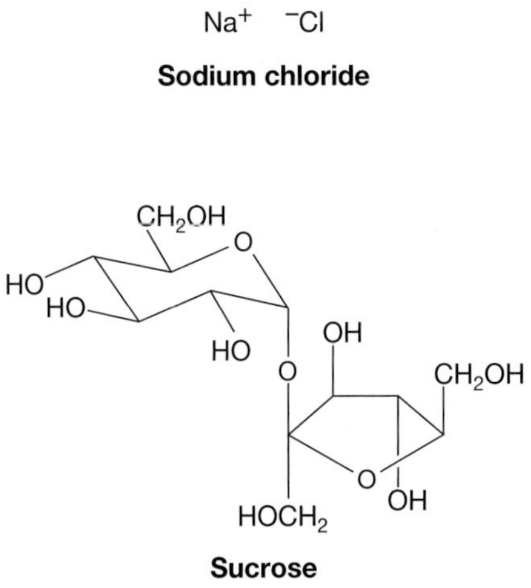

6.5b Solubility: "Like Dissolves Like" Oil spreads on water, but very little dissolves. Two solids, salt (the prototypal ionic solid sodium chloride) and the sugar sucrose ($C_{12}H_{22}O_{11}$, an organic material bristling with OH groups), both dissolve completely in water (Fig. 6.24). What is going on?

Sodium chloride, Na^+Cl^-, dissolves in the polar solvent water because both the positive sodium ion and the negative chloride ion are well solvated by the protic, polar solvent water (Fig. 6.25). Nonpolar solvents cannot solvate ions efficiently and do not dissolve sodium chloride.

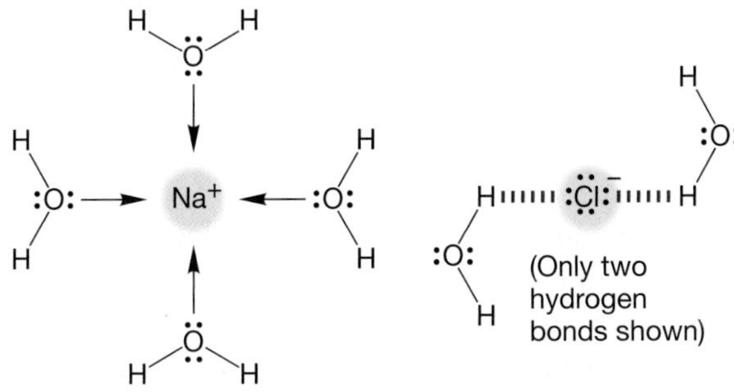

FIGURE 6.24 Two very different solids that dissolve easily in water.

FIGURE 6.25 Solvation of the positive sodium cation, and hydrogen bonding to the negative chloride ion (only two hydrogen bonds shown).

Protic solvents are particularly good at dissolving molecules capable of hydrogen bonding. Formation of hydrogen bonds in solution is highly stabilizing. Thus sucrose, a molecule with many OH groups, dissolves easily in water. The polar solvent water stabilizes sucrose both through formation of many hydrogen bonds and through stabilization of the polarized O—H groups of sucrose. Figure 6.26 shows a schematic picture of hydrogen bonding (only one of many shown) and dipolar stabilization of the polyhydroxy compound sucrose by water.

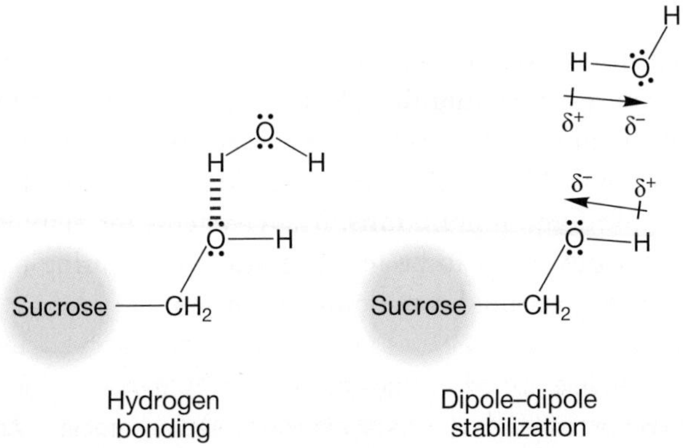

FIGURE 6.26 Solvation of an O—H bond by water.

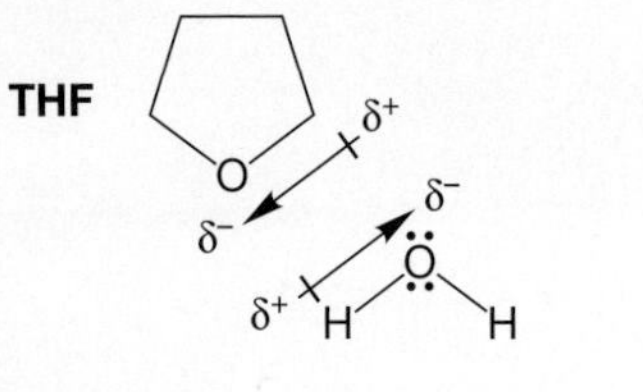

THF

FIGURE 6.27 Stabilization (solvation) of water by tetrahydrofuran, THF.

Polar aprotic solvents can help disperse the partial charges in other polar aprotic molecules and thus dissolve them well. Although they cannot act as proton donors, such solvents can align their negative ends toward the positive charges in a polar molecule and their positive ends toward the negative charges (Fig. 6.27). For example, Figure 6.27 shows the polar solvent THF solvating a water molecule by aligning dipoles in a stabilizing way. Water and THF are miscible.

By contrast, hydrocarbons cannot form hydrogen bonds (they are neither proton donors nor acceptors) and are not very polar. Typical ε values for hydrocarbons are around 2. The more nonpolar and "greasy" the solvent, the more easily the similarly nonpolar hydrocarbon molecules dissolve in it. Thus, hydrocarbons (oil is a mixture of many hydrocarbons) are almost completely insoluble in water, but very soluble in other hydrocarbons. The polar molecule acetone [dimethyl ketone, $(CH_3)_2CO$], with an ε value of 21, is completely soluble in water, but the alkene isobutene, with the same number of nonhydrogen atoms, is almost completely insoluble in water. Like does dissolve like.

Summary

In the practical world of making chemical reactions work, we need to allow molecules to "find each other" and react. Dissolving molecules in a fluid medium provides an opportunity for molecules to move around and do just that—find other reactive molecules. The choice of solvent is critical because we need to dissolve the molecules we want to react together. In making that choice, the rather simple notion that "like dissolves like" really works! As we work through chemical reactions in the next few chapters, we will have to take account of solvents and solvation.

6.6 Diols (Glycols)

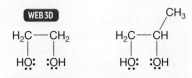

1,2-Ethanediol   **1,2-Propanediol**
or ethylene glycol   or propylene glycol

FIGURE 6.28 1,2-Ethanediol and 1,2-propanediol are called ethylene glycol and propylene glycol, respectively.

Although molecules containing two OH groups are logically enough called **diols**, there is an often-used common name as well, **glycol**.

Some 1,2-diols or 1,2-glycols are familiar and are often important molecules. 1,2-Ethanediol, or ethylene glycol, is the commonly used antifreeze, for example. Note that in naming these compounds, the final "e" of the alkane parent compound is not dropped, as it is in naming simple alcohols, and that the almost universally used common name can be very misleading. *Ethylene glycol* implies the presence of an alkene, and *propylene glycol* (1,2-propanediol) suggests a three-carbon unsaturated chain. In neither case is the implied unsaturation present (Fig. 6.28). Be aware that the real structures are based on *saturated* hydrocarbon chains.

6.7 Amines

6.7a Nomenclature We first saw **amines** in Chapter 1, albeit very briefly. Amines are derivatives of **ammonia**, :NH_3 (Fig. 6.29). Successive replacement of the hydrogens of ammonia leads to **primary, secondary,** and **tertiary amines,** :NRH_2, :NR_2H, and :NR_3. Quaternary nitrogen compounds are positively charged and are called **ammonium ions**. Many systems for naming amines exist, and the chemical world seems to be resisting attempts to bring order out of this minichaos by keeping to the old common names.

FIGURE 6.29 Substituted amines.

:NH_3	:NH_2R	:NHR_2	:NR_3	$\overset{+}{N}R_4$
Ammonia	Primary amine	Secondary amine	Tertiary amine	Ammonium ion

Primary amines are commonly named by using the name of the substituent, R, and appending the suffix "amine." Secondary and tertiary amines in which the R groups are all the same are simply named as di- and trialkylamines. Amines bearing different R groups are named by ordering the groups alphabetically. Ethylmethylamine is correct, methylethylamine is not, although both are surely intelligible (Fig. 6.30).

WEB3D
CH3NH2
Methylamine
CH3CH2NH2
Ethylamine
PhCH2NH2
Benzylamine
(CH3)2NH
Dimethylamine
(CH3CH2)3N:
Triethylamine
NH
CH2CH3
N
H
CH3
Ethylmethylamine,
(not methylethylamine)
CH2CH3
N
PhCH2
CH3
Benzylethyl-
methylamine
(CH3)3CNH2
tert-Butylamine
CH3CH2CH2NH2
Propylamine
NH2
Cyclohexylamine
Dicyclopentylamine
WEB3D

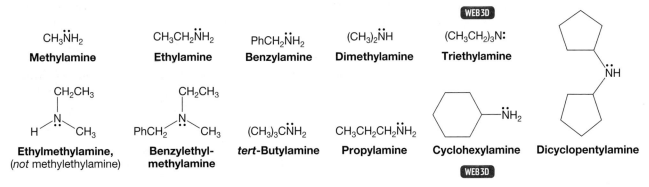

FIGURE 6.30 A naming scheme for primary, secondary, and tertiary amines.

CH3NH2
Methanamine
NH2
Cyclohexanamine

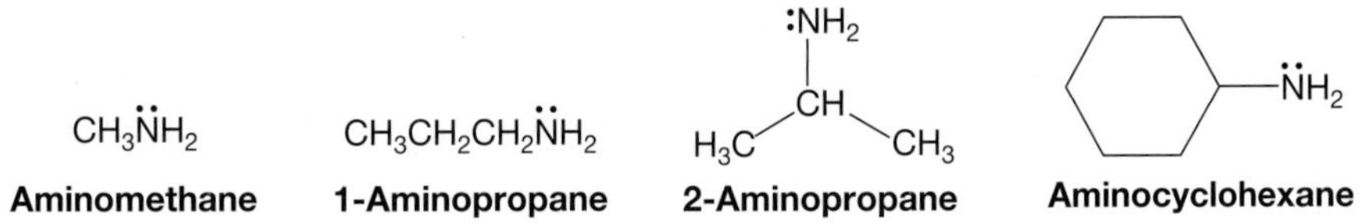

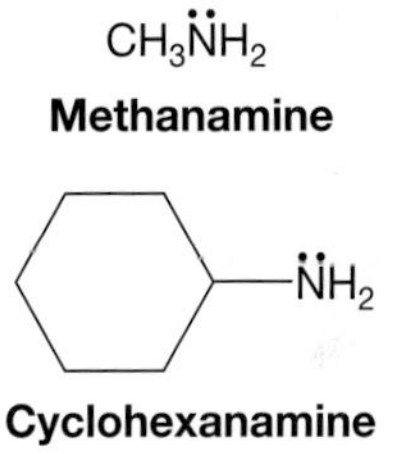

FIGURE 6.31 Two systematically named amines.

The IUPAC system names amines analogously to alcohols. If CH_3OH is officially known as methan<u>ol</u>, then CH_3NH_2 is methan<u>amine</u>. Figure 6.31 gives two examples.

Still another method names amines as substituted alkanes. Now CH_3NH_2 becomes aminomethane. Figure 6.32 gives some examples of amines named this way.

CH3NH2
Aminomethane
CH3CH2CH2NH2
1-Aminopropane
:NH2
CH
H3C
CH3
2-Aminopropane
NH2
Aminocyclohexane

FIGURE 6.32 Still another method for naming amines.

The amino group is just below the alcohol group in the naming protocol priority system. Figure 6.33 gives some correct and incorrect names of polyfunctional compounds containing amino groups.

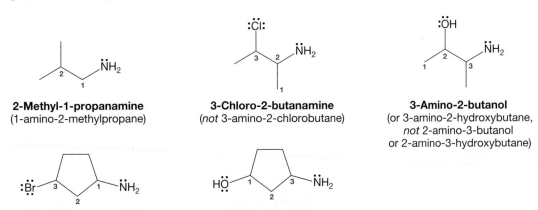

NH2
2
1
2-Methyl-1-propanamine
(1-amino-2-methylpropane)
:Cl:
NH2
3
2
1
3-Chloro-2-butanamine
(not 3-amino-2-chlorobutane)
:OH
NH2
2
1
3
3-Amino-2-butanol
(or 3-amino-2-hydroxybutane,
not 2-amino-3-butanol
or 2-amino-3-hydroxybutane)
:Br
3
1
NH2
2
3-Bromocyclopentanamine
(neither 3-aminobromocyclopentane
nor 4-bromoaminocyclopentane)
HO
1
3
NH2
2
3-Aminocyclopentanol
(not 3-hydroxycyclopentylamine)

FIGURE 6.33 Some polyfunctional amines and their names.

Simple, cyclic amino compounds are also widely seen in nature, and are almost invariably known by their common names. These compounds are often named as "aza" analogues of an all-carbon system. Thus **aziridine** becomes azacyclopropane. Methylamine could be called azaethane, although in point of fact this system is almost never used for acyclic compounds. Unless there is an oxygen atom in the ring, the numbering scheme places the nitrogen at 1 and proceeds so as to minimize the numbering of any substituents. Figure 6.34 gives a few important examples. Once again there is no way out other than to learn these names.

Aziridine (azacyclopropane) | **Azetidine** (azacyclobutane) | **Pyrrolidine** (azacyclopentane) | **Piperidine** (azacyclohexane) | **Pyridine** (azabenzene)

Morpholine | **3-Methylpiperidine** (*not* 5-methylpiperidine) | **3-Bromopyrrolidine** (*not* 4-bromopyrrolidine) | **4-Benzylmorpholine** (*N*-benzylmorpholine) | **Pyrrole**

FIGURE 6.34 Cyclic amines and their names.

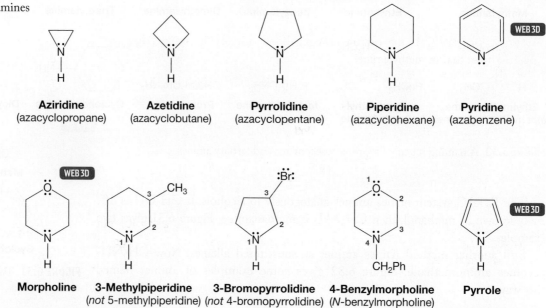

Despite this seeming excess of nomenclature, another aspect of naming amines must be mentioned. Substituted amines can also be named as *N*-substituted amines. In this system, the substituent on nitrogen is designated with the locating prefix *N*. Some examples are shown in Figure 6.35.

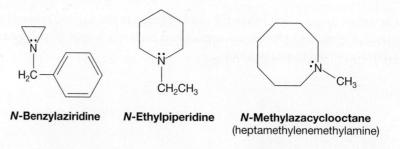

***N*-Benzylaziridine** | ***N*-Ethylpiperidine** | ***N*-Methylazacyclooctane** (heptamethylenemethylamine)

FIGURE 6.35 Still another naming protocol for amines.

Ammonium ions are named by alphabetically attaching the names of the substituents. Don't forget to append the name of the negatively charged counterion (Fig. 6.36).

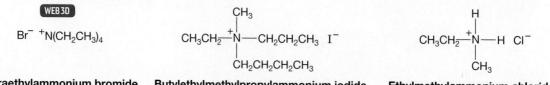

Tetraethylammonium bromide | **Butylethylmethylpropylammonium iodide** | **Ethylmethylammonium chloride**

FIGURE 6.36 Names for three ammonium ions.

6.7b Structure and Physical Properties of Amines In simple amines, the carbon–nitrogen bond distance is a little shorter than the corresponding carbon–carbon bond distance in alkanes. This phenomenon is similar to that found in the alcohols and ethers (Section 6.8) (Fig. 6.37).

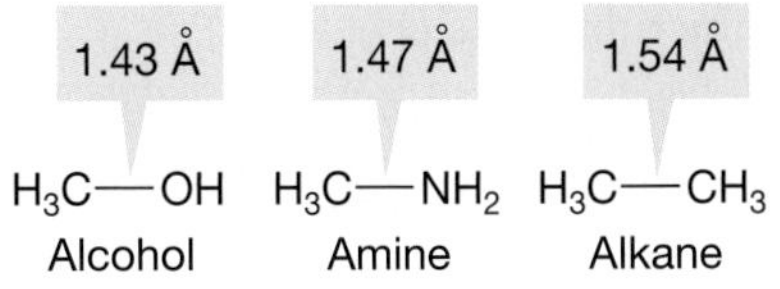

1.43 Å   1.47 Å   1.54 Å

H_3C—OH H_3C—NH_2 H_3C—CH_3
Alcohol   Amine   Alkane

FIGURE 6.37 The carbon–nitrogen bond in amines is shorter than the carbon–carbon bond in alkanes.

PROBLEM 6.11 Why is the carbon–nitrogen bond distance in aniline slightly shorter than that in methylamine?

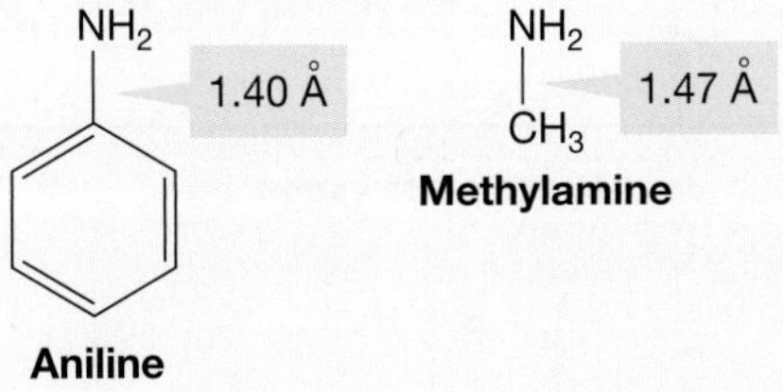

NH_2 1.40 Å

NH_2 1.47 Å
CH_3
Methylamine

Aniline

Simple amines are hybridized approximately sp^3 at nitrogen and thus are pyramidal. The bond angles are close to the tetrahedral angle of 109.5° but, of course, cannot be *exactly* the tetrahedral angle.

PROBLEM 6.12 Why not?

For amines in which all three R groups are identical, there will be only a single R—N—R angle, but for less symmetrical amines, the different R—N—R angles cannot be the same (Fig. 6.38).

The most important structural feature of amines comes from the presence of the pair of nonbonding electrons acting as the fourth substituent on the pyramid. In alkanes, where there are four substituents, there is nothing more to consider once the pyramid is described. In amines there is, because the pyramid can undergo an "umbrella flip," **amine inversion**, forming a new, mirror-image pyramid (Fig. 6.39).

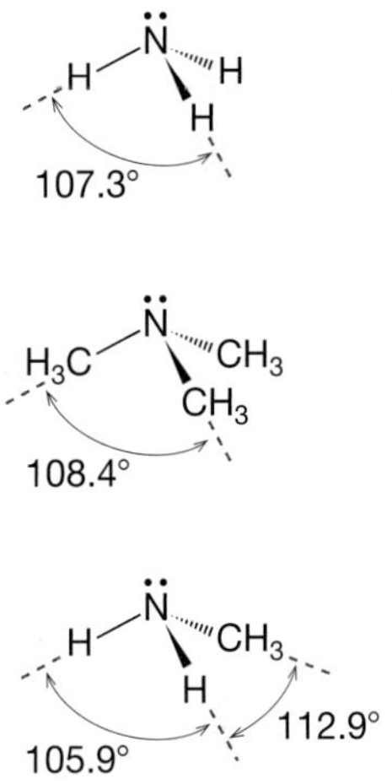

107.3°

108.4°

105.9°   112.9°

FIGURE 6.38 Some bond angles in simple amines.

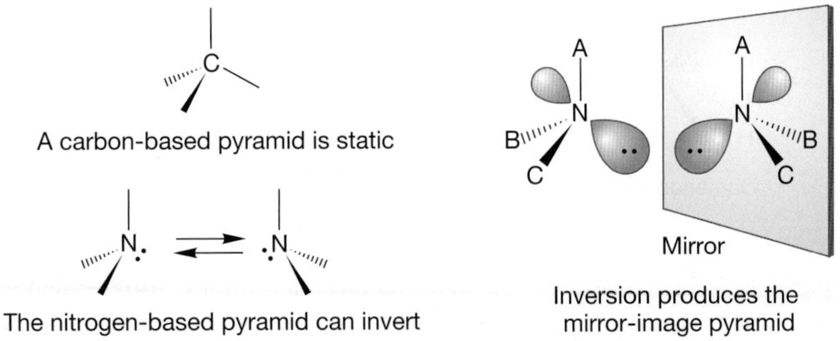

A carbon-based pyramid is static

The nitrogen-based pyramid can invert

Inversion produces the mirror-image pyramid

Mirror

FIGURE 6.39 Amine inversion interconverts enantiomeric amines.

This inversion process is sometimes confusing when first encountered, so let's work through it, making comparisons with the alkanes as we go along. At the beginning of the process, the pyramid begins to flatten and the R—N—R bond angles increase.

At the halfway point along the path to the inverted pyramid, the transition state is reached. Here the molecule is planar, with 120° R—N—R bond angles. At this point, the hybridization of the central nitrogen is sp^2! The inversion process maintains a mirror plane of symmetry. In the alkanes, we could certainly imagine the beginning of the flattening process, but there is no way to force the fourth substituent through the molecule to the other side (Fig. 6.40).

FIGURE 6.40 The transition state for amine inversion is planar.

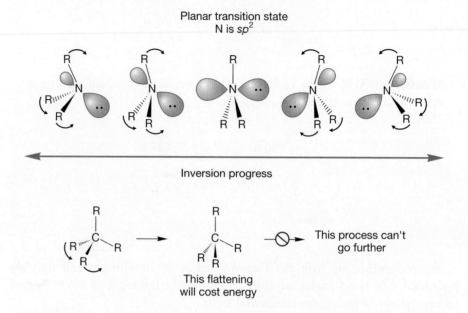

PROBLEM 6.13 In general terms, what is the hybridization of nitrogen at some point between the starting pyramidal amine and the transition state?

An important concern is the rate of this inversion process. If it is slow at room temperature, then we should be able to isolate appropriately substituted optically active amines. If inversion is fast, however, racemization will also be fast as the inversion process interconverts enantiomers. In practice, this is what happens. Barriers to inversion in simple amines are very low, about 5–6 kcal/mol, and isolation of one enantiomer, even at very low temperature, is difficult. Inversion is simply too facile, and the rapid inversion interconverts enantiomeric amines (Fig. 6.41).

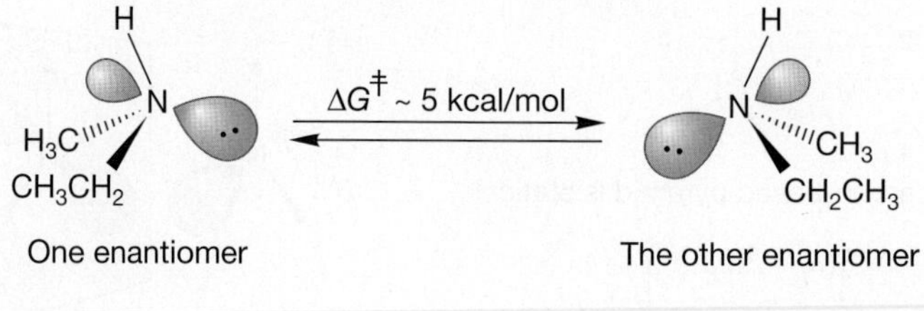

FIGURE 6.41 Rapid inversion racemizes optically active amines.

PROBLEM 6.14 In contrast to simple amines, aziridines (three-membered rings containing a nitrogen) can often be separated into enantiomers. For example, the activation energy ($\Delta G^{\ddagger}$) for the inversion of 1,2,2-trimethylaziridine is about 18.5 kcal/mol, much higher than that for simple amines. Explain.

(continued)

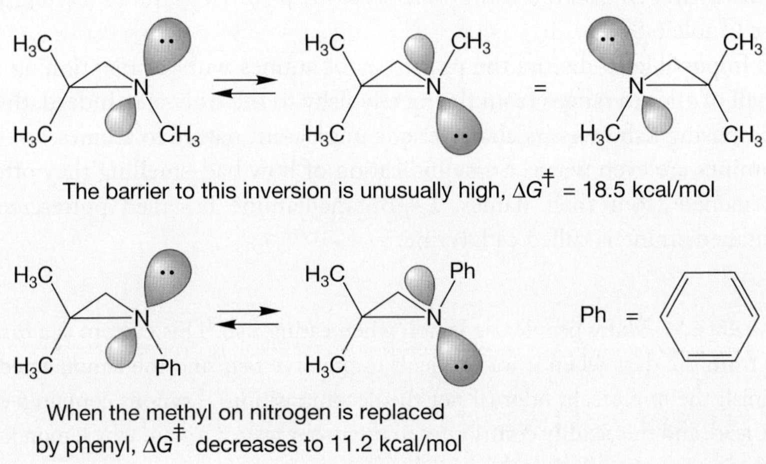

The barrier to this inversion is unusually high, $\Delta G^{\ddagger} = 18.5$ kcal/mol

When the methyl on nitrogen is replaced by phenyl, $\Delta G^{\ddagger}$ decreases to 11.2 kcal/mol

---

PROBLEM 6.15 If a phenyl group (Ph in the figure above) is attached to the nitrogen atom of the aziridine, the barrier to inversion decreases (see figure in Problem 6.14). Explain.

Like alcohols, amines are much higher boiling than hydrocarbons of similar molecular weight, although the effect is not as great as with alcohols (Table 6.8; compare to Table 6.3). The reasons for the increased boiling points are the same as

TABLE 6.8 Some Physical Properties of Amines and a Few Related Alkanes

Compound	bp (°C)	mp (°C)	Density (g/cm^3)
NH$_3$ Ammonia	−33.4	−77.7	0.77
CH$_3$NH$_2$ Methylamine	−6.3	−93.5	0.70
CH$_3$CH$_3$ Ethane	−88.6	−183.3	0.57
CH$_3$CH$_2$NH$_2$ Ethylamine	16.6	−81	0.68
CH$_3$CH$_2$CH$_3$ Propane	−43.1	−189.7	0.59
(CH$_3$)$_2$NH Dimethylamine	7.4	−93	0.68
(CH$_3$)$_3$N Trimethylamine	2.9	−117.2	0.64
⬡—NH$_2$ **WEB 3D** Aniline	184	−6.3	1.02

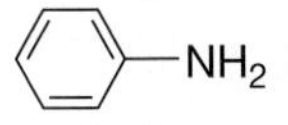

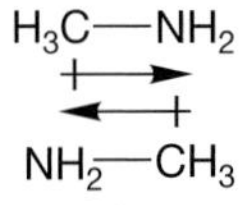

for alcohols. Amines are polar compounds and aggregate in solution. Like alcohols, small amines are miscible with water. Boiling requires overcoming the intermolecular attractive forces between the dipoles. Moreover, amines form hydrogen-bonded oligomers in solution, effectively increasing the molecular weight and further increasing the boiling point (Fig. 6.42).

Association because of favorable dipole alignment

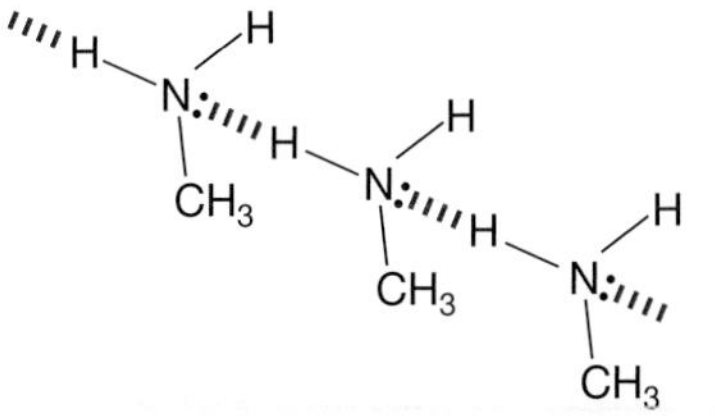

Oligomer formation through hydrogen bonding

FIGURE 6.42 The effective molecular weight of amines is increased through association and hydrogen bonding.

When hydrogen bonding is impossible, as with tertiary amines, boiling points decrease (Table 6.8).

It is impossible to discuss the properties of amines without mentioning smell! The smell of amines ranges from the merely fishy to the truly vile. Indeed, the odor of decomposing fish owes its characteristic unpleasant nature to amines.

Diamines are even worse. Some indication of how bad-smelling they often are can be gained from their names. 1,4-Butanediamine is called putrescine, and 1,5-pentanediamine is called cadaverine.

PROBLEM 6.16 Many people use lemon when eating fish. This custom is a carry-over from the days when it was difficult to preserve fish, and the lemon acted to diminish the unpleasant odor (if not the decomposition). Lemons contain 5–8% citric acid, and this acidity contributes to their sour taste. Explain why lemon juice is effective at reducing the odor of fish.

6.7c Acid and Base Properties of Amines Like water and alcohols, amines are Brønsted bases (proton acceptors). Primary and secondary amines are weak Brønsted acids (proton donors) as well. Brønsted basicity results in the formation of ammonium ions through proton transfer (Fig. 6.43).

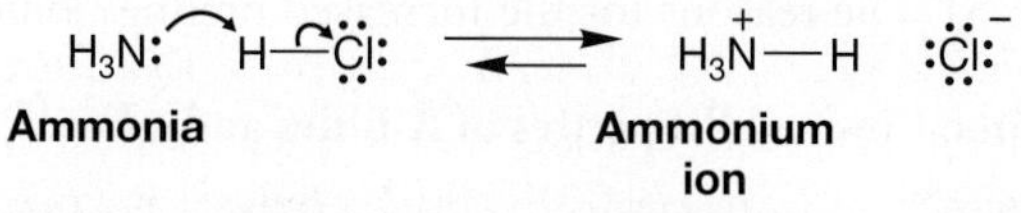

FIGURE 6.43 Ammonia acting as a Brønsted base in proton transfer.

$$H_3N: \quad H-\overset{\cdot\cdot}{\underset{\cdot\cdot}{Cl}}: \quad \rightleftharpoons \quad H_3\overset{+}{N}-H \quad :\overset{\cdot\cdot}{\underset{\cdot\cdot}{Cl}}:^-$$

Ammonia   **Ammonium ion**

The better competitor an amine is in the proton-transfer reaction, the stronger is its Brønsted basicity. A way of turning this statement around is to focus not on the Brønsted basicity of the amine, but on the Brønsted acidity of its conjugate acid, the ammonium ion. If the ammonium ion is a strong Brønsted acid, the related amine must be a weak Brønsted base. If it is easy to remove a proton from the ammonium ion to give the amine, the amine itself must be a poor competitor in the proton-transfer reaction, which is exactly what we mean by a weak Brønsted base. Strongly basic amines give ammonium ions from which it is difficult to remove a proton, ammonium ions with high pK_a values. So the pK_a of the ammonium ion has come to be a common measure of the basicity of the related amine. Ammonium ions with high pK_a values (weak acids) are related to strongly basic amines, and ammonium ions with low pK_a values (strong acids) are related to weakly basic amines (Fig. 6.44).

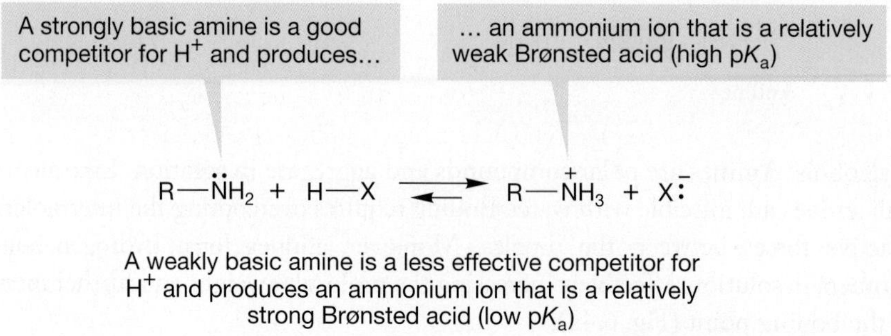

A strongly basic amine is a good competitor for H$^+$ and produces…

… an ammonium ion that is a relatively weak Brønsted acid (high pK_a)

FIGURE 6.44 Ammonium ions with high pK_a values are related to strongly basic amines, and ammonium ions with low pK_a values are related to weakly basic amines.

$$R-\overset{\cdot\cdot}{N}H_2 + H-X \rightleftharpoons R-\overset{+}{N}H_3 + X:^-$$

A weakly basic amine is a less effective competitor for H$^+$ and produces an ammonium ion that is a relatively strong Brønsted acid (low pK_a)

Table 6.9 relates the basicity of ammonia and some very simple alkylamines by tabulating the pK_a values of the related ammonium ions.

TABLE 6.9 Some pK_a Values for Simple Ammonium Ions in Solution

Amine	Formula	Ammonium Ion	pK_a (in aqueous solution)
Ammonia	NH_3	$^+NH_4$	9.24
Methylamine	CH_3NH_2	$CH_3\overset{+}{N}H_3$	10.63
Dimethylamine	$(CH_3)_2NH$	$(CH_3)_2\overset{+}{N}H_2$	10.78
Trimethylamine	$(CH_3)_3N$	$(CH_3)_3\overset{+}{N}H$	9.80

The general trend of the first three entries in Table 6.9 is understandable if we make the analogy between ammonium ions and carbocations. *Remember* (Chapter 3, p. 138): The more substituted a carbocation, the more stable it is. Similarly, the more substituted an ammonium ion, the more stable it is. The more stable the ammonium ion, the less readily it loses a proton and the higher its pK_a. At least for the first three entries of Table 6.9, increasing substitution of the ammonium ion carries with it a decreasing ease of proton loss.

Actually, even the first three entries in Table 6.9 should make you suspicious. Why is there a much bigger change between ammonia and methylamine (1.4 pK_a units) than between methylamine and dimethylamine (only 0.15 pK_a units)? The structural change is the same in each case, the replacement of a hydrogen with a methyl group. When we get to the fourth entry, we find that things have truly gone awry; trimethylammonium ion is a stronger acid than the dimethylammonium ion and the methylammonium ion, and almost as strong as the ammonium ion itself, which implies that trimethylamine is a weaker base than dimethylamine or methylamine. There seems to be no continuous trend in these data.

Indeed there isn't, and it takes a look at matters in the gas phase to straighten things out. In the gas phase, the trend *is* regular. The basicity of amines increases with substitution, and the acidity of ammonium ions decreases with substitution (Fig. 6.45).

There must be some effect present in solution that disappears in the gas phase, and this effect must be responsible for the irregularities in Table 6.9. The culprit is the solvent itself. Ions in solution are strongly stabilized by solvation, by interaction of the solvent molecules with the ion. These interactions can take the form of electrostatic stabilization or of partial covalent bond formation, as in hydrogen bonding (Fig. 6.46).

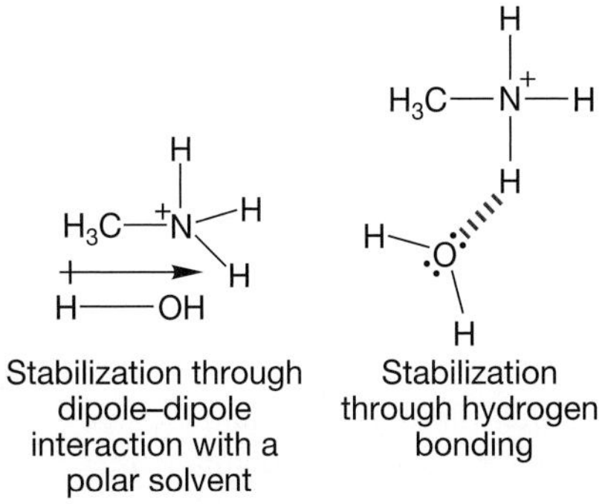

In solution, an alkyl group has two effects on the stability of an ammonium ion. It stabilizes by helping the ammonium ion to disperse the charge, but it destabilizes the ion by interfering with solvation, by making intermolecular solvation (charge dispersal) more difficult. These two effects operate in opposite directions! There is little steric interference with solvation when we replace one hydrogen of the ammonium ion with a methyl group, and the pK_a change is more than a full pK_a unit. This change reflects the stabilizing effect of the methyl group on the ammonium ion. However, when a second hydrogen is replaced by a methyl, the stabilization and interference with solvation nearly balance. The result is practically no change in the overall stability of the ammonium ion. When a third hydrogen is replaced with methyl, the destabilizing effects outweigh the stabilizing forces, and the result is a less stable, more acidic ammonium ion. In the gas phase, where there can be no solvation, only the stabilizing effects remain, and each replacement of a hydrogen with a methyl is stabilizing.

$^+NH_4 < {}^+NRH_3 < {}^+NR_2H_2 < {}^+NR_3H$
Order of increasing gas-phase pK_a
Lowest pK_a Highest pK_a

FIGURE 6.45 The gas-phase acidity of ammonium ions.

Stabilization through dipole–dipole interaction with a polar solvent

Stabilization through hydrogen bonding

FIGURE 6.46 Stabilization of amines in solution.

In addition, as we progress along the series $^{+}NH_4$, $^{+}NRH_3$, $^{+}NR_2H_2$, $^{+}NR_3H$, $^{+}NR_4$, there are fewer hydrogens available for hydrogen bonding and therefore less stabilization of the ammonium ion in solution.

PROBLEM 6.17 Amines can be stabilized by other factors. Explain the following pK_a data.

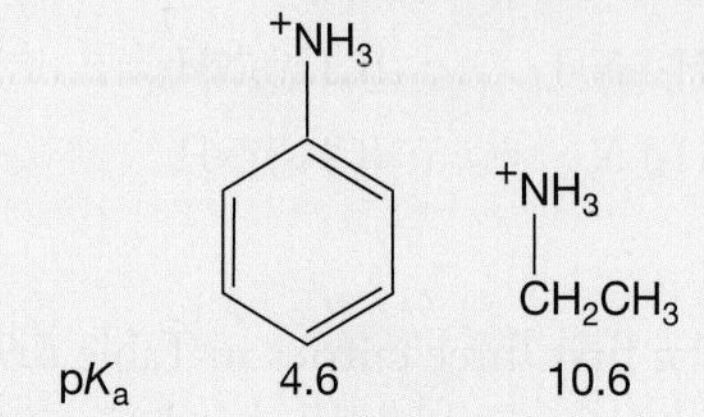

+NH3
+NH3
CH2CH3
pKa
4.6
10.6

It must be admitted that this traditional discussion of base strength of amines in terms of the pK_a of the related ammonium ion is indirect. There is another, less common but undeniably more direct method. That is to frame the discussion in terms of the pK_b ($-\log K_b$), where K_b is the basicity constant of the amine (Fig. 6.47).

FIGURE 6.47 The expression for pK_b.

$$\text{For } RNH_2 + H_2O \underset{}{\overset{K}{\rightleftharpoons}} R\overset{+}{N}H_3 + HO^-$$

$$K_b = \frac{[R\overset{+}{N}H_3][HO^-]}{[RNH_2]}$$

The higher the pK_b, the weaker is the base, just as the higher the pK_a, the weaker is the acid. Typical amines have pK_b values of about 4, which makes them quite strong bases.

Amines are much stronger bases than alcohols are. For example, it is much harder to deprotonate $^{+}NH_4$ than $^{+}OH_3$. The pK_a value for the ammonium ion (9.2) is much higher than that for the oxonium ion (–1.7) (Fig. 6.48).

$^{+}NH_4$	$^{+}OH_3$
Ammonium ion	**Oxonium ion**
pK_a = 9.2	pK_a = −1.7
$^{+}N(CH_3)_4$ Cl^-	$^{+}O(CH_3)_3$ BF_4^-
Tetramethylammonium chloride, a stable solid, easily isolable	**Trimethyloxonium fluoborate**, not easily isolated, and very reactive

FIGURE 6.48 Ammonium ions are weaker acids than oxonium ions.

Why is the oxonium ion much more acidic than the ammonium ion? Oxygen is a more electronegative atom than nitrogen and bears the positive charge less well. We can see the results of the increased stability of the ammonium ion in practical terms. Stable ammonium salts are common, but salts of oxonium ions, though known, are rare and usually unstable (Fig. 6.48).

PROBLEM 6.18 The stability of oxonium ions depends on the nature of the negatively charged counterion. Fluoborate ($^-BF_4$) is an especially favorable counterion. Why? *Hint*: Lewis bases, $B:^-$, can transfer methyl (and other) groups through a process called the S_N2 reaction. You will learn all about it in Chapter 7, but you might want to use it here in this problem.

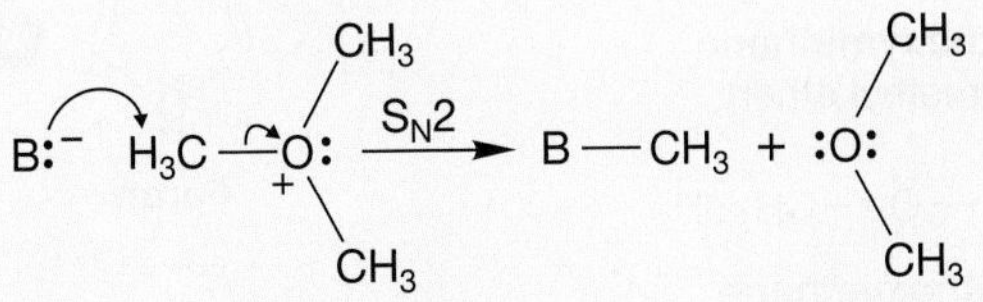

$$B:^- \quad H_3C \overset{\curvearrowleft}{-} \overset{CH_3}{\underset{\underset{CH_3}{|}}{\overset{|}{\overset{+}{O}}:} \quad \xrightarrow{S_N2} \quad B-CH_3 \;+\; \overset{CH_3}{\underset{\underset{CH_3}{|}}{\overset{|}{:O:}}}$$

Primary and secondary amines are Brønsted acids, but only very weak ones. Amines are much weaker acids than alcohols. Removal of a proton from an alcohol gives an alkoxide in which the negative charge is borne by oxygen. An amine forms an **amide**[3] in which the less electronegative nitrogen carries the negative charge (Fig. 6.49).

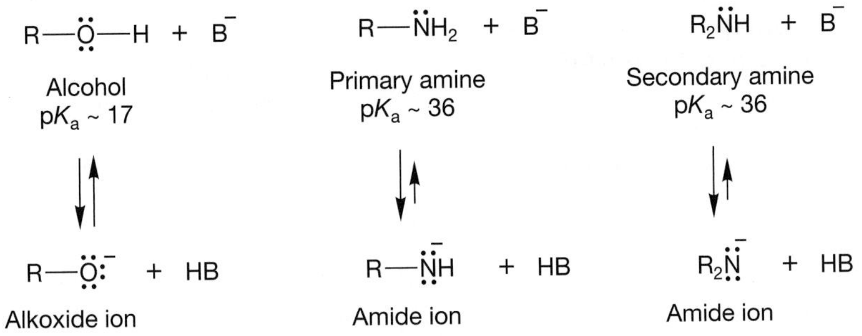

$$R-\ddot{O}-H \;+\; B^- \qquad R-\ddot{N}H_2 \;+\; B^- \qquad R_2\ddot{N}H \;+\; B^-$$

Alcohol
$pK_a \sim 17$

Primary amine
$pK_a \sim 36$

Secondary amine
$pK_a \sim 36$

$$R-\ddot{\underset{..}{O}}:^- \;+\; HB \qquad R-\ddot{\underset{..}{N}}H \;+\; HB \qquad R_2\ddot{\underset{..}{N}} \;+\; HB$$

Alkoxide ion

Amide ion

Amide ion

FIGURE 6.49 Primary and secondary amines are weak Brønsted acids.

It takes a very strong base indeed to remove a proton from an amine to give the amide ion, which is itself a very strong base. Alkyllithium reagents (p. 228) are typically used (Fig. 6.50).

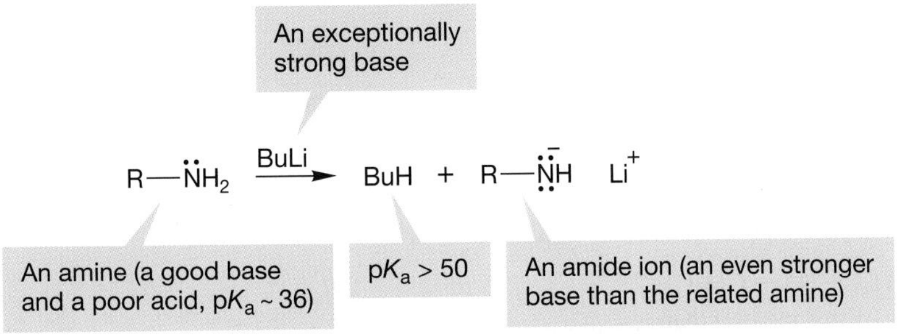

An exceptionally strong base

$$R-\ddot{N}H_2 \xrightarrow{\text{BuLi}} BuH \;+\; R-\ddot{\underset{..}{N}}H \;\; Li^+$$

An amine (a good base and a poor acid, $pK_a \sim 36$)

$pK_a > 50$

An amide ion (an even stronger base than the related amine)

FIGURE 6.50 Very strong bases, such as alkyllithium reagents, can remove a proton from an amine to give an amide ion (Bu = $CH_2CH_2CH_2CH_3$).

So, amines are good bases, but relatively poor acids. Amides, the conjugate bases of amines, are even stronger bases than the amines themselves (Fig. 6.50).

6.8   Ethers

Alcohol (ROH) and amine (RNH$_2$) chemistry depends on both the OH or NH and the OR or NR parts of the molecule. In this section, we look at molecules that no longer have any OH groups, but retain the R group. These compounds are called **ethers**, and have the structure R—O—R or R—O—R′.

---

[3]Be careful—there is another kind of amide that has the structure R—CO—NH$_2$. There is no written or verbal distinction made, so you need to know the context before you know which kind of amide is meant.

6.8a Nomenclature Naming ethers is simple. In the widely used common system, the two groups joined by oxygen are named in alphabetical order, and the word *ether* appended. In the IUPAC system, ethers are named as alkoxy (RO) alkanes. Figure 6.51 gives some simple examples of both usages, IUPAC first.

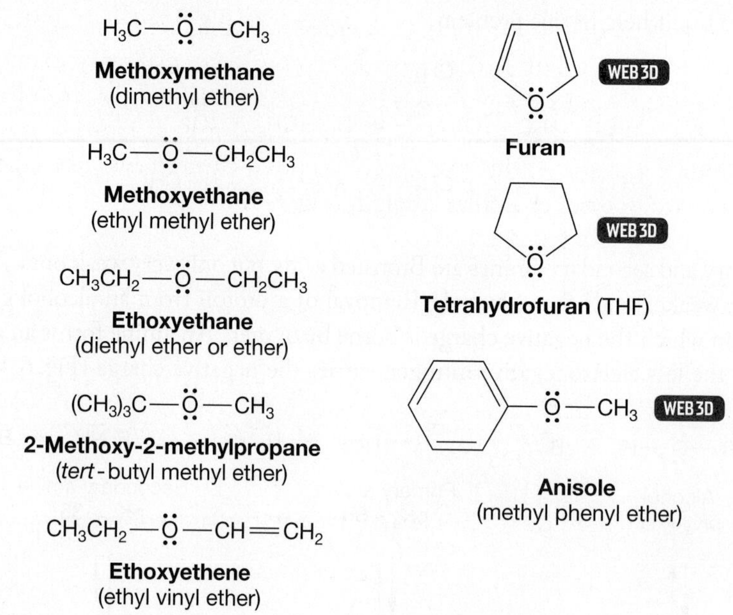

FIGURE 6.51 Some ethers and their common names.

H_3C—$\ddot{O}$—CH_3
Methoxymethane
(dimethyl ether)

H_3C—$\ddot{O}$—CH_2CH_3
Methoxyethane
(ethyl methyl ether)

CH_3CH_2—$\ddot{O}$—CH_2CH_3
Ethoxyethane
(diethyl ether or ether)

$(CH_3)_3C$—$\ddot{O}$—CH_3
2-Methoxy-2-methylpropane
(*tert*-butyl methyl ether)

CH_3CH_2—$\ddot{O}$—CH=CH_2
Ethoxyethene
(ethyl vinyl ether)

Furan WEB 3D

Tetrahydrofuran (THF) WEB 3D

Anisole
(methyl phenyl ether) WEB 3D

6.8b Physical Properties Ethers are polar, but only for the smallest members of the class does this polarity strongly affect physical properties. Other ethers are sufficiently hydrocarbon-like so as to behave as their all-carbon relatives. For example, diethyl ether has nearly the same boiling point as pentane and is only modestly soluble in water. Table 6.10 gives some physical properties for a few common ethers.

TABLE 6.10 Some Physical Properties of Ethers

Compound	bp (°C)	mp (°C)	Density (g/mL)
H_3C—O—CH_3 Dimethyl ether	−23	−138.5	0.668
CH_3CH_2—O—CH_2CH_3 Diethyl ether	34.6	−119.2	0.725
H_3C—O—CH_2CH_3 Ethyl methyl ether	7.6	−139	0.714
$(CH_3)_3C$—O—CH_3 *tert*-Butyl methyl ether	55.2	−109	0.740
CH_3CH_2—O—CH=CH_2 Ethyl vinyl ether	35.5	−115.8	0.759
 Furan	31.4	−85.6	0.951
 THF	67	−108	0.889
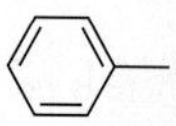 Anisole	155.0	−37.5	0.996

6.8c Structure The carbon–oxygen bond length in ethers is similar to that in alcohols, and the opening of the angle as we go from H—O—H (104.5°) to R—O—H (~109°) to R—O—R (~112°) continues. As the angle in ethers is approximately 112°, the oxygen is hybridized approximately sp^3.

6.8d Acidity and Basicity Of course, ethers lack the OH group of their relatives water and the alcohols and so are not Brønsted acids. The oxygen atom has two pairs of nonbonding electrons, however, and they make ethers weak Brønsted and Lewis bases. Ethers can be protonated, and the protonated form has a pK_a in the same range as those of the protonated alcohols, which means that ethers and alcohols are strong bases of similar strength (Fig. 6.52).

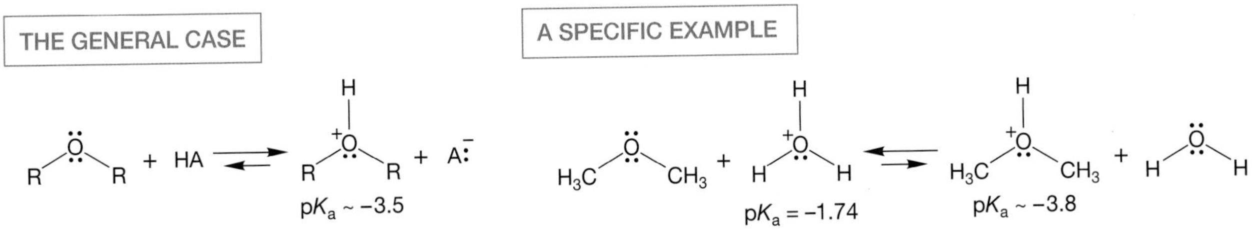

THE GENERAL CASE	A SPECIFIC EXAMPLE

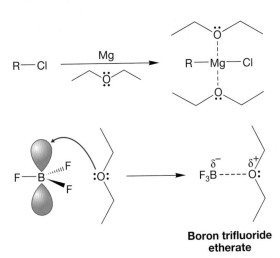

FIGURE 6.52 Ethers are weak Brønsted bases and can be protonated. Protonated ethers (and protonated alcohols) are strong acids.

We have already seen one instance in which the ability of ethers to donate electrons, that is, to act as a Lewis base, was a critical property. Recall from p. 228 that an ether solvent is vital to success in the formation of the Grignard reagent (Fig. 6.53). Ethers are stabilizing to Lewis acidic species. Boron trifluoride etherate is a commercially available complex of BF_3 and ether that can even be distilled. It is commonly used as a source of BF_3. For similar reasons, the cyclic ether THF (Fig. 6.51) is used to stabilize the highly reactive molecule BH_3.

FIGURE 6.53 Ethers will react with Lewis acids as well as Brønsted acids. The BF_3–etherate complex is a stable source of boron trifluoride.

Boron trifluoride etherate

6.9   Special Topic: Thiols (Mercaptans) and Thioethers (Sulfides)

Thiols (RSH) and **thioethers** (RSR′) are the sulfur-containing counterparts of alcohols and ethers, and their chemistries are generally similar to those of their oxygen-containing relatives. These sulfur compounds have a poor reputation, because some are quite extraordinarily evil-smelling. The skunk uses a variety of four-carbon thiols in its defense mechanism, for example. But this bad press is not entirely deserved; not all sulfur compounds are malodorous. Garlic derives its odor from small sulfur-containing compounds. You may never need to ward off a vampire with it, but garlic contains many sulfur compounds that have quite remarkably positive qualities. Figure 6.54 shows some of the sulfur-containing molecules that can be found in garlic.

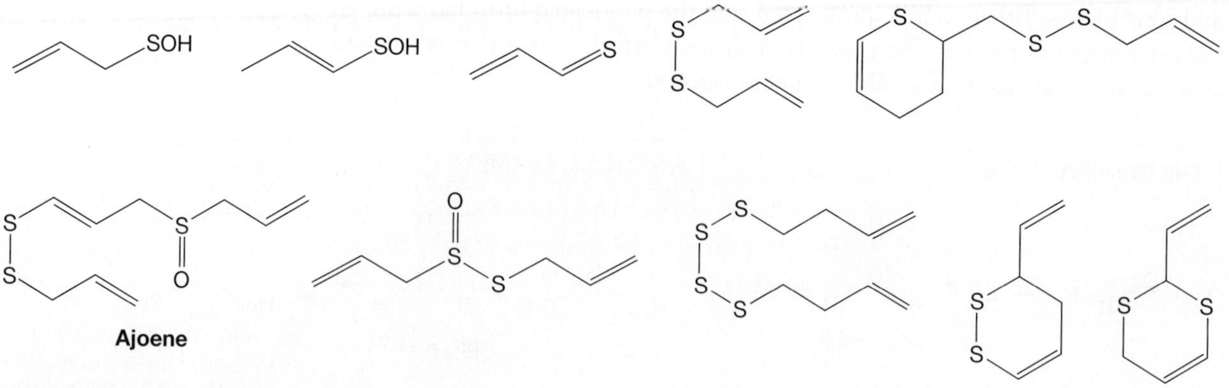

Ajoene

FIGURE 6.54 Some of the sulfur-containing molecules in garlic.

AJOENE

Ajoene is only one of the beneficial compounds present in garlic. Ajoene is lethal to certain tumor-prone cells, promotes the antiaggregative action of some molecules on human blood platelets, and seems to be effective against viruses (including HIV) as well. It appears to reduce repeat heart attacks among people who have already suffered an initial attack. Garlic contains many other compounds that also appear to have beneficial medicinal properties. And, of course, if that vampire is hot on your trail. . . .

6.9a Nomenclature Thiols, also called **mercaptans**, are named by adding the suffix "thiol" to the parent hydrocarbon name. Note that the final "e" is not dropped, as it is, for example, in alcohol nomenclature (Fig. 6.55).

FIGURE 6.55 Some thiols, or mercaptans.

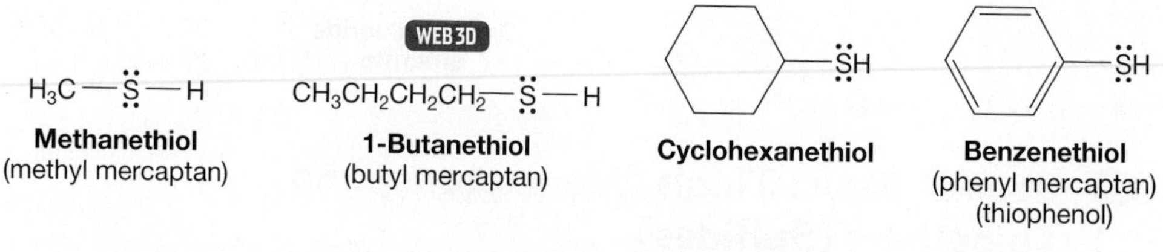

H_3C — $\ddot{S}$ — H

Methanethiol
(methyl mercaptan)

$CH_3CH_2CH_2CH_2$ — $\ddot{S}$ — H

1-Butanethiol
(butyl mercaptan)

Cyclohexanethiol

Benzenethiol
(phenyl mercaptan)
(thiophenol)

Thioethers, the sulfur counterparts of ethers, are also called **sulfides**, and are named in the same way as ethers. The two groups attached to the sulfur atom are followed by the word *sulfide*. Disulfides (R—S—S—R) are the counterparts of peroxides, R—O—O—R, and have a significant biological role, which we will discuss later. Peroxides are nastily unstable and tend to explode, but disulfides are relatively benign. Disulfides are named in a similar fashion as are sulfides (Fig. 6.56).

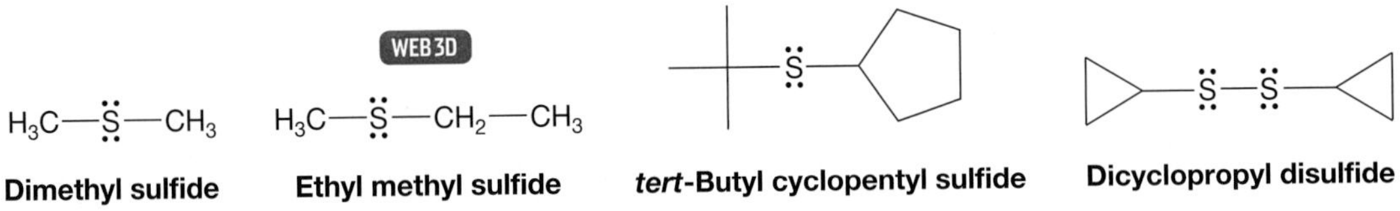

Dimethyl sulfide   **Ethyl methyl sulfide**   ***tert*-Butyl cyclopentyl sulfide**   **Dicyclopropyl disulfide**

FIGURE 6.56 Some thioethers (sulfides) and disulfides.

6.9b Acidity Thiols (pK_a = 9–12) are stronger acids than alcohols and form **mercaptides**, the sulfur counterparts of alkoxides, when treated with base. The relatively high acidity of thiols makes formation of the conjugate base more favorable than formation of alkoxides from alcohols (Fig. 6.57).

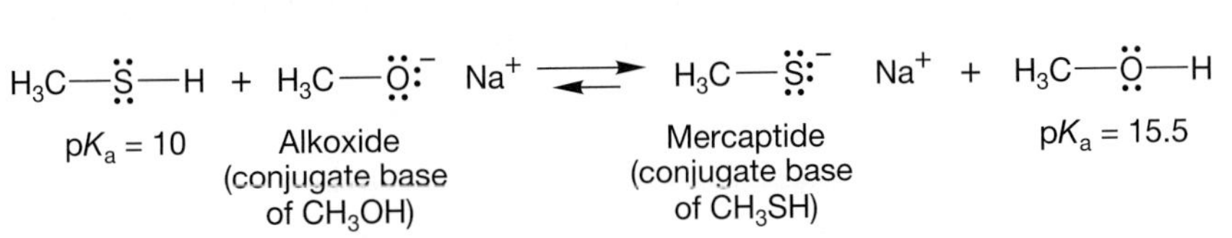

pK_a = 10   Alkoxide   Mercaptide pK_a = 15.5
                 (conjugate base   (conjugate base
                 of CH₃OH)   of CH₃SH)

FIGURE 6.57 Mercaptides are easier to form than alkoxides because thiols are much more acidic than alcohols.

6.9c Reduction of Sulfur Compounds with Raney Nickel: A New Alkane Synthesis Thiols and thioethers are reduced by a catalyst called **Raney nickel** (essentially just hydrogen adsorbed on finely divided nickel) to give hydrocarbons (Fig. 6.58). This synthetic method, called desulfuration, gives you a second way to make alkanes.

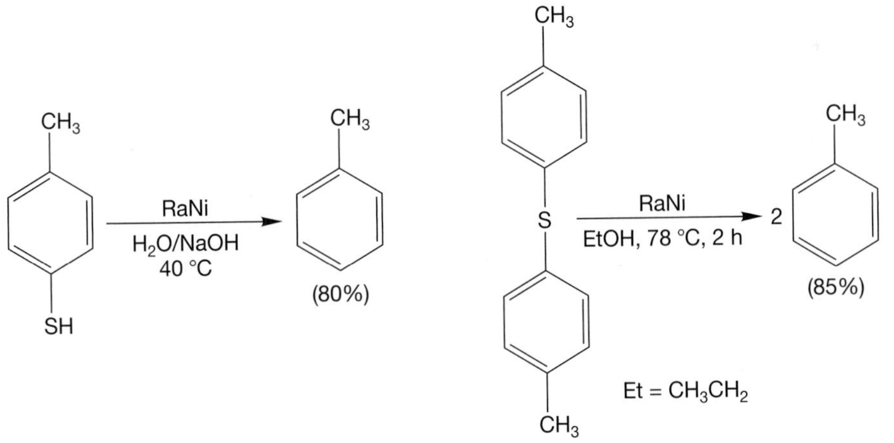

FIGURE 6.58 Desulfuration with Raney nickel (RaNi).

PROBLEM 6.19 What reagents would you use to convert the indicated starting materials into cyclohexane?

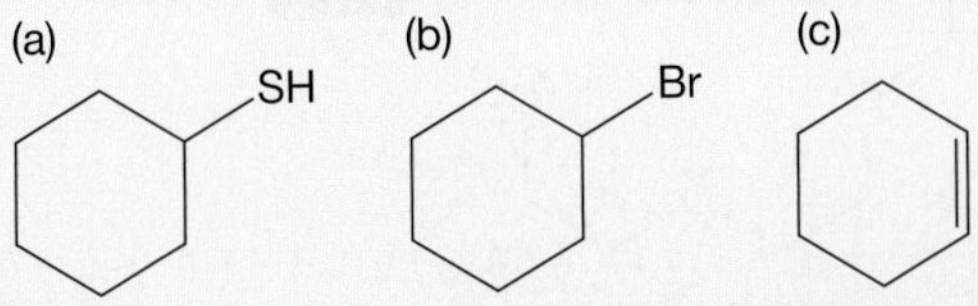

(a)   (b)   (c)

6.10 Special Topic: Crown Ethers

Cyclic ethers are common solvents and, as we will see in Section 7.10, can often be prepared rather easily (Williamson ether synthesis, p. 315). It is possible to make cyclic polyethers, compounds in which there is more than one ether linkage. Figure 6.59 gives some examples of cyclic ethers and cyclic polyethers.

FIGURE 6.59 Some cyclic ethers and polyethers.

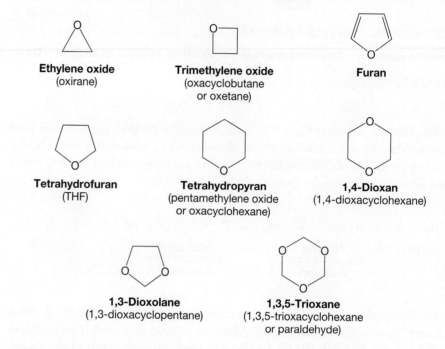

Ethylene oxide
(oxirane)

Trimethylene oxide
(oxacyclobutane
or oxetane)

Furan

Tetrahydrofuran
(THF)

Tetrahydropyran
(pentamethylene oxide
or oxacyclohexane)

1,4-Dioxan
(1,4-dioxacyclohexane)

1,3-Dioxolane
(1,3-dioxacyclopentane)

1,3,5-Trioxane
(1,3,5-trioxacyclohexane
or paraldehyde)

We have mentioned more than once the remarkable solvating powers of ethers. Ethers are Lewis bases and can donate electrons to Lewis acids, thus stabilizing them. One example of this stabilization is the requirement for an ether, usually diethyl ether or THF, in the formation of a Grignard reagent. Ethers are also polar compounds and therefore are able to stabilize other polar molecules through non-covalent dipole–dipole interactions.

This stabilization was carried to extremes in work that ultimately led to the 1987 Nobel Prize in Chemistry for Charles J. Pedersen (1904–1989) of du Pont, Donald J. Cram (1919–2001) of UCLA, and Jean-Marie Lehn (b. 1939) of Université Louis Pasteur in Strasbourg for opening the field of "host–guest" chemistry. Pedersen discovered that certain cyclic polyethers (the hosts) had a remarkable affinity for metal cations (the guests). Molecules were constructed whose molecular shapes created different-sized cavities into which different metal ions fit well. Because of their vaguely crown-shaped structures, these molecules came to be called **crown ethers**. Figure 6.60 shows two of them.

FIGURE 6.60 Two crown ethers. The first number in the name shows the number of atoms in the ring, and the final number shows the number of heteroatoms in the ring. In these molecules the heteroatoms are all oxygen, but others are possible.

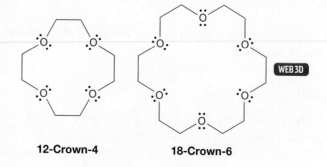

WEB 3D

12-Crown-4

18-Crown-6

The molecule 12-crown-4 is the proper size to capture lithium ions, and 18-crown-6 has a remarkable affinity for potassium ions. These crown ethers can stabilize positive ions of different sizes, depending on the size of the cavity (Fig. 6.61).

Three-dimensional versions of these essentially two-dimensional molecules have now also been made, using both oxygen and other heteroatoms, such as sulfur and nitrogen, as the Lewis bases. These host molecules are called **cryptands** and can incorporate positive ions into the roughly spherical cavity within the cage (Fig. 6.62).

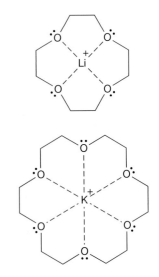

FIGURE 6.61 Two crown ethers stabilizing cations.

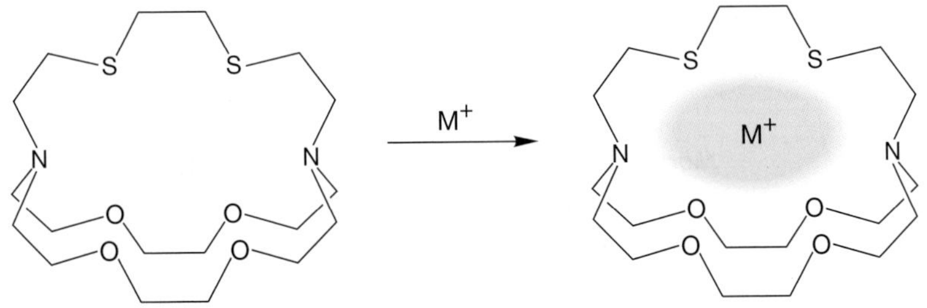

M^+ ⟶

A cryptand

FIGURE 6.62 Three-dimensional cryptands can incorporate metal ions into a central cavity.

There are chemical and larger implications of this work. Potassium permanganate, $KMnO_4$, is an effective oxidizing agent. Its great insolubility in organic reagents limits its use, however. For example, the deep purple $KMnO_4$ is completely insoluble in benzene. It simply forms a solid suspension when added to benzene, which remains a colorless liquid. The addition of a little 18-crown-6 results in the dissolution of the permanganate, and the benzene takes on the deep purple color of the permanganate ion. The potassium ion has been effectively solvated by the crown ether and now dissolves in the organic solvent. The permanganate counterion, though not solvated by the crown compound, accompanies the positive ion into solution, as it must to preserve electrical neutrality. The purple benzene solution can be used as an effective oxidizing agent.

Schemes have been envisioned in which specially designed crown compounds are used to scavenge poisonous heavy metal ions from water supplies. There have been less humanitarian, though admittedly clever, schemes for extracting the minute amounts of gold ions present in seawater by passing vast amounts of water over a crown compound.

6.11 Special Topic: Complex Nitrogen-Containing Biomolecules—Alkaloids

Alkaloids are naturally occurring amines that are powerfully active biologically. They are among the most useful medicinal agents known. The painkiller morphine and the antimalarial agent quinine are typical examples. At the same time, some of the most destructively addictive substances known are also alkaloids. Heroin and cocaine are examples from our own time, and anyone at all familiar with popular literature knows of the notorious poison strychnine (Fig. 6.63).

FIGURE 6.63 Some alkaloids.

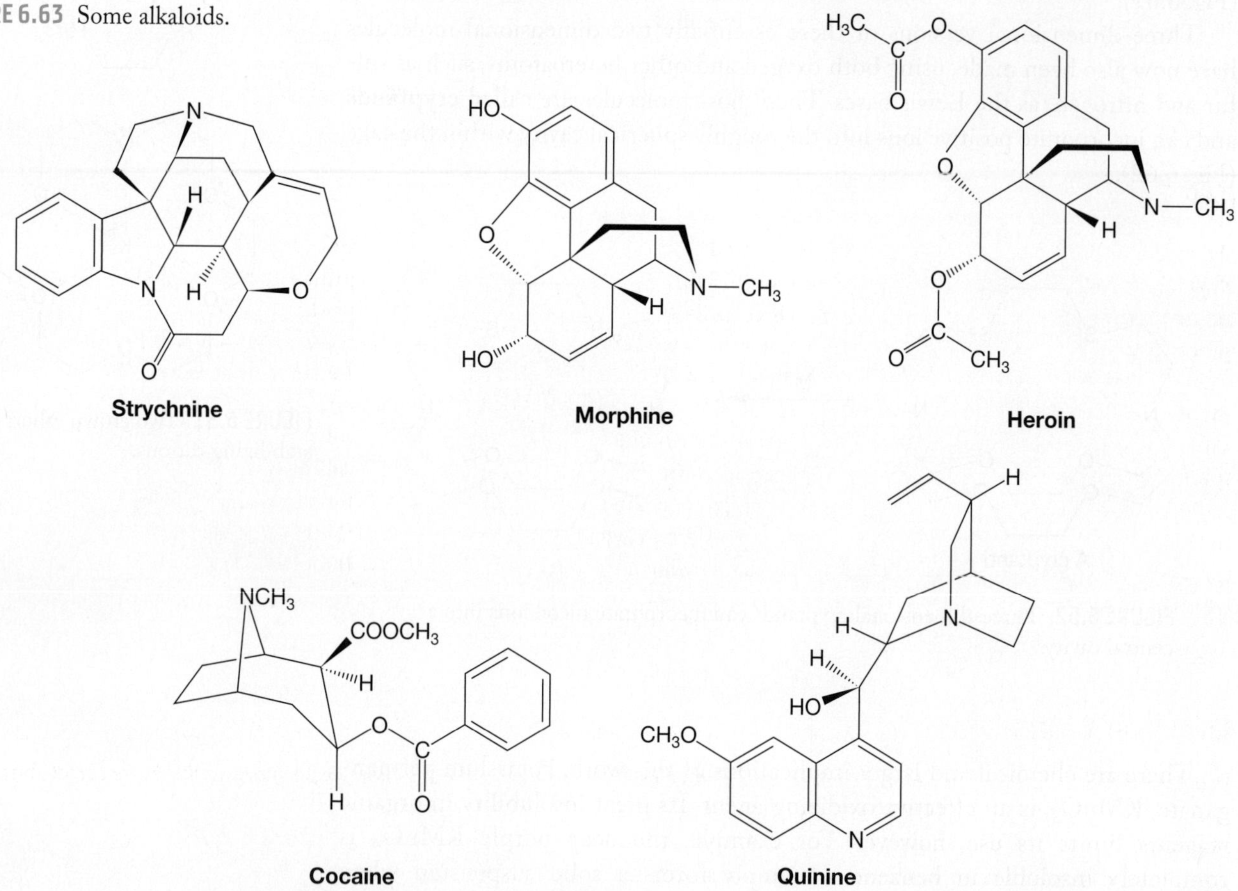

Strychnine

Morphine

Heroin

Cocaine

Quinine

We have already seen strychnine and its close relative brucine in Figure 4.44, where they served as reagents for the resolution of optically active carboxylic acids into separate (R) and (S) enantiomers (p. 171). As we work through the book, alkaloids will reappear now and again. The important thing to "take home" from this picture is not the detail but rather a broad view of the great structural diversity—molecular architecture—of these bioactive molecules.

6.12 Summary

New Concepts

A measure of a compound's acidity—its pK_a value—is introduced. The lower the pK_a value, the more acidic is the compound in question.

Solvents appear as important players in questions of stability. For example, both the conjugate bases of alcohols (alkoxides) and the conjugate acids of amines (ammonium ions) depend greatly on solvation for stabilization. In assessing basicity or acidity, one must take account of the solvent.

A general discussion of solvents and solvation also appears. Protic solvents that can participate in hydrogen bonding

dissolve other protic, polar molecules well. By contrast, polar solvents do a poor job of dissolving "greasy" nonpolar organic molecules such as hydrocarbons, which are dissolved by nonpolar, similarly "greasy" solvents. Like dissolves like—polar solvents dissolve polar molecules, and nonpolar solvents dissolve nonpolar molecules.

Organic halogenated molecules can be used to form organometallic reagents. Many of the important synthetic reactions of organic chemistry use organometallic reagents. In this chapter, we see them only as very strong bases and as sources of hydrocarbons through their reaction with water.

Amines, like alkanes with sp^3-hybridized carbons, are tetrahedral. Unlike alkanes, amines invert their "umbrellas" through an sp^2-hybridized transition state.

The hydrogen bond is introduced. This kind of bond is nothing more than a partially completed Brønsted acid–base reaction in which a partial bond is formed between a pair of electrons on one atom and a hydrogen on another (Fig. 6.64).

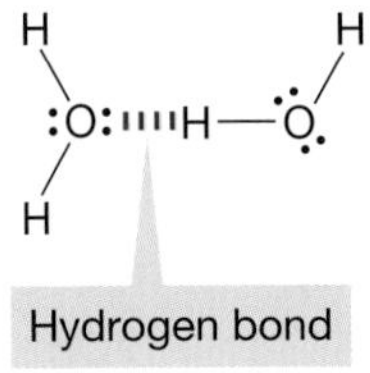

FIGURE 6.64 A hydrogen bond.

Key Terms

alkoxide ion (p. 236)
alkyl halide (p. 225)
amide (p. 249)
amine (p. 240)
amine inversion (p. 243)
ammonia (p. 240)
ammonium ion (p. 240)
aprotic solvent (p. 238)
aziridine (p. 242)
conjugate acid (p. 234)
conjugate base (p. 234)
crown ether (p. 254)

cryptand (p. 255)
diol (p. 240)
ether (p. 249)
free radical (p. 228)
glycol (p. 240)
Grignard reagent (p. 228)
hydrogen bonding (p. 232)
mercaptan (p. 252)
mercaptide (p. 253)
organolithium reagent (p. 228)
organometallic reagent (p. 227)
peptide (p. 224)

pK_a (p. 235)
primary amine (p. 240)
protein (p. 224)
protic solvent (p. 238)
Raney nickel (p. 253)
secondary amine (p. 240)
solvation (p. 237)
sulfide (p. 252)
tertiary amine (p. 240)
thioether (p. 251)
thiol (p. 251)

Common Errors

People sometimes get the connection between pK_a and acid strength wrong. A strong acid has a low pK_a; a weak acid has a high pK_a.

It is easy to get confused when using the pK_a of an ammonium ion as a measure of the acidity of the related amine.

Review the discussion (p. 246) or use the more direct (but less common) pK_b values.

Syntheses

1. Grignard Reagents

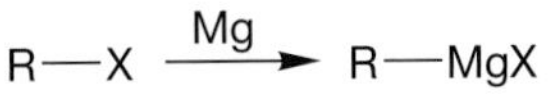

$$R-X \xrightarrow{\text{Mg}} R-MgX$$

R can be almost any organic structure. X is I, Br, or Cl

2. Hydrocarbons

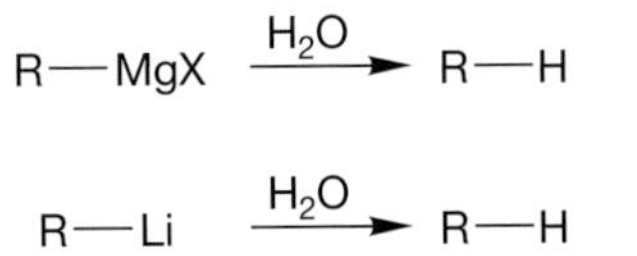

$$R-MgX \xrightarrow{\text{H}_2\text{O}} R-H$$

$$R-Li \xrightarrow{\text{H}_2\text{O}} R-H$$

If D$_2$O is used instead of H$_2$O, deuterated hydrocarbons can be made

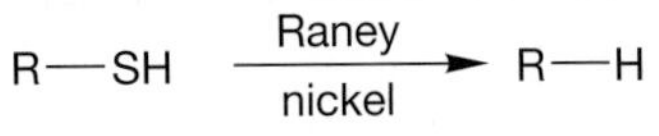

$$R-SH \xrightarrow[\text{nickel}]{\text{Raney}} R-H$$

Reduction of thiols by Raney nickel

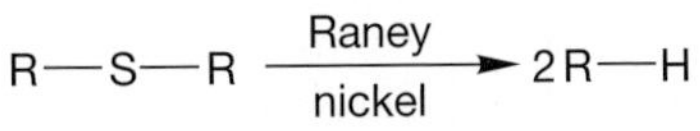

$$R-S-R \xrightarrow[\text{nickel}]{\text{Raney}} 2R-H$$

Reduction of sulfides by Raney nickel

3. Organolithium Reagents

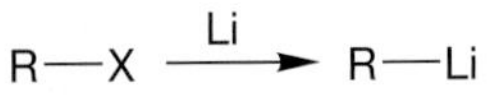

$$R-X \xrightarrow{\text{Li}} R-Li$$

R can be almost any organic structure. X is I, Br, or Cl

6.13 Additional Problems

PROBLEM 6.20   Name the following compounds:

(a)

(b)

OH

(c)

OH

(d)

OH

(e)

OH

(f)

OH

(g)

OH

OH

(h)

OH

(i)

OH

(j)

OCH₂CH₃

OH

(k)

OH   OH

(l)

SH

(m)

S

PROBLEM 6.21   Draw structures for the following compounds:

(a) 1,4-butanediol
(b) 3-amino-1-pentanol
(c) *cis*-4-*tert*-butoxycyclohexanol
(d) 2,2,4,4-tetramethyl-3-pentanol
(e) (*R*)-*sec*-butyl alcohol
(f) *trans*-3-chlorocyclohexanol
(g) *tert*-butyl cyclopentyl ether

PROBLEM 6.22   Draw structures for the following compounds:

(a) neopentyl alcohol
(b) 18-crown-6
(c) tetrahydropyran
(d) THF
(e) furan

PROBLEM 6.23   Draw structures for the following compounds:

(a) (*R*)-3-methyl-2-butanamine
(b) *cis*-2-ethylcyclopentanamine
(c) 1-hexanamine
(d) 3-methoxy-2-heptanamine
(e) *N*-methyl-3-hexanamine

PROBLEM 6.24   Draw structures for the following compounds:

(a) *N*-ethyl-1-ethanamine
(b) *N*,*N*-diethyl-1-ethanamine
(c) (*E*)-2-penten-1-amine
(d) (*Z*)-2-penten-2-amine
(e) 2,3-pentanediamine

PROBLEM 6.25   Name the following compounds according to the official naming system (IUPAC):

(a)

NH₂

(b)

NH₂

(c)

H₂N͵͵͵   NH₂

(d)

NH₂

OH

(e)

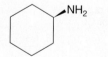

OCH₃   NH₂

PROBLEM 6.26   Indicate the hybridization of the carbons, nitrogens, and oxygens in each of the following compounds:

(a)

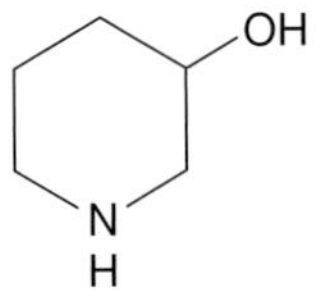

(b)

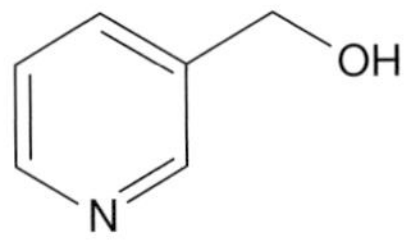

(c)

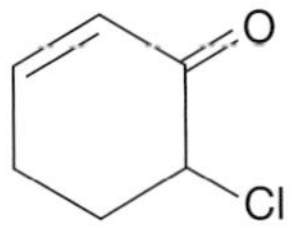

(d)

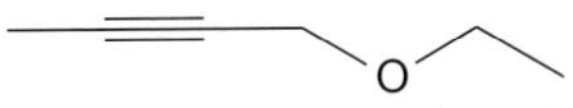

(e)

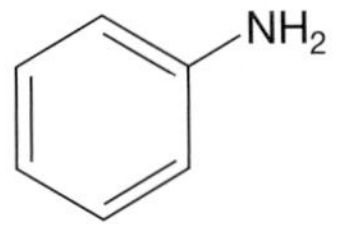

PROBLEM 6.27   Which of the molecules in Problem 6.26 are able to hydrogen bond with water? Which are able to hydrogen bond with another molecule of the same structure (i.e., with themselves)?

PROBLEM 6.28   In the following molecules, there exists the possibility of forming more than one alkoxide or thiolate. In each case, explain which will be formed preferentially, and why.

(a)

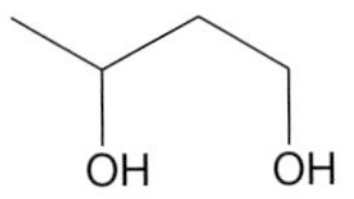

OH   OH
(solution)

(b)

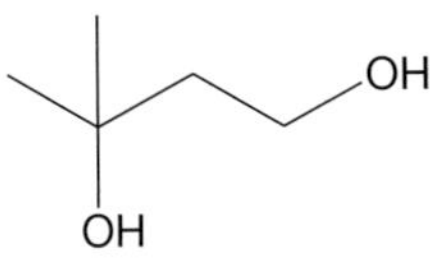

OH
(gas phase)

(c)

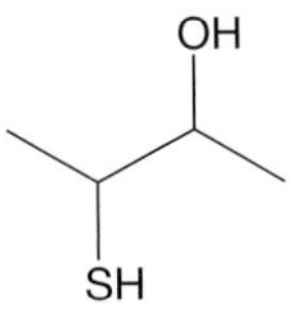

OH

SH
(solution)

PROBLEM 6.29   Predict the relative acidities of the following three alcohols:

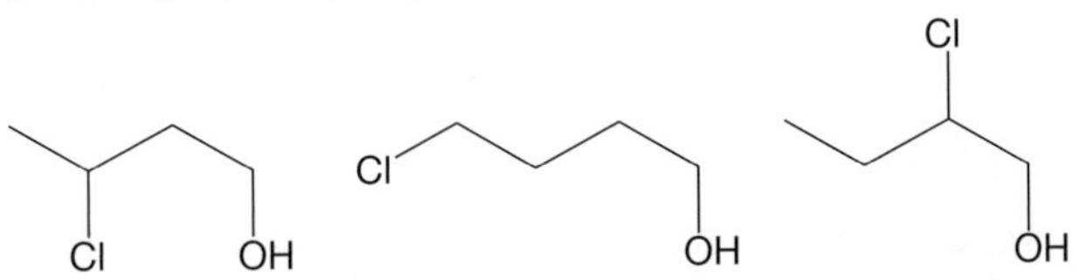

PROBLEM 6.30   Write all the isomers of the alcohols of the formula $C_6H_{14}O$.

PROBLEM 6.31   Name the following compounds (in more than one way, if possible). Identify each as a primary, secondary, or tertiary amine, or as a quaternary ammonium ion.

(a)

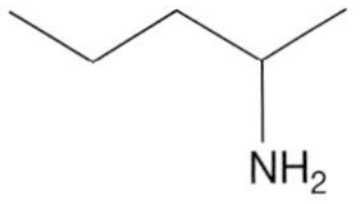

NH$_2$

(b)

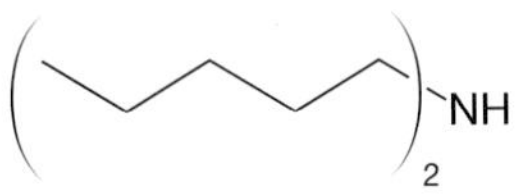

NH

(c)

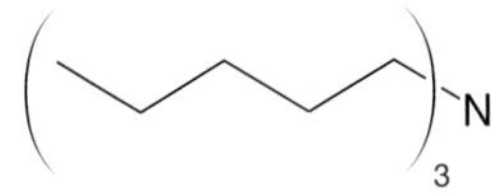

N

(d)

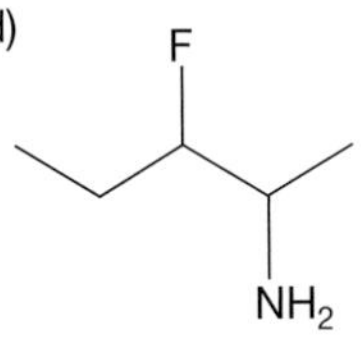

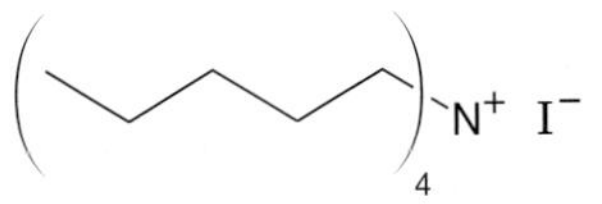

F

NH$_2$

(e)

N$^+$ I$^-$

(f)

NH$_2$

NH$_2$

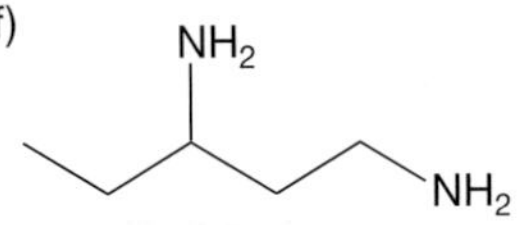

PROBLEM 6.32   Name the following compounds (in more than one way, if possible). Identify each as a primary, secondary, or tertiary amine.

(a)

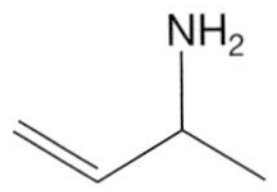

NH$_2$

(b)

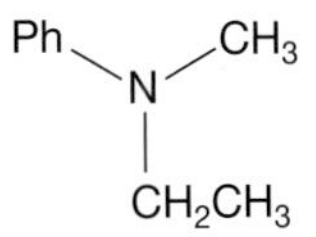

Ph   N   CH$_3$

CH$_2$CH$_3$

(c)

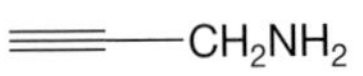

—CH$_2$NH$_2$

(d)

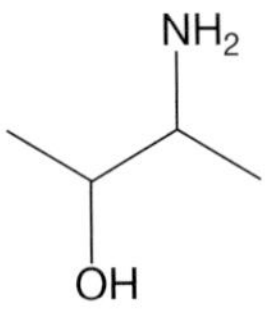

NH$_2$

OH

PROBLEM 6.33   Explain why the compound below shows five signals in the ^{13}C NMR spectrum at low temperature but only three at higher temperature.

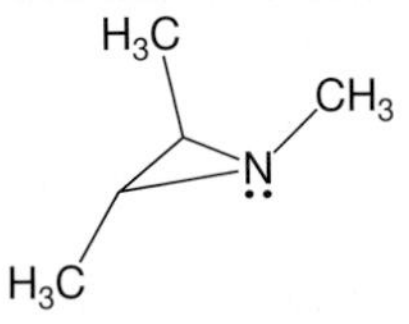

H$_3$C

CH$_3$

N

H$_3$C

PROBLEM 6.34   Write the expressions for K_a and K_b and show that $pK_a + pK_b = 14$.

PROBLEM 6.35   The pK_b values for a series of simple amines are given below. Explain the relationship between pK_b and base strength, and then rationalize the nonlinear order.

Amine	:NH_3	:NH_2CH_3	:$NH(CH_3)_2$	:$N(CH_3)_3$
pK_b	4.76	3.37	3.22	4.20

Use Organic Reaction Animations (ORA) to answer the following questions:

PROBLEM 6.36   Select the animation of the reaction titled "Hofmann elimination." Observe the reagents that are involved. What type of amine is required for the reaction? Is it primary, secondary, or tertiary?

PROBLEM 6.37   How would a polar solvent impact the Hofmann elimination reaction? Would it slow the reaction down or make the reaction faster? Why? Use an energy diagram to help explain your answer.

PROBLEM 6.38   There are three products formed in the Hofmann elimination reaction. They can be seen moving off the screen at the end at the reaction. Stop the animation while all three are still on the screen. You should be able to see water, an alkene, and an amine. Which do you suppose is the best nucleophile? Select the HOMO representation for the animation. Remember that the HOMO represents the most available electrons, in other words, the most nucleophilic electrons. Based on the calculated HOMO, which of the three products is the most nucleophilic?

PROBLEM 6.39   Select the animation titled "Carbocation rearrangement: E1." There are several intermediates in this reaction. Which one is the most stable? Which one is the least stable? What process is occurring in the first step of the reaction? Use Table 6.4 to explain why the first step occurs.

Substitution and Elimination Reactions: The S_N2, S_N1, E1, and E2 Reactions

7.1 Preview

7.2 Review of Lewis Acids and Bases

7.3 Reactions of Alkyl Halides: The Substitution Reaction

7.4 Substitution, Nucleophilic, Bimolecular: The S_N2 Reaction

7.5 The S_N2 Reaction in Biochemistry

7.6 Substitution, Nucleophilic, Unimolecular: The S_N1 Reaction

7.7 Summary and Overview of the S_N2 and S_N1 Reactions

7.8 The Unimolecular Elimination Reaction: E1

7.9 The Bimolecular Elimination Reaction: E2

7.10 What Can We Do with These Reactions? How to Do Organic Synthesis

7.11 Summary

7.12 Additional Problems

INVERSION In this chapter we will learn that some carbons are like umbrellas. Both can undergo inversion.

Don't look back, something might be gaining on you.

—LEROY ROBERT "SATCHEL" PAIGE[1]

7.1 Preview

We are about to embark on an extensive study of chemical change. The last six chapters mostly covered various aspects of structure, and still more analysis of structure will appear as we go along. Without this structural underpinning no serious exploration of chemical reactions is possible. The key to all studies of reactivity is an initial, careful analysis of the structures of the molecules involved. We have been learning parts of the grammar of organic chemistry. Now we are about to use that grammar to write some sentences and paragraphs. Let's begin with a reaction central to most of organic and biological chemistry, the replacement of one group by another (Fig. 7.1).

When one group is replaced by another in a chemical reaction, the result is a **substitution reaction**. Conversion of one alkyl halide into another is one example of the substitution reaction (Fig. 7.2). This reaction looks simple, and in some ways it is. The only change is the replacement of one halogen by another.

THE GENERAL CASE

$$R-L + Nu \rightleftharpoons R-Nu + L$$

$$R = \begin{array}{cc} H_3C & (CH_3)_2CH \\ \text{Methyl group} & \text{Isopropyl group} \\ CH_3CH_2 & (CH_3)_3C \\ \text{Ethyl group} & \textit{tert-}\text{Butyl group} \end{array}$$

FIGURE 7.1 This generic reaction presents an overall picture of the substitution process. A leaving group, L, is replaced by a new group, Nu, to give R—Nu. A standard set of alkyl groups that will be used throughout this chapter is presented.

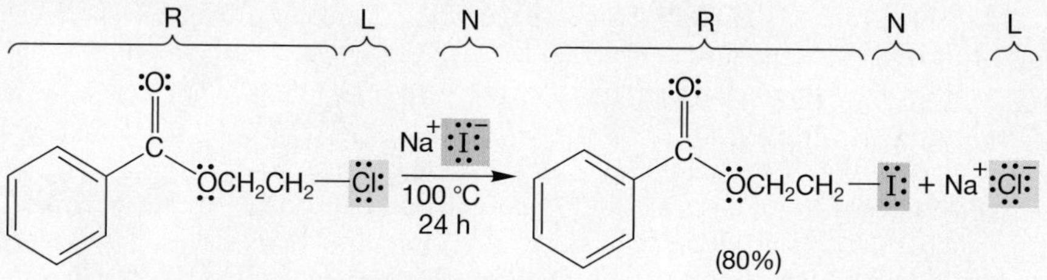

FIGURE 7.2 The replacement of one halogen by another is an example of a substitution reaction.

A SPECIFIC EXAMPLE

Yet as we examine the details of this process, we find ourselves looking deeply into chemical reactivity. This reaction is an excellent prototype that nicely illustrates general techniques used to determine reaction mechanism.

One reason that we spend so much time on the substitution reaction is that it *is* so general; what we learn about it and from it is widely applicable. Not only will we apply what we learn here to many other chemical reactions, substitution reactions have vast practical consequences as well. Substitution reactions, commonly called displacement reactions, are at the heart of many industrial processes and are central to the mechanisms of many biological reactions including some forms of carcinogenesis, the cancer-causing activity of many molecules. Biomolecules are bristling with substituting agents. It seems clear that when some groups on our DNA (deoxyribonucleic acid, see Chapter 23) are modified in a substitution reaction, tumor production is initiated or facilitated. Beware of substituting agents.

This chapter is long, important, and possibly challenging. In it, we cover four of five building block reactions, two of them substitution reactions (S_N1 and S_N2) and two of them elimination reactions (E1 and E2). We have encountered another fundamental reaction, additions, already in Chapter 3, and it will reappear shortly, in Chapter 9. From now on, you really cannot memorize your way through the material;

---

[1] Satchel Paige (1906–1982) was a pitcher for the Cleveland Indians and later the St. Louis Browns. He was brought to the major leagues by Bill Veeck after a long and spectacular career in the Negro Leagues. Some measure of his skill can be gained from realizing that Paige pitched in the major leagues after his 50th birthday.

you must try to generalize. Understanding concepts and applying them in new situations is the route to success. This chapter contains lots of problems, and it is important to try to work through them as you go along. We know this warning is repetitious, and we don't want to insult your intelligence, but it would be even worse not to alert you to the change in the course that takes place now or not to suggest ways of getting through this new material successfully. You cannot simply read this material and hope to get it all. You must work with the text and the various people in your course, the professor and the teaching assistants, in an interactive way. The problems try to help you do this. When you hit a snag or can't do a problem, find someone who can and get the answer. The answer is more than a series of equations or structures and arrows; it also involves finding the right approach to the problem. There is much technique to problem solving in organic chemistry; it can be learned, and it gets much easier with practice. *Remember*: Organic chemistry must be read with a pencil in your hand.

ESSENTIAL SKILLS AND DETAILS

1. We have seen it before, as early as Chapter 1, but in this chapter the curved arrow formalism becomes more important than ever. If you are at all uncertain of your ability to "push" electrons with arrows, now is definitely the time to solidify this skill. In the arrow formalism convention, arrows flow *from* electrons. But be careful about violating the rules of valence; arrows either displace another pair of electrons or fill a "hole"—an empty orbital.

2. It is really important to be able to generalize the definition of Lewis bases and Lewis acids. All electron-pair donors are Lewis bases, and nearly any pair of electrons can act as a Lewis base under some conditions. Similarly, there are many very different appearing Lewis acids—acceptors of electrons.

3. The four fundamental reactions we study in this chapter, S_N2, S_N1, E2, and E1, are affected by the reaction conditions as well, of course, as by the structures of the molecules themselves. These four reactions often compete with each other, and it is important to be able to select conditions and structures that favor one reaction over the others. You will rarely be able to find conditions that give specificity, but you will usually be able to choose conditions that give selectivity.

4. At the end of this chapter, we will sum up what we know about making molecules—synthesis. ALWAYS attack problems in synthesis by doing them backward. Do not attempt to see forward from starting material to product. In a multistep synthesis, this approach is often fatal. The way to do it is to look at the ultimate product and ask yourself what molecule might lead to it in a single step. That technique will usually produce one or more candidates. Then apply the same question to those candidates, and before you know it, you will be at the indicated starting material. This technique works, and can make what is often a tough, vexing task easy.

5. Alcohols are important starting materials for synthesis—they are inexpensive and easily available. Alcohols can be made even more useful by learning how to manipulate the OH group to make it more reactive—to make it a better "leaving group." Several ways to do that are outlined in Section 7.4f.

6. Keeping track of your growing catalog of useful synthetic reactions is not easy—it is a subject that grows rapidly. A set of file cards on which you collect ways of making various types of molecules is valuable. See page 320 for an outline of how to make this catalog.

7.2   Review of Lewis Acids and Bases

7.2a The Curved Arrow Formalism   Recall that there is a most important bookkeeping technique called the curved arrow formalism (p. 23), which helps to sketch the broad outlines of what happens during a chemical reaction. However, it does not constitute a full **reaction mechanism**, a description of how the reaction occurs.

If a reaction mechanism is not an arrow formalism, what is it? A short answer would be a description, in terms of structure and energy, of the intermediates and transition states separating starting material and product. Implied in this description is a picture of how the participants in the reaction must approach each other. The arrow formalism is a powerful device of enormous utility in helping to keep track of bond makings and breakings. In this formalism, chemical bonds are shown as being formed from pairs of electrons flowing from a donor toward an acceptor, often, but not always, displacing other electron pairs. Figure 7.3 shows two examples. In the first, hydroxide ion, acting as a Brønsted base, displaces the pair of electrons in the hydrogen–chlorine bond. The reaction of ammonia ($:NH_3$) with hydrogen chloride is another example in which the electrons on neutral nitrogen are shown displacing the electrons in the hydrogen–chlorine bond.

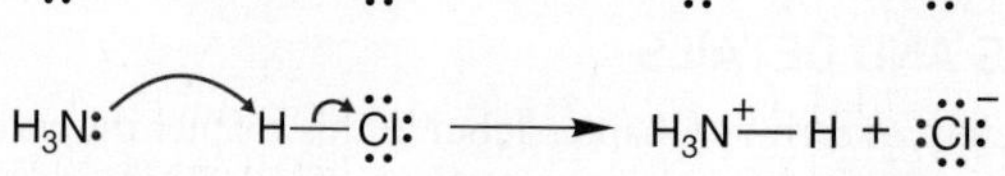

FIGURE 7.3 Two demonstrations of the application of the curved arrow formalism. We will see many other examples.

WORKED PROBLEM 7.1 Draw arrow formalisms for the reversals of the two reactions in Figure 7.3.

ANSWER This problem may seem to be mindless, but it really is not. The key point here is to begin to see all reactions as proceeding in two directions. Nature doesn't worry about "forward" or "backward." Thermodynamics dictates the direction in which a reaction will run.

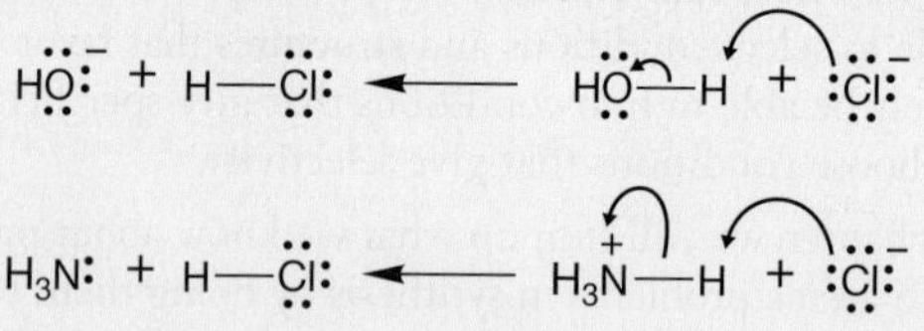

7.2b Lewis Acids and Bases

In Sections 1.7 and 2.15, we began a discussion of acids and bases that will last throughout this book. Then we spent some time in Chapter 3 reviewing Brønsted acids and bases in the context of the addition reaction and extended the subject of acidity and basicity to Lewis acids and Lewis bases. These concepts will hold together much of the material in the remaining chapters.

Consider two conceptually related processes: the reaction of boron trifluoride with fluoride ion, F^-, and the reaction of the methyl cation with hydride, $H:^-$ (Fig. 7.4).

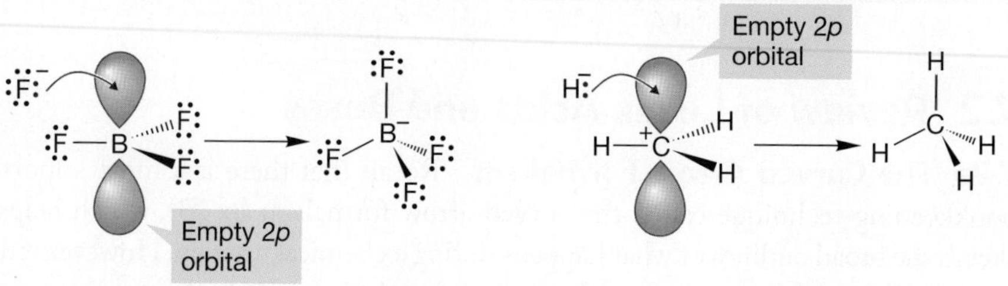

FIGURE 7.4 The fluoride and hydride ion are Lewis bases reacting with the Lewis acids BF_3 and $^+CH_3$.

In these two reactions there is no donating or accepting of a proton, and so by a strict definition of Brønsted acids and bases we are not dealing with acid–base chemistry. As we already know, however, there is a more general definition of acids and bases that deals with this situation. A little review may be helpful: A Lewis base is defined as any species with a reactive pair of electrons (an electron donor). This definition may seem very general and indeed it is. It gathers under the "base" category all manner of odd compounds that don't really look like bases. The key word that confers all this generality is *reactive*. Reactive under what circumstances? Almost everything is reactive under some set of conditions. And that's true; compounds that are extraordinarily weak Brønsted bases can still behave as Lewis bases.

If the definition of a Lewis base seems general and perhaps vague, the definition of a Lewis acid is even worse. A Lewis acid is anything that will react with a Lewis base (thus, an electron acceptor). Therein lies the power of these definitions. They allow us to generalize—to see as similar reactions that on the surface appear very different.

To describe Lewis acid–Lewis base reactions in orbital terms, we need only point out that a stabilizing interaction of orbitals involves the overlap of a filled orbital (Lewis base) with an empty orbital (Lewis acid) (Fig. 7.5).

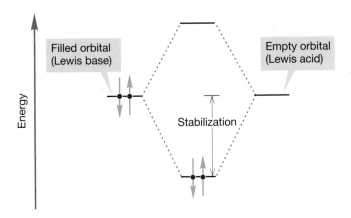

FIGURE 7.5 The interaction of a filled orbital with an empty orbital is stabilizing.

In the examples in Figure 7.4, the Lewis bases are the fluoride and hydride ions. In fluoride, the electrons occupy a $2p$ orbital and in hydride, a $1s$ orbital. The Lewis acids are boron trifluoride and the methyl cation, each of which bears an empty $2p$ orbital. A schematic picture of these interactions is shown in Figure 7.6.

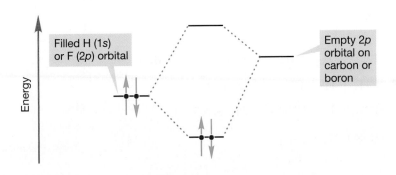

FIGURE 7.6 A schematic interaction diagram for the reactions of Figure 7.4.

7.2c HOMO–LUMO Interactions   It is possible to elaborate a bit on the generalization that "Lewis acids react with Lewis bases" by realizing that the strongest interactions between orbitals are between the orbitals closest in energy. For stabilizing filled–empty overlaps, this will almost always mean that we must look at the interaction of the highest occupied molecular orbital (HOMO) of one molecule (**A**) with the lowest unoccupied molecular orbital (LUMO) of the other (**B**) (Fig. 7.7).

FIGURE 7.7 In two typical molecules, the strongest stabilizing interaction is likely to be between the HOMO of one molecule (**A**) and the LUMO of the other molecule (**B**), because the strongest interactions are between the orbitals closest in energy.

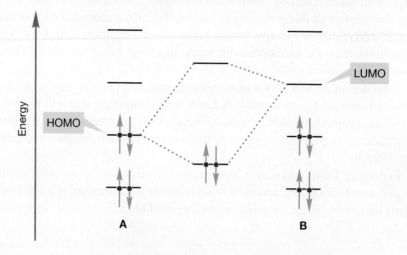

In the reaction of BF_3 with fluoride, it is the filled fluorine $2p$ orbital that is the HOMO and the empty $2p$ orbital on boron that is the LUMO (Fig. 7.8). *All* Lewis acid–Lewis base reactions can be described in similar terms.

FIGURE 7.8 The interaction of the LUMO for BF_3 (an empty $2p$ orbital on boron) and the HOMO of F^- (a filled $2p$ orbital on fluorine) produces two new molecular orbitals: an occupied bonding σ orbital and an unoccupied antibonding σ^* orbital.

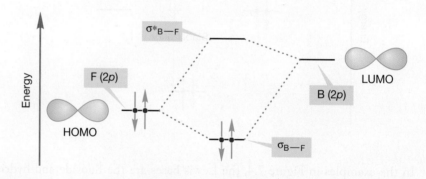

PROBLEM 7.2 Identify the HOMO and LUMO in the reaction of the methyl cation with hydride in Figure 7.4.

WORKED PROBLEM 7.3 Point out the HOMO and LUMO in the following reactions:

(a) $H\ddot{\overset{..}{O}}:^- + Li^+ \rightarrow H\ddot{\overset{..}{O}}Li$

(b) $H_2C{=}CH_2 + BH_3 \rightarrow {}^+CH_2{-}CH_2{-}{}^-BH_3$

(c) $:CH_3^- + H{-}\overset{..}{\underset{..}{Cl}}: \rightarrow CH_4 + :\overset{..}{\underset{..}{Cl}}:^-$

(d) $H_3N: + H{-}\overset{..}{\underset{..}{Cl}}: \rightarrow {}^+NH_4 + :\overset{..}{\underset{..}{Cl}}:^-$

(e) $H\ddot{\overset{..}{O}}:^- + H_3\overset{+}{\ddot{O}}: \rightarrow 2\ H_2\ddot{\overset{..}{O}}:$

(continued)

ANSWER

(a) HOMO: Filled nonbonding orbital on HO^-
LUMO: Empty $2s$ orbital on Li^+
(b) HOMO: Filled π orbital of ethylene
LUMO: Empty $2p$ orbital on BH_3
(c) HOMO: Filled sp^3 nonbonding orbital of methyl anion
LUMO: Empty σ^* orbital of H—Cl
(d) HOMO: Filled sp^3 nonbonding orbital of ammonia, $:NH_3$
LUMO: Empty σ^* orbital of H—Cl
(e) HOMO: Filled nonbonding orbital of hydroxide
LUMO: Empty σ^* orbital of an O—H bond of H_3O^+

7.3   Reactions of Alkyl Halides: The Substitution Reaction

A great many varieties of substitution reactions are known, but all involve a competition between a pair of Lewis bases for a Lewis acid. Figure 7.9 shows five examples with the substituting group, the nucleophile (from the Greek, "nucleus loving"), and displaced group, the **leaving group**, identified. *Both* the nucleophile (usually Nu) and leaving group (usually L) are Lewis bases.

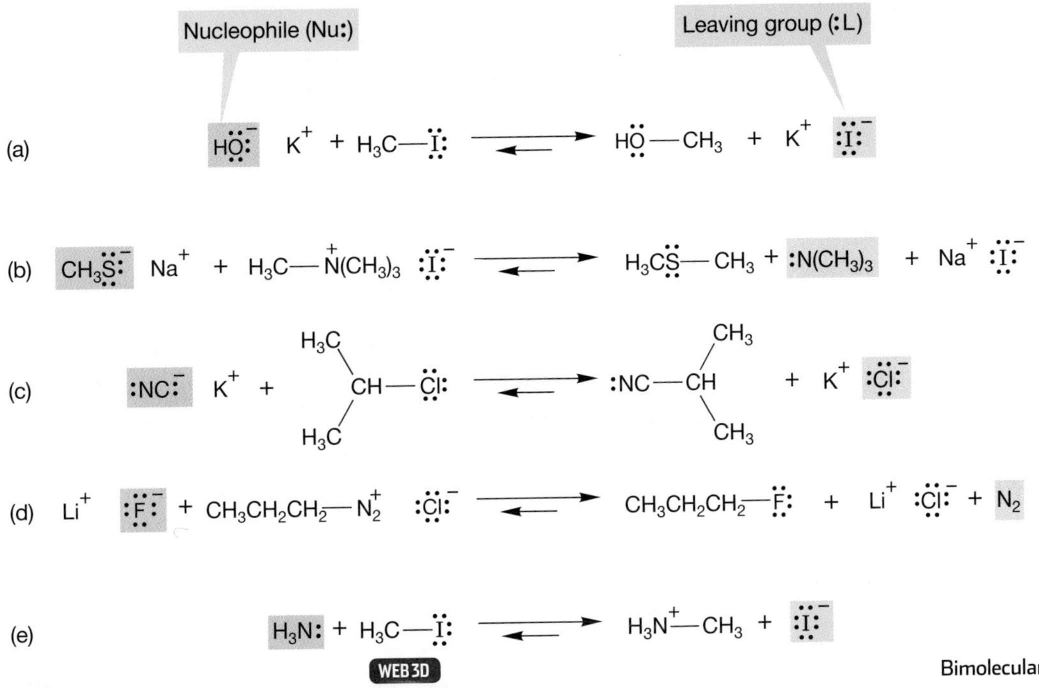

Nucleophile (Nu:)   Leaving group (:L)

(a) $HO:^-$ K^+ + $H_3C—I:$ ⇌ $HO—CH_3$ + K^+ $:I:^-$

(b) $CH_3S:^-$ Na^+ + $H_3C—\overset{+}{N}(CH_3)_3$ $:I:^-$ ⇌ $H_3CS—CH_3$ + $:N(CH_3)_3$ + Na^+ $:I:^-$

(c) $:NC:^-$ K^+ + $\underset{H_3C}{\overset{H_3C}{>}}CH—Cl:$ ⇌ $:NC—\underset{CH_3}{\overset{CH_3}{|}}CH$ + K^+ $:Cl:^-$

(d) Li^+ $:F:^-$ + $CH_3CH_2CH_2—\overset{+}{N_2}$ $:Cl:^-$ ⇌ $CH_3CH_2CH_2—F:$ + Li^+ $:Cl:^-$ + N_2

(e) $H_3N:$ + $H_3C—I:$ ⇌ $H_3\overset{+}{N}—CH_3$ + $:I:^-$

[WEB 3D]

Bimolecular nucleophilic substitution: S_N2

FIGURE 7.9 Five examples of the substitution reaction. The nucleophiles for the reactions reading left to right are (a) hydroxide (HO^-), (b) methyl mercaptide (CH_3S^-), (c) cyanide (NC^-), (d) fluoride (F^-), and (e) ammonia ($:NH_3$). The leaving groups are iodide (I^-), trimethylamine [$N(CH_3)_3$], chloride (Cl^-), nitrogen (N_2), and iodide (I^-) again.

PROBLEM 7.4 Identify the HOMOs and LUMOs in the reactions of Figure 7.9.

Notice that the reactions in Figure 7.9 are all equilibrium reactions. We will consider equilibrium in detail in Chapter 8, but we need to see here that these competition reactions are all, in principle at least, reversible. A given reaction may not be reversible in a practical sense, but that simply means that one of the competitors has been overwhelmingly successful. There are many possible reasons why this might be so. One of the equilibrating molecules might be much more stable than the other. If one of the nucleophiles were much more powerful than the other, the reaction would be extremely exothermic *in one direction or the other*. Reactions a–c in Figure 7.9 are examples of this. We are used to thinking of reactions running from left to right, but this prejudice is arbitrary and even misleading. Reactions run *both ways*. There are other reasons one of these reactions might be irreversible. Perhaps the equilibrium is disturbed by one of the products being a solid that crystallizes out of the solution, or a gas that bubbles away. If so, the reaction can be driven all the way to one side, even against an unfavorable equilibrium constant. Under such circumstances, an endothermic reaction can go to completion. In reaction d of Figure 7.9 both propyl fluoride (bp 2.5 °C) and nitrogen are gases, and their removal from the solution will drive the equilibrium to the right.

The second point to notice is that there is no obligatory charge type in these reactions. The displacing nucleophile is not always negatively charged (reaction e), and the group displaced—the leaving group—does not always appear as an anion (reactions b and d).

If we examine a large number of substitution reactions, two categories emerge. The boundaries between them are not absolutely fixed, as we will see, but there are two fairly clear-cut limiting classes of reaction. One is a kinetically second-order process, common for primary and secondary substrates (R—L, Section 7.4). The other is a kinetically first-order reaction, common for tertiary substrates (Section 7.6). We take up these two reactions in turn, examining each in detail.

7.4 Substitution, Nucleophilic, Bimolecular: The S$_N$2 Reaction

7.4a Rate Law   The rate law tells us the number and kinds of molecules involved in the transition state—it tells you how the reaction rate depends on the concentrations of reactants and products. For many substitution reactions, the rate v is proportional to the concentrations of both the nucleophile, [Nu:$^-$], and the substrate, [R—L]. This bimolecular reaction is first order in Nu:$^-$ and first order in substrate, R—L. The rate constant, k, is a fundamental constant of the reaction (Fig. 7.10). Here, R stands for a generic alkyl group. Don't confuse this R with the (R) used to specify one enantiomer of a chiral molecule. This bimolecular reaction is second order overall, which leads to the name **S**ubstitution, **N**ucleophilic, **bi**molecular, or **S$_N$2 reaction**.

CONVENTION ALERT

FIGURE 7.10 A bimolecular reaction.

$$:Nu + R—L \rightleftharpoons Nu—R + L:^-$$

$$\text{Rate } (v) = k\,[R—L]\,[:Nu]$$

7.4b Stereochemistry of the S$_N$2 Reaction   One of the most powerful of all tools for investigating the mechanism of a reaction is a stereochemical analysis. In this case, we need first to see what's possible and then to design an experiment that will enable us to determine the stereochemical results. If we start with a single enantiomer (Chapter 4, p. 151) (R) or (S), there are several possible outcomes of a substitution reaction. We could get **retention of configuration**, in which the stereochemistry of the starting molecule is preserved. The entering nucleophile occupies the same stereochemical position as did the departing leaving group, and the product has the same handedness as the starting material (Fig. 7.11).

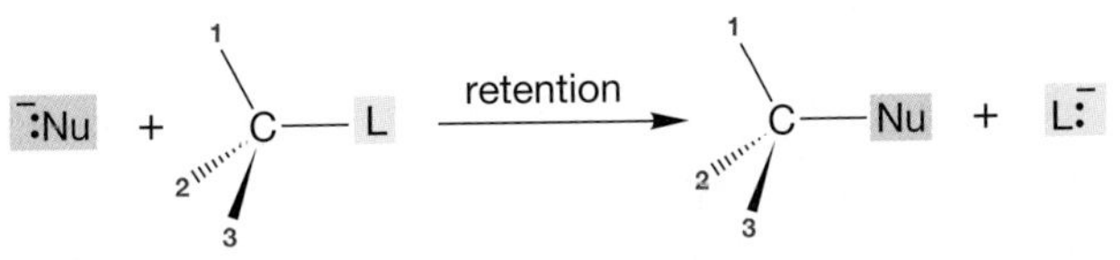

FIGURE 7.11 One possible stereochemical outcome for the S$_N$2 reaction. In this scenario, the product is formed with *retention of configuration*.

A second possibility is **inversion of configuration**, in which the stereochemistry of the starting material is reversed: (R) starting material going to (S) product and (S) starting material going to (R) product. In this kind of reaction, the pyramid of the starting material becomes inverted in the reaction like an umbrella in the wind (Fig. 7.12). In this and Figure 7.11, notice how the numbers enable us to keep track of inversions and retentions.

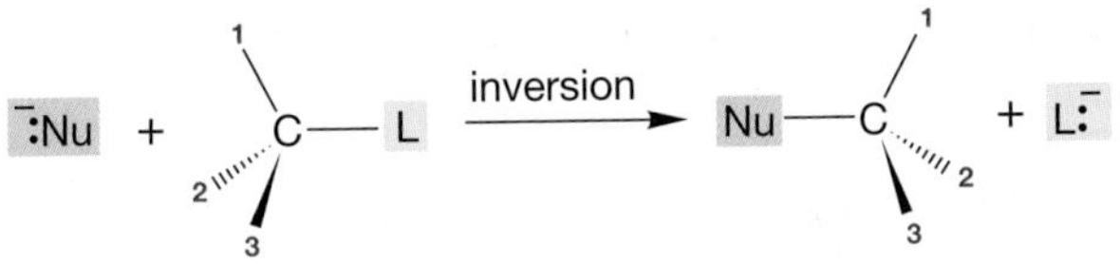

FIGURE 7.12 Another possible stereochemical result for the S$_N$2 reaction. The handedness of the starting material could be reversed in the product, in what is called *inversion of configuration*.

It is also possible that optical activity could be lost. In such a case, (R) or (S) starting material goes to a racemic mixture—a 50:50 mixture of (R) and (S) product (Fig. 7.13).

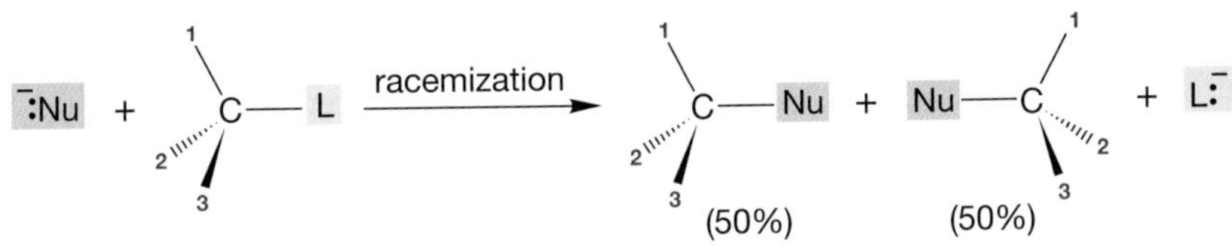

FIGURE 7.13 A third possible stereochemical result for the S$_N$2 reaction.

Finally, it is possible that some dreadful mixture of the (R) and (S) forms is produced. In such a case, it is very difficult to analyze the situation, as several reaction mechanisms may be operating simultaneously.

WORKED PROBLEM 7.5 Sometimes matters appear more complicated than they really are. It is possible, for example, that an (R) starting material could produce an (R) product by a process involving inversion. Here is a conceivable example. Explain.

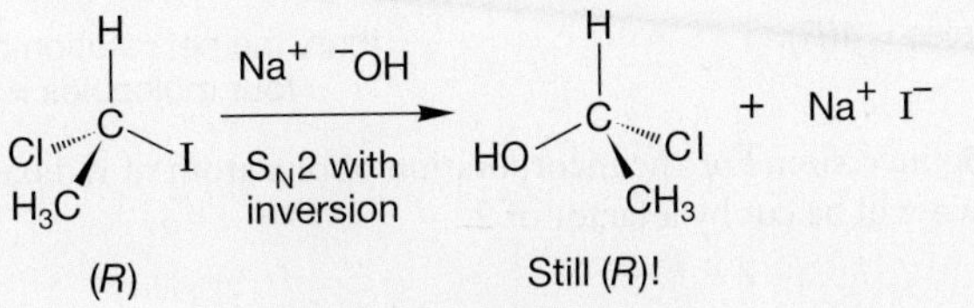

(continued)

ANSWER The designations (R) and (S) are determined for each molecule through the Cahn–Ingold–Prelog priority system (Chapter 3, p. 111; Chapter 4, p. 152). If the groups attached to carbon change, as in the example shown, there is no guarantee that inversion will change an (R) carbon to an (S) carbon. In the example shown here, the "umbrella" is inverted, but at the same time the substituents attached to the stereogenic carbon have also changed from (C, Cl, H, I) to (C, Cl, H, OH). Application of the priority system shows that both molecules are (R) despite the inversion.

For most secondary and primary substrates, substitution occurs with complete inversion of configuration. We now are faced with three questions: (1) How do we know the reactions go with inversion? (2) What does this result tell us about the reaction mechanism? (3) Why is inversion preferred to retention or racemization?

The answer to the first question is easy: We measure it. Suppose we start with the optically active iodide R—I in Figure 7.14 and monitor the reaction with radioactive iodide ion ($^-$I*).[2] We measure two things: the incorporation of radioactive iodide and the optical activity of R—I*. If radioactive iodide reacts with R—I with retention of configuration, there will be no loss of rotation at all [I* and I are not significantly different in their contributions to the rotation of plane-polarized light (Fig. 7.14)].

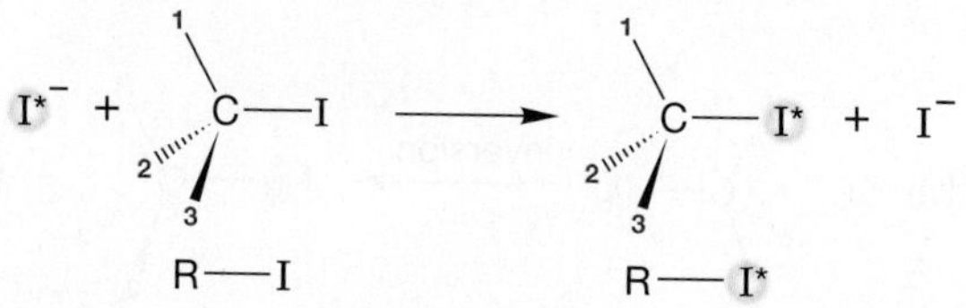

FIGURE 7.14 A mechanism involving retention predicts that incorporation of radioactive iodide will induce no change in optical rotation.

What happens if the reaction goes with inversion? The observed optical rotation of R—I is certainly going to decrease as, for example, (R)-iodide is converted into (S)-iodide. But it is usually not obvious how the rate of loss of optical activity compares with the rate of incorporation of I*. If every replacement occurs with inversion, the rate of loss of optical activity must be twice the rate of incorporation of I*. Every conversion of an (R)-iodide (R—I) into an (S)-iodide (I*—R) generates a product molecule (I*—R) whose rotation exactly cancels that of a molecule of starting R—I. In effect, the rotation of two molecules goes to zero for every inversion (Fig. 7.15).

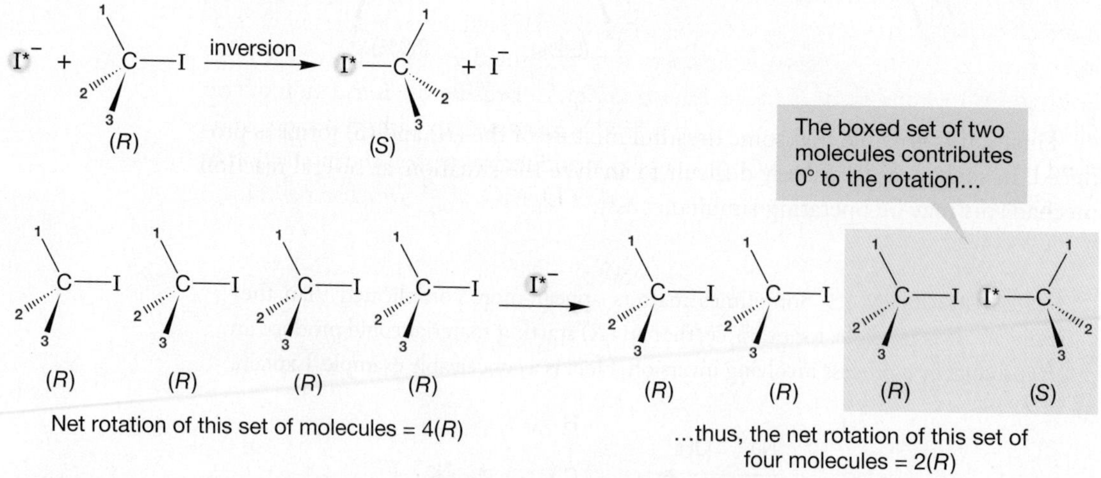

FIGURE 7.15 A mechanism involving inversion. For the incorporation of one atom of radioactive iodide, the rotation of the set of molecules shown will be cut by a factor of 2.

[2]In this discussion, we indicate radioactive iodine with an asterisk in the text and with the asterisk and a yellow circle in the figures.

This change is exactly what is observed experimentally. Every incorporation of I^* is accompanied by the inversion of one (R)-iodide to an (S)-iodide.

Now, what does this result tell us about the mechanism? The observation of inversion tells us that the nucleophile, the incoming atom or molecule, must be approaching the substrate from the rear of the departing atom. Only this path can lead to the observed inversion. Figure 7.16 shows the reaction, along with an arrow formalism description of the bonds that are forming and breaking. The arrows show the path of approach of the incoming nucleophile and the departing leaving group. The tetrahedral "umbrella" is inverted during the reaction.

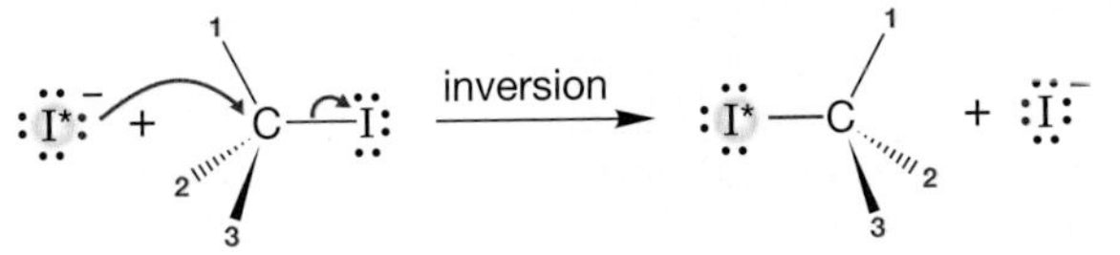

FIGURE 7.16 The observation of inversion in the S_N2 reaction means that the entering nucleophile, radioactive (or labeled) iodide, must enter the molecule from the side opposite the departing leaving group, here unlabeled iodide.

Had the reaction occurred with retention, a frontside approach would have been demanded. The incoming nucleophile would have had to enter the molecule from the same side as the departing leaving group. The tetrahedral "umbrella" would have retained its configuration. This result is *not* observed (Fig. 7.17).

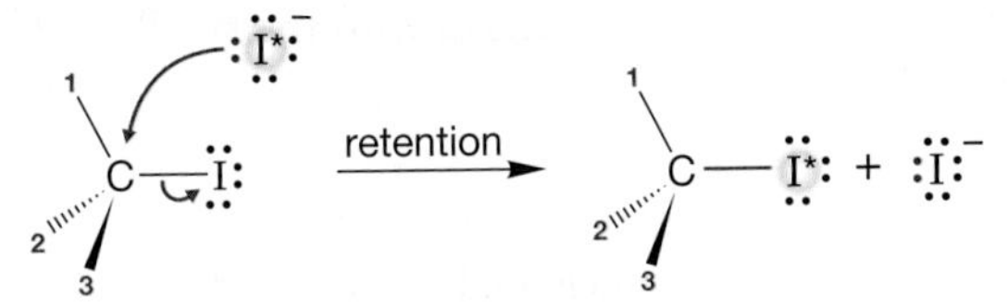

FIGURE 7.17 If retention had been the experimentally observed result, the nucleophile would have added from the same side that the leaving group departed.

All this analysis is useful; we now know the path of approach for the nucleophile in the S_N2 reaction, and so we know more about the mechanism of this process. But our newfound knowledge serves only to generate the inevitable next question. *Why* is inversion preferred? Why is frontside displacement—retention—never observed? Here comes part of the payoff for all the work we did in earlier chapters on molecular orbitals. One of the key lessons of those early chapters was that interactions of *filled* and *empty* orbitals are stabilizing, because the two electrons originally in the filled orbital can be accommodated in the new bonding orbital, and no electrons need occupy the antibonding molecular orbital. Now let's look at the S_N2 reaction in orbital terms. A filled nonbonding orbital "n" on the nucleophile interacts with an orbital on the substrate, R—L, involved in the bond from R to the leaving group, L. Because the interaction of two filled orbitals is not stabilizing, the filled orbital n must overlap with the empty, antibonding σ^* orbital of R—L. This overlapping is nothing more than the interaction of the nucleophile's HOMO, n, with the sigma bond's LUMO, σ^*, shown schematically in Figure 7.18.

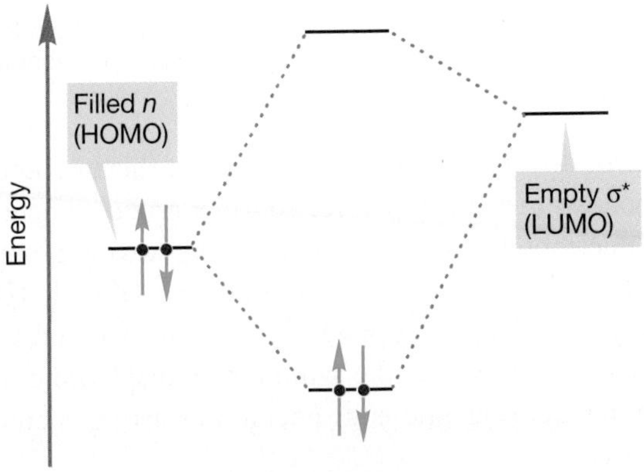

FIGURE 7.18 An orbital interaction scheme for the stabilizing interaction of a filled nonbonding orbital (n) on the nucleophile overlapping with the empty, antibonding (σ^*) orbital on the substrate.

(a) Stabilizing interaction of filled and empty orbitals (strong)   (b) Destabilizing interaction of two filled orbitals (weak)

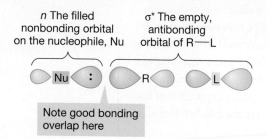

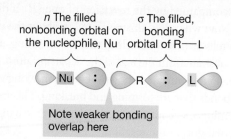

n The filled nonbonding orbital on the nucleophile, Nu   σ* The empty, antibonding orbital of R—L

n The filled nonbonding orbital on the nucleophile, Nu   σ The filled, bonding orbital of R—L

Note good bonding overlap here

Note weaker bonding overlap here

FIGURE 7.19 (a) The overlap of the filled nonbonding orbital (*n*) with the "outside" lobe of the empty σ* orbital of R—L takes advantage of the good overlap available with this lobe of σ*. (b) Contrast (a) with the destabilizing interaction of *n* and σ, the two filled orbitals. The interaction of two filled orbitals is destabilizing, but the overlap of the lobes involved, one fat, the other thin, is poor. Therefore, this interaction is not important, and the stabilizing interaction shown in (a) dominates.

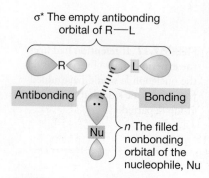

σ* The empty antibonding orbital of R—L

Antibonding   Bonding

n The filled nonbonding orbital of the nucleophile, Nu

FIGURE 7.20 Frontside substitution (retention) requires overlap with the inside part of σ*. There is poor overlap with the small lobes of σ* as well as offsetting bonding and antibonding interactions.

What does σ* look like? Figure 7.19 shows us. Note that σ* has one of its large lobes pointing to the *rear* of R—L. That is where overlap with the filled *n* orbital will be most efficient.

Now look what happens if the orbitals overlap in a way that would lead to retention through a frontside substitution (Fig. 7.20). Not only is the magnitude of interaction reduced by the relatively poor overlap with the smaller lobe of σ*, but there is both a bonding and an antibonding interaction, and these will tend to cancel each other. Frontside displacement is a very unlikely process indeed.

We can now map out the progress of the bond making and breaking in the S_N2 reaction. The reaction starts as the nucleophile Nu:⁻, approaches the rear of the substrate. Because the HOMO, the nonbonding orbital *n* on Nu:⁻, overlaps with the LUMO, the antibonding σ* orbital of C—L, the bond from C to L is weakened as the antibonding orbital begins to be occupied. The bond length of C—L increases as the new bond between Nu and C begins to form. As C—L lengthens, the other groups attached to C bend back, in umbrella fashion, ultimately inverting (just like the windblown umbrella) to form the product. For a symmetrical displacement such as the reaction of radioactive ⁻I* with R—I discussed before, the midpoint of the reaction will be the point at which the groups attached to C are exactly halfway inverted (Fig. 7.21).

FIGURE 7.21 In a symmetrical inversion process, the midpoint, called the transition state, is the point at which the umbrella is exactly half-inverted.

$$I^{\cdot-} + {}^{\cdots}C\!-\!I \longrightarrow \left[I^* \quad C \quad I \right]^{-} \longrightarrow I^*\!-\!C^{\cdots} + I^{-}$$

sp²

In the transition state for this symmetrical S_N2 reaction, the umbrella is halfway inverted and the carbon is *sp²* hybridized

What is the hybridization of the central carbon at this midpoint? It must be *sp²*, because this carbon is surrounded by three coplanar groups. There are partial bonds to the incoming nucleophile, ⁻:I*, and the departing leaving group, ⁻:I. No picture shows the competition between the two nucleophiles (⁻:I* and ⁻:I) as clearly as this one. What Lewis acid are they competing for? A carbon 2*p* orbital, as the figure shows. What is this structure? At least in the symmetrical displacement of ⁻:I by ⁻:I*, it is the halfway point between the equi-energetic starting material and product.

This structure is the transition state for this reaction. Is this transition state something one could potentially catch and isolate? No! It occupies not an energy minimum, as stable compounds do, but it sits on an energy maximum along the path from starting material to product. We have seen such structures before: Recall, for example, the eclipsed form of ethane or the transition state for the interconversion of two pyramidal methyl radicals (Fig. 7.22).

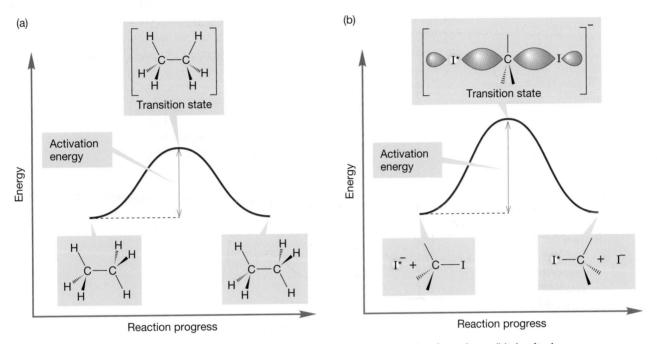

FIGURE 7.22 Two analogous diagrams. (a) The rotation about the carbon–carbon bond in ethane; (b) the displacement of $^-$I by $^-$I*. In each case, two equi-energetic molecules are separated by a single energy barrier, the top of which is called the *transition state*, shown in brackets.

Now, let's consider the change in energy as the S$_N$2 reaction proceeds (Fig. 7.22b). The starting material and products are separated by the high-energy point of the reaction, the transition state. The energy difference between starting material and transition state is related to the rate of the reaction. The larger this difference, the higher is the barrier, and the slower is the reaction. So, a quantity of vital importance to any study of a chemical reaction is the energy difference between starting material and the transition state, called the activation energy. The reaction progress diagram is symmetrical only if the substitution reaction is symmetrical, that is, if starting material and product are of equal energy.[3]

This specific picture is transferable to a general second-order substitution reaction, but some differences will appear. First of all, usually the starting material and product will not be at the same energy, and the transition state cannot be *exactly* halfway between starting material and product. It will lie closer to one than the other.

---

[3]Actually, even our picture of the displacement of $^-$:I by $^-$:I* is ever so slightly wrong. Normal iodine (I) is not *exactly* the same as radioactive iodine (I*), and even this reaction is not truly symmetrical. We can ignore the tiny differences in isotopes, however, as long as we keep the principle straight. If you want to be more correct, insert the words "except for a tiny isotope effect" wherever appropriate.

The transition state cannot be exactly sp^2, only *nearly* sp^2 (notice its slightly pyramidal shape) (Fig. 7.23).

Figure 7.23 shows both an exothermic and endothermic S_N2 substitution reaction. If we read the figure conventionally from left to right, we see the exothermic reaction, but if we read it "backward," from right to left, we see the endothermic reaction. Notice that the activation energy for the forward process must be smaller than that for the reverse reaction. Thermodynamics will determine where the equilibrium settles out, and there may be practical consequences if one partner is substantially favored over the other, but one diagram will suffice to examine both directions. Many of us are prisoners of our language, which is read from left to right, and we tend to do the same for these energy diagrams. Nature is more versatile, or perhaps less prejudiced, than we are.

FIGURE 7.23 A general S_N2 reaction. If the nucleophile (Nu) and the leaving group (L) are different, the transition state is not exactly at the midpoint of the reaction, and the hybridization of carbon is only approximately sp^2.

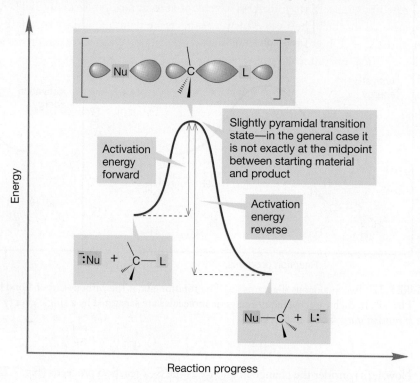

Slightly pyramidal transition state—in the general case it is not exactly at the midpoint between starting material and product

Activation energy forward

Activation energy reverse

Energy

Reaction progress

MALIC ACID

The chiral malic acid figures strongly in the original discovery of inversion of configuration in the S_N2 reaction. Malic acid is sometimes called "apple acid" because of its high concentration in apples, nectarines, and some other fruits. In fact, it was first isolated from apple juice as early as 1785. It functions as a molecular carrier of the carbon dioxide absorbed by plants. The CO_2 appears in the CH_2COOH group of malic acid. The principal actor in the early mechanistic work on the S_N2 reaction was Paul Walden, a chemist born in 1863 in what is

now Latvia and who died in Tübingen, Germany, in 1957. Indeed, the inversion now known to occur universally in the S_N2 reaction is sometimes called Walden inversion.

MJ's teacher of organic chemistry, William Doering, told a wonderful story about Paul Walden. Professor Doering was visiting Tübingen in 1956 to talk about his work and overheard a now-famous German chemist lamenting the fact that Tübingen had not been bombed during the war. It seems that most of his colleagues in other "more fortunate" cities had newly built laboratories courtesy of Allied bombing. There was an old gentleman sitting in the corner who had been wheeled into the department from his nursing home to hear Doering's seminar. Paul Walden—for indeed it was he—was 92 at the time. He overheard the remark and replied, scathingly, "Es ist nicht der Käfig, sondern der kleine Vogel darin!" (It is not the cage that matters, but the little bird inside!)

The take-home lesson here is an important one: There is a strong preference for inversion in the S_N2 reaction. Indeed, there is *no* authenticated example of retention of configuration in this process, despite a great deal of searching by some very clever people.

Now let's examine the effects of structural change in the various participants in the S_N2 reaction, the R group, the nucleophile, the leaving group, and the solvent.

7.4c Effects of Substrate Structure: The R Group

The structure of the R group makes a huge difference in the rate of the S_N2 reaction. We anticipated this result when we mentioned earlier that the practically useful S_N2 reaction was restricted to methyl, primary, and secondary substrates. By implication, the rate of the S_N2 reaction with tertiary substrates is zero, or at least negligibly small (Table 7.1). Why should this be? The simple answer seems to be that in a tertiary substrate the rear of the C—L bond is guarded by three alkyl groups, and the incoming nucleophile can find no unhindered path along which to approach the fat lobe of σ^* (Fig. 7.24). So the S_N2 reaction is disfavored for tertiary substrates, for all of which steric hindrance to the approaching nucleophile is prohibitively severe. Another substitution mechanism, favorable for tertiary substrates, becomes possible. It is called the S_N1 reaction, and we will deal with its mechanism in Section 7.6.

If this steric argument is correct, secondary substrates should react more slowly than primary substrates, and primary substrates should be slower than methyl compounds. In general, this is the case (Table 7.1). In practice, the S_N2 reaction is usually useful as long as there is at least one hydrogen attached to the same carbon as the leaving group. Thus, the S_N2 reaction works only for methyl, primary, and secondary substrates, all of which have at least one hydrogen attached to the carbon at which the substitution is occurring. The small size of hydrogen opens a path at the rear for the incoming nucleophile.

PROBLEM 7.6   Explain the following change in rate for the S_N2 reaction:

$$R\text{—}O^- + CH_3CH_2\text{—}I \rightarrow R\text{—}O\text{—}CH_2CH_3 + I^-$$

Rate for R = CH_3 is much faster than for R = $(CH_3)_3C$

This picture of the S_N2 reaction, which emphasizes steric effects, allows us to make a prediction. In principle, there must be some primary group so gigantic that the S_N2 reaction would be unsuccessful.

In practice, it is rather easy to find such groups. Even the neopentyl group, $(CH_3)_3CCH_2$, is large enough to slow the bimolecular displacement reaction severely, because the *tert*-butyl group blocks the best pathway for rearside displacement of the leaving group (Fig. 7.25; Table 7.1).

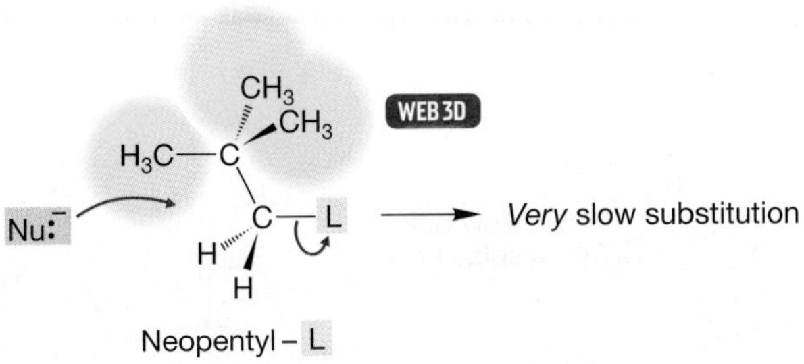

TABLE 7.1 Average Rates of S_N2 Substitution Reactions for Different Groups

R	Average Relative Rate
$CH_2\text{=}CHCH_2$	1.3
CH_3	1
CH_3CH_2	0.033
$CH_3CH_2CH_2$	0.013
$(CH_3)_2CH$	8.3×10^{-4}
$(CH_3)_3CCH_2$	2×10^{-7}
$(CH_3)_3C$	~0

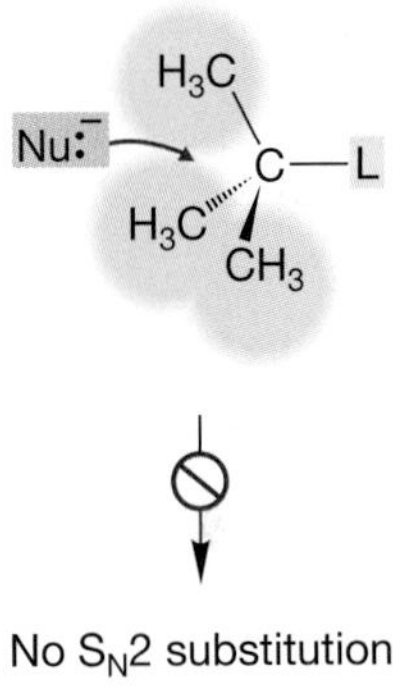

No S_N2 substitution

FIGURE 7.24 For tertiary substrates, approach from the rear is hindered by the alkyl groups, here all shown as methyls. This steric effect makes the S_N2 reaction impossible.

FIGURE 7.25 Even neopentyl compounds, in which a *tert*-butyl group shields the rear of the C—L bond, are hindered enough so that the rate of the S_N2 reaction is extremely slow.

Neopentyl—L

TABLE 7.2 Relative Reactivities of Cycloalkyl Bromides in the S$_N$2 Reaction

Compound	Relative Rate
Cyclopropyl bromide	$<10^{-4}$
Cyclobutyl bromide	8×10^{-3}
Cyclopentyl bromide	1.6
Cyclohexyl bromide	1×10^{-2}
Isopropyl bromide	1.0

The S$_N$2 reaction also takes place when ring compounds are used as the substrates, although there are some interesting effects of ring size on the rate of the reaction (Table 7.2; Fig. 7.26).

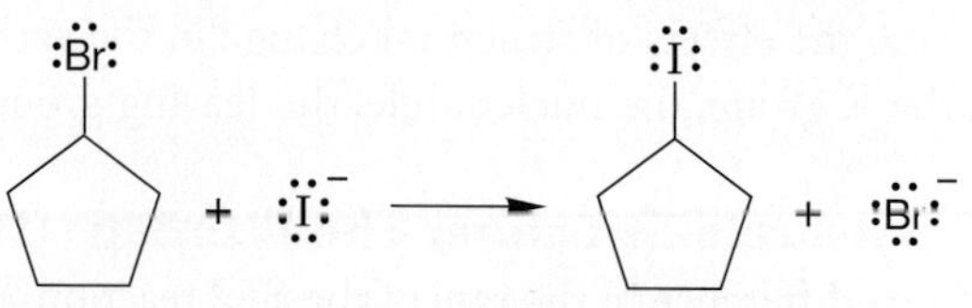

FIGURE 7.26 The S$_N$2 reaction takes place in a normal fashion with cyclic substrates.

PROBLEM 7.7 Use a ring compound to design a test of the stereochemistry of the S$_N$2 reaction.

Why should small rings be so slow in the S$_N$2 reaction? As with any question involving rates, we need to look at the structures of the transition states for the reaction to find an answer. In the transition state, the substrate is hybridized approximately sp^2, and that requires bond angles close to 120°. The smaller the ring, the smaller are the C—C—C angles. For a three-membered ring, the optimal 120° must be squeezed to 60°, and for a cyclobutane, squeezed to 90°. This contraction introduces severe angle strain in the transition state, raises its energy, and slows the rate of reaction (Fig. 7.27).

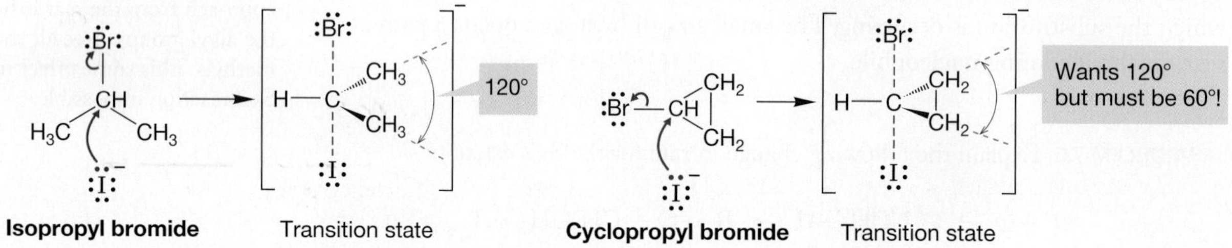

FIGURE 7.27 A comparison of S$_N$2 transition states for isopropyl bromide and cyclopropyl bromide.

PROBLEM 7.8 Wait! The argument just presented ignores angle strain in the starting material. Cyclopropane itself is strained. Won't that strain raise the energy of the starting material and offset the energy raising of the transition state? Comment. *Hint*: Consider angle strain in both starting material and transition state—in which will it be more important?

In cyclohexane, the reaction rate is apparently slowed by steric interactions with the axial hydrogens. There can be no great angle strain problem here (Fig. 7.28).

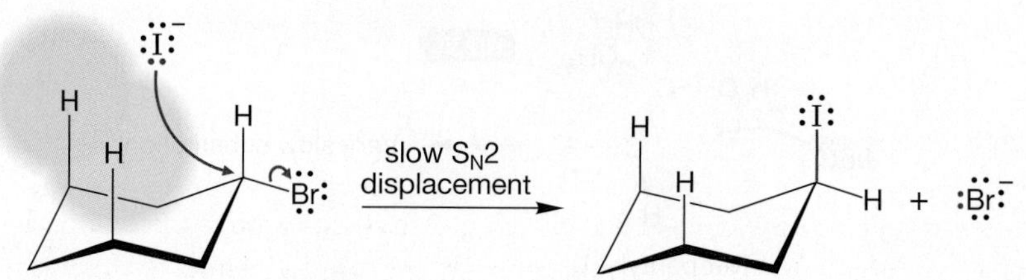

FIGURE 7.28 The slow S$_N$2 displacement of bromide by iodide in cyclohexyl bromide. Incoming iodide is apparently blocked somewhat by the axial carbon–hydrogen bonds.

WORKED PROBLEM 7.9 The axial cyclohexyl iodide shown below reacts more quickly in S$_N$2 displacements than the equatorial cyclohexyl iodide. Draw the transition states for displacement of iodide from axial cyclohexyl iodide by iodide ion and for the analogous reaction of equatorial cyclohexyl iodide. What is the relation between these transition states? Why does the axial iodo compound react faster than the equatorial compound? **Caution!** Hard, tricky question.

ANSWER A tricky problem indeed. The transition states for the two reactions are the same! There is no energy difference between the two transition states:

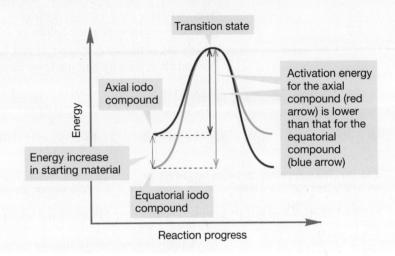

The transition state is the same for displacement of axial and equatorial cyclohexyl iodide from 4-iodo-*tert*-butylcyclohexane

Therefore, the reason that axial cyclohexyl iodide reacts more quickly than equatorial cyclohexyl iodide cannot lie in the transition state energies. However, the axial iodo group raises the energy of the *starting material*, thus lowering the activation energy.

Transition state

Axial iodo compound

Activation energy for the axial compound (red arrow) is lower than that for the equatorial compound (blue arrow)

Energy increase in starting material

Equatorial iodo compound

Energy

Reaction progress

7.4d Effect of the Nucleophile   Some displacing agents are more effective than others. They are better players in the competition for the carbon 2p orbital, the Lewis acid. In this section, we examine what makes a good nucleophile, that is, what makes a powerful displacing agent.

Remember: Size can be important. We have already seen how the S$_N$2 reaction can be slowed, or even stopped altogether, by large R groups (Fig. 7.25). Presumably, large nucleophiles will also have a difficult time in getting close enough to the substrate to overlap effectively with σ* (Problem 7.6, p. 275). Figure 7.29 shows two nucleophiles, carefully chosen to minimize all differences except size.

WEB 3D

CH$_2$

H$_2$C   H

H$_2$C   C

H$_2$C   CH$_2$

        N

H$_2$C   CH$_2$

Azabicyclooctane

CH$_3$

H$_2$C

H$_3$C—N   CH$_3$

H$_2$C   CH$_2$

Triethylamine

C—C bond rotations

WEB 3D

CH$_3$

H$_2$C

H$_2$C   N   CH$_2$

H$_3$C   CH$_3$

FIGURE 7.29 In both azabicyclooctane and triethylamine, the nucleophilic nitrogen atom is flanked by three two-carbon chains. In the bicyclic cage compound azabicyclooctane, they are tied back by the CH group shown in red and are not free to rotate. In triethylamine, they are free to rotate and effectively increase the bulk of the nucleophile.

The bicyclic molecule, in which the three ethyl-like groups are tied back out of the way, is a more effective displacing agent than is triethylamine, in which the three ethyl groups are freely rotating. Triethylamine is effectively larger than the cage compound in which the alkyl groups are tied back (Fig. 7.30).

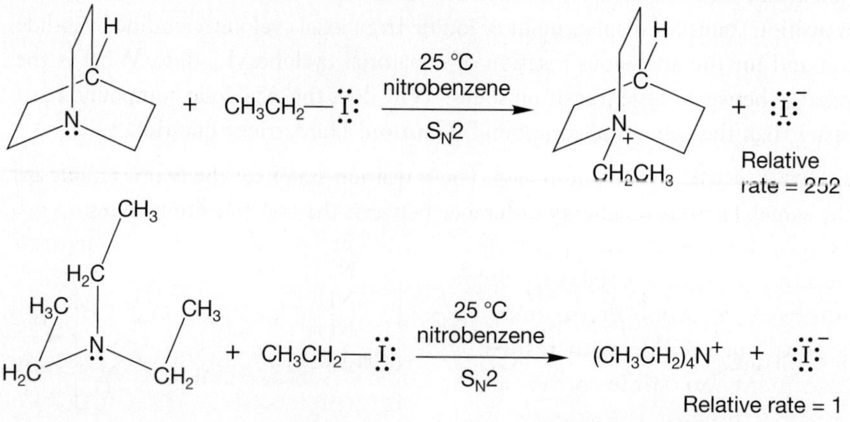

FIGURE 7.30 The tied-back bicyclic compound reacts faster with ethyl iodide than does triethylamine, which is effectively the larger compound. The larger nucleophile has a more difficult time in attacking the rear of the carbon–iodine bond in ethyl iodide.

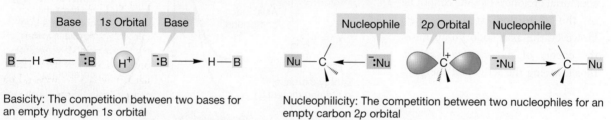

FIGURE 7.30 The tied-back bicyclic compound reacts faster with ethyl iodide than does triethylamine, which is effectively the larger compound. The larger nucleophile has a more difficult time in attacking the rear of the carbon–iodine bond in ethyl iodide.

Nucleophilicity, Lewis basicity, is a measure of how well a nucleophile competes for an empty carbon $2p$ orbital, and Brønsted basicity is a measure of how well a base competes for an empty hydrogen $1s$ orbital. One would expect there to be a general correlation of nucleophilicity with base strength although the two are not exactly the same thing, because $1s$ and $2p$ orbitals are different in energy and shape. Nevertheless, basicity and nucleophilicity are related phenomena (Fig. 7.31).

Basicity: The competition between two bases for an empty hydrogen $1s$ orbital

Nucleophilicity: The competition between two nucleophiles for an empty carbon $2p$ orbital

FIGURE 7.31 Brønsted basicity and nucleophilicity are different, but related, phenomena.

Actually, it is simple to refine our discussion of good nucleophiles. What do we mean by "a good competitor for a carbon $2p$ orbital"? We are talking about the overlap of a filled orbital and an empty orbital, and the resulting stabilization is a measure of how strong the interaction is. *Remember*: The strongest orbital interactions, and hence the greatest stabilizations, come from the overlap of orbitals close in energy (Fig. 7.32).

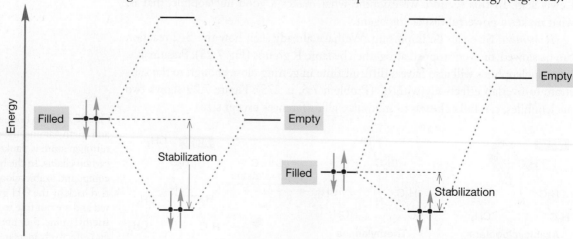

FIGURE 7.32 Stabilization is greater when filled and empty orbitals of equal or nearly equal energy interact than it is when orbitals that differ greatly in energy interact. The further apart two orbitals are in energy, the smaller is the stabilization resulting from their overlap.

So we can anticipate that energy matching between the orbital on the nucleophile and an empty carbon orbital will be important in determining nucleophilicity.

Table 7.3 and Figure 7.33 show some nucleophiles segregated into general categories. It is not possible to do much better than this, because nucleophilicity is a property that depends on the reaction partner. A good nucleophile with respect to carbon may or may not be a good nucleophile with respect to displacement on some other atom. Energy matching is critical in the reaction. If we change the reaction partner, and thus the energy of the orbital involved, we change the stabilization involved as well (Fig. 7.32).

TABLE 7.3 Relative Nucleophilicities of Some Common Species

Species	Name	Relative Nucleophilicity
Excellent Nucleophiles		
NC^-	Cyanide	126,000
HS^-	Mercaptide	126,000
I^-	Iodide	80,000
Good Nucleophiles		
HO^-	Hydroxide	16,000
Br^-	Bromide	10,000
N_3^-	Azide	8,000
NH_3	Ammonia	8,000
NO_2^-	Nitrite	5,000
Fair Nucleophiles		
Cl^-	Chloride	1,000
CH_3COO^-	Acetate	630
F^-	Fluoride	80
CH_3OH	Methyl alcohol	1
H_2O	Water	1

Based on charge	Based on electronegativity
$^-NH_2 > NH_3$	$^-NH_2 > {}^-OR > {}^-OH$
$^-SH > SH_2$	$H_2Se > H_2S > H_2O$
$^-OH > OH_2$	$R_3P > R_3N$
	$I^- > Br^- > Cl^- > F^-$

FIGURE 7.33 Some relative nucleophilicities. Be careful! Nucleophilicity (Lewis basicity) is a hard-to-categorize quantity. Relative nucleophilicity depends, for example, on the identity of the reaction partner, which is always a Lewis acid, as well as the nature of the solvent.

Table 7.3 and Figure 7.33 reveal some spectacular exceptions to the "a good Brønsted base is a good nucleophile" rule. Look at the halide ions, for example. As we would expect from the pK_a values of the conjugate acids, HX (where X is a halide), the basicity order is $F^- > Cl^- > Br^- > I^-$. The nucleophilicity order is *exactly opposite* the basicity order (Fig. 7.34). Iodide is the *weakest* Brønsted base but the *strongest* nucleophile. Fluoride is the *strongest* Brønsted base but the *weakest* nucleophile. Why?

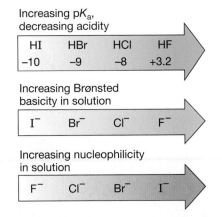

FIGURE 7.34 For the halide ions in solution, the orders of Brønsted basicity and nucleophilicity are opposite.

To answer this question, we need more data. It turns out that the effectiveness of these (and other) nucleophiles depends on the so-often-neglected solvent. It is easy to forget that most chemical reactions are run in "oceans" of solvent, and it is perhaps not too surprising that these oceans of other molecules are not always without substantial effects on the reaction! Figure 7.35 examines rates for typical displacement reactions by halides in solvents of different polarities and in the total absence of solvent (the gas phase). Note that iodide is a particularly effective nucleophile, relative to other halides, in water, a protic solvent. Recall our discussion in Chapter 6 (p. 238): Protic solvents are polar solvents containing hydrogens (e.g., water or alcohols). In polar solvents that do not contain hydroxylic hydrogen, such as acetone, $(CH_3)_2C—O$, the reactivity order is reversed. In the absence of solvent, in the gas phase, the order is also reversed.

FIGURE 7.35 In the highly polar and protic solvent water, iodide is the best nucleophile (fastest reacting). In the polar but aprotic solvent $(CH_3)_2C═O$ (acetone), the order of nucleophilicity is reversed. In the gas phase, where there is no solvent, fluoride emerges as the best nucleophile! Read the table across, not down. The dashed line means "not measured."

$$X^- + CH_3—I \xrightarrow[H_2O]{S_N2} X—CH_3 + I^-$$

$$X^- + CH_3—Br \xrightarrow[\substack{acetone \\ solvent}]{S_N2} X—CH_3 + Br^-$$

$$X^- + CH_3—Br \xrightarrow[\substack{gas\ phase \\ no\ solvent}]{S_N2} X—CH_3 + Br^-$$

X = halogens

	Relative Rates		
I	Br	Cl	F
160	14	1	---
1	5	11	---
---	<0.015	0.02	1

In a protic solvent, a halide can act as a *nucleophile* (Lewis base) and react with the substrate through interaction with σ^* of the C—X bond. Alternatively, a halide can behave as a Brønsted *base* by interacting with one of the protonic solvent molecules. Chloride and fluoride are by far the better bases and thus hydrogen bond to protonic solvents much more strongly than iodide does. The halides become highly solvated in protonic solvents and their relative sizes increase dramatically. They become *more* encumbered, and therefore *less* effective nucleophiles. Iodide, the poorer base, is less hydrogen bonded to protonic solvents and thus more free to do the S_N2 reaction (Fig. 7.36).

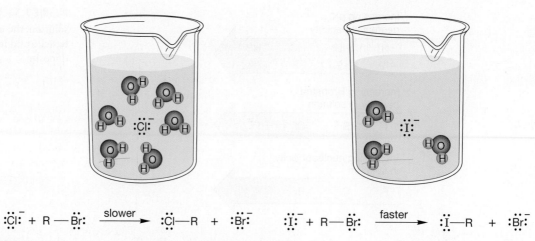

FIGURE 7.36 In a hydroxylated solvent, the highly solvated chloride ion reacts more slowly with R—Br than the less encumbered iodide ion does.

As the polarity of the solvent decreases, fluoride becomes more competitive with iodide. What's the limit of this effect? A reaction with *no* solvent, of course. Pure nucleophilicity, with no competing effects of solvent, can be examined only in the absence of other bulk molecules. We can do that by running the reaction in the gas phase. It is no trivial matter to examine reactions of ions in the gas phase, as the stability of ions depends heavily on solvation. Yet, since the 1990s such experiments have become common. In the gas phase, fluoride, the better base, is also the better nucleophile (Fig. 7.35).

7.4e Effect of the Leaving Group   Many leaving groups depart as anions.

$$\text{Nu:}^- + \text{R—L} \rightarrow \text{Nu—R} + \text{L:}^-$$

It seems obvious that the more stable L:$^-$ is, the easier it will be to displace it. The stability of the anion L:$^-$ is related to how easily its conjugate acid, H—L, dissociates and thus to the pK_a of H—L. Low-pK_a acids (strong acids) are related to good leaving groups, L:$^-$ (weak bases). High pK_a acids (weak acids) are related to poor leaving groups (strong bases). Table 7.4 shows some good and poor leaving groups and the pK_a values of their conjugate acids, H—L.

$$\text{H—L} + \text{H}_2\text{O} \rightleftharpoons \text{H}_3\text{O}^+ + \text{L:}^-$$

TABLE 7.4 Some Good and Poor Leaving Groups and Their Conjugate Acids

Acid	pK_a	Leaving Group	Name
		Good Leaving Groups	
HI	-10	$^-$I	Iodide
HBr	-9	$^-$Br	Bromide
HCl	-8	$^-$Cl	Chloride
HOSO$_2$R	-3	$^-$OSO$_2$R	Sulfonate
H$_3$O$^+$	-1.7	OH$_2$	Water
		Poor Leaving Groups	
HF	$+3.2$	$^-$F	Fluoride
H$_2$S	$+7.0$	$^-$SH	Thiolate
HCN	$+9.4$	$^-$CN	Cyanide
H$_2$O	$+15.7$	$^-$OH	Hydroxide
HOCH$_2$CH$_3$	$+15.9$	$^-$OCH$_2$CH$_3$	Ethoxide
HOR	$+16$–18	$^-$OR	Alkoxide

7.4f How to Play "Change the Leaving Group"   Many times it would be convenient to convert an alcohol (R—OH), which is generally cheap and available (in the chemical industry, those two terms are usually synonymous), into another compound, R—Nu. One naturally considers an S$_N$2 substitution reaction of R—OH (Fig. 7.37) with some nucleophile Nu:$^-$.

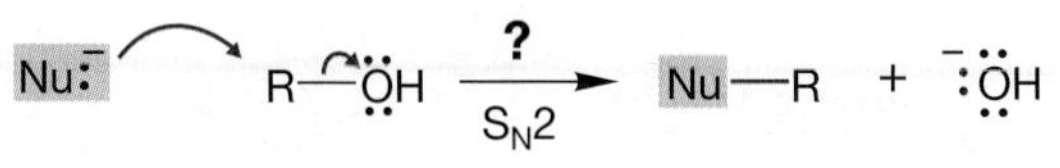

FIGURE 7.37 A hypothetical S$_N$2 reaction converting an alcohol, R—OH, into a new compound, Nu—R.

The problem is that $^-$OH is an extraordinarily poor leaving group (Table 7.4). The pK_a of water is 15.7. Water is a rather weak acid, and its conjugate base, hydroxide ion, is a strong base. There are several ways to circumvent the problem, the simplest of which is to transfer a proton to the alcohol (protonate the alcohol) with a strong acid to give the conjugate acid of the alcohol, R—$\overset{+}{O}H_2$. Protonation converts the leaving group from a very poor leaving group, hydroxide, into an excellent leaving group, water (Fig. 7.38).

FIGURE 7.38 The first step in the reaction of an alcohol with the acid H—B is protonation of the Lewis base oxygen.

$$R—\ddot{O}H + H—B \rightleftharpoons R—\overset{+}{\ddot{O}}H_2 + B\overset{-}{:}$$

> **PROBLEM 7.10** What *will* happen when an alcohol is treated with a strong base? Think "simple."

Water is a weak base. Its conjugate acid, H_3O^+, is a strong acid (pK_a −1.7). Therefore, water is easily displaced by a good nucleophile. In the reaction of an alcohol with HBr, the strongest available nucleophile is Br^-, the conjugate base of the HBr used to protonate the alcohol. So, if we want to convert an alcohol into a bromide we need only treat it with HBr. The reaction and its mechanism are shown for the conversion of ethyl alcohol into ethyl bromide in Figure 7.39. Note especially the technique of changing the leaving group from the very poor hydroxide to the excellent water.

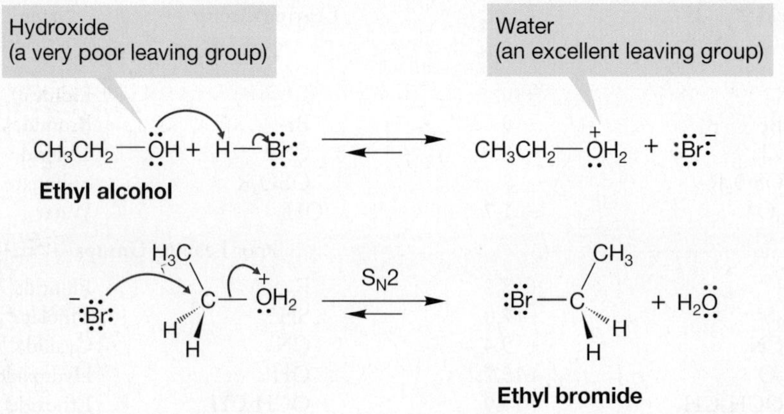

FIGURE 7.39 Playing "change the leaving group" with ethyl alcohol.

Hydroxide (a very poor leaving group)

Water (an excellent leaving group)

$$CH_3CH_2—\ddot{O}H + H—\ddot{B}r: \rightleftharpoons CH_3CH_2—\overset{+}{\ddot{O}}H_2 + :\ddot{B}r\overset{-}{:}$$

Ethyl alcohol

Ethyl bromide

A variant of the "change the leaving group" game is very closely related to the reactions of alcohols in halogenated acids to give halides. It is called ether cleavage, and it follows essentially the same mechanism. In strong halogenated acids, such as HI or HBr, ethers can be cleaved to the corresponding alcohol and halide. In the presence of excess HBr, the alcohols formed in this reaction are often converted into the bromides (Fig. 7.40).

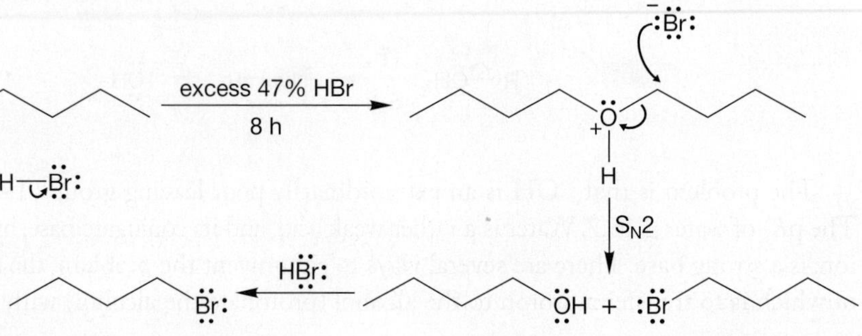

FIGURE 7.40 The mechanism of the cleavage of ethers by strong haloacids.

excess 47% HBr

8 h

In this reaction, the ether oxygen is first protonated, turning the potential leaving group in an S_N2 reaction from alkoxide, ^-OR, into the much more easily displaced alcohol, HOR. The displacement itself is accomplished by the halide ion formed when the ether is protonated. This reaction is quite analogous to the reaction of alcohols with haloacids. The halide must be a nucleophile strong enough to do the displacement, and the reaction is subject to the limitations of any S_N2 reaction (Fig. 7.40).

PROBLEM 7.11 Clearly, an unsymmetrical ether can cleave in two ways. Yet often only one path is followed. Explain the specificity shown in the following reaction.

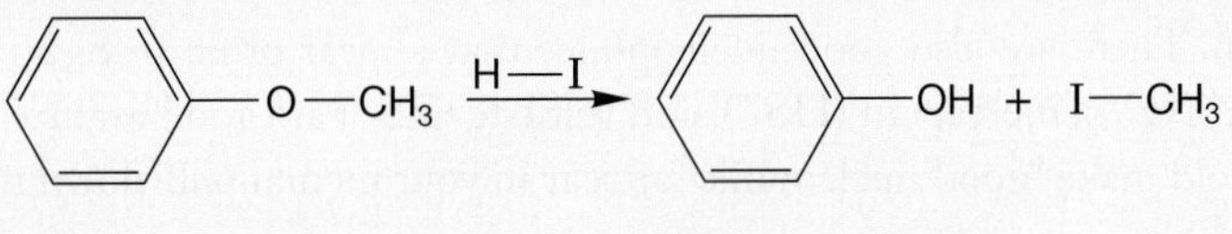

PROBLEM 7.12 Provide a mechanism for the following reaction:

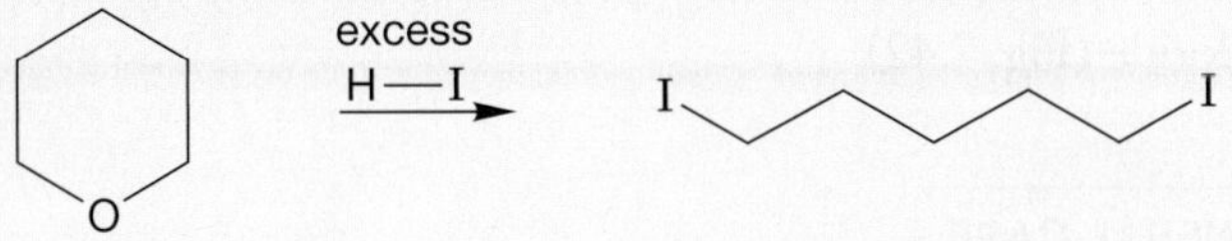

Another technique for changing the leaving group uses a two-step process in which the alcohol is first converted into a sulfonate. Sulfonates are excellent leaving groups (Table 7.4) and can be displaced by all manner of nucleophiles to give desirable products (Fig. 7.41).

THE GENERAL CASE

$$CH_3CH_2—OH + Cl—SO_2—R \longrightarrow CH_3CH_2—O—SO_2—R + H—Cl$$

Ethyl alcohol An ethyl sulfonate

A SPECIFIC EXAMPLE

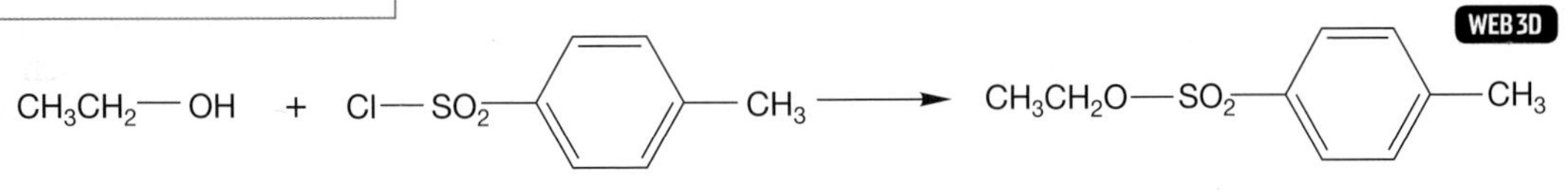

Ethyl tosylate
(an ethyl sulfonate)

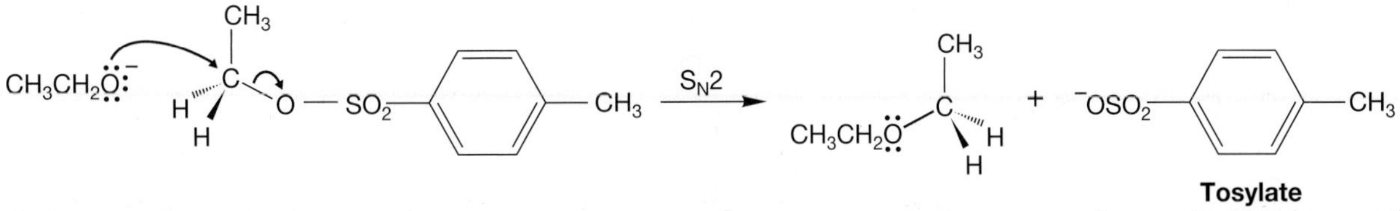

Tosylate
(an excellent leaving group)

FIGURE 7.41 The conversion of ethyl alcohol into another ethyl derivative. In this sequence of reactions the poor leaving group OH is first converted into the good leaving group sulfonate. The leaving group is then displaced through an S_N2 reaction with a nucleophile, here $CH_3CH_2O^-$.

A variation on the sulfonate reaction of Figure 7.41 involves the treatment of an alcohol with **thionyl chloride** ($SOCl_2$). The end result is the conversion of an alcohol into a chloride (Fig. 7.42).

FIGURE 7.42 Treatment of an alcohol with thionyl chloride results in the formation of an alkyl chloride.

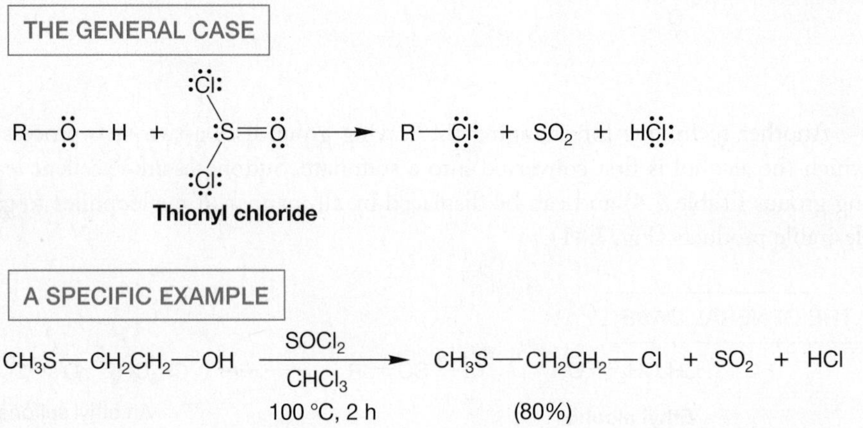

THE GENERAL CASE

$$R—\overset{..}{\underset{..}{O}}—H \ + \ \underset{\overset{|}{:\overset{..}{Cl}:}}{\overset{:\overset{..}{Cl}:}{:S=\overset{..}{\underset{..}{O}}}} \longrightarrow R—\overset{..}{\underset{..}{Cl}}: \ + \ SO_2 \ + \ H\overset{..}{\underset{..}{Cl}}:$$

Thionyl chloride

A SPECIFIC EXAMPLE

$$CH_3S—CH_2CH_2—OH \ \xrightarrow[\substack{CHCl_3 \\ 100\,°C,\ 2\ h}]{SOCl_2} \ CH_3S—CH_2CH_2—Cl \ + \ SO_2 \ + \ HCl$$
(80%)

The initial product in this reaction is a chlorosulfite ester. Mechanisms of breakdown of the intermediate chlorosulfite ester involve both S_N2 displacement by chloride ion to give the alkyl chloride, sulfur dioxide (SO_2), and a chloride ion (Fig. 7.43a), as well as direct decomposition to SO_2 and a pair of ions that recombine (Fig. 7.43b).

FIGURE 7.43 Mechanisms of chloride formation from reaction of an alcohol with thionyl chloride. (a) The S_N2 pathway. (b) Direct decomposition.

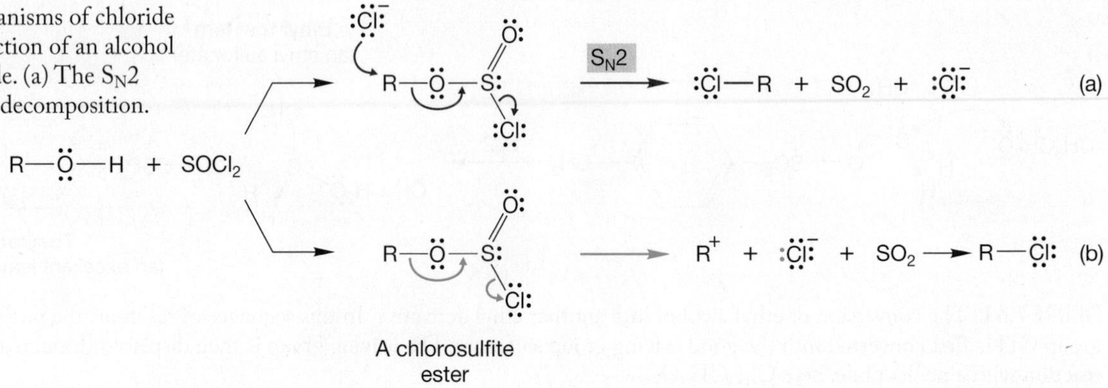

$$R—\overset{..}{\underset{..}{O}}—H \ + \ SOCl_2$$

$$R—\overset{..}{\underset{..}{O}}—\underset{:\overset{..}{Cl}:}{\overset{\overset{\displaystyle ..}{\underset{..}{O}:}}{S}} \ \xrightarrow{\ S_N2\ } \ :\overset{..}{\underset{..}{Cl}}—R \ + \ SO_2 \ + \ :\overset{..}{\underset{..}{Cl}}:^- \quad (a)$$

$$R—\overset{..}{\underset{..}{O}}—\underset{\overset{..}{Cl}:}{\overset{\overset{\displaystyle ..}{\underset{..}{O}:}}{S}} \ \longrightarrow \ R^+ \ + \ :\overset{..}{\underset{..}{Cl}}:^- \ + \ SO_2 \ \longrightarrow \ R—\overset{..}{\underset{..}{Cl}}: \quad (b)$$

A chlorosulfite ester

In either case, the end product is the chloride. Once again, the strategy in this reaction has been to convert a poor leaving group ($^-$OH) into a good one (SO$_2$ and Cl$^-$).

PROBLEM 7.13 Treatment of an alkoxide with dimethyl sulfate leads to methyl ethers. Write a mechanism for this reaction.

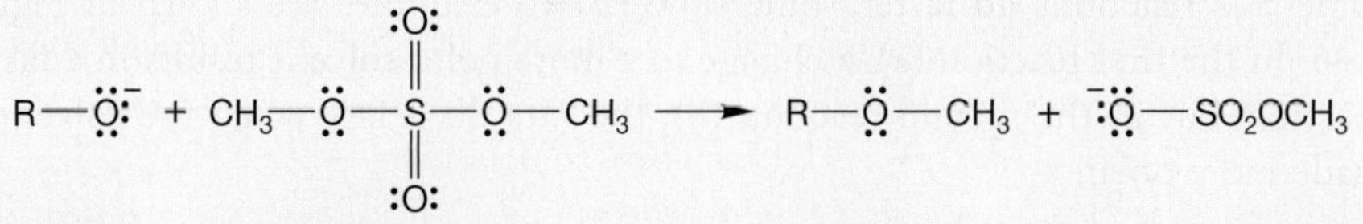

R—Ö:$^-$ + CH$_3$—Ö—S—Ö—CH$_3$ ⟶ R—Ö—CH$_3$ + $^-$:Ö—SO$_2$OCH$_3$

Alkoxide   **Dimethyl sulfate**   Methyl ether

Several related reactions involve the use of phosphorus reagents. Alkyl halides can be made from the treatment of an alcohol with phosphorus halides, PX$_3$ or PX$_5$, as shown in Figure 7.44. Many of these phosphorus-containing reagents cause rearrangement, especially in secondary systems, and the stereochemical outcome, retention or inversion, depends on solvent and other reaction conditions. Accordingly, modifications have been worked out in recent years in order to avoid such stereochemical problems.

Halide formation

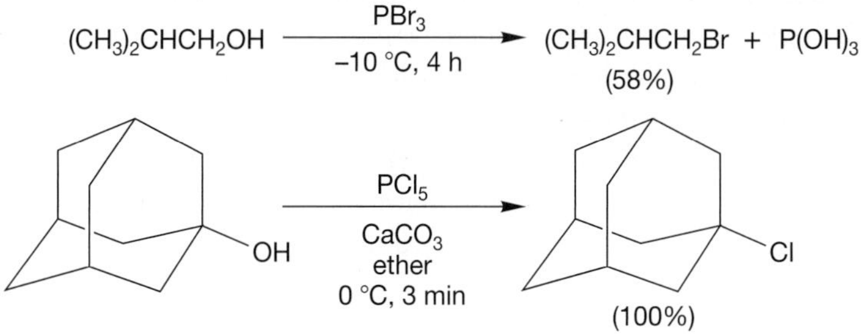

(CH$_3$)$_2$CHCH$_2$OH $\xrightarrow[\text{–10 °C, 4 h}]{\text{PBr}_3}$ (CH$_3$)$_2$CHCH$_2$Br + P(OH)$_3$
(58%)

$\xrightarrow[\substack{\text{CaCO}_3 \\ \text{ether} \\ \text{0 °C, 3 min}}]{\text{PCl}_5}$

(100%)

FIGURE 7.44 Formations of alkyl halides with phosphorus reagents.

For example, alcohols can be converted into the related bromides and chlorides through treatment with triphenylphosphine and a carbon tetrahalide. The alkyl halide is generally formed with inversion, and a rough mechanism involving a series of S$_N$2 reactions is sketched out in Figure 7.45.

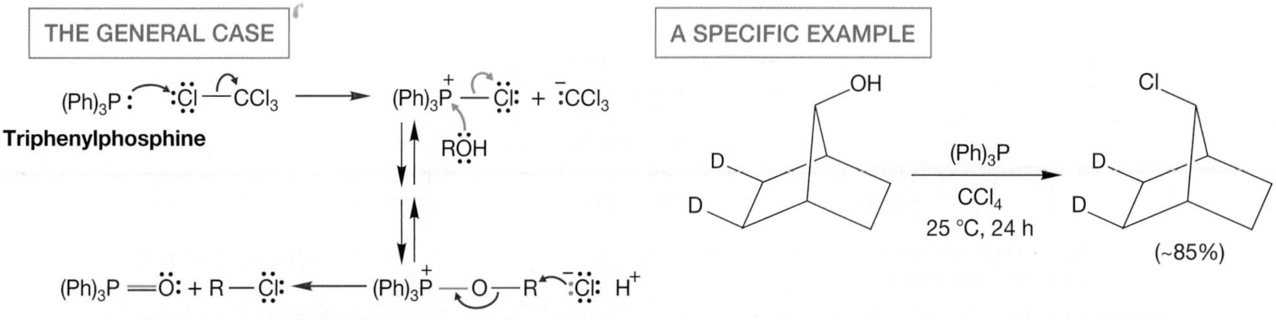

THE GENERAL CASE

(Ph)$_3$P: :Ċl—CCl$_3$ ⟶ (Ph)$_3$P$^+$—Ċl: + $^-$:CCl$_3$

Triphenylphosphine

⇅ RÖH

(Ph)$_3$P=Ö: + R—Ċl: ⟵ (Ph)$_3$P$^+$—O—R :Ċl: H$^+$

A SPECIFIC EXAMPLE

$\xrightarrow[\substack{\text{CCl}_4 \\ \text{25 °C, 24 h}}]{\text{(Ph)}_3\text{P}}$

(~85%)

FIGURE 7.45 Halides are formed from the reaction of alcohols with the intermediate produced from triphenylphosphine and a carbon tetrahalide.

PROBLEM 7.14 Explain why sulfonates should give rather stable anions (weak nucleophiles) and are therefore good leaving groups.

7.4g Effect of Solvent At first, the behavior of the S_N2 reaction as the solvent polarity is changed is perplexing. As the solvent polarity is increased, some S_N2 reactions go faster, some slower. Two examples are shown in Figure 7.46. In the first reaction (a), a change to a more polar solvent results in a faster reaction, but in the second reaction (b), the rate decreases when the solvent is made more polar.

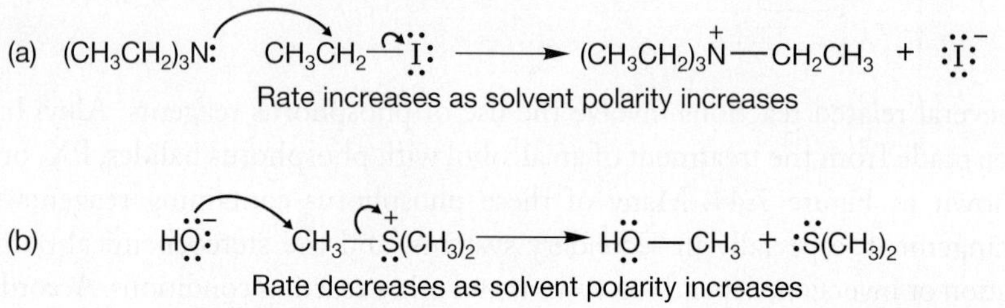
(a) $(CH_3CH_2)_3N$: CH_3CH_2—I: → $(CH_3CH_2)_3\overset{+}{N}$—$CH_2CH_3$ + :I:⁻
Rate increases as solvent polarity increases
(b) HO:⁻ CH_3—$\overset{+}{S}(CH_3)_2$ → HO—CH_3 + :$S(CH_3)_2$
Rate decreases as solvent polarity increases

FIGURE 7.46 Opposite effects of a change in solvent polarity on the rates of two S_N2 reactions. Reaction a goes faster when the solvent is made more polar, whereas reaction b goes more slowly.

There is a way to attack almost every problem that poses a question having to do with rates. The rate of a reaction is determined by the energy of the transition state—the high point in energy as the reaction proceeds from starting material to products. In order to solve a question involving rates, we must know the structure of the transition state of the reaction.

Figure 7.47 shows the structures of the transition states for the two S_N2 reactions of Figure 7.46. In reaction a, a more polar solvent will act to stabilize the charged products and the polar, charge-separated transition state. There will be very

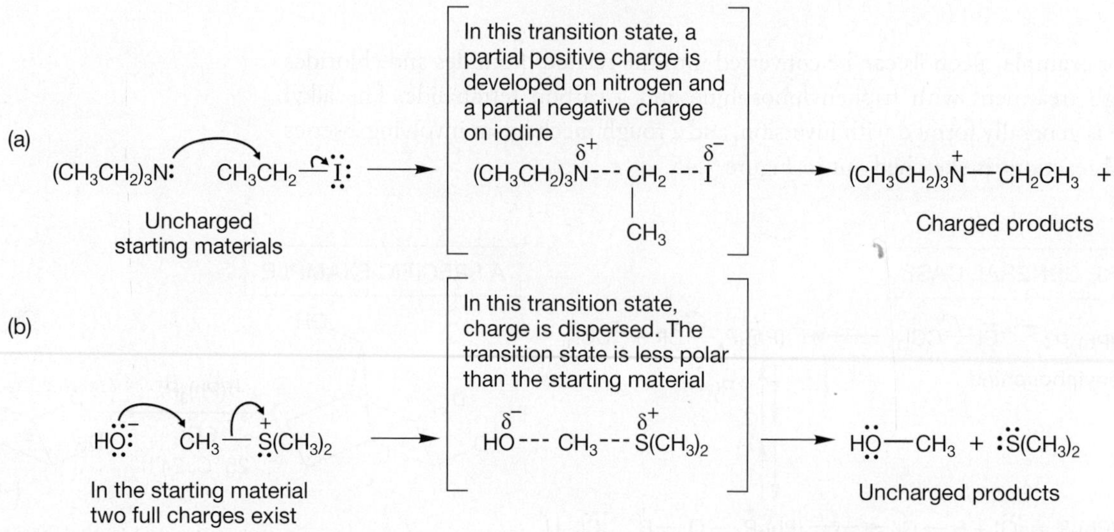
(a)
$(CH_3CH_2)_3N$: CH_3CH_2—I:
Uncharged starting materials
In this transition state, a partial positive charge is developed on nitrogen and a partial negative charge on iodine
$(CH_3CH_2)_3\overset{\delta+}{N}$--- CH_2---$\overset{\delta-}{I}$
CH_3
$(CH_3CH_2)_3\overset{+}{N}$—$CH_2CH_3$ + :I:⁻
Charged products
(b)
HO:⁻ CH_3—$\overset{+}{S}(CH_3)_2$
In the starting material two full charges exist
In this transition state, charge is dispersed. The transition state is less polar than the starting material
$\overset{\delta-}{HO}$--- CH_3---$\overset{\delta+}{S}(CH_3)_2$
HO—CH_3 + :$S(CH_3)_2$
Uncharged products

FIGURE 7.47 The transition state for reaction a is more polar than the starting material. The transition state for reaction b is less polar than the starting material.

little effect on the uncharged starting materials in reaction a. The energy diagram in Figure 7.48 shows what this differential solvation does to the rate of this reaction. The height of the transition state is decreased relative to the starting material and the reaction goes faster. It takes less energy for the molecules to traverse the barrier when the polarity of the solvent increases.

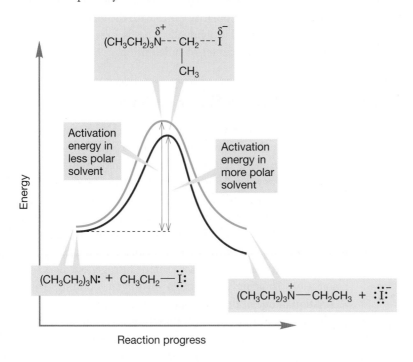

FIGURE 7.48 The difference between the energy of the starting materials and the energy of the transition state is smaller when the reaction is run in a more polar solvent (red curve) and larger when the reaction is run in a less polar solvent (blue curve). The result is a decreased activation energy for the reaction and a faster rate.

Reaction b is different. Here the starting materials bear full charges. This charge is dispersed as the reaction goes on. A change to a more polar solvent will stabilize the fully charged starting materials more than the transition state or products. Therefore, the activation energy increases and the reaction slows (Fig. 7.49).

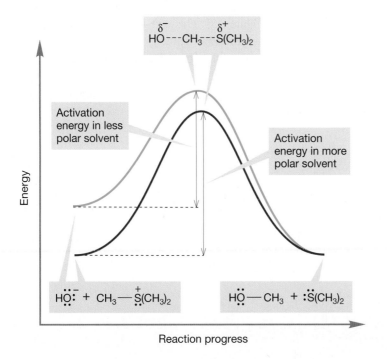

FIGURE 7.49 The difference between the energy of the starting materials and the energy of the transition state is larger when the reaction is run in a more polar solvent (red curve) and smaller when the reaction is run in a less polar solvent (blue curve). The result is an increased activation energy for the reaction and a slower rate.

Summary

The S$_N$2 reaction—Substitution, Nucleophilic, bimolecular—is common for methyl, primary, and secondary substrates. It is disfavored by steric hindrance in either the substrate (the "R" group) or the entering nucleophile. It is favored by a strong Lewis base (nucleophile) and a good leaving group. The response of the S$_N$2 reaction to changes in solvent polarity varies with the charge type of the reaction. Each case must be examined individually. The S$_N$2 reaction always involves inversion of configuration at the center of substitution.

7.5 The S$_N$2 Reaction in Biochemistry

As mentioned in Section 7.1, your vital biomolecules—DNA, for example—are covered with nucleophiles, each of them ready to do the S$_N$2 reaction if presented with a suitable alkylating agent. Once that alkylation takes place, two important changes are induced. First of all, the shape of the molecule is changed, and the critical fit, on which proper bioaction depends, is altered. Second, alkylation often removes the possibility for hydrogen bonding, and in turn that changes the ability of DNA to associate properly with other molecules. As we will see in Chapter 23, DNA exists as pairs of associated "nitrogenous bases" and that association is achieved through hydrogen bonding (Fig. 7.50). Alkylation of one of the nucleophilic groups in DNA can disrupt base pairing and lead to destructive mutations.

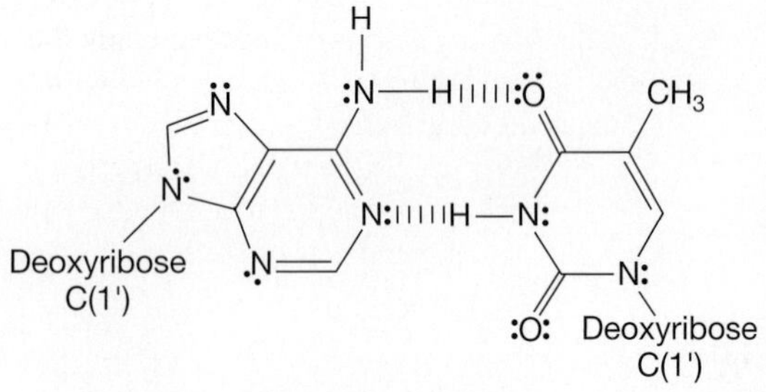

FIGURE 7.50 Replacement of the N—H hydrogens by an alkyl group would hinder—or even eliminate—hydrogen bonding. Thus, proper association of the bases would no longer be possible.

Remember the methyl transfer from *S*-adenosylmethionine (p. 142)? Heinz Floss (b. 1934) and his co-workers have demonstrated that a similar transfer occurs through an S$_N$2 displacement. They did this experiment by showing that transfer takes place with complete inversion of configuration—the hallmark of the S$_N$2 reaction. How can one show inversion in a simple methyl group, CH$_3$? That's not easy! Floss and his co-workers used two isotopes of hydrogen, deuterium (D) and tritium (T), so that their methyl was not CH$_3$ but CHDT. Use of a single enantiomer revealed the diagnostic inversion in the transfer shown in Figure 7.51. The work involved in synthesizing that single enantiomer and in analyzing the product was prodigious, but these prototypal techniques have been borrowed several times in similar studies of biochemical transformations.

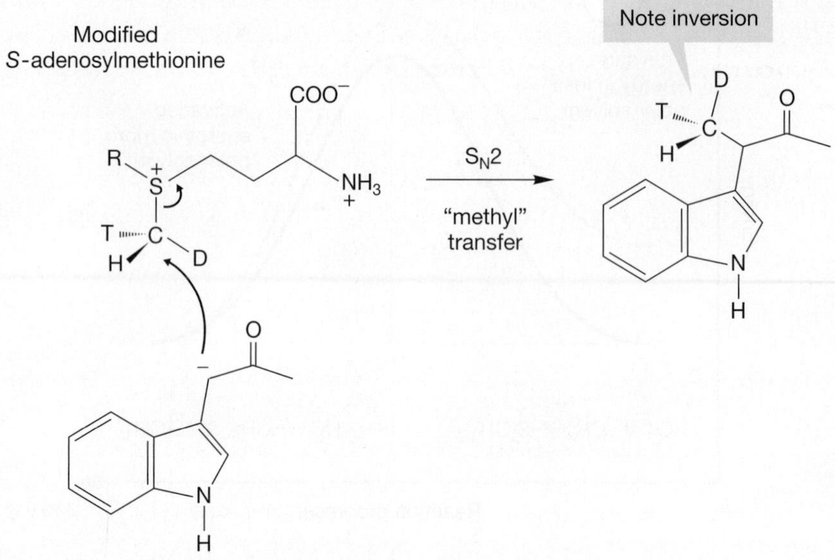

FIGURE 7.51 Methyl transfer involves an S$_N$2 reaction with inversion.

The S$_N$2 reaction is by no means irrelevant to you—your life literally depends on it happening properly, as in the example studied by Floss, and not in the wrong place— as demonstrated by the sometimes lethal mutations induced by alkylating agents.

7.6 Substitution, Nucleophilic, Unimolecular: The S$_N$1 Reaction

7.6a Rate Law   *tert*-Butyl bromide does not undergo a substitution reaction when treated with the good nucleophiles listed in Table 7.3. The rate of the second-order S$_N$2 displacement on *tert*-butyl bromide by hydroxide ion—or any nucleophile—is effectively zero. Yet, *tert*-butyl bromide does react with the much less powerfully nucleophilic molecule water to form the product of substitution, *tert*-butyl alcohol (Fig. 7.52).

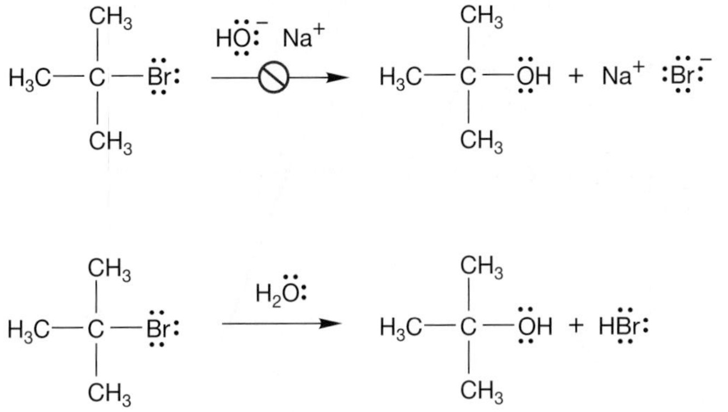

FIGURE 7.52 Despite the far greater nucleophilicity of hydroxide ion, the reaction of *tert*-butyl bromide with water is successful in producing *tert*-butyl alcohol, whereas the reaction with hydroxide ion is not.

PROBLEM 7.15 Although *tert*-butyl bromide does not give *tert*-butyl alcohol on reaction with hydroxide ion, an alkene product (C$_4$H$_8$) is produced. Suggest a structure for the product and an arrow formalism for this bimolecular reaction.

The first clue to what's happening in Figure 7.52 comes from an examination of the rate law for the reaction. It turns out that the formation of *tert*-butyl alcohol in this reaction is a first-order process in which the concentration of added nucleophile is irrelevant to the rate. The rate law leads to the name **S**ubstitution, **N**ucleophilic, **uni**molecular, or **S$_N$1 reaction**, where k is the rate constant, a fundamental constant of the reaction.

Unimolecular nucleophilic substitution: S$_N$1

$$\text{Rate} = \nu = k\,[\textit{tert}\text{-butyl bromide}]$$

At this point, a reasonable mechanistic hypothesis might simply view the S$_N$1 reaction as beginning with the unimolecular ionization shown in Figure 7.53.

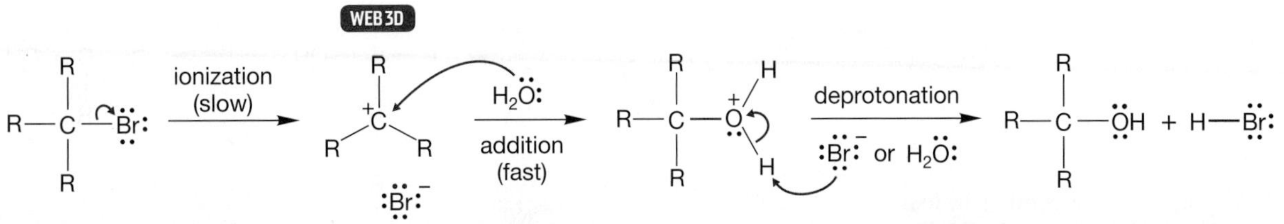

FIGURE 7.53 A schematic mechanism for the S$_N$1 reaction. A slow ionization is followed by a relatively fast capture of the carbocation by a nucleophile, water in this case.

The reaction continues as the carbocation is captured by the solvent, often water, as in Figure 7.53, in what is called a **solvolysis reaction**. Solvolysis simply means that the solvent, here water, plays the role of nucleophile in the reaction. Next, any base present, here likely to be bromide or water, can remove a proton (deprotonate the protonated alcohol) to give the alcohol itself (Fig. 7.53).

We can describe the S_N1 reaction with an energy diagram, as we did for the S_N2 reaction (pp. 273–274). The S_N1 diagram is more complicated because of the formation of the intermediate carbocation. The presence of the intermediate means that there will be two transition states separating the starting material and the product. The transition state for the slow step, the initial endothermic ionization, is higher than the transition state for the second, faster step, the capture of the ion by the nucleophile (Fig. 7.54).

FIGURE 7.54 A generic energy diagram for the S_N1 reaction. Notice that the transition state for the slow step in the reaction—the ionization—is higher in energy than the transition state for the capture of the ion by a nucleophile.

Transition state for ionization—higher

Transition state for ion capture by nucleophile—lower

Energy

R^+ Br^-

R — Nu

R — Br

Reaction progress

This second step determines the structure of the product *but not the rate of the reaction*. If there are several nucleophiles present, they will compete with one another in the capture of the carbocation to produce different products, but the rate of the reaction will not be affected (Fig. 7.55). The nucleophile determines the structures of the products, but not the rate of their formation. Note that the reverse of the ionization, recapture of the carbocation by bromide, is nothing more than another example of a reaction of this intermediate with a nucleophile.

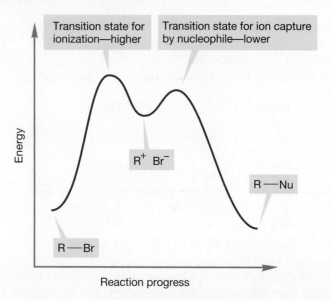

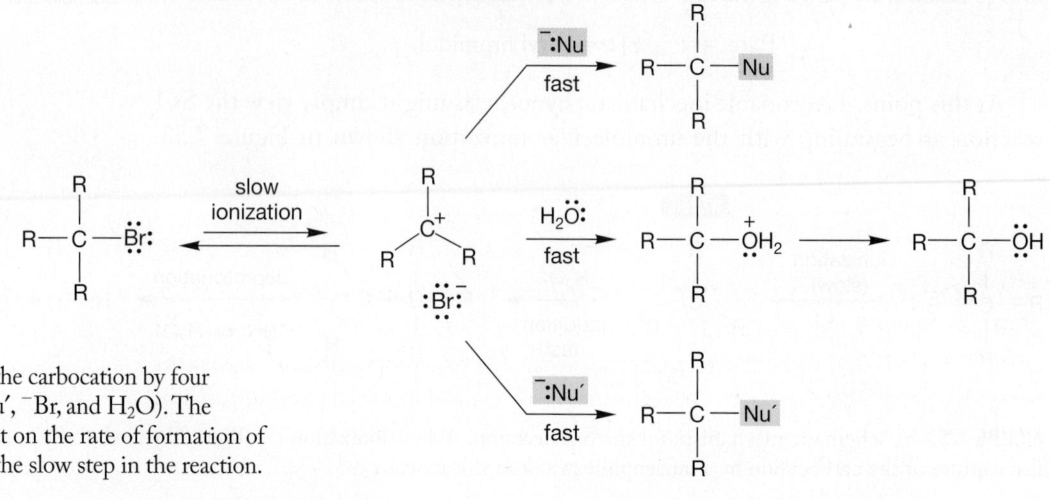

FIGURE 7.55 Capture of the carbocation by four nucleophiles ($^-$:Nu, $^-$:Nu′, $^-$Br, and H_2O). The nucleophiles have no effect on the rate of formation of the carbocation, which is the slow step in the reaction.

We will now see if this simple hypothesis allows a reasonable prediction of the stereochemical course of the reaction.

7.6b Stereochemistry of the S$_N$1 Reaction   If the S$_N$1 reaction were as straightforward as suggested by Figures 7.53–7.55, what would be expected if an optically active starting material were used? The crucial intermediate in the reaction is the planar carbocation, an achiral molecule! Once the free, planar carbocation is formed, optical activity is inevitably lost. Addition of the nucleophile must take place at both faces of the cation, forming equal amounts of (*R*) and (*S*) product. This simple mechanism demands racemization, as Figure 7.56 shows.

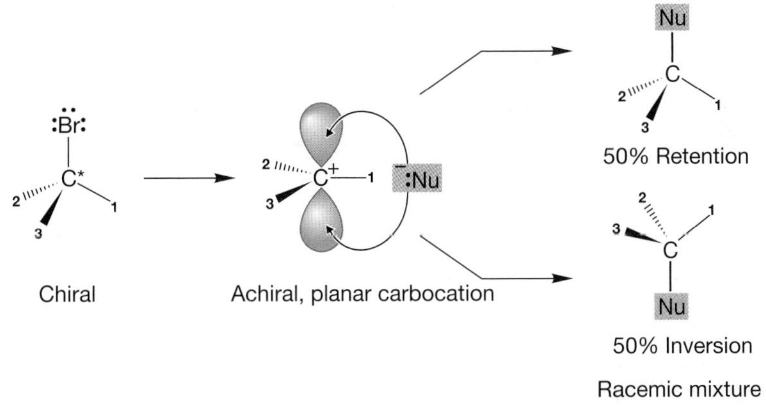

50% Retention

Racemic mixture

50% Inversion

Chiral   Achiral, planar carbocation

FIGURE 7.56 Loss of optical activity occurs as soon as the free, planar carbocation is formed. Addition of any nucleophile will occur equally at the two available lobes of the 2p orbital to give a racemic mixture of products.

Sometimes racemization *is* the actual result, and there is 50% inversion and 50% retention in the product. Usually, however the stereochemical results are not so clean, and an excess of inversion is found. For example, the two optically active molecules of Figure 7.57 give 21 and 18% excess inversion in the S$_N$1 reaction with water.

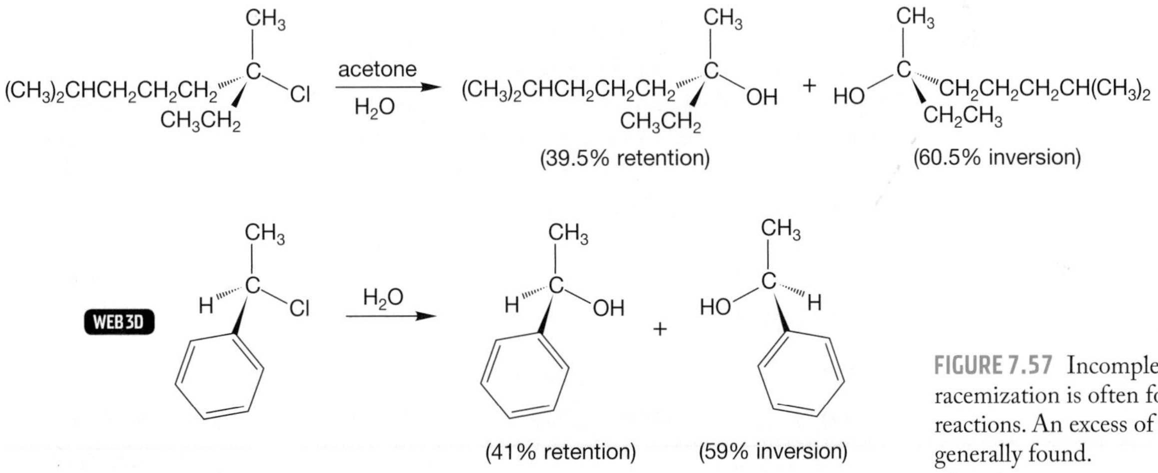

(39.5% retention)

(60.5% inversion)

(41% retention)

(59% inversion)

FIGURE 7.57 Incomplete racemization is often found in S$_N$1 reactions. An excess of inversion is generally found.

Why is the product not racemic, and why is excess inversion (as opposed to retention) observed? These two questions are vitally important, as satisfactory answers must be found if we are not to be forced to abandon the proposed mechanism for the S$_N$1 reaction. The answers turn out to be simple. In the mechanism outlined in Figures 7.53–7.55, we assumed that the leaving group had diffused away from the carbocation, leaving behind a *symmetrical* ion. What if reaction with solvent occurs before the leaving group diffuses completely away? What if reaction occurs with the *pair* of ions shown in Figure 7.58? The leaving group guards the side of the carbocation from which it departed, and makes addition by solvent water more difficult at this "retention" side. The result is usually an excess of inversion.

FIGURE 7.58 Until the leaving group, here bromide ion, completely diffuses away, it shields the carbocation along the retention pathway, and this results in an excess of inversion in most S_N1 reactions.

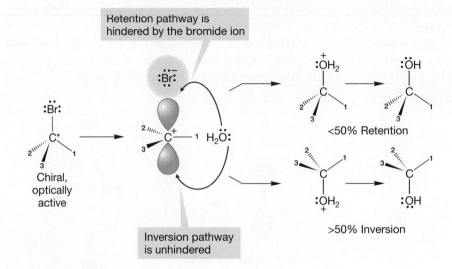

PROBLEM 7.16 The second reaction of Figure 7.57 shows an S_N1 reaction of a secondary halide. Explain why the S_N1 reaction should be especially favorable for this secondary halide.

PROBLEM 7.17 Consider the role of solvent on the stereochemistry of the S_N1 reaction shown in Figure 7.58. A more polar solvent favors racemization. Why?

7.6c Effects of Substrate Structure: The R Group   The rate of the S_N1 reaction is much greater for tertiary substrates than for secondary substrates, and simple primary alkyl and methyl halides do not react at all by the S_N1 process. Our mechanistic hypothesis shows us why. If carbocation formation is the key step, only those compounds that can form relatively stable ions can react in S_N1 fashion.

Modern measurements show that the stability of carbocations increases with substitution (p. 137). Table 7.5 shows the heats of formation of carbocations in the gas phase. *Remember*: The more positive its heat of formation, the less stable a species is (p. 115). We can't make too much of the *quantitative* data in Table 7.5,

TABLE 7.5 Heats of Formation for Some Carbocations

Entry	Cation	Substitution	ΔH_f° (kcal/mol)
1	$^+CH_3$	Methyl	261.3 Least stable
2	$^+CH_2CH_3$	Primary	215.6
3	$^+CH_2CH_2CH_3$	Primary	211
4	$(CH_3)_2\overset{+}{C}H$	Secondary	190.9
5	$^+CH_2CH_2CH_2CH_3$	Primary	203
6	$H_3C\overset{+}{C}HCH_2CH_3$	Secondary	183
7	$(CH_3)_3C^+$	Tertiary	165.8 Most stable

because these gas-phase measurements cannot take into account the effects of solvent, which are very large indeed. Nonetheless, the data do give a good qualitative picture of carbocation stability: Entries 5, 6, and 7 show that tertiary carbocations (three alkyl groups) are more stable than secondary ones (two alkyl groups), and secondary carbocations are more stable than primary ones (one alkyl group). The methyl cation ($^+CH_3$), with no alkyl groups, is the least stable of all. Do not confuse these relative stabilities with absolute stabilities. Carbocations are almost *all* very unstable species. A small atom such as carbon does not bear a positive charge with any grace, and a carbocation will react with available Lewis bases (e.g., halide ion) very quickly. Do not say, "The *tert*-butyl cation is very stable," but rather, "*For a carbocation*, the *tert*-butyl cation is quite stable."

PROBLEM SOLVING

There are many reactions in which tertiary and secondary carbocations are intermediates but no reactions for which a simple primary or methyl cation is required. These intermediates are simply too unstable. In practice, this instability means that when you find yourself writing a mechanism using a primary carbocation, think again, it's almost certainly wrong!

WORKED PROBLEM 7.18 We have again avoided (p. 138) a detailed discussion of *why* more substituted carbocations are more stable than less substituted ones.[4] Nevertheless you can now see the outlines of why this is so. Draw a three-dimensional picture of the ethyl cation. (*Remember*: The positively charged carbon and the three surrounding atoms are coplanar—this part of the ethyl cation is flat.) Now consider the overlap of one of the carbon–hydrogen bonds of the methyl group with the empty $2p_z$ orbital. Is this interaction stabilizing? Why?

ANSWER Interactions between filled and empty orbitals are stabilizing. This fundamental tenet of chemistry can be applied in this case. If the filled orbital of the carbon–hydrogen bond overlaps with the empty $2p$ orbital of the carbocation, the result is the stabilizing interaction called hyperconjugation.

(continued)

---

[4]If avoiding the question drives you crazy, and we think it should at least make you a little uncomfortable, be sure to do this problem—the answer falls right out. See also Chapter 9.

This interaction can occur only if the appropriate carbon–hydrogen bonds exist. The more alkyl groups, the more stabilization there is. For example, in the methyl cation no such stabilization is possible because there are no alkyl groups to supply the needed C—H bonds.

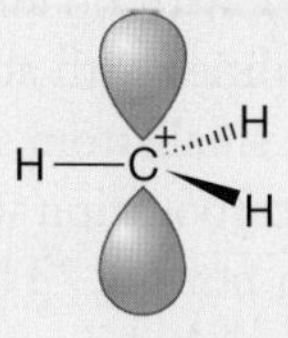

So, tertiary compounds, which form the most stable, *tertiary* carbocations, commonly react through the S_N1 mechanism, and methyl and primary compounds, which would have to form extremely unstable carbocations, do not. Naturally enough, secondary compounds react faster than primary ones in the S_N1 reaction but more slowly than tertiary compounds (Fig. 7.59).

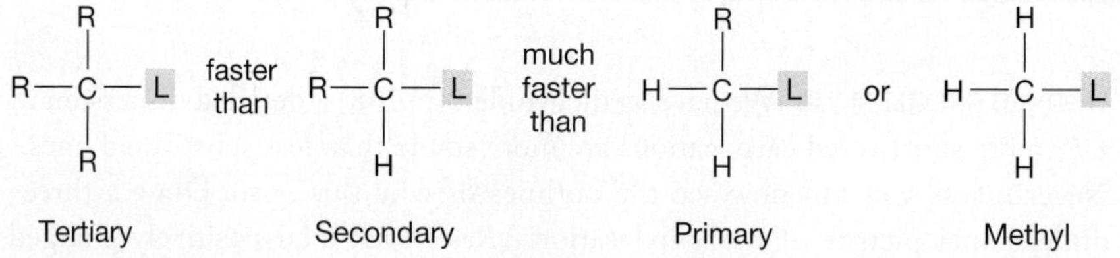

FIGURE 7.59 Relative reactivities in the S_N1 reaction.

PROBLEM 7.19 The heat of formation for cation **A** is 201 kcal/mol, and that of cation **B** is 218.6 kcal/mol. Draw the two cations carefully and explain why **A** is more stable than **B**.

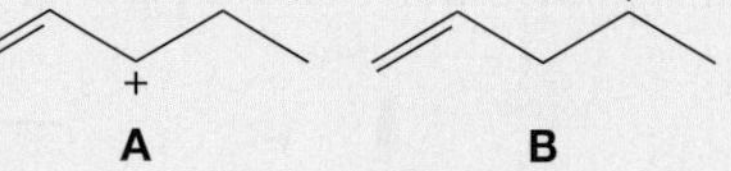

7.6d Effects of the Nucleophile   We have already seen that the nucleophile does not affect the rate of the S_N1 reaction—the reaction is first order in substrate. Yet, the nucleophile does determine the product structure. Note the distinction between the **rate-determining step** (or rate-limiting step) and the **product-determining step**. For example, reaction of *tert*-butyl bromide in water containing a small amount of an added nucleophilic ion, such as cyanide, leads to substantial amounts of *tert*-butyl cyanide. Even though water is present as solvent and thus has an immense advantage over cyanide ion because of concentration, the more powerfully nucleophilic cyanide still competes. Cyanide is many times more reactive toward the carbocation than the solvent water is (Table 7.3, p. 279).

In this reaction, the rate-determining step (the slow ionization) and the product-determining step (the faster capture of the intermediate ion by nucleophiles) are sharply separated (Fig. 7.60).

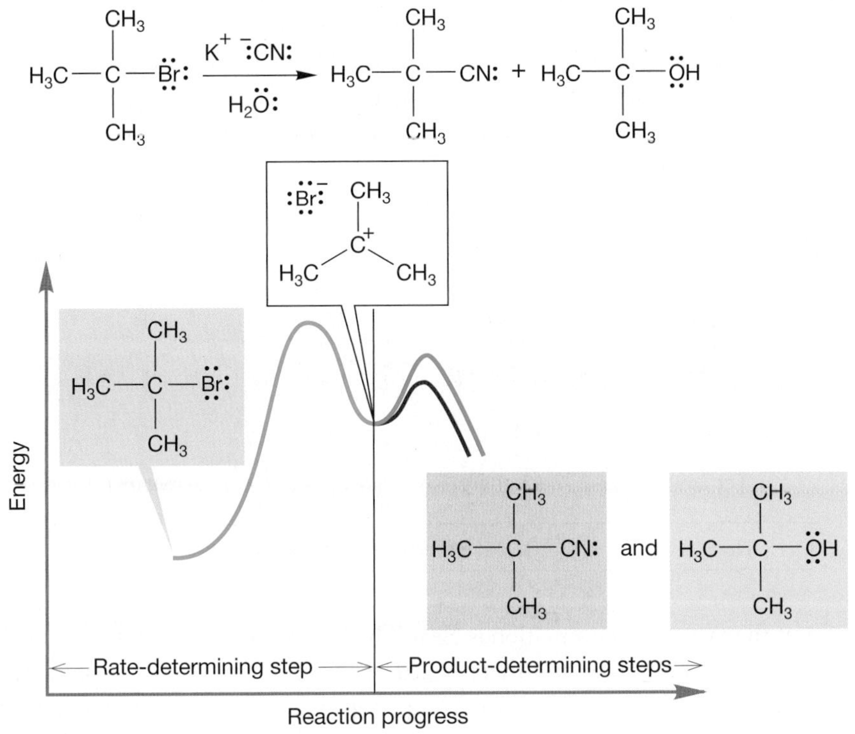

FIGURE 7.60 The rate-determining step and product-determining step are sharply distinguished in the S$_N$1 reaction.

7.6e Effect of the Leaving Group

The identity of the leaving group has a large influence on the rate of the S$_N$1 reaction because the structure of the departing group affects the ease of ionization. We have already discussed how the structure—and hence stability—of the carbocation affects the rate of ionization, but obviously the stability of the negatively charged counterion (L:$^-$) will be important as well. The more stable the anion, the more easily it will be lost (Fig. 7.61). Some good and poor leaving groups are shown in Table 7.4 (p. 281).

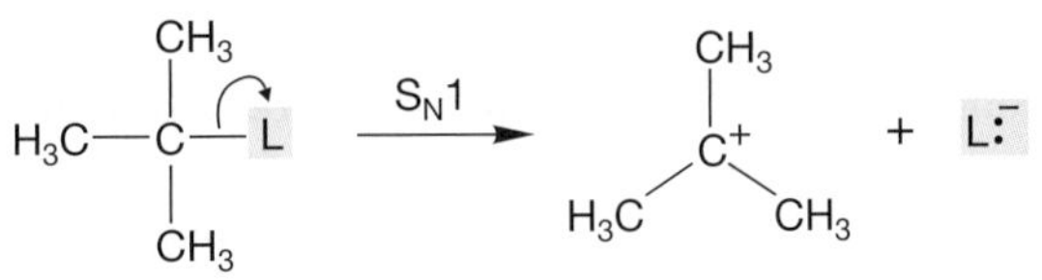

FIGURE 7.61 The rate of ionization is affected by the stability of the leaving group, L:$^-$. The more stable the leaving group, the faster is the ionization.

7.6f Effect of Solvent

For the S$_N$2 reaction, the effect of solvent is rather complicated and depends on the charge type of the reaction. Some S$_N$2 reactions are accelerated by a change to a more polar solvent; others are slowed down by such a change (p. 286). For most S$_N$1 reactions the situation is simpler. Ions are usually formed in the slow step of the reaction. In a polar solvent, represented by the dipole arrows in Figure 7.62, the charged species formed in the S$_N$1 reaction are stabilized. A nonpolar solvent cannot stabilize as well. One can expect that the S$_N$1 reaction will usually be favored by a polar solvent.

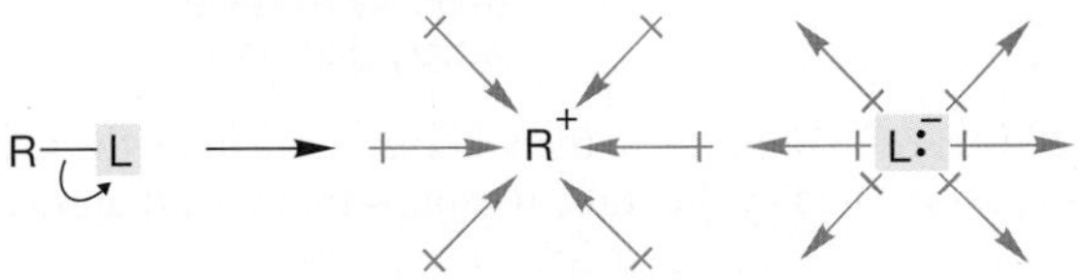

FIGURE 7.62 A polar solvent stabilizes the ions formed in an S$_N$1 reaction.

7.7 Summary and Overview of the S$_N$2 and S$_N$1 Reactions

These two substitution reactions are interconnected by more than the ultimate formation of a substitution product. They complement each other—when one is ineffective, the other is likely to work well. For example, the S$_N$2 reaction works best for methyl and primary substrates, which is exactly the point at which the S$_N$1 reaction is least effective. When the S$_N$2 reaction fails, as with tertiary substrates, the S$_N$1 reaction is most effective (Fig. 7.63).

FIGURE 7.63 The S$_N$2 and S$_N$1 reactions are complementary.

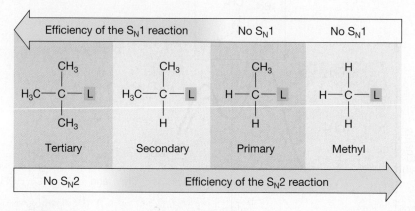

So, at the extremes the situation is clear. The S$_N$2 reaction dominates for methyl and primary substrates because there is little steric hindrance to the displacement of the leaving group and because the competing S$_N$1 reaction is disfavored by the necessity of forming the highly unstable methyl or primary carbocations (Fig. 7.64).

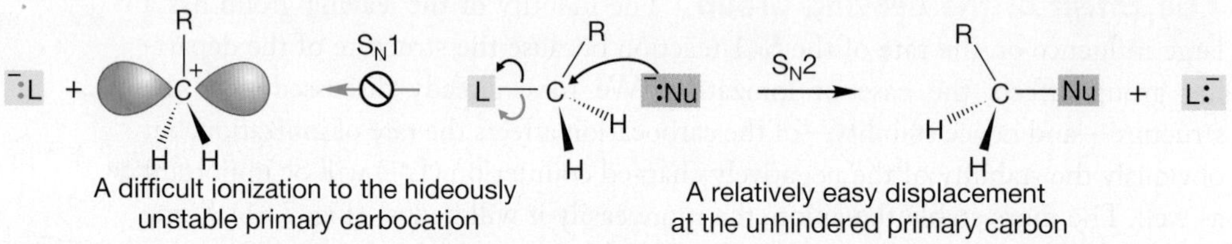

A difficult ionization to the hideously unstable primary carbocation

A relatively easy displacement at the unhindered primary carbon

FIGURE 7.64 A comparison of the S$_N$1 and S$_N$2 reactions for primary substrates.

The S$_N$1 reaction dominates for tertiary substrates because ionization can produce the relatively stable tertiary carbocation and because the competing S$_N$2 reaction is strongly disfavored by the steric hindrance to the nucleophile approaching from the rear (Fig. 7.65).

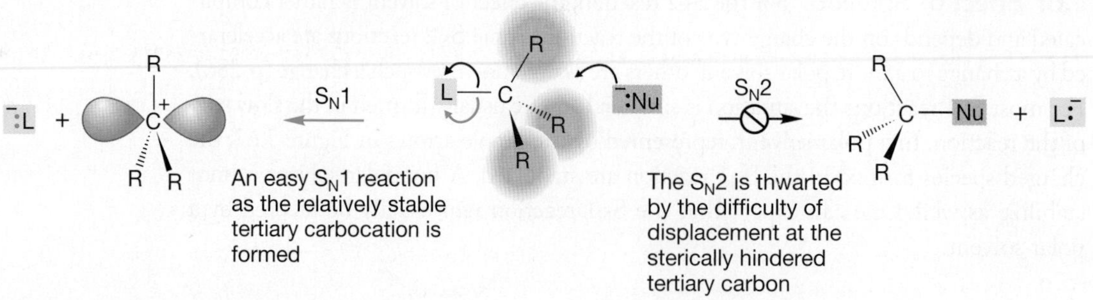

An easy S$_N$1 reaction as the relatively stable tertiary carbocation is formed

The S$_N$2 is thwarted by the difficulty of displacement at the sterically hindered tertiary carbon

FIGURE 7.65 For tertiary systems, the S$_N$1 reaction dominates; the ionization produces the relatively stable tertiary carbocation and the S$_N$2 reaction is hindered by the three R groups that thwart approach of the nucleophile from the rear.

In the middle ground—secondary substrates—we might expect to be able to find both reactions. Small R groups, use of a powerful nucleophile, and a solvent of proper polarity for the charge type of the reaction favor the S$_N$2 reaction. Large R groups, a weak nucleophile, and a polar solvent favor the S$_N$1 reaction (Fig. 7.66).

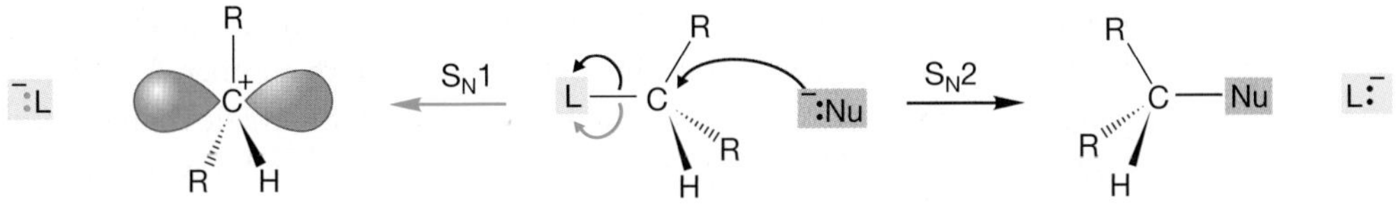

FIGURE 7.66   For secondary substrates, both S$_N$1 and S$_N$2 reactions may be possible.

PROBLEM 7.20   We saw at least one S$_N$1 reaction before, although it was not labeled as such because we didn't yet know the name. Where was it? If you get this one, you have a great eye and memory for detail. *Hint*: It is in this chapter.

We have given a picture of these reactions in which extremes are emphasized, in which an "either S$_N$2 or S$_N$1" view is developed. However, the situation is messier than this simple explanation would suggest, especially for secondary compounds. Imagine beginning to ionize a bond to a leaving group. As the bond stretches, positive charge begins to develop on carbon (δ^+) and rehybridization toward sp^2 begins as the groups attached to the carbon flatten out (Fig. 7.67).

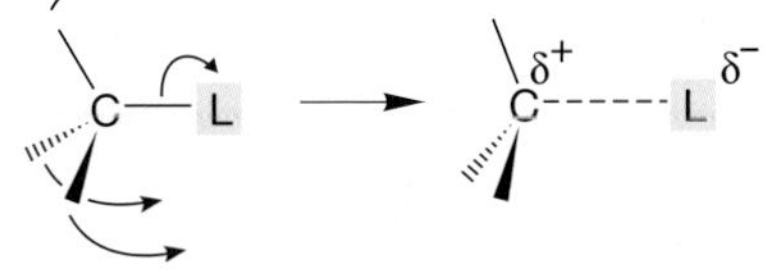

If we simply continue the process of ionization, we have the pure S$_N$1 reaction. If we let a nucleophilic solvent molecule interact with the rear of the departing bond and ultimately form a bond to carbon, we have the S$_N$2 reaction with inversion (Fig. 7.68).

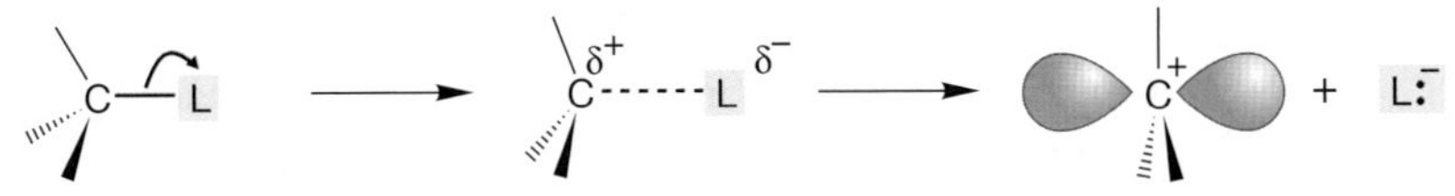

FIGURE 7.67   The beginning of an S$_N$1 reaction. The bond to the leaving group :L$^-$ has lengthened, and the tetrahedron has begun to flatten out. The hybridization of the central carbon is changing from nearly sp^3 to the sp^2 of the planar carbocation.

In the classic S$_N$1 reaction, this ionization continues to produce the planar carbocation. Notice, however, that the carbocation when first formed is not symmetrical. The leaving group lurks on one side, partially shielding the cation from solvent and other nucleophiles

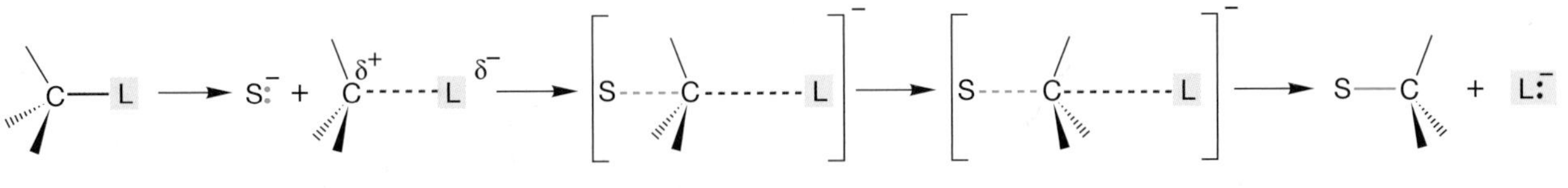

However, solvation at the side away from the old leaving group can be converted into nucleophilic attack (S$_N$2) to give inverted product

FIGURE 7.68   Especially for secondary substrates, the difference between S$_N$2, displacement by solvent, and preferential solvation at the rear of the departing bond is a subtle one.

How strong an interaction with the rear lobe of the orbital is needed before we call it bonding? In polar solvents, ions are highly solvated, and in an S$_N$1 reaction solvation will begin to be important as soon as a δ^+ charge begins to build up on carbon. Solvation will be especially easy at the rear of the departing group as the leaving group shields the frontside. We have already seen that the leaving group can hinder reaction at one side of the developing cation and produce some net inversion in the reaction (Fig. 7.58).

Now we begin to see the difficulty of telling the two mechanisms apart in some situations. When does "preferential solvation from the rear" become "displacement from the rear"? It becomes painfully hard to sort out in some cases. Organic chemistry is frustratingly, if delightfully, messy.

This messiness raises its head in other ways. Almost all substitution reactions are accompanied by some formation of alkene. In organic chemistry few reactions proceed in 100% yield, quantitatively producing one molecule from another. A large part of the business of organic synthesis is the search for specificity. Synthetic chemists seek reagents and reaction conditions that greatly favor one reaction pathway over others. In particular, attempts to synthesize molecules through substitution reactions must take account of competitive reactions called eliminations. A hydrogen atom and a leaving group (L) are lost (eliminated) from the substrate to give an alkene (Fig. 7.69). Sections 7.8 and 7.9 examine two general mechanisms for these elimination reactions.

FIGURE 7.69 In almost all S_N1 and S_N2 reactions, substitution is accompanied by some elimination of H—L to give an alkene.

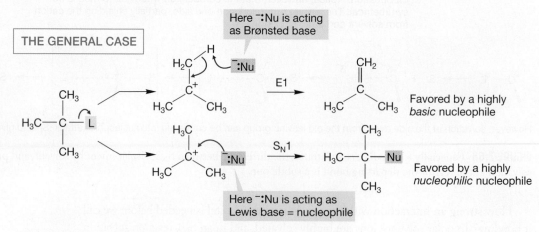

Substitution reaction   Elimination reaction

7.8 The Unimolecular Elimination Reaction: E1

7.8a Rate Law If we determine the rate law for the formation of the alkene accompanying an S_N1 reaction, we find that this **elimination reaction** is also a first-order process, $v = k[\text{R—L}]$. As in the S_N1 reaction, the slow step in this reaction is the ionization of the substrate to give a carbocation, which can do two things: It can continue the S_N1 reaction by adding a nucleophile (Nu:^-) or, if a proton is available, an alkene can be formed in what is called the **E1 reaction**. Note that in the elimination phase of the reaction, the loss of hydrogen, the nucleophile is acting as a Brønsted base. The carbocation is an intermediate in *both* processes (Fig. 7.70).

Unimolecular elimination: E1

THE GENERAL CASE

Here ⁻:Nu is acting as Brønsted base

E1 — Favored by a highly *basic* nucleophile

S_N1 — Favored by a highly *nucleophilic* nucleophile

Here ⁻:Nu is acting as Lewis base = nucleophile

FIGURE 7.70 Competing E1 and S_N1 reactions.

The ratio of E1 and S_N1 products depends primarily on the nucleophile's basicity. Strong Brønsted bases are more effective at removing the proton than weak Brønsted bases. That's exactly what we mean by a strong Brønsted base—something that has a high affinity for a proton. Good nucleophiles will be less effective in the elimination reaction, which requires removal of a proton, but more effective in the S_N1 reaction (Fig. 7.71).

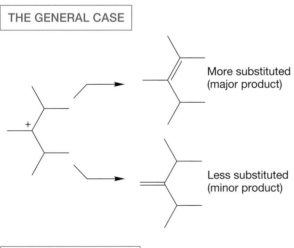

SOME SPECIFIC EXAMPLES

FIGURE 7.71 The elimination and S_N1 reactions compete. In pure ethyl alcohol, the ratio of alkene (2-methylpropene) to S_N1 product (*tert*-butyl ethyl ether) is 0.25. In the presence of the much stronger base sodium ethoxide, this ratio increases to about 13. A strong base (ethoxide) favors the elimination reaction.

7.8b Selectivity in the E1 Reaction

In many E1 reactions, more than one hydrogen can be removed from the carbocationic intermediate to give an alkene. When there is a choice, the major product is the more substituted, more stable alkene (Chapter 3, p. 115). Of course, statistical effects will also influence the product mix, but the overwhelming factor is alkene stability. The more substituted alkene is favored strongly in the E1 reaction. For example, 2-bromo-2-methylbutane reacts in a solution of ethyl alcohol and water to give a mixture of 2-methyl-2-butene and 2-methyl-1-butene in which the more substitued alkene predominates by a factor of 4 (Fig. 7.72).

THE GENERAL CASE

FIGURE 7.72 The more substituted (more stable) alkene is favored in the E1 reaction, even when statistical factors favor the less substituted isomer. This reaction is known as Saytzeff elimination.

A SPECIFIC EXAMPLE

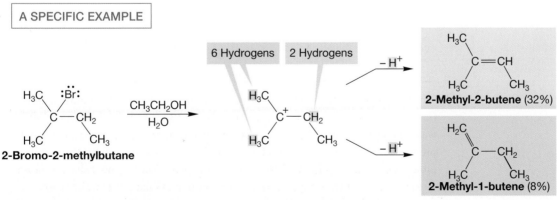

PROBLEM 7.21 Draw arrow formalisms for the reactions of Figure 7.72.

PROBLEM 7.22 The specific example in Figure 7.72 also produces two products of the S_N1 reaction: one $C_5H_{12}O$ and the other $C_7H_{16}O$. What are these products and how are they formed?

To see why the more substituted alkene is favored, look at the transition state for proton removal. In the transition state the carbon–hydrogen bond is lengthened and the double bond of the alkene is partially formed (Fig. 7.73).

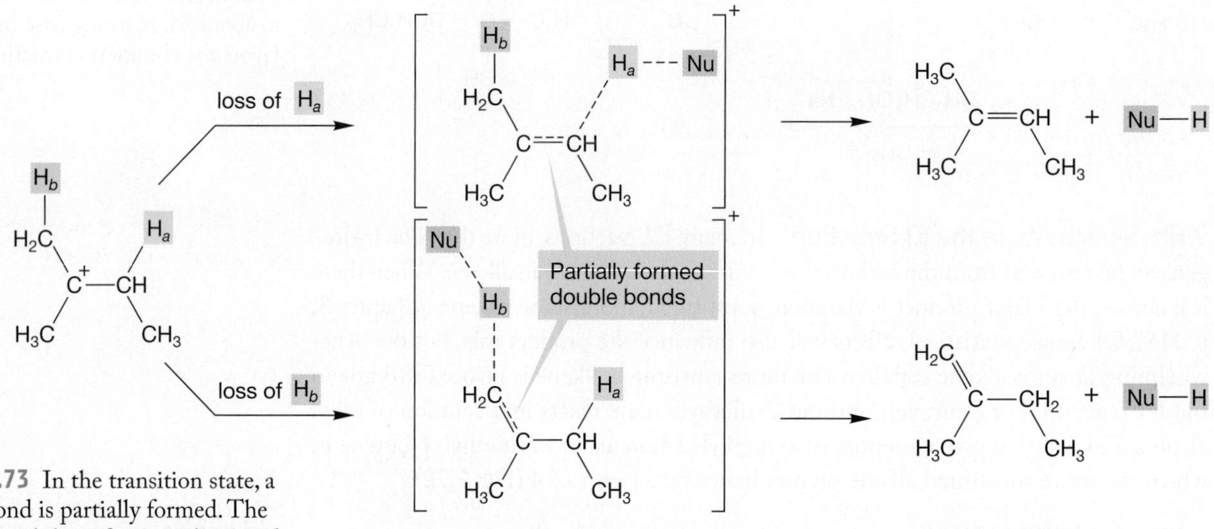

FIGURE 7.73 In the transition state, a double bond is partially formed. The increased stability of more substituted double bonds will be felt, and the transition state for formation of the more substituted alkene (loss of H_a) will be lower in energy than the transition state for formation of the less substituted alkene (loss of H_b).

Transition states for proton removal by a base :Nu to give the alkene and NuH

The factors that operate to stabilize alkenes with fully formed double bonds are also operative in the transition states in which there are partially formed double bonds. So, the more alkyl groups, the better, and the E1 reaction preferentially produces the most substituted alkene possible. This phenomenon is called **Saytzeff elimination**, after Alexander M. Saytzeff (1841–1910), the Russian chemist who discovered it.[5]

PROBLEM 7.23 Predict the products of the E1/S_N1 reactions of the following molecules in water:

(a)

$(CH_3CH_2)_3C$—Br

(b)

(c)

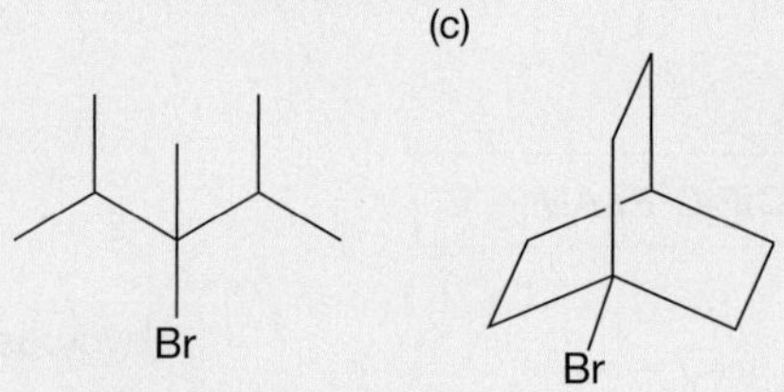

PROBLEM 7.24 It was noticed as early as 1862 that treatment of isobutyl halides, $(CH_3)_2CHCH_2$—X, with silver acetate, Ag^+ ^-OAc (^-OAc is the acetate ion, a weak nucleophile, and Ag^+ is a strong Lewis acid, especially toward chloride), produced *tert*-butyl acetate, $(CH_3)_3C$—OAc, and isobutene. Use the stability order of carbocations (p. 292) to provide an explanation for this behavior. *Hint*: Reason backward from the structure of the product. What ion must be present to form this product?

---

[5] Saytzeff can be transliterated from the Cyrillic as Zaitzev, Saytzev, or Saytseff.

7.9 The Bimolecular Elimination Reaction: E2

As with the substitution reaction, there are two common mechanisms for the elimination reaction. We have just explored the first-order process, the E1 reaction. Now let's look at the bimolecular reaction, the **E2 reaction**.

7.9a Rate Law   The S_N2 reaction is also complicated by alkene formation. Just as we saw for the S_N2 reaction, the typical rate law for the bimolecular E2 reaction shows that the alkene is formed in a second-order process that is first order in both substrate and nucleophile:

$$v = k[\text{substrate}]\,[\text{nucleophile}]$$

Bimolecular elimination: E2

The mechanism involves attack by the base at a hydrogen attached to a carbon adjacent to the carbon bearing the leaving group. Figure 7.74 shows the arrow formalisms for the competing E2 and S_N2 reactions.

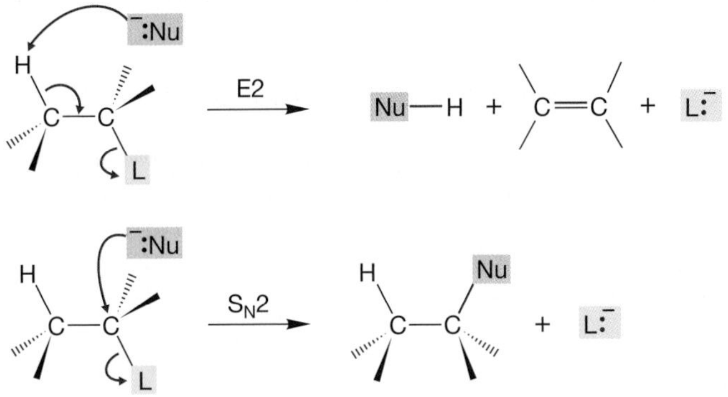

FIGURE 7.74 The competing E2 and S_N2 reactions.

This picture gives us an immediate clue as to how to influence the product mixture from competing E2 and S_N2 reactions. As in the E1–S_N1 competition (p. 298), strong Brønsted bases (high affinity for hydrogen $1s$ orbital) should favor the E2 reaction and strong nucleophiles (high affinity for carbon $2p$ orbital) should favor the S_N2 reaction (Fig. 7.75).

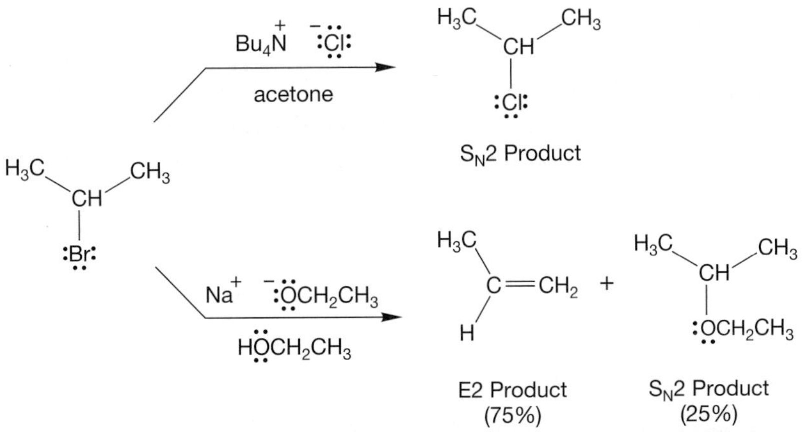

FIGURE 7.75 The E2 and S_N2 reactions compete. In the examples shown, the relatively good nucleophile chloride gives exclusively the S_N2 reaction, whereas the much more basic ethoxide ion gives mostly the E2 reaction. Bu means butyl.

7.9b Effect of Substrate Branching on the E2–S_N2 Mix   The rate of the S_N2 reaction depends strongly on the structure of the substrate. The more highly branched the substrate, the more hindered is the obligatory approach from the rear of the leaving group and the slower is the S_N2 reaction. This effect reaches a limit with tertiary substrates, in which the S_N2 rate is effectively zero. The slower the S_N2 reaction, the more important is the competing E2 reaction. With a typical tertiary substrate such as *tert*-butyl bromide, the E2 reaction is the only effective reaction under

E2–S_N2 conditions. The steric requirements for deprotonation are far less severe than for attack at the rear of the carbon–bromine bond of *tert*-butyl bromide (Fig. 7.76).

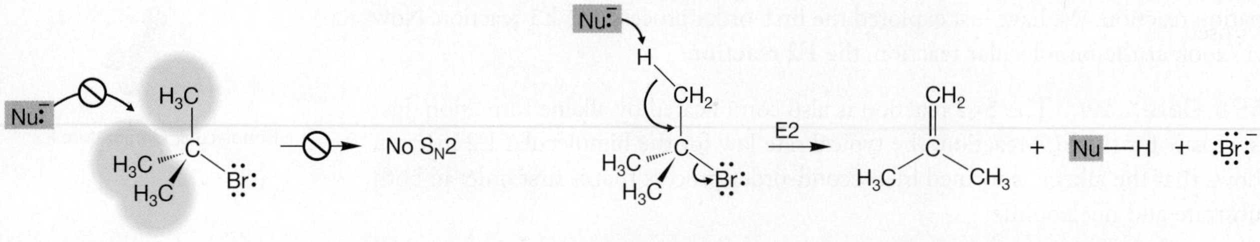

FIGURE 7.76 For tertiary substrates, the S_N2 reaction does not proceed, but the E2 reaction is relatively unhindered. Attack from the rear in S_N2 fashion is blocked by the three methyl groups. The E2 reaction is relatively easy.

With less hindered substrates, the S_N2 reaction can compete much more effectively with the E2 reaction, and with primary substrates the S_N2 reaction is dominant (Fig. 7.77).

FIGURE 7.77 For secondary substrates, the E2 and S_N2 reactions are competitive. For primary substrates, the S_N2 reaction dominates.

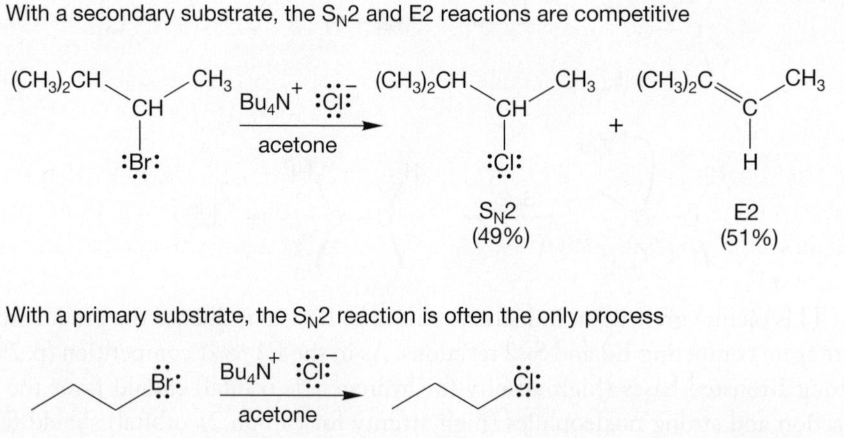

With a secondary substrate, the S_N2 and E2 reactions are competitive

S_N2 (49%)   E2 (51%)

With a primary substrate, the S_N2 reaction is often the only process

7.9c Stereochemistry of the E2 Reaction When we explored the mechanism of the S_N1 and S_N2 reactions, an examination of the stereochemical outcome of the reactions was crucial to our development of their mechanisms. Stereochemical experiments are common in organic chemistry and are one of the reasons we spent so much time and effort earlier on stereochemical analysis. In the E2 reaction, there are four possible extreme orientations. They can be classified by noting the dihedral angle between the two groups that come off, the hydrogen and the leaving group L (Fig. 7.78).

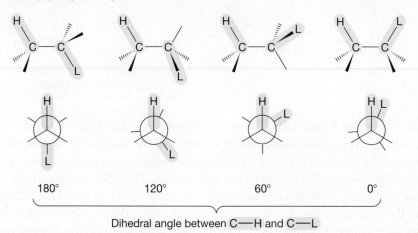

FIGURE 7.78 Four possible arrangements for the E2 reaction. The Newman projections emphasize the dihedral angle between the C—H and C—L bonds involved in the reaction.

180°   120°   60°   0°

Dihedral angle between C—H and C—L

As the C—H and C—L bonds lengthen in this reaction, the molecule flattens out as the $2p$ orbitals that will constitute the π system of the alkene are formed. *Remember* (p. 105): In the product alkene the $2p$ orbitals are perfectly lined up, eclipsed, with a 0° dihedral angle (Fig. 7.79).

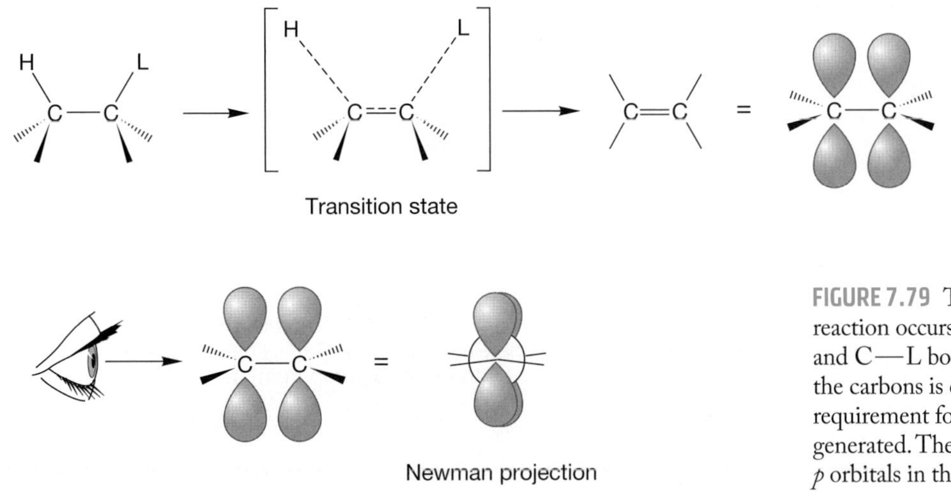

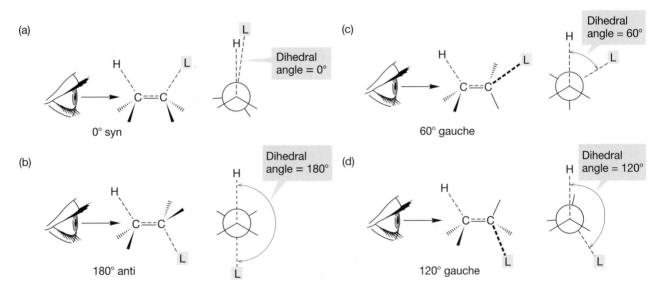

Transition state

Newman projection

FIGURE 7.79 The generation of a π bond as the E2 reaction occurs. In the E2 reaction, the C—H and C—L bonds lengthen as the π bond between the carbons is created. Note especially the requirement for orbital lineup as the π bond is generated. The dihedral angle between the two p orbitals in the product alkene is 0°.

Note that only the 0° (syn) and 180° (anti) arrangements shown in Figure 7.78 can lead smoothly to an alkene with perfectly overlapping $2p$ orbitals. The 120° and 60° forms cannot produce alkene without an additional rotation to line up the orbitals (Fig. 7.80).

(a)

Dihedral angle = 0°

0° syn

(c)

Dihedral angle = 60°

60° gauche

(b)

Dihedral angle = 180°

180° anti

(d)

Dihedral angle = 120°

120° gauche

FIGURE 7.80 Four views of the transition state for the E2 reaction. In examples (a) and (b), where the dihedral angle is 0° or 180°, the π bond can be generated without an extra rotation. In examples (c) and (d), where the dihedral angle is 60° or 120°, an extra rotation must take place in order for the π bond to be made.

Thus, we can anticipate that the 60° and 120° gauche arrangements for elimination should be disfavored. It is much more difficult to decide which of the remaining two possibilities, 0° (syn) or 180° (anti) will be preferred in the E2 reaction.

The E2 is a kind of extended S_N2 reaction. The Lewis base (Nu:$^-$) attacks the hydrogen, and the pair of electrons in the carbon–hydrogen bond does the displacing of the leaving group. If we keep this analogy in mind we can see that the 180° elimination involves backside displacement, as does the S_N2 reaction, but the 0° elimination requires frontside displacement. We might therefore expect the 180° E2 to be favored (Fig. 7.81).

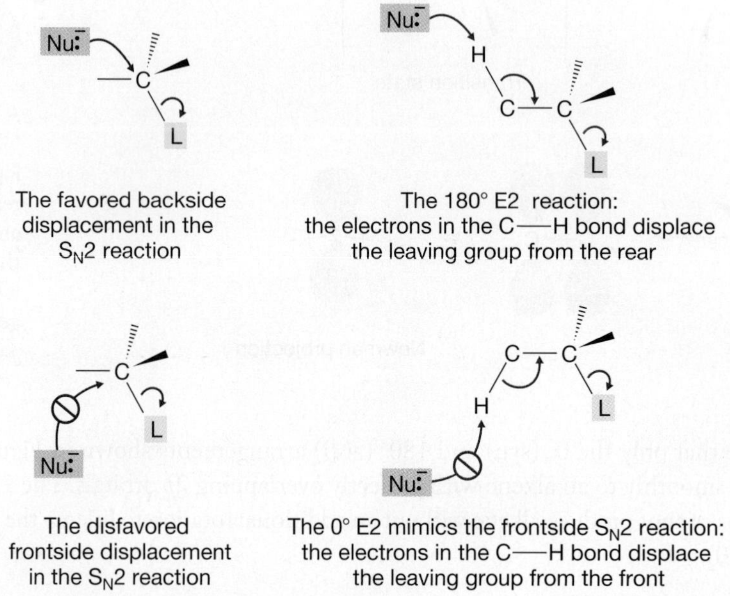

FIGURE 7.81 The 0° E2 reaction resembles a frontside S_N2 reaction, whereas the 180° E2 reaction looks like the favored, backside S_N2 displacement reaction.

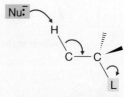

The favored backside
displacement in the
S_N2 reaction

The 180° E2 reaction:
the electrons in the C—H bond displace
the leaving group from the rear

The disfavored
frontside displacement
in the S_N2 reaction

The 0° E2 mimics the frontside S_N2 reaction:
the electrons in the C—H bond displace
the leaving group from the front

Now the problem is to find an experiment that will let us see this level of detail. As usual, we will use a stereochemical analysis, and this recquires a certain complexity of substitution in the substrate molecule. For example, there is no use in examining the stereochemistry of the E2 reaction of a molecule as simple as 2-chloro-2,3-dimethylbutane because there is only one tetrasubstituted alkene possible (Fig. 7.82).

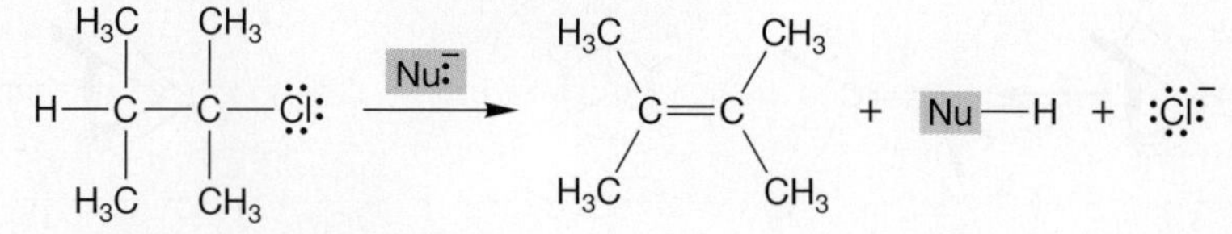

FIGURE 7.82 In this reaction there is only one possible tetrasubstituted alkene product.

2-Chloro-2,3-dimethylbutane

PROBLEM 7.25 What other alkene could be formed in the reaction of Figure 7.82?

Molecules more highly substituted than 2-chloro-2,3-dimethylbutane can reveal the stereochemical preference, however. For example, consider the molecules of

Figure 7.83. In diastereomer **A**, **syn** (0°) **elimination** would lead to a product in which the methyl groups are cis [this is, however, the (*E*) alkene], whereas **anti** (180°) **elimination** would give the compound with trans methyls [the (*Z*) alkene]. In **B**, these preferences are reversed: syn elimination would produce the (*Z*) alkene, and anti elimination would give the (*E*) alkene. *Remember*: The syn form is an eclipsed, energy-maximum structure and the anti form is a staggered, energy-minimum structure. Many tests of this kind have been carried out, and it is certain that in acyclic molecules anti (180°) elimination is favored over the syn process (0°) by *very* large amounts, typically greater than 10^4 (Fig. 7.83).

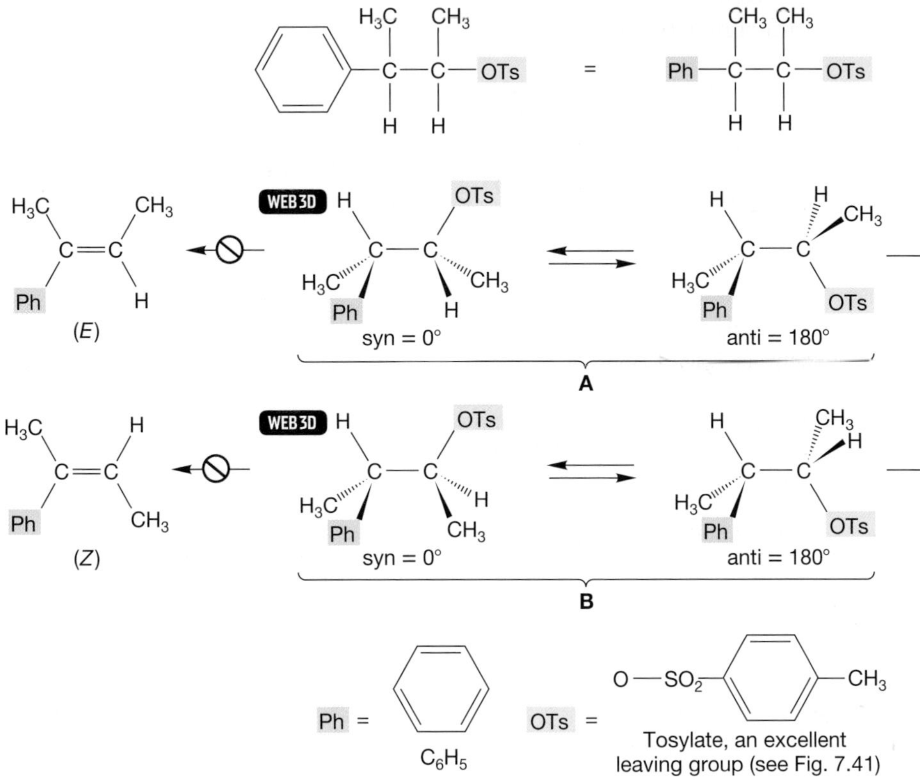

FIGURE 7.83 The possibilities for 0° and 180° E2 reactions for the two diastereomers **A** and **B** shown.

WORKED PROBLEM 7.26 In chemistry, it is customary to report one's research results by submitting papers to the chemical literature. As a kind of quality control, these papers are sent to experts in the field for comment and review before publication. Imagine that you are such a referee looking over the experiment described in Figure 7.83. Do the data absolutely require the conclusion that anti (180°) eliminations are preferred to syn (0°)? Obviously, we think not. What do you think?

ANSWER There is a problem. The analysis of Figure 7.83 shows a 180° elimination proceeding from a staggered, energy-minimum form. By contrast, the 0°, syn elimination of Figure 7.83 proceeds from a fully eclipsed arrangement, which is a transition state, not an energy minimum (p. 67; Fig. 7.22, p. 273). This comparison is scarcely fair! In order to make a real comparison of the syn and anti E2 processes, we will have to be more clever.

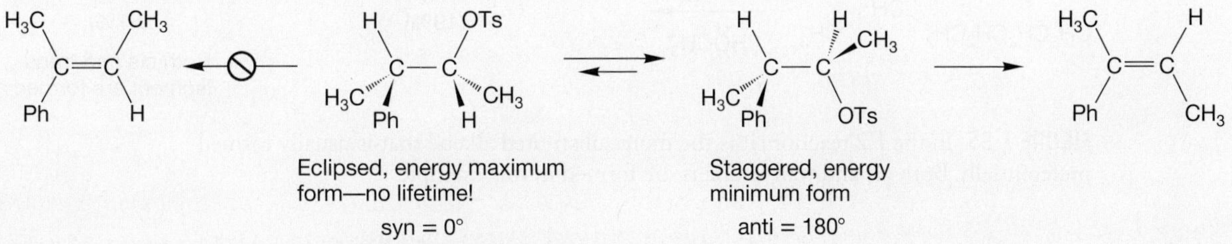

It is important to remember that orbital overlap is vital in the E2 elimination reaction, and a syn elimination in which the orbitals overlap well may compete effectively with an anti elimination in which the perfect 180° dihedral angle cannot be attained. For example, some polycyclic molecules give syn elimination. In the example shown in Figure 7.84, the 2-bicyclo[2.2.1]heptyl, or 2-norbornyl, system, the *ideal* anti (180°) elimination (loss of HOTs) is impossible, because only a 120° arrangement of the departing groups can be achieved. By contrast, the syn elimination (loss of DOTs) has a perfect 0° dihedral angle between the two departing groups. The result is that elimination is very slow and the perfect syn elimination is preferred to the 120° anti elimination (Fig. 7.84).

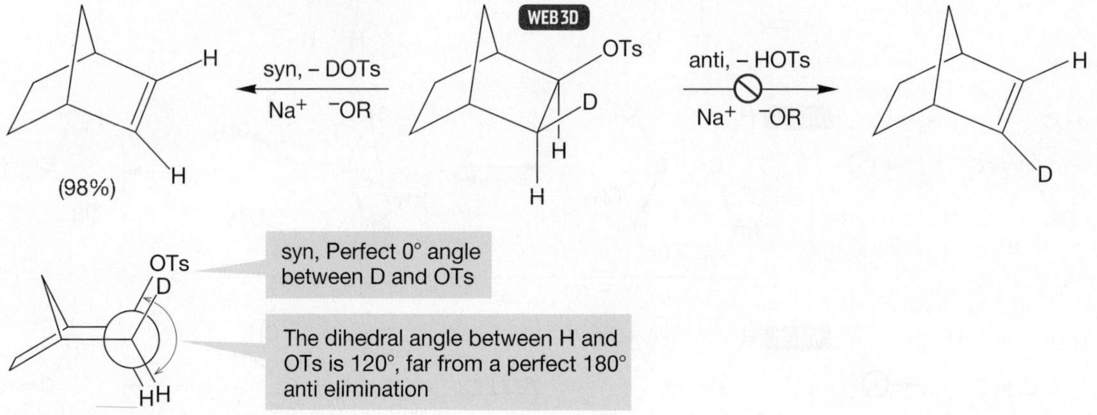

FIGURE 7.84 Sometimes syn elimination can be preferred.

PROBLEM 7.27 Draw all nine possible stereoisomers of 1,2,3,4,5,6-hexachlorocyclohexane. First, draw the molecules flat using some schematic device to indicate the stereochemistry of the chlorine atoms. Next, make three-dimensional drawings of these molecules.

PROBLEM 7.28 One of the isomers you drew in Problem 7.27 reacts much more slowly in the E2 reaction than all the others. Which one is it, and why is it so slow to undergo an E2 reaction? *Hint*: Remember the steric requirements for the E2 reaction.

7.9d Selectivity in the E2 Reaction   Generally, Saytzeff elimination is the rule for the E2 elimination. The most substituted, and therefore most stable, alkene is the major product in this regioselective reaction (Fig. 7.85). Recall that

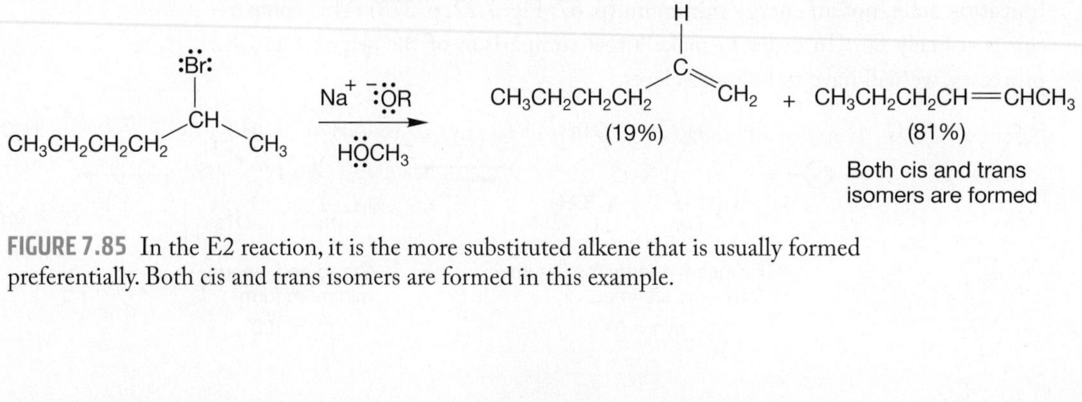

FIGURE 7.85 In the E2 reaction, it is the more substituted alkene that is usually formed preferentially. Both cis and trans isomers are formed in this example.

regiochemistry simply refers to the outcome of a reaction in which more than one orientation of substituents is possible in the product, and in a **regioselective reaction**, one product predominates.

In the transition state for alkene formation, the π bond is partially developed. The factors that make more substituted alkenes more stable than less substituted alkenes will operate on the partially developed alkenes in the transition states as well. The more substituted the partially developed alkene, the more stable it is, and this effect inevitably favors the more substituted alkene as the major product (Fig. 7.86).

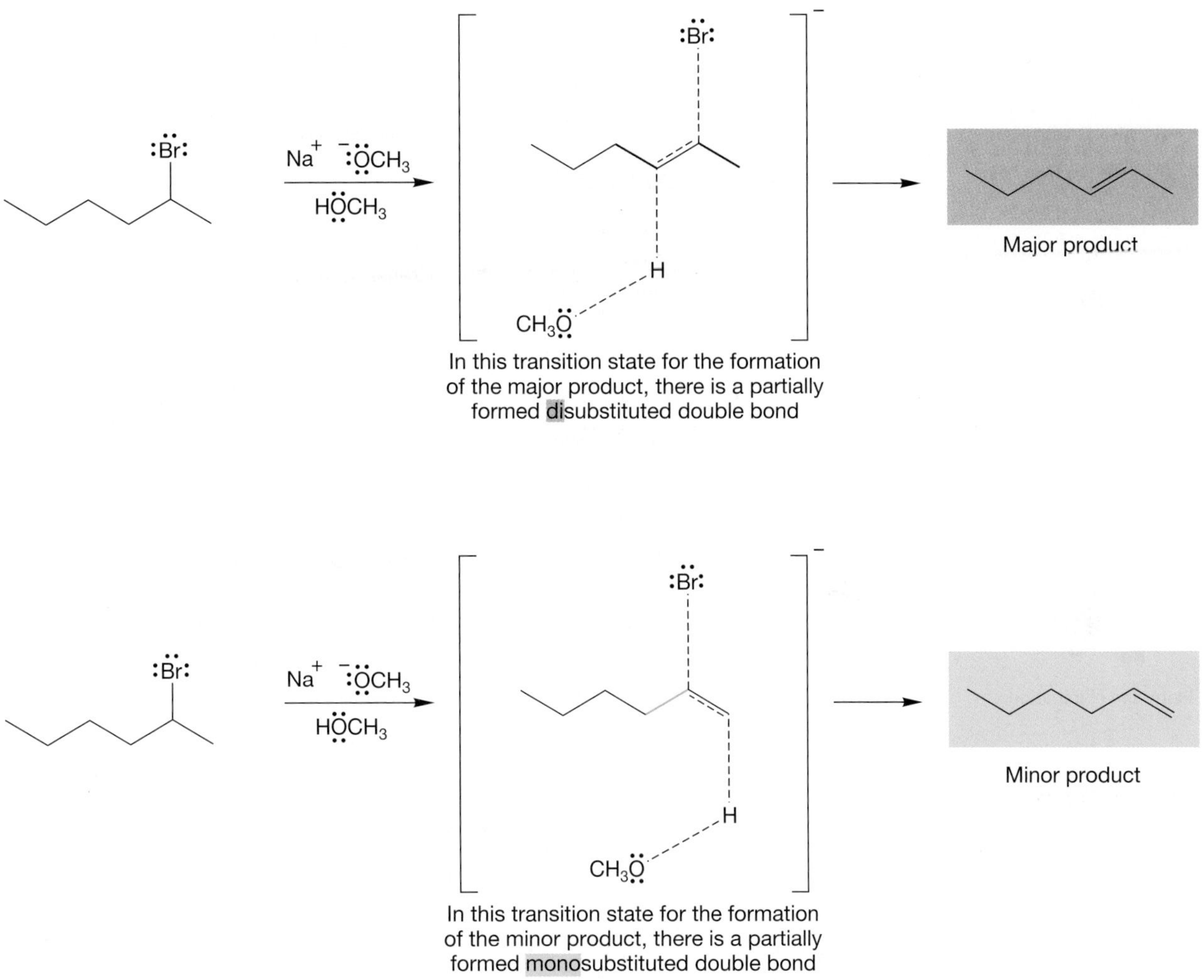

In this transition state for the formation
of the major product, there is a partially
formed disubstituted double bond

Major product

In this transition state for the formation
of the minor product, there is a partially
formed monosubstituted double bond

Minor product

FIGURE 7.86 In the E2 reaction, the more stable transition state leads to the major product, the more substituted alkene.

Many factors influence the ratio of alkene products, however, and some are quite surprising at first. The size of the base makes a small difference, with larger bases favoring the less stable alkene. Although there is only a very small change in the product distribution when the base is changed from methoxide to *tert*-butoxide in the reaction of 2-iodobutane with alkoxide (Fig. 7.87), when the base is made truly enormous these effects can be magnified. In order to form the more substituted alkene, the base must penetrate farther into the middle of the substrate. Formation of the less stable alkene requires attack by base farther toward the periphery of the molecule and thus has lower steric requirements.

FIGURE 7.87 The effect of increasing base size is to increase the amount of the less stable "Hofmann" product somewhat.

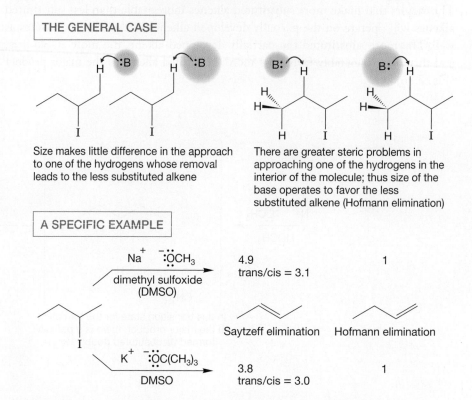

THE GENERAL CASE

Size makes little difference in the approach to one of the hydrogens whose removal leads to the less substituted alkene

There are greater steric problems in approaching one of the hydrogens in the interior of the molecule; thus size of the base operates to favor the less substituted alkene (Hofmann elimination)

A SPECIFIC EXAMPLE

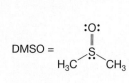

Na^+ $\ddot{\text{:}}\ddot{\text{O}}CH_3$
dimethyl sulfoxide (DMSO)

4.9
trans/cis = 3.1

1

Saytzeff elimination   Hofmann elimination

K^+ $\ddot{\text{:}}\ddot{\text{O}}C(CH_3)_3$
DMSO

3.8
trans/cis = 3.0

1

Hofmann elimination

Formation of the less stable alkene is called **Hofmann elimination** after the man who discovered this preference, August W. von Hofmann (1818–1892). Hofmann elimination contrasts with the more usual Saytzeff process, in which the more stable alkene is the major product. Hofmann elimination is favored by certain leaving groups, however. Typical examples of Hofmann leaving groups are fluoride and the 'onium ions, such as ammonium and sulfonium (Fig. 7.88).

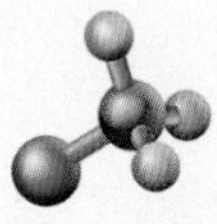

L	2-Alkene (%)	1-Alkene (%)
I	81	19 (Saytzeff preferred)
Br	72	28 (Saytzeff preferred)
Cl	67	33 (Saytzeff preferred)
F	30	70 (Hofmann preferred)

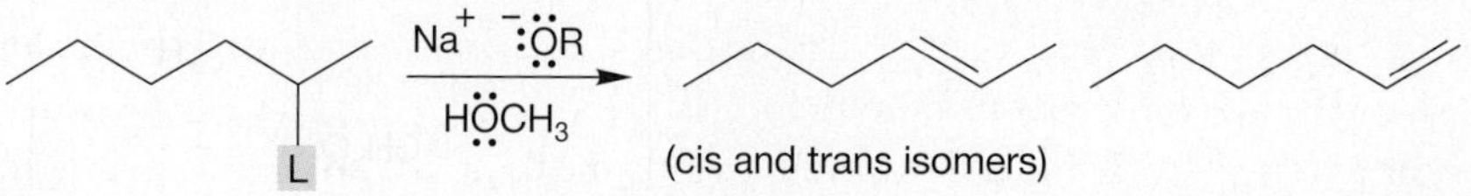

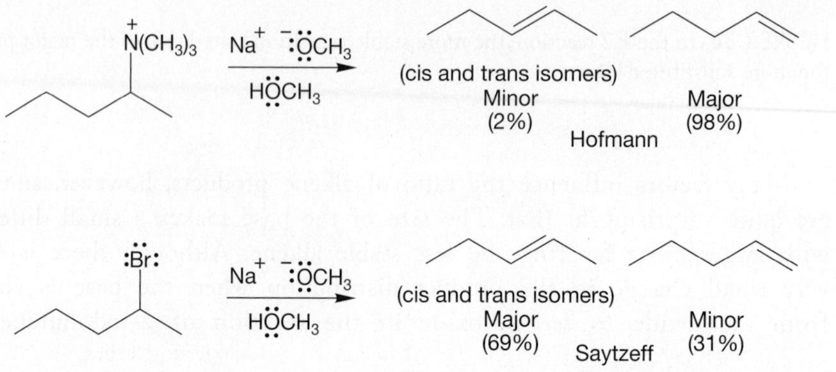

Na^+ $\ddot{\text{:}}\ddot{\text{O}}R$
$H\ddot{\text{O}}CH_3$

(cis and trans isomers)

$\overset{+}{N}(CH_3)_3$ Na^+ $\ddot{\text{:}}\ddot{\text{O}}CH_3$
$H\ddot{\text{O}}CH_3$

(cis and trans isomers)
Minor (2%)   Major (98%)
Hofmann

FIGURE 7.88 The regioselectivity of the E2 reaction depends on the identity of the leaving group. Leaving groups such as fluoride, ammonium (R_3N^+), and sulfonium (R_2S^+) lead to predominant Hofmann elimination, in which the less stable isomer is the major product.

$\ddot{\text{:}}\ddot{Br}\text{:}$ Na^+ $\ddot{\text{:}}\ddot{\text{O}}CH_3$
$H\ddot{\text{O}}CH_3$

(cis and trans isomers)
Major (69%)   Minor (31%)
Saytzeff

Why is there a preference for the less-substituted alkene in a Hofmann elimination? And why does the leaving group make a difference? To answer these questions, we first have to mention another kind of loss of H—L, the **E1cB reaction** (elimination, unimolecular, conjugate base). In this reaction, in essence the opposite of the E1 elimination, the hydrogen is lost first to generate an anion. Subsequent ejection of the leaving group as an anion generates the alkene. In the E1 reaction, the leaving group is lost first to give the carbocation, and a proton lost in the second step to give the alkene (Fig. 7.89).

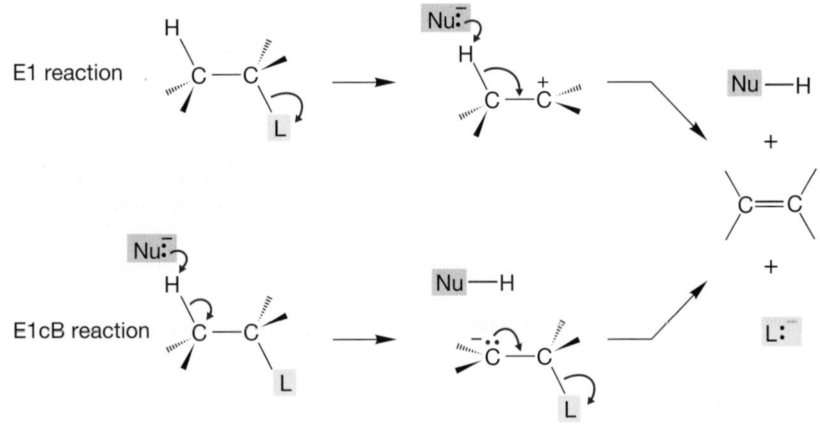

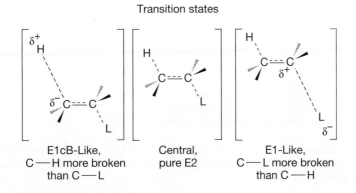

FIGURE 7.89 The E1 and E1cB reactions contrasted.

The E1cB reaction is relatively uncommon, but some examples are known. The requirements for the E1cB reaction are a poor leaving group and a rather easily lost proton.

PROBLEM 7.29 Design a good E1cB reaction. *Hint*: What kind of leaving group do you need? What kind of carbon–hydrogen bond?

The E2 reaction of alkyl fluorides, ammonium ions, and sulfonium ions does not go by the E1cB mechanism but the transition state *is* E1cB-like. In the transition state for any E2 reaction, the bonds to hydrogen and the leaving group are both breaking. In the E2 reaction, the transition state may or may not have the bonds to H and L breaking roughly simultaneously. If breaking of the C—H bond is further advanced in the transition state than is breaking of the C—L bond, the transition state will resemble the E1cB reaction. If the C—L bond is more broken in the transition state than is C—H bond, the transition state is E1-like (Fig. 7.90).

Transition states

$$\left[\begin{array}{c} \delta^+ \\ H \end{array} \right] \quad \left[\begin{array}{c} H \end{array} \right] \quad \left[\begin{array}{c} H \end{array} \right]$$

E1cB-Like,	Central,	E1-Like,
C—H more broken	pure E2	C—L more broken
than C—L		than C—H

FIGURE 7.90 Three transition states for an E2 elimination.

In an E1cB-like transition state, the carbon adjacent to the carbon bearing the leaving group becomes partially negatively charged (Fig. 7.90). When the leaving group is highly electronegative, as with fluoride, sulfonium, and ammonium, there will be a partial positive charge on the carbon to which the leaving group is attached, adjacent to that partial negative charge. For an E1cB-like transition state, this situation is favorable (Fig. 7.91).

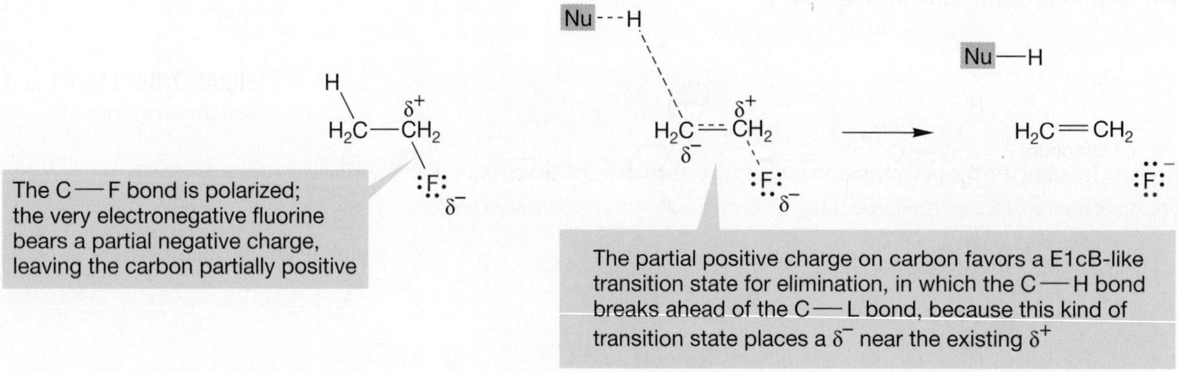

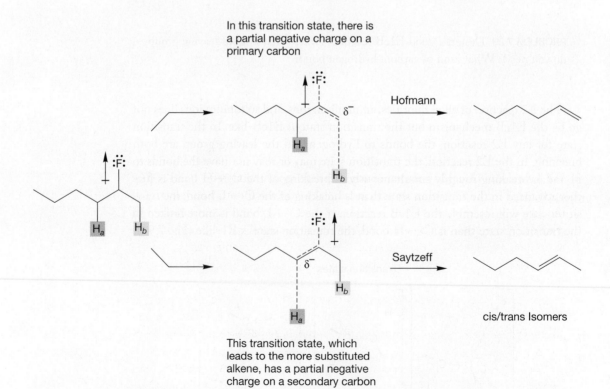

The C—F bond is polarized; the very electronegative fluorine bears a partial negative charge, leaving the carbon partially positive

The partial positive charge on carbon favors a E1cB-like transition state for elimination, in which the C—H bond breaks ahead of the C—L bond, because this kind of transition state places a δ^- near the existing δ^+

FIGURE 7.91 An E1cB-like transition state is favored by strongly electron-withdrawing groups, such as fluorine and 'onium ions.

In the example shown in Figure 7.92, the transition state for Hofmann elimination has a partial negative charge on a primary carbon, whereas the transition state for Saytzeff elimination has a partial negative charge on a secondary carbon (Fig. 7.92).

In this transition state, there is a partial negative charge on a primary carbon

This transition state, which leads to the more substituted alkene, has a partial negative charge on a secondary carbon

cis/trans Isomers

FIGURE 7.92 In the E1cB-like transition state for Hofmann elimination, a partial primary carbanion (less substituted, more stable) is formed. In the E1cB-like transition state for Saytzeff elimination, it is a partial secondary carbanion (more substituted, less stable).

In contrast to carbocations and radicals, less substituted carbanions are more stable than more substituted carbanions. Methyl and primary carbanions are more stable than secondary carbanions, and secondary are more stable than tertiary (Fig. 7.93).

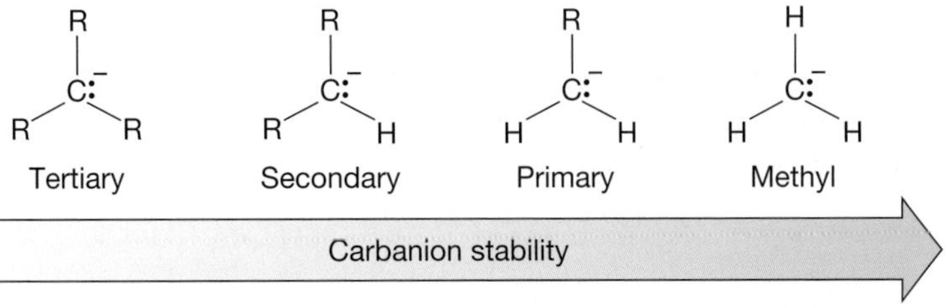

FIGURE 7.93 The order of carbanion stability is the opposite of that for carbocations and radicals, partly because of the greater ease of solvation for the smaller carbanions.

So, the more favorable transition state is the one with the developing negative charge on the primary carbon, the one leading to the less substituted, less stable alkene (Fig. 7.92).

PROBLEM 7.30   How else can the nature of the leaving group determine the structure of the alkene? With the exception of F, the "Hofmann" leaving groups are all large. Consider the possible products of an E2 reaction from the compound in the figure at the right. First show what the possible products are and identify the different hydrogens that must be lost to produce them. Then draw Newman projections for the arrangements leading to the two products. Remember the preference for 180° anti elimination in the E2 process. Finally, use the Newman projections to explain why a large leaving group might favor Hofmann elimination.

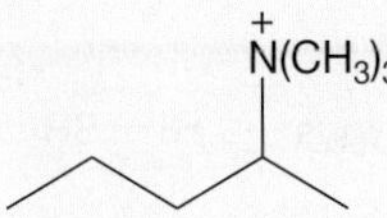

PROBLEM 7.31   Predict the products of reaction when each of the following compounds reacts with hydroxide ion. Be careful with the molecule in (c). The answer here is tricky.

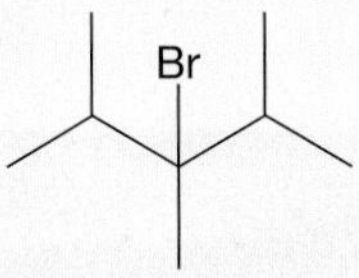

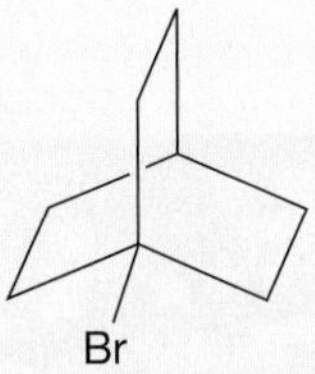

Summary

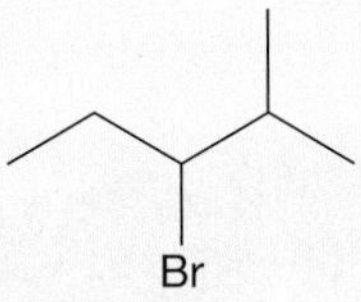

We have already been introduced to addition reactions in Chapter 3 (p. 132). The opposite of addition is elimination, in which two groups or atoms are lost with the introduction of a π bond. There are two general mechanisms for elimination reactions. First, there is a bimolecular E2 process in which starting material goes to product over a single transition state. The E2 reaction resembles (and competes with) the bimolecular S_N2 reaction but is favored by a strong Brønsted base rather than a strong Lewis base. The E2 elimination "prefers" an anti arrangement of the groups being eliminated. The regiochemistry of the product alkene depends on the nature of the leaving group. Most leaving groups favor the more substituted alkene (Saytzeff elimination), but a few large and/or highly electronegative leaving groups lead mainly to the less substituted alkene (Hofmann elimination).

Second, there is a unimolecular E1 elimination reaction that parallels (and competes with) the S_N1 substitution reaction. The S_N1 and E1 reactions share the same intermediate, the carbocation formed by ionization of the leaving group. Once this carbocation is formed it can be captured by nucleophiles (S_N1) or deprotonated to give an alkene (E1). Saytzeff elimination is the rule.

Acetylide addition

7.10 What Can We Do with These Reactions? How to Do Organic Synthesis

Your ability to make molecules has just increased enormously. From rather easily accessible compounds, the alkyl halides and inorganic materials, myriad other structural types are suddenly available. Figure 7.94 summarizes the synthetic potential of the substitution reaction. Not all reactions will work for all R—L molecules and elimination reactions will surely also occur.

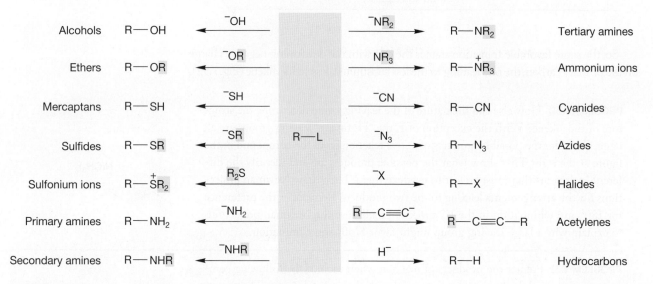

Alcohols	R—OH	⁻OH ←	⁻NR₂ →	R—NR₂	Tertiary amines
Ethers	R—OR	⁻OR ←	NR₃ →	R—N⁺R₃	Ammonium ions
Mercaptans	R—SH	⁻SH ←	⁻CN →	R—CN	Cyanides
Sulfides	R—SR	⁻SR ←	⁻N₃ →	R—N₃	Azides
Sulfonium ions	R—S⁺R₂	R₂S ←	⁻X →	R—X	Halides
Primary amines	R—NH₂	⁻NH₂ ←	R—C≡C⁻ →	R—C≡C—R	Acetylenes
Secondary amines	R—NHR	⁻NHR ←	H⁻ →	R—H	Hydrocarbons

FIGURE 7.94 The synthetic potential of the substitution reaction.

Beginning in the 1940s, the term "organic" in connection with food took on the extra meaning of "lacking use of man-made pesticides, herbicides, fertilizers, growth regulators, or genetically modified organisms." This is clearly a misuse of the word "organic" because all food is "organic." "Organic synthesis" deals with the formation of organic molecules, whether man-made or Nature-made.

Figure 7.94 adds several reactions to the ones we have discussed specifically. For example, it sneaks in a brand-new synthesis of substituted alkynes by using the acetylide ion as a nucleophile (usually effective only with primary R—L compounds). We mentioned the acidity of acetylenes in Chapter 3 (p. 129) and even warned that more was coming about this reaction. Monosubstituted (terminal) alkynes are decent acids ($pK_a \sim 25$), which means that treatment of a terminal alkyne with a strong base can result in removal of a proton and formation of the acetylide. A favorite base for doing this is the amide ion, $^-NH_2$ (Fig. 7.95). There are some other new reactions in Figure 7.94. Go over each of them carefully and be sure you see how each works.

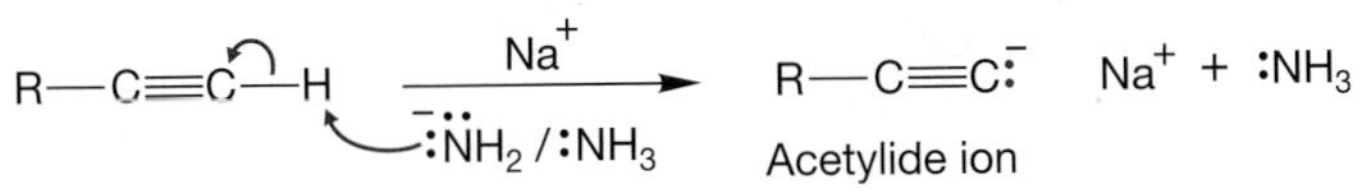

$$R-C\equiv C-H \xrightarrow[\ :NH_2 \ / \ :NH_3]{Na^+} R-C\equiv C:^- \quad Na^+ + :NH_3$$
Acetylide ion

FIGURE 7.95 Formation of an acetylide from a terminal alkyne.

It is *relatively* easy (note that we didn't say it was easy!) to keep track of the available reactions and therefore not too difficult to think up one-step transformations. Once synthetic chemistry goes beyond one-step reactions, however, it gets much harder. Now, the intellectual challenge of synthetic chemistry increases sharply, and synthesis problems, whether in the real world or in this book, become not only more challenging, but much more fun. One can begin to see why the art of making molecules fascinates so many people. Let's look at a few examples.

7.10a Sulfur as Nucleophile   Suppose we are set the task of making a thiol from an alkyl halide. We will use the nucleophilicity of sulfur to accomplish this transformation. The mercaptide ion itself (^-SH) and alkyl mercaptides (^-SR) are powerfully nucleophilic and can be used to great effect in the S_N2 reaction. This nucleophilicity leads to the most common synthesis of thiols and sulfides. The only serious restriction is that the alkyl halide must be active in the S_N2 reaction (Fig. 7.96).

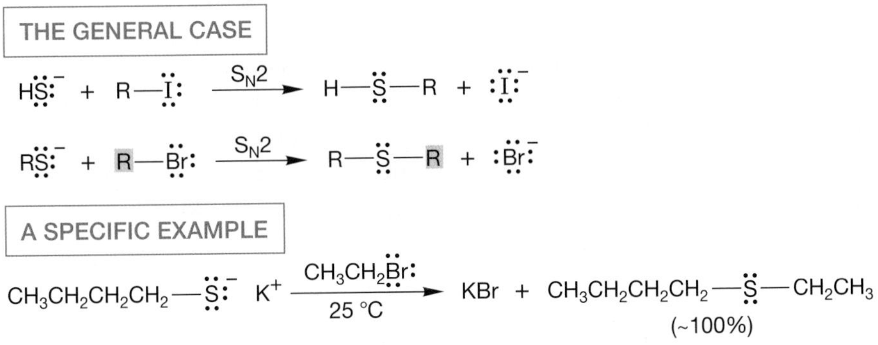

THE GENERAL CASE

$$H\ddot{S}:^- + R-\ddot{I}: \xrightarrow{S_N2} H-\ddot{S}-R + :\ddot{I}:^-$$

$$R\ddot{S}:^- + R-\ddot{B}r: \xrightarrow{S_N2} R-\ddot{S}-R + :\ddot{B}r:^-$$

A SPECIFIC EXAMPLE

$$CH_3CH_2CH_2CH_2-\ddot{S}:^- \ K^+ \xrightarrow[25\ ^\circ C]{CH_3CH_2\ddot{B}r:} KBr + CH_3CH_2CH_2CH_2-\ddot{S}-CH_2CH_3$$
(~100%)

FIGURE 7.96 The S_N2 reactions of mercaptide and substituted mercaptides lead to thiols and sulfides.

Sulfides have two lone pairs of electrons, and these relatives of ethers are quite nucleophilic. S_N2 reactions of sulfides give **sulfonium ions** (Fig. 7.97).

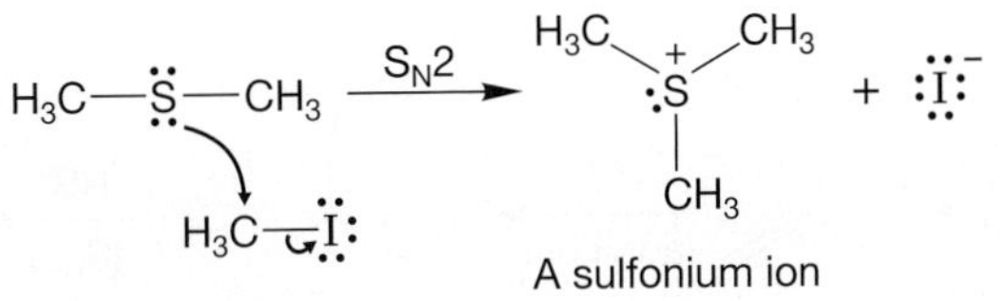

$$H_3C-\ddot{S}-CH_3 \xrightarrow{S_N2} \begin{matrix} H_3C \ \ \ \ CH_3 \\ \diagdown \ \ \ \ \diagup \\ :\overset{+}{S} \\ | \\ CH_3 \end{matrix} + :\ddot{I}:^-$$

$$H_3C-\ddot{I}:$$

A sulfonium ion

FIGURE 7.97 Sulfides can also be alkylated to give sulfonium ions.

Sulfonium ions and the related, less stable, oxonium ions are used as alkylating agents. Dimethyl sulfide and dimethyl ether are excellent leaving groups and can be displaced by nucleophiles in S_N2 reactions (Fig. 7.98). As we have already seen (p. 288), Nature uses a variant of this reaction, displacement on the amino acid derivative S-adenosylmethionine, to do many methylation reactions.

FIGURE 7.98 Alkylations by sulfonium and oxonium ion.

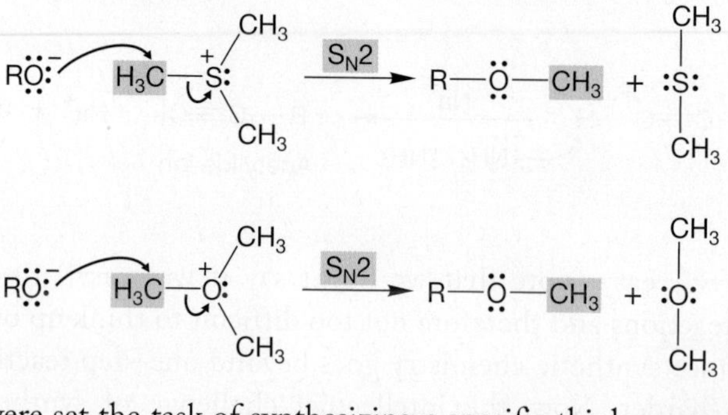

So, if we were set the task of synthesizing a specific thiol, say, propane-2-thiol, we have an answer right at hand: Just treat isopropyl bromide with thiolate, HS^-. You might get a little elimination product as well, but the reaction should work quite well. That's an easy problem, at least if you know the reaction in Figure 7.99.

FIGURE 7.99 The conversion of 2-bromopropane into propane-2-thiol.

2-Bromopropane   Propane-2-thiol

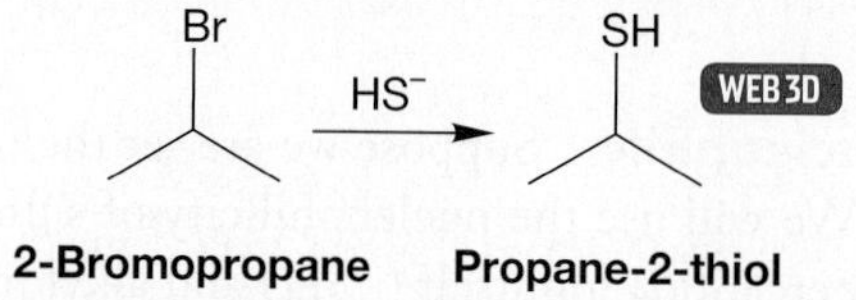

Now, however, suppose we had to make propane-2-thiol not from an alkyl halide, but from propene. We have no direct way to do this transformation, and Figure 7.94 is no help. But we do have a way to convert propene into 2-bromopropane (Chapter 3, p. 132). So, a two-step synthesis is possible (Fig. 7.100), but it is *much* harder to conjure up than the simple one-step method. By far the best way to find it is to work backward from the desired product, asking yourself the question, From what immediate precursor molecule or molecules can I get the ultimate product? In the reaction we are looking at here, the answer to that question should lead you to 2-bromopropane, among others. Now ask again, From what immediate molecule or molecules can I get 2-bromopropane? The answer to that question should include propene plus HBr. Working backward in this way is the only successful route to solving problems in synthesis, and *all* successful chemists, both beginners and pros, do it this way.

FIGURE 7.100 The two-step conversion of propene into propane-2-thiol.

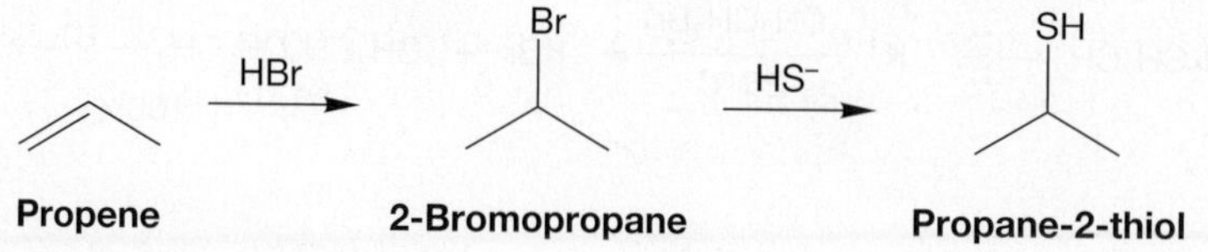

Propene   2-Bromopropane   Propane-2-thiol

Now let's make the sequence even harder and ask how to make propane-2-thiol from 1-bromopropane. Now a three-step sequence is necessary, and one suggestion is shown in Figure 7.101. Be sure you go through the process of working backward for Figure 7.101.

FIGURE 7.101 A three-step sequence for converting 1-bromopropane into propane-2-thiol.

1-Bromopropane   Propene   2-Bromopropane   Propane-2-thiol

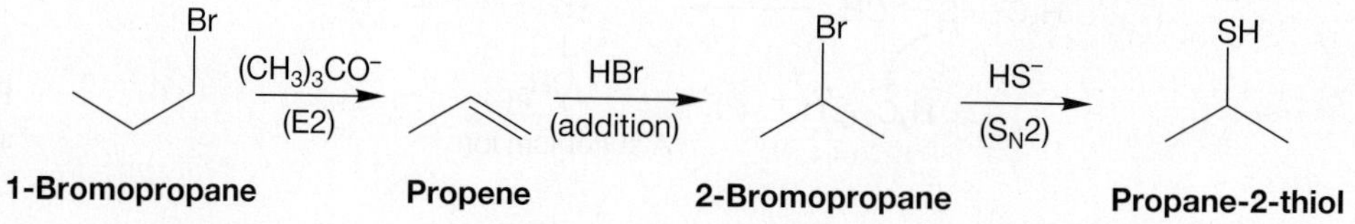

7.10b Oxygen as Nucleophile: The Williamson Ether Synthesis

Here's another synthetic challenge that uses propyl bromide: Make methyl propyl ether. Given our discussion of Figure 7.99, you have almost certainly come up with the idea of displacing bromide ion from 1-bromopropane with methoxide in a straightforward S_N2 sequence (Fig. 7.102). Good idea! In fact, you have just invented a good way to make ethers—well, not exactly, because A. W. Williamson (1824–1904) discovered this process, now called the **Williamson ether synthesis**, over 100 years ago (Fig. 7.103). There are both limitations and opportunities in this synthesis, and they rather nicely illustrate some of the kinds of thinking that synthetic chemists have to go through. So let's explore the Williamson synthesis a bit.

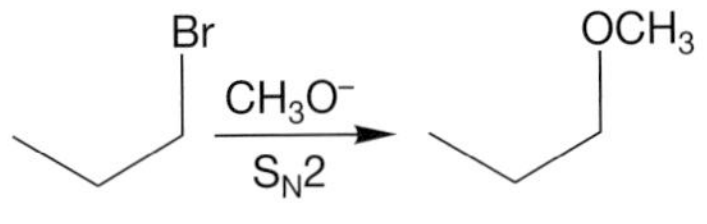

FIGURE 7.102 The Williamson ether synthesis.

THE GENERAL CASE

A SPECIFIC EXAMPLE

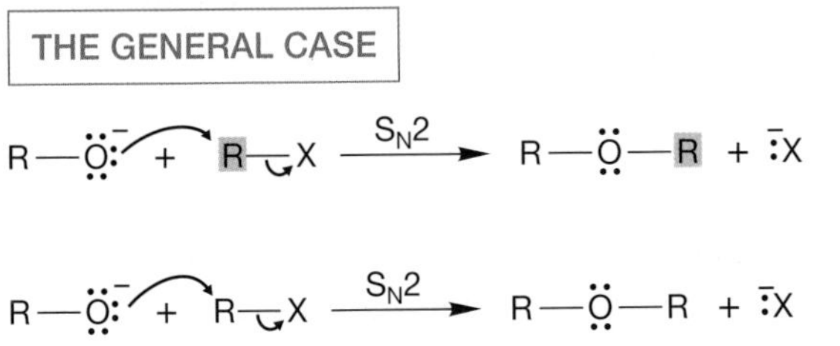

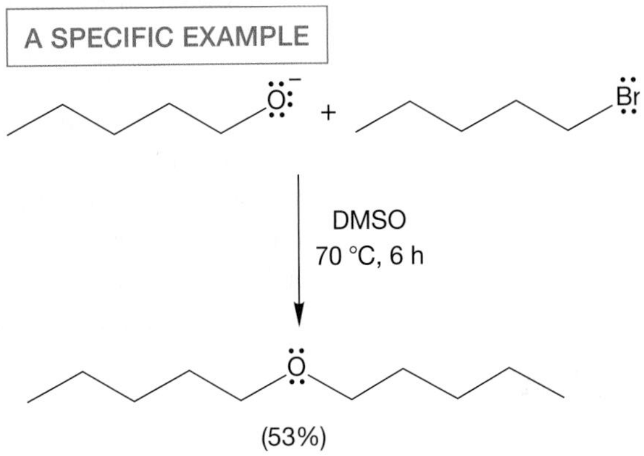

FIGURE 7.103 Alkoxides can displace halides in an S_N2 reaction to make ethers. This reaction is the Williamson ether synthesis.

First of all, we need to be able to make the necessary alkoxide. As we saw in Chapter 6 (p. 236), when an alcohol is treated with a much stronger base, it can be converted entirely into the alkoxide. Note that the base used must be much stronger than the alkoxide. If that is not the case, an equilibrium mixture of the base and the alkoxide will result. A favorite reagent for forming alkoxides is sodium hydride (Fig. 7.104). Sodium hydride has the advantage of producing hydrogen gas, which can easily be removed from the reaction mixture, thus driving the equilibrium toward the alkoxide.

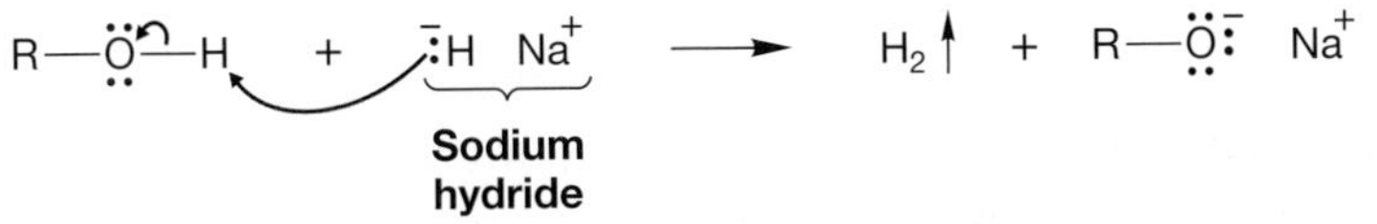

Sodium hydride

FIGURE 7.104 Sodium hydride irreversibly removes a proton to give the alkoxide, the conjugate base of the alcohol.

Just as water reacts with metallic sodium or potassium to give the corresponding metal hydroxide, alkoxides can be made through a reaction in which sodium or potassium metal is dissolved directly in the alcohol. Hydrogen is liberated in what can be a most vigorous reaction indeed. Potassium is particularly active in all cases, and the reaction of sodium metal with smaller alcohols, such as methyl and ethyl alcohol, is also rapid and exothermic (Fig. 7.105).

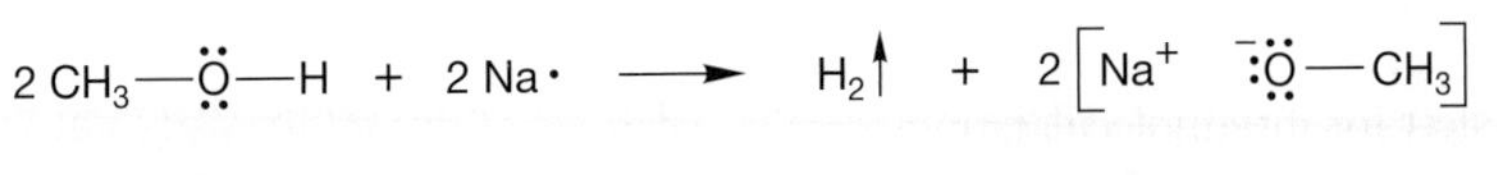

FIGURE 7.105 Alkoxides can be made through the reaction of metallic sodium or potassium with alcohols.

> **WORKED PROBLEM 7.32** Is a molar equivalent amount of sodium hydroxide an effective reagent for forming sodium ethoxide from ethyl alcohol? Explain your answer using Table 7.4.
>
> $CH_3CH_2—\ddot{O}—H + Na^+\ ^-:\ddot{O}H \rightleftarrows CH_3CH_2—\ddot{O}:^-\ Na^+ + H_2\ddot{O}:$
>
> *(continued)*

ANSWER The pK_a values for water (15.7) and ethyl alcohol (15.9) are very close, and the equilibrium concentration of sodium ethoxide will be slightly less than that of hydroxide ion. Hydroxide ion is a slightly weaker base than ethoxide ion in solution. Thus, hydroxide will not be effective in making large amounts of ethoxide. At equilibrium, both species will be present:

$$CH_3CH_2-O-H + {}^-OH \rightleftharpoons CH_3CH_2-O^- + H-O-H$$

$$pK_a = 15.9 \qquad\qquad\qquad\qquad pK_a = 15.7$$

Now let's use alkoxides to make ethers—the Williamson ether synthesis. Several modifications of the original procedure have made this venerable method quite useful, although there are some restrictions. The reaction works only for alkyl halides that are active in the S_N2 reaction. Therefore, tertiary halides cannot be used. They are too hindered to undergo the crucial S_N2 reaction. Sometimes there is an easy way around this problem, but sometimes there isn't. For example, *tert*-butyl methyl ether cannot be made from *tert*-butyl iodide and sodium methoxide, but it can be made from *tert*-butoxide and methyl iodide (Fig. 7.106). However, there is no way to use the Williamson ether synthesis to make di-*tert*-butyl ether.

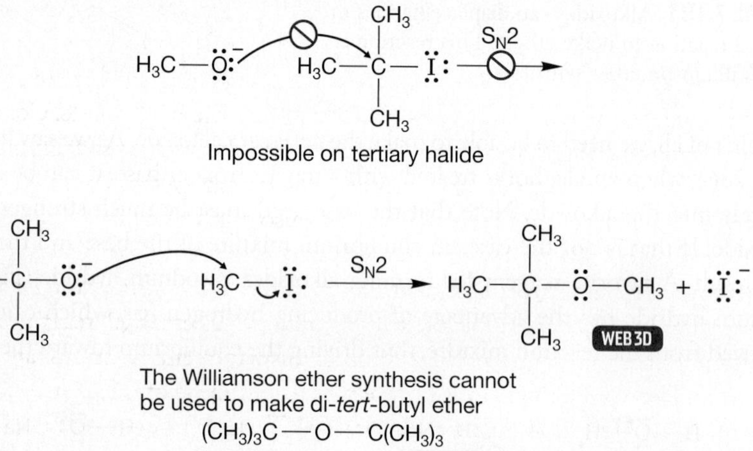

FIGURE 7.106 How to use the Williamson ether synthesis.

Impossible on tertiary halide

The Williamson ether synthesis cannot be used to make di-*tert*-butyl ether
$(CH_3)_3C-O-C(CH_3)_3$

WEB 3D

Even secondary halides are of little use in the Williamson reaction. Alkoxides are strong bases (high affinity for a hydrogen 1s orbital) and the E2 reaction is a prominent side reaction when a secondary halide is used (Fig. 7.107).

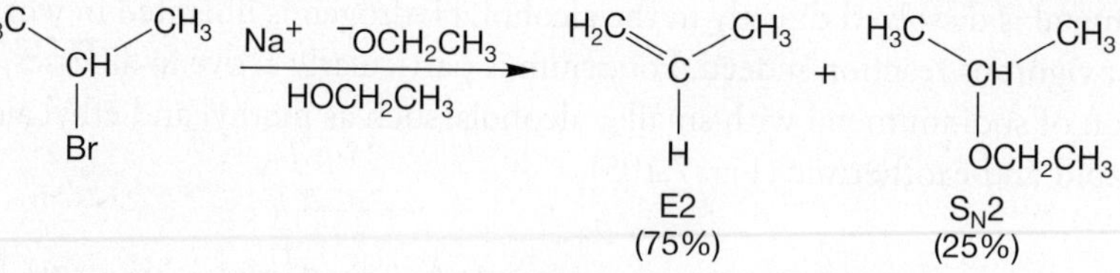

FIGURE 7.107 Secondary halides undergo substantial E2 elimination with alkoxides.

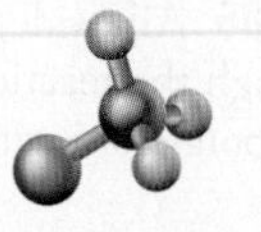

E2
(75%)

S_N2
(25%)

Be alert for intramolecular, ring-forming versions of the Williamson ether synthesis. Cyclic ethers can be formed by intramolecular S_N2 reactions. There is no mystery in this reaction; it is simply the acyclic process applied in an intramolecular way (Fig. 7.108).

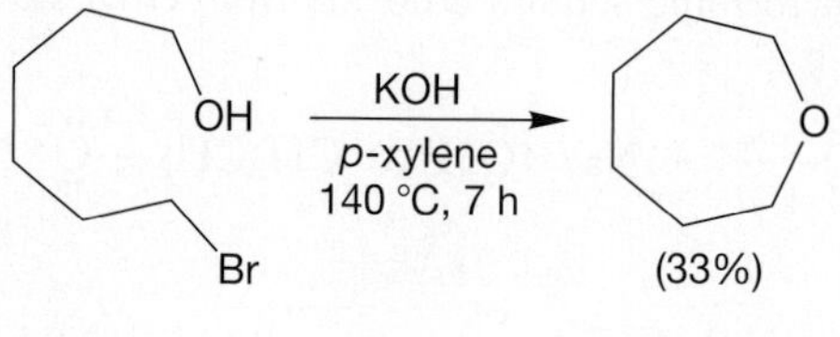

Intramolecular S_N2

FIGURE 7.108 An intramolecular Williamson ether synthesis.

KOH
p-xylene
140 °C, 7 h

(33%)

An important use of the intramolecular Williamson ether synthesis is in the making of **oxiranes**, ethers with the oxygen atom in a three-membered ring, also called **epoxides**. A 2-halo-1-hydroxy compound (a halohydrin) is treated with base to form the haloalkoxide, which then undergoes an intramolecular backside S_N2 displacement to give the three-membered ring (Fig. 7.109).

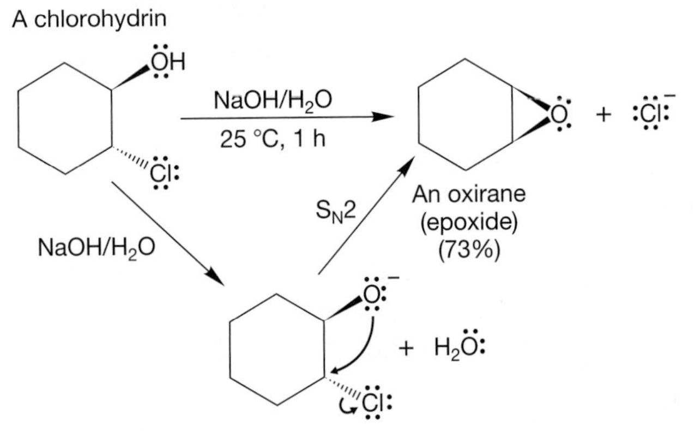

FIGURE 7.109 A common epoxide synthesis takes advantage of an intramolecular S_N2 reaction.

7.10c Nitrogen as Nucleophile: Alkylation of Amines   We have just examined a series of synthetically useful reactions that uses oxygen as the nucleophile. Nitrogen has a lone pair of electrons, and therefore is also nucleophilic. Let's look here at some synthetic reactions in which nitrogen is the nucleophile. Conceptually, the following reactions are rather closely related to the Williamson ether synthesis. Amines are good nucleophiles and can displace leaving groups in the S_N2 reaction. Of course, the limitations of the S_N2 reaction also apply—there can be no displacements at tertiary carbons. Successive displacement reactions yield primary, secondary, and tertiary amines. The initial products are alkylammonium ions that are subsequently deprotonated by the amine present. These S_N2 reactions produce amines that are also nucleophilic. Further displacement reactions are possible, which can lead to problems. Indeed, the product amine is often a better participant in the S_N2 reaction than the starting material. Therefore it can be difficult to stop the reaction at the desired point, even if an excess of the starting amine is used to minimize overalkylation (Fig. 7.110).

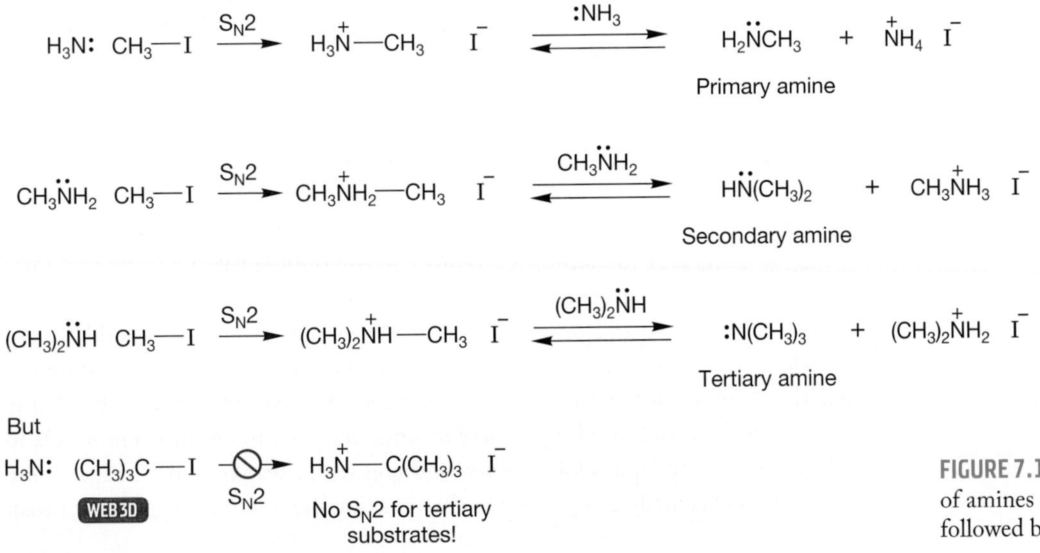

FIGURE 7.110 Alkylation reactions of amines through S_N2 displacement followed by deprotonation.

Tertiary amines are also strong nucleophiles, and a final displacement reaction can occur to give a stable quaternary ammonium ion (Fig. 7.111). Thus, the alkylation process can be pushed to its logical conclusion in an effective synthesis of quaternary ammonium ions.

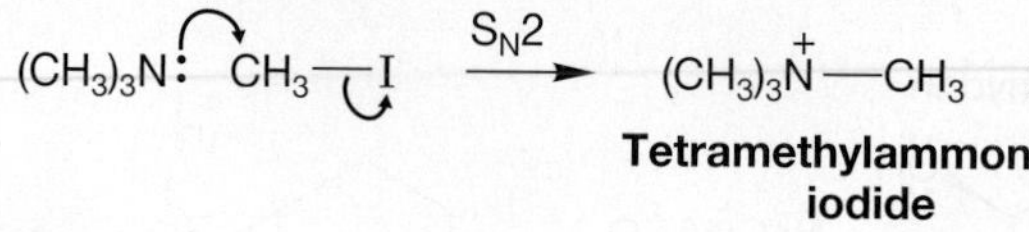

FIGURE 7.111 The formation of an ammonium ion through alkylation of a tertiary amine.

$$(CH_3)_3N\colon \quad CH_3 - I \quad \xrightarrow{S_N2} \quad (CH_3)_3\overset{+}{N} - CH_3 \quad I^-$$

Tetramethylammonium iodide

These alkylation reactions of amines are so efficient that it is difficult to stop at monoalkylation. However, Theodore Cohen (b. 1929) and his co-workers at the University of Pittsburgh worked out a method that uses ammonia in excess under pressure to generate primary amines efficiently. The vast excess of ammonia means that it is difficult for the initial alkylated product, methylamine, to compete with ammonia for a molecule of methyl iodide (Fig. 7.112).

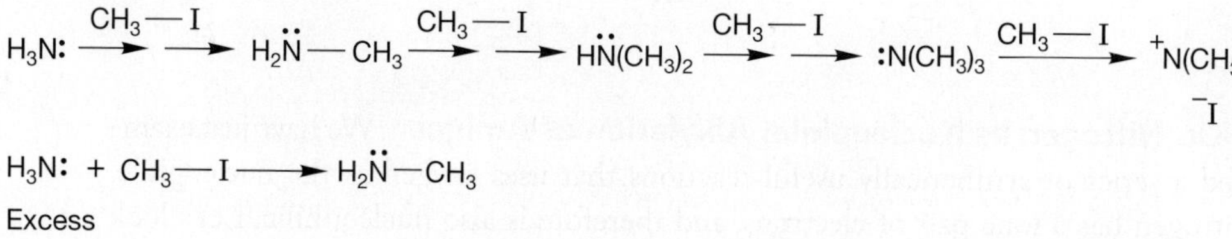

$$H_3N\colon \xrightarrow{CH_3 - I} H_2\ddot{N} - CH_3 \xrightarrow{CH_3 - I} H\ddot{N}(CH_3)_2 \xrightarrow{CH_3 - I} \colon N(CH_3)_3 \xrightarrow{CH_3 - I} {}^{+}N(CH_3)_4$$
$$I^-$$

$$H_3N\colon + CH_3 - I \longrightarrow H_2\ddot{N} - CH_3$$
Excess

FIGURE 7.112 The synthesis of alkylamines through alkylation reactions (S_N2 reactions).

PROBLEM 7.33 What product do you expect from an attempt to make *tert*-butylamine from the reaction of the amide ion ($^-NH_2$) with *tert*-butyl bromide?

PROBLEM 7.34 Another nucleophilic synthesis of amines involves the reaction of simple amines with epoxides. Write a mechanism for the reaction below.

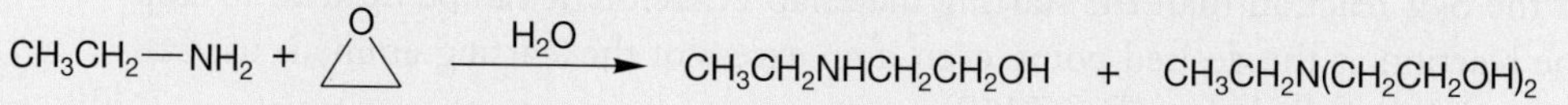

$$CH_3CH_2 - NH_2 + \overset{O}{\triangle} \xrightarrow{H_2O} CH_3CH_2NHCH_2CH_2OH + CH_3CH_2N(CH_2CH_2OH)_2$$

7.10d "Real-World" Difficulties   In the real world, the challenges of synthesis are great—one needs not only to find a reaction or, more likely, a sequence of reactions that will make the desired molecule, but one must do so specifically and, in industry, economically. Specificity becomes increasingly important these days as we recognize how critical chirality is in biological action. And, of course, the more stereogenic centers (p. 152) a molecule has, the more difficult the synthesis. Imagine making a molecule with only a modest number of stereogenic atoms, say 4. There are 16 possible isomers and you almost certainly can use only 1. Every reaction you use must lead to the correct stereochemistry and no others.

Although it is easy to look down one's nose at questions of cost, one would be very wrong to do so. Chemists in industry must find sophisticated ways to make those stereochemically rich molecules without using superexpensive reagents that might be fine in very small-scale work but utterly useless in work designed to produce a molecule to be used by millions of people. Doing clever syntheses with cheap reagents is much more difficult, both intellectually and practically, than making the molecule regardless of cost.

PROBLEM 7.35 Predict the products of the following reactions:

(a) $HO^- + CH_3Br \longrightarrow$

(b) $HO^- + (CH_3)_3CBr \longrightarrow$

(c) $H_2O + (CH_3)_3CBr \longrightarrow$

PROBLEM 7.36 Devise syntheses for the following molecules. You may use any inorganic reagent, organic iodides containing up to four carbons, and the special reagents potassium cyanide and propyne.

(a)   (b) 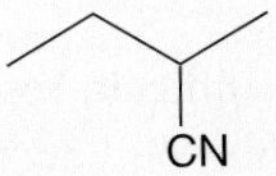   (c) 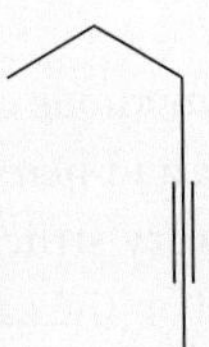   (d)   (e)

$(CH_3)_3C-OCH_3$ $(CH_3)_3C-N_3$

PROBLEM 7.37 Indicate which member of the following pairs of compounds will react faster in (1) the S_N1 reaction and (2) the S_N2 reaction. Explain your reasoning.

(a) 1-Bromobutane and 2-bromobutane

(b) 1-Chloropentane and cyclopentyl chloride

(c) 1-Chloropropane and 1-iodopropane

(d) *tert*-Butyl iodide and isopropyl iodide

(e) *tert*-Butyl iodide and *tert*-butyl chloride

PROBLEM 7.38 Devise synthetic routes that will produce the following compounds as the major products from each of the indicated starting materials.

(a)

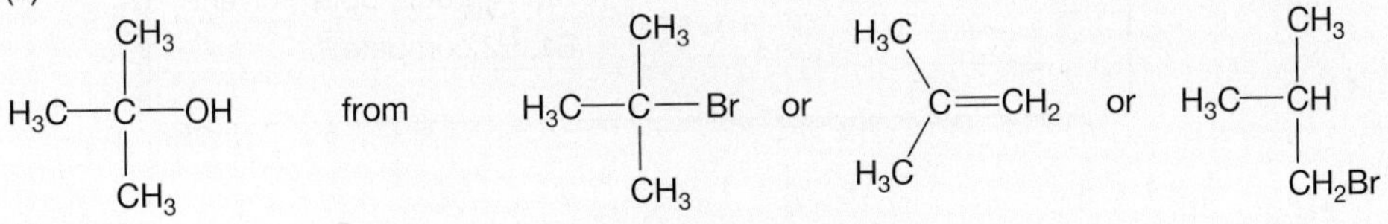

(b)

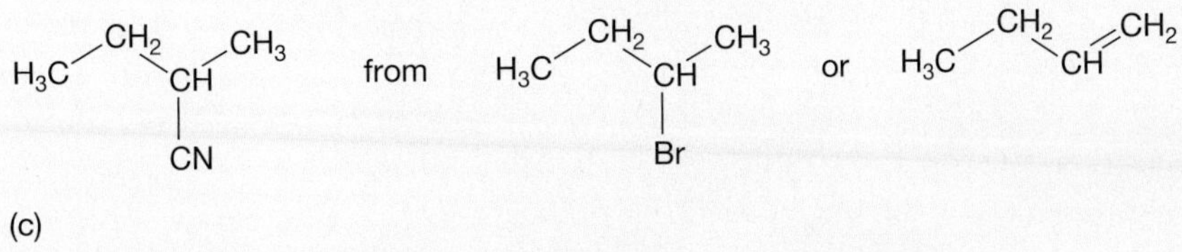

(c)

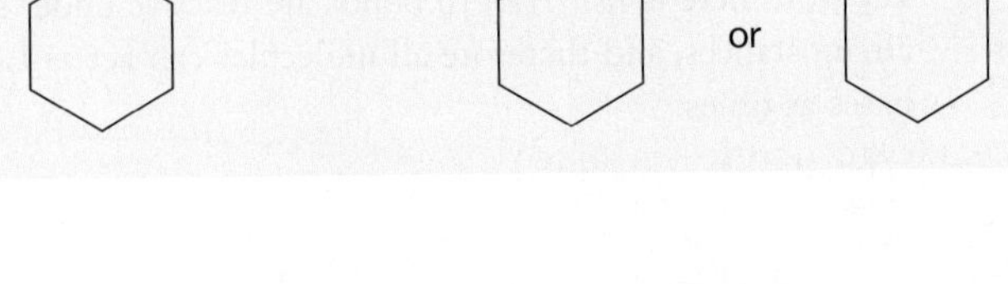

Figure 7.94 fails to show the potential complications we have talked about. *Remember*: There is always a competition between the substitution reactions, S_N1 and S_N2, and the E1 and E2 elimination reactions. Under some circumstances the major products of the reactions summarized in Figure 7.94 will be alkenes. Your file card catalog should include the synthesis of alkenes through the E1 and E2 reactions. The identity of the leaving group, so conveniently avoided by using the symbol L in the figure, makes a difference, and the structure of R, also deliberately vague in the figure, is vital. In order to design a successful synthesis, you have to think about all the components of the reaction.

PROBLEM SOLVING

Keeping track of synthetic methods can be difficult. You are right at the beginning of synthetic chemistry now, and here is a hint on how to keep track of it. Make a 4×6-inch file card for every structural type and then keep a short list of the available synthetic reactions on the card. To help in ordering things, and to keep as far as possible from the dreaded mindless memorization, you might make a mechanistic notation for each reaction as well. The reaction of acetylides with alkyl halides to make new acetylenes might get an "S_N2 with primary substrates" beside it, for example. You can start your catalog with the reactions in Figure 7.94. Be sure to keep up to date; synthetic ability is the kind of thing that sneaks up on you, like whatever it was that worried Mr. Paige, as quoted at the beginning of this chapter. A sample file card for alkyl bromides is shown in Figure 7.113.

FIGURE 7.113 A sample file card for alkyl bromides.

Alkyl Bromides		
Reaction	Mechanism	Notes
R — L + Br⁻	S_N2	Needs good leaving group
		Primary or secondary R
		E2 competes
R — L + Br⁻	S_N1	tert R group, polar solvent
		E1, E2 compete

PROBLEM 7.39 Use the following pK_a data to decide whether sodium amide would be effective at producing vinyl anions ($R_2C=\bar{C}H$). pK_a: ammonia, 38; alkene, ~50.

7.11 Summary

New Concepts

The most important idea in this chapter is the continuing expansion of the concept of Brønsted acidity and basicity to embrace the species known as Lewis acids and Lewis bases. The relatively narrow notion that acids and bases compete for a proton (Brønsted acids and bases) is extended to define all species with reactive pairs of electrons as Lewis bases and all molecules that will react with Lewis bases as Lewis acids. The key word here is *reactive*. All bonds are reactive under some circumstances, and therefore all molecules can act as Lewis bases at times.

Another way to describe the reaction of Lewis bases with Lewis acids is to use the graphic device showing the stabilizing interaction of a filled orbital (Lewis base) with an empty orbital (Lewis acid) (Fig. 7.5). This idea can be generalized if you recall that the strength of orbital interaction depends on the energy difference between the orbitals involved. The closer the orbitals are to each other in energy, the stronger is their interaction and the greater is the resulting stabilization. Therefore the strongest interactions between orbitals are likely to be between the highest occupied molecular orbital (HOMO) of one molecule (**A**, a Lewis base) and the lowest unoccupied molecular orbital (LUMO) of the other (**B**, a Lewis acid) (Fig. 7.7). This notion is used in this chapter to explain the overwhelming preference for backside attack in the S_N2 reaction.

Stereochemical experiments are used here for the first time to probe the details of reaction mechanisms. The observation of inversion of configuration in the S_N2 reaction shows that the incoming nucleophile must approach the leaving group from the rear, the side opposite the leaving group. Similarly, if the stereochemistry of the E2 reaction is monitored, it can be shown that there is a strong preference for the anti (180°) E2 reaction in acyclic molecules.

The curved arrow formalism is used extensively in this chapter. This bookkeeping device is extremely useful in keeping track of the pairs of electrons involved in the making and breaking of bonds in a reaction.

The terms *transition state* (the structure representing the high-energy point separating starting material and products) and *activation energy* (the height of the transition state above the starting material) appear several times in this chapter. Chapter 8 will continue a detailed discussion of these terms (Fig. 7.114).

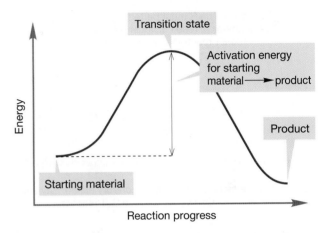

FIGURE 7.114 A typical Energy versus Reaction progress diagram.

Key Terms

anti elimination (p. 305)
E1cB reaction (p. 309)
E1 reaction (p. 298)
E2 reaction (p. 301)
elimination reaction (p. 298)
epoxide (p. 317)
Hofmann elimination (p. 308)
inversion of configuration (p. 269)
leaving group (p. 267)

nucleophilicity (p. 278)
oxirane (p. 317)
product-determining step (p. 294)
rate-determining step (p. 294)
reaction mechanism (p. 263)
regioselective reaction (p. 307)
retention of configuration (p. 269)
Saytzeff elimination (p. 300)
S_N1 reaction (p. 289)

S_N2 reaction (p. 268)
solvolysis reaction (p. 290)
substitution reaction (p. 262)
sulfonium ion (p. 313)
syn elimination (p. 305)
thionyl chloride (p. 284)
Williamson ether synthesis (p. 315)

Reactions, Mechanisms, and Tools

Four exceptionally important prototypal reactions are introduced in this chapter: the S_N1, S_N2, E1, and E2 reactions. The E1 and S_N1 reactions involve the same carbocationic intermediate. Addition of a nucleophile completes the S_N1 reaction; removal of a proton finishes the E1 (Fig. 7.70).

The S_N2 reaction and the E2 reaction are also competitive processes. In the S_N2 reaction, a nucleophile attacks the substrate at the rear of a carbon–leaving group bond. In the E2 reaction, it is the adjacent hydrogen that is removed along with the leaving group (Fig. 7.74).

For primary substrates, it is the S_N2 and E2 reactions that compete. For tertiary substrates in polar solvents with weak bases, it is the S_N1 and E1 reactions that are important. If strong bases are employed in nonionizing solvents, E2 reactions are most important. For secondary substrates all four reactions compete.

It is absolutely vital to have these four building block reaction mechanisms under complete control.

Syntheses

1. Substituted Alkanes

$$R-L \xrightarrow[\text{Nu:}^-]{\text{strong nucleophile}} Nu-R$$

S_N2 reaction. The R must be methyl, primary, or secondary
Always goes with inversion. The E2 reaction is competitive

$$R-L \xrightarrow[\text{HOS}]{\text{polar solvent}} SO-R$$

S_N1 reaction. The R must be tertiary or secondary.
Stereochemical result is largely racemization.
The E1 reaction competes. HOS is
a polar, protic solvent

2. Alkenes

E1 reaction. Must be a tertiary or secondary halide. Saytzeff
elimination. Competes with the S_N1 reaction

E2 reaction. Competes with S_N2, which dominates for
primary substrates. anti Elimination favored, regiochemistry
depends on leaving group

3. Alkoxides

$$R-OH \xrightarrow{Na^+ \ ^-H} R-O^- Na^+ + H_2$$

$$R-OH \xrightarrow{Na} R-O^- Na^+ + H_2$$

4. Alkyl Halides

$$R-OH \xrightarrow{HX} R-X$$ This reaction can be S_N1 or S_N2, but the intermediate is always the protonated alcohol

$$R-OH \xrightarrow{SOCl_2} R-Cl$$ This reaction can be S_N1 or S_N2, but the intermediate is a chlorosulfite ester

$$R-OH \xrightarrow{CCl_4 \ (Ph)_3P} R-Cl$$ Inversion is the usual result

$$R-OH \xrightarrow{PCl_5} R-Cl$$

$$R-OH \xrightarrow{PBr_3} R-Br$$

5. Alkyl Sulfonates

$$R-OH \xrightarrow{Cl-SO_2-R} R-O-SO_2-R$$

These sulfonates are good substrates for further
conversions using S_N1, S_N2, E1, and E2 reactions

6. Amines

$$H_3N + CH_3-I \xrightarrow{S_N2} H_3\overset{+}{N}-CH_3 \ \bar{I} \ \rightleftarrows H_2NCH_3$$
Primary
amine

$$CH_3NH_2 + CH_3-I \xrightarrow{S_N2} CH_3\overset{+}{N}H_2-CH_3 \ \bar{I} \ \rightleftarrows HN(CH_3)_2$$
Secondary
amine

$$(CH_3)_2NH + CH_3-I \xrightarrow{S_N2} (CH_3)_2\overset{+}{N}H-CH_3 \ \bar{I} \ \rightleftarrows N(CH_3)_3$$
Tertiary
amine

Restrictions on S_N2 reaction apply;
overalkylation is a serious problem

7. Ethers

$$R-O^- + R-X \longrightarrow R-O-R$$

Williamson ether synthesis.
R must be able to undergo
the S_N2 reaction—it cannot
be tertiary

The S_N1 and S_N2 reactions allow the conversion of alkyl
halides into a host of new structures. These reactions were
summarized earlier in Figure 7.94. Remember that the S_N2
reaction works only for primary and secondary halides and
the S_N1 reaction works only for tertiary and secondary
molecules.

Alcohols can also be used as starting materials in the S_N1
and S_N2 reactions, but the OH group is a very poor leaving
group and must be transformed either by protonation into
water or by other reactions into some other good leaving
group.

Common Errors

Students of organic chemistry long for certainty. They have not yet learned to love the messiness of the subject or at least to accommodate to it. That the answer to the question, What happens if we treat A with B? might be "All sorts of things" can be profoundly unsettling. In this chapter, we meet four basic building blocks: the S_N1, S_N2, E1, and E2 reactions. It is a sad fact that in many instances they compete with one another. We can usually adjust reaction conditions to favor the reaction we desire, but it will be a rare joyous occasion when we achieve perfect selectivity.

There is much detail in this chapter, but that is not usually a problem for most students. Good nucleophiles can be distinguished from good bases, good leaving groups can be identified, and so on. What is hard is to learn to adjust reaction conditions (reagents, temperature, and solvent) so as to achieve the desired selectivity, to favor the product of just one of these four reactions. That's hard in practice as well as in answering a question on an exam or problem set. To a certain extent, we all have to learn to live with this problem!

7.12   Additional Problems

This chapter expands the topic of organic synthesis. We still have very few reactions for making molecules at our disposal, basically just additions and the E1, E2, S_N1, and S_N2 reactions. Nonetheless, it is possible to devise simple routes to target molecules and even to craft a few multistep sequences. Organic synthesis is one of the most creative (and fun) parts of this science, and so it is important to start early. Here we group several synthesis-related problems before going to more mechanistic material. In just a few chapters, we will have many more reactions under control, and be able to do much fancier work. Stay tuned.

PROBLEM 7.40   The S_N2 reaction occurs between an alkyl halide and a nucleophile. It is a very important process in organic chemistry. For this reason it is necessary to recognize the nucleophilicity of various reagents. In the following pairs of compounds, indicate the more nucleophilic reagent. Explain your reasoning.

(a) $(CH_3CH_2)_3N$ $(CH_3CH_2)_3P$
(b) $(CH_3CH_2)_3N$ $(CH_3CH_2)_2N^-$
(c) CH_3CH_2SNa CH_3CH_2OK

PROBLEM 7.41   Which nucleophiles would serve to effect the following conversions of 1-iodopropane?

PROBLEM 7.42   Explain with crystal clarity why treatment of propyl alcohol with bromide ion fails to give propyl bromide, but treatment with HBr succeeds.

PROBLEM 7.43   Propose a synthesis of each of the following compounds starting with 1-butene:

(a) 2-chlorobutane
(b) 2-methoxybutane
(c) 2-butanamine
(d) 2-butene (mixture of cis and trans)
(e) butane

PROBLEM 7.44   Propose a synthesis of each of the following compounds starting with 2-butanol:

(a) 2-chlorobutane
(b) 2-methoxybutane
(c) 2-butanamine
(d) 2-butene (mixture of cis and trans)
(e) 2-butanethiol

PROBLEM 7.45 Predict the product(s) for each of the reactions shown. Indicate the pathway (S_N1, S_N2, E1, or E2) for formation of each product.

(a)

$$\xrightarrow[\text{THF}]{\text{NaSCH}_3}$$

(b)

$$\xrightarrow[\Delta]{\begin{array}{c}\text{catalytic}\\ \text{H}_2\text{SO}_4\end{array}}$$

(c)

$$\xrightarrow[\text{CH}_3\text{OH}]{\text{CH}_3\text{ONa}}$$

(d)

$$\xrightarrow[\text{H}_2\text{O}]{\Delta}$$

(e)

$$\xrightarrow[\text{H}_2\text{O}]{\Delta}$$

PROBLEM 7.46 The contents of four flasks containing the compounds shown in (a), (b), (c), and (d) are treated with $(CH_3)_3C{-}O^-/(CH_3)_3C{-}OH$. In two cases, a product C_7H_{12} is formed that shows only five signals in its ^{13}C NMR spectrum. In two other cases, a product of the same formula, C_7H_{12}, is formed that shows seven signals in the ^{13}C NMR spectrum. What are the products, and which of the compounds, (a), (b), (c), and (d), leads to which product? Explain.

(a)   (b)   (c)   (d)

PROBLEM 7.47 Why are sulfonates (p. 283) such good participants in the S_N1 and S_N2 reactions?

PROBLEM 7.48 Write a mechanism for the formation of alkyl bromides from the reaction of alcohols and PBr_3.

PROBLEM 7.49 You may be familiar with the compound *tert*-butyl methyl ether. Its common acronym is MTBE. It has been an additive for gasoline since 1979, although its use has declined in the United States since 2003. How would you make MTBE if you have bottles of *tert*-butyl alcohol and methyl alcohol? You may use any other inorganic reagents needed. Your proposed route should avoid mixtures containing undesired products.

PROBLEM 7.50 The reaction shown below was used to help clarify the S_N2 nature of the alkylation reactions of *S*-adenosylmethionine. Explain what the results tell us about the mechanism.

^3H = Tritium (T)
^2H = Deuterium (D)

PROBLEM 7.51 (a) Reaction of 1,2-dimethylpyrrolidine (**1**) with ethyl iodide leads to two salts, $C_8H_{18}IN$. Explain.

(a)

$$\xrightarrow{\text{CH}_3\text{CH}_2\text{I}} \begin{array}{c}\text{Two}\\ C_8H_{18}IN\end{array}$$

1

(b) However, when 2-methylpyrrolidine (**2**) undergoes a similar reaction with ethyl iodide, followed by treatment with weak base, analysis of the product by ^1H NMR spectroscopy reveals only one set of signals for $C_7H_{15}N$. Explain.

(b)

$$\xrightarrow[\text{2. weak base}]{\text{1. CH}_3\text{CH}_2\text{I}} \begin{array}{c}\text{One}\\ C_7H_{15}N\end{array}$$

2

PROBLEM 7.52 Treatment of 1,2-dibromoethane with the dithiolate shown below leads to two products, $C_4H_8S_2$ and $C_6H_{12}S_2Br_2$. Write structures for these products and explain how they are formed.

A dithiolate

$$C_4H_8S_2$$
$$+$$
$$C_6H_{12}S_2Br_2$$

PROBLEM 7.53 Devise a synthetic route that will convert (R)-*sec*-butyl alcohol into (S)-*sec*-butyl alcohol.

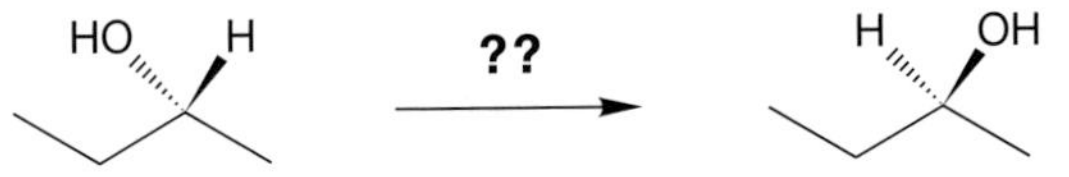

PROBLEM 7.54 Provide reagents that will effect the following changes:

(a)

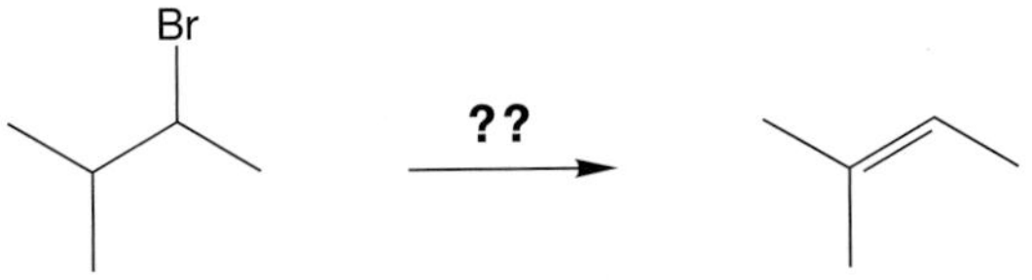

(b)

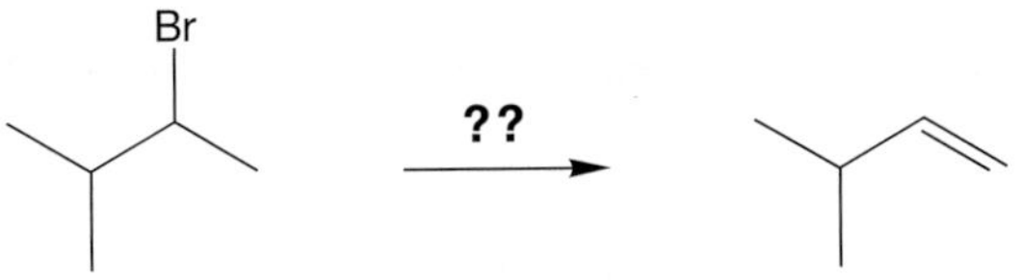

(c)

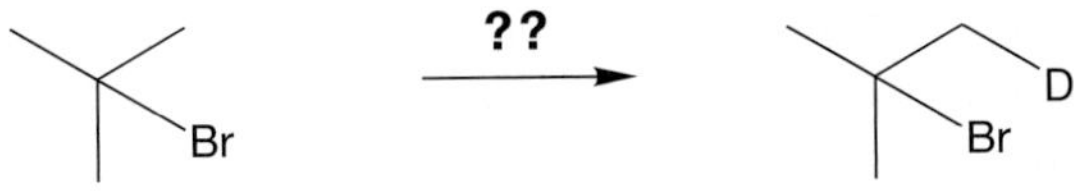

PROBLEM 7.55 Devise S$_N$2 reactions that would lead to the following products:

(a)

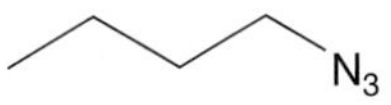

(b)

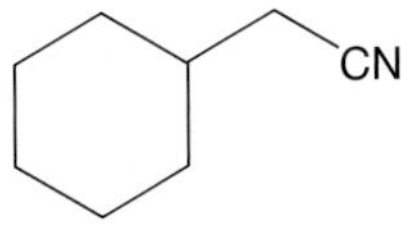

(c)

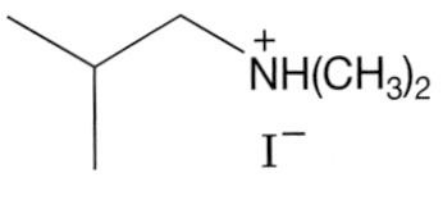

(d)

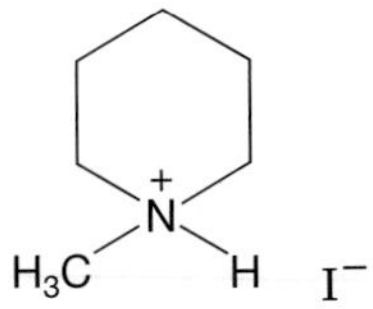

(e)

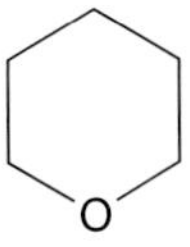

PROBLEM 7.56 Predict the products of the following S$_N$2 reactions:

(a)

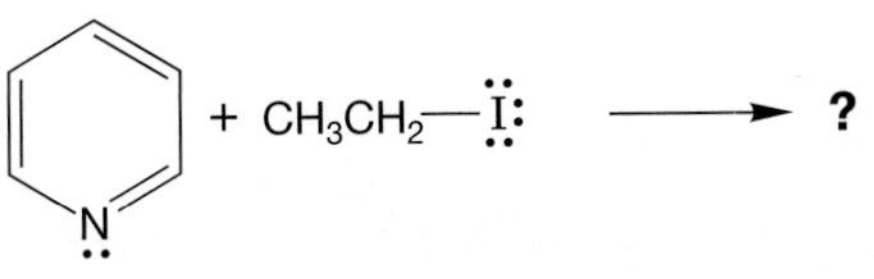

(b)

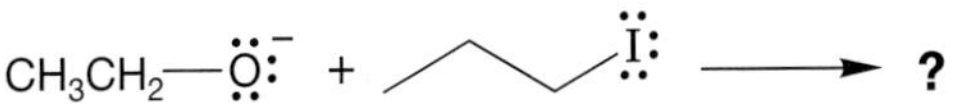

$$CH_3CH_2 \!-\! \ddot{\text{O}}\!:^- \ + \ \text{(propyl iodide)} \longrightarrow \ ?$$

(c)

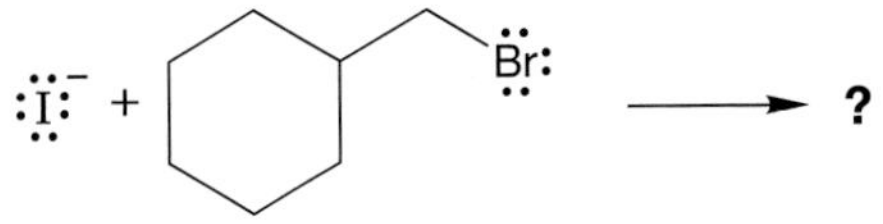

(d)

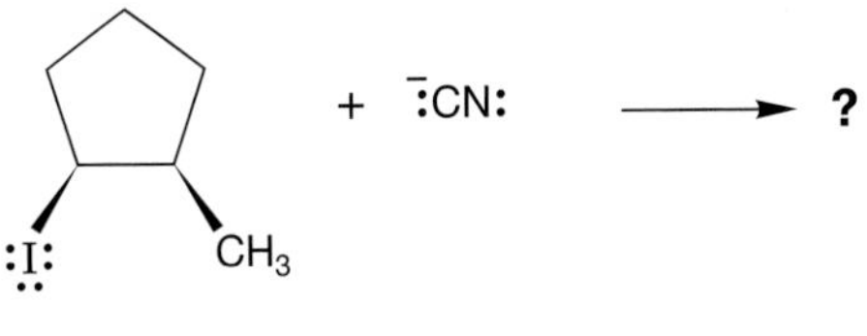

(e)

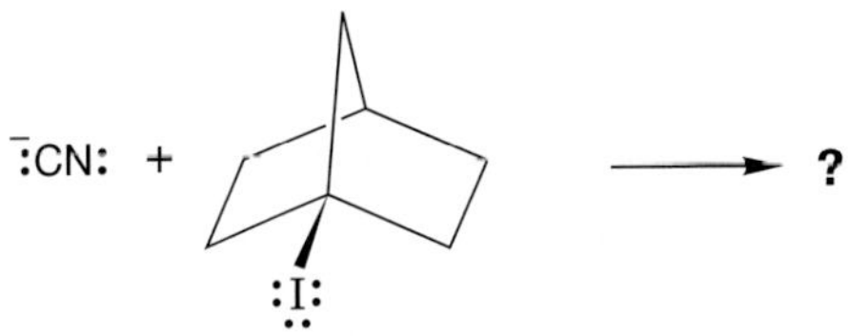

PROBLEM 7.57 Treatment of butyl ethyl ether with HI leads to both butyl iodide and ethyl iodide, along with the related alcohols. By contrast, when *tert*-butyl ethyl ether is treated the same way, only ethyl iodide and *tert*-butyl alcohol are formed. *tert*-Butyl iodide and ethyl alcohol are not produced. Explain.

$$CH_3CH_2CH_2CH_2\!-\!O\!-\!CH_2CH_3$$

$$\Big\downarrow HI$$

$$CH_3CH_2CH_2CH_2I \ + \ CH_3CH_2I$$
and
$$CH_3CH_2CH_2CH_2OH \ + \ CH_3CH_2OH$$

$$(CH_3)_3C\!-\!O\!-\!CH_2CH_3 \xrightarrow{\ HI\ } (CH_3)_3COH \ + \ CH_3CH_2I$$

PROBLEM 7.58 Write all the possible products of the reaction of alcohol **1** with HCl/H$_2$O.

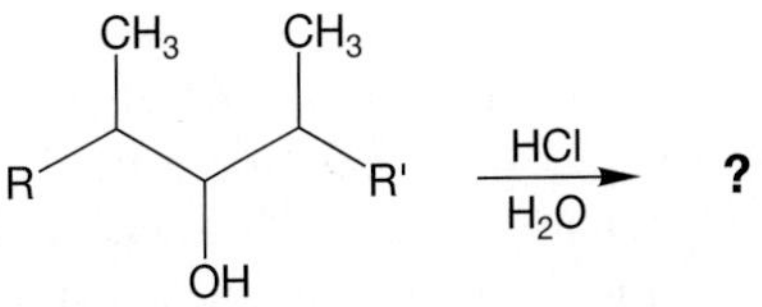

PROBLEM 7.59 Predict the products of the following S_N1 reactions:

(a)

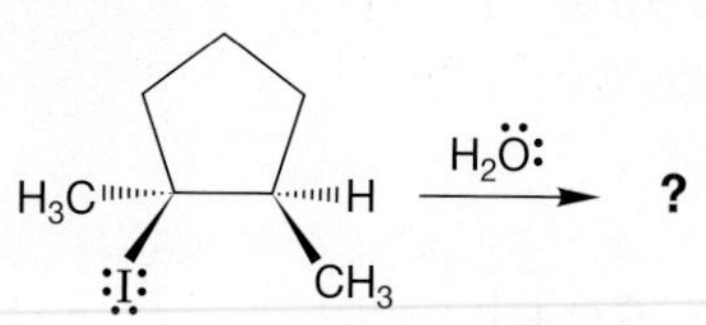

(b)

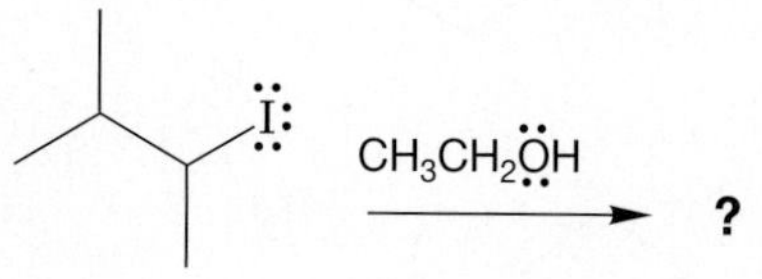

(c)

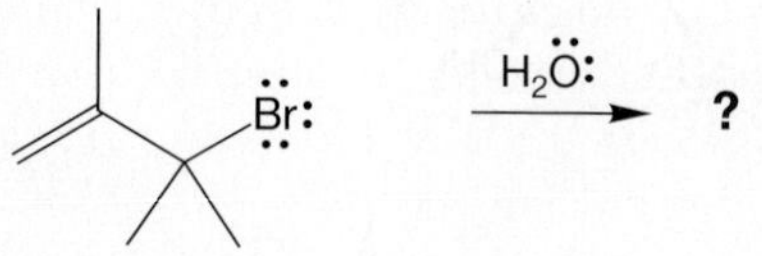

(d)

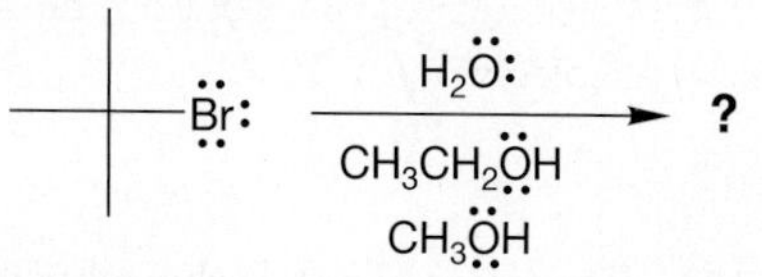

PROBLEM 7.60 You are given a supply of ethyl iodide, *tert*-butyl iodide, sodium ethoxide, and sodium *tert*-butoxide. Your task is to use the S_N2 reaction to make as many different ethers as you can. In principle, how many are possible? In practice, how many can you make?

PROBLEM 7.61 The following molecules can form conjugate bases by loss of a proton from more than one position (arrows). Start by drawing a good Lewis structure (electron dots are deliberately left out) and then choose which proton will be lost more easily. Explain your choice.

(a)

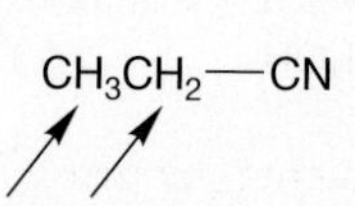

CH_3CH_2—CN

(b)

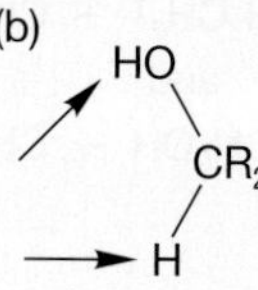

(c)

CH_3CH_2—C

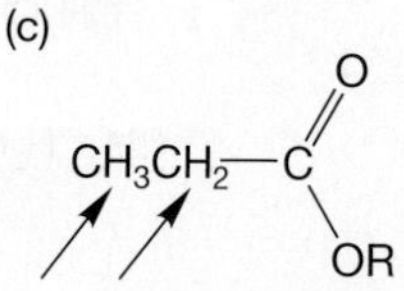

PROBLEM 7.62 The following molecules can accept a proton to form their conjugate acids by addition of a proton at more than one possible place (arrows). Choose the position at which protonation will be preferred and explain your choice. Once again, it makes sense to start by drawing a good Lewis structure for the molecule in question.

(a)   (b) 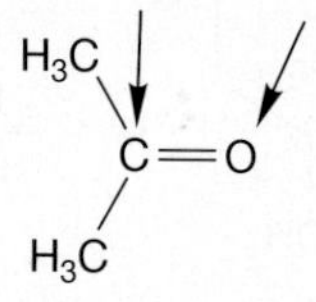   (c)

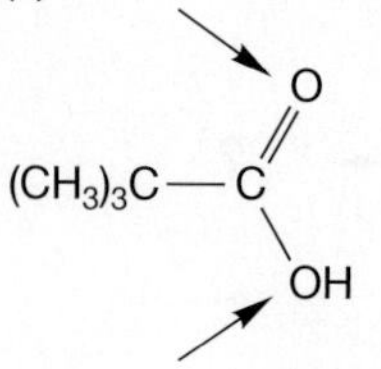

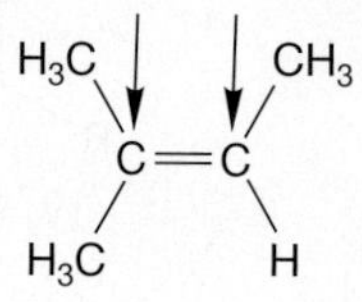

PROBLEM 7.63 Write arrow formalisms for the following reactions. Charges and electron dots are deliberately left out of the reagents and products, so you must start by putting them in.

(a)

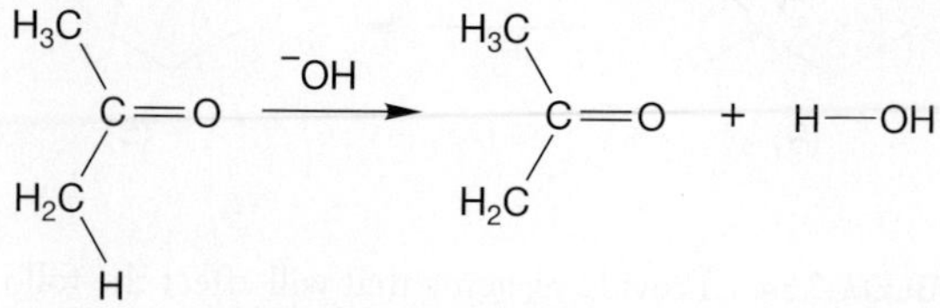

(b)

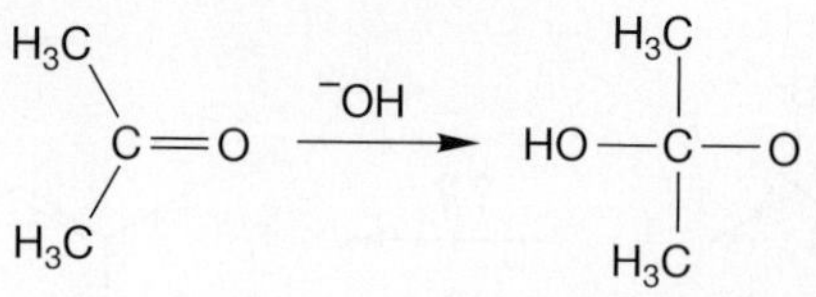

(c)

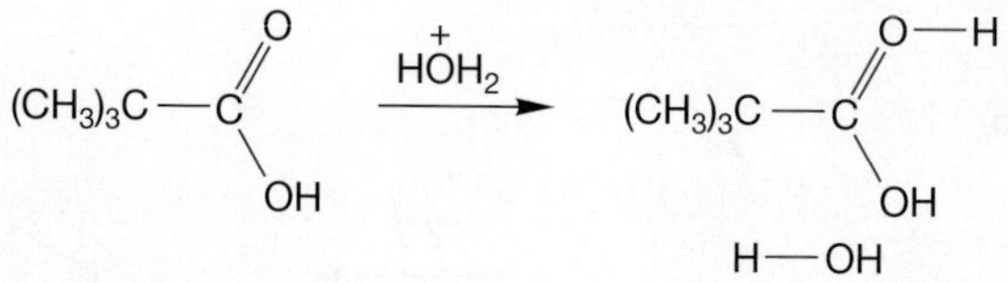

PROBLEM 7.64 The S_N2 reactions at the allyl position are especially facile. For example, allyl compounds are even more reactive than methyl compounds (Table 7.1, p. 275). Explain this rate difference. Why are allyl compounds especially fast? *Hint*: This, like all "rate" questions, can be answered by examining the structures of the transition states involved.

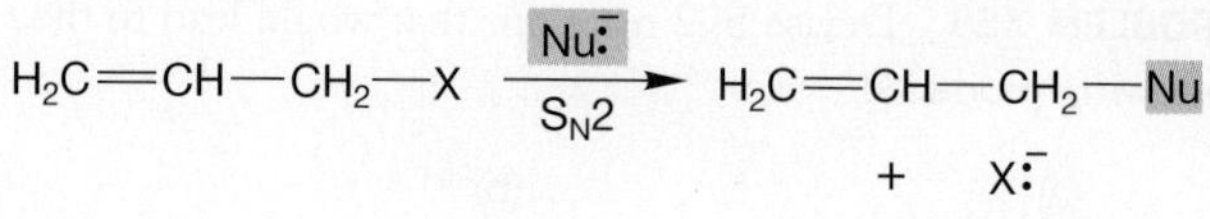

PROBLEM 7.65 In the following pairs of compounds, rank the molecules in order of rate of reaction in the S_N2 process. Explain your reasoning.

(a)

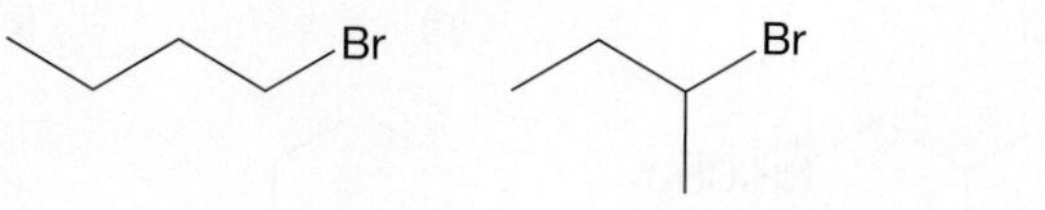

(b)

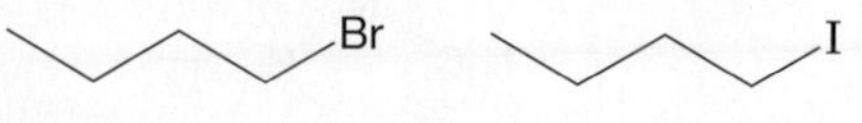

(c)

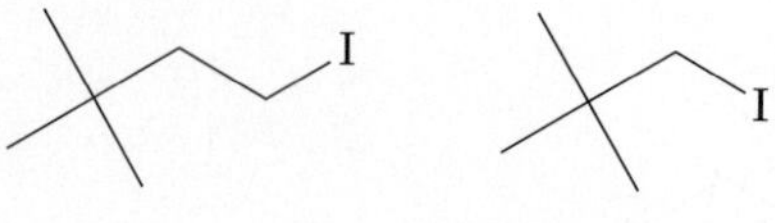

(d)

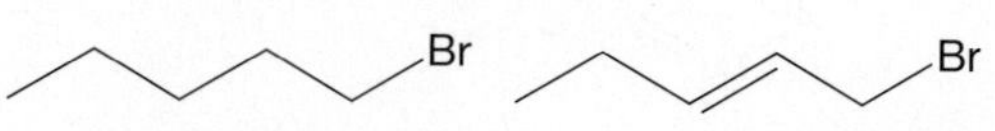

PROBLEM 7.66   In the following pairs of compounds, rank the molecules in order of rate of reaction in the S_N1 process. Explain your reasoning.

(a)

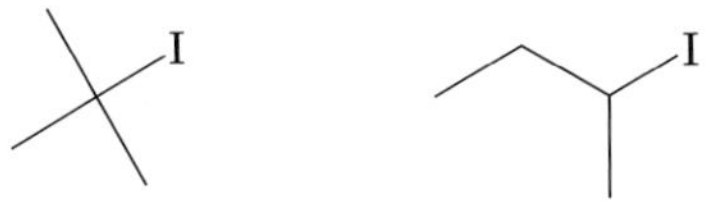

(b)

(c)

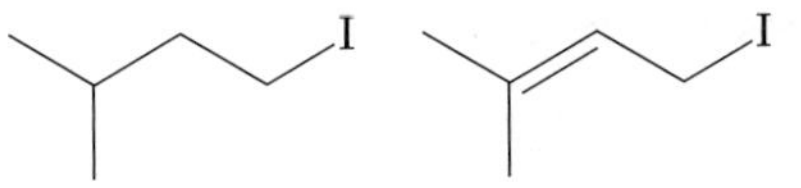

PROBLEM 7.67   Each of the following molecules contains two halogens of different kinds. They are either in different positions in the molecule or are different halogens entirely. In each case, determine which of the two halogens will be the more reactive in the S_N2 reaction. Explain.

(a)

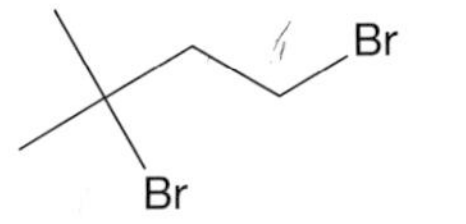

(b)

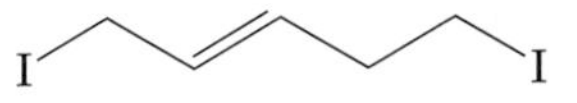

(c)

PROBLEM 7.68   Each of the following molecules contains two halogens at different positions in the molecule. In each case, determine which of the two halogens will be the more reactive in the S_N1 reaction. Explain.

(a)

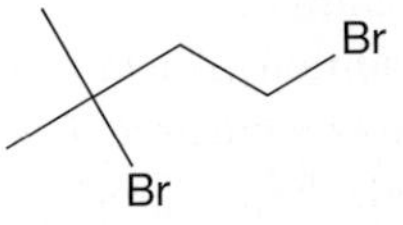

(b)

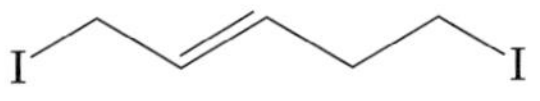

(c)

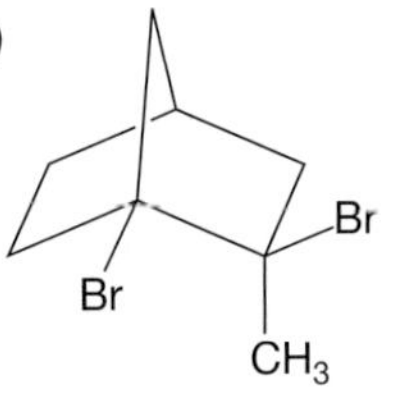

PROBLEM 7.69   You have two bottles containing the two possible diastereomers of the compound shown below ($C_8H_{11}IO_2$). When the salts of these two carboxylic acids are made, only one of them is converted into a new compound, $C_8H_{10}O_2$. What is the structure of the new compound, and how does its formation allow you to determine the stereochemistry of the original two compounds? The squiggly bond indicates both stereoisomers.

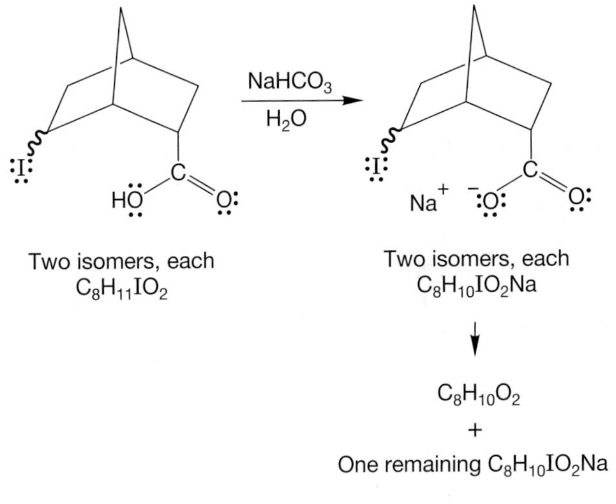

PROBLEM 7.70   You have two bottles containing the two possible diastereomers of the compound shown below (C_5H_9IO). When the molecules are treated with a strong base, such as sodium hydride, only one of them is converted into a new compound, C_5H_8O. What is the new compound, and how does its formation allow you to determine the stereochemistry of the original two compounds?

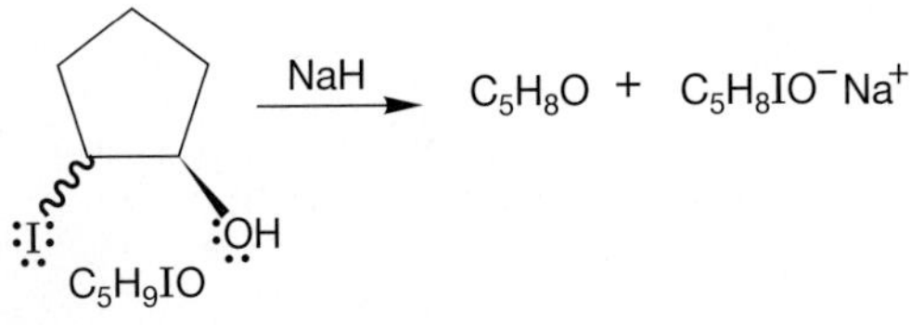

PROBLEM 7.71   Predict the products of the following E2 reactions. If more than one product is expected, indicate which will be the major compound formed.

(a)

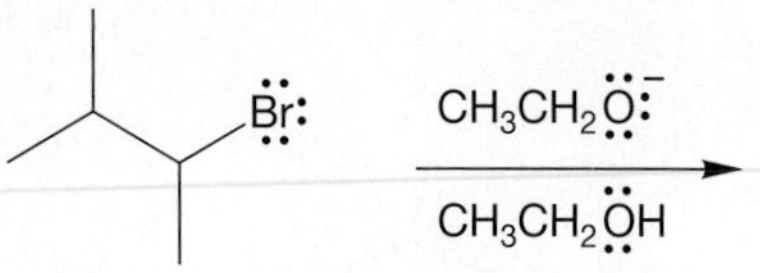

(b)

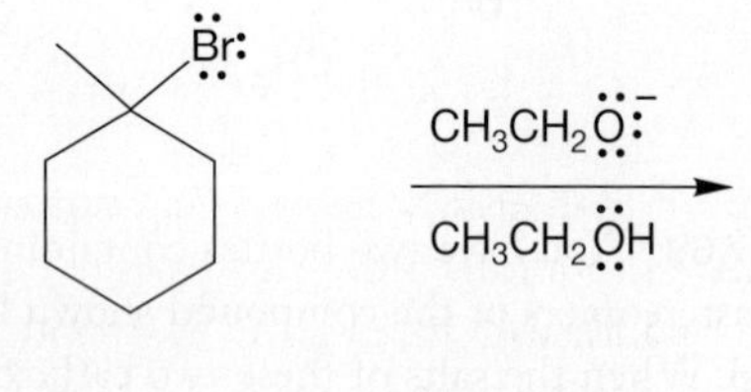

(c)

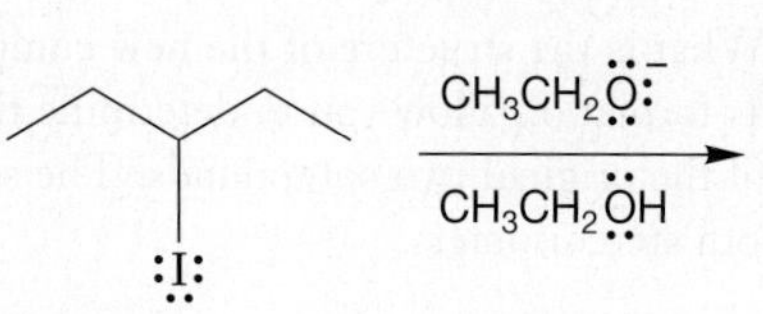

(d)

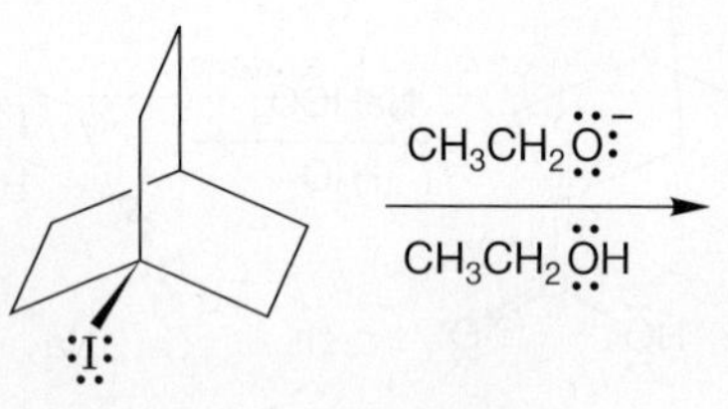

(e)

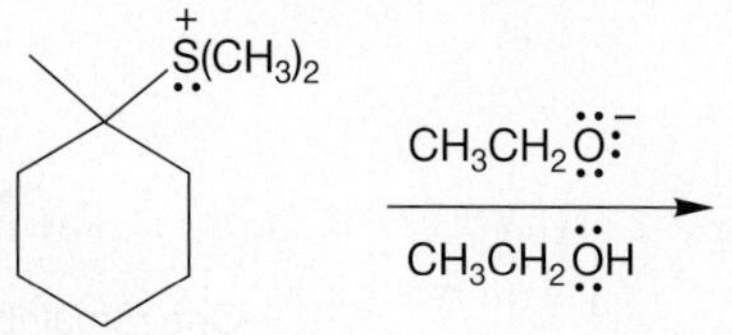

PROBLEM 7.72   Predict the products of the following E1 reactions. If more than one product is expected, indicate which will be the major compound formed.

(a)

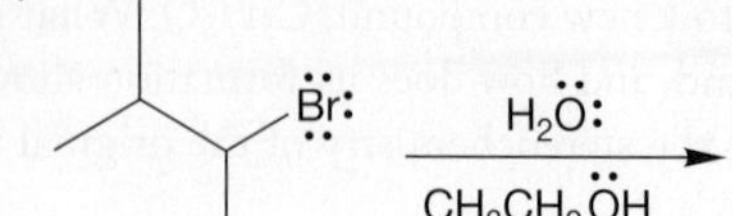

(b)

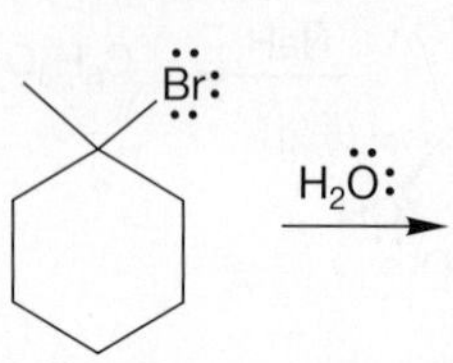

(c)

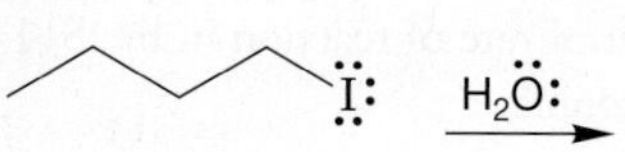

(d)

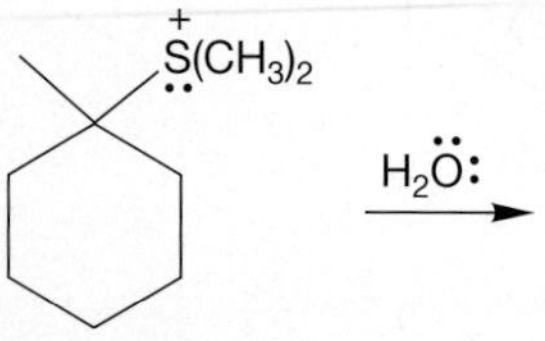

PROBLEM 7.73   (a) Write the "obvious" arrow formalism for the following transformation. Of course it is going to be wrong; you are being set up here. But don't mind, it's all a learning experience.

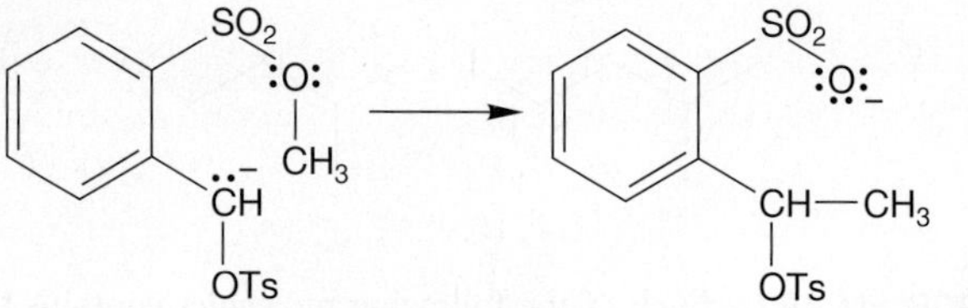

(b) OK, what you wrote is almost certainly wrong. To find the correct mechanism, you measure the kinetics of this process and discover that the reaction is bimolecular. Now write another arrow formalism.

(c) What's wrong with the mechanism suggested by the original arrow formalism? *Hint*: Think about the geometry of the transition state for the S_N2 reaction.

PROBLEM 7.74   (a) Menthyl chloride reacts with sodium ethoxide in ethyl alcohol to give a single product as shown. Analyze this reaction mechanistically to explain why this is the only alkene produced. *Hint*: Think in three dimensions.

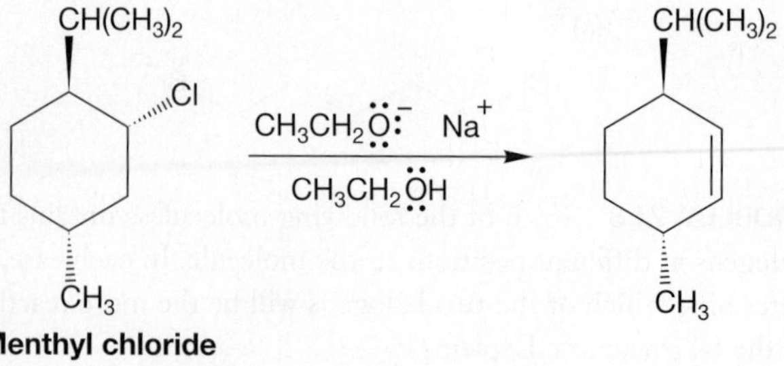

Menthyl chloride

By contrast, neomenthyl chloride treated in the same fashion gives an additional, major, product. Moreover, menthyl chloride reacts much more slowly than neomenthyl chloride. Explain these observations mechanistically.

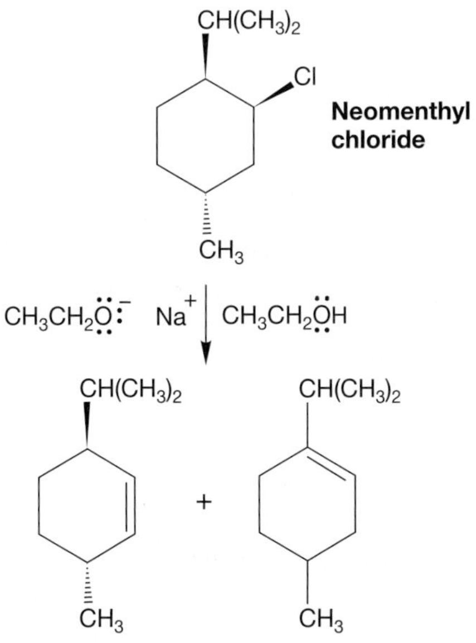

Neomenthyl chloride

$CH_3CH_2\ddot{O}\!:^-$ Na$^+$ | $CH_3CH_2\ddot{O}H$

(25%)   (75%)

(b) When menthyl chloride is treated with 80% aqueous ethyl alcohol with no ethoxide present, once again, two products are formed in the ratio shown. Explain mechanistically.

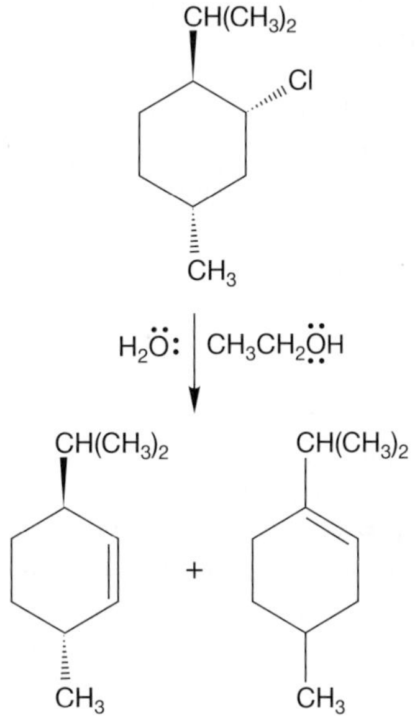

$H_2\ddot{O}\!:$ | $CH_3CH_2\ddot{O}H$

(32%)   (68%)

Steric effects are important in solvolysis reactions. Problems 7.75 and 7.76 illustrate this point.

PROBLEM 7.75   Explain the following rates for S$_N$1 solvolysis. Construct an Energy versus Reaction progress diagram showing the two reactions.

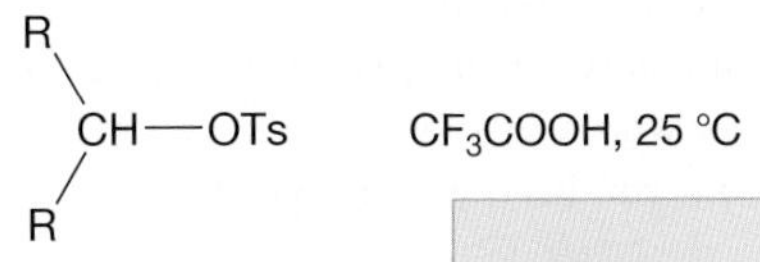

CF$_3$COOH, 25 °C

R	Rate
CH$_3$	1
(CH$_3$)$_3$C	10^5

PROBLEM 7.76   Explain the relative rates of the following compounds in the S$_N$1 reaction:

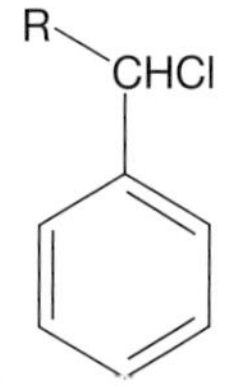

R	Relative Rate
CH$_3$	540
CH$_3$CH$_2$	125
(CH$_3$)$_2$CH	27
(CH$_3$)$_3$C	1

PROBLEM 7.77   Rationalize the following curious experimental results. When the optically active chlorosulfite **1** is heated in toluene, the chloride with inverted configuration is obtained. However, when **1** is decomposed in 1,4-dioxane (**2**), retention of configuration in the resulting chloride is observed. *Hint:* See Figure 7.43, and also think about the differences between toluene and 1,4 dioxane. A good Lewis structure for 1,4-dioxane may also be helpful.

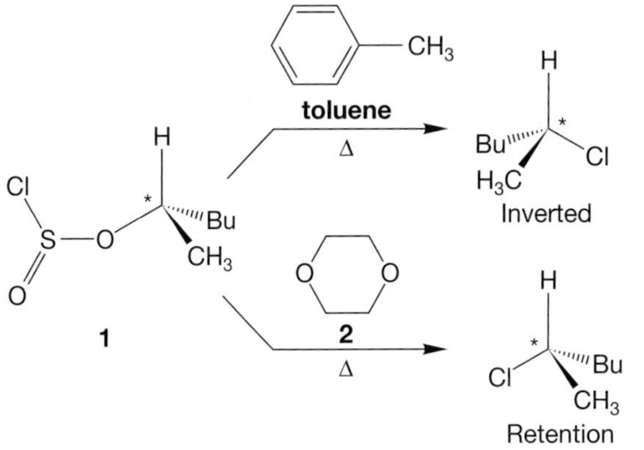

For Problems 7.78 and 7.79 you need to know that treatment of amines, R—NH$_2$, with nitrous acid, HONO, gives diazonium ions, R—N$_2^+$.

PROBLEM 7.78   Provide arrow formalism mechanisms for the following reactions:

(a)

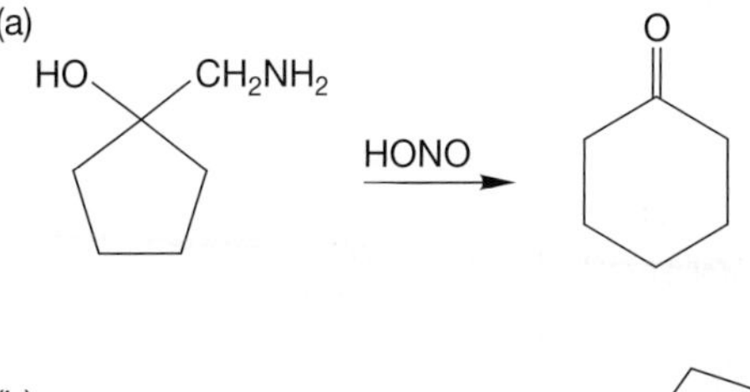

HONO

(b)

HONO

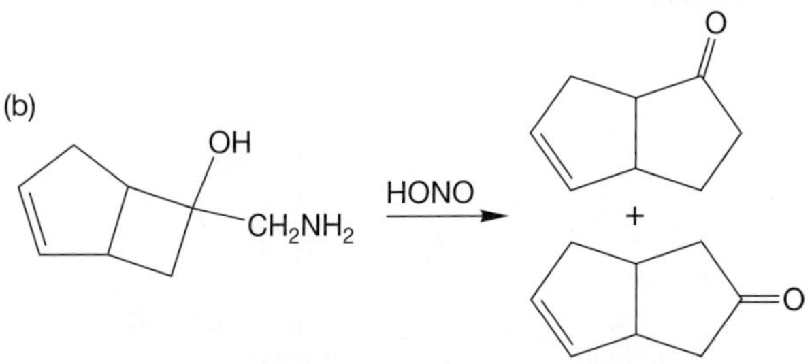

PROBLEM 7.79 Rationalize the different products formed on treatment of the stereoisomeric aminocyclohexanols **1–4** with nitrous acid.

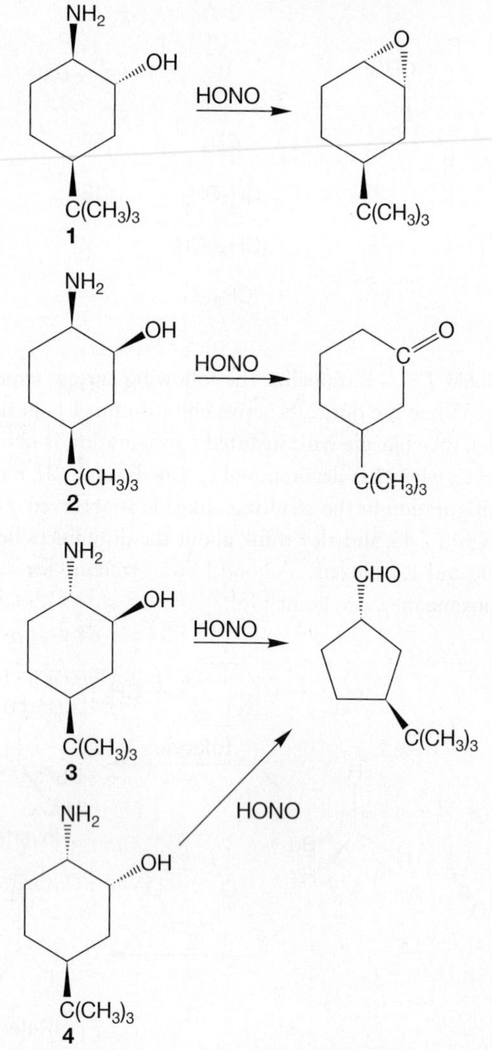

PROBLEM 7.80 You are given a supply of 2-bromo-1-methoxybicyclo[2.2.2]octane. Your task is to prepare 1-methoxybicyclo-[2.2.2]oct-2-ene (**A**) from this starting material. In your first attempt, you treat the bromide with sodium ethoxide in ethyl alcohol. Although a small amount of the desired alkene is formed, most of the product is 2-ethoxy-1-methoxybicyclo[2.2.2]octane (**B**).

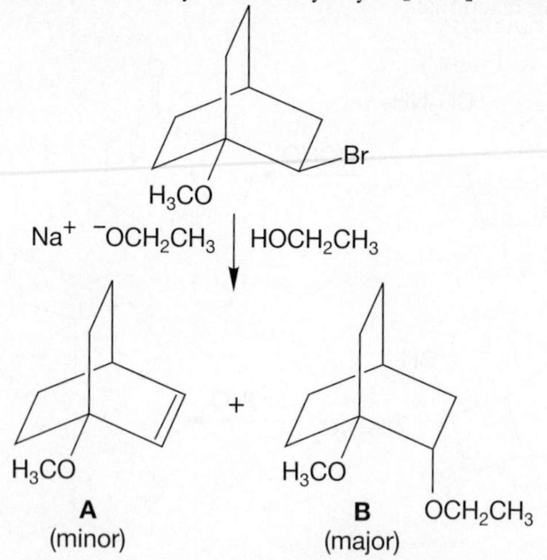

Suggest a simple experimental modification that should result in an increase in the amount of the desired alkene, and a decrease in the amount of the unwanted substitution product, **B**.

PROBLEM 7.81 A common procedure to increase the rate of an S_N2 displacement is to add a catalytic (very small) amount of sodium iodide. Notice that even though the rate increases, the structure of the product is not the iodide. Explain the function of the iodide. How does it act to increase the rate, and why is only a catalytic amount necessary?

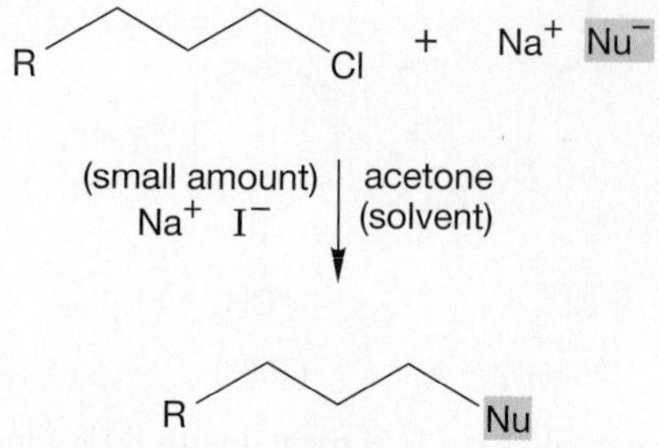

Use Organic Reaction Animations (ORA) to answer the following questions:

PROBLEM 7.82 Select the animation titled "Halide formation" in the panel of reactions labeled Semester 2A. Write out the reaction (see Fig. 7.45). How many intermediates are included in the animation? According to the energy diagram shown for the reaction, which step is the fastest?

PROBLEM 7.83 The "Halide formation" reaction has several intermediates formed that aren't shown in the animation. Predict what they are. It might help to know that one of the final products in the reaction is $P(OH)_3$.

PROBLEM 7.84 After you watch the "Halide formation" reaction, you should be able to describe the mechanism of the second step. What kind of reaction would you call it? Do you suppose a tertiary alcohol goes through the same mechanism?

PROBLEM 7.85 Select the "Bimolecular elimination: E2" reaction. You know about the stereochemistry of the reaction from Section 7.9c. Select the LUMO track for the reaction and observe the starting alkyl halide. Which hydrogens have LUMO character? Why?

PROBLEM 7.86 If you were to add the methoxide base to propane, would propane have a hydrogen acidic enough to react with the base? In other words, would you predict a reaction? See Table 6.4 (p. 235) if you need pK_a information.

Equilibria

8.1 Preview

8.2 Equilibrium

8.3 Entropy in Organic Reactions

8.4 Rates of Chemical Reactions

8.5 Rate Constant

8.6 Energy Barriers in Chemical Reactions: The Transition State and Activation Energy

8.7 Reaction Mechanism

8.8 The Hammond Postulate: Thermodynamics versus Kinetics

8.9 Special Topic: Enzymes and Reaction Rates

8.10 Summary

8.11 Additional Problems

ENERGY BARRIERS Reactions have energy barriers that must be crossed to go from reactants to products. A reasonable analogy is the mountain range that must be crossed to go from your starting point to the other side of the mountain. There may be intermediates in the reaction path, just like the valleys that separate the peaks.

"I hope," the Golux said, "that this is true. I make things up, you know."

—JAMES THURBER,[1] *THE 13 CLOCKS*

8.1 Preview

The study of how reactions occur depends on an analysis of the structures of the starting materials, the products, and the transition states separating them. We have spent much time on structure in the last several chapters. However, there are other, nonstructural factors that are important. To understand chemical reactivity, and to be able to use our knowledge in a predictive way, we also need to know something of the energetics involved in the reaction. Which are more stable, starting materials or products? Of course those designations—starting materials and products—are arbitrary. Nature doesn't care about such synthetic labels! How high is the energy of the transition state separating "starting materials and products"? If the transition-state energy is too high, no reaction will occur. If there are several available low-energy pathways—several accessible transition states—there will be a mixture of products, and the synthetic chemist's longed-for specificity will be impossible. So we need to know about energy.

In Chapter 7, we saw some examples of chemical equilibria. For instance, the S_N2 reaction involves an equilibrium in which two nucleophiles ($Nu:^-$ and $L:^-$) compete for the R group (Fig. 8.1).

FIGURE 8.1 The S_N2 reaction is an equilibrium in which two nucleophiles (the Lewis bases $Nu:^-$ and $L:^-$) compete for a Lewis acid.

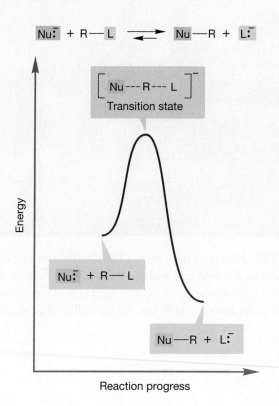

[1]James Grove Thurber (1894–1961) was an American writer and cartoonist, much of whose work first appeared in *The New Yorker* magazine. *The 13 Clocks* is a fairy tale of surpassing charm and wit written in Bermuda.

All chemical reactions can be regarded as equilibria. Multistep reactions are simply series of equilibria in which species occupying energy minima are separated by barriers, the tops of which are transition states. The S_N1 solvolysis reaction of *tert*-butyl iodide in water makes a good example. Here we have three equilibria. The first is between *tert*-butyl iodide and the *tert*-butyl cation and iodide ion. This equilibrium is highly unfavorable, because the cation–anion pair is far less stable than the covalent iodide. The second equilibration is the exothermic capture of the cation by water to give the protonated alcohol, an oxonium ion. Finally, the oxonium ion is deprotonated to give the ultimate product, *tert*-butyl alcohol (Fig. 8.2).

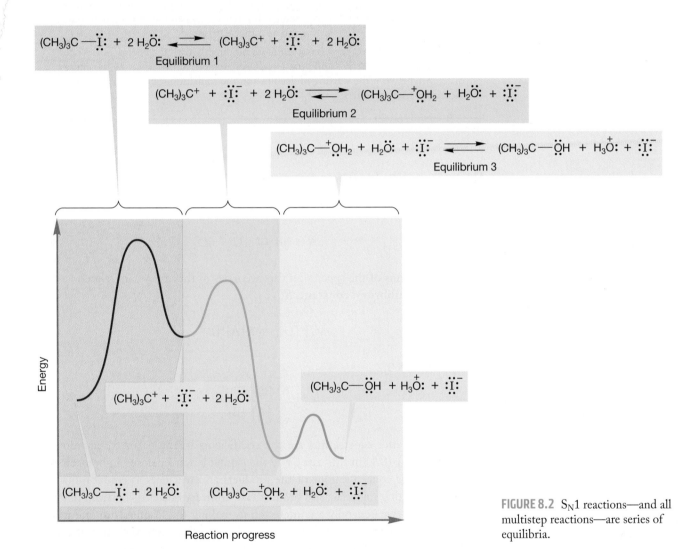

FIGURE 8.2 S_N1 reactions—and all multistep reactions—are series of equilibria.

In practice, thermodynamics may dictate that one side of an equilibrium be overwhelmingly favored over the other. Nevertheless, we can still apply techniques for studying equilibria to the reaction, deriving information about the position of the equilibrium (thermodynamics) and the rate of reaction (kinetics). The concepts we will develop in this short chapter are extremely general and very important.

ESSENTIAL SKILLS AND DETAILS

1. Thermodynamics. Understanding the importance of the relative energies of starting materials and products, as well as the heights of the barriers separating these species, is critical to understanding reactivity. The relative energies of starting materials and products determine the ratio of products under conditions that allow all energy barriers to be passed.

2. Kinetics. The heights of the barriers separating starting materials from products determine the rates of formation of products. If there is not sufficient energy to return to starting material from products, it is the heights of energy barriers—topped by transition states—that will determine the ratio of products.

3. A little bit of energy goes a long way at equilibrium. The outcome of a reaction at equilibrium is determined by the energy difference between the starting materials and products: $\Delta G = -RT \ln K$ is the operative equation.

4. Rates. The rate of a reaction has nothing to do directly with the energy difference between starting materials and products. Rates are determined solely by the energy difference between the starting molecules and the transition states separating those molecules from products.

8.2 Equilibrium

For a general chemical reaction at equilibrium,

$$aA + bB \rightleftarrows cC + dD \tag{8.1}$$

the concentrations of the species on the two sides of the arrows are related to each other by an **equilibrium constant, K**:

$$[C]^c [D]^d = K[A]^a [B]^b \tag{8.2}$$

or, more familiarly:

$$K = \frac{[C]^c [D]^d}{[A]^a [B]^b} \tag{8.3}$$

The small letter superscripts are the coefficients in the balanced equation for the reaction [Eq. (8.1) in this case], and the capital letters enclosed in brackets represent the molar concentrations of the products and reactants at equilibrium. The concepts of starting materials and products are arbitrary ones, arising from practical considerations—we generally are interested in making the reaction run one way or the other, and by convention the desired reaction is written left to right. But it need not be; it is just as valid to look at the reaction in the other direction.

A reaction will be a successful one if the compounds we seek, the products, are strongly favored at equilibrium. So we are very interested in what determines the magnitude of the equilibrium constant (K): the difference in the free energies ($\Delta G° = \Delta H° - T \Delta S°$; see below for a fuller discussion) of the compounds involved in the reaction. The exact relationship between K and the relative free energies of the starting materials and products is shown in Eq. (8.4):

$$\Delta G° = -RT \ln K \tag{8.4}$$

or, if we want to use base 10 rather than natural logarithms, in Eq. (8.5):

$$\Delta G° = -2.3\,RT\log K \qquad (8.5)$$

R is the gas constant, 1.986×10^{-3} kcal/deg·mol, T is the absolute temperature, and $\Delta G°$, the **Gibbs free energy change**, is the difference in energy between starting materials and products in their standard states (Figs. 8.3 and 8.4).

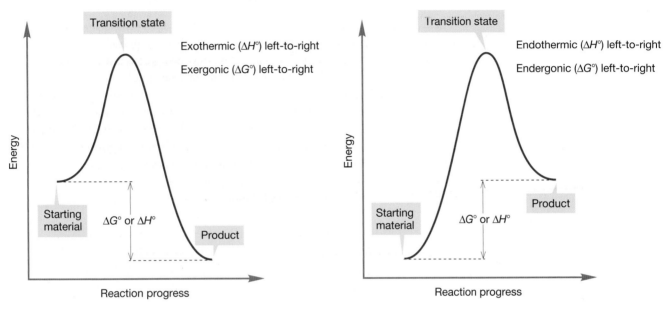

FIGURE 8.3 An exothermic (exergonic) reaction.   FIGURE 8.4 An endothermic (endergonic) reaction.

Up to now, we have more or less equated energy with $H°$, the enthalpy. The enthalpy change $\Delta H°$, or the heat of reaction, is a measure of the bond energy changes in a reaction. The Gibbs free energy $(\Delta G°)$ is more complicated than this, as we will see in Section 8.3, and includes not only $\Delta H°$, but **entropy change ($\Delta S°$)** as well. The entropy of a reaction is a measure of the change in amount of order in starting materials and products. The higher the degree of ordering of reactants necessary, the greater are the energy requirements for the reaction. Conversely, an increase in *dis*order will require less energy for the reaction.

A number of times in Chapter 7 we drew diagrams that described the course of a reaction. We can now label those diagrams a bit more completely. The diagrams are plots of energy versus something called, most often, "reaction progress." Figures 8.3 and 8.4 show an exothermic reaction (products more stable than starting materials) and an endothermic reaction (starting materials more stable than products).

The terms "exothermic" and "endothermic" really refer to the enthalpy change, $\Delta H°$, which may or may not be in the same direction as the free energy change, $\Delta G°$, which involves entropy, $\Delta S°$, as well as enthalpy, $\Delta H°$. If we really mean free energy, $\Delta G°$, the proper terms are **exergonic** and **endergonic**.

For an exothermic reaction, the difference in enthalpy between products and starting materials is negative and the equilibrium constant (K) is greater than 1. For an endothermic reaction (read the first reaction backward to produce an endothermic process), the enthalpy difference is positive and K is less than 1.

More stable products are favored at equilibrium. By how much? Note that the relationship between the difference in free energy between starting material and products ($\Delta G°$) and the equilibrium constant K is logarithmic. We can rewrite Eqs. (8.4) and (8.5) to emphasize that point:

$$\Delta G° = -RT \ln K \quad \text{or} \quad K = e^{-\Delta G°/RT} \tag{8.4}$$

or, in base 10:

$$\Delta G° = -2.3RT \log K \quad \text{or} \quad K = 10^{-\Delta G°/2.3RT} \tag{8.5}$$

These equations show that a small difference in energy between starting materials and products results in a large excess of one over the other at equilibrium. Let's do a few calculations to show this. First, let's take a case in which products are more stable than starting materials by the seemingly tiny amount of 1 kcal/mol, $\Delta G° = -1$ kcal/mol,[2] so $K = 10^{1/(1.364)} = 10^{0.7331} = 5.4$.

An equilibrium constant of 5.4 translates into 84.4% product at equilibrium! One would very often be quite happy to find a reaction that gives about 85% product, and a mere 1-kcal/mol suffices to ensure this. Conversely, if we are "fighting" a 1-kcal/mol endothermic process, it will be a futile fight indeed because equilibrium will settle out at 84.4% starting material. *A little bit of energy goes a long way at equilibrium.* Table 8.1 summarizes a number of similar calculations.

TABLE 8.1 The Relationship between $\Delta G°$ and K at 25 °C

$\Delta G°$ (kcal/mol)	K	Equilibrium Ratio
−0.1	1.2	54.5/45.5
−0.5	2.3	69.7/30.3
−1	5.4	84.4/15.6
−2	29.3	96.7/3.3
−5	4631	99.98/0.02
−10	2.1×10^7	99.999996/0.000004

PROBLEM 8.1 Calculate the energy difference between starting material and products at 25 °C given equilibrium constants of 7 and 14.

In practice, there often is something we can do to drive a reaction in the direction we want. **Le Châtelier's principle**[3] states that a system at equilibrium responds to stress in such a way as to relieve that stress. This statement means that if we increase the concentration of one of the starting materials, an equilibrium process will be driven toward products. In Eqs. (8.1–8.3), if we increase $[A]^a$, the quantity $[C]^c[D]^d/[B]^b$ must also increase to maintain K, and we will have more product. Similarly, if we can remove one product as it is formed, the equilibrium will be driven toward products. Perhaps either C or D is a solid that crystallizes out of solution or is a gas that can be driven off. As the concentration of this compound drops, the equilibrium will be disturbed and more C or D will be formed.

---

[2] At 25 °C, $2.3RT = 1.364$ kcal/mol. This number is useful to remember because with it one can do calculations easily at odd moments without having to look up the value of R and multiply it by T.

[3] The principle is named for Henri Louis Le Châtelier (1850–1936). The mathematics of this principle had been worked out earlier by the American Josiah Willard Gibbs (1839–1903), who donated the G in Gibbs to ΔG. Le Châtelier knew of Gibbs's contributions and was a great popularizer of Gibbs's work in France.

In a multistep reaction, it often does not matter that an early step involves an unfavorable equilibrium, as long as the overall reaction is exergonic. As long as all barriers can be traversed, the ratio of C and A will be governed by the $\Delta G°$ between C and A. So, A → B is endergonic, but, overall, A → C is exergonic ($\Delta G° < 0$). Figure 8.5 shows such a process for the conversion of A into C via an intermediate B.

As long as some B is present in equilibrium with starting material A, the reaction will be successful. As the equilibrium A ⇌ B is established, B will be converted into the much more stable final product C. As the small amount of B present in equilibrium with A is depleted, the A ⇌ B equilibrium will be reestablished and more B will be formed from A. In the long run, A will be largely converted into C, and as long as there is enough energy to reach the intermediate B, the final mixture of A and C will depend on the $\Delta G°$ between these two compounds.

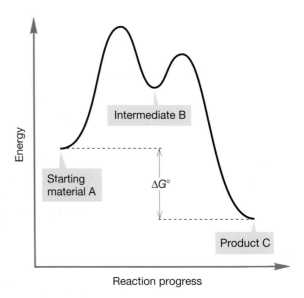

FIGURE 8.5 As long as equilibrium is reached between A and B, it does not matter that formation of B from A is endergonic, compound C will be the major product.

8.3 Entropy in Organic Reactions

The parameter $\Delta G°$ is composed of enthalpy ($\Delta H°$, the change in bond strengths in the reaction) and entropy ($\Delta S°$, the change in freedom of motion in the reaction). The exact relationship is

$$\Delta G° = \Delta H° - T\Delta S° \tag{8.6}$$

where T is the absolute temperature. We can make rather accurate estimates of the $\Delta H°$ term by knowing the bond dissociation energies for a variety of bonds. We know a few of these values already from our discussion of alkanes. Table 8.2 gives several more.

TABLE 8.2 **Some Homolytic Bond Dissociation Energies (BDE)**

Bond	BDE (kcal/mol)	Bond	BDE (kcal/mol)
I—I	36.1	H—CH_2CH_3	101.1
Br—Br	46.1	H—$CH(CH_3)_2$	98.6
Cl—Cl	59.0	H—$C(CH_3)_3$	96.5
F—F	38.0	I—CH_3	57.6
HO—OH	51	Br—CH_3	72.1
$(CH_3)_3CO$—$OC(CH_3)_3$	38	Cl—CH_3	83.7
H—H	104.2	F—CH_3	115
H—I	71.3	HO—CH_3	92.1
H—Br	87.5	CH_3O—CH_3	83.2
H—Cl	103.2	H_2N—CH_3	85.2
H—F	136.3	H_3C—CH_3	90.1
H—OH	118.8	H_3C—CH_2CH_3	89.0
H—OCH_3	104.6	H_2C══CH_2 (π bond only)	66
H—NH_2	107.6	H_2C══CH_2 (total)	174.1
H—CH_3	105.0	HC☰CH (total)	230.7
		H_2C══O (total)	178.8

PROBLEM 8.2 Estimate the $\Delta H°$ for the following reactions:

(a) $CH_3OH + H_2C{=}CH_2 \rightarrow H_3C{-}CH_2OCH_3$

(b) $H_2 + H_2C{=}CH_2 \rightarrow H_3C{-}CH_3$

(c) $Br_2 + H_2C{=}CH_2 \rightarrow BrCH_2{-}CH_2Br$

(d) $HCl + H_2C{=}CH_2 \rightarrow ClCH_2{-}CH_3$

PROBLEM 8.3 The bond dissociation energies for the carbon–hydrogen bonds in Table 8.2 show a steady decline as the breaking carbon–hydrogen bond becomes more substituted. What can you infer from these data about the stability of the neutral species, R•, called "free radicals"?

The entropy term ($\Delta S°$) is related to freedom of movement. The more restricted or "ordered" the product(s) of a reaction, the more negative is the entropy change in the reaction. Since it is the negative of $\Delta S°$ that is related to $\Delta G°$ [Eq. (8.6)], the more negative the entropy change, the higher is the free energy change ($\Delta G°$) of the reaction. Note also that the entropy term is temperature-dependent. Entropy becomes more important at high temperature. Imagine the reaction in Figure 8.6, in which strong carbon–carbon σ bonds are broken and weaker carbon–carbon π bonds are made (Table 8.2). The $\Delta H°$ term will surely favor the left-hand side. At the same time, one molecule is made into two other, smaller molecules, and this change will be favored by the entropy change, $\Delta S°$. This reaction, in which entropy change favors the right-hand side, can be driven to the right by using a high temperature, at which the $T\Delta S°$ term is large. An unfavorable reaction can even become favorable at high temperature where the entropy term becomes more important in determining the value of $\Delta G°$.

FIGURE 8.6 This reaction is a process in which one direction is favored by enthalpy and the other by entropy. Although $\Delta H°$ favors cyclohexene, at very high temperature the formation of two compounds from one can drive this reaction to the right.

WORKED PROBLEM 8.4 Use $\Delta H°$ values from Table 8.2, as well as the value for the BDE of $H_3C{-}CH_2CH{=}CH_2 \rightarrow H_3C\cdot\ \cdot CH_2CH{=}CH_2$ of 80 kcal/mol, to estimate $\Delta H°$ for the reaction in Figure 8.6.

ANSWER Compare the bonds that are breaking with those that are being made. Two carbon–carbon single bonds are broken. As the problem says, each is worth 80 kcal/mol. In addition, a π bond (66 kcal/mol) is broken in the reaction. So, the bonds broken are worth a total of 226 kcal/mol. Three π bonds are made, each worth 66 kcal/mol, for a total of 198 kcal/mol. The left-to-right reaction is endothermic by about 28 kcal/mol, because the bonds made are weaker than those broken.

$$\begin{aligned} 2 \times C{-}C &= +160 \text{ kcal/mol (bonds broken)} \\ C{=}C\ (\pi) &= +66 \text{ kcal/mol (bond broken)} \\ 3 \times C{=}C\ (\pi) &= -198 \text{ kcal/mol (bonds made)} \\ \hline \Delta H° &= +28 \text{ kcal/mol} \end{aligned}$$

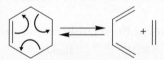

Summary

Now we know something of the factors that affect equilibrium. Enthalpy change, ΔH, is important. Basically, the side of the equilibrium with the stronger bonds will be favored by enthalpy. Also important is the temperature-dependent entropy change, ΔS, term. The two terms may act in concert or oppose each other.

We have said nothing so far of the factors influencing the *rate* of establishing equilibrium.

8.4   Rates of Chemical Reactions

Just because a reaction is exothermic does not mean that it will be a rapid reaction or even that it will proceed at all under normal conditions. First of all, an exothermic reaction can be endergonic if entropy plays an especially important role. Entropy does not raise its ugly head in this way often, but there is another, much more frequently encountered reason why an *exergonic* reaction might not be a *fast* reaction. As we have seen, there are energy barriers to chemical reactions. In principle, a very stable product can be separated from a much less stable starting material by a barrier high enough to stop all reaction (Fig. 8.7).

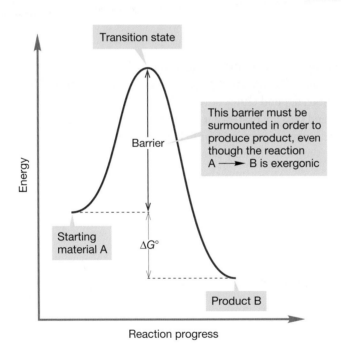

FIGURE 8.7 Rates of reactions are determined not by the $\Delta G°$ between starting material and product, but by the height of the energy barrier separating starting material from product. Some very unstable materials are protected by high barriers and thus can be isolated.

PROBLEM SOLVING

Whenever you see the word "rate" in a problem, or when you see words such as "faster" that talk about rates, you are very likely to have to answer the question by drawing the transition state for the reaction. Remember that the rate of a reaction is determined by $\Delta G^{\ddagger}$, the energy difference between starting material and the transition state, and not by the energy difference between starting material and product. You need a pull-down menu that says, "Think about the transition state" when the word "rate" appears in a problem.

At a given temperature, collections of molecules exist in ranges of energies called Boltzmann distributions. In a **Boltzmann distribution**, the energies of most molecules cluster about an average value. As energy is provided to the molecules, usually through the application of heat, the average energy increases. Although there is always a small number of highly energetic molecules present (the high-energy tail of the Boltzmann distribution), in practice there may not be enough molecules of sufficient energy to make the reaction proceed at an observable rate if the barrier is high enough. The application of energy in the form of heat produces more molecules with sufficient energy to traverse the barrier (Fig. 8.8). A reaction rate approximately doubles for every increase in temperature of 10 °C.

FIGURE 8.8 Boltzmann distributions of molecules at two temperatures. The blue line is the distribution at higher temperature, and the red line is the distribution at lower temperature.

As we already saw in our discussion of the S_N2 and S_N1 reactions, rates of chemical reactions also depend on the concentrations of reactants. The rates of most organic reactions depend either only on the concentration of a single reactant (S_N1, for example) and thus are **first-order reactions** or on the concentrations of two species (S_N2) and are **second-order reactions** (Fig. 8.9). There are examples of higher-order reactions, but they are relatively rare.

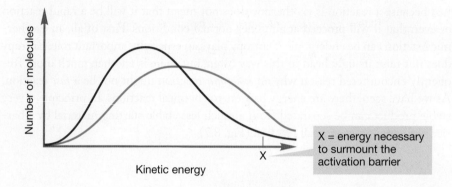

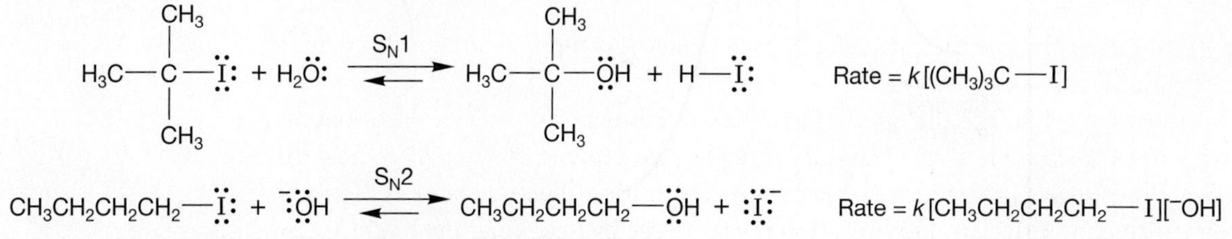

FIGURE 8.9 Rate laws for the S_N1 and S_N2 reactions.

Sometimes a reaction we expect to be second order appears not to be. The S_N2 solvolysis reaction of methyl iodide with solvent (therefore in great excess) ammonia is an example. This reaction is first order in substrate, methyl iodide, because the concentration of one of the reactants, ammonia, does not change appreciably during the reaction. Such reactions are called **pseudo-first-order reactions** (Fig. 8.10).

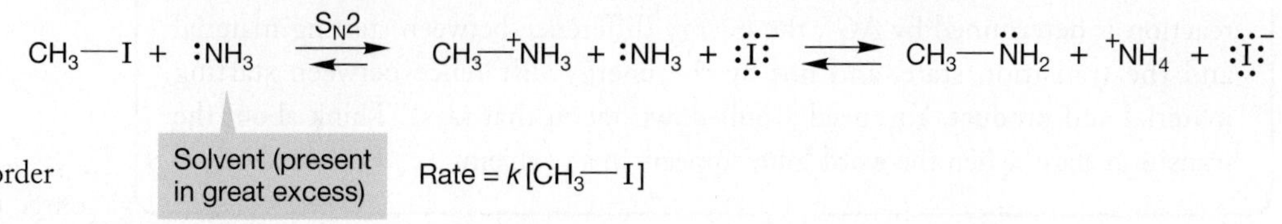

$$CH_3-I + :NH_3 \xrightleftharpoons{S_N2} CH_3-^+NH_3 + :NH_3 + :\ddot{\underset{..}{I}}:^- \rightleftharpoons CH_3-\ddot{N}H_2 + ^+NH_4 + :\ddot{\underset{..}{I}}:^-$$

Solvent (present in great excess)   Rate = $k[CH_3-I]$

FIGURE 8.10 A pseudo-first-order reaction.

8.5 Rate Constant

In the rate laws mentioned so far, there has always been a proportionality constant, the **rate constant (k)**, which is a fundamental property of any given chemical reaction. A rate must have the units of concentration per unit time, often moles per liter per second. For a first-order process such as the S_N1 reaction, the rate constant k must have the units of reciprocal seconds (s^{-1}) (if we use the second as our time unit, Fig. 8.11).

$$\text{Rate} - k[A]$$

$$\text{in units, } \left(\frac{mol}{L}\right) \times (s)^{-1} = k \times \left(\frac{mol}{L}\right)$$

Rate = concentration/time = (rate constant) (concentration)

$$\text{rearranging we have: } \left(\frac{mol}{L}\right) \times (s)^{-1} \times \left(\frac{mol}{L}\right)^{-1} = k$$

$$\text{so, } (s)^{-1} = k \quad \text{or} \quad \frac{1}{s} = k$$

FIGURE 8.11 For a first-order reaction, the rate constant (k) has the units of time^{-1} = s^{-1}.

For a second-order reaction, k must have units of reciprocal moles per liter per second $[(mol/L)^{-1}s^{-1}]$ (Fig. 8.12).

$$\text{Rate} = k[A][B]$$

$$\text{in units, } \left(\frac{mol}{L}\right) \times (s)^{-1} = k \times \left(\frac{mol}{L}\right) \times \left(\frac{mol}{L}\right)$$

Rate = concentration/time = (rate constant) (concentration) (concentration)

$$\text{rearranging we have: } \left(\frac{mol}{L}\right) \times (s)^{-1} \times \left(\frac{mol}{L}\right)^{-1} \times \left(\frac{mol}{L}\right)^{-1} = k$$

$$\text{so, } \left(\frac{mol}{L}\right)^{-1} \times (s)^{-1} = k \quad \text{or} \quad \frac{L}{(mol \times s)} = k$$

FIGURE 8.12 For a second-order reaction, the rate constant (k) has the units of $(mol/L)^{-1}s^{-1}$.

We can determine the value of k by measuring how the reaction rate varies with the concentration(s) of the reactants.

It is easy to confuse the *rate* of a reaction, which is simply how fast the concentrations of the reactants are changing, with the *rate constant*, which tells us how the rate will change as a function of reactant concentration. The rate of a reaction will vary with concentration. This variation is easy to see by imagining the limiting case in which one reactant is used up. When its concentration goes to zero, the reaction rate must also be zero. On the other hand, the rate constant is an intrinsic property of the reaction. It will vary with temperature, pressure, and solvent, but does not depend on the concentrations of reactants.

Only when the concentrations of all reactants are 1 M are the rate and the rate constant the same. The rate constant is thus a measure of the rate under standard conditions (Fig. 8.13).

$$\text{Rate} = k[A][B][C][D] \dots [X]$$

if all [ ] = 1 M

$$\text{Rate} = k[1][1][1][1] \dots [1]$$

and therefore, rate = k

FIGURE 8.13 If the concentrations of the reactants are all 1 M, the rate constant and the rate are the same.

8.6 Energy Barriers in Chemical Reactions: The Transition State and Activation Energy

We have already mentioned that even exothermic reactions may be very slow, indeed immeasurably slow. And it's a good thing. Combustion is a very exothermic chemical reaction, and yet we and our surroundings exist more or less tranquilly in air without bursting into flame. Somehow we are protected by an energy barrier. For example, methane and oxygen do not react unless energy, called activation energy, is supplied. If we are considering free energy, the activation energy is $\Delta G^{\ddagger}$; if we are speaking of enthalpy, the activation energy is $\Delta H^{\ddagger}$. The double dagger (‡) is always used when we are referring to activation parameters involving the energy of a transition state. A match will suffice to provide the energy in our methane plus oxygen example. This highly exothermic process (p. 68) then provides its own energy as very stable molecules are produced from less stable ones (Fig. 8.14).

CONVENTION ALERT

FIGURE 8.14 An exergonic reaction can provide the energy necessary to pass over an activation barrier. Once such a reaction is started, it continues until one reactant is used up.

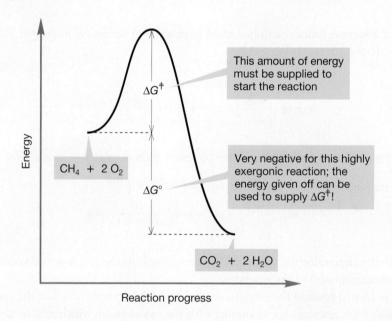

This amount of energy must be supplied to start the reaction

Very negative for this highly exergonic reaction; the energy given off can be used to supply $\Delta G^{\ddagger}$!

$\Delta G^{\ddagger}$

ΔG°

$CH_4 + 2 O_2$

$CO_2 + 2 H_2O$

Energy

Reaction progress

PROBLEM 8.5 Draw an arrow to show the activation energy for the reverse of the reaction in Figure 8.14, the formation of methane and oxygen from carbon dioxide and water.

We are protected from spontaneous combustion in the same way we are protected from hungry bears and tigers during a visit to the zoo. At the zoo the barrier is tangible and operates in a clear way to ensure our survival. Activation barriers are less visible but no less necessary for the continuance of our lives. In this important section, we will discuss these energy barriers to chemical reactions.

In a process such as the S_N1 reaction, a bond must be broken. Consider the ionization of *tert*-butyl iodide. The first step in this reaction is the ionization of the substrate, and surely this step will involve an energy cost in breaking the bond from carbon to the leaving group. The enthalpy change $\Delta H°$ (and $\Delta G°$) is positive for this reaction (Fig. 8.15).

WEB 3D   WEB 3D

$$(CH_3)_3C \overset{\curvearrowright}{—} \ddot{\underset{..}{I}}: \quad \underset{\longleftarrow}{\overset{S_N1}{\rightleftharpoons}} \quad (CH_3)_3C^+ + :\ddot{\underset{..}{I}}:^-$$

FIGURE 8.15 The enthalpy change $\Delta H°$ (and $\Delta G°$) is positive for this endothermic reaction.

This energy cost is partially offset by an increase in entropy as we form two particles out of one. But we have to be careful when we estimate entropy changes because there are invisible participants in most reactions, including this one. These invisible participants are the solvent molecules. They will have to be disordered and reordered in any ionic reaction in solution, and the net change in entropy may not be at all easy to estimate.

WORKED PROBLEM 8.6 (a) Construct an Energy versus Reaction progress diagram for the reaction of *tert*-butyl iodide and water (see Fig. 8.2), showing clearly the activation energies and transition states for all steps. A discussion of this reaction is coming, but try to anticipate it here. (b) Show the activation energies for the reverse reactions involved in the formation of *tert*-butyl iodide from *tert*-butyl alcohol and hydrogen iodide (HI).

ANSWER

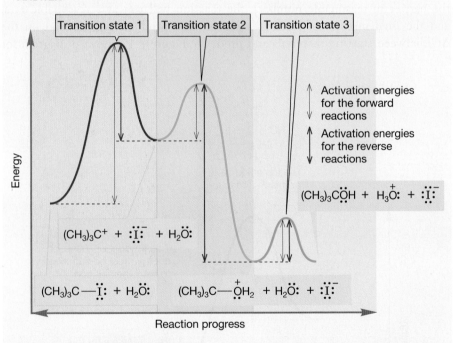

PROBLEM 8.7 Show that at equilibrium the equilibrium constant (K) is given by the ratio of the rate constants (k) for the forward and reverse reactions. *Hint*: At equilibrium the rates of the forward and reverse reactions are equal.

The situation is a little different for the S_N2 reaction. Here we must also break the bond from carbon to the leaving group, but this energy cost is partially offset by the formation of the bond from carbon to the entering nucleophile. We might imagine that there must be a substantial entropy cost in this reaction as we must order the reactants in a special way for the reaction to proceed. The S_N2 reaction can occur only if the nucleophile attacks from the rear of the departing leaving group, and this requirement demands a specific arrangement of the molecules and ions involved in the reaction (Fig. 8.16).

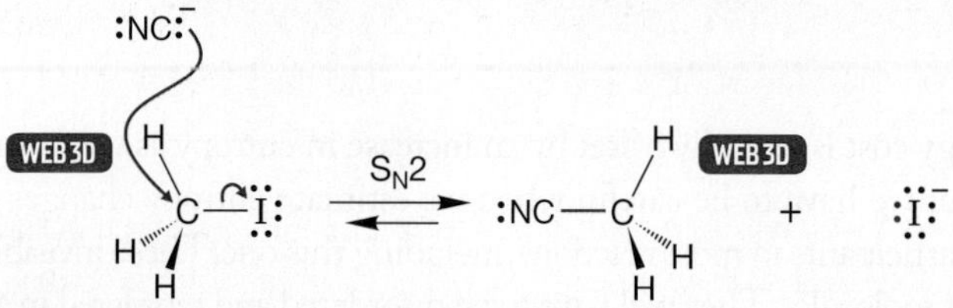

FIGURE 8.16 The transition state for the S_N2 reaction has severe ordering requirements. Attack of the nucleophile must be from the rear for the reaction to succeed.

We can follow the course of both these reactions in diagrams that plot Energy versus Reaction progress. We usually don't specify reaction progress in detail because there are many possible measures of this quantity, and for this kind of qualitative discussion it doesn't much matter what we pick. In the S_N2 reaction, for example, we might choose to plot energy versus the length of the bond between carbon and the leaving group (which increases in the reaction), or the length of the bond between carbon and the nucleophile, which decreases through much of the process. The overall picture is the same.

If we have enough energy for molecules to reach the transition state, to cross the barrier in reasonable numbers, the reaction will run at a reasonable rate, and the final mixture of starting materials and products will depend on the $\Delta G°$ between starting materials and products. Figure 8.17 shows a diagram for

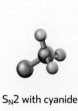

S_N2 with cyanide

FIGURE 8.17 An energy diagram for the S_N2 displacement of iodide by cyanide (and its reverse reaction, the much less favorable displacement of the strong nucleophile cyanide by iodide). The activation energy for the exergonic displacement of iodide is shown.

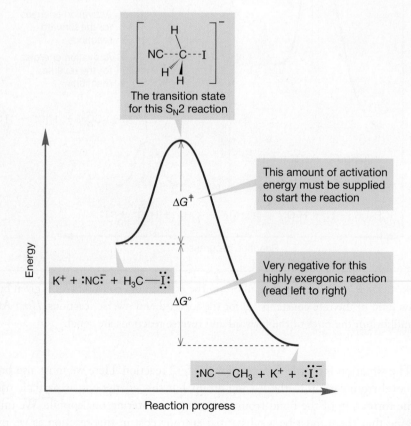

The transition state for this S_N2 reaction

This amount of activation energy must be supplied to start the reaction

Very negative for this highly exergonic reaction (read left to right)

the exothermic formation of methyl cyanide (CH_3CN) and potassium iodide (KI) from methyl iodide (CH_3I) and potassium cyanide (KCN). The activation energy is the height of the transition state above the starting material. As there is enough energy to pass over the transition state in both directions, the ratio of product and starting material will depend on $\Delta G°$, the difference in free energy between starting material and product. The reaction shown in Figure 8.17 has a negative $\Delta G°$, and will be successful in producing product.

In Figure 8.18, this same reaction is considered from the reverse point of view: the endothermic formation of methyl iodide from methyl cyanide and iodide ion. The activation energy $\Delta G^{\ddagger}$ is much higher than that for the forward reaction, and $\Delta G°$ is positive for the endergonic formation of methyl iodide. The transition state is exactly the same as that for the forward reaction. This reaction will not be very successful. It not only requires overcoming a high activation energy but is "fighting" an unfavorable equilibrium because $\Delta G°$ is positive.

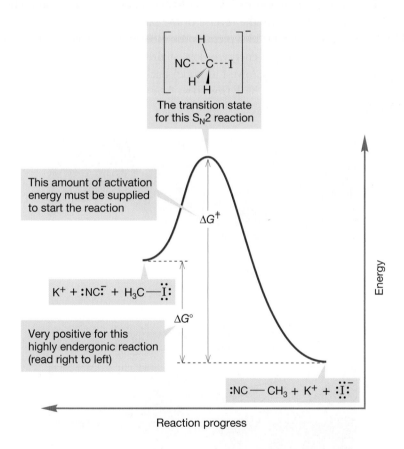

FIGURE 8.18 This energy diagram for the same reaction shown in Figure 8.17 emphasizes the reversal. Here the activation energy for displacement of cyanide by iodide is shown.

Of course, this separation of the two reactions is absolutely artificial—we are really describing a single equilibrium between methyl iodide and cyanide ion on one side and methyl cyanide and iodide ion on the other (Fig. 8.19).

$$K^+ + :NC^- + H_3C-\ddot{\underset{\cdot\cdot}{I}}: \quad \rightleftharpoons \quad :NC-CH_3 + K^+ + :\ddot{\underset{\cdot\cdot}{I}}:^-$$

FIGURE 8.19 The competition between cyanide and iodide for a Lewis acidic carbon.

The exergonic formation of methyl cyanide and the endergonic formation of methyl iodide are one and the same reaction. The reactions share exactly the same transition state; only the ease of reaching it is different ($\Delta G^{\ddagger}_{\text{forward}} < \Delta G^{\ddagger}_{\text{reverse}}$). The energy difference between the starting material and the product (ΔG°) is the difference between the two activation energies.

The rates of the forward and reverse reactions are determined by the magnitudes of the activation energies for the two halves of the reaction. They are intimately related to the equilibrium constant for the reaction K, determined by ΔG° [Eq. (8.7)].

$$\Delta G^{\ddagger}_{\text{reverse}} - \Delta G^{\ddagger}_{\text{forward}} = \Delta G^{\circ} = -RT\ln K \qquad (8.7)$$

The principle of **microscopic reversibility** says that if we know the mechanism of a forward reaction, we know the mechanism of the reverse reaction. One diagram suffices to tell all about both forward and reverse reactions. Figures 8.17 and 8.18 are in reality a single diagram describing both reactions, in this case one "forward" and exergonic, the other "reverse" and endergonic. There is no lower-energy route from methyl cyanide and iodide ion back to methyl iodide and cyanide ion. The lowest-energy path back is exactly the reverse of the path forward.

The classic analogy for this idea pictures the molecules as crossing a high mountain pass. The best path (lowest energy, or lowest altitude) from village A to village B is the best path (lowest energy, lowest altitude) back from village B to village A. If there were a lower-energy route back from B to A, it would have been the better route from A to B as well (Fig. 8.20).

FIGURE 8.20 Molecules, unlike explorers, take the best (lowest-energy) path. You might take the red path from A to B and the orange path back, but a molecule wouldn't.

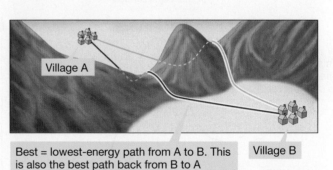
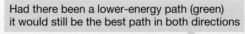

The Energy versus Reaction progress diagram for the S_N1 reaction is more complicated than that for the S_N2 reaction, because there are more steps than in the S_N2 reaction. We dealt with the description of the S_N1 reaction as a sequence of equilibria earlier (Fig. 8.2). Let's again take the reaction of *tert*-butyl iodide with solvent water as an example. The first step is the formation of an intermediate car-bocation. We expect this first reaction to be quite endothermic as a carbon–leav-ing group bond must be broken and ions must be formed. The high-energy point in this reaction, transition state 1, occurs as the carbon–iodine bond stretches (Fig. 8.21).

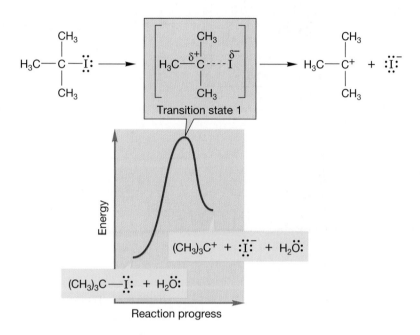

FIGURE 8.21 The first step in the S_N1 reaction of *tert*-butyl iodide in water. The formation of the cation is highly endothermic.

The second step involves the reaction of the nucleophilic solvent water with the carbocation, and is highly exothermic because a bond is made and none is broken. There is a second barrier, and therefore another transition state (2) for this second step of the reaction (Fig. 8.22).

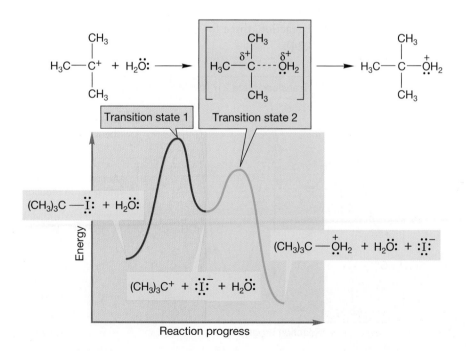

FIGURE 8.22 By contrast, the second step in this S_N1 reaction, the capture of the cation by water, is highly exothermic. Transition state 2 is lower than transition state 1; therefore, capture of the carbocation by water is favored.

The third and final step in this particular S_N1 reaction is removal of a proton from the oxonium ion by a base to give the ultimate product, *tert*-butyl alcohol. So there is a third barrier and a third transition state (Fig. 8.23).

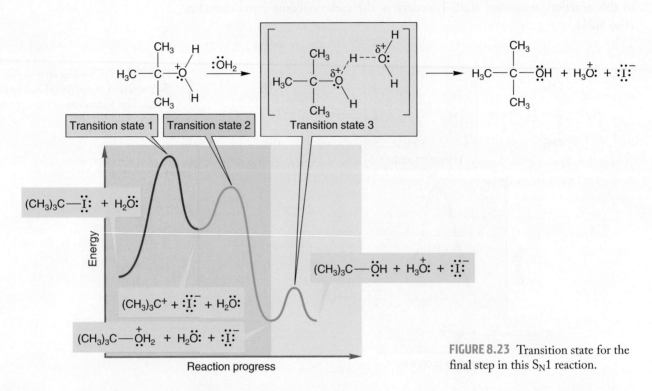

FIGURE 8.23 Transition state for the final step in this S_N1 reaction.

PROBLEM 8.8 Draw an Energy versus Reaction progress diagram for the S_N1 reaction of *tert*-butyl iodide in H_2O/KOH containing an additional nucleophile, $Nu:^-$. Note that the major product is not necessarily *tert*-butyl alcohol (Fig. 7.60, p. 295).

As you saw in Worked Problem 8.6, there are six activation energies in the completed diagram for this S_N1 reaction, three for the forward reactions and three for the reverse reactions (Fig. 8.24).

FIGURE 8.24 The three black double-headed arrows represent activation energies for the forward reactions and the three red double-headed arrows represent activation energies for the reverse reactions.

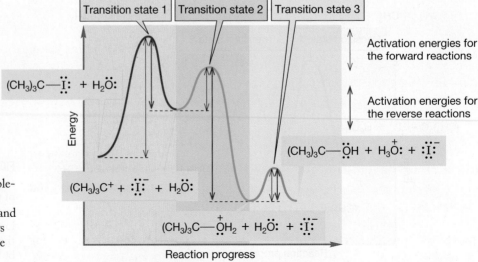

Let us now concentrate on these energy barriers separating the various intermediates. The barriers are made up of the same quantities, enthalpy and entropy, that define the energy of any chemical species. Recall that $\Delta G° = \Delta H° - T\Delta S°$ [Eq. (8.6)]. Transition state theory assumes that we can treat a transition state as if it were a real molecule occupying a potential energy minimum. It is important to emphasize that it is *not*, however. It occupies an energy maximum and can never be bottled and examined as real molecules can. Nonetheless it has an energy, as indeed does every other point in the energy diagrams on the preceding pages. The energy of the transition state, and thus the height of the energy barrier in the reaction, can be calculated using Eq. (8.8).

$$\Delta G^\ddagger = \Delta H^\ddagger - T\Delta S^\ddagger \tag{8.8}$$

Remember: The double daggers (‡) simply indicate that we are referring to a transition state, not to an energy minimum. The energy barrier, $\Delta G^\ddagger$, is the Gibbs free energy of activation, $\Delta H^\ddagger$ is the enthalpy of activation, and $\Delta S^\ddagger$ is the entropy of activation.

Clearly, the rate constant for any reaction will depend on the height of the barrier, and thus on $\Delta G^\ddagger$. The theoretical study of reaction rates gives us the exact equation relating the rate constant k and $\Delta G^\ddagger$ [Eq. (8.9)].

$$k = v^\dagger\, e^{-\Delta G^\ddagger/RT} \qquad \text{or} \qquad k = v^\ddagger\, e^{-\Delta H^\ddagger/RT}\, e^{\Delta S^\dagger/R} \tag{8.9}$$

It is worth dissecting this equation a bit further. The quantity $v^\ddagger$ is a frequency—its units are reciprocal seconds (s^{-1}) in this case. In qualitative terms, v is the frequency with which molecules with enough energy to cross the barrier actually do so. So, the rate constant k is equal to $e^{-\Delta G^\ddagger/RT}$ (the fraction of molecules with enough energy to cross the barrier) times $v^\ddagger$ (the frequency with which such molecules actually do cross the barrier).

Summary

Reaction rates are determined not by the relative energies of the starting material and product, but by the height of the transition state separating them. Even though the transition state is an energy maximum, a transient species that cannot be isolated, its energy is determined by enthalpy and entropy factors, just as is the energy of any stable compound.

8.7 Reaction Mechanism

Throughout this book we will frequently be concerned with the determination of reaction mechanisms. A mechanism is nothing less than the description of the structures and energies of the starting materials and products of a reaction, as well as of any reaction intermediates. In addition, all of the transition states (energy maxima) separating the stable molecules lying in energy minima must be described. It is relatively easy to deal with energy minima—we can often isolate the molecules themselves, take their spectra, and measure their properties. As we can by definition never isolate a transition state, it is exceptionally difficult to make a direct measurement on it. Yet if we are to have a good picture of a reaction, we must arrive at good descriptions of the transition states involved.

We can usually measure the rate law, which tells us the number and kinds of molecules involved in the transition state. The rate law does not tell us anything about the orientation of the molecules in the transition state, however. Usually, stereochemical experiments do that. Recall the S_N2 reaction (p. 268), in which inversion of configuration is observed, thus specifying the direction of approach of the entering nucleophile in the transition state (Fig. 8.25).

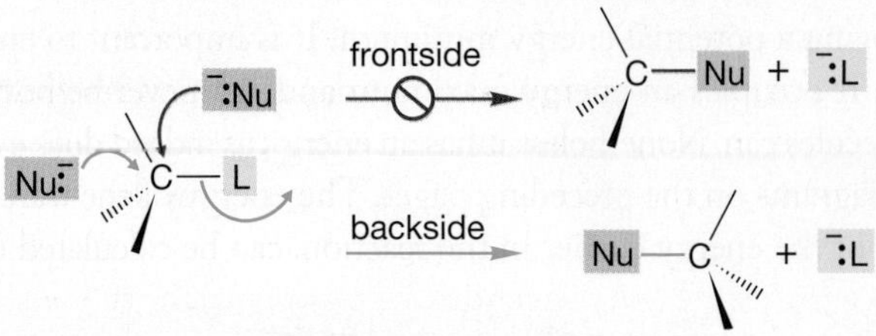

FIGURE 8.25 In the S_N2 reaction, kinetics cannot tell frontside from backside displacement. It takes a stereochemical experiment to do that.

This observation of inversion in the S_N2 reaction tells us that the incoming nucleophile must approach from the rear of the departing leaving group. The rate law, rate = k[Nu:⁻][R—L], does not tell us that. It shows only that both the nucleophile and the substrate are involved in the transition state. As we work through the important reactions of organic chemistry, we will see many similar experiments designed to determine the stereochemistry of the reaction. The energy of the transition state can be measured if good quantitative kinetic experiments can be performed. If the rate constant for the reaction is determined at a variety of temperatures, the height of the transition state can be calculated. But we can do this only for the step that has the highest-energy transition state. In a multistep reaction, we can only measure the rate for the slowest step in the sequence.

This slowest step is the rate-determining step, sometimes called the rate-limiting step. It is always the step in the reaction with the highest-energy transition state. We cannot measure the rates of faster steps in the reaction. In the S_N1 reaction, for example, the slowest step is the ionization of the substrate to a carbocation. This ion may then be captured by a host of nucleophiles in following faster steps. There will be a different product for every capturing nucleophile (p. 295; Fig. 8.26).

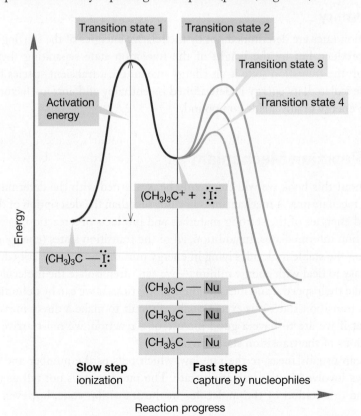

FIGURE 8.26 In this S_N1 reaction, the rate-determining step is the ionization of the starting iodide to give the *tert*-butyl cation. The faster, product-determining steps follow, but their rates cannot be measured kinetically (unless the intermediate carbocation can be isolated and investigated independently).

If we determine the rate of this process by measuring the disappearance of *tert*-butyl iodide over time, we can ultimately find the height of transition state 1. The rates of the faster captures of the carbocation *do not affect the rate of ionization of tert-butyl iodide*. The carbocation is gobbled up by the nucleophiles at a rate faster than that at which it is formed. In the limit, each molecule of the *tert*-butyl cation is captured before another is produced from *tert*-butyl iodide. The rate of formation of product depends on how fast the *tert*-butyl cation is produced in the slow, rate-determining step but not on the rate of the fast capture by a nucleophile.

Imagine a cage of especially ravenous rats into which we are slowly dropping incredibly delicious food pellets. At first, the pellets are gobbled up by the hungry rats at a rate far faster than they are delivered to the cage. The slowest step in this reaction, the step that determines the rate of disappearance of the food, is the rate at which the pellets are delivered to the cage. This process is analogous to a reaction in which the first step (food dropping into the cage) is slower than the second step (the rats eating the food). Later, however, a different situation obtains. Even these rats can become satiated, and the food pellets accumulate in the cage as the no-longer-starving rodents largely ignore them, only taking a nibble now and then. The second step is now slower than the first step. Now the step that determines the rate of disappearance of the food is the rate at which the rats eat the pellets. This second scenario has a different rate-determining step than the first scenario.

8.8   The Hammond Postulate: Thermodynamics versus Kinetics

We are used to saying, "Product Y is more stable than product X and therefore is formed faster." This statement has a comfortable sound to it and intuitively makes sense. But it need not be true! The first half of the statement describes **thermodynamics**, the relationship of the energies of product X and product Y. The second half speaks of the *rates* of formation of X and Y, which is a statement about **kinetics**. Despite the good intuitive feel of the original statement, there need be no direct connection between thermodynamic stabilities and kinetics.

Figure 8.27 shows the relative energies of two products, X and Y, being formed from a starting material. One might expect that it is the $\Delta G°$ between the two

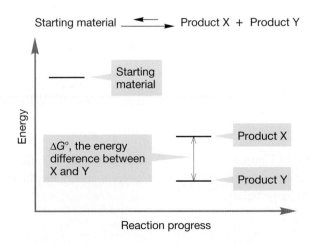

FIGURE 8.27 The energy difference between X and Y does not necessarily affect their rates of formation.

products X and Y that determines the amount of the two compounds formed, but this notion is true only if X and Y are in equilibrium via the starting material (Fig. 8.28).

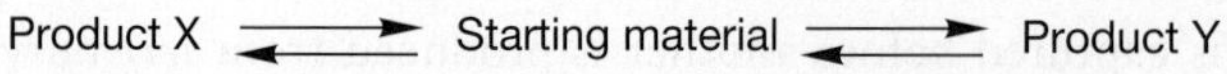

FIGURE 8.28 Only if X and Y are in equilibrium does $\Delta G°$ determine their ratio.

$$\text{Product X} \rightleftharpoons \text{Starting material} \rightleftharpoons \text{Product Y}$$

It is conceivable that the products are *not* in equilibrium with each other. For instance, there might not be enough energy for the products to go back to the starting material. Under these conditions, the amounts of X and Y are determined not by their relative energies but only by the heights of the transition states leading to them. Figure 8.29a shows a conventional picture in which the transition state leading to the more stable product, Y, is lower in energy than the transition state leading to the higher-energy product, X. Under such conditions, the major product will be the more stable compound, Y. We have seen many examples of reactions like this.

The scenario of Figure 8.29a is not the only possibility. It could be the case that the transition state leading to the more stable compound, Y, is actually higher in energy than the transition state leading to the less stable product, X (Fig. 8.29b).

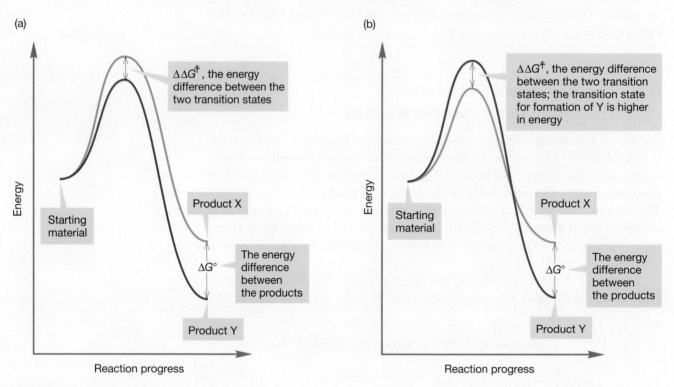

FIGURE 8.29 (a) The usual picture is this: The higher transition state leads to the less stable product. The lower transition state leads to the more stable product. (b) The situation could be otherwise; the higher transition state could lead to the lower-energy product, and the lower-energy transition state to the higher-energy product. This second scenario is rare, but it does happen.

PROBLEM 8.9 In Figure 8.29a what is the activation energy for the reaction of X going to starting material? What is the activation energy for the reaction of Y going to starting material?

This inversion will not matter if X and Y equilibrate via the starting material. In such a case, the greater stability of Y will win out eventually, and the ratio of X and Y depends only on the $\Delta G°$ between them. Such a reaction is said to be under **thermodynamic control**. *But what if the two products do not equilibrate?* What if there is enough energy available so that the forward reactions are possible but the backward reactions from X or Y to starting material are not? In the exothermic processes shown, the forward reaction is easier—has a lower activation barrier—than the backward reaction (Fig. 8.30).

Under these circumstances, the relative amounts of X and Y depend only on the relative heights of the transition states leading to the two products. The product for which the transition state is lower is the product that is formed in greater amount. Such a reaction is said to be under **kinetic control**.

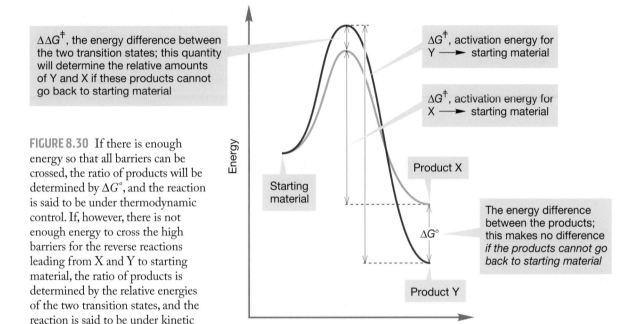

$\Delta\Delta G^{\ddagger}$, the energy difference between the two transition states; this quantity will determine the relative amounts of Y and X if these products cannot go back to starting material

$\Delta G^{\ddagger}$, activation energy for Y ⟶ starting material

$\Delta G^{\ddagger}$, activation energy for X ⟶ starting material

Product X

Starting material

$\Delta G°$

The energy difference between the products; this makes no difference *if the products cannot go back to starting material*

Product Y

Reaction progress

Energy

FIGURE 8.30 If there is enough energy so that all barriers can be crossed, the ratio of products will be determined by $\Delta G°$, and the reaction is said to be under thermodynamic control. If, however, there is not enough energy to cross the high barriers for the reverse reactions leading from X and Y to starting material, the ratio of products is determined by the relative energies of the two transition states, and the reaction is said to be under kinetic control.

There is no *necessary* connection between thermodynamics and kinetics. In our original statement, "Product Y is more stable than product X and *therefore* is formed faster," we are equating thermodynamics and kinetics. We tacitly assume that it will always be the case that the product of greater stability (lower energy) will be formed at a faster rate via the lower-energy (more stable) transition state (Fig. 8.29a).

In principle, it could be that the transition state for the reaction leading to the more stable product is actually higher in energy than the transition state leading to the less stable product. Such a situation can lead to the preferred formation of the less stable product, if the reaction is under kinetic control (Fig. 8.29b).

Although there are a few reactions for which the scenario of Figure 8.29b applies, fortunately they are rare because there are factors that operate to preserve a parallelism between the energies of products and the transition states leading to them. The rest of this section will try to show you why the parallelism between product energy and transition state energy exists for many reactions. Consider, for a start, Worked Problem 8.10.

WORKED PROBLEM 8.10 As we saw in Chapter 3, alkenes can be protonated to give carbocations. Unsymmetrical alkenes, such as 2-methyl-2-butene, could be protonated at either end of the double bond to give two different carbocations. Predict the direction of protonation of 2-methyl-2-butene and explain your answer.

ANSWER The choice is between the formation of secondary carbocation **B** and tertiary carbocation **A**. It is surely tempting to predict that the more stable tertiary carbocation will be formed faster than the less stable secondary carbocation. However, this comfortable reasoning equates thermodynamics (more stable) with kinetics (will be formed faster), and this equation is not necessarily valid. Read on.

Carbocation **A**   2-Methyl-2-butene   Carbocation **B**

Before we accept the comfortable reasoning laid out in the answer to Worked Problem 8.10, we had better think it through. Look first at the Energy versus Reaction progress diagram (Fig. 8.31).

FIGURE 8.31 The more substituted cation is at lower energy than the less substituted cation. Note also that the transition state for formation of the tertiary cation is lower than the transition state for formation of the secondary cation. Why?

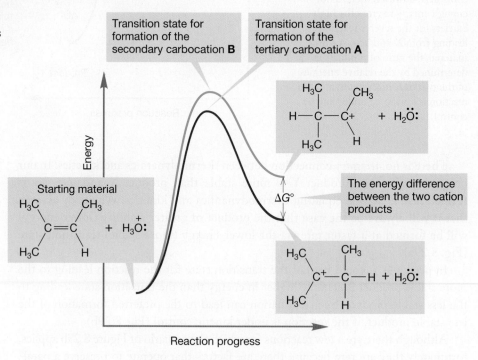

The tertiary cation is more substituted than the secondary cation and thus is more stable by virtue of hyperconjugative overlap of a sigma bond with the empty $2p$ orbital (p. 292). Substitution stabilizes the positive charge on carbon in these species.

Therefore, a small amount of energy difference ($\Delta G°$) has a great influence on the equilibrium constant K.

The ratios of products formed in a reaction depend either on the energy difference between the products $\Delta G°$ (thermodynamic control) or on the relative heights of the transition states leading to products (kinetic control).

Differences in $G°$, the Gibbs free energy, whether they are $\Delta G°$, the difference in energy between starting material and product, or $\Delta G^{\ddagger}$, the difference in energy between starting material and the transition state (activation energy), are made up of entropy ($\Delta S°$) and enthalpy ($\Delta H°$) terms ($\Delta G° = \Delta H° - T\Delta S°$ or $\Delta G^{\ddagger} = \Delta H^{\ddagger} - T\Delta S^{\ddagger}$).

The rate of a reaction depends on temperature, the concentration of the reactant(s), and the rate constant k for the reaction. The rate constant is a property of the reaction and depends on temperature, pressure, and solvent, but not on the concentrations of the reactants.

The rate-determining step of a reaction is the step with the highest-energy transition state. It may or may not be the same as the product-determining step. For example, in the S_N2 reaction it is the same, but in the S_N1 reaction it is not.

The principle of microscopic reversibility says that the lowest-energy path in one direction of an equilibrium reaction is the lowest-energy path for the reverse reaction.

The Hammond postulate says that the transition state for an endothermic reaction resembles the product, or, equivalently, that the transition state for an exothermic reaction resembles the starting material.

Key Terms

Boltzmann distribution (p. 340)
endergonic (p. 335)
entropy change ($\Delta S°$) (p. 335)
equilibrium constant (K) (p. 334)
exergonic (p. 335)
first-order reaction (p. 340)

Gibbs free energy change ($\Delta G°$) (p. 335)
Hammond postulate (p. 355)
kinetic control (p. 353)
kinetics (p. 351)
Le Châtelier's principle (p. 336)
microscopic reversibility (p. 346)

pseudo-first-order reaction (p. 340)
rate constant (k) (p. 341)
second-order reaction (p. 340)
thermodynamic control (p. 353)
thermodynamics (p. 351)

Common Errors

The most common problem experienced with the material in this chapter is the confusion between thermodynamics and kinetics. Rates of reactions (kinetics) are determined by the relative heights of the available transition states leading to products.

If there is insufficient energy available for the products to re-form starting materials, it will be the relative heights of the transition states that determine the mixture of products, no matter what the energy differences among those products (Fig. 8.36).

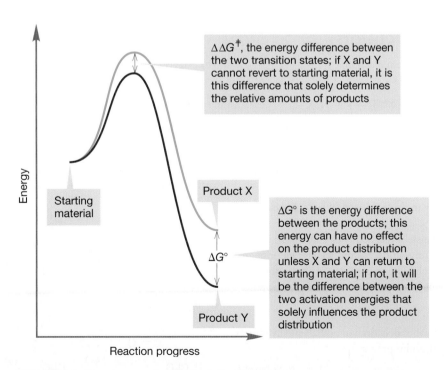

$\Delta\Delta G^{\ddagger}$, the energy difference between the two transition states; if X and Y cannot revert to starting material, it is this difference that solely determines the relative amounts of products

$\Delta G°$ is the energy difference between the products; this energy can have no effect on the product distribution unless X and Y can return to starting material; if not, it will be the difference between the two activation energies that solely influences the product distribution

FIGURE 8.36 Kinetic control of product distribution.

On the other hand, if equilibrium is established, if the products can revert to starting materials, it will be the relative energies of the product molecules that determine how much of each is formed (Fig. 8.37).

Remember also to be careful to distinguish between the *rate* of a reaction and the *rate constant* for that reaction. The rate depends on the concentration(s) of the reacting molecules; the rate constant does not, and is a fundamental property of the reaction.

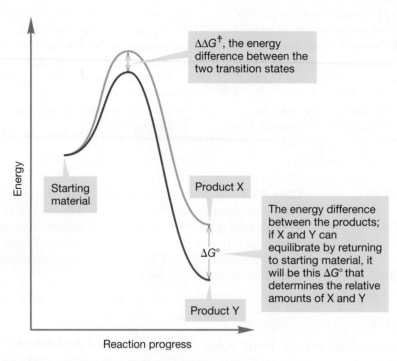

FIGURE 8.37 Thermodynamic control of product distribution.

8.11 Additional Problems

PROBLEM 8.11 The Finkelstein reaction:

$$R{-}X + NaI \rightleftarrows R{-}I + NaX$$

$$X = Cl \text{ or } Br$$

is often used to prepare alkyl iodides from alkyl chlorides or bromides through an S_N2 reaction.

(a) Suggest a simple way to drive this equilibrium toward the desired iodide.

(b) In practice, the Finkelstein reaction is usually run in acetone solvent, because sodium iodide, but not sodium chloride or sodium bromide, is soluble in acetone. Explain carefully how this difference in solubilities drives the equilibrium toward the desired iodide.

PROBLEM 8.12 The alcohol dehydration and alkene hydration shown below is an equilibrium process.

(a) Suggest experimental conditions that will favor cyclohexene.

(b) Suggest experimental conditions that will favor cyclohexanol.

PROBLEM 8.13 Calculate the energy differences between the carbonyl (C=O) compounds and their hydrates at 25 °C.

(a)

$$K = 1.4 \times 10^{-3}$$

(b)

$$K = 2.8 \times 10^{4}$$

PROBLEM 8.14 Under certain highly acidic conditions, the three bicyclooctanes shown below can be isomerized. Given the ratios at equilibrium (3.66:32.95:63.35), calculate the relative energies of the three isomers at 25 °C.

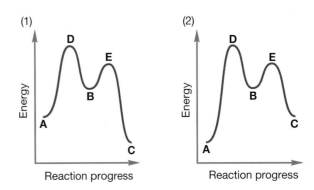

Bicyclo[2.2.2]octane
(3.66%)

Bicyclo[3.3.0]octane
(32.95%)

AlBr₃ | 25 °C

Bicyclo[3.2.1]octane
(63.35%)

PROBLEM 8.15 It is possible to equilibrate the following two ethers. Compound **B** is favored over **A** by 1.51 kcal/mol. What is the equilibrium constant for this interconversion at 25 °C?

CH₃O

A

Fe(CO)₅

CH₃O

B

PROBLEM 8.16 Use the BDE of Table 8.2 (p. 337) to estimate $\Delta H°$ for the following reactions. Reaction mechanisms are not necessary!

(a)

F—F + H₂C=CH₂ ⇌ H₂C—CH₂
                                           | |
                                           F   F

(b)

I—I + H₂C=CH₂ ⇌ H₂C—CH₂
                                        | |
                                        I   I

(c)

I· + (CH₃)₂C=CH₂ ⇌ (CH₃)₂C—CH₂
                                               |
                                               I

(d)

(CH₃)₂C—CH₂ + H—Cl ⇌ (CH₃)₂C—CH₂ + Cl·
           | | |
           Cl   H   Cl

PROBLEM 8.17 Use the data in Table 8.2 (p. 337) to design an endothermic reaction. Don't worry about mechanism! This problem does not ask you to design a *reasonable* reaction—it merely asks you to find reactions that would be endothermic if they occurred.

PROBLEM 8.18 For the equilibrium A + B ⇌ C, calculate:
(a) $\Delta G°$ at 25 and 200 °C given $\Delta H° = -14.0$ kcal/mol and $\Delta S° = -15.8$ cal/deg·mol.
(b) $\Delta G°$ at 25 and 200 °C given $\Delta H° = -6.0$ kcal/mol and $\Delta S° = -15.8$ cal/deg·mol.

PROBLEM 8.19 Draw the complete reaction mechanism for the E1 and S_N1 reactions of *tert*-butyl alcohol with a catalytic amount of H₂SO₄ and solvent ethanol.

PROBLEM 8.20 Because the E1 and S_N1 reactions are reversible, the selection between the two can often be driven by increasing the temperature. In this fashion the more stable product is favored. Use data in Table 8.2 to choose the more stable products in the reaction of *tert*-butyl alcohol in catalytic H₂SO₄ and ethanol.

PROBLEM 8.21 Draw Energy versus Reaction progress diagrams for E1 and E2 reactions that are overall (a) exothermic, (b) endothermic. Use a tertiary iodide as a sample substrate molecule.

PROBLEM 8.22 For the following two reactions, described by the two Energy versus Reaction progress diagrams, (1) and (2):

(1)

Energy vs Reaction progress

(2)

Energy vs Reaction progress

(a) In each case, which compounds will be present at the end of the reaction?
(b) In each case, which compound will be present in the largest amount? Which will be present in the smallest amount?
(c) What is the rate-determining step for the left-to-right reaction in each case?
(d) Use the diagrams to point out in each case the activation energy for the following reactions:

$$A \rightarrow B, A \rightarrow C, C \rightarrow B, \text{ and } C \rightarrow A$$

(e) In each case, what is the rate-determining step for the reaction $A \rightarrow C$?

PROBLEM 8.23   Explain in painstaking detail why one is justified in saying, "The reaction of 2-methyl-2-butene with HCl leads to the tertiary chloride because a tertiary carbocation is more stable than a secondary carbocation." You might start with the construction of Energy versus Reaction progress diagrams and good drawings for the transition states.

PROBLEM 8.24   Find the errors in each diagram.

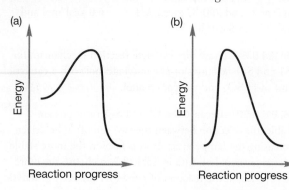

Use Organic Reaction Animations (ORA) to answer the following questions:

PROBLEM 8.25   Select the animation titled "Unimolecular elimination: E1" and observe the energy diagram. Do you suppose this reaction is reversible? Use the energy diagram to answer this question. Use the nature of the products to answer the same question.

PROBLEM 8.26 Compare the E1 reaction with the "Bimolecular elimination: E2" in terms of the energy diagram. Do you suppose the E2 reaction is more or less reversible than the E1 reaction? Use the energy diagram to answer this question. Use the nature of the products to answer the same question.

PROBLEM 8.27   If you needed to make an alkene, would you prefer to use the E1 reaction or the E2 reaction? Why?

PROBLEM 8.28   Observe the "Hofmann elimination" reaction. Draw the energy diagram shown for the reaction. Draw a new reaction curve for the same reaction run in a more polar solvent. Draw the curve for a reaction run in a less polar solvent. Which solvent would likely give the faster Hofmann elimination?

Additions to Alkenes 1

9.1 Preview

9.2 Mechanism of the Addition of Hydrogen Halides to Alkenes

9.3 Effects of Resonance on Regiochemistry

9.4 Brief Review of Resonance

9.5 Resonance and the Stability of Carbocations

9.6 Inductive Effects on Addition Reactions

9.7 HX Addition Reactions: Hydration

9.8 Dimerization and Polymerization of Alkenes

9.9 Rearrangements during HX Addition to Alkenes

9.10 Hydroboration

9.11 Hydroboration in Synthesis: Alcohol Formation

9.12 Special Topic: Rearrangements in Biological Processes

9.13 Summary

9.14 Additional Problems

TWO PATHS Depending on the nature of the alkene, addition of an electrophile to the alkene can follow more than one path. The path taken and the final product depend on the nature of the starting alkene and the reagents encountered along the way.

In the last third of his life, there came over Laszlo Jamf—so it seemed to those who from out in the wood lecture halls watched his eyes slowly granulate, spots and wrinkles grow across his image, disintegrating it towards old age—a hostility, a strangely *personal* hatred, for the covalent bond.

—THOMAS PYNCHON,[1] *GRAVITY'S RAINBOW*

9.1 Preview

In this chapter, we revisit one of the building-block reactions of organic chemistry—addition to alkenes. We first saw this useful process in Chapter 3, where we used it to introduce reactivity. Here and in Chapter 10, we review very briefly the material from Chapter 3 and then expand on the addition process, using the style of theme and variations. By all means, go back and reread the material in Chapter 3 (pp. 132–139) if this introduction is not completely obvious to you.

In Chapter 3, we restricted ourselves to a few examples of the addition of HX reagents to alkenes. Now we expand to a variety of old and new HX reagents, and in Chapter 10, we add X_2 and XY reagents (Fig. 9.1). Our growing catalog of reactions will continue to add to your expertise in synthesis.

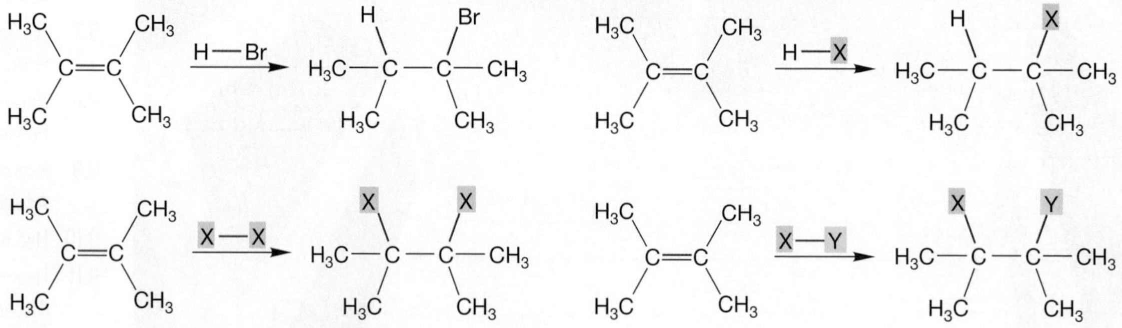

FIGURE 9.1 A variety of additions to alkenes. Many HX, X_2, and XY reagents undergo the addition reaction.

The kinds of molecules we know how to make will increase sharply as a result of material in Chapters 9 and 10. Addition of hydrogen bromide and hydrogen chloride to alkenes provides access to alkyl halides. We already have the S_N1 and S_N2 reactions available for transforming the alkyl halides into other compounds (Fig. 7.94, p. 312).

ESSENTIAL SKILLS AND DETAILS

1. Many addition reactions, especially HX additions, proceed through formation of a carbocation that is captured in a second step by X^-. You must be able to analyze the possibilities for cation formation and pick out the most stable one possible. In practice, this task is not especially difficult.

[1]Thomas Pynchon is an American author born in 1937.

2. One of the important factors influencing cation stability is delocalization of charge through the phenomenon we call resonance. Writing resonance forms well is an essential skill, and this chapter provides a review of how to do it properly.

3. Hydration and hydroboration/oxidation give you the ability to add water across an unsymmetrical double bond in either a Markovnikov or an anti-Markovnikov fashion, respectively. Although the mechanistic details of the hydroboration process are complex, the synthetic outcome is decidedly not. Be sure you know how to use these reactions to good effect in making alcohols.

4. Less stable carbocations rearrange to more stable ones through hydride (H:⁻) shifts. Stay alert—from now on, every time you see a carbocation, you have to consider whether it will rearrange into a lower-energy carbocation. Once again, you have to be able to judge and predict stability.

9.2 Mechanism of the Addition of Hydrogen Halides to Alkenes

Consider once again the reaction of 2,3-dimethyl-2-butene with hydrogen chloride to give 2-chloro-2,3-dimethylbutane (Fig. 9.2). This process is called an **alkene hydrohalogenation** because it is an addition of hydrogen and halogen across a π bond. The arrow formalism maps out the process for us and develops the picture of a two-step mechanism. In the first step, the alkene, with the filled π orbital acting as base, is protonated by hydrogen chloride to give a carbocation and a chloride ion. In the second step, chloride acts as a nucleophile and adds to the strongly Lewis acidic cation to give the product.

Alkene hydrohalogenation

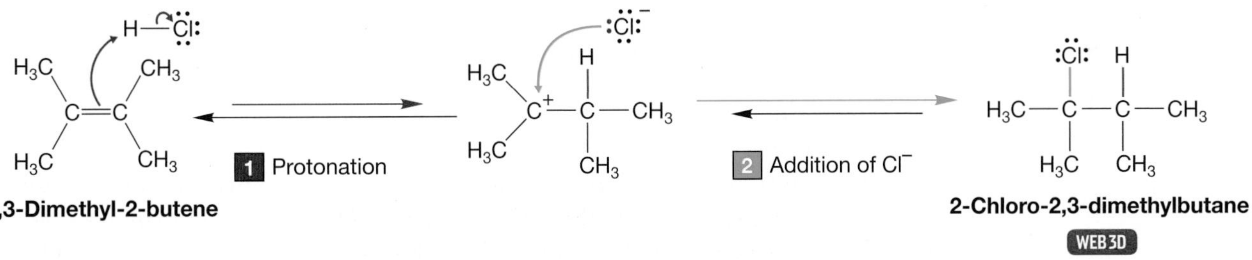

2,3-Dimethyl-2-butene   **1** Protonation   **2** Addition of Cl⁻   **2-Chloro-2,3-dimethylbutane**

WEB 3D

FIGURE 9.2 The first step in this two-step reaction is the protonation of the alkene to give a carbocation. In the second step, a chloride ion adds to the cation to give the final product.

PROBLEM 9.1 Use the bond dissociation energies of Table 8.2 (p. 337) to estimate the exothermicity or endothermicity of the reaction in Figure 9.2.

The protonation of the alkene to give a carbocation is the slow step in the reaction. The rate-determining step, the one with the highest-energy transition state, is this addition of a proton. Because this step is endothermic, the transition state will resemble the carbocation intermediate, and it will have substantial positive charge developed on carbon (Fig. 9.3). That notion leads directly to the next topic.

FIGURE 9.3 In the addition of hydrogen chloride to an alkene, the first step, the endothermic formation of a carbocation, is the slow, rate-determining step. In the transition state for this step, the positive charge accumulates on one of the carbons of the starting alkene.

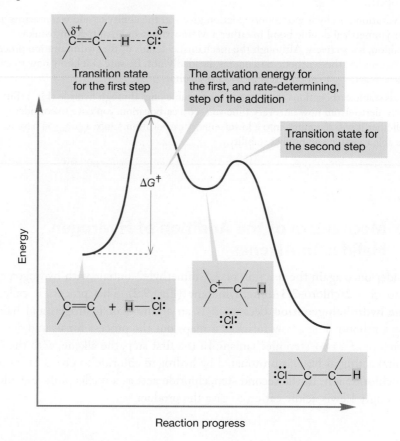

9.3 Effects of Resonance on Regiochemistry

The reaction of hydrogen chloride with a symmetrically substituted alkene such as 2,3-dimethyl-2-butene is relatively simple. Protonation is followed by chloride addition to give the alkyl chloride. There are no choices to be made about the position of protonation, and once the carbocation is made, chloride formation is easy to rationalize. What happens if the ends of the alkene are differently substituted? In such an instance, two products are possible (Fig. 9.4). We discussed this question briefly in Chapter 3 (p. 137), and the answer is that it depends on *how* the ends of

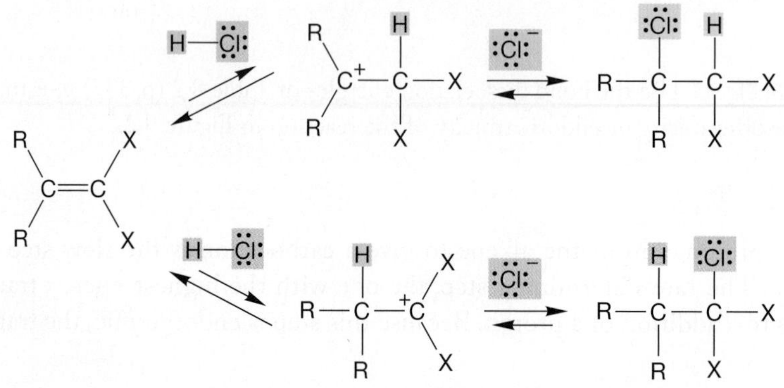

FIGURE 9.4 For an unsymmetrical alkene, there are two possible intermediate cations, and therefore two possible chloride products.

the alkene differ from each other. The preferential formation of one isomer in those cases where a choice is possible is called regioselectivity. The next few sections discuss factors influencing regiochemistry.

WORKED PROBLEM 9.2 Chloride formation isn't the only result possible in the reaction of Figure 9.2. In the early stages of this reaction, a small amount of an alkene isomeric with the 2,3-dimethyl-2-butene is produced. What is this other alkene, and what is the mechanism of its formation?

ANSWER Remember the E1 reaction! Just because we are starting a new subject does not mean that old material disappears. Cations will undergo the E1 and S_N1 reactions as well as the addition reactions encountered in this chapter. The cation formed on protonation of 2,3-dimethyl-2-butene can undergo an E1 elimination in two ways. Path a simply reverses to give starting compounds back, whereas path b leads to the isomeric alkene, 2,3-dimethyl-1-butene.

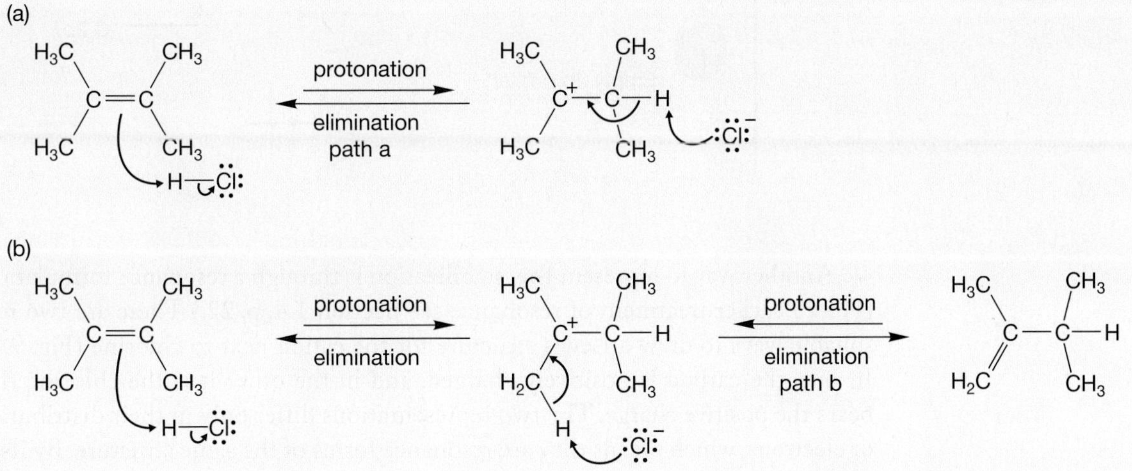

(a)

(b)

When hydrogen chloride adds to vinyl chloride (chloroethylene), the major product is 1,1-dichloroethane, *not* 1,2-dichloroethane (Fig. 9.5). The best way to rationalize the regiospecificity of this reaction is to examine the two possible intermediate carbocations

Vinyl chloride
(chloroethylene)

1,1-Dichloroethane *not* **1,2-Dichloroethane**

FIGURE 9.5 Addition of hydrogen chloride to vinyl chloride gives 1,1-dichloroethane as the major product, *not* 1,2-dichloroethane.

(Fig. 9.6) and see which is more stable. *Remember:* The molecule has an empty *p* orbital on the carbon bearing the positive charge. In the first carbocation shown in Figure 9.6, this empty orbital overlaps with a *p* orbital on chlorine, which contains two electrons.

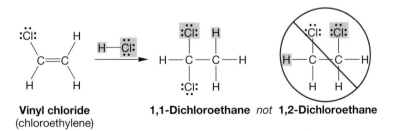

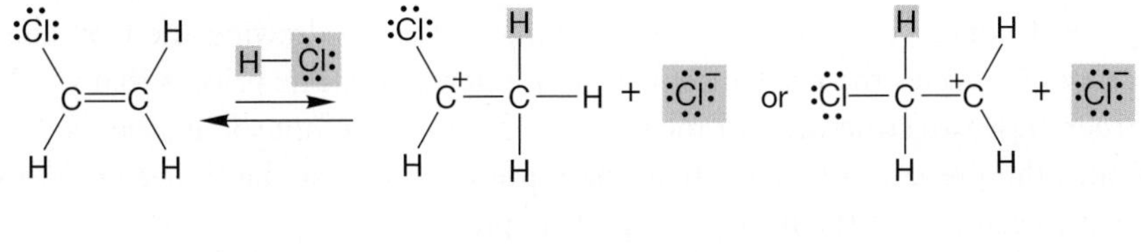

FIGURE 9.6 The two cations that could be formed by protonation of vinyl chloride.

This overlap of filled and empty orbitals produces strong stabilization (Fig. 9.7). In this ion, the chlorine atom helps bear the positive charge. Only when the charge is adjacent to the chlorine is there stabilization through overlap of the empty $2p$ orbital on carbon with the filled $3p$ orbital on chlorine. Because this stabilization is not available to the second carbocation shown in Figure 9.6, it is much harder to form.

FIGURE 9.7 Stabilization of a cation through overlap of a filled and empty orbital.

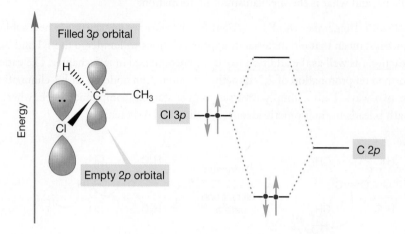

Another way to represent this stabilization is through a resonance formulation. (For an earlier treatment of resonance see Section 1.4, p. 22.) There are two reasonable ways to draw a Lewis structure for the cation next to chlorine (Fig. 9.8). In one, the carbon is positively charged, and in the other it is the chlorine that bears the positive charge. The two representations differ only in their distribution of electrons, which means they are resonance forms of the same structure. By itself neither is exactly right, as you learned in Section 1.4. Neither by itself shows that two atoms share the charge. Taken together they do. The real structure of this cation is not well represented by either form alone, but the two combined do an excellent job.

FIGURE 9.8 Resonance stabilization of a carbocation.

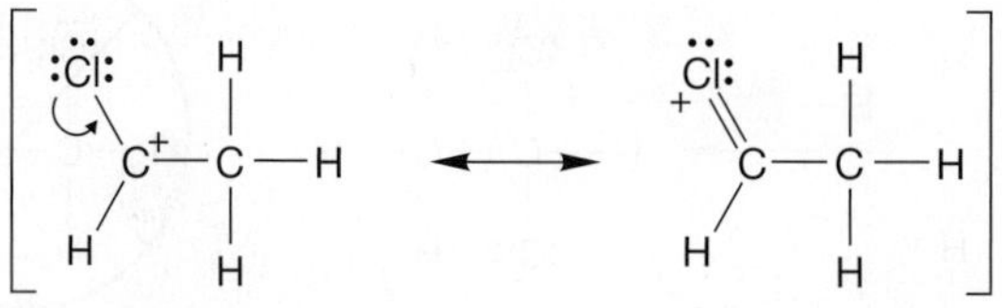

In this resonance form, the carbon bears the positive charge

In this resonance form, the chlorine bears the positive charge

The molecular orbital picture we developed in Figure 9.7 does a good job of showing the stabilization, but it is not particularly well suited for bookkeeping purposes. In practice, the ease of bookkeeping—of easily drawing the next chemical reaction—makes the resonance picture simpler to use. The price is that you have to either draw two structures (or more in other cases) or adopt some other code that indicates the presence of more than one representation for the molecule. We will see many examples in the next pages and chapters.

Figure 9.9 illustrates a reaction that we will return to in some detail later but can examine quickly now. Hydrogen chloride adds to 1,3-butadiene very easily to yield two products. One product you may be able to predict, but the other will probably not be immediately obvious (Fig. 9.9).

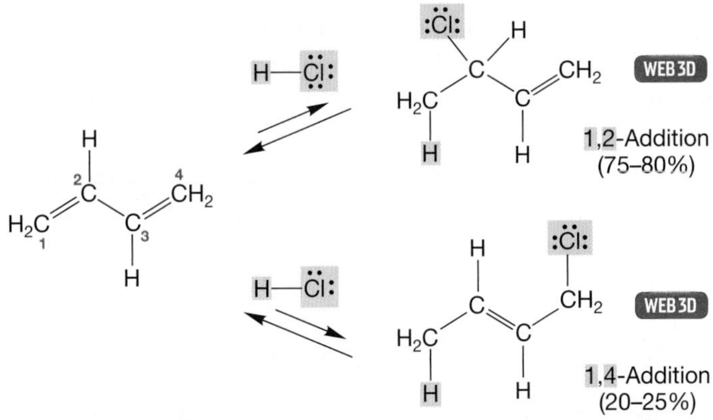

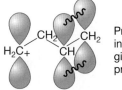

FIGURE 9.9 1,2- and 1,4-Addition of HCl to 1,3-butadiene.

Look at the possibilities. There are two cations that can arise from protonation of 1,3-butadiene by hydrogen chloride (Fig. 9.10). In the allylic cation, the positive charge is on the carbon adjacent to the double bond; in the other, it is isolated on a terminal

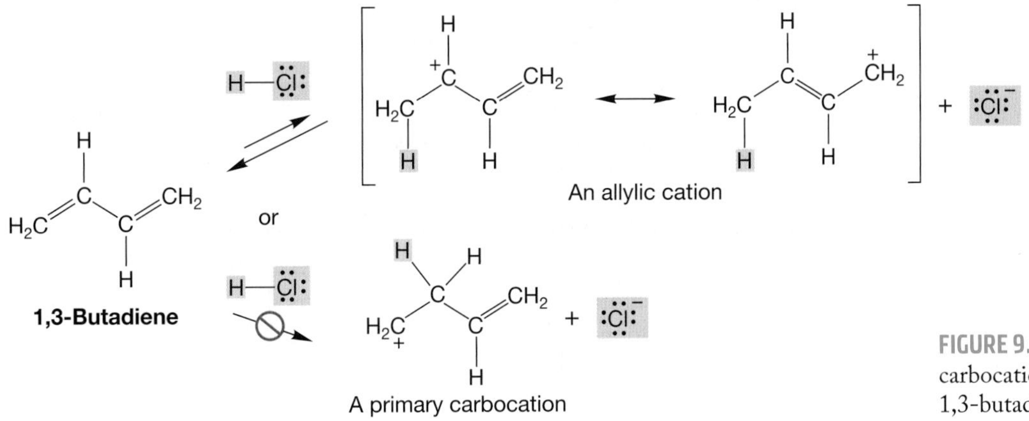

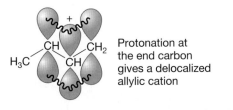

FIGURE 9.10 The two possible carbocations from protonation of 1,3-butadiene.

primary carbon. The choice between these two ions becomes clear if we draw orbital pictures of the two carbocations. In the allylic cation, the p orbital on the positively charged carbon atom overlaps the π orbitals of the double bond (Fig. 9.11). We have

Allyl cation

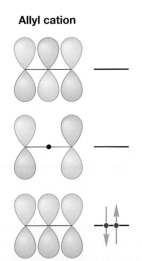

FIGURE 9.11 In the allylic cation, the positive charge is shared by two carbons; in the primary cation it is localized on a single carbon.

seen the system of three parallel overlapping p orbitals before (Problems 1.16c and 1.61). Here we have an allylic cation again, and the molecular orbital description is the same as before. The three $2p$ atomic orbitals can be combined to yield three molecular orbitals. There are only two electrons in an allylic cation and they will occupy the lowest, most stable molecular orbital (Fig. 9.12). The presence of the double bond is highly stabilizing for this cation, because the charge is delocalized over the allyl system.

FIGURE 9.12 The molecular orbitals of the allylic cation.

PROBLEM 9.3 Derive the π orbitals of the allyl system shown in Figure 9.12. *Hint*: Do this task by placing a carbon 2*p* orbital between the two *p* orbitals of ethylene. *Another Hint*: Watch out for "net zero," orthogonal interactions!

For the primary carbocation in the reaction shown in Figure 9.10, the positive charge is localized at the end of the molecule and has no resonance stabilization. Protonation at the end carbon gives an ion (the allylic cation) that can be represented by more than one resonance structure. The choice is easy once we have looked carefully at the structures. Formation of the lower energy delocalized allylic cation will be favored over formation of the higher energy localized primary cation (Fig. 9.13).

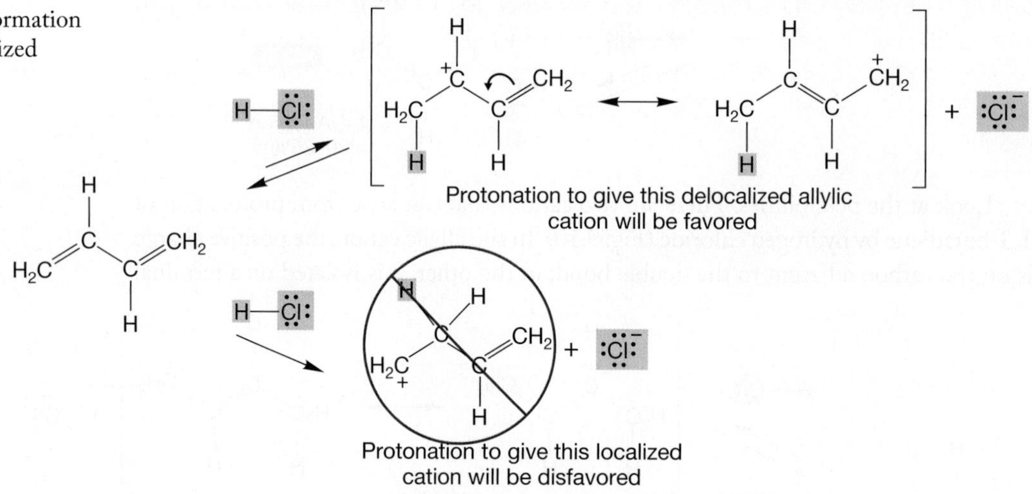

FIGURE 9.13 Preferential formation of the lower energy, delocalized cation.

The real structure of the delocalized cation is best represented as a combination of the two resonance forms. This way of looking at the allylic cation is especially helpful because it points out simply and clearly which carbons share the positive charge. When the chloride ion approaches the allylic cation it can add at either of the two carbons that share the positive charge to give the two products shown in Figure 9.14.

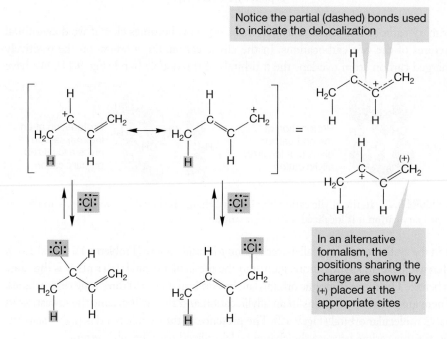

FIGURE 9.14 In the allylic cation, the positive charge is shared by two carbons. Note the summary structures in which dashed bonds or charges in parentheses are used to indicate delocalization without writing out all the resonance forms.

Notice the partial (dashed) bonds used to indicate the delocalization

In an alternative formalism, the positions sharing the charge are shown by (+) placed at the appropriate sites

Be careful not to think of this resonance-stabilized structure as spending part of its time with the charge on carbon 1 and part of the time with the charge on carbon 3! The cation has a *single* structure in which two carbons of the allylic cation share the positive charge.

This single, summary resonance structure is often shown using dashed bonds to represent the bonds that are double in one resonance form and single in the other. The charge is placed at the midpoint of the dashes as shown in the structure at the top right of Figure 9.14. Alternatively, one resonance form is drawn out in full and the other atoms sharing the charge are shown with a (+) or (−), as also shown in Figure 9.14.

CONVENTION ALERT

Now let's ask if the molecular orbital description can predict both products. Of course it can, or it would hardly be very useful. What happens as the chloride ion, with its filled nonbonding orbitals, begins to interact with the allylic cation? Stabilizing interactions occur between filled orbitals and empty orbitals. Chloride bears the filled orbital; therefore we must look for the lowest *unoccupied* molecular orbital (LUMO) of allyl. Figure 9.15 shows it, Φ_2. There are two points at which chloride can add to Φ_2, and they lead to the two observed products. Note that the middle carbon, through which the node passes, cannot be attacked. So the molecular orbital description also explains the regiochemistry of the addition.

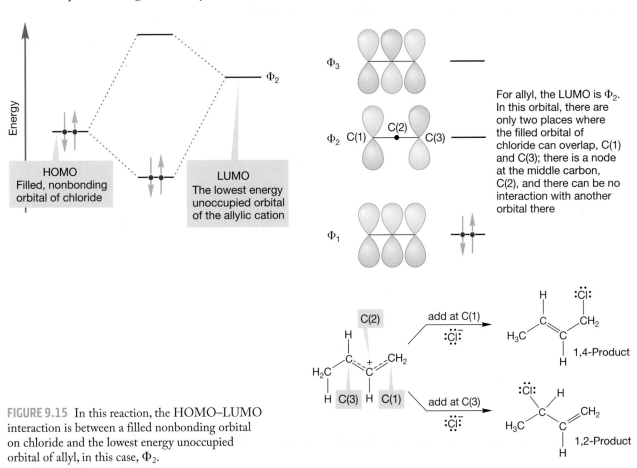

For allyl, the LUMO is Φ_2. In this orbital, there are only two places where the filled orbital of chloride can overlap, C(1) and C(3); there is a node at the middle carbon, C(2), and there can be no interaction with another orbital there

FIGURE 9.15 In this reaction, the HOMO–LUMO interaction is between a filled nonbonding orbital on chloride and the lowest energy unoccupied orbital of allyl, in this case, Φ_2.

Whether one uses resonance forms or molecular orbital descriptions is largely a matter of taste. Resonance forms are the traditional way, and they are extremely good at letting you see where the charge resides. We will use them extensively. It's worth taking time now to review the whole resonance system.

Summary

Addition of HX acids to alkenes really is straightforward. An initial protonation to give the most stable carbocation possible is followed by capture by X^-. The only difficulty is figuring out which carbocation intermediate is the lowest energy. Here, an analysis of the "usual suspects" is necessary. You have to consider the substitution pattern, resonance effects, and, as we shall soon see, inductive effects as well.

9.4 Brief Review of Resonance

This subject was treated in detail in Chapter 1. Here we only pick up the highlights of that treatment. If anything seems unfamiliar or strange go back to Chapter 1! A good way to check that this important material is under control is to see if you can do Problems 1.18–1.21.

Remember that a molecule that is best described as a resonance hybrid, *does not*, repeat *not*, spend part of its time as one form and part as another. That is chemical equilibrium not resonance. The two are very different phenomena. Resonance is always indicated by the special double headed arrow $\longleftrightarrow$, whereas equilibrium is shown by two arrows pointing in opposite directions $\rightleftarrows$.

As we pointed out as early as Chapter 1, the real structure of a resonance-stabilized molecule is a hybrid, or combination, of the resonance forms contributing to the structure. These forms represent different electronic descriptions of the molecule. *Resonance forms differ only in the distributions of electrons and never in the positions of atoms.* If you have moved an atom, you have written a chemical equilibrium.

It is important to be able to estimate the relative importance of resonance forms in order to get an idea of the best way to represent a molecule. To do so, we can assign a weighting factor, c, to each resonance form. The weighting factor indicates the percent contribution of each resonance form to the overall structure. Some guidelines for assigning weighting factors are summarized below.

1. Equivalent resonance forms contribute equally.

2. The more bonds in a resonance form, the more stable the form is.

3. Resonance forms must have the same number of paired and unpaired electrons.

4. Separation of charge is bad.

5. In ions, delocalization is especially important. As we have mentioned a number of times, small charged atoms are usually most unstable. Delocalization of electrons that allows more than one atom to share a charge is stabilizing. The protonation of vinyl chloride is a good example. Vinyl chloride is protonated to give the cation in which the charge is shared by carbon and chlorine. The resonance forms show this sharing clearly (Fig. 9.16).

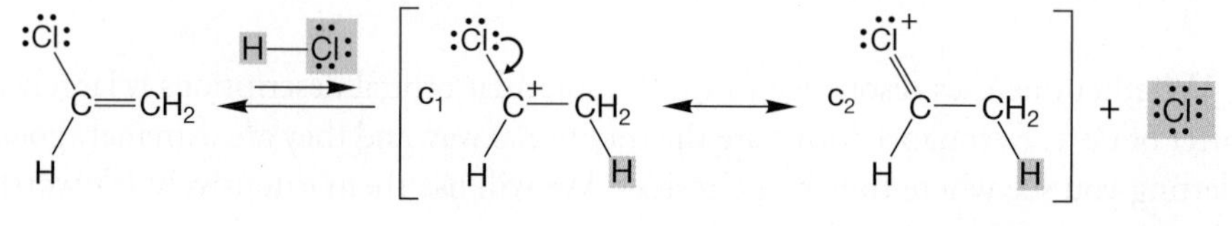

FIGURE 9.16 In the chlorine-substituted carbocation, the charge is shared by carbon and chlorine.

6. Electronegativity is important. In the enolate anion, which has two resonance forms, the two do not contribute equally because they are not equivalent (Fig. 9.17). Each has the same number of bonds, so we cannot choose the better representation that way. However, form **A** has the charge on the relatively electronegative oxygen while form **B** has it on carbon. Form **A** is the better representation for the enolate anion, although both forms contribute. Mathematically, we would say that the weighting factor for **A**, c_1, is larger than that for **B**, c_2.

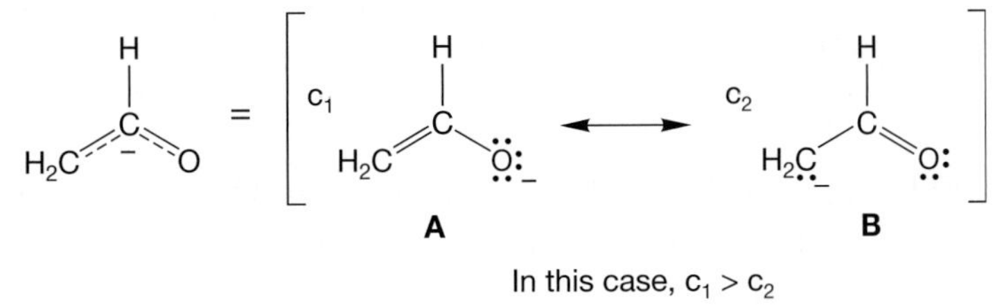

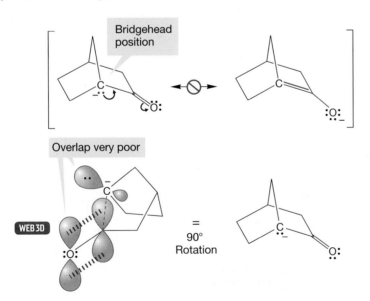

In this case, $c_1 > c_2$

FIGURE 9.17 Two resonance forms for the enolate anion.

7. Now that we have a good deal of structure under our belts, we can look at one more phenomenon relevant to evaluating good (low energy) resonance forms. We can tell you to "Watch out! Geometry is important." Is the "enolate" anion in Figure 9.18 resonance stabilized? On paper it certainly looks like it is, and there is at least a typographical relationship to the enolate anion of Figure 9.17. If you answered yes, however, you have been fooled by the two-dimensional surface. The three-dimensional structure of the cage molecule prevents substantial overlap of the orbitals. Even though the paper doesn't protest when you draw the second resonance form, in reality, it does not contribute. The negative charge is borne on only one carbon—the bridgehead—and is not shared by any other atom. Build a model and you will see it easily.

Bridgehead position

Overlap very poor

WEB 3D

=
90°
Rotation

FIGURE 9.18 In this enolate-like anion, the orbital containing the pair of electrons does not overlap well with the p orbitals of the C=O. There is no resonance, no matter what the deceptive two-dimensional surface says.

PROBLEM 9.4 Are the two structures shown below resonance forms? Explain why or why not?

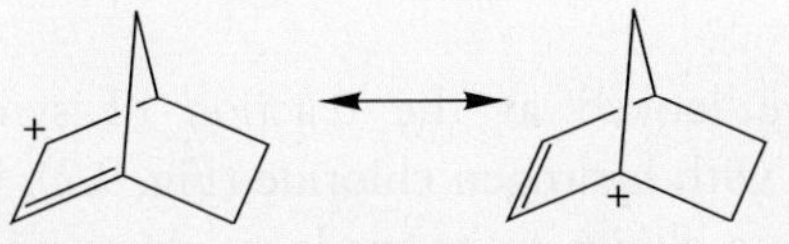

PROBLEM 9.5 How many signals will the following species show in their ^{13}C NMR spectra? In fact, not all of these species are stable enough to be detected by NMR, but in each case it is possible to predict what would be (will be?) seen.

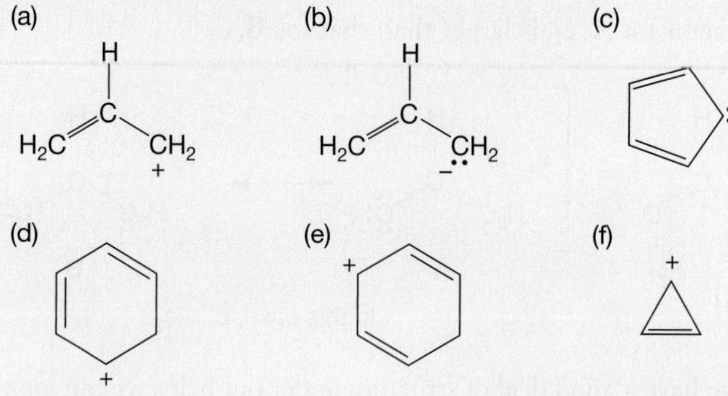

(a)

(b)

(c)

(d)

(e)

(f)

9.5 Resonance and the Stability of Carbocations

We saw in Figure 9.5 one example of how an atom capable of delocalizing electrons can help share a charge and influence the regiochemistry of an addition reaction. Addition of hydrogen chloride to vinyl chloride gives 1,1-dichloroethane, *not* 1,2-dichloroethane. Protonation produces a carbocation adjacent to the chlorine, which is far more stable than the alternative in which the positive charge sits localized on a primary carbon (Fig. 9.19).

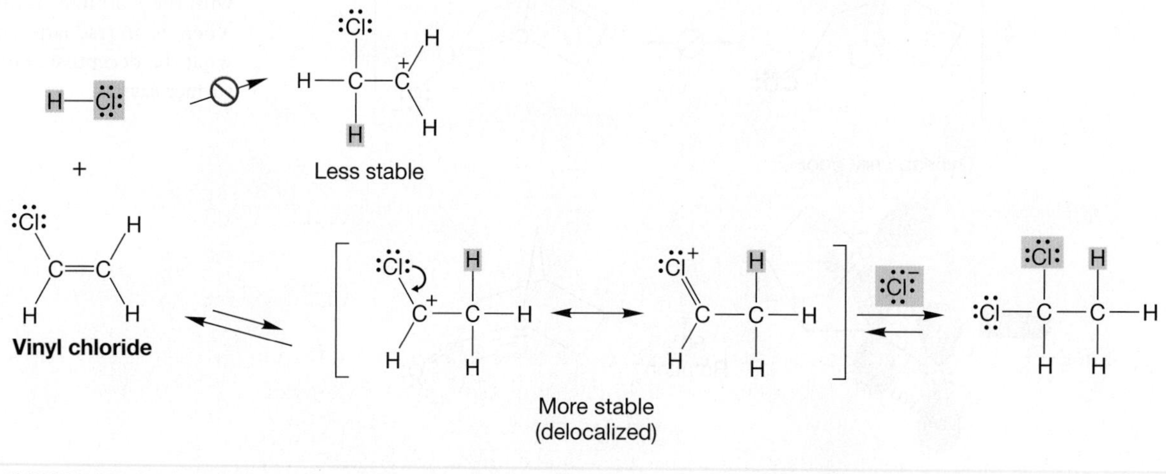

Less stable

Vinyl chloride

More stable
(delocalized)

FIGURE 9.19 In the addition of hydrogen chloride to vinyl chloride, it is formation of the more stable, resonance-stabilized carbocation that determines the product.

In Section 9.2, we looked at the reaction of symmetrically substituted 2,3-dimethyl-2-butene with hydrogen chloride (Fig. 9.2). In the formation of the carbocation, there was no choice to be made—only one cation could be formed. When the less symmetrical alkene 2-methyl-1-butene reacts with hydrogen chloride, there are two possible carbocations (Fig. 9.20). The only chloride formed is the one produced from the more stable, tertiary carbocation. This product results from the hydrogen adding to the less substituted end of the alkene and the chlorine to the more substituted end (Fig. 9.20).

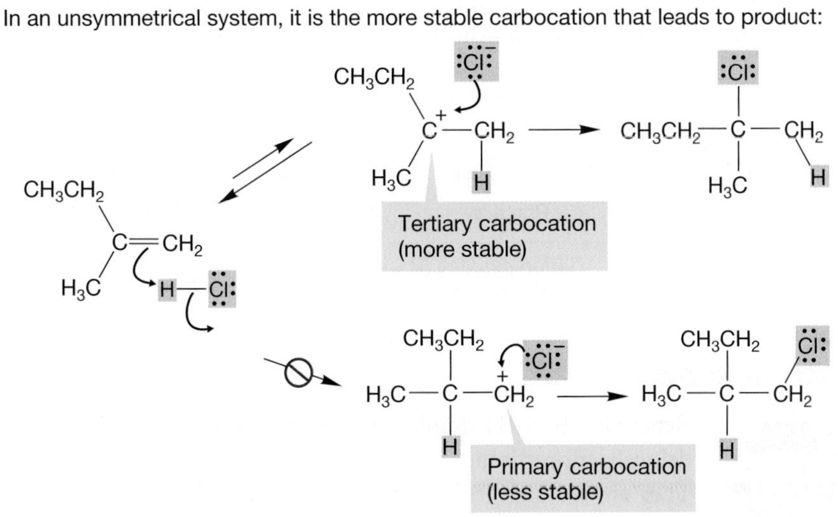

In an unsymmetrical system, it is the more stable carbocation that leads to product:

FIGURE 9.20 The regiospecific addition of HCl to 2-methyl-1-butene. It is the more stable carbocation that leads to product.

Unsymmetrical alkenes such as 2-pentene, from which two secondary carbocations of roughly equal stability can be formed, do give two products in comparable amounts (Fig. 9.21).

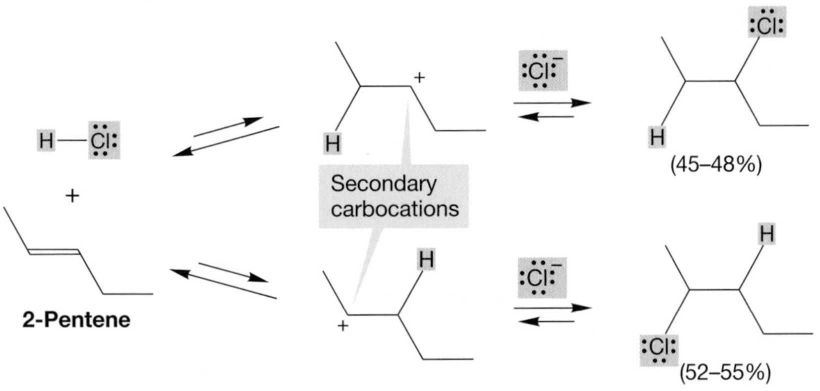

FIGURE 9.21 Two products are formed from the reaction of HCl and 2-pentene.

These phenomena have been known a long time and were first correlated by the Russian chemist Vladimir V. Markovnikov (1838–1904) in 1870. The observation that the more substituted halide is formed now bears his name in **Markovnikov's rule**.[2] Rules are useless (or worse) unless we understand them, so let's see why Markovnikov's rule works.

---

[2]Markovnikov states in his 1870 paper (*Ann.* **1870**, *153*, 228): "wenn ein unsymmetrisch constituirter Kohlenwasserstoff sich mit einer Haloïdwasserstoffsäure verbindet, so addirt sich das Haloïd an das weniger hydrogenisirte Kohlenstoffatom, d.h. zu dem Kohlenstoff, welcher sich mehr unter dem Einflusse anderer Kohlenstoffe befindet." Which is to say "when an unsymmetrical alkene combines with a hydrohalic acid, the halogen attaches itself to the carbon atom containing the fewer hydrogen atoms—that is to say, the carbon that is more under the influence of other carbons."

Figures 9.19–9.21 show that the mechanism of HX addition must involve formation of the more substituted cation, because the less substituted cation would lead to a product that is either not observed or is only formed in minor amounts. In turn, a reasonable presumption is that the more substituted a carbocation is, the more stable it is (Fig. 9.22). *Remember*: These differences in stability are large. In practice, reactions in which tertiary and secondary carbocations are intermediates are common, but primary and methyl carbocations cannot be formed.

FIGURE 9.22 Stability order for carbocations.

PROBLEM SOLVING

Remember (p. 141) simple primary and methyl carbocations (no delocalization through resonance) are mechanistic "stop signs." They are too high in energy to be formed, and are not known as intermediates under "normal" conditions.

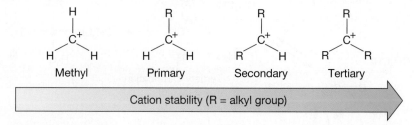

How does the difference in energy of these two carbocations, the tertiary and secondary cationic intermediates in the addition reaction, become translated into a preference for one product over the other? The relative rates of the two possible protonation steps will not depend directly on the relative energies of the two carbocations themselves, but on the energy difference between the two transition states leading to them ($\Delta\Delta G^{\ddagger}$ in Fig. 9.23). Both protonation steps are endothermic and it is likely that the

$\Delta G^{\ddagger}$ Activation energy for formation of more stable tertiary carbocation

$\Delta G^{\ddagger}$ Activation energy for formation of less stable secondary carbocation

$\Delta\Delta G^{\ddagger}$ Difference in energies of transition states is the difference in activation energies

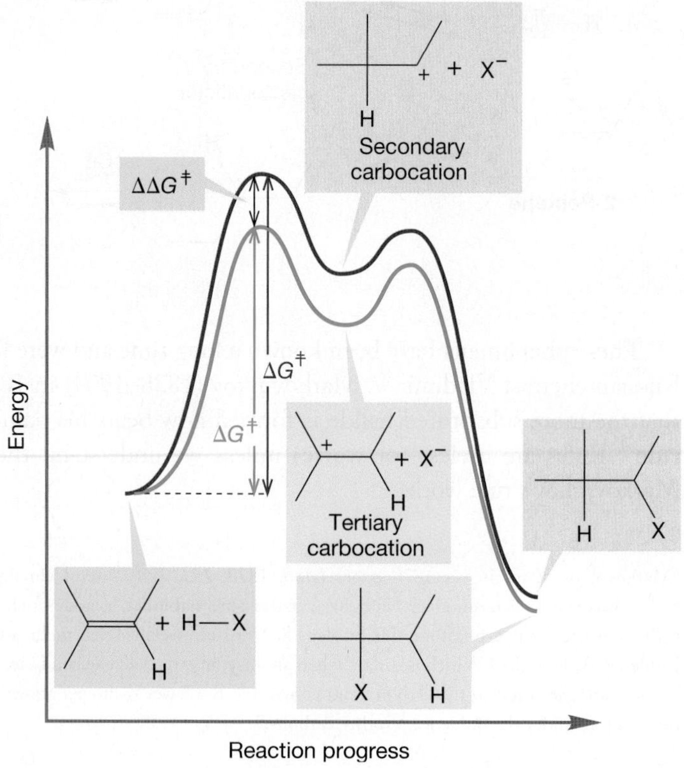

FIGURE 9.23 The transition states for formation of the high-energy carbocations will contain partial positive charges (δ^+). The same order of stability that holds for full positive charges applies to partial positive charges. A δ^+ on a tertiary carbon is more stable than a δ^+ on a secondary carbon, and so on.

transition states will bear a strong resemblance to the cations. (Remember the Hammond postulate, "For an endothermic reaction the transition state will resemble the product," p. 355.) In these two transition states, positive charge is building up on carbon. There are partial positive charges in the transition states. *A tertiary partial positive charge is more stable than a secondary partial positive charge, just as a full positive charge on a tertiary carbon is more stable than a full positive charge on a secondary carbon.* Once the carbocation intermediate is formed, the structure of the product is determined. Capture of the tertiary carbocation leads to the observed product, in which X is attached to the more substituted carbon of the starting alkene (the site of the tertiary positive charge in the cation). Similarly, the secondary carbocation would lead inevitably to a product in which X is attached to the less substituted carbon (Fig. 9.23).

In Chapters 3 and 7, we avoided the issue, but it is now time to deal with the question of why alkyl groups stabilize carbocations. Our mechanistic hypothesis is consistent with the idea that the more substituted halide is produced because it is the result of the initial formation of the more substituted (and hence more stable) carbocation (Fig. 9.20). But this consistency doesn't answer anything; we have just pushed the question further back.

Why are more substituted carbocations more stable than less substituted ones? To answer such questions, we start with structure. Carbocations are flat, and the central carbon is sp^2 hybridized and therefore trigonal. Figure 9.24 compares the methyl cation ($^+CH_3$) with the ethyl cation ($^+CH_2CH_3$). Note that in the ethyl cation there is a filled carbon–hydrogen bond that can overlap with the empty p orbital. This phenomenon has the absolutely grotesque name of **hyperconjugation**, but there is nothing particularly "hyper" about it. It is a simple effect and can be rationalized in familiar molecular orbital or resonance terms. The orbital mixing is stabilizing (Fig. 9.24) and nothing like it is available to the methyl cation. The more interactions of this kind that are possible, the more stable the ion. Thus the *tert*-butyl cation is most stable, the isopropyl cation next, and so on.

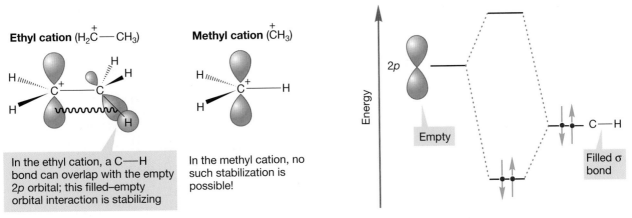

Ethyl cation ($H_2\overset{+}{C}$—CH_3)

In the ethyl cation, a C—H bond can overlap with the empty $2p$ orbital; this filled–empty orbital interaction is stabilizing

Methyl cation ($\overset{+}{C}H_3$)

In the methyl cation, no such stabilization is possible!

Energy

$2p$

Empty

Filled σ bond

C—H

FIGURE 9.24   The ethyl cation is stabilized by a filled orbital overlapping with an empty orbital.

Alternatively, we could write the following kind of resonance form, which shows the charge delocalization from carbon to hydrogen especially well (Fig. 9.25). It's important to see that the second carbon–carbon bond in the newly drawn

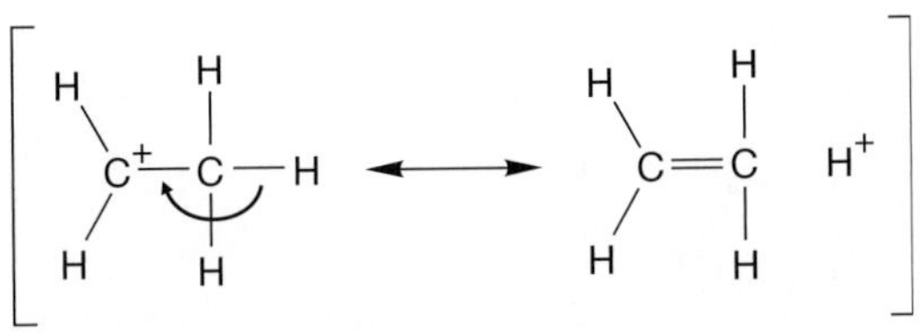

FIGURE 9.25   A resonance description of the ethyl cation ($^+CH_2CH_3$).

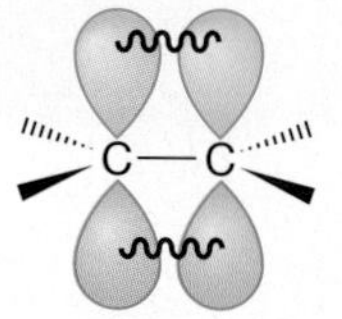

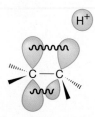

The π bond in ethylene is made up of 2p/2p overlap

In the "hyperconjugative" resonance form for the ethyl cation, the "double" part of the double bond is not made up of 2p/2p overlap, but of 2p/sp³ overlap

FIGURE 9.26 A detailed look at hyperconjugative stabilization of the ethyl cation.

resonance form of Figure 9.25 is *not* identical to a π bond of an alkene. It is not formed from 2p/2p overlap but rather from 2p/sp³ overlap (Fig. 9.26). Be sure you see the difference between these two fundamentally different, but similar-looking structures.

PROBLEM 9.6 It is often claimed that the *tert*-butyl cation has 10 resonance forms, 1 in which the carbon bears the positive charge and 9 in which each hydrogen bears the positive charge. Is this claim true?

PROBLEM 9.7 On page 115 we described the stabilization of alkenes by attached alkyl groups in different terms, focusing on the σ bond system. We could construct a similar treatment here. Provide a justification for the relative stability of differently substituted carbocations using an analysis of the σ system.

PROBLEM 9.8 Do you think the π part of the double bond in the resonance form in Figure 9.25 is stronger or weaker than a normal π bond? Explain why.

We have just seen that carbocations are stabilized by delocalization through the overlap of filled and empty orbitals. One example of this effect is fairly straightforward; a lone pair of electrons on an adjacent atom interacts with an empty orbital. Less obvious is the last phenomenon, the overlap of filled σ orbitals of carbon–hydrogen bonds with empty orbitals, called hyperconjugation. Now we proceed to other factors influencing the stability of carbocations.

9.6 Inductive Effects on Addition Reactions

Polar bonds in a molecule can have a major effect on both the rate and the regiochemistry of an addition reaction. Let's look first at the series of ethylenes in Figure 9.27,

FIGURE 9.27 Addition reaction to three different alkenes. HX = HOH in this case.

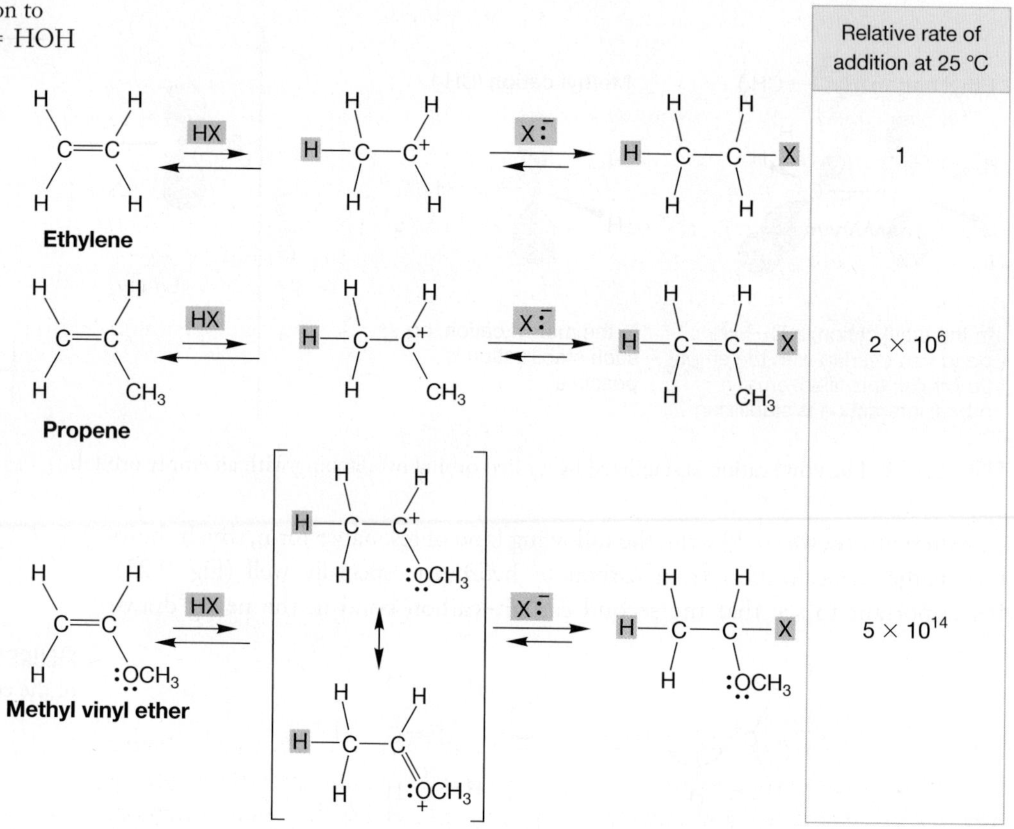

	Relative rate of addition at 25 °C
Ethylene	1
Propene	2×10^6
Methyl vinyl ether	5×10^{14}

which shows typical rates of addition for a series of substituted ethylenes. The reaction is, in fact, a hydration reaction in which HOH adds across the double bond.

Ethylene itself will protonate only very slowly. Propene will surely protonate so as to give the more stable secondary cation, and the methoxyl group of methyl vinyl ether will provide resonance stabilization to an adjacent positive charge. Hyperconjugative stabilization by the methyl group of propene and resonance stabilization by the methoxyl group of methyl vinyl ether, respectively, have a great accelerating effect on the rate of protonation, the rate-determining step in this reaction. Figure 9.27 shows the rates of protonation compared to that of unsubstituted ethylene itself.

PROBLEM 9.9 Why is the rate of protonation of ethylene especially slow?

Substituents can slow reactions as well, especially if they are not in a position to provide resonance stabilization. Figure 9.28 shows typical rates for addition to some substituted alkenes. To see why bromine and chlorine so dramatically slow the rate of addition, we must again look more closely at the carbon–halogen bond (see p. 225 for an earlier look).

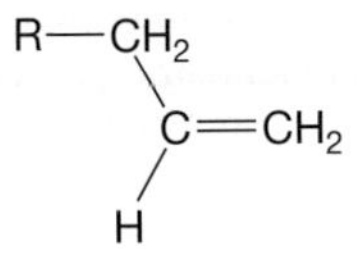

R = H	Br	Cl
Rate 1	10^{-4}	10^{-4}

FIGURE 9.28 Rates of addition to three alkenes.

The two atoms involved in the bond, carbon and halogen, are different, so the bond between them must be polarized. Therefore, the electrons in the bond cannot be equally shared. Bromine and chlorine are far more electronegative than carbon (see Table 1.8, p. 15) and they will attract electrons strongly. The carbon–halogen bond is polarized as shown in Figure 9.29. The carbon will bear a partial positive charge (δ^+) and the more electronegative halogen a partial negative charge (δ^-).

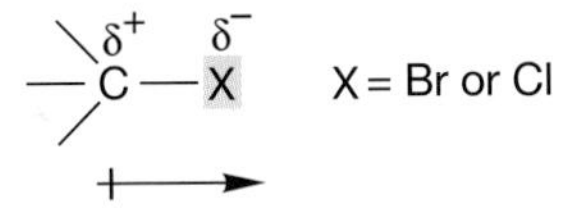

For the molecules in Figure 9.28, protonation in the Markovnikov sense places a full positive charge on the carbon adjacent to the already partially positive carbon (Fig. 9.30). This interaction will surely be destabilizing and accounts for the decrease in rate for addition reactions to the bromine- and chlorine-substituted molecules shown in Figure 9.28. Effects such as these, which are transmitted through the σ bonds, are called **inductive effects**.

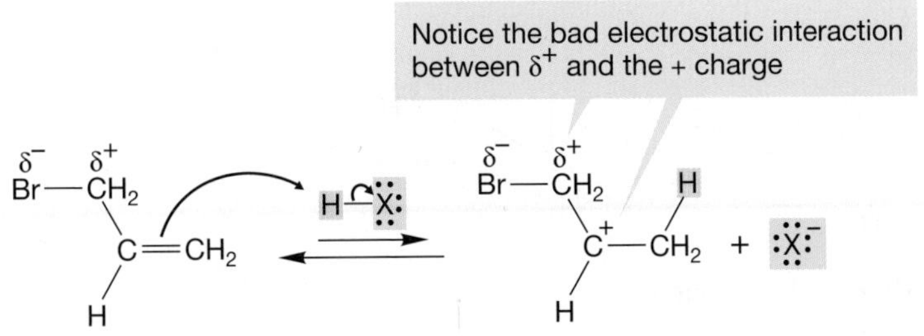

FIGURE 9.29 The polar C—X bond.

Notice the bad electrostatic interaction between δ^+ and the + charge

FIGURE 9.30 Protonation gives an intermediate destabilized by the electrostatic interaction between the newly formed full positive charge and the partial positive charge (δ^+).

Wouldn't the methoxyl group of methyl vinyl ether act in the same way? Oxygen is also more electronegative than carbon, and wouldn't this difference in electronegativity act to discourage any addition that puts more positive charge on the already partially positive carbon adjacent to oxygen? Why does addition to methyl vinyl ether occur so rapidly (Fig. 9.27) in the Markovnikov sense? Figure 9.31 frames the question, and answers it as well. The answer is simply that the resonance effect of oxygen overwhelms the inductive effect. The two effects *are* in competition, and the stronger resonance effect wins out. Now let's summarize, then look at a number of related addition reactions.

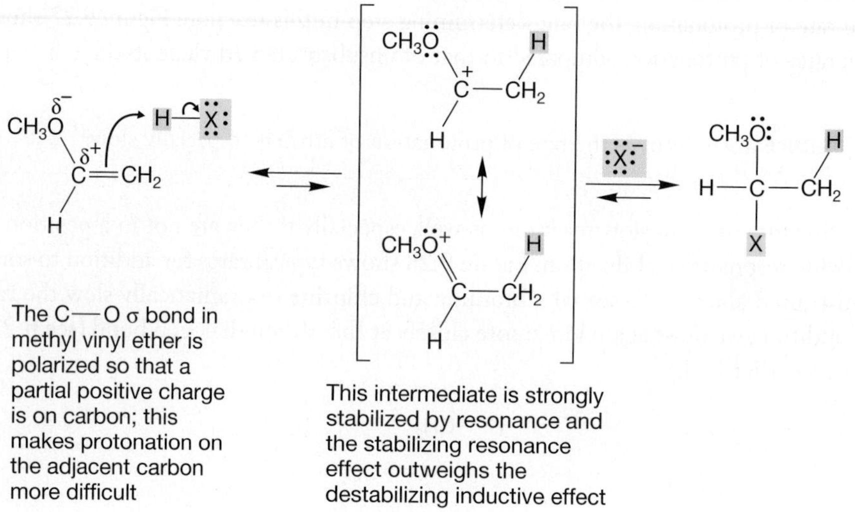

FIGURE 9.31 Stabilizing resonance effects and destabilizing inductive effects can "fight" each other. Generally, resonance stabilization is more important.

The C—O σ bond in methyl vinyl ether is polarized so that a partial positive charge is on carbon; this makes protonation on the adjacent carbon more difficult

This intermediate is strongly stabilized by resonance and the stabilizing resonance effect outweighs the destabilizing inductive effect

Summary

Protonation (or addition of any Lewis acid) to an unsymmetrical double bond leads preferentially to the more stable carbocation. The energy of carbocations is strongly influenced by stabilizing resonance effects. Substituents bearing a lone pair of electrons are especially effective at stabilization, but even alkyl groups can donate electrons to the empty $2p$ orbital of a carbocation. In cases in which inductive and resonance effects oppose each other, resonance usually wins.

9.7 HX Addition Reactions: Hydration

If HBr and HCl add to double bonds, it should not be surprising that other acidic HX compounds undergo similar addition reactions. Hydrogen iodide is a simple example, and HI addition rather closely resembles the reactions we have already studied in this chapter. Markovnikov's rule is followed, for example (Fig. 9.32).

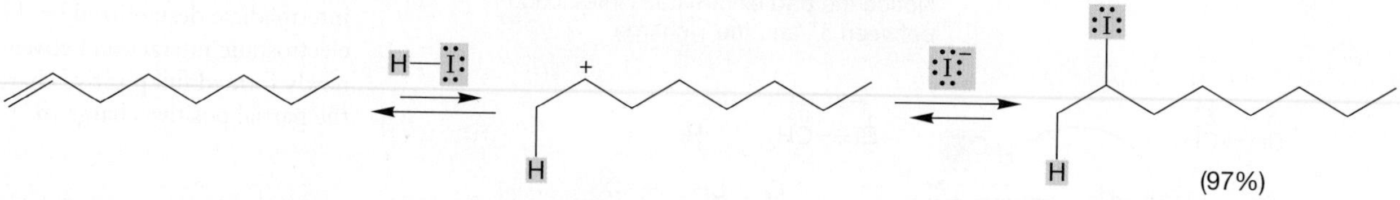

FIGURE 9.32 Addition of HI to 1-octene in Markovnikov fashion.

Water (HOH) is an HX reagent, but, as we saw on p. 139 it is not a strong enough acid to undergo addition to simple alkenes at a useful rate. If one mixes water and a typical alkene, no reaction occurs. The difference between HOH and HX is that HCl and HBr are strong enough acids to protonate the alkene and start things going. Water is not. The pK_a values of HBr, HCl, and HOH are -9, -8, and $+15.7$, respectively, which means that water is a weaker acid than HBr and HCl by factors of about 10^{25} and 10^{24}! Acid must be added to protonate the alkene and thus generate a carbocation that can continue the reaction. The acid catalyst in this reaction sets things going. The catalyst has no effect on the position of the final equilibrium, but it does lower the energy requirement for generating product. As Figure 9.33 illustrates, the catalyst provides a new mechanistic pathway for product formation. It is able to carve a new pathway along a lower slope of the mountain we first encountered in Figure 8.20 (p. 346).

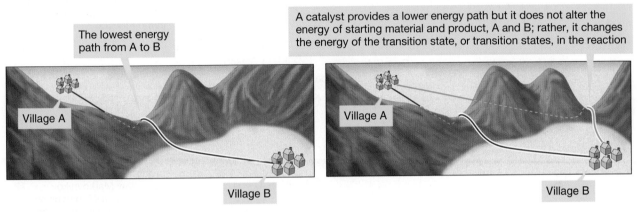

FIGURE 9.33 A catalyst does not change the energy of either starting material or product but does provide a lower energy path between them.

As can be seen in Figure 9.34, the catalyst H_3O^+ is used up in the first step of the reaction and is regenerated in the last step. A number of acids can be used in the catalytic role; sulfuric (H_2SO_4 or $HOSO_2OH$) or nitric (HNO_3 or $HONO_2$) are commonly used to promote the addition reaction. In water containing a little sulfuric acid for instance, 2,3-dimethyl-2-butene is protonated to generate the tertiary carbocation. This intermediate is then attacked by HOH (water, not hydroxide, ^-OH) to give the oxonium ion, $R-OH_2^+$, which can be deprotonated by water to regenerate the catalyst, H_3O^+, and produce a molecule of an alcohol, $R-OH$. Thus, the outcome of the **hydration** reaction is addition of HOH across a π bond.

Alkene hydration

FIGURE 9.34 The mechanism of hydration of 2,3-dimethyl-2-butene.

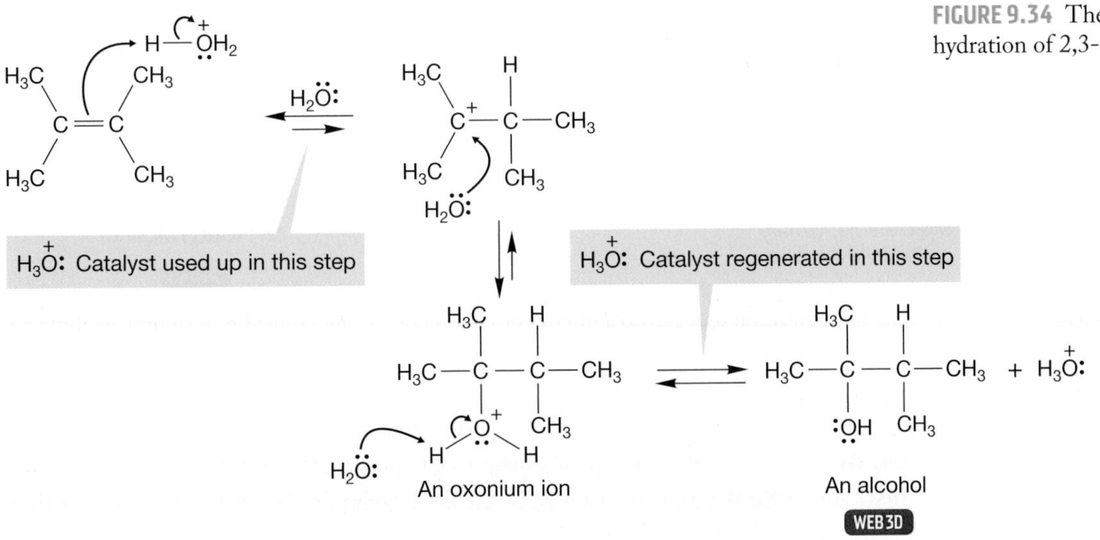

The mechanism shown in Figure 9.34 predicts, indeed requires, that the Markovnikov rule be followed, and that sets up a mechanistic test. We will examine the hydration of 2-methylpropene to probe the regiochemistry of the reaction. Before we even start, it is worth thinking a bit about the results. What will we know after we have done the experiment? There are only two general results possible: Either the Markovnikov rule is followed or it is not (Fig. 9.35). If it is not, we can say with certainty that the mechanism we have suggested for this reaction is wrong.

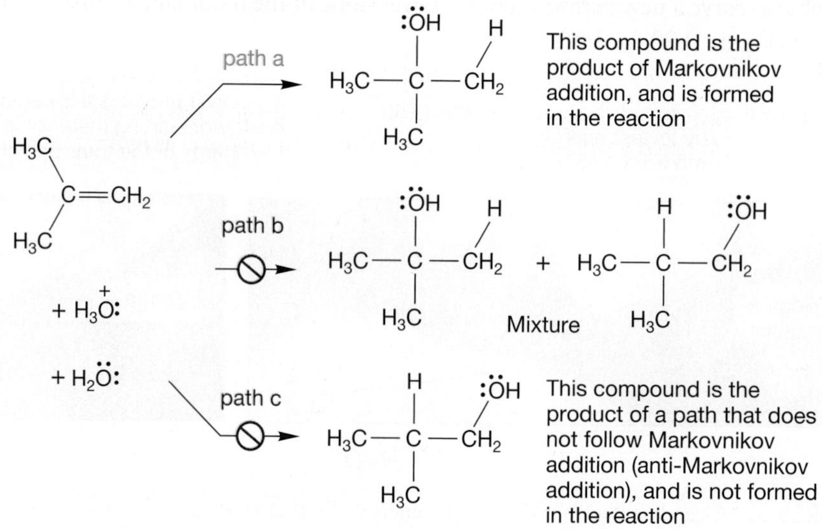

FIGURE 9.35 Three possible regiochemical results from the hydration of 2-methylpropene.

A crucial step in our mechanism involves the formation of a carbocation. When there is a choice, as there certainly is with 2-methylpropene, the more stable carbocation must be formed. Once formed, this cation must ultimately give the product in which the hydroxyl group is attached to the more substituted carbon. If our mechanism postulated in Figure 9.36 is correct, there is no other possible result. So if we get either the "wrong" alcohol or a mixture of alcohols (paths b and c in Fig. 9.35), we know the mechanism is wrong. It may need only a minor adjustment, or it could require complete scrapping, but it cannot be correct as written.

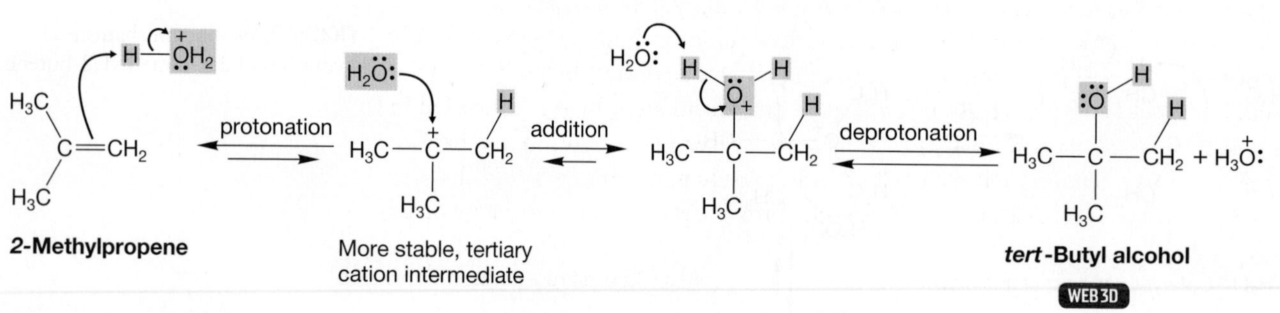

FIGURE 9.36 The mechanism for Markovnikov addition of water to 2-methylpropene. The product must be *tert*-butyl alcohol if this is indeed the mechanism.

More interesting, and less obvious, is what we know if our prediction turns out to be correct. Have we proved the mechanism? Absolutely not. Our mechanistic hypothesis remains just a hypothesis: a guess at how the reaction goes that is consistent with all experiments that have been done so far. There may be some other test experiment that would not work out as well. The sad fact is that no number of experiments we can do will suffice to *prove* our mechanism. We are doomed to be forever in doubt, and our mechanism to be forever suspect. There is no way out. Indeed, all the mechanisms in this book are "wrong," at least in the sense that they are broad-brush treatments that do not look closely enough at detail.

Organic chemistry has come a long way since the days of the rectangular hypothesis, which was little more than a bookkeeping device. In olden times it was common practice to write a reaction such as the S_N1 reaction of *tert*-butyl iodide with water in the way shown in Figure 9.37. To use the rectangular hypothesis, one simply surrounds the appropriate groups with a rectangle, which somehow removes the captured atoms and reconstitutes them, in this case as hydrogen iodide.

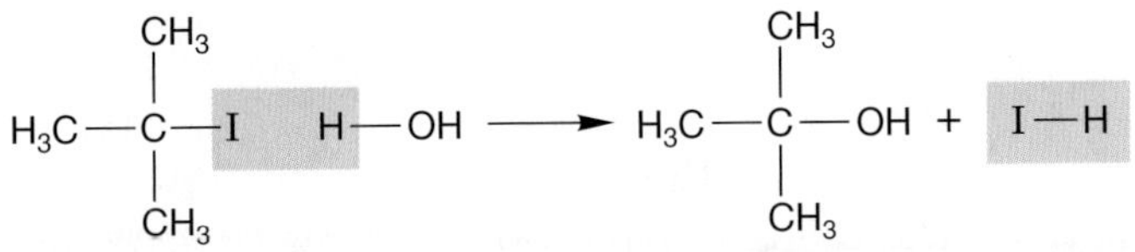

FIGURE 9.37 The rectangular hypothesis applied to the reaction of *tert*-butyl iodide with water.

This archaic device is scarcely a reasonable description of how the reaction occurs and has almost no predictive value. If you know one rectangle, you do not know them all! But let's not get too smug. We do know more than chemists of other generations. We know, for example, that carbocations are involved in the reactions we are currently studying, and molecular orbital theory gives us additional insight into the causes of these reactions. But we are still very ignorant. We are certain that much of what we "know" is wrong, at least in detail. Chemists of the future will regard us as primitives, and they will be quite right. It doesn't matter, of course. Our job is to do the best we can; to add what we can to the fabric of knowledge and to provide some shoulders for those future chemists to stand on, as our earlier colleagues did for us.

In this case, things turn out well for the mechanistic hypothesis shown in Figure 9.36, and the Markovnikov rule is followed in hydration reactions.

There are variations on this reaction. Ethylene, for example, is hydrated only under more forcing conditions—concentrated sulfuric acid must be used—and the initial product is not the alcohol, but ethyl hydrogen sulfate, which is converted in a second step into the alcohol (Fig. 9.38). It is not easy to be sure of the mechanism of this reaction. The concentrated sulfuric acid may be required to produce the very unstable primary carbocation, or a different, perhaps cyclic cation may be involved (see Problem 9.9).

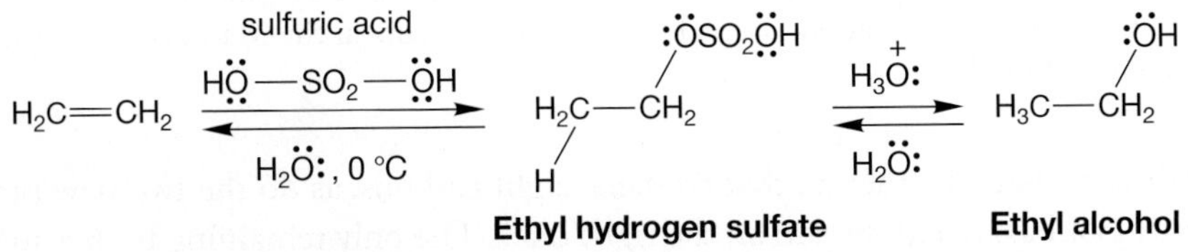

FIGURE 9.38 Ethyl hydrogen sulfate is an intermediate in the hydration of ethylene.

Summary

Water is not a strong enough acid to protonate a simple alkene. However, if a trace of acid catalyst is present, an initial protonation will be followed by addition of water to produce an oxonium ion. A final deprotonation gives the alcohol. As usual, you have to be careful to analyze which carbocation is formed, and to consider structural, resonance, and inductive effects in making that determination.

9.8 Dimerization and Polymerization of Alkenes

Let's look again at the hydration of 2-methylpropene. What if we change the conditions a bit from what they were in Figure 9.36? What if both the temperature and concentration of sulfuric acid are increased? Figure 9.39 shows that two new products, each containing eight carbons, are formed. These are dimeric products (C_8H_{16}), formed somehow by the combining of two molecules of 2-methylpropene (C_4H_8).

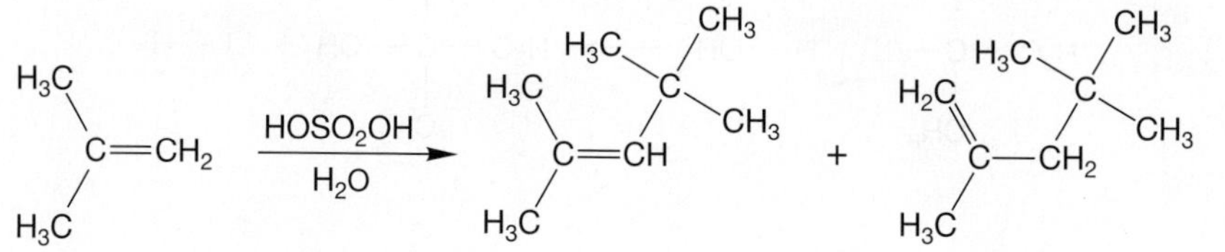

FIGURE 9.39 Dimeric products formed from 2-methylpropene.

Surely, increasing the acid concentration can only make protonation of the alkene more probable, so the first step of the reaction is again formation of the tertiary carbocation intermediate shown in Figure 9.36. Now, however, the initially formed carbocation finds itself in the presence of fewer water molecules than it did in the reaction we looked at before. Accordingly, formation of the alcohol through reaction with water is less likely. We might ask what other Lewis bases are available for the strongly Lewis acidic carbocation. The most obvious one is the starting alkene itself. The *tert*-butyl cation intermediate can react with it, just as the proton did. Both the proton and the carbocation are Lewis acids and can add to the double bond of 2-methylpropene to give a new cation (Fig. 9.40). There is no essential difference in the two reactions, and once more it is the more stable, tertiary carbocation intermediate that is formed.

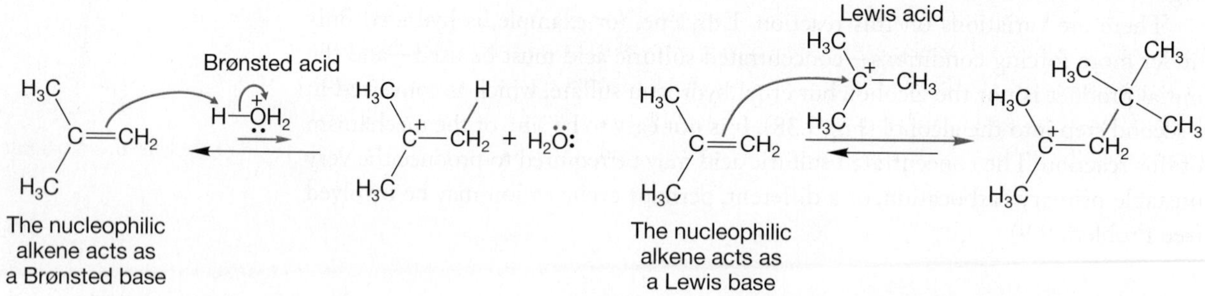

FIGURE 9.40 Many different acids add to 2-methylpropene (and other alkenes). Here we see H_3O^+ and the *tert*-butyl cation adding to 2-methylpropene to give tertiary carbocations. In each case, addition occurs in the Markovnikov sense, as the more stable carbocation is formed preferentially.

Now we have a structure that contains eight carbons, as do the two new products. It looks as though we are on the right track. The only remaining trick is to see how to lose the "extra" proton and form the new alkenes. Losing H_a as shown in

Figure 9.41, will lead to the first product and losing H$_b$ gives the second. We have seen similar deprotonations before in our discussion of the E1 reaction (p. 298), and we have also seen the reverse reaction. Loss of these protons is just the reverse of protonation of an alkene (Fig. 9.41).

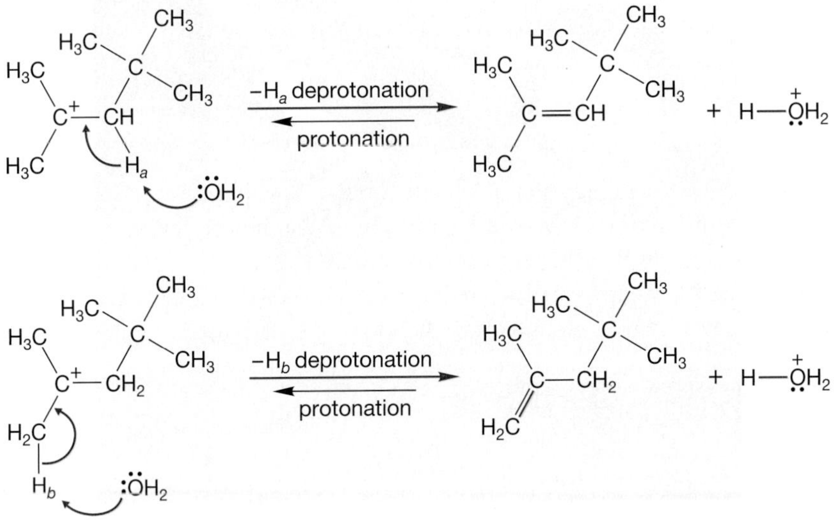

FIGURE 9.41 The two products of Figure 9.39 are formed by deprotonation of the intermediate carbocation.

The proton can't just leave, however. It must be removed by a base. In concentrated sulfuric acid, strong bases do not abound. There are some bases present, however, and the carbocation is a high-energy species, and thus rather easily deprotonated. Under these conditions, both water and bisulfate ion (HSO$_4^-$) are capable of assisting the deprotonation.

Note the exact correspondence of the steps involved in alkene protonation and deprotonation. The two reactions are just the two sides of an equilibrium (Fig. 9.42).

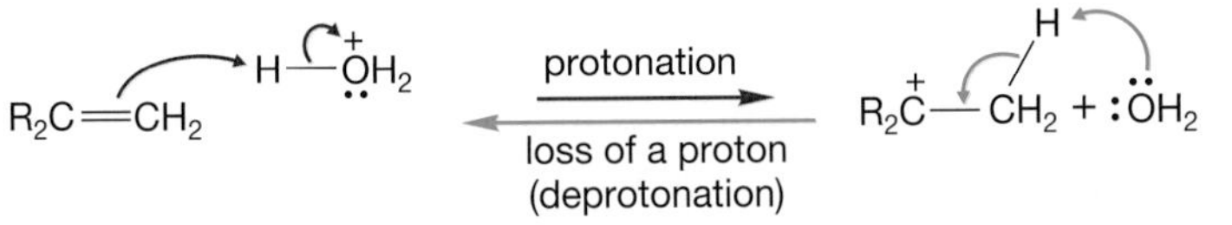

FIGURE 9.42 Protonation and deprotonation are just two sides of an equilibrium.

If the concentration of acid is even higher, the base concentration will be even lower and both alcohol formation and deprotonation will be slowed (both require water to act as Lewis base). There is no reason why the dimerization process cannot continue indefinitely under such conditions, leading first to a trimer (a molecule formed from three molecules of 2-methylpropene) and, eventually, to a polymer (Fig. 9.43).

Alkene polymerization

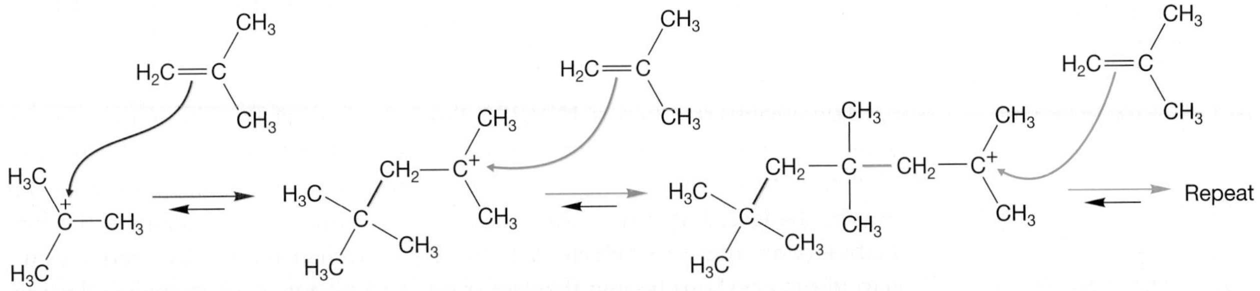

FIGURE 9.43 The beginning stages of the cationic polymerization of 2-methylpropene.

The cation-induced polymerization is called, logically enough, **cationic polymerization** and ends only when the alkene concentration decreases sufficiently to slow the addition step.

Teflon is the polymer formed from polymerization of tetrafluoroethene. Here are several useful applications of this wonderful material.

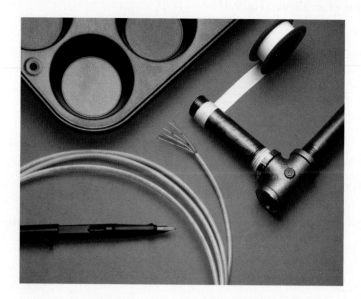

PROBLEM 9.10 Treating 2-ethyl-1-butene with H_3O^+/H_2O leads to three products each having the formula $C_{12}H_{24}$. Draw the three structures and provide a mechanism for their formation.

9.9 Rearrangements during HX Addition to Alkenes

When HCl adds to isopropylethylene (3-methyl-1-butene), a quite surprising thing happens. We do get some of the product we expect, 2-chloro-3-methylbutane, but the major compound formed is 2-chloro-2-methylbutane (Fig. 9.44).

FIGURE 9.44 Reaction of HCl with 3-methyl-1-butene gives a new product in addition to the expected one.

3-Methyl-1-butene

2-Chloro-3-methylbutane
(40%)

2-Chloro-2-methylbutane
(60%)

The obvious question is, Why? Will our mechanism for addition have to be severely changed to accommodate these new data? In fact, we will be able to graft a small modification onto our general mechanism for addition reactions and avoid drastic changes. The way to attack a problem of this kind—explaining the formation of an unusual product—is to write the mechanism for the reaction as we have developed it so far and then see if we can find a way to rationalize the new product. In this case, protonation of 3-methyl-1-butene can give either a secondary or primary carbocation (Fig. 9.45). There is no real choice here, because the secondary carbocation is of far lower energy than the hideously unstable primary carbocation and will surely be the one formed.

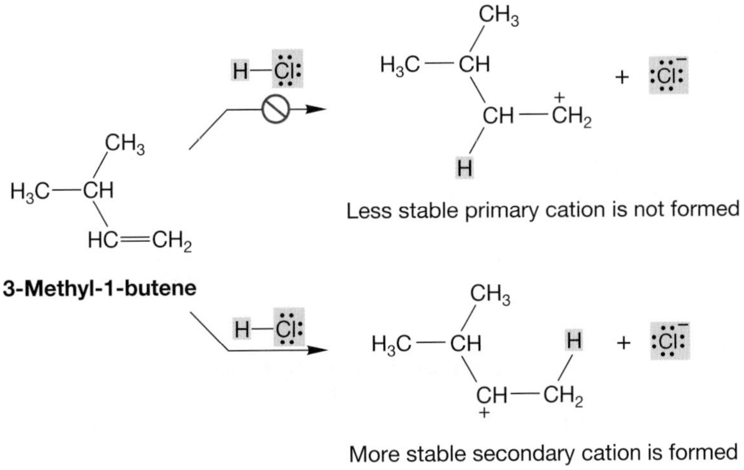

3-Methyl-1-butene

Less stable primary cation is not formed

More stable secondary cation is formed

FIGURE 9.45 No real choice here—the more stable secondary carbocation is formed.

If the chloride ion adds to the positively charged carbon of the secondary carbocation, we get the minor product of the reaction, and the one we surely expected (path a, Fig. 9.46). To get the major product, we must have a **rearrangement**, which is a word describing the relocation of an atom (or atoms) in a molecule as a result of electrons shifting. In this case, a hydrogen must move, with its pair of electrons, from the carbon adjacent to the cation to the positively charged carbon itself (path b, Fig. 9.46). Such migrations of hydrogen with a pair of electrons are called **hydride shifts**.

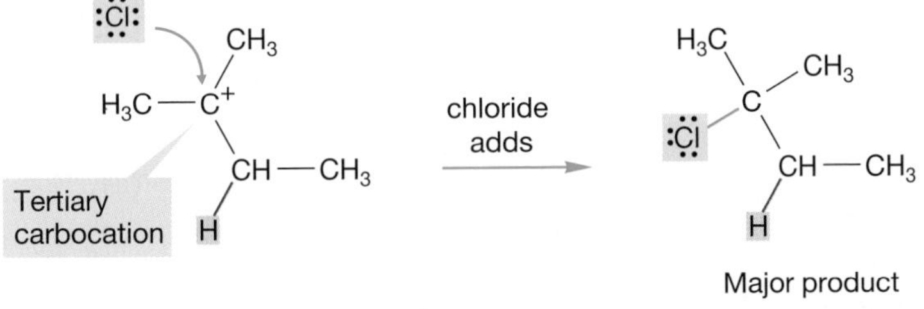

Secondary carbocation

chloride adds

path a

shift of hydride, H:⁻

path b

Minor product

Tertiary carbocation

chloride adds

Major product

FIGURE 9.46 In the reaction between 3-methyl-1-butene and HCl the major product is formed through a rearrangement mechanism (path b) involving the migration of hydrogen with its pair of electrons (hydride, H:⁻) to give the more stable tertiary carbocation from the less stable secondary ion.

Why should this rearrangement happen? Note that in such a process we have generated a lower-energy tertiary carbocation from a higher-energy secondary carbocation. The cation is certainly "better off" (lower in energy) thermodynamically if the hydride moves. In practice, such rearrangements are quite common. Hydride shifts to produce more stable cations from less stable ones are easy. Indeed, not only are hydride shifts common, but so are alkyl shifts, called **Wagner–Meerwein rearrangements** (Georg Wagner, 1849–1903; Hans Lebrecht Meerwein, 1879–1965).

For example, 3,3-dimethyl-1-butene adds hydrogen chloride to give not only 2-chloro-3,3-dimethylbutane but also 2-chloro-2,3-dimethylbutane. The new, rearranged compound results from an alkyl shift—in this case the migration of a methyl group with its pair of electrons (Fig. 9.47). Why is this reaction called an "alkyl shift," when by analogy to "hydride shift" it should be "alkide shift"? We don't know, but it is!

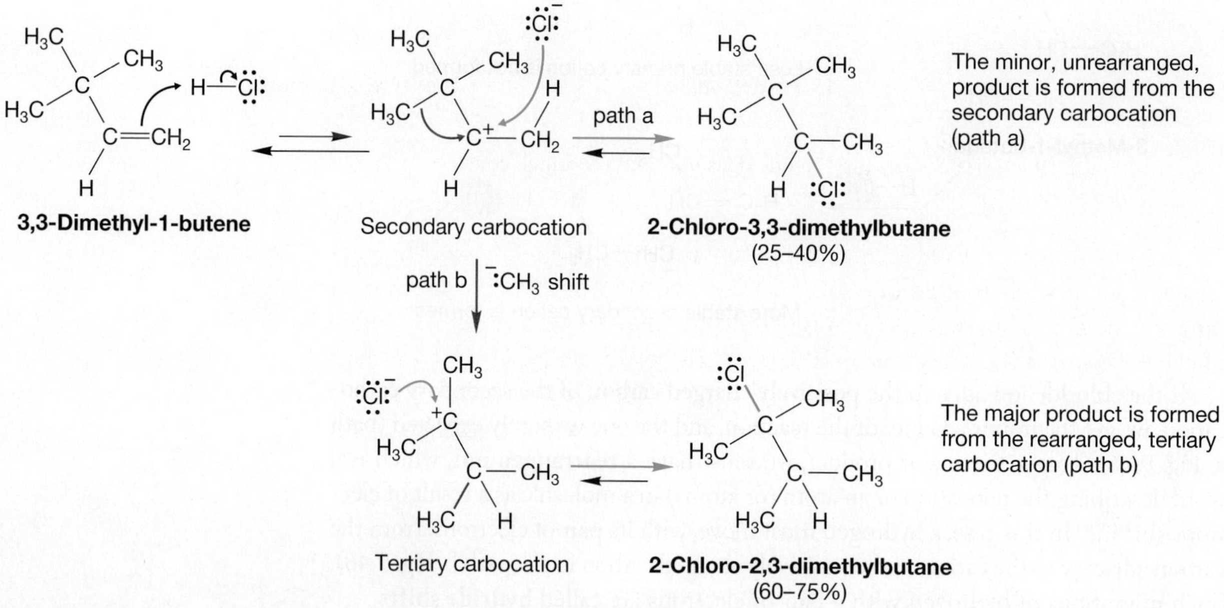

FIGURE 9.47 Formation of the major product involves a rearrangement mechanism in which a shift of a methyl group with its pair of electrons generates a more stable tertiary carbocation from a less stable secondary carbocation.

PROBLEM 9.11 Write mechanisms that account for the products in the following reaction:

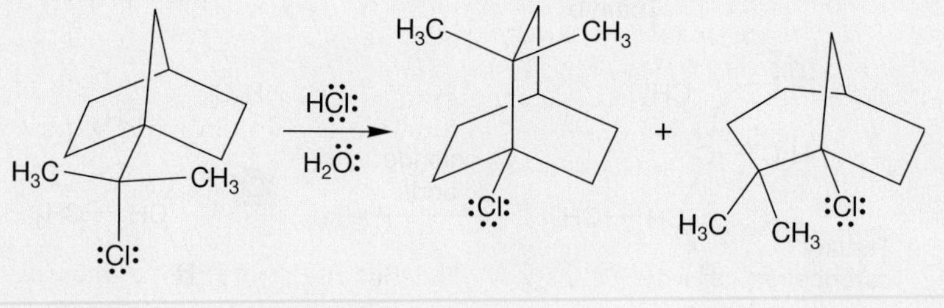

PROBLEM 9.12 A hydride shift to a positively charged carbon is just another example of a Lewis base–Lewis acid reaction. Although it looks very different from the more obvious electrophile–nucleophile reactions we have studied, it really is quite similar. In the reaction shown in the previous problem, there is a very strong and strategically located electrophile and a weak, but nonetheless reactive enough nucleophile. Identify the nucleophile and electrophile in this reaction.

Such rearrangements are so common that they have come to be diagnostic for reactions involving carbocations. Molecules prone to such rearrangements are used to test for carbocation involvement. Here is an example: Attempted S_N1 solvolysis of neopentyl iodide (1-iodo-2,2-dimethylpropane) in water leads not to neopentyl alcohol, as might be anticipated, but instead to a rearranged alcohol, 2-methyl-2-butanol (Fig. 9.48).

Carbocation rearrangement E1

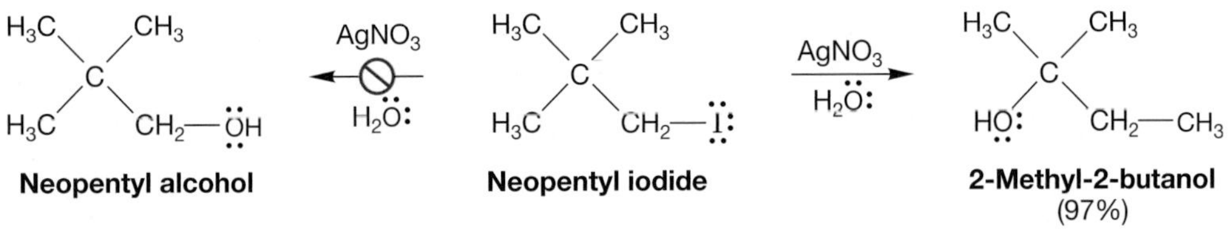

Neopentyl alcohol   **Neopentyl iodide**   **2-Methyl-2-butanol**
(97%)

FIGURE 9.48 The solvolysis of neopentyl iodide does not take a "normal" course.

These results imply the scenario outlined in path a of Figure 9.49, in which an unstable primary carbocation rearranges to a much more stable tertiary carbocation through the shift of a methyl group. In fact, the presence of the primary ion is so unsettling—they are most unstable and not likely to be formed—that a variant of this mechanism has been proposed in which the methyl group migrates in a concerted fashion as the iodide departs (path b, Fig. 9.49). A **concerted reaction** is one that has no intermediates.

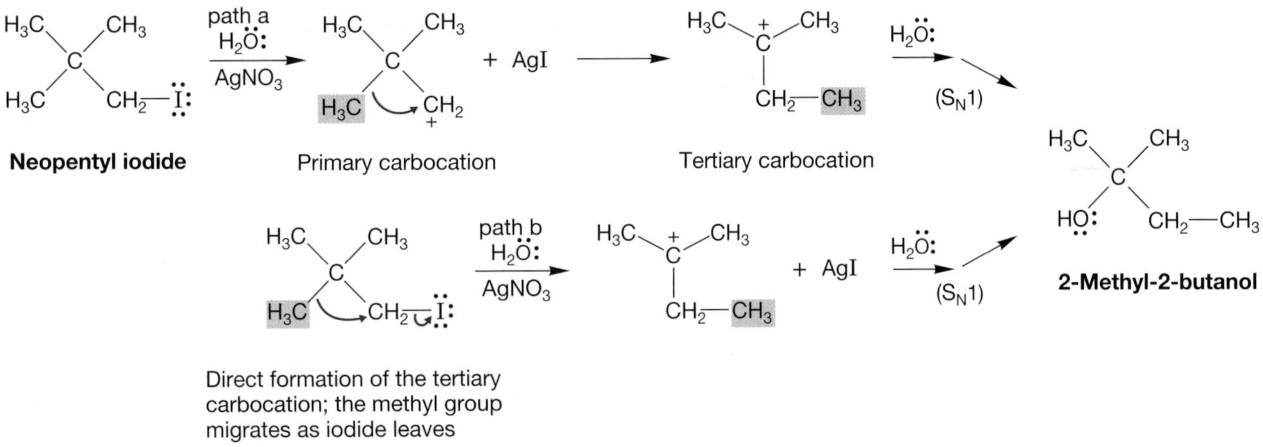

Neopentyl iodide   Primary carbocation   Tertiary carbocation   2-Methyl-2-butanol

Direct formation of the tertiary
carbocation; the methyl group
migrates as iodide leaves

FIGURE 9.49 Rearrangements are common in many reactions in which carbocations are involved. Path a shows the stepwise process and path b is the concerted process.

At very low temperature in highly polar but nonnucleophilic solvents, some carbocations can be observed spectroscopically. Under such conditions it is possible to determine that many carbocations are rapidly rearranging as we observe them. For example, the two tertiary carbocations in Figure 9.50 rapidly interconvert at −60 °C.

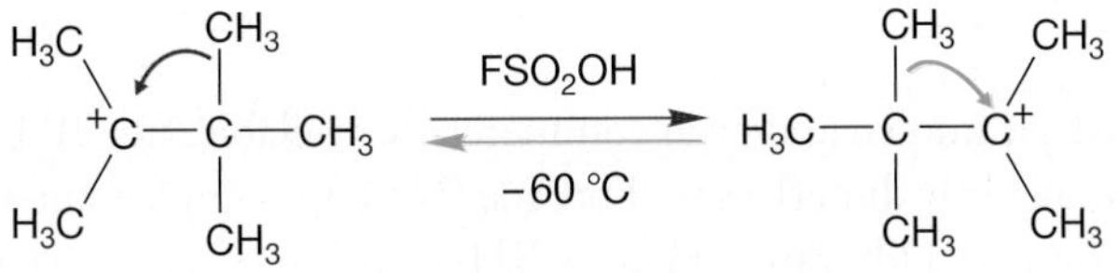

FIGURE 9.50 Rearranging tertiary carbocations.

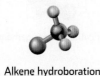

Alkene hydroboration

9.10 Hydroboration

Now we pass on to another addition reaction, **hydroboration**, which is the addition of hydrogen and boron across a π bond. The mechanism is not simple, and it will take some effort for you to become comfortable with all the details. The mechanistic discussion does make several worthwhile points, so it merits your attention for that reason alone. But there is another, more compelling reason to master the hydroboration reaction—it is one of the most useful of synthetic reactions. Indeed H. C. Brown (1912–2004) won a Nobel Prize in 1979 for the development of hydroboration. Consider the following seemingly simple synthetic task outlined in Problem 9.13.

WORKED PROBLEM 9.13 Try to provide a synthesis of 2-methyl-1-propanol. You may start from any alkene.

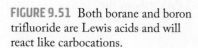

2-Methyl-1-propanol

ANSWER Well, there isn't one (yet). At this point in your study of organic chemistry, you have no way to make this simple alcohol. If we start from the obvious alkene, 2-methylpropene, the hydration reaction we learned in this chapter (p. 380) can only give the product of Markovnikov addition, *tert*-butyl alcohol. This nonanswer to such a simple question points out how limited our synthetic skills are so far.

The hydroboration reaction solves both Problem 9.13 and the general difficulty, so let's see how it works.

In the previous sections, we saw both the protonation of alkenes and the addition of a carbocation to an alkene to generate a new carbocation. The familiar theme of overlap between filled (alkene π) and empty (carbon 2*p*) orbitals was recapitulated: "Lewis bases react with Lewis acids." Other reagents containing empty *p* orbitals, all good Lewis acids, might also be expected to add to alkenes, and so they do.

We know another kind of molecule that has an empty 2*p* orbital, the trigonal boranes, BF_3 and BH_3 (Fig. 9.51). There is no positive charge on these compounds, but they should still be strong Lewis acids (electrophiles).

FIGURE 9.51 Both borane and boron trifluoride are Lewis acids and will react like carbocations.

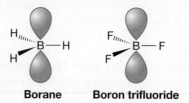

Borane   **Boron trifluoride**

PROBLEM 9.14 Verify that boron in BF_3 and BH_3 is neutral and that each boron atom has an empty 2*p* orbital.

Although BF_3 is known and even commercially available, free BH_3 is unavailable because it spontaneously dimerizes to diborane (B_2H_6), an unpleasant-smelling, flammable, and toxic gas. Diborane (H_3B—BH_3) does not have a structure similar to ethane, H_3C-CH_3. An ethane-like structure requires 14 electrons to make 7 boron–hydrogen and boron–boron σ bonds. The 6 hydrogens in diborane contribute 6 electrons and the 2 borons another 6 to give a total of 12, so we are 2 electrons shy of the required number. The real structure is shown in Figure 9.52, along with a resonance formulation. The molecule is a hybrid of the two resonance forms **A** and **A′**. In this molecule, partial bonds (shown as dashed bonds in Fig. 9.52) are used to connect the atoms. This kind of bonding is more common than once thought. It abounds in inorganic chemistry, and in the organic chemistry of electron-deficient species, such as carbocations.

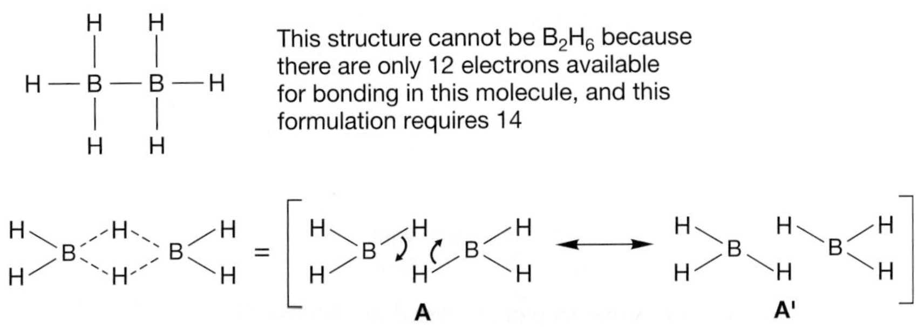

This structure cannot be B_2H_6 because there are only 12 electrons available for bonding in this molecule, and this formulation requires 14

FIGURE 9.52 The real structure of diborane.

When diborane dissolves in diethyl ether ($CH_3CH_2-O-CH_2CH_3$) a borane–ether complex is formed. This complex can be used as a source of BH_3. Often the cyclic ether tetrahydrofuran is used. Some borane–ether complexes are so stable that they can even be distilled.

WORKED PROBLEM 9.15 Propose a structure for the ether–borane complex, and suggest a mechanism for its formation.

ANSWER Borane is a Lewis acid because the boron atom has an empty $2p$ orbital. An ether, here tetrahydrofuran (THF), is a Lewis base. The reaction between the Lewis base and Lewis acid produces the borane–THF complex.

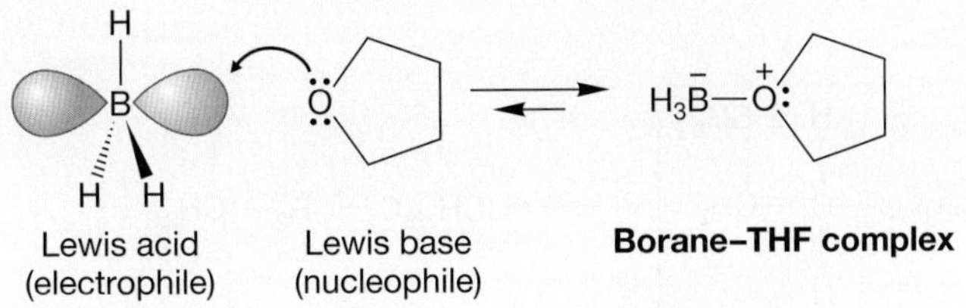

Lewis acid (electrophile) Lewis base (nucleophile) **Borane–THF complex**

When the borane–ether complex is allowed to react with an alkene, there is a rapid addition of the borane across the double bond. The initial product is an alkylborane (Fig. 9.53). As in the addition of a carbocation to a double bond, the initial interaction is between the empty $2p$ orbital (in this case on neutral boron,

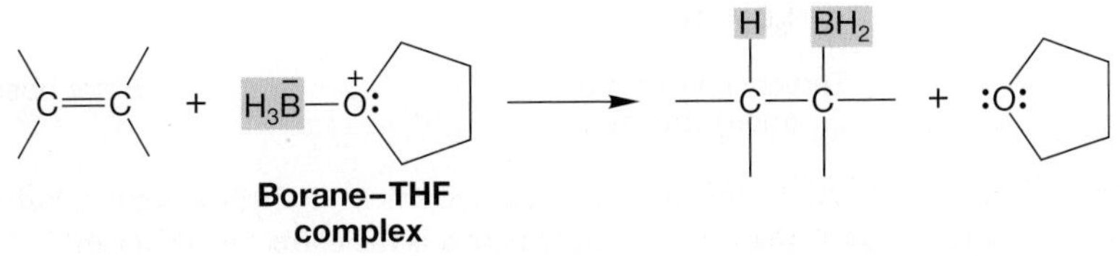

Borane–THF complex

FIGURE 9.53 A schematic hydroboration. The elements of hydrogen and boron have been added across the double bond.

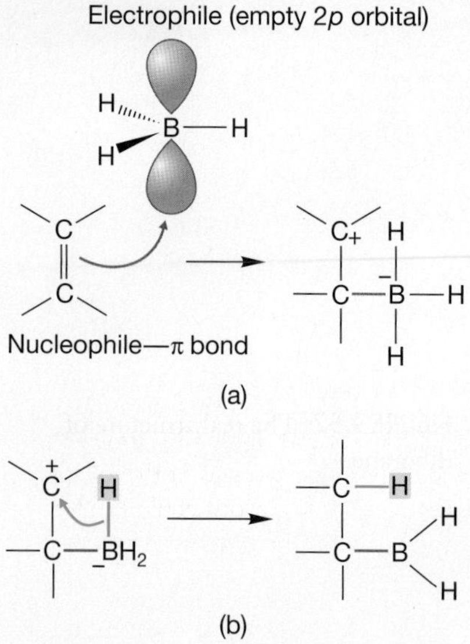

Electrophile (empty 2*p* orbital)

Nucleophile—π bond

(a)

(b)

FIGURE 9.54 (a) A possible first step in addition of BH_3 to an alkene. The π bond of the alkene acts as nucleophile and reacts with the strong Lewis acid, BH_3. (b) A possible second step in the hydroboration reaction is transfer of a hydride ($H:^-$).

not positive carbon) and the filled π orbital of the alkene (Fig. 9.54a). If the mechanism were directly parallel to that of cation addition, we would form the charge-separated molecule shown in Figure 9.54a. A hydride ($H:^-$) could then be delivered from the negative boron to the positively charged carbon to complete the addition (Fig. 9.54b). However, the mechanism is *not* exactly parallel to what we saw before. In this case, the hydride is transferred as the carbon–boron bond is made (Fig. 9.55). This reaction occurs in a single step. There is no intermediate. In the transition state, partial bonds from carbon to boron and carbon to hydrogen are formed. Let's see how we know this mechanism satisfies the experimental data and, as we do this, why the more conventional mechanism is inadequate.

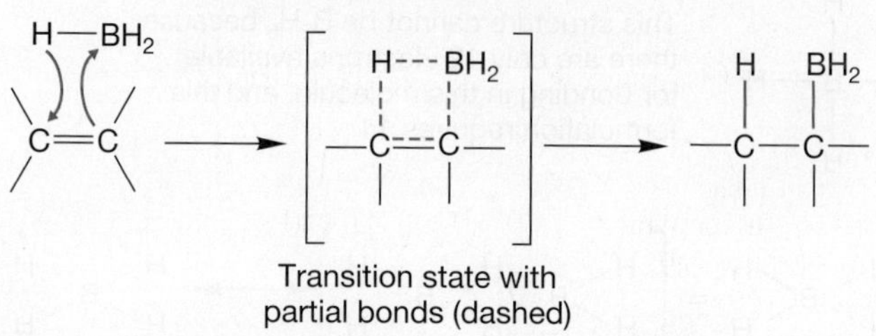

Transition state with
partial bonds (dashed)

FIGURE 9.55 A one-step or concerted mechanism for hydroboration.

First of all, there are no rearrangements during hydroboration. The concerted mechanism outlined in Figure 9.55 cannot lead to rearrangements, but the intermediate carbocation of Figure 9.54 would surely do so. For example, the hydroboration of 3,3-dimethyl-1-butene leads exclusively to borane **A**, shown in path a of Figure 9.56. Were a carbocation intermediate involved, rearrangement of the secondary carbocation to the more stable tertiary carbocation would seem inevitable, and product **B** should be observed (path b of Fig. 9.56). But the reaction yields a regiospecific hydroboration. Only one of two possible product boranes is formed. This phenomenon is general—in hydroboration the boron becomes attached to the less substituted carbon of the double bond. Our one-step mechanism must be able to explain this regiochemical preference.

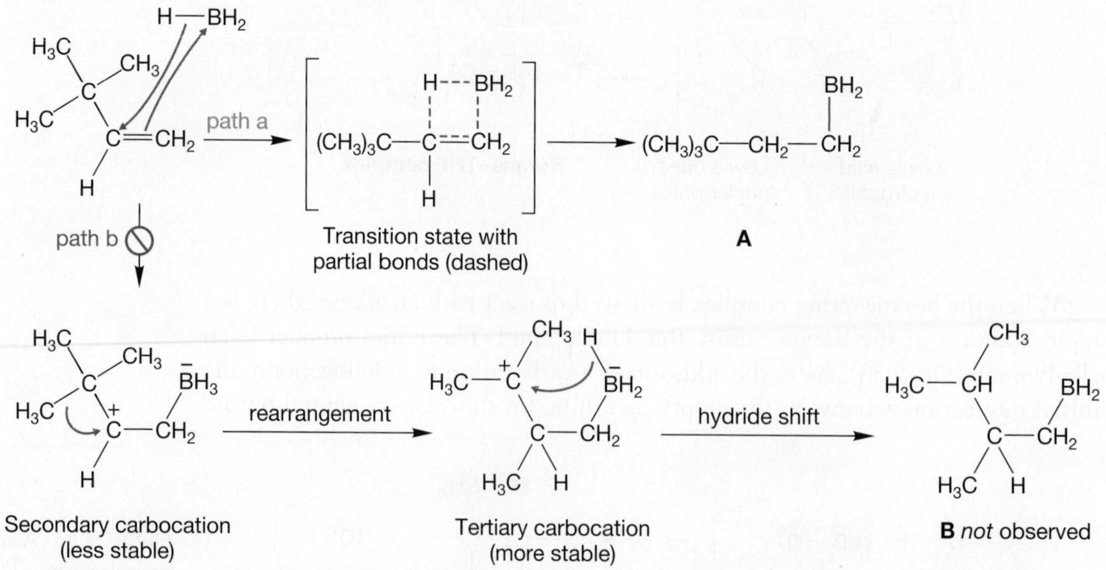

FIGURE 9.56 An intermediate carbocation of path b should lead to rearrangements in hydroboration, but no evidence of such can be found, which suggests that the concerted path a is the correct mechanism.

There are at least two factors, one simple, the other more complicated, that explain the regiospecific addition in the hydroboration reaction. The simple analysis is a steric argument. Consider the reaction of BH_3 with 2-methyl-1-butene. There are steric differences in the two possible concerted additions. In the addition leading to the observed product (Fig. 9.57a), the larger end of the borane molecule, the BH_2 portion, is nearer to the smaller hydrogens than to the larger alkyl groups. In the path leading to the product that is not observed (Fig. 9.57b), the larger end opposes the relatively large methyl and ethyl groups.

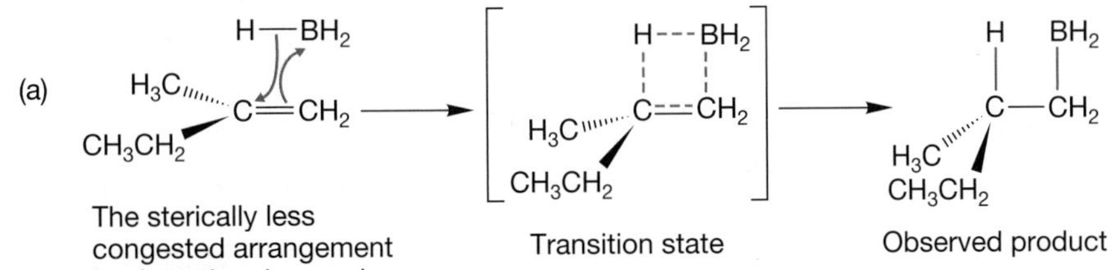

(a) The sterically less congested arrangement leads to the observed product

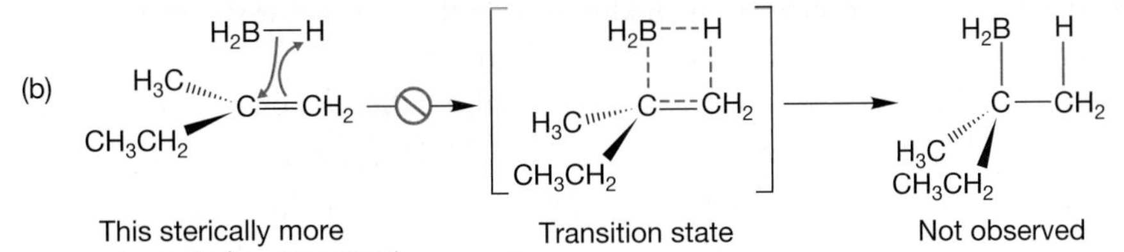

Transition state

Observed product

(b) This sterically more congested arrangement leads to the product that is not observed

Transition state

Not observed

FIGURE 9.57 (a) Steric factors favor addition of the larger BH_2 group to the less congested end of the alkene. (b) Steric interactions could hinder formation of the nonobserved product.

The more complicated factor has to do with the way in which the two new bonds are made in the transition state for the hydroboration reaction. Two bonds are formed in the same step, without a cationic intermediate, but there is no need for the new bonds to develop to the same extent as the reaction proceeds. Indeed, in a philosophical sense, they cannot, because they are different. In the transition state, one bond will be further along in its formation than the other. This differential progress can be symbolized by bonds of different lengths, as in Figure 9.58.

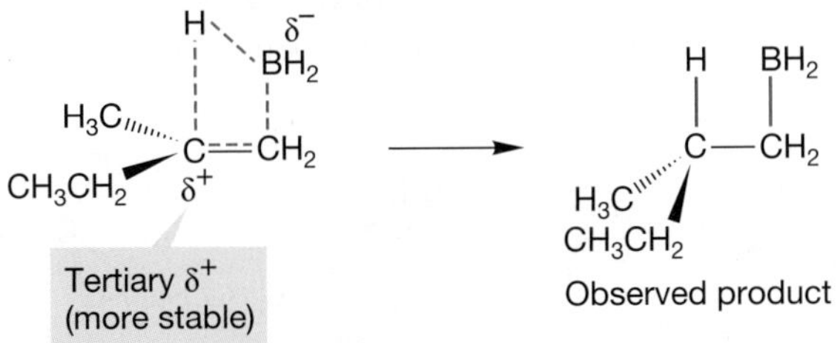

Tertiary δ^+ (more stable)

Observed product

Primary δ^+ (less stable)

This product is not formed

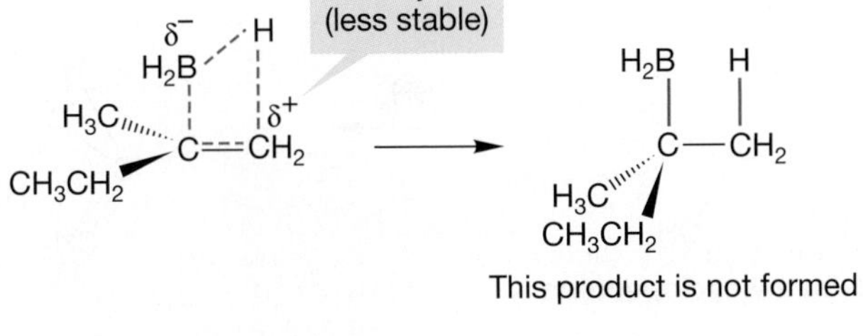

FIGURE 9.58 In the two possible transition states for the concerted addition, there will be partial positive charge developed on carbon. This δ^+ will be more stable at the more substituted position.

In such a mechanism, a full positive charge is not developed on carbon, *but a partial one is*. All the factors that operate to stabilize a full positive charge also operate to stabilize a partial positive charge. A more-substituted partial positive charge is more stable than a less-substituted partial positive charge, and this difference favors the observed product. So, the initial product of hydroboration of an alkene is the monoalkylborane formed by addition in which boron becomes attached to the less substituted end of the double bond (Fig. 9.58). Does this result violate Markovnikov's rule? No!

Here is an instance where rules can be confusing. When we say "Markovnikov addition," we are used to seeing the hydrogen attached to the less substituted carbon of the double bond and the X group to the other carbon (Fig. 9.59).

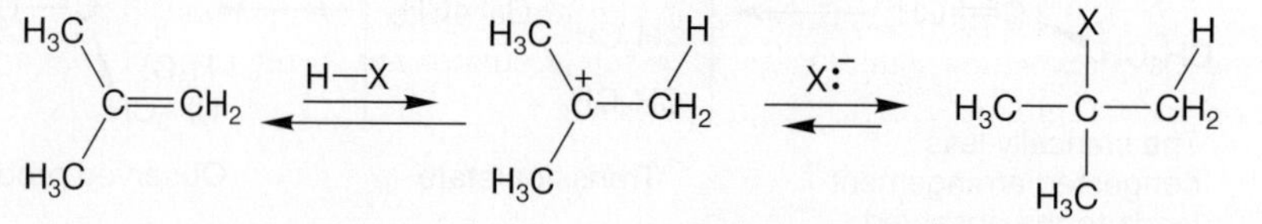

FIGURE 9.59 Typical Markovnikov addition in which the more stable cation is formed as an intermediate to give the product in which H is attached to the less substituted carbon of the original double bond, and X is attached to the more substituted end.

When H—BH$_2$ adds across a π bond, it is the boron that is the positive end of the dipole, the electrophile in the addition. It is the boron, not the hydrogen, that becomes attached to the less substituted end of the double bond. As Figure 9.60 tries to show, it is the terminology that gets complicated, not the reaction mechanisms. There is danger in learning rules; it is much safer to understand the reaction mechanism.

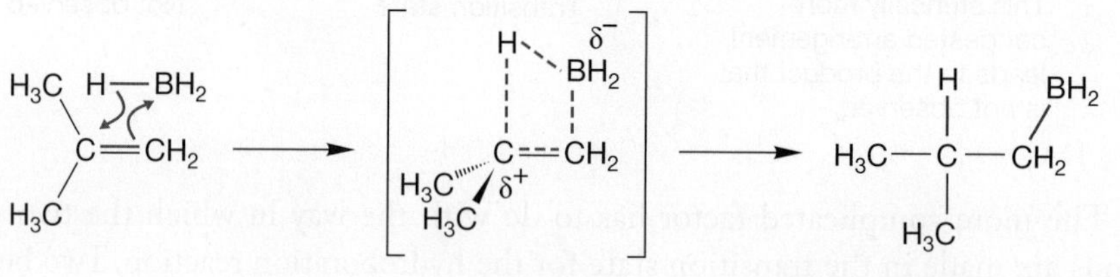

FIGURE 9.60 This reaction is still Markovnikov addition. The electrophilic boron in this case adds to the less substituted end of the alkene. Positive charge develops on the more substituted carbon.

Transition state

There is another reason we know that the concerted mechanism for hydroboration is correct, and it involves the stereochemistry of the addition reaction.

Hydroboration can be shown to proceed in syn fashion. A **syn addition** describes the result when two pieces (H and BH$_2$ in this case) are delivered to the same side of an alkene. Figure 9.61 shows the hydroboration of *cis*-1,2-dideuterio-1-hexene and of 1-methylcyclopentene. These reactions exclusively produce the products of syn addition.

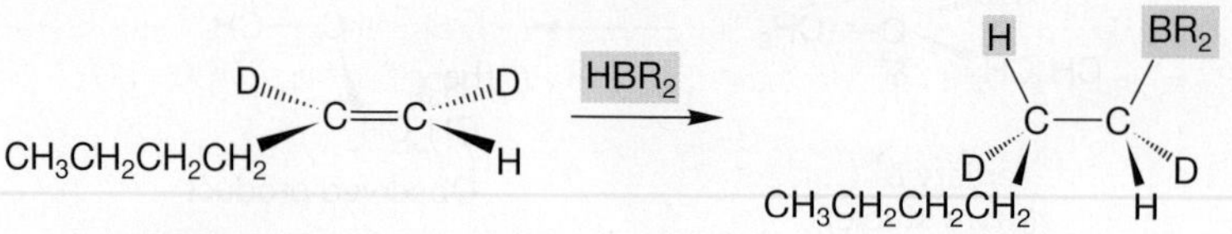

cis-1,2-Dideuterio-1-hexene

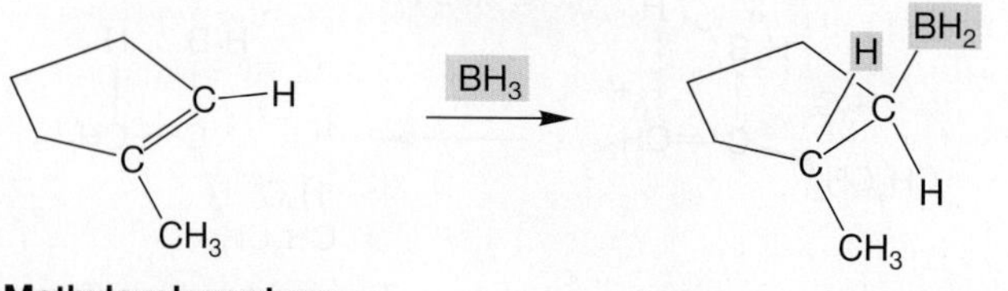

FIGURE 9.61 These two stereochemical experiments show that addition of BH$_3$ proceeds in a syn fashion. The H and BH$_2$ groups are delivered to the same side of the alkene (R = cyclohexyl).

1-Methylcyclopentene

Of course, the one-step addition process must involve syn addition because both new bonds, the one to boron and the one to hydrogen, are made at the same or nearly the same time. Figure 9.62 shows the observed syn addition to deuterated 1-hexene.

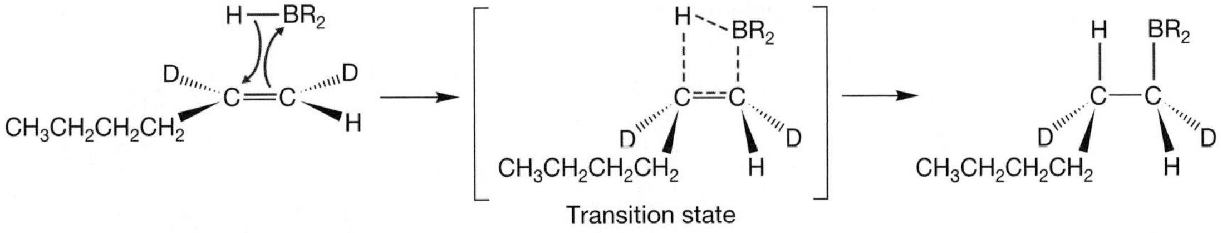

FIGURE 9.62 Syn addition is accommodated by a mechanism in which both new bonds are forming in the transition state. There are no intermediates and therefore no chance for rotation and scrambling of stereochemistry.

A mechanism involving an intermediate carbocation has great problems accounting for syn addition. If a planar cation were formed on addition to the deuterated 1-hexene, we would expect the rapid (~10^{11} s^{-1}) rotation about carbon–carbon single bond to produce the two diastereomeric products shown in Figure 9.63.

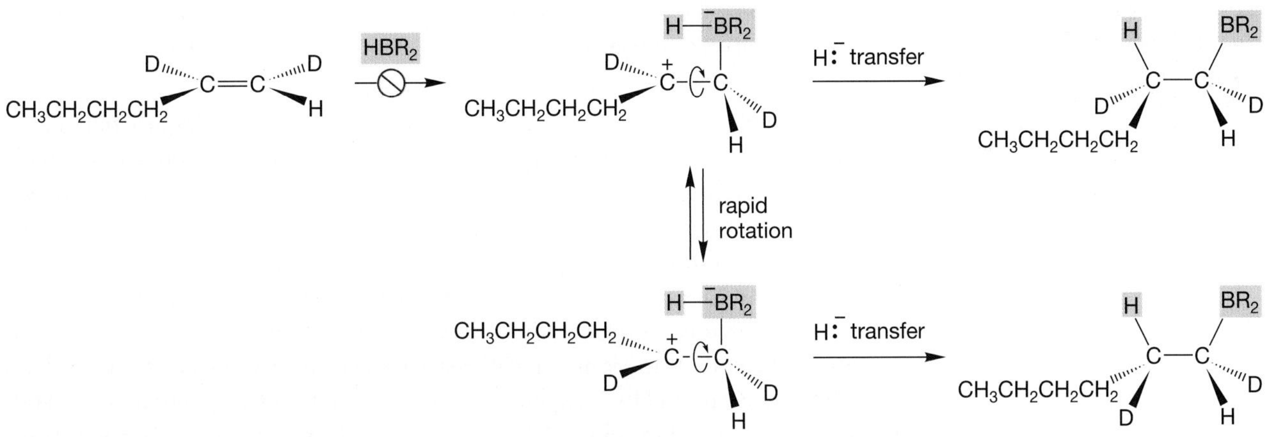

FIGURE 9.63 Rotation about carbon–carbon single bonds is very fast. A two-step mechanism predicts that two products should be formed from the hydroboration of *cis*-1,2-dideuterio-1-hexene. Both syn and anti addition products should be formed. However, only the product of syn addition is found, indicating that there is no carbocation intermediate.

Summary

The current mechanistic hypothesis for hydroboration, a concerted, one-step syn addition to alkenes in which the two new bonds are partially formed to different extents in the transition state, nicely rationalizes all the experimental data. A mechanism involving formation of an open cation followed by hydride transfer fails to account for the observed syn stereochemistry of addition, which is demanded by the one-step mechanism.

WORKED PROBLEM 9.16 Draw Energy versus Reaction progress diagrams for the concerted and two-step mechanisms for the addition of BH_3 to alkenes.

ANSWER The crucial difference is the presence of an intermediate in the two-step process. The one-step (concerted) reaction simply passes from starting material to product over a single transition state.

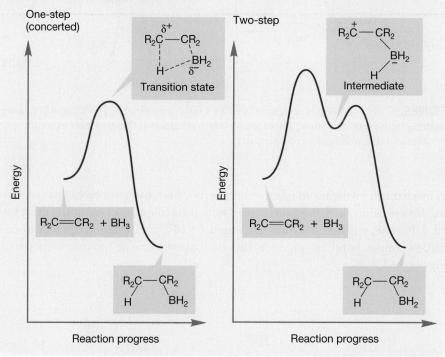

This example is typical of the use of stereochemistry in determining reaction mechanism. One reason we spent so much time on the stereochemical relationships of molecules of various shapes (diastereomers, enantiomers, and meso forms) was that we will use these relationships over and over again as we dig into questions of reaction mechanism. If it is not obvious what we are talking about in the last few paragraphs, and if Figures 9.61–9.63 are at all unclear, be sure to go back to Chapter 4 and work over the stereochemical arguments again. Otherwise, you run the risk of being lost in further discussion. The key points to see are the difference between the two hypothetical products shown in Figure 9.63 and their required formation from the rotating cation. A good test is to see if you can do Problems 9.17 and 9.18 easily. If you can, you are in good shape.

PROBLEM 9.17 Bromine adds to alkenes to give dibromo compounds. Based on the products from the reaction of bromine with the 2-butenes, determine if the addition is syn or anti. This problem does not pose a mechanistic question—you do not have to draw arrow formalisms. You only need to determine from the structure of the products the direction from which the two bromines have added.

PROBLEM 9.18 Devise a way to determine the stereochemistry of the addition of hydrogen bromide to alkenes. Assume you have access to any necessary starting materials.

A single addition isn't the end of the hydroboration story. The initially produced monoalkylborane still has a trigonal boron; therefore, it has an empty $2p$ orbital (Fig. 9.64). It's still a Lewis acid, and it can participate in a second hydroboration reaction to give the product of two hydroborations of the alkene, a dialkylborane.

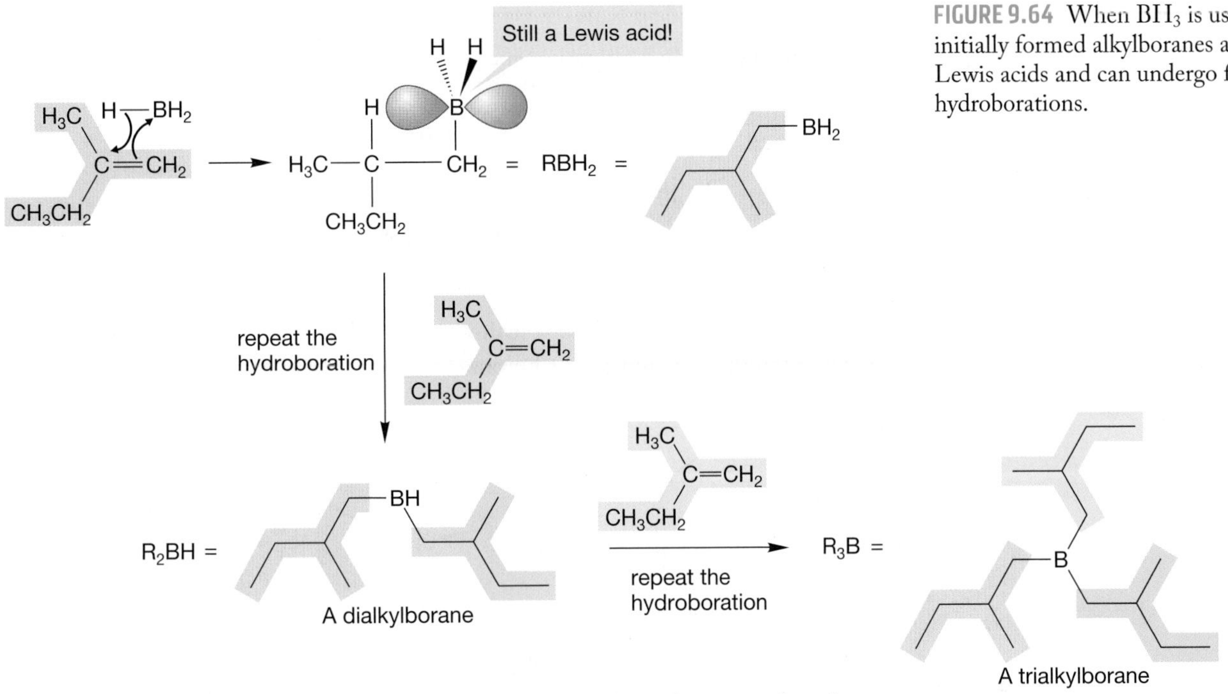

FIGURE 9.64 When BH_3 is used, the initially formed alkylboranes are Lewis acids and can undergo further hydroborations.

Nor must the reaction stop here. The dialkylborane is also a Lewis acid, and it can do one more hydroboration to give a trialkylborane. Now, however, all the original hydrogens that were on the boron are used up and there can be no further hydroborations. In practice, the number of hydroborations depends on the size of the alkyl groups in the alkene. When the groups are rather large, further reaction is retarded by steric effects—the alkyl groups just get in the way. For example, 2-methyl-2-butene hydroborates only twice, and 2,3-dimethyl-2-butene only once (Fig. 9.65).

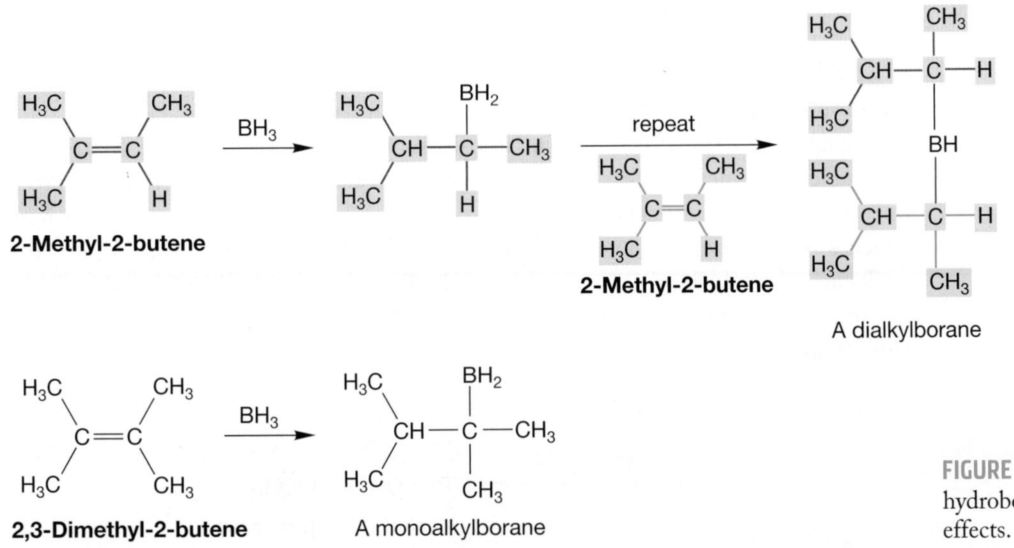

FIGURE 9.65 The number of hydroborations depends on steric effects.

WORKED PROBLEM 9.19 Provide a mechanism for the following reaction:

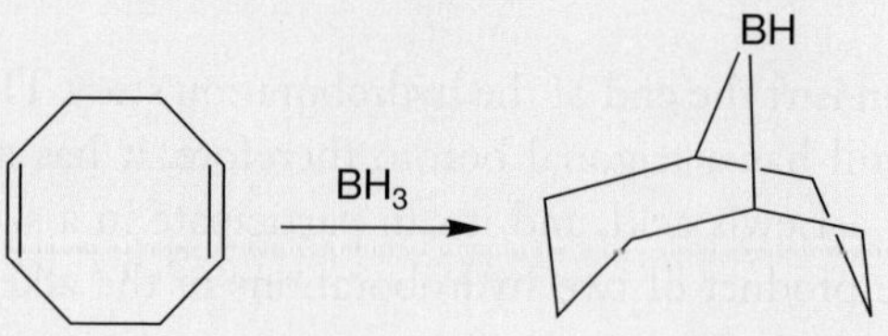

ANSWER The first hydroboration goes in normal fashion.

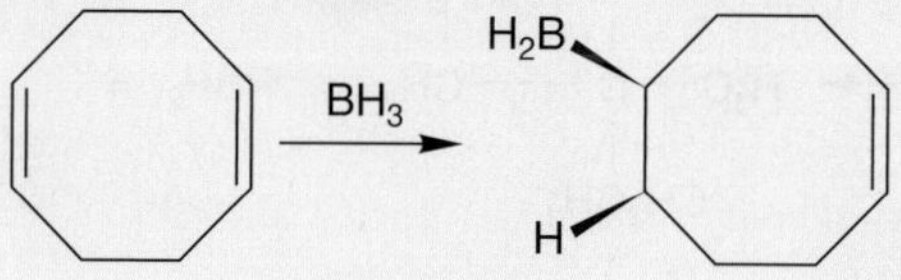

Now a second, intramolecular hydroboration takes place as the remaining double bond in the ring attacks the boron.

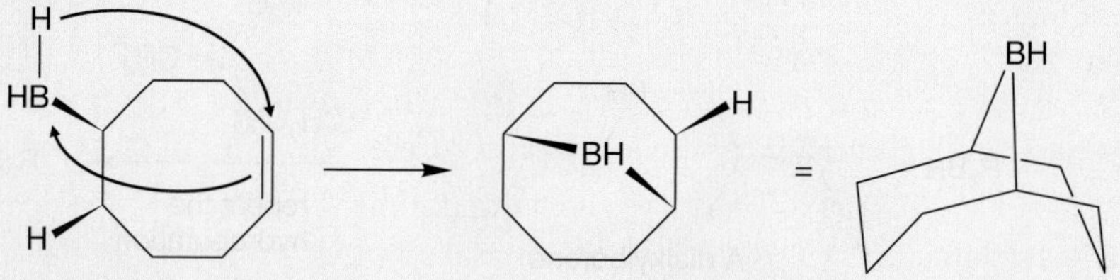

It takes some unraveling to see the structure of the final product. It is important to begin to hone your skills at this kind of thing, and this problem provides practice.

9.11 Hydroboration in Synthesis: Alcohol Formation

It would be presumptuous to attempt to summarize fully the utility of the hydroboration reaction here. There are many variations of this reaction, each designed to accomplish a specific synthetic transformation. A book could be written on the subject. In fact, Professor H. C. Brown has written just such a book. We'll only mention one especially important reaction here, one that leads to a new process for the synthesis of alcohols (recall Problem 9.13, p. 390). When one molecule of trialkylborane is treated with hydrogen peroxide (H_2O_2) and hydroxide ion (HO^-), three molecules of an alcohol are formed along with boric acid (Fig. 9.66).

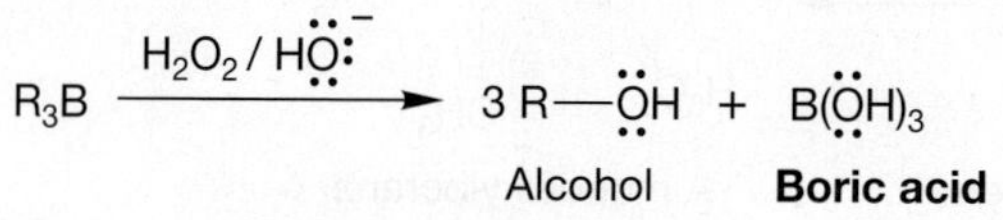

FIGURE 9.66 Formation of an alcohol from a borane.

$$R_3B \xrightarrow{\;H_2O_2\,/\,HO{:}^-\;} 3\ R\!-\!\ddot{O}H\ +\ B(\ddot{O}H)_3$$

$$\text{Alcohol} \qquad \textbf{Boric acid}$$

We are not yet up to a full discussion of the mechanism of this reaction, which is outlined in Figure 9.67, but a number of important points are within our powers.

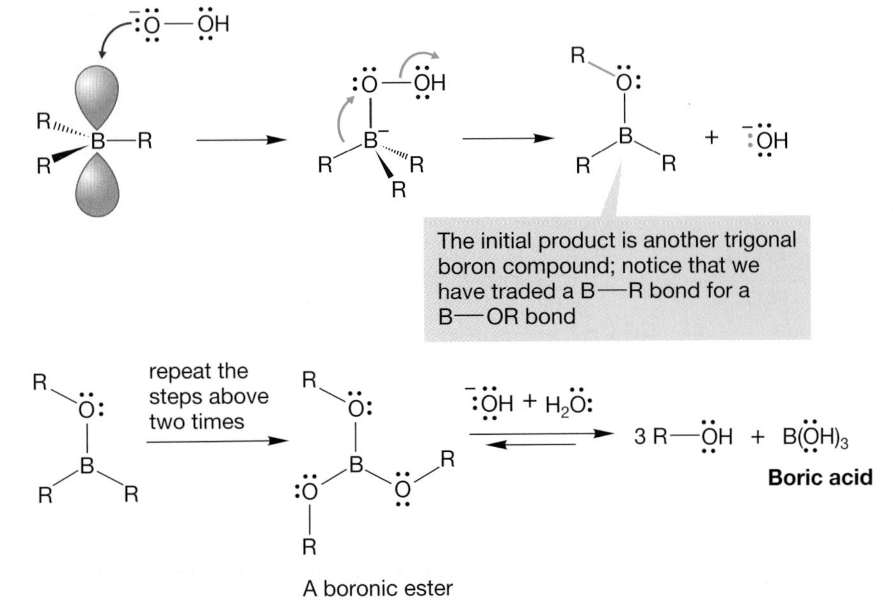

The initial product is another trigonal boron compound; notice that we have traded a B—R bond for a B—OR bond

A boronic ester

FIGURE 9.67 The mechanism of alcohol formation from an alkylborane.

The first step, addition of peroxidate ion (H—O—O⁻) to the borane, is just the reaction of a filled orbital on oxygen with the empty $2p$ orbital on boron. It's another nucleophile–electrophile reaction. Notice that the boron becomes oxidized in this process. A discussion of the later steps is deferred for now, but at least you can see that the boron–oxygen bonds in the $B(OR)_3$ (boronic ester) should be especially strong ones and the reaction will be favored thermodynamically. Finally, in excess hydroxide the boronic ester equilibrates with boric acid and the alcohol.

PROBLEM 9.20 Why are boron–oxygen bonds especially strong?

PROBLEM 9.21 Write a mechanism for the formation of boric acid, $B(OH)_3$, from hydroxide ion, HO^-, and a boronic ester, $B(OR)_3$.

In this case, questions of mechanism are secondary to the utility of this reaction. Note one very important thing: In the overall reaction we are able to produce the *less* substituted alcohol from the alkene. The hydroxyl group enters at the position to which the boron was attached: the less substituted end of the double bond. This reaction is an overall **anti-Markovnikov addition** (Fig. 9.68) because the regioselectivity of the product is the opposite of the Markovnikov product. But Markovnikov's rule was not broken.

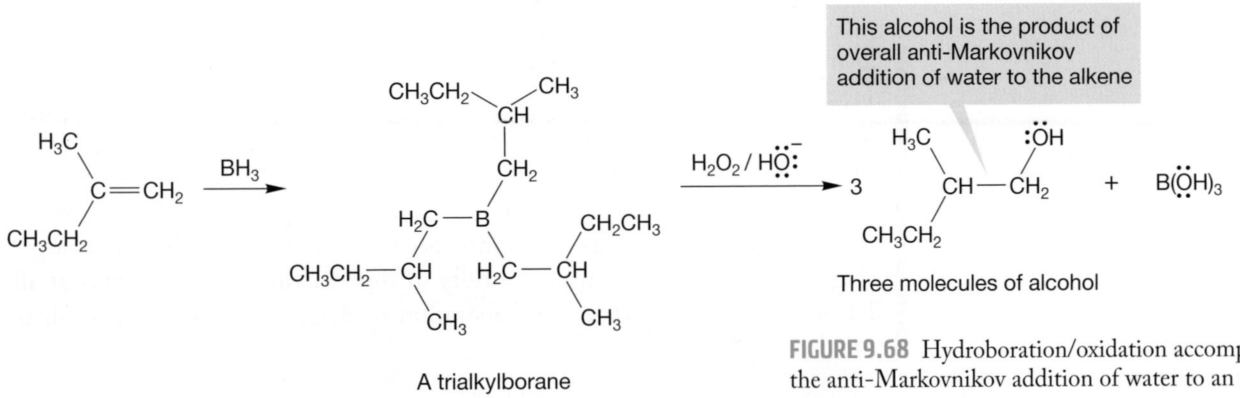

This alcohol is the product of overall anti-Markovnikov addition of water to the alkene

A trialkylborane

Three molecules of alcohol

FIGURE 9.68 Hydroboration/oxidation accomplishes the anti-Markovnikov addition of water to an alkene.

Our previous method for alcohol synthesis, hydration of alkenes, necessarily produced the *more* substituted alcohol. So now we have complementary synthetic methods for producing both possible alcohols from a given alkene (Fig. 9.69). Direct hydration gives the more substituted product (Markovnikov addition) and the indirect hydroboration/oxidation method gives the less substituted alcohol (anti-Markovnikov addition). Be sure to add these reactions to your file-card collection of synthetically useful reactions.

FIGURE 9.69 Two addition reactions that can produce alcohols with different regiochemistry.

ETHYL ALCOHOL

Ethyl alcohol, or ethanol, can be produced by several reactions described in this chapter. Although it has other uses, most ethyl alcohol is made for the production of alcoholic beverages, a use deeply rooted in history. Indeed, the rise of civilization has been attributed to the discovery of beer, and the necessity to give up a nomadic life to attend to the cultivation of the ingredients. In 2005, the average adult American consumed slightly more than 31 gal of beer and more than 4.5 gal of wine and spirits.

Ethyl alcohol acts as a depressant in humans, interfering with neurotransmission. It works by binding to one side of the synapse, changing its shape slightly and making it able to bind γ-aminobutyric acid (GABA) more efficiently. Bound GABA widens the synaptic channel, allowing chloride ions to migrate in, thus changing the voltage across the synapse. The nerve cell is less able to fire, and neurotransmission is inhibited. Your body cleanses itself of alcohol by converting it into acetaldehyde with the enzyme alcohol dehydrogenase. A second enzyme, aldehyde dehydrogenase, completes the conversion into acetic acid. There are individuals, indeed whole races, who are short of aldehyde dehydrogenase and thus overly susceptible to the deleterious effects of alcohol. By contrast, alcoholics have been found to contain elevated levels of this enzyme.

Ethyl alcohol

$$CH_3CH_2OH \xrightarrow[\text{dehydrogenase}]{\text{alcohol}} H_3C-\overset{\displaystyle O}{\underset{\displaystyle H}{C}} \xrightarrow[\text{dehydrogenase}]{\text{aldehyde}} H_3C-\overset{\displaystyle O}{C}-OH$$

Acetaldehyde   **Acetic acid**

Summary

Hydroboration/oxidation is a process that yields an alcohol that is the product of overall anti-Markovnikov addition. The mechanism of hydroboration is complex, but several lines of evidence have led to the picture we have of a concerted reaction with an unsymmetrical transition state in which one of the alkene's carbon atoms becomes partially positively charged (Figs. 9.55–9.60). The synthetic utility of this reaction is not complex at all. For unsymmetrical alkenes, hydroboration/oxidation leads to the less substituted alcohol.

PROBLEM 9.22 Devise syntheses for the following molecules from 2-methyl-2-butene. You may also use any inorganic reagent as well as organic compounds containing no more than one carbon.

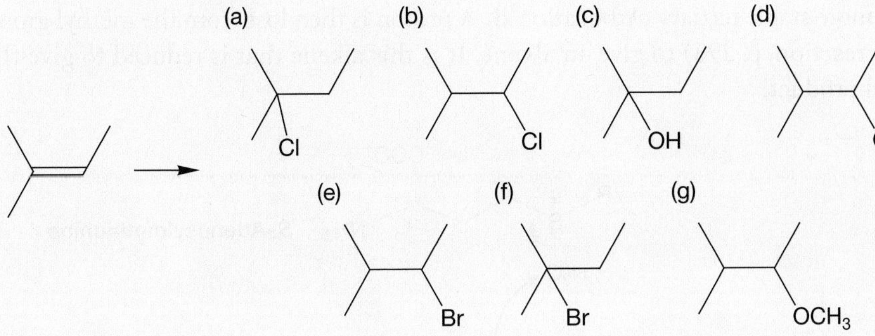

9.12   Special Topic: Rearrangements in Biological Processes

The hydride shifts and alkyl migrations discussed in Section 9.9 are not arcane examples fit only for the nether regions of an organic text (and for especially vexing problems). They are going on all the time as part of many of the enzyme-driven reactions in your body. So, as you read this, hydrides are migrating somewhere deep inside you.

Here is an example of a biological reaction in which an intramolecular hydride shift must be occurring. Remember the section in Chapter 3 (p. 142) on methyl transfer to an alkene? We are a bit further along now and can flesh out the rather sketchy mechanistic description given earlier. In particular, Figure 9.70 shows the ultimate product of the reaction introduced in Figure 3.82. It looks as

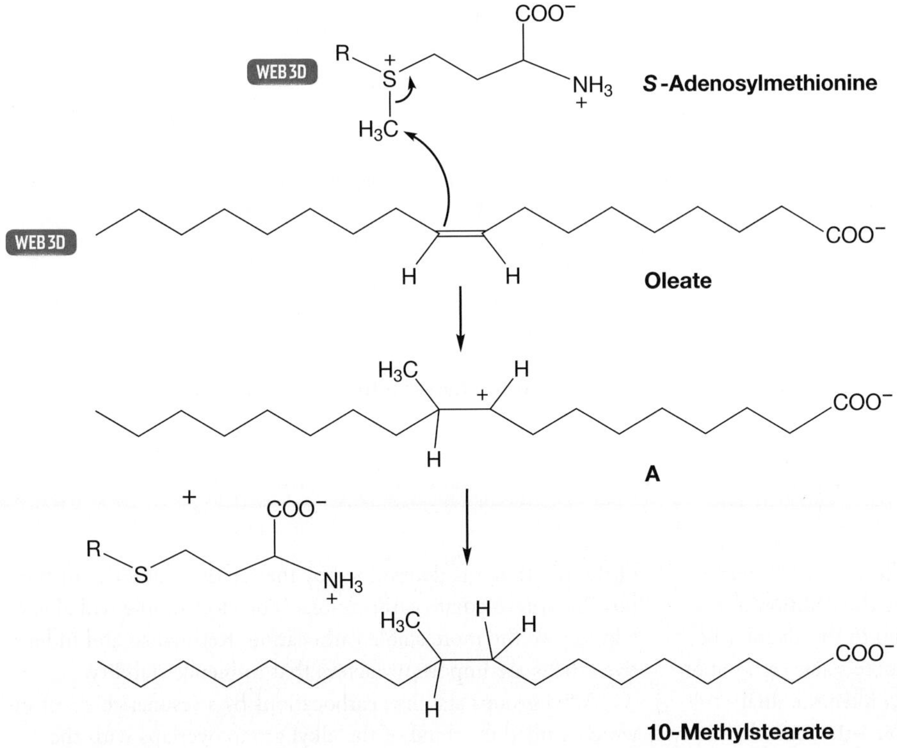

FIGURE 9.70 The final product of the methylation of oleate is 10-methylstearate.

if intermediate **A** has somehow been reduced to give the final product. But a clever labeling experiment (Fig. 9.71) shows that the process is far more complex and that a hydride shift is involved. It should be no surprise to you that the secondary carbocationic intermediate **A′** undergoes an intramolecular hydride shift to give the more stable tertiary carbocation **B**. A proton is then lost from the methyl group (E1 reaction, p. 298) to give an alkene. It is this alkene that is reduced to give the final product.

FIGURE 9.71 This deuterium labeling experiment shows that there is a "hydride" shift at the center of this reaction.

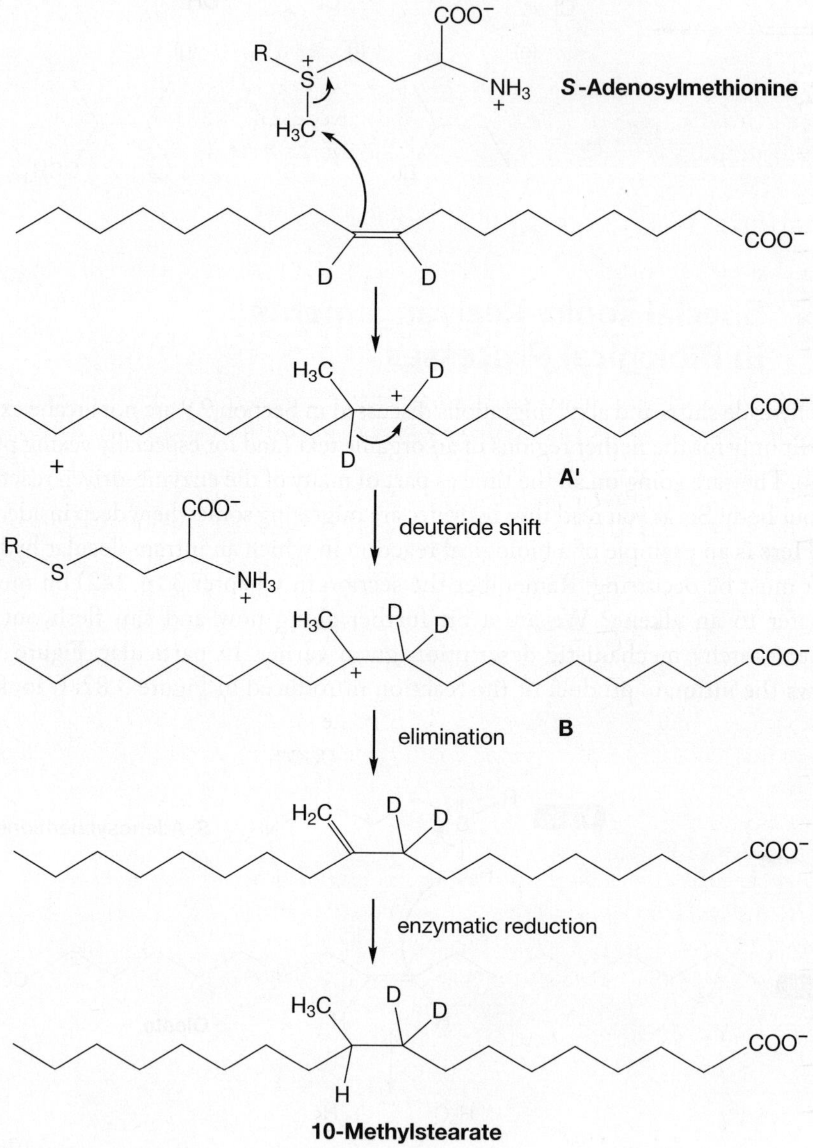

10-Methylstearate

9.13 Summary

New Concepts

The central topic of this chapter is the addition of HX molecules to alkenes. These reactions begin by the addition of the positive end of the dipole (the electrophile) to the alkene (the nucleophile) in such a way as to form the more stable carbocation. The anion (X^-) then adds to the cation to form the final addition product. The direction of addition—the regiochemistry of the reaction—is determined by the relative stabilities of the possible intermediate carbocations. The electrophile will always add to give the more stable carbocation. Resonance and inductive effects are important factors that influence stability.

Alkyl groups stabilize carbocations by a resonance effect in which a filled σ orbital of the alkyl group overlaps with the empty *2p* orbital on carbon. This effect is known by the curious name of "hyperconjugation" (Figs. 9.24–9.26).

In this chapter, we continue to use stereochemical experiments to determine reaction mechanisms. Generally, one-step (concerted) reactions preserve the stereochemical relationships of groups in the reacting molecules. Multistep processes intro-

duce the possibility of rotation around carbon–carbon bonds and loss of the initial stereochemistry.

Remember: No experiment can prove a mechanism. We are always at the mercy of the next experiment, the results of which might contradict our current mechanistic ideas. Experiments can certainly disprove a mechanism, but they can never prove one.

Key Terms

alkene hydrohalogenation (p. 365)
anti-Markovnikov addition (p. 399)
cationic polymerization (p. 386)
concerted reaction (p. 389)
hydration (p. 381)

hydride shift (p. 387)
hydroboration (p. 390)
hyperconjugation (p. 377)
inductive effects (p. 379)
Markovnikov's rule (p. 375)

rearrangement (p. 387)
syn addition (p. 394)
Wagner–Meerwein rearrangement
(p. 387)

Reactions, Mechanisms, and Tools

Many acids (HBr, HCl, HI, HOSO$_2$OH, generally, HX) will add directly to a π bond. The first step is addition of a proton to give the more stable carbocation. In the second step of the reaction, the negative end of the original HX dipole adds to complete the addition process. The regiochemistry of the addition is determined by the formation of the more stable carbocation in the original addition (Fig. 9.20).

Other HX molecules are not strong enough acids to protonate alkenes. Water (HOH) is an excellent example. However, such molecules will add across the double bond if the reaction is acid catalyzed. Enough acid catalyst is added to give the protonated alkene, which is then attacked by water. The catalyst is regenerated in the last step and recycles to carry the reaction further (Fig. 9.34).

If the concentration of alkene is high enough, the nucleophilic alkene can compete with other nucleophiles in the

system (for example, X$^-$), and dimerization or even polymerization of the alkene can take place. This reaction is not mysterious; the alkene is merely acting as a nucleophile toward the electrophilic carbocation (Fig. 9.43).

In hydroboration, the boron atom of "BH$_3$" adds to an alkene to give an alkylborane. The mechanism involves a single step in which the new carbon–hydrogen and carbon–boron bonds are both made. Subsequent further hydroborations can give dialkyl- and trialkylboranes. Treatment of these boranes with basic hydrogen peroxide leads to alcohols (Figs. 9.66 and 9.67). Hydroboration is especially important because it gives access to the less substituted alcohol (anti-Markovnikov addition).

Additions that go through carbocationic intermediates can be complicated by rearrangements of hydride (H:$^-$) or alkyl groups (R:$^-$) to produce more stable carbocations from less stable ones (Figs. 9.46 and 9.47).

Syntheses

In this chapter, we see the synthesis of alkyl halides and alcohols from alkenes through addition reactions. Hydroboration allows us to do addition reactions in the anti-Markovnikov sense. The S$_N$2 and S$_N$1 reactions from earlier chapters allow us to do further transformations of the alcohols and halides. The new synthetic methods are summarized below.

1. Alcohols

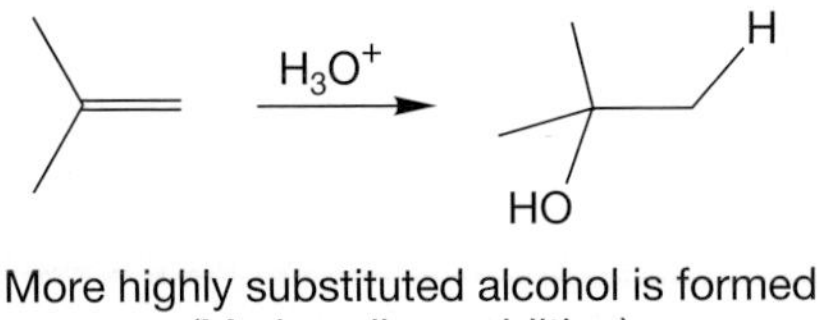

More highly substituted alcohol is formed
(Markovnikov addition)

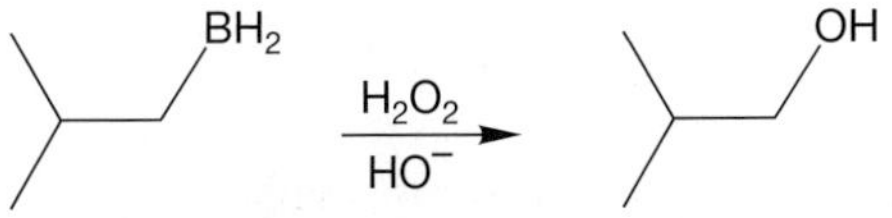

Less highly substituted alcohol is formed
(anti-Markovnikov addition)

2. Alkylboranes

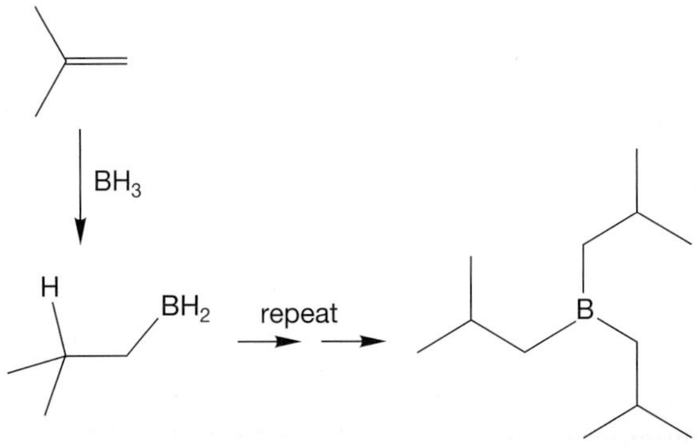

The number of hydroborations depends on the bulk of the alkene; boron becomes attached to the less substituted end of the alkene

3. Alkyl Halides

More highly substituted halide is formed
(Markovnikov addition)
(X = Br, Cl, or I)

4. Allyl Halides

1,2- and 1,4-Addition compete

Common Errors

Keeping clear the difference between resonance and equilibrium is a constant difficulty for many students. Resonance forms are simply different electronic descriptions for a single molecule. The key word is "electronic." In resonance, only electrons are allowed to move to produce the various representations of the molecule. Atoms may not change their positions. If they do, we are not talking about resonance, but about equilibrium. Be careful! This point is trickier than it sounds.

Remember also that the two-dimensional paper surface can fool us. Molecules are three-dimensional. To have delocalization of electrons, orbitals must overlap. Sometimes it looks in two dimensions as though resonance forms exist when, in fact, they do not in the real world of three dimensions. Two excellent examples appear in Figure 9.18 (p. 373) and in Problem 9.4 (p. 373).

The grammar of chemistry—our use of arbitrary conventions—is important because we need to be precise in communicating with each other. The double-headed resonance arrow is reserved for resonance and never used for anything else. There are several schemes for representing resonance forms, ranging from drawing them all out in full, to the summary structures of Figure 9.14 (p. 370).

Both mechanistic analysis and synthesis are gaining in complexity. There is much stereochemical detail to keep track of in many mechanistic analyses, for example. Perhaps the most complex mechanism considered so far is that for hydroboration. If you understand why it was necessary to modify the standard mechanism for the addition of Lewis acids to alkenes to accommodate the experimental observations, you are in fine shape so far.

There are no really complex synthetic procedures yet. However, even very simple steps taken in sequence can lead to difficulty. This area will rapidly proliferate and become more difficult, so do not be lulled by the deceptive simplicity of the reactions so far.

9.14 Additional Problems

This chapter and Chapter 10 continue our cataloging of the standard reactions of organic chemistry. To the S_N1, S_N2, E1, and E2 reactions we now add a variety of alkene addition reactions. Although there are several different mechanisms for additions, many take place through a three-step sequence of protonation, addition, and deprotonation. The following new problems allow you to practice the basics of addition reactions and to extend yourself to some more complex matters. Even simple additions become complicated when they occur in intramolecular fashion, for example. These problems also allow you to explore the influence of resonance and inductive effects, and to use the regiochemistry and stereochemistry of addition to help work out the probable mechanisms of reactions.

Your sophistication in synthesis is also growing, and the variety of addition reactions encountered in this chapter adds to the ways you have available to make differently substituted molecules. You are still not quite ready to undertake multistep syntheses, but you are now very close. In anticipation of these tougher problems, be sure to start working synthetic questions backward. Always ask the question, What molecule is the immediate precursor of the target? Don't start, even with simple problems, thinking of how an ultimate starting material might be transformed into product. That approach will work in one- or two-step synthesis, but becomes almost impossible to do efficiently when we come to longer multistep syntheses.

PROBLEM 9.23 See if you can write the π molecular orbitals of allyl from memory. Show the electronic configuration (orbital occupancy) for the allyl cation, radical, and anion.

PROBLEM 9.24 The molecular orbitals for pentadienyl are shown on the next page.

Pentadienyl

Compare these molecular orbitals with those of allyl. What rules can you develop to predict the molecular orbitals of any odd-carbon, fully conjugated (2*p* orbital on each carbon) chain? It is probably easiest to work from the schematic top views at the right of the figure.

Views from the top of pentadienyl; beneath every (+) is a (–) and beneath every (–) is a (+); nodes are shown as (0)

$= + - + - +$

$= + - 0 + -$

$= + 0 - 0 +$

$= + + 0 - -$

$= + + + + +$

PROBLEM 9.25 Apply your rules to the molecular orbitals of heptatrienyl.

Heptatrienyl

PROBLEM 9.26 Draw a third resonance form for the allyl cation that involves a three-membered ring. Be careful not to move any atoms! What is the relative importance of this resonance form? Explain.

PROBLEM 9.27 Write resonance forms for the following species. You will first have to write good Lewis structures, as the lone pairs have been deliberately left out. You will also have to write a full Lewis structure for the nitro (NO$_2$) group.

(a)

(b)

(c)

(d)

PROBLEM 9.28 How many signals would appear in the ^{13}C NMR spectra of the compounds in Problem 9.27 (a) and (b)?

PROBLEM 9.29 How many signals would appear in the ^{13}C NMR spectra of the following compounds?

(a)   (b)

PROBLEM 9.30 Which of the following pairs represent resonance forms and which do not? Explain your choices.

(a)

(b)

$$H_2\overset{+}{C}-CH_2 \longleftrightarrow H_2\overset{..}{C}-\overset{+}{C}H_2$$

(c)

(d)

$$:C{=}\overset{..}{O}: \longleftrightarrow {}^{-}:C{\equiv}\overset{+}{O}:$$

PROBLEM 9.31 Analyze the relative importance of the following two resonance forms. What would you say to someone who claimed that form **B** must be unimportant because the positive charge is on the more electronegative atom? You will have to add electron dots to make good Lewis structures first.

A   B

PROBLEM 9.32 Predict the regiochemistry of the following addition reactions of a generic acid, HX. Explain your predictions.

(a)

(b)

(c)

(d)

(e)

PROBLEM 9.33 What would be the structure of the polymer produced from each of the following two monomers through cationic polymerization?

PROBLEM 9.34 At −78 °C, hydrogen bromide adds to 1,3-butadiene to give the two products shown. Carefully write a mechanism for this reaction, and rationalize the formation of two products.

$$+ \text{HBr} \xrightarrow{-78\,°C}$$

(81%)   +   (18%)

PROBLEM 9.35 Reaction of *cis*-2-butene with hydrogen bromide gives racemic 2-bromobutane.
(a) Show carefully the origin of the two enantiomers of 2-bromobutane.
(b) In addition, small amounts of two alkenes isomeric with *cis*-2-butene can be isolated. What are these two alkenes and how are they formed?

PROBLEM 9.36 Indene can be hydrated to give a different product (**A** or **B**) under the following two conditions.

$$\longrightarrow C_9H_{10}O$$

Indene

Reagents for the first route: H_3O^+/H_2O
Reagents for the second route: (1) BH_3, (2) $HOOH/HO^-$

One of these two routes leads to a compound, **A** ($C_9H_{10}O$), which has a ^{13}C NMR spectrum consisting of nine signals. The other route leads to a different compound, **B**, which also has the formula $C_9H_{10}O$, and has a ^{13}C NMR spectrum consisting of only five signals. What are the structures of the two compounds **A** and **B**, and which route leads to which compound? Explain your reasoning in mechanistic terms for each compound and make very clear how you used the ^{13}C data in the structure determination. (You do not have to sketch a mechanism for the oxidation step of route 2.)

In the following four problems, reactions appear that are quite similar to those discussed in this chapter, but are not exactly the same. Each raises one or more slightly different mechanistic points. In this book, we aim to develop an ability to understand new observations in the light of old knowledge. In a sense, you are being put in the place of the research worker presented with new experimental results. In each case, explain the products mechanistically.

PROBLEM 9.37

(a)

$$\xrightarrow[CH_3OH]{H_3O^+} (CH_3)_3C-OCH_3$$

(b)

$$\xrightarrow[CH_3OH]{H_3O^+}$$

PROBLEM 9.38

$$\xrightarrow[H_2S]{H_3O^+}$$

PROBLEM 9.39

$$\xrightarrow[H_2O]{H_3O^+} \qquad not$$

PROBLEM 9.40

(a)

$$+$$

$(CH_3)_2CH-C-BH_2$ with CH_3 groups $\longrightarrow$

(b)

$$\xrightarrow{BH_3}$$

PROBLEM 9.41   Given the following experimental observation and the knowledge that the initial hydroboration step involves a syn addition, what can you say about the overall stereochemistry of the migration and oxidation steps of the organoborane (see Fig. 9.67, p. 399). Is there overall retention or inversion of configuration?

PROBLEM 9.42   Show the reactions you would use to synthesize 1-butanethiol from 1-butene.

PROBLEM 9.43   Show how you would make 1-azidobutane if your only source of carbon is 1-butene.

PROBLEM 9.44   Show how you would make 1-isopropoxyhexane starting with 1-hexene and using any other reagents you need.

PROBLEM 9.45   See the syntheses of 2-chloro-3-methylbutane and 2-bromo-3-methylbutane suggested in the answer to Problem 9.22 (b) and (e), given in the Study Guide. Carefully consider the last step in the proposed syntheses, the reaction of 3-methyl-2-butanol with hydrogen bromide or hydrogen chloride. Do you see any potential problems with this step? What else might happen?

PROBLEM 9.46   Propose syntheses of the following molecules starting from methylenecyclohexane, alcohols, iodides containing no more than four carbon atoms, and any inorganic materials (no carbons) you want.

PROBLEM 9.47   Predict the major product(s) for the hydroboration/oxidation reaction (1. BH_3/THF; 2. H_2O_2/NaOH) with each of the following alkenes:

(a) 1-pentene   (b) 2-methyl-2-pentene
(c) cyclopentene   (d) 3-hexene

PROBLEM 9.48   Which of the products in the previous problem are chiral and which are achiral?

PROBLEM 9.49   Predict the possible products in the reaction between HBr with the following alkenes:

(a) cyclopentene   (b) *trans*-2-butene
(c) *cis*-2-butene   (d) 3-methylcyclohexene
(e) 2-hexene

PROBLEM 9.50   Predict the major product(s) for the reaction between (*E*)-2-pentene and the following reagents:

(a) H_3O^+/H_2O (b) 1. BH_3/THF; 2. H_2O_2/NaOH
(c) HCl   (d) HBr
(e) H_3O^+/CH_3CH_2OH

PROBLEM 9.51   Predict the major product(s) for the reaction between (*E*)-3-methyl-2-pentene and the following reagents:

(a) H_3O^+/H_2O (b) 1. BH_3/THF; 2. H_2O_2/NaOH
(c) HCl   (d) HBr
(e) H_3O^+/CH_3CH_2OH

In each of the following reactions, more than a simple addition takes place. Keep your wits about you, think in simple stages, and work out mechanisms.

PROBLEM 9.52   Provide a mechanism for the following change:

PROBLEM 9.53   Provide a mechanism for the following change:

PROBLEM 9.54   Provide a mechanism for the following change: *Hint for the hard part and for the next few problems*: Think Wagner–Meerwein, and use the CH_3 groups as markers (p. 387).

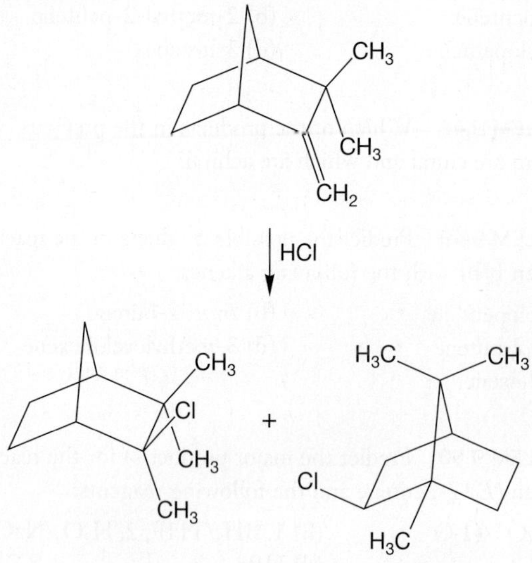

Easy   Hard!!

PROBLEM 9.55   Provide mechanisms for the following transformations. Once again, use the CH_3 groups to track where the atoms go.

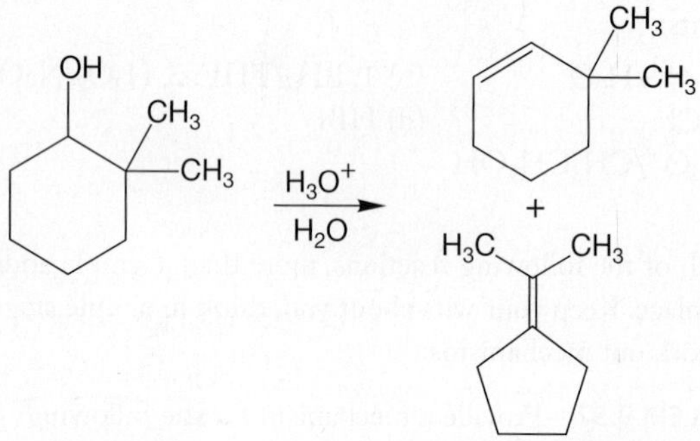

PROBLEM 9.56   What two products do you anticipate for the reaction shown below? Apply what you already know about the chemistry of alcohols in acid. There are two reasonable products, each with the formula $C_6H_{12}O$, but with very different structures. In fact, these compounds are formed in only minor amounts. The major product is pinacolone, shown in Problem 9.57.

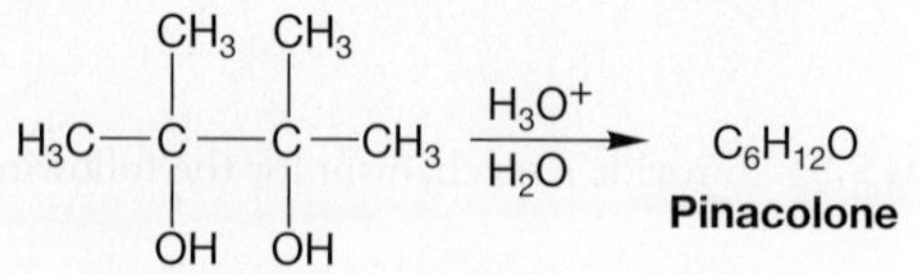

$$H_3C-\overset{\underset{OH}{|}}{\underset{\underset{CH_3}{|}}{C}}-\overset{\underset{OH}{|}}{\underset{\underset{CH_3}{|}}{C}}-CH_3 \xrightarrow[H_2O]{H_3O^+} C_6H_{12}O$$
Pinacolone

PROBLEM 9.57   What is the mechanism of pinacolone formation in the reaction in Problem 9.56?

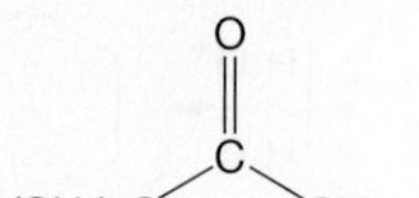

Pinacolone

$(CH_3)_3C \overset{\overset{O}{\|}}{C} CH_3$

PROBLEM 9.58   Provide a mechanism for the following transformation of morphine into apomorphine. *Caution*: This problem is very hard.

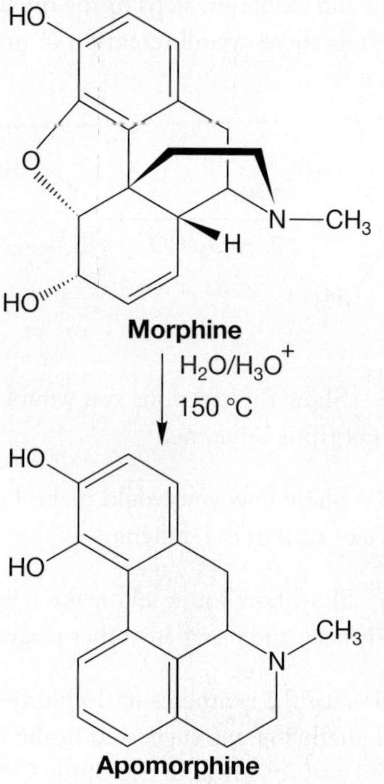

Morphine

H_2O/H_3O^+
150 °C

Apomorphine

Use Organic Reaction Animations (ORA) to answer the following questions:

PROBLEM 9.59   Select the animation titled "Alkene hydrohalogenation" and observe the intermediate in the reaction. What is the hybridization of the central carbon? Select the LUMO representation of the intermediate. What do we learn from this calculated information? What options are available for the nucleophile in the second step of this reaction?

PROBLEM 9.60   Observe the "Alkene polymerization" animation. What experimental conditions are necessary for polymerization to be the major pathway? The animation shows formation of the dimer and the trimer. How many alkene molecules need to add before we can call the product a polymer?

PROBLEM 9.61   The "Alkene polymerization" animation does not show a final product. Do you suppose an alkene polymer would have a charge? Milk jugs are made from alkene polymers. What do you know about the properties of a milk jug that help you answer the question? What are possible final steps in the formation of the polymer?

PROBLEM 9.62   The "Alkene hydroboration" reaction was a challenge to calculate. Observe the first part of the hydroboration animation, the reaction of the alkene with the borane. Why is it difficult to calculate this part of the reaction? Observe the second part of the animation, the oxidation of the boron. Notice that there is a 1,2-alkyl shift. What molecular orbital relationship do you think is the key for that shift?

Additions to Alkenes 2 and Additions to Alkynes

10.1 Preview

10.2 Addition of H_2 and X_2 Reagents

10.3 Hydration through Mercury Compounds: Oxymercuration

10.4 Other Addition Reactions Involving Three-Membered Rings: Oxiranes and Cyclopropanes

10.5 Dipolar Addition Reactions: Ozonolysis and the Synthesis of Carbonyl ($R_2C{=\!=}O$) Compounds

10.6 Addition Reactions of Alkynes: HX Addition

10.7 Addition of X_2 Reagents to Alkynes

10.8 Hydration of Alkynes

10.9 Hydroboration of Alkynes

10.10 Hydrogenation of Alkynes: Alkene Synthesis through syn Hydrogenation

10.11 Reduction by Sodium in Ammonia: Alkene Synthesis through anti Hydrogenation

10.12 Special Topic: Three-Membered Rings in Biochemistry

10.13 Summary

10.14 Additional Problems

NATURAL DYES The red dragonfly may be one of several insects that produce the red cochineal dye, a tricyclic compound with many multiple bonds.

A red dragonfly hovers above a backwater of the stream, its wings moving so fast that the eye sees not wings in movement but a probability distribution of where the wings might be, like electron orbitals: a quantum-mechanical effect that maybe explains why the insect can apparently teleport from one place to another, disappearing from one point and reappearing a couple of meters away, without seeming to pass through the space in between. There sure is a lot of bright stuff in the jungle. Randy figures that, in the natural world, anything that is colored so brightly must be some kind of serious evolutionary badass.

—NEAL STEPHENSON,[1] *CRYPTONOMICON*

10.1 Preview

We continue the discussion of additions to π systems in this chapter. Most reactions we see here follow the pattern seen in Chapter 9, in which a Lewis acid (usually a cation) is formed, and subsequently captured by a nucleophile, often an anion. Other reactions follow a different path. We begin with one of these, hydrogenation, the addition of H_2. We then move on to reactions involving X_2 reagents, which often proceed through the formation of an intermediate nonisolable, three-membered ring. Next, we take up additions in which the three-membered ring is stable, and finish with reactions forming different-sized rings. We will also discuss the related additions to alkynes. Throughout this chapter, attention is paid to what all these reactions allow us to do in the way of synthesis, which is rather a lot.

ESSENTIAL SKILLS AND DETAILS

1. Addition reactions have stereochemical consequences. In Chapters 3 and 9, where we first encountered additions, we focused mainly on regiochemistry, the direction of addition. Here, when we describe additions of X_2 and XY reagents, the stereochemistry, syn or anti, of the reaction can also be followed easily, and it is essential that you keep this factor in mind.

2. Addition reactions can take place through either open or cyclic intermediates, and you will learn to assess the factors that control which way the reaction goes. If an open cation is well stabilized, usually by resonance, it is likely to be favored. If not, the cyclic intermediate generally prevails.

3. Open cations permit addition of a nucleophile to each lobe of the empty orbital. Cyclic ions require opening in S_N2 fashion, leading to overall trans addition.

4. Epoxides open differently in acid and base. In base, the nucleophile adds to the less hindered position, whereas in acid the nucleophile becomes attached to the more substituted, more hindered side.

10.2 Addition of H_2 and X_2 Reagents

In this section, we see two very different reactions of diatomic reagents, the addition of H_2 and of X_2. In the first reaction, the concerted addition of hydrogen to the double bond of an alkene, a catalyst is required to get things going. The second reaction, halogen addition, fits better into the pattern seen in Chapter 9, in which a two-step process is followed.

---

[1] Neal Stephenson (b. 1959) is the author of the brilliant *Snow Crash*, as well as *The Diamond Age, Zodiac: The Eco-Thriller*, and, of course, *Cryptonomicon*.

10.2a Hydrogenation   When an alkene is mixed with hydrogen gas, nothing happens (unless one is careless with oxygen and a match). If the two reagents are mixed in the presence of any of a large number of metallic catalysts, however, an efficient addition of H_2 converts the alkene into an alkane. This reaction is called **hydrogenation**, and it can be achieved by either of two catalytic processes. First, there is **heterogeneous catalysis**, in which an insoluble catalyst, often palladium (Pd) on charcoal, is used. Many other insoluble metallic catalysts work as well in heterogeneous catalysis. A second process is **homogeneous catalysis**, in which a soluble catalyst, often the rhodium-based Wilkinson's catalyst, after Geoffrey Wilkinson (1921–1996), is employed (Fig. 10.1).

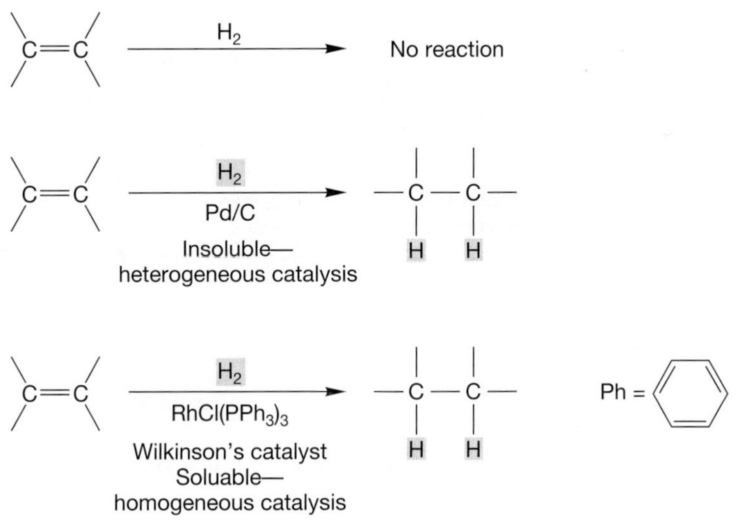

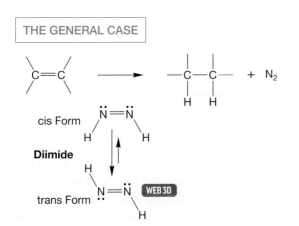

FIGURE 10.1 Hydrogen gas and an alkene react very quickly in the presence of a metal catalyst such as palladium adsorbed on charcoal (Pd/C). The rhodium-based Wilkinson's catalyst is typical of soluble molecules that also catalyze hydrogenation.

PROBLEM 10.1 Use bond dissociation energies (BDE) (Table 8.2, p. 337) to calculate the exothermicity or endothermicity of the hydrogenation of ethylene to ethane.

Nonmetallic reducing agents can be used in alkene hydrogenation as well. Diimide (HN=NH) is a good example. It is not stable under normal conditions and is made only as it is needed. Diimide exists in two forms, cis and trans, but only the less stable cis form is active in hydrogenation (Fig. 10.2).

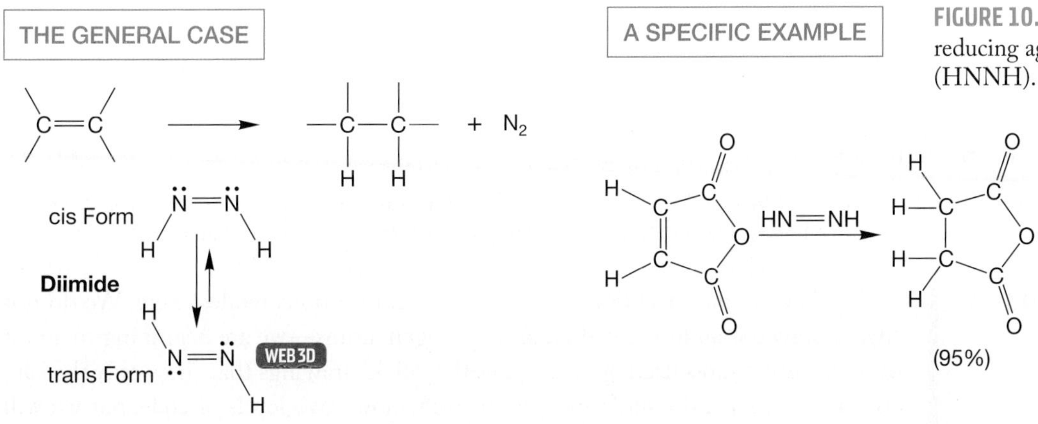

FIGURE 10.2 An effective nonmetallic reducing agent for alkenes is diimide (HNNH).

If hydrogenation is carried out with a cycloalkene, transfer of hydrogen takes place in predominately syn fashion. It is a syn addition because both new hydrogens are delivered to the same side of the alkene. In Figure 10.3, D_2 is used so we can see the stereochemical outcome of the reaction.

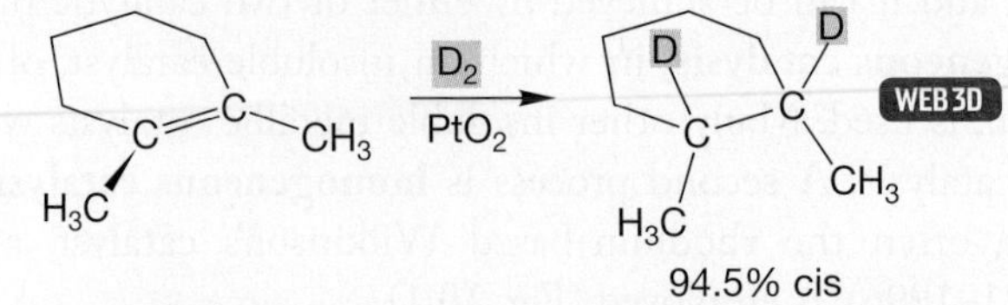

FIGURE 10.3 Metal-catalyzed addition to a cycloalkene shows that deuterium is added in syn fashion.

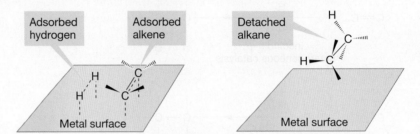

94.5% cis

The mechanistic details of hydrogenation remain vague. Presumably, the metal surface binds both hydrogen and alkene, weakening the π bond of the alkene and the σ bond of hydrogen. The alkane is formed by addition of a weakened hydrogen molecule to the coordinated alkene. As the transfer is completed, the product alkane detaches from the metal surface (Fig. 10.4).

FIGURE 10.4 Partial bonding to the metal surface is followed by hydrogen transfer to the alkene.

Adsorbed hydrogen   Adsorbed alkene   Detached alkane

Metal surface   Metal surface

There are many kinds of selectivity in the hydrogenation reaction. Not all carbon–carbon double bonds are hydrogenated at the same rate, and the differences in rate often allow hydrogenation of one double bond in the presence of another. Generally, the more sterically hindered a double bond, the more slowly it is hydrogenated. This rate change is probably related to ease of adsorption on the catalyst surface. Large groups hinder adsorption and therefore slow the rate of hydrogenation. One can often hydrogenate a less substituted double bond in the presence of more substituted double bonds, as shown in Figure 10.5.

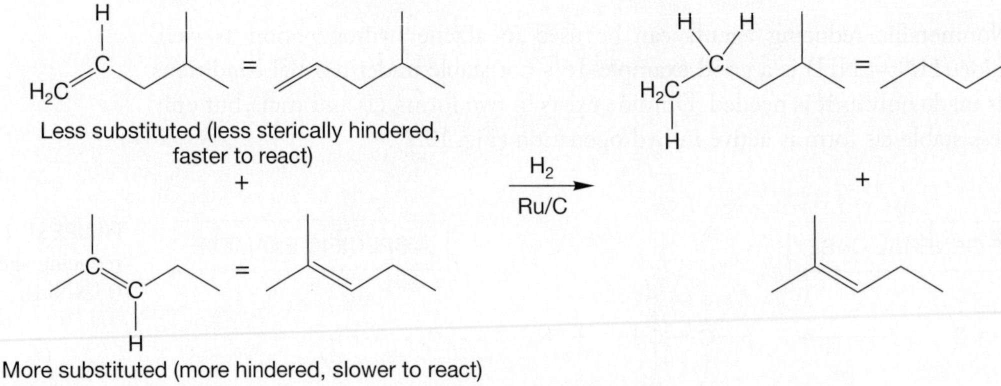

Less substituted (less sterically hindered, faster to react)

+

More substituted (more hindered, slower to react)

H_2
Ru/C

+

FIGURE 10.5 It is possible to hydrogenate less-substituted double bonds in the presence of more-hindered, more-substituted double bonds. The catalyst in this example is a ruthenium–carbon complex.

CONVENTION ALERT   We have to pay attention to the drawing conventions (codes) here. We do not always draw out all of the carbon and hydrogen atoms—we are beginning to adopt the convention most used by chemists—the "stick" drawings that show only the carbon framework of the molecule. Figure 10.5 shows two levels of code, but we will often lapse into the most schematic representations of molecules.

PROBLEM 10.2 Hydrogenation of β-pinene gives only one of the possible diastereomeric products, as shown below. Explain why.

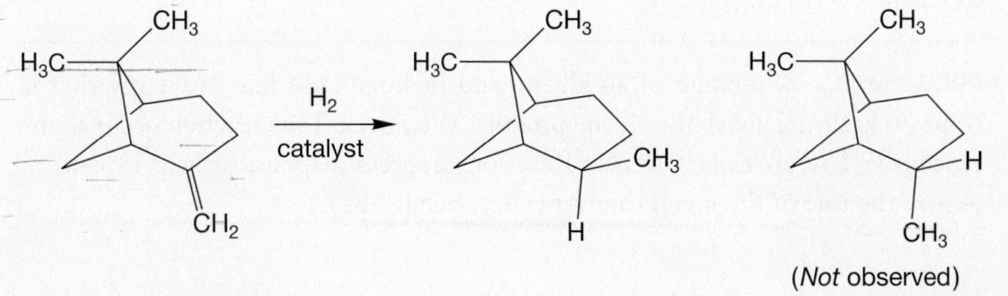

(*Not* observed)

Now let's consider the energetics of hydrogenation. Is it likely to be exothermic or endothermic? As shown in Problem 10.1, a simple and common technique for making such estimates uses bond energies. In the hydrogenation reaction, two bonds are breaking: the hydrogen–hydrogen σ bond and the π bond of the alkene. Two bonds are also being made in this reaction: the two new σ bonds in the alkane product. If we know bond energies (Table 8.2, p. 337), we can estimate $\Delta H°$ for this reaction. We made this estimate in Problem 10.1, and Figure 10.6 repeats the calculation. The process is exothermic by about 30 kcal/mol.

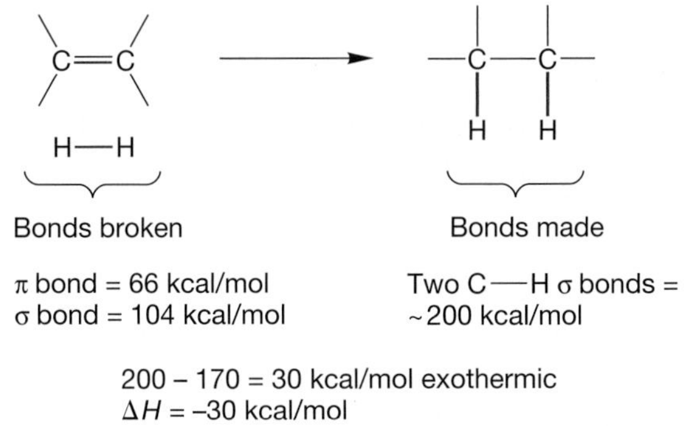

Bonds broken

π bond = 66 kcal/mol
σ bond = 104 kcal/mol

Bonds made

Two C—H σ bonds =
~200 kcal/mol

200 – 170 = 30 kcal/mol exothermic
$\Delta H = -30$ kcal/mol

FIGURE 10.6 In a typical hydrogenation, the π bond of the alkene and the σ bond of hydrogen are broken. Two new carbon–hydrogen bonds are made. The hydrogenation of an alkene is exothermic by approximately 30 kcal/mol.

Table 10.1 collects some measured heats of hydrogenation. We can see that the crude bond energy calculation of Figure 10.6 has actually done quite well; the heats of hydrogenation of simple alkenes are quite close to our calculation of 30 kcal/mol.

TABLE 10.1 Some Heats of Hydrogenation

Alkene	$\Delta H°_{H_2}$ (kcal/mol)
$H_2C{=}CH_2$	−32.7
$H_2C{=}CH{-}CH_3$	−30.0
$H_2C{=}CH{-}CH_2CH_3$	−30.3
cis $CH_3{-}CH{=}CH{-}CH_3$	−28.6
trans $CH_3{-}CH{=}CH{-}CH_3$	−27.6
$(CH_3)_2C{=}CH_2$	−28.4
$(CH_3)_2C{=}CHCH_3$	−26.9
$H_2C{=}CH{-}CH_2CH_2CH_2CH_3$	−29.6
cis $CH_3CH_2{-}CH{=}CH{-}CH_2CH_3$	−27.3
trans $CH_3CH_2{-}CH{=}CH{-}CH_2CH_3$	−26.4
$H_2C{=}\overset{\overset{\displaystyle CH_3}{\displaystyle \vert}}{C}{-}CH_2CH_2CH_3$	−28.0
$(CH_3)_2C{=}C(CH_3)_2$	−26.6

PROBLEM 10.3 Use the values in Table 10.1 to estimate the relative energies of the hexenes. Compare your estimates with those generated from heats of formation in Chapter 3 (p. 115).

PROBLEM 10.4 A mixture of an alkene and hydrogen gas has an energy that is some 30 kcal/mol above the alkane product. Why doesn't the reaction occur spontaneously? Use an Energy versus Reaction progress diagram to help explain *in general* the role of the metal catalyst (*Hint*: See p. 381).

10.2b Additions of Halogens

Alkene halogenation is the reaction between X_2 and an alkene and it results in addition of a halogen atom to each of the carbons of the double bond. Molecular bromine and chlorine (Br_2 and Cl_2) do not need a catalyst to add to alkenes to give these **vicinal** dihalides (Fig. 10.7). Groups are described as vicinal when they are on adjacent atoms.

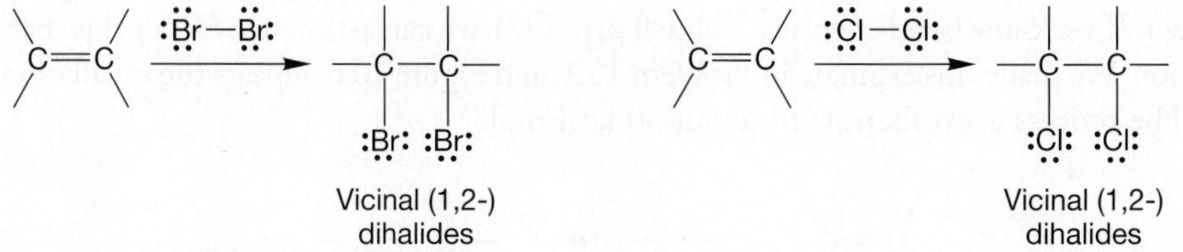

FIGURE 10.7 Addition of Br_2 and Cl_2 to yield vicinal dihalides.

Halogen addition is stereospecifically anti; the groups add to opposite sides of the alkene. For example, addition to cyclopentene gives only *trans*-1,2-dibromo-cyclopentane (Fig. 10.8).

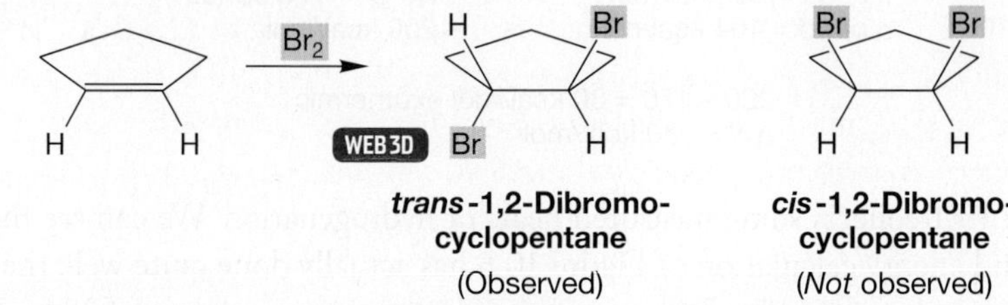

FIGURE 10.8 Anti addition of Br_2 to cyclopentene to give the trans product.

trans-1,2-Dibromo-cyclopentane
(Observed)

cis-1,2-Dibromo-cyclopentane
(*Not* observed)

If we take the addition of hydrogen bromide to an alkene as a model for the alkene halogenation reaction, we can build a mechanism quite quickly. Just as the alkene reacts with hydrogen bromide to produce a carbocation and bromide (Br^-), so reaction with bromine–bromine might give a brominated cation and bromide (Fig. 10.9). The π bond of the alkene acts as a nucleophile, displacing Br^- from bromine. Addition of bromide to the carbocation would give the dibromide product.

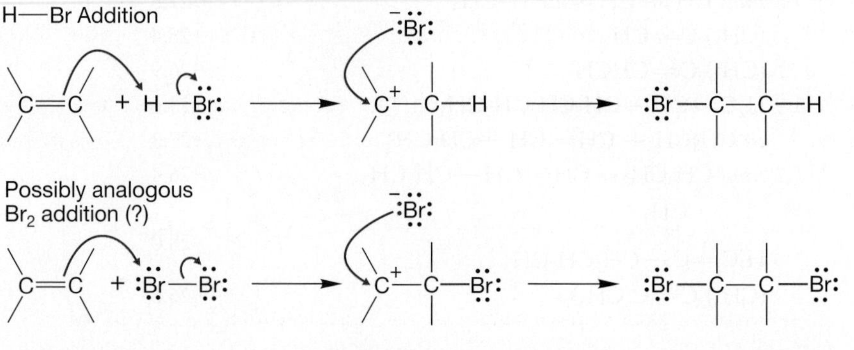

Alkene halogenation

FIGURE 10.9 A first-guess mechanism for the reaction of an alkene and bromine. For analogy, it draws on the mechanism of hydrogen bromide addition to an alkene.

H—Br Addition

Possibly analogous Br₂ addition (?)

There are cases of bromination in which the mechanism shown in Figure 10.9 *is* exactly correct, but for most alkenes it is not quite right. First, carbocation rearrangements (p. 386) are not observed in most bromination and chlorination reactions. For example, addition of bromine to 3,3-dimethyl-1-butene gives only 1,2-dibromo-3,3-dimethylbutane, *not* 1,3-dibromo-2,3-dimethylbutane, as would be expected if an open cation were involved (Fig. 10.10). We already know that if a cycloalkene is used, the product is exclusively the *trans*-1,2-dibromide. The cis compound cannot be detected (Fig. 10.8).

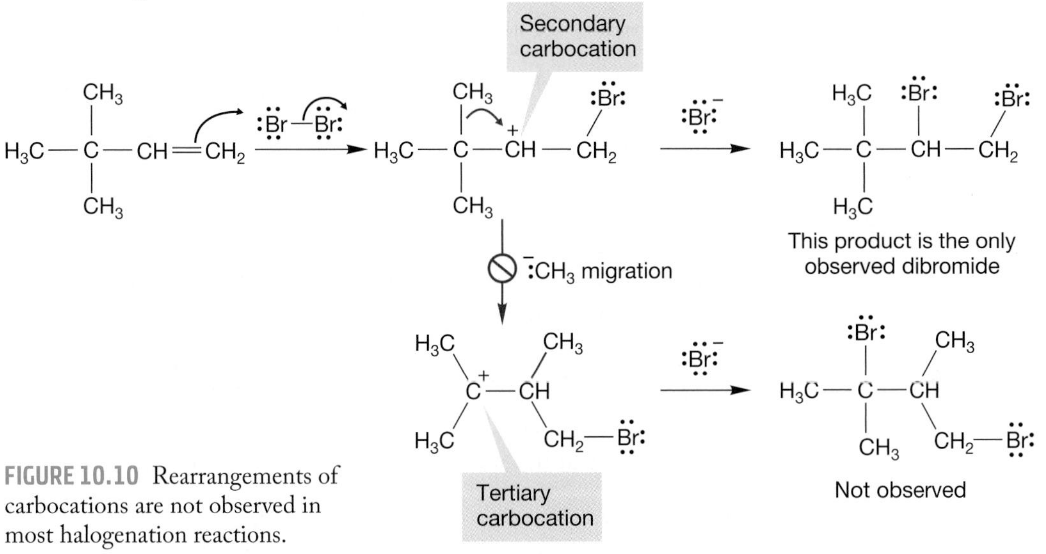

FIGURE 10.10 Rearrangements of carbocations are not observed in most halogenation reactions.

If the addition involved an open cation, we would expect the bromide ion to add from both sides to produce a mixture of products. The yields of cis and trans compounds would not be equal, but both compounds should be formed (Fig. 10.11). Because only anti addition is observed, our initial proposed mechanism cannot be completely correct. So our mechanism needs modification. How do we explain the simultaneous preference for anti addition and absence of rearrangement? It seems that an open carbocation, which would lead to cationic rearrangements and permit both syn and anti addition, must be avoided in some way.

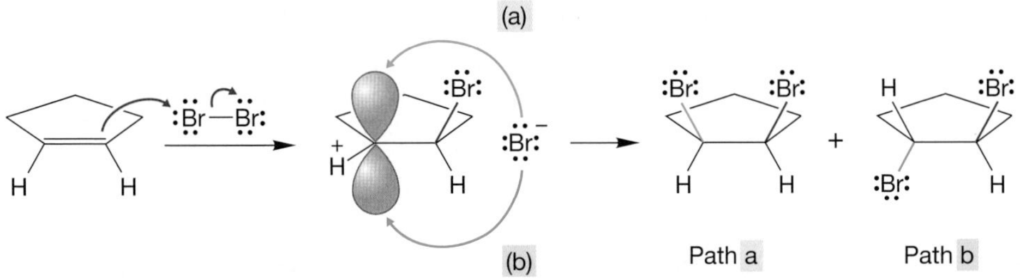

FIGURE 10.11 An open carbocation must lead to both syn and anti addition.

What if the initial displacement of Br^- from Br_2 were to take place in a symmetrical fashion (Fig. 10.12)? Once again, the filled π orbital of the alkene acts as nucleophile in an S_N2 reaction in which Br^- is displaced from Br_2. As the figure shows, a bromine-containing three-membered ring, a **bromonium ion** (brom = bromine; onium = positive), is formed. An open carbocation is avoided

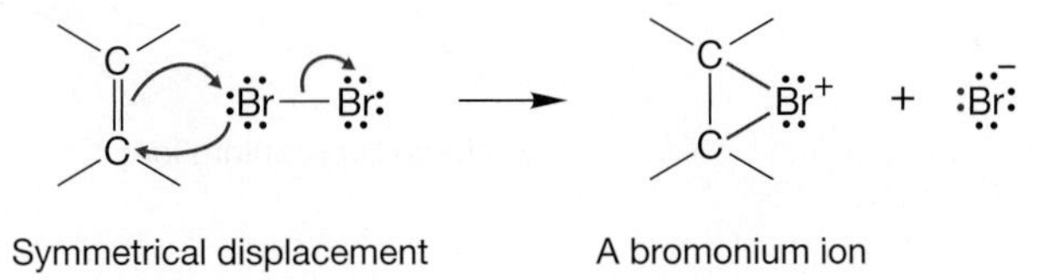

Symmetrical displacement   A bromonium ion

FIGURE 10.12 A symmetrical attack of the alkene on bromine yields a cyclic bromonium ion along with bromide, Br^-. The π bond is the nucleophile in this S_N2 reaction.

and rearrangements cannot occur. But what of the preference for anti addition? Look closely at the second step of the reaction, the opening of the bromonium ion by the nucleophile Br⁻ (Fig. 10.13). This reaction is an S_N2 displacement, and therefore must take place with inversion, through addition from the rear as we saw so often in Chapter 7. Overall anti addition of the two bromine atoms is assured.

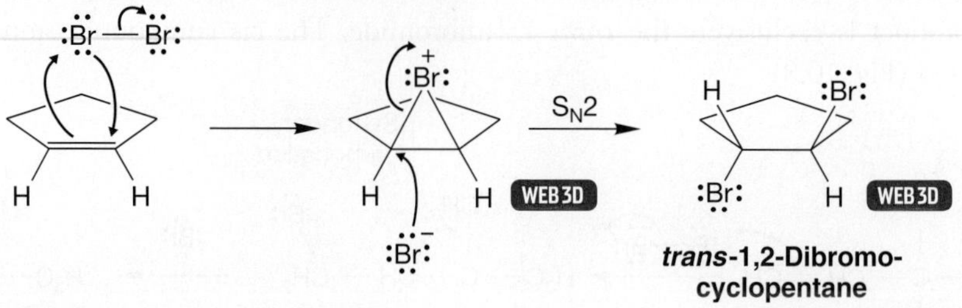

FIGURE 10.13 This anti addition of the two bromine atoms requires the formation of *trans*-1,2-dibromocyclopentane.

trans-1,2-Dibromo-cyclopentane

WORKED PROBLEM 10.5 In Figure 10.13, the reaction starts with achiral molecules. Yet the product, as drawn in the figure, is a single enantiomer. Clearly something is awry. What is it? *Hint*: Convention Alert! (p. 157).

CONVENTION ALERT

ANSWER You were warned about this! Of course there can be no optical activity produced from a reaction of two achiral reagents. In this reaction, the bromonium ion must open in two equivalent ways to produce a racemic pair of enantiomers. In situations such as this one, it is the custom not to draw both enantiomers.

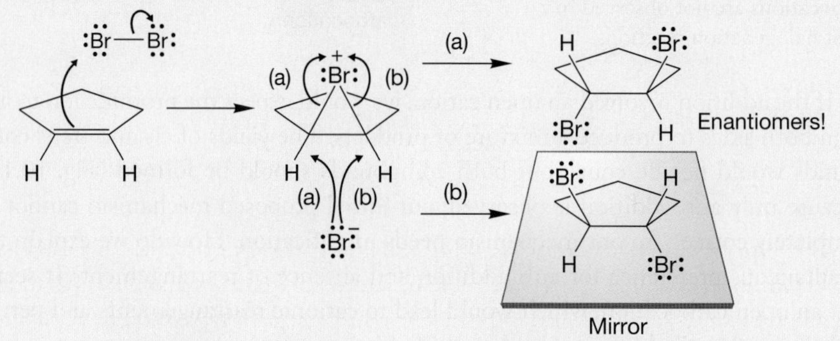

So, a modification of our earlier addition mechanism suffices to explain the reaction. In contrast to the addition to alkenes we saw in Chapter 9 (HX, carbocations, and boranes), in alkene halogenation (bromination and chlorination) a three-membered ring is produced as an intermediate. In some, very specialized cases the bromonium ion can even be isolated, and it seems its intermediacy in most brominations is assured.

PROBLEM 10.6 One example in which the bromonium ion is stable to further addition of nucleophiles is the reaction of adamantylideneadamantane with bromine (see below). Explain why this cyclic ion might be stable.

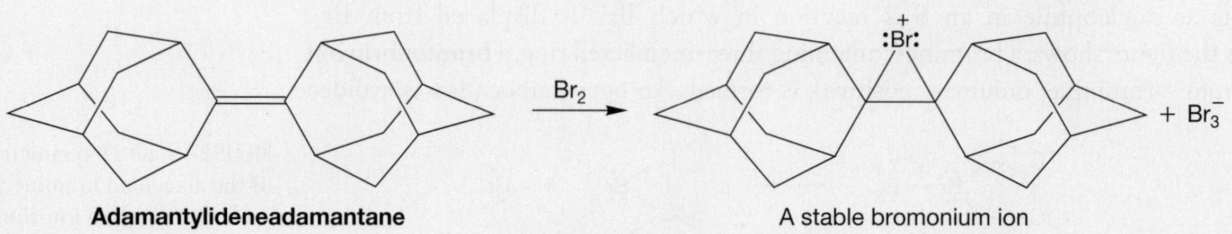

Adamantylideneadamantane   A stable bromonium ion

PROBLEM 10.7 Addition of Br₂ to acenaphthylene gives large amounts of the product of cis addition along with the usual trans dibromide. Give a mechanism for this bromination and, more important, explain why the mechanism of the reaction shown below is different from the typical selectivity for only trans addition. Why is cis product formed in this case?

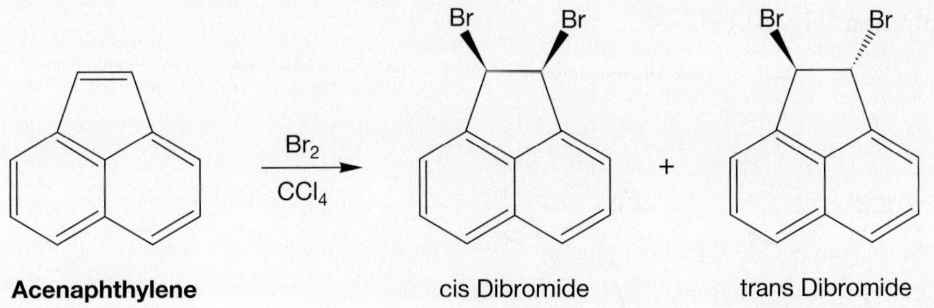

Stabilized alkene halogenation

Acenaphthylene Br₂ / CCl₄ → **cis Dibromide** + **trans Dibromide**

In the bromonium ion, there is the full complement of six electrons for the three σ bonds binding the two carbons and the bromine. The bromonium ion is a normal compound even though it may look odd at first. To see this clearly, imagine forming the bromonium ion from an open cation.

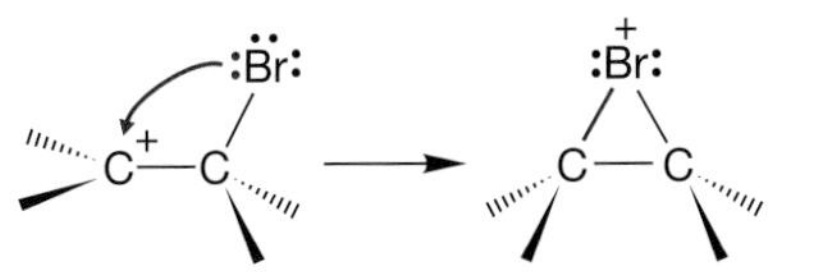

Bromonium ion

Normally, the nonnucleophilic solvent carbon tetrachloride (CCl₄) is used for the bromination or chlorination of alkenes. But the reaction will also work in protic solvents such as water and simple alcohols. In these reactions, new products appear that incorporate molecules of the solvent (Fig. 10.14). How do the OH or OR groups get into the product molecule? To see the answer, write out the mechanism and look for an opportunity to make the new products. The first step

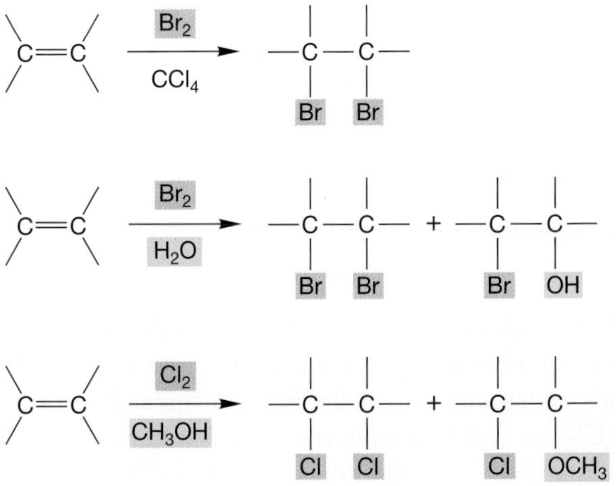

FIGURE 10.14 In the presence of nucleophilic solvents, addition reactions of bromine and chlorine give products incorporating solvent molecules as well as dihalides.

is the same as in Figure 10.12, in this case formation of the cyclic chloronium ion (Fig. 10.15). Normally, the chloronium ion is opened in S_N2 fashion by the nucleophilic Cl^- to give the dichloride. This opening must be the path for dichloride formation in these examples. But in these reactions there are other nucleophiles present: molecules of water or alcohol. Of course, they are weaker nucleophiles than the negatively charged Cl^-, but they are present as solvent and have an enormous numerical advantage over the more powerfully nucleophilic chloride, which is present only in relatively low concentration (Fig. 10.15).

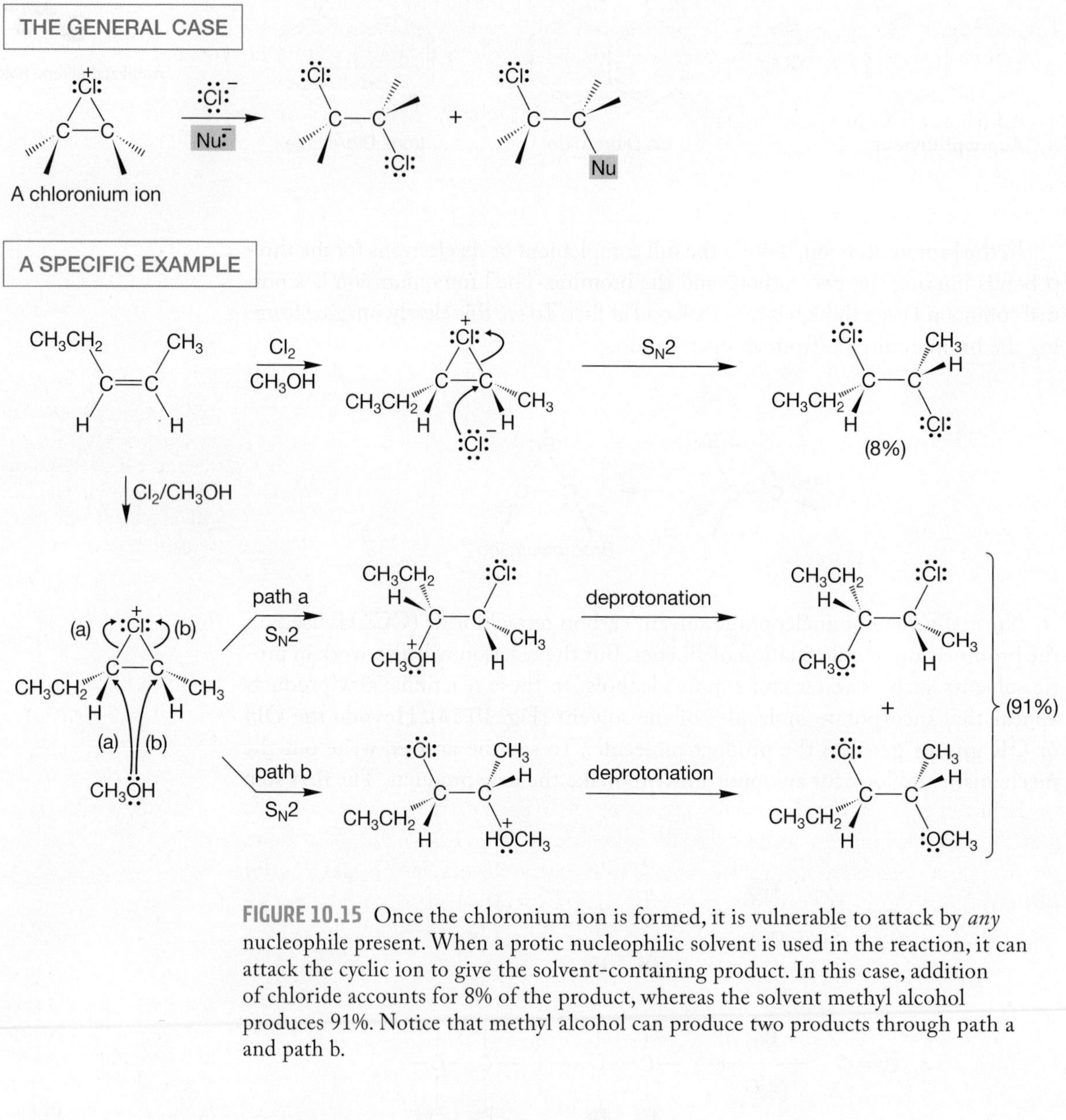

THE GENERAL CASE

A chloronium ion

A SPECIFIC EXAMPLE

(8%)

(91%)

FIGURE 10.15 Once the chloronium ion is formed, it is vulnerable to attack by *any* nucleophile present. When a protic nucleophilic solvent is used in the reaction, it can attack the cyclic ion to give the solvent-containing product. In this case, addition of chloride accounts for 8% of the product, whereas the solvent methyl alcohol produces 91%. Notice that methyl alcohol can produce two products through path a and path b.

Halohydrin formation

If this mechanism is correct, we can predict that the required inversion in the opening of the cyclic ion must produce a trans stereochemistry in the products. So, we have the opportunity to test the mechanism, and it turns out that addition of Br_2 (or Cl_2) to cyclohexene in the solvent water exclusively produces the compound resulting from anti addition in which the bromine (or chlorine) and hydroxyl groups have a trans relationship. Such compounds are called **halohydrins**. The bromonium ion mechanism fits all the data (Fig. 10.16).

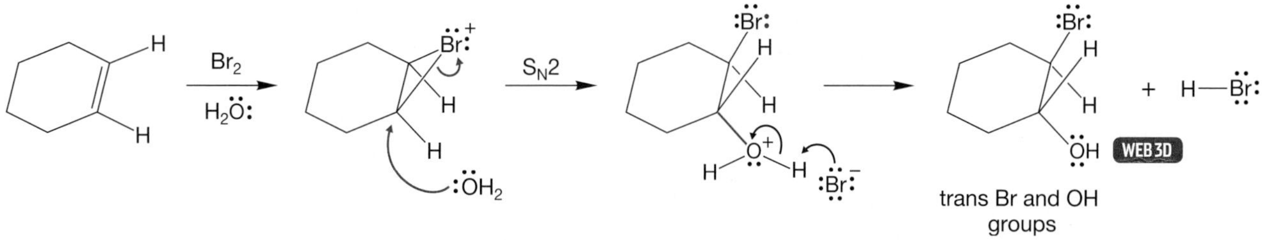

FIGURE 10.16 The bromonium ion mechanism demands that the products be formed by anti addition, as they are. This stereochemical experiment uses cyclohexene to show that the bromine and solvent molecule become attached from different sides of the ring through the anti addition enforced by the presence of the bromonium ion.

Addition of X_2 in nucleophilic solvents presents us with another opportunity. We can measure the regioselectivity of the reaction. For example, in the bromination of 2,3,3-trimethyl-1-butene in carbon tetrachloride, there is no way to determine the regiochemistry of the bromide attack of the bromonium ion. Bromination in water allows us to see where the hydroxyl group goes as it opens the bromonium ion. There are two possibilities (Fig. 10.17). The hydroxyl group can attach to the more substituted or to the less substituted carbon of the original double bond.

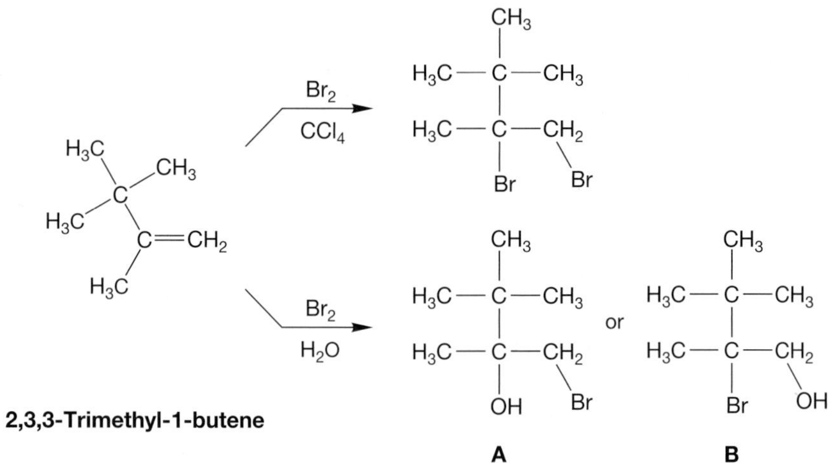

2,3,3-Trimethyl-1-butene

FIGURE 10.17 A test of the regiochemistry of the addition of Br_2/H_2O to alkenes. Halohydrin **A** would result from water attacking the more substituted carbon of the bromonium ion and halohydrin **B** would result from attack at the less substituted carbon.

In practice, the hydroxyl group becomes attached to the more substituted carbon. It is as if the reaction were proceeding through a carbocation (Fig. 10.18). But we are quite certain that it does not! The problem now is to reconcile the observed regioselectivity with the intermediacy of a bromonium ion (recall trans addition and the lack of carbocation rearrangements).

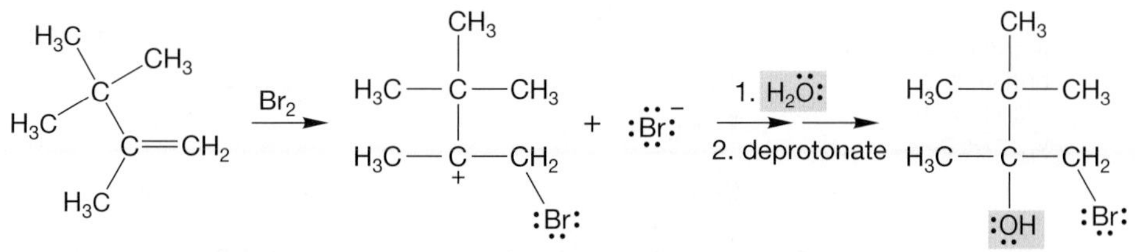

FIGURE 10.18 The bromine becomes attached to the less substituted carbon and the nucleophilic solvent to the more substituted carbon of the alkene. It is *as if* the reaction went through a carbocationic intermediate.

The bromonium ion is *symmetrical* only when formed from an alkene in which the two carbons of the double bond are *the same*. In all other cases, the two carbon–bromine bonds must be of different strength. In resonance terms, we say that the weighting factor (coefficient) for the more stable form is greater than that for the less stable form ($c_1 > c_2$, Fig. 10.19). The two carbons share the positive charge, but not equally. The charge is more stable on the more substituted carbon. Therefore, the more substituted carbon–bromine bond is longer and weaker than the less substituted bond in this positively charged intermediate. For this reason, the transition state for

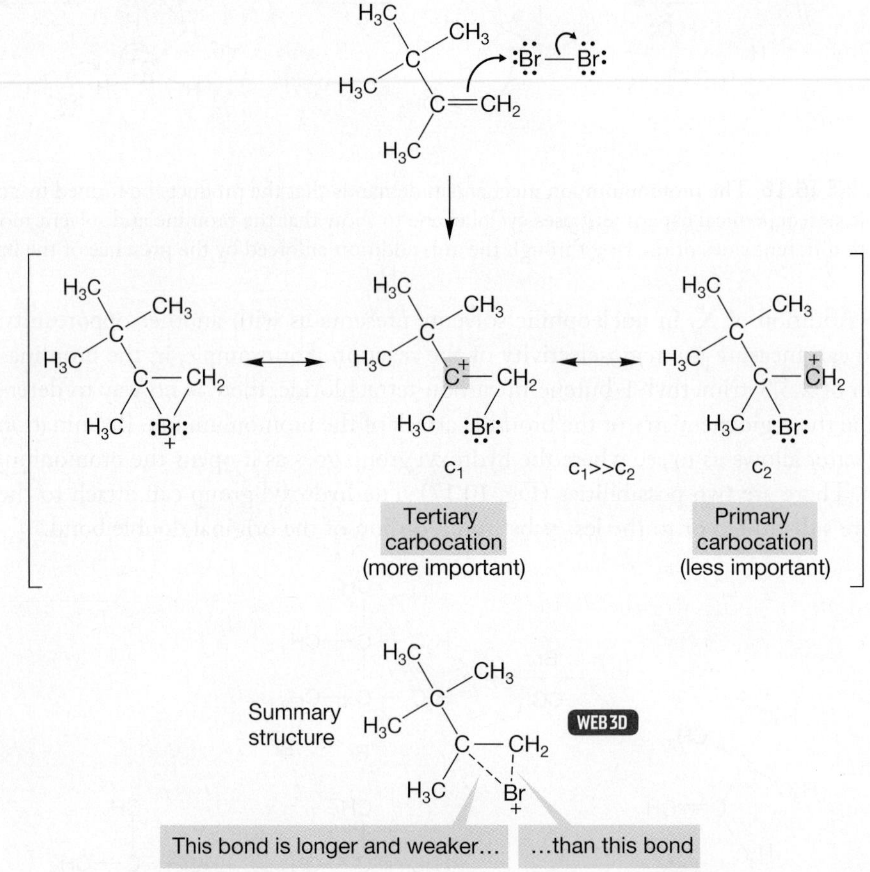

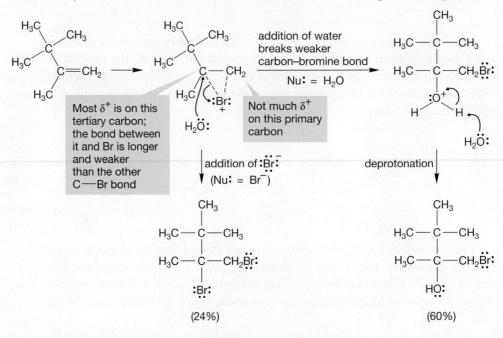

FIGURE 10.19 A resonance description of the bromonium ion.

breaking the bond to the more substituted carbon will be lower in energy than that for breaking the bond to the less substituted carbon. Therefore nucleophiles such as methyl alcohol, water, and halide add at the more substituted position (Fig. 10.20).

FIGURE 10.20 Addition of nucleophiles breaks the longer, weaker C—Br bond.

Summary

The critical difference between addition of halogen (X_2) and the many HX additions we saw in Chapter 9 is the presence of the intermediate halonium (bromonium or chloronium) ion. This intermediate is demanded by the observed preference for anti addition. However, occasionally the cyclic halonium ion is *less* stable than the corresponding open ion and the open ion will be favored. One way to stabilize the open ion is through resonance. If the open ion is the intermediate, the products of both cis and trans addition will be observed. When a protic nucleophilic solvent is used in the reaction, the solvent as well as the halide will add to the intermediate. In such cases halohydrins or halogenated ethers are formed along with dihalides.

10.3 Hydration through Mercury Compounds: Oxymercuration

For small-scale hydrations, it is convenient to use mercury salts, such as mercuric acetate, $Hg(OAc)_2$, to carry out the addition of H—OH across a double bond. This process is called **oxymercuration**. In this two-step process, an alkene is first treated with mercuric acetate, then the initial alkylmercury compound formed is reduced with sodium borohydride (Na^+ $^-BH_4$).

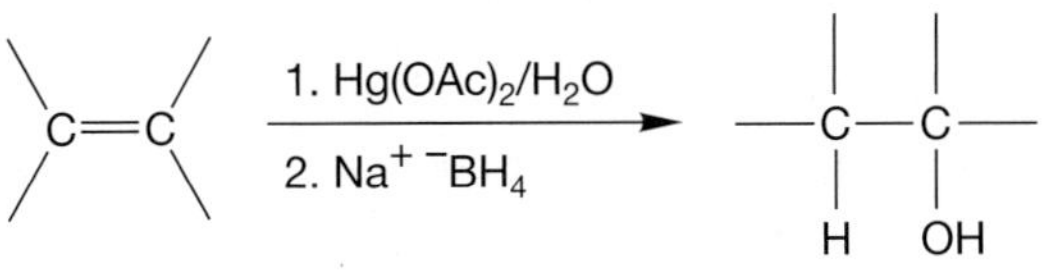

The abbreviation AcO^- (or ^-OAc) represents the acetate ion (CH_3COO^-), the conjugate base of acetic acid, CH_3COOH (often abbreviated AcOH). Lewis structures are given in Figure 10.21, which shows the relationship between the acetate ion and its conjugate acid (AcOH). Note the resonance stabilization of the acetate ion—the two oxygen atoms are equivalent.

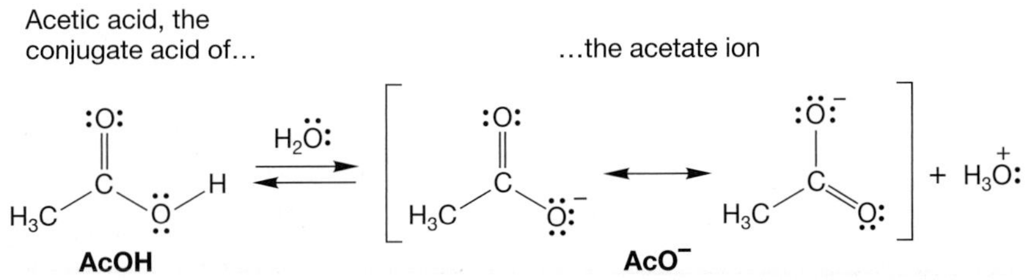

CONVENTION ALERT

Acetic acid, the conjugate acid of...   ...the acetate ion

FIGURE 10.21 Note the resonance stabilization of the acetate ion.

PROBLEM 10.8 The pK_a of acetic acid is 4.8. The corresponding two-carbon alcohol, ethyl alcohol (CH_3CH_2OH), is a much weaker acid, as its pK_a of 15.9 demonstrates. Explain why acetic acid is much more acidic than ethyl alcohol.

The first step in the oxymercuration reaction is the formation of a cyclic mercurinium ion (Fig. 10.22). Open carbocations are unlikely intermediates, as they would surely undergo carbocationic rearrangements, and these are not seen even in systems normally prone to rearrangement.

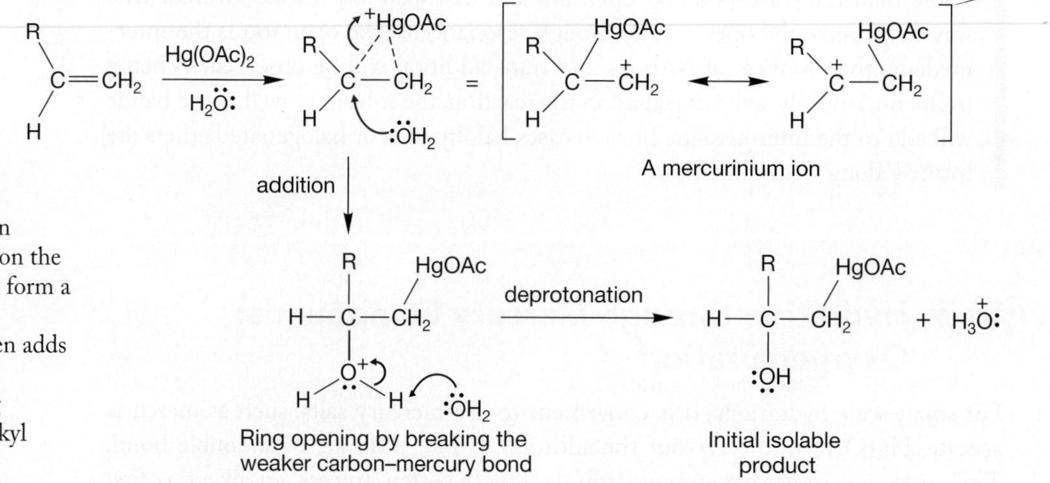

FIGURE 10.22 Oxymercuration begins by attack of the alkene on the Lewis acid mercuric acetate to form a cyclic mercurinium ion. The nucleophilic solvent, water, then adds in S_N2 fashion to give the first product, a mercury-containing alcohol, where R is a simple alkyl group.

WORKED PROBLEM 10.9 Design an experiment to test for carbocationic rearrangements in the oxymercuration reaction.

ANSWER It's simple. Just use any alkene in which migration of hydride ($H:^-$) or migration of $R:^-$ would give a more thermodynamically stable cation. 3-Methyl-1-butene would work well. If addition gave an open carbocation rather than the cyclic mercurinium ion, rearrangements would surely occur. That they are *not* observed provides strong evidence that open cations are not involved in oxymercuration.

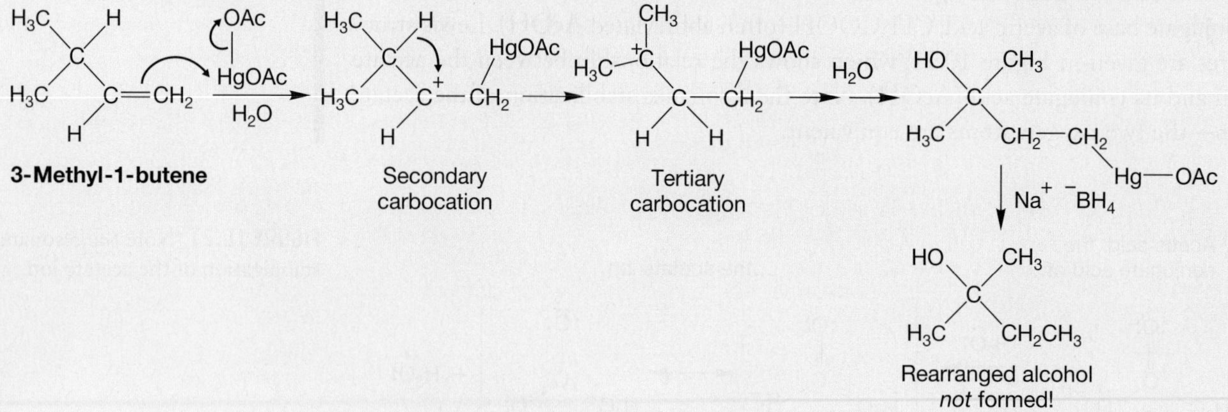

With unsymmetrically substituted alkenes, the two bonds to mercury will not be equally strong. (Recall our discussion of asymmetrical bromonium ions just a few pages ago.) The bond between mercury and the more substituted carbon will be longer and weaker than the bond to the less substituted carbon of the mercurinium ion. It is at the more substituted carbon that water opens the ring. So, the first product of this reaction is the mercury-containing alcohol shown in Figure 10.22.

Oxymercuration is synthetically useful because the reagent sodium borohydride (Na^+ $^-BH_4$) efficiently replaces the mercury with hydrogen. Figure 10.23 shows the reaction with sodium borohydride and sodium borodeuteride (Na^+ $^-BD_4$). The use of the deuterated molecule allows us to see with certainty the position of the entering deuterium.

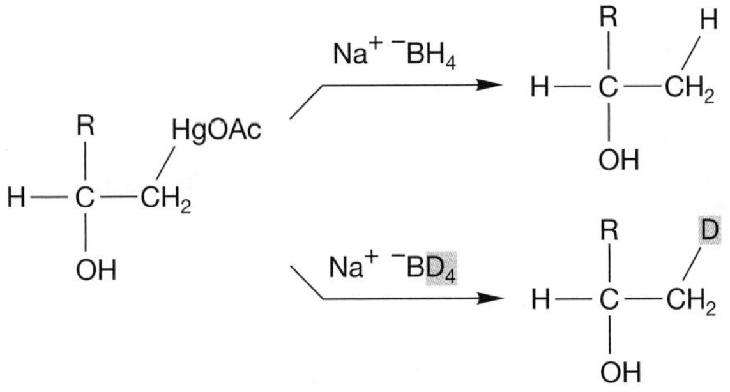

FIGURE 10.23 Reduction of the mercury-containing products gives alcohols.

So we now have a new hydration reaction, one that proceeds through a cyclic mercurinium ion but ultimately gives the product of Markovnikov addition, the more substituted alcohol.

Summary

Here are our three methods of hydration:

1. *Direct hydration* (p. 380) proceeds through an intermediate carbocation that is captured by water to give the product of Markovnikov addition. The reaction is limited in utility because rearrangements of the initially formed cation to more stable species can lead to undesired, rearranged products.

2. *Hydroboration/oxidation* (p. 390) first generates an alkylborane that subsequently reacts with peroxide in base to give the product of anti-Markovnikov addition.

3. *Oxymercuration* proceeds through a cyclic mercury-containing ion and also gives the product of Markovnikov addition. No rearrangements take place, which is sometimes a great advantage.

PROBLEM 10.10 Work out the consequences of applying the three procedures outlined above to 3-methyl-1-butene.

10.4 Other Addition Reactions Involving Three-Membered Rings: Oxiranes and Cyclopropanes

10.4a Oxiranes We have seen that many addition reactions begin with formation of a three-membered ring. These intermediates have almost always been rapidly converted into the final products through addition of a nucleophile. This state of affairs will now change dramatically, as you learn about the addition reactions that are the best routes to two types of stable three-membered rings: the oxiranes (epoxides, p. 317) and cyclopropanes.

An organic acid has the formula R—COOH and is called a **carboxylic acid**. Its structure is shown in Figure 10.24. We will study this functional group in detail in Chapter 18. A *per*acid has the formula R—COO̲OH. Figure 10.24 shows the difference in more detail. The "per" indicates an "extra" oxygen is present. Thus hydrogen *per*oxide is HOO̲H, whereas the normal oxide of hydrogen is water, HOH.

FIGURE 10.24 The prefix per- means extra. So a "peroxide" and a "peracid" each contain an extra oxygen atom, one more than the oxide or acid.

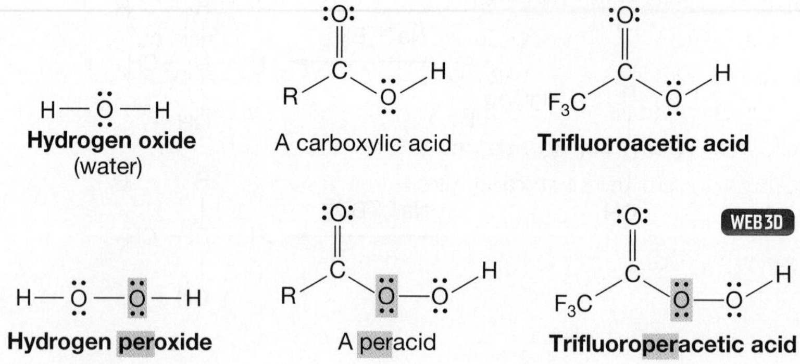

H—O̲—H
Hydrogen oxide
(water)

A carboxylic acid

Trifluoroacetic acid

H—O̲—O̲—H
Hydrogen peroxide

A peracid

Trifluoroperacetic acid

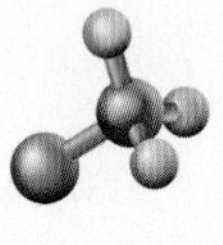

WEB 3D

Alkene epoxidation

The **epoxidation** reaction resembles the formation of a bromonium ion. In both reactions, a leaving group is displaced by the alkene acting as nucleophile in an S_N2 reaction (Fig. 10.25). Simultaneous transfer of hydrogen to the carbonyl (C=O) oxygen gives a stable three-membered ring, an epoxide.

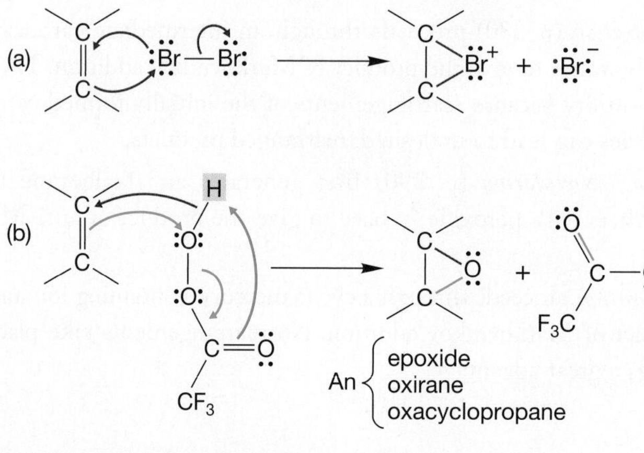

(a)

(b)

An { epoxide
oxirane
oxacyclopropane }

FIGURE 10.25 (a) Bromination of an alkene. (b) Epoxidation of an alkene. There is an additional transfer of hydrogen to give the isolable, stable epoxide.

It is important to be clear why we are so sure that epoxidation occurs in a single step. Why not just form the carbocation and then close up to the oxirane (Fig. 10.26)?

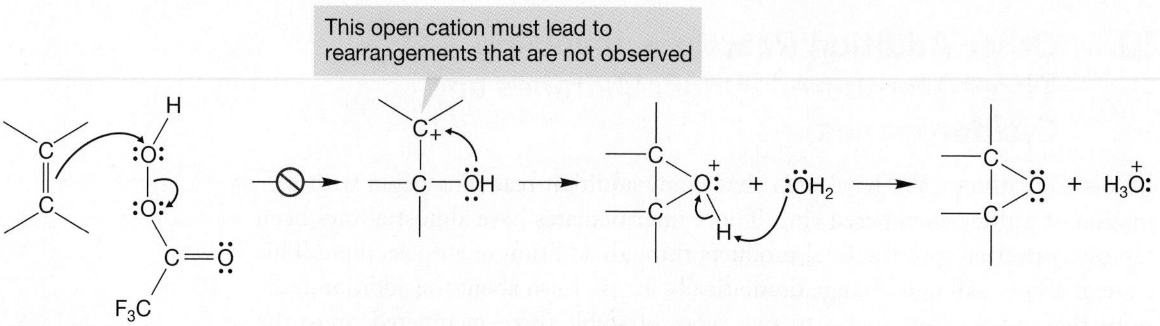

This open cation must lead to rearrangements that are not observed

FIGURE 10.26 A stepwise (nonconcerted) mechanism for epoxidation does not happen.

If a free cation were involved, carbocation rearrangements would be expected, and they are not found. Moreover it is easy to monitor the stereochemistry of this reaction. Concerted (no intermediates) formation of the oxirane explains the observed syn addition very well. A reaction in which the two new bonds are formed at the same time cannot change the stereochemical relationships of the groups in the original alkene. On the other hand, a mechanism involving an open carbocation would very likely lead to a mixture of stereoisomeric oxiranes (Fig. 10.27). If only one bond were made, there would be a carbocationic intermediate in which rapid rotation about carbon–carbon single bonds would take place. These rotations would scramble the original stereochemical relationship present in the alkene, and this result is not observed. Instead, oxirane formation occurs with retention of the stereochemical relationships present in the starting alkene.

THE GENERAL CASE

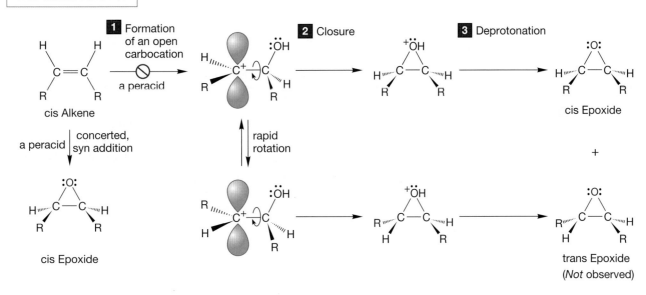

A SPECIFIC EXAMPLE

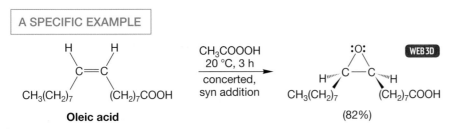

Oleic acid

(82%)

FIGURE 10.27 A test of the mechanism of epoxidation. A stepwise mechanism predicts scrambling of stereochemistry, whereas a concerted mechanism must retain the stereochemistry of the starting alkene. Because cis alkenes give only cis epoxides, the reaction must be concerted.

PROBLEM SOLVING

The mechanistic technique outlined in Figure 10.27 above is a general one, often used to distinguish one-step from two-step mechanisms. Retention of stereochemistry (cis → cis and trans → trans) implicates a one-step process, whereas scrambling of stereochemistry (cis → cis and trans, or trans → cis and trans) is indicative of a stepwise process proceding through an intermediate (or intermediates) capable of rotation. We'll see it again in a few pages. (Can you spot it?) You should be alert for it, as it forms the basis for many mechanism-based problems.

10.4b Asymmetric (Sharpless) Epoxidation

In recent years, an enormously useful synthetic technique for producing enantiomeric oxiranes (epoxides) has appeared. Discovered by Barry Sharpless (b. 1941), now at the Scripps Research Institute, it uses a witches' brew of titanium isopropoxide, *tert*-butyl peroxide, and one enantiomer of a tartaric ester to react with an allylic alcohol ($R—CH=CH—CH_2OH$). Part of the great utility of this procedure comes from the observation that the two enantiomers of tartaric ester lead to two different stereochemistries of product oxirane (Fig. 10.28). Professor Sharpless shared the Nobel prize in chemistry in 2001 for his discovery of chiral catalysts.

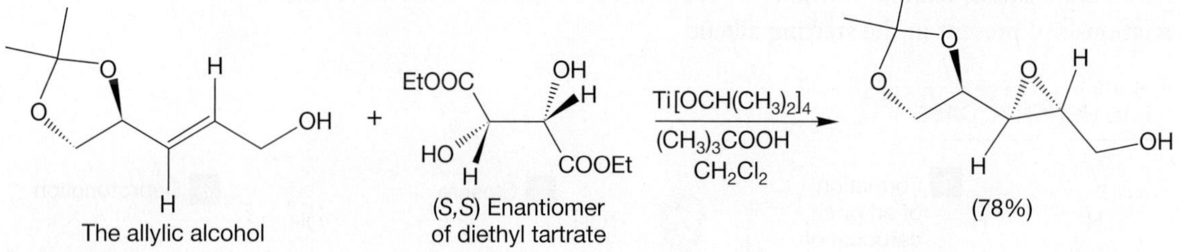

FIGURE 10.28 Asymmetric (Sharpless) epoxidation of alkenes ($Et = CH_3CH_2$).

The mechanism of this complex reaction involves the titanium compound acting as a clamp, holding the alkene, peroxide, and (S,S)-tartaric ester together. Because the ester is asymmetric, the clamped combination of molecules is also asymmetric. In one cluster, the oxirane oxygen is delivered from one side, whereas in the enantiomeric cluster formed from (R,R)-tartaric ester, the oxygen comes from the other side. These and other epoxides can then be transformed into all manner of compounds, as we will now see.

> **PROBLEM 10.11** The (R,R) enantiomer of diethyl tartrate gives a diastereomer of the product shown in Figure 10.28. Draw this diastereomer.

10.4c Further Reactions of Oxiranes

Unlike the closely related bromonium ions, oxiranes can be isolated under many reaction conditions. The bromonium ion is doomed to bear a positive charge and is therefore a more powerful Lewis acid than the neutral oxirane. Oxiranes will react, however, when treated in a second step with either acids or bases. For example, reaction with either H_3O^+/H_2O or HO^-/H_2O leads to opening of the three-membered ring and formation of a 1,2-diol (Fig. 10.29).

FIGURE 10.29 Epoxides (oxiranes) open in either acid or base.

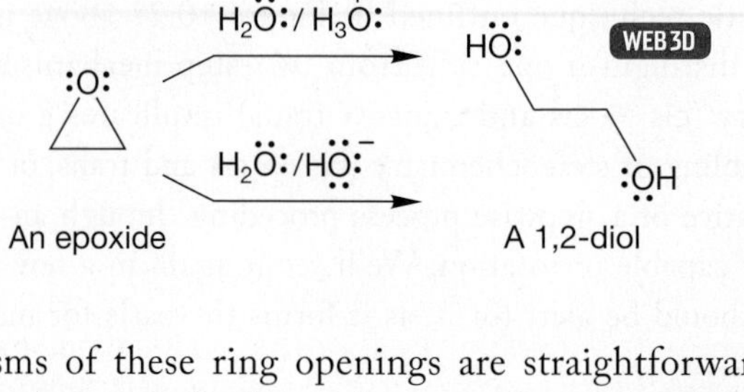

Basic epoxide ring opening

The mechanisms of these ring openings are straightforward extensions of reactions you already know. In base, the strongly nucleophilic hydroxide ion attacks the oxirane and displaces the oxygen atom from one carbon in an S_N2

reaction (Fig. 10.30). In acid, there is no strong nucleophile, but protonation of the oxygen atom of the oxirane creates a good Lewis acid. Water is a strong enough nucleophile to open the protonated oxirane to give, after deprotonation, the 1,2-diol (Fig. 10.30).

The mechanism of epoxide opening in base

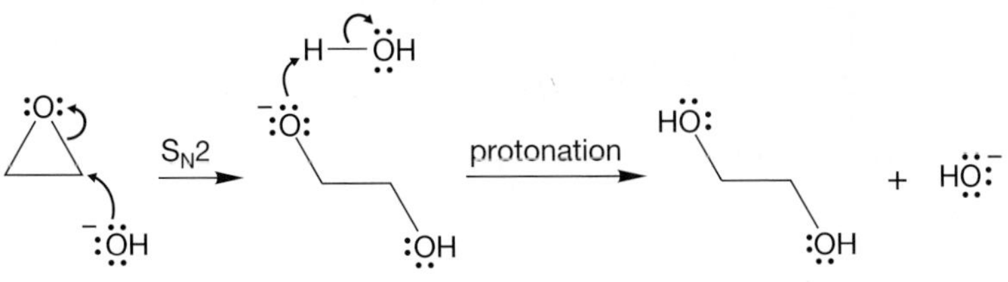

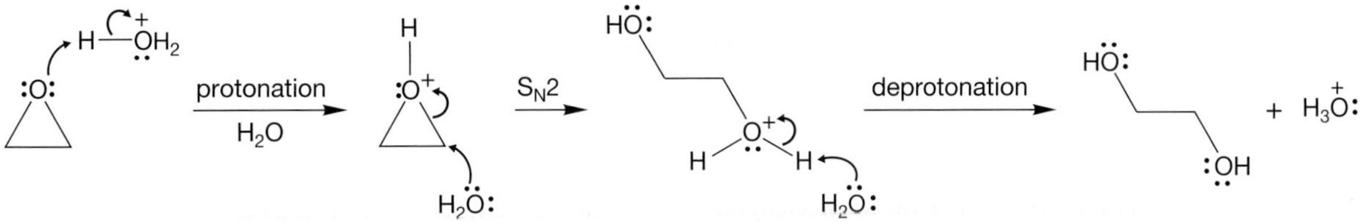

FIGURE 10.30 Mechanisms for epoxide opening in base and acid.

The mechanism of epoxide opening in acid

There is a strong parallel here to reactions we have already seen. Hydroxide is a poor leaving group. Treatment with acid or formation of sulfonate esters converts hydroxide into water or another good leaving group, and many reactions require this transformation (see p. 281 for examples). The opening of the oxirane in acid involves exactly the same kind of transformation; protonation facilitates the opening by making the leaving group ROH rather than RO⁻ (Fig. 10.31).

Acidic epoxide ring opening

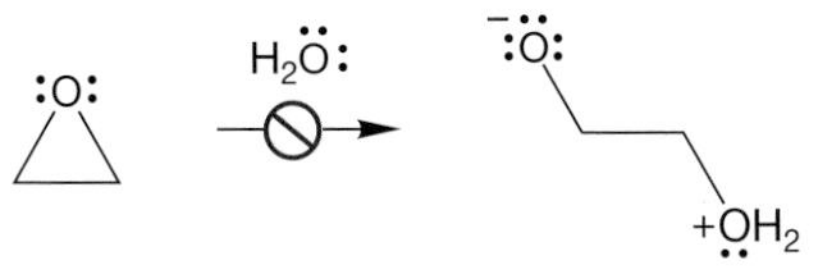

(*Not* observed)

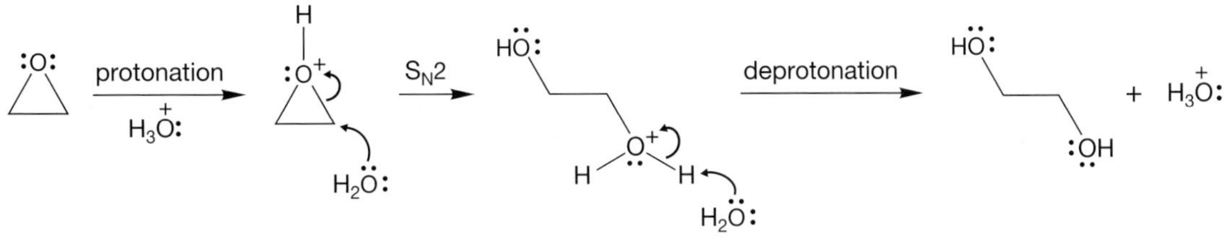

FIGURE 10.31 Water is not a strong enough nucleophile to open the unprotonated oxirane, because RO⁻ is too poor a leaving group. Acid catalysis changes the leaving group to ROH, which is much easier to displace.

Unsymmetrical oxiranes can open in two ways and the regiochemical result is generally different in acid and base (Fig. 10.32). How does the difference in regiochemistry arise? Why should the two openings be different? In base, the ring opening is an

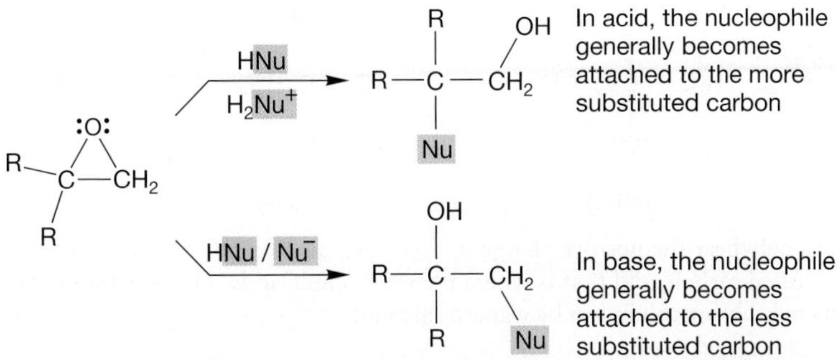

In acid, the nucleophile generally becomes attached to the more substituted carbon

In base, the nucleophile generally becomes attached to the less substituted carbon

FIGURE 10.32 The regiochemistry of epoxide opening is different in acid and base.

S_N2 displacement by hydroxide on the unprotonated oxirane. As in any S_N2 reaction, steric matters are important, and the ring opens at the sterically more accessible, less substituted carbon (Fig. 10.33). In acid, the first step is the protonation of the oxygen

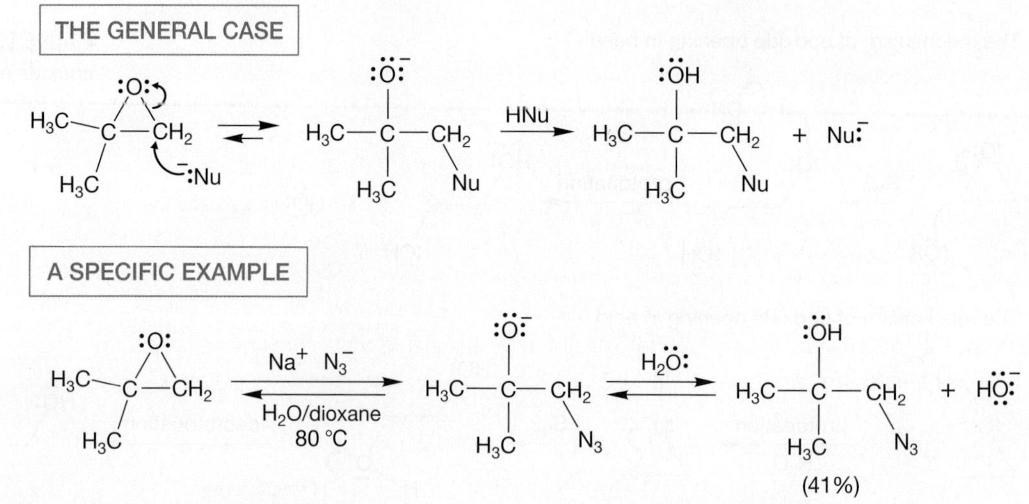

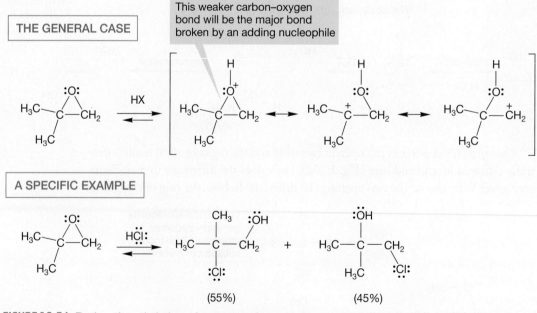

FIGURE 10.33 In base, the nucleophile adds at the sterically less encumbered carbon.

of the three-membered ring (Fig. 10.34). Recall the discussion of the regiochemistry of the opening of unsymmetrical bromonium ions (p. 419) and mercurinium ions (p. 422). The positive charge is borne not only by the oxygen but also by the two carbons of the oxirane. Most of the positive charge on carbon will reside at the more substituted position. The more substituted carbon–oxygen bond will be longer and weaker than the other, less substituted carbon–oxygen bond (Fig. 10.34). As in unsymmetrical bromonium ions, the transition state for addition at the more substituted position will be lower in energy than that for addition at the less substituted position.

FIGURE 10.34 Both carbons help bear the positive charge in the protonated oxirane. Most of the positive charge is on the more substituted carbon, and that is where the nucleophile adds. This reaction is much like the opening of an unsymmetrical bromonium ion by water or alcohol.

PROBLEM 10.12 If hydroxide is such a poor leaving group, one would expect RO⁻ would also be a bad leaving group, and it is. In fact, simple ethers (R—O—R) are not generally cleaved in strong base. Why, then, do oxiranes open so easily in base (Fig. 10.32)? What is special about the three-membered ring that facilitates the displacement?

We have seen that the apparently mysterious change in regiochemistry of the opening in acid and base can be reconciled easily. The trick is to learn to see new reactions in terms of what you already know. Books will always be organized in this way. Questions are set up by recently discussed phenomena. Nature is not always so cooperative and organized, however.

In Chapter 6, we met two organometallic reagents, the Grignard reagent, RMgX, and organolithium reagents, RLi. These compounds are very strong bases (p. 227) and are nucleophiles as well. Like other nucleophiles, organometallic reagents react with oxiranes to open the strained ring. The initial product is an alkoxide, which is protonated in a second step to give the alcohol. This reaction is quite useful for synthesizing alcohols in which the initial carbon atom frame of the organometallic reagent has been augmented by the two carbons originally in the oxirane (Fig. 10.35).

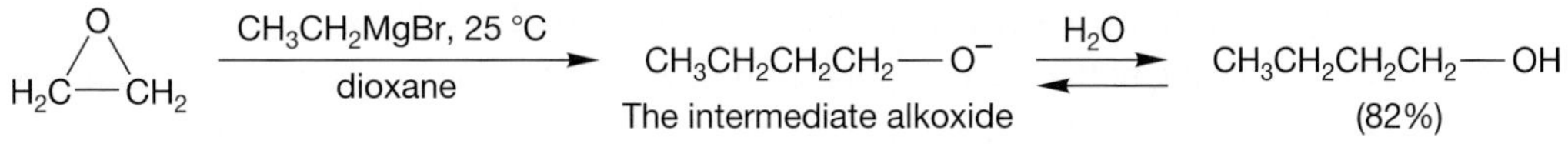

FIGURE 10.35 Opening of an epoxide by an organometallic reagent. The initial product is an alkoxide, which is protonated in a second step to give the alcohol.

The strongly basic organometallic reagents inevitably attack at the less substituted, less sterically hindered carbon of the oxirane, as does any strong nucleophile (Fig. 10.36).

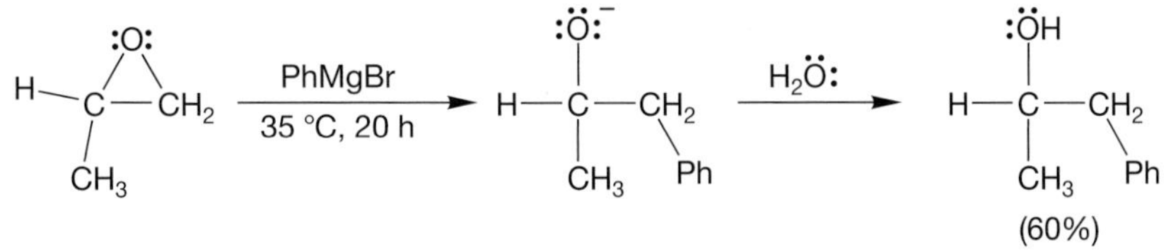

FIGURE 10.36 The S_N2 ring opening occurs at the less sterically demanding carbon of the epoxide.

The Grignard reagent and organolithium reagents are sources of "R⁻." Because they are not free carbanions, we call them carbanion equivalents. There are also reagents that are hydride (H:⁻) equivalents. We will see several in Chapter 16 and beyond, but it is useful to introduce lithium aluminum hydride (LiAlH₄, "LAH") now. Like the organometallic reagents, lithium aluminum hydride opens oxiranes by nucleophilic addition, and this reaction provides you with another synthesis of alcohols. Addition of hydride is predominantly from the less hindered side of the epoxide.

Once again, an initially produced alkoxide is protonated by water in a second step (Fig. 10.37).

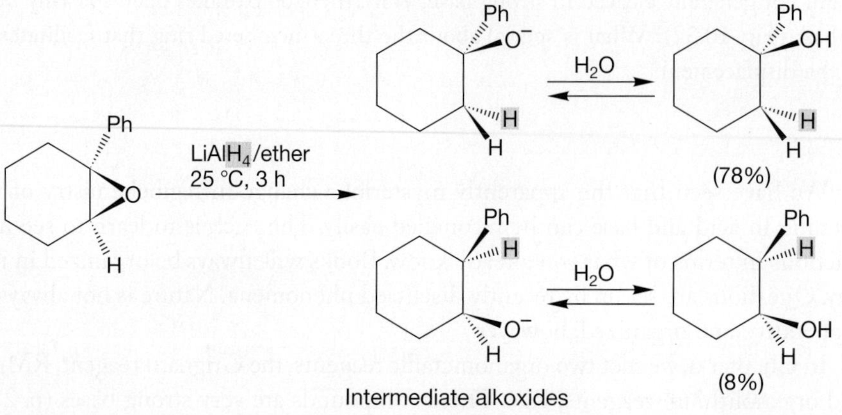

FIGURE 10.37 Lithium aluminum hydride reduction of epoxides, followed by hydrolysis, is yet another pathway for alcohol synthesis.

Intermediate alkoxides

Organometallic reagents have drawbacks, however. Reduction of the oxirane by hydride transfer from the organometallic reagent sometimes occurs and rearrangements can be induced, especially by organolithium reagents. Moreover, Grignard and organolithium reagents inevitably add to any carbonyl groups present in the molecule (Chapter 16). By contrast, $R_2Cu(CN)Li_2$, the **organocuprates** which are formed from RLi and CuCN, open epoxides smoothly without rearrangement (Fig. 10.38) and are not reactive enough to attack a carbon–oxygen double bond. As in other S_N2 displacements, opening of the three-membered ring is from the rear and occurs predominately at the less hindered position.

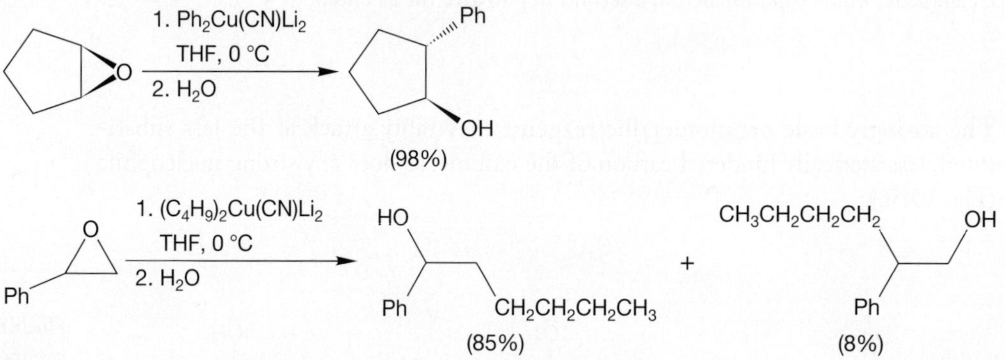

FIGURE 10.38 Organocuprates also open epoxides.

Summary

Epoxides are typically synthesized through peracid-mediated additions to alkenes. Opening of epoxides in either acid or base is stereospecifically trans (here is the S_N2 reaction at work), but the regiochemistry of opening depends on the conditions. In base, steric factors lead to addition of the nucleophile at the less hindered position. In acid, the nucleophile adds predominantly to the protonated epoxide at the more sterically hindered carbon.

Openings of oxiranes by organometallic reagents or hydride reagents such as lithium aluminum hydride are valuable synthetic tools for making alcohols.

10.4d Carbenes   Additions to alkenes lead to cyclopropanes through intermediates called **carbenes**. Carbenes are neutral compounds containing divalent carbon, and they are often formed from nitrogen-containing molecules called **diazo compounds** (Fig. 10.39). Heat (Δ) or light (*h*ν) liberates carbenes from their parent diazo compounds through loss of nitrogen, and they react rapidly with alkenes (which are generally used as solvent for the reaction) to give cyclopropanes. Another common source of carbenes is halocarbons. Treatment of chloroform (CHCl₃) or bromoform (CHBr₃) with strong bases such as hydroxide or potassium *tert*-butoxide gives dichlorocarbene or dibromocarbene (Fig. 10.39).

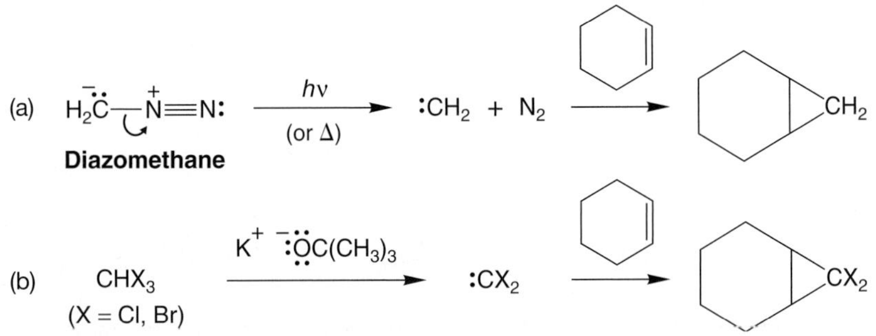

FIGURE 10.39 (a) Loss of nitrogen from a diazo compound liberates a carbene that can add to an alkene to give a cyclopropane. (b) Dihalocarbenes can be made by treatment of halogenated alkanes with strong base. Dihalocarbenes also add to alkenes.

Diazo compounds have the attractive feature (aside from being sources of carbenes) of often being quite handsomely colored. Diazomethane (CH₂N₂) is a bright yellow gas, diazocyclopentadiene an iridescent orange liquid, and 4,4-dimethyldiazocyclohexa-2,5-diene an almost blue liquid (Fig. 10.40). However, diazo compounds have a downside: They are certainly poisonous, demonstrably explosive, and probably carcinogenic. So one must be careful (and brave, and perhaps a little crazy) to do carbene chemistry! But this care is worth it because carbenes are fascinating species.

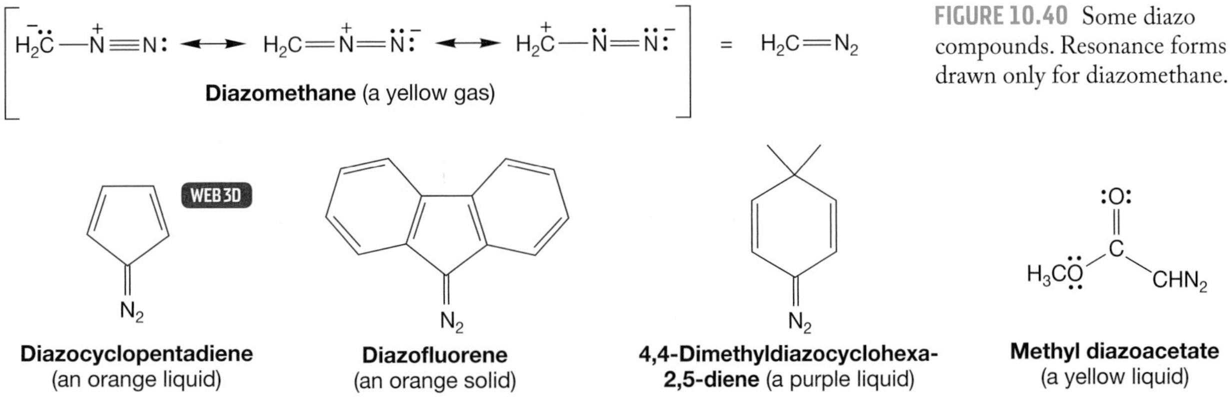

FIGURE 10.40 Some diazo compounds. Resonance forms are drawn only for diazomethane.

PROBLEM 10.13 Write resonance forms for 4,4-dimethyldiazocyclohexa-2,5-diene shown in Figure 10.40.

PROBLEM 10.14 Propose a mechanism for the conversion of chloroform (CHCl₃, pK_a = 29) into dichlorocarbene. *Hint*: When the reaction is carried out in D₂O/DO⁻, recovered chloroform is partially deuterated (CDCl₃).

Most carbenes add to the double bond of an alkene in stereospecifically syn fashion, which means that cis alkenes react with carbenes to give only cis cyclopropanes and trans alkenes give the corresponding trans cyclopropanes (Fig. 10.41).

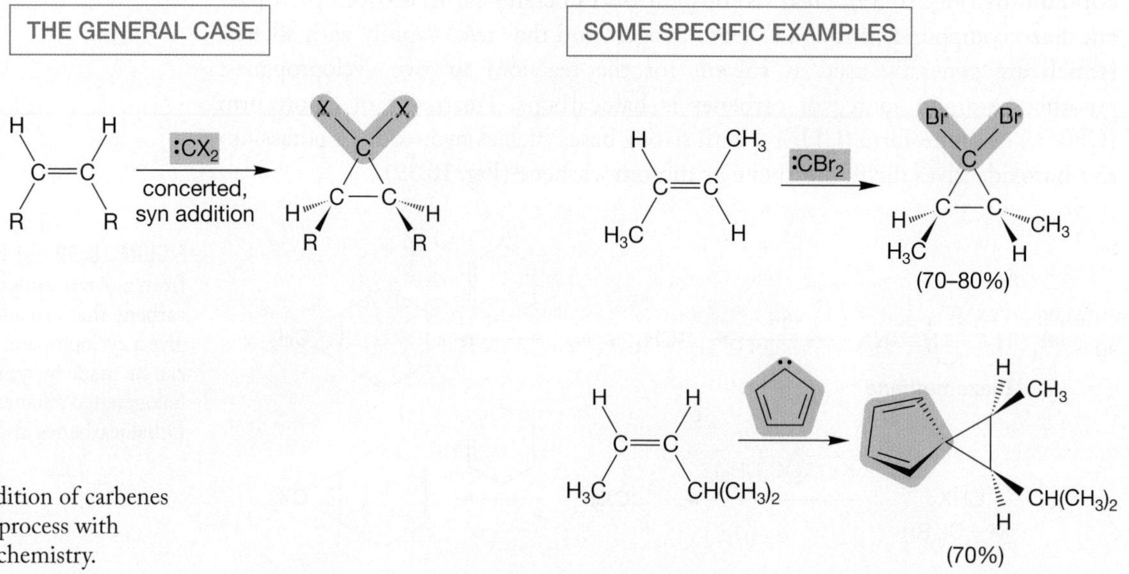

FIGURE 10.41 Addition of carbenes to alkenes is a syn process with retention of stereochemistry.

This stereospecificity tells us that there can be no intermediates capable of rotation about carbon–carbon bonds in the ring-forming reaction. The mechanism involves a single step; in the addition both new ring bonds are formed in a concerted fashion. A two-step addition would lead to formation of both cis and trans cyclopropanes (Fig. 10.42).

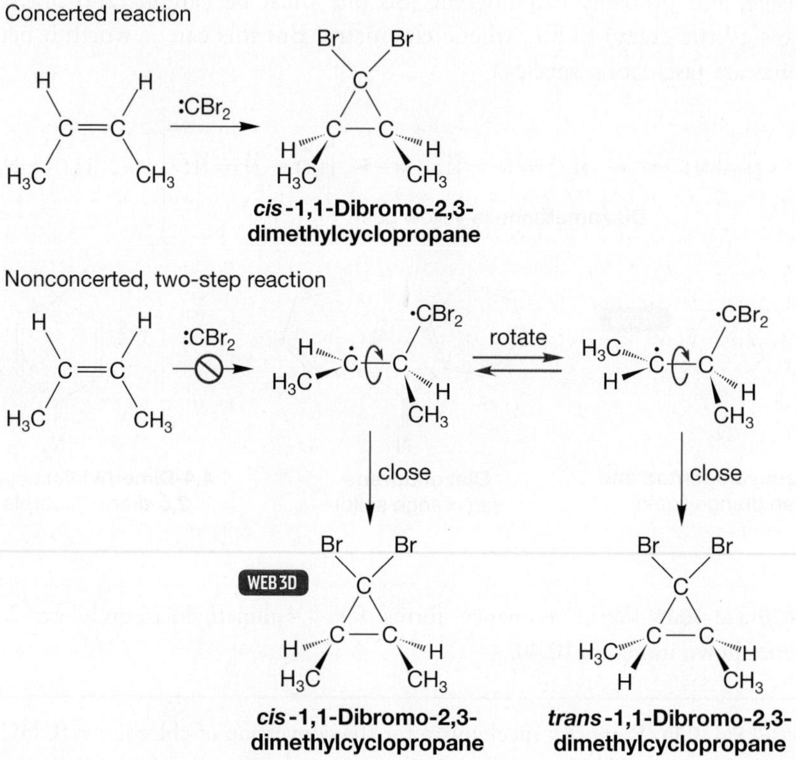

FIGURE 10.42 The concerted and nonconcerted mechanisms contrasted.

Regardless of the mechanism, the reactions of carbenes with alkenes provide the most widely used synthetic route to cyclopropanes.

10.4e Triplet Carbenes   One of the reasons carbenes are so interesting is that, unlike almost all other reactive intermediates we know, they have two reactive states, and these states are quite close to each other in energy. The reactions considered in Figure 10.41 were reactions of a carbene species in which two spin-paired electrons occupied the relatively low-energy hybrid orbital and the higher energy $2p$ orbital was empty, as shown in Figure 10.43. In fact, this species, called a **singlet carbene**, generally lies a few kilocalories per mole *higher* in energy than another form of divalent carbon called a **triplet carbene**. In the triplet species, one electron occupies the hybrid orbital and the other has been promoted to the previously empty $2p$ orbital. Of course, promotion incurs an energy cost—the approximately sp^2 hybrid orbital has more s character than the $2p$ orbital, and thus an electron in the sp^2 orbital is more stable than an electron in a pure $2p$ orbital. But there are energetic compensations. For example, the two nonbonding electrons are farther apart in the triplet and thus electron–electron repulsion is lowered. With some exceptions (R = halogen, for example), the triplets are somewhat more stable than the corresponding singlets. For many carbenes the two species are in equilibrium, and sorting out the chemistry can be complicated because both species can react.

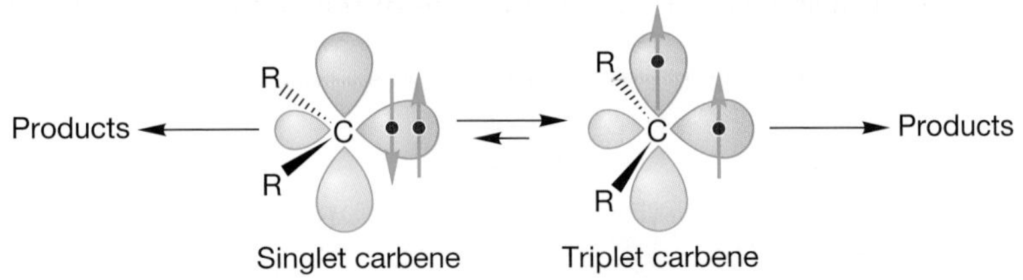

FIGURE 10.43 Both spin states of a carbene may form products.

Products ←   Singlet carbene   ⇌   Triplet carbene   → Products

PROBLEM 10.15 Singlet $:CF_2$ is much more stable than the corresponding triplet. Consider the structure of this intermediate and explain why the singlet is especially favored in this case.

Normally, even though the singlet is disfavored at equilibrium, reactions of the singlet are the ones that are observed. The singlet is usually *much* more reactive than the triplet. Even though there is more triplet present at equilibrium, its reactions cannot compete with those of the small amount of the much more reactive singlet. This point is important. A species does not have to be present in great quantity for its chemical reactions to dominate those of a more thermodynamically stable (lower energy) and relatively unreactive equilibrium partner. The activation barriers surrounding the higher energy partner may be substantially lower than those surrounding the lower energy partner. Figure 10.44 shows this situation in an energy diagram.

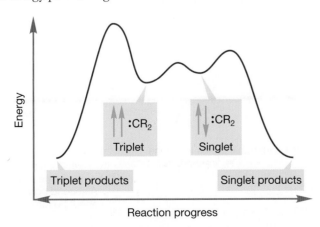

Triplet :CR₂
Singlet :CR₂
Triplet products
Singlet products
Energy
Reaction progress

FIGURE 10.44 Even though the triplet carbene is usually lower in energy and thus favored at equilibrium, most of the chemistry comes from the singlet. The triplet is protected by quite high activation barriers. As long as some of the singlet is formed in equilibrium with the triplet, the singlet accounts for most or all of the chemistry.

Sometimes the barriers to triplet reactions are not effectively higher than those for singlets. In these cases, triplet carbene reactions can be observed. Like their singlet counterparts, triplet carbenes add to alkenes to give cyclopropanes (Fig. 10.45).

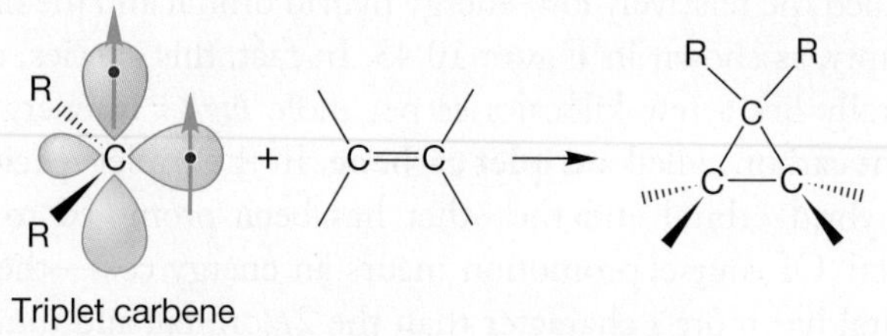

FIGURE 10.45 Triplet carbenes also add to alkenes to give cyclopropanes.

Triplet carbene

Now let's look at electron spin. For a singlet carbene, the two nonbonding electrons have paired (opposite) spins and can produce the two new cyclopropane carbon–carbon bonds without any problem by combining with the two paired electrons in the π bond of the alkene (Fig. 10.46). As we saw earlier (Fig. 10.41), the reaction of a singlet carbene with an alkene takes place in a concerted fashion, preserving the original stereochemical relationships of the groups on the alkene.

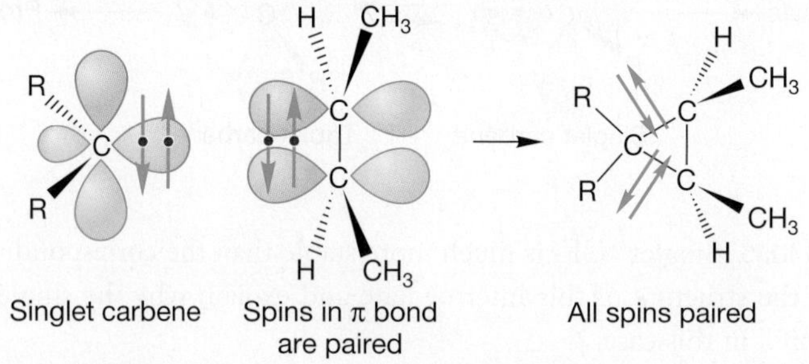

FIGURE 10.46 In the addition of a singlet carbene to an alkene, two new bonds are formed in a concerted fashion.

Singlet carbene   Spins in π bond are paired   All spins paired

What do we expect of the triplet? In particular, what will be the stereochemistry of the addition reaction? The first step in the addition of a triplet carbene to an alkene is shown in Figure 10.47. In contrast to the singlet, the triplet cannot form two new bonds in a single step. A new species, called a **diradical**, is produced, with two nonbonding electrons on different carbons. Now what can happen? Why not simply close up to the three-membered ring and be done with it? Look closely in Figure 10.47 at the spins of the electrons involved in the reaction. The second bond cannot close! The two electrons that would make up the new bond have the same spin and therefore cannot occupy the same orbital.

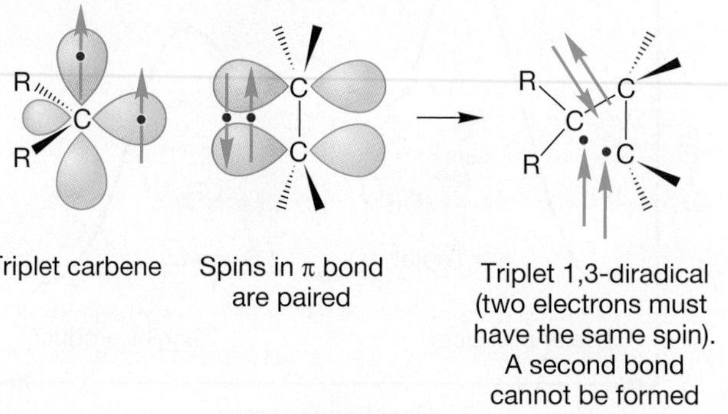

Triplet carbene   Spins in π bond are paired   Triplet 1,3-diradical (two electrons must have the same spin). A second bond cannot be formed

FIGURE 10.47 Only one C—C bond can be formed in the reaction of a triplet carbene with an alkene. The first step in triplet addition is reaction with the π bond of the alkene to give a 1,3-diradical.

The triplet carbene, with its two nonbonding electrons of the same spin, must produce an intermediate that cannot close to a cyclopropane until a spin is somehow changed. Such spin "flips" are not impossible, but they are usually slow on the time scale of molecular processes. In particular, they are slow compared with the very rapid rotations about carbon–carbon single bonds. Therefore the stereochemistry of the original alkene need not be maintained in the eventual product of the reaction, the cyclopropane (Fig. 10.48). If rotation about the carbon–carbon bond is faster than the rate at which the spin quantum number of one electron changes (spin flip), it will not matter whether cis or trans alkene is used in the reaction with triplet methylene. The product will be a mixture of cis and trans 1,2-dialkylcyclopropanes in either case.

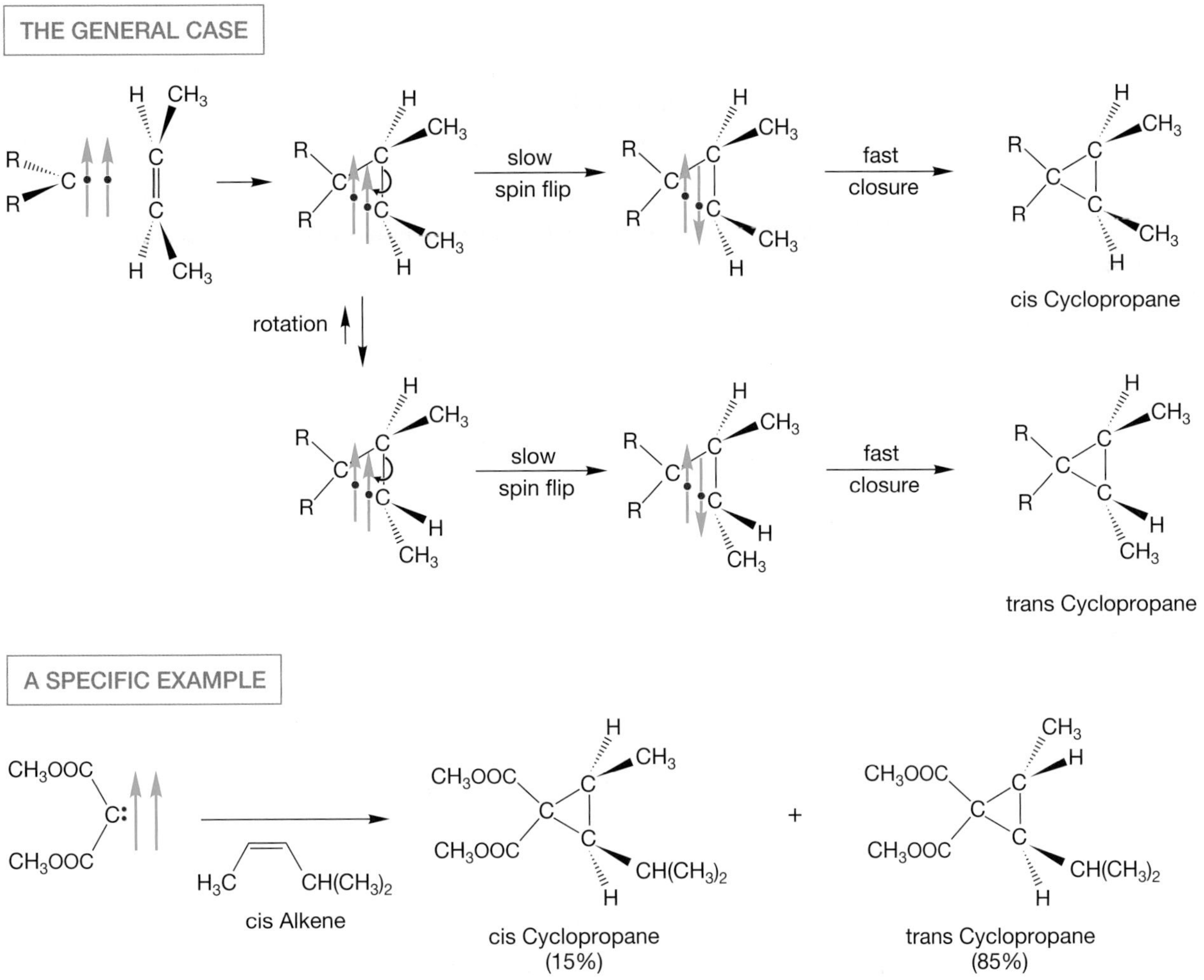

THE GENERAL CASE

A SPECIFIC EXAMPLE

FIGURE 10.48 Triplet carbene addition to alkenes. The formation of an intermediate 1,3-diradical induces scrambling of stereochemistry. Unlike singlets, which add in a single step to generate cyclopropanes stereospecifically, triplets give both cis and trans cyclopropanes.

It turns out that triplets do give mixtures of the cis and trans cyclopropanes, and the examination of the stereochemistry of the addition reaction of carbenes has become the method of choice for determining the spin state of the reacting carbene.

This discussion of triplet carbenes and diradicals leads directly to Chapter 11, which will concentrate on the chemistry of radicals: atoms with a single nonbonding electron.

Summary

Carbenes come in two "flavors"—spin-paired singlets and spin-unpaired triplets. Often the triplet is the more stable form, but the singlet state is usually more reactive. Both singlet and triplet carbenes add to alkenes to give cyclopropanes. In the concerted reaction of the singlets, any stereochemistry present in the alkene is preserved. In the nonconcerted additions of triplets, stereochemistry is lost.

10.5 Dipolar Addition Reactions: Ozonolysis and the Synthesis of Carbonyl (R₂C=O) Compounds

So far, we have seen a number of addition reactions leading to three-membered rings, either transient (bromonium ions, p. 414) or stable (oxiranes and cyclopropanes, p. 423). Now we look at other additions leading to different-sized rings. These reactions lead in new directions, and their products are not at all similar. Nonetheless they are tied together and to earlier material by a mechanistic thread—they involve an initial addition to an alkene to give a cyclic intermediate.

10.5a The Mechanism of Ozonolysis   The precursors of carbenes, diazo compounds (Fig. 10.39), are members of a class of molecules called **1,3-dipoles**, or **1,3-dipolar reagents** (Fig. 10.49). They are so called because good neutral resonance forms cannot be written for them. An important characteristic of 1,3-dipoles is that they add to alkenes to give five-membered rings.

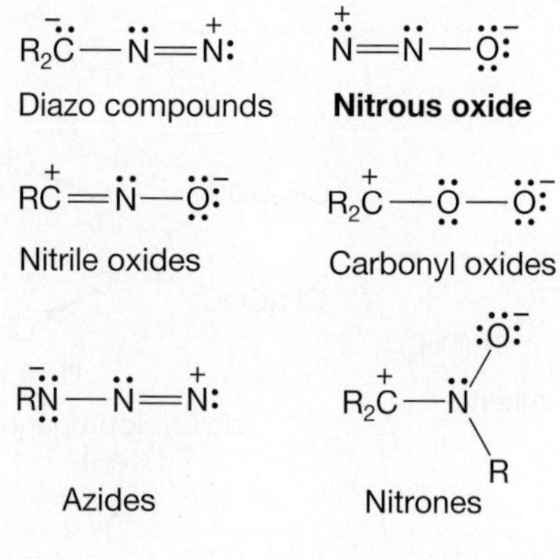

FIGURE 10.49 Some 1,3-dipolar reagents.

PROBLEM 10.16 Draw more resonance forms for each 1,3-dipole in Figure 10.49.

PROBLEM 10.17 Use the arrow formalism to show the compounds formed from reaction of the azides, nitrile oxides, and carbonyl oxides shown in Figure 10.49 with a typical alkene and alkyne.

Ozone (O$_3$) is a particularly important 1,3-dipole. The ozone molecule is often in the news these days. There is too little of this gas in the upper atmosphere, where it functions to screen the surface of the earth from much dangerous ultraviolet (UV) radiation. Yet, in all too many places near the surface, there is too much of it. Ozone is a principal component of photochemical smog.

The process called **ozonolysis** involves the reaction of ozone with an alkene to give a five-membered ring containing three oxygen atoms adjacent to one another, as shown in Figure 10.50. Such a ring is called a **primary ozonide** or a **molozonide**.

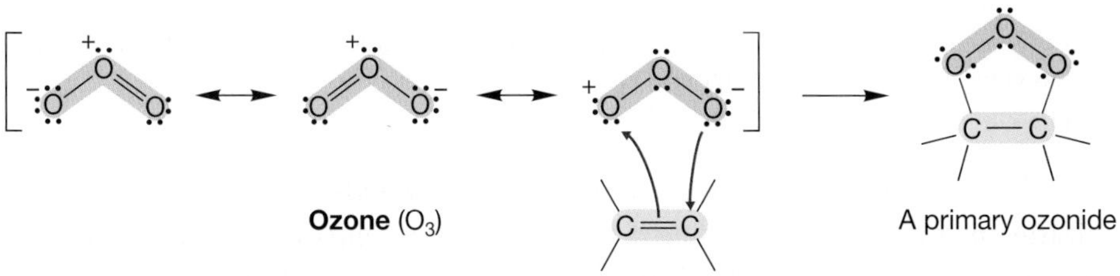

FIGURE 10.50 A 1,3-dipolar addition of ozone to an alkene gives a primary ozonide, a five-membered ring containing three oxygen atoms in a row.

The arrow formalism lets us map out the formation of the ozonide. The π bond of the alkene reacts with ozone to produce the five-membered ring. These primary ozonides are extremely difficult to handle, and it took great experimental skill on the part of Rudolf Criegee (1902–1975) and his co-workers at the University of Karlsruhe in Germany to isolate the ozonide produced in the reaction between ozone and *trans*-di-*tert*-butylethylene. A concerted mechanism predicts that the stereochemical relationship of the alkyl groups in the original alkene will be preserved in the ozonide, and this is what happens (Fig. 10.51).

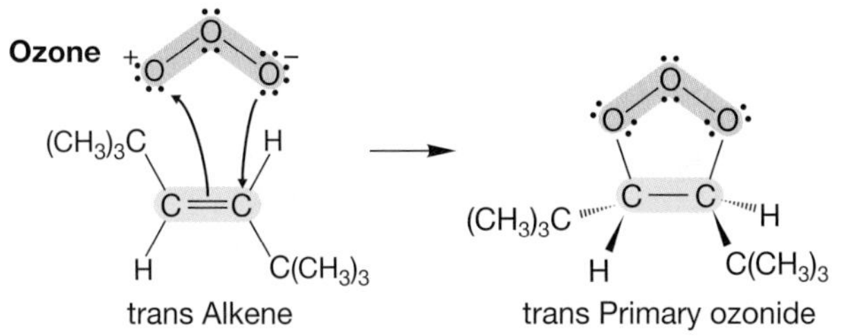

FIGURE 10.51 The stereospecific formation of a primary ozonide.

As implied previously, unless great care is taken, these initially formed primary ozonides are not stable, but rearrange to another compound, known simply as an **ozonide** (Fig. 10.52). The rearrangement of primary ozonides to ozonides involves a

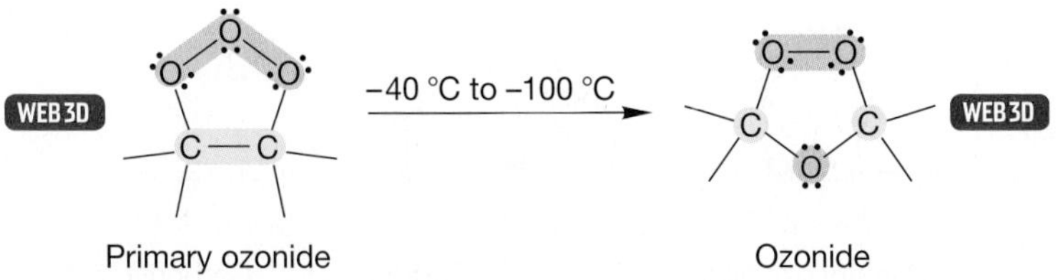

FIGURE 10.52 The rearrangement of a primary ozonide into an ozonide.

pair of reverse and forward 1,3-dipolar additions (Fig. 10.53). The first step in the rearrangement is the opening of the primary ozonide to give a molecule containing a carbon–oxygen double bond called a **carbonyl compound**, along with a carbonyl oxide, a new 1,3-dipole (see Fig. 10.49). These two species can recombine to give the ozonide.

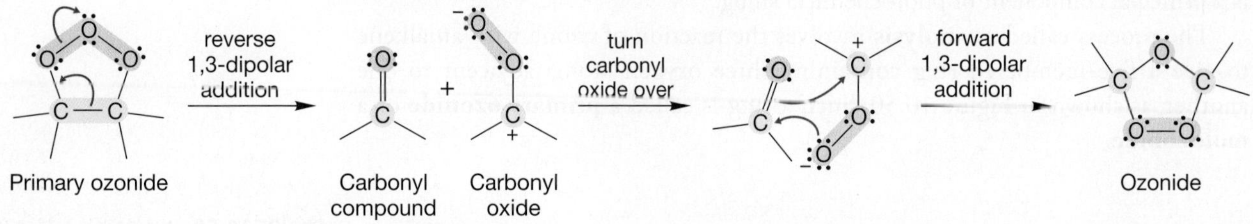

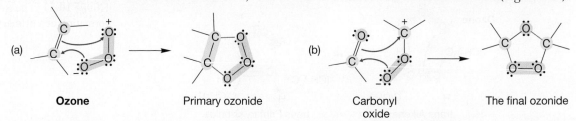

FIGURE 10.53 The mechanism of rearrangement by which a primary ozonide becomes an ozonide involves two steps. In the first, a reverse 1,3-dipolar addition generates a carbonyl compound and a carbonyl oxide. The carbonyl oxide then re-adds to the carbonyl in the opposite sense to give the new ozonide.

The mechanism of rearrangement from primary ozonide to ozonide may look a bit bizarre, and there can be no hiding the fact that it's not simple. It is not really far from things we already know, however. More frustrating is our distance from being able to predict the course of complicated reactions. At this point, you have little choice but to learn the course of the ozonolysis reaction, and that is not intellectually satisfying. It would be a far better situation if you were able to predict what must happen in this multistep, multi-intermediate process. Perhaps it is some comfort to know that you are not far behind the current frontier. Organic chemists have only recently emerged from the "gee whiz" phase to the point where advances in experimental skill (mostly in analytical techniques, if one is to be honest) and theoretical methods have begun to allow reliable predictions to be made in risky cases. All too often we are reduced to making explanations after the fact; to rationalizing, as we are about to do with the rearrangement of one ozonide to the other.

We will take it one step at a time, working backward. In the final step, the addition of the carbonyl oxide to a carbon–oxygen double bond is very little removed from the addition of ozone to an alkene (Fig. 10.54a). Like ozone, the carbonyl oxide can add to π bonds. In this case, such an addition leads to the observed ozonide (Fig. 10.54b).

FIGURE 10.54 Like diazo compounds and ozone, a carbonyl oxide is a 1,3-dipole and adds to π systems to give five-membered ring compounds.

The first step in the conversion of the primary ozonide into the ozonide is another 1,3-dipolar addition, but this time in reverse, and it is therefore a bit harder to see (Fig. 10.55). The reaction is completed when the carbonyl oxide turns over and re-adds to the carbonyl compound (Fig. 10.54b).

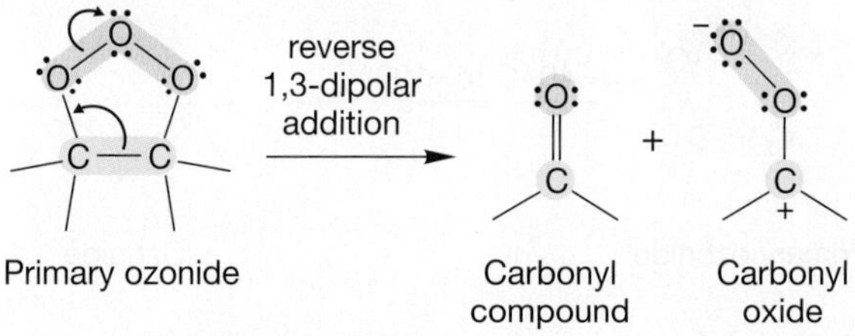

FIGURE 10.55 The first step in the formation of an ozonide from a primary ozonide is a reverse 1,3-dipolar addition to generate a carbonyl compound and a carbonyl oxide.

The overall reaction is a series of three 1,3-dipolar additions (Fig. 10.56). First, ozone adds to the alkene to give the primary ozonide. Then the primary ozonide undergoes a reverse 1,3-dipolar addition to give a carbonyl compound and a new 1,3-dipole, the carbonyl oxide. Finally, the carbonyl oxide turns over and re-adds to the carbonyl compound in the opposite sense to give the new ozonide. The reaction is a sequence of three reversible 1,3-dipolar additions driven by thermodynamics toward the relatively stable ozonide.

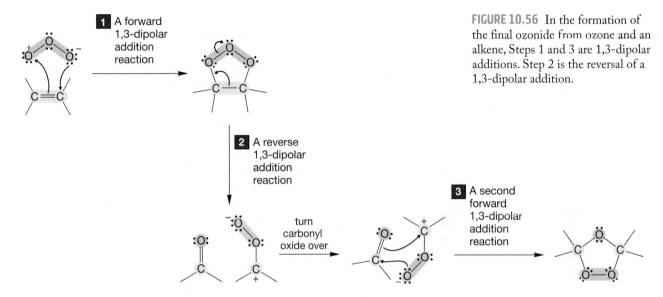

FIGURE 10.56 In the formation of the final ozonide from ozone and an alkene, Steps 1 and 3 are 1,3-dipolar additions. Step 2 is the reversal of a 1,3-dipolar addition.

Now that we have described the reaction, however, the tougher part is to explain *why* it runs the way it does. Why does thermodynamics favor the final, more stable ozonide over the primary ozonide (Fig. 10.57)?

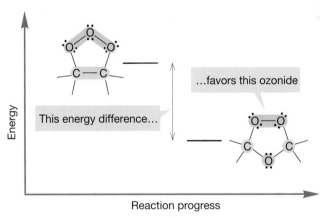

FIGURE 10.57 The ozonide contains fewer weak oxygen–oxygen bonds than the primary ozonide and is therefore more stable.

The oxygen–oxygen bond is quite weak (~40 kcal/mol). Much more stable, though, is the carbon–oxygen bond (85–90 kcal/mol). The most important factor that makes an ozonide more stable than a primary ozonide is the difference in the numbers of weak oxygen–oxygen bonds and strong carbon–oxygen bonds. As long as the kinetic barriers to the forward and reverse 1,3-dipolar additions are not too high, thermodynamics favors formation of the stronger carbon–oxygen bonds and the ultimate result is the final, more stable, ozonide.

PROBLEM 10.18 Use Table 8.2 (p. 337) to do a full calculation of the bond energy difference between a primary ozonide and an ozonide.

PROBLEM 10.19 Why are oxygen–oxygen sigma bonds so weak? *Hint*: It is not just electron–electron repulsion. Do a quick molecular orbital analysis.

PROBLEM 10.20 Complete the Energy versus Reaction progress diagram of Figure 10.57 to show the conversion of the primary ozonide into the ozonide. Assign the intermediates to the appropriate points in the reaction pathway.

10.5b The Synthetic Potential of Ozonolysis: Carbonyl Compounds

Why do we spend so much time with this complicated reaction? We have already mentioned generality. A vast number of 1,3-dipolar agents has been made, and many of them will add to π bonds to give five-membered rings.

Second, and more important, the ozonides are very useful compounds. Their further transformations give us entry into new classes of compounds. Ozonides can be reduced (Fig. 10.58) and, depending on the structure of the starting alkene, the products can be **ketones** ($R_2C\!=\!O$), which are compounds that have two R groups attached to the carbonyl carbon; **aldehydes** ($RCH\!=\!O$), which are compounds that have an R group and a H attached to the carbonyl; or a mixture of the two. Many reducing agents are known, among them H_2/Pd, Zn, and $(CH_3)_2S$ (dimethyl sulfide).

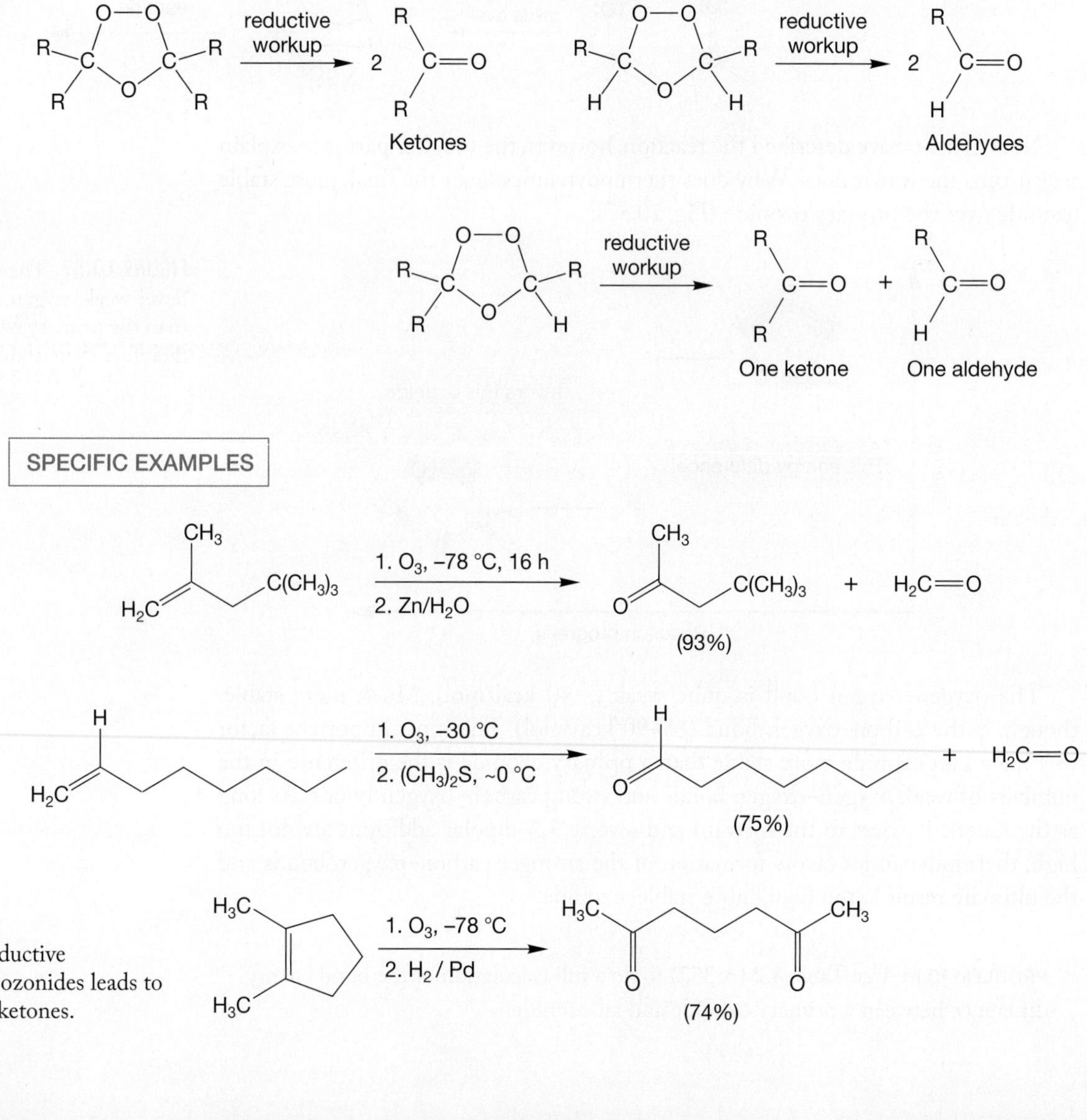

FIGURE 10.58 Reductive decomposition of ozonides leads to aldehydes and/or ketones.

Oxidation of the ozonides, usually with hydrogen peroxide (H_2O_2), leads either to ketones or carboxylic acids (see Fig. 10.24). The structure of the product again depends on the structure of the ozonide, which itself depends on the structure of the original alkene (Fig. 10.59).

GENERAL CASES

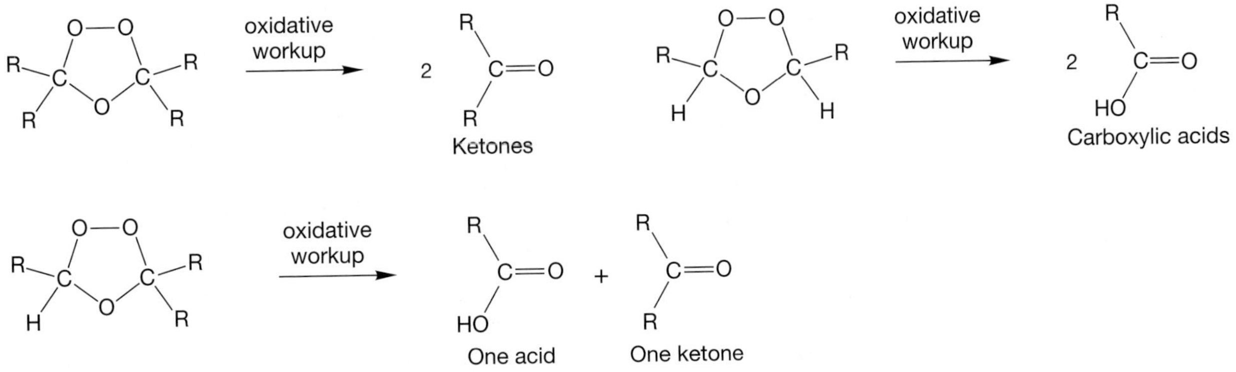

SPECIFIC EXAMPLES

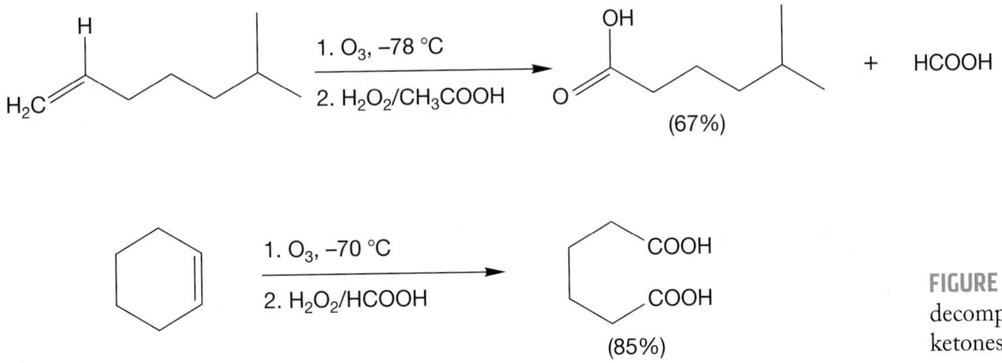

FIGURE 10.59 Oxidative decomposition of ozonides leads to ketones and/or carboxylic acids.

In addition to providing a synthetic pathway for numerous carbonyl compounds, ozonolysis has another use for us: The structures of the products of an ozonolysis reaction can be used to reason out the composition of an unknown alkene. To see ozonolysis used this way, try Problem 10.21.

PROBLEM SOLVING

Whenever you see ozone (O_3) in a problem, you can bet the farm that a double bond is going to be converted into a pair of carbonyl compounds. The question may be mechanistic or synthetic in nature, but that processs is going to be involved. It is not a bad idea at this early point to jot down the general scheme shown below whenever you see ozone.

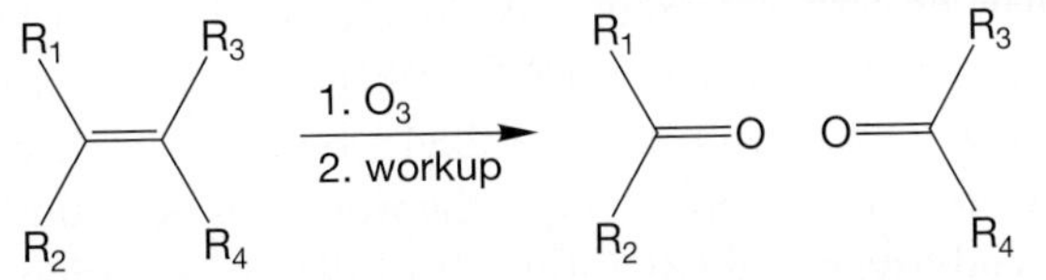

WORKED PROBLEM 10.21 Ozonolysis of alkenes can give the carbonyl products shown below. Supply an alkene structure that can produce the carbonyl compounds shown. Specify the nature of the workup step in the process (oxidation or reduction).

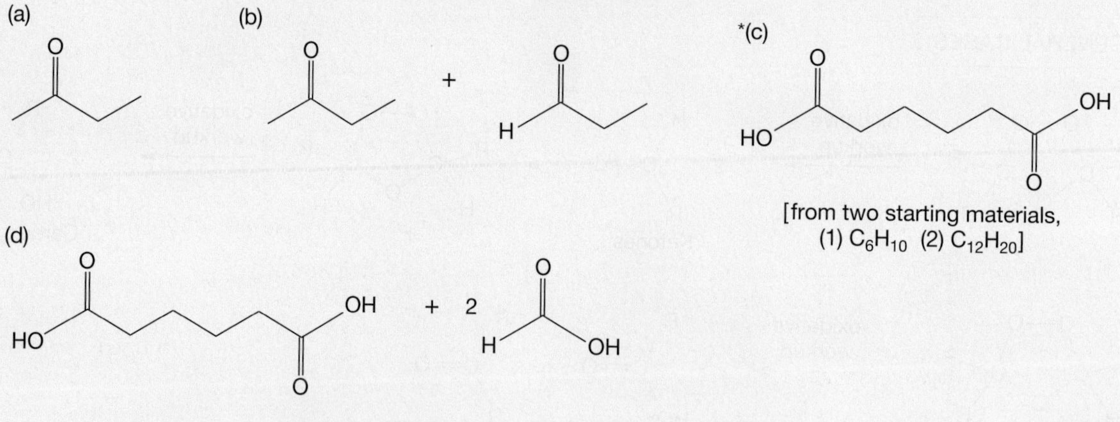

(a)

(b)

+

*(c)

[from two starting materials, (1) C_6H_{10} (2) $C_{12}H_{20}$]

(d)

OH   + 2

ANSWER (c) Ozonolysis converts a carbon–carbon double bond into two carbon–oxygen double bonds.

$$\text{C=C} \quad \xrightarrow[\substack{\text{2. oxidative} \\ \text{or reductive} \\ \text{workup}}]{\text{1. } O_3} \quad \text{C=O + O=C}$$

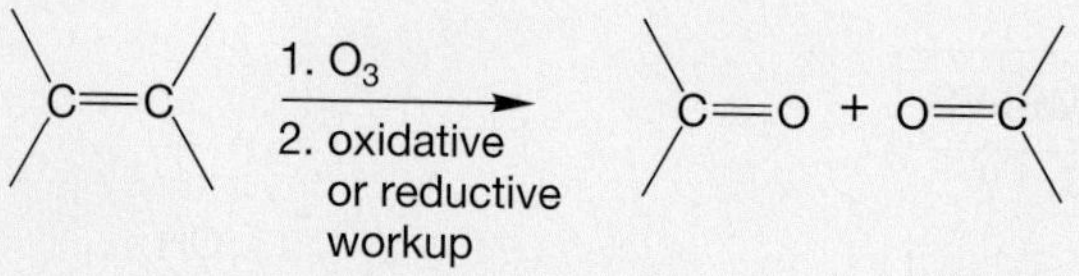

In this case, both carbon–oxygen double bonds are in the same product molecule, which means that the starting material must be a cyclic alkene. The product contains carboxylic acids, so we know that an oxidative workup is required. A reductive workup would give an aldehyde.

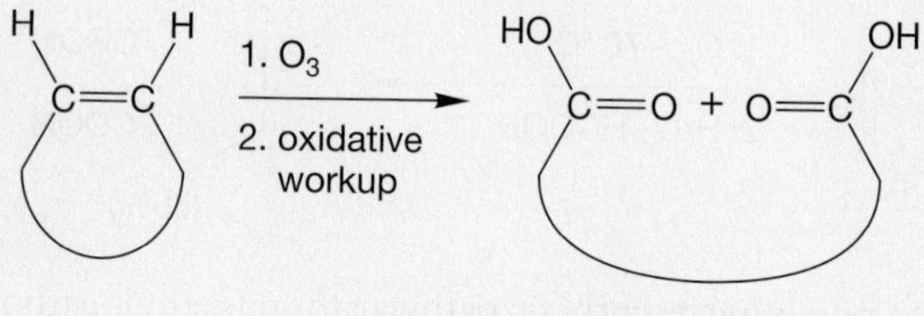

$$\text{H C=C H} \quad \xrightarrow[\substack{\text{2. oxidative} \\ \text{workup}}]{\text{1. } O_3} \quad \text{HO C=O + O=C OH}$$

There are several ways to construct an appropriate starting material. Two possibilities are

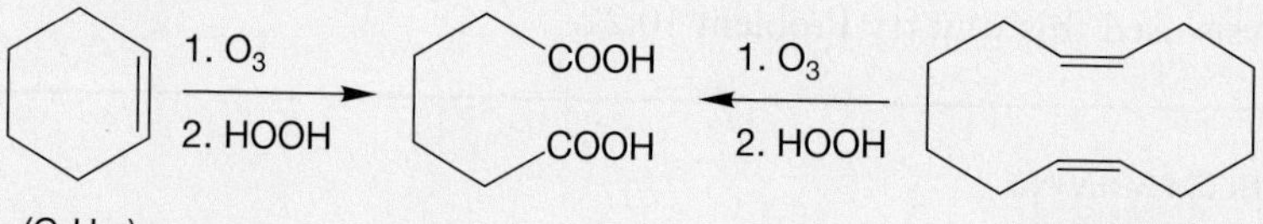

1. O_3
2. HOOH

COOH
COOH

1. O_3
2. HOOH

(C_6H_{10}) $(C_{12}H_{20})$

Summary

Ozonolysis of alkenes breaks both bonds in the carbon–carbon double bond and generates a pair of carbonyl groups. A double bond carbon bearing two alkyl groups is transformed into a ketone (R_2C=O) regardless of workup conditions. When a double bond carbon bears one alkyl group and one hydrogen it is transformed into an aldehyde (RHC=O) if the workup is done under reducing conditions and a carboxylic acid (RCOOH) if oxidizing conditions are used.

10.5c Oxidation with Permanganate or Osmium Tetroxide

Although they are not 1,3-dipolar reagents, both potassium permanganate ($KMnO_4$) and osmium tetroxide (OsO_4) add to alkenes to form five-membered rings in reactions that are related to the additions of ozone and other 1,3-dipoles (Fig. 10.60).

Dihydroxylation of alkenes

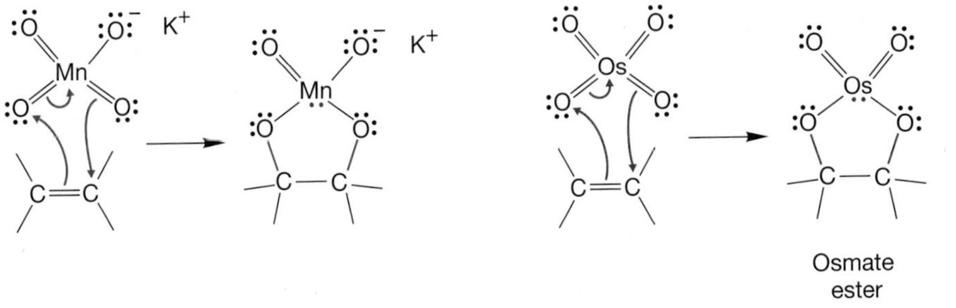

FIGURE 10.60 Addition of potassium permanganate or osmium tetraoxide to alkenes gives five-membered rings.

Osmate ester

The cyclic intermediate in the permanganate reaction cannot be isolated and is generally decomposed as it is formed to give vicinal diols (Fig. 10.61a). Good yields of cis 1,2-diols can be isolated from the treatment of alkenes with basic permanganate. The cyclic osmate ester can be isolated, but it, too, is generally transformed directly into a diol product by treatment with aqueous sodium sulfite (Na_2SO_3) as shown in Figure 10.61b. In both of these reactions it is the metal–oxygen bonds, not the carbon–oxygen bonds, that are broken. There are many variations of these reactions, which are among the best ways of synthesizing vicinal diols.

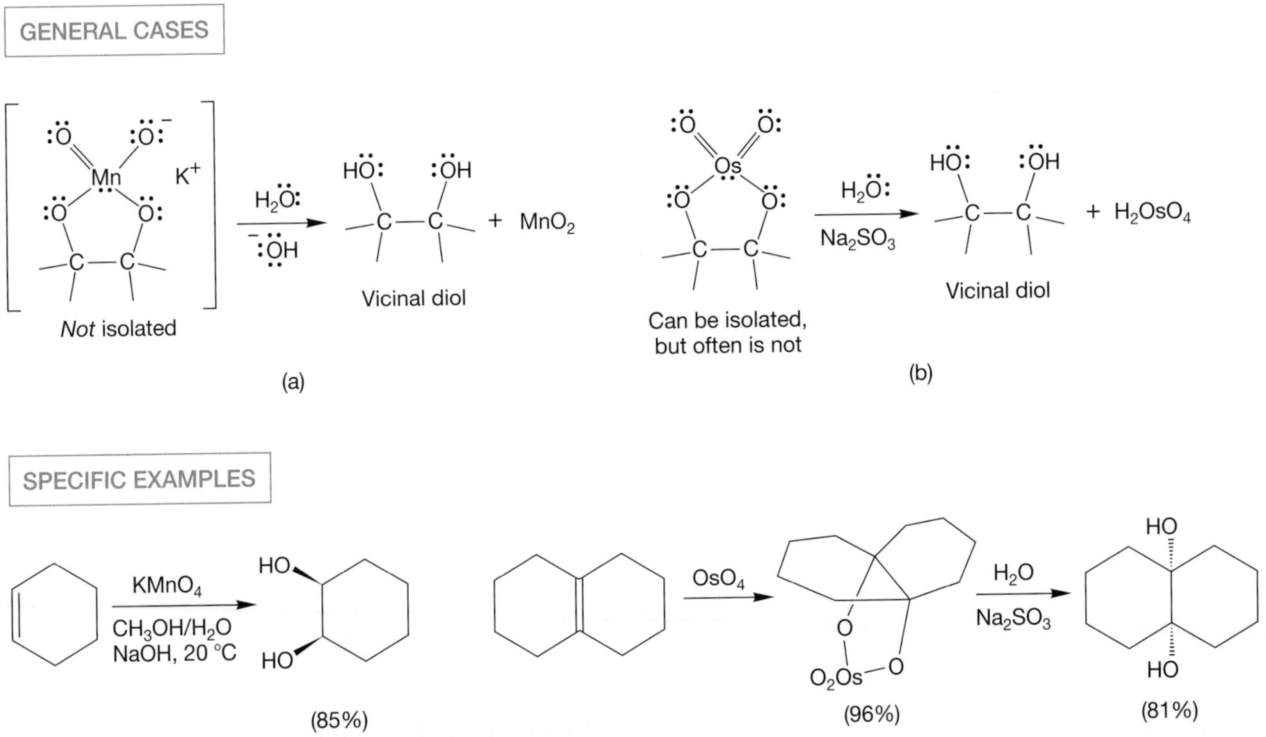

GENERAL CASES

Not isolated → Vicinal diol + MnO_2

Can be isolated, but often is not → Vicinal diol + H_2OsO_4

(a)   (b)

SPECIFIC EXAMPLES

KMnO₄, CH₃OH/H₂O, NaOH, 20 °C (85%)

OsO₄ (96%) → H₂O, Na₂SO₃ (81%)

FIGURE 10.61 A five-membered ring containing either (a) Mn or (b) Os can react further to generate vicinal diols.

PROBLEM 10.22 What mechanistic conclusions can be drawn from the observation that reaction of OsO_4 with *cis*-2-butene gives this single product?

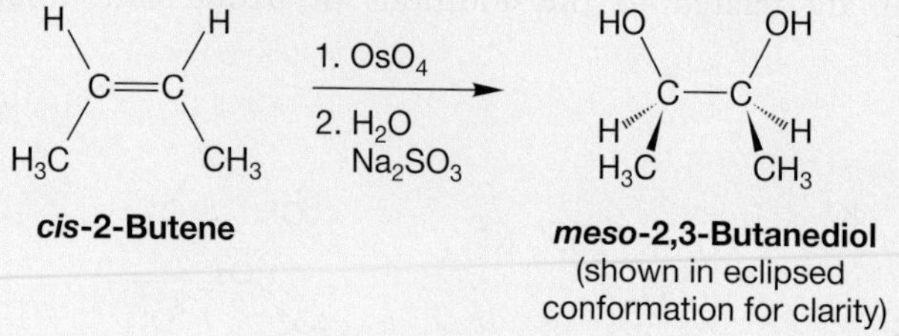

cis-**2-Butene**

1. OsO_4
2. H_2O
   Na_2SO_3

meso-**2,3-Butanediol**
(shown in eclipsed
conformation for clarity)

10.6 Addition Reactions of Alkynes: HX Addition

Much of Chapters 9 and 10 has been devoted to the addition of various reagents to carbon–carbon double bonds. Alkynes contain two π bonds and it would be quite astonishing if they did not undergo similar addition reactions. If we keep in mind what we have learned about alkene additions, it is reasonably easy to work through alkyne additions. The presence of the second double bond will add mechanistic complications, however, and some synthetic opportunities as well.

Addition of HBr or HCl to 3-hexyne gives the corresponding vinyl halides. It is tempting to begin with a protonation of the π system to give a carbocation and a halide ion as shown in Figure 10.62. In this case, the positive ion would be a vinyl cation. Addition of the halide to this cation would give the vinyl halide.

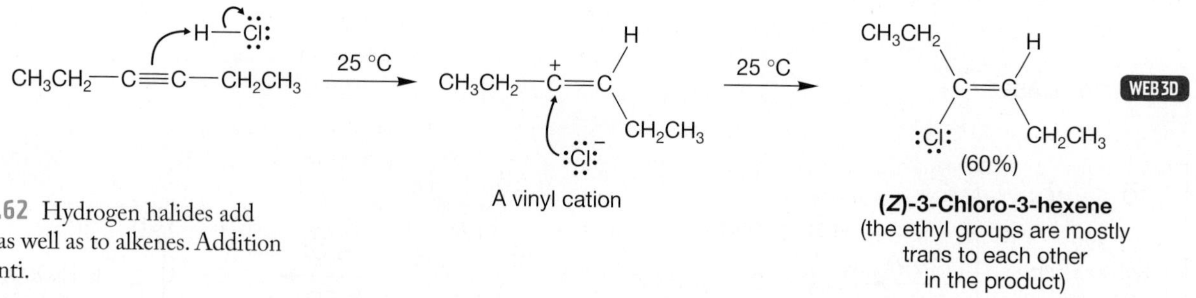

A vinyl cation

(Z)-3-Chloro-3-hexene
(the ethyl groups are mostly
trans to each other
in the product)

FIGURE 10.62 Hydrogen halides add to alkynes as well as to alkenes. Addition is mostly anti.

The stereochemistry of this addition is mixed, as would be expected from an open carbocation, although the trans diethyl compound is favored (Fig. 10.62).

However, there are problems with this simple extension to alkynes of the mechanism for additions to alkenes. Vinyl cations are unusual species and we must stop for a moment to consider their structures. The positive carbon in a vinyl cation is attached to two groups: the other carbon in the original triple bond and an R group. Accordingly, we would expect *sp* hybridization, and that leads to the structure in Figure 10.63.

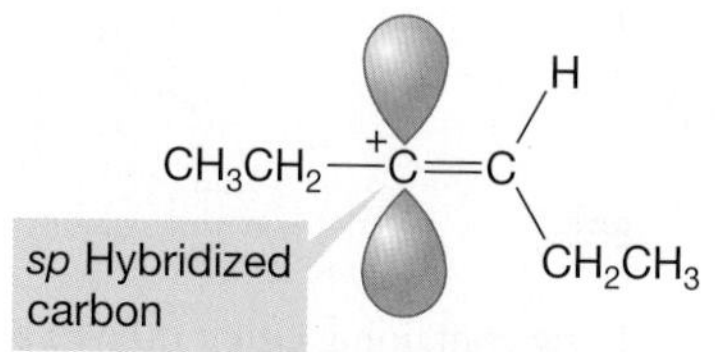

sp Hybridized
carbon

FIGURE 10.63 A vinyl cation contains an *sp* hybridized carbon.

Measurements in the gas phase show that vinyl carbocations are very unstable. Remember that the gas-phase heat of formation values do not include the solvent, which is very important to any cationic species. Remember also that direct comparisons can be made only between isomeric species. Nonetheless, one can get an approximate idea of the great instability of vinyl carbocations from Table 10.2. The absolute values of the energy differences are not important, but the relative stability order of the various carbocations is.

TABLE 10.2 Heats of Formation for Some Carbocations

Cation	Substitution	ΔH_f° (kcal/mol)
$^+CH{=}CH_2$	Primary vinyl	285
$^+CH_3$	Methyl	261.3
$CH_3{-}\overset{+}{C}{=}CH_2$	Secondary vinyl	231
$^+CH_2CH_3$	Primary	215.6
$^+CH_2CH_2CH_3$	Primary	211
$^+CH_2CH_2CH_2CH_3$	Primary	203
$H_3C\overset{+}{C}HCH_2CH_3$	Secondary	183
$(CH_3)_2\overset{+}{C}H$	Secondary	190.9
$CH_2{=}CH{-}\overset{+}{C}H_2$	Primary allyl	226
$(CH_3)_3C^+$	Tertiary	165.8

Table 10.2 shows that, just as with alkyl carbocations, secondary vinyl cations are more stable than primary vinyl cations. However, it also shows that even a secondary vinyl cation is not even as stable as a primary alkyl carbocation. Primary carbocations serve as a kind of mechanistic stop sign: They are generally too unstable to be viable intermediates in most reactions. Presumably, we should treat unstabilized vinyl cations the same way.

What intermediate could we use to replace the highly unstable vinyl cations in addition reactions involving alkynes? Perhaps cyclic intermediates are involved. Moreover, alkynes are known to form complexes with HX acids. A cyclic protonium ion or complex could accommodate the complicated kinetics, which show that more than one molecule of halide is involved, and also account for the generally observed predominance of trans addition (Fig. 10.64). We will write cyclic ions in the answers to problems, but you should know that the intermediacy of vinyl cations in these reactions is not a fully resolved issue. There is still lots to do in organic chemistry.

The modified mechanism shown in Figure 10.64 predicts that Markovnikov addition should be observed: that the more substituted secondary vinyl halide should be

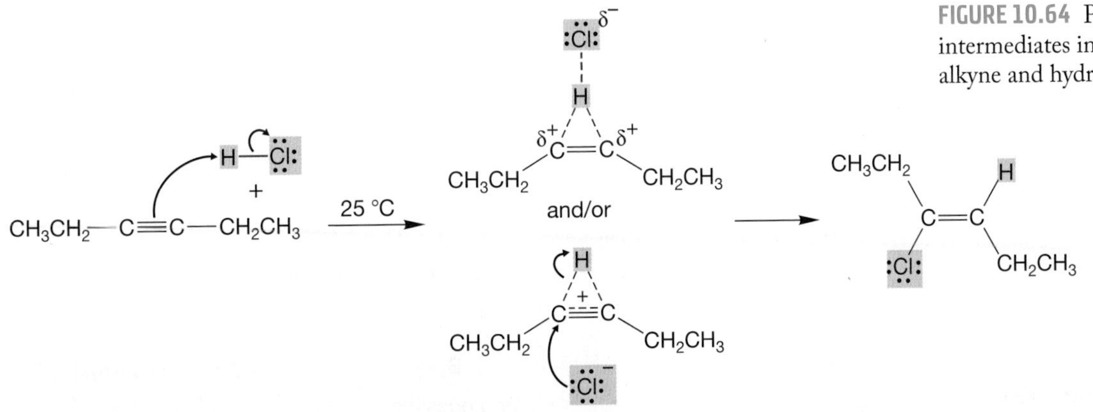

FIGURE 10.64 Possible cyclic intermediates in the reaction of an alkyne and hydrogen chloride.

formed whenever there is a choice (Fig. 10.65). In the cyclic intermediate, there is a partial positive charge on the carbons of the original triple bond. The partial bond to hydrogen from the *more* substituted carbon of an unsymmetrical alkyne is weaker than the bond from the *less* substituted carbon because the partial positive charge is more stable at the more substituted position. Accordingly, addition of halide would be expected to take place more easily at the more substituted position, leading to Markovnikov addition.

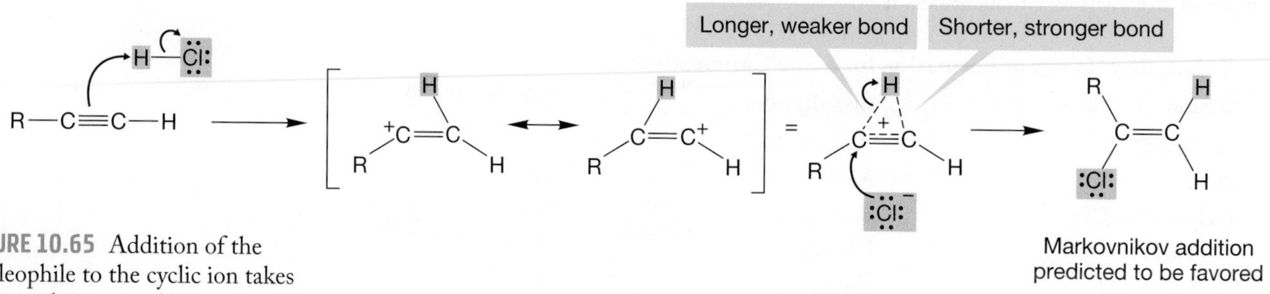

FIGURE 10.65 Addition of the nucleophile to the cyclic ion takes place at the more substituted position.

Let's look at the experimental data. There is no difference in the two carbons of the triple bond in a molecule such as 3-hexyne, so let's examine the addition to a terminal alkyne—a 1-alkyne—to see if Markovnikov addition occurs. Our mechanism predicts that the more substituted halide should be the product, and it is. The products shown in Figure 10.66 are the only addition products isolated.

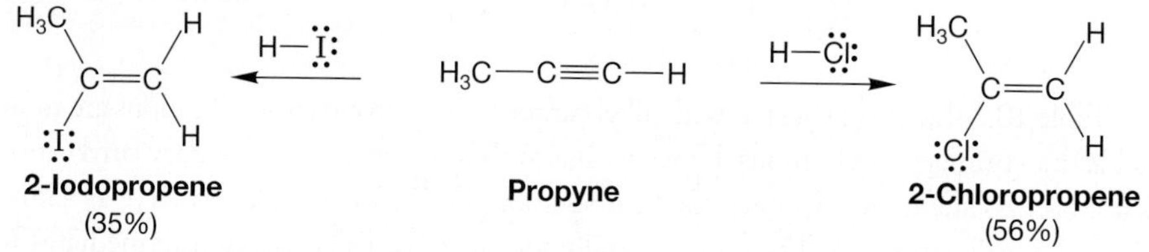

FIGURE 10.66 Markovnikov addition generally predominates in reactions of HX with alkynes.

Additions of HX to alkynes are mechanistically complex. Also, they are usually not practical sources of vinyl halides because the vinyl halides also contain π bonds and often compete favorably with the starting alkyne in the addition reaction. The vinyl halide products are not symmetrical and two products of further reaction are possible (p. 367).

It makes a nice problem to figure out which way the second addition of HX should go. As always in a mechanistic problem of this sort, the answer only appears through an analysis of the possible pathways for further reaction. Draw out both mechanisms, compare them at every point, and search each step for differences. We will use the reaction of propyne with hydrogen chloride as a prototypal example. The initial product is 2-chloropropene, as shown in Figure 10.66. There are two possible ions from the reaction of this vinyl halide with additional hydrogen chloride. Structure **A** in Figure 10.67 has the positive charge adjacent to a chlorine, but structure **B** does not.

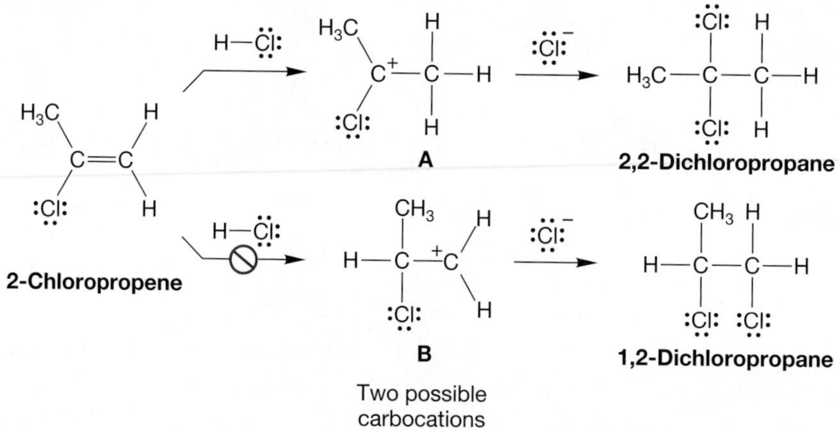

FIGURE 10.67 2-Chloropropene can protonate in two ways.

Two possible carbocations

There is a way for the chlorine in structure **A** to share the positive charge. The charge is shared by the two atoms through $2p/3p$ overlap as shown in Figure 10.68. In molecular orbital terms, we see that a filled $3p$ orbital on chlorine overlaps with the empty $2p$ orbital on carbon, and this orbital overlap is stabilizing. The primary carbocation **B** in Figure 10.67 is much less stable, and products from it are not observed.

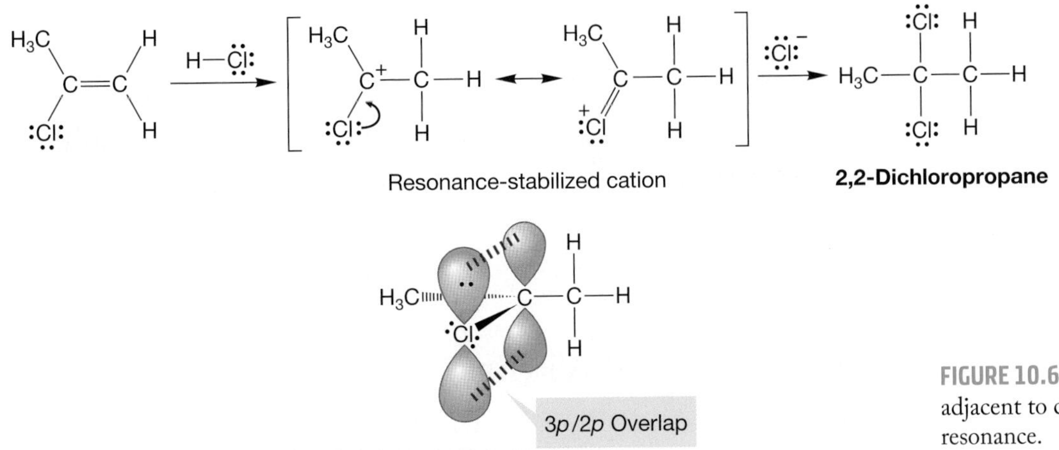

Resonance-stabilized cation

2,2-Dichloropropane

$3p/2p$ Overlap

FIGURE 10.68 The carbocation adjacent to chlorine is stabilized by resonance.

So, addition of a second halogen will preferentially result in formation of the **geminal** dihalide (geminal means groups substituted on the same carbon) not the vicinal (groups substituted on adjacent carbons) compound because the cation with chlorine attached is lower in energy, and leads inevitably to the observed geminal product. The reactions of Figure 10.66 are complicated by the formation of compounds arising through double addition in the presence of excess hydrogen halide. The double addition always gives the geminal dihalide (Fig. 10.69).

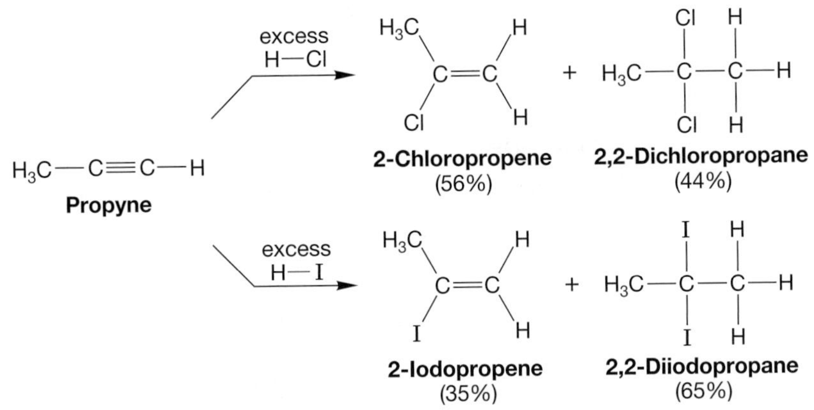

2-Chloropropene (56%)   **2,2-Dichloropropane** (44%)

Propyne

2-Iodopropene (35%)   **2,2-Diiodopropane** (65%)

FIGURE 10.69 Addition of excess hydrogen halide to an alkyne gives a mixture of products of mono- and di-addition.

10.7 Addition of X₂ Reagents to Alkynes

The pattern of alkene addition is also followed when X₂ reagents add to alkynes, although the mechanism of the alkyne reaction is not well worked out. Both ionic and neutral intermediates (i.e., radicals, which are discussed in Chapter 11) are involved. Alkynes add Br_2 or Cl_2 to give vicinal dihalides. A second addition can follow to give the tetrahalides (Fig. 10.70).

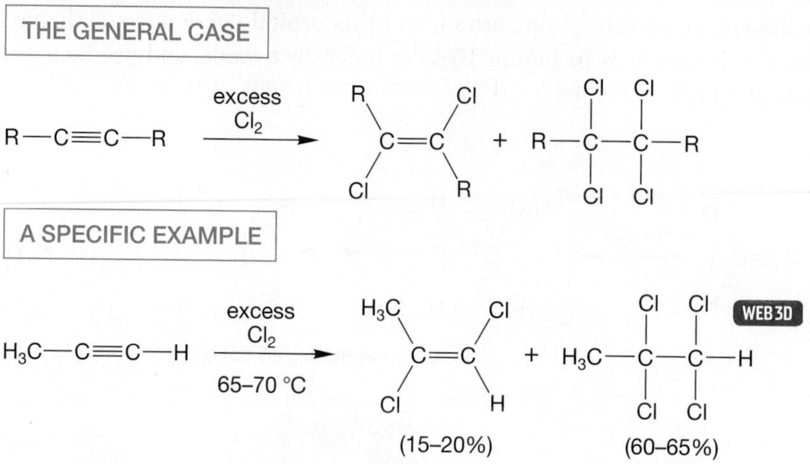

FIGURE 10.70 Addition of Cl_2 (or Br_2) to alkynes gives both vicinal dihalides and tetrahalides.

THE GENERAL CASE

A SPECIFIC EXAMPLE

10.8 Hydration of Alkynes

Like alkenes, alkynes can be hydrated. The reaction is generally catalyzed by mercuric ions in an oxymercuration process (Fig. 10.71), although simple acid catalysis is also known. In contrast to the oxymercuration of alkenes, no second, reduction step is required in this alkyne hydration. By strict analogy to the oxymercuration of alkenes, the product should be a hydroxy mercury compound, but the second double bond exerts its influence and further reaction takes place. The double bond is protonated and mercury is lost to generate a species called an **enol**. An enol is part alk*ene* and part alcoh*ol*, hence the name.

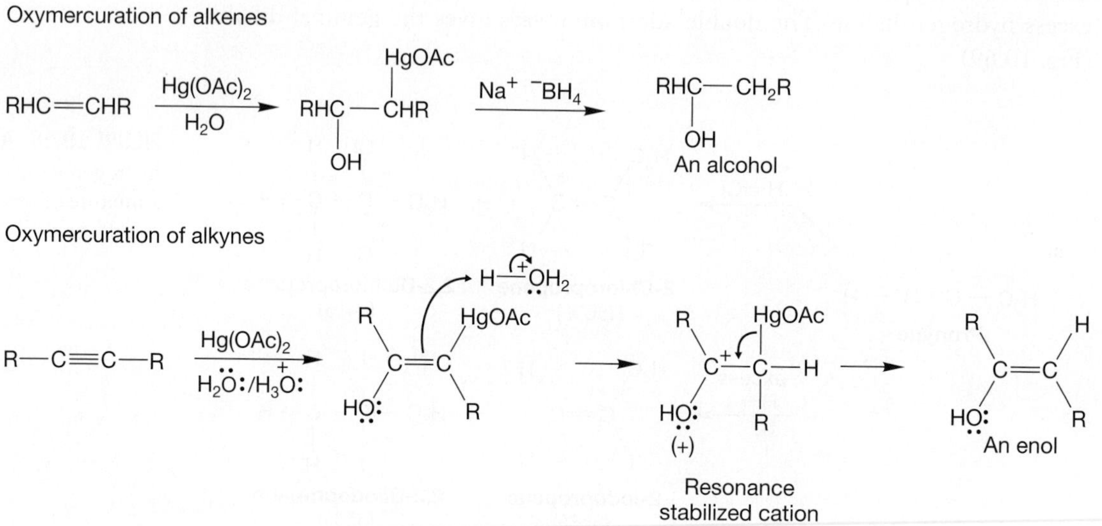

Oxymercuration of alkenes

Oxymercuration of alkynes

FIGURE 10.71 The oxymercuration of alkynes resembles the oxymercuration of alkenes. In the alkyne case, the product is an enol, which can react further.

Enols are extraordinarily important compounds, and more than one chapter will mention their chemistry. Here their versatility is exemplified by their conversion into ketones. We have already seen a synthesis of ketones in this chapter through ozonolysis of alkenes (p. 436). Here, ketones are the final products of the hydration reactions of alkynes (Fig. 10.72). The mechanism of their formation from enols is simple: The alkene part is protonated on carbon and the oxygen of the enol is deprotonated to generate the compound containing the carbon–oxygen double bond. Note that this process gives you another synthetic pathway to carbonyl compounds. We will see this process many times in the future.

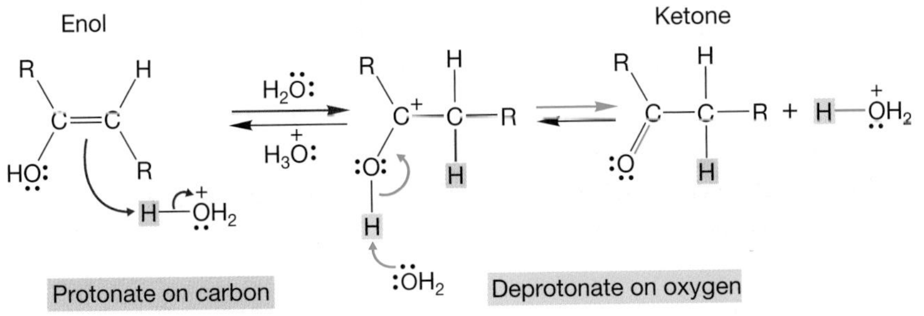

FIGURE 10.72 Protonation of carbon, followed by deprotonation at oxygen, generates the carbonyl compound. This sequence is a general reaction of enols.

Nearly all ketones and aldehydes containing a hydrogen on the carbon adjacent to the carbonyl carbon are in equilibrium with the related enol forms. How much enol is present at equilibrium is a function of the detailed structure of the molecule, but there is almost always some enol present. In simple compounds, the ketone form is greatly favored. This interconversion is called keto-enol **tautomerization**.

PROBLEM 10.23 Draw a mechanism for the acid-catalyzed hydration of an alkyne.

WORKED PROBLEM 10.24 Why were we so sure the enol in Figure 10.72 would protonate to give the cation shown? There is another cation possible. Why is it not formed?

ANSWER Protonation as shown in Figure 10.72 leads to a resonance-stabilized carbocation. The adjacent oxygen shares the positive charge. Protonation on the other carbon of the double bond leads to a carbocation that is not resonance stabilized, which is therefore of much higher energy.

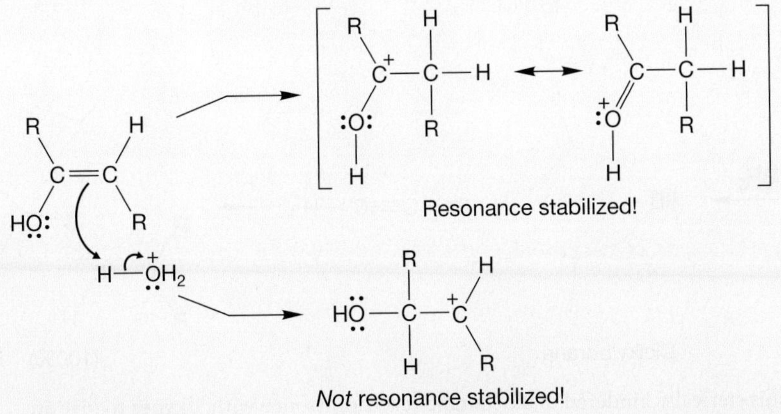

Resonance stabilized!

Not resonance stabilized!

PROBLEM 10.25 In general, hydration of unsymmetrical alkynes such as 2-pentyne is not a practical source of ketones. Why not?

PROBLEM 10.26 Terminal alkynes are certainly unsymmetrical, yet in these cases hydration is a useful process. In principle, either ketones or aldehydes could be formed, but in practice only ketones are produced. Write mechanisms for both ketone and aldehyde formation and explain why the product is always the ketone. You may write open vinyl carbocations in your answer for simplicity, but be aware that the real intermediates may be more complex cyclic species.

PROBLEM 10.27 Write a mechanism for the acid-catalyzed conversion of diethyl ketone into its enol form.

10.9 Hydroboration of Alkynes

Like alkenes, alkynes can be hydroborated by boranes. The mechanism is very similar to that for alkenes (p. 390); the stereochemistry of addition of HBH_2 is syn, and it is the boron that becomes attached to the less substituted end of a terminal alkyne (Fig. 10.73).

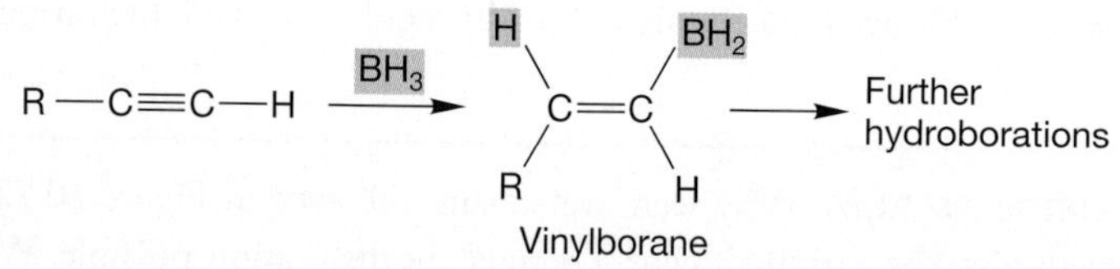

FIGURE 10.73 Hydroboration of alkynes gives vinylboranes that can react further (R = alkyl).

Vinylborane

The initial products of hydroboration are prone to further hydroboration reactions. This complication can be avoided by using a sterically hindered borane so that further reaction is slowed. A favorite suitably hindered reagent is the borane formed by reaction of two equivalents of 2-methyl-2-butene with BH_3 (Fig. 10.74).

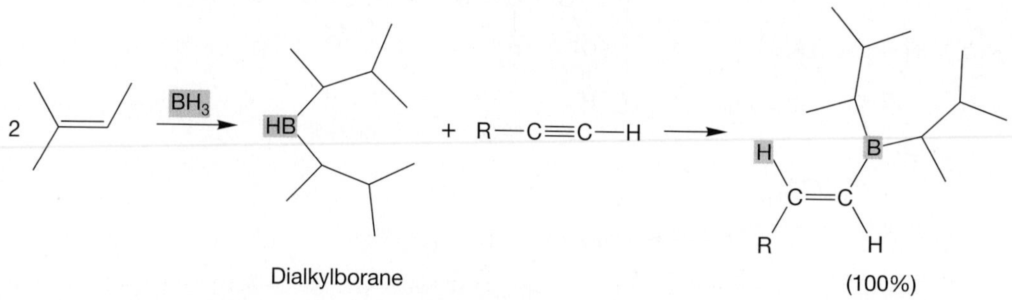

Dialkylborane   (100%)

FIGURE 10.74 This sterically hindered dialkylborane reacts only once with alkynes to give an isolable vinylborane (R = alkyl).

Like alkylboranes, vinylboranes can be converted into alcohols by treatment with basic peroxide (Fig. 10.75). The products are enols in this case, and they are in equilibrium with the corresponding carbonyl compounds. Here's an important point: The regiochemistry of this reaction is the opposite of the regiochemistry of the hydration reaction of alkynes. This situation is exactly the same as the one that exists with the alkenes: Hydration of an alkene gives overall Markovnikov addition, whereas the hydroboration/oxidation sequence gives the anti-Markovnikov product.

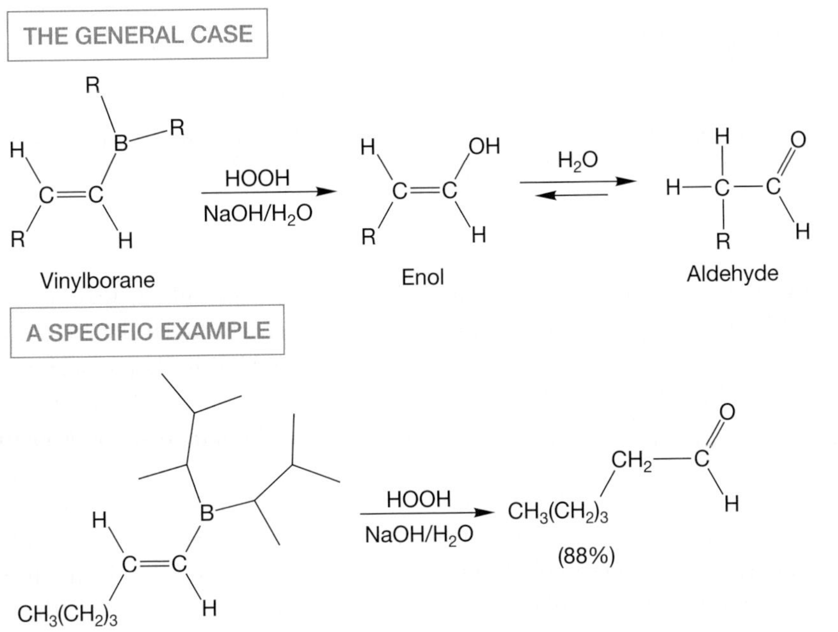

FIGURE 10.75 Oxidation of a vinylborane leads to an enol that can be converted into the aldehyde.

The ultimate formation of a carbonyl compound disguises the matter, but if we look back one step to the enols, we can see that in alkyne reactions the same regioselectivity holds (Fig. 10.76). Hydration goes through Markovnikov addition to the alkyne and, ultimately, gives the ketone. The hydroboration/oxidation reaction goes through anti-Markovnikov addition to the alkyne and, ultimately, gives the aldehyde.

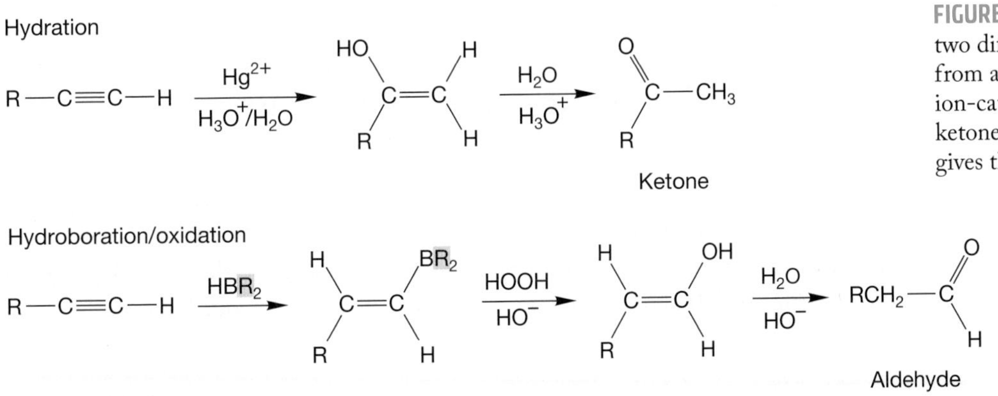

FIGURE 10.76 It is possible to make two different carbonyl compounds from a terminal alkyne. Mercuric ion-catalyzed hydration gives the ketone, and hydroboration/oxidation gives the aldehyde.

Be sure you see the similarities here as well as the mechanistic underpinnings of the different paths followed. There is too much to memorize. Be sure to annotate your file cards as you update your catalog of synthetic methods.

10.10 Hydrogenation of Alkynes: Alkene Synthesis through syn Hydrogenation

Like alkenes, alkynes can be hydrogenated in the presence of many catalysts. Alkenes are the first products, but they are not stable under the reaction conditions and are usually hydrogenated themselves to give alkanes (Fig. 10.77). This reaction is your second synthetic pathway to alkanes.

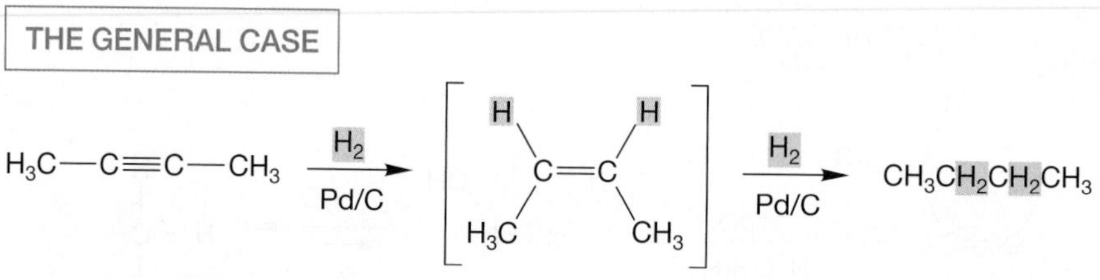

FIGURE 10.77 Hydrogenation of an alkyne proceeds all the way to the alkane stage. The intermediate alkene cannot usually be isolated.

THE GENERAL CASE

It would be extremely convenient to be able to stop the alkyne hydrogenation at the alkene stage, and special, "poisoned" catalysts have been developed to do just this. A favorite is the Lindlar catalyst (Fig. 10.78), which is Pd on calcium carbonate that has been treated with lead acetate in the presence of certain amines (H. H. M. Lindlar, b. 1909). For many years it was thought that the lead treatment poisoned the palladium catalyst through some sort of alloy formation, rendering it less active and the reaction more selective. It now appears that the treatment forms no alloys, but does modify the surface of the palladium. Regardless, the reaction can be stopped at the alkene stage, and the alkenes formed by hydrogenation with Lindlar catalyst have cis stereochemistry because the usual syn addition of hydrogen has occurred.

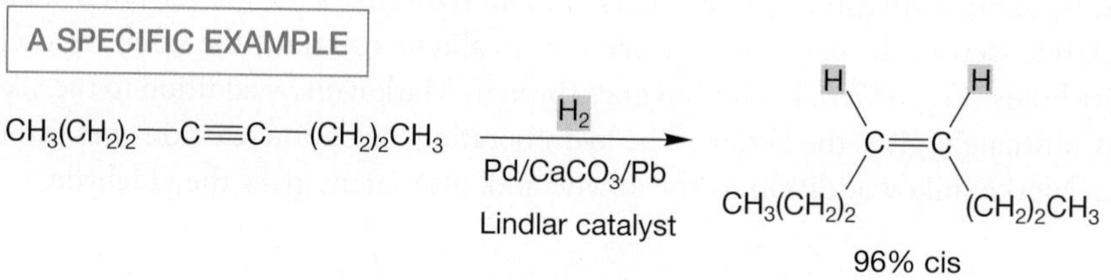

FIGURE 10.78 If the Lindlar catalyst is used in hydrogenation of a alkyne, the cis alkene can be obtained.

A SPECIFIC EXAMPLE

10.11 Reduction by Sodium in Ammonia: Alkene Synthesis through anti Hydrogenation

In a reaction that must at first seem mysterious, an alkyne dissolved in a solution of sodium in liquid ammonia forms the trans alkene (Fig. 10.79). The practical consequences of this are that we now have a way of synthesizing either a cis (Fig. 10.78) or trans (Fig. 10.79) alkene from a given alkyne. So, update your file cards.

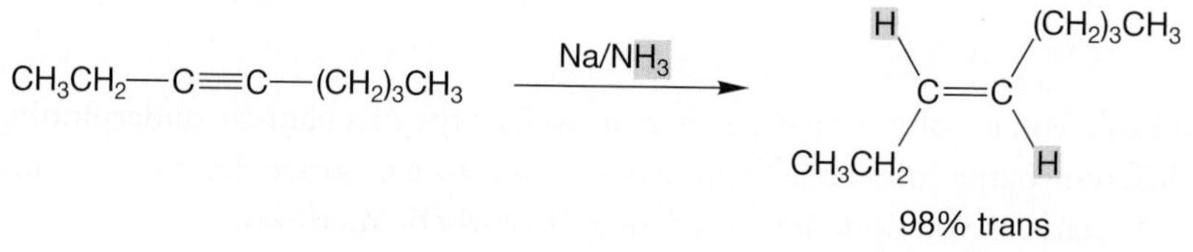

FIGURE 10.79 Reduction of an alkyne with sodium in ammonia gives the trans alkene.

When sodium is dissolved in ammonia, the sodium atom loses its odd electron to give a sodium ion and a **solvated electron**, which is an electron surrounded by solvent molecules:

$$Na\cdot + NH_3 \longrightarrow Na^+ + e^-[NH_3]_n$$

Solvated
electron

This complexed electron can add to an alkyne to give an intermediate called a **radical anion** (Fig. 10.80). A radical anion is a negatively charged molecule that has an unpaired electron.

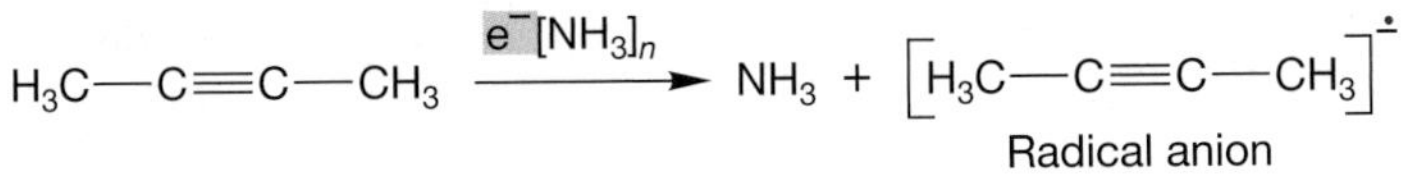

$$H_3C-C\equiv C-CH_3 \xrightarrow{e^-[NH_3]_n} NH_3 + \left[H_3C-C\equiv C-CH_3\right]^{\cdot-}$$

Radical anion

FIGURE 10.80 A solvated electron can add to an alkyne to generate a radical anion.

I (MJ) vividly recall being quite uncertain about the structure of the radical anion when I was first studying organic chemistry. I suspect I solved the problem by memorizing this reaction and hoping that no one would ever ask me to explain what was happening. But it's not so hard and you should not repeat my error. First of all, where is the "extra" electron in the radical anion? It is most certainly not in one of the bonding π orbitals of the alkyne, because they are each filled with two electrons already. If it is not there, it must be in one of the antibonding π^* orbitals (Fig. 10.81).

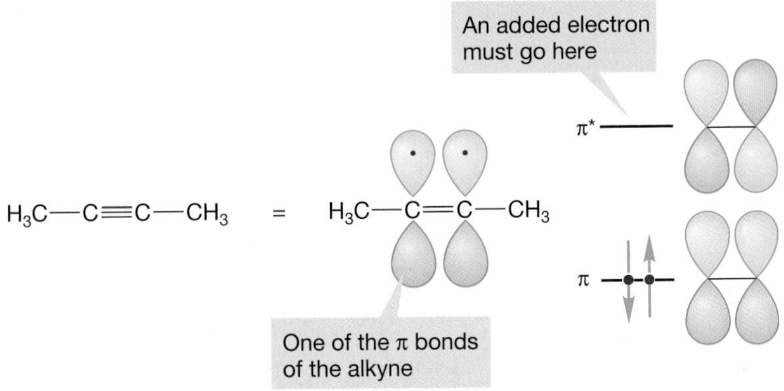

FIGURE 10.81 The electron must go into an antibonding π^* orbital of the alkyne.

An added electron must go here

$$H_3C-C\equiv C-CH_3 = H_3C-C=C-CH_3$$

One of the π bonds of the alkyne

Why should the radical anion put up with this electron in an antibonding orbital? The answer is twofold. First of all, the intermediate is held together by many electrons occupying bonding orbitals, and a single electron in a high-energy orbital is not destabilizing enough to overcome all the bonding interactions. Figure 10.82 gives both molecular orbital and resonance pictures of the radical anion.

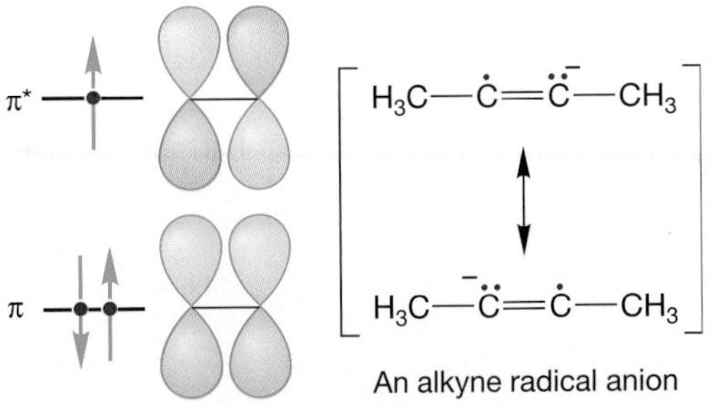

$$\left[H_3C-\overset{\cdot}{C}=\overset{..}{\overset{-}{C}}-CH_3 \longleftrightarrow H_3C-\overset{-}{\overset{..}{C}}=\overset{\cdot}{C}-CH_3\right]$$

An alkyne radical anion

FIGURE 10.82 The alkyne radical anion.

Second, the radical anion *is* quite unstable and not much of it is likely formed at equilibrium. Its very instability contributes to its great base strength, however, and the small amount present at equilibrium is rapidly protonated by ammonia to give the amide ion ($^-NH_2$) and the vinyl radical (Fig. 10.83). There are two forms of the

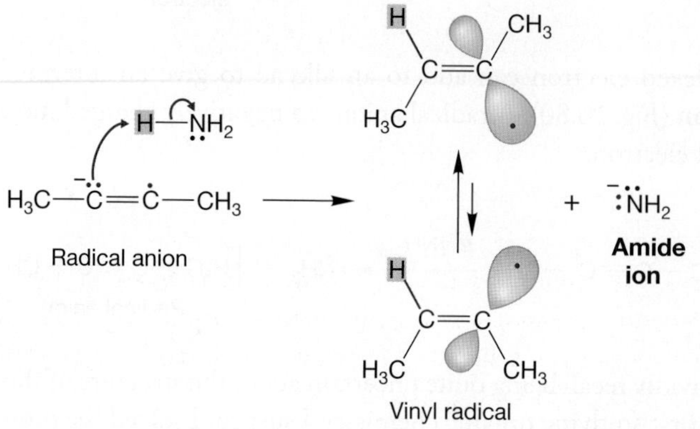

FIGURE 10.83 The radical anion can abstract a proton from ammonia to generate a vinyl radical and the amide ion. The trans form is favored.

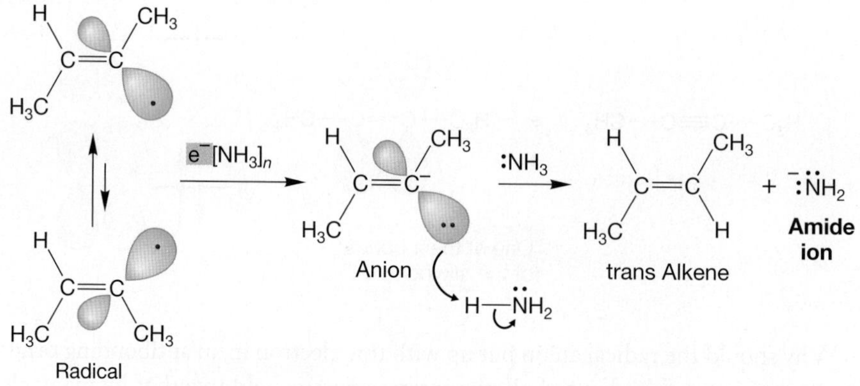

vinyl radical, and it will be the trans form, with the alkyl groups as far from each other as possible, that is the more stable version. A solvated electron can be donated to this radical forming the vinyl anion, which is rapidly protonated by ammonia to give the final product, the trans alkene (Fig. 10.84).

FIGURE 10.84 Transfer of another electron and protonation completes the overall trans hydrogenation of an alkyne.

PROBLEM 10.28 Devise a synthetic pathway for each of the following compounds, starting in each case with 2-butyne:

(a)

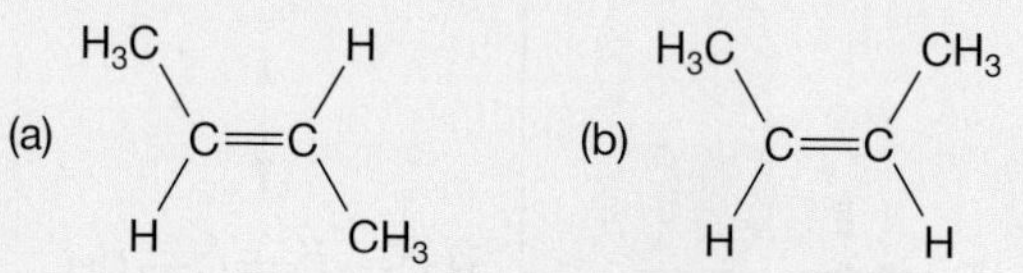

(b)

(continued)

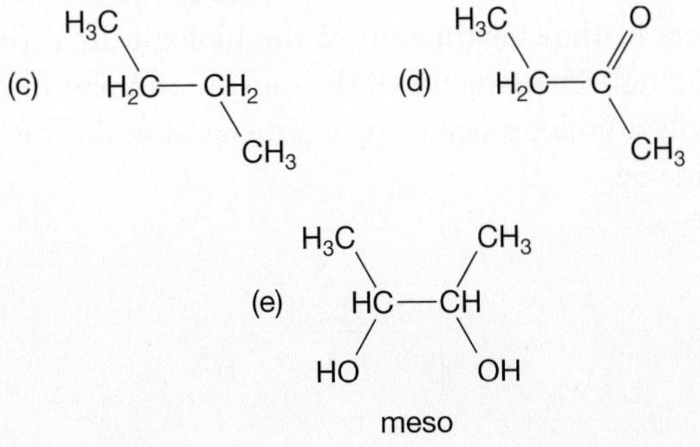

(c)   H₃C, H₂C—CH₂, CH₃

(d)   H₃C, H₂C—C, O, CH₃

(e)   H₃C, CH₃, HC—CH, HO, OH
meso

Summary

Normal hydrogenation of an alkyne adds two equivalents of H_2 to give the alkane. Use of a "poisoned" Lindlar catalyst leads to absorption of one equivalent of H_2 and formation of the cis alkene. The trans alkene can be made by reducing the alkyne with Na in ammonia.

10.12 Special Topic: Three-Membered Rings in Biochemistry

Despite the strain inherent in cyclopropanes, Nature finds ways to make them. Bacteria, in particular, contain surprising amounts of cyclopropanated fatty acids. Even more remarkable is the conversion of such molecules into cyclopropenes. The source of the single "extra" carbon is *S*-adenosylmethionine, the same agent involved in the methyl transfers discussed in Chapters 3 and 7 (pp. 142, 288; Fig. 10.85). The mechanism of this intriguing change hasn't been worked out yet.

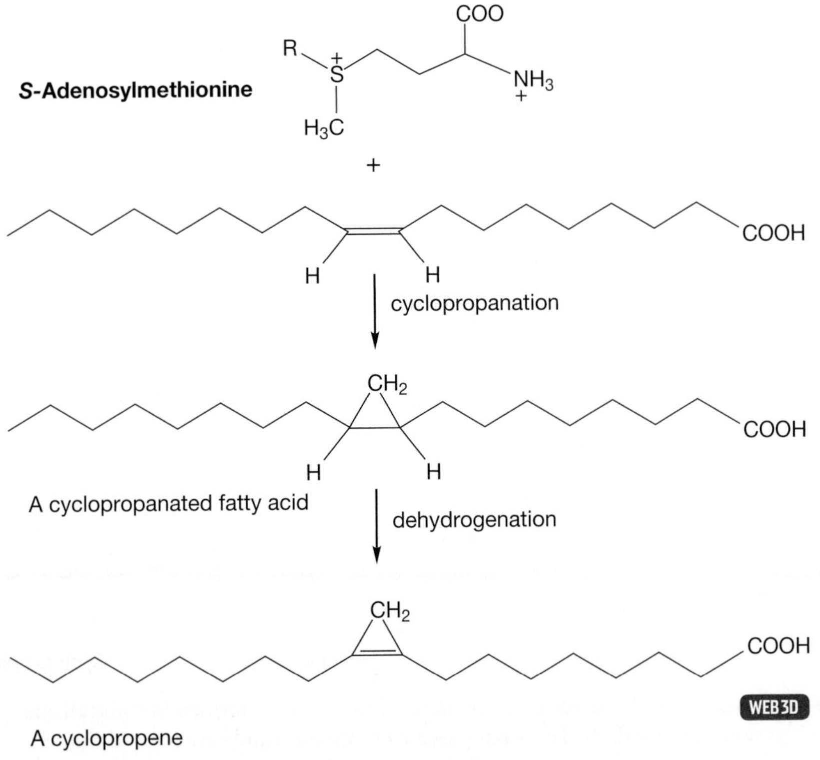

FIGURE 10.85 Cyclopropanation of a fatty acid is followed by dehydrogenation.

In Chapter 12 we will see other three-membered rings found in Nature. Epoxides are used as triggers in the construction of the biologically active molecules called steroids, and, amazingly enough, one of the most prominent molecules in interstellar space is not only a cyclopropene, but a carbene as well. This molecule is known as cyclopropenylidene:

Cyclopropenylidene

PYRETHRINS

Cyclopropanes are occasionally found in Nature, and one class of naturally occurring cyclopropanes, the pyrethroids, are useful insecticides. These compounds are highly toxic to insects but not to mammals, and because they are rapidly biodegraded, cyclopropanes do not persist in the environment. The naturally occurring pyrethroids are found in members of the chrysanthemum family and are formally derived from chrysanthemic acid. Many modified pyrethrins not found in Nature have been made in the laboratory and several are widely used as pesticides. The molecule known as pyrethrin I has the systematic name 2,2-dimethyl-3-(2-methyl-1-propenyl)cyclopropanecarboxylic acid 2-methyl-4-oxo-3-(Z-2,4-dipentadienyl)-2-cyclopenten-1-yl ester. See why the shorthand is used?

Pyrethrin I

10.13 Summary

New Concepts

This chapter continues our discussion of addition reactions, and many of the mechanistic ideas of Chapter 9 apply. For example, many polar addition reactions start with the formation of the more stable carbocation. The most important new concept in this chapter is the requirement for overall anti addition introduced by three-membered ring intermediates. This concept appears most obviously in the trans bromination and chlorination of alkenes. The openings of the intermediate bromonium and chloronium ions by nucleophiles are S_N2 reactions in which inversion is always required. This process insures anti addition (Fig. 10.13).

Remember also the formation of epoxides, stable three-membered rings containing an oxygen atom, and cyclopropanes, formed by carbene additions to alkenes (Fig. 10.86).

Alkenes and alkynes react in similar fashion in most addition reactions, although additional complications and opportunities are introduced by the second π bond in the alkynes.

FIGURE 10.86 Stable three-membered rings can be formed in some alkene addition reactions. Two examples are epoxidation and carbene addition.

Key Terms

aldehydes (p. 440)
alkene halogenation (p. 414)
bromonium ion (p. 415)
carbene (p. 431)
carbonyl compound (p. 438)
carboxylic acid (p. 424)
diazo compounds (p. 431)
1,3-dipolar reagents (p. 436)
1,3-dipoles (p. 436)
diradical (p. 434)
enol (p. 448)

epoxidation (p. 424)
geminal (p. 447)
halohydrin (p. 419)
heterogeneous catalysis (p. 411)
homogeneous catalysis (p. 411)
hydrogenation (p. 411)
ketone (p. 440)
molozonide (p. 437)
organocuprate (p. 430)
oxymercuration (p. 421)
ozonide (p. 437)

ozonolysis (p. 437)
primary ozonide (p. 437)
radical anion (p. 453)
singlet carbene (p. 433)
solvated electron (p. 453)
tautomerization (p. 449)
triplet carbene (p. 433)
vicinal (p. 414)

Reactions, Mechanisms, and Tools

In this chapter, a series of ring formations is described: the three-membered oxiranes, cyclopropanes, bromonium, chloronium, and mercurinium ions, as well as the five-membered rings formed through 1,3-dipolar additions, ozonolysis, and the additions of osmium tetroxide and potassium permanganate. Many of these compounds react further to give the final products of the reaction (Fig. 10.87).

Cyclopropanes and oxiranes can be isolated, but the other rings react further under the reaction conditions to give the

final products. Oxiranes can be opened in either acid or base to give new products. Additions to alkynes resemble those of alkenes, but the second π bond adds complications, and further addition reactions or rearrangements occur. Alkynes can be reduced in either cis or trans fashion (Fig. 10.78 and Fig. 10.79).

Both homogeneous and heterogeneous catalytic hydrogenation are syn additions, because both added hydrogens are transferred from the metal surface to the same side of the alkene.

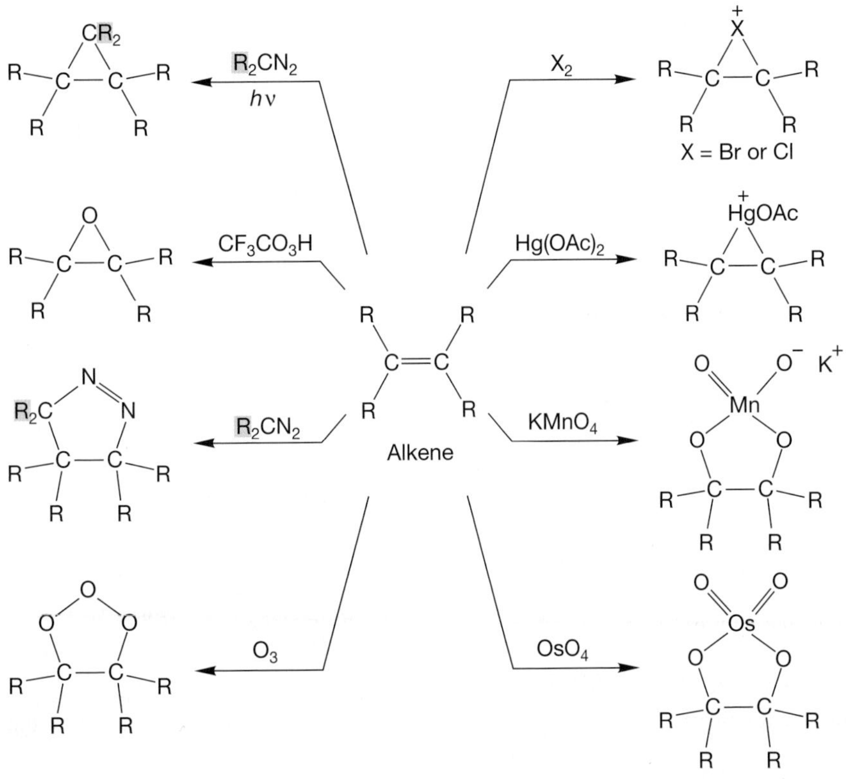

FIGURE 10.87 Ring formation through addition across the double bond.

Syntheses

This chapter provides many new synthetic methods.

1. Carboxylic Acids

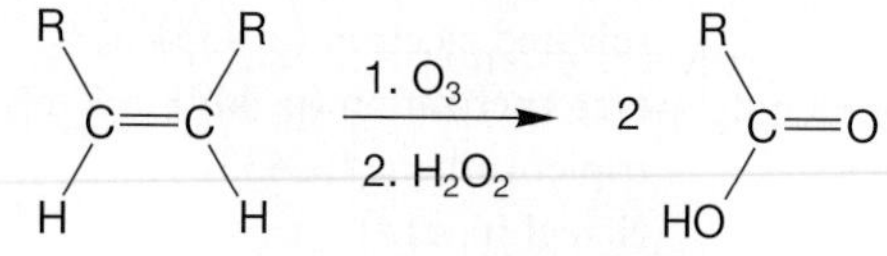

Via ozonide intermediates

2. Alcohols

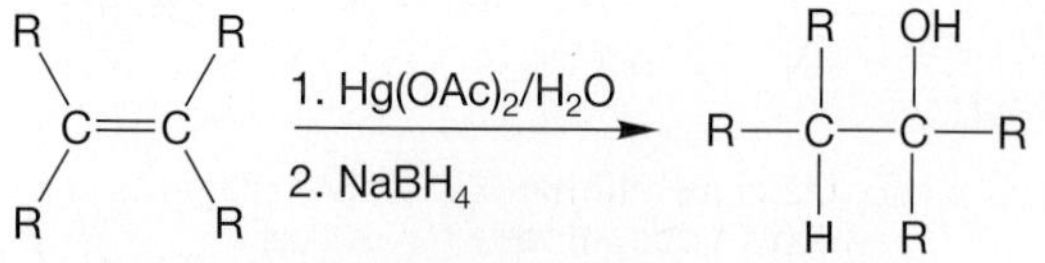

Markovnikov addition, no rearrangements, anti addition

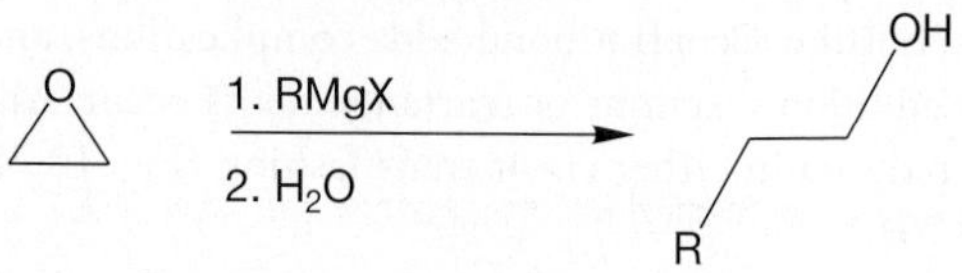

Organocopper reagents also work

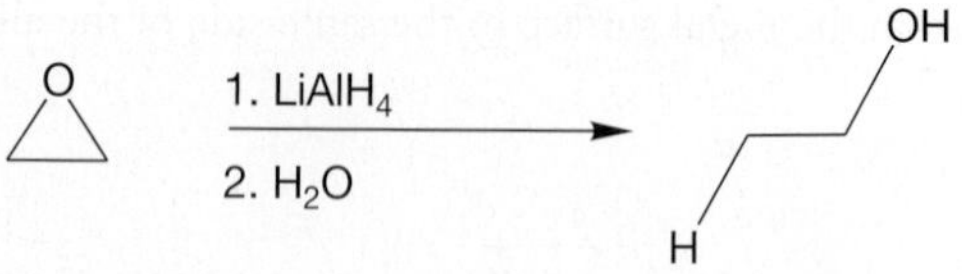

3. Aldehydes

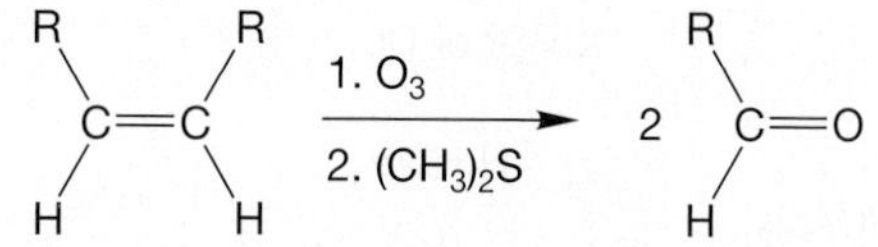

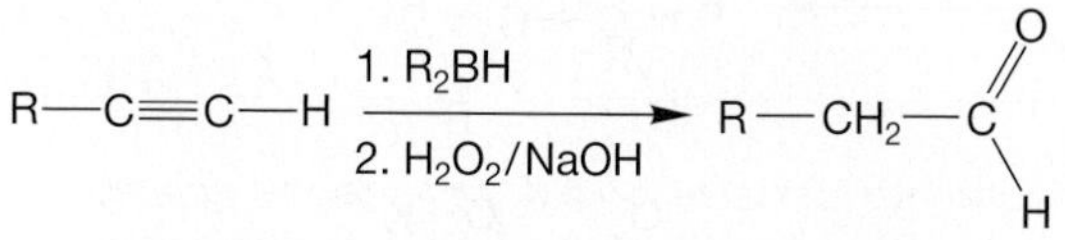

A vinyl borane and an enol are intermediates

4. Alkanes

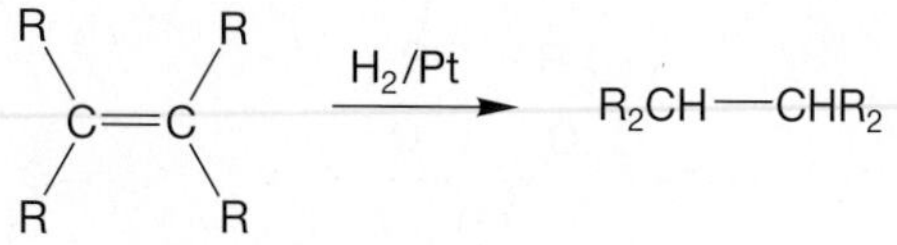

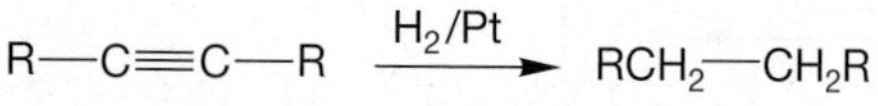

syn Addition, many catalysts possible

5. Alkenes

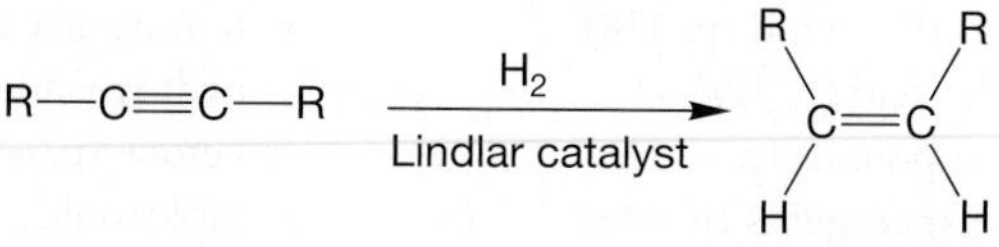

syn Addition

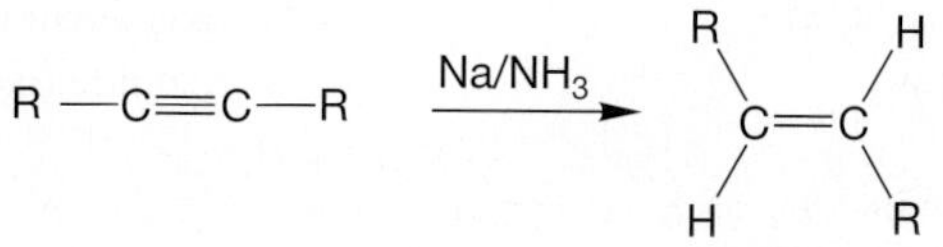

anti Addition

6. Halohydrins

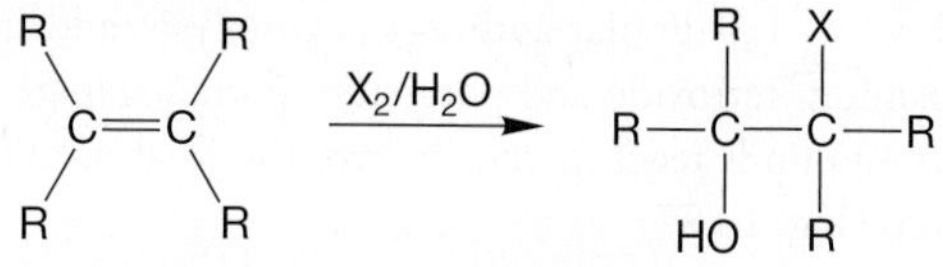

anti Addition, X = Br or Cl

7. Bromo Ethers and Chloro Ethers

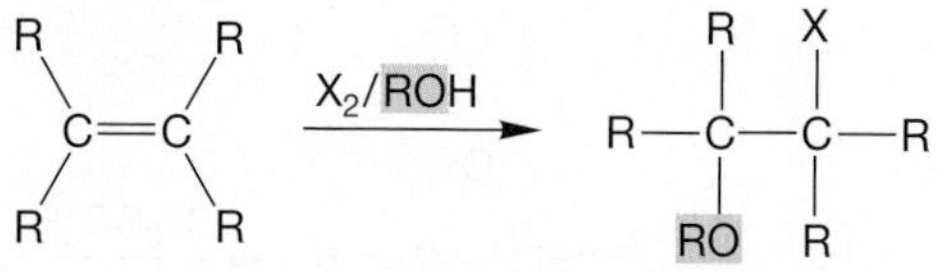

anti Addition, X = Br or Cl

8. Cyclopropanes

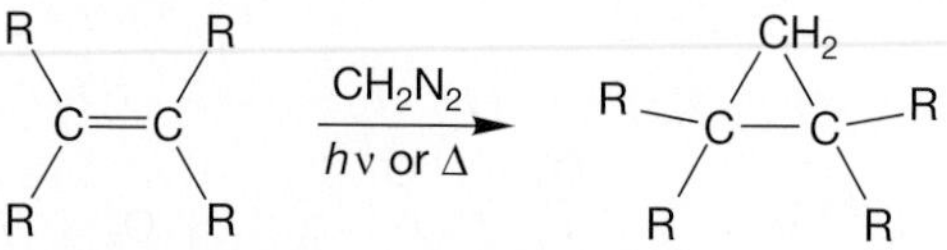

cis Addition for singlet carbene, but both
cis and trans addition for triplet carbene

9. Dihalides

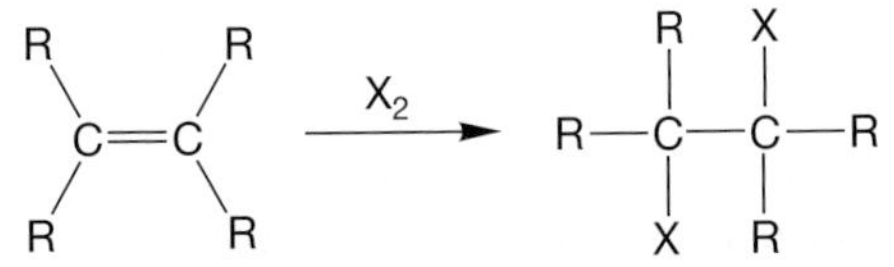

Vicinal dihalide, anti addition, X = Br or Cl

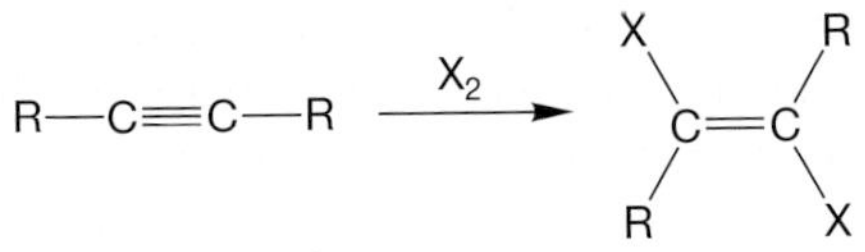

Vicinal dihalide, mostly anti addition,
X = Br or Cl, further reaction likely

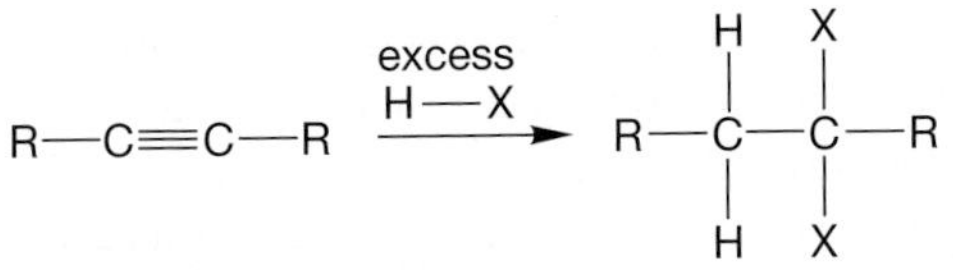

Geminal dihalide, X = Br, Cl, or I,
double HX addition

10. Vicinal Diols

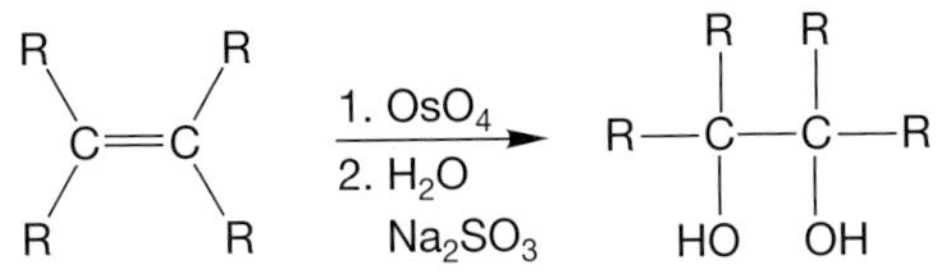

syn Addition

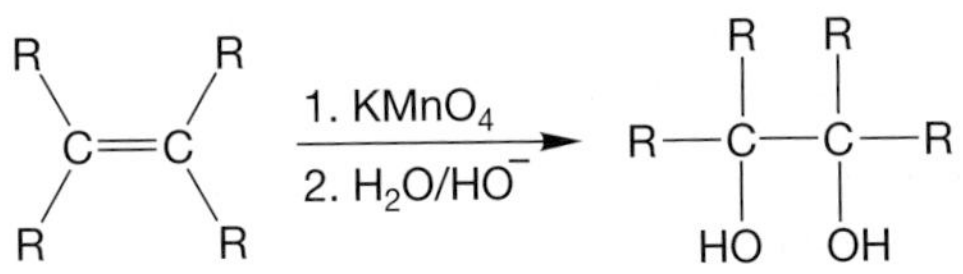

syn Addition

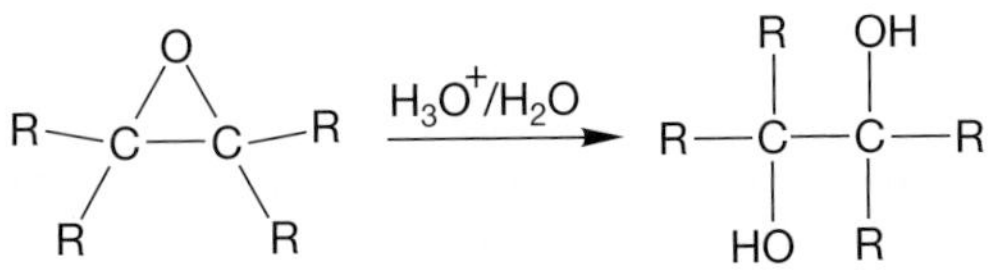

anti Addition

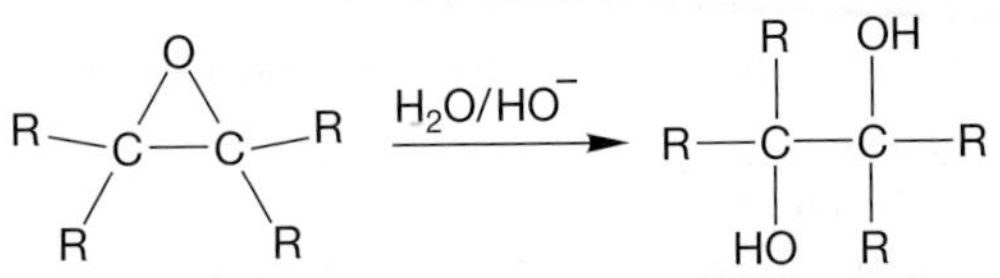

anti Addition

11. Heterocyclic Five-Membered Rings

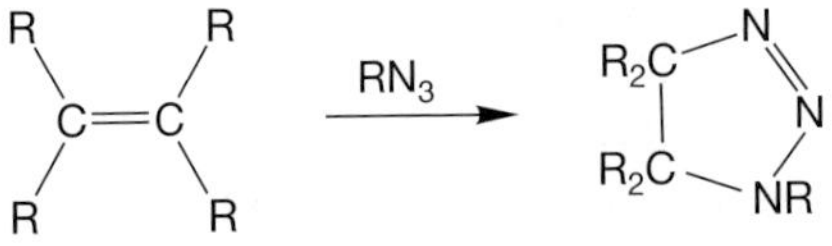

This is only one example; there is a different
product for every 1,3-dipole; alkynes work too

12. Ketones

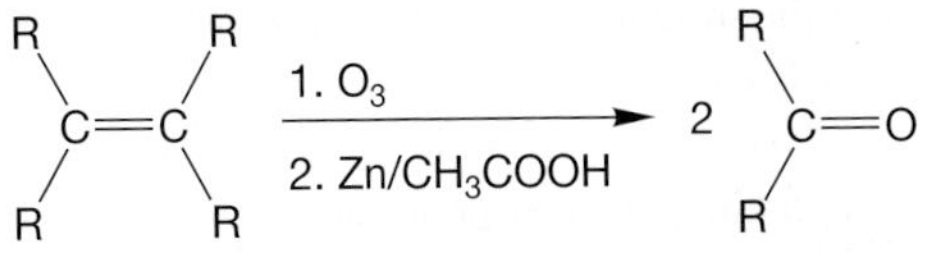

An ozonide is an intermediate; other reagents
will also decompose the ozonide

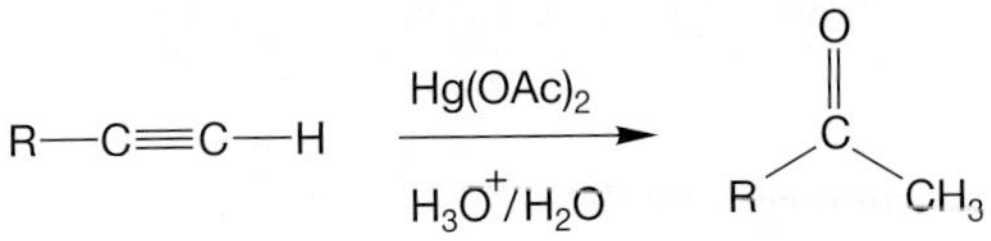

Forms methyl ketones, never the aldehyde

13. Epoxides (Oxiranes)

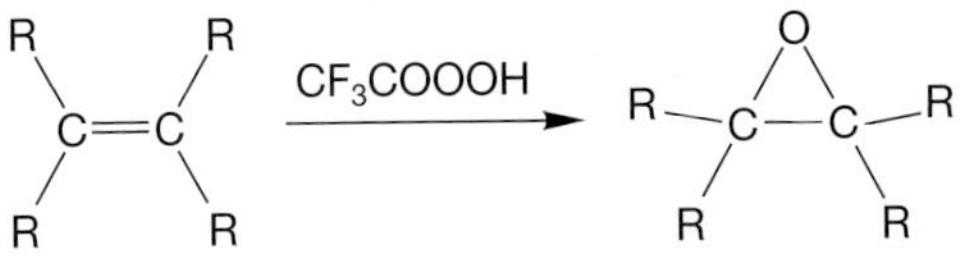

syn Addition

14. Tetrahalides

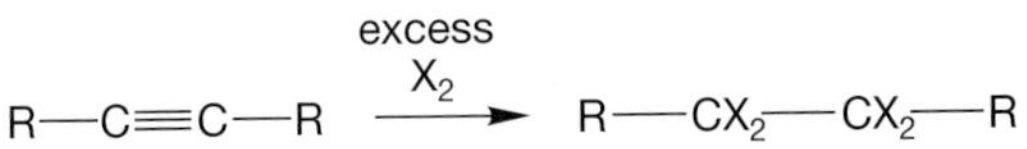

The intermediate vicinal dihalide can be isolated

15. Vinyl Halides

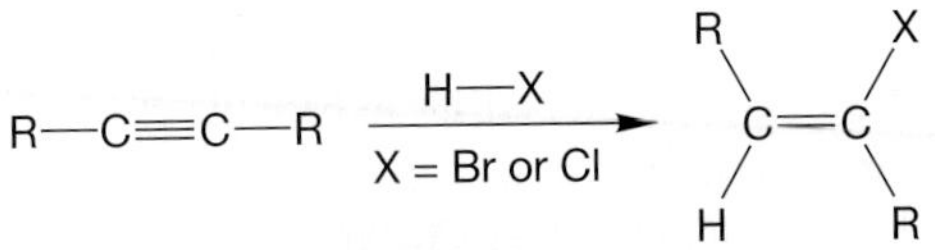

Markovnikov addition and it's hard to stop here;
a second addition of H—X occurs

Common Errors

What psychopathology exists in this chapter comes not so much from "something everyone always gets wrong" or from an overwhelmingly difficult concept, but from the mass of detail. There have been those who have succumbed to despair at sorting out all the stereochemical nuances, at dealing with the many new mechanisms and synthetic methods—at just making rational sense of all the material in this chapter and Chapter 9. There certainly is a lot of information, and you must be very careful to keep your new synthetic methods properly cataloged. There is a unifying principle that may help you to keep from getting lost in the mechanism jungle: "Lewis acids (electrophiles) react with Lewis bases (nucleophiles)."

The Lewis base or nucleophile in most of the reactions in this chapter and Chapter 9 is the π bond of an alkene or an alkyne. The Lewis acids or electrophiles are numerous, but most add to give either an intermediate, or a stable three-membered ring (not all, however; hydrogenation does not, and 1,3-dipolar addition leads to a host of five-membered rings).

The three-membered rings are themselves often prone to further reaction with nucleophiles, and the final products of reaction may be quite far removed in structure from the starting material! Moreover, not all the mechanistic details are known about all these processes. There are reactions about which we need much more information. The reactions of alkynes with halogens and HX are examples, and even hydrogenation has a complicated mechanism, still somewhat obscure to organic chemists.

A good technique that helps one not to get too overwhelmed or lost is to anchor oneself in one or two specific reactions and then to generalize; to relate other reactions to the anchor reaction. For example, the polar addition of hydrogen chloride and hydrogen bromide to alkenes (p. 365) is within anyone's ability to master. Extensions to hydration reactions of alkenes and alkynes and to hydroboration become easier if the analogy with the anchor is always kept in mind. Similarly, use the reaction of alkenes with Br_2 as an anchor on which to hang the other ring-forming addition reactions. It is useful to start a set of mechanism cards to go along with your synthesis cards in order to keep track of the detail.

10.14 Additional Problems

PROBLEM 10.29   Show the reaction you would use to synthesize bromoethene if your only source of carbon is acetylene.

PROBLEM 10.30   Show how you would make 2-azidobutane if your only source of carbon is 1-butene.

PROBLEM 10.31   Show how you would make *trans*-2-methoxycyclohexanol starting with cyclohexene.

PROBLEM 10.32   Show how you would make *cis*-1-bromo-2-methoxycyclohexane from cyclohexene and any other reagents you might need.

PROBLEM 10.33   Show the major organic product(s) expected when 1-methylcyclohexene reacts with the following reagents. Pay close attention to stereochemistry and regiochemistry where appropriate.

(a) D_2/Pd/C
(b) Br_2/CCl_4
(c) Br_2/CH_3OH
(d) $Hg(OAc)_2$, H_2O, then $NaBD_4$ in base (no stereochemical preference in this reaction)
(e) 1. B_2H_6 (BH_3)/THF; 2. H_2O_2/HO^-
(f) CF_3COOH
(g) CH_2N_2, *hv*

(h) 1. O_3; 2. $(CH_3)_2S$
(i) 1. OsO_4; 2. $NaHSO_3$/H_2O
(j) $HN{=}NH$

PROBLEM 10.34   Show in detail how both enantiomers of product are formed in Problem 10.33a.

PROBLEM 10.35   Predict the major product(s), including stereochemistry where relevant, for the bromination reaction (Br_2/CCl_4) with each of the following alkenes:

(a) 1-pentene
(b) *cis*-2-pentene
(c) cyclopentene
(d) *cis*-3-hexene
(e) *trans*-3-hexene

PROBLEM 10.36   Which of the products in the previous problem are chiral and which are achiral?

PROBLEM 10.37   Predict the possible products in the reaction between bromine in water with the following alkenes:

(a) cyclobutene
(b) *trans*-2-butene
(c) 2-methyl-2-pentene
(d) *cis*-2-hexene

PROBLEM 10.38   Show the major organic products expected when the following acetylenes react with the reagents shown. Pay attention to stereochemistry and regiochemistry where appropriate.

(a)

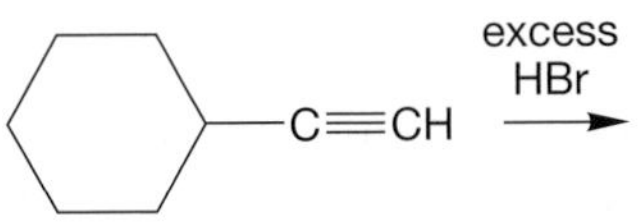

excess
HBr

(b)

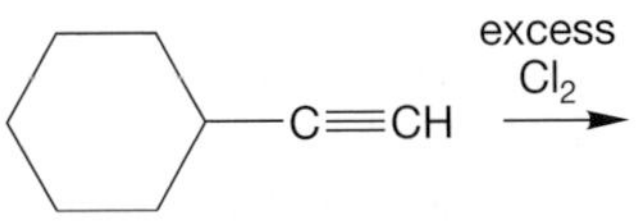

excess
Cl_2

(c)

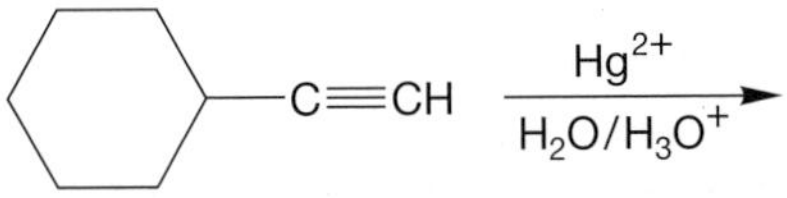

Hg^{2+}
$\overline{H_2O/H_3O^+}$

(d)

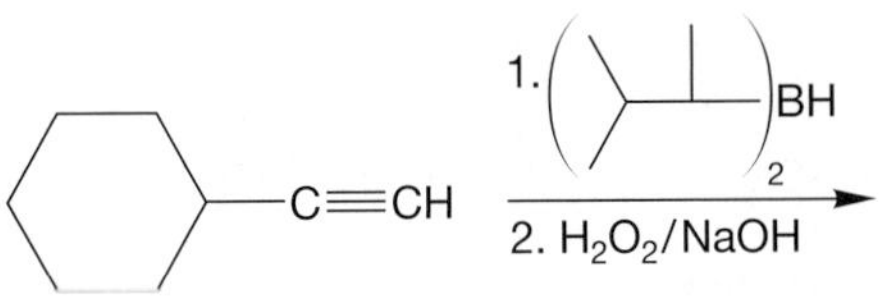

1. BH
2. H_2O_2/NaOH

(e)

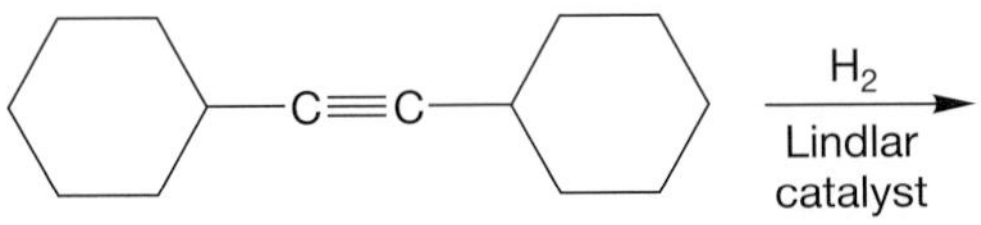

H_2
Lindlar
catalyst

(f)

Na
NH_3

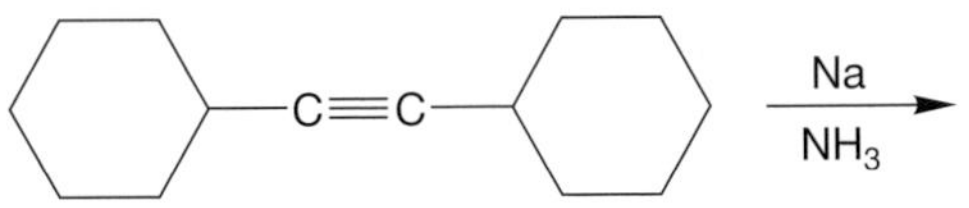

PROBLEM 10.39   Devise syntheses for the following molecules starting from 3,3-dimethyl-1-butene (**1**). You may use any inorganic reagent (no carbons), and the following "special" organic reagents: diazomethane, carbon tetrachloride, *tert*-butyl alcohol, *tert*-butyl iodide, Hg(OAc)$_2$, carbon, (CH$_3$)$_2$S, trifluoroperacetic acid, strychnine, and Igor Likhotvorik's famous boiled goose-in-a-kettle. Your answers may be very short, and no mechanisms are required.

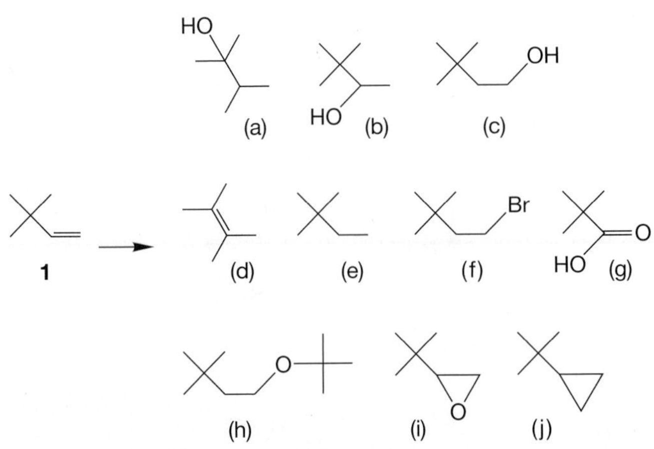

(a)   (b)   (c)

1 →

(d)   (e)   (f)   (g)

(h)   (i)   (j)

PROBLEM 10.40   What reagents would you use to convert the following starting materials into the products shown? There are no restrictions on reagents, but you must start from the molecule shown.

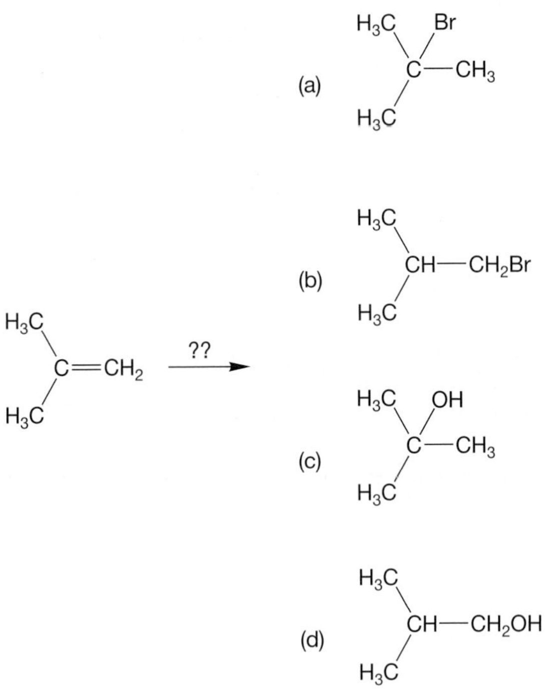

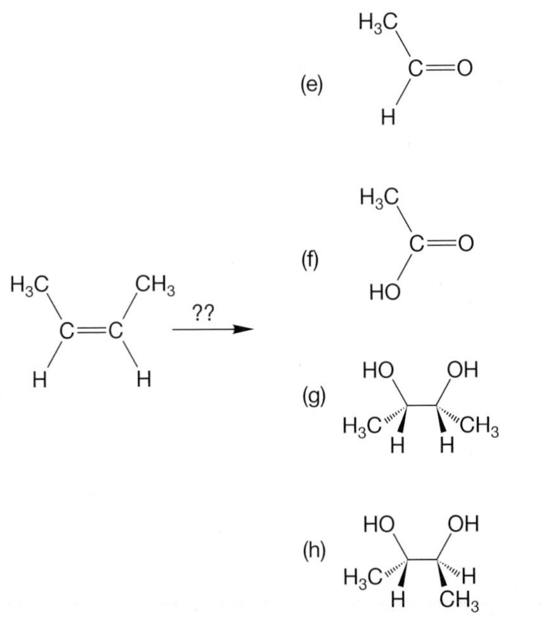

(a)

(b)

(c)

(d)

??

- -

(e)

(f)

??

(g)

(h)

PROBLEM 10.41   Devise syntheses for the following molecules starting from methylenecyclopentane (**1**). You may use any inorganic reagent (no carbons), and the following "special" organic reagents: diazomethane, carbon tetrachloride, *tert*-butyl alcohol, methyl alcohol, ethyl alcohol, *tert*-butyl iodide, $Hg(OAc)_2$, carbon, $(CH_3)_2S$, trifluoroperacetic acid, and the secret contents of Zhou Enlai's favorite veggie dumplings. Your answers may be very short, and no mechanisms are required.

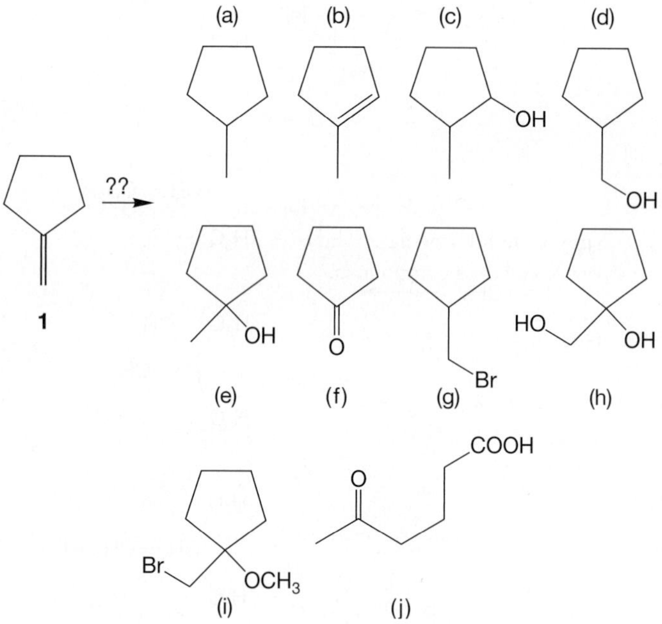

(a)   (b)   (c)   (d)

1

(e)   (f)   (g)   (h)

(i)   (j)

PROBLEM 10.42   Provide syntheses for the following molecules starting from the indicated compound. You may use any appropriate reagents you need, including Mrs. Tao's incredible baby eels in white pepper sauce.

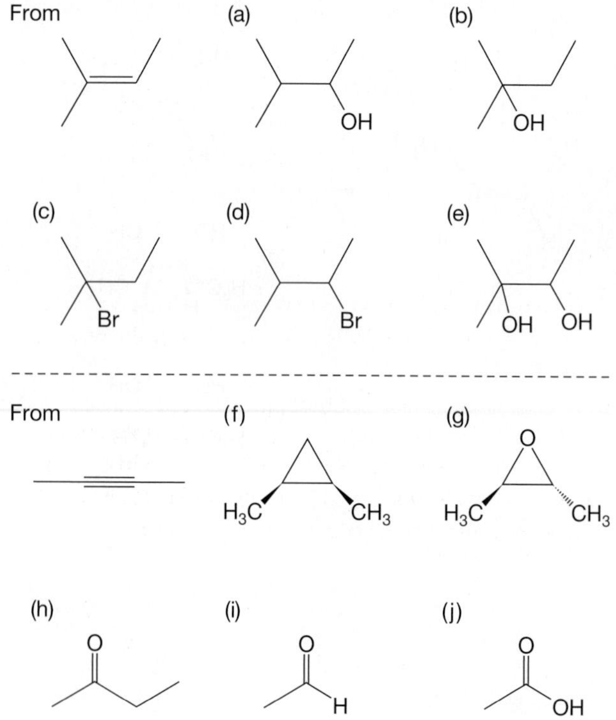

From   (a)   (b)

(c)   (d)   (e)

From   (f)   (g)

(h)   (i)   (j)

PROBLEM 10.43   Draw four possible products for the reaction of (*R*)-3-methylcyclopentene with bromine in methanol. Show the reaction pathway for the compound you think would be the major product and explain why you think it would be the major product.

PROBLEM 10.44   Which of the products in the previous problem are chiral and which are achiral?

PROBLEM 10.45   In Section 10.2b (p. 414), we considered two possible mechanisms for the addition of Br_2 to alkenes. Ultimately, a stereochemical experiment that used a ring compound was used to decide the issue in favor of a mechanism in which addition proceeded through a bromonium ion rather than an open carbocation. Use a detailed stereochemical analysis to show how the experimental results shown below are accommodated by an intermediate bromonium ion but not by an open carbocation.

$$cis\text{-}2\text{-Butene} \xrightarrow[\text{CCl}_4]{\text{Br}_2} \text{Racemic dibromide}$$

$$trans\text{-}2\text{-Butene} \xrightarrow[\text{CCl}_4]{\text{Br}_2} \text{meso Dibromide}$$

PROBLEM 10.46   Explain the following regiochemical results mechanistically.

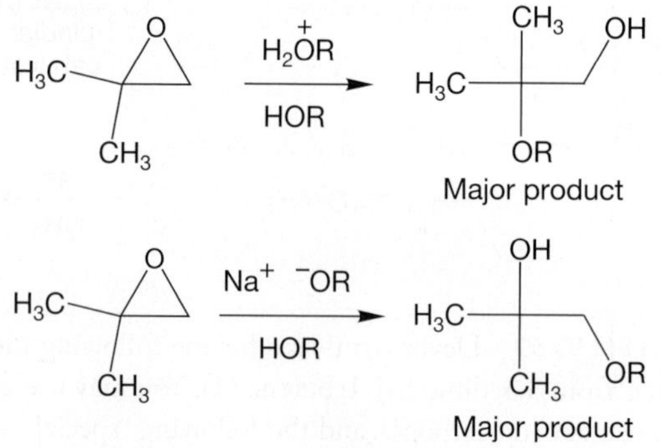

Major product

Major product

PROBLEM 10.47   Predict the major product for the following reactions:

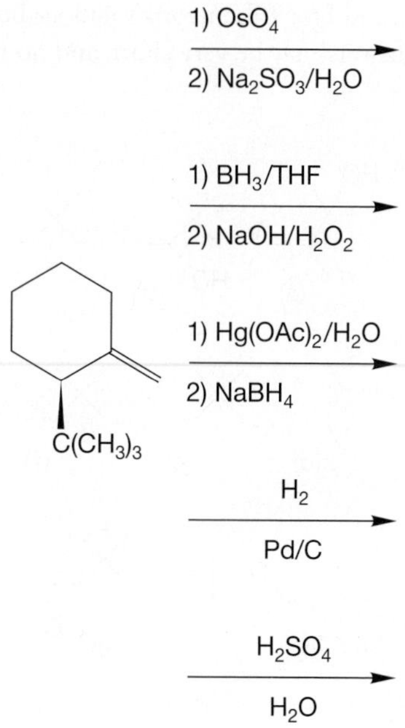

1) OsO_4
2) Na_2SO_3/H_2O

1) BH_3/THF
2) $NaOH/H_2O_2$

1) $Hg(OAc)_2/H_2O$
2) $NaBH_4$

$C(CH_3)_3$

H_2
Pd/C

H_2SO_4
H_2O

PROBLEM 10.48 In Section 10.4a (p. 423), you saw that epoxidation of an alkene with a peracid such as trifluoroperacetic acid results in addition of an oxygen atom to the alkene in such a way as to preserve in the product epoxide the stereochemical relationships originally present in the alkene. That is, cis alkene leads to cis epoxide, and trans alkene leads to trans epoxide. There is another route to epoxides that involves cyclization of halohydrins. Use *cis*-2-butene as a substrate to analyze carefully the stereochemical outcome of this process. What happens to the original alkene stereochemistry in the product?

A bromohydrin

PROBLEM 10.49 Treatment of *cis*-2-butene with Br_2/H_2O gives a product (C_4H_9OBr), that reacts with sodium hydride to give a meso compound of the formula C_4H_8O. With the same sequence of reagents, *trans*-2-butene gives a different compound (a racemic mixture) of the same formula, C_4H_8O. Explain mechanistically, paying close attention to stereochemical relationships.

PROBLEM 10.50 The difference in nucleophilicity between differently substituted alkenes can lead to selectivity in the epoxidation reaction. Rationalize the position of faster epoxidation in the 1,4-cyclohexadiene below and explain your reasoning.

(73%)   (Not observed)

PROBLEM 10.51 There is one alcohol that can be synthesized in one or two easy steps from each of the following precursors. What is the alcohol, and how would you make it from each starting material?

(a)   (b)   (c)

(d)   (cis or trans)

PROBLEM 10.52 Show the final products of the reactions below. Pay attention to stereochemistry.

(a)

CF_3COOOH H_2O / NaOH

(b)

$KMnO_4$ / $NaOH/H_2O$

PROBLEM 10.53 Glycols react with periodic acid to form a cyclic periodate intermediate. The periodate then decomposes to a pair of carbonyl compounds.

Periodate intermediate

If the product from (b) in Problem 10.52 is cleaved with periodic acid, what would be the final product of the reaction sequence?

From a synthetic point of view, this two-step sequence—the treatment of an alkene with basic permanganate followed by periodate cleavage—is equivalent to what other direct method of cleaving carbon–carbon double bonds? What other way do you know to convert an alkene into a pair of carbonyl compounds?

PROBLEM 10.54 Why doesn't the product of Problem 10.52a react with periodate?

PROBLEM 10.55   Incomplete catalytic hydrogenation of a hydrocarbon, **1** (C_5H_8), gives a mixture of three hydrocarbons, **2**, **3**, and **4**. Ozonolysis of **3**, followed by reductive workup, gives formaldehyde and 2-butanone. When treated in the same way, **4** gives formaldehyde and isobutyraldehyde. Provide structures for compounds **1–4** and explain your reasoning.

$R = CH_3$

PROBLEM 10.56   α-Terpinene (**1**) and γ-terpinene (**2**) are isomeric compounds ($C_{10}H_{16}$) that are constituents of many plants. Upon catalytic hydrogenation, they both afford 1-isopropyl-4-methylcyclohexane. However, on ozonolysis followed by oxidative workup, each compound yields different products. Provide structures for **1** and **2** and explain your reasoning.

PROBLEM 10.57   In practice, it is often very difficult to isolate ozonides from ozonolysis of tetrasubstituted double bonds. The product mixture depends strongly on reaction conditions, but the products shown in the next column can all be isolated. Construct reasonable arrow formalisms for all of them.

PROBLEM 10.58   Ozonolysis of **1** in solvent acetone (dimethyl ketone) leads to **2** as the major product. Explain.

PROBLEM 10.59   *trans*-β-Methylstyrene reacts with phenyl azide to give a single product, triazoline (**1**). What other stereoisomeric products might have been produced? Draw an arrow formalism for this reaction and explain what mechanistic conclusions can be drawn from the formation of a *single* isomer of **1**.

$Ph = $

PROBLEM 10.60   Triazoline (**1**) (formed in Problem 10.59) decomposes upon photolysis or heating to give aziridine (**2**). Write two mechanisms for the conversion of **1** into **2**; one a concerted, one-step reaction, and the other a nonconcerted, two-step process. How would you use the stereochemically labeled triazoline (**1**) to tell which mechanism is correct?

PROBLEM 10.61 Explain the formation of aziridine (**2**) in the following reaction of azide (**1**). *Hints*: Draw a full Lewis structure for the azide, and see Section 10.4d (p. 431).

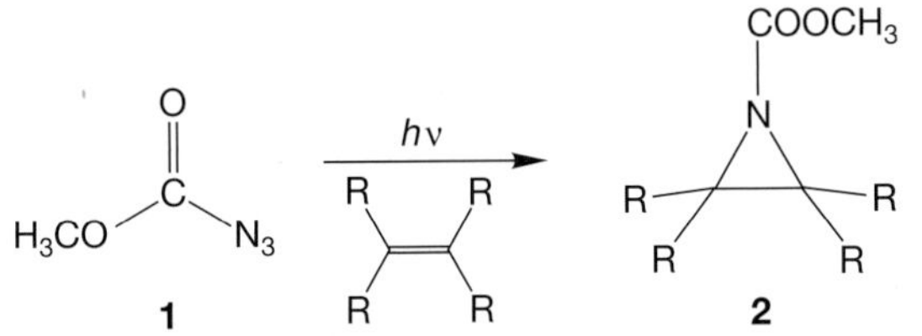

PROBLEM 10.62 What does the formation of only cis aziridine (**2**) from irradiation of azide (**1**) in *cis*-2-butene tell you about the nature of the reacting species in Problem 10.61?

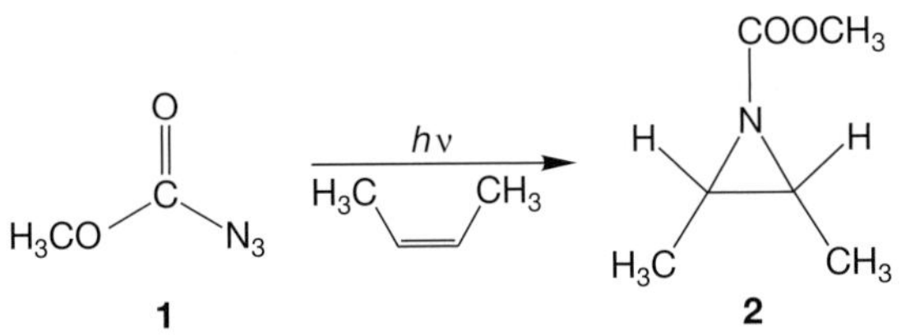

PROBLEM 10.63 A reaction of carbenes not mentioned in the text is called "carbon–hydrogen insertion." In this startling reaction, a reactive carbene ultimately places itself between the carbon and hydrogen of a carbon–hydrogen bond.

(a) Write two mechanisms for this reaction, one with a single-step and the other having two steps.

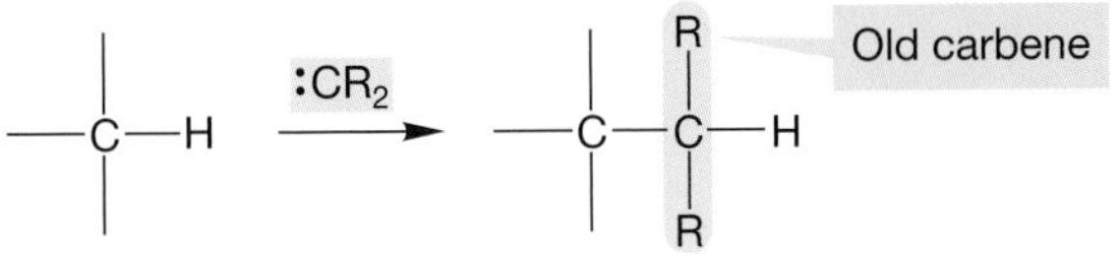

(b) In 1959, in a classic experiment in carbene chemistry, Doering and Prinzbach used ^{14}C-labeled isobutene to distinguish the two mechanisms. Explain what their results mean. The figure shows only one product of the many formed in the reaction. Focus on this compound only.

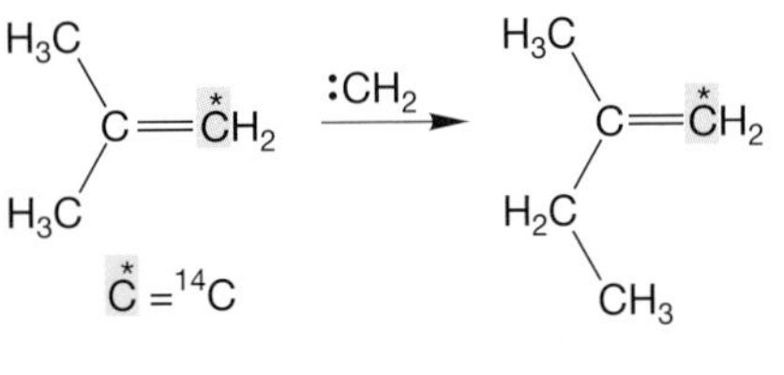

PROBLEM 10.64 Contrast the results of hydroboration/oxidation and mercury (Hg^{2+})-catalyzed hydration for 2-pentyne and 3-hexyne. Would any of these procedures be a practical preparative method? Explain.

PROBLEM 10.65 Recall that terminal alkynes are among the most acidic of the hydrocarbons (p. 129), and that the acetylide ions can be used in S_N2 alkylation reactions with an appropriate alkyl halide. For example,

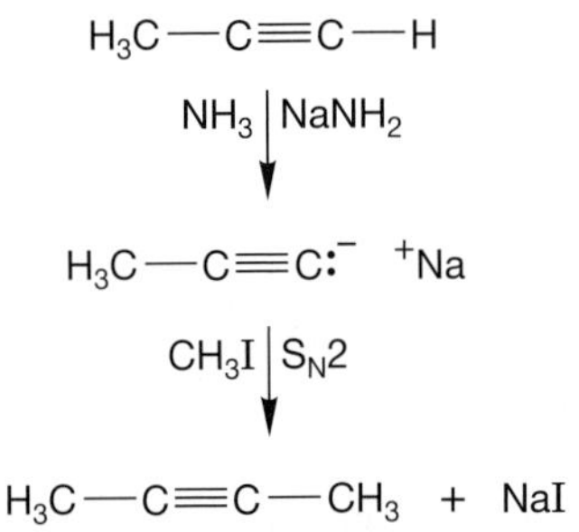

Provide syntheses for the following molecules, free of other isomers. You must use alkynes containing no more than four carbon atoms as starting materials. You may use inorganic reagents of your choice, and other organic reagents containing no more than two carbon atoms. Mechanisms are not required.

(a)

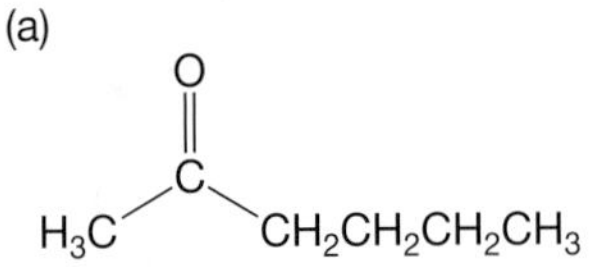

(b)

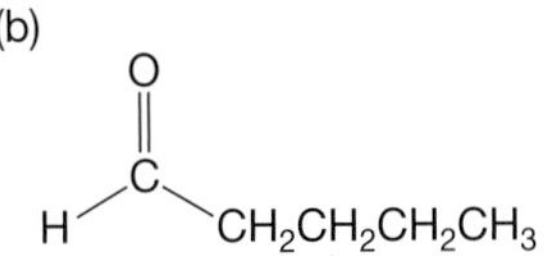

(c)

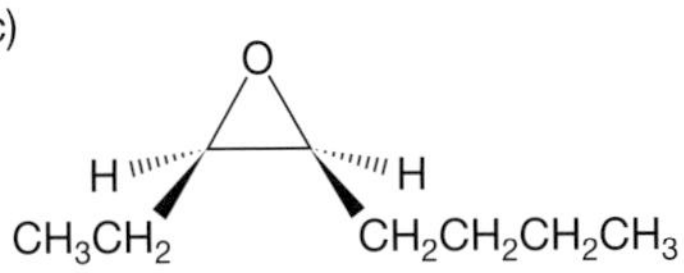

(d)

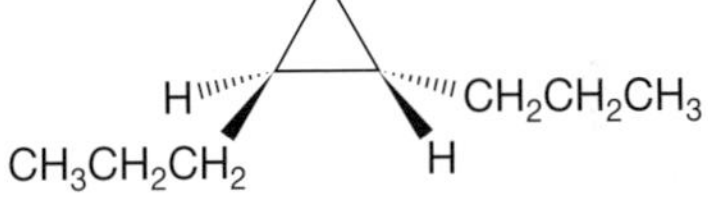

PROBLEM 10.66 Many additions of bromine to acetylenes give only trans dibromide intermediates. However, there are exceptions. Here is one. Explain why phenylacetylene gives both cis and trans dibromides on reaction with bromine.

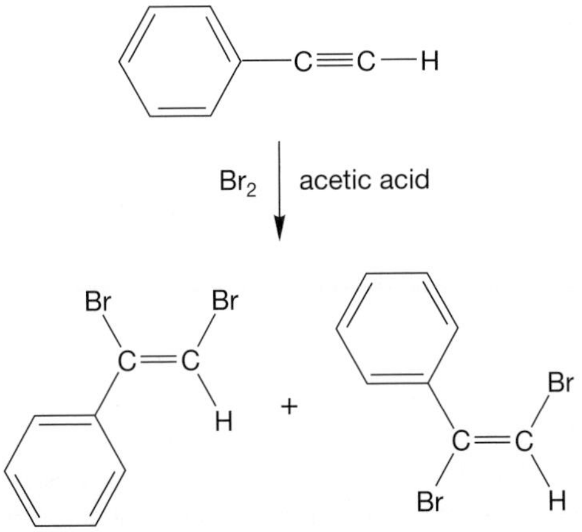

Use Organic Reaction Animations (ORA) to answer the following questions:

PROBLEM 10.67   The second page of reactions on the ORA CD has several reactions that are discussed in Chapter 10. Select the "Stabilized alkene halogenation" animation. This animation shows the calculated pathway for the bromination of acenaphthylene (see Problem 10.7). Notice the energy diagram for the reaction. What does the energy diagram tell you about the intermediate? Stop the animation at the intermediate and click on the LUMO track. Notice the interesting pattern of LUMO (or cation) density. Draw all the resonance structures that are consistent with this calculated representation of the intermediate. Based on the LUMO picture, which of the resonance structures that you have drawn contribute most to the overall species? Which contribute least? Can you guess why?

PROBLEM 10.68   The "Halohydrin formation" animation was calculated using a symmetrical alkene. The symmetry of the alkene means that nucleophilic addition to the bromonium intermediate will not be regioselective. Select play for this reaction. Stop at the first intermediate and observe the calculated LUMO. Based on this data, where is the positive charge located? Show four products that would be formed through nucleophilic attack at each of the places of LUMO density.

PROBLEM 10.69   Epoxidation of alkenes is thought to be a concerted process. Select the "Alkene epoxidation" animation. Does the energy diagram indicate this reaction is concerted? Why do the methyl groups on the *cis*-2-butene rotate during the reaction? This reaction occurs because of the weak oxygen–oxygen bond of a peracid. Hydrogen peroxide (H_2O_2) also has an oxygen–oxygen bond. Would this reaction mechanism work with hydrogen peroxide?

Radical Reactions

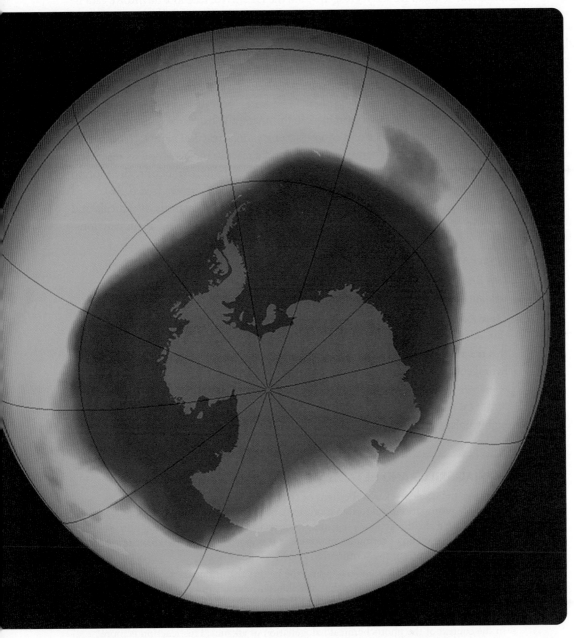

11.1 Preview

11.2 Formation and Simple Reactions of Radicals

11.3 Structure and Stability of Radicals

11.4 Radical Addition to Alkenes

11.5 Other Radical Addition Reactions

11.6 Radical-Initiated Addition of HBr to Alkynes

11.7 Photohalogenation

11.8 Allylic Halogenation: Synthetically Useful Reactions

11.9 Special Topic: Rearrangements (and Nonrearrangements) of Radicals

11.10 Special Topic: Radicals in Our Bodies; Do Free Radicals Age Us?

11.11 Summary

11.12 Additional Problems

RADICALS IN THE AIR It is radical initiation and propagation steps like those shown in Figure 11.40 that are thought to be the cause of ozone depletion. This chart shows ozone concentrations over the Antarctic in 2006.

Knowledge may have its purposes,
But guessing is always
More fun than knowing.

—W. H. AUDEN[1]

11.1 Preview

Although there have been many variations in the details, with few exceptions all of the reactions we have studied so far have been polar ones. The S_N1 and S_N2 substitutions, the addition reactions of HX and X_2 reagents, and the E1 and E2 reactions all involve cationic electrophiles and anionic nucleophiles, or at least have obviously polar transition states.

Now we come to a series of quite different processes involving the decidedly nonpolar, neutral intermediates called radicals or, sometimes, free radicals. We will also encounter **chain reactions**, in which a small number of radicals can set a repeating reaction in motion. In such a case, a few starter molecules can determine the course of an entire chemical process.

The question of selectivity is prominent in this chapter. What factors influence the ability of a reactive species to pick and choose among various reaction pathways? In particular, what is the connection between the energy of a reactive molecule and its reactivity?

CONVENTION ALERT

Don't forget—throughout our discussion of these radical reactions, the motion of single electrons is shown with single-barbed, "fishhook" arrows (see p. 38).

We begin with a section on the formation and simple reactions of radicals and then move on to structure before considering the more complicated chain reactions.

WORKED PROBLEM 11.1 Find an exception to the generalization stated in this chapter's first paragraph. What reaction have we studied that involves no polar intermediates and doesn't have a highly polar transition state? *Hint*: Think "recent."

ANSWER One example is the reaction studied in Section 10.4e (p. 433), the addition of a triplet carbene to an alkene. Only neutral intermediates are involved in this reaction. Another example is hydrogenation, the conversion of an alkene and hydrogen into an alkane (p. 411).

PROBLEM 11.2 Name a reaction that involves no intermediates but certainly has a highly polar transition state.

ESSENTIAL SKILLS AND DETAILS

1. You have to be able to sketch the steps of chain reactions. This chapter introduces initiation, propagation, and termination steps, which are quite common in radical reactions. In a chain reaction, a repeating process keeps the reaction going until starting materials are used up or termination occurs. It is important to be able to write initiation, propagation, and termination steps.

2. Another important skill is understanding the factors that influence selectivity. In synthesis, selectivity is everything, because the synthetic chemist seeks a single product over all others. In order to maximize the formation of that single product, we need to understand the factors that control selectivity. In this chapter, selectivity is discussed in the context of the photochemical halogenation of alkanes. This reaction is more important for understanding selectivity than for its use in synthesis.

3. *N*-Bromosuccinimide (NBS) is a standard reagent for inducing bromine into the allylic and benzylic positions.

---

[1]Wystan Hugh Auden (1907–1973) was one of the best known and most influential British poets of his generation.

11.2 Formation and Simple Reactions of Radicals

We saw radicals in Chapters 1 and 2 when we considered the formation of molecules such as hydrogen, methane, and ethane from their constituent parts: the hydrogen atom, which is the smallest radical, and the methyl radical (see Sections 1.5, 2.4 and 2.5; Fig. 11.1).

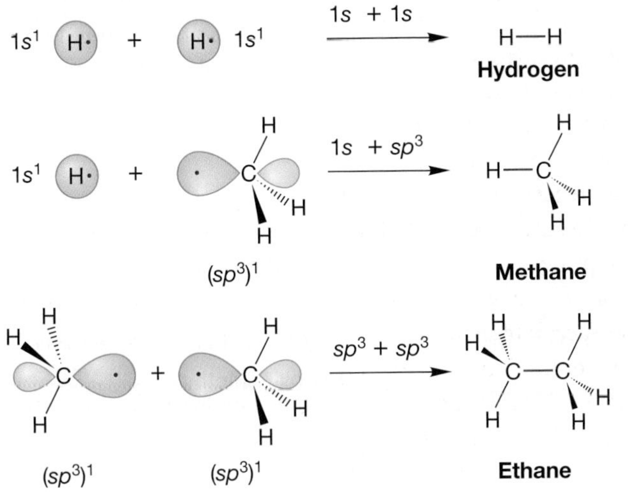

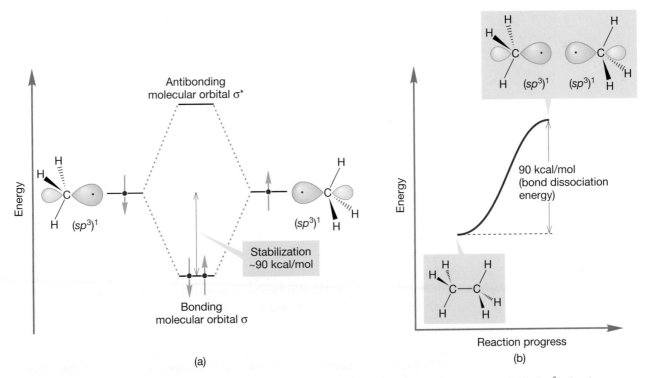

FIGURE 11.1 The formation of hydrogen, methane, and ethane through reactions of the radicals, H· and ·CH₃.

Can we reverse the bond-forming process to produce radicals from these molecules? Energy will surely have to be added in the form of heat, because bond forming is highly exothermic and bond breaking is highly endothermic. Figure 11.2a

FIGURE 11.2 (a) A schematic picture of the formation of a covalent bond through overlap of two half-filled sp^3 orbitals. (b) To reverse the process, energy must be supplied. For example, methyl radicals can be formed by supplying enough energy (90 kcal/mol) to cause homolytic cleavage of the carbon–carbon bond in ethane.

shows schematically the stabilization gained when two singly occupied sp^3 orbitals overlap to form a bond. In order to reverse the bond-making process and form two radicals, energy must be provided up to the amount of the stabilization achieved in the bond making. For example, we can apply heat to ethane in a process called **pyrolysis** or **thermolysis**, eventually providing 90 kcal/mol and homolytically breaking the carbon–carbon bond (Fig. 11.2b). Carbon–carbon bond cleavage is the lowest energy process easily detectable in ethane.

Recall that the energy required to break the sigma bond homolytically is called the bond dissociation energy (BDE, p. 337). Note that the bond breaks to give two neutral species, two radicals, and not to give two polar species, a cation and an anion (Fig. 11.3).

FIGURE 11.3 Two possible bond breakings in ethane. Homolytic cleavage requires less energy because no charges are generated.

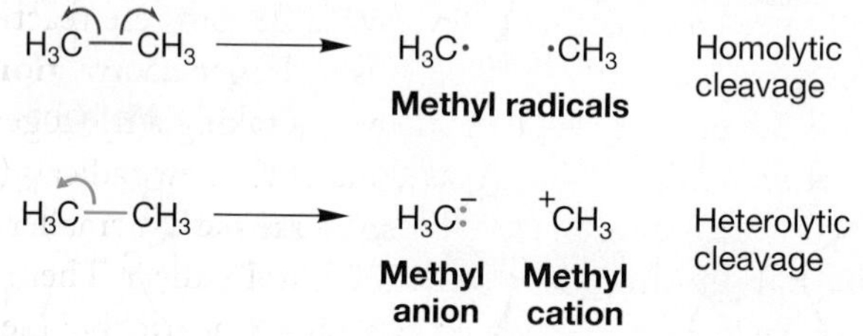

$$H_3C\text{—}CH_3 \longrightarrow H_3C\cdot \quad \cdot CH_3$$

Methyl radicals   Homolytic cleavage

$$H_3C\text{—}CH_3 \longrightarrow H_3C\overset{\text{..}}{\cdot}{}^- \quad {}^+CH_3$$

Methyl anion **Methyl cation**   Heterolytic cleavage

WORKED PROBLEM 11.3 In addition to bond vibration, what hard-to-see process is going on in ethane as we heat it?

ANSWER Don't forget rotation about the carbon–carbon bond! No new product is formed, but this process is still occurring.

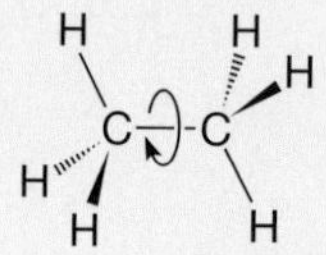

Although adding heat is effective in producing methyl radicals from ethane, it is not a generally useful method of making radicals. The carbon–carbon sigma bond in hydrocarbons is strong, and therefore harsh conditions are required to break it, even in a homolytic fashion. Other, more delicate functional groups will not survive such treatment. In practice, this means that if we want to generate radicals in the presence of other functional groups we should not attempt to do so by heating simple hydrocarbons.

PROBLEM 11.4 When ethane is heated, why does the carbon–carbon bond break rather than one of the carbon–hydrogen bonds?

The pyrolysis of alkanes can be effective in decreasing the overall chain length of the molecule and in introducing unsaturation. Let's consider the pyrolysis of butane. First, there are two possible homolytic cleavages of carbon–carbon bonds in butane. The methyl, ethyl, and propyl radicals can all be formed from breaking carbon–carbon bonds (Fig. 11.4).

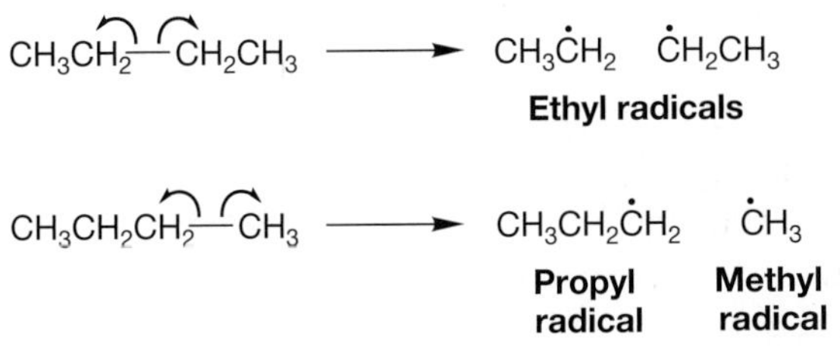

$$CH_3CH_2 \quad CH_2CH_3 \longrightarrow CH_3\dot{C}H_2 \quad \dot{C}H_2CH_3$$
Ethyl radicals

$$CH_3CH_2CH_2 \quad CH_3 \longrightarrow CH_3CH_2\dot{C}H_2 \quad \dot{C}H_3$$
**Propyl   Methyl
radical   radical**

FIGURE 11.4 There are two possible homolytic cleavages of carbon–carbon bonds in butane.

Now we need to evaluate the pathways available for these reactive intermediates. One reaction possible for these radicals is hydrogen **abstraction**. Abstraction is used to describe the process of a radical species taking a hydrogen atom (not a proton) from another molecule. In this case, one of the new radicals (methyl, ethyl, and propyl) can abstract hydrogen from butane to make the smaller hydrocarbons methane, ethane, and propane, along with the butyl radical. There are two positions from which hydrogen abstraction can take place in butane, the primary C—H of the methyl group or the secondary C—H of the methylene group, to give the primary radical or the secondary radical (Fig. 11.5). We will consider the question of selectivity, of choosing between the two different positions, a little later in this chapter.

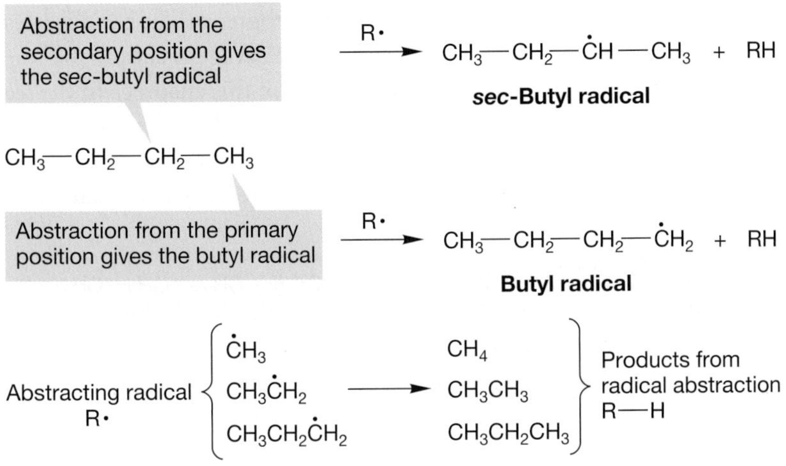

Abstraction from the secondary position gives the *sec*-butyl radical

$$\xrightarrow{R\cdot} CH_3—CH_2—\dot{C}H—CH_3 \quad + \quad RH$$
***sec*-Butyl radical**

$$CH_3—CH_2—CH_2—CH_3$$

Abstraction from the primary position gives the butyl radical

$$\xrightarrow{R\cdot} CH_3—CH_2—CH_2—\dot{C}H_2 \quad + \quad RH$$
Butyl radical

Abstracting radical
R·
$$\begin{cases} \dot{C}H_3 \\ CH_3\dot{C}H_2 \\ CH_3CH_2\dot{C}H_2 \end{cases} \longrightarrow \begin{array}{l} CH_4 \\ CH_3CH_3 \\ CH_3CH_2CH_3 \end{array} \left.\begin{array}{l} \\ \\ \\ \end{array}\right\} \begin{array}{l} \text{Products from} \\ \text{radical abstraction} \\ \text{R—H} \end{array}$$

FIGURE 11.5 Hydrogen abstraction from butane by a radical ultimately leads to the formation of smaller hydrocarbons and two possible butyl radicals.

PROBLEM 11.5 Draw the transition states for the two possible abstractions of hydrogen from butane by a methyl radical.

It is also possible for one radical to abstract a hydrogen from the vicinal (adjacent or β) carbon of another radical in a process called **disproportionation**, which produces an alkane and an alkene. In the propyl radical, for example, a carbon–hydrogen bond could be broken by a methyl radical to give propene and methane (Fig. 11.6). In the butane pyrolysis example, any of the radicals can do the abstracting so an R· is used in Figure 11.6.

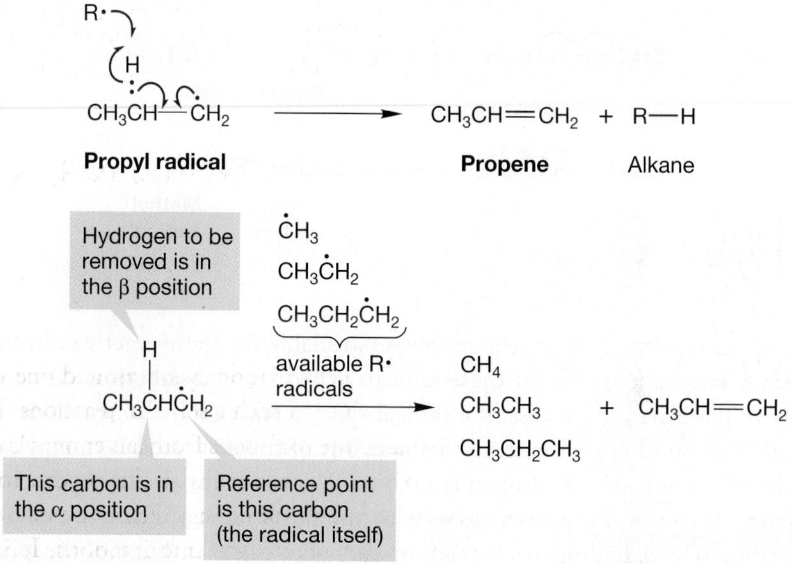

FIGURE 11.6 The disproportionation reaction produces an alkane and an alkene from a pair of radicals. A hydrogen atom of one radical is abstracted by another radical.

PROBLEM 11.6 Draw the transition state for hydrogen abstraction in a disproportionation reaction.

Another reaction possible for many radicals is **β cleavage** (Fig. 11.7). In the terminology of radical chemistry, the point of reference is the radical itself; the adjacent atom is α and the next atom is β. If an α–β carbon–carbon bond is present, it can break to give a new radical and an alkene. This cleavage is simply the reverse of the addition of a radical to an alkene (discussed later in this chapter). In the *sec*-butyl radical, for example, cleavage between the methylene and methyl groups produces a methyl radical and propene. The β cleavage reaction is favored at high temperature and can be quite important in high-temperature reactions of alkanes.

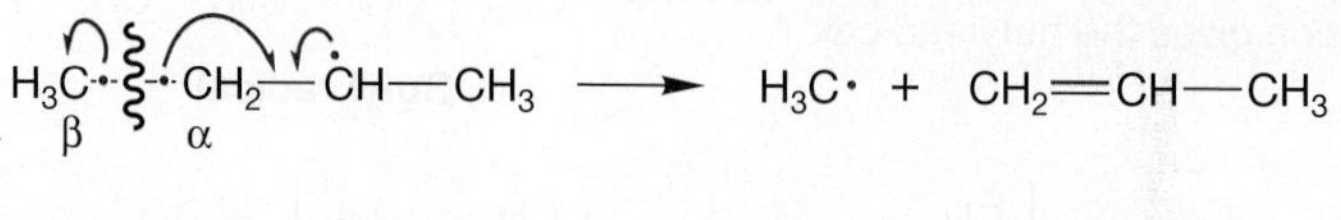

FIGURE 11.7 β Cleavage of the *sec*-butyl radical.

PROBLEM 11.7 Why will β cleavage be favored at high temperature?

To make the issue even more complex, two radicals can occasionally react with each other (dimerize) in a very exothermic formation of new hydrocarbons (dimers) that can reenter the reaction sequence and provide new species. For example, two propyl radicals might combine to form hexane (Fig. 11.8). Further pyrolysis of hexane would lead to *many* other products.

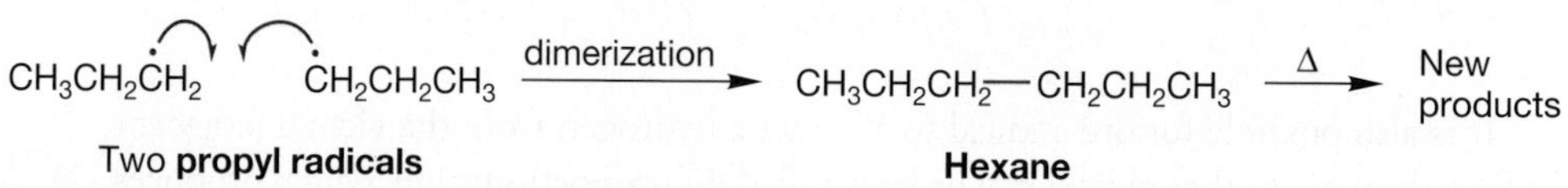

FIGURE 11.8 Two radicals can form a bond in a process called dimerization.

Summary

The reactions available to an alkyl radical include hydrogen abstraction to make a saturated compound and a new radical, disproportionation (hydrogen abstraction from another radical to give an alkene and an alkane), β cleavage (the fragmentation of a radical into a new radical and an alkene), and dimerization. (Fig. 11.9).

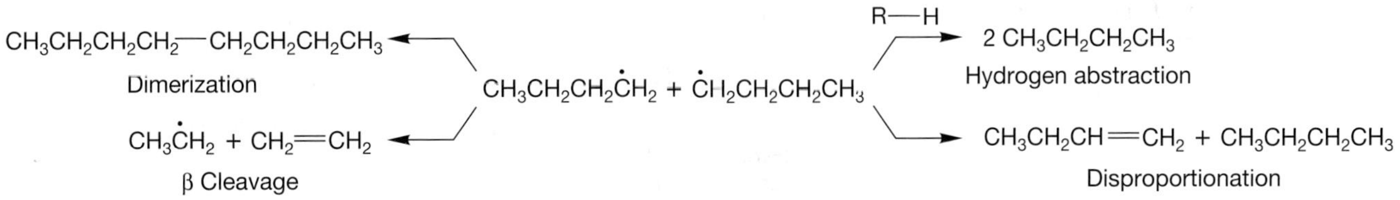

$CH_3CH_2CH_2CH_2$—$CH_2CH_2CH_2CH_3$

Dimerization

$CH_3CH_2CH_2\dot{C}H_2 + \dot{C}H_2CH_2CH_2CH_3$

$\dot{C}H_3\dot{C}H_2 + CH_2\!\!=\!\!CH_2$

β Cleavage

R—H

2 $CH_3CH_2CH_2CH_3$

Hydrogen abstraction

$CH_3CH_2CH\!\!=\!\!CH_2 + CH_3CH_2CH_2CH_3$

Disproportionation

FIGURE 11.9 Simple reactions of radicals.

Clearly, little specificity is possible in hydrocarbon pyrolysis, and one might at first expect that there would be little utility in such a series of reactions. That's not quite right. Using heat to form smaller hydrocarbons from larger ones is called **hydrocarbon cracking**, and the petroleum industry relies on this process to convert the high molecular weight hydrocarbons making up much of crude oil into the much more valuable lower molecular weight gasoline fractions. It is also possible to do the pyrolysis in the presence of hydrogen to help produce smaller alkanes from the radicals that are produced initially.

PROBLEM 11.8 What function does the hydrogen perform in hydrocarbon cracking?

WORKED PROBLEM 11.9 What products do you expect from the thermal cracking of hexane?

ANSWER The point of this problem is to show you how incredibly complicated this "simple" process becomes even in a relatively small alkane such as hexane.

First of all, carbon–carbon bond breakings give the methyl, ethyl, propyl, butyl, and pentyl radicals, which can abstract hydrogen to give methane, ethane, propane, butane, and pentane:

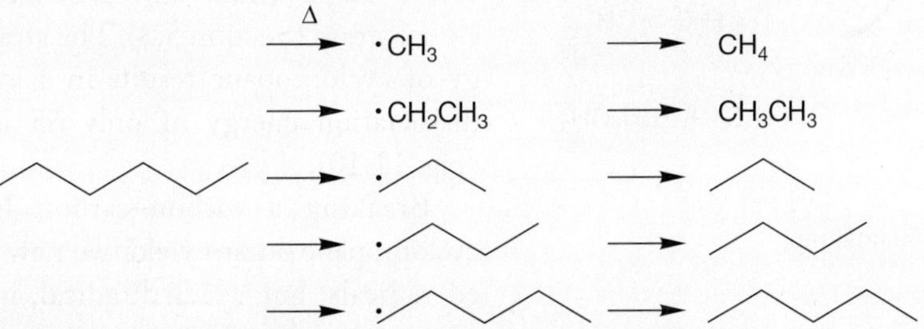

Δ

$\cdot CH_3$ → CH_4

$\cdot CH_2CH_3$ → CH_3CH_3

Disproportionation gives the alkanes already mentioned, in addition to ethylene, propene, 1-butene, and 1-pentene:

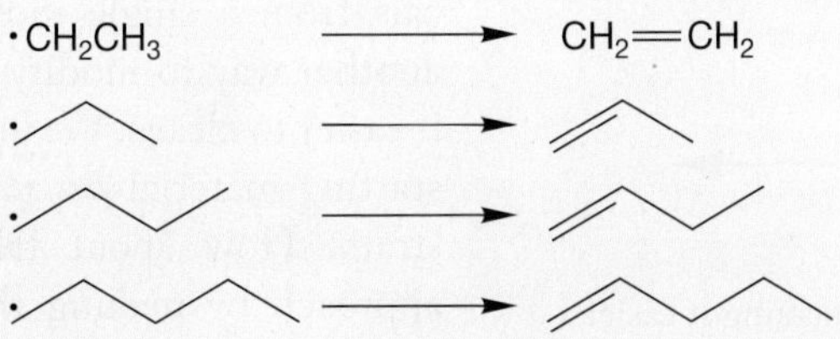

$\cdot CH_2CH_3$ → $CH_2\!\!=\!\!CH_2$

(continued)

β Cleavage of these radicals gives more ethylene and the methyl, ethyl, and propyl radicals:

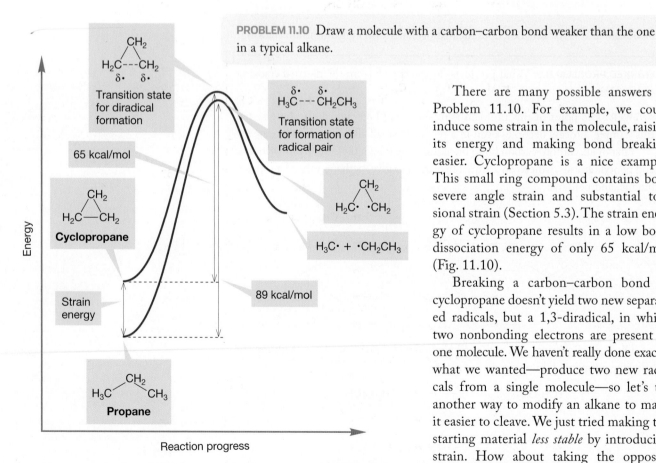

Of course, the alkanes and alkenes formed in these reactions can react further. Propane, butane, and pentane, as well as the alkenes formed in disproportionation and β cleavage, can participate in similar radical reactions. Moreover, dimerization reactions have the potential of producing larger hydrocarbons, only two of which are shown here. These dimers can go on to produce other molecules as they, too, participate in the radical reactions. It's an incredible mess.

Now comes a big question. How can an alkane be modified in order to make the carbon–carbon bond weaker and thus more easily broken?

PROBLEM 11.10 Draw a molecule with a carbon–carbon bond weaker than the one in a typical alkane.

There are many possible answers to Problem 11.10. For example, we could induce some strain in the molecule, raising its energy and making bond breaking easier. Cyclopropane is a nice example. This small ring compound contains both severe angle strain and substantial torsional strain (Section 5.3). The strain energy of cyclopropane results in a low bond dissociation energy of only 65 kcal/mol (Fig. 11.10).

Breaking a carbon–carbon bond in cyclopropane doesn't yield two new separated radicals, but a 1,3-diradical, in which two nonbonding electrons are present in one molecule. We haven't really done exactly what we wanted—produce two new radicals from a single molecule—so let's try another way to modify an alkane to make it easier to cleave. We just tried making the starting material *less stable* by introducing strain. How about taking the opposite approach by making the products of the reaction, the two radicals, *more stable*?

FIGURE 11.10 Strain can raise the energy of a hydrocarbon, making it easier to break a carbon–carbon bond.

PROBLEM 11.11 Construct an Energy versus Reaction progress diagram that illustrates this second general approach of making the product of the reaction more stable.

The question now becomes: How can we make a radical more stable than the methyl radical? We know that delocalization of electrons is stabilizing, so one approach would be to use a starting material that will yield resonance-stabilized radicals. That change should lower the bond strength of the crucial carbon–carbon bond. For example, introduction of a single vinyl group (as in 1-butene) lowers the bond dissociation energy by about 13 kcal/mol to a value of 76 kcal/mol. As shown in Figure 11.11, propane breaks a carbon–carbon bond to give a methyl radical and an ethyl radical and 1-butene breaks a carbon–carbon bond to give a methyl radical and an allyl radical. The difference in the energy required for the two bond breakings is reasonably attributed mostly to the stabilization of the allyl radical, and, of course, of the transition state leading to it. Introduction of a second vinyl group, as in 1,5-hexadiene, reduces the BDE by approximately an additional 13 kcal/mol (Fig. 11.11). The key to lowering the bond dissociation energy is the formation of allyl radicals with their stabilizing delocalization of the nonbonding electrons.

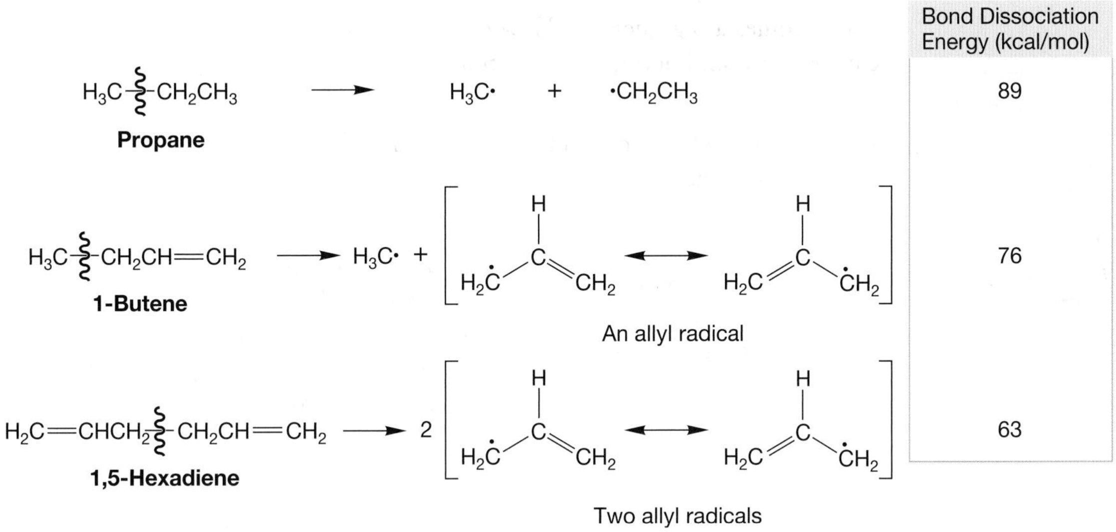

FIGURE 11.11 Stabilization of the products of bond breaking through delocalization can make the bond cleavage process easier.

PROBLEM 11.12 There is an even lower energy process occurring in 1,5-hexadiene, although no *new* molecule appears! Can you draw an arrow formalism for the conversion of one molecule of 1,5-hexadiene into another?

Hint: 1,1,6,6-Tetradeuterio-1,5-hexadiene equilibrates on heating with 3,3,4,4-tetradeuterio-1,5-hexadiene:

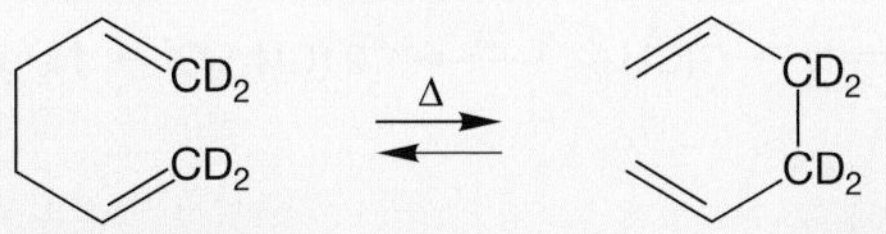

1,1,6,6-Tetradeuterio-1,5-hexadiene **3,3,4,4-Tetradeuterio-1,5-hexadiene**

Another technique for producing radicals is to start with molecules containing especially weak, and therefore easily broken bonds. Examples are peroxides and acyl peroxides (Fig. 11.12). The oxygen–oxygen bond in simple peroxides is already quite weak (~38 kcal/mol; Table 8.2, p. 337). The addition of the carbon–oxygen double bond of the **acyl group (R—C=O)**, helps delocalize the nonbonding electron in the carboxyl radical and lowers the BDE to about 29 kcal/mol.

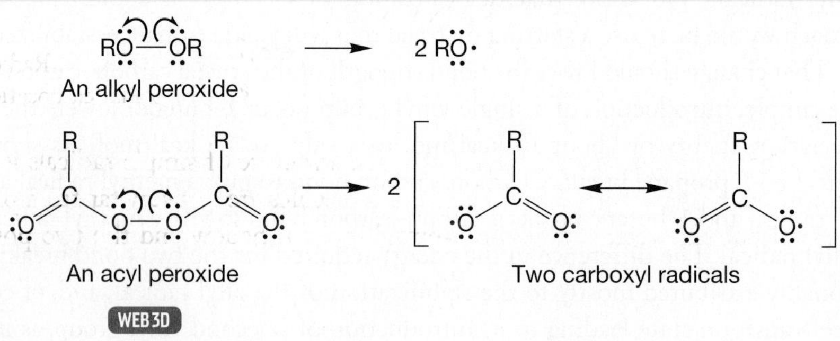

FIGURE 11.12 In alkyl peroxides, the weak oxygen–oxygen bond is easily cleaved. In acyl peroxides, the product radicals are stabilized and as a result the oxygen–oxygen bond is even weaker.

Some molecules give stable gases (often nitrogen or carbon dioxide) on heating and thus produce radicals irreversibly. **Azo compounds**, which have the structure R—N=N—R, and acyl peroxides are good examples (Fig. 11.13).

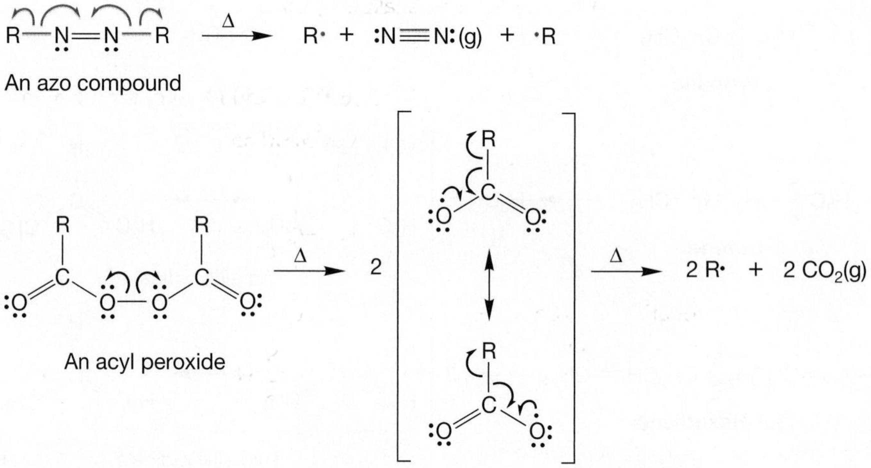

FIGURE 11.13 Some molecules give free radicals irreversibly through the formation of gases.

A particularly useful source of radicals is azobisisobutyronitrile (AIBN), a molecule that decomposes much more easily than simpler azo compounds (Fig. 11.14a).

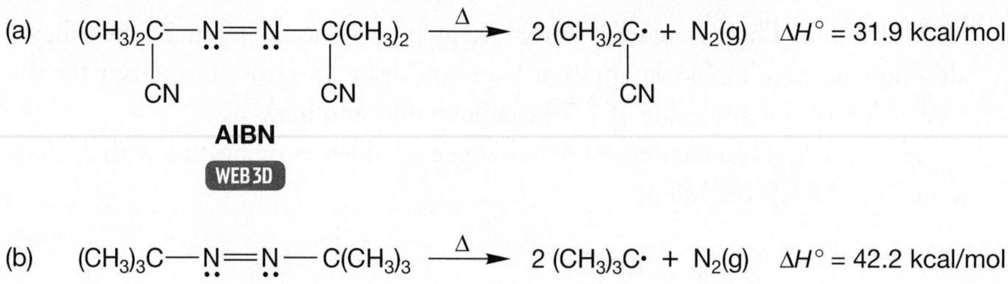

(a) $(CH_3)_2C—N=N—C(CH_3)_2 \xrightarrow{\Delta} 2(CH_3)_2C\cdot + N_2(g)$ $\Delta H° = 31.9$ kcal/mol
with CN groups; **AIBN**

FIGURE 11.14 Azobisisobutyronitrile is an azo compound that is especially easy to cleave.

(b) $(CH_3)_3C—N=N—C(CH_3)_3 \xrightarrow{\Delta} 2(CH_3)_3C\cdot + N_2(g)$ $\Delta H° = 42.2$ kcal/mol

PROBLEM 11.13 Why does AIBN decompose much more easily than a simple azo compound such as the molecule shown in Figure 11.14b?

11.3 Structure and Stability of Radicals

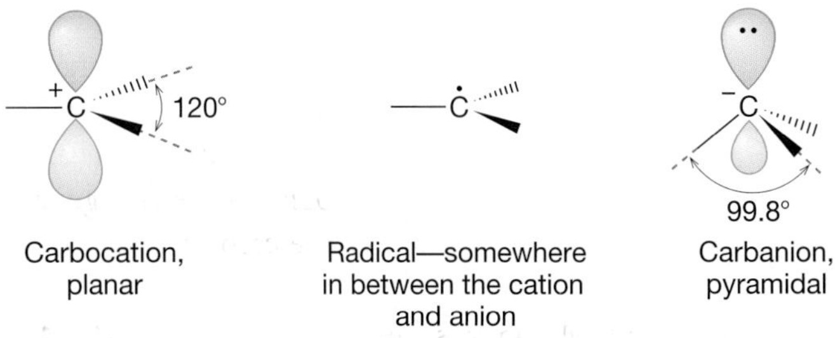

We have already encountered both carbocations and carbanions, and spent some time discussing their structures (Section 2.4, p. 62). The methyl cation, and carbocations in general, are flat, sp^2-hybridized molecules with an empty $2p$ orbital extending above and below the plane of the three substituents on carbon (Fig. 11.15). By contrast, the methyl anion is pyramidal. One might well guess that the methyl radical, with a single nonbonding electron, would have a structure intermediate between that of the two ions.

Carbocation, planar

Radical—somewhere in between the cation and anion

Carbanion, pyramidal

FIGURE 11.15 Radicals are intermediate in structure between the planar carbocations and the pyramidal carbanions.

Both theory and experiment agree that the structure of simple radicals is hard to determine! However, it is clear that these molecules are not very far from planar. If the molecules are not planar, the pyramid is *very* shallow and the two possible forms are rapidly inverting (Fig. 11.16).

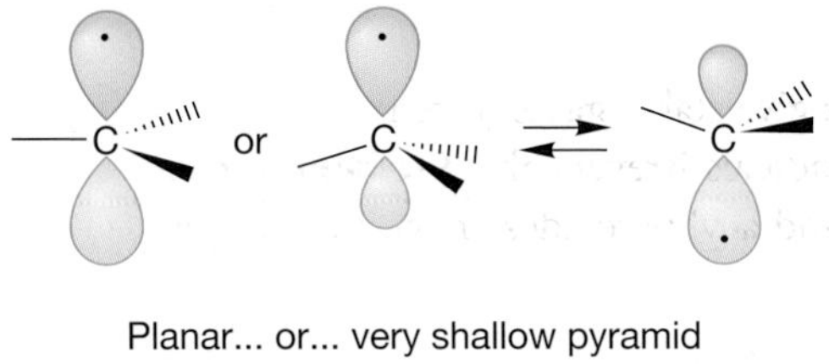

Planar... or... very shallow pyramid

FIGURE 11.16 Alkyl radicals are either flat or very shallow, rapidly interconverting pyramids.

PROBLEM 11.14 Draw the transition state for the inversion process in Figure 11.16.

As with carbocations, more substituted radicals are more stable than less substituted ones. The first four entries of Table 11.1 show BDEs of four hydrocarbons that lead to a methyl, primary, secondary, and tertiary radical. The more substituted the radical, the easier it is to form, and the easier it is to break the bond that produces it. The data presented in the table are slightly suspect, because we are forming the radicals from different starting materials. Might not the different energies of the different starting materials influence the BDEs? Yes they might, and do. But this effect isn't large, and the data in the table can be used to show that more substituted radicals are more stable than less substituted radicals.

TABLE 11.1 Bond Dissociation Energies for Some Hydrocarbons

Hydrocarbon	Radical	BDE (kcal/mol)
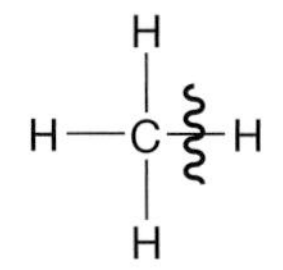	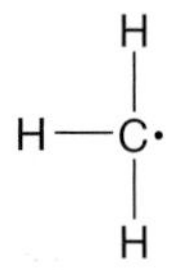	105.0
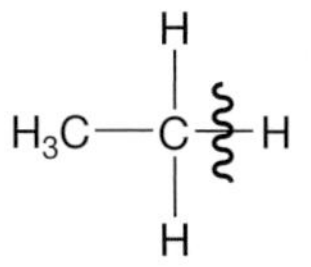	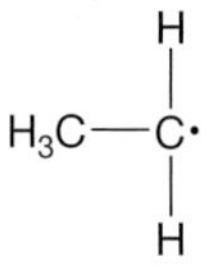	101.1
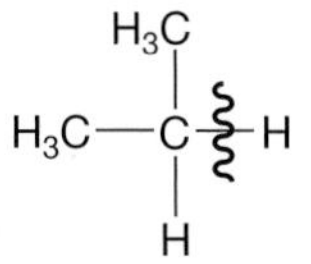	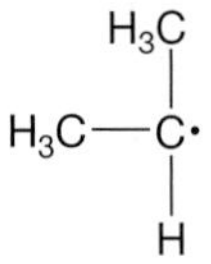	98.6
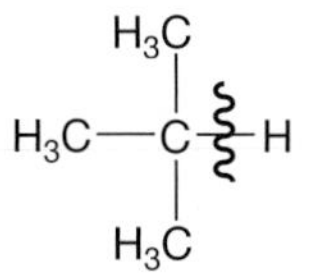	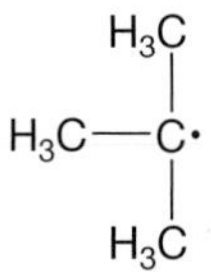	96.5
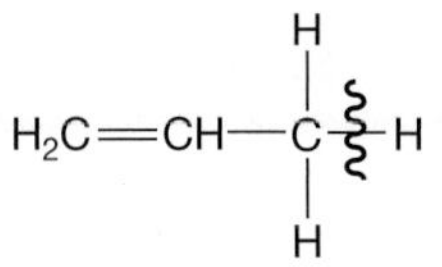	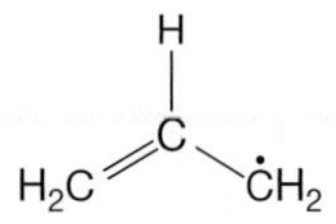	89.8
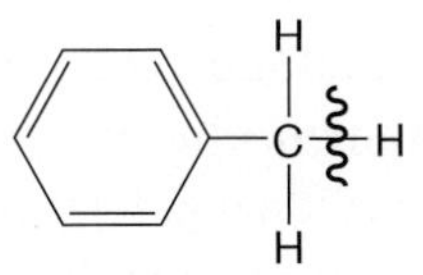	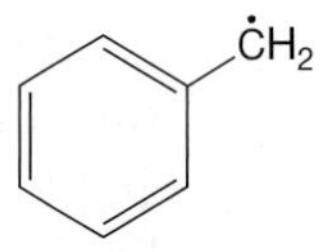	88.8

Why should more substituted radicals be more stable than less substituted radicals? Let's start with an analogy. In explaining the order of stability for carbocations (tertiary > secondary > primary > methyl), we drew resonance forms in which a filled carbon–hydrogen bond overlapped with the empty $2p$ orbital in a process called hyperconjugation (Fig. 11.17).

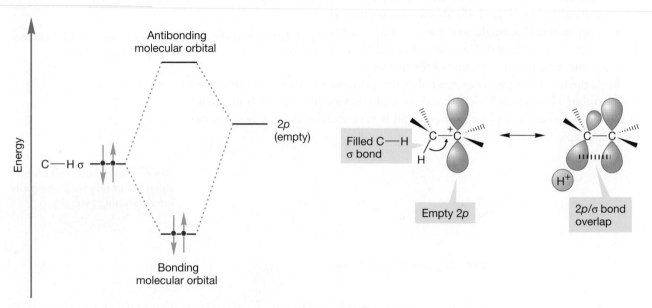

FIGURE 11.17 Resonance stabilization of a carbocation through hyperconjugation, shown both by an interaction energy diagram and by an orbital overlap picture.

We can draw similar resonance forms for radicals. As before, the more alkyl groups attached to the carbon bearing the nonbonding electron, the more resonance forms are possible. But there is an important difference: Stabilization of the carbocation involves overlap of a filled carbon–hydrogen bond with an *empty 2p* orbital (Fig. 11.17), whereas in the radical, overlap is between a filled carbon–hydrogen bond and a *half-filled 2p* orbital (Fig. 11.18).

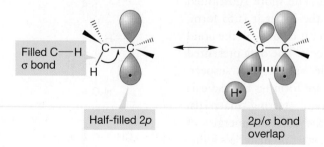

FIGURE 11.18 Resonance stabilization of an alkyl free radical.

Why should this orbital mixing be stabilizing? The evident problem is that this is a *three*-electron system. One electron must occupy the antibonding orbital, and this is certainly destabilizing. Nevertheless, the *overall* system is still stabilized because two electrons are able to occupy the low-energy bonding orbital.

Recall He_2^+ in which a similar orbital system was encountered (p. 41). Figure 11.19 compares hyperconjugation of a carbocation, hyperconjugation of a radical, and the He_2^+ molecule. As we saw in Chapter 1, He_2^+ is held together (bound) by over 60 kcal/mol.

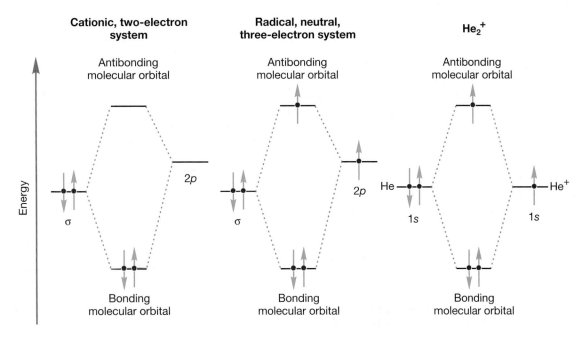

FIGURE 11.19 A comparison of the stabilizations of carbocations, radicals, and He_2^+. Notice the electron in the antibonding orbital of He_2^+ and the radical.

Consider the bond strengths in the carbon radicals, as we did earlier in discussing alkenes and carbocations (p. 378, Problem 9.7). In the 2-propyl radical we have two sp^2/sp^3 carbon–carbon bonds, whereas in the less stable 1-propyl radical there is one sp^2/sp^3 bond and one less strong sp^3/sp^3 carbon–carbon bond (Fig. 11.20).

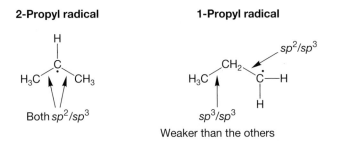

FIGURE 11.20 The 2-propyl radical contains two relatively strong sp^2/sp^3 bonds, whereas the 1-propyl radical has one strong sp^2/sp^3 bond and one less strong sp^3/sp^3 bond.

Of course, we should analyze the carbon–hydrogen bonds as well. If we do this, we find that the differences in the strengths of the C—H bonds should favor the 1-propyl radical. As with carbocations, it is the C—C bonds that dominate and determine the issue. As a practical matter, it appears that we can "get away with" an analysis of the C—C bonds alone.

PROBLEM 11.15 Analyze the difference in bond strength between the C—H bonds in the 2-propyl and 1-propyl radical.

Radicals substituted with electron-delocalizing groups, such as the vinyl or phenyl groups, are even more stable. As Table 11.1 shows, breaking a carbon–hydrogen bond to give an allyl radical or benzyl radical is relatively easy. The allyl and benzyl radicals are even more stable than a tertiary radical (Fig. 11.21).

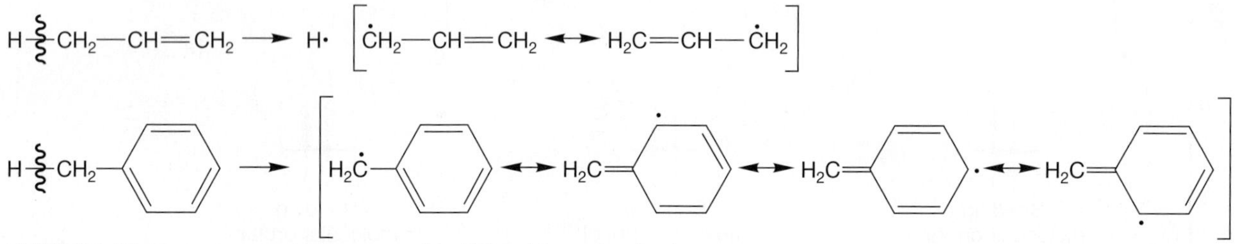

FIGURE 11.21 Resonance stabilization in the allyl and benzyl radicals.

If they are sufficiently stabilized, radicals can be made and studied spectroscopically at low temperature. A few highly delocalized radicals are even stable under normal, room temperature conditions. For example, the triphenyl-methyl radical happily exists in solution as an equilibrium mixture with its dimer (Fig. 11.22).

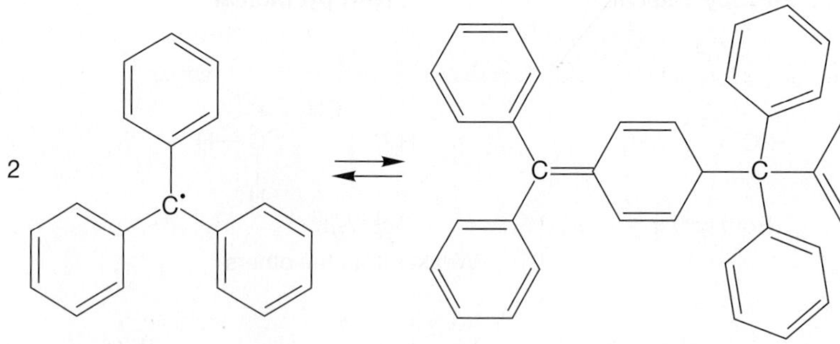

FIGURE 11.22 The triphenylmethyl radical exists in equilibrium with its dimer.

Triphenylmethyl radical   Dimer

PROBLEM 11.16 As you can see from Figure 11.22, the dimer is not the "obvious" one, hexaphenylethane, but a much more complicated molecule. Can you see why hexaphenylethane is a most unstable molecule, and draw an arrow formalism mechanism for the formation of the real dimer?

Summary

Radical stability parallels carbocation stability: Benzylic and allylic are more stable than tertiary, tertiary more stable than secondary, secondary more stable than primary, and primary more stable than methyl. The determining factors are resonance and hyperconjugative stabilization.

11.4 Radical Addition to Alkenes

In Section 9.3 (p. 366), the addition of hydrogen chloride to alkenes was presented with the reaction always leading to Markovnikov addition. Chapter 9 stressed the initial formation of the more stable carbocation followed by addition of the residual halide ion. The product of Markovnikov addition must be produced in this reaction (Fig. 11.23). Of course, HBr reacts the same way as HCl.

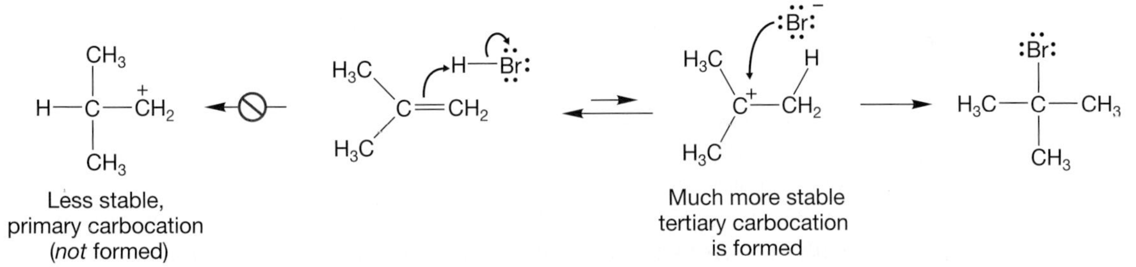

Less stable, primary carbocation (*not* formed)

Much more stable tertiary carbocation is formed

FIGURE 11.23 The addition of hydrogen bromide to an unsymmetrical alkene follows the Markovnikov rule because protonation of the alkene must give the more substituted (more stable) carbocation.

The historical reality was far from that simple. The regiochemistry of the reaction appeared strangely dependent on the source of the starting materials and even on the place in which the reaction was carried out. In some worker's hands, the reaction did follow the Markovnikov path; in other places, with other investigators, anti-Markovnikov addition was dominant. Confusion and argument were the rule of the day in the late 1800s. Oddly enough, the problem was restricted to hydrogen bromide addition. Both hydrogen chloride and hydrogen iodide addition remained securely Markovnikov in everyone's hands (Fig. 11.24).

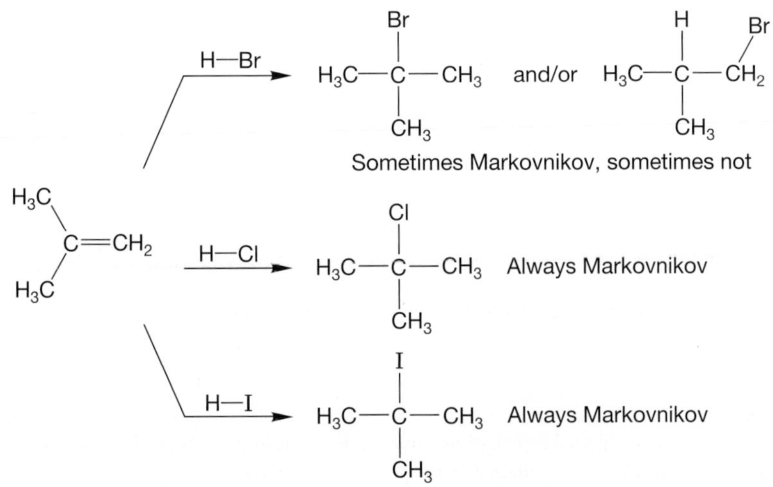

Sometimes Markovnikov, sometimes not

Always Markovnikov

Always Markovnikov

FIGURE 11.24 Both hydrogen chloride and hydrogen iodide addition always follow the Markovnikov rule, but hydrogen bromide addition is strangely capricious.

In the 1930s, largely through the efforts of M. S. Kharasch,[2] it was discovered that the purity of the compounds used was vital. Pure materials led to Markovnikov addition, and we can remain content with the mechanism for the polar reaction. Impure reagents led to predominately anti-Markovnikov addition. A special culprit was found to be peroxides (Fig. 11.25). Breaking the weak oxygen–oxygen bond in peroxides provides a relatively easy route to free radicals. The mechanism of addition in the presence of peroxides is different from the familiar polar one and can be triggered by these small amounts of radicals.

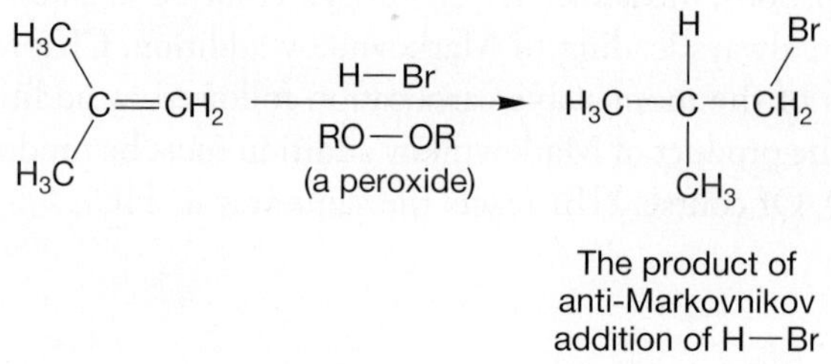

FIGURE 11.25 Addition of hydrogen bromide to alkenes in the presence of peroxide is anti-Markovnikov.

The product of anti-Markovnikov addition of H—Br

11.4a The Mechanism of Radical HBr Addition   As noted earlier (p. 476), peroxides contain a very weak oxygen–oxygen bond and are easily cleaved by heat (Δ) or light ($h\nu$) to give two R—O· radicals (Fig. 11.26, Step 1). The radical can react with H—Br by hydrogen abstraction to form R—OH and a new radical, a bromine atom (Step 2). Addition of the bromine atom to an alkene gives still another radical (Step 3), which can react with H—Br to produce an alkyl bromide and another bromine atom (Step 4). The bromine atom can recycle to Step 3 and start the reaction over again with another alkene molecule. The overall result is the addition of hydrogen bromide to the alkene.

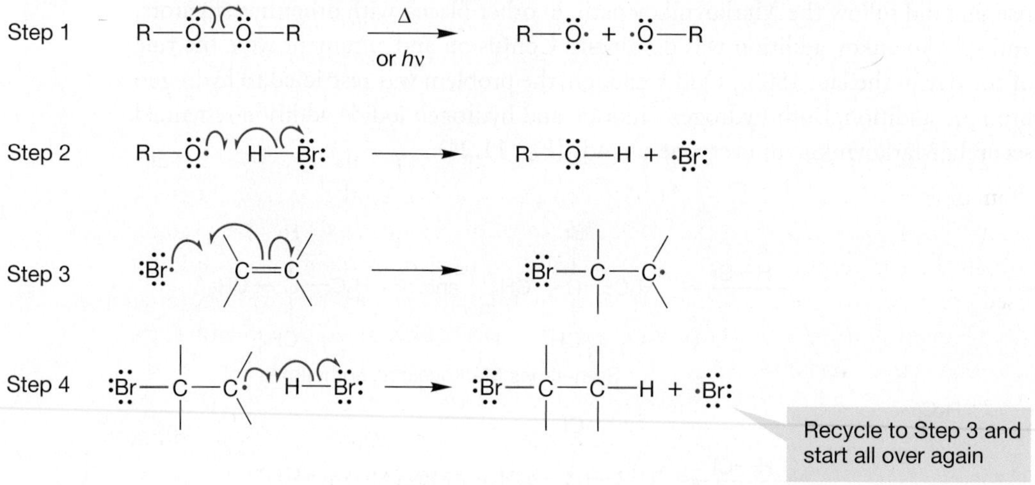

Recycle to Step 3 and start all over again

FIGURE 11.26 The mechanism for the radical addition of HBr to an alkene.

---

[2]Kharasch, 1895–1957, is reported by one of his students, F. R. Mayo (b. 1908), to have suggested that the direction of addition may have been affected by the phases of the moon.

PROBLEM 11.17 Why is abstraction of H from H—Br preferred to abstraction of Br? *Hint*: The bond dissociation energy of HO—Br is 56 kcal/mol.

Other products can be formed in this reaction. When two radicals meet they can react in a very exothermic fashion to form a new bond and end the sequence, called a chain reaction. There are many possibilities for this kind of termination reaction using the radicals in Figure 11.26. A few ways are shown in Figure 11.27.

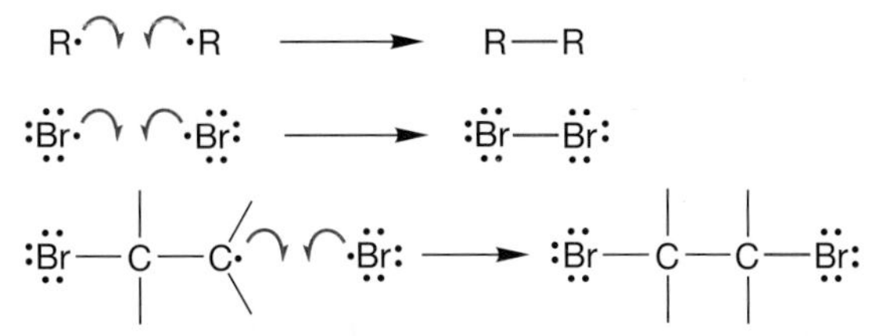

FIGURE 11.27 Whenever two radicals come together they can react to form a bond. The bond formation removes two radicals and terminates this chain reaction.

We have just outlined the radical-induced addition of hydrogen bromide to an alkene. The steps of this chain reaction naturally divide into three kinds:

Initiation. The reaction cannot start without the presence of a radical. A radical must be formed to start the process, in a reaction known as an initiation step. Peroxides are good generators of radicals because of the weakness of their oxygen–oxygen bond (Fig. 11.26, Step 1). Once a radical is created in the presence of hydrogen bromide, a second initiation step, abstraction of hydrogen (Fig. 11.26, Step 2), takes place to form a bromine atom, which is the radical that propagates the chain reaction.

Propagation. Once a bromine atom has been formed in the initiation steps, the chain reaction can begin. The reactions making up the chain reaction are called propagation steps. In this example, the two propagation steps are the addition of the bromine atom to the alkene to give a new, carbon-centered alkyl radical and the subsequent abstraction of hydrogen from hydrogen bromide by the alkyl radical (Fig. 11.26, Steps 3 and 4).

The two radicals involved in the propagation steps, the bromine atom and the alkyl radical, are called chain-carrying radicals. The chain reaction continues as the newly formed bromine atom adds to another alkene to repeat the process over and over again.

Termination. Any time one of the chain-carrying radicals (in this case a bromine atom or an alkyl radical) is annihilated by combination with another radical, the chain reaction is stopped (Fig. 11.27). If such a step does not happen, a single radical could initiate a chain reaction that would continue until starting material was completely used up. In practice, a competition is set up between the propagation steps carrying the chain reaction and the steps ending it. Such reactions are called termination steps and there are many possibilities.

For a chain reaction to produce large amounts of products effectively, the propagation steps must be much more successful than the termination steps. Such is often the case, because in order for a termination step to occur, two radicals must wander through the solution until they find each other. Typically, the propagation steps require reaction with a reagent present in relatively high concentration. So, even though termination steps often can take place very easily once the reactants find each other, propagation steps can compete successfully, and reaction cycles of many thousands are common. Under such circumstances, the small amounts of product formed by the occasional termination step are inconsequential.

11.4b Inhibitors   Small amounts of substances that react with the chain-carrying radicals can stop chain reactions by removing the chain-carrying radicals from the process. These substances are called **inhibitors**. They act as artificial terminators of the propagation steps (Fig. 11.28).

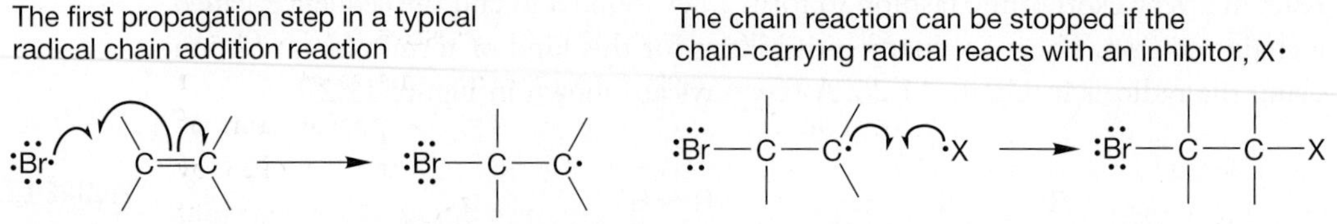

The first propagation step in a typical radical chain addition reaction

The chain reaction can be stopped if the chain-carrying radical reacts with an inhibitor, X·

FIGURE 11.28 An inhibitor of radical chain reactions can act by intercepting one of the chain-carrying radicals.

A few molecules can have an impact out of proportion to their numbers, because stopping one chain reaction can mean stopping the formation of thousands of product molecules. Oxygen (O_2), itself a diradical, can act in this way, and many radical reactions will not proceed until the last oxygen molecules are scavenged from the system, thereby allowing the radical chains to carry on. Radical inhibitors are often added to foods to retard radical reactions causing spoilage. Figure 11.29 shows a few typical radical inhibitors. Both butylated hydroxy toluene (BHT) and butylated hydroxy anisole (BHA) are widely used in the food industry. It is not important that you memorize these structures. What is worth remembering is the general structural features. You should also think a bit about how these molecules might function in intercepting radicals.

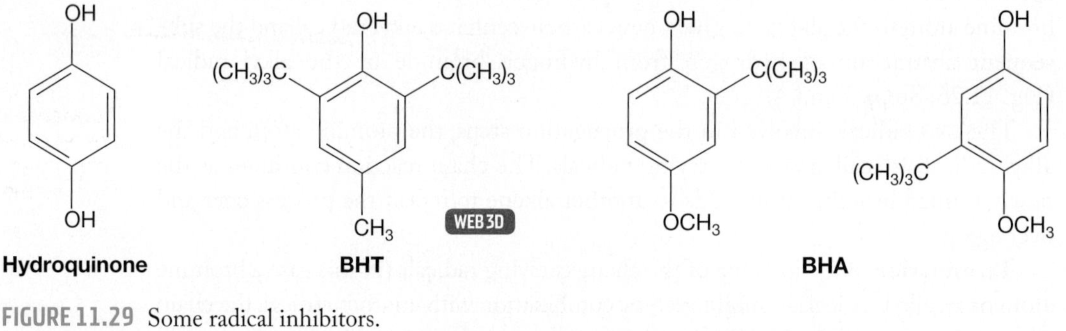

Hydroquinone   BHT   BHA

FIGURE 11.29 Some radical inhibitors.

11.4c Regiospecificity of Radical Additions: Anti-Markovnikov Addition
When a radical adds to an unsymmetrical alkene, there are two possible outcomes: formation of a more substituted or less substituted alkyl radical (Fig. 11.30).

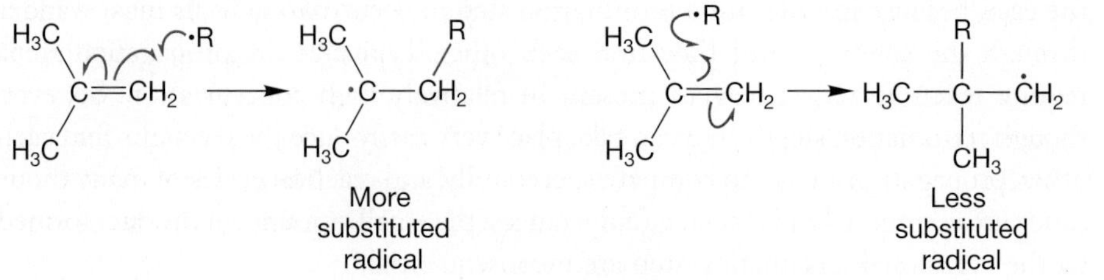

More substituted radical

Less substituted radical

FIGURE 11.30 When a radical adds to an unsymmetrical alkene, there are two possible radical products.

So, the regiospecificity of radical addition to an unsymmetrical alkene such as 2-methylpropene (isobutene) is determined by the direction of radical attack on the alkene, and thus by the stability of the alkyl radical formed. In the case of alkene radical hydrohalogenation, the bromine atom adds to give the more substituted, more stable radical. The reaction continues when the newly formed alkyl radical abstracts a hydrogen atom from hydrogen bromide to give, ultimately, the compound in which the bromine is attached to the *less* substituted end of the original alkene, the anti-Markovnikov product (Fig. 11.31a). This reaction is in contrast to polar hydrohalogenation, in which the first step is protonation of the alkene to give the more substituted carbocation, which is then attacked by bromide ion to give the *more* substituted bromide, the Markovnikov product (Fig. 11.31b).

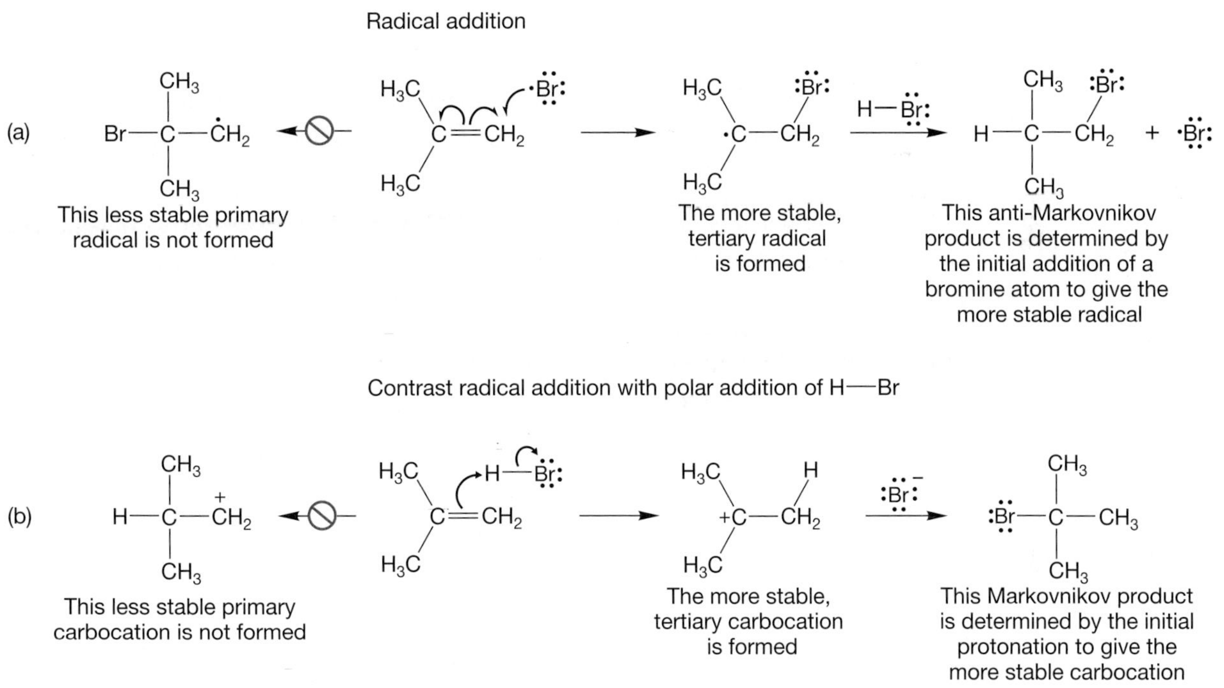

Radical addition

(a)   [structure diagram]

This less stable primary
radical is not formed

The more stable,
tertiary radical
is formed

This anti-Markovnikov
product is determined by
the initial addition of a
bromine atom to give the
more stable radical

Contrast radical addition with polar addition of H—Br

(b)   [structure diagram]

This less stable primary
carbocation is not formed

The more stable,
tertiary carbocation
is formed

This Markovnikov product
is determined by the initial
protonation to give the
more stable carbocation

FIGURE 11.31 (a) The first step in a radical addition to an alkene gives the more substituted, and therefore more stable radical. (b) The first step in the polar addition gives the more stable intermediate, which leads to the Markovnikov product. Notice, however, that in the polar reaction it is the hydrogen atom that adds first, and in the radical reaction it is the bromine atom that adds first.

The historical difficulty in determining the regiospecificity of hydrogen bromide addition had to do only with recognizing that radical initiators such as peroxides had to be kept out of the reaction. If even a little peroxide were present, the chain nature of the radical addition of hydrogen bromide could overwhelm the nonchain polar reaction and produce large amounts of anti-Markovnikov product. This understanding of the causes of the regiochemical outcome of these reactions presents an opportunity for synthetic control. We now have ways to do Markovnikov addition of hydrogen chloride and hydrogen iodide, and either Markovnikov or anti-Markovnikov addition of hydrogen bromide. Of course, the halides formed in these reactions can be used for further synthetic work.

Radical alkene hydrohalogenation

PROBLEM 11.18 Show how to carry out the following conversions. You need not write mechanisms in such problems unless they help you to see the correct synthetic pathways.

Devise two syntheses for each

Neither hydrogen chloride nor hydrogen iodide will add in anti-Markovnikov fashion, even if radical initiators are present. Now we have to see why.

11.4d Thermochemical Analysis of Radical HX Additions to Alkenes

Why does the chain reaction of Figure 11.26 succeed with HBr but not for other hydrogen halides? We can *write* identical radical chain mechanisms for HCl and HI, but these must fail for some reason.

We can understand the difference if we examine the thermochemistry of the two propagation steps for the different reactions.

The energetics of the first propagation step are shown in Table 11.2. These values are estimated by noting that in this first reaction a π bond is broken and a X—C bond is made. The overall exothermicity or endothermicity of the reaction is a combination of these two bond strengths. A π bond is worth 66 kcal/mol, and we approximate the X—C bond strength in the product by taking the values for ethyl halides.

TABLE 11.2 The Thermochemistry of the First Propagation Step: Addition of Halogen Radicals to Ethylene

| | X· + H$_2$C=CH$_2$ → X—CH$_2$CH$_2$· | | |
X	Bond Strength (π bond)	Bond Strength (X—C in X—CH$_2$CH$_2$·)	ΔH (kcal/mol)
Cl	66	85	−19
Br	66	72	−6
I	66	57	+9

We are making all sorts of assumptions, but the errors involved turn out to be relatively small. For both bromine and chlorine this first step is exothermic—the product is more stable than starting material and is favored at equilibrium. For iodine, the reaction is substantially *endothermic*. The product is strongly disfavored and the reaction proceeds very slowly. So now we know why the addition of HI fails; the first propagation step of the reaction is highly endothermic and does not compete with the termination steps. But both the bromine case, which we know succeeds, and the chlorine case, which we know fails, have exothermic first steps.

Let's look at the second propagation step in Table 11.3.

TABLE 11.3 **The Thermochemistry of the Second Propagation Step:**
Abstraction of a Hydrogen Atom from H—X

	$H-X + \cdot CH_2-CH_2X \rightarrow \cdot X + H-CH_2-CH_2-X$		
X	Bond Strength (H—X)	Bond Strength (H—C in $H-CH_2-CH_2-X$)	ΔH (kcal/mol)
Cl	103	98	+5
Br	87	98	−11
I	71	98	−27

The second step is exothermic for Br and I, but *endothermic for Cl*. So, we can expect this reaction to succeed for Br and I, but fail for Cl. If we sum up the analysis for the two propagation steps we see that it is only for HBr addition that *both* steps are exothermic. For HCl and HI, one of the two propagation steps is endothermic, and the radical addition reaction cannot successfully compete with the termination steps. The polar addition of HX wins out.

Summary

The polar addition of HBr to alkenes leads to the more substituted bromide, whereas the radical-initiated addition of HBr gives the less substituted bromide. Thus, you have the ability to make either regioisomer when the alkene is unsymmetrically substituted. The addition of HCl and HI always gives the more substituted halide.

11.5 Other Radical Addition Reactions

Hydrogen bromide addition to alkenes is only one example of a vast number of radical chain reactions. Another example involves carbon tetrahalides, such as CCl_4 and CBr_4, which add to alkenes in the presence of initiators. The steps in the peroxide-initiated reaction are shown in Figure 11.32.

Step 1 $ROOR \longrightarrow RO\cdot + \cdot OR$ ⎫
                                                                        ⎬ Initiation
Step 2 $RO\cdot + CCl_4 \longrightarrow ROCl + \cdot CCl_3$ ⎭

Step 3 $\cdot CCl_3 + CH_2=CH_2 \longrightarrow \cdot CH_2CH_2CCl_3$ ⎫
                                                                                              ⎬ Propagation
Step 4 $\cdot CH_2CH_2CCl_3 + CCl_4 \longrightarrow ClCH_2CH_2CCl_3 + \cdot CCl_3$ ⎭
                                                                                    (recycle
                                                                                    to Step 3)

Step 5   There are many possible termination steps

FIGURE 11.32 The mechanism for the radical-induced addition of carbon tetrachloride to an alkene.

PROBLEM 11.19 Write as many termination steps as you can imagine for the reaction of Figure 11.32.

Under some conditions, alkenes can be polymerized by free radicals. In the examples we have discussed, radical addition to an alkene produced an alkyl radical, which then abstracted hydrogen (or halogen in the case of CX_4) in a second propagation step to give the product of overall addition to the alkene. What if Step 4 is not easy? If the concentration of alkene is very high, for example, or if the addition step (Step 3) is especially favorable, the abstraction reaction (Step 4) might not compete effectively with it. Under such conditions the alkyl radical can undergo addition to the alkene to form a new alkyl radical (the dimer). The dimer radical can react with alkene to form a trimer radical and so forth, until either a termination step intercedes or the supply of alkene is exhausted (Fig. 11.33).

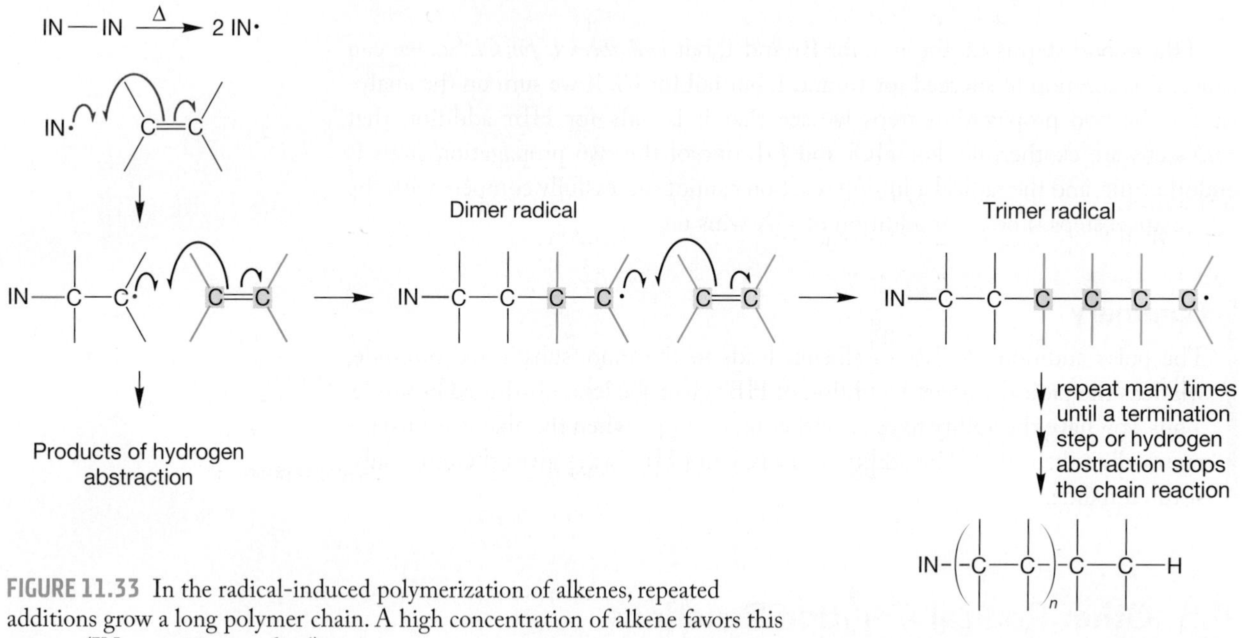

FIGURE 11.33 In the radical-induced polymerization of alkenes, repeated additions grow a long polymer chain. A high concentration of alkene favors this process (IN· = initiator radical).

Such polymerization reactions are common and the products often have very long chain lengths. Very high molecular weight products can be formed. Table 11.4 shows a number of common polymers and the monomers from which they are built up.

TABLE 11.4 Some Monomers and Polymers

Monomer	Polymer Formula	Name	Monomer	Polymer Formula	Name
$H_2C{=}CH_2$ Ethylene	$(CH_2)_n$	Polyethylene, polymethylene	$F_2C{=}CF_2$ Tetrafluoroethylene	$(CF_2)_n$	Poly(tetrafluoro-ethylene), Teflon
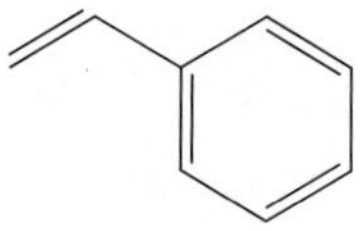 Styrene	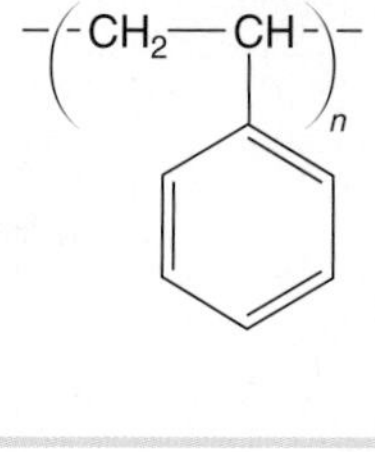 $-(CH_2-CH)_n-$	Polystyrene	$ClCH{=}CH_2$ Vinyl chloride	$-(CH-CH_2)_n-$ Cl	Poly(vinyl chloride) (PVC)
			$CH_3OOCCH{=}CH_2$ Methyl acrylate	$-(CH-CH_2)_n-$ COOCH$_3$	Polyacrylate

PROBLEM 11.20 Styrene (vinylbenzene) is an example of a molecule in which the addition of a radical is especially favorable. Explain why.

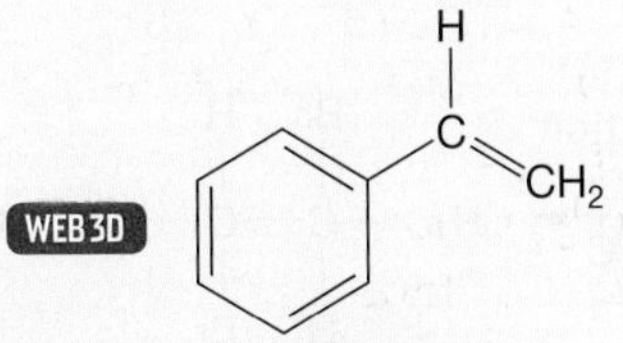

Styrene
(vinylbenzene)

WEB 3D

PROBLEM 11.21 Write a mechanism for the formation of polystyrene, the product of the polymerization of styrene (vinylbenzene).

11.6   Radical-Initiated Addition of HBr to Alkynes

Just as with alkenes, hydrogen bromide adds to alkynes in the anti-Markovnikov sense when free radicals initiate the reaction. The light-initiated reaction of H—Br with propyne gives *cis*- and *trans*-1-bromopropene, *not* 2-bromopropene (Fig. 11.34).

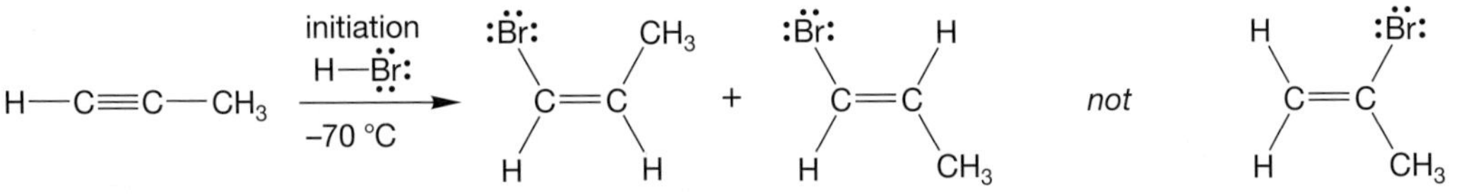

cis- and *trans*-**1-Bromopropene**   **2-Bromopropene**

FIGURE 11.34 The radical-induced addition of HBr to propyne.

The free-radical chain reaction for addition to alkynes is quite analogous to that operating in additions to alkenes. The crucial step is the formation of the more stable vinyl radical rather than the less stable vinyl radical (Fig. 11.35).

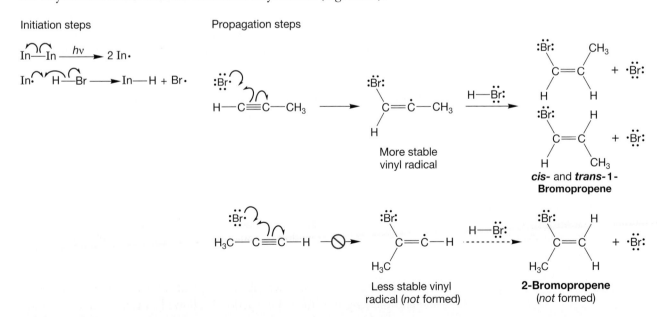

FIGURE 11.35 Addition of a bromine atom to 1-propyne gives the more stable vinyl radical, which then abstracts the hydrogen atom from hydrogen bromide to give the anti-Markovnikov product.

In principle, a second radical addition of hydrogen bromide could form either a geminal or vicinal dihalide. In contrast to ionic reactions, it is the vicinal dibromide that is preferred (Fig. 11.36).

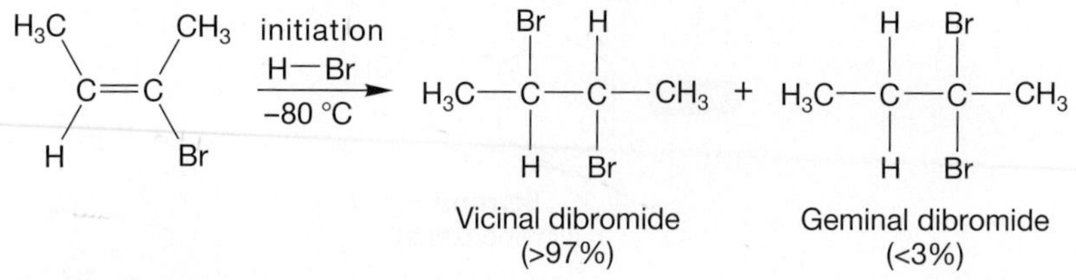

FIGURE 11.36 A second radical addition of hydrogen bromide to a vinyl bromide generates the vicinal dibromo compound as the major product, *not* the geminal dibromo compound. Problem 11.22 asks you to explain why.

Vicinal dibromide
(>97%)

Geminal dibromide
(<3%)

PROBLEM 11.22 Explain in resonance and molecular orbital terms why the vicinal dihalide is formed in Figure 11.36. Be careful! This question is more subtle than it looks at first. Consider the stabilization of an adjacent radical (a half-filled orbital) by an adjacent bromine atom. Why is this a stabilizing interaction? *Hint*: Recall the molecule He_2^+ from p. 41, and see p. 479 in this chapter.

PROBLEM 11.23 Now predict whether R· will abstract hydrogen from the methylene or methyl position of diethyl ether ($CH_3CH_2OCH_2CH_3$). Explain your answer.

11.7 Photohalogenation

So far, we have seen a variety of reactions of radicals with alkenes. Alkanes are much less reactive than alkenes, and we have not seen ways to induce these notoriously unreactive species into reaction in a specific way. We have seen pyrolysis, the thermal cracking of saturated hydrocarbons (p. 470), but this method is anything but useful in making specific molecules. Here we find a way to use alkanes in synthesis. There still will be little specificity here, and these reactions will not find a prominent place in your catalog of synthetically important processes. Yet they are occasionally useful, and provide a nice framework on which to base a discussion of selectivity.

11.7a Halogenation of Methane

A mixture of chlorine and methane is stable indefinitely at room temperature in the dark, but reacts vigorously to give a mixture of methyl chloride, methylene chloride, chloroform (trichloromethane), and carbon tetrachloride when the mixture is either irradiated or heated (Fig. 11.37).

FIGURE 11.37 Photolysis or pyrolysis of chlorine in methane gives a mixture of chlorinated methanes.

$$CH_4 + Cl_2 \xrightarrow[\text{or } \Delta]{h\nu} CH_3Cl + CH_2Cl_2 + CHCl_3 + CCl_4$$

Methyl chloride   **Methylene chloride**   **Chloroform**   **Carbon tetrachloride**

The composition of the product mixture depends on the amount of chlorine available and the length of time the reaction is allowed to run. If we use a sufficient amount of chlorine and allow the reaction to run long enough, all the methane can be converted into carbon tetrachloride. However, measuring the product composition as a function of time shows that the initial product of the reaction is methyl chloride. Methylene chloride and chloroform are formed on the way to carbon tetrachloride (Fig. 11.38).

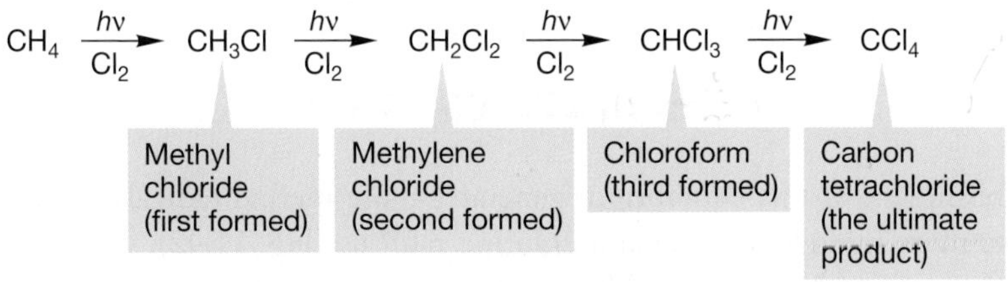

$$CH_4 \xrightarrow[\text{Cl}_2]{h\nu} CH_3Cl \xrightarrow[\text{Cl}_2]{h\nu} CH_2Cl_2 \xrightarrow[\text{Cl}_2]{h\nu} CHCl_3 \xrightarrow[\text{Cl}_2]{h\nu} CCl_4$$

| Methyl chloride (first formed) | Methylene chloride (second formed) | Chloroform (third formed) | Carbon tetrachloride (the ultimate product) |

FIGURE 11.38 The chlorinated products are formed by sequential chlorinations of methane.

Let's first examine the formation of the initial product, methyl chloride. The reaction is exothermic by about 23 kcal/mol (Fig. 11.39).

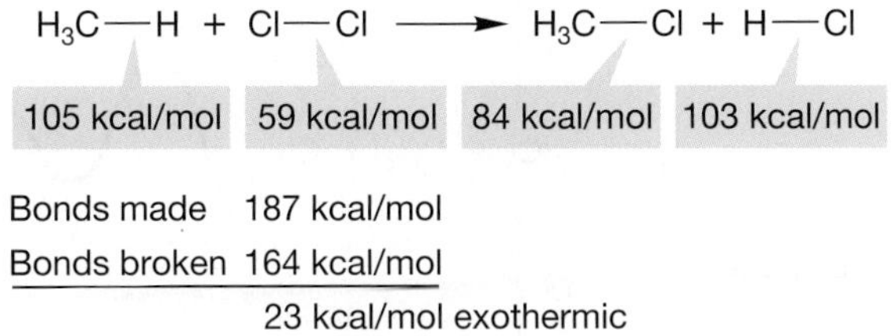

$$H_3C\text{—}H + Cl\text{—}Cl \longrightarrow H_3C\text{—}Cl + H\text{—}Cl$$

105 kcal/mol   59 kcal/mol   84 kcal/mol   103 kcal/mol

Bonds made   187 kcal/mol
Bonds broken 164 kcal/mol
_____
              23 kcal/mol exothermic

FIGURE 11.39 The formation of methyl chloride is exothermic.

The first step is the thermal or photochemical breaking of the chlorine–chlorine bond. This initiation reaction is followed by the first propagation step (Fig. 11.40), the abstraction of a hydrogen atom from methane by a chlorine atom to produce a methyl radical and hydrogen chloride. In the second propagation step, the methyl radical abstracts a chlorine atom from a chlorine molecule to give methyl chloride and another chlorine atom that can carry the chain reaction forward. There are many possible termination reactions; Figure 11.40, which shows the overall mechanism, includes only one.

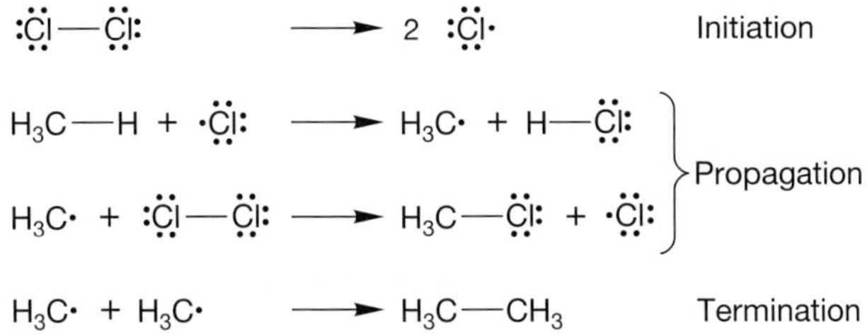

:Cl—Cl: $\longrightarrow$ 2 :Cl·   Initiation

$H_3C\text{—}H + \cdot Cl:$ $\longrightarrow$ $H_3C\cdot + H\text{—}Cl:$ ⎫
                                                                      ⎬ Propagation
$H_3C\cdot + :Cl\text{—}Cl:$ $\longrightarrow$ $H_3C\text{—}Cl: + \cdot Cl:$ ⎭

$H_3C\cdot + H_3C\cdot$ $\longrightarrow$ $H_3C\text{—}CH_3$ Termination

FIGURE 11.40 The mechanism of methyl chloride formation.

Now let's look at the thermochemistry of the propagation steps. The second propagation step is exothermic by 25 kcal/mol and thus is very favorable (Fig. 11.41).

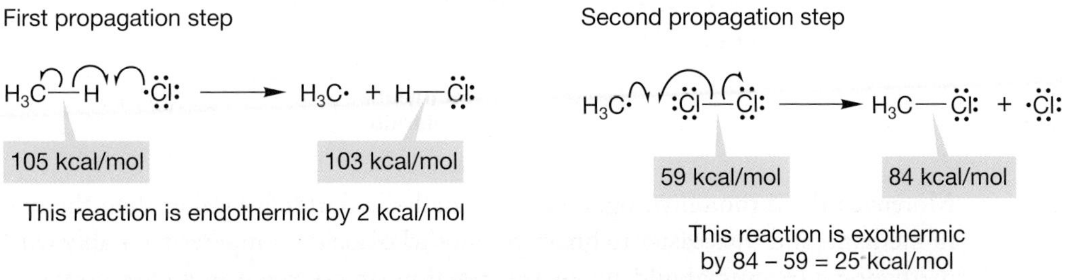

First propagation step

$H_3C\text{—}H \quad \cdot Cl:$ $\longrightarrow$ $H_3C\cdot + H\text{—}Cl:$

105 kcal/mol   103 kcal/mol

This reaction is endothermic by 2 kcal/mol

Second propagation step

$H_3C\cdot \quad :Cl\text{—}Cl:$ $\longrightarrow$ $H_3C\text{—}Cl: + \cdot Cl:$

59 kcal/mol   84 kcal/mol

This reaction is exothermic by 84 − 59 = 25 kcal/mol

FIGURE 11.41 The thermochemistry of the two propagation steps for methyl chloride formation.

The first propagation step is endothermic by about 2 kcal/mol. Will this endothermic step be enough to stop the reaction? No. Although an energy difference of 2 kcal/mol produces an unfavorable equilibrium mixture, the second highly exothermic step results in the overall reaction being favored. The few methyl radicals formed react rapidly to give methyl chloride, thus disrupting the slightly unfavorable equilibrium

$$Cl\cdot + CH_4 \underset{\longleftarrow}{\overset{\longrightarrow}{}} Cl-H + \cdot CH_3$$

Reestablishment of this equilibrium generates more methyl radicals, which go on to make more methyl chloride and chlorine radicals (Fig. 11.42), which continue the chain reaction. Note that a chlorine radical is just a chlorine atom.

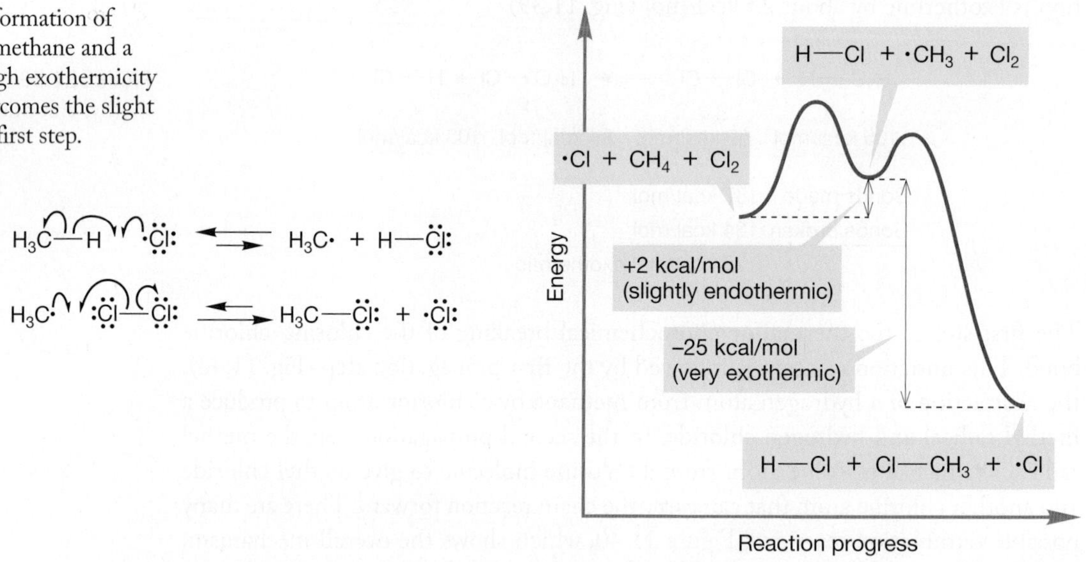

FIGURE 11.42 In the formation of methyl chloride from methane and a chlorine radical, the high exothermicity of the second step overcomes the slight endothermicity of the first step.

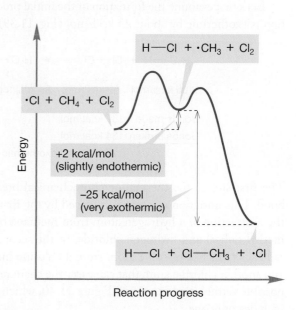

As methyl chloride builds up in the reaction chamber, it, too, begins to react with chlorine atoms in another chain reaction (Fig. 11.43).

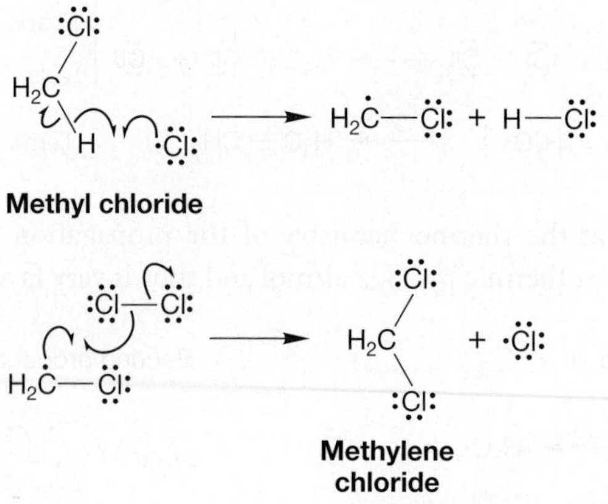

Methyl chloride

Methylene chloride

FIGURE 11.43 The formation of methylene chloride (H_2CCl_2) from methyl chloride and a chlorine atom.

Moreover, the carbon–hydrogen bond in methyl chloride is weaker than the one in methane, and thus easier to break. So methyl chloride competes favorably with methane and cannot build up in the reaction unless there is a vast excess of methane.

PROBLEM 11.24 Explain why the carbon–hydrogen bond in methyl chloride is easier to break than the carbon–hydrogen bond in methane itself.

Methylene chloride is more reactive than either methyl chloride or methane, and so it reacts to form chloroform as it builds up. Chloroform is more reactive still, and rapidly produces carbon tetrachloride through abstraction of its single hydrogen (Fig. 11.44).

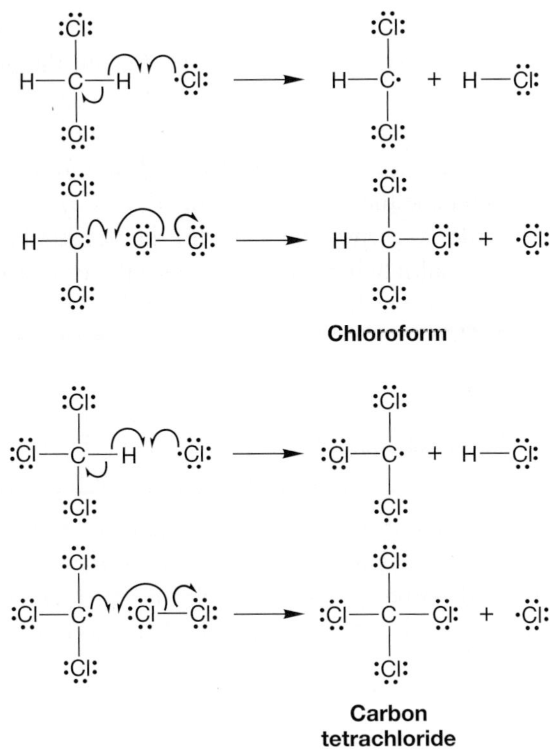

FIGURE 11.44 The formation of chloroform from methylene chloride and of carbon tetrachloride from chloroform.

Methane can also be photohalogenated with bromine, but the overall reaction is much less exothermic than is chlorination. Moreover, the first propagation step is endothermic by the substantial amount of 17 kcal/mol (Fig. 11.45). As we will see, this strongly endothermic step will have important consequences for the selectivity of the reaction when a choice of carbon–hydrogen bonds exists.

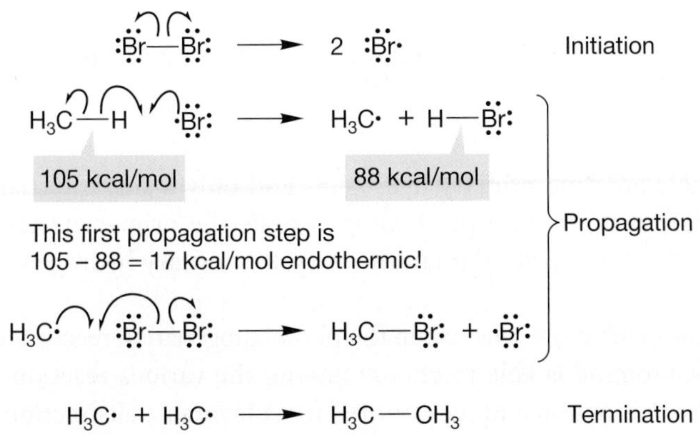

FIGURE 11.45 The formation of methyl bromide from bromine and methane.

11.7b Halogenation of Other Alkanes   If we irradiate a mixture of chlorine and a hydrocarbon more complicated than methane or ethane, there is usually more than one possible initial product. For example, butane can give either 1-chlorobutane or 2-chlorobutane. Figure 11.46 shows that the hydrogens on the secondary

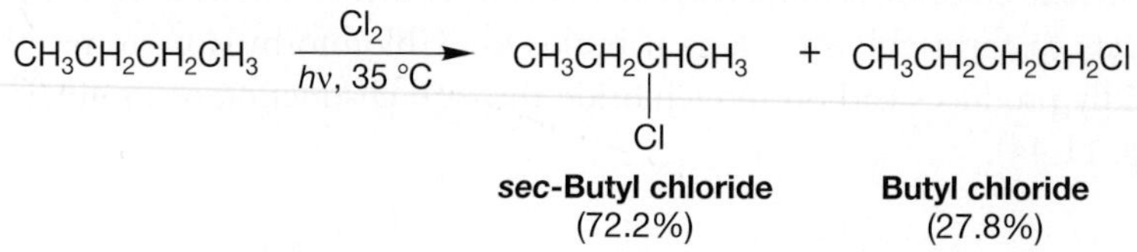

FIGURE 11.46 On a per hydrogen basis, abstraction of a secondary hydrogen is favored over abstraction of a primary hydrogen by a factor of $(72.2/27.8) \times 6/4$.

$$CH_3CH_2CH_2CH_3 \xrightarrow[hv,\ 35\ °C]{Cl_2} CH_3CH_2CHCH_3 \quad + \quad CH_3CH_2CH_2CH_2Cl$$

$$\underset{\underset{(72.2\%)}{\textbf{\textit{sec}-Butyl chloride}}}{\overset{|}{Cl}} \qquad \underset{(27.8\%)}{\textbf{Butyl chloride}}$$

carbons of butane are more reactive than the hydrogens on the primary carbons. Abstraction by a chlorine atom occurs more easily at the secondary position than at the primary position. The percentages of products shown in Figure 11.46 indicate that the preference is 72.2:27.8, or a factor of 2.6. However, this simple analysis ignores the fact that in butane there are more primary hydrogens (6) than secondary hydrogens (4). In order to correct this value for the statistical advantage of the primary bonds, we must multiply it by 6/4 to produce the true factor of 3.9. On a per hydrogen basis, secondary hydrogens are abstracted more easily than primary hydrogens by a factor of 3.9.

PROBLEM 11.25 Write a reaction mechanism for the photochlorination of ethane.

PROBLEM 11.26 Analyze the thermochemistry of the two propagation steps for the photochlorination of ethane. The bond dissociation energies of the chlorine–chlorine and carbon–chlorine bonds of ethyl chloride and the carbon–hydrogen bond of ethane are 59, 84.8, and 101.1 kcal/mol, respectively.

Selectivity is increased when the possibilities include the formation of a tertiary radical. Photochlorination of 2-methylpropane, for example, gives a 64:36 mixture of 1-chloro-2-methylpropane (isobutyl chloride) and 2-chloro-2-methylpropane (*tert*-butyl chloride). The primary:tertiary ratio is 64:36 = 1.78 (Fig. 11.47).

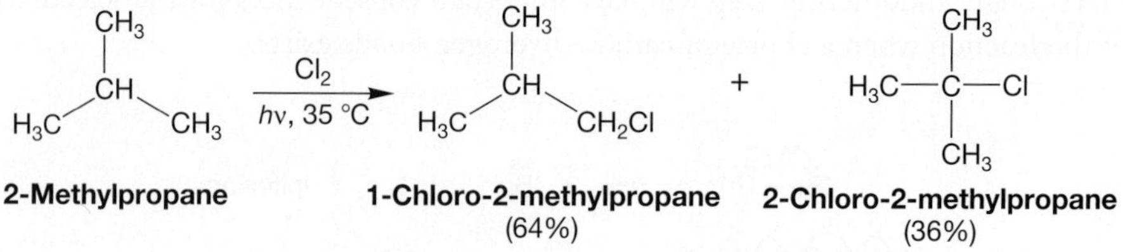

FIGURE 11.47 For a chlorine radical, abstraction of a tertiary hydrogen is favored over abstraction of a primary hydrogen by a factor of about 5 on a per hydrogen basis.

2-Methylpropane $\xrightarrow[hv,\ 35\ °C]{Cl_2}$ **1-Chloro-2-methylpropane** (64%) + **2-Chloro-2-methylpropane** (36%)

However, there are nine primary hydrogens and only a single tertiary hydrogen in 2-methylpropane, so on a per hydrogen basis the primary:tertiary reactivity ratio is $1.78 \times 1/9 = 0.198$. Put another way, the tertiary hydrogen is favored by $1:0.198 = 5.1$.

Bromine is much more selective in the photohalogenation reaction than is chlorine. That is, bromine is able to choose among the various reaction possibilities with more discrimination and pick out the most favorable abstraction pathway by a wider margin than chlorine. For example, a bromine atom prefers secondary to

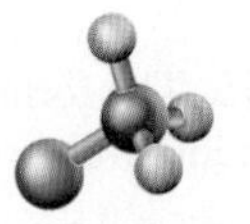

Alkane halogenation

primary hydrogens by a factor of 82, and tertiary hydrogens to primary hydrogens by a factor of about 1600 (Fig. 11.48; Table 11.5).

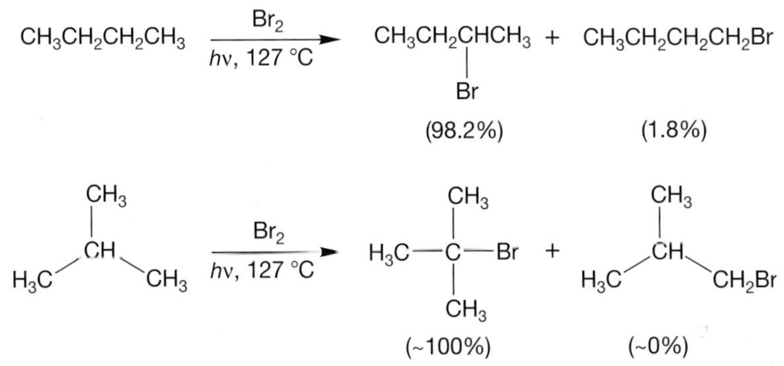

$$CH_3CH_2CH_2CH_3 \xrightarrow[h\nu,\ 127\ °C]{Br_2} \underset{\underset{(98.2\%)}{\underset{|}{Br}}}{CH_3CH_2CHCH_3} + \underset{(1.8\%)}{CH_3CH_2CH_2CH_2Br}$$

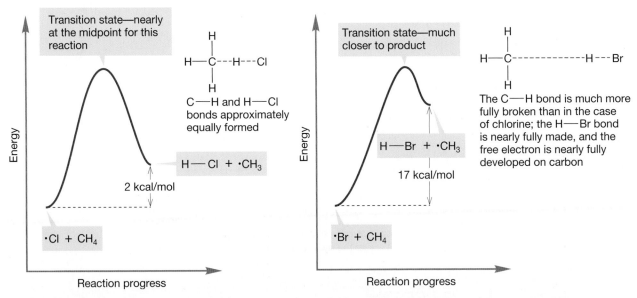

FIGURE 11.48 Photobromination is far more selective than photochlorination.

TABLE 11.5 Selectivities of Halogen Atoms in Carbon–Hydrogen Bond Abstraction

Halogen	Primary	Secondary	Tertiary
F	1	1.2	1.4
Cl	1	3.9	5.1
Br	1	82	1600

PROBLEM 11.27 Use the bond dissociation energies of Table 8.2 (p. 337) to estimate the thermochemistry of the propagation steps for photochemical fluorination and iodination of methane.

Neither fluorine nor iodine undergoes a useful photohalogenation reaction. The reaction of fluorine with carbon–hydrogen bonds is overwhelmingly exothermic and can even occur explosively. Almost no selectivity is possible (Table 11.5). In direct contrast to the reaction of fluorine atoms, reaction of iodine atoms with hydrocarbons is highly endothermic. It is so slow as to be of no synthetic utility.

What is the reason for bromine being more selective than chlorine? Recall that the abstraction reaction is close to thermoneutral for chlorine (Fig. 11.49), but very

FIGURE 11.49 For the nearly thermoneutral abstraction of hydrogen by a chlorine atom, the transition state is almost halfway between starting material and products. For the highly endothermic abstraction of hydrogen by the bromine atom, the transition state will be quite product-like.

endothermic for bromine (p. 493, Fig. 11.45).[3] The Hammond postulate (the transition state for an endothermic process resembles the product) tells us that the transition state for the abstraction of hydrogen by chlorine is nearly halfway between starting material and product (the radical intermediate) (Fig. 11.50a), but the transition state is very product-like for bromine (Fig. 11.50b). Because the reaction is more endothermic, the transition state for abstraction of hydrogen by bromine is product-like. There is a strongly developed hydrogen–bromine bond and a large degree of radical character on carbon in the transition state (Fig. 11.50b).

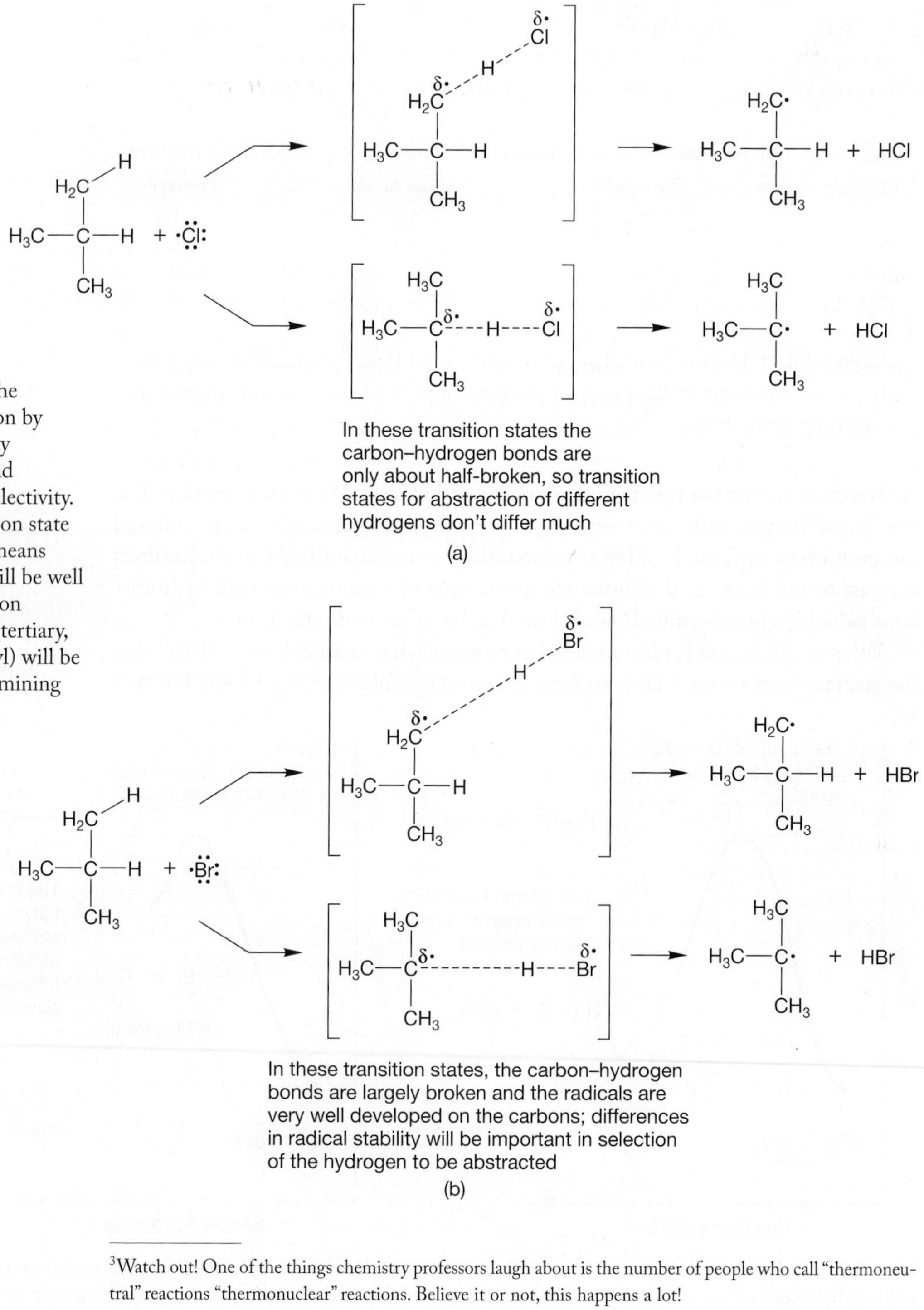

In these transition states the carbon–hydrogen bonds are only about half-broken, so transition states for abstraction of different hydrogens don't differ much

(a)

FIGURE 11.50 (a) Because the transition state for abstraction by chlorine lies roughly half way between starting material and product there can be little selectivity. (b) The product-like transition state for abstraction by bromine means that the radical on carbon will be well developed and the substitution pattern of the carbon atom (tertiary, secondary, primary, or methyl) will be relatively important in determining the ultimate major product.

In these transition states, the carbon–hydrogen bonds are largely broken and the radicals are very well developed on the carbons; differences in radical stability will be important in selection of the hydrogen to be abstracted

(b)

----

[3] Watch out! One of the things chemistry professors laugh about is the number of people who call "thermoneutral" reactions "thermonuclear" reactions. Believe it or not, this happens a lot!

The transition states for abstraction of hydrogen from an alkane by chlorine and bromine radicals are very different from each other. In the transition states for abstractions by bromine, the radical on carbon is far more developed than it is in the abstractions by chlorine, which means that differences in the stabilities of different radicals will be far more important in determining the major product of the bromine abstractions than in the chlorine abstractions (Fig. 11.50).

Summary

When questions of selectivity arise, it is always important to consider what the transition state looks like. Selective radicals will have more product-like transition states in any abstraction reaction, and therefore the factors influencing the stability of those products will be important in the transition state as well.

11.8 Allylic Halogenation: Synthetically Useful Reactions

One of the goals of synthetic organic chemists is to increase selectivity—to be able to do chemical transformations at specific sites in a molecule without affecting the rest of the molecule. Selectivity is useful—even necessary—when one is dealing with a complicated molecule. We must have ways of inducing change at one place in a molecule without affecting others if we are to be able to manipulate molecules rationally. It's quite clear that the photochlorination reaction of alkanes is not a good way of doing this because this reaction is far too unselective. Chlorination takes place all too easily at the various positions in the hydrocarbon.

Photobromination more closely approaches the goal of perfect selectivity because the bromine atom abstracts tertiary hydrogens much more readily than secondary or primary hydrogens. But even this reaction is not without its problems. It is a slow process and the side products of di- and polybromination are formed.

There are some instances where the chlorination or bromination reaction can be used to good effect, however. Alkenes that have *allylic* hydrogens can sometimes be halogenated specifically at an allylic position in a process called **allylic halogenation**, another free radical chain reaction. For example, when low concentrations of bromine are photolyzed in the presence of the complicated cyclohexene shown in Figure 11.51, the product is exclusively brominated in one allylic position.

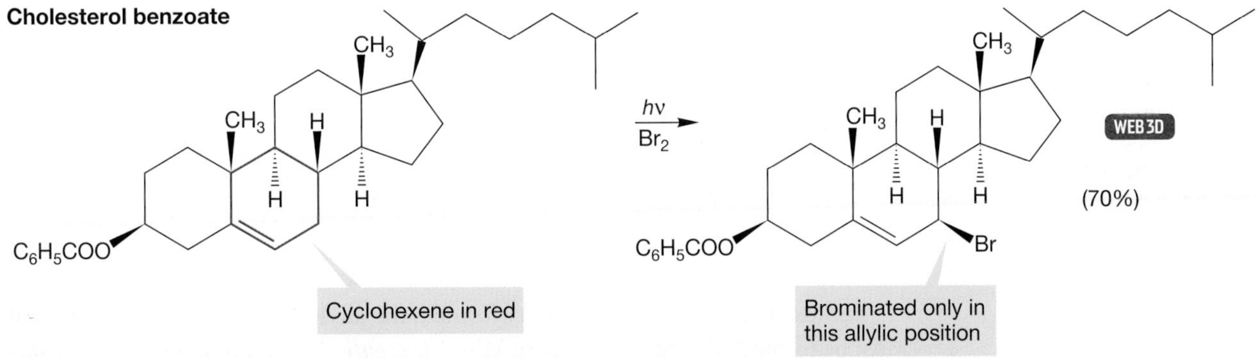

FIGURE 11.51 When low concentrations of bromine are photolyzed in the presence of the substituted cyclohexene, the product is exclusively the allylic bromide shown.

The mechanism is a typical radical chain reaction in which a low concentration of bromine atoms is first produced by breaking of the weak bromine–bromine bond (initiation step). Abstraction of hydrogen from the cyclohexene is the first propagation step, but there are three possible kinds of hydrogen that could be removed: H$_v$ (vinylic), H$_a$ (allylic), and H$_m$ (methylene) (Fig. 11.52).

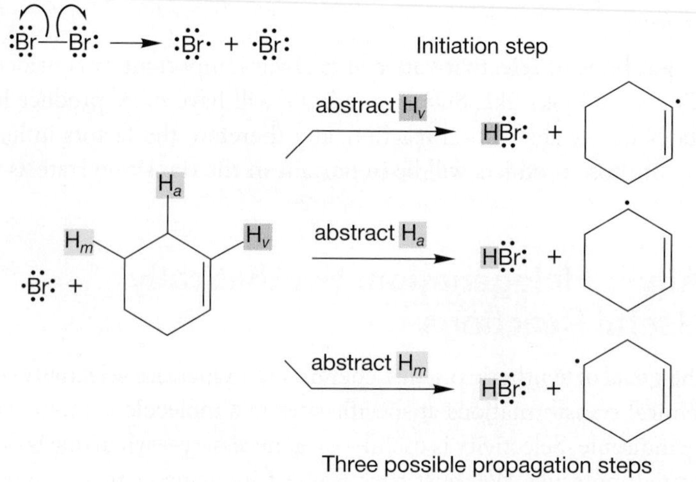

FIGURE 11.52 In principle, a bromine atom could abstract a vinylic (H$_v$), allylic (H$_a$) or methylene (H$_m$) hydrogen.

Three possible propagation steps

As shown in Figure 11.53, each of these radicals could lead to a bromocyclohexene through reaction with bromine in the second propagation step. The propagation is then carried forward by the new bromine atom formed in this step. The formation of only the allylic bromide tells us that only the allylic radical is produced originally.

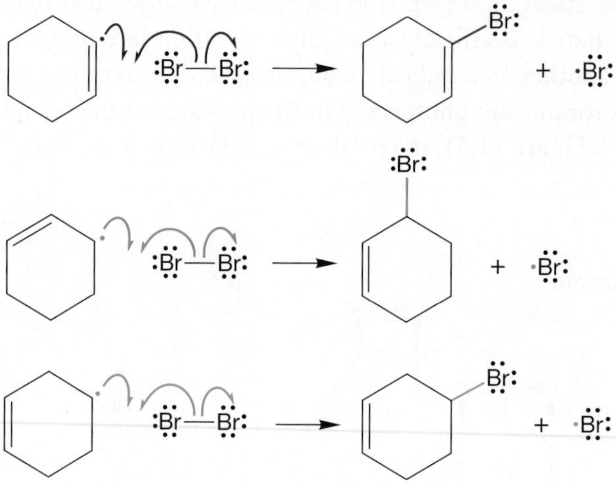

FIGURE 11.53 In a propagation step, any of the carbon radicals could react with bromine to give a bromocyclohexene and a chain-carrying bromine atom.

Because there is only one product formed, the allylic bromide, the allylic radical must be formed with high selectivity. Why? *It is only this radical that is resonance stabilized.* There are two equivalent resonance forms for the allylic radical and this delocalization makes it far more stable than the other two possibilities (Table 11.1).

Further reaction with bromine leads to the product of allylic bromination and a new chain-carrying bromine atom (Fig. 11.54).

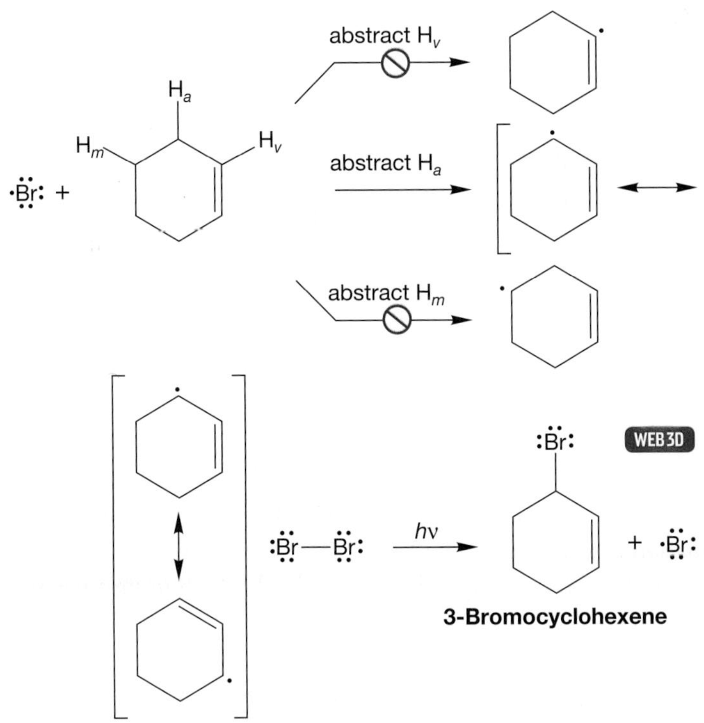

FIGURE 11.54 Only the allylic bromide is formed.

One might hope that the photohalogenation reaction would be general, that photochlorination would be as useful as photobromination. Alas it is not, and Table 11.5 shows why. A chlorine atom is far less selective than a bromine atom. Utility in synthesis comes from selectivity—from doing one reaction, not several. Photobromination is useful because the bromine atom picks and chooses among the various possible carbon–hydrogen bonds with fairly high discrimination. A chlorine atom does not. For example, the allylic hydrogens in cyclohexene (H_a) are only slightly more reactive toward a chlorine atom than the H_m methylene hydrogens (Fig. 11.55).

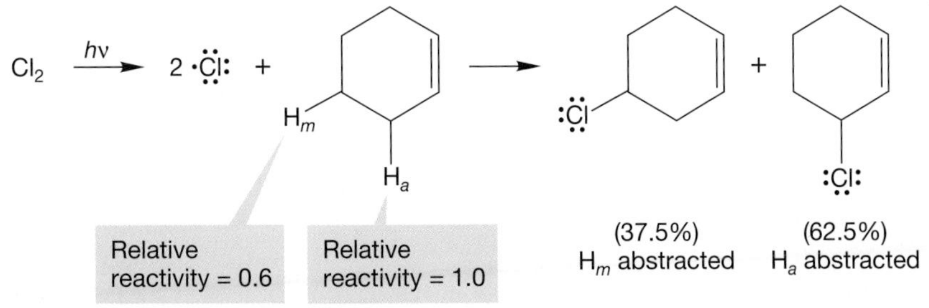

FIGURE 11.55 Chlorine is much less selective than bromine. The allylic chloride is not the only product formed in this reaction.

In photohalogenation reactions, keeping the concentration of X_2 low is essential to success. If there is too much halogen, polar addition to give vicinal dihalides competes with formation of the allylic halide, and the desired selectivity is lost.

Methods have been developed to ensure that this low concentration is maintained and to lower the temperature necessary for the initiation step. A common and very effective method of allylic bromination involves the molecule *N*-bromosuccinimide, generally known as NBS, in the solvent carbon tetrachloride (CCl_4). The overall reaction is shown in Figure 11.56.

FIGURE 11.56 Allylic bromination with NBS.

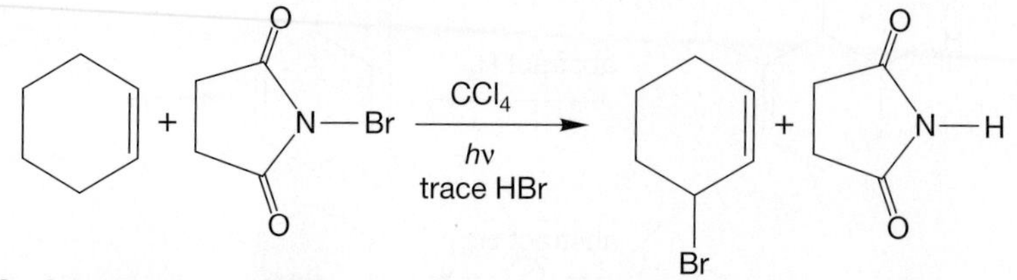

Cyclohexene   NBS   3-Bromocyclohexene   NHS

WEB 3D

The initiation steps in this reaction are formation of succinimide (NHS), followed by formation of bromine atoms (Fig. 11.57). The bromine radical abstracts the allylic hydrogen to produce the cyclohexenyl radical, which in turn reacts with

FIGURE 11.57 The first two steps in bromination of cyclohexene using NBS.

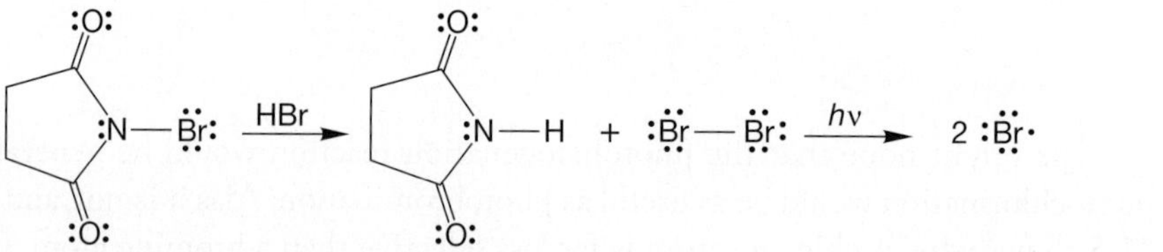

the Br_2 to give the allylic bromide and a new, chain-carrying bromine atom. The NBS is only slightly soluble in carbon tetrachloride, thus ensuring the low concentration of bromine necessary for the success of this reaction (Fig. 11.58).

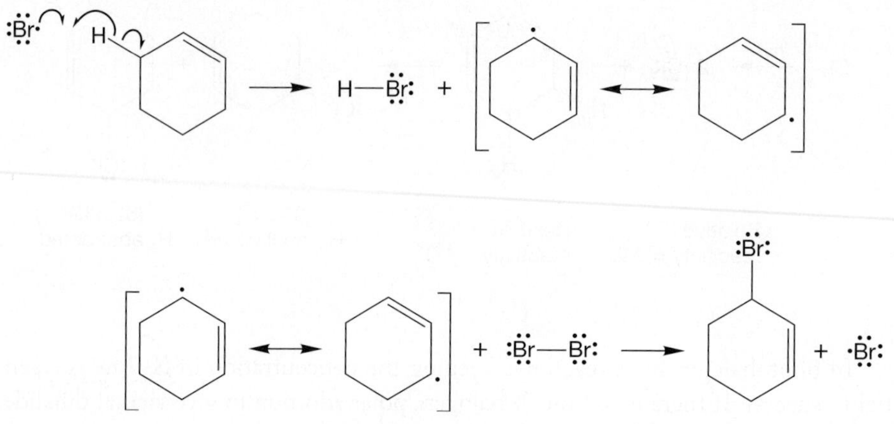

FIGURE 11.58 The bromine radical abstracts a hydrogen to form the cyclohexenyl radical. Independently, a molecule of bromine is formed from the NBS. The Br_2 then reacts with the cyclohexenyl radical to give the product and a new bromine atom.

PROBLEM 11.28 Suggest a mechanism for the formation of bromine from NBS and hydrogen bromide. *Hint*: The oxygen of a carbon–oxygen double bond is a base.

PROBLEM 11.29 Selectivity is not possible in a photochemical reaction of Br$_2$ with a disubstituted alkene such as *trans*-2-pentene. Explain.

PROBLEM SOLVING

NBS is a reagent that comes with its own pull-down menu that says, "watch out for allylic bromination." When you see NBS in a problem, allylic bromination is very likely to be involved in the answer.

11.9 Special Topic: Rearrangements (and Nonrearrangements) of Radicals

We saw in Chapters 9 and 10 that polar additions to alkenes are bedeviled by hydride or alkyl shifts when there is a possibility of forming a more stable cation from a less stable one. A typical example, the polar addition of hydrogen chloride to 3,3-dimethyl-1-butene, is shown again in Figure 11.59. Yet free radical additions

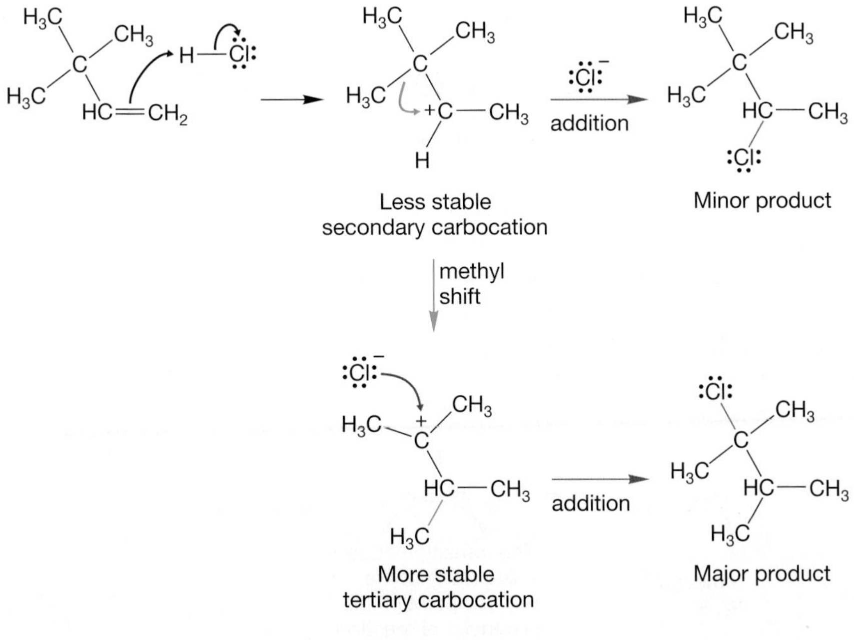

Less stable
secondary carbocation

methyl
shift

Minor product

More stable
tertiary carbocation

addition

Major product

FIGURE 11.59 Rearrangements occur in the polar addition of hydrogen chloride to 3,3-dimethyl-1-butene.

are usually uncomplicated by similar rearrangements. For example, the peroxide-initiated addition of hydrogen bromide to 3,3-dimethyl-1-butene gives only the unrearranged product (Fig. 11.60).

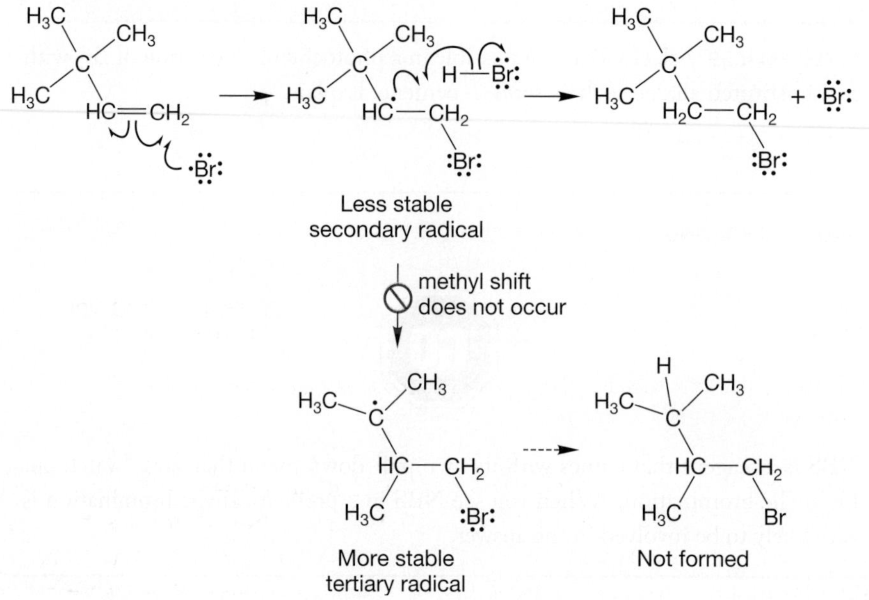

FIGURE 11.60 The related radical addition gives no rearranged product.

Less stable secondary radical

methyl shift does not occur

More stable tertiary radical

Not formed

We might have expected rearrangement through the shift of a methyl group to give the more stable tertiary radical from the less stable secondary one, yet this reaction doesn't happen. In fact, the situation is quite dramatic: 1,2-Hydride shifts (migration of H:⁻) are among the fastest known organic reactions but 1,2-shifts of a hydrogen atom or an alkyl group are unknown (Fig. 11.61).

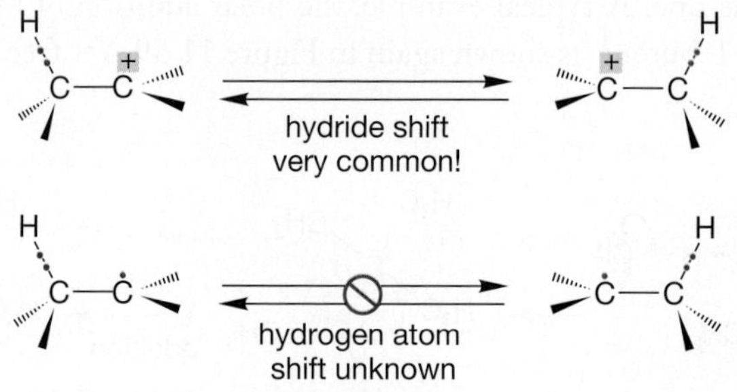

FIGURE 11.61 The simple 1,2-shift of a hydrogen atom (H·) is unknown.

hydride shift very common!

hydrogen atom shift unknown

This situation seems mysterious, but a look at the molecular orbitals involved in the transition state for the reaction shows the reason. There is more than one way to approach this problem, but we think an analysis of the transition state for the rearrangement is the easiest. In the transition state for any symmetrical 1,2-shift of hydrogen, the hydrogen atom is half-transferred from one carbon to the other (Fig. 11.62).

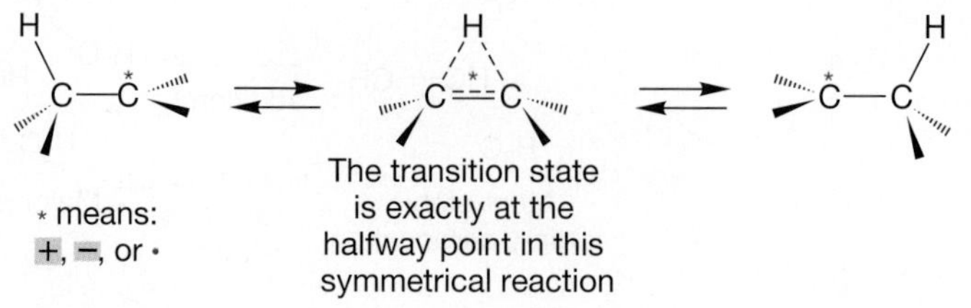

FIGURE 11.62 In the transition state for *any* symmetrical 1,2-shift, the hydrogen is halfway between the two carbons. Here the symbol * = +, −, or ·.

* means: +, −, or ·

The transition state is exactly at the halfway point in this symmetrical reaction

What do the molecular orbitals of the transition state look like? To answer this question, we need only construct the molecular orbitals for the system of a hydrogen 1*s* orbital interacting with the π and π* orbitals of an alkene. The combination of 1*s*, π, and π* will yield three new molecular orbitals **A**, **B**, and **C**. They are constructed in Figure 11.63.

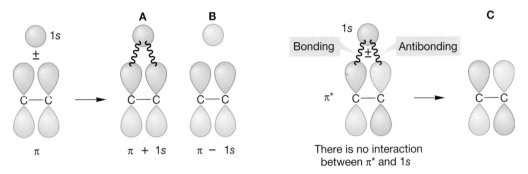

FIGURE 11.63 The three molecular orbitals for the transition state in this reaction can be easily constructed from H (1*s*), π, and π*.

The 1*s* orbital interacts with the π orbital to form two new molecular orbitals (**A** and **B**), but there is no interaction between 1*s* and π* in this geometry. Accordingly, π* becomes **C**, the third molecular orbital for the transition state. To order the energy of **A**, **B**, **C**, we need only count nodes. Molecular orbital **A** has no new nodes, but **B** and **C** each have one additional node (Fig. 11.64).

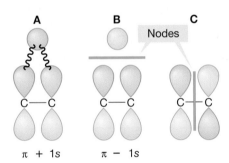

FIGURE 11.64 The new molecular orbitals (**A–C**) can be ordered in energy by counting nodes.

The ordering of the new molecular orbitals is shown in Figure 11.65. Now, how many electrons do we need to put into the orbital system? For the cationic reaction of Figure 11.61, there are only two and they will go nicely into the new stabilized

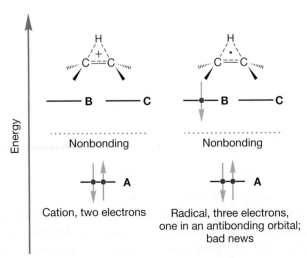

FIGURE 11.65 The transition state will have a single bonding molecular orbital (**A**) and two, equi-energetic antibonding molecular orbitals (**B** and **C**). For the cationic rearrangement, the two electrons involved nicely fill the bonding molecular orbital (**A**). In the radical case, there is another electron and it must occupy an antibonding molecular orbital (**B** or **C**) in the transition state. Occupancy of this antibonding orbital is unfavorable.

bonding molecular orbital **A**. However, for the radical reaction of Figure 11.61 there is another electron, and it must be placed in one of the antibonding molecular orbitals, **B** or **C**. This high-energy electron is evidently destabilizing enough to completely shut off the reaction (Fig. 11.65).

PROBLEM 11.30 Analyze the shift of hydrogen in the related anionic system.

WORKED PROBLEM 11.31 1,2-Shifts of vinyl groups are known in radicals. Describe how this reaction might differ from the 1,2-shift of a hydrogen atom or a methyl radical, both of which are unknown.

ANSWER A vinyl group, or any other group with p orbitals, is not restricted to using the σ bond in the migration. In this case, the radical on C(2) could add to the vinyl group to give the symmetrical intermediate **A**. Opening of intermediate **A** can occur in two directions (a and a′) either to regenerate the starting material (a′) or complete the vinyl migration (a).

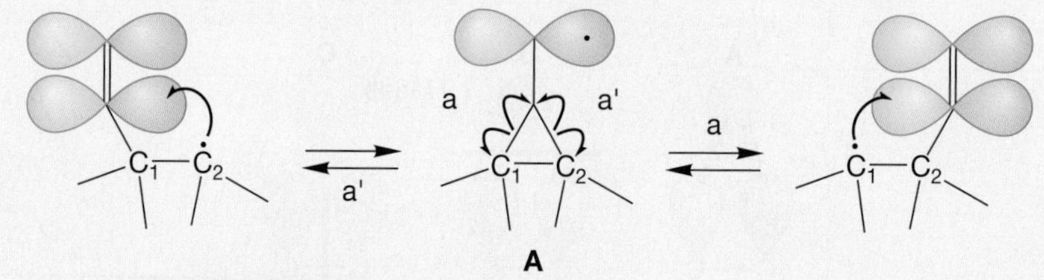

WORKED PROBLEM 11.32 We have already seen a reaction much like the opening of the cyclic intermediate shown in the answer to Problem 11.31. What is the name of this process?

ANSWER It is just a β cleavage (p. 472).

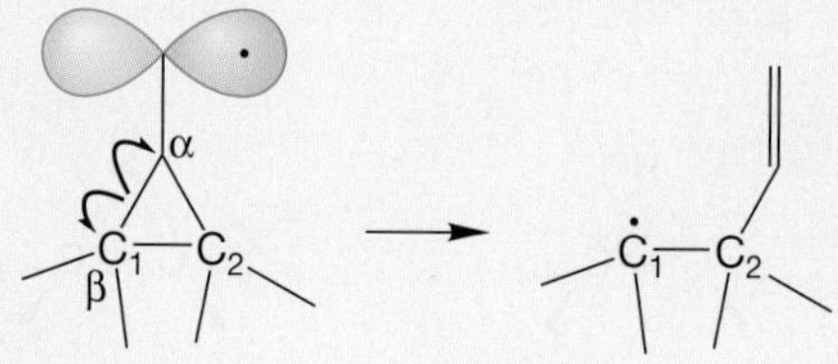

11.10 Special Topic: Radicals in Our Bodies; Do Free Radicals Age Us?

Free radicals are strongly implicated in the aging process. Cell membranes, which are made up of long-chain hydrocarbon carboxylic acids called fatty acids, are particularly vulnerable to oxidation through free radical attack. Once a cell's membrane has been oxidized, the cell is damaged. In time, the immune system comes to recognize damaged cells as "foreign" and attacks them. This leads to cell death and ages the organ that the cell is in. If radicals are involved in aging, then radical trapping agents might be effective in arresting the aging process. One such potential trapping agent is Vitamin E (Fig. 11.66):

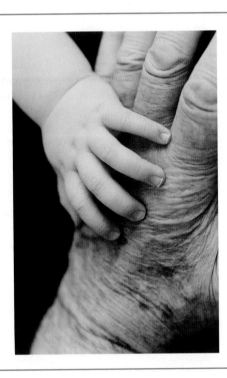

FIGURE 11.66 Vitamin E.

Vitamin E

VITAMIN E

In rats, vitamin E is essential for proper liver function as well as nerve and muscle condition. At least one observer has pointed out that because rats and humans are so alike, vitamin E might be vital for humans as well. Despite the obvious truth of the observation, there is no proven need for vitamin E in humans. Still, that lack is not sufficient to stop speculation, and it has been suggested that the ability of vitamin E to act as a free radical interceptor, an antioxidant, might make it effective in delaying aging. Vitamin E traps radicals through the hydroxyl group highlighted in red on the six-membered ring (Fig. 11.66). A radical (RO·) plucks a hydrogen atom from this hydroxyl to make ROH and destroys itself.

The long hydrocarbon chain of vitamin E also has its function. It aids in the delivery of vitamin E to the cell membrane by making the molecule soluble in fatty acids, which also contain long greasy hydrocarbon chains. *Remember*: "Like dissolves like" (p. 239).

11.11 Summary

New Concepts

One of the most important concepts of this chapter is that of selectivity. It is not really new because you have already seen many selective reactions. A simple example is the protonation of isobutene to give the more stable tertiary carbocation rather than the less stable primary carbocation (Fig. 11.23).

In this chapter, you also see selectivity when a radical abstracts one of several possible hydrogens from an alkane. The ease of hydrogen abstraction is naturally determined by the energies of the transition states leading to the products

(Figs. 11.49 and 50). A tertiary radical is more likely to be formed than a primary radical. In the transition state for radical formation, the factors that make a tertiary radical more stable than a primary radical will operate to stabilize the partially formed tertiary radical over the partially formed primary radical, as noted in Figures 11.49 and 11.50.

What is more subtle, and at least somewhat new, is the idea that the degree of selectivity in a reaction will be influenced by *how product-like* the transition state is. The more the transition state is like the product (the radical intermediate), the more important the relative stabilities of two possible products will be. Abstraction of a hydrogen by chlorine is endothermic by about 2 kcal/mol. The same abstraction reaction by bromine is endothermic by about 17 kcal/mol. The transition state for the much more endothermic abstraction reaction by bromine will be more product-like than that for abstraction by chlorine. So we expect bromine to be more selective than chlorine, because the carbon–hydrogen bond will be more broken in the transition state. The stability of the product radical will be more important in the transition state for hydrogen abstraction by a bromine than it will in the transition state for abstraction by a chlorine.

There are a few other points to keep in mind. Radical stability parallels carbocation stability. Tertiary radicals are more stable than secondary, secondary more stable than primary,

and primary more stable than methyl. The differences aren't as great for radicals as for cations. Delocalization helps stabilize radicals by spreading the odd electron over two or more atoms, so allylic and benzylic radicals are more stable than tertiary radicals.

Radical reactions are often cyclic: Radical chain reactions produce a chain-carrying radical in their last, product-making step, and the new radical can begin another series of propagation steps leading to another molecule of product. Chain reactions only stop when starting material is exhausted or a chain is terminated by bond formation by a pair of radicals.

Chains can be started by small amounts of radicals released by radical initiators. In this way, a small number of chain-initiating radicals can have an impact far out of proportion to their numbers.

Key Terms

abstraction (p. 471)

acyl group (R—C=O) (p. 476)

allylic halogenation (p. 497)

azo compounds (p. 476)

chain reaction (p. 468)

β cleavage (p. 472)

disproportionation (p. 471)

hydrocarbon cracking (p. 473)

inhibitors (p. 484)

initiation (p. 483)

propagation (p. 483)

pyrolysis (p. 470)

termination (p. 483)

thermolysis (p. 470)

Reactions, Mechanisms, and Tools

The reactions in this chapter involve neutral free radicals, in contrast to the ionic species of earlier chapters. Chain reactions are characteristic of radical chemistry and consist of three steps: (1) Initiation—a small number of radicals serves to start the reaction; (2) Propagation—radical reactions are carried out that generate a product molecule and a new, chain-carrying radical, which recycles to the beginning of the propagating steps and starts a new series of product-forming steps; and

(3) Termination—radical recombinations destroy chain-carrying species and end the chain reaction. The success of a chain reaction depends on the relative success of the chain-carrying and termination steps. Figure 11.26 gives an example of a typical chain reaction.

Important reactions of radicals include abstraction, additions to alkenes and alkynes, β cleavage, and disproportionation (Fig. 11.9), but not simple 1,2-rearrangements.

Syntheses

The new synthetic reactions of this chapter are shown below.

1. Alkanes

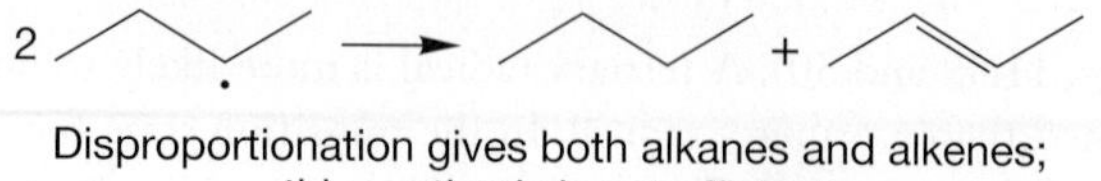

Disproportionation gives both alkanes and alkenes; this synthesis is not efficient

2. Alkenes

Disproportionation gives both alkanes and alkenes; this synthesis is not efficient

3. Alkyl Bromides

$$R{-}H + Br_2 \xrightarrow{h\nu} R{-}Br + HBr$$

Also works for Cl_2; photobromination is more selective than photochlorination

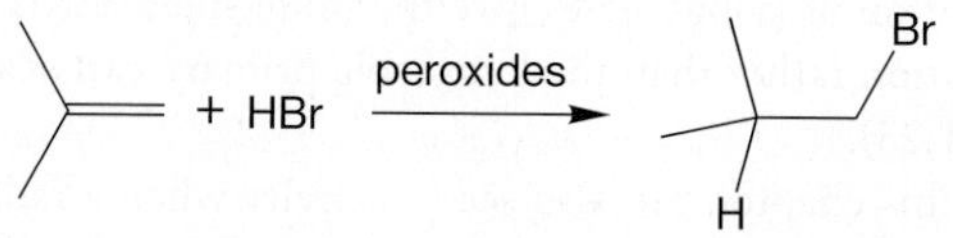

A chain reaction that works only for H—Br, not other H—X molecules; note the anti-Markovnikov regiochemistry

4. Allylic Halides

NBS
hv/CCl₄

Bromination is specifically at the allylic position

5. Radicals

$$R—R \xrightarrow{\Delta} 2\,R\cdot$$

Pyrolysis of hydrocarbons is an unselective method for making radicals; many different radicals are typically produced

$$R—N{=}N—R \xrightarrow{\Delta} R\cdot + N_2 + R\cdot$$

Pyrolysis of azo compounds

6. Vicinal Dihalides

Vinyl halides (see 7) can add a second molecule of HBr to give vicinal (not geminal) dihalides

7. Vinyl Bromides

$$HC{\equiv}C—R \xrightarrow[\text{peroxides}]{\text{HBr}} BrHC{=}CHR$$

Radical-induced addition to alkynes gives anti-Markovnikov addition; cis and trans products are formed

Common Errors

A very common mistake is to confuse the propagation step in a chain reaction with the termination step. In the photobromination of an alkane such as methane, the product is formed in a chain reaction in which bromine and methyl radicals carry the chain (Fig. 11.45). It is all too tempting to imagine that the product, methyl bromide, is formed by the combination of one methyl radical and one bromine radical! In fact, although this step does produce a molecule of product, it destroys two chain-carrying radicals. This reaction would terminate the chain process and so we know it does not contribute significantly to product formation.

11.12   Additional Problems

PROBLEM 11.33   Let's build right away on Problem 11.30 and Section 11.9. Suppose you are walking down the street, and a stranger comes up to you and asks, Which is more stable, linear or triangular H₃⁺? We maintain that you should be able to give him or her a quick answer to this seemingly bizarre question. *Obvious hint*: Look at Section 11.9.

Here's an outline of how to do this problem: First, build the molecular orbitals for linear and cyclic H₃. You did the linear molecule long ago in Chapter 1 (Problem 1.60, p. 48). Next, put the appropriate number of electrons in. Finally, look at the orbital containing those electrons in each species. Which is lower in energy?

PROBLEM 11.34   Show the reactions you would use to synthesize propene if your only source of carbon is propane.

PROBLEM 11.35   Show how you would make 2-butene if your only source of carbon is butane.

PROBLEM 11.36   Show how you would make *trans*-1,2-dibromocyclohexane starting with cyclohexane.

PROBLEM 11.37   Predict the major product(s) for the radical hydrohalogenation reaction (HBr, peroxides, heat) with each of the following alkenes:

(a) 1-pentene
(b) 2-pentene
(c) cyclopentene
(d) *cis*-3-hexene
(e) *trans*-3-hexene

PROBLEM 11.38   Draw the monochlorinated products formed by radical chlorination of the following compounds:

(a) cyclobutane
(b) cyclopentane
(c) methylcyclopentane
(d) pentane

PROBLEM 11.39   Draw four dibromo products that could be obtained by radical bromination of propane. Show the reaction pathway for the compound you think would be the major product.

PROBLEM 11.40   Which of the products in the previous problem are chiral and which are achiral?

PROBLEM 11.41 Give the major product(s) expected for each of the following reactions. Pay attention to regiochemistry and stereochemistry where appropriate. Mechanisms are not required, although they may be useful in finding the answers.

(a)

$$\xrightarrow[\text{ROOR}]{\text{HBr}}$$

(b)

$$\xrightarrow{\text{HBr}}$$

(c)

$$\xrightarrow[\text{ROOR}]{\text{HCl}}$$

(d)

$$+ \text{CCl}_4 \xrightarrow{h\nu}$$

(e)

$$\xrightarrow[h\nu]{\text{NBS, CCl}_4}$$

(f)

$-\text{C}\equiv\text{CH}$

$$\xrightarrow{\text{HBr}}$$

(g)

$-\text{C}\equiv\text{CH}$

$$\xrightarrow[\text{ROOR}]{\text{HBr}}$$

PROBLEM 11.42 Predict the products of hydrogen bromide addition to 1-butyne. Consider both ionic (a) and radical-initiated (b) reactions. Rationalize your predictions with arrow formalism descriptions. Remember that a second molecule of hydrogen bromide can add to the initial product(s) in both cases.

(a)

$$\text{CH}_3\text{CH}_2-\text{C}\equiv\text{CH} \xrightarrow{\text{HBr}}$$

(b)

$$\text{CH}_3\text{CH}_2-\text{C}\equiv\text{CH} \xrightarrow[\text{ROOR}]{\text{HBr}}$$

PROBLEM 11.43 The relative rates of recombination of methyl and isopropyl radicals are quite different. Give two possible reasons for this difference.

	Relative Rate
$2\ \text{H}_3\text{C}\cdot \longrightarrow \text{H}_3\text{C}-\text{CH}_3$	22
$2\ \begin{array}{c}\text{H}_3\text{C}\\ \dot{\text{C}}\text{H}\\ \text{H}_3\text{C}\end{array} \longrightarrow \begin{array}{c}\text{H}_3\text{C}\quad\quad\text{CH}_3\\ \text{CH}-\text{CH}\\ \text{H}_3\text{C}\quad\quad\text{CH}_3\end{array}$	1

PROBLEM 11.44 Decomposition of the following azo compounds at 300 °C occurs at quite different rates. Give two possible explanations for these differences.

R_1	R_2	R_3	Relative Rate
CH_3,	CH_3,	CH_3	13,000
CH_3,	CH_3,	H	111
CH_3,	H,	H	11
H,	H,	H	1

PROBLEM 11.45 Predict the stereochemical results of the photochlorination of the optically active compound **1**. The (*) indicates a single enantiomer, here (R). It is the tertiary carbon–hydrogen bond that is chlorinated.

No stereochemistry implied in this drawing

PROBLEM 11.46 In Problem 11.25, we examine the photochlorination of ethane to give ethyl chloride. As was the case in the photochlorination of methane (p. 490), products containing more than one chlorine are also isolated in the photochlorination of ethane. What is the structure of the major dichloroethane produced in this reaction? Rationalize your prediction with an analysis of the mechanism of the reaction.

PROBLEM 11.47 Write a detailed mechanism for the bromination of ethyl phenylacetate (**1**) with NBS.

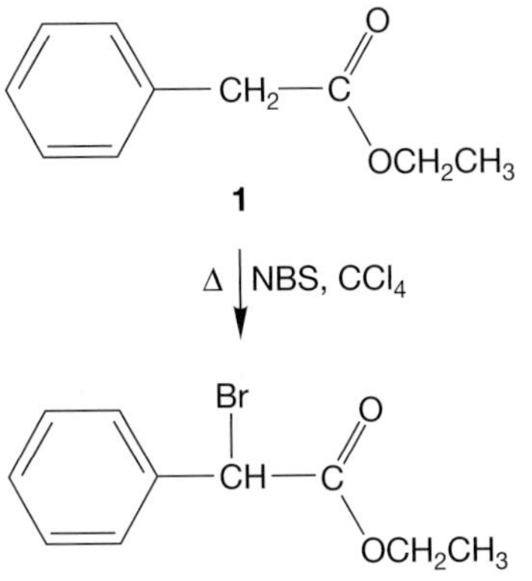

1

Δ | NBS, CCl₄

PROBLEM 11.48 Draw each of the resonance structures for the radical intermediate formed in Problem 11.47. Focus on the radical part.

PROBLEM 11.49 Predict the major product(s) for the reaction between NBS (allylic halogenation) and the following alkenes:
(a) propene
(b) 1-butene
(c) 2-butene
(d) cyclopentene
(e) 1-methylcyclohexene

PROBLEM 11.50 Write a detailed mechanism for the photochlorination of **1** to give **2**. Why is compound **1** reactive even though it contains no π bond?

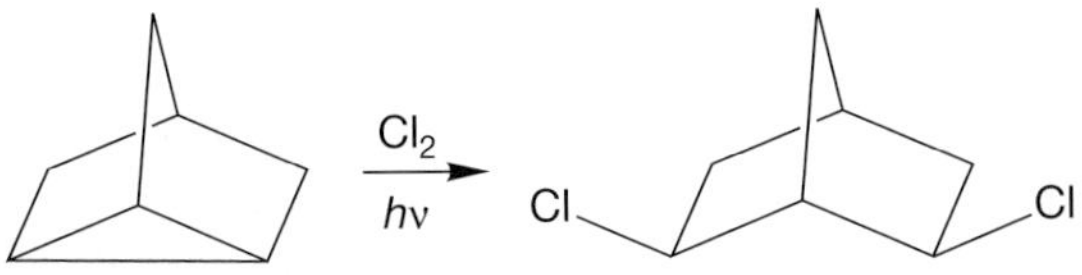

PROBLEM 11.51 Here is a more complicated photochlorination of a compound containing a three-membered ring. Explain the formation of the following products from spiropentane (**1**).

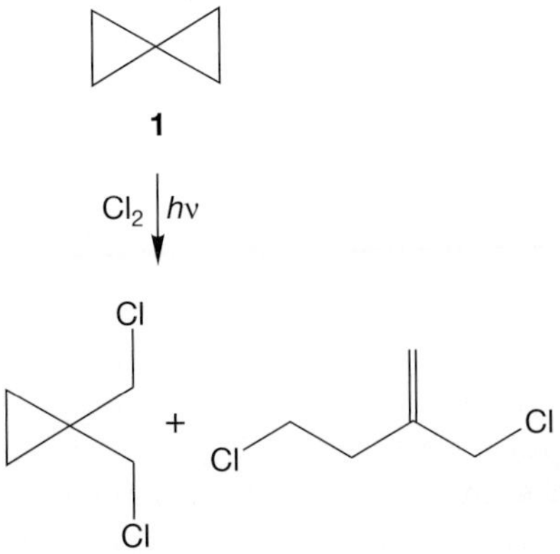

PROBLEM 11.52 Photolysis of iodine in the presence of cis-2-butene leads to isomerization of the alkene to a mixture of cis- and trans-2-butene. Explain.

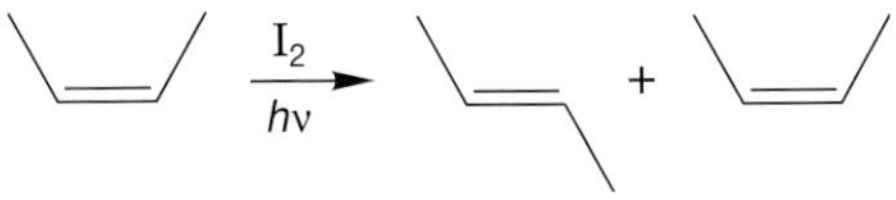

PROBLEM 11.53 Because of their high-potency, low-mammalian toxicity, and photostability, certain esters of halovinylcyclo-propanecarboxylic acids (pyrethroids) are promising insecticides (see Chapter 10, p. 456). An important precursor to this class of insecticides is the carboxylic acid (**1**), prepared as follows:

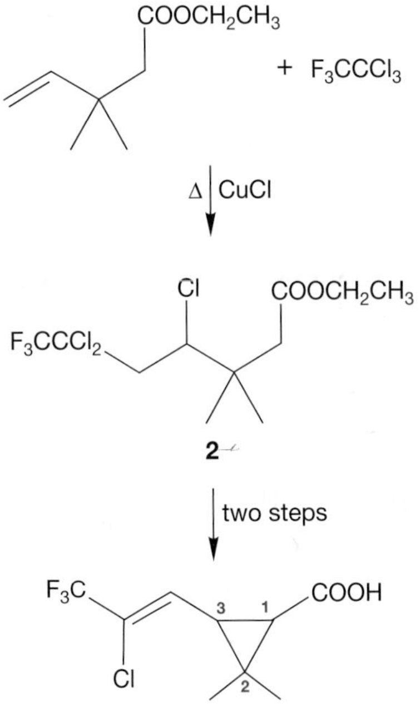

(a) Write a mechanism for the formation of **2**. Be sure to account for the observed regiochemistry. *Hint*: The CuCl can abstract a chlorine atom.
(b) What transformations must be accomplished in the **2 → 1** conversion? Don't worry about mechanisms.
(c) Given the stereochemistry about the carbon–carbon double bond in **1**, how many stereoisomers of **1** are possible? The most potent insecticides are derived from the (1R,3S) isomer of **1**. Draw this isomer.

PROBLEM 11.54 Write a mechanism for the following reaction of 1-octene:

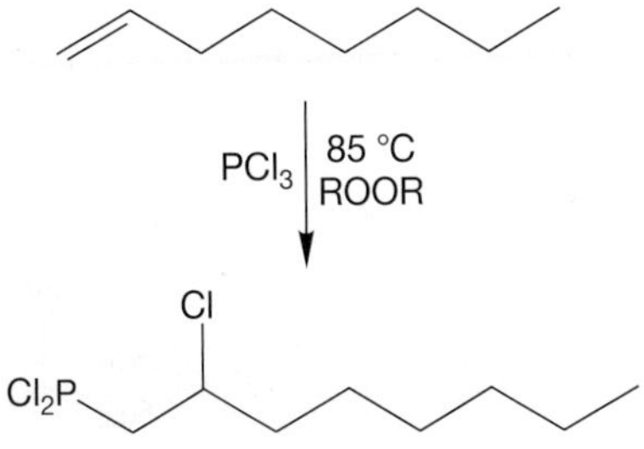

PROBLEM 11.55   Provide mechanisms for the formation of the following products:

77 °C | Ph—CH₃

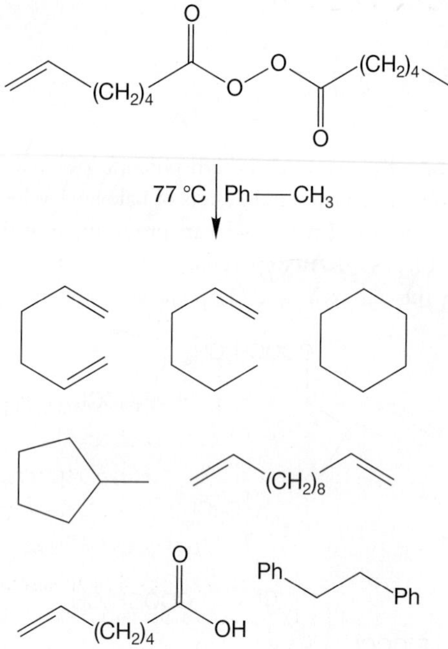

PROBLEM 11.56   Provide mechanisms for the formation of the following products:

$$CH_3(CH_2)_4 - C \equiv CH$$

CCl₄ | ROOR, 77 °C

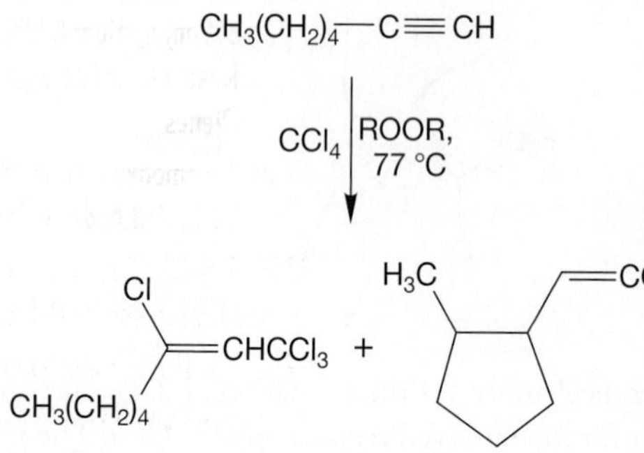

PROBLEM 11.57   Propose a mechanism for the following reaction. Be sure to rationalize the preferred formation of diene **1**.

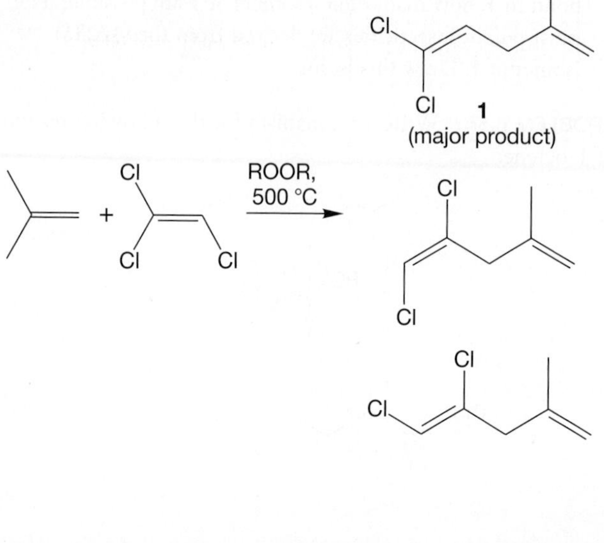

PROBLEM 11.58   Write a mechanism for the formation of compound **1**.

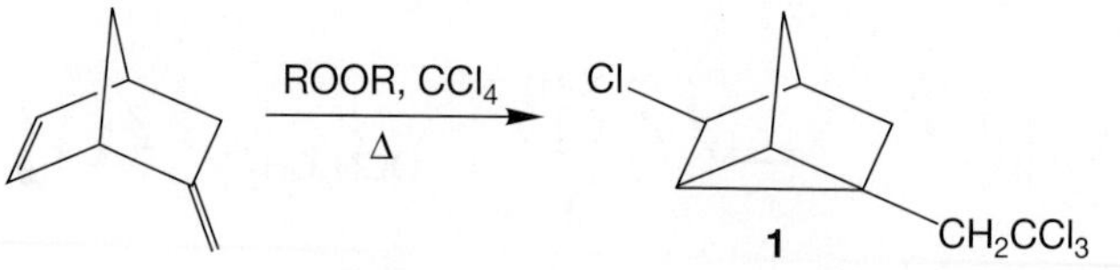

Use Organic Reaction Animations (ORA) to answer the following questions:

PROBLEM 11.59   There are three radical processes included in ORA. All three are on the second page of reactions. Select the "Radical alkene hydrohalogenation" reaction. The animation shows initiation steps and propagation steps for the overall process. Write out the steps of the reaction and indicate which are initiation steps and which are propagation steps.

PROBLEM 11.60   No termination steps are shown in the animation. Write out possible termination steps for the reaction.

PROBLEM 11.61   The "Alkane halogenation" reaction is also a radical process. Select this reaction and watch it several times. What is the hybridization of the central carbon of the intermediate? Select the SOMO (singly occupied molecular orbital) track. Why has the calculation come up with this hybridization? Is the hybridization fixed or does it change?

PROBLEM 11.62   A third radical animation is "Benzylic oxidation." This reaction is shown using oxygen (O_2) in the initiation step. Why can oxygen remove a hydrogen from isopropylbenzene? Is the benzylic hydrogen acidic? Is oxygen basic? What is the hybridization of the benzylic carbon in the intermediate? Is the hybridization of the benzylic carbon fixed or does it change?

Dienes and the Allyl System: 2*p* Orbitals in Conjugation

12

12.1 Preview

12.2 Allenes

12.3 Related Systems: Ketenes and Cumulenes

12.4 Allenes as Intermediates in the Isomerization of Alkynes

12.5 Conjugated Dienes

12.6 The Physical Consequences of Conjugation

12.7 Molecular Orbitals and Ultraviolet Spectroscopy

12.8 Polyenes and Vision

12.9 The Chemical Consequences of Conjugation: Addition Reactions of Conjugated Dienes

12.10 Thermodynamic and Kinetic Control of Addition Reactions

12.11 The Allyl System: Three Overlapping 2*p* Orbitals

12.12 The Diels–Alder Reaction of Conjugated Dienes

12.13 Special Topic: Biosynthesis of Terpenes

12.14 Special Topic: Steroid Biosynthesis

12.15 Summary

12.16 Additional Problems

CONJUGATED ALKENES The different colors in fruits and vegetables arise from the different conjugated alkenes that are produced by plants.

> Since chemists are not physicians, we shall scarcely benefit by their art,
> except by making the physician a chemist.
>
> —GEORGE REES[1]

12.1 Preview

Sometimes the whole really is greater than, or at least different from, the sum of the parts. In this chapter, we see just such a situation. One cannot always generalize from the properties of alkenes to those of molecules with multiple double bonds. Dienes, containing two double bonds separated only by a single bond (Fig. 12.1), often have properties that are quite different from those of isolated, individual alkenes. This difference depends on a phenomenon called conjugation. **Conjugated double bonds** are alkenes connected by a single bond. A molecule containing widely separated double bonds will generally react as would the individual alkenes. Double bonds in conjugation react quite differently. A conjugated diene constitutes a separate functional group. **Cumulated alkenes (cumulenes)** are molecules that have three or more atoms attached only by double bonds.

Unconjugated dienes

(E)-1,5-Decadiene   **1,4-Pentadiene**   **1,4-Cyclooctadiene**   **1,4-Cyclohexadiene**

Conjugated dienes

1,3-Butadiene   **2-Methyl-1,3-butadiene**   **(E,E)-2,4-Hexadiene**   **1,3-Cyclohexadiene**
                           (isoprene)

FIGURE 12.1 Some conjugated and unconjugated double bonds.

First we will cover the cumulenes, then we will generalize our discussion of the phenomenon of conjugation and its effect on chemical reactions. Next we will develop further the chemistry of a molecule we already encountered, allyl. Figure 12.2 shows the neutral allyl radical, a system of three parallel 2*p* orbitals.

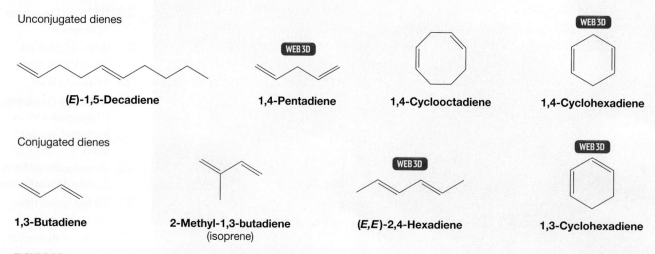

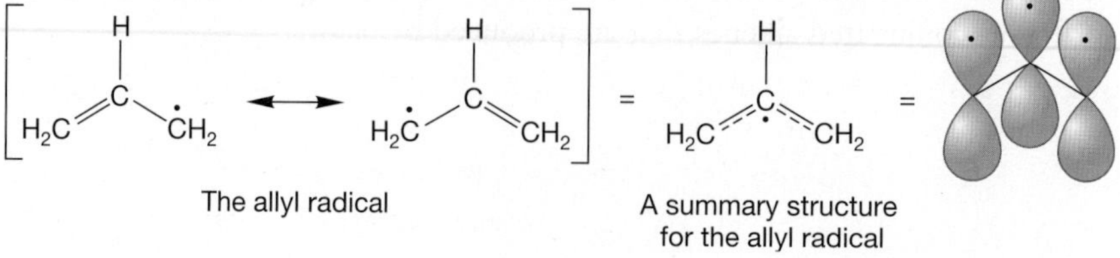

The allyl radical   A summary structure for the allyl radical

FIGURE 12.2 Allyl has a 2*p* orbital on each carbon atom.

---

[1]George Owen Rees was a nineteenth-century clinician who specialized in the analysis of blood and urine. A student of MJ's, Jordan M. Cummins, found this nice quote in the book by Stanley J. Reiser, *Medicine and the Reign of Technology*, Cambridge University Press, Cambridge, 1978.

Finally, we will see the outlines of how large molecules are constructed in Nature from a rather simple building block, the diene isoprene (2-methyl-1,3-butadiene), shown in Figure 12.1.

ESSENTIAL SKILLS AND DETAILS

1. The overall skill that you will take away from this chapter is the ability to think about the structural and chemical consequences of conjugation. Conjugation affects reactivity markedly, and an ability to see how and why is essential before progressing to the following chapters.

2. The difference between kinetic control and thermodynamic control must be understood. Kinetic control means that the products are determined by the relative heights of the transition states connecting starting materials and products, not by the energies of the products themselves. Under conditions that supply enough energy to go over *all* the transition states, thermodynamic control will operate. Under these conditions, the product mixture is determined by the relative energies of the products themselves.

3. The Diels–Alder reaction has been characterized as the most important synthetic reaction in organic chemistry. It is surely among the most important, because it makes the very common six-membered rings in a single step. Every cyclohexene is the potential product of a Diels–Alder reaction and the potential source of a diene and a dienophile through a reverse Diels–Alder reaction. Be sure you can play the game of "find the cyclohexene"!

4. The number of *sp* hybridized carbons in an allene or cumulene has stereochemical consequences: An even number of such carbons leads to cis/trans isomerism, whereas an odd number produces a pair of enantiomers.

5. Allyl resonance is important in reactions of cations, radicals, and anions. Be sure you can manipulate the allyl system easily.

12.2 Allenes

The simplest diene possible is propadiene, almost always known as allene (Fig. 12.3).[2] Allene is the general term for any molecule containing two double bonds that share a carbon atom. Allenes are systematically named as dienes or commonly named as derivatives of allene (Fig. 12.3).

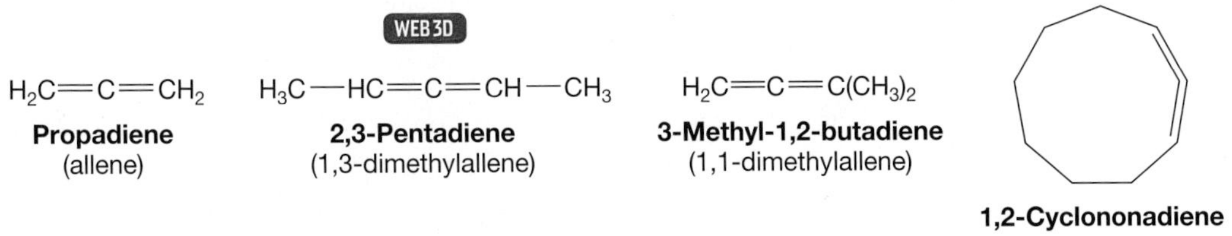

$H_2C{=}C{=}CH_2$

Propadiene
(allene)

$H_3C{-}HC{=}C{=}CH{-}CH_3$

2,3-Pentadiene
(1,3-dimethylallene)

$H_2C{=}C{=}C(CH_3)_2$

3-Methyl-1,2-butadiene
(1,1-dimethylallene)

1,2-Cyclononadiene

FIGURE 12.3 Some allenes.

The first problem is to work out a structure for allene. The key position in this cumulated diene is certainly the central carbon atom that is shared by the two double bonds, and the bonding involved has important structural consequences. The central carbon in any allene is attached to only two groups and, therefore, *sp* hybridization is appropriate. The central carbon uses two *sp* hybrid orbitals to form a pair of σ bonds with the flanking methylene groups, leaving the central carbon's remaining two *p* orbitals available for π bonding (Fig. 12.4).

2p Orbitals

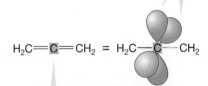

This carbon is attached to only two other groups (*sp* hybridization is appropriate)

FIGURE 12.4 The central carbon of allene is *sp* hybridized.

[2]There is a chance for confusion here between the system of three parallel 2*p* orbitals called "allyl" and the "allene" in which two double bonds are attached to one carbon. Be careful; these species are very different from each other!

Each end carbon, though, is attached to three groups and is therefore hybridized sp^2. This easy approach leads immediately to a structure for allene (shown in Fig. 12.5). Notice how each $2p$ orbital of the sp^2 hybridized end carbons overlaps perfectly with one of the $2p$ orbitals of the central, sp hybridized carbon. The important thing is not to be panicked by the odd central carbon and to keep in mind how we dealt with new structures before. Assign an appropriate hybridization, be sure to pay attention to the geometrical requirements of the hybrid, and the structure will always work out.

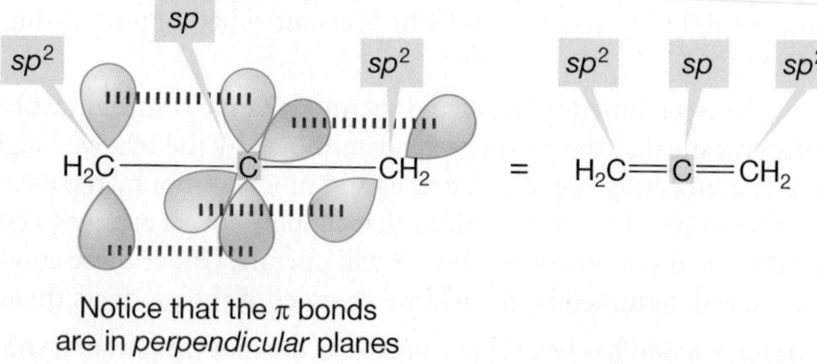

FIGURE 12.5 An orbital picture of allene. Notice the perpendicular π systems.

Notice that the π bonds
are in *perpendicular* planes

We have to be careful and pay attention to detail. Finding a reasonable structure for allene is a classic example of a problem asked at this point in almost every course in organic chemistry.

WORKED PROBLEM 12.1 It would appear that there could be cis and trans isomers of 1,3-dimethylallene, but there aren't. Explain.

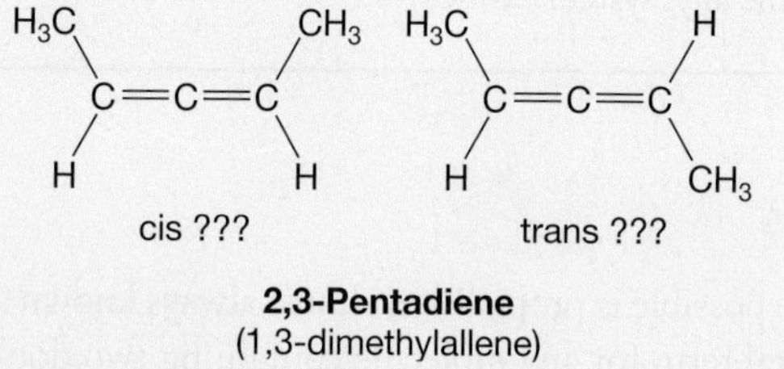

cis ???   trans ???

2,3-Pentadiene
(1,3-dimethylallene)

ANSWER The two-dimensional paper will fool us all the time if we do not think in three dimensions. The end groups of an allene are not in the same plane, but lie in perpendicular planes. The words cis (same side) and trans (opposite sides) have absolutely no meaning for an allene.

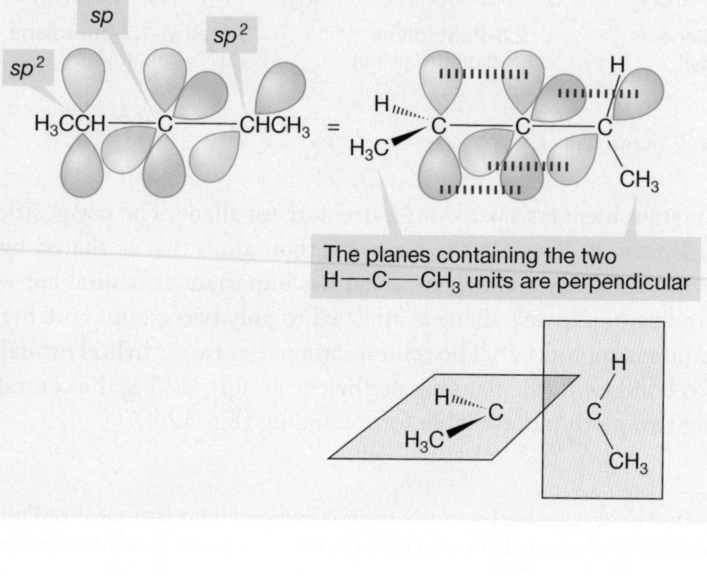

The planes containing the two
H—C—CH₃ units are perpendicular

WORKED PROBLEM 12.2 Two isomers of 1,3-dimethylallene do exist. What are they?

ANSWER There are two isomers of 1,3-dimethylallene, because it is a chiral molecule. The mirror image of 1,3-dimethylallene is not superimposable on the original; therefore a pair of enantiomers exists.

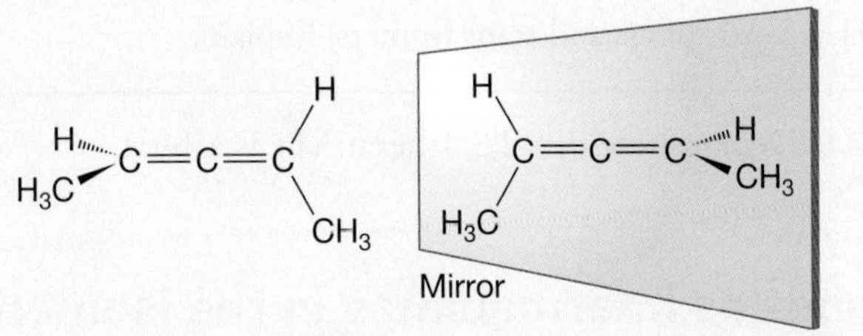

Allenes are high-energy molecules, just as are alkynes. The sp hybridization of the central carbon of this functional group is not an efficient way of using orbitals to make bonds, because π bonds are much weaker than σ bonds. The π bond energy of about 66 kcal/mol is substantially lower than most σ bond energies. This inefficiency is reflected in relatively positive heats of formation (ΔH_f° values). Table 12.1 gives a few heats of formation for allenes and alkynes. Remember that heats of formation are strictly comparable only among isomers. As you can see, allenes are roughly of the same energy as alkynes. This similarity is not surprising because the two molecules contain the same number of π bonds.

TABLE 12.1 Heats of Formation for Some Small Hydrocarbons

Alkynes	ΔH_f° (kcal/mol)	Allenes	ΔH_f° (kcal/mol)
Propyne	44.6	Propadiene $H_2C{=}C{=}CH_2$	47.7
1-Butyne	39.5	1,2-Butadiene $H_2C{=}C{=}CH{-}CH_3$	38.8
2-Butyne	34.7	1,2-Pentadiene $H_2C{=}C{=}CH{-}CH_2CH_3$	33.6
		2,3-Pentadiene $H_3C{-}CH{=}C{=}CH{-}CH_3$	31.8

12.3 Related Systems: Ketenes and Cumulenes

In allenes, the double bonds share a central carbon atom. There are other cumulated molecules in which heteroatoms (noncarbon atoms such as oxygen or nitrogen) take the place of one or more carbons of an allene and/or in which there are more than two double bonds in a row. When one end carbon of allene is replaced with an oxygen, the compound is called **ketene** (Fig. 12.6). If both end atoms are oxygens, the molecule is carbon dioxide. Cumulated hydrocarbons with three double bonds attached in a row are called cumulenes, although they can also be named as derivatives of 1,2,3-butatrienes (Fig. 12.6). Cumulenes are very reactive and most difficult to handle.

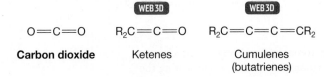

$O{=}C{=}O$ $R_2C{=}C{=}O$ $R_2C{=}C{=}C{=}CR_2$

Carbon dioxide   Ketenes   Cumulenes
(butatrienes)

FIGURE 12.6 Some molecules containing cumulated double bonds.

PROBLEM 12.3 Draw Lewis structures for the molecules in Figure 12.6 as well as for the related molecules: carbon disulfide (S=C=S), isocyanates (R—N=C=O), carbodiimides (RN=C=NR), and carbon suboxide (O=C=C=C=O).

PROBLEM 12.4 Unlike 1,3-dimethylallene, 2,3,4-hexatriene (1,4-dimethylbutatriene) exists as a pair of cis and trans isomers. Explain.

PROBLEM 12.5 Demonstrate that 2,3,4-hexatriene is achiral.

12.4 Allenes as Intermediates in the Isomerization of Alkynes

Now that we have the structures of allenes and cumulenes under control, it is time to look at reactivity. Our brief look at allenes permits us to understand one of the most astonishing reactions of alkynes: the base-catalyzed isomerization of an internal triple bond to the terminal position (Fig. 12.7).

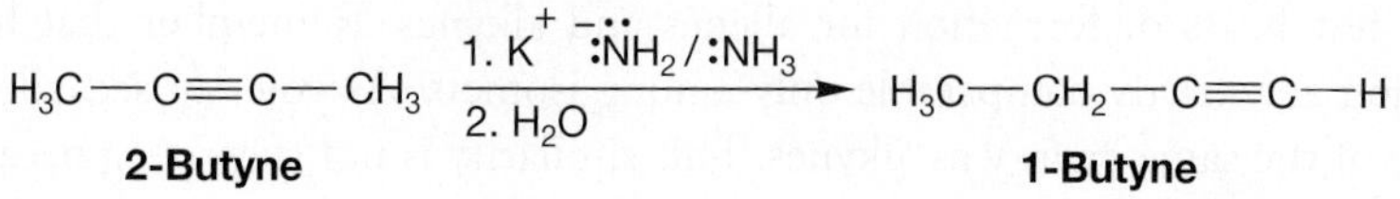

FIGURE 12.7 The isomerization of 2-butyne to 1-butyne can be carried out in strong base.

$$H_3C—C≡C—CH_3 \xrightarrow[\text{2. } H_2O]{\text{1. } K^+ \ ^-:NH_2 / :NH_3} H_3C—CH_2—C≡C—H$$

2-Butyne   **1-Butyne**

Let's look at this isomerization in detail. We do not begin with an allene, rather we encounter one as an intermediate in this reaction. When 2-butyne is treated with a very strong base such as an amide ion ($^-NH_2$), a proton can be removed from one of the methyl groups to give a resonance-stabilized carbanion (Fig. 12.8).

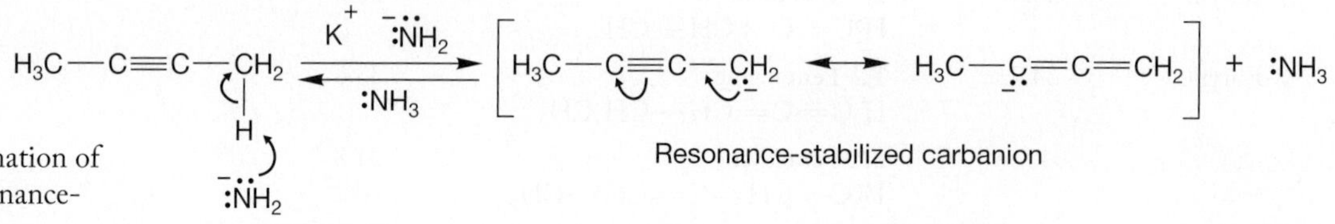

FIGURE 12.8 Deprotonation of 2-butyne to give a resonance-stabilized anion.

Resonance-stabilized carbanion

Ammonia and the carbon–hydrogen bonds of the methyl groups of 2-butyne both have pK_a values of about 38. Therefore, we can expect reasonable amounts of the carbanion to be produced on reaction with the amide ion. 2-Butyne is much more acidic than the parent hydrocarbon butane because the anion formed from 2-butyne is resonance stabilized, and that from butane is not (Fig. 12.8).

What can happen to this carbanion? It can be reprotonated by ammonia to regenerate 2-butyne, but this reaction surely gets us nowhere, even though it is reasonable and must happen frequently. Suppose, however, that the anion reprotonates at the *other* carbon sharing the negative charge? Now the product is 1,2-butadiene, an allene (Fig. 12.9).

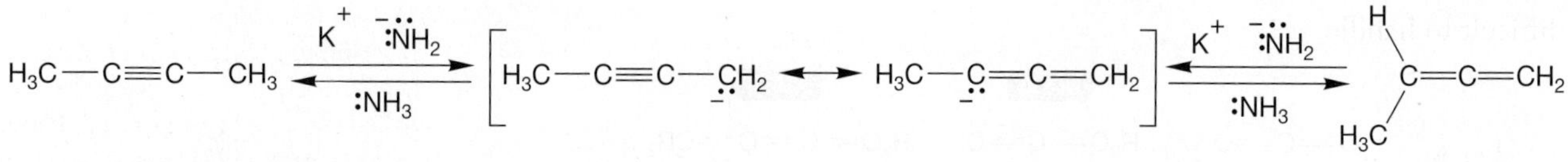

FIGURE 12.9 This resonance-stabilized anion can reprotonate in two ways. Reattachment of a proton at the end re-forms 2-butyne, but protonation at the other carbon sharing the negative charge produces an allene.

1,2-Butadiene
(methylallene)

The pK_a of the allene at the terminal methylene position is also about 38, so the amide ion is strong enough to remove the terminal proton to generate a new, resonance-stabilized anion (Fig. 12.10). Reprotonation at one of the carbons sharing the

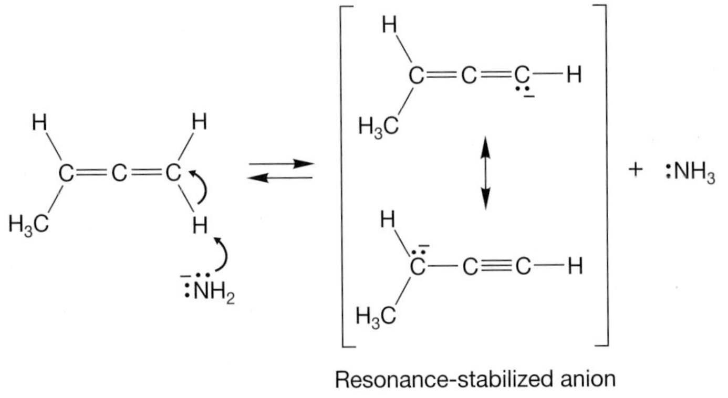

FIGURE 12.10 Deprotonation of the allene at the terminal position leads to a new resonance-stabilized anion.

Resonance-stabilized anion

negative charge regenerates the allene, but reprotonation at the other carbon yields a new, terminal alkyne, 1-butyne (Fig. 12.11).

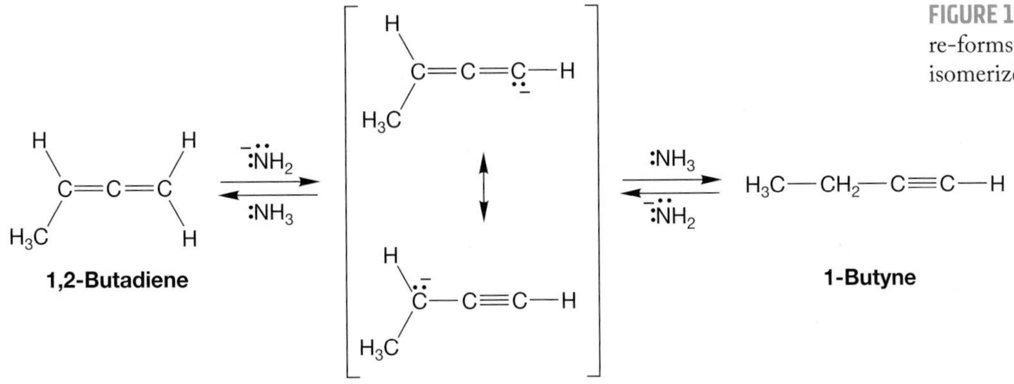

FIGURE 12.11 Reprotonation either re-forms the allene or gives the isomerized alkyne, 1-butyne.

1,2-Butadiene

1-Butyne

1-Butyne (pK_a = 25) is a *much* stronger acid than any other molecule present in the mixture and will be rapidly and, in a practical sense, irreversibly deprotonated to give the acetylide (Section 3.14). This deprotonation is the key step in this reaction (Fig. 12.12).

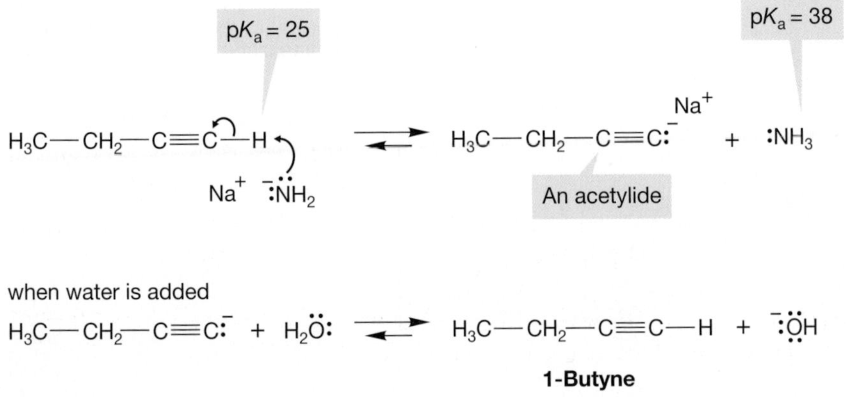

FIGURE 12.12 1-Butyne is by far the strongest acid in this equilibrating system. Removal of the acetylenic hydrogen gives an acetylide, the thermodynamic energy minimum in this reaction. When the reaction mixture is quenched by the addition of water, the acetylide protonates to give 1-butyne. It is critical to understand why it is not possible to re-enter the equilibrium at this point. After acidification there is no base present strong enough to remove the acetylenic hydrogen!

pK_a = 25

pK_a = 38

An acetylide

when water is added

1-Butyne

No matter where the equilibria in the earlier steps of this reaction settle out, the thermodynamic stability of the acetylide anion will drag the *overall* equilibrium far to the right. When water is added to neutralize the solution and end the reaction, the acetylide will be protonated and 1-butyne can be isolated. The overall reaction is summarized in Figure 12.13.

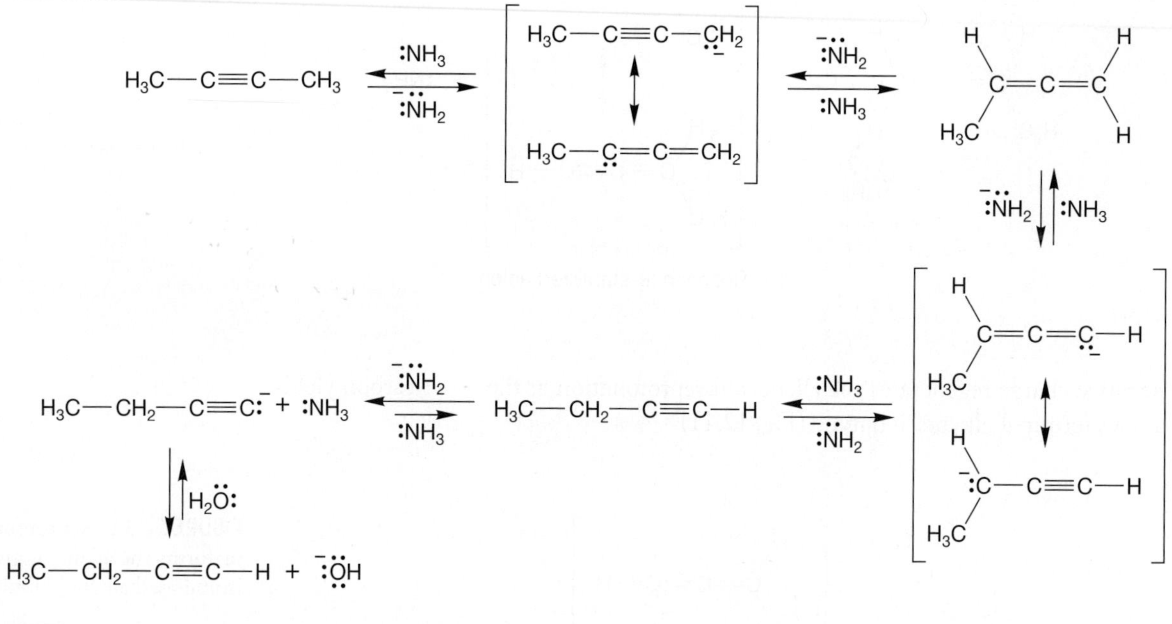

FIGURE 12.13 The full mechanism for the isomerization of 2-butyne to 1-butyne in strong base.

PROBLEM 12.6 Write a mechanism for the isomerization of 3-hexyne to 1-hexyne in the presence of sodium amide (NaNH$_2$) and NH$_3$.

WORKED PROBLEM 12.7 Di-*sec*-butylacetylene (3,6-dimethyl-4-octyne) fails to isomerize to another alkyne when treated with NaNH$_2$/NH$_3$. Explain.

ANSWER Isomerization is blocked by the alkyl groups. An allene can be formed, but the reaction can go no further because there is no allenic hydrogen that can be removed to produce a new resonance-stabilized anion.

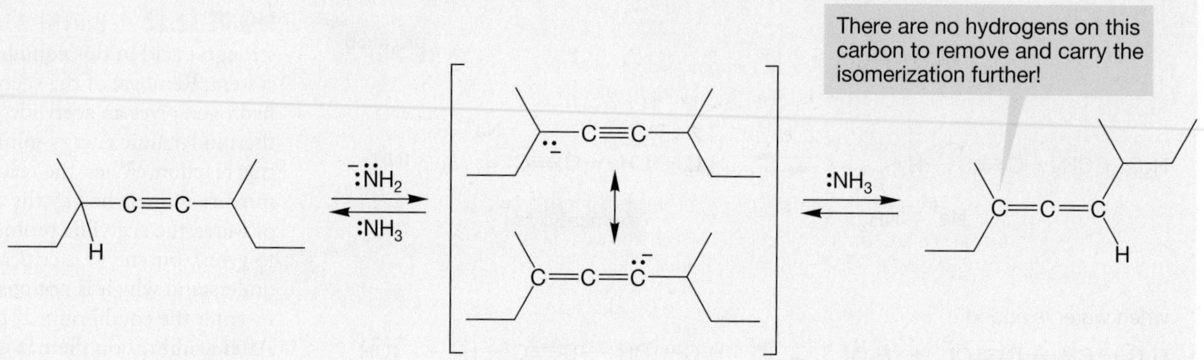

There are no hydrogens on this carbon to remove and carry the isomerization further!

12.5 Conjugated Dienes

The bonding in conjugated dienes is not so obviously exotic as it is in cumulated dienes, such as the allenes. Nevertheless, there are some subtle effects with far-reaching implications. The source of these effects can be seen from simple, schematic orbital drawings of 1,3-butadiene and 1,4-pentadiene. In a conjugated diene, the orbitals of the two double bonds overlap; in an unconjugated diene, they do not (Fig. 12.14).

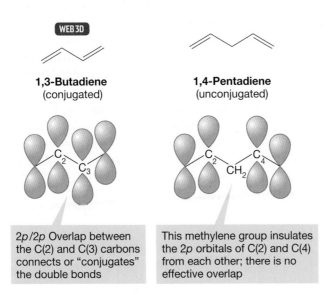

WEB 3D

1,3-Butadiene
(conjugated)

1,4-Pentadiene
(unconjugated)

2p/2p Overlap between the C(2) and C(3) carbons connects or "conjugates" the double bonds

This methylene group insulates the 2p orbitals of C(2) and C(4) from each other; there is no effective overlap

FIGURE 12.14 A conjugated and an unconjugated diene.

In the conjugated diene, the two double bonds are connected through 2p/2p overlap. By contrast, in the 1,4-diene the two double bonds are insulated by a methylene group. These bonds are "not conjugated," or "unconjugated," and will behave much like two separate alkenes. The question we now have to examine is how the chemical behavior of a conjugated diene is influenced by the overlap between the 2p orbitals on C(2) and C(3). To start, let's take a closer look at the structure of 1,3-butadiene, giving special attention to the π system.

> **PROBLEM 12.8** Sketch the σ system of 1,3-butadiene. What orbitals overlap to form the carbon–carbon and carbon–hydrogen bonds?

First, let's look at 1,3-butadiene from a resonance point of view. 1,3-Butadiene does have resonance forms that contribute to the overall electronic structure of the molecule (Fig. 12.15).

The best resonance form, and therefore the best single description of the molecule, is the diene structure we conventionally write. This structure has one more bond than the others and involves no charge separation. Thus, it wins by almost all of our criteria for estimating the relative energies of resonance forms (p. 372). Nevertheless, the three other resonance forms in Figure 12.15 show that there is some "double-bond character" to the C(2)—C(3) bond.

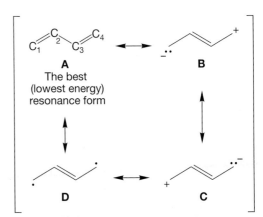

Note the double-bond character between C(2) and C(3) as shown in resonance forms **B**, **C**, and **D**

FIGURE 12.15 Resonance forms contributing to the structure of 1,3-butadiene.

PROBLEM 12.9 For each of the resonance forms **B**, **C**, and **D** in Figure 12.15 indicate what makes it less stable than the conventional structure, **A**.

Conjugation can be seen, perhaps even more clearly, if we examine the π molecular orbitals of 1,3-butadiene. In butadiene, we have a system of four 2*p* orbitals on four adjacent carbons. These four atomic 2*p* orbitals will overlap to produce four π molecular orbitals. We can get the symmetries of the four molecular orbitals in a number of equivalent ways. It is always easiest to build up something new out of things we know already, so it's productive to construct the molecular orbitals of butadiene out of two sets of the π molecular orbitals of ethylene, π and π*. One way to accomplish this is to allow the π molecular orbitals of ethylene to overlap side by side as indicated in Figure 12.16. In producing the π molecular orbitals in this way, we have made an important approximation. As often mentioned before, the closer in energy two orbitals are, the stronger the stabilization resulting from their overlap. In practice, all orbital overlap except for the strongest interactions can be ignored, at least at this level of theory. This approximation means that we need to look only at the result of the π ± π and π* ± π* interactions and do not have to consider π ± π*.

The symmetries of the butadiene molecular orbitals can be easily derived from the symmetries of the starting ethylene orbitals (Fig. 12.16).

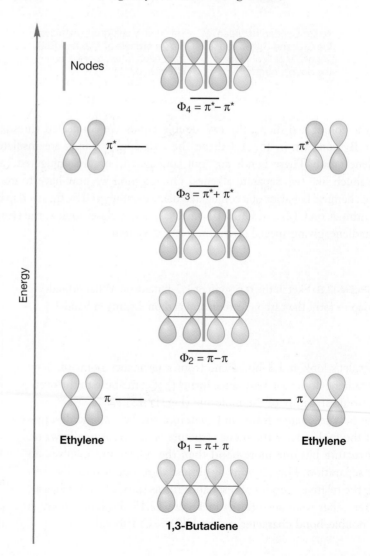

FIGURE 12.16 The schematic formation of the π molecular orbitals of 1,3-butadiene from the π molecular orbitals of ethylene.

To order the butadiene molecular orbitals in terms of energy, count the new nodes. Orbital Φ_1 in Figure 12.16 has no new nodes, Φ_2 has one, Φ_3 has two, and Φ_4 has three. As the number of nodes in an orbital goes up, so does the energy of the orbital (p. 32). Not all methods of construction will give the correct molecular orbitals *in the correct order* of relative energy. So you must always go through the procedure of counting nodes to do an energy ordering.

PROBLEM 12.10 Construct the π molecular orbitals of 1,3-butadiene from those of ethylene by putting one ethylene "inside" the other.

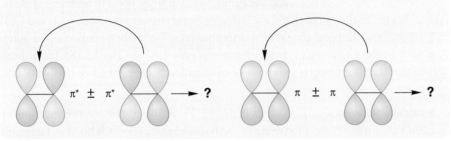

12.6 The Physical Consequences of Conjugation

In 1,3-butadiene, or any other conjugated diene, there are four electrons to accommodate in the π molecular orbital system. The lowest two bonding molecular orbitals are occupied and the two higher energy, antibonding molecular orbitals are empty. The HOMO for butadiene is Φ_2 (see Fig. 12.17), and to a first approximation Φ_2 will control the reactivity of the molecule. It is Φ_2 that looks most like the traditional representation of 1,3-butadiene. In the resonance formulation it is the traditional picture of 1,3-butadiene that is surely the lowest-energy resonance form, and thus the best representation of the molecule. So we see the usual (and necessary) convergence between the resonance and molecular orbital way of looking at structures. Each method generates a picture of 1,3-butadiene that looks mostly like a molecule containing a pair of double bonds (Fig. 12.17). Both methods warn us, however, that there are likely to be some consequences of conjugation in this molecule. This is shown

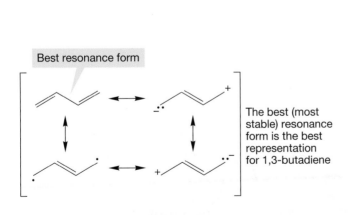

Best resonance form

The best (most stable) resonance form is the best representation for 1,3-butadiene

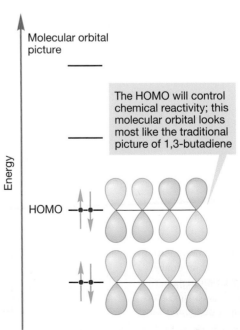

Molecular orbital picture

The HOMO will control chemical reactivity; this molecular orbital looks most like the traditional picture of 1,3-butadiene

HOMO

FIGURE 12.17 Both the resonance view and the molecular orbital view of 1,3-butadiene lead to the conclusion that 1,3-butadiene is a molecule containing two traditional double bonds. However, both pictures make it clear that there will be some consequences of conjugation between the two double bonds.

by the C(1)—C(2)—C(3)—C(4) connection of Φ_1 in the molecular orbital picture of Figure 12.16, and by the existence of resonance forms in which there is double-bond character between C(2) and C(3).

What are the effects of the C(2)—C(3) overlap on the structure of 1,3-butadiene? Carbon–carbon single bonds are relatively long; for example, in butane the carbon–carbon bond is 1.53 Å. The carbon–carbon double bond in 2-butene is only 1.32 Å, reflecting the presence of two bonds between the carbons, and the carbon–carbon triple bond in acetylene, 1.20 Å, is even shorter. To the extent that there is double-bond character between C(2) and C(3) in 1,3-butadiene, the bond should be shorter than an sp^3/sp^3 carbon–carbon single bond. And so it is; the C(2)—C(3) bond distance is only 1.47 Å (Fig. 12.18). It has been argued that the shortening is indicative of conjugation between the two double bonds.

FIGURE 12.18 The central carbon–carbon bond length in 1,3-butadiene lies between that for a single bond and double bond.

WORKED PROBLEM 12.11 Life is never simple. There is a persuasive counterargument that attributes the shortening of the butadiene C(2)—C(3) bond to factors other than conjugation. Can you come up with another explanation? *Hint for this tough question*: We have ignored the σ system in this argument. What kind of a σ bond joins C(2) and C(3)? So what? See Problem 12.8.

ANSWER The two inner carbons of 1,3-butadiene are joined by an sp^2/sp^2 σ bond, in contrast to the sp^3/sp^3 σ bonds of butane. An sp^2 orbital has more s character than an sp^3 orbital, and bonds made from sp^2/sp^2 overlap are stronger and shorter than those made from sp^3/sp^3 overlap. Accordingly, some of the shortening of the central σ bond in 1,3-butadiene may be attributed to the σ system and not to π orbital overlap.

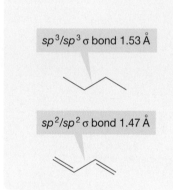

sp^3/sp^3 σ bond 1.53 Å

sp^2/sp^2 σ bond 1.47 Å

Now the question becomes: How strong is the C(2)—C(3) overlap? What consequences will this overlap have for the structure of conjugated dienes? Overlap between 2p orbitals on adjacent atoms has profound consequences for the structure of alkenes, for example, and the same might be true for dienes. It is the overlap between 2p orbitals that makes cis and trans isomers of alkenes possible (Fig. 12.19).

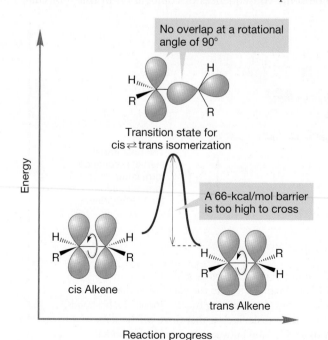

No overlap at a rotational angle of 90°

Transition state for cis ⇌ trans isomerization

A 66-kcal/mol barrier is too high to cross

Energy

cis Alkene

trans Alkene

Reaction progress

FIGURE 12.19 In alkenes, rotation about the carbon–carbon double bond requires overcoming a barrier of 66 kcal/mol. This barrier is too high to be surmounted under most conditions.

Might the same be true for conjugated dienes? Will overlap between the *2p* orbitals at the C(2) and C(3) positions create a large enough barrier so that isomers, called the **s-cis** and **s-trans** conformations, can be isolated? The "s" nomenclature in dienes stands for "single" because these are two conformations involving stereochemistry about the single bond of the diene (Fig. 12.20).

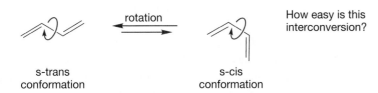

rotation

s-trans
conformation

s-cis
conformation

How easy is this
interconversion?

FIGURE 12.20 Interconversion of the s-trans and s-cis conformations of 1,3-butadiene.

As we might guess, there is a barrier to the rotation, but it's not very large. Overlap of *2p* orbitals between C(2) and C(3) is not nearly as important as that between the *2p* orbitals of the two adjacent carbons in ethylene. At the transition state for the rotation of 1,3-butadiene about the central, C(2)—C(3) bond, overlap between the *2p* orbitals on C(2) and C(3) vanishes, but C(1)—C(2) and C(3)—C(4) overlap is undisturbed (Fig. 12.21a).

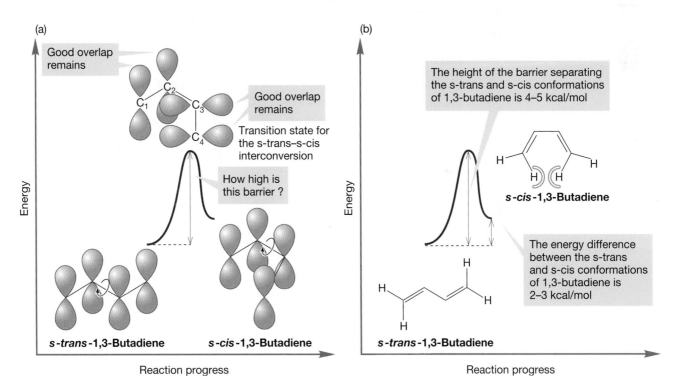

FIGURE 12.21 The barrier for the interconversion of the s-trans and s-cis conformations of 1,3-butadiene.

The barrier to rotation in 1,3-butadiene turns out to be 4–5 kcal/mol, only a bit larger than the barrier to rotation about the carbon–carbon bond in ethane. It is the s-trans form that is favored at equilibrium, by 2–3 kcal/mol. There is a simple steric effect favoring the s-trans conformation. In the s-cis conformation the two inside hydrogens are quite close to each other (make a model!) and in the s-trans conformation they are not. In order to make sense of these structures, we have to watch out not only for the locations of the double bonds, but for the positions of the hydrogens attached to them as well (Fig. 12.21b).

PROBLEM 12.12 Calculate the amount of s-*cis*-1,3-butadiene present at equilibrium at 25 °C, assuming an energy difference between s-*cis*- and s-*trans*-1,3-butadiene of 2.5 kcal/mol.

PROBLEM 12.13 There is another kind of rotation possible in 1,3-butadiene, that about the C(1)—C(2) or C(3)—C(4) bonds. We might guess that the barrier to this rotation would be little different from that for rotation in a typical double bond but, as a former president of the United States once said, "that would be wrong." In 1,3-butadiene, it takes only 52 kcal/mol to do this rotation, some 14 kcal/mol less than the barrier of 66 kcal/mol for rotation in ethylene. Explain. *Hint*: Look at the transition state for C(1)—C(2) bond rotation in 1,3-butadiene.

PROBLEM 12.14 Design an experiment to determine whether rotation about the C(1)—C(2) bond in 1,3-butadiene is possible. Assume you have the ability to make any labeled molecules you need.

As both the resonance and molecular orbital pictures indicate, there is $2p/2p$ overlap between C(2) and C(3), but this overlap does not dominate the structure of 1,3-butadiene the way $2p/2p$ overlap does for alkenes. We can get another idea of the magnitude of this interaction from measuring the heats of hydrogenation (p. 413) of a series of dienes.

We begin with 1-butene. When it is hydrogenated to butane, the reaction is exothermic by 30.3 kcal/mol (Fig. 12.22). A reasonable assumption would be that a molecule containing two double bonds would be hydrogenated in a reaction twice as exothermic as the hydrogenation of 1-butene. And this idea is just right; the heat of hydrogenation is -60.8 kcal/mol for 1,4-pentadiene and -60.5 kcal/mol for 1,5-hexadiene.

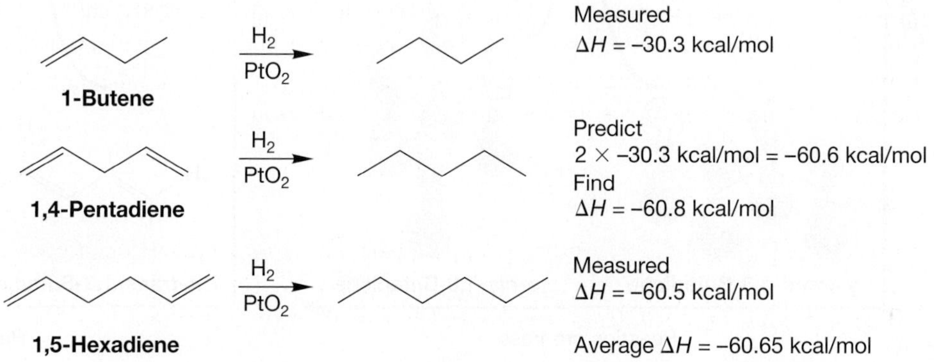

FIGURE 12.22 Hydrogenation of 1-butene is exothermic by 30.3 kcal/mol. Hydrogenation of unconjugated dienes is exothermic by almost exactly twice the heat of hydrogenation of 1-butene.

1-Butene

$$\xrightarrow[\text{PtO}_2]{\text{H}_2}$$

Measured
$\Delta H = -30.3$ kcal/mol

1,4-Pentadiene

$$\xrightarrow[\text{PtO}_2]{\text{H}_2}$$

Predict
2×-30.3 kcal/mol $= -60.6$ kcal/mol
Find
$\Delta H = -60.8$ kcal/mol

1,5-Hexadiene

$$\xrightarrow[\text{PtO}_2]{\text{H}_2}$$

Measured
$\Delta H = -60.5$ kcal/mol

Average $\Delta H = -60.65$ kcal/mol

Both of our sample dienes in Figure 12.22 were unconjugated. Now let's look at the heat of hydrogenation of the conjugated diene, 1,3-butadiene; ΔH is only -57.1 kcal/mol (Fig. 12.23).

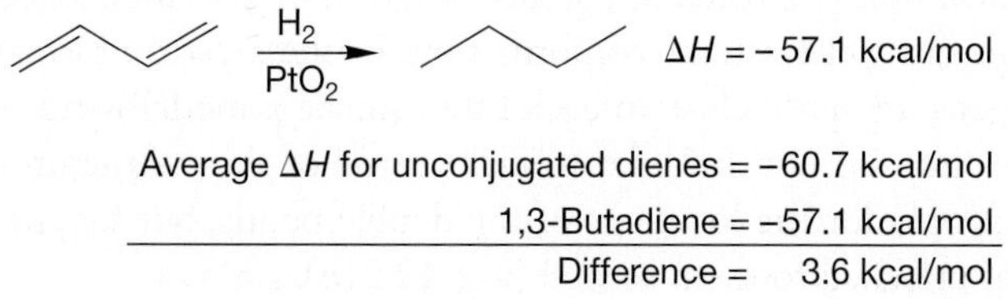

$$\xrightarrow[\text{PtO}_2]{\text{H}_2}$$ $\Delta H = -57.1$ kcal/mol

Average ΔH for unconjugated dienes $= -60.7$ kcal/mol

1,3-Butadiene $= -57.1$ kcal/mol
Difference $=   3.6$ kcal/mol

FIGURE 12.23 The heat of hydrogenation of 1,3-butadiene is 57.1 kcal/mol, smaller than expected by about 3.6 kcal/mol.

The heat of hydrogenation of the conjugated diene is about 3.6 kcal/mol lower than we would expect by analogy to the unconjugated dienes in Figure 12.22. This amount of energy can be attributed to the energy-lowering effects of conjugation. The situation is summarized in Figure 12.24.

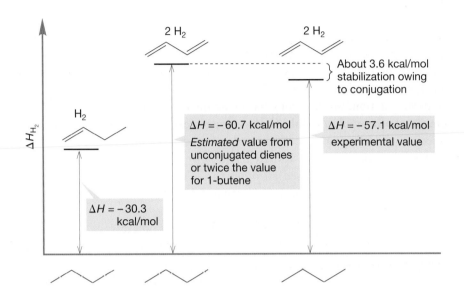

FIGURE 12.24 A summary of the heat of hydrogenation data.

About 3.6 kcal/mol stabilization owing to conjugation

$\Delta H = -60.7$ kcal/mol
Estimated value from unconjugated dienes or twice the value for 1-butene

$\Delta H = -57.1$ kcal/mol
experimental value

$\Delta H = -30.3$ kcal/mol

So now we have two measures of the energetic value of conjugation in 1,3-butadiene, and they are remarkably close to each other. The first is the height of the barrier to rotation about the C(2)—C(3) bond (4–5 kcal/mol), and the second is the lowering of the heat of hydrogenation of a conjugated diene relative to that of an unconjugated diene (~3.6 kcal/mol).

Two conjugated double bonds are not utterly unconnected; the whole is somewhat different from the sum of the parts. Conjugation, the orbital overlap between middle carbons in a conjugated diene, results in a shortening of the single bond and in lower heats of hydrogenation for conjugated dienes than for unconjugated systems. Now we move on to the consequences of further conjugation, and to an examination of the reactivity of these molecules. Here, too, we will see that conjugation influences the physical properties and chemical reactivity of these molecules.

Summary

Overlap between middle carbons of a conjugated diene has structural consequences. The central sigma bond is slightly shorter than a normal carbon–carbon single bond. The magnitude of the conjugative stabilization is small, about 4 kcal/mol.

12.7 Molecular Orbitals and Ultraviolet Spectroscopy

In the early days of organic chemistry, there were two extraordinarily difficult general problems. One was to separate mixtures into pure compounds, and the other was to determine the structures of the compounds once pure samples were obtained. How simple these problems seem! Yet, much of the mechanistic and synthetic organic chemistry of the last decades would have been quite impossible without the progress in solving these two old difficulties. We will have little to do with separation techniques in this book (your laboratory will take another approach), but we must certainly spend some time here on structure determination.

About 70 years ago, structure determination was an entire branch of chemistry. People spent whole careers working out the structures of complicated molecules by running chemical reactions. The products of many reactions of an unknown compound were analyzed and inferences were made regarding the structure of the unknown. This task was time-consuming and extraordinarily demanding of experimental technique, intellectual rigor, and the development, over years of work, of an intuitive sense of what was right, of what a given experiment might mean. In a way, this work had all the satisfactions and charms of handwork, of a craft. Nowadays, the work of years can usually be done in hours, and chemists attack structural problems that would have been beyond any chemical determination. Spectroscopy has made the difference. First came the development of **ultraviolet/visible (UV/vis) spectroscopy**. We will deal with that subject in this chapter. Then followed infrared (IR) spectroscopy and finally nuclear magnetic resonance (NMR). The power of NMR is so great that this technique has largely eclipsed all other forms of structure determination. We introduced NMR spectroscopy in Section 2.14, and we will devote much of Chapter 15 to this subject.

The development of spectroscopy certainly brought progress, and no reasonable person would want to turn back the clock to the days of hand-crafted structure determination. Yet, one can sympathize with the feeling that there has been some loss with the progress. We are somehow not as close to the molecules as we once were, even though we know much more about them. There is a psychological distance involved here. One tends to believe what spectrometers say and to read too much into the data. In chemistry, nothing is more important than honing one's critical sense. It is somehow easier to ignore some small, but ultimately vital glitch on a piece of chart paper than to ignore the results of an experiment gone in an unexpected direction. The former somehow melts into the noise of the baseline, while the latter teases the imagination.

One can also sympathize with those practitioners of the art of structure determination who saw the future when the first NMR machine was delivered, and realized that for them, "NMR" meant "No More Research." They were right.

Ultraviolet/visible spectroscopy is called **electronic spectroscopy** because it deals with the absorption of energy by electrons in molecules. The energy absorbed (ΔE) is used to promote an electron from one orbital to a higher-energy orbital (Fig. 12.25). The magnitude of ΔE is proportional to the frequency of the light, ν, and inversely proportional to wavelength of the light, λ, as shown by

$$\Delta E = h\nu = hc/\lambda \qquad (12.1)$$

where h is Planck's constant and c is the speed of light. Because electrons are restricted to orbitals of specific energy, ΔE is also quantized and may have only certain values.

The distinction between ultraviolet and visible spectroscopy is an artificial one and is based solely on the fact that the detectors built into our bodies, our eyes, are sensitive only to radiation of wavelength 400–800 nm (nm = nanometers; 1 nm = 10^{-9}m), which we call the visible spectrum. Spectrometers are less limited in this respect than we are, and UV spectrometers easily detect radiation from 200–400 nm, as well as the entire visible range and beyond. At higher energy, below 200 nm, the atmosphere begins to absorb strongly, and spectroscopy in this range is more difficult, although by no means impossible.

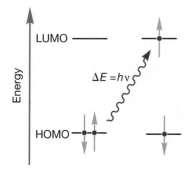

FIGURE 12.25 Absorption of UV light promotes an electron from the HOMO to the LUMO.

A schematic spectrometer is shown in Figure 12.26. It consists of a source of radiation, a monochromator that is capable of scanning the range of light emitted by the source, a sample cell, and a detector.

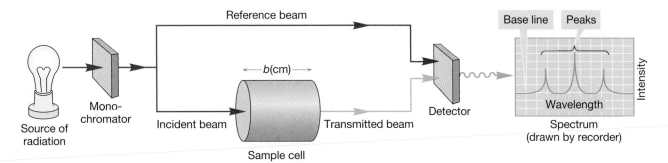

FIGURE 12.26 A schematic picture of a UV/vis spectrometer.

Generally, the spectrometer determines A, the absorbance, and plots A versus wavelength (λ in nm). The absorbance is calculated according to the Beer–Lambert law

$$A = \log I_0/I = \varepsilon bc \tag{12.2}$$

where the intensity of the light entering the sample cell is I_0, the intensity of the light exiting the sample cell is I, the concentration in moles per liter is c (not to be confused with the c for speed of light), the length of the sample cell in centimeters is b, and ε is a proportionality constant known as the **extinction coefficient**. Every compound has a characteristic ε, which depends not only on A, b, and c as shown in Equation 12.2, but also on wavelength, solvent, and temperature.

It is important to have a feel for the magnitudes of the energies involved in spectroscopy. Equation 12.1 shows the familiar relationship between energy, Planck's constant h, and the frequency of light. Because $v = c/\lambda$, and because we are interested in the energy per mole, we obtain

$$E = Nhc/\lambda \tag{12.3}$$

where $N = 6.02 \times 10^{23}$ molecules/mol, $h = 1.583 \times 10^{-37}$ kcal $\cdot$ s/molecule, and $c = 3.00 \times 10^{17}$ nm/s. The quantity Nhc is equal to 28.6×10^3 (nm $\cdot$ kcal/mol).

PROBLEM 12.15 The wavelength of light used in UV/vis spectroscopy ranges from 200 nm to 800 nm. Calculate the energy (in kcal/mol) corresponding to these two wavelengths.

For a given molecule there will be many possible UV (or vis) transitions, because there are many different energies of electrons even in a simple molecule. For ethylene there are several possible absorptions, as σ or π electrons are promoted to antibonding σ^* or

π^* orbitals. Figure 12.27 ignores the carbon–hydrogen bonds and shows only $\sigma \rightarrow \sigma^*$ and $\pi \rightarrow \pi^*$ transitions. There are other possible absorptions, even without including the carbon–hydrogen σ and σ^* orbitals: $\sigma \rightarrow \pi^*$ and $\pi \rightarrow \sigma^*$ are two examples. Such transitions between orbitals of different symmetry are usually relatively weak, however.

FIGURE 12.27 A $\pi \rightarrow \pi^*$ and $\sigma \rightarrow \sigma^*$ transition for ethylene.

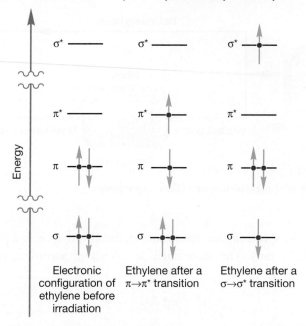

The starting point and final destination of the electron are used to denote the transition; hence $\sigma \rightarrow \sigma^*$, $\sigma \rightarrow \pi^*$, $\pi \rightarrow \pi^*$, and $\pi \rightarrow \sigma^*$ in this case. The energies, and therefore the wavelengths of light involved, depend on the energy separating the orbitals. For ethylene, even the $\pi \rightarrow \pi^*$ separation is great enough so that detection is difficult with common spectrometers. The $\pi \rightarrow \pi^*$ absorption occurs at about 165 nm, which corresponds to 173 kcal/mol, a high energy indeed. Little change is induced by alkyl substitution, and neither simple alkenes nor unconjugated dienes absorb strongly above 200 nm.

When a diene is conjugated, however, substantial changes occur in orbital energies, and therefore in the possible UV transitions. In general, the lowest energy (longest wavelength λ) transition possible for a polyene, a molecule containing multiple double bonds, will be from the HOMO to the LUMO. In ethylene or butadiene, this transition is $\pi \rightarrow \pi^*$. Figure 12.28 shows the relative energies of the π

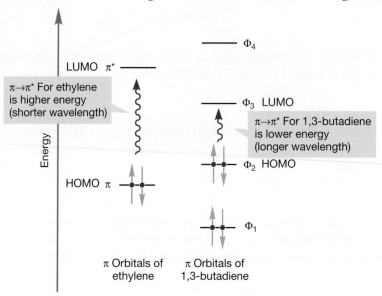

FIGURE 12.28 The HOMO–LUMO transition in ethylene is higher in energy (at shorter wavelength) than the HOMO–LUMO transition in 1,3-butadiene.

molecular orbitals of ethylene and 1,3-butadiene. You can see that there is a much lower energy $\pi \rightarrow \pi^*$ (HOMO–LUMO) transition possible in the conjugated molecule 1,3-butadiene than there is in ethylene. Spectroscopy bears this out, and 1,3-butadiene absorbs at 217 nm, which is a much longer wavelength (lower energy) than the absorption in ethylene.

WORKED PROBLEM 12.16 How much energy is contained in light of wavelength 217 nm?

ANSWER $E = Nhc/\lambda$ and the quantity $Nhc = 28.6 \times 10^3$ (p. 527). So, $E = (28.6 \times 10^3)/217 = 132$ kcal/mol at 217 nm.

Further conjugation does two things. First, it reduces the energy gap between the HOMO and LUMO, and this pushes the position of the lowest energy absorption to a longer wavelength (lower energy). Eventually, the HOMO–LUMO gap becomes small enough so that absorption enters the visible range (400–700 nm), and as a result we perceive highly conjugated polyenes as colored. The first polyene to absorb in the visible range appears yellow, because it is violet light (400–420 nm) that is first absorbed as the HOMO–LUMO gap shrinks. Remember that perceived color is a result of reflected light, the wavelengths of light that are not absorbed. β-Carotene (Fig. 12.29), an important precursor to vitamin A, absorbs blue light at 453 and 483 nm, and appears orange.

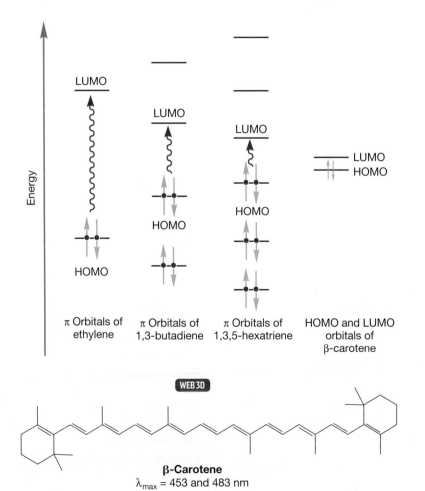

β-Carotene
λ_{max} = 453 and 483 nm

FIGURE 12.29 As conjugation increases, the HOMO–LUMO gap decreases and the position of the $\pi \rightarrow \pi^*$ absorption shifts to longer wavelength (lower energy).

The second impact of increased conjugation is that the number of possible UV transitions increases and, therefore, the spectra become complex. Figure 12.30 shows the spectrum of one moderately complicated conjugated molecule, which surely will resist a simple analysis, and which would be most difficult to predict in detail. However, many important molecules contain simple diene or triene systems, either acyclic or contained in rings, and UV spectroscopy remains a useful tool for structure determination in such molecules.

FIGURE 12.30 A complicated UV spectrum.

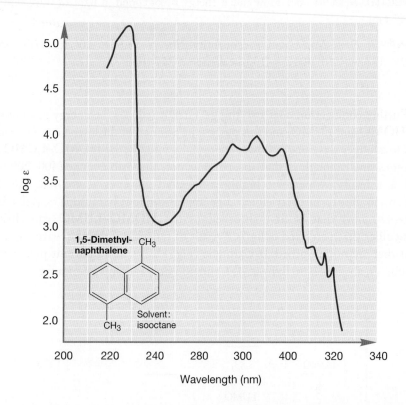

The position of longest wavelength absorption (λ_{max}) depends in a predictable way on the substitution pattern of these simple molecules. A set of empirical correlations was collated by Robert Burns Woodward (1917–1979) and was known initially as "Woodward's rules" and now sometimes as "Woodward's first rules." These rules were extended by Louis Fieser (1899–1977) and Mary Fieser (1909–1997) to include cyclic dienes in the polycyclic compounds called steroids. Table 12.2 summarizes Woodward's and the Fiesers' rules. Woodward's very different second set of rules will appear in Chapter 20.

TABLE 12.2 Woodward's and Fieser's Rules for Ultraviolet Spectra of Dienes

Woodward's Rules for Acyclic Dienes	(nm)	Fieser's Rules for Steroid Dienes	(nm)
Base diene value	217	Base diene with both double bonds in the same ring	253
Each added alkyl group	+5	Base diene with double bonds in different rings	214
Each double-bond exocyclic in a six-membered ring	+5	Each additional conjugated double bond	+30
		Each alkyl substituent	+5
		Each exocyclic double bond	+5

R. B. Woodward was acknowledged as the greatest American organic chemist, and was certainly one of the greatest chemists of all time. There is remarkable agreement on the subjective evaluation of Woodward's preeminence among organic chemists, and it's worth taking a little time to see what set this man apart from his merely excellent colleagues. First of all, he was devoted to the subject and worked unbelievably hard at it. Although the main themes of his professional life were the elucidation of structure and the synthesis of molecules of ever more breathtaking complexity, his work is also notable for the attention he paid to mechanistic detail, and to the broad implications of the work. Working hard is not a matter of mere hours, but of maintaining focus as well. Woodward was able to *concentrate* and either had or developed a prodigious memory for minutiae. In another person that might amount to mere obsession, but Woodward was able to extract from the sea of detail the important facts that allowed him time and again to turn his work from the specific to the general. Early on he was able to propose correct structures for such disparate molecules as penicillin, tetracycline, and ferrocene (Fig. 12.31).

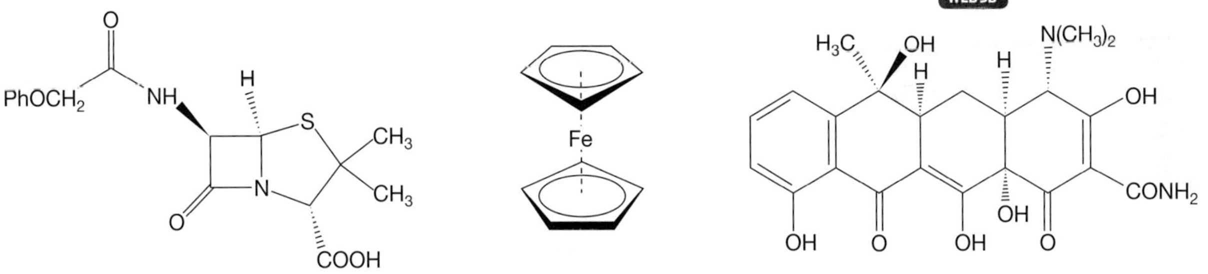

FIGURE 12.31 Some complex molecules whose structures were proposed by Woodward and his co-workers.

Woodward was able to pull obscure facts out of remote chemical hats and apply them to the matter at hand in a way that often transformed the very nature of the problem. The molecules he considered worthy of synthesis grew in complexity throughout his career, but it was more than the difficulty of the target that made his work so universally impressive. Woodward's work led places—it had "legs." For example, Woodward, primarily a synthetic chemist, made the observations [with Roald Hoffmann (b. 1937)] that led to the development and exploitation of what is arguably the most important *theoretical* construct of recent times. Hoffmann and Kenichi Fukui (1918–1998) of Kyoto University received the Nobel prize for this work in 1981. Many believe Woodward would have shared this Nobel prize had he lived (it would have been his second; his first was awarded in 1965 for "his outstanding achievements in the art of organic synthesis"). Some of the molecules synthesized by Woodward and his co-workers are shown in Figure 12.32. Woodward said that the synthesis of quinine was conceived when he was 12 years old.

Should you memorize Woodward's rules? Frankly, we think not—although there are those who strongly disagree, and some of these people write examination questions. These rules were an important part of structure determination but are restricted in utility to relatively simple polyenes. You should know they exist, know what systems they apply to, and look them up when you need them. If you wind up in steroid research, you'll learn them.

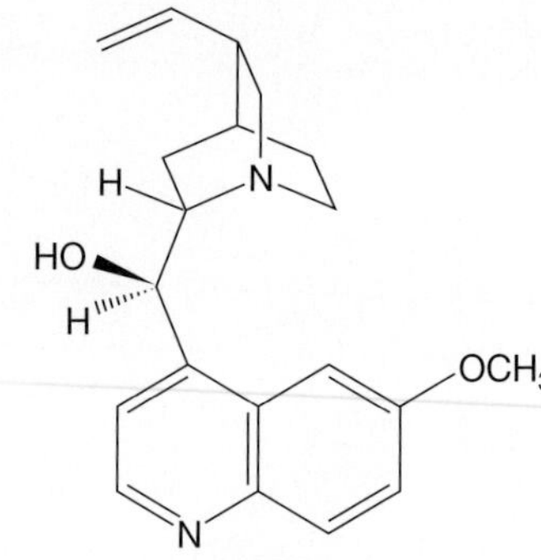

Quinine,
with W. von E. Doering, 1944

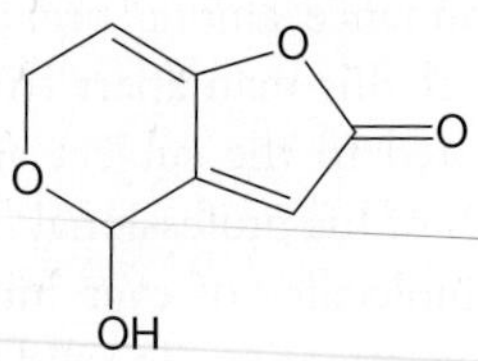

Patulin, 1950

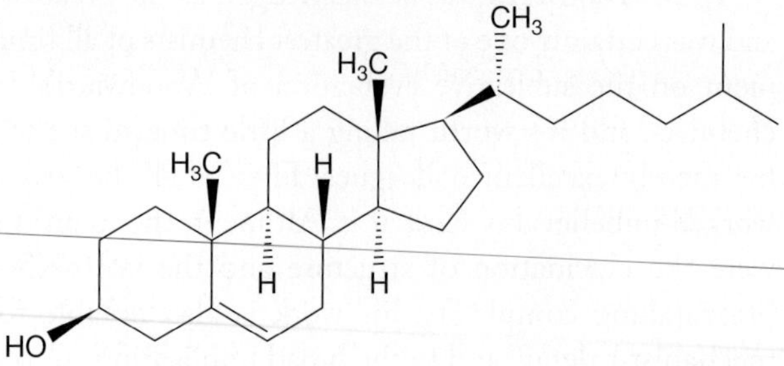

Cholesterol, 1951

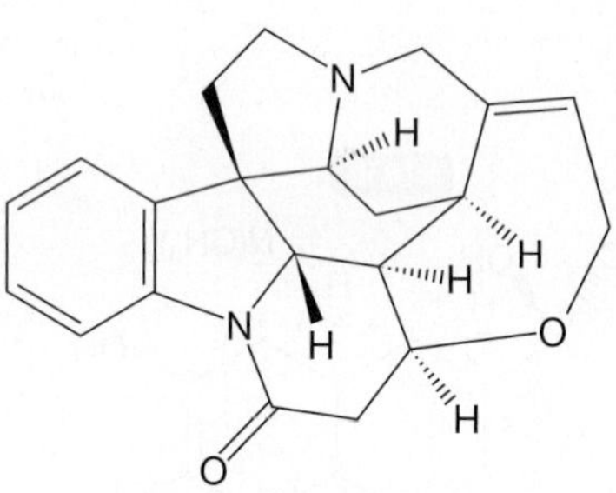

Strychnine, 1954

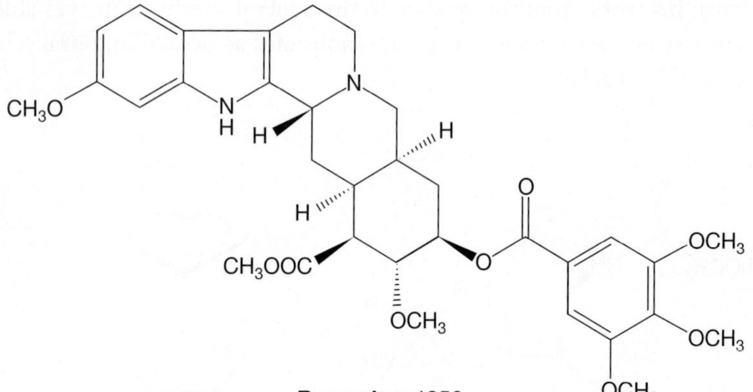

Reserpine, 1956

WEB 3D

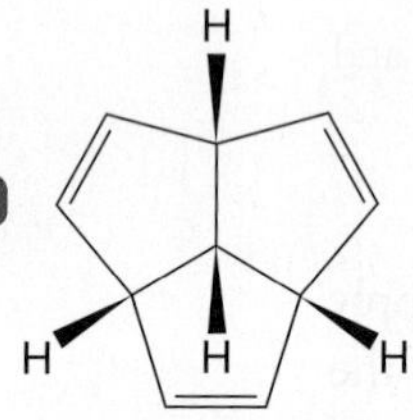

Triquinacene, 1964

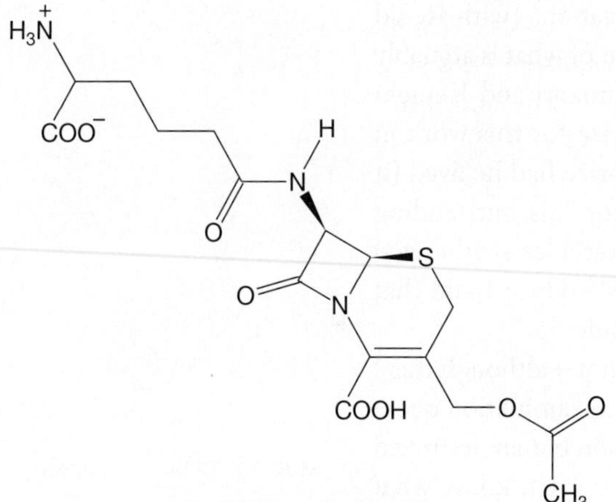

Cephalosporin C, 1966

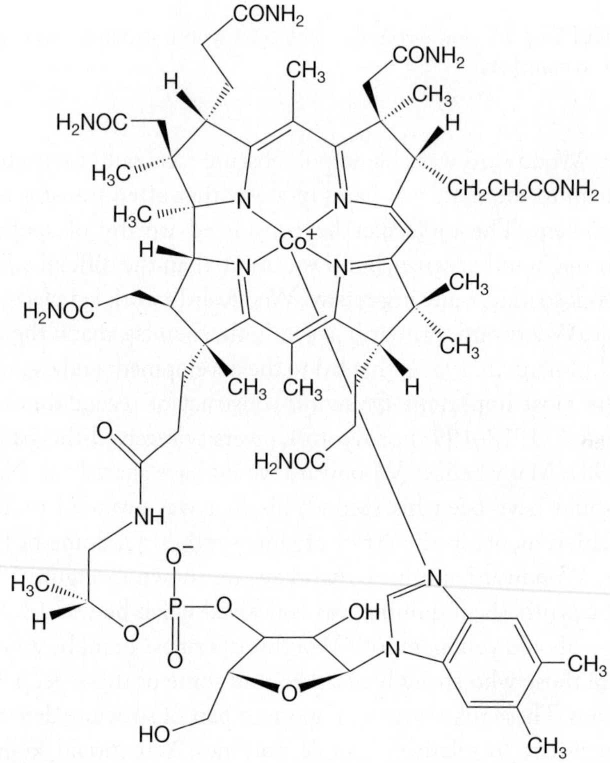

Vitamin B$_{12}$ with A. Eschenmoser and a cast of thousands, 1972

FIGURE 12.32 Some molecules synthesized by the Woodward group.

12.8 Polyenes and Vision

When we hear or read the word "spectroscopy," we tend to imagine rooms full of expensive machines measuring away. It's quite an inhuman vision. Yet we are spectrometers ourselves. All of our sensory experiences, most obviously vision, are examples of spectroscopic measurements. Those machines in our imaginations are mere extensions of our biological, in-house equipment. When you see a red apple among a basket of green ones, and decide that it is the ripe red one you will eat, you are acting on a spectroscopic measurement. Let's examine vision a little bit here, because it depends on polyene chemistry.

Many polyenes are important in biological processes, and as we might imagine, polyenes that absorb in the visible spectra are important in human vision. Vitamin A is *trans*-retinol, for example (Fig. 12.33). We have two enzymes, one of which, retinol dehydrogenase, oxidizes *trans*-retinol to *trans*-retinal and another, retinal isomerase, which isomerizes one double bond of *trans*-retinal to produce *cis*-retinal.

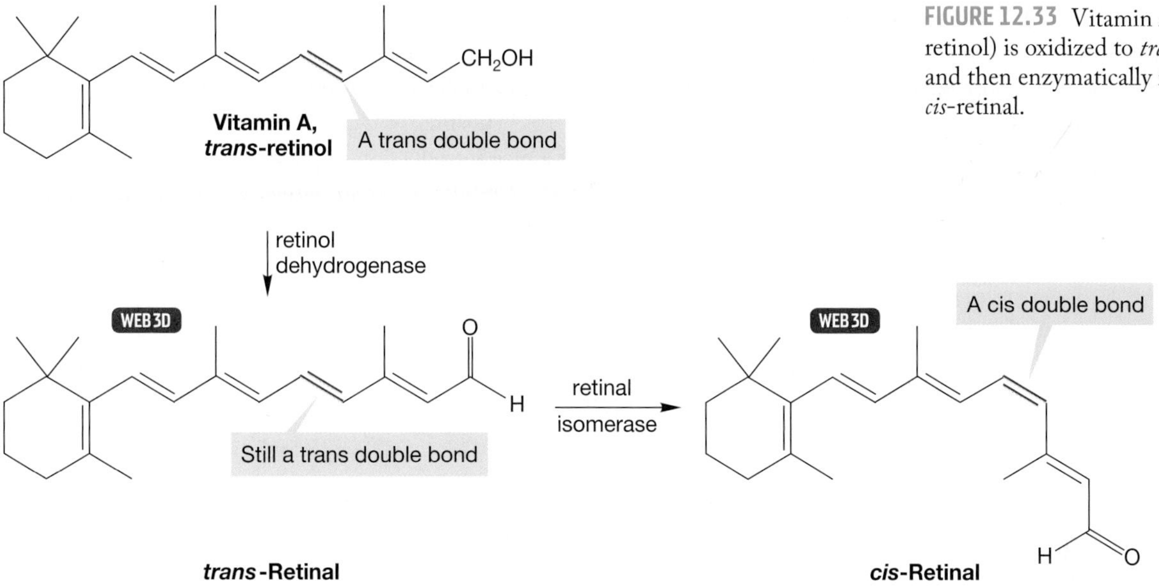

FIGURE 12.33 Vitamin A (*trans*-retinol) is oxidized to *trans*-retinal and then enzymatically isomerized to *cis*-retinal.

cis-Retinal (but *not trans*-retinal) has the proper shape to react with a protein called opsin to form rhodopsin (Fig. 12.34). Rhodopsin absorbs strongly in the visible spectrum, which results in isomerization of the cis double bond back to trans. This absorption of visible light changes the shape of the molecule severely, destroying the crucial fit of molecule and protein. The attachment point is exposed, and the retinal is detached from the protein. As this process occurs, a nerve impulse is generated and perceived by our brains as light. The chemistry involved in the attachment and detachment of retinal and opsin is covered in Chapter 16.

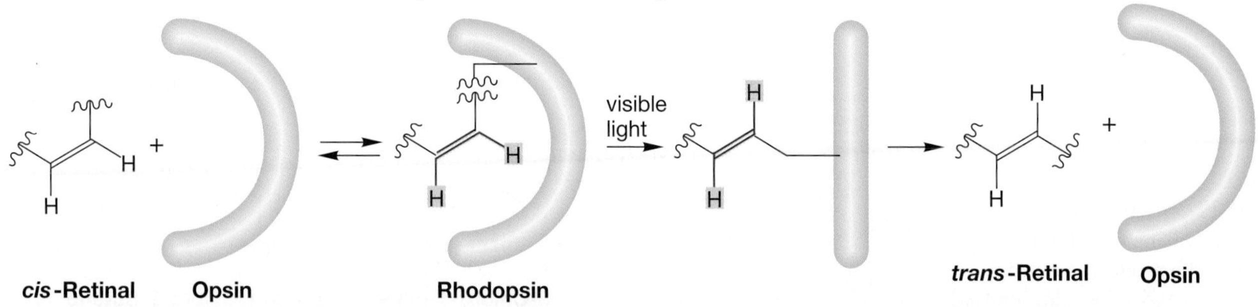

FIGURE 12.34 *cis*-Retinal has the correct shape to react with opsin to form rhodopsin. The squiggly lines indicate the rest of the molecule.

BOMBYKOL

Most insects (and other animals, including humans) "talk" to each other by releasing chemicals that carry meaning. Such molecules are called "pheromones." Some

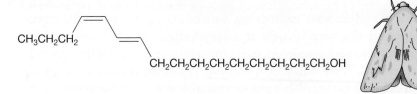

are sexual attractants, others are trail markers, and still others carry an alarm signal. Here is bombykol, a simple cis, trans dienol that serves as the female silkworm moth's attractant for the male. The female of the species, *Bombyx mori*, releases bombykol from two glands located at her abdomen. The structure of bombykol was worked out by the German chemist Adolph Butenandt over a long period, ending in 1956. He collected 6.4 mg of the pure pheromone from 500,000 moth abdomens. Such work was extraordinarily difficult in those days, as he did not have the advantage of modern spectroscopy to help with structure determination.

Eventually, the correct structure was determined and the molecule was made independently. The male moth's

receptor for bombykol also recognizes the other stereoisomers, but the proper *Z, E* (cis, trans) isomer is far more active than the *Z, Z* (cis, cis) or *E, E* (trans, trans) versions.

As you might imagine, if bombykol is to function effectively as an attractant in the open air, reception must be very efficient; a little bombykol must go a long way if the silkworm moths are to have a satisfactory love life. In fact this is so; the male is staggeringly efficient at detecting this molecule. It is effective in doses as small as 1×10^{-12} μg/mL.

For a nice introduction to pheromones, see William C. Agosta, *Chemical Communication*, Scientific American Library, New York, 1992.

12.9 The Chemical Consequences of Conjugation: Addition Reactions of Conjugated Dienes

1,2-Hydrohalogenation of dienes

1,4-Hydrohalogenation of dienes

Now we move on to reactions (at last!). Conjugation affects not only the physical properties of molecules but their reactions as well. Although the basic chemistry of conjugated dienes does resemble that of isolated alkenes, there are detailed, but important differences introduced by the presence of the *pair* of connected double bonds.

12.9a Addition of HBr and HCl When hydrogen chloride adds to 1,3-butadiene at −78 °C, two major products of hydrohalogenation are formed, 3-chloro-1-butene and *trans*-1-chloro-2-butene (Fig. 12.35). There is a very small amount of *cis*-1-chloro-2-butene also formed. The main product is produced through a pathway of 1,2-addition and the minor products by a 1,4-addition.

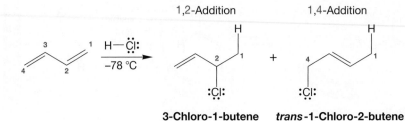

FIGURE 12.35 Hydrohalogenation of 1,3-butadiene with HCl produces the products resulting from a 1,2- and a 1,4-addition (red numbers not IUPAC).

We saw this reaction briefly in the discussion of resonance in Chapter 9 (p. 369). The source of the two pathways becomes clear as we work through the mechanism step by step. The first reaction must be protonation of butadiene by hydrogen chloride (Fig. 12.36). There is a choice of two cations, one of which is a resonance-stabilized allylic cation, and the other an unstabilized primary carbocation. The stabilized cation is much lower in energy and its formation is greatly preferred.

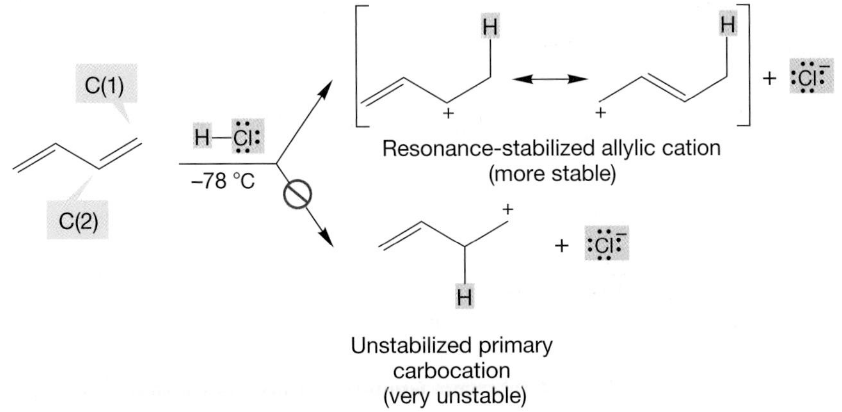

FIGURE 12.36 Formation of a resonance-stabilized allylic cation is preferred.

The resonance description of the allylic cation shows the source of the two products, because chloride can add to either of the two carbons sharing the positive charge to give the two observed alkyl chloride products (Fig. 12.37). Hydrohalogenation of conjugated dienes with HBr also results in 1,2- and 1,4-addition.

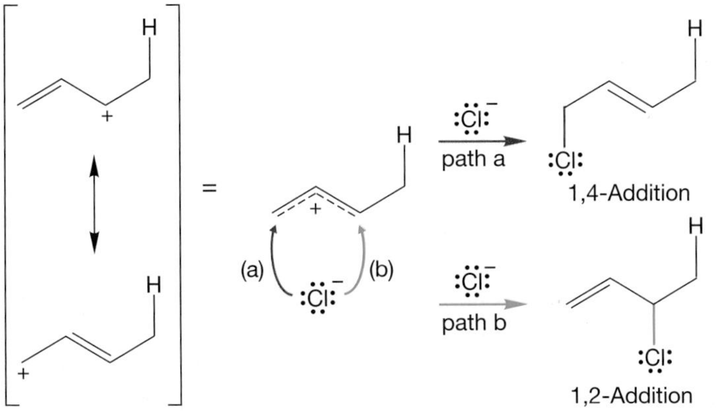

FIGURE 12.37 Addition of the nucleophilic chloride ion at the two positions (a and b) sharing the positive charge gives the products of 1,4- and 1,2-addition.

PROBLEM 12.17 When treated under S_N1 conditions, the two allylic chlorides shown here give the same pair of alcohols in the same ratio. Draw structures for the two alcohols and show the mechanism leading to each product.

12.9b Addition of Br₂ and Cl₂   Other addition reactions of conjugated dienes also produce two products. Like an isolated alkene, conjugated dienes react with bromine or chlorine to give dihalides. As in hydrohalogenation addition, products of both 1,2- and 1,4-addition are formed (Fig. 12.38).

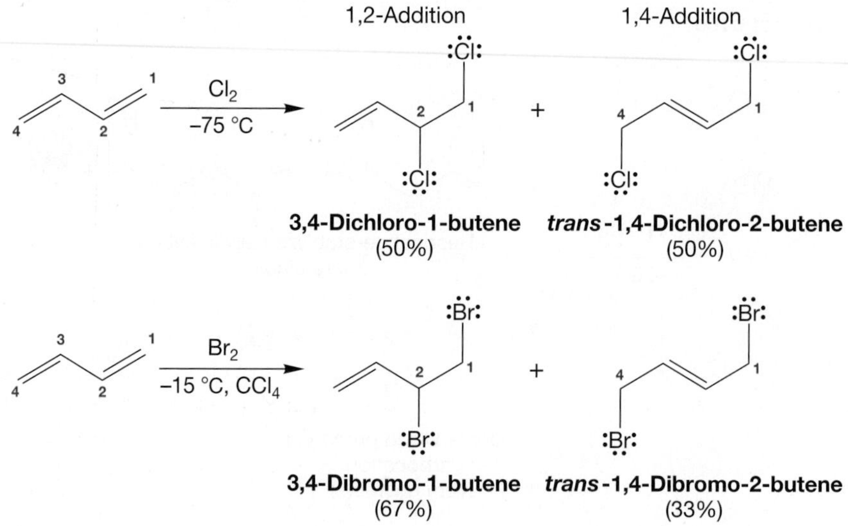

FIGURE 12.38 Addition of bromine or chlorine to conjugated dienes gives two products.

WORKED PROBLEM 12.18 One potential source of the 1,4-addition product in the bromination of a conjugated diene is the bromonium ion shown below. Look carefully at the products of the reaction of 1,3-butadiene and bromine in Figure 12.38 and then explain why the mechanism shown is certainly wrong.

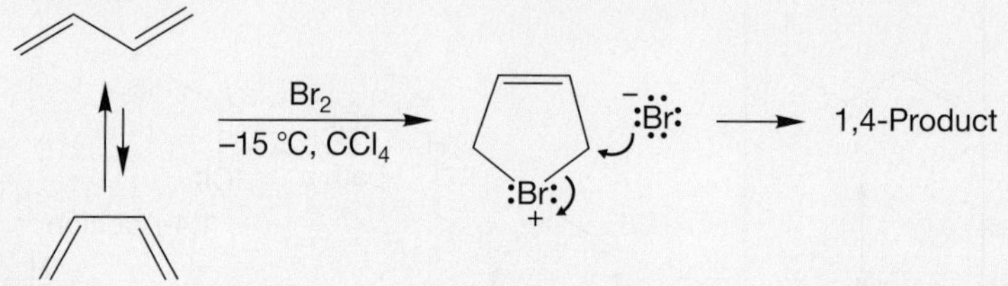

ANSWER Although there is nothing wrong structurally or electronically with the five-membered ring bromonium ion, it cannot be involved in this reaction. Opening of this ring must lead to an alkene containing a cis double bond, and the 1,4-product is, in fact, only trans (Fig. 12.38). The technique for solving this problem is to use the five-membered bromonium ion as the intermediate, see what the structure of the product must be, and compare that to what is actually observed. If we pay careful attention to stereochemistry, we can see that the five-membered ring cannot be involved.

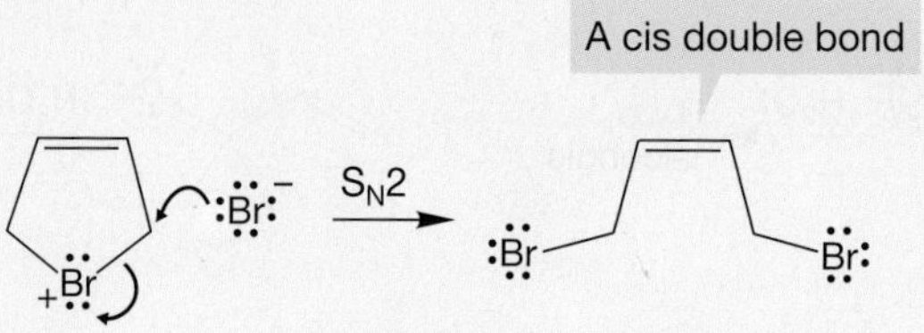

12.10 Thermodynamic and Kinetic Control of Addition Reactions

The addition reactions in the previous sections allow us to look at general properties of chemical reactions and to address fundamental issues of reactivity. Just what are the factors that influence the direction a reaction takes? Why is one product formed more than another? Is it because of the energies of the products themselves, or something else?

There are curious effects of temperature and time on the addition reactions of HX and X_2 with conjugated dienes. If the reactions are run at low temperature, it is the product of 1,2-addition that is generally favored. The warmer the reaction conditions, the more product of 1,4-addition is formed. At high temperature (25 °C) the product of 1,4-addition is favored (Fig. 12.39).

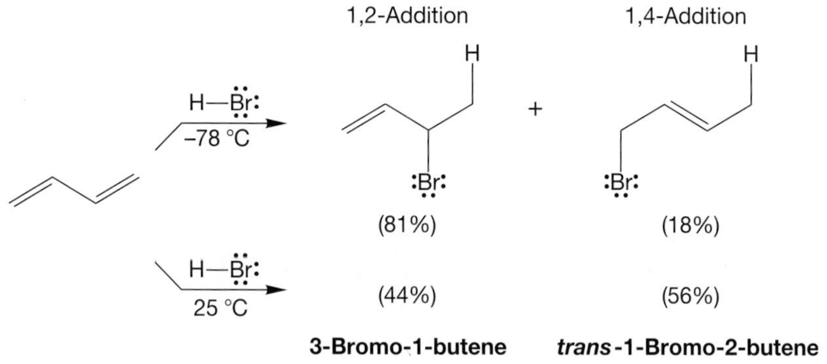

FIGURE 12.39 The ratio of 1,2- to 1,4-addition is temperature dependent.

The effect of time is seen in the observation that the products shown in Figure 12.39 will equilibrate if allowed to stand in solution. As Figure 12.40 shows, the same mixture of 1,2- and 1,4-addition products is formed from each compound, and the product of 1,4-addition is favored.

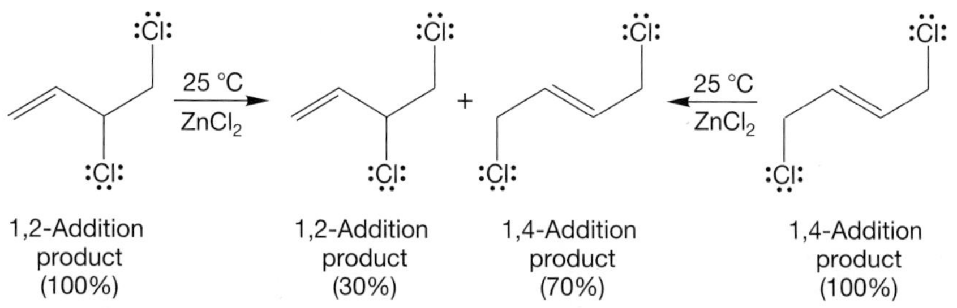

FIGURE 12.40 Equilibration of the products of 1,2- and 1,4-addition at 25 °C.

Several questions now confront us: (1) Why is the product of 1,4-addition favored? (2) How do the products of 1,2- and 1,4-additions equilibrate? (3) Why is the 1,2-addition product favored at low temperatures?

The first question is easy. The product of 1,4-addition contains a disubstituted double bond, whereas the 1,2-product has only a monosubstituted double bond, and is therefore less stable (Fig. 12.41). *Remember*: The more substituted an alkene, the more stable it is (p. 115).

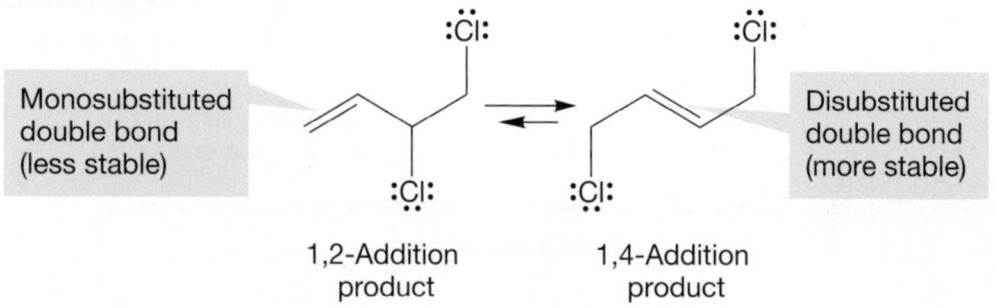

FIGURE 12.41 The compound containing a disubstituted double bond is more stable than the compound with a monosubstituted double bond.

It is also not difficult to see how the two products equilibrate. For example, S_N1 solvolysis of either of the two products in Figure 12.40 leads to the same, resonance-stabilized allylic cation (Fig. 12.42). The chloride ion can then add at either of the two positions sharing the positive charge and at equilibrium it will be the more stable product of 1,4-addition that dominates.

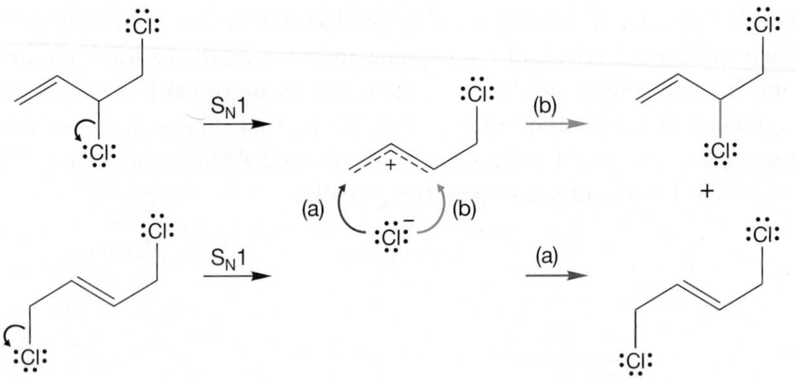

FIGURE 12.42 The 1,2- and 1,4-addition products equilibrate through the formation of a resonance-stabilized allylic cation.

The last question is tougher. Why should one product be favored at low temperature and another at higher temperature and at longer reaction times? With longer reaction times and higher temperatures, thermodynamics takes over and the more stable product prevails. At shorter reaction time, or under mild conditions, it is not the *more stable* 1,4-compound that is favored, but the *less stable* 1,2-addition product. This change can be explained if the 1,2-addition product is formed more easily. The rate of formation of the 1,2-product is then by definition faster than that of the 1,4-product. Therefore, the transition state for formation of the less stable 1,2-product must be lower in energy than the transition state for formation of the more stable 1,4-product. The *lower energy* product is formed by passing over the *higher energy* transition state and the *higher energy* product is formed by passing over the *lower energy* transition state (Fig. 12.43). In Chapter 8, you were warned that such cases would occur; here is one.

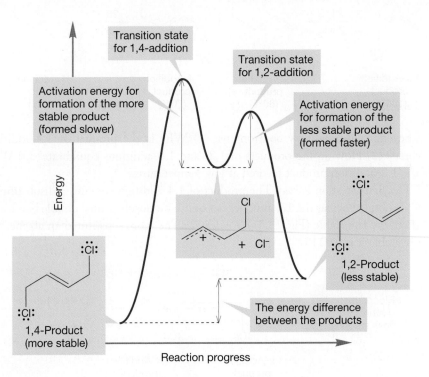

FIGURE 12.43 An energy diagram that summarizes formation of 1,2- and 1,4-addition products from a common intermediate.

The 1,2-addition product is favored under conditions of *kinetic control* (short time, lower temperature), and the 1,4-product is favored under conditions of *thermodynamic control* (long time, higher temperature). The 1,2-product is the kinetic product, and results from taking a pathway over a lower-energy transition state. The 1,4-product results from conditions that favor formation of the more stable product; it is the thermodynamic product in this case.

Now we have a fourth question, and it's a nice one. For both Cl_2 and HCl addition, we know why thermodynamics favors the 1,4-product; a disubstituted alkene is more stable than a monosubstituted alkene. But why does kinetic control favor the product of 1,2-addition?

Let's look at HCl addition because the best experiments have been done in this area. The simplest explanation focuses on distance. Where is the chloride ion after the initial protonation at C(1)? At the moment of its birth, it is much closer to C(2) than to C(4). Perhaps addition of chloride is faster at the 2-position than at the 4-position by virtue of simple proximity (Fig. 12.44).

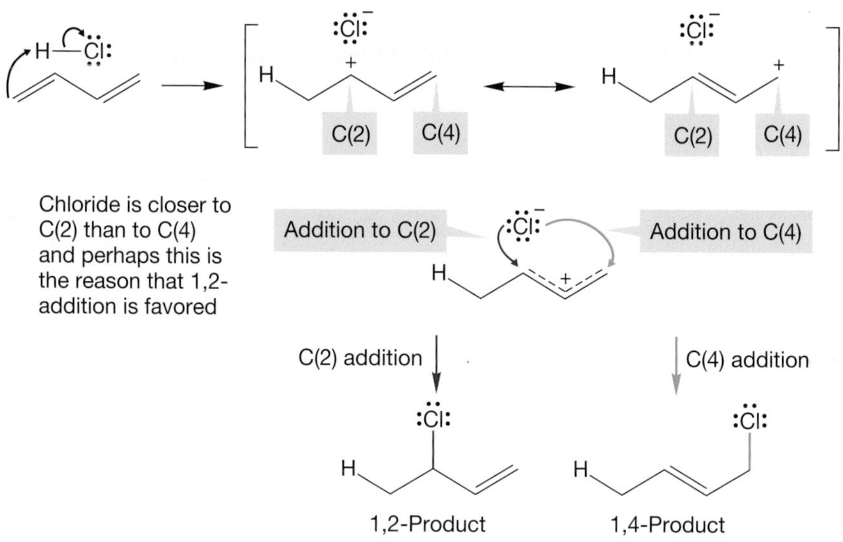

FIGURE 12.44 After protonation, the chloride ion is closer to C(2) than to C(4). Perhaps this proximity accounts for the preference for 1,2-addition.

A second possibility looks at the two resonance forms for the allylic cation (Fig. 12.44), which are not equivalent. The one on the left is a secondary carbocation, the one on the right is a primary carbocation. The first will be weighted more heavily than the second, and most of the positive charge in the ion will be at the secondary position. Perhaps it is this more positive position that is attacked by the anion under kinetic control to give the 1,2-addition product. It is this factor that is almost always cited as the reason 1,2-addition is preferred kinetically.

There is a brilliant experiment that settles the issue, done by Eric Nordlander (1934–1984) at Case Western Reserve University. He studied the addition of deuterium chloride (D—Cl), not to 1,3-butadiene, but to 1,3-pentadiene. In this molecule, addition of D^+ yields a resonance-stabilized diene in which both resonance forms are secondary cations. Indeed these resonance forms are identical except for the remote isotopic deuterium label (Fig. 12.45).

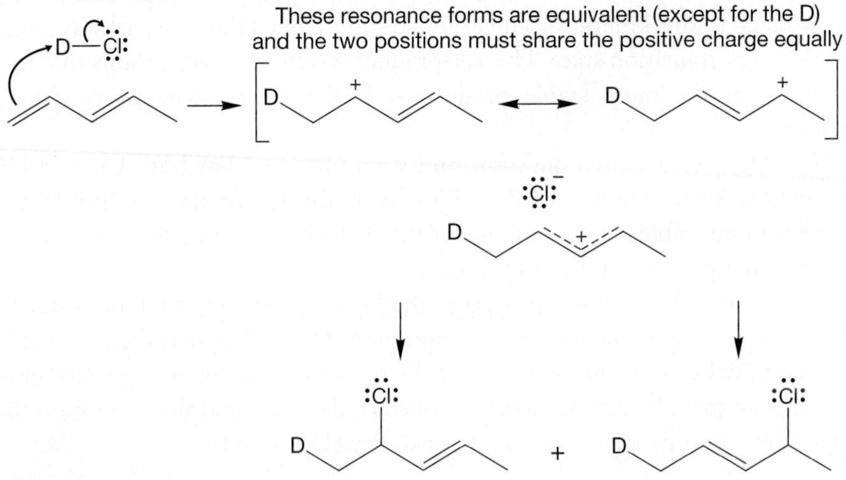

FIGURE 12.45 Protonation (D⁺ in this case) of 1,3-pentadiene gives a resonance-stabilized allylic carbocation in which both contributing resonance forms are secondary. Addition of chloride at the two positions must take place to give equal amounts of 1,2- and 1,4-addition.

If the preference for 1,2-addition depends on the difference between a secondary and a primary carbocation, that preference should disappear in the reaction with 1,3-pentadiene. If proximity is at the root of the effect, 1,2-addition should still be favored. For once simplicity wins out, and 1,2-addition is still preferred (Fig. 12.46).

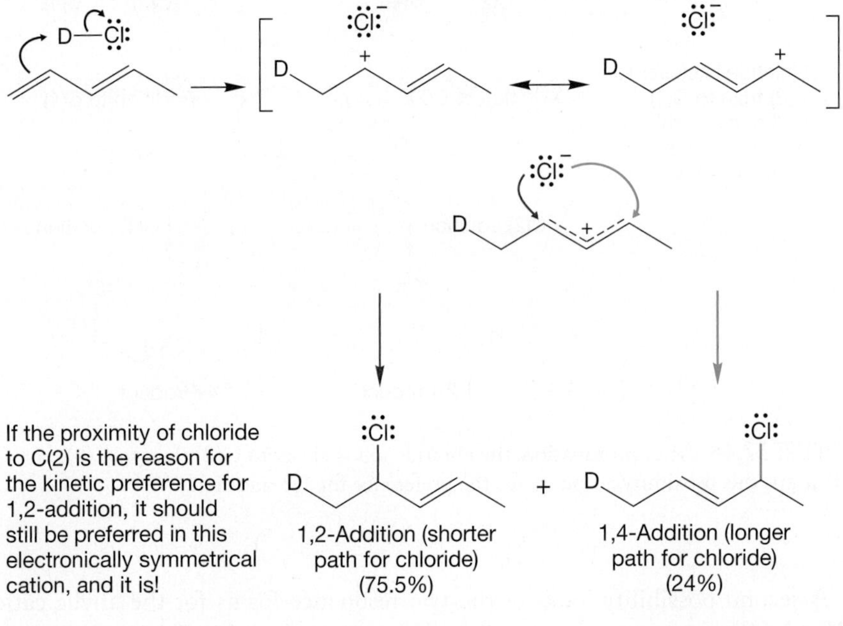

FIGURE 12.46 1,2-Addition is still preferred in 1,3-pentadiene. The kinetic preference for 1,2-addition is a simple proximity effect.

If the proximity of chloride to C(2) is the reason for the kinetic preference for 1,2-addition, it should still be preferred in this electronically symmetrical cation, and it is!

1,2-Addition (shorter path for chloride) (75.5%)

1,4-Addition (longer path for chloride) (24%)

PROBLEM 12.19 Predict the kinetic and thermodynamic products for HCl addition to 4-methyl-1,3-pentadiene. Be careful, thermodynamic products are not always a result of 1,4-addition.

Summary

Under thermodynamic (relatively high energy) conditions, the more stable product is the major compound formed in addition reactions with conjugated dienes. Under kinetic conditions, the major product comes from 1,2-addition—even if it is the less stable product. Why? Proximity! In the formation of the allyl cation, the nucleophile is "born" closer to the 2-position than to the 4-position.

We will now continue with an examination of a few more reactions in which allylic stabilization is important.

12.11   The Allyl System: Three Overlapping 2p Orbitals

We have met the structure of the allyl system several times already (Problems 1.61, 9.3), but Figure 12.47 again summarizes both the molecular orbital treatment and the resonance treatment of this system of three contiguous 2p orbitals.

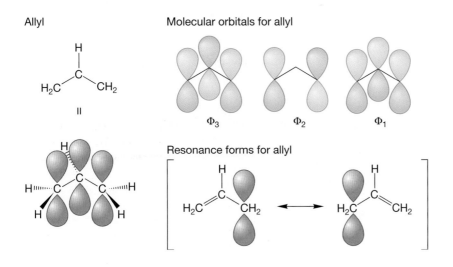

FIGURE 12.47 The resonance and molecular orbital treatment of the allyl system.

> **PROBLEM 12.20** Add electrons to both the resonance and molecular orbital descriptions in Figure 12.47 to form the allyl anion, radical, and cation.

12.11a The Allylic Cation: S_N1 Solvolysis of Allylic Halides

12.11a The Allylic Cation: S_N1 Solvolysis of Allylic Halides   The resonance stabilization of the allylic cation makes allylic halides relatively reactive under S_N1 conditions (Fig. 12.48).

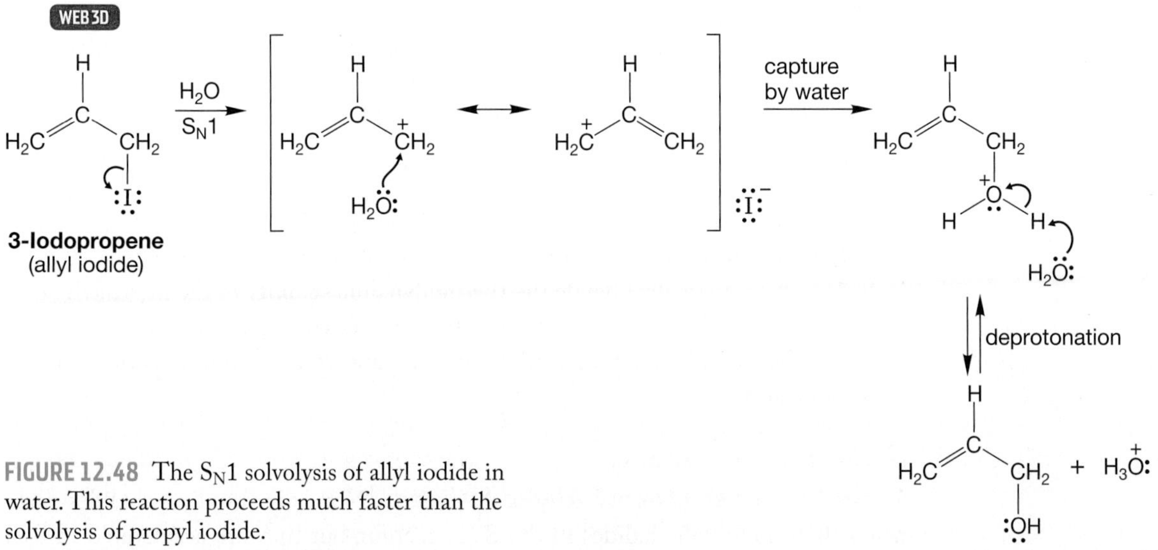

FIGURE 12.48 The S_N1 solvolysis of allyl iodide in water. This reaction proceeds much faster than the solvolysis of propyl iodide.

As mentioned earlier, the allyl cation can be captured at either carbon sharing the positive charge. Table 12.3 gives some average rates of S_N1 solvolysis for a few common structural types. Primary allylic halides react much faster than primary alkyl halides. Secondary and tertiary allylic halides also ionize faster than their non-allylic counterparts. Resonance stabilization does have an accelerating effect on the ionizations.

TABLE 12.3 Relative Rates of S_N1 Solvolysis of RCl in 50% Ethyl Alcohol at 45 °C

Molecule	Ion	Relative Rate	Molecule	Ion	Relative Rate
	Primary [a]	1.0 [a]		Allyl: one resonance form is primary, one secondary	1157
	Secondary	1.7		Tertiary	3×10^4
	Allyl: both resonance forms are primary	14.3		Allyl: one resonance form is primary, one tertiary	1.9×10^6
	Allyl: both resonance forms are primary	21.4		Allyl: one resonance form is primary, one tertiary	7.9×10^6
	Allyl: one resonance form is primary, one secondary	1300			

[a] The value for the primary substrate is suspect, and may include some contribution from S_N2 reactions.

PROBLEM 12.21 Explain why the last two entries in Table 12.3 react at nearly the same rate and are so much faster than the others. Can you devise a molecule that would react even faster than the last two entries? Draw it.

Remember: We are doing a dangerous thing here. We are again equating thermodynamic stability (a secondary allylic cation is more stable than a non-resonance-stabilized secondary cation) with kinetics (therefore secondary allylic halides form cations faster than do typical secondary halides). Although justified here, we must be clear on the reason why thermodynamics and kinetics are closely related in this example. Work Problem 12.22.

PROBLEM 12.22 Explain carefully, with the use of an energy diagram, why in an S_N1 reaction of an alkyl halide the thermodynamic stability of the intermediate cation is related to the ease of ionization of the corresponding alkyl halide. Remember the Hammond postulate (p. 351), and consider the transition state for ionization.

12.11b S_N2 Reactions of Allylic Halides
Not only do allylic halides react faster than non-allylic halides in the S_N1 reaction, but they are accelerated in S_N2

reactions as well. Table 12.4 gives some average rates of S$_N$2 reactions for a variety of halides.

TABLE 12.4 Relative Rates of S$_N$2 Reaction with Ethoxide in Ethyl Alcohol at 45 °C

Molecule	Relative Rate	Molecule	Relative Rate
(propyl chloride)	1.0	(but-2-enyl chloride)	97
(allyl chloride)	37.0	(3-chlorobut-1-ene)	1.85
(methallyl chloride)	33.0	(prenyl chloride)	556

When we are considering a question involving reaction rates, the answer can always be found through an analysis of the transition states. The rate is dependent on the activation energy for the reaction, which is the height of the transition state relative to the starting halide. Let's compare the transition state for the reaction of a 1-propyl halide with that for the reaction of an allyl halide (Fig. 12.49).

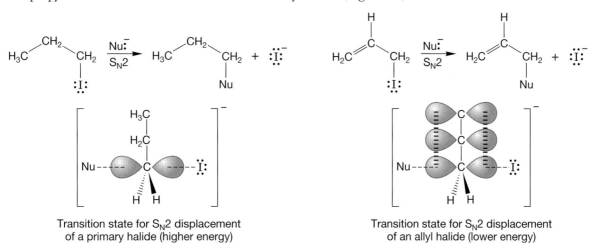

Transition state for S$_N$2 displacement of a primary halide (higher energy)   Transition state for S$_N$2 displacement of an allyl halide (lower energy)

FIGURE 12.49 A comparison of the transition states for S$_N$2 displacement of a primary iodide and allyl iodide.

In the transition state for the S$_N$2 reaction of the allyl halide, there is overlap between the 2p orbital developed on the carbon at which displacement occurs and the 2p orbitals of the double bond (Fig. 12.49). Therefore, this transition state contains an allylic system. It will benefit energetically from the delocalization of electrons, and will lie at lower energy than the transition state for S$_N$2 displacement in the 1-propyl system, in which there is no delocalization.

12.11c The Allyl Radical It is much easier to form a resonance-stabilized allylic radical than an undelocalized radical. We saw the effects of delocalization in reactions involving allylic radicals in Chapter 11 (p. 497). For example, the regiospecific bromination of the allylic position of alkenes with NBS is possible because of the relative ease of forming the delocalized allyl radical (Fig. 11.56).

12.11d The Allyl Anion Now let's look at the acidity of allylic hydrogens to see if the formation of a resonance-stabilized allylic anion has any effect on that acidity. We would surely expect it to. If delocalization of electrons is important, then removal of a proton at the allylic position should be easier than at a position not leading to a delocalized anion. A comparison of the acidities of propane and propene should test this idea (Fig. 12.50).

FIGURE 12.50 Propene is far more acidic than propane.

In fact, the pK_a of propene is 43, some 7–17 pK_a units lower than that for propane ($pK_a \sim 50$–60). The uncertainty arises from the difficulty of measuring the pK_a of an acid as weak as a saturated hydrocarbon. In any event, propene is a much stronger acid than propane, and it is reasonable to suppose that the formation of the delocalized allyl anion is responsible for the increased acidity of the alkene.

12.12   The Diels–Alder Reaction of Conjugated Dienes

In 1950, Otto Diels (1876–1954) and his student Kurt Alder (1902–1958) received the Nobel prize for their work on what has appropriately come to be known as the **Diels–Alder reaction**, which they discovered in 1928. (So much for timeliness; at least they lived long enough to outlast the Nobel committee's renowned conservatism.) We will have more to say about its mechanism in Chapter 20, but the Diels–Alder reaction is characteristic of conjugated dienes and is of enormous synthetic utility. Therefore we must outline it here.

When energy is supplied to a mixture of a conjugated diene and an alkene, a ring-forming reaction takes place to produce a cyclohexene. The alkene in this reaction is called a **dienophile**, because it has demonstrated an affinity for the diene. The product of the reaction is often called the adduct, which is another word for product in a reaction between two molecules. Our discussion of the Diels–Alder mechanism begins with the arrow formalism, which points out which bonds are broken and shows where the new bonds are made in the forward reaction (Fig. 12.51).

Which orbitals are involved? The Diels–Alder reaction begins with the overlap of the π systems of the dienophile and the diene. The two reaction partners approach in parallel planes and as the bonds form, the end carbons of both the diene and alkene rehybridize from sp^2 to sp^3 (Fig. 12.52).

Diels–Alder reaction

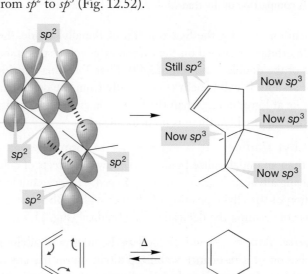

FIGURE 12.51 The arrow formalism for the Diels–Alder reaction.

FIGURE 12.52 In the Diels–Alder reaction, the two participants approach each other *in parallel planes*. As the reaction occurs, the end atoms of the diene and the dienophile rehybridize from sp^2 to sp^3.

PROBLEM 12.23 Curiously, it matters just how energy is supplied to the molecules in the Diels–Alder reaction. If we heat the reactants up, the reaction occurs, but if we attempt to supply energy photochemically, it generally fails. We will examine such matters in detail in Chapter 20, but we might start to consider them now. When we think about reactions between atoms, we do not consider every electron in the atom, just those most loosely held, the valence electrons. Similarly, in molecular reactions it is the most weakly held electrons, those in the HOMO, that control reactivity. The electrons in the HOMO might well be thought of as the valence electrons of molecules. Remember what Figure 12.25 tells us about the interaction of light and molecules. First, consider the phase relationships in the possible HOMO–LUMO interactions in the transition state for a simple Diels–Alder reaction. Then think about how the absorption of a *single* light photon might change those interactions and offer an explanation for why light energy does not promote the Diels–Alder reaction.

PROBLEM 12.24 Write an arrow formalism for the reverse Diels–Alder reaction.

It is possible to estimate the thermochemistry in the Diels–Alder reaction by comparing the bonds made and broken in the reaction. Three π bonds in the starting material are converted into one π bond and two σ bonds in the product. Accordingly, the reaction is calculated to be exothermic by approximately 28 kcal/mol (Fig. 12.53).

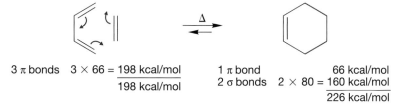

3 π bonds   3 × 66 = 198 kcal/mol
            _____
                 198 kcal/mol

1 π bond   66 kcal/mol
2 σ bonds   2 × 80 = 160 kcal/mol
                     _____
                          226 kcal/mol

The reaction is 226–198 = 28 kcal/mol exothermic;
$\Delta H = -28$ kcal/mol

FIGURE 12.53 The simplest Diels–Alder reaction is exothermic by about 28 kcal/mol.

PROBLEM 12.25 Why is the bond strength of the new σ C—C bonds in cyclohexene (80 kcal/mol each) used in Figure 12.53 lower than the C—C bond in ethane (90 kcal/mol)?

As shown in the previous figures, the Diels–Alder reaction requires the s-cis form of the diene. However, it is the s-trans arrangement that is the more stable one, by approximately 3 kcal/mol (p. 523). If equilibrium favors the s-trans form, why is the Diels–Alder reaction observed at all? Even though there is little s-cis form at equilibrium, reaction with the dienophile disturbs the equilibrium between the s-trans and s-cis form by depleting the s-cis partner. Reestablishment of equilibrium generates more s-cis molecules, which can react in turn. Eventually, all the diene can be converted into Diels–Alder product, even though at any moment there is very little of the active s-cis form present (Fig. 12.54).

FIGURE 12.54 Even though the unreactive s-trans form of 1,3-butadiene is favored at equilibrium, the small amount of the s-cis form present can lead to product. The vertical axis of the figure is not to scale. The squiggly lines indicate gaps.

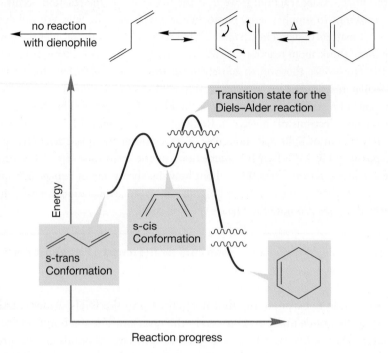

When there is no possibility of any s-cis form, there can be no Diels–Alder reaction. For example, although 1,2-dimethylenecyclohexane reacts normally with dienophiles, 3-methylenecyclohexene does not. The product of this reaction would contain a double bond far too strained to be formed (Fig. 12.55).

FIGURE 12.55 When there can be no s-cis form of the diene, the Diels–Alder reaction fails.

WORKED PROBLEM 12.26 Identify the source of strain in the putative product shown for Diels–Alder addition to 3-methylenecyclohexene in Figure 12.55. Use your models.

ANSWER Although the paper puts up with the double bond drawn in Figure 12.55, Nature won't. The orbitals making up this bridgehead double bond do not, in reality, overlap.

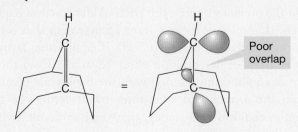

Another serious mechanistic question involves the timing of the formation of the two new σ bonds in the cyclohexene. Are they made simultaneously in a concerted fashion, or are they formed in two separate steps? In principle, the mechanism of the Diels–Alder reaction could either be concerted or involve two steps (Fig. 12.56).

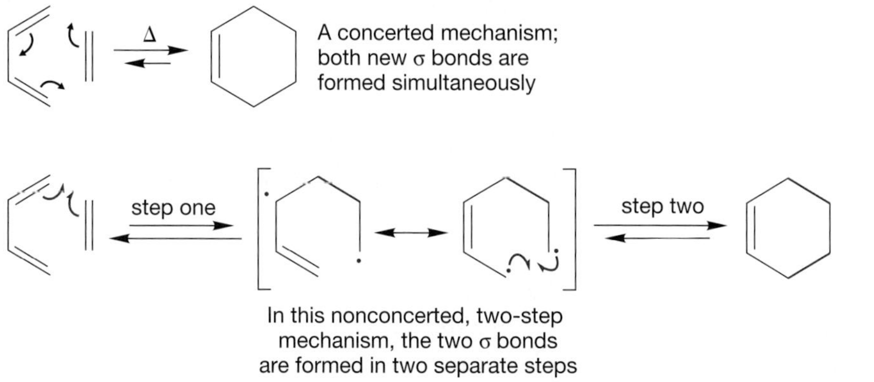

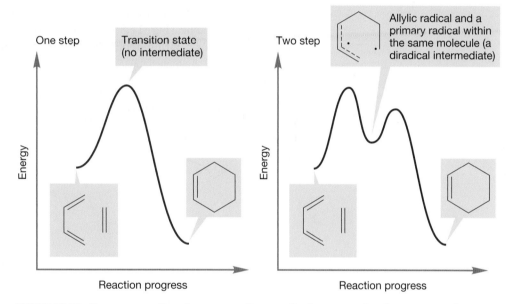

A concerted mechanism; both new σ bonds are formed simultaneously

In this nonconcerted, two-step mechanism, the two σ bonds are formed in two separate steps

FIGURE 12.56 Arrow formalisms for concerted and two-step mechanisms for the Diels–Alder reaction.

This question asks if there is an intermediate in the reaction or not. Figure 12.56 shows the two arrow formalisms and Figure 12.57 gives Energy versus Reaction progress diagrams for both the concerted and two-step processes.

We might immediately be suspicious of the two-step mechanism. As it involves an intermediate allylic radical (Figs. 12.56 and 12.57), shouldn't we observe closure of the diradical at both possible positions? Should there not be a vinylcyclobutane produced as well as the cyclohexene (Fig. 12.58)?

One step

Energy

Transition state (no intermediate)

Reaction progress

Two step

Energy

Allylic radical and a primary radical within the same molecule (a diradical intermediate)

Reaction progress

FIGURE 12.57 Energy versus Reaction progress diagrams for the concerted and two-step mechanisms.

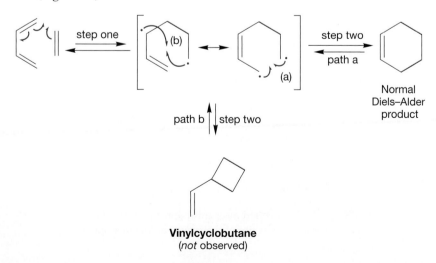

Normal Diels–Alder product

Vinylcyclobutane
(*not* observed)

FIGURE 12.58 The two-step reaction could lead to vinylcyclobutane. This product is not observed.

In fact, Diels–Alder reactions do not lead to vinylcyclobutanes, but give cyclohexenes exclusively. Although this argument would seem at first to favor the concerted mechanism, which can only give cyclohexenes, it could be countered that the transition state for closing the four-membered ring will contain much of the strain present in all cyclobutanes, and will therefore be high in energy.

PROBLEM 12.27 Write an Energy versus Reaction progress diagram for the two-step Diels–Alder reaction that includes formation of the vinylcyclobutane.

As is so often the case, the mechanistic question of the timing of new bond formation is finally answered with a stereochemical experiment. A one-step reaction must preserve the stereochemical relationships present in the original alkene. For example, in a concerted reaction, a cis alkene must give a cis disubstituted cyclohexene. If the reaction is concerted, which means the two new σ bonds form simultaneously, there can be no change in the stereochemical relationship of groups on the alkene (Fig. 12.59). If the bonds are formed in two steps, there should be time for rotation about the carbon–carbon σ bond in the dienophile.

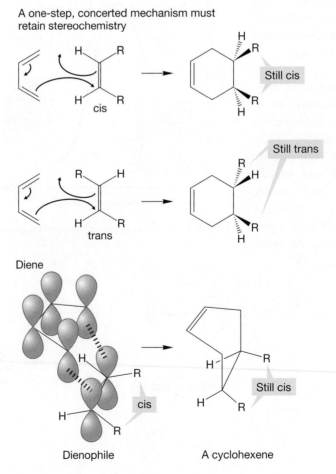

FIGURE 12.59 A concerted mechanism must preserve the stereochemical relationships of the groups on the dienophile.

If the mechanism is stepwise, it should lead to stereochemical scrambling of the groups originally attached to the dienophile (Fig. 12.60).

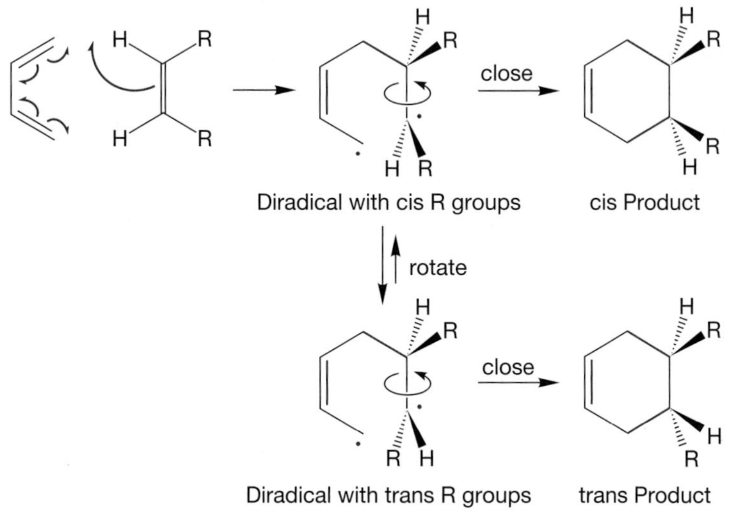

FIGURE 12.60 A two-step mechanism predicts a mixture of stereoisomers.

Diradical with cis R groups   cis Product

rotate

Diradical with trans R groups   trans Product

In fact, retention of stereochemistry is observed. The reaction is a concerted, one-step process (Fig. 12.61).

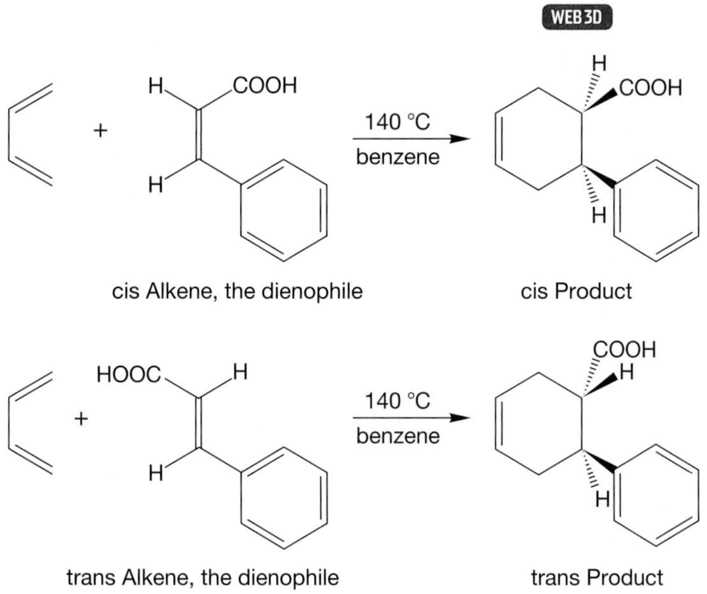

FIGURE 12.61 The preservation of stereochemistry in the Diels–Alder reaction favors the concerted mechanism.

cis Alkene, the dienophile   cis Product

trans Alkene, the dienophile   trans Product

PROBLEM 12.28 The experiment described in Figures 12.59–12.61 is usually easy to understand, but there is another stereochemical experiment possible. It tests the same thing, but is harder to see without models. So, if this problem is difficult to follow, by all means work it through with models. Let's follow the stereochemical fate of groups attached to the ends of the diene, not the alkene. Consider the Diels–Alder reaction between a trans,trans-conjugated diene and ethylene. Only a single cyclohexene product is formed and it has the R groups cis. Explain how this result shows that the mechanism of cycloaddition cannot be a two-step process.

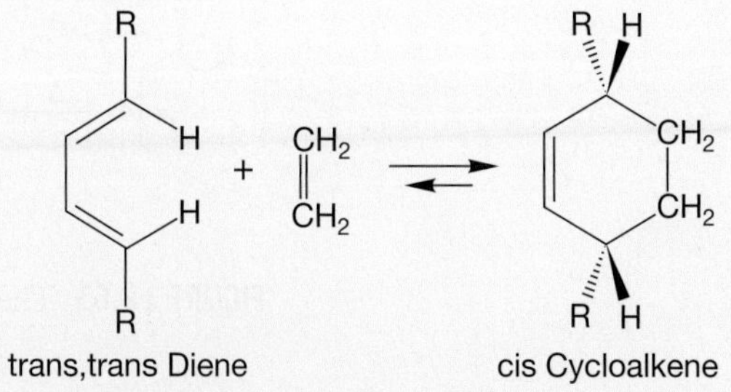

trans,trans Diene   cis Cycloalkene

In the preceding pages, mechanistic points have been illustrated using very simple examples. Actually, the simplest Diels–Alder reaction, ethylene and 1,3-butadiene, has a rather high activation enthalpy of about 27 kcal/mol and requires high pressure and temperature for success. Even 200 °C and a reaction pressure of 4500 psi yields only 18% cyclohexene after 17 hours. Other Diels–Alder reactions proceed under milder conditions. Reactions are more facile when the dienophile bears one or more electron-withdrawing groups and if the diene can easily achieve the required s-cis conformation. Two examples of these relatively fast Diels–Alder reactions are shown in Figure 12.62.

FIGURE 12.62 Two efficient Diels–Alder reactions.

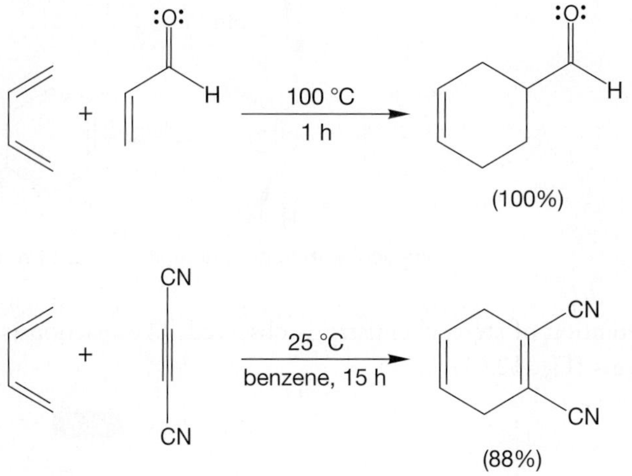

Why is the Diels–Alder reaction so important and so useful? Six-membered rings are very common, and the reaction builds six-membered rings from relatively simple building blocks. Moreover, the reaction is very general. Most dienes react with most dienophiles to give Diels–Alder products. From now on, every cyclohexene you see must trigger the thought "can be made by a Diels–Alder reaction" in your mind. That's an important concept for both synthesis and problem solving.

Like most reactions in organic chemistry, the Diels–Alder reaction is not a one-way street. Reverse Diels–Alder reactions are common, especially at high temperature. *Remember*: $\Delta G = \Delta H - T\Delta S$ (Eq. 8.6, p. 337), and at high temperature the temperature-dependent ΔS term can overwhelm the temperature-independent ΔH term. The reverse Diels–Alder reaction makes two molecules from one and therefore is favored by entropy, ΔS. Cyclohexenes are formed in a reaction of a diene and a dienophile, but they can be used as sources of dienes through the reverse Diels–Alder reaction (Fig. 12.63).

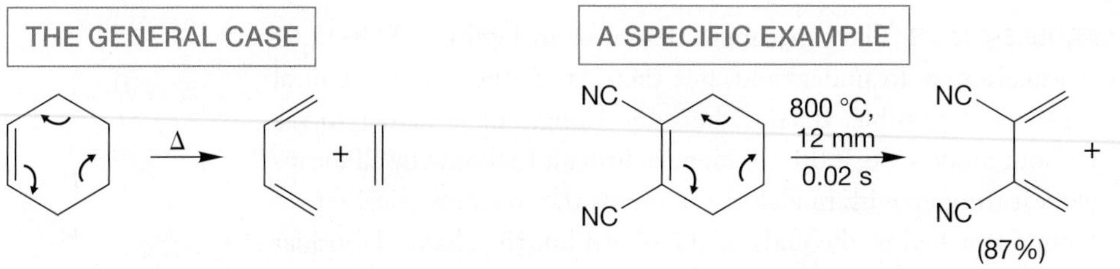

FIGURE 12.63 The reverse Diels–Alder reaction.

Now let's press on to look at other uses of the Diels–Alder reaction, which is surely one of the most synthetically useful of all the reactions in organic chemistry.

Chemists often use bicyclic molecules to investigate effects requiring a fixed geometrical arrangement of orbitals or groups. Consider the simple Diels–Alder reaction
between ethylene or acetylene and a common cyclic diene, cyclopentadiene (Fig. 12.64).

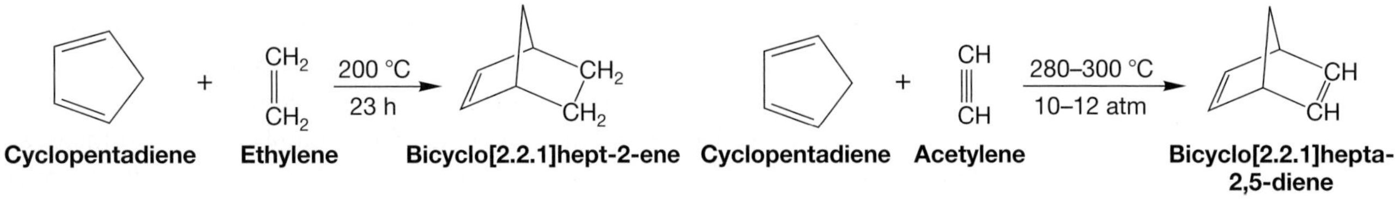

FIGURE 12.64 The Diels–Alder reaction between cyclic dienes and alkenes (or alkynes) gives
bicyclic molecules.

Presto! Bicyclic compounds! But this instant synthetic expertise is not without its price,
because now there is a new stereochemical effect to be considered. When a substituted alkene reacts with a cyclic diene, there are two possible stereoisomeric products even
though the two new σ bonds are formed simultaneously. The substituents on the
alkene must retain their stereochemical relationship during the reaction, but there are
still two possible products. The substituents on the alkene can point toward the newly
formed double bond or away from it (Fig. 12.65). In this molecule, the isomer with

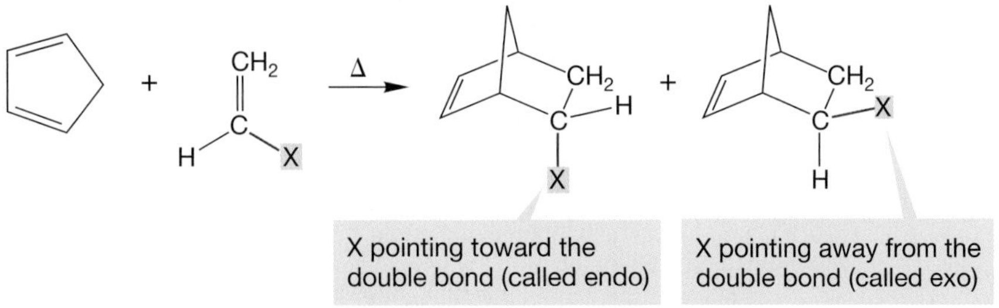

FIGURE 12.65 In principle, this kind
of Diels–Alder reaction can give
either endo or exo products.

the substituents aimed toward the double bond is called **endo** and that with the groups
aimed away from the double bond is called **exo**. More generally, a substituent that is
on the same side as the smaller bridge of a bicyclic system is exo. The exo isomers are
usually more stable, but the endo compounds are kinetically favored in the Diels–Alder
reaction (Fig. 12.66).

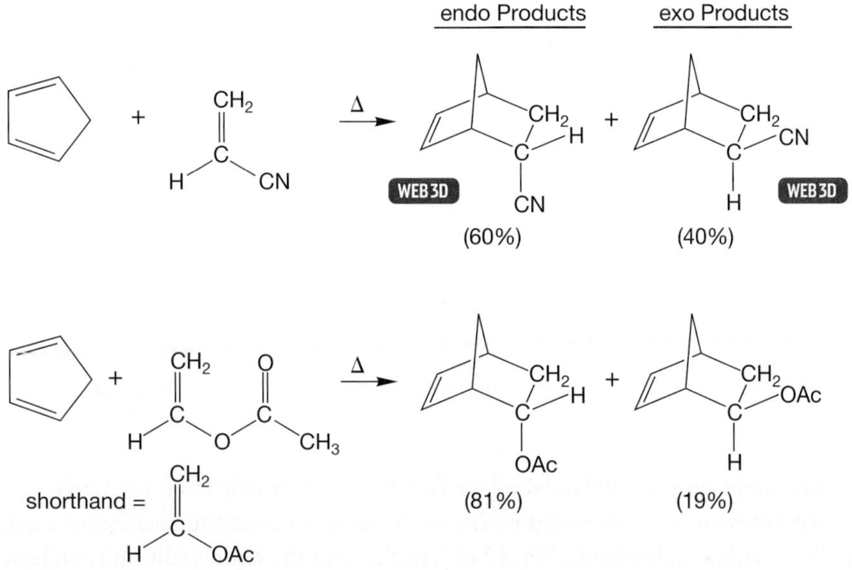

FIGURE 12.66 It is the endo product
that is favored in most Diels–Alder
reactions. In each case, the percentage
of endo product is greater than the
percentage of exo product.

The exo compounds formed from a substituted dienophile are more stable for steric reasons: The substituent points away from the more congested part of the molecule. Indeed, the transition state for exo product formation must also be less crowded than the transition state for endo product formation. But the transition state for the endo product benefits from other factors that overwhelm steric considerations. Chemists have performed high-level molecular orbital calculations and have found that in the endo transition state, but not in the exo transition state, there can be "secondary orbital overlap." This fancy-sounding phrase simply means that there can be interaction between the "back" diene orbitals and orbitals on the substituent X only in the endo transition state. Figure 12.67 shows this effect for the first example

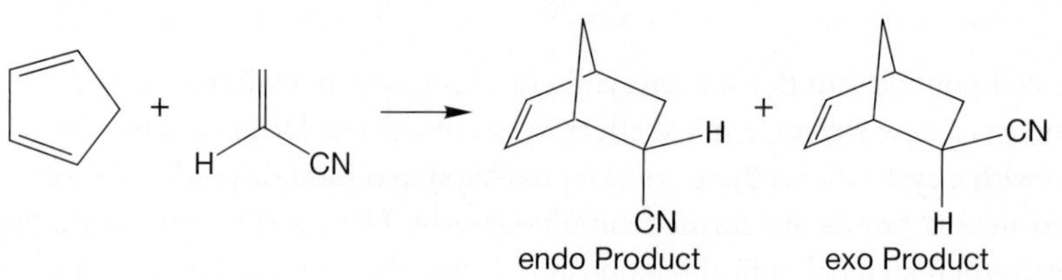

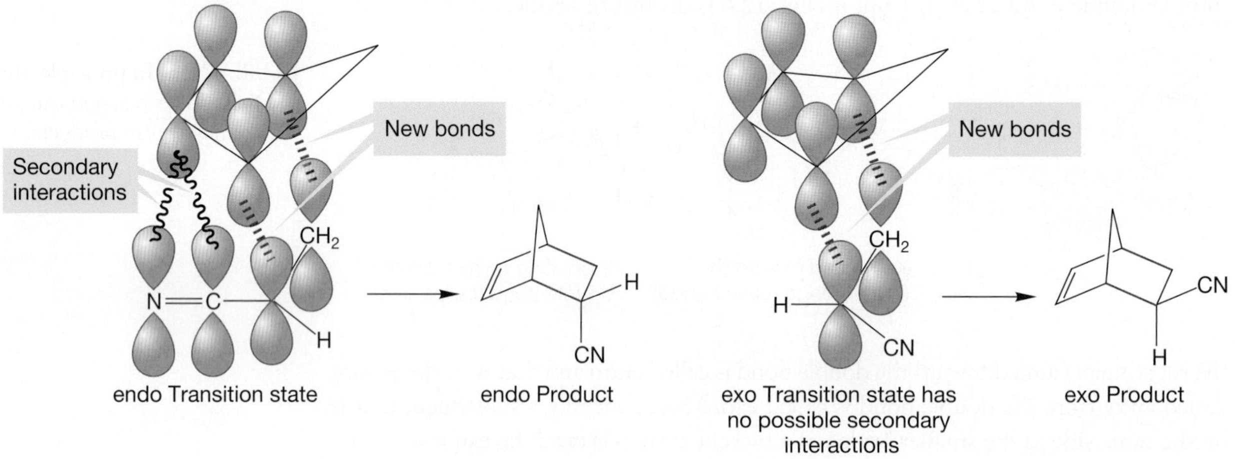

FIGURE 12.67 Secondary orbital interactions stabilize the endo transition states.

of Figure 12.66. Another explanation that has been suggested for the endo selectivity relates to solvent interactions. Whatever the cause, the endo product is usually formed preferentially and sometimes with very high selectivity (Fig. 12.68).

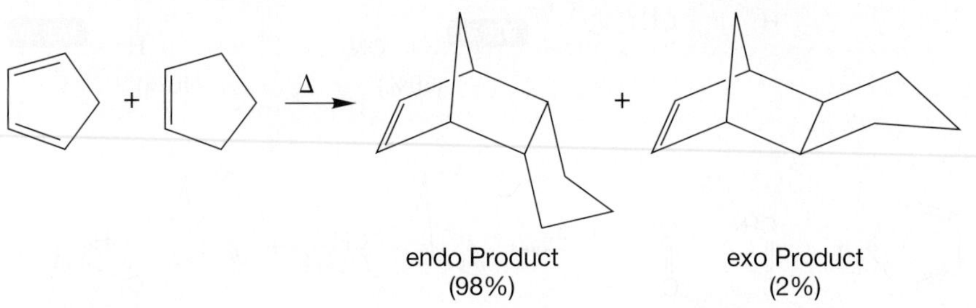

FIGURE 12.68 The Diels–Alder reaction between a cyclic diene and a cyclic dienophile gives a polycyclic product. Once again, it is the endo form that is favored.

Even more complicated polycyclic structures can be made from the Diels–Alder reaction between two ring compounds. In the simplest case, the reaction of a cyclic 1,3-diene with a cycloalkene (Fig. 12.68), notice that the endo addition is preferred.

Two cyclic dienes can also react in Diels–Alder fashion. The self-condensation of cyclopentadiene is the classic example. We'll illustrate this reaction with a pair of problems.

WORKED PROBLEM 12.29 You attempt to prepare 1,3-cyclopentadiene from the base-catalyzed loss of hydrogen bromide (E2 reaction) from 3-bromocyclopentene, but instead of isolating a compound having the formula C_5H_6, you get a high yield of a compound with the formula $C_{10}H_{12}$.

Fame, fortune, and the Nobel prize await you if you can explain this result (at least if you were explaining it in 1928). Be careful—although this question is obviously set up in the text, it is not so easy to write a structure correct in every detail.

ANSWER The formula of $C_{10}H_{12}$ is a clue. Clearly, two C_5H_6 molecules have joined to make the $C_{10}H_{12}$ product. The reaction is, in fact, successful. The problem is that cyclopentadiene dimerizes in a Diels–Alder reaction very quickly. The molecule you isolate is the Diels–Alder dimer. Note formation (mainly) of the endo adduct. Anytime you see a 1,3-diene, especially a cyclic one, think Diels–Alder.

WORKED PROBLEM 12.30 In a fit of frustration or inspiration you decide to boil (reflux) your unanticipated $C_{10}H_{12}$, and, in order to do that, you attach a long condenser to a receiving vessel that is kept very cold. Your $C_{10}H_{12}$ boils at about 165 °C, but you collect a product of the formula C_5H_6 boiling at 40 °C. Moreover, although this C_5H_6 product is stable at low temperature for a long time, at room temperature it rapidly reverts to your starting $C_{10}H_{12}$. Explain.

ANSWER At 165 °C, the Diels–Alder reaction reverses, and the low-boiling cyclopentadiene can be collected and kept at low temperature. If it is allowed to warm up, it again forms the Diels–Alder dimer.

12.13 Special Topic: Biosynthesis of Terpenes

When we discussed cationic addition reactions in Chapter 9, we included a section on the polymerization reactions of alkenes (p. 384). Dienes form quite stable intermediates in addition reactions and are especially prone to polymerization.

PROBLEM 12.31 Write a mechanism for the radical-induced polymerization of 1,3-butadiene.

Nature uses related polymerization reactions (the cart is way before the horse here; we make approximations of Nature, not the other way round!). One form of natural rubber is polyisoprene, which is polymerized **isoprene** (2-methyl-1,3-butadiene).

Nature!

Isoprene   **Polyisoprene**

PROBLEM 12.32 Notice the *Z* configuration of the alkenes in natural rubber. There is another, stereoisomeric form of natural rubber, also a polymer of isoprene, called by the wonderful name gutta percha.[3] What do you suppose it is?

Polyisoprene is a limiting case—a molecule composed of a vast number of isoprene units. Many important natural products called **terpenes** are constructed of smaller numbers of isoprene units put together in various ways. Terpenes are widespread in nature and are still often extracted from plants. The smaller terpenes are widely used as flavors and fragrances. Indeed, terpene chemistry is the very basis of the fragrance industry.

---

[3]Gutta percha comes from the Malay, *getah percha* (gum of the percha tree).

Most of the reactions used to make terpenes from diene units can be largely understood in terms of reaction mechanisms we know well. For example, 3-methyl-3-buten-1-ol is converted first into the corresponding 3-methyl-3-buten-1-ol pyrophosphate (isopentenyl pyrophosphate) by reaction with pyrophosphoric acid (Fig. 12.69).

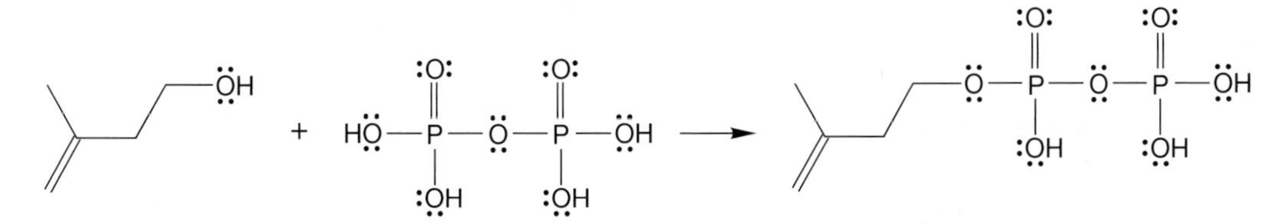

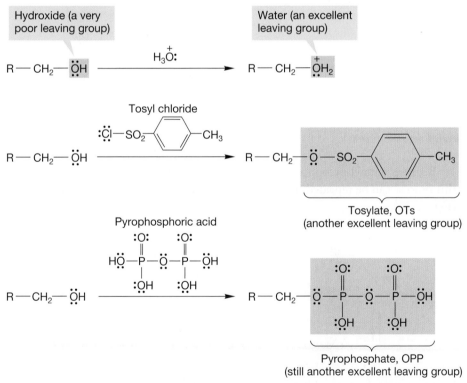

3-Methyl-3-buten-1-ol   **Pyrophosphoric acid**   **3-Methyl-3-buten-1-ol pyrophosphate**
(isopentenyl pyrophosphate)

FIGURE 12.69 In the biosynthesis of terpenes, 3-methyl-3-buten-1-ol is first converted into the corresponding pyrophosphate.

Recall the discussion of various ways to change OH, a very poor leaving group, into much better leaving groups (p. 281). We saw examples of protonation to make the leaving group water, and of formation of a sulfonate ester (reviewed in Fig. 12.70). Formation of the pyrophosphate ester serves exactly the same purpose and is widely used in Nature to generate a good leaving group from a poor one.

FIGURE 12.70 Three ways of converting OH, a poor leaving group, into a good leaving group.

Hydroxide (a very poor leaving group)

Water (an excellent leaving group)

$R-CH_2-\ddot{O}H \xrightarrow{H_3\overset{+}{O}:} R-CH_2-\overset{+}{\ddot{O}}H_2$

Tosyl chloride

$:\ddot{Cl}-SO_2-\!\!\!\!\bigcirc\!\!\!\!-CH_3$

$R-CH_2-\ddot{O}H \longrightarrow R-CH_2-\ddot{O}-SO_2-\!\!\!\!\bigcirc\!\!\!\!-CH_3$

Tosylate, OTs
(another excellent leaving group)

Pyrophosphoric acid

$R-CH_2-\ddot{O}H \longrightarrow R-CH_2-\ddot{O}-\overset{\overset{\displaystyle :O:}{\|}}{P}-\ddot{O}-\overset{\overset{\displaystyle :O:}{\|}}{P}-\ddot{O}H$

Pyrophosphate, OPP
(still another excellent leaving group)

PROBLEM 12.33 Explain why the pyrophosphate ester in Figure 12.70 is an excellent leaving group. The abbreviation OPP is used for the pyrophosphate group.

There is an enzyme that equilibrates 3-methyl-3-buten-1-ol pyrophosphate with an isomeric pyrophosphate, 3-methyl-2-buten-1-ol pyrophosphate (Fig. 12.71).

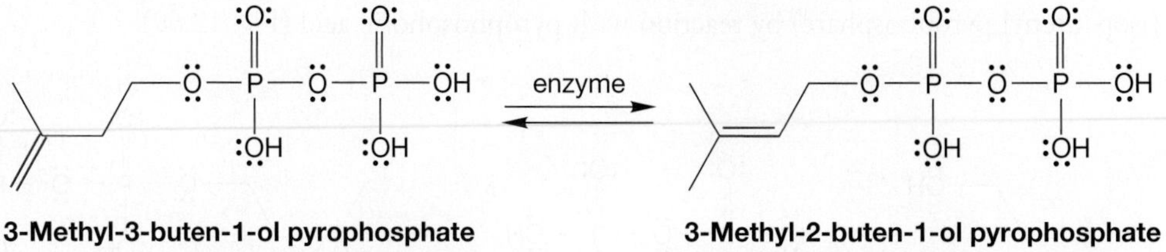

FIGURE 12.71 An enzyme converts 3-methyl-3-buten-1-ol pyrophosphate into a new, allylic pyrophosphate.

3-Methyl-3-buten-1-ol pyrophosphate   **3-Methyl-2-buten-1-ol pyrophosphate**

Note that this new pyrophosphate now has a good leaving group *in an allylic position*. The combination of a good leaving group and the allylic position makes ionization to form an allylic cation easy. The cation is attacked by 3-methyl-3-buten-1-ol pyrophosphate to produce ultimately the C_{10} compound, geranyl pyrophosphate (Fig. 12.72).

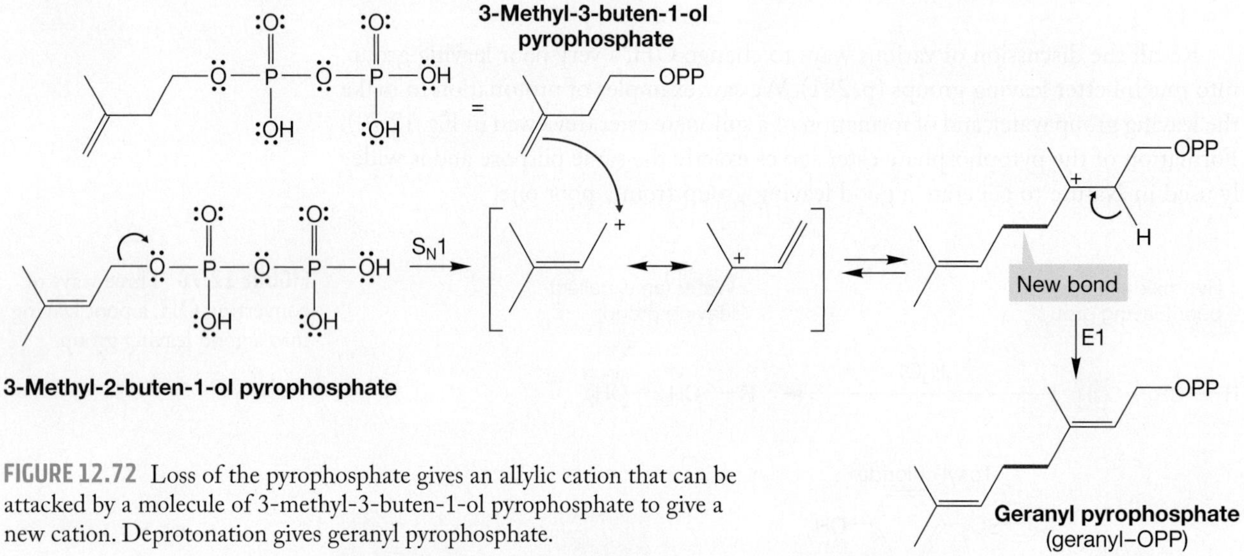

FIGURE 12.72 Loss of the pyrophosphate gives an allylic cation that can be attacked by a molecule of 3-methyl-3-buten-1-ol pyrophosphate to give a new cation. Deprotonation gives geranyl pyrophosphate.

Because geranyl pyrophosphate has a good leaving group in an allylic position, this process can be repeated to produce the C_{15} compound, farnesyl pyrophosphate (Fig. 12.73).

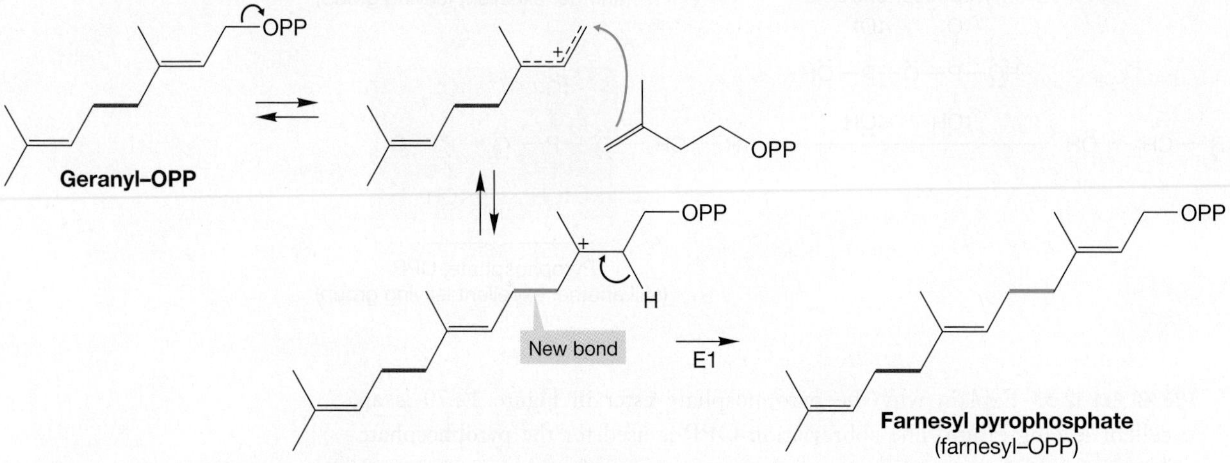

FIGURE 12.73 The addition process can be repeated to give farnesyl pyrophosphate.

Hydrolysis of geranyl pyrophosphate and farnesyl pyrophosphate produces the terpenes geraniol and farnesol, respectively (Fig. 12.74).

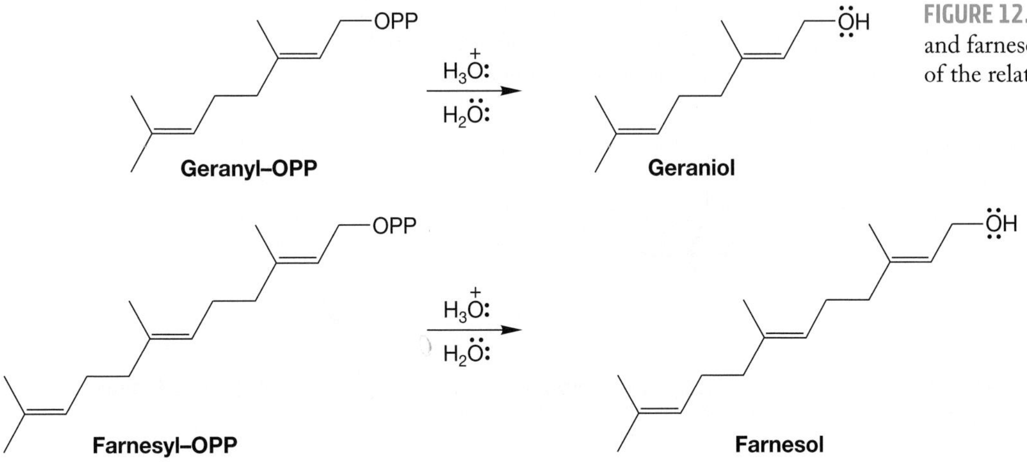

FIGURE 12.74 The alcohols geraniol and farnesol are formed by hydrolysis of the related pyrophosphates.

Isoprene units can be characterized as having a *head* and a *tail* as shown in Figure 12.75. Notice that both geranyl and farnesyl pyrophosphate are formed by head-to-tail connections of isoprenes.

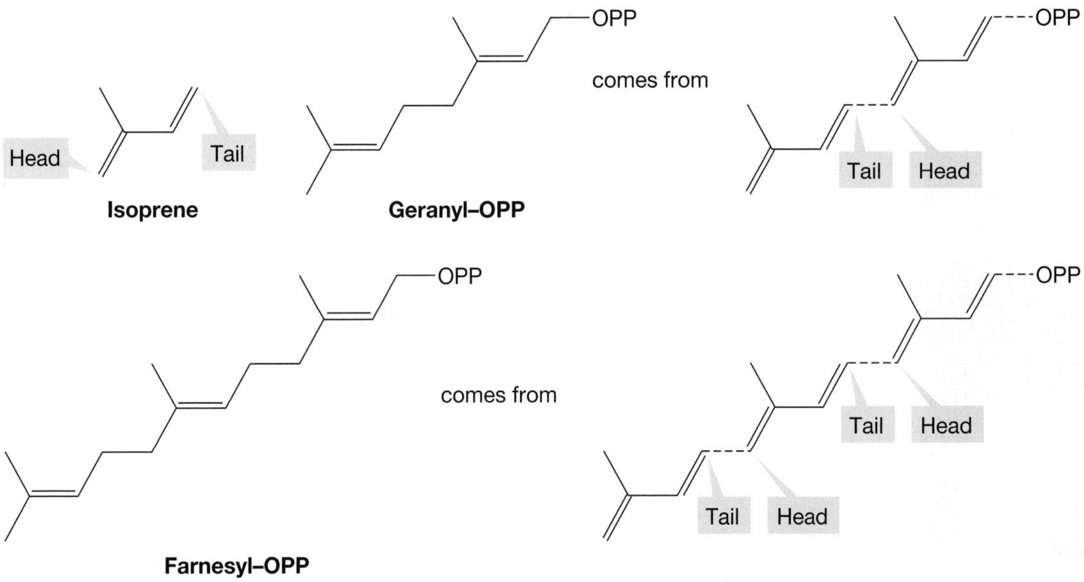

FIGURE 12.75 Head-to-tail attachment of isoprene units gives geranyl and farnesyl pyrophosphates.

The limit of head-to-tail combining is polyisoprene, natural rubber (p. 554). Your drawing of the polymer of isoprene called gutta percha (Problem 12.32) should also have head-to-tail connections.

Molecules formed from isoprene units are made of multiples of five carbons. Different names are used corresponding to the number of isoprene units in the molecule.

Two units of isoprene combine to form terpenes, C_{10} compounds such as geraniol. Three isoprene units form sesquiterpenes, C_{15} compounds such as farnesol. Diterpenes are C_{20} compounds, triterpenes are C_{30} compounds, and so on. Myriad terpenoid compounds are known, and most can be traced to derivatives of isoprene as starting material. Most terpenes follow the **isoprene rule**, which dictates the head-to-tail formation described earlier. All of the terpenes shown in Figure 12.76 do not strictly follow the pattern, but each has at least one head-to-tail connection.

Limonene   **Menthol**   **(1*R*)-(+)-Camphor**   **(1*S*)-(−)-β-Pinene**

Thujone   **(1*S*)-(−)-Camphor**   **(1*R*)-(+)-α-Pinene**

FIGURE 12.76 Some terpenes.

If you have experienced the rich odor of a pine tree forest, then this photo will remind you of that pleasant smell. It comes from volatile terpenes that fill the air.

WORKED PROBLEM 12.34 Dissect limonene and (1*R*)-(+)-camphor (Fig. 12.76) into isoprene units.

ANSWER The isoprene units can usually be found by searching for the methyl groups first. One of these problems is easy, the other is more complicated. Don't worry about oxygen atoms or double bonds, just look for carbon framework. The two isoprenes for each compound are shown in red.

PROBLEM 12.35 Devise a mechanism for converting geranyl pyrophosphate into racemic limonene (Fig. 12.76). Such processes are usually enzyme-mediated in Nature. When you try this problem you should encounter a roadblock, because the process can't really be done without a cis/trans isomerization, very likely carried out in Nature by the enzyme.

12.14   Special Topic: Steroid Biosynthesis

There is an extraordinarily important class of molecules called **steroids**. The carbon framework common to all steroids is three cyclohexanes and a cyclopentane fused together (Fig. 12.77). This arrangement is often referred to as "three rooms and a bath." Steroids are formed in Nature from isoprene units. We will start our summary of the biosynthesis with two units of farnesyl pyrophosphate, which can be enzymatically coupled in a tail-to-tail fashion to give the symmetrical triterpene called squalene (Fig. 12.78).

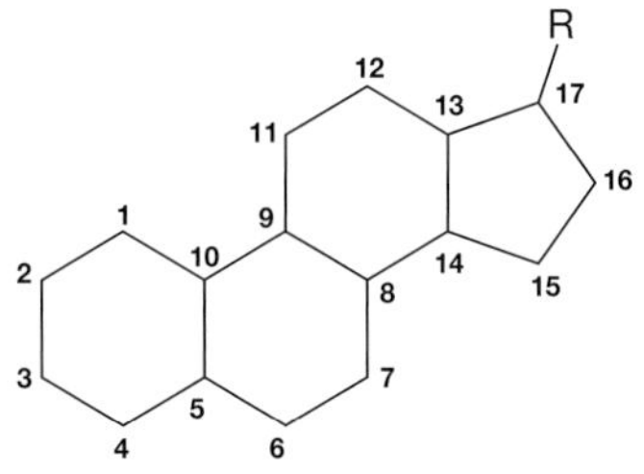

FIGURE 12.77 Basic steroid ring system.

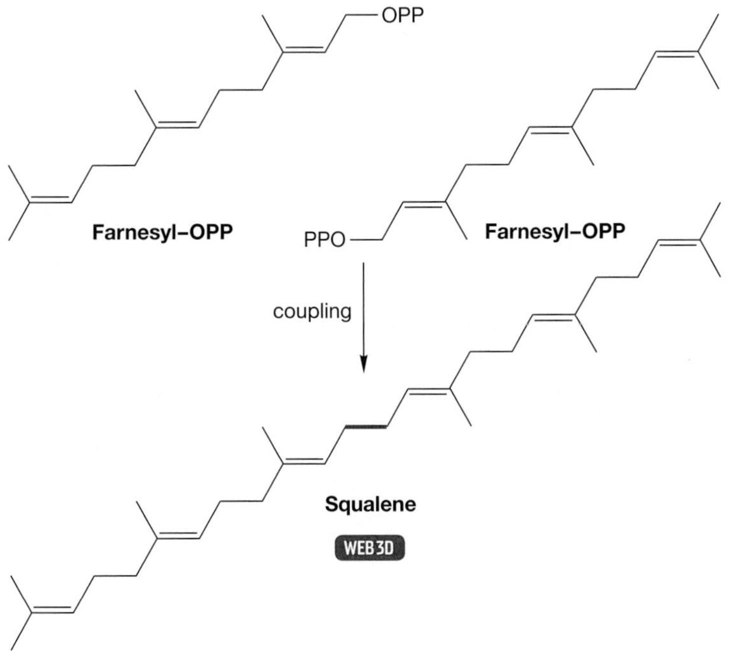

FIGURE 12.78 Coupling of two molecules of farnesyl pyrophosphate gives squalene, a triterpene.

Squalene can be epoxidized to give squalene oxide, which is written in a suggestive way in Figure 12.79. Protonation of squalene epoxide leads to a relatively

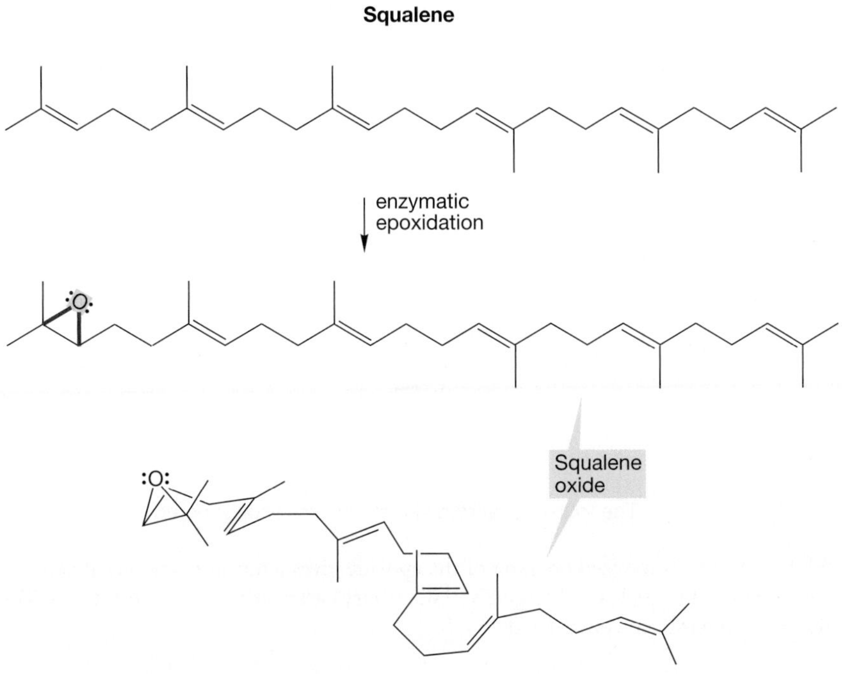

FIGURE 12.79 Enzymatic epoxidation of squalene gives squalene oxide, redrawn in a suggestive fashion at the bottom.

stable tertiary carbocation that is attacked by the proximate nucleophilic double bond to give a new cation. This cation is attacked in turn by another nearby alkene. The process continues until the molecule is sewn up in a four-ring steroid skeleton as shown in Figure 12.80. The series of rearrangements summarized in

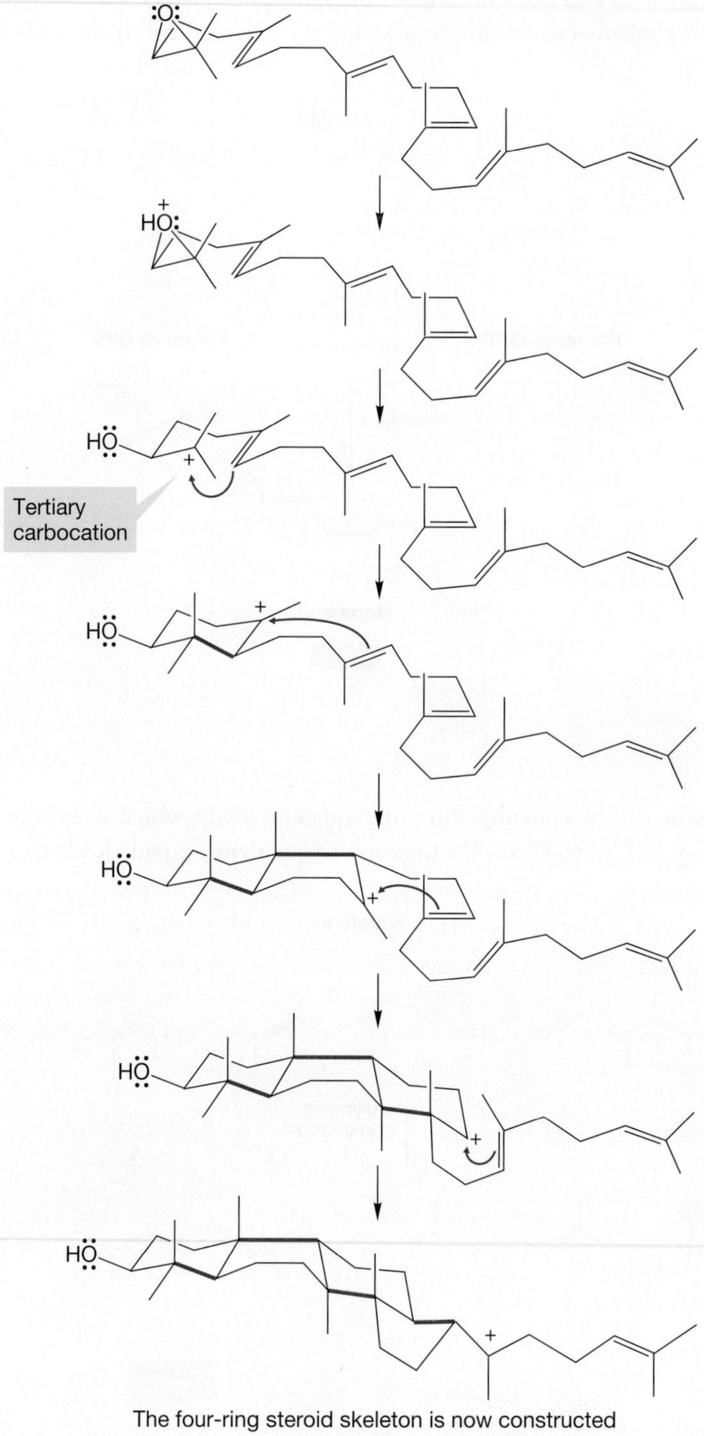

The four-ring steroid skeleton is now constructed

FIGURE 12.80 Acid-catalyzed opening of the epoxide gives a tertiary cation that can undergo a series of ring closings to give a new, tertiary carbocation in which the four-ring steroid skeleton has been constructed.

Figure 12.81 ensues and a proton is finally lost to give the compound known as lanosterol.

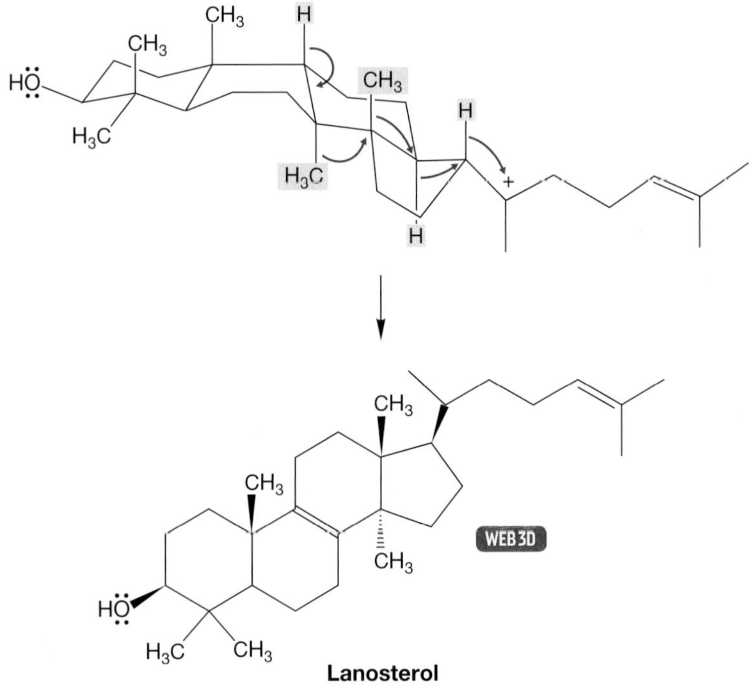

FIGURE 12.81 A series of skeletal rearrangements, followed by removal of a proton, gives lanosterol.

PROBLEM 12.36 Starting at the right-hand end of the top molecule in Figure 12.81, write a stepwise mechanism for the conversion to lanosterol. Take the several rearrangements one at a time and draw out each reaction intermediate.

Steroids are commonly found in Nature and are considered important primarily because of their powerful biological activities. They often function as regulators of human biology. The four rings are designated A, B, C, and D, and the carbons are numbered as shown in Figure 12.82. As in lanosterol, steroids usually have axial methyl groups attached to C(10) and C(13), and often carry an oxygen atom at C(3) and a carbon chain at C(17). The rings are usually, but not always, fused in trans fashion.

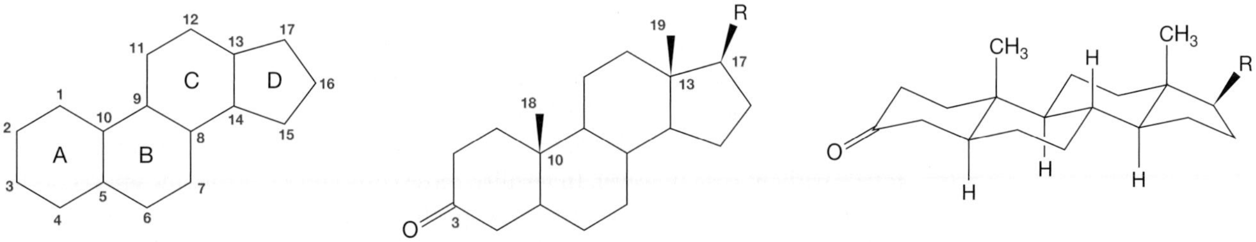

FIGURE 12.82 The steroid ring system.

Steroids were, and still can be, isolated from natural sources. Such important biomolecules quickly became the subject of attempts at chemical modification and total synthesis in the hopes of uncovering routes to new molecules of greater and different bioactivities. Many naturally occurring steroids have been constructed in the laboratory and numerous "unnatural" steroids have been created as well. Figure 12.83 shows a few important steroids found in Nature.

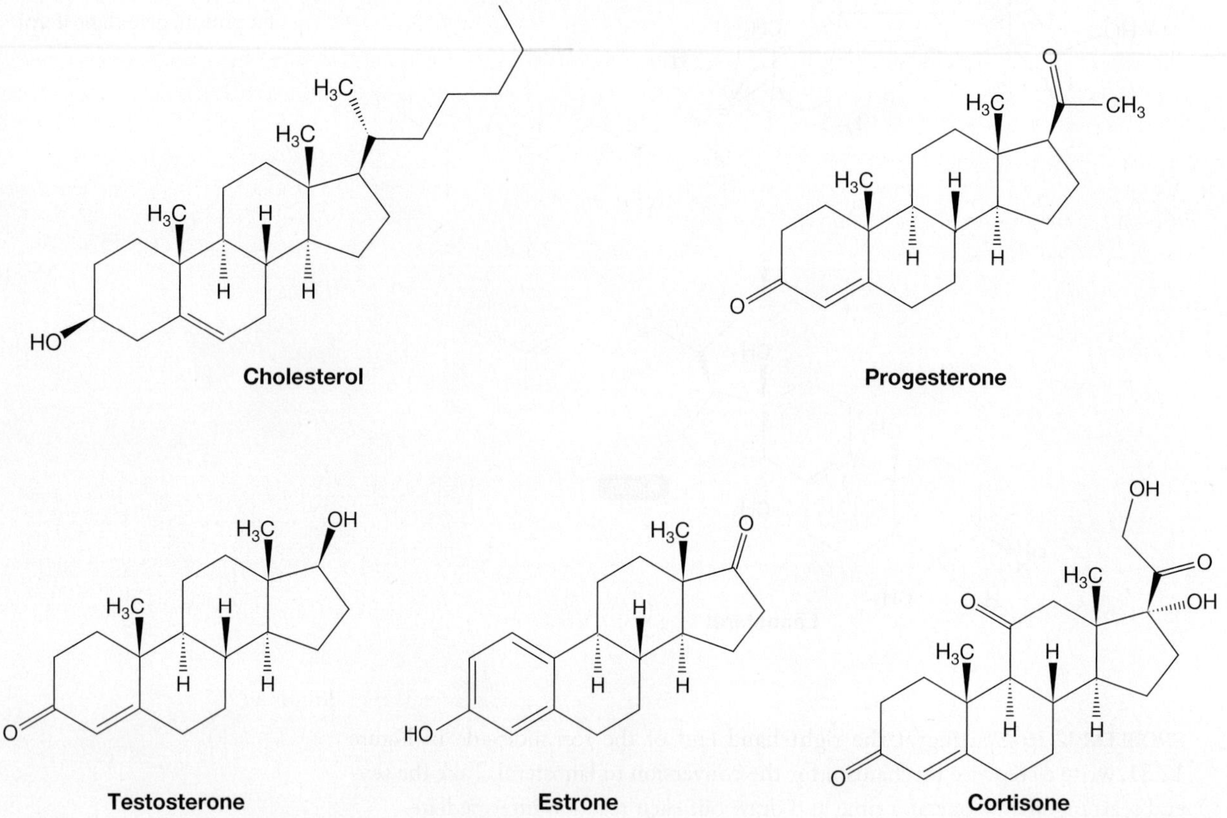

FIGURE 12.83 Some naturally occurring steroids.

Whole industries have been derived from the laboratory syntheses of steroids. Consider, for example, the steroids used in our attempts to control human fertility. "The pill" used by women to avoid pregnancy or to control hormonal imbalance contains synthetic steroids that mimic the action of the body's natural steroids, the estrogens and progesterone, which are involved in regulation of ovulation. The body is essentially tricked into behaving as if it were pregnant, so ovulation does not take place. It should be no surprise that it has been difficult to use such powerfully biologically active molecules in a completely specific way. The goal of this work is to interfere specifically with ovulation without doing *anything* else. Considering the activity of the molecules involved and the complexity of the mechanisms we are trying to regulate, it is remarkable that we have been so successful. This control is not totally without cost, however. In fertility, those costs appear as the side effects of the pill, which occur in some women. Perhaps the most difficult of human problems is to find the balance between benefits and risks in cases such as this. Although such problems have been with us throughout our history, questions of biology are particularly vexing, it seems. It is easier to judge that the benefits of the automobile outweigh the societal costs of the certain accidents than to make a decision regarding the steroids used in the birth-control pill. This problem is certain to become much worse as our ability to manipulate complicated molecules continues to grow.

12.15 Summary

New Concepts

In this chapter, you learned about the cumulated dienes. These molecules are characterized by at least one *sp* hybridized internal carbon atom.

The central topics of this chapter are the chemical and physical consequences of the overlap of 2*p* orbitals (conjugation). In structural terms, conjugation appears as a short C(2)—C(3) bond, about which the barrier to rotation is small (Fig. 12.18).

Conjugation leads to the formation of allyl systems in addition reactions to dienes, and is a requirement for the Diels–Alder reaction (Fig. 12.84).

An important fundamental concept in this chapter involves the difference between thermodynamic and kinetic control of a reaction. Conjugated dienes will react with HX or X_2 to give 1,2-addition products when kinetic conditions are used. The use of thermodynamic conditions will favor formation of the most stable product, usually the 1,4 addition product.

In UV/vis spectroscopy, absorption of light results in the promotion of an electron from the HOMO to the LUMO.

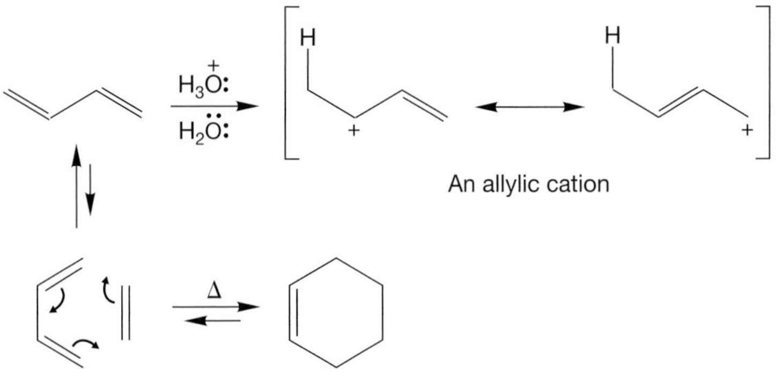

FIGURE 12.84 Two typical reactions of conjugated dienes: addition to give an allyl cation and the Diels–Alder reaction.

Key Terms

s-cis (p. 523)
conjugated double bonds (p. 512)
cumulated alkene (cumulene) (p. 512)
Diels–Alder reaction (p. 544)
dienophile (p. 544)
electronic spectroscopy (p. 526)

endo (p. 551)
exo (p. 551)
extinction coefficient (p. 527)
isoprene (p. 555)
isoprene rule (p. 558)
ketene (p. 515)

steroid (p. 559)
terpenes (p. 554)
s-trans (p. 523)
ultraviolet/visible (UV/vis) spectroscopy
(p. 526)

Reactions, Mechanisms, and Tools

The base-catalyzed isomerization of disubstituted to terminal alkynes takes place through allene intermediates (Fig. 12.13). Addition reactions of HX or X_2 to conjugated dienes can give products of both 1,2- and 1,4-additions (Fig. 12.38). The Diels–Alder reaction forms cyclohexenes and cyclohexadienes from the thermal reaction of conjugated dienes and alkenes or alkynes. Stereochemical labeling experiments show that both new σ bonds are formed simultaneously (Figs. 12.61 and the figure in Problem 12.28). Many natural products (terpenes, polyisoprene, steroids) are produced through the combination of various numbers of isoprene units.

Syntheses

The most important synthetic methods described in this chapter are a route to allylic halides through additions to conjugated dienes and the Diels–Alder synthesis of cyclohexenes and cyclohexadienes. Be especially careful about the Diels–Alder reaction, because there are many structural possibilities for the products depending on the complexity of the diene and dienophile used as starting materials. See the following page.

1. Allylic Halides

The 1,2- and 1,4-addition; watch out, these allylic halides can be transformed further though S$_N$2 reactions

Both 1,2- and 1,4-addition take place

2. Cyclohexadienes

The dienophile is an alkyne; watch out for structural variation and the reverse reaction

3. Cyclohexenes

Diels–Alder reaction; watch out for structural variation; the reverse reaction is a synthesis of 1,3-dienes and alkenes from cyclohexenes

4. Dienes

The reverse Diels–Alder reaction; watch out for structural variation; this picture gives only the simplest version

Common Errors

Be sure to understand how the structures of allenes and cumulenes can be derived from an analysis of the hybridizations of the atoms involved. Although the structures all *look* flat on a two-dimensional page, they are not all planar by any means.

Although the Diels–Alder reaction mechanism can be quickly drawn using only three little curved arrows, and although it is not difficult to see that all cyclohexenes can be constructed on paper from the reaction of a diene and a dienophile, very complex structures can be built very quickly using Diels–Alder reactions. The simplicity can quickly disappear in the wealth of detail! Try to see to the essence of the problem; locate the cyclohexene, and then take it apart to find the building-block diene and dienophile.

It is even harder to find a reverse Diels–Alder reaction in a mechanism problem or a chemical synthesis. Not only can

cyclohexenes be constructed through Diels–Alder reactions, but they can serve as sources of dienes and dienophiles as well.

It is time to mention again another very common conceptual error that often traps students. Look, for example, at Figure 12.44. Protonation of 1,3-butadiene leads to an allyl cation in which two different carbons share the positive charge. Addition of chloride at these two positions leads to the two products. It is very easy to think of the resonance forms of the allyl cation as each having a separate existence; of chloride adding to one resonance form to give one product and addition to the other resonance form to give the other product. This notion is absolutely wrong! The allyl cation is a single species that has two partially positive carbon atoms. Chloride adds to the allyl cation, not to one resonance form or the other. The figure tries to help out by showing a summary structure, but this is not always done, and you must be careful.

12.16 Additional Problems

PROBLEM 12.37   Use the rules given in Table 12.2 (p. 530) to calculate λ_{max} for the following steroids:

(a)

(b)

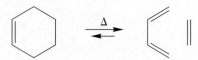

PROBLEM 12.38 Write resonance forms for the following ions:

(a)

(b)

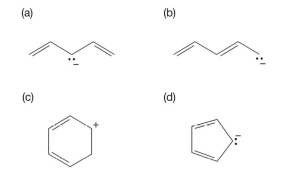

(c)

(d)

PROBLEM 12.39 Are any of the following compounds chiral? Explain with good drawings.

PROBLEM 12.40 Show the reaction conditions you would use to make 3-azido-1-butene if your only source of carbon is 1,3-butadiene.

PROBLEM 12.41 What are the possible nonhalogenated products in the reaction between *cis*-1,2-dibromocyclohexane and excess strong base? Show the mechanism for the formation of each product. Which one do you suppose would be the major product? Explain why.

PROBLEM 12.42 Predict the major product for the hydrohalogenation reaction (HBr, cold, short time) with each of the following dienes:

(a) 1,3-butadiene   (b) 1,3-cyclopentadiene
(c) (Z)-1,3-pentadiene   (d) (E)-2-ethyl-1,3-pentadiene
(e) 1-methyl-1,3-cyclohexadiene

PROBLEM 12.43 Predict the major product for the hydrohalogenation reaction (HBr, reflux, long time) with each of the following dienes:

(a) 1,3-butadiene   (b) 1,3-cyclopentadiene
(c) (Z)-1,3-pentadiene   (d) (E)-3-ethyl-1,3-pentadiene
(e) 1-methyl-1,3-cyclohexadiene

PROBLEM 12.44 Show the main products of the reactions of *trans,trans*-2,4-hexadiene under the following conditions. There may be more than one product in some cases.

(a) HCl   (b) H₂/Pd   (c) Cl₂/CCl₄
(d) Δ, H₃COOC —C≡C— COOCH₃

PROBLEM 12.45 Suggest a plausible arrow formalism mechanism to account for the products in the following reaction:

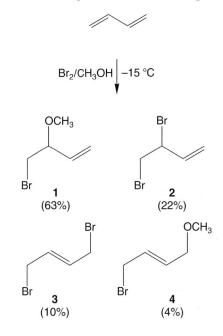

PROBLEM 12.46 Predict the products of the reaction of hydrogen chloride with isoprene under kinetic and thermodynamic control.

Isoprene

PROBLEM 12.47 Solvolysis of 3-chloro-4,4-dimethylcyclohexene proceeds through an intermediate carbocation. Draw that ion. Does an alkyl shift occur? Why or why not? *Hint*: Table 12.3.

PROBLEM 12.48 Show the reactions you would use to synthesize *cis*-4,5-dimethylcyclohexene if your only source of carbon is *cis*-2-butene.

PROBLEM 12.49 Devise syntheses for the following molecules. You may use 1-butyne and ethyl iodide as your only sources of carbon, as well as any inorganic reagents you need.

PROBLEM 12.50   Conversion of 2-butyne into 1-butyne is easy. The reverse reaction is more difficult. Suggest routes for doing both these conversions. Mechanisms are not required.

$$H_3C-C\equiv C-CH_3 \xrightarrow{?} CH_3CH_2-C\equiv C-H$$

$$CH_3CH_2-C\equiv C-H \xrightarrow{?} H_3C-C\equiv C-CH_3$$

PROBLEM 12.51   You have a bottle of C_6H_{10}. You know it contains a terminal alkyne. Which alkyne does it need to be in order to make 2,2-dimethyl-3-octyne most directly?

PROBLEM 12.52   Predict the major product for the Diels–Alder reaction (toluene, reflux) between 2-methyl-1,3-butadiene and the following dienophiles:

(a) cyclopentene
(b) diphenylacetylene
(c) (*E*)-4-octene
(d) (*Z*)-4-octene

PROBLEM 12.53   Predict the products of the following Diels–Alder reactions:

(a)

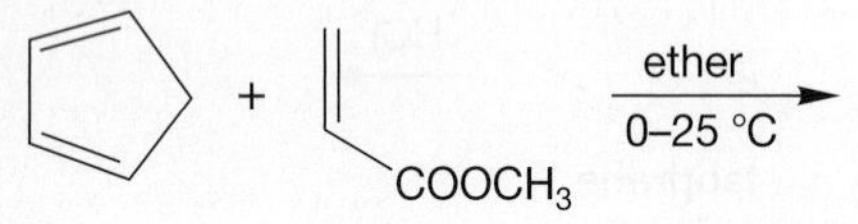

$$\xrightarrow[0-25\ °C]{ether}$$

(b)

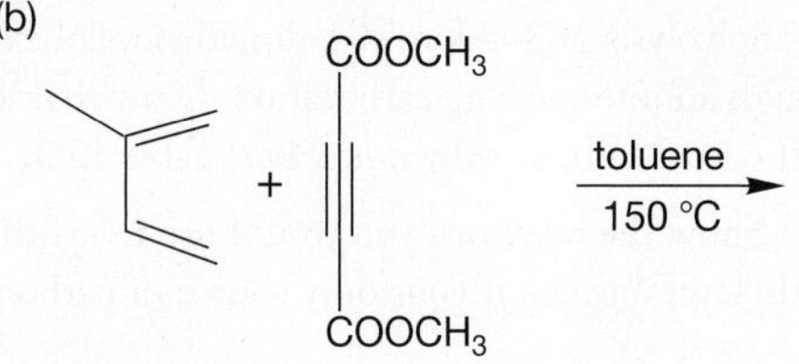

$$\xrightarrow[150\ °C]{toluene}$$

(c)

$$\xrightarrow[0\ °C]{alcohol}$$

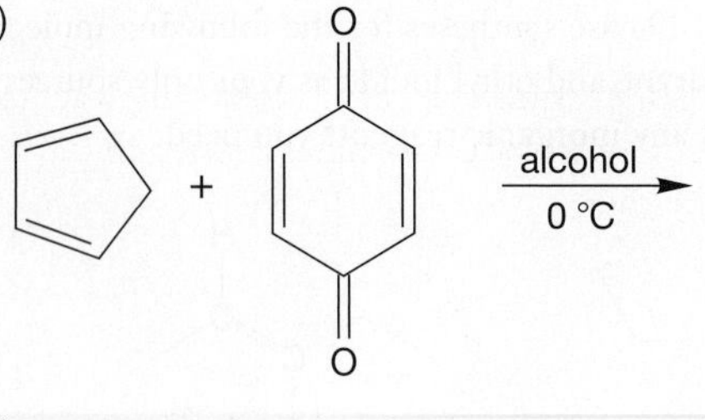

(d)

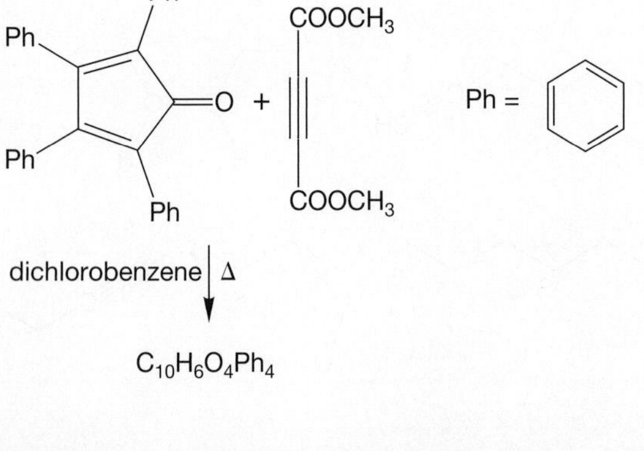

$$\text{dichlorobenzene} \Big\downarrow \Delta$$

$$C_{10}H_6O_4Ph_4$$

(e)

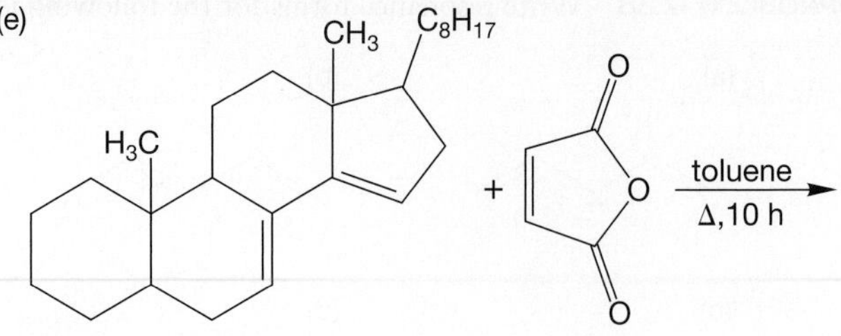

$$\xrightarrow[\Delta,\ 10\ h]{toluene}$$

(f)

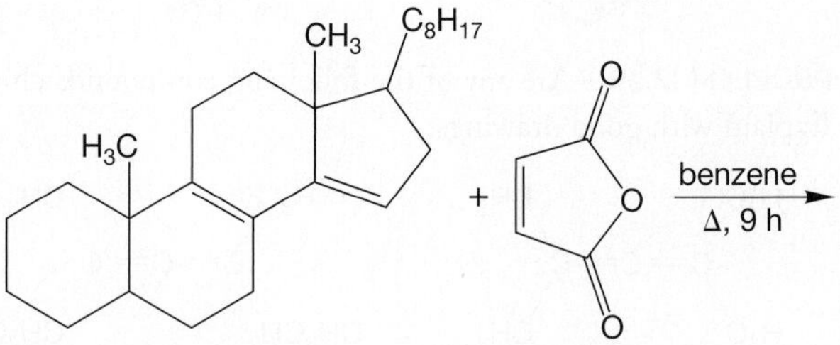

$$\xrightarrow[\Delta,\ 9\ h]{benzene}$$

(g)

$$\xrightarrow[25\ °C]{dioxane} C_8H_8N_2O_4$$

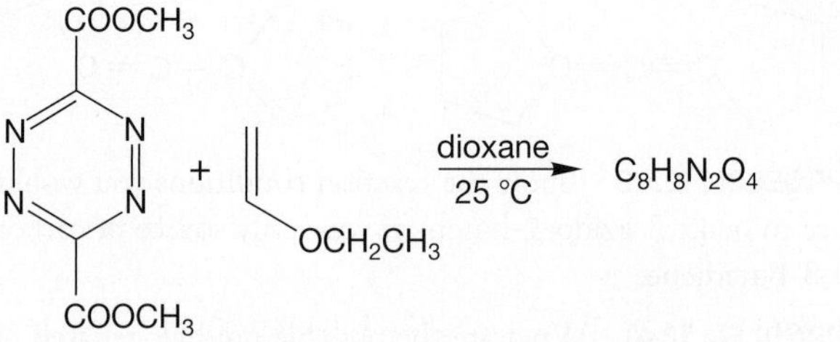

PROBLEM 12.54   Although simple alkyl and aryl nitriles are poor dienophiles, nitriles substituted with electron-withdrawing groups are more reactive. Provide mechanisms for the formations of the two products in the following reaction:

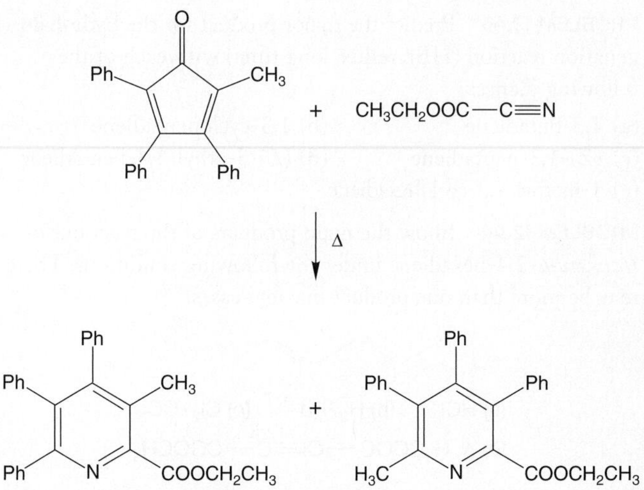

$$+ \quad CH_3CH_2OOC-C\equiv N$$

$$\Big\downarrow \Delta$$

$+$

PROBLEM 12.55 Although the Diels–Alder reaction is concerted, an analysis of a polar, stepwise mechanism can explain the regiochemical preferences often observed. Use such an analysis to explain the following data:

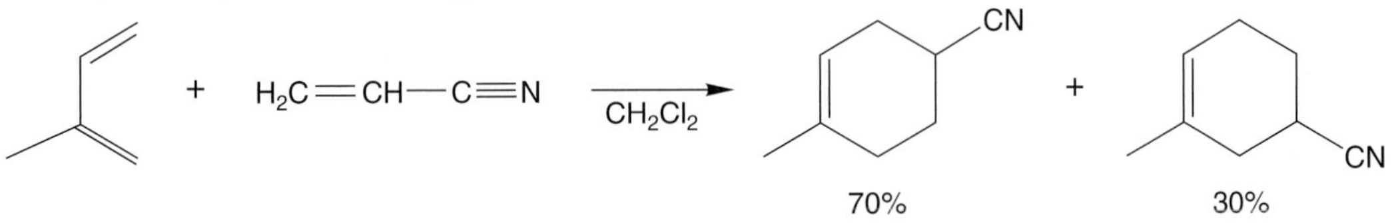

70%   30%

PROBLEM 12.56 In Problem 12.53c, the initially formed adduct can be closed in a photochemical reaction to give a complex cage structure (shown below). Write an arrow formalism for this process and explain how the observation of this reaction was helpful in determining the stereochemistry (exo or endo) of the original Diels–Alder adduct.

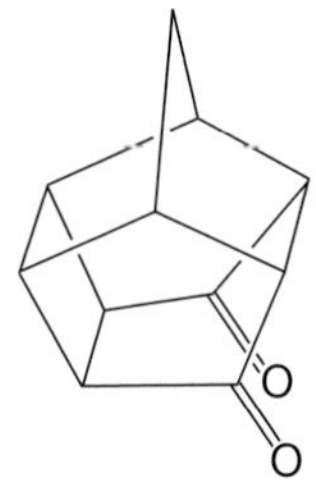

PROBLEM 12.57 Use a carefully constructed Energy versus Reaction progress diagram to rationalize the following data. Be sure to explain why the product at low temperature is mainly endo and why the product at high temperature is mainly exo.

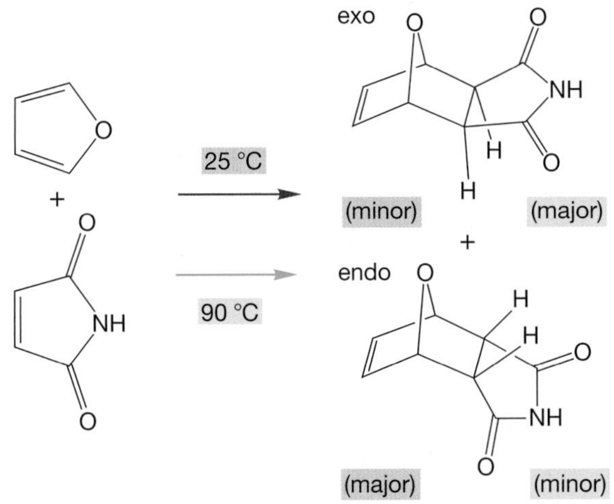

25 °C

90 °C

exo   (minor)   (major)
+
endo   (major)   (minor)

PROBLEM 12.58 In each of the following three reactions, two products are possible. Could ^{13}C NMR spectroscopy distinguish between the isomeric products formed? Explain. Can you remember or predict which products are actually formed in these reactions?

(a)

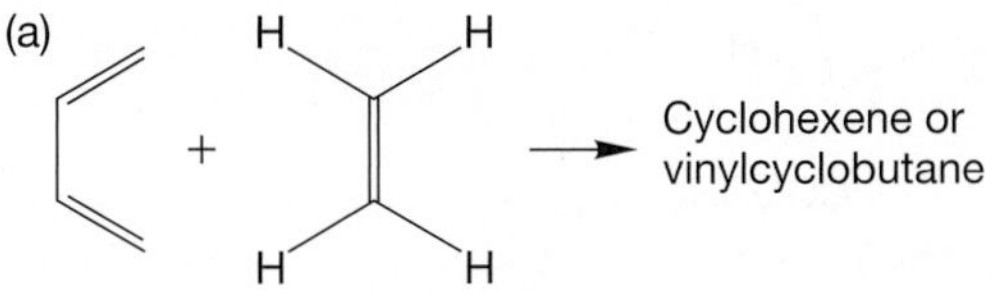

Cyclohexene or vinylcyclobutane

(b)

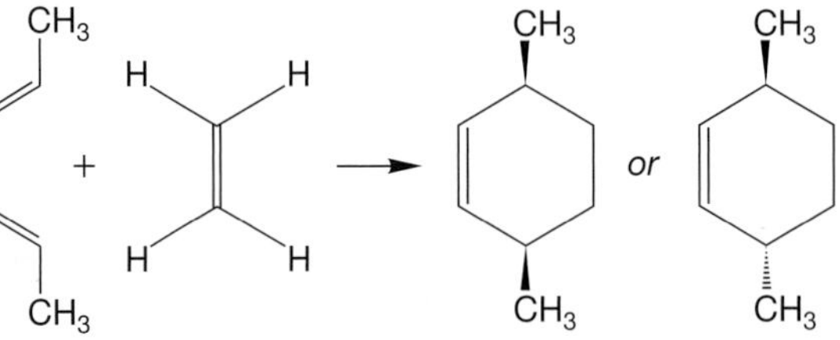

or

(c)

endo or exo Product

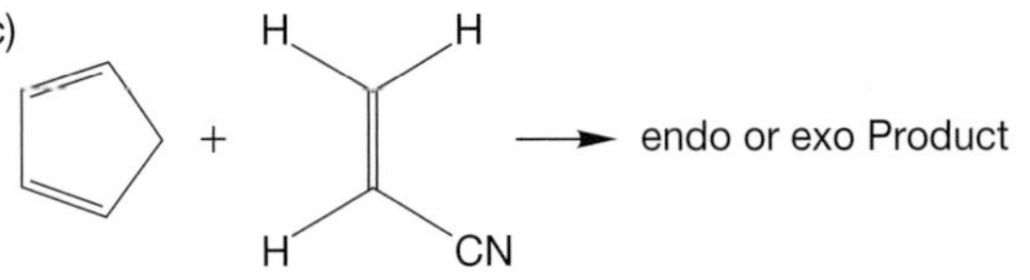

PROBLEM 12.59 Here are two reactions of allyl systems that are related to the Diels–Alder reaction. Use a HOMO–LUMO molecular orbital analysis to show why reaction a succeeds whereas reaction b fails.

(a)

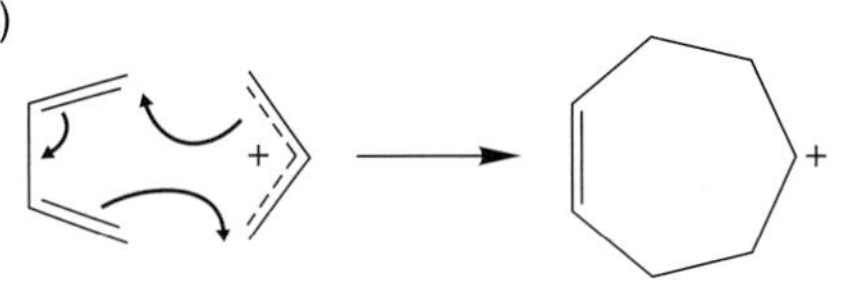

(b)

PROBLEM 12.60 In principle, one of the following reaction sequences could give two products, while the other must produce only one. Explain.

(a)

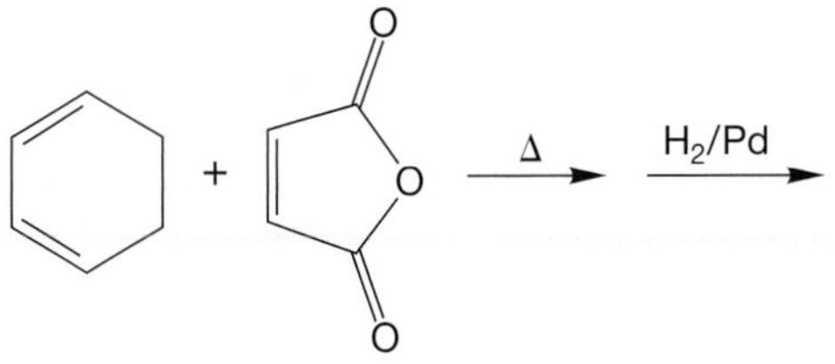

Δ H_2/Pd

(b)

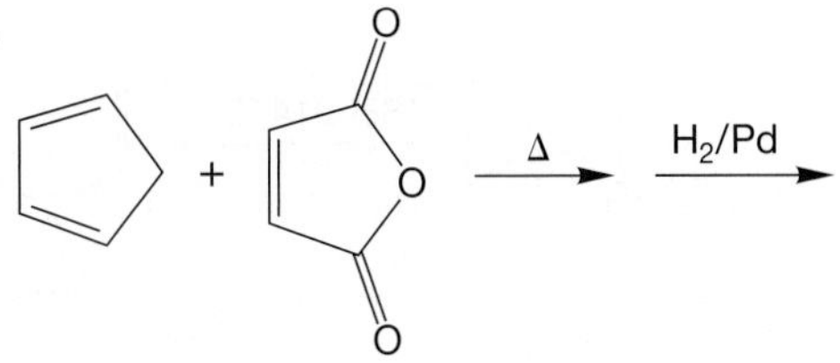

Δ H_2/Pd

PROBLEM 12.61 In 1951, the great Kurt Alder himself examined the reactions of a mixture of dienes **1** and **2** with maleic anhydride. He discovered that diene **1** reacted at 35 °C to give a single adduct (**A**). Diene **2** reacted only at 150 °C to give a single, different, adduct (**B**). Write structures for **A** and **B** and explain the difference in the ease of reaction. Why is reaction of **1** so much easier than reaction with **2**? Alder never actually looked at the reaction with diene **3**, but if he had, we predict that no reaction would have occurred at 150 °C. Explain.

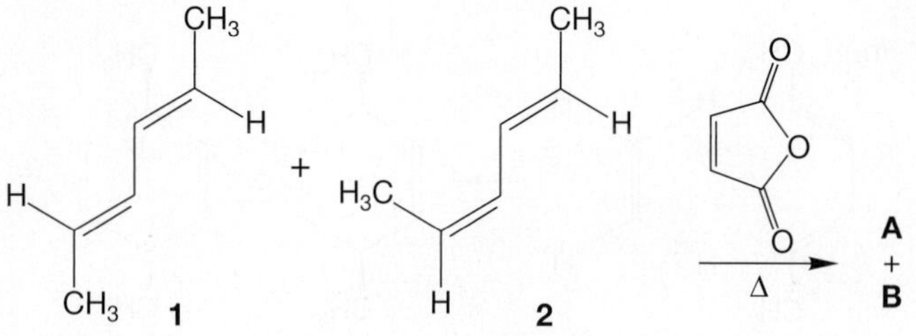

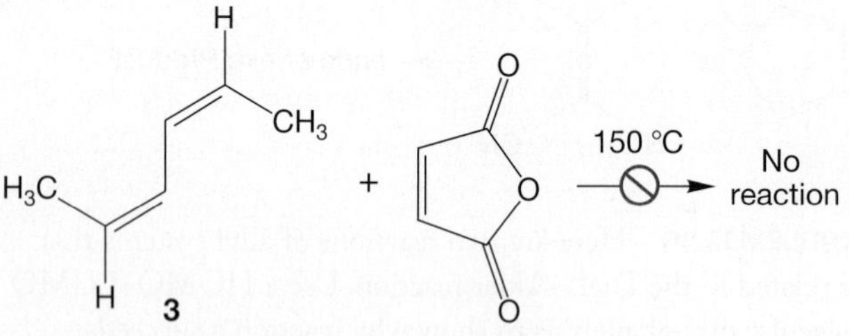

PROBLEM 12.62 Propose a synthesis of the powerful vesicant (blister inducer) cantharidin, the active ingredient in the putative aphrodisiac "Spanish fly." You may assume that organic starting materials containing no more than six carbon atoms are available, along with any inorganic materials you may need. *Hint*: The thermodynamically more stable exo Diels–Alder adducts of furan are generally isolable.

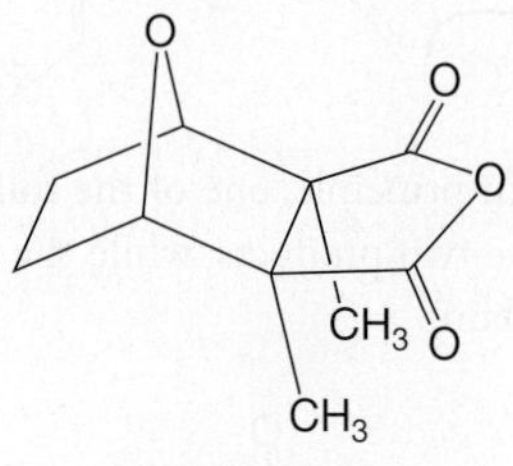

Cantharidin

PROBLEM 12.63 Write arrow formalism mechanisms for the following transformations:

(a)

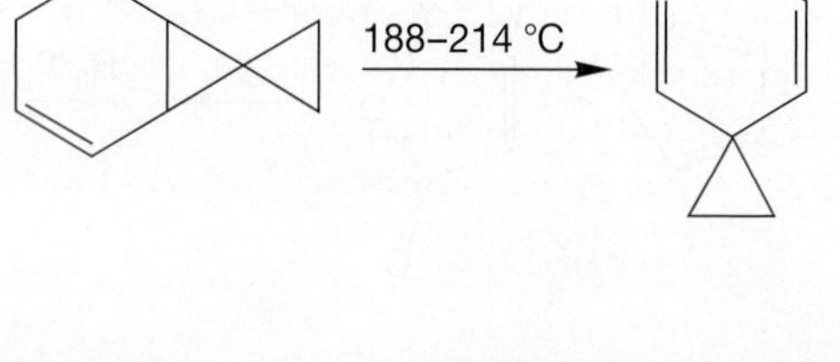

(b)

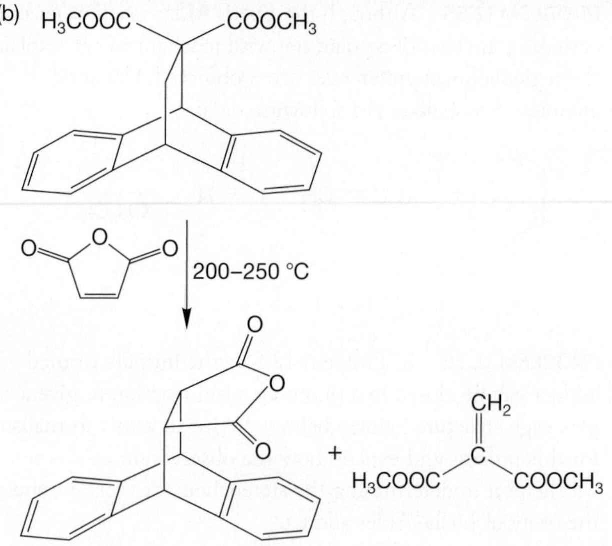

(c)

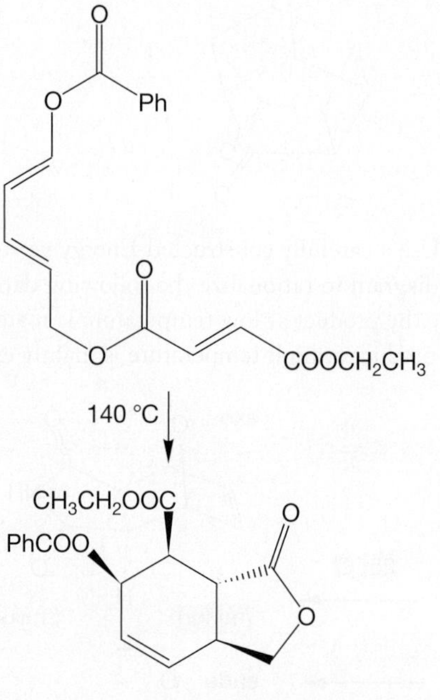

(d)

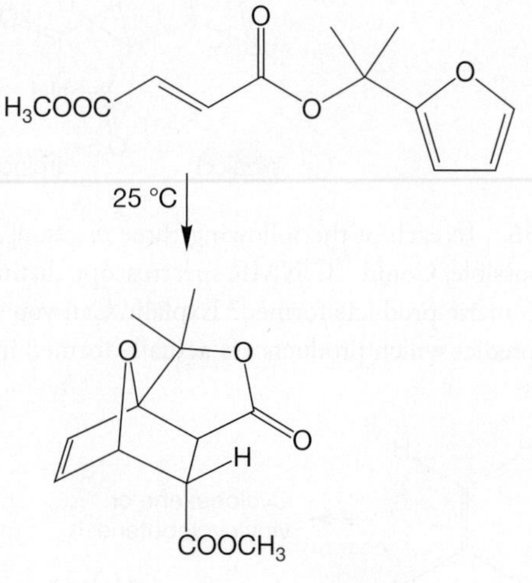

PROBLEM 12.64 Allylic halides undergo the S_N2 reaction much faster than saturated halides (Section 12.11b). There is another substitution reaction pathway open to allylic halides called the S_N2' reaction.

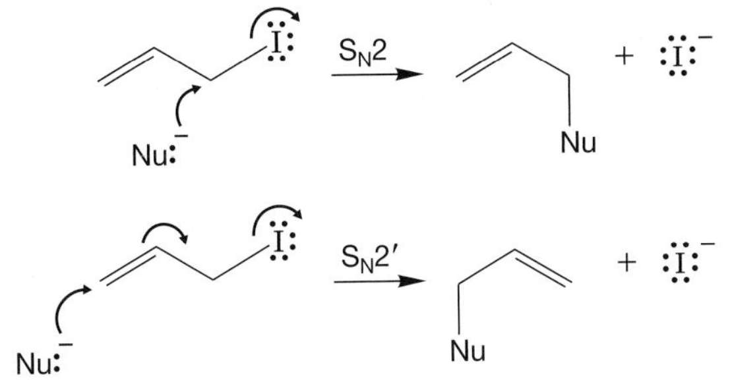

As the figure shows, in this simple example it is not possible to distinguish the two reactions. Devise an experiment that would allow you to tell if the product were formed through the S_N2 or S_N2' reaction.

PROBLEM 12.65 Devise an experiment using isotopic labeling that would determine if the S_N2 or S_N2' reaction were occurring in the reaction of Problem 12.64.

PROBLEM 12.66 Now here's a much harder problem. The stereochemistry of the S_N2 reaction has been well worked out and has been described in detail in Section 7.4b. The S_N2' reaction has also been investigated, and this problem asks you to put yourself in the place of R. M. Magid (b. 1938) and his co-workers. They used the following experiment to determine the stereochemistry of the S_N2' reaction. Use their data to determine whether the nucleophile approaches the double bond from the same side as the leaving group or from the opposite side.

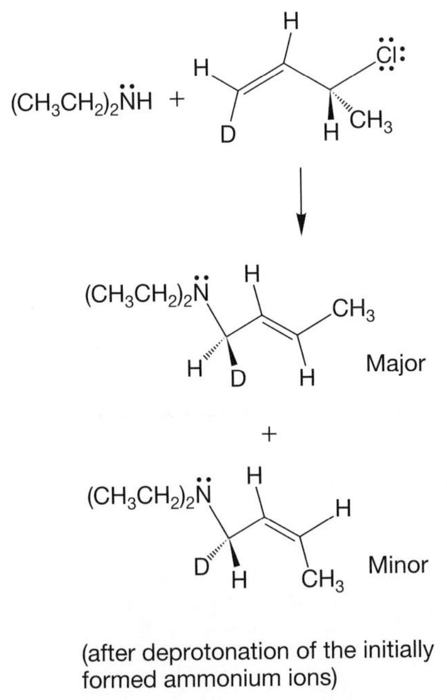

(after deprotonation of the initially formed ammonium ions)

PROBLEM 12.67 Write an arrow formalism mechanism for the following bimolecular process. Be sure to account for the stereochemistry of the product.

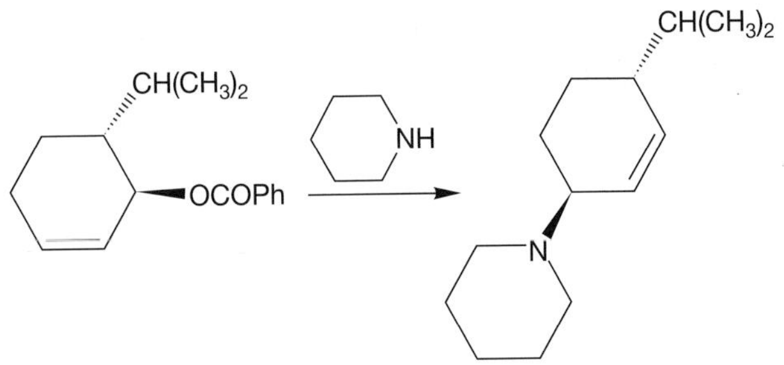

PROBLEM 12.68 Prostaglandins are a class of very important natural products found in all animals. They are responsible for a wide range of physiological events such as dilation of smooth cells, aggregation of platelets, pain sensation, cell growth, hormone regulation, and inflammatory mediation. The central compound in the formation of prostaglandins is called prostaglandin H_2 (PGH$_2$). PGH$_2$ is an endoperoxide. Endoperoxides are formed through a Diels–Alder reaction in which oxygen (O$_2$) is the dienophile. Based on the structure of PGH$_2$ predict the structure of the organic compound that leads to PGH$_2$. *Hint*: Find the six-membered ring, then think of the reverse Diels–Alder-like reaction that would give O$_2$.

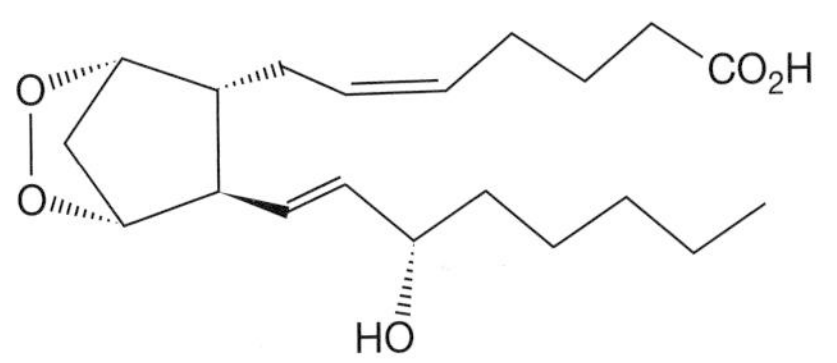

Prostaglandin H$_2$ (PGH$_2$)

Use Organic Reaction Animations (ORA) to answer the following questions:

PROBLEM 12.69 Select the reaction "1,2-Hydrohalogenation of diene" that is on the Semester 1B panel. Observe the reaction several times. When the electrophilic HBr reacts with the nucleophilic diene, an intermediate carbocation is formed. The animation shows the end carbon being protonated. Draw the intermediates that could result from protonation of each of the other carbons. Convince yourself that protonation at the end is preferred and that protonation at either end is equivalent in this case. Now go through this exercise with 2-methyl-1,3-butadiene. Draw each possible intermediate and decide which is most stable. Don't forget resonance structures!

PROBLEM 12.70 Chose the LUMO track of the "1,2-Hydrohalogenation of diene" and stop the animation at the intermediate (there is only one intermediate in the energy diagram). Notice the nature of this molecular orbital. It is the same as the LUMO of an allyl cation. Does there appear to be any more LUMO density on one carbon than the other?

PROBLEM 12.71   Look closely at the energy diagrams for the "1,2-Hydrohalogenation of diene" and "1,4-Halogenation of diene" reactions. Notice there is a faint pathway in the second half of the reaction in each case. Why is the product of the 1,4 addition lower in energy than the product of 1,2 addition? Will the 1,4-addition always be lower? Consider the HBr addition to 1,3-pentadiene. Is the product of 1,4-addition lower in this example?

PROBLEM 12.72   Select the "Diels–Alder reaction." Observe the reaction several times. Pause the animation just as the bottom molecule has completely entered the screen. This is the point on the "Reaction Progress" slide of the energy diagram (the x-axis) where the reaction is about one-fourth of the way along. The red ball has not moved. With the reaction paused in this position, click on LUMO and then HOMO. Based on the information that you gain from this visualization, which of the two reagents is the nucleophile? Which is the electrophile? Why?

Conjugation and Aromaticity

13.1 Preview

13.2 The Structure of Benzene

13.3 A Resonance Picture of Benzene

13.4 The Molecular Orbital Picture of Benzene

13.5 Quantitative Evaluations of Resonance Stabilization in Benzene

13.6 A Generalization of Aromaticity: Hückel's $4n + 2$ Rule

13.7 Substituted Benzenes

13.8 Physical Properties of Substituted Benzenes

13.9 Heterobenzenes and Other Heterocyclic Aromatic Compounds

13.10 Polynuclear Aromatic Compounds

13.11 Introduction to the Chemistry of Benzene

13.12 The Benzyl Group and Its Reactivity

13.13 Special Topic: The Bio-Downside, the Mechanism of Carcinogenesis by Polycyclic Aromatic Compounds

13.14 Summary

13.15 Additional Problems

BAD AIR Forest fires are a significant source of polycyclic aromatic hydrocarbons, which make up the solid, cancer-causing particulates that come from combustion.

> Curious, though, isn't it, um, Patwardhan, that the number, er, six should be, um, embodied in one of the most, er, er, beautiful, er, shapes in all nature: I refer, um, needless to say, to the, er, benzene ring with its single, and, er, double carbon bonds. But is it, er, truly symmetrical, Patwardhan, or, um, asymmetrical? Or asymmetrically symmetrical, perhaps. . . .
>
> —VIKRAM SETH,[1] *A SUITABLE BOY*

13.1 Preview

Chapter 12 was devoted to the consequences of conjugation—the overlap of $2p$ orbitals in acyclic systems. Now we will see what happens when conjugated systems are cyclized, turned into rings. Nothing much changes if the atoms of the rings do not maintain orbital overlap throughout the ring—if they are not fully conjugated. 1,3-Cyclohexadiene is not especially different from 1,3-hexadiene, for example (Fig. 13.1). There are small differences in chemical and physical properties, but the two molecules clearly belong to the same chemical family. Hydrogenation is easy, addition reactions abound, the Diels–Alder reaction is common, and so on.

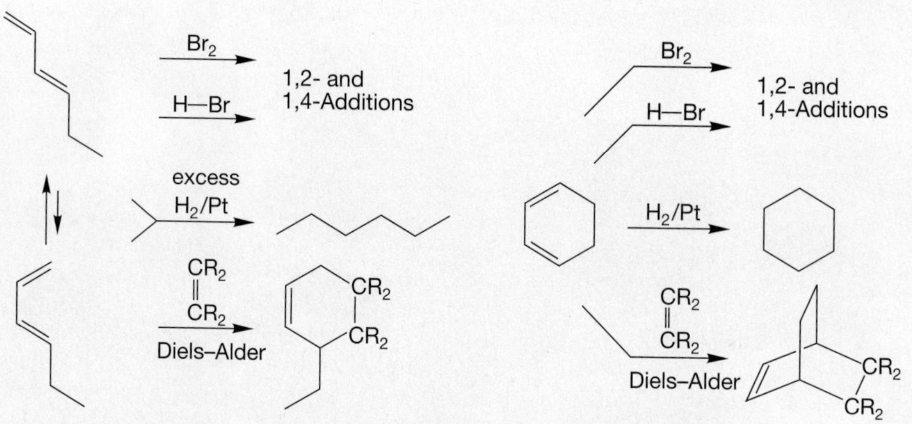

FIGURE 13.1 Acyclic and cyclic dienes have similar chemical properties.

When the conjugation is continued around the 1,3-cyclohexadiene ring to give what we might call "1,3,5-cyclohexatriene," matters change (Fig. 13.2). None of the reactions shown in Figure 13.1 is easy, and this molecule appears to be remarkably unreactive. It is clearly in a different class from 1,3-cyclohexadiene and the acyclic dienes.

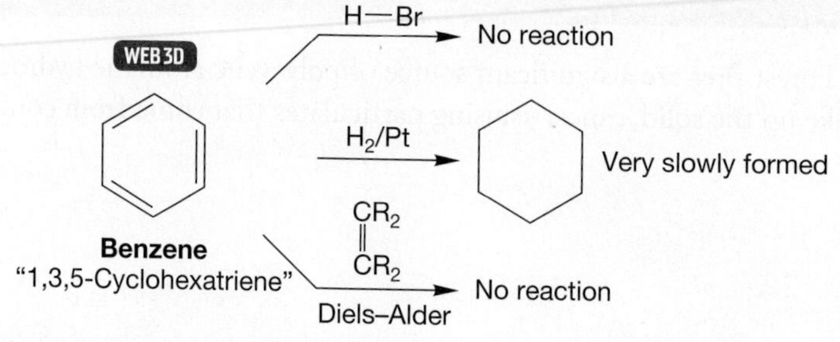

FIGURE 13.2 Benzene ("1,3,5-cyclohexatriene") does not undergo most of the reactions of normal alkenes.

---

[1]Vikram Seth was born in Calcutta in 1952 and educated in India, Britain, and the United States. This brilliant novel gets the MJ 10-star recommendation.

But not all ring compounds containing double bonds behave in this way. Some cycloalkenes react as expected; cycloheptatriene and cyclooctatetraene are examples of this kind of cyclic molecule. Others are exceptionally unstable; cyclobutadiene is the archetype of this kind of cyclic polyene (Fig. 13.3).

1,3,5-Cycloheptatriene   **1,3,5,7-Cyclooctatetraene**

These are normal polyenes (for example, they undergo addition reactions and hydrogenate easily)

Cyclobutadiene

This molecule is extraordinarily unstable—it can only be detected at very low temperature

FIGURE 13.3   Many cyclic polyenes are either normal in their chemical properties or exceptionally unstable.

In this chapter, we will uncover the sources of these differences in stability between cyclic polyenes and begin to see the chemical and physical consequences of this special kind of conjugative interaction of 2*p* orbitals. We will encounter the molecule **benzene** (Fig. 13.2), in which the overlap of six carbon 2*p* orbitals in a ring has great consequences for both structure and reactivity. We will also see a generalization of the properties that make benzene so stable, and will learn how to predict which cyclic polyenes should share benzene's stability and which should not.

ESSENTIAL SKILLS AND DETAILS

1. The properties that determine whether or not a molecule is properly termed "aromatic" are simple to write down but are a bit more difficult to apply. It is very important that you be able to think about these qualities, to see why they apply to certain molecules and why they do not apply to others. Understanding the hard-to-define quality called aromaticity is critical.

   Chemists have argued for years about the special stability called aromaticity that attends cyclic, planar polyenes with the proper numbers of π electrons. The wonderful quote that begins this chapter is not so far from an accurate representation of remarks heard at many a chemical conference! However, even though it is not easy to define this slippery concept precisely, especially in marginal cases, it is not difficult to understand the simple examples, to see why the number of π electrons is important, and to learn which molecules are likely to exhibit this property.

2. This chapter introduces the generic aromatic substitution reaction. We will see this reaction again many, many times in Chapter 14, but it would be a very good idea if you arrived at that chapter knowing the basics.

3. The Frost circle allows you to determine quickly the relative energies of the π molecular orbitals for any planar, cyclic, fully conjugated polyene. *Remember*: The polygon must be inscribed vertex down.

4. Aromaticity is not magic! It is just another stabilizing effect that can be augmented or opposed by other effects.

13.2   The Structure of Benzene

The molecule in Figure 13.2 that might be called "1,3,5-cyclohexatriene" was isolated from the thermal rendering of whale blubber in the nineteenth century. Although the formula was established as C_6H_6 [or better, $(CH)_6$], there remained a serious problem: What was its structure? There are many possibilities, among them the four shown in Figure 13.4. Two of these, **Dewar benzene** and Ladenburg

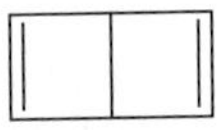

Bicyclo[2.2.0]hexa-2,5-diene
(Dewar benzene)

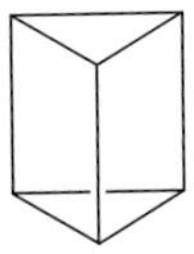

Tetracyclo[2.2.0.0^{2,6}.0^{3,5}]hexane
(Ladenburg benzene or prismane)

Tricyclo[3.1.0.0^{2,6}]hex-3-ene
(benzvalene)

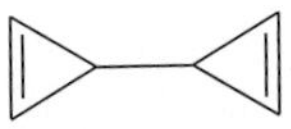

3,3'-Bicyclopropenyl

FIGURE 13.4   Four possible structures for benzene.

benzene, are named for the people who suggested them, Sir James Dewar (1842–1923) and Albert Ladenburg (1842–1911). The third, benzvalene, is more cryptically named. The last compound in Figure 13.4, 3,3′-bicyclopropenyl, was first made only in 1989 by a group headed by W. E. Billups (b. 1939) at Rice University.

The German chemist Friedrich August Kekulé von Stradonitz (1829–1896) suggested the correct ring structure. Legend has it that Kekulé was provoked by a dream of a snake biting its tail. Kekulé describes dozing and seeing long chains of carbon atoms twisting and turning until one "gripped its own tail and the picture whirled scornfully before my eyes." Kekulé's suggestion for the structure of benzene is shown in cyclohexatriene form in Figure 13.5.

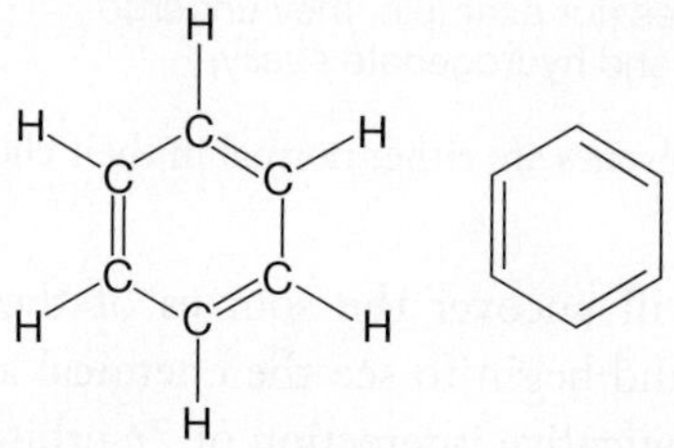

FIGURE 13.5 The structure of benzene, drawn as 1,3,5-cyclohexatriene.

PROBLEM 13.1 Suppose Dewar, Ladenburg, and Kekulé had had a ^{13}C NMR spectrometer. How many signals would the various candidates shown in Figure 13.4 show in their spectra? How many would cyclohexatriene show?

But there were problems with the Kekulé formulation[2] as well as with the four shown in Figure 13.4. Benzene didn't react with halogens or hydrogen halides like any self-respecting polyene, for example. Hydrogenation was much slower than for

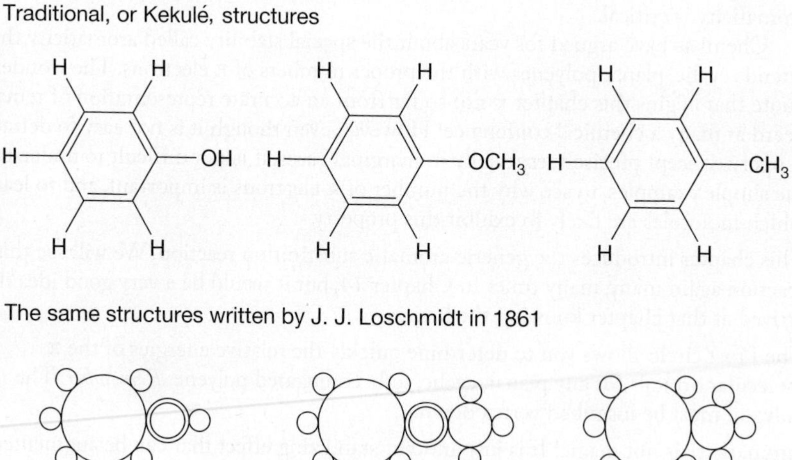

FIGURE 13.6 Loschmidt's formulations of benzene structures (see footnote).

Traditional, or Kekulé, structures

The same structures written by J. J. Loschmidt in 1861

---

[2]Kekulé gets the lion's share of the credit, but in the mid-1800s there were others hot on the trail of structural theory in general and a decent representation of benzene in particular. One such chemist was Alexandr M. Butlerov (1828–1886) from Kazan, Russia, who may well have independently conceived of the ring structure for benzene, and who surely contributed greatly, if quite invisibly in the West, to the development of much of modern chemistry. Another, even more obscure contributor, was the Austrian chemist–physicist Johann Josef Loschmidt (1821–1895) who, four years before Kekulé's proposal in 1865, clearly wrote a ring structure for benzene. Figure 13.6 shows how forward-looking were the structures proposed by this too-little-known Austrian schoolteacher.

other alkenes and required severe conditions. Benzene was clearly a remarkably stable compound, and the cyclohexatriene structure couldn't account for this.

Moreover, it eventually became known that the structure of benzene was that of a *regular* hexagon (i.e., equal sided). Our 1,3,5-cyclohexatriene is immediately ruled out as a structural possibility, because it must contain alternating long single bonds and short double bonds. We might expect normal double bonds of about 1.33 Å and single bonds of about the length of the C(2)—C(3) bond in 1,3-butadiene: 1.47 Å. Instead, benzene has a uniform carbon–carbon bond distance of 1.39 Å. Note that the measured bond distance, 1.39 Å, is just about the average of the double-bond distance, 1.33 Å, and the single-bond distance, 1.47 Å (Fig. 13.7). At this point, you can probably see the beginnings of the answer to these structural problems. Kekulé's static cyclohexatriene formulation doesn't take account of resonance stabilization.

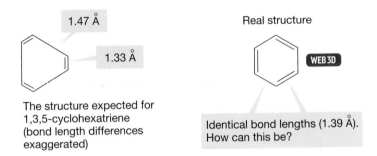

1.47 Å

1.33 Å

The structure expected for 1,3,5-cyclohexatriene (bond length differences exaggerated)

Real structure

WEB 3D

Identical bond lengths (1.39 Å). How can this be?

FIGURE 13.7
Any 1,3,5-cyclohexatriene structure must contain alternating long single bonds and short double bonds (shown here with exaggerated short and long bonds). Such is not the case for benzene, which is a regular hexagon with a uniform carbon–carbon bond distance of 1.39 Å.

PROBLEM 13.2 One early problem with the Kekulé structure was that only one 1,2-disubstituted isomer of benzene could be found for any given substituent. That is, there is only one 1,2-dimethylbenzene (*o*-xylene). Explain why this observation would have been hard for Kekulé to explain.

13.3 A Resonance Picture of Benzene

Perhaps the remarkable stability of benzene can be explained just by realizing that delocalization of electrons is strongly stabilizing. There are really *two* **Kekulé forms** for benzene. They are resonance forms, differing from each other only in the distribution of electrons but not in the positions of atoms. An orbital picture for cyclohexatriene should have influenced us in this direction. Each carbon is hybridized sp^2 and there is, therefore, a $2p$ orbital on every carbon. The structure in Figure 13.8a

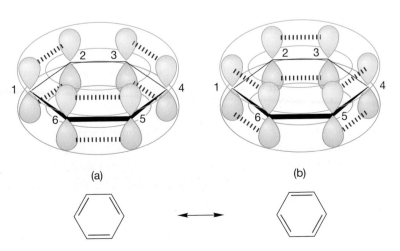

(a)   (b)

FIGURE 13.8 These two Kekulé forms together provide a real structure for benzene. The dashed lines represent the overlap implied in the line structures.

stresses overlap between the $2p$ orbitals on C(1) and C(2), C(3) and C(4), and C(5) and C(6), but not between C(2) and C(3), C(4) and C(5), or C(6) and C(1). This specificity is suspicious, because there is a $2p$ orbital on every carbon atom and the drawing of Figure 13.8a takes no account of the symmetry of the situation. If there is overlap between the orbitals on C(1) and C(2), then there must be equivalent overlap between the $2p$ orbitals on C(2) and C(3) as well, as shown in Figure 13.8b. Drawing both Kekulé forms emphasizes this, and produces a regular hexagon as the combination of the two Kekulé resonance forms.

So, our picture of benzene has now been elaborated somewhat to show the orbitals. The $2p$ orbitals on the six carbons, extending above and below the plane containing the carbons and their attached hydrogens, overlap to form a circular cloud of electron density above and below the plane of the ring.

CONVENTION ALERT

A convention has been adopted that attempts to show the cyclic overlap of the six $2p$ orbitals. Instead of drawing out each double bond as in the Kekulé forms, a circle is inscribed inside the ring (Fig. 13.9). The cyclic overlap of $2p$ orbitals is nicely evoked by this formulation, but a price is paid. The Kekulé forms are easier for bookkeeping purposes: for keeping track of bonds making and breaking in chemical reactions. We will use the Kekulé form most often in this book.

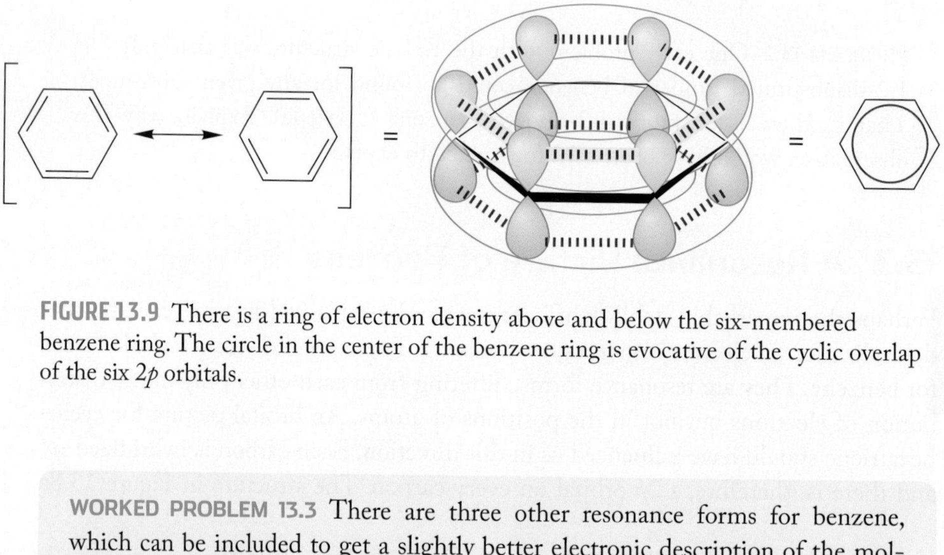

FIGURE 13.9 There is a ring of electron density above and below the six-membered benzene ring. The circle in the center of the benzene ring is evocative of the cyclic overlap of the six $2p$ orbitals.

WORKED PROBLEM 13.3 There are three other resonance forms for benzene, which can be included to get a slightly better electronic description of the molecule than is provided by the pair of Kekulé forms alone. In these other resonance forms, overlap between two p orbitals on "across the ring" carbons is taken into account. Draw these three resonance forms, called **Dewar forms**, showing the orbitals involved in the "cross-ring" bond. Be careful! This problem uses bonding that you probably haven't seen before.

ANSWER Any resonance form represents one electronic description for a molecule and ignores all others. For example, in one resonance form for the allyl cation, overlap between the $2p$ orbitals on C(1) and C(2) is emphasized, and the $2p$ orbital on C(3) is written as if it did not overlap. This error is taken into

(continued)

account in the other resonance form that shows overlap between C(3) and C(2), and ignores any overlap with the *2p* orbital on C(1).

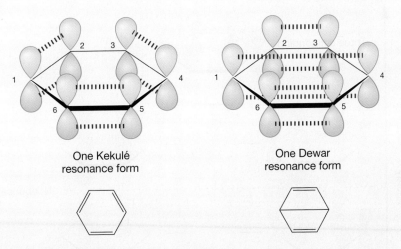

The two Kekulé forms of benzene shown in Figure 13.8 make similar approximations and do a good job of giving the structure of benzene only when taken in combination.

In the Dewar forms, 1,4-overlap is explicitly emphasized. There are three Dewar forms that differ in which 1,4-overlap (1,4; 3,6; or 2,5) is counted.

Notice that this business is tricky. By now, your eye is trained to see this central bond as a σ bond, *and it isn't!* As in the Kekulé forms, it is *2p/2p*, π overlap:

One Kekulé
resonance form

One Dewar
resonance form

For the central bond to be a σ bond, atoms would have to move, and that cannot be done in resonance forms. Only electrons can move.

PROBLEM 13.4 Do you think the Dewar forms will contribute strongly to the benzene structure? Why or why not?

PROBLEM 13.5 There is a real compound $(CH)_6$ called Dewar benzene containing a pair of fused (they share an edge) cyclobutene rings (Dewar benzene is correctly named bicyclo[2.2.0]hexa-2,5-diene). This molecule is not flat, and therefore *not* a resonance form of benzene, but a separate molecule potentially in equilibrium with benzene (see figure below). Make a good three-dimensional drawing of Dewar benzene.

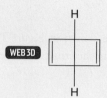

13.4 The Molecular Orbital Picture of Benzene

The raw material for a calculation of the molecular orbitals of benzene is six $2p$ orbitals in a ring. Linear combination of these atomic orbitals leads to a set of six molecular orbitals: three bonding and three antibonding. Their symmetries and relative energies are shown in Figure 13.10.

There are six electrons to go into the system, one from each carbon, and these electrons naturally occupy the three lowest-energy, bonding molecular orbitals. The antibonding molecular orbitals are empty. If we compare benzene to a system of three double bonds not tied together into a ring (1,3,5-hexatriene is a good model) we can see that there are substantial energetic savings involved in the ring structure.

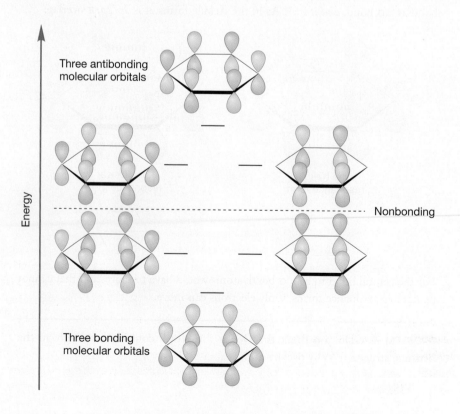

FIGURE 13.10 The π molecular orbitals of benzene.

One can calculate the energy difference in terms of a quantity (β) whose value is approximately −18 kcal/mol (Fig. 13.11). If we sum all the energies of the electrons in Figure 13.11, we find an advantage for benzene over 1,3,5-hexatriene of 1β. The energetic advantage over three separated ethylenes is even greater: 2β (Fig. 13.11).

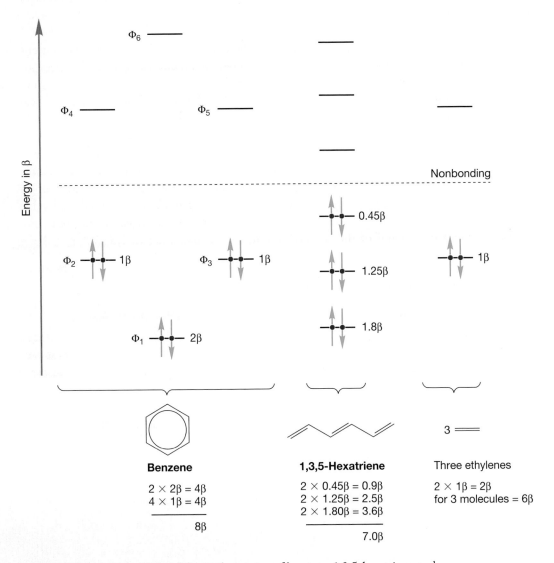

FIGURE 13.11 A comparison of the total π energy of benzene, 1,3,5-hexatriene, and three ethylenes. The calculations are based on the number of electrons and the energy of each π electron.

Look carefully at how well the orbitals are used in benzene. There is maximum occupancy of bonding molecular orbitals and no electrons are placed in destructive antibonding molecular orbitals or wasted in nonbonding molecular orbitals. The molecular orbital picture gives a stronger hint of the special stability of benzene than does the resonance formulation. The stability comes not from a large number of resonance contributors, but from an especially good, especially low-energy, set of orbitals. We will soon see other examples. Now the question is: Just how good is this electronic arrangement? *How much* more stable is benzene than we might have expected from a cyclohexatriene model?

13.5   Quantitative Evaluations of Resonance Stabilization in Benzene

Although we can appreciate that delocalization of electrons is an energy-lowering phenomenon, it is not easy to see intuitively whether this effect is big or small. In this section, we see two experimental procedures for determining the magnitude of resonance stabilization.

13.5a Heats of Hydrogenation

The heat of hydrogenation of *cis*-2-butene is -28.6 kcal/mol (Table 10.1, p. 413). There is no change induced by incorporating the double bond into a six-membered ring, as the heat of hydrogenation of cyclohexene is also -28.6 kcal/mol (Fig. 13.12). A reasonable guess about the heat of hydrogenation of a compound containing two double bonds in a ring would be twice the value for cyclohexene, or $2 \times -28.6 = -57.2$ kcal/mol. This guess is very close to being correct; the heat of hydrogenation of 1,3-cyclohexadiene is -55.4 kcal/mol. There is no special effect apparent from confining two double bonds in a six-membered ring. The difference between the observed heat of hydrogenation for 1,3-cyclohexadiene (-55.4 kcal/mol) and the predicted value (-57.2 kcal/mol) is approximately what one would expect, because our prediction takes no account of the stabilizing effects of conjugation in 1,3-cyclohexadiene (p. 522).

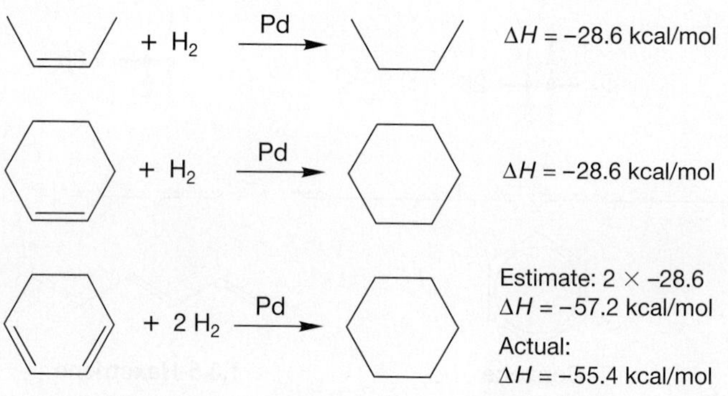

FIGURE 13.12 Heats of hydrogenation for some alkenes and cycloalkenes.

Now let's estimate the heat of hydrogenation for 1,3,5-cyclohexatriene, the hypothetical molecule with three double bonds in a six-membered ring. We might make this estimate by assuming that the third double bond will increase the heat of hydrogenation of 1,3-cyclohexadiene by as much as the addition of a second double bond did for cyclohexene [$-55.4 - (-28.6) = -26.8$ kcal/mol]. This increment would give a predicted value of $-55.4 + (-26.8) = -82.2$ kcal/mol for the heat of hydrogenation of the hypothetical 1,3,5-cyclohexatriene. Now we can measure the heat of hydrogenation of the real molecule, benzene. Hydrogenation is extremely slow under normal circumstances but there are special catalysts that can speed the reaction up a bit, and eventually even the stable benzene will hydrogenate to give cyclohexane. You will learn about these catalysts in Chapter 14. The important point is that the measured heat of hydrogenation for benzene, -49.3 kcal/mol, is much lower than the calculated value for 1,3,5-cyclohexatriene.

Figure 13.13 combines all these numbers and shows graphically how this procedure yields the amount by which benzene is more stable than the hypothetical 1,3,5-cyclohexatriene, -32.9 kcal/mol (Fig. 13.13). This difference represents the amount of energy the special stabilization of benzene is worth. It is more than 30 kcal/mol and is called the **resonance energy** or **delocalization energy** of benzene.

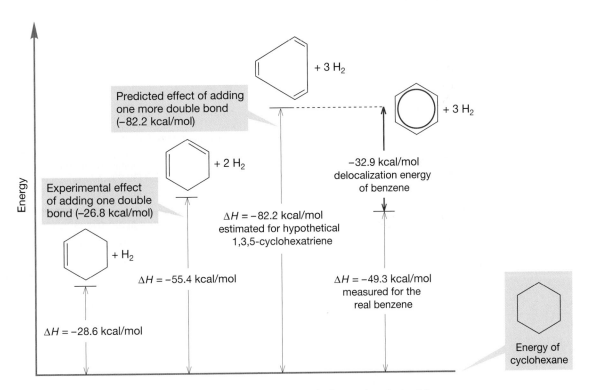

Predicted effect of adding one more double bond (-82.2 kcal/mol)

Experimental effect of adding one double bond (-26.8 kcal/mol)

$+ 3$ H$_2$

$+ 3$ H$_2$

-32.9 kcal/mol delocalization energy of benzene

$+ 2$ H$_2$

$\Delta H = -82.2$ kcal/mol estimated for hypothetical 1,3,5-cyclohexatriene

$+$ H$_2$

$\Delta H = -55.4$ kcal/mol

$\Delta H = -49.3$ kcal/mol measured for the real benzene

$\Delta H = -28.6$ kcal/mol

Energy of cyclohexane

Energy

FIGURE 13.13 The value for the special stability of benzene (red arrow) as derived from measured and calculated heats of hydrogenation. The 1,3,5-cyclohexatriene is drawn with exaggerated bond lengths.

Summary

We calculated the heat of hydrogenation of the unreal, unknown, hypothetical molecule 1,3,5-cyclohexatriene by using values for the heats of hydrogenation of the real molecules cyclohexene and 1,3-cyclohexadiene. The heat of hydrogenation of 1,3-cyclohexadiene was calculated quite accurately, so we can be rather confident that our estimate for 1,3,5-cyclohexatriene won't be far off. That's important, because we can never make the cyclic triene to check the calculation. Any synthetic route will inevitably produce the much more stable real molecule, benzene. The heat of hydrogenation for the real molecule, benzene, is far lower than our estimate.

13.5b Heats of Formation   The delocalization energy of benzene can be estimated in another way, using heats of formation (ΔH_f°). The chemistry of 1,3,5,7-cyclooctatetraene tells us there is no special stabilization in this molecule. Hydrogenation is fast and addition reactions occur with ease. The ΔH_f° for cyclooctatetraene is $+71.23$ kcal/mol, or $+8.9$ kcal/mol per CH unit.

The ΔH_f° for benzene is $+19.82$ kcal/mol, or 3.3 kcal/mol per CH unit (Fig. 13.14). Therefore, benzene is about 5.6 kcal/mol more stable per CH unit than a cyclic polyene with no special stability ($8.9 - 3.3 = 5.6$ kcal/mol). Because there are six CH units in benzene we can estimate the special stability (resonance or delocalization energy) to be $6 \times 5.6 = 33.6$ kcal/mol. Note that the two methods, which use the related heats of hydrogenation and heats of formation, are in rather good agreement.

FIGURE 13.14 The delocalization energy of benzene as calculated from heats of formation.

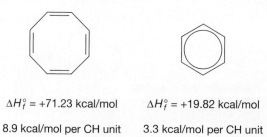

$\Delta H_f^\circ = +71.23$ kcal/mol $\Delta H_f^\circ = +19.82$ kcal/mol

8.9 kcal/mol per CH unit   3.3 kcal/mol per CH unit

$6 \times (8.9 - 3.3) = 33.6$ kcal/mol delocalization energy—this value compares well with that calculated from heat of hydrogenation data (32.9 kcal/mol)

This special stability has come to be known as benzene's **aromatic character** or **aromaticity**, a name derived in an obvious way from the pleasant (?) odor of many benzene derivatives. In the next few pages, we will expand on the theme of aromaticity, eventually ending up with a general method for identifying aromatic compounds.

VANILLIN

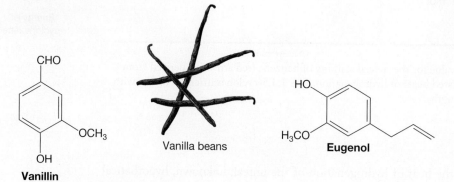

Vanilla beans

Vanillin

Eugenol

The term "aromatic" was coined because many such compounds were odoriferous. Vanillin is a good example, although there are a great many more. Vanillin is found in nature attached to a sugar molecule, glucose, in the seed pods of the vanilla orchid, *Vanilla fragrens*. It appears as well in many other natural sources found in the tropical world. It is now made synthetically from eugenol, which, itself has a pleasant clove-like smell.

13.6 A Generalization of Aromaticity: Hückel's 4n + 2 Rule

It would certainly be surprising if there were not other compounds sharing the special stability we have just called aromaticity. To begin our search for other aromatic compounds, and, ultimately, for a general way of finding such compounds, let's review the structural features of the prototype, benzene, that lead to special stability.

1. The molecule is cyclic. Aromaticity is a property of ring compounds. Recall the discussion of the difference between the molecular orbitals for the cyclic molecule, benzene, and the acyclic molecule, 1,3,5-hexatriene (Fig. 13.11).

2. The molecule is *fully conjugated*, which simply means that there is a 2*p* orbital on every atom in the ring, and that connectivity between adjacent 2*p* orbitals is maintained throughout. 1,3,5-Cycloheptatriene, for instance, is not aromatic, because the 2*p* orbitals of the complete cycle do not overlap. Its properties resemble those of the acyclic 1,3,5-hexatriene (Fig. 13.15).

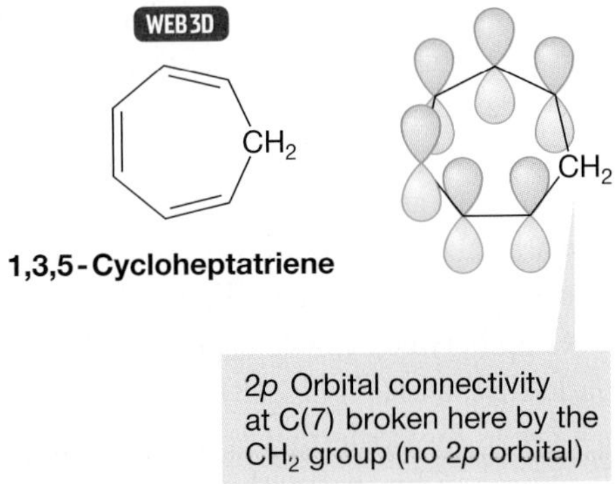

WEB 3D

1,3,5-Cycloheptatriene

2*p* Orbital connectivity at C(7) broken here by the CH₂ group (no 2*p* orbital)

FIGURE 13.15 In 1,3,5-cycloheptatriene, 2*p* orbital connectivity is broken by the CH₂ group. This molecule does not have benzene's aromaticity.

3. The molecule is planar. Why should this make a difference? It ensures optimal overlap of the 2*p* orbitals. If the ring is not planar, or at least quite close to planar, the 2*p* orbitals do not overlap well and some connectivity between orbitals is lost. Large deviations from planarity can nearly or completely uncouple the cycle of orbitals. An example is the tub-shaped molecule cyclooctatetraene (Fig. 13.16).

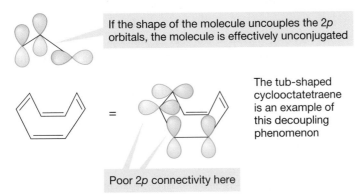

If the shape of the molecule uncouples the 2*p* orbitals, the molecule is effectively unconjugated

The tub-shaped cyclooctatetraene is an example of this decoupling phenomenon

Poor 2*p* connectivity here

FIGURE 13.16 The nonplanar cyclooctatetraene is not aromatic.

One might now be tempted to try a generalization. Might not all molecules that are cyclic, planar, and fully conjugated share benzene's aromatic character and stability? Apparently not. Recall from Figure 13.3 that cyclobutadiene, a molecule that meets all three criteria mentioned above, is most *unstable* (Fig. 13.17).

There must be more to it, and in our search for further criteria, we must take into account the difference between benzene and cyclobutadiene. The crucial generalization was made by the German Erich Hückel (1896–1980) in the early 1930s. His observation can be added to our previous list of three criteria (also recognized by Hückel). It is called **Hückel's rule:**

4. Aromatic systems will contain 4*n* + 2 π electrons, where *n* is an integer, 0, 1, 2, 3, . . . *Please note carefully that "n" has nothing to do with the number of atoms in the ring. We will now discuss the significance of "n."*

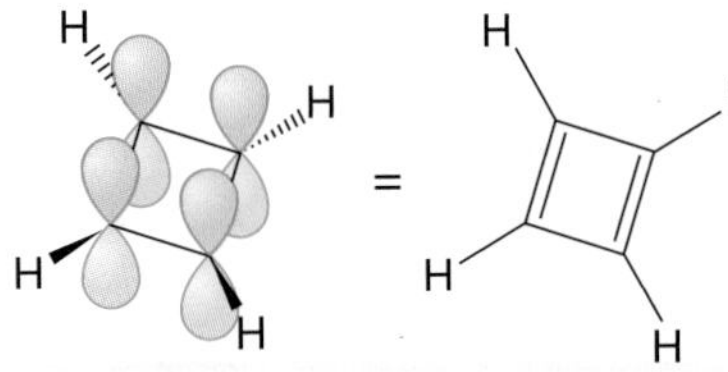

A most unstable molecule, even though it is planar, cyclic, and fully conjugated

FIGURE 13.17 Cyclobutadiene is cyclic, planar, and fully conjugated, but it is not aromatic and it is very unstable.

We are now faced with the most important question of this section: Why should the number of π electrons make a difference? Why does Hückel's rule work? Briefly stated, it works because it is *only* those molecules containing $4n + 2$ π electrons (and satisfying the other three criteria for aromaticity) that will have molecular orbitals arranged like those of benzene. These molecules will have a single orbital at the lowest energy and equal energy pairs of orbitals above it. Orbitals that are equal in energy are called degenerate. When the lowest molecular orbital and all the bonding degenerate pair(s) above it are filled, the molecule is aromatic. In "$4n + 2$", the "2" is that lowest molecular orbital, and the "$4n$" refers to the electrons filling the degenerate pairs. The value of n is simply the number of filled degenerate pairs. An aromatic molecule is stable as a result of the filled molecular orbital system much as the noble gases are stable as a result of their filled shells. It's best to look at some examples now.

cis-1,3,5-Hexatriene can serve as an example for all acyclic molecules. Although it is fully conjugated (a $2p$ orbital on every carbon), and can be planar without inducing great strain, it most certainly is not cyclic. The lack of a ring means that there is no chance of strong overlap between the $2p$ orbitals on C(1) and C(6) and, therefore, no chance of aromatic character (Fig. 13.18). 1,3,5-Cycloheptatriene, which we saw in Figure 13.15, is a cyclic molecule that is not fully conjugated. The connectivity of the $2p$ orbitals is broken by the methylene group at C(7). This molecule cannot be aromatic. Cyclobutadiene is a planar, cyclic molecule that is fully conjugated. It fails only the fourth criterion because it has 4 π electrons, not a "Hückel number" of $4n + 2$.

Before we can go on, we need a reliable way of finding the molecular orbital patterns for these molecules. Consider the device called a **Frost circle**. It is named for the American Arthur A. Frost (1909–2002) who devised it in 1953. It is a method of finding, mercifully without the use of mathematics, the relative energies of the molecular orbitals for planar, cyclic, fully conjugated molecules. One merely inscribes a regular polygon, having the same number of sides as the cyclic molecule, into a circle of radius 2β, *vertex down*, and the intersections of the ring with the circle will mark the positions of the molecular orbitals.

For example, to locate the molecular orbitals of benzene, inscribe a hexagon in a circle, vertex down as shown in Figure 13.19, with the nonbonding line dividing the circle exactly in half.

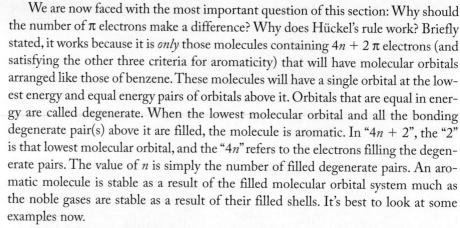

cis-**1,3,5-Hexatriene**

No strong overlap here; the molecule is not aromatic

FIGURE 13.18 *cis*-1,3,5-Hexatriene is fully conjugated and can be planar, but the lack of a ring structure means that overlap between the $2p$ orbitals on C(1) and C(6) is essentially zero.

CONVENTION ALERT

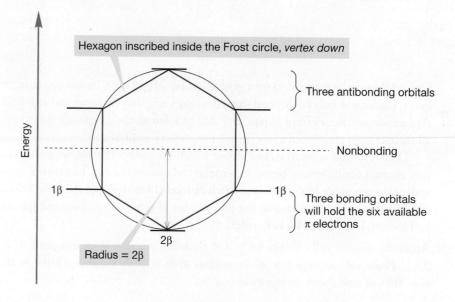

Hexagon inscribed inside the Frost circle, *vertex down*

Three antibonding orbitals

Nonbonding

Energy

1β   1β

Three bonding orbitals will hold the six available π electrons

2β

Radius = 2β

FIGURE 13.19 The relative energies of the molecular orbitals of benzene derived from a Frost circle.

The familiar (p. 578) arrangement of molecular orbitals appears, and with some plane geometry we can accurately get the energies of the molecular orbitals in terms of β. For benzene, the available six π electrons completely fill the three available bonding molecular orbitals with no offending electrons remaining to be placed in the antibonding set. There are no nonbonding molecular orbitals. Most important, the degenerate pair of bonding orbitals is filled.

If we follow this procedure for cyclobutadiene, a set of four molecular orbitals appears (Fig. 13.20). One is bonding, one is antibonding, and two are degenerate, nonbonding molecular orbitals. Four π electrons are available and two of them will surely go into the bonding molecular orbital. The remaining two electrons will have to occupy the nonbonding degenerate pair, and will be most stable in this square arrangement if the spins are unpaired and a single electron is put into each orbital of the pair (Fig. 13.20). This arrangement is surely not particularly stable, and we would not expect the properties of cyclobutadiene to resemble those of benzene, as they most emphatically do not. It is also true that cyclobutadiene must pay a price in ring strain as the internal angles are constrained at 90° but "want to be" 120°.

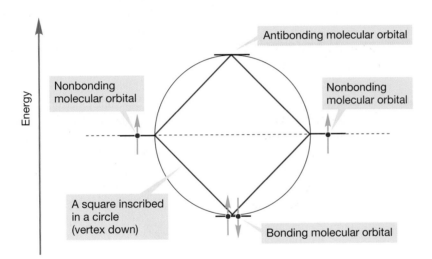

FIGURE 13.20 The relative energies of the molecular orbitals of cyclobutadiene derived from a Frost circle. Electrons have been added to this construct after the energies of the orbitals have been determined.

But it is not this strain that makes cyclobutadiene so especially unstable. There are many compounds with considerable ring strain that are isolable and stable at room temperature. Its instability is due to the fact that it has 4*n* π electrons. The degenerate orbitals have a single electron each, essentially making square cyclobutadiene a diradical species. Remember from Chapter 11 that radicals are very reactive. Cyclobutadiene can be isolated and observed at very low temperatures (its IR spectrum was measured at 8 K = −265 °C), as long as it is segregated from other reactive molecules, including itself. Theory and experiment agree that cyclobutadiene distorts from the square arrangement to form a rectangle with short carbon–carbon double bonds and longer carbon–carbon single bonds. This distortion to a rectangle minimizes the conjugation of the *p* orbitals (Fig. 13.21).

Square, delocalized cyclobutadiene, not observed   Rectangular cyclobutadiene

FIGURE 13.21 In rectangular cyclobutadiene, shown here with bond lengths exaggerated, the orbital connectivity is minimized. Square cyclobutadiene has never been observed.

1,3,5,7-Cyclooctatetraene is another example of a 4*n* molecule. But here we find neither the special stability of benzene, nor the special instability of cyclobutadiene. The molecule behaves like four separated, normal alkenes. Let's first examine planar cyclooctatetraene. The Frost circle again determines the relative positions of the molecular orbitals (Fig. 13.22).

FIGURE 13.22 The molecular orbitals of planar cyclooctatetraene. Because there are eight π electrons, two must occupy nonbonding orbitals.

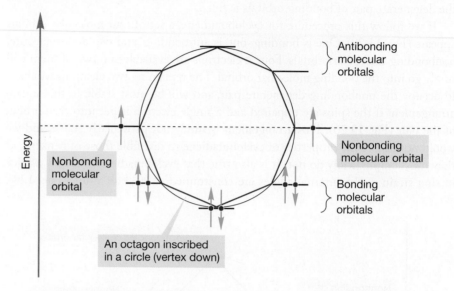

Like cyclobutadiene, planar cyclooctatetraene must have one electron in each of the degenerate nonbonding molecular orbitals, making it a diradical. Unlike cyclobutadiene, however, there is a way for cyclooctatetraene to get out of its energetic misery. Cyclobutadiene must be planar, or very nearly so. By contrast, cyclooctatetraene can distort easily into a tublike form in which most of the angle strain and the diradical nature of the planar form is avoided (Fig. 13.23).

FIGURE 13.23 In planar cyclooctatetraene, there is both angle strain and diradical reactivity. Both can be avoided by distorting into a tub shape.

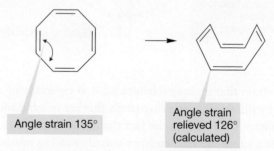

Kinetic measurements reveal that the energy difference between the tub and the planar, delocalized species is 17 kcal/mol. The tub is more stable than the planar, symmetrical, fully conjugated octagon (Fig. 13.24). This example is a rare case of conjugation being avoided because it destabilizes the molecule.

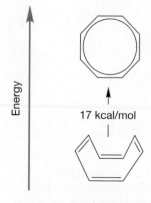

FIGURE 13.24 The tub-shaped cyclooctatetraene is 17 kcal/mol more stable than the planar, delocalized structure.

PROBLEM 13.6 The interconversion of cyclooctatetraene forms is a bit more complicated. There are actually two tub forms that interconvert rather easily. The two localized tubs (alternating single and double bonds, not conjugated) equilibrate without the double bonds changing position. The activation energy for this process is 14.6 kcal/mol. The conversion into the planar, delocalized form requires an additional 2.4 kcal/mol. Neither the localized planar form nor the delocalized planar form is an energy minimum. In other words, they can't be isolated. Based on this information, sketch an Energy versus Reaction progress diagram for these processes.

PROBLEM 13.7 Which of the following molecules are properly termed "aromatic" and which are not? For those that are not, explain why not.

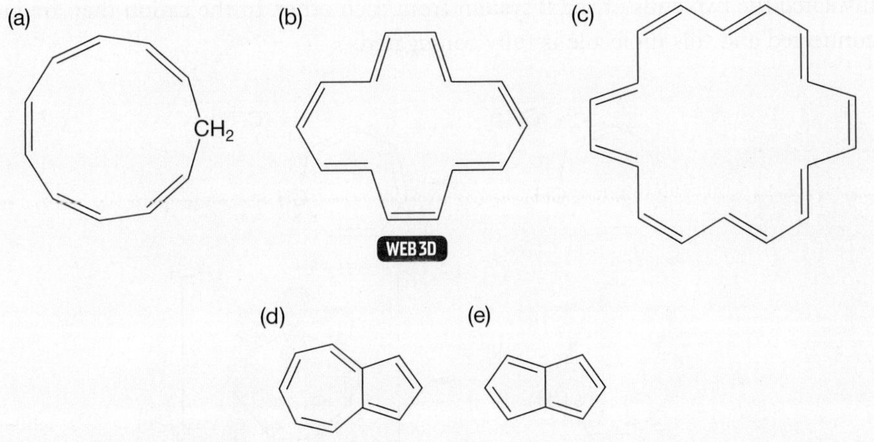

(a)

(b)

CH$_2$

WEB 3D

(c)

(d)

(e)

We have now discussed the lack of aromaticity in 4n molecules and we have seen a few examples of aromaticity (benzene and some of the molecules in Problem 13.7); it would seem to be time to look at some other potentially aromatic 4n + 2 molecules. It is in the stability of certain ions that the most spectacular examples have appeared. As we have stressed over and over, small carbocations are most unstable. But there are a few exceptions to this generality. One of them is the **cycloheptatrienylium ion**, or **tropylium ion** (C$_7$H$_7^+$), the ion derived from the loss of hydride (H:$^-$) from 1,3,5-cycloheptatriene, also called tropilidene (Fig. 13.25). Hydride cannot be simply lost from cycloheptatriene, but it can be *transferred* to the trityl cation, a carbocation attached to three benzenes, which is itself a quite stable carbocation.

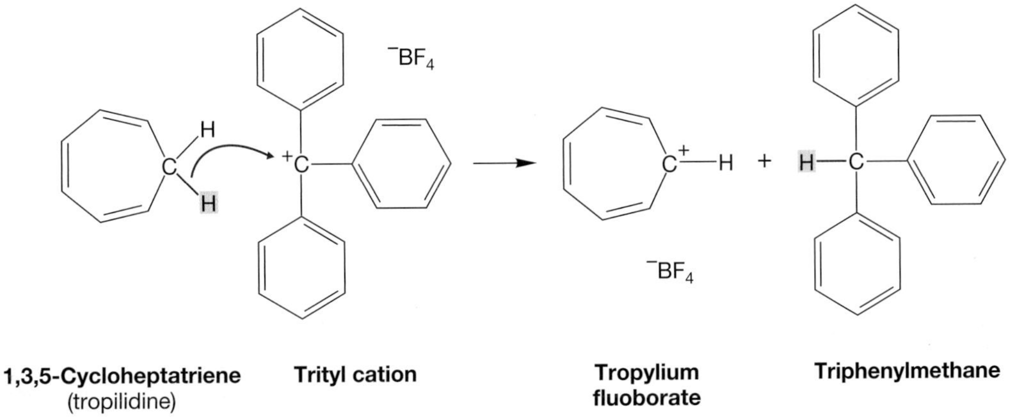

$^-$BF$_4$

$^-$BF$_4$

1,3,5-Cycloheptatriene   **Trityl cation**   **Tropylium**   **Triphenylmethane**
(tropilidine)   **fluoborate**

FIGURE 13.25 The tropylium ion (C$_7$H$_7^+$) can be made through transfer of hydride (H:$^-$) from 1,3,5-cycloheptatriene to the trityl cation.

Tropylium fluoborate appears to be stable indefinitely under normal conditions. It sits uncomplaining on the desktop, for all the world resembling table salt. This stuff is no ordinary carbocation!

PROBLEM 13.8 Why is the trityl cation relatively stable?

Note first that the tropylium ion has one important difference from its parent, 1,3,5-cycloheptatriene. The ion is fully conjugated (Fig. 13.26). The removal of a hydride from the cyclic triene generates a $2p$ orbital at the 7-position, which once insulated the two ends of the π system from each other. In the cation they are now connected and this molecule is fully conjugated.

FIGURE 13.26 In the tropylium ion (the cycloheptatrienylium ion) orbital connectivity is continuous around the ring.

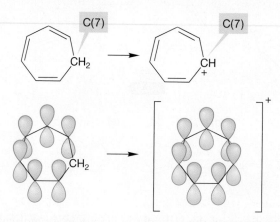

1,3,5-Cycloheptatriene is not fully conjugated   The cycloheptatrienylium (tropylium) ion is fully conjugated

The Frost circle analysis lets us see the relative energies of the molecular orbitals for this system (Fig. 13.27). Notice the degenerate pairs of orbitals. How many electrons must go into the π system? In the cation, there is an empty $2p$ orbital on C(7) and the only π electrons are the six from the three original double bonds. These electrons can fit nicely into the three bonding molecular orbitals of the π system. Note how this system resembles the arrangement of benzene—the occupied degenerate molecular orbitals are full (Fig. 13.27). This ion qualifies as aromatic, and the increased stability *relative to other carbocations* is dramatic indeed. Notice that the number of atoms involved in the ring—here seven—is independent of the number of π electrons—here

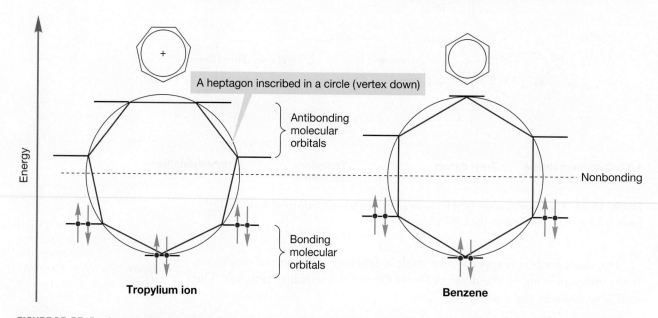

FIGURE 13.27 In the tropylium ion, the lowest single orbital and the bonding degenerate molecular orbitals are fully occupied. The molecular orbital picture of the tropylium ion closely resembles that of benzene.

six $= (4n + 2)$, $n = 1$. The value of n is simply the number of filled degenerate pairs. Hückel's rule is $4n + 2$ because aromaticity requires a filled degenerate pair of orbitals (the $4n$ part) and the lowest single orbital is filled first (the 2 part).

There is an anionic counterpart to the stable tropylium cation, the **cyclopentadienyl anion**. The pK_a of propene is 43, and the pK_a of propane is 50–60. The added delocalization in the allyl anion makes propene a stronger acid than propane by a factor of about 10^{20} (Fig. 13.28). But the pK_a of 1,3-cyclopentadiene is 15!

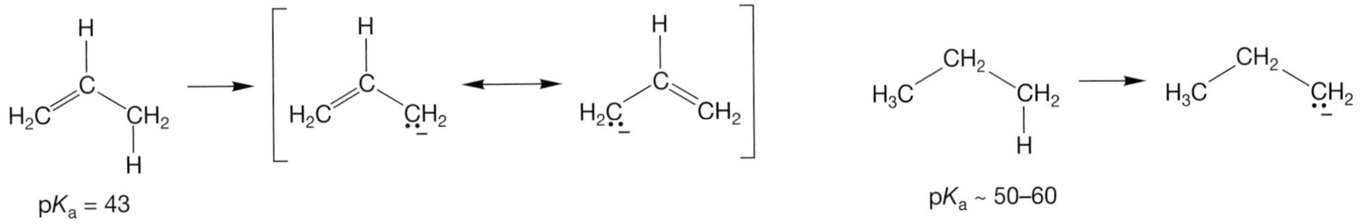

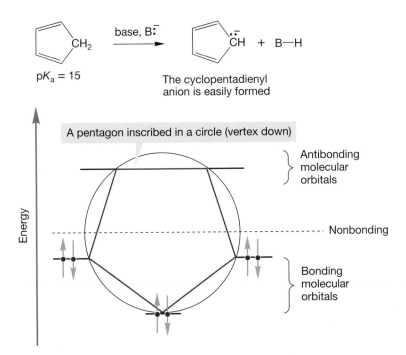

FIGURE 13.28 Resonance stabilization of its conjugate base makes propene far more acidic than propane.

Cyclopentadiene is a *much* stronger acid than propene. The difference in acidity is enormous. Look carefully at the structure of the cyclopentadienyl anion. Here too, we have a planar, cyclic, and fully conjugated system. The molecular orbitals can be derived from a Frost circle (Fig. 13.29). There are six electrons to put into the molecular orbitals, and, as in the tropylium ion or benzene, they fully occupy the lowest molecular orbital and the set of degenerate bonding molecular orbitals. The cyclopentadienyl anion can be described as aromatic, and *for an anion*, this species is remarkably stable. Do not fall into the trap of expecting this anion to be as stable as benzene. Benzene is a neutral molecule in which all of carbon's valences are satisfied. The anion is charged and contains unsatisfied valence. It is very stable, but only in comparison with the rest of the family of carbon anions.

FIGURE 13.29 Cyclopentadiene is quite a strong acid. The cyclopentadienyl anion is an aromatic species. Notice the filled degenerate orbitals. For an anion, it is extremely stable.

But wait. We have been at pains to equate stability with delocalization of electrons, which results in a distribution of charge to several atoms in the molecule.

Both the tropylium ion and the cyclopentadienyl anion can be described by a most impressive array of resonance forms (Fig. 13.30).

FIGURE 13.30 Resonance descriptions of the tropylium (cycloheptatrienylium) cation and the cyclopentadienyl anion.

Cycloheptatrienylium ion (tropylium ion)

Cyclopentadienyl ion

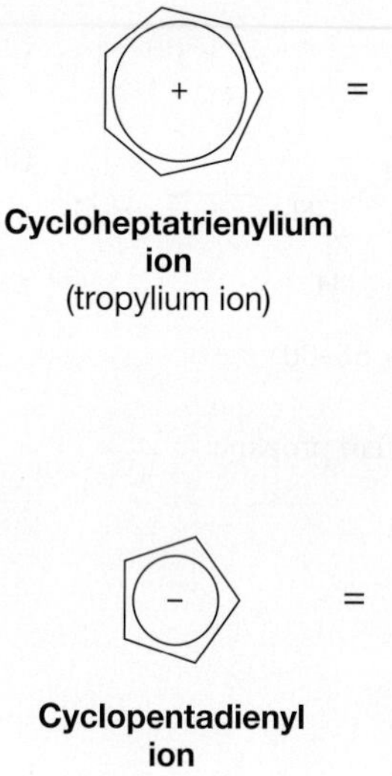

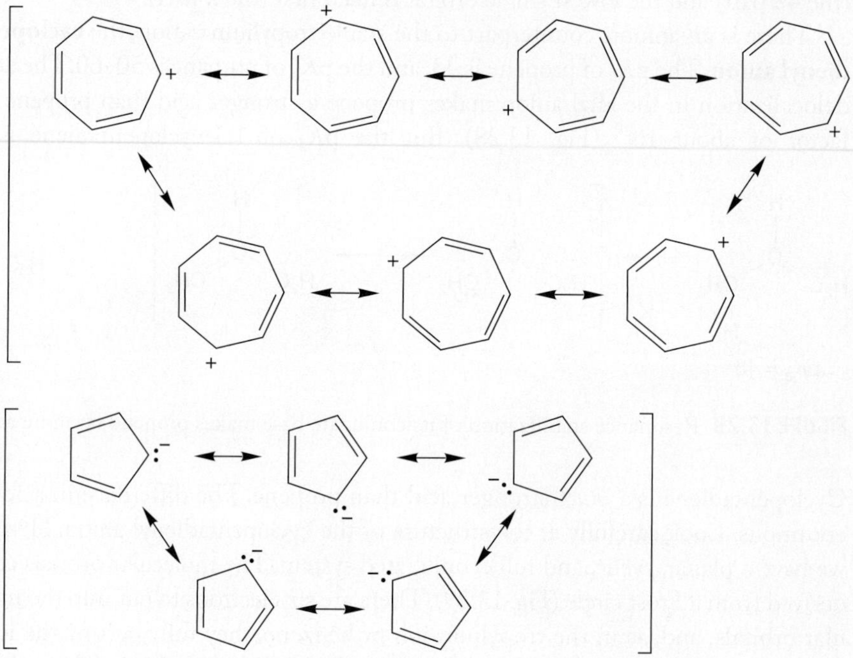

Perhaps the aromaticity concept is unnecessary. Why attribute the stability of these species to aromaticity when simple delocalization of electrons may do as well? That question is a good one, and needs an answer. It is not necessary to invoke special effects if simpler concepts will suffice. The principle of "Ockham's razor"[3] says that when faced with a variety of equally attractive explanations, pick the simplest.

Wielding Ockham's razor, we might say that the cyclopentadienyl anion is very stable because of the five resonance forms so apparent in Figure 13.30. What then would we expect of the cycloheptatrienyl anion, whose seven resonance forms are shown in Figure 13.31? If an ion with five resonance forms is stable, surely one

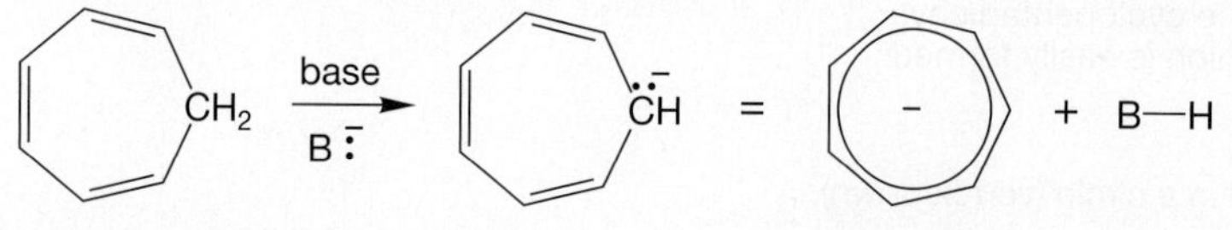

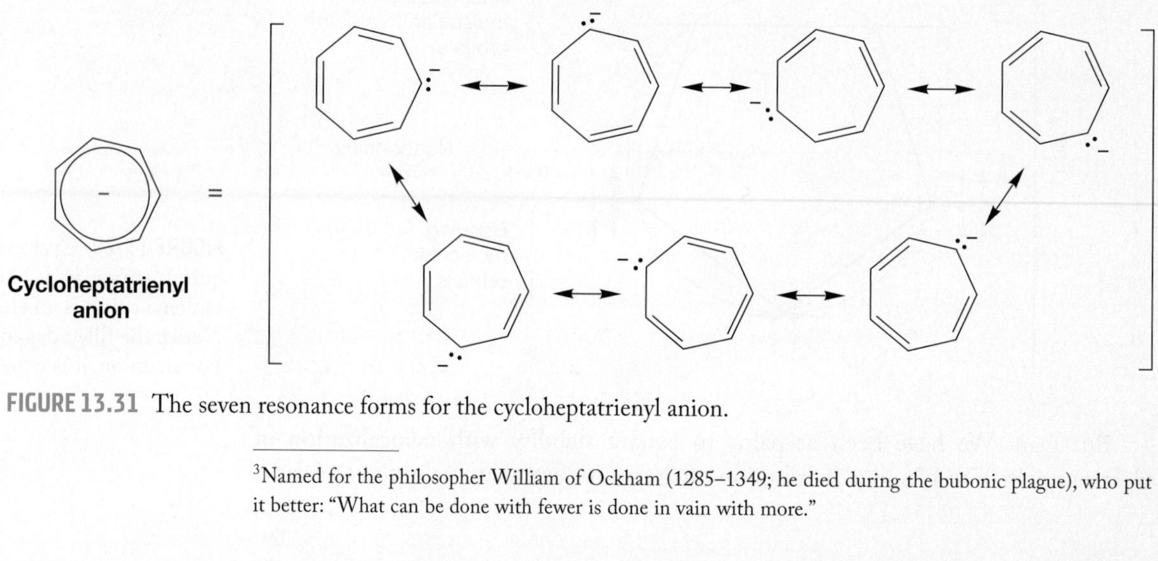

Cycloheptatrienyl anion

FIGURE 13.31 The seven resonance forms for the cycloheptatrienyl anion.

---

[3]Named for the philosopher William of Ockham (1285–1349; he died during the bubonic plague), who put it better: "What can be done with fewer is done in vain with more."

with seven forms will be even more stable. Our explanation predicts a formidable acidity for 1,3,5-cycloheptatriene; in particular, it must be more acidic than cyclopentadiene. Its pK_a must be <15. However, the pK_a of cycloheptatriene is about 39. Cycloheptatriene is a *weaker* acid than cyclopentadiene by a factor of 10^{24} (Fig. 13.32)!

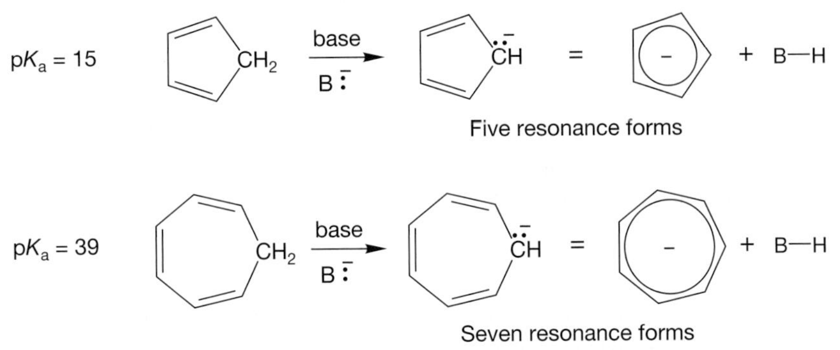

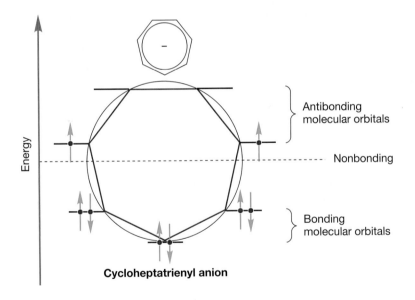

FIGURE 13.32 Cycloheptatriene is a weaker acid than cyclopentadiene by a factor of 10^{24}.

So, the number of resonance forms is *not* always a predictor of stability. No longer are we faced with equally attractive explanations. The aromaticity concept is necessary to make sense of the data. Aromaticity trumps resonance. A look at the molecular orbitals of cycloheptatrienyl anion shows the source of the instability. The cycloheptatrienyl anion must contain unpaired electrons in degenerate antibonding orbitals (Fig. 13.33)! It is not aromatic, it is a diradical.

FIGURE 13.33 In the cycloheptatrienyl anion, two electrons must occupy degenerate antibonding orbitals.

PROBLEM 13.9 How many signals do the ^{13}C NMR spectra of the cyclopentadienyl anion and the tropylium cation show?

Now let's search for other neutral compounds exhibiting aromaticity. Of course, all manner of substituted benzenes exist, but such aromatic compounds hardly constitute exciting new examples. It is possible to combine benzene rings so that they share an edge in what is called a fused relationship (p. 211). These molecules are also aromatic. Atoms other than carbon can be introduced into the ring while maintaining the sextet of electrons to produce **heterobenzenes**, which are fully conjugated, six-membered rings containing one or more noncarbon atoms (N, O, S, etc.).

Some smaller rings also contain a sextet of π electrons; **pyrrole** is an example. Molecules of these various types are shown in Figure 13.34, and will be considered further in later sections.

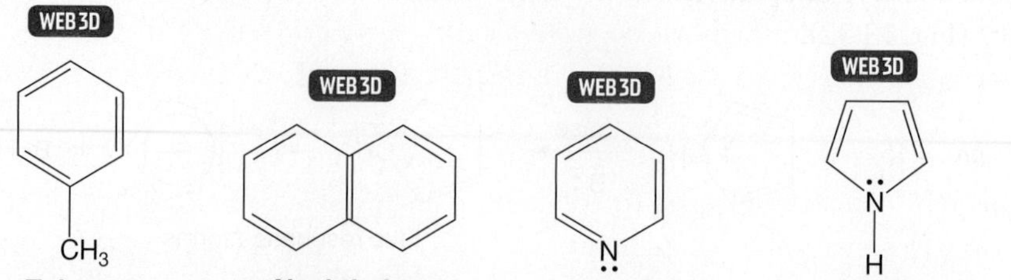

FIGURE 13.34 Some simple extensions of aromaticity: substituted, fused, and heterobenzene compounds, as well as a five-membered ring aromatic compound.

WEB 3D

WEB 3D   WEB 3D   WEB 3D

Toluene
(a simple substituted benzene)

Naphthalene
(a fused, two-ring aromatic compound)

Pyridine
(a six-membered heterobenzene)

Pyrrole
(a five-membered aromatic compound)

PROBLEM 13.10 It is startlingly easy to form the cyclopropenyl cation. Explain why. For this system, what is the value of n in Hückel's formula?

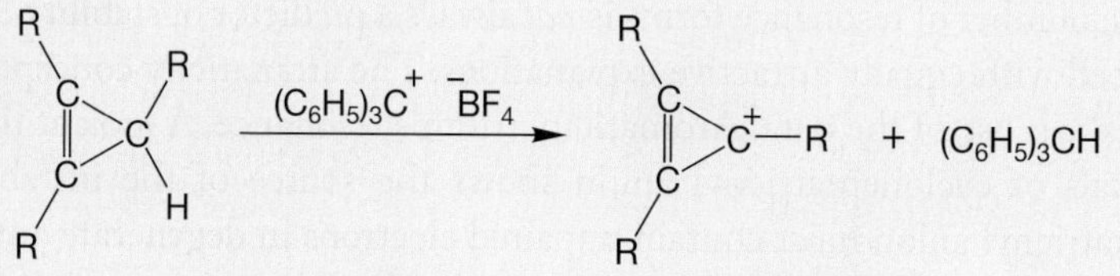

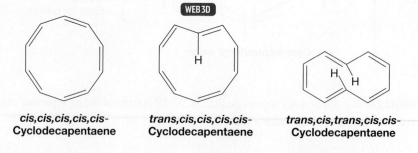

A cyclopropenium ion

PROBLEM 13.11 Do you expect the corresponding cyclopropenyl anion to be stable? Explain.

What we are looking for here are not examples of relatively straightforward extensions, but new $4n + 2$ molecules. You have learned that n can be any integer. We have seen many examples of $n = 1$. In Problems 13.7 and 13.10 we have seen $n = 0, 2, 3,$ and 4.

Let's consider the cyclodecapentaenes, cyclic molecules with 10 π electrons ($4n + 2$, $n = 2$) (Fig. 13.35). Are they as stable as the six-electron molecules we have been examining?

FIGURE 13.35 Some cyclodecapentaenes. These molecules have 10 π electrons ($4n + 2 = 10$, $n = 2$). There are two pairs of degenerate orbitals that are filled.

WEB 3D

cis,cis,cis,cis,cis-
Cyclodecapentaene

trans,cis,cis,cis,cis-
Cyclodecapentaene

trans,cis,trans,cis,cis-
Cyclodecapentaene

PROBLEM 13.12 The three isomeric cyclodecapentaenes of Figure 13.35 differ in the stereochemistries of the double bonds. Why isn't there more than one isomer of benzene?

PROBLEM 13.13 How many signals would be seen in the ^{13}C NMR spectra of the three cyclodecapentaenes in Figure 13.35?

A remarkable amount of effort was expended on the synthesis of the cyclodecapentaenes before they were successfully made. The magnitude of this effort should make us suspicious; aromatic compounds are everywhere and extraordinarily easy to make. Moreover, the known cyclodecapentaenes are not only not especially stable, they are instead strikingly *unstable*. So, we have some explaining to do. We have been relying on a theory that predicts special stabilization conferred by the aromatic 4*n* + 2 number of π electrons, and the cyclodecapentaenes are members of the 4*n* + 2 club. If we cannot explain why these molecules are unstable, Hückel's rule must be abandoned. It turns out to be not too difficult to resolve the problem. Aromatic stability is not some magical force. It is just one kind of stabilizing effect, which in the face of stronger destabilizing effects will not prevail. Put simply, the planar cyclodecapentaenes are destabilized by an amount greater than aromaticity can overcome. What might those destabilizing effects be? Look first at the molecule on the right in Figure 13.35, the trans,cis,trans,cis,cis isomer. For the molecule to be flat, the two hydrogens inside the ring must occupy the same space. This crowding induces extraordinary strain that is greater than any resonance stabilization. The molecule cannot be flat. Aromaticity cannot compensate for this strong destabilizing effect.

The trans,cis,cis,cis,cis isomer, the middle molecule in Figure 13.35, has some of the same problems, although here only one hydrogen is inside the ring. In this molecule, however, there is severe angle strain, and the planar form, required for aromaticity, is badly destabilized.

The all-cis isomer, the molecule on the left in Figure 13.35, has even more angle strain, and it too can be planar, and therefore aromatic, only at a prohibitively high energy cost. The benefits of delocalization are overwhelmed by the angle strain.

A good set of models will show this strain easily. Construct the planar, all-cis cyclodecapentaene and the strain will be apparent. If you are lucky, the plastic will mimic the molecule and the model will snap into a pretzel shape, which approximates the energy minimum form of the molecule. These molecules are best described as strained polyenes. They fail the test of planarity for aromatic molecules because the planar structures are much more strained than the nonplanar forms. In this case, strain trumps aromaticity.

PROBLEM 13.14 The all-cis cyclodecapentaene forms a new, cross-ring bond (see below). As the molecule warms up, a bicyclo[4.4.0]decatetraene is formed. Write an arrow formalism for this reaction.

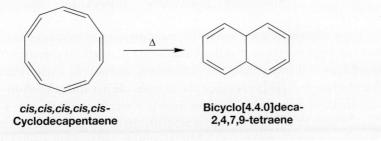

cis,cis,cis,cis,cis-
Cyclodecapentaene

Bicyclo[4.4.0]deca-
2,4,7,9-tetraene

So we haven't really judged the limits of the theory of aromaticity by using the cyclodecapentaenes. We might be able to if the offending strain could somehow be removed. On paper, this transformation is simple. For example, we might just erase the two inside hydrogens in one of the molecules (Fig. 13.36). Unfortunately, this erasure doesn't give a real molecule because this structure has two trivalent carbons at the positions marked with dots in Figure 13.36.

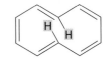

trans,cis,trans,cis,cis-
Cyclodecapentaene
(strain caused by the
inside red hydrogens)

magic
hydrogen
eraser

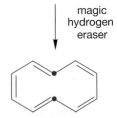

(much less strain! But…
what is going on at the
red-dot positions?)

FIGURE 13.36 The strain caused by two inside hydrogens in one cyclodecapentaene can be eliminated by removing the offending hydrogens.

A particularly clever German chemist, Emanuel Vogel (b. 1927), found a way to do this erasure chemically. He used a remarkable synthesis to replace the offending inside hydrogens with a bridging methylene group (Fig. 13.37). The removal of the offending hydrogens eliminates the strain that comes from their "bumping" into each other.

FIGURE 13.37 Vogel's bridged cyclodecapentaene.

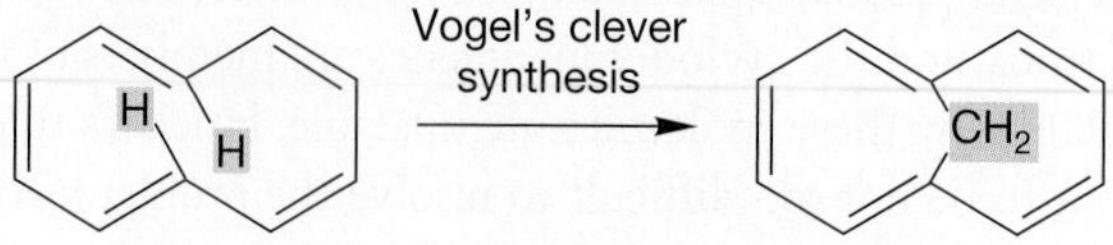

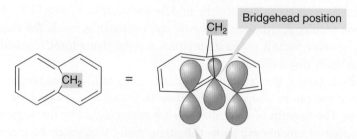

The resulting molecule, a bridged system containing 10 π electrons, isn't a perfect cyclodecapentaene. The overlap between the $2p$ orbitals at the bridgehead and adjacent positions isn't optimal, for example (Fig. 13.38). Still, the molecule maintains enough orbital overlap to remain fully conjugated and satisfy the criteria for aromaticity (cyclic, flat, conjugated, and $4n + 2$ π electrons). Both its chemical and physical properties allow it to be classified as aromatic.

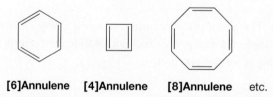

Bridgehead position

Overlap here is not ideal (make a model!), but it is good enough to allow substantial electron delocalization

FIGURE 13.38 Vogel's molecule is fully conjugated, but orbital overlap is not optimal, as it is, for example, in benzene. Nonetheless, this molecule is aromatic.

The generic term for monocyclic, fully conjugated molecules is **annulene**. Benzene could be called [6]annulene, square cyclobutadiene is [4]annulene, and planar cyclooctatetraene is [8]annulene (Fig. 13.39).

FIGURE 13.39 Some simple annulenes.

[6]Annulene   **[4]Annulene**   **[8]Annulene**   etc.

Some other annulenes, besides [6]annulene, appear to be aromatic. These annulenes are large enough so that the destabilizing steric effects present in the cyclodecapentaenes ([10]annulenes) are diminished, and severe angle strain is avoided. A nice example is [18]annulene, a planar molecule in which all carbon–carbon bond lengths are very similar. It contains $4n + 2$ π electrons, where $n = 4$ (Fig. 13.40). The description of this molecule as aromatic derives largely from an examination of its nuclear magnetic resonance (NMR) spectrum, and that must wait for Chapter 15, but [18]annulene does appear to be aromatic.

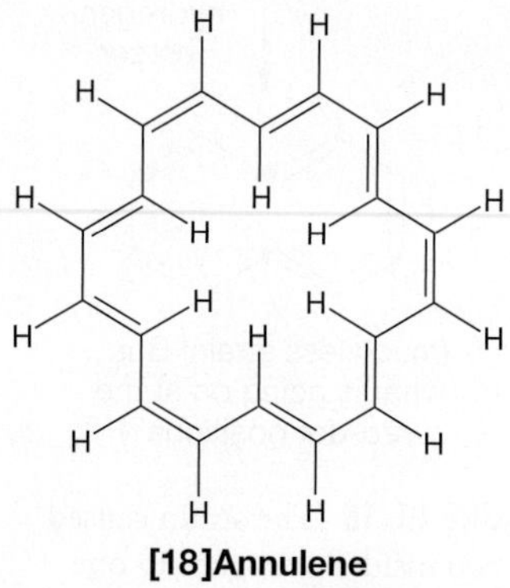

[18]Annulene

FIGURE 13.40 [18]Annulene.

Summary

Planar, cyclic, fully conjugated molecules containing $4n + 2$ π electrons are especially stable and are called "aromatic." The "$4n + 2$ rule" works because such molecules have a set of filled degenerate molecular orbitals, in a sense a closed shell.

We have concentrated on the phenomenon of aromaticity as applied to benzene and larger and smaller polyenes. We saw two experimental methods of evaluating the stabilization of aromaticity quantitatively, and with the aid of Hückel's insight, we found a generalization of the phenomenon. Now we move on to ways of varying the structure of our prototypal example, benzene, and then take a quick look at reactivity.

13.7 Substituted Benzenes

13.7a Mono- and Disubstitution Monosubstituted benzenes are usually named as derivatives of benzene, and the names are written as single words. There are numerous common names that have survived a number of attempts to systemize them out of existence. You really do have to know a number of these names in order to be literate in organic chemistry. We don't think the systematic "benzenol" will replace "phenol" in your lifetime, for example. At least we hope it doesn't. Figure 13.41 shows some simple benzenes with both the systematic and common names.

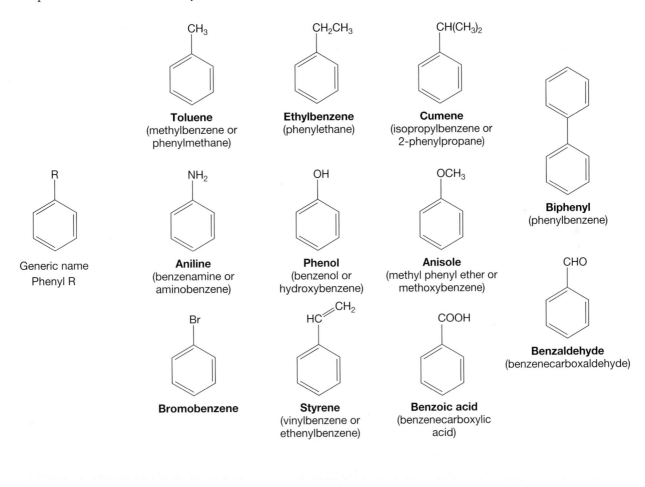

FIGURE 13.41 Some simple monosubstituted benzenes. Common names are widely used (boldface).

CONVENTION ALERT

Two of the most often used common names are **phenyl** (sometimes even **Ph**), which stands for a benzene ring with an open valence (the ring is attached to an R group, for example), and **benzyl**, which stands for the Ph—CH$_2$ group that has an open valence. Both are shown in Figure 13.41.

When one Kekulé structure is drawn, you know that you must (mentally or physically) draw the other form as well for a full picture.

There are three possible isomers of disubstituted benzenes (Fig. 13.42). 1,2-Substitution is known as **ortho** (*o*-), 1,3-substitution as **meta** (*m*-), and 1,4-substitution as **para** (*p*-).

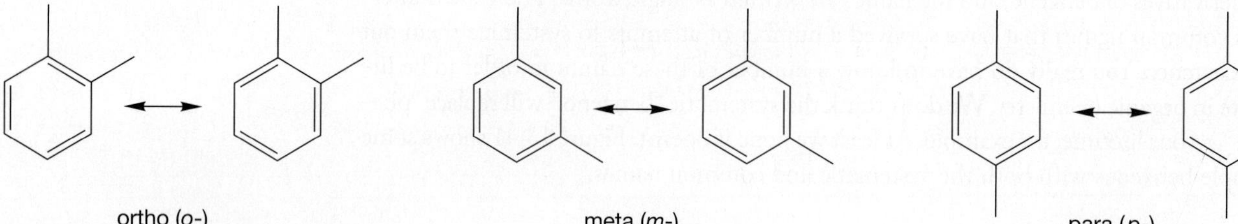

ortho (*o*-)   meta (*m*-)   para (*p*-)
1,2-Disubstituted   1,3-Disubstituted   1,4-Disubstituted

FIGURE 13.42 The three possible substitution patterns for disubstituted benzenes.

In practice, either the numbers or the "ortho, meta, para" designations are used. If there is a widely used common name, it is used as a base for the full name. For example, the first and third molecules in Figure 13.43 are quite properly named as *o*-bromotoluene and *p*-nitrophenol, respectively.

FIGURE 13.43 Four disubstituted benzenes.

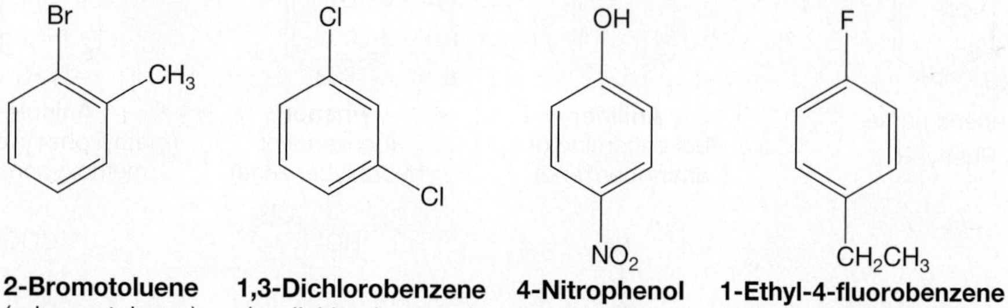

2-Bromotoluene   **1,3-Dichlorobenzene**   **4-Nitrophenol**   **1-Ethyl-4-fluorobenzene**
(*o*-bromotoluene)   (*m*-dichlorobenzene)   (*p*-nitrophenol)   (*p*-ethylfluorobenzene)

PROBLEM 13.15 The isolation of three, and only three isomers of disubstituted benzenes was a crucial factor in the assignment of the Kekulé form as the correct structure of benzene. Use the fact that there are three and only three achiral isomers of a disubstituted benzene (C$_6$H$_4$R$_2$) to criticize the suggestions that benzene might have the following structures:

13.7b Polysubstitution There are disubstituted benzenes that are also known by common names. Figure 13.44 shows a sampling of these molecules.

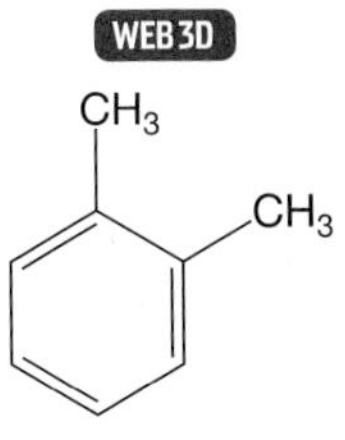

o-Xylene
(1,2-dimethylbenzene)

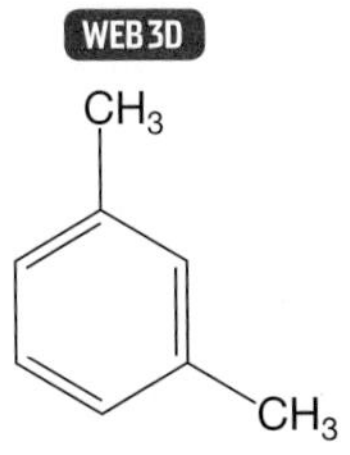

m-Xylene
(1,3-dimethylbenzene)

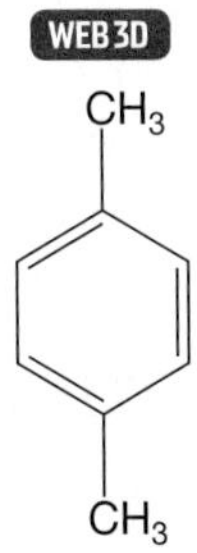

p-Xylene
(1,4-dimethylbenzene)

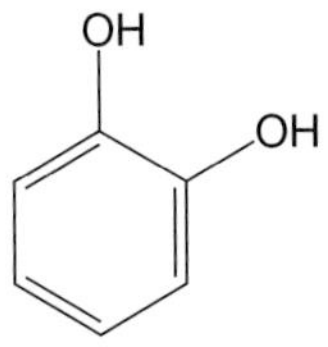

Catechol
(o-dihydroxybenzene)

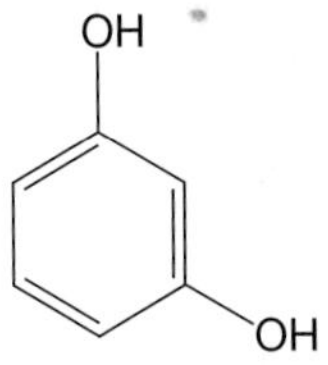

Resorcinol
(m-dihydroxybenzene)

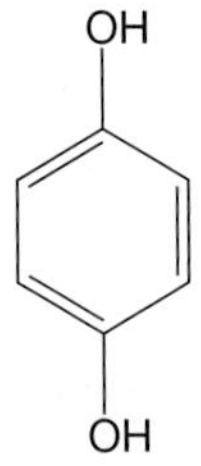

Hydroquinone
(p-dihydroxybenzene)

FIGURE 13.44 More common names for disubstituted benzenes.

PROBLEM 13.16 How many signals will appear in the ^{13}C NMR spectra of the three dihydroxybenzenes at the bottom of Figure 13.44?

The substituents on polysubstituted benzenes are numbered and named alphabetically (Fig. 13.45). Notice that a benzene with a methyl-, a hydroxy-, or an amino-group will have the root word of toluene, phenol, or aniline respectively. Because it is the root word, the carbon attached to that group is automatically C(1).

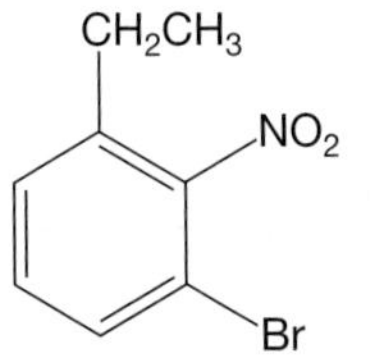

1-Bromo-3-ethyl-2-nitrobenzene

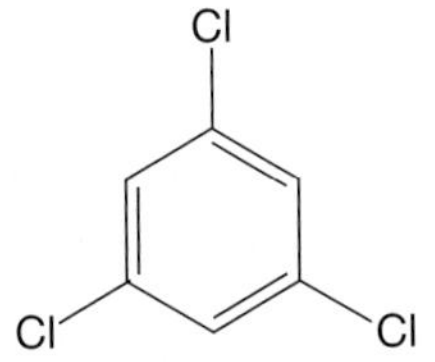

1,3,5-Trichlorobenzene

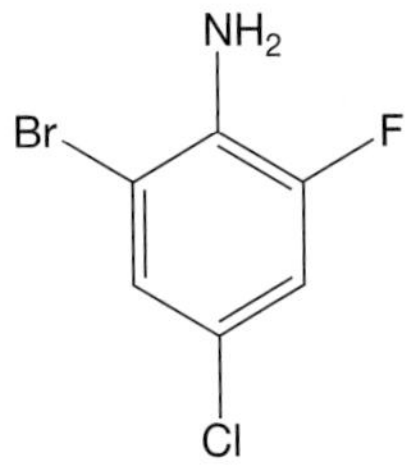

2-Bromo-4-chloro-6-fluoroaniline

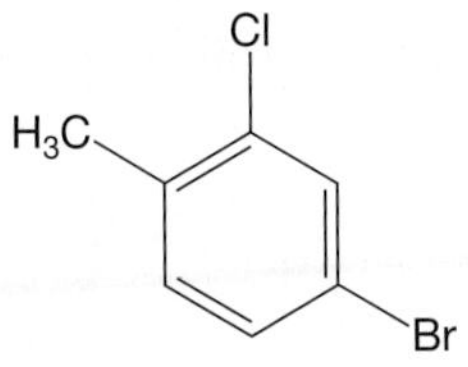

4-Bromo-2-chlorotoluene

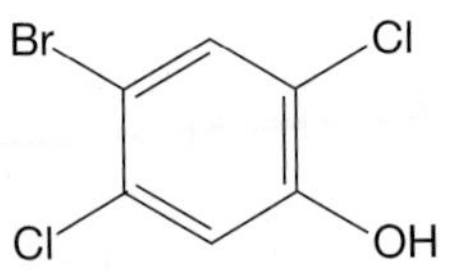

4-Bromo-2,5-dichlorophenol

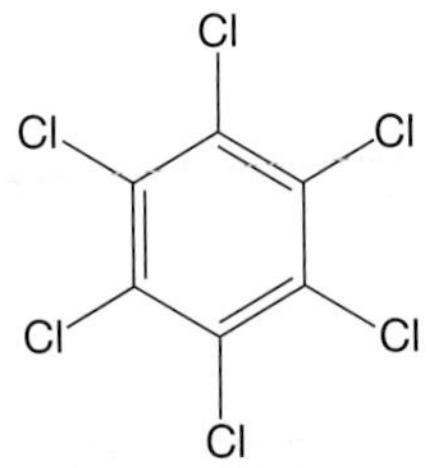

Hexachlorobenzene

FIGURE 13.45 Some named polysubstituted benzenes.

13.8   Physical Properties of Substituted Benzenes

The physical properties of substituted benzenes resemble those of alkanes and alkenes of similar shape and molecular weight. Table 13.1 collects some physical properties for a number of common substituted benzenes. Notice the effects of symmetry. *p*-Xylene melts at a much higher temperature than the ortho or meta isomer, for example. Many para isomers have high melting points, and crystallization can sometimes be used as a means of separating the para regioisomer from the others.

TABLE 13.1 Physical Properties of Some Substituted Benzenes

Name	bp (°C)	mp (°C)	Density (g/mL)
Cyclohexane	80.7	6.5	0.78
Benzene	80.1	5.5	0.88
Methylcyclohexane	100.9	−126.6	0.77
Toluene	110.6	−95	0.87
o-Xylene	144.4	−25.2	0.88
m-Xylene	139.1	−47.9	0.86
p-Xylene	138.3	13.3	0.86
Aniline	184.7	−6.3	1.02
Phenol	181.7	43	1.06
Anisole	155	−37.5	1.0
Bromobenzene	156.4	−30.8	1.5
Styrene	145.2	−30.6	0.91
Benzoic acid	249.1[a]	122.1	1.1
Benzaldehyde	178.6	−26	1.0
Nitrobenzene	210.8	5.7	1.2
Biphenyl	255.9	71	0.87

[a]At 10 mm pressure.

You have learned that aromatic compounds are particularly stable. This stability is reflected in the chemical and physical properties of substances that contain aromatic compounds, such as graphite and coal. Aromatic molecules are hardy.

Another important physical property of aromatic compounds is their involvement in π stacking. These flat molecules lie easily on top of each other, and this property is a critical feature of enzyme structure–activity and in DNA double helix formation.

CONVENTION ALERT

Aromatic compounds are called **arenes** and the term for an arene with an open valence is aryl. A general abbreviation for aryl is Ar, in analogy to the R used for a general alkyl species.

13.9   Heterobenzenes and Other Heterocyclic Aromatic Compounds

13.9a Contrasting Structures of Pyridine and the Five-Membered Rings Pyrrole, Furan, and Thiophene   If a CH unit of benzene is replaced with a nitrogen, the result is **pyridine**, another aromatic six-membered ring (Fig. 13.46). From the provincial view of organic chemistry all atoms except

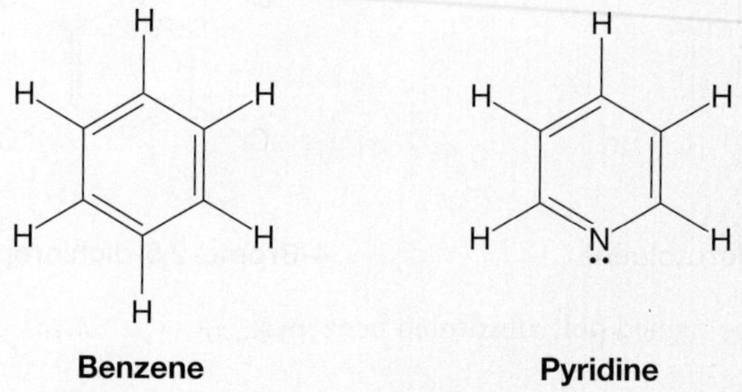

FIGURE 13.46 Pyridine, a heterobenzene.

Benzene   **Pyridine**

carbon are called "heteroatoms," and pyridine is a heterobenzene, which we saw in Figure 13.34. It is a member of the general class of compounds called **heteroaromatic compounds**, defined as aromatic molecules containing at least one heteroatom.

The electron counting for pyridine can be confusing at first because of the extra pair of electrons on nitrogen. Doesn't pyridine have eight π electrons and thus violate the $4n + 2$ rule? Put a better way, we have to worry whether antibonding orbitals are occupied in pyridine. A good orbital drawing shows that the "extra" two electrons are not in the π system, and pyridine is not in violation of the Hückel rule (Fig. 13.47).

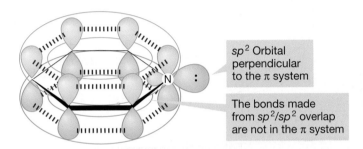

sp^2 Orbital perpendicular to the π system

The bonds made from *sp*^{2}/*sp*^{2} overlap are not in the π system

FIGURE 13.47 Pyridine has only 6 π electrons. The lone-pair electrons on nitrogen are in an *sp*^{2} orbital perpendicular to the π system, not in the π system of the ring.

Other six-membered rings containing one heteroatom are known, but none is very stable. Oxabenzene must be cationic and the positive charge contributes to making it a reactive molecule (Fig. 13.48). Aromaticity must overcome the destabilizing influences of the charged atom and doesn't completely succeed in this case, in part because the electronegative oxygen is forced to carry the positive charge.

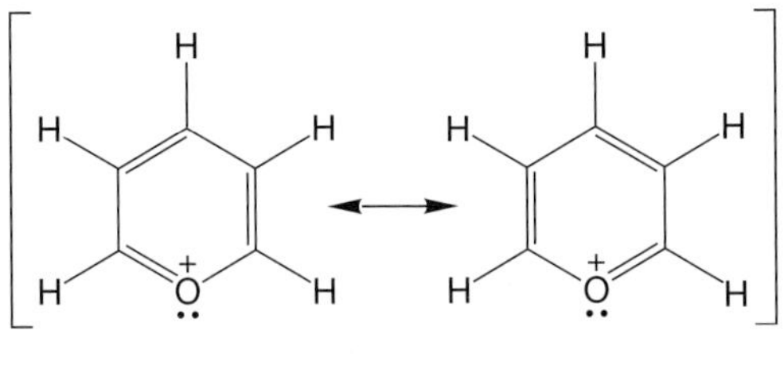

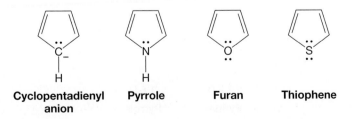

FIGURE 13.48 Oxabenzene must be charged and is not very stable.

The neutral, five-membered heterocyclic ring compounds, pyrrole and **furan**, also show aromatic character. Even thiophene is considered aromatic (Fig. 13.49). Furan and pyrrole can complete an aromatic sextet of π electrons by using a pair of nonbonding electrons in a *2p* orbital. These molecules have the same number of π electrons as the cyclopentadienyl anion, but none of the problems induced by

Cyclopentadienyl anion   **Pyrrole**   **Furan**   **Thiophene**

FIGURE 13.49 Four five-membered aromatic rings. There are $4n + 2$ π electrons in the three neutral molecules, which are isoelectronic (same number of electrons) with the cyclopentadienyl anion.

the charge on the all-carbon molecule. In thiophene and furan, there is a second pair of nonbonding electrons remaining in an sp^2 orbital perpendicular to the π orbital system (Fig. 13.50).

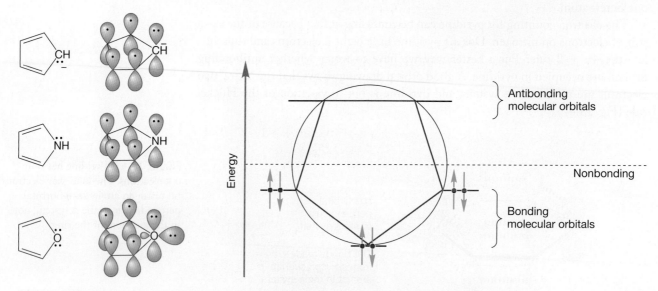

FIGURE 13.50 Orbital pictures of the isoelectronic aromatic compounds that are five-membered rings.

Pyridine and pyrrole have contrasting resonance descriptions (Fig. 13.51), and this difference results in some very different physical properties. For example, the direction of the dipole moment is different in the two aromatic molecules.

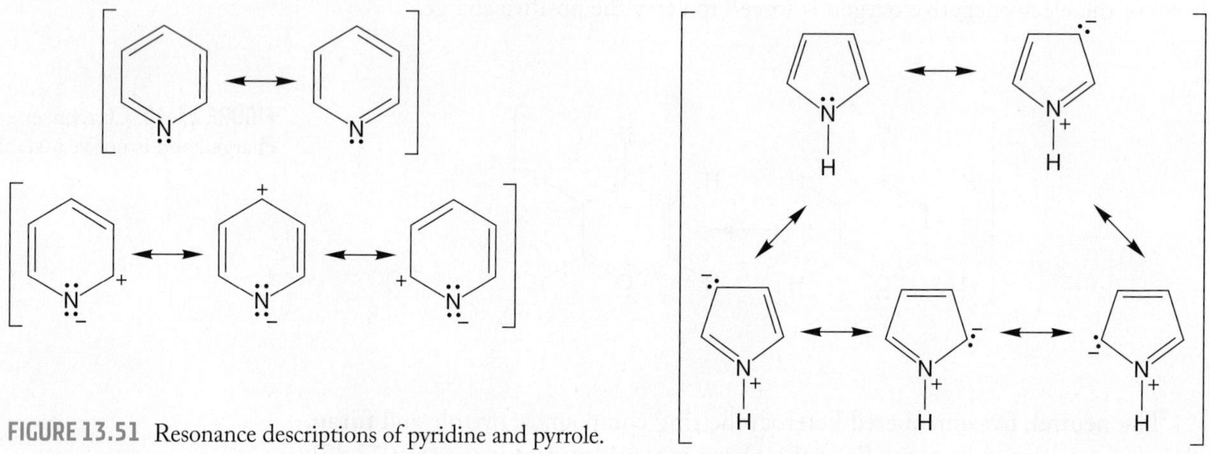

FIGURE 13.51 Resonance descriptions of pyridine and pyrrole.

In pyridine, the electronegative nitrogen is the negative end of the dipole, but in pyrrole, nitrogen is the positive end (Fig. 13.52).

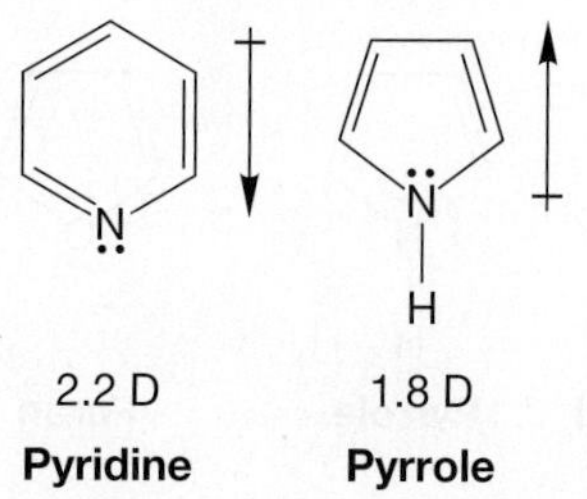

FIGURE 13.52 The dipole moments of pyridine and pyrrole are in different directions.

2.2 D   1.8 D
Pyridine   **Pyrrole**

13.9b Acid and Base Properties Pyridine can be protonated in acid, but pyridine is a weaker base than a simple cyclic secondary amine such as piperidine (Fig. 13.53). The pK_a values of the conjugate acid ammonium ions show this clearly. Protonated pyridine (the pyridinium ion) is a much stronger acid ($pK_a = 5.2$) than protonated piperidine, an ammonium ion formed from a typical secondary amine ($pK_a \sim 11$). The stronger acid corresponds to a weaker conjugate base. Alternatively, we might make this argument in terms of pK_b values. The pK_b of pyridine is 8.8, whereas that for a simple secondary amine is about 3. These values also tell us that the secondary amine is a stronger base. Most organic chemists use pK_a values to analyze base strength.

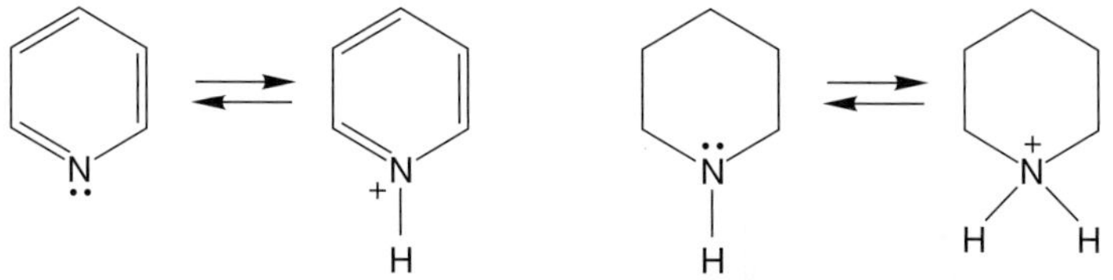

| Pyridine
$pK_b = 8.8$ | Pyridinium ion
$pK_a = 5.2$ | Piperidine
$pK_b \sim 3$ | Piperidinium ion
$pK_a \sim 11$ |

FIGURE 13.53 Pyridine is a weaker base than the similar, but nonaromatic piperidine.

One reason for the decreased basicity of pyridine is that the lone pair of electrons of the sp^2 hybridized nitrogen of pyridine are held more tightly than those of the approximately sp^3 hybridized nitrogen of a simple secondary amine like piperidine, and therefore are less basic.

Pyrrole is even less basic than pyridine (Fig. 13.54). One can estimate the pK_a of the conjugate acid at about -4! When pyrrole is protonated, aromaticity is lost because the aromatic sextet is disrupted (*not* true for pyridine). The lone pair of electrons is tied up in the new bond to hydrogen and no longer contributes to the completion of the aromatic sextet.

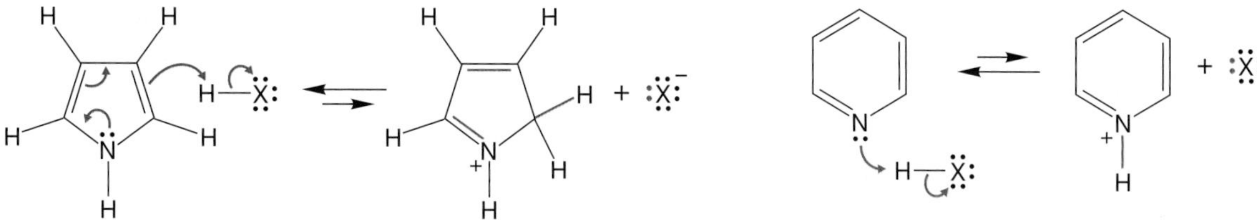

| Aromatic | *Not* aromatic
$pK_a \sim -4$ | Aromatic | Still aromatic |

FIGURE 13.54 Pyrrole is a very weak base because protonation on nitrogen destroys aromaticity. Protonation on carbon also destroys aromaticity, but is better than protonation on nitrogen.

PROBLEM 13.17 Notice in Figure 13.54 that pyrrole is protonated on one of the carbons, not nitrogen. Explain why.

For an organic compound, pyrrole is a rather strong Brønsted acid, $pK_a \sim 23$; but it is much weaker than cyclopentadiene, $pK_a = 15$ (Fig. 13.55). This difference may seem curious at first, because removal of a proton from pyrrole leaves a negative

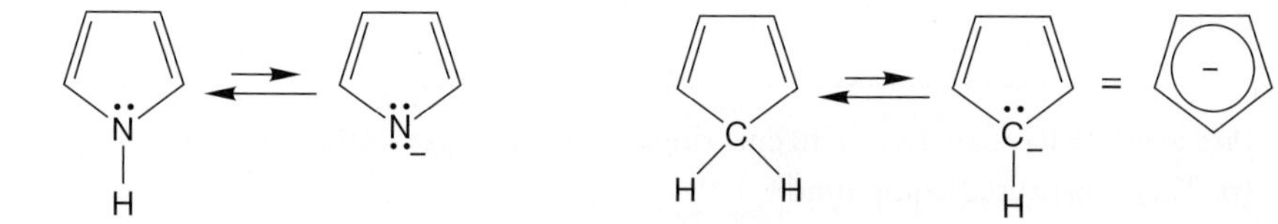

| Already aromatic
$pK_a = 23$ | Still aromatic | *Not* aromatic
$pK_a = 15$ | Aromatic! |

FIGURE 13.55 Pyrrole is a relatively weak acid compared to cyclopentadiene.

charge on a relatively electronegative nitrogen, whereas cyclopentadiene gives a carbon anion, generally a much less happy event.

However, cyclopentadiene *becomes* aromatic when a proton is removed, and pyrrole is *already* aromatic. Removal of a proton from cyclopentadiene is made easier by the gain in aromaticity, which does not happen in pyrrole.

13.9c Pyridine as Nucleophile   Like other amines, pyridine can be alkylated in an S_N2 reaction in which the lone pair electrons on nitrogen displace a leaving group (p. 317). Treatment of pyridine with a primary or secondary alkyl iodide leads to alkyl pyridinium ions. Pyridine gives pyridine *N*-oxide when treated with hydrogen peroxide (Fig. 13.56).

FIGURE 13.56 Pyridine can act as a nucleophile in the formation of pyridinium salts and pyridine *N*-oxides.

13.10 Polynuclear Aromatic Compounds

13.10a Polynuclear Aromatic Compounds   At this point, we can see benzene as a building block from which more complicated, polycyclic molecules can be constructed, as we mentioned in Chapter 4 (p. 179). For example, benzene and other aromatic rings can share edges to form fused, **polynuclear aromatic compounds** (also called polyaromatic hydrocarbons or polycyclic aromatic compounds). The simplest of these is naphthalene, in which two benzene rings share an edge. In quinoline and isoquinoline, it is a benzene ring and a pyridine that share an edge:

Naphthalene   Quinoline   Isoquinoline

PROBLEM 13.18 Draw schematic orbital pictures of naphthalene and isoquinoline. Put in the π electrons as dots. It is not necessary to draw the carbon–hydrogen σ bonds.

PROBLEM 13.19 Can two benzene rings share a single carbon to form a spiro (p. 211) substituted compound?

Clearly, we can continue this building process. Three fused benzene rings produce either anthracene or phenanthrene:

WEB 3D

WEB 3D

Anthracene   **Phenanthrene**

WORKED PROBLEM 13.20 Why is the following molecule not stable? *Hint*: Kekulé forms are often useful.

ANSWER The circles are misleading. Drawing the molecule in Kekulé form allows you to see that there is no way for each carbon to be hybridized sp^2 if each carbon has four valences.

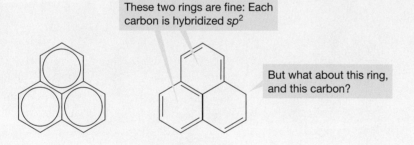

These two rings are fine: Each carbon is hybridized sp^2

But what about this ring, and this carbon?

PROBLEM SOLVING

Watch out for those circles! They evoke the MO description of benzene nicely, but they can be dangerous, as Problem 13.20 demonstrates. In polycyclic systems, it is probably a good idea not to use circles. Draw out Kekulé forms instead.

PROBLEM 13.21 Draw all the fused polynuclear aromatic compounds made from four benzene rings. Be careful. This question is much harder than it seems.

The limit of the ring-fusing business is reached in graphite, a polymer composed of fused benzene rings extended in what amounts to an infinite plane (Fig. 13.57). Of course, the plane is not really infinite, but it is large enough so that the edges do not affect the physical or chemical properties of graphite. Graphite's lubricating properties are related to the ease of sliding the planar layers of graphite over each other.

In the mid-1980s, R. F. Curl (b. 1933), H. W. Kroto (b. 1939), R. E. Smalley (1943–2005), and their co-workers were attempting to explain the striking abundance

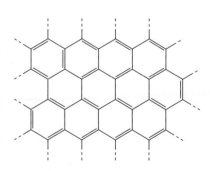

FIGURE 13.57 Graphite, an infinitely extended planar array of fused benzene rings.

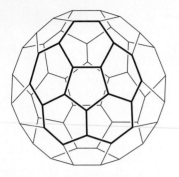

FIGURE 13.58 Buckminsterfullerene, C_{60}, a soccer ball–shaped form of carbon. The double bonds (or circles) are left out for clarity.

of certain fragments observed when a carbon rod was vaporized. In the face of much skepticism in the chemical community, they proposed that the major fragment, C_{60}, had the structure of a soccer ball of pure carbon (Fig. 13.58). They turned out to be absolutely right, and named this molecule "buckminsterfullerene" after R. Buckminster Fuller (1895–1983), the inventor of the geodesic dome. This compound, which can now be made easily in multigram lots through the vaporization of carbon rods, is a testimony to the proposition that organic chemistry is hardly played out as a provider of exciting new molecules. Curl, Kroto, and Smalley shared the 1996 Nobel prize in chemistry for this discovery.

Fuller's geodesic dome was the inspiration for the name of C_{60}, buckminsterfullerene.

PROBLEM 13.22 The ring-fusing process produces some interesting molecules along the way to graphite. "Hexahelicene" and "twistoflex" are chiral, for example. Why?

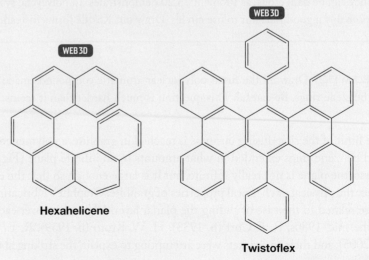

WEB 3D

Hexahelicene

WEB 3D

Twistoflex

Expanding this building block process to generate ever larger and more varied poly-cyclic aromatic compounds may be fun, but there is a downside to the process. Many of the polycyclic aromatic compounds (and some of the simple ones as well) are carcino-genic. Some of the worst are shown in Figure 13.59. A prudent person avoids exposure to the known strong carcinogens (it is impossible to avoid exposure to all carcinogens), and minimizes contact with related aromatic compounds. It is harder to avoid such com-pounds than it sounds. It is clearly simple to avoid bathing in benzene, and usually (but not always) possible to substitute other molecules when benzene is called for in a chem-ical recipe. However, one of the bad carcinogens, benz[*a*]pyrene, is a product of the inter-nal combustion engine as well as combustion (cigarettes, charcoal-broiled meat) in general. The control of such molecules is not only physically difficult, but intersects with politics and economics as well. This situation creates serious problems of social policy because, beyond the questions of public health, there are fortunes to be made and elec-tions to be won. Such matters seem often to complicate the purely scientific questions!

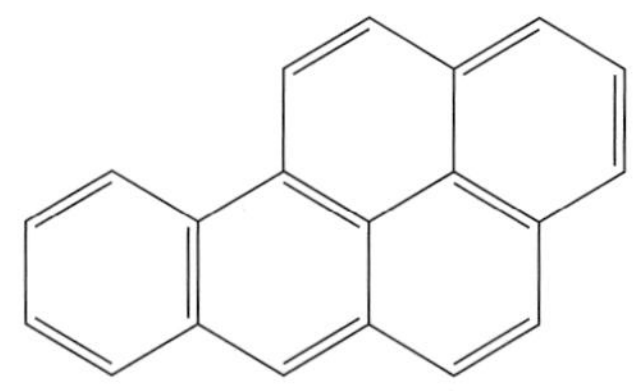

13.10b   Related Heterocycles: Indole, Benzofuran, and Benzothiophene

The benzo-fused counterparts of the simple heterocycles pyrrole, furan, and thiophene are indole, benzofuran, and benzothiophene. These compounds are related to the two-ring, all-carbon aromatic compound naphthalene:

WEB 3D

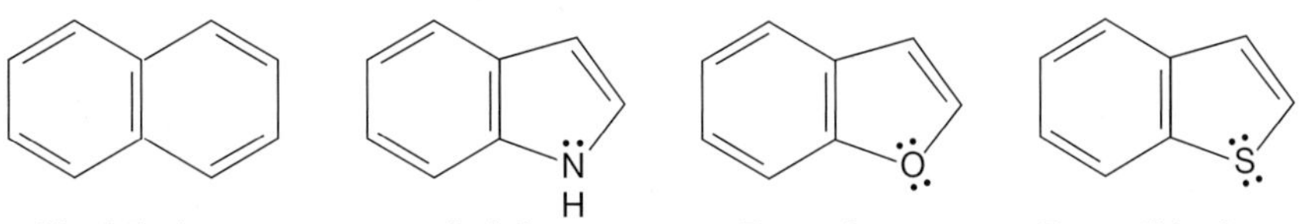

Naphthalene   **Indole**   **Benzofuran**   **Benzothiophene**

Benz[*a*]pyrene

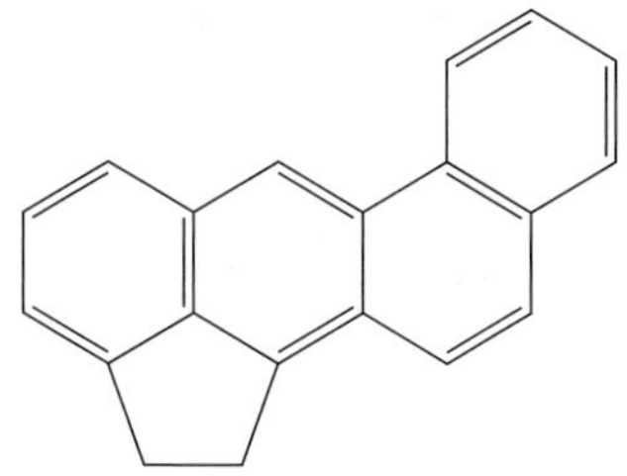

Cholanthrene

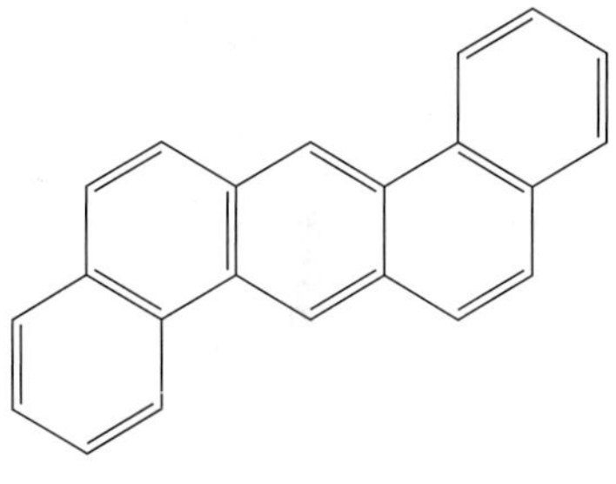

Dibenz[*a,h*]anthracene

FIGURE 13.59 Some carcinogenic aromatic hydrocarbons.

They are important chiefly because of the large variety of natural products contain-ing these ring systems. Some examples of natural products containing simple and complex five-membered heterocyclic rings are given in Figure 13.60.

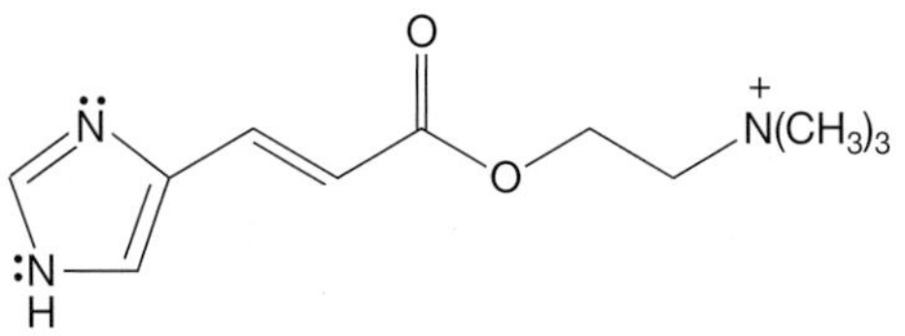

Murexine
(neuromuscular blocker)

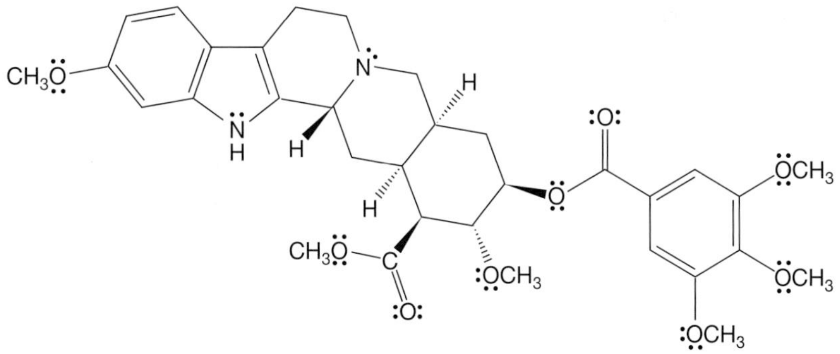

Reserpine
(antihypertensive)

FIGURE 13.60 Two natural products containing five-membered heterocyclic rings.

Summary

We have seen that both pyridine and pyrrole qualify as aromatic. However, we have also seen sharp differences. For example, the lone-pair electrons on nitro-gen are in the σ system in pyridine, whereas in pyrrole they are in the π system, completing the aromatic sextet. We have also learned about the Brønsted–Lowry basicity and nucleophilicity (Lewis basicity) of these heteroaromatic compounds. Our eyes have been opened to the wonderful, but sometimes hazardous, world of polyaromatic compounds.

13.11 Introduction to the Chemistry of Benzene

We will take a quick look here at reactions. We will first introduce a general reaction that preserves the aromatic character of benzene. It is risky even to suggest that molecules have desires, but the imperative in benzene chemistry definitely is, Preserve the Aromatic Sextet! A less flamboyant way of saying this is to point out that thermodynamics will surely favor very stable molecules, and therefore benzene chemistry is likely to lead to more very low energy aromatic compounds. However, occasionally there are useful reactions in which the aromatic stabilization is lost, and we will see one of them in Section 13.11b.

13.11a Aromatic Substitution
We have already mentioned that benzene does not react with Br_2 or HBr, as do almost all alkenes (p. 572), to give addition products (Fig. 13.61). This apparent lack of reactivity results from the stabilization of

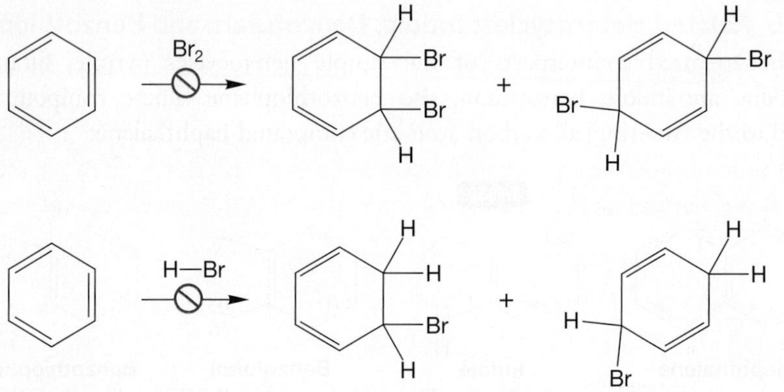

FIGURE 13.61 Benzene does not react with Br_2 or HBr. Neither 1,2- nor 1,4-addition reactions occur.

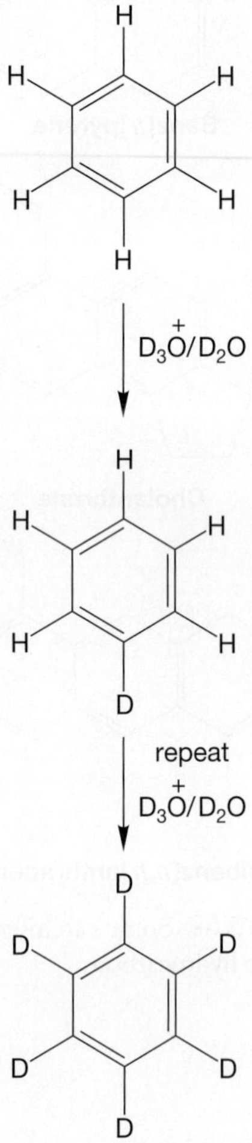

benzene by aromaticity, which produces activation barriers that are too high for the addition reactions. If we use a deuterated acid we can see that there is a reaction taking place, at least in strong acid, but the product is merely the exchanged, deuterated benzene, not an addition product (Fig. 13.62).

What is happening in this deuteration? And why is the reaction diverted from ordinary addition? The answers to these questions will serve to summarize much of the chemistry of aromatic compounds, which will be encountered in detail in Chapter 14. In deuterio acid, addition of a deuteron gives a resonance-stabilized carbocation, but aromaticity is lost (Fig. 13.63).

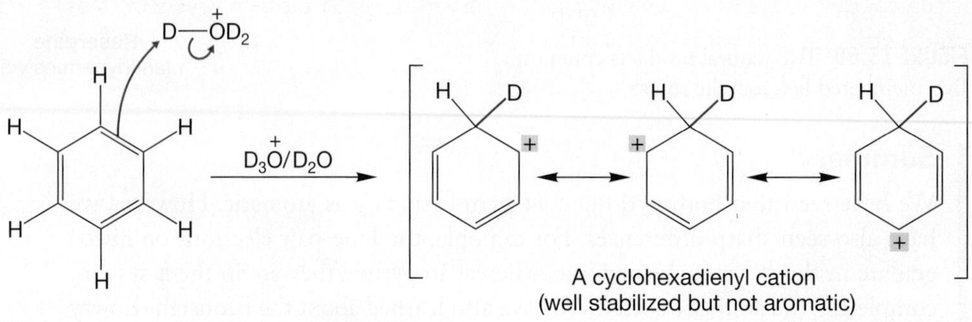

FIGURE 13.62 The hydrogens of benzene can be exchanged for deuterium in deuterated acid.

A cyclohexadienyl cation
(well stabilized but not aromatic)

FIGURE 13.63 Addition of a deuteron to benzene gives a resonance-stabilized, but nonaromatic, carbocation.

WORKED PROBLEM 13.23 Be sure you see that the ion in Figure 13.63 is not aromatic. Why is aromaticity lost?

ANSWER

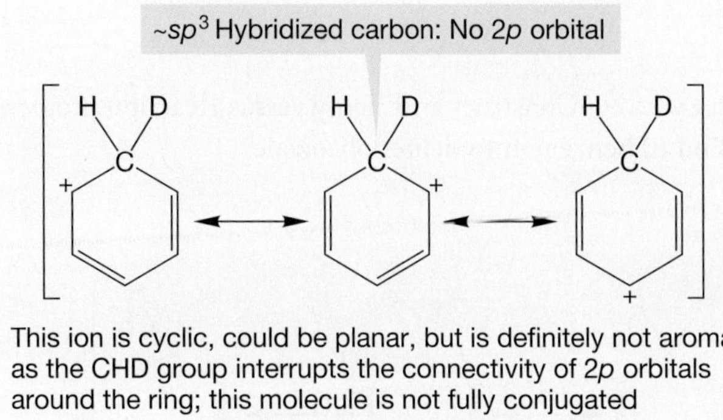
~sp3 Hybridized carbon: No 2p orbital
H D C
This ion is cyclic, could be planar, but is definitely not aromatic, as the CHD group interrupts the connectivity of 2p orbitals around the ring; this molecule is not fully conjugated

In the normal addition reaction of alkenes or dienes, a nucleophile such as bromide would attack the carbocation (Fig. 13.64) to give the final product. In this case, we would anticipate the two products of 1,2- and 1,4-addition. But aromaticity, the more than 30 kcal/mol of delocalization energy, has been lost in each of these products.

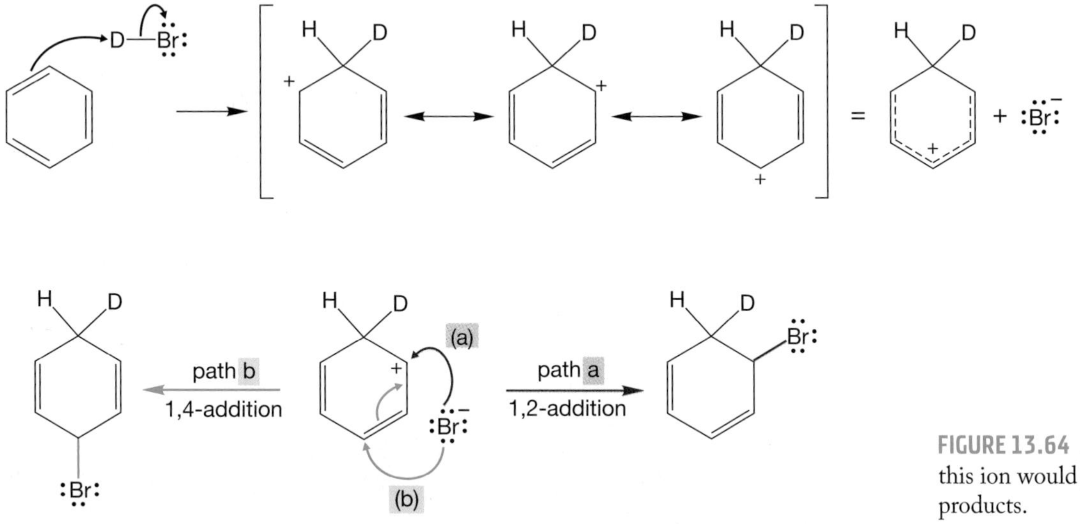
D—Br:
H D
+ :Br:
path b
1,4-addition
(a)
(b)
path a
1,2-addition

FIGURE 13.64 Addition of Br^- to this ion would give two nonaromatic products.

Their formation from benzene is generally endothermic. A far better reaction that regenerates the benzene ring is possible. Loss of a proton (or deuteron) regenerates the aromatic sextet of π electrons (Fig. 13.65).

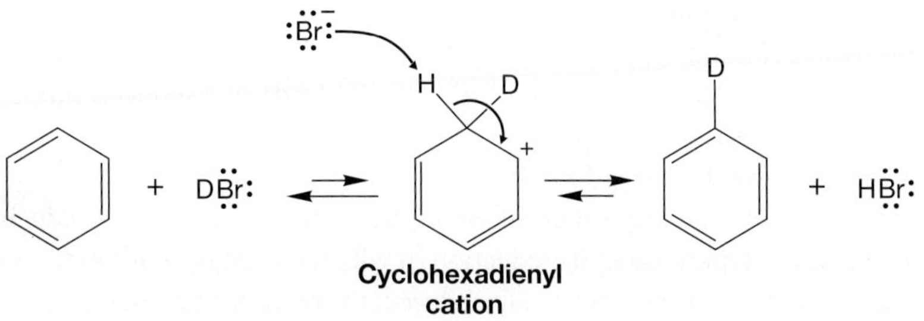
:Br:
H D
+ DBr:
+ HBr:
Cyclohexadienyl cation

FIGURE 13.65 Loss of a proton or deuteron from the carbocation regains aromaticity.

If a deuteron is lost, we see no reaction, as benzene is simply regenerated. If a proton is lost, we see the product of an exchange reaction in which deuteriobenzene is formed (Fig. 13.65). If a large excess of deuterio acid is used, all of the hydrogens can be exchanged to form hexadeuteriobenzene. We will elaborate on this reaction in Chapter 14.

WORKED PROBLEM 13.24 Construct an Energy versus Reaction progress diagram for the conversion of benzene into deuteriobenzene.

ANSWER

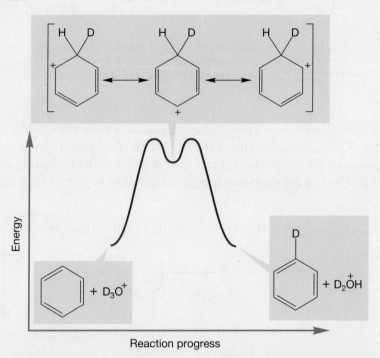

The first step is protonation of the benzene ring (using a deuteron, not a proton) to give the resonance-stabilized cyclohexadienyl cation. This process is surely very endothermic, because aromaticity is lost. The second half of this symmetrical reaction (except for the difference between the isotopes H and D) involves the deprotonation of the intermediate to produce monodeuterated benzene. If there is a sufficient source of deuterium, the reaction can continue until all H atoms are replaced with D atoms. The second step is the reverse of the first, and simply is the exothermic deprotonation of the intermediate cation to regenerate the aromatic ring.

13.11b Birch Reduction of Aromatic Compounds
Now we come to a reaction in which aromaticity *is* lost, the **Birch reduction**. Remember that alkynes can be treated with sodium in liquid ammonia to give, ultimately, trans alkenes (p. 452). Catalytic hydrogenation of benzene is difficult (p. 572), but treatment of benzene with sodium in liquid ammonia and a little alcohol leads to a reduced product—in this case, 1,4-cyclohexadiene (Fig. 13.66). The reaction is named after the Australian chemist Arthur J. Birch (1915–1995).

Immediately note the important difference between this method of reducing the benzene ring and a typical catalytic reduction in which all carbon–carbon π bonds are reduced. In this case, the ring is not totally reduced; the reaction stops at the diene stage.

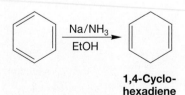

1,4-Cyclo-hexadiene

FIGURE 13.66 Birch reduction of aromatic compounds leads to 1,4-cyclohexadienes.

The mechanism of the Birch reduction is similar to that for dissolving metal reduction of alkynes (p. 452), and starts in the same way, with a transfer of an electron from the metal to one of the empty antibonding orbitals of benzene. The product is a resonance-stabilized radical anion (Fig. 13.67).

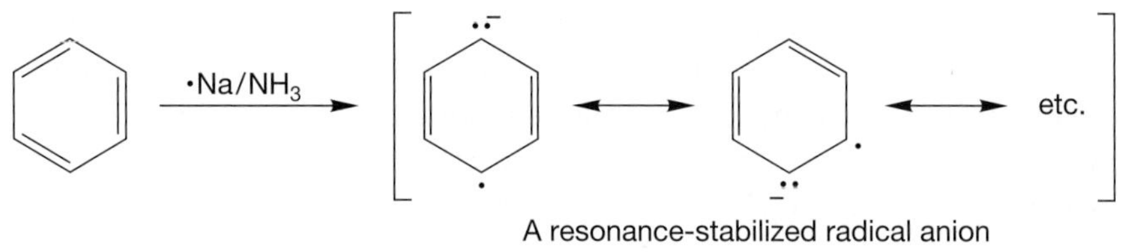

A resonance-stabilized radical anion

FIGURE 13.67 Transfer of an electron from sodium to benzene leads to a resonance-stabilized radical anion.

Protonation by alcohol gives a radical that can undergo a second electron transfer–protonation sequence to complete the reduction (Fig. 13.68).

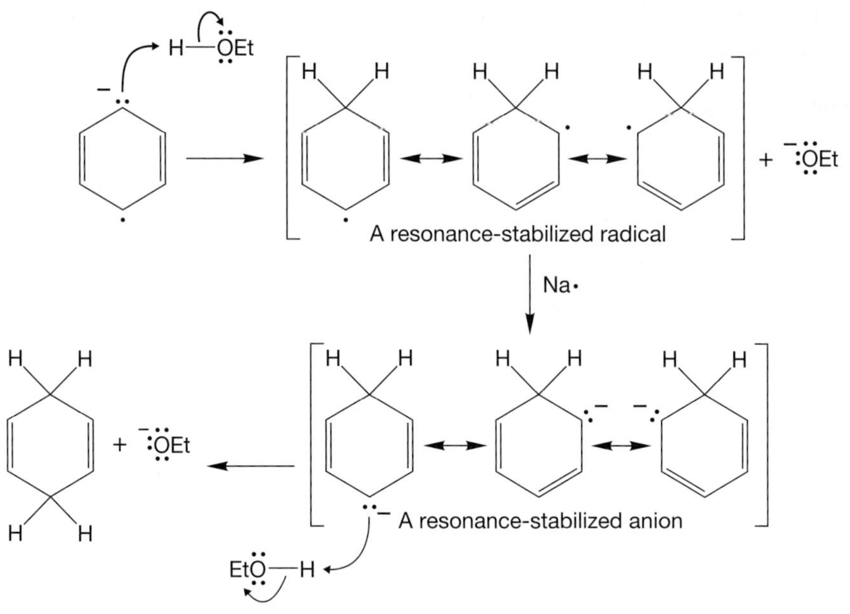

A resonance-stabilized radical

A resonance-stabilized anion

FIGURE 13.68 The radical anion is protonated to give a resonance-stabilized cyclohexadienyl radical that goes on to produce the 1,4-cyclohexadiene through another reduction–protonation sequence (Et = CH_2CH_3).

Substituted benzenes can also be reduced using these conditions. Electron-releasing groups such as alkyl or alkoxy always wind up on one of the double bonds, but electron-withdrawing groups, such as a carboxylic acid (COOH), invariably appear on the methylene positions of the product dienes (Fig. 13.69).

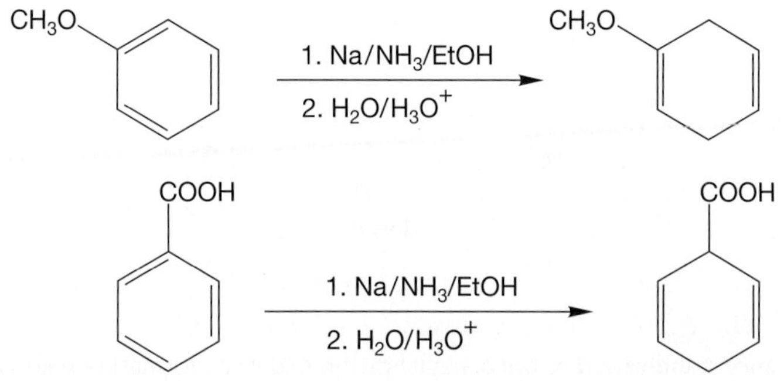

1. $Na/NH_3/EtOH$
2. H_2O/H_3O^+

FIGURE 13.69 In the Birch reduction of substituted benzenes, electron-releasing groups appear on one of the double bonds, but electron-withdrawing groups are located at the methylene position.

13.12 The Benzyl Group and Its Reactivity

Thus far we have been looking at processes directly involving the benzene ring of an aromatic compound. Here we sketch out some reactions in which benzene plays a major role, but in which the action takes place on the outskirts of the ring. The carbon adjacent to a benzene ring is called the benzylic carbon, or the benzylic position. Remember from p. 596, the benzyl group is Ph—CH_2. Any group attached to a benzylic carbon is referred to as being in a benzylic position. In many ways, the reactivity of the benzylic position resembles that of the allylic position. Indeed, an orbital at the benzylic position is stabilized by overlap with adjacent orbitals in much the same way as is an orbital at the allylic position (Fig. 13.70).

The benzyl group

The benzylic position

The benzylic position

The allyl group

The allylic position

FIGURE 13.70 The position adjacent to a benzene ring is quite analogous to the allylic position.

13.12a Substitution Reactions of Benzyl Compounds   Benzyl compounds are unusually active in both the S_N1 and S_N2 reactions. Clearly, the benzene ring helps enormously to stabilize the carbocation formed in the rate-determining ionization of an S_N1 reaction. For example, the benzylic chloride in Figure 13.71 reacts more than 100,000 times faster than isopropyl chloride in a typical S_N1 process. Both involve secondary cations as intermediates, but the secondary cation that is also benzylic is significantly more stable, as is the transition state leading to it.

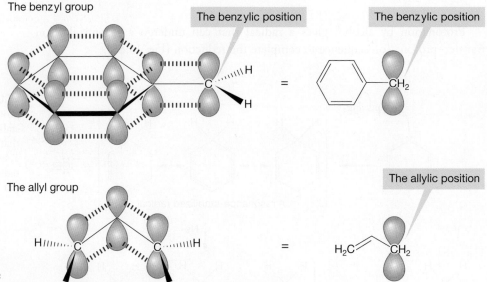

Relative rate = 1.37×10^5 Resonance-stabilized benzylic cation

Relative rate = 1

FIGURE 13.71 The resonance stabilization of the benzylic cation makes its formation relatively easy. Accordingly, S_N1 reactions of benzylic compounds containing good leaving groups are quite fast.

In the benzyl cation itself (Fig. 13.72), the positive charge is borne by 4 carbons, in the benzhydryl cation (Ph_2C^+H) (the **benzhydryl group** is Ph_2CH) by 7 carbons, and in the trityl cation (Ph_3C^+) by a whopping 10 carbons.

Benzyl chloride   The benzyl cation—4 resonance forms

A composite picture in which the positions sharing the positive charge are shown as (+)

Benzhydryl chloride   The benzhydryl cation—7 resonance forms

In this composite representation of the benzhydryl cation, the positions sharing the positive charge are shown as (+)

Trityl chloride   The trityl cation—10 resonance forms

In this composite representation of the trityl cation, the positions sharing the positive charge are shown as (+)

FIGURE 13.72 Resonance stabilization of the benzyhydryl and trityl cations.

PROBLEM 13.25 Why is triptycenyl chloride not nearly as reactive in the S_N1 reaction as trityl chloride?

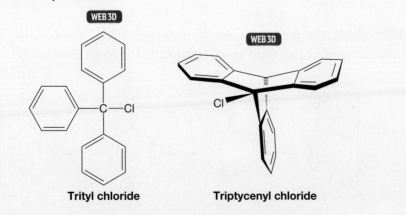

Trityl chloride   **Triptycenyl chloride**

PROBLEM SOLVING

In a Problem like 13.25, in which the parts of two molecules are just about the same (here, three benzene rings and a chlorine bonded to a tertiary carbon are common to the two molecules), it is a good bet that the answer lies in geometry. In most such problems it is the geometrical relationship between the various parts that leads to the crucial difference.

Benzyl halides are also especially reactive in the S_N2 reaction. The reason is the same as that for the enhanced reactivity of allyl chloride (p. 543). The transition state for S_N2 displacement is delocalized, and therefore especially stable. If the transition state for a reaction is at relatively low energy, the activation energy for the reaction will also be relatively low and the reaction will be fast (Fig. 13.73).

FIGURE 13.73 The transition state for S_N2 displacement at the benzylic position is stabilized through orbital overlap with the ring.

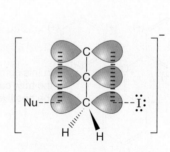
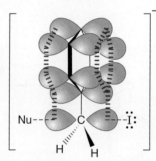

Transition state for S_N2 displacement of an allyl halide   Transition state for S_N2 displacement of a benzyl halide

TABLE 13.2 Some Relative Rates of S_N2 Displacements in the Reaction of R—I with Iodide Ion at 50 °C in Ethyl Alcohol Solvent

R Group	Relative Rate
Ethyl	1.0
Propyl	0.6
Butyl	0.4
Allyl	33
Benzyl	78

Table 13.2 compares the rates of reaction for benzyl iodide and other alkyl iodides in an S_N2 reaction.

13.12b Radical Reactions at the Benzylic Position   A benzene ring will also stabilize an adjacent half-filled orbital (a free radical) through resonance (Fig. 13.74).

FIGURE 13.74 The resonance forms for benzyl radical.

In Chapter 11 (p. 497), we discussed radical reactivity at the allylic position, and very little need be added to get a description of benzylic reactivity. Toluene, for example, can be brominated by *N*-bromosuccinimide (NBS) to give benzyl bromide (Fig. 13.75).

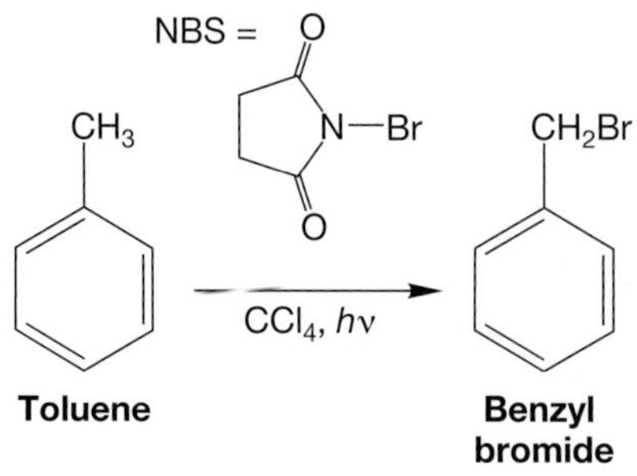

NBS =

CH₃   CH₂Br

Toluene   Benzyl
                       bromide

CCl₄, *h*ν

FIGURE 13.75 The radical bromina-tion of toluene.

PROBLEM 13.26 Write a mechanism for the reaction between toluene and NBS. *Hint*: See Chapter 11 (p. 500) for a description of the reaction of NBS with allylic compounds.

PROBLEM 13.27 Ethylbenzene yields only the product of bromination at the α position (the position adjacent to the ring) when treated with NBS in CCl₄. Explain.

As you should be able to deduce from Problem 13.27, abstraction of hydrogen is especially easy at the α position of ethylbenzene. This abstraction occurs easily because of the resonance stabilization of the benzylic radical (Fig. 13.74). Because bromine will not attack the benzene ring without a catalyst, benzylic positions can often be brominated by simply heating or photolyzing bromine in their presence (Fig. 13.76).

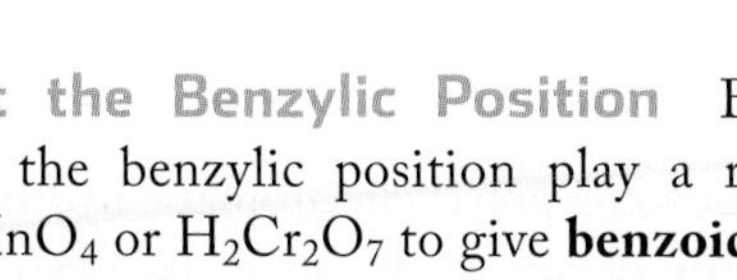

$\overset{\alpha}{C}H_2\overset{\beta}{C}H_3$

Br
|
CHCH₃

Br₂
———→
Δ or *h*ν

FIGURE 13.76 The thermal or photochemical decomposition of bromine in ethylbenzene leads to 1-bromo-1-phenylethane through radical bromination at the benzylic position.

13.12c Oxidation at the Benzylic Position Both free radicals and the special activity of the benzylic position play a role in the reaction of alkylbenzenes with KMnO₄ or H₂Cr₂O₇ to give **benzoic acid**, which is benzene with a carboxylic acid (COOH) attached. Side chains are "chewed down" to carboxylic acids no matter what their length. The details are vague, but the key observation is that tertiary side chains are untouched by oxidizing agents.

Benzylic oxidation

There must be at least one hydrogen in the benzylic position for the reaction to succeed (Fig. 13.77).

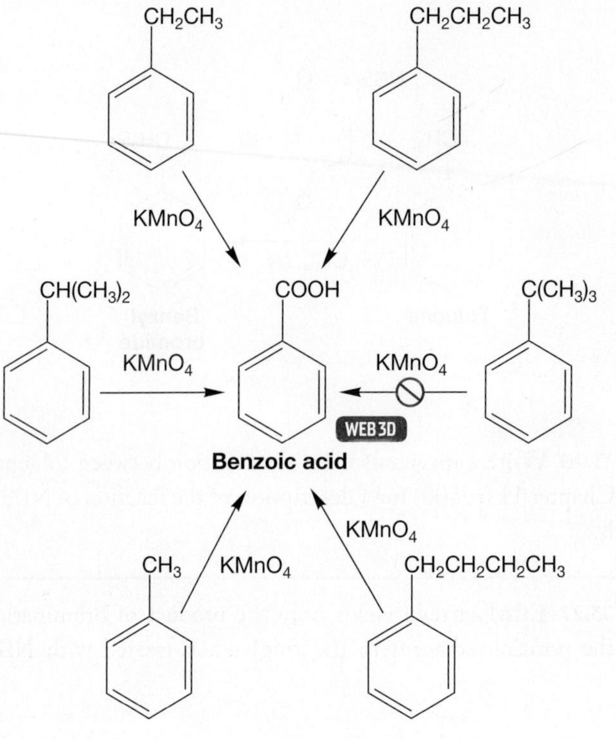

FIGURE 13.77 Oxidizing agents convert most alkyl side chains on benzene rings into the acid group, COOH. However, for the oxidation to succeed there must be at least one benzylic hydrogen. Thus, *tert*-butylbenzene does not react.

Summary

The chemistry of benzenes is summed up by the imperative: "Preserve the circle!" In plain English, that statement means that there is a strong thermodynamic driving force to preserve the delocalization energy—more than 30 kcal/mol—that the circle represents. In this chapter, the classic aromatic substitution reaction in which an electrophile replaces (substitutes for) one of the hydrogens on the ring is introduced. In Chapter 14, this prototypal reaction will be expanded in many ways.

Other aromatic chemistry features a few reactions in which aromaticity is lost (the Birch reduction is one example) and the special influence of the aromatic ring on the benzylic position.

13.13 Special Topic: The Bio-Downside, the Mechanism of Carcinogenesis by Polycyclic Aromatic Compounds

We mentioned earlier that some polycyclic aromatic hydrocarbons are strongly carcinogenic. Although all the details of this carcinogenesis are not worked out, the broad outlines are known for some molecules. Deoxyribonucleic acid (DNA) and ribonucleic acid (RNA) are huge biomolecules that contain and help transcribe genetic information (Chapter 23). Both DNA and RNA are phosphate-linked sequences of nucleotides, which are sugar molecules attached to heterocyclic bases.

These molecules bear large numbers of nucleophilic groups that can be alkylated by other molecules (electrophiles) in what seem to be simple S_N2 reactions (Fig. 13.78).

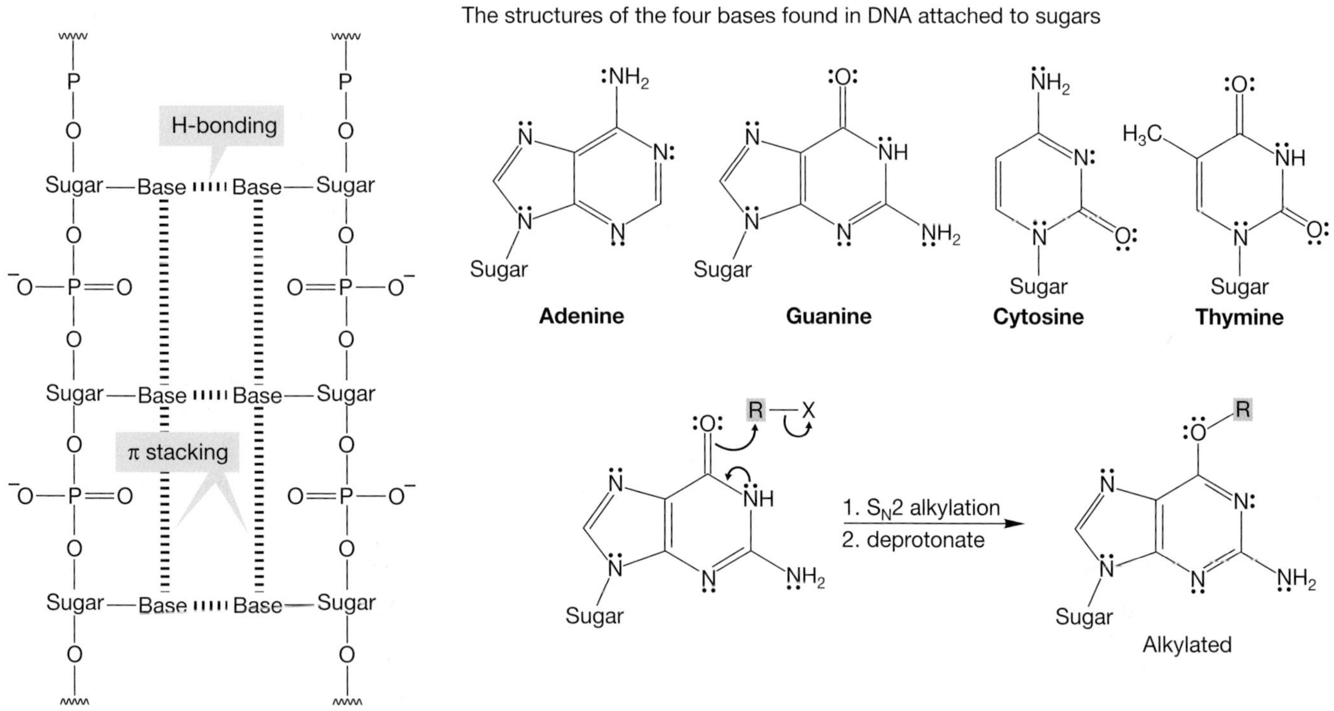

The structures of the four bases found in DNA attached to sugars

FIGURE 13.78 The DNA polymer is composed of phosphate-linked sugar-base units. Two strands of DNA are held together by hydrogen bonding between pairs of bases and by π stacking between the aromatic bases. Alkylation of the bases changes the size and shape of one of the units. This change can interfere with hydrogen bonding and base pairing. In turn, this interference can lead to mutations and cancer.

Alkylation changes the size and shape of the DNA molecule and makes mismatched base pairing through hydrogen bonding more likely (see Chapter 23 for more details of base pairing). This mismatch can lead to errors in replication and to mutations. Most mutations are harmless to the organism, because they lead to the death of only the individual cell. On rare occasions, however, a mutation can lead to uncontrolled cell division and cancer.

The polyaromatic hydrocarbons (PAHs) are only indirectly responsible for the chemical processes that lead to cancer. Aromatic hydrocarbons are not very reactive and there is no obvious way for them to react chemically with the nucleophilic bases in DNA or RNA (Fig. 13.79).

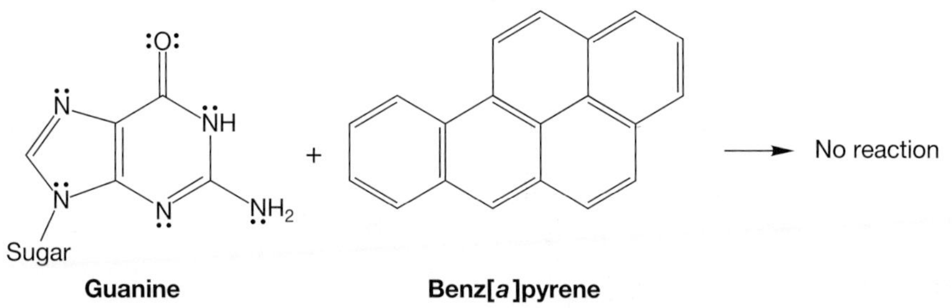

FIGURE 13.79 Aromatic hydrocarbons do not react with nucleophiles, and there is no way for a molecule like benz[*a*]pyrene to alkylate one of the bases directly.

However, there are enzymes that can modify the PAH rings. Nonpolar molecules such as aromatic hydrocarbons accumulate in fat cells and natural methods have been evolved to purge our bodies of such molecules. These methods generally involve making the hydrocarbons more water soluble by adding polar groups.

For example, benz[*a*]pyrene is epoxidized by the enzyme P-450 monooxygenase and the product epoxide is then opened enzymatically to give a trans diol (Fig. 13.80).

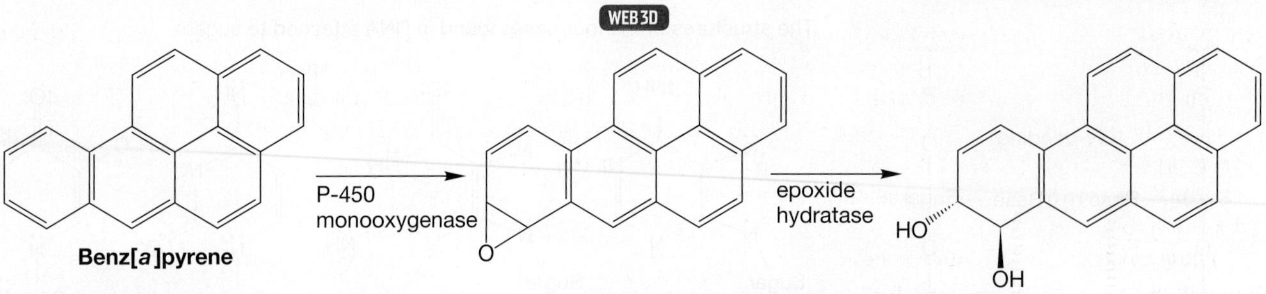

FIGURE 13.80 Benz[*a*]pyrene is enzymatically epoxidized and ring opened to give a polar trans 1,2-diol.

A second enzymatic epoxidation, again by P-450 monooxygenase, gives a molecule that reacts with guanine, one of the DNA bases, to give DNA that is severely encumbered by the large, alkylated guanine (Fig. 13.81). This modified base has difficulty in maintaining normal hydrogen bonding with another complementary base, and this mismatch can lead to mutations.

FIGURE 13.81 A second epoxide is formed. This epoxide is opened by guanine to give a highly encumbered molecule, **A**, that has problems maintaining proper hydrogen bonding.

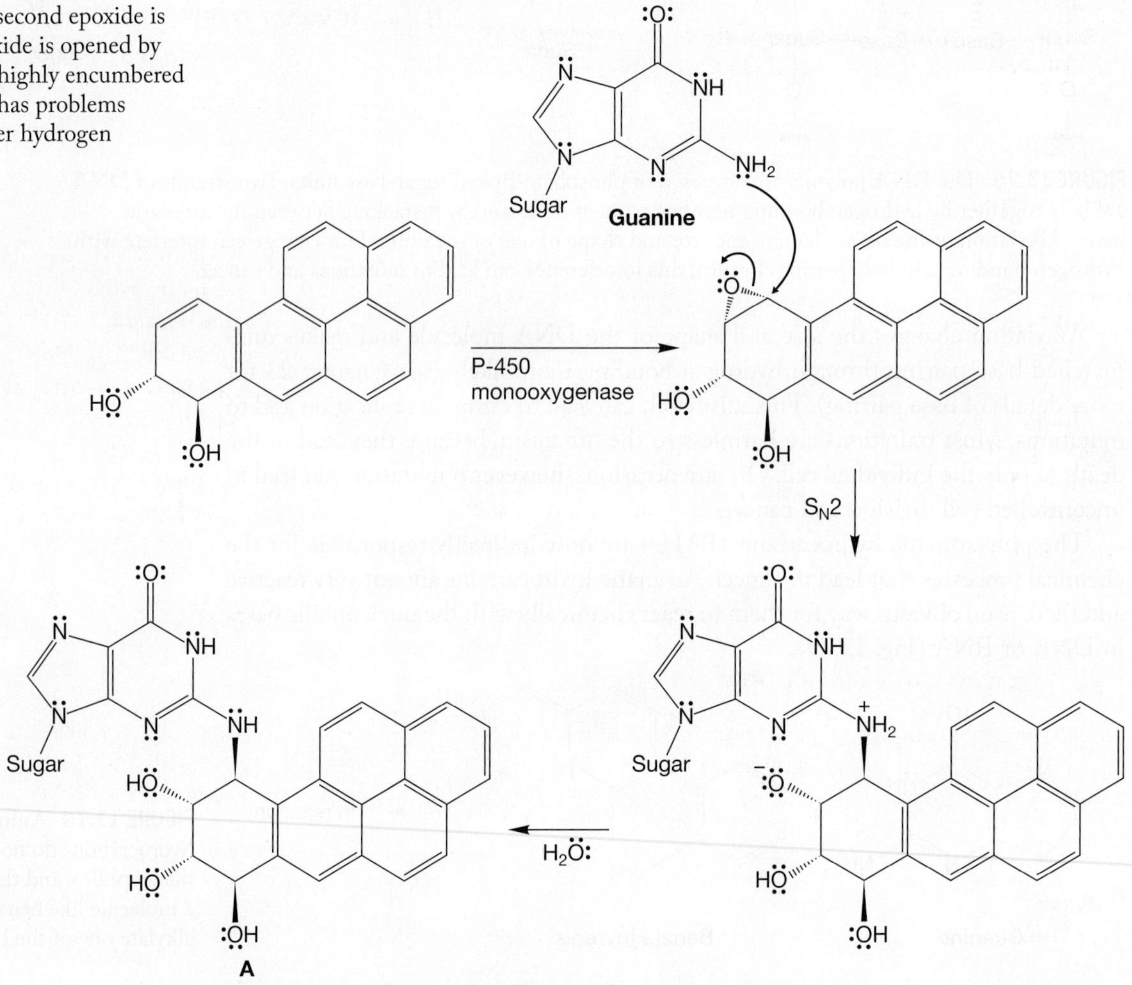

This alkylation of guanine is not a new reaction. You already know of the opening of epoxides in both base and acid (p. 426), and this process is nothing more than a complicated version of that simple reaction.

13.14 Summary

New Concepts

The material in this chapter is composed almost entirely of concepts. There are few new reactions or synthetic procedures. We concentrate here on the special stability of some planar, cyclic, and fully conjugated polyenes. The special stability called *aromaticity* is encountered when the cyclic polyene has a molecular orbital system in which all degenerate bonding molecular orbitals are completely filled.

These especially stable molecular orbital systems are found in planar, cyclic, fully conjugated polyenes that contain $4n + 2$ π electrons (Hückel's rule).

Heats of hydrogenation or heats of formation can be used to calculate the magnitude of the stabilization. For benzene, the delocalization energy or resonance energy amounts to more than 30 kcal/mol.

It's vital to keep clear the difference between resonance forms and molecules in equilibrium. In this chapter, that difference is exemplified by Dewar benzene, bicyclo[2.2.0]hexa-2,5-diene (Problem 13.5), and the Dewar resonance forms contributing slightly to the structure of benzene. As always, resonance forms are related only by the movement of electrons, whereas equilibrating molecules have different geometries—different arrangements of atoms in space. Benzene and Dewar benzene are different molecules; the Kekulé and Dewar resonance forms contribute to the real structure of benzene (Fig. 13.82).

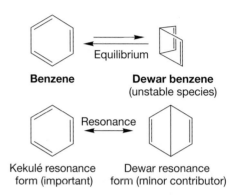

FIGURE 13.82 Equilibrium versus resonance.

Key Terms

annulene (p. 594)
arene (p. 598)
aromatic character (p. 582)
aromaticity (p. 582)
benzene (p. 573)
benzhydryl group (p. 611)
benzoic acid (p. 613)
benzyl (p. 596)
Birch reduction (p. 608)
cycloheptatrienylium ion (p. 587)

cyclopentadienyl anion (p. 589)
delocalization energy (p. 581)
Dewar benzene (p. 573)
Dewar forms (p. 576)
Frost circle (p. 584)
furan (p. 599)
heteroaromatic compound (p. 599)
heterobenzene (p. 591)
Hückel's rule (p. 583)
Kekulé forms (p. 575)

meta (p. 596)
ortho (p. 596)
para (p. 596)
phenyl (Ph) (p. 596)
polynuclear aromatic compound (p. 602)
pyridine (p. 598)
pyrrole (p. 592)
resonance energy (p. 581)
tropylium ion (p. 587)

Reactions, Mechanisms, and Tools

In this chapter, the first reaction mechanism encountered is the important and general electrophilic substitution of benzene. A host of aromatic substitution reactions will be studied in Chapter 14 and are exemplified here by deuterium exchange. The aromatic ring is destroyed by an endothermic addition of D^+, but reconstituted by an exothermic loss of H^+ (Fig. 13.65).

Syntheses

There is little in the way of new synthetic procedures in this chapter. You might remember the formation of the tropylium ion by hydride abstraction and the Birch reduction of benzenes to 1,4-cyclohexadienes. You also have a method of synthesizing deuteriobenzene through the acid-catalyzed exchange reaction of benzene. This reaction will serve as the prototype of many similar substitution reactions to be found in Chapter 14.

1. Benzylic Compounds

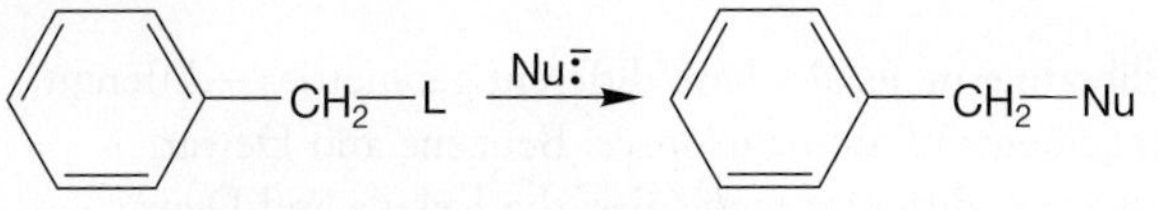

S_N1 and S_N2 reactions

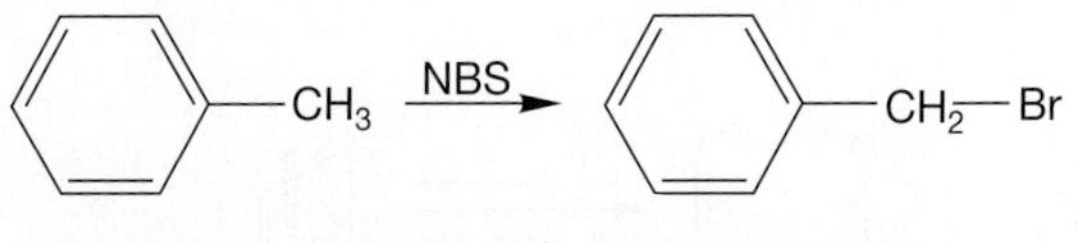

Radical bromination at the benzyl position

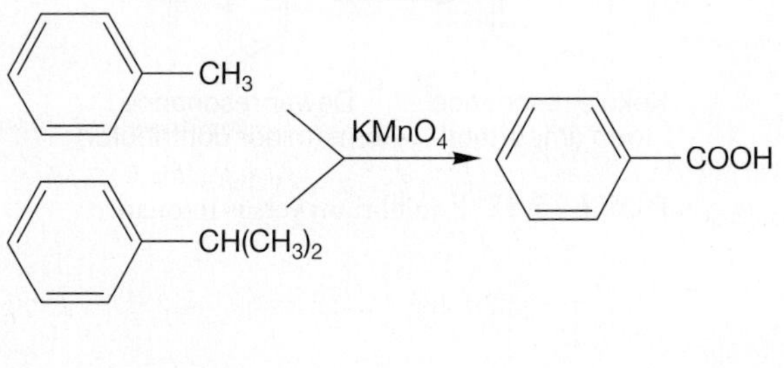

Oxidation of benzyl positions bearing at least one hydrogen

2. Cyclohexadienes

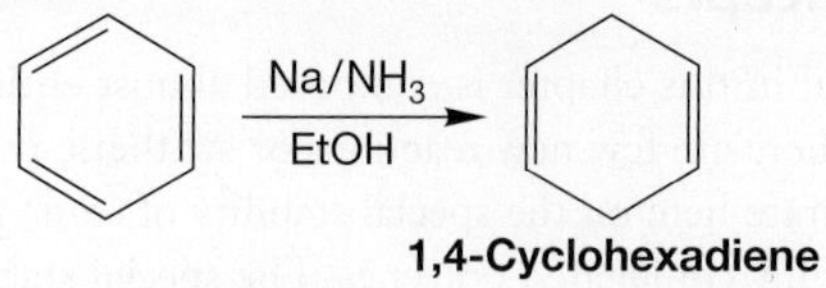

1,4-Cyclohexadiene

Birch reduction of benzene

3. Deuteriobenzenes

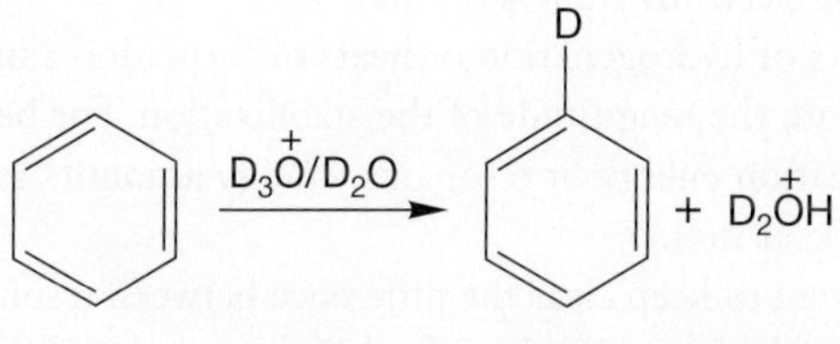

Acid-catalyzed exchange of hydrogens on a benzene ring; all hydrogens can eventually be exchanged

4. Tropylium Ions

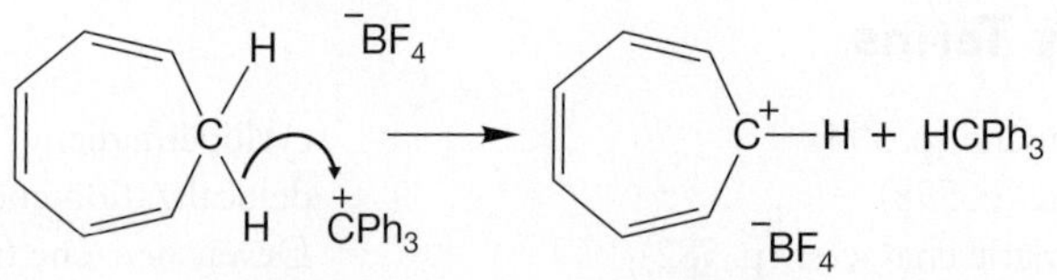

Tropylium fluoborate

Hydride transfer from cycloheptatriene to the trityl cation ($^+CPh_3$)

Common Errors

Certainly the most common error is to overgeneralize Hückel's rule. We often forget to check each potentially aromatic molecule to be certain it satisfies all the criteria. The molecule must be cyclic or there cannot be the fully connected set of p orbitals necessary (Fig. 13.18). The molecule must be planar so that the p orbitals can overlap effectively (Fig. 13.16). There may be no positions in the ring at which p/p overlap is interrupted (Fig. 13.15). Finally, there must be $4n + 2$ π electrons. Be certain you are counting only π electrons. Do not include in your count electrons not in the π system. The classic example is pyridine, which contains a pair of electrons in an sp^2 orbital that are often mistakenly counted as π electrons (Fig. 13.47). If *all four* of these criteria are satisfied we find aromaticity.

Do not misapply the Frost circle device. The most common error is to forget to inscribe the circle with the vertex of the appropriate polygon down. This requirement is hard to remember because at this level of discussion this point is arbitrary— there is no easy way to figure it out if you don't remember the proper convention.

Do not confuse absolute and relative stability. For example, the cyclopentadienyl anion is aromatic, and therefore exceptionally stable, *but only within the family of anions*. The negatively charged cyclopentadienyl anion is not as stable as benzene.

As always, you must be clear about the difference between resonance and equilibrium. Be certain that you know the difference between the two Kekulé resonance forms for benzene and the hypothetical molecule 1,3,5-cyclohexatriene.

13.15 Additional Problems

PROBLEM 13.28   Give names for the following compounds:

(a)

(b)

(c)

(d)

(e)

(f)

PROBLEM 13.29   Write structures for the following compounds:

(a) 2,4,6-trinitrotoluene (TNT)
(b) 4-bromotoluene
(c) *p*-bromotoluene
(d) *m*-diiodobenzene
(e) 3-ethylphenol
(f) *m*-ethylphenol
(g) 1,2,4,5-tetramethylbenzene (durene)
(h) *o*-deuteriobenzenesulfonic acid (Note the "i" in "deuter*io*." It is not "deutero.")
(i) *p*-aminoaniline
(j) *m*-chlorobenzoic acid

PROBLEM 13.30   Write arrow formalisms for the conversions of Dewar benzene, Ladenburg benzene (prismane), benzvalene, and 3,3′-bicyclopropenyl (Fig. 13.4) into benzene. Some of these tasks may be quite challenging.

PROBLEM 13.31   Which of the following heterocyclic compounds might be aromatic? Explain. It will help if you start by drawing good Lewis structures for all the compounds.

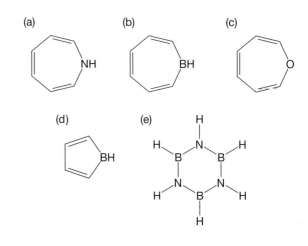

PROBLEM 13.32   Pyridine could be called azabenzene. There are three diazabenzenes. Draw these compounds. Are they aromatic?

PROBLEM 13.33   Which of the following hydrocarbons might be aromatic? Explain.

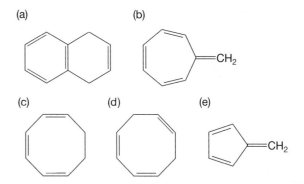

PROBLEM 13.34   Which of the following ions might be aromatic? Explain.

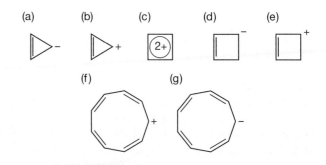

PROBLEM 13.35 Design an aromatic heterocyclic compound that
(a) contains one boron atom.
(b) contains one boron and one oxygen atom in a six-membered ring.
(c) contains one oxygen and one nitrogen atom in a five-membered ring.
(d) contains two nitrogen atoms in a five-membered ring.
(e) contains three boron and three nitrogen atoms in a six-membered ring.

PROBLEM 13.36 Try to construct an aromatic heterocycle in which there is an oxygen and a nitrogen in a neutral six-membered ring.

PROBLEM 13.37 The molecule 5-bromo-1,3-cyclohexadiene is very difficult to isolate, but 1-bromo-1,3-cyclohexadiene is much easier to obtain. Explain why there is so much difference.

PROBLEM 13.38 Carefully explain why azirines of structure **1** have never been isolated, whereas the isomeric azirines of structure **2** are well known.

1   **2**

PROBLEM 13.39 The dipole moment of THF is 1.7 D, but that of furan is only 0.7 D. Explain.

PROBLEM 13.40 The heat of hydrogenation of 1,3,5,7-cyclooctatetraene (COT) is about 101 kcal/mol. The heat of hydrogenation of cyclooctene is about 23 kcal/mol. Is COT aromatic? Explain.

PROBLEM 13.41 As early as 1891 the German chemist G. Merling showed that the bromination of 1,3,5-cycloheptatriene leads to a liquid dibromide. When the dibromide is heated, HBr is evolved and a yellow solid of the formula C_7H_7Br (mp 203 °C) can be isolated. This compound is soluble in water, but cannot be reisolated from water solution. Instead, ditropyl ether is obtained. Merling could not explain what was happening, but perhaps you can.

Ditropyl ether

PROBLEM 13.42 Most diazo compounds are exceedingly, and sometimes disastrously unstable. They are notorious explosives. By contrast, the brilliant, iridescent orange compound diazocyclopentadiene is quite stable. Explain.

Diazocyclopentadiene

PROBLEM 13.43 Why is it that (E)-1-phenyl-2-butene reacts only once with OsO_4? Draw the product that is formed.

PROBLEM 13.44 Predict the major product(s), if any, for the bromination (Br_2, CH_2Cl_2) of the following compounds:
(a) styrene
(b) 1-ethyl-3-nitrobenzene
(c) (Z)-2-phenyl-2-butene
(d) 3-phenyl-1-propene
(e) 3-phenylheptane

PROBLEM 13.45 The major product in the bromination (1 equivalent of Br_2 in CH_2Cl_2) of (E,E)-1-phenyl-1,3-pentadiene is (E)-3,4-dibromo-1-phenyl-1-pentene regardless of the reaction conditions (thermodynamic or kinetic control). Show the mechanism of the reaction and account for this selectivity.

PROBLEM 13.46 Predict the major product(s), if any, for the reaction of HBr in THF (solvent) with each of the following:
(a) styrene
(b) (Z)-2-phenyl-2-hexene
(c) 2-methyl-1-phenyl-1-propene
(d) 3-phenyl-1-propene
(e) naphthalene

PROBLEM 13.47 Predict the product of the reaction with NBS (CCl_4, $h\nu$) for each of the following compounds:
(a) 3-phenyloctane
(b) 3-methyl-1-phenylbutane
(c) (Z)-1-phenyl-2-butene
(d) isopropylbenzene
(e) phenylcyclohexane

PROBLEM 13.48 Predict the major product in the reaction between 1 equivalent of NaI and 1,5-dichloro-1-phenylhexane. Explain the regioselectivity.

PROBLEM 13.49 Construct the π molecular orbitals for benzene (p. 578) from the π molecular orbitals of allyl. *Remember*: Only combinations of orbitals of comparable energy need be considered.

PROBLEM 13.50 Vogel and Roth's synthesis of the bridged cyclodecapentaene (**1**) is outlined below. Supply the missing reagents (a, b, c, and d) and an arrow formalism for the last step.

1

PROBLEM 13.51 Pyridine and imidazole are modest Brønsted bases at nitrogen, whereas pyrrole is not.

Pyridine   Pyrrole   C(2)   Imidazole

In fact, pyrrole is protonated only in strong acid, and protonation occurs at C(2), not on nitrogen. First, explain why pyridine and imidazole are basic and pyrrole is not. Then, explain why eventual protonation of pyrrole is at carbon, not nitrogen. *Caution!* You will have to write full Lewis structures to answer this question.

PROBLEM 13.52 Draw cyclopropene. How many types of hydrogens are there in this molecule? Which hydrogen is more acidic? Explain why. Use the table inside the back cover of this book to estimate the pK_a for each type of hydrogen.

PROBLEM 13.53 Explain why the 1,2-bond of anthracene is shorter than the 2,3-bond.

1.37 Å

1.42 Å

Anthracene

PROBLEM 13.54 The nitronium ion ($^+NO_2$) is known and can react with simple benzenes to give nitrobenzenes. Write an arrow formalism mechanism for this reaction.

$^+NO_2$

$^-BF_4$

NO_2

PROBLEM 13.55 The molecule shown below undergoes reaction with cupric nitrate in acetic anhydride (a source of $^+NO_2$) to give a high yield of the substituted product shown. Write a mechanism for this reaction.

CH_3

CH_3

$Cu(NO_3)_2$
acetic anhydride
0 °C

NO_2

CH_3

CH_3

PROBLEM 13.56 The compound shown below was synthesized by Professor I. M. Dimm, who was astonished to find that it was exceedingly stable. Prof. Dimm had expected world renown for making what he thought would be a most unstable compound. Dimm reasoned that with six double bonds and a triple bond in the molecule, there would be 16 π electrons and thus, no aromatic character ($4n$, $n = 4$; not $4n + 2$). Where was his reasoning wrong?

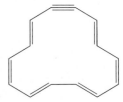

PROBLEM 13.57   Provide a mechanism for the following reaction sequence:

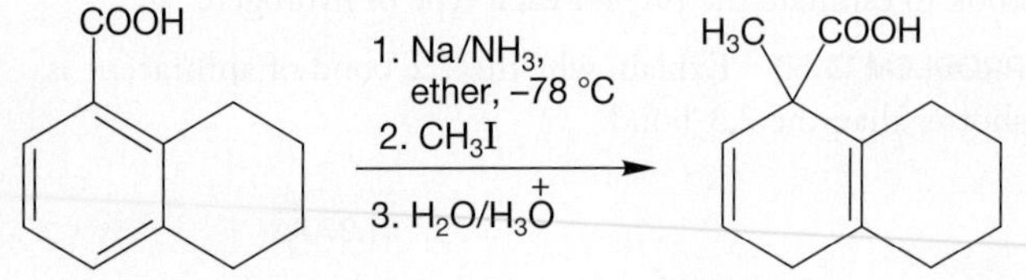

PROBLEM 13.58   There are many aromatic rings in nature. You have learned about the nucleotide bases. Draw adenine, guanine, cytosine, and thymine (p. 615). Circle any aromatic rings in these molecules. You might need to draw resonance structures to see the aromaticity.

PROBLEM 13.59   Even though naphthalene and anthracene are polynuclear aromatic compounds, they undergo the Diels–Alder reaction. For example, anthracene readily reacts with maleic anhydride. However, the reaction of naphthalene with maleic anhydride proceeds efficiently only under high pressure.

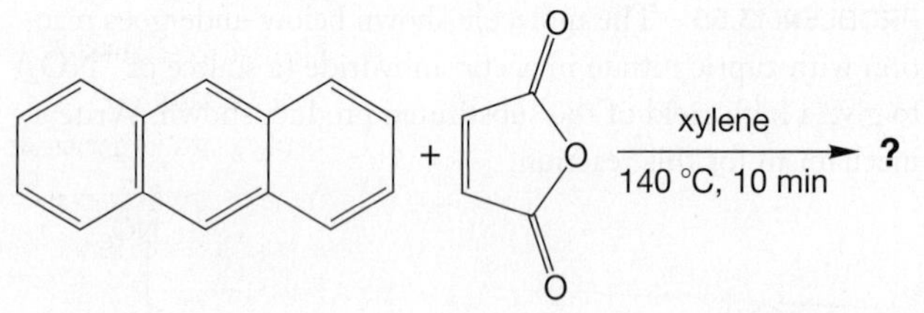

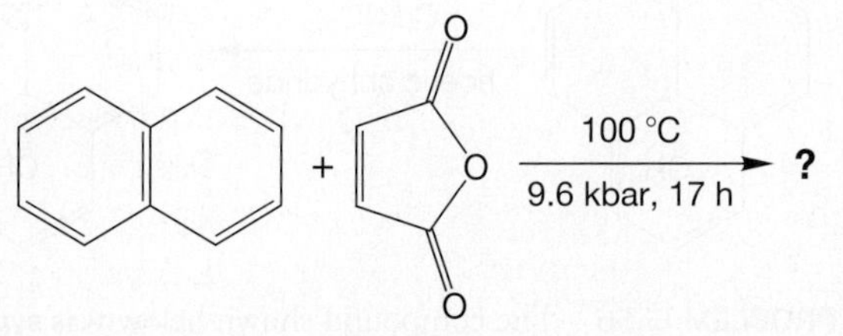

(a) Draw the structures of the Diels–Alder adducts formed in these reactions.
(b) Explain why anthracene is so much more reactive than naphthalene in the Diels–Alder reaction. *Hint*: The calculated resonance energies of naphthalene and anthracene are 61 kcal/mol and 84 kcal/mol, respectively.

PROBLEM 13.60   A certain alkylbenzene was known to have one of the isomeric structures shown below.

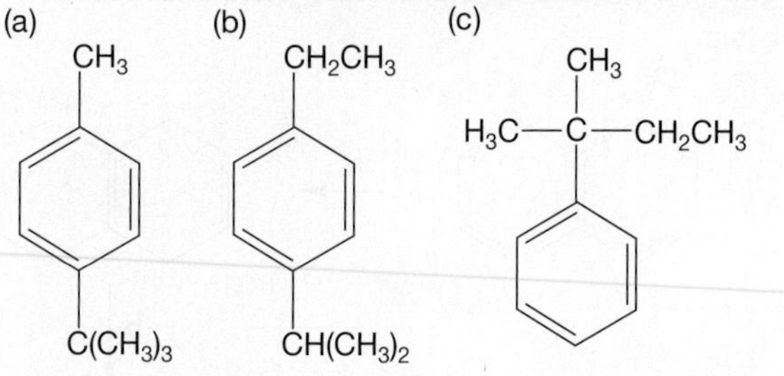

Explain how the structure of the unknown compound could be determined by oxidation with $KMnO_4$.

Use Organic Reaction Animations (ORA) to answer the following questions:

PROBLEM 13.61   Select the reaction "Benzylic oxidation" on the Semester 1B panel. Observe the reaction several times. The animation uses oxygen (O_2) as the initiator of the reaction, which also occurs in Nature. One of the reasons for the common practice of packing organic foods over nitrogen (N_2) is to avoid the reaction of O_2 with benzylic (and allylic) positions in natural products present in food. Draw each step of the reaction between O_2 and 2-phenylpropane. Include a termination reaction. Label your steps as initiation, propagation, or termination.

PROBLEM 13.62   Watch the "Benzylic oxidation" reaction using the SOMO track. What does SOMO stand for? Stop the reaction as O_2 is approaching the benzylic hydrogen. Notice the aromatic ring. What positions are showing SOMO character? Why these carbons? Does this help you understand the resonance structures? Explain.

PROBLEM 13.63   One of the final products from the benzylic oxidation of 2-phenylpropane is phenol. See if you can show a mechanism for the formation of phenol from the peroxide product shown in the animation. Remember that the oxygen–oxygen bond is very weak. The mechanism will require a rearrangement and a second radical oxidation process.

PROBLEM 13.64   We answered questions about the "Stabilized alkene halogenation" reaction in Chapter 10. This reaction is at the top of the Semester 1B page. Watch it again to remind yourself about the pathway. Bromination in this animation is shown to occur via a syn addition. Bromination usually proceeds through an anti addition. Why would syn addition be favored over anti addition for this reaction? Notice the benzylic-like intermediate in the reaction. Do you suppose the napthylmethyl cation (like the one in this animation) is more stable than a benzyl cation? Why?

In such a
again reca
small activ
plete ener
another is
and 14.2.

14

Substitution Reactions of Aromatic Compounds

In this
pounds in
an electron
Note that
retained in
tion of hov
make a sec
tions that
is lost.

H

H

14.1 Preview
14.2 Hydrogenation of Aromatic Compounds
14.3 Diels–Alder Reactions
14.4 Substitution Reactions of Aromatic Compounds
14.5 Carbon–Carbon Bond Formation: Friedel–Crafts Alkylation
14.6 Friedel–Crafts Acylation
14.7 Synthetic Reactions We Can Do So Far
14.8 Electrophilic Aromatic Substitution of Heteroaromatic Compounds
14.9 Disubstituted Benzenes: Ortho, Meta, and Para Substitution
14.10 Inductive Effects in Aromatic Substitution
14.11 Synthesis of Polysubstituted Benzenes
14.12 Nucleophilic Aromatic Substitution
14.13 Special Topic: Stable Carbocations in "Superacid"
14.14 Special Topic: Benzyne
14.15 Special Topic: Biological Synthesis of Aromatic Rings; Phenylalanine
14.16 Summary
14.17 Additional Problems

AROMATIC SUBSTITUTION The most common reaction for aromatic compounds is substitution. This process is much like a soccer substitution; one group (in this case, a person) replaces another.

ESSENTIA

1. The gre
   ultimate
   reaction
   of electr
   synthesi

2. Some substituents increase the rate of a subsequent substitution and generally direct it to the ortho and para positions. Other substituents slow the rate and direct further substitution to the meta position. Understanding the dependence of substitution position and rate on the structure of the initial substituent is essential.

3. The resonance-stabilized cyclohexadienyl cation is an intermediate in the central reaction of this chapter. Be sure you understand why three—and only three—of the carbons in the ring share the positive charge.

4. The nitrobenzene–aminobenzene (aniline) pair is one important key to doing aromatic syntheses. Be sure you exit this chapter being able to interconvert these two compounds and being able to use both nitrobenzene and aniline in synthesis.

5. Making bonds between carbon and the aromatic ring generally requires either Friedel–Crafts alkylation or Friedel–Crafts acylation. There are subtle differences between the two reactions, and you'll need to know these differences to take advantage of Friedel–Crafts reactions in synthesis.

14.2 Hydrogenation of Aromatic Compounds

If one were to try to predict the reactions of benzene from first principles, one would probably start by examining the reactions of other π systems and applying them to aromatic compounds. For example, we know that hydrogenation of cyclohexene is exothermic by 28.6 kcal/mol (p. 580). It is no great surprise to find that absorption of 3 mol of hydrogen by benzene to give cyclohexane is exothermic as well; indeed we examined this reaction in some detail in Chapter 13. Although the reaction is very exothermic, it is far less exothermic than a guess based on the value of −28.6 kcal/mol for a simple alkene would suggest. The difference between the estimated value and the measured value was taken as a measure of the delocalization energy of benzene, approximately 33 kcal/mol (Fig. 14.5).

FIGURE 14.5 The determination of the delocalization, or resonance energy, of benzene. Note the exaggerated short double bonds and long single bonds in cyclohexatriene.

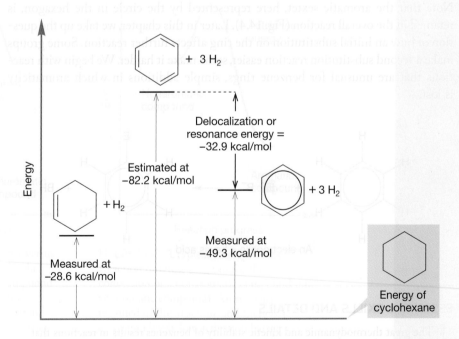

Often, conditions under which alkenes are swiftly transformed into alkanes are quite without effect on benzene and other simple aromatic compounds (Fig. 14.6a). Indeed, benzene can be used as solvent in many hydrogenation reactions. We can eventually

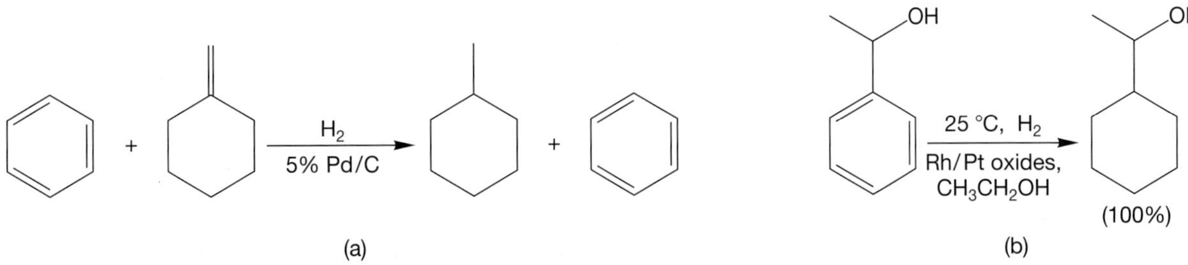

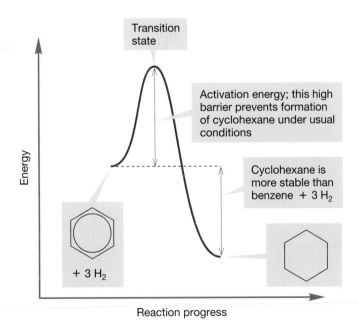

FIGURE 14.6 (a) Benzene does not hydrogenate under the same conditions as alkenes. (b) More active catalysts can be used to hydrogenate the aromatic ring.

make the reaction occur by using high temperature and pressure, or by using especially active catalysts, such as rhodium, ruthenium or platinum oxide (Fig. 14.6b).

The hydrogenation reaction is highly exothermic, by 49.3 kcal/mol in the case of the parent benzene. Even though the combination of benzene and three hydrogen molecules is higher in energy than cyclohexane by a substantial amount, benzene is protected by high activation barriers. Heat, pressure, and/or special catalysts are needed to help overcome these barriers. Benzene rests in a valley, separated from other valleys—even lower energy valleys—by especially high mountains.

In the language of chemistry, we say that benzene is both thermodynamically and kinetically very stable. The valley containing benzene by itself is low (thermodynamics), and the mountains surrounding it (the activation barriers, kinetics) are very high. Benzene is even protected by these barriers from undergoing reactions that are, ultimately, exothermic. The example of benzene plus hydrogen nicely illustrates this point, as the combination of these molecules lies higher in energy than the product of hydrogenation, cyclohexane (Fig. 14.7).

FIGURE 14.7 The activation energy barrier for cyclohexane formation is high and makes this reaction difficult under normal conditions.

PROBLEM 14.1 Explain why the only product of hydrogenation of benzene is cyclohexane. Cyclohexadienes and cyclohexene must be formed first. Why can they never be isolated, even if the reaction is interrupted before completion?

14.3 Diels–Alder Reactions

Benzene and other aromatic compounds contain a conjugated π system. Such molecules are known to undergo the Diels–Alder reaction (p. 544). Yet, it is extremely difficult to induce the Diels–Alder reaction for simple benzenes, and yields are often poor. Harsh conditions (temperature as high as 200 °C, for example) and very reactive dienophiles such as hexafluoro-2-butyne are necessary (Fig. 14.8). The reason for the lack of reactivity is simple: Benzene is too stable (lies in a deep energy well) and the activation energies for this reaction are significant.

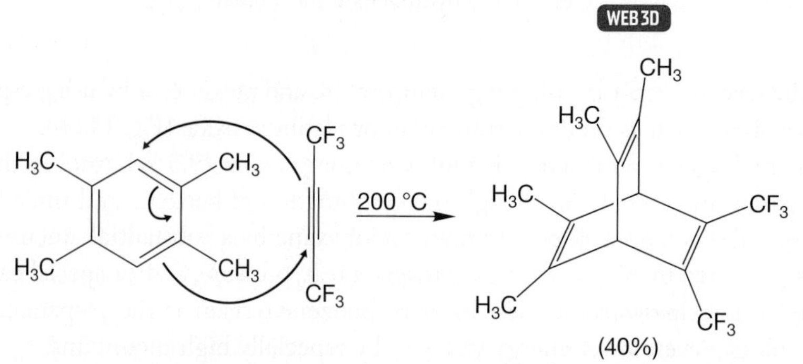

FIGURE 14.8 Benzenes will undergo the Diels–Alder reaction, but only with reactive dienophiles under forcing conditions.

To see if all this makes sense, let's try a thought experiment. Suppose we were set the task of finding a way to make the Diels–Alder reaction possible for benzenes. We might first try to use as reactive (high energy) a dienophile as possible, as in Figure 14.8; but there is another way. We might try instead to destabilize the benzene so that surmounting the activation energy barriers for Diels–Alder reaction becomes easier. Figure 14.9 shows this idea in graphic form.

FIGURE 14.9 If benzene is destabilized, the energy barrier for reaction should be reduced.

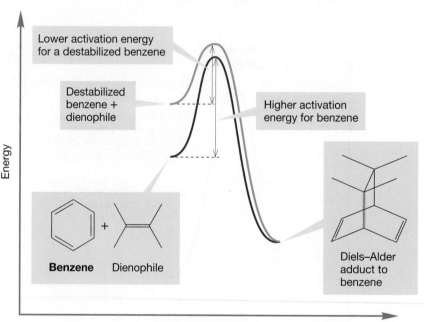

There are several nice practical examples of this idea. Imagine a benzene ring bridged by a chain of methylene groups between the 1- and 4-positions (Fig. 14.10). Such a molecule is called an **[*n*]paracyclophane,** where *n* gives the number of methylene groups in the bridge. As long as there are enough methylene groups (more than 7) the molecule behaves like a simple benzene. But as *n* diminishes (molecules with *n* = 5, 6, and 7 are known and the *n* = 4 isomer

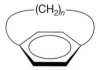

[*n*]Paracyclophanes

For large *n*, the ring can be flat   A small *n* forces the ring into a boat form

FIGURE 14.10 1,4-Bridged benzenes: [*n*]paracyclophanes.

has been detected), the ring becomes warped, and must assume a boat shape (Fig. 14.10), which severely destabilizes the molecule. In this case, the models tell all. Make a model of a paracyclophane with $n = 10$ and then reduce the number of bridging methylenes one by one. The benzene ring will increasingly deform as this happens. The source of the destabilization should also become apparent. Try Problem 14.2.

PROBLEM 14.2 Why are boat-shaped benzenes less stable than flat ones?

As anticipated, these destabilized benzenes are much more reactive in the Diels–Alder reaction than their undistorted relatives (Fig. 14.11).

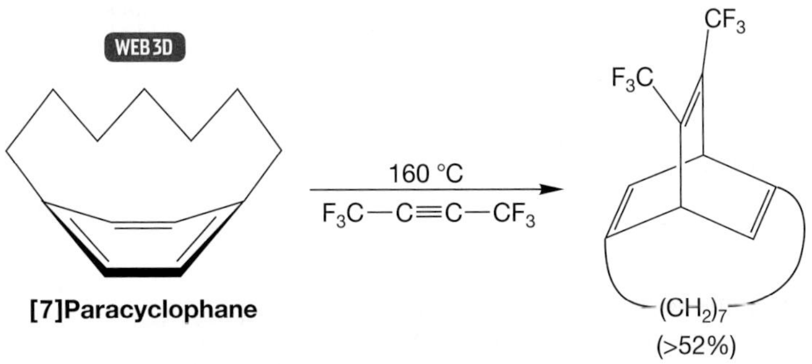

FIGURE 14.11 [7]Paracyclophane undergoes the Diels–Alder reaction at a lower temperature than benzene.

PROBLEM 14.3 Use an arrow formalism to show how [7]paracyclophane forms the product shown in Figure 14.11 through a Diels–Alder reaction.

We really needn't go to the trouble of making paracyclophanes (unless one is interested in what happens to benzene's π system as it deforms), because there are naturally occurring molecules that are less aromatic than benzene and thus should be more reactive. Recall how we reached a quantitative determination of the resonance energy for benzene through measurements of the heats of hydrogenation (or combustion) for the compounds cyclohexene and cyclohexadiene. We then used these data to estimate the heat of hydrogenation of the *hypothetical* molecule 1,3,5-cyclohexatriene. Finally, we compared our estimate with the experimental value determined by hydrogenating the *real* molecule benzene (p. 581). Table 14.1 gives similarly determined values for a variety of heteroaromatic compounds.

There is less delocalization energy in furan than benzene (Table 14.1) and preservation of aromaticity is less important. Diels–Alder reactions are common for furans, which make excellent trapping agents for reactive π systems (Fig. 14.12).

TABLE 14.1

Resonance Energies of Some Heteroaromatic Compounds and Benzene

Compound	Resonance Energy (kcal/mol)
Benzene	33
Thiophene	29
Pyridine	23
Pyrrole	21
Furan	16

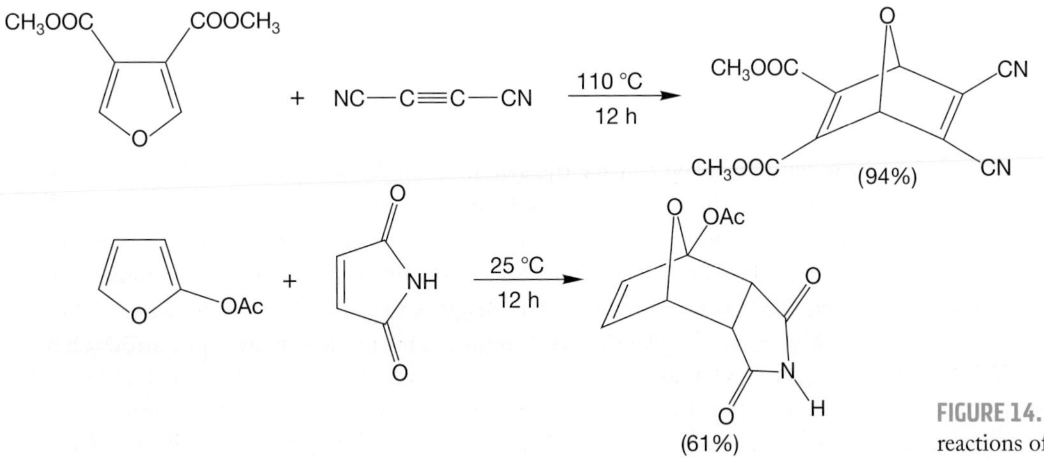

FIGURE 14.12 Two Diels–Alder reactions of furans.

Pyrroles, like furans, are also 1,3-dienes and one might suspect that the Diels–Alder reaction would occur as with furan. This suspicion is correct, but the aromaticity of pyrrole introduces complications. Such reactions are often slow, and aromaticity is sometimes regained through a retro Diels–Alder reaction. With those hints, you should be able to tackle Problem 14.4.

WORKED PROBLEM 14.4 Write a mechanism for the following reaction:

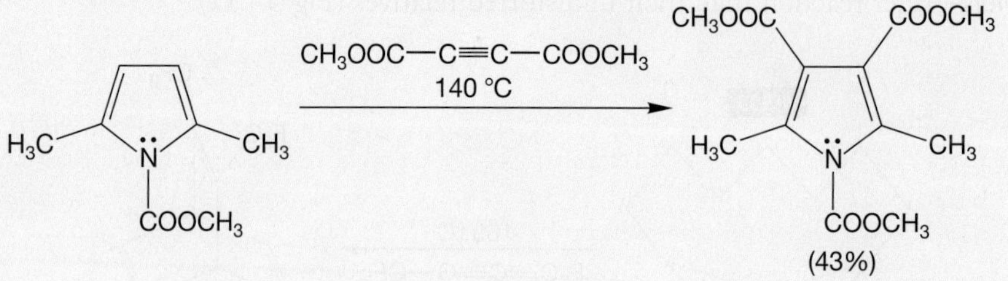

ANSWER All those hints in the text surely point directly at a Diels–Alder reaction. In this case, that leads to intermediate **A**.

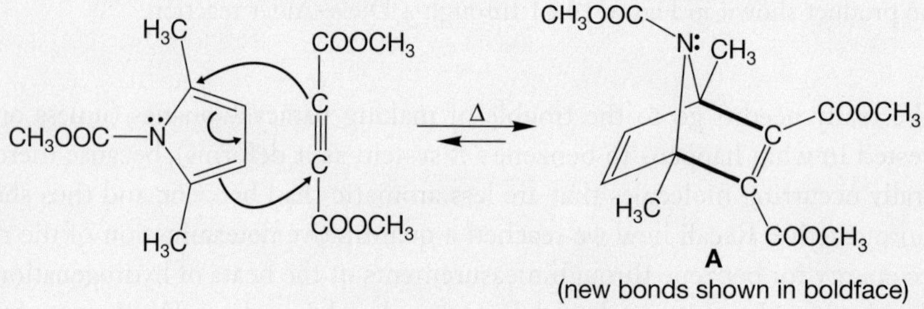

A
(new bonds shown in boldface)

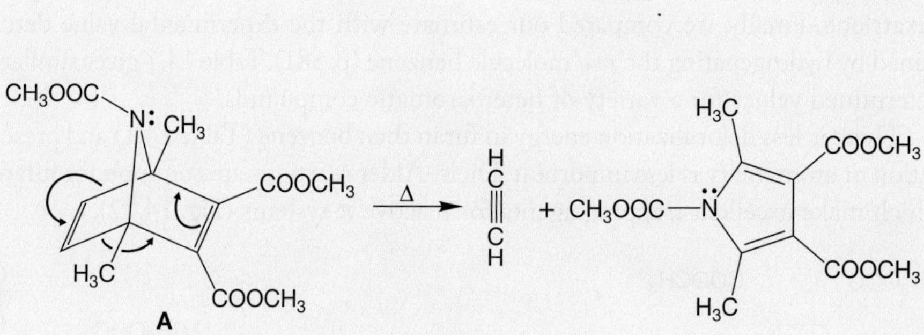

Now, with luck, you can see the symmetry in **A**. Because Diels–Alder reactions are reversible, there are two possible reverse Diels–Alder routes for **A** to take. In one, the reaction simply reverses to give the starting materials. In the other, the reverse Diels–Alder reaction proceeds in the other direction, and acetylene is lost to give the observed product. One imagines that it is the irreversible loss of the acetylene gas that drives the reaction toward the product shown. The arrows show the product-forming pathway.

Summary

It is not easy to disrupt the aromatic sextet, and few reactions of benzene do so permanently. It takes a very reactive reagent to overcome the high activation energy barriers surrounding the benzene ring and usually there is a path available that regains the lost aromaticity. In this section, you saw two reactions that do "dearomatize" benzene; in the next sections you will see many that don't.

14.4 Substitution Reactions of Aromatic Compounds

14.4a Why Benzene Doesn't Undergo Simple Electrophilic Addition Reactions
We are first going to do some thought experiments, beginning with a "What if" question. This process will lead us to an understanding of why benzene reacts so differently from simple alkenes. In particular, we are going to see why benzene does not undergo simple addition of HX (hydrohalogenation) as alkenes do. Then we will go on to see why benzene reacts as it does.

What if benzene reacted as the simple alkenes do to give addition products with HX reagents? The answer to this hypothetical question is more complicated than it appears at first. Let's work through a prediction for the addition of HBr to benzene, a reaction that in fact does not occur. By analogy to alkene reactions, the first step should be the addition of a proton to the π system of benzene to give an intermediate cyclohexadienyl carbocation. Note that aromaticity is lost (Fig. 14.13).

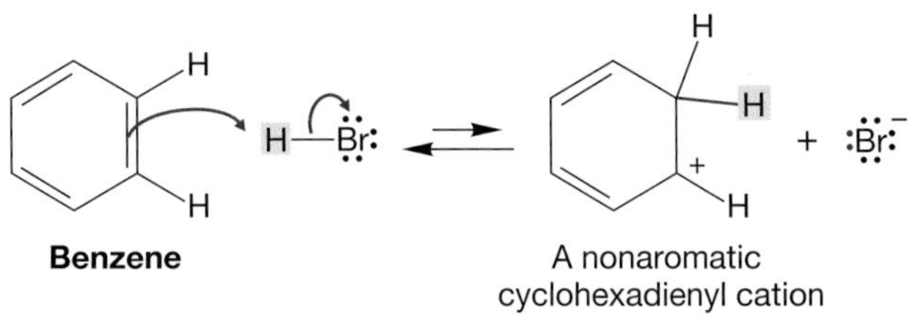

Benzene   A nonaromatic cyclohexadienyl cation

FIGURE 14.13 Protonation of benzene to give a cyclohexadienyl cation.

Simple addition of the bromide ion to the cation, again by analogy to the addition of HBr to an alkene, would generate a cyclohexadienyl bromide (Fig. 14.14).

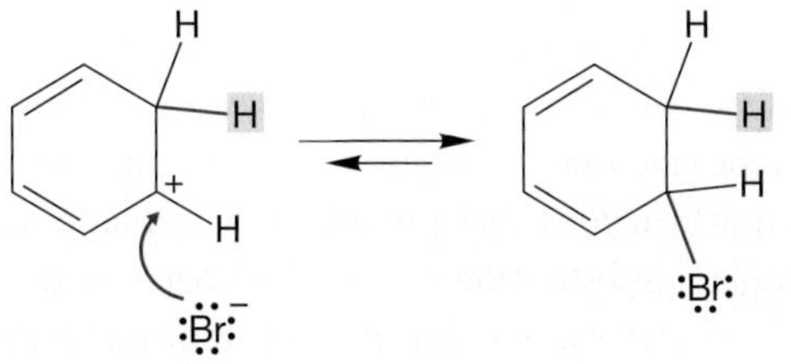

FIGURE 14.14 Addition of bromide would give a cyclohexadienyl bromide.

But we have forgotten to look closely at the structure of the carbocation formed by the addition of a proton. Just as is the case with simple dienes, the positive charge is not localized on one atom, but instead is shared by two other carbons. The resonance picture of Figure 14.15 shows a better representation of the cyclohexadienyl cation.

CONVENTION ALERT

Look carefully again at the shorthand used to indicate the delocalized species. The dashed loop of Figure 14.15 indicates delocalization quite clearly, but it does not do a good job of showing which atoms share the positive charge. To get this information, you must either draw out the separate resonance forms, or use a single resonance form with the other positive positions labeled as (+). Keep in mind that the dashed loop does not mean that *all* the atoms surrounding the loop are partially positive. In Figure 14.15, the positions sharing the positive charge are shown with dots.

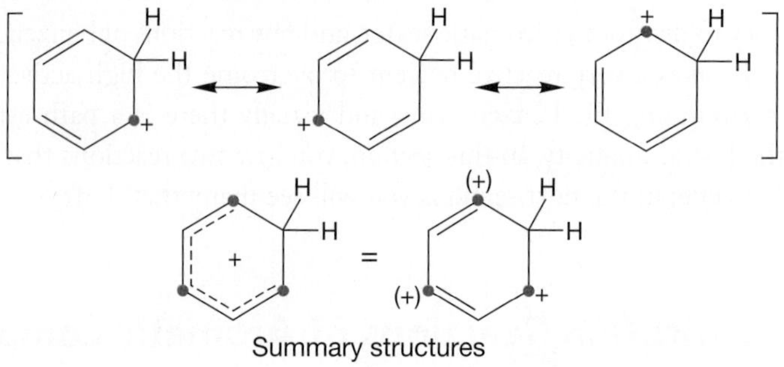

Summary structures

FIGURE 14.15 The cyclohexadienyl cation is resonance stabilized. The positive charge is shared by the three carbons marked with dots.

So, addition of bromide could take place at any one of the three carbons sharing the positive charge to produce two different products (Fig. 14.16). Yet, neither of these compounds is the product of the reaction! In fact, there is no apparent reaction at all, and benzene is recovered unchanged from the attempted addition reaction. One searches in vain for the predicted cyclohexadienyl bromides.

FIGURE 14.16 Capture of the cyclohexadienyl cation at the three possible positions bearing positive charge would give two cyclohexadienyl bromides.

| 5-Bromo-1,3-cyclohexadiene | 3-Bromo-1,4-cyclohexadiene | 5-Bromo-1,3-cyclohexadiene |

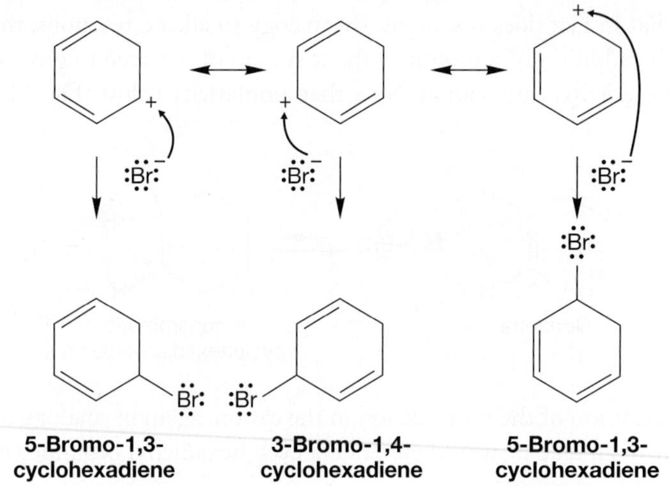

14.4b Electrophilic Deuteration of Benzene   In this section, we see how benzene *does* react. What we will see is the result of an energetic imperative to preserve the aromatic sextet. Benzene reacts not by addition, which would destroy aromatic stabilization, but by substitution reactions that preserve the aromatic sextet and stabilization. The prototypal example is substitution of each hydrogen by deuterium. If sufficiently acidic conditions are used, we can see that the apparent nonreaction of HX with benzene is an illusion. As we saw in Chapter 13, when benzene is treated with D_2SO_4, the recovered benzene has incorporated deuterium, with the amount of incorporation depending on the strength of acid used and the time of the reaction. Ultimately, excess deuteriosulfuric acid will produce fully exchanged benzene (Fig. 14.17).

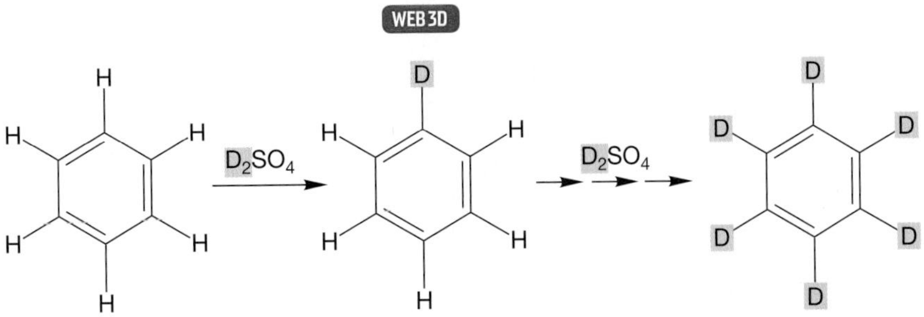

FIGURE 14.17 Hydrogen–deuterium exchange in benzene.

Because deuterium must become incorporated onto the ring, perhaps the proposed reaction shown in Figure 14.13 is not wrong after all. If benzene is protonated to produce the cyclohexadienyl cation shown in Figure 14.18, there are two possible further reactions.

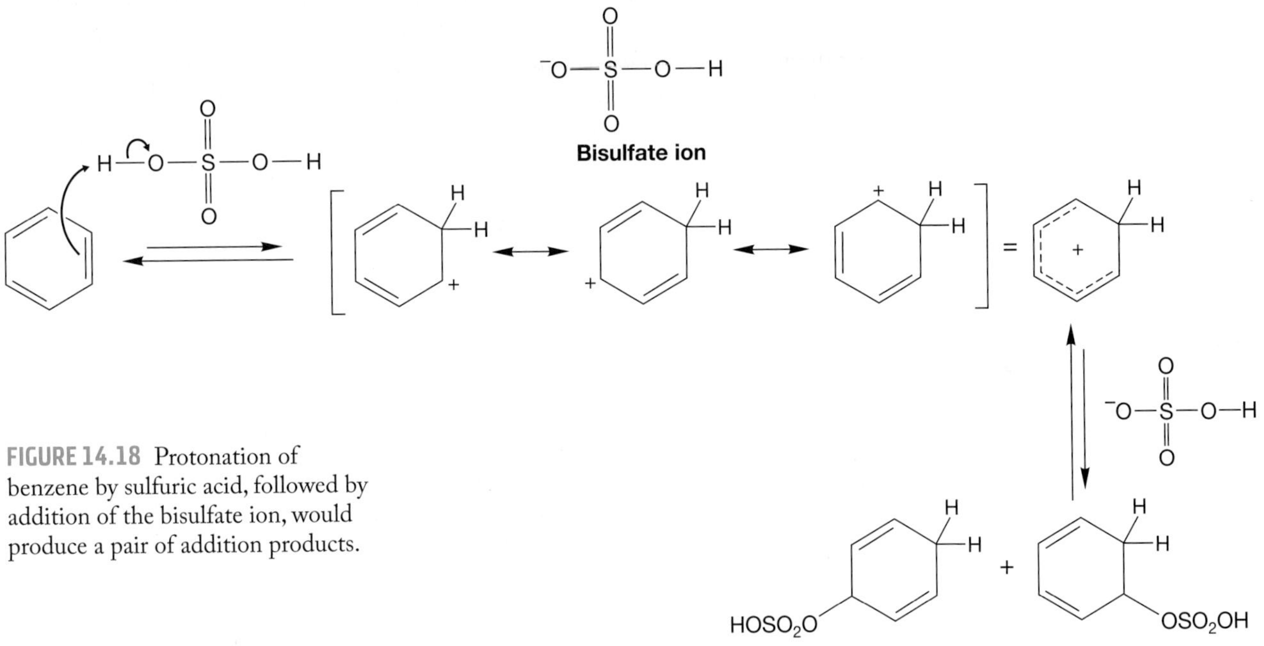

FIGURE 14.18 Protonation of benzene by sulfuric acid, followed by addition of the bisulfate ion, would produce a pair of addition products.

As shown in Figure 14.18, the bisulfate anion produced in the first step of the reaction could add to give two cyclohexadienyl compounds. We imagined a related reaction in Section 14.4a, where the Lewis base was bromide, Br^-. In this reaction, aromaticity is lost, and we would not expect this process to be favorable energetically. Alternatively, a proton might be lost to reverse the initial addition reaction as shown in Figure 14.19. In this scenario, the aromatic sextet is regenerated. In the absence of an isotope such as deuterium, this reaction is invisible because it simply regenerates starting material.

FIGURE 14.19 The protonation reaction is reversible. Bisulfate could remove a proton to regenerate benzene.

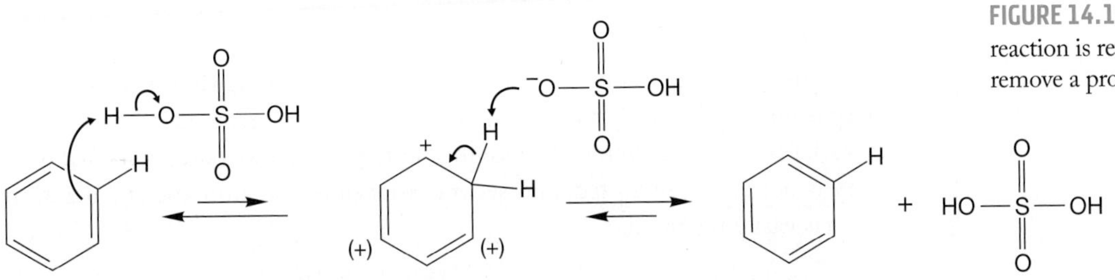

If it is a deuteron that is added to produce an isotopically labeled cyclohexadienyl cation, either a deuteron or a proton could be lost to regenerate benzene (Fig. 14.20).

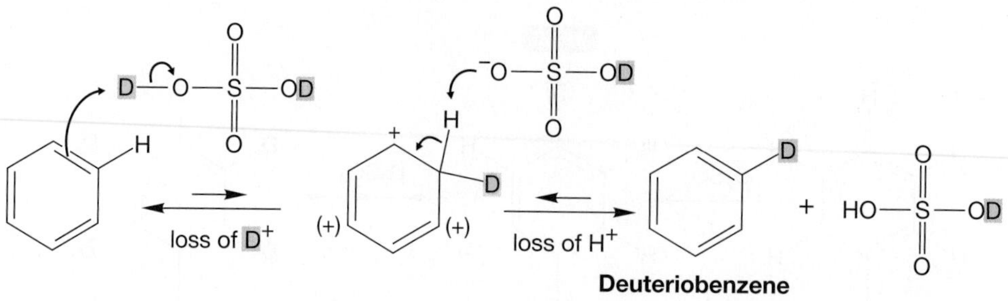

FIGURE 14.20 If deuteriosulfuric acid is used, the initial addition must generate a deuterated cyclohexadienyl cation. Reversal of the reaction will regenerate benzene, but loss of a proton will give deuteriobenzene. In an excess of D_2SO_4, all the protons of benzene can eventually be exchanged for deuterons.

If the proton is lost, deuteriobenzene will appear as product, because one hydrogen will have been exchanged for a deuterium. Repetition of the reaction can serve to exchange all the hydrogens for deuterons. If the cyclohexadienyl cation is formed, why aren't normal addition products, such as the dienes in Figure 14.18, observed? The answer is that loss of a proton (or deuteron) regenerates an aromatic system in a highly exothermic reaction. Addition to the relatively high-energy carbocation would still be exothermic, but not nearly as much as the substitution reaction (green); no aromatic molecule would be generated. The addition reaction (red) is much less exothermic than the rearomatization pathway and cannot compete with it (Fig. 14.21).

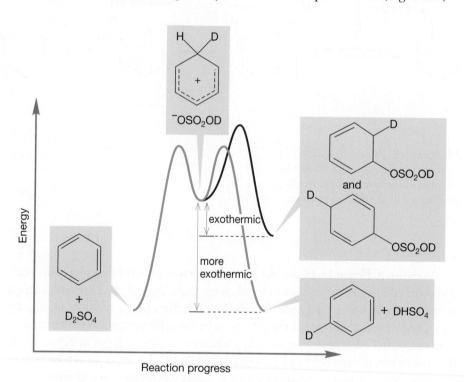

FIGURE 14.21 Both proton removal and addition are exothermic reactions for the cyclohexadienyl cation, but the regaining of aromaticity by proton removal is much more so.

Summary

We have just seen one example of substitution of a benzene ring in which aromaticity is initially lost only to be regained later. Next we generalize, discovering many variations on this theme. The main points to remember are, first, that aromaticity is not easily disrupted, and second, that once the aromatic stabilization is lost, it is easily regained.

14.4c Electrophilic Aromatic Substitution: The General Reaction

The exchange reaction of benzene in deuterated acid serves as a prototype for a host of similar reactions in which an electrophile, E^+, is substituted for a hydrogen on the ring:

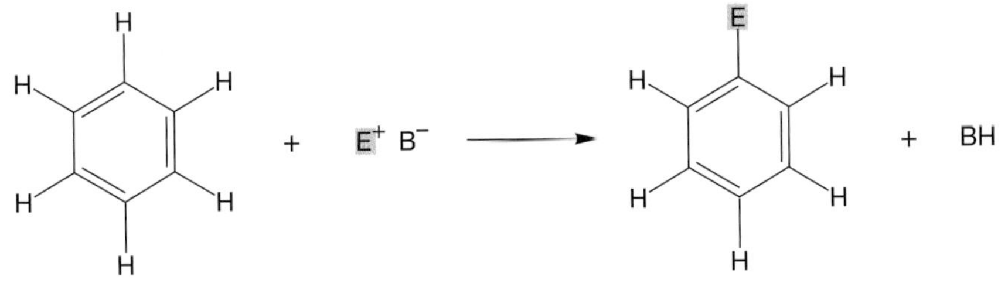

The reaction is always referred to as **electrophilic aromatic substitution**, because the emphasis is on the agent doing the substituting, the electrophile. Fair enough, but in every electrophilic reaction there must be a nucleophile as well. In this case, the π system of the benzene ring acts as the nucleophile. In orbital terms, it is one of the degenerate filled molecular orbitals of benzene that overlaps with the empty orbital on the electrophile (Fig. 14.22).

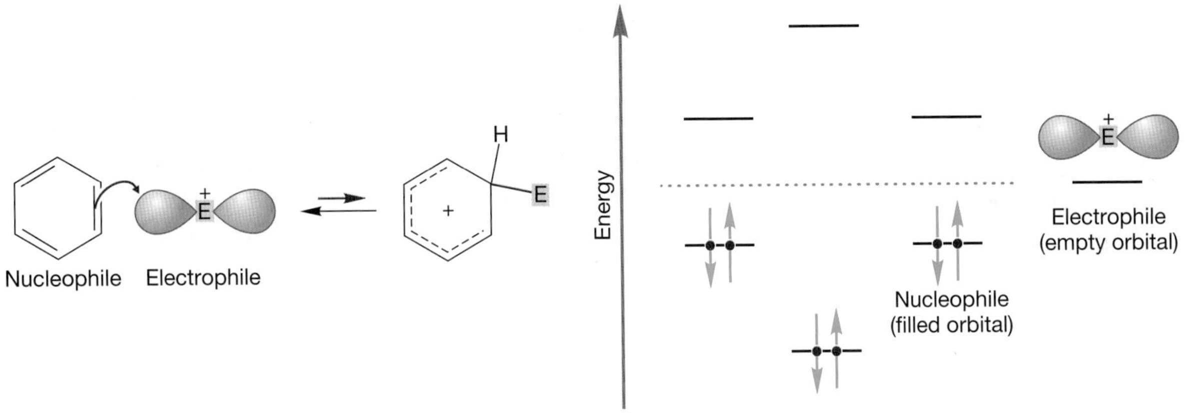

FIGURE 14.22 One of benzene's degenerate HOMOs overlaps with the empty orbital on E^+ in a Lewis base–Lewis acid reaction.

As in the substitution reaction of D for H on the aromatic ring, the general reaction is completed by the removal of a proton from the intermediate cyclohexadienyl cation by a base, regenerating the aromatic benzene ring (Fig. 14.23).

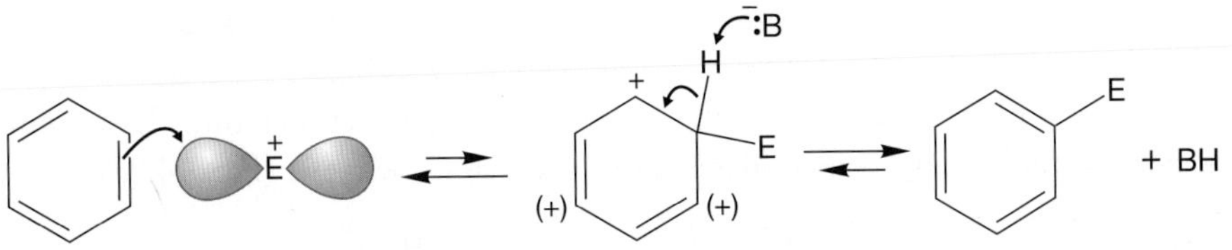

FIGURE 14.23 The general aromatic substitution mechanism involves formation of a cyclohexadienyl cation, followed by proton removal, regenerating the aromatic ring.

14.4d Sulfonation   When benzene is treated either with very concentrated sulfuric acid or with a solution of sulfur trioxide in sulfuric acid, called oleum, benzenesulfonic acid (Fig. 14.24) is the product. When oleum is used, the sulfur trioxide is protonated to produce the electrophile, $H\overset{+}{O}SO_2$.

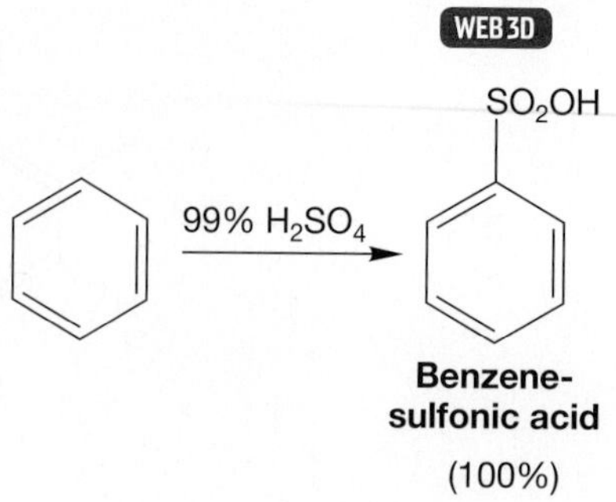

FIGURE 14.24 Sulfonation of benzene.

Benzene-
sulfonic acid

(100%)

PROBLEM 14.5 Write a mechanism for the formation of benzenesulfonic acid when benzene is treated with SO_3/H_2SO_4. *Hint*: Sulfur trioxide (SO_3) can be written as a sulfur atom doubly bonded to each of the three oxygens.

The resonance-
stabilized
cyclohexadienyl
cation intermediate

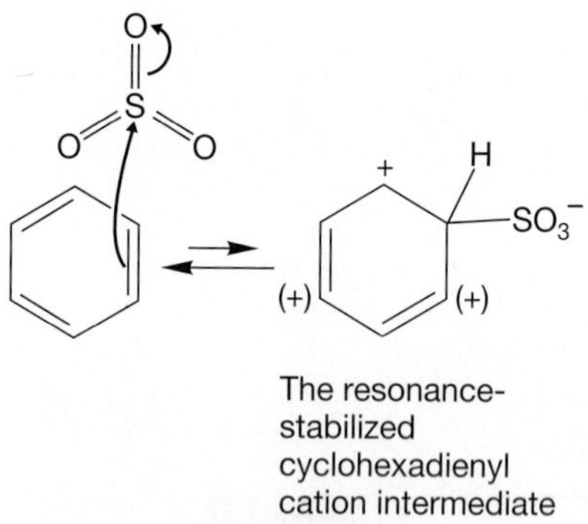

FIGURE 14.25 The SO_3 molecule acts as an electrophile and adds to benzene.

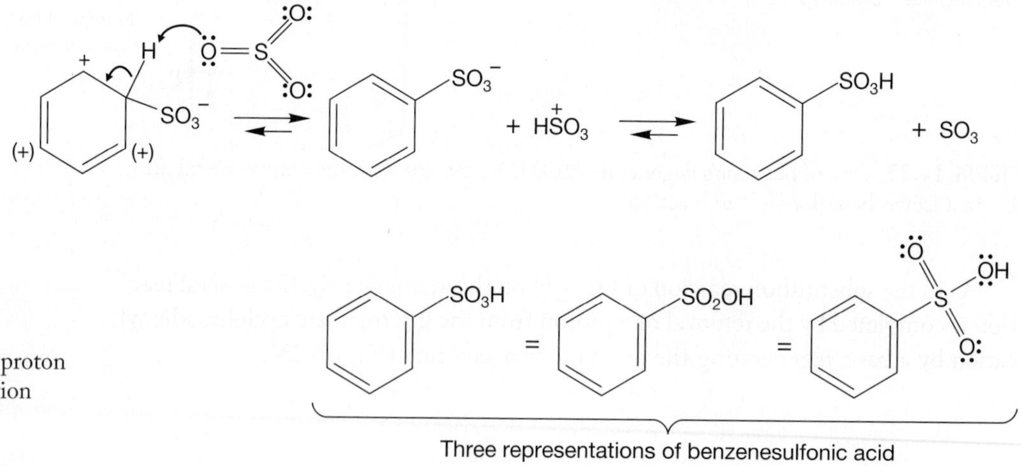

When used in an aprotic medium, SO_3 is a strong enough electrophile to react with benzene without protonation (Fig. 14.25). Under these circumstances the sulfonation reaction looks a little different. Addition leads to a cyclohexadienyl cation, but now there is also a negative charge in the molecule. As in the general mechanism for electrophilic aromatic substitution, a base now removes a proton to give the aromatic product. The SO_3 in this particular reaction is the base that will remove the very acidic hydrogen to restore aromaticity. When the reaction is run in H_2SO_4 we find that even sulfuric acid is basic enough to deprotonate the cyclohexadienyl cation. The structure of the product of this reaction, benzenesulfonic acid, gives students problems because the formulas, SO_3H or SO_2OH hide the detail. The full structure is drawn out in Figure 14.26.

FIGURE 14.26 Removal of a proton from the cyclohexadienyl cation intermediate regenerates the aromatic ring.

Three representations of benzenesulfonic acid

Sulfonation of benzene is reversible, especially at high temperature. Treatment of benzenesulfonic acid in steam regenerates the parent benzene (Fig. 14.27).

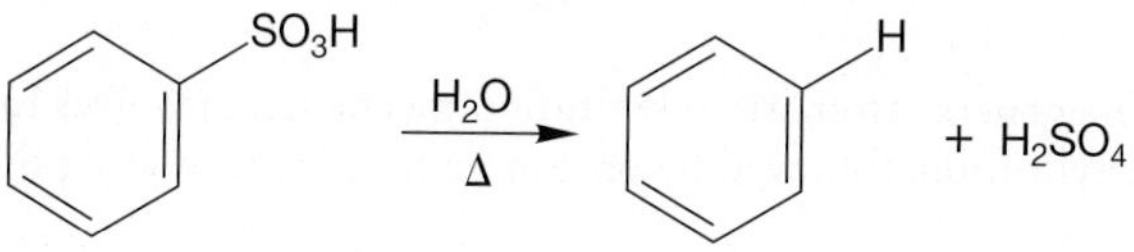

FIGURE 14.27 Sulfonation is reversible.

WORKED PROBLEM 14.6 Write a mechanism for the reaction in Figure 14.27.

ANSWER The mechanism for the transformation of benzenesulfonic acid into benzene is the reverse of the mechanism for formation of benzenesulfonic acid from benzene (Figs. 14.25 and 14.26). Protonation gives a cyclohexadienyl cation that can either lose a proton to revert to benzenesulfonic acid, or lose the sulfonic acid group to give benzene and protonated sulfuric acid, which will lose its proton to water.

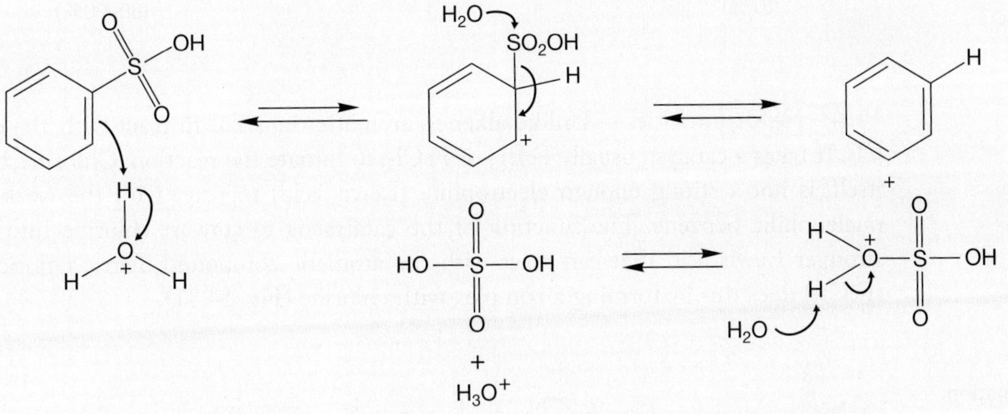

14.4e Nitration

In concentrated nitric acid, often in the presence of sulfuric acid, benzene is converted into nitrobenzene (Fig. 14.28). In strong acid, nitric acid

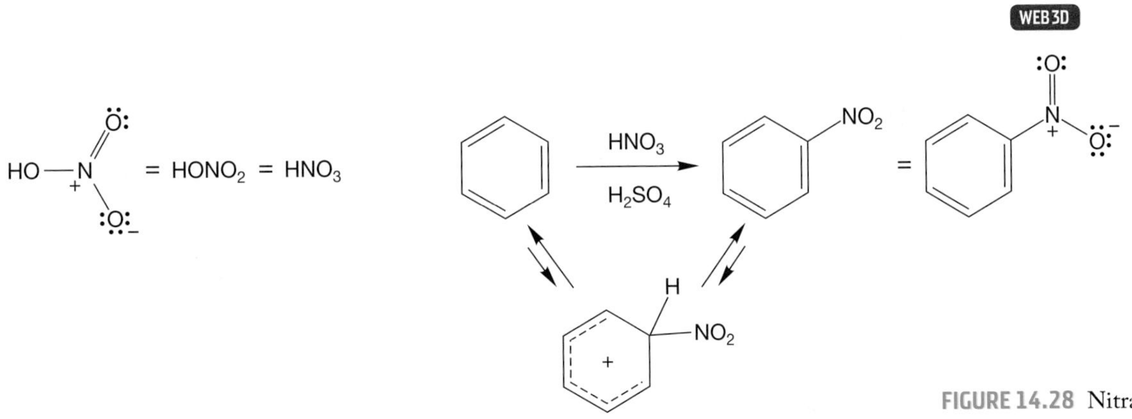

FIGURE 14.28 Nitration of benzene.

is protonated to give $H_2ONO_2^+$ (Fig. 14.29). Loss of water can generate a powerful Lewis acid, the nitronium ion, $^+NO_2$, that can act as the electrophile in an aromatic

$$HO-NO_2 + H_3O^+ \rightleftharpoons H_2\overset{+}{O}-NO_2 + H_2O \rightleftharpoons H_2O + \overset{+}{N}O_2$$

Nitronium ion

FIGURE 14.29 In strong acid, nitric acid becomes protonated, then loses water to form the nitronium ion.

substitution reaction (Fig. 14.30). Indeed, stable nitronium salts are known and can also effect nitration. As usual, the reaction is completed by deprotonation of the cyclohexadienyl cation by one of the weak Lewis bases present in the mixture (water or even nitric acid).

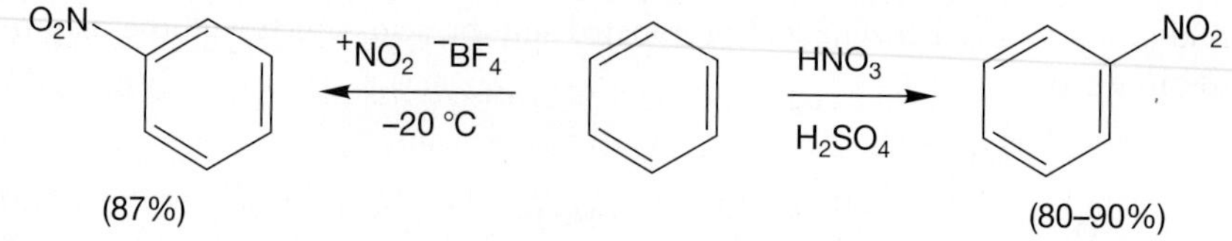

FIGURE 14.30 Two nitration reactions. In each case the electrophile is the nitronium ion, $^+NO_2$.

$^+NO_2\ ^-BF_4$
$-20\ °C$

HNO_3
H_2SO_4

(87%)

(80–90%)

14.4f Halogenation

Unlike alkenes, aromatic rings fail to react with Br_2 or Cl_2. It takes a catalyst, usually $FeBr_3$ or $FeCl_3$, to initiate the reaction. Chlorine, by itself, is not a strong enough electrophile (Lewis acid) to react with the weakly nucleophilic benzene. The function of the catalyst is to convert chlorine into a stronger Lewis acid that can react with the aromatic compound. Ferric chloride ($FeCl_3$) does this by forming a complex with chlorine (Fig. 14.31).

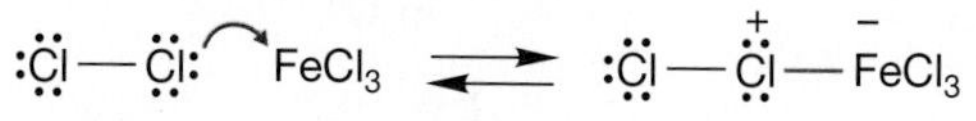

FIGURE 14.31 Complex formation from Cl_2 and $FeCl_3$.

Benzene is nucleophilic enough to displace the good leaving group $FeCl_4^-$ from the Cl_2—$FeCl_3$ complex (Fig. 14.32). Displacement generates a cyclohexadienyl cation that can be deprotonated to give chlorobenzene. The first step of this reaction may look strange, but it has a most familiar analogy in the protonation of an alcohol to convert OH into the good leaving group OH_2 (Fig. 14.32). The $FeCl_3$ plays the role of the proton in the alcohol reaction by converting the Cl into the good leaving group Cl—$\bar{Fe}$—Cl_3. The mechanism of bromination using Br_2 and $FeBr_3$ is similar.

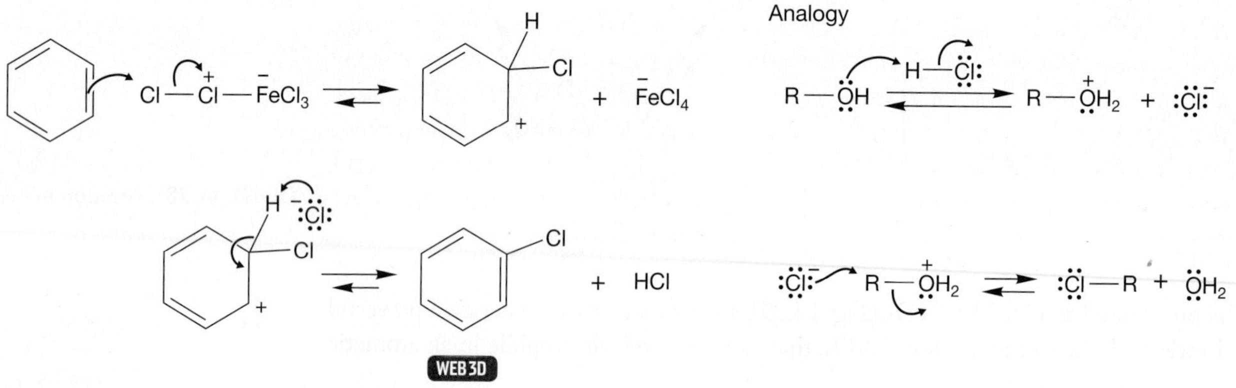

Analogy

FIGURE 14.32 Chlorination of benzene.

PROBLEM 14.7 Write a mechanism for the bromination of benzene using $Br_2/FeBr_3$.

Summary

Benzene fails to react with most HX reagents under conditions that easily result in addition to alkenes and dienes. The X_2 reagents, such as bromine and chlorine, are similarly unreactive. Benzene lies in too deep a well, and it is protected by activation energy barriers that are too high for these reactions to occur.

Sulfonation, nitration, chlorination, and bromination all serve to underscore the general principles of electrophilic aromatic substitution. In each reaction, an electrophile powerful enough to react with the weakly nucleophilic benzene ring must be generated. The aromatic ring and the powerful electrophile together will have a high enough energy that the reaction can proceed over the energy barrier. The addition reaction first generates a resonance-stabilized (but not aromatic) cyclohexadienyl cation that can be deprotonated to regenerate an aromatic system. The general mechanism is shown again in Figure 14.33.

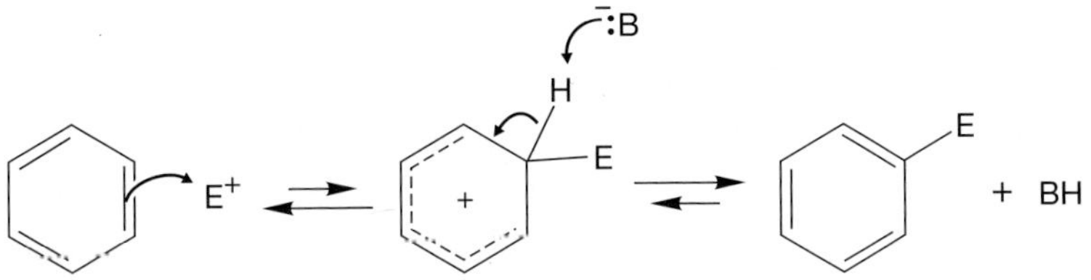

FIGURE 14.33 Once again, here is the general mechanism for aromatic substitution, where $E^+ = D^+$, $^+SO_2OH$, $^+NO_2$, ^+Cl, or ^+Br.

14.5 Carbon–Carbon Bond Formation: Friedel–Crafts Alkylation

We are now able to make a few substituted aromatic compounds, and this capability will shortly prove quite useful. However, much of the interest in synthetic chemistry involves the construction of carbon–carbon bonds. We have made no progress at all in finding ways to make bonds from benzene to carbon. Here and in the next section we do just that by making use of another electrophilic aromatic substitution reaction quite closely related to the bromination and chlorination reactions we have just seen.

The **Friedel–Crafts alkylation** reaction is an electrophilic aromatic substitution that attaches an alkyl group to the aromatic ring. It is named for Charles Friedel (1832–1899) and James M. Crafts (1839–1917), who discovered it by accident when they tried to synthesize pentyl chloride from pentyl iodide through reaction with aluminum chloride in an aromatic solvent. Instead of the hoped-for chloride, substituted aromatic hydrocarbons appeared. The Friedel–Crafts reaction is closely related to bromination and chlorination of benzene. If we treat benzene with isopropyl bromide in the hope that a nucleophilic attack of benzene on isopropyl bromide will produce isopropylbenzene (cumene), we are sure to be disappointed (Fig. 14.34). Benzene is by no means a strong nucleophile and isopropyl bromide can scarcely be described as a powerful electrophile. The proposed reaction is utterly hopeless, and fails to give product.

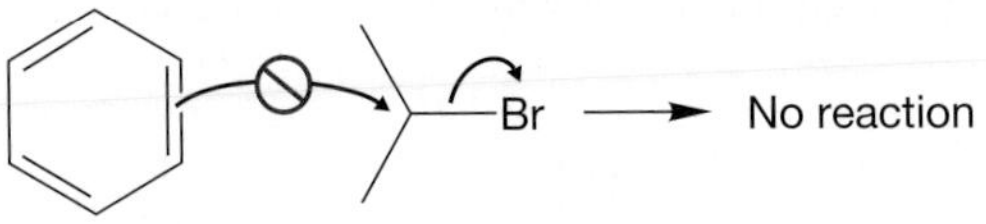

FIGURE 14.34 Benzene fails to react with isopropyl bromide.

So why even bring the subject up? Notice that this process does look a bit like another reaction that fails to give product—the attempt to form chlorobenzene or bromobenzene from chlorine or bromine and benzene. Those halogenations are transformed into a success by the addition of a catalyst, either $FeCl_3$ or $FeBr_3$, which converts the halogen, a weak electrophile, into a much stronger Lewis acid

(Figs. 14.31 and 14.32). Perhaps a related technique will work here. Indeed it does, as the addition of a small amount of the catalyst AlBr₃ or AlCl₃ to a mixture of benzene and isopropyl bromide or chloride results in the formation of isopropyl-benzene (Fig. 14.35). The mechanism should be easy to write now. A complex is first formed between isopropyl bromide and AlBr₃. This complex can be attacked by benzene to give a resonance-stabilized cyclohexadienyl cation. Deprotonation of the intermediate cyclohexadienyl cation gives isopropylbenzene.

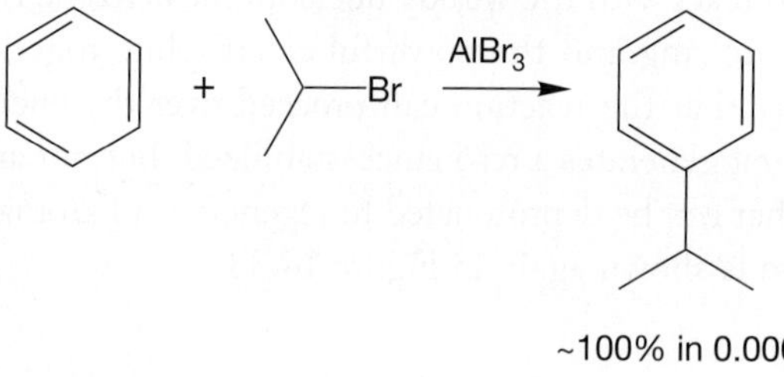

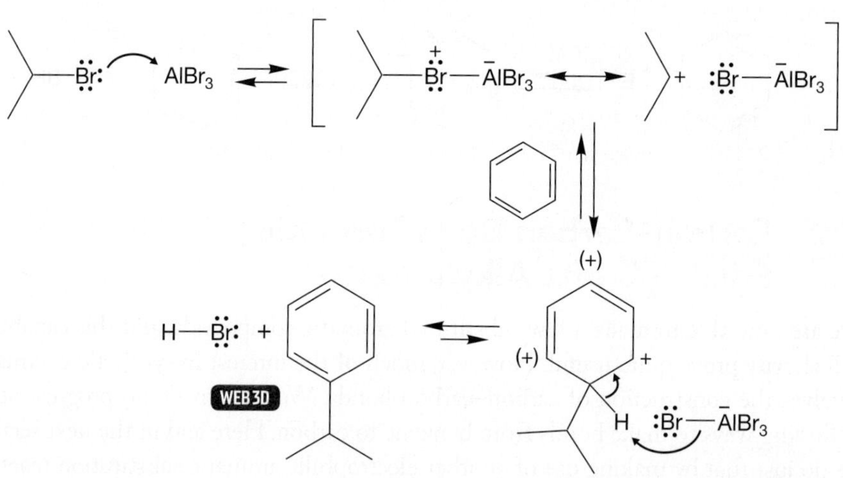

FIGURE 14.35 The Friedel–Crafts alkylation of benzene.

~100% in 0.0005 s

Whether or not the complex in Figure 14.35 can actually form a free carbocation depends on the alkyl halide used in the reaction. Friedel–Crafts alkylation succeeds with a variety of alkyl halides, including methyl-, ethyl-, isopropyl-, and *tert*-butyl chlorides and bromides (Fig. 14.36).

We would not expect the methyl or ethyl cation to be formed in these reactions, and surely these alkylations take place by attack of benzene on the complexed halide. *tert*-Butyl chloride, on the other hand, can produce the relatively stable *tert*-butyl cation, not an unlikely intermediate at all. So, the detailed mechanism of the Friedel–Crafts alkylation depends on the nature of the R group.

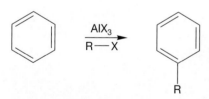

FIGURE 14.36 Friedel–Crafts alkylation succeeds for a wide variety of alkyl halides.

PROBLEM 14.8 Why is attack by benzene on complexed *tert*-butyl bromide to give a cyclohexadienyl cation an unlikely mechanistic step?

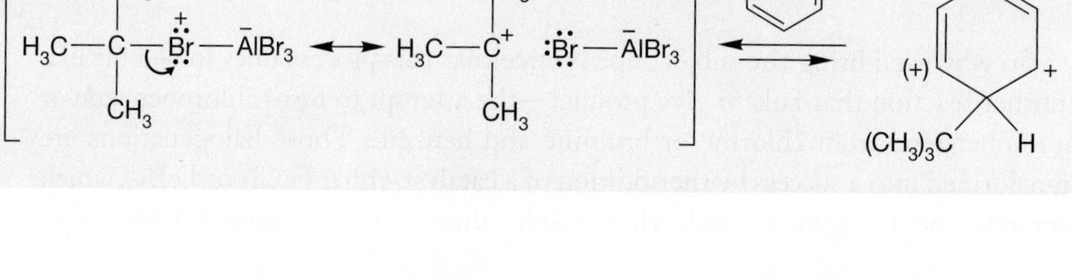

PROBLEM 14.9 The synthesis of toluene by the aluminum chloride–catalyzed Friedel–Crafts alkylation of benzene with methyl chloride is badly complicated by the formation of di-, tri-, and polymethylated benzenes. It appears that the initial product of the reaction, toluene, is more reactive in the Friedel–Crafts reaction than is benzene. Analyze the mechanism of electrophilic aromatic substitution to see why toluene is more reactive than benzene. *Hint*: Look carefully at substitution in the position directly across the ring from the methyl group (the para position) for toluene, and look for differences from the reaction with benzene.

PROBLEM 14.10 How would you minimize the overalkylation mentioned in Problem 14.9 and achieve a reasonable synthesis of toluene from benzene?

Not only do the mechanistic details of the Friedel–Crafts alkylation depend on the alkyl group (R), but even the gross outcome of the reaction depends on the structure of R. Although the alkyl halides we have mentioned so far all succeed in making the related alkylated benzenes, there are many that do not. Suppose we set out to make propylbenzene by the Friedel–Crafts alkylation of benzene using 1-chloropropane, benzene, and aluminum chloride. As we run this reaction, all looks well at first. The products formed are of the correct molecular weight, but a closer look uncovers a problem. Some propylbenzene is formed, but the major product is isopropylbenzene (Fig. 14.37).

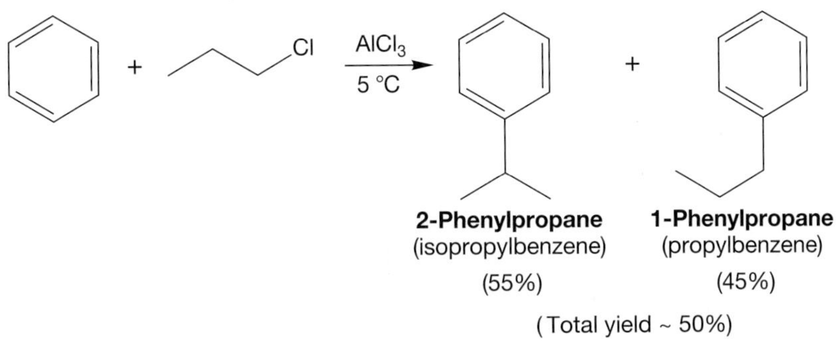

2-Phenylpropane
(isopropylbenzene)
(55%)

1-Phenylpropane
(propylbenzene)
(45%)

(Total yield ~ 50%)

FIGURE 14.37 An attempt to make propylbenzene by Friedel–Crafts alkylation gives isopropylbenzene as the major product.

A careful look at the mechanism of the reaction shows the problem. In the complex between the 1-chloropropane and aluminum chloride (AlCl₃), the carbon–chlorine bond is weakened and the positive charge is shared by the chlorine and the adjacent carbon. The resonance form shown in Figure 14.38 shows this clearly.

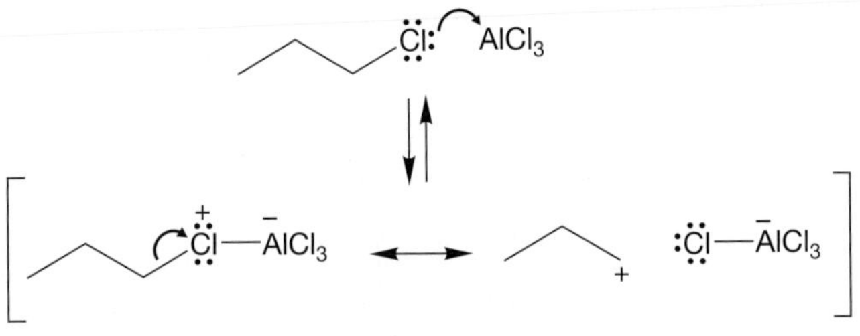

FIGURE 14.38 In the complex formed between 1-chloropropane and aluminum chloride, the positive charge is partially on the primary carbon adjacent to the chlorine.

The complex contains a partial positive charge on a primary carbon, a most unhappy state of affairs. When this happens in reactions of methyl chloride or ethyl chloride, there is nothing that can be done about it, but propyl chloride has a way out. A rearrangement of hydride generates the much more stable complex, or even a free cation in which a secondary carbon bears the positive charge. Straightforward alkylation now gives the observed major product, isopropylbenzene (Fig. 14.39).

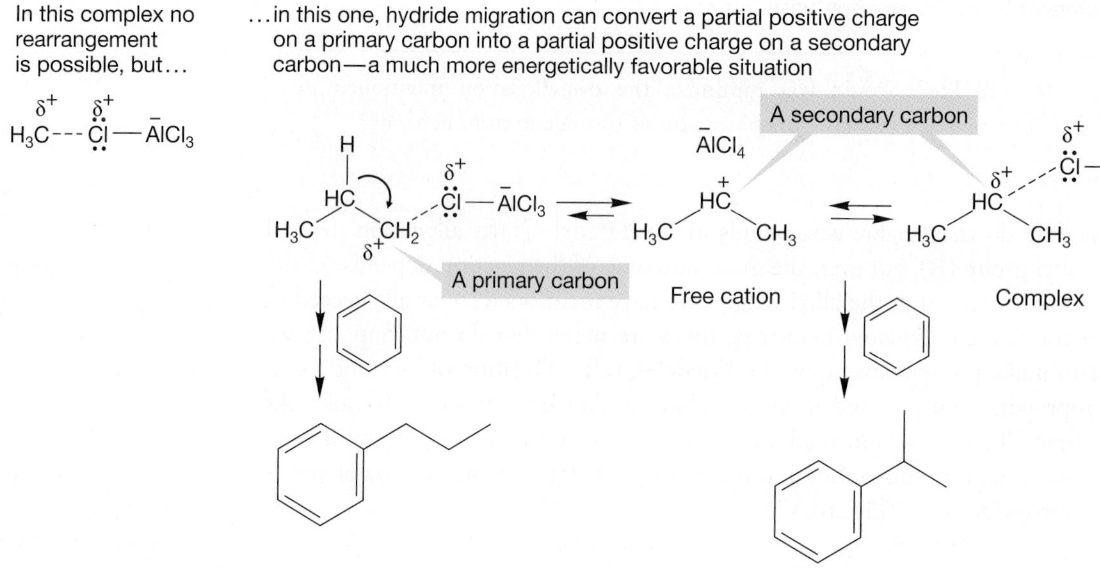

In this complex no rearrangement is possible, but...

...in this one, hydride migration can convert a partial positive charge on a primary carbon into a partial positive charge on a secondary carbon—a much more energetically favorable situation

FIGURE 14.39 A hydride shift can convert a less stable complex with a partial positive charge on a primary carbon into a more stable complex in which the partial positive charge is on a secondary carbon. The substitution reaction follows.

These rearrangements are very common, and are encountered whenever a hydride or alkyl shift will generate a more stable cation, or complex. This phenomenon is not new. Recall, for example, the rearrangements of hydride and alkyl groups encountered in reactions involving carbocations (Fig. 14.40; also p. 388).

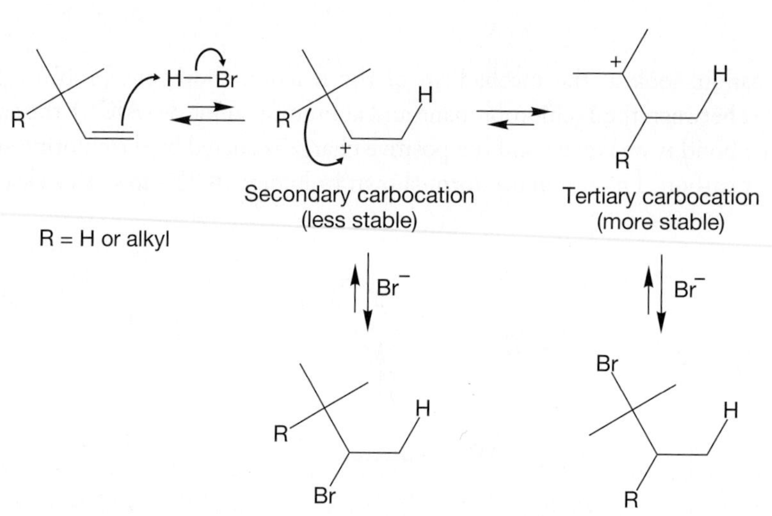

FIGURE 14.40 Hydride shifts are commonly encountered in addition reactions whenever a more stable carbocation can be generated from a less stable one.

PROBLEM 14.11 *tert*-Butylbenzene can be obtained from treatment of 1-bromo-2-methylpropane with aluminum chloride. Write a mechanism for this reaction.

PROBLEM 14.12 *tert*-Butylbenzene is converted into benzene on treatment with AlCl$_3$ and a trace of HCl. Write a mechanism for this reaction.

14.6 Friedel–Crafts Acylation

The **Friedel–Crafts acylation** reaction places an acyl group (RC=O) onto an aromatic ring (Fig. 14.41). We will take up the chemistry of all manner of acyl derivatives in Chapter 18, but here we need to use acyl chlorides (RCOCl), which are more commonly known as **acid chlorides**. In practice, acid chlorides are rather easy to make from the relatively available carboxylic acids, RCOOH, by treatment with thionyl chloride (SOCl$_2$) (Fig. 14.42).

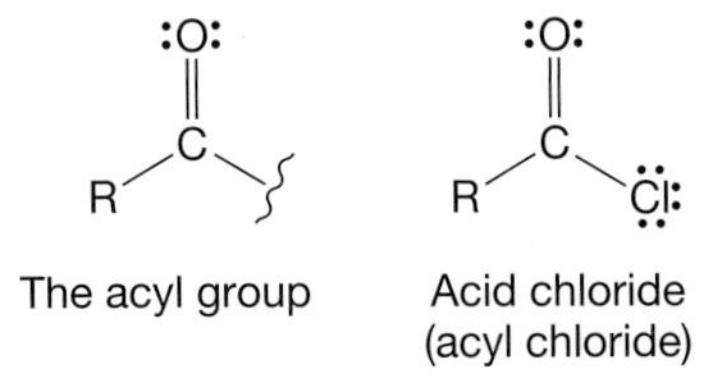

The acyl group   Acid chloride
                      (acyl chloride)

FIGURE 14.41 The acyl group.

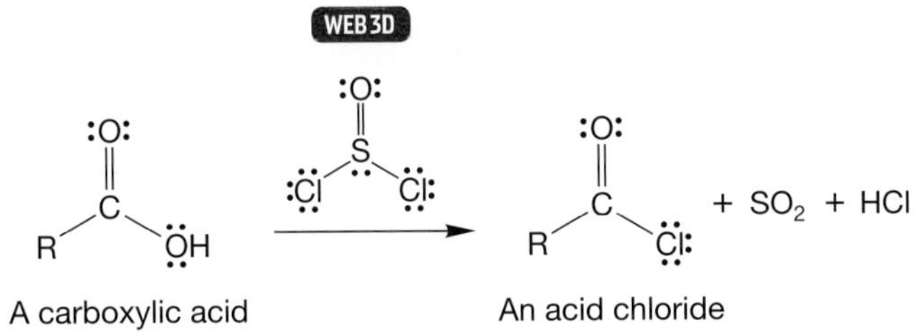

A carboxylic acid   An acid chloride

FIGURE 14.42 Acid chlorides can be made by the reaction of carboxylic acids with thionyl chloride.

Acid chlorides react with benzene under the influence of an equivalent (not a catalytic amount) of AlCl$_3$. A molecule of AlCl$_3$ is needed for each molecule of product. By now it should be easy to write a general mechanism. A complex is first formed with AlCl$_3$. But which complex? Unlike alkyl chlorides, acid chlorides contain two nucleophilic sites, the oxygen and chlorine atoms. The evidence is that both complexes are formed, and that the two are in equilibrium. The resonance-stabilized **acylium ion** can be formed by dissociation of the complex (Fig. 14.43).

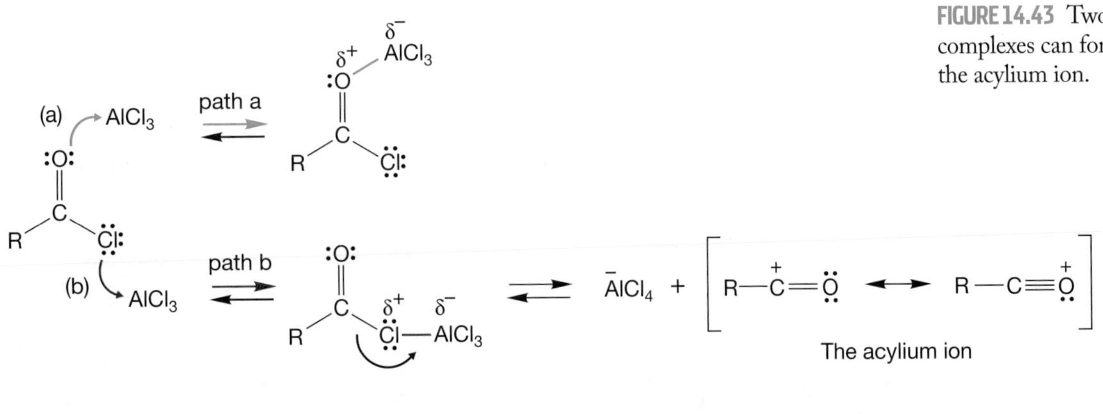

FIGURE 14.43 Two acid chloride–AlCl$_3$ complexes can form; dissociation gives the acylium ion.

The acylium ion

The acylium ion can then add to a benzene ring to generate an intermediate cyclohexadienyl cation. Deprotonation by a Lewis base completes the reaction and produces

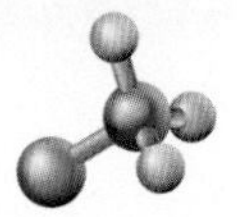

Friedel–Crafts acylation

acyl benzenes (Fig. 14.44). As mentioned above, unlike Friedel–Crafts alkylation, Friedel–Crafts acylation is not a catalytic reaction. The acyl benzene is also reactive toward AlCl₃ and forms its own 1:1 complex as it is produced. So, for every molecule of acyl benzene formed in the reaction, one molecule of AlCl₃ is used up in complex formation. A stoichiometric amount, or one equivalent, of AlCl₃ is therefore required in this reaction. At the end of the reaction, addition of water destroys the complex and generates the free, acylated benzene, as shown in the last step in Figure 14.44.

FIGURE 14.44 The mechanism of Friedel–Crafts acylation. Aromatic ketones can form complexes with AlCl₃, which is why a full equivalent of AlCl₃ is necessary for Friedel–Crafts acylation.

A final hydrolysis step liberates the ketone, an acyl benzene

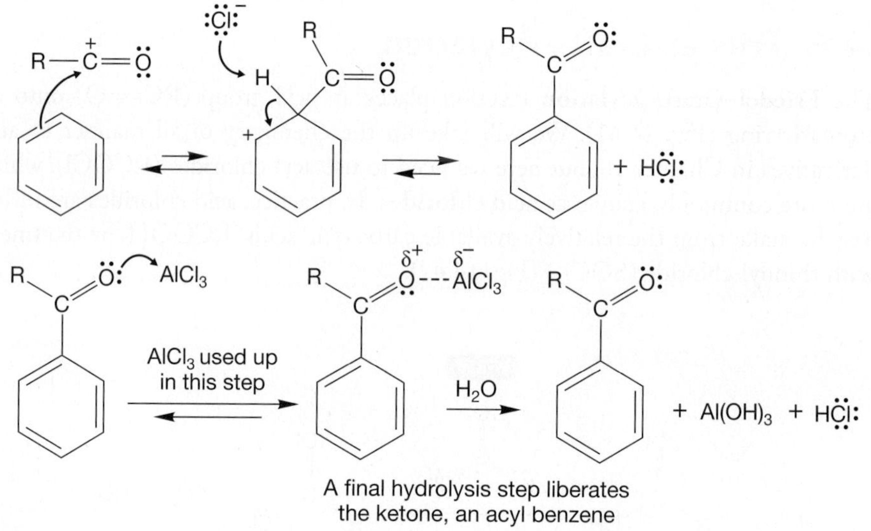

Despite the mechanistic complexity, in practice the Friedel–Crafts acylation reaction works very well, and a variety of phenyl-substituted ketones can be produced in good yield. Be certain to add this reaction to your collection of synthesis file cards (Fig. 14.45).

FIGURE 14.45 Two typical Friedel–Crafts acylations.

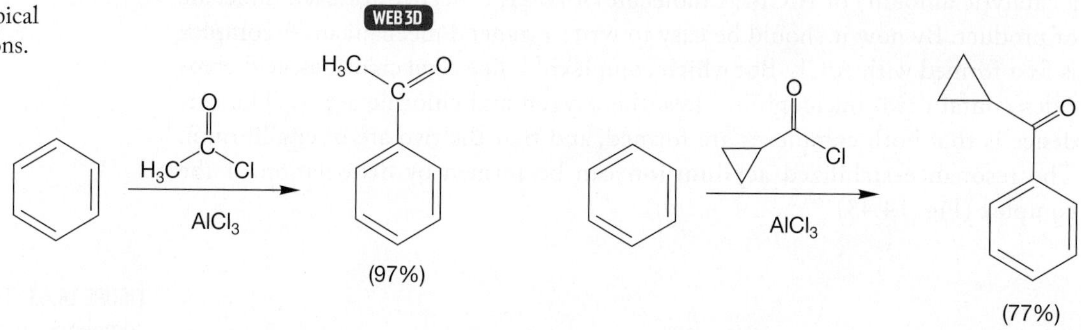

(97%)

(77%)

Figure 14.46 shows another successful Friedel-Crafts acylation, but this reaction contains a bonus. The reaction of the four-carbon acid chloride with benzene yields the unrearranged ketone. This result implies that acyl complexes are not prone to rearrangement (Fig. 14.46).

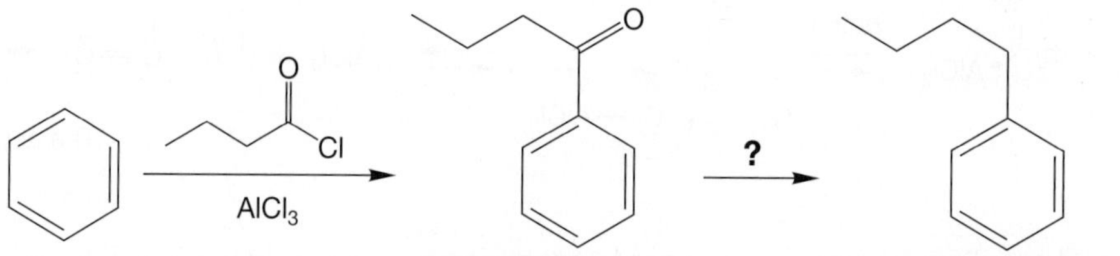

FIGURE 14.46 Friedel–Crafts acylation is not complicated by rearrangements.

1-Phenylbutanone
(phenyl propyl ketone)
(butyrophenone)

1-Phenylbutane
(butylbenzene)

The four-carbon chain is intact in the product, and if we had a way to transform the carbon–oxygen double bond into a methylene group, we would have a way to make butylbenzene without the complications of rearrangement that come from the Friedel–Crafts alkylation.

WORKED PROBLEM 14.13 Explain why an acylium ion (or acyl complex) is not likely to rearrange. Two reasons should be apparent. First, what differentiates the acylium ion from alkyl carbocations? Next, look closely at the product of rearrangement and consider the polarity of the carbon–oxygen double bond.

ANSWER The first reason has to do with the stability of the acylium ion, which is well stabilized by resonance and is therefore less likely to rearrange than alkyl carbocations.

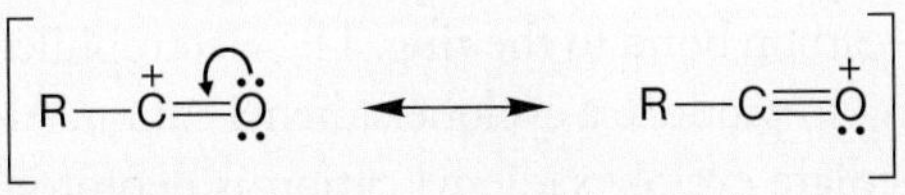

An acylium ion

Second, consider the structure of the putative rearranged cation.

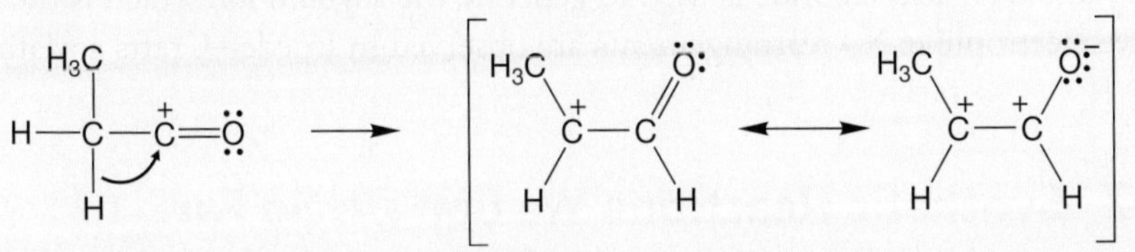

The carbonyl group is strongly polarized in the sense shown in the drawing. The carbon is δ^+ and the oxygen is δ^-. Rearrangement generates a full and partial positive charge on adjacent atoms. This situation is very bad energetically and will act to retard migration of hydrogen or an alkyl group.

In fact, there are many such ways to convert a carbonyl into a methylene group, two of which are shown in Figure 14.47. The Clemmensen reduction, named for its discoverer E. C. Clemmensen (1876–1941), takes place under strongly acidic conditions, whereas the Wolff–Kishner reduction (Ludwig Wolff, 1857–1919; N. Kishner, 1867–1935) requires strong base.

Clemmensen reduction

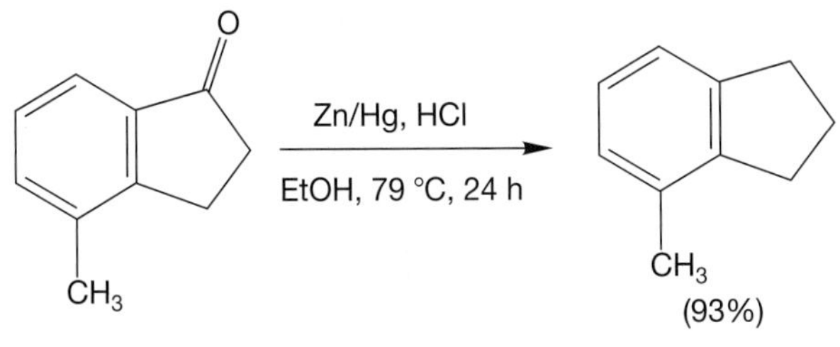

Wolff–Kishner reduction

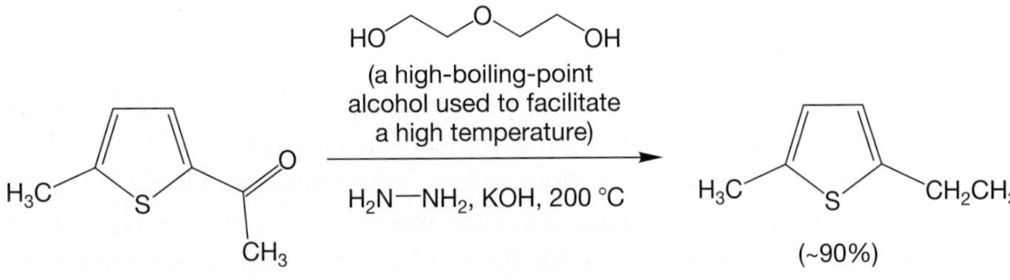

FIGURE 14.47 Two methods for converting a carbonyl group into a methylene group.

PROBLEM 14.14 Other reagents besides acid chlorides can lead to acylation reactions. Write a mechanism for the following reaction:

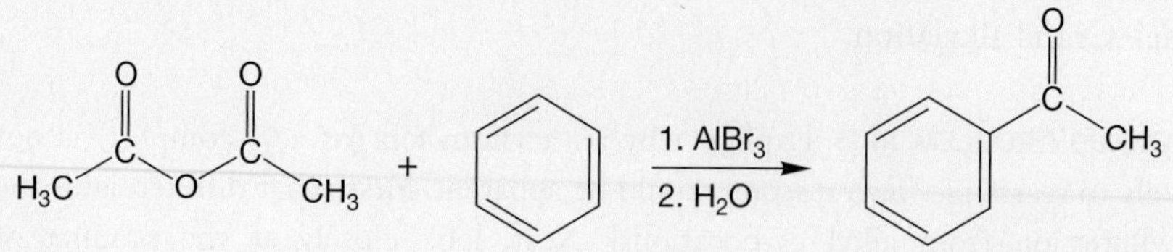

Summary

A Friedel–Crafts alkylation is an electrophilic aromatic substitution reaction that attaches a carbon–carbon bond to the ring. The electrophile is R^+, which will add to the aromatic ring to produce a cyclohexadienyl cation. Aromaticity is regained when that intermediate cyclohexadienyl cation is deprotonated. That's all there is to it—all the rest is details. Remember to watch out for rearrangements, because this Friedel–Crafts alkylation is especially prone to them. In the Friedel–Crafts acylation, an acid chloride is used to generate the acylium ion which is the reactive electrophile. No rearrangements are observed in Friedel–Crafts acylation.

14.7 Synthetic Reactions We Can Do So Far

We have uncovered seven synthetically important reactions in this chapter so far. We can (1) brominate and (2) chlorinate benzene. We can (3) nitrate and reversibly (4 and 5) sulfonate benzene. The two Friedel–Crafts reactions (6 and 7) allow the construction of carbon–carbon bonds to give some alkyl benzenes and give you another synthetic route to compounds containing carbon–oxygen double bonds. Figure 14.48 summarizes the situation.

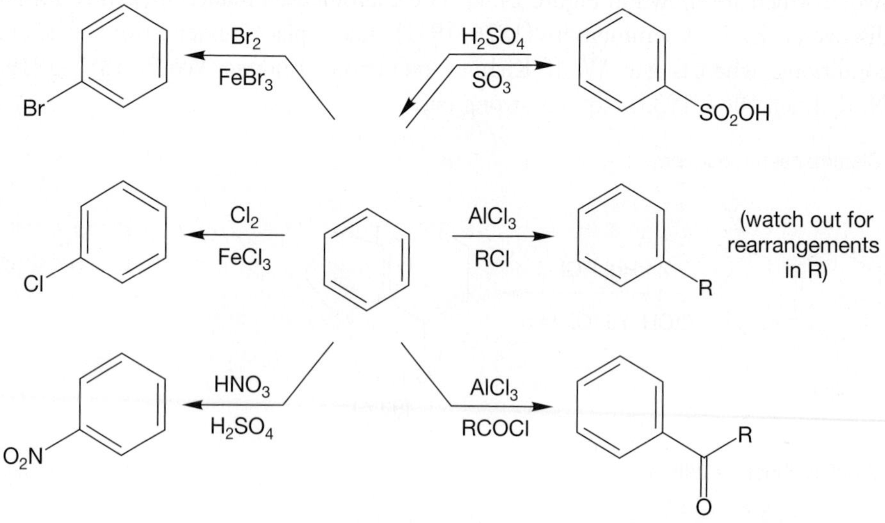

FIGURE 14.48 Aromatic substitution reactions encountered so far.

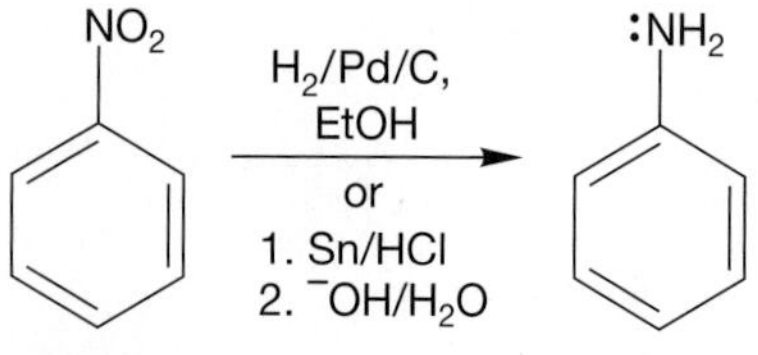

Nitro-benzene   **Aniline**

FIGURE 14.49 The reduction of nitrobenzene to aniline.

Although there are relatively few groups we can attach to a benzene ring directly, we can often transform one group into another through chemical reactions. It is nitrobenzene that is the key to many of these transformations. Nitrobenzene itself is relatively unreactive, but the nitro group can be reduced in many ways to the amino group (Fig. 14.49). This transformation opens the way for the synthesis of a host of new molecules.

Aminobenzene, generally known by its common name aniline (p. 243), can be treated with nitrous acid (HONO) or its sodium salt (Na$^+$ $^-$ONO) and HCl, to produce benzenediazonium chloride, which is explosive when dry, but a relatively safe material when wet (Fig. 14.50). A molecule of the structure R—N$_2^+$ is called a **diazonium ion**. We will soon use them in a number of important reactions.

Aniline derivatives are used to make azo dyes. Almost all dyes, including azo dyes and these natural dyes from the Fez Tannery in Morocco, are colored because they contain highly conjugated aromatic amines.

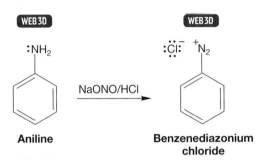

FIGURE 14.50 Reaction of aniline with nitrous acid to give benzenediazonium chloride.

PROBLEM 14.15 Draw resonance forms for benzenediazonium chloride.

The mechanism for the reaction that converts aniline into the diazonium salt contains many steps, but is really within our range now. A critical intermediate is dinitrogen trioxide, formed in equilibrium with nitrous acid by loss of water. When nitrous acid is protonated, a good leaving group, water, is created and can be displaced by a nitrite ion (Fig. 14.51).[2]

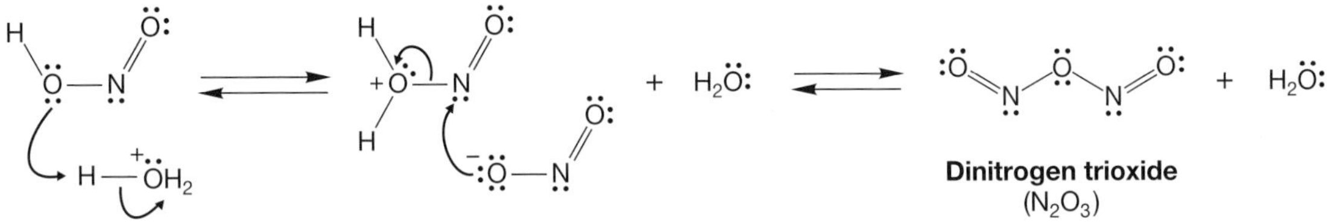

Dinitrogen trioxide (N$_2$O$_3$)

FIGURE 14.51 The formation of dinitrogen trioxide, N$_2$O$_3$.

The nucleophilic nitrogen of aniline can displace the good leaving group nitrite ($^-$ONO) from N$_2$O$_3$, which leads, after removal of a proton, to *N*-nitrosoaniline (Fig. 14.52).

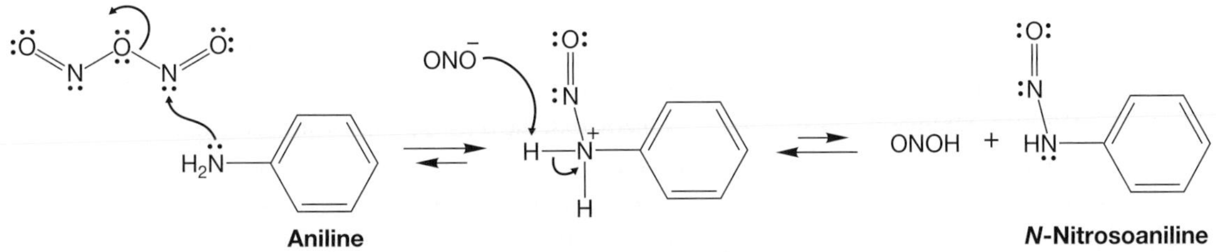

Aniline   ***N*-Nitrosoaniline**

FIGURE 14.52 Formation of a nitrosamine by displacement of nitrite from N$_2$O$_3$.

[2]This reaction (Fig. 14.51) is actually a bit more complicated than is shown in the figure. It involves addition of nitrite to the nitrogen–oxygen double bond followed by loss of water. We will explore this kind of mechanism in detail beginning in Chapter 16.

The *N*-nitrosoaniline equilibrates with a molecule called a diazotic acid, which loses water in acid to generate the diazonium ion (Fig. 14.53). These transformations are probably bewilderingly complex right now, even though they are made up of very simple reactions in sequence. It certainly takes some practice to get them right. However, many reaction mechanisms involve similar sequences of protonations, deprotonations, and additions and losses of water molecules. Now is the time to start practicing them. We will encounter many more when we discuss carbonyl chemistry in Chapters 16 through 19.

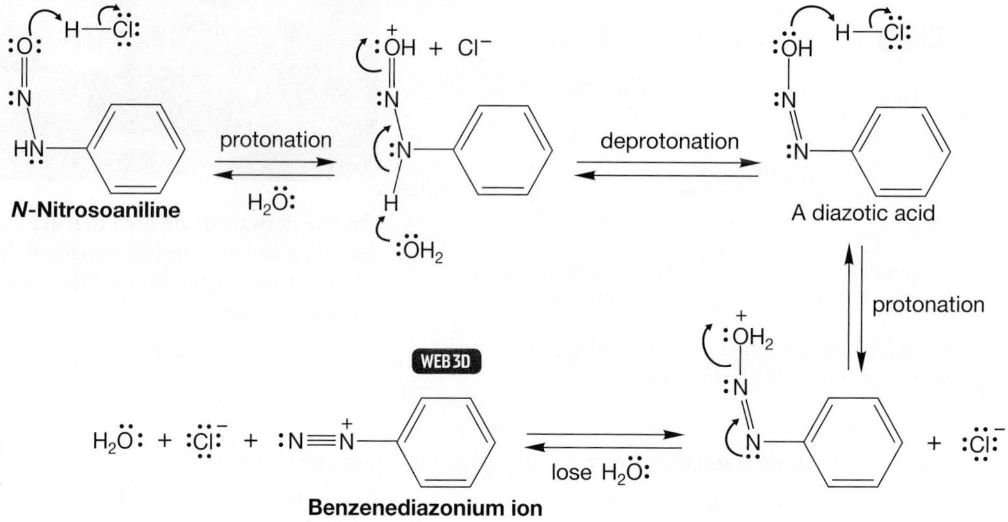

FIGURE 14.53 Loss of water from the protonated diazotic acid gives the diazonium ion, benzenediazonium chloride.

Benzenediazonium ion

ANILINE

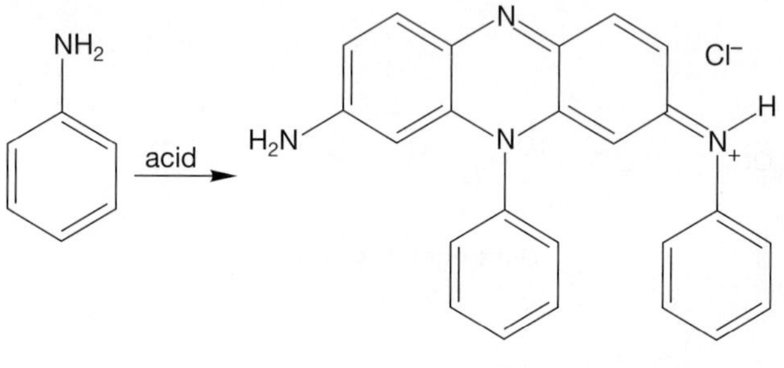

Pseudomauvine

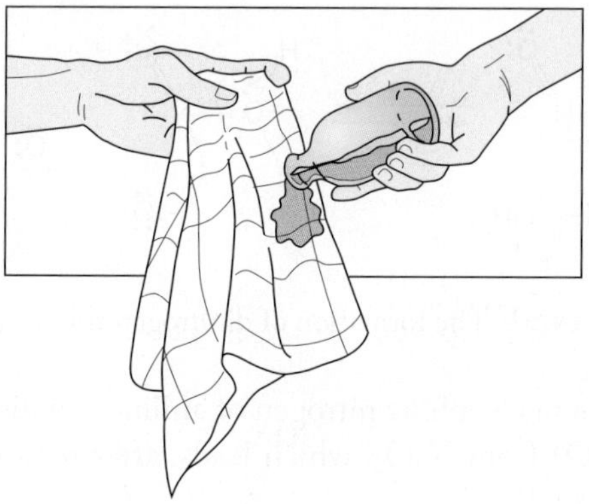

Aniline featured prominently in the development of the synthetic dye industry. In 1856, the 17-year-old William Henry Perkin, on holiday from the Royal College of Chemistry where he was studying with August Wilhelm Hofmann, was working in his home lab (isn't this what you do on your vacations?). As an unexpected outcome of a failed attempt to make quinine, Perkin obtained a black goop that stained his laboratory rag a beautiful mauve. This dye is the molecule shown above, pseudomauvine. Over Hofmann's objections, Perkin and his family set out to commercialize this dye. It was not a vast financial success, as colorfastness became a problem, and

mauve was very expensive, being roughly the same price per pound as platinum at the time. It was overtaken by other, related dyes made in similar ways from substituted anilines. Nonetheless, the serendipitous synthesis of mauve marks the beginnings of a great chemical industry.

It shouldn't be difficult to find the four molecules of aniline within the structure of pseudomauvine. It is harder to map out a general mechanism through which the four molecules come together; keep in mind that the necessary oxidations would be easy to accomplish under Perkin's crude reaction conditions.

The diazonium ion is important because it is the critical intermediate in a number of transformations lumped under the heading of the **Sandmeyer reaction**, named for Traugett Sandmeyer (1854–1922). Cuprous salts convert a diazonium ion into the related cyanide or halide. In recent times, numerous variations on the basic theme of the Sandmeyer reaction have improved yields and reduced side products. Figure 14.54 shows three typical Sandmeyer reactions.

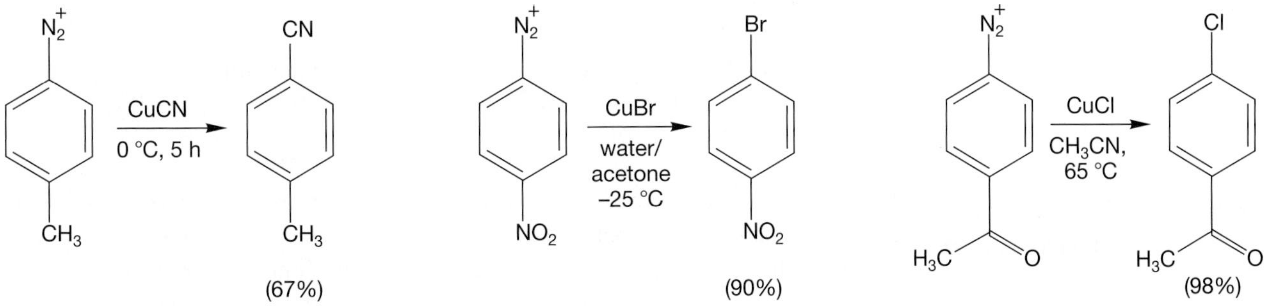

FIGURE 14.54 Three Sandmeyer reactions. Decomposition of the diazonium ion with cuprous salts leads to substitution.

Reagents other than cuprous salts are also effective. Figure 14.55 shows four such reactions. Treatment with potassium iodide (KI) or fluoboric acid (HBF$_4$) gives the iodide or fluoride, respectively (the Schiemann reaction; G. Schiemann 1899–1969), and hydrolysis gives phenols (p. 231). Heating with hypophosphorous acid (H$_3$PO$_2$) removes the diazonium group and replaces it with hydrogen.

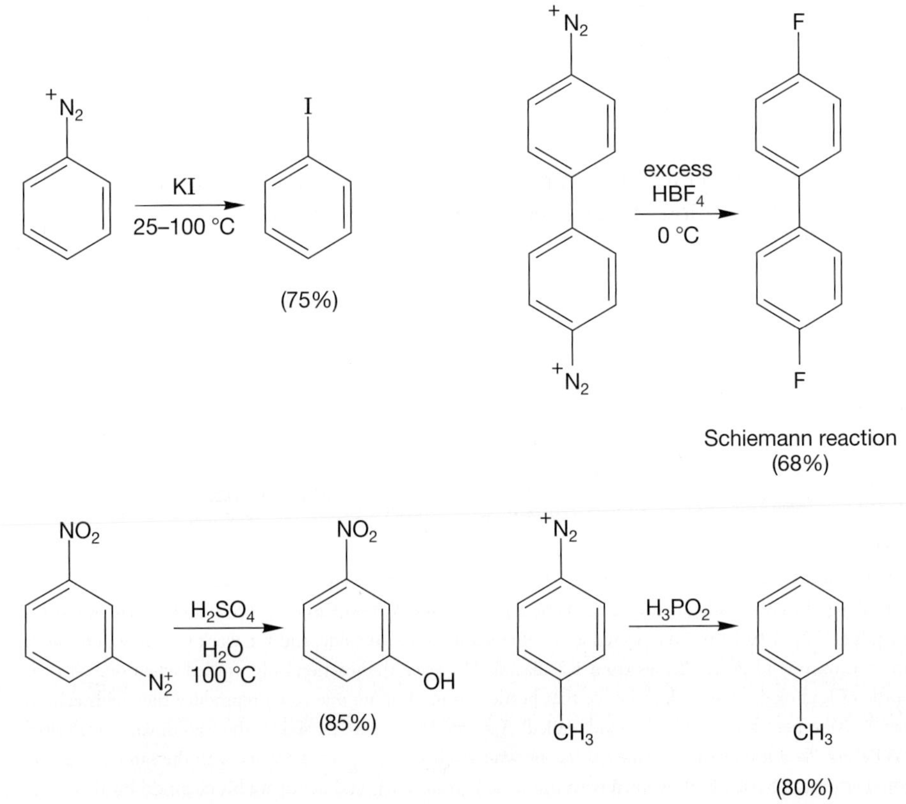

FIGURE 14.55 Other substitution reactions of diazonium ions.

Of course, you are now clamoring for the mechanisms for all these changes. Sadly, only the outlines of the mechanisms are known, and much remains to be filled in. The details of many reactions are imperfectly understood, and there remains a delightful (to some) atmosphere of magic about parts of organic chemistry. What are we to make of the admonition in one procedure for the Sandmeyer reaction to "stir the mixture with a lead rod." What happens if we use a glass rod? Although one would think that the reaction might still succeed, there are examples to the contrary, and perhaps microscopic pieces of lead do affect the success of the reaction. We'd use lead.[3]

PROBLEM 14.16 We reject out of hand the notion that any of these net displacement reactions could go by something as simple as an S_N2 reaction. Why?

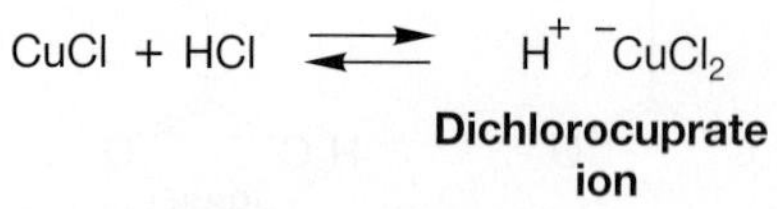

$$CuCl + HCl \rightleftharpoons H^+ \ ^-CuCl_2$$

Dichlorocuprate ion

FIGURE 14.56 Formation of the dichlorocuprate ion from CuCl and HCl.

In the Sandmeyer reactions involving copper, electron-transfer-mediated radical reactions are certainly involved. In HCl, cuprous chloride (CuCl) is in equilibrium with the dichlorocuprate ion $^-CuCl_2$ (Fig. 14.56). The dichlorocuprate ion can transfer an electron to the diazonium ion, which can then lose molecular nitrogen to give a phenyl radical. In turn, the phenyl radical can pluck a chlorine atom from a copper chloride molecule to give the chlorobenzene and a molecule of cuprous chloride. The mechanism for the formation of chlorobenzene in the Sandmeyer reaction is shown in Figure 14.57.

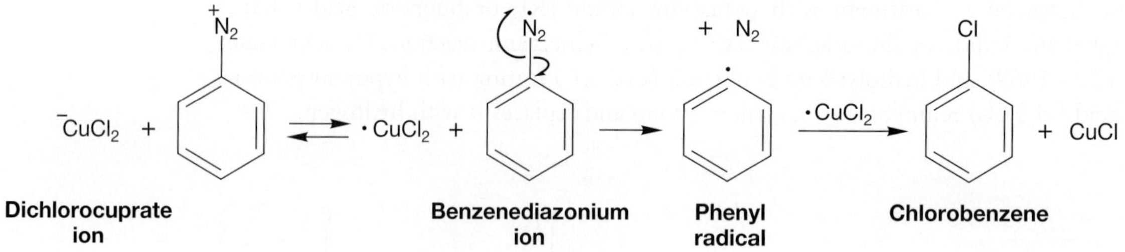

FIGURE 14.57 The mechanism of chlorobenzene formation.

Some Sandmeyer reactions apparently involve a phenyl cation formed by the irreversible S_N1 ionization of benzenediazonium chloride (Fig. 14.58).

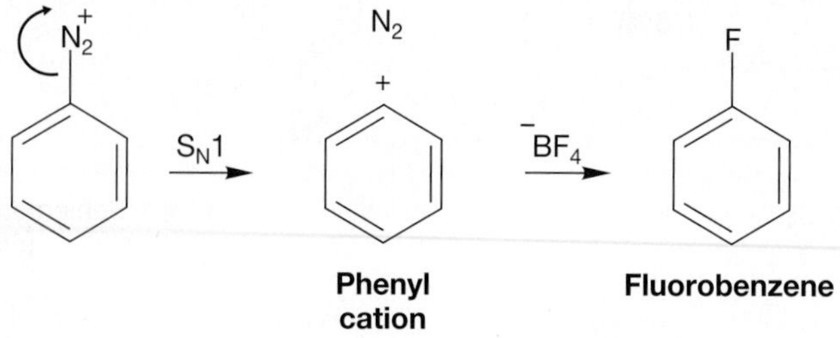

FIGURE 14.58 Some reactions of benzenediazonium chloride involve a phenyl cation.

---

[3] Speaking of lead and magic, two of MJ's chemical relatives, Milton Farber (b.1925) and Adnan Abdul-rida Sayigh (~1922–1980), once discovered a useful reaction of lead dioxide, and the work was duly published as an effective way to make alkenes from 1,2-diacids. However, the reaction only worked when one particular bottle of lead oxide was used, and once that bottle was used up no one could reproduce the yields achieved earlier. All other sources of lead dioxide failed! Apparently, the particle size of the lead dioxide was critical. (What's a "chemical relation"? That's someone who did his or her graduate work with the same professor as you. Farber and Sayigh both worked with the same person as MJ and hence are his chemical brothers.)

It doesn't matter that some mechanistic details are sketchy; the formation of the diazonium ion from aniline (and therefore ultimately from the easily made nitrobenzene) and its further transformations give you the ability to make a great number of substituted benzenes from very simple precursors.

PROBLEM 14.17 Draw resonance forms for the phenyl radical, Ph• (be careful).

PROBLEM 14.18 In the formation of diazonium ions from aromatic amines, side products called azo compounds are occasionally encountered, as shown below. Give a mechanism for the formation of these compounds.

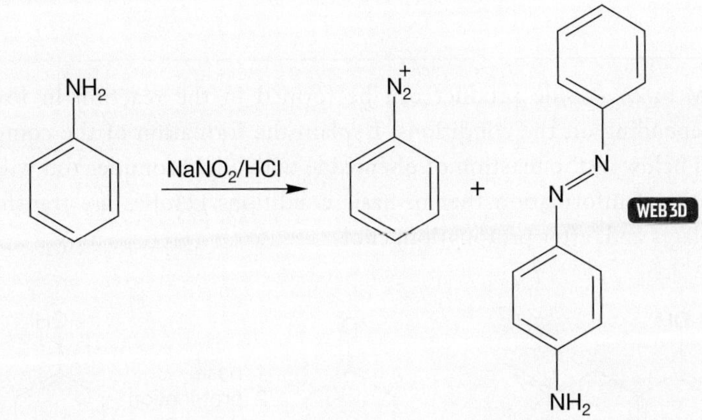
NH2
NaNO2/HCl
N2 +
+
N
N
WEB 3D
NH2

An azo compound

PROBLEM 14.19 Suggest synthetic procedures leading to each of these compounds. You may start with benzene, any inorganic reagent, and organic compounds containing no more than four carbon atoms.

(a)
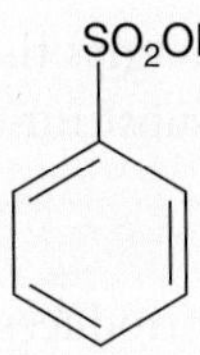
SO2OH

(b)
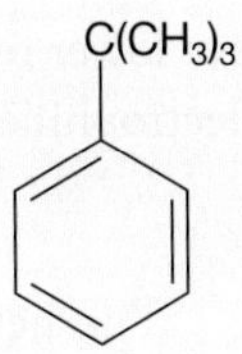
C(CH3)3

(c)
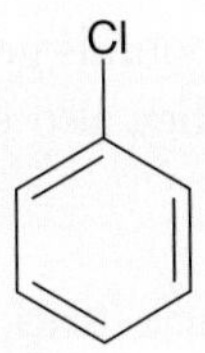
Cl

(d)
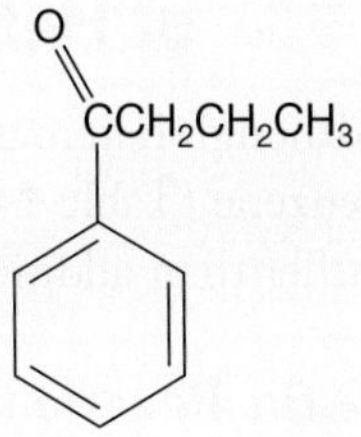
O
CCH2CH2CH3

(e)
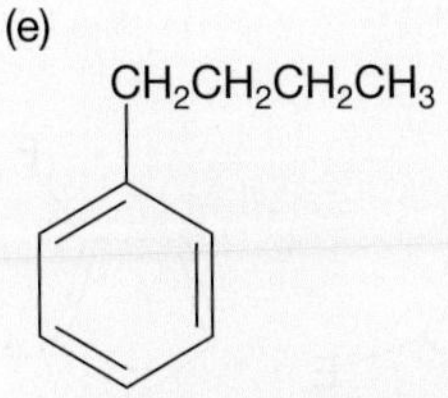
CH2CH2CH2CH3

(f)
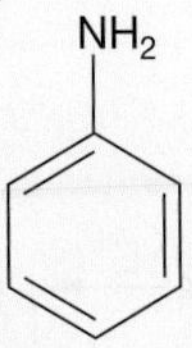
NH2

(g)
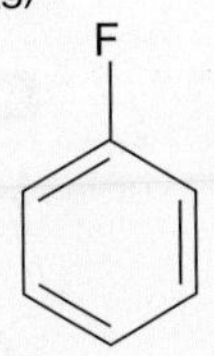
F

(h)
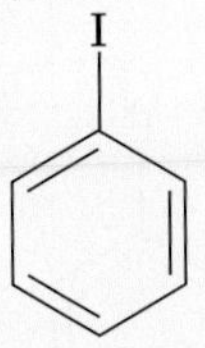
I

(i)
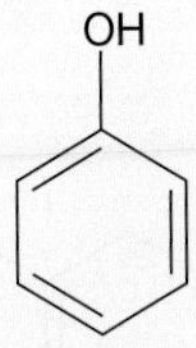
OH

PROBLEM 14.20 Phenols made by the Sandmeyer reaction can be used further in synthesis. First, explain why phenol ($pK_a = 10$) is a much stronger acid than cyclohexanol ($pK_a = 16$).

PROBLEM 14.21 Aryl ethers can be made by the following reaction. Write an arrow formalism mechanism for each step.

OH → (NaH, 0 °C, THF) O⁻ → (27 °C, CH₃OCH₂CH₂OCH₃, with allyl bromide) O-allyl

Phenol   **Phenoxide**   **3-Phenoxypropene**
(allyl phenyl ether)
(100%)

PROBLEM 14.22 A side product can be formed in the reaction in Problem 14.21, depending on the conditions. Explain the formation of the compound **A** shown below in the reaction of phenoxide with allyl bromide. You will need the additional information that in basic conditions ketones are transformed into enolates and, after protonation, enols.

A   **Ketone**   1. base   2. protonation   **Enol**

14.8 Electrophilic Aromatic Substitution of Heteroaromatic Compounds

Although heteroaromatic compounds generally have lower resonance energies than benzene (Table 14.1, p. 629), they also undergo electrophilic aromatic substitution rather than addition.

14.8a Reactions of Pyridine and Pyrrole The heteroatom introduces a complication in aromatic substitution not present in benzene. Benzene has six equivalent positions, but substitution could occur at two positions in pyrrole and three in pyridine (Fig. 14.59).

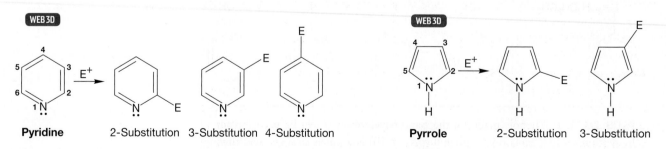

Pyridine   2-Substitution 3-Substitution 4-Substitution   **Pyrrole**   2-Substitution 3-Substitution

FIGURE 14.59 Possible positions for electrophilic aromatic substitution in pyridine and pyrrole.

As with benzene, electrophilic attack on pyridine gives an intermediate in which three atoms share the positive charge (Fig. 14.60). Substitution at the 2- or 4-position (ortho or para to the nitrogen in pyridine) places a partial positive charge on the relatively electronegative nitrogen atom. Substitution at the 3-position (meta to the nitrogen) does not and is accordingly favored.

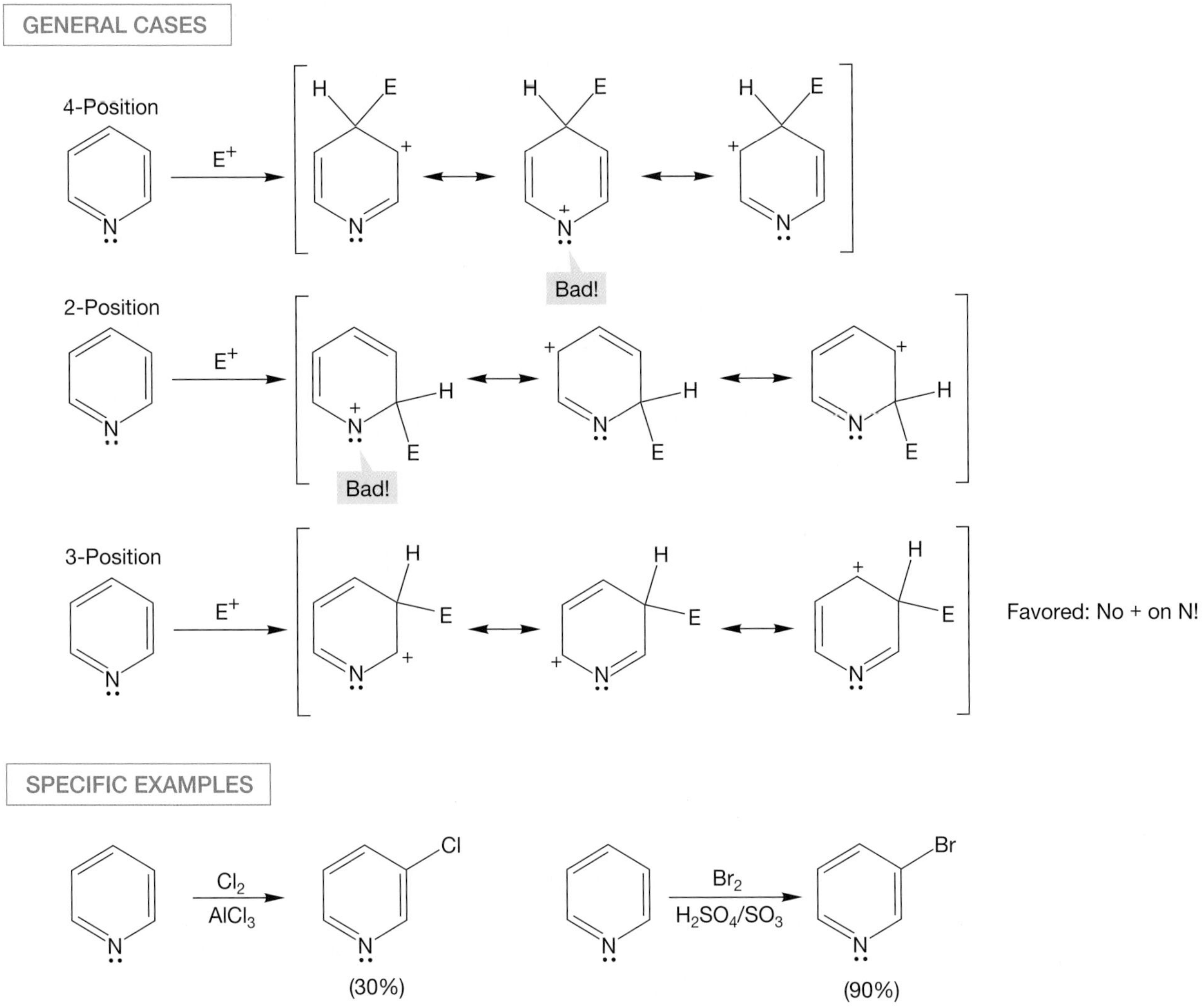

GENERAL CASES

SPECIFIC EXAMPLES

FIGURE 14.60 In pyridine, substitution at the 3-position is favored, although the reaction is slower than with benzene.

This rate-diminishing effect is accentuated if the nitrogen is protonated, as it very often is under the strongly acid conditions necessary to carry out substitution on pyridine. Therefore it is not surprising that substitution of pyridine takes place more slowly than analogous reactions involving benzene. Pyridine itself substitutes about as rapidly as nitrobenzene, 10^6 times more slowly than benzene. The pyridinium ion substitutes an impressive 10^{18} times more slowly than benzene.

A similar analysis serves to locate the more favorable position for substitution on pyrrole. Addition at either carbon 2 or 3 (Fig. 14.59) allows the nitrogen to help bear the positive charge. However, in the intermediate produced by substitution at carbon 2, two ring carbons share the charge with the nitrogen, whereas substitution at carbon 3 leads to an intermediate in which only one carbon shares the charge with the nitrogen. Addition to the 2-position is favored, although the formation of mixtures is commonly observed (Fig. 14.61).

Pyrrole is more reactive than benzene or pyridine, because pyrrole is less well stabilized by resonance than the six-membered ring molecules (Table 14.1).

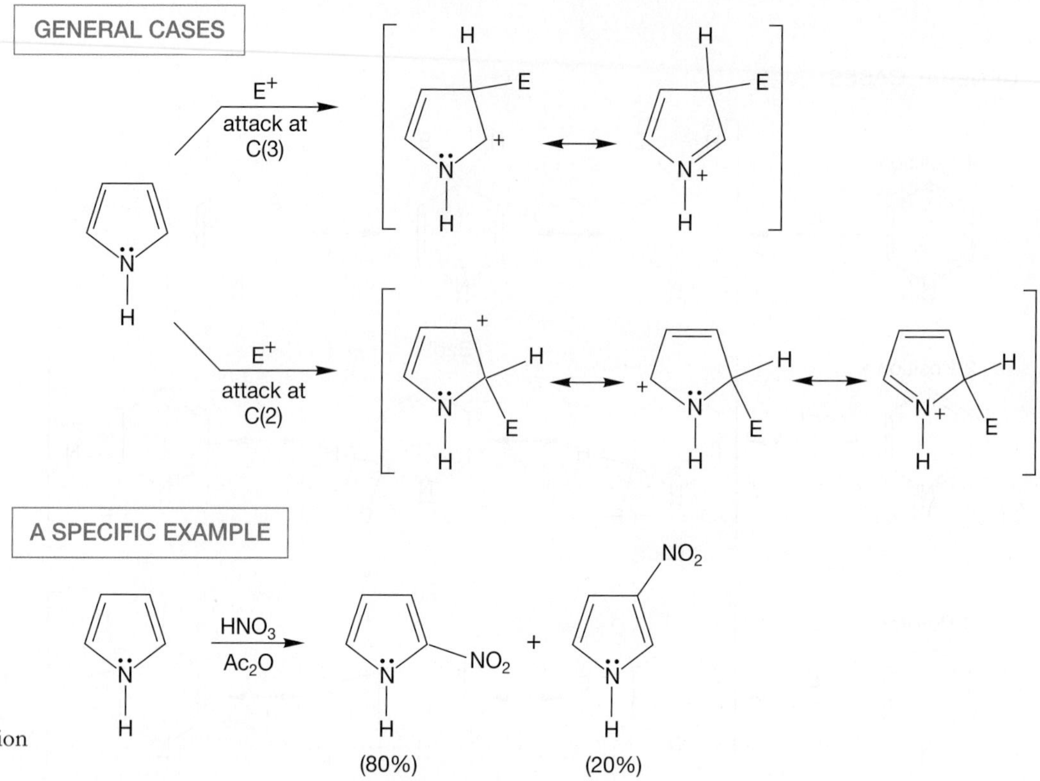

GENERAL CASES

A SPECIFIC EXAMPLE

FIGURE 14.61 In pyrrole, substitution at the 2-position is favored.

14.8b Reactions of Furan and Thiophene   These compounds are very reactive toward electrophiles, E^+. As in pyrrole, aromatic substitution is directed first to the 2-position (Fig. 14.62), although mixtures are sometimes obtained in which some 3-substitution appears.

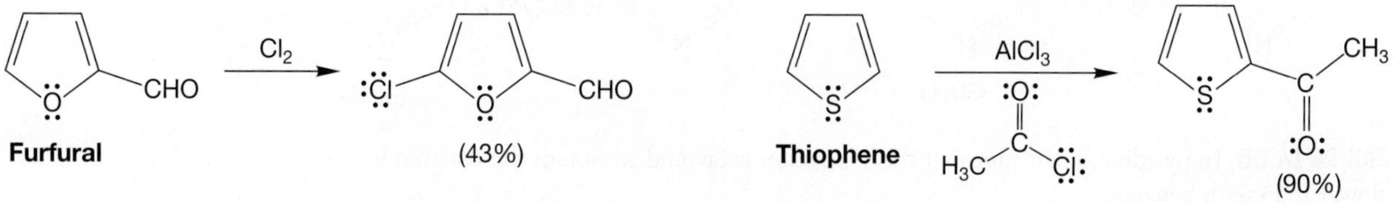

FIGURE 14.62 Electrophilic substitution reactions of furan and thiophene.

PROBLEM 14.23 Explain why 2-substitution is preferred to 3-substitution in electrophilic aromatic substitution of furan.

PROBLEM SOLVING

All problems of this kind ask you to make comparisons, but they usually require you to write mechanisms so that you can see the molecules that must be compared. It may seem obvious, but attacking such problems is best done by carefully writing out the mechanisms in parallel so that you can see the possible points of comparison. Don't just think, write.

PROBLEM 14.24 In addition to the mechanism following the path shown for pyrrole in Figure 14.61, there is another reasonable mechanism for aromatic substitution of furan. Suggest one. *Hint*: Think simple! What are the possible reactions between electrophiles and double bonds (Chapters 9 and 10)? Use the chlorination of furan as an example.

PROBLEM 14.25 There is a difficulty with the answer to Problem 14.24. Find what is wrong (or unlikely) and suggest an alternative mechanism. *Hint*: Recall what you know of the stereoelectronic requirements of the E2 reaction (Section 7.9, p. 301).

14.9 Disubstituted Benzenes: Ortho, Meta, and Para Substitution

We have already mentioned (Problem 14.9) that the synthesis of toluene from benzene is made difficult because once a significant amount of toluene is formed, it competes successfully with the unsubstituted benzene in further methylation reactions. Unless one is careful to use a vast excess of benzene, the result is a horrible mixture of benzene, toluene, dimethylbenzenes (xylenes), and even tri- and tetra-substituted molecules.

WORKED PROBLEM 14.26 If one attempts to methylate benzene completely with $CH_3Cl/AlCl_3$, the product is a greenish solid of the formula $C_{13}H_{21}AlCl_4$. Propose a structure for the compound and a mechanism for its formation.

ANSWER This problem involves a sequence of Friedel–Crafts alkylations. Toluene is made first, then the xylenes, then trimethylbenzenes, and so on, until hexamethylbenzene is formed.

What happens if another Friedel–Crafts reaction occurs? The product is the heptamethylcyclohexadienyl cation tetrachloroaluminate salt, and this material is stable under the reaction conditions. Why? The resonance forms contributing

(continued)

to the structure are all tertiary carbocations, and there is no proton that can be lost to give a new aromatic system. All positions are occupied by methyl groups.

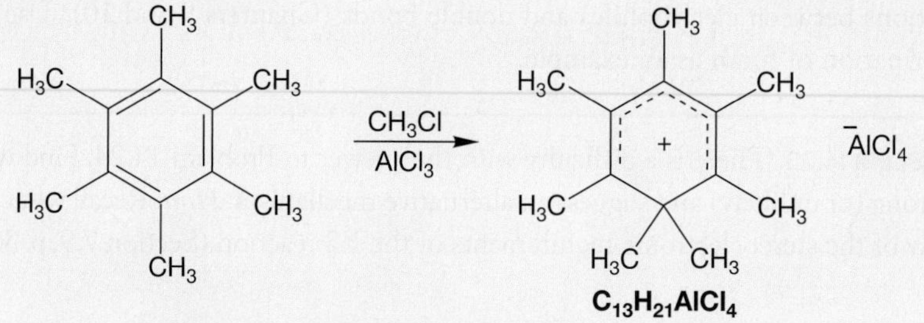

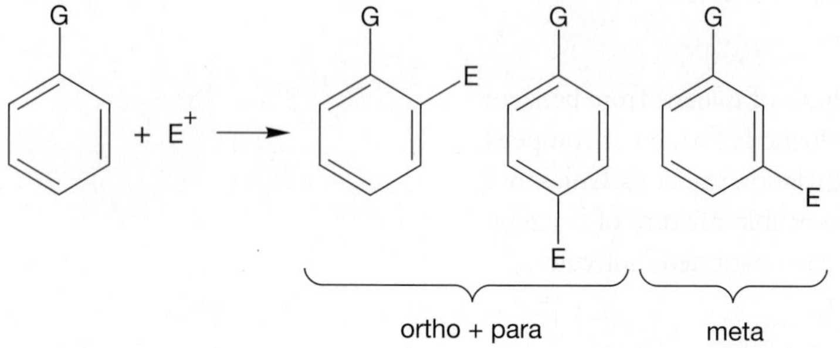

$C_{13}H_{21}AlCl_4$

FIGURE 14.63 The three possible disubstituted benzenes.

TABLE 14.2 Patterns of Further Reactions of Substituted Benzenes

G	Position[a]	Relative Rate
NH_2, NHR, NR_2	o/p	Very fast
OH	o/p	Very fast
OR	o/p	Very fast
R (alkyl)	o/p	Fast
NH—C(=O)R′	o/p	Fast
H (benzene)		1
I, Cl, Br, F	o/p	Slow
C(=O)R′ (R′ = OH, OR, NH2, Cl, alkyl, aryl)	m	Slow
S(=O)(=O)—OH	m	Slow
N^+(=O)(O^-)	m	Slow
C≡N	m	Slow
$^+NR_3$	m	Slow

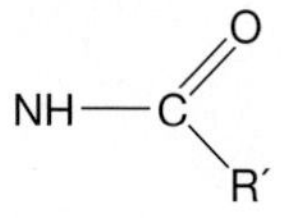

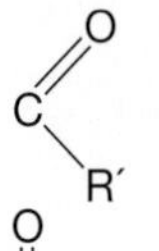

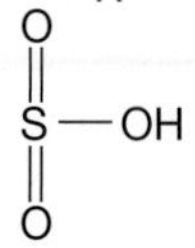

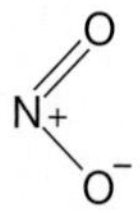

[a]Ortho = o, para = p, meta = m.

The general question now arises of how the substituted benzenes we have learned to make will react with electrophilic reagents. What are the possibilities? There are three isomers of disubstituted benzene, ortho, meta, and para, as shown in Figure 14.63.

Substituents influence the distribution of the three isomeric disubstituted products dramatically. Further substitution of monosubstituted benzenes by no means proceeds to give a statistical distribution of two parts ortho product, two parts meta product, and one part para product. In fact, if one looks at the reactions of a number of monosubstituted benzenes, a pattern emerges. Some groups (G) substitute predominantly ortho and para, while for other groups meta is dominant. There is also a correlation between the position of further substitution and the rate of the reaction. Monosubstituted benzenes that give mainly ortho and para products usually react faster than benzene. Monosubstituted benzenes that give mainly meta product react more slowly than benzene (Table 14.2).

A look at the general mechanism of substitution shows how different groups influence the position of subsequent reaction. We can also see how these orientation (regiochemical) effects are connected to the rates of the reactions. Let's look first at methoxybenzene (anisole). Methoxy (CH_3O) is one of the groups most effective at directing further substitution to the ortho and para positions and at increasing the rate of reaction. We will work through the mechanism for meta and para substitution and look for a point of difference. Figure 14.64 shows the mechanism for meta substitution of anisole by a generic E^+ reagent, and compares the reaction to that of benzene. As usual, a resonance-stabilized cyclohexadienyl cation is formed first, and then deprotonated to give the substituted product.

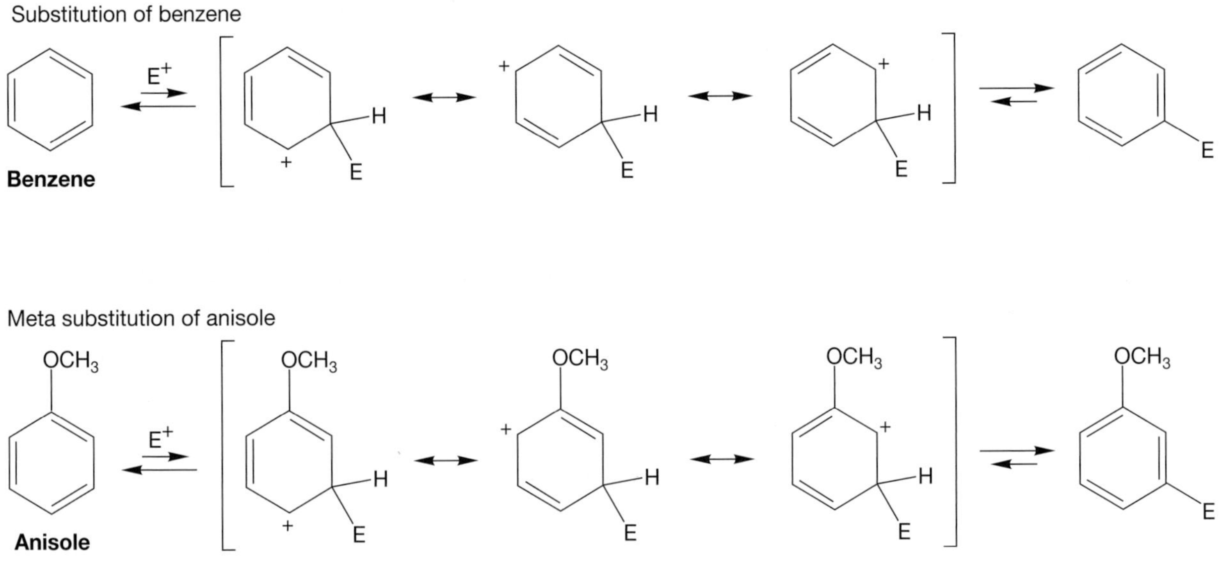

FIGURE 14.64 The intermediate cyclohexadienyl cations involved in the substitution of benzene and the meta substitution of anisole.

At first glance, substitution at the para position of anisole seems similar to meta substitution (Fig. 14.65). But look carefully at the intermediate in para substitution and remember our discussion of the stabilization of carbocations that are adjacent to an oxygen (p. 378). In the intermediate from para substitution, there is a fourth resonance form in which oxygen shares the positive charge

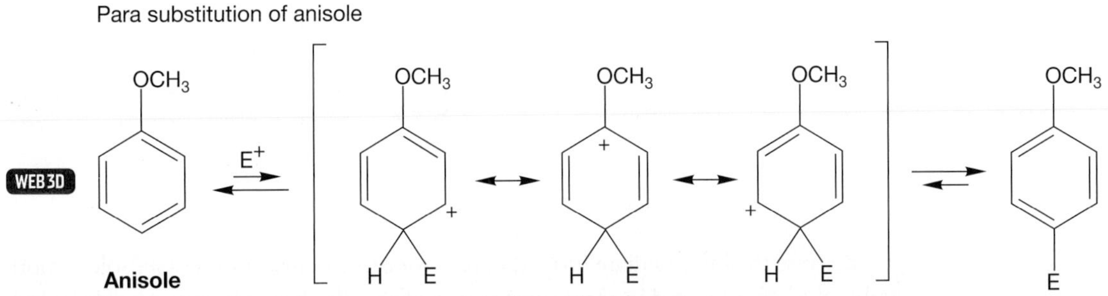

FIGURE 14.65 The intermediate cyclohexadienyl cation involved in para substitution of anisole.

(Fig. 14.66). This fourth form stabilizes the cation relative to that formed in the meta substitution or in the substitution of benzene.

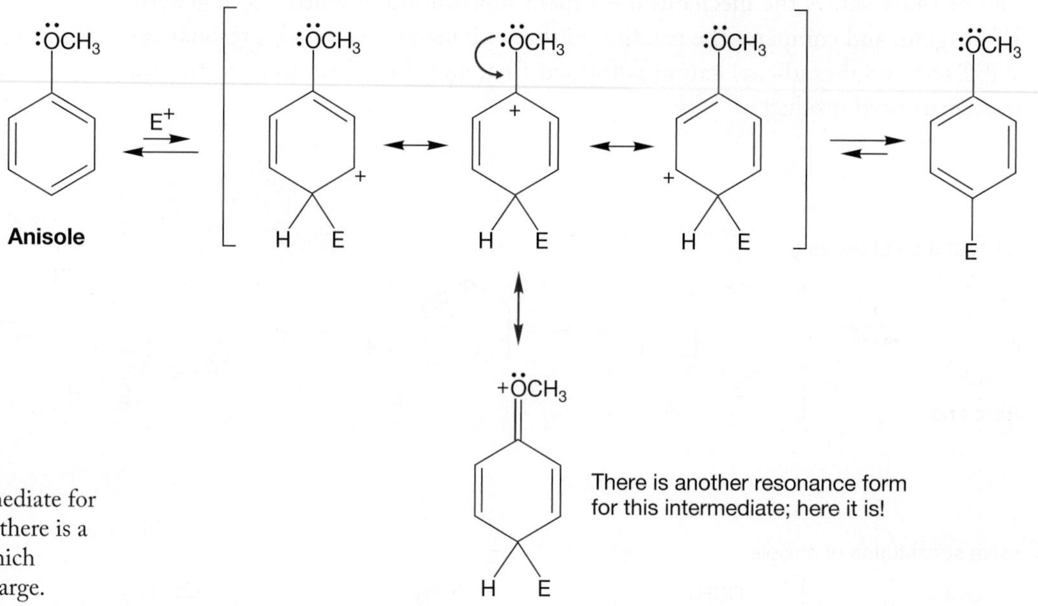

FIGURE 14.66 In the intermediate for para substitution of anisole, there is a fourth resonance form in which oxygen bears the positive charge.

There is another resonance form for this intermediate; here it is!

WORKED PROBLEM 14.27 Draw the resonance structures for the intermediate formed by substitution of anisole at the ortho position.

ANSWER The figure shows the four resonance forms of the cyclohexadienyl cation intermediate. As in the para case, the methoxy group is in a position to help share the positive charge.

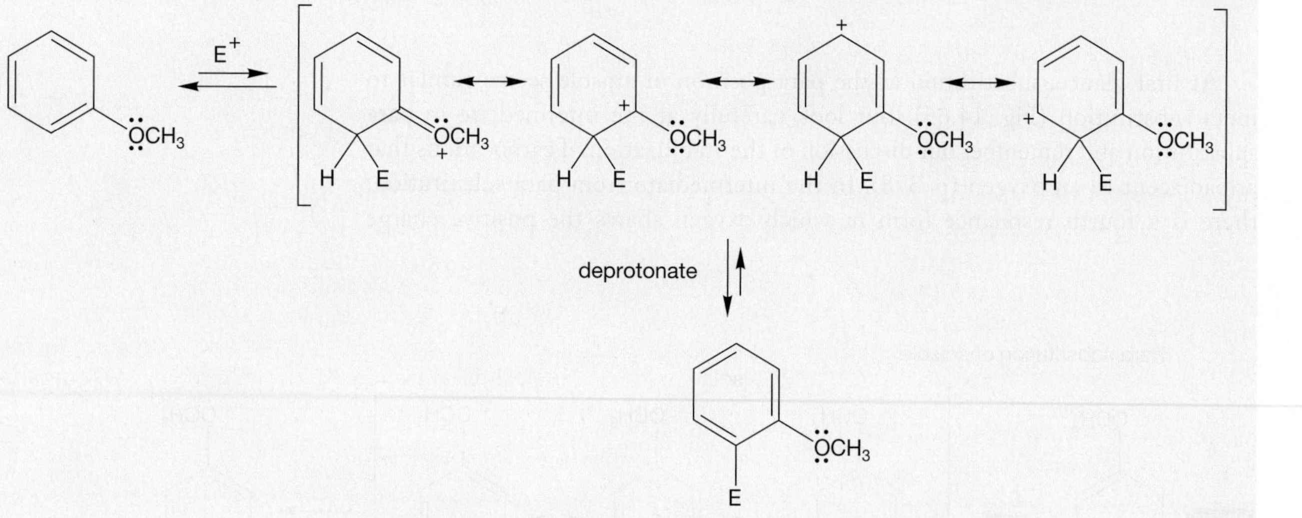

Because the intermediate formed in para (or ortho) substitution of anisole is more stable than that formed in meta substitution of anisole or in substitution of benzene, the transition state leading to the para (or ortho) intermediate is more stable than the transition states for meta substitution of anisole or substitution of benzene itself.

As a result, we can expect para (or ortho) substitution to occur more rapidly than the other two reactions (Fig. 14.67). In fact, the chlorination of anisole gives only

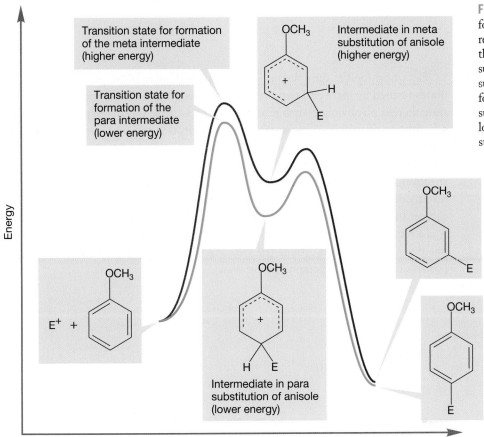

FIGURE 14.67 The intermediate for para substitution, with four resonance forms, is lower in energy than the intermediate for meta substitution (three forms) or that for substitution of benzene (also three forms). The transition state for para substitution of anisole will also be lower in energy than the transition state for meta substitution.

o-chloroanisole and *p*-chloroanisole as products (Fig. 14.68). The relatively low energy of the intermediate formed by addition to anisole at the ortho or para substitution (four resonance forms, Fig. 14.66) means that this reaction will prevail over meta attack (the intermediate has only three resonance forms, Fig. 14.64). Moreover, if we compare the rate of formation of the lower-energy intermediate to the formation of the benzene intermediate (also only three resonance forms), we would expect the lower-energy intermediate from anisole to be formed faster, as is indeed observed.

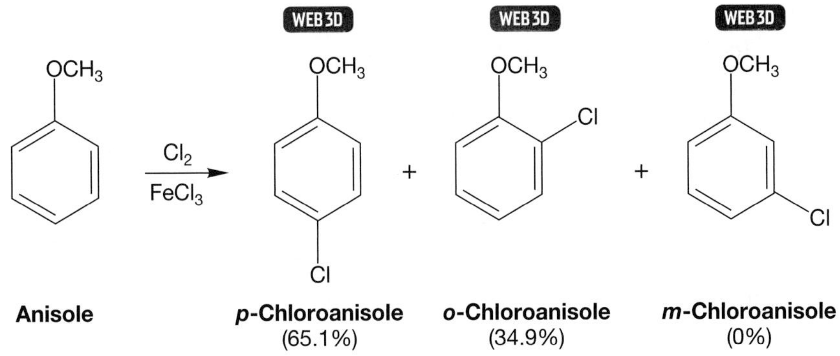

FIGURE 14.68 The chlorination of anisole.

PROBLEM 14.28 Provide a possible explanation for why ortho substitution is often less favorable than para substitution, even though ortho substitution is favored statistically by a 2:1 factor?

PROBLEM 14.29 Hold it! Explain *exactly why* we "expect" that attack on anisole by an electrophilic reagent to form a relatively low energy, resonance-stabilized intermediate will be faster than attack on benzene? Are we not confusing thermodynamics with kinetics?

Let's now look at the substitution of toluene. Figure 14.69 shows the intermediates formed by meta and para substitution using a generic electrophile, E⁺. This time there is no obvious fourth resonance form, as there was in the para (and ortho) substitution of anisole. Why is ortho/para substitution still preferred, and why does toluene react faster than unsubstituted benzene (Table 14.2 p. 656)? Look at the positions sharing the positive charge. In the intermediate formed from meta substitution, the three positions sharing the positive charge are all secondary. In the intermediate formed by para substitution, one of the three positions is a tertiary carbon and is therefore more effective in stabilizing a positive charge. So there really is a fourth resonance form! Recall the description of the stabilization of adjacent positive charge by alkyl groups (p. 376). An alkyl group can stabilize an adjacent positive charge in a hyperconjugative way.

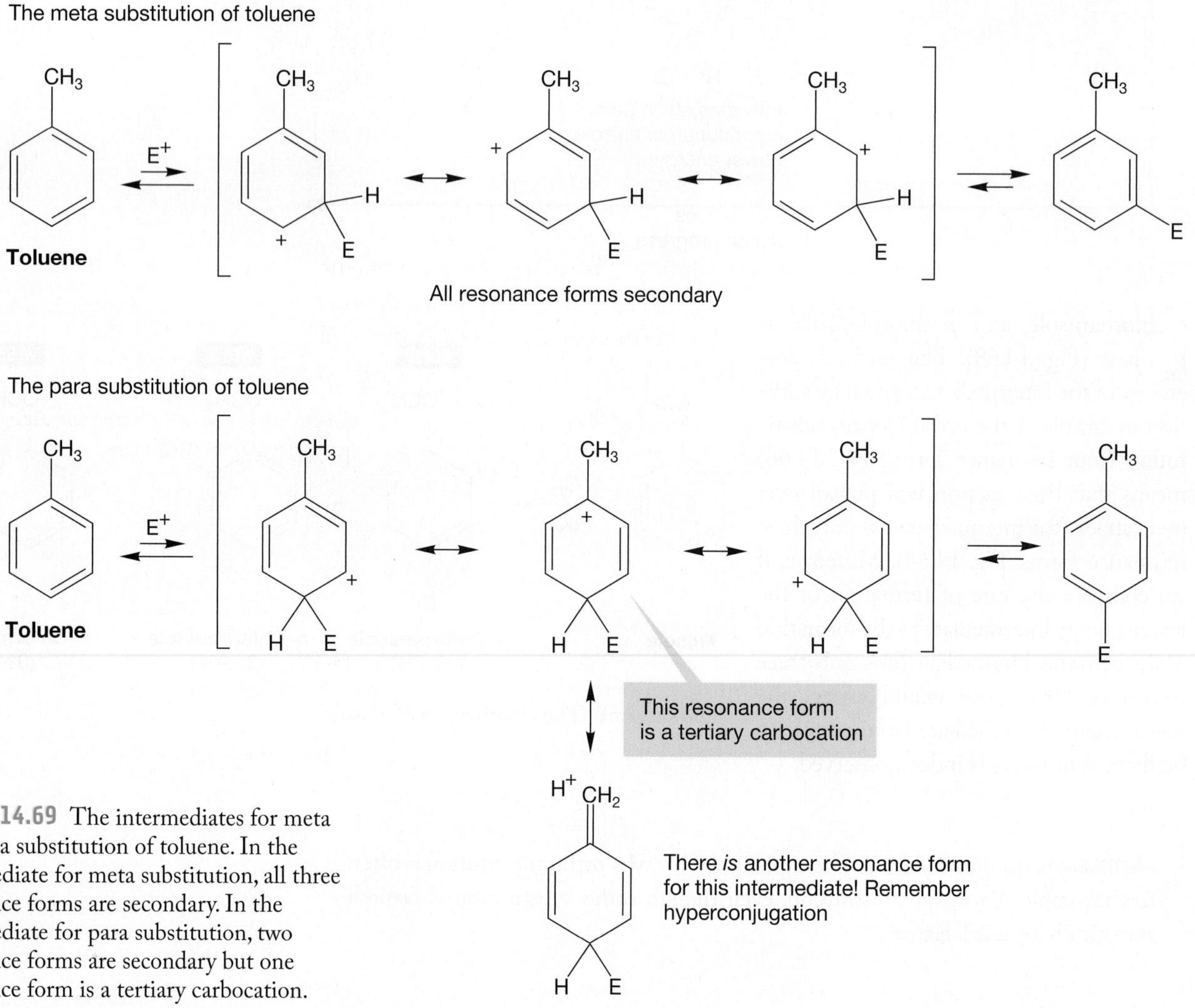

The meta substitution of toluene

All resonance forms secondary

The para substitution of toluene

This resonance form is a tertiary carbocation

There *is* another resonance form for this intermediate! Remember hyperconjugation

FIGURE 14.69 The intermediates for meta and para substitution of toluene. In the intermediate for meta substitution, all three resonance forms are secondary. In the intermediate for para substitution, two resonance forms are secondary but one resonance form is a tertiary carbocation.

The intermediate for para substitution, *and the transition state leading to it*, are lower in energy than those involved in meta substitution (Fig. 14.70). The transition state

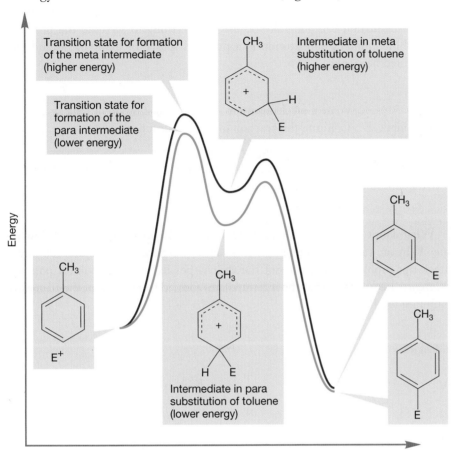

FIGURE 14.70 The intermediate and transition states for para substitution of toluene are lower in energy than those for meta substitution.

is also lower than the transition state for substitution of benzene. When a group on an aromatic ring lowers the transition state compared to that for benzene, the ring is said to be activated. This theoretical analysis is borne out in practice. Figure 14.71 shows some typical substitution reactions of toluene.

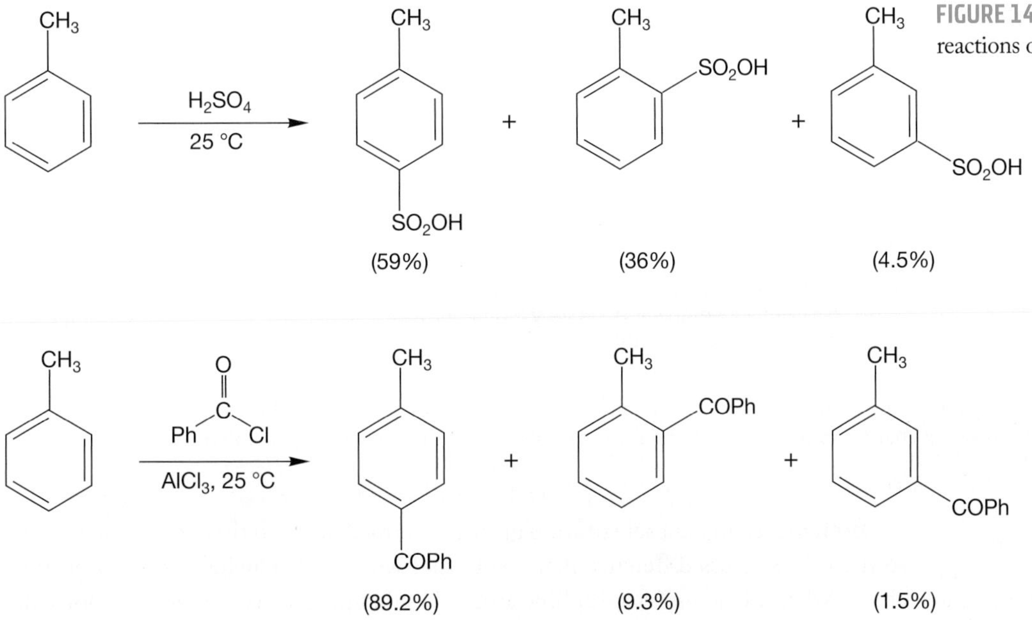

FIGURE 14.71 Some substitution reactions of toluene.

So, for many groups, meta substitution does not present a problem. However, separation of ortho and para products is a problem, especially because the two are often formed in comparable amounts. Happily, physical methods such as crystallization and distillation can generally be used successfully to separate the products. Electrophilic aromatic substitution is a practical source of many ortho and para disubstituted benzenes.

PROBLEM 14.30 Write a mechanism for the ortho substitution of toluene by an electrophile, E^+. Compare this reaction to meta substitution.

PROBLEM 14.31 Explain why bromobenzene and aniline direct electrophilic aromatic substitution reactions ortho/para.

What about groups that direct substitution to the meta position and slow the rate? Why are the effects we have seen so far not general? Once more, the way to attack this problem is to write out the various possibilities and look for points of difference. Let's look first at the intermediates formed by reaction of the trimethylanilinium ion in Figure 14.72.

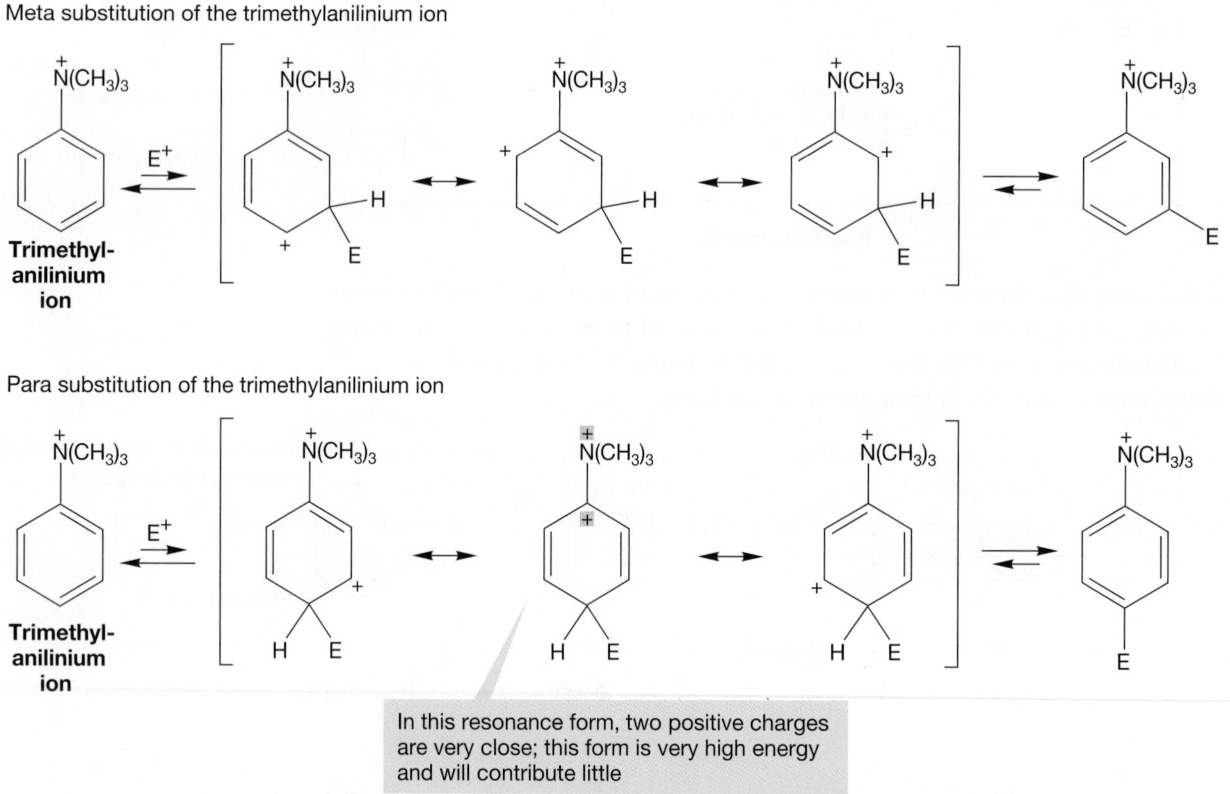

Meta substitution of the trimethylanilinium ion

Para substitution of the trimethylanilinium ion

In this resonance form, two positive charges are very close; this form is very high energy and will contribute little

FIGURE 14.72 Meta and para substitution of the trimethylanilium ion. Both intermediates have three resonance forms, but the adjacent positive charges in the intermediate for para substitution are especially destabilizing.

Both meta and para substitution give an intermediate with three resonance forms, so there is no gross difference here. Note that we are introducing a second positive charge when we do *any* electrophilic aromatic substitution reaction on this molecule.

This second charge is bound to slow the reaction relative to substitution of benzene itself. In this case the aromatic ring is strongly deactivated with respect to electrophilic substitution. Now we know why this substitution reaction is especially slow, but we can't yet explain why the meta position is preferred. The difficulty in para substitution is that the reaction places the two positive charges especially close together. In para substitution, the positive charge produced by reacting with the electrophile resides partly on the carbon bearing the original substituent. When this substituent can stabilize the charge through resonance, as in anisole or toluene, para substitution will be favored. When it brings two positive charges together, as in the trimethylanilinium ion, para substitution is decidedly disfavored for electrostatic reasons.

PROBLEM 14.32 Draw all of the resonance structures of the intermediate formed by ortho substitution of the trimethylanilinium ion.

PROBLEM 14.33 Use an Energy versus Reaction progress diagram to compare ortho and meta substitution on the trimethylanilinium ion.

So the situation is clear for the trimethylanilinium ion. Meta substitution, which does not place two like charges on adjacent carbons, is favored over para (or ortho) substitution, which does. Figure 14.73 shows one example: bromination of the trimethylanilinium ion.

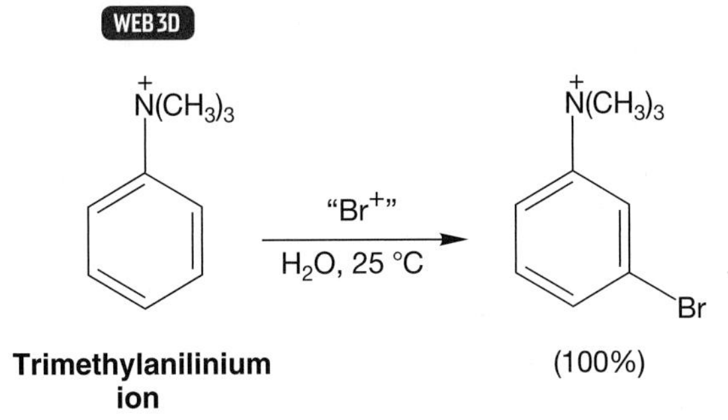

FIGURE 14.73 Bromination of the trimethylanilinium ion.

What about the other meta-directing groups? They do not bear obvious positive charges. Nitrobenzene is a good example. Nitrobenzene is so deactivated with respect to electrophilic substitution that it can be used as a solvent for Friedel–Crafts reactions of other aromatic compounds and not suffer attack itself. When substitution does occur on nitrobenzene, it is invariably largely in the meta position (Table 14.2; Fig. 14.74).

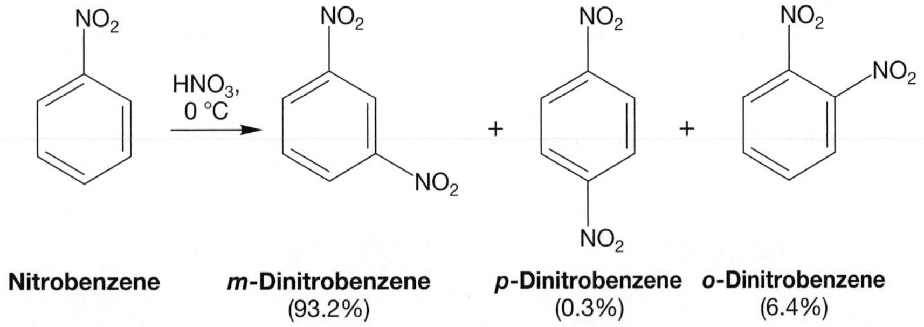

FIGURE 14.74 Nitrobenzene nitrates very slowly at 0 °C, and almost entirely in the meta position.

Yet there is no obvious positive charge in the starting material, as there is in the trimethylanilinium ion we analyzed before. However, a close look at the structure of the nitro group reveals the problem, which has been hidden by the obfuscation of nitro as NO_2. If you draw out a nitro group carefully, you will see (1) that there is no good uncharged structure, and (2) that the nitrogen always bears a positive charge:

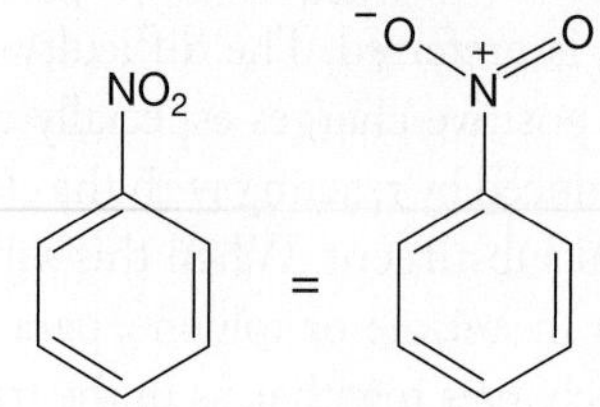

In nitrobenzene, there is a positive charge adjacent to the ring

So there really is a positive charge adjacent to the ring in nitrobenzene. Therefore the same factors that lead to meta substitution and a slow rate in the trimethylanilinium ion will operate for nitrobenzene as well (Fig. 14.75).

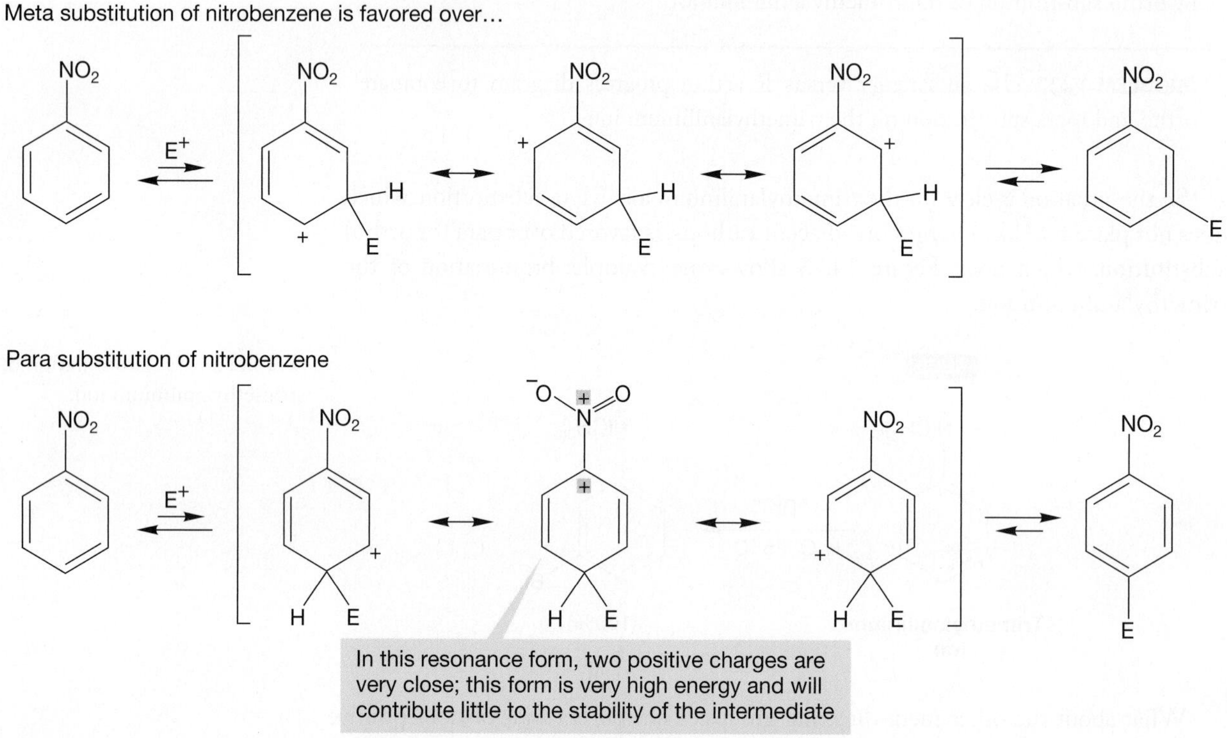

Meta substitution of nitrobenzene is favored over...

Para substitution of nitrobenzene

In this resonance form, two positive charges are very close; this form is very high energy and will contribute little to the stability of the intermediate

FIGURE 14.75 The resonance description for the nitro group reveals that nitrogen bears a full positive charge. Electrophilic aromatic substitution will be directed to the meta position. The ring is deactivated.

We still need to discuss the other meta directors in Table 14.2, those with a carbonyl group attached to the ring. Ethyl benzoate ($C_6H_5COOCH_2CH_3$) is a good example. It too substitutes slowly, and predominantly in the meta position (Fig. 14.76).

WEB 3D

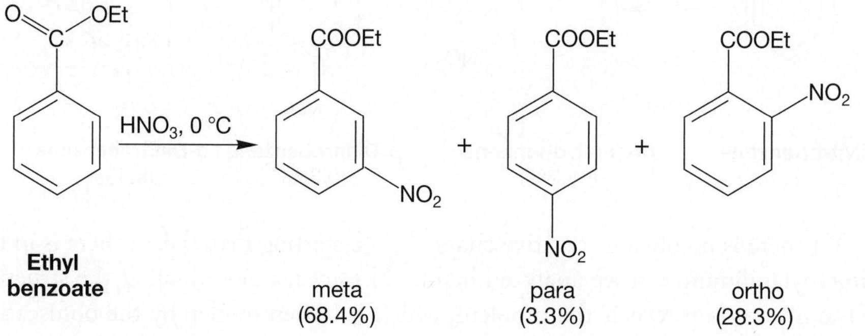

FIGURE 14.76 The nitration of ethyl benzoate gives mainly meta product.

Ethyl benzoate

meta (68.4%)

para (3.3%)

ortho (28.3%)

The explanation for this behavior again comes from a close look at the structure of the group substituting the benzene ring, in this case, the carbonyl group. Although there are no hidden full positive charges as in the nitro group, the carbon–oxygen double bond is highly polar. Because oxygen is much more electronegative than carbon, the dipole in the bond will place a partial positive charge on the carbonyl carbon attached to the ring (Fig. 14.77).

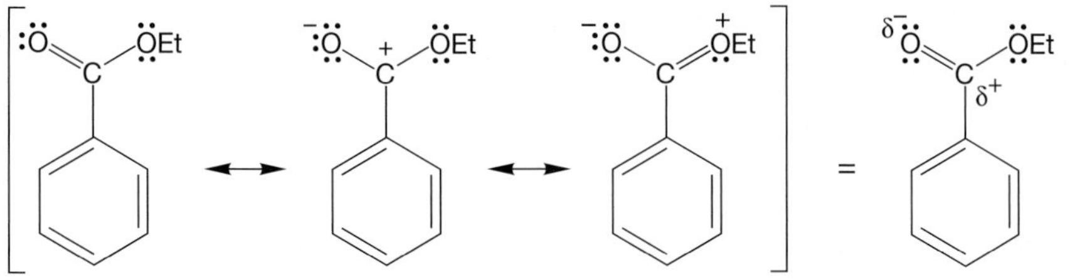

FIGURE 14.77 The carbonyl group is polar, with the carbon bearing a partial positive charge.

So, once again, ortho or para substitution of an electrophile results in a less stable intermediate. This time it is not two full positive charges that must be adjacent, but a full and partial positive charge. No matter—meta substitution, which keeps the charges further apart, will still be favored (Fig. 14.78). The rate of the reaction will be slow relative to that for benzene. An aromatic ring with a carbonyl group attached is deactivated.

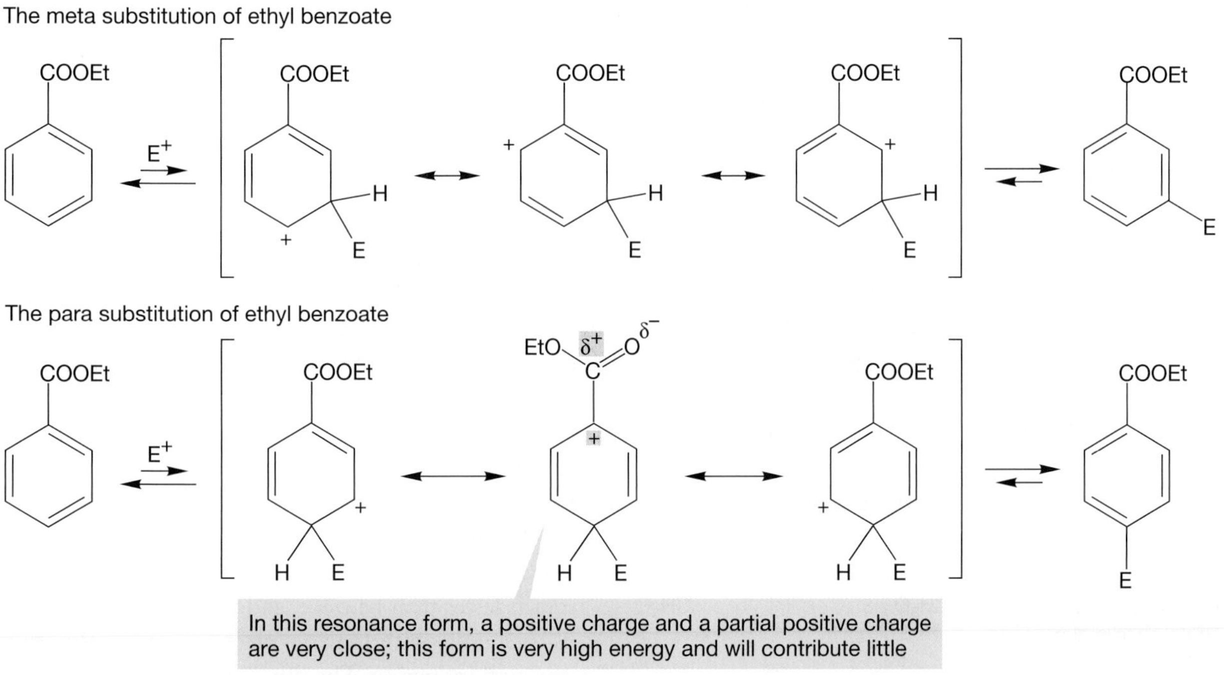

The meta substitution of ethyl benzoate

The para substitution of ethyl benzoate

In this resonance form, a positive charge and a partial positive charge are very close; this form is very high energy and will contribute little

FIGURE 14.78 Meta substitution is preferred to para substitution. In the intermediate for para (or ortho) substitution, a full and partial positive charge are opposed on adjacent carbons.

PROBLEM 14.34 Explain why cyanobenzene (benzonitrile, C_6H_5CN) substitutes preferentially in the meta position.

Summary

A group (G) on the benzene ring determines the position of the next electrophilic aromatic substitution reaction. Generally, groups that direct ortho/para also increase the rate of substitution. In such cases, the ring is said to be activated. If substitution is in the ortho or para position (but not the meta position), G provides additional stabilization to the cyclohexadienyl cation intermediate, and thus increases the rate. Meta-directing groups make the best of an energetically bad set of choices. Meta substitution is slow and wins by default—it is the least-bad alternative for some substituents. In this case the ring is said to be deactivated with respect to electrophilic aromatic substitution.

14.10 Inductive Effects in Aromatic Substitution

So far, we have mainly considered the way in which the group attached to a benzene ring can affect further reaction through either resonance stabilization or electrostatic destabilization of the intermediate cyclohexadienyl cation. We should also consider the inductive effects of groups, which act by withdrawing or donating electrons through the σ bonds. For example, both the trimethylammonium group and the nitro group are strongly electron-withdrawing by induction (Fig. 14.79).

FIGURE 14.79 In the trimethylanilinium ion and nitrobenzene, the σ bond from the ring to the substituent is polarized by the electron-withdrawing group. Further substitution is difficult.

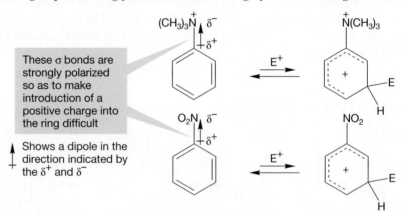

These σ bonds are strongly polarized so as to make introduction of a positive charge into the ring difficult

Shows a dipole in the direction indicated by the δ^+ and δ^-

These subsituents will reduce the electron density in the ring, making the aromatic compound a weaker nucleophile, and thus a less active participant in any reaction with an electrophilic reagent. More important is the destabilizing effect of electron withdrawal on the charged intermediate and the partially charged transition state leading to the intermediate. So, it is no surprise to find that both trimethylanilinium ion and nitrobenzene are very slow to react in electrophilic aromatic substitution. Similar rate-retarding effects might be expected to operate whenever an atom more electronegative than carbon is attached to the ring.

If that is the case, what is one to make of the rate *accelerating* effect of the methoxy group? Anisole reacts much faster than benzene, and yet the methoxy group is certainly withdrawing electrons from the ring through induction. The dipole in the carbon–oxygen σ bond surely points toward the electronegative oxygen atom (Fig. 14.80).

The answer to this conundrum is simply that the resonance effect must substantially outweigh the inductive effect. The fourth resonance form in Figure 14.66 helps more in energy terms than the inductive withdrawal of σ electrons by methoxy hurts.

This σ bond is also strongly polarized so as to make introduction of a positive charge into the ring difficult; yet, the reactions of anisole are very fast

FIGURE 14.80 The carbon–oxygen bond in anisole is also electron-withdrawing, yet substitution is directed ortho/para and the reactions are fast. The resonance effect outweighs this inductive effect.

Thus, often we have two effects operating in opposition. In anisole, the resonance effect stabilizes the intermediate for ortho or para substitution and the transition states leading to them, accelerating the reaction and directing ortho/para. The inductive effect is decelerating for any substitution reaction. For anisole and most groups, the resonance effect wins out, but we can surely find examples where the two effects are largely in balance; where substitution is still ortho/para because of a resonance effect, but where an inductive effect slows the reaction relative to unsubstituted benzene. A look at Table 14.2 shows that there are such cases. For example, halobenzenes direct substitution to the ortho/para positions but substitution proceeds at a slower rate than that for the reaction of benzene. In halobenzenes, all electrophilic aromatic substitution is retarded by the strong electron-withdrawing effect of the electronegative halogens (Fig. 14.81). As a result, the ring is deactivated.

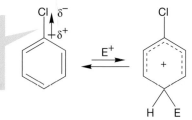

This bond is strongly polarized and the δ+ on carbon acts to retard any substitution reaction that would place more positive charge on the ring carbons

FIGURE 14.81 Chlorine and the other halogens are strongly electron-withdrawing and slow the rate of electrophilic substitution. The net result is an overall slowing of the rate.

However, ortho/para substitution is still favored over meta because the halogen can stabilize an adjacent positive charge by resonance (Fig. 14.82). The rate of meta

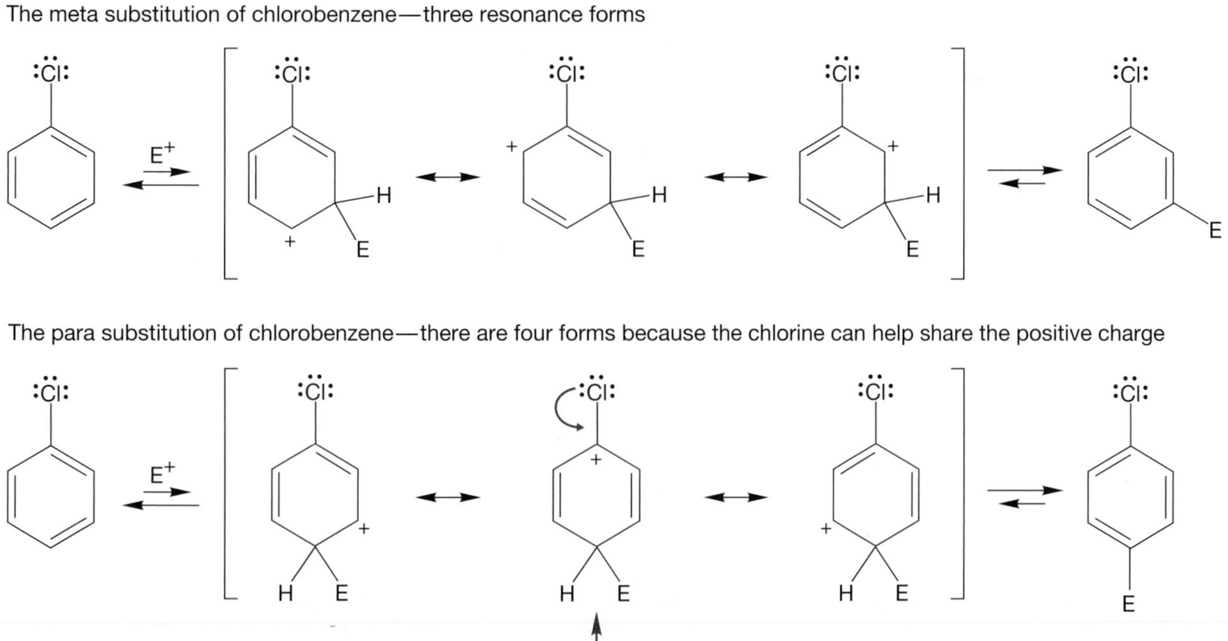

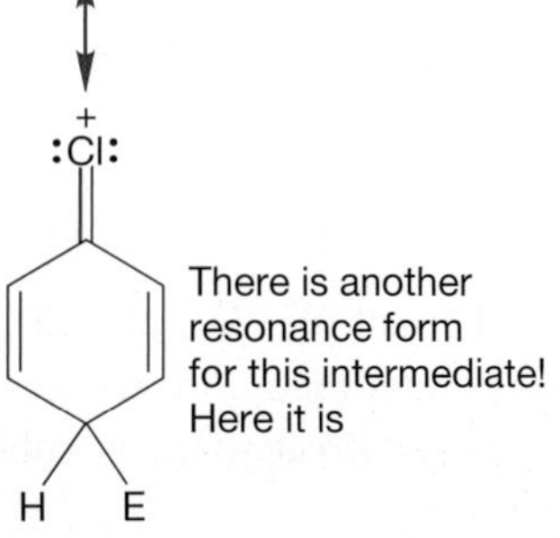

The meta substitution of chlorobenzene—three resonance forms

The para substitution of chlorobenzene—there are four forms because the chlorine can help share the positive charge

There is another resonance form for this intermediate! Here it is

FIGURE 14.82 Chlorine and other halogens can help stabilize the intermediate for para (and ortho) substitution through a fourth resonance form. The halogens strongly withdraw electrons from the ring through an inductive effect, however, and substitution is slow relative to benzene.

substitution, in which the stabilizing resonance effect cannot operate, is slowest of all, and so meta product is greatly disfavored (Fig. 14.83).

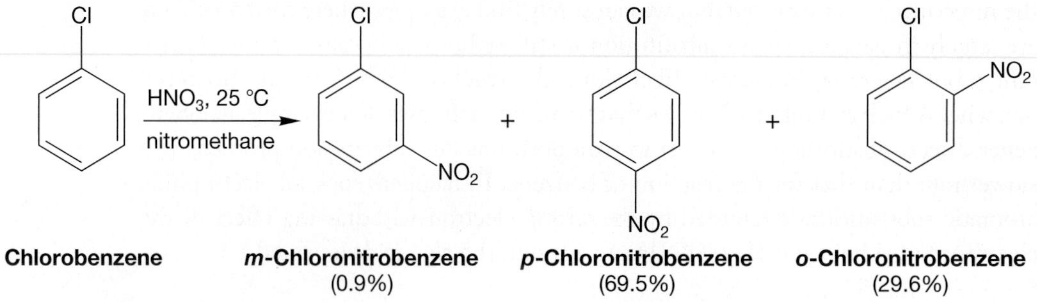

Chlorobenzene **m-Chloronitrobenzene** (0.9%) **p-Chloronitrobenzene** (69.5%) **o-Chloronitrobenzene** (29.6%)

FIGURE 14.83 The nitration of chlorobenzene gives almost entirely ortho and para products.

14.11 Synthesis of Polysubstituted Benzenes

Sometimes it is easy to predict the position of electrophilic attack on a polysubstituted benzene ring. For example, if we have a ring substituted 1,3 with meta-directing groups, further substitution will surely occur at the position meta to both groups, although the rate will be severely retarded by the presence of a pair of deactivating groups (Fig. 14.84).

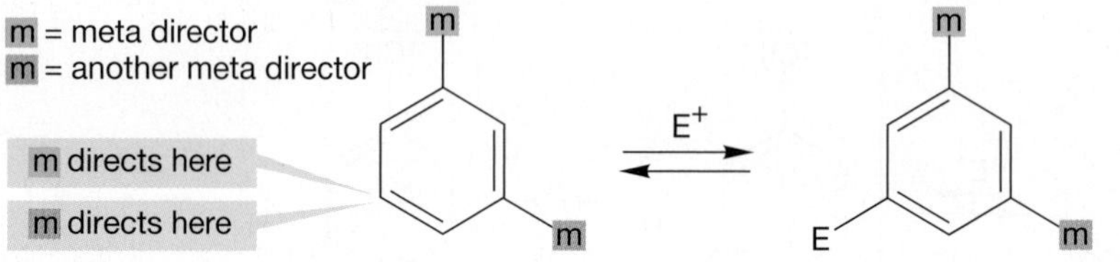

FIGURE 14.84 On a benzene ring substituted 1,3 with a pair of meta-directing groups, both substituents direct further substitution to the same position.

Accordingly, nitration of 1,3-dinitrobenzene gives only meta substitution (Fig. 14.85).

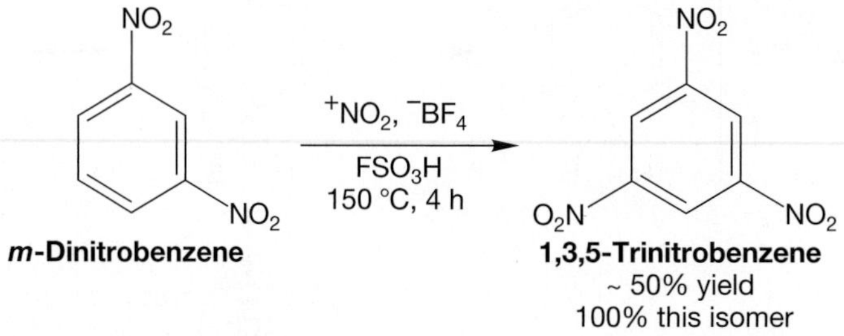

m-Dinitrobenzene **1,3,5-Trinitrobenzene** ~ 50% yield 100% this isomer

FIGURE 14.85 Nitration of 1,3-dinitrobenzene gives a single product, 1,3,5-trinitrobenzene.

Further substitution of a disubstituted aromatic ring containing an ortho/para director and a meta director at the 1- and 4-positions is similarly easy to predict. Both groups direct the next substitution ortho to the ortho/para-directing group.

An example of this kind of reaction is the nitration of 4-nitrotoluene (Fig. 14.86).

In the example in Figure 14.86, we really didn't have to worry much about the directing effect of the nitro group on the disubstituted ring, because it is a rate-retarding substituent that will be overwhelmed by the activating and ortho/para-directing methyl group. However, there are less obvious situations. If the benzene ring bears ortho/para-directing groups at both the 1- and the 4-positions, the position of further substitution is not immediately clear. Sometimes steric effects determine the outcome. For example, Friedel–Crafts acylation of 4-isopropyltoluene gives reaction only at the less encumbered position adjacent to the methyl group (Fig. 14.87). In other cases, electronic factors of one of the groups may determine the outcome, and further reaction will be at the position activated by the more strongly activating group. For example, the Friedel–Crafts alkylation of 4-methylphenol gives almost entirely substitution ortho to the strongly ortho/para-directing hydroxyl group, and only traces of substitution ortho to the less activating methyl group (Fig. 14.87).

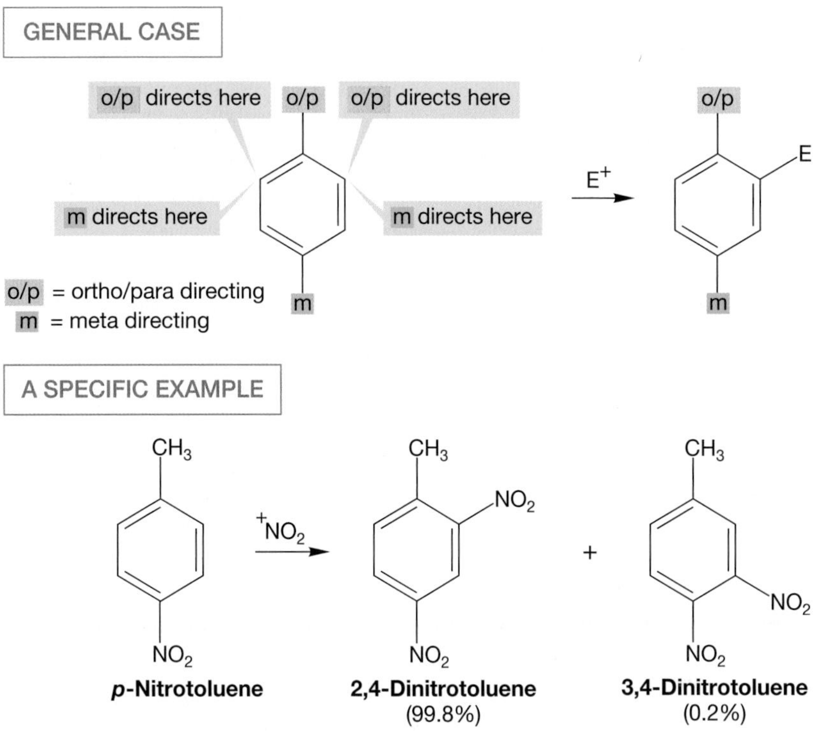

GENERAL CASE

o/p = ortho/para directing
m = meta directing

A SPECIFIC EXAMPLE

p-Nitrotoluene

2,4-Dinitrotoluene
(99.8%)

3,4-Dinitrotoluene
(0.2%)

FIGURE 14.86 Substitution is directed to a single position if a benzene ring is substituted 1,4 with one ortho/para-directing group and one meta-directing group.

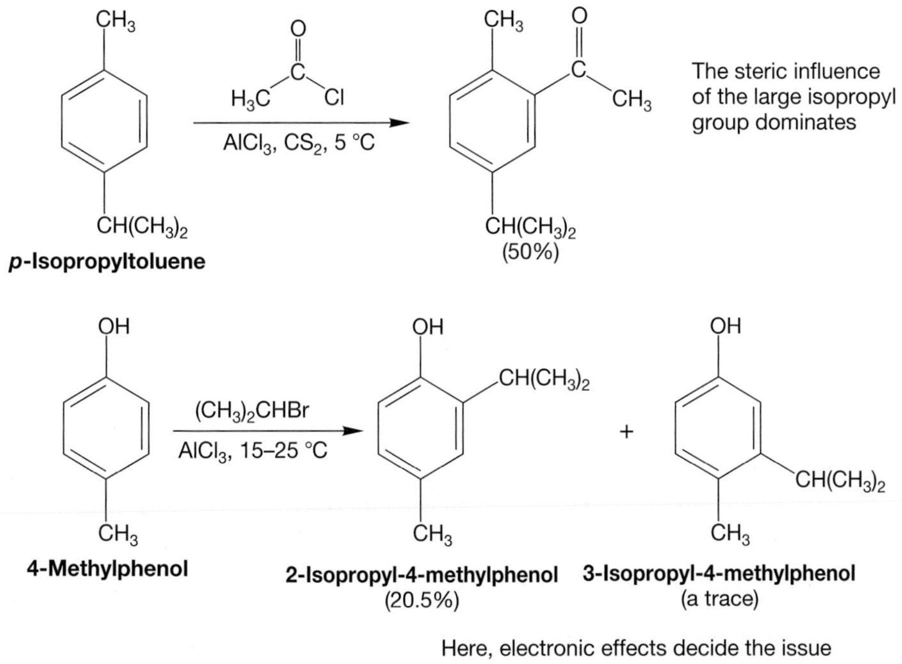

p-Isopropyltoluene

(50%)

The steric influence of the large isopropyl group dominates

4-Methylphenol

2-Isopropyl-4-methylphenol
(20.5%)

3-Isopropyl-4-methylphenol
(a trace)

Here, electronic effects decide the issue

FIGURE 14.87 Two examples of reactions of benzenes substituted 1,4 with both groups ortho/para-directing. Steric effects decide the first case, electronic effects the second.

Other situations are more difficult to predict. Generally, the result will be determined by the sum of the steric and electronic effects of the groups already present on the ring.

Some substituents are so strongly activating that no catalyst is needed, and it is often difficult to stop substitution before more than one new substituent has been added to the ring. For example, many anilines are brominated with Br_2 alone to give 2,4,6-tribromoanilines (Fig. 14.88). Phenols are also highly activated toward further substitution to give polysubstituted products. Mild conditions are needed to restrict the reaction to monosubstition.

NH₂ + excess Br₂, H₂O, 40–50 °C → 3-Amino-2,4,6-tribromo- ; OH + excess Br₂, CCl₄, 0 °C → 2,4,6-Tribromophenol

3-Aminobenzoic acid ; **3-Amino-2,4,6-tribromo-benzoic acid** (~100%) ; **Phenol** ; **2,4,6-Tribromophenol** (97%)

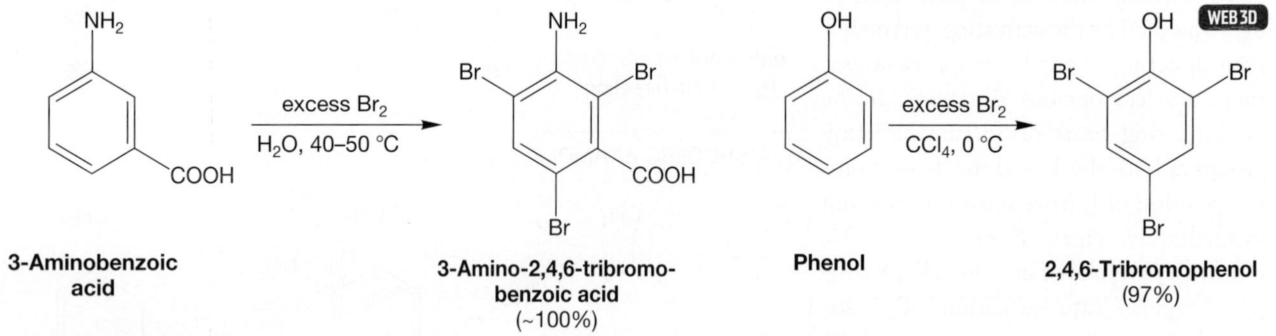

FIGURE 14.88 The bromination of anilines and phenols often leads to multiple additions of several bromines. The strongly activating groups, amino and hydroxyl, direct substitution to the ortho and para positions.

Arene halogenation

The hydroxyl group of phenol or the amino group of aniline are so strongly activating that neither monobromo nor dibromo compounds can be isolated unless great care is taken. Note that when the first bromine is added in the para position, we have the general situation of Figure 14.87. In this case, the OH (or NH_2) and the Br direct further substitution to different positions. The more powerfully activating OH or NH_2 groups win out (bromine is actually *de*activating), and the second and third bromines add ortho to the amino or hydroxyl group as shown in Figure 14.88.

The overly active amino group can be deactivated, as shown in Figure 14.89, by acylating it through reaction with acetyl chloride to give acetamidobenzene.

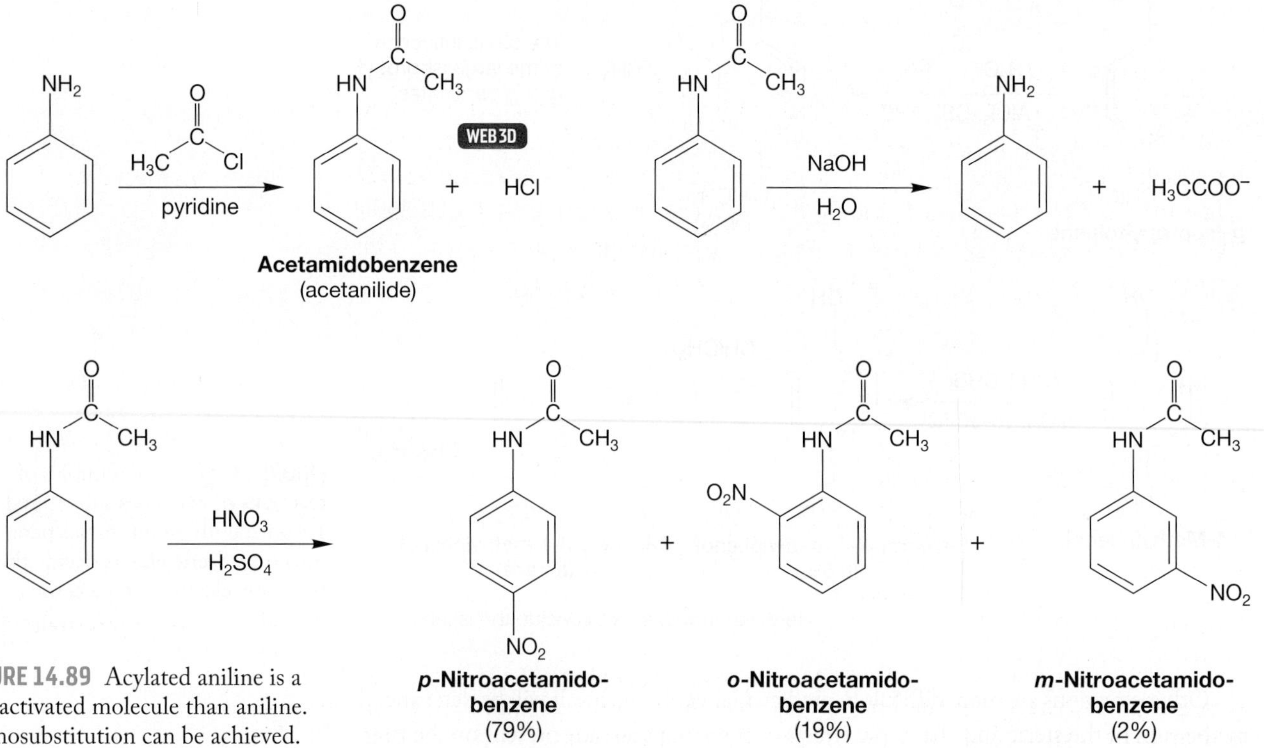

NH₂ + H₃C–C(=O)–Cl, pyridine → **Acetamidobenzene** (acetanilide) + HCl ; acetanilide + NaOH / H₂O → NH₂ + H₃CCOO⁻

acetanilide + HNO₃ / H₂SO₄ → **p-Nitroacetamido-benzene** (79%) + **o-Nitroacetamido-benzene** (19%) + **m-Nitroacetamido-benzene** (2%)

FIGURE 14.89 Acylated aniline is a less activated molecule than aniline. Monosubstitution can be achieved.

N-Acylated aromatic amines still direct ortho/para, but substitution is slower than for the unsubstituted aniline, and the reaction can be stopped after a single new substitution. The protecting acyl group can be removed later by treatment with acid or base. We will see the mechanisms for these reactions in Chapter 18.

PROBLEM 14.35 Through a mechanistic analysis of the electrophilic aromatic substitution of acetamidobenzene, explain why this compound substitutes mainly para. Suggest a reason why the reaction rate of acetamidobenzene with electrophiles is slower than the rate of the reaction of aniline with electrophiles.

Effective use can sometimes be made of removable blocking groups on the ring. Suppose you are set the task of making pure *o*-bromophenol. Simple bromination of phenol will fail because of the dominant formation of the para isomer or of polybrominated compounds (Fig. 14.90). A solution to this problem is first to sulfonate

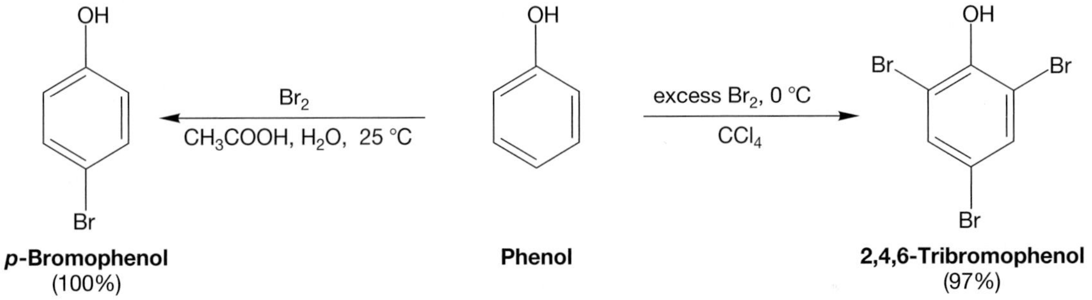

p-Bromophenol
(100%)

Br_2
$CH_3COOH, H_2O,$ 25 °C

Phenol

excess Br_2, 0 °C
CCl_4

2,4,6-Tribromophenol
(97%)

FIGURE 14.90 Direct bromination of phenol cannot give pure *o*-bromophenol.

the ring to give the 2,4-disulfonic acid of phenol (Fig. 14.91). The deactivating effect of the sulfonic acid groups prevents further sulfonation. This compound can now be brominated at the position ortho to the activating hydroxyl group and meta to the deactivating groups. The sulfonic acids have acted to block substitution at the other ortho and para positions. Removal of the blocking sulfonate groups with acidic steam completes the reaction. (Note that the SO_2OH group is an acid.) This approach allows the formation of a single isomer of bromophenol.

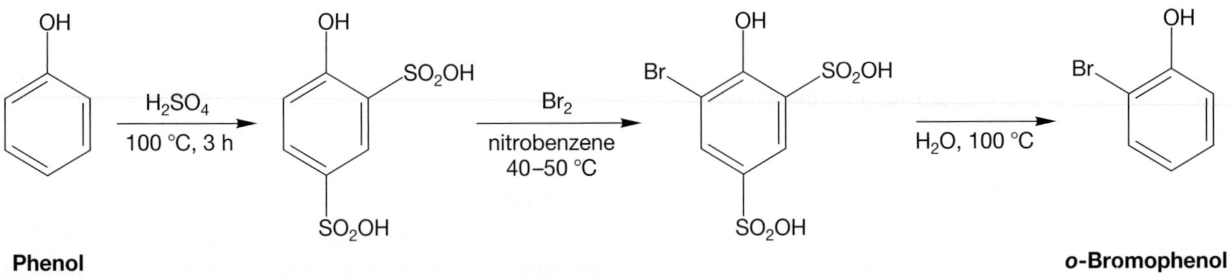

Phenol

H_2SO_4
100 °C, 3 h

SO_2OH

SO_2OH

Br_2
nitrobenzene
40–50 °C

SO_2OH

SO_2OH

H_2O, 100 °C

o-Bromophenol

FIGURE 14.91 The para and one ortho position can be reversibly blocked by the sulfonic acid groups. Once these groups are attached, only one ortho position remains for further substitution. After bromination, the sulfonic acid groups can easily be removed to give the pure ortho isomer.

The ability to interconvert nitro and amino groups is also useful in synthesis of aromatic compounds. The synthesis of *p*-bromonitrobenzene from nitrobenzene is an example. Nitrobenzene cannot be brominated in the para position because the meta-directing nitro group will force formation of only the meta isomer (Fig. 14.92).

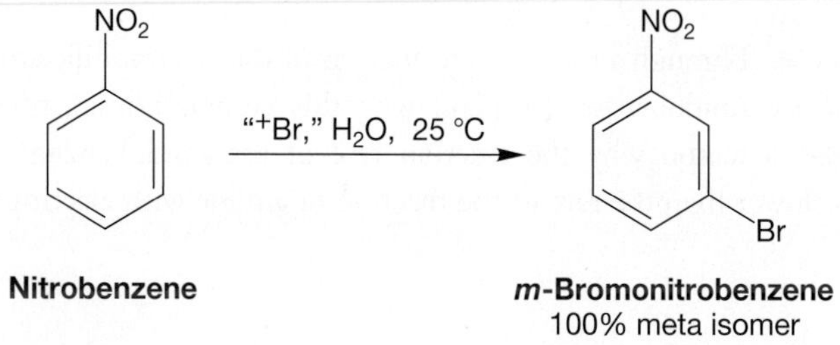

FIGURE 14.92 Nitrobenzene cannot be brominated in the ortho or para positions. Substitution is entirely meta.

The solution is to convert nitrobenzene into aniline by a reduction step, acetylate, and then brominate. Bromination now goes 100% in the para position. Now the amino group can be regenerated by removal of the acyl group and then oxidized to a nitro group (Fig. 14.93). The initial reduction of nitro to amino can be carried out using catalytic hydrogenation, or by a variety of other reducing agents. The oxidation step is usually accomplished with trifluoroperacetic acid or *m*-chloroperbenzoic acid.

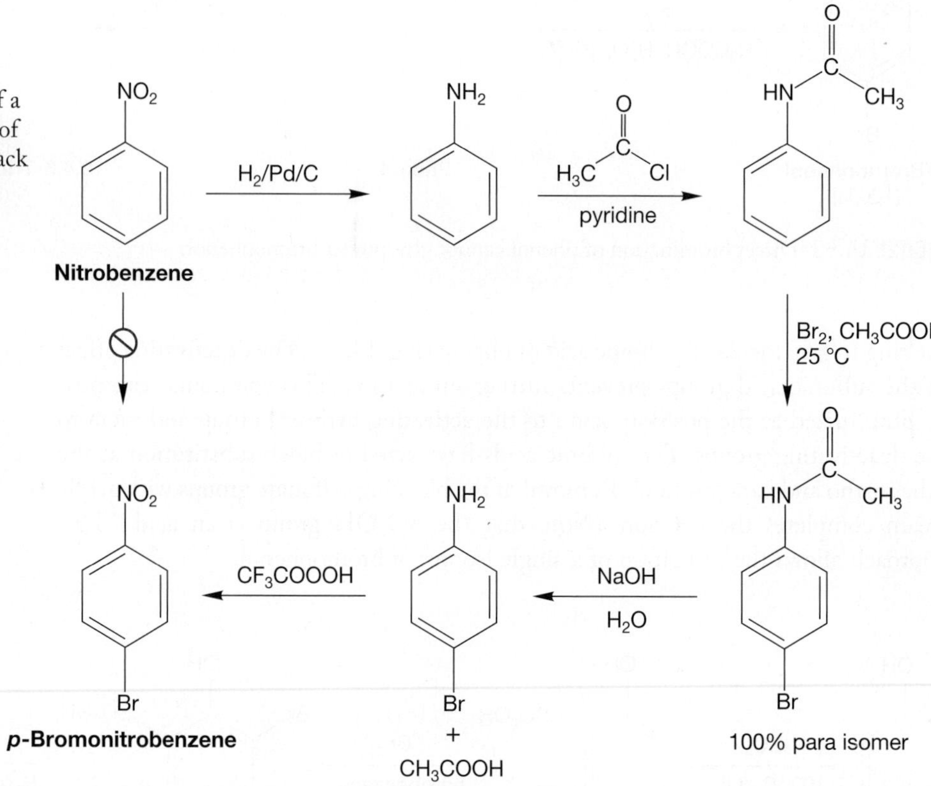

FIGURE 14.93 Conversion of the nitro group into the ortho/para-directing amino group followed by acylation allows the introduction of a single para bromine. Regeneration of the amine, followed by oxidation back to nitro, completes the reaction sequence.

The synthesis of a disubstituted benzene in which two meta-directing groups are substituted ortho or para to each other constitutes a general synthetic problem. To solve such a problem, the ability to convert a meta-directing group into an ortho/para director *and back again* is crucial. A similar difficulty is encountered in the construction of disubstituted benzenes bearing ortho/para-directing groups in meta positions.

WORKED PROBLEM 14.36 The nitration of aniline takes place in acid. Under acidic conditions the amino group must be largely protonated to give an anilinium ion, a meta-directing group. In fact, if the amino group is not acetylated, the product does contain substantial amounts of *m*-nitroaniline. Why is the product not entirely *m*-nitroaniline? *Hint*: Think about the relative rates of the reactions involved.

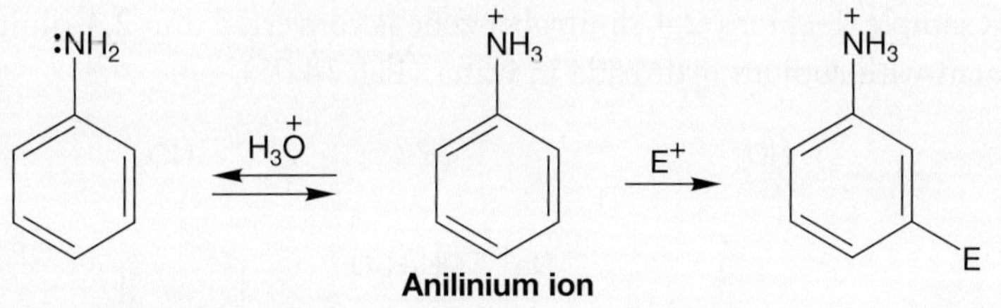

Anilinium ion

ANSWER Aniline will protonate in acid, and the anilinium ion will substitute meta. However, as long as there is some unprotonated aniline present at equilibrium, it will continue to substitute ortho and para. It is estimated that some anilines react at 10^{19} times the rate of the deactivated anilinium ions! So, even if there is only a little free amine present there can still be a great deal of product formed from it.

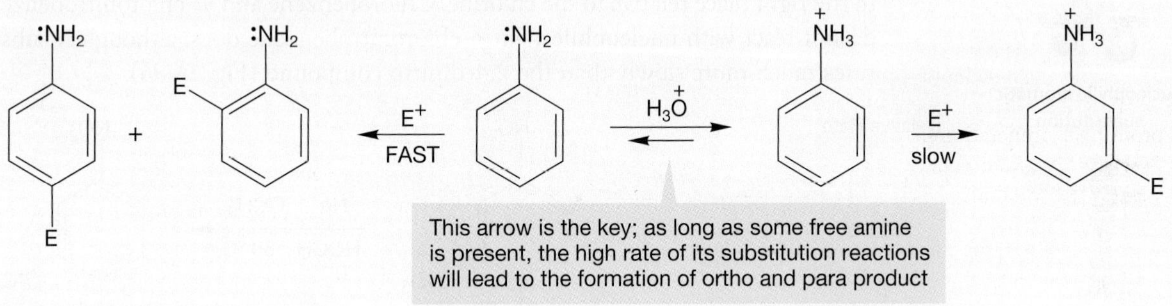

This arrow is the key; as long as some free amine is present, the high rate of its substitution reactions will lead to the formation of ortho and para product

PROBLEM 14.37 Devise syntheses of the following substituted benzenes. You may start from benzene, organic reagents containing no more than two carbons, and any inorganic reagents you may need.

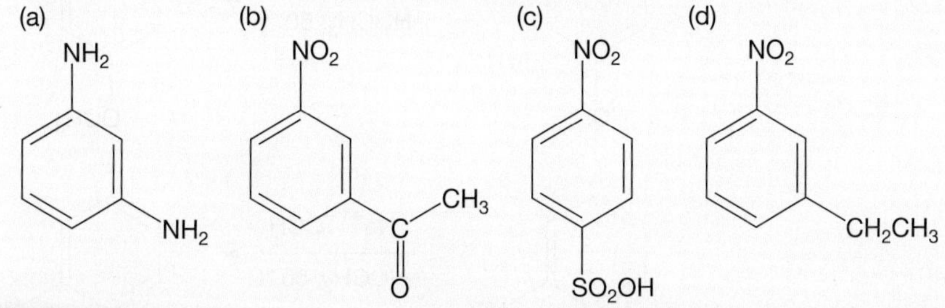

(a)

(b)

(c)

(d)

Summary

When the reacting benzene ring is di- or polysubstituted, the original substituents may reinforce each other or compete. One has to determine the relative influences, resonance, inductive, and steric of each substituent in assessing where the next substitution will occur. Resonance effects usually dominate inductive effects.

14.12 Nucleophilic Aromatic Substitution

14.12a Nucleophilic Additions to Benzenes Normally, alkenes are unaffected by treatment with nucleophiles, and it is no surprise to find that benzene also survives treatment with strong nucleophiles. In particular, no reaction occurs when benzene or halobenzenes are treated with a nucleophile (Fig. 14.94).

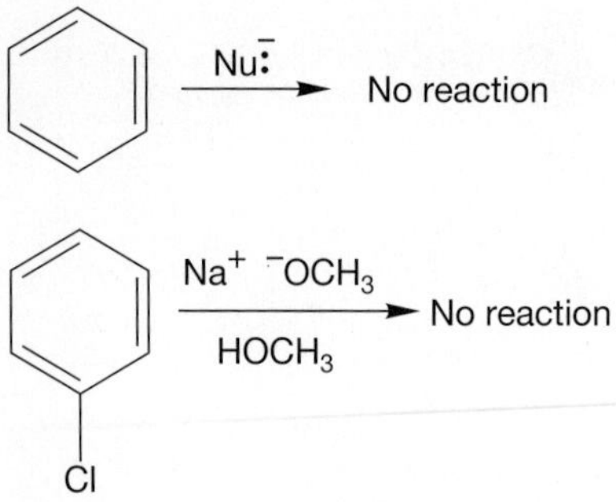

Nu:⁻
No reaction
Na+ ⁻OCH3
HOCH3
No reaction
Cl

FIGURE 14.94 Benzenes and halobenzenes do not normally react with nucleophiles to give substituted products.

However, a few specially substituted benzenes do undergo nucleophilic substitution in base. Electron-withdrawing groups on the ring are necessary for this reaction. For example, 1-chloro-2,4-dinitrobenzene is converted into 2,4-dinitrophenol by treatment with sodium hydroxide in water (Fig. 14.95).

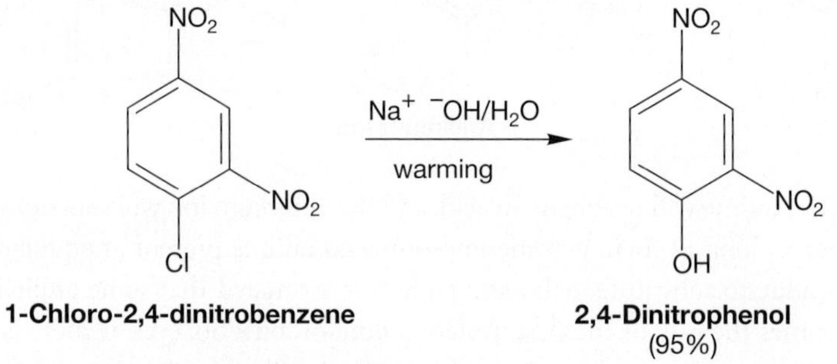

NO2
NO2
Cl
Na+ ⁻OH/H2O
warming
NO2
NO2
OH
1-Chloro-2,4-dinitrobenzene
2,4-Dinitrophenol
(95%)

FIGURE 14.95 Some substituted benzenes are reactive toward nucleophiles. 1-Chloro-2,4-dinitrobenzene can be swiftly transformed into the corresponding phenol by hydroxide ion in water.

Nucleophilic aromatic substitution

At least one nitro group is essential for this substitution reaction, and it must be in the right place relative to the chlorine. Chlorobenzene and *m*-chloronitrobenzene do not react with nucleophiles, but *p*-chloronitrobenzene does, although it substitutes much more slowly than the 2,4-dinitro compound (Fig. 14.96).

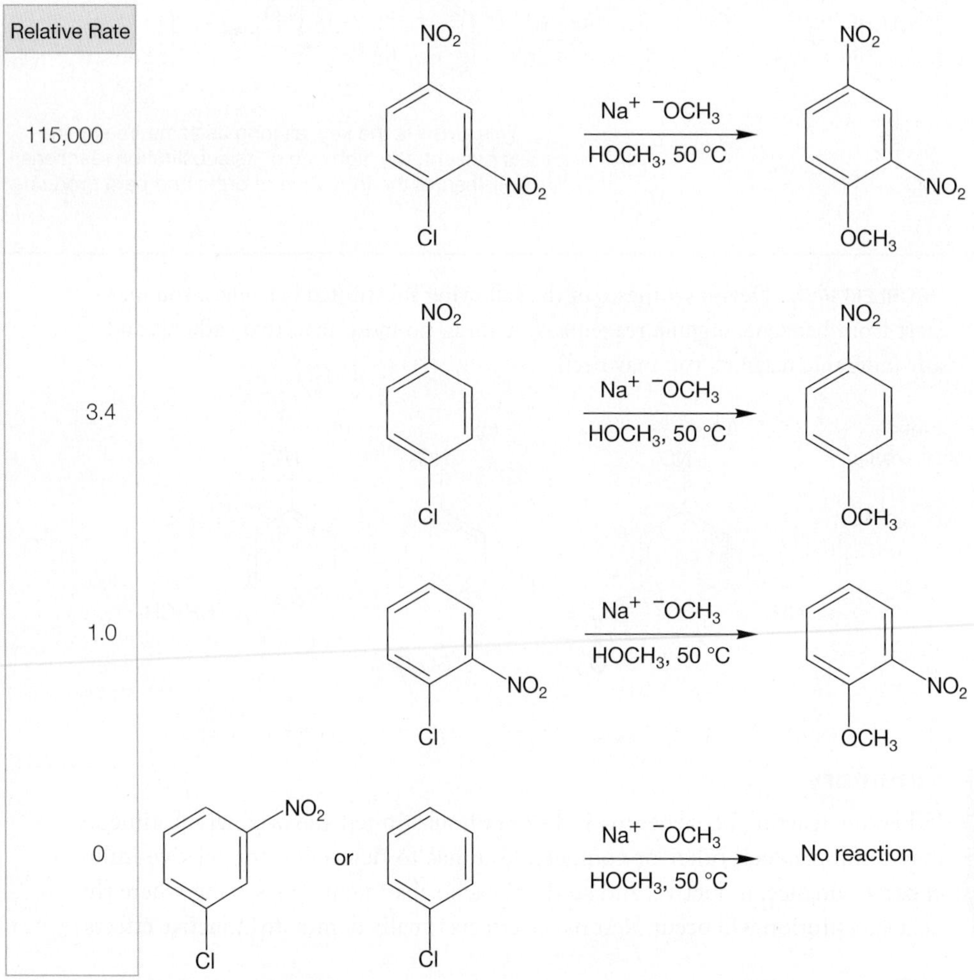

Relative Rate
115,000
3.4
1.0
0
NO2
NO2
Cl
Na+ ⁻OCH3
HOCH3, 50 °C
NO2
NO2
OCH3
NO2
Cl
Na+ ⁻OCH3
HOCH3, 50 °C
NO2
OCH3
Cl
NO2
Na+ ⁻OCH3
HOCH3, 50 °C
NO2
OCH3
NO2
Cl
or
Cl
Na+ ⁻OCH3
HOCH3, 50 °C
No reaction

FIGURE 14.96 For nucleophilic aromatic substitution to succeed, at least one nitro group must be present. 2,4-Dinitro compounds (represented by the first molecule in this figure) react much faster than mononitro compounds.

Neither an S_N2 nor an S_N1 reaction seems likely. Attack from the rear is impossible, as the ring blocks the path of any entering nucleophile, so an S_N2 reaction is not involved. Ionization would produce a most unstable carbocation (Fig. 14.97), which precludes the S_N1 reaction path.

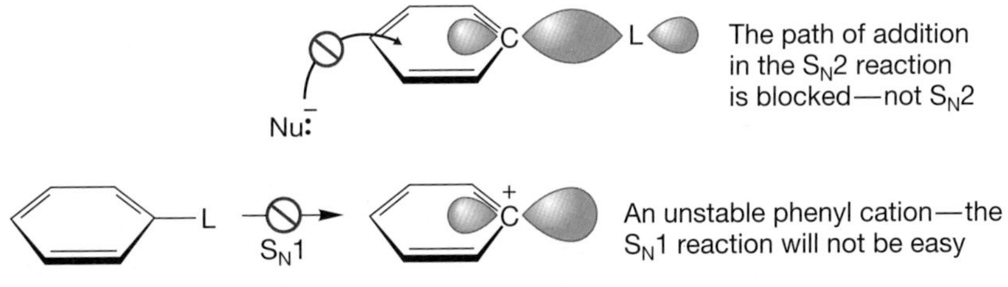

The mechanism of this reaction involves addition to the benzene ring by methoxide (or any nucleophile) to generate a resonance-stabilized anionic intermediate, called a **Meisenheimer complex**, after Jakob Meisenheimer (1876–1934) (Fig. 14.98). This mechanism is called the S_NAr reaction, because it is a **Nucleophilic Aromatic Substitution**.

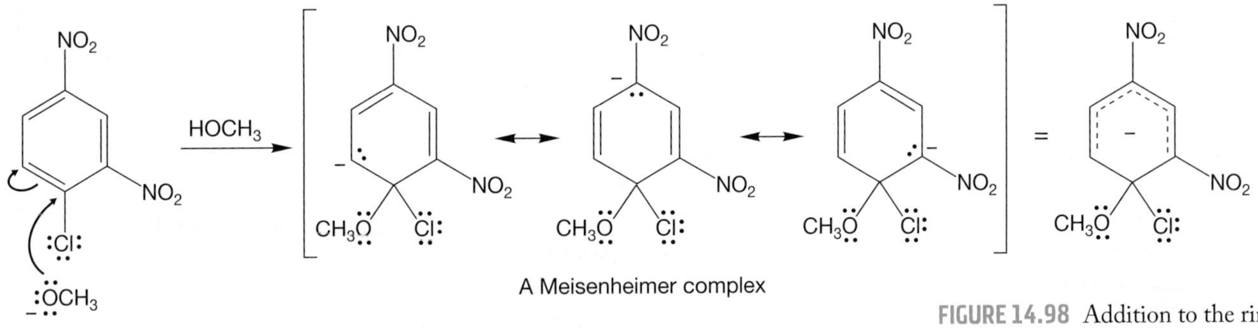

FIGURE 14.97 Neither an S_N2 nor S_N1 reaction occurs in substitution of the halobenzenes.

A Meisenheimer complex

FIGURE 14.98 Addition to the ring by methoxide leads to a highly resonance-stabilized anion. Both nitro groups help bear the negative charge.

The nitro groups stabilize the negative charge through resonance, and therefore exert their effect *only* when they are substituted at a carbon that helps to bear the negative charge. This stabilization by nitro is the reason that *o*- or *p*-chloronitrobenzene undergoes the reaction, but the meta isomer does not (Fig. 14.99). The aromatic

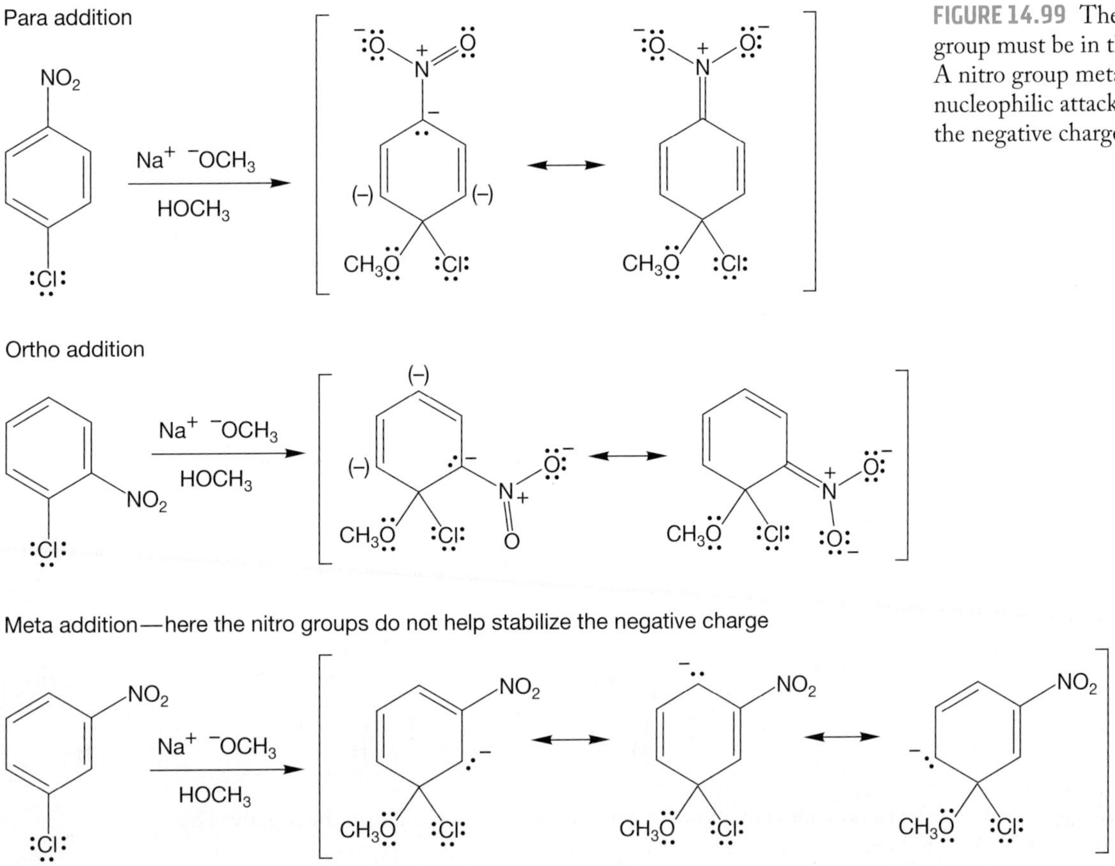

Para addition

FIGURE 14.99 The stabilizing nitro group must be in the proper position. A nitro group meta to the point of nucleophilic attack cannot stabilize the negative charge by resonance.

Ortho addition

Meta addition—here the nitro groups do not help stabilize the negative charge system can be regenerated by reversal of the attack of nucleophile or by elimination of chloride to give 2,4-dinitroanisole (Fig. 14.100). This is the addition–elimination pathway of the S_NAr reaction.

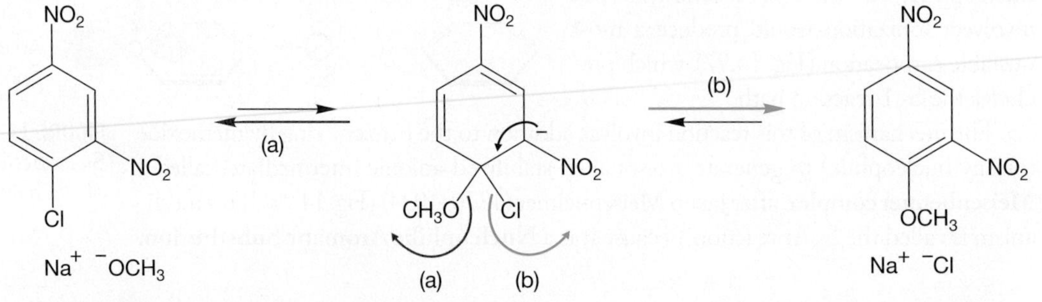

FIGURE 14.100 The addition–elimination S_NAr mechanism starts with addition of the nucleophile; the reaction is completed by elimination of chloride and regeneration of the aromatic ring. Loss of methoxide only leads back to starting material.

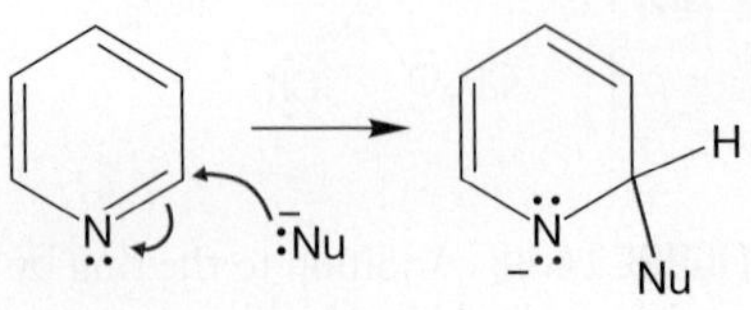

FIGURE 14.101 Nucleophilic attack on pyridine.

14.12b S_NAr Reactions of Heteroaromatics   Pyridine is more active in nucleophilic aromatic substitution than benzene. Not only is pyridine less resonance stabilized than benzene (Table 14.1), but the electronegative nitrogen atom provides an attractive place for an incoming negative charge (Fig. 14.101).

The classic example of nucleophilic aromatic substitution of pyridine is the **Chichibabin reaction**, named for Alexei E. Chichibabin (1871–1945), in which pyridine is converted into 2-aminopyridine through treatment with potassium amide in liquid ammonia (Fig. 14.102).

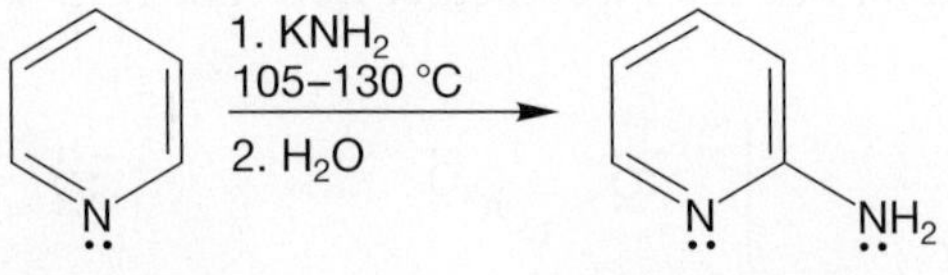

FIGURE 14.102 The Chichibabin reaction.

The first step of the mechanism is consistent with the addition–elimination S_NAr process. The strongly nucleophilic $^-NH_2$ adds to pyridine. Addition at the 2- or 4-position makes it possible for the negative charge to be shared by nitrogen and so is preferred to attack at the 3-position (Fig. 14.103).

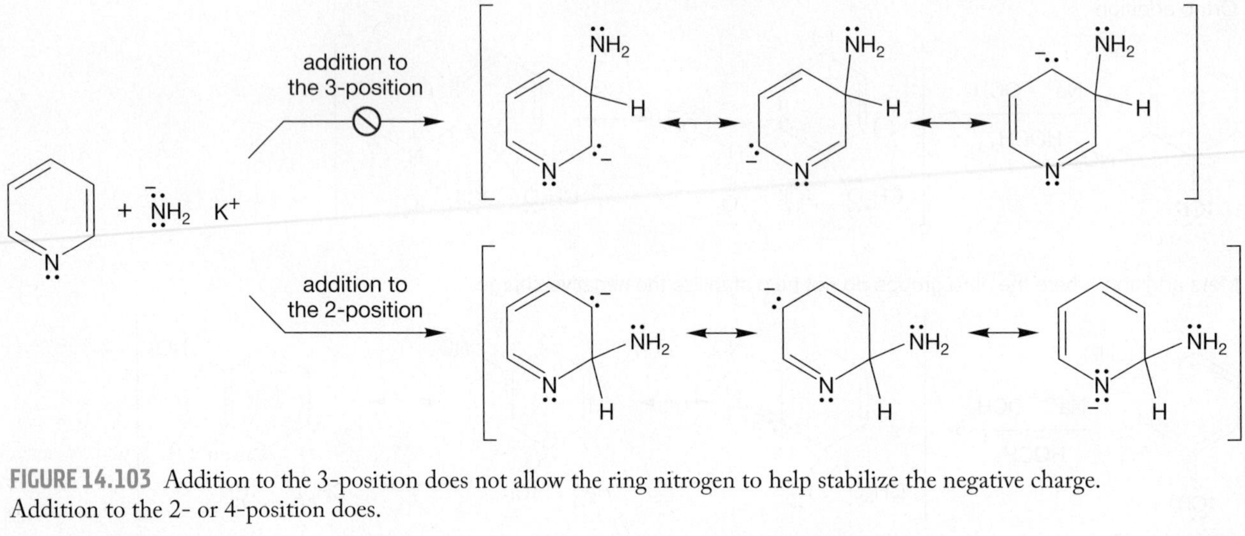

FIGURE 14.103 Addition to the 3-position does not allow the ring nitrogen to help stabilize the negative charge. Addition to the 2- or 4-position does.

This conventional beginning is followed by a rather unconventional conclusion, as aromaticity is regained through elimination of hydride (Fig. 14.104). This step is unconventional because loss of hydride from an organic compound is rare. Aminopyridine and potassium hydride are the initially formed products. These react rapidly and irreversibly to form hydrogen gas and a nicely resonance-stabilized amide ion. A final hydrolysis step regenerates aminopyridine. It is presumably the irreversible nature of hydrogen gas formation that helps drive this reaction to completion.

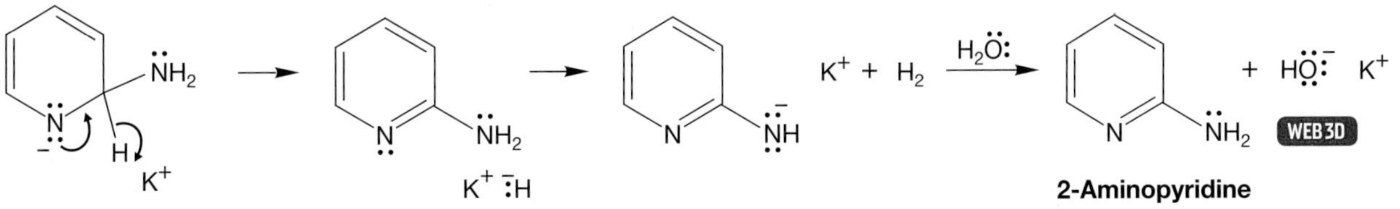

FIGURE 14.104 In this S_NAr reaction there is a loss of hydride followed by formation of H_2. Hydrolysis regenerates 2-aminopyridine.

Other strong nucleophiles such as organolithium reagents also undergo the S_NAr reaction to form substituted pyridines. In these reactions, hydrogen gas cannot be formed until the final hydrolysis step (Fig. 14.105).

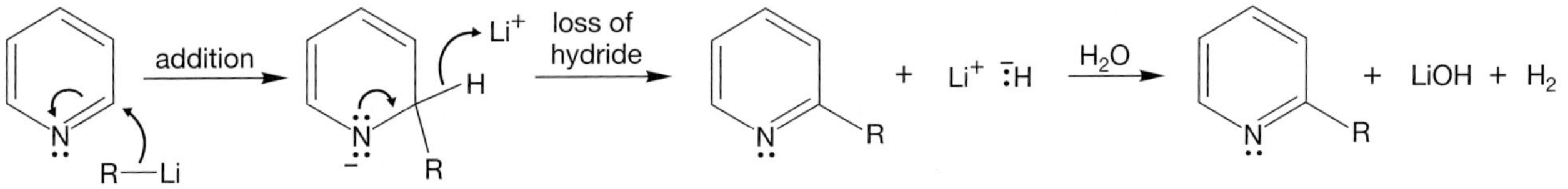

FIGURE 14.105 Addition of alkyllithium reagents also leads to substituted pyridines.

Nucleophilic substitution provides a means of converting pyridines bearing a leaving group at the 2- or 4-position into differently substituted molecules. Addition to the position bearing the leaving group (called **ipso attack**) leads to an intermediate that can either revert to starting material or eliminate the leaving group to give a new compound (Fig. 14.106).

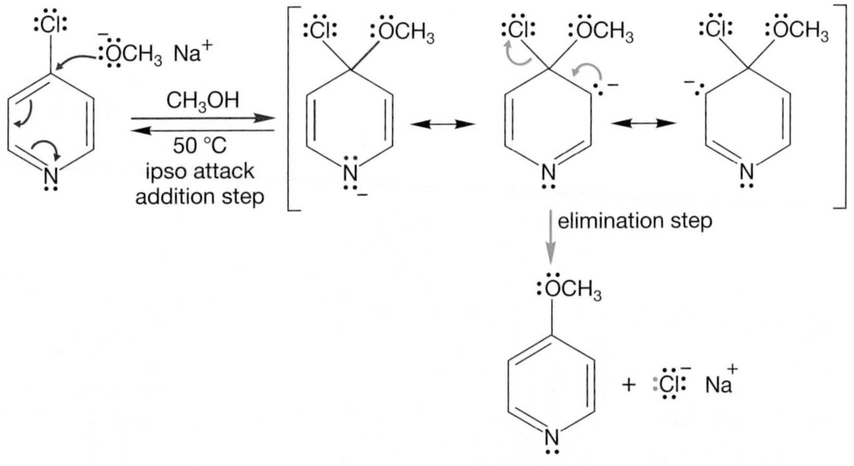

FIGURE 14.106 S_NAr reaction of a pyridine through ipso attack. A leaving group is displaced through the addition–elimination process.

WORKED PROBLEM 14.38 Explain why the S_NAr reaction works only for halopyridines substituted at the 2- or 4-position and fails for 3-substituted halopyridines.

ANSWER The key to the relatively easy nucleophilic additions to pyridines is the placement of the negative charge on nitrogen. If the leaving group is attached to the 3-position, ipso addition of a nucleophile does not place the negative charge on the electronegative nitrogen atom. Only if the leaving group is at the 2- or 4-position does this happen.

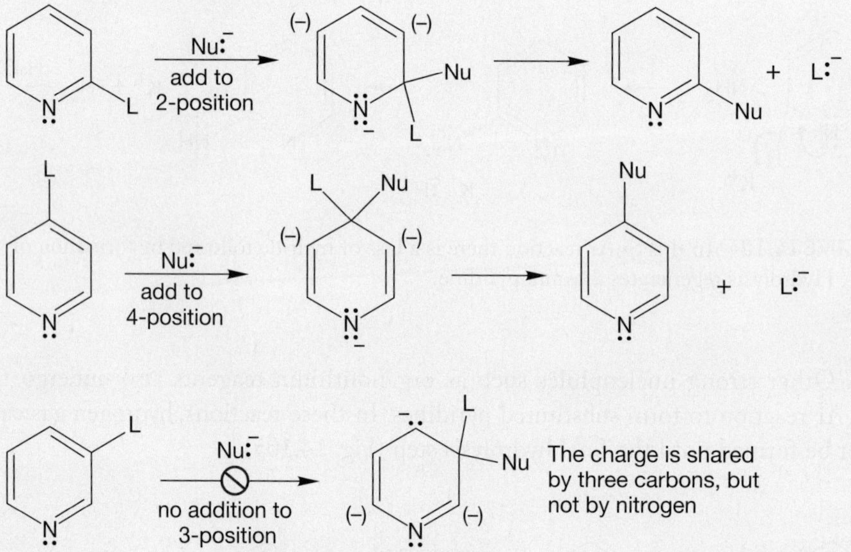

N-oxides are especially active in nucleophilic aromatic substitution because the incoming nucleophile annihilates the positive charge on nitrogen as it adds. Addition destroys aromaticity, but for the *N*-oxides, loss of a leaving group regenerates the aromatic sextet and the aromatic *N*-oxide. An interesting twist to this reaction is shown in Figure 14.107 where the S_NAr reaction is carried out in

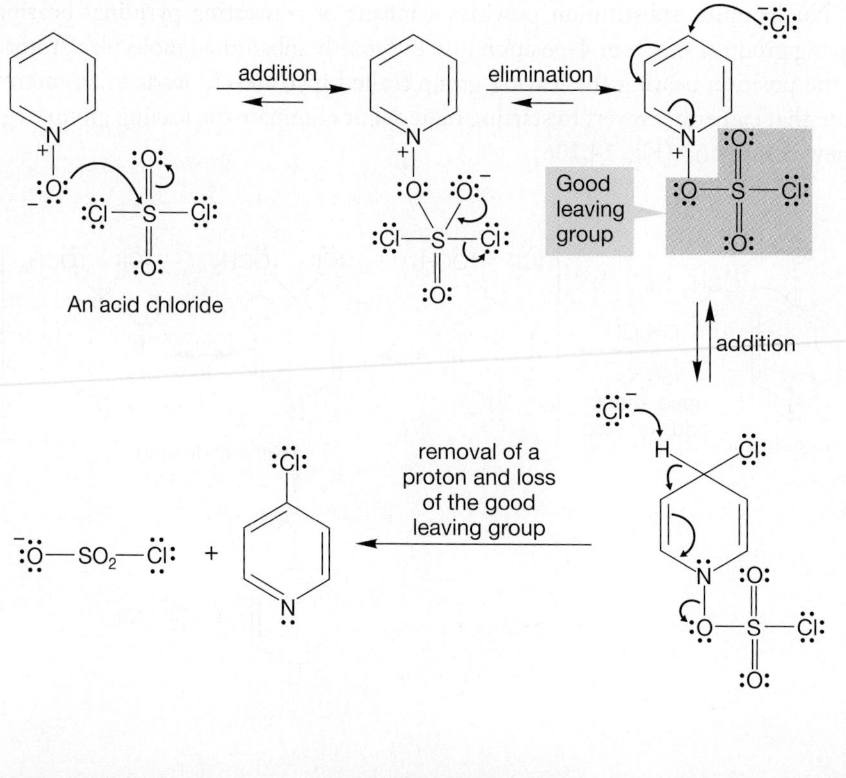

FIGURE 14.107 Pyridine *N*-oxides are relatively easily attacked by nucleophiles. Loss of a leaving group leads to a substituted pyridine.

the presence of an acid chloride (here, SO_2Cl_2), in order to produce a good leaving group. This reaction converts the oxidized pyridine into the substituted nonoxidized pyridine.

Nature uses a similar reaction in biological oxidation, the transfer of hydride to the para position of the pyridinium ion, NAD^+, the oxidized form of nicotinamide adenine dinucleotide. We will see more of this reaction in Chapter 16 (Fig. 14.108).

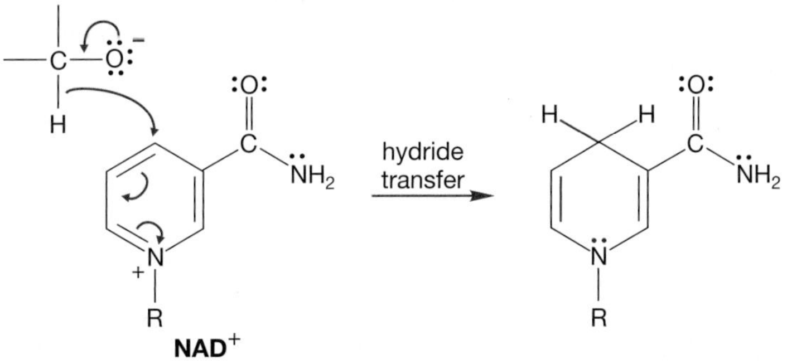

FIGURE 14.108 Reduction of NAD^+ through hydride transfer.

Summary

There are similarities between nucleophilic aromatic substitution (S_NAr) and its more usual counterpart, electrophilic aromatic substitution. Each involves the formation of a resonance-stabilized intermediate, and each involves a temporary loss of aromaticity that is regained in the final step of the reaction. But the similarities are only so deep. The electrophilic reaction involves cationic intermediates; the nucleophilic involves anionic intermediates. Use the differing effects of a nitro group, strongly deactivating in the electrophilic substitution and strongly activating in the nucleophilic substitution, to keep the two mechanisms distinct in your mind.

14.13 Special Topic: Stable Carbocations in "Superacid"

The Friedel–Crafts reactions succeed because relatively stable carbocations can be formed by the reaction of aluminum chloride (an exceptionally strong Lewis acid with which an alkyl halide will react) in solvents that are nonnucleophilic. The strongest nucleophile available in the reaction vessel is benzene, which attacks the cation to start the Friedel–Crafts reaction. If stable carbocations cannot be formed, as with methyl chloride or ethyl chloride, the $AlCl_3$ complexes can also alkylate benzenes.

How would we modify the reaction conditions so as to maximize the possibility of generating the carbocation and minimize the chances for its further reaction? Antimony pentafluoride (SbF_5) is an even stronger Lewis acid than $AlCl_3$ toward halogens, particularly fluorine. When SbF_5 removes a fluoride ion from an alkyl compound the anion antimony hexafluoride (SbF_6^-) is produced (Fig. 14.109). This anion is nonnucleophilic, and fluoride ion is a poor nucleophile as well.

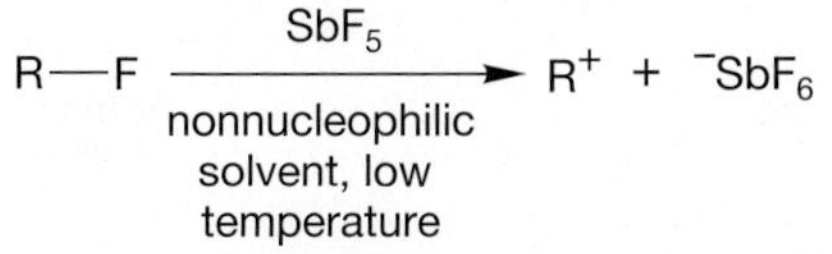

FIGURE 14.109 Cation formation through reaction of an alkyl fluoride with SbF_5.

PROBLEM 14.39 Explain why SbF_6^- is nonnucleophilic.

Treatment of alkyl fluorides with SbF_5 at very low temperatures in highly polar, nonnucleophilic solvents such as SO_2, SO_2ClF, or FSO_3H (so-called **superacid** conditions) can generate many stable carbocations. This process is Friedel–Crafts chemistry carried to extremes. Methyl and primary carbocations cannot be generated this way, but tertiary and secondary cations can be as long as they are not prone to rearrangement to more stable cations (Fig. 14.110).

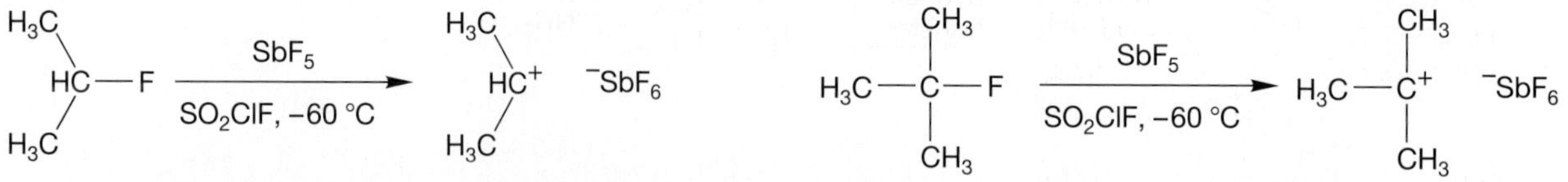

FIGURE 14.110 Stable secondary and tertiary carbocations can be formed in superacid.

Many carbocations, once thought to be impossible species even as reaction intermediates, can be generated and studied under superacid conditions. This work was pioneered by George A. Olah (b. 1927) and earned Olah the Nobel Prize in Chemistry in 1994. In future chapters, we will see a number of examples of the utility of looking at cations in superacid.

PROBLEM 14.40 When 1-fluorobutane is dissolved in superacid at low temperature, it is possible to observe a stable species. However, the ^{13}C NMR spectrum of that stable species shows only two signals. (a) Why doesn't the "expected" spectrum appear? What is the species that is observed? (b) Although part (a) is quite easy, this reaction actually poses a vexing problem. Try to write a mechanism for formation of the stable species. What problems appear when you do?

14.14 Special Topic: Benzyne

In Section 14.12, we mentioned the lack of reactivity of halobenzenes with nucleophiles. It took the activation of a nitro group on the ring for substitution to occur (see Figs. 14.94–14.96). In truth, if the nucleophile is basic enough, even chlorobenzene will react. In the very strongly basic medium potassium amide ($K^+\ {}^-NH_2$) in liquid ammonia (NH_3), chlorobenzene is converted into aniline (Fig. 14.111).

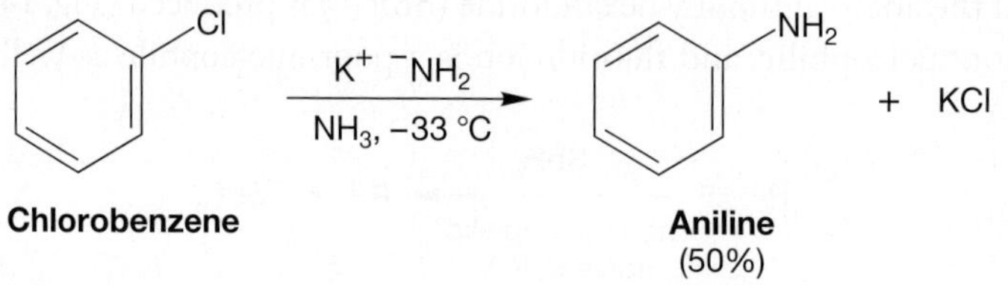

FIGURE 14.111 Aniline can be formed from chlorobenzene by reaction with very strong base, potassium amide in ammonia.

The mechanism is surely not an S_N2 displacement, and there is no reason to think S_NAr addition–elimination is possible with an unstabilized benzene ring. This difficult mechanistic problem begins to unravel with the observation by John D. Roberts (b. 1918) and his co-workers that labeled chlorobenzene produces aniline *labeled equally in two positions* (Fig. 14.112).

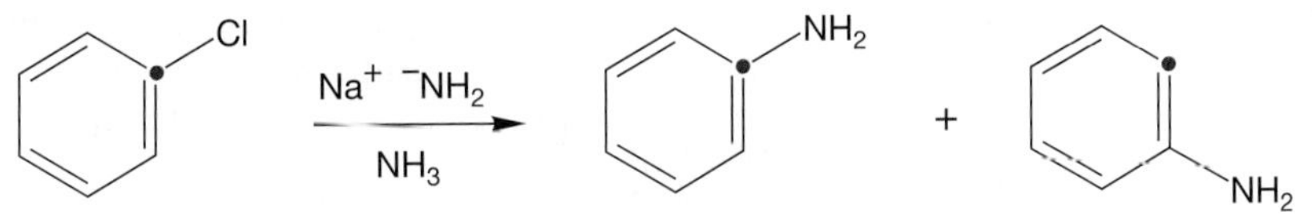

FIGURE 14.112 A clue to the mechanism for this reaction is provided by Roberts' observation of isotopic scrambling when labeled chlorobenzene is used. The red dot represents ^{14}C, an isotopic label.

Benzyne formation

What reactions of halides do you know besides displacement? The E1 and E2 reactions compete with displacement reactions (Chapter 7). If HCl were lost from chlorobenzene, a cyclic alkyne, called **benzyne** (or dehydrobenzene), would be formed. This symmetrical bent alkyne might be reactive enough to undergo an addition reaction with the amide ion. If this were the case, the labeled material must produce two differently labeled products of addition (Fig. 14.113), because the $^-NH_2$ can attack either carbon of the alkyne. This mechanism is the elimination–addition version of the S_NAr reaction.

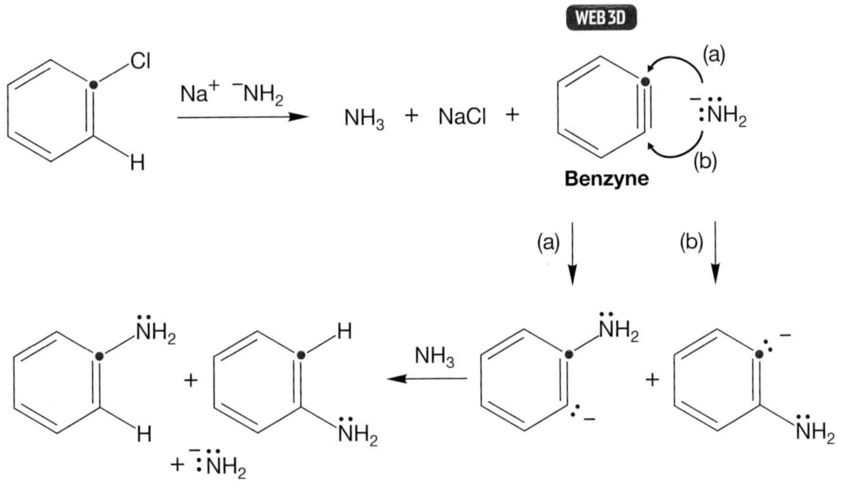

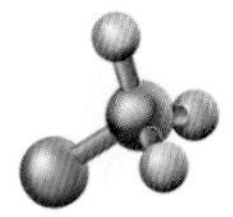

FIGURE 14.113 The elimination–addition variation of the S_NAr reaction begins with a halobenzene. The symmetrical intermediate benzyne is formed. Addition of the amide ion to benzyne produces the observed labeling results.

PROBLEM SOLVING

Almost all "magical migration" reactions, such as the mysterious movement of the red label in Figure 14.112, involve the formation of a symmetrical intermediate—here benzyne (Fig. 14.113). The Special Topic in Chapter 21 is full of such problems, but they have come up elsewhere as well. Problems 14.67, 20.16, and 20.42b are nice examples.

Benzyne is surely an unusual species and deserves a close look. Although this intermediate retains the aromatic sextet, the triple bond is badly bent, and there must be very severe angle strain indeed. Remember, the optimal angle in triple bonds is 180°. Moreover, the "third" bond is not composed of $2p/2p$ overlap like a normal alkyne, but rather of overlap of two hybrid orbitals (Fig. 14.114).

FIGURE 14.114 The structure of benzyne.

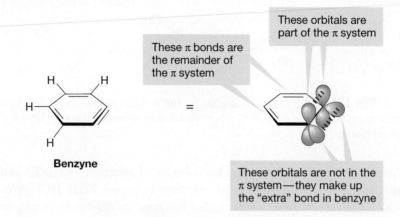

One would therefore not be too surprised to find that this molecule is able to add strong bases such as the amide ion. In fact, one would expect very high reactivity for this strained alkyne. Problem 14.41 gives you a chance to see some of the reactions of benzyne.

PROBLEM 14.41 Provide arrow formalisms for the formation of these four products from benzyne:

14.15 Special Topic: Biological Synthesis of Aromatic Rings; Phenylalanine

Some of the amino acids from which our tissues are constructed contain benzene rings, and so nature must have a way of making them. You might actually set yourself that problem right now; you have at your disposal the means to generate benzene from simple hydrocarbon precursors.

PROBLEM 14.42 Design a synthesis of benzene starting from inorganic compounds, and organic compounds containing only hydrogen and no more than four carbon atoms. *Hint*: Your primary goal is to construct the six-membered ring. You have only one easy way to do that (so far).

Nature can't use the method you designed in Problem 14.42, because too much energy is needed to do the crucial reaction that constructs the six-membered ring. Nature starts from simple sugars, which contain the carbons necessary for the ring, along with many hydroxyl groups. (For much more on carbohydrates, wait for Chapter 22.) The overall strategy, if one may speak of Nature having a strategy, is to build a cyclohexane ring, then use the OH groups in various ways to do elimination reactions and put in the necessary double bonds.

Here is how the process works for phenylalanine, one of your amino acids. A sugar is transformed enzymatically into dehydroquinate (**A**), which undergoes an elimination reaction, also enzyme mediated, to give dehydroshikimate (**B**) (Fig. 14.115).

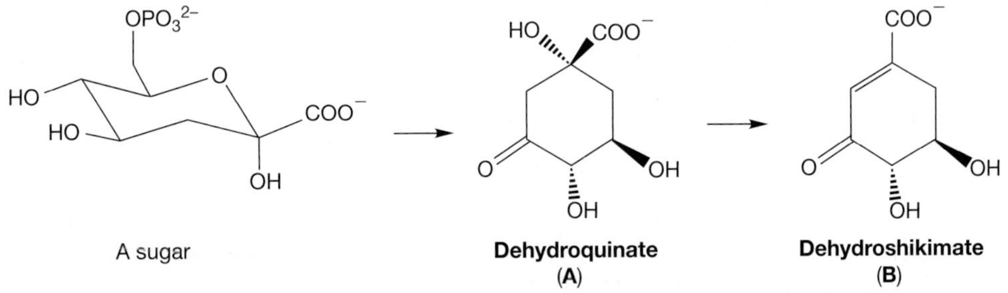

FIGURE 14.115 In the biosynthesis of phenylalanine, enzymes transform a sugar into dehydroquinate (**A**) and then dehydroshikimate (**B**).

The carbon–oxygen double bond in **B** is reduced and phosphorylated to give shikimate-3P (**C**) (Fig. 14.116). Another enzyme adds a three-carbon fragment to one of the hydroxyl groups (you'll see why this is critical in a moment) to give **D**, which undergoes a second elimination to place the second double bond in the ring in **E**, chorismate.

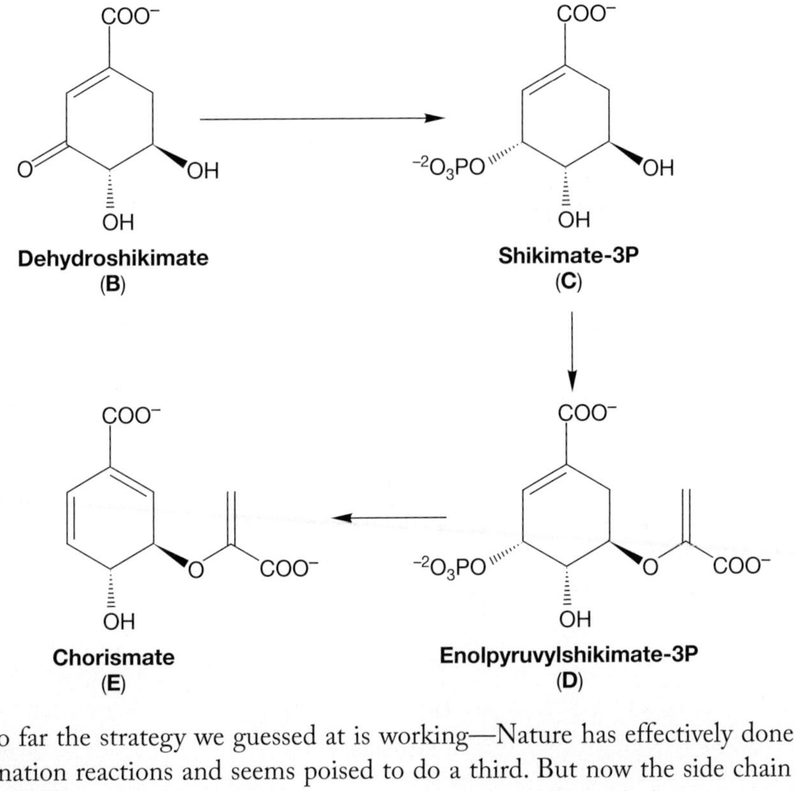

FIGURE 14.116 Dehydroshikimate (**B**) is transformed into chorismate (**E**).

So far the strategy we guessed at is working—Nature has effectively done two elimination reactions and seems poised to do a third. But now the side chain that was added to make **D** in Figure 14.116 comes into play, and chorismate mutase induces a transformation of **E** to prephenate, **F** (Fig. 14.117). You can see this process in some detail in Chapter 20.

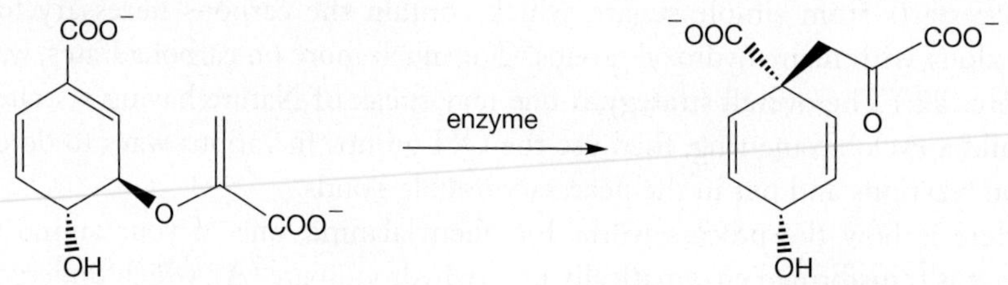

FIGURE 14.117 Chorismate (**E**) is made into prephenate (**F**).

Chorismate (E)   enzyme   **Prephenate (F)**

PROBLEM 14.43 Write an arrow formalism for the transformation of compound **E** into **F**. Don't worry about the role of the enzyme.

The final elimination occurs in prephenate, with the carboxylate (COO⁻) directly attached to the ring acting as trigger. Elimination of this group and the hydroxide leads to phenylpyruvate (**G**) (Fig. 14.118). Now the three-carbon chain is elaborated into the side chain of the amino acid phenylalanine.

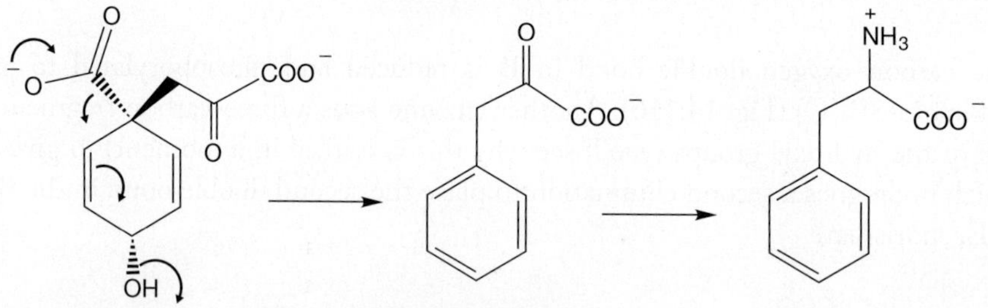

FIGURE 14.118 Finally, prephenate (**F**) is transformed into phenylalanine.

Prephenate (F)   **Phenylpyruvate (G)**   **Phenylalanine**

This sequence is far from simple, but even though it is laden with bells and whistles, some of which we surely could not anticipate, the general adherence to our early strategy—build the six-membered ring and then use the hydroxyls as leaving groups—is apparent.

PROBLEM 14.44 There is another amino acid, tyrosine, that is formed from prephenate (Fig. 14.117). At first this transformation seems surprising, perhaps even impossible. Analyze what must be done and explain why it will not be easy.

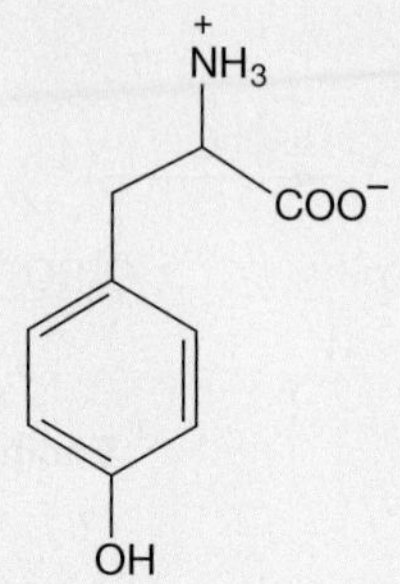

Tyrosine

14.16 Summary

New Concepts

This chapter involves the chemistry of aromatic compounds and is dominated by two fundamental interrelated notions. First, it is simply difficult to disturb the aromatic sextet, and second, once aromaticity is disrupted, it is easily regained. In chemical terms, this means that reactions of benzene and other simple aromatic compounds generally involve a highly endothermic first step with formation of a high-energy intermediate. In electrophilic aromatic substitution, the most common reaction of aromatic compounds, this intermediate is the cyclohexadienyl cation. Because this intermediate lies far above benzene in energy, the transition state leading to it is also high in energy, and the reaction is difficult. Reactions of the cyclohexadienyl cation that regenerate aromatic systems are highly exothermic, and therefore have low activation energy barriers.

Further substitution reactions of an already substituted benzene are controlled by the properties of the first sub-stituent. Both resonance and inductive effects are involved, but the resonance effect usually dominates. Many substituents can stabilize an intermediate cyclohexadienyl cation through resonance if substitution is in the ortho or para position. The extra stabilization afforded by the substituent leads to relatively high rates of reaction. Positively charged substituents generally substitute meta, because ortho or para substitution places two positive charges on adjacent positions. The introduction of a second positive charge means that these substitutions are relatively slow.

At low temperature, in sufficiently polar, but nonnucleophilic solvents, some carbocations are stable. Even under these superacid conditions, rearrangements through hydride or alkyl shifts to the most stable possible carbocation are common.

Key Terms

acid chloride (p. 643)
acylium ion (p. 643)
benzyne (p. 681)
Chichibabin reaction (p. 676)
diazonium ion (p. 647)

electrophilic aromatic substitution (p. 635)
Friedel–Crafts acylation (p. 643)
Friedel–Crafts alkylation (p. 639)
ipso attack (p. 677)

Meisenheimer complex (p. 675)
nucleophilic aromatic substitution (p. 675)
[n]paracyclophane (p. 628)
Sandmeyer reaction (p. 649)
superacid (p. 680)

Reactions, Mechanisms, and Tools

By far the most important reaction in this chapter is electrophilic aromatic substitution by a variety of electrophiles (E^+). This reaction involves the initial formation of a resonance-stabilized, but not aromatic, cyclohexadienyl cation. Deprotonation regenerates an aromatic system (Fig. 14.119).

Nucleophilic aromatic substitution occurs only when the ring carries substituents that can stabilize the negative charge introduced through addition of a nucleophile. Departure of a leaving group regenerates the aromatic system (Fig. 14.100).

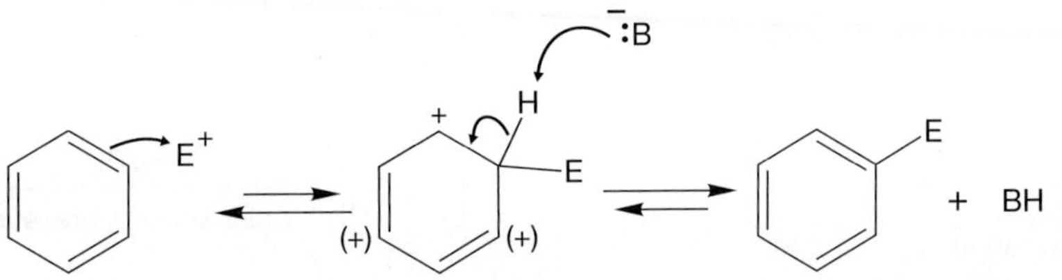

FIGURE 14.119 The general mechanism for electrophilic aromatic substitution.

Syntheses

This chapter provides many new syntheses of aromatic compounds. Although we can now make all sorts of substituted benzenes, there is still a unifying theme: Each product ultimately derives from a reaction of benzene with an electrophile.

1. Alkanes

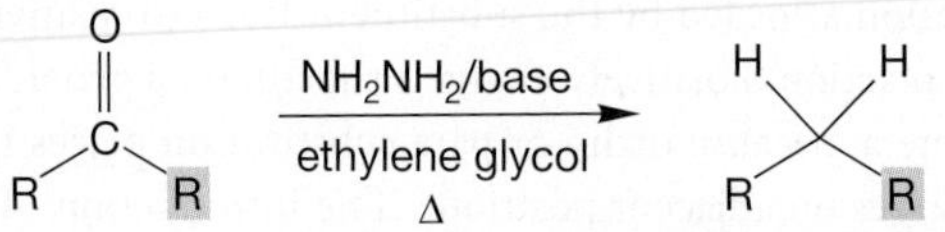

Wolff–Kishner reduction

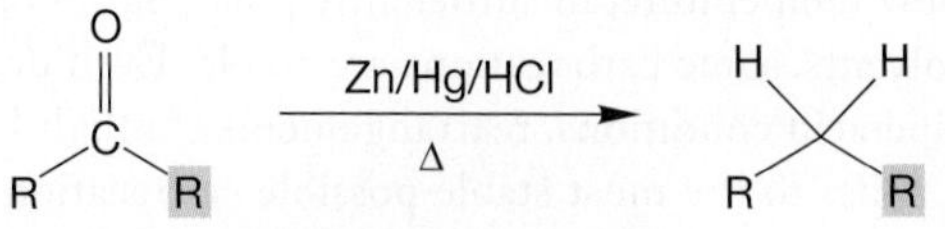

Clemmensen reduction

2. Alkylbenzenes

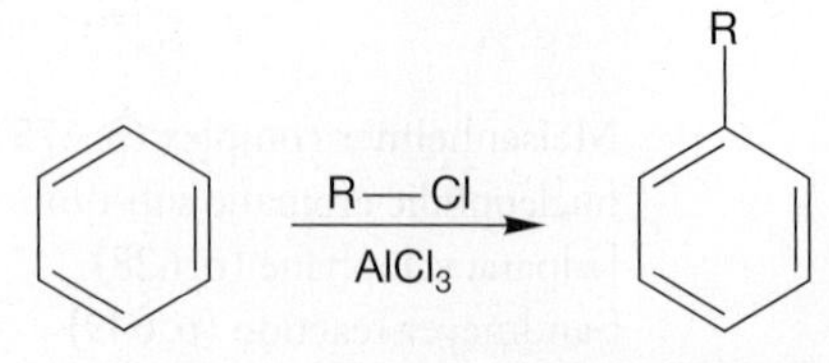

Friedel–Crafts alkylation; carbocationic rearrangements occur to give the most stable ion; the possible structures of R are therefore limited

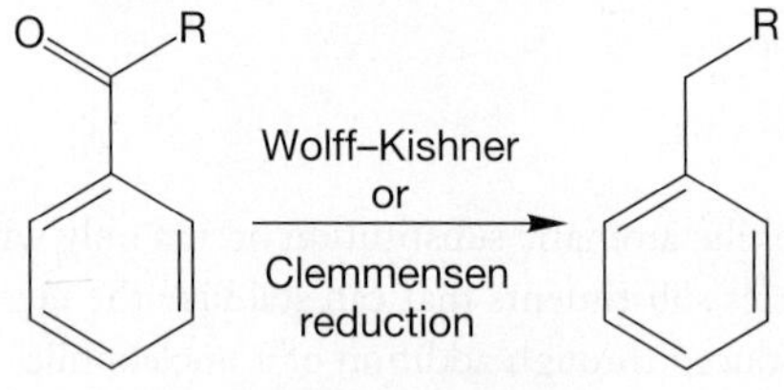

Wolff–Kishner
or
Clemmensen
reduction

3. Anilines

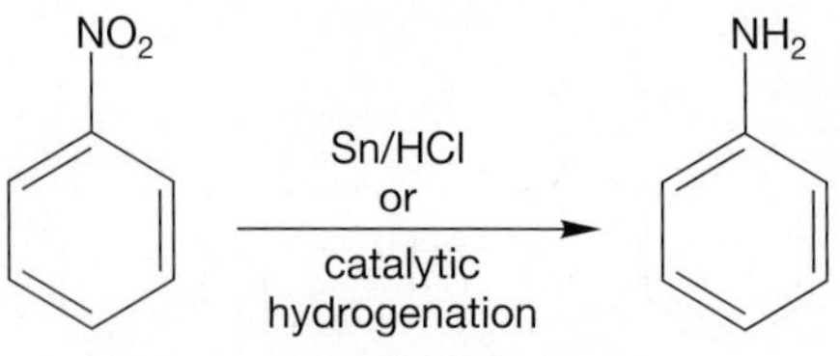

Sn/HCl
or
catalytic
hydrogenation

Reduction of nitro groups

4. Benzenes

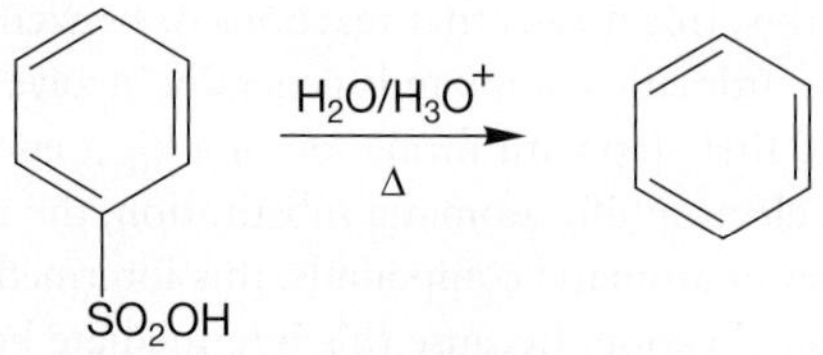

Reversal of sulfonation reaction

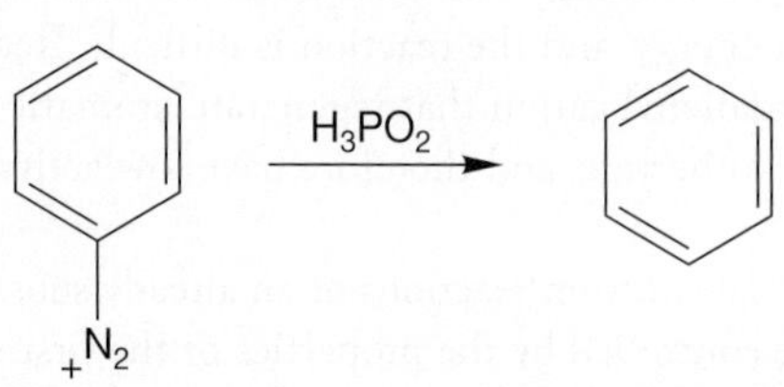

The diazonium ion can be made from nitrobenzene

5. Benzenesulfonic Acids

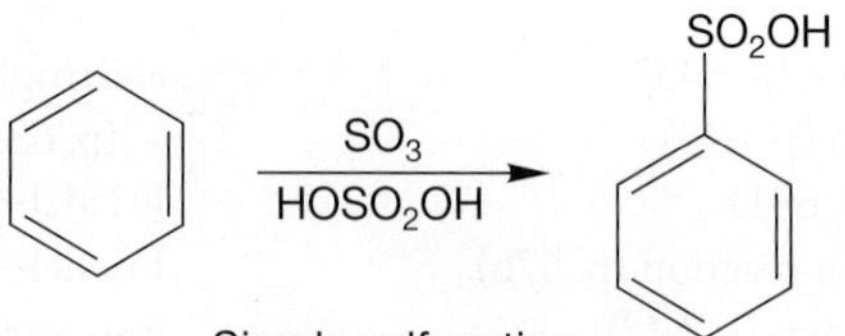

Simple sulfonation

6. Cyanobenzenes

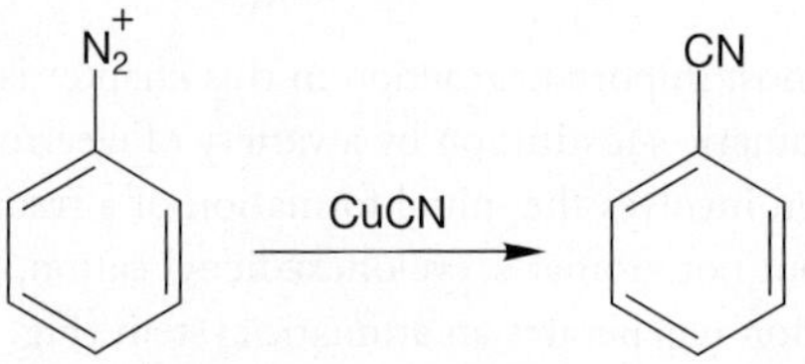

Sandmeyer reaction

7. Cyclohexanes

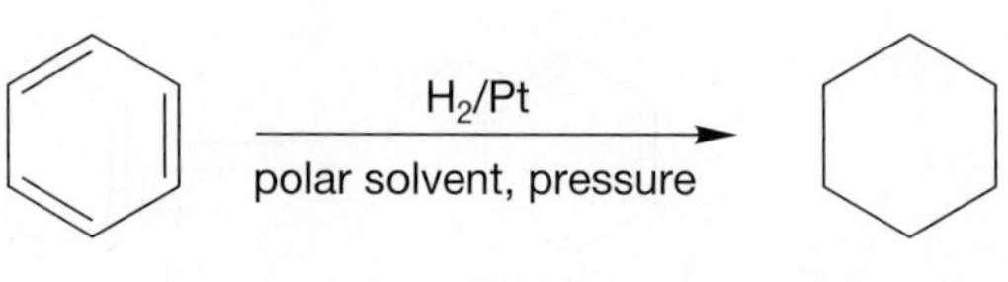

High-pressure catalytic hydrogenation

8. Deuterated Benzenes

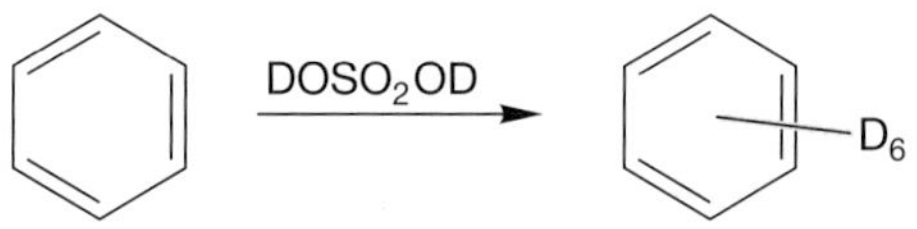

See also methods for making benzenes in item 4

9. Diazonium Ions

10. Halobenzenes

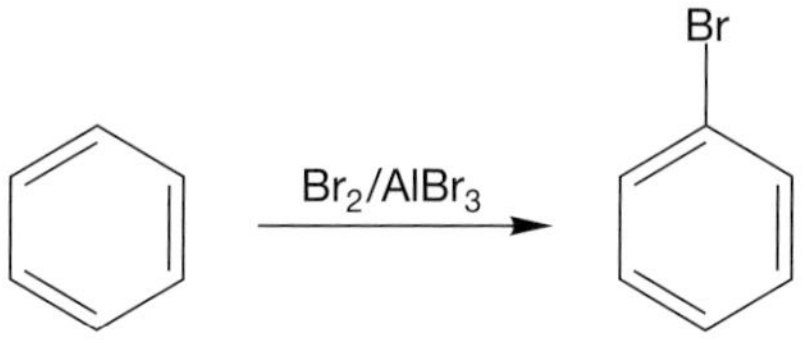

Aromatic halogenation

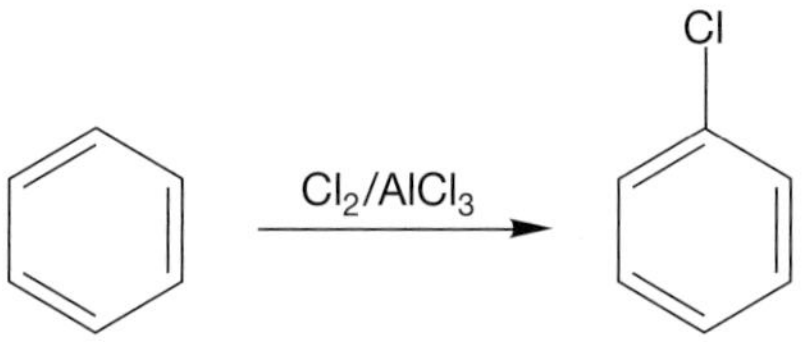

Aromatic halogenation

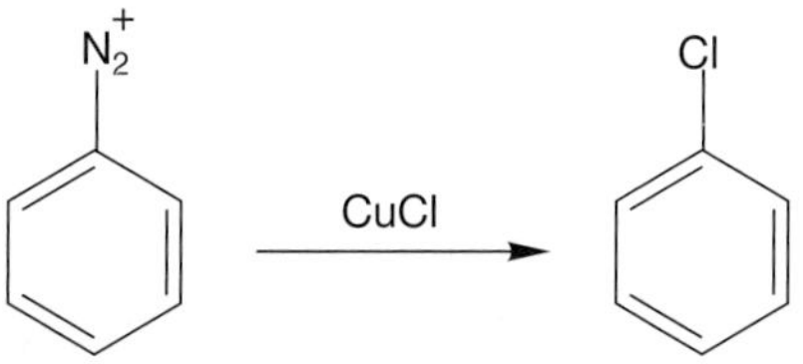

Sandmeyer reaction

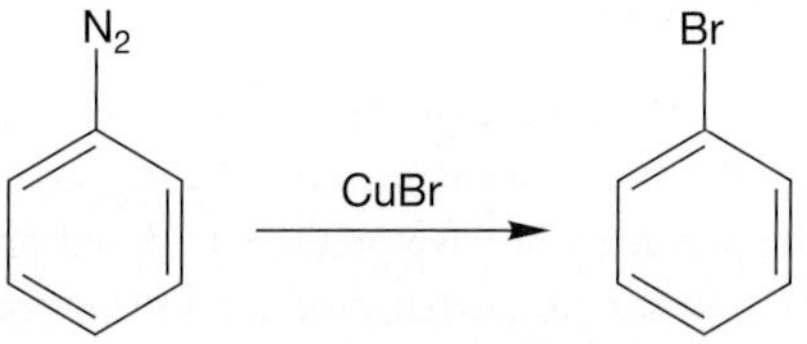

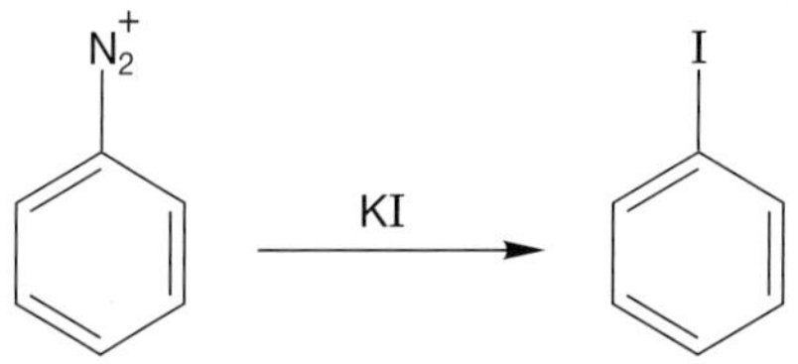

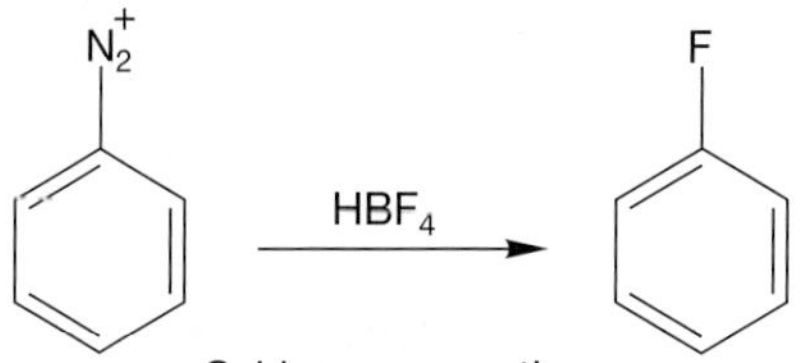

Sandmeyer reaction

Schiemann reaction

11. Nitrobenzenes

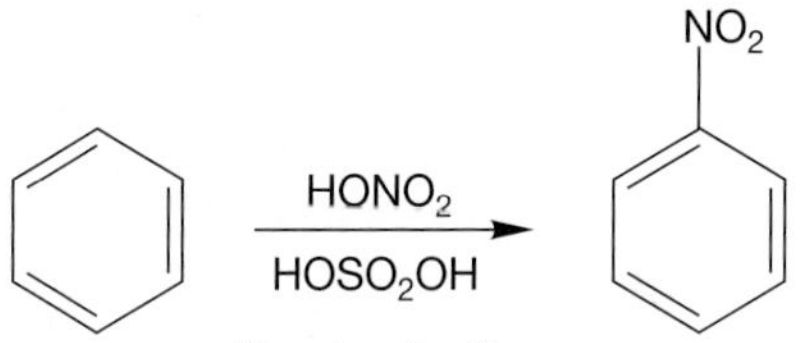

Simple nitration

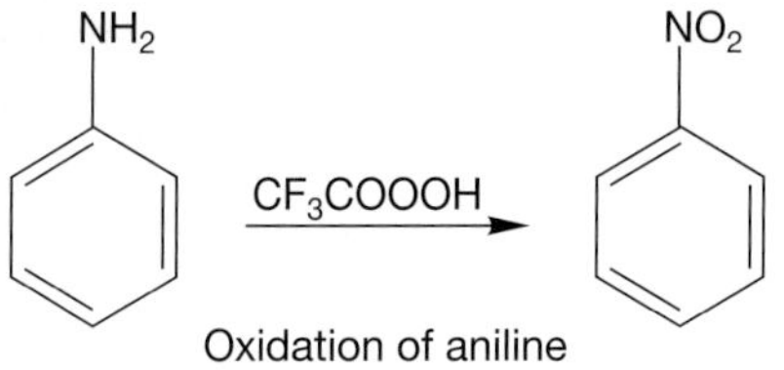

Oxidation of aniline

12. Phenols

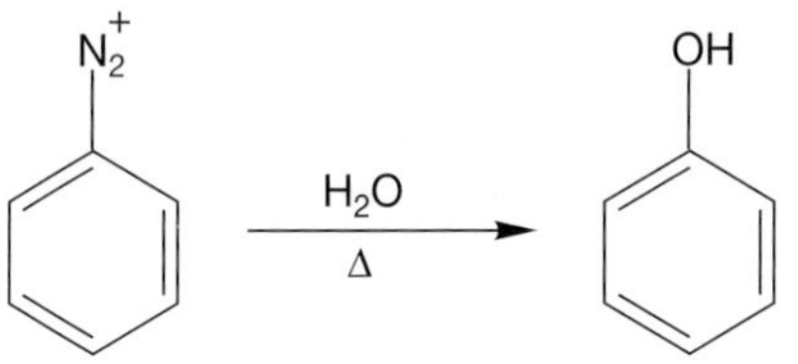

13. Phenyl Ketones

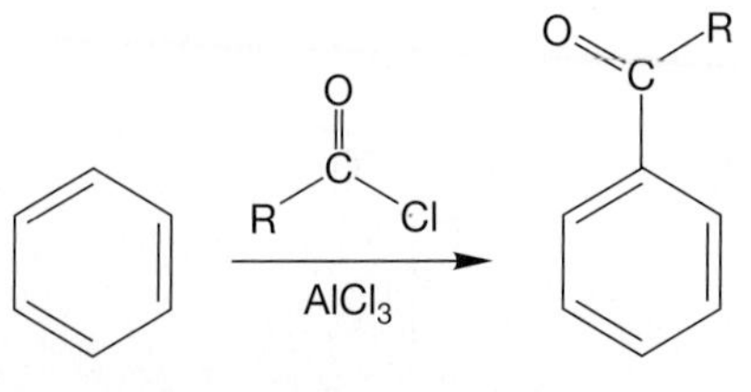

Friedel–Crafts acylation; no rearrangements occur

Common Errors

The most common conceptual error is to attempt to analyze the course of aromatic substitution reactions by looking at the starting materials and products. The various resonance and inductive effects discussed in this chapter are far more important in the charged intermediates and in the transition states leading to those intermediates. When analyzing one of these reactions, always look first at the cyclohexadienyl intermediate (a cation in electrophilic aromatic substitution and an anion in the less common nucleophilic reactions), and then consider the transition state leading to it.

Another potential error is failing to keep up with the proliferating synthetic detail. Be sure to look at each new reaction in this and subsequent chapters with an eye to using it in synthesis. Keep up your file cards!

14.17 Additional Problems

PROBLEM 14.45 Explain why the trifluoromethyl group is meta-directing in electrophilic aromatic substitution. Would you expect CF_3 to be activating or deactivating? Why?

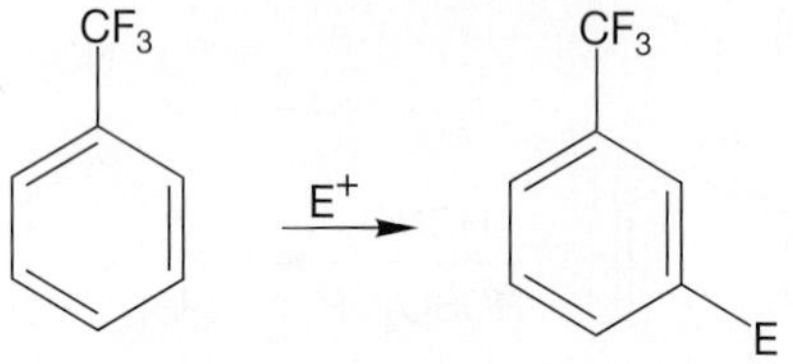

PROBLEM 14.46 Show where the following compounds would undergo further substitution in electrophilic aromatic substitution. There may be more than one position for further substitution in some cases. Explain your reasoning.

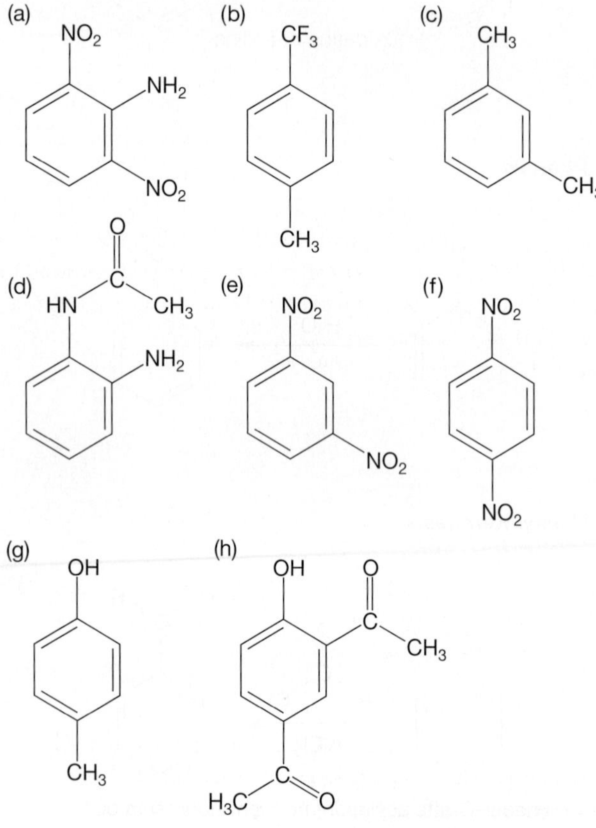

PROBLEM 14.47 Predict the position of further substitution in biphenyl. Explain your reasoning.

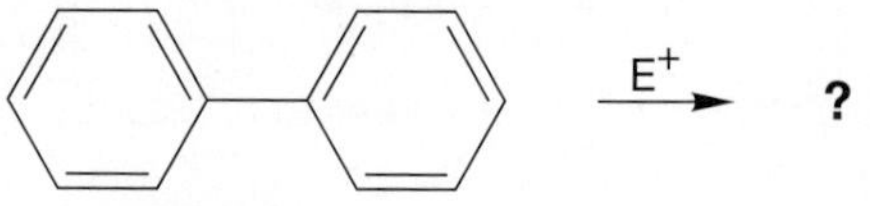

Biphenyl

PROBLEM 14.48 Show specific reagent(s) and product(s) for the following reactions with anisole (methoxybenzene):

(a) nitration
(b) alkylation
(c) sulfonation
(d) monobromination
(e) acylation

PROBLEM 14.49 Show the product(s) for chlorination (Cl_2, $FeCl_3$) of the following compounds:

(a) toluene
(b) chlorobenzene
(c) 4-nitrotoluene
(d) 2,4-dinitrophenol
(e) diphenylether

PROBLEM 14.50 Diclofenac is a nonsteroidal antiinflammatory drug (NSAID) that is effective for treating pain. In an effort to better understand its enzyme selectivity, it has been nitrated. A single product was obtained. Predict the product and explain your choice.

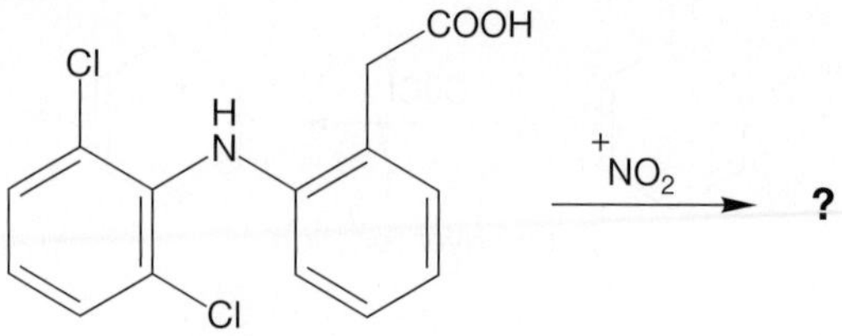

Diclofenac

PROBLEM 14.51 Predict the product for the first reaction shown on the next page. What mechanism do you suppose is involved in making the product you have predicted? Based on your answer to the first reaction, what product do you predict for the second reaction shown?

First reaction:

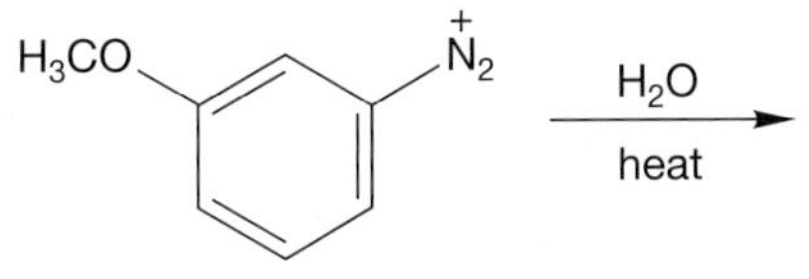

Second reaction:

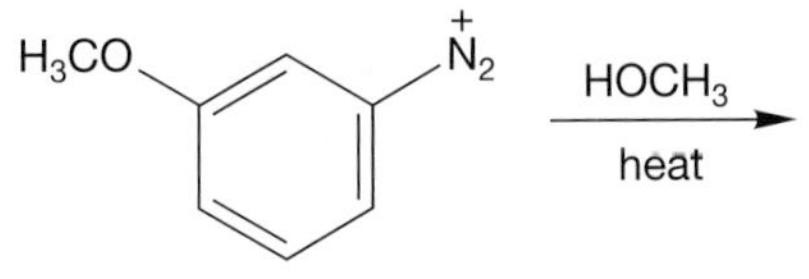

PROBLEM 14.52   Problem 14.18 describes the formation of an azo compound. The dye industry is dependent on the reaction between diazonium ions and activated aromatic rings to form azo compounds. What aspect of the azo compounds makes them good dyes? What diazonium salt would you use to make Para Red (shown below)? Para Red is reported to be the first azo dye synthesized, and has been used as a dye for both clothing and food.

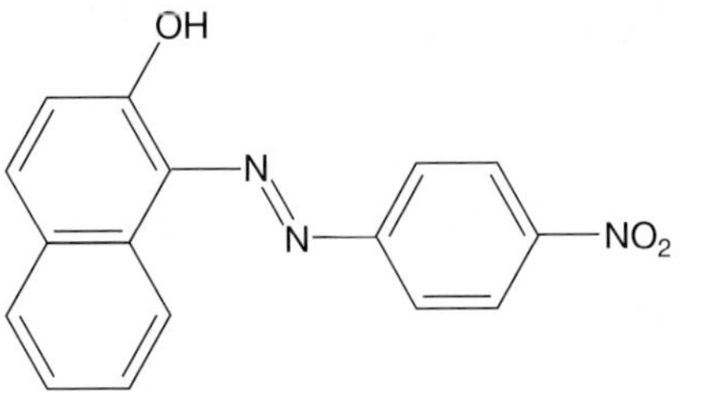

PROBLEM 14.53   In the presence of very strong acid, phenol is brominated in the meta position. Why do you suppose this is the case? When phenol is brominated in slightly basic conditions, tribromophenol is the major product. Why would basic conditions favor polybromination?

PROBLEM 14.54   Provide syntheses for the following compounds, as free of other isomers as possible. You do not need to write mechanisms, and you may start from benzene, inorganic reagents of your choice, organic reagents containing no more than four carbons, and pyridine.

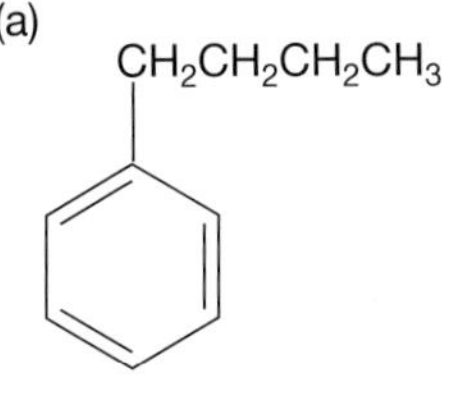

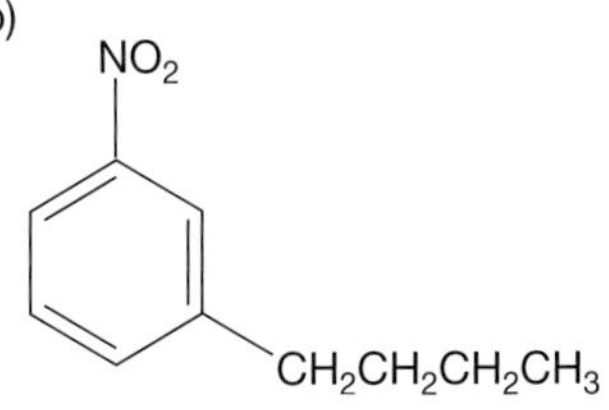

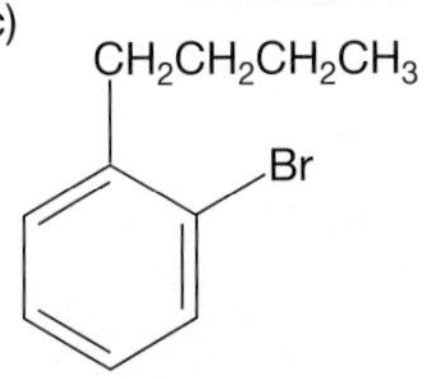

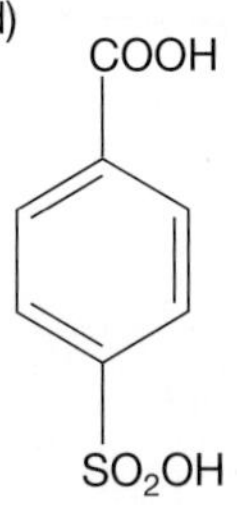

PROBLEM 14.55   Provide syntheses for the following compounds, free of other isomers. You do not need to write mechanisms, and you may start from benzene, inorganic reagents of your choice, organic reagents containing no more than two carbons, and pyridine.

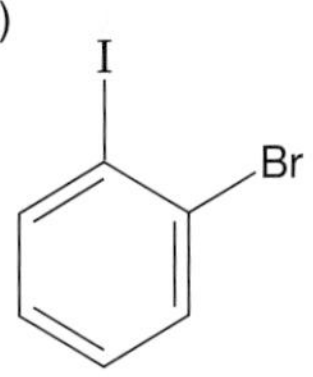

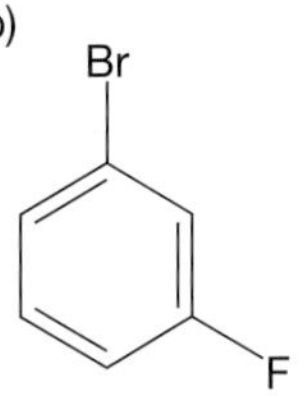

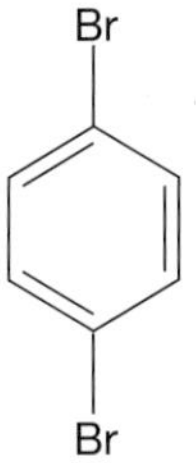

PROBLEM 14.56   How many signals would appear in the ^{13}C NMR spectra of the following four compounds? Try to use an analysis of symmetry in your answer.

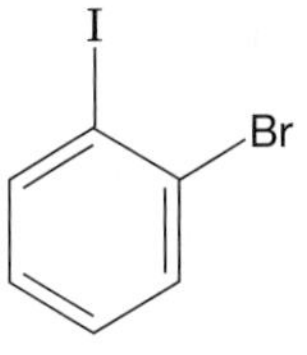

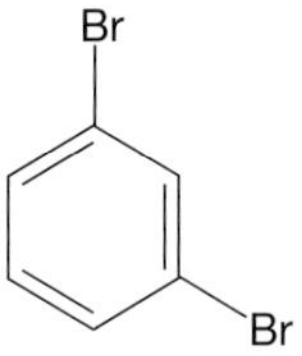

PROBLEM 14.57   Provide syntheses for the following compounds, as free of other isomers as possible. You do not need to write mechanisms, and you may start from benzene, inorganic reagents of your choice, organic reagents containing no more than three carbons, and pyridine.

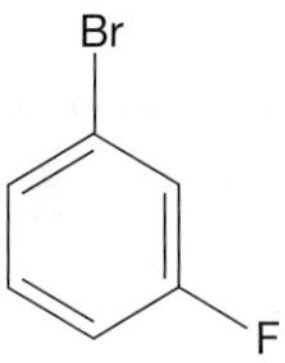

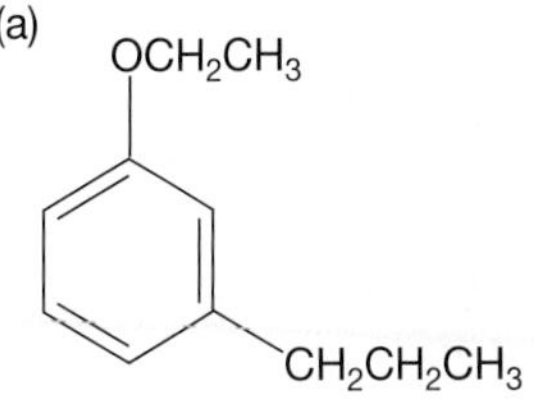

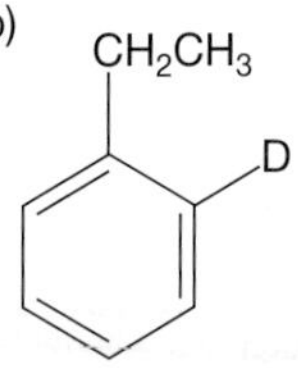

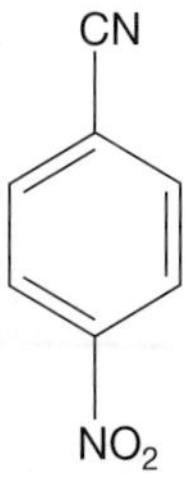

PROBLEM 14.58   Provide structures for compounds **A–E**. Mechanisms are not required, but a mechanistic analysis of the various reactions will probably prove helpful in deducing

the structures. *Hint*: You might look at the figure in Problem 14.14 (p. 646).

PROBLEM 14.59 One substituent you do not know how to attach to a benzene ring is the aldehyde group, HC=O. Here is a way. The price of this knowledge is that you must write a mechanism for the following aromatic substitution reaction.

PROBLEM 14.60 Lies! Lies! Lies! We can think of at least one way (other than the reaction in Problem 14.59) you can put an aldehyde group on a benzene ring with your current knowledge. It is a roundabout process, and so makes a good "roadmap" problem. Try this one. Provide structures for compounds **A–D**.

PROBLEM 14.61 Explain mechanistically why it is possible to alkylate a phenol selectively in the presence of an alcohol.

PROBLEM 14.62 Write structures for compounds **A**, **B**, and **C**.

PROBLEM 14.63 How would you convert aniline into each of the following molecules? Mechanisms are not necessary.

PROBLEM 14.64   Write mechanisms for the following two related reactions:

(43%)

75–80%
H₂SO₄

(75%)

PROBLEM 14.65   Rationalize the following data. An Energy versus Reaction progress diagram will be useful. Be sure to explain why the low-temperature product is mainly naphthalene-1-sulfonic acid, and why the high-temperature product is mainly naphthalene-2-sulfonic acid. Note that naphthalene-2-sulfonic acid is more stable than naphthalene-1-sulfonic acid. You do not have to explain the relative energies of the two naphthalene sulfonic acids, although you might guess, and the answer will tell you.

concd H₂SO₄ | 40 °C   concd H₂SO₄ | 160 °C

SO₂OH   SO₂OH

(major)   (minor)
+   +
SO₂OH   SO₂OH

(minor)   (major)

PROBLEM 14.66   Write a mechanism for the following chloromethylation reaction:

H₂C=O
HCl

CH₂Cl

(75%)

PROBLEM 14.67   The following wonderful question appeared on a graduate cumulative examination at Princeton in January, 1993. With a little care you can do it easily. Write a mechanism for this seemingly bizarre change. *Hint*: It is not bizarre at all. Think simple! First consider what HCl might do if it reacts with the starting material.

AlCl₃/H₂O/HCl

benzene,
80 °C

● = ¹³C

PROBLEM 14.68   At low temperature, only a single product (**A**) is formed from the treatment of *p*-xylene with isopropyl chloride and AlCl₃, which is surely no surprise, as there is only one position open on the ring. However, at higher temperature some sort of rearrangement occurs, as two products are formed in roughly equal amounts. First, write a mechanism for formation of both products. *Hint*: Note that HCl is a product of the first reaction to give **A**. Second, explain why product **A** rearranges to product **B**. What is the thermodynamic driving force for this reaction?

(CH₃)₂CHCl
AlCl₃, 70 °C,
2 h

CH₃
CH(CH₃)₂
A
(48%)
CH₃

H₃C
CH(CH₃)₂
B
(52%)
CH₃

p-Xylene

PROBLEM 14.69   Provide mechanisms for the following reactions:

(a)

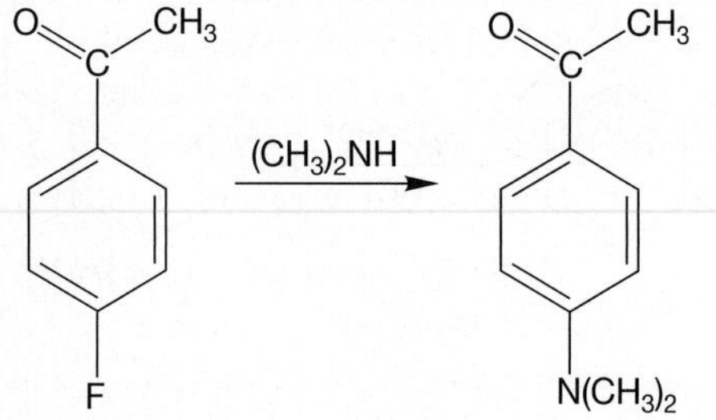

(b)

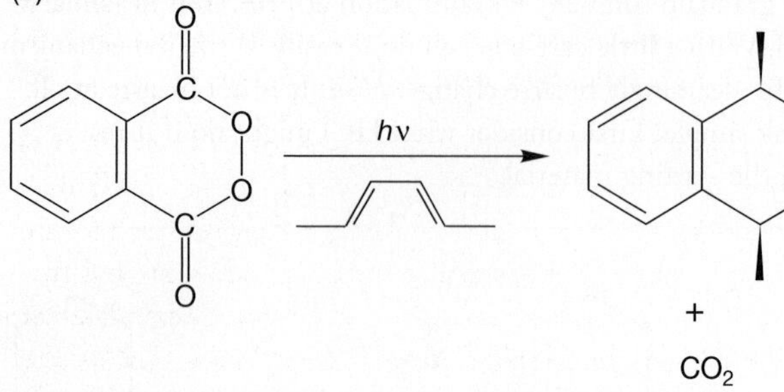

$+$

CO_2

PROBLEM 14.70   Construct an Energy versus Reaction progress diagram for the reaction of 1-chloro-2,4-dinitrobenzene with sodium methoxide.

PROBLEM 14.71   When 1,2-dichloro-4-nitrobenzene is treated with an excess of sodium methoxide, three different substitution products (two monosubstituted, one disubstituted) could, in principle, be formed. In fact, only a single product is obtained.

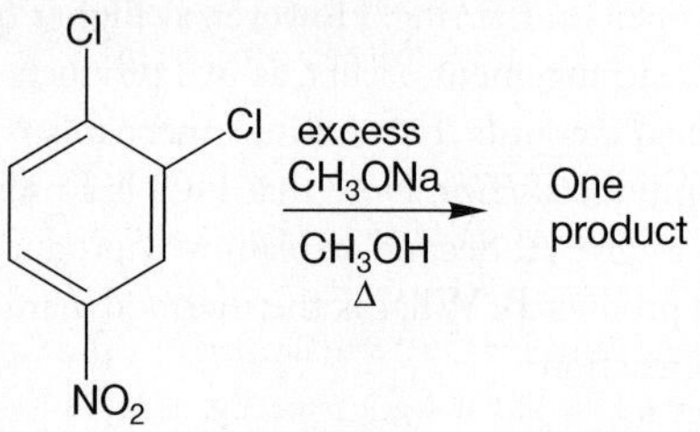

(a) Draw the three possible products.
(b) Carefully analyze the mechanism for this nucleophilic aromatic substitution reaction. What single product will be formed and why?

PROBLEM 14.72   A useful benzyne precursor is benzenediazonium-2-carboxylate (**1**), a molecule readily formed from anthranilic acid. Upon heating, **1** affords benzyne, which can be trapped by anthracene to give the architectural marvel, triptycene (**2**).

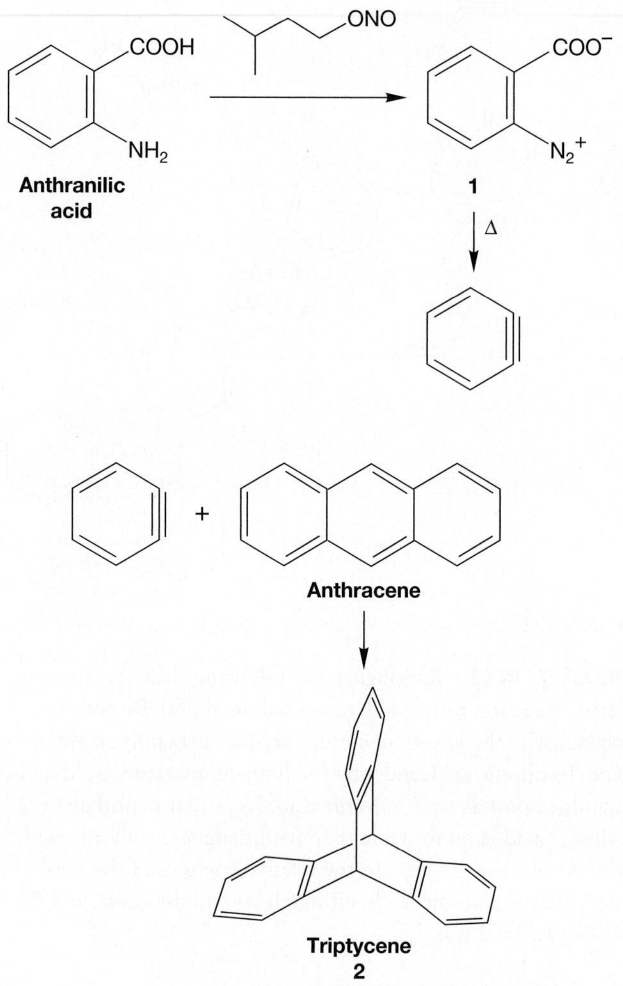

Anthracene

Triptycene
2

(a) Provide an arrow formalism for the formation of benzyne on decomposition of benzenediazonium-2-carboxylate.
(b) Similarly, provide an arrow formalism for the formation of triptycene from benzyne and anthracene.

PROBLEM 14.73   Years before J. D. Roberts' classic labeling experiment (p. 681), it was known that nucleophilic aromatic substitution occasionally resulted in what was termed *cine* (from the Greek *kinema* meaning "movement") substitution. In this type of substitution reaction, the nucleophile becomes attached to the carbon adjacent to the carbon bearing the leaving group. Here are two examples of this kind of substitution:

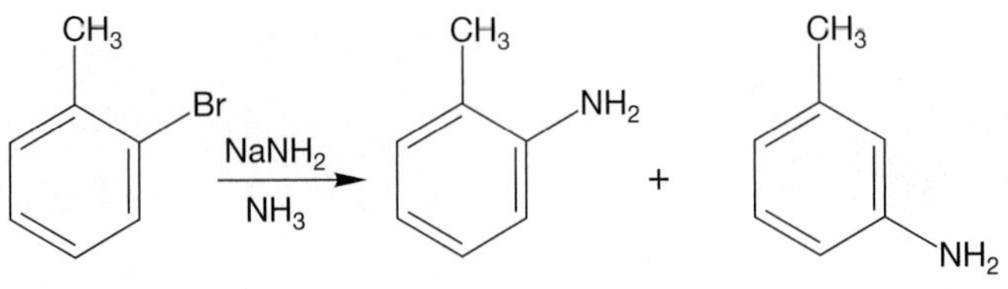

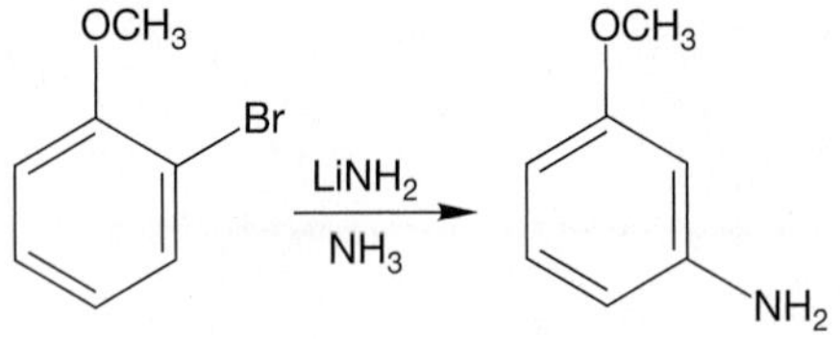

Only isomer formed

Rationalize these *cine* substitutions through use of a substituted benzyne as an intermediate. Explain why only a single isomer is formed in the second experiment.

PROBLEM 14.74   In phenylalanine biosynthesis, the OH group is a leaving group (see Fig. 14.118). That's not normal. Why do you suppose it happens in this case? How could an enzyme make it easier?

PROBLEM 14.75   2,4-Dinitrofluorobenzene is a better electrophile for S_NAr reactions than the corresponding chloride (see Fig. 14.95). 2,4-Dinitrofluorobenzene reacts with amines at room temperature to give orange-colored adducts. Show the mechanism of the reaction between 2,4-dinitrofluorobenzene and dimethylamine. Predict the product. Why is the fluoride more reactive than the chloride? Why is the product highly colored?

Use Organic Reaction Animations (ORA) to answer the following questions:

PROBLEM 14.76   Select the reaction "Arene halogenation" on the Semester 1B panel. Observe the reaction several times, then select the LUMO track. As the reaction reaches the intermediate, pause the animation. Notice that the LUMO shows the locations sharing cationic character. Draw the resonance forms for this intermediate. Which site has the lowest cationic character? Why?

PROBLEM 14.77   Watch the "Friedel–Crafts acylation" reaction. What is the hybridization of the carbon of the acylium ion? What species is animated as the weak base that deprotonates the nonaromatic intermediate? Consider the hydrogen that is being deprotonated in this last step of the reaction. Give a pKa value that would reflect how acidic that hydrogen is. (*Hint*: It is the most acidic hydrogen we have encountered.)

PROBLEM 14.78   We have animated the addition–elimination version of the S_NAr reaction for your viewing pleasure. This reaction has much to teach us. Select the LUMO track for the reaction. Notice that the starting aromatic compound has the largest LUMO character on one of the atoms. Which one is it? Where does the nucleophile attack? The animation shows the amine being deprotonated before the chloride is kicked out. Why do you suppose deprotonation might be necessary prior to the chloride leaving? Look at the final product. What is the hybridization of the amine nitrogen? Finally, look at the HOMO track for the second intermediate in the reaction. It is the clearest picture of the resonance available in the anionic intermediate. Using this visual insight, draw each of the resonance forms.

PROBLEM 14.79   The Chapter 14 benzyne reaction is also animated. Benzyne is an intermediate in the elimination–addition version of the S_NAr reaction, and is a very reactive intermediate. Observe the reaction several times. What are the most acidic hydrogens in the reaction mixture? What are the second most acidic hydrogens? The animation shows benzene without double bonds. Why are the double bonds left out? The animation shows the amide ion ($^-NH_2$) attacking one carbon of the benzyne. Observe the LUMO of the benzyne. Could the $^-NH_2$ attack either carbon of the benzyne? Remember that the LUMO tells us about cationic character. How can both carbons of the benzyne have equal cationic character?

15

Analytical Chemistry: Spectroscopy

15.1 Preview

15.2 Chromatography

15.3 Mass Spectrometry (MS)

15.4 Infrared Spectroscopy (IR)

15.5 ¹H Nuclear Magnetic Resonance Spectroscopy (NMR)

15.6 NMR Measurements

15.7 ¹³C NMR Spectroscopy

15.8 Problem Solving: How to Use Spectroscopy to Determine Structure

15.9 Special Topic: Dynamic NMR

15.10 Summary

15.11 Additional Problems

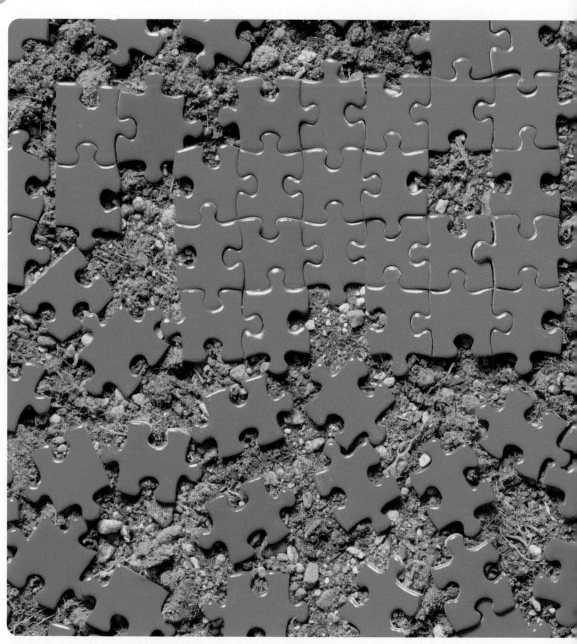

PUZZLES Using spectroscopy to determine the structure of an unknown organic molecule is very much like putting together the pieces of a jigsaw puzzle.

After a great expenditure of life and treasure a Daring Explorer had succeeded in reaching the North Pole, when he was approached by a Native Galeut who lived there.

"Good morning," said the Native Galeut. "I'm very glad to see you, but why did you come here?"

"Glory," said the Daring Explorer, curtly.

"Yes, yes, I know," the other persisted; "but of what benefit to man is your discovery? To what truths does it give access which were inaccessible before?—facts, I mean, having scientific value?"

"I'll be Tom scatted if I know," the great man replied frankly; "you will have to ask the Scientist of the Expedition."

But the Scientist of the Expedition explained that he had been so engrossed with the care of his instruments and the study of his tables that he had found no time to think of it.

—AMBROSE BIERCE,[1] "AT THE POLE," FROM *FANTASTIC FABLES*

15.1 Preview

Years ago, all universities gave one or several courses in the subject of analytical chemistry. Sadly, in recent times this material has sometimes been folded into courses in physical and organic chemistry, and the subject "lost." Sometimes also lost, we fear, is the sense of how important the subject is. Nothing has influenced the development of organic chemistry more strongly over the last 50 years than progress in analytical chemistry. After a chemical reaction is run, the experimentalist is faced with two general problems before he or she can begin to think about what the results mean: separating the products of the reaction from one another, then identifying those products.

Separation was a severe problem in the early days of organic chemistry. For many compounds, the only available methods were distillation or, for solids, fractional crystallization. Fractional crystallization works well (if done properly, which it rarely is), but this method requires reasonably large amounts of material and, of course, works only for compounds that are solids at or near room temperature. If you are taking an organic lab along with this course, then by now your laboratory experience has likely revealed that distillation is not often the final answer to securing large amounts of very pure materials. Distillation works well if there is a large difference between the boiling points of all the desired components of a mixture, and if substantial amounts of material are available. It seems that reactions in the research laboratory are all too prone to give several isomeric products, and distillation and fractional crystallization are often impractical methods. The vexing problem of separation was relieved somewhat by the advent of column chromatography, then more or less solved, at least in the research lab, by the discovery of gas and liquid chromatography. Other chromatographic techniques include thin-layer, preparative-layer, and flash chromatography. Nowadays, it is the chemist's realistic expectation that any mixture can be separated. Even enantiomers are now often separable by chromatography.

---

[1] Ambrose Bierce (1842–1914?) was an American journalist and writer of bitter, sardonic tales who vanished during the Mexican civil war at the start of the last century.

> **PROBLEM 15.1** How would a separation of enantiomers be possible? Are not all physical properties (except, of course, for the direction of rotation of the plane of plane-polarized light) identical for enantiomers? Can you guess the principle on which the chromatographic separation of enantiomers must rest? *Hint*: What is true for enantiomers is not necessarily true for diastereomers (p. 169).

Even given a successful separation of a reaction mixture, the chemist is now faced with the daunting task of working out the structures of the various components. Just how do we find out what's in that bottle? In the old days, structure determination was a respected profession, and many careers were built through the chemical determination of complex molecular structures. Diagnostic reactions were run, and the presence or absence of functional groups inferred from the results. Gradually, the results from many chemical experiments allowed the formulation of a reasonable structural hypothesis that could be tested through further chemical reactions. The intellectual complexity of such work was substantial, and the experimental skill required was prodigious. These days, the correlation of the results of chemical reactions with chemical structure has been largely replaced with the spectroscopic identification of compounds, and *every* chemist is an analytical chemist. The profession of analytical chemist has by no means disappeared, but rather has expanded to include us all. Ultraviolet/visible (UV/vis) spectroscopy (p. 525) was the earliest form of spectroscopy to gain widespread use. It was followed by mass spectrometry (MS), which is capable of determining molecular weights, and infrared (IR) spectroscopy, in which the vibrations of atoms in bonds are detected. At the end of the 1950s, nuclear magnetic resonance (NMR) spectroscopy appeared,[2] and the real revolution in the determination of the structures of organic molecules began. In this chapter, we will briefly discuss chromatography, MS, and IR spectroscopy, and then proceed to a more detailed investigation of NMR spectroscopy.

ESSENTIAL SKILLS AND DETAILS

1. "All" there is to understanding nuclear magnetic resonance (NMR) is to know how the chemical shift (δ) arises, how coupling between atoms works, and how intensities of signals are compared. That's only three things, but each requires a lot of experience before using it becomes second nature. Understanding the chemical shift δ requires no more than realizing that "different" hydrogens will resonate at different frequencies. That's not a difficult concept, but understanding what makes two hydrogens "different" is less simple! The coupling between atoms means that hydrogens on adjacent carbons "talk" to each other. The state of one influences the properties of the other. There is a simple method for translating the electron–nuclear "talk" but, again, it requires an

---

[2] Professor Martin Saunders (b. 1931) told MJ of the arrival of the long and eagerly awaited first NMR spectrometer at Yale in the late 1950s. The magnet arrived in a plywood carton labeled in bold letters, "2.5 tons." Unfortunately, the workman who was to bring the machine into the laboratory didn't believe that such a heavy object could be in such a flimsy box and attempted to use his forklift, rated at 1 ton, to bring it in. The forklift did its best and labored mightily to lift the object into the air. It succeeded at this task, but then, apparently exhausted, gave out and dropped the magnet to the stone floor, where it burst its packing and lay blocking the entrance. The workman then attached a chain to the magnet and dragged it about 30 feet across the cobblestones to the side. Naturally Professor Saunders, whose professional fate depended to a fair extent on the performance of this instrument, was less than happy about this state of affairs. However, the story has a happy ending. The machine functioned brilliantly after minor repairs, and Professor Saunders has had a long and outstanding career at Yale. He claims that he advocated to Varian Associates, the manufacturer of the instrument, that they add a 5-foot drop to a cobblestone floor to the manufacturing process for their magnets, in order to bring all their magnets up to the standard of this one. Perhaps understandably, they ignored his advice.

understanding of difference. The intensity of signals is determined by the technique of integrating the signals in the spectrum.

2. Problem solving. Good spectroscopy problem solvers are good at solving spectroscopy problems because they have solved lots of spectroscopy problems. There is no way out! To be a decent spectroscopist you have to do many problems; the skill is highly experience-intensive and there is only one way to gain that experience. Section 15.8 will give you one way to approach spectroscopy problems, but it is certainly appropriate to develop your own.

15.2 Chromatography

In Chapter 4 (p. 172), we briefly encountered column chromatography. All chromatography works in much the same way. The components of a mixture are forced to equilibrate between an immobile material, the stationary phase, and a moving phase. The more time one component spends adsorbed on the stationary phase, the more slowly it moves. If it is possible to monitor the moving phase as it emerges from the chromatograph, each component of the original mixture can be detected and collected as it appears. The efficiency of column chromatography can be greatly increased by a technique called **high-performance liquid chromatography (HPLC)**. In this variation, high-pressure pumps force the liquid phase through a column of densely packed uniform microspheres, which provide an immense surface area for adsorption. As the liquid elutes, a detector, often a small UV/vis spectrometer, monitors the composition of the liquid. Figure 15.1 shows a schematic instrument as well as a nice separation of a series of organic compounds.

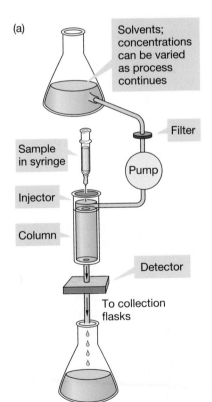

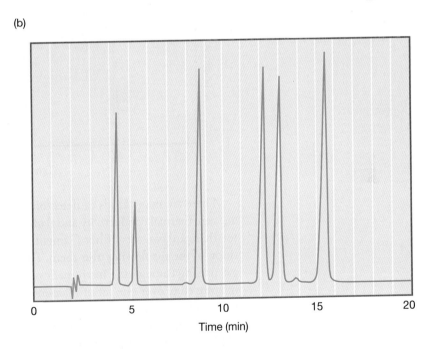

FIGURE 15.1 (a) A schematic view of a liquid chromatograph. (b) A chromatogram showing the separation of a mixture of organic compounds into its components.

For separating many organic molecules, **gas chromatography (GC)** is the technique of choice. Gas chromatographs are remarkably simple and it is still possible to build one on your own that will function well in a research setting. The stationary phase consists of a high-boiling material adsorbed on a solid support. High molecular weight silicone oil polymers are often used, as well as many other organic chemicals. The moving phase is a gas, usually helium or hydrogen. Typically, the solid support is packed in a coiled column through which the carrier gas flows. The mixture to be separated is injected at one end. As the carrier gas moves the mixture through the column, the various components are differentially adsorbed by the stationary phase and exist in an equilibrium between the gas phase and solution. The more strongly a component is adsorbed, the more slowly it moves through the column. At the end of the column, there is a detector that plots the amount of material emerging as a function of time. A schematic instrument and a typical chromatogram are shown in Figure 15.2.

FIGURE 15.2 (a) A schematic view of a gas chromatograph. (b) A chromatogram showing the separation of a mixture into its components.

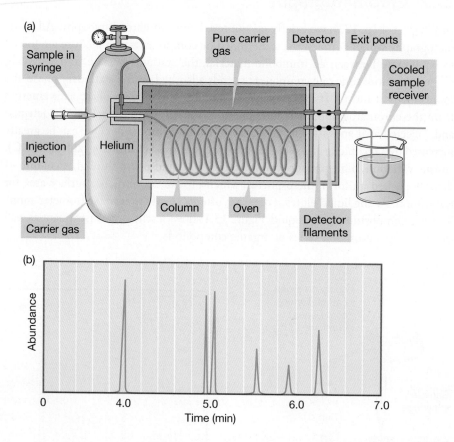

There are many kinds of detectors used for gas chromatography; two common ones are thermal conductivity and flame ionization. In a thermal conductivity detector, there are two sets of hot filaments whose temperature depends on the composition of the gas flowing over them. The detector can be set to zero for pure carrier gas passing through the instrument over both filaments. As the components of the mixture elute over one set of filaments, their temperature will change and the current needed to bring them back to the zero point can be measured and plotted. A flame ionization detector measures the current induced by ions created as the effluent is burned. Again, a zero point for pure carrier gas is determined and deviations from zero are detected and plotted as the components of the mixture elute.

All sorts of fancy bells and whistles are possible, and the art of creating efficient GC columns has progressed immensely from the days when one packed large-bore columns by hanging them down a stairwell, and pouring in the stationary phase. The resulting columns were then tested for efficiency in the sometimes vain hope that one would do a good enough job of separating the mixture of products. One especially effective modification is the construction of capillary columns. Such columns can be several hundred feet long and contain only a tiny amount of stationary phase coated on the walls. They often have immense separating powers. Yet, for the practicing chemist, the old-style simple packed GC columns are still very important because they allow for the collection of the components of the mixture as they pass out of the column. Not only can the mixtures be separated, but the pure components can be isolated for identification by the simple means of attaching a cooled receiver to the end of the chromatograph (Fig. 15.2).

A particularly useful modification of a gas chromatograph is to attach it to a mass spectrometer (**GC/MS**) or infrared spectrometer (**GC/IR**) so that the various components can be analyzed directly as they emerge from the column. In the following sections, we will see what kinds of information MS and IR spectroscopy can give about molecules.

15.3 Mass Spectrometry (MS)

First of all, never make the mistake of calling it "mass spectroscopy." Spectroscopy involves the absorption of electromagnetic radiation and as we will see later, **mass spectrometry (MS)** is different. The mass spectrometrists sometimes get upset if you confuse the issue.

15.3a The MS Spectrometer
In MS, a compound is bombarded with very high-energy electrons. Typically, electrons of 70 eV are used.

> **WORKED PROBLEM 15.2** How many kilocalories per mole does 70 eV represent? See page 4.
>
> **ANSWER** One electron volt (eV) is 23.06 kcal/mol, so 70 eV is 1614 kcal/mol. The point of this problem is to show that the incoming electron is carrying a *lot* of energy!

One thing these very high-energy electrons can do is eject an electron from a molecule, M, to give the corresponding **radical cation**, $M^{\bullet+}$ (Fig. 15.3). Now what can happen to the newly formed radical cation? In a typical mass spectrometer, it is first accelerated toward a negatively charged plate (Fig. 15.4). There is a slit in the plate, and some of the accelerated ions pass through the slit to form a beam.

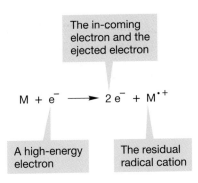

The in-coming electron and the ejected electron

$$M + e^{-} \longrightarrow 2\,e^{-} + M^{\bullet+}$$

A high-energy electron

The residual radical cation

FIGURE 15.3 When bombarded with high-energy electrons, molecules can lose an electron to give a radical cation.

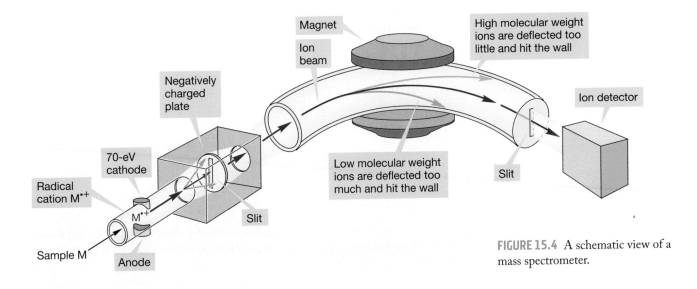

Magnet

Ion beam

High molecular weight ions are deflected too little and hit the wall

Negatively charged plate

Ion detector

70-eV cathode

Low molecular weight ions are deflected too much and hit the wall

Slit

Radical cation $M^{\bullet+}$

$M^{\bullet+}$

Slit

Sample M

Anode

FIGURE 15.4 A schematic view of a mass spectrometer.

The beam of ions then follows a curved path between the poles of a magnet. In a magnetic field, a moving charged particle, the radical cation in this case, is deflected by an amount proportional to the strength of the magnetic field. Too little deflection (too weak a magnetic field) results in the ion hitting the far wall of the tunnel and in loss of the ion. If there is too much deflection (too strong a magnetic field), the ion impinges on the near wall and is also lost. If, however, the strength of the magnetic field is just right, the ion will follow a curved path to the detector and be "seen." The magnetic field can be tuned so that ions of different mass are curved just the right amount to be detected. For ions of low mass, a relatively low magnetic field will be necessary; for ions of higher mass, the field will have to be stronger to do the required bending of the path.

Mass spectra are reported as graphs of the mass-to-charge ratio, m/z, versus intensity. As long as we are dealing with singly charged ions, the positions of the signals will correspond to the molecular weights of the ions detected because the charge, z, will be 1.

So far, we have a device that can detect $M^{\cdot +}$ ions formed by bombardment of a molecule M. This species, $M^{\cdot +}$, is called the **molecular ion**. In our detector, we should see an ion with the mass m corresponding to the molecular weight of M. For cyclohexane, we should see a molecular ion of 84 mass units, because the molecular weight of cyclohexane (C_6H_{12}) is 84 (Fig. 15.5). In the mass spectrum of cyclohexane, there is a small peak above the molecular ion, at $m/z = 85$. It is only about 7% of the molecular ion, and therefore one must look closely in order to see it.

FIGURE 15.5 Many different ions can have a molecular weight of 84. Two possibilities are the molecular ions of cyclohexane and 3,4-dihydro-2*H*-pyran.

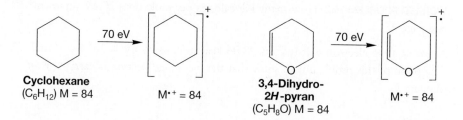

Cyclohexane
(C_6H_{12}) M = 84 $M^{\cdot +}$ = 84

3,4-Dihydro-
2*H*-pyran
(C_5H_8O) M = 84 $M^{\cdot +}$ = 84

This **M + 1 peak** is not the result of an impurity, however; it is always there no matter how carefully the cyclohexane is purified. This M + 1 peak exists because the carbon and hydrogen have naturally occurring isotopes, and the mass spectrometer can detect molecules containing these heavy isotopes. Table 15.1 gives the relative natural abundance of some common isotopes.

TABLE 15.1 Some Common Isotopes and Their Abundances in Nature

Isotope	Abundance (%)	Isotope	Abundance (%)
^1H	99.985	^{17}O	0.038
^2H (D)	0.015	^{18}O	0.200
^{12}C	98.90	^{35}Cl	75.77
^{13}C	1.10	^{37}Cl	24.23
^{14}N	99.63	^{79}Br	50.69
^{15}N	0.37	^{81}Br	49.31
^{16}O	99.762	^{127}I	100

For cyclohexane, there is a 1.1% probability that *each* carbon will be a ^{13}C, and therefore about 7% probability that the mass of cyclohexane will be one unit high (M + 1).

Deuterium (^{2}H or D) has a natural abundance of only 0.015%, and therefore contributes relatively little to the M + 1 peak. Sometimes the presence of isotopes is useful in the diagnosis of structure. An analysis of isotope peaks is particularly useful for organic chlorides and organic bromides, as we will see at the end of this section.

But how do we know that a mass spectrum with $m/z = 84$ is really cyclohexane, C_6H_{12}? If the starting material is of unknown structure, we probably do not know the formula of the molecule. The detection of a molecular ion of $m/z = 84$ only tells us that the molecular weight of the unknown is 84, nothing more. Any other molecule of $m = 84$ could give the same molecular ion. In this case, we might be looking at 3,4-dihydro-2*H*-pyran, for example ($C_5H_8O = 84$, Fig. 15.5).

PROBLEM 15.3 Find other possible structures that might fit the observed m/z of 84.

These days, mass spectrometers are so accurate that we can actually tell C_6H_{12} from C_5H_8O. The two species do not have the same molecular weight *if we measure molecular weights accurately enough* (Fig. 15.6). Table 15.2 gives the accurate atomic weights for a variety of isotopes as well as the average values.

C_6H_{12}

$6\ ^{12}C = 6 \times 12.000 = 72.000$
$12\ ^1H = 12 \times 1.008 = \underline{12.096}$
84.096
 M of C_6H_{12}

C_5H_8O

$5\ ^{12}C = 5 \times 12.000 = 60.000$
$8\ ^1H = 8 \times 1.008 = 8.064$
$1\ ^{16}O = 15.995 = \underline{15.995}$
$\phantom{1\ ^{16}O = 15.995 = }84.059$
$\phantom{1\ ^{16}O = 15.9}$ M of C_5H_8O

FIGURE 15.6 If molecular weights are measured carefully enough, mass spectrometry can differentiate between different molecules of the nominal molecular weight of 84.

TABLE 15.2 Some Atomic Weights

Element	Atomic Weight	Element	Atomic Weight
^{1}H	1.008	^{19}F	19.998
^{2}H	2.010	^{28}Si	27.977
H (av)	1.008	^{29}Si	28.975
^{10}B	10.013	^{30}Si	29.974
^{11}B	11.009	Si (av)	28.086
B (av)	10.810	^{31}P	30.974
^{12}C	12.000	^{32}S	31.972
^{13}C	13.003	^{33}S	32.972
C (av)	12.011	^{34}S	33.968
^{14}N	14.003	^{36}S	35.967
^{15}N	15.000	S (av)	32.066
N (av)	14.007	^{35}Cl	34.969
^{16}O	15.995	^{37}Cl	36.966
^{17}O	16.999	Cl (av)	35.453
^{18}O	17.999	^{79}Br	78.918
O (av)	15.999	^{81}Br	80.916
		Br (av)	79.904
		^{127}I	126.905

15.3b MS Measurements When a molecular formula is calculated from combustion data, it is the average values in Table 15.2 that must be used, because unless there has been a deliberate introduction of an isotope during the synthesis of

CHAPTER 15 Analytical Chemistry: Spectroscopy

702

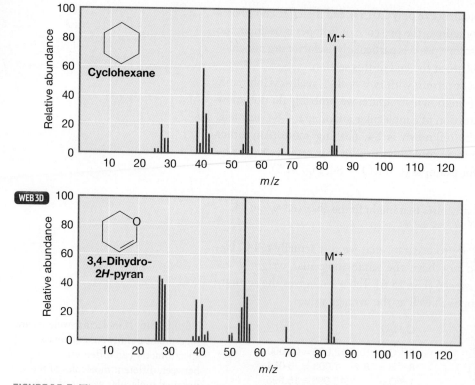

FIGURE 15.7 The mass spectra of cyclohexane and 3,4-dihydro-2*H*-pyran.

the material, it will be the natural isotopic mixture that appears in the compound. The mass spectrometer, on the other hand, is able to separate ions containing different isotopes. Each peak represents a particular set of isotopes, for example, $^{12}C_6{}^1H_{12}$ or $^{12}C_5{}^1H_8{}^{16}O$. So, a good mass spectrometer can give us the molecular formula of an unknown, as long as we can detect the molecular ion. The full mass spectra of our two $m = 84$ molecules are shown in Figure 15.7. One can see that the mass spectra are not simple composites of molecular ion and M + 1 peaks. There are many additional lower molecular weight ions detected that give a characteristic array of ions known as the **fragmentation pattern** for each molecule.

The original radical cation, $M^{\bullet+}$, can undergo many fragmentation reactions before it is accelerated into the magnetic field and passed on to the detector. The fragmenting ion is called the **parent ion (p)**. It can react to give many **daughter ions**, and these ions appear as peaks in the mass spectrum—the fragmentation pattern. Often the most intense or largest peak is not the molecular ion. In all cases, the intensities of mass spectral peaks are given as percentages of the largest peak in the spectrum, which may be the molecular ion, but usually is not. The largest peak is called the **base peak** of the spectrum. You can see in Figure 15.7 that the base peak for cyclohexane is 56 and the base peak for 3,4-dihydro-2*H*-pyran is 55. Remember that the molecular ion, $M^{\bullet+}$, is the furthest to the right on the spectrum, except for the small isotope peaks.

Is there any way to tell how a given molecular ion will fragment? Yes and no. There is no way to predict the detailed pattern of daughter ions. Yet we can make general predictions by applying what we know about reaction mechanisms and the stabilities of ions. Fragmentations usually take place to give the most stable possible cations, and we are now able to make some reasonable guesses as to what these might be.

First of all, note that there are two ways for a typical parent radical cation to produce two fragments (Fig. 15.8). When the radical cation A—B$^{\bullet+}$ produces A and B by fragmentation, the charge may go with either A or B. The detector can see only positively charged ions, and so any neutral species formed are invisible. In the fragmentation process, the charge will generally wind up where it is more stable, and we are often able to predict where it will go. For example, *tert*-butyl chloride gives large amounts of the *tert*-butyl cation, *m/z* = 57, in its mass spectrum, not Cl⁺, *m/z* = 35 (Fig. 15.8). The charge is more easily borne on the tertiary cation than on the relatively electronegative chlorine. In fact, the cleavage of *tert*-butyl chloride is so facile, the molecular ion is not even detected. Nevertheless, it must be admitted that predicting is not always a straightforward process. High energies are involved, and there is plenty of energy available to do a lot of bond breaking. Rearrangements are also common, especially if a particularly stable ion can be reached.

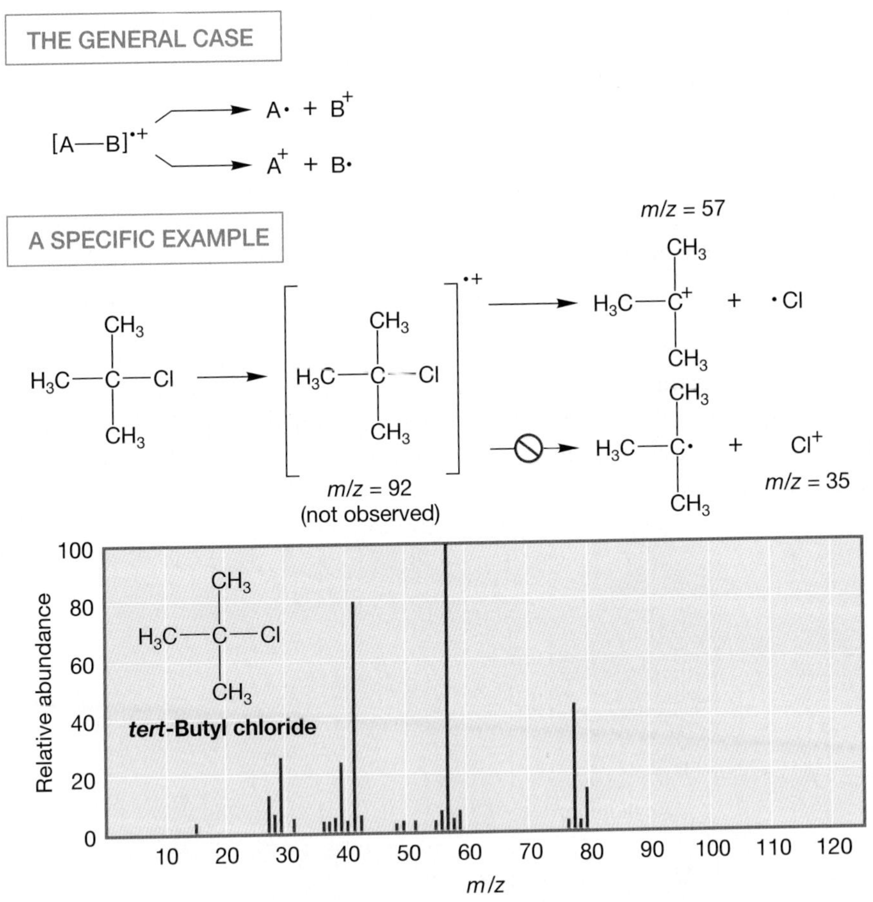

FIGURE 15.8 Two possible fragmentations for a general molecule A—B and the mass spectrum of *tert*-butyl chloride.

THE GENERAL CASE

$$[A{-}B]^{\bullet+} \nearrow A\bullet + B^+$$
$$\searrow A^+ + B\bullet$$

A SPECIFIC EXAMPLE

$m/z = 57$

$$\text{H}_3\text{C}{-}\overset{\overset{\text{CH}_3}{|}}{\underset{|}{\text{C}}}{-}\text{Cl} \longrightarrow \left[\text{H}_3\text{C}{-}\overset{\overset{\text{CH}_3}{|}}{\underset{|}{\text{C}}}{\cdots}\text{Cl} \right]^{\bullet+}$$

$$\longrightarrow \text{H}_3\text{C}{-}\overset{\overset{\text{CH}_3}{|}}{\underset{\underset{\text{CH}_3}{|}}{\text{C}^+}} + \bullet\text{Cl}$$

$$\xcancel{\longrightarrow} \text{H}_3\text{C}{-}\overset{\overset{\text{CH}_3}{|}}{\underset{\underset{\text{CH}_3}{|}}{\text{C}\bullet}} + \text{Cl}^+ \quad m/z = 35$$

$m/z = 92$
(not observed)

tert-Butyl chloride

Some other examples will probably be useful. Figure 15.9 shows the mass spectrum of toluene. The molecular ion is visible as a strong peak at $m/z = 92$, but it is not the base peak. The largest peak in the spectrum is the M − 1 peak at $m/z = 91$. What can this be?

FIGURE 15.9 The mass spectrum of toluene.

Toluene

$M^{\bullet+}$

A little thought brings us quickly to a reasonable answer. If a hydrogen atom (not a proton, which would be positively charged and leave a neutral fragment behind) is lost from the parent ion, what is left behind is a resonance-stabilized benzyl ion (Fig. 15.10).

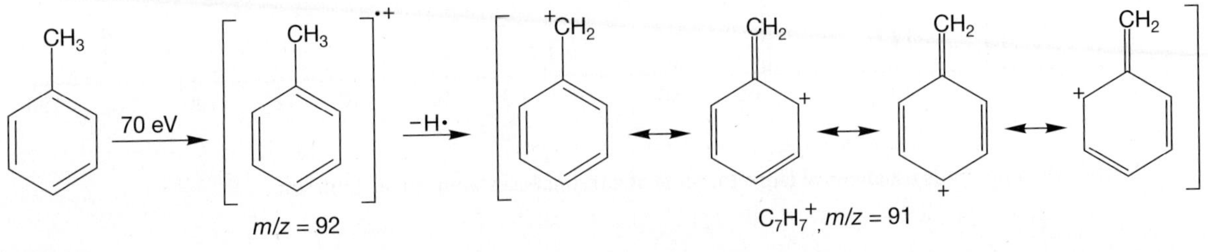

FIGURE 15.10 The molecular ion of toluene loses a hydrogen atom to give the benzyl cation.

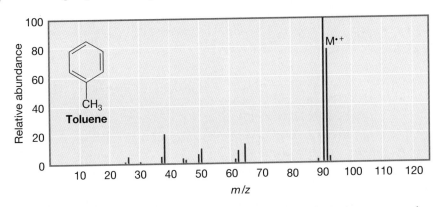

$m/z = 92$ $\qquad\qquad\qquad\qquad$ $C_7H_7^+, m/z = 91$

Accordingly, loss of the hydrogen atom from the first-formed radical cation is relatively easy, and the largest peak in the spectrum belongs to the ion $C_7H_7^+$, $m/z = 91$, the benzyl cation.

PROBLEM 15.4 There is some evidence that the $m/z = 91$ peak may not be the benzyl cation but another species. Can you (a) think of a $C_7H_7^+$ species that might be even more stable than the benzyl cation, and (b) write an arrow formalism for the conversion of the benzyl cation into this new species? *Hint*: See p. 587.

The mass spectra of compounds containing carbon–oxygen double bonds (carbonyl compounds) often have large peaks for the resonance-stabilized acylium ions (Fig. 15.11). For example, the molecular ion of 2,2,4,4-tetramethylpentan-3-one (di-*tert*-butyl ketone) ionizes to give large amounts of the acylium ion of $m/z = 85$. Acylium ions can break down further to give carbon monoxide and a carbocation.

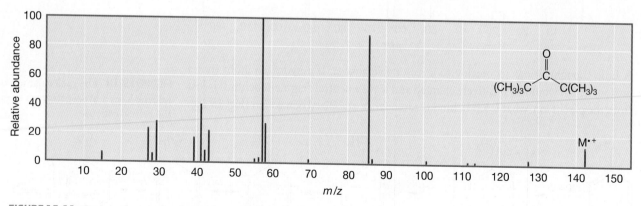

FIGURE 15.11 Carbonyl compounds usually give large amounts of a resonance-stabilized acylium ion.

Alkanes cleave so as to give the most stable possible carbocation (Fig. 15.12). For example, 2,2-dimethylpropane (neopentane) fragments in a mass spectrometer to give an ion of $m/z = 57$, the *tert*-butyl cation, much more easily than it gives the far less stable methyl cation, $m/z = 15$.

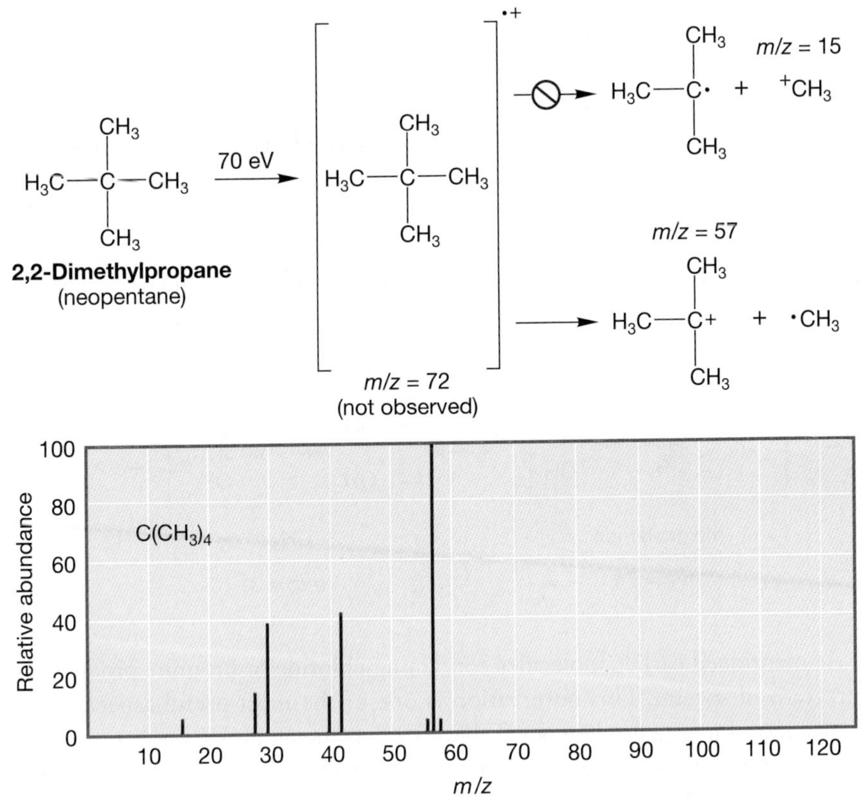

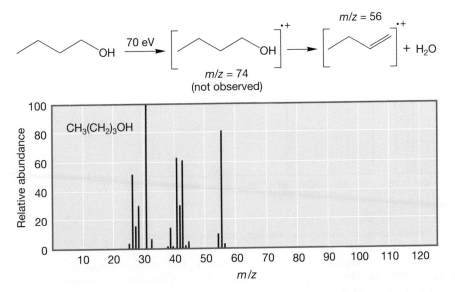

FIGURE 15.12 As illustrated by the mass spectrum of 2,2-dimethylpropane, alkanes fragment to give the most stable carbocation possible.

There are other fragmentation reactions that resemble solution chemistry. For example, an important peak in the mass spectrum of 1-butanol is at $m/z = 56$. High-resolution analysis of this peak shows that its formula is C_4H_8. Apparently, the parent ion eliminates water to give a butene radical cation (Fig. 15.13). This behavior is typical for alcohols and has clear roots in the solution chemistry described in our sections on elimination reactions (p. 298).

FIGURE 15.13 In the mass spectrometer, alcohols such as 1-butanol can eliminate water to give alkene radical cations.

PROBLEM 15.5 Draw a schematic molecular orbital diagram for the π orbitals of the butene radical cation, $C_4H_8^{\bullet+}$.

WORKED PROBLEM 15.6 What do you suggest as a structure for the ion that corresponds to the base peak ($m/z = 31$) in the mass spectrum of 1-butanol shown in Figure 15.13?

ANSWER The base peak is $m/z = 31$. The mass of 1-butanol ($C_4H_{10}O$) is 74, so some quite serious fragmentation must have taken place. A mass of 31 can only have two heavy (nonhydrogen) atoms, so it must be either a C_2 ($m = 24$) or CO ($m = 28$) compound. If the ion contained two carbons, it would have to be C_2H_7, an impossible formula, so it can only be CH_3O. The problem now is to find a reasonably stable ion of this formula that can be made from a fragmentation of 1-butanol. A simple carbon–carbon bond breaking leads to a resonance-stabilized cation that certainly is a good candidate.

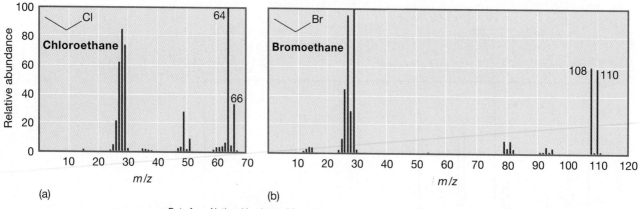

As mentioned earlier, molecules containing chlorine or bromine produce easily characterized spectra. This observation is one of the most useful aspects of mass spectrometry for organic chemists. Table 15.1 lists the natural isotopic ratio for ^{35}Cl to ^{37}Cl (75.77:24.23), which is essentially a 3:1 ratio. The ratio for ^{79}Br to ^{81}Br (50.69:49.31) is close to a 1:1 ratio. Chlorine and bromine are the only elements frequently found in organic molecules that have significant M + 2 isotopes. There are other elements (osmium and mercury, for example) that have M + 2 isotopes, but we are less likely to deal with mass spectra of such compounds. Therefore, if a mass spectrum shows an M + 2 peak that is one-third the intensity of the molecular ion ($M^{\bullet+}$) as shown in Figure 15.14a, then we know there is a chlorine in the molecule. If a spectrum has an M + 2 peak that is the same intensity as the molecular ion, we can conclude that the molecule has a bromine (see Fig. 15.14b).

Data from National Institute of Standards and Technology, http://webbook.nist.gov

FIGURE 15.14 (a) The mass spectrum of chloroethane. If there is a chlorine in the molecule, then an M + 2 peak with one-third the height of the M peak will be observed. (b) The mass spectrum of bromoethane. When there is bromine in a molecule there is a 1:1 ratio of the M and the M + 2 peaks.

There are still plenty of mysteries in mass spectrometry, however. For example, the Jones research group first made the molecule bicyclo[4.2.2]deca-2,4,7,9-tetraene ($m = 130$; Fig. 15.15) over 40 years ago, and were startled then to find that the base peak in the mass spectrum was at $m/z = 115$. It seemed likely that this ion ($m - 15$) represented the loss of a methyl group from the parent ion. We

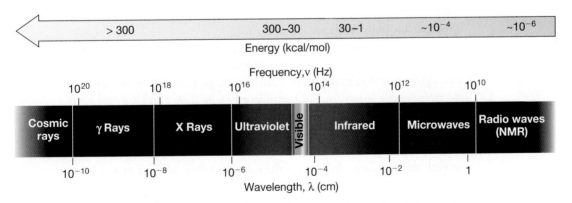

WEB 3D

70 eV

$m/z = 130$

$m/z = 115$???

Bicyclo[4.2.2]deca-2,4,7,9-tetraene

FIGURE 15.15 Bicyclo[4.2.2]deca-2,4,7,9-tetraene apparently loses a methyl group in the mass spectrometer. How?

have been thinking about this process (not constantly!) over the ensuing three decades and still have no reasonable explanation for the loss of a methyl group. Any ideas?

Despite occasional mysteries, mass spectrometry is still highly useful, even in the bare-bones form described here. It gives us a way to determine molecular composition through detection of a molecular ion, and can give hints as to structure from the fragment ions produced. Even though it is difficult to predict the fragmentation patterns in more than a general way, it is possible to compare a fragmentation pattern with that of a known compound in order to prove identity.

Summary

Mass spectrometry usually provides the molecular weight of an unknown compound. If run at sufficiently high resolution, MS can even allow you to determine the molecular formula of a compound. Some information can be gathered about structure from the ions formed on fragmentation.

15.4   Infrared Spectroscopy (IR)

Infrared (IR) spectroscopy is true spectroscopy—it informs us about the electromagnetic radiation absorbed by a molecule. In Chapter 12, we discussed UV/vis spectroscopy, and IR spectroscopy is qualitatively similar. A schematic picture of the electromagnetic spectrum (Fig. 15.16) identifies the regions used for UV/vis, IR, and the spectroscopy we will study in the second half of this chapter, nuclear magnetic resonance.

> 300   300–30   30–1   ~10^{-4} ~10^{-6}

Energy (kcal/mol)

Frequency, ν (Hz)

10^{20} 10^{18} 10^{16} 10^{14} 10^{12} 10^{10}

| Cosmic rays | γ Rays | X Rays | Ultraviolet | Visible | Infrared | Microwaves | Radio waves (NMR) |

10^{-10} 10^{-8} 10^{-6} 10^{-4} 10^{-2} 1

Wavelength, λ (cm)

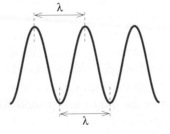

λ

λ

FIGURE 15.16 The electromagnetic spectrum.

We begin our discussion of IR spectroscopy by recalling certain properties of light and energy (p. 526). The wavelength of light, λ, is the length of a wave as measured from peak to peak or from trough to trough (Fig. 15.16). In IR spectroscopy we will express wavelength in centimeters (cm). Wavelength is related to frequency as shown in Eq. (15.1). The frequency (v) is the number of cycles (or waves) passing a given point per unit time, usually seconds. So, the units of frequency (v) are cycles per second (usually written as reciprocal seconds, s^{-1}), or more simply, hertz (Hz).

$$v = \frac{c}{\lambda}, \quad c = \text{the speed of light in centimeters per second (cm/s)} \tag{15.1}$$

The quantity $1/\lambda = \bar{v}$ is called the **wavenumber** and its units are reciprocal centimeters (cm^{-1}). Physically, $\bar{v}$ is the number of wavelengths contained in one centimeter (cm^{-1}) and is related to the more familiar frequency, v, by Eq. (15.2). Signals from IR spectra are reported in wavenumbers (cm^{-1}).

$$v = \frac{c}{\lambda} \quad \text{or} \quad v = c\bar{v} \tag{15.2}$$

Absorption of IR light involves energies of roughly 1–10 kcal/mol, and is associated with molecular vibrations. A chemical bond can be approximated by a picture of two objects connected by a vibrating spring. Hooke's law, which is shown in Eq. (15.3), describes this situation.

$$v = k\sqrt{f(m_1 + m_2)/m_1 m_2} \tag{15.3}$$

The frequency of vibration, v, is related by a constant k to the masses of the objects (m_1 and m_2) and another constant, f, which is the **force constant** of the spring. Chemical bonds do behave rather like classical springs. So, like a real spring, the frequency of vibration in a bond depends on the masses of the two atoms involved and the strength of the bond between them. Bond strength turns out to be the dominant feature, although the masses of the atoms do make a difference. We would expect strong bonds, with higher force constants, to absorb at higher wavenumbers. Carbon–carbon single bonds ($\sim$90 kcal/mol, $\sim$1450 cm^{-1}), double bonds ($\sim$168 kcal/mol, $\sim$1650 cm^{-1}), and triple bonds ($\sim$230 kcal/mol, $\sim$2180 cm^{-1}) illustrate this point well.

Not all molecular vibrations can be detected. For a vibration to be infrared active, it must produce a net change in the dipole moment of the molecule, which means that symmetrical vibrations are either weak or invisible in the IR region of the electromagnetic spectrum. For example, ethylene shows no IR signal for the carbon-carbon double-bond stretch. However, Raman spectroscopy, which is related to IR spectroscopy, does detect symmetrical vibrations, and shows a signal at 1623 cm^{-1} for ethylene. Signals in IR and Raman are usually broad and actually extend over several wavenumbers. As a result the signal is often referred to as a band.

15.4a The IR Spectrometer

In practice, an IR spectrum is obtained by passing IR radiation (light) through a sample. If no light is absorbed at a given wavelength, transmission is 100%. If light is absorbed by the sample at a certain wavelength, transmission is less than 100%. A plot of absorption versus wavenumber (cm^{-1}) is thus generated and constitutes the IR spectrum of the sample. Spectra of gases are measured in cells with nonabsorbing windows. For liquids, spectra can be obtained either as pure samples or in solution. Pure samples are usually described as "neat." To measure the spectrum of a solid, the sample is usually compressed into a disk with KBr and placed in the beam of IR radiation; alternatively, it can be dissolved in Nujol®, which is a petroleum oil.

Spectrometers are routinely attached to computers that can search for matches between the spectrum of an unknown and a library of known spectra. As with mass spectrometry, gas chromatographs can be attached to IR spectrometers and spectra can be determined as the individual components of a mixture elute from a column. As noted in Section 15.2, this technique is called gas chromatography/infrared spectroscopy, or GC/IR.

15.4b Characteristic Infrared Absorptions
From the foregoing discussion, one might expect IR spectroscopy to be a simple matter of noting what characteristic absorptions appear in the spectrum. An IR spectrum might be a very simple phenomenon, composed only of signals for each kind of bond present in the sample molecule. There certainly are characteristic regions of absorption for certain bonds, but the practical situation is not quite so simple because connected vibrating springs interact. Two springs attached to the same object (or atom) do not vibrate independently—try it. If you set one spring in motion, the other moves as well. In fact, this phenomenon occurs even if the two springs (bonds) are attached to different atoms within the same molecule. The closer in energy the two vibrations are, the more strongly they affect each other. A molecule is like a collection of interacting springs, and it is often not possible to make a simple assignment of all the bands in an IR spectrum, although computational methods are becoming more proficient at this task. Most IR spectra are complicated series of bands from which we can extract information, but which we cannot completely rationalize in more than a general way. Figure 15.17 shows the IR spectrum of a typical organic molecule and includes a few band assignments.

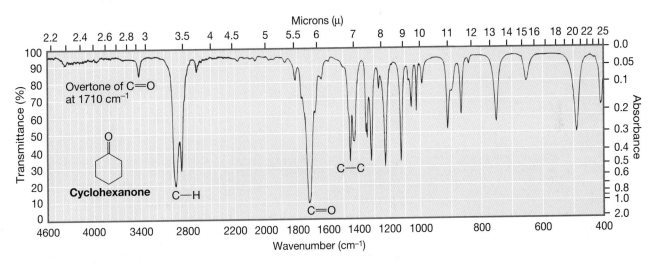

FIGURE 15.17 A typical IR spectrum.

Although this complexity may seem confusing at first, it is usually possible to gain a lot of information about the functional groups present in a molecule, even if we cannot assign all the bands, or draw a complete structure of the sample molecule. It is a great help in structure assignment to know which types of bonds are present in a molecule. There is also an important side benefit to these complicated IR spectra. Their very complexity means that the IR spectrum of every molecule is different. Each spectrum serves as a fingerprint of the molecule. If two IR spectra are identical (not similar—MJ's research advisor insisted on there being no difference greater than the width of the pen line drawn by the recorder) the compounds must be the same.

Table 15.3 gives the general positions of absorptions of a variety of functional groups.

TABLE 15.3 Typical Infrared Absorptions of Functional Groups[a]

Functional Group	Position (cm^{-1})	Intensity[b]
Alkanes		
C—H	2980–2850	m–s (C—H stretch)
C—C	1480–1420	m (C—C bend)
Alkenes (conjugation generally lowers C=C double-bond stretching vibrations by about 20 cm^{-1})		
C=C—H	3150–3000	m (C—H stretch)
C=C	1680–1620	m–w (C=C stretch)
C=C (conj)	1630–1600	m–w (C=C stretch)
R₂C=CH₂ (with H)	995–985 / 915–905	s (C—H out-of-plane bends)
RHC=CHR (trans)	980–960	s (C—H out-of-plane bend)
R₂C=CHR	730–665	s (C—H out-of-plane bend) (br, variable)
R₂C=CH₂	895–885	s (C—H out-of-plane bend)
R₂C=CR(H)	840–790	m (C—H out-of-plane bend)
Alkynes		
C≡C—H	3350–3300	s (C—H stretch)
C≡C	2260–2100	m–w (C≡C stretch)
Alcohols		
RO—H free	3650–3580	m (O—H stretch)
RO—H hydrogen bonded	3550–3300	br, s (O—H stretch)
R—OH	1260–1000 / 1150–1050	s (C—O stretches)
Amines		
RNH₂	3500–3100 (two bands)	br, m (N—H stretches)
R₂NH	3500–3100 (one band)	br, m (N—H stretch)
RNH₂, R₂NH, or R₃N	~1200	m (C—N stretch)

TABLE 15.3 Typical Infrared Absorptions of Functional Groups[a] *(continued)*

Functional Group	Position (cm^{-1})	Intensity[b]	
Aromatic Compounds			
Ar—H	3080–3020	m–w	(C—H stretch)
	1600–1580	m–w	(C=C stretch)
Monosubstituted	770–730	s	(C—H out-of-plane bends)
	710–690		
Ortho disubstituted	770–735	s	(C—H out-of-plane bend)
Meta disubstituted	900–860	m	(C—H out-of-plane bend)
	810–750	s	(C—H out-of-plane bend)
	725–680	m	(C—H out-of-plane bend)
Para disubstituted	860–800	s	(C—H out-of-plane bend)
Carbonyl Compounds (subtract 20–30 cm^{-1} for a conjugated carbonyl)			
RCHO	2900–2700 (two bands)	w	(C—H stretch)
R$_2$C=O	1730–1700 (higher in strained cyclic molecules)	s	(C=O stretch)
![ester]	1750–1735	s	(C=O stretch)
	1300–1000	s	(C—O stretch)
![acid]	1730–1700	s	(C=O stretch)
	3200–2800	br, s	(O—H stretch)
![acid chloride]	1820–1770	s	(C=O stretch)
![anhydride]	1820–1750 (two bands)	s	(C=O stretch)
	1150–1000	s	(C—O stretch)
Imines			
R$_2$C=NR or R$_2$C=NH	1680–1640	m	(C=N stretch)
Nitriles			
RC≡N	~2250	s	(C≡N stretch)

[a] CAUTION! There certainly is some subjectivity in this table, and the values represent average positions for "normal" compounds.

[b] Medium = m, strong = s, weak = w, broad = br.

Once again, we come to the question of memorization. Should one learn this chart by heart? We think not. It is important to know that this kind of general correlation exists, and you should have a rough idea of where some important functional groups absorb. If you come to use IR often, you will automatically learn the relevant details of the chart, as you work out what the signals in your IR spectra tell you.

WORKED PROBLEM 15.7 Use the data in Table 15.3 to identify the functional groups in the IR spectra below. Perhaps, given the molecular formulas, you can make some guesses as to structure.

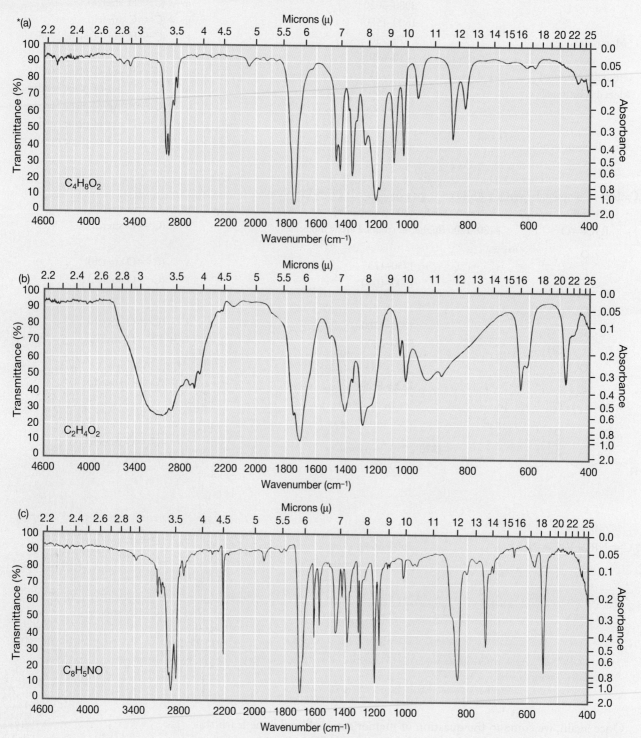

ANSWER (a) In the C—H stretch region (3350–2850 cm^{-1}) almost all the bands are below 3000 cm^{-1}, indicating that there are only alkane carbon–hydrogen bonds. No hydrogens are attached to double or triple bonds, because such carbon–hydrogen

(continued)

bonds would absorb above 3000 cm^{-1}. There is a strong absorption in the carbonyl region at about 1740 cm^{-1}. A 1740 cm^{-1} band is too high to be a ketone or an acid. The compound must be an ester, or perhaps an aldehyde. Finally, there is a strong band at 1200 cm^{-1}, appropriate for the C—O stretch for an ester. Given the formula of $C_4H_8O_2$, one can make two guesses at the structure.

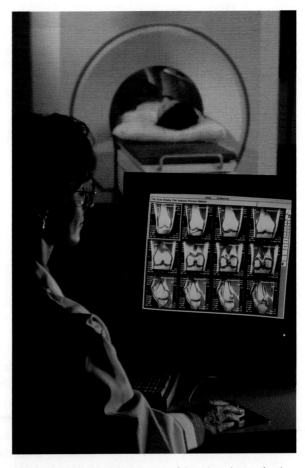

Methyl propanoate   **Ethyl acetate**

   This problem tries to show that without further information there is no reliable way to choose between these two possibilities. Given the information so far, either answer is possible. We could either compare the IR spectrum with those of known samples, or obtain further spectra. An NMR spectrum would distinguish the two possibilities easily, as you will soon see. In fact, this compound is methyl propanoate. For (b) and (c), see the *Study Guide*.

Summary

The IR spectrum of a compound provides information on the kinds of functional groups in the molecule. In addition, if standard spectra of known compounds are available, comparison of the highly detailed IR spectra can be used as a proof of identity—two identical IR spectra can only come from identical compounds.

15.5 ¹H Nuclear Magnetic Resonance Spectroscopy (NMR)

In contrast to IR spectroscopy, which gives an overall view of the functional groups in a molecule, nuclear magnetic resonance (NMR) spectra can give a wealth of detail. It is the general feeling among organic chemists that given good NMR spectra of a molecule, the structure must follow. Today, the frontier of structure determination by NMR lies in immense biomolecules, and even these structures are yielding to an increasing variety of sophisticated NMR techniques.

   Nuclear magnetic resonance was introduced very early on in Section 2.14 (p. 88). There we only developed a bare-bones outline—just enough to let you work out the numbers of different carbon or hydrogen atoms in a molecule and to have you get used to working with symmetry. Here, we go on to many of the gory details of this incredibly powerful spectroscopic technique. We will start with ¹H NMR (hydrogen nuclear magnetic resonance), and then pick up ¹³C NMR (carbon nuclear magnetic resonance) again.

An increasingly useful and powerful tool in the medical field is NMR spectroscopy, called Magnetic Resonance Imaging (MRI) in this setting.

As you can see from Figure 15.16, NMR spectroscopy involves energies much smaller than those in IR, UV, or visible spectroscopy. What happens when radiation that has a wavelength of 1–100 m ($\sim 10^{-6}$ kcal/mol) is absorbed? What change can such a tiny amount of energy induce? Molecular vibrations require much higher energies (1–10 kcal/mol). Molecular rotations demand much less energy than vibrations ($\sim 10^{-4}$ kcal/mol), but even these motions require about 100 times more energy than that of the radio waves used for NMR. To see what happens when radio waves are absorbed by molecules, we must first go back to the beginning of our discussion of atomic structure, and learn a bit more about the structure of the atomic nucleus.

Like the electron, the nucleus has spin. For some nuclei ($^{1}H, {}^{13}C, {}^{15}N, {}^{19}F, {}^{29}Si$), the value of the nuclear spin (I) is $\frac{1}{2}$. For others, the spin can take different values, such as $0 ({}^{12}C, {}^{16}O)$, $1 ({}^{2}H, {}^{14}N)$, $\frac{3}{2} ({}^{11}B, {}^{35}Cl)$, or $\frac{5}{2} ({}^{17}O)$. A nonzero spin is a requirement for the NMR phenomenon. It is convenient that some very common isotopes have zero spin ($^{12}C, {}^{16}O$) and therefore are NMR inactive. However, the most common nucleus in organic chemistry, hydrogen, is one of the nuclei with a spin of $\frac{1}{2}$ and is NMR active. Keep in mind that hydrogens in molecules are sometimes referred to as protons. The NMR effect arises as follows: Any spinning charged particle generates a magnetic field, and therefore we can think of the proton as a bar magnet. In the absence of a magnetic field, these bar magnets will be oriented randomly (Fig. 15.18a). However, when we apply a magnetic field (B_0), the proton can align either with the field or against it (Fig. 15.18b).

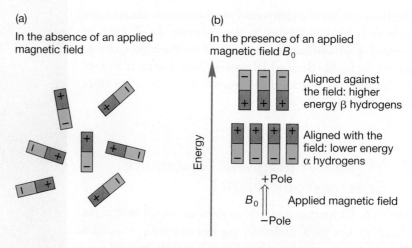

FIGURE 15.18 (a) In the absence of an applied magnetic field, nuclear spins will be randomly oriented. (b) When an external magnetic field is applied, alignment with the field will be slightly favored energetically over alignment against the field; there will be a small excess of molecules aligned with the field.

Alignment with the field ($I = +\frac{1}{2}$, α hydrogen) will be slightly more favorable energetically than alignment against the field ($I = -\frac{1}{2}$, β hydrogen), and there will be an excess of nuclei in the lower energy, more favorable orientation. Because the energy difference between the two orientations is very small, there will only be a slight excess, but it will exist, and there is the possibility of inducing transitions between the two orientations if the proper amount of energy is supplied. The function of the radio waves is to supply the energy necessary to change the orientation of the nuclear spin (often called "flipping" the spin).

The disturbed equilibrium is reestablished when this absorbed energy is returned to the environment as heat (Fig. 15.19).

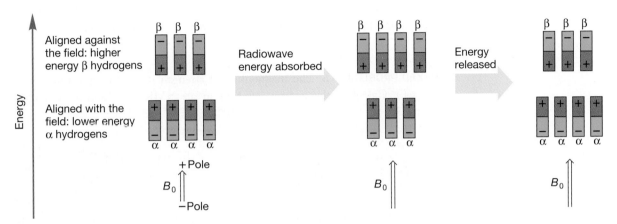

FIGURE 15.19   Absorption of energy can "flip" a nuclear spin, moving the low energy proton into the higher energy level.

The magnitude of the separation of the two states—alignment with and against the applied magnetic field—depends on the strength of that field, B_0, as shown in Eq. (15.4) and Figure 15.20. The higher the field strength, the greater the energy difference between hydrogens aligned with and opposed to the applied magnetic field. For values of B_0 in the range commonly used, the frequencies ν for typically studied isotopes are in the range of 60–750 MHz. The gyromagnetic ratio, γ, is characteristic of a given nucleus, which means there is a different γ for every NMR active nucleus.

$$\nu = \gamma B_0/2\pi \qquad\qquad (15.4)$$

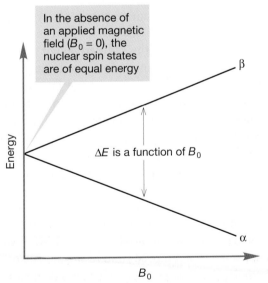

FIGURE 15.20   The energy difference between the two spin states of hydrogen depends on the strength of the applied magnetic field.

If we can build an instrument capable of delivering the required radio waves and of detecting the absorption of tiny amounts of energy, we should be able to find a signal from any hydrogen-containing material. But won't there be problems from the other nuclei? Many common nuclei are NMR active, and there might be too many overlapping signals for us to make sense of the spectrum. We are saved from this fate by the gyromagnetic ratios γ, which are sufficiently different so that for the usual field strengths, there is a wide separation of signals from different nuclei. For example, signals from ^{13}C or ^{19}F nuclei appear far from those for ^{1}H.

WORKED PROBLEM 15.8 The magnetic field strength is measured in units of Tesla (T). Calculate the energy involved in an NMR transition at a field strength of 4.7 T. For H, γ is 2.7×10^8 T^{-1} s^{-1}, and Planck's constant is 9.5×10^{-14} (kcal/mol)(s).

ANSWER Energy equals Planck's constant (h) $\times$ frequency (ν) or $E = h\nu$. The frequency is given by Equation (15.4), $\nu = \gamma B_0 / 2\pi$, so the useful relationship for this problem is

$$E = h\gamma B_0 / 2\pi$$

$$E = (9.5 \times 10^{-14})(2.7 \times 10^8)(4.7)/2(3.14) = 1.9 \times 10^{-5}$$

1.9×10^{-5} whats? Figure out the units using the same equation.

$$E = (kcal/mol)(s)(T^{-1} s^{-1})(T) = kcal/mol$$

The answer is 1.9×10^{-5} kcal/mol or 0.019 cal/mol. Notice that this last value is *calories* per mole! These NMR transitions involve *tiny* energies.

In the early NMR spectrometers, the magnetic field was varied, while the frequency of the radio wave was held constant. The magnetic field was swept over the appropriate range for the nucleus in question, most often hydrogen or carbon. As the chemically different nuclei absorbed energy (came into resonance) they would be detected one by one. It took several minutes to acquire the spectrum and rather large samples—several milligrams—were necessary. Today a different and more sensitive spectrometer is used. These days, superconducting magnets supply far higher magnetic fields than did the old permanent magnets. The magnetic field is not varied; instead, the sample is irradiated with a very short pulse of radio frequency that extends over the entire range for the nucleus in question. All hydrogen atoms or all carbon atoms absorb energy at once. They return to equilibrium over time, and a computer transforms the intensity versus time data into intensity versus resonance frequency data through what is called a Fourier transform (FT). An FT spectrum can be obtained in a few seconds on very small samples, and the results from many pulses are averaged to increase the signal-to-noise ratio.

Resonance frequencies are always referenced to the resonance frequency of a standard signal, and given in parts per million (ppm) of the magnetic field. The standard chosen for ^{1}H NMR is **tetramethylsilane (TMS)**, which shows a sharp signal at a frequency somewhat removed from resonance frequencies of hydrogens in typical organic molecules. The position of TMS is defined as the zero point of the **ppm scale**. Why is the somewhat odd ppm scale used? Let's assume that we are using a 60-MHz spectrometer and that the resonance frequency of the sample molecule appears at 120 Hz from TMS. If we measured the NMR spectrum of the same sample on one of the higher field instruments common today, say,

300 MHz, the resonance frequency would be different because the gap between the two nuclear spin states depends on the strength of the applied magnetic field (Fig. 15.20). At 300 MHz, our sample hydrogen signal would appear at 600 Hz from TMS. Although the measured resonance position is different for every applied magnetic field, *the position is always the same in ppm*. In the example above, it appears at 2.0 ppm each time: 120 Hz/(60 × 10⁶ Hz), or 600 Hz/(300 × 10⁶ Hz) (Fig. 15.21). The ppm scale was devised so that chemists could report the *same* values for resonance positions regardless of the magnetic field strength of their spectrometer.

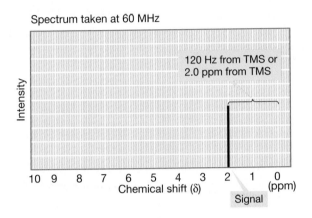

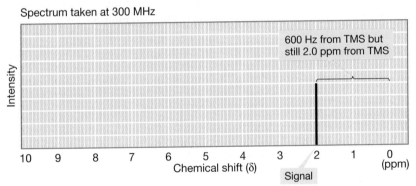

FIGURE 15.21 The ppm scale ensures that peak positions will be the same regardless of external field strength.

15.6 NMR Measurements

Up to this point, NMR spectroscopy might seem to be merely an expensive way of seeing if a compound contained hydrogen. We pay some hundreds of thousands of dollars for a 300-MHz machine, put a sample of an organic compound in it, and if we see absorption of radio waves at an appropriate frequency, we know the sample contains hydrogen. That's usually not very surprising if we are dealing with organic molecules. So what's the big deal? It turns out that we don't see just one big signal from a hydrogen-containing molecule. Instead, there is a signal at a different position for every different hydrogen in the molecule. The resonance frequency for each hydrogen in the molecule is critically dependent on its precise local environment—any difference in environment will product a different resonance frequency.

The NMR spectrum is obtained by dissolving approximately 10 mg of the compound to be analyzed in about 1 mL of a deuterated solvent, usually deuteriochloroform ($CDCl_3$), with 0.05% TMS added. The spectrum is then taken and the signals analyzed.

WORKED PROBLEM 15.9 The higher the external magnetic field of the NMR spectrometer, the greater the separation between the different signals, and therefore the greater the resolving power of the instrument. This resolution is one of the reasons that chemists are willing to spend extra thousands of dollars to get extra megahertz. For example, two signals appearing at 120 and 126 Hz in a spectrum taken on a 60-MHz machine will be much better separated if the spectrum is obtained at 300 MHz. Show that this last statement is true.

ANSWER At 60 MHz, the signals at 120 and 126 Hz are at 2.0 and 2.1 ppm, respectively, $[120 \text{ Hz}/(60 \times 10^6 \text{ Hz})$ and $126 \text{ Hz}/(60 \times 10^6 \text{ Hz})]$. At 300 MHz, these signals are at 600 and 630 Hz, respectively. So the two signals are separated by 6 Hz at 60 MHz, but by 30 Hz at 300 MHz. That's quite an improvement in separation.

There are three pieces of data associated with every NMR signal. The first is called the **integral**. One can determine the relative number of hydrogens giving rise to the signal by measuring the integral, or the area under the peak. The second piece of data is the **chemical shift**, δ. The chemical shift is the location of the NMR signal along the x-axis (see Fig. 15.21). It is reported in ppm; for example, as $\delta = 4$ ppm. The chemical shift of most organic molecules lies between 0.8 and 10 ppm. The third piece of data is the coupling, or multiplicity—the number of lines in each signal. Coupling changes the signal from a single line into multiple lines. The coupled signals are called "doublets" (d, two lines), "triplets" (t, three lines), "quartets" (q, four lines), and so on. The distance between the lines of the multiplet (the **coupling constant**, J) can be measured in Hz. A typical coupling constant J is about 7 Hz, but the range is 1 to 16 Hz.

15.6a The Integral   The ^{1}H NMR spectrum of 4,4-dimethyl-2-pentanone is shown in Figure 15.22. Three signals appear, and the integration shows that they are in the ratio 2:3:9. These days, integrations are calculated electronically and printed out as numbers below or next to the signal. In older spectra, one sees a line traced by a pen, as in Figure 15.22. The integral is determined by measuring the vertical (y-axis) distance the pen traverses as it moves across the signal along the x-axis.

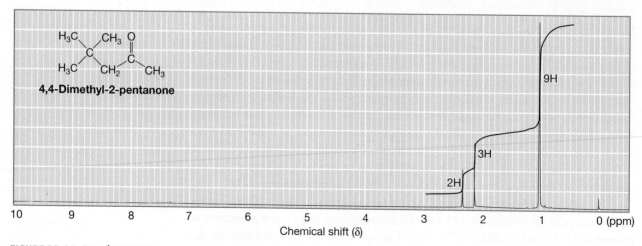

FIGURE 15.22 The ^{1}H NMR spectrum of 4,4-dimethyl-2-pentanone.

In this spectrum, the assignment is easy; the nine hydrogens of the *tert*-butyl group give rise to a signal around δ 1 ppm, the methyl group produces the signal for three hydrogens, and the methylene group gives the peak for two hydrogens. From now on, integrations in the figures will often be indicated by a number followed by H, as in 1H, 2H, and so on.

WORKED PROBLEM 15.10 Which of the molecules below give rise to which spectra?

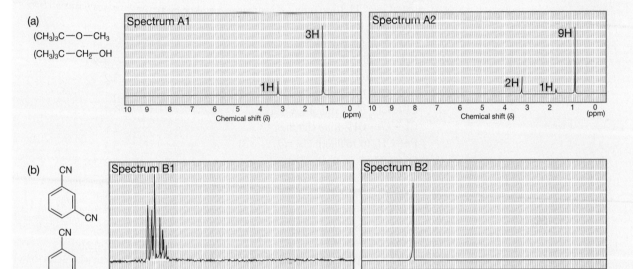

ANSWER (a) First of all, notice that this problem can be solved just by counting signals. *tert*-Butyl methyl ether has only two kinds of hydrogen whereas neopentyl alcohol (2,2-dimethylpropanol) has three. Thus, spectrum A1 belongs to the ether and spectrum A2 belongs to the alcohol. That's all there is to it. Why is the ratio of hydrogens in spectrum A1 1:3 not 3:9? Remember that integration only gives the relative number of hydrogens—it is not an absolute measure. There are other clues here. A nine-hydrogen signal? That's certain to be a *tert*-butyl group. In spectrum A1 there is a signal at approximately δ = 3.3 ppm. That region is something of a desert in ¹H NMR spectra, and almost always indicates a hydrogen attached to a carbon that is also attached to an oxygen.

(b) Here symmetry also allows a quick assignment. 1,4-Dicyanobenzene has only one kind of hydrogen and must give rise to the spectrum B2. 1,3-Dicyanobenzene has three different hydrogens and has the far less symmetrical spectrum B1.

The examples in Problem 15.10 are straightforward, and are designed to ease you into analyzing NMR spectra. In no area of organic chemistry is practice so necessary in the development of skill. You simply cannot become proficient at this kind of structure determination by reading about how it is done and studying

Table 15.4, which gives approximate values of δ for many kinds of hydrogen. Working problems is an absolute must. Do the problems at the end of this chapter and seek out others. Other textbooks have good ones, and NMR problems can also be found on-line; don't be reluctant to dig them out.

TABLE 15.4 Chemical Shifts of Various Hydrogens[a]

Hydrogen	δ (ppm)
CH_3 (methyl hydrogens)	0.8–1.0
CH_2 (methylene hydrogens)	1.2–1.5
CH (methine hydrogen)	1.4–1.7
$C{=}C{-}CH_x$ (allylic hydrogens, $x = 1, 2,$ or 3)	1.7–2.3
$O{=}C{-}CH_x$ (α to carbonyl hydrogens, $x = 1, 2,$ or 3)	2.0–2.7
$Ph{-}CH_x$ (benzylic hydrogens, $x = 1, 2,$ or 3)	2.3–3.0
${\equiv}C{-}H$ (alkynyl hydrogen)	2.5
$R_2N{-}CH_x$ (α to amine hydrogens, $x = 1, 2,$ or 3)	2.0–2.7
$I{-}CH_x$ (α to iodine, $x = 1, 2,$ or 3)	3.2
$Br{-}CH_x$ (α to bromine, $x = 1, 2,$ or 3)	3.4
$Cl{-}CH_x$ (α to chlorine, $x = 1, 2,$ or 3)	3.5
$Cl{-}C{-}CH_2$ (β to chlorine)	1.8
$F{-}CH_x$ (α to fluorine, $x = 1, 2,$ or 3)	4.4
$O{-}CH_x$ (α to ether or alcohol oxygen, $x = 1, 2,$ or 3)	3.2–3.8
$C{=}CH$ (vinylic hydrogens)	4.5–7.5
$Ar{-}H$ (aromatic hydrogens)	6.5–8.5
$O{=}CH$ (aldehydic hydrogen)	9.0–10.0
ROH (hydroxyl hydrogens)	1.0–5.5
ArOH (phenolic hydrogens)	4.0–12.0
RNH_x (amine hydrogens, $x = 1$ or 2)	0.5–5.0
$CONH_x$ (amide hydrogens, $x = 1$ or 2)	5.0–10.0
RCOOH (carboxylic hydrogens)	10–13

[a] These values are approximate. There will surely be examples that lie outside the ranges indicated. Use them as guidelines, not "etched in stone" inviolable numbers.

15.6b The Chemical Shift

We need to know how the dependence of the chemical shift (δ) on the local environment arises. In a sense, this is a nonquestion. If two hydrogens are different, they *must* give different NMR signals. It is in fact just a question of whether our ability to detect the difference is great enough, because the difference must be there. Electrons in molecules occupy regions of space defined by the molecular orbitals of the molecule. The chemical shift depends on the local environment because, when an external magnetic field (B_0) is applied, the electrons will circulate within those regions of space so as to create an induced magnetic field (B_i) that opposes the applied field B_0 (Fig. 15.23).

The net magnetic field ($B_{net} = B_0 \pm B_i$) must be slightly different at every different hydrogen in the molecule, and therefore the resonance frequency will also be different for every different hydrogen. Relative to the hydrogens in TMS, for most (not quite all) hydrogens in a sample molecule, the circulation of electrons produces an induced magnetic field that augments the applied magnetic field. The net magnetic field (B_{net}) felt by most hydrogens is actually greater than the applied magnetic field. Such hydrogens are said to be "deshielded" by the induced field and they appear downfield of TMS, that is, to the left of TMS, in the NMR spectrum. In a very few instances, we find that hydrogens are "shielded" and appear upfield of TMS. In these cases, the magnetic field, B_0, is opposed by the induced field, B_i, and it takes less energy to flip the nuclear

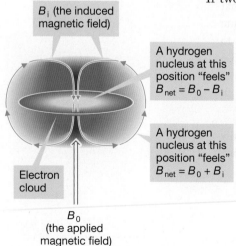

B_i (the induced magnetic field)

A hydrogen nucleus at this position "feels" $B_{net} = B_0 - B_i$

A hydrogen nucleus at this position "feels" $B_{net} = B_0 + B_i$

Electron cloud

B_0 (the applied magnetic field)

FIGURE 15.23 In an applied magnetic field B_0, electrons circulate so as to generate an induced magnetic field (B_i) that opposes the applied field.

spin. Because there are so few molecules that have signals upfield of TMS, we can set δ for TMS as 0, with the spectrum usually appearing entirely to the left of TMS.

We can expect to see a signal for each different hydrogen in a molecule, and this idea leads to the another sometimes thorny question, What constitutes different hydrogens? There is a simple scheme for finding identical (chemically and magnetically equivalent) hydrogens. First, carry out mental substitutions of the hydrogens in question with a phantom group X one at a time. If the resulting molecules are identical, the hydrogens are also identical, or **homotopic**. Figure 15.24 gives an example using methylene chloride. Replacement of either hydrogen (red or green in the figure) with X gives the same CHCl$_2$X. The two hydrogens of methylene chloride are homotopic and equivalent in all chemical and spectroscopic operations.

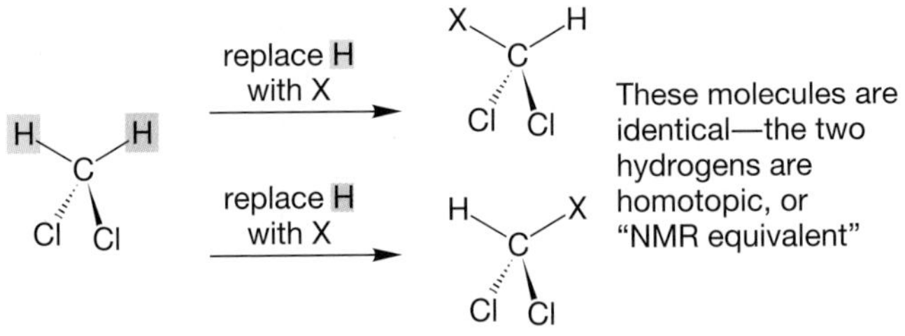

Hydrogens in more complicated molecules are usually not homotopic. Chlorofluoromethane is a good example. Replacement of the two hydrogens by the phantom X in this case does not yield identical molecules (Fig. 15.25). The resulting compounds are enantiomers. When this mental replacement exercise gives enantiomers, the hydrogens are said to be **enantiotopic**. Enantiotopic hydrogens are equivalent in chemical reactions and NMR spectra, as long as chiral reagents or solvents are not involved.

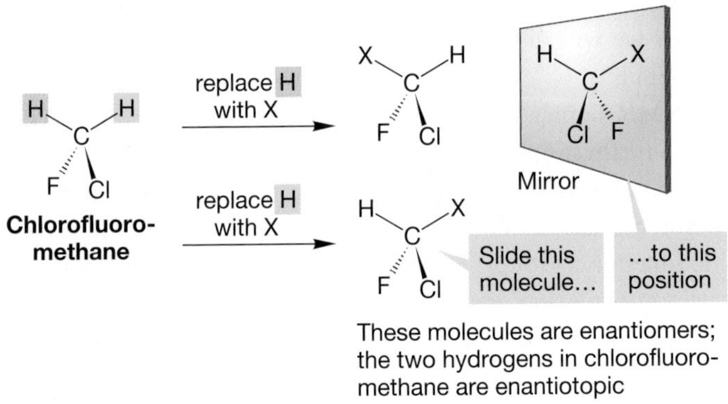

FIGURE 15.24 Substituting either of the two hydrogens gives identical molecules; therefore the hydrogens in methylene chloride are homotopic. Homotopic hydrogens are equivalent.

These molecules are enantiomers; the two hydrogens in chlorofluoromethane are enantiotopic

FIGURE 15.25 The two molecules obtained by substituting one of the two hydrogens are enantiomers; therefore the hydrogens are enantiotopic. Enantiotopic hydrogens are equivalent in typical NMR experiments.

PROBLEM 15.11 Explain why the enantiotopic hydrogens of chlorofluoromethane will show different signals if the NMR spectrum is obtained with an optically active additive (often the solvent).

Finally, consider 1,2-dichlorofluoroethane (Fig. 15.26). Now replacement of the two methylene hydrogens gives neither identical molecules nor enantiomers, but diastereomers (p. 164). These hydrogens are **diastereotopic**. Diastereotopic hydrogens are chemically different and will always give different signals in the NMR

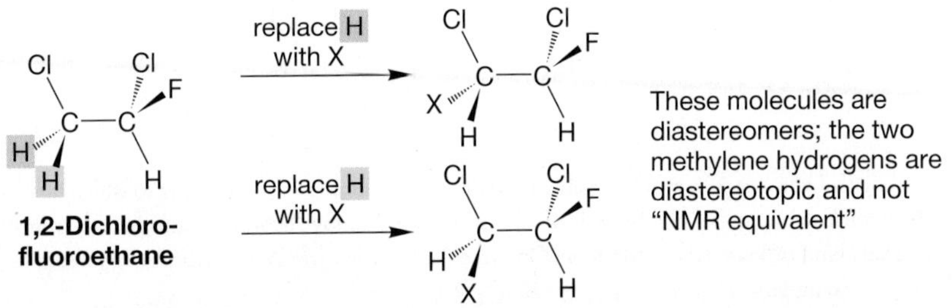

These molecules are diastereomers; the two methylene hydrogens are diastereotopic and not "NMR equivalent"

FIGURE 15.26 The two molecules obtained by substituting one of the two hydrogens are diastereomers; therefore the hydrogens are diastereotopic. Diastereotopic hydrogens are not equivalent in NMR analysis.

spectrum. Of course, diastereotopic hydrogens can be so similar that the spectrometer cannot resolve the signals. There are many such examples, especially in old spectra in which low-field spectrometers were used.

PROBLEM 15.12 Identify the underlined hydrogens in each of the following molecules as homotopic, enantiotopic, or diastereotopic.

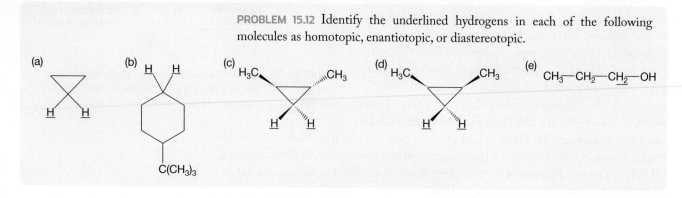

Now let's return to the issue of chemical shifts. There is a general correlation between the resonance frequency in an NMR spectrum of hydrogens that are in similar enviroments in different molecules. For example, all methyl groups attached to double bonds appear at roughly the same position, all aromatic hydrogens at another, and so on.[3] Table 15.4 gives a correlation chart for the peak positions of many different kinds of hydrogen. Use this table when working problems and you will gradually become familiar with the chemical shifts of commonly encountered hydrogens. Now we will discuss several of the factors that influence the chemical shift in more detail.

Alkanes. NMR signals from saturated alkanes (and the simplest cycloalkanes) are typically the furthest upfield. Because alkane hydrogens are unperturbed by any substituents, we can think of the region where they appear (around 1 ppm) as a beginning point. Signals for all organic compounds would appear at about 1 ppm if there were no functional groups. There are exceptions of course. For example, hydrogens on a cyclopropane ring are at unusually high field, sometimes even appearing upfield (to the right) of TMS (Fig. 15.27). A useful trend for analyzing NMR

FIGURE 15.27 The ^{1}H NMR chemical shifts of alkanes.

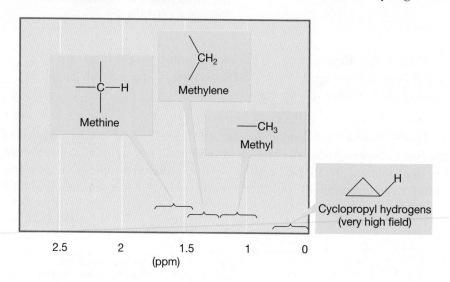

[3]To say "aromatic hydrogen" is certainly wrong; the hydrogen isn't aromatic, it is the ring to which it is attached that is. Nonetheless, one must admit that it is slightly awkward to say "hydrogens attached to aromatic systems," and this kind of loose talk is common. One refers to vinylic hydrogens, allylic hydrogens, cyclopropyl hydrogens, for example.

spectra of alkanes is that in any given environment a methyl group will be about 0.3 ppm upfield from a methylene group and a methylene group will be about 0.3 ppm upfield of a methine. This trend is the reason why several entries of Table 5.4 have a range of chemical shift values that depends on the number of hydrogens on the carbon.

Electronegative Groups. Electron-withdrawing groups will reduce the electron density at neighboring hydrogens through an inductive effect (p. 237), which decreases electron-shielding of the nucleus. As a result of this deshielding, the chemical shifts appear further downfield than those of alkanes. This effect is directly related to the electronegativity of the substituent. Notice, for example, that because fluorine is the most electronegative halogen, alkyl fluorides appear furthest downfield of all the halogen-substituted molecules (Fig. 15.28). Chlorine and oxygen have almost the same electronegativity and, as a result, the signals for the hydrogens of CH_2—O and CH_2—Cl both appear at about 3.5 ppm (Table 5.4).

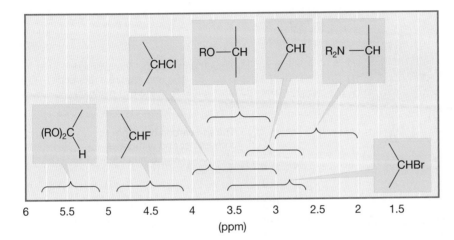

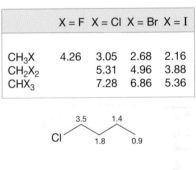

	X = F	X = Cl	X = Br	X = I
CH_3X	4.26	3.05	2.68	2.16
CH_2X_2		5.31	4.96	3.88
CHX_3		7.28	6.86	5.36

FIGURE 15.28 The ^{1}H NMR chemical shifts of alkanes substituted with electron-withdrawing groups.

The effect of electronegative groups is understandably additive. Thus, increasing the number of electronegative groups in the neighborhood of a hydrogen increases the overall electron-withdrawing effect, and the resonance position shifts further downfield. A clear example of this additivity is the comparison between CH_3Cl (δ 3.1 ppm), CH_2Cl_2 (δ 5.3 ppm), and $CHCl_3$ (δ 7.3 ppm). The chlorine has a deshielding effect of about 2.2 ppm (remember that unperturbed methyl hydrogens appear about 0.9 ppm). The effect of the second chlorine in CH_2Cl_2 is an additional 2.2 ppm, which produces the observed chemical shift of 5.3 ppm. To predict the chemical shift of $CHCl_3$, one adds the effect of another chlorine (2.2 ppm), giving 7.5 ppm, which is close to its observed chemical shift of 7.3 ppm. The inductive effect is also felt to a small extent by the hydrogens on the β carbon. The α carbon is the carbon that is directly attached to the electronegative group. The hydrogens on the α carbon are significantly shifted as shown in Table 5.4. The β carbon is attached to the α carbon (there can be more than one β carbon). The hydrogens on the β carbon in 1-chlorobutane (see Fig. 15.28) appear at δ 1.8 ppm. If the chlorine were not present, then the chemical shift would be about 1.3 ppm. So the effect of the chlorine on the β position is about 0.5 ppm, compared to the 2.2 ppm effect on the α position. As we move further away from the electronegative group there is little to no impact on the chemical shift.

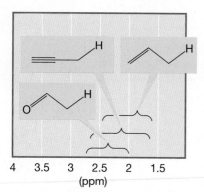

FIGURE 15.29 The ¹H NMR chemical shifts of hydrogens adjacent to double and triple bonds.

Allylic Hydrogens. A carbon–carbon double or triple bond is electron-withdrawing, and therefore exerts a deshielding effect, shifting the signal downfield for hydrogens in the allylic position. The carbon–oxygen double bond exerts a similar, but greater effect (Fig. 15.29).

Hybridization. Hydrogens attached directly to a carbon–carbon double bond are called **vinylic hydrogens**. Vinylic hydrogens appear downfield from hydrogens on sp^3 hybridized carbons. Two effects are operating to deshield such hydrogens and produce the downfield signals. In an external magnetic field, B_0, the electrons in the π bond will circulate so as to create a magnetic field B_i that adds to B_0 in the region where the hydrogens reside (Fig. 15.30). In addition, the sp^2 carbon of the double bond has high s character and attracts electrons, thereby removing electrons from the vicinity of the hydrogen and deshielding it. As a result of these two phenomena, vinylic hydrogens appear around δ 5.5 ppm.

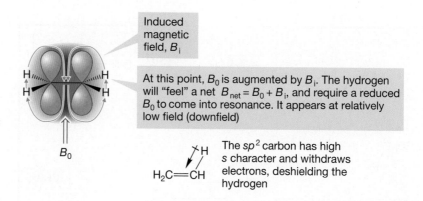

FIGURE 15.30 A carbon–carbon double bond acts in two ways to deshield attached hydrogens. Such deshielded hydrogens appear relatively downfield.

There are important differences between the various kinds of hydrogens attached to double bonds. For example, terminal methylene hydrogens appear further upfield than most vinylic hydrogens the same way that a methylene group appears upfield of a methine. Based on the hybridization trend, a hydrogen directly attached to a triply bonded carbon (alkynyl hydrogen) ought to have a chemical shift around δ 9 ppm. However, it appears at δ 2.2–2.8, at substantially higher field than hydrogens attached to double bonds (Fig. 15.31).

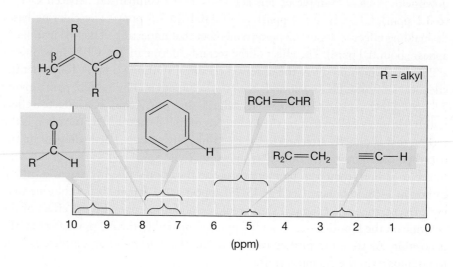

FIGURE 15.31 The ¹H NMR chemical shifts of alkenes and alkynes.

The upfield shift results from a strong shielding effect induced by circulation of electrons in the triple bond (Fig. 15.32).

Delocalization. Hydrogens attached to a double bond, and β to a carbonyl group, have a chemical shift surprisingly far downfield, and the hydrogens of aldehydes are even further to the left (Fig. 15.31). These two examples show the effect that delocalization can have on chemical shifts. Groups that withdraw electrons by resonance significantly deshield hydrogens on carbons at the β position. Alternatively, a group that can donate electrons by resonance shifts hydrogens on the adjacent carbon upfield (Problem 15.13). Such hydrogens are the furthest upfield of all vinylic hydrogens.

PROBLEM 15.13 Predict the chemical shifts for the hydrogens in methoxyethene (methyl vinyl ether).

WORKED PROBLEM 15.14 (a) A hydrogen attached directly to the carbon of a carbonyl group (aldehyde hydrogen) appears at δ 9–10 ppm, and must be strongly deshielded. Explain why. (b) Vinylic hydrogens that are on the β position of a conjugated carbonyl compound are also substantially downfield. Use resonance structures to explain why.

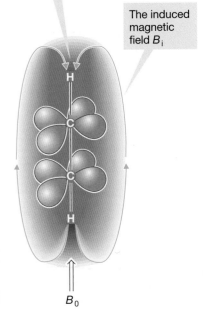

At this point the induced magnetic field B_i opposes the applied field B_0; a hydrogen here "feels" a net magnetic field, $B_{net} = B_0 - B_i$. The alkynyl hydrogen is shielded, and a relatively high B_0 will have to be applied to bring this shielded hydrogen into resonance

The induced magnetic field B_i

B_0

FIGURE 15.32 Acetylenic hydrogens are strongly shielded by the induced magnetic field.

ANSWER (a) In the absence of other effects, we would expect to see the hydrogen of an aldehyde group at about the position of hydrogens attached to carbon–carbon double bonds. However, the carbonyl group is very polar. The oxygen atom is much more electronegative than the carbon and will attract electrons strongly, becoming partially negative (δ^-) and leaving the carbon partially positive (δ^+). The same effect can be seen from a resonance formulation. The electron density around carbon is depleted and the attached hydrogen is the most deshielded hydrogen in neutral organic molecules. Accordingly, a weaker magnetic field is necessary to reach the resonance frequency, and the hydrogen appears downfield.

(b) Conjugated carbonyl compounds have important resonance forms in which the β carbon is positively charged. The resulting deshielding leads to an especially downfield chemical shift for the β hydrogens.

Aromaticity. Hydrogens on aromatic rings are found downfield from most hydrogens attached to double bonds. In fact, so-called aromatic hydrogens can be quite reliably diagnosed by the existence of signals in the range δ 6.5–8.0 ppm. The reason for the downfield chemical shift is similar to that for hydrogens attached to double bonds. The external magnetic field induces a circulation of electrons in the ring (a "ring current") that creates its own magnetic field B_i opposing B_0 in the center of the ring. At the edge of the ring where the hydrogens are, the induced magnetic field augments B_0. Therefore, such hydrogens require a weaker applied magnetic field to bring them into resonance (Figs. 15.31 and 15.33).

FIGURE 15.33 The field induced (B_i) by the ring current of an aromatic ring in an applied magnetic field (B_0).

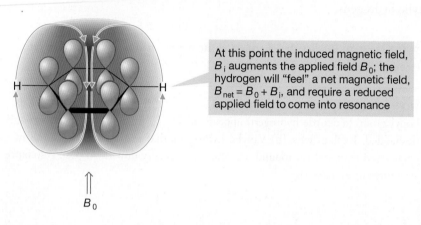

At this point the induced magnetic field, B_i augments the applied field B_0; the hydrogen will "feel" a net magnetic field, $B_{net} = B_0 + B_i$, and require a reduced applied field to come into resonance

PROBLEM 15.15 Explain the unusual position (far upfield, above TMS) of the hydrogens indicated in the following molecules:

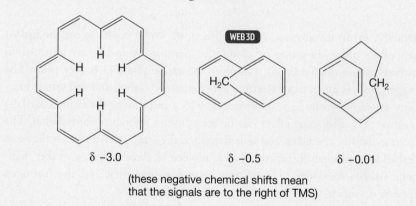

δ −3.0 δ −0.5 δ −0.01

(these negative chemical shifts mean that the signals are to the right of TMS)

Hydrogens Attached to Oxygen or Nitrogen. Hydrogens attached to a carbon bearing the electronegative oxygen or nitrogen atom of an alcohol or a primary or secondary amine absorb, as we might expect, in the same region as the related hydrogens in ethers and tertiary amines (Fig. 15.28). But what about the OH and NH hydrogens themselves? Here we do not see what a simple analysis would lead us to expect. First of all, the chemical shifts of OH and NH vary greatly from sample to sample. The problem is that these molecules are extensively hydrogen bonded. The extent of hydrogen bonding depends on the concentration of the alcohol or amine, the nature of the solvent, the pH of the solution, and the temperature. The chemical shift depends strongly on the chemical environment and thus on the extent of hydrogen bonding. As a result, it is very difficult to predict where a hydrogen attached to oxygen or nitrogen will appear. Generally, the less hydrogen bonding, the further upfield the resonance appears. In the gas phase, where hydrogen bonding does not occur, the chemical shifts of these hydrogens are about 1 ppm.

15.6c Spin–Spin Coupling: The Coupling Constant, _J_ The utility of NMR spectroscopy goes far beyond that of a simple hydrogen-detecting machine. We can determine the number of different kinds of hydrogen in a molecule, and gain a general idea of what chemical environments those hydrogens might occupy. But there is much more information available.

Most NMR spectra are not as simple as those of Figure 15.22 or Problem 15.10. The signals in typical NMR spectra are not all single peaks, but are often multiplets containing much fine structure. Ethyl iodide makes a nice example. Our initial expectation might be that the spectrum would contain only two signals, one for the three equivalent hydrogens of the methyl group and another for the two hydrogens of the methylene group. A look at Table 15.4 would allow us to estimate the general position for each peak. What we see, however, is somewhat different (Fig. 15.34). There are signals at the positions we estimated, but each signal is composed of several lines. They are not singlets as were most of the signals of the molecules in Problem 15.10. The signal for the methyl hydrogens shows three lines (a triplet) in a 1:2:1 ratio centered at δ 1.85 ppm, and the methylene appears as four lines (a quartet) in the ratio 1:3:3:1, centered at δ 3.2 ppm. The spacings between the lines are all the same.

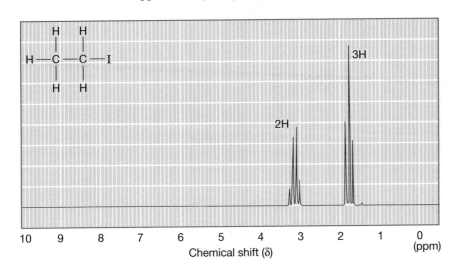

FIGURE 15.34 The ¹H NMR spectrum of ethyl iodide. The methyl signal is a triplet and the methylene signal is a quartet.

Our job is first to see how these multiple signals arise, and then to see how we can use that information.

The (_n_ + 1) Rule. The net magnetic field experienced by a hydrogen will be significantly modified by adjacent (vicinal) hydrogens. Let's start by examining the triplet for the methyl hydrogens in Figure 15.34. To do this, we look first at the neighboring methylene hydrogens. Each of the two equivalent methylene hydrogens has a spin of either $+\frac{1}{2}$ or $-\frac{1}{2}$. Therefore, there are four different combinations of these spins, $(+\frac{1}{2} +\frac{1}{2})$, $(+\frac{1}{2} -\frac{1}{2})$, $(-\frac{1}{2} +\frac{1}{2})$, and $(-\frac{1}{2} -\frac{1}{2})$ (Fig. 15.35). The methyl hydrogens will experience an applied magnetic field (B_0) modified by each

To explain the three-line signal for the methyl hydrogens of ethyl iodide, look first at the neighboring methylene hydrogens

The possible combinations for the methylene hydrogens are

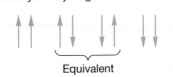

↓ Means spin = $-\frac{1}{2}$

↑ Means spin = $+\frac{1}{2}$

Equivalent

FIGURE 15.35 For two hydrogens there are four possible combinations of the nuclear spins. Two of these combinations are equivalent.

of the different spin combinations of the neighboring methylene hydrogens. So we now expect to see four lines, one for each possible spin combination. But two of these combinations, $(+\frac{1}{2} -\frac{1}{2})$ and $(-\frac{1}{2} +\frac{1}{2})$, are equivalent and will give rise to the same net magnetic field, which is the reason we see only a three-line signal in approximately a 1:2:1 ratio for the methyl hydrogens (Fig. 15.36). The methyl hydrogens are said to be coupled to the adjacent methylene hydrogens. The distance between any two of the lines, as mentioned earlier, is the coupling constant (J), and is measured in hertz. In ethyl iodide the methyl signal is split by 7.6 Hz (Fig. 15.36).

FIGURE 15.36 The methyl group adjacent to the two methylene hydrogens will be split into three lines in the ratio 1:2:1. The splitting between the lines is called the coupling constant, J.

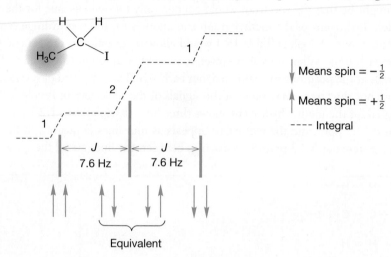

Now let's rationalize the four-line signal for the methylene hydrogens of ethyl iodide. There are eight possible combinations of the nuclear spins of the three equivalent hydrogens that are vicinal to the methylene group (Fig. 15.37). Note that there are now two sets of three equivalent spin combinations (two up and one down or one up and two down). So, the methylene hydrogens feel an applied magnetic field modified by the four different spin combinations of the adjacent methyl hydrogens. Accordingly, the spectrum consists of four lines in a 1:3:3:1 ratio, a quartet. The methylene hydrogens are coupled to the methyl hydrogens with the same coupling constant, $J = 7.6$ Hz, by which the methyl hydrogens are coupled to the methylene hydrogens (Fig. 15.38).

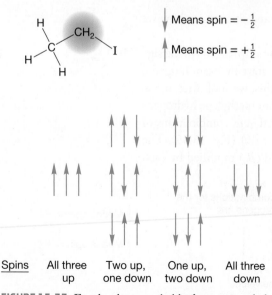

FIGURE 15.37 For the three methyl hydrogens in ethyl iodide there are eight possible spin combinations, but only four different energy combinations.

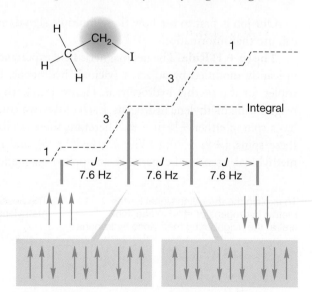

FIGURE 15.38 The signal for the methylene group of ethyl iodide will be split into four lines in the ratio 1:3:3:1.

In the general situation, for a given hydrogen, H_a, the NMR signal will consist of $n + 1$ lines, where n is the number of equivalent adjacent (vicinal) hydrogens, H_b. This observation is called the **$n + 1$ rule**. The lines for the signal are separated from each other by a frequency of J Hz. The intensity of the lines can be determined by analyzing the various possible spins of the coupled hydrogens as we have been doing, or by using a device called "Pascal's triangle." In Pascal's triangle, the value (intensity) of a number is the sum of the two numbers above it. Figure 15.39 explains it better than these words. Pascal's triangle allows the rapid determination of the intensities of lines in NMR signals. Note that we are speaking here of the relative intensities of the lines *within a given signal*, not of the intensities of signals for different hydrogens in the molecule (the integral).

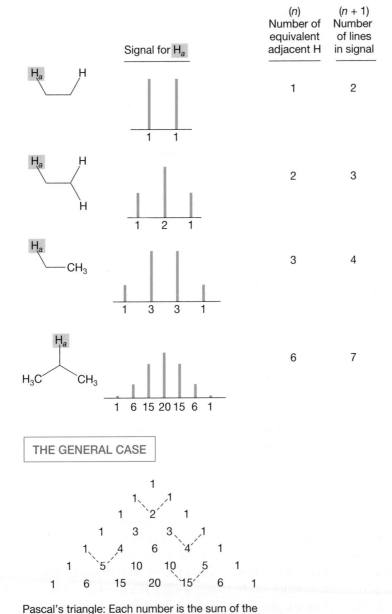

Pascal's triangle: Each number is the sum of the numbers above (see examples shown with dashed lines); the intensities of NMR signals follow this pattern

FIGURE 15.39 In the general case, there will be $n + 1$ lines for a hydrogen adjacent (vicinal) to n equivalent hydrogens.

Multiple Couplings. What happens when a hydrogen is flanked by more than one kind of hydrogen, a very common situation in complex organic molecules? The answer is quite simple—just apply the $n + 1$ rule for each set of equivalent adjacent hydrogens. Consider the situation in Figure 15.40 in which H_a is adjacent to two different hydrogens, H_b and H_c. The NMR signal for H_a will be split into two lines by H_b and each of these lines will be further split into two by coupling to H_c. Let's say that J_{ab} is equivalent to 4 squares and J_{ac} is equivalent to 2 squares. The result is shown in Figure 15.40a: four lines, a doublet of doublets. But does it matter which way we do it? In Figure 15.40a we started with J_{ab}. In Figure 15.40b, we do it the other way, starting with J_{ac}. As you can see, the order makes no difference, we get the same four lines either way.

FIGURE 15.40 The pattern for a single hydrogen, H_a, is determined by coupling to two different adjacent single hydrogens, H_b and H_c. Note that the ultimate pattern of lines, a double doublet, does not depend on the order in which we write the coupling constants J_{ab} and J_{ac}.

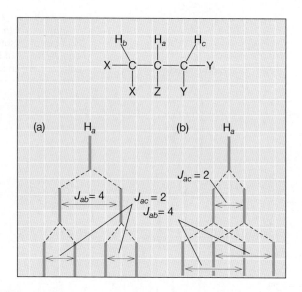

Similarly, Figure 15.41 shows what happens if hydrogen H_a has one adjacent hydrogen H_b and two adjacent hydrogens H_c, with $J_{ab} = 3$ squares and $J_{ac} = 1$ square. A doublet of triplets results.

> **PROBLEM 15.16** What happens in general if hydrogen H_a is flanked by one hydrogen H_b and three hydrogens H_c?
>
> **PROBLEM 15.17** What happens to the NMR spectrum of the molecule in Figure 15.41 if $J_{ab} = 2$ squares and $J_{ac} = 1$ square?

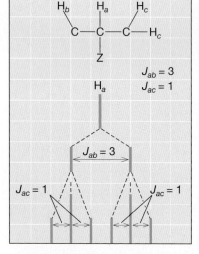

FIGURE 15.41 The splitting pattern for a single hydrogen, H_a, is determined by coupling to a single adjacent hydrogen, H_b, and two equivalent adjacent hydrogens, H_c.

Summary

We have learned about the three important NMR phenomena: the integral, the chemical shift, and the coupling constant. A signal will appear for each set of equivalent hydrogens. The integration will tell us the relative number of hydrogens in that signal. The chemical shift of the signal is dependent on the chemical environment in which the hydrogen resides. The number of lines in an NMR signal results from coupling with vicinal hydrogens. Multiple couplings can be analyzed by constructing "tree" diagrams such as those in Figures 15.40 and 15.41 in which the different couplings are worked out in sequence. The order in which one applies the different couplings never matters.

The Magnitude of _J_: The Karplus Curve. There is no observable coupling between equivalent nuclei. But why is coupling observed only to hydrogens on adjacent atoms and not to all other hydrogens in a molecule? Usually, only "two-bond" or "three-bond coupling" is large enough to be detectable (Fig. 15.42). Two-bond coupling only occurs when hydrogens on the same carbon are diastereotopic (different). Such coupling can be very small (<2 Hz), and thus hard to detect, or it can be relatively large (20 Hz). In some cases, **long-range coupling** can be observed, especially if the hydrogens are connected in allylic fashion.

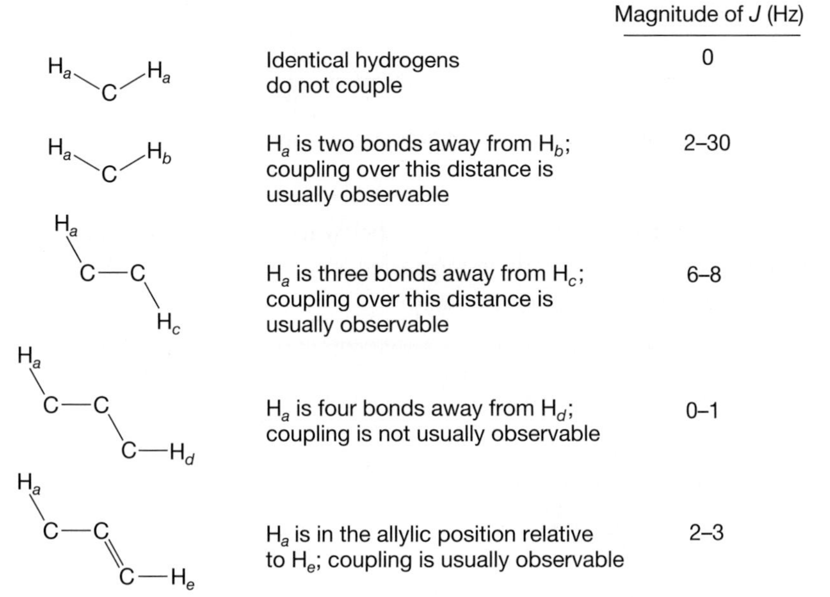

		Magnitude of _J_ (Hz)
H_a—C—H_a	Identical hydrogens do not couple	0
H_a—C—H_b	H_a is two bonds away from H_b; coupling over this distance is usually observable	2–30
H_aC—CH_c	H_a is three bonds away from H_c; coupling over this distance is usually observable	6–8
H_aC—C—CH_d	H_a is four bonds away from H_d; coupling is not usually observable	0–1
H_aC—C=CH_e	H_a is in the allylic position relative to H_e; coupling is usually observable	2–3

FIGURE 15.42 Coupling can be measured over a three-bond distance, but is not usually easily detectable over longer distances.

The coupling constant between hydrogens on adjacent carbons is most sensitive to the dihedral angle between those hydrogens, and this dependence can often be used in assigning structure. Figure 15.43a shows the **Karplus curve,** named for

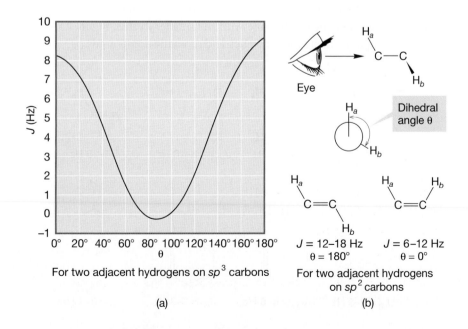

(a)

(b)

For two adjacent hydrogens on sp^3 carbons

For two adjacent hydrogens on sp^2 carbons

J = 12–18 Hz J = 6–12 Hz
θ = 180° θ = 0°

FIGURE 15.43 Coupling between vicinal hydrogens is sensitive to the dihedral angle between them. (a) The Karplus curve shows the calculated relationship between _J_ and the dihedral angle. (b) Dihedral angles can be determined by using a Newman projection and taking the angle between the C—H_a bond and the C—H_b bond.

Martin Karplus (b. 1930) of Harvard University, which shows the dependence of J on the dihedral angle for two hydrogens on adjacent carbons. Figure 15.43b shows the dependence of the coupling constant on the angle between hydrogens attached to sp^2 hydridized carbons.

PROBLEM 15.18 How many lines do you expect to see in the NMR signal for the central hydrogen in 2-methoxypropane? How about 1,3-dichloro-2-methoxypropane? *Hint*: Be careful! See page 721.

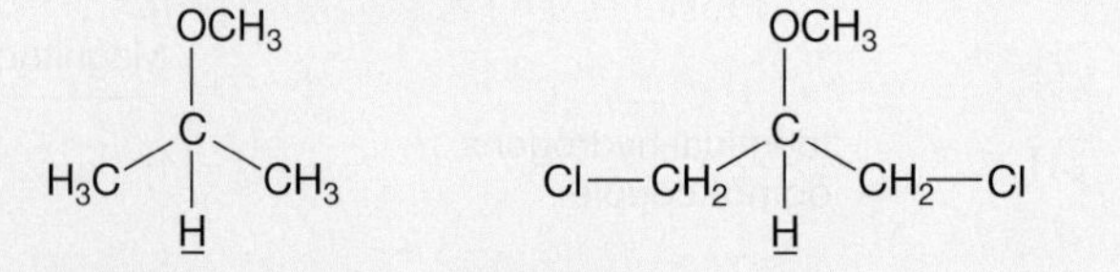

2-Methoxypropane   **1,3-Dichloro-2-methoxypropane**

PROBLEM 15.19 The two molecules shown below have coupling constants for the hydrogens attached to the three-membered ring of 8.4 and 5.3 Hz. Which compound has which coupling constant?

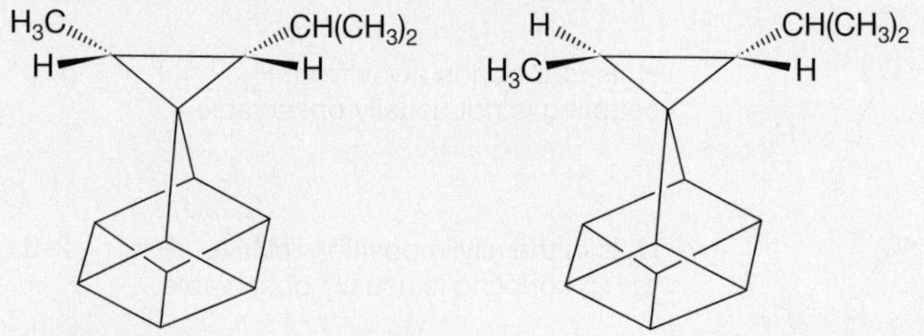

The cis coupling constant, J_{cis}, for alkenes and rigid rings is smaller than the trans coupling constant, J_{trans}. In alkenes, J_{cis} is in the range 6–12 Hz and J_{trans} is in the range 12–18 Hz. Notice that the ranges slightly overlap. One cannot always assign stereochemistry from just one stereoisomer. What would you make of a coupling constant of 12 Hz, for example? But if both isomers are available, the cis stereochemistry can be reliably assigned to the one with the smaller J value. Figure 15.44 gives some typical coupling constants in alkenes and aromatic compounds.

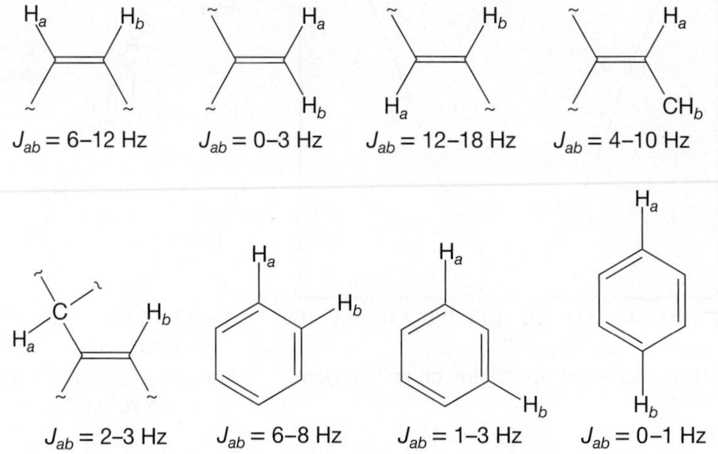

$J_{ab} = 6$–12 Hz $J_{ab} = 0$–3 Hz $J_{ab} = 12$–18 Hz $J_{ab} = 4$–10 Hz

FIGURE 15.44 Some coupling constants for hydrogens attached to double bonds.

$J_{ab} = 2$–3 Hz $J_{ab} = 6$–8 Hz $J_{ab} = 1$–3 Hz $J_{ab} = 0$–1 Hz

Chemical Exchange. The NMR splitting patterns of alcohols and amines are often not what we would expect. Consider the common alcohol, ethyl alcohol, whose ¹H NMR spectrum is shown twice in Figure 15.45. Notice that these two spectra, taken from different catalogs in the chemical literature, show a different chemical shift for the OH signal. Remember there are several factors that can influence the hydrogen bonding, and thus the chemical shift of a hydrogen on oxygen. These include concentration, pH, solvent, and temperature. The methyl and methylene signals are in the same place in the two spectra. The spectra show the expected triplet for the methyl group, but the methylene and hydroxyl hydrogens do not appear as our analysis would predict. The most accurate description of them might be "blobs." What is going on?

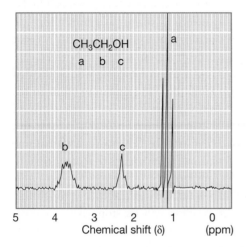 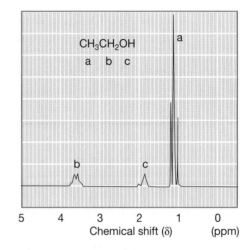

FIGURE 15.45 Two ¹H NMR spectra of ethyl alcohol show different shifts for the OH signal.

Here the issue is one of chemical exchange. In the presence of either an acidic or a basic catalyst, or even under neutral conditions, alcohols and amines exchange the OH or NH hydrogens. Figure 15.46 gives the mechanism for base-catalyzed exchange of an alcohol and acid-catalyzed exchange of an amine, using deuterated materials in order to detect the exchange reaction. This exchange reaction can be used to locate the signals for O—H and N—H groups. The NMR spectrum is taken, then a drop of D_2O is added to the sample tube.

FIGURE 15.46 The base-catalyzed exchange reaction of alcohols, and the acid-catalyzed exchange reaction of amines.

The tube is shaken and the spectrum taken again. Hydrogens on carbon will not exchange in this experiment, so their signals will not shift significantly. However, the RO—H hydrogens exchange and the molecule becomes RO—D. The signals for deuterium, which has a different magnetogyric ratio γ from hydrogen, will not be observed because they appear in a different NMR range, so the signals for the exchangeable hydrogens effectively vanish from the spectrum and are replaced by an HOD signal.

PROBLEM 15.20 Write a mechanism for acid-catalyzed exchange of ethyl alcohol and the base-catalyzed exchange of diethylamine.

In IR spectroscopy, the absorption of energy is essentially instantaneous. The IR spectrometer takes a stop-action picture of the molecule. Nuclear magnetic resonance spectroscopy is different. Acquisition of the signal is much slower, about 10^{-3} s, and the exchange reaction is fast compared to the time it takes to make the NMR measurement. During this time the alcohol oxygen of a given molecule will be bonded to many different hydrogens. This is why OH and the NH signals are often broad. In addition, the two methylene hydrogens of ethyl alcohol "see" only an average of all spins of the hydroxyl hydrogen, not the two different $+\frac{1}{2}$ and $-\frac{1}{2}$ states. Exchange can be so rapid that there is no coupling with the OH hydrogen. The same phenomenon serves to remove coupling between alkyl hydrogens that are vicinal to the NH hydrogens of primary and secondary amines.

If this explanation is correct, the coupling should be detectable in scrupulously purified ethyl alcohol because the exchange reaction will be stopped or, at least, greatly slowed. It is very hard to remove the last vestiges of acidic or basic catalysts from these polar molecules, but if the effort is made, the coupling returns, as it should. Absolutely pure ethyl alcohol does give the anticipated triplet for the OH hydrogen and a well-defined doublet of quartets for the methylene signal (Fig. 15.47).

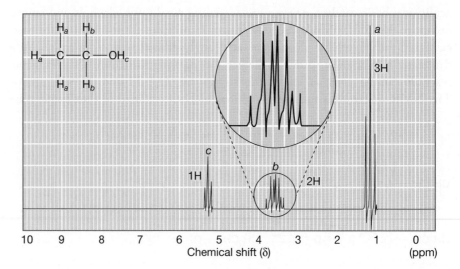

FIGURE 15.47 The ^{1}H NMR spectrum of absolutely pure ethyl alcohol. See Problem 15.60.

This section gives the first hint that NMR spectra are dependent on the rates of reactions. If the rate of the exchange reaction were slow on the NMR time scale, we would not see the averaged spectra that we do for alcohols and amines. We will return to this notion in Section 15.9.

Summary

In the best of all worlds (which means in this case that your budget includes enough money to buy the highest magnetic field spectrometer available), we will see a different chemical shift for every different hydrogen in our molecule, and we will be able to determine the number of equivalent adjacent hydrogens from the $n + 1$ rule. In practice, things may not be quite so simple. We can see some possible complications right away. Perhaps the signals will not be sufficiently separated, and the spectrum may consist of complicated overlapping multiplets. Here is an example that shows why chemists are so anxious to spend taxpayers' money on ever higher field spectrometers. The higher the field, the better the dispersion of the signals (recall Problem 15.9, p. 718). Remember that chemical shifts (in hertz) are dependent on the strength of the applied field and the higher the field, the more separated the signals.

15.6d More Complicated NMR Spectra Spectra with coupling that clearly fits the $n + 1$ rule are called **first-order spectra**. Even at high field, NMR spectra are often not the simple first-order collections of lines we might expect. A classic example of this phenomenon occurs in the two-spin system shown in Figure 15.48. In the NMR spectrum for this molecule, we expect to see a pair of doublets because each hydrogen is different, and coupled to only one hydrogen on the adjacent carbon. The COOH hydrogen is not shown in this spectrum (it is at δ 12 ppm).

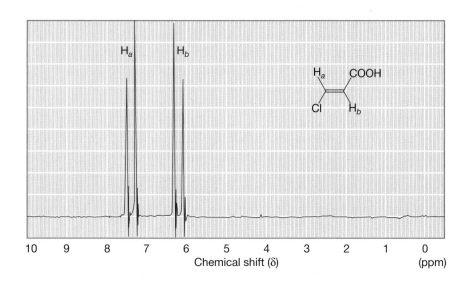

FIGURE 15.48 Two different hydrogens on adjacent carbons will give rise to a pair of doublets as long as the hydrogens are very different. The signal at δ 12 ppm for the COOH hydrogen is not shown.

Remember that a doublet should appear as two lines in a 1:1 ratio. The signals in Figure 15.48 are indeed doublets. But the doublets are not simple 1:1 pairs of lines; the doublet at δ 7.4 ppm is leaning to the right and the doublet at δ 6.2 ppm is leaning to the left. In fact, we see the simple spectrum of 1:1 doublets only when H_a and H_b are *very* different from each other: when they have *very* different chemical shifts. The more alike the two hydrogens, the more the spectrum deviates from first order, the spectrum predicted by the $n + 1$ rule. As the chemical environments of the two hydrogens become more similar, the chemical shifts for the corresponding NMR signals of course also become more alike.

The more alike the hydrogens are, the closer the signals are and the more pronounced the leaning of the peaks toward the center (Fig. 15.49).

FIGURE 15.49 As two hydrogens in a molecule become more similar, the inner lines grow at the expense of the outer lines. The signals increasingly lean toward each other. In the limit, the two hydrogens become identical, and there is only a single line.

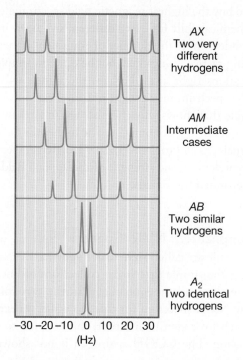

In the limit, when H_a and H_b become the same, they have the same chemical shift. Because there is no coupling between equivalent hydrogens, the signal for them collapses to a single sharp line. Figure 15.49 shows the gradual transition between the case in which the two hydrogens are very different and that in which they are identical. In the language of NMR, the very different system is called *AX*, the identical system A_2, and an intermediate case either *AB* or *AM*.

As we saw earlier (p. 730) it is often simple to derive the expected multiplicity of an NMR signal from the $n + 1$ rule. For example, in a molecule such as 2-furoic acid (Fig. 15.50), the two different hydrogens on C(3) and C(5) (H_A and H_X) are each coupled to the single hydrogen on C(4) (H_M). This system of three very different hydrogens is called an *AMX* system. We would predict the signals for H_A and H_X to be a pair of doublets. The actual signals are more complicated because H_A and H_X are coupled to each other through long-range coupling (Fig. 15.44). Hydrogen H_M in furoic acid is coupled differently to H_A and H_X by the coupling constants J_{AM} and J_{MX}. How can we figure out the pattern for H_M that should be observed in the

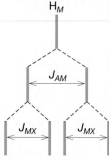

2-Furoic acid

Predicted spectrum for H_M, four lines

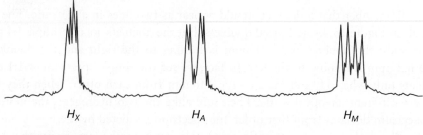

Here is the real spectrum of 2-furoic acid; the COOH hydrogen is off-scale

FIGURE 15.50 In 2-furoic acid, we expect H_A and H_X to appear as doublets. Each is strongly coupled to a single adjacent hydrogen, H_M. However, both hydrogens are further split by long-range coupling (J_{AX}) into doublets of doublets.

spectrum? The first-order pattern can be determined by working out the number of lines from each coupling one at a time (p. 730). We first determine the number of lines produced by coupling H_M to H_A, and then apply the coupling of H_M to H_X to each of the lines in a "tree diagram." For example, coupling of H_M to H_A produces a doublet, separated by J_{AM}. Each of these two lines is further split into two by coupling to H_X. The result is four lines, a doublet of doublets. Figure 15.50 shows the theoretical prediction for H_M and the real spectrum of 2-furoic acid.

First-order spectra are quite easy to predict. For example, according to the $n + 1$ rule, the two different methylene groups of a generic molecule X—CH_2—CH_2—Y should appear as a pair of 1:2:1 triplets. 2-Chloropropanoic acid fits the pattern and, as Figure 15.51 shows, satisfactorily produces the expected pair of methylene triplets in its ^1H NMR spectrum at δ 2.89 and 3.78 ppm. However, when the molecules become more complex we can anticipate complicated NMR spectra, even for first-order systems.

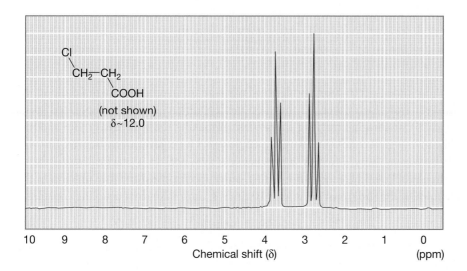

FIGURE 15.51 In 2-chloropropanoic acid, each methylene group appears as a simple triplet. Notice that the signals for these coupled hydrogens lean toward each other.

Methyl butanoate is a simple molecule, and we can predict its first-order spectrum quite easily (Fig. 15.52). The methyl group attached to oxygen has no neighboring hydrogens and appears as a singlet at δ 3.68 ppm. The other methyl group is adjacent to a pair of methylene hydrogens and appears as a triplet at δ 0.94 ppm. The methylene group adjacent to the carbon–oxygen double bond is flanked by the same methylene group. It, too, is a triplet, this time at δ 2.30 ppm. But what about the other methylene group? It is split by three methyl hydrogens, and then split again

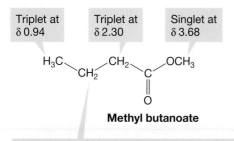

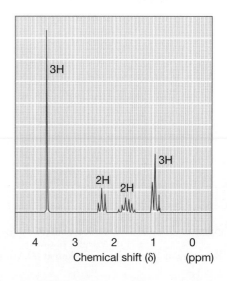

FIGURE 15.52 The multiplet at δ 1.66 ppm results from the splitting of one methylene group by the adjacent methyl and methylene groups. The result is 12 lines that overlap in a complex pattern to give a sextet.

Triplet at δ 0.94

Triplet at δ 2.30

Singlet at δ 3.68

Methyl butanoate

These hydrogens are split into a triplet by the two adjacent methylene hydrogens, and into a quartet by the three adjacent methyl hydrogens; the result is a sextet at δ 1.66, an overlapping set of 12 lines by two methylene hydrogens. The result should be a quartet of triplets, or, equivalently, a triplet of quartets. Either way you work it out, the result is a 12-line pattern, a dodectuplet! Fortunately, the coupling constants are nearly identical for a freely rotating alkane and the signal collapses to a sextet (1:5:10:10:5:1). We can analyze the coupling by noting that this methylene sees five very nearly equivalent hydrogens and therefore the $n + 1$ rule correctly predicts the sextet.

PROBLEM 15.21 In Figure 15.52, the expected dodectuplet has simplified to six lines. The methylene group giving rise to that signal is coupled to both the adjacent methyl and methylene groups. Use a "tree" diagram such as the one in Figure 15.50 to show that, if the coupling constants J to the methyl group and the methylene are fortuitously the same, the observed six lines should appear. Graph paper is very useful!

PROBLEM 15.22 Explain why the methylene group adjacent to the carbonyl group in methyl butanoate is downfield of the methylene group adjacent to the methyl group.

First-order spectra are only observed when all the hydrogens in a molecule are very different from each other. As they become less different, more complicated spectra appear. If we imagine carrying the process of making the hydrogens more and more alike to its logical conclusion, we get a single signal for equivalent hydrogens. But when are hydrogens "very different" and when are they not? In a practical sense, for two hydrogens H_a and H_b to give rise to first-order spectra, the difference in hertz in their chemical shifts ($\Delta\delta$) must be at least 10 times greater than their coupling constant, J_{ab}

$$\Delta\delta > 10 J_{ab} \tag{15.5}$$

This requirement is another reason why high-field NMR spectrometers are so much more useful than lower field instruments. Although the coupling constant J does not depend on the field strength, the value (in hertz) of the chemical shifts does (p. 717). The greater the field strength, the greater the chemical shift difference ($\Delta\delta$) in hertz, and therefore the greater the likelihood that the condition of Eq. (15.5) will be satisfied. Figure 15.53 shows the ^1H NMR spectrum of 1-bromobutane at 60, 90, and 300 MHz. Even the modest change from 60 to 90 MHz results in substantial improvement in the spectrum. You get what you pay for.

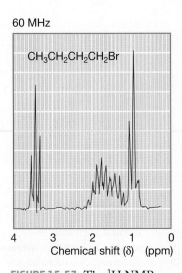
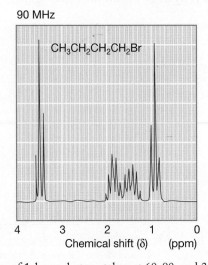
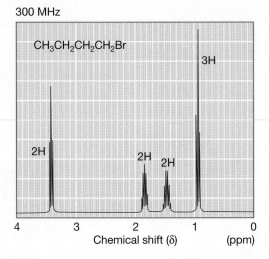

FIGURE 15.53 The ^1H NMR spectra of 1-bromobutane taken at 60, 90, and 300 MHz.

15.6e Decoupled Spectra   When we discussed the NMR spectra of alcohols and amines, we encountered the phenomenon of chemical exchange of the OH and NH hydrogens (Fig. 15.46). Rapid exchange of these hydrogens resulted in an averaging of the coupling to adjacent hydrogens. Exchange effectively **decoupled** the OH and NH hydrogens from hydrogens at adjacent positions. This same trick can be accomplished electronically, and it can be very useful, or even essential in the interpretation of NMR spectra.

Suppose we are set the task of determining the stereochemistry of ethyl crotonate from its ^{1}H NMR spectrum (Fig. 15.54). We might hope to make this distinction by

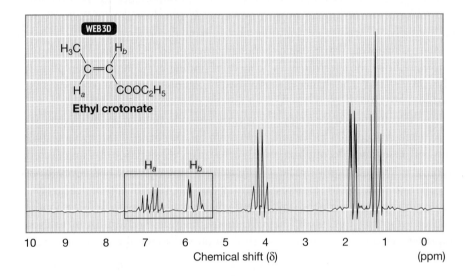

FIGURE 15.54 The ^{1}H NMR spectrum of ethyl crotonate. The coupling between H_a and H_b is obscured by the coupling to the adjacent methyl groups.

measuring the coupling constant between the two vinyl hydrogens. If the compound is trans, we would expect J_{ab} to be 12–18 Hz; if it is cis, J_{ab} would be smaller, 6–12 Hz (Fig. 15.43). However, the rather complex spectrum makes the determination of J_{ab} difficult. The problem is that both vinylic hydrogens are coupled to the methyl group, and this coupling leads to the complex signals we see. If we could somehow get rid of the coupling to the methyl group, the spectrum would surely simplify, and it might become possible to pick out the couplings between H_a and H_b. It is possible to accomplish this simplification through electronic decoupling. The spectrometer is modified to allow a second application of radio-frequency waves during the NMR experiment. While the spectrum is being measured, a radio wave whose frequency matches the resonance frequency of the methyl hydrogens is applied. The result is to induce rapid transitions between the two possible spin states for each methyl hydrogen. As in chemical exchange of ethyl alcohol, the spin of the methyl hydrogens is averaged to zero and so there is no observable coupling to H_a or H_b. Figure 15.55 shows a blow-up of the signals for the two vinyl hydrogens with and without irradiation at the methyl group. It is now easy to pick out the coupling between H_a and H_b and the stereochemistry of the molecule can be determined with ease from the observed J_{ab} of 17 Hz. The compound is trans.

There are many similar applications, and modern NMR spectrometers are all equipped to do decoupling experiments routinely.

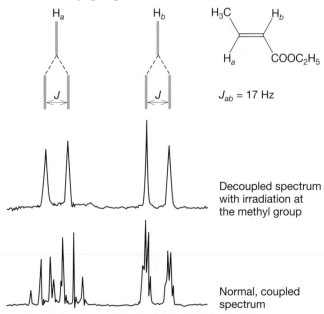

J_{ab} = 17 Hz

Decoupled spectrum with irradiation at the methyl group

Normal, coupled spectrum

FIGURE 15.55 Irradiation at the resonance frequency of the allylic methyl group averages the spins of the methyl hydrogens and decouples them from H_a and H_b. The coupling constant J between H_a and H_b is now easier to determine.

15.7 ^{13}C NMR Spectroscopy

Several other nuclei are of interest to the organic chemist, mainly ^{19}F, ^{15}N, ^2H (deuterium), and ^{13}C, an isotope of carbon, which is present in 1.1% natural abundance. Even though all organic compounds contain carbon, the natural abundance of ^{13}C is low enough so that coupling to ^{13}C is not observed in hydrogen spectra. These other nuclei can be used either in natural abundance or in enriched samples to provide their own NMR spectra. Recall that the constant γ in Eq. 15.4 will be different for ^{19}F, ^2H, ^{15}N, or ^{13}C, and their NMR signals will not overlap with the hydrogen spectra.

For organic chemists, only ^{13}C is comparable in importance to ^1H. Like hydrogen, ^{13}C has a spin of ½, and coupling between ^{13}C and ^1H will complicate ^{13}C spectra. Normally, ^{13}C spectra are run using a broad-band decoupling so that all ^{13}C—^1H coupling is removed. What about coupling between two ^{13}C atoms? Won't this coupling complicate the spectrum? The chance of finding a ^{13}C at any given position is only 1.1% or 0.011. The chance of finding two ^{13}C atoms on adjacent positions is therefore $0.011 \times 0.011 = 0.00012$! At least for natural abundance spectra we can ignore ^{13}C—^{13}C coupling. Accordingly, ^{13}C spectra are collections of sharp singlets. If the broad-band decoupler is turned off, a spectrum can be obtained through a technique called **off-resonance decoupling**, which allows the ^{13}C signals to be split according to the $n + 1$ rule, but only by the *directly attached* (not adjacent) hydrogens. Because they have no attached hydrogens, quaternary carbons will still appear as singlets, but methine carbons (one attached hydrogen) will be doublets, methylene carbons (two attached hydrogens) triplets, and methyl carbons (three attached hydrogens) will appear as quartets. Figure 15.56 summarizes this notion and shows the decoupled and off-

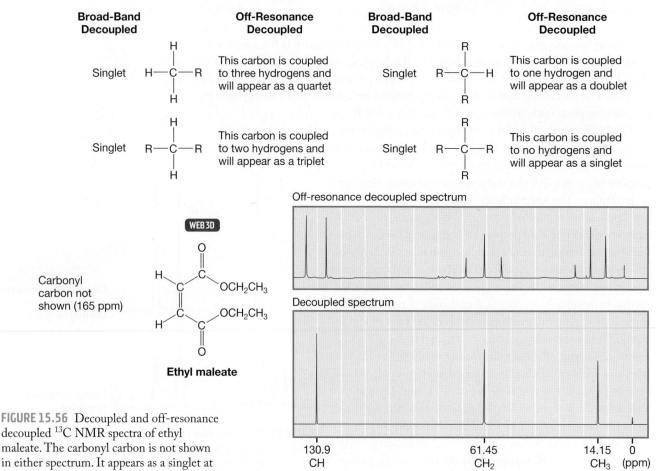

FIGURE 15.56 Decoupled and off-resonance decoupled ^{13}C NMR spectra of ethyl maleate. The carbonyl carbon is not shown in either spectrum. It appears as a singlet at δ 165 ppm.

resonance decoupled spectra of ethyl maleate. Notice that ^{13}C NMR spectra are not integrated. Only heights of similar carbons (methyls, methylenes, or methines) can be compared to determine the relative number of carbons giving rise to a signal.

Several other techniques now allow for the easy determination of the number of hydrogens attached to a carbon. One of the most convenient techniques goes by the acronym **DEPT (distortionless enhancement with polarization transfer)**. In this technique, several separate NMR spectra are determined for a molecule under conditions that allow for the appearance of only methine (CH), methylene (CH$_2$), or methyl (CH$_3$) carbons. Quaternary carbons can be determined by difference: Any signal in the full spectrum that does not appear in the separate spectra for CH, CH$_2$, and CH$_3$ carbons must belong to a quaternary carbon.

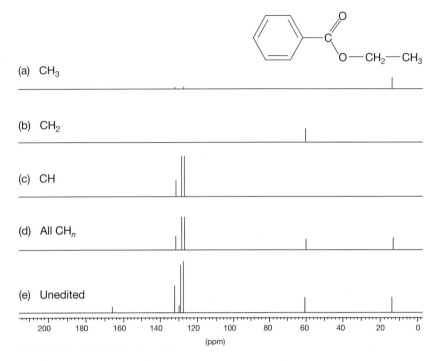

FIGURE 15.57 The DEPT technique applied to ethyl benzoate.

Figure 15.57 shows the DEPT technique applied by Kathryn Williams and Roy W. King of the University of Florida to ethyl benzoate. The full ^{13}C spectrum is shown on the bottom line, (e). The line (d) shows the signals for all carbons attached to any number of hydrogens, note the absence of the signals for the quaternary carbons at 130 and 166 ppm. The spectra on lines (c), (b), and (a) show the methine, methylene, and methyl carbons, respectively.

Table 15.5 gives a series of chemical shift correlations for ^{13}C NMR. Notice the greatly expanded range of chemical shifts extending over 200 ppm, as opposed to the mere 10 ppm for ^{1}H (Table 15.4). The factors that give rise to chemical shifts in ^{13}C NMR are the same as in ^{1}H NMR (electronegativity, hybridization, aromaticity, and delocalization). Dispersion is much greater for ^{13}C NMR than for ^{1}H NMR.

TABLE 15.5 Some ^{13}C Chemical Shifts

Type of Carbon	Chemical Shift (δ)[a]	Type of Carbon	Chemical Shift (δ)[a]
Alkanes		Alcohols, ethers	
Methyl	0–30	C—O	50–90
Methylene	15–55	Amines	
Methine	25–55	C—N	40–60
Quaternary	30–40	Halogens	
Alkenes		C—F	70–80
C=C	80–145	C—Cl	25–50
Alkynes		C—Br	10–40
C≡C	70–90	C—I	−20–10
Aromatics	110–170	Carbonyls, C=O	
Benzene	128.7	R$_2$C=O	190–220
		RXC=O (X = O or N)	150–180

[a] The chemical shift δ is in parts per million (ppm) from TMS.

Once again, the question of memorization arises. Use Table 15.5 to work problems, and you will automatically memorize the information you really need (if you do enough problems!).

15.8 Problem Solving: How to Use Spectroscopy to Determine Structure

The task set here is an almost impossible one—to show in a practical way how to use the various kinds of spectral data to come up with a structure. Every compound is different, of course, and so there can be no absolutely general method. Even worse, there are many different approaches; what works for you may not be the optimal way for others. Yet, there are some general techniques that are useful, and we'll try to set them out here. Remember, though, as the old lady in New York answered, when asked for directions on how to get to Carnegie Hall, "practice, practice, practice." For you, that ancient joke means, "do lots and lots of problems." Nowhere else is that advice so essential.

First of all, when confronted by a set of spectral charts or data derived from them, try to determine the molecular formula. Sometimes it will be given, sometimes you can work it out. Other times, it may be obvious or at least likely from the chemical reactions involved. Sometimes you may have only a molecular weight, from the molecular ion of a MS for example. Either way, having a formula or molecular weight lets you determine the degrees of unsaturation (p. 131) and see what's left once you have determined part of the structure.

Next, use any available IR data, along with NMR chemical shift data, to get an idea of what functional groups are present. The most useful data usually come from the NMR spectrum. The convention for reporting signals is to list the chemical shift followed by parenthetical listing of the coupling and the integration. So the NMR spectrum of 1-bromobutane (Fig. 15.53) would be δ 0.9 (t, 3H), 1.5 (sextet, 2H), 1.8 (quintet, 2H), 3.4 (t, 2H). Coupling information is given by using s for singlet, d for doublet, t for triplet, q for quartet, dd for doublet of doublets, and so on. If the coupling is indecipherable, then we use m for multiplet (or mess, if you want). There are many items in the Additional Problems section that let you practice this technique.

Surprisingly, symmetry is often all that is needed to solve a structural problem—a simple counting of signals may be sufficient to make a choice between several alternatives, so you should always look hard at symmetry *before* starting on any detailed analysis.

WORKED PROBLEM 15.23 You have bottles containing three isomeric compounds, **A**, **B**, and **C**, each of the formula $C_{10}H_{14}$. The only significant bands in the IR spectra of all three compounds are at about 3050, 2950, and 1610 cm^{-1}. The ^1H NMR spectrum of **A** shows signals at δ 2.13 (s, 6H) and 6.88 (s, 1H); **B** shows signals at δ 2.18 (s, 3H), 2.25 (s, 3H), and 6.89 (s, 1H); and **C** shows signals at δ 2.11 (s, 3H), 2.19 (s, 3H), 2.22 (s, 6H), and 6.80 (s, 2H). The ^{13}C NMR spectrum of **A** shows lines at δ 131.02, 133.52, and 19.04 ppm. Compound **B** shows signals at δ 133.76, 133.79, 126.88, 20.65, and 15.78 ppm. Compound **C** shows signals at δ 136.15, 134.43, 131.69, 128.29, 20.74, 20.20, and 14.86 ppm. What are the structures?

(continued)

ANSWER First of all, these compounds seem to be aromatic hydrocarbons. The formula $C_{10}H_{14}$ establishes that there are four degrees of unsaturation (p. 131), and the IR spectra show both aromatic and aliphatic CH stretches, as well as a band at 1610 cm^{-1} appropriate for an aromatic double-bond stretch. An aromatic ring has three double bonds and a ring, which adds up to four degrees of unsaturation. If six of the available carbons are tied up in an aromatic ring, the other four must be attached in some way. A look at the ^{1}H NMR spectra shows only singlets, and so the four remaining carbons are certainly methyl groups. The chemical shift for each of the methyl groups is about δ 2.2, which is consistent with a methyl group on an aromatic ring. Thus, putting the data together leads to the idea that **A**, **B**, and **C** are the three possible tetramethylbenzenes. But which is which?

    To answer this question, there is no need to do anything more elaborate than look at the number of signals. There is more than one way to do this, but let's use the methyl signals in the ^{1}H NMR spectrum. Compound **A** has only one methyl signal, so all four methyl groups must be identical. This compound must be the molecule on the left. Compound **B** has two methyl signals, and must be the structure in the middle. Of course, we now know that the remaining isomer must be **C**, but it is gratifying that it has the appropriate three methyl signals. There was no need to consider many details, only to use symmetry and count.

PROBLEM 15.24 Could we have used the aromatic CH signals in Problem 15.23 to determine the answer?

PROBLEM 15.25 Could we have used the ^{13}C NMR spectra in Problem 15.23 to determine the answer?

PROBLEM SOLVING

In almost all spectroscopy problems in which NMR is used to distinguish between various possible structures, the very first thing to do is count carbons and/or hydrogens. Many otherwise quite tough problems can be shortcut, and made much easier, by determining how many signals there will be. Problem 15.23 is a good example. Moreover, problem writers aren't always aware that simple counting can solve otherwise difficult problems. By counting, you may be able to solve easily a problem that was meant to be very hard, and to require specific detailed knowledge of spectroscopy. It happens all the time.

Chemical shift data are often decisive. After considering symmetry, that is what we usually look at next. Table 15.4 will let you make many structural decisions.

WORKED PROBLEM 15.26 You have four bottles, each containing one of the four compounds shown below, **E**, **F**, **G**, and **H**. From the spectral data given, determine which bottle holds which compound.

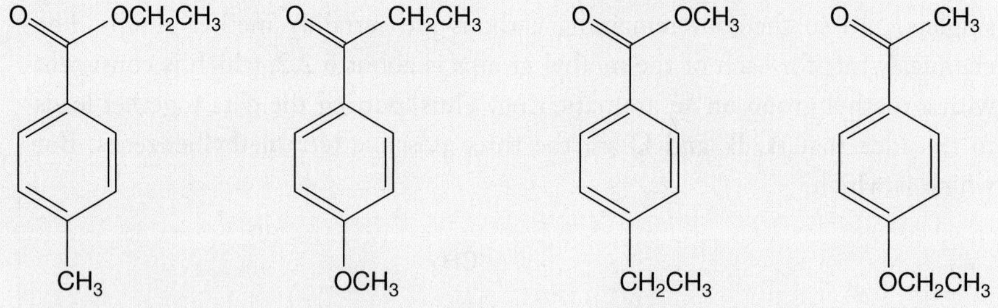

All four compounds have signals for 4H in the aromatic region of the ^{1}H NMR spectrum at δ 7–8 ppm.

Bottle 1   IR: 1720 cm^{-1} ^{1}H NMR: δ 1.38 (t, 3H), 2.40 (s, 3H), 4.36 (q, 2H)

Bottle 2   IR: 1688 cm^{-1} ^{1}H NMR: δ 1.12 (t, 3H), 2.81 (q, 2H), 3.76 (s, 3H)

Bottle 3   IR: 1725 cm^{-1} ^{1}H NMR: δ 1.21 (t, 3H), 2.65 (q, 2H), 3.85 (s, 3H)

Bottle 4   IR: 1680 cm^{-1} ^{1}H NMR: δ 1.40 (t, 3H), 2.50 (s, 3H), 4.07 (q, 2H)

ANSWER First, use Table 15.3 to determine that the two esters are in bottles 1 and 3 (IR: conjugated C=O stretch at about 1720 cm^{-1}) and the two ketones are in bottles 2 and 4 (IR: conjugated C=O stretch much lower, at about 1680 cm^{-1}).

Now the chemical shift data allow a quick determination of the rest of the problem. Again, there are several ways to do this last part, but the easiest is to look at the chemical shift of the singlet methyl group (you could also use the methylene quartet). Every signal has three pieces of data: integral, chemical shift, and coupling. If we look at the chemical shift (δ 2.40) of the methyl (3H) singlet in bottle 1, we see that it cannot be a methyl group attached to an oxygen. Methyl groups on oxygen are at δ 3.2 or even further downfield. A methyl group on an aromatic ring should be around δ 2.2 ppm. If the compound in bottle 1 must be an ester based on IR, then it must contain compound **E**. The ester in bottle 3 has a 3H singlet at δ 3.85 ppm and is consistent with compound **G**. A similar analysis of the ketones leads us to assign the 3H singlet at δ 2.50 ppm in bottle 4 to the methyl ketone of compound **H**, and the 3H singlet at δ 3.76 ppm in bottle 2 must be assigned to the methoxy group on the aromatic ring in compound **F**.

Sometimes the coupling constant J is the key to solving the puzzle, and often it is the magnitude of J that is the key. The molecules in Problem 15.19 are a perfect example of this use of J. More often, it is the splitting pattern that solves the problem. An examination of J reveals who is adjacent to whom in the hydrogen world, and thus provides deep insight into structure. Here's another problem, this time one that requires you to think about J.

WORKED PROBLEM 15.27 Again you have bottles 1 and 2, this time containing **I** and **J**. Explain how you would use ^{1}H NMR spectroscopy to tell which molecule is in which bottle.

I   **J**

ANSWER In compound **J**, the hydrogen adjacent to the phenyl group and the hydrogen adjacent to the chlorine will couple. Each will appear as a doublet, with a small coupling constant of a few hertz (Hz). The hydrogen adjacent to the phenyl ring and the hydrogen adjacent to the chlorine in compound **I** are too far from each other to couple; they will appear as singlets because identical hydrogens do not couple. Be sure you see that there is no other easy way to make this determination.

Now we pass on to some "bells and whistles" of NMR.

MAITOTOXIN

These days sophisticated NMR techniques seem capable of determining any structure. This box is almost pure "gee-whiz." We are trying to dazzle you with the complexity of the molecular structures that can be solved by the application of spectroscopy (and a lot of hard work). The example is maitotoxin.

Maitotoxin, a molecule for which one of your authors maintains a certain affection, is extraordinarily lethal (for half the mice treated the lethal dose is only 50 ng/kg) and is one of the largest natural products known. In 1993, a group of Japanese chemists reported a variety of complex NMR studies that established the general structure, and in 1996, chemists at Harvard used much spectroscopy, along with organic synthesis, to establish the complete relative stereochemistry. That same year the absolute stereochemistry was determined in Japan by K. Tachibana (b. 1949) and his research group

at the University of Tokyo. How many stereogenic carbons are there in maitotoxin? How many possible stereoisomers are there? To solve this structure is a remarkable accomplishment.

Why show this molecule? No one is ever going to ask you to reproduce the structure, we hope, and there are not likely to be any immediate practical applications. We think that the answer is simply that the sheer difficulty of the enterprise is overwhelming. We humans can do some quite amazing things. There are times when that gee-whiz factor must be confronted directly, when we are allowed to sit back and marvel at some accomplishment, and this is one of those moments.

15.9 Special Topic: Dynamic NMR

We have already seen that rates of hydrogen exchange are important in NMR spectroscopy. Recall that fast exchange of hydroxyl and amine hydrogens generally results in a decoupling of the OH and NH hydrogens (p. 733). An adjacent hydrogen "feels" only an averaged perturbation of the applied magnetic field if exchange is fast enough. If exchange is slowed by careful removal of all acidic or basic catalysts, the coupling can be detected. The *rate* of exchange is important to the spectrum. If it is fast, the NMR experiment detects an averaged spectrum; if it is slow, it sees a static picture.

The classic example of this phenomenon involves the spectrum of cyclohexane. As you know, there are two kinds of hydrogen in cyclohexane, axial and equatorial, and these are interconverted by a ring flip, with an activation energy of about 11 kcal/mol (p. 198). Despite the existence of two kinds of hydrogen, the room-temperature spectrum of cyclohexane shows only a single line at δ 1.4 ppm (Fig. 15.58).

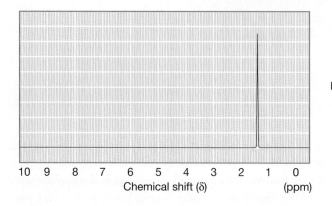

FIGURE 15.58 The ^{1}H NMR spectrum of cyclohexane at 25 °C.

This spectrum appears simple because the rate of interconversion of axial and equatorial hydrogens is so fast, with respect to the time necessary for the NMR measurement, that the spectrum appears as an average. Suppose we decrease the energy available to the system by lowering the temperature. If we can decrease the rate of the ring flip sufficiently, the spectrometer will no longer detect an average of the two possible positions but will be able to see both axial and equatorial hydrogens. Figure 15.59 shows the ^{1}H NMR spectrum of undecadeuteriocyclohexane (cyclohexane-d_{11}) at −89 °C. Two equal signals appear, centered on the position of the room temperature, averaged signal.

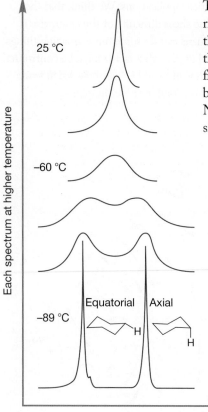

FIGURE 15.59 At low temperature, ring flipping is slowed and the axial and equatorial hydrogens of cyclohexane can be resolved in the NMR spectrum. As the temperature is raised, the −89 °C spectrum approaches the 25 °C spectrum.

In practice, for a signal to be detectable for a given species, it must have a lifetime of about 1 s. As the temperature is raised to room temperature, the lifetime of individual cyclohexanes decreases as the rate of ring flipping increases, and an averaged spectrum appears.

Now, what happens at an intermediate temperature? Figure 15.59 shows you. A common-sense answer would be correct here. If two signals are found at -89 °C, and a single signal at 25 °C, at intermediate temperatures the two kinds of signal must approach each other. The two sharp peaks of the -89 °C spectrum will slowly merge into the single peak of the 25 °C spectrum. In fact, it is possible to derive kinetic information from the temperature dependance of NMR spectra, and this is still another use of NMR spectroscopy.

More molecules than you might at first expect give this kind of averaged spectrum. It takes a detailed look at structure to see why. Consider ethyl alcohol. If we ignore the OH signal, which we know to be dependent on concentration, pH, solvent, and temperature, we expect to see only one signal for the three equivalent methyl hydrogens and another for the two equivalent methylene hydrogens. But wait—look at the Newman projection in Figure 15.60, which shows the most stable staggered conformation of ethyl alcohol. This perspective shows that there are three kinds of hydrogen (H_a, H_b, and H_c), not two. The three methyl hydrogens are *not* all the same. The two that are gauche to the alcohol are in a different environment from the hydrogen that is anti to the alcohol. Yet, we do not see separate signals for these hydrogens. By now you should be able to see the answer to this conundrum. As shown in Figure 15.60, rotation about the carbon–carbon bond of ethyl alcohol interconverts the three methyl hydrogens, rendering them equivalent. Moreover, we know that the low barrier for rotation, about 3 kcal/mol, ensures a rapid rate for this simple process. In Figure 15.60, the hydrogens are individualized with colored squares. Note how each methyl hydrogen occupies one of the three possible positions as rotation occurs. The NMR spectrometer detects only the averaged signal.

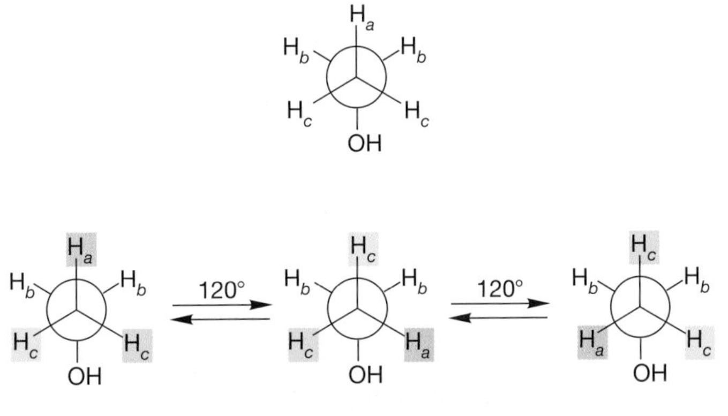

FIGURE 15.60 In static ethyl alcohol, there are three kinds of hydrogen attached to carbon. However, rapid rotation around the carbon–carbon bond interconverts the methyl hydrogens and makes them equivalent. The two methylene hydrogens are enantiotopic.

What about the two methylene hydrogens, H_b? The Newman projection of Figure 15.60 shows that even in the absence of rapid rotation they are equivalent, and therefore must *always* give rise to one signal. These hydrogens are enantiotopic. As we saw earlier, the origin of the name lies in the mirror-image relationship of the two to each other.

So, as long as rotation is fast, neither the individual methyl hydrogens nor the individual methylene hydrogens of ethyl alcohol are distinguishable by NMR spectroscopy.

PROBLEM 15.28 In a solvent consisting of a single enantiomer, the two enantiotopic hydrogens of ethyl alcohol (H_b in Fig. 15.60) appear as separate signals. Explain.

WORKED PROBLEM 15.29 Use the device of replacing the methylene hydrogens (H_b in Fig. 15.60) with an X group to show that they are enantiotopic (p. 721).

ANSWER

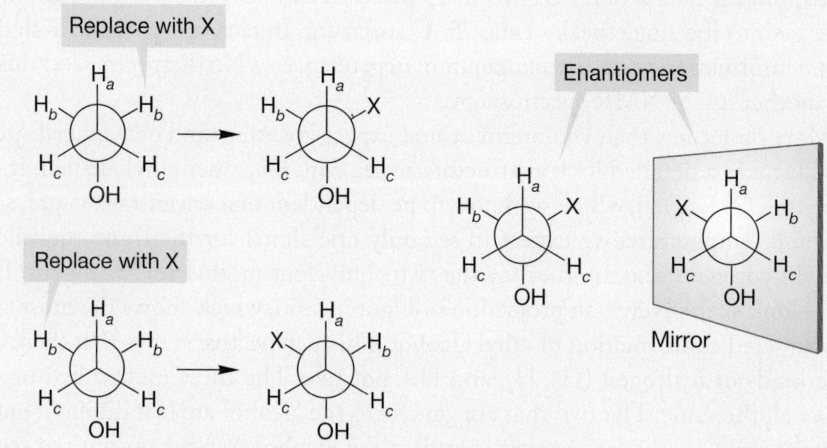

Now let's change some things about ethyl alcohol. Let's first convert it into 1-propanol by replacing one methyl hydrogen with a methyl group (Fig. 15.61). There are now three sets of hydrogens: the methyl hydrogens and two different sets of methylene hydrogens. Rapid rotation ensures that the three methyl hydrogens will be equivalent, and each pair of methylene hydrogens consists of two magnetically equivalent, enantiotopic hydrogens.

FIGURE 15.61 In propyl alcohol, the methyl hydrogens are homotopic; there are two pairs of enantiotopic methylene hydrogens.

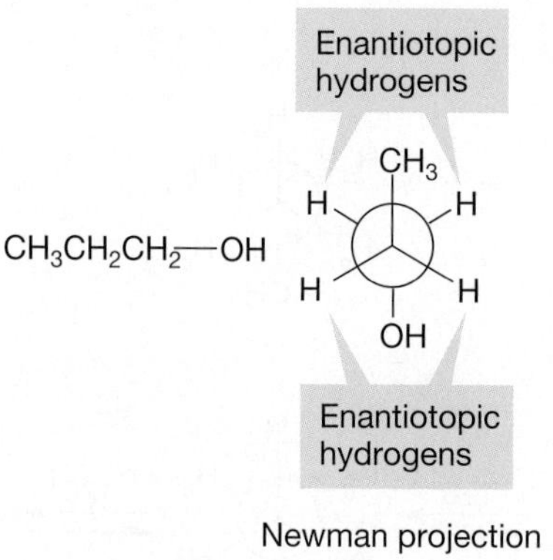

Next let's replace a hydrogen of 1-propanol with a chlorine to make 2-chloro-1-propanol. Look at the Newman projection of this molecule in Figure 15.62.

FIGURE 15.62 In 2-chloro-1-propanol, the two methylene hydrogens, H_x and H_y, are neither equivalent (homotopic) nor enantiotopic.

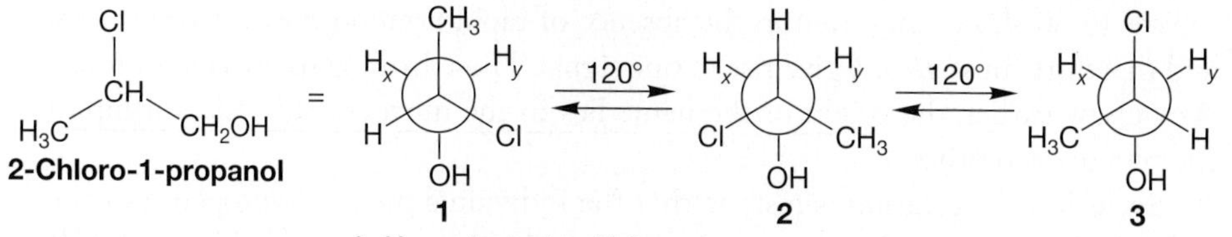

Are the two methylene hydrogens equivalent? No. The Newman projections of Figure 15.62 show the three rotational isomers and in *all* of them the two hydrogens, H_x and H_y are different. *Never* do they become equivalent. They are forever different no matter how fast rotation about the carbon–carbon bond is, and so must give rise to separate signals. These hydrogens are diastereotopic, and will give rise to distinct signals. This point is particularly difficult, so let's anticipate the Common Errors section a little bit and deal with the problem right here. Is not H_y in Newman projection **2** of Figure 15.62 in the same position as H_x in drawing **1**? Are the two not equivalent? Both are 120° from the hydroxyl group, and both lie in between a methyl group and a hydrogen. The answer is no, and to see why, we must look at the OH groups. A short demonstration shows that the two OH groups in drawings **1** and **2** are not identical. One is between hydrogen and chlorine, the other is between methyl and chlorine. Thus, H_x and H_y are 120° from *different* OH groups in the two forms. A more elaborate demonstration takes advantage of the thought process mentioned earlier (p. 721). Replace H_x with a group X and determine the relationship between this compound and the one produced by replacing H_y with X (Fig. 15.63). These two compounds are different, neither identical nor enantiotopic. There is no way out, H_x and H_y are diastereotopic.

These two structures are neither identical nor enantiomers; they are diastereomers, and H_x and H_y are diastereotopic hydrogens

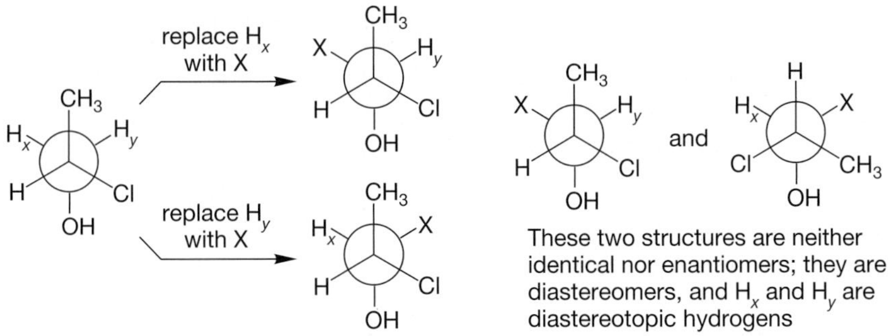

FIGURE 15.63 The two hydrogens H_x and H_y are diastereotopic and will give different signals in an NMR spectrum.

PROBLEM 15.30 The methyl group of *trans*-1-*tert*-butyl-4-methylcyclohexane appears as a doublet, but in the spectrum for *cis*-1-isopropyl-3-phenylcyclohexane there are two methyl doublets. Explain.

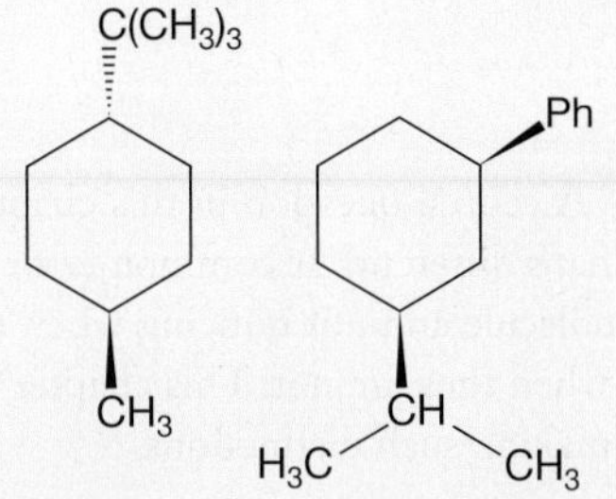

15.10 Summary

New Concepts

This chapter recapitulates the ideas introduced in Chapter 4 in the discussion of column chromatography, and introduces the much more versatile separation methods of gas chromatography and high-performance liquid chromatography. All chromatography works in much the same way: The components of a mixture are forced to partition themselves between a moving phase and a stationary phase. The more a given component exists in the moving phase, the faster it moves through the column; the more strongly it is absorbed by the stationary phase, the longer it remains on the column.

In mass spectrometry, a molecule is first ionized by bombardment with high-energy electrons. The molecular ion so formed can undergo numerous fragmentation reactions before it passes, along with the daughter ions formed through fragmentation, into a region between the poles of a magnet. Ions are separated by mass, and focused on a detector. The mass spectrometer gives the molecular weight of a molecule, provided that the molecular ion can be detected, and high-resolution mass spectrometry can determine the molecular formula of an ion. Structural information can often be gained through an analysis of the important daughter ions in the fragmentation

pattern. The presence of chlorine or bromine in a molecule can be deduced from the appearance of signals from M + 2 isotopes.

Infrared spectroscopy detects vibrational excitation of chemical bonds. Functional groups have characteristic vibrational frequencies in the IR region that depend on the masses of the atoms in the bond and on the force constant of the bond. Analysis of an IR spectrum gives information on the kinds of functional groups present.

Nuclear magnetic resonance spectroscopy detects the energy absorbed in the transition from a low-energy state in which the nuclear spin is aligned with an external magnetic field to a slightly higher energy state in which it is aligned against the external field. The ^{1}H NMR spectra can be integrated to reveal the relative number of hydrogens in each peak. Hydrogens, ^{13}C, and some other nuclei have characteristic absorptions (the chemical shift, δ), which depend critically on the surrounding chemical environment. Hydrogen nuclei are coupled to adjacent hydrogens through a coupling constant, J. The number of lines observed for a particular hydrogen depends on the number of other nuclei to which it is coupled. For first-order spectra, the number of lines equals $n + 1$, where n is the number of equivalent coupled nuclei.

Key Terms

base peak (p. 702)
chemical shift (δ) (p. 718)
coupling constant (J) (p. 718)
daughter ion (p. 702)
decoupling (p. 739)
DEPT (distortionless enhancement with polarization transfer) (p. 741)
diastereotopic (p. 721)
enantiotopic (p. 721)
first-order spectrum (p. 735)
force constant (p. 708)
fragmentation pattern (p. 702)
gas chromatography (GC) (p. 697)
GC/IR (p. 699)

GC/MS (p. 699)
high-performance liquid chromatography (HPLC) (p. 697)
homotopic (p. 721)
infrared (IR) spectroscopy (p. 707)
integral (p. 718)
Karplus curve (p. 731)
long-range coupling (p. 731)
M + 1 peak (p. 700)
mass spectrometry (MS) (p. 699)
molecular ion (p. 700)
$n + 1$ rule (p. 729)
off-resonance decoupling (p. 740)
parent ion (p) (p. 702)

ppm scale (p. 716)
radical cation (p. 699)
tetramethylsilane (TMS) (p. 716)
vinylic hydrogen (p. 724)
wavenumber (p. 708)

Common Errors

Practice, practice, practice! The successful interpretation of spectra requires it. The most common error is not to realize this fact. The memorization of tables of data will not make you into a spectroscopist, but working lots and lots of problems will. See Section 15.8 for some hints on solving problems.

A central question in this chapter, and in all of chemistry, is, What's different? A common error is missing the symmetry of a molecule and not noticing when two hydrogens are equivalent or when they are not. This chapter offers a device that is useful in making such distinctions.

It is often difficult to keep track of the interplay of the integration, the chemical shift, δ, and the coupling constant, *J*. A common error is confusing a hydrogen's signal with those of the hydrogens that are vicinal to it. For example, if the signal for H_a is a doublet then it "sees" one hydrogen. But H_a itself might be a 3H methyl, a 2H methylene, or a 1H methine.

15.11   Additional Problems

General advice: Although this section is longer than most problem sections, it cannot do more than illustrate some of the difficulties involved in going from a set of lines on chart paper, or a table of numbers, to a structure. Moreover, in the real world of chemistry there is close interaction between the reactions involved and the interpretation of spectra. The chemist has some idea (usually!) of what is likely to be present. Some of the later problems must be solved in an interactive way. The chemistry involved will suggest structures, and you will have to see if they are consistent with the data available. All of the following problems involve manipulating spectral data. You should start well-equipped with a table of IR frequencies and ^{1}H NMR and ^{13}C NMR correlation charts for chemical shifts. A table of coupling constants will also be essential, and a copy of the Karplus curve is often useful. Do not be shy about looking values up. Most chemists analyze their "real world" spectra that way, and *everyone* starts that way.

   Specific advice: In Chapter 3 we first encountered Ω, the number of "degrees of unsaturation." In this chapter, we see nitrogen-containing compounds and will have to modify our formula a bit. Here is the protocol for nitrogen and some other heteroatoms. Be sure to review Ω in Chapter 3 (p. 131).

1. Oxygen and divalent sulfur can be ignored.
2. Halogens (F, Cl, Br, and I) are counted as if they were hydrogens.
3. Treat nitrogen as if it were a carbon. However, as nitrogen is trivalent, not tetravalent, subtract one hydrogen per nitrogen from the total number of hydrogens computed for the saturated hydrocarbon. A formula that can be used is

$$\frac{[2(\#C's) + 2 - (\#H's \text{ and halogens}) + (\#N's)]}{2} = \Omega$$

We begin with some relatively straightforward problems largely involving infrared spectroscopy and mass spectrometry, and then move on.

PROBLEM 15.31   Here is a practical problem of no intellectual content whatsoever. In doing problems involving IR, it is often much easier to read the values on the chart paper in microns (μ), rather than wavenumber ($\bar{v}$). Unfortunately, most of us have learned to think in wavenumbers. There is a simple conversion, $1\mu = 10^{-4}$ cm. Convert 3.00, 5.00, 7.00, 9.00, and 11.00 μ into wavenumbers. *Hint*: See p. 708.

PROBLEM 15.32   Calculate the degrees of unsaturation in the following molecules:

PROBLEM 15.33   The IR spectrum of a molecule known to have structure **A, B, C,** or **D** is shown below. Which structure is most consistent with the spectrum? Explain your reasoning.

PROBLEM 15.34   Each of compounds **1–5** is either an aldehyde, an anhydride, a carboxylic acid, an ester, or a ketone. Identify to which class each compound belongs based on the IR spectrum shown. Explain your reasoning by assigning characteristic and/or distinguishing absorptions. Each class is represented by one compound.

PROBLEM 15.35   For each of the following IR spectra (**1–5**) there is a choice of three possible types of compounds. For each spectrum, choose the most appropriate class of compound. Explain your reasoning by noting the presence or absence of characteristic bands in the spectrum.

1

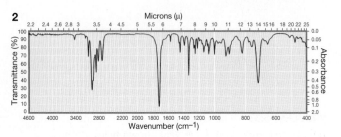

1 Alcohol
Carboxylic acid
Phenol

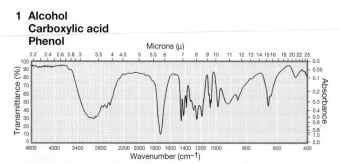

2

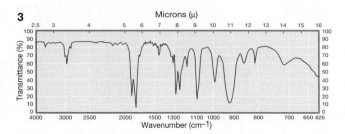

2 Aldehyde
Ester
Ketone

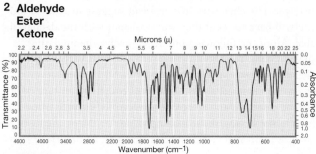

3

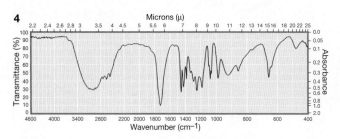

3 1-Alkyne
Symmetrical disubstituted alkyne
Nitrile

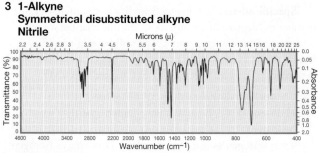

4

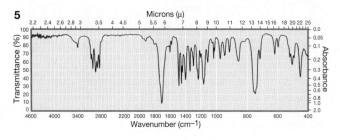

4 Primary amide
Primary amine
Nitro compound

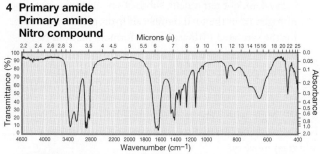

5

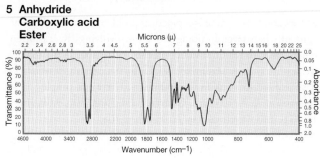

5 Anhydride
Carboxylic acid
Ester

PROBLEM 15.36 Assign the IR spectra below to the correct octene isomer. Explain your reasoning.

1-Octene

***trans*-2-Octene**

***cis*-2-Octene**

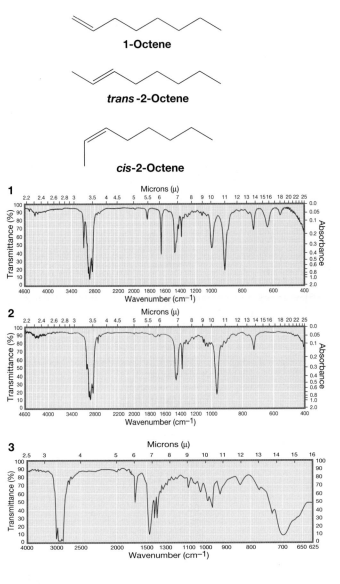

PROBLEM 15.37 The following IR spectra belong to 5-phenyl-1-pentyne and 6-phenyl-2-hexyne. Which is which? Explain your reasoning.

$$PhCH_2CH_2CH_2C{\equiv}CH \qquad PhCH_2CH_2CH_2C{\equiv}CCH_3$$

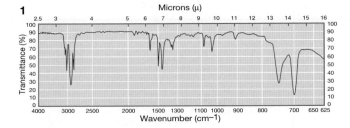

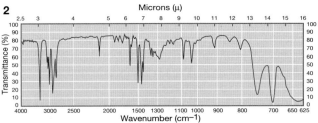

PROBLEM 15.38 Assign the IR spectra below to the correct chlorotoluene isomer.

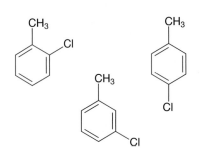

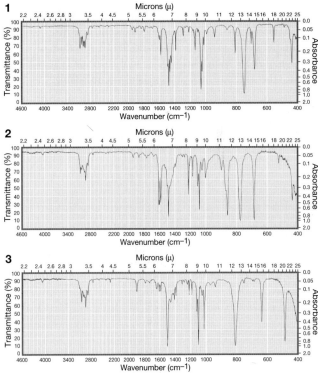

PROBLEM 15.39 Compound **A** has the formula $C_8H_7BrO_2$. The two most intense peaks in the mass spectrum are at $m/z = 216$ and 214, and they appear in almost equal amounts. What ion gives rise to each of these peaks? Compound **A** is an aldehyde. Explain why the next most intense peaks in the mass spectrum come at m/z values of 215 and 213.

PROBLEM 15.40   Compound **A**, containing only C, H, and O, gave 80.00% C and 6.70% H upon elemental analysis. Summaries of the mass spectrum and the IR spectrum of compound **A** are shown below. The molecular ion is at *m/z* 120. Deduce the structure of **A** and explain your reasoning.

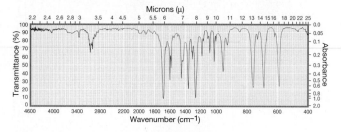

m/z	Relative Abundance
105	100
77	88
51	40
120	29
50	21
43	17
78	10
39	9

PROBLEM 15.41   Compound **A**, containing only C, H, and O, gave 70.60% C and 5.90% H upon elemental analysis. Summaries of the mass spectrum and the IR spectrum of compound **A** are shown below. The molecular ion is at *m/z* 136. Deduce the structure of **A** and explain your reasoning.

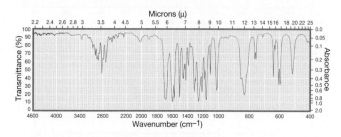

m/z	Relative Abundance
135	100
136	71
77	28
92	15
107	15
39	8
65	8
63	8

PROBLEM 15.42   Compounds **A**, **B**, and **C** all gave 71.17% C, 5.12% H, and 23.71% N upon elemental analysis. Summaries of the mass spectra and the IR spectra of these compounds are shown in the next column. Each one has a molecular ion of *m/z* 118. Deduce the structures of **A–C** and explain your reasoning.

A

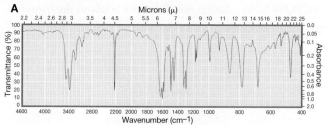

m/z	Relative Abundance
118	100
96	20
119	9
64	8
90	7
63	6
39	5
117	5

B

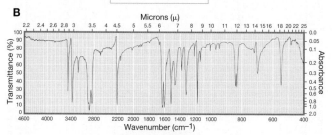

m/z	Relative Abundance
118	100
96	40
119	22
64	22
90	19
63	14
39	14
117	12

C

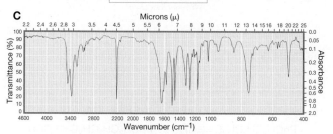

m/z	Relative Abundance
118	100
96	35
119	12
64	7
90	7
63	7
39	6
117	4

PROBLEM 15.43 Extraction of black Hemiptera bugs affords a colorless oil **A** of the formula $C_7H_{12}O$ possessing a sickening odor. The IR spectrum of **A** displays peaks at 2833 (w), 2725 (w), 1725 (s), 1380 (m), and 975 (m) cm^{-1}. Upon catalytic hydrogenation, the peak at 975 cm^{-1} disappears. Ozonolysis of compound **A**, followed by an oxidative workup, gives succinic acid ($HOOC—CH_2CH_2—COOH$) and propionic acid ($HOOC—CH_2CH_3$). Deduce a structure for **A** and explain your reasoning.

PROBLEM 15.44 The O—H portion of the IR spectrum changes with concentration. As the solution becomes more dilute, the broad band centered at roughly 3300 cm^{-1} sharpens. Explain.

PROBLEM 15.45 The O—H stretching IR band for *tert*-butyl alcohol is much sharper than that for methyl alcohol at the same concentration. Explain.

PROBLEM 15.46 Explain the difference between the IR spectra of the starting material and the product that you would expect to see in the reaction of benzyl bromide with NaOH.

PROBLEM 15.47 The IR stretching frequency for the O—H group in the syn unsaturated molecule shown here is sharp at all concentrations and shows no change on dilution, whereas that for the related saturated molecule or the anti unsaturated molecule is broad and sharpens with dilution. Explain. *Hint*: Oxygen is not the only base that can interact with an electrophilic hydrogen through hydrogen bonding.

HO␣␣R R␣␣OH HO␣␣R

syn-Unsaturated   anti-Unsaturated   Saturated

Now we move on to NMR spectroscopy. Again, we start with simple exercises:

PROBLEM 15.48 Explain why TMS appears so far upfield in the ^1H NMR spectrum.

PROBLEM 15.49 List the factors that can influence the chemical shift of a hydrogen. Which of these factors do you suppose accounts for the fact that cyclopropane hydrogens appear at unusually high field, near δ 0.2 ppm.

PROBLEM 15.50 Explain carefully why the following two ^1H NMR spectra of 1-bromobutane look so different:

(a) Spectrum taken at 60 MHz

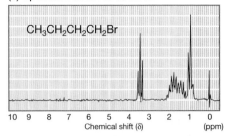

CH₃CH₂CH₂CH₂Br

(b) Spectrum taken at 300 MHz

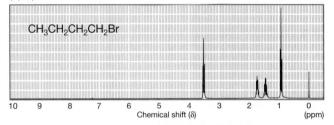

CH₃CH₂CH₂CH₂Br

PROBLEM 15.51 Are the underlined hydrogens in the following molecules homotopic, enantiotopic, or diastereotopic? Justify your answers.

(a) H₃C␣␣CH₃ / H̲ H̲

(b) H₃C␣␣CH₂CH₃ / H̲ H̲

(c) H₃C␣␣CH(CH₃)₂ / H̲ H̲

(d) H / H₃C␣␣CH₃ , CH₂CH₃ / H̲ H̲

PROBLEM 15.52 Predict the ^1H NMR spectrum (chemical shift and coupling for each signal) of 2-chloropropane.

PROBLEM 15.53 Predict the ^1H NMR spectrum (chemical shift and coupling for each signal) of 1,2-dichloropropane.

PROBLEM 15.54 Assign the following chemical shifts to the hydrogens shown in parts (a) through (d). Use the ($n + 1$) rule to predict the multiplicities of the peaks.

(a) $Cl_2CH—CHCl—CH_3$
δ (ppm)
5.88
4.37
1.70

(b) $ClCH_2—CH(OCH_3)_2$
δ (ppm)
4.53
3.53
3.43

(c) $(CH_3)_2CH—CH_2I$
δ (ppm)
3.16
1.72
1.03

(d) $BrCH_2CH_2CH_2Br$
δ (ppm)
3.59
2.36

PROBLEM 15.55 The two isotopes of bromine are ^{79}Br and ^{81}Br. These isotopes occur naturally in a 1:1 ratio. Based on this information, predict the ratios of the molecular ion and related isotopes of 1,1-dibromoethane.

PROBLEM 15.56   Predict the first-order spectra of the indicated hydrogens in the molecules shown in parts (a) through (c).

PROBLEM 15.57   Nitrobenzene shows three signals in its ^{1}H NMR spectrum at $\delta = 7.8$ ppm (2H), $\delta = 7.5$ ppm (1H), and $\delta = 7.2$ ppm (2H). Assign these signals and explain your reasoning.

PROBLEM 15.58   In the nitration of toluene, two products are formed. Predict the products and describe how you would use ^{1}H NMR spectral data to distinguish between the two.

PROBLEM 15.59   Cyclobutanes are relatively rigid molecules. As a result, a cis 1,2-disubstituted cyclobutane has significant steric interactions that translate into a much higher energy than the trans isomer. Spectral data for four cyclobutanediol isomers are given. Assign each set of ^{1}H NMR data to one of the isomers and explain your reasoning. It might be helpful to use your model set.

Compound 1: This diol is the least polar; it is the first isomer eluted from a chromatography column.
^{1}H NMR: δ 7.6–7.2 (m, 10H)
    3.78 (q, $J = 8$ Hz, 1H)
    3.20 (br s, 2H, signal shifts when D_2O is added)
    2.70 (q, $J = 8$ Hz, 1H)
    1.28 (d, $J = 8$ Hz, 3H)
    0.40 (d, $J = 8$ Hz, 3H)

Compound 2: This diol is the most polar; it comes off the column last.
^{1}H NMR: δ 7.5–7.2 (m, 5H)
    3.07 (q, $J = 8$ Hz, 1H)
    2.62 (br s, 1H, signal shifts when D_2O is added)
    1.18 (d, $J = 8$ Hz, 3H)

Compound 3: This compound is the second off the column.
^{1}H NMR: δ 7.4–7.0 (m, 5H)
    3.40 (br s, 1H, signal shifts when D_2O is added)
    3.10 (m, 1H)
    1.24 (d, $J = 8$ Hz, 3H)

Compound 4: This compound is the third isomer off the column.
^{1}H NMR: δ 7.4–7.0 (m, 10H)
    4.20 (br s, 2H, signal shifts when D_2O is added)
    2.58 (m, 1H)
    2.39 (m, 1H)
    1.38 (d, $J = 8$ Hz, 3H)
    0.71 (d, $J = 8$ Hz, 3H)

PROBLEM 15.60   Figure 15.47 shows an enlarged view of the multiplet centered at roughly δ 3.6 ppm in the ^{1}H NMR spectrum of pure ethyl alcohol. Use a tree diagram to analyze this complex multiplet.

PROBLEM 15.61   The ^{1}H NMR spectra of the two scrupulously dry alcohols shown below have signals (among others) at δ 2.60 (d, $J = 3.8$ Hz, 1H) (**A**) and δ 2.26 (t, $J = 6.1$ Hz, 1H) (**B**). Assign the spectra, and explain your reasoning.

PROBLEM 15.62   When a drop of aqueous acid is added to the samples described in Problem 15.61, the signals described collapse to broad singlets. Explain.

PROBLEM 15.63   In this chapter, we claimed that the NMR signals for ^{2}H (deuterium) and ^{13}C will not overlap ^{1}H spectra. Verify that this statement is true. Show that the frequency (ν) will be different for ^{1}H, ^{2}H, and ^{13}C at a field strength (B_0) of 4.7 T. The gyromagnetic ratios for ^{1}H, ^{2}H, and ^{13}C are 2.7×10^8, 0.41×10^8, and 0.67×10^8 rad T^{-1} s^{-1}, respectively.

PROBLEM 15.64   Match the following ^{1}H NMR absorptions to the compounds shown below, each of which has a one-signal spectrum: δ 0.04, 0.90, 1.42, 2.25, 7.3 ppm.

PROBLEM 15.65 Match the following ^{13}C NMR absorptions to the following compounds, each of which has a one-signal spectrum: δ −2.9, 26.9, 59.7, 128.5 ppm.

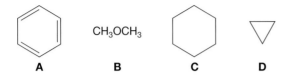

CH₃OCH₃

A   B   C   D

PROBLEM 15.66 Explain why the hydrogens α to an amino group ($R—CH_2—NH_2$) appear further upfield in the ^1H NMR spectrum than those α to a hydroxyl group ($R—CH_2—OH$).

PROBLEM 15.67 Describe the signals for the methylene hydrogens (underlined) of CH_3CH_2NHPh (**1**) in its ^1H NMR spectrum. Analyze **1** exactly as drawn in the figure.

H₃C   Ph
         H
    N
H   H
    H
    1

PROBLEM 15.68 The carbon-bound hydrogens of pyrrole appear at δ 6.05 and 6.62 ppm in the ^1H NMR spectrum, upfield of the typical aromatic signal region. Does this observation mean that pyrrole is not aromatic?

PROBLEM 15.69 At room temperature the pyridinophane shown below has a signal in its ^1H NMR spectrum at about δ −0.2 ppm. What relevance does this observation have to the question of the aromaticity of pyridine? *Hint*: See p. 720.

—(CH₂)₈
N
··

PROBLEM 15.70 Use the partial spectral data to predict a structure for the product. Outline the steps in this reaction.

OH
            H₃O⁺
          ————→   (C₇H₁₂O)   IR: (CCl₄) 1729 cm⁻¹
            H₂O ^1H NMR δ 9.5 (1H)
OH

PROBLEM 15.71 The following spectrum appears in an old catalog of ^1H NMR spectra of organic compounds. Is the assigned structure correct? What's wrong?

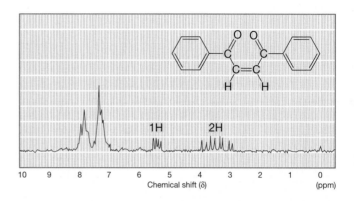

PROBLEM 15.72 In the synthesis of the compound described in Problem 15.71, HCl is produced. Use this information to postulate a reasonable structure for the compound whose ^1H NMR spectrum is shown above.

PROBLEM 15.73 2-Methylpropene can be hydrated in acid, or under hydroboration–oxidation conditions to give two different alcohols. Give structures for these alcohols and assign the peaks in the following ^1H NMR spectra (**1** and **2**). Which peaks would vanish if a drop of D_2O were added?

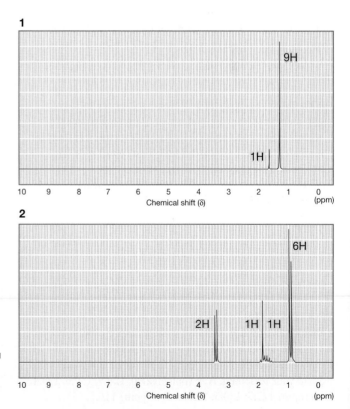

PROBLEM 15.74   2-Methyl-2-butene can be hydrated in acid, or under hydroboration–oxidation conditions to give two different alcohols. Give structures for these alcohols and assign the following ^{1}H NMR spectra (**1** and **2**) to these isomers. Explain your answers.

1

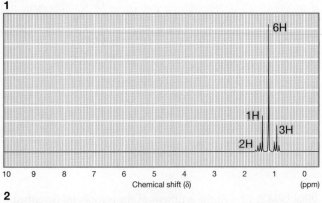

2

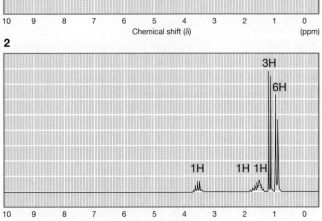

PROBLEM 15.75   On careful examination, it can be seen that the upfield signal in spectrum **2** of Problem 15.74 (at δ 0.90 ppm) which appears to be a doublet, is actually two doublets. Explain.

PROBLEM 15.76   The ^{1}H NMR spectra for the following eight-membered ring compounds are summarized below (remember m means multiplet). Assign structures, and explain your reasoning. **A**: δ (ppm) 1.54 (s); **B**: δ (ppm) 5.74 (s); **C**: δ (ppm) 2.39 (m, 8H), 5.60 (m, 4H); **D**: δ (ppm) 1.20–1.70 (m, 4H), 1.85–2.50 (m, 4H), 5.35–5.94 (m, 4H).

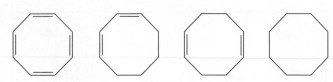

PROBLEM 15.77   You are given three bottles containing *o*-dichlorobenzene, *m*-dichlorobenzene, and *p*-difluorobenzene, along with broad-band decoupled ^{13}C NMR spectra of the three isomers. Assign the three spectra to the three compounds and explain your reasoning. **A**: δ (ppm) 127.0, 128.9, 130.6, 135.1. **B**: δ (ppm) 127.7, 130.5, 132.6. **C**: δ (ppm) 116.5, 159.1.

PROBLEM 15.78   Could looking at symmetry allow you to assign the spectra of 4-methylcyclohexanone, 2-methylcyclohexanone, and 2,3-dimethylcyclopentanone, **A**, **B**, and **C**? If not, suggest an experiment, involving only ^{13}C NMR spectroscopy, that would let you finish the job. (*Hint*: See p. 741.)

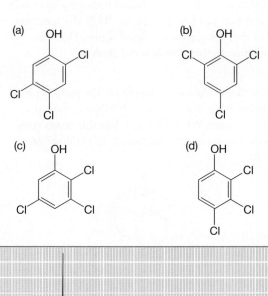

PROBLEM 15.79   The three isomers of pentane, C_5H_{12}, have the following ^{13}C NMR spectra. Assign the spectra of these three compounds. **A**: δ (ppm) 13.9, 22.8, 34.7; **B**: δ (ppm) 22.2, 31.1, 32.0, 11.7; **C**: δ (ppm) 31.7, 28.1.

PROBLEM 15.80   Assign the following ^{1}H NMR spectra to the isomers of trichlorophenol shown below and explain your reasoning.

1

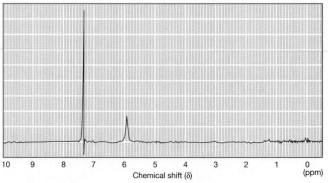

2

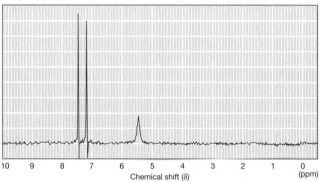

3

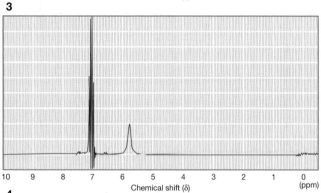

4

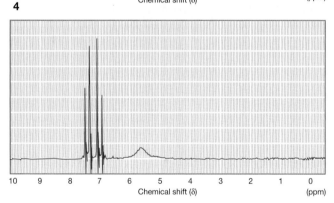

PROBLEM 15.81 Isomeric compounds **A** and **B** have the composition $C_{11}H_{12}O_4$. Spectral data are summarized below. Deduce structures for **A** and **B** and explain your reasoning.

Compound A

IR (KBr): 1720 (s), 1240 (s) cm^{-1}

^{1}H NMR (CDCl$_3$): δ 2.46 (s, 3H)
3.94 (s, 6H)
8.05 (d, J = 2 Hz, 2H)
8.49 (t, J = 2 Hz, 1H)

Compound B

IR (Nujol®): 1720 (s), 1245 (s) cm^{-1}

^{1}H NMR (CDCl$_3$): δ 2.63 (s, 3H)
3.91 (s, 6H)
7.28 (d, J = 8 Hz, 1H)
8.00 (dd, J = 8 Hz, 2 Hz; 1H)
    dd = doublet of doublets
8.52 (d, J = 2 Hz, 1H)

PROBLEM 15.82 Compounds **A** and **B**, containing only C, H, and O, gave 63.15% C and 5.30% H upon elemental analysis. Their mass spectra do not show a molecular ion, but freezing point depression data indicate a molecular weight of 150 g/mol. Catalytic hydrogenation of both **A** and **B** leads to **C**. Spectral data for compounds **A**, **B**, and **C** are summarized below. Assign structures for compounds **A**, **B**, and **C** and explain your reasoning.

Compound A

IR (Nujol®): 1845 (m), 1770 (s) cm^{-1}

^{1}H NMR (CDCl$_3$): δ 2.03–2.90 (m, 2H)
3.15–3.70 (m, 1H)
5.65–6.25 (m, 1H)

Compound B

IR (Nujol®): 1825 (m), 1755 (s) cm^{-1}

^{1}H NMR (CDCl$_3$/DMSO-d_6): δ 1.50–2.10 (m, 1H)
2.15–2.70 (m, 1H)

Compound C

IR (neat): 1852 (m), 1786 (s) cm^{-1}

^{1}H NMR (CDCl$_3$/DMSO-d_6): δ 1.10–2.30 (m, 4H)
3.05–3.55 (m, 1H)

PROBLEM 15.83 Spectral data for isomeric compounds **A** and **B** are summarized below. Assign structures for compounds **A** and **B**, and explain your reasoning.

Compound A

Mass spectrum: m/z = 148 (M, 7%), 106 (8%), 105 (100%), 77 (29%), 51 (8%)

IR (neat): 1675 (s), 1220 (s), 980 (s), and 702 (s) cm^{-1}

^{1}H NMR (CDCl$_3$): δ 1.20 (d, J = 7 Hz, 6H)
3.53 (septet, J = 7 Hz, 1H)
7.20–7.60 (m, 3H)
7.80–8.08 (m, 2H)

Compound B

IR (neat): 1705 (s), 738 (m), and 698 (s) cm^{-1}

^{1}H NMR (CDCl$_3$): δ 0.95 (t, J = 7 Hz, 3H)
2.35 (q, J = 7 Hz, 2H)
3.60 (s, 2H)
7.20 (s, 5H)

PROBLEM 15.84 Upon direct photolysis or heating at 220 °C, the dimer of 2,5-dimethyl-3,4-diphenyl-2,4-cyclopentadien-1-one (**A**) yields compound **B**, mp 166–168 °C. Spectral data for compound **B** are summarized on the next page. Deduce the structure of compound **B** and explain your reasoning. (You may have to make an educated guess about the actual stereochemistry of **B**.)

Compound **B**

Mass spectrum: $m/z = 492$ (M)

IR (CCl$_4$): 1704 cm^{-1}

^{1}H NMR (CDCl$_3$): δ 0.73 (s, 3H)
0.92 (s, 3H)
1.51 (s, 3H)
1.88 (s, 3H)
6.6–7.5 (m, 20H)

PROBLEM 15.85 An appreciation of the concepts of electronegativity (inductive effects) and of electron delocalization (resonance effects), combined with an understanding of ring current effects, permits both rationalization and prediction of many chemical shifts.

(a) The carbons of ethylene have a ^{13}C NMR chemical shift of δ 123.3 ppm. Rationalize the ^{13}C NMR chemical shifts of the β carbons of 2-cyclopenten-1-one (**A**) and methyl vinyl ether (**B**). (*Hint*: Think resonance!)

(b) The ^{1}H NMR chemical shifts of hydrogens ortho, meta, and para to a substituent on a benzene ring can also be correlated in this way. For reference sake, note that the hydrogens in benzene appear at about δ 7.3 ppm. Aniline (**C**) exhibits three shielded aromatic hydrogens (i.e., δ < 7.0), whereas the other two aromatic hydrogens appear in the normal range. On the other hand, two of the aromatic hydrogens in acetophenone (**D**) are strongly deshielded (δ 7.9 ppm), one is somewhat deshielded (δ 7.5 ppm), and the other two aromatic hydrogens appear at δ 7.4. Finally, all five aromatic hydrogens in chlorobenzene (**E**) appear at δ 7.3 ppm. Assign the ^{1}H NMR chemical shifts of the aromatic hydrogens for **C**, **D**, and **E** by considering inductive, resonance, and ring current effects. Additionally, the ^{13}C NMR chemical shifts for the aromatic carbons of **C** and **E** are shown in color in the next column. (The ^{13}C NMR chemical shift for benzene is δ 128.5 ppm.) Are these ^{13}C NMR chemical shifts consistent with the ^{1}H NMR chemical shifts for **C** and **E**?

C   D   E

PROBLEM 15.86 Compound **A** was isolated from the bark of the sweet birch (*Betula lenta*). Compound **A** is soluble in 5% aqueous NaOH solution but not in 5% aqueous NaHCO$_3$ solution. The spectral data for compound **A** are summarized below. Deduce the structure of compound **A**. (Be sure to assign the aromatic resonances in the ^{1}H NMR spectrum. Note that J_{meta} and J_{para} were not observed.) *Hint*: The pK_a of phenol is about 10, and that of benzoic acid is about 4.2.

Compound **A**

Mass spectrum: $m/z = 152$ (M, 49%), 121 (29%), 120 (100%), 92 (54%)

IR (neat): 3205 (br), 1675 (s), 1307 (s), 1253 (s), 1220 (s), and 757 (s) cm^{-1}

^{1}H NMR (CDCl$_3$, 300 MHz): δ 3.92 (s, 3H)
6.85 (t, $J = 8$ Hz, 1H)
7.00 (d, $J = 8$ Hz, 1H)
7.44 (t, $J = 8$ Hz, 1H)
7.83 (d, $J = 8$ Hz, 1H)
10.8 (s, 1H)

^{13}C NMR (CDCl$_3$): δ 52.1 (q), 112.7 (s), 117.7 (d), 119.2 (d),
(hydrogen coupled)   130.1 (d), 135.7 (d), 162.0 (s), 170.7 (s)

PROBLEM 15.87 Dehydration of α-terpineol (**A**) affords a mixture of at least two isomers, one of which is **B**.

Isomer **B**, which exhibits the pleasant odor of lemons and can be isolated from the essential oils of cardamom, marjoram, and coriander, reacts with maleic anhydride to yield compound **C**, mp 64–65 °C. Spectral data for compound **C** are summarized on the next page. Deduce the structures of compounds **B** and **C**. (You may have to make an educated guess about the stereochemistry of compound **C**.)

Compound **C**

Mass spectrum: $m/z = 234$ (M)

IR (Nujol®): 1870 (shoulder), 1840 (m), and 1780 (s) cm^{-1}

^{1}H NMR (CDCl$_3$): δ 0.98 (d, $J = 6.8$ Hz, 3H)

1.08 (d, $J = 6.8$ Hz, 3H)

1.36 (m, 4H)

1.46 (s, 3H)

2.60 septet ($J = 6.8$ Hz, 1H)

2.80 (d, $J = 9$ Hz, 1H)

3.20 (d, $J = 9$ Hz, 1H)

6.01 (d, $J = 8.5$ Hz, 1H)

6.10 (d, $J = 8.5$ Hz, 1H)

PROBLEM 15.88   Isoquinoline (**1**) undergoes the Chichibabin amination reaction when treated with potassium amide in liquid ammonia to yield 1-aminoisoquinoline (**2**) (p. 676).

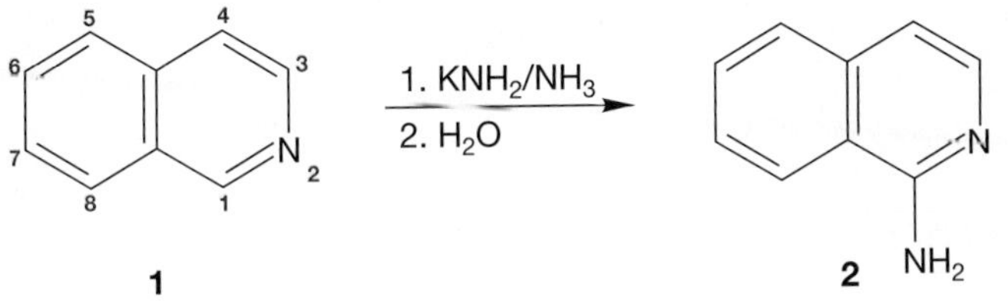

When the reaction mixture containing **1**, potassium amide, and liquid ammonia was examined at -10 °C by ^{1}H NMR spectroscopy, the following signals were observed (solvent hydrogens and NH$_2$ hydrogens not reported):

δ 4.87 (d, $J = 5.5$ Hz, 1H), 5.34 (t, $J = 7.0$ Hz, 1H; signal collapses to a singlet when amide ion is present in excess), 6.35–7.3 (br m, 5H). Rationalize this low-temperature ^{1}H NMR spectrum. *Hint*: For reference sake, the ^{1}H NMR spectrum of isoquinoline itself (**1**) is summarized below:

δ 7.3–8.2 (m, H$_4$, H$_5$, H$_6$, H$_7$, H$_8$), 8.61 (d, $J = 6$ Hz, H$_3$), 9.33 (s, H$_1$).

PROBLEM 15.89   When reactions are run in THF, it is often possible to isolate traces of an impurity, compound **1**. Compound **1** gives 81.76% C and 10.98% H upon elemental analysis. Spectral data for **1** are summarized below. Deduce the structure of compound **1** and explain your reasoning.

Compound **1**

Mass spectrum: $m/z = 220$ (M, 23%), 205 (100%), 57 (27%)

IR (Nujol®): 3660 (m, sharp) cm^{-1}

^{1}H NMR (CDCl$_3$): δ 1.43 (s, 18H), 2.27 (s, 3H), 5.00 (s, 1H, shifts with D$_2$O), 6.98 (s, 2H)

^{13}C NMR (CDCl$_3$): δ 21.2 (q), 30.4 (q), 34.2 (s), 125.5 (d), 128.2 (s), 135.8 (s), 151.5 (s)

Use Organic Reaction Animations (ORA) to answer the following questions:

PROBLEM 15.90   Select the reaction "Nucleophilic aromatic substitution." Predict the ^{1}H NMR spectra of the starting material (1-chloro-2,4-dinitrobenzene) and the product (*N*-methyl-2,4-dinitroaniline). How would the IR data differ for the starting material and the product?

PROBLEM 15.91   View the "Alkene epoxidation" reaction. Predict the ^{1}H NMR spectra for the starting material and the product. Compare the ^{1}H NMR spectra of the starting material and the product for the same reaction using (*Z*)-2-butene.

PROBLEM 15.92   The animated "Acidic epoxide ring opening" and "Basic epoxide ring opening" show isomeric products. What analytical tool (^{1}H NMR, ^{13}C NMR, IR, MS) could you use to distinguish between the 2-methoxy-1-propanol and the 1-methoxy-2-propanol products? What specific differences would you look for and what methods would you use?

16 Carbonyl Chemistry 1: Addition Reactions

16.1 Preview

16.2 Structure of the Carbon–Oxygen Double Bond

16.3 Nomenclature of Carbonyl Compounds

16.4 Physical Properties of Carbonyl Compounds

16.5 Spectroscopy of Carbonyl Compounds

16.6 Reactions of Carbonyl Compounds: Simple Reversible Additions

16.7 Equilibrium in Addition Reactions

16.8 Other Addition Reactions: Additions of Cyanide and Bisulfite

16.9 Addition Reactions Followed by Water Loss: Acetal Formation

16.10 Protecting Groups in Synthesis

16.11 Addition Reactions of Nitrogen Bases: Imine and Enamine Formation

16.12 Organometallic Reagents

16.13 Irreversible Addition Reactions: A General Synthesis of Alcohols

16.14 Oxidation of Alcohols to Carbonyl Compounds

16.15 Retrosynthetic Alcohol Synthesis

16.16 Oxidation of Thiols and Other Sulfur Compounds

16.17 The Wittig Reaction

16.18 Special Topic: Biological Oxidation

16.19 Summary

16.20 Additional Problems

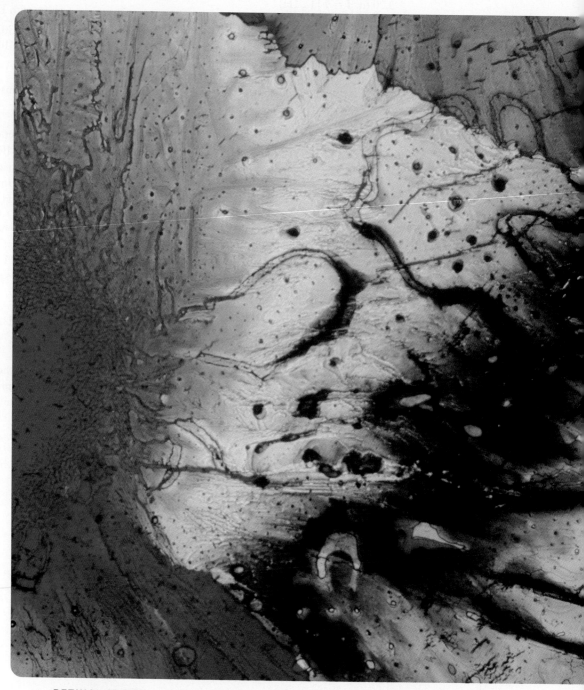

RETINAL CRYSTAL This is a polarized light photo of retinal (vitamin A), which is a nutritionally important aldehyde found in green vegetables and fruit (see p. 533).

"You have two choices," Jamf cried, his last lecture of the year: outside were the flowery strokings of wind, girls in pale colored dresses, oceans of beer, male choruses intensely, movingly lifted as they sang *Semper sit in flores! Semper sit in flo-ho-res* . . . "stay behind with carbon and hydrogen, take your lunch-bucket in to the works every morning with the faceless droves who can't wait to get in out of the sunlight—or move *beyond*. Silicon, boron, phosphorus—these can replace carbon, and can bond to nitrogen instead of hydrogen.—" a few snickers here, not unanticipated by the playful old pedagogue, be he always in flower: his involvement in getting Weimar to subsidize IG's Stickstoff Syndikat was well known—"move beyond life, toward the inorganic. Here is no frailty, no mortality—here is Strength, and the Timeless." Then his well-known finale, as he wiped away the scrawled C—H on his chalkboard and wrote, in enormous letters, Si—N.

—THOMAS PYNCHON,[1] GRAVITY'S RAINBOW

16.1 Preview

In Chapters 3, 9, and 10 we examined the structure of alkenes and the addition reactions of their carbon–carbon double bonds. Indeed, we have been exploring various aspects of carbon–carbon double-bond chemistry since Chapter 9. Now it is time to extend what we know to the chemistry of carbon–oxygen double bonds, the carbonyl group. Although it bears many similarities to carbon–carbon double bonds, the carbon–oxygen double bond is not "just another double bond," because there are many important differences. The chemistry of the carbonyl group is the cornerstone of synthetic chemistry. Much of our ability to make molecules depends on manipulation of carbonyl groups because they allow us to make carbon–carbon bonds. There are several centers of reactivity in a molecule containing a carbon–oxygen double bond, and this versatility contributes to the widespread use of carbonyl compounds in synthesis. There is the π bond itself (don't forget the antibonding π^* orbital), and the nonbonding electrons on oxygen. The centerpiece reaction of this chapter is the acid- or base-induced addition to the carbon–oxygen double bond. In addition, there is special reactivity associated with the hydrogens on the carbon in the position adjacent (α) to the carbon–oxygen double bond (Fig. 16.1). These hydrogens are called α hydrogens. A ketone can have α hydrogens on either side of its carbonyl group. The chemistry of the α position is very important, but will wait until Chapter 19.

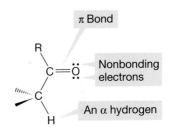

FIGURE 16.1 A generic carbonyl group has three loci of reactivity: the nonbonding electrons on the oxygen, the π bond, and its α hydrogens.

ESSENTIAL SKILLS AND DETAILS

1. Almost this entire chapter involves additions of nucleophiles to carbon–oxygen double bonds. Many of these reactions are reversible, others are not. It is the reversible reactions that trouble students most, because there seems to be a bewildering array of additions, proton transfers, and eliminations. So there is, but keep your wits about you, and you'll learn to push those protons and waters around properly.

2. In this chapter, you will learn a lot about synthesis, especially alcohol synthesis. It is not hard to do these syntheses, but you have to learn the basic reaction—the irreversible addition of organometallic reagents and metal hydrides to carbonyl compounds—very well.

---

[1] Our first Pynchon quote was in Chapter 9. Same man, same book.

3. A compound does not have to be favored at equilibrium to be the reactive ingredient in a chemical reaction. Even if it is disfavored, it can still have low barriers to further reaction. In such a case, it is used up, then regenerated at equilibrium. In such a way, all of the product can come from the minor partner in an equilibrium.

4. It is very useful to become comfortable with the redox interconversion of carbonyl compounds (aldehydes and ketones) and the related alcohols. There is a lot of detail here, and a none-too-simple mechanism as well.

5. Protecting groups are introduced in this chapter. A reactive functional group can be converted into a nonreactive group and stored in this fashion. Later, when the reactivity of the functional group is required, it can be regenerated as needed. There are many protecting groups, and it is very helpful to be able to use them in synthesis.

16.2 Structure of the Carbon–Oxygen Double Bond

By analogy to the carbon–carbon double bond, we might expect sp^2 hybridization for the carbon atom in the carbonyl group. The $2p$ orbitals on carbon and oxygen form the π bond, and two lone pairs of electrons remain on oxygen. Figure 16.2 gives a schematic orbital picture of both the σ and π systems for the simplest carbonyl compound, formaldehyde.

FIGURE 16.2 One orbital picture of the simplest carbonyl compound, formaldehyde.

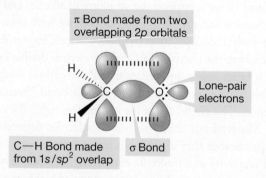

π Bond made from two overlapping 2p orbitals

Lone-pair electrons

C—H Bond made from 1s/sp² overlap

σ Bond

PROBLEM 16.1 Make an orbital interaction diagram for each different kind of bond in the molecule formaldehyde, shown in Figure 16.2.

The structures of some known carbonyl compounds show that this picture is a reasonable one. Bond angles are close to the 120° required by pure sp^2 hybridization, and of course we would not expect a molecule of less than perfect trigonal symmetry to have exactly sp^2 hybridization. Bond lengths are similar to those in simple alkenes, although the carbon–oxygen double bond is a bit shorter and stronger than its carbon–carbon counterpart (Fig. 16.3).

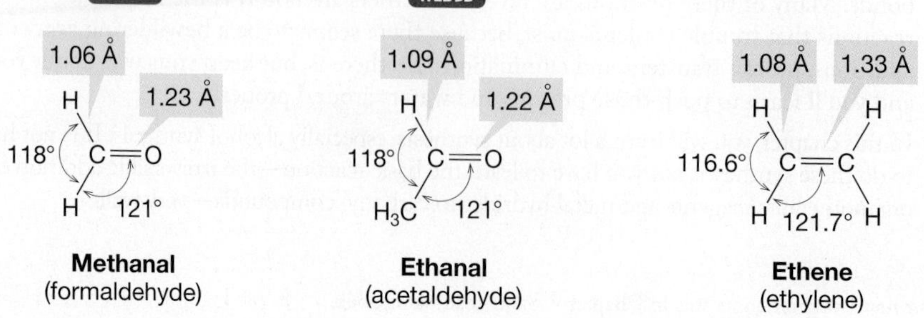

FIGURE 16.3 A comparison of the structures of formaldehyde, acetaldehyde, and ethylene.

WEB 3D

1.06 Å
1.23 Å
H
118°   C=O
H   121°

Methanal
(formaldehyde)

WEB 3D

1.09 Å
1.22 Å
H
118°   C=O
H₃C   121°

Ethanal
(acetaldehyde)

1.08 Å   1.33 Å
H   H
116.6°   C=C
H   121.7° H

Ethene
(ethylene)

Formaldehyde is the smallest and simplest aldehyde. It can be found in substantial concentrations in new buildings as commonly used wood products age.

The carbonyl π bond is constructed in the same way as the related carbon–carbon π bond. The $2p$ orbitals on the adjacent carbon and oxygen atoms are allowed to interact in a constructive way to form a bonding π molecular orbital, and in a destructive, antibonding way to form π^*. But there is a difference. In an alkene, the constituent $2p$ orbitals are of equal energy, and therefore contribute equally to the formation of π and π^*. In a carbonyl group, the two $2p$ orbitals are not identical. An electron in the $2p$ orbital on the more electronegative oxygen atom is lower in energy than one in the $2p$ orbital on the carbon atom, so the oxygen atom contributes more to π than the more energetically remote carbon $2p$ orbital (Fig. 16.4). The bonding π molecular orbital will be constructed from more than 50% of the oxygen $2p$ atomic orbital and less than 50% of the carbon $2p$ atomic orbital. In a similar way, we can see that π^* will be made up of more than 50% of the energetically close carbon $2p$ orbital and less than 50% of the energetically more remote $2p$ orbital on oxygen.

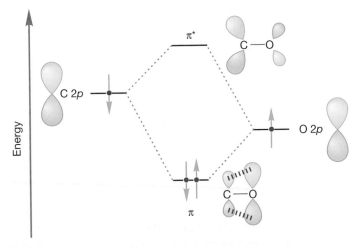

FIGURE 16.4 A graphical construction of π and π^* for the carbonyl group. The π orbital is constructed from more than 50% of the oxygen $2p$ orbital, and the π^* orbital from more than 50% of the carbon $2p$ orbital.

So the carbonyl π and π^* orbitals bear a general resemblance to the π and π^* orbitals of the alkenes, but are less symmetrical. The two electrons naturally occupy the lower energy bonding molecular orbital. In the π molecular orbital, there is a greater probability of finding an electron in the neighborhood of the relatively electronegative oxygen atom than near the more electropositive carbon, and this unequal sharing of electrons contributes to the dipole moment in molecules containing carbonyl groups. As can be seen from Figure 16.5, carbonyl compounds do have substantial dipole moments and are quite polar molecules.

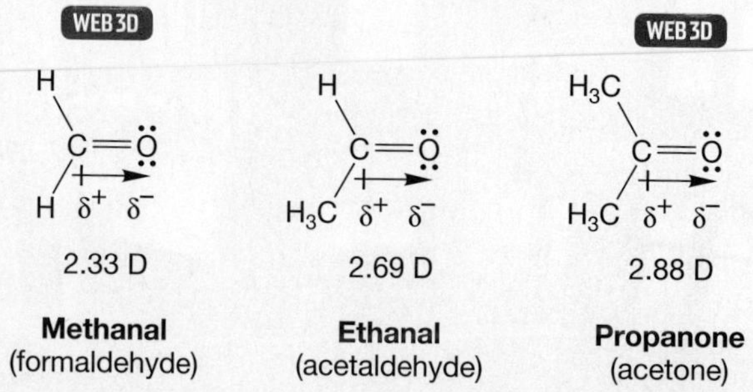

FIGURE 16.5 Carbonyl compounds are polar molecules with substantial dipole moments [measured in debye (D) units]. Notice the small bond dipole arrow that points from the positive end toward the negative end of the dipole.

| 2.33 D | 2.69 D | 2.88 D |
| **Methanal** (formaldehyde) | **Ethanal** (acetaldehyde) | **Propanone** (acetone) |

PROBLEM 16.2 It is not just the π bonds that are unsymmetrical in a carbonyl group. Use the hybrid orbitals of oxygen and carbon to construct the C–O σ and σ* orbitals.

PROBLEM 16.3 One way to construct a carbon–oxygen double bond using the hybridization formalism is to use *sp*-hybridized oxygen. Make a drawing of this model in the style of Figure 16.2.

A GENERAL CARBONYL GROUP

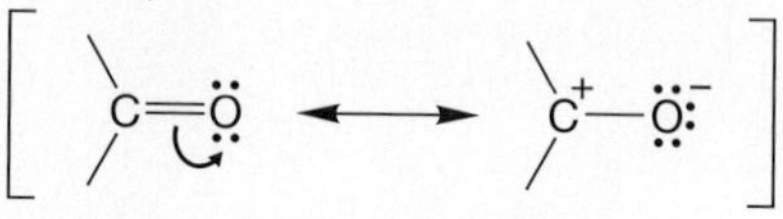

FIGURE 16.6 A resonance formulation of a carbonyl group.

In resonance terms, we can describe the carbonyl group as a combination of the two major contributing forms shown in Figure 16.6. This picture emphasizes the Lewis acid (electrophilic) character of the carbon of the carbon–oxygen double bond.

WORKED PROBLEM 16.4 There is another possible polar resonance form in addition to the one shown in Figure 16.6. What is it, and why is it not considered a major contributor to the carbonyl group?

ANSWER The other polar resonance form has the positive charge on oxygen and the negative charge on carbon. Oxygen is more electronegative than carbon and the negative charge will be more stable on oxygen than on carbon. Similarly, the relatively electropositive carbon should have the positive charge, not the negative. For these reasons, this form is not a major contributor to the structure.

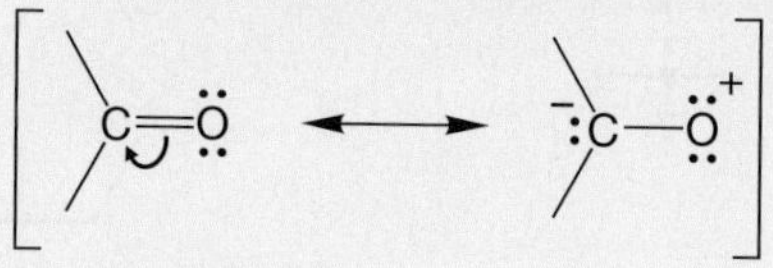

WORKED PROBLEM 16.5 Why is the dipole moment in acetone larger than that in acetaldehyde (Fig. 16.5)?

ANSWER Because the negative end of the dipole is the same for each species, there can be no answer there. However, the two methyl groups in acetone stabilize the positive charge much better than the methyl and hydrogen of acetaldehyde do. The polar form is more stable in acetone than it is in acetaldehyde, and as a result, acetone has a larger dipole moment.

16.3 Nomenclature of Carbonyl Compounds

In a generic carbonyl compound $R_2C=O$, there is a vast array of possible substitution patterns, and there is a separate name for each one of them. Simple compounds in which both R groups are hydrocarbons such as alkyl, alkenyl, or aryl are called ketones. If one R group is hydrogen, the compound is an aldehyde, and the molecule in which both R groups are hydrogen is called formaldehyde.

Common names are often used for the aldehydes containing four carbons or fewer, but thereafter the IUPAC system takes over, and the naming process becomes rational. The "e" of the parent alkane's name is dropped and the suffix "al" is added. In the IUPAC numbering system the aldehyde group is always number 1. When aldehydes are attached to cycloalkanes, the term carboxaldehyde is used, as in "cyclopropanecarboxaldehyde." Some aldehydes and their IUPAC names are given in Figure 16.7.

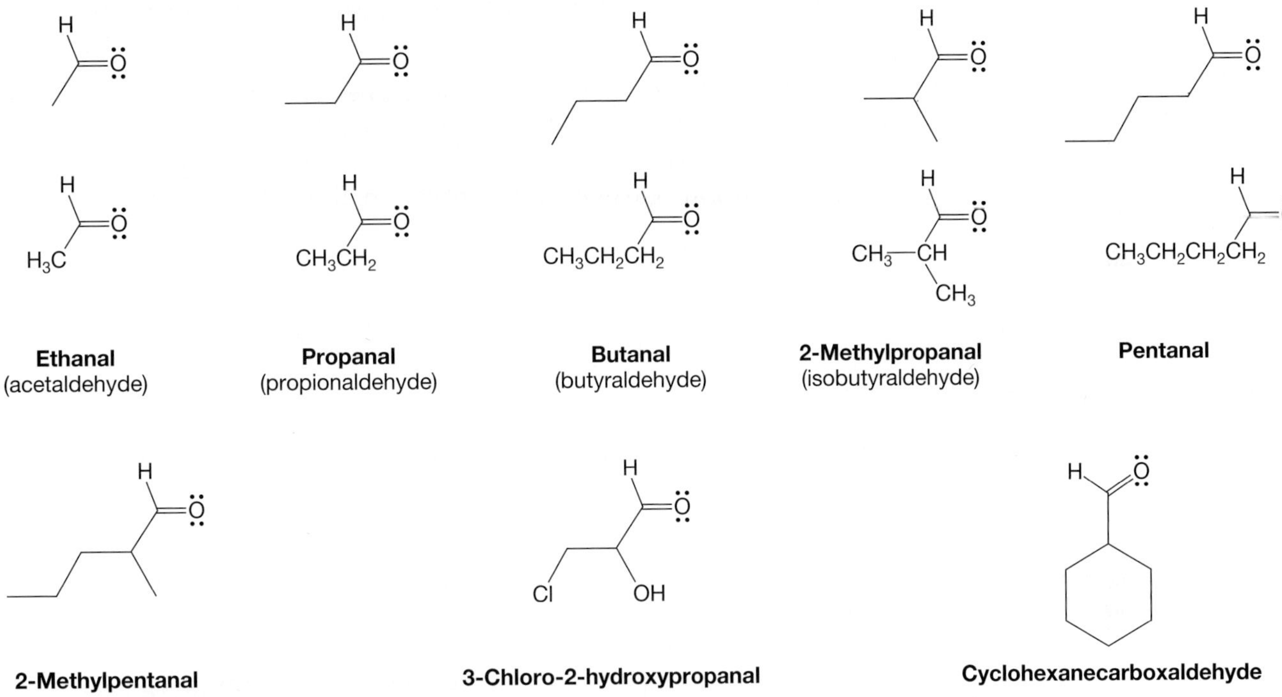

Ethanal
(acetaldehyde)

Propanal
(propionaldehyde)

Butanal
(butyraldehyde)

2-Methylpropanal
(isobutyraldehyde)

Pentanal

2-Methylpentanal

3-Chloro-2-hydroxypropanal

Cyclohexanecarboxaldehyde

FIGURE 16.7 A few aldehydes and their IUPAC names are shown. Common names are given in parentheses.

Compounds containing two aldehydes are **dials**. In this case, the final "e" of the parent name is not dropped before adding "dial" (Fig. 16.8).

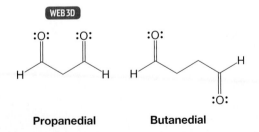

WEB 3D

Propanedial

Butanedial

FIGURE 16.8 Dialdehydes are named as dials.

Ketones, too, have common names, but most have been discarded. An exception is the name for the simplest member of the class, dimethyl ketone, which is still called acetone. Ketones are named in several ways. For example, acetone is also known as

2-oxopropane, propanone, or dimethyl ketone (Fig. 16.9). In using the IUPAC system, the final "e" of the parent alkane is dropped, and the suffix "one" is appended. The number of the carbon atom bearing the carbon–oxygen double bond is placed before the parent name.

FIGURE 16.9 Simple ketones can be named in many ways. The IUPAC names are listed first.

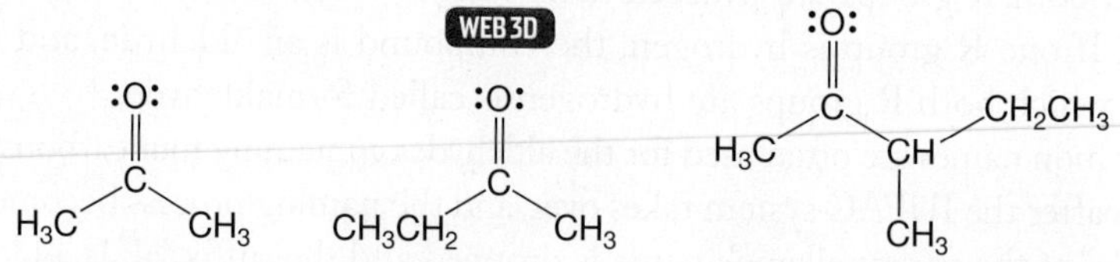

WEB 3D

Propanone
(acetone)
(dimethyl ketone)

Butanone
(ethyl methyl ketone)

3-Methyl-2-pentanone
(*sec*-butyl methyl ketone)

PROBLEM 16.6 Why is acetone properly named propanone and *not* 2-propanone?

Compounds containing two ketones are **diones**. As with dials, the final "e" of the parent alkane is retained, and the name begins with the position numbers of the two carbonyl groups (Fig. 16.10). Ketoaldehydes are named as aldehydes, not ketones, because the aldehyde group has a higher priority than a keto group (Fig. 16.10).

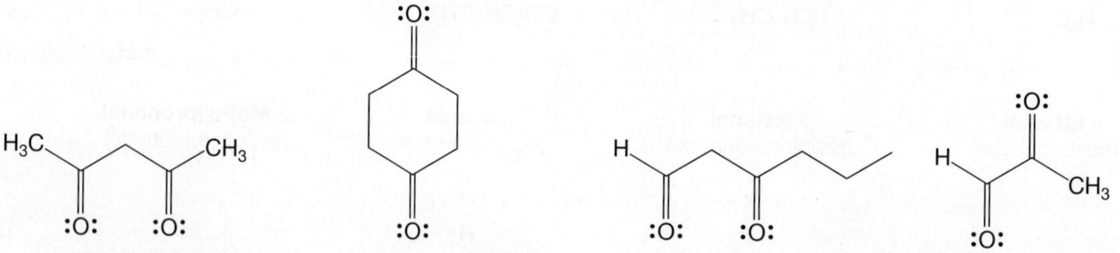

2,4-Pentanedione

1,4-Cyclohexanedione

3-Oxohexanal

2-Oxopropanal
(pyruvic aldehyde)

FIGURE 16.10 Diketones are named as diones. The final "e" of the parent hydrocarbon is retained. Ketoaldehydes are named as aldehydes, not ketones.

The rules are fairly simple: in the presence of a higher priority group the IUPAC system treats the carbonyl oxygen atom as a substituent that is described by the prefix "oxo-" and given a number like any other attached group. A great many common names are retained for aromatic aldehydes and ketones. For example, the simplest aromatic aldehyde is not generally called "benzenecarboxaldehyde" but **benzaldehyde**. That one's quite easy to figure out, but most are not. A few commonly used names are given in Figure 16.11.

FIGURE 16.11 Some common names for aromatic aldehydes are shown. The IUPAC accepted names are listed first.

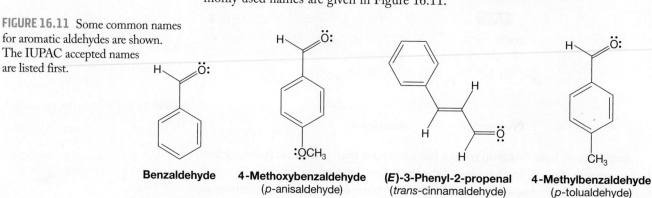

Benzaldehyde

4-Methoxybenzaldehyde
(*p*-anisaldehyde)

(*E*)-3-Phenyl-2-propenal
(*trans*-cinnamaldehyde)

4-Methylbenzaldehyde
(*p*-tolualdehyde)

CIVETONE

This simple, but unusual cycloalkenone is called civetone, as it is isolated from a gland present in both male and female civet cats, *Viverra civetta* and *Viverra zibetha*. The cats use it as a sexual attractant and marking agent. It has a strong musky odor, overwhelming when concentrated (to humans, that is, probably not to civet cats). In high dilution the odor becomes quite pleasant, and civetone and the related cyclic 15-membered ring ketone, muskone, are important starting materials in the perfumery industry. In the old days, you had to catch yourself a mess o' civet cats, described as "savage and unreliable," to make your perfumes. Nowadays, this compound can be made quite easily in the laboratory, a fact presumably reassuring to the civet cat population.

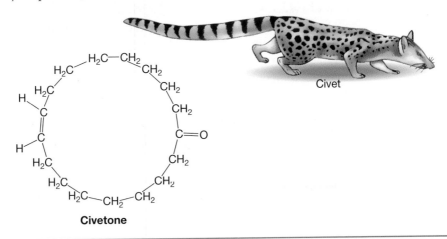

Civet

Civetone

Aromatic ketones are often named by treating the benzene ring as a phenyl group, as in isopropyl phenyl ketone (Fig. 16.12). A few can be named through the IUPAC system in which the framework Ph—C=O is indicated by the term "ophenone" and the remainder of the molecule designated by one of the alternative names in Figure 16.12. Thus, diphenyl ketone is benzophenone, methyl phenyl ketone is acetophenone, ethyl phenyl ketone is propiophenone, and phenyl propyl ketone is butyrophenone (Fig. 16.12).

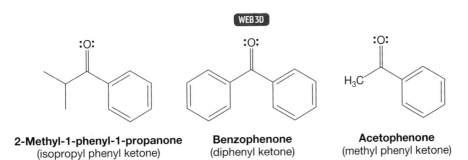

2-Methyl-1-phenyl-1-propanone
(isopropyl phenyl ketone)

Benzophenone
(diphenyl ketone)

Acetophenone
(methyl phenyl ketone)

FIGURE 16.12 Ketones containing benzene rings are phenyl ketones. This figure also shows the "ophenone" method of naming these molecules.

Figure 16.13 gives the names of some compounds containing carbon–oxygen double bonds in which one of the R groups is neither hydrogen nor a hydrocarbon group. We met these compounds in earlier chapters and we will see them again in later chapters and deal with their nomenclature more extensively there.

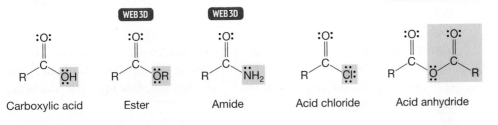

Carboxylic acid   Ester   Amide   Acid chloride   Acid anhydride

FIGURE 16.13 Some names for other carbonyl compounds in which one R group is neither hydrogen nor a hydrocarbon group.

16.4 Physical Properties of Carbonyl Compounds

Carbonyl compounds are all somewhat polar, and the smaller, less hydrocarbon-like molecules are all at least somewhat water soluble. Acetone and acetaldehyde are miscible with water, for example. Their polarity means that they can profit energetically by associating in solution, and therefore their boiling points are considerably higher than those of hydrocarbons containing the same number of carbons and oxygens. For example, propane and propene are gases with boiling points of -42.1 °C and -47.4 °C, respectively. By contrast, acetaldehyde boils at $+20.8$ °C, about 63 degrees higher. Acetone boils more than 60 degrees higher than 2-methylpropene (isobutene).

Table 16.1 gives some physical properties of a number of common carbonyl-containing compounds (and some related hydrocarbons in parentheses).

TABLE 16.1 Some Physical Properties of Carbonyl Compounds

Compound	bp (°C)	mp (°C)	Density (g/mL)	Dipole Moment (D)
Formaldehyde	-21	-92	0.815	2.33
Acetaldehyde	20.8	-121	0.783	2.69
Propionaldehyde	48.8	-81	0.806	2.52
(Propane)[a]	-42.1			
(Propene)[a]	-47.4			
Benzaldehyde	178.6	-26	1.04	
Acetone	56.2	-95.4	0.789	2.88
(Isobutene)[a]	-6.9			
Ethyl methyl ketone	79.6	-86.3	0.805	
Cyclohexanone	155.6	-16.4	0.948	
Methyl phenyl ketone	202.6	20.5	1.03	
Diphenyl ketone	305.9	48.1	1.15	2.98

[a]Hydrocarbons for comparison purposes.

16.5 Spectroscopy of Carbonyl Compounds

16.5a Infrared Spectroscopy Carbonyl compounds are ubiquitous in nature. The C=O bond is one of the most widely found of the "functional groups." Spectroscopy, especially IR spectroscopy, played an important role in identifying many medicinally and otherwise important compounds. The carbon–oxygen double bond has a diagnostically useful and quite strong stretching frequency in the IR, generally near 1700 cm^{-1} (p. 711). This absorption is at substantially higher frequency than that of carbon–carbon double bonds. This shift should be no surprise, because we already know that carbon–oxygen double bonds are stronger than carbon–carbon double bonds, and therefore require more energy for stretching. A useful diagnostic for aldehydes is the pair of bands observed in the C—H stretching region at about 2850 cm^{-1} and 2750 cm^{-1}. Table 16.2 gives some IR, NMR, and UV data for a few aldehydes and ketones.

TABLE 16.2 **Some Spectral Properties of Carbonyl Compounds**

Compound	IR (CCl$_4$, cm^{-1})	^{13}C NMR (δ C=O)	UV [nm (ε)]
Formaldehyde	1744 (gas phase)		270
Acetaldehyde	1733	199.6	293.4 (11.8)
Propanal	1738	201.8	289.5 (18.2)
Acetone	1719	205.1	279 (14.8)
Cyclohexanone	1718	207.9	285 (14)
Cyclopentanone	1751	213.9	299 (20)
Cyclobutanone	1788	208.2	280 (18)
Butanone	1723	206.3	277 (19.4)
3-Butenone	1684	197.2	219 (3.6)
2-Cyclohexenone	1691	197.1	224.5 (10,300) ($\pi \to \pi^*$)
Benzaldehyde	1710	191.0	244 (15,000) ($\pi \to \pi^*$)
			328 (20) ($n \to \pi^*$)
Benzophenone	1666	195.2	252 (20,000) ($\pi \to \pi^*$)
			325 (180) ($n \to \pi^*$)
Acetophenone	1692	196.8	240 (13,000) ($\pi \to \pi^*$)
			319 (50) ($n \to \pi^*$)

Note that conjugated carbonyl compounds are always found at lower frequency than their unconjugated relatives. For example, cyclohexanone appears at 1718 cm^{-1} and 2-cyclohexenone at 1691 cm^{-1} (Fig. 16.14). Thus there is reduced double-bond character in the C=O group in conjugated molecules. The resonance formulation of 2-cyclohexenone clearly shows the partial single-bond character of the carbon–oxygen bond in these compounds.

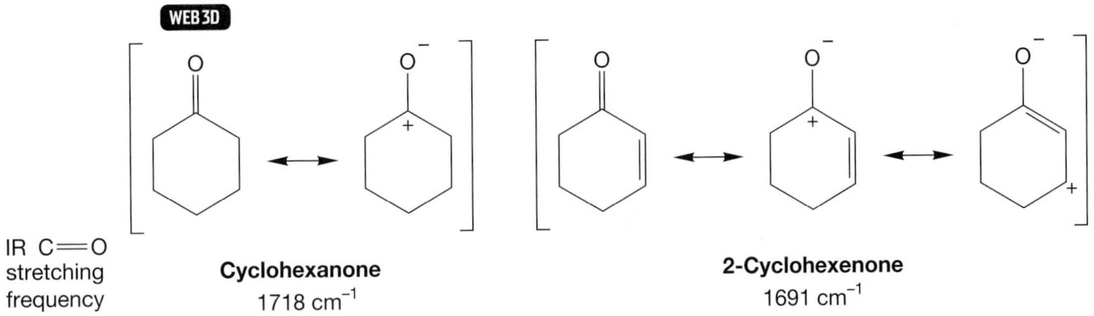

WEB 3D

IR C=O stretching frequency

Cyclohexanone
1718 cm^{-1}

2-Cyclohexenone
1691 cm^{-1}

FIGURE 16.14 Conjugation reduces the double-bond character of a carbonyl group. The result is a shift of the IR stretching band to lower frequency.

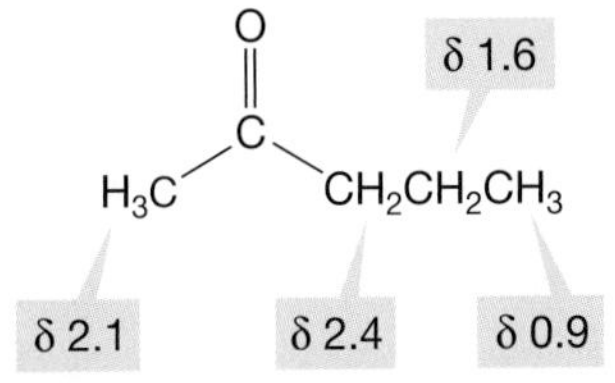

16.5b Nuclear Magnetic Resonance Spectroscopy

The carbonyl group inductively withdraws electrons from the nearby positions, and this inductive effect results in a deshielding of adjacent hydrogens. The further away from the carbonyl group, the less effective the deshielding. This effect is well illustrated by 2-pentanone (Fig. 16.15).

In aldehydes, hydrogens directly attached to the carbonyl group resonate far downfield, $\delta \sim 8.5$–10 ppm. Not only does the carbon bear a partial positive charge that deshields the attached hydrogen, but the applied magnetic field B_0 creates an induced field B_i in the carbonyl group that augments the applied field in the region

FIGURE 16.15 Hydrogens on the α carbons are affected most by the electron-withdrawing carbonyl group. Hydrogens on the β carbon are only slightly shifted. Hydrogens that are further away are not significantly affected (δ in ppm).

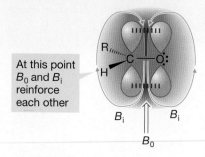

FIGURE 16.16 Aldehyde hydrogens are in a strongly deshielded region in which the induced magnetic field B_i augments the applied field B_0.

of aldehydic hydrogens (Fig. 16.16). Accordingly, a relatively weak applied magnetic field is required to reach the resonance frequency, and they appear unusually far downfield (p. 720).

The carbon atoms of carbonyl groups carry partial positive charges and are greatly deshielded by their relative lack of surrounding electrons. They, too, resonate especially far downfield in the ^{13}C NMR. Table 16.2 shows typical resonance positions.

16.5c Ultraviolet Spectroscopy

Simple aldehydes and ketones, like simple alkenes, do not have a $\pi \rightarrow \pi^*$ absorption in the region of the UV spectrum accessible to most spectrometers. However, unlike alkenes, carbonyl compounds have another possible absorption which, though weak, is detectable. This absorption involves promotion of a nonbonding, or n, electron to the antibonding π^* orbital, the $n \rightarrow \pi^*$ absorption. Table 16.2 gives wavelengths for UV absorptions of a few compounds.

As for alkenes (p. 528), conjugation of the carbonyl group greatly reduces the energy necessary to promote an electron from the highest occupied π molecular orbital to the LUMO. So, the strong $\pi \rightarrow \pi^*$ transition in unsaturated carbonyl compounds can be detected because of the conjugation-induced shift to the longer wavelengths (lower energy) observable by normal UV/vis spectrometers (Table 16.2; Fig. 16.17).

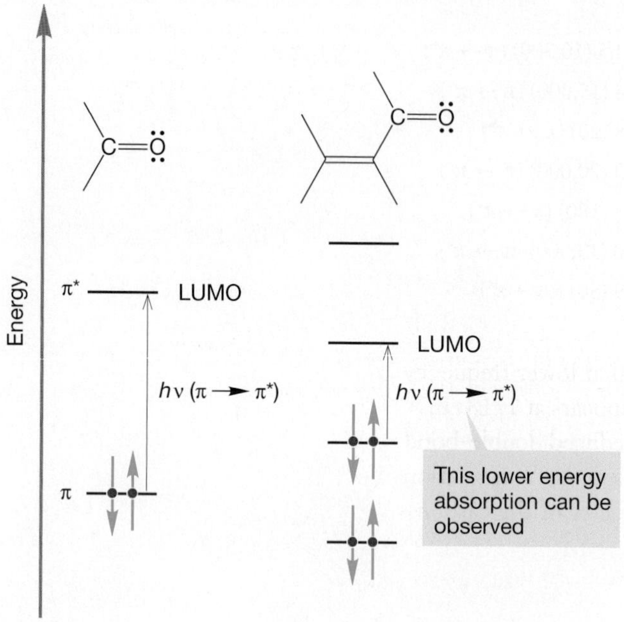

16.5d Mass Spectrometry

There is a diagnostic cleavage reaction that occurs for many simple carbonyl-containing compounds. It involves breaking the bond attached directly to the carbonyl carbon, with the charge remaining on the carbonyl group. Cleavage of this bond generates a resonance-stabilized acylium ion. It is this resonance stabilization that accounts for the prominence of this particular fragmentation. Figure 16.18 shows the mass spectrum for acetone, the archetypal methyl ketone.

FIGURE 16.17 Conjugation leads to lower energy (longer wavelength) $\pi \rightarrow \pi^*$ absorptions that are observable by UV/vis spectrometers.

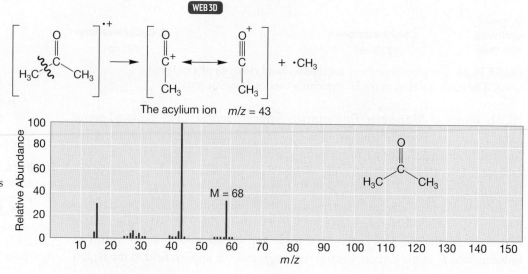

The acylium ion $m/z = 43$

FIGURE 16.18 In the mass spectrum of acetone, the base peak results from formation of the resonance-stabilized acylium ion.

16.6 Reactions of Carbonyl Compounds: Simple Reversible Additions

In simple addition reactions, carbonyl compounds can behave as either Lewis acids or Lewis bases depending on the other reagents present. As we might expect, there is some correspondence between addition reactions of carbon–oxygen double bonds and those of the carbon–carbon double bonds we saw in Chapters 9 and 10. As we might also expect, additional complexity is introduced by the presence of two *different* atoms in the carbon–oxygen double bond. Figure 16.19 compares the reaction between water and a generic carbonyl compound (the formation of a **hydrate**) with the hydration of an alkene.

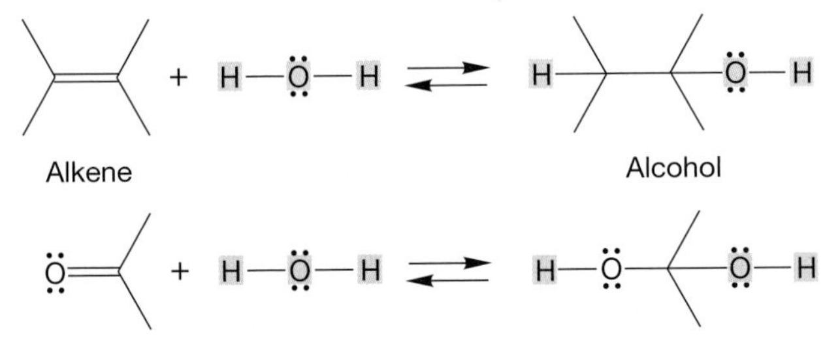

FIGURE 16.19 Addition of water to an alkene to give an alcohol is analogous to addition to a ketone or aldehyde to give a hydrate.

If water acts as a nucleophile and adds to the Lewis acid carbonyl group, there is a choice to be made. In principle, water could add to either the carbon or the oxygen of the carbonyl group (Fig. 16.20).

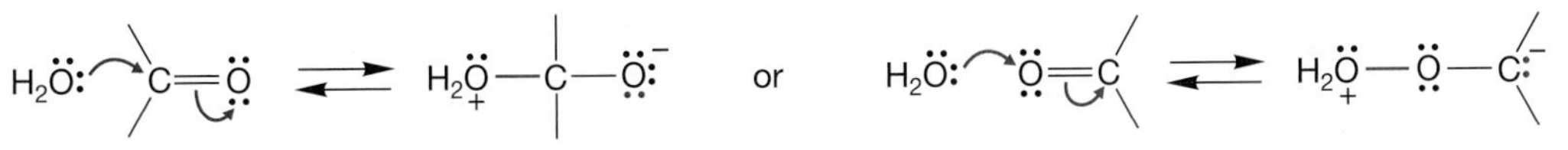

FIGURE 16.20 In principle, addition of water to a carbonyl group could occur in two ways.

Three factors must be considered in predicting which way the addition will go. First of all, addition to the carbon of the carbon–oxygen double bond places the negative charge on the highly electronegative oxygen atom, whereas addition at oxygen generates a much less stable carbanion (Fig. 16.21). Second, addition to oxygen forms a weak oxygen–oxygen bond, whereas addition to the carbon makes a stronger oxygen–carbon bond.

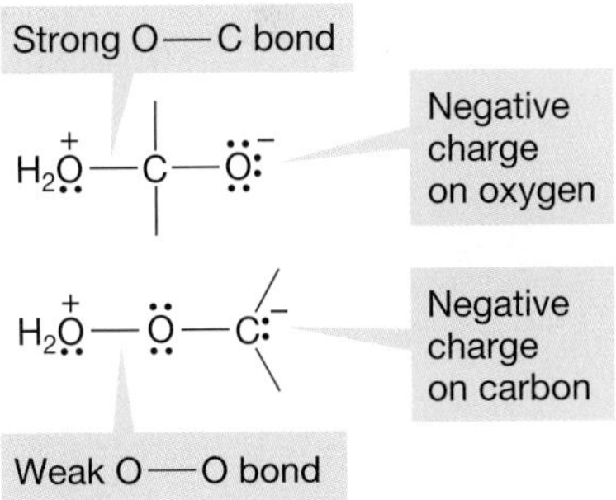

FIGURE 16.21 Addition to the carbon end, which puts the negative charge on the relatively electronegative oxygen atom, will be greatly preferred.

WORKED PROBLEM 16.7 Explain how the thermodynamic preference for placing a negative charge on oxygen, rather than carbon, can influence the rates of addition to the two ends of the carbon–oxygen double bond.

ANSWER The rates of reaction depend, of course, on the heights of the transition states, not on the energies of the products in which a full negative charge resides on either oxygen or carbon. In the transition states for addition, there will be a partial negative charge on either oxygen or carbon. The factors that favor a full negative charge on oxygen rather than carbon (electronegativity) will also favor a partial negative charge (δ^-) on oxygen.

(continued)

Therefore, in the transition state, a partial negative charge will be more stable on oxygen than on carbon.

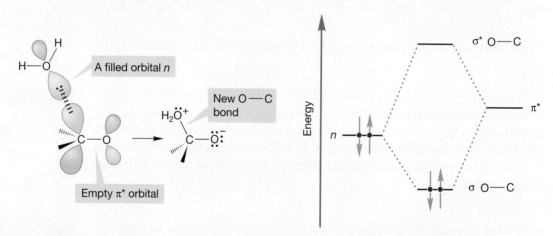

The third factor to consider involves the orbitals making up the carbonyl π bond. If we look at the molecular orbitals involved, we can see how they influence the direction in which the reaction occurs. As in all organic reactions, we look for the combination of a nucleophile (here, water) and an electrophile. The electrophile in this case must be the *empty* π^* orbital of the carbonyl group. Figure 16.22 shows the orbital interactions graphically, and gives pictures of the orbital lobes involved.

FIGURE 16.22 In the addition of water to a carbonyl group, a filled n orbital of water overlaps with the empty π^* orbital of the carbonyl group. This interaction between a filled and an empty orbital is stabilizing.

As we saw in our earlier section on structure (p. 764), neither the π nor the π^* of the carbonyl group is symmetric with respect to the midpoint of the carbon–oxygen bond. In particular, the π^* is much larger in the region of the carbon atom than it is near the oxygen. Because the magnitude of stabilization depends on the amount of orbital overlap, it is no surprise to see that addition at the carbon, where the π^* is largest, dominates the reaction. So, not only is the direction of the addition affected by the electronegativities and the bond strengths, but it's affected by the shapes of the orbitals involved as well.

After the addition of water to form the alkoxide, a proton must be removed and a proton added to give the hydrate, not necessarily in that order (Fig. 16.23). It is not necessary that the proton lost from one molecule be re-added to the same molecule. The protonation–deprotonation process is very likely to be intermolecular, as shown in Figure 16.23.

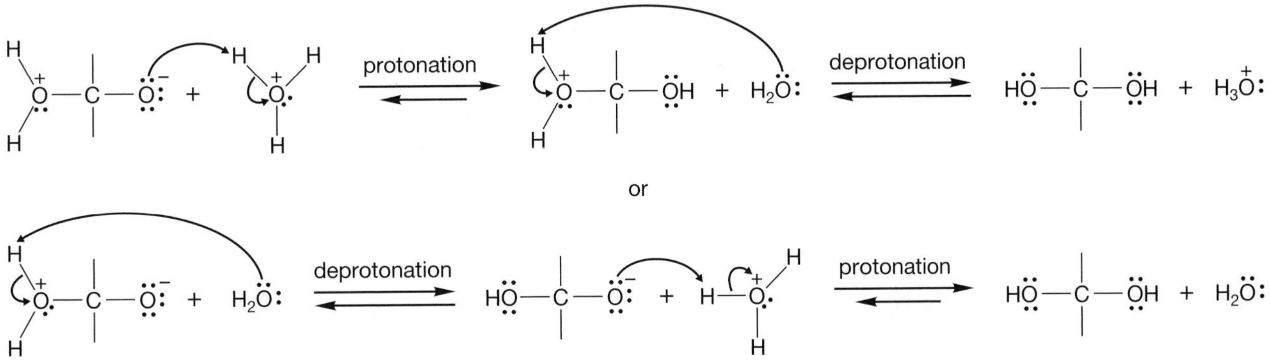

FIGURE 16.23 The carbonyl hydration reaction is completed by two proton transfers. The order of the two steps is not certain.

Now consider how we might make the carbonyl addition reaction easier. Two ways seem possible—we could make either the electrophile or the nucleophile stronger. Let's start with the acid-catalyzed reaction in which the carbonyl group is converted into a stronger electrophile. If we take as our model for the mechanism the acid-catalyzed additions to alkenes, a reasonable first step for hydration of a carbonyl would be protonation. But which end is protonated? Again, there are two possiblities (Fig. 16.24).

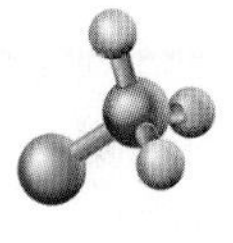

Carbonyl hydration

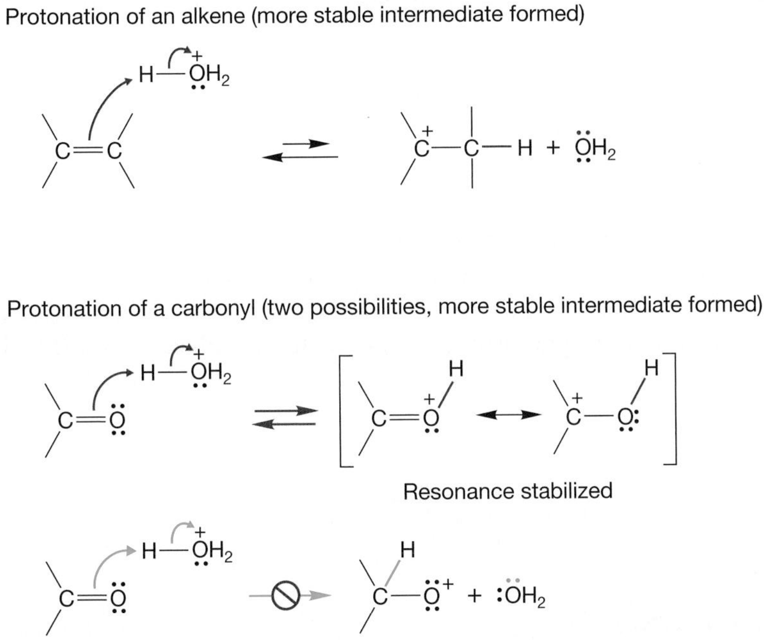

FIGURE 16.24 Protonation of the carbonyl group can occur in two ways. The placement of a proton on oxygen to give a resonance-stabilized cation will be greatly preferred.

Protonation of an alkene (more stable intermediate formed)

Protonation of a carbonyl (two possibilities, more stable intermediate formed)

Resonance stabilized

The choice is easy this time. First, the carbonyl oxygen atom is a good Brønsted base because it has two lone pairs of nonbonding electrons. Moreover, it is surely better to protonate on oxygen, and thus place the positive charge on the relatively electropositive carbon rather than to protonate on carbon, which would place the positive charge on the highly electronegative oxygen atom. Finally, protonation at oxygen gives a resonance-stabilized cation, whereas protonation at carbon does not, as shown in

Figure 16.24. So there is not really much choice in the direction of the protonation step, and the result is a very electrophilic species, the protonated carbonyl compound.

In the second step of the reaction, water acts as a nucleophile and adds to the Lewis acid (the protonated carbonyl) to generate a protonated diol (Fig. 16.25). In the final step of this acid-catalyzed reaction, the protonated diol is deprotonated by a water molecule to generate the neutral diol and regenerate the acid catalyst, H_3O^+. A diol with both OH groups on the same carbon is called a ***gem*-diol** (gem stands for geminal) or a hydrate. Note how closely related are the acid-catalyzed additions of water to an alkene and to a carbonyl group. Each reaction is a sequence of three steps: protonation, addition of water, and deprotonation.

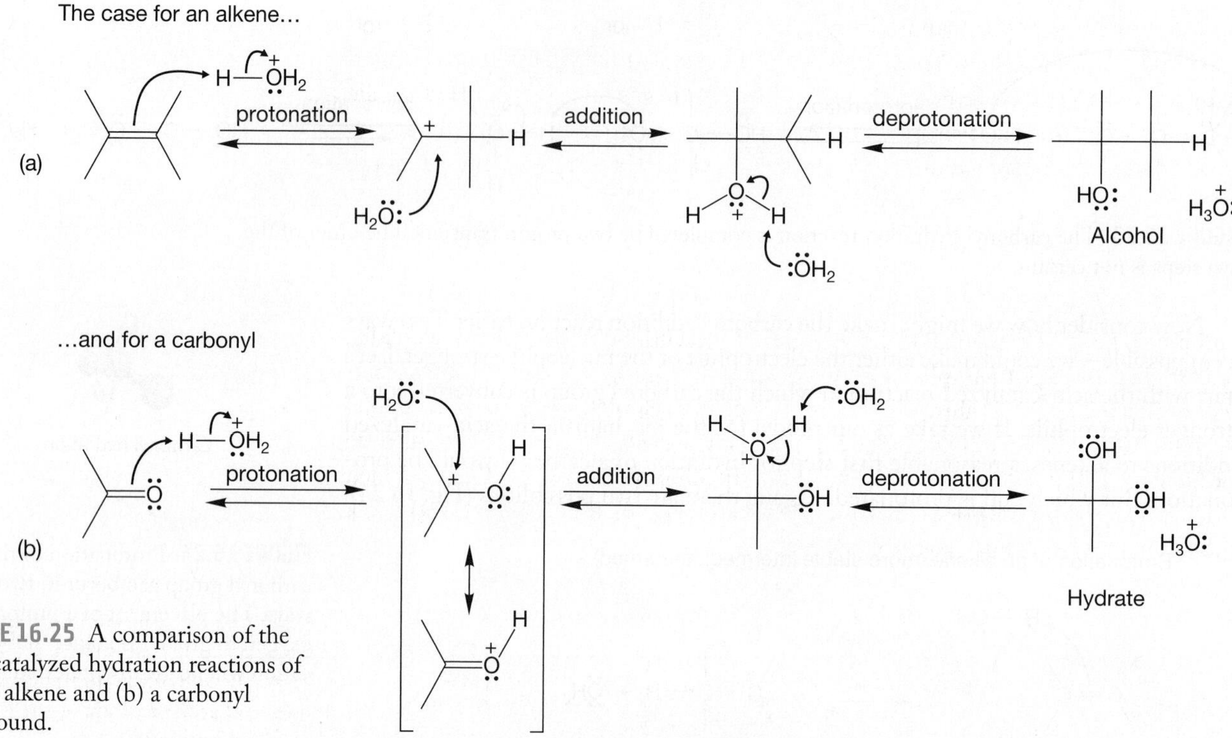

The case for an alkene...

(a)

...and for a carbonyl

(b)

FIGURE 16.25 A comparison of the acid-catalyzed hydration reactions of (a) an alkene and (b) a carbonyl compound.

What about making the nucleophile more reactive? There is another mechanism for hydration and it operates under basic conditions. Here the active agent is the hydroxide ion, $^-$OH. The first step in the base-catalyzed hydration reaction is the addition of the strong nucleophile hydroxide to the carbonyl group (Fig. 16.26).

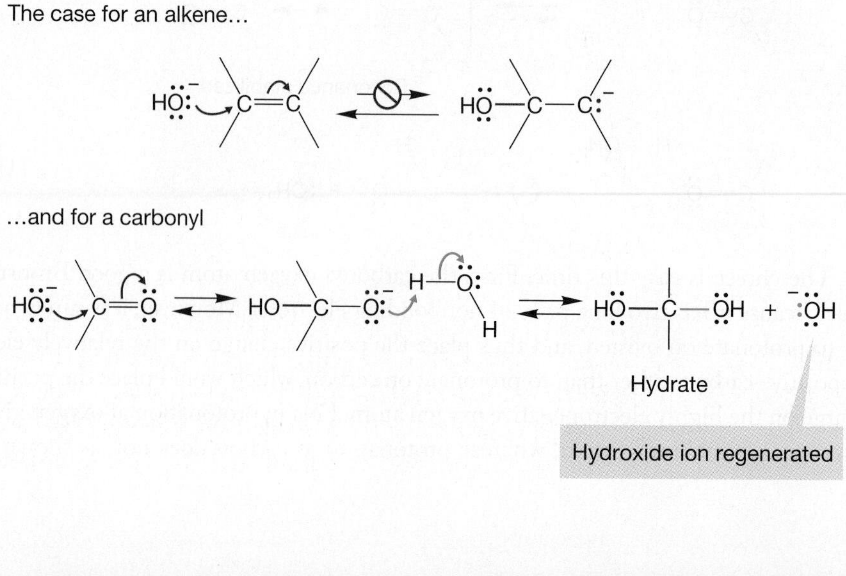

The case for an alkene...

...and for a carbonyl

FIGURE 16.26 Although strong bases such as hydroxide ion do not add to alkenes, they do attack carbonyl groups. The final step in the base-catalyzed hydration reaction is protonation of the newly formed alkoxide ion by water. Hydroxide is regenerated.

Hydrate

Hydroxide ion regenerated

Unlike protonation, the first step in the acid-catalyzed reaction, this reaction has no counterpart in the chemistry of simple alkenes. After addition, a proton is transferred from water to the newly formed alkoxide, generating the hydrate and regenerating the hydroxide ion.

PROBLEM 16.8 Explain why bases such as the hydroxide ion do not add to the oxygen of the carbon–oxygen double bond, nor to alkenes.

Summary

The carbonyl group is hydrated under acidic (Fig. 16.25), basic (Fig. 16.26), and neutral conditions (Figs. 16.20 and 16.23). In acid, the mechanism is accelerated by protonation of the carbon–oxygen double bond to generate a strong electrophile (Lewis acid). The relatively weak nucleophile, water, then adds. In base, the strong nucleophile hydroxide adds to the relatively weak Lewis acid, the carbonyl group.

16.7 Equilibrium in Addition Reactions

Carbonyl hydration reactions consist of several consecutive equilibria. In basic, acidic, or even neutral water, simple carbonyl compounds are in equilibrium with their hydrates, as Figure 16.27 shows. Even for molecules as similar as acetone and ethanal (acetaldehyde), these equilibria settle out at widely different points. If we include formaldehyde, the difference is even more striking. For these three molecules, which differ only in the number of methyl groups attached to the carbonyl, there is a difference in K of about a factor of 10^6.

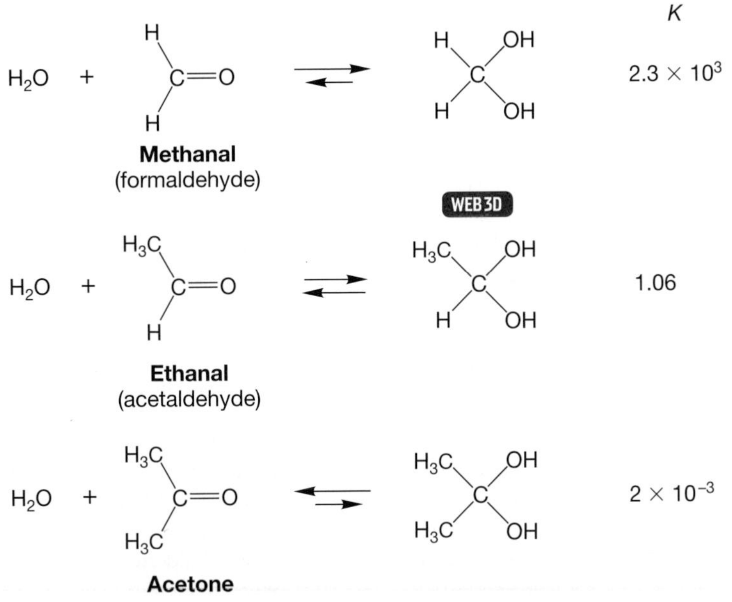

FIGURE 16.27 The equilibrium constant for hydration of a carbonyl-containing molecule can favor either the hydrate or the carbonyl compound. The equilibrium constants shown here are for pH = 7 conditions.

Why should there be such great differences among these rather similar compounds? First of all, remember that a little difference in energy goes a long way at equilibrium (p. 336). Recall also that alkyl groups stabilize double bonds (Table 3.1). The more highly substituted an alkene, the more stable it is. The same phenomenon holds for carbon–oxygen double bonds. So, the stability of the carbonyl compound increases along the series: formaldehyde, acetaldehyde, acetone as the substitution of the carbon–oxygen double bond increases (Fig. 16.28, left side). If we now look at the molecule at the other side of the equilibrium, the hydrate, we can see that increased methyl substitution introduces steric destabilization.

FIGURE 16.28 Like alkenes, more substituted carbonyl groups are more stable than their less substituted counterparts. In the hydrates, increasing substitution results in increasing destabilization through steric interactions.

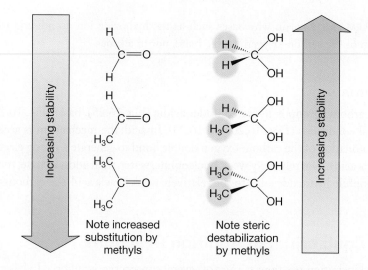

The more substituted the hydrate, the less stable it is (Fig. 16.28, right side). We can immediately see the sources of the difference in K of 10^6 (Fig. 16.29). They are the increased stabilization of the starting carbonyl compounds and destabilization of the hydrates, both of which are induced by increased substitution by methyl groups.

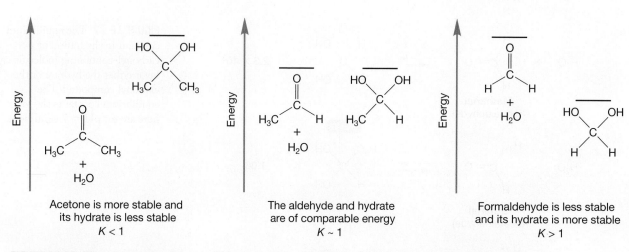

Acetone is more stable and its hydrate is less stable $K < 1$

The aldehyde and hydrate are of comparable energy $K \sim 1$

Formaldehyde is less stable and its hydrate is more stable $K > 1$

FIGURE 16.29 The magnitude of the equilibrium constant (K) depends on the relative energies of the carbonyl compounds and their hydrates. We know that for acetaldehyde the energies of hydrate and aldehyde are comparable, because $K \sim 1$. Acetone is more stable than acetaldehyde and formaldehyde is less stable. Steric factors make the acetone hydrate less stable than the acetaldehyde hydrate. By contrast, the formaldehyde hydrate is more stable than the acetaldehyde hydrate. The equilibrium constants reflect these changes.

We can see other striking effects on this equilibrium process in Table 16.3. For example, benzene rings, which can interact with the carbon–oxygen double bond by resonance, decrease the amount of hydrate present at equilibrium. However, electron-withdrawing groups on the α position adjacent to the C=O strongly favor the hydrate. There also appears to be a steric effect. Large alkyl groups favor the carbonyl compound, as can be seen from the sequence of decreasing equilibrium constants in Table 16.3: propanal > butanal > 2,2-dimethylpropanal. Each of these phenomena can be understood by considering the relative stabilities of the starting materials and products.

TABLE 16.3 Equilibrium Constants for Carbonyl Hydration

	Compound	
Structure	Name	K
(structure: m-chlorobenzaldehyde) —CHO	m-Chlorobenzaldehyde	0.022
CH_3CH_2CHO	Propanal	0.85
$CH_3CH_2CH_2CHO$	Butanal	0.62
$(CH_3)_3CCHO$	2,2-Dimethylpropanal	0.23
CF_3CHO	Trifluoroethanal	2.9×10^4
CCl_3CHO	Trichloroethanal	2.8×10^4
$ClCH_2CHO$	Chloroethanal	37

Resonance stabilization of the carbonyl group is generally not accompanied by a resonance stabilization of the product hydrate (Fig. 16.30). For example, the aromatic aldehyde, m-chlorobenzaldehyde (Table 16.3), is more strongly stabilized by resonance than acetaldehyde. There is no corresponding difference in the hydrates, and therefore the aromatic aldehyde is less hydrated at equilibrium than acetaldehyde.

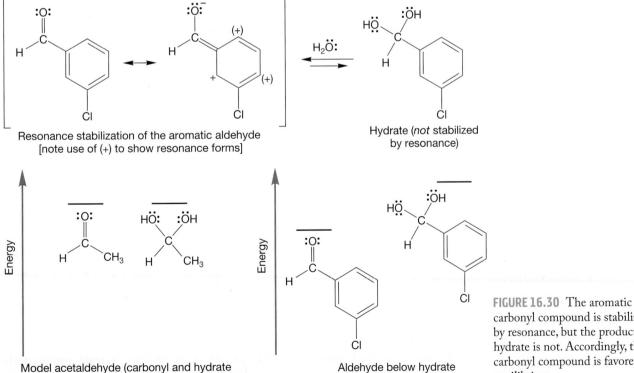

Resonance stabilization of the aromatic aldehyde
[note use of (+) to show resonance forms]

Hydrate (*not* stabilized by resonance)

Model acetaldehyde (carbonyl and hydrate are of comparable stability $K \sim 1$)

Aldehyde below hydrate

FIGURE 16.30 The aromatic carbonyl compound is stabilized by resonance, but the product hydrate is not. Accordingly, the carbonyl compound is favored at equilibrium.

The reason electron-withdrawing groups such as chlorine or fluorine sharply destabilize carbonyl groups when attached at the α position is that the dipoles oppose each other. The carbon atom is already electron deficient, and further inductive removal of electrons by the halogen on the adjacent carbon is costly in energy terms. There is relatively little effect in the product hydrate, although there is some destabilization in this molecule as well. So, as Figure 16.31 shows, the strong destabilization of the carbonyl group favors the hydrate at equilibrium.

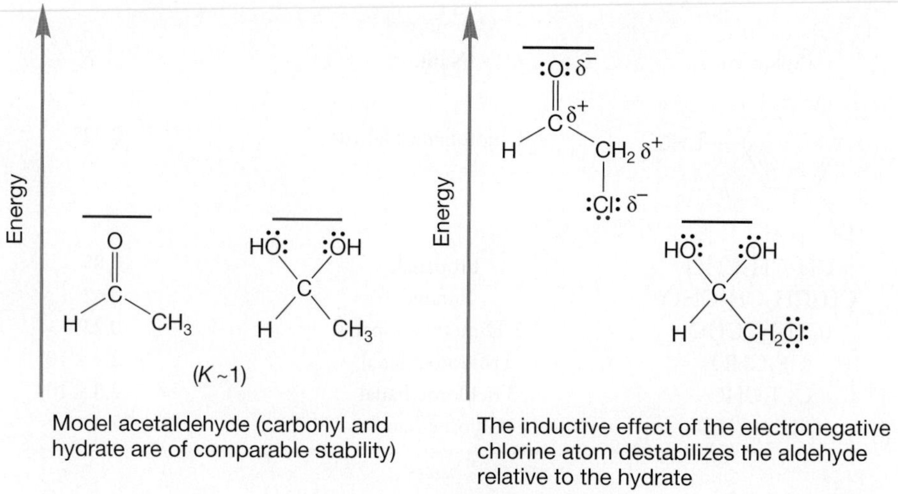

FIGURE 16.31 Addition of a chlorine in the α position of acetaldehyde raises the energy of the carbonyl compound.

Model acetaldehyde (carbonyl and hydrate are of comparable stability)

The inductive effect of the electronegative chlorine atom destabilizes the aldehyde relative to the hydrate

The steric effect is slightly more subtle. In the carbonyl group, the carbon is hybridized sp^2 and the bond angles are therefore close to 120° (Fig. 16.32). In the product hydrate, the carbon is nearly sp^3 hybridized, and the bond angles must be close to 109°. So, in forming the hydrate, groups are moved closer to each

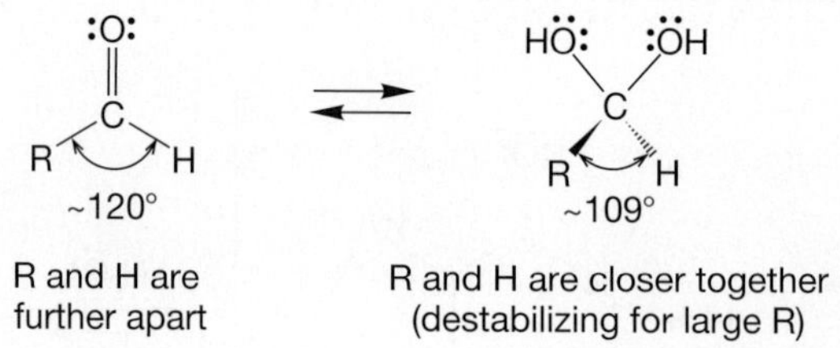

FIGURE 16.32 In the hydrate, groups are closer together than they are in the carbonyl compound. This proximity destabilizes the hydrate relative to the carbonyl compound.

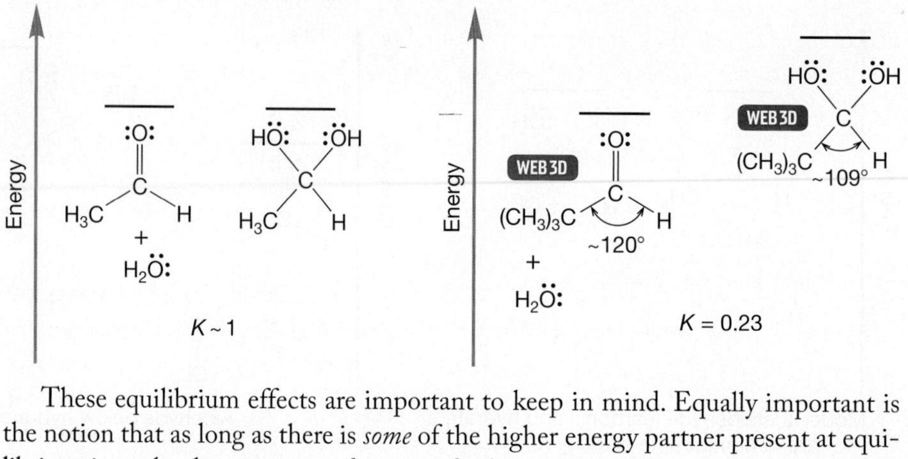

R and H are further apart

R and H are closer together (destabilizing for large R)

other, and for a large group that is relatively destabilizing. It is less costly in energy terms to move a methyl group closer to a hydrogen than to move a *tert*-butyl group closer to a hydrogen. Accordingly, acetaldehyde is more strongly hydrated than is 2,2-dimethylpropanal (pivaldehyde) (Fig. 16.33).

FIGURE 16.33 The larger the R group, the more important the steric effect. For example, 2,2-dimethylpropanal is less hydrated at equilibrium than is acetaldehyde.

These equilibrium effects are important to keep in mind. Equally important is the notion that as long as there is *some* of the higher energy partner present at equilibrium, it can be the active ingredient in a further reaction. For example, even though it is a minor component of the equilibrium mixture (Fig. 16.34a), the hydrate of acetone could go on to other products (X), with more hydrate being formed as the small amount present at equilibrium is used up. Similarly, even though it is extensively hydrated in aqueous solution, reactions of formaldehyde to give Y can still be detected (Fig. 16.34b). It is also worth mentioning that *gem*-diols (hydrates) are very difficult to isolate. Even though we know they exist in aqueous environment, in order to isolate the diol we must remove the water and that drives the equilibrium toward the carbonyl species.

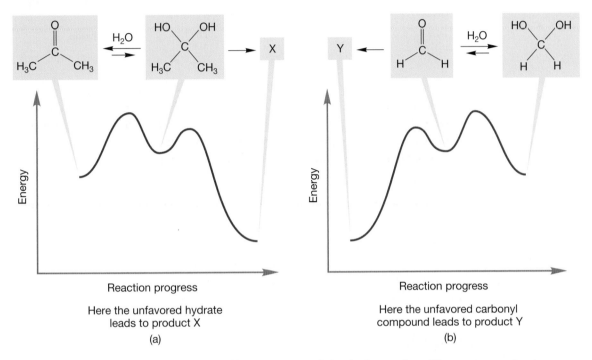

FIGURE 16.34 Molecules not favored at equilibrium may nonetheless lead to products. These two examples show a pair of carbonyl compound–hydrate equilibria in which it is the higher energy, *unfavored* partner that leads to product. As the small amount of the higher energy molecule is used up, equilibrium is reestablished and more of the unfavored compound is formed. (a) Reaction of the acetone hydrate is possible even though the hydrate is not the favored species. (b) Reaction of the formaldehyde carbonyl group is possible even though the carbonyl is not the favored species.

As usual, mechanisms stick in the mind most securely if we are able to write them backward. "Backward" is an arbitrary term anyway, because by definition an equilibrium runs both ways. Carbonyl addition reactions are about to grow more complex, and you run the risk of being overwhelmed by a vast number of proton additions and losses in the somewhat more complicated reactions that follow. Be sure you are completely comfortable with the acid-catalyzed and base-catalyzed hydration reactions *in both directions* before you go on. That advice is important.

16.8   Other Addition Reactions: Additions of Cyanide and Bisulfite

Many other addition reactions follow the pattern of carbonyl hydration. Two typical examples are **cyanohydrin** formation and bisulfite addition. Both of these reactions are generally base catalyzed. Because of an understandable and widespread aversion to working with the volatile and notorious HCN, cyanide additions are generally carried out with cyanide ion. Potassium cyanide ($K^+ {}^- CN$) is certainly poisonous, but it's not volatile, and one knows where it is at all times, at least as long as basic conditions are maintained.

Cyanide ion is a good nucleophile and attacks the electrophilic carbonyl compound to produce an alkoxide (Fig. 16.35). This alkoxide is protonated by solvent, often water, to give the final product, a cyanohydrin. The cyanohydrin is often not stable under basic conditions and reverses to the carbonyl compound and cyanide.

FIGURE 16.35 Cyanohydrin formation.

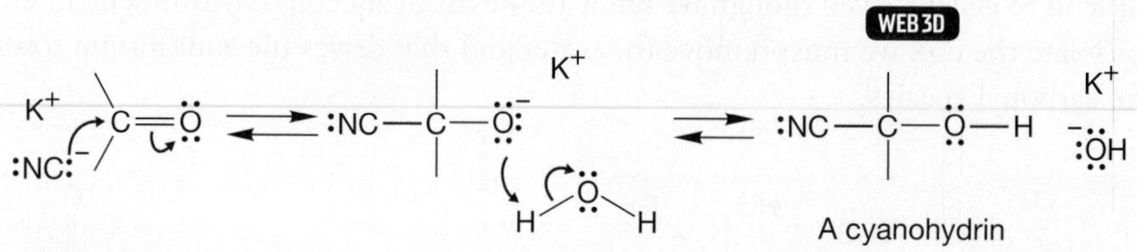

PROBLEM 16.9 Why does acetaldehyde in aqueous solution form an excellent yield of cyanohydrin even though acetaldehyde is substantially hydrated in water ($K \sim 1$)? Might it not be argued that, because about one-half of the aldehyde is present in hydrated form at equilibrium, no more than about 50% cyanohydrin can be formed. Why is this argument wrong? *Hint*: See Figure 16.34.

Bisulfite addition to an aldehyde or ketone is similar to cyanide addition, although the nucleophile in this case is sulfur. Sodium bisulfite adds to many aldehydes and ketones to give an addition product, often a nicely isolable solid (Fig. 16.36).

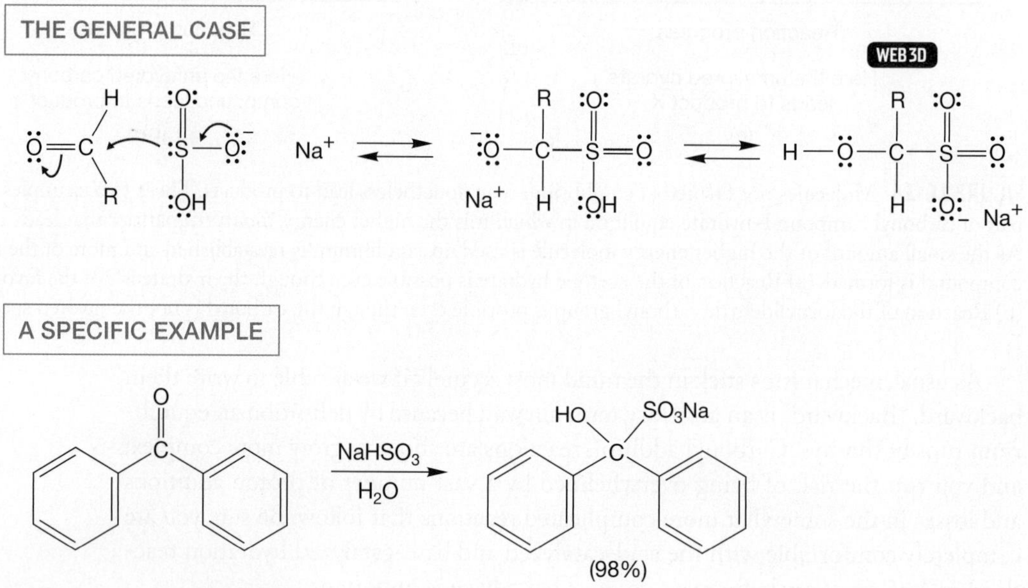

THE GENERAL CASE

A SPECIFIC EXAMPLE

FIGURE 16.36 Bisulfite addition.

The reactions we have seen so far have all been simple additions of a generic HX to the carbon–oxygen bond of a carbonyl group. Now we move on to related reactions in which there is one more dimension. In these new reactions, addition of HX is accompanied by the loss of some other group.

PROBLEM 16.10 Bisulfite addition is generally far less successful with ketones than it is with aldehydes. Explain why.

16.9 Addition Reactions Followed by Water Loss: Acetal Formation

If aldehydes and ketones form hydrates in water (H—OH), shouldn't there be a similar reaction in the presence of alcohols (R—OH)? Indeed there should be, and there is. However, the reactions are slightly more complicated than simple hydration, at least under acidic conditions. In base, the two reactions are entirely analogous, so let's look at that reaction first.

Figure 16.37 contrasts the base-catalyzed reaction of a carbonyl compound in water to give a hydrate with the related formation of a **hemiacetal** in an alcohol. The hemiacetal is similar to a *gem*-diol, except that the central carbon has an OH and an OR group rather than two OH groups.

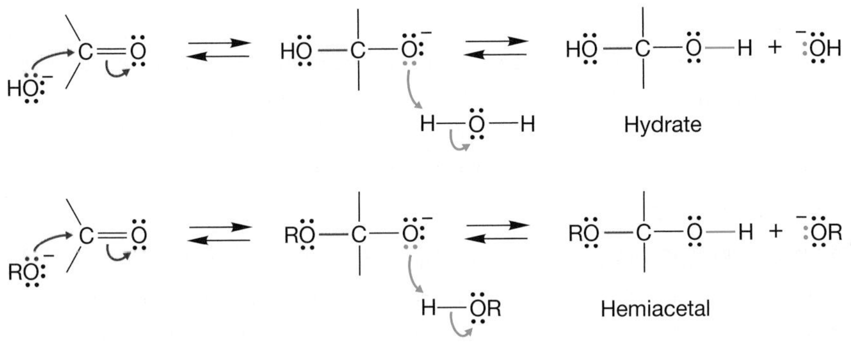

FIGURE 16.37 The reaction of a carbonyl group with hydroxide ion in water parallels the reaction of a carbonyl with an alkoxide ion in alcohol. In water, a *gem*-diol (or hydrate) is formed; in alcohol, the analogous product is called a hemiacetal.

Hydration and hemiacetal formation are comparable, so it is no surprise that the discussion in Section 16.7 on equilibrium effects is quite directly transferable to this reaction. Aldehydes, for example, are largely in their hemiacetal forms in basic alcoholic solutions (in acid the reaction goes further, as we will see), whereas ketones are not.

Most simple hemiacetals, like *gem*-diols, are not isolable. However, the cyclic versions of these compounds often are. Consider combining the carbonyl and hydroxyl components of the hemiacetal reaction in Figure 16.37 within the same molecule (Fig. 16.38). Now hemiacetal formation is an intramolecular process. If the ring size is neither too large nor too small, the hemiacetals are often more stable than the open-chain molecules.

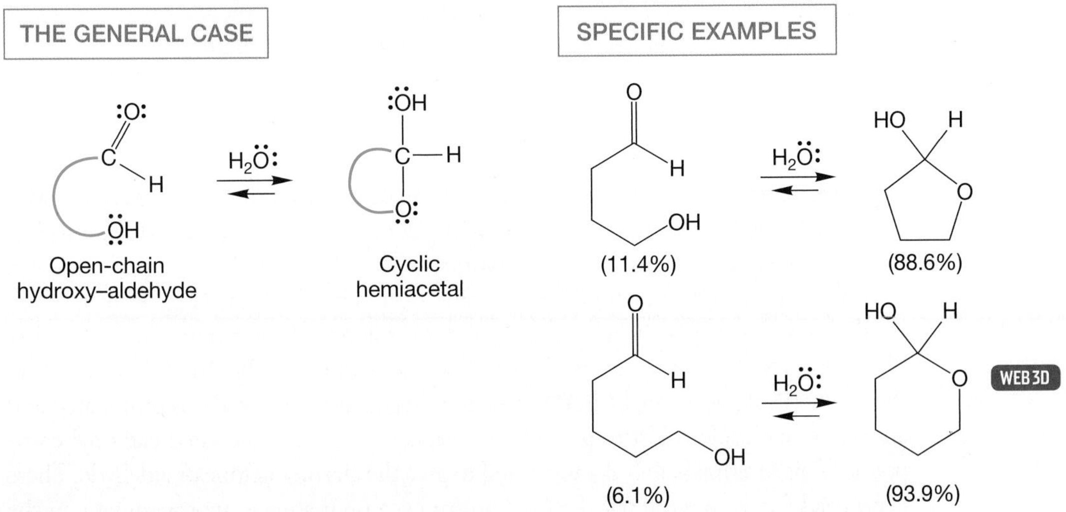

FIGURE 16.38 Cyclic hemiacetals are often more stable than their open-chain counterparts.

The most spectacular examples of this effect are the sugars, polyhydroxylated compounds that exist largely, though not quite exclusively, in their cyclic hemiacetal forms (Fig. 16.39). We will explore the chemistry of sugars in Chapter 22.

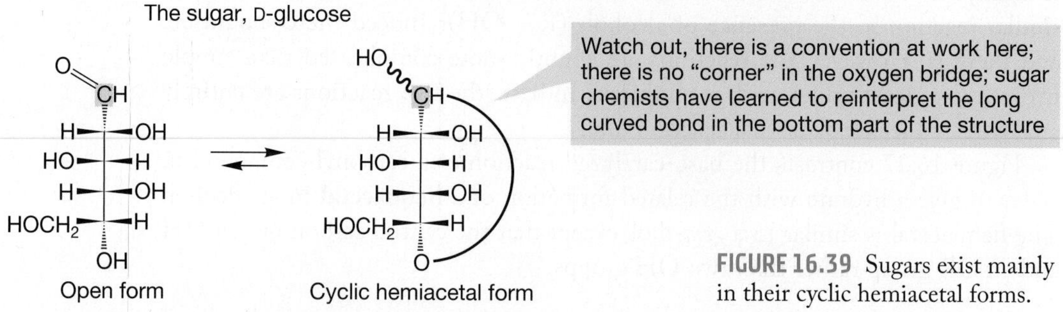

The sugar, D-glucose

Open form   Cyclic hemiacetal form

Watch out, there is a convention at work here; there is no "corner" in the oxygen bridge; sugar chemists have learned to reinterpret the long curved bond in the bottom part of the structure

FIGURE 16.39 Sugars exist mainly in their cyclic hemiacetal forms.

PROBLEM 16.11 Write a mechanism for the intramolecular formation of one of the hemiacetals of Figure 16.38.

PROBLEM SOLVING

Problems that ask you to write a mechanism for the transformation of a starting material into a product are common, and they will appear increasingly throughout the remaining chapters. This particular problem should be relatively easy, but it serves as an example of how to analyze what must happen. That kind of analysis is absolutely critical to success in solving hard mechanistic problems. It is very, very important to get into the habit right now of defining goals in mechanism problems. For example, in the first part of this problem, what happens? A carbon–oxygen bond is made. A carbonyl group becomes an alcohol. So, we have two goals (write them down!):

1. Close the ring by making a C—O bond.
2. Break the C=O double bond.

It is easy to see that the two goals are connected—this *is* an easy problem—but defining and paying attention to those goals is critical practice, and a great habit to get into.

PROBLEM 16.12 Note how odd the cyclic structure in Figure 16.39 looks. Construct a three-dimensional picture of D-glucose that better represents its structure in the cyclic hemiacetal form. Note that the position of the OH on the carbon marked in red in Figure 16.39 can be either axial or equatorial.

Now, what about the acid-catalyzed version of the reaction of alcohols and aldehydes or ketones? The process begins with hemiacetal formation, and the mechanism parallels that of acid-catalyzed hydration (Fig. 16.25b). A three-step sequence of protonation, addition, and deprotonation occurs in each case (Fig. 16.40).

Can anything happen to a hemiacetal in acidic solution? Does the reaction stop there? One thing you know happens is the set of reactions indicated by the reverse arrows in the equilibria of Figure 16.40. In this case, the oxygen atom of the OR is protonated and an alcohol molecule is lost to give the resonance-stabilized protonated carbonyl compound. This intermediate is deprotonated to give the starting ketone or aldehyde. There is no need for us to write out the mechanism because it already appears exactly in the reverse steps of Figure 16.40. It is simply the forward reaction run directly backward.

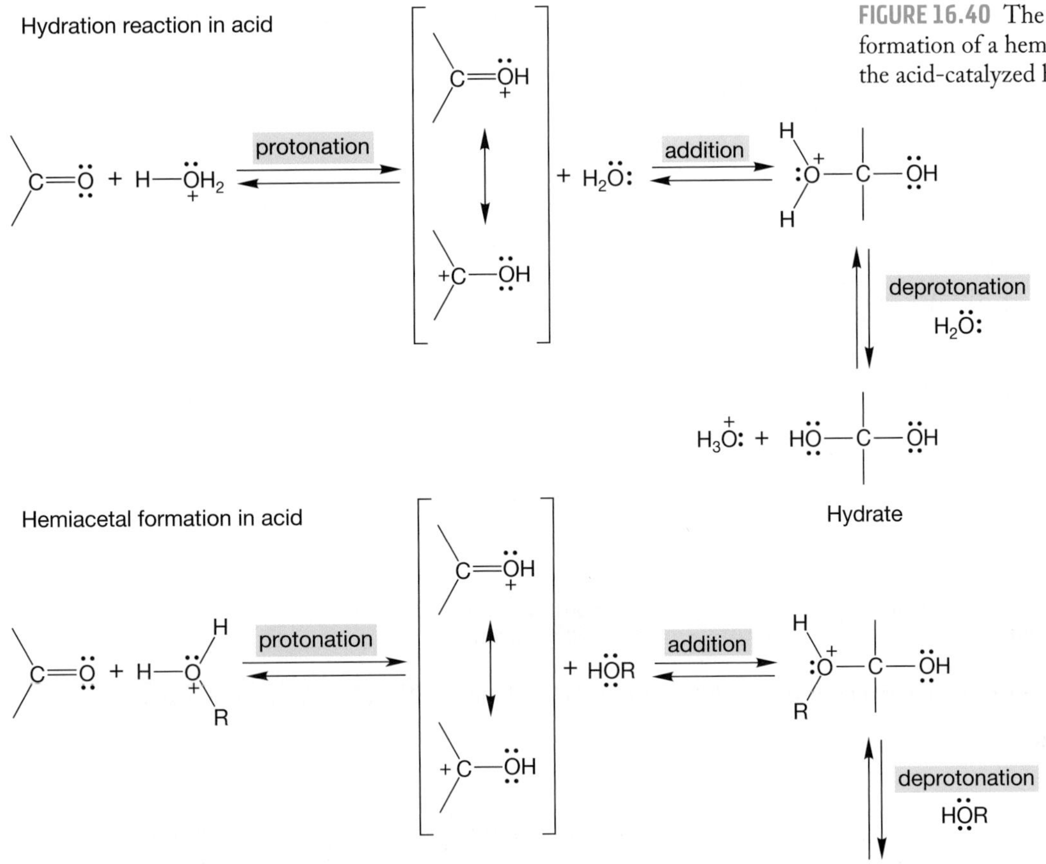

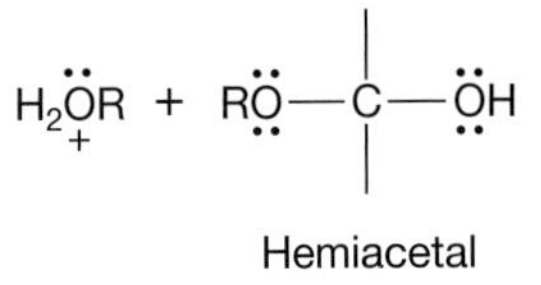

Hydration reaction in acid

FIGURE 16.40 The acid-catalyzed formation of a hemiacetal parallels the acid-catalyzed hydration reaction.

Hydrate

Hemiacetal formation in acid

Hemiacetal

But suppose it is not the OR oxygen that is protonated, but instead the OH oxygen? Surely, if one of these reactions is possible, so is the other. If we keep strictly parallel to the reverse process, we can now lose water to give a resonance-stabilized species (Fig. 16.41), which is quite analogous to the

FIGURE 16.41 Protonation of the OH in the hemiacetal, followed by water loss, leads to a resonance-stabilized species that is analogous to the protonated carbonyl.

A protonated carbonyl

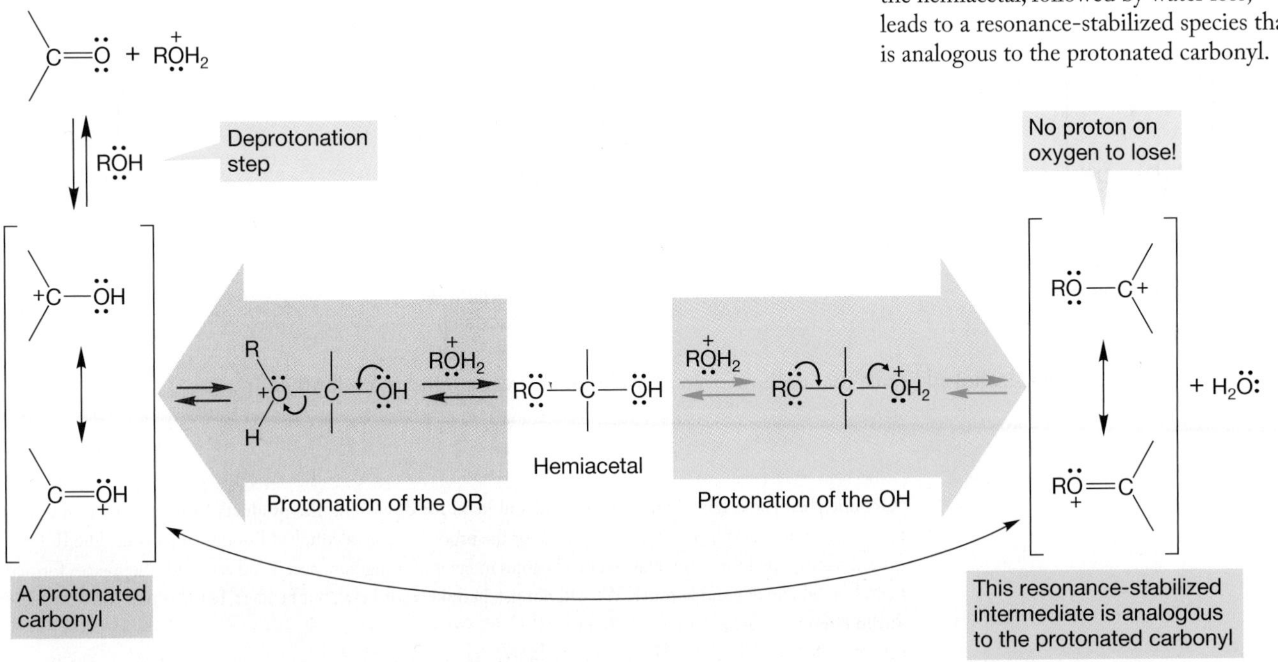

This resonance-stabilized intermediate is analogous to the protonated carbonyl

resonance-stabilized protonated carbonyl compound of Figure 16.40. But now deprotonation is not possible—*there is no proton on oxygen to lose.* Another reaction occurs instead, the addition of a second molecule of alcohol. Deprotonation gives a full **acetal**[2] (Fig. 16.42). An acetal contains an sp^3 hybridized carbon that has two OR groups attached.

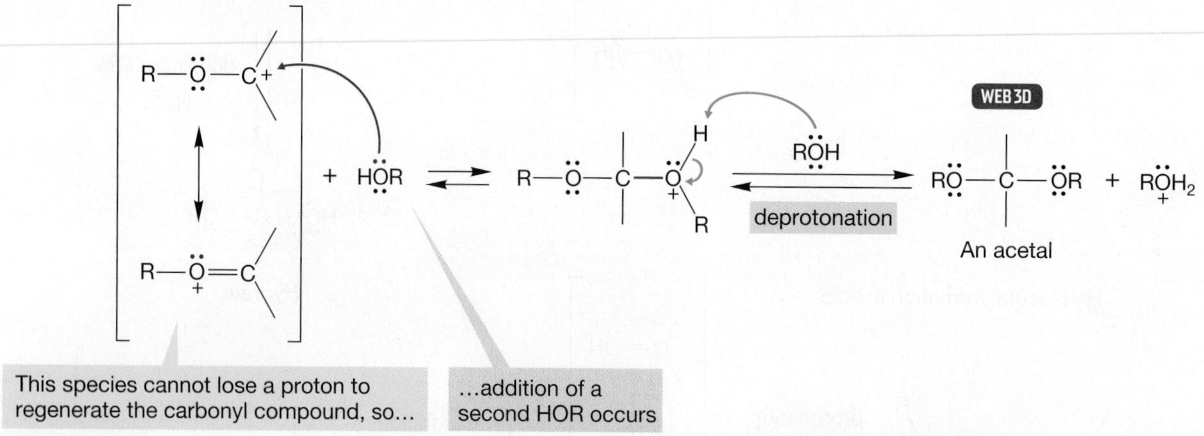

This species cannot lose a proton to regenerate the carbonyl compound, so...

...addition of a second HOR occurs

WEB 3D

An acetal

FIGURE 16.42 An acetal is formed by addition of a second molecule of alcohol to the resonance-stabilized intermediate.

Acetal formation

Notice that the second addition of alcohol in acetal formation is just like the first; the only difference is the presence of an H or R group in the species to be attacked (Fig. 16.43). Notice also that acetal formation is completely reversible.

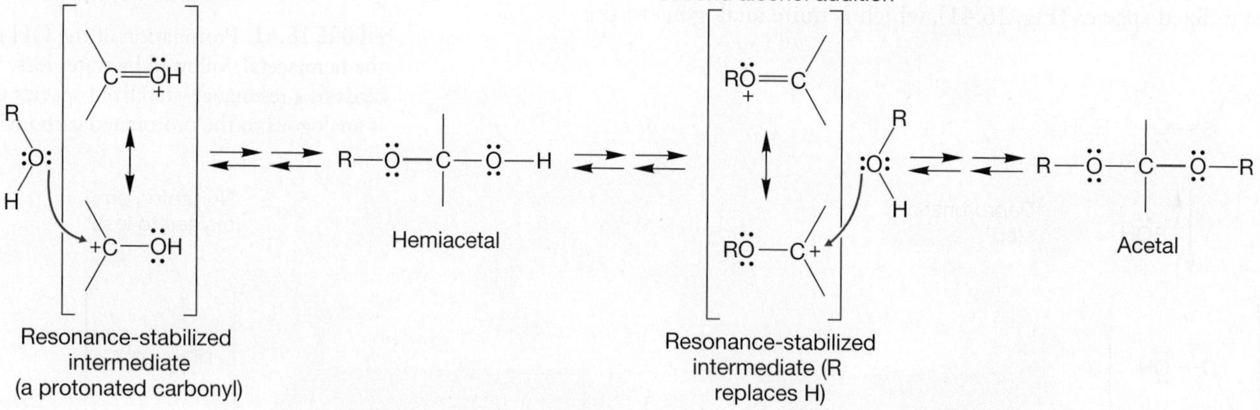

First alcohol addition

Resonance-stabilized intermediate (a protonated carbonyl)

Hemiacetal

Second alcohol addition

Resonance-stabilized intermediate (R replaces H)

Acetal

FIGURE 16.43 The two alcohol additions that occur in acetal formation are very similar to each other.

---

[2]Organic chemists originally used hemiketal and ketal for describing the products from alcohol addition to ketones; and hemiacetal and acetal were used for the products from alcohol addition to aldehydes. The IUPAC system has dropped the hemiketal and ketal names in favor of using hemiacetal and acetal for the species formed from both aldehydes and ketones. We will not use hemiketal or ketal, but be alert, because one still sees these sensible terms occasionally.

PROBLEM 16.13 Write the mechanism for the formation of the acetal from treatment of cyclopentanone in ethylene glycol (1,2-ethanediol) with an acid catalyst.

PROBLEM 16.14 Write the mechanism for the reverse reaction, the regeneration of the starting ketone on treatment of the acetal with aqueous acid.

WORKED PROBLEM 16.15 Explain why hemiacetals, but not full acetals, are formed under basic conditions.

ANSWER If we work through the mechanism we can see why. The first step is addition of an alkoxide to the carbonyl group to form the alkoxide intermediate. Protonation then gives the hemiacetal and regenerates the base.

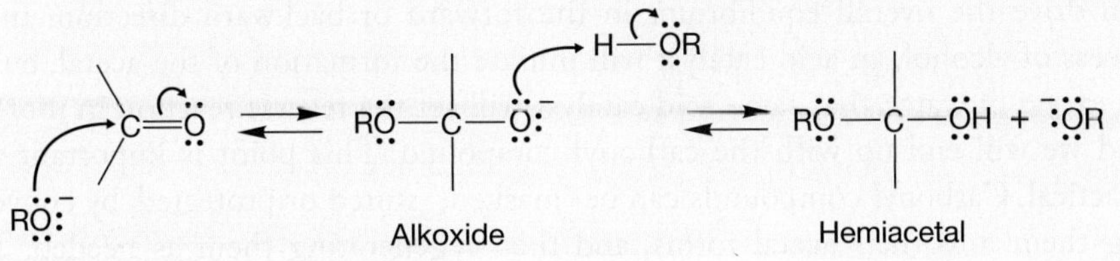

Now what? There is no acid available to protonate the OH oxygen. In order to get an acetal, RO⁻ must displace hydroxide! This reaction is most unfavorable. The more favorable reaction between RO⁻ and the hemiacetal is removal of the proton to re-form the intermediate alkoxide. In base, the reaction cannot go beyond hemiacetal formation.

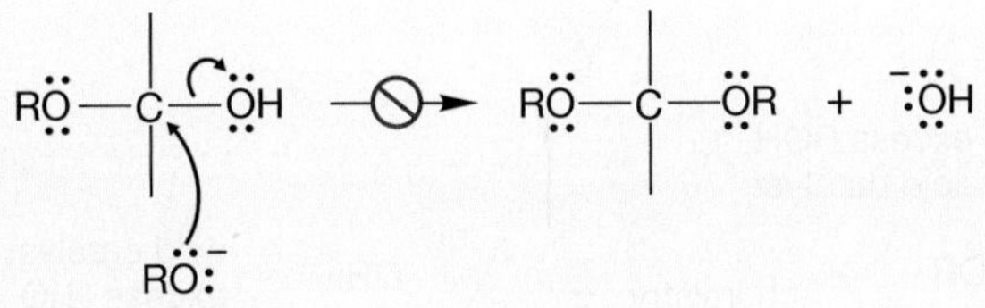

PROBLEM 16.16 When acetone is treated with $^{18}OH_2/^{18}\overset{+}{O}H_3$, acetone is found to be labeled with ^{18}O. Explain.

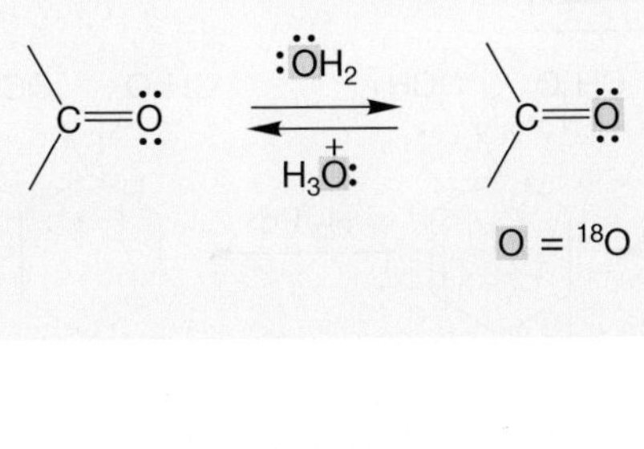

$O = {}^{18}O$

Summary

By now, you may be getting the impression that almost any nucleophile will add to carbonyl compounds. If so, you are quite right. So far, we have seen water, cyanide, bisulfite, and alcohols act as nucleophiles toward the Lewis acidic carbon of the carbon–oxygen double bond. Addition can take place under neutral, acidic, or basic conditions. All of the addition reactions are similar, with only the details being different. Those details are important though, as well as highly effective at camouflaging the essential similarity of the reactions. We'll now see some examples of acetal chemistry.

16.10 Protecting Groups in Synthesis

16.10a Acetals as Protecting Groups for Aldehydes and Ketones

By all means work out the mechanism of acetal formation by setting yourself the problem of writing the mechanism for the *reverse* reaction (go back and do Problem 16.14!), the formation of a ketone or aldehyde from the reaction of an acetal in acidic water. Acetal formation is an equilibrium process, and conditions can drive the overall equilibrium in the forward or backward direction. In an excess of alcohol, an acid catalyst will initiate the formation of the acetal, but in an excess of water, that same acid catalyst will set the reverse reaction in motion, and we will end up with the carbonyl compound. This point is important and practical. Carbonyl compounds can be "masked," stored or protected, by converting them into their acetal forms, and then regenerating them as needed. The acetal constitutes your first **protecting group**. For example, it might be desirable to convert group X, in the carbonyl-containing molecule shown in Figure 16.44,

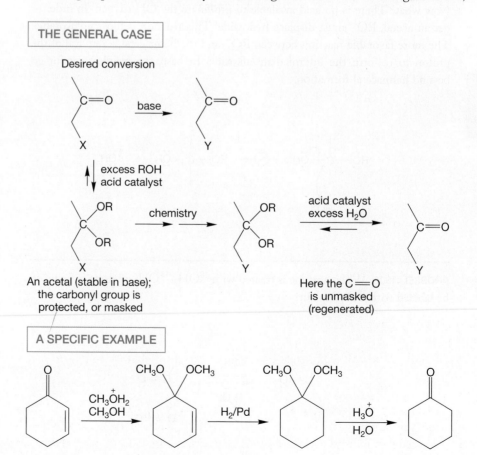

FIGURE 16.44 A protecting group can be used to mask a sensitive functional group while reactions are carried out on other parts of the molecule. When these reactions are completed, the original functionality can be regenerated.

into group Y through a base-catalyzed reaction. But we know that carbonyl compounds are sensitive to base, and the starting material may be destroyed in the attempted reaction. The solution is first to convert the carbonyl group into an acetal (which is unreactive in base), then carry out the conversion of X to Y, and finally regenerate the carbonyl group.

16.10b Protecting Groups for Alcohols It is often desirable to protect a hydroxyl group so that further reactions in base can take place without complication. One widely used method is to convert the hydroxyl group into a **tetrahydropyranyl (THP) ether** through reaction with dihydropyran (Fig. 16.45). The mechanism involves protonation of dihydropyran to give the resonance-stabilized cation, which is then captured by the alcohol. The result is an acetal, a molecule stable to base, but not to acid (p. 783). Regeneration of the alcohol can be accomplished through acid-catalyzed hydrolysis.

THE GENERAL CASE

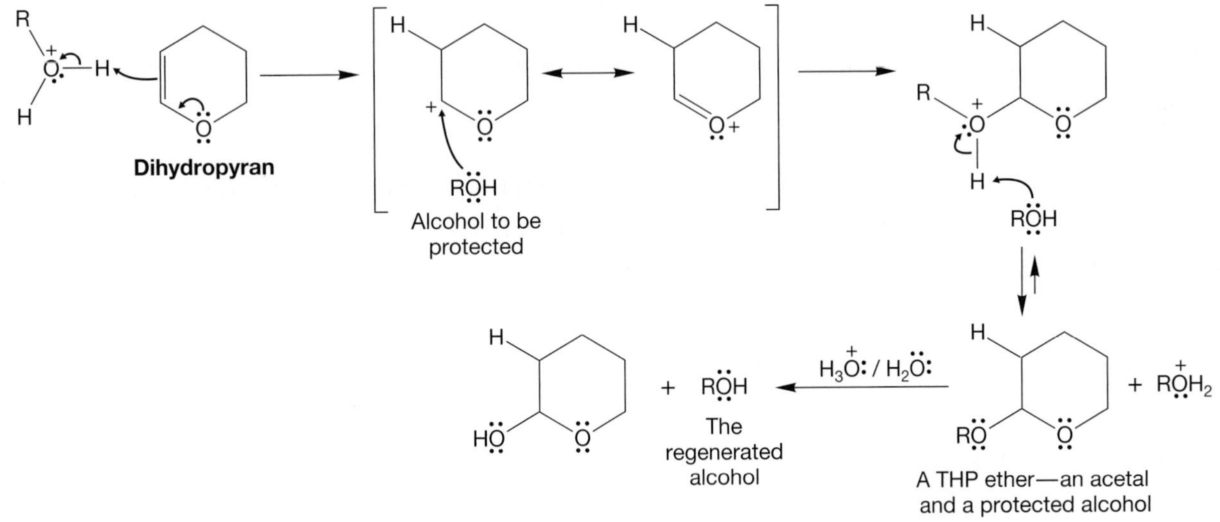

A SPECIFIC EXAMPLE

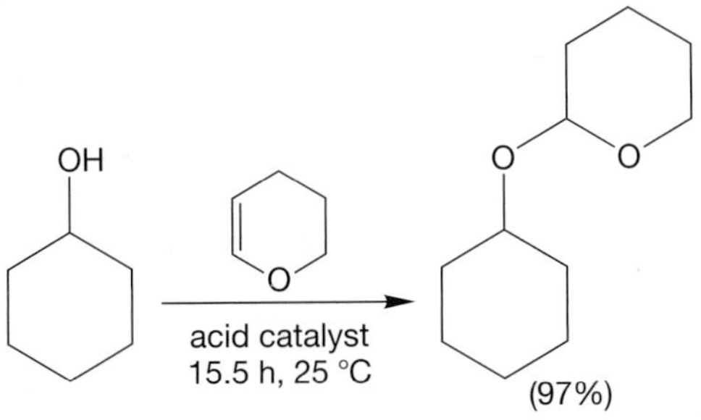

FIGURE 16.45 The use of THP ethers as protecting groups for alcohols.

Another type of protecting group for alcohols is a **silyl ether**. A silyl ether has the general formula of R′—O—SiR_3. The reaction of an alcohol with a trialkylsilyl halide leads to a trialkylsilyl ether in near-quantitative yield

(Fig. 16.46). The reaction depends for its success on the enormous strength of the silicon–oxygen bond (~110 kcal/mol). The silyl group can be removed through reaction with fluoride ion or hydrogen fluoride. The silicon–fluorine bond is even stronger (~140 kcal/mol) than the silicon–oxygen bond, and the formation of the trialkylsilyl fluoride provides the thermodynamic driving force for the reaction.

So, alcohols can be protected by converting them into THP or silyl ethers, and regenerating them when needed.

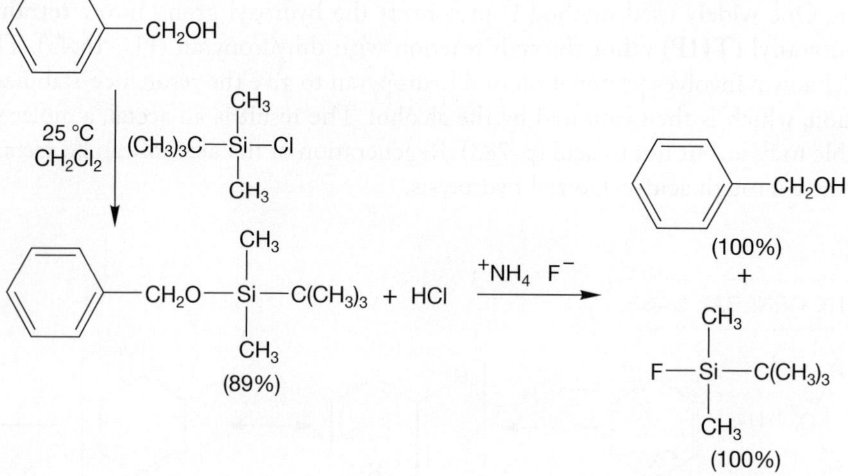

FIGURE 16.46 The use of trialkylsilyl ethers as protecting groups for alcohols.

16.11 Addition Reactions of Nitrogen Bases: Imine and Enamine Formation

The nitrogen atom of an amine (RNH_2) is a nucleophile, and like the related oxygen nucleophiles, it can add to carbonyl compounds. Indeed, the acid-catalyzed reaction between a carbonyl compound and an amine to give a **carbinolamine** is quite analogous to the reactions of water and alcohols we have just seen (Fig. 16.47).

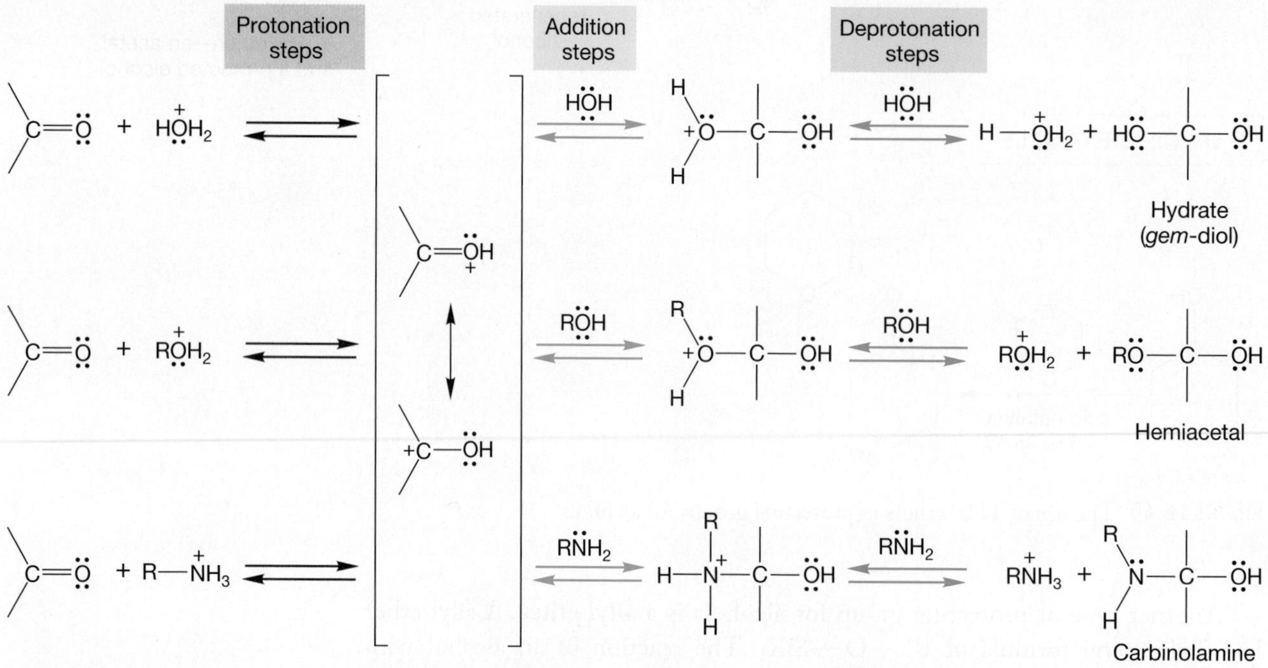

FIGURE 16.47 Analogous reactions of carbonyl compounds give hydrates (called *gem*-diols), hemiacetals, and carbinolamines. The usual sequence of protonation, addition, and deprotonation steps is followed.

Although the amine is more basic than the carbonyl oxygen, there still must be sufficient protonation of carbonyl and availability of the free amine under the acid-catalyzed conditions.

PROBLEM 16.17 Write the base-catalyzed versions of the reactions in Figure 16.47.

Like the hemiacetal, the carbinolamine is not a final product in acid. Further steps in the two reactions are all similar. An OH is protonated, and then lost as water (note that ⁻OH is *not* the leaving group) to give a resonance-stabilized cation (Fig. 16.48).

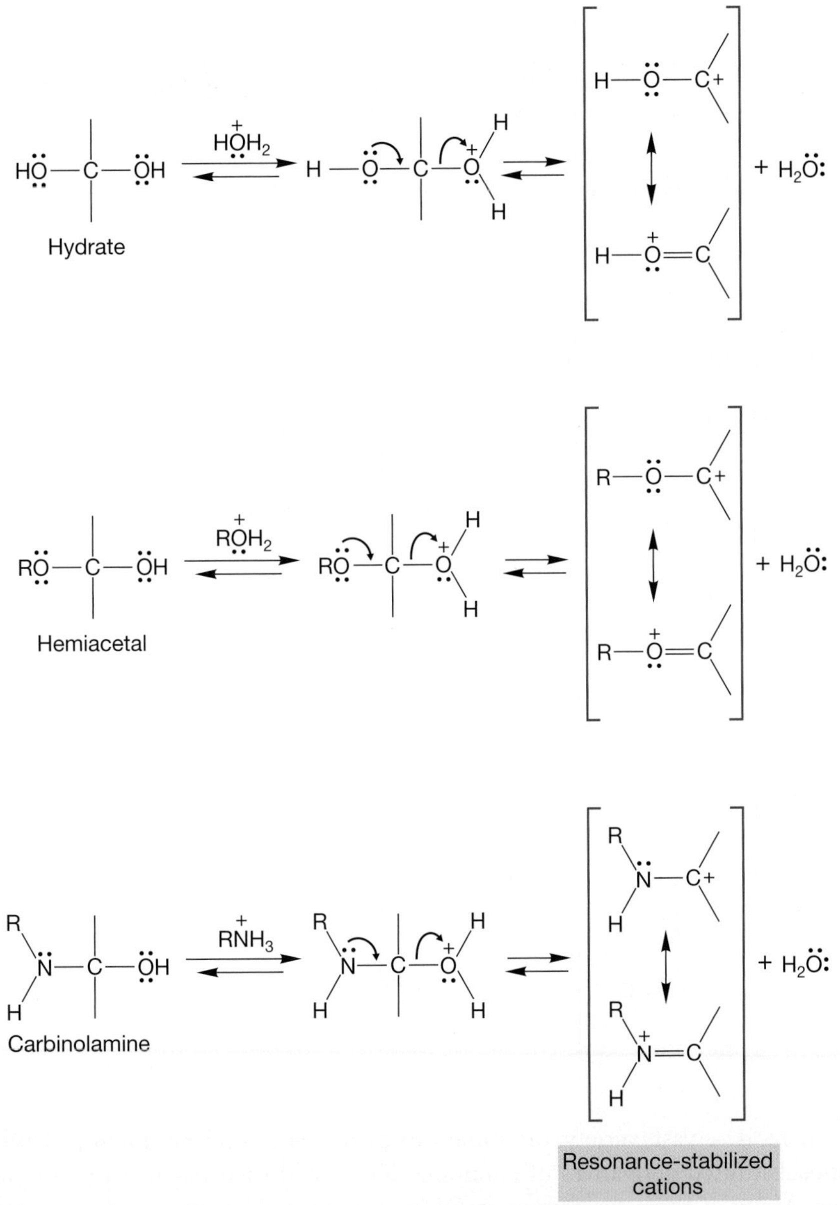

FIGURE 16.48 Continuation of the reactions from Figure 16.47. Loss of water leads to a resonance-stabilized intermediate in all three reactions.

In the acid-catalyzed reaction with alcohol shown in Figures 16.47 and 16.48, the resonance-stabilized cation reacts by adding a nucleophile, ROH, because R^+ cannot usually be lost. Instead, the cation goes on to acetal formation by a second addition of alcohol. For the water and amine additions, however, there remains an available hydrogen on the resonance-stabilized cation that can be lost as a proton, as shown in Figure 16.49. In the hydration reaction, this loss simply completes the re-formation of the carbonyl group of the starting material, but in the amine example it leads to the formation of a carbon–nitrogen double bond, called either a **Schiff base** or an **imine** (Fig. 16.49).

FIGURE 16.49 Continuation of reactions from Figure 16.48. In the reaction with water, loss of a proton leads to the recovered carbonyl compound. In the alcohol reaction, there is no proton to be lost, and a second molecule of alcohol adds to give an acetal. In the reaction with amine, loss of a proton leads to an imine.

Imine formation

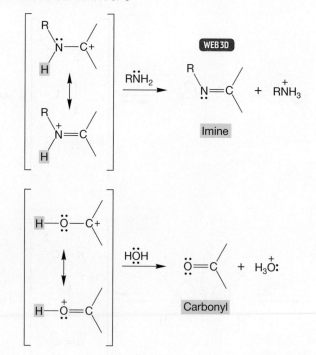

There is a vast variety of imines formed from primary amines (RNH_2, monosubstituted derivatives of ammonia, NH_3), and each has its own common name. Many of them were once useful in analytical chemistry because they are generally solids, and therefore could be characterized by their melting points.

For example, formation of a 2,4-dinitrophenylhydrazone or a semicarbazone (Fig. 16.50) from an unknown was not only diagnostic for the presence of a carbonyl group, but it yielded a specific derivative that could be compared to a similar sample made from a compound of known structure. Figure 16.50 gives a few of these substituted imines. Be sure to note that all of these reactions are the same— the only difference is the identity of the R group. If you know one, you know them all—but you have to *know* that one.

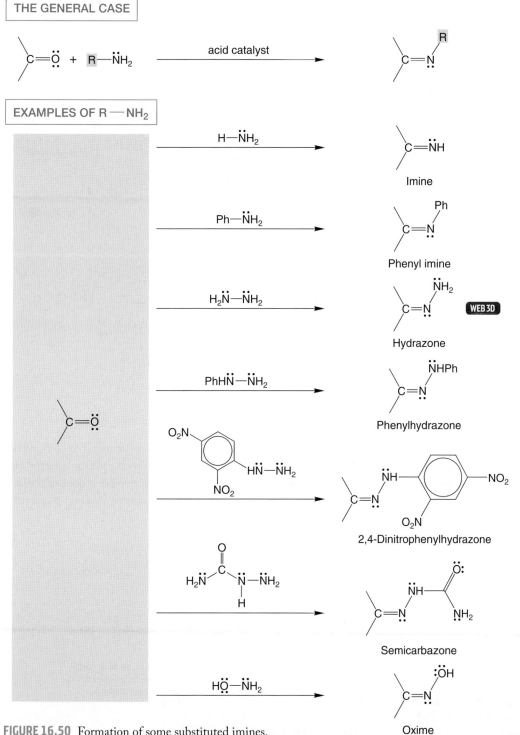

FIGURE 16.50 Formation of some substituted imines.

PROBLEM 16.18 Write a mechanism for the general reaction of Figure 16.50, the acid-catalyzed reaction of a carbonyl compound with RNH_2.

Note also that the success of the imine-forming reaction relies on the presence of at least two hydrogens in the starting amine. One is lost as a proton in the first step of the reaction, the formation of the carbinolamine. The second proton is lost in a deprotonation step as the final imine is produced (Fig. 16.51).

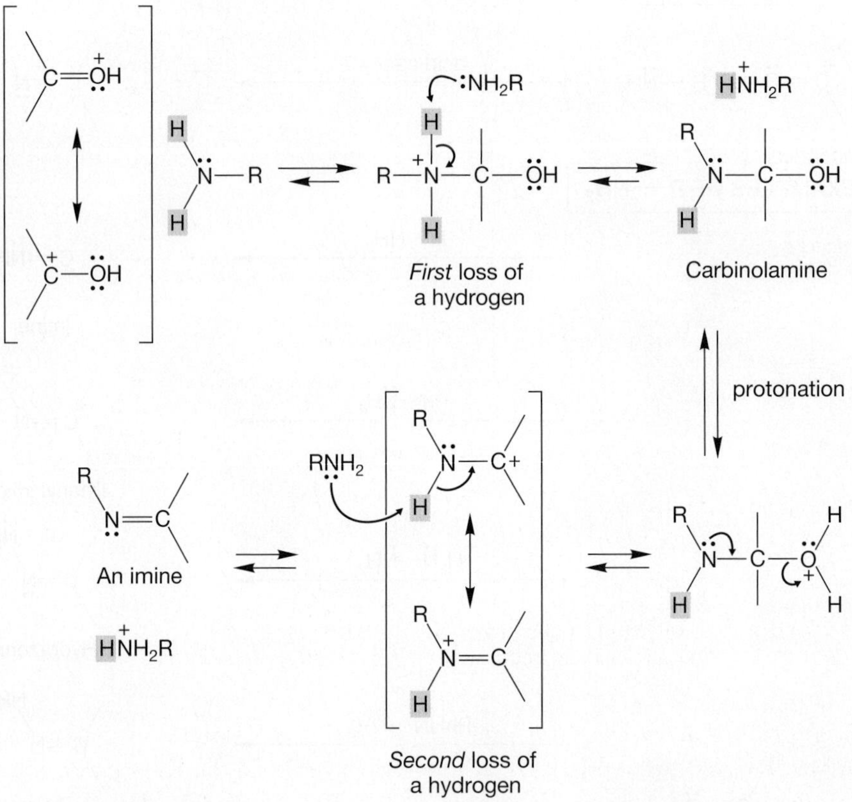

FIGURE 16.51 In order to form an imine, there must be *two* hydrogens (red) on the starting amine. That is, the amine must be either primary (RNH_2) or ammonia (NH_3).

First loss of a hydrogen

Carbinolamine

protonation

An imine

Second loss of a hydrogen

What happens if one or both of these hydrogens is missing? Tertiary amines (R_3N) carry no hydrogens and undergo no visible reaction with carbonyl groups. But this masks reality. In fact, addition to the carbon–oxygen double bond does take place, but the unstable intermediate has no available reaction except reversal to starting material (Fig. 16.52).

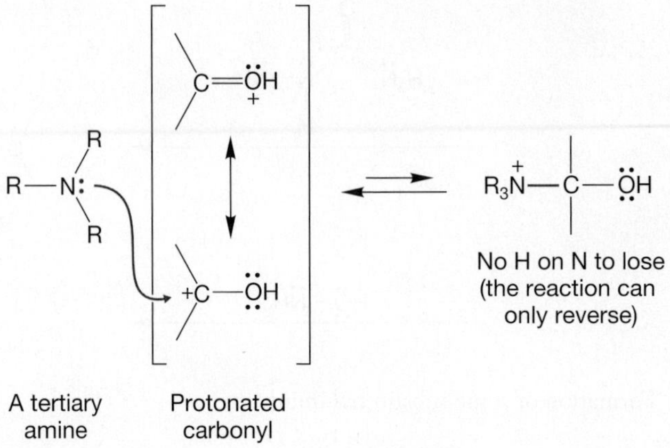

FIGURE 16.52 Tertiary amines (R_3N) can add to carbonyl groups, but the only available further reaction is reversion to starting materials.

A tertiary amine

Protonated carbonyl

No H on N to lose (the reaction can only reverse)

Secondary amines (R_2NH) have one hydrogen, and so carbinolamine formation is possible (Fig. 16.53). This reaction can be followed by protonation and water loss to give a resonance-stabilized cation, called an **iminium ion**. An iminium ion has the general structure $R_2\overset{+}{N}=CR_2$. Deprotonation of the iminium ion on nitrogen is *not* possible as there remains no second proton on nitrogen to be lost.

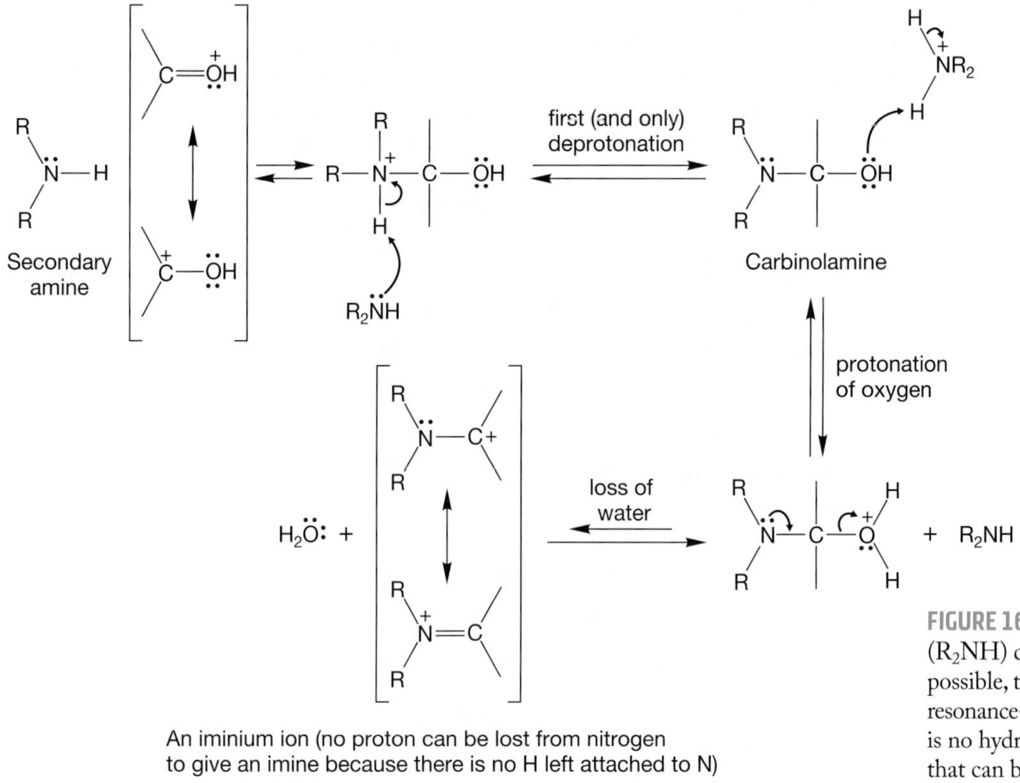

An iminium ion (no proton can be lost from nitrogen to give an imine because there is no H left attached to N)

FIGURE 16.53 For secondary amines (R_2NH) carbinolamine formation is possible, then water is lost to give a resonance-stabilized iminium ion. There is no hydrogen remaining on nitrogen that can be lost to give an imine.

Think now about what you know of the possibilities for proton loss by carbocations. The E1 reaction (p. 298) is a perfect example. In the E1 reaction, a leaving group departs to give a carbocation, which reacts further by deprotonating at all possible adjacent positions. If an adjacent carbon-attached hydrogen is available, the iminium ion can do this same reaction. Figure 16.54 makes the analogy. This reaction gives a molecule called an **enamine**, which is the final product of the reaction of secondary amines and simple aldehydes and ketones. Enamine is the name used to describe a compound that has an alkene attached to an amine.

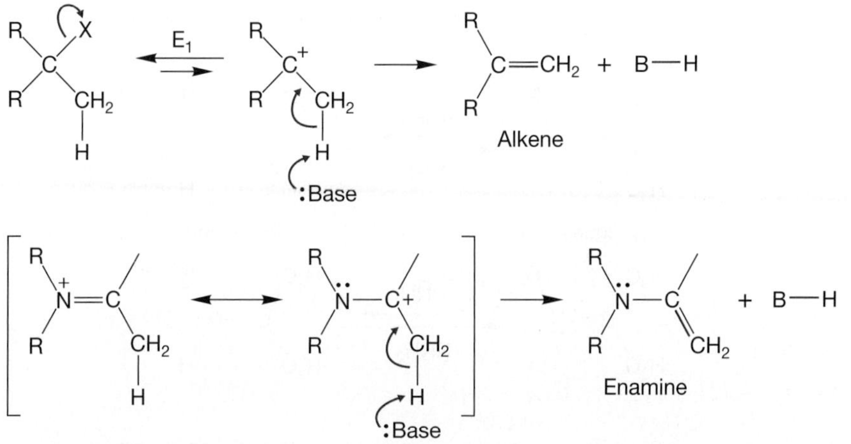

FIGURE 16.54 Deprotonation of an iminium ion to give an enamine is exactly like deprotonation of a carbocation to give an alkene.

Notice that enamines are the nitrogen counterpart of enols (p. 448). Now comes an important question: Why doesn't the iminium ion formed from primary amines or ammonia (Fig. 16.49) also lose a proton from carbon to give an enamine? Why is imine formation preferred when there is a choice (Fig. 16.55)? The answer is extraordinarily simple: Just as the ketone tautomer is more stable than the enol, the imine is more stable than the enamine and so is preferred when possible. It is only in cases where imine formation is not possible, as when secondary amines are used, that enamines are formed. We hope you have noticed that the imine- and enamine-forming reactions are reversible. The reactions work best under slightly acidic conditions with heat applied to drive off water. If water is added to an imine or enamine, the reverse reaction occurs to yield the carbonyl compound and an amine.

FIGURE 16.55 Why don't iminium ions formed from primary amines give enamines rather than imines?

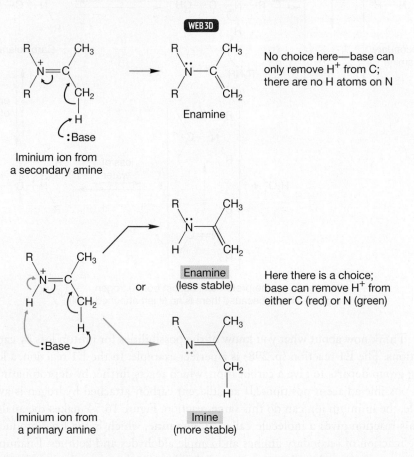

PROBLEM 16.19 Many enamines and imines are in equilibrium. Write a mechanism for the forward and the reverse pathways in the acid- and base-catalyzed reactions below.

Summary

All manner of oxygen and nitrogen bases will add reversibly to carbonyl compounds. The ultimate fate of the reaction and the ultimate structure of the product depend on concentration, and, for nitrogen bases, on the number of hydrogens available on nitrogen.

16.12 Organometallic Reagents

Not all additions to carbonyl compounds are reversible. Some of the irreversible reactions are of great importance in synthetic chemistry, as they involve the creation of carbon–carbon bonds. In order to discuss these, we must first talk a bit about the formation of organometallic reagents.

We first met organometallic reagents in Section 6.3 (p. 227). It would be a very good idea to go back and re-read those few pages carefully. You will recall that we can make Grignard reagents (RMgX) and organolithium reagents (RLi) from the corresponding alkyl halides. We observed that these reagents will react with water or D_2O to produce R—H or R—D, respectively (p. 229). We have also noted that organometallic compounds add to epoxides (p. 429).

PROBLEM 16.20 Given inorganic reagents of your choice (including D_2O), devise syntheses of the following molecules from the indicated starting materials:

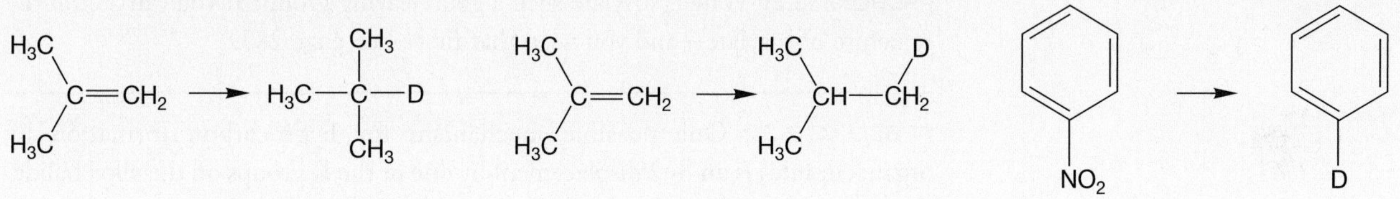

Neither Grignard reagents nor organolithium reagents are good nucleophiles toward organic halides (Fig. 16.56). If they were, the syntheses we use to generate the organometallic reagents in the presence of the starting organic halide would be quite ineffective, because hydrocarbons would be the major products. Small amounts of coupling are sometimes observed, especially with highly reactive halides such as allyl bromides.

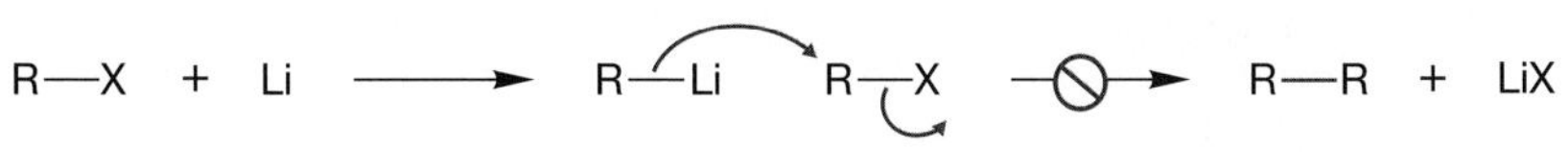

FIGURE 16.56 Organometallic reagents are strong bases but not strong nucleophiles.

There are organometallic reagents that can be used to displace halides to good effect. The most common examples are the organocuprates formed from organolithium compounds and copper iodide. The organocuprates are quite versatile reagents, and undergo a variety of synthetically useful reactions. We will meet them again a number of times, but for now, note that they are able to displace chloride, bromide, and iodide ions, as well as other good leaving groups, from primary alkyl halides and often from secondary alkyl halides (Fig. 16.57). This capability gives you another quite effective synthesis of hydrocarbons.

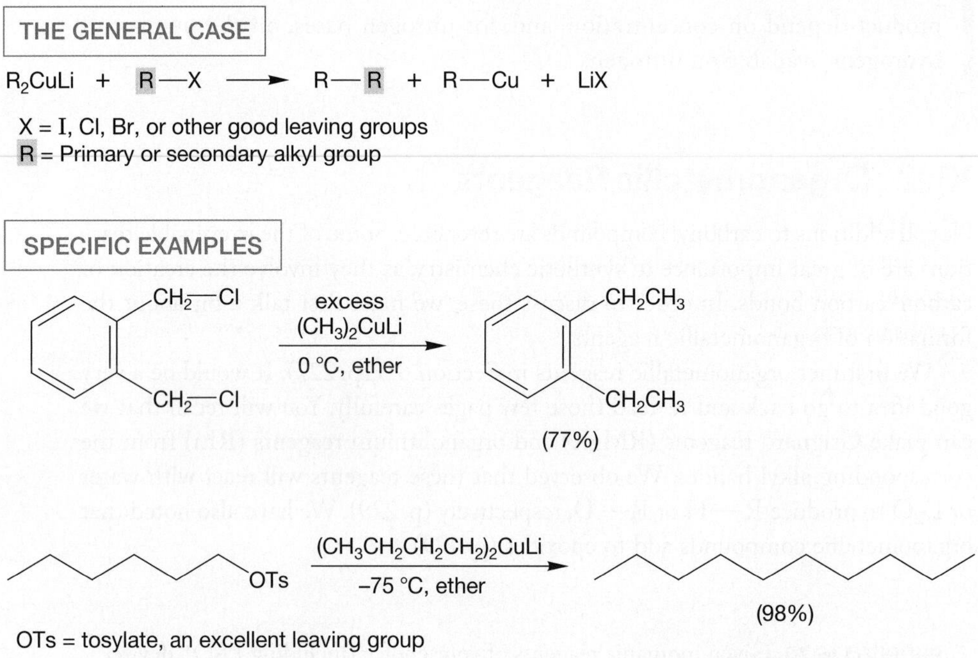

FIGURE 16.57 Lithium organocuprates will react with primary and secondary alkyl halides to give hydrocarbons.

THE GENERAL CASE

$$R_2CuLi \ + \ \boxed{R}-X \ \longrightarrow \ R-\boxed{R} \ + \ R-Cu \ + \ LiX$$

X = I, Cl, Br, or other good leaving groups
R = Primary or secondary alkyl group

SPECIFIC EXAMPLES

(77%)

OTs = tosylate, an excellent leaving group

(98%)

PROBLEM 16.21 Why is tosylate such a good leaving group? If you can't recall the structure of tosylate—and you need that first—see page 283.

PROBLEM 16.22 One possible mechanism for hydrocarbon formation by organocuprates is an S$_N$2 displacement by one of the R groups on the alkyl halide. Can you devise an experiment to test this possibility?

Many metal hydrides, such as lithium aluminum hydride (LiAlH$_4$), undergo reactions that are similar to those of organometallic reagents. Many hydride reagents react violently with moisture to liberate hydrogen gas. Lithium aluminum hydride and some other metal hydrides are strong enough nucleophiles to displace halides from most organic halides. Lithium triethylborohydride, LiHB(Et)$_3$, is an especially effective H:$^-$ source that can be used as a displacing agent (Fig. 16.58).

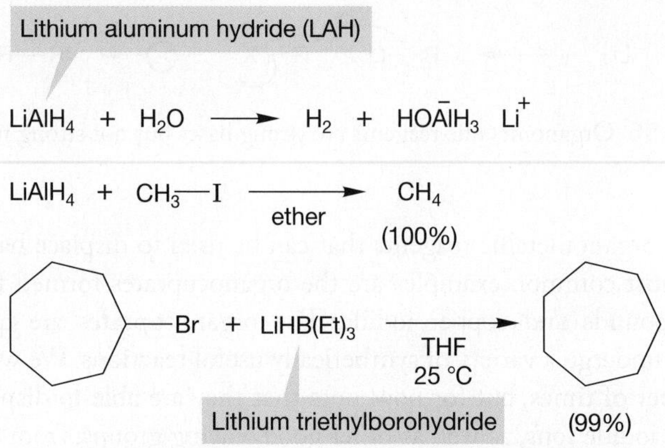

Lithium aluminum hydride (LAH)

$$LiAlH_4 \ + \ H_2O \ \longrightarrow \ H_2 \ + \ HO\bar{A}lH_3 \ Li^+$$

$$LiAlH_4 \ + \ CH_3-I \ \xrightarrow{\text{ether}} \ CH_4$$
(100%)

FIGURE 16.58 Metal hydrides react with water to generate hydrogen, and are often strong enough nucleophiles to displace halides and generate hydrocarbons.

Lithium triethylborohydride

(99%)

Like other Lewis bases, Grignard reagents, organolithium reagents, and metal hydrides are also able to undergo addition to carbonyl-containing compounds. These additions lead directly to the next subject.

16.13 Irreversible Addition Reactions: A General Synthesis of Alcohols

Now that we know something of the structures of these organometallic reagents, it is time to look at their reactivity with ketones and aldehydes. This subject is important, because these reactions form carbon–carbon bonds, a central goal in synthetic chemistry. Like oxygen- and nitrogen-centered nucleophiles, organometallic reagents (and metal hydrides) add to carbonyl compounds. Although there is no free R:$^-$ (or H:$^-$) in these reactions, the organometallic reagents are reactive enough to deliver an R:$^-$ group, acting as a nucleophile, to the Lewis acidic carbon of the carbonyl group (Fig. 16.59). The initial product is a metal alkoxide, which is protonated

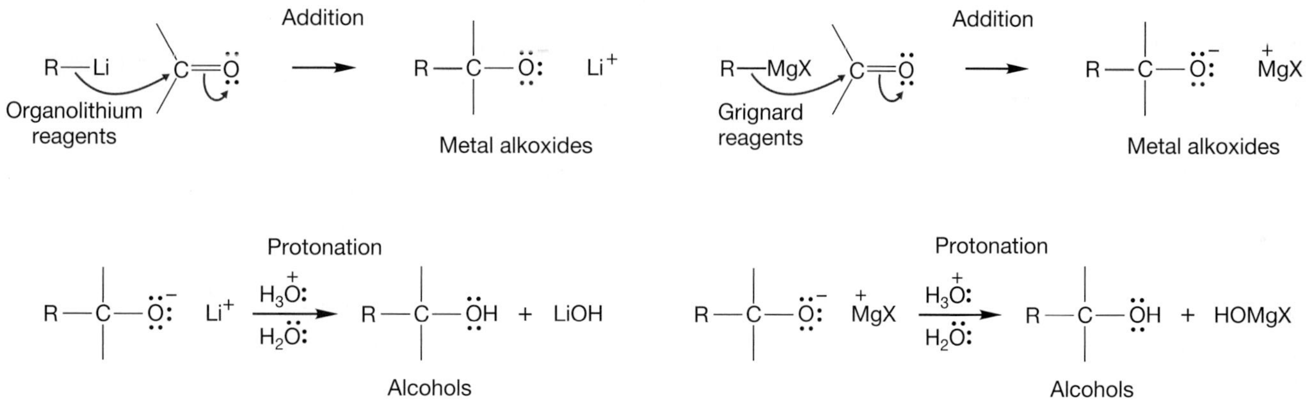

FIGURE 16.59 Organometallic reagents are strong enough nucleophiles to add to carbonyl compounds. When water is added in a second step, alcohols are produced.

when the solution is neutralized by addition of aqueous acid in a second step. Water cannot be present at the start of the reaction because it would destroy the organometallic reagent (Fig. 16.60).

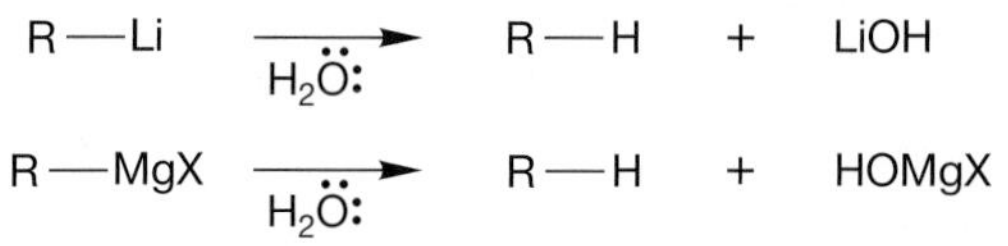

FIGURE 16.60 If water is present at the beginning of the reaction, the organometallic reagent is destroyed through hydrocarbon formation.

It is easy to make the mistake of writing this reaction sequence as shown in Figure 16.61a, which means, "Treat the ketone with a mixture of alkyllithium reagent and water." This procedure is impossible, because the water would destroy the alkyllithium immediately (Fig. 16.60). The correct way to write the sequence is shown in

Figure 16.61b and means, "First allow the ketone to react with the alkyllithium reagent and then, after some specified time, add water."

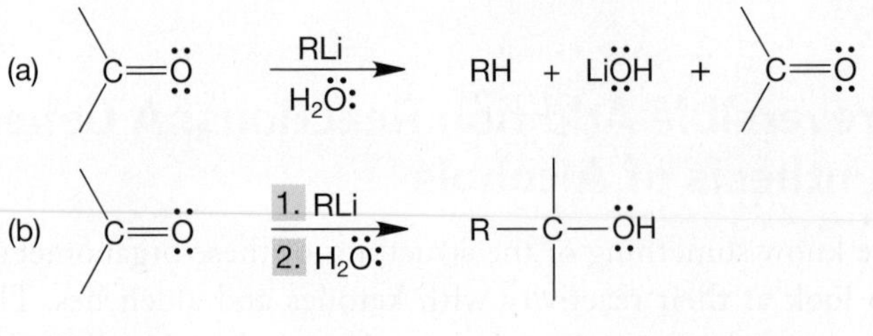

FIGURE 16.61 Reaction (a) means, Add a mixture of R—Li and water to the carbonyl compound. Under such conditions the organometallic reagent will be destroyed. Reaction (b) means, First add the organometallic reagent to the carbonyl compound and then, in a second step, add water.

This addition of organometallic to carbonyl compounds gives us a new and most versatile alcohol synthesis. Primary alcohols can be made through the reaction of an organometallic reagent with formaldehyde (Fig. 16.62). Figure 16.62 also shows the shorthand form of the reaction, which emphasizes the synthetic utility.

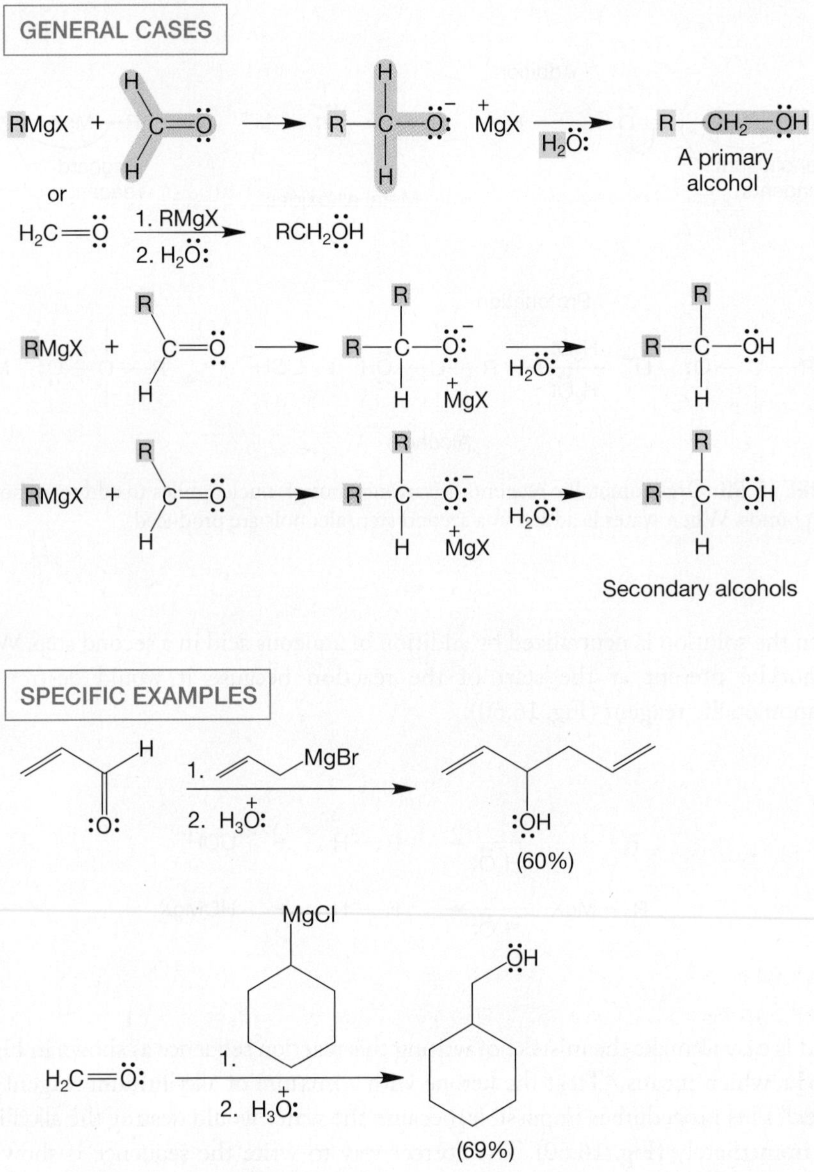

FIGURE 16.62 Reaction of organometallic reagents with formaldehyde, followed by hydrolysis, leads to primary alcohols. The reaction of other aldehydes with organometallic reagents, followed by hydrolysis, gives secondary alcohols.

Similarly, reactions of organometallic reagents (RLi or RMgX) with aldehydes other than formaldehyde give secondary alcohols. The R of the organometallic reagent can either be the same as that of the aldehyde or different (Fig. 16.62).

Ketones react with Grignard reagents or organolithium reagents to give tertiary alcohols. Three different substitution patterns are possible depending on the number of different R groups in the reaction (Fig. 16.63).

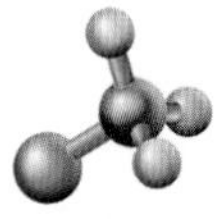

Grignard reaction

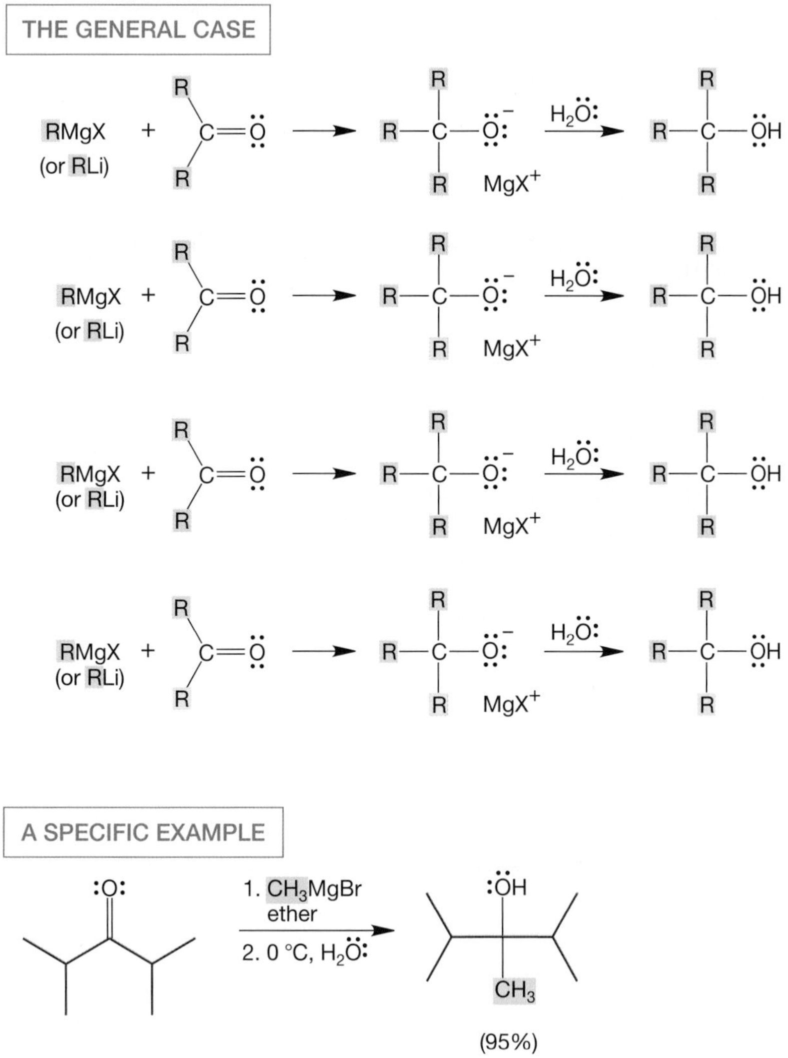

THE GENERAL CASE

A SPECIFIC EXAMPLE

(95%)

FIGURE 16.63 The reaction of ketones with organometallic reagents, followed by hydrolysis, yields tertiary alcohols.

Metal hydrides (LiAlH$_4$, NaBH$_4$, and many others) react in a similar fashion to deliver hydride (H:$^-$) as the nucleophile (Fig. 16.64). Alcohols are formed when water is added in a second, quenching step. Notice that this reaction is formally a reduction. Aldehydes are reduced to primary alcohols and ketones are reduced to secondary alcohols. There is no way to make a tertiary alcohol through this kind of reduction (Fig. 16.64).

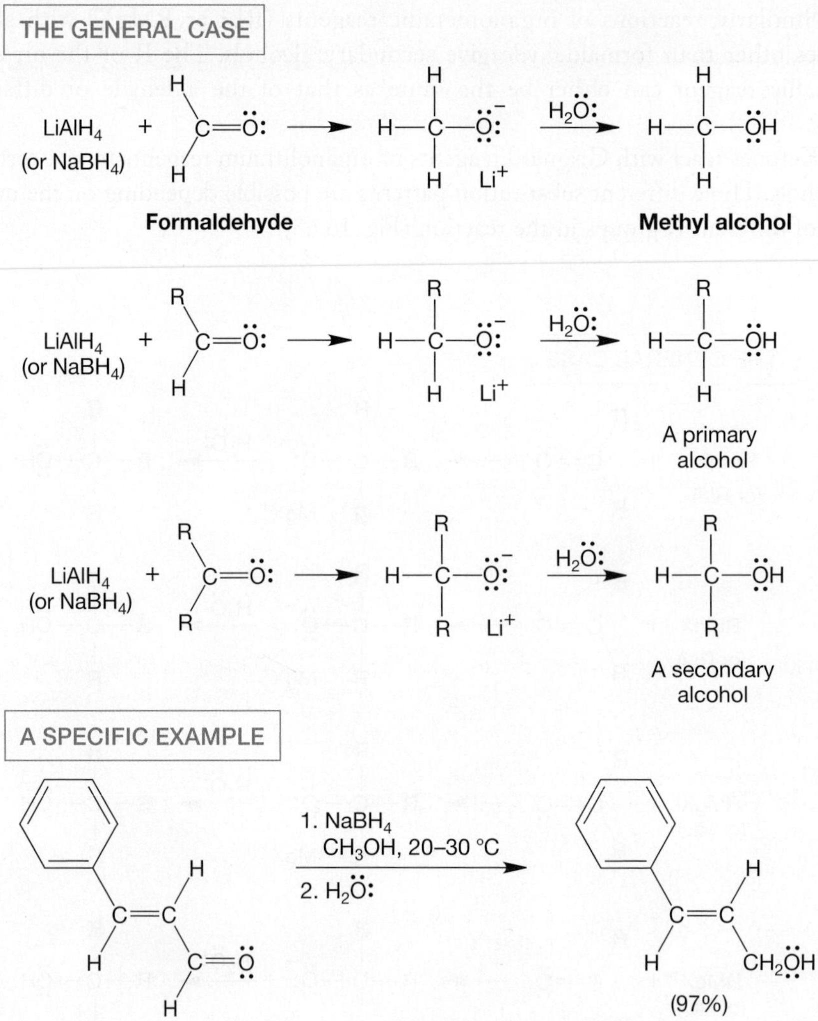

FIGURE 16.64 Metal hydride addition, followed by hydrolysis, gives primary alcohols from aldehydes and secondary alcohols from ketones.

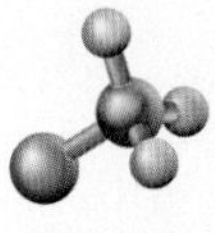

THE GENERAL CASE

$$LiAlH_4 \text{ (or } NaBH_4) + \underset{\text{H}}{\overset{\text{H}}{C}}=\ddot{O}: \longrightarrow H-\underset{\underset{\text{H}}{|}}{\overset{\overset{\text{H}}{|}}{C}}-\ddot{O}:^{-} Li^{+} \xrightarrow{H_2\ddot{O}:} H-\underset{\underset{\text{H}}{|}}{\overset{\overset{\text{H}}{|}}{C}}-\ddot{O}H$$

Formaldehyde   **Methyl alcohol**

$$LiAlH_4 \text{ (or } NaBH_4) + \underset{\text{H}}{\overset{\text{R}}{C}}=\ddot{O}: \longrightarrow H-\underset{\underset{\text{H}}{|}}{\overset{\overset{\text{R}}{|}}{C}}-\ddot{O}:^{-} Li^{+} \xrightarrow{H_2\ddot{O}:} H-\underset{\underset{\text{H}}{|}}{\overset{\overset{\text{R}}{|}}{C}}-\ddot{O}H$$

A primary alcohol

$$LiAlH_4 \text{ (or } NaBH_4) + \underset{\text{R}}{\overset{\text{R}}{C}}=\ddot{O}: \longrightarrow H-\underset{\underset{\text{R}}{|}}{\overset{\overset{\text{R}}{|}}{C}}-\ddot{O}:^{-} Li^{+} \xrightarrow{H_2\ddot{O}:} H-\underset{\underset{\text{R}}{|}}{\overset{\overset{\text{R}}{|}}{C}}-\ddot{O}H$$

A secondary alcohol

A SPECIFIC EXAMPLE

$$\text{(cinnamaldehyde)} \xrightarrow[\text{2. } H_2\ddot{O}:]{\begin{array}{c}\text{1. } NaBH_4 \\ CH_3OH, 20\text{–}30\,°C\end{array}} \text{(cinnamyl alcohol, } CH_2\ddot{O}H\text{)}$$

(97%)

Carbonyl reduction

We have seen reduction of carbonyl compounds before. In Chapter 14 (p. 645), we encountered the Clemmensen and Wolff–Kishner reductions in which a carbon–oxygen double bond is reduced all the way to a methylene group.

Summary

The irreversible addition of organometallic reagents and metal hydrides provides a simple synthesis of primary, secondary, and tertiary alcohols. But you have to be careful to choose your reagents properly, so that the desired substitution pattern is achieved.

16.14 Oxidation of Alcohols to Carbonyl Compounds

16.14a Oxidation of Monoalcohols

So far, we have generated two new and very general alcohol syntheses, the reaction of aldehydes and ketones with organometallic reagents, and the reduction of carbonyl compounds by metal hydrides. A complement to the reduction reaction is the oxidation of alcohols to aldehydes and ketones (Fig. 16.65). These oxidations extend the utility of the alcohol syntheses we have just learned, and complicate your life by increasing the complexity of the molecules you are able to make.

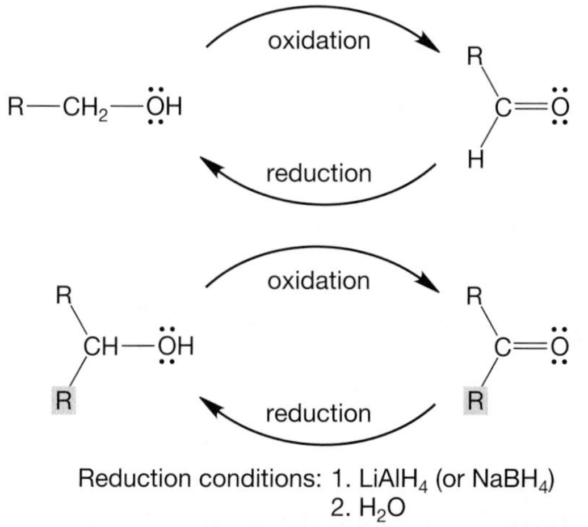

FIGURE 16.65 Alcohols can be oxidized to carbonyl compounds. This reaction is the reverse of what you have just learned—the reduction of carbonyl compounds to alcohols through reactions with metal hydride reagents.

Reduction conditions: 1. LiAlH$_4$ (or NaBH$_4$)
2. H$_2$O

Now your collection of syntheses can be extended by using the alcohols as starting materials in making carbonyl compounds using the oxidation reaction. Tertiary alcohols cannot be easily oxidized, but primary and secondary ones can be. As shown in Figure 16.66, secondary alcohols can only give ketones, but primary alcohols can give either aldehydes or carboxylic acids (RCOOH) depending on the oxidizing agent used.

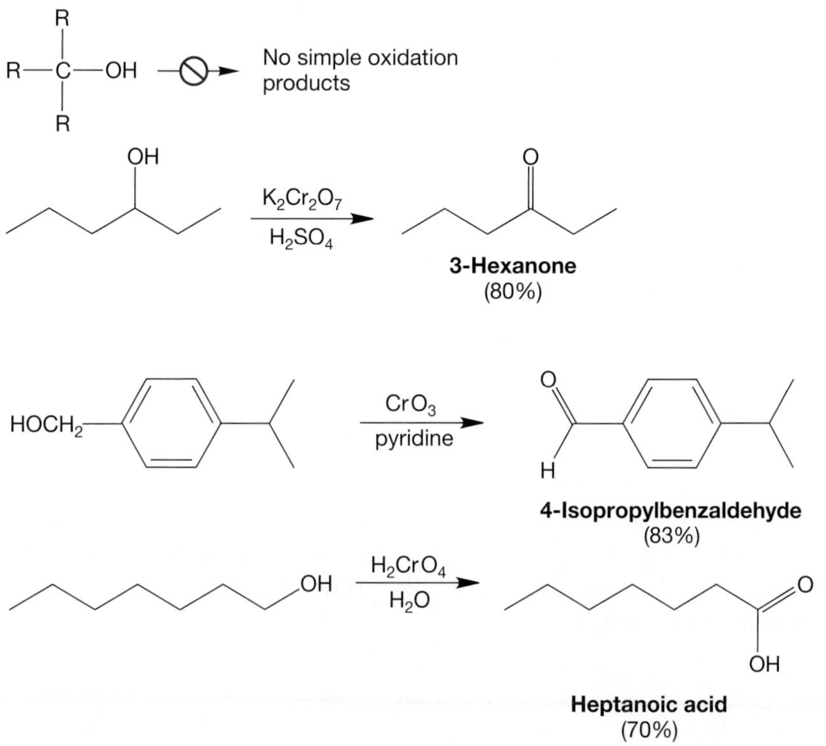

FIGURE 16.66 Tertiary alcohols cannot be easily oxidized, but secondary and primary alcohols can be. It is logical to infer that in order for the oxidation reaction to succeed there must be a carbon–hydrogen bond available on the alcohol carbon.

3-Hexanone
(80%)

4-Isopropylbenzaldehyde
(83%)

Heptanoic acid
(70%)

How do these oxidation reactions work? A typical oxidizing agent for primary alcohols is chromium trioxide (CrO$_3$) in pyridine. As long as the reagent is kept dry, good yields of aldehydes can often be obtained (Fig. 16.66). Although a metal–oxygen bond is more complicated than a carbon–oxygen bond, the double bonds in CrO$_3$ and

$R_2C=O$ react in similar ways. For example, an alcohol oxygen is a nucleophile and can attack the chromium–oxygen bond to give, ultimately, a chromate ester intermediate (Fig. 16.67). This reaction of a $Cr=O$ with an alcohol is analogous to the formation of a hemiacetal from the reaction of a $C=O$ with an alcohol.

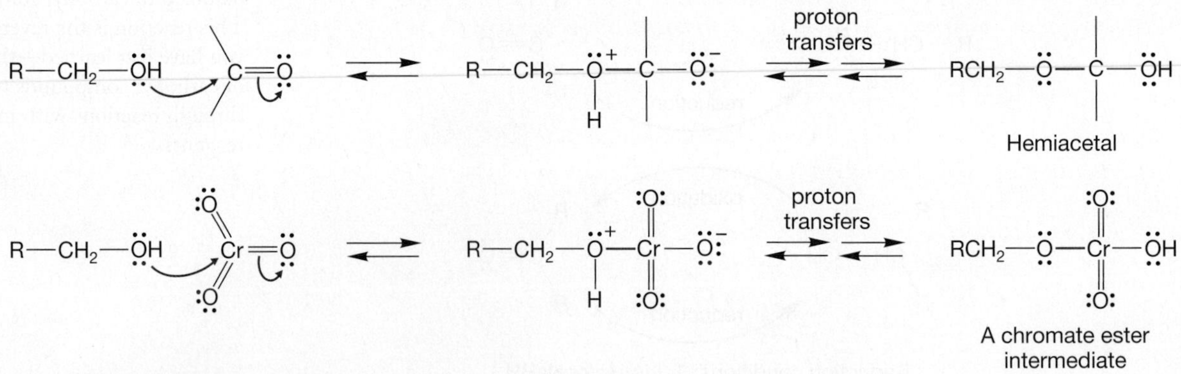

FIGURE 16.67 Nucleophilic alcohols can add to $Cr=O$ bonds just as they do to $C=O$ bonds. This reaction is the first step in the oxidation process, and in this case gives a chromate ester intermediate.

Alcohol oxidation

Once the chromate intermediate is formed, pyridine (or any other base) initiates an E2 reaction to give an aldehyde (an oxidized alcohol) and $HOCrO_2^-$ (a reduced chromium species) (Fig. 16.68).

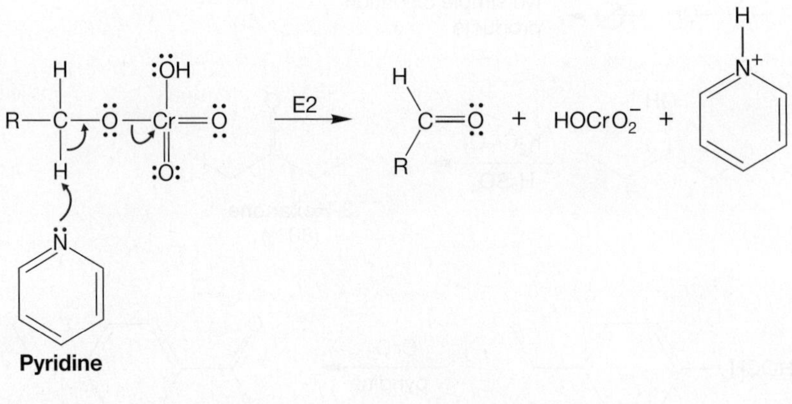

FIGURE 16.68 The second step in the oxidation reaction is a simple E2 reaction to generate the new $C=O$ double bond.

Pyridine

Chromium trioxide (CrO_3) will also oxidize secondary alcohols to ketones. It should now be clear why tertiary alcohols resist oxidation by this reagent. The initial addition reaction can proceed, but there is no hydrogen to be removed in the elimination process (Fig. 16.69).

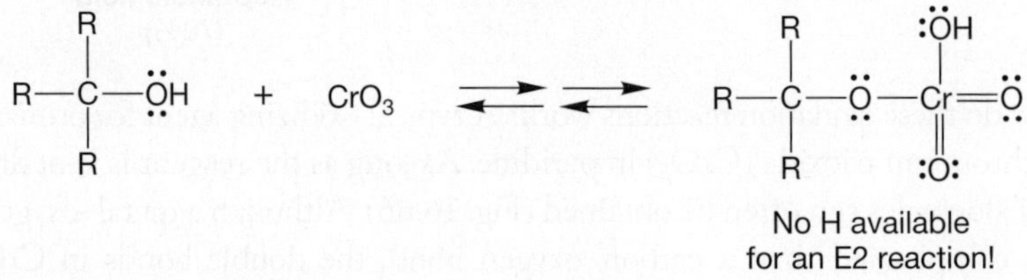

FIGURE 16.69 A tertiary alcohol can add to the $Cr=O$ double bond, but there is no hydrogen in the chromate intermediate available for an E2 reaction.

No H available for an E2 reaction!

Notice how this seemingly strange reaction, oxidation of alcohols with CrO_3, is really made up of two processes we have already encountered. The first step is addition of an oxygen base to a chromium–oxygen double bond. The second step is nothing more than a garden-variety elimination. The lesson here is that you should not be bothered by the identity of the atoms involved. When you see a reaction that looks strange, rely on what you already know. Make analogies, and more often than not you will come out all right. That is exactly what all chemists do when we see a new reaction. We think first about related reactions and try to find a way to generalize.

Another useful oxidizing reagent is sodium chromate (Na_2CrO_4) or sodium dichromate ($Na_2Cr_2O_7$) in aqueous strong acid (Fig. 16.70). Under these aqueous conditions, it is difficult to stop oxidation of primary alcohols at the aldehyde stage.

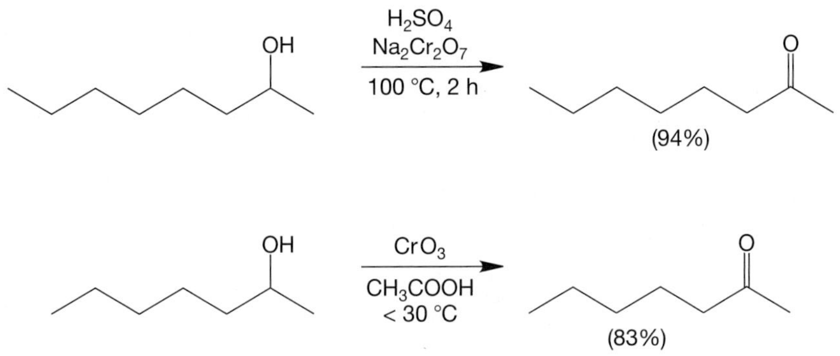

FIGURE 16.70 A typical reagent for oxidation of secondary alcohols to ketones is sodium dichromate in acid. Other oxidizing agents also work.

Overoxidation to the carboxylic acid almost always occurs. The difficulty is that aldehydes are hydrated in water (p. 773). These hydrates are alcohols and can be oxidized further (Fig. 16.71). If one wants to convert a primary alcohol into a carboxylic acid, $Na_2Cr_2O_7$ in H_2SO_4 and H_2O is the reagent of choice.

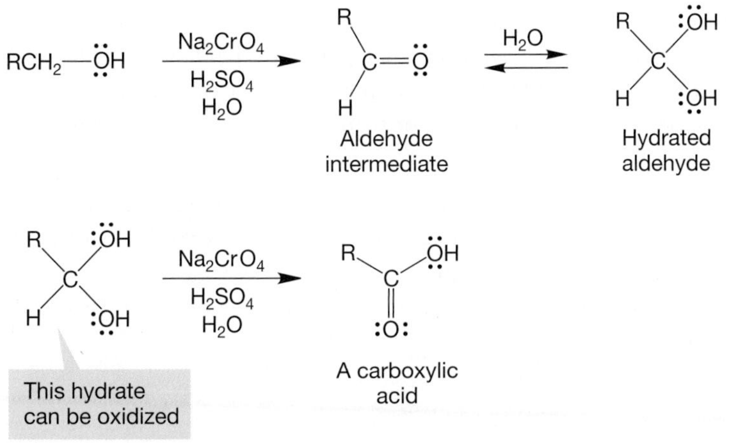

FIGURE 16.71 Oxidation of a primary alcohol leads to an aldehyde. Aldehydes are hydrated in the presence of water, and the hydrates are alcohols. Further oxidation of the hydrate leads to the carboxylic acid.

PROBLEM 16.23 Write a mechanism for the oxidation of a secondary alcohol to a ketone by chromic oxide.

In addition to the chromium reagents, there are many other oxidizing agents of varying strength and selectivity (Fig. 16.72). Manganese dioxide (MnO_2) is effective at making aldehydes from allylic or benzylic alcohols, nitric acid (HNO_3) oxidizes primary alcohols to carboxylic acids, and potassium permanganate ($KMnO_4$) is an excellent oxidizer of secondary alcohols to ketones and of primary alcohols to carboxylic acids, although it can be difficult to control. One of the mildest and most useful oxidizing reagents is dimethyl sulfoxide (DMSO), which can be used in combination with one of a variety of co-reactants to convert secondary alcohols into ketones and primary alcohols into aldehydes.

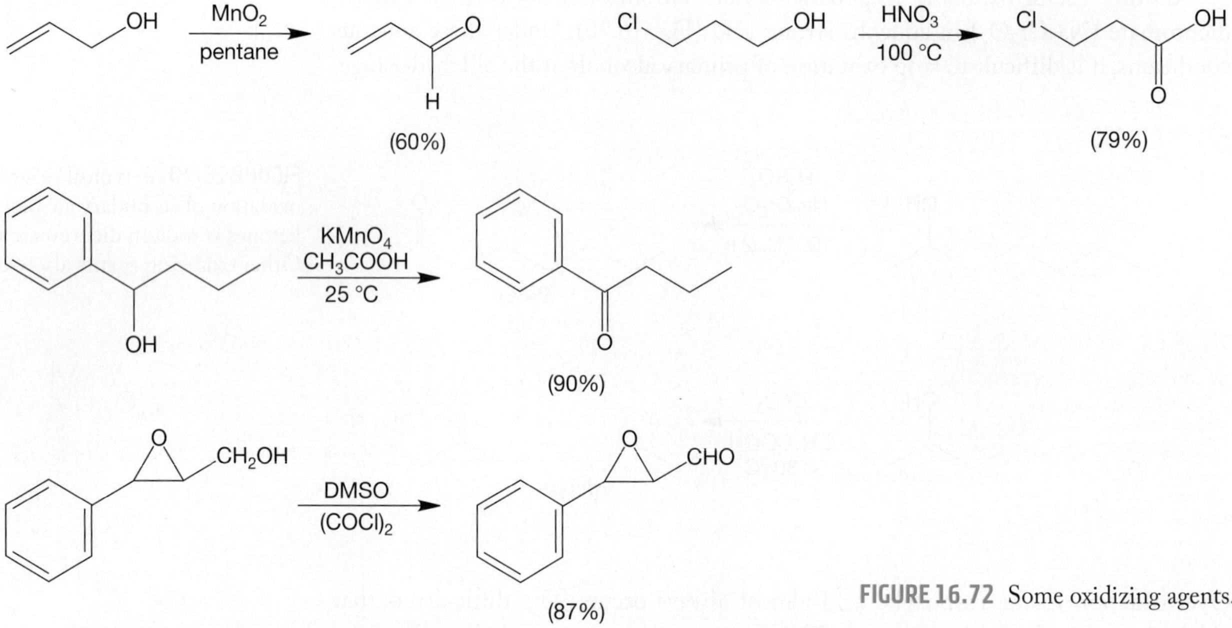

FIGURE 16.72 Some oxidizing agents.

Summary

Now we can reduce carbonyl compounds to the alcohol level with hydride reagents, and oxidize them back again using the various reagents of this section. Although one probably wouldn't want to spend one's life cycling a compound back and forth between the alcohol and carbonyl stage, these reactions are enormously useful in synthesis. Many alcohols are now the equivalent of aldehydes or ketones, because they can be oxidized, and carbonyl compounds are the equivalent of alcohols because they can be reduced with hydride or treated with an organometallic reagent.

PROBLEM 16.24 Starting from alcohols containing no more than three carbon atoms and inorganic reagents of your choice, devise syntheses of the following molecules:

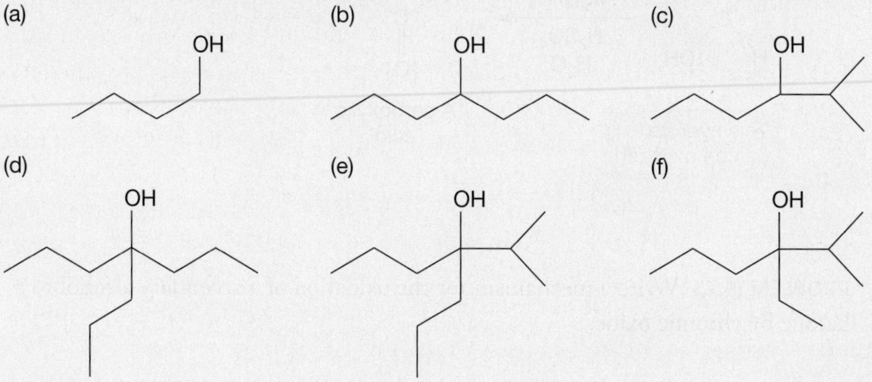

16.14b Oxidative Cleavage of Vicinal Diols   There is an oxidation reaction that works only for vicinal diols (1,2-diols). When a vicinal diol is oxidized by periodic acid (HIO_4) the diol is cleaved to a pair of carbonyl compounds (Fig. 16.73). This reaction, which has long been used as a test for 1,2-diols, involves the formation of a cyclic periodate intermediate.

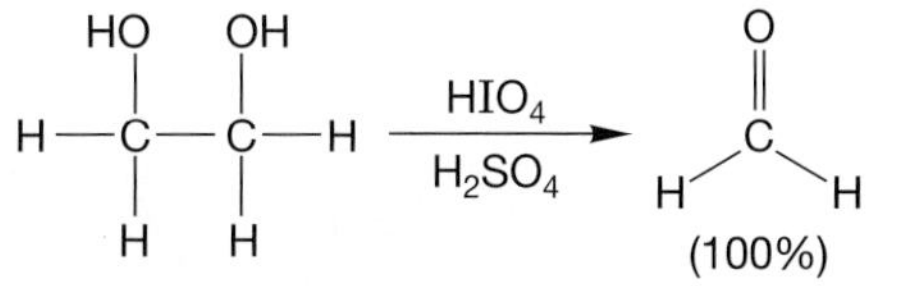

FIGURE 16.73 The cleavage of vicinal diols by periodic acid.

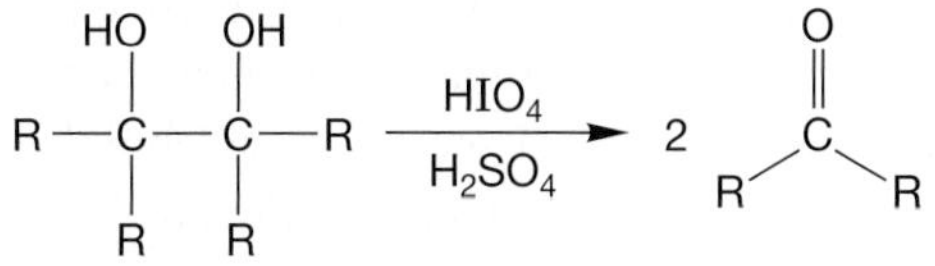

THE GENERAL CASE

A SPECIFIC EXAMPLE

This reaction looks much like the formation of a cyclic acetal from a 1,2-diol and a ketone. The intermediate breaks down to form the two carbonyl compounds. As you can see from the mechanism (Fig. 16.74), only vicinal diols can undergo this cleavage reaction.

Cleavage of vicinal diols

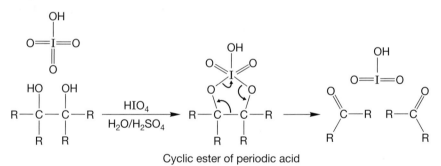

Cyclic ester of periodic acid

FIGURE 16.74 The vicinal diol cleavage by periodic acid involves a cyclic intermediate that breaks down to give a pair of carbonyl compounds.

> **PROBLEM 16.25** *trans*-1,2-Cyclohexanediol is much more reactive toward periodic acid than the cis isomer. Explain why.

16.15 Retrosynthetic Alcohol Synthesis

It's worth taking a moment to say a word again about strategy in doing synthesis problems, and to reinforce an old idea. It is almost never a good idea to try to see all the way back from the target molecule to the initial starting materials. Chemists have learned to be much less ambitious and to work back one step at a time. This process has been dignified with the title **retrosynthetic analysis**, which makes the idea seem far more complicated than it is. Let's assume you want to make molecule **A**. The idea behind this retrosynthesis strategy is simply that you can almost always see the synthetic step immediately leading to the product **A**. That is, compound **A** could come from **B**. Sometimes the problem is one of sorting among various possibilities of reactions that would lead to the product. Each possibility might come from a different compound, so perhaps **B**, **C**, or **D** are possible precursors to compound **A**. Now we just ask ourselves for a set of reactions leading to the new target **B**, or **C**, or **D**. The idea is that eventually this technique will lead back to starting materials that are readily available. Don't be too proud to work back step by step, even in simple problems.

As an example of using retrosynthetic analysis, consider synthesizing a general tertiary alcohol **A** (Fig. 16.75). It could be made in three ways. Your thoughts should be led to each of the three ketones **B**, **C**, or **D** shown in the figure. You know the retrosynthesis for the RMgX. It comes from an RX and Mg. To carry the analysis further, you need to decide which ketone to use. Whichever one you choose, you need to devise a synthesis for it. Fortunately, we have just covered this topic. You know that a ketone can come from a secondary alcohol. So, **B** could be obtained from the secondary alcohol **E**, which can come from an aldehyde **F** and RMgX. The beauty of it all is that aldehyde **F** comes from a primary alcohol **G**. So you will inevitably find your way back to simple alcohols and alkyl halides in this analysis. All that remains is to write out the forward reactions.

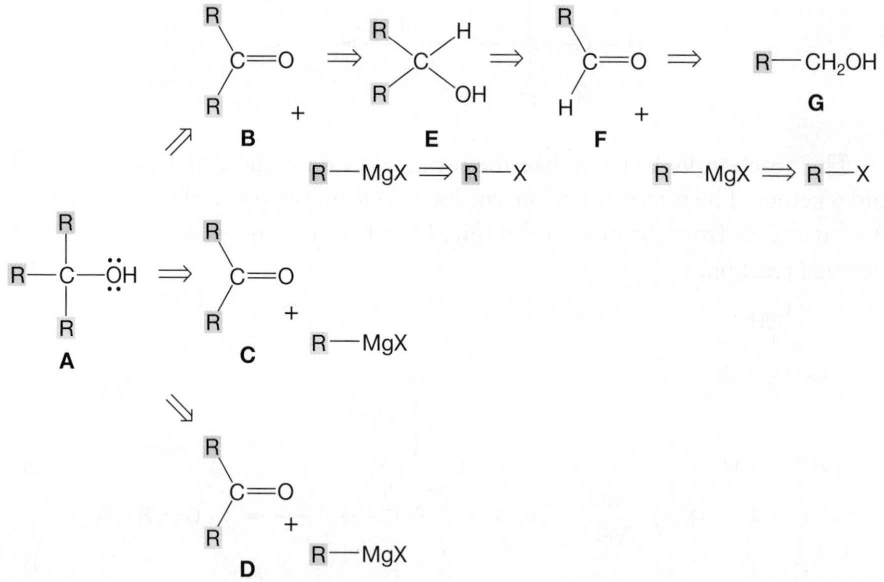

FIGURE 16.75 A retrosynthetic analysis of ways of making a tertiary alcohol.

CONVENTION ALERT Note in Figure 16.75 that retrosynthetic analysis has appropriated a special arrow for its own use. The double-lined arrow always points *from* the target molecule *to* the immediate precursor. So a double-lined arrow pointing from **A** to **B** means that **A** comes from **B**.

WORKED PROBLEM 16.26 Here is a specific example of using retrosynthetic analysis. Provide a retrosynthetic synthesis of 2-phenyl-2-butanol. Assume that you have benzene, ethanol, pyridine, and access to any inorganic reagents you might need.

ANSWER Here is the retrosynthetic analysis:

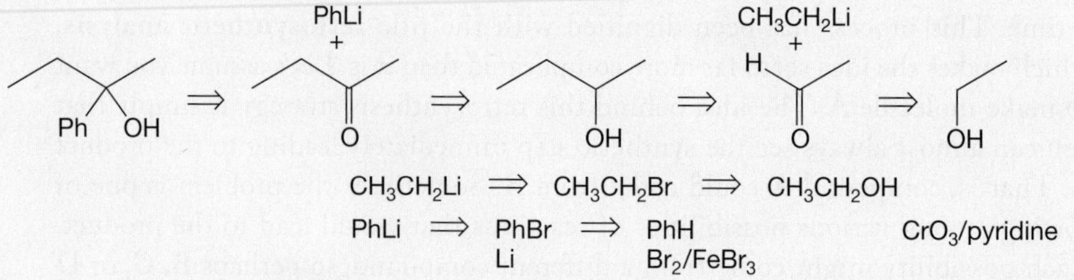

PROBLEM 16.27 Starting from inorganic reagents, tosyl chloride, formaldehyde, acetaldehyde, acetone, and butyl alcohol, devise syntheses of the following molecules:

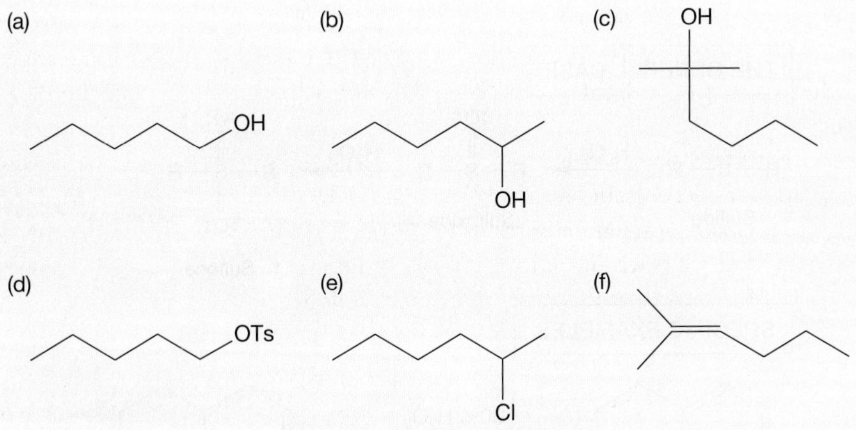

(a)

(b)

(c)

(d)

(e)

(f)

16.16 Oxidation of Thiols and Other Sulfur Compounds

Like alcohols, thiols (mercaptans) can be oxidized, but oxidation usually takes place not at carbon, as with alcohols, but on sulfur to give sulfonic acids (Fig. 16.76).

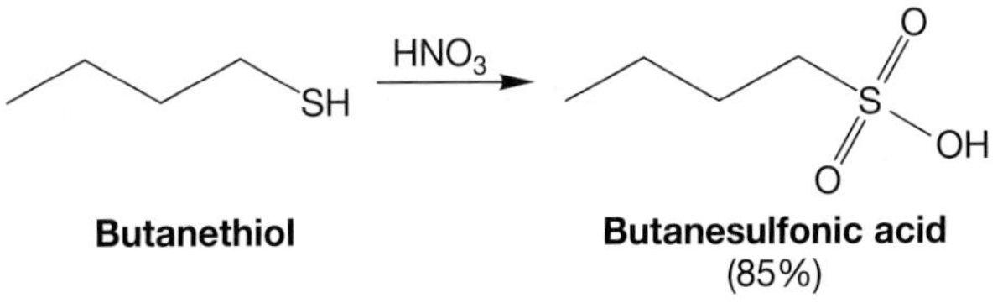

Butanethiol **Butanesulfonic acid**
(85%)

FIGURE 16.76 Oxidation of thiols (mercaptans) with nitric acid gives sulfonic acids.

Milder oxidation, using halogens (I_2 or Br_2) in base, is a general source of disulfides. This reaction probably involves a sequence of displacement reactions (Fig. 16.77).

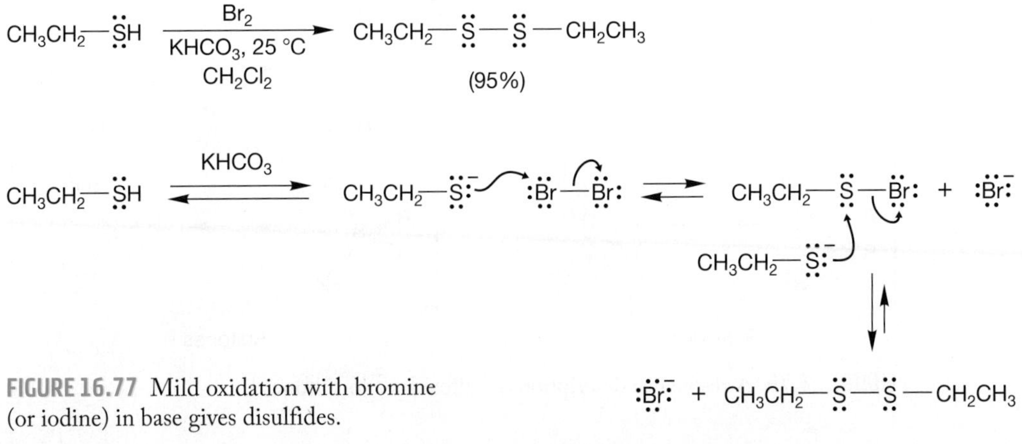

$$CH_3CH_2{-}\overset{..}{\underset{..}{S}}H \xrightarrow[\substack{KHCO_3,\ 25\ °C \\ CH_2Cl_2}]{Br_2} CH_3CH_2{-}\overset{..}{\underset{..}{S}}{-}\overset{..}{\underset{..}{S}}{-}CH_2CH_3$$

(95%)

$$CH_3CH_2{-}\overset{..}{\underset{..}{S}}H \underset{\xrightarrow{KHCO_3}}{\rightleftharpoons} CH_3CH_2{-}\overset{..}{\underset{..}{S}}\!:^{-} \quad :\overset{..}{\underset{..}{Br}}{-}\overset{..}{\underset{..}{Br}}: \rightleftharpoons CH_3CH_2{-}\overset{..}{\underset{..}{S}}{-}\overset{..}{\underset{..}{Br}}: \ + \ :\overset{..}{\underset{..}{Br}}\!:^{-}$$

$$CH_3CH_2{-}\overset{..}{\underset{..}{S}}\!:^{-}$$

FIGURE 16.77 Mild oxidation with bromine (or iodine) in base gives disulfides.

$$:\overset{..}{\underset{..}{Br}}\!:^{-} \ + \ CH_3CH_2{-}\overset{..}{\underset{..}{S}}{-}\overset{..}{\underset{..}{S}}{-}CH_2CH_3$$

Unlike ethers, sulfides (thioethers) can easily be oxidized to **sulfoxides**, usually with hydrogen peroxide, H_2O_2. Sulfoxides have the general formula R_2SO. Further or more vigorous oxidation gives a **sulfone**, a compound of the formula R_2SO_2 (Fig. 16.78).

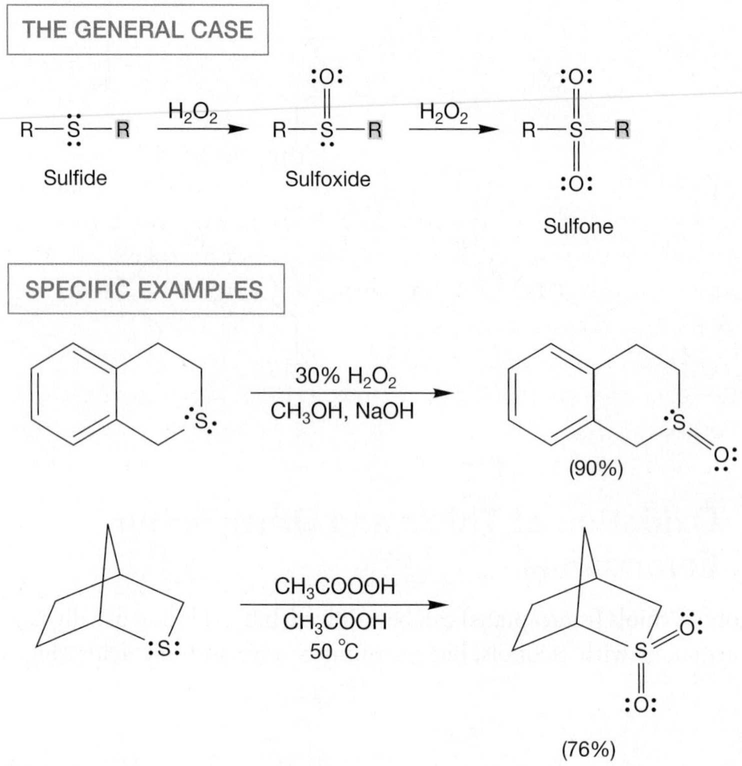

FIGURE 16.78 Sulfides can be oxidized to sulfoxides with hydrogen peroxide. Further oxidation gives sulfones.

Sulfoxides and sulfones are more complicated structurally than they appear. First, they are not planar, but approximately tetrahedral. Second, the uncharged resonance forms are not the major contributors to the structures. Better representations involve charge-separated resonance forms in which the sulfur octet is not expanded (Fig. 16.79).

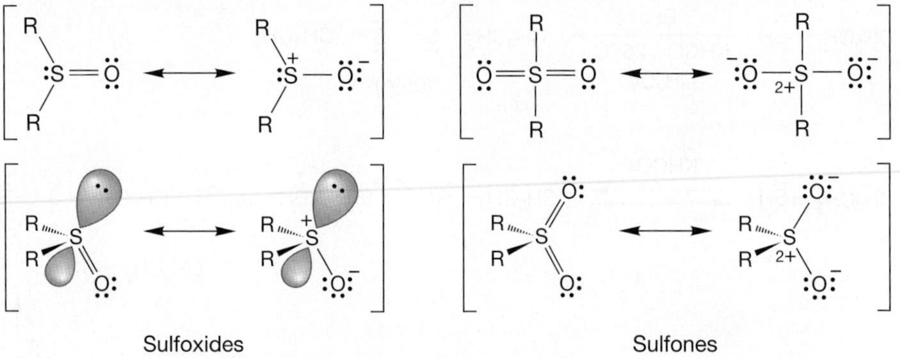

FIGURE 16.79 A resonance description of sulfoxides and sulfones.

Summary

Thiol and sulfide chemistry resembles alcohol and ether chemistry. There are differences, but many of the differences are quantitative, not fundamental. For example, we have learned that mercaptides (RS^-) are much better nucleophiles than alkoxides (RO^-), but even an alkoxide can undergo the S_N2 reaction given the right conditions and reagents. There are areas in which substantial differences do appear; for example, sulfur is far more prone to oxidation than oxygen.

16.17   The Wittig Reaction

As we have seen, all manner of nucleophiles add to carbon–oxygen double bonds, often leading to useful compounds such as alcohols. Now we will learn about a synthetically useful carbonyl addition reaction that produces certain kinds of alkenes. This reaction was discovered in the laboratories of Georg Wittig (1897–1987) and is called the **Wittig reaction**. Wittig shared the 1979 Nobel prize in chemistry largely for his work on this reaction. The process begins with the reaction between phosphines, (R_3P), and alkyl halides to give phosphonium halides (Fig. 16.80). Although we have not encountered reactions of phosphorus yet, it sits right below nitrogen in the periodic table and, like nitrogen, is a good nucleophile. Formation of the phosphonium halide takes place through an S_N2 displacement.

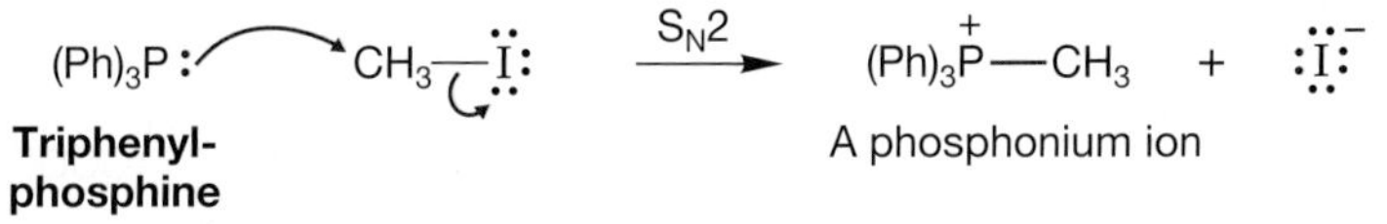

FIGURE 16.80 Displacement of iodide by the nucleophilic phosphorus atom of triphenylphosphine leads to phosphonium ions.

The phosphonium ion contains acidic hydrogens that can be removed by strong bases such as alkyllithium reagents. The product is an **ylide** (pronounced *ill-id*), a compound containing opposite charges on adjacent atoms (Fig. 16.81).

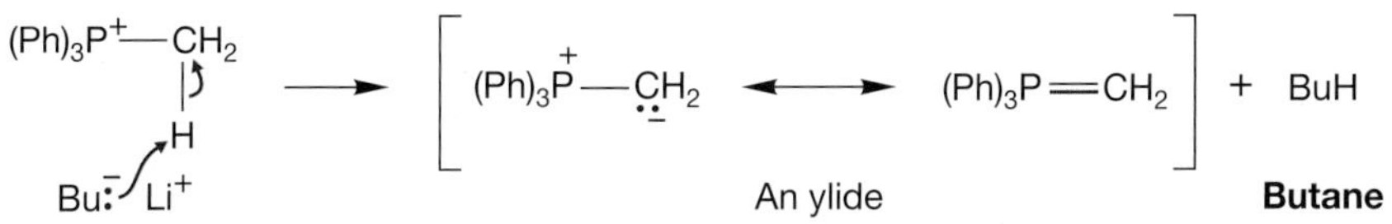

FIGURE 16.81 Protons adjacent to the positively charged phosphorus atom can be removed in strong base to give ylides.

The carbon of the ylide is a nucleophile, and like other nucleophiles, adds to carbon–oxygen double bonds (Fig. 16.82). Intramolecular closure of the intermediate

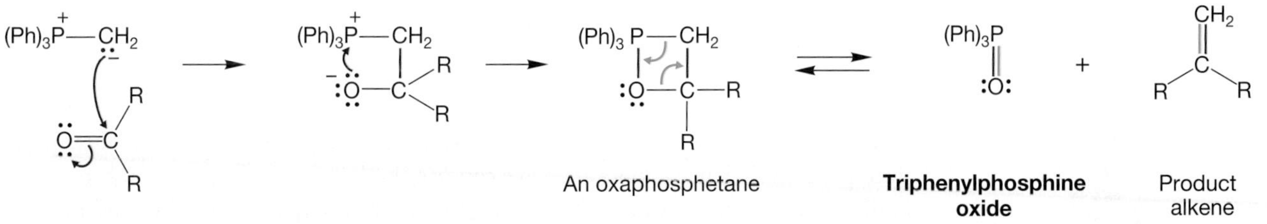

FIGURE 16.82 Ylides are nucleophiles and will add to carbonyl compounds to give intermediates that can close to oxaphosphetanes. These four-membered ring compounds can open to give triphenylphosphine oxide and the product alkenes.

leads to a four-membered ring, an **oxaphosphetane**. Oxaphosphetanes are strained and unstable relative to their quite stable constituent parts, phosphine oxides and alkenes. Therefore, it is easy for these intermediates to fragment.

The Wittig synthesis does contain some unfamiliar intermediates. You have not seen either an ylide or an oxaphosphetane before, for example. Yet the key reactions, the S_N2 displacement to form the phosphonium halide (Fig. 16.80) and the addition to the carbonyl group (Fig. 16.82), are simple variations on basic processes. In a practical sense, it is worth working through the new parts of the Wittig reaction because it is so useful synthetically. Note that the Wittig reaction is, in a formal sense, the reverse of ozonolysis (p. 436). Figure 16.83 shows some specific examples.

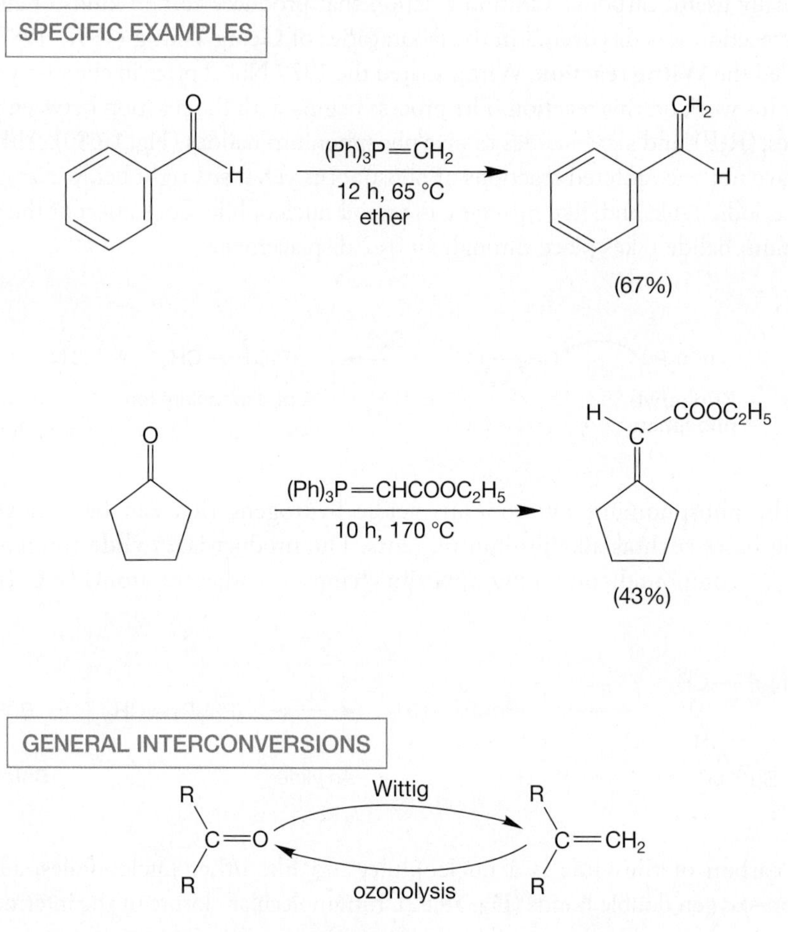

SPECIFIC EXAMPLES

$$\text{PhCHO} \xrightarrow[\substack{\text{12 h, 65 °C} \\ \text{ether}}]{(Ph)_3P{=}CH_2} \text{PhCH=CH}_2 \quad (67\%)$$

$$\text{cyclopentanone} \xrightarrow[\substack{\text{10 h, 170 °C}}]{(Ph)_3P{=}CHCOOC_2H_5} \quad (43\%)$$

GENERAL INTERCONVERSIONS

$$\underset{R}{\overset{R}{{}}}C{=}O \underset{\text{ozonolysis}}{\overset{\text{Wittig}}{\rightleftarrows}} \underset{R}{\overset{R}{{}}}C{=}CH_2$$

FIGURE 16.83 Some typical synthetic uses of the Wittig reaction. It converts an aldehyde or ketone into an alkene.

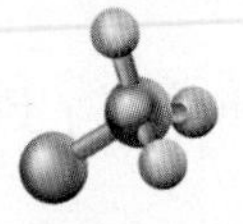

Wittig reaction

The Wittig reaction is very useful. How else could we convert cyclohexanone into pure methylenecyclohexane? We can use the reactions of this chapter to devise a different synthesis. Some thought might lead to a sequence in which cyclohexanone reacts with methyllithium to give, after hydrolysis, a tertiary alcohol. An acid-catalyzed elimination reaction would give some of the desired product, but there is no easy way to avoid the predominant formation of the undesired isomer, 1-methylcyclohexene (Fig. 16.84). The Wittig reaction solves this synthetic problem and is an important constituent of any synthetic chemist's bag of tricks.

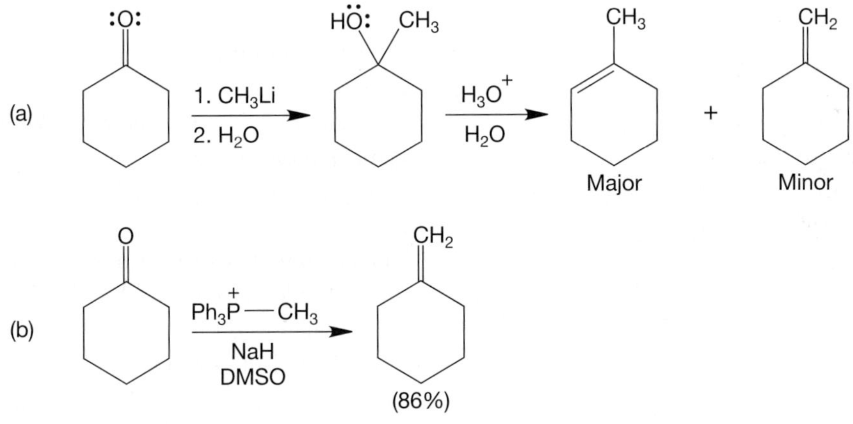

FIGURE 16.84 (a) The dehydration approach fails to give very much of the desired product. (b) The Wittig reaction is the preferred alternative.

PROBLEM 16.28 Explain why the acid-catalyzed dehydration reaction of Figure 16.84a will give mostly the undesired 1-methylcyclohexene.

PROBLEM 16.29 Starting with cycloheptanone, cyclopentanone, triphenylphosphine, propyl iodide, butyllithium, and inorganic reagents of your choice, devise synthesis of the following molecules:

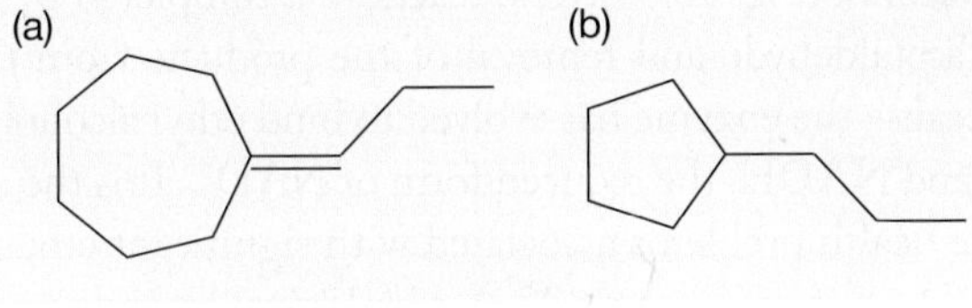

16.18   Special Topic: Biological Oxidation

We have seen several examples of redox reactions in the last few sections. Related processes are of vital importance in biological systems. We humans, for example, derive much of our energy from the oxidation of ingested sugars and fats. However, we do not use chromium reagents or nitric acid as oxidizing agents to accomplish these reactions, for these reagents are far too unselective and harsh. They would surely destroy most of our constituent molecules; a quite uncomfortable process one imagines. Instead, a series of highly selective oxidizing biomolecules has evolved, each dedicated to a particular purpose. One of these, nicotinamide adenine dinucleotide (NAD$^+$),[3] is the subject of this section. Recall that we first looked at this molecule in Chapter 14 (p. 679).

[3] A glance at the structure of this molecule, and some brief thoughts on how complicated its systematic name must be, quickly reveal the obvious reasons why biochemists are addicted to acronyms.

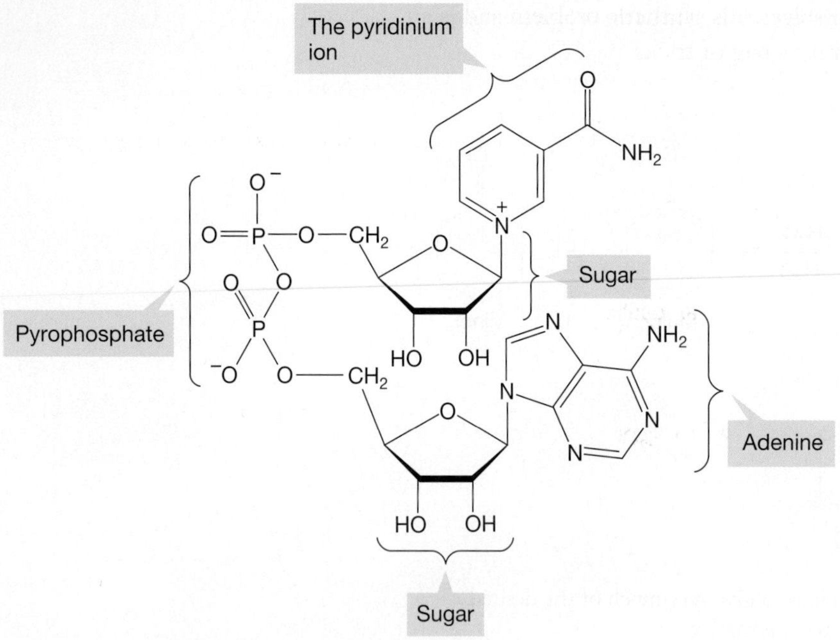

FIGURE 16.85 Nicotinamide adenine dinucleotide, NAD^+.

Nicotinamide adenine dinucleotide is a pyridinium ion, and its full structure is shown in Figure 16.85. The business end of the molecule, the pyridinium ion, is connected through a sugar (ribose) and a pyrophosphate linkage to a second ribose which is, in turn, attached to the base adenine. The molecule is a **coenzyme**, a molecule needed by an enzyme to carry out a reaction. In this case, the enzyme is alcohol dehydrogenase. The enzyme binds the alcohol, which is the molecule to be oxidized, and NAD^+, which will be reduced. Ethanol is a common alcohol in our diet (it is in bread, fruit, flavoring, and often in our drinks) and it is on this alcohol that much of the research on the mechanism of NAD^+ has been performed.

In this reaction, the NAD^+ behaves as a Lewis acid, and accepts a hydride ion. Ethanol is a Lewis base in this setting, because it donates the H^-. The enzyme serves to bring the NAD^+ and the ethyl alcohol together, but this bringing together is of immense importance to the reaction, as it allows the redox process to take place without the requirements that molecules in solution have for finding each other and orienting properly for reaction.

The NAD^+ is a strong acid by virtue of its positively charged quaternary nitrogen. It can be reduced by transfer of a hydrogen *with its pair of electrons* (a hydride reduction), from ethyl alcohol (Fig. 16.86). The reaction is completed by deprotonation to give the product acetaldehyde and removal of the products from the enzyme. This removal is easy because the enzyme has evolved to bind ethyl alcohol and NAD^+, and not acetaldehyde and NADH, the reduced form of NAD^+. It is the acetaldehyde that causes most of the health problems associated with significant ethanol consumption. We will see NADH again in Chapter 23 as a biological hydride source.

This hydride reduction may seem strange, but it is not. There is a close relative in the intramolecular hydride shifts in carbocation chemistry. These, too, are redox

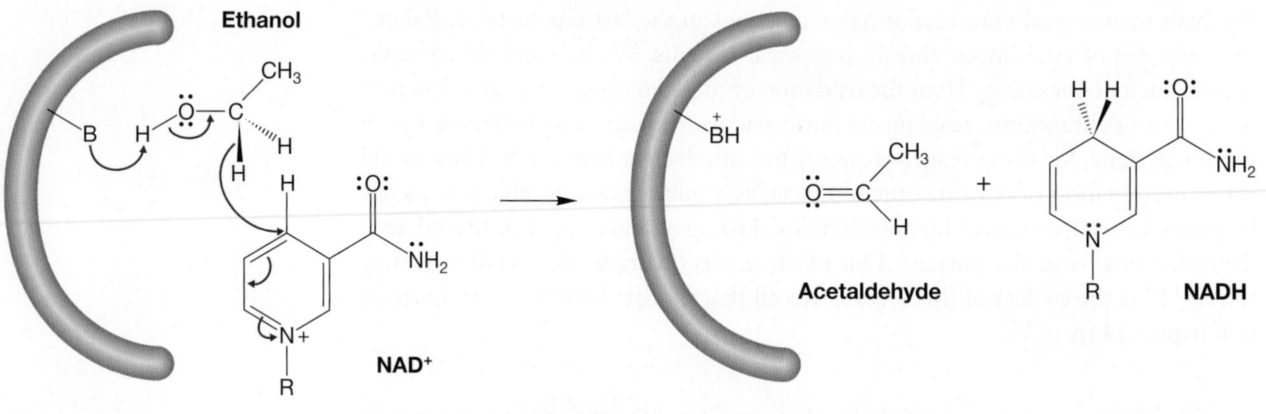

FIGURE 16.86 The NAD^+ is reduced by transfer of hydride (H^-) from ethanol.

reactions. The originally positive carbon is reduced and the carbon from which the hydride moves is oxidized (Fig. 16.87).

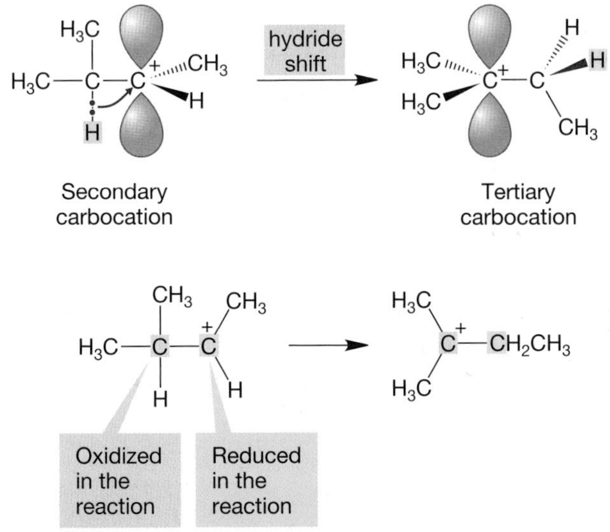

FIGURE 16.87 Hydride shifts in carbocations are also redox reactions.

WORKED PROBLEM 16.30 (a) Each methylene hydrogen of ethanol can be replaced by deuterium to give a pair of chiral deuterioethanols. Show this process and indicate which of the new compounds is (*R*) and which is (*S*). (b) Oxidation of (*R*)-1-deuterioethanol produces deuterio-NADH (NADD) and acetaldehyde. Predict the products of the oxidation of (*S*)–1–deuterioethanol with NAD^+. Explain carefully.

ANSWER (a) Replacement of one methylene H with D gives the (*S*) enantiomer; replacement of the other gives the (*R*) enantiomer. The hydrogens are enantiotopic.

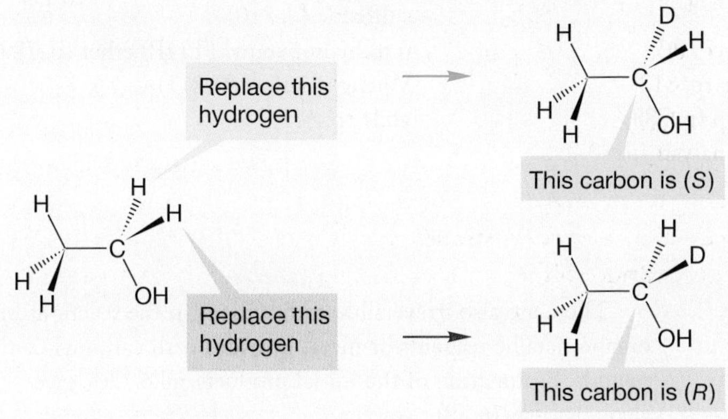

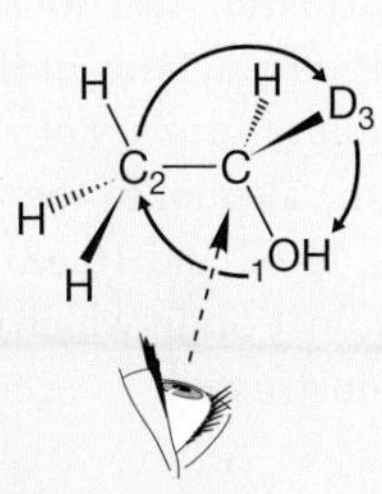

This structure shows the priorities for the (*R*) compound; the 1 ... 2 ... 3 arrow is clockwise

(b) The product will be 1-deuterioacetaldehyde. The binding site of the enzyme is chiral and therefore the enantiotopic hydrogens are distinguishable.

16.19 Summary

New Concepts

The centerpiece of this chapter and its unifying theme is the addition reaction in which nucleophiles of all kinds add to carbonyl groups. This process can be either acid- or base-catalyzed, and starts with the overlap of a filled orbital on the nucleophilic Lewis base with the empty π^* orbital of the carbonyl to give an alkoxide in which the highly electronegative oxygen atom bears the negative charge (Fig. 16.22). The π system of the carbonyl group is constructed through combinations of carbon and oxygen $2p$ orbitals. This process leads to the π orbitals of Figure 16.4.

Many addition reactions are equilibria, and in these cases the position of the equilibrium depends on the relative stabilities of the starting materials and products. These stabilities depend, as always, on a variety of factors including electronic structure, substitution pattern, and steric effects.

The close relationship between alcohols and carbonyls has become apparent in this chapter. Oxidation of alcohols produces the carbonyl group and reduction of the carbonyl, by hydrides or organometallic reagents, produces the alcohol. Protecting groups for alcohols, aldehydes, and ketones were introduced in Section 16.10.

The concept of retrosynthetic analysis appears in this chapter. Retrosynthetic analysis simply means that it is best to approach problems of synthesis by searching not for the *ultimate* starting material, but rather to undertake the much easier task of finding the *immediate* precursor of the product. This process can be repeated until an appropriate level of simplicity is reached (Fig. 16.88).

FIGURE 16.88 Retrosynthetic analysis.

Key Terms

acetal (p. 786)
benzaldehyde (p. 768)
carbinolamine (p. 790)
coenzyme (p. 814)
cyanohydrin (p. 781)
dial (p. 767)
dione (p. 768)
enamine (p. 795)

gem-diol (p. 776)
hemiacetal (p. 783)
hydrate (p. 773)
imine (p. 792)
iminium ion (p. 795)
oxaphosphetane (p. 812)
protecting group (p. 788)
retrosynthetic analysis (p. 807)

Schiff base (p. 792)
silyl ether (p. 789)
sulfone (p. 810)
sulfoxide (p. 810)
tetrahydropyranyl (THP) ether (p. 789)
Wittig reaction (p. 811)
ylide (p. 811)

Reactions, Mechanisms, and Tools

As mentioned in the New Concepts section, there is a single, central reaction in this chapter, the addition of a nucleophile to the Lewis acid carbonyl group. The details of the mechanism vary, depending on whether the reaction is acid-catalyzed, base-catalyzed, reversible, or irreversible. Here are some general examples.

A simple case is the acid- or base-catalyzed reversible addition reaction (Fig. 16.35). Examples are hydration or cyanohydrin formation.

A slightly more complicated reaction involves an addition followed by loss of water. An example is the reaction of primary amines with carbonyl groups to give substituted imines (Fig. 16.51).

If there is no proton that can be removed after loss of water, a second addition reaction can occur, as in acetal formation (Fig. 16.43).

There are also irreversible additions, as in the reactions of organometallic reagents or metal hydrides with carbonyl compounds. Protonation of the initial products, alkoxides, gives alcohols (Fig. 16.59).

This chapter also introduces the oxidation reactions of alcohols. These reactions involve addition reactions in their first steps. For example, in the oxidation of primary alcohols by CrO_3, the first step is addition of the alcohol to the chromium–oxygen double bond (Figs. 16.67 and 16.68). The oxidation is completed by an elimination in which the new carbon–oxygen double bond is constructed.

Syntheses

This chapter contains lots of new syntheses.

1. Acetals

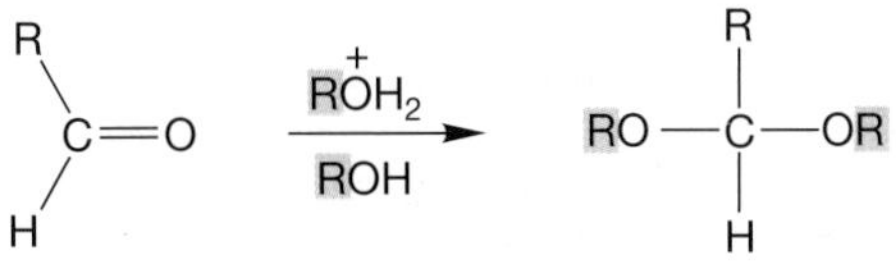

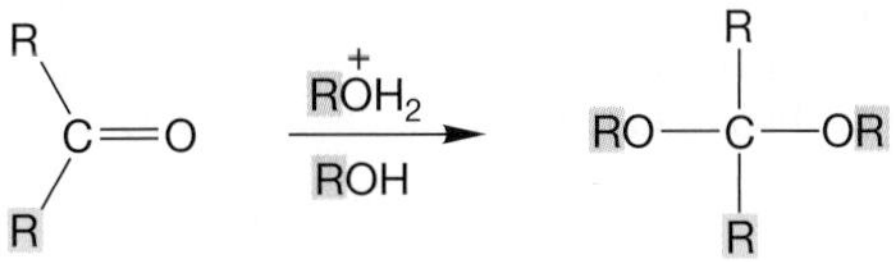

Acetals can be used as protecting groups for aldehydes and ketones; treatment with H_2O/H_3O^+ regenerates the aldehyde. The hemiacetal is an intermediate

2. Acids

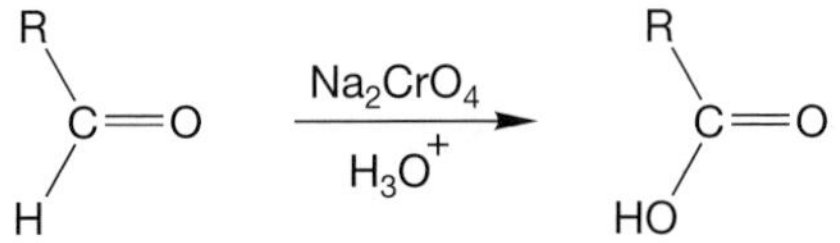

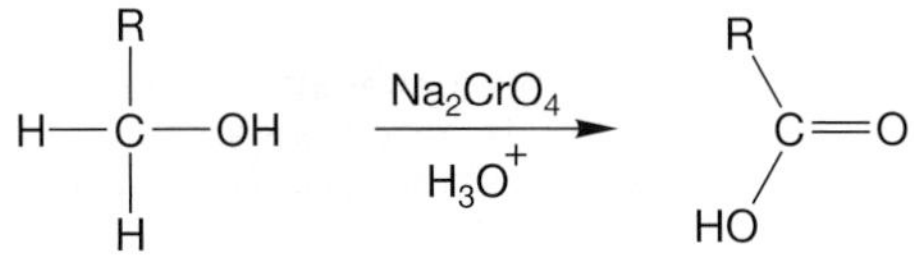

Other oxidizing agents such as $KMnO_4$ work also. The aldehyde and its hydrate are intermediates

3. Alcohols

Primary alcohols

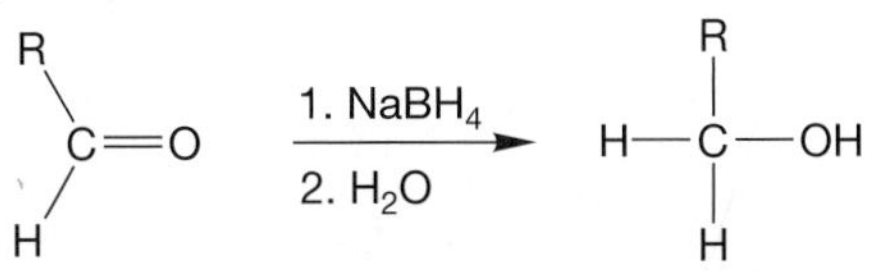

The alkoxide is an intermediate; $LiAlH_4$ can also be used

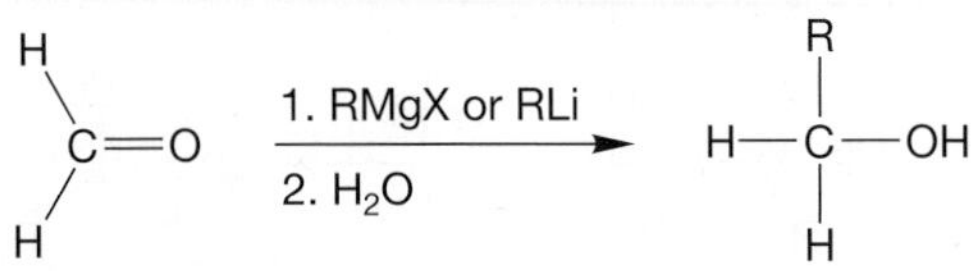

The alkoxide is an intermediate

Secondary alcohols

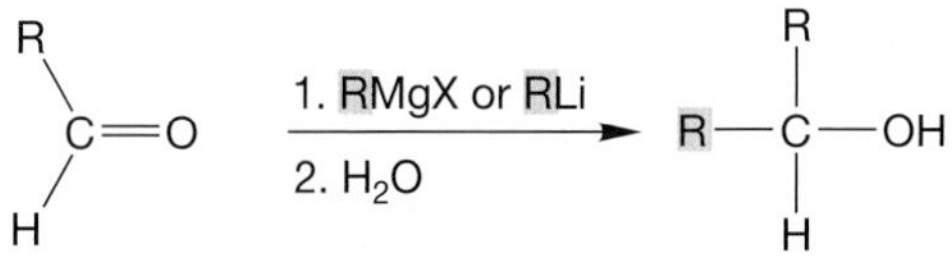

The alkoxide is an intermediate; the R groups can be the same or different

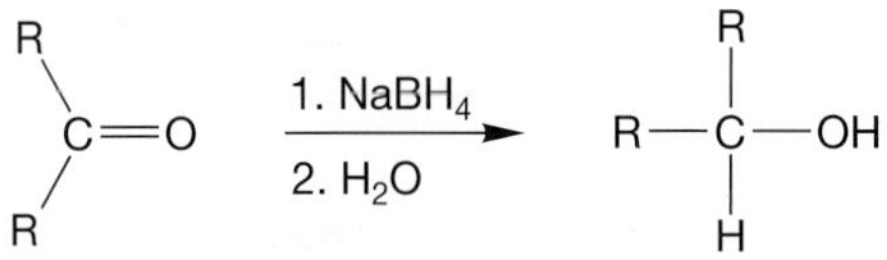

The alkoxide is an intermediate; the R groups can be the same or different, depending on the structure of the ketone; $LiAlH_4$ can also be used

Tertiary alcohols

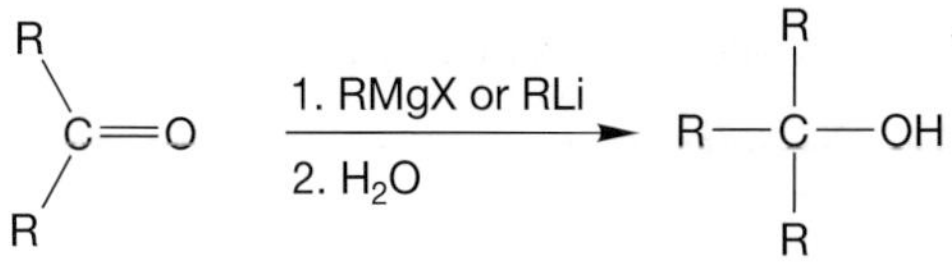

The product can be R_3COH, R_2RCOH, or $RRRCOH$, depending on the structures of the starting ketone and organometallic reagent

4. Aldehydes

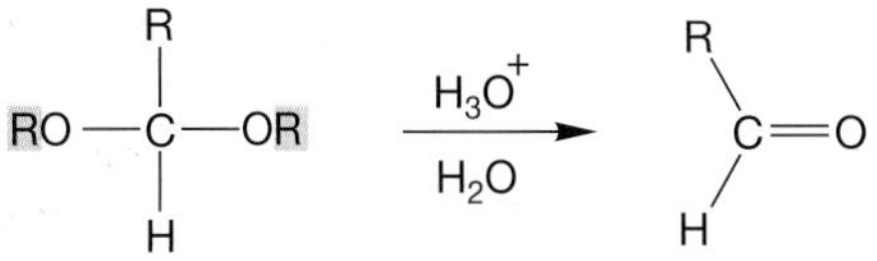

Hydrolysis of an acetal

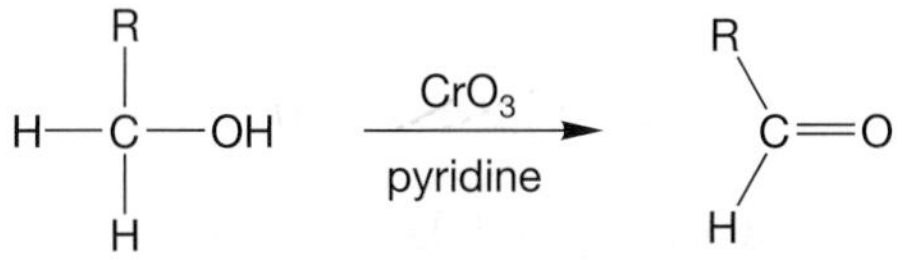

Oxidation of a primary alcohol; water must be absent

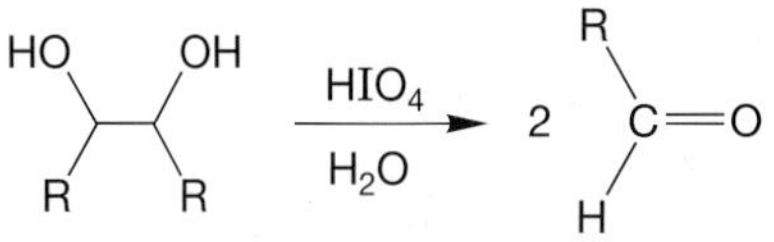

Periodate cleavage of a vicinal diol; reaction involves a cyclic intermediate

5. Alkenes

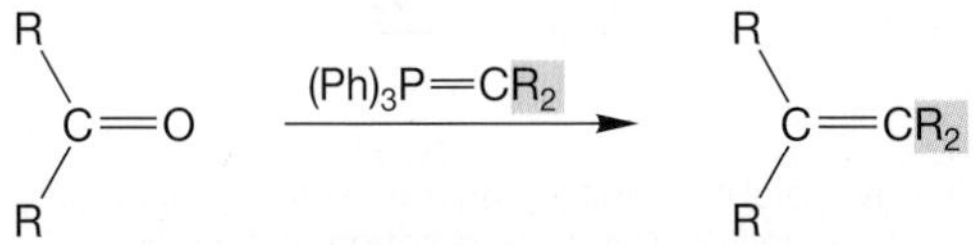

The Wittig reaction

6. Alkylithium Reagents

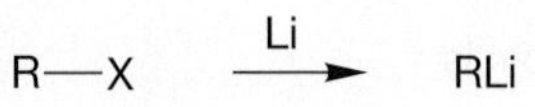

$$R-X \xrightarrow{\text{Li}} RLi$$

X = Cl, Br, or I

RLi is a simplistic picture of the reagent, which is *not* monomeric

7. Bisulfite Addition Products

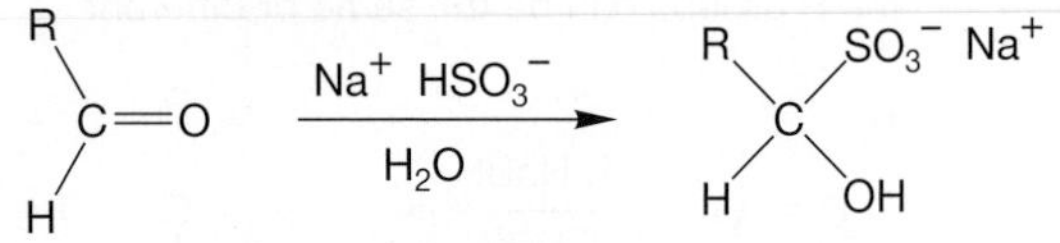

Also works with some ketones

8. Cyanohydrins

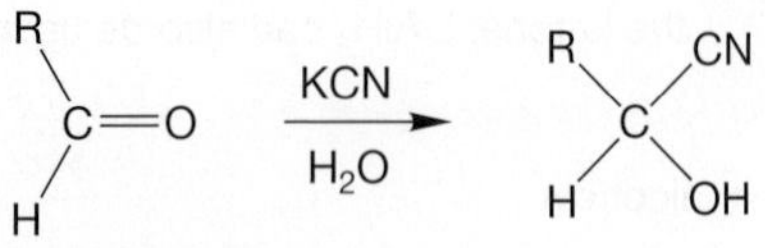

Cyanohydrin formation is more favorable for aldehydes than for ketones

9. Disulfides

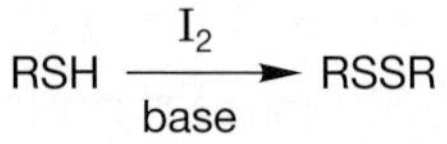

$$RSH \xrightarrow[\text{base}]{I_2} RSSR$$

10. Enamines

α-Position

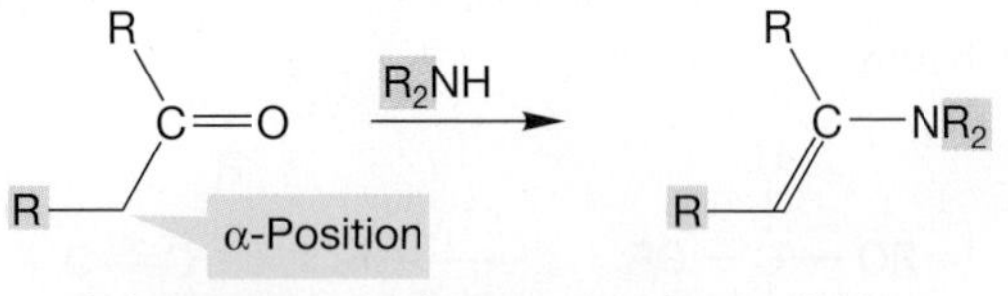

At least one α-hydrogen must be available; the amine must be secondary

11. Grignard Reagents

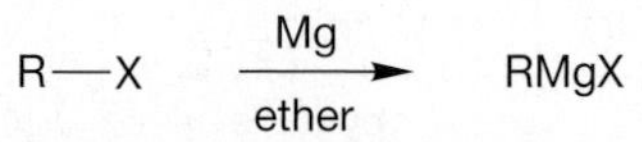

$$R-X \xrightarrow[\text{ether}]{\text{Mg}} RMgX$$

The structure of the organometallic reagent is complicated, as RMgX is in equilibrium with other molecules; the ether solvent is also critical to the success of the reaction

12. Hemiacetals

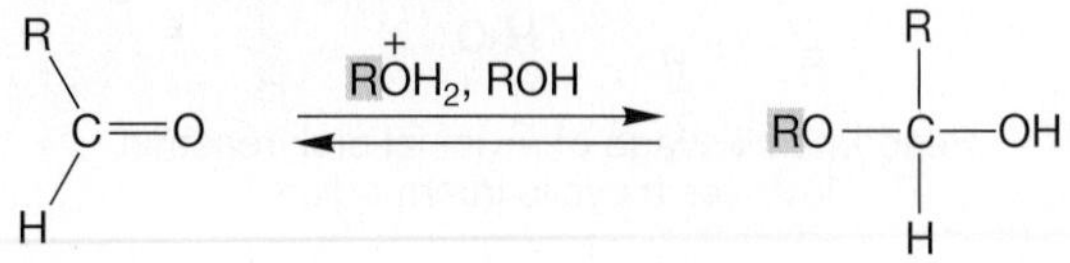

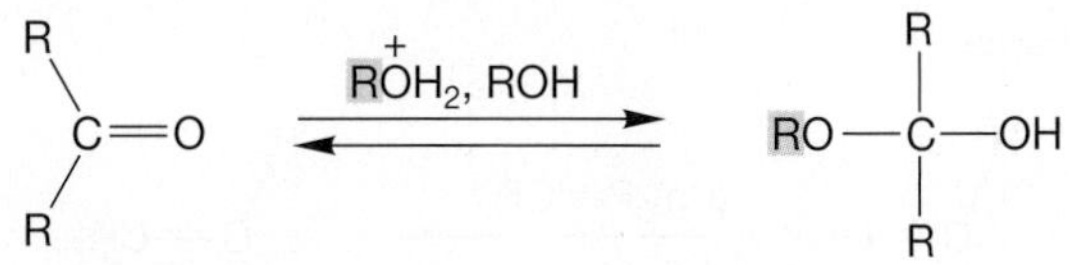

Rarely isolable; exceptions are cyclic hemiacetals, especially sugars; the reaction usually goes further to give the full acetal

13. Hydrates

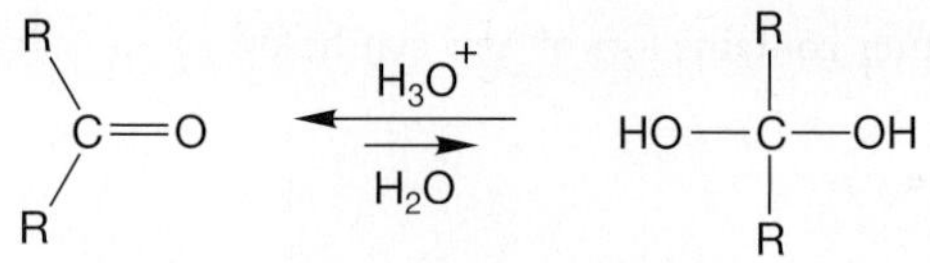

Unstable; for ketones the starting material is usually favored

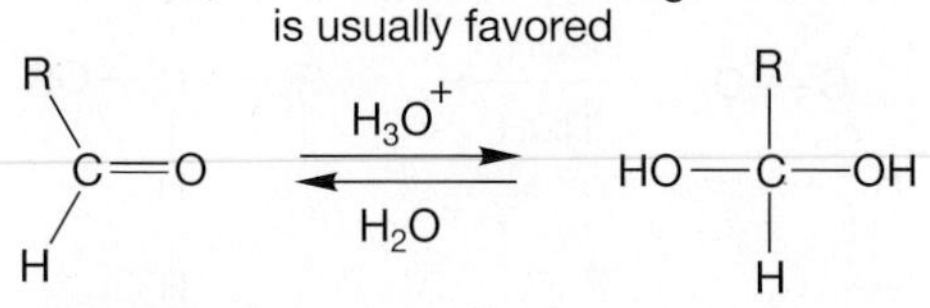

More likely to be favored at equilibrium than the hydrated ketones; not usually isolable

14. Hydrocarbons

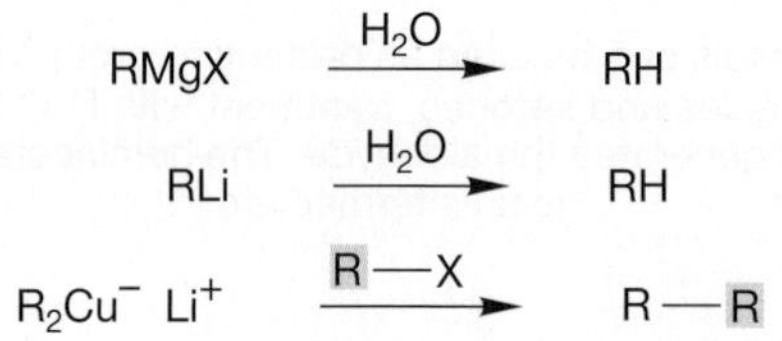

$$RMgX \xrightarrow{H_2O} RH$$

$$RLi \xrightarrow{H_2O} RH$$

$$R_2Cu^- Li^+ \xrightarrow{R-X} R-R$$

R—X must be primary or secondary

15. Imines

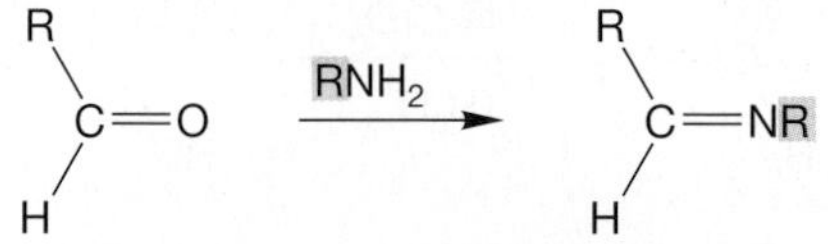

This reaction works for ketones as well; many varieties of imine are known, depending on the structure of R in the amine, which must be primary

16. Ketones

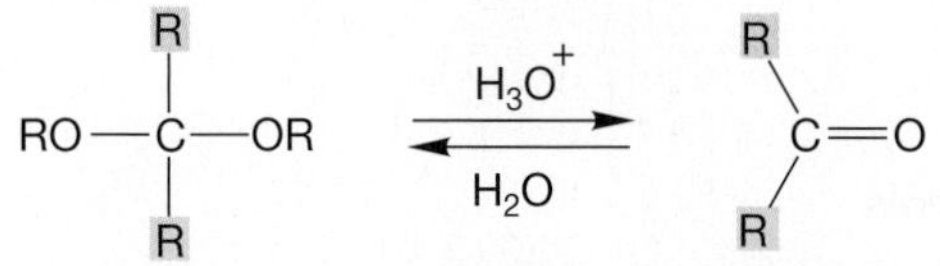

The hydrolysis of an acetal

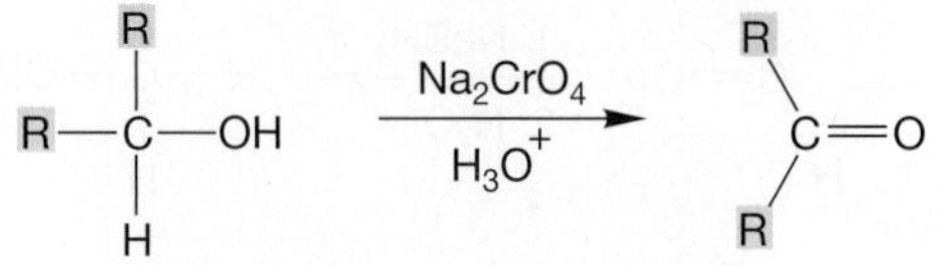

Many other oxidizing agents will oxidize secondary alcohols to ketones

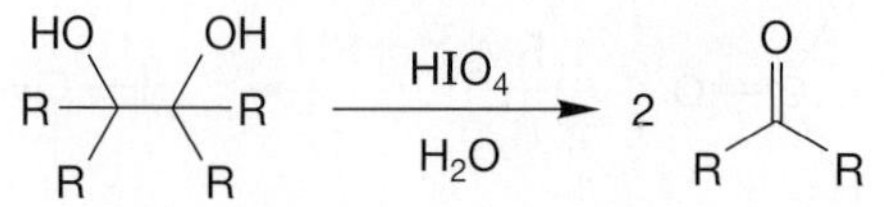

Periodate cleavage of a vicinal diol involves a cyclic intermediate

17. Lithium Organocuprates

$$RLi + CuX \longrightarrow R_2Cu^- \; Li^+ + LiX$$

$$X = I, Br, or Cl$$

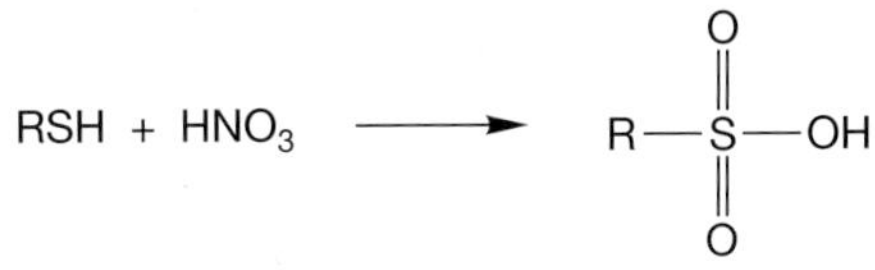

18. Sulfones

excess

$$RSR + H_2O_2 \longrightarrow$$

The sulfoxide is an intermediate

19. Sulfonic Acids

$$RSH + HNO_3 \longrightarrow$$

20. Sulfoxides

$$RSH + H_2O_2 \longrightarrow$$

Further oxidation gives the sulfone

Common Errors

By far the hardest thing about this chapter is the all-too-plentiful detail. To get this material under control it is absolutely necessary to be able to generalize. The easiest mistake to make is to memorize the details and lose the broad picture. Although there is one general principle in this chapter (nucleophiles add to carbon–oxygen double bonds), it is easy to get lost in the myriad protonation, addition, and deprotonation steps. The reversible addition reactions of this chapter are the hardest to keep straight. The following analysis is a model for the sorting process necessary to focus on the fundamental similarities, while being mindful of the small, but important differences introduced by structural changes in the molecules involved in these reactions.

In base, hydroxide, alkoxides, and amide ions all add to the carbonyl group to give alkoxides that can be protonated to produce hydrates, hemiacetals, and carbinolamines. The three reactions are very similar (Fig. 16.89).

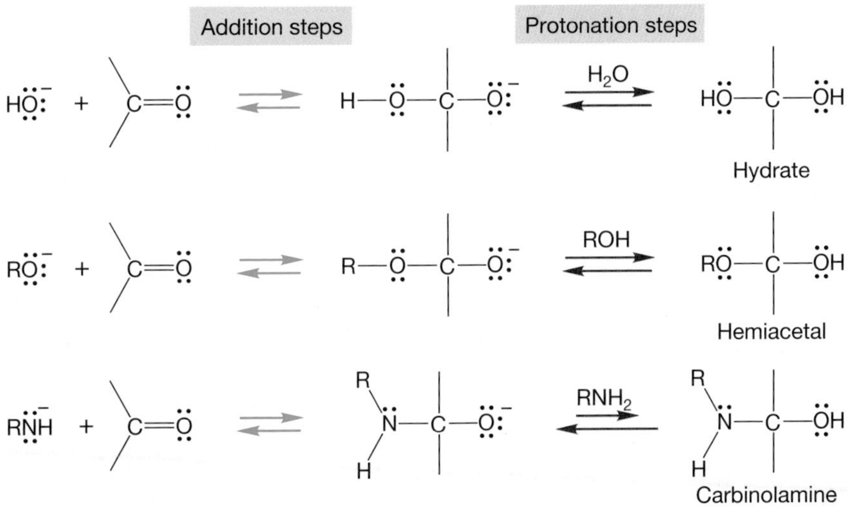

FIGURE 16.89 Base-catalyzed additions to carbonyl groups.

In acid, similar reactions occur, and the three processes begin with closely related sequences of protonation, addition, and deprotonation (Fig. 16.90).

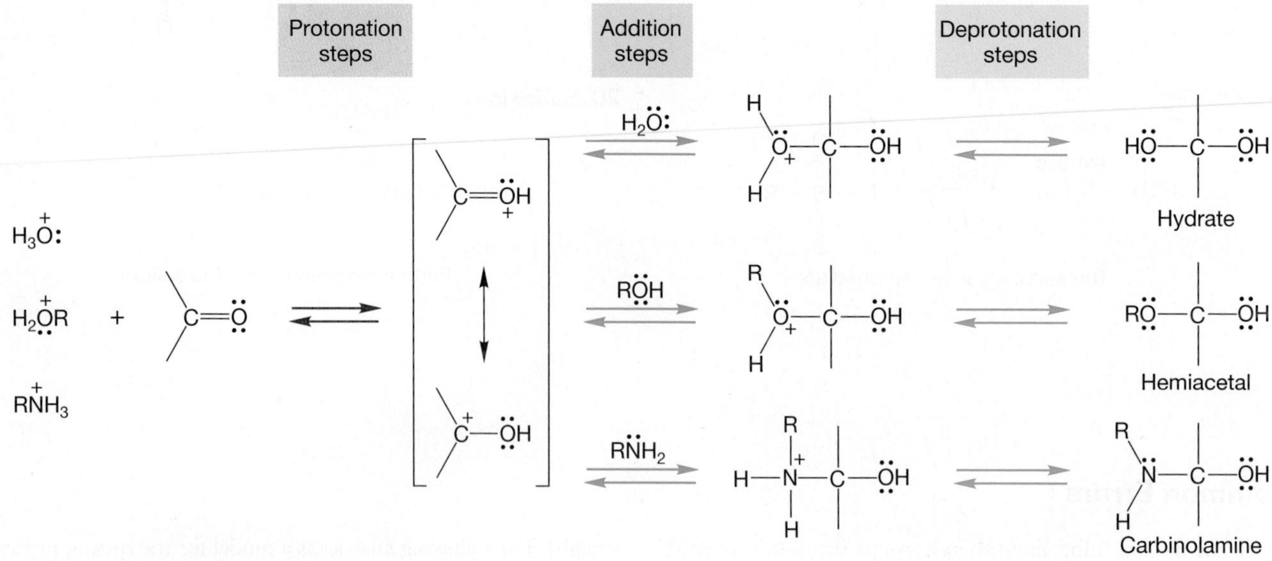

FIGURE 16.90 Acid-catalyzed additions to carbonyl groups.

In the acid-catalyzed reactions, water can be lost from the hydrate, hemiacetal, and carbinolamine to give new, resonance-stabilized intermediates. In the hydrate, this results only in re-formation of the original carbonyl group, but new compounds can be produced in the other two cases. The carbinolamines formed from primary amines follow a similar path that leads ultimately to imines. Hemiacetals can lose water but do not have a second proton that can be lost. Instead, they add a second molecule of alcohol to give acetals (Fig. 16.91).

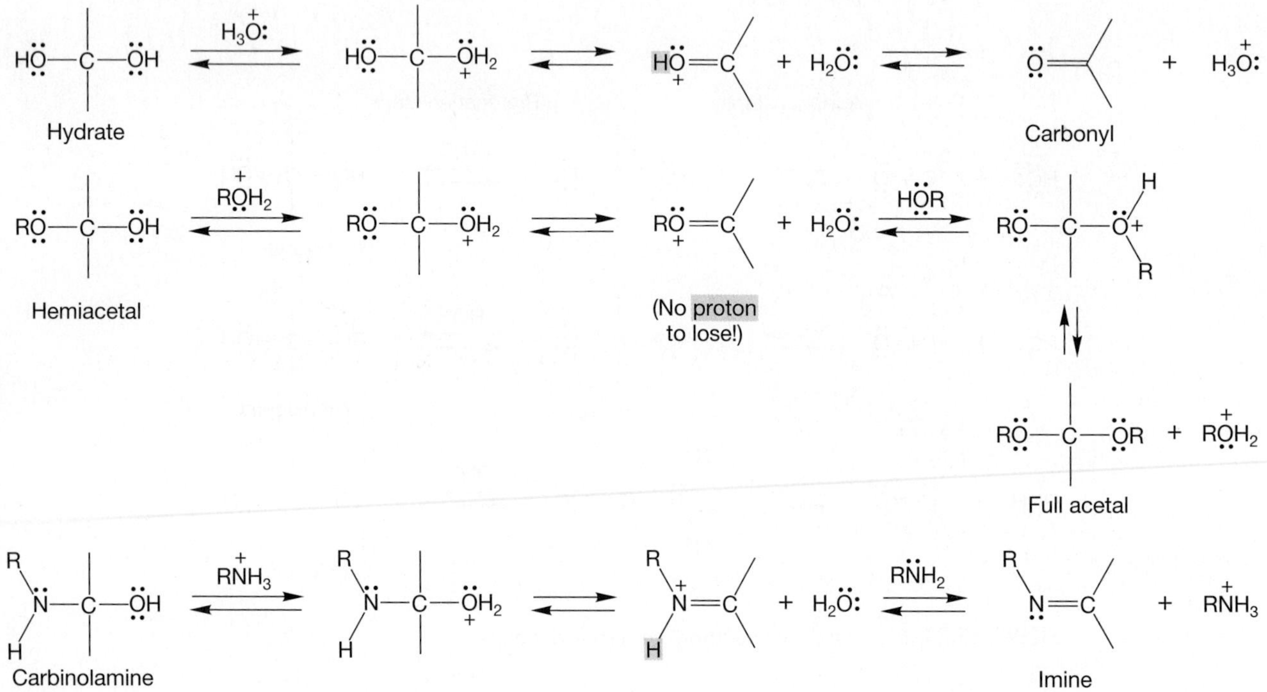

FIGURE 16.91 Further reactions in acid of hydrates, carbinolamines, and hemiacetals.

The carbinolamines formed from secondary amines and carbonyl compounds also have no second proton to lose and instead are deprotonated at carbon to give enamines (Fig. 16.92).

Carbinolamine

FIGURE 16.92 Further reactions of carbinolamines formed from secondary amines.

16.20 Additional Problems

PROBLEM 16.31 Name the following molecules:

(a)

(b)

(c)

(d)

(e)

(f)

PROBLEM 16.32 Draw structures for the following compounds:

(a) *p*-nitrobenzaldehyde
(b) 4-methylhexanal
(c) 2-methylpropiophenone (isobutyrophenone)
(d) *cis*-2-bromocyclopropanecarboxaldehyde

PROBLEM 16.33 Draw structures for the following compounds:

(a) (*R*)-2-amino-3-heptanone
(b) 4-oxopentanal
(c) 3-hydroxycyclopentanone
(d) phenylacetone
(e) (*E*)-2-octenal

PROBLEM 16.34 Name the following compounds according to the official naming system (IUPAC):

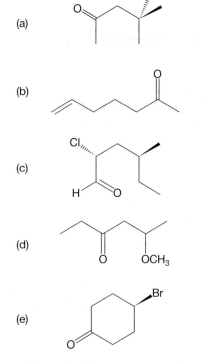

(a)

(b)

(c)

(d)

(e)

PROBLEM 16.35 Name the following compounds:

(a)

(b)

(c)

PROBLEM 16.36 Draw structures for the following compounds:

(a) methyl propyl ketone

(b) 2-pentanone

(c) 2,3-hexanedione

(d) butyrophenone

(e) 3-bromo-4-chloro-5-iodo-2-octanone

PROBLEM 16.37 Draw structures for the following compounds:

(a) 3,5-di-*tert*-butyl-4-hydroxybenzaldehyde

(b) 2-acetylcyclopentanone

(c) 2,4′-dichloro-4-nitrobenzophenone

(d) 6-methyl-2-pyridinecarboxaldehyde

(e) (*E*)-4-phenyl-3-buten-2-one

(f) aminoacetaldehyde dimethyl acetal

PROBLEM 16.38 Use the indicated spectroscopy to distinguish the following pairs of isomers:

(a)

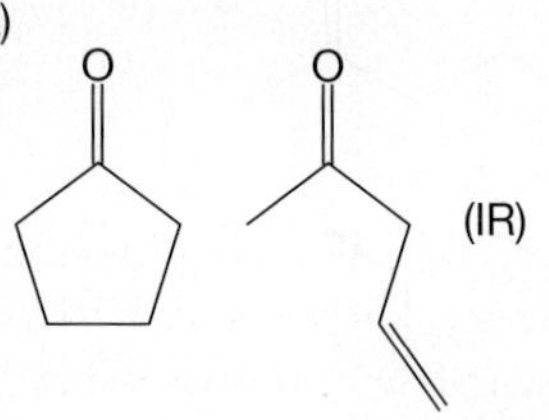

O
O
(IR)

(b)

O
O
(IR)

(c)

O
O
(^{13}C NMR)

(d)

CH3
H3C—C—CH2
H
C=O
H
CH3 O
H3C—C—C
H
CH3
(IR, ^{1}H NMR)

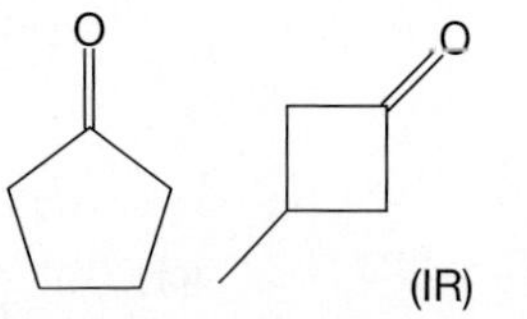

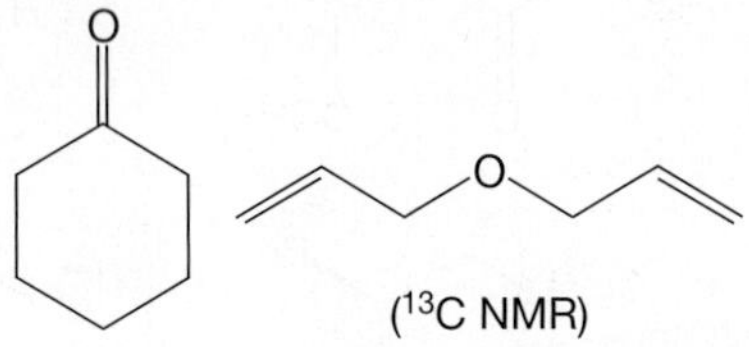

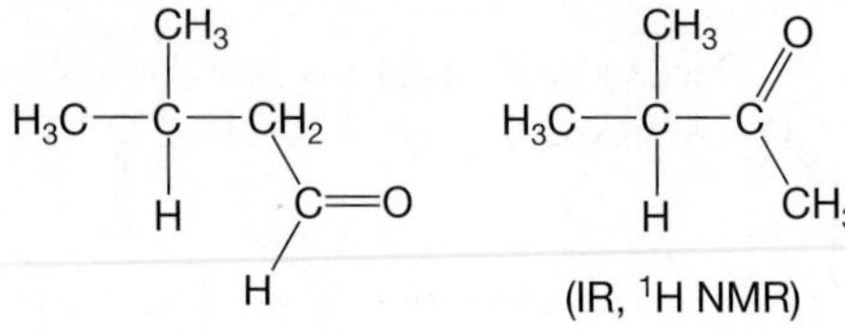

PROBLEM 16.39 Give the major organic products expected in each of the reactions in the next column. *Hint*: Part (k) involves an intramolecular reaction.

(a)

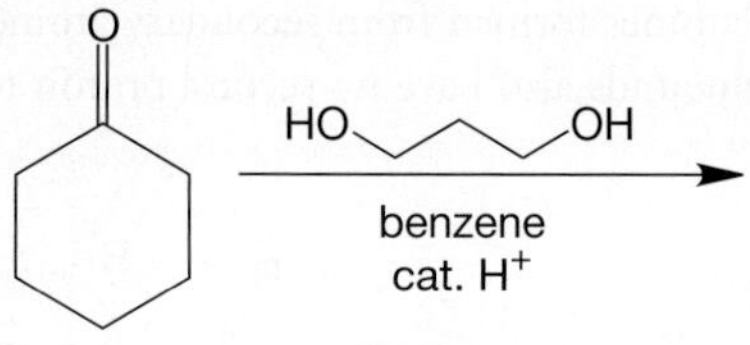

O
HO OH
benzene
cat. H+

(b)

O
PhNH2
cat. H+

(c)

O
NH
cat. H+

(d)

O
C
Ph CH3
1. PhMgBr
2. H2O/H3O+

(e)

O
C
Ph Ph
1. CH3Li
2. H2O/H3O+

(f)

O
C
Ph CH3
NaBH4
CH3OH

(g)

O
C
Ph Ph
1. LiAlH4
2. H2O/H3O+

(h)

OH
CH3
Na2Cr2O7
H2SO4

(i)

PhCH2OH
H2CrO4
H2O

(j)

Ph
OH
CrO3
pyridine

(k)

O
C
Ph (CH2)4CH2I
1. Ph3P
2. BuLi
3. Δ

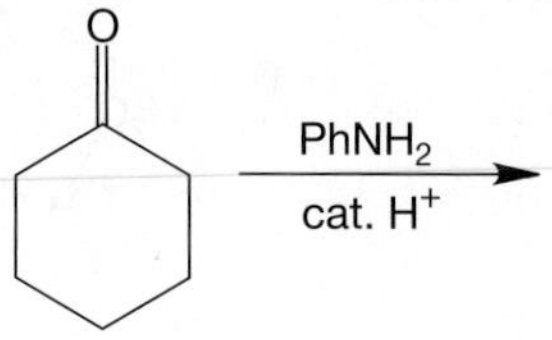

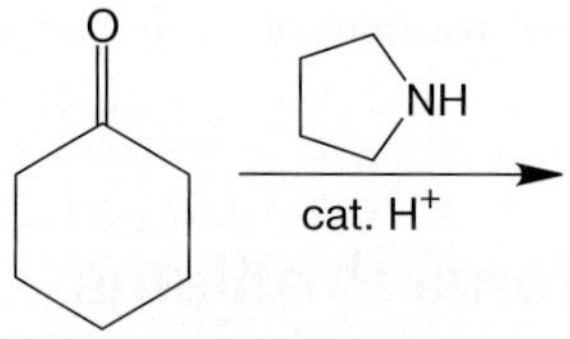

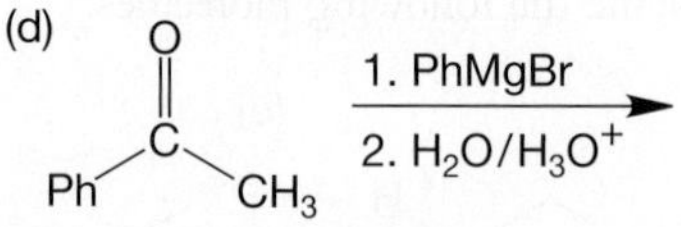

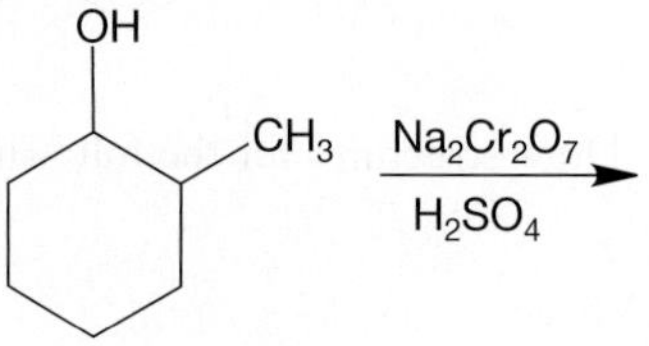

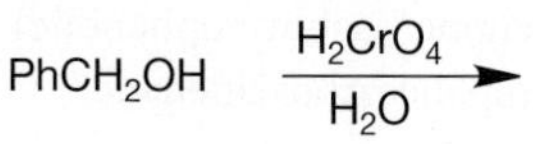

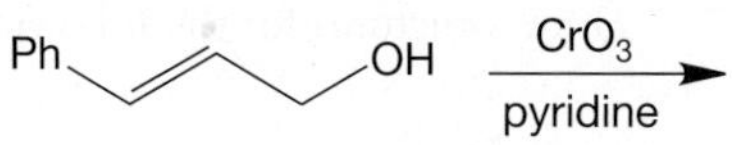

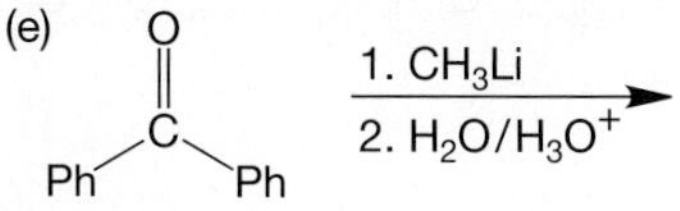

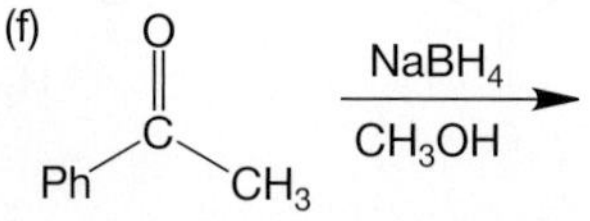

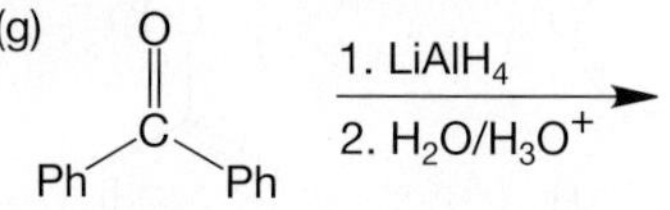

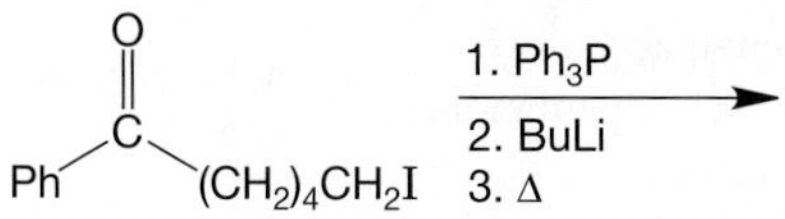

PROBLEM 16.40 Determine the product for each of the following reactions:

(a)
$$\xrightarrow[\text{toluene, }\Delta]{\substack{\text{catalytic H}^+ \\ \text{NH}_2\text{OH}}}$$

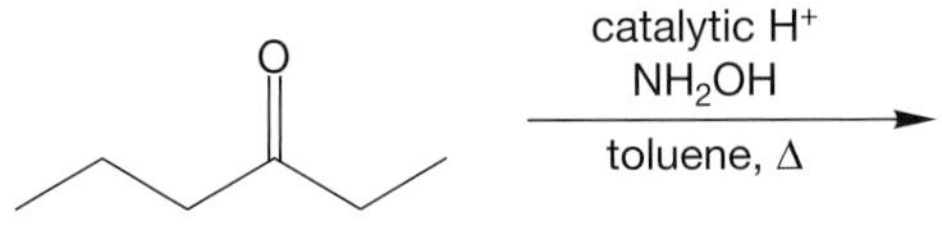

(b)
$$\xrightarrow[\text{toluene, }\Delta]{\substack{\text{catalytic H}^+ \\ \text{NH}_2\text{CH}_2\text{Ph}}}$$

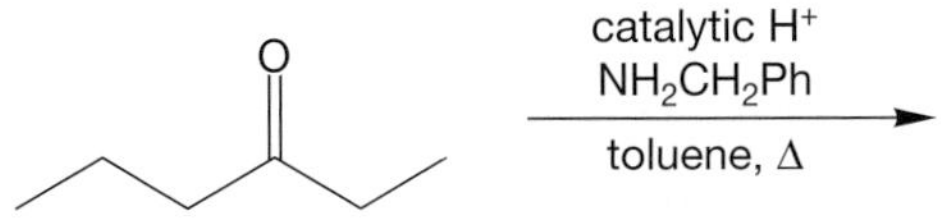

(c)
$$\xrightarrow[\text{toluene, }\Delta]{\substack{\text{catalytic H}^+ \\ \text{NH}_2\text{NH}_2}}$$

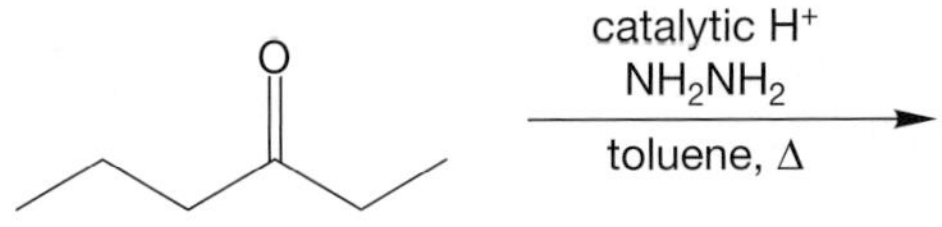

(d)
$$\xrightarrow[\text{toluene, }\Delta]{\substack{\text{catalytic H}^+ \\ \text{NH}_2\text{NHPh}}}$$

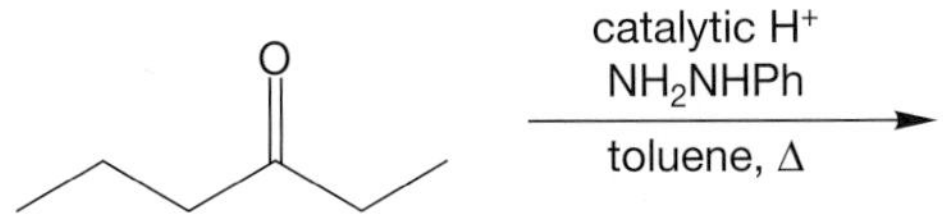

(e)
$$\xrightarrow[\text{toluene, }\Delta]{\substack{\text{catalytic H}^+ \\ \text{NH}_2\text{NHCONH}_2}}$$

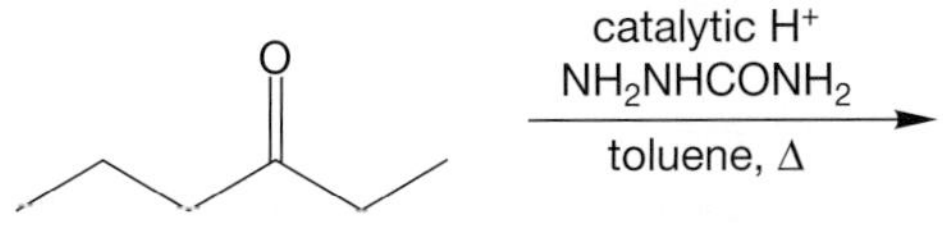

PROBLEM 16.41 Provide an argument not involving molecular orbitals to explain why a carbon–oxygen double bond protonates on O, not C.

PROBLEM 16.42 Write products and arrow formalism mechanisms for the following simple changes (indicate NR if no reaction is expected):

(a)
$$\xrightarrow{\text{H}_2\text{O/H}_3\text{O}^+}$$

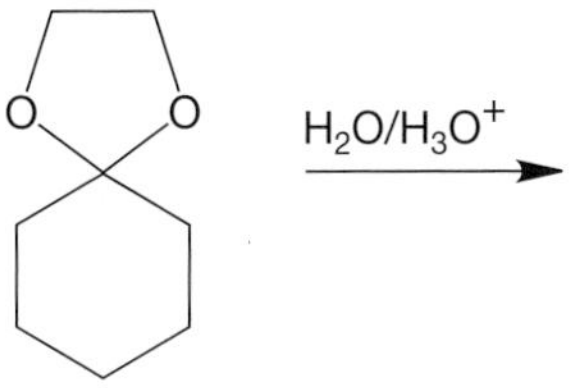

(b)
$$\xrightarrow{\text{H}_2\text{O}}$$

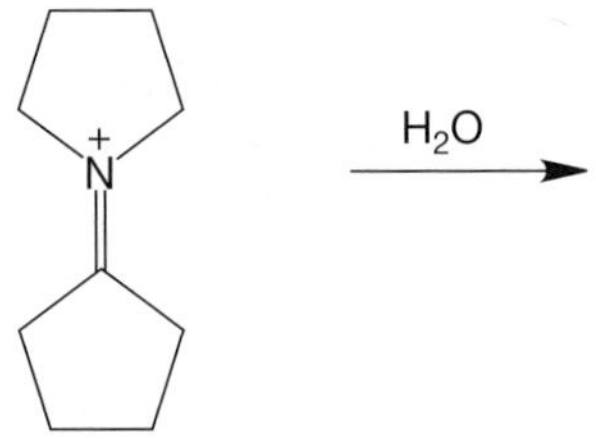

(c)
$$\xrightarrow[\substack{\text{CH}_3\text{OH}_2 \\ +}]{\text{CH}_3\text{OH}}$$

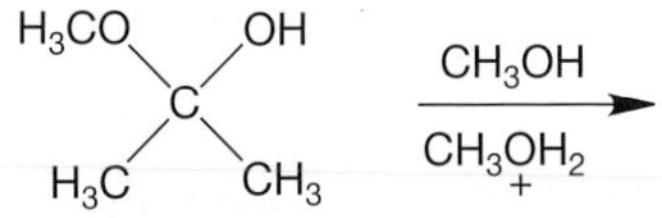

(d)
$$\xrightarrow{\text{KOH/H}_2\text{O}}$$

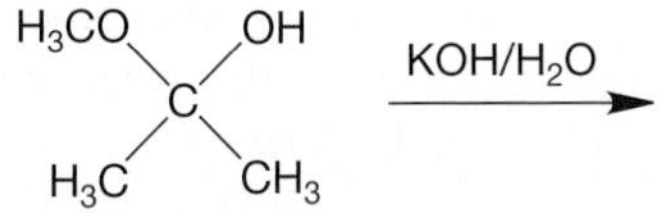

(e)
$$\xrightarrow{\text{KOH/H}_2\text{O}}$$

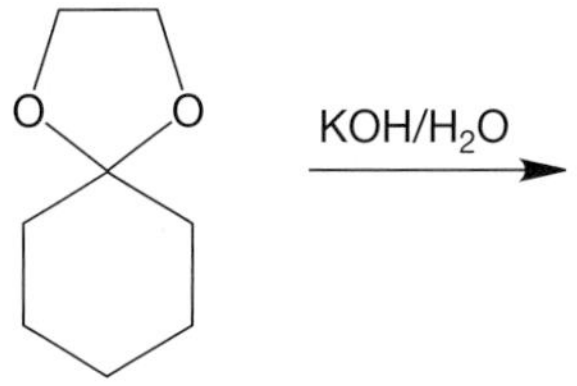

(f)
$$\xrightarrow[\text{NH}_2\text{OH}]{^+\text{NH}_3\text{OH}}$$

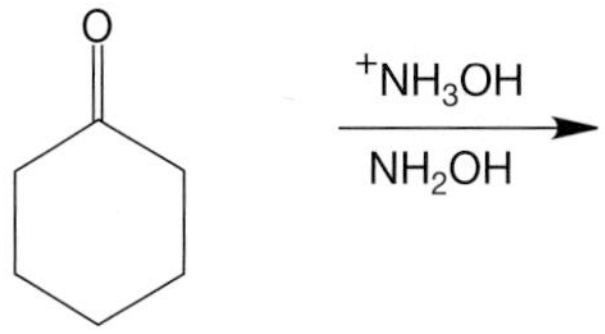

PROBLEM 16.43 The reduction of 4-*tert*-butylcyclohexanone with lithium aluminum hydride affords two isomeric products. What are the structures of these two products? Which isomer do you expect to predominate?

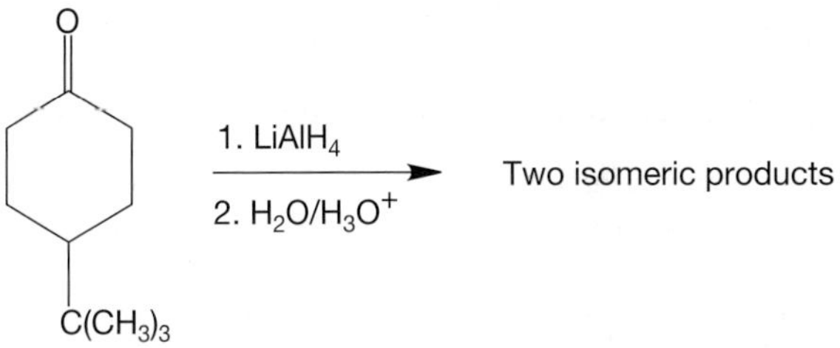

$$\xrightarrow[\text{2. H}_2\text{O/H}_3\text{O}^+]{\text{1. LiAlH}_4} \text{Two isomeric products}$$

C(CH₃)₃

PROBLEM 16.44 As we saw on page 798, benzyl halides can be reduced by metal hydrides. Why can compound **A** not be reduced successfully with lithium aluminum hydride? What reaction will complicate matters?

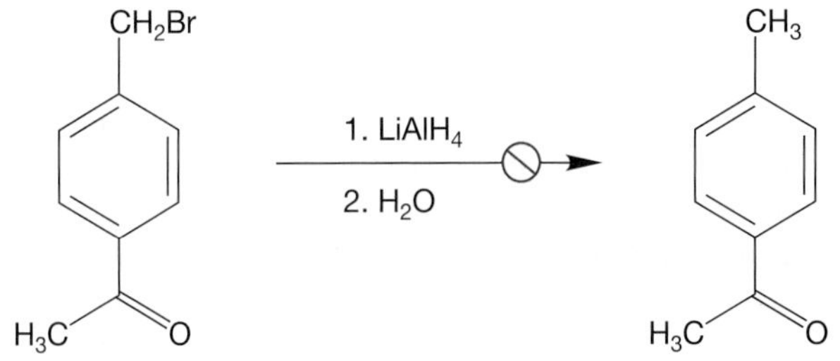

CH₂Br

$$\xrightarrow[\text{2. H}_2\text{O}]{\text{1. LiAlH}_4}$$

CH₃

H₃C
A

H₃C

PROBLEM 16.45 Devise a scheme for avoiding the problem you discovered in your answer to Problem 16.44. *Hint*: See page 788.

PROBLEM 16.46 Write an arrow formalism for the following reaction:

Br
CH₃
$$\xrightarrow[\text{2. H}_2\text{O/H}_3\text{O}^+]{\text{1. Mg/THF}}$$
CH₃
OH

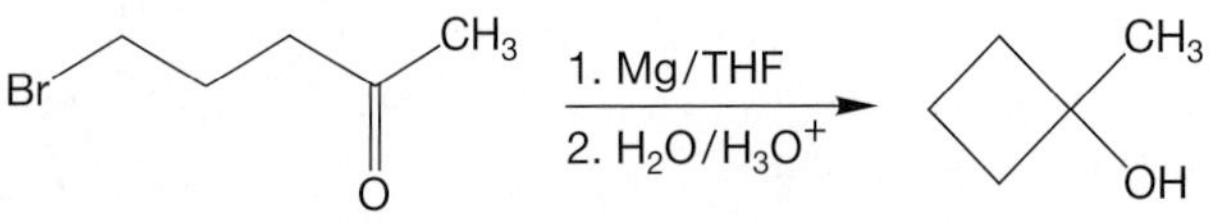

PROBLEM 16.47 Starting from isopropyl alcohol as your primary source of carbon, incidental organic reagents such as bases and solvents, as well as inorganic reagents of your choice (including labeled materials as needed), devise syntheses of the following labeled compounds. It is not necessary to write mechanisms.

(a)   (b) 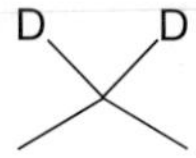   (c)

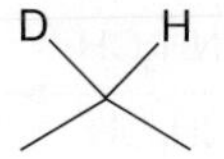

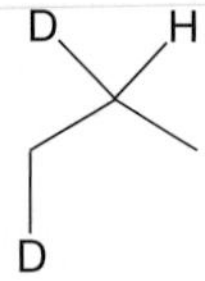

(d) 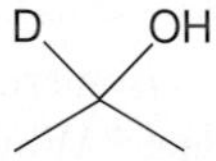   (e)

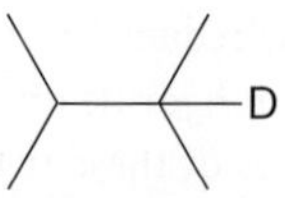

PROBLEM 16.48 Starting from propyl alcohol, organic reagents containing no more than one carbon, and incidental organic reagents such as bases and solvents, as well as inorganic materials, devise syntheses of the following compounds. Some of the necessary transformations will require conversions covered in the earlier chapters. See, for example, figures on page 283.

(a)   (b) 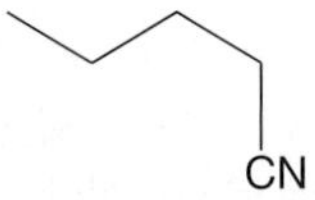   (c)

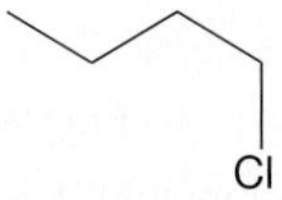

(d)

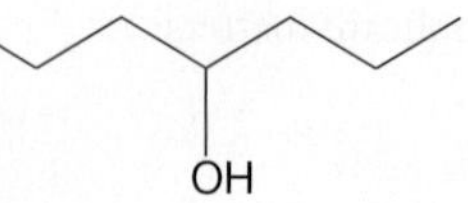

PROBLEM 16.49 Starting from alcohols containing no more than four carbon atoms, inorganic reagents of your choice, and "special" organic reagents (such as ether and pyridine) provide syntheses of the following molecules:

(a)   (b)

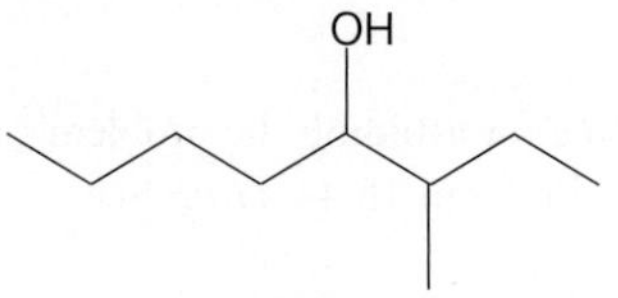

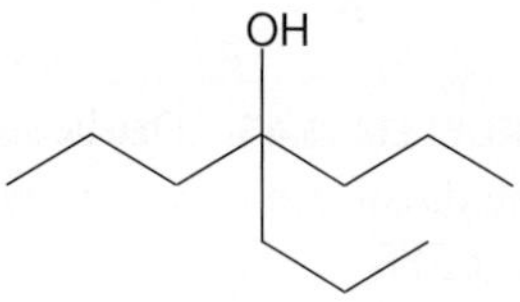

(c) 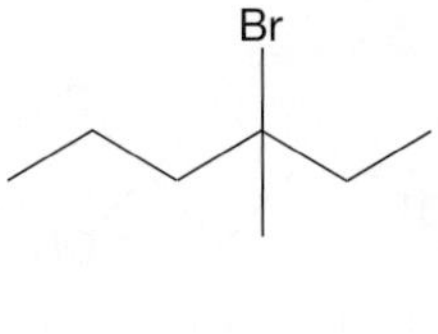   (d)

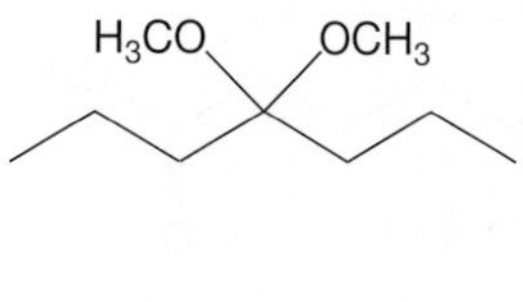

PROBLEM 16.50 Propose an arrow formalism mechanism for the following reaction:

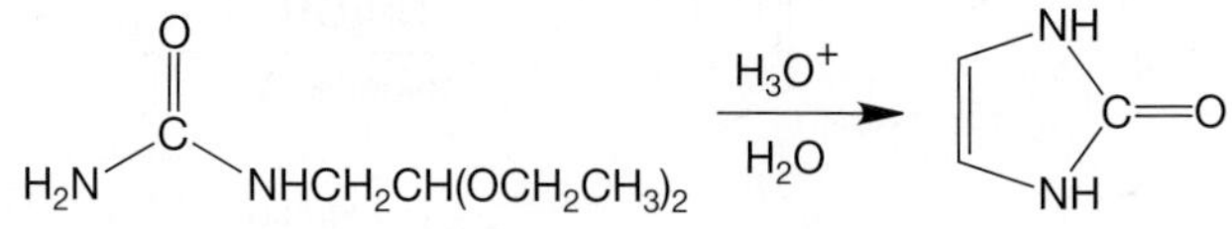

PROBLEM 16.51 Propose an arrow formalism mechanism for the following reaction. What is the driving force for this reaction? Why does it go from the seven-membered ring to the five-membered ring?

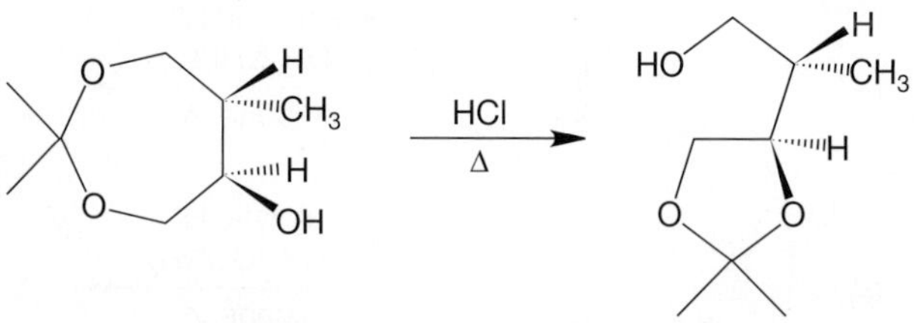

PROBLEM 16.52 Propose an arrow formalism mechanism for the following reaction:

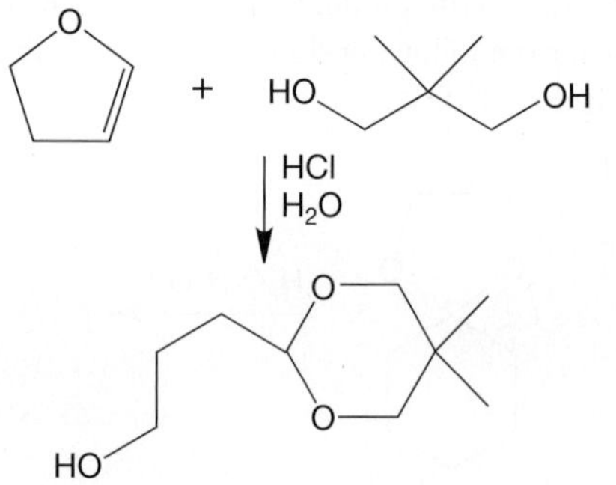

PROBLEM 16.53 In Section 16.10 we learned about protecting groups for alcohols. One of the advantages for using a silyl ether as a protecting group is that it can be selectively removed by treating with fluoride ion. The fluoride ion attacks the silicon atom and breaks the silicon-oxygen bond. You know that the Si—O bond is very strong. Why is it the Si—O bond that is broken rather than one of the Si—C bonds in this reaction? If you are familiar with the process of etching glass, you know this chemistry is related. What acid do you suppose is used to etch glass?

PROBLEM 16.54 There are many reagents that have been developed to oxidize alcohols to ketones or aldehydes. The chromium reagents work very well, but other procedures are preferred. Why do you suppose that other reagents are favored? What are the hazards of using Cr(VI)?

PROBLEM 16.55   Rationalize the following somewhat surprising result. *Hint*: This problem is not as difficult as it looks. First, consider what the reaction of the alcohol starting material with the chromium reagent will give. Second, what will happen when this compound reacts with more alcohol? Finally, use the chromium reagent again.

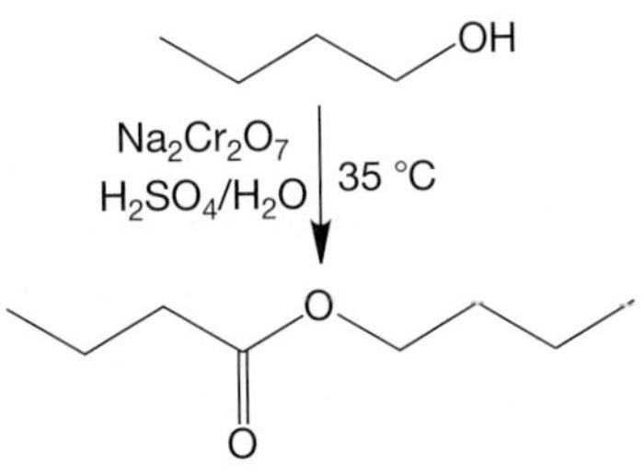

PROBLEM 16.56   The reaction of acetone with methylamine reaches equilibrium with a carbinolamine and, eventually, an imine. There is another participant in the equilibrium called an aminal. Its formula is $C_5H_{14}N_2$. Provide a structure and a mechanism.

PROBLEM 16.57   Treatment of aromatic compound **A** as shown below leads to two products of the same formula, C_9H_{10}, **B** and **C**. The ^1H NMR spectra of **B** and **C** are very similar except that the coupling constant between two hydrogens at δ 6.3 is 8 Hz in **B** and 14 Hz in **C**. Treatment of both **B** and **C** with ozone followed by oxidative workup leads to benzoic acid and acetic acid. Provide structures for **B** and **C** and explain your reasoning.

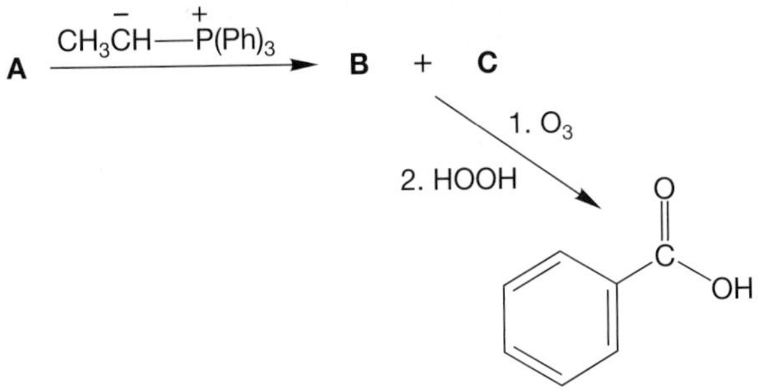

PROBLEM 16.58   Les Gometz, a professor from New York who is very often disappointed in September, was interested in investigating the Diels–Alder reaction of methyl coumalate **A** and the enamine **B**. The professor attempted to prepare **B** by allowing propionaldehyde and morpholine **C** to react.

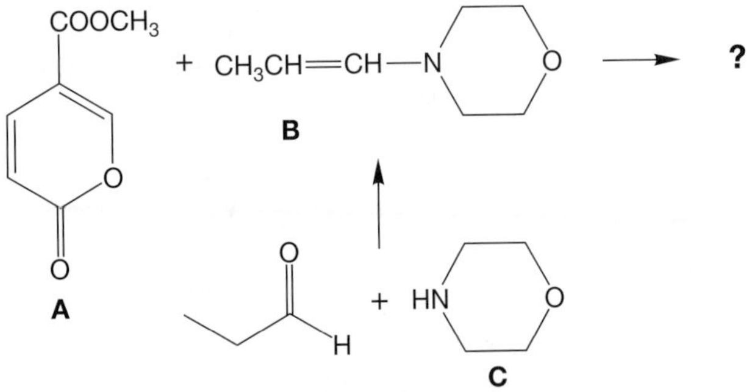

After an appropriate time, Professor Gometz assayed the reaction mixture by ^1H NMR spectroscopy and was disappointed to find that only a small portion (5–10%) of the desired enamine **B** had formed. Ever the optimist, he ran the Diels–Alder reaction with diene **A** anyway, using the reaction mixture containing what he knew to be only a small amount of **B**. He was delighted and astonished to obtain an 80% yield of cycloadducts. Your task is to explain how Professor Gometz could get such a good yield of products when only a small amount of enamine **B** was present in the reaction mixture. *Hint*: See Problem 16.9.

PROBLEM 16.59   Hydrogens adjacent to carbonyl groups, C=O, are relatively acidic ($pK_a \sim 20$). That is, they are relatively easily removed by a base to leave an anion. Explain briefly, then do Problem 16.60.

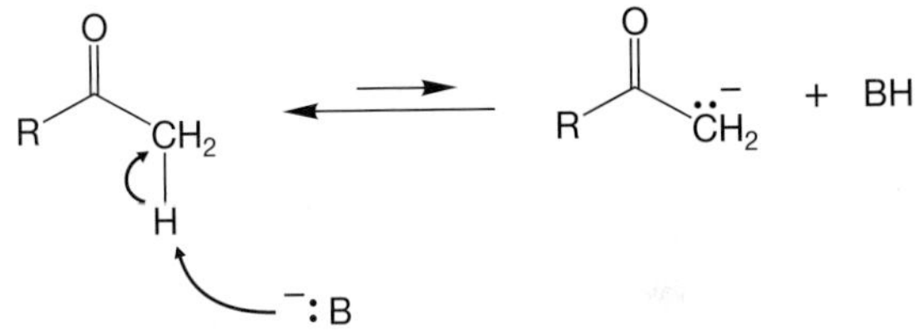

PROBLEM 16.60   Remember that most nucleophiles add to carbonyl groups. Given that notion, and your answer to Problem 16.59, provide a mechanism for the following reaction:

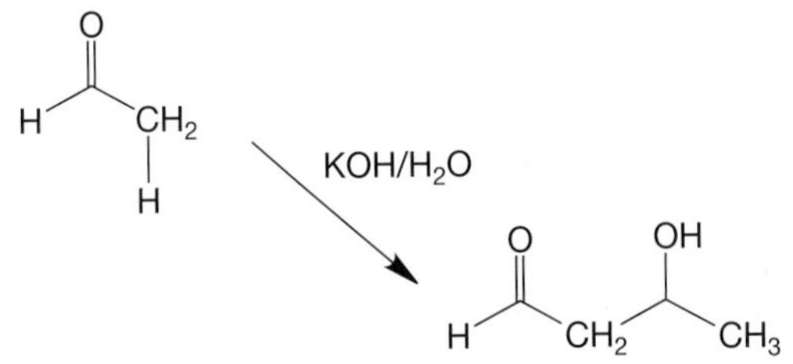

PROBLEM 16.61   Predict the product for the following reaction:

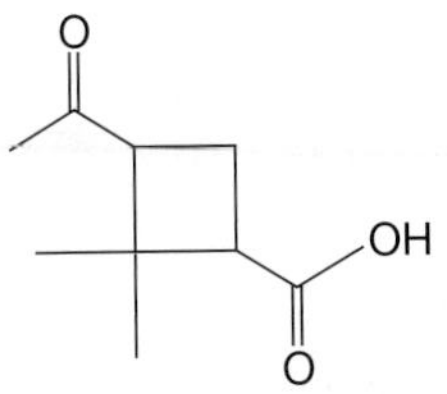

PROBLEM 16.62   Provide structures for compounds **A–D**. Spectral data for compound **D** are shown below.

Compound **D**

Mass spectrum: $m/z = 332$ (p, 82%), 255 (85%), 213 (100%), 147 (37%), 135 (43%), 106 (48%), 77 (25%), 43 (25%)

IR (KBr): 3455 (s) and 1655 (s) cm^{-1}

^{1}H NMR (CDCl$_3$): δ 2.58 (s, 3H), 2.85 (s, 1H, vanishes when a drop of D$_2$O is added), 3.74 (s, 3H), 6.77–7.98 (m, 13H)

^{13}C NMR (acetone-d_6): δ 27.0, 55.8, 82.0, 128.1–159.9 (12 lines), 197.8

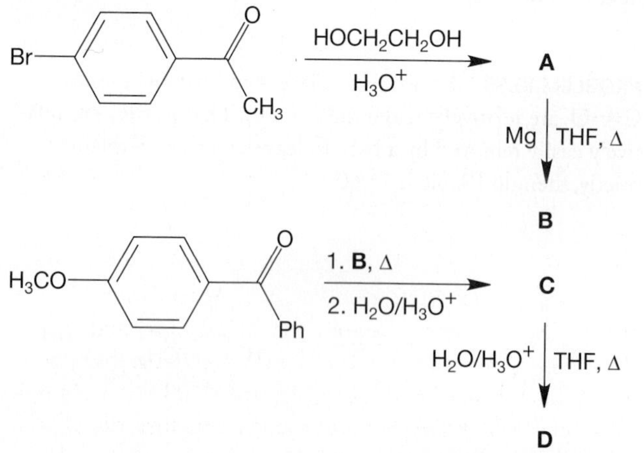

PROBLEM 16.63   Provide structures for compounds **A–D**. Spectral data for compound **D** are shown below. Although mechanisms are not required, a mechanistic analysis may prove helpful in deducing the structures.

Compound **D**

IR (neat): 2817 (w), 2717 (w), 1730 (s) cm^{-1}

^{1}H NMR (CDCl$_3$): δ 0.92 (t, $J = 7$ Hz, 6H), 1.2–2.3 (m, 5H), 9.51 (d, $J = 2.5$ Hz, 1H)

^{13}C NMR (CDCl$_3$): δ 11.4, 21.5, 55.0, 205.0

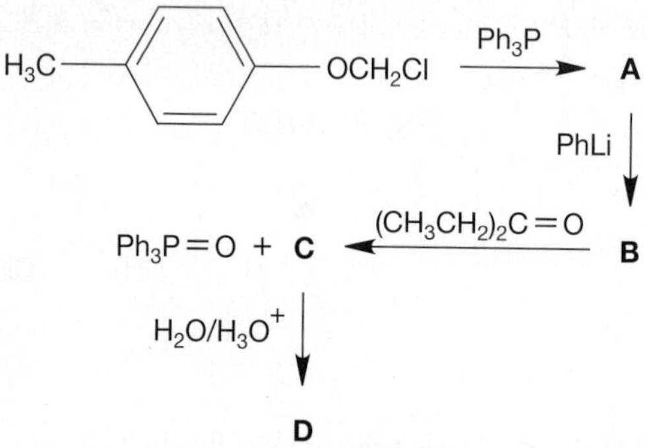

PROBLEM 16.64   Write a mechanism for the acid-catalyzed transformation of glycerol into acrolein.

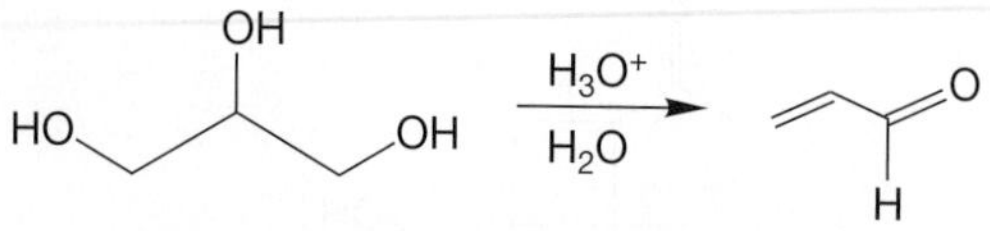

PROBLEM 16.65   This problem introduces another alcohol protecting group, methoxymethyl (MOM), particularly popular in the month of May. The MOM protecting group is stable to base, but it can be cleaved upon treatment with mild acid. Provide an arrow formalism for the introduction of the MOM protecting group. Also explain why this protecting group is readily cleaved upon treatment with aqueous acid.

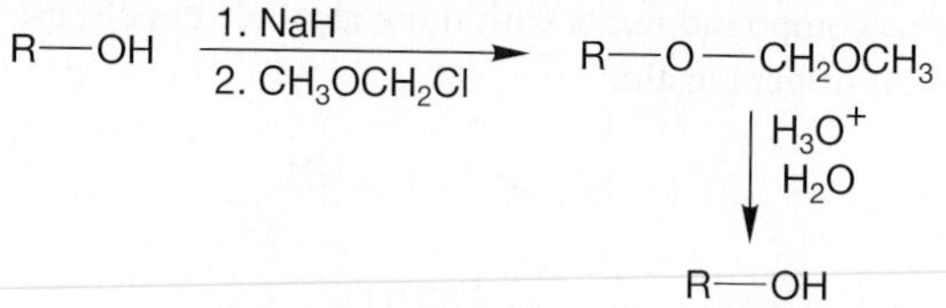

PROBLEM 16.66   Thiols are important participants in determining enzyme shapes and reactivity. Nature has selected for this thiol involvement because the sulfur-sulfur bond of the disulfide is formed reversibly. The reaction is

$$RSH + RSH \rightleftharpoons RS{-}SR$$

What is the name for the type of reaction that converts thiols into disulfides? What type of reaction is the reverse of this process? What extraordinarily common biological reagent will convert a thiol into a disulfide? What biological reagent will convert a disulfide into a thiol? *Hint*: see p. 814.

PROBLEM 16.67   Metaproterenol (orciprenaline, **1**) is a β-adrenoreceptor stimulant used therapeutically as a bronchodilator. In the synthesis of the hydrobromide salt of racemic **1**, outlined below, supply the appropriate reagents.

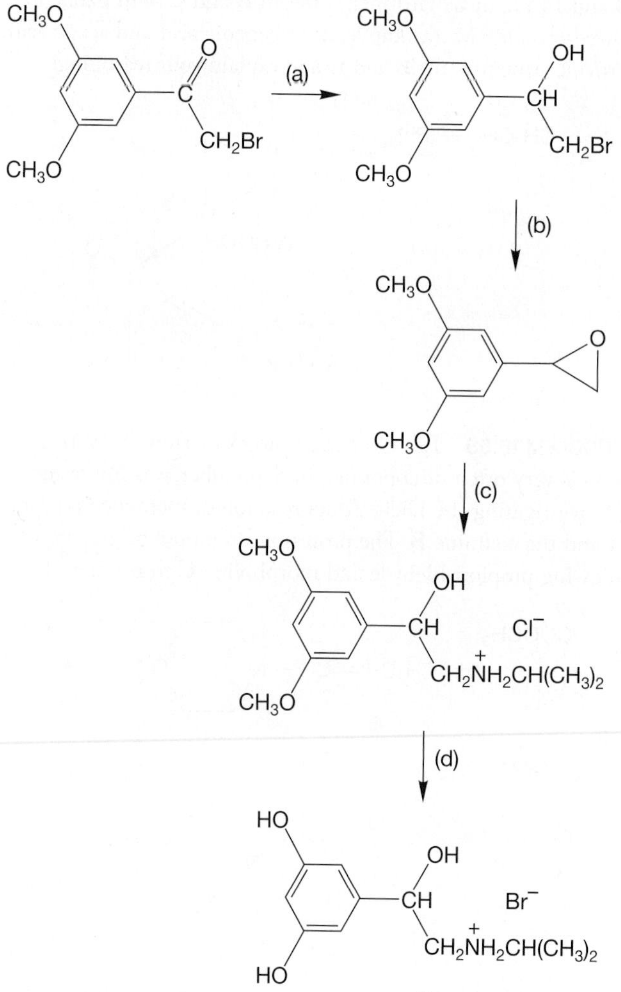

PROBLEM 16.68 Treatment of 2-thiabicyclo[2.2.1]-heptane (**1**) with 1 equivalent of hydrogen peroxide affords a mixture of diastereomers **2** and **3** of molecular formula $C_6H_{10}OS$. Both diastereomers **2** and **3** yield the same product **4** ($C_6H_{10}O_2S$) upon treatment with peracetic acid. Compound **4** is also produced when **1** reacts with 2 equivalents of peracetic acid. Propose structures for compounds **2**, **3**, and **4**. *Hint*: Start by drawing a good Lewis structure for the starting material **1**.

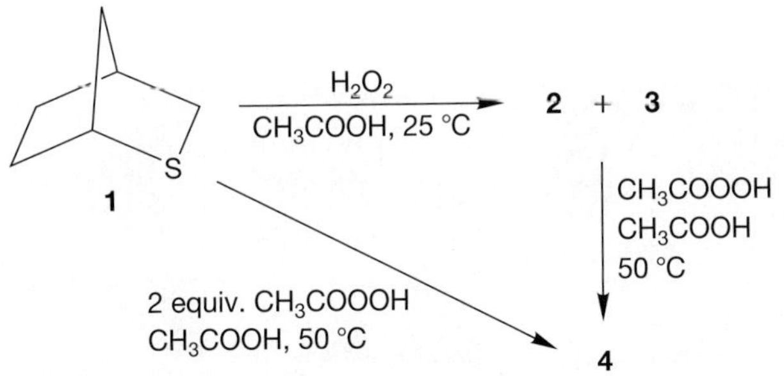

PROBLEM 16.69 Provide structures for compounds **A–C**.

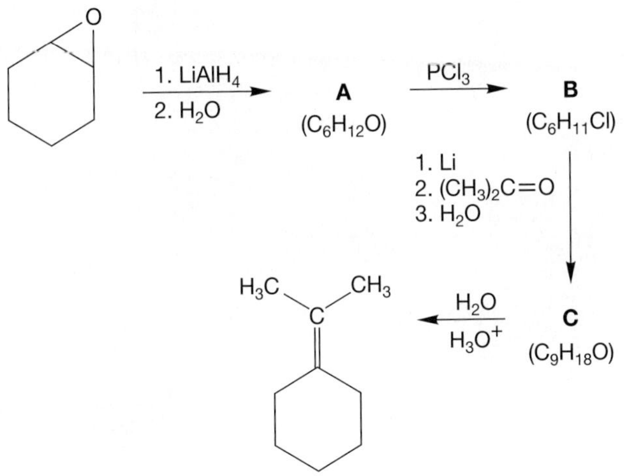

PROBLEM 16.70 Provide structures for compounds **A–C**.

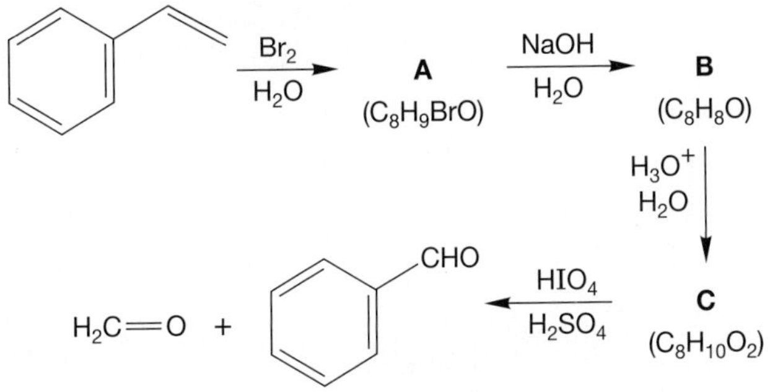

Use Organic Reaction Animations (ORA) to answer the following questions:

PROBLEM 16.71 Select the "Carbonyl hydration" reaction and view the animation several times. Notice that there are two intermediates in the reaction. The first intermediate is protonated acetone and the second is the protonated *gem*-diol. The final product in this reaction is higher in energy than the starting material. Why is this the case? Provide a one-word description for the first step in the reaction. Provide a one-word description of the second step. Is this reaction reversible?

PROBLEM 16.72 View the "Acetal formation" reaction. There are many intermediates in this reaction. How many? Which of them is the most stable intermediate? Is the acetal or the ketone favored in this reaction? Explain why. Remember that every step in this reaction is reversible. If you want to drive the reaction to the acetal, what would you do?

PROBLEM 16.73 The "Imine formation" reaction shows the methanamine attacking acetone in the first step. The alkoxide intermediate is then protonated by an acidic ammonium species in a proton transfer process. It is known that imine formation occurs faster under slightly acidic conditions (pH ~ 5). Why not show the acetone being protonated in the first step as we did in Carbonyl hydration and Acetal formation? Select the LUMO track for this animation and examine this MO on acetone by pausing the reaction at the beginning. There are several important facts that are confirmed by this calculated image. Which atom of the molecule has the most LUMO density? Can you see the effect of hyperconjugation? Where? Why?

PROBLEM 16.74 This chapter also covered the reactions that are animated in "Grignard reaction," "Carbonyl reduction," "Alcohol oxidation," "Wittig reaction," and "Diol cleavage." Observe each of these reactions. Why is the Grignard reaction so unusual? Why does the Wittig reaction start at such a high energy?

17

Carboxylic Acids

17.1 Preview

17.2 Nomenclature and Properties of Carboxylic Acids

17.3 Structure of Carboxylic Acids

17.4 Infrared and Nuclear Magnetic Resonance Spectra of Carboxylic Acids

17.5 Acidity and Basicity of Carboxylic Acids

17.6 Syntheses of Carboxylic Acids

17.7 Reactions of Carboxylic Acids

17.8 Special Topic: Fatty Acids

17.9 Summary

17.10 Additional Problems

NONSTEROIDAL ANTIINFLAMMATORY DRUGS Most nonsteroidal antiinflammatory drugs (NSAIDs) are carboxylic acids. Aspirin, ibuprofen, and naproxen are the three most common NSAIDs. NSAIDs are frequently used to relieve muscle and back pain, which improves our flexibility.

If I mix CH$_2$ with NH$_4$ and boil the atoms in osmotic fog, I should get speckled nitrogen!

—DONALD DUCK[1]

He's talking chemical talk!

—HUEY

But he knows nothing

—DEWEY

About chemicals!

—LOUIE

17.1 Preview

In this chapter, we explore the properties and chemistry of carboxylic acids (general formula R—COOH). As with other functional groups, we will begin by learning the IUPAC naming system, and then move on to structure. Carboxylic acids have a rich chemistry; they are both acids (hence the name) and bases, and we saw in Chapter 16 how the carbonyl group is a locus of reactivity. Although many new reactions will appear in this chapter, most important will be the emergence of a general reaction mechanism, addition–elimination, that we have already seen briefly in Chapter 14. Chapter 18 will extend the exploration of this process.

WORKED PROBLEM 17.1 What are the reactions we have learned for making carboxylic acids?

ANSWER Carboxylic acids can be made from oxidative ozonolysis of alkenes, by oxidation of primary alcohols using dichromate under aqueous conditions, and by oxidation of aldehydes.

ESSENTIAL SKILLS AND DETAILS

1. This chapter's centerpiece is the Fischer esterification–acid hydrolysis equilibrium. If you know this pair of reactions well, you will have much of this chapter under control. The two mechanisms are exactly the reverse of each other—one is the "forward" reaction, the other is the "backward" reaction.

2. The reaction of a carboxylic acid with two equivalents of an alkyllithium reagent, followed by hydrolysis, yields a ketone. This reaction is a very effective way to make these carbonyl compounds. Be sure you understand why two equivalents of alkyllithium are necessary.

3. The reaction of LiAlH$_4$ with a carboxylic acid results in formation of a primary alcohol through addition of two equivalents of hydride.

[1]This quotation comes from a 1944 issue of *Walt Disney's Comics and Stories*, "Donald Duck the Mad Chemist." It was discovered by Peter P. Gaspar, then a postdoctoral fellow at Caltech, now a professor at Washington University. Those of us who work with CH$_2$ have still not reproduced Professor Duck's experiment with osmotic fog, but a few of us are still trying.

17.2 Nomenclature and Properties of Carboxylic Acids

In the IUPAC systematic nomenclature, the final "e" of the parent alkane is dropped, the suffix "oic" is added, and the separate word "acid" is added. Diacids are named similarly, except that the suffix is "dioic," and the final "e" is not dropped. Cyclic acids are named as "cycloalkanecarboxylic acids." Table 17.1 shows some common mono- and diacids along with some of their physical properties and common names, many of which are still used.

TABLE 17.1 Some Carboxylic Acids and Their Properties

Structure	Systematic Name	Common Name	bp (°C)	mp (°C)	pK_a
WEB 3D HCOOH	Methanoic acid	Formic acid	100.7	8.4	3.77
CH_3COOH	Ethanoic acid	Acetic acid	117.9	16.6	4.76
CH_3CH_2COOH	Propanoic acid	Propionic acid	141	−20.8	4.87
$CH_3CH_2CH_2COOH$	Butanoic acid	Butyric acid	165.5	−4.5	4.81
$CH_3CH_2CH_2CH_2COOH$	Pentanoic acid	Valeric acid	186	−33.8	4.82
$CH_3CH_2CH_2CH_2CH_2COOH$	Hexanoic acid	Caproic acid	205	−2	4.83
$CH_2{=}CHCOOH$	Propenoic acid	Acrylic acid	141.6	13	4.25
⬠—COOH	Cyclopentanecarboxylic acid		216	−7	4.91
⬡—COOH	Cyclohexanecarboxylic acid		232	31	4.88
WEB 3D ⬡(benzene)—COOH	Benzenecarboxylic acid	Benzoic acid	249	122	4.19
HOOC—COOH	Ethanedioic acid	Oxalic acid		190	1.23 / 4.19[a]
HOOC—CH_2—COOH	Propanedioic acid	Malonic acid		136	2.83 / 5.69[a]
HOOC—$(CH_2)_2$—COOH	Butanedioic acid	Succinic acid		188	4.16 / 5.61[a]
HOOC—$(CH_2)_3$—COOH	Pentanedioic acid	Glutaric acid	~300	99	4.31 / 5.41[a]
HOOC—$(CH_2)_4$—COOH	Hexanedioic acid	Adipic acid	>300	156	4.43 / 5.41[a]
HOOC—CH=CH—COOH cis Isomer	*cis*-Butenedioic acid	Maleic acid		140	1.92 / 6.23[a]
HOOC—CH=CH—COOH trans Isomer	*trans*-Butenedioic acid	Fumaric acid		~300	3.02 / 4.38[a]

[a] These are the pK_a values for loss of the second proton.

For substituted acids, the numbering of the longest chain begins with the acid carbon itself, which is given the number "1" (Fig. 17.1). In cyclic compounds it is the carbon attached to the "COOH" that is carbon "1" in naming. Because carboxylic acids are the highest priority functional group, all functional groups except alkenes and

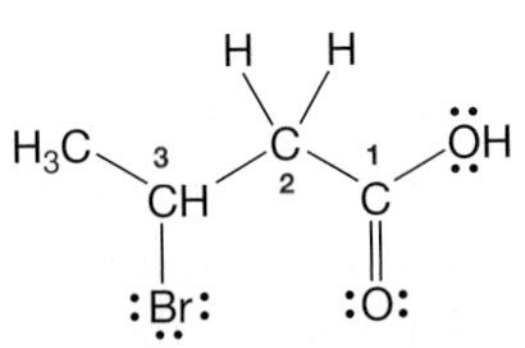

FIGURE 17.1 Some carboxylic acids and their names.

3-Bromobutanoic acid
(3-bromobutyric acid)

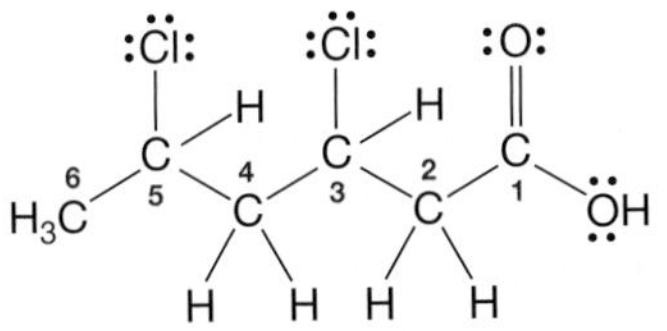

3,5-Dichlorohexanoic acid

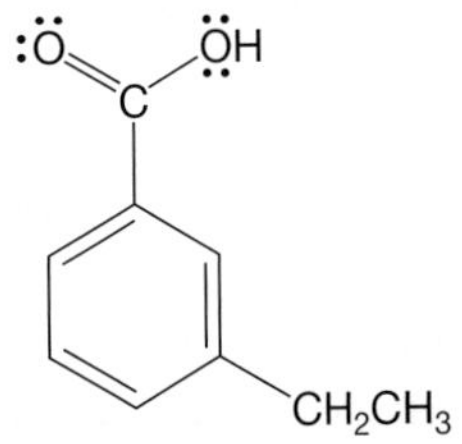

3-Ethylbenzoic acid
(*m*-ethylbenzoic acid)

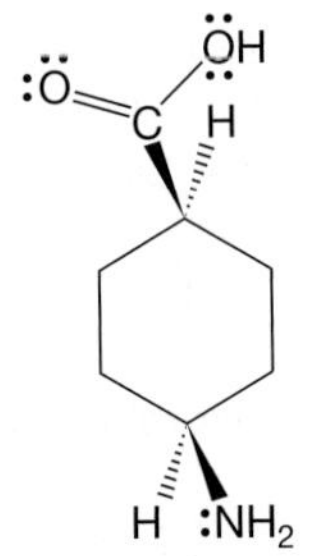

cis-**4-Aminocyclohexane-carboxylic acid**

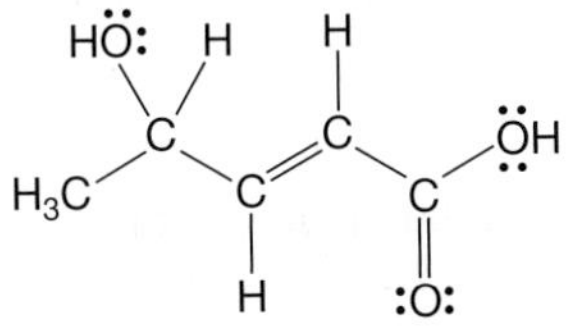

(*E*)-4-Hydroxy-2-pentenoic acid

alkynes are listed as substituents in the prefix. Alkenes and alkynes are indicated in the parent name. You will need to be able to recognize the condensed formula for a carboxylic acid, which is RCOOH, often written as RCO_2H.

PROBLEM 17.2 Name the following compounds:

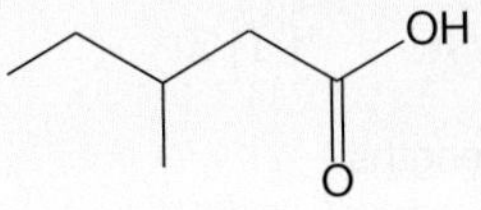

(a)

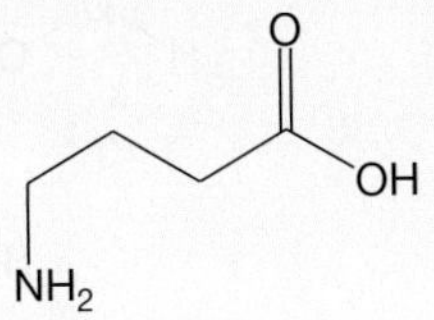

(b)

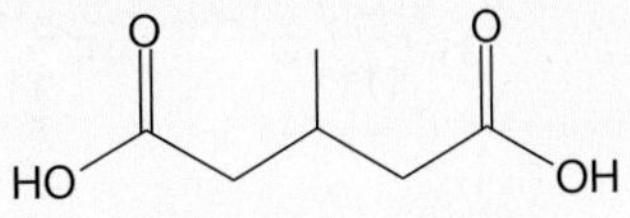

(c)

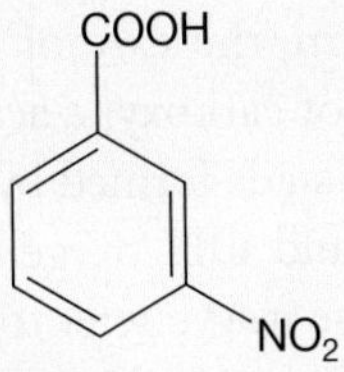

(d)

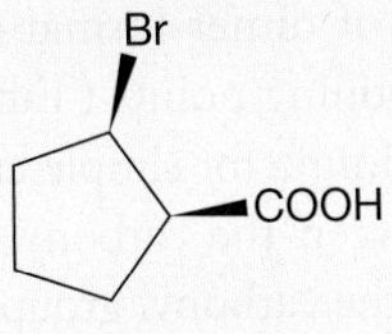

(e) 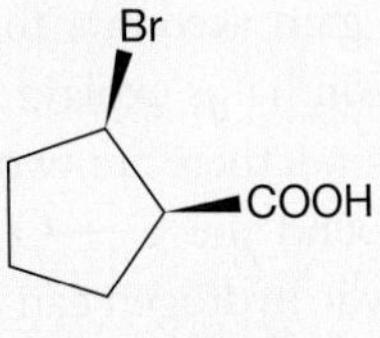
(racemic)

(f)
This enantiomer

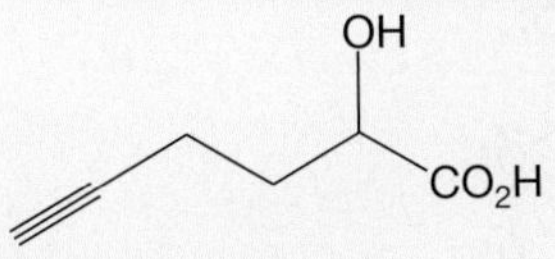

(g)

A **carboxylate anion** is the conjugate base of a carboxylic acid. Carboxylate anions are encountered frequently in any study of organic chemistry. They are also often found listed as contents of soaps and shampoos. So, it is worth learning the IUPAC system for naming the carboxylates. We refer to $RCOO^-$ and its counterion (Li^+, Na^+, or K^+, for example) as a salt because the $RCOO^- M^+$ is a result of an acidic hydrogen replaced by a metal, which is the definition of a salt. The salt is named as if it were a carboxylic acid, the suffix being changed from *–ic acid* to *–ate*. The name of the metal counterion is added before the parent organic name as a separate word (Fig. 17.2). An important property of carboxylate salts is that they are often soluble in water.

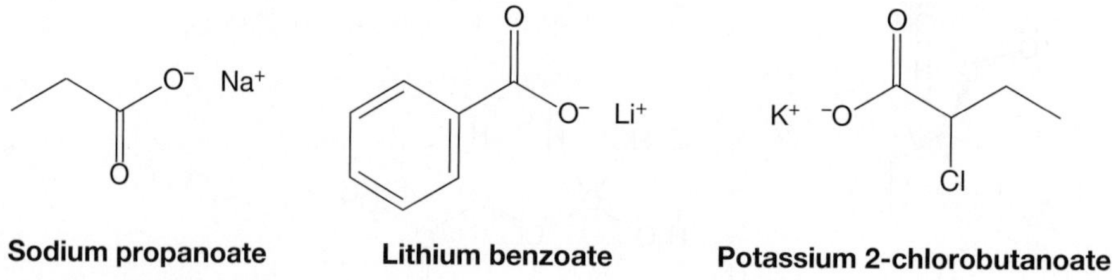

FIGURE 17.2 Some carboxylate salts and their names.

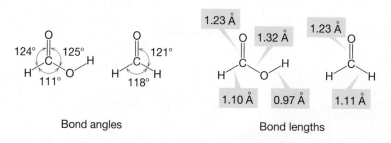

Sodium propanoate   **Lithium benzoate**   **Potassium 2-chlorobutanoate**

17.3 Structure of Carboxylic Acids

In Figure 17.3 we compare the structure of formic acid and the structure of formaldehyde. The structure of acids is essentially what one would expect by analogy to other carbonyl compounds. The carbonyl carbons are sp^2 hybridized, and therefore carbonyl compounds are planar. The strong carbon–oxygen double bonds are appropriately short, about 1.23 Å.

FIGURE 17.3 The structure of formic acid compared to that of formaldehyde.

124° 125°
111°

121°
118°

Bond angles

1.23 Å   1.32 Å   1.23 Å

1.10 Å   0.97 Å   1.11 Å

Bond lengths

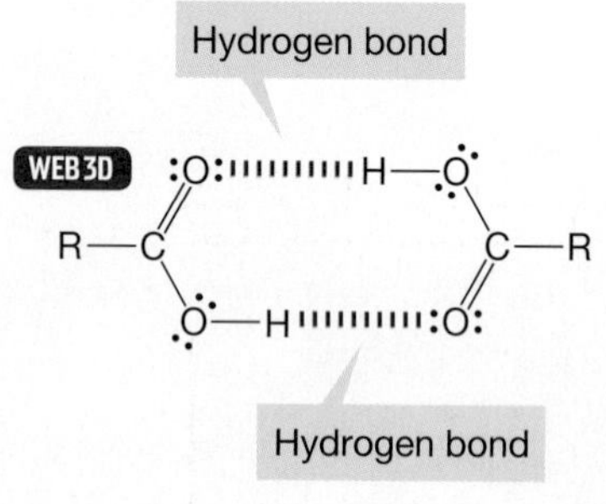

FIGURE 17.4 A dimeric carboxylic acid.

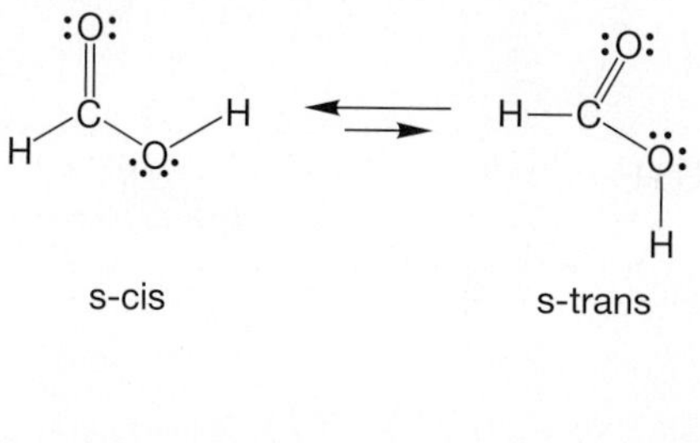

Hydrogen bond

WEB 3D

Hydrogen bond

There are two complications with the structural picture. First, in solution the simplest carboxylic acids are substantially dimerized (Fig. 17.4). It is possible to form two rather strong hydrogen bonds (~7 kcal/mol, each) in the dimeric form, and this energy gain accounts for the ease of dimer formation. In turn, the ease of dimer formation helps explain the high boiling points (Table 17.1) of carboxylic acids.

Second, there are two energy minima for simple carboxylic acids formed by rotation around the C—O bond between the carbonyl carbon and OH oxygen. The hydroxylic hydrogen can be anti to the carbonyl group, or eclipse it. We refer to these conformations as s-trans and s-cis, respectively (p. 523) as shown in Figure 17.5. These conformations are real energy minima separated in the gas phase by a barrier of about 13 kcal/mol. The s-cis form for formic acid is more stable by about 6 kcal/mol.

FIGURE 17.5 The s-cis and s-trans forms of a simple carboxylic acid.

s-cis   s-trans

PROBLEM 17.3 Explain why the s-cis form of the carboxylic acid is more stable than the s-trans form.

PROBLEM 17.4 Calculate the equilibrium constant for the equilibrium between s-cis and s-trans formic acid at 25 °C.

WORKED PROBLEM 17.5 Interconversion of the s-cis and s-trans forms can occur very easily in acidic water solution. Devise a mechanism not involving simple rotation about a bond.

ANSWER A series of protonations and deprotonations will do the job.

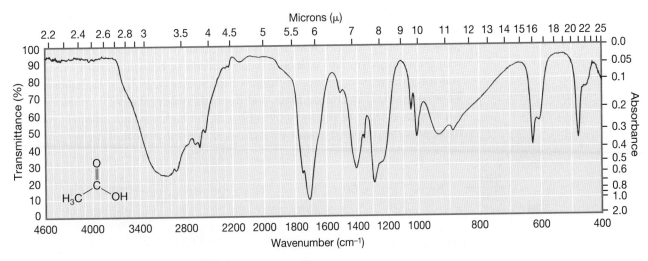

17.4 Infrared and Nuclear Magnetic Resonance Spectra of Carboxylic Acids

The most prominent features of the IR spectra of carboxylic acids are the strong absorption for the C=O stretch at about 1710 cm^{-1} and the strong and very broad band centered at 3100 cm^{-1} for the O—H stretching frequencies. Figure 17.6 shows the IR spectrum of acetic acid. As usual, conjugation of the carbonyl shifts the C=O stretch to a lower frequency by about 20 cm^{-1}. The O—H stretching band is broad for the same reasons that the O—H stretching bands of alcohols are broad (p. 710). There are many hydrogen-bonded dimers and oligomers with different O—H bond strengths present in any carboxylic acid sample.

The NMR spectra for a carboxylic acid are also informative. Hydrogens in the position α to the carbonyl group of carboxylic acids are deshielded by both the

FIGURE 17.6 The IR spectrum of acetic acid.

electron-withdrawing nature of the carbonyl group and by local magnetic fields set up by the circulation of electrons in the π bond. These hydrogens appear in the same region as other hydrogens α to carbonyl groups at δ 2–2.7 ppm (see Table 15.4). The acidic hydrogen of the RCOOH is the most deshielded hydrogen of all the typical organic functional groups. It usually appears in the region of δ 10–13 ppm as a broad singlet. Like the hydrogen in the OH group of an alcohol, the hydrogen of a carboxylic acid is exchangeable, and the signal vanishes when D_2O is added to the sample and the OH becomes OD. Alcohol hydrogens rarely appear as low as δ 10–13 ppm.

The carbon of the carboxylic acid group is also strongly deshielded and appears downfield, as do other carbonyl carbons in ^{13}C NMR spectra. The position is about 180 ppm, slightly upfield of the chemical shifts of the carbonyl carbons of aldehydes and ketones.

17.5 Acidity and Basicity of Carboxylic Acids

Because carboxylic acids are both acids and bases, we might well expect that a rather diverse chemistry would be found. Carboxylic acids wouldn't be called acids if they were not strong Brønsted acids. In fact, carboxylic acids are the organic functional group that Nature uses as an acid, just as amines are the organic functional group that Nature uses as its organic base. Table 17.1 gave the pK_a values for some mono- and dicarboxylic acids.

Organic acids (pK_a = 3–5) are much stronger acids than alcohols (pK_a = 15–17). Reasons for this increased acidity are not hard to find, but assessing their relative importance is more difficult. One explanation focuses on the formation of a resonance-stabilized carboxylate anion after proton loss (Fig. 17.7).

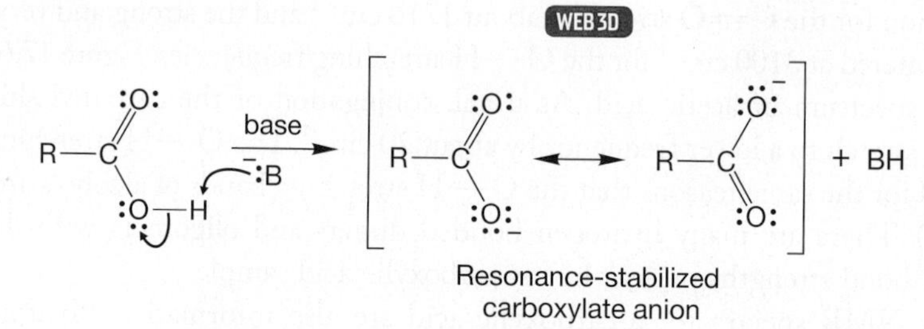

FIGURE 17.7 The formation of a resonance-stabilized carboxylate anion through removal of a proton from the hydroxyl group of a carboxylic acid.

Resonance-stabilized carboxylate anion

Deprotonation of an alcohol gives an oxygen anion, an alkoxide, but the ion is not stabilized by resonance (Fig. 17.8). It is reasonable to accept the notion that the formation of the highly stabilized carboxylate is responsible for the high acidity of carboxylic acids. Recently, however, this traditional view has been challenged by at least three research groups with both theoretical and experimental arguments. In various ways, the research groups of Andrew Streitwieser (b. 1927) at Berkeley, Darrah Thomas (b. 1932) at Oregon State, and Kenneth Wiberg (b. 1927) at Yale pointed out the importance of the inductive effect of the highly polar carbonyl group. In their view, the increased acidity of acetic acid over ethyl alcohol derives not from

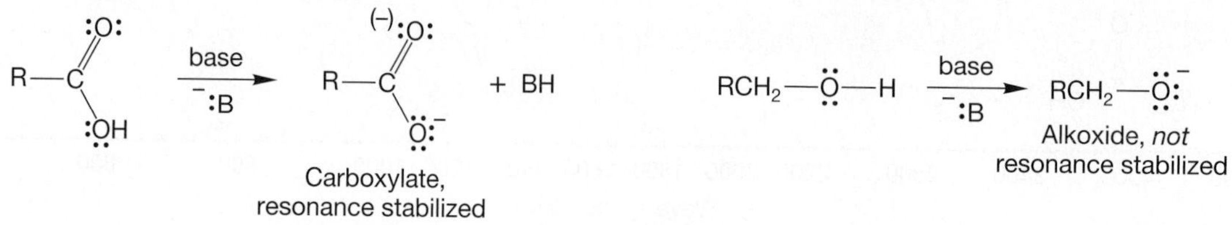

Carboxylate, resonance stabilized

Alkoxide, *not* resonance stabilized

FIGURE 17.8 A comparison of carboxylate and alkoxide anions.

resonance stabilization of the carboxylate anion, but from the electrostatic stabilization afforded the developing negative charge by the adjacent polar carbonyl group in which the carbon bears a partial positive charge (Fig. 17.9). They reckon that resonance could account for no more than about 15% of the total stabilization.

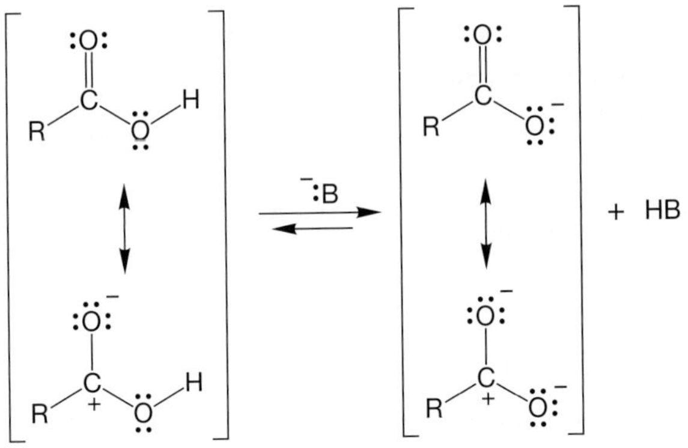

FIGURE 17.9 Electrostatic stabilization may account for the acidity of carboxylic acids. The highly polar carbonyl group stabilizes the carboxylate anion.

Now we should ask *why* resonance might not be so important. The carboxylate anion is surely delocalized, whereas the alkoxide ion is not. The key point is that the carbonyl group of the acid is *already* so polar that little further delocalization can occur as the anion is formed. The negative charge on the carbonyl oxygen and the positive charge on the carbonyl carbon aren't developed as the oxygen–hydrogen bond breaks; the charges are already largely there! Both factors, the resonance stabilization of the carboxylate anion and the polarity of the carbonyl group, surely contribute to the acidity of carboxylic acids. The question is only over their relative importance. We will use the resonance stabilization argument, keeping in mind that there may be more to the story.

PROBLEM SOLVING

Watch out! Carboxylic *acids* are called *acids* because they are *acidic*. They will deprotonate in base before they do any other reactions typical of compounds containing carbonyl groups. "Everyone" forgets that seemingly simple fact, and problem writers may try to trap you. Always remember: A carboxylic acid deprotonates in base.

WORKED PROBLEM 17.6 Explain why fluoroacetic acid has a lower pK_a than acetic acid (2.66 vs. 4.76).

ANSWER The structure of the anion formed by removal of a proton tells the story. The dipole in the carbon–fluorine bond stabilizes the fluoroacetate anion, which makes removal of the acidic proton easier.

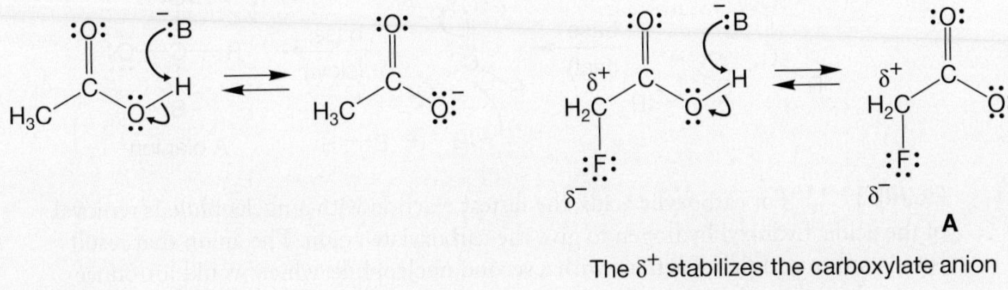

The δ^+ stabilizes the carboxylate anion

The stabilization by electron-withdrawing groups pointed out in Problem 17.6 is a general phenomenon. As Table 17.2 shows, acids bearing electron-withdrawing groups are stronger acids than their parent compounds. The further the electron-withdrawing group is from the acid, the smaller the effect on the pK_a (Fig. 17.10).

TABLE 17.2 Acidities of Some Substituted Carboxylic Acids

Acid	pK_a
Acetic	4.76
α-Chloroacetic	2.86
α,α-Dichloroacetic	1.29
α,α,α-Trichloroacetic	0.65
α-Fluoroacetic	2.66
α,α-Difluoroacetic	1.24
α,α,α-Trifluoroacetic	−0.25
α-Bromoacetic	2.86
α-Iodoacetic	3.12
α-Nitroacetic	1.68

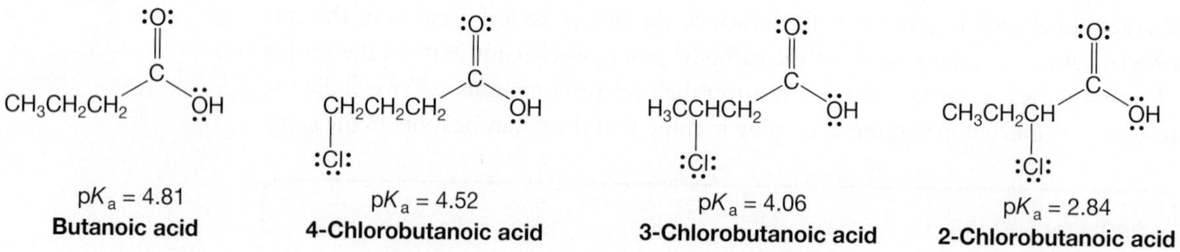

FIGURE 17.10 The greater the distance (number of bonds) between an electron-withdrawing group, here chlorine, and the point of ionization, the less effect an electron-withdrawing group has on the acidity.

Carboxylic acids are electrophiles (Lewis acids) as well as Brønsted acids. The presence of the carbonyl group ensures that. Remember all the addition reactions of carbonyl groups encountered in Chapter 16. However, expression of the Lewis acidity of the carbonyl carbon is often thwarted because the easiest reaction with a nucleophile is not addition to the carbonyl group, but removal of the acid's OH hydrogen to give the carboxylate anion. Once it is formed, the carboxylate anion is far more resistant to addition than an ordinary carbonyl because, in this case, addition would introduce a *second* negative charge (Fig. 17.11).

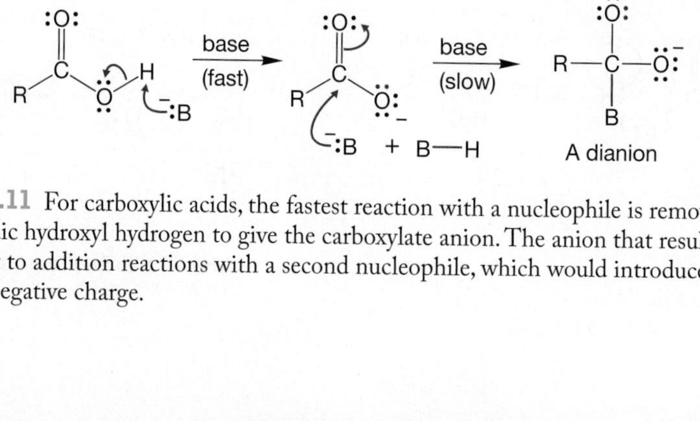

FIGURE 17.11 For carboxylic acids, the fastest reaction with a nucleophile is removal of the acidic hydroxyl hydrogen to give the carboxylate anion. The anion that results is resistant to addition reactions with a second nucleophile, which would introduce a second negative charge.

Nevertheless, some especially strong nucleophiles are able to do this second addition. An example is the organolithium reagent, RLi, and shortly we will see the synthetic consequences of this reaction. The formation of the dianion of Figure 17.12 is not as bad as it looks because the O—Li bond is substantially covalent. It is not completely ionic.

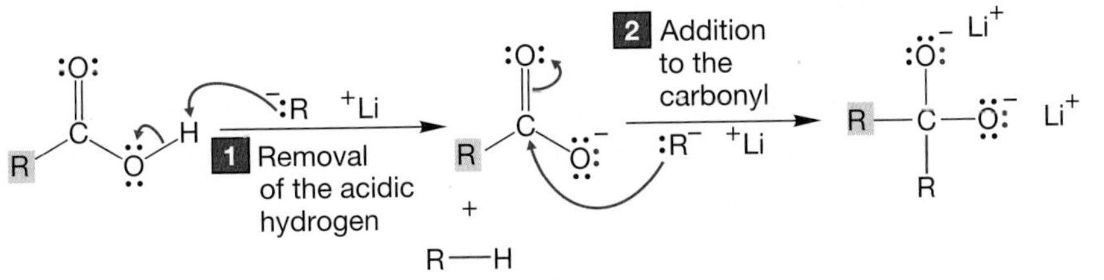

FIGURE 17.12 Some strong nucleophiles can add to the carboxylate anion. The organolithium reagent is one example.

WORKED PROBLEM 17.7 Anticipate a little. What synthetic use can you see for the reaction in Figure 17.12? *Hint*: What will happen when water is added once the dianion is formed?

ANSWER When water is added, a hydrate will be formed. You know from Chapter 16 that simple hydrates are generally unstable relative to a carbonyl compound and water. So, this reaction should be a good synthesis of ketones. It is, as we will see on page 856.

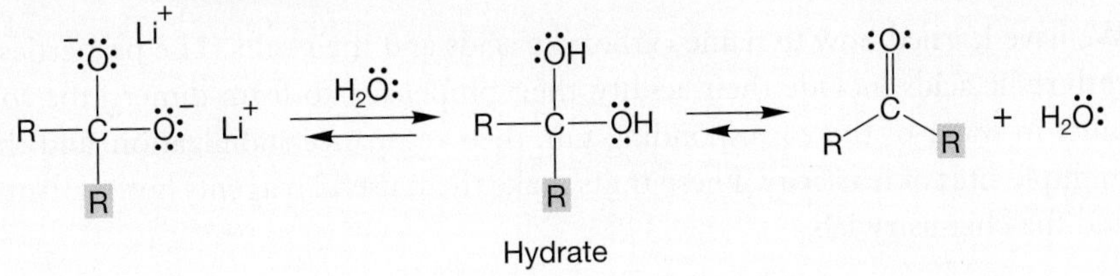

Hydrate

Finally, organic acids are nucleophiles (Lewis bases), and can react with electrophiles (Lewis acids). The simplest reaction is the protonation of a carboxylic acid. There are two possible sites for protonation, the carbonyl oxygen and the hydroxyl oxygen (Fig. 17.13). Which will it be?

The intermediate in which protonation has taken place at the carbonyl oxygen is resonance stabilized, whereas the species in which the hydroxyl oxygen is protonated is not (Fig. 17.14). Moreover, protonation of the hydroxyl oxygen is destabilized by the dipole in the carbon–oxygen double bond, which places a partial positive charge on carbon. It will not be energetically favorable to introduce a positive charge adjacent to this already partially positive carbon. The more stable cation, in which the carbonyl oxygen is protonated, is preferred.

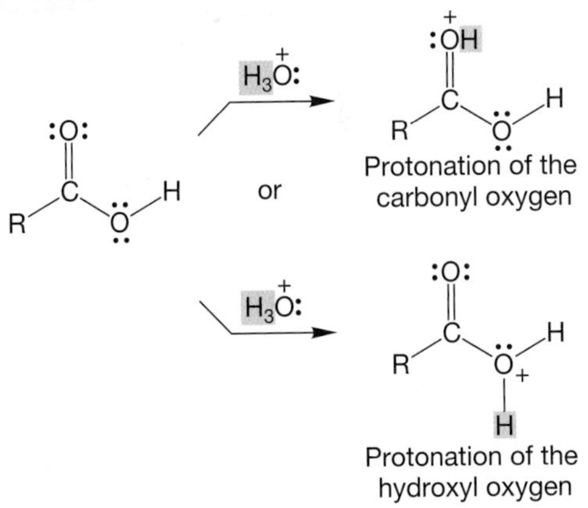

Protonation of the carbonyl oxygen

Protonation of the hydroxyl oxygen

FIGURE 17.13 The two possible sites for protonation of a carboxylic acid.

Resonance stabilization of the product of protonation of the carbonyl group

Unstabilized by resonance, destabilized by the C═O dipole

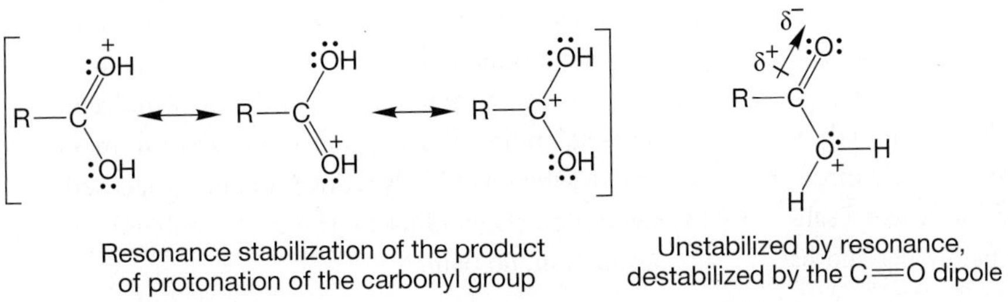

FIGURE 17.14 Delocalization stabilizes the intermediate resulting from protonation of the carbonyl oxygen. The carbon–oxygen dipole destabilizes the cation formed from protonation of the hydroxyl group.

For these reasons, carboxylic acids are more strongly basic at the carbonyl oxygen than at the hydroxyl oxygen. Does this mean that the hydroxyl group is never protonated? Certainly not, but reaction at the more basic site is favored and will be faster.

PROBLEM 17.8 Explain why the thermodynamic stability of the protonated carbonyl should influence the rate (a kinetic parameter) of protonation.

With so many possible sites for reaction we might expect a rich chemistry of carboxylic acids. That idea would be exactly right. Figure 17.15 summarizes the sites of reactivity we have discussed so far.

FIGURE 17.15 The various sites of reactivity for a carboxylic acid.

Lewis basicity (greater) (nucleophile)

Lewis basicity (lesser) (nucleophile)

Brønsted acidity (electrophile)

Lewis acidity (electrophile)

Summary

We have learned how to name carboxylic acids and their salts. The properties of carboxylic acids include their acidity, their propensity to form dimers, the solubility in water of the corresponding salt, their resonance stabilization, and their multiple sites of reactivity. These traits make them useful reagents both in Nature and the chemistry lab.

SALICYLIC ACID

The simple aromatic carboxylic acid, salicylic acid, is a plant hormone and is involved in many of the marvelously complex interactions between plants and animals that makes the study of biology so fascinating. Here's a typical example involving the voodoo lily. The flower of the voodoo lily emits foul odors that attract flies, and the flies are used to transmit pollen from the male reproductive organs to the female organs of another lily. In the late afternoon, salicylic acid triggers the first of two surges of heat, some 10–20 °C above normal. This first surge releases the odor and attracts the flies, which become trapped and coated with pollen. A second heat wave the next morning opens the flower, awakens the pollen-covered flies, which escape until the afternoon when they are attracted to another voodoo lily and deposit the pollen.

Salicylic acid is also an analgesic. Indeed, the chewing of willow bark, which contains salicylic acid, has been used to control pain for thousands of years. You are probably most familiar with salicylic acid in its acetylated form, acetyl salicylic acid, or aspirin. Aspirin is a versatile pain killer and has

Salicylic acid

Voodoo lily

myriad uses. It was synthesized as early as 1853, and its analgesic properties were recognized by a group at Farbenfabriken Baeyer in 1897. It apparently works by inhibiting the production of an enzyme, prostaglandin cyclooxygenase, that catalyzes the synthesis of molecules called prostaglandins. Prostaglandins are active in many ways, one of which is to help transmit pain signals across synapses. No prostaglandins, no signal transmission; no signal transmission, no pain.

17.6   Syntheses of Carboxylic Acids

We already know a number of routes to carboxylic acids, and this section will add an important new one, the reaction of organometallic reagents with carbon dioxide.

17.6a Oxidative Routes   As you know, there are several oxidative reactions that produce carboxylic acids (Fig. 17.16). Primary alcohols and aldehydes can be oxidized to acids by a variety of oxidants including HNO_3, $KMnO_4$, CrO_3, $K_2Cr_2O_7$, or RuO_4 (p. 802). In Figure 17.16, these reagents are generalized by "[O]", which stands for oxidant. Alkenes with a hydrogen attached to one or both of the doubly bonded carbons can undergo ozonolysis using an oxidative workup to give acids (p. 436). In addition, the side chains of alkyl aromatic compounds can be oxidized to acid groups with $KMnO_4$ (p. 613).

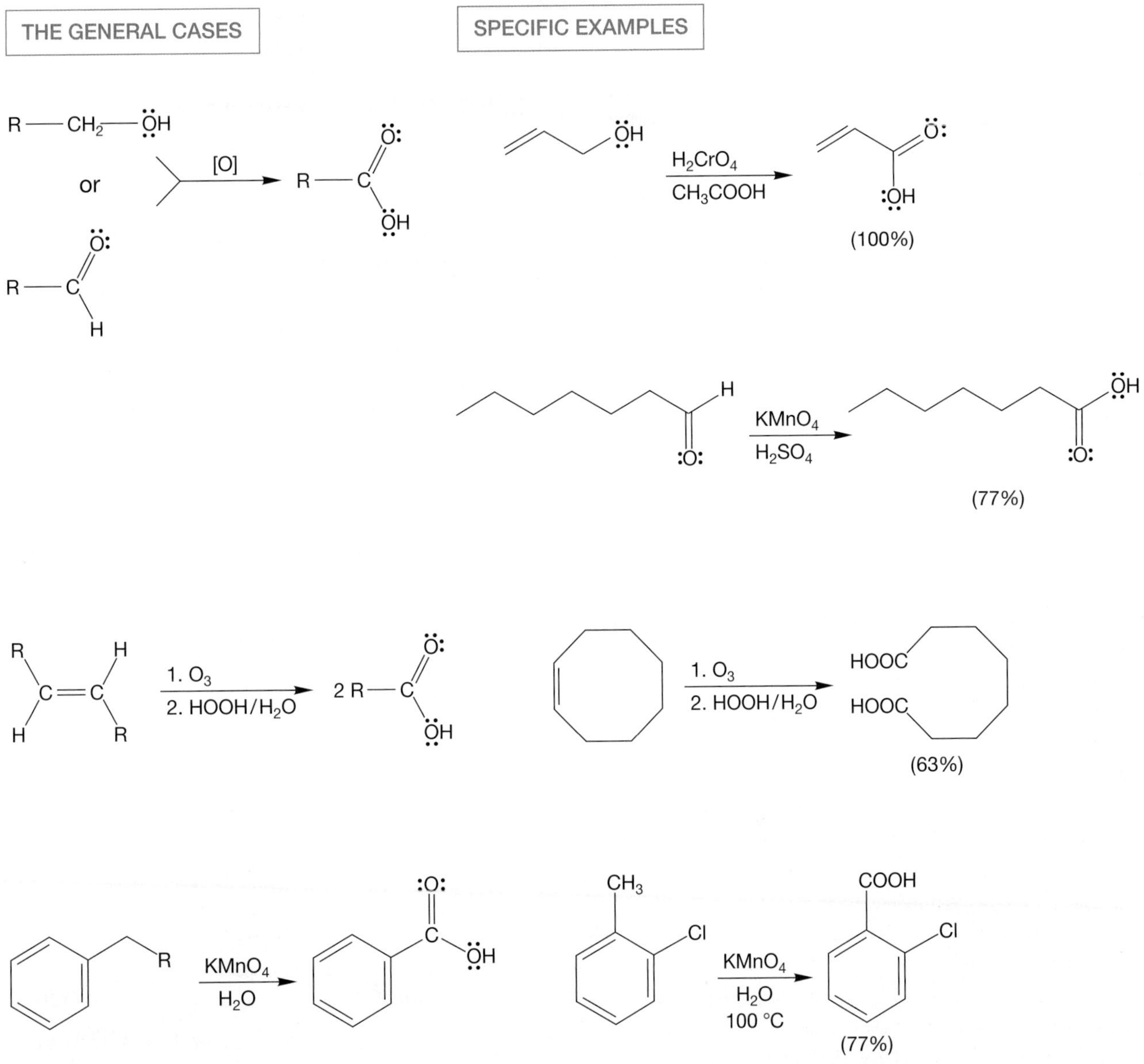

THE GENERAL CASES   SPECIFIC EXAMPLES

FIGURE 17.16 Oxidative routes to acids.

Alkenes can also be oxidatively cleaved by potassium permanganate to give carboxylic acids (Fig. 17.17). We already know that alkenes form vicinal diols when treated with basic permanganate or OsO_4 (p. 443). The reaction with permanganate can go further to form two acids. The trick is to use the polyether 18-crown-6 (p. 254) to make $KMnO_4$ soluble in benzene. The crown ether has a great ability to bind potassium ions, and this property allows the negatively charged permanganate ion (MnO_4^-) to follow the bound cation K^+ (crown) into solution, where it is more available for reaction with the organic substrate.

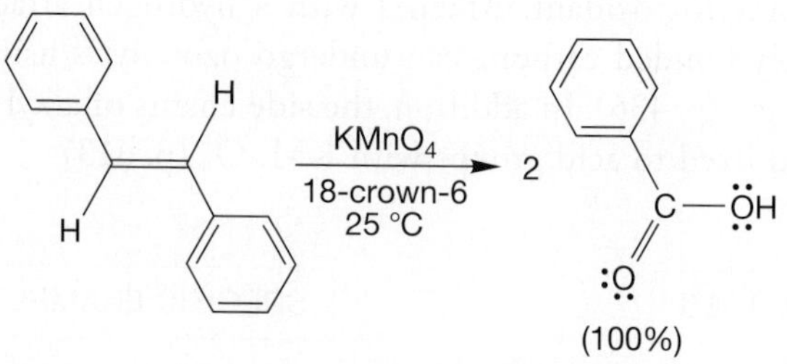

FIGURE 17.17 Another oxidative route to carboxylic acids.

17.6b Reaction of Organometallic Reagents with Carbon Dioxide

Carboxylic acids can be obtained from the Grignard reaction. Like any other compound containing a carbonyl group, carbon dioxide reacts with organometallic reagents through an addition reaction (Fig. 17.18). The initial product is a carboxylate salt that is converted into the acid when the reaction mixture is acidified. This reaction is very general and a source of many acids. The typical procedure for these reactions is simply addition of dry ice to the Grignard reagent.

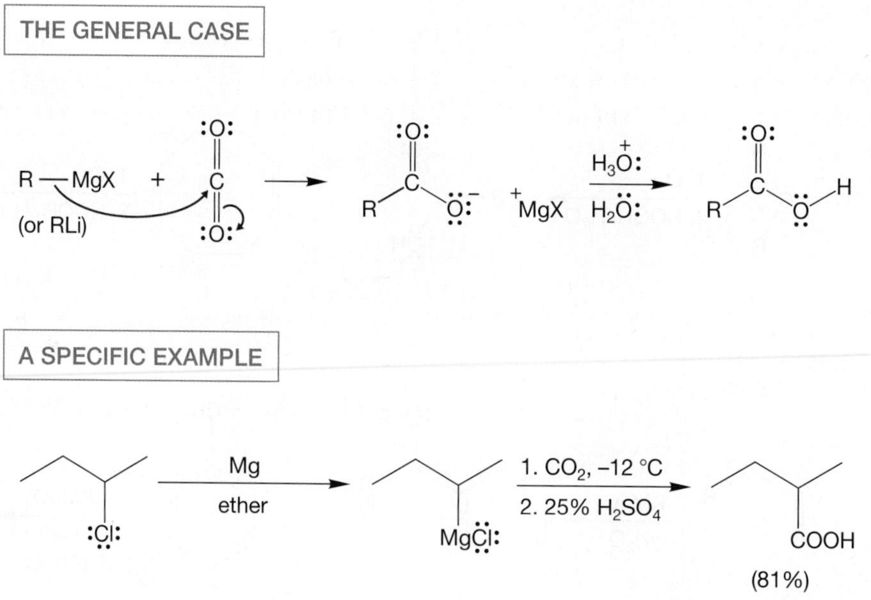

FIGURE 17.18 A general route to carboxylic acids uses the carboxylation of Grignard reagents with carbon dioxide.

PROBLEM 17.9 Starting from inorganic reagents, 1-butanol, CO_2, and tosyl chloride as your only sources of carbon, devise syntheses of the following molecules:

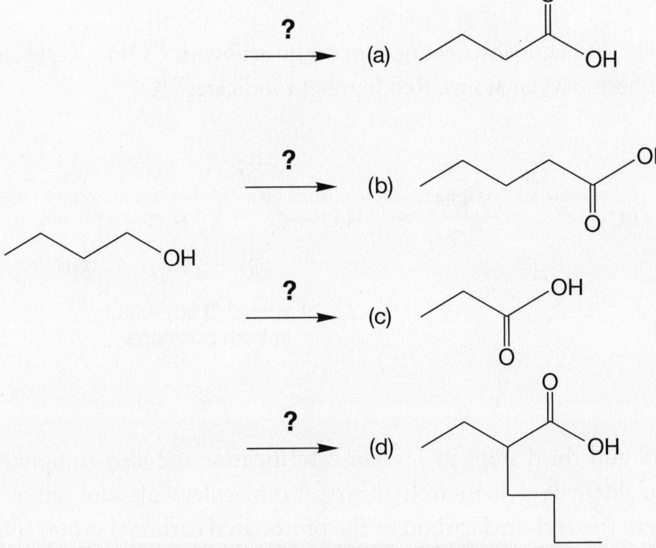

17.7 Reactions of Carboxylic Acids

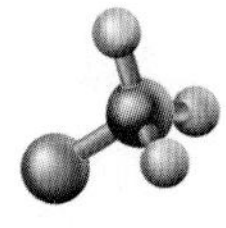

17.7a Formation of Esters: Fischer Esterification

An ester is a derivative of a carboxylic acid in which the hydrogen of the hydroxyl group has been replaced with an R group. We will examine the nomenclature, properties, and chemistry of esters in detail in Chapter 18, but one of the most important syntheses of these species involves the acid-catalyzed reaction of carboxylic acids with excess alcohol, and must be given here. The reaction is called **Fischer esterification** after the great German chemist Emil Fischer (1852–1919) (Fig. 17.19).

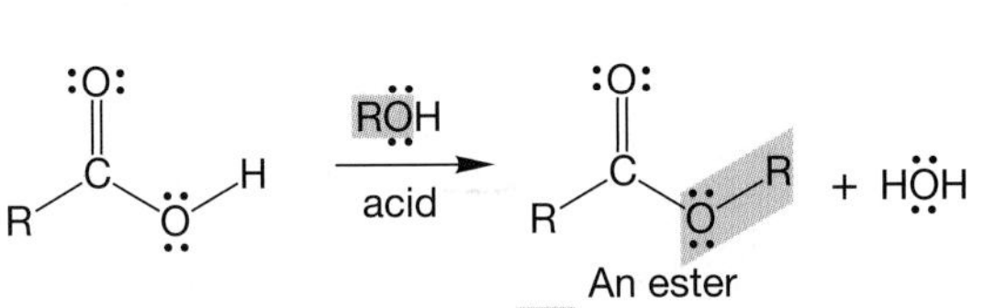

We can write the mechanism for this reaction rather easily, because its important steps are quite analogous to reactions of other carbonyl-containing compounds (Fig. 17.20). As in the reaction of

Fischer esterification

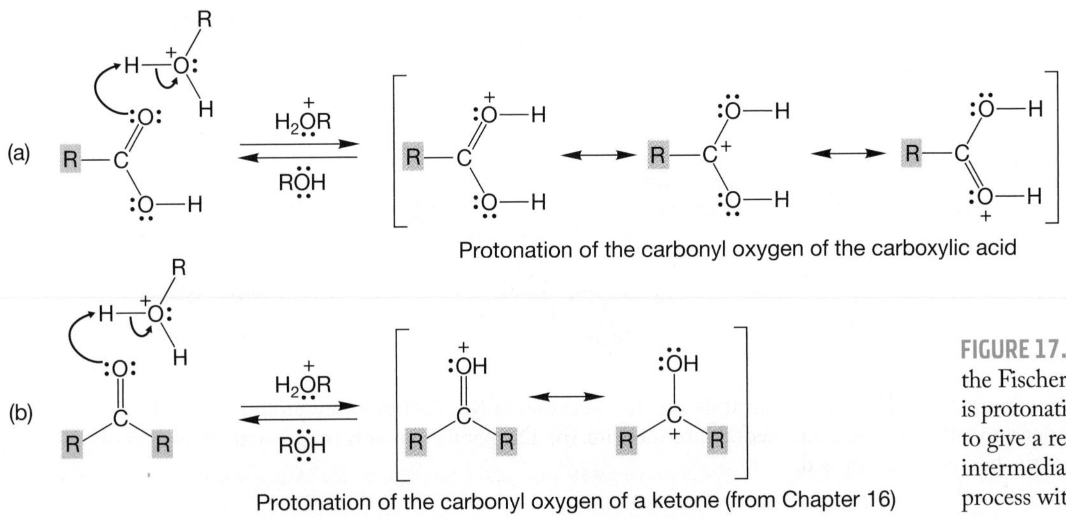

FIGURE 17.19 Fischer esterification, the acid-catalyzed reaction of a carboxylic acid with an alcohol, yields an ester.

An ester
(RO replaces HO)

(a) Protonation of the carbonyl oxygen of the carboxylic acid

(b) Protonation of the carbonyl oxygen of a ketone (from Chapter 16)

FIGURE 17.20 (a) The first step in the Fischer esterification mechanism is protonation of the carbonyl oxygen to give a resonance-stabilized intermediate. (b) The analogous process with an aldehyde or ketone.

an aldehyde or ketone in acid, the first step is surely protonation of the carbonyl group by the acid catalyst, to form a resonance-stabilized intermediate. Sulfuric acid is often used as the acid catalyst.

PROBLEM 17.10 Explain why treatment of acetic acid with $^{18}OH_3^+/^{18}OH_2$ leads to exchange of both oxygen atoms. Red highlight indicates ^{18}O.

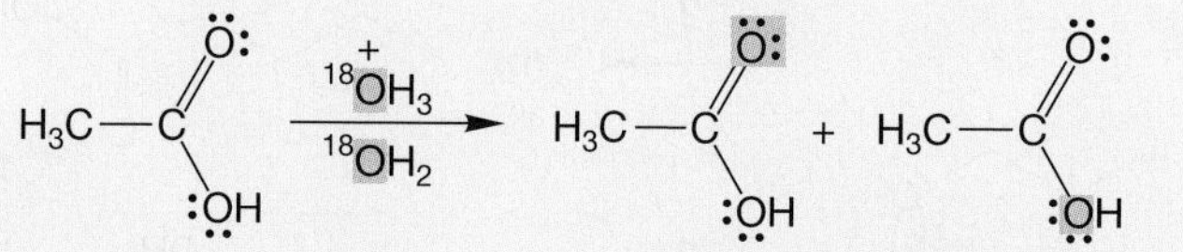

Labeled O appears
in both positions

The second and third steps in Fischer esterification are also completely analogous to parts of aldehyde or ketone chemistry. A molecule of alcohol acts as a nucleophile and adds to the carbonyl carbon of the protonated carbonyl group (Fig. 17.21). For a ketone, this addition is followed by deprotonation, and gives the hemiacetal; for the acid, a somewhat more complicated intermediate with one more hydroxyl group is formed. In both reactions, however, a planar, sp^2 hybridized carbon has been converted into an intermediate with a tetrahedral carbon, called the **tetrahedral intermediate**.

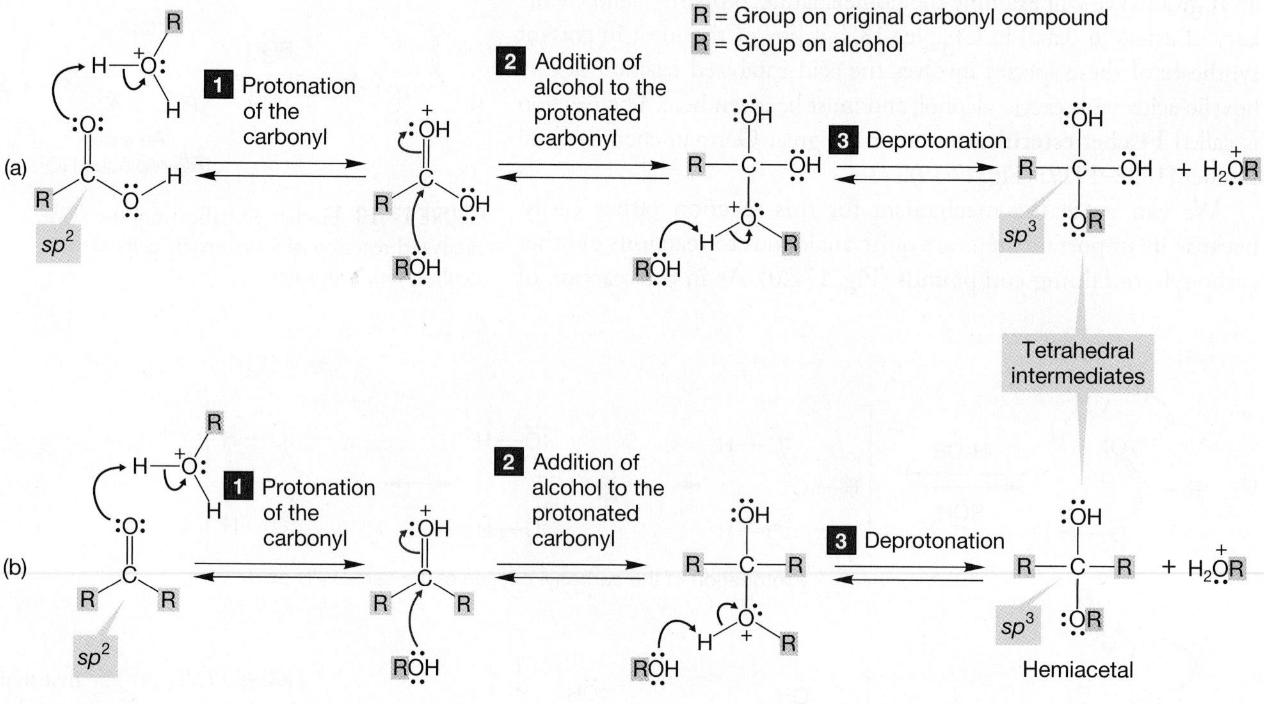

FIGURE 17.21 (a) The second step in Fischer esterification involves addition of the alcohol to the protonated acid, which is followed by deprotonation to give a tetrahedral intermediate. (b) The parallel process occurs with aldehydes or ketones, in which case the hemiacetal is formed.

Now what can these tetrahedral intermediates do in acid? The hemiacetal has only two options. It can reprotonate the OR group and then revert to the starting ketone, or it can protonate the OH group and proceed to the full acetal (Fig. 17.22).

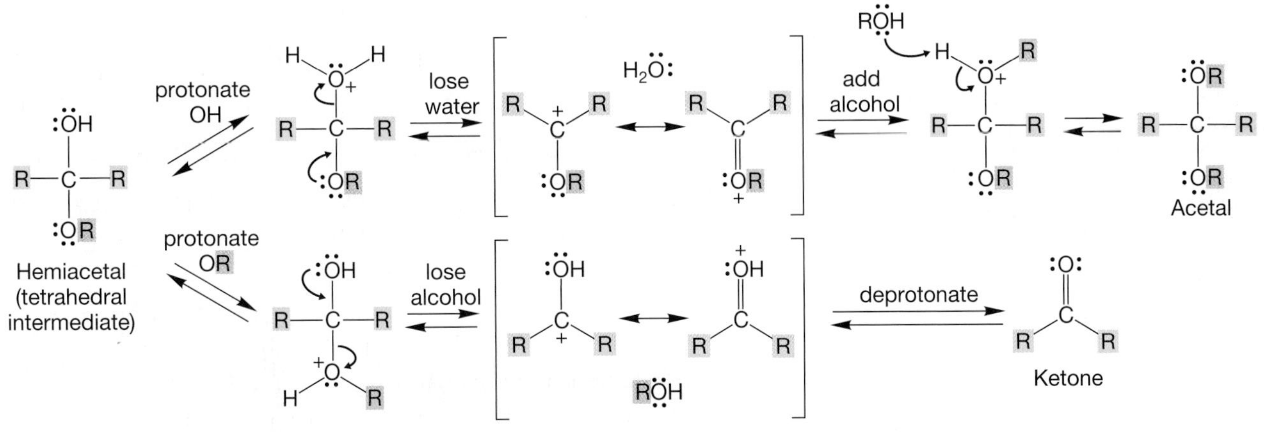

FIGURE 17.22 For the hemiacetal intermediate, there are two further possibilities: Protonation of the OH leads to water loss and formation of the full acetal, whereas protonation of the OR leads back to the original starting ketone.

The tetrahedral intermediate formed from the acid has more options, because it has one OR and two OH groups. Protonation of the OR simply takes us back along the path to the starting acid. Protonation of one of the OH groups leads to an intermediate that can lose water, deprotonate, and give the ester (Fig. 17.23). In excess alcohol, the reaction is driven in this direction.

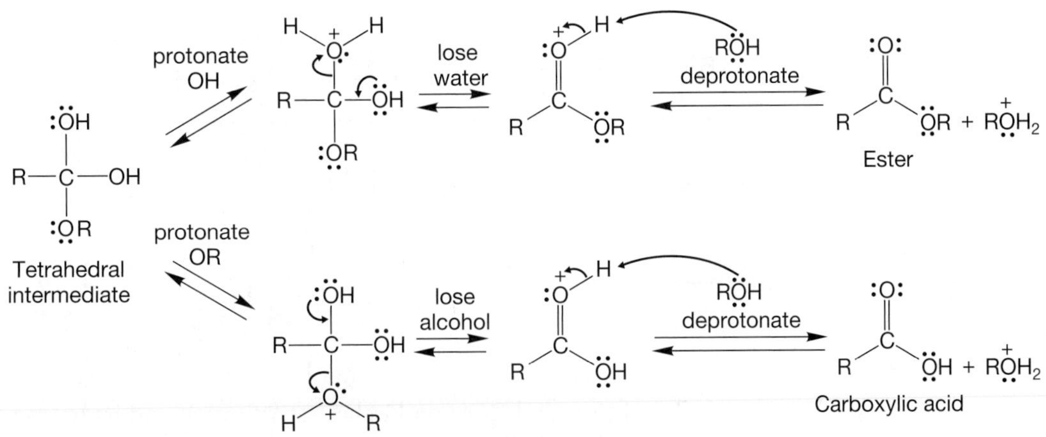

FIGURE 17.23 For the tetrahedral intermediate from Figure 17.21a, there are also two further possibilities: Protonation of the OR leads back to the starting acid, whereas protonation of either of the OH groups leads to water loss and formation of the ester.

The complete mechanism for Fischer esterification is shown in Figure 17.24. Note the symmetry of the process about the tetrahedral intermediate **A**. Note also the similarity of **B** and **C**, the intermediates on either side of **A**. In intermediate **B**, the OR is protonated. This pathway leads back to the starting carboxylic acid. In intermediate **C**, it is an OH that is protonated. This path leads to the ester. To the left of **B** and **C** are **D** and **E**, the two resonance-stabilized intermediates resulting from protonation of the acid or ester carbonyl.

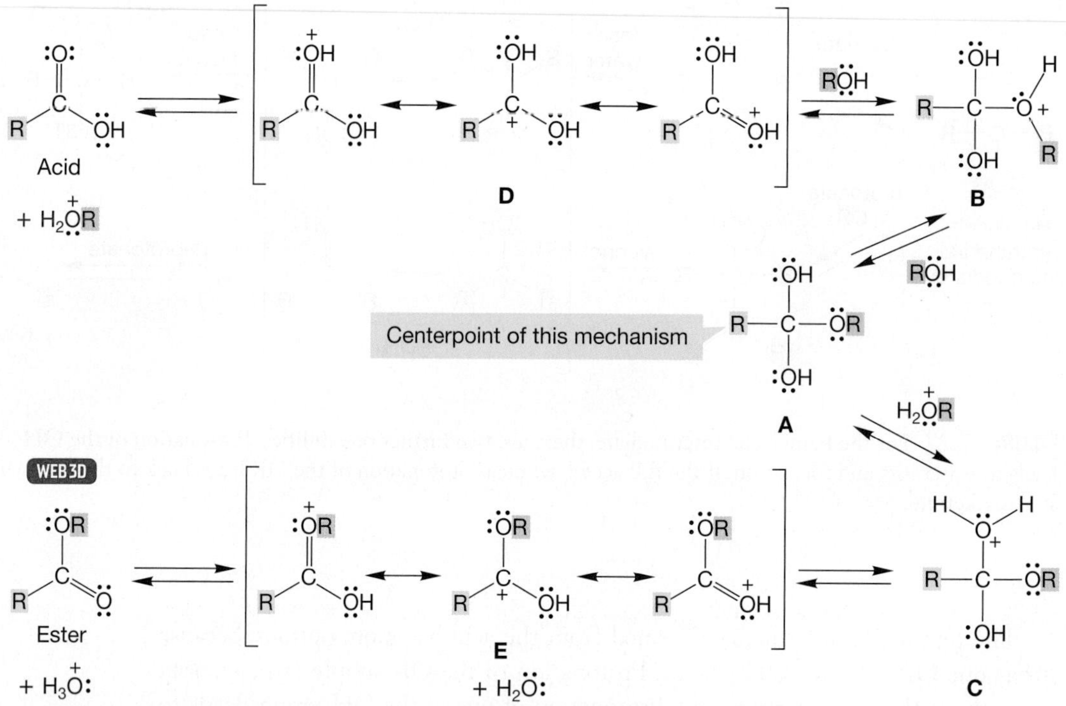

FIGURE 17.24 The full mechanism for Fischer esterification as well as for the reverse reaction, hydrolysis of an ester to a carboxylic acid.

PROBLEM 17.11 Write an arrow formalism mechanism, showing the electron-pushing, for the acid-catalyzed formation of an organic acid from the reaction of an ester with excess water.

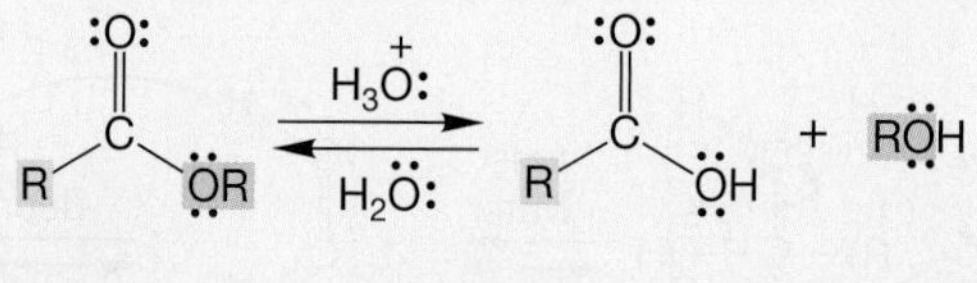

PROBLEM 17.12 Ester formation is not new to you. Give products and mechanisms for the following reactions, which appear in earlier chapters.

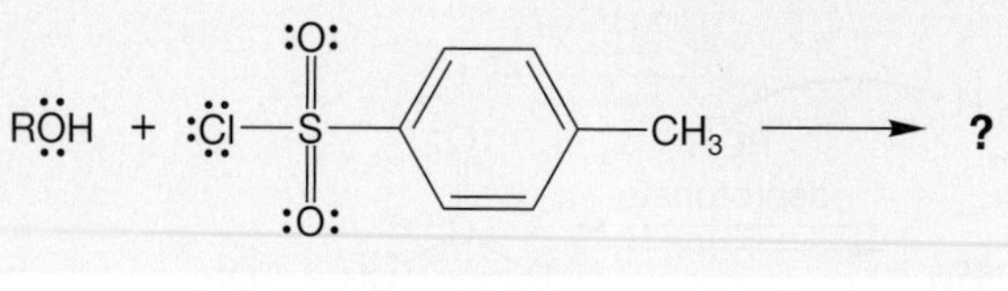

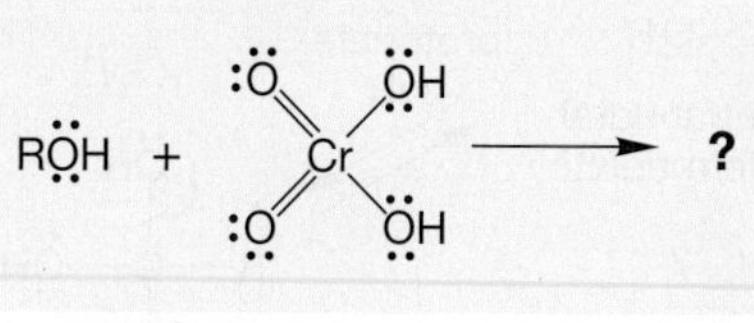

We have spent a lot of time and space on Fischer esterification and its reverse, acid hydrolysis of the ester. We paid this close attention because this reaction is prototypal and an anchor for further discussion. As for other anchor reactions (for example, the S_N2 reaction and addition of Br_2 to alkenes), knowing this reaction, quite literally backward and forward, allows you to generalize and to organize the many following reactions that show a similar pattern, although the details may be different. Fischer esterification is definitely a reaction to know well.

Note here what makes the chemistry of aldehydes and ketones different from that of acids (or esters). In the aldehydes and ketones there is no leaving group present, whereas in an acid (or ester) there is (Fig. 17.25). The chemistry of aldehydes and ketones on the one hand, and the chemistry of acids and esters on the other, diverge only after the nucleophilic addition of the alcohol to the carbonyl (see Fig. 17.21). In the acids and esters there is a group that can be lost, providing a route to other molecules; in the aldehydes and ketones there isn't, which means that the acids and esters can undergo an addition–elimination reaction that the aldehydes and ketones can't.

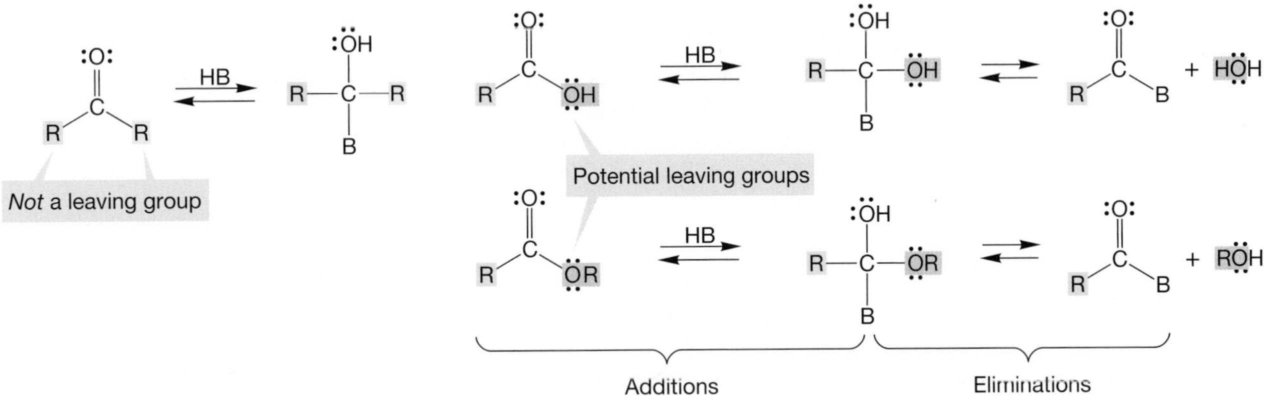

FIGURE 17.25 Unlike aldehydes and ketones, acids and esters bear OH and OR groups that are potential leaving groups.

But we should wait a moment. Mechanisms other than the one shown in Figure 17.24 for Fischer esterification can be imagined, and it is not fair to dismiss them without evidence. For example, why not use the protonated alcohol as the electrophile and the carbonyl oxygen of the carboxylic acid as the nucleophile rather than the protonated carboxylic acid as the electrophile and the alcohol as the nucleophile? Displacement of water using the acid as nucleophile could give the ester (Fig. 17.26).

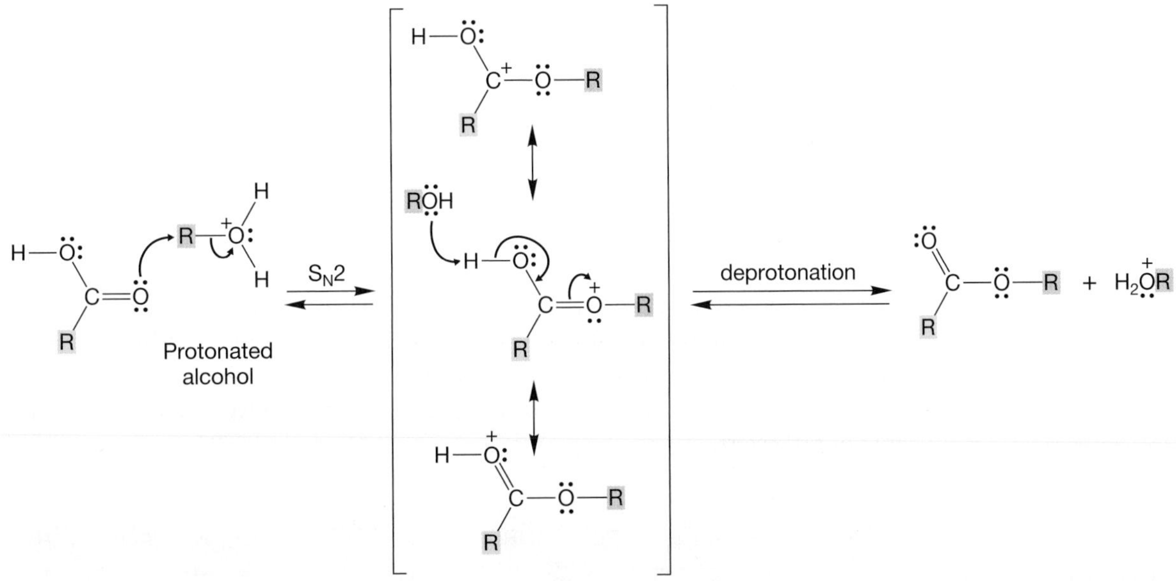

FIGURE 17.26 A potential mechanism for Fischer esterification. Here the acid is the nucleophile and the protonated alcohol is the electrophile. Watch out for conventions! The figure shows proton removal from one resonance form, which is purely for convenience sake—the resonance form has no individual existence.

The key difference between the mechanism for Fischer esterification shown in Figure 17.24 (which is correct) and this hypothetical one in Figure 17.26 is the site of carbon–oxygen bond breaking. In the first (correct) mechanism, it is a carbon–oxygen bond in the *carboxylic acid* that is broken; in the new (incorrect) mechanism, it is the carbon–oxygen bond in the *alcohol* that breaks.

WORKED PROBLEM 17.13 Design an experiment to tell these two mechanisms (Fig. 17.24 and Fig. 17.26) apart. Assume you have access to any isotopically labeled compounds you may need.

ANSWER Using labeled alcohol will do the trick. If the mechanism involved carbon–oxygen bond breaking in the alcohol (it does not), the use of ^{18}O-labeled alcohol would not lead to ^{18}O incorporation in the product ester.

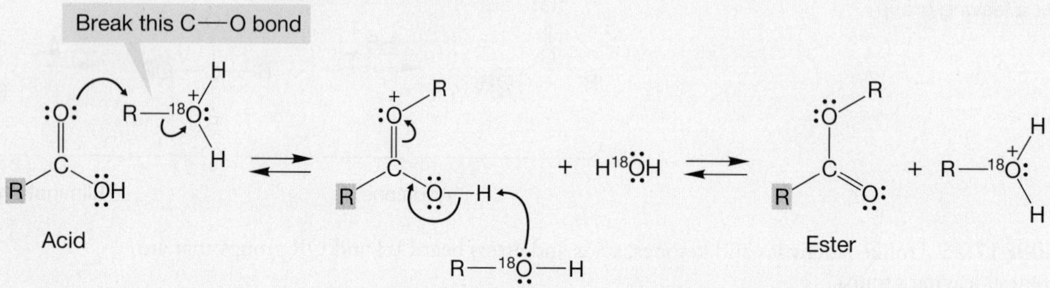

However, if the mechanism involves breaking of a carbon–oxygen bond in the original acid (it does) ^{18}O will be incorporated, which it is.

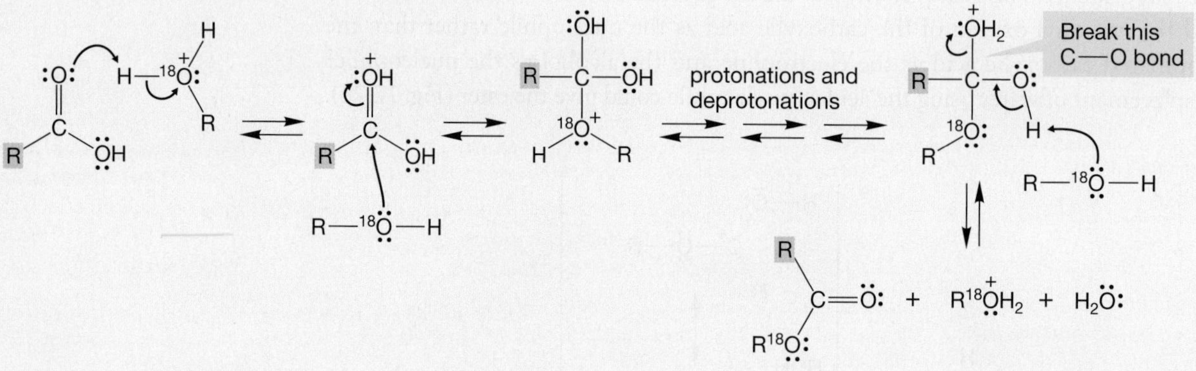

Why don't esters react further under these conditions? Aldehydes and ketones form acetals when treated with an acid catalyst and excess alcohol. Why don't esters go on to form **ortho esters** (Fig. 17.27)? Ortho ester is the rather misleading name used for the functional group that has the general structure of $RC(OR)_3$. Ortho esters are not real esters.

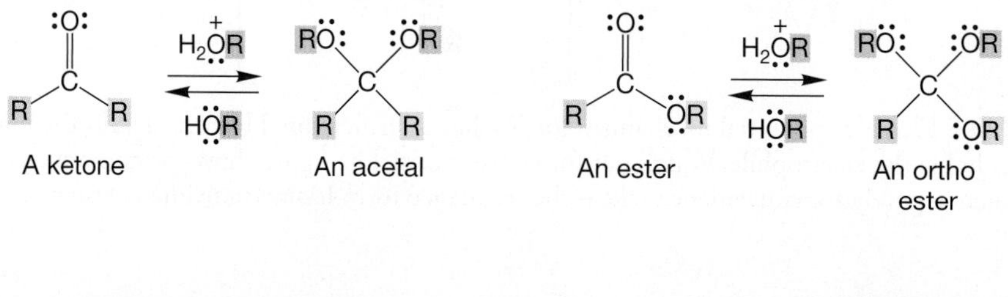

FIGURE 17.27 The conversion of an ester into an ortho ester is analogous to the formation of acetals from aldehydes and ketones.

A ketone   An acetal   An ester   An ortho ester

A reasonable mechanism for the conversion of an ester into an ortho ester is outlined in Figure 17.28. Ortho esters are known compounds, and there is nothing fundamentally wrong with the steps outlined in Figure 17.28. The steps are quite analogous

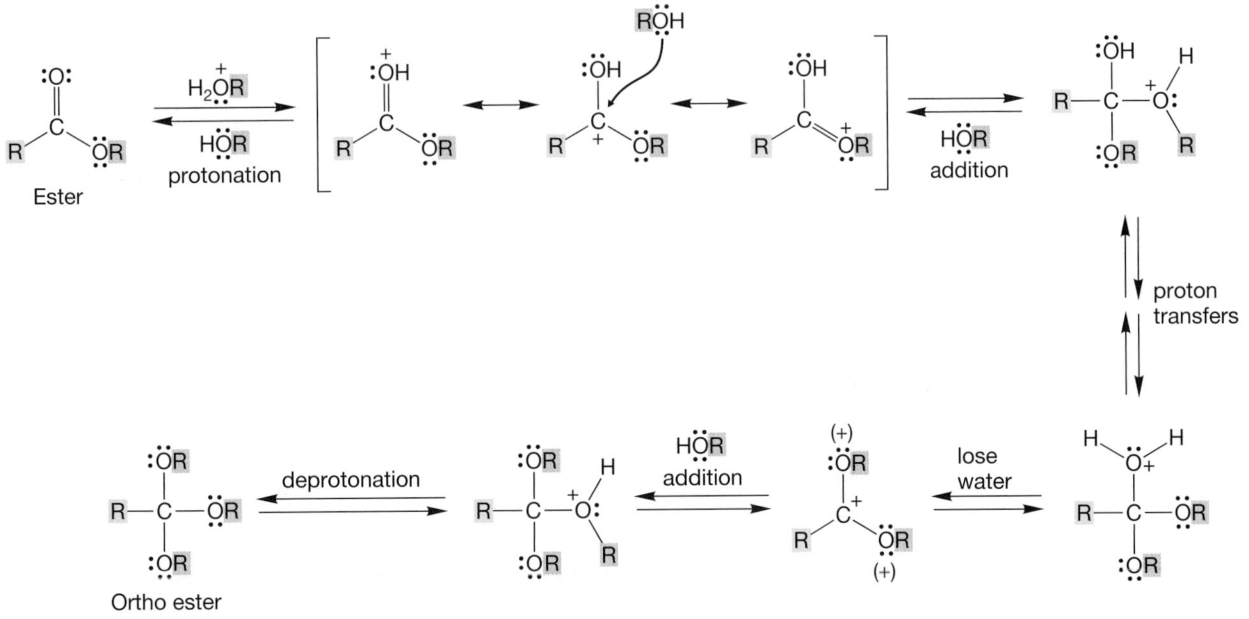

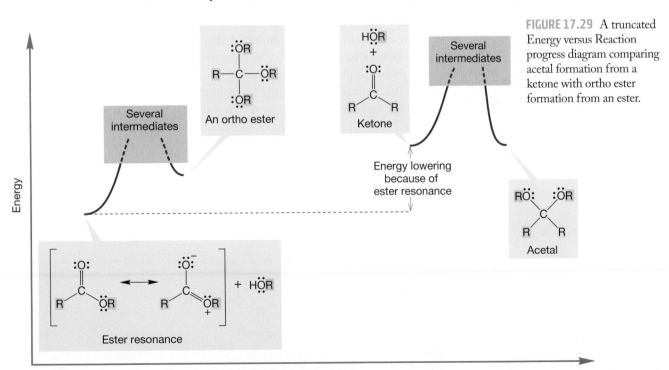

FIGURE 17.28 A reasonable
mechanism for the conversion of an
ester into an ortho ester.

to those for acetal formation. However, stable ortho esters cannot be made this way.
The problem is one of thermodynamics. The sequence of Figure 17.28 is a series of equilibria, and at equilibrium it is the ester that is greatly favored, not the ortho ester. Note
that the starting material in this reaction, the ester, is stabilized by resonance but the
product, the ortho ester, is not. The ortho ester is analogous to the thermodynamically
unstable intermediate **A** in Figure 17.24. Aldehydes and ketones lack this ester resonance,
and are therefore less stable, and more prone to further addition reactions. Figure 17.29

FIGURE 17.29 A truncated
Energy versus Reaction
progress diagram comparing
acetal formation from a
ketone with ortho ester
formation from an ester.

makes this point with a partial Energy versus Reaction progress diagram. The acetal and the ortho ester are roughly equivalent in energy, but the ester lies well below the ketone. The result is that acetals are thermodynamically accessible from ketones, but ortho esters are disfavored relative to esters.

There are other mechanisms for ester formation, and some of them do not involve breaking a carbon–oxygen bond in the starting carboxylic acid. Sometimes the carboxylate anion is nucleophilic enough to act as the displacing agent in an S_N2 reaction. The partner in the displacement reaction must be especially reactive. In practice, this means that a primary halide or even more reactive species such as a methyl halide must be used (Fig. 17.30).

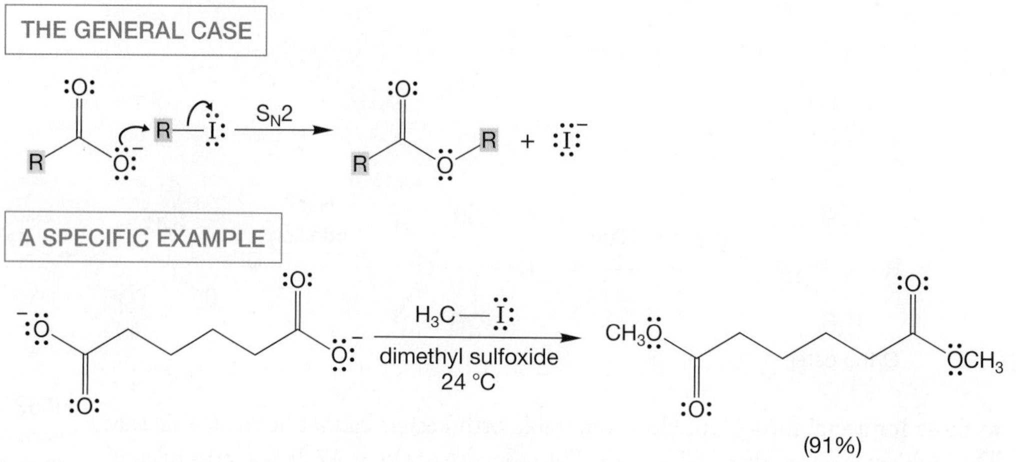

FIGURE 17.30 Carboxylate anions can be used to displace very reactive halides in an ester-forming S_N2 reaction.

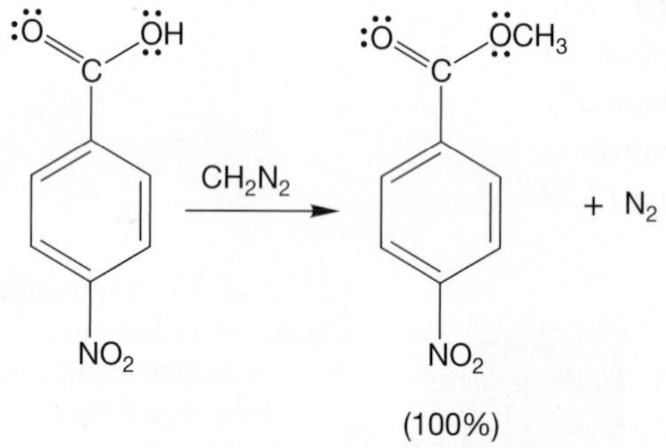

FIGURE 17.31 Diazomethane can be used to make methyl esters from acids in yields as high as 100%.

Another example of a reaction in which the carboxylate acts as a nucleophile is the ester-forming reaction of acids with diazomethane, CH_2N_2. Diazomethane, though easily made, is quite toxic and a powerful explosive. So, this method is generally used only when small amounts of the methyl esters are needed (Fig. 17.31). Don't try this reaction at home!

The first step in this reaction is deprotonation of the acid by the basic carbon of the diazo compound (Fig. 17.32). The result is a carboxylate anion and a diazonium ion, an extraordinarily reactive alkylating agent. Next, an alkylation step similar to that in Figure 17.30 occurs between the carboxylate and the diazonium ion. This reaction takes place because nitrogen (N_2) is a superb leaving group, perhaps the world's best.

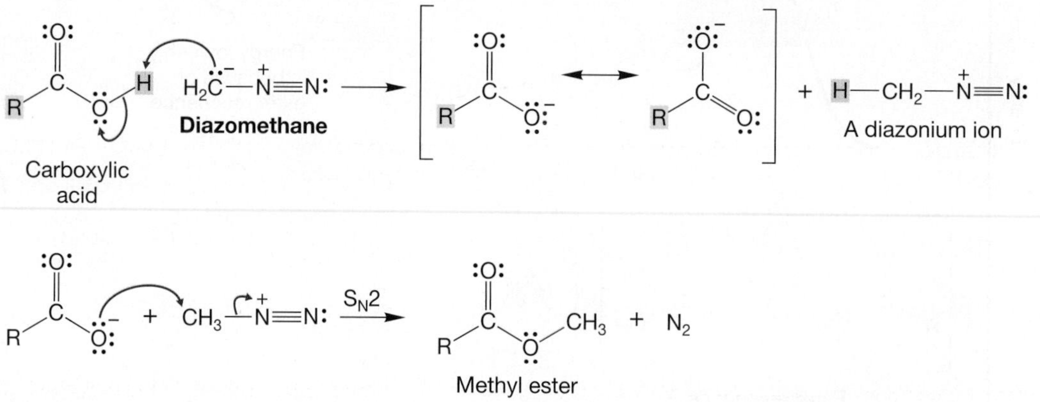

FIGURE 17.32 In the first step of this methyl ester–forming reaction, diazomethane acts as a Brønsted base and deprotonates the carboxylic acid. In the second step, the carboxylate does an S_N2 substitution on the very reactive methyl diazonium species.

In practice, this method of esterification is restricted to the synthesis of methyl esters because other diazo compounds are relatively unstable.

A final note about ester formation addresses the case when an alcohol and a carboxylic acid in the same molecule undergo esterification. The result is a cyclic ester, known as a **lactone**. As usual, the relatively unstrained five- and six-membered rings are formed more easily than other sized rings. Figure 17.33 gives an example of lactone formation.

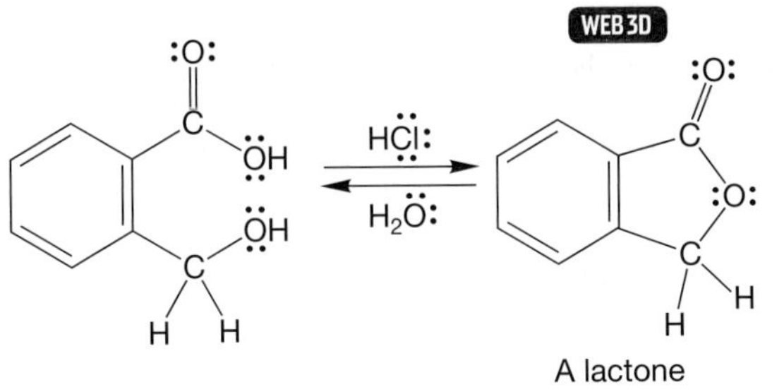

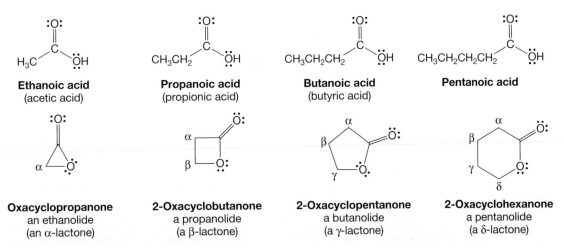

FIGURE 17.33 A lactone can be formed via an intramolecular Fischer esterification.

A lactone

PROBLEM 17.14 Write a detailed mechanism for the reaction in Figure 17.33.

PROBLEM 17.15 Treatment of the compound in the figure below with I_2/KI and Na_2CO_3 in water leads to a neutral molecule of the formula $C_8H_{11}IO_2$. Propose a structure for the product and a mechanism for its formation.

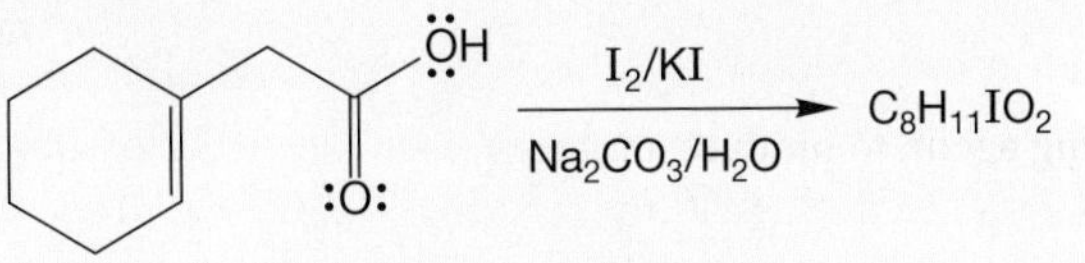

$$\xrightarrow[Na_2CO_3/H_2O]{I_2/KI} C_8H_{11}IO_2$$

Lactones are commonly named in reference to the acid containing the same number of carbon atoms, with "olide" added as a suffix. Thus a butanolide has four carbons, a pentanolide has five, and so on. A Greek letter is used to designate the size of the ring. An α-lactone has one carbon joining the oxygen and carbonyl carbon, a β-lactone has two carbons, a γ-lactone three carbons, and so on. In the IUPAC systematic naming protocol, lactones are called oxacycloalkanones (Fig. 17.34).

Ethanoic acid
(acetic acid)

Propanoic acid
(propionic acid)

Butanoic acid
(butyric acid)

Pentanoic acid

Oxacyclopropanone
an ethanolide
(an α-lactone)

2-Oxacyclobutanone
a propanolide
(a β-lactone)

2-Oxacyclopentanone
a butanolide
(a γ-lactone)

2-Oxacyclohexanone
a pentanolide
(a δ-lactone)

FIGURE 17.34 The naming options for lactones.

17.7b Formation of Amides   Although carboxylic acids do react with primary or secondary amines to form **amides**,[2] in general, this reaction is not a very useful process. The dominant reaction between the basic amine and the carboxylic acid is proton transfer to give the ammonium salt of the carboxylic acid. Heating of the salts has been used as a source of amides (Fig. 17.35). However, the amount of heat needed is more than 200 °C. Such conditions are often too harsh for the organic compounds and result in degradation of the material.

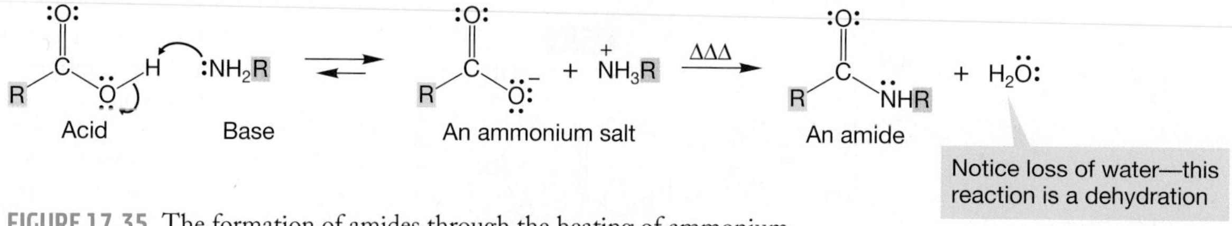

FIGURE 17.35 The formation of amides through the heating of ammonium salts of carboxylic acids.

Amide formation from the carboxylic acid is much easier if the acid is first activated. Several **activating agents** have been developed that greatly facilitate amide formation. Dicyclohexylcarbodiimide (DCC) is used most often (Fig. 17.36).

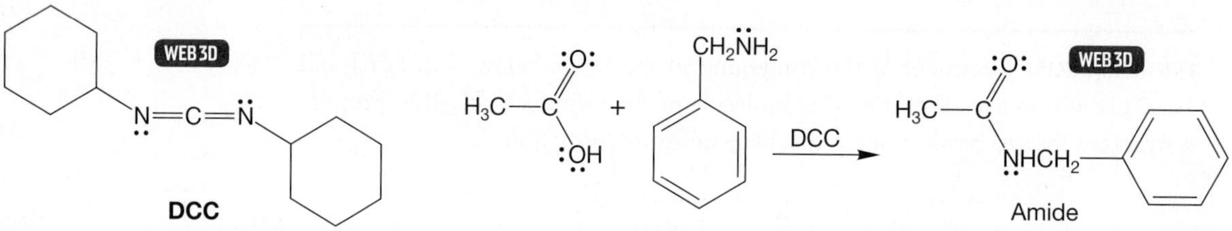

FIGURE 17.36 The use of DCC, an activating agent, to produce amides.

The strategy is to convert the poor leaving group, OH, into a better one. We used this technique extensively in Chapter 7 (p. 282), when we discussed the transformation of alcohols. Here, the carboxylate adds to one carbon–nitrogen double bond of DCC, accomplishing the transformation of the leaving group (Fig. 17.37).

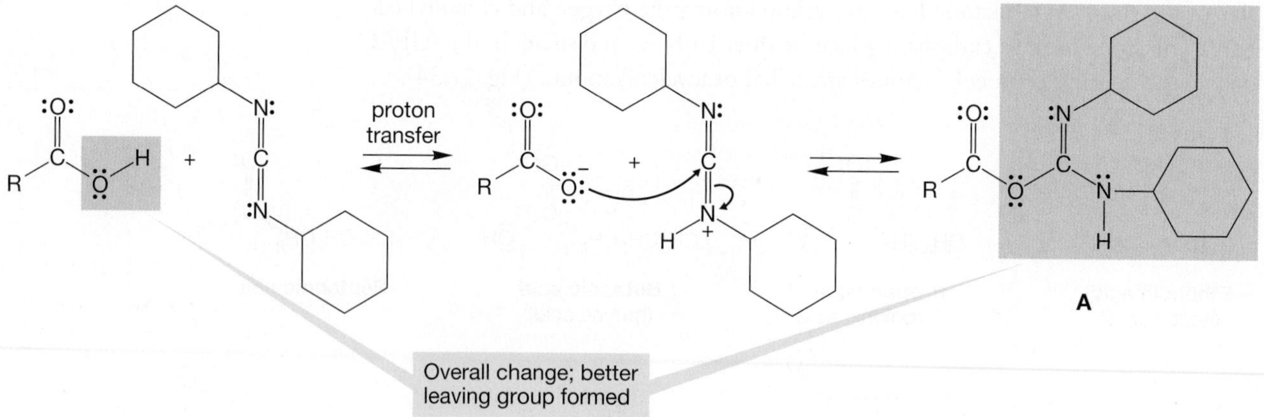

FIGURE 17.37 The first step in the DCC mechanism is proton transfer. This reaction is followed by addition of the carboxylate to one of the carbon–nitrogen double bonds. The transformation involves a change of leaving group from OH to OR to give the activated acid **A**.

---

[2]As noted on p. 249, there is potential for confusion here. There are two kinds of amides, the carbonyl derivative shown here (R—CO—NHR) and the negatively charged ions, ⁻NR₂. You need to know the context of the discussion to know which is meant. In this case we are referring to R—CO—NHR.

Two mechanistic pathways are now possible. In the simpler one, the primary or secondary amine adds to the carbon–oxygen double bond to give a tetrahedral intermediate that expels a relatively stable ion to generate the amide (Fig. 17.38).

If the amide is made from an amine and a carboxylic acid in the same molecule, then a cyclic amide is formed. Cyclic amides are called **lactams**.

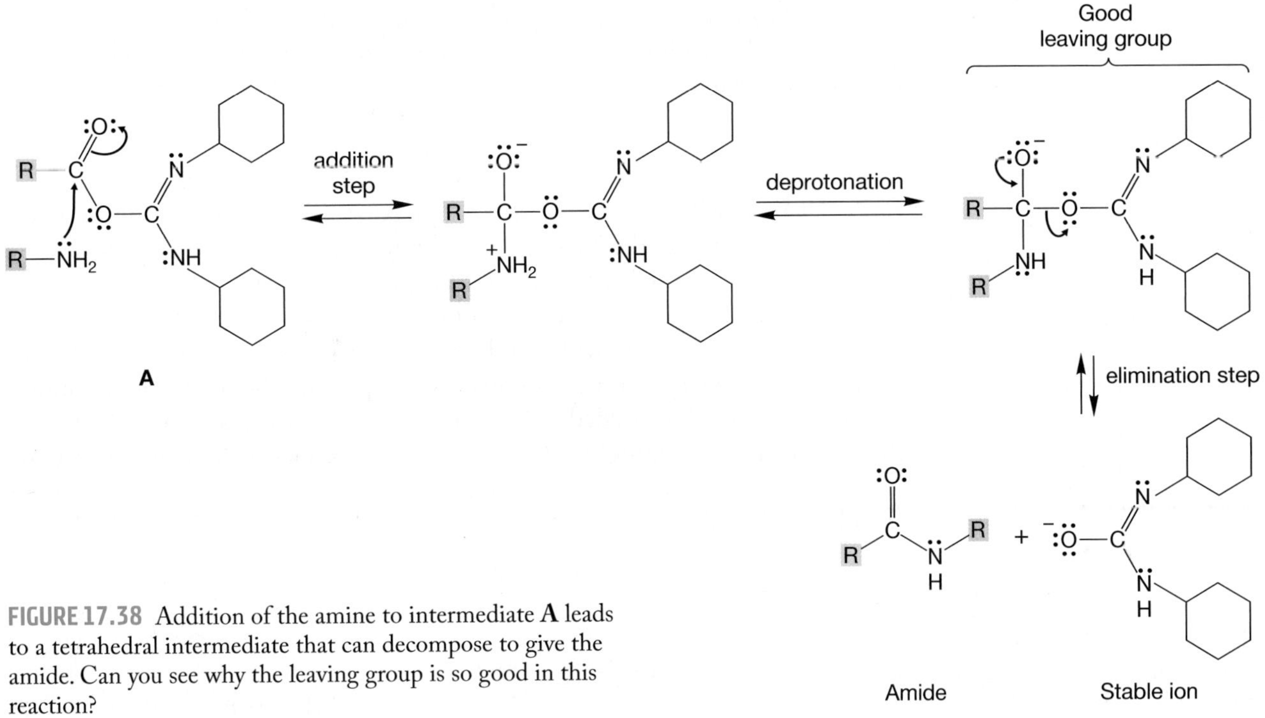

FIGURE 17.38 Addition of the amine to intermediate **A** leads to a tetrahedral intermediate that can decompose to give the amide. Can you see why the leaving group is so good in this reaction?

PROBLEM 17.16 There is another, more complicated (and accepted), mechanism for the reaction of the activated acid **A**. The initially formed activated acid reacts not with the amine, as in Figure 17.38, but with another molecule of carboxylic acid. The result is an anhydride. The anhydride is the actual reagent that reacts with the amine to give the product amide. Sketch mechanisms for anhydride and amide formation.

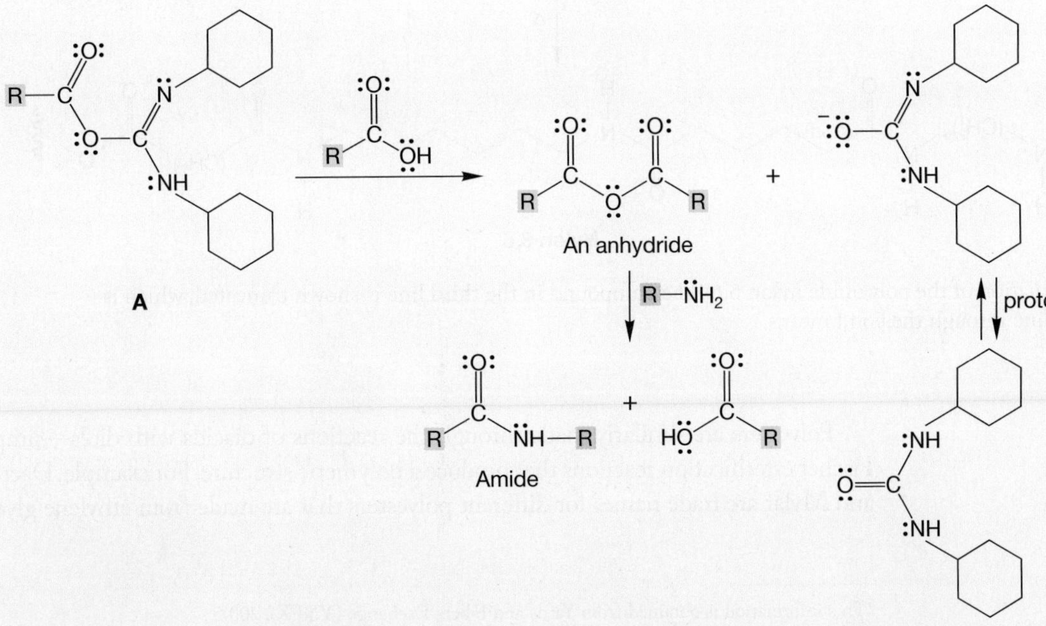

Summary

You have learned several methods to make carboxylic acids. We have also discussed how to make esters and amides from organic acids. Notice the similarity between amide formation and ester formation. Amide formation is in fact merely a collection of steps closely resembling those in Fischer esterification. We are beginning to see the generality of the addition–elimination process. The addition–elimination reaction will continue to be prominent throughout this and other chapters.

17.7c Polyamides (Nylon) and Polyesters   The reaction shown at the start of Section 17.7b, the heating of the ammonium salt of a carboxylic acid to give an amide, is at the heart of the synthesis of a set of polyamides collectively known as nylons. This reaction is certainly not insignificant, as evidenced by the fact that the annual worldwide production of nylon is 4 million tons.[3] In one example of nylon synthesis, a double-headed amine, hexamethylenediamine, reacts with a double-headed acid, adipic acid, to form a salt (Fig. 17.39). Pyrolysis (extreme heating) of this salt at 275 °C produces the long-chain polymer known as nylon 6,6. Nylons of other chain lengths are also produced by the chemical industry.

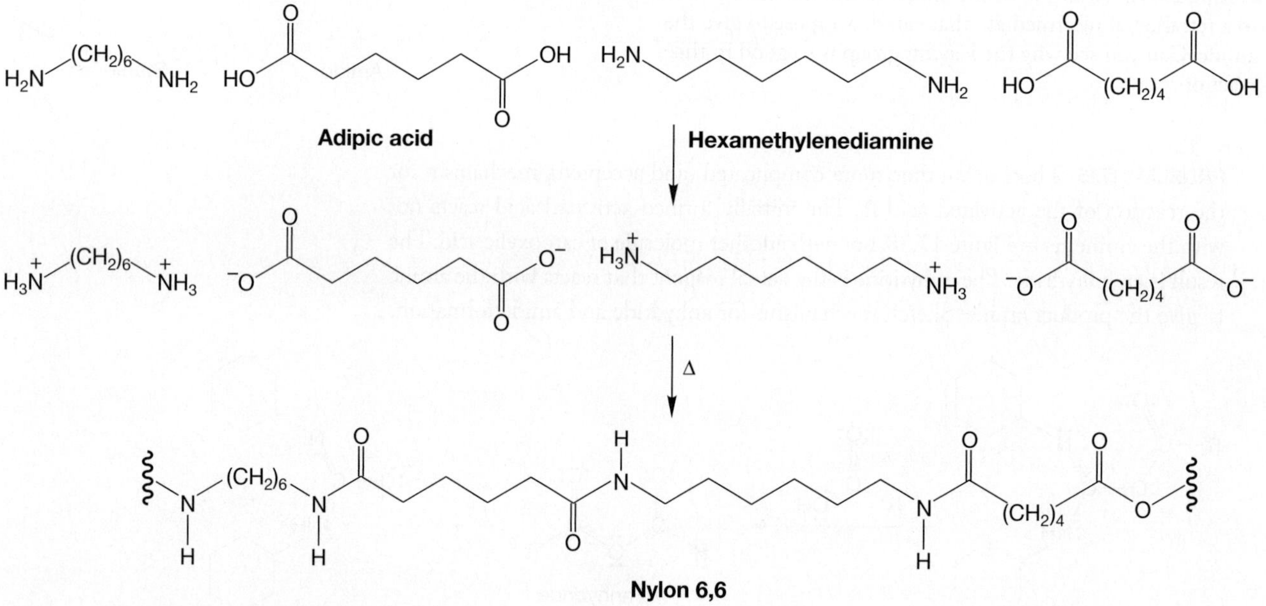

Adipic acid   **Hexamethylenediamine**

Nylon 6,6

FIGURE 17.39 Formation of the polyamide nylon 6,6. The compound in the third line is shown truncated, which is what the squiggly line through the bond means.

Polyesters are similarly made through the reactions of diacids with diols—simple Fischer esterification reactions that produce a polymeric structure. For example, Dacron and Mylar are trade names for different polyesters that are made from ethylene glycol

---

[3]This information is obtained from Yarns and Fibers Exchange (YNFX), 2007.

and terephthalic acid (Fig. 17.40). This type of polyester, polyethylene terephthalate (PET), is also used to make clear recyclable plastic bottles. It is more recyclable and therefore more environmentally friendly than most other kinds of polymers.

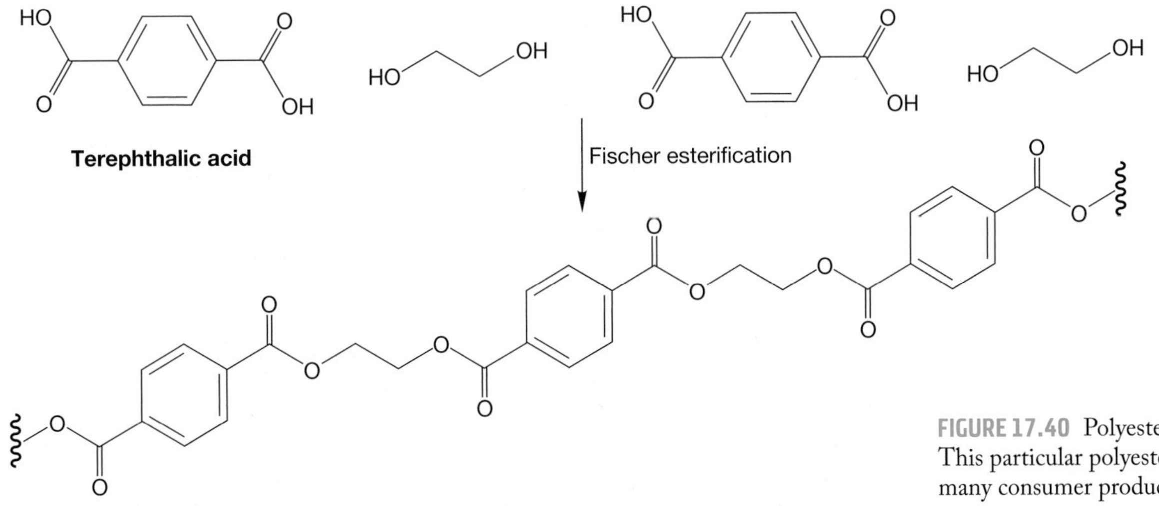

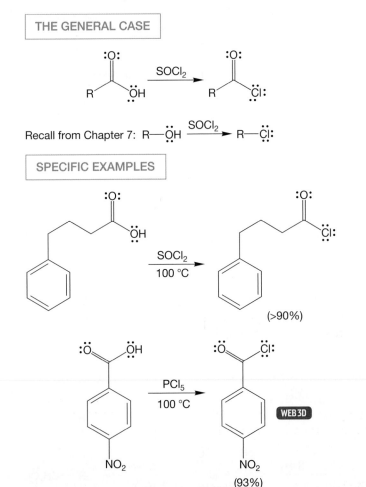

Terephthalic acid
Fischer esterification

FIGURE 17.40 Polyester synthesis. This particular polyester is used in many consumer products.

17.7d Formation of Acid Chlorides We saw acid chlorides in our study of Friedel–Crafts acylation (p. 643). One can make an acid chloride in good yield by treatment of a carboxylic acid with thionyl chloride or phosphorus pentachloride (Fig. 17.41). This reaction closely resembles the formation of chlorides by the reaction

THE GENERAL CASE
$SOCl_2$
Recall from Chapter 7:
SPECIFIC EXAMPLES
$SOCl_2$
100 °C
(>90%)
PCl_5
100 °C
WEB 3D
NO_2
NO_2
(93%)

FIGURE 17.41 The reaction of carboxylic acids with thionyl chloride ($SOCl_2$) or phosphorus pentachloride (PCl_5) gives acid chlorides. The reaction of alcohols with $SOCl_2$ is similar.

Acid chloride formation

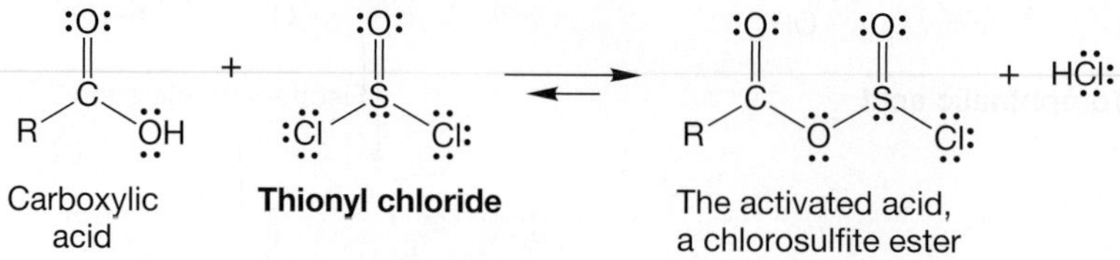

FIGURE 17.42 An activated acid is formed in the first step of acid chloride formation.

of alcohols with thionyl chloride (p. 284). Once again, the key concept in acid chloride formation is activation of the carboxylic acid by the transformation of a very poor leaving group, OH, into a better one. In the first step of the acid chloride formation using $SOCl_2$, a chloride is displaced to form an activated acid intermediate, in this case a chlorosulfinic acid derivative (Fig. 17.42).

Carboxylic   **Thionyl chloride**   The activated acid,
acid   a chlorosulfite ester

PROBLEM 17.17 Write a mechanism for the formation of the product in Figure 17.42. Watch out for the details! Almost everyone makes a small error the first time on this problem.

The hydrogen chloride generated in the formation of the activated acid protonates the carbonyl oxygen, and the chloride ion adds to the strong electrophile. The tetrahedral intermediate then breaks down into chloride, sulfur dioxide, and the acid chloride. Notice the change in leaving group induced by the overall transformation of OH into OSOCl (Fig. 17.43).

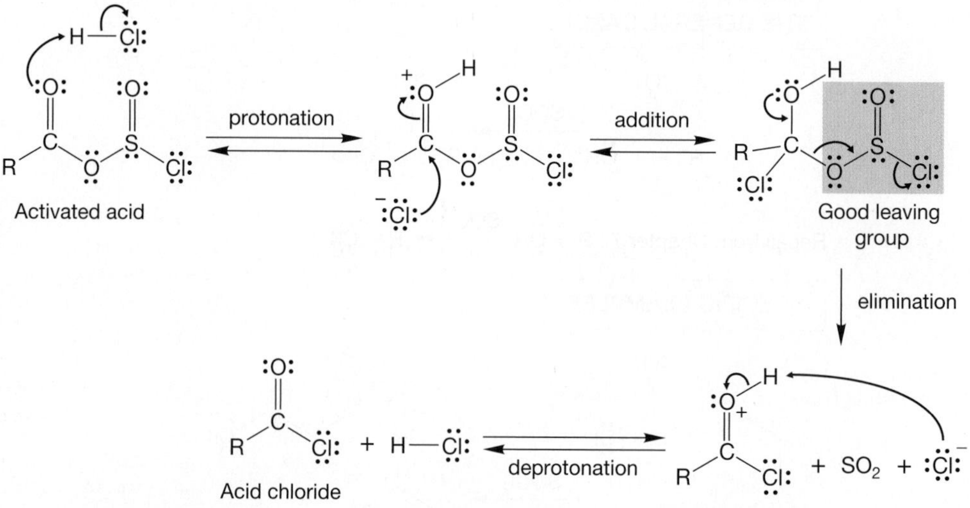

FIGURE 17.43 Hydrogen chloride protonates the carbonyl group, and chloride attacks this strong electrophile. In the crucial step, sulfur dioxide and chloride ion are lost as the acid chloride is formed.

Activated acid   protonation   addition   Good leaving group

elimination

Acid chloride   deprotonation

Acid chlorides contain a good potential leaving group, the chloride; and the addition–elimination process can lead to all manner of acyl derivatives, R—CO—X, as we will see in Chapter 18 (Fig. 17.44).

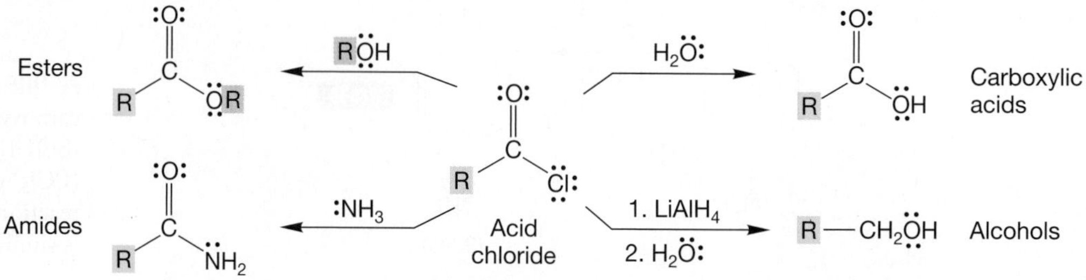

Acid chloride aminolysis

FIGURE 17.44 Some compounds formed from addition–elimination reactions between different nucleophiles and acid chlorides. See Problem 17.18.

Esters   Amides   Acid chloride   Carboxylic acids   Alcohols

PROBLEM 17.18 Write a mechanism for the transformation of an acid chloride into a primary alcohol as shown in Figure 17.44.

PROBLEM 17.19 Another excellent synthesis of acid chlorides from carboxylic acids uses either phosgene (Cl—CO—Cl) or oxalyl chloride (Cl—CO—CO—Cl) as the reactive agent. Suggest a mechanism for the reaction with phosgene.

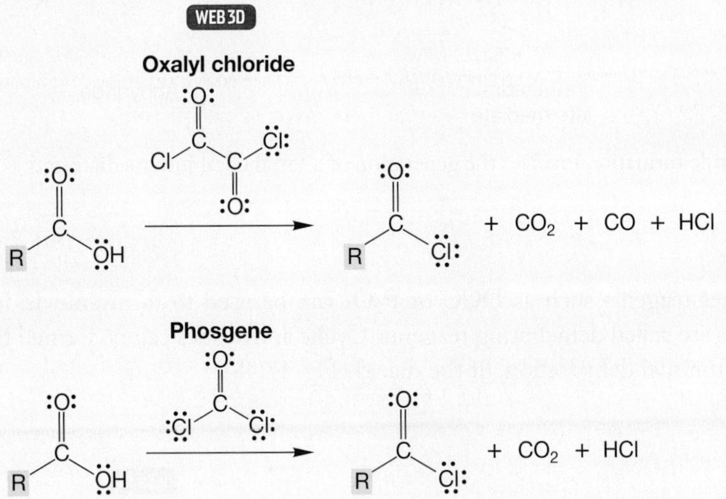

PROBLEM 17.20 Phosgene (see Problem 17.19) is a most effective poison. Can you guess its mode of action? What happens when phosgene is absorbed by moist lung tissue?

17.7e Anhydride Formation   Acid anhydrides are, as the name suggests, related to carboxylic acids by a formal loss of water. An anhydride is "two carboxylic acids less a molecule of water" (Fig. 17.45).

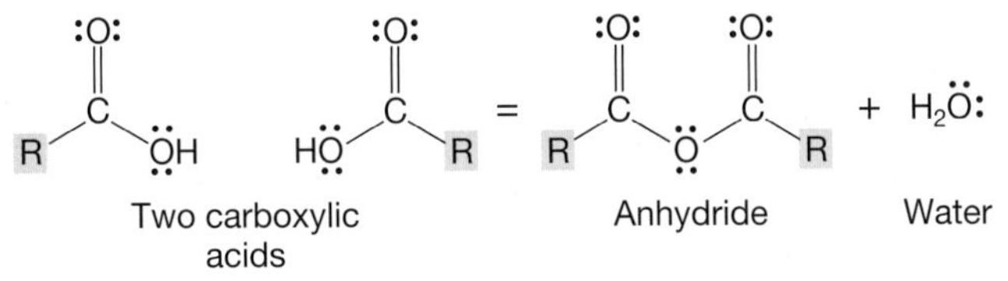

FIGURE 17.45 A comparison between two carboxylic acids and acid anhydrides.

Carboxylic acids and their conjugate bases, the carboxylate anions, react with acid halides to give acid anhydrides (Fig. 17.46).

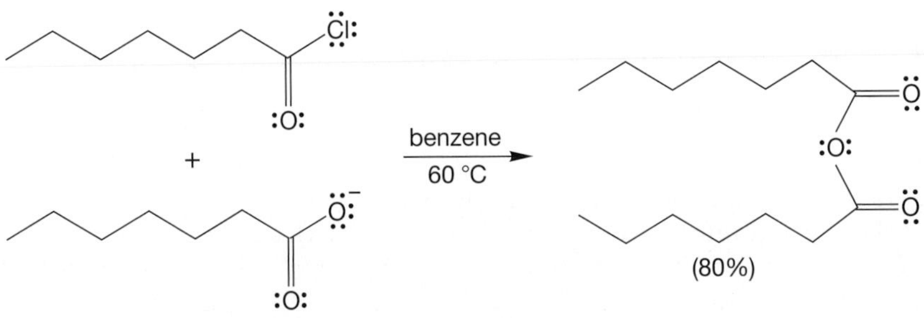

FIGURE 17.46 Anhydride formation through the reaction of a carboxylate salt with an acid chloride.

The mechanisms for these reactions involve addition of the carboxylate anion to the carbonyl group of the acid chloride to give a tetrahedral intermediate that can lose a chloride ion, which is a good leaving group (Fig. 17.47).

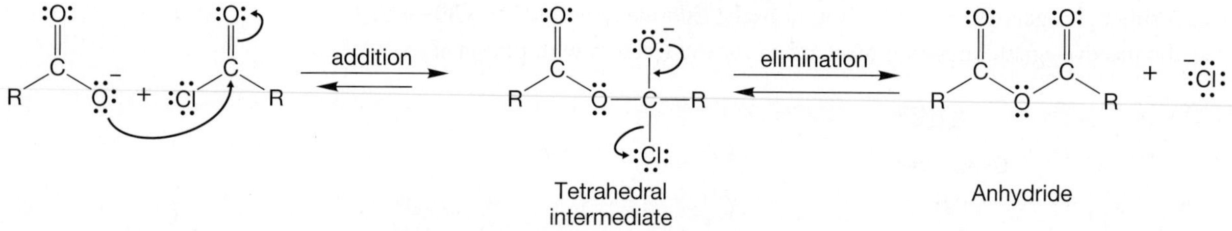

FIGURE 17.47 The mechanism of anhydride formation involves the generation of a tetrahedral intermediate and loss of chloride ion.

Other reagents such as DCC or P_2O_5 can be used to form anhydrides. Such reagents are called dehydrating reagents. Cyclic anhydrides can be formed by pyrolysis, or thermal dehydration, of the diacid (Fig. 17.48).

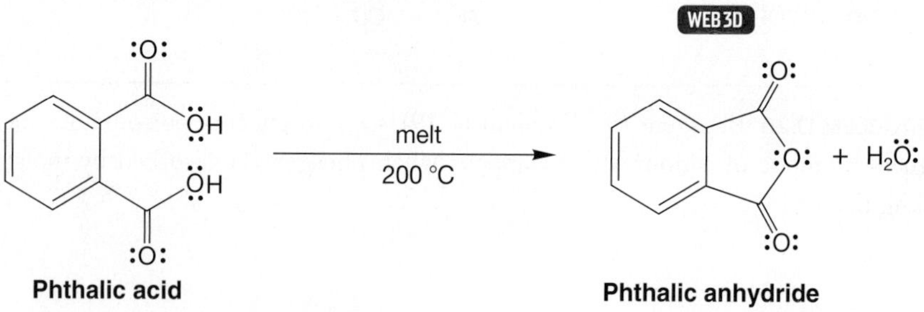

FIGURE 17.48 Dehydration of a dioic acid can form a cyclic anhydride.

17.7f Reactions with Organolithium Reagents and Metal Hydrides

In a useful synthetic reaction that was mentioned on p. 837, organic acids react with two equivalents of an organolithium reagent to give the corresponding ketone (Fig. 17.49).

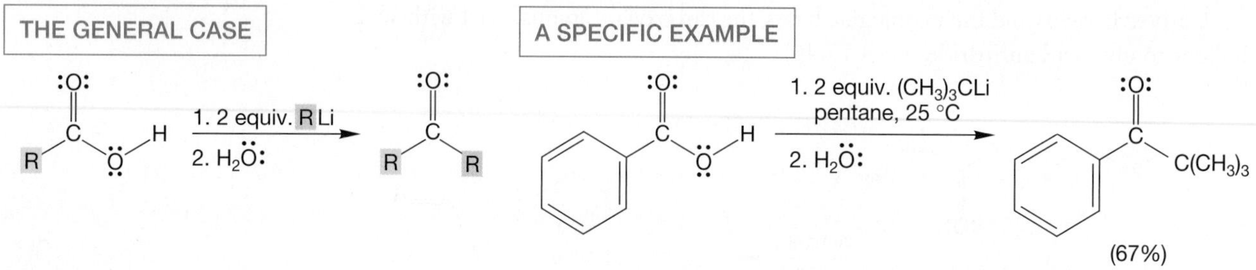

FIGURE 17.49 A general ketone synthesis from a carboxylic acid and an organolithium reagent. Note again the convention that differentiates sequential reactions (written 1. RLi; 2. H_2O) from "dump 'em all together" reactions (written RLi, H_2O).

The first step in the reaction is the formation of the lithium salt of the carboxylic acid. Although this species already bears a negative charge, the organolithium reagent is a strong enough nucleophile to add to it to give the "dianion" (Fig. 17.50).

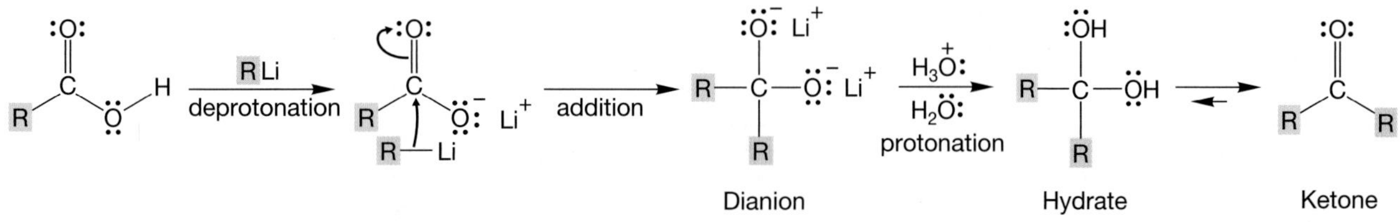

FIGURE 17.50 The first step in the reaction is formation of the carboxylate anion. In a second step, the organolithium reagent is a strong enough nucleophile to add to the carboxylate anion to give a lithium-stabilized dianion. When water is added to the reaction, a hydrate is formed. Hydrates are generally unstable compared to ketones. The equilibrium favors the final product of the reaction, the ketone.

What can happen to the product dianion in this basic solution? This question is one of the rare instances when the answer really is, "nothing." There is no possible leaving group, because neither R^- nor O^{2-} can be lost. So, the dianion remains in solution until water is added to the reaction mixture, at which point a hydrate is formed. Hydrates are unstable relative to their ketone precursors (p. 777), and so the end result is the ketone.

PROBLEM 17.21 What reagent(s) would you use to synthesize acetophenone from acetic acid?

The key to the formation of the ketone is the inability of the dianionic intermediate to expel a leaving group. Compare this nonreaction with a general process in which a leaving group is present (Fig. 17.51). We have already seen many reactions of this type, and in Chapter 18 we will see many more.

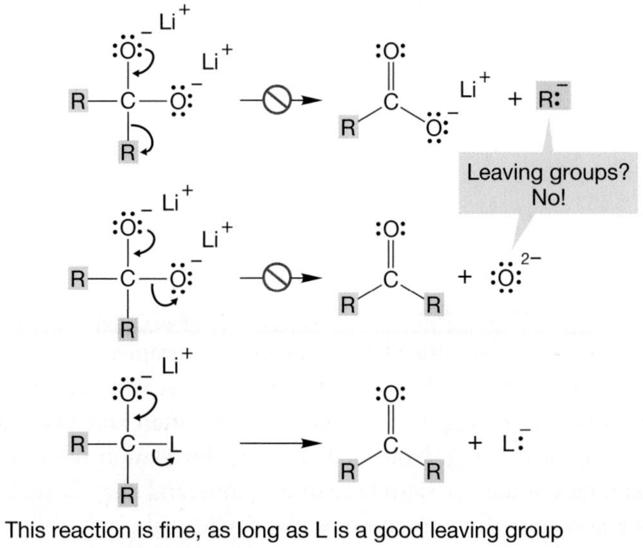

This reaction is fine, as long as L is a good leaving group

FIGURE 17.51 The dianion from Figure 17.50 is foiled in any attempt to expel a negative charge because of the absence of a possible leaving group.

Reduction of carboxylic acids by lithium aluminum hydride, followed by hydrolysis, gives the corresponding primary alcohols (Fig. 17.52). There is no unanimity on the details of the reaction mechanism, but the first step must be formation of the carboxylate anion and hydrogen. Anion formation is followed by addition of hydride to the carbonyl group. The result is another "dianion." In this case, the aluminum is undoubtedly tightly bound to one of the oxygen anions.

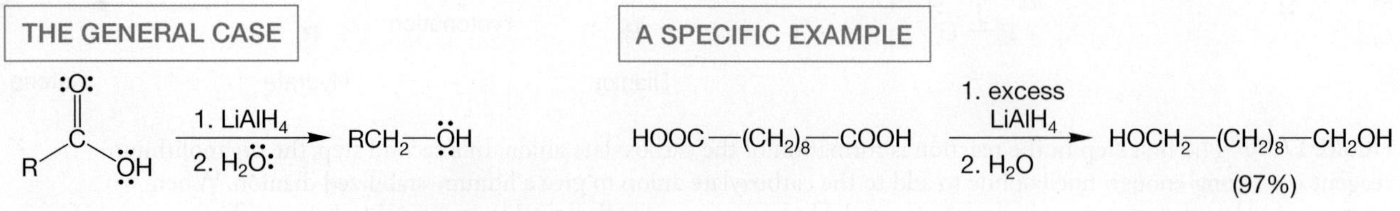

THE GENERAL CASE	A SPECIFIC EXAMPLE

FIGURE 17.52 The reduction of carboxylic acids to alcohols.

Now, unlike the "dianion" in Figure 17.50, a molecule of aluminum oxide can be lost to generate the aldehyde (Fig. 17.53). Because the $LiAlH_4$ can continue to provide hydride (some experiments show 3.8 equivalents of hydride come from $LiAlH_4$), the newly formed aldehyde cannot survive for long (p. 801). Another addition of hydride gives the alkoxide. The alkoxide is stable in the basic environment and subsequent addition of water forms the primary alcohol.

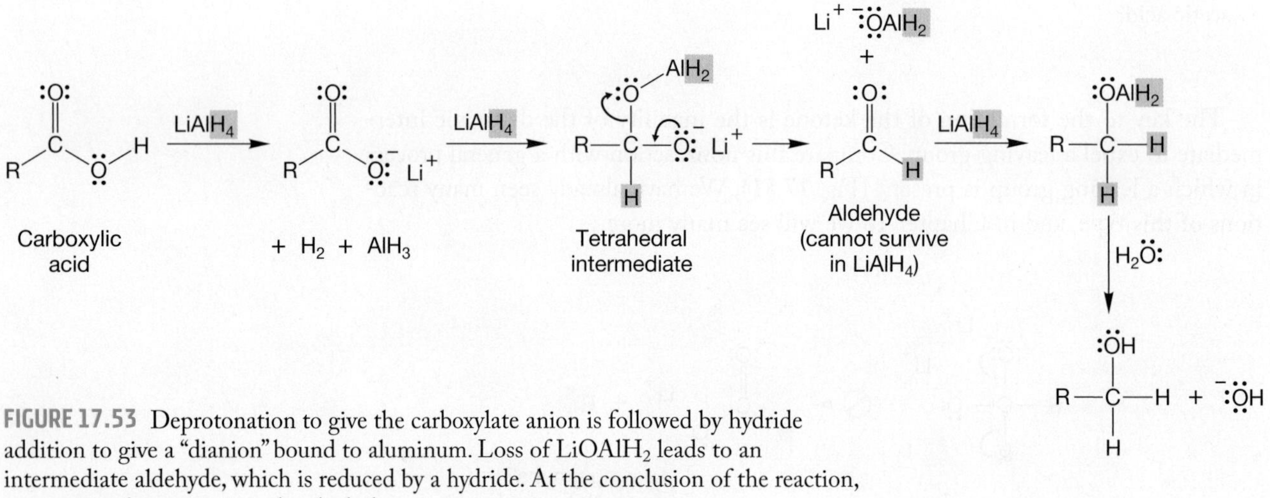

FIGURE 17.53 Deprotonation to give the carboxylate anion is followed by hydride addition to give a "dianion" bound to aluminum. Loss of $LiOAlH_2$ leads to an intermediate aldehyde, which is reduced by a hydride. At the conclusion of the reaction, protonation by water gives the alcohol.

17.7g Decarboxylation   Some carboxylic acids easily lose carbon dioxide in a reaction logically called **decarboxylation**. The best examples are β-keto acids, and 1,3-diacids such as malonic acid (propanedioic acid) in which there is also a carbonyl group β to the acid (Fig. 17.54). The reactions often take place at room temperature or on gentle heating. Note that the decarboxylation reaction of a β-keto acid provides a new route for syntheses of a ketone and the decarboxylation of a 1,3-diacid is a new route for formation of carboxylic acids.

GENERAL REACTIONS

SPECIFIC EXAMPLES

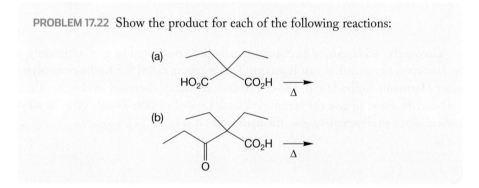

FIGURE 17.54 Three kinds of carboxylic acid that decarboxylate (lose CO_2) easily.

PROBLEM 17.22 Show the product for each of the following reactions:

(a)

(b)

We will defer discussion of the decarboxylation of β-keto acids until we learn the utility of such compounds in Chapter 19, but carbonates and malonic acids will be mentioned here.

Carbonic acid (H_2CO_3 or HOCOOH) is itself unstable, although its salts are isolable and familiar as sodium bicarbonate and sodium carbonate:

Carbonic acid
(unstable)

Sodium bicarbonate
(stable)

Sodium carbonate
(stable)

Monoesters of carbonic acid lose carbon dioxide easily. The loss of carbon dioxide is irreversible, because the gaseous carbon dioxide escapes. The diesters of carbonic acid (carbonates) are stable (Fig. 17.55).

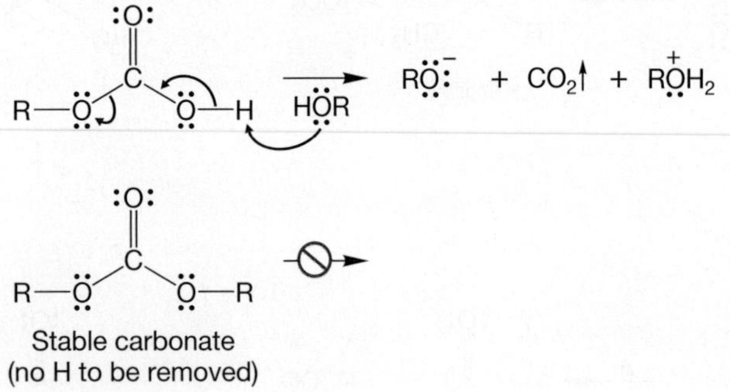

FIGURE 17.55 Monoesters of carbonic acid decarboxylate easily, but carbonates (acid diesters) do not.

Similarly, **carbamic acid**, the amide derivative of carbonic acid, is also unstable, and loses carbon dioxide to give amines (Fig. 17.56). The diamide, **urea**, and **carbamates**, which cannot lose carbon dioxide, are stable.

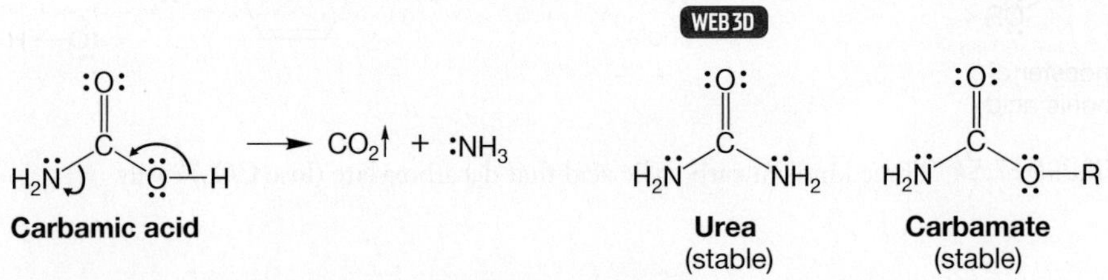

FIGURE 17.56 Carbamic acids decarboxylate, but urea and carbamates are stable.

Carbamic acid

Urea
(stable)

Carbamate
(stable)

Carboxylic acids can be electrochemically decarboxylated to give, ultimately, a hydrocarbon composed of two R groups. This reaction, called the **Kolbe electrolysis** after Hermann Kolbe (1818–1884), involves an electrochemical oxidation of the carboxylate anion to give the carboxyl radical. Loss of carbon dioxide gives an alkyl radical that can dimerize to give the hydrocarbon (Fig. 17.57).

THE GENERAL CASE

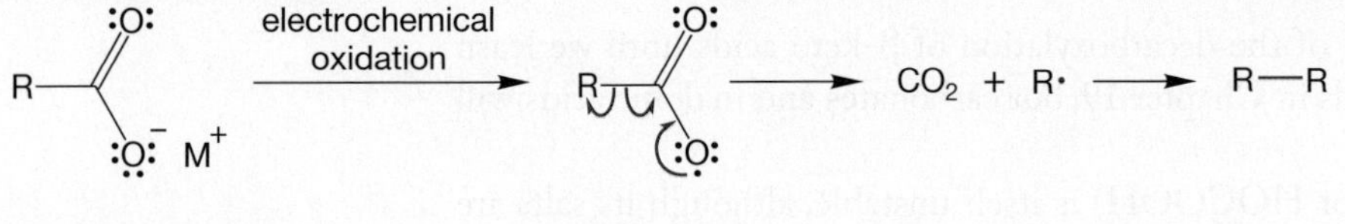

A SPECIFIC EXAMPLE

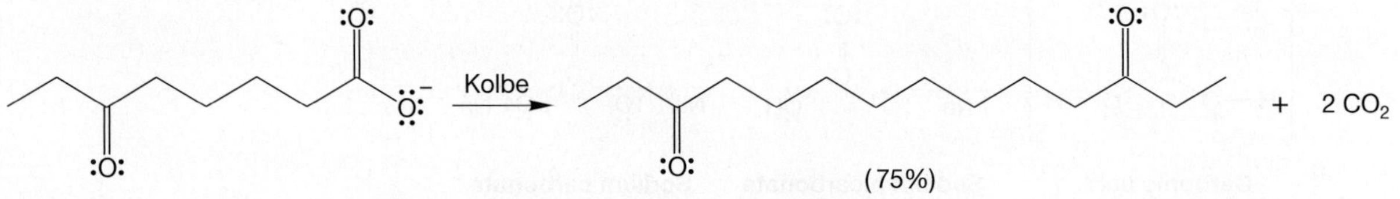

FIGURE 17.57 The Kolbe electrolysis is a source of molecules, R—R, made up of two R groups from R—COOH.

17.7h Formation of Alkyl Bromides: The Hunsdiecker Reaction   The carboxyl radical encountered in Figure 17.57 is also the active ingredient in the **Hunsdiecker reaction** in which alkyl bromides are formed from acids (Heinz Hunsdiecker, 1904–1981; Cläre Dieckmann Hunsdiecker, 1903–1995). Although this reaction is named for the Hunsdieckers, this reaction was actually discovered by Alexander Borodin, the Russian composer-physician-chemist. Borodin is far more famous for his somewhat schmaltzy music rather than his excellent chemistry. This reaction was first done by Borodin in 1861, 81 years before the Hunsdieckers! Typically, the silver salt of the acid is used and allowed to react with bromine. The first step is the formation of the hypobromite (Fig. 17.58). On heating, the weak oxygen–bromine bond breaks homolytically to give the carboxyl radical and a bromine atom. As in the Kolbe reaction, the carboxyl radical undergoes decarboxylation to give an alkyl radical. In this case, however, the alkyl radical can carry the chain reaction further by reacting with the hypobromite to produce an alkyl bromide and another molecule of carboxyl radical as shown in Figure 17.58. For a review of radical chain reactions, see Chapter 11, pp. 481ff.

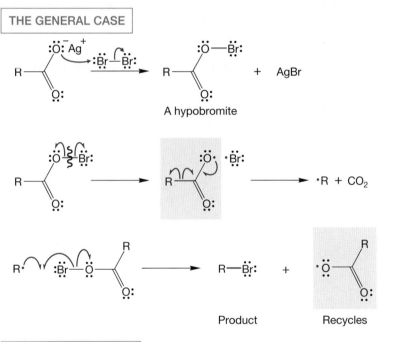

FIGURE 17.58 The decarboxylation of the carboxylate radical is also involved in the Hunsdiecker reaction, a synthesis of alkyl bromides from carboxylic acids. The recycling, chain-propogating carboxyl radical is highlighted.

Summary

That's a lot of material. Many new kinds of compounds have appeared and there may seem to be an overwhelming number of new reactions. Don't be dismayed! Most involve versions of the addition–elimination reaction. Keep this generalization in mind and you will be able to figure out what is happening in these reactions relatively easily. It is not necessary, or productive, to try to memorize all this material. What is far more effective is to learn the basic process well, and then practice applying it in different circumstances.

In this chapter we learned that carboxylic acids can be used to make polyamides and polyesters. We also learned the reactions that use organic acids to make acid chlorides, acid anhydrides, esters, amides, ketones, and primary alcohols. Finally, we've learned about a few decarboxylation reactions of carboxylic acids. Perhaps the most important of the reactions in this chapter are acid chloride formation and Fischer esterification. You certainly will encounter these two reactions often in your study of organic chemistry.

17.8 Special Topic: Fatty Acids

We have spent many pages looking at the reactions of carboxylic acids. Now let's consider a class of molecules found extensively in Nature, called **fatty acids**. These compounds contain long unbranched hydrocarbon chains with a single carboxylic acid at one end. The properties of such molecules are strongly influenced by the presence of both the long-chain and the acid group. Most fatty acids in Nature are even-numbered long-chain acids because they are formed by a process that puts the hydrocarbon chain together two carbons at a time using an acetic acid derivative.

The fatty acids are usually found in plants and animals as the corresponding fat or oil (Fig. 17.59), a fatty acid triester of glycerol. The fatty acid triesters that are solids at room temperature are called fats and those that are liquid at room temperature are known as oils. The base-induced hydrolysis of these esters is called **saponification**. We will learn more about this reaction in Chapter 18.

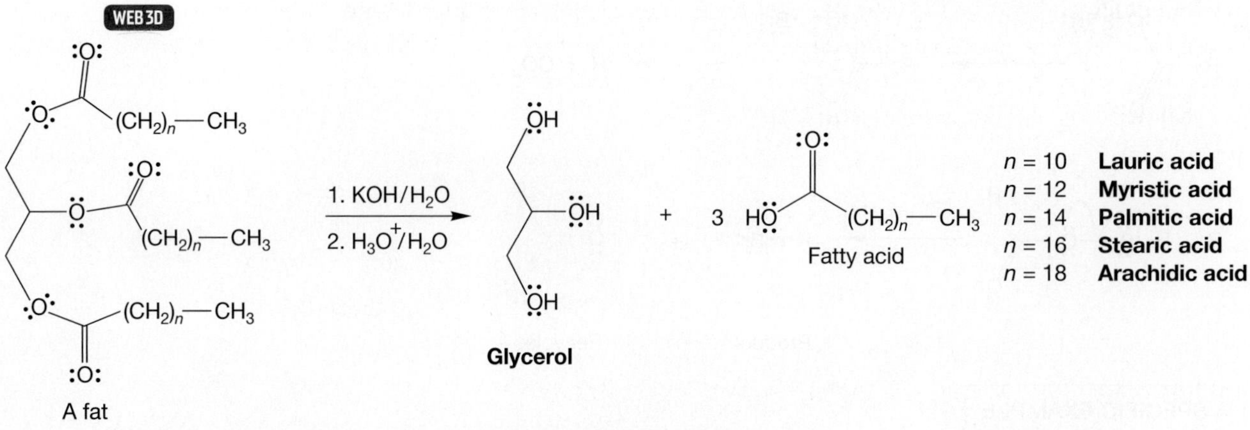

FIGURE 17.59 Some fatty acids that can be made from the hydrolysis of fats.

PROBLEM 17.23 Suggest a mechanism for the hydroxide ion–induced hydrolysis (saponification) of a fat to a fatty acid.

Fatty acids are most important molecules. For example, part of our body's energy supply comes from a pathway that starts with the hydrolysis of a fat to fatty acids, such as palmitic acid, which is then converted into a thioester (R—CO—SCoA[4]).

[4]CoA stands for coenzyme A, one of many molecules necessary to promote an enzymatic reaction.

The thioester is repetitively cleaved by enzymes to give multiple two-carbon units (Fig. 17.60). These acetyl groups are called acetyl-CoA, and are fed into the Krebs cycle, a series of reactions in which acetyl-CoA is oxidized to CO_2.

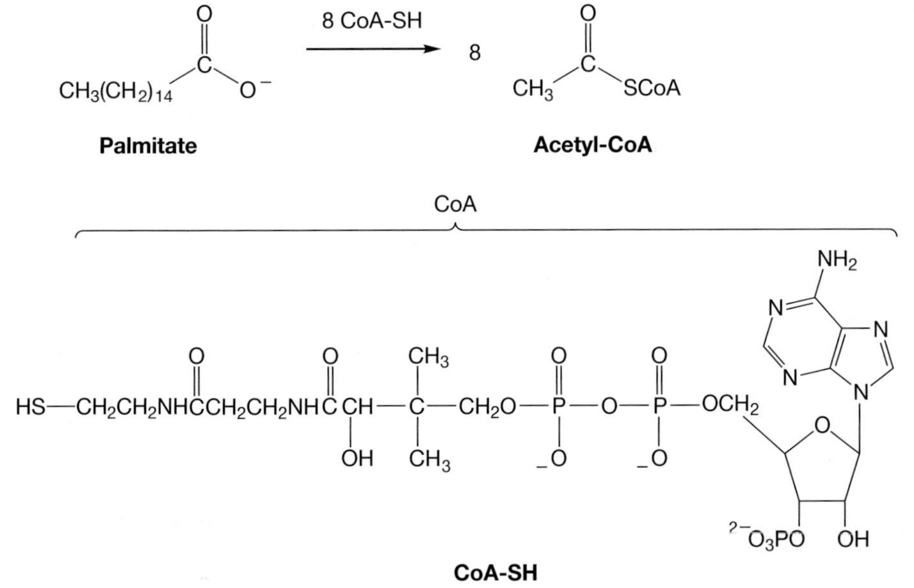

FIGURE 17.60 Enzymes ultimately convert fatty acids such as palmitate, a 16-carbon fatty acid, into eight molecules of acetyl-CoA.

Thus, it seems that the methyl group of acetyl-CoA must be the α carbon of the original fatty acid and the β carbon of the fatty acid becomes the carbonyl carbon of myristate (Fig. 17.61). We will come back to this reaction in Chapter 19, but you should be able to appreciate both the extent and difficulty of the task that Nature faces here—How to break the α–β bond in palmitate? That's no simple task, and we shall need the material in Chapter 19 to solve it.

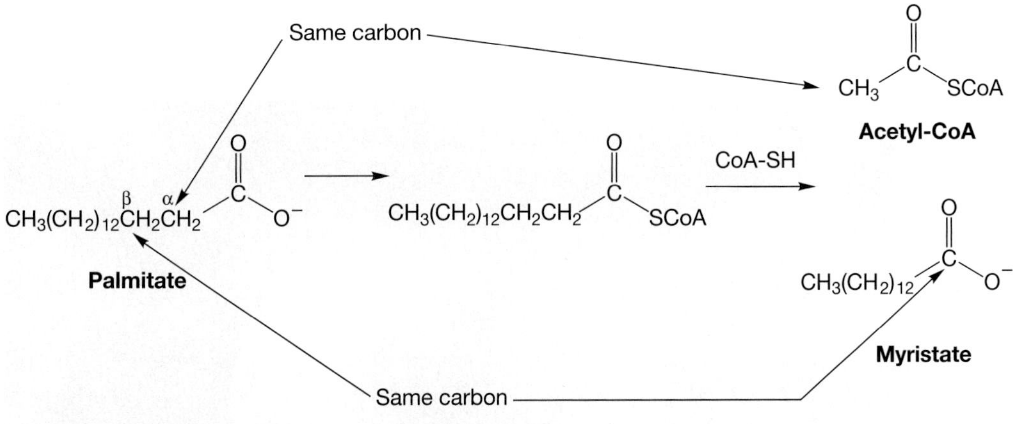

FIGURE 17.61 Fatty acids are deconstructed two carbons at a time. Note that the β carbon of the palmitic acid becomes the carbonyl carbon of the 14-carbon acid.

Fatty acids often have double bonds in the carbon chain. The presence of double bonds in the triester fat affects their solubility and melting points. Saturated fats (those having no double bonds) are solids, and there are well-documented health concerns with diets that are high in saturated fat content. Fatty acids with trans double bonds are also likely to be solid, one of the reasons for recent "trans-fat" health concerns.

When fatty acids have cis double bonds, their corresponding triesters are oils, liquids at room temperature. Because the cis double bond results in poorer stacking between the chains, these molecules do not solidify as easily as their trans counterparts (Fig. 17.62).

FIGURE 17.62 The triesters with no double bonds can stack much better than the triesters with a cis double bond. Trans fats also line up well and have higher melting points than cis fats.

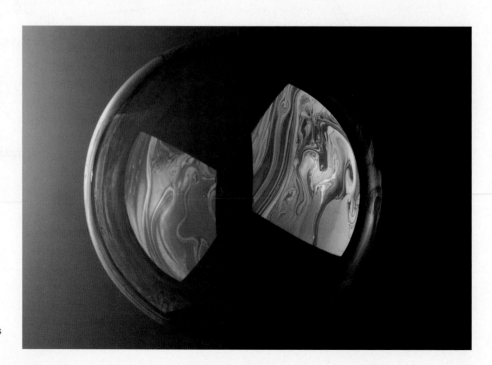

Saturated fat

cis Unsaturated fat (oil)

trans Unsaturated fat (trans fat)

Soap bubbles are thin films made from long-chain carboxylic acid salts that are aligned with each other.

There are many fatty acids that are necessary for proper cellular function. Humans make most of the fatty acids that are needed. However, the omega-3 fatty acids (Fig. 17.63) are an essential part of our diet. Omega-3 fatty acids belong to a family of long-chain carboxylic acids that have a cis double bond at the third carbon from the end of the chain. The specific molecules differ in chain length and in the number of additional double bonds in the chain. The FDA has reported data that suggest omega-3 fatty acids reduce the risk of coronary heart disease.

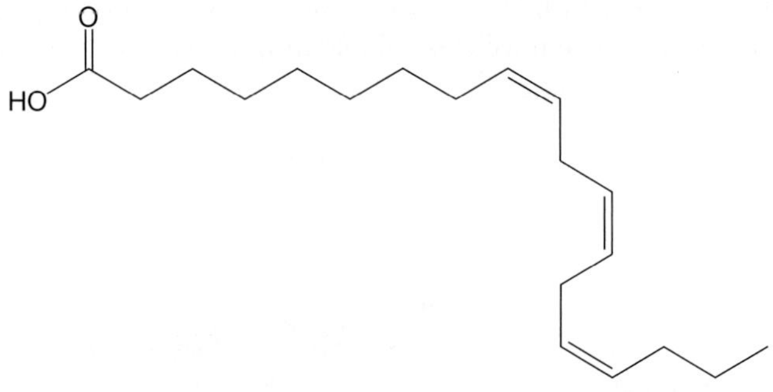

FIGURE 17.63 The omega-3 fatty acids involved in cellular function: α-linolenic acid (ALA), eicosapentaenoic acid (EPA), and docosahexaenoic acid (DHA).

α-Linolenic acid (ALA)
(Z, Z, Z)-9,12,15-Octadecatrienoic acid

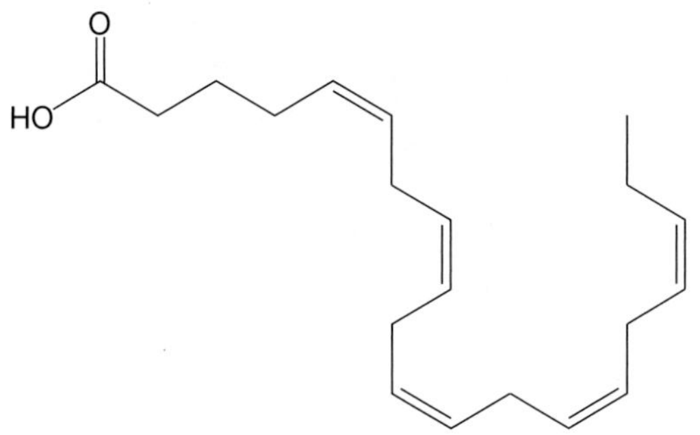

(Z, Z, Z, Z, Z)-5,8,11,14,17-
Eicosapentaenoic acid (EPA)

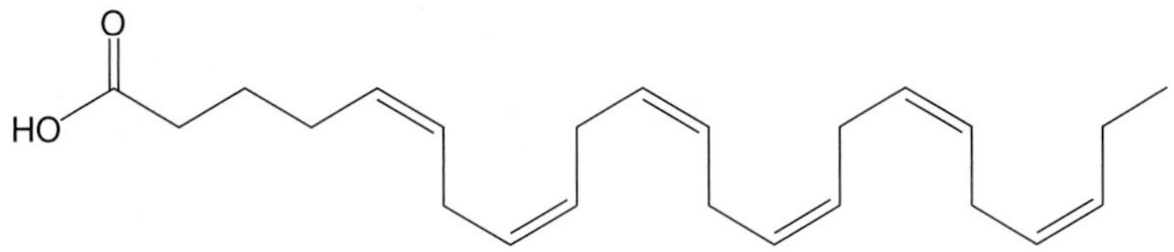

(all Z)-4,7,10,13,16,19-
Docsahexaenoic acid (DHA)

Omega-3 fatty acids are found in high concentrations in fish oil. Fish do not actually synthesize these compounds, but obtain them from algae. Even higher concentrations of omega-3 fatty acids are found in flaxseed oil. Lingonberry and kiwi are also rich in these compounds.

There is another topic of importance related to fatty acids. **Soaps** are the sodium salts of long-chain fatty acids, and work in the following way. These compounds contain both a highly polar end, the carboxylic acid salt, and a most nonpolar end, the long-chain hydrocarbon end (Fig. 17.64). The polar end is soluble in polar solvents such as water, but the nonpolar, hydrocarbon end is not. In aqueous solution, the **hydrophobic** (water-hating) hydrocarbon ends cluster together at the inside of a sphere, protected from the aqueous environment by the **hydrophilic** (water-loving) polar acid-salt groups forming the interface with the polar water molecules. Such aggregates are called **micelles**. The cleaning ability of soaps comes from the oil-capturing property of the hydrophobic portion of the micelle, which can remove oil and grease from a surface. The micelle is soluble in water and so it washes away the non-water-soluble oil and grease.

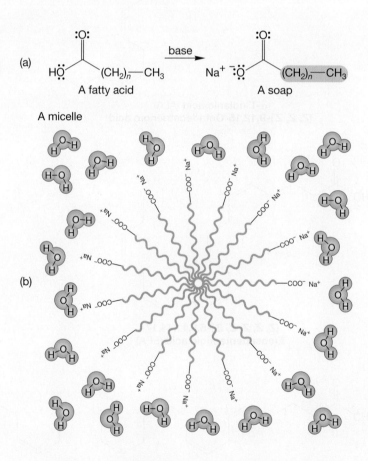

FIGURE 17.64 (a) The formation of a soap from a fatty acid. (b) In polar media, soaps form micelles, with the hydrophobic hydrocarbon chains directed toward the center of a sphere, away from the polar solvent. The outside of the sphere contains the polar carboxylate anions and their positive counterions.

Similar micellar species are formed from **detergents**, which are synthetic soaps. One of the most common synthetic soaps is a long-chain sulfonic acid salt (Fig. 17.65). Because detergents are industrially synthesized, they haven't always been biodegradable. Perhaps not surprisingly, the key to chemists finding an environmentally friendly soap was recognizing that Nature can oxidatively cleave long-chain carbon compounds, but does not do as well with branched alkyl groups. Using this "green" recognition has given us biodegradable detergents.

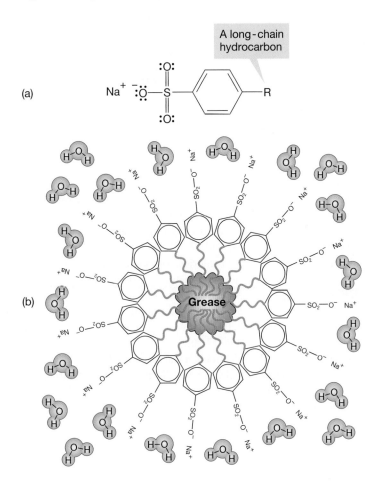

FIGURE 17.65 (a) A synthetic soap, a detergent, and (b) a micelle of the detergent solubilizing a grease glob.

PROBLEM 17.24 What process do you suppose Nature uses to degrade the long-chain detergent?

17.9 Summary

New Concepts

Carboxylic acids exhibit several kinds of reactivity. So far, we have seen many kinds of molecules (aldehydes and ketones, for example) that can be both electrophiles and nucleophiles. But carboxylic acids are even more polyreactive. They are Brønsted acids through proton loss from the hydroxyl group. They are also electrophiles (Lewis acids) through reaction at the carbonyl carbon.

Carboxylic acids act as Brønsted bases and nucleophiles (Lewis bases) through reaction at the more basic carbonyl oxygen.

The Brønsted acidity of carboxylic acids, long thought to rest largely upon the delocalization of the carboxylate anion produced by ionization, is also dependent on the inductive effect of the polarized carbonyl group.

Key Terms

acid anhydride (p. 855)
activating agent (p. 850)
amide (p. 850)
carbamates (p. 860)
carbamic acid (p. 860)
carboxylate anion (p. 832)
decarboxylation (p. 858)
detergent (p. 866)

fatty acids (p. 862)
Fischer esterification (p. 841)
Hunsdiecker reaction (p. 861)
hydrophilic (p. 866)
hydrophobic (p. 866)
Kolbe electrolysis (p. 860)
lactam (p. 851)
lactone (p. 849)

micelle (p. 866)
ortho esters (p. 846)
saponification (p. 862)
soap (p. 866)
tetrahedral intermediate
  (p. 842)
urea (p. 860)

Reactions, Mechanisms, and Tools

Most of the new reactions in this chapter are examples of addition–elimination processes. This mechanism has appeared occasionally before, but it is emphasized in this chapter for the first time. Any carbonyl compound bearing a leaving group can be attacked by a nucleophile to give a tetrahedral intermediate (addition) that can then expel the leaving group (elimination) (Fig. 17.66).

For carboxylic acids, the success of an addition–elimination reaction depends on the transformation of the carboxyl hydroxyl group into a better leaving group. This conversion is often done through protonation in acid-catalyzed reactions. The best example is Fischer esterification, and its cyclic variation, lactone formation (Fig. 17.33).

The carboxyl OH can also be made into a better leaving group through other chemical conversions. An excellent exam-ple is acid chloride formation in which a chlorosulfite ester is first produced. Addition of chloride to the carbonyl group of the ester leads to an intermediate that can eliminate sulfur dioxide and chloride, thus producing the acid chloride (Figs. 17.42 and 17.43).

The acid chlorides contain an excellent leaving group, and addition–elimination reactions are common (Fig. 17.44).

Occasionally, the carboxylate anion can be used as a displacing agent in an S_N2 reaction. In order for this reaction to be successful, the Lewis acid partner in the reaction must be very reactive (Fig. 17.30).

The electrochemical reaction of carboxylate anions leads to hydrocarbons (Kolbe electrolysis). Other reactions involving decarboxylations are briefly mentioned (Fig. 17.54).

FIGURE 17.66 The addition–elimination reaction.

Syntheses

1. Acid Halides

2. Acids

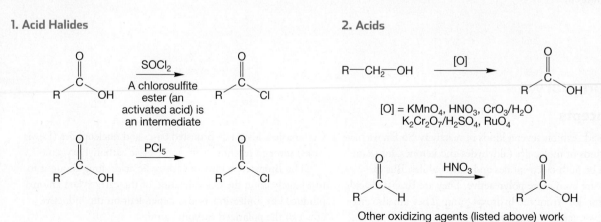

$[O] = KMnO_4, HNO_3, CrO_3/H_2O$
$K_2Cr_2O_7/H_2SO_4, RuO_4$

Other oxidizing agents (listed above) work

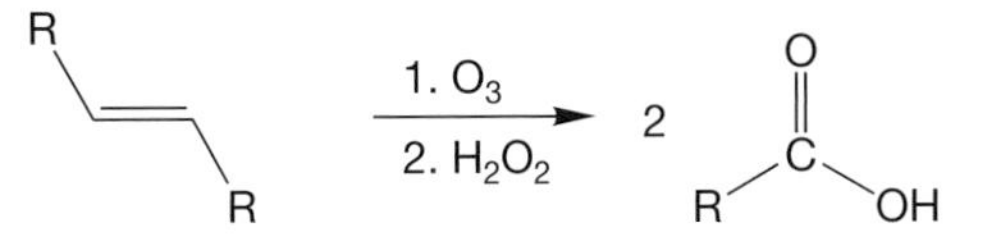

Note the oxidative workup; the aldehyde can be obtained by using dimethyl sulfide or H_2/Pd

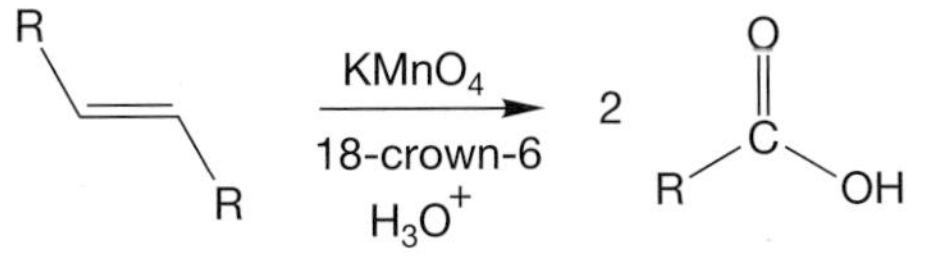

The crown ether is essential, as it brings the $KMnO_4$ into solution

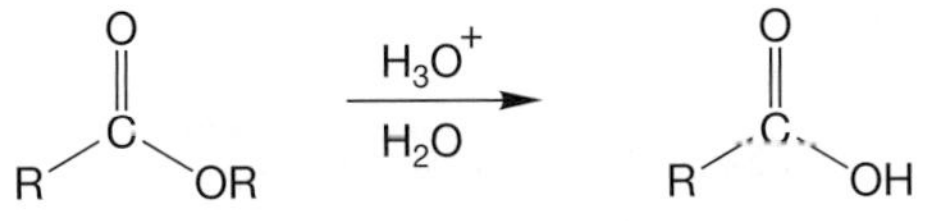

Acid-catalyzed ester hydrolysis

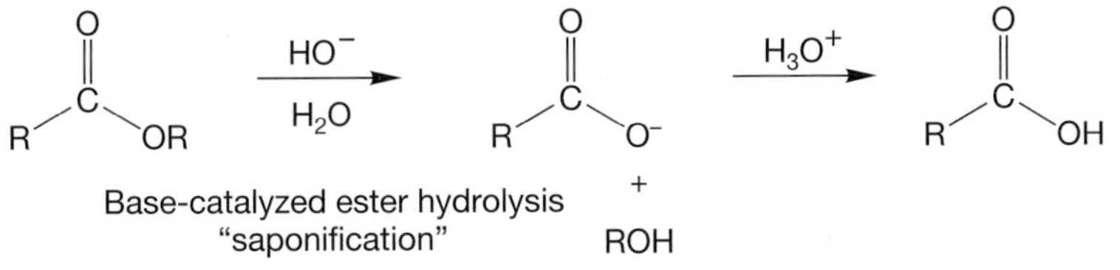

Base-catalyzed ester hydrolysis "saponification"

3. Alcohols

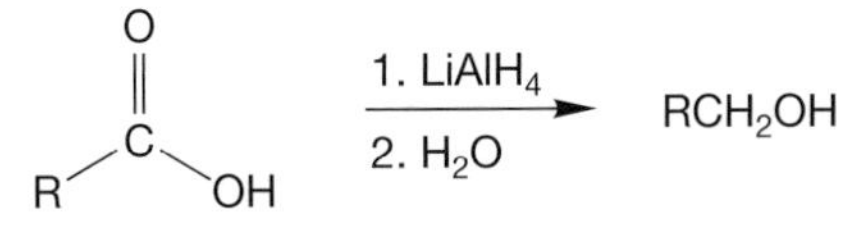

An aluminum alkoxide is a leaving group in this reaction

Decarboxylation of a bicarbonate ester

See also other "hydroxy" compounds in this section

4. Alkyl Halides

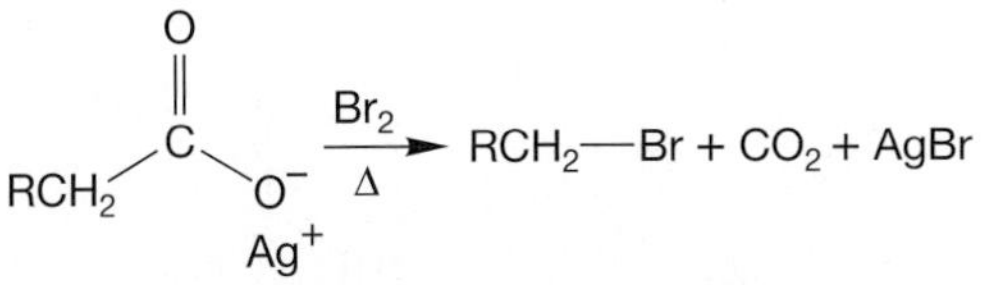

The Hunsdiecker reaction is used almost exclusively for bromides

5. Amides (for cyclic amides, see "Lactams")

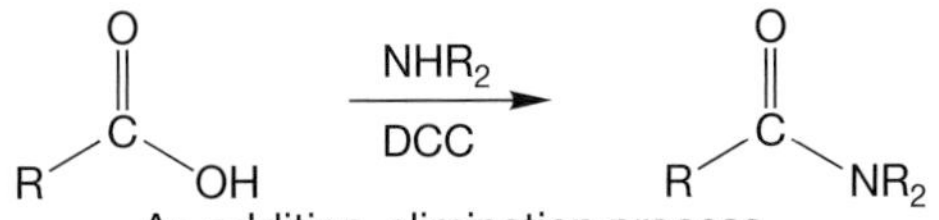

An addition–elimination process

6. Amines

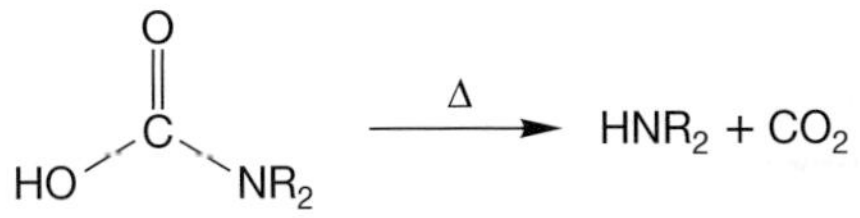

Decarboxylation of a carbamic acid

7. Anhydrides

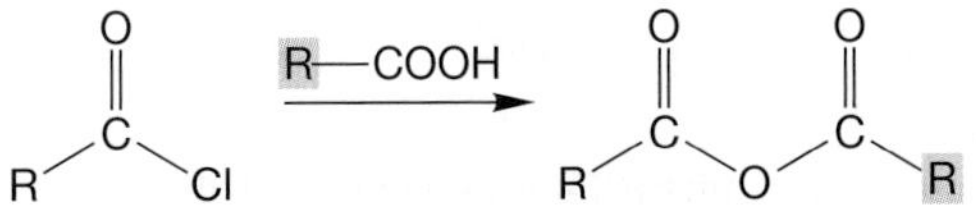

This addition–elimination process works even better with the carboxylate anion as nucleophile

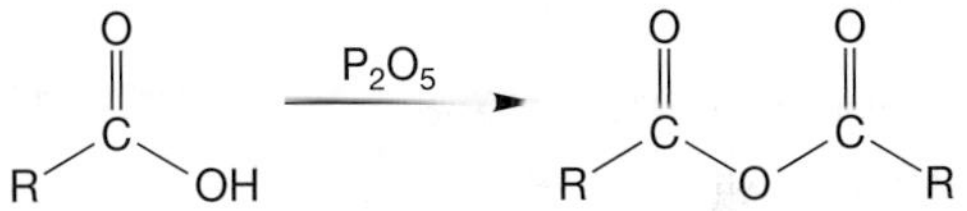

Other dehydrating agents also work; intramolecular reactions can often be accomplished by simple heating

8. Esters (for cyclic esters, see "Lactones")

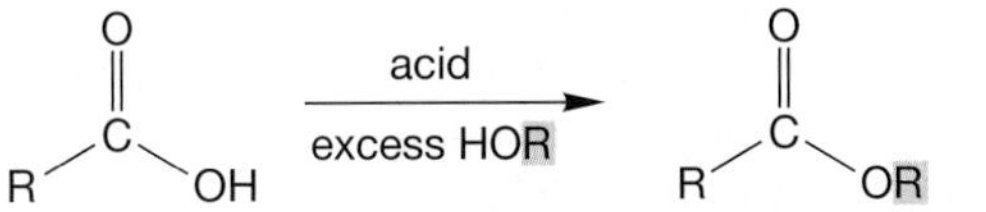

Fischer esterification

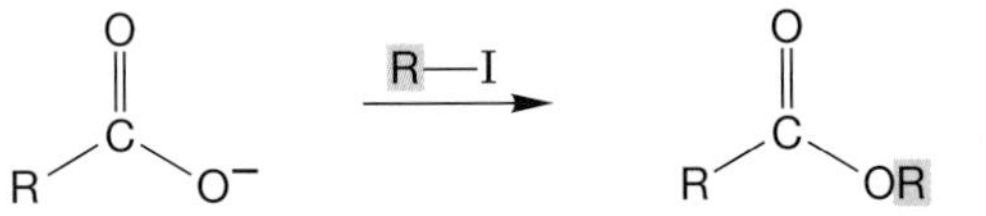

For this S_N2 reaction to succeed, R—I must be very reactive

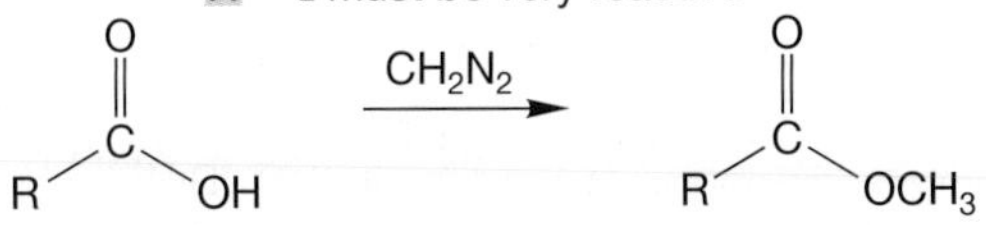

Only common for methyl esters

9. Hydrocarbons

The Kolbe electrolysis, a radical dimerization process

10. Ketones

The hydrate is an intermediate,
2 moles of RLi are required

Decarboxylation of β-keto acids

11. Lactams

This formation of cyclic amides works
best for unstrained rings

12. Lactones

Intramolecular Fischer esterification

Common Errors

Addition–elimination reactions are straightforward: A nucleophile adds to the carbonyl carbon, then the carbonyl is reconstituted as a leaving group is displaced. But there is such variety! That's where the difficulty lies—in seeing through all the "spinach" hanging on the nucleophile and hiding the carbonyl—to find the simple mechanism at the heart of matters.

Also, there are vast numbers of small steps—proton transfers mostly—in these mechanisms, and that makes it more challenging to get them all right.

Finally, there is a small error that all of us seem to make on occasion. This mistake is to add an electrophile to the wrong oxygen of a carboxylic acid. It is the carbonyl oxygen that is the more nucleophilic and basic site, not the hydroxyl oxygen.

17.10 Additional Problems

PROBLEM 17.25   Write names for the following carboxylic acids:

PROBLEM 17.26   Write names for the following carboxylate salts:

(a)

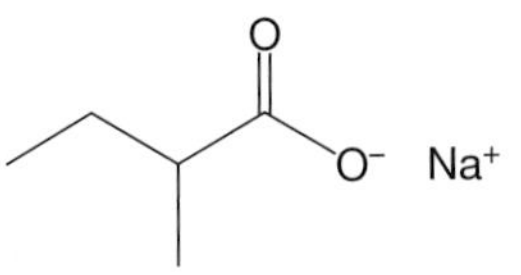

(b)

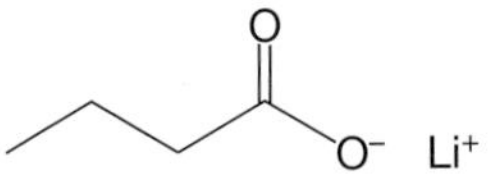

(c)

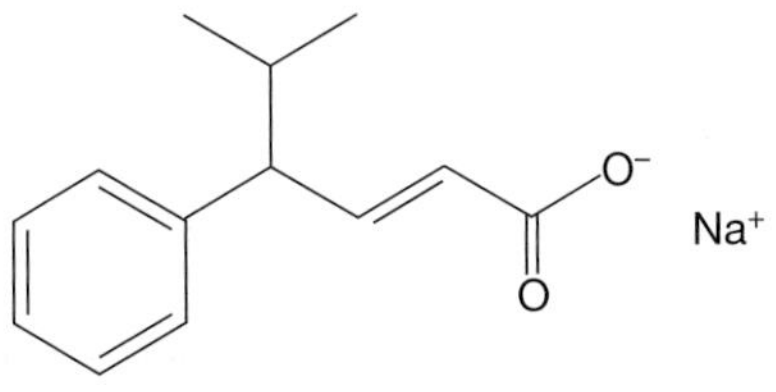

(d)

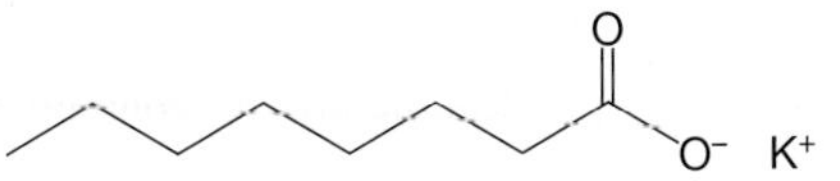

PROBLEM 17.27   Provide the IUPAC name for the following compounds:

(a)

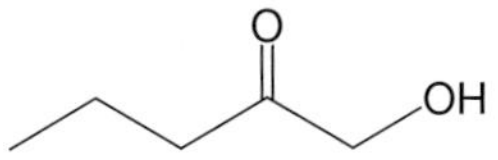

(b)

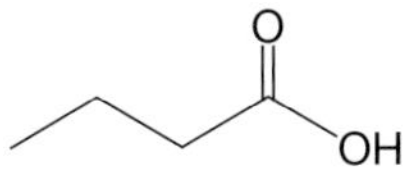

(c)

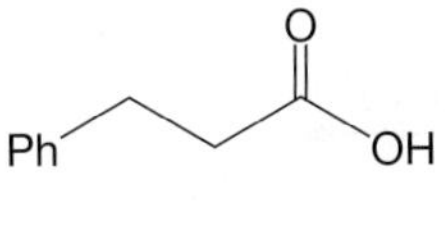

(d)

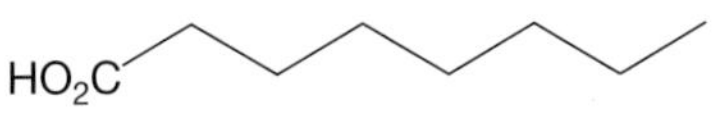

(e)

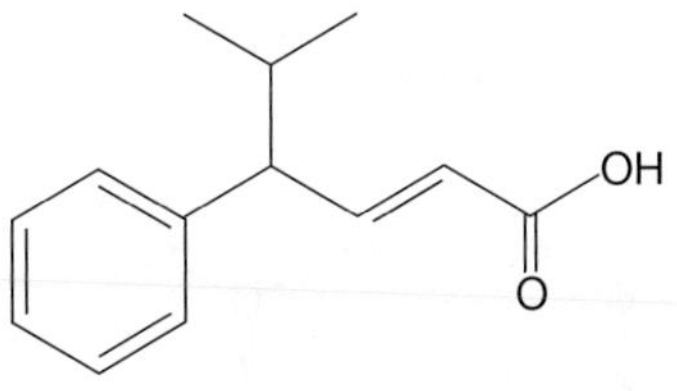

PROBLEM 17.28   Draw a structure for each of the following molecules:

(a) 2-aminopropanoic acid (alanine)
(b) 2-hydroxypropanoic acid (lactic acid)
(c) (Z)-2-methyl-2-butenoic acid
(d) 2-hydroxy-1,2,3-propanetricarboxylic acid (citric acid)
(e) 2-hydroxybenzoic acid (o-hydroxybenzoic acid or salicylic acid)
(f) 2-oxopropanoic acid (pyruvic acid)

PROBLEM 17.29   Draw a structure for each of the following compounds:

(a) lithium 2-hydroxy-3-methylpentanoate
(b) sodium benzoate
(c) sodium 2-amino-3-phenylpropanoate
(d) potassium acetate
(e) potassium 3-butenoate

PROBLEM 17.30   Give reagents for converting butanoic acid into the following compounds:

(a)

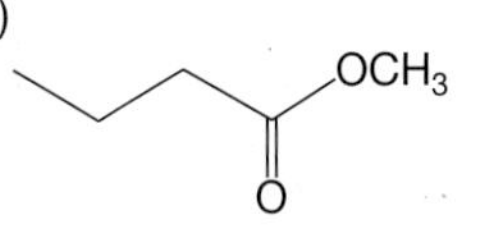

(b)

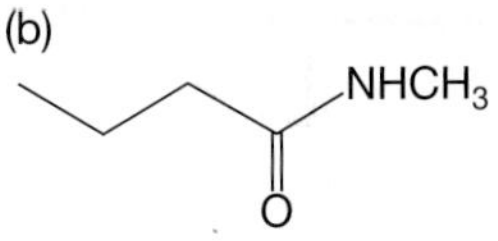

(c)

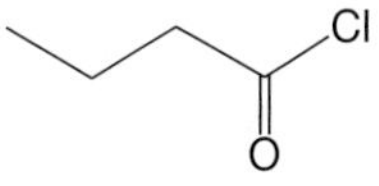

(d)

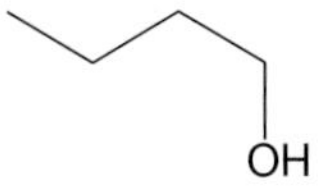

(e)

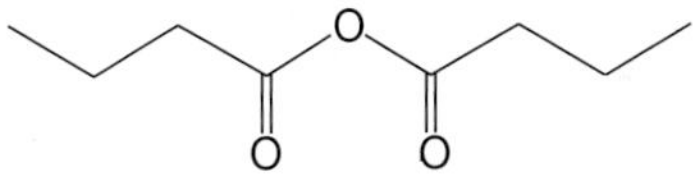

(f)

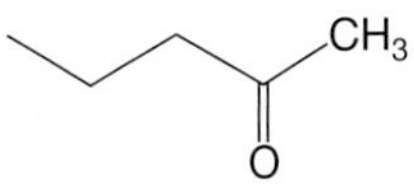

(g)

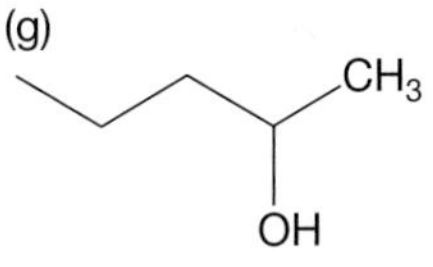

PROBLEM 17.31   Give the major organic products expected in each of the following reactions:

(a)

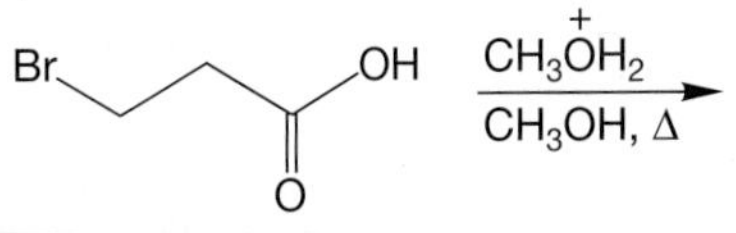

$$Br \quad \text{---} \quad OH \quad \xrightarrow[\text{CH}_3\text{OH, } \Delta]{\overset{+}{\text{CH}_3\text{OH}_2}}$$

(b)

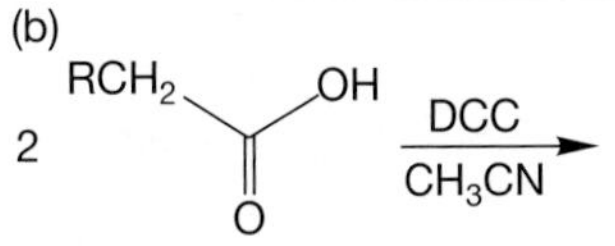

$$2 \quad RCH_2 \quad \text{---} \quad OH \quad \xrightarrow[\text{CH}_3\text{CN}]{\text{DCC}}$$

(continued)

(c)

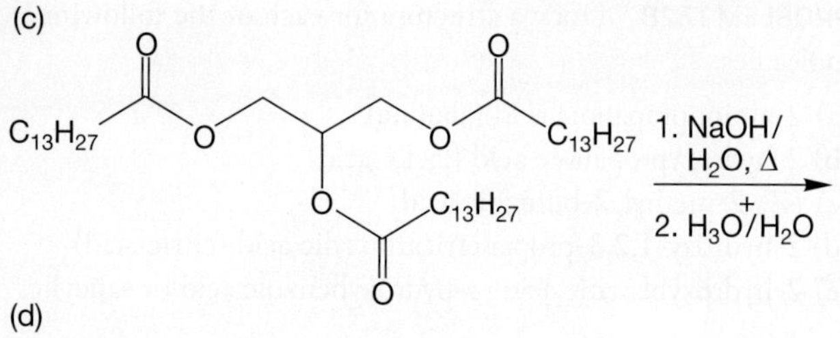

1. NaOH/
   H$_2$O, Δ

2. H$_3$O$^+$/H$_2$O

(d)

RCH$_2$—C(OH)(=O) + RRCHNH$_2$ $\xrightarrow{\text{DCC} \atop \text{CH}_3\text{CN}}$

(e)

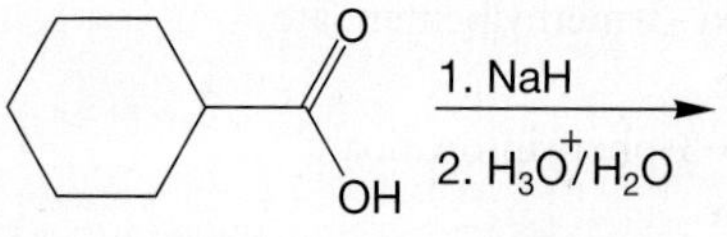

1. NaH

2. H$_3$O$^+$/H$_2$O

(f)

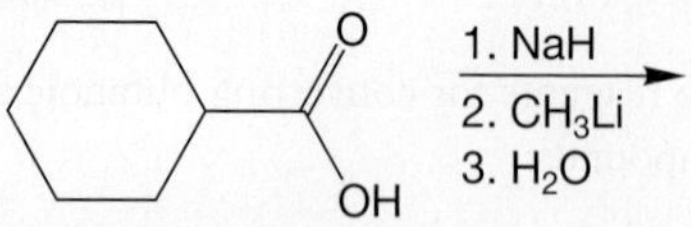

1. NaH

2. CH$_3$Li

3. H$_2$O

(g)

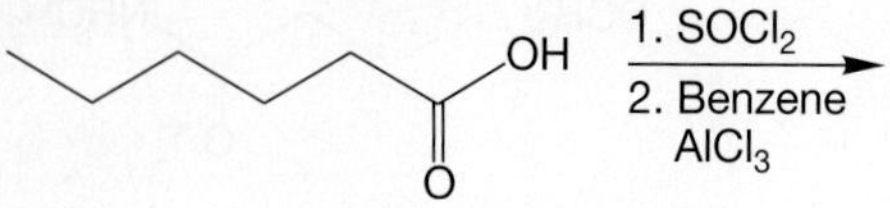

1. SOCl$_2$

2. Benzene
   AlCl$_3$

(h)

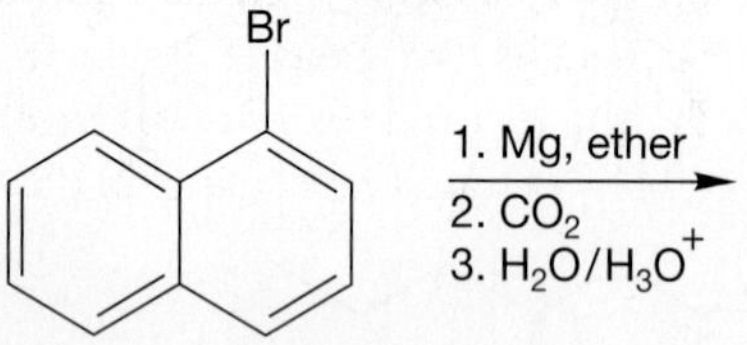

1. Mg, ether

2. CO$_2$

3. H$_2$O/H$_3$O$^+$

PROBLEM 17.32 Provide structures for compounds **A–F**. Mechanisms are not required, but may be helpful in your analysis.

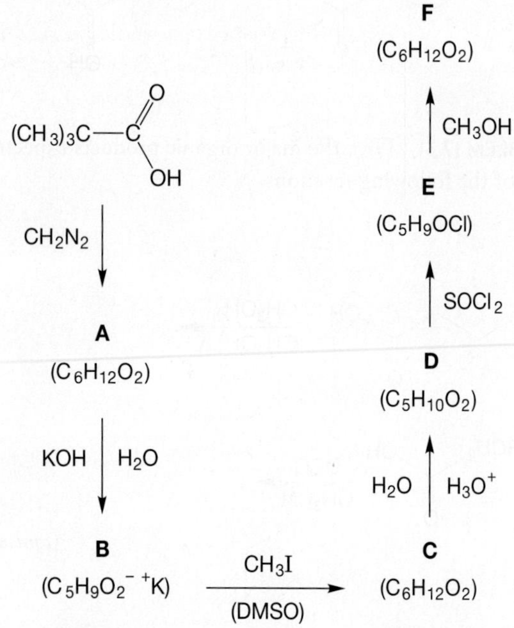

F
(C$_6$H$_{12}$O$_2$)

↑ CH$_3$OH

E
(C$_5$H$_9$OCl)

↑ SOCl$_2$

(CH$_3$)$_3$C—C(OH)(=O)

CH$_2$N$_2$ ↓

A
(C$_6$H$_{12}$O$_2$)

KOH | H$_2$O ↓

B
(C$_5$H$_9$O$_2^-$ $^+$K)

$\xrightarrow[\text{(DMSO)}]{\text{CH}_3\text{I}}$

C
(C$_6$H$_{12}$O$_2$)

D
(C$_5$H$_{10}$O$_2$)

↑ H$_2$O | H$_3$O$^+$

PROBLEM 17.33   We begin this problem with cyclohexanol. Your job is to determine the structures of the compounds **A–D**.

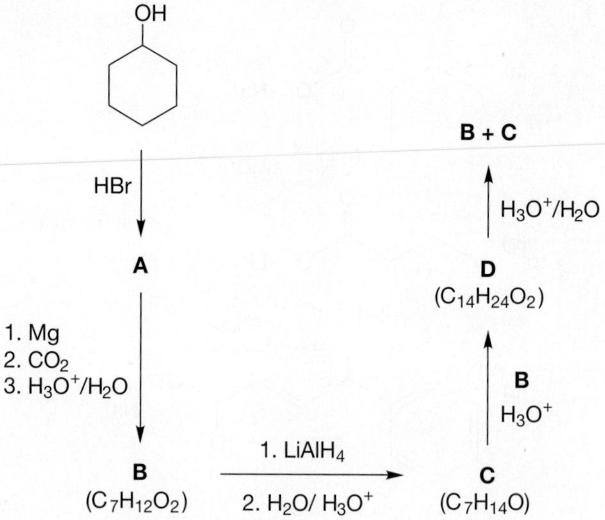

B + C

↑ H$_3$O$^+$/H$_2$O

HBr ↓

A

1. Mg
2. CO$_2$
3. H$_3$O$^+$/H$_2$O

↓

B
(C$_7$H$_{12}$O$_2$)

$\xrightarrow[\text{2. H}_2\text{O/ H}_3\text{O}^+]{\text{1. LiAlH}_4}$

C
(C$_7$H$_{14}$O)

D
(C$_{14}$H$_{24}$O$_2$)

↑ **B**
   H$_3$O$^+$

PROBLEM 17.34   Provide structures for compounds **A–D**.

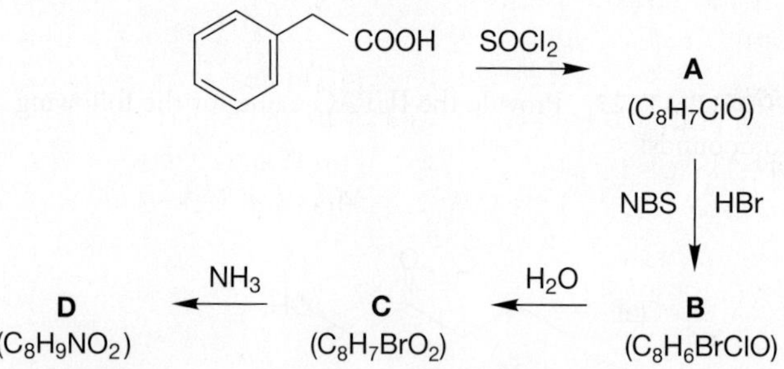

COOH $\xrightarrow{\text{SOCl}_2}$ **A**
(C$_8$H$_7$ClO)

NBS | HBr ↓

B
(C$_8$H$_6$BrClO)

D
(C$_8$H$_9$NO$_2$) $\xleftarrow{\text{NH}_3}$ **C**
(C$_8$H$_7$BrO$_2$) $\xleftarrow{\text{H}_2\text{O}}$ **B**

PROBLEM 17.35   Provide a mechanism for this unusual esterification process.

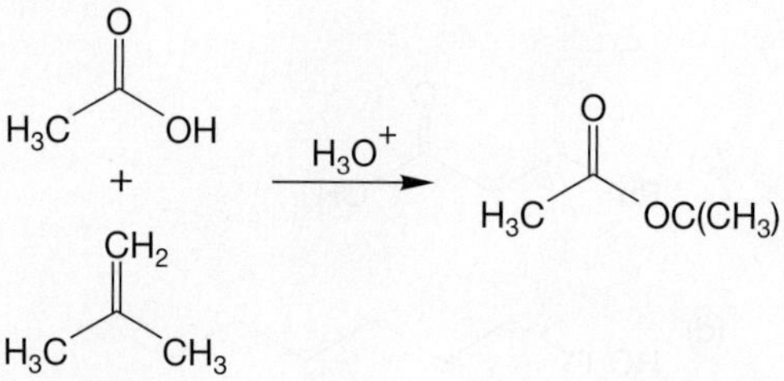

PROBLEM 17.36   Provide brief explanations for the pK_a differences noted in the following:

(a)

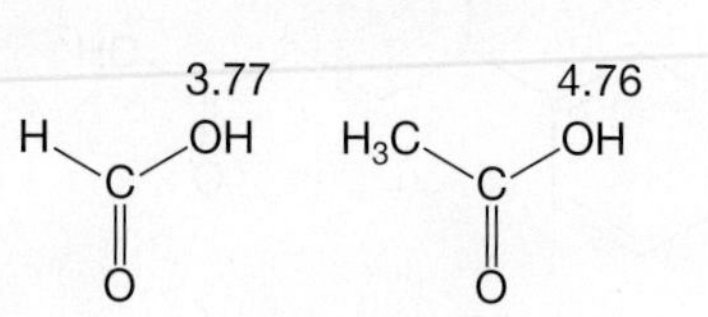

3.77   4.76

(b)

4.76   1.68

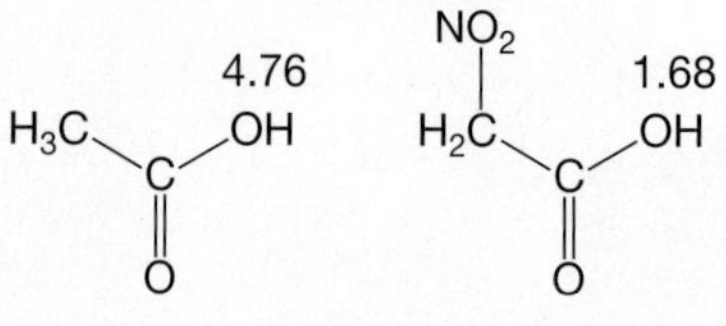

(c)

4.19
—COOH

4.88
—COOH

(d)

1.92
HOOC   COOH   HOOC

3.02
COOH

(e)

6.23
HOOC   COO⁻

4.38
HOOC

COO⁻

PROBLEM 17.37   In Section 17.7b, we saw how dicyclohexyl-carbodiimide (DCC) could be used as a dehydrating agent for the formation of amides from carboxylic acids and amines. Another reagent that can be used to good effect is the phosgene derivative N,N′-carbonyldiimidazole (CDI). Carboxylic acids and CDI react under mild conditions to form imidazolides, **1**, which then easily react with amines to give amides. Propose mechanisms for the formation of **1** from carboxylic acids and CDI, and for the formation of amides from **1** and amines.

CDI

THF | 25 °C

1

PROBLEM 17.38   Dicyclohexylcarbodiimide (DCC) can also be used to activate a carboxylic acid in the synthesis of esters. Show how you would use DCC to synthesize cyclohexyl acetate starting with any alcohols.

PROBLEM 17.39   Show the mechanism for the esterification reaction between pentanoic acid and ethanol using DCC.

PROBLEM 17.40   Another method of forming an anhydride is to allow a dicarboxylic acid to react with another anhydride such as acetic anhydride, Ac₂O. Propose a mechanism for the reaction in the next column.

PROBLEM 17.41   In Problem 17.14, you worked out the mechanism for lactone formation from an hydroxy acid. Here is the related reaction, the formation of a cyclic amide, a lactam. Provide a mechanism.

H_2N COOH   →   acid catalyst
25 °C, 5 h

A lactam
(97%)

PROBLEM 17.42   There is a compound called the Vilsmeier reagent (we will ask you about its formation in Chapter 18) that reacts with carboxylic acids to give acid chlorides and the molecule dimethylformamide (DMF). Provide a mechanism for this process.

Cl^-
$(CH_3)_2\overset{+}{N}=C$ + $R-C$ → $R-C$ + $(CH_3)_2N-C$
          Cl OH Cl H

Vilsmeier reagent   **DMF**

PROBLEM 17.43   Show the product that would be obtained from the decarboxylation of each of the following compounds:

(a)
HO_2C

(b)
Ph   OH

(c)
HO_2C CO_2H

(d)
CO_2H

PROBLEM 17.44   Predict the product for each of the following reactions:

(a)
SOCl₂
benzene

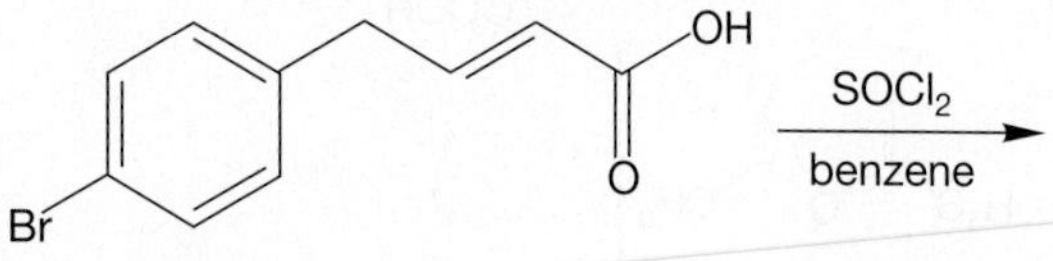

(b)
SOCl₂
benzene

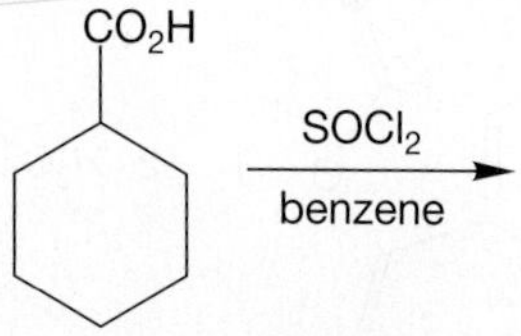

(c)
SOCl₂
benzene

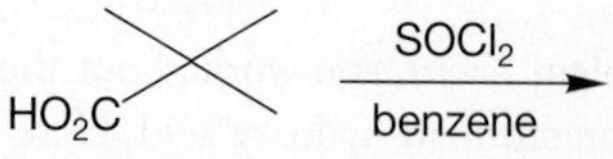

(d)
SOCl₂
benzene

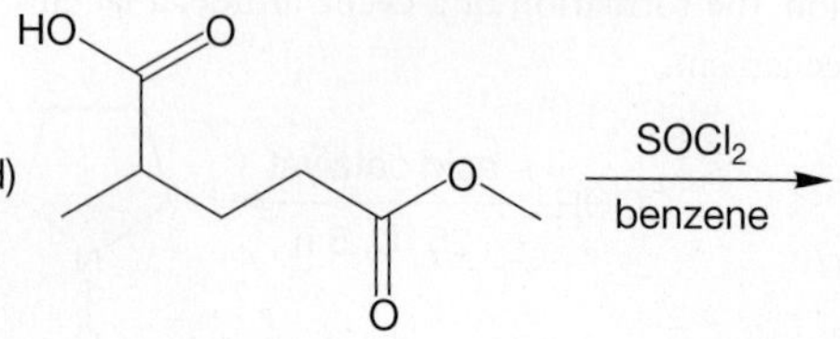

(e)
excess
SOCl₂
benzene

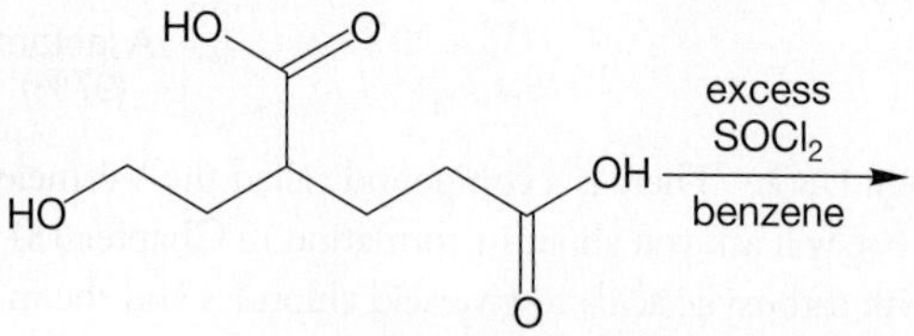

PROBLEM 17.45   An imide is a functional group similar to an acid anhydride. Instead of an oxygen between two carbonyls, an imide has a nitrogen between two carbonyls. Suggest a method for making the imide shown below starting with pentanedioic acid and benzyl amine.

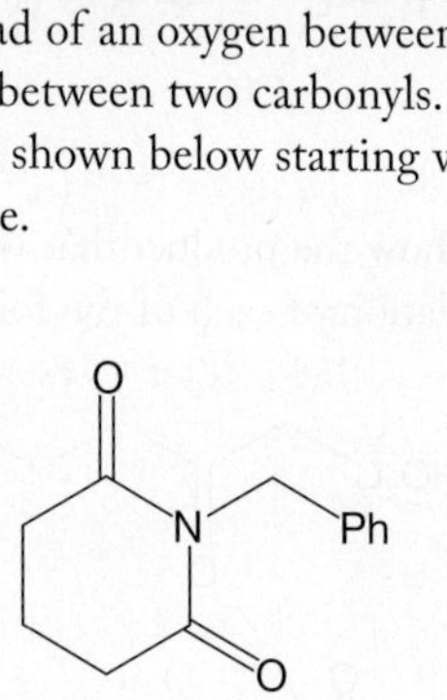

PROBLEM 17.46   Propose syntheses of the following molecules from the readily available fatty acid, lauric acid (**1**):

(a) $CH_3(CH_2)_{11}Br$

$CH_3(CH_2)_{10}COOH$ ⟶ (b) $CH_3(CH_2)_{10}Br$

1

(c) $CH_3(CH_2)_{11}COOH$

PROBLEM 17.47   Propose an arrow formalism mechanism for the following reaction. *Hint*: See Section 17.7g and Figure 17.57.

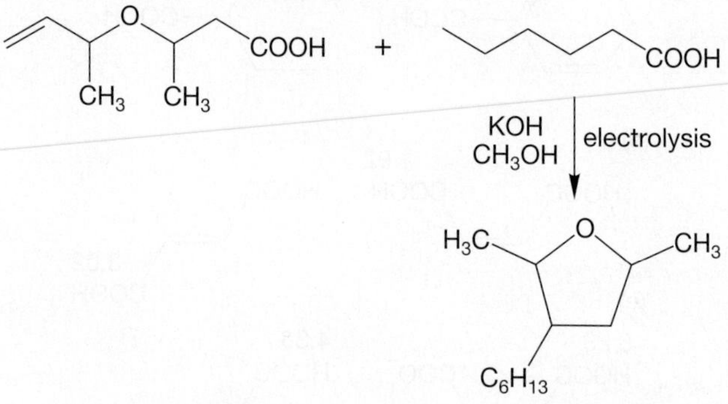

KOH
CH₃OH   electrolysis

PROBLEM 17.48   Explain the following observations. Note that the only difference in the starting materials is the stereochemistry of the carbon–carbon double bond.

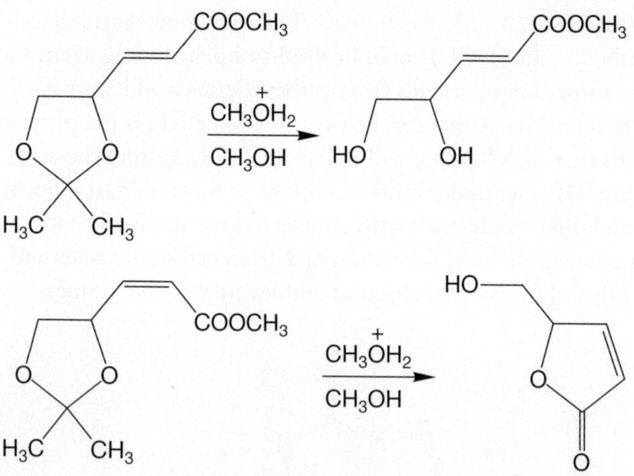

$CH_3\overset{+}{O}H_2$
CH₃OH

$CH_3\overset{+}{O}H_2$
CH₃OH

PROBLEM 17.49   Here are some somewhat more complicated syntheses. Devise ways to make the compounds shown below. You may use cyclopentanecarboxylic acid as your only source of carbon.

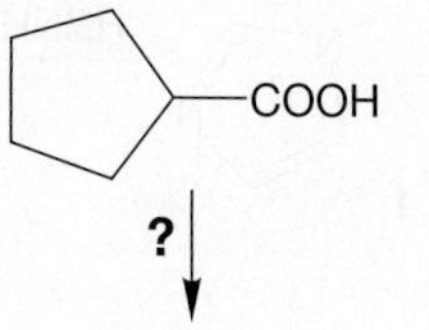

?

(a)

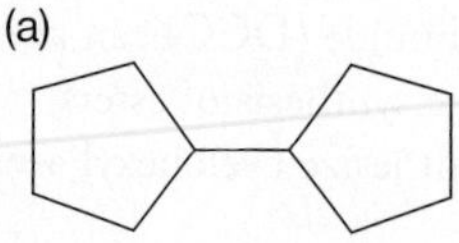

(b)

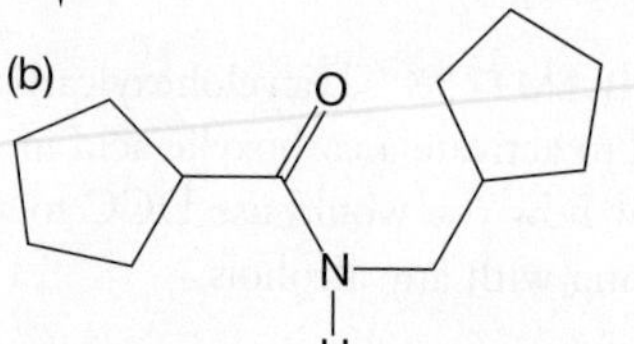

(c)

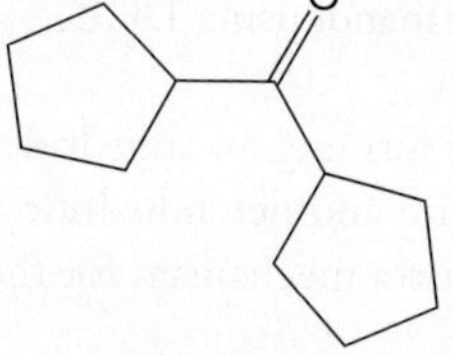

(d)

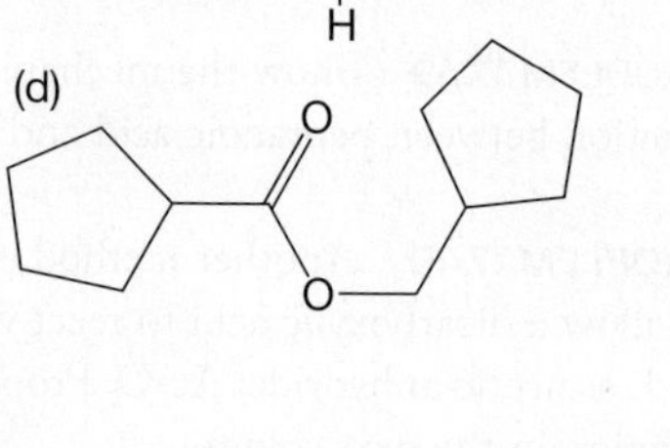

PROBLEM 17.50   Estragole (**1**) is a major constituent of tarragon oil. Treatment of **1** with hydrogen bromide in the presence of peroxides affords **2**. Compound **2** reacts with Mg in ether, followed by addition of carbon dioxide and acidification, to give **3**. Spectral data for compound **3** are summarized below. Deduce the structure of compound **3** and propose structures for estragole and compound **2**.

Compound **3**

Mass spectrum: m/z = 194 (p)
IR (Nujol): 3330–2500 (br, s), 1696 (s) cm^{-1}
^{1}H NMR (CDCl$_3$): δ 1.95 (quintet, J = 7 Hz, 2H)
2.34 (t, J = 7 Hz, 2H)
2.59 (t, J = 7 Hz, 2H)
3.74 (s, 3H)
6.75 (d, J = 8 Hz, 2H)
7.05 (d, J = 8 Hz, 2H)
11.6 (s, 1H)

PROBLEM 17.51   Reaction of cyclopentadiene and maleic anhydride affords compound **1**. Hydrolysis of **1** leads to **2**, which gives an isomeric compound **3** upon treatment with concentrated sulfuric acid. Spectral data for compounds **1–3** are shown below. Propose structures for compounds **1–3** and provide mechanisms for their formations.

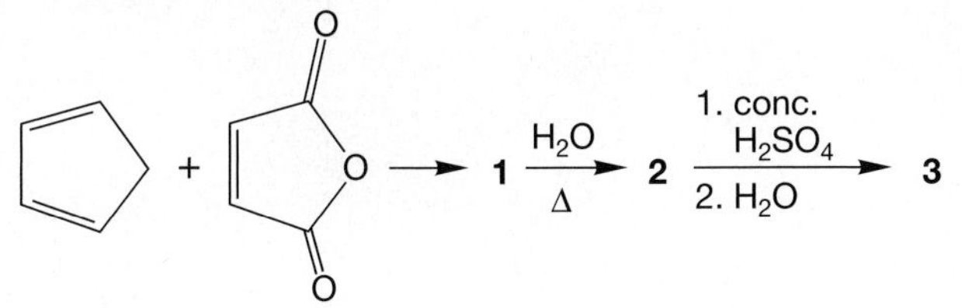

**Cyclo-   Maleic
pentadiene   anhydride**

Compound **1**

Mass spectrum: m/z = 164 (p, 3%), 91 (23%), 66 (100%), 65 (22%)
IR (Nujol): 1854 (m) and 1774 (s) cm^{-1}
^{1}H NMR (CDCl$_3$): δ 1.60–1.70 (m, 1H)
3.40–3.50 (m, 1H)
3.70–3.80 (m, 1H)
6.20–6.30 (m, 1H)

Compound **2**

IR (Nujol): 3450–2500 (br, s), 1710 (s) cm^{-1}
^{1}H NMR (CDCl$_3$): δ 0.75–0.95 (m, 1H)
2.50–2.60 (m, 1H)
2.70–2.80 (m, 1H)
5.60–5.70 (m, 1H)
11.35 (s, 1H)

Compound **3**

IR (KBr): 3450–2500 (br, s), 1770 (s), 1690 (s) cm^{-1}
^{1}H NMR (CDCl$_3$): δ 1.50–1.75 (m, 3H)
1.90–2.05 (m, 1H)
2.45–2.55 (m, 1H)
2.65–2.75 (m, 1H)
2.90–3.05 (m, 1H)
3.20–3.30 (m, 1H)
4.75–4.85 (m, 1H)
12.4 (s, 1H)

PROBLEM 17.52   When *N*-nitroso-*N*-methylurea (NMU) is treated with a two-phase solvent system, H$_2$O/KOH and ether, the yellow color of diazomethane (p. 848) rapidly appears in the ether layer. Show the mechanism.

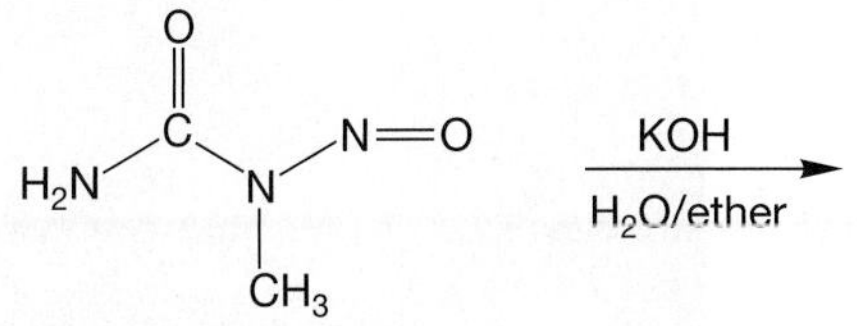

***N*-Nitroso-*N*-methylurea   Diazomethane
(NMU)   (as a yellow solution in ether)**

Use Organic Reaction Animations (ORA) to answer the following questions:

PROBLEM 17.53   Select the reaction "Fischer esterification" and click on the play button. Explain why the first step of the reaction, protonation of acetic acid, occurs at the oxygen that is shown. How does the LUMO track of the first intermediate help explain this selectivity.

PROBLEM 17.54   Observe the "Acid chloride formation" animation. What is the shape of thionyl chloride? What is the shape of the SO$_2$ that is formed at the end of the reaction? This reaction converts a carboxylic acid into a much more reactive acid chloride. Why does the energy diagram show the reaction as being exothermic?

PROBLEM 17.55   The reaction titled "Ester hydrolysis" is also known as saponification. How many intermediates are there in this reaction? Why does the tetrahedral intermediate eliminate the methoxy group so rapidly in this animation?

18

Derivatives of Carboxylic Acids: Acyl Compounds

18.1 Preview

18.2 Nomenclature

18.3 Physical Properties and Structures of Acyl Compounds

18.4 Acidity and Basicity of Acyl Compounds

18.5 Spectral Characteristics

18.6 Reactions of Acid Chlorides: Synthesis of Acyl Compounds

18.7 Reactions of Anhydrides

18.8 Reactions of Esters

18.9 Reactions of Amides

18.10 Reactions of Nitriles

18.11 Reactions of Ketenes

18.12 Special Topic: Other Synthetic Routes to Acid Derivatives

18.13 Special Topic: Thermal Elimination Reactions of Esters

18.14 Special Topic: A Family of Concerted Rearrangements of Acyl Compounds

18.15 Summary

18.16 Additional Problems

TART! The flavors and fragrances of fruit come from volatile esters that we taste and smell.

Strange shadows from the flames will grow,
Till things we've never seen seem familiar . . .

—JERRY GARCIA AND ROBERT HUNTER,[1] *TERRAPIN STATION*

18.1   Preview

In this chapter, we focus on things that should seem familiar even though we have never seen them in detail. Many of the molecules we saw in Chapter 17 will reappear, for example. Acid halides, anhydrides, esters, and amides are all **acyl compounds** of the general structure R—CO—L. These compounds are also known as **acid derivatives**, because historically they were first derived from carboxylic acids. Many are shown in Figure 18.1 along with the related nitriles and ketenes. This chapter will deal with all six of these functional groups (acid halides, anhydrides, esters, amides, nitriles, and ketenes). The nitriles and ketenes are included in this discussion because, like acid derivatives, they undergo reaction with water to form a carboxylic acid. These six functional groups are at the same oxidation level.

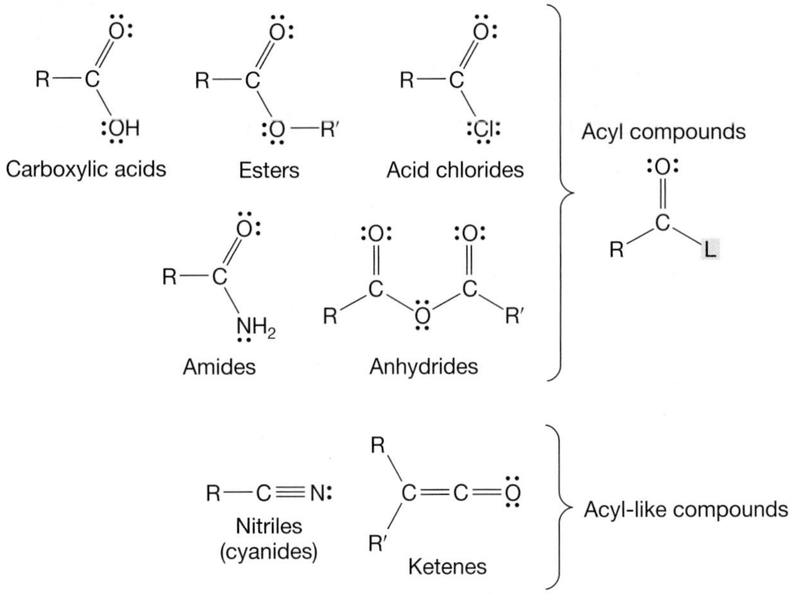

FIGURE 18.1 Some acyl and related derivatives of carboxylic acids.

Not only are these molecules related by having similar structures, but their chemistries are also closely intertwined—a reaction of one acyl compound is generally a synthesis of another, for example. The unifying properties of these seemingly disparate compounds allow us to learn their chemistry as a group. The structures of the acyl compounds are determined largely by the interaction between the L group and the carbonyl, and the chemistry of these acid derivatives is dominated by addition–elimination reactions in which the L group is replaced with some other substituent (Fig. 18.2).

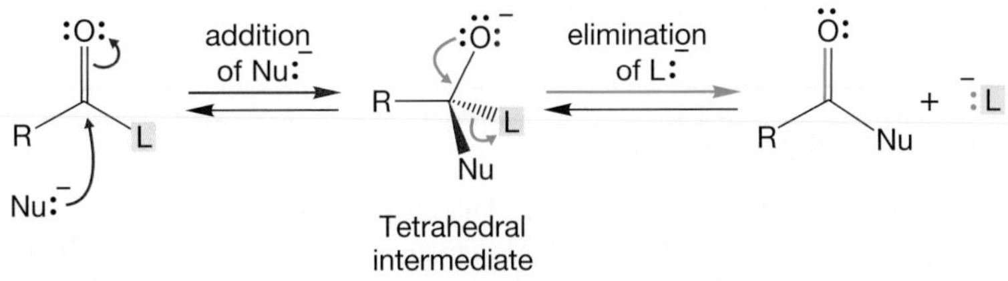

FIGURE 18.2 The most important mechanism for reactions of acyl compounds is the addition–elimination process.

[1]Jerry Garcia (1942–1995) played lead guitar with the Grateful Dead, and Robert Hunter (b. 1941) was a frequent songwriting collaborator.

Note the tetrahedral intermediate at the center of this generic mechanism. This addition–elimination process is an equilibrium, and can proceed in both the "forward" and "backward" directions. The old leaving group L$^-$ is also a nucleophile.

The acid derivatives deserve our full attention. They are all around us in life. Esters are of utmost importance to the fragrance and flavoring industry. The sweet odors of fruits and perfumes are usually a result of volatile esters. Amides are found throughout biochemistry. It is the amide group that defines enzyme structure, which in turn defines us. We will cover more of the bio-organic chemistry of amides in Chapter 23.

In this chapter we will first go over some common ground—nomenclature, structure, and spectra—and then go on to examine reactions and syntheses for the different derivatives.

The chemistry in the next two chapters can be either daunting or exhilarating. Much depends on how well you have the basic reactions understood. If you do understand them, you will be able to apply them in the new situations we are about to explore. If not, the next chapters will seem crammed with new and complicated reactions, each of which requires an effort to understand. Like anything else, learning chemistry can have a strong psychological component. When you are able to think, "That's not so bad, it's just an application of . . . ," you gain confidence and what you see on the page is likely to stick in your mind.

At this point in your study of organic chemistry, it is possible to ask questions that stretch your knowledge and talent. It's hard to teach this kind of problem solving, because success has a lot to do with having worked lots of related problems in the past. Nevertheless, there are some techniques that we can try to pass on. Don't worry if you don't get every problem! Our colleagues can write problems that we will have difficulty with. It's no disgrace to have trouble with a problem. There are several favorite problems that we've been working on (not continuously) for many years. You'll see some of them. Success in the chemistry business has relatively little to do with how *fast* you can solve problems, at least not problems that require thought and the bringing together of information from different areas. People think in different ways and therefore answers to problems appear at different rates. There are those who are blindingly fast at this kind of thing and others who reach answers more slowly. There is nothing wrong with chewing over a problem several times, approaching it in several ways. Letting a particularly vexing problem rest for a while is often a good idea. Of course, this is the bane of exam writers (and even more so of exam answerers). How do you write an exam that allows for thought and doesn't simply favor the student who can answer quickly? It's hard, especially if you don't have the luxury of asking a small number of questions and giving lots of time.

What does make for success in the real world of organic chemistry? Nothing correlates better than love of the subject, and by "subject" we do not necessarily mean what you are doing now, but rather really *doing* organic chemistry. Doing organic chemistry means not only liking the manipulative aspects of the subject, although that is a great help, but also loving the *questions themselves*. At any rate, it is not the number of A's on your transcript that correlates best with your later success in life!

ESSENTIAL SKILLS AND DETAILS

1. As always, we need to be able to generalize. This chapter describes the chemistry of a set of related acyl compounds. It will be very hard to memorize the vast array of addition–elimination processes, but it should be relatively easy to see them all as the same reaction repeated over and over. Only the detailed structure changes—the overall reaction does not.

2. It will be helpful to keep in mind the order of reactivity that acid derivatives have with respect to nucleophiles: acid chlorides > anhydrides > esters > amides.

3. Amides are not like amines. Students often equate the two functional groups, but they are very dissimilar. You should be able to recognize their differences and explain why these compounds are so different.

4. Carbon–nitrogen triple bonds (nitriles) undergo addition reactions, as do carbonyl compounds.

18.2 Nomenclature

The names of acyl compounds are based on the parent carboxylic acid. The IUPAC systematic naming protocol for esters (R—COOR′) has the R group named as the carboxylic acid and the *ic acid* is replaced with the suffix *ate*. So the root word becomes *alkanoate*. The R′ group is named as the appropriate *alkyl* group and it is written as a separate word in front of the alkanoate. Therefore, the generic ester is *alkyl alkanoate*. The carbonyl carbon of the ester gets top priority in the naming scheme and substituents are numbered with respect to the carbonyl carbon as position 1 (Fig. 18.3). The common names for methanoate and ethanoate are formate and acetate, respectively, so we will frequently encounter alkyl (methyl, ethyl, etc.) formates and acetates.

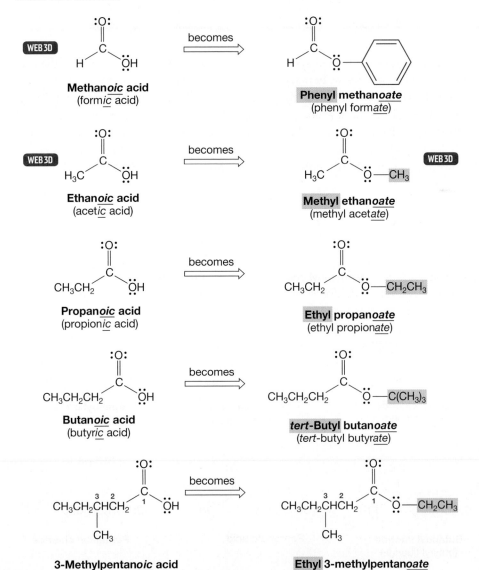

FIGURE 18.3 Some examples of the naming protocol for esters. For each example, the R′ group in the drawing and in the name are highlighted.

As we learned in Chapter 17 (p. 849), cyclic esters are called lactones. The IUPAC system names these compounds as "oxa" cycloalkanones (Fig. 18.4). Alternatively, the ring size of a lactone can be described by starting at the carbonyl carbon and designating the other carbons in the ring with Greek letters until the oxygen atom is reached. A three-membered ring is an α-lactone, a four-membered ring is a β-lactone, and so on.

FIGURE 18.4 Cyclic esters are known as lactones or oxacycloalkanones.

WEB 3D

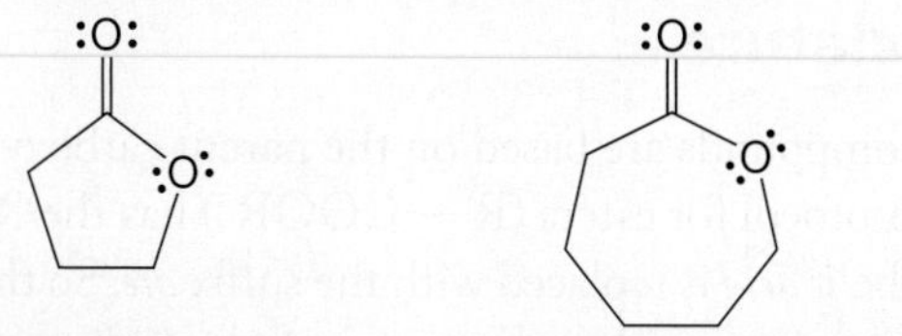

2-Oxacyclopentanone   **2-Oxacycloheptanone**

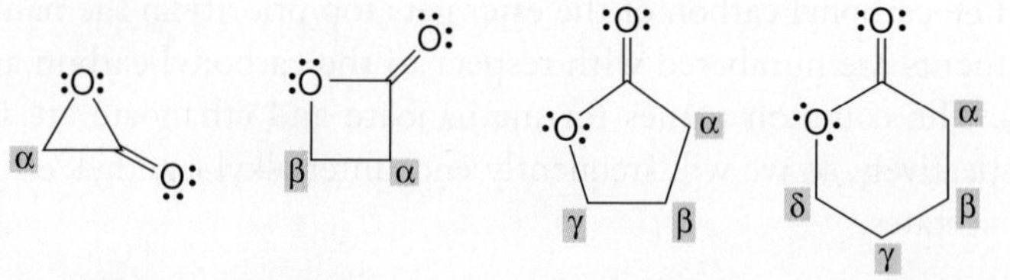

α-Lactone   β-Lactone   γ-Lactone   δ-Lactone

Acid halides are named systematically by replacing the final *ic acid* of the carboxylic acid with *yl halide*. In practice, the names for the smaller acid halides are based upon the common names for the related acids. Thus, acetyl chloride and formyl fluoride would be understood everywhere. Formyl chloride is most unstable and efforts to make it under most conditions are rewarded only with the formation of HCl and CO. The diacid chloride of carbonic acid (Cl—CO—Cl) is known, however, and this deadly poison is called phosgene (Fig. 18.5). Throughout this

FIGURE 18.5 Examples of the systematic and common naming protocols for acid halides.

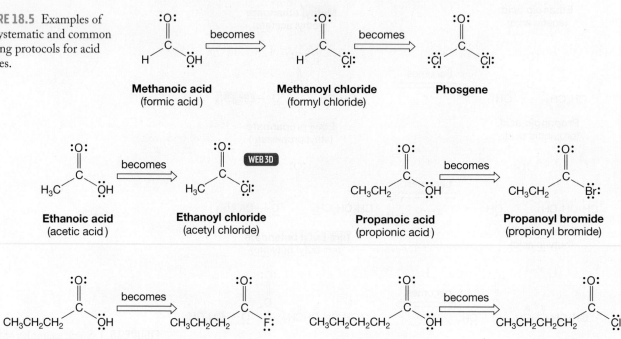

Methanoic acid (formic acid) *becomes* **Methanoyl chloride** (formyl chloride) *becomes* **Phosgene**

Ethanoic acid (acetic acid) *becomes* **Ethanoyl chloride** (acetyl chloride)   WEB 3D

Propanoic acid (propionic acid) *becomes* **Propanoyl bromide** (propionyl bromide)

Butanoic acid (butyric acid) *becomes* **Butanoyl fluoride** (butyryl fluoride)

Pentanoic acid *becomes* **Pentanoyl chloride**

chapter we will use acid chlorides as the example for acid halides. The other acid halides are less stable, less accessible, and less studied.

Acid anhydrides are named by reference to the carboxylic acid from which they are formally derived by loss of water. The *acid* in the name is replaced by *anhydride*. Both systematic and common names are used in this process (Fig. 18.6).

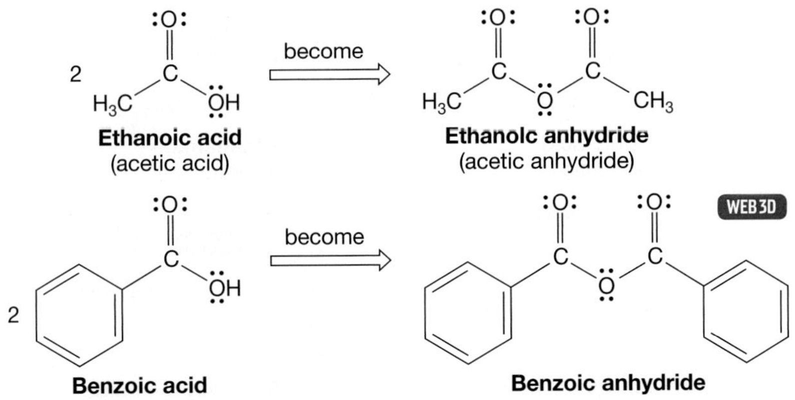

FIGURE 18.6 Examples of the naming protocol for symmetrical anhydrides.

Mixed anhydrides present more problems. Here, both constituent acids must be named and they are listed in alphabetical order. Two examples appear in Figure 18.7.

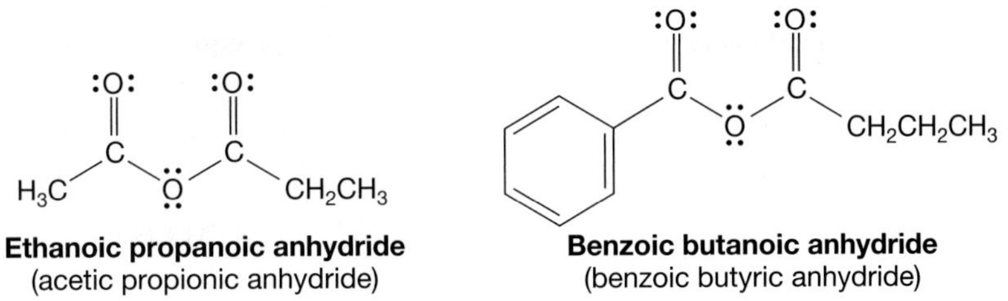

FIGURE 18.7 Named unsymmetrical anhydrides.

Cyclic anhydrides are named as derivatives of the related carboxylic acids (Fig. 18.8).

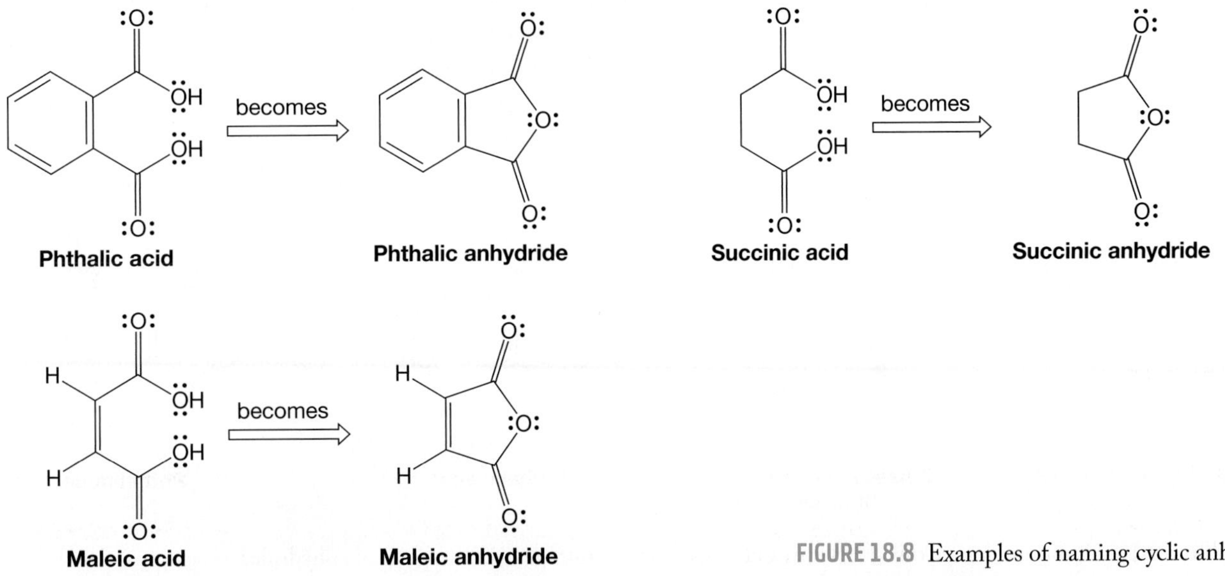

FIGURE 18.8 Examples of naming cyclic anhydrides.

Amides, the nitrogen counterparts of esters, are systematically named by dropping the final *oic acid* of the corresponding carboxylic acid and adding the suffix *amide*. Substitution on nitrogen is indicated with a one-word prefix *N*-alkyl, as in *N*-methyl, *N,N*-diethyl, and so on (Fig. 18.9). Thus, an amide that has one alkyl group on the nitrogen is named as *N-alkylalkanamide*. An amide with two hydrogens on the nitrogen is called a primary amide. A secondary amide has only one hydrogen on the amide nitrogen.

FIGURE 18.9 Examples of the naming protocol for amides.

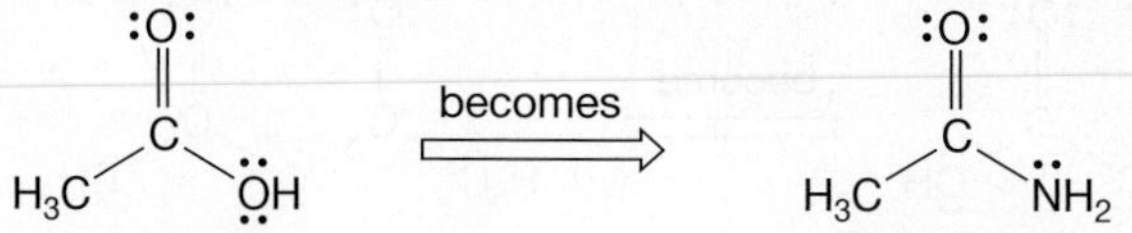

Ethanoic acid (acetic acid) → **Ethanamide** (acetamide)

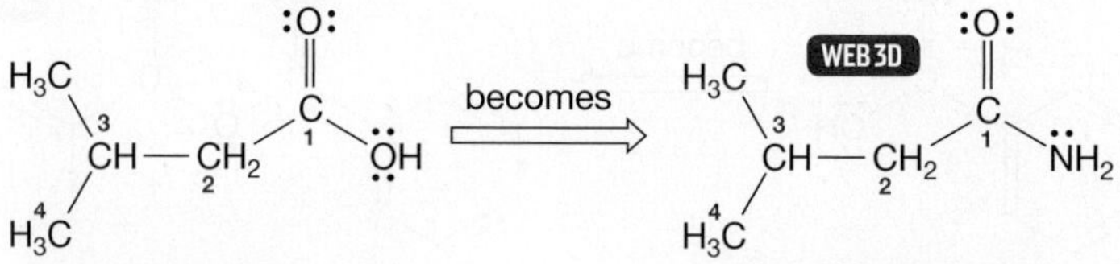

3-Methylbutanoic acid → **3-Methylbutanamide**

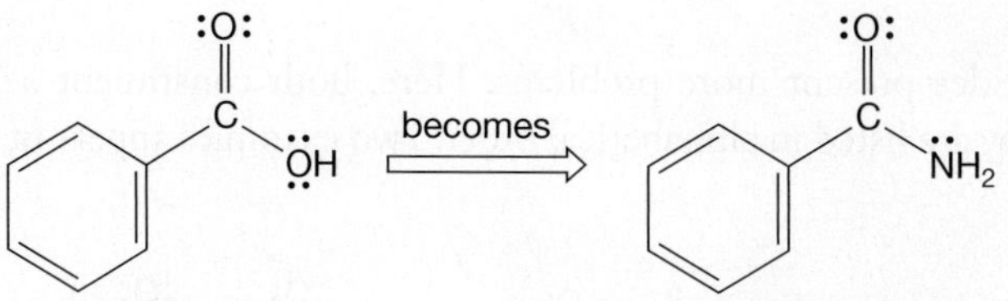

Benzoic acid → **Benzamide**

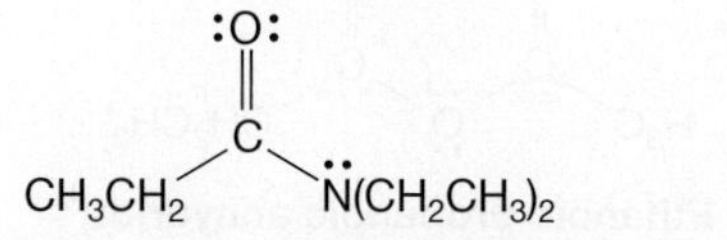

***N,N*-Diethylpropanamide**

A tertiary amide has no hydrogens on the nitrogen. Substituents on the nitrogen are listed alphabetically together with the other substituents from the carboxylic acid portion of the chain. As usual, the smaller members of the class have retained their common names, which are based on the common names for the smaller carboxylic acids. For example, you will often encounter *N,N*-dimethylformamide (abbreviated DMF) because it is a useful polar, high-boiling solvent.

Cyclic amides are known as lactams (p. 851), and the naming protocol closely follows that for lactones, the cyclic esters. Lactams are named systematically as azacycloalkanones. **Imides** are the nitrogen counterparts of anhydrides (Fig. 18.10).

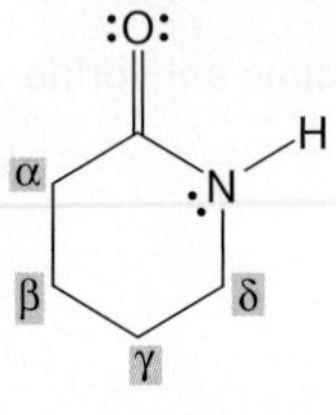

2-Azacyclohexanone (δ-lactam)

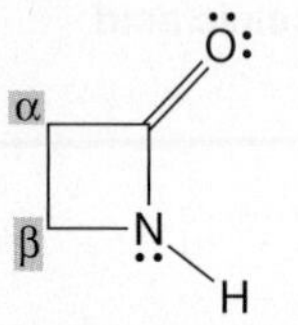

2-Azacyclobutanone (β-lactam)

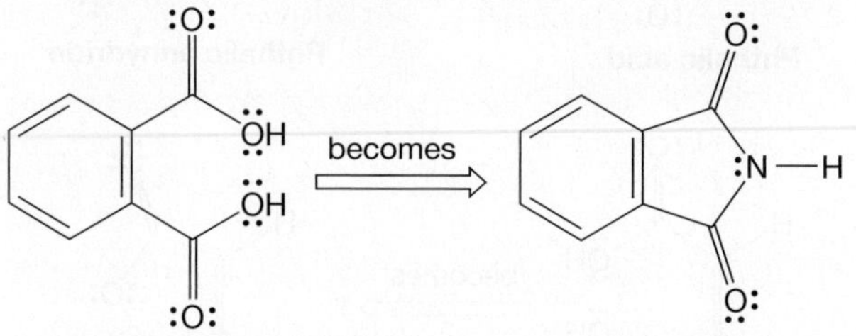

Phthalic acid → **Phthalimide**

FIGURE 18.10 Cyclic amides are called lactams, and imides are the nitrogen counterparts of anhydrides.

Compounds containing carbon–nitrogen triple bonds with the general formula of RCN are named as **nitriles**. Systematically, these compounds are alkanenitriles. Notice that the final "e" of the alkane root word is not dropped. The root word is determined by counting the number of carbons in the longest chain, including the carbon of the nitrile (Fig. 18.11). If the nitrile group is not the highest priority group, the CN substituent is identified as a cyano- and listed as a prefix. Small nitriles are commonly named as derivatives of the parent carboxylic acid or as cyanides. Ethanenitrile, for example, is almost always called acetonitrile. It is a common solvent.

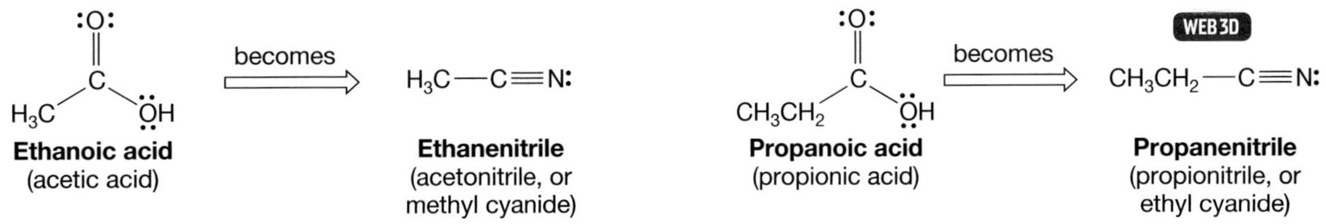

Ethanoic acid
(acetic acid)

becomes

H₃C—C≡N:

Ethanenitrile
(acetonitrile, or
methyl cyanide)

Propanoic acid
(propionic acid)

becomes

CH₃CH₂—C≡N:

WEB 3D

Propanenitrile
(propionitrile, or
ethyl cyanide)

FIGURE 18.11 A few examples of the naming protocol for nitriles.

Ketenes (p. 515) are named using the IUPAC system as alkene-substituted ketones (Fig. 18.12). Common names for ketenes are often used and are obtained by simply describing the substituents on the ketene. Ketenes are related to allenes, compounds with two directly attached carbon–carbon double bonds (Fig. 12.3, p. 513).

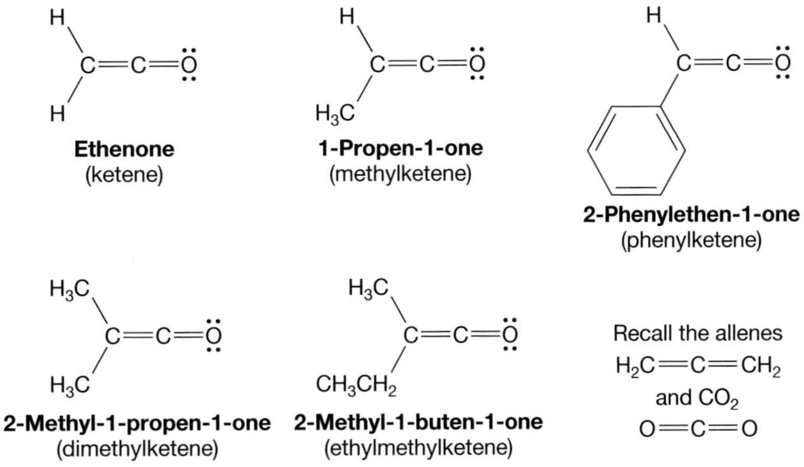

Ethenone
(ketene)

1-Propen-1-one
(methylketene)

2-Phenylethen-1-one
(phenylketene)

2-Methyl-1-propen-1-one
(dimethylketene)

2-Methyl-1-buten-1-one
(ethylmethylketene)

Recall the allenes
H₂C=C=CH₂
and CO₂
O=C=O

FIGURE 18.12 Ketenes are named as alkenyl ketones or as derivatives of the parent unsubstituted ketene.

PROBLEM 18.1 Draw structures for the following compounds: cyclopropyl 2-methylbutanoate, 3-chlorobutanoyl chloride, benzoic propanoic anhydride, *N,N*-diethyl-4-phenylpentanamide, and ethylpropylketene.

Acyl compounds and nitriles are Brønsted bases and Lewis bases as well, with the carbonyl oxygen or the nitrile nitrogen being the center of basicity (Fig. 18.16).

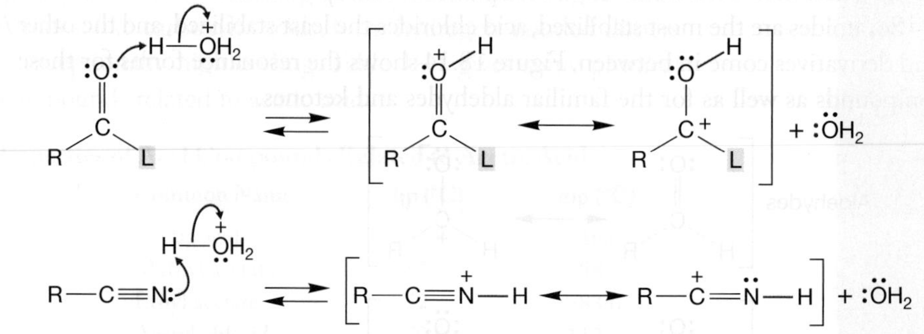

FIGURE 18.16 As these protonation reactions show, acyl compounds and nitriles can act as Brønsted bases.

As we have seen, all these acid derivatives are stabilized by dipolar resonance forms in which the L group bears a positive charge and the carbonyl oxygen a negative charge (Fig. 18.14). Because the acid chlorides are the least stabilized by this resonance, they have the least basic carbonyl oxygen atoms. Esters and carboxylic acids are stabilized by resonance to a similar extent and therefore are roughly equally basic. In amides it is a nitrogen atom, not an oxygen atom, that bears the positive charge. As was mentioned earlier, nitrogen is less electronegative than oxygen and therefore bears this positive charge with more grace than oxygen. The result is a more polar carbonyl group for the amide and a more basic carbonyl oxygen atom (Fig. 18.17).

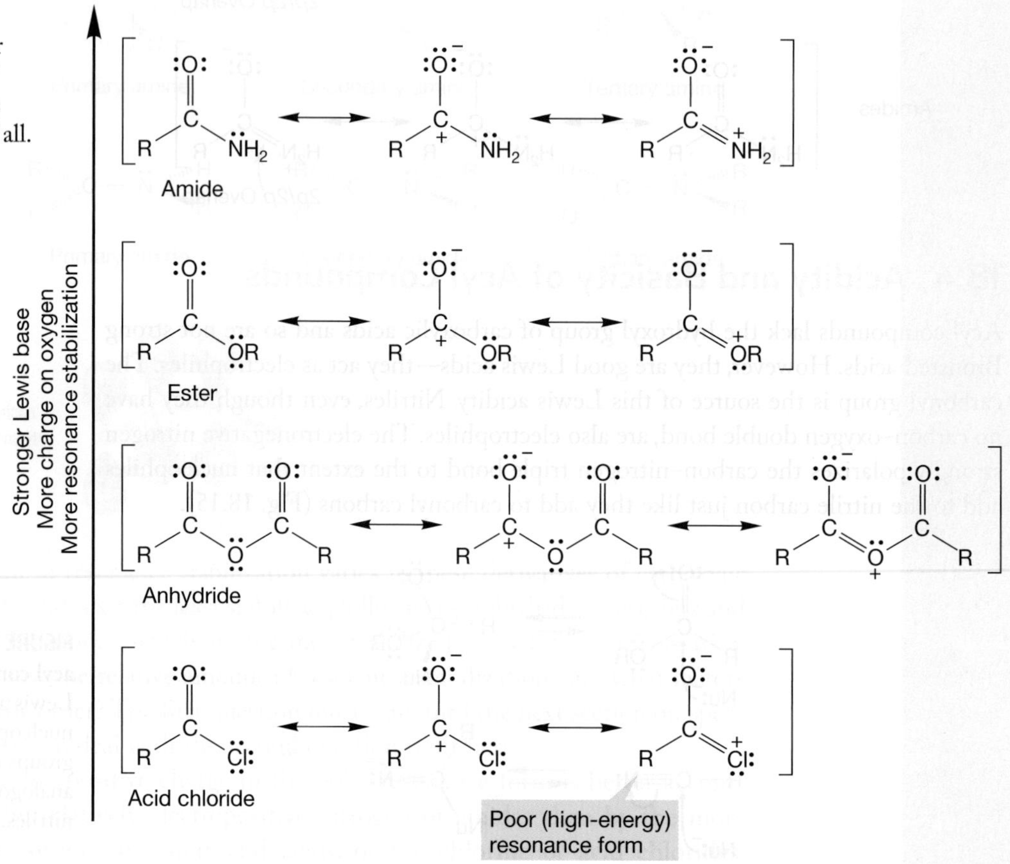

FIGURE 18.13 A... by resonance.

FIGURE 18.17 The carbonyl oxygens of amides are stronger nucleophiles than are those of esters. Acid chloride carbonyls are the weakest nucleophiles of all.

Resonance stabilization influences the bond lengths in these compounds as well (Fig. 18.18). Although the carbon–chlorine bond distance is little changed in acid chlorides from that in alkyl chlorides, the carbon–oxygen and carbon–nitrogen bond distances in esters and amides are shortened compared to alcohols and amines, reflecting some double-bond character in these compounds.

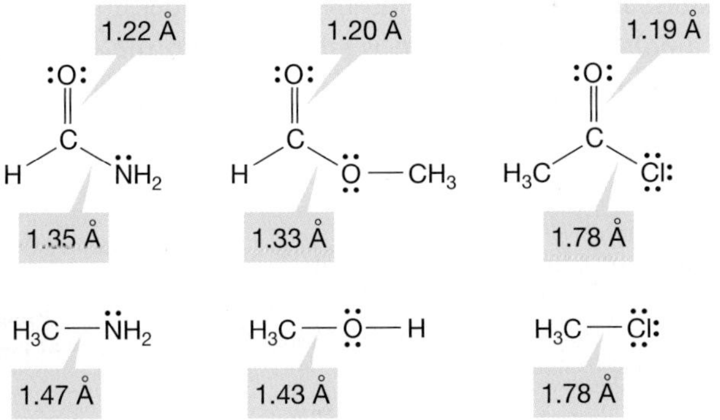

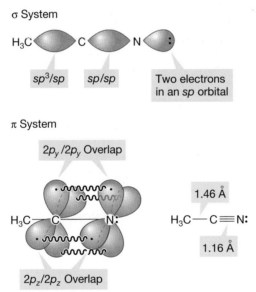

FIGURE 18.18 The structures of some simple acyl compounds. Notice how the carbon–oxygen and carbon–nitrogen single bonds are shortened by the double-bond character that results from the resonance shown in Figure 18.17.

Nitriles are linear molecules with short carbon–nitrogen bond distances. The sp hybridization of the carbon and nitrogen atoms ensures this. Figure 18.19 gives orbital pictures of both the σ and π systems, which closely resemble those of the alkynes (p. 125).

σ System

H_3C C   N   :

sp^3/sp sp/sp Two electrons in an sp orbital

π System

$2p_y/2p_y$ Overlap

H_3C—C≡N:

1.46 Å

1.16 Å

$2p_z/2p_z$ Overlap

FIGURE 18.19 The σ and π orbital systems of acetonitrile and the bond lengths.

18.5 Spectral Characteristics

18.5a Infrared Spectra   Characteristic strong carbonyl stretching frequencies are observed for all these compounds (except, of course, for the nitriles, which contain no carbonyl group). The contribution of the dipolar resonance forms can be seen in the position of the C=O stretch. Ketenes have a very strong carbonyl bond, as the inner carbon is hybridized sp, not the usual sp^2 (Fig. 18.20).

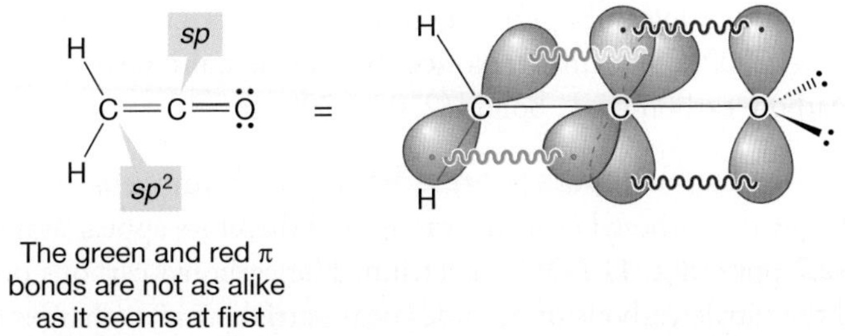

sp

H
  \
   C=C=Ö   =
  /
H

sp^2

The green and red π bonds are not as alike as it seems at first

FIGURE 18.20 The bonding scheme in ketene.

The more *s* character in a bond, the stronger it is. In addition, the ketene carbonyl has no substituent to feed electrons into its π system. Therefore, ketenes absorb in the IR at very high frequency (~2150 cm^{-1}).

The acid chloride carbonyl is the strongest of the acid derivatives, with the most double-bond character, so its C=O stretch appears at the highest frequency. Anhydrides, esters, and carboxylic acids are next. Amides, in which the contribution from the dipolar resonance form is strongest, and the *single*-bond character of the carbonyl group is greatest, have the lowest carbonyl stretching frequencies (Table 18.2).

TABLE 18.2 Infrared Stretching Frequencies of Some Carbonyl Compounds

Compound	Frequency of C=O in CCl₄ (cm⁻¹)	Compound	Frequency of C=O in CCl₄ (cm⁻¹)
$H_2C{=}C{=}O$ Ketene	2151	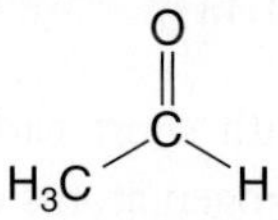 Acetaldehyde	1733
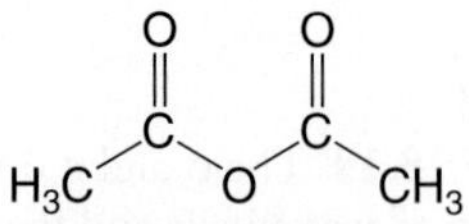 Acetic anhydride	1833, 1767	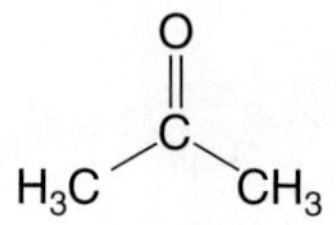 Acetone	1719
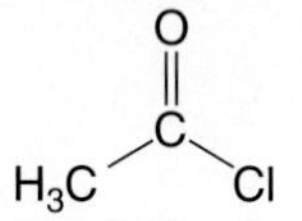 Acetyl chloride	1799	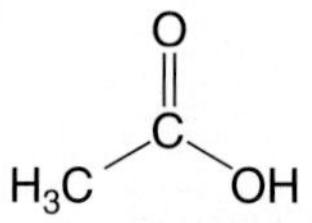 Acetic acid	1717
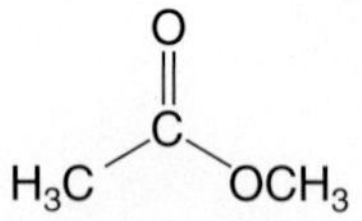 Methyl acetate	1750	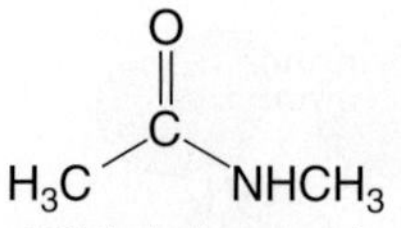 *N*-Methylacetamide	1688

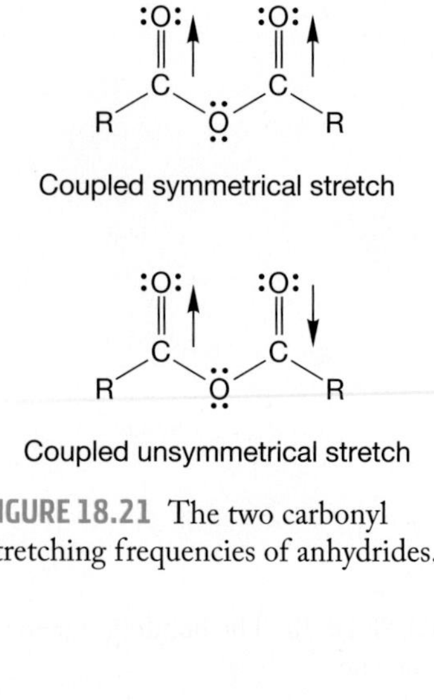

Coupled symmetrical stretch

Coupled unsymmetrical stretch

FIGURE 18.21 The two carbonyl stretching frequencies of anhydrides.

In the anhydrides, which contain two carbonyl groups, there are usually two carbonyl stretching frequencies. This doubling is not the result of independent stretching of the separate carbonyls, because symmetrical anhydrides as well as unsymmetrical anhydrides show the two IR bands. Instead, the two bands are caused by the coupled symmetrical and unsymmetrical stretching modes of these compounds (Fig. 18.21). Remember, vibrations of bonds in molecules are no more independent of each other than are the vibrations of attached real springs (p. 708).

Nitriles show a strong carbon–nitrogen triple-bond stretch at characteristically high frequency (2200–2300 cm^{-1}). Notice that this is quite near the frequency of the related carbon–carbon triple bonds (2100–2300 cm^{-1}).

18.5b Nuclear Magnetic Resonance Spectra   Hydrogens in the α position are deshielded by the carbonyl or nitrile group, and therefore appear at relatively low field, δ 2.0–2.7 ppm in a ^{1}H NMR spectrum. The carbonyl carbons of acyl compounds, and the nitrile carbons of cyanides bear partial positive charges and appear downfield in the ^{13}C NMR spectra (Table 18.3). Carbon NMR spectroscopy is a useful tool for differentiating between a ketone (or aldehyde) and an acid derivative because the carbonyl carbon of a ketone appears near δ 200 ppm and the acid derivative appears near δ 170 ppm.

TABLE 18.3 Some NMR Properties of Carbonyl Compounds and CH₃CN (ppm, CDCl₃)

Compound	^{13}C δ (C=O, C≡N)	δ (H$_α$)	Compound	^{13}C δ (C=O, C≡N)	δ (H$_α$)
Acetaldehyde	199	2.25	Ethyl acetate	169	2.05
Acetone	205	2.18	N,N-Dimethylacetamide	170	2.09
Acetyl chloride	169	2.66	Acetonitrile	117	2.05

PROBLEM 18.2 In the room temperature ^{1}H NMR spectrum of N,N-dimethyl-formamide (DMF), *two* methyl signals are observed. As the temperature is raised they merge into a single signal. Explain.

N,N-Dimethylformamide (DMF)

PROBLEM SOLVING

The words "NMR spectrum" and a change in the spectrum with temperature always mean that some groups in the molecule are interchanging positions through some sort of molecular motion, typically bond rotation (as here) or ring flipping in cyclohexanes.

18.6   Reactions of Acid Chlorides: Synthesis of Acyl Compounds

As mentioned before, all acyl compounds participate in the addition–elimination process. Acid chlorides are especially reactive toward nucleophiles. Their carbonyl groups, being the least stabilized by resonance, have the highest energy and are the most reactive. So, an initial addition reaction with a nucleophile is relatively easy. The chloride atom of acid chlorides is an excellent leaving group, and sits poised, ready to depart once the tetrahedral intermediate has been formed

Acid chloride aminolysis

(Fig. 18.22). The result is an exceedingly facile and common example of the addition–elimination mechanism.

THE GENERIC ADDITION–ELIMINATION REACTION

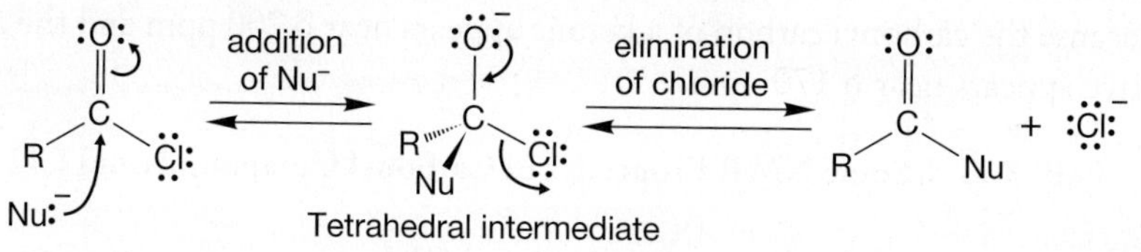

GENERAL REACTIONS

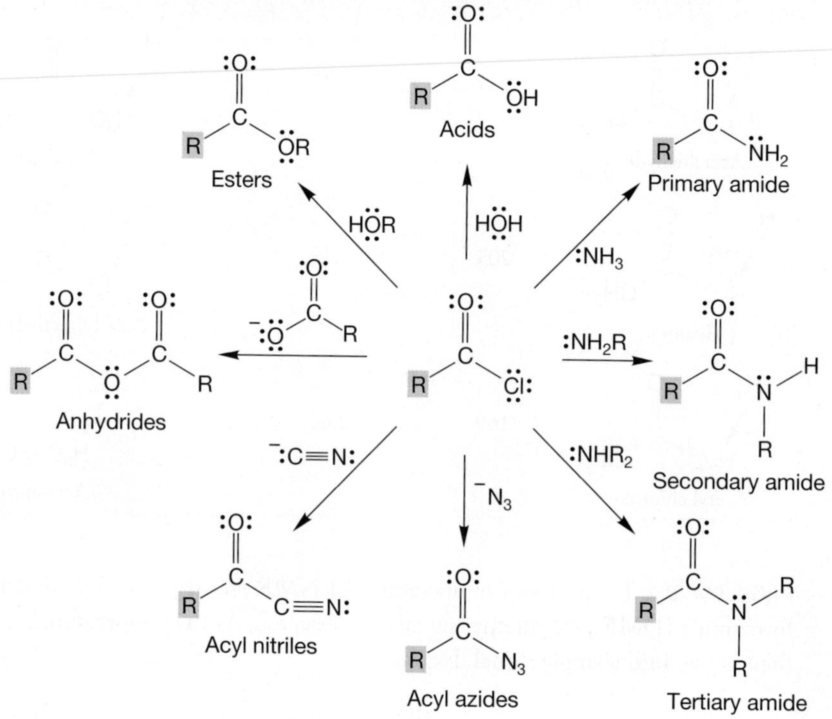

FIGURE 18.22 Addition–elimination reactions of acid chlorides. Note the synthetic potential.

SOME SPECIFIC EXAMPLES

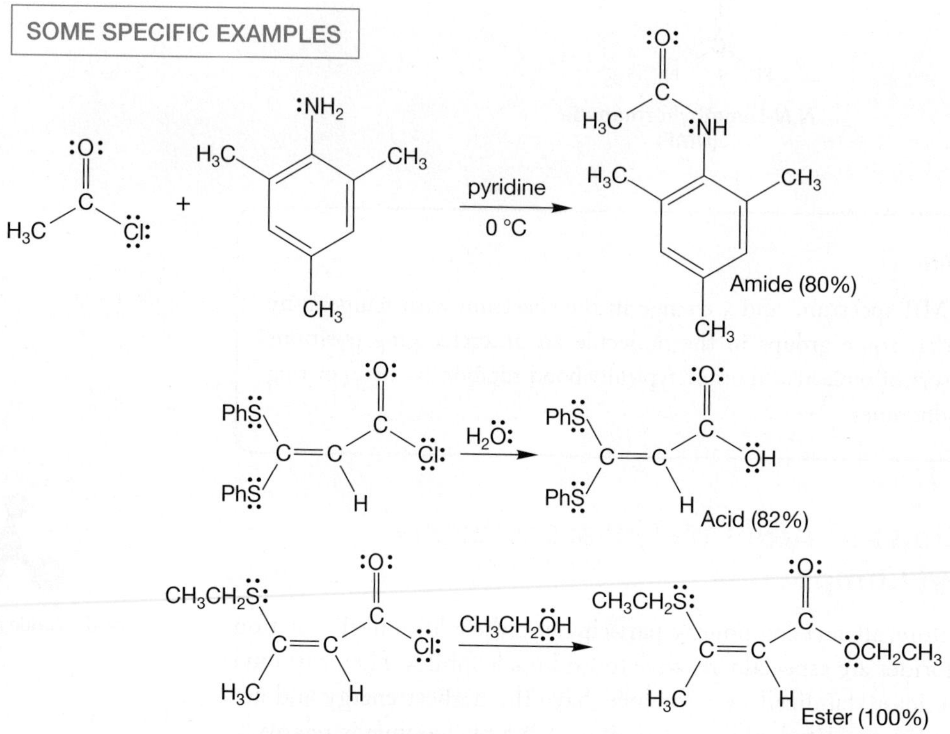

PROBLEM 18.3 Acid chlorides often smell like hydrochloric acid. Explain.

Many nucleophiles are effective in the addition–elimination reaction of acid chlorides, and a great many acyl compounds can be made using acid chlorides as starting materials.

Of course, strong nucleophiles such as organolithium compounds, Grignard reagents, and the metal hydrides also react rapidly with acid chlorides. The problem here is one of controlling secondary reactions. For example, reduction of an acid chloride with lithium aluminum hydride initially gives an aldehyde. But the newly born aldehyde finds itself in the presence of a reducing agent easily strong enough to reduce it further to the alkoxide, and hence to the primary alcohol after water is added (Fig. 18.23).

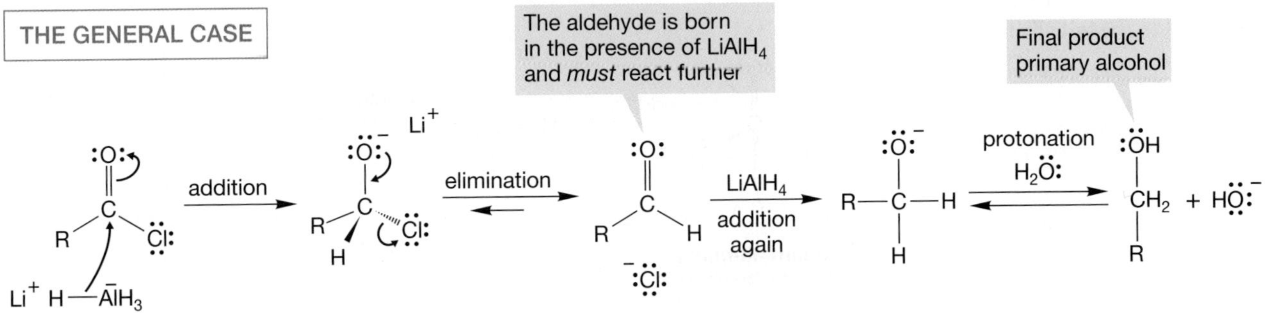

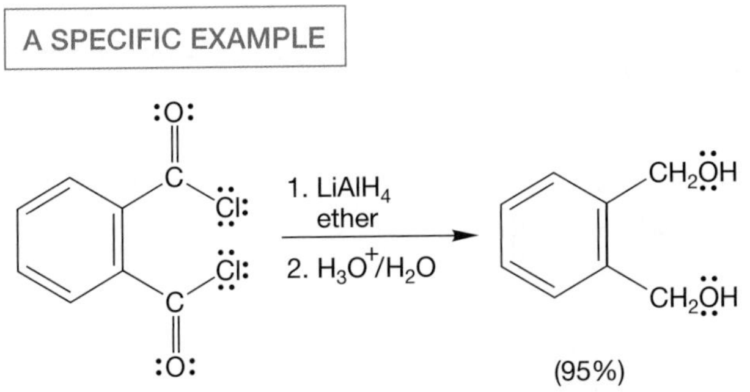

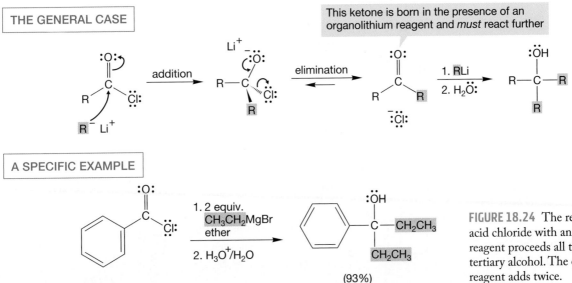

FIGURE 18.23 The reaction of an acid chloride with lithium aluminum hydride leads to a primary alcohol. The hydride adds twice.

Similarly, reaction of an acid chloride with an organometallic reagent, R—M, initially gives a ketone that usually reacts with a second equivalent of R—M to give the tertiary alcohol (Fig. 18.24). Note that this reaction must give a tertiary alcohol in which at least two R groups are the same.

FIGURE 18.24 The reaction of an acid chloride with an organometallic reagent proceeds all the way to the tertiary alcohol. The organometallic reagent adds twice.

All is fine if we are trying to make these particular alcohols, but often we are not. How to stop the process at the aldehyde or ketone stage is the problem, and several solutions have been found over the years. Less reactive organometallic reagents or metal hydrides allow the isolation of the intermediate aldehydes or ketones. The reactivity of lithium aluminum hydride can be attenuated by replacing some of the hydrogens with other groups. For example, use of lithium aluminum tri-*tert*-butoxyhydride allows isolation of the aldehyde (Fig. 18.25). Presumably, the bulky metal hydride is unable to reduce the aldehyde, which is less reactive than the original acid chloride.

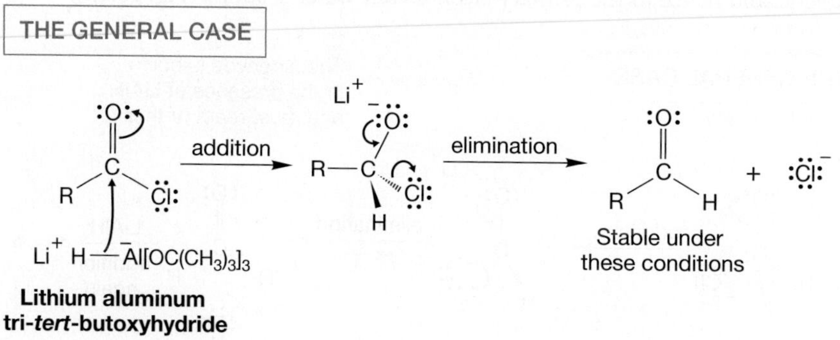

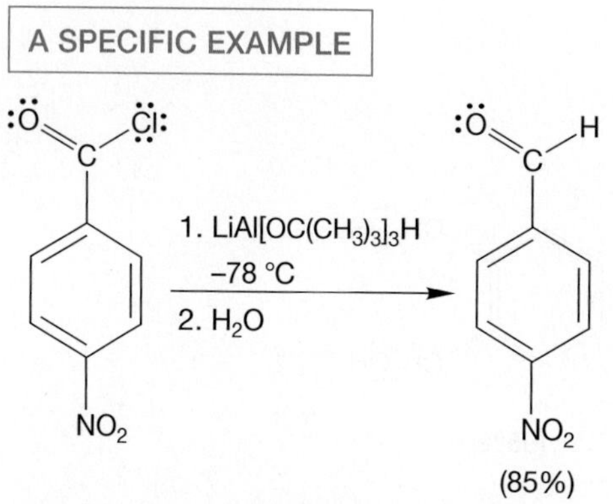

FIGURE 18.25 Lithium aluminum tri-*tert*-butoxyhydride is not reactive enough to add to the initially formed aldehyde.

Catalytic reduction of an acid chloride using a deactivated, or "poisoned" catalyst is called the **Rosenmund reduction** after Karl W. Rosenmund (1884–1964) (Fig. 18.26). In this reaction, only the more reactive acid chloride is reduced.

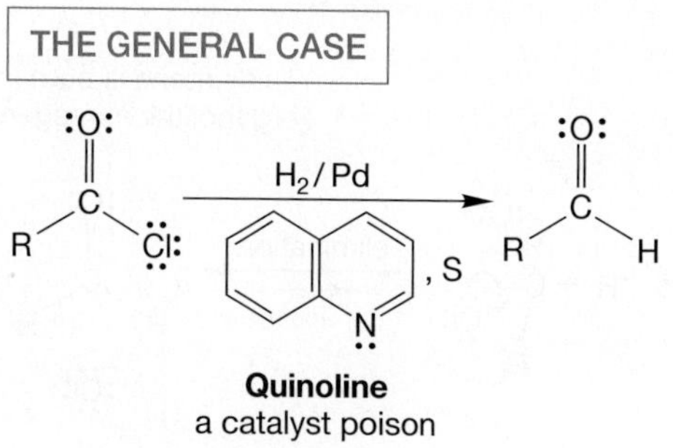

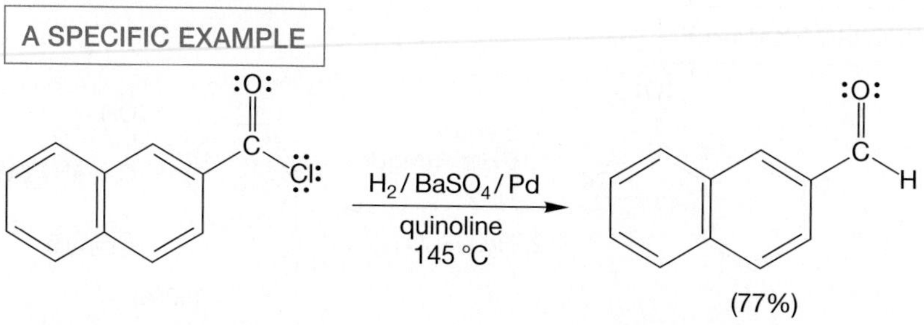

FIGURE 18.26 The Rosenmund reduction of an acid chloride gives an aldehyde.

The more stable, less reactive aldehyde can be isolated. Although this is not the same poisoned catalyst used for hydrogenation of an alkyne to give the cis alkene (Lindlar catalyst, p. 452), they are very similar.

Organocuprates react with acid chlorides but are not reactive enough to add to the product ketones (Fig. 18.27).

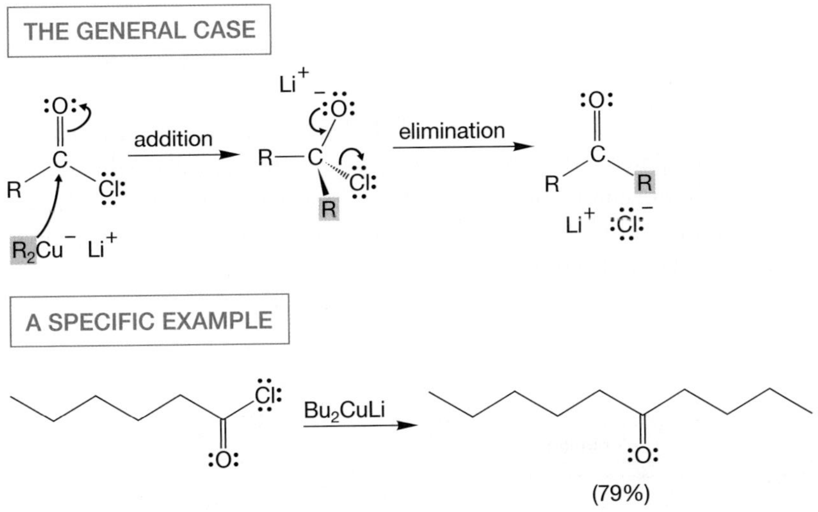

FIGURE 18.27 Cuprates are reactive enough to add R⁻ to the carbonyl group of the acid chloride, but not reactive enough to attack the product ketones.

Finally, don't forget that acid chlorides are used in the synthesis of aromatic ketones through Friedel–Crafts acylation (p. 643). This very useful reaction is shown in Figure 18.28.

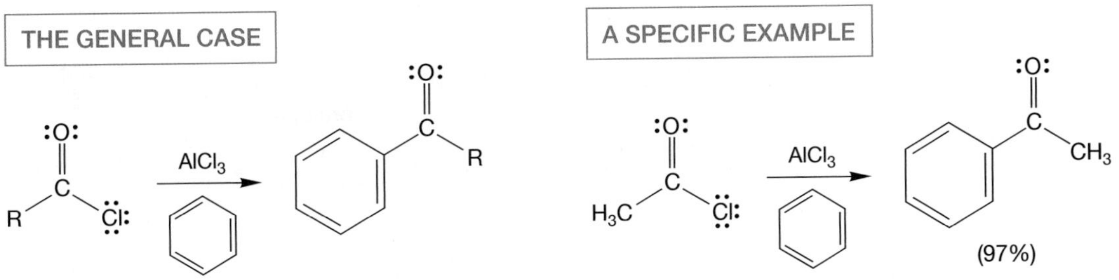

FIGURE 18.28 Friedel–Crafts acylation of benzene produces an aromatic ketone.

PROBLEM 18.4 Write a mechanism for the formation of the product of Figure 18.28.

PROBLEM 18.5 Starting with acetyl chloride, devise a synthesis of each of the following molecules.

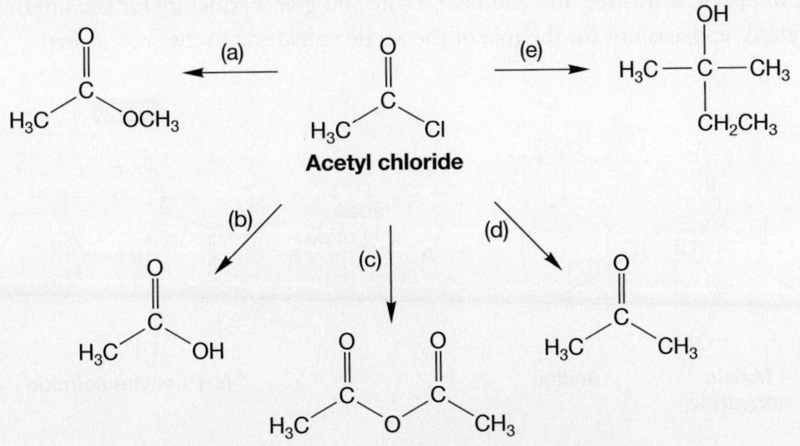

Summary

The presence of the good leaving group (chloride) attached directly to the carbon–oxygen double bond makes all manner of addition–elimination reactions possible for acid chlorides. The acid chloride can be used to make anhydrides, esters, carboxylic acids, amides, aldehydes, ketones, and alcohols.

18.7 Reactions of Anhydrides

The reactivity of anhydrides is similar to that of acid chlorides. A carboxylate anion is the leaving group in a variety of syntheses of acyl derivatives, all of which are examples of the addition–elimination process (Fig. 18.29). One reaction of this kind is the basic hydrolysis of phthalic anhydride to phthalic acid. Here the leaving group is an internal carboxylate anion.

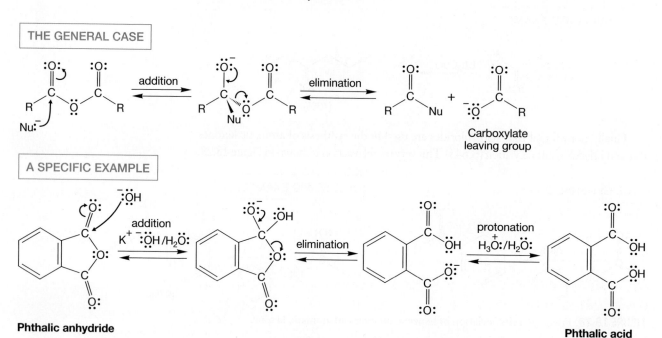

FIGURE 18.29 Addition–elimination reactions of an anhydride.

PROBLEM 18.6 Provide a mechanism for the formation of *N*-phenylmaleimide from maleic anhydride and aniline. Be sure you give a structure for the intermediate **A**, and account for the role of the acetic anhydride in the second step.

Like acid chlorides, anhydrides can be used in the Friedel–Crafts reaction (Fig. 18.30). Once again, a strong Lewis acid catalyst such as aluminum chloride is used to activate the anhydride. The mechanism follows the pattern of the Friedel–Crafts reactions we saw earlier (p. 643).

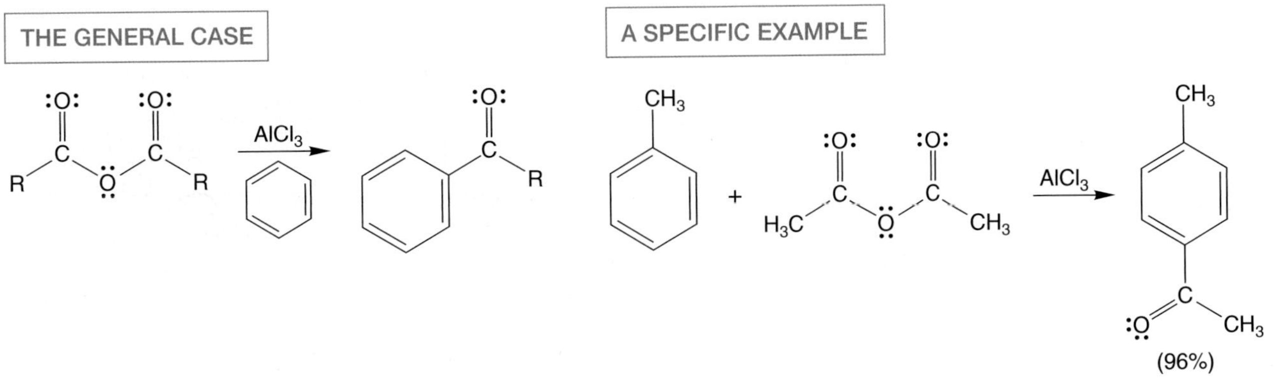

THE GENERAL CASE   A SPECIFIC EXAMPLE

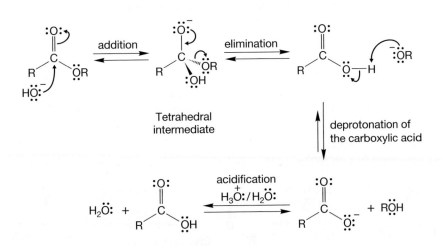

FIGURE 18.30 A Friedel–Crafts reaction using an anhydride as the source of the acyl group.

PROBLEM 18.7 Write a mechanism for the reaction of Figure 18.30.

18.8 Reactions of Esters

Esters are less reactive than acid chlorides and anhydrides in addition reactions, but more reactive than amides. Esters can be converted into their parent carboxylic acids under either basic or acidic aqueous conditions in a process called, logically enough, **ester hydrolysis**. In base, the mechanism is the familiar addition–elimination one (Fig. 18.31). Hydroxide ion attacks the carbonyl group to form a tetrahedral intermediate. Loss of alkoxide then gives the acid, which is rapidly deprotonated to the carboxylate anion in basic solution. Notice that this reaction, saponification (p. 862), is not catalytic. The hydroxide ion used up in the reaction is not regenerated at the end. To get the carboxylic acid itself, a final acidification step is necessary.

Ester hydrolysis

FIGURE 18.31 The base-induced ester hydrolysis reaction, saponification.

Esters can also be converted into carboxylic acids in acid (Fig. 18.32). The carbonyl oxygen is first protonated to give a resonance-stabilized cation to which water adds. Water is by no means as nucleophilic as hydroxide, but the protonated carbonyl is a very strong electrophile and the overall reaction is favorable. The reaction concludes with proton transfers and loss of alcohol.

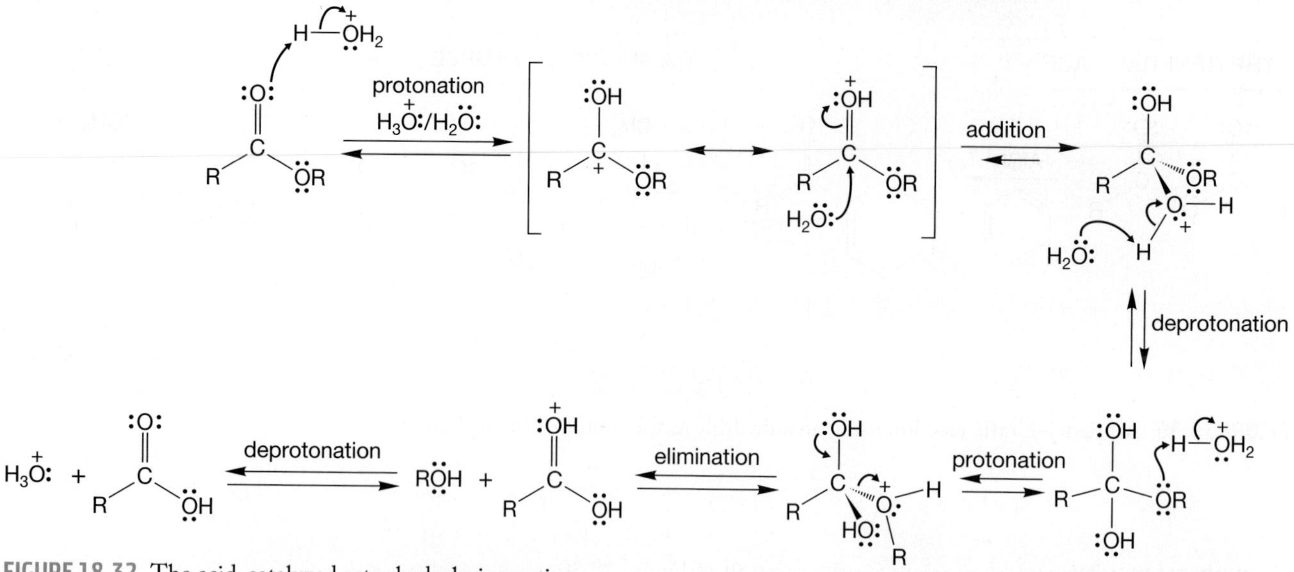

FIGURE 18.32 The acid-catalyzed ester hydrolysis reaction.

Have you seen this reaction before? You certainly have. It is the exact reverse of the mechanism for Fischer esterification (p. 841), so there is nothing new here at all. What determines where the overall equilibrium settles out? The structure of the ester is an important factor, but much more important are the reaction conditions. Excess water favors the acid; excess alcohol favors the ester. Le Châtelier's principle is at work here, and the chemist has the ability to manipulate this equilibrium (Fig. 18.33) to favor either side. In practice, most acids and esters are easily interconvertible.

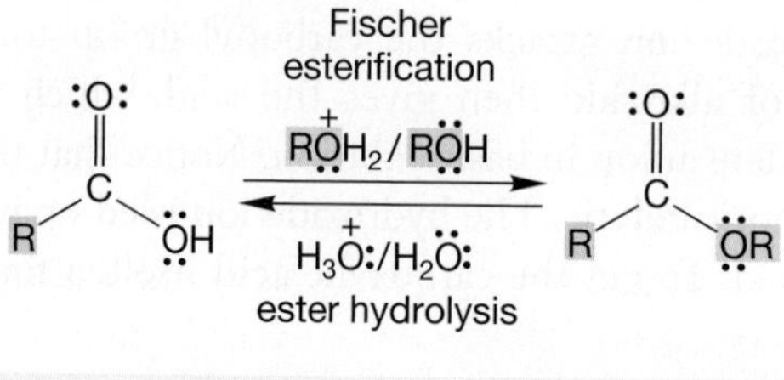

FIGURE 18.33 Acids and esters are usually interconvertible.

PROBLEM 18.8 Where appropriate, draw resonance forms for the intermediate species in the reactions of Figure 18.33. You will have to write mechanisms for the reactions first.

A reaction closely related to acid-catalyzed ester hydrolysis is acid-catalyzed **transesterification**. In this reaction, an ester is treated with an excess of an alcohol and an acid catalyst. The result is replacement of the ester OR group with the alcohol OR group (Fig. 18.34). Like ester hydrolysis, transesterification can be carried out under either acidic or basic conditions. The mechanisms are extensions of those you have already seen in the ester hydrolysis reactions or Fischer esterification.

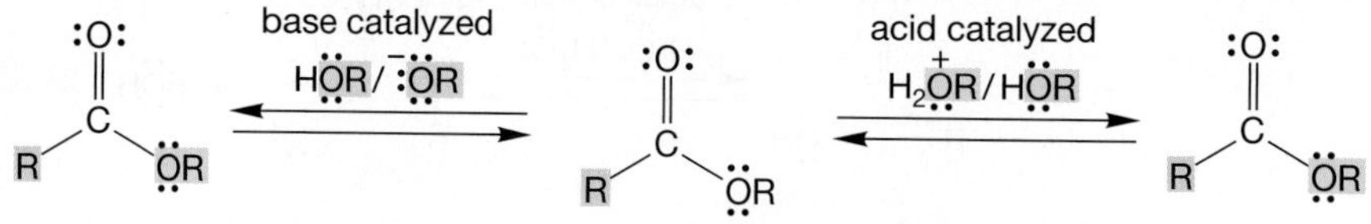

FIGURE 18.34 Base- and acid-catalyzed transesterification.

PROBLEM 18.9 Write mechanisms for acid- and base-induced transesterification.

WORKED PROBLEM 18.10 There is an important difference between the reaction of an ester with hydroxide ($^-$OH) and of an ester with alkoxide ($^-$OR). The reaction with hydroxide is neither catalytic nor reversible, whereas the reaction with alkoxide is both catalytic and reversible. Analyze the mechanisms of these reactions and explain these observations.

ANSWER Look at the final products of the two reactions. Reaction with hydroxide leads to a carboxylic acid ($pK_a \sim 4.5$) and an alkoxide ion. These two species *must* react very rapidly to make the more stable carboxylate anion (resonance stabilized) and the alcohol ($pK_a \sim 17$). The hydroxide reagent is consumed in this reaction, and the overall process is so thermodynamically favorable that it is irreversible in a practical sense.

With an alkoxide reagent, things are different. Now the product is not a carboxylic acid but another ester. There is no proton that can be removed, and there is a new molecule of alkoxide generated. The reaction is approximately thermoneutral. The reaction is reversible.

WORKED PROBLEM 18.11 Given that there are acid- and base-catalyzed transesterification reactions, should there not be both acid- and base-catalyzed Fischer esterifications as well? Explain clearly why there is no base-catalyzed version of Fischer esterification (i.e., RCOOH + RO$^-$ → RCOOR + HO$^-$ does not occur).

ANSWER The reaction of a carboxylic acid with an alkoxide can't proceed by addition–elimination to give an ester because there is another much easier reaction available; that reaction is simple removal of the carboxylic acid hydroxyl proton to give the resonance-stabilized carboxylate anion. They don't call these compounds "acids" for nothing! The lesson in this problem is that you have to "think simple." Look first for "trivial" reactions (loss of the proton) before proceeding on to more complicated processes (addition–elimination).

$pK_a \sim 4.5$ $pK_a \sim 17$

PROBLEM 18.12 Esters can also react with amines to give amides. Write a mechanism for the reaction of methyl acetate (CH_3COOCH_3) and ammonia to form acetamide (CH_3CONH_2).

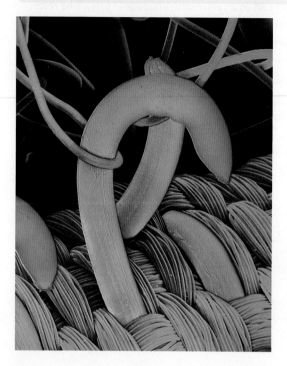

Esters react with amines to make amides. One of the most important synthetic polymers is nylon. It can be made through the reaction of a diester with a diamine. The resulting polymer is a polyamide (p. 852). This electron microscope photo illustrates the binding mechanism for Velcro, which is a polyamide.

Summary

Anhydrides react with water to generate carboxylic acids, with alcohols to give esters, and with amines to form amides. Esters behave similarly. There are both acid-catalyzed and base-induced reactions of esters with water (ester hydrolysis) to give carboxylic acids. There are both acid- and base-catalyzed versions of the reaction of esters with alcohols (transesterification) that generate new esters. Esters also react with amines to form amides.

Organometallic reagents and metal hydrides react with esters in still another example of the addition–elimination process. Usually, these strong nucleophiles react further with the ketones that are the initial products of the reactions (Fig. 18.35). The ketones are not as well stabilized by resonance as are the esters and so are more reactive in the addition reaction.

The reaction shown in Figure 18.35 is another general synthesis of complex tertiary alcohols (compare with Fig. 18.24). Here are the possibilities for the

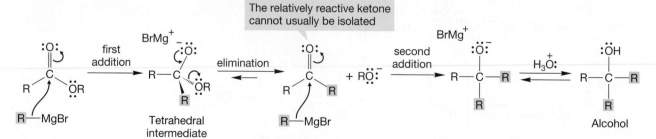

FIGURE 18.35 The reaction of an ester with a Grignard reagent to form a tertiary alcohol.

addition of two equivalents of an organometallic reagent with an ester. First, tertiary alcohols in which all three R groups are the same can be made if the R group of the organometallic reagent is the same as that of the ester alkyl group. Second, the R of the organometallic reagent can be different from that of the alkyl R part of the ester. In this case, the tertiary alcohol will contain two different R groups (Fig. 18.36).

Because two R groups must come from the organometallic reagent, there is no way to use this ester-based synthesis to make a tertiary alcohol with three different R groups.

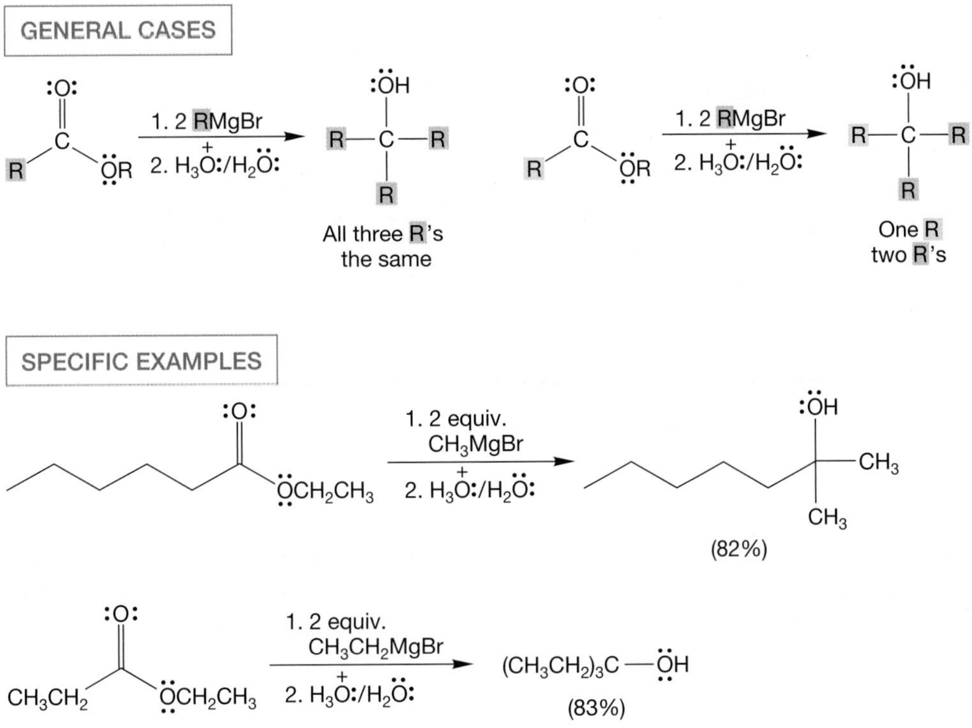

GENERAL CASES

All three R's
the same

One R
two R's

SPECIFIC EXAMPLES

(82%)

(83%)

FIGURE 18.36 Tertiary alcohols in which all three R groups are the same, or molecules in which only two R groups are the same can be made, but the structural type in which all three R groups are different cannot be made this way. Alkyllithium reagents, RLi, can also be used.

PROBLEM 18.13 Dialkyl carbonates can be made from the reaction of phosgene with alcohols. These carbonates are converted into tertiary alcohols with three identical R groups on reaction with three equivalents of a Grignard reagent or an organolithium reagent. Provide mechanisms for the formation of carbonates and tertiary alcohols in these reactions.

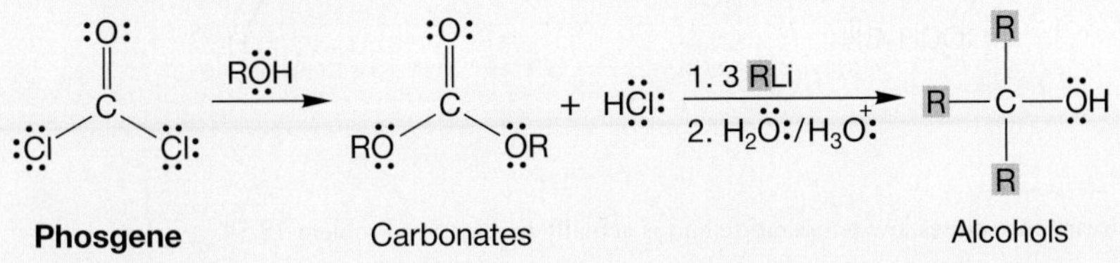

Phosgene   Carbonates   Alcohols

Metal hydride reduction of esters is difficult to stop at the intermediate aldehyde stage, and the usual result is further reduction to the primary alcohol (Fig. 18.37). Either $LiAlH_4$ or $LiBH_4$ (but not $NaBH_4$, which will not reduce esters) is the reagent of choice.

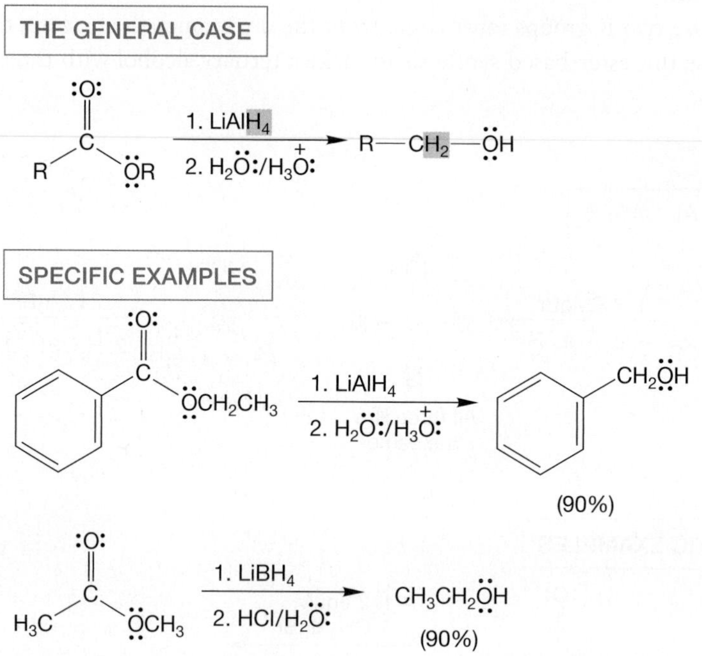

FIGURE 18.37 The reduction of esters with the powerful hydride donor lithium aluminum hydride leads to primary alkoxides and then, after acidification, to primary alcohols.

As with acid chlorides, other metal hydrides that permit isolation of the intermediate aldehydes have been developed. For example, diisobutylaluminum hydride (DIBAL-H) reduces ketones to alcohols, but can be used to prepare the intermediate aldehydes from esters (Fig. 18.38).[2]

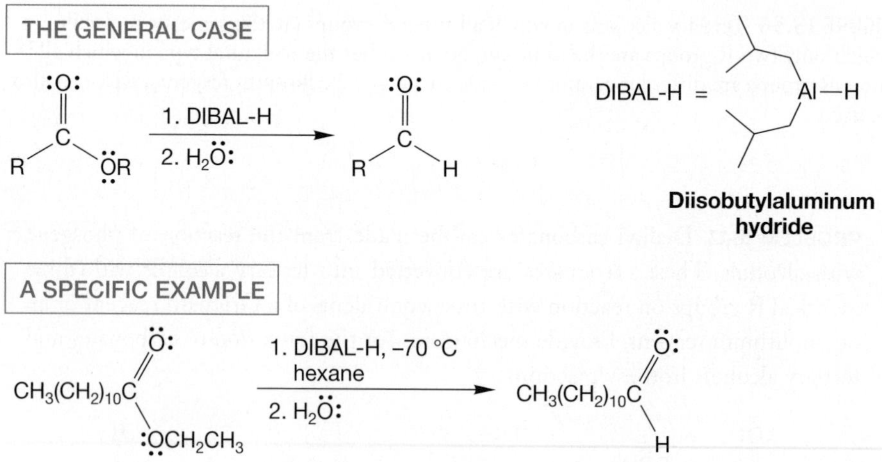

FIGURE 18.38 If DIBAL-H is used as the hydride donor, the reduction of esters can often be stopped at the aldehyde stage.

[2]This reaction requires low temperature and is actually finicky. See Problem 18.54.

Esters also have a substantial chemistry involving the related enolate anions. We will develop this subject in detail in Chapter 19.

> **PROBLEM 18.14** Suggest one reason why DIBAL-H might be slower than LiAlH₄ to react with the aldehyde of Figure 18.38.

18.9 Reactions of Amides

Even though amides are the least reactive of the acyl compounds, they can nevertheless be hydrolyzed in either acid or base. The mechanisms of these reactions are directly related to those of acid- or base-induced ester hydrolysis. In base, the carbonyl carbon of an amide is attacked to give tetrahedral intermediate **A** (Fig. 18.39).

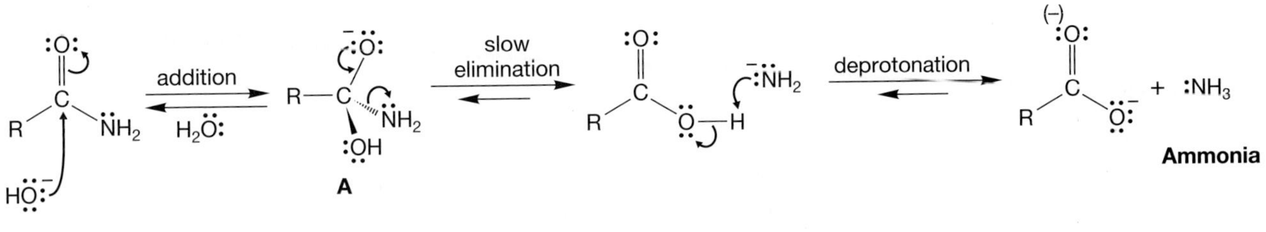

FIGURE 18.39 The base-induced hydrolysis of amides gives carboxylate anions. Acidification of the carboxylate at the end of the reaction would give the carboxylic acid.

Loss of the amide ion (⁻NH₂) gives the carboxylic acid, and deprotonation of the acid gives the carboxylate anion. Loss of amide ion is surely difficult, because ⁻NH₂ is a much stronger base than hydroxide. We know the relative base strengths by comparing the pK_a's of ammonia and water. So the reverse reaction that regenerates the starting material is much faster. However, when an amide ion *is* lost, a carboxylic acid is formed, and is quickly converted into the carboxylate. So, loss of the amide ion, though slow, is essentially irreversible. A final acidification (not shown in Fig. 18.39) regenerates the carboxylic acid in isolable form. Overall, it's addition–elimination once again.

> **PROBLEM 18.15** In strong base, the hydrolysis of an amide is second order in hydroxide. In other words, two molecules of HO⁻ are required to hydrolyze one amide molecule. Write a modified mechanism to account for the need for two molecules of hydroxide.

In acid, the carbonyl oxygen is protonated creating a Lewis acid strong enough to react with the relatively weak nucleophile water (Fig. 18.40). The leaving group

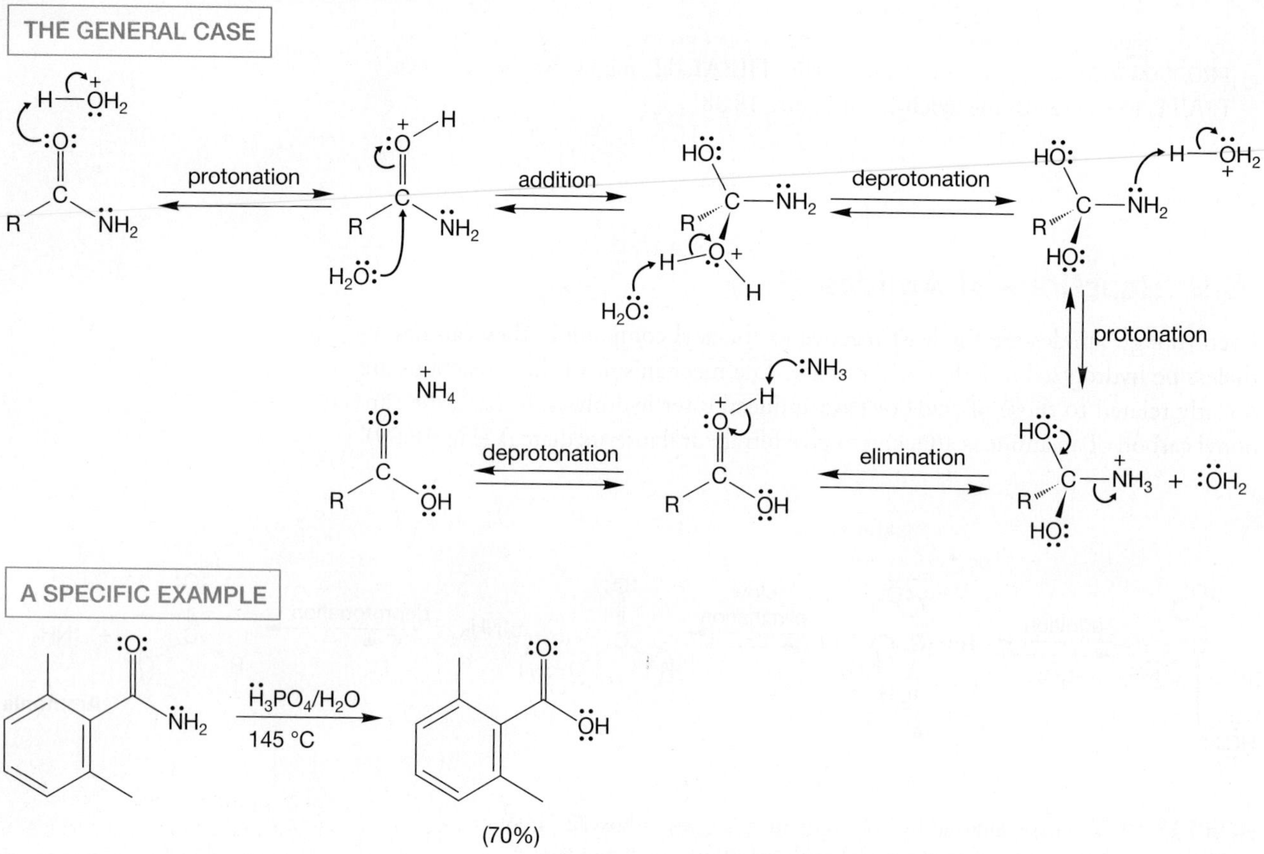

THE GENERAL CASE

A SPECIFIC EXAMPLE

(70%)

FIGURE 18.40 The acid-induced hydrolysis of amides gives the carboxylic acid directly.

in the acid-induced reaction is the amine (Fig. 18.41), which makes this reaction much easier than the base-induced hydrolysis. NH_3 is a better leaving group than $^-NH_2$.

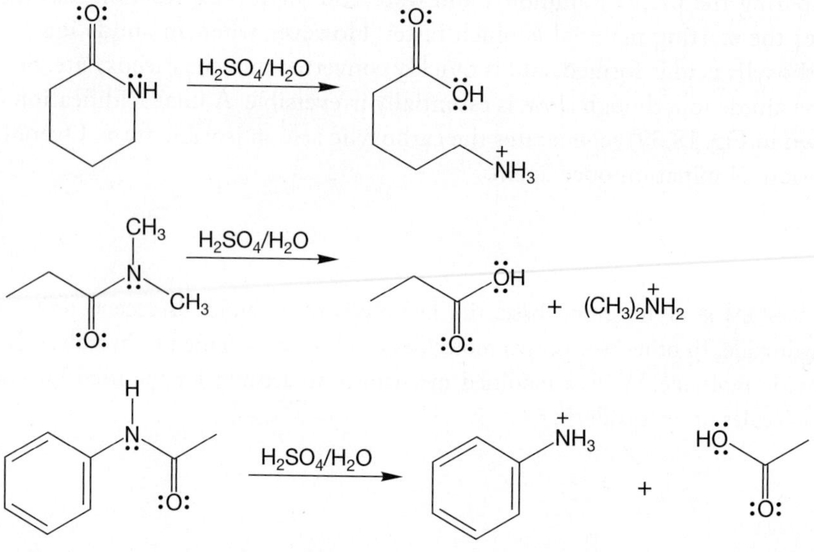

FIGURE 18.41 Examples of amide hydrolysis.

WORKED PROBLEM 18.16 Write resonance forms for the structures in Figure 18.40 (The General Case) where appropriate.

ANSWER This problem is very similar to Problem 18.8. In this case, it is the protonated amide and protonated carboxylic acid that are most stabilized by resonance.

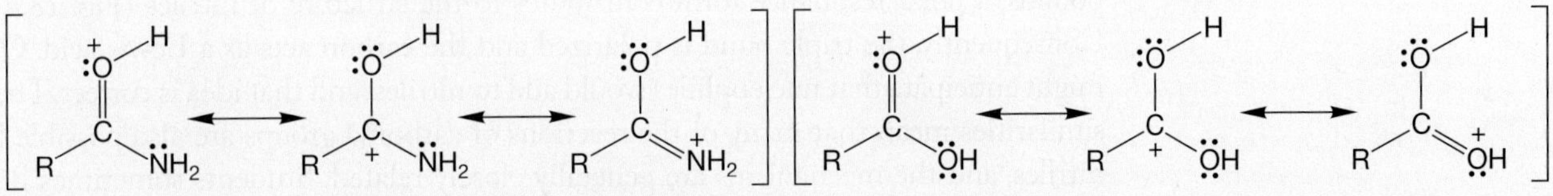

The starting amide and acid are also resonance stabilized, but resonance stabilization is more important for the charged intermediates in the reaction.

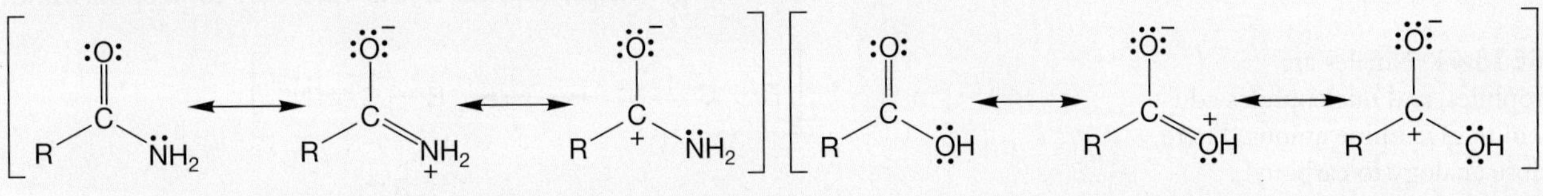

PROBLEM 18.17 Predict the organic products obtained from the acid-induced hydrolysis of *N*-phenylpentanamide.

Metal hydrides reduce amides to amines (Fig. 18.42). Clearly, something new is happening here, because analogy to the reactions of acids, esters, and acid chlorides would lead us to expect alcohols as products.

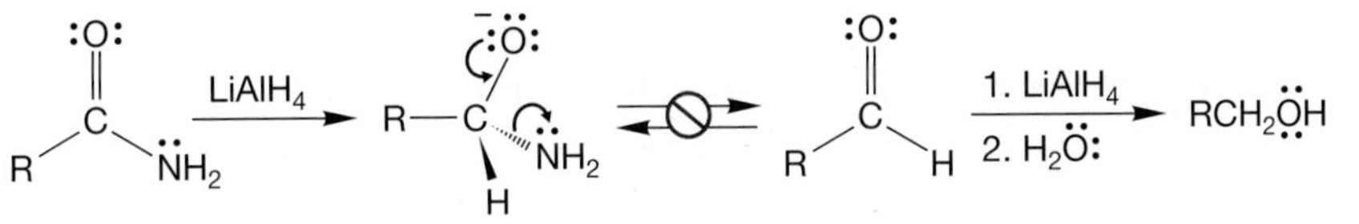

FIGURE 18.42 By analogy to what we know of other acyl compounds, we might well expect the treatment of an amide with a metal hydride to result in the formation of an alcohol. In fact, this reaction does *not* take place. An amine is formed instead.

The mechanism of this reaction is complex and may vary with the number of alkyl groups on the amide nitrogen. However, the first step is certainly the same as in the other reductions by metal hydrides; the hydride adds to the carbonyl to give an alkoxide that is coordinated to a metal (Fig. 18.43). Now, the metal oxide is lost with formation of an iminium ion. There are several possible ways to describe the formation of the iminium ion, but once it is produced, reduction by the metal hydride to give the amine is sure to be rapid.

THE GENERAL CASE

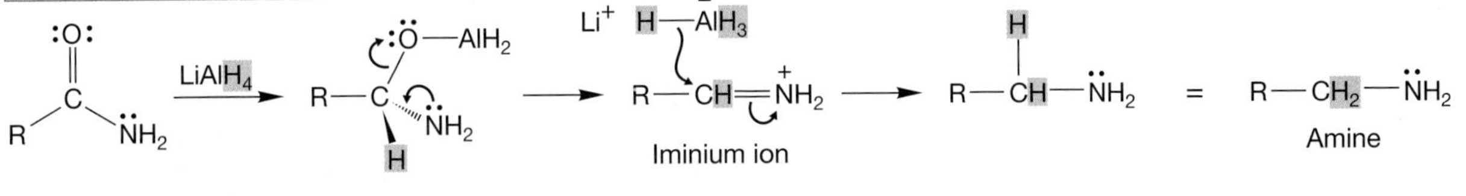

A SPECIFIC EXAMPLE

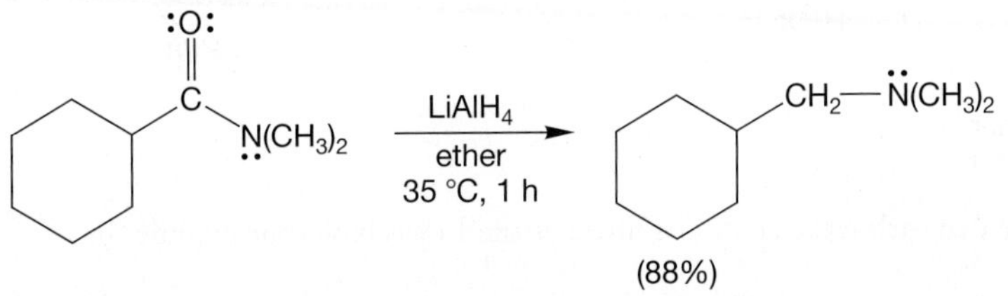

(88%)

FIGURE 18.43 Amines are the actual products of the reduction of amides with metal hydrides.

18.10   Reactions of Nitriles

Of course a nitrile is not a carbonyl. But nitrile and carbonyl chemistries are similar, and an examination of nitrile reactions naturally belongs in this section. Like carbonyl compounds, nitriles are both electrophiles and nucleophiles. As in carbonyl compounds, a polar resonance form contributes to the structure of nitriles (Fig. 18.44). Consequently, the triple bond is polarized and the carbon acts as a Lewis acid. One might anticipate that nucleophiles would add to nitriles, and that idea is correct. These similarities mean that many of the reactions of carbonyl groups are also possible for nitriles, and the mechanisms are generally closely related. Students sometimes have trouble with the nitrile reactions, however, so it is important not to dismiss them with only a cursory look. For some reason, the analogy between the carbon–oxygen double bond and the carbon–nitrogen triple bond is not always easy to keep in mind.

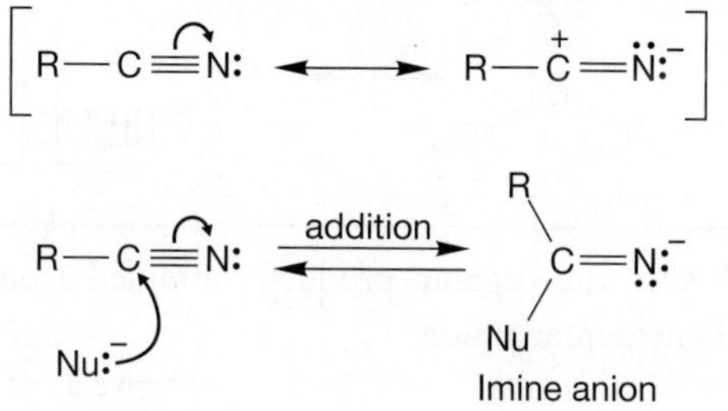

FIGURE 18.44 Nitriles are electrophiles, and nucleophiles add to them to give imine anions. There is a close analogy to carbonyl chemistry here.

The nitrogen atom of the nitrile bears a pair of nonbonding electrons and is therefore the center of nucleophilicity and Brønsted basicity. The nitrogen is *sp*-hybridized, however, and is a relatively weak base.

Nitrile hydrolysis

PROBLEM 18.18 Explain why an *sp*-hybridized nitrogen should be a weaker base than an sp^2-hybridized nitrogen. What about sp^3-hybridized nitrogens?

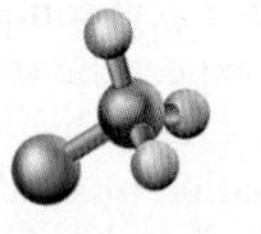

Both acid- and base-induced hydrolysis of a nitrile gives the amide, but rather severe conditions are required for the reactions, and further hydrolysis of the intermediate amides gives the carboxylic acids (p. 902). In acid, the first step is protonation of the nitrile nitrogen to give a strong Lewis acid (**A**, Fig. 18.45), which is attacked by water. A series of proton shifts then gives an amide that is hydrolyzed to the carboxylic acid.

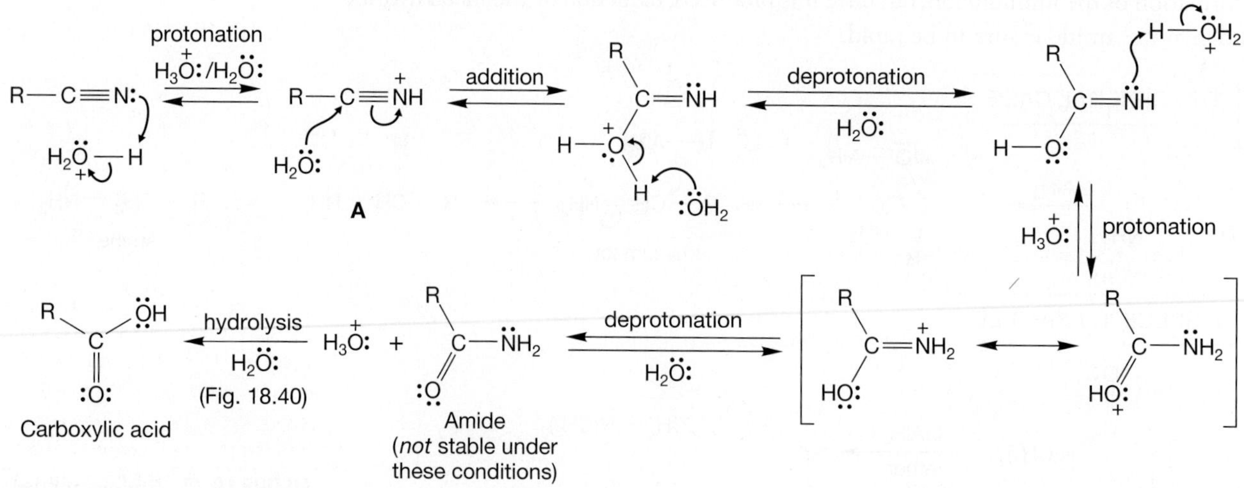

FIGURE 18.45 The acid-induced hydrolysis of a nitrile to a carboxylic acid. The nitrile is the Lewis base (nucleophile) in the first step.

In base, the nucleophilic hydroxide ion adds to the carbon–nitrogen triple bond (Fig. 18.46). Protonation on nitrogen and deprotonation from oxygen leads to the amidate anion, which is protonated to give an intermediate amide. Basic hydrolysis of the intermediate amide leads to the carboxylate salt.

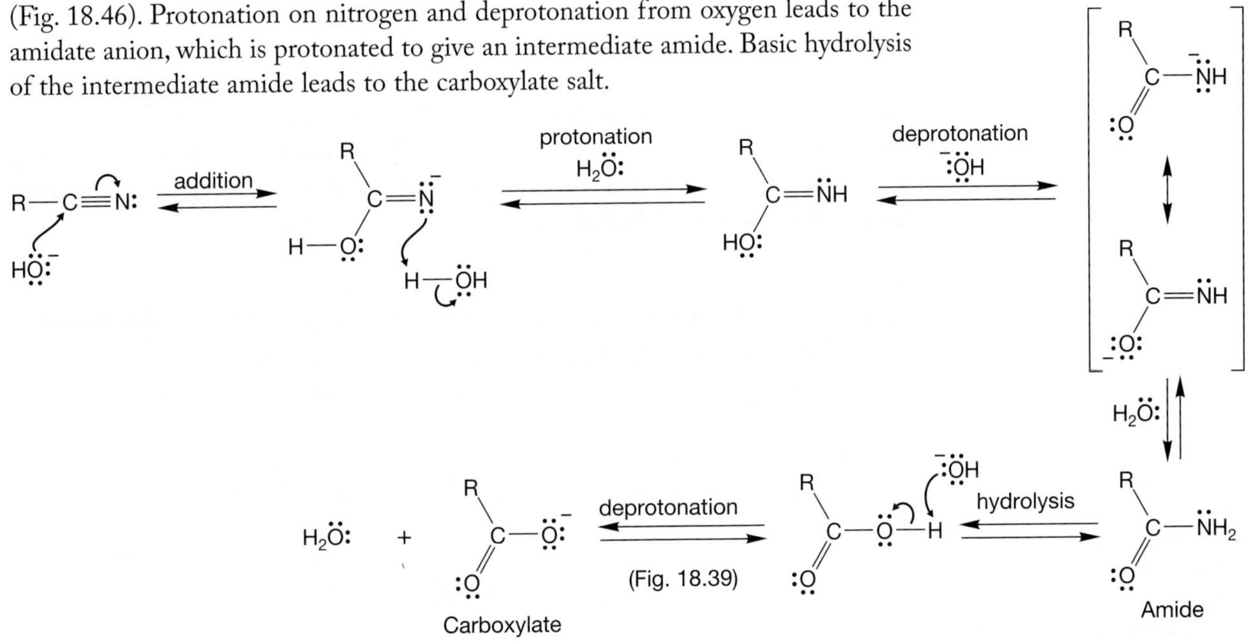

FIGURE 18.46 The base-induced hydrolysis of a nitrile to give a carboxylate anion. The nitrile is the Lewis acid (electrophile) in the first step.

Stronger bases also add to nitriles. Organometallic reagents give ketones, although the product before the hydrolysis step is the imine (Fig. 18.47a). Addition to the nitrile first takes place in normal fashion to give the imine salt.

THE GENERAL CASE

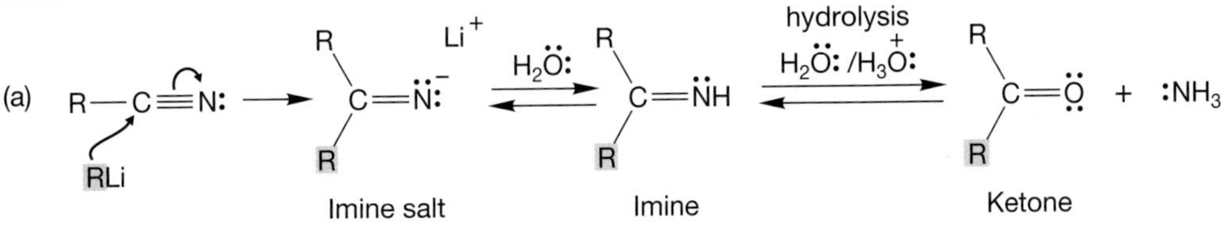

A SPECIFIC EXAMPLE

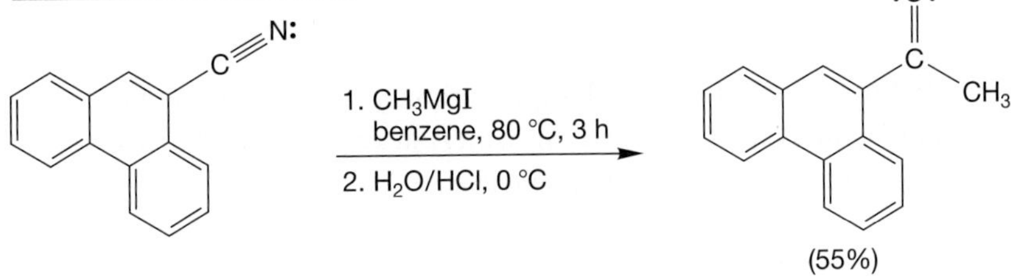

(b) Recall reactions of esters and acid chlorides

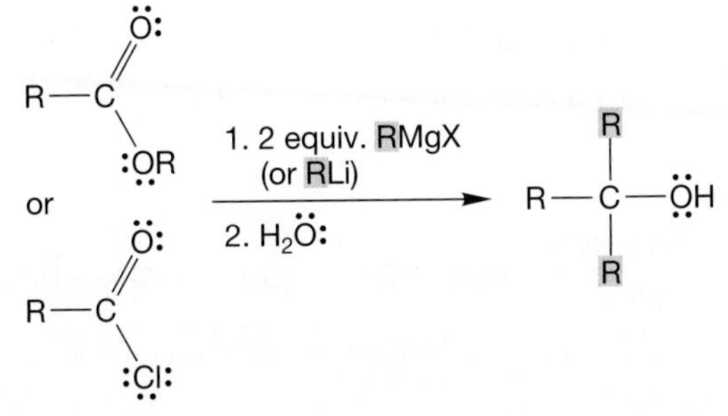

FIGURE 18.47 (a) Reaction of a nitrile with an organometallic reagent gives an imine salt. Acidification first generates the imine and then the ketone. The organometallic reagent adds once. (b) Reaction of an ester or acid chloride with organometallic reagents gives an alcohol. The organometallic reagent adds twice.

When the reaction is worked up with water, protonation gives the imine, which in acid undergoes further hydrolysis to the ketone. Note the difference between this reaction and those of alkyllithium or Grignard reagents with acid chlorides or esters, which go all the way to alcohols. With nitriles the reaction can be stopped at the carbonyl stage (Fig. 18.47a). With acid chlorides and esters the intermediate carbonyl compounds are born in the presence of the organometallic reagent and must react further. In the reaction with nitriles, there is no leaving group present and the imine salt is stable until hydrolysis.

> **PROBLEM 18.19** Write a mechanism for the hydrolysis of imines under neutral conditions to give a ketone (Fig. 18.47).

Metal hydrides reduce nitriles to give primary amines (Fig. 18.48). Two equivalents of hydride are delivered, and subsequent addition of water yields the amine.

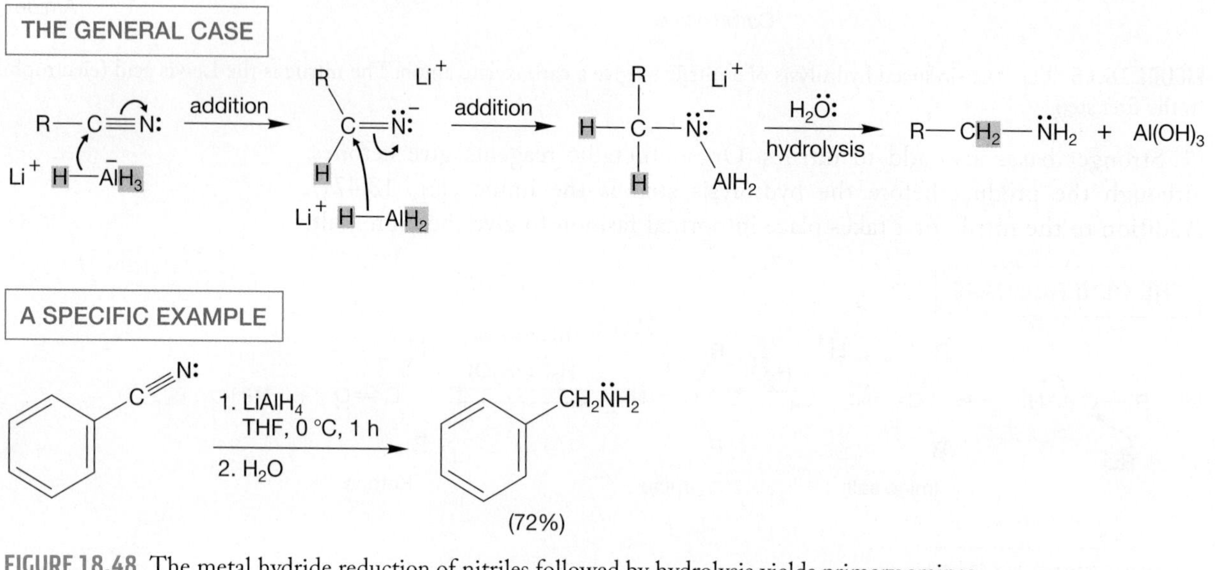

THE GENERAL CASE

A SPECIFIC EXAMPLE

(72%)

FIGURE 18.48 The metal hydride reduction of nitriles followed by hydrolysis yields primary amines.

Catalytic hydrogenation of nitriles also gives primary amines (Fig. 18.49). Notice that only primary amines can be formed this way. Alkenes are more reactive toward typical hydrogenation conditions than are nitriles.

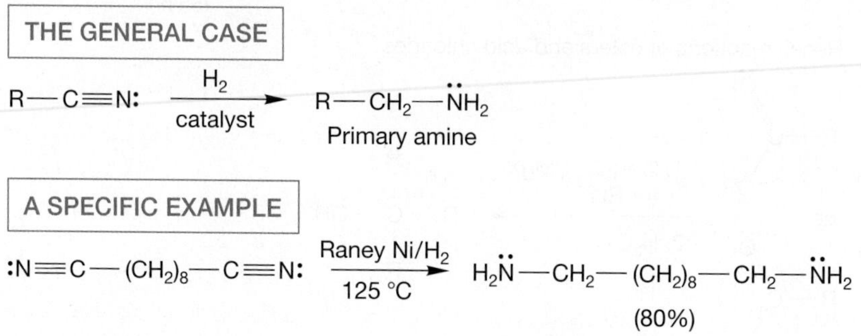

THE GENERAL CASE

$$R-C\equiv N: \xrightarrow[\text{catalyst}]{H_2} R-CH_2-\overset{..}{N}H_2$$
Primary amine

A SPECIFIC EXAMPLE

$$:N\equiv C-(CH_2)_8-C\equiv N: \xrightarrow[125\,°C]{Raney\ Ni/H_2} H_2\overset{..}{N}-CH_2-(CH_2)_8-CH_2-\overset{..}{N}H_2$$

(80%)

FIGURE 18.49 Catalytic hydrogenation of nitriles gives primary amines.

18.11 Reactions of Ketenes

Ketenes have a very electrophilic *sp*-hybridized carbon and are therefore reactive even at low temperature. Most ketenes are so reactive that they must be carefully protected from atmospheric moisture. Nucleophiles add to the carbonyl group to give acyl derivatives. This addition is another reaction that gives students trouble, but shouldn't. The ketene reacts with nucleophiles like any other carbonyl compound to give an addition product (Fig. 18.50). In this case, the addition product is an enolate (p. 373), and protonation gives the stable carbonyl compound. Figure 18.50 gives some examples.

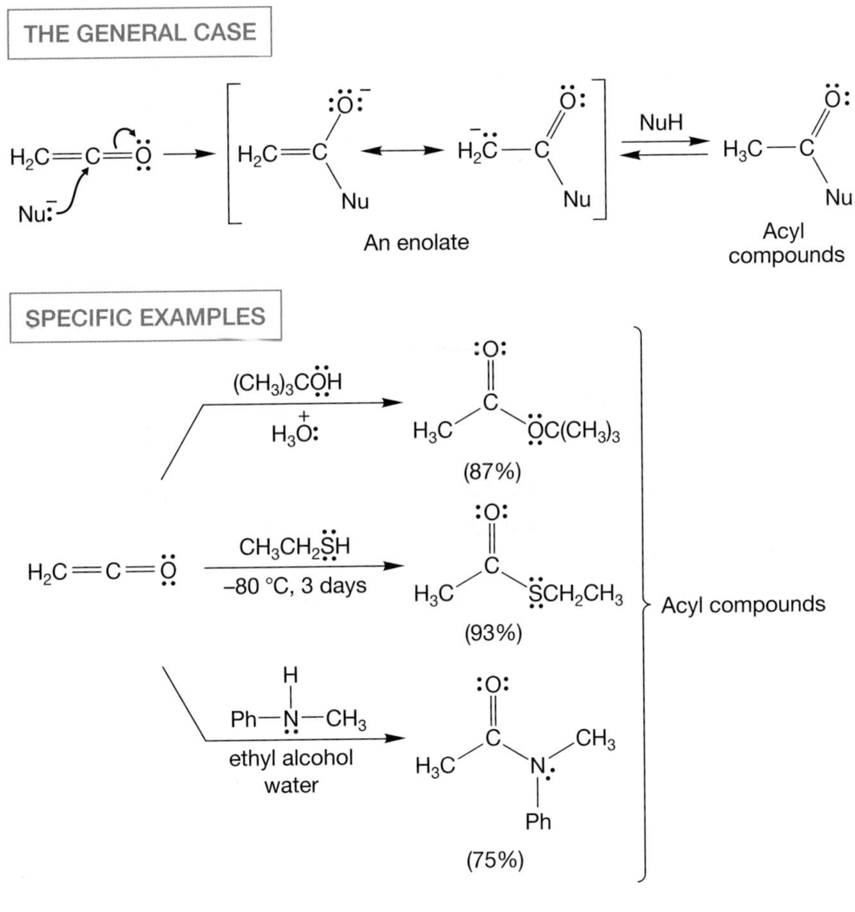

FIGURE 18.50 Ketenes react with nucleophiles to give acyl compounds.

THE GENERAL CASE

SPECIFIC EXAMPLES

18.12 Special Topic: Other Synthetic Routes to Acid Derivatives

As mentioned in the introduction to this chapter, reactions and syntheses are even more intertwined than usual with these acid derivatives. A reaction of one is a synthesis of another. The various interconversions of acyl compounds discussed in this chapter (and in Chapter 17) are summarized in Section 18.15. Here we first discuss two rearrangement reactions that are useful as syntheses of esters and amides, respectively, then move on to a brief discussion of routes to nitriles and ketenes.

18.12a The Baeyer–Villiger Reaction
Named by Kurt Mislow (p. 152) for Adolf von Baeyer (p. 187) and his student Victor Villiger (1868–1934), the **Baeyer–Villiger reaction** is an oxidation of aldehydes and ketones, usually with a

Baeyer–Villiger reaction

peracid (p. 424), that results in the insertion of an oxygen atom next to the carbonyl carbon (Fig. 18.51). Seen in its bare form, the reaction appears quite remarkable—how can that "extra" oxygen atom insert itself into a carbon–carbon bond? Fleshed out in a

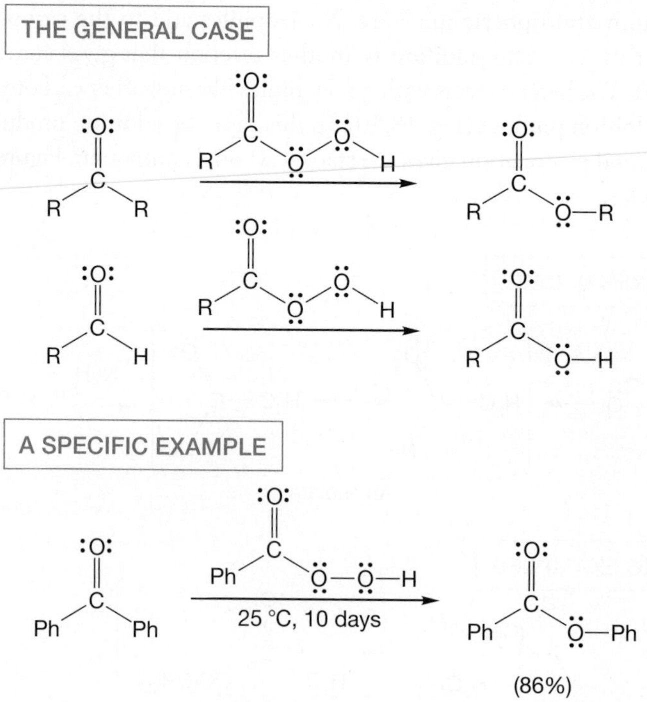

FIGURE 18.51 Baeyer–Villiger oxidation of aldehydes and ketones.

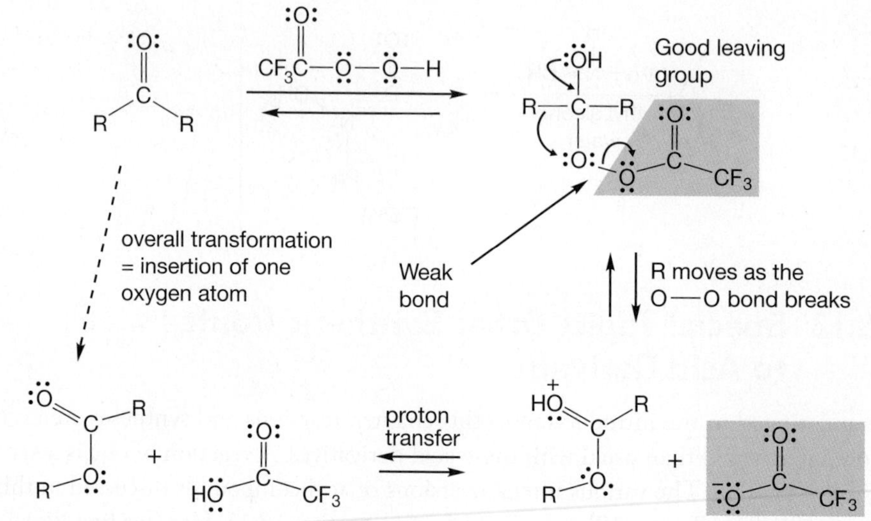

real mechanism, however, the reaction becomes intelligible. The first stage is a straight-forward addition of the peracid to the carbonyl group (Fig. 18.52). Now look at that the weak oxygen–oxygen bond, and at the good leaving group formed when it breaks. Migration of an R group as the oxygen–oxygen bond breaks gives the product directly.

FIGURE 18.52 The mechanism of the Baeyer–Villiger reaction.

Of course when there are two different R groups in the ketone, two products can result, because either R group can migrate. However, there is a well-established order of migration for different groups: H > tertiary alkyl > secondary alkyl > primary alkyl > methyl, and there is quite substantial selectivity in product formation. That is why aldehydes give the acid, as shown in Figure 18.51. The hydrogen migrates in preference to the alkyl or aryl group.

When the ketone is cyclic, the Baeyer–Villiger reaction becomes a synthesis of lactones (p. 849). Watch out for this reaction. The cyclic ketone adds apparent complexity, and this reaction is a favorite of problem writers (Fig. 18.53). Nevertheless, the mechanism is the same for the cyclic ketone as you will see in Problem 18.20.

:O:
Ph
25 °C, CHCl3
(71%)

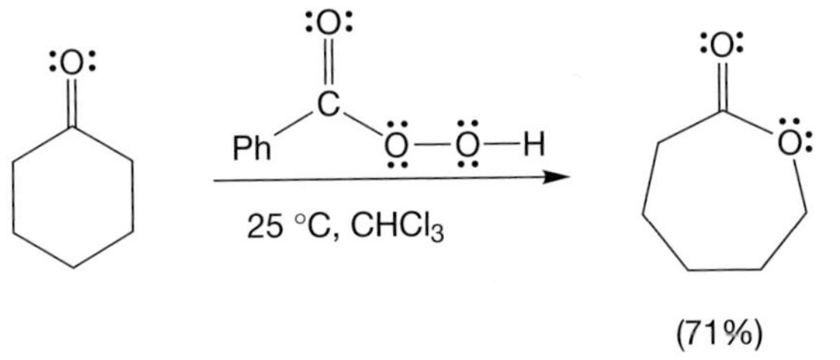

FIGURE 18.53 When a cyclic ketone is used, the Baeyer–Villiger reaction becomes a synthesis of lactones.

PROBLEM 18.20 Write a mechanism for the reaction in Figure 18.53.

PROBLEM 18.21 In the paper that named the Baeyer–Villiger reaction, Mislow and his student Joseph Brenner showed that in the reaction below, the product was formed with *retention* of optical activity. What does that observation tell you about the details of the mechanism?

H3C
Ph
H
CH3
25 °C, 65 h, CHCl3
(S)
(S)

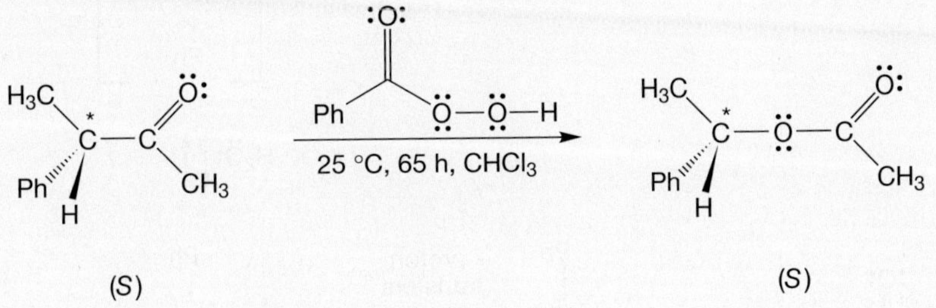

18.12b The Beckmann Rearrangement

Here is a related reaction called the **Beckmann rearrangement**, named for Ernst Otto Beckmann (1853–1923), in which the end product is not an ester, but an amide. The first step is formation of an oxime (p. 793), the product of the addition of hydroxylamine to a ketone (Fig. 18.54). Treatment of the oxime with any of a variety of strong acids leads to the amide.

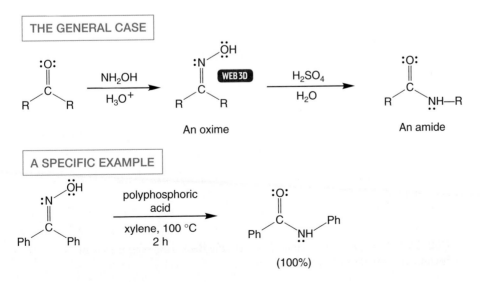

THE GENERAL CASE
NH2OH
H3O+
WEB3D
H2SO4
H2O
NH—R
An oxime
An amide
A SPECIFIC EXAMPLE
polyphosphoric acid
xylene, 100 °C
2 h
Ph
(100%)

FIGURE 18.54 The Beckmann rearrangement converts a ketone into an amide.

The mechanism of the Beckmann rearrangement (Fig. 18.55) involves a migration much like the one in the Baeyer–Villiger reaction, but the reaction is not triggered by the breaking of a weak oxygen–oxygen bond. Instead, it is initiated by the displacement of a good leaving group, water. That good leaving group is formed as the OH of the oxime is protonated by acid. Intramolecular rearrangement leads to a nicely resonance-stabilized intermediate that is then captured by water. Proton shifts lead to the final product, the amide.

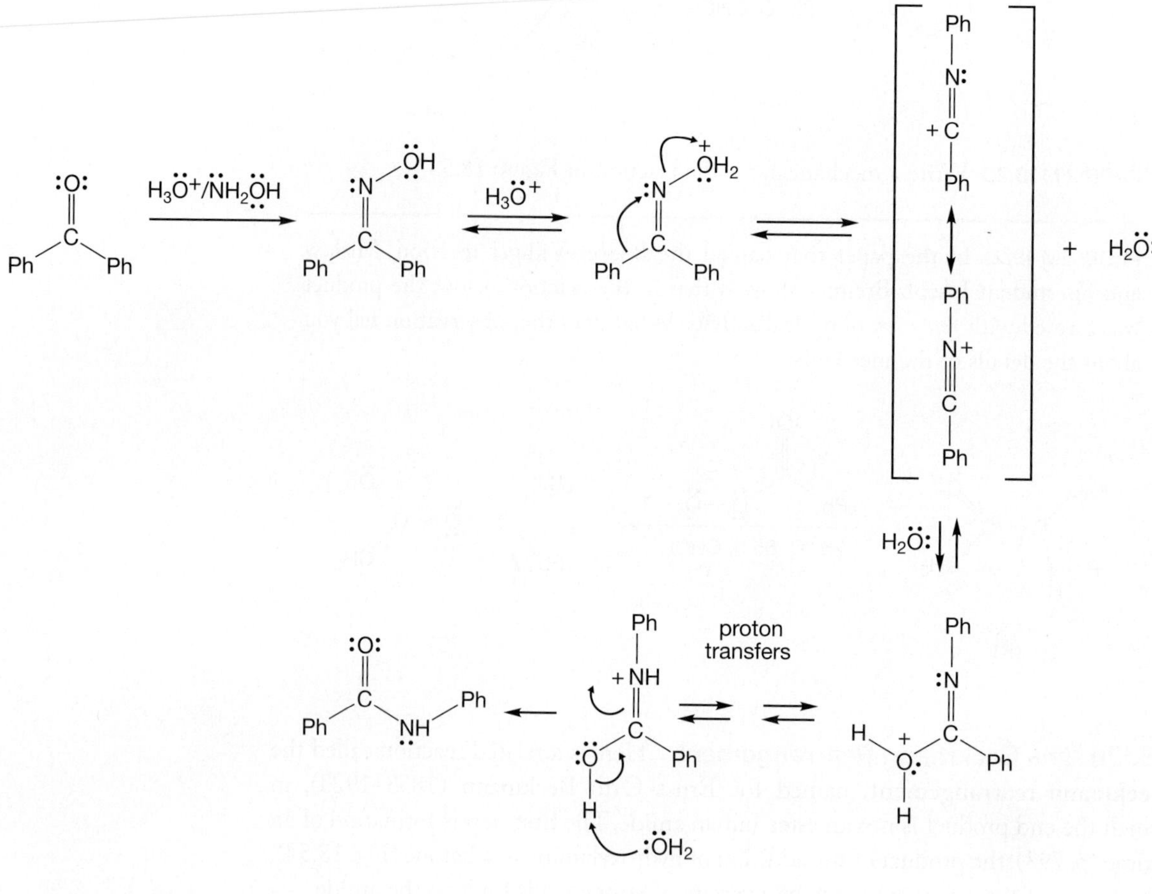

FIGURE 18.55 The mechanism of the Beckmann rearrangement.

As shown in Figure 18.56, in its cyclic version, the Beckmann rearrangement leads to a lactam (p. 851).

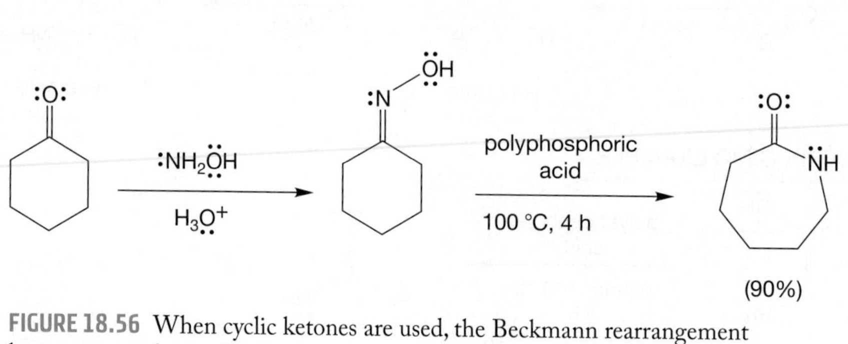

FIGURE 18.56 When cyclic ketones are used, the Beckmann rearrangement becomes a synthesis of lactams.

PROBLEM 18.22 Provide a mechanism for the reaction shown in Figure 18.56.

PROBLEM 18.23 Oximes can be obtained as the *E* or the *Z* isomers, and the two stereoisomers react differently. Explain the results shown here:

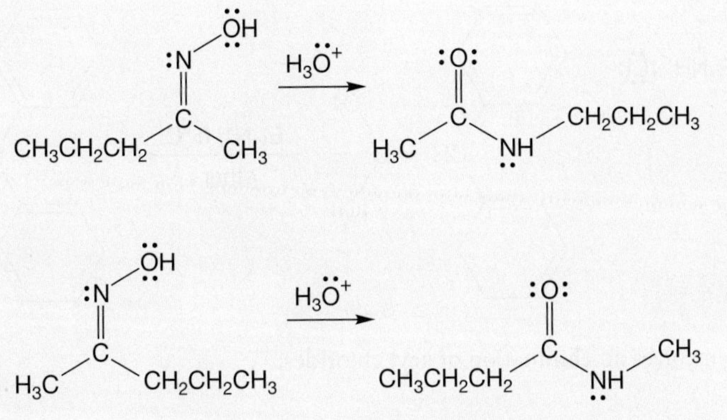

18.12c Nitrile Formation   Nitriles are most often made by displacement reactions of alkyl halides by cyanide ion or by dehydration of amides, generally with P_2O_5 (Fig. 18.57).

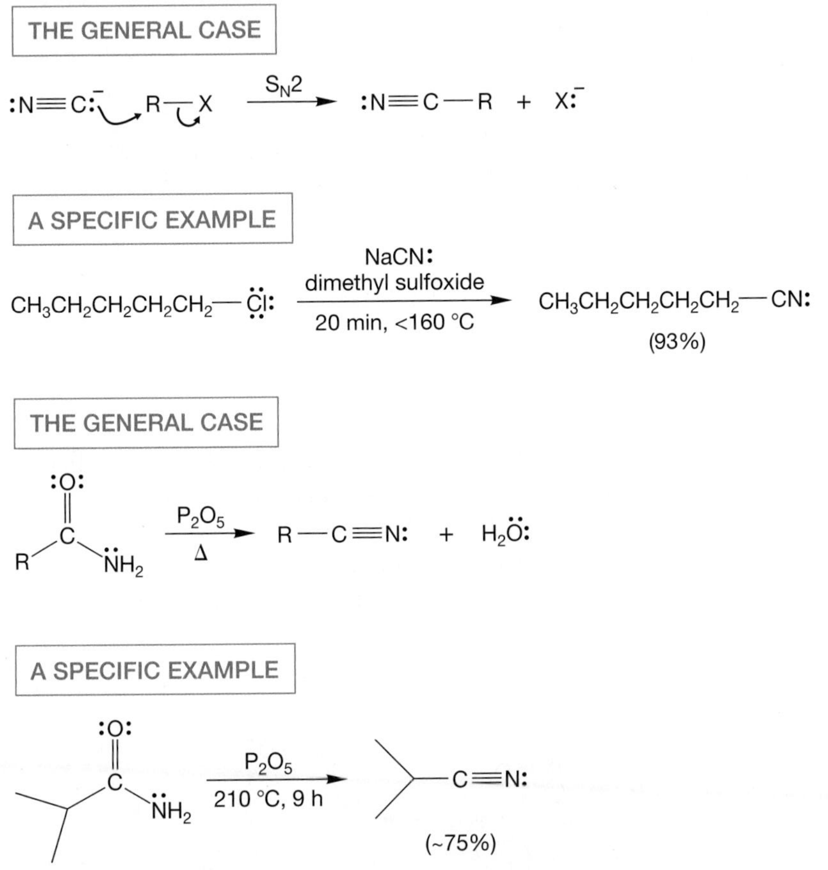

THE GENERAL CASE

$:N\equiv C:^- \quad R-X \xrightarrow{S_N2} \quad :N\equiv C-R \ + \ X:^-$

A SPECIFIC EXAMPLE

$CH_3CH_2CH_2CH_2CH_2-\ddot{C}l: \xrightarrow[\substack{\text{dimethyl sulfoxide} \\ \text{20 min, <160 °C}}]{\text{NaCN}:} CH_3CH_2CH_2CH_2CH_2-CN:$

(93%)

THE GENERAL CASE

R—C(=O)—NH₂ $\xrightarrow[\Delta]{P_2O_5}$ $R-C\equiv N:$ + $H_2\ddot{O}:$

A SPECIFIC EXAMPLE

$\xrightarrow[\text{210 °C, 9 h}]{P_2O_5}$ C≡N:

(~75%)

FIGURE 18.57 Nitriles are usually made through S_N2 reactions using cyanide as the nucleophile. Nitriles can also be formed by dehydration of amides.

18.12d Ketene Formation   Ketenes are synthesized through elimination reactions of acyl chlorides using tertiary amines as bases (Fig. 18.58).

THE GENERAL CASE

A SPECIFIC EXAMPLE

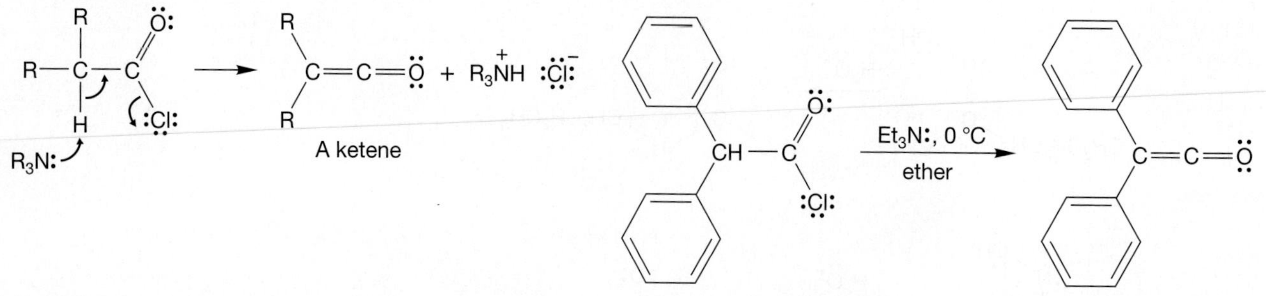

FIGURE 18.58   One common synthesis of ketenes involves an elimination of acyl chlorides.

18.13 Special Topic: Thermal Elimination Reactions of Esters

Esters that contain a hydrogen in the β position of the OR group give alkenes and carboxylic acids when heated to high temperatures (>400 °C) (Fig. 18.59).

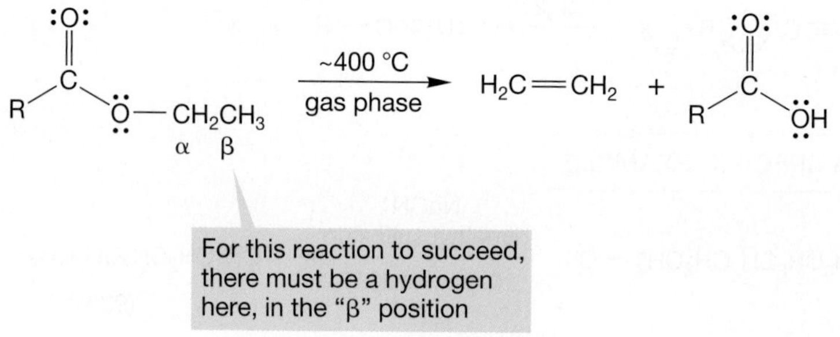

FIGURE 18.59 The thermal elimination reaction of esters containing a hydrogen in the β position.

For this reaction to succeed, there must be a hydrogen here, in the "β" position

In this reaction, the carbonyl group acts as a base, attacking the β hydrogen intramolecularly. At the same time the ester acts as a leaving group, and a carboxylic acid is lost in a single, concerted step (Fig. 18.60). In contrast to most E2 reactions, this pyrolytic, intramolecular version proceeds through a syn elimination. The reason is simple: Only the syn hydrogen is within reach of the carbonyl group.

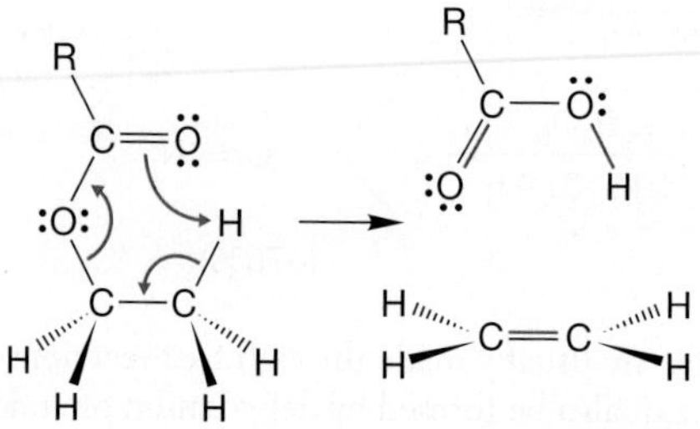

FIGURE 18.60 An arrow formalism description of the thermal elimination reaction of esters.

There is no possibility of doing an anti elimination in this reaction. The elimination reactions of the deuterium-labeled compounds shown in Figure 18.61 provide an example of the stereospecificity of this elimination reaction. In diastereomer **A**, it is only the cis hydrogen that is lost as *trans*-stilbene is formed. By contrast, in isomer **B** it is the deuterium atom that must occupy the cis position in the formation of *trans*-stilbene, and this time it is D, not H, that is lost.

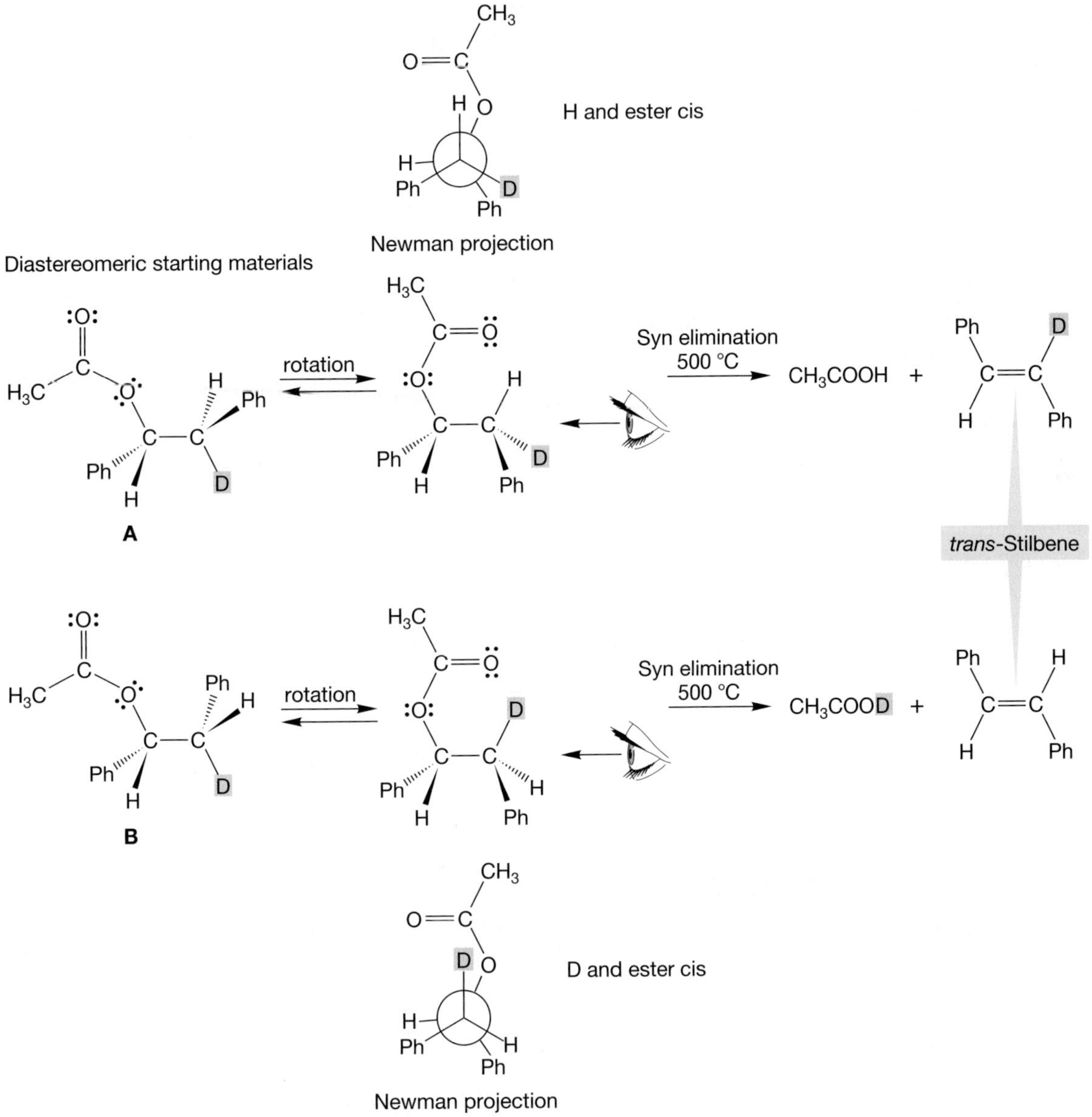

FIGURE 18.61 In these intramolecular elimination reactions, only a cis hydrogen is removed. Because the phenyl groups are required to stay as far from each other as possible for steric reasons, syn elimination demands that H be lost from stereoisomer **A** and D lost from stereoisomer **B**.

There are other versions of this new elimination process, some of which proceed under much milder conditions than those necessary for ester pyrolysis.

For example, the sulfur-containing **xanthate esters** undergo elimination especially easily (Fig. 18.62). Xanthate esters can be prepared as shown in the figure by addition of alkoxide to CS_2 (recall the addition of nucleophiles to carbon dioxide, p. 840). These intramolecular thermal elimination reactions provide a new route to alkenes, so update your file cards.

THE GENERAL CASE

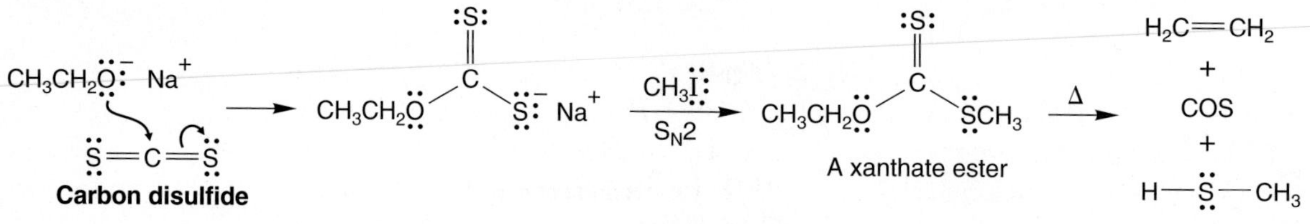

Carbon disulfide

A xanthate ester

A SPECIFIC EXAMPLE

(62%)

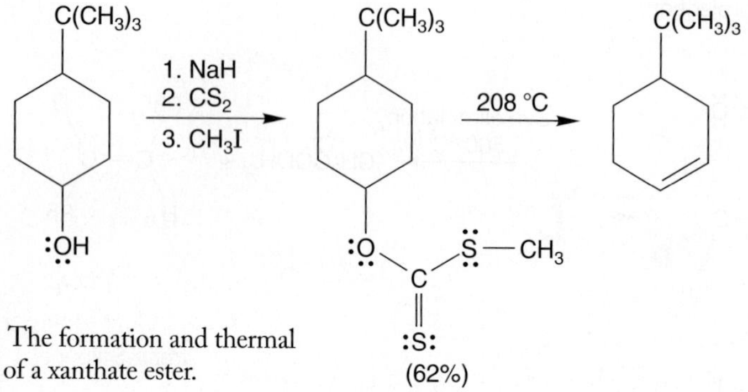

FIGURE 18.62 The formation and thermal fragmentation of a xanthate ester.

PROBLEM 18.24 Write arrow formalisms for the general reaction of Figure 18.62.

PROBLEM 18.25 Provide mechanisms for the following reactions. The first reaction (the thermal elimination of an *N*-oxide) is called the Cope elimination, after Arthur C. Cope (1909–1966).

(a) Cope elimination

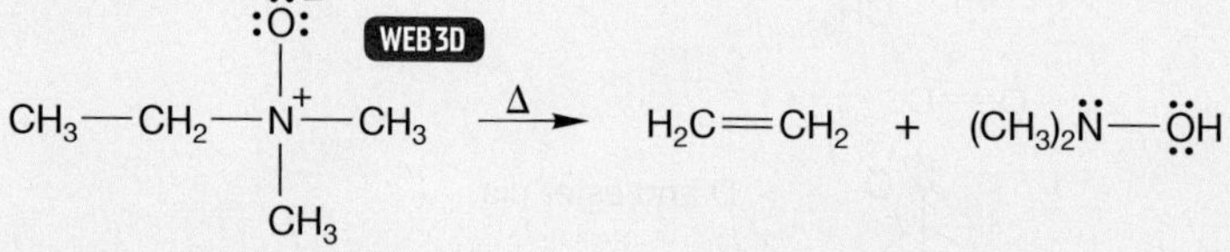

(b) Ester pyrolysis

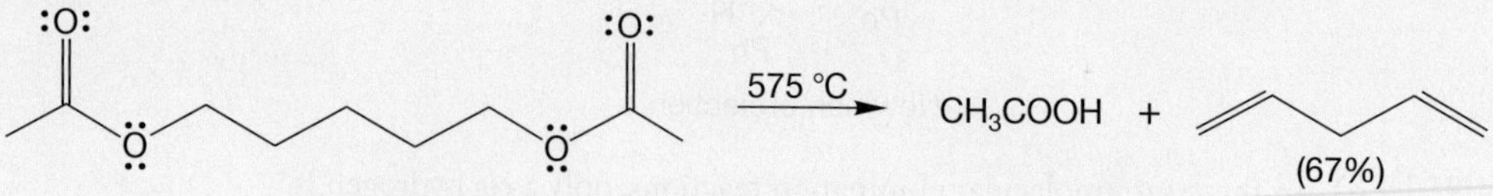

(67%)

18.14 Special Topic: A Family of Concerted Rearrangements of Acyl Compounds

In this section, we take up a few rearrangement reactions of some new kinds of molecules related to acyl compounds. The mechanisms are not always fully worked out, but the reactions are of interest and are often synthetically useful.

18.14a Diazo Ketones: The Wolff Rearrangement

When an acid chloride is allowed to react with a diazo compound, the result is formation of a **diazo ketone** (Fig. 18.63).

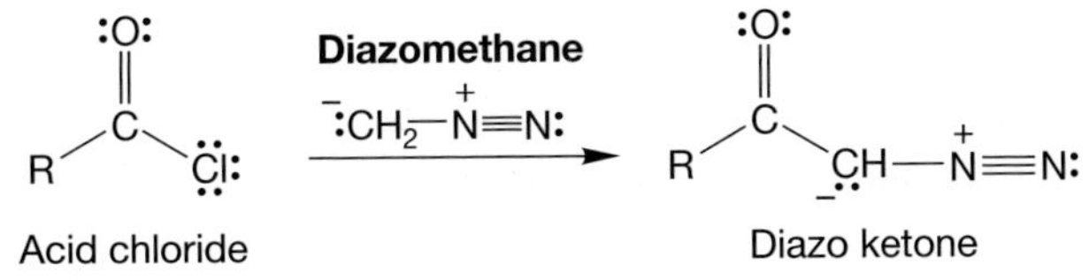

FIGURE 18.63 Diazo ketone formation.

The mechanism of this reaction is only slightly more complicated than the usual reaction of nucleophiles with acid chlorides (p. 854). Diazo compounds are nucleophiles at carbon and add to the carbonyl group of the reactive acid chloride. Chloride ion is expelled in the elimination phase of this normal addition–elimination process. In the only new step of the reaction, chloride removes the newly acidic hydrogen to give the diazo ketone (Fig. 18.64). This hydrogen is acidic because the conjugate base is stabilized by resonance.

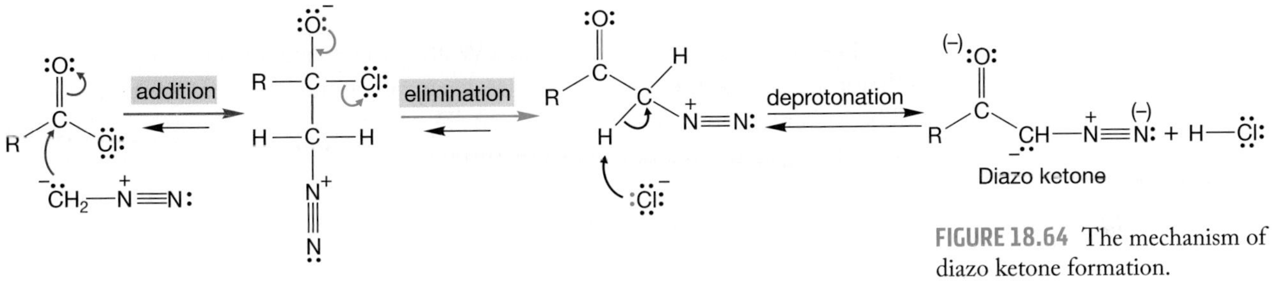

FIGURE 18.64 The mechanism of diazo ketone formation.

Diazo compounds are generally rather dangerous. They are not only poisonous but explosive as well. The resonance-stabilized diazo ketones are far less explosive than simple diazo compounds, but they must still be treated with the greatest respect.

PROBLEM 18.26 Draw resonance forms for a diazo ketone.

Like other diazo compounds, diazo ketones are sensitive to both heat and light. Upon exposure to either, nitrogen is lost and a carbene intermediate (p. 431), in this case, a ketocarbene, is formed (Fig. 18.65).

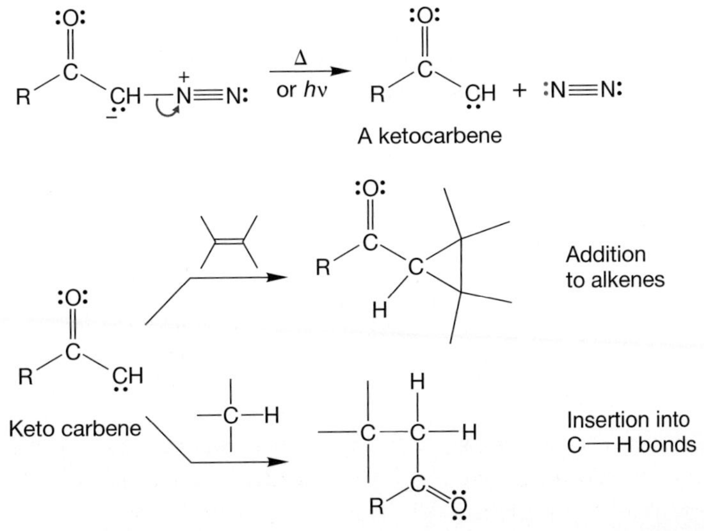

FIGURE 18.65 Diazo ketones decompose to give ketocarbenes on heating or photolysis. The reactions of ketocarbenes include additions to π systems and carbon–hydrogen insertion.

Ketocarbenes are capable of the usual carbene reactions, addition to multiple bonds and insertion into carbon–hydrogen bonds chief among them (Fig. 18.65). In addition, it is possible for ketocarbenes to undergo an intramolecular carbon–carbon insertion reaction. This reaction is often catalyzed by silver, and involves a migration of the R group (Fig. 18.66). The product of this reaction is a ketene. This reaction is known as the **Wolff rearrangement**, after Ludwig Wolff (1857–1919; remember the Wolff–Kishner reduction, p. 645? Same Wolff).

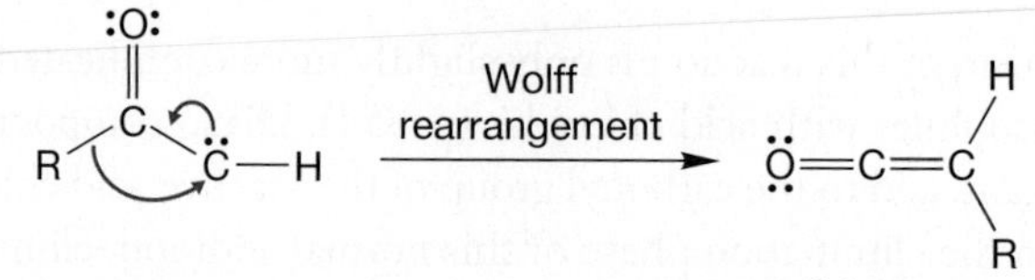

FIGURE 18.66 Intramolecular ketene formation from a ketocarbene, through a migration of an R group.

The mechanistic complication in a Wolff rearrangement comes from the notion that it may not be the carbene, but the diazo compound itself that gives the ketene. The carbene is not necessarily involved (Fig. 18.67). The question isn't completely sorted out yet, but the best evidence currently favors at least some direct formation of the ketene from most diazo ketones. However the evidence doesn't exclude the involvement of the carbene in all cases.

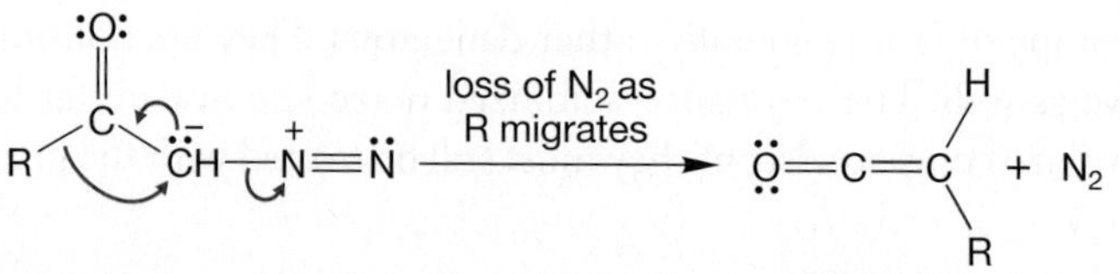

FIGURE 18.67 In the Wolff rearrangement, the C—R bond can break as the nitrogen departs. Perhaps there is no ketocarbene intermediate.

WORKED PROBLEM 18.27 Much experimentation indicates that only one of the two possible stereoisomers of a diazo ketone can give ketene directly. What are the two stereoisomers of the diazo ketone?

ANSWER There are s-cis and s-trans isomers of diazo ketones.

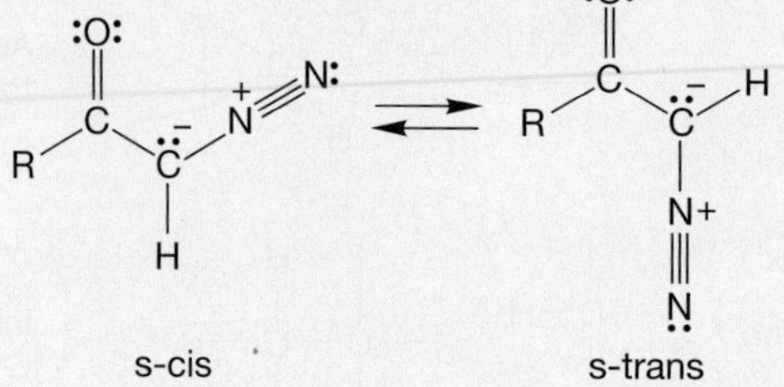

s-cis   s-trans

WORKED PROBLEM 18.28 Explain why only one of the isomers in the previous problem can rearrange easily to a ketene.

ANSWER The s-cis isomer is set up for migration of R with displacement from the rear in an intramolecular S_N2 reaction. By contrast, the s-trans isomer would have to do a frontside S_N2 for R to migrate as the nitrogen leaves.

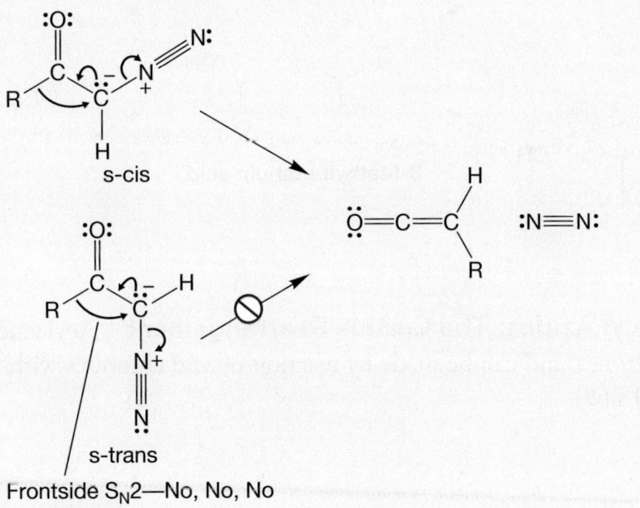

If the mechanistic questions still are not completely resolved, the synthetic applications certainly are. The Wolff rearrangement generates a ketene in the absence of nucleophiles, and therefore the ketene can be isolated in this reaction. If the product ketene is subsequently hydrolyzed to the carboxylic acid, the overall sequence is called the **Arndt–Eistert reaction** after Fritz Arndt (1885–1969) and Bernd Eistert (1902–1978). It constitutes a most useful chain-lengthening reaction (Fig. 18.68). The acid chloride that was used at the outset (Fig. 18.63) must come from a carboxylic acid, because that is the only way we have for making an acid chloride. So the overall process starting with the original acid is as follows: make the acid chloride, convert the acid chloride into the diazo ketone, rearrange the ketone to the ketene, and hydrolyze the ketene to give the new carboxylic acid.

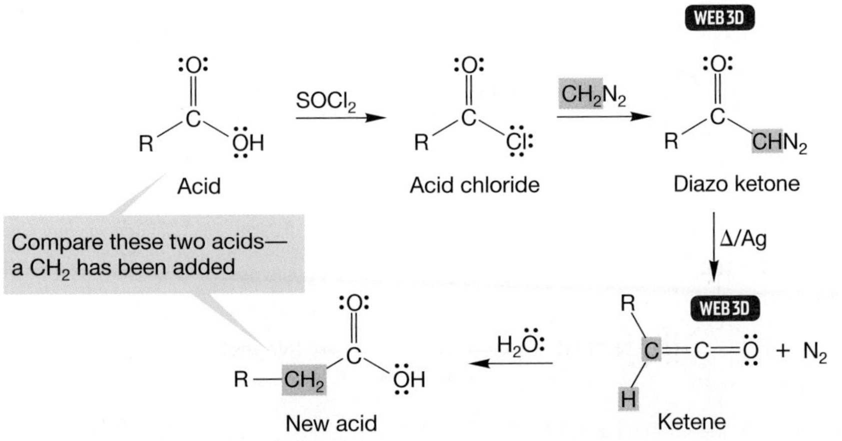

FIGURE 18.68 The Arndt–Eistert synthesis adds a methylene group to a carboxylic acid.

PROBLEM 18.29 Devise a synthetic scheme for converting each of the starting materials shown into 3-methylbutanoic acid.

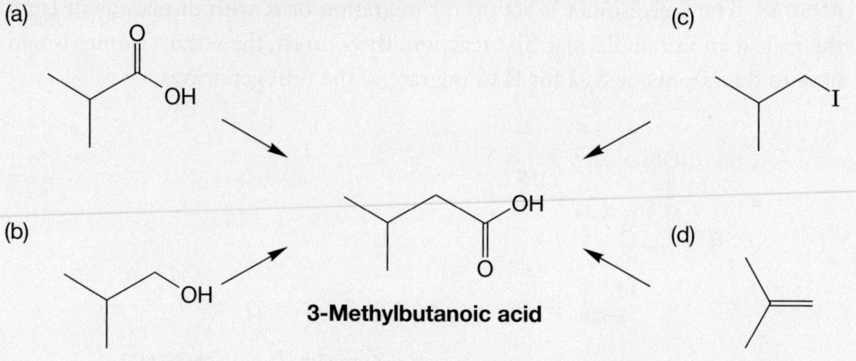

(a)

(c)

(b)

(d)

3-Methylbutanoic acid

18.14b Acyl Azides: The Curtius Rearrangement

Acyl azides are related to diazo ketones, and can be made by reaction of acid chlorides with the azide ion, $^-N_3$ (Fig. 18.69).

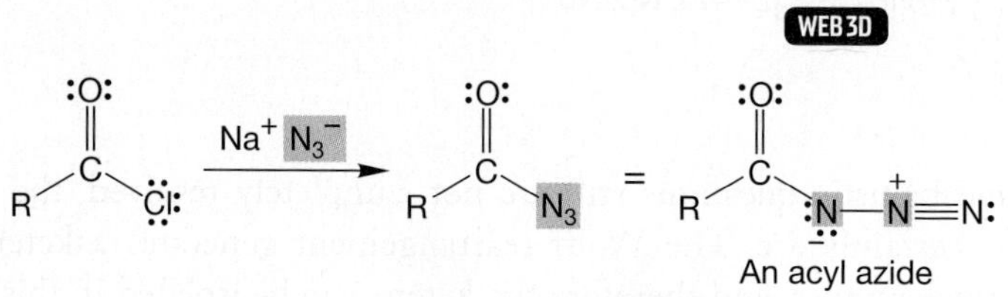

WEB 3D

An acyl azide

FIGURE 18.69 Acid chlorides react with azide ion ($^-N_3$) to give acyl azides.

Like diazoketones, acyl azides lose nitrogen when heated or irradiated. The end result is an **isocyanate**, formed either directly from the acyl azide, in a reaction called the **Curtius rearrangement** after Theodore Curtius (1857–1928), or by intramolecular rearrangement of a **nitrene**, which is the nitrogen analogue of a carbene (Fig. 18.70). For most acyl azides, it is known that the isocyanate is formed by direct rearrangement of the acyl compound and that the nitrene, though also formed from the azide, is not involved in isocyanate production.

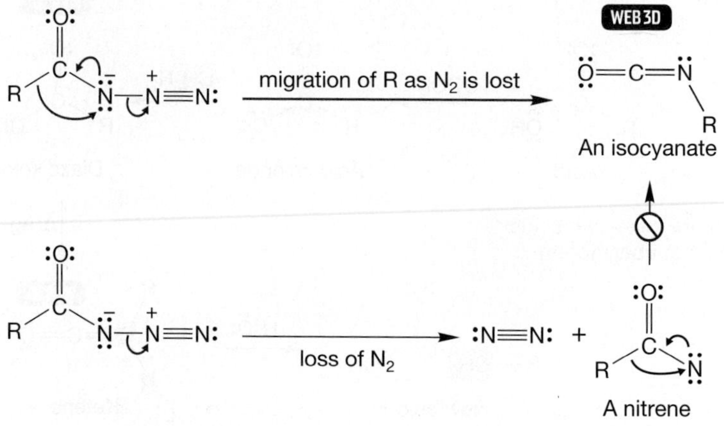

migration of R as N₂ is lost

WEB 3D

An isocyanate

loss of N₂

A nitrene

FIGURE 18.70 The Curtius rearrangement leads to isocyanates. The mechanism usually involves a concerted, one step rearrangement.

PROBLEM 18.30 Can you guess what nitrene chemistry will be like? What products do you anticipate from the reaction of a nitrene with the carbon–carbon double bond? What product would you get from reaction with a carbon–hydrogen bond? Remember that a nitrene is the nitrogen analogue of a carbene (p. 431).

Isocyanates, like ketenes, are very sensitive to nucleophiles. For example, alcohols add to isocyanates to give carbamate esters (Fig. 18.71).

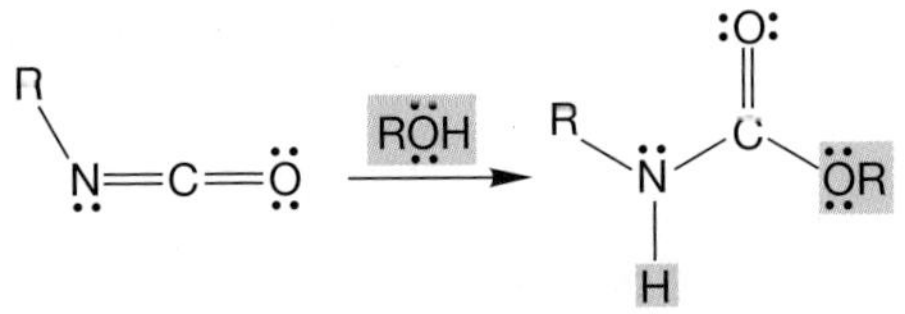

An isocyanate   A carbamate ester

FIGURE 18.71 Isocyanates react with alcohols to give carbamate esters.

PROBLEM 18.31 Write a mechanism for the addition of an alcohol to an isocyanate to give a carbamate ester (Fig. 18.71).

PROBLEM 18.32 Write a mechanism for the formation of the amine in the following reaction:

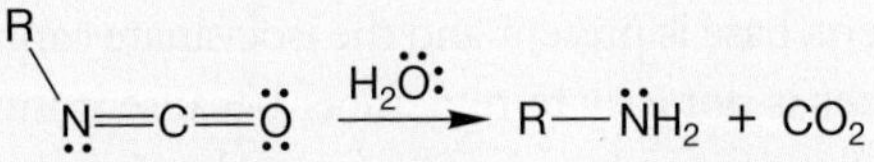

18.14c Hofmann Rearrangement of Amides

Amides are able to take part in a reaction similar to the Wolff and Curtius rearrangements. The reaction is called the **Hofmann rearrangement**[3] and it allows you to convert an amide into an amine with the loss of the carbonyl carbon. This process occurs in two steps. The amide is first brominated to give the *N*-bromoamide (Fig. 18.72). Base then removes the very acidic α hydrogen and rearrangement to the isocyanate ensues. The question of whether a nitrene is involved in this rearrangement is not completely settled, but we would bet on the direct rearrangement shown in Figure 18.72.

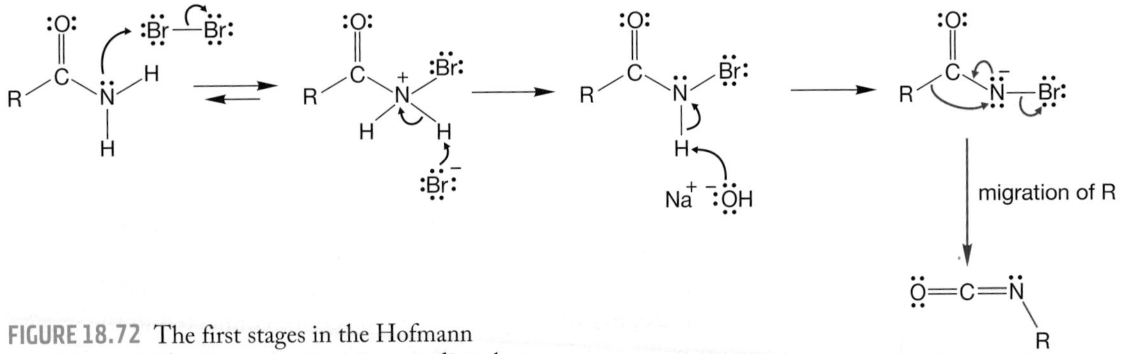

FIGURE 18.72 The first stages in the Hofmann rearrangement. An isocyanate is an intermediate, but cannot be isolated.

An isocyanate
(but *not* isolable under these conditions)

migration of R

---

[3]In earlier, more colorful times, this reaction was known as the "Hofmann degradation."

The isocyanate is not isolable, as it is in the Curtius rearrangement. Instead, in this reaction, it is born *in the presence of base* and is hydrolyzed to the carbamic acid, which is unstable and decarboxylates (Fig. 18.73, recall Problem 18.32).

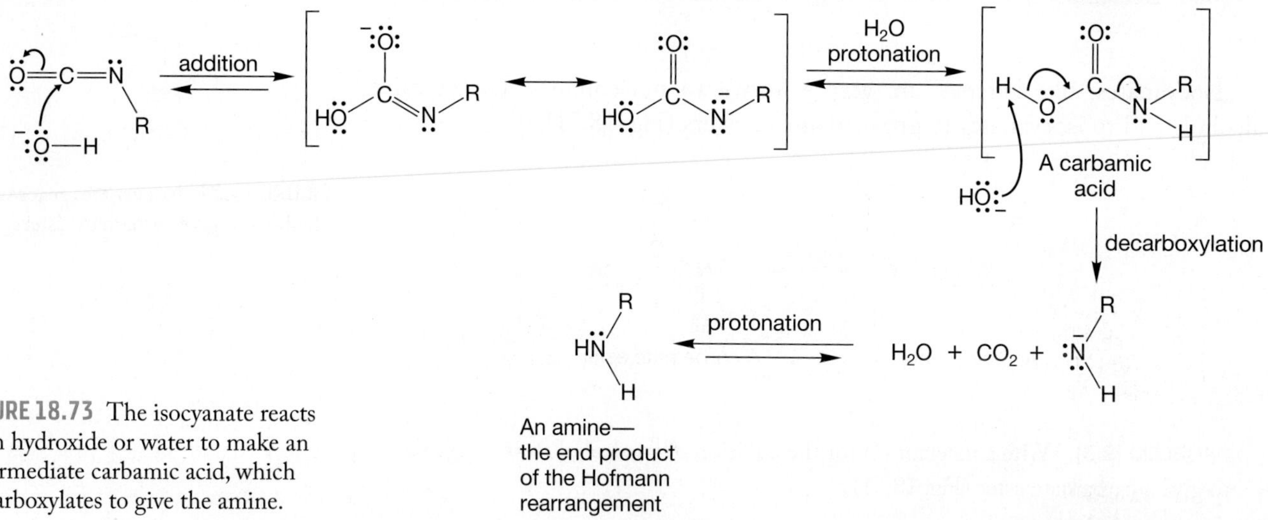

FIGURE 18.73 The isocyanate reacts with hydroxide or water to make an intermediate carbamic acid, which decarboxylates to give the amine.

An amine—
the end product
of the Hofmann
rearrangement

The end product is the amine. There is an essential difference between the Curtius rearrangment and the Hofmann rearrangement. Both involve the formation of isocyanates, but only in the Curtius rearrangement is this intermediate isolable. In the Hofmann rearrangement, base is present and the isocyanate cannot survive (Fig. 18.74). This synthesis of amines is not easy to remember because it involves many steps, thus making it a great favorite of problem writers (open-book, of course).

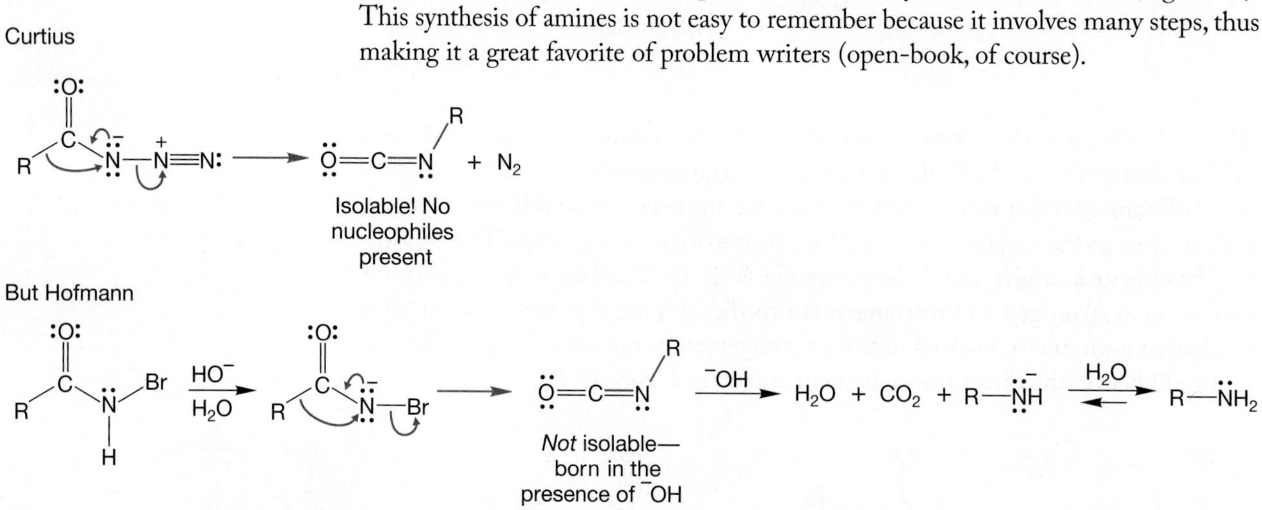

FIGURE 18.74 The Curtius and Hofmann rearrangements contrasted.

Summary

In this section we have explored a number of rearrangement reactions that involve ketenes or ketene-like intermediates. Notice that each of these rearrangements is a name reaction (Wolff, Arndt-Eistert, Hofmann,[4] and Curtius). Don't panic about keeping the specifics of the reactions connected to the names. Most instructors will not ask you to reproduce the reactions by name. However, if your future includes organic chemistry, you will see them again and become familiar with them. Of course if your future does not include organic chemistry, at least you can appreciate the predictability and trends these reactions illustrate.

---

[4]Do you remember the Hofmann elimination (p. 308)? It's the same Hofmann.

EAT YOUR BROCCOLI!

The last section showed a set of reactions unified by formation of an intermediate containing two cumulated (adjacent) double bonds. Their common ancestor is allene, $H_2C{=}C{=}CH_2$. Are these molecules mere curiosities, chemical oddities of no real importance outside a textbook or exam? Not at all. In 1992, broccoli was shown to contain a sulfur-containing isocyanate, an iso*thio*cyanate, that induces formation of something called a "phase II detoxication enzyme," which is involved in the metabolism of carcinogens. Eat your broccoli and study your orgo.

$$H_2C{=}C{=}CH_2$$
Allene

$$\ddot{O}{=}C{=}CH_2$$
Ketene

$$R{-}N{=}C{=}\ddot{O}$$
Isocyanate

$$R{-}N{=}C{=}\ddot{S}$$
Isothiocyanate

1-Isothiocyanato-(4R)-(methylsulfinyl)butane
(an anticancer agent found in broccoli)

18.15 Summary

New Concepts

This chapter is filled with detail, but there are no new fundamental concepts. Most important, although not really new, is the continuation of the exploration of the addition–elimination reaction of carbonyl compounds begun in Chapter 17. When nucleophiles add to carbonyl compounds that do not bear a leaving group (for example, aldehydes or ketones) reversal of the addition or protonation are the only two possible reactions. When there is a leaving group attached to the carbonyl, as there is with most acid derivatives, another option is loss of the leaving group (Fig. 18.75).

FIGURE 18.75 Carbonyl groups bearing good leaving groups can undergo the addition–elimination reaction; others cannot.

Key Terms

acid derivative (p. 877)	Curtius rearrangement (p. 918)	nitrene (p. 918)
acid halide (p. 880)	diazo ketone (p. 915)	nitrile (p. 883)
acyl compound (p. 877)	ester hydrolysis (p. 895)	Rosenmund reduction (p. 892)
Arndt–Eistert reaction (p. 917)	Hofmann rearrangement (p. 919)	transesterification (p. 896)
Baeyer–Villiger reaction (p. 907)	imide (p. 882)	Wolff rearrangement (p. 916)
Beckmann rearrangement (p. 909)	isocyanate (p. 918)	xanthate ester (p. 914)

Reactions, Mechanisms, and Tools

The addition–elimination reaction, introduced in Chapter 17, dominates the reaction mechanisms discussed in this chapter.

The chemistry of nitriles is similar to that of carbonyl compounds. The carbon–nitrogen triple bond can act as an acceptor of nucleophiles in addition reactions.

There is a class of intramolecular thermal elimination reactions that provides a new route to alkenes. Esters, xanthates, and amine oxides are commonly used in this reaction. The reactions are concerted (one-step), and steric requirements dictate that a syn elimination must occur in the reaction, as the carbonyl group cannot reach a hydrogen in an anti position (Fig. 18.61).

Syntheses

Here are the many important synthetic reactions found in this chapter. A few have been touched upon already in Chapter 17, but most are new to you, at least in a formal sense. Many of the following reactions share a common (addition–elimination) mechanism.

1. Acids

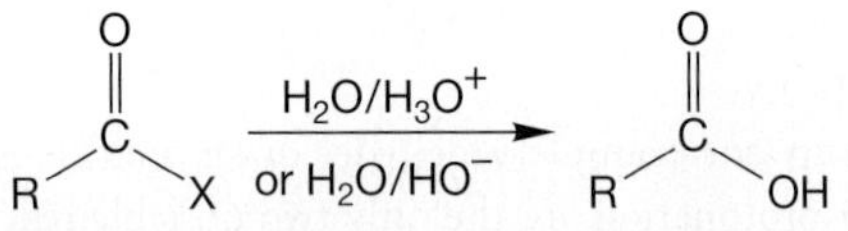

Acid- or base-induced hydrolysis of acid halides, anhydrides, esters, and amides; X = Cl, Br, I, O—CO—R, OR, NH$_2$

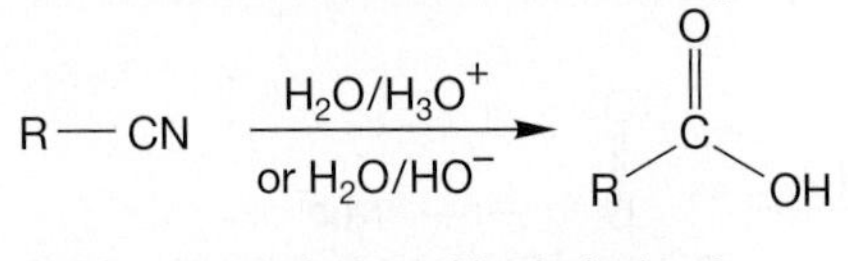

Acid- or base-induced hydrolysis of nitriles; the amide is an intermediate

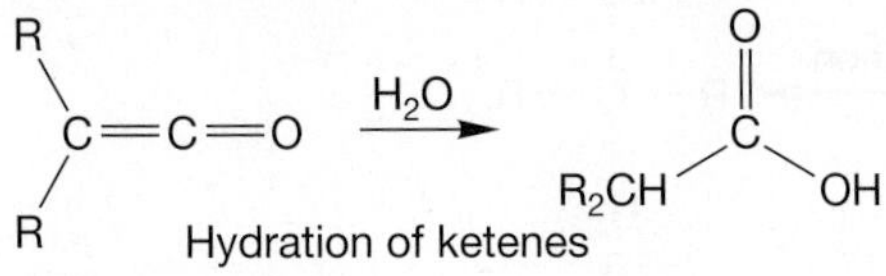

Hydration of ketenes

2. Acyl Azides

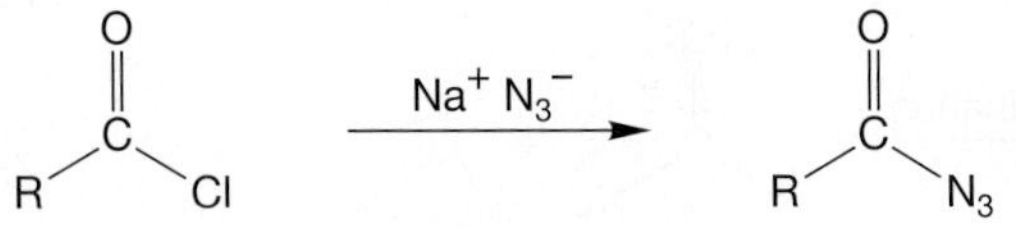

Addition–elimination reaction

3. Alcohols

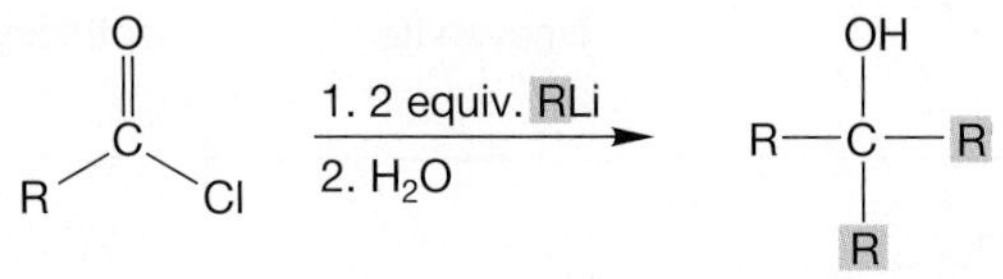

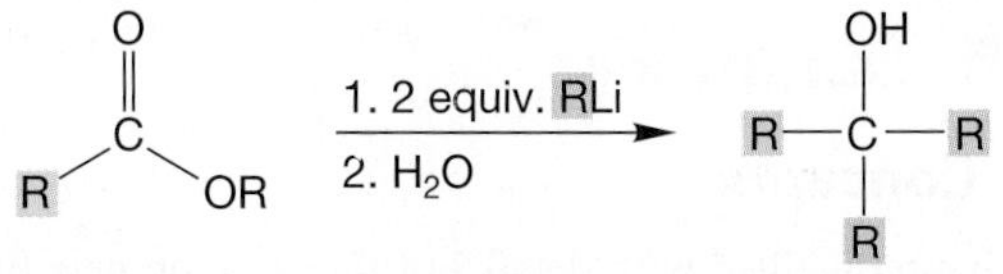

The ketone is an intermediate but cannot be isolated; all three R groups may be the same, but all three R's cannot be different, RMgX also works

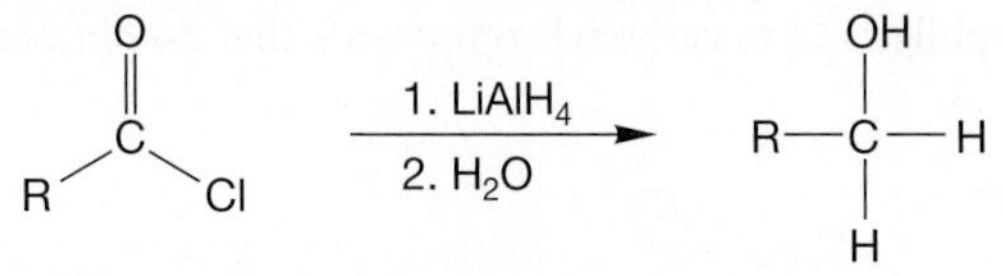

Other metal hydride donors also work; the aldehyde is an intermediate, but cannot be isolated

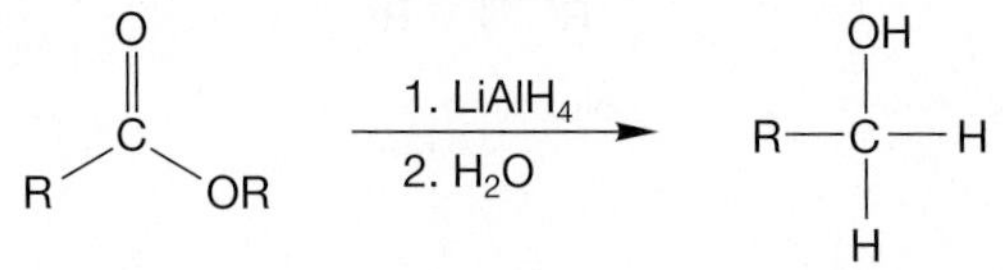

Other metal hydride donors also work; the aldehyde is an intermediate, but cannot be isolated

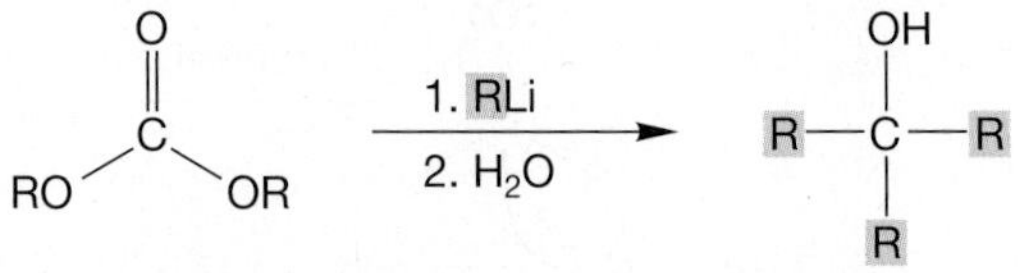

The ester and ketone are intermediates, but cannot be isolated; all three R groups must be the same, RMgX also works

4. Aldehydes

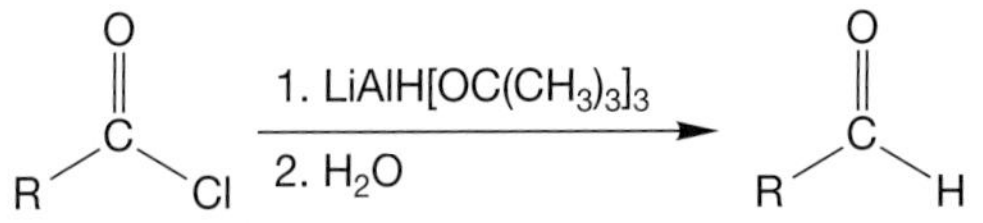

Other encumbered metal hydrides also work

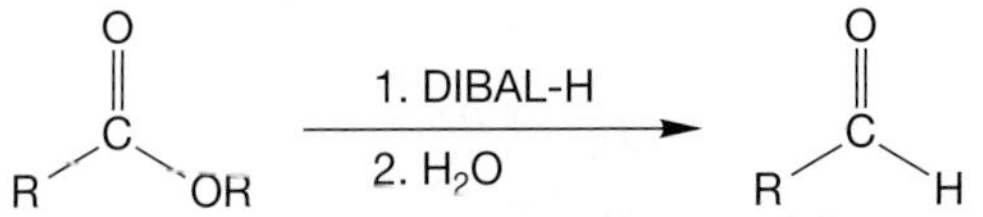

Diisobutylaluminum hydride (DIBAL-H) will not reduce the aldehyde if the reaction is carefully run

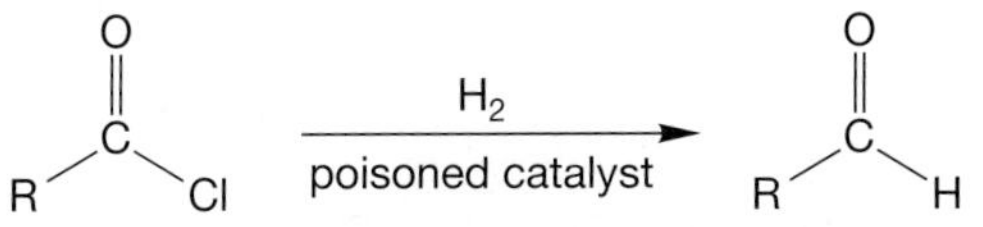

Rosenmund reduction

5. Alkenes

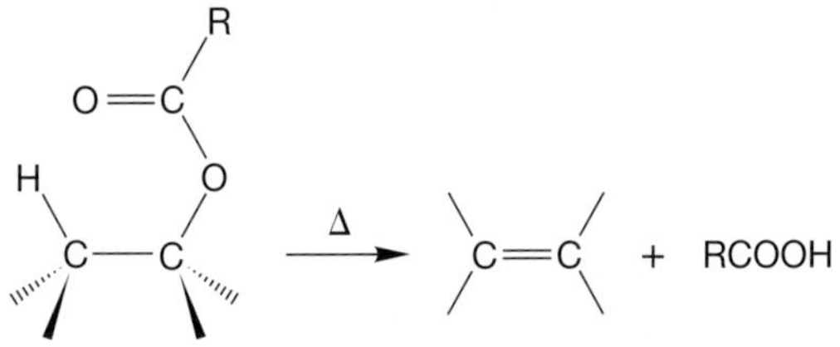

Note that only a cis hydrogen can be removed

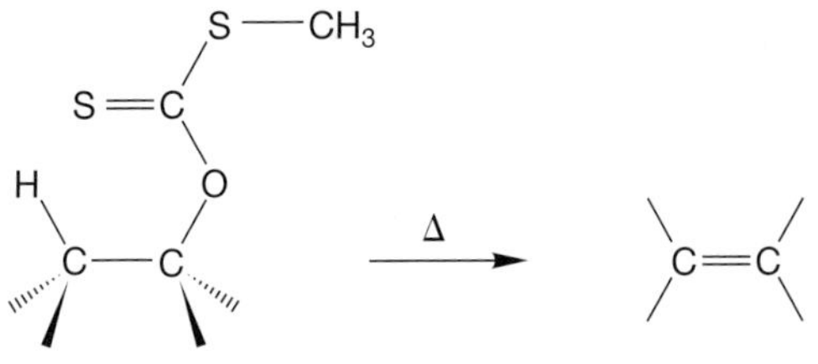

Xanthate ester pyrolysis requires lower temperature than ester pyrolysis

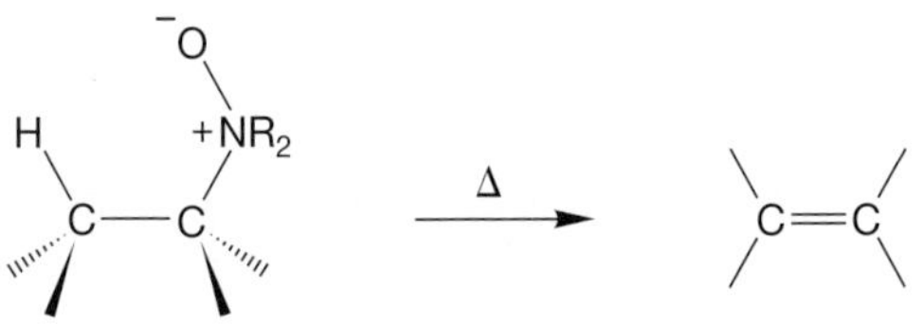

A Cope elimination

6. Amides

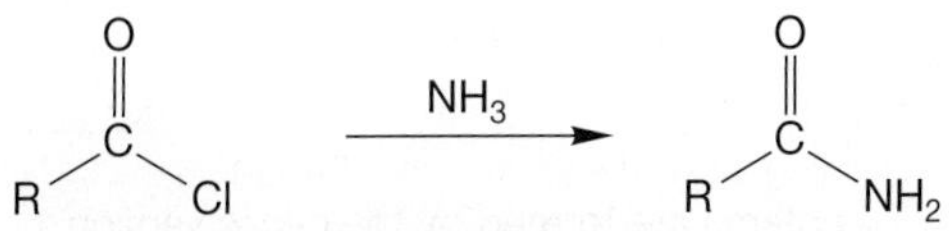

Substituted amines give substituted amides

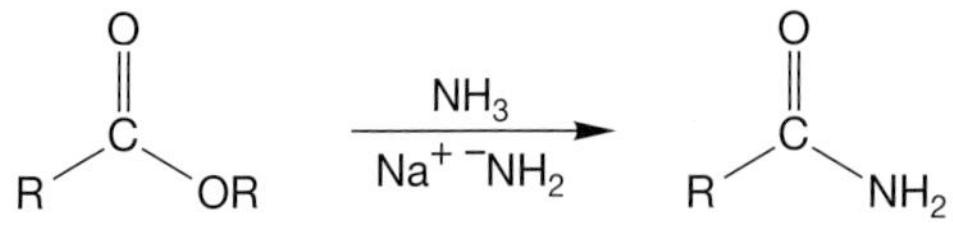

Acid catalysis also works

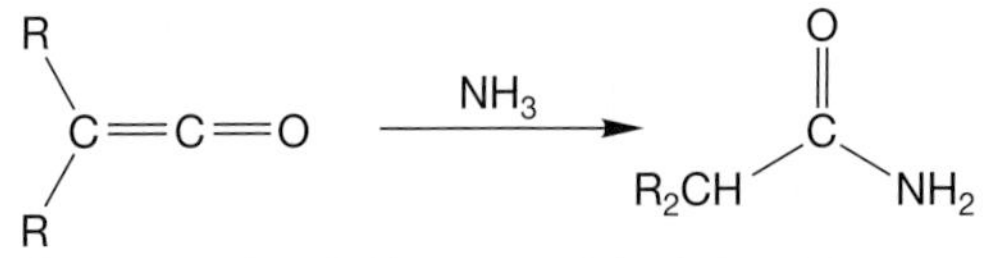

Substituted amines give substituted amides

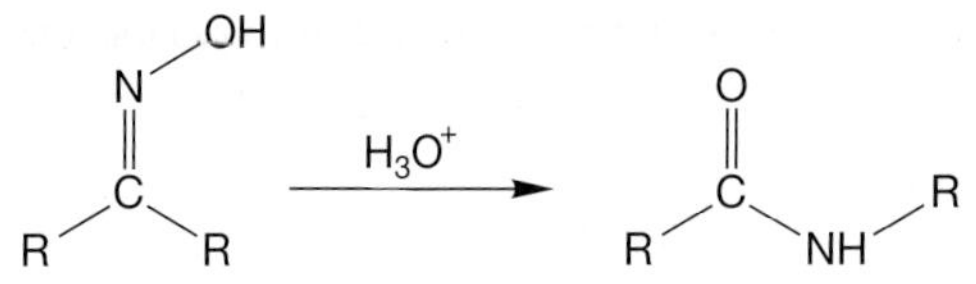

The Beckmann rearrangement

7. Amines

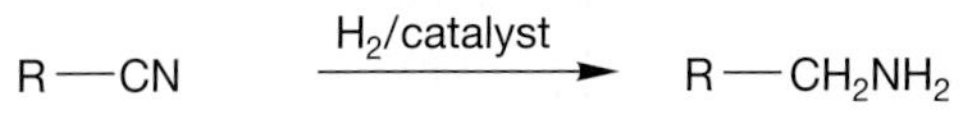

Catalytic hydrogenation

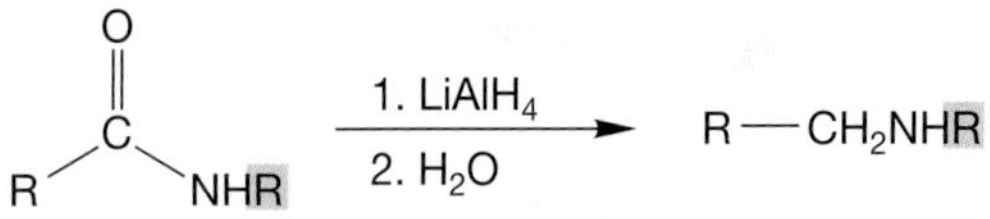

Loss of a metal oxide from the initially formed intermediate is the key to this reaction

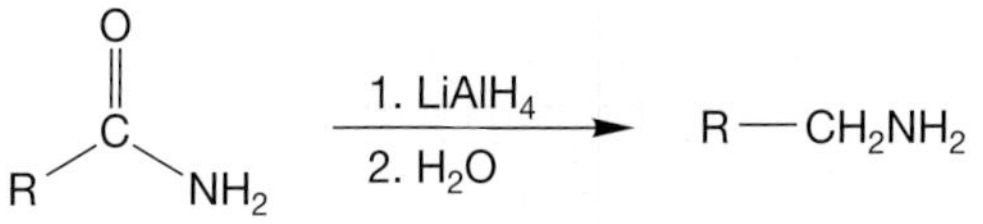

Loss of a metal oxide from the initially formed intermediate is the key to this reaction, too

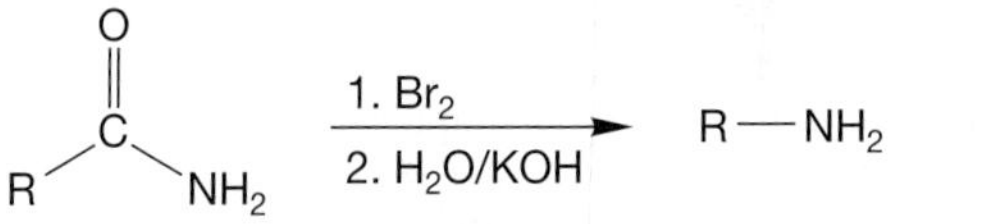

The Hofmann rearrangement; an isocyanate is an intermediate

8. Anhydrides

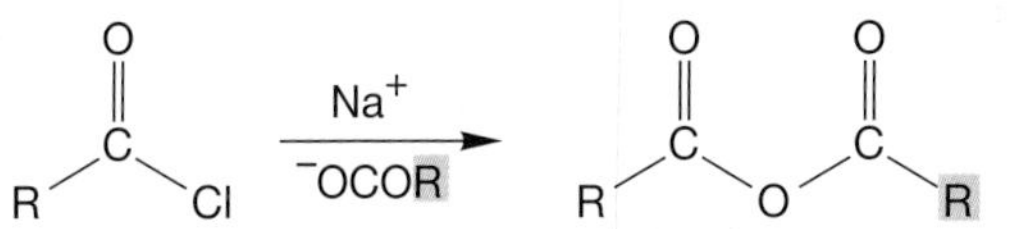

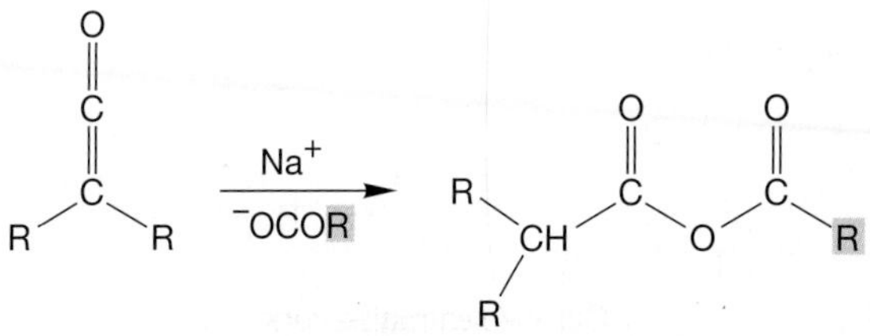

9. Carbamates

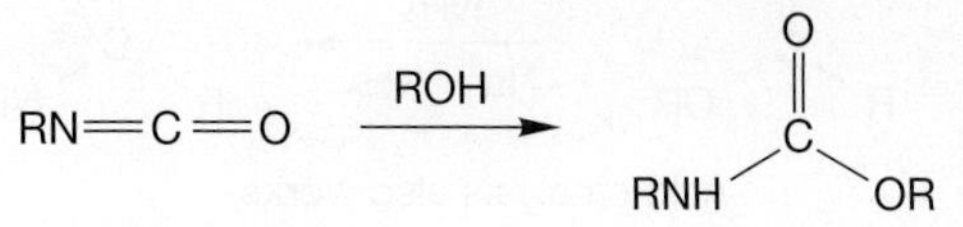

$$RN{=}C{=}O \xrightarrow{ROH}$$

(product: carbamate with RNH and OR groups)

10. Carbonates

$$\xrightarrow{ROH}$$

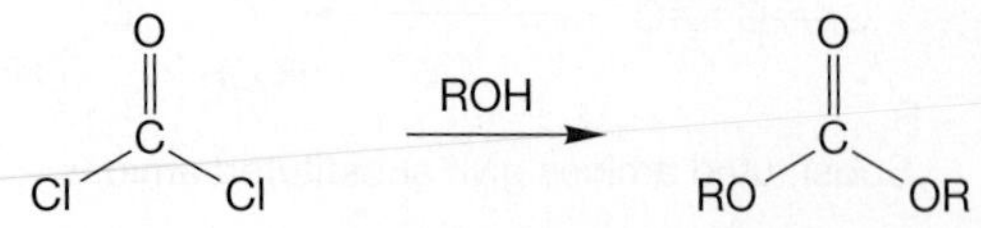

The mixed carbonate cannot be made this way

11. Diazo Ketones

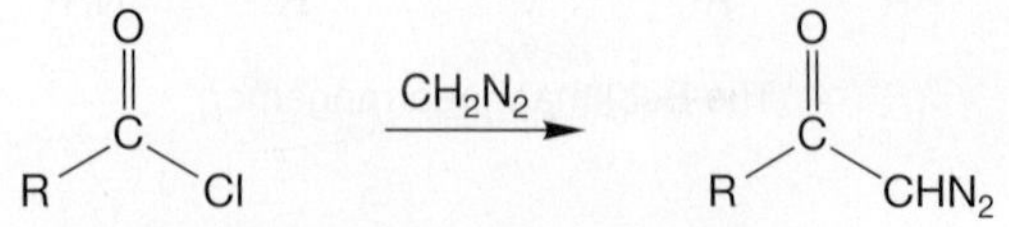

$$\xrightarrow{CH_2N_2}$$

Addition–elimination mechanism followed by removal of an acidic hydrogen

12. Esters

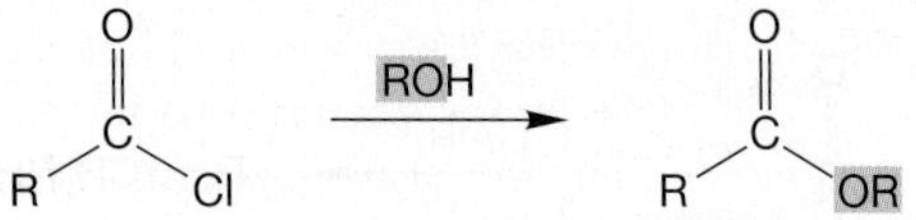

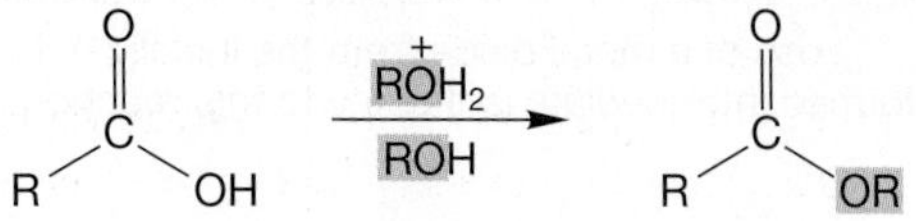

$$\xrightarrow{ROH}$$

$$\xrightarrow[ROH]{ROH_2^+}$$

Fischer esterification

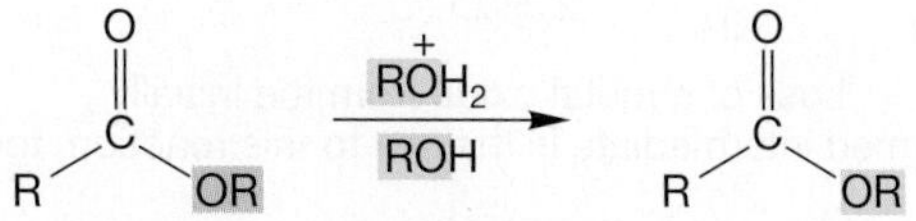

$$\xrightarrow[ROH]{ROH_2^+}$$

Acid-catalyzed transesterification; this transformation also works with base catalysis

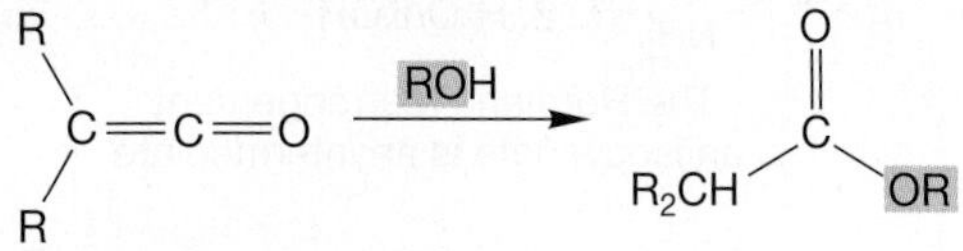

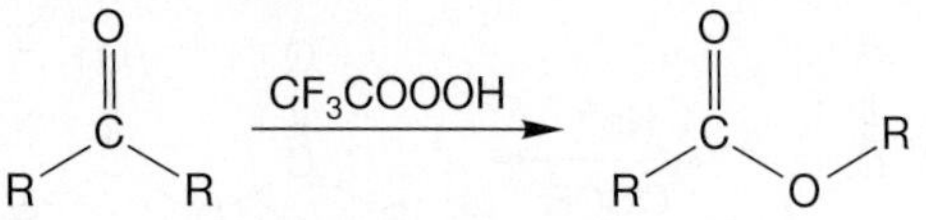

$$\xrightarrow{ROH}$$

$$\xrightarrow{CF_3COOH}$$

The Baeyer–Villiger reaction

13. Isocyanates

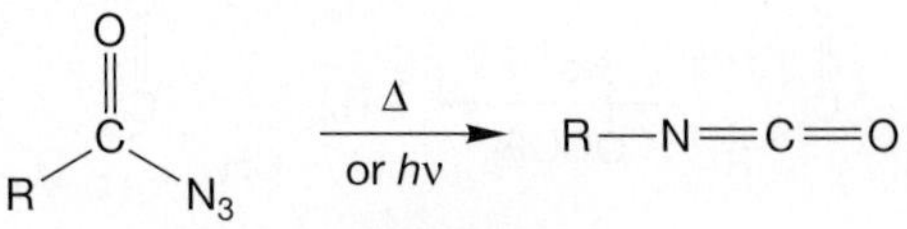

$$\xrightarrow[\text{or } h\nu]{\Delta} R{-}N{=}C{=}O$$

Curtius rearrangement

14. Ketenes

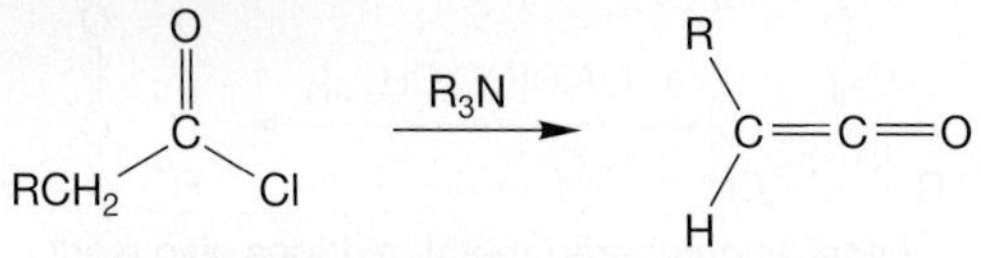

$$\xrightarrow{R_3N}$$

Elimination using a tertiary amine as base

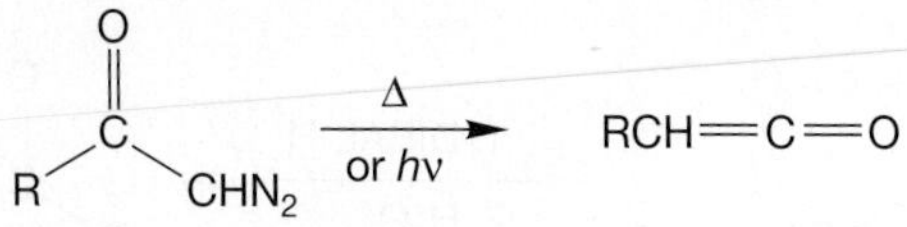

$$\xrightarrow[\text{or } h\nu]{\Delta} RCH{=}C{=}O$$

Wolff rearrangement

15. Ketones

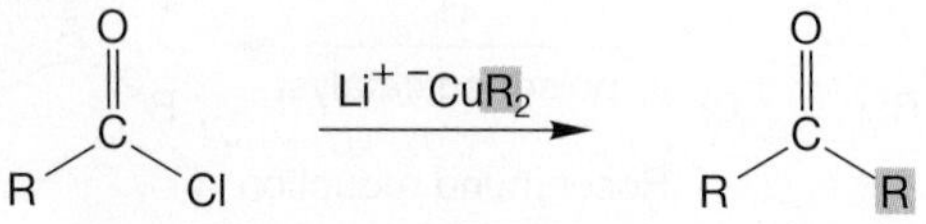

$$\xrightarrow{Li^+ \ ^-CuR_2}$$

The cuprate will react with the acid chloride but not with the product ketone

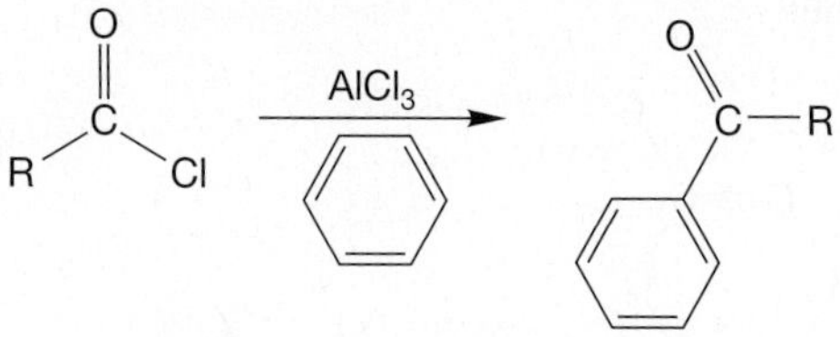

$$\xrightarrow{AlCl_3}$$

Friedel–Crafts acylation; this reaction works with the anhydride as well

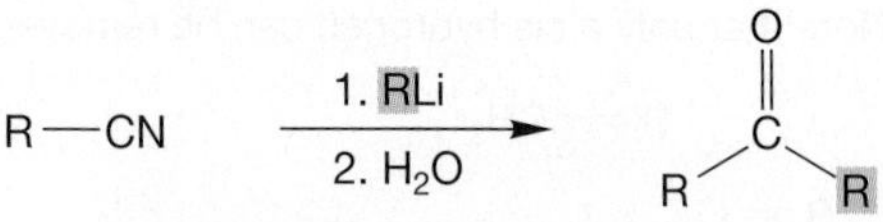

$$R{-}CN \xrightarrow[\text{2. } H_2O]{\text{1. } RLi}$$

The imine is an intermediate

16. Lactams

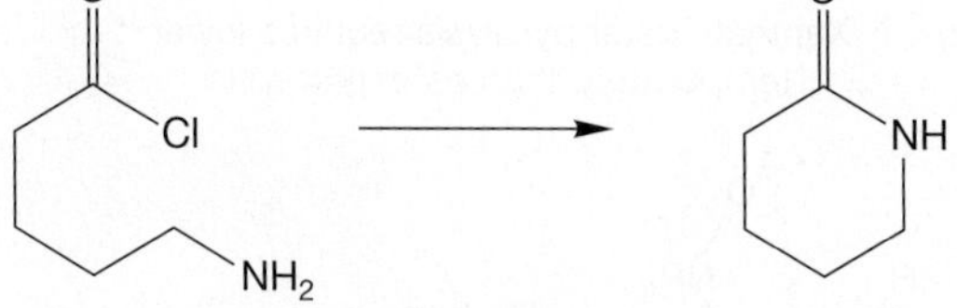

Amide-forming reactions applied in an intramolecular way will work

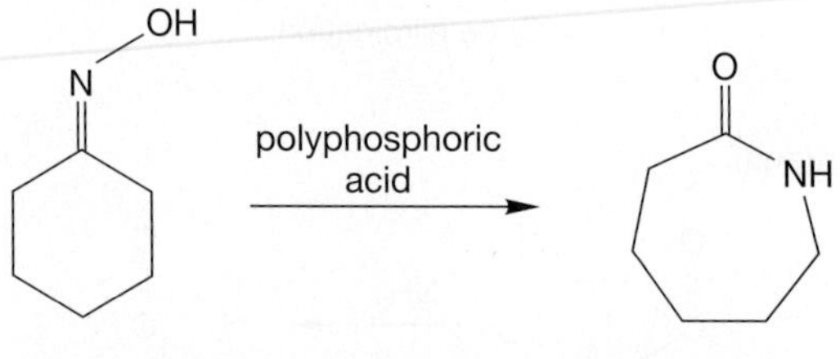

$$\xrightarrow{\text{polyphosphoric acid}}$$

Lactams are formed by the cyclic version of the Beckmann rearrangement

17. Lactones

Ester-forming reactions applied in an
intramolecular way will work

Lactones are formed by the cyclic version
of the Baeyer–Villiger reaction

18. Nitriles

$$NC^- + R-X \longrightarrow NC-R$$

S_N2 displacement

Dehydration of amides

19. Xanthate Esters

Common Errors

Analyze problems before starting. Do not be too proud to do the obvious. Use the molecular formula to see what molecules must be combined. In acid (H_3O^+, HCl, etc.), carbocations are the likely intermediates. In base ($^-$OH, $^-$OR, etc.), carbanions are probably involved. Try to identify what must be accomplished in a problem. Although sometimes it will be hard to plan, and you may have to try all possible intermediates, usually a problem will tell you much of what you must do.

Ask yourself questions and set goals. What rings must be opened? What rings must be closed? What atoms become incorporated in the product? Pay attention to "stop signs." Primary carbocations are stop signs. Unstabilized carbanions are stop signs. If you are forced to invoke such a species in an answer, it is almost certainly wrong, and you must backtrack. Hard problems become easier if you keep such questions in mind. All of this advice sounds obvious, but it's not. Thinking a bit about what must be done in a problem saves much time in the long run and avoids the generally hopeless random arrow pushing that can trap you into a wrong answer.

18.16 Additional Problems

Let's warm up with some simple problems, just to be certain that the basic (and acidic) mechanisms of this chapter are under control. Then we can go on to other things. Please be sure you can do the "easy" problems before you try the more difficult ones at the end of this section.

PROBLEM 18.33 Write mechanisms for the following conversions:

(a)

(b)

PROBLEM 18.34 Write mechanisms for the following conversions in acid:

(a)

(b)

PROBLEM 18.35 Now write the mechanism for a slightly more complicated acid-catalyzed hydrolysis.

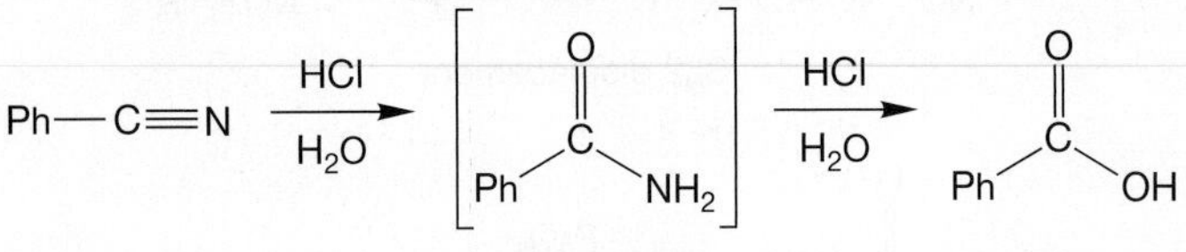

PROBLEM 18.36 Write a mechanism for the acid-catalyzed formation of ethyl hexanoate from ethanol and hexanoic acid.

PROBLEM 18.37 Show a synthetic route to methyl benzoate starting with methanol and toluene. Assume you have access to any other needed reagents.

PROBLEM 18.38 Write Lewis structures for the following compounds:

(a) propanoic acid
(b) propanoyl chloride
(c) *N,N*-diethyl-2-phenylpropanamide
(d) methylketene
(e) propyl cyclopropanecarboxylate
(f) phenyl benzoate

PROBLEM 18.39 Give the IUPAC name for each of the following molecules:

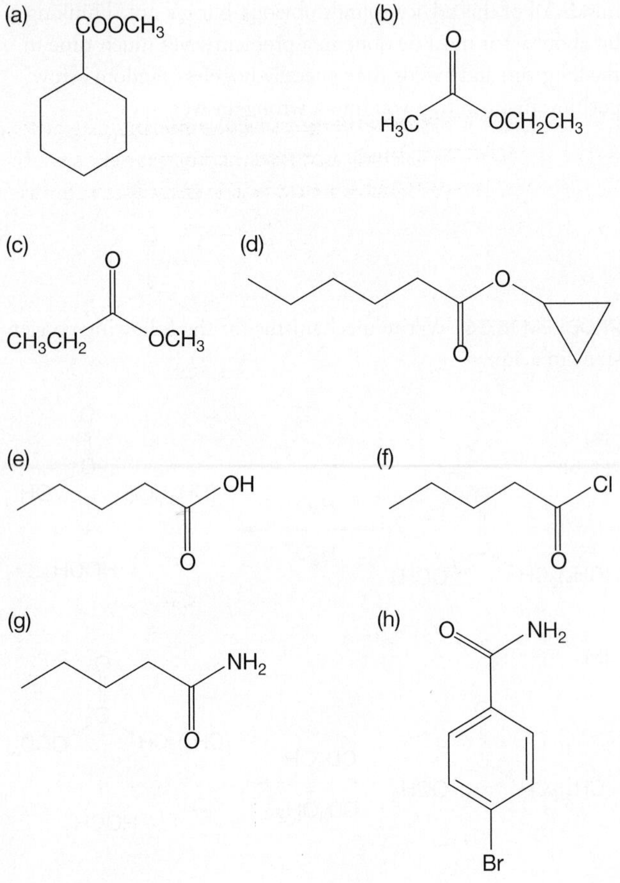

PROBLEM 18.40 Show how you would synthesize *N*-phenyl-1-pentanamine from 1-pentanol, aniline, tosylchloride, and any necessary inorganic reagents.

PROBLEM 18.41 Amine oxides can be formed by the reaction of a tertiary amine with a peroxy acid. Show the synthetic steps you would use to make cyclohexene from cyclohexanamine using any necessary reagents.

PROBLEM 18.42 Provide structures for compounds **A–F** in this series of reactions.

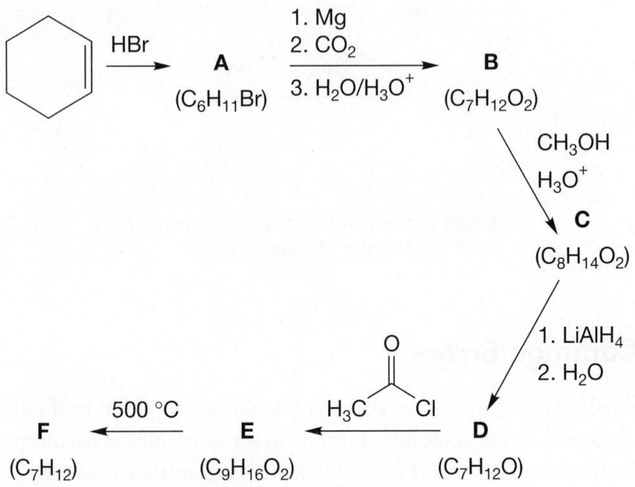

PROBLEM 18.43 Analysis of the IR stretching frequencies of carbonyl compounds involves an assessment of resonance and inductive effects. This assessment is a somewhat risky business, and it must be admitted that near-circular reasoning is sometimes encountered. Nonetheless, see if you can make sense of the following observations taken from Table 18.2. In each case, it will be profitable to think of all the important resonance forms of the acyl compound.

(a) Acetone absorbs at 1719 cm^{-1}, whereas acetaldehyde absorbs at 1733 cm^{-1}.
(b) Explain why methyl acetate absorbs at higher frequency (1750 cm^{-1}) than acetone (1719 cm^{-1}).
(c) *N*-Methylacetamide absorbs at much lower frequency (1688 cm^{-1}) than methyl acetate (1750 cm^{-1}).

PROBLEM 18.44 Is your explanation in Problem 18.43 (b) consistent with the observation that the carbonyl carbon of methyl acetate appears at δ 169 ppm in the ^{13}C NMR spectrum, whereas the carbonyl carbon of acetone is far downfield at δ 205 ppm? Explain.

PROBLEM 18.45 Give the major organic products expected in each of the following reactions or reaction sequences.

(a)

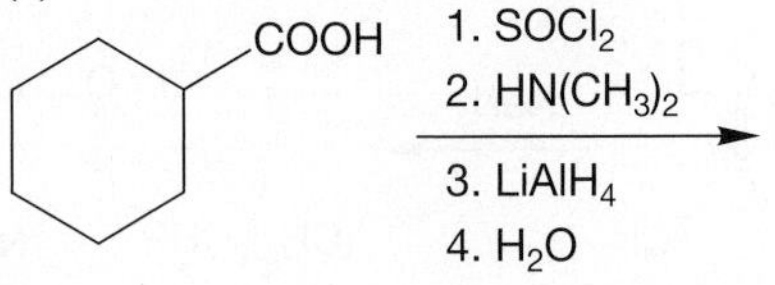

1. $SOCl_2$
2. $HN(CH_3)_2$
3. $LiAlH_4$
4. H_2O

(b)

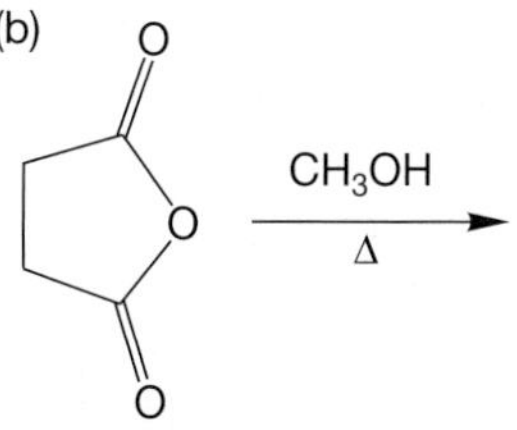

CH₃OH
Δ

(c)

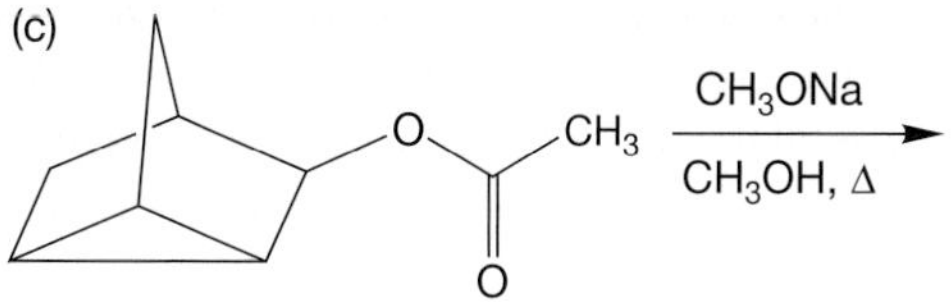

CH₃ONa
CH₃OH, Δ

(d)

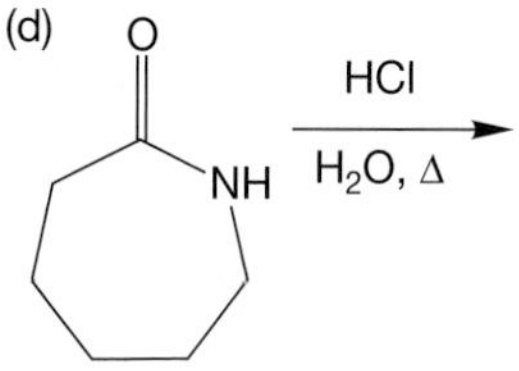

HCl
H₂O, Δ

(e)

1. NaH
2. CS₂
3. CH₃I
4. 210–235 °C

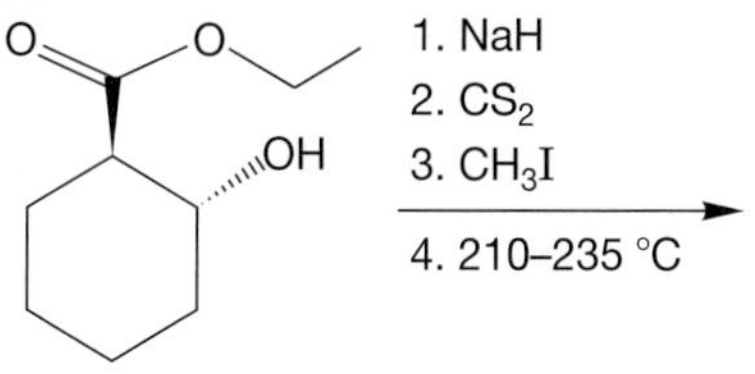

PROBLEM 18.46   We learned in Chapter 17 that most β-keto acids decarboxylate very easily (p. 858). The compound below is an exception, as it survives heating to very high temperature without losing carbon dioxide. Explain.

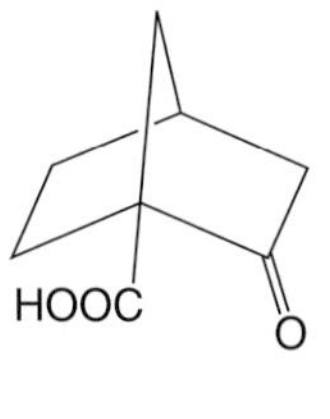

PROBLEM 18.47   In Section 18.12c, we saw that primary amides could be dehydrated to nitriles (cyanides) with P₂O₅. Nitriles can also be prepared by treating aldoximes (**1**) with dehydrating agents such as acetic anhydride. Propose a mechanism for the formation of benzonitrile from the reaction of **1** with acetic anhydride. How are oximes prepared?

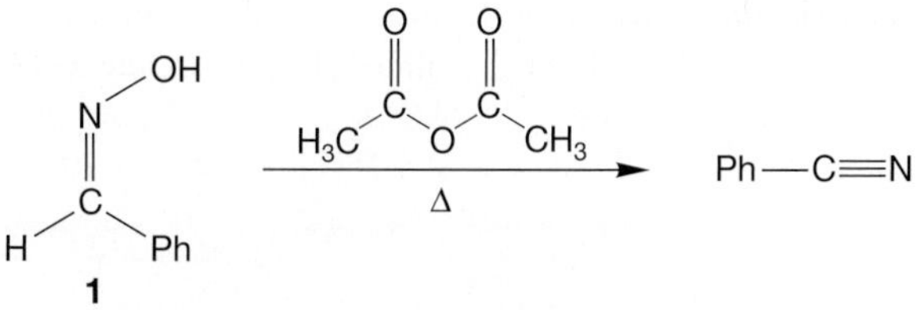

PROBLEM 18.48   Reaction of benzonitrile (**1**) and *tert*-butyl alcohol in the presence of concentrated sulfuric acid, followed by treatment with water, gives *N-tert*-butylbenzamide (**2**). Provide an arrow formalism mechanism for this reaction.

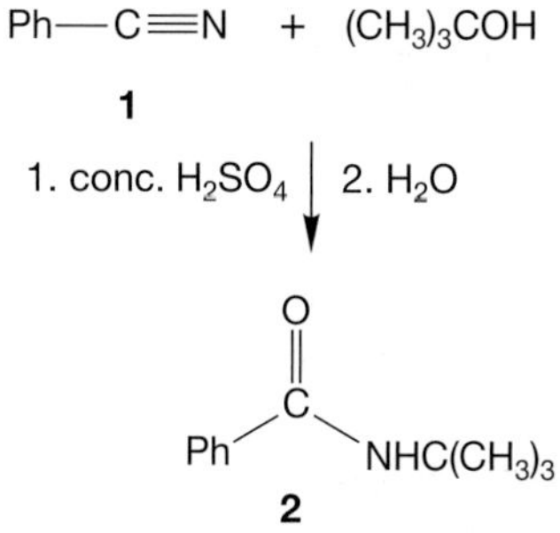

Ph—C≡N   +   (CH₃)₃COH

1

1. conc. H₂SO₄ | 2. H₂O

2

PROBLEM 18.49   Devise syntheses for the following molecules. You may start with benzene, methyl alcohol, sodium methoxide, ethyl alcohol, sodium ethoxide, butyl alcohol (BuOH), phosgene, and propanoic acid as your sources of carbon. You may also use any inorganic reagent and solvents as needed.

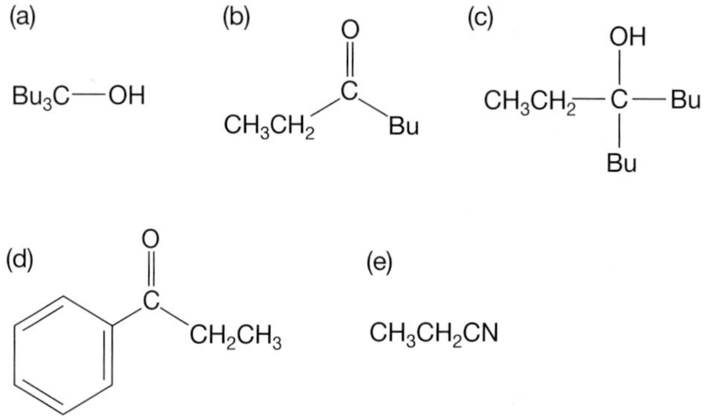

(a)   Bu₃C—OH

(b)

(c)

(d)

(e)   CH₃CH₂CN

PROBLEM 18.50   Propose syntheses for the following target molecules starting from the indicated material. You may use any other reagents that you need. Use a retrosynthetic analysis in each case.

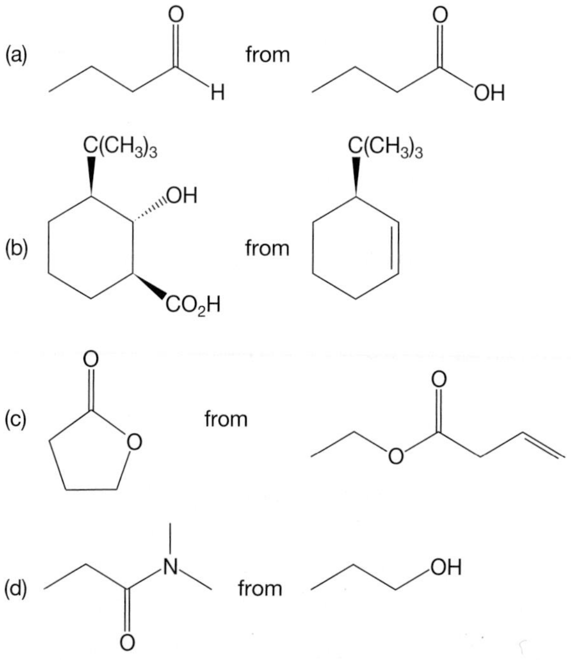

(a) from

(b) from

(c) from

(d) from

PROBLEM 18.51   Propose syntheses of the following molecules starting from cyclopentanone:

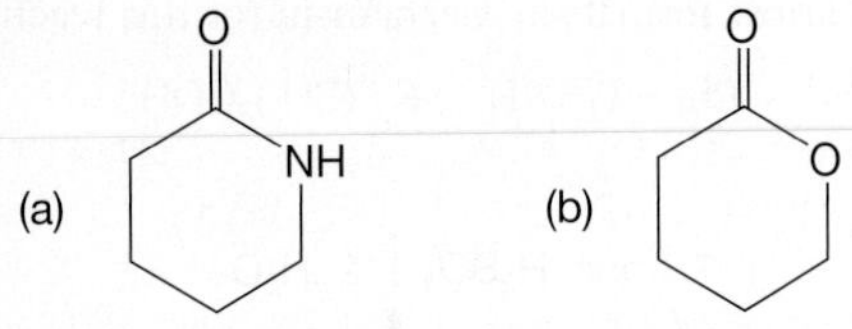

(a)   (b)

PROBLEM 18.52   Provide an arrow formalism mechanism for the following reaction:

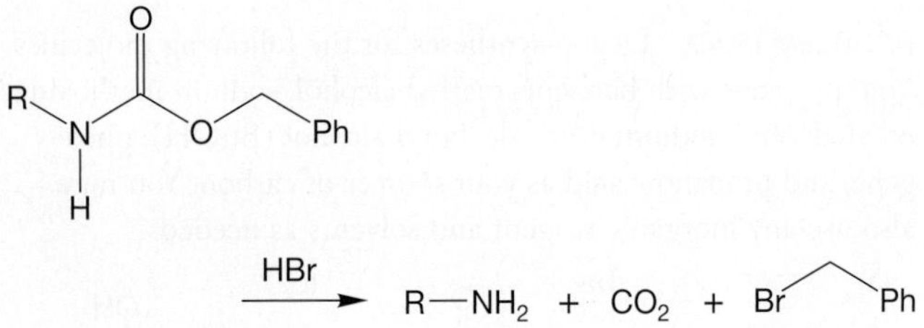

$$\xrightarrow{\text{HBr}} \text{R}-\text{NH}_2 \; + \; \text{CO}_2 \; + \; \text{Br}\!\!\!\diagup\!\!\!\diagup\text{Ph}$$

PROBLEM 18.53   Here is the problem concerning the formation of the Vilsmeier reagent we promised you in Chapter 17. As we saw in Section 17.7d, acid chlorides can be prepared from carboxylic acids and reagents such as thionyl chloride, phosphorus pentachloride, phosgene, and oxalyl chloride. The conditions for these reactions are sometimes quite severe. It is possible to prepare acid chlorides under appreciably milder conditions by using the Vilsmeier reagent (Problem 17.42). The Vilsmeier reagent is prepared from DMF and oxalyl chloride. Write an arrow formalism mechanism for its formation. *Hint*: DMF acts as a nucleophile, but not via the nitrogen atom.

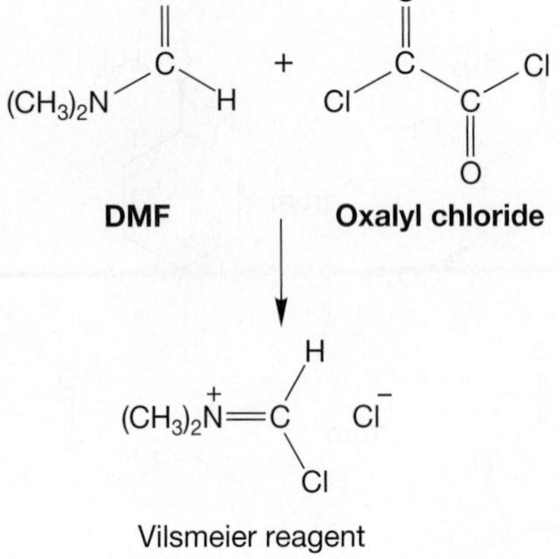

DMF   **Oxalyl chloride**

Vilsmeier reagent

PROBLEM 18.54   In Section 18.8, we saw that aldehydes are available from the reduction of esters at low temperature ($-70\ ^\circ$C) by diisobutylaluminum hydride (DIBAL-H; see Fig. 18.38). If the reaction mixture is allowed to warm before hydrolysis, the yield of aldehyde drops appreciably. For example, reduction of ethyl butyrate with DIBAL-H at $-70\ ^\circ$C followed by hydrolysis at $-70\ ^\circ$C, affords a 90% yield of butyraldehyde. However, if the reaction mixture is allowed to warm to $0\ ^\circ$C before hydrolysis, the yield of aldehyde is less than 20%. Among the products under these conditions are ethyl butyrate (recovered starting material) and butyl alcohol ($\text{CH}_3\text{CH}_2\text{CH}_2\text{CH}_2\text{OH}$), formed from hydrolysis of $\text{CH}_3\text{CH}_2\text{CH}_2\text{CH}_2\text{O}-\text{AlR}_2$ (R = isobutyl). Rationalize these observations with arrow formalism mechanisms.

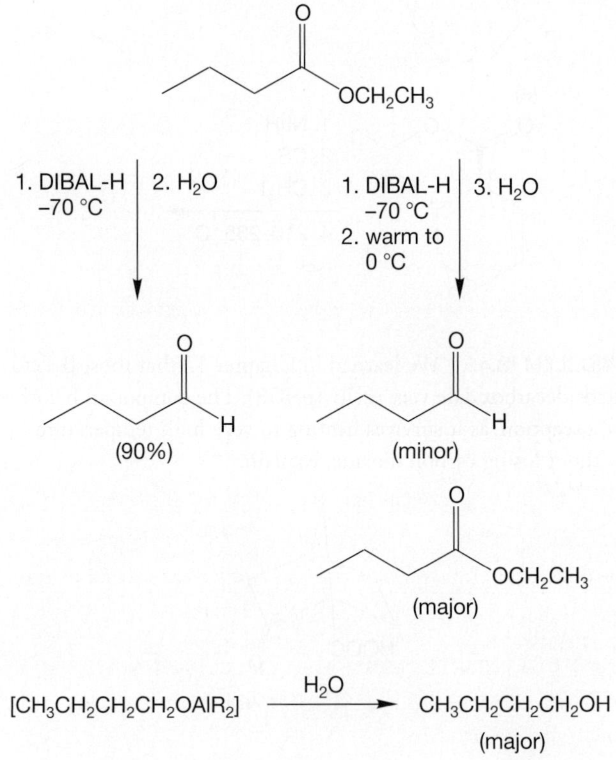

1. DIBAL-H   2. H_2O
$-70\ ^\circ$C

1. DIBAL-H   3. H_2O
$-70\ ^\circ$C
2. warm to $0\ ^\circ$C

(90%)   (minor)

(major)

$$[\text{CH}_3\text{CH}_2\text{CH}_2\text{CH}_2\text{OAlR}_2] \xrightarrow{\text{H}_2\text{O}} \text{CH}_3\text{CH}_2\text{CH}_2\text{CH}_2\text{OH}$$
(major)

PROBLEM 18.55   γ-Aminobutyrate transaminase (GABA-T) is the key enzyme controlling the metabolism of γ-aminobutyric acid (GABA), a mammalian inhibitory neurotransmitter. In the following synthesis of the hydrochloride salt of the putative GABA-T inhibitor, 3-amino-1,4-cyclohexadiene-1-

carboxylic acid (**G**), propose structures for intermediates **A**–**F**. Mechanisms are not necessary, but may be helpful in some cases.

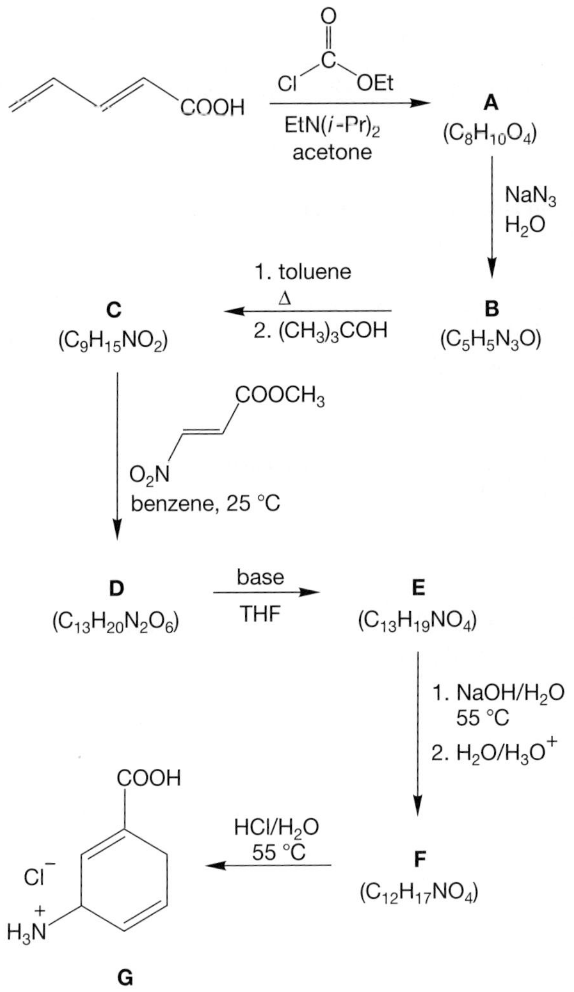

PROBLEM 18.56 Treatment of phthalimide (**1**) with bromine in aqueous sodium hydroxide, followed by acidification with acetic acid, affords compound **2**. A summary of spectral data for **2** follows. Deduce the structure of **2**, and then propose a mechanism for its formation.

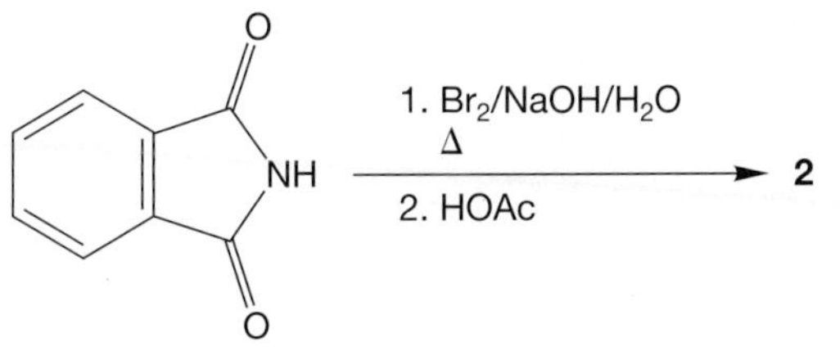

1

Compound **2**

Mass spectrum: m/z = 137 (M, 59%), 119 (100%), 93 (79%), 92 (59%)

IR (KBr): 3490 (m), 3380 (m), 3300–2400 (br), 1665 (s), 1245 (s), and 765 (m) cm^{-1}

^{1}H NMR (DMSO-d_6):[5] δ 6.52 (t, J = 8 Hz, 1H)
6.77 (d, J = 8 Hz, 1H)
7.23 (t, J = 8 Hz, 1H)
7.72 (d, J = 8 Hz, 1H)
8.60 (br s, 3H, vanishes with D$_2$O)

^{13}C NMR (DMSO-d_6): δ 109.5 (s), 114.5 (d), 116.2 (d), 131.1 (d), 133.6 (d), 151.4 (s), 169.5 (s)

PROBLEM 18.57 Sketch the transition state for the Cope elimination (p. 914) shown below.

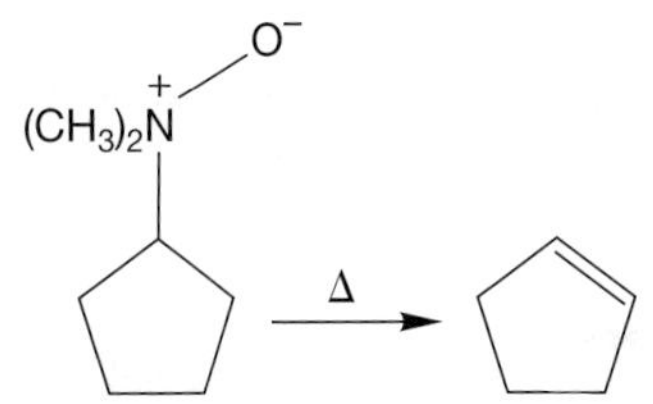

PROBLEM 18.58 Design an experiment to test the stereochemistry of the Cope elimination. How would one determine that it is a cis hydrogen that is lost? Assume you can make any labeled compounds you need.

PROBLEM 18.59 Write a mechanism for the Schmidt reaction shown below. *Hint*: The first step is probably the formation of an acylium ion, R—$\overset{+}{\text{C}}$=O. Note also that the reagent is HN$_3$, not NH$_3$.

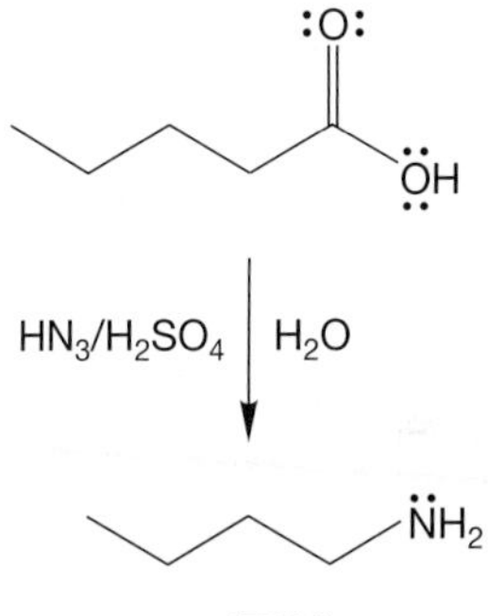

(72%)

---

[5] Note that J_{meta} and J_{para} were not observed under these conditions.

PROBLEM 18.60   Provide structures for compounds **A–E**. Mechanisms are not necessary.

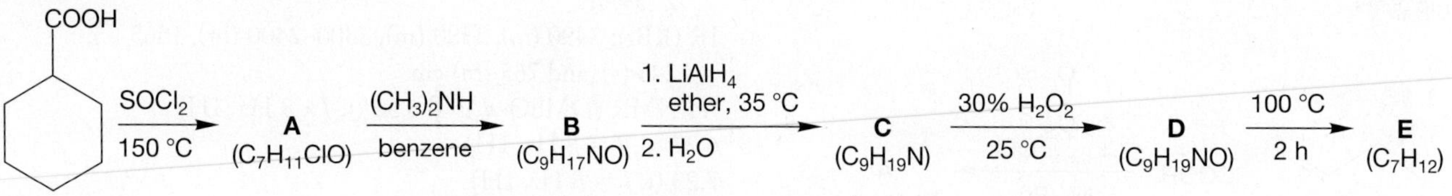

PROBLEM 18.61   Reaction of 1,3-diphenylisobenzofuran (**1**) and vinylene carbonate (**2**) gives compound **3**. Acid hydrolysis of **3** affords **4**. Propose a structure for **3** and write a mechanism for its formation and hydrolysis to **4**.

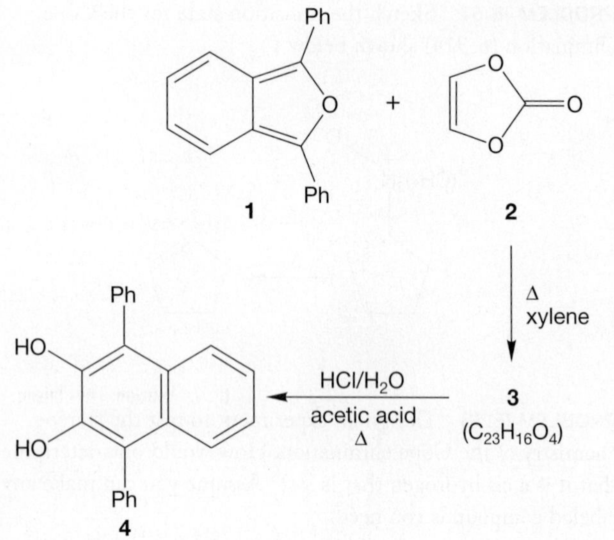

PROBLEM 18.62   Write an arrow formalism mechanism for the following reaction:

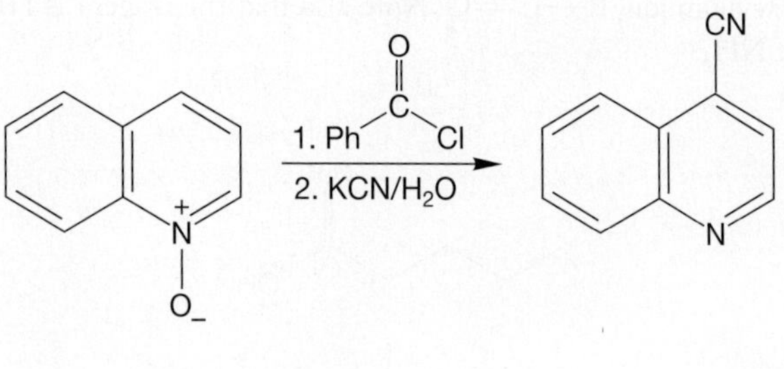

Use Organic Reaction Animations (ORA) to answer the following questions:

PROBLEM 18.63   Select the reaction "Ester hydrolysis" and click on the play button. Note the products of the reaction. Is this reaction catalyzed by base, or does this reaction require a stoichiometric amount of base?

PROBLEM 18.64   Observe the LUMO animation of the "Ester hydrolysis." Describe the LUMO of the starting ester (methyl methanoate). On what atom is the LUMO density concentrated? Where else is there LUMO density? Why?

PROBLEM 18.65   Select the reaction "Nitrile hydrolysis" and watch the LUMO animation. Stop the animation at the first intermediate. Notice that the first molecular orbital animated is along the C—N axis. Then the animation shifts to a molecular orbital that is perpendicular to the C—N axis. These orbitals are very close in energy. Why does the animation shift to the second LUMO? Does this mean that a nucleophile (a water molecule in this case) has more than one option for reaction with the first intermediate?

Carbonyl Chemistry 2: Reactions at the α Position

MAKING CARBON–CARBON BONDS All plants and animals such as these fish, the coral, and the diver use enol chemistry in the citric acid cycle as a way to make carbon–carbon bonds.

19.1 Preview

19.2 Many Carbonyl Compounds Are Weak Brønsted Acids

19.3 Racemization of Enols and Enolates

19.4 Halogenation in the α Position

19.5 Alkylation in the α Position

19.6 Addition of Carbonyl Compounds to the α Position: The Aldol Condensation

19.7 Reactions Related to the Aldol Condensation

19.8 Addition of Acid Derivatives to the α Position: The Claisen Condensation

19.9 Variations on the Claisen Condensation

19.10 Special Topic: Forward and Reverse Claisen Condensations in Biology

19.11 Condensation Reactions in Combination

19.12 Special Topic: Alkylation of Dithianes

19.13 Special Topic: Amines in Condensation Reactions, the Mannich Reaction

19.14 Special Topic: Carbonyl Compounds without α Hydrogens

19.15 Special Topic: The Aldol Condensation in the Real World, an Introduction to Modern Synthesis

19.16 Summary

19.17 Additional Problems

> I want to beg you, as much as I can, dear sir, to be patient towards all that
> is unsolved in your heart and try to love the *questions themselves*.
>
> —RAINER MARIA RILKE,[1] *LETTERS TO A YOUNG POET*

19.1 Preview

In this chapter, we explore the chemistry of the α position in the carbonyl compounds we met in Chapters 16–18. Our ability to understand complex reactions
and to construct complicated molecules will greatly increase. There really is little fundamentally new chemistry—much of what we find from now on will be
applications of reactions we already know, placed in more complicated settings.
For example, in Chapter 16 we saw a vast array of additions of nucleophiles to
carbon–oxygen double bonds. Here that same reaction appears, when a new nucleophile, an enolate, adds to carbonyl groups to yield products of increased complexity. The reaction is the same old addition, but the context makes it look
complicated (Fig. 19.1).

FIGURE 19.1 Additions of
nucleophiles to carbonyl compounds.

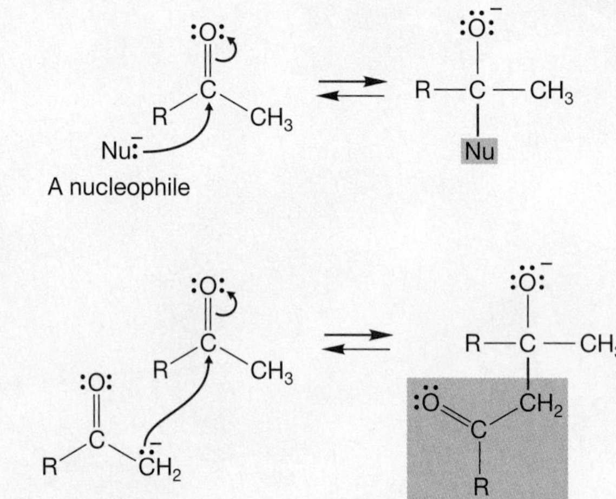

ESSENTIAL SKILLS AND DETAILS

1. Two reactions critical to understanding the material in this chapter are enol formation
   in acid and enolate formation in base. Be sure you understand these two basic reactions
   completely.

2. The central reactions of this chapter are the aldol and Claisen condensations. Both
   acid- and base-catalyzed aldol reactions exist. The Claisen condensation only succeeds
   under basic conditions.

3. Although both the acid- and base-catalyzed aldol condensations lead to the same
   initial product, a β-hydroxy carbonyl compound, the reaction usually goes further
   in acid (sometimes in base) to give an α,β-unsaturated carbonyl compound. All
   β-hydroxy carbonyl compounds and all α,β-unsaturated carbonyl compounds can,
   in principle, be made through aldol or closely related condensations. This observation
   is an important problem-solving tool. Always look for the α,β-unsaturated carbonyl
   compound.

---

[1] Rilke (1875–1926) was a German lyric poet much influenced by the French sculptor, Rodin. The italics are
Rilke's.

4. Watch out for the addition of nucleophiles to α,β-unsaturated carbonyl compounds to give enolates. This process, called the Michael reaction, is very common in Nature, in organic synthesis, and in organic chemistry problems.

5. The synthetic procedures known as the acetoacetate synthesis and the malonic ester synthesis are very useful. Acetoacetate can be used to make ketones and malonic esters can be used to make carboxylic acids.

6. Alkylation of enolates provides an excellent route for making carbon–carbon bonds.

19.2 Many Carbonyl Compounds Are Weak Brønsted Acids

One of the remarkable things about carbonyl compounds is that they are not only electrophiles (see the addition reactions throughout Chapters 16–18 and Fig. 19.1), but are proton donors as well.

19.2a Enolates of Ketones and Aldehydes
The pK_a values of typical aldehydes and ketones are generally in the high teens (Table 19.1). It is the hydrogen at the **α position**, the position adjacent to the carbon–oxygen double bond, that is lost to a base as a proton (Fig. 19.2). Our first task is to see why these compounds are weak acids at their α positions, and then to see what new chemistry results from this acidity.

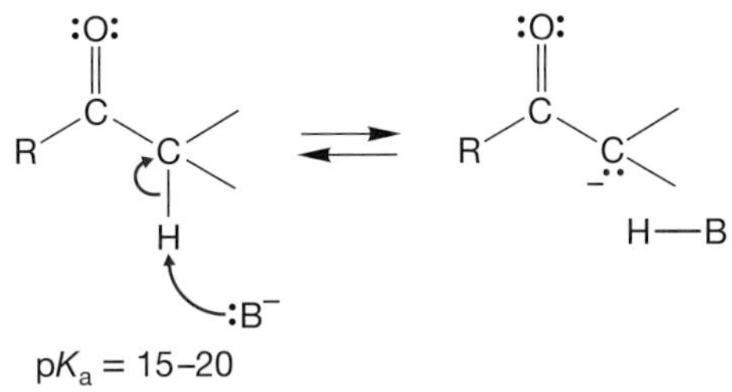

TABLE 19.1 Some pK_a Values for Simple Ketones and Aldehydes

Compound[a]	pK_a
$CH_3CH_2COCH_2CH_3$	19.9
CH_3COCH_3	19.3
$PhCOCH_3$	18.3
$PhCH_2COCH_3$	18.3
$PhCH_2COCH_3$	15.9
Cyclohexanone (α hydrogen)	18.1
CH_3CHO	16.7
$(CH_3)_2CHCHO$	15.5

[a]The proton to be lost is underlined when there is a choice.

$$pK_a = 15\text{--}20$$

FIGURE 19.2 Carbonyl compounds bearing hydrogen at the α position are weak acids, with pK_a values in the high teens.

We'll start with butanal ($pK_a = 16.7$) and look at the three possible anions we might form from it by breaking an $sp^3/1s$ carbon–hydrogen bond (Fig. 19.3). First of all, the dipole in the carbon–oxygen bond will stabilize an adjacent anion more than a more remote anion. Loss of the α hydrogen will lead to a more stable anion

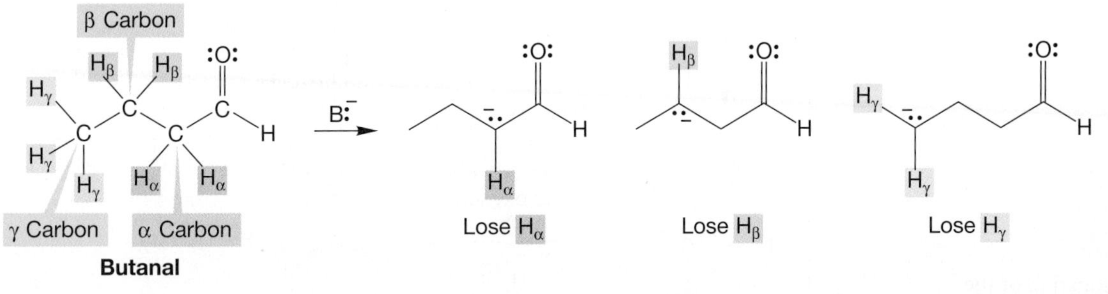

FIGURE 19.3 There are three possible anions that can be formed from butanal by breaking an $sp^3/1s$ carbon–hydrogen bond.

than loss of either the β or γ hydrogen (Fig. 19.4). Moreover, one of these anions, but only one of them, is stabilized by the resonance resulting from the oxygen atom

FIGURE 19.4 The dipole in the carbon–oxygen bond will stabilize an adjacent anion more than a more distant anion.

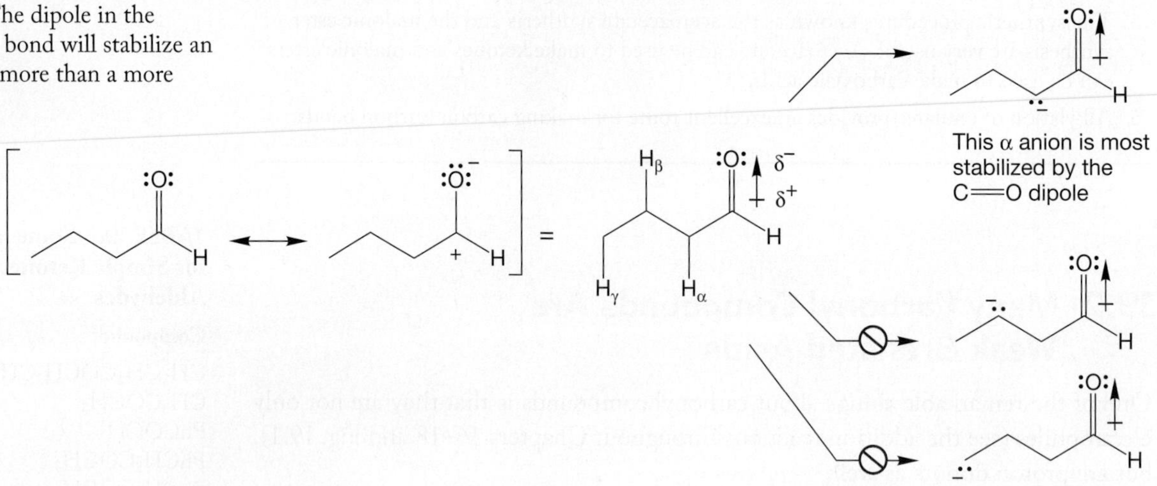

This α anion is most stabilized by the C=O dipole

sharing the negative charge (Fig. 19.5). It is this stabilization that makes the α position, but only the α position, a proton source. The intermediate is called an **enolate** anion, because it is part alkene (*ene*), part alcohol (*ol*), and an anion (*ate*). Please note again that the enolate anion is *not* "part of the time an alkoxide" and "part of the time a carbanion," but all of the time a single species, the resonance-stabilized enolate.

FIGURE 19.5 Loss of the α hydrogen leads to a resonance-stabilized enolate anion.

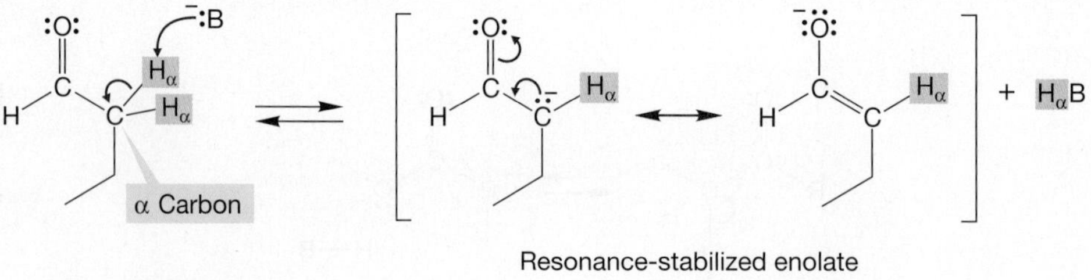

Resonance-stabilized enolate

The enolate anion can reprotonate in two ways. If it reacts with a proton source at carbon, the original carbonyl compound is regenerated (Fig. 19.6a). If it protonates at

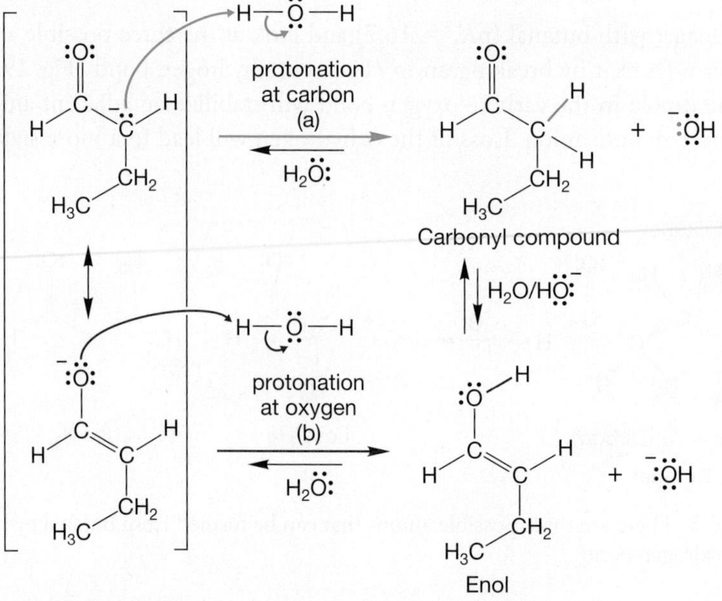

FIGURE 19.6 Reprotonation of the enolate can lead either to (a) the original carbonyl compound or to (b) the related enol.

oxygen, the neutral enol (p. 448) is formed (Fig. 19.6b). In aqueous base, carbonyl compounds are in equilibrium with their enol forms.

Watch out! Arrows can emanate from either one of a pair of resonance forms. One can always write an acceptable arrow formalism from any legitimate resonance form. You should be able to write an arrow formalism using either resonance form.

WORKED PROBLEM 19.1 Write a mechanism for the base-catalyzed equilibration of the carbonyl and enol forms of acetone.

ANSWER This problem is not hard, but it needs to be done right now. To be able to do the more complicated material that will appear later, these simple interconversions must become second nature to you. This problem is drill, but it is nevertheless very important.

Loss of either the α hydrogen from acetone or the hydroxyl hydrogen of the enol form of acetone leads to the same enolate. Protonation at carbon gives the ketone; protonation at oxygen gives the enol. It is formation of the enolate that allows the equilibration to take place.

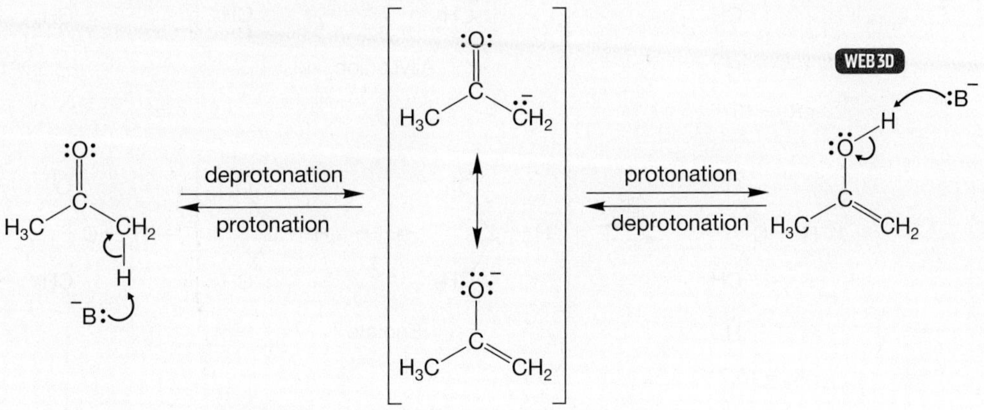

The orbital picture of Figure 19.7 clearly shows the source of the stabilization of the enolate. The three $2p$ orbitals overlap to form an allyl-like system (p. 373).

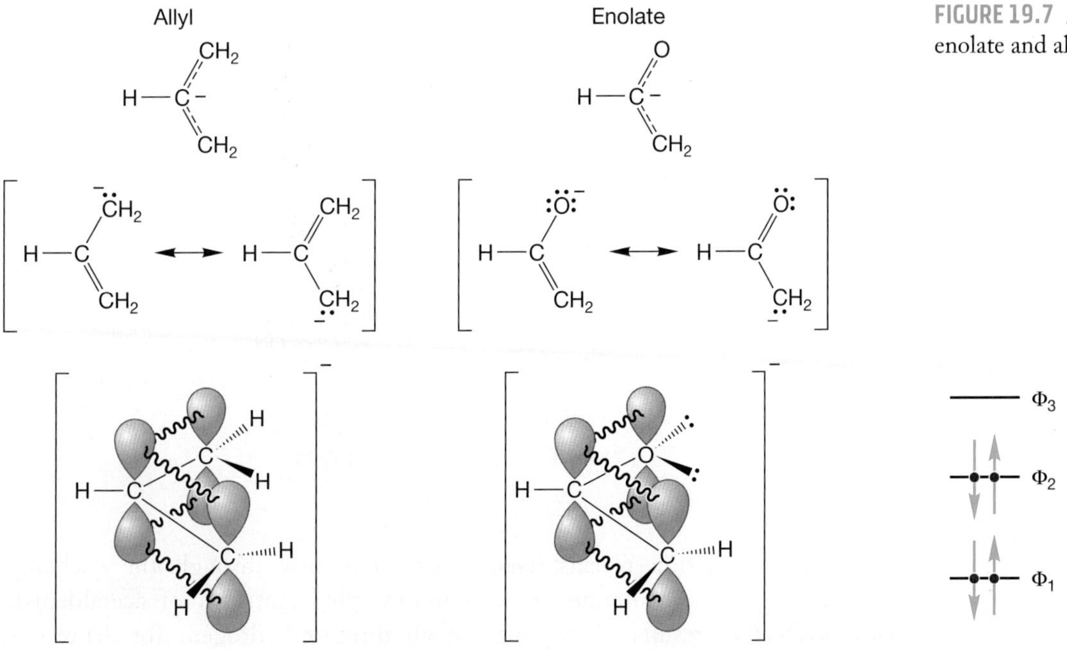

FIGURE 19.7 A comparison of the enolate and allyl anions.

Two of the four π electrons are accommodated in the lowest bonding molecular orbital, Φ_1, and two in the nonbonding orbital, Φ_2. The presence of the oxygen atom makes the enolate less symmetrical than an allyl system formed from three carbon $2p$ orbitals. The bonding orbital (Φ_1) of the enolate has its highest electron density on the oxygen atom. The electron density resides mostly on the carbon of the enolate in the Φ_2 orbital, the HOMO. The carbon atom is the usual nucleophilic site because it is the electrons in Φ_2 that react with an electrophile.

How important is the electronegative oxygen atom to the stabilization of the enolate? The pK_a of propene is 43 and that of acetaldehyde is 16.7. There is a 10^{26} difference in acidity in the two molecules, which can be attributed to the influence of the oxygen (Fig. 19.8).

FIGURE 19.8 The oxygen atom of the enolate plays a crucial role in promoting the acidity at the α position. Acetaldehyde is much more acidic than propene.

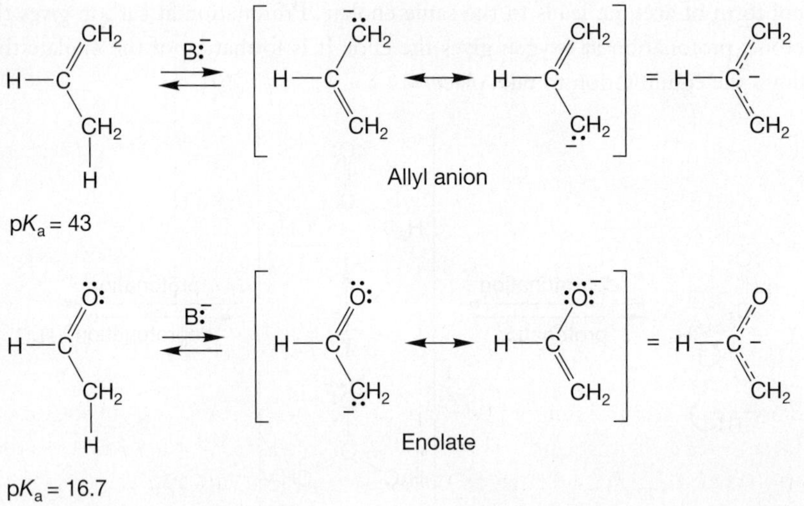

WORKED PROBLEM 19.2 Propanal can form two enols. What are they?

ANSWER There are stereochemical differences between the two enols. Like any other unsymmetrical alkene, this enol can exist in Z or E forms.

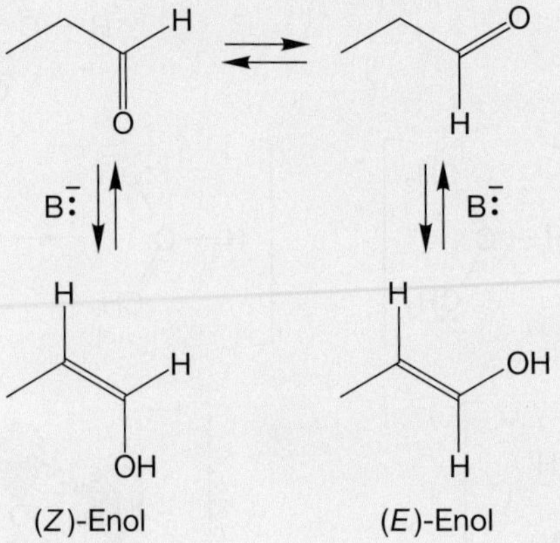

Enolate formation reveals itself most commonly through the exchange of α hydrogens in deuterated base. For example, treatment of acetaldehyde with D_2O/DO^- results in exchange of all three α hydrogens for deuterium

(Fig. 19.9). In order to understand this exchange process we need to consider the reversibility of the deprotonation reaction. In hydroxide (here deuteroxide), not all of the acetaldehyde will be transformed into the enolate. The pK_a of acetalde-

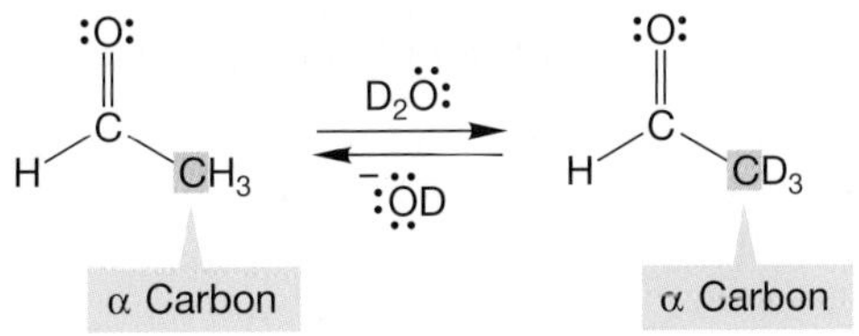

FIGURE 19.9 In D_2O/DO^-, the three α hydrogens of acetaldehyde are exchanged for deuterium.

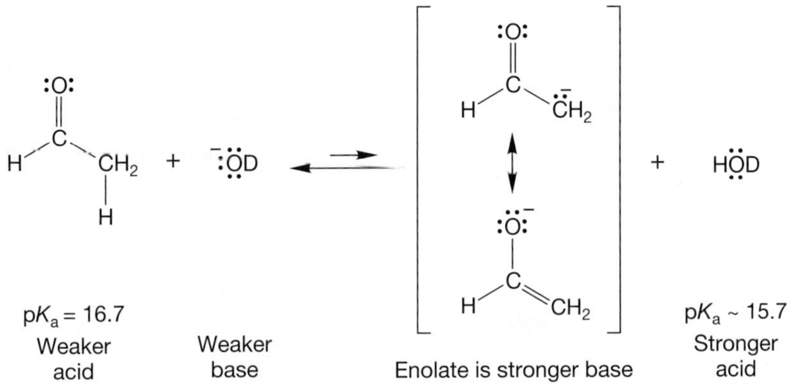

hyde is 16.7 and that of water is 15.7. So, hydroxide is a *weaker* base than the enolate anion and at equilibrium only a small amount of the enolate anion will be present. Enolate formation is endothermic in this case (Fig. 19.10).

FIGURE 19.10 Enolate formation is an equilibrium reaction and is endothermic in the case of acetaldehyde.

$pK_a = 16.7$
Weaker acid

Weaker base

Enolate is stronger base

$pK_a \sim 15.7$
Stronger acid

These exchange reactions are typically run using catalytic amounts of base, DO^-, in a large excess of D_2O. Under such conditions only a small amount of enolate can be formed. However, any enolate present will react with the available water, in this case D_2O, to remove a deuteron and produce a molecule of exchanged acetaldehyde (Fig. 19.11). A new molecule of DO^- that can carry the reaction further is also formed. Eventually all the available α hydrogens will be exchanged for deuterons.

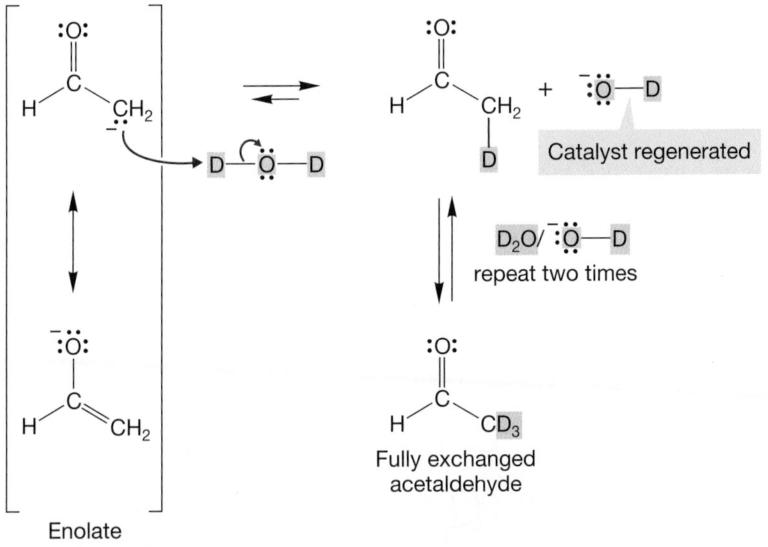

Catalyst regenerated

$D_2O/ \; ^-\ddot{O}-D$
repeat two times

Enolate

Fully exchanged acetaldehyde

FIGURE 19.11 The exchange reaction is a catalytic process, with deuteroxide ion (^-OD) acting as the catalyst.

PROBLEM 19.3 Explain why the aldehydic hydrogen in acetaldehyde (the one bonded to the carbonyl carbon) does not exchange in D_2O/DO^-.

PROBLEM 19.4 Explain why the bicyclic ketone in the following reaction exchanges only the two α hydrogens shown (H$_\alpha$) and not the bridgehead hydrogen, which is also "α." *Hint*: Use models and examine the relationship between the bridgehead carbon–hydrogen bond and the π orbitals of the carbonyl group.

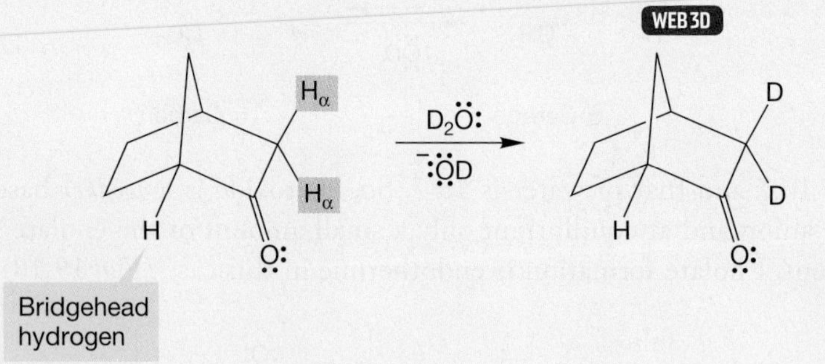

Bridgehead hydrogen

WEB 3D

The α hydrogens of acetaldehyde can also be exchanged under acidic conditions (Fig. 19.12). This result illustrates a general principle: In carbonyl chemistry, there will usually (not quite always) be an acid-catalyzed version for every base-catalyzed reaction.

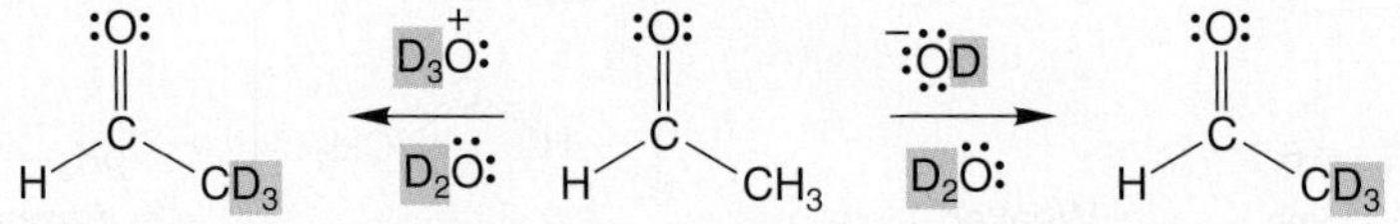

FIGURE 19.12 Exchange of the α hydrogens of acetaldehyde can also be carried out in deuterated acid, D_3O^+/D_2O.

The mechanisms of the two reactions will be different, however, even though the end results are similar or even identical. For example, in this acid-catalyzed exchange reaction there is no base strong enough to remove an α hydrogen from the starting acetaldehyde. The only reasonable reaction is protonation of the Brønsted basic oxygen by D_3O^+ to give a resonance-stabilized cation (Fig. 19.13).

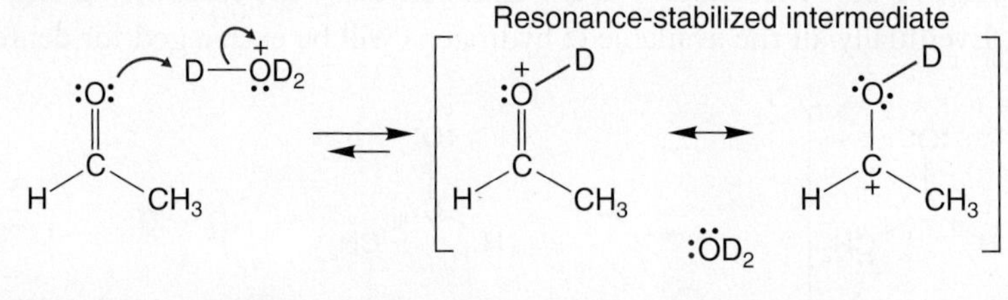

FIGURE 19.13 The first step in the acid-catalyzed exchange is addition of a deuteron to the carbonyl oxygen. A resonance-stabilized cation results.

Now we have created a powerfully acidic species, the protonated carbonyl compound. Removal of a deuteron from the oxygen simply reverses the reaction to regenerate acetaldehyde and D_3O^+, but deprotonation at the α carbon generates the enol (Fig. 19.14). Note that the product is not an anionic *enolate*, but the neutral *enol*. If this enol regenerates the carbonyl compound in D_3O^+, exchanged acetaldehyde will result.

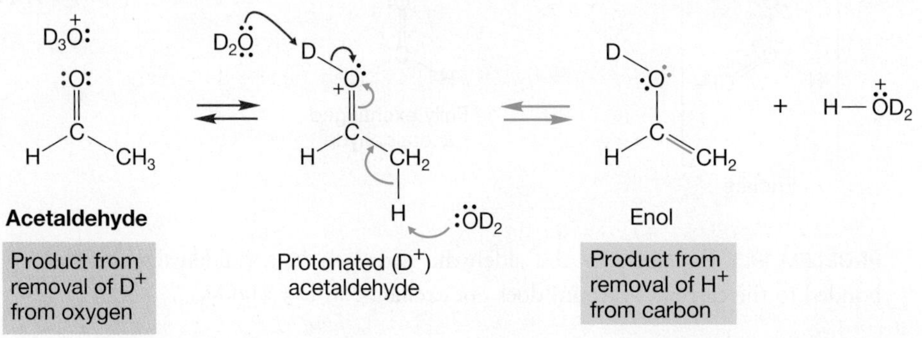

FIGURE 19.14 Removal of a proton from carbon (green) generates the neutral enol form. Removal of a deuteron from oxygen (red) regenerates starting material.

Notice that the steps in the enol → acetaldehyde reaction are simply the reverse of the acetaldehyde → enol reaction (Fig. 19.15). Note also that in acid, as in base, aldehydes and ketones that have α hydrogens are in equilibrium with their enol forms. We will soon see that although enols are in equilibrium with the related keto forms, it is usually the keto forms that are favored. This equilibrium is called the **keto–enol tautomerization**. The carbonyl compound and its associated enol are called tautomers.

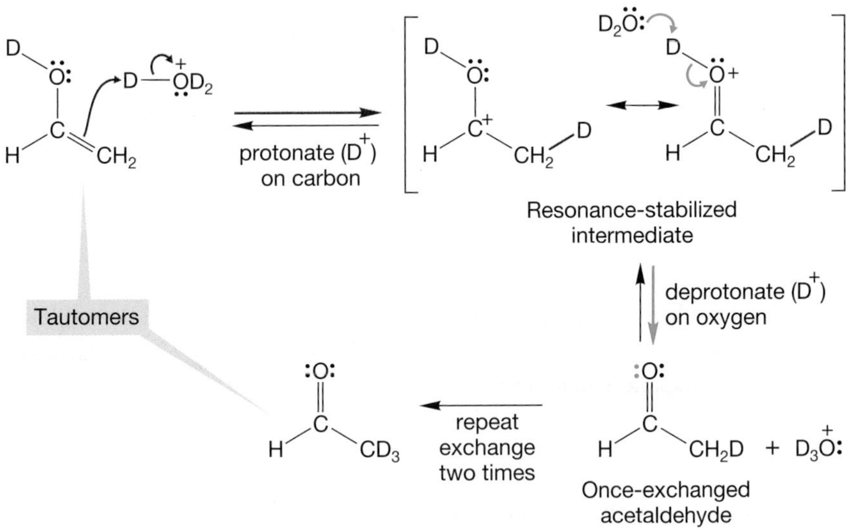

FIGURE 19.15 Reaction of the enol with D₃O⁺ will generate deuterated acetaldehyde.

WORKED PROBLEM 19.5 Write a mechanism for the equilibration of acetaldehyde and its enol form in acid. This problem may seem mindless, and it certainly has little intellectual content at this point, but it is an important drill problem. Doing this problem is part of "reading organic chemistry with a pencil."

ANSWER This process must become part of your chemical vocabulary from now on. First the carbonyl oxygen is protonated and then the resulting cation is deprotonated at carbon to give the enol. This process is a keto–enol tautomerization even though we are dealing with the aldehyde in this particular case.

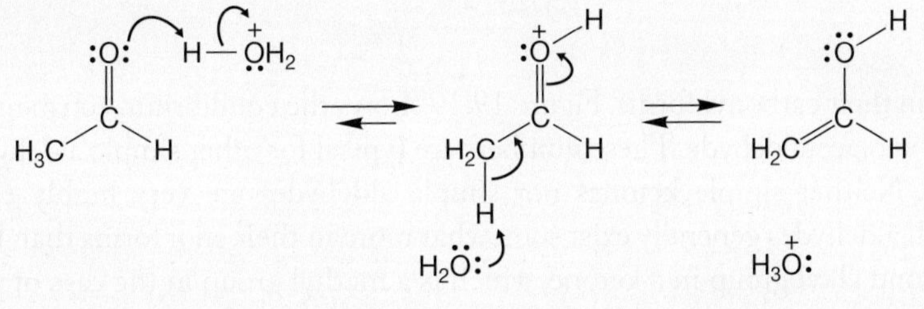

Recently, it has become possible to generate the simplest enol, vinyl alcohol (hydroxyethene), in such a fashion that the acid or base catalysts of the previous reactions are absent (Fig. 19.16). As long as it is protected from acids or bases, the enol

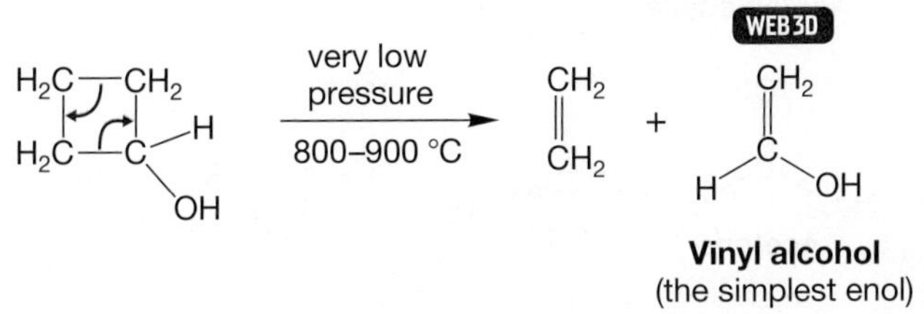

FIGURE 19.16 The parent of all enols, vinyl alcohol, is stable as long as it is protected from acid or base catalysts.

is stable and can be isolated and studied. The detailed structure of vinyl alcohol provides no surprises. Bond lengths and angles are not unusual (Fig. 19.17).

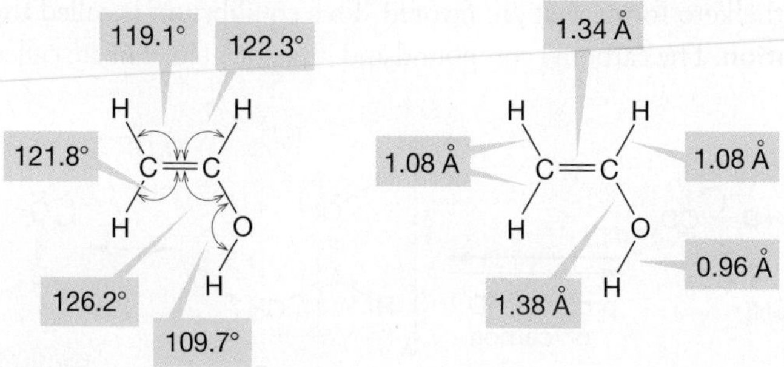

FIGURE 19.17 The structure of vinyl alcohol.

PROBLEM 19.6 It is impossible to protect vinyl alcohol from *all* acidic and basic materials, and so it eventually reverts to the carbonyl form, acetaldehyde. Explain what the unavoidable acid and base catalysts are.

So far we have seen that α hydrogens can be removed from carbonyl compounds, and that carbonyl compounds are often in equilibrium with their enol forms. The next question to address is the position of the equilibrium between a carbonyl compound and its tautomeric form, the enol. Where does the equilibrium lie, and how does it depend on structure (Fig. 19.18)? Simple aldehydes and ketones are predom-

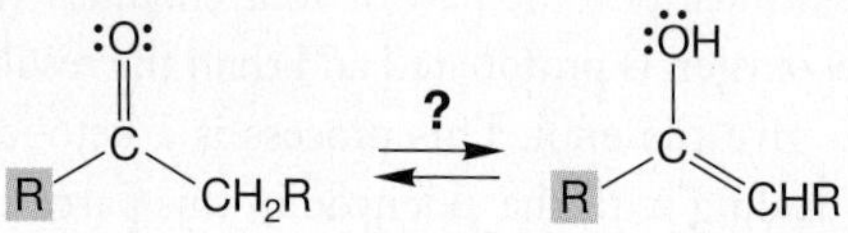

FIGURE 19.18 Where does the equilibrium between carbonyl compound and enol lie? Which tautomer is more stable?

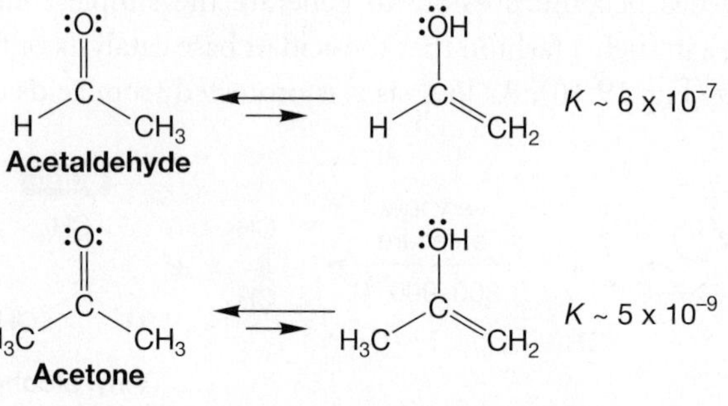

inantly in their carbonyl forms. Figure 19.19 shows the equilibrium constants, K, for acetone and acetaldehyde. These numbers are typical for other simple aldehydes and ketones. Neither simple ketones nor simple aldehydes are very highly enolized, although aldehydes generally exist somewhat more in their enol forms than ketones. The second alkyl group in a ketone, which is a methyl group in the case of acetone,

FIGURE 19.19 For simple aldehydes and ketones, it is the carbonyl form that is favored at equilibrium.

stabilizes the carbon–oxygen double bond more than it does a carbon–carbon double bond (Fig. 19.20).

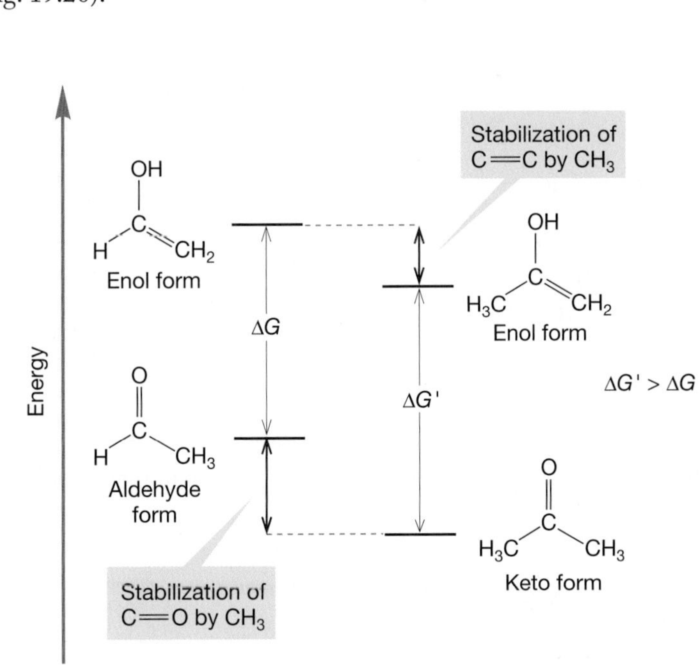

FIGURE 19.20 Acetone is less enolized than acetaldehyde because of the greater stabilization provided to the keto form by the second methyl group.

PROBLEM 19.7 From a bond strength analysis (ΔH), decide whether the keto or enol form of acetone is more stable. Use the data from Table 8.2 (p. 337). These molecules both contain sp^2/sp^3 carbon–carbon bonds. Use a value of 101.4 kcal/mol for these bonds.

More complex carbonyl compounds can be much more strongly enolized. β-Dicarbonyl compounds (1,3-dicarbonyl compounds) exist largely in their enol forms, for example (Fig. 19.21). In the enol forms of these compounds, it is possible to form an intramolecular hydrogen bond, thus forcing the equilibrium position away from the diketo forms. There is also conjugation between the carbon–carbon

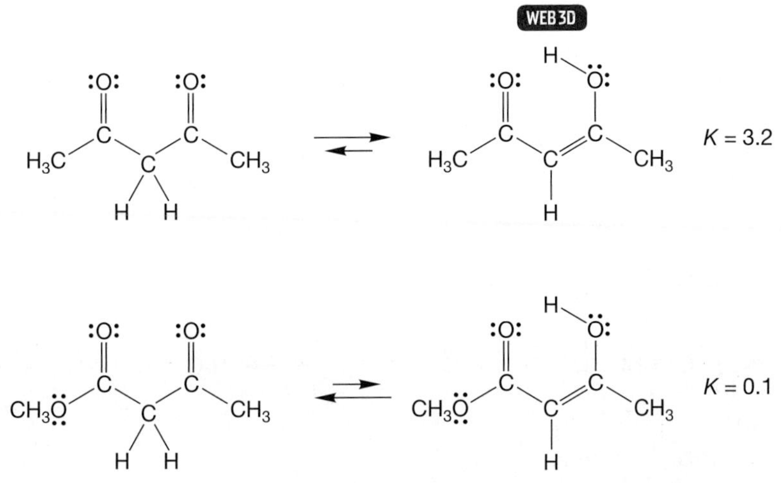

FIGURE 19.21 β-Dicarbonyl compounds are more enolized than are monoketones.

double bond and the remaining carbonyl group in these enols (Fig. 19.22). We will revisit the β-dicarbonyl compounds in Section 19.5c.

FIGURE 19.22 The formation of an intramolecular hydrogen bond, as well as conjugation between the carbon–carbon double bond and the carbonyl group, contributes to the increased stability of enols in β-dicarbonyl compounds.

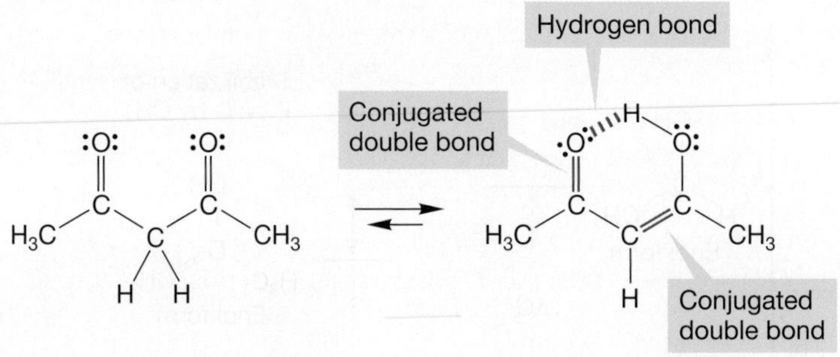

The ultimate stabilized enol is phenol. It is estimated that the equilibrium constant for formation of the aromatic enol form from the nonaromatic ketone is greater than 10^{13} (Fig. 19.23).

FIGURE 19.23 The equilibrium constant for enolization of cyclohexadienone is estimated to be greater than 10^{13}.

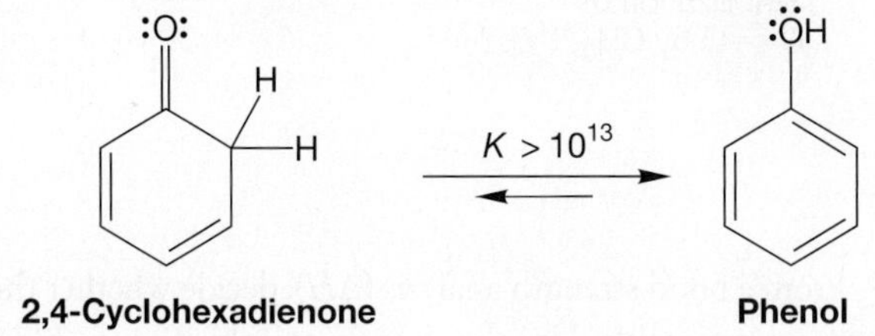

19.2b Enolates of Acid Derivatives

Hydrogens in the α position of acid derivatives are acidic for the same reasons that α hydrogens are acidic in aldehydes and ketones. The conjugate bases are resonance stabilized by the adjacent carbonyl or nitrile group (Fig. 19.24). The ease of deprotonation at the α position varies,

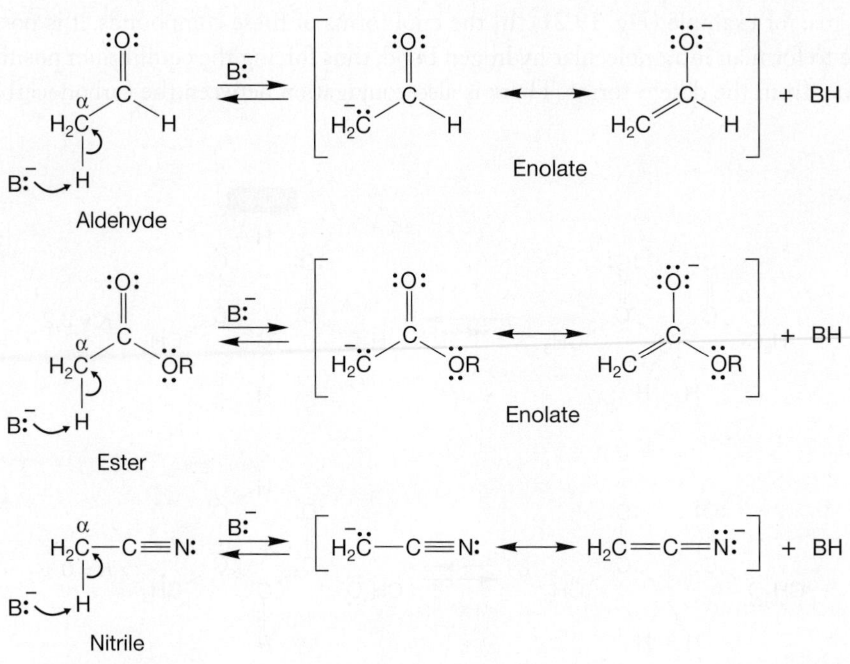

FIGURE 19.24 A hydrogen at the α position of an acyl compound or nitrile is acidic and can be removed by base to give an enolate, or, in the case of nitriles, a resonance-stabilized species much like an enolate.

however, depending on the ability of the adjacent carbonyl group to provide stabilization (Table 19.2).

Esters and amides are more stabilized by resonance than are aldehydes and ketones. Therefore their carbonyl groups have less double-bond character, and they are weaker acids at the α position. Notice the connection between resonance stabilization and acidity. The more resonance stabilized the starting material, the more difficult it is to remove a proton from it.

Amides are the most resonance stabilized of acyl compounds and therefore the weakest acids. In an amide's polar resonance form it is the relatively electropositive nitrogen atom that bears the positive charge (Table 19.2; Fig. 19.25). Thus the polar form is relatively stable. There is no substantial chemistry of amide enolates.

TABLE 19.2 The Acidity of Some Acyl Compounds[a]

Compound	pK_a
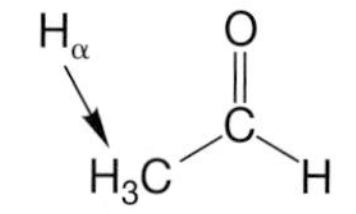 Acetaldehyde	16.7
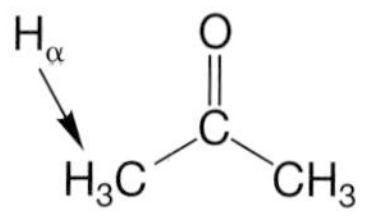 Acetone	19.3
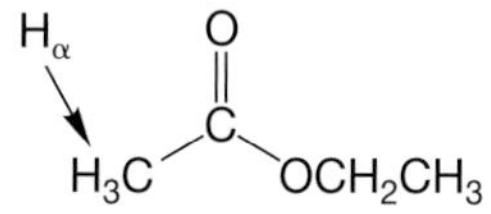 Ethyl acetate	24
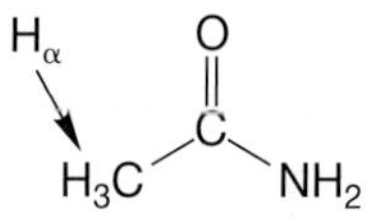 Acetamide	25
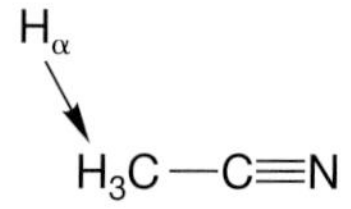 Acetonitrile	24

[a] The arrow shows the proton lost.

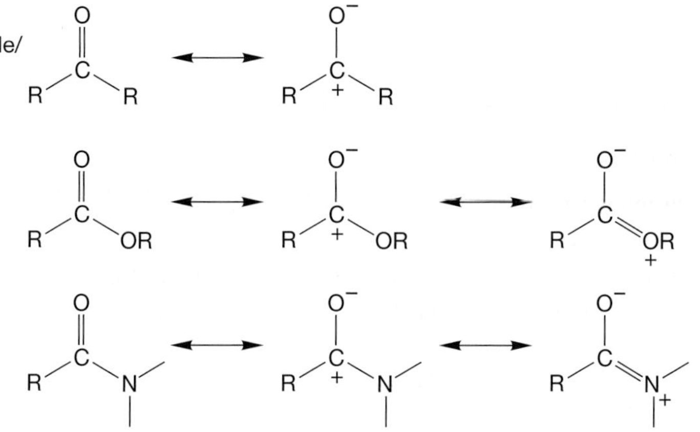

FIGURE 19.25 Esters and amides are more stabilized by resonance than aldehydes and ketones. It is harder to remove an α hydrogen from the R group of these more stable species. They are weaker acids as shown by their higher pK_a values.

The problem is that the α hydrogens on the amide nitrogen are more acidic than the α hydrogens on carbon, and therefore are more easily removed by base to give anions (Fig. 19.26).

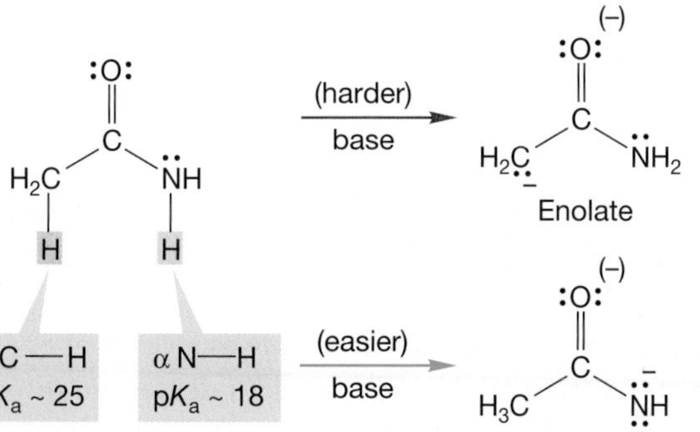

FIGURE 19.26 There are two kinds of α hydrogen in most amides. It is much easier to remove the proton attached to nitrogen than the proton attached to carbon.

PROBLEM 19.8 Explain why α hydrogens on the nitrogen of amides are more acidic than α hydrogens on carbon (Fig. 19.26).

If there are no hydrogens on the nitrogen, it becomes possible to form enolates. Strong, hindered bases such as **lithium diisopropylamide** (**LDA**, Fig. 19.27) are generally used because they are sufficiently basic and nonnucleophilic.

FIGURE 19.27 Amides in which there are no hydrogens attached to nitrogen can form enolates.

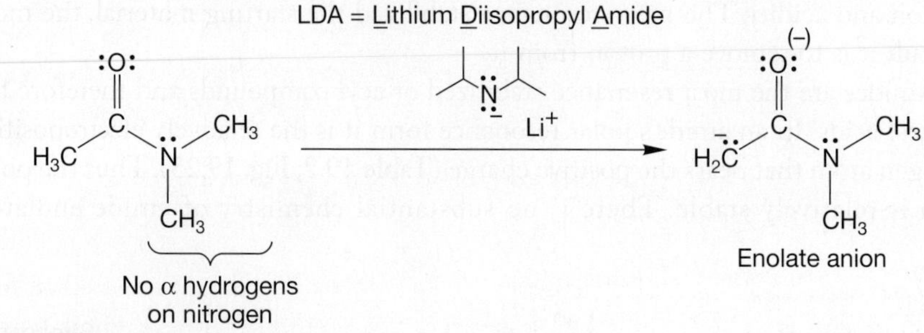

LDA = Lithium Diisopropyl Amide

No α hydrogens on nitrogen

Enolate anion

PROBLEM SOLVING

In many problems there are clues in the very question itself! Here's one: Whenever you see LDA on a problem—every time—think "deprotonation." LDA, a strong, very hindered base, is going to remove the most sterically available hydrogen to give an anion. That anion must be stabilized by resonance, however.

Summary

Ketones and aldehydes bearing α hydrogens are in equilibrium with their enol forms, although for simple ketones and aldehydes the carbonyl forms are greatly favored. This equilibrium is the keto–enol tautomerization. Equilibration with the enol form can be either acid- or base-catalyzed. The enol form can be favored in special cases. Esters and other acid derivatives also have acidic α hydrogens. LDA is a strong base that can be used to drive ketones, aldehydes, or esters completely to their corresponding enolates.

19.3 Racemization of Enols and Enolates

Now that we know something of the origins of enols and enolates, and we have seen that enols are in equilibrium with carbonyl compounds, it is time to move on to a look at reactions of these species.

We saw in the previous sections that as long as an enol or enolate can be formed, hydrogens in the α position can be exchanged for deuterium through treatment with deuterated acid or base (Fig. 19.28). Treatment of a simple optically active aldehyde or ketone with acid or base results in loss of optical activity, racemization, *as long as*

FIGURE 19.28 Exchange reactions of carbonyl compounds bearing α hydrogens can be either acid or base catalyzed.

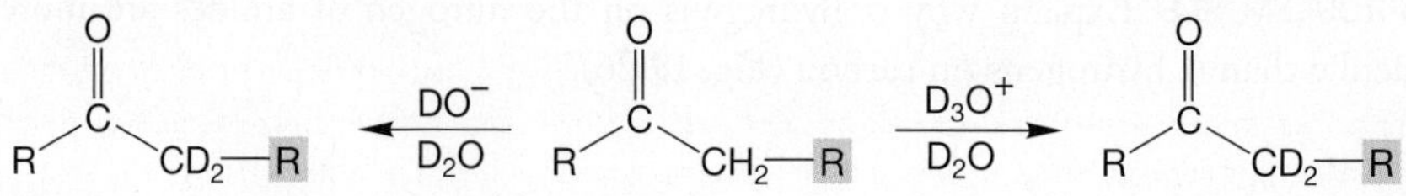

the stereogenic carbon is in the position α to the carbon–oxygen double bond, and as long as there is an α hydrogen (Fig. 19.29).

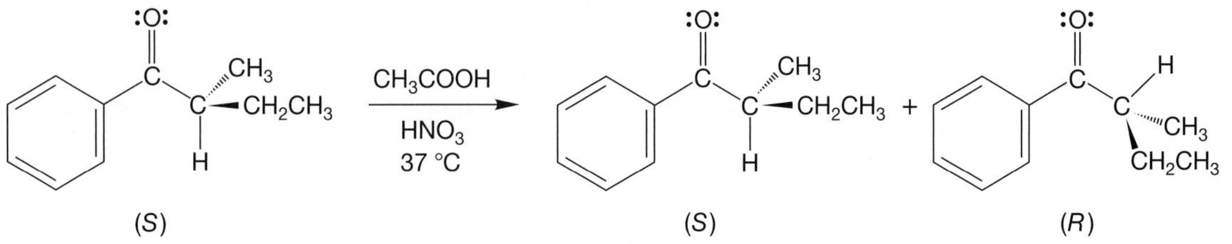

FIGURE 19.29 Optically active carbonyl compounds are racemized in acid or base, as long as an α hydrogen is present on the stereogenic carbon.

Once again the mechanism depends on enol formation in acid, or enolate formation in base. In acid, an achiral, planar enol is formed (Fig. 19.30). Re-formation of the keto form occurs by equal protonation from either side of the enol. In base, removal of the α hydrogen generates a planar enolate anion that is also achiral (Fig. 19.31).

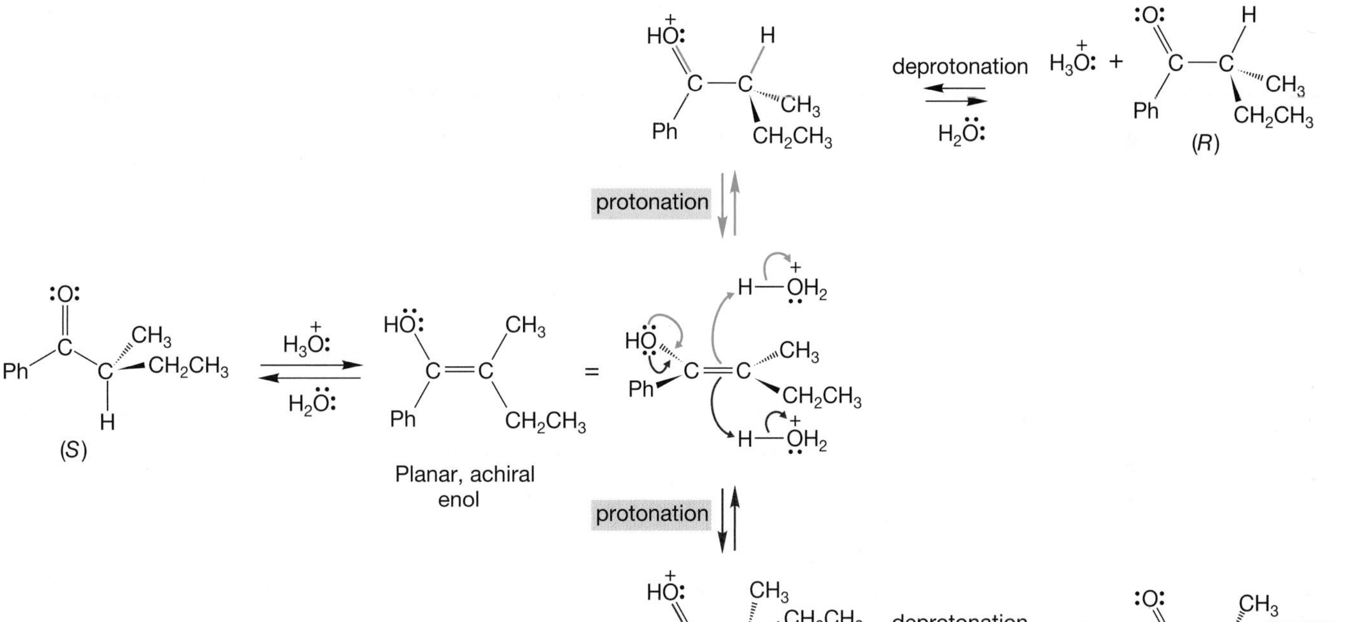

FIGURE 19.30 Protonation of the planar enol must result in formation of equal amounts of two enantiomers. An optically active carbonyl compound is racemized on enol formation.

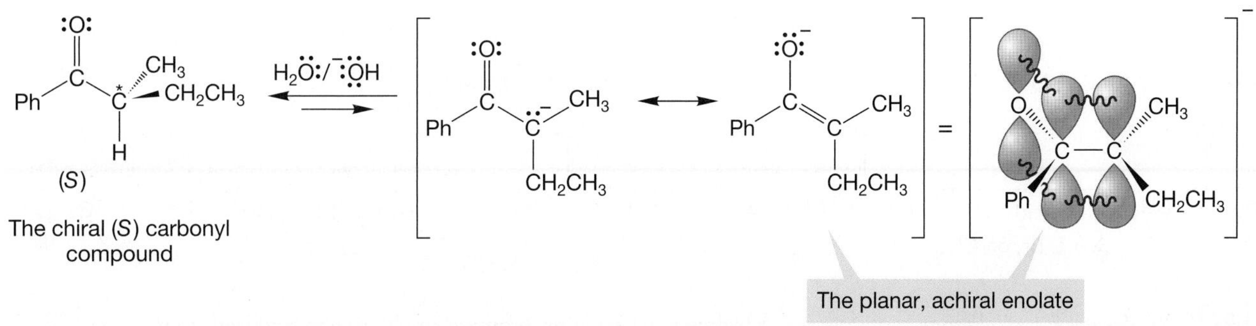

The chiral (*S*) carbonyl compound

The planar, achiral enolate

FIGURE 19.31 Resonance stabilization of the enolate depends on overlap of the 2*p* orbitals on carbon and oxygen. Maximum overlap requires planarity, and the planar enolate is necessarily achiral. Racemization has been accomplished at the enolate stage. Remember that the asterisk tells you that the structure is a single enantiomer.

For overlap between the $2p$ orbitals on carbon and oxygen to be maximized, the enolate must be planar, and the planar structure has a nonstereogenic carbon at the α position. A racemic mixture is formed when the enolate reprotonates (Fig. 19.32). Addition of a proton *must* take place with equal probability at the two equivalent faces ("top" = green; "bottom" = red) of the planar molecule. Racemization will also occur with esters and other acid derivatives that have a hydrogen on a stereogenic α carbon.

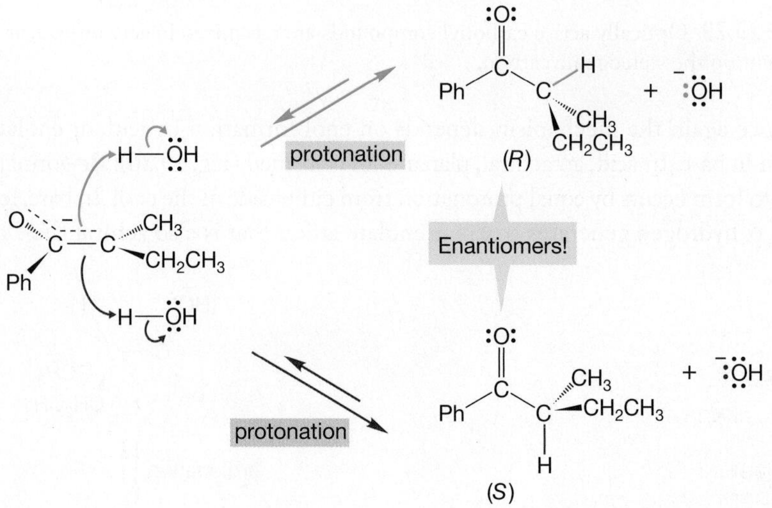

FIGURE 19.32 Reprotonation of the enolate from the two equivalent faces must give a pair of enantiomers.

19.4 Halogenation in the α Position

We have already learned that halogens (X_2) are able to react with alkenes, alkynes, and (with a little help) aromatic rings. Halogenation is also a useful tool for adding functionality to ketones, aldehydes, carboxylic acids, and esters.

19.4a Halogenation of Ketones and Aldehydes   Halogenation at the α position of ketones and aldehydes is one of the relatively rare examples of a reaction that takes a different course in its acid- and base-catalyzed versions. As you will see, though, the two reactions are closely related.

Treatment of a ketone or an aldehyde containing an α hydrogen with halogen (X_2, X = I, Cl, or Br) in acid results in the replacement of an α hydrogen with a halogen atom (Fig. 19.33). For the specific example shown in Figure 19.33, the

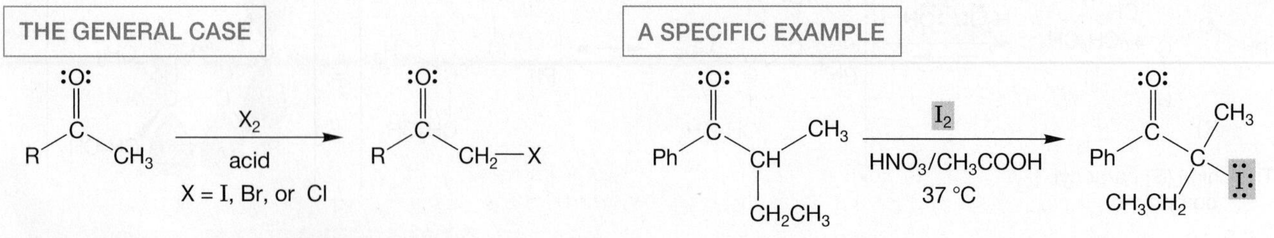

THE GENERAL CASE

A SPECIFIC EXAMPLE

FIGURE 19.33 Treatment of a ketone containing an α hydrogen with iodine, bromine, or chlorine in acid leads to α halogenation.

enol is formed first and, like most alkenes, reacts with iodine to displace iodide and form a resonance-stabilized cation (Fig. 19.34). The iodide ion deprotonates the intermediate cation generating the α-iodo carbonyl compound, and ending the reaction.

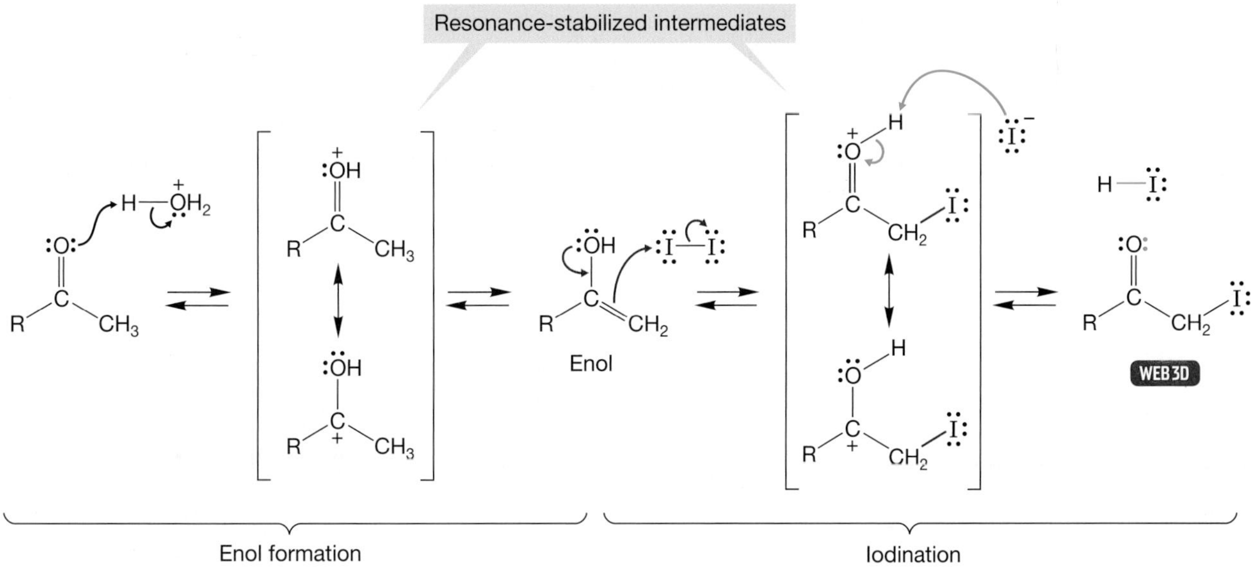

Resonance-stabilized intermediates

Enol formation

Iodination

FIGURE 19.34 Under acidic conditions the enol is formed and then reacts with iodine to give the resonance-stabilized cation. Deprotonation leads to the α-iodide.

Earlier, we saw how deuterium replaces *all* α hydrogens in acetaldehyde. Why doesn't the α-iodo compound react further to add more iodines? The way to answer a question such as this one is to work through the mechanism of the hypothetical reaction and see if you can find a point at which it is disfavored. In this case, that point appears right away. It is enol formation itself that is slowed by the introduction of the first halogen. The introduced iodine inductively withdraws electrons and disfavors the first step in enol formation, protonation of the carbonyl group (Fig. 19.35).

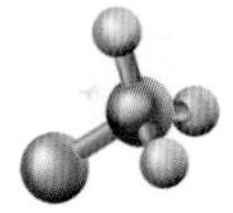

Enol halogenation

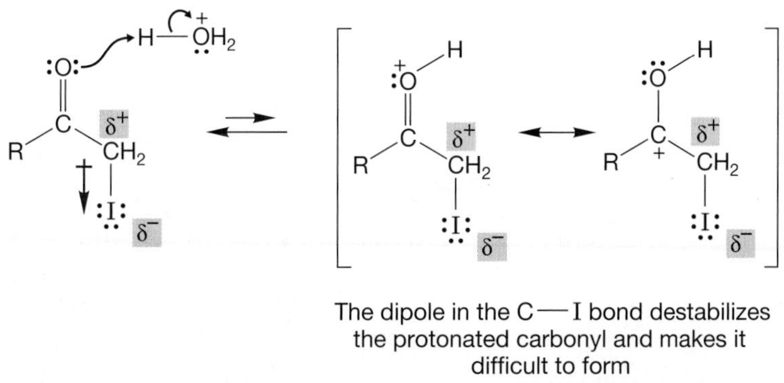

The dipole in the C—I bond destabilizes
the protonated carbonyl and makes it
difficult to form

FIGURE 19.35 Protonation of the α-iodoketone is disfavored by the electron-withdrawing character of the halogen.

It is not easy to *predict* the extent of this sequence of enol formations and halogenations. It is not reasonable to expect you to anticipate that further enol formation would be sufficiently slowed so as to make the monoiodo compound the end product. On the other hand, it is not so difficult to *understand* this effect. Rationalization of results is much easier than predicting new reactions (or in this case, no reaction).

Under basic conditions matters are quite different. In fact, it is very difficult to stop the reaction at the monoiodo stage. The general result is that all α hydrogens are replaced with halogen. Note that ketones might have α hydrogens on either side of the carbonyl. For molecules with three α hydrogens on one side, methyl ketones, the reaction goes even further in what is called the **haloform reaction**. In base, it is not the enol that is formed, but the enolate (Fig. 19.36).

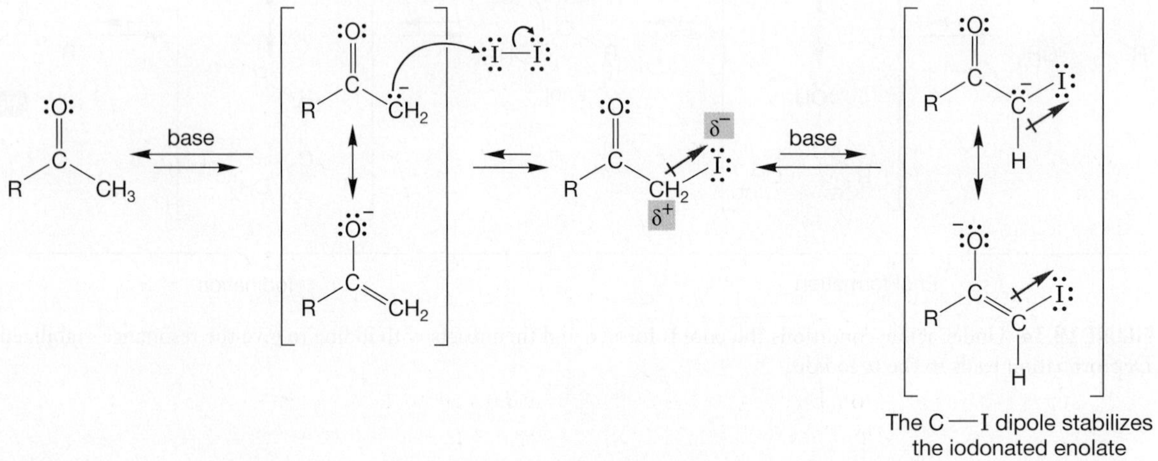

The C—I dipole stabilizes
the iodonated enolate

FIGURE 19.36 The initially formed α-iodo carbonyl compound is a stronger acid than the carbonyl compound itself. The introduced iodine makes enolate formation easier.

This anion reacts in S$_N$2 fashion with iodine to generate the α-iodo carbonyl compound. Now, however, the electron-withdrawing inductive effect of iodine makes enolate formation easier, not harder, as for enols. The negative charge of the enolate is *stabilized* by the electron-withdrawing effect. A second and third displacement reaction on iodine generates the α,α,α-triiodo carbonyl compound (Fig. 19.37). Similar reactions are known for bromine and chlorine.

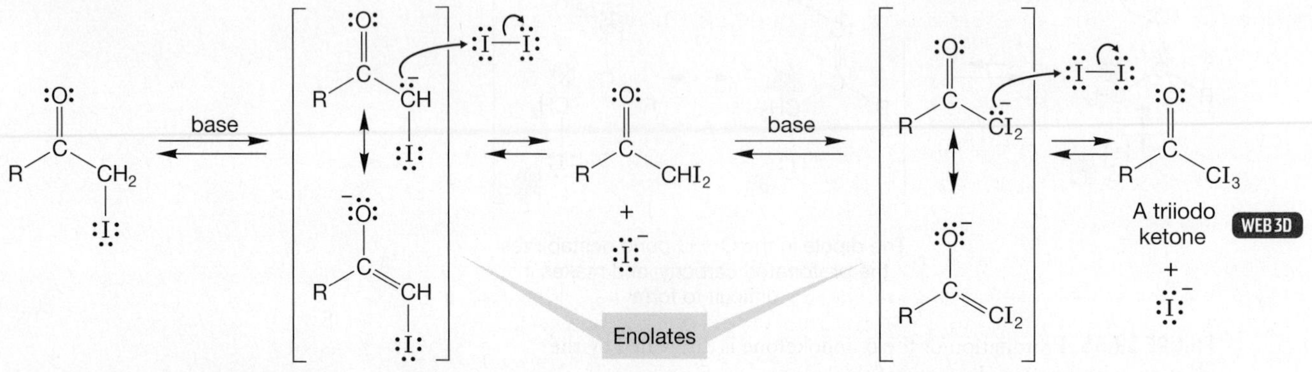

Enolates

FIGURE 19.37 Sequential enolate formations and iodinations lead to the α,α,α-triiodo carbonyl compound.

In NaOH/H$_2$O, these trihalocarbonyl compounds react further because hydroxide addition can lead to elimination of the stabilized trihalocarbanion. After these addition–elimination steps, the carbanion can gain a proton from solvent or the carboxylic acid to form a molecule of a **haloform**, which is a common name for a trihalomethane (Fig. 19.38). Under these basic conditions the carboxylic acid will remain deprotonated.

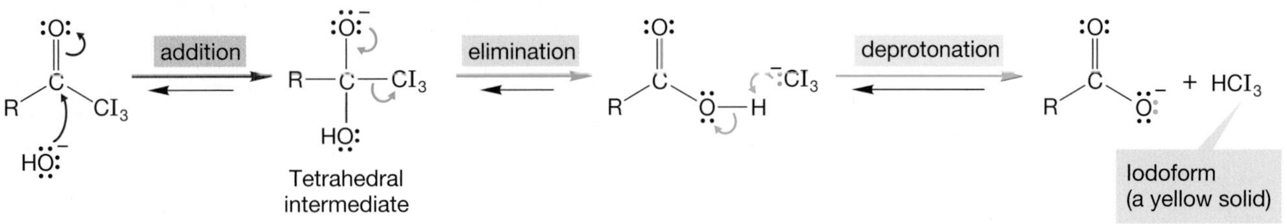

FIGURE 19.38 Addition of hydroxide to the carbonyl group leads to a tetrahedral intermediate that can lose triiodomethide anion to generate the carboxylic acid. Transfer of a proton completes the reaction.

Essentially all nucleophilic reagents add to carbonyl groups, often in a reversible fashion. In this case, the addition reaction is strongly favored by the halogen substitution (p. 780)! So, the addition of hydroxide to the carbonyl looks like a reaction that is almost certain to occur. Once the addition reaction has taken place, there is an opportunity to generate the acid and the haloform in an elimination step if the triiodomethyl anion can be lost as the carbonyl group re-forms. Protonation of the carbanion completes the reaction.

Is this mechanism reasonable? Is the triiodomethyl anion a good enough leaving group to make this step a sensible one? The iodines are strongly electron withdrawing, and that property will stabilize the anion. As a check we might look up the pK_a of iodoform to see how easily the molecule is deprotonated to form the anion (Fig. 19.39). The value is about 14, and so iodoform is a relatively strong acid, at least compared to water (pK_a = 15.7).

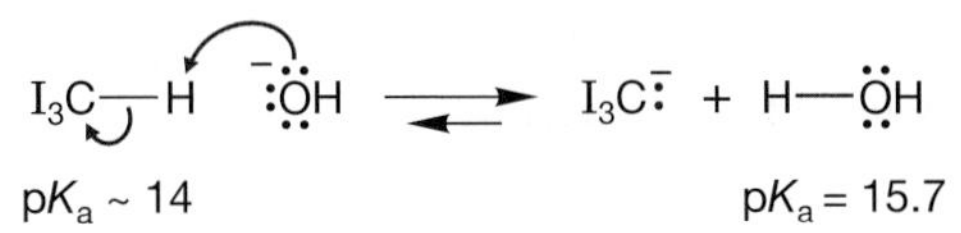

$$I_3C \!\!-\!\! H \quad :\ddot{O}H^- \quad \rightleftharpoons \quad I_3C:^- \ + \ H\!\!-\!\!\ddot{O}H$$

pK_a ~ 14   pK_a = 15.7

FIGURE 19.39 The pK_a of iodoform is about 14. Iodoform is a relatively strong acid, and the loss of $^-$CI$_3$ as a leaving group is a reasonable step.

So the loss of $^-$CI$_3$ in Figure 19.38 does look reasonable. In fact, the iodoform reaction serves as a diagnostic test for methyl ketones. Formation of iodoform, a yellow solid, is a positive test for a molecule containing a methyl group attached to a carbonyl carbon.

PROBLEM 19.9 Write out the mechanism for the reaction between acetophenone and excess chlorine in aqueous sodium hydroxide.

PROBLEM 19.10 If we measure the rates of three reactions of the ketone shown on the next page, exchange of the α hydrogen for deuterium in D$_2$O/DO$^-$, racemization in H$_2$O/HO$^-$, and α bromination using Br$_2$/H$_2$O/HO$^-$, we find that they are identical.

(continued)

How can the rates of three such different reactions be the same? Explain, using an Energy versus Reaction progress diagram.

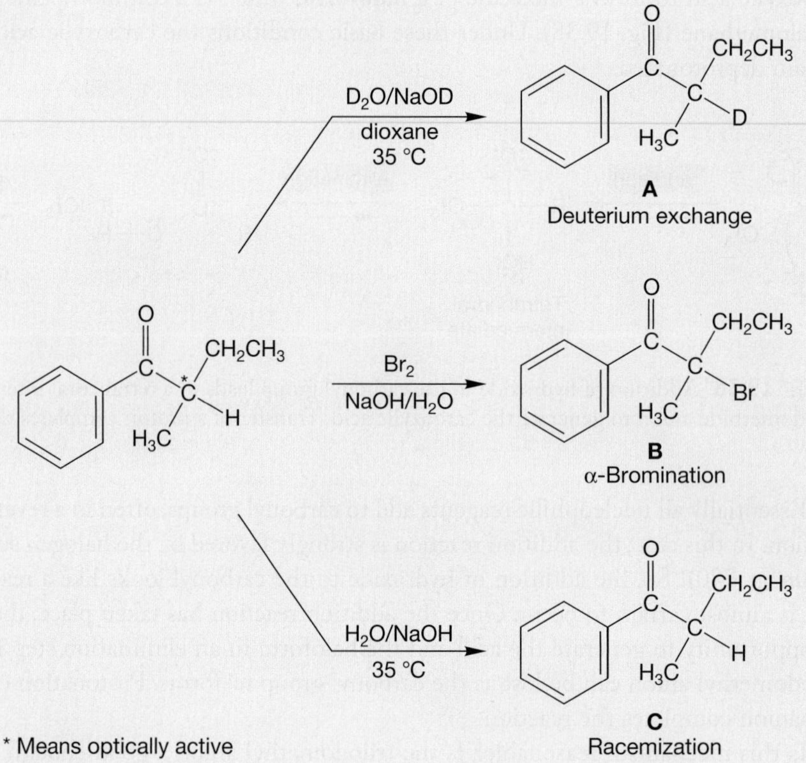

A
Deuterium exchange

B
α-Bromination

C
Racemization

* Means optically active

19.4b Halogenation of Carboxylic Acids

Halogenation can occur at the α position of carboxylic acids. Treatment of carboxylic acids with Br_2 and PBr_3, or the equivalent, a mixture of phosphorus and bromine, leads ultimately to formation of the α-bromo acid (Fig. 19.40). The process is known as the **Hell–Volhard–Zelinsky (HVZ) reaction** after Carl M. Hell (1849–1926), Jacob

THE GENERAL CASE

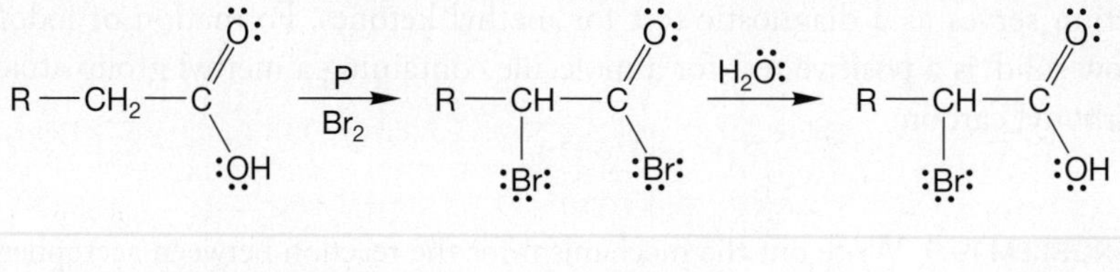

A SPECIFIC EXAMPLE

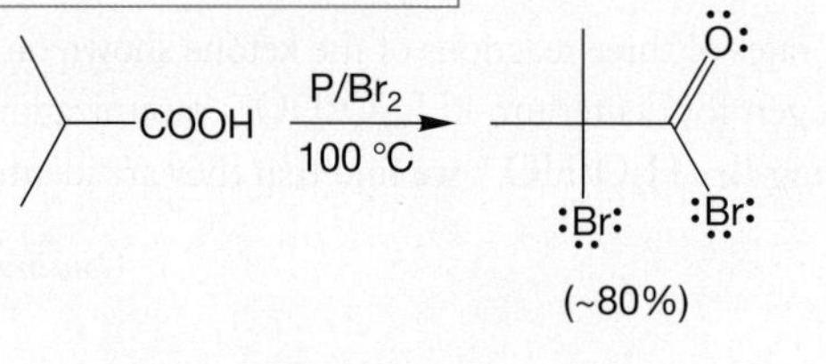

(~80%)

FIGURE 19.40 The Hell–Volhard–Zelinsky reaction.

Volhard (1834–1910), and Nicolai D. Zelinsky (1861–1953). The first step is formation of the acid bromide through reaction with PBr₃. The acid bromide is in equilibrium with its enol form (Fig. 19.41). Bromination of the enol form with Br₂ gives an isolable α-bromo acid bromide.

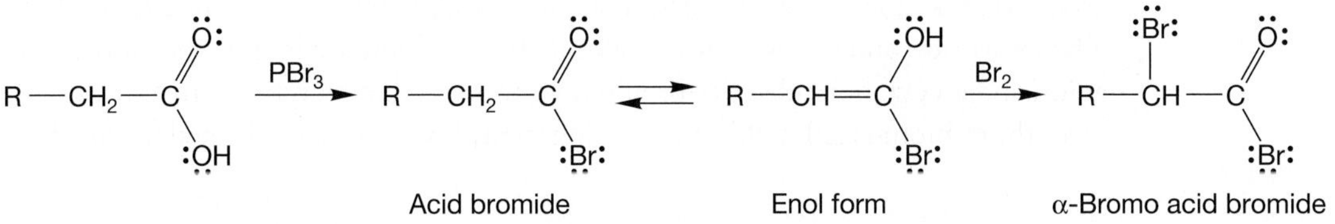

FIGURE 19.41 An intermediate in the Hell–Volhard–Zelinsky reaction is the α-bromo acid bromide. This compound can be isolated if a full equivalent of PBr₃ is used in the reaction.

The α-bromo acid halide reacts like any other acid halide, and it can be used to generate the α-bromo acid, ester, and amide, for example (Fig. 19.42).

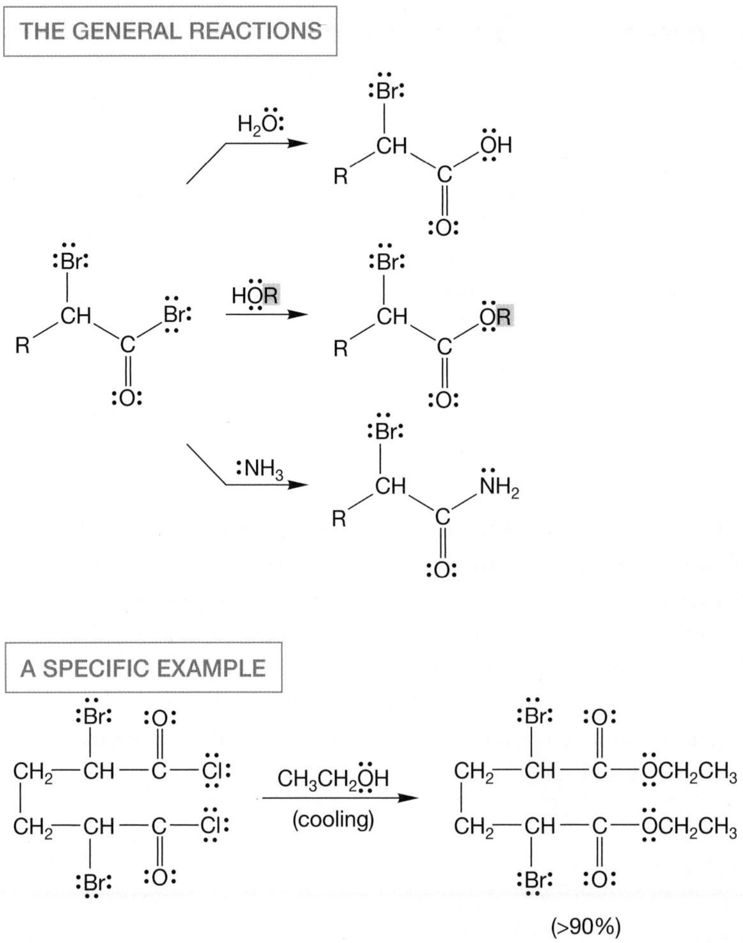

FIGURE 19.42 Compounds available from reactions of an α-bromo acid halide.

PROBLEM 19.11 Write a general mechanism that accounts for the reactions of Figure 19.42.

If a catalytic amount of PBr$_3$ is used, and not the full equivalent as shown in the reaction in Figure 19.41, the product of the reaction is the α-bromo acid, not the α-bromo acid bromide (Fig. 19.43a). The α-bromo acid bromide is still an intermediate, but only a small amount of it can be made at any one time, because its formation depends on PBr$_3$. The small amount of α-bromo acid bromide produced is attacked by a molecule of the acid in an anhydride-forming reaction (Fig. 19.43b). The anhydride undergoes addition–elimination by bromide to give the carboxylate anion and a new molecule of α-bromo acid bromide that recycles. A final hydrolysis gives the α-bromo acid itself. This reaction surely involves a complicated mechanism.

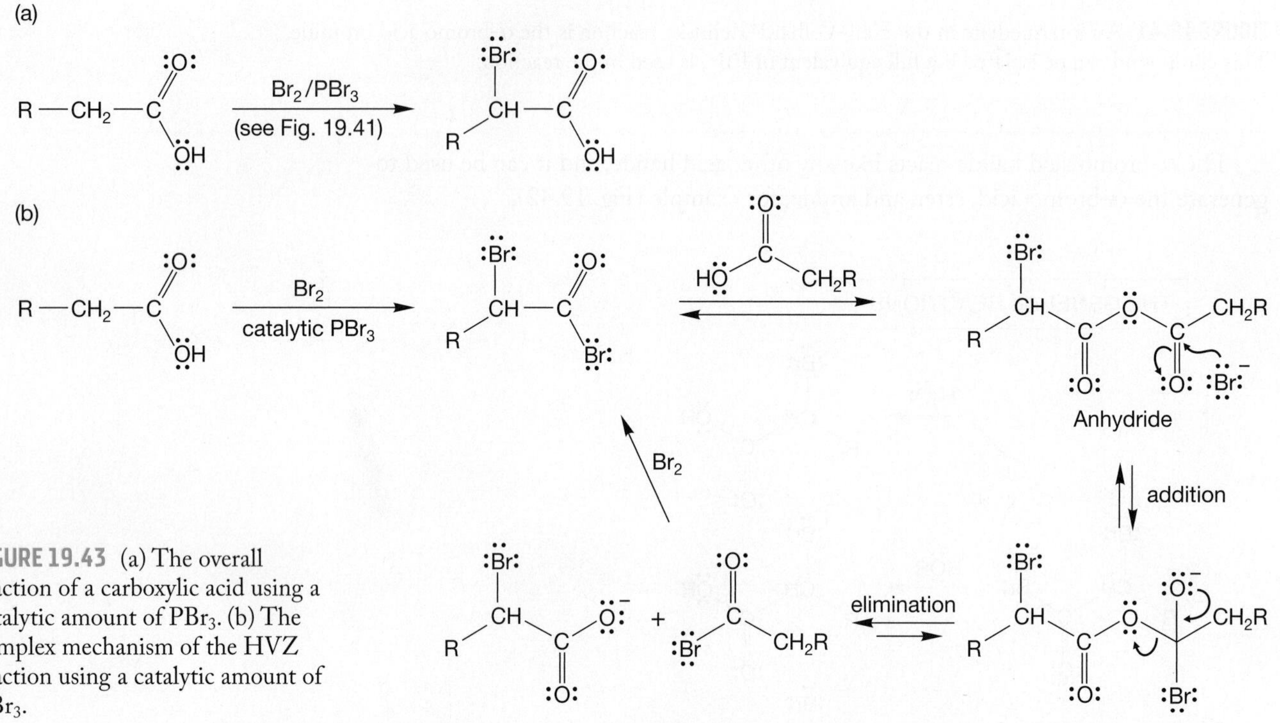

(a)

(b)

FIGURE 19.43 (a) The overall reaction of a carboxylic acid using a catalytic amount of PBr$_3$. (b) The complex mechanism of the HVZ reaction using a catalytic amount of PBr$_3$.

The HVZ reaction is limited to the formation of α-bromo compounds and is, in truth, sometimes awkward to carry out. The reagents, bromine and phosphorus, are noxious; reaction times are often long, and reaction conditions are harsh. For these reasons, methods have evolved to extend and replace the classic HVZ process. For example, David Harpp (b. 1937) and his colleagues at McGill University use the reaction of acid chlorides and N-bromosuccinimide (NBS, p. 613), in the presence of a catalytic amount of HBr, to form α-bromo acid chlorides conveniently and in excellent yields (Fig. 19.44).

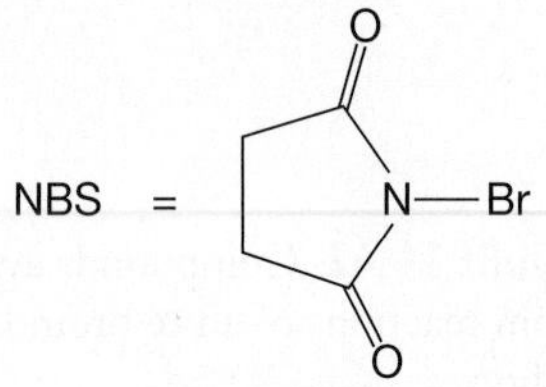

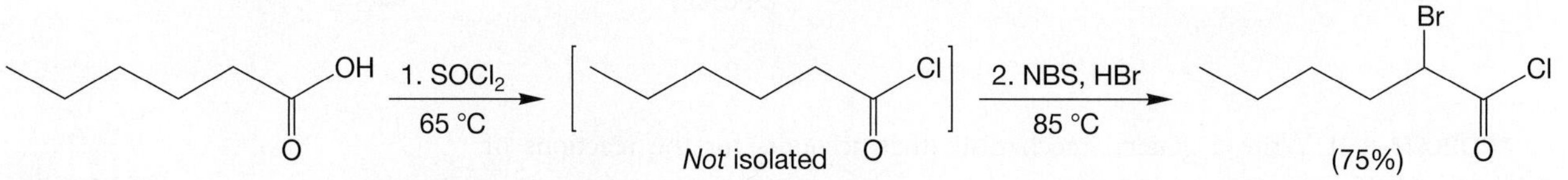

FIGURE 19.44 The Harpp modification of the HVZ reaction.

The mechanism of Harpp's modification is an ionic one, and uses many of the steps we have seen already. The HBr catalyst protonates the carbonyl oxygen of the acid halide (Fig. 19.45). Then NBS deprotonates the α carbon to form an enol.

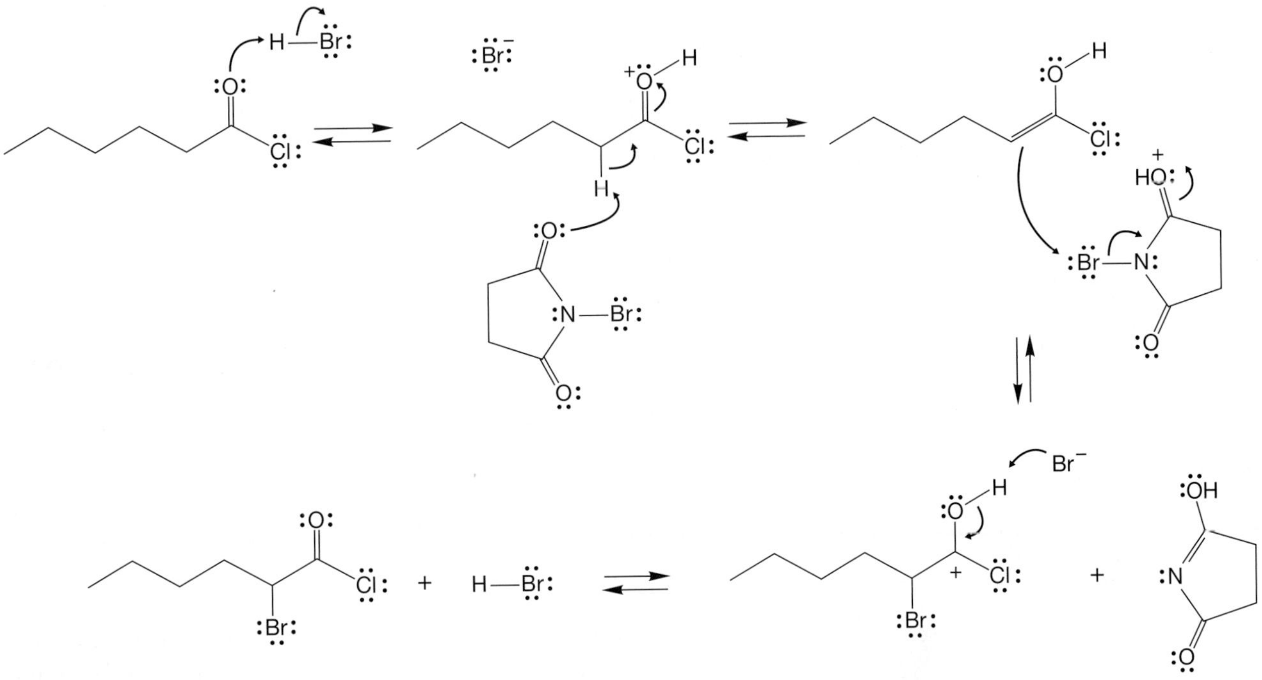

FIGURE 19.45 The mechanism of the Harpp modification of the HVZ reaction.

The enol now functions as a base to remove a Br from NBS. Finally deprotonation of the protonated carbonyl gives the product and a new molecule of acid catalyst.

Like the α-bromo aldehydes and ketones, the α-bromo acids are very reactive in displacement reactions and therefore serve as sources of many α-substituted carboxylic acids (Fig. 19.46). A particularly important example is the formation of α-amino acids through reaction with ammonia.

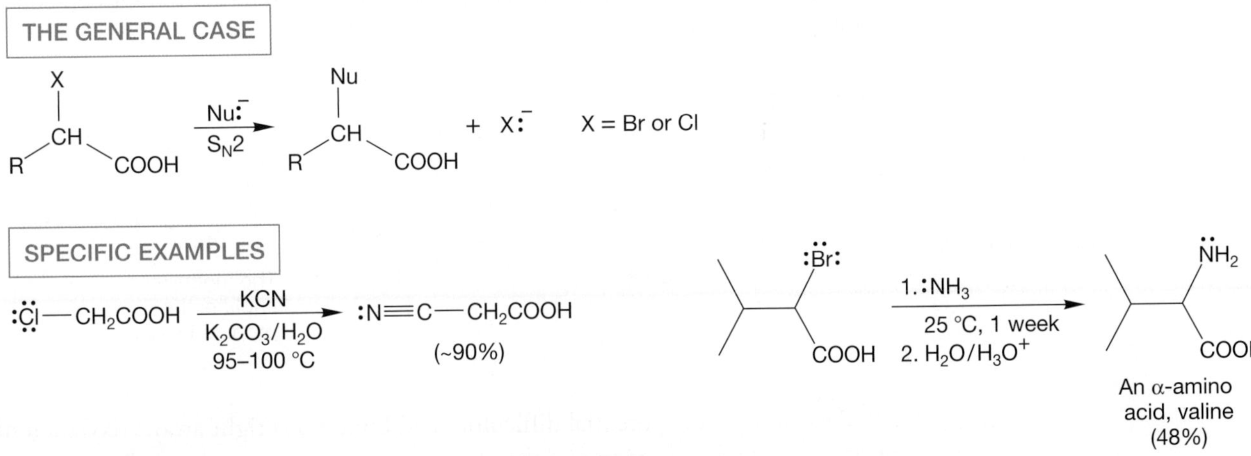

FIGURE 19.46 Some reactions of α-halo acids.

19.4c Halogenation of Esters   In theory, one should be able to halogenate an ester in the α position. However, there are very few examples of this reaction. Here's one (Fig. 19.47).

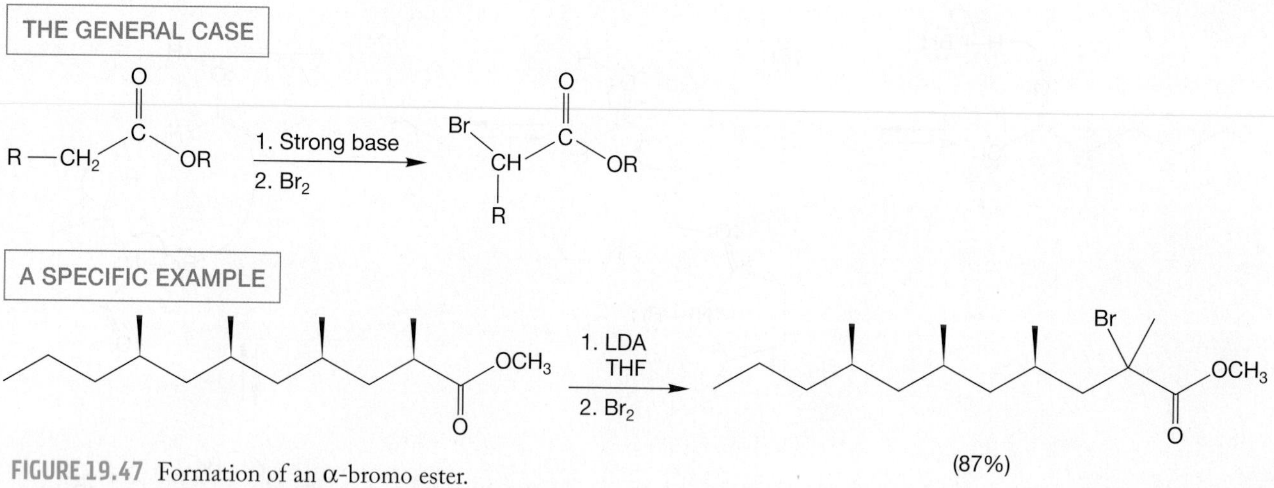

THE GENERAL CASE

A SPECIFIC EXAMPLE

FIGURE 19.47 Formation of an α-bromo ester.

Summary

We have now discussed two reactions that take place at the α position of carbonyl compounds. We will use these as prototypes on which to base further, more complicated processes. The enolate can react with an electrophilic source of H^+ or the electrophile X_2. The unifying theme is formation of an enolate ion (in base) or an enol (in acid) that can act as a nucleophile. With what other electrophiles might such a nucleophile react?

19.5 Alkylation in the α Position

In the previous sections, we saw several examples of enolate anions acting as nucleophiles. A logical extension would be to attempt to use enolates as nucleophiles in S_N2 displacements. If we could do this reaction, we would have a way to alkylate the α position of carbonyl compounds, and a new and most useful carbon–carbon bond-forming reaction would result (Fig. 19.48).

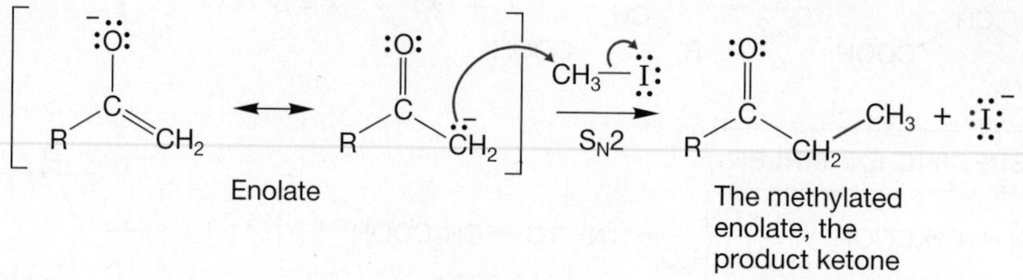

FIGURE 19.48 If the enolate could act as a nucleophile in the S_N2 reaction, we would have a way of alkylating at the α position, thus forming a new carbon–carbon bond.

We can see some potential difficulties and limitations right away. Alkylation of an enolate is an S_N2 reaction, and that means we cannot use a tertiary halide, a species too hindered for participation in the S_N2 reaction.

The resonance formulation of enolate anions clearly shows that the negative charge is shared between an oxygen and a carbon atom (Fig. 19.49). At which atom will alkylation be faster? If there is little or no selectivity in the alkylation reaction, it will surely be of limited use.

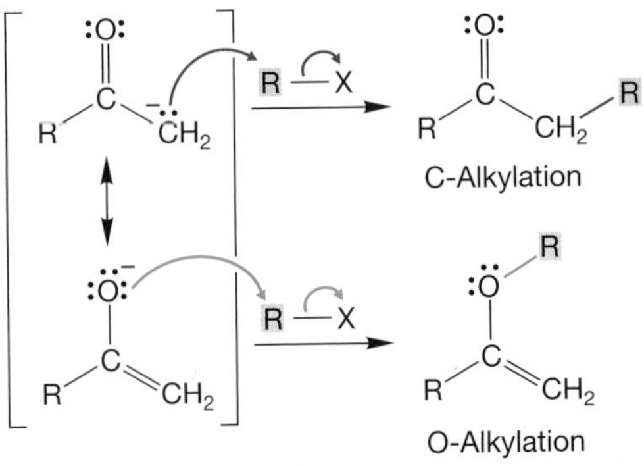

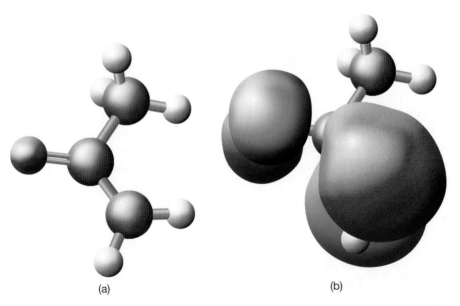

C-Alkylation

O-Alkylation

FIGURE 19.49 In principle, alkylation of the enolate could take place at either carbon or oxygen.

It turns out that most enolates are better nucleophiles at carbon. So, alkylation takes place faster at carbon than at oxygen (Fig. 19.50). One explanation for this fact is that the highest occupied molecular orbital of the nucleophile has more electron density at the carbon than at oxygen. Selectivity might also be a result of counter ion (Li⁺, for example) coordination between the enolate and the electrophile. If the counterion is complexed with the site of overall higher electron density, the oxygen in this case, it sterically hinders alkylation at that position.

(a)   (b)

FIGURE 19.50 (a) The enolate of acetone. (b) The highest occupied molecular orbital of the enolate of acetone is represented. Note the greater contribution of electron density on the carbon of the enolate. In practice, alkylation generally takes place at carbon because of the greater electron density of the orbital of the nucleophile involved in the reaction.

19.5a Alkylation of Ketone and Aldehyde Enolates

Even though alkylation at oxygen is not common, there are other problems with the alkylation reaction of simple ketones using hydroxide base. Consider the alkylation of 2-methylcyclohexanone in H_2O/HO^-. Both α positions are active, and two enolates will be produced (Fig. 19.51). Moreover, once alkylated, the compounds can undergo further alkylation. Reactions of the initial products compete with the desired single

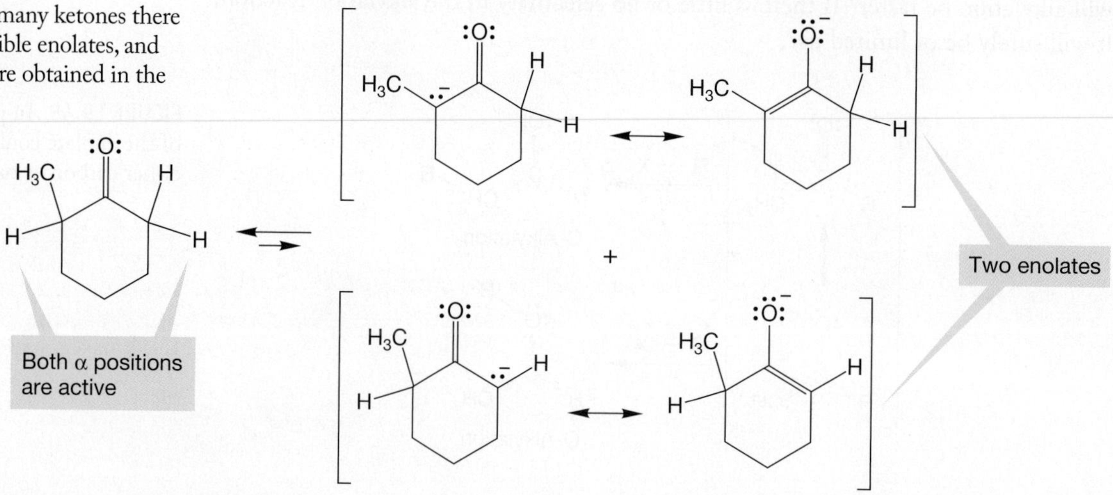

FIGURE 19.51 For many ketones there are at least two possible enolates, and therefore mixtures are obtained in the alkylation reaction.

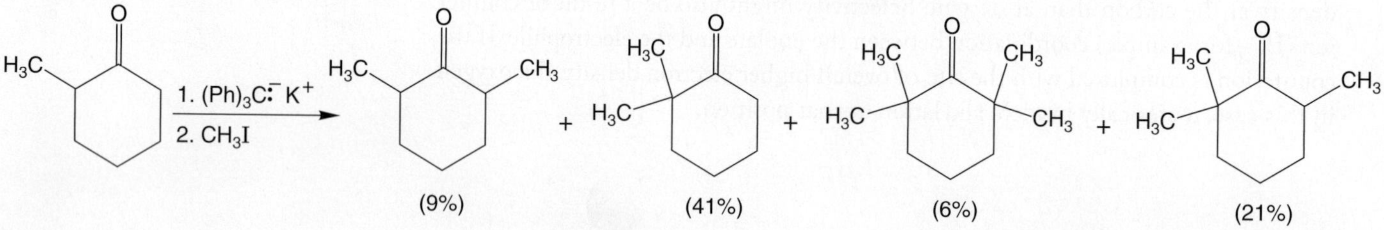

Both α positions are active

Two enolates

alkylation reaction. Mixtures of products can be formed, and that is not a very useful situation (Fig. 19.52). Most aldehydes encounter similar problems.

FIGURE 19.52 For ketones, there can often be more than one alkylation. Mixtures of products can be formed.

1. (Ph)₃C⁻ K⁺
2. CH₃I

(9%)   +   (41%)   +   (6%)   +   (21%)

However, if a strong, poorly nucleophilic base such as LDA is used to form the enolate at very low temperatures (Fig. 19.53), then one can usually control the reaction. LDA is strong enough to convert all available carbonyl compound into an enolate. This technique has become a useful tool for the synthesis of organic molecules.

CONVENTION ALERT

Note again the use in Figure 19.53 of the shorthand notation for resonance forms. Instead of always drawing out each important resonance form, one summary form is drawn. The positions sharing the charge or electron are indicated with parentheses, (+), (−), or (·).

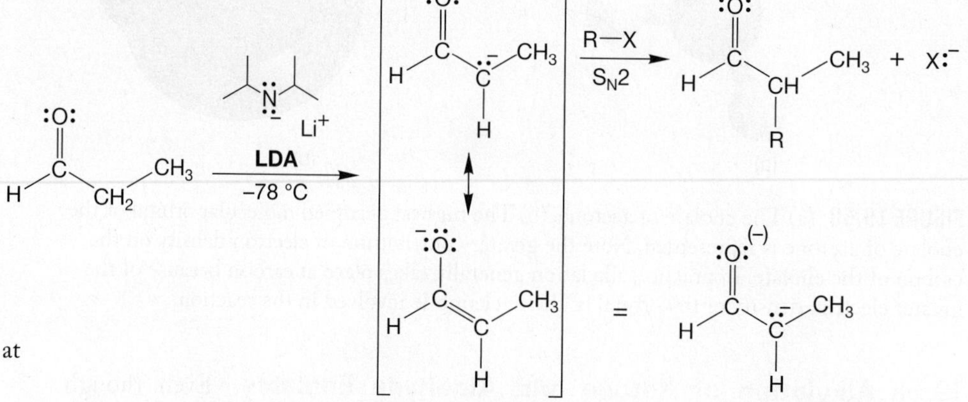

FIGURE 19.53 Strong base (LDA) and low temperatures are effective at forming enolates.

LDA
−78 °C

R—X
SN2

PROBLEM 19.12 LDA is a nonnucleophilic base. It does not add to the carbonyl carbon. Explain why LDA is a poor nucleophile.

19.5b Alkylation of Carboxylic Acids   Carboxylic acids can also be alkylated, but two equivalents of strong base are necessary, because this reaction proceeds through a dianion. The first equivalent of base removes the carboxyl hydrogen to give the carboxylate salt. If the base used is a strong *base* and a weak *nucleophile*, a second hydrogen can be removed, this time from the α position. The typical base used is LDA. The dianion can now be brominated or alkylated at the α position, as long as the alkylating agent is reactive (Fig. 19.54). After the alkylation step the resulting carboxylate anion can be reprotonated to obtain the neutral product. Primary halides work well in these S$_N$2 alkylation reactions, but more substituted halides lead mostly to the products of an E2 reaction. Steric hindrance around the carbon–halogen antibond blocks the S$_N$2 path. As a result, the enolate reacts as a base to give the alkene product in the E2 elimination.

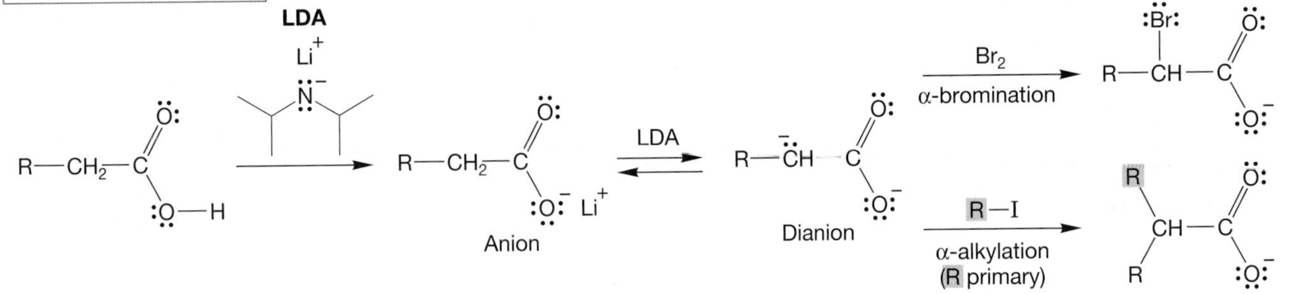

FIGURE 19.54 Dianions of carboxylic acids can function as enolates. They can be alkylated in the α position with primary or other reactive halides.

THE GENERAL CASE

A SPECIFIC EXAMPLE

(63%)

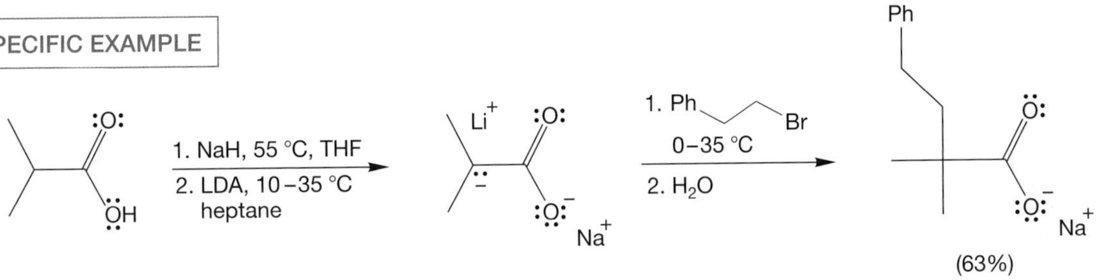

PROBLEM 19.13 Draw the resonance forms for the dianion formed from butanoic acid. Why do you suppose the electrophile adds to the α carbon?

PROBLEM 19.14 Provide mechanisms for the following reactions that show related electrophiles that can react at the α position of carboxylic acids:

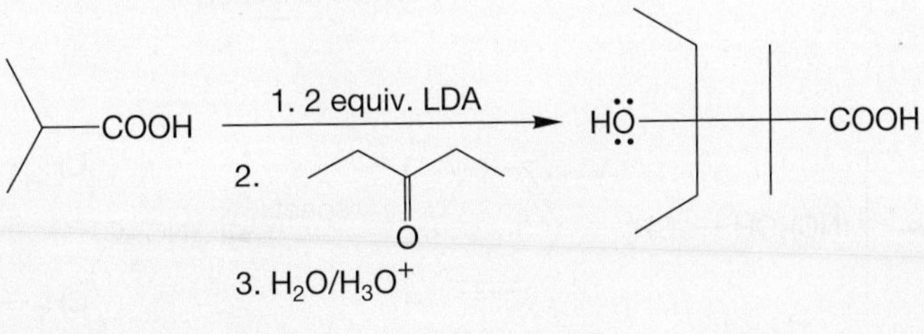

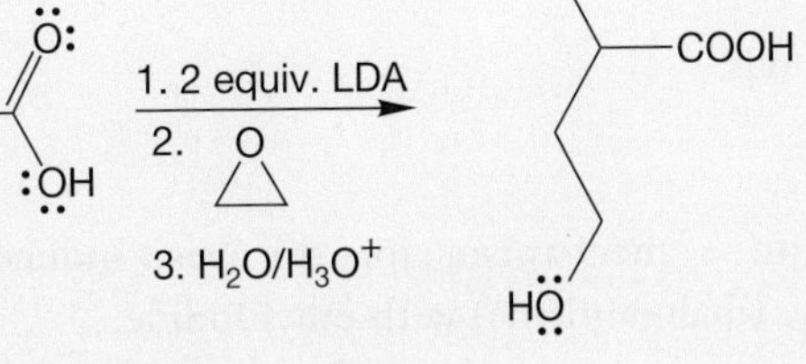

TABLE 19.3 pK$_a$ Values for Some Dicarbonyl Compounds

Compound	Conjugate Base (Enolate)	pK$_a$
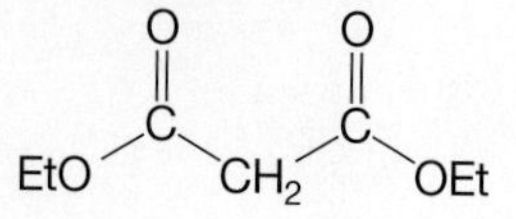	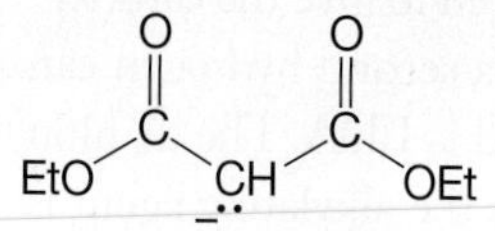	13.3
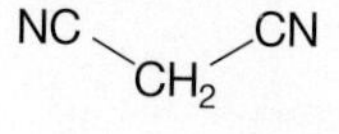	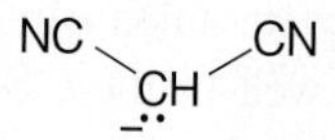	11
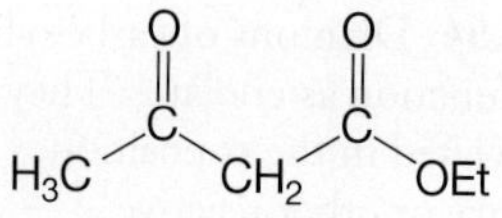	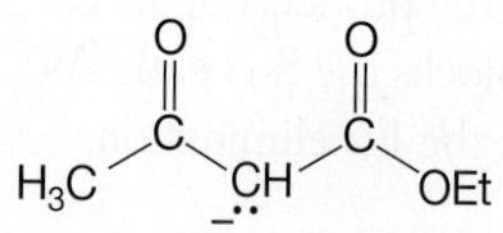	10.7
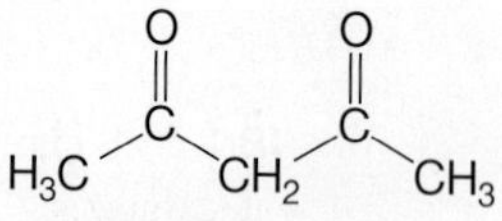	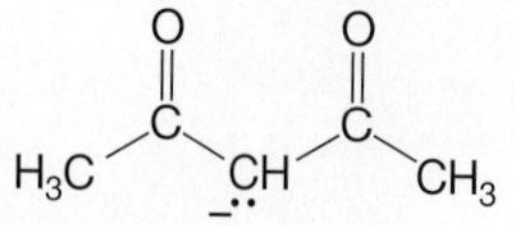	8.9
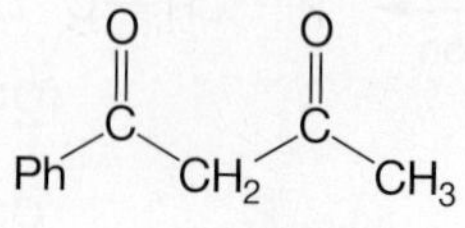	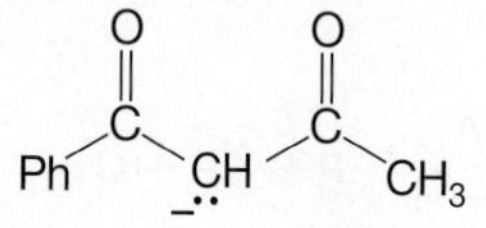	8.5
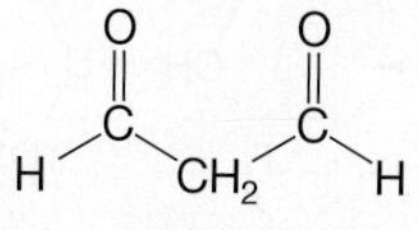	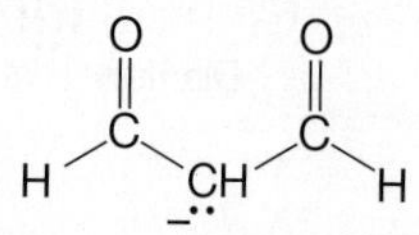	5

19.5c Alkylation of β-Dicarbonyl Compounds β-Dicarbonyl compounds are quite strong acids. The pK$_a$ of diethyl malonate is about 13 and other β-dicarbonyl compounds, β-cyano carbonyl compounds, and dicyano compounds are similarly acidic (Table 19.3). The anions formed from these compounds can be alkylated in S$_N$2 reactions (Fig. 19.55). The once-alkylated compounds still contain one doubly α hydrogen, and therefore remain acidic. A second alkylation reaction is often possible, as long as the requirements of the S$_N$2 reaction are kept in mind. These β-dicarbonyl compounds are sufficiently acidic so that alkylation through the enolate using an alkoxide base is easy. Both mono- and dialkylation reactions of β-keto esters and related diesters are common.

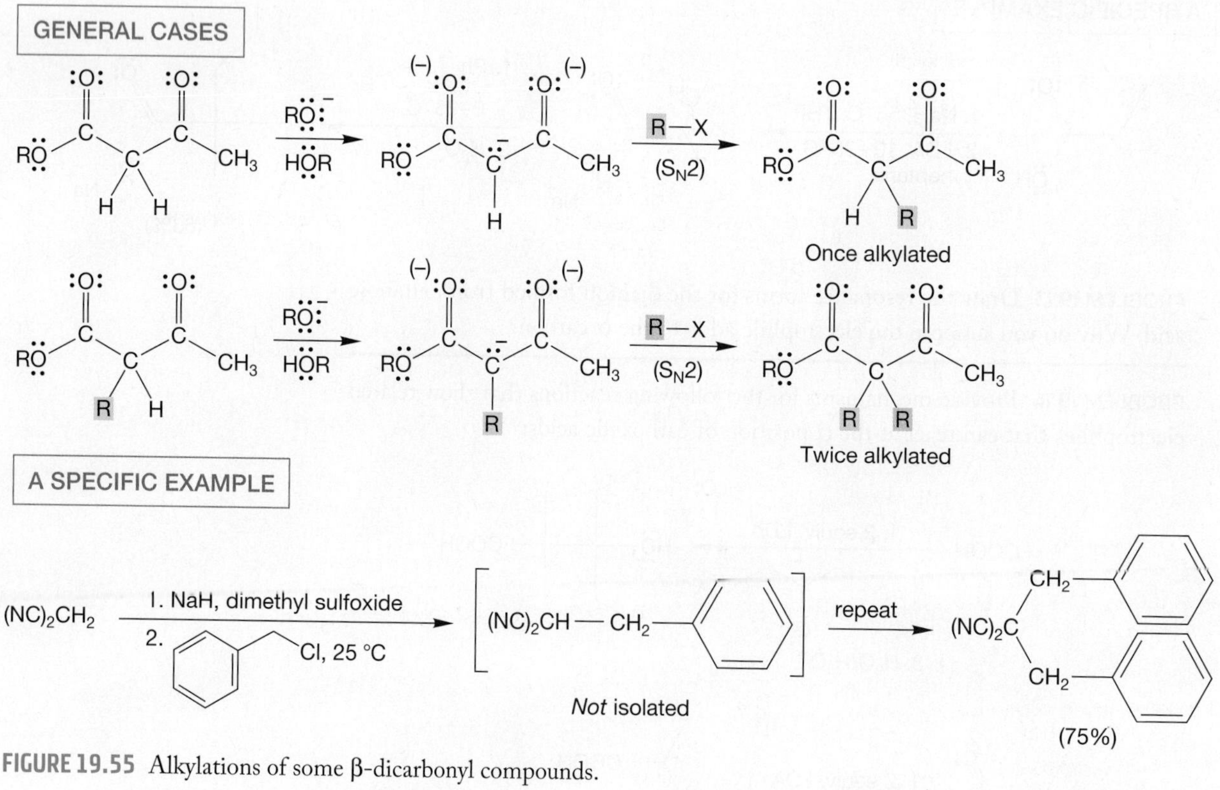

GENERAL CASES

A SPECIFIC EXAMPLE

FIGURE 19.55 Alkylations of some β-dicarbonyl compounds.

PROBLEM 19.15 Write a mechanism for the base-induced alkylation of 1, 1-dicyanomethane (malononitrile) with ethyl iodide.

19.5d Hydrolysis and Decarboxylation of β-Dicarbonyl Compounds

When a β-keto ester such as ethyl acetoacetate is hydrolyzed, the resulting β-keto acid is most unstable and decarboxylates easily (Fig. 19.56). The end result is one

THE GENERAL CASE

molecule of a ketone and one of carbon dioxide. This reaction is a roundabout method of making simple ketones. However, the combination of the alkylation reactions of Figure 19.55 followed by acid or base hydrolysis and decarboxylation is a useful source of more complex ketones (Fig. 19.57).

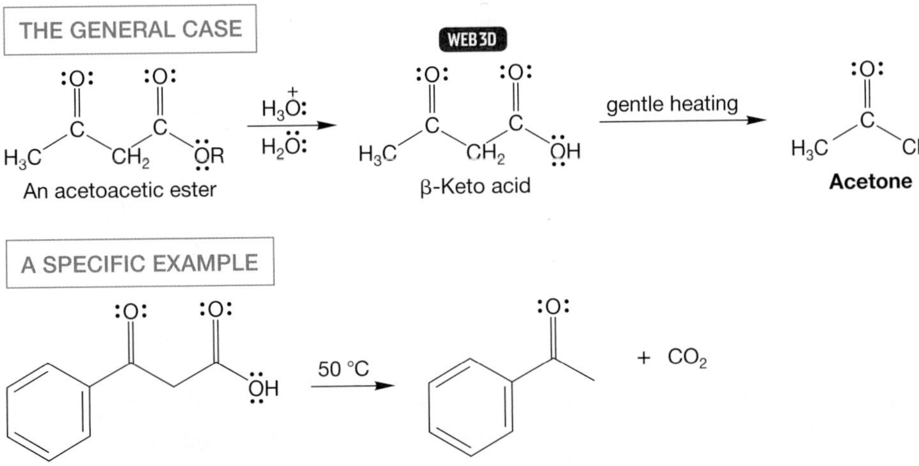

FIGURE 19.56 The acid-catalyzed hydrolysis and decarboxylation of acetoacetic esters.

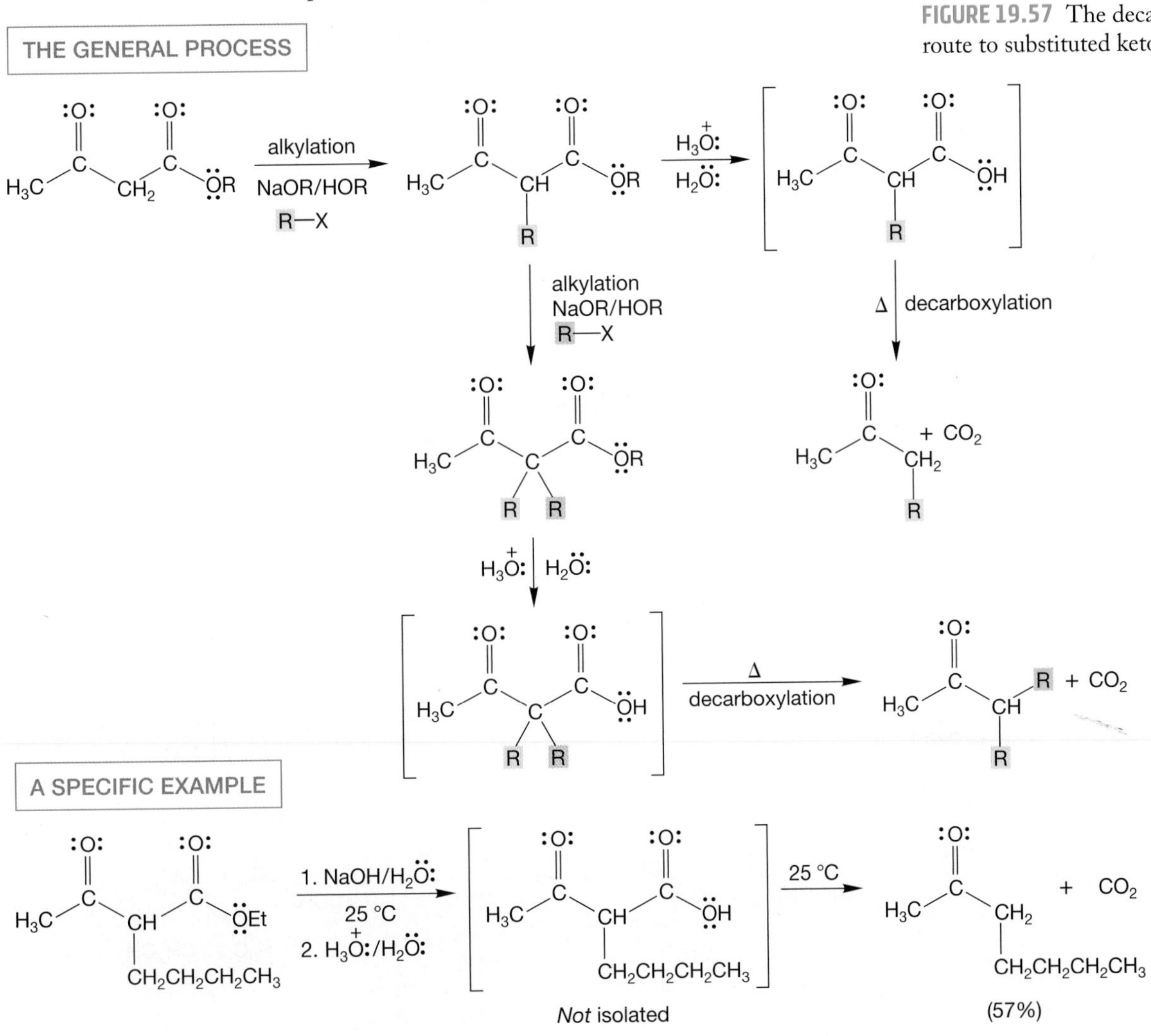

FIGURE 19.57 The decarboxylation route to substituted ketones.

In the decarboxylation reaction, the carbonyl oxygen of the keto group acts as a base and removes the acidic hydrogen of the carboxylic acid group. Notice the nicely organized six-membered ring transition state for this reaction (Fig. 19.58). Carbon dioxide is lost and an enol is initially formed. Normal tautomerization of the enol to the more stable ketone gives the final product.

FIGURE 19.58 The mechanism for the decarboxylation of β-keto acids.

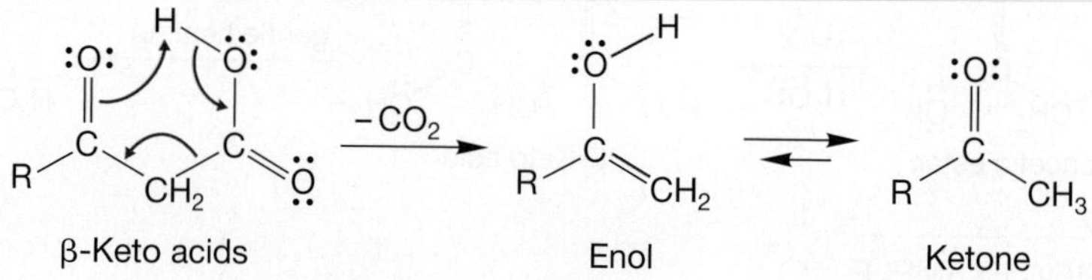

Decarboxylation

β-Keto acids $-CO_2$ Enol   Ketone

Be sure to update your file cards for ketone syntheses with this sequence of reactions, which is called the **β-keto ester** or **acetoacetate synthesis**. This reaction is not easy to see, as it requires a number of steps. In practice, you must dissect a potential target molecule into its parts to determine if it is available through this route. You must identify the R groups to be attached through alkylation, then piece together the original β-keto ester (we will learn how to make β-keto esters in Section 19.8b).

Some examples are useful here, and as usual they will appear as problems. For instance, see Problem 19.16.

WORKED PROBLEM 19.16 Show how the following ketones can be made from suitable β-keto esters through the acetoacetate synthesis. Mechanisms are not required.

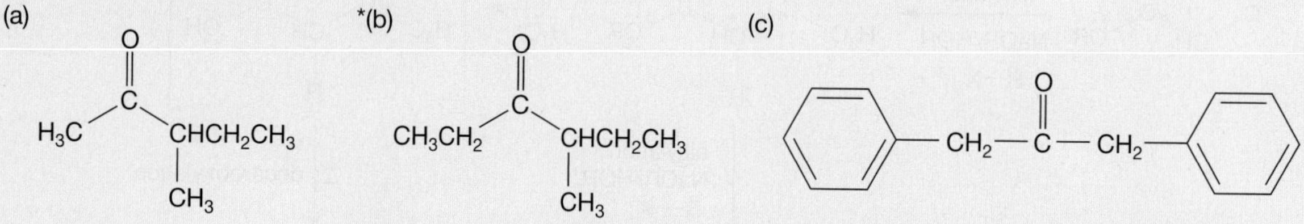

(a)   *(b)   (c)

ANSWER (b) Ethyl 3-oxopentanoate can be alkylated twice, first with methyl iodide and then with ethyl iodide. Hydrolysis and decarboxylation give the desired product.

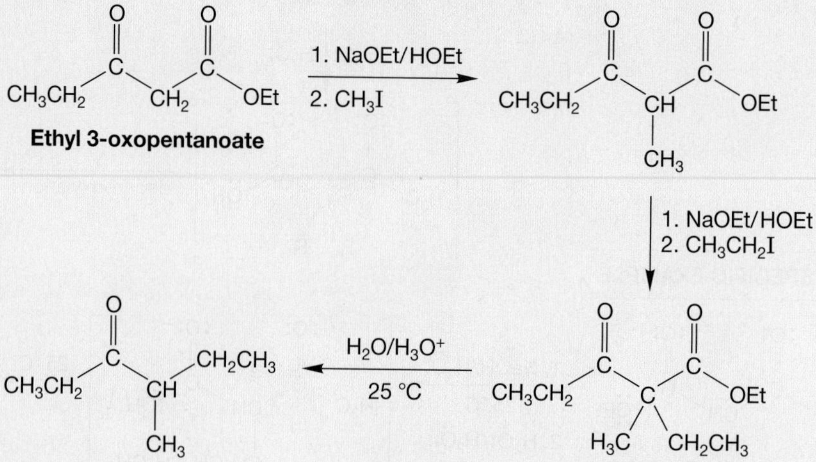

Ethyl 3-oxopentanoate

1. NaOEt/HOEt
2. CH_3I

1. NaOEt/HOEt
2. CH_3CH_2I

H_2O/H_3O^+

25 °C

Diesters such as malonic ester can also be used in this kind of synthetic sequence. The diester is first hydrolyzed in acid or base to the diacid, and then heated to induce loss of carbon dioxide (Fig. 19.59). This process is known as the **malonic ester synthesis**. The end products are carboxylic acids, not the ketones formed in the related reactions of β-keto esters.

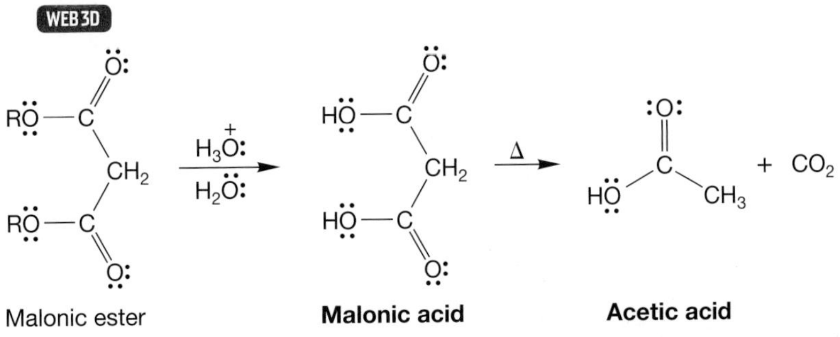

FIGURE 19.59 The formation of acetic acid from malonic acid.

Malonic ester   Malonic acid   Acetic acid

Note an intuitive point here. The β-keto esters, which contain a ketone, are the sources of ketones after decarboxylation, and the diesters, which contain only ester carbonyls, are the source of carboxylic acids after decarboxylation (Fig. 19.60).

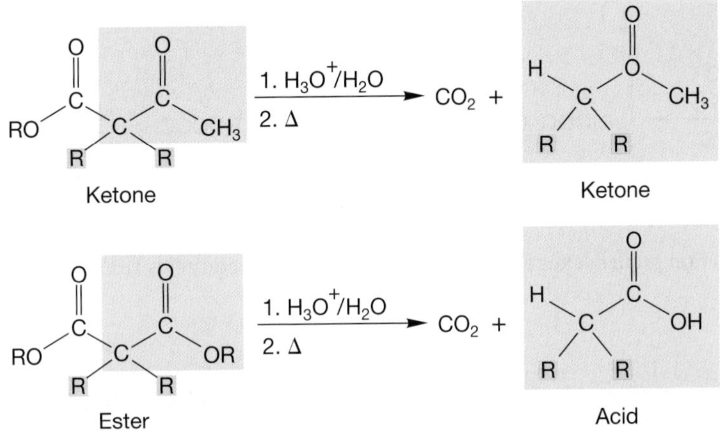

FIGURE 19.60 Hydrolysis and decarboxylation of acetoacetic esters gives substituted *ketones*, whereas the same sequence applied to malonic esters gives substituted *acids*.

Here is how the reaction works. The hydrolysis converts the diester into a diacid (Fig. 19.61). These compounds are similar to β-keto acids and will decarboxylate on heating to give the enols of acetic acids. Notice again the six-membered ring in the transition state for this decarboxylation. The enols are in equilibrium with the more stable carbonyl compounds, which are the isolated carboxylic acid products.

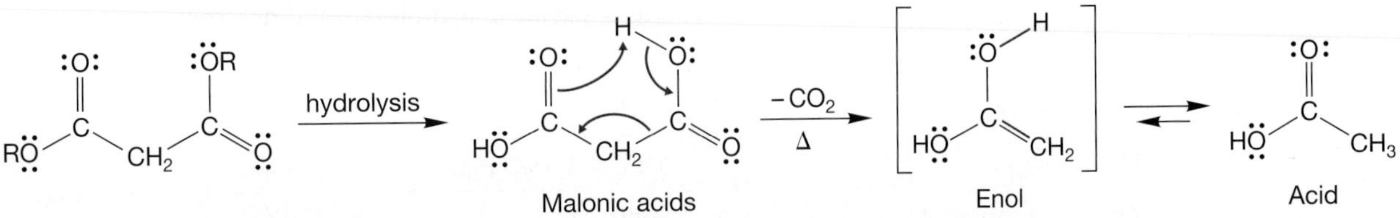

FIGURE 19.61 Hydrolysis and decarboxylation of a diester is shown. The mechanism of decarboxylation has a six-membered ring transition state. The initially formed enol is not isolable. It undergoes keto–enol tautomerization to give the carboxylic acid.

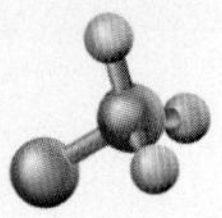

Malonate alkylation

When this reaction is carried out on the parent diethyl malonate, acetic acid results, and the process is of no practical synthetic utility. As with the synthesis of ketones starting from β-keto esters, this process is far more useful when combined with the alkylation reaction. One or two alkylations of malonic ester give substituted diesters that can be hydrolyzed and decarboxylated to give carboxylic acids (Fig. 19.62). Substituted acetic acids are potential sources of esters, acid halides, and any other compound that can be made from a carboxylic acid.

THE GENERAL SEQUENCE

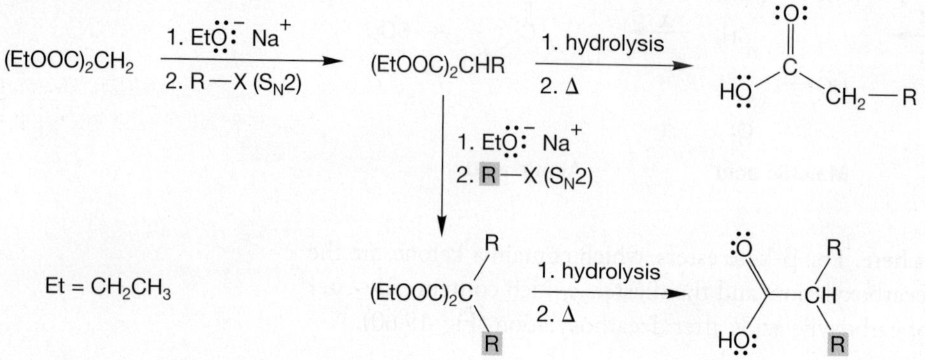

$(EtOOC)_2CH_2$ $\xrightarrow[\text{2. R—X (S}_N2)]{\text{1. EtÖ:}^- \text{Na}^+}$ $(EtOOC)_2CHR$ $\xrightarrow[\text{2. }\Delta]{\text{1. hydrolysis}}$ (structure: HO–C(=O)–CH$_2$–R)

↓ 1. EtÖ:$^-$ Na$^+$
  2. R—X (S$_N$2)

Et = CH$_2$CH$_3$

$(EtOOC)_2C$ (with R above and R below) $\xrightarrow[\text{2. }\Delta]{\text{1. hydrolysis}}$ (structure: HO–C(=O)–CH with R groups)

A SPECIFIC EXAMPLE

$(EtOOC)_2CH—CH(CH_3)(CH_2CH_3)$ $\xrightarrow[\text{2. HCl, 15 °C}]{\substack{\text{1. KOH/H}_2\text{O} \\ \text{100 °C, 5 h}}}$ $[(HOOC)_2CH—CH(CH_3)(CH_2CH_3)]$ *Not* isolated $\xrightarrow{\Delta}$ $HOOCCH_2—CH(CH_3)(CH_2CH_3)$ + CO_2 (64%)

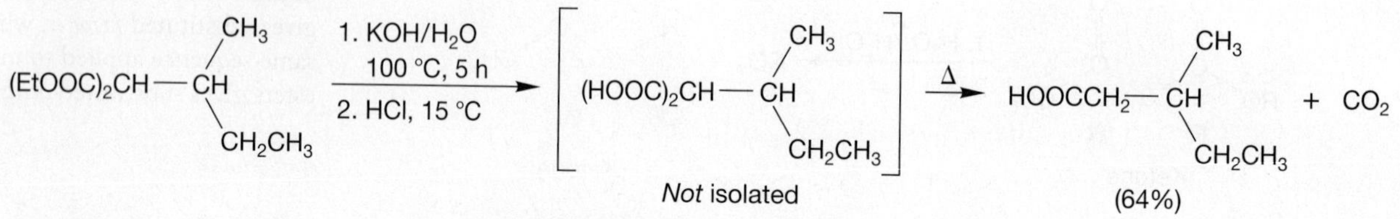

FIGURE 19.62 The alkylation–decarboxylation sequence for the synthesis of substituted acetic acids from diethyl malonate.

WORKED PROBLEM 19.17 Show how the following molecules can be made through a malonic ester synthesis.

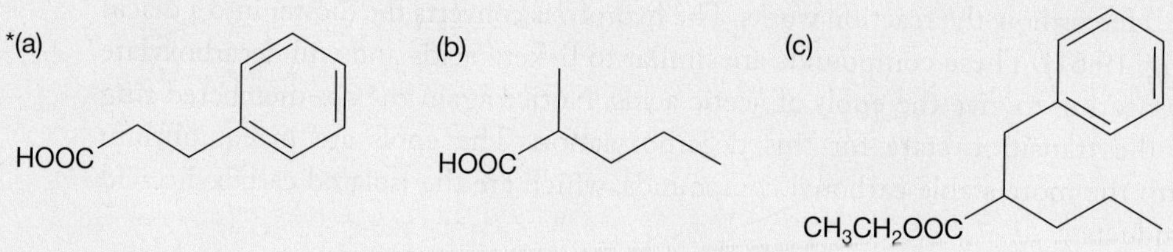

*(a)   (structure: HOOC–CH$_2$CH$_2$–phenyl)

(b)   (structure: HOOC–CH$_2$–CH(CH$_3$)–CH$_2$CH$_3$)

(c)   (structure: CH$_3$CH$_2$OOC–CH with benzyl and propyl groups)

ANSWER (a) Diethyl malonate is alkylated, hydrolyzed, and decarboxylated.

(structure: diethyl malonate) $\xrightarrow[\text{2. PhCH}_2—\text{I}]{\substack{\text{1. NaOEt} \\ \text{(one equiv.)} \\ \text{HOEt}}}$ (structure: benzyl-substituted diethyl malonate, Ph) $\xrightarrow[\text{2. }\Delta]{\text{1. H}_2\text{O/H}_3\text{O}^+}$ (structure: carboxylic acid with OH and Ph) + CO_2

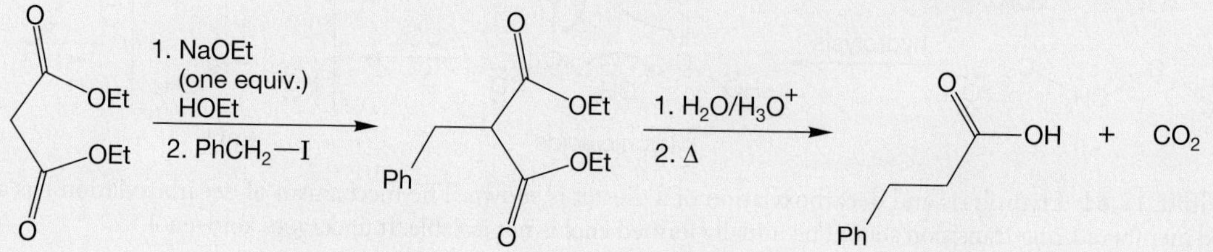

19.5e Direct Alkylation of Esters   It is possible to alkylate esters directly and thereby avoid the extra steps of the malonic ester or acetoacetate synthesis. Very strong bases such as LDA can be used to remove the α hydrogens of esters (Fig. 19.63). When esters are added to these bases at low temperature, an α hydrogen is removed to give the enolate anion. This anion can then be alkylated by any alkyl halide able to participate in the S$_N$2 reaction. The end result is an alkylated ester. This method is usually preferred for small-scale, research laboratory syntheses. The large-scale, industrial syntheses are more likely to use a β-keto ester or malonic ester synthesis, because these methods employ less expensive reagents and can be performed using milder conditions.

THE GENERAL CASE

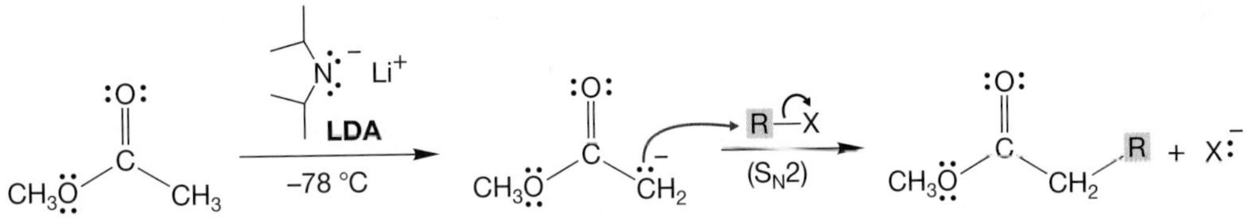

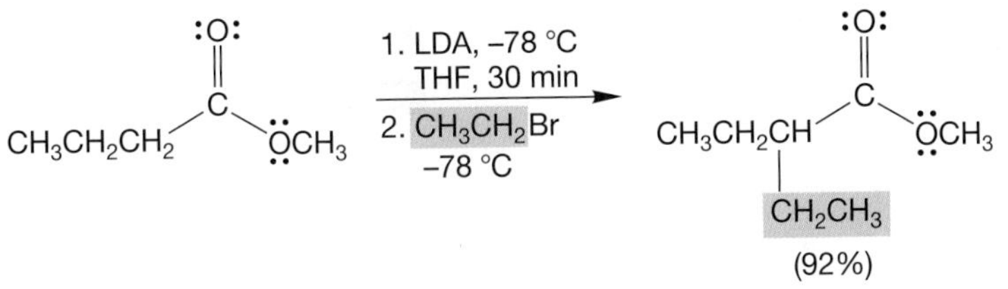

FIGURE 19.63 The low-temperature alkylation of an ester using LDA as base.

A SPECIFIC EXAMPLE

(92%)

Summary

This section gives you new ways of making complex molecules by alkylating ketones, dicarbonyl compounds, and esters. One can efficiently add an alkyl group to a β-dicarbonyl compound, which can then be converted into a ketone or carboxylic acid through the acetoacetate or malonic ester syntheses. Although these routes are somewhat roundabout—the acetoacetic ester or malonic ester must be made, alkylated, and finally decarboxylated—in a practical setting they are efficient and easily performed. A more direct method for alkylating esters involves using LDA at low temperatures.

19.5f Alkylation of Enamines   Recall that α alkylation of ketones or aldehydes using hydroxide base resulted in multiple products. Perhaps not surprisingly, a way around this problem has been found, and it uses enamines (p. 795) as substitutes for carbonyl compounds. Remember that enamines can be formed by the addition of secondary amines (not primary or tertiary amines) to carbonyl compounds. Pyrrolidine is a secondary amine that is often used for this reaction.

PROBLEM 19.18 Why can't primary or tertiary amines be used to make enamines?

Enamines are electron-rich and nucleophilic and will attack S$_N$2-active halides (no tertiary halides!) to form iminium ions. The reaction is ended by adding water to hydrolyze the iminium ions to the corresponding ketone or aldehyde (Fig. 19.64).

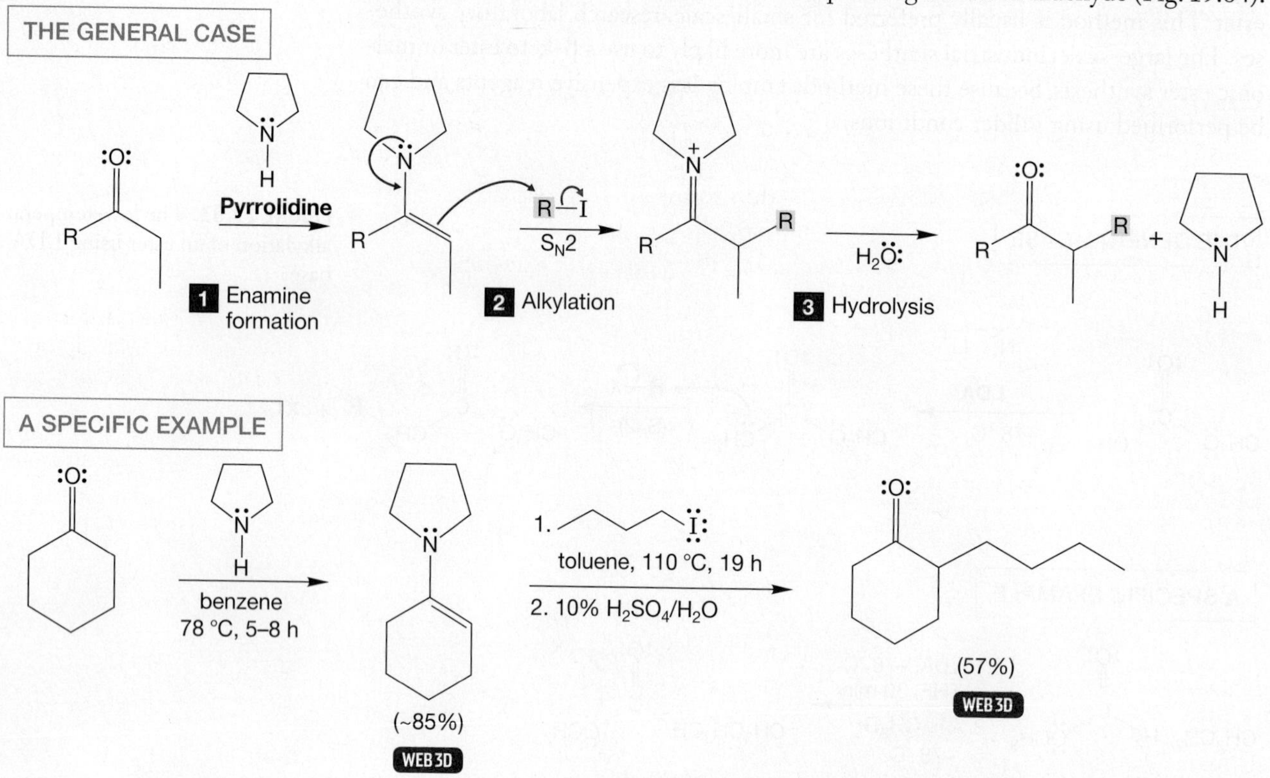

FIGURE 19.64 Alkylation of an enamine initially gives an iminium ion. The iminium ion can be hydrolyzed to give the alkylated carbonyl compound.

PROBLEM 19.19 Write the mechanism for the hydrolysis of the iminium ion to the ketone. *Hint*: You have seen this reaction before, in reverse, in the mechanism for acid-catalyzed enamine formation (p. 795).

There are still some problems, however. It is not easy to use alkylations of enamines made from ketones where more than one possible enamine can be formed. If the two α positions are similar, then the double bond of the enamine can be formed in either direction and α alkylation will take place on both sides. There are occasionally difficulties with overalkylation or substitution on nitrogen. If the two α positions of the ketone are different, then the alkene of the enamine can sometimes be formed selectively on the more hindered side.

Summary

We have observed several bases employed in the formation of enolates, which can be alkylated in S$_N$2 fashion. A useful alternative involves reactions of enamines. Figure 19.64 shows the overall procedure: The ketone or aldehyde is first converted into the enamine, the enamine is alkylated, and then the ketone or aldehyde is regenerated by hydrolysis.

19.6   Addition of Carbonyl Compounds to the α Position: The Aldol Condensation

So far, we have seen the reactions of enolates and enols with a variety of electrophiles, for example, D_2O, H_2O, Br_2, Cl_2, I_2, and alkyl halides. What happens if these acids are absent or, as is the case for water, simply regenerate starting material? Suppose, for example, we treat acetaldehyde with KOH/H_2O (Fig. 19.65). First of all, the answer to the question of what happens when a base and a carbonyl compound are combined is almost never, "nothing." Nucleophiles add to carbonyl groups, and hydroxide will surely initiate hydrate formation. The hydrate is usually not isolable, but it is formed in equilibrium with the aldehyde. Moreover, enolate formation is also possible and the enolate, as well as the hydrate, will be in equilibrium with the starting aldehyde. Although hydrate formation does not lead directly to an isolable product, formation of the enolate does. Be very careful about "nothing happens" answers. They are almost always wrong.

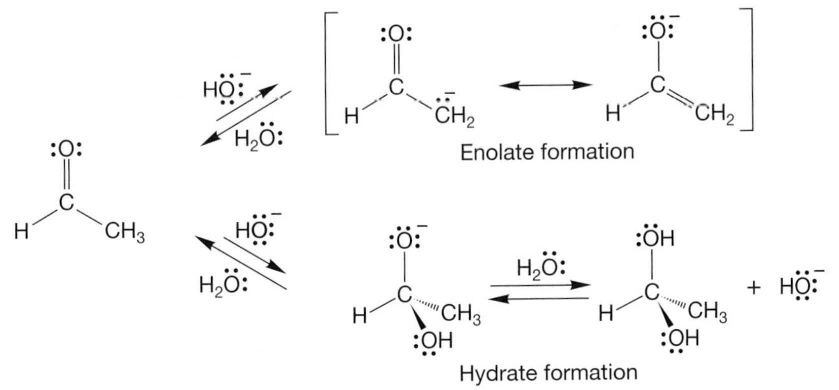

Enolate formation

Hydrate formation

FIGURE 19.65 Two reactions of acetaldehyde with hydroxide ion: addition (hydrate formation) and enolate formation.

19.6a Ketones and Aldehydes React with Ketone or Aldehyde Enolates

In the solution of acetaldehyde and KOH/H_2O there is another electrophile in addition to water, the carbonyl group of acetaldehyde. Nucleophiles add to carbonyl groups, and the enolate anion is surely a nucleophile. There is no essential difference between the addition of hydroxide to the electrophilic carbonyl group of acetaldehyde and the addition of the enolate anion (Fig. 19.66). Nor is there much structural

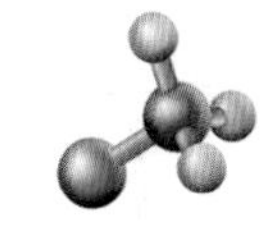

Aldol condensation

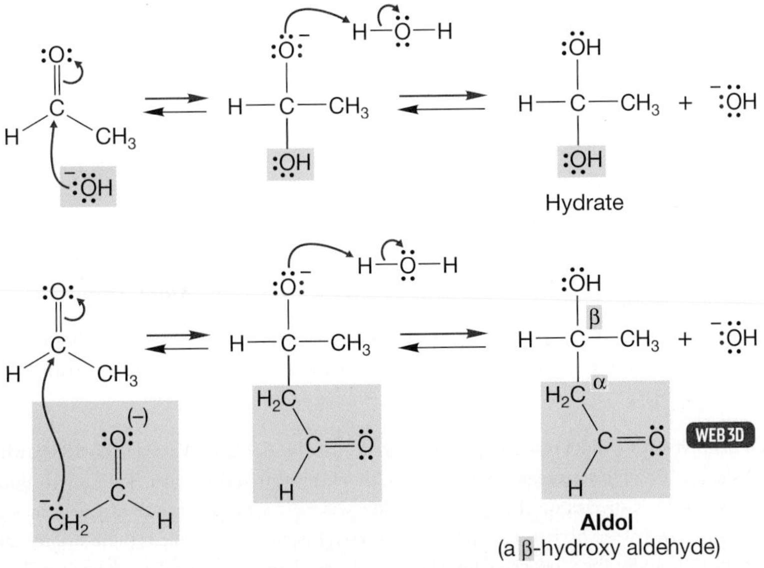

Hydrate

Aldol
(a β-hydroxy aldehyde)

WEB 3D

FIGURE 19.66 Addition of hydroxide and the enolate anion to the carbonyl group are simply two examples of the reaction of nucleophiles with electrophilic carbonyl groups. In the second step, protonation gives the hydrate in the top sequence, or aldol in the lower sequence.

difference between the hydrate formed by protonation of the hydroxide addition product, and the molecule known as aldol (*ald* because of the aldehyde and *ol* because of the alcohol) formed by protonation of the enolate addition product. Note that hydroxide ion, the catalyst, is regenerated in the final step of both hydrate formation or aldol formation. But the aldol is surely more complex than the relatively simple hydrate! The conversion of an aldehyde or ketone into a β-hydroxy carbonyl compound is called the **aldol condensation** or aldol reaction[2] after the product of the reaction of acetaldehyde, and is quite general for carbonyl compounds with α hydrogens.

As usual, this base-catalyzed reaction has its acid-catalyzed counterpart. For the acid-catalyzed reaction the catalyst is not hydroxide, but H_3O^+, and the active ingredient is not the enolate anion, but the enol itself. The first step in the reaction is acid-catalyzed enol formation, the keto–enol tautomerization we have come to know (Fig. 19.67).

FIGURE 19.67 The acid-catalyzed aldol condensation begins with enol formation.

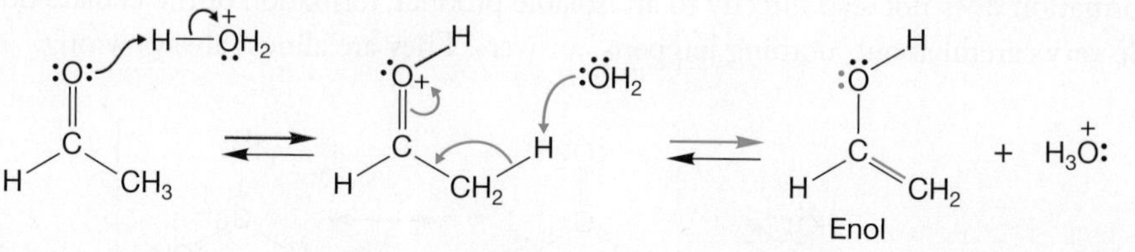

The enol, though nucleophilic, is not nearly as strong a nucleophile as hydroxide. However, the Lewis acid present is the protonated carbonyl, and it is a far stronger electrophile than the carbonyl group itself. The reaction between the enol and the protonated carbonyl group is analogous to the reaction of the enol with H_3O^+ (Fig. 19.68). Once again, the final step of the reaction regenerates the catalyst, this time H_3O^+. The acid- and base-catalyzed aldol reactions of acetaldehyde give the same product, the β-hydroxy carbonyl compound called aldol. The aldol reaction is a general synthesis of β-hydroxy ketones and aldehydes.

Protonation/deprotonation of the enol to regenerate acetaldehyde

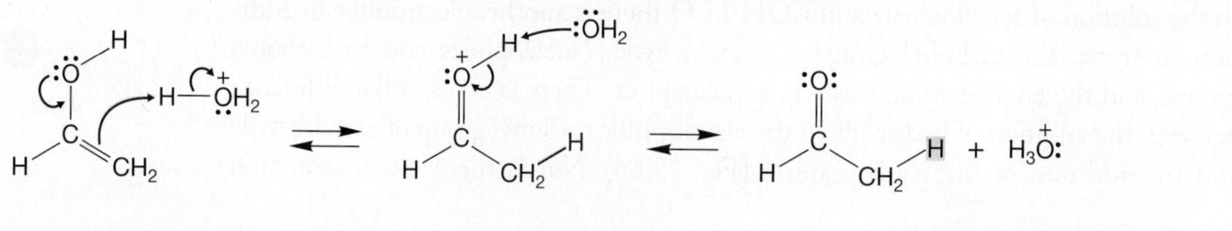

Reaction of the enol with the protonated carbonyl to give aldol

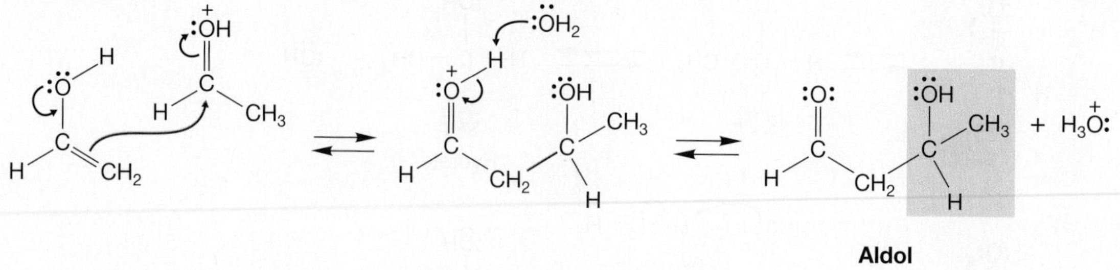

Aldol

FIGURE 19.68 Two reactions of the weakly nucleophilic enol with Lewis acids. In the first case, it is protonated to regenerate acetaldehyde; in the second, the enol adds to the strongly Lewis acidic protonated carbonyl group to give aldol.

---

[2]Credit for discovery of the aldol reaction generally goes to Charles Adolphe Wurtz (1817–1884), who coined the word "aldol" in 1872. However, equal billing at least should also go to Alexander Porfir'yevich Borodin, the Russian composer–physician–chemist (1833–1887). Borodin was the illegitimate son of Prince Gedianov, and actually spent his formative years as his father's serf. Happily, he was given his freedom and became an eminent chemist who is far better known for the products of his avocation, music, than for his excellent chemistry. Borodin also discovered the Hunsdiecker reaction (p. 861).

WORKED PROBLEM 19.20 Write the products of the aldol condensations of the following compounds. Write both an acid- and a base-catalyzed mechanism for (b).

(a)   *(b)   (c)   (d)

ANSWER (b) As usual, in base the enolate is formed first and then adds to the carbonyl group of another molecule. A final protonation generates the condensation product and regenerates the catalyst, here hydroxide.

Watch out! In these and many other figures, arrows emanate from one of a pair of resonance forms, which is done for simplicity's sake only. You should be able to write the arrow formalism using the other resonance form.

In acid, the protonated carbonyl group reacts with the enol form of the ketone to give a resonance-stabilized intermediate. Deprotonation gives the product and regenerates the acid catalyst, H_3O^+.

In acid, a second reaction, the dehydration of the β-hydroxy carbonyl compound to the α,β-unsaturated aldehyde or ketone is very common, and it is generally the dehydrated products that are isolated (Fig. 19.69). In the overall reaction, two carbonyl compounds condense to produce a larger carbonyl compound and give off a small molecule—water in this case. Thus, the aldol reaction is often called the aldol condensation.

FIGURE 19.69 One possible mechanism for the acid-catalyzed dehydration of an aldol.

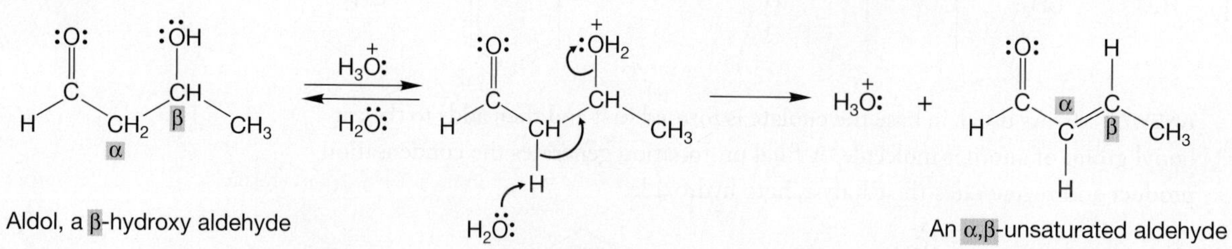

Aldol, a β-hydroxy aldehyde

An α,β-unsaturated aldehyde

PROBLEM SOLVING

Success in solving what are often called "fun in base" problems has a lot to do with recognition, of seeing the clues that are present in almost every problem. We will have more to say on this subject later, but for now remember that "all" β-hydroxy aldehydes and ketones and "all" α,β-unsaturated aldehydes and ketones come from aldol condensations. The quotes around "all" are there just to remind you that there will inevitably be exceptions.

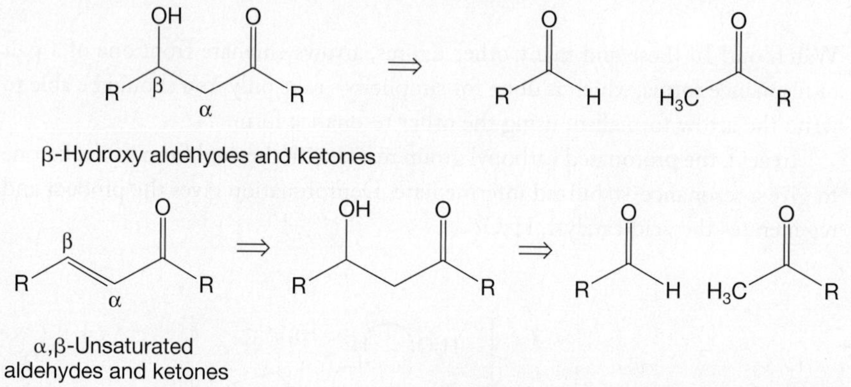

β-Hydroxy aldehydes and ketones

α,β-Unsaturated aldehydes and ketones

However, when you are dealing with a condensation problem, one of the first things you should do is scan for β-hydroxy aldehydes and ketones, and α,β-unsaturated aldehydes and ketones. If you find one, and you will very often, you can be almost certain that an aldol condensation is involved.

A second point is to remember that in the aldol condensation that forms these compounds, it is the α,β bond that is made.

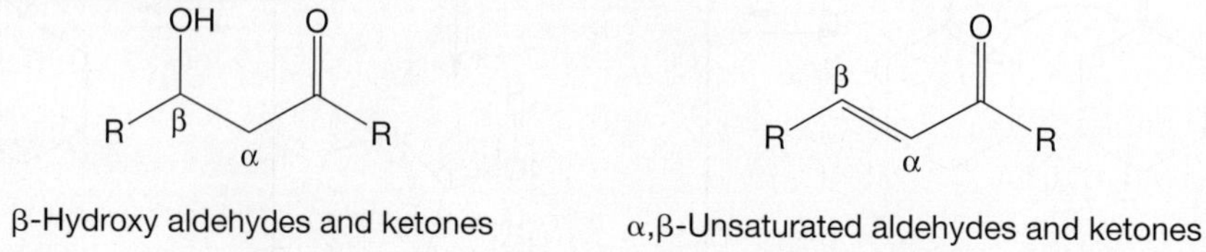

β-Hydroxy aldehydes and ketones   α,β-Unsaturated aldehydes and ketones

Let's consider the mechanism for the loss of water. Hydroxide is a poor leaving group. Therefore, dehydration is more difficult in base and the β-hydroxy compounds can often be isolated. However, even under basic conditions dehydration can be achieved at relatively high temperature, forming the conjugated carbonyl compound.

But the mechanism is not a simple E2 reaction (Fig. 19.70). Removal of the α hydrogen is relatively easy, as this process gives a resonance-stabilized enolate anion. In a second step, hydroxide is lost and the α,β-unsaturated compound is formed. This kind of elimination mechanism, called the E1cB reaction, was the subject of discussion in Chapter 7 (p. 309).

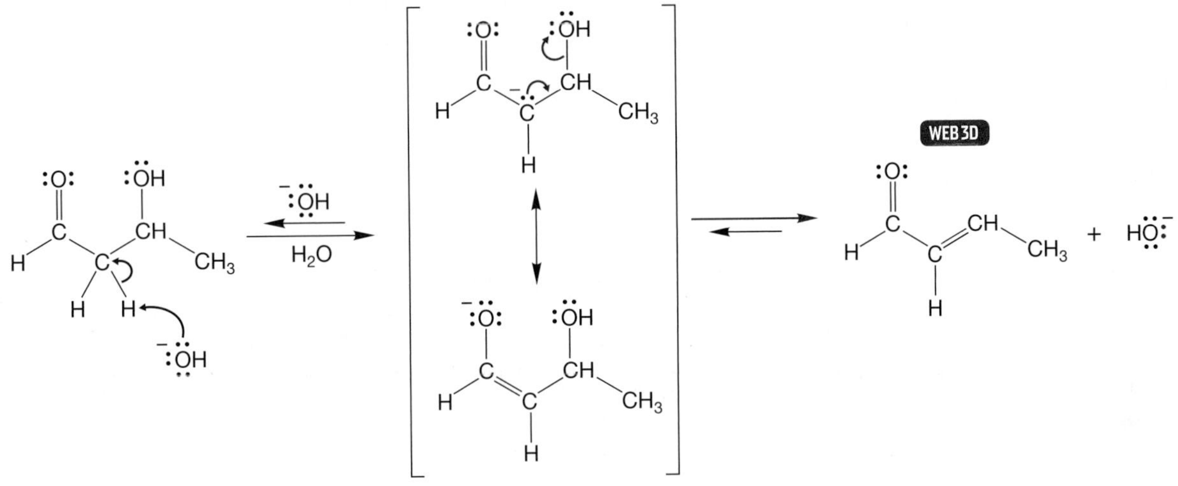

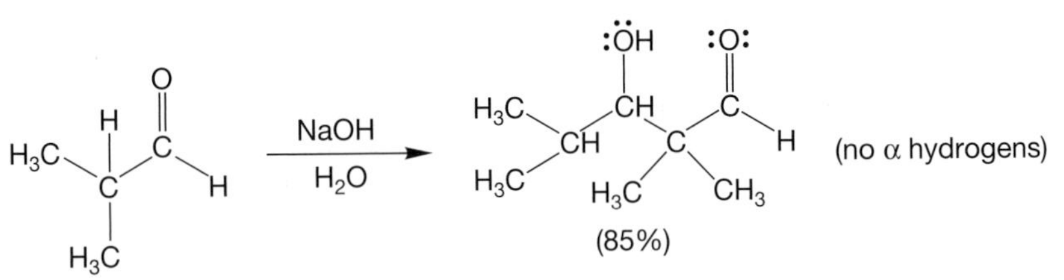

Resonance-stabilized enolate

(85%)

(no α hydrogens)

FIGURE 19.70 The base-catalyzed elimination of water from aldol. The reaction mechanism is E1cB. If no α hydrogen is available for elimination, then the initial aldol product can be isolated as shown in the specific example.

PROBLEM 19.21 What are the products of dehydration for the reactions of Problem 19.20?

PROBLEM 19.22 Figure 19.69 shows protonation of the hydroxyl oxygen, not the carbonyl oxygen. That picture is correct as the pK_a of a protonated alcohol is about -2, and the pK_a of a protonated aldehyde is about -10. First explain the relevance of the pK_a data, and then explain why the hydroxyl oxygen is protonated in preference to the aldehyde. *Hint*: Think hybridization.

Like many reactions, the aldol condensation is a series of equilibria and it is not always clear what is the slow step, the rate-determining step. For example, consider the reaction of acetaldehyde in D_2O/DO^- (Fig. 19.71). In the aldol formed, deuterium incorporation at carbon is a function of acetaldehyde concentration. At high acetaldehyde concentration the aldol product initially formed is not deuterated on any carbon, but at low acetaldehyde concentration it is deuterated at two positions. The reason for the concentration dependence is that the rates of the first two steps of the reaction are similar. Once the enolate is formed it can either revert to acetaldehyde, which will result in exchange in D_2O, or go on to the aldol product in a bimolecular reaction with a molecule of acetaldehyde. At high acetaldehyde concentrations the enolate is more likely to form the aldol.

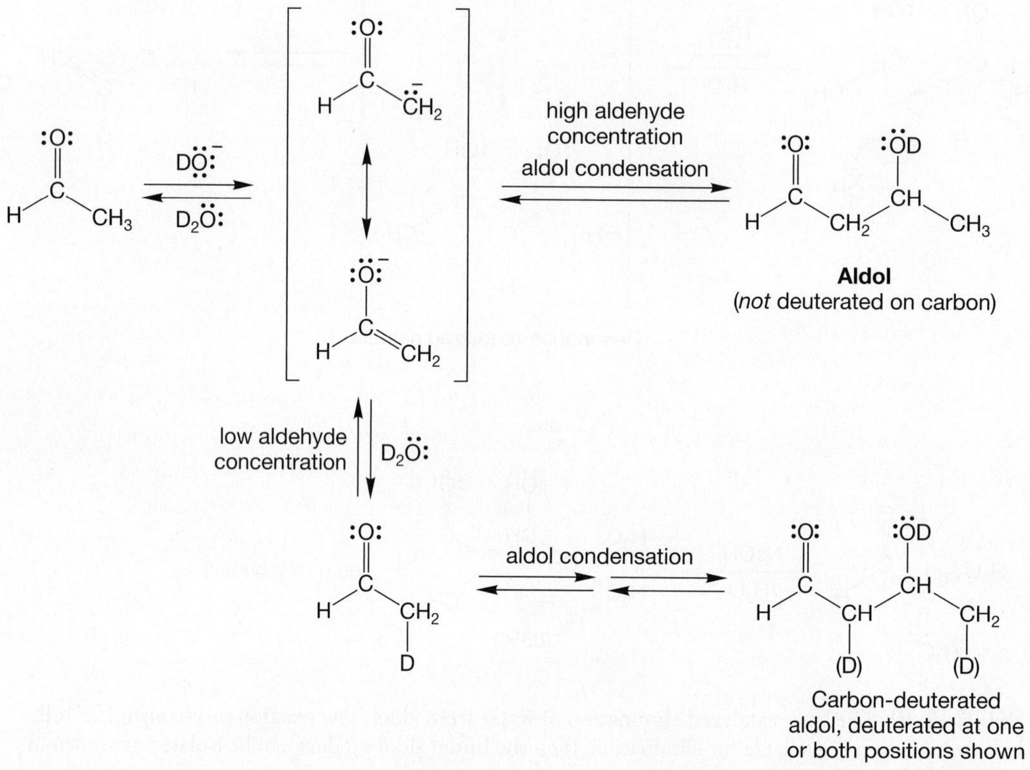

FIGURE 19.71 A high concentration of acetaldehyde results in a relatively rapid aldol reaction and little or no exchange at carbon. If, however, the concentration of acetaldehyde is low, the bimolecular reaction will be slowed and the enolate can revert to acetaldehyde. In D_2O, this reaction leads to exchange of hydrogens attached to carbon. The deuterium can appear in either or both of the two positions and is shown as (D).

If the concentration of acetaldehyde is high, the rate of the bimolecular reaction is faster (rate = k[enolate][acetaldehyde]), and acetaldehyde is not re-formed from the enolate. Under these conditions, it is the formation of the enolate that is the slow step in the reaction, the rate-determining step (p. 350). If the concentration of acetaldehyde is low, the enolate is reconverted to acetaldehyde, and in D_2O this reversal results in exchange. When exchanged acetaldehyde goes on to form aldol, the product also contains deuterium attached to carbon. At low acetaldehyde concentrations it is the addition step itself that is the slower one, and therefore the rate-determining step.

Ketones can undergo the aldol reaction too, and the mechanism is similar to that for the aldol condensation of aldehydes. We will use acetone as an example (Fig. 19.72). In base, the enolate is formed first, and adds to the electrophilic carbonyl compound. Protonation by water yields a molecule once known as diacetone alcohol, 4-hydroxy-4-methyl-2-pentanone.

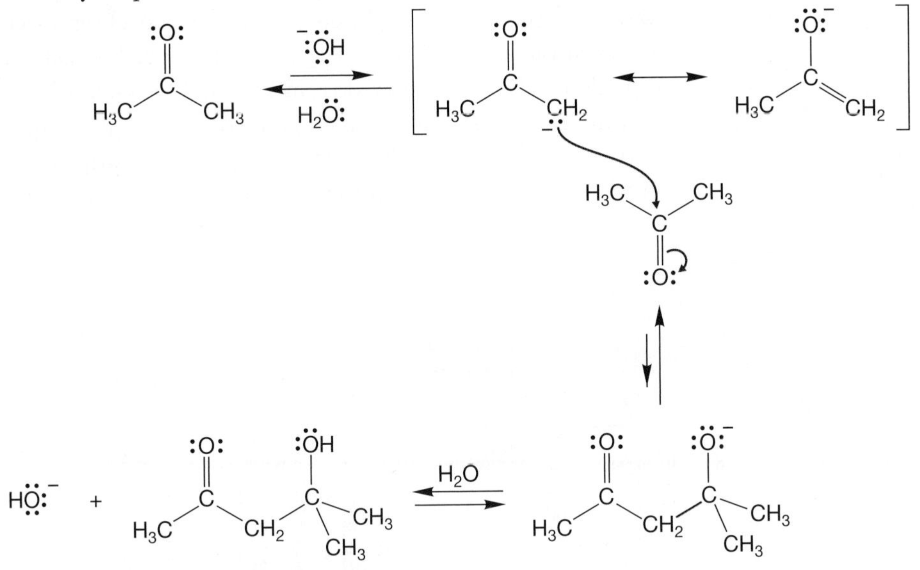

FIGURE 19.72 The base-catalyzed aldol reaction of acetone.

4-Hydroxy-4-methyl-2-pentanone
(diacetone alcohol)

In the acid-catalyzed condensation of acetone, the enol is the intermediate, and it adds to the very strong Lewis acid, the protonated carbonyl group, to give initially the same product as the base-catalyzed reaction, diacetone alcohol (Fig. 19.73). Dehydration generally follows to give the α,β-unsaturated ketone, 4-methyl-3-penten-2-one (mesityl oxide).

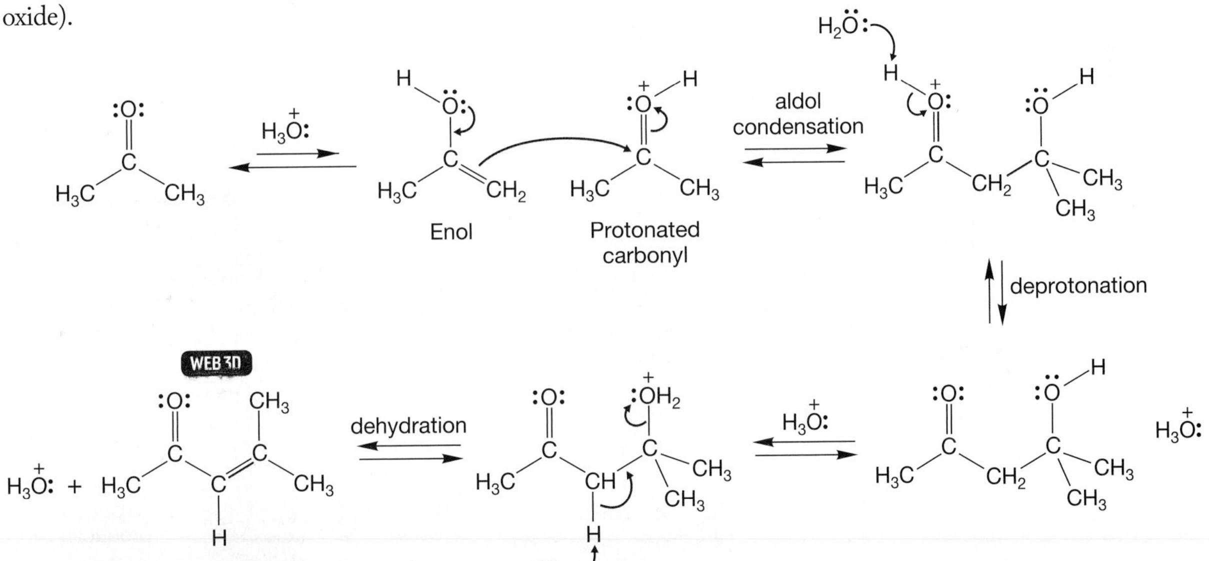

4-Methyl-3-penten-2-one
(mesityl oxide)

FIGURE 19.73 The acid-catalyzed aldol condensation of acetone. The first product, diacetone alcohol, is generally dehydrated in acid to give 4-methyl-3-penten-2-one.

We have been at pains to point out the similarities between these aldol reactions and other addition reactions of carbonyl compounds. Like simple additions, such as hydration, the aldol condensations are series of equilibria. They are affected by structural factors in the same way as the simple reactions. Product formation is unfavorable in base-catalyzed aldol reactions of ketones, just as it is in hydration

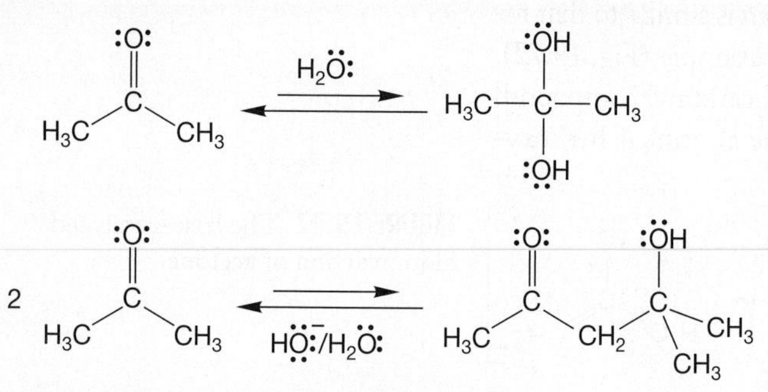

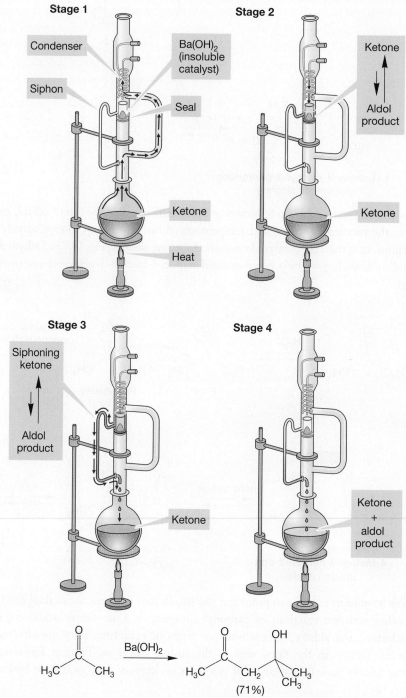

FIGURE 19.74 In the hydration of ketones, the hydrate is usually not favored at equilibrium. Similarly, in the aldol reaction of ketones the initial product molecule is not usually favored, but the dehydrated product is.

(Fig. 19.74). In acid, dehydration generally occurs and formation of the α,β-unsaturated ketone pulls the equilibrium to the right.

How can the aldol condensation of ketones be useful? If equilibrium favors starting material, how can the product be isolated? Ingenious ways have been found to thwart the dictates of thermodynamics. In one, a spectacularly clever piece of apparatus called a Soxhlet extractor is used. The ketone is boiled in a flask separated from the catalyst, usually barium hydroxide, Ba(OH)$_2$, which is insoluble in organic materials. The ketone distills to the condenser, and drips from it to come in contact with the catalyst (stage 1, Fig. 19.75).

FIGURE 19.75 The operation of a Soxhlet extractor.

Now the aldol condensation takes place to give an equilibrium mixture of starting ketone (favored) and the product of aldol condensation (unfavored) (stage 2). As the starting ketone continues to boil, and the level of the liquid in the trap rises, the whole solution eventually is siphoned from the chamber into the flask that previously contained only ketone (stage 3). There is still no catalyst in the flask, and so the aldol product *cannot revert to starting material*. The mixture in the flask continues to boil, the remaining ketone, which has a much lower boiling point than the product, continues to come in contact with the catalyst in the isolated chamber, and slowly the concentration of product in the flask increases to a useful amount (stage 4).[3]

All β-hydroxy carbonyl compounds are potential products of an aldol reaction. Whenever you see one, your thoughts about synthesis must turn first to the aldol reaction. The same is true for the dehydration products, the α,β-unsaturated carbonyl compounds. It is important to be able to go quickly to the new bond, to find the one formed during the aldol condensation, and to be able to dissect the molecule into its two halves. In doing this operation, you are merely carrying out a retrosynthetic analysis and following the mechanism backward (Fig. 19.76).

FIGURE 19.76 A retrosynthetic analysis for the product of an aldol condensation. Note the retrosynthesis arrows that mean "can be formed from."

PROBLEM 19.23 Write mechanisms for the acid- and base-catalyzed reverse aldol reaction of diacetone alcohol shown in Figure 19.72.

WORKED PROBLEM 19.24 Perform retrosynthetic analyses on the following three molecules:

(a)   *(b)   (c)

ANSWER (b) In any aldol, or aldol-like condensation, the carbon–carbon π bond is formed through an elimination reaction (Fig. 19.73). So, the first part of our retrosynthetic analysis recognizes that the precursor to the final product is β-hydroxy ketone **A**.

(continued)

---

[3] One would be wrong to think that the day of clever apparatus has passed. See, for example, Charlie Biggs' (age 5) creation here. It's not totally clear exactly what it's for, but there must be a good use for it.

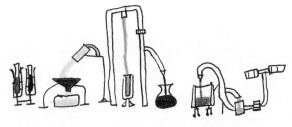

Now, where does **A** come from? All β-hydroxy carbonyl compounds are aldol products, so the σ bond joining the two rings must come from the following aldol reaction:

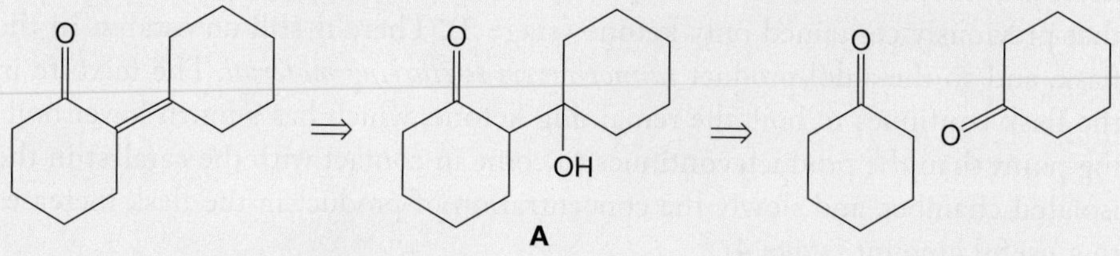

Summary

Base- and acid-catalyzed aldol reactions of aldehydes generally lead to useful amounts of aldol product, the β-keto alcohol, or the product of dehydration, the α,β-unsaturated carbonyl compound. Aldol reactions of ketones are less useful because equilibrium does not usually favor product. Special techniques can be used (Soxhlet extractor) to thwart thermodynamics. If dehydration to the α,β-unsaturated carbonyl compound occurs, as it often does in acid-catalyzed aldols, the reaction may be practically successful.

19.6b Ester Enolates React with Ketones or Aldehydes Any compound that can support an anion through resonance stabilization can attack a ketone or aldehyde. As we have seen, malonate esters and similar compounds readily form resonance-stabilized anions when treated with bases.

PROBLEM 19.25 Draw resonance forms for the anions of each compound in Table 19.3 (p. 958).

Bases, often amines, can generate significant amounts of enolates from the β-dicarbonyl compounds in Table 19.3, and similar molecules. In the presence of an acceptor ketone or aldehyde, condensation occurs to give, ultimately, the α,β-unsaturated diester shown in Figure 19.77. This reaction is called the **Knoevenagel condensation** after Emil Knoevenagel (1865–1921).

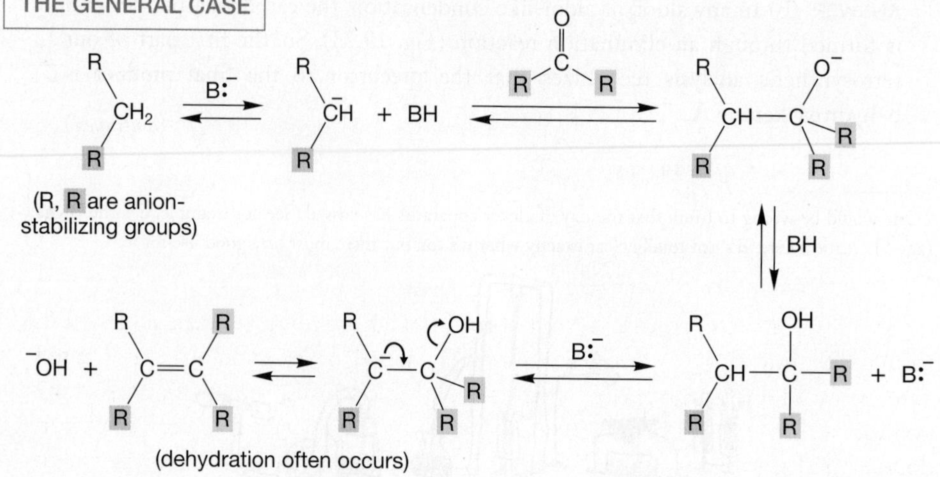

FIGURE 19.77 The Knoevenagel condensation.

SPECIFIC EXAMPLES

WEB 3D

FIGURE 19.77 *(continued)*

(89%)

(83%)

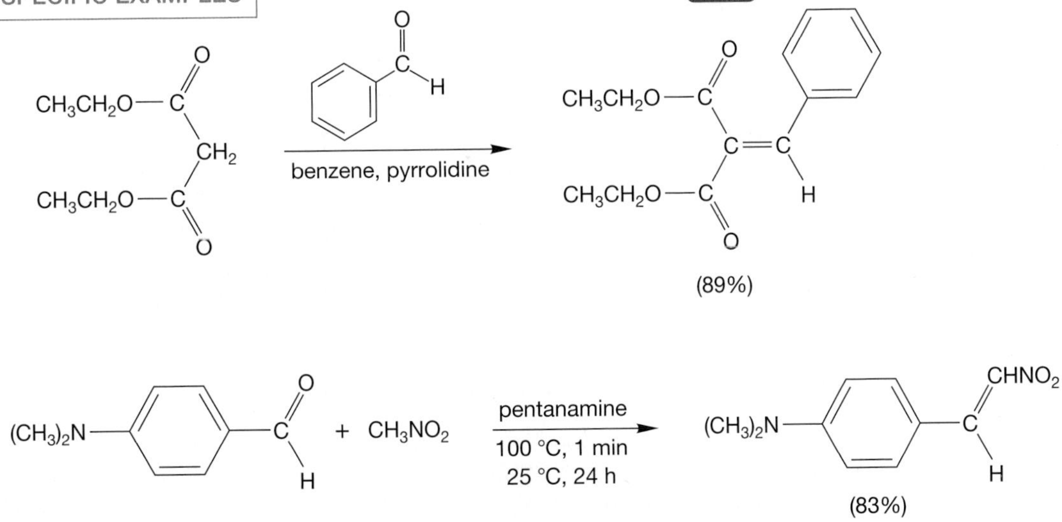

Many other reactions, some with names, many without, are conceptually similar, in that they involve the addition of a complex anion to a ketone or aldehyde. But, in these other reactions the source of the nucleophilic anionic partner is not always an enolate. For example, cyclopentadiene (pK_a = 15) is a strong enough acid to give an anion when treated with a base, and the cyclopentadienide ion (p. 589) can add to ketones. The end result is a synthesis of the beautifully colored compounds known as fulvenes (Fig. 19.78).

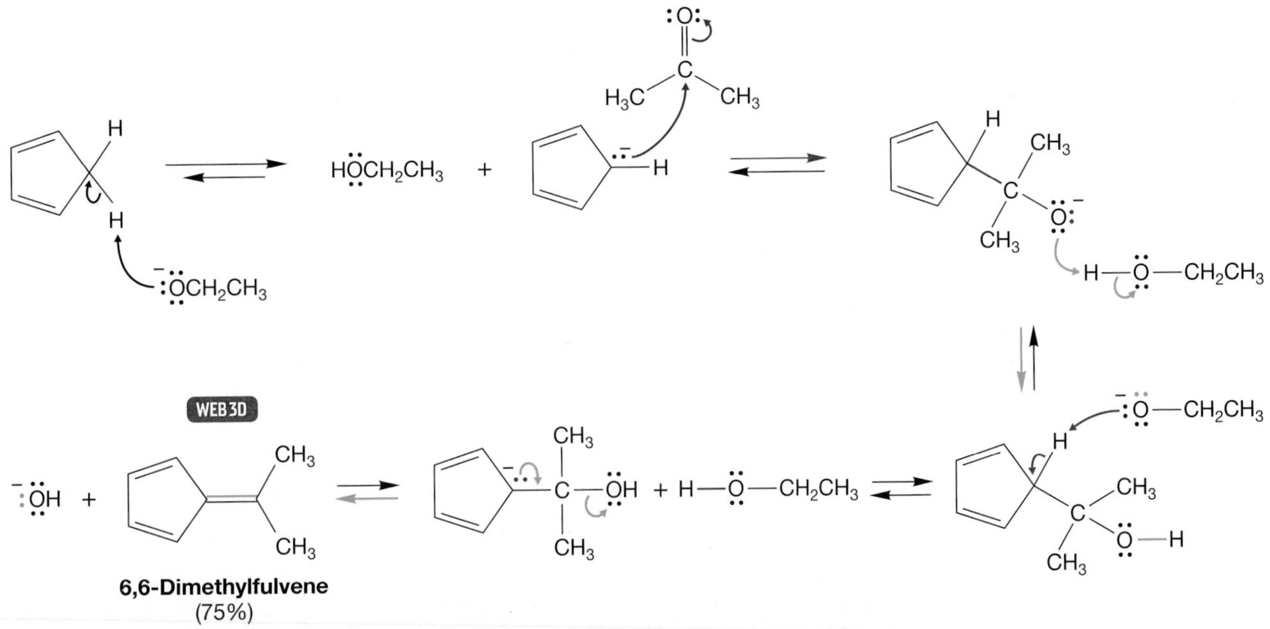

WEB 3D

6,6-Dimethylfulvene
(75%)

FIGURE 19.78 A condensation reaction leading to fulvenes.

With Knoevenagel reactions, your analytical task is more complicated than with aldols. It is quite easy to learn to dissect β-hydroxy ketones or α,β-unsaturated ketones into their components, but how is one to deal with the less obvious compounds like fulvenes? There is no easy answer to this question. A carbon–carbon double bond is always the potential result of a condensation reaction. You must learn to look on them that way, and to evaluate the pieces thus created as parts of a possible synthesis (Fig. 19.79).

The following problem gives some practice at devising syntheses of molecules formed by condensation reactions.

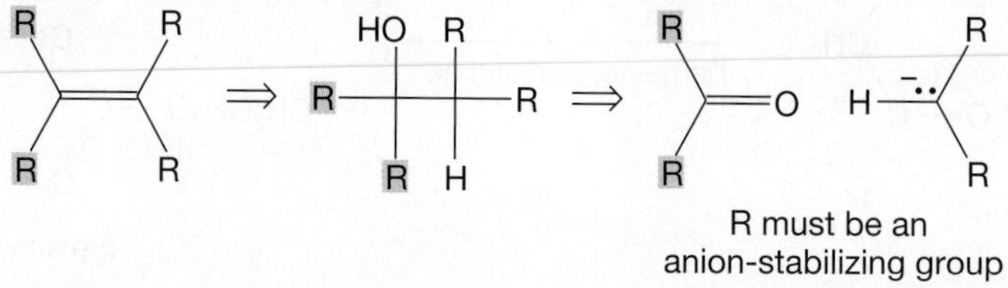

FIGURE 19.79 Any carbon–carbon double bond is the formal result of a condensation reaction.

PROBLEM 19.26 Propose syntheses for the following molecules:

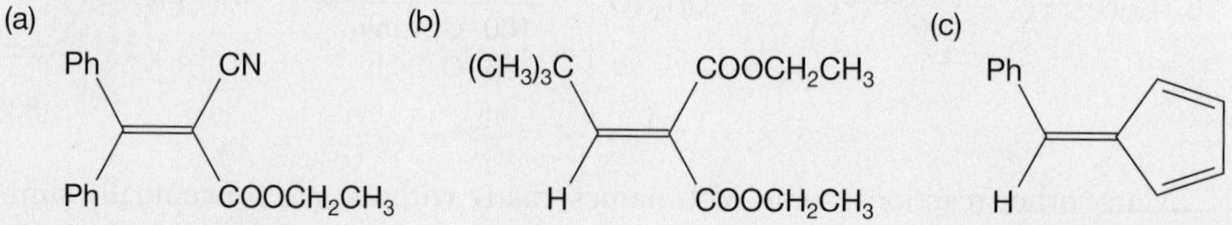

(a)

Ph   CN

Ph   COOCH₂CH₃

(b)

(CH₃)₃C   COOCH₂CH₃

H   COOCH₂CH₃

(c)

Ph

H

Michael addition

19.6c Ester Enolates Add to Enones: The Michael Addition

We now have a synthetic route to α,β-unsaturated carbonyl compounds, the aldol (and related) condensation reactions. There is an especially important process associated with these compounds, called the **Michael reaction**, after its discoverer, the American chemist Arthur Michael (1853–1942).

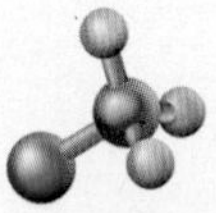

Ordinary carbon–carbon double bonds are unreactive toward nucleophiles. In particular, addition is not observed, because it would require formation of an unstabilized anion, the conjugate base of an alkane. Alkanes are extraordinarily weak acids. Their pK_a values are difficult to measure, but numbers as high as 70 have been suggested (Fig. 19.80). When working on problems in organic chemistry, an unstabilized carbanion is the functional equivalent of a methyl or primary carbocation. It is a stop sign, a signal that you are almost certainly wrong!

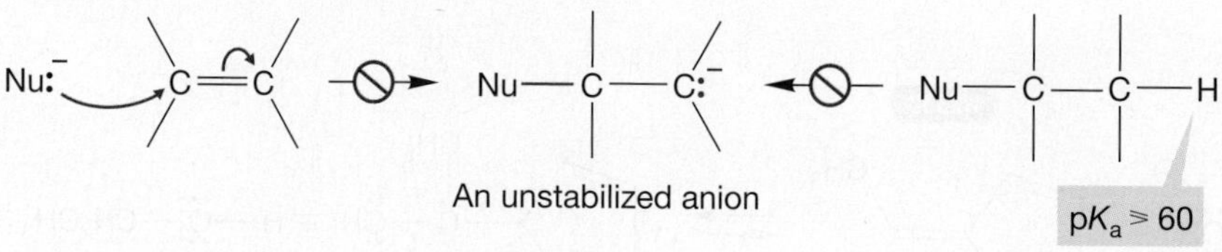

FIGURE 19.80 Addition of a base such as alkoxide to a simple alkene would yield an unstabilized anion. A measure of the difficulty of this reaction can be gained by examining the acidity of the related hydrocarbon. Such species are extraordinarily weak acids and their pK_a values are very difficult to determine.

An unstabilized anion

p$K_a \geqslant 60$

By contrast, many nucleophiles add easily to the double bonds of α,β-unsaturated carbonyl compounds, also known as **enones**. This process is called the Michael reaction. Notice the resonance stabilization of the resulting enolate anion (Fig. 19.81).

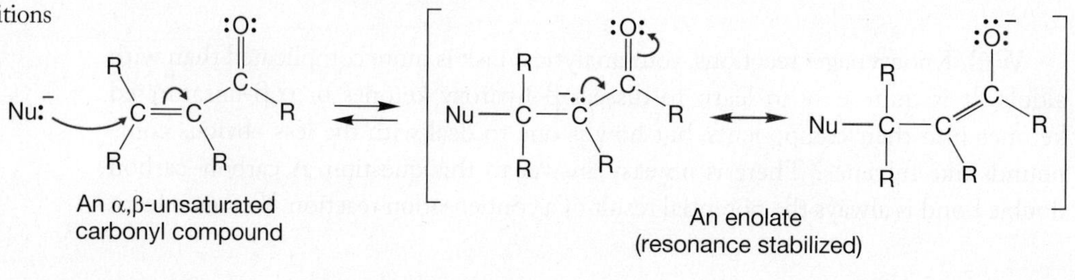

FIGURE 19.81 Nucleopilic additions to α,β-unsaturated carbonyl compounds are common.

An α,β-unsaturated carbonyl compound

An enolate (resonance stabilized)

The reason is the initial formation, not of a bare, unstabilized carbanion as was the case in Figure 19.80, but of a nicely resonance-stabilized species. The carbonyl group stabilizes any α anion, no matter how it is formed. The Michael addition is another way of generating an enolate.

When the species adding to the α,β-unsaturated compound is itself an enolate, the reaction is properly called the Michael reaction, but related additions are often also lumped under this heading (Fig. 19.82).

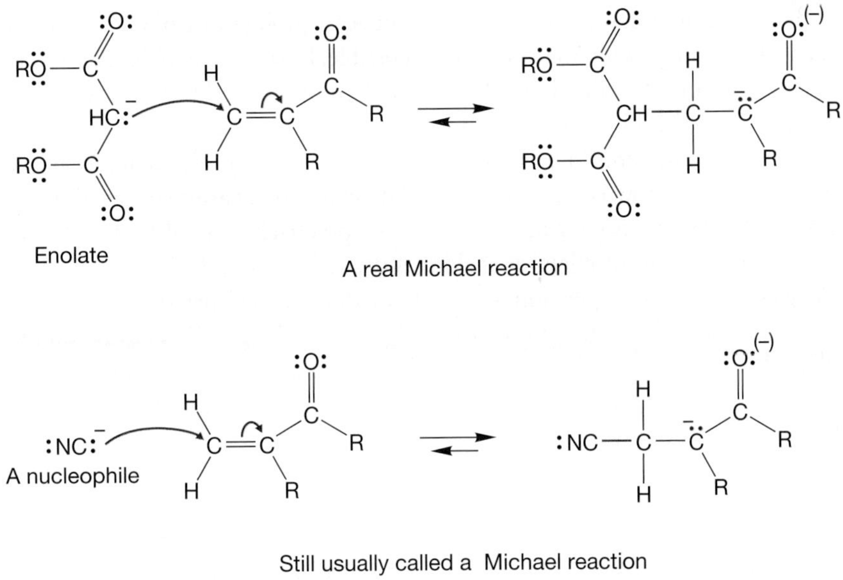

Enolate

A real Michael reaction

FIGURE 19.82 Two Michael reactions.

A nucleophile

Still usually called a Michael reaction

There is also an acid-catalyzed version of the Michael reaction. In this reaction the carbonyl group is initially protonated, and the carbon–carbon double bond is then attacked by a nucleophile to give the enol (Fig. 19.83). The enol equilibrates with the more stable carbonyl compound to produce the final product of addition.

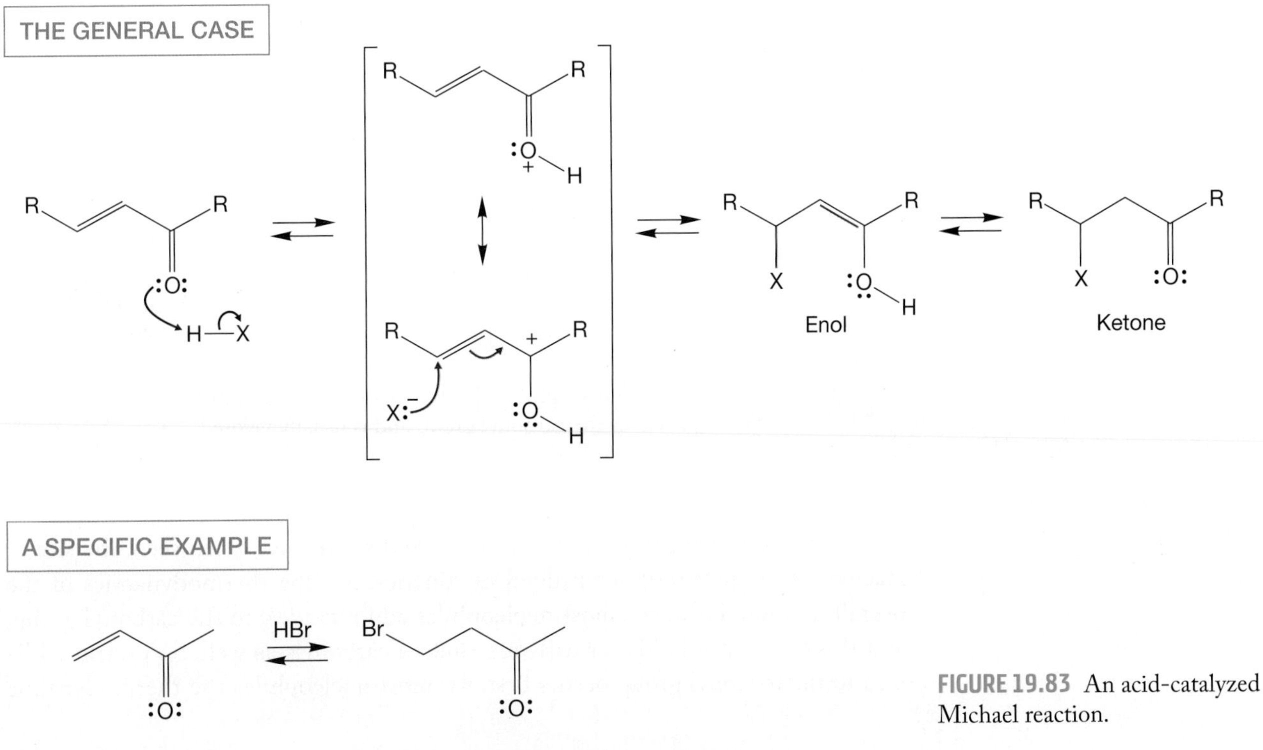

THE GENERAL CASE

Enol

Ketone

A SPECIFIC EXAMPLE

HBr

FIGURE 19.83 An acid-catalyzed Michael reaction.

PROBLEM 19.27 Write a mechanism for the following reaction:

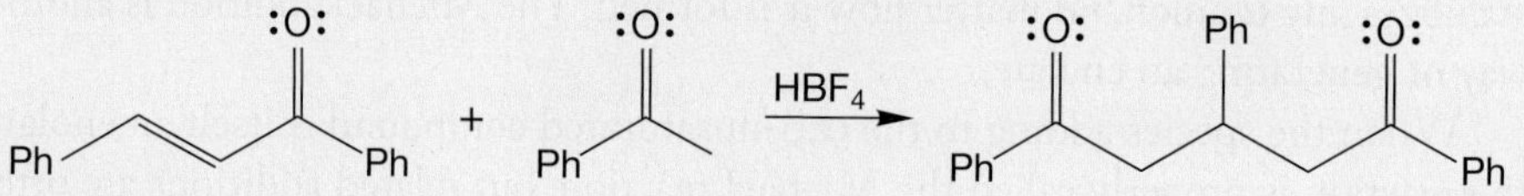

α,β-Unsaturated compounds present a general problem in predicting reaction products. There are just too many reactions possible. How, for example, can we decide whether a nucleophile will add to the carbonyl group (1,2-addition) or to the double bond (Michael)? This question captures in microcosm the difficulties presented by condensation problems in particular, and by the chemistry of polyfunctional compounds in general. How is it possible to decide which of a variety of reactions is the best one (either in energy terms or in the more practical sense of finding the route to the end of an exam problem or making a desired compound)?

At least the specific question above about the Michael reaction has a rational answer. First, notice that the product of Michael addition is likely to be more stable than the product of addition to the carbonyl group (Fig. 19.84). The product of Michael addition retains the strong carbon–oxygen double bond, and the simple addition product does not.

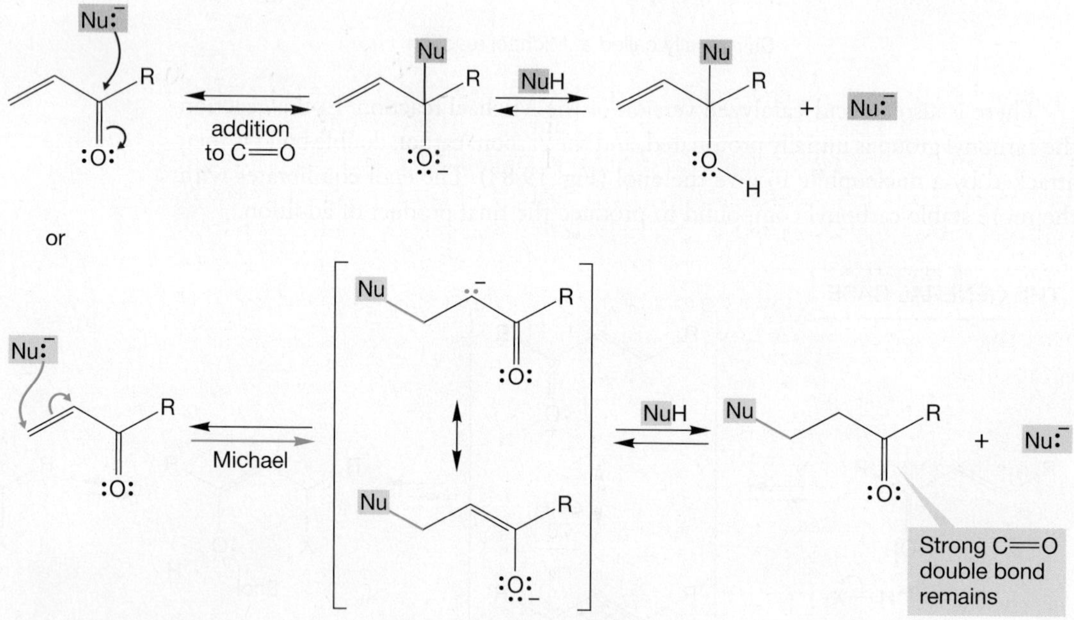

FIGURE 19.84 The two possible reactions of a nucleophile (Nu⁻) with an α,β-unsaturated carbonyl compound. Michael addition preserves the carbonyl group and is usually favored thermodynamically.

But this is a thermodynamic argument, and as we have often seen, the site of reactivity is many times determined by kinetics, not the thermodynamics of the overall reaction. However, most nucleophiles add *reversibly* to the carbonyl group, and this preference holds for α,β-unsaturated carbonyls as well. So even if addition to the carbonyl group occurs first, for most nucleophiles the thermodynamic advantage of Michael addition should ultimately win out. This notion suggests that strong nucleophiles such as hydride (H⁻) and alkyllithium compounds, which surely add *irreversibly* to carbonyls, might be found to give the products of attack at the carbonyl group, and such is the case (Fig. 19.85).

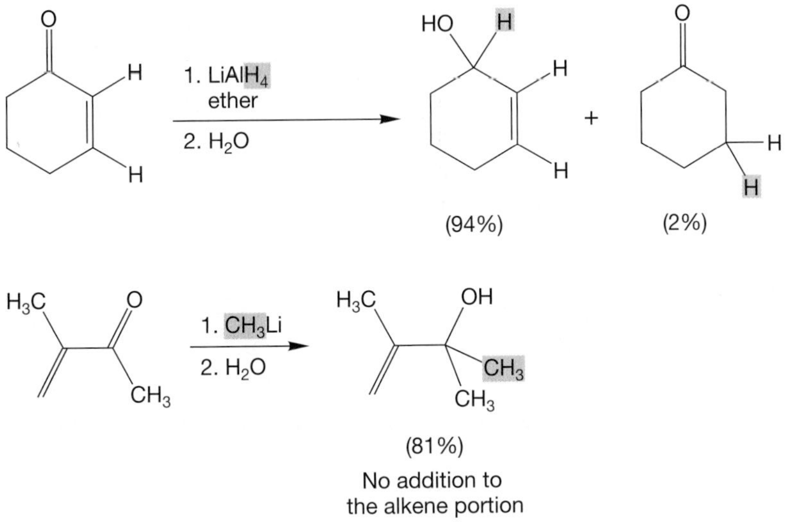

(94%)   (2%)

(81%)
No addition to
the alkene portion

FIGURE 19.85 Two irreversible addition reactions to the carbonyl group of an α,β-unsaturated carbonyl compound.

A very common reaction in organic synthesis is the addition of a cuprate (p. 798) to an enone. Although the organometallic cuprate is a strong base and therefore might be expected to add to the carbonyl carbon, this reagent is found to give Michael addition products (Fig. 19.86). Grignard reagents usually give 1,2-addition with α,β-unsaturated aldehydes, but the outcome of Grignard addition to α,β-unsaturated ketones is difficult to predict. Steric factors are thought to control the selectivity of such Grignard additions. However, Grignard reactions that are copper catalyzed dependably undergo Michael addition to α,β-unsaturated compounds.

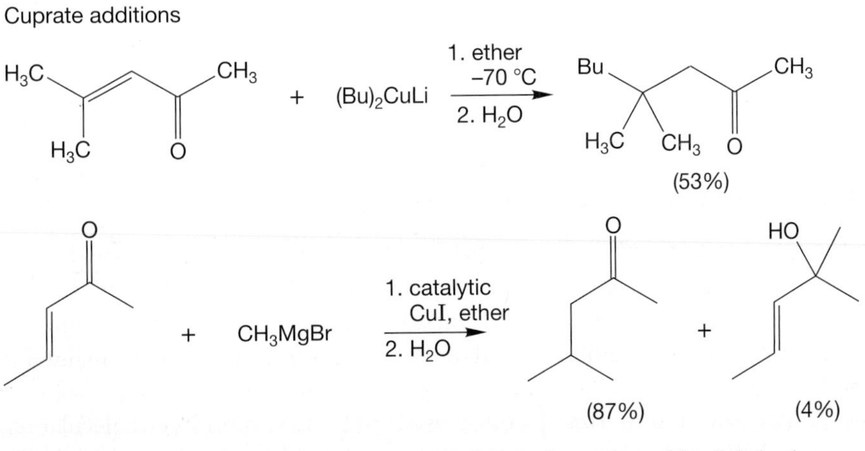

Cuprate additions

(53%)

(87%)   (4%)

FIGURE 19.86 Cuprates and copper-catalyzed Grignard reagents add in Michael fashion to α,β-unsaturated carbonyl compounds.

The Michael reaction is critical to the operation of the anticancer drug calicheamicin (Fig. 19.87). In the first step of its operation, the trisulfide bond in **A** is broken. The nucleophilic sulfide then adds in Michael fashion to give the enolate (**B**). This addition changes the shape of the molecule, bringing the ends of the two acetylenes closer together. A cyclization occurs to give a diradical (**C**), and this diradical abstracts hydrogen from the cancer cell's DNA, which ultimately kills the cell. Calicheamicin depends for its action on the change of shape. Before the Michael reaction, the ends of the two acetylenes are too far apart to cyclize. They are freed to do so only when the sulfur adds to the α,β-unsaturated carbonyl.

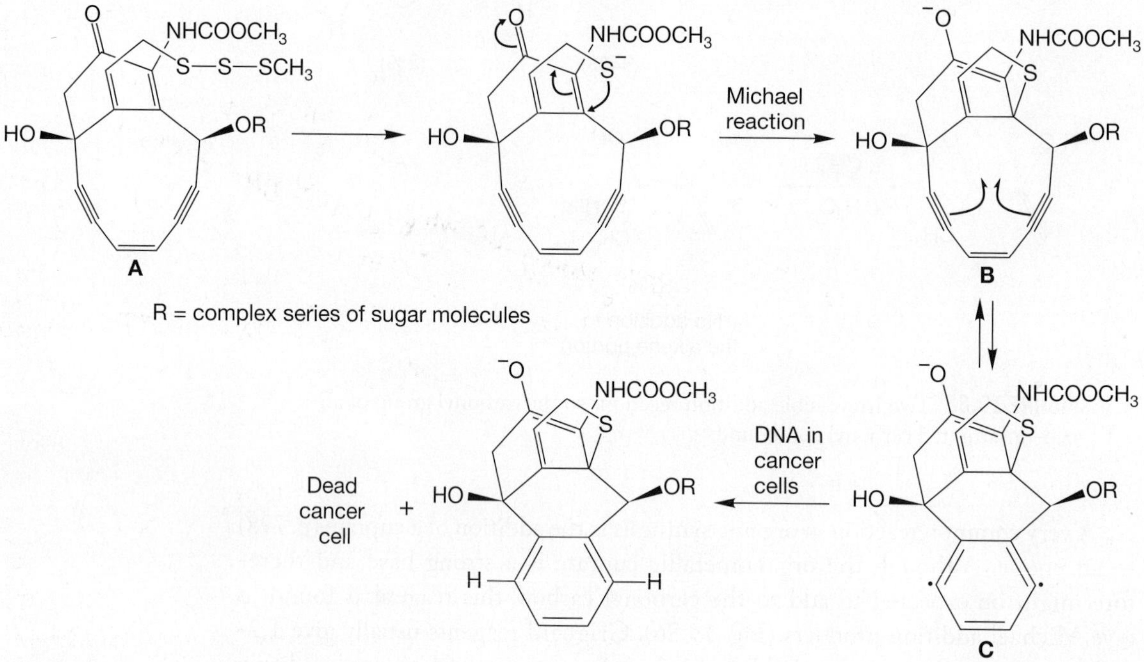

FIGURE 19.87 The operation of the antitumor agent calicheamicin.

Summary

In this section, we have seen two variations on the aldol theme, the Knoevenagel condensation and the Michael reaction. In all of these reactions, a single nucleophile adds to a single electrophile. The reactions are fundamentally the same as the aldol condensation; the variations arise from the structural differences of the nucleophiles.

19.7 Reactions Related to the Aldol Condensation

In the next few pages, we will use what we know of the simple aldol condensation to understand some related reactions. We will start with reactions that are very similar to simple aldol reactions, and slowly increase the complexity. The connection to the simple, prototypal acid- and base-catalyzed aldol reactions will always be maintained.

19.7a Intramolecular Aldol Condensations   Like most intermolecular reactions, the aldol reaction has an intramolecular version. If a molecule contains both an enolizable hydrogen and a receptor carbonyl group, an intramolecular addition will be theoretically possible. Particularly favorable are intramolecular cyclizations that

form the relatively strain-free five- and six-membered rings. The products are still β-hydroxy ketones (or the dehydration products, α,β-unsaturated ketones), and the mechanism of the reaction remains the same (Fig. 19.88).

THE GENERAL
BASE-CATALYZED CASE

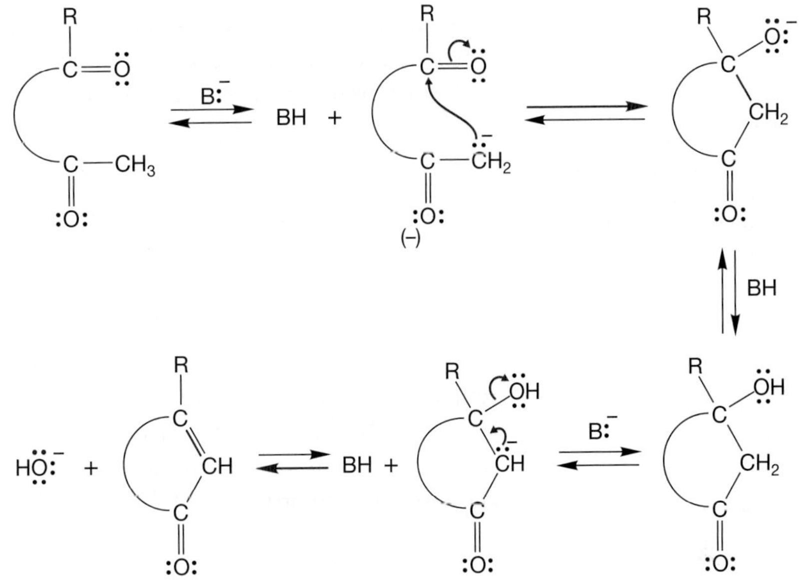

SPECIFIC BASE- AND
ACID-CATALYZED EXAMPLES

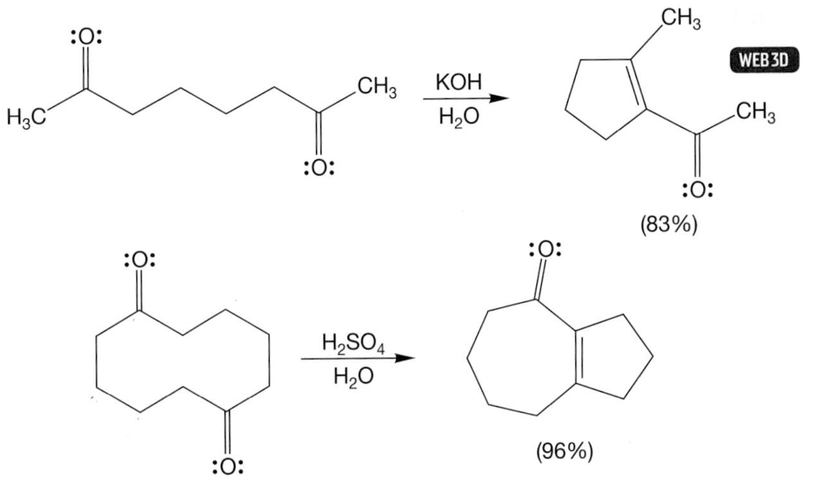

(83%)

(96%)

FIGURE 19.88 The mechanism for intramolecular aldol reactions.

PROBLEM 19.28   Write the mechanisms for the acid- and base-catalyzed aldol condensations of the molecule at the bottom of Figure 19.88.

PROBLEM 19.29   Write the mechanisms for the acid- and base-catalyzed reverse aldol reaction for 3-hydroxy-3-methylcycloheptanone. *Hint*: See Problem Solving, p. 968.

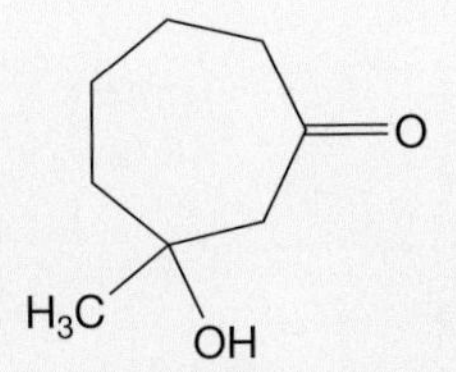

3-Hydroxy-3-methylcycloheptanone

19.7b Crossed Aldol Condensations

In practice, not all β-hydroxy ketones can be formed efficiently by aldol reactions. Suppose we were set the task of synthesizing 5-hydroxy-4,5-dimethyl-3-hexanone (Fig. 19.89). We recognize it as a potential aldol product through the dissection shown in the figure, which suggests that an aldol condensation of diethyl ketone and acetone should be a good route to the molecule. An aldol condensation between two different carbonyl compounds is called a **crossed** (or **mixed**) **aldol condensation**.

FIGURE 19.89 A retrosynthetic analysis suggests that a crossed aldol reaction between diethyl ketone and acetone should give 5-hydroxy-4,5-dimethyl-3-hexanone.

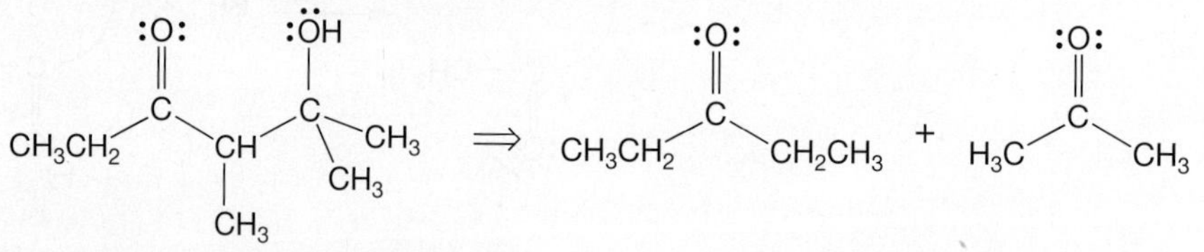

5-Hydroxy-4,5-dimethyl-3-hexanone

However, a little thought about the mechanism reveals potential problems. Two enolates can be formed and each enolate has two carbonyl groups to attack (Fig. 19.90). Thus, four products, **A**, **B**, **C**, and **D**, are possible (more if any of the β-hydroxy ketones are dehydrated to α,β-unsaturated ketones), and all are likely to be formed in comparable yield. The enolate of diethyl ketone can add to the carbonyl of acetone or diethyl ketone to give **A** and **B**. Similarly, the enolate of acetone gives **C** and **D** by reaction with the two carbonyl compounds.

FIGURE 19.90 When this synthetic route is attempted, four β-hydroxy ketones, **A**, **B**, **C**, and **D**, are likely to be produced.

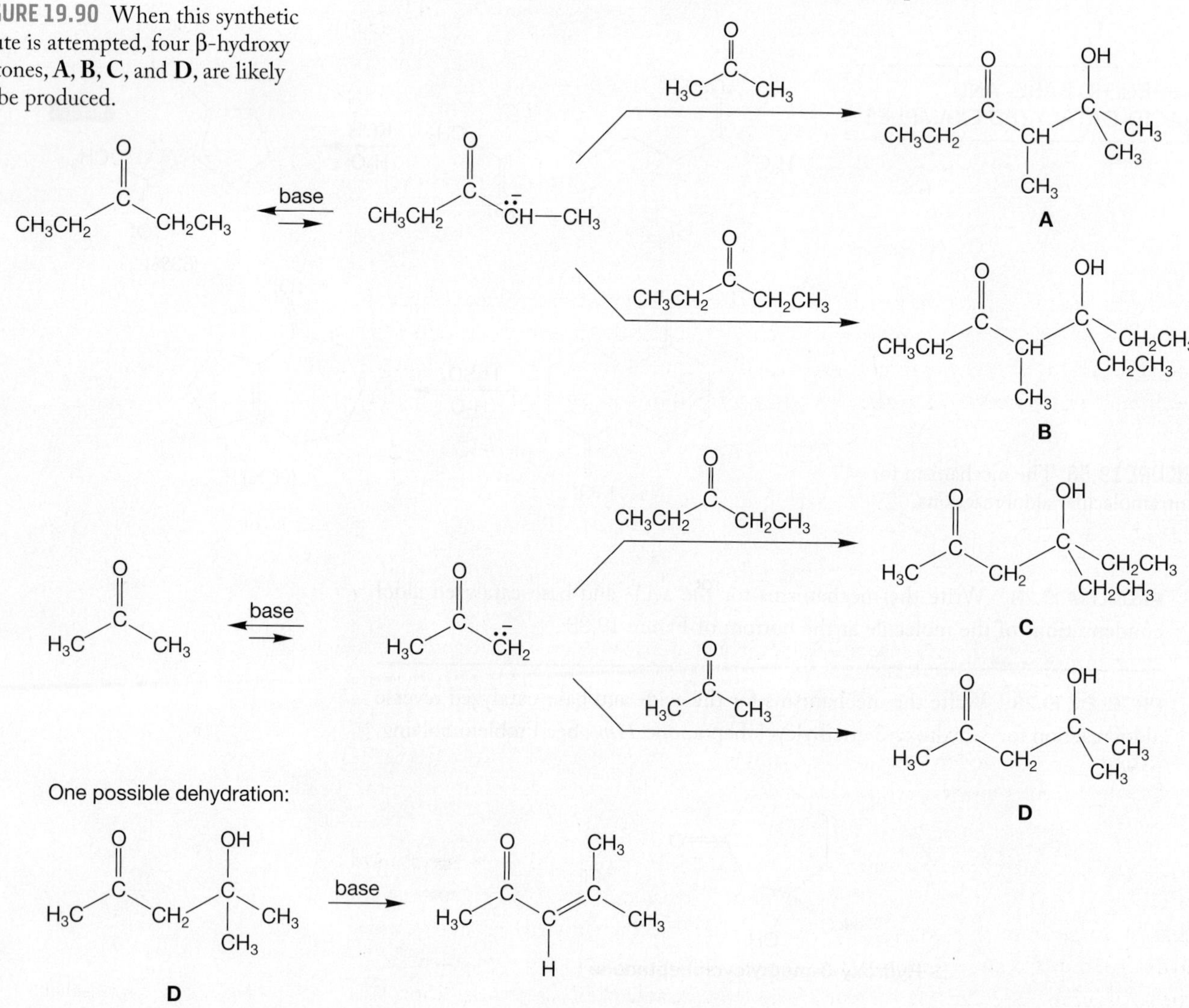

One possible dehydration:

There are some simple ways to limit the possibilities in a crossed aldol reaction. If one partner has no α hydrogens, it can function only as an acceptor, and not as the nucleophilic enolate partner in the reaction (Fig. 19.91). For example, we might hope

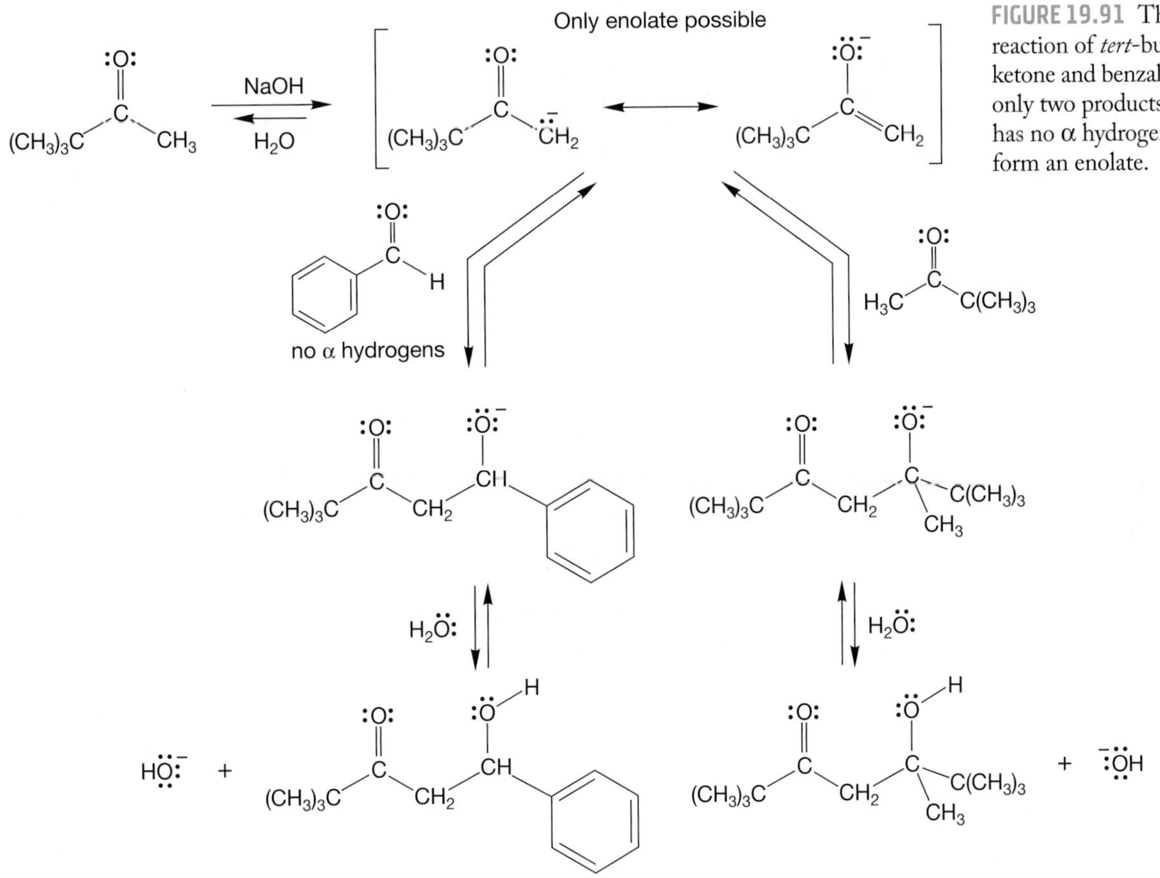

Only enolate possible

FIGURE 19.91 The crossed aldol reaction of *tert*-butyl methyl ketone and benzaldehyde can give only two products. Benzaldehyde has no α hydrogens and cannot form an enolate.

no α hydrogens

for better luck in a reaction pairing *tert*-butyl methyl ketone and benzaldehyde. In principle, however, there are still two possibilities, because the enolate can add to either the ketone or the aldehyde to generate two different β-hydroxy ketones. When this reaction is run, however, only one product is formed in substantial yield (Fig. 19.92).

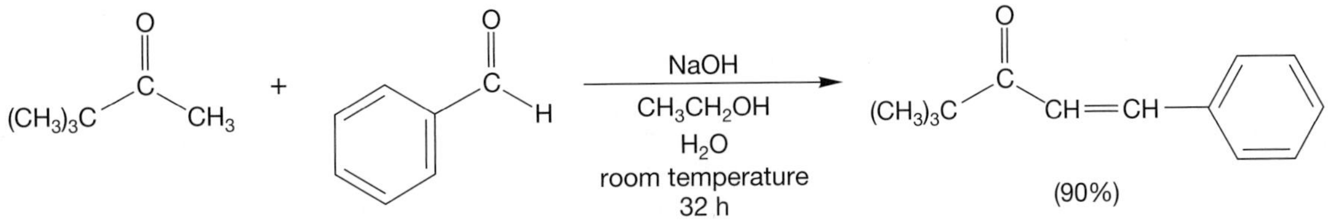

NaOH
CH₃CH₂OH
H₂O
room temperature
32 h

(90%)

FIGURE 19.92 In fact, there is only one major product in the aldol condensation of *tert*-butyl methyl ketone and benzaldehyde.

PROBLEM 19.30 There is a reaction between benzaldehyde and hydroxide ion. What is it, and why does it not interfere with the aldol condensation?

There are several reasons that this crossed aldol is successful. First, the rate of addition of the enolate to the carbonyl group of benzaldehyde is much greater than that of addition to *tert*-butyl methyl ketone because the carbonyl groups of aldehydes are more reactive than those of ketones. Second, the equilibrium constant for the addition to an aldehyde carbonyl is more favorable than that for addition to a

Mixed aldol condensation

ketone carbonyl (Fig. 19.93). Third, because the addition and dehydration steps are reversible, if elimination occurs, thermodynamics will favor the most stable product, which in this case is the phenyl-substituted enone.

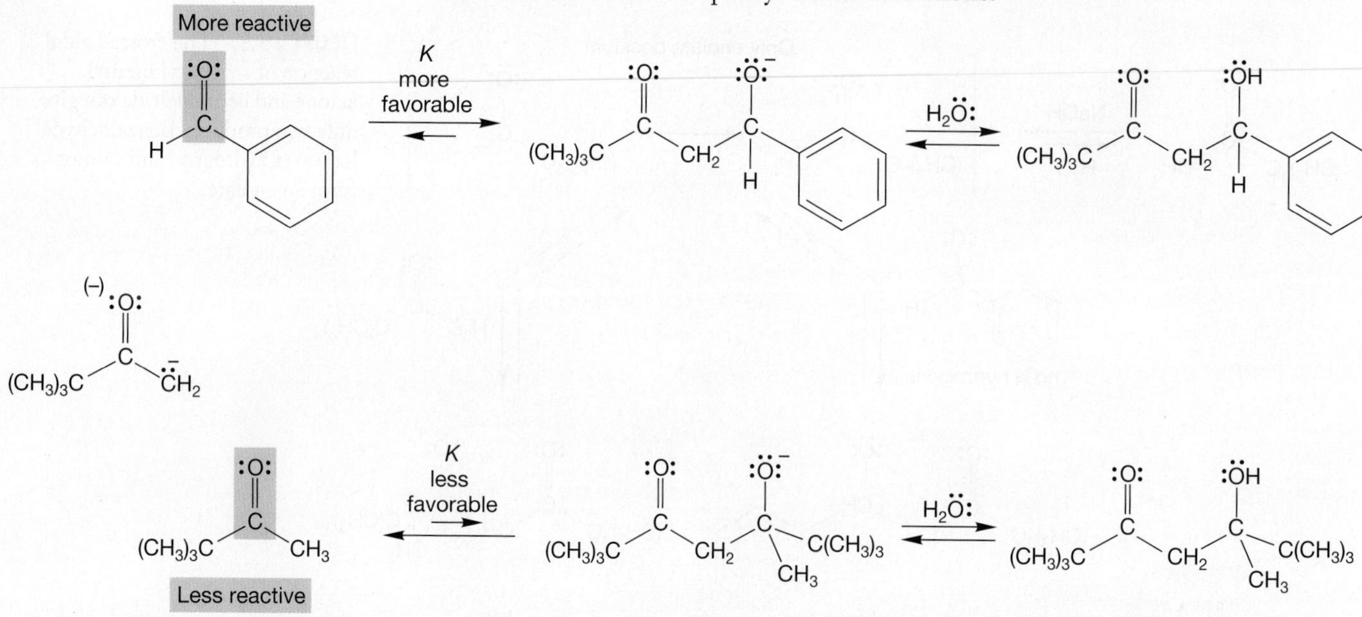

FIGURE 19.93 The carbonyl group of benzaldehyde is more reactive than that of *tert*-butyl methyl ketone, and equilibrium favors the product for the reaction with benzaldehyde but not for the reaction with *tert*-butyl methyl ketone.

The crossed aldol reaction is important enough to have been given its own name, the **Claisen–Schmidt condensation** (Ludwig Claisen, 1851–1930).

Earlier in this chapter we saw LDA, a base that is effective in producing enolates without the complications of addition to the carbonyl group. Because LDA is such a strong base, all the initial carbonyl compound is driven to its enolate (Fig. 19.94).

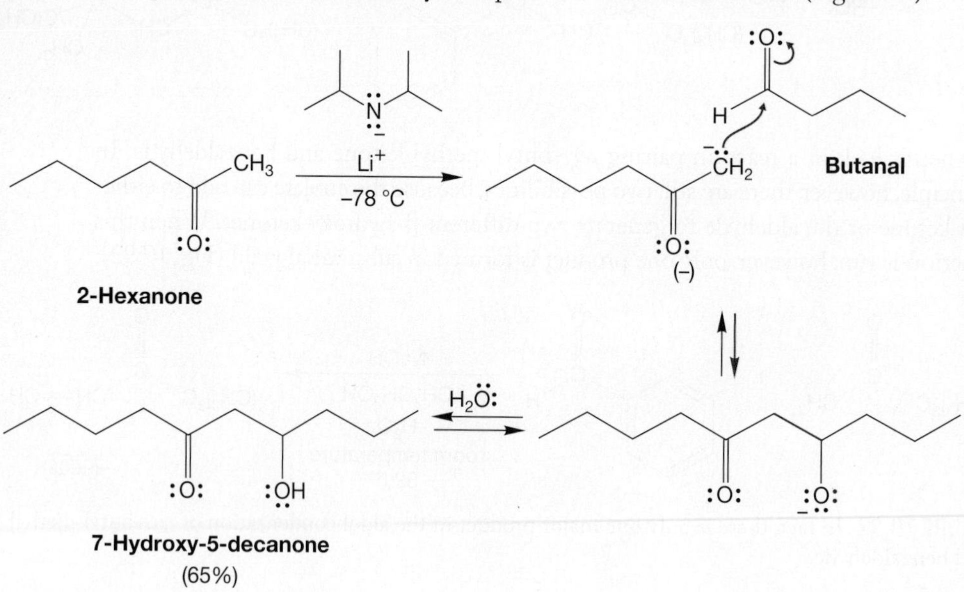

FIGURE 19.94 A crossed aldol reaction with LDA as base. Because of its size, LDA selectively deprotonates the less substituted side of the initial ketone making the kinetic enolate. Because of its strength, LDA completely deprotonates the carbonyl compound. A second carbonyl compound can then be added and undergo the aldol reaction. The alkoxide is protonated at the end of the reaction when water is added.

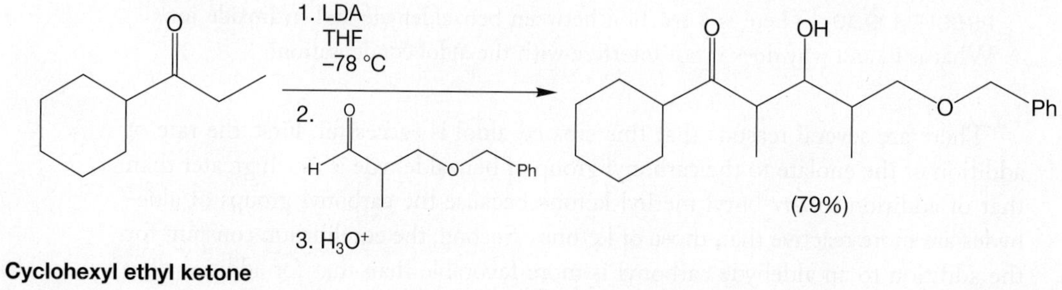

No aldol reaction is possible until the LDA is consumed and a second carbonyl compound is added. This procedure is a convenient way to do crossed aldol reactions in a controlled fashion. When there is a choice, LDA forms the more accessible, less substituted enolate. Lithium diisopropylamide is a large, sterically encumbered base and removes a proton from the sterically less hindered position to give what is called the **kinetic enolate**. Notice that this reaction can be used to generate two new stereocenters adjacent to each other. This procedure is very useful and has received considerable attention in the past 30 years.

19.8 Addition of Acid Derivatives to the α Position: The Claisen Condensation

The reaction between enolates and esters is an addition–elimination process in which the enolate is the nucleophile that replaces the alkoxy group of the ester. This reaction is similar to many you saw in Chapter 18.

19.8a Condensations of Ketone Enolates with Esters   Many reactions can occur between an ester and a ketone under basic conditions. It is unlikely that condensation of the ketone enolate with another molecule of ketone will be a problem, because aldol reactions of ketones are reversible, and usually endothermic. So even though the product of the aldol reaction *will* be formed, starting ketone will be favored. If ketone is drained away through reaction with the ester, the reversibly formed aldol product merely serves as a storage point for the ketone. The condensation of an ester enolate with an ester is usually not a problem because the ketone is a much stronger acid ($pK_a \sim 19$) than the ester ($pK_a \sim 24$, see Table 19.2), and formation of the ketone enolate will be much preferred to formation of the ester enolate. In the reaction between the ketone enolate and an ester, the enolate adds to the ester carbonyl, alkoxide is lost, and a β-dicarbonyl compound is formed (Fig. 19.95). If a

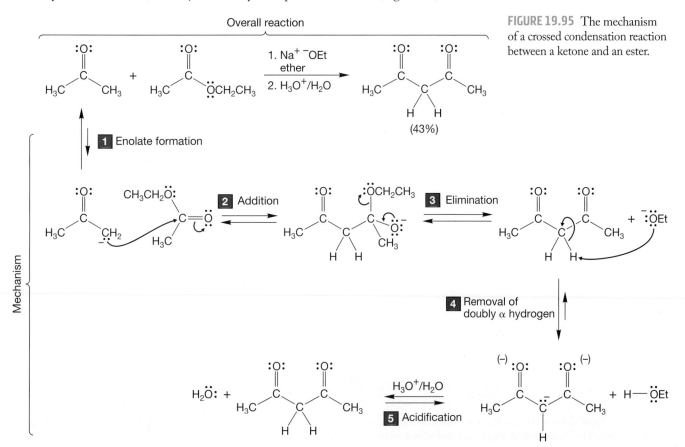

FIGURE 19.95 The mechanism of a crossed condensation reaction between a ketone and an ester.

full equivalent of base is used, the doubly α hydrogen of the β-dicarbonyl compound is removed to give a nicely resonance-stabilized anion. The reaction is completed by an acidification step that regenerates the β-dicarbonyl compound in the absence of base.

PROBLEM 19.31 Why are esters less reactive in addition reactions than aldehydes and ketones?

Frequently, the strong base sodamide ($NaNH_2$) or LDA is used in condensing esters with ketones. These conditions favor removal of the kinetically more accessible hydrogen when there is a choice of possible enolates (Fig. 19.96).

THE GENERAL CASE

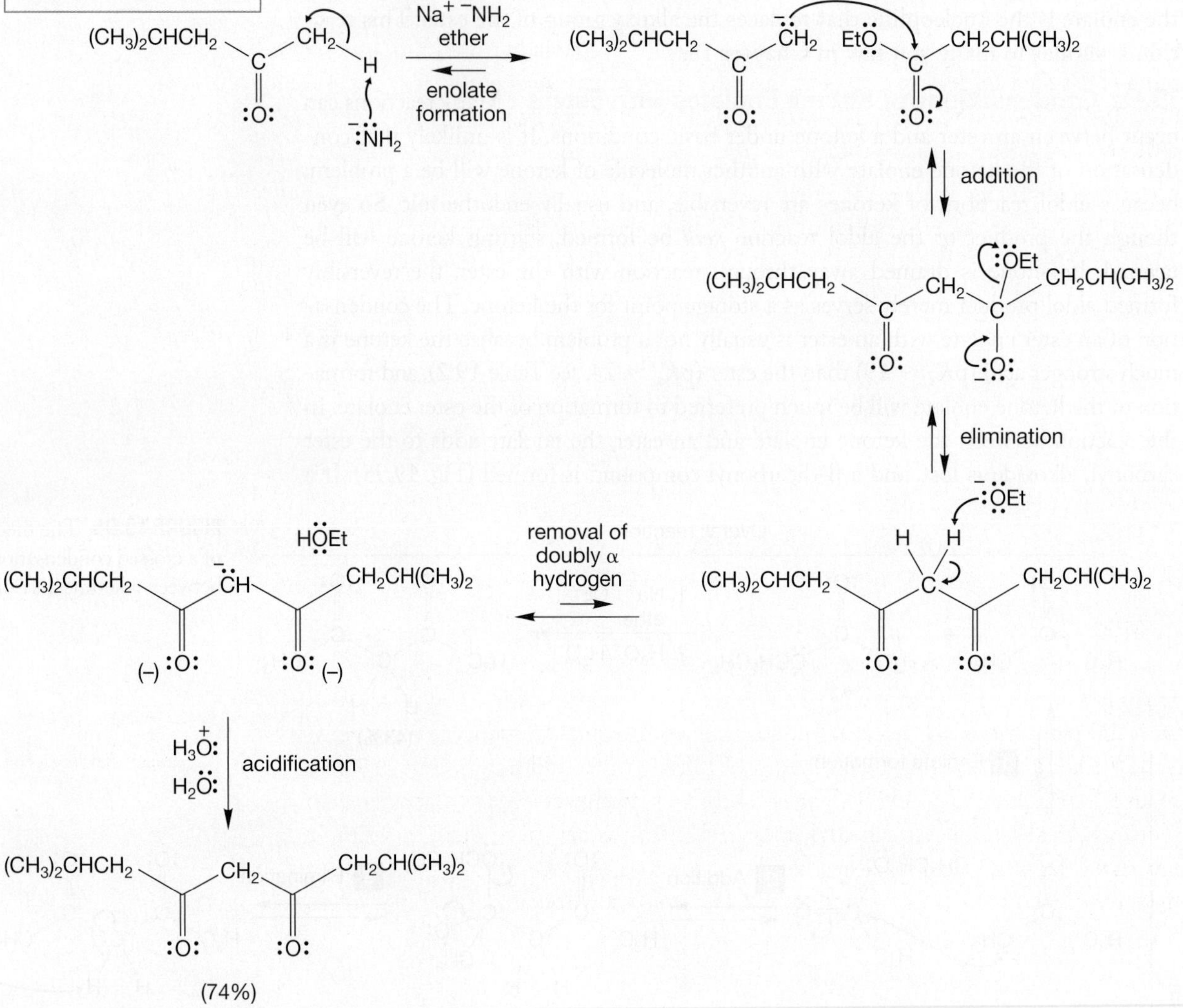

A SPECIFIC EXAMPLE

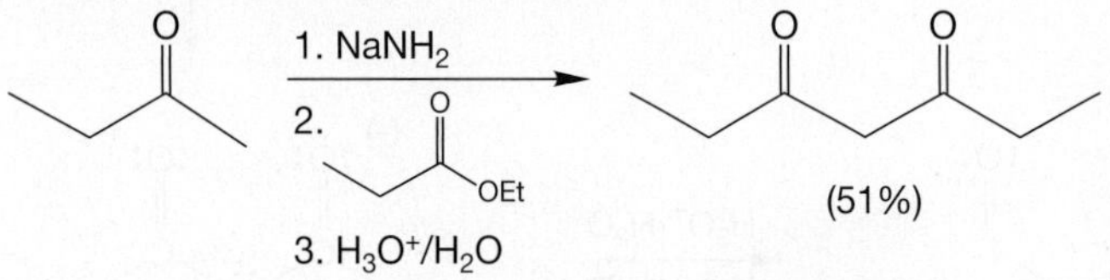

FIGURE 19.96 Amide bases are often used in crossed aldol condensations of esters and ketones. They lead to formation of the kinetic enolate through removal of the most accessible hydrogen.

19.8b Ester Enolates React with Esters   Esters are approximately seven pK_a units less acidic than alcohols. Nevertheless, it is possible for some ester enolate to be formed in the presence of a base such as an alkoxide ion, even though the reaction must be substantially endothermic (Fig. 19.97).

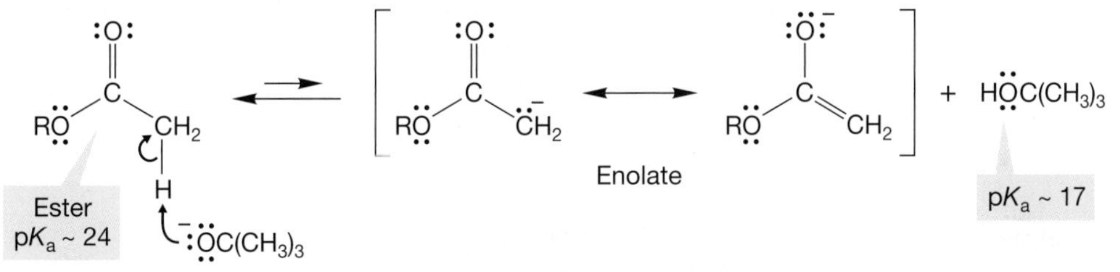

FIGURE 19.97 Formation of an enolate anion from an ester. Because *tert*-butyl alcohol has a pK_a of about 17, and the ester a pK_a of about 24, this reaction must be highly endothermic.

Why doesn't the alkoxide in Figure 19.97 add to the ester carbonyl group and start a base-catalyzed transesterification reaction? It does! If one is not careful to use the same OR group in the alkoxide and the ester, then mixtures of products are found (Fig. 19.98). As long as the OR groups match, however, the base-catalyzed transesterification doesn't change anything—the product is the same as the starting material.

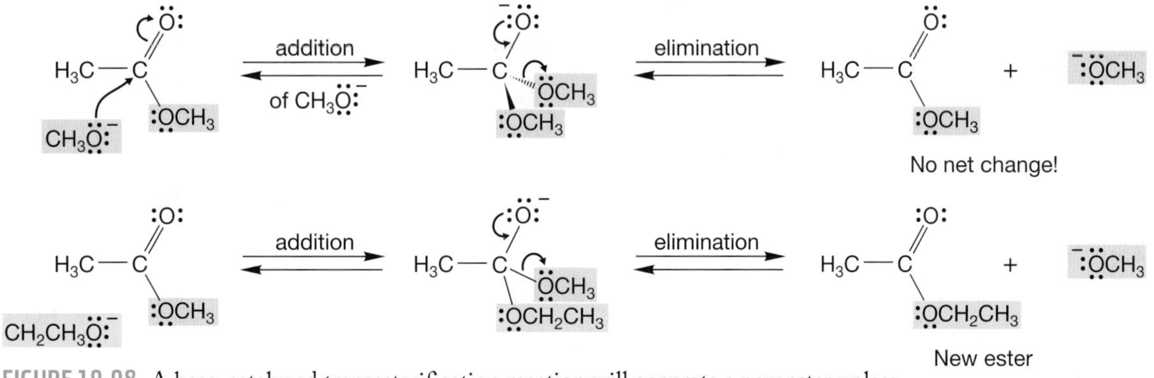

FIGURE 19.98 A base-catalyzed transesterification reaction will generate a new ester unless the OR group of the alkoxide is identical to the OR group of the ester.

What can the ester enolate do? It can react with the acidic alcohol to reprotonate, regenerating the starting ester. However, it can also react *with any other electrophile in the system*. The carbonyl group of the ester is just such a species, and it can react with the ester enolate in an addition reaction. Although the product is more complex, the addition reaction is no different from any other addition of a nucleophile to an ester carbonyl (Fig. 19.99). The second step of the reaction is also analogous to previous reactions. The tetrahedral intermediate can expel the enolate in a simple reverse of the original addition reaction, or it can lose alkoxide to give a molecule of β-keto ester. This reaction is called the **Claisen condensation** after Ludwig Claisen,

Claisen condensation

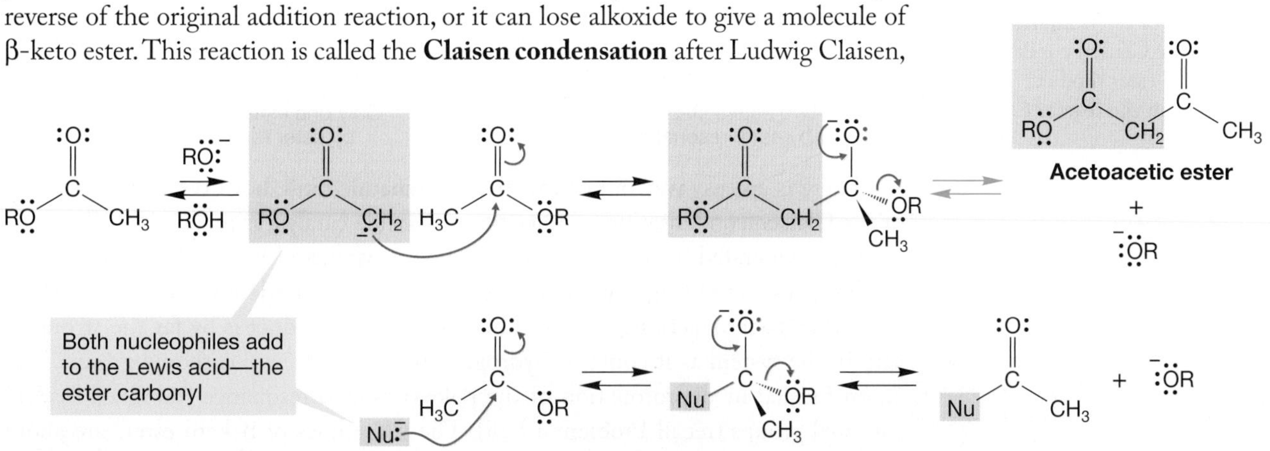

FIGURE 19.99 In the first step of the reaction between an ester enolate and an ester, the nucleophilic ester enolate adds to the electrophilic carbonyl carbon of the ester. In the second step the alkoxide ion is lost from the tetrahedral intermediate.

who has many reactions named for him. Recall the Claisen–Schmidt condensation, for example (p. 984). There are more to come.

The Claisen condensation bears the same relationship to the aldol condensation as the addition of a nucleophile to an ester bears to the addition of a nucleophile to an aldehyde or ketone. Whereas the tetrahedral alkoxide formed in both aldehyde (or ketone) reactions can only revert to starting material or protonate, both ester reactions have the added option of losing the ester alkoxide. What is added in ester chemistry is the possibility of the elimination phase of the addition–elimination process (Fig. 19.100).

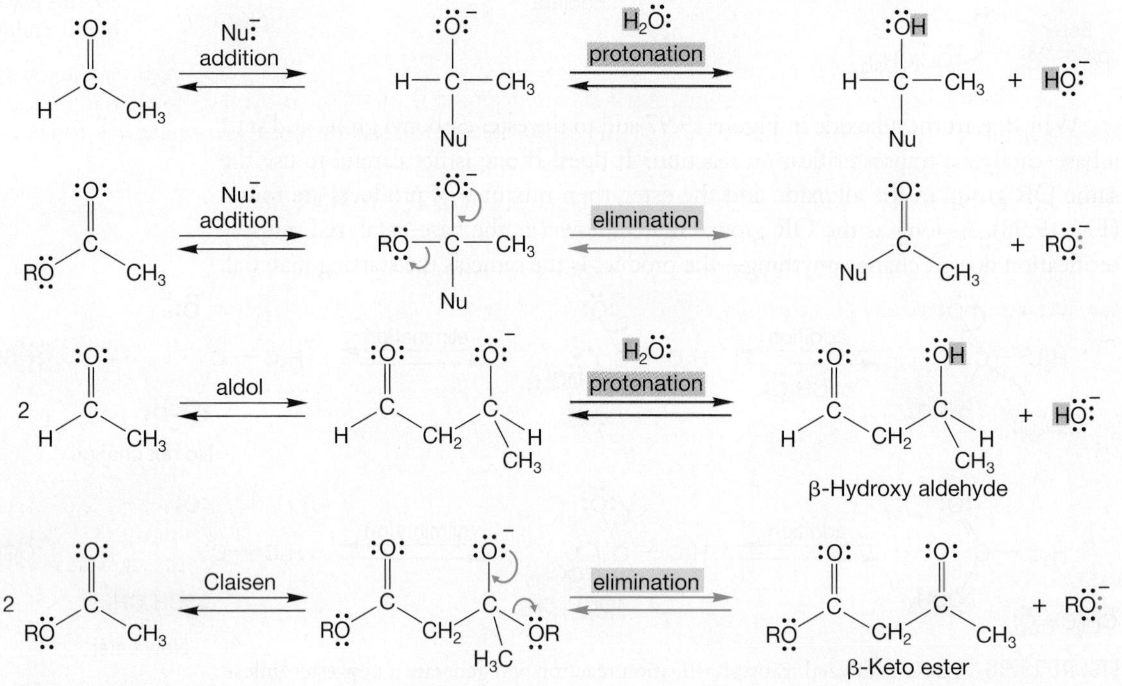

FIGURE 19.100 The aldol condensation is just one example of the addition of a nucleophile to a carbonyl group followed by protonation. The Claisen condensation is just one example of the addition–elimination sequence available to esters.

As in aldol condensations of ketones, in the Claisen condensation it is the starting materials that are favored thermodynamically (Fig. 19.101). Esters are stabilized by resonance and there are two ester groups in the starting materials but only one in the product. The result is a thermodynamic favoring of the two separated esters.

FIGURE 19.101 Thermodynamics favors the pair of esters (starting material) in the Claisen condensation, not the β-keto ester product. The reaction to form the β-keto ester is endothermic.

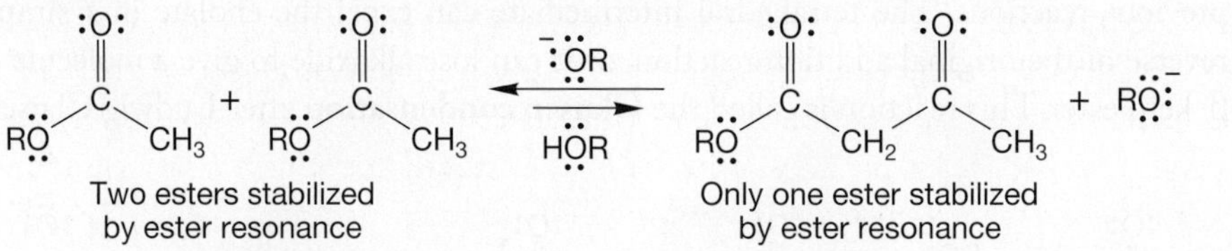

Two esters stabilized by ester resonance

Only one ester stabilized by ester resonance

There is an easy way out of this thermodynamic bind, however. In the last step of the Claisen condensation a molecule of alkoxide catalyst is regenerated. The catalyst is regenerated in the aldol condensation as well, and in the aldol reaction the catalyst goes on to form another enolate. In the Claisen condensation, however, there is another, faster reaction possible. The β-keto ester product is by far the strongest acid in the system as it contains hydrogens that are α to *two* carbonyl groups. The anion formed by deprotonation of the β-keto ester is resonance stabilized by *both* carbonyl groups (recall Problem 19.25). The pK_a values of β-keto esters are about 11 (Table 19.3), and therefore these molecules are about 10^{13} more acidic than the starting esters! By far the best reaction for the alkoxide regenerated at the end of the condensation step is removal of one of the doubly α hydrogens to give the highly resonance-stabilized salt of the β-keto ester (Fig. 19.102).

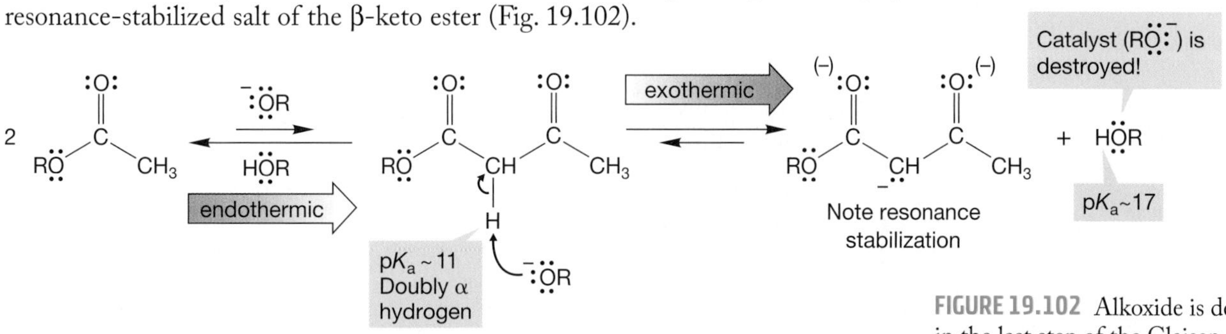

FIGURE 19.102 Alkoxide is destroyed in the last step of the Claisen condensation as a doubly α proton is removed to give a stable anion.

If the doubly α proton is removed, all is well thermodynamically; now the product is more stable than the starting material, *but the catalyst for the reaction has been destroyed*! A Claisen condensation using a catalytic amount of base must fail, because it can only generate an amount of product equal to the original amount of base. The simple remedy is to use a full equivalent of alkoxide, not a catalytic amount. A thermodynamically favorable Claisen condensation is not run with catalytic base, but requires a full equivalent of base. Be certain you are clear as to the reason why the aldol condensation requires only a catalytic amount of base but the Claisen needs a full equivalent (Fig. 19.103).

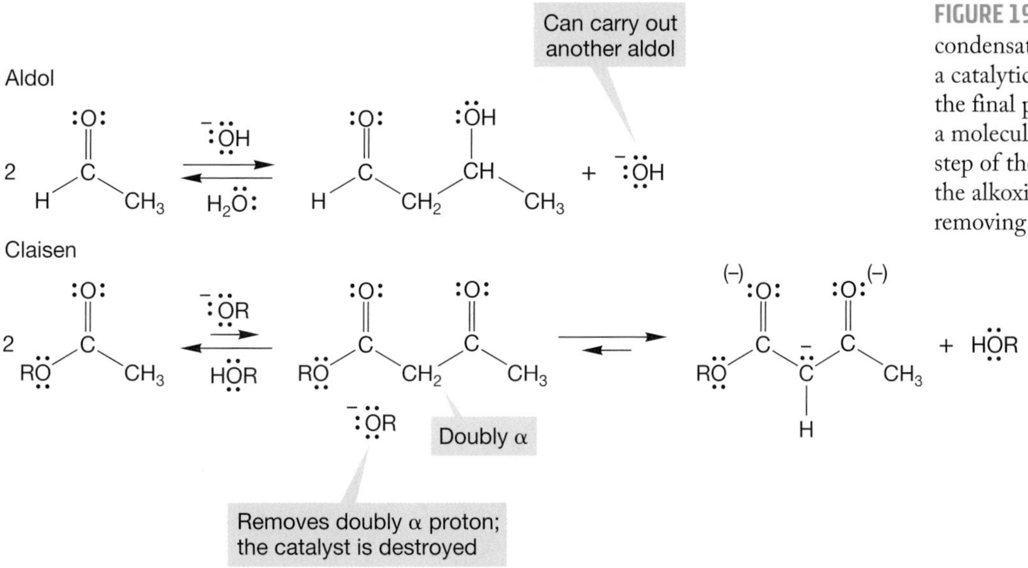

FIGURE 19.103 The aldol condensation can be carried out using a catalytic amount of base because the final protonation step regenerates a molecule of hydroxide. In the last step of the Claisen condensation, the alkoxide ion is consumed by removing the doubly α proton.

Because we are left with a carbanion after the Claisen condensation, acidification of the reaction mixture is required in order to form the neutral β-keto ester. What prevents this product, thermodynamically unstable with respect to the starting ester molecules, from re-forming the starting esters? Certainly the Claisen condensation is reversible. But the β-keto ester that is formed in the basic conditions is much more likely to form the resonance-stabilized enolate than go back to starting material. And once the stable enolate is formed it is not likely to react with alkoxide, a nucleophile + nucleophile reaction. And after acidification there is no base present! Without the base, there can be no reverse Claisen condensation, and the product ester is obtained (Fig. 19.104).

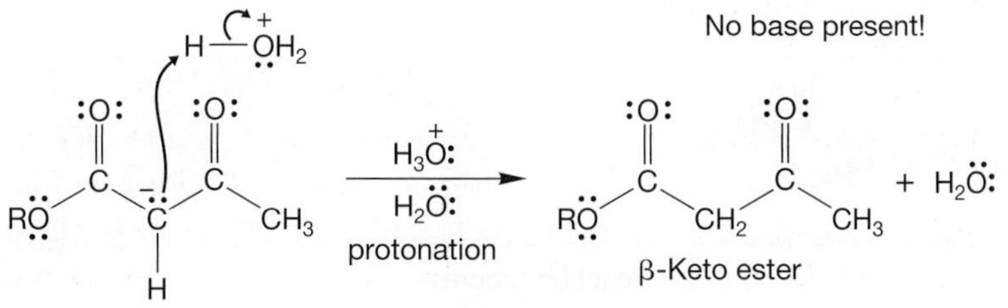

FIGURE 19.104 Acidification of the Claisen condensation produces a β-keto ester *in the absence of base.*

PROBLEM 19.32 Draw the mechanism for the Claisen condensation of ethyl phenyl-acetate using the appropriate base. Include the acidification process at the end.

Figure 19.105 shows Energy versus Reaction progress diagrams for the aldol and Claisen condensations.

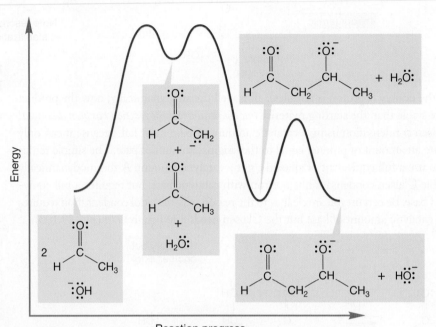

The aldol condensation for acetaldehyde is approximately thermoneutral

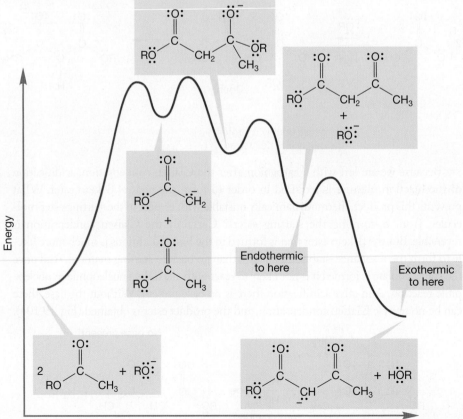

FIGURE 19.105 Energy versus Reaction progress diagrams for the aldol and Claisen condensations.

Summary

A full equivalent of base makes the Claisen condensation a practical source of acetoacetic esters (β-keto esters). The five steps of the process are summarized in Figure 19.106:

1. Enolate formation.
2. The condensation itself (addition).
3. Loss of alkoxide from the tetrahedral intermediate (elimination).
4. Removal of the doubly α hydrogen.
5. Acidification to produce the β-keto ester.

THE GENERAL CASE

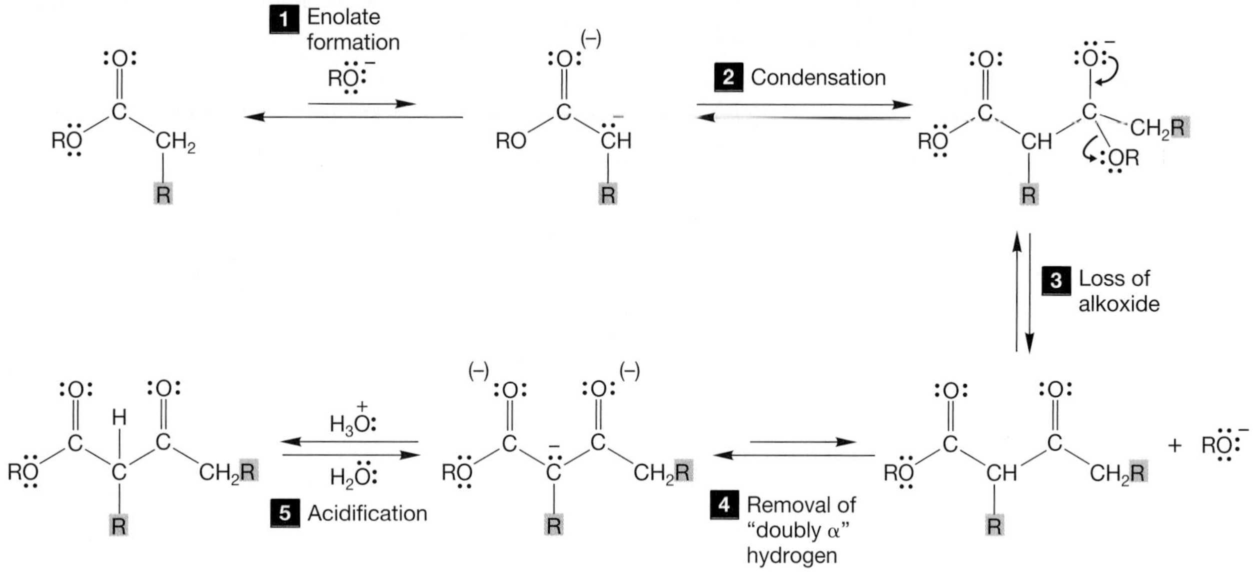

A SPECIFIC EXAMPLE

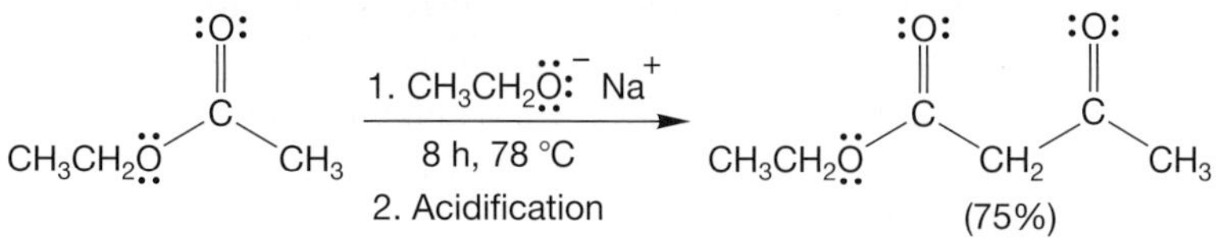

FIGURE 19.106 The Claisen condensation.

PROBLEM SOLVING

GO

"All" β-keto esters are Claisen products or products of closely related reactions. Every time you see a β-keto ester, think—"What Claisen condensation would have formed it?"

19.8c Reverse Claisen Condensations

As mentioned, formation of the initial condensation product in the Claisen condensation is usually endothermic. If no doubly α hydrogens exist for anion formation, β-keto esters will revert to their component ester starting materials when treated with alkoxide base.

WORKED PROBLEM 19.33 Even when a full equivalent of sodium ethoxide is used, ethyl 2-methylpropanoate fails to give useful amounts of a condensation product. Explain with a careful mechanistic analysis.

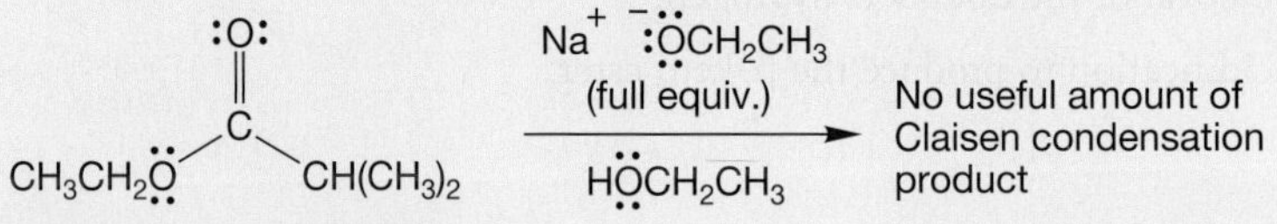

Ethyl 2-methylpropanoate

ANSWER In this classic problem, Claisen condensation leads to product **A** through the usual addition–elimination scheme:

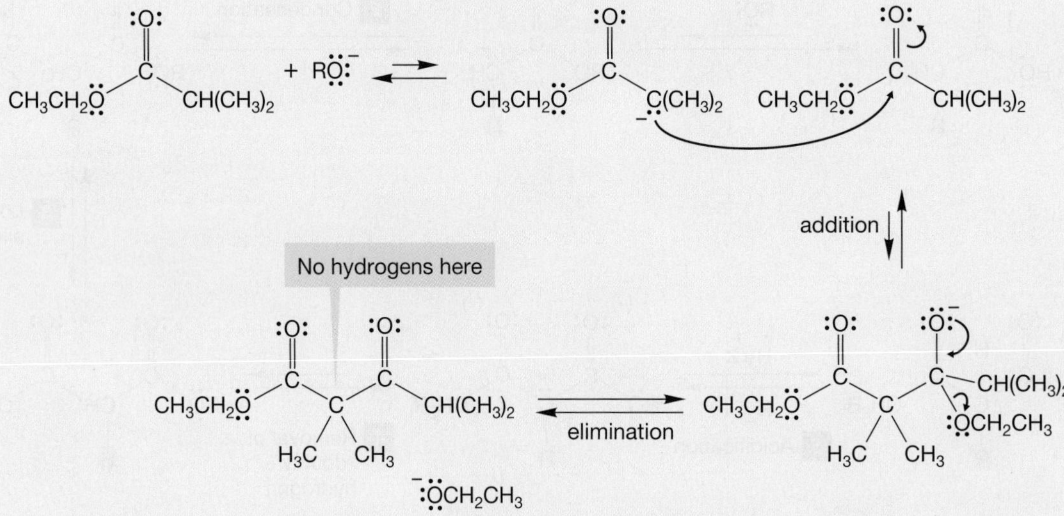

However, the product **A** contains no doubly α hydrogens! There can be no formation of the stable, highly resonance-stabilized anion that accounts for the success of the Claisen condensation. Thermodynamics dictates that the starting materials be more stable than the product, and the reaction settles out at a point at which there is little product. The mechanism of the reverse reaction is written for you: Just read the scheme backward.

19.9 Variations on the Claisen Condensation

So, that's the Claisen condensation: Two esters combine with an equivalent of base to make a β-keto ester. Now it's time to look at how changes in structure of the starting materials affect the structure of the product and the mechanism of its formation.

19.9a Intramolecular Claisen Condensations

The **Dieckmann condensation**, the cyclic version of the Claisen condensation, is named after Walter Dieckmann (1869–1925). The reaction transforms an acyclic diester into a cyclic β-keto ester. Formation of five- and six-membered rings is favored, and there is little beside the intramolecular nature of the reaction to distinguish the Dieckmann from the standard

Claisen condensation (Fig. 19.107). Like the Claisen, the Dieckmann condensation requires a full equivalent of base. Unless there is sufficient base available to remove the doubly α hydrogen and drive the reaction to a thermodynamically favorable conclusion, the reaction is not successful. A final acidification liberates the β-keto ester in the absence of base.

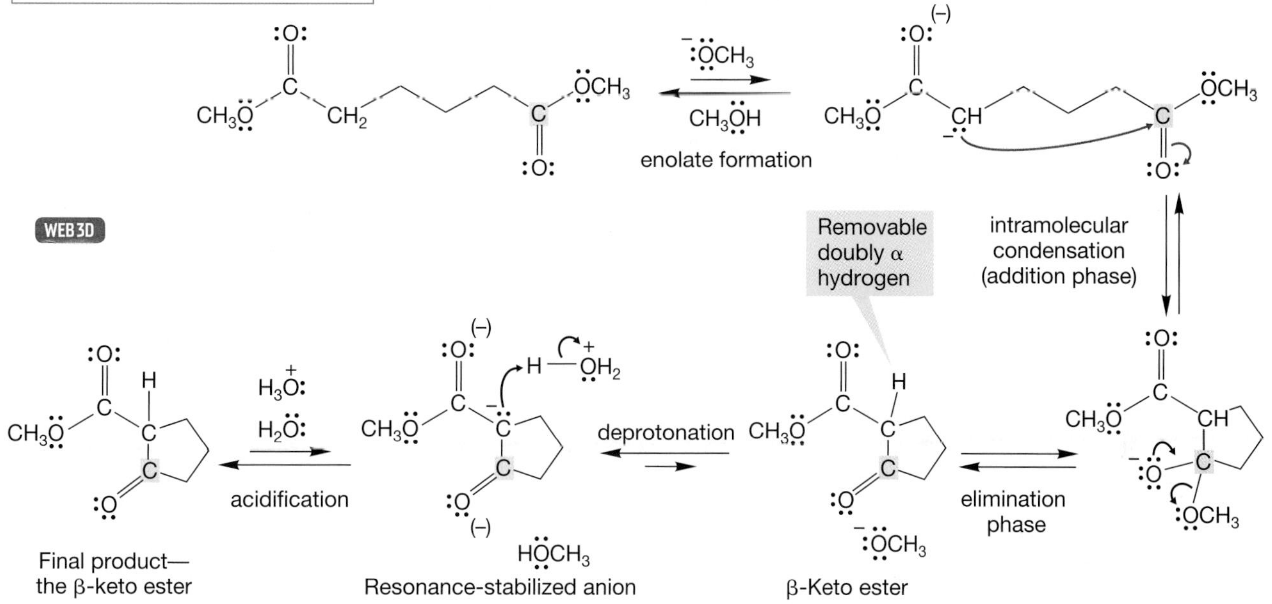

FIGURE 19.107 The Dieckmann condensation—an intramolecular Claisen condensation.

THE STANDARD DIECKMANN

A MORE COMPLEX EXAMPLE

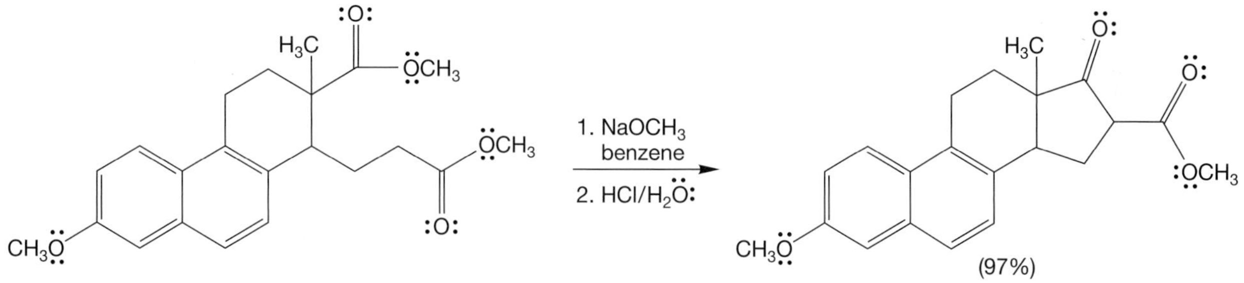

19.9b Crossed Claisen Condensations A naively designed **crossed** (or **mixed**) **Claisen condensation** is doomed to the same kind of failure as is the crossed aldol condensation. There are four possible products (Fig. 19.108), and therefore no easy source of the selectivity we need for a practical synthetic reaction.

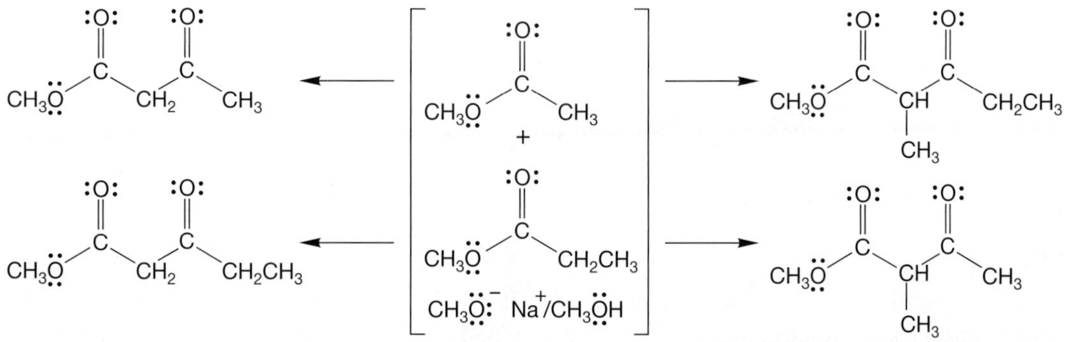

FIGURE 19.108 There are four possible—and likely—products from a crossed Claisen condensation.

PROBLEM 19.34 Show in a general way how these products are produced and write a mechanism for the formation of one of the four products shown in Figure 19.108.

As in the crossed aldol condensation, there are some simple steps we can take to improve matters. If one of the starting esters has no α hydrogen, there can be no enolate formation from it, and it can act only as an electrophile, never as a nucleophile. It is helpful to add the ester with an α hydrogen to a mixture of the partner without an α hydrogen and the base. As the enolate is formed it is more likely to react with the ester that has no α hydrogen since it is present in excess. Using this technique, a successful crossed Claisen can be achieved. Benzoates, formates, and carbonates are examples of esters without α hydrogens and are often used (Fig. 19.109).

THE "CROSSED CLAISEN"—THE GENERAL MECHANISM

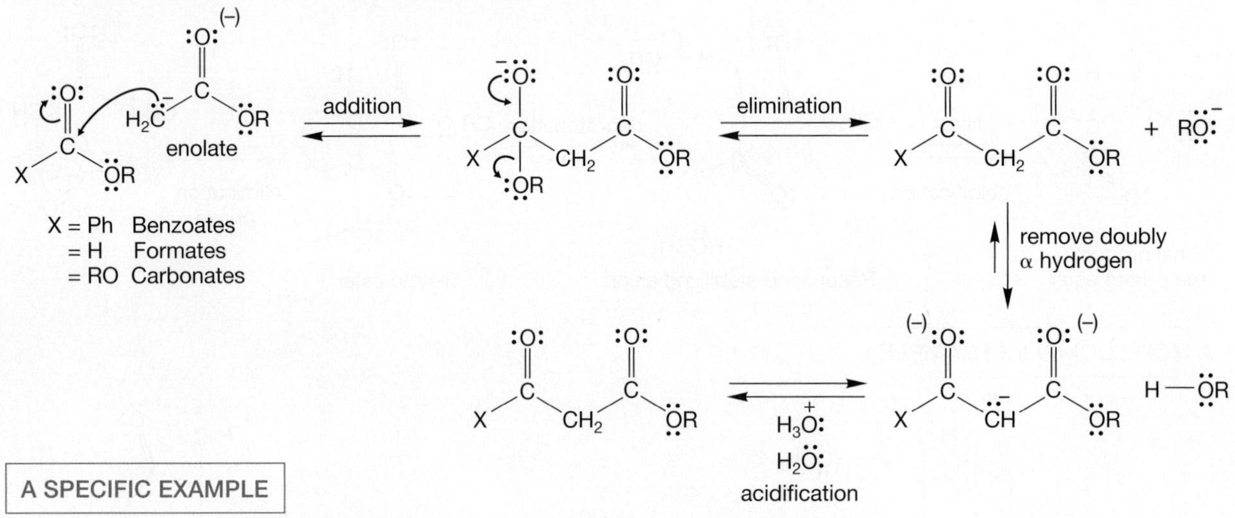

A SPECIFIC EXAMPLE

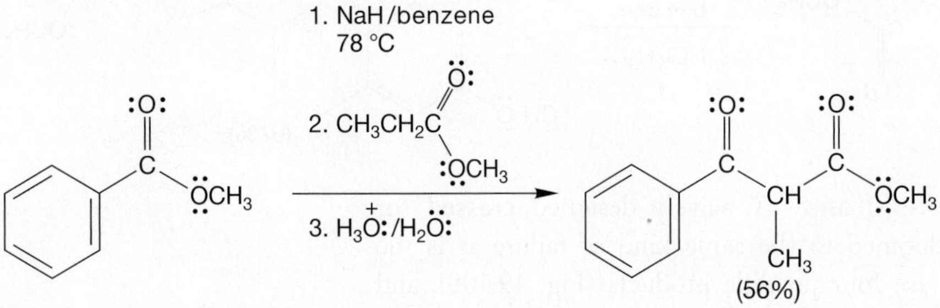

A RELATED EXAMPLE

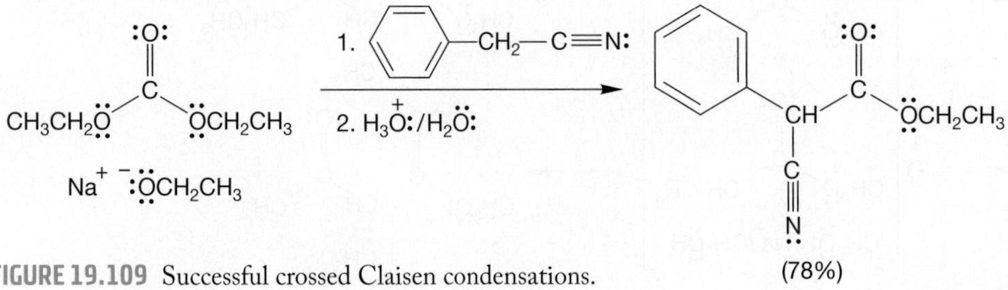

FIGURE 19.109 Successful crossed Claisen condensations.

We now have some new reactions to use in problems. Put another way, your lives are now even more complicated than before, as even more sadistically difficult problems are within your abilities. There really aren't too many new things to think about. You only have to add the Claisen-like processes that involve the loss of a leaving group from a carbonyl carbon that has been attacked by an enolate. In other words, you always have to keep the addition–elimination reaction in mind.

But be careful—reactions are driven by thermodynamics, not our concepts of forward and backward. The reverse Claisen is much harder to detect in problems than the forward version, but no less reasonable mechanistically. As always in matters such as these, experience counts a lot, and there is no way to get it except to work a lot of problems.

PROBLEM 19.35 Provide mechanisms for the following reactions:

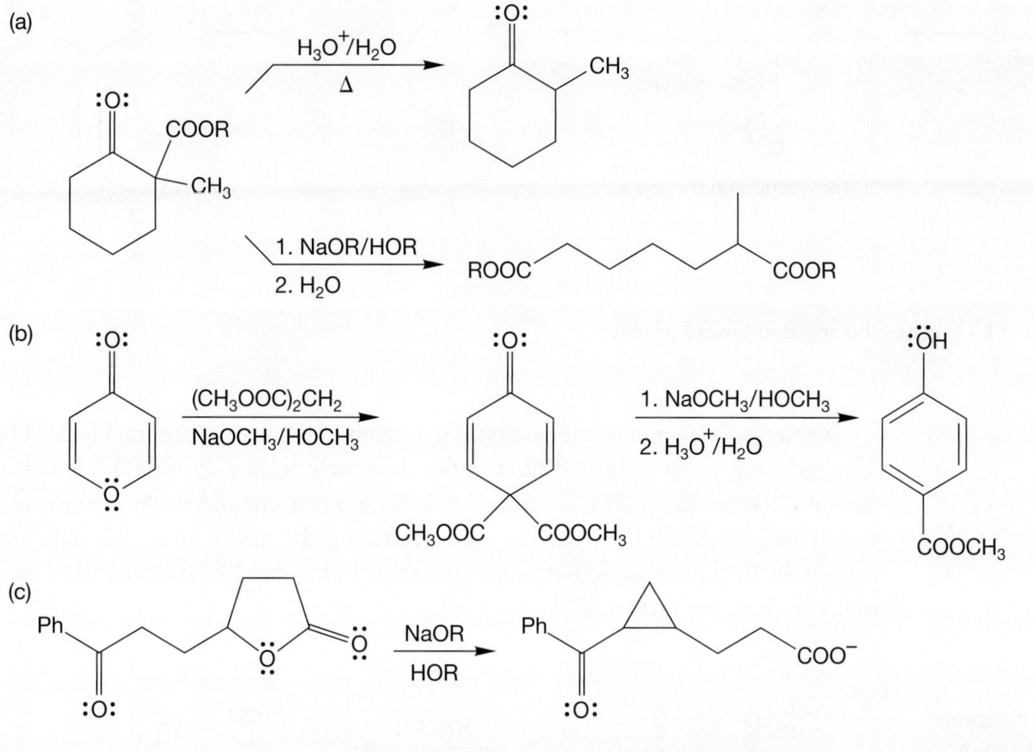

(a)

(b)

(c)

PROBLEM 19.36 Provide a mechanism for the following transformation. Be careful. This problem is *much* harder than it looks at first. *Hint*: The mechanism involves a lactone intermediate.

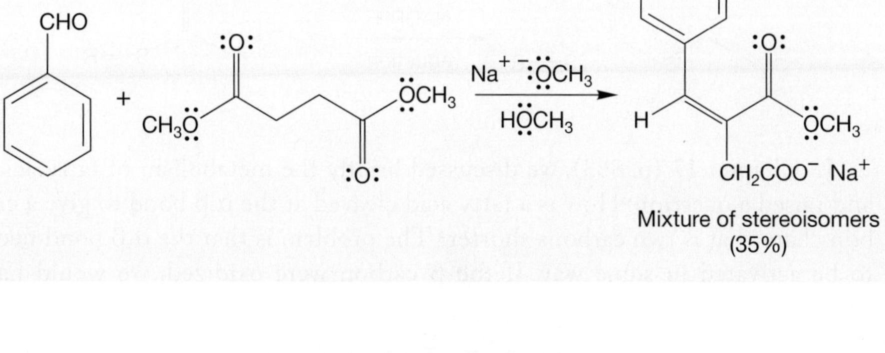

Mixture of stereoisomers
(35%)

19.10 Special Topic: Forward and Reverse Claisen Condensations in Biology

Both forward and reverse Claisen condensations feature prominently in the biochemical synthesis and degradation of fatty acids (p. 862). A critical step in the construction of these long-chain carboxylic acids involves enzyme-mediated Claisen condensations in which activated two-carbon fragments are sewn together. Of course, Nature has a thermodynamic problem here—how is the endothermicity of the Claisen condensation to be overcome? The trick is to use an activated malonate in the condensation. Loss of carbon dioxide is used to drive the equilibrium toward the product (Fig. 19.110).

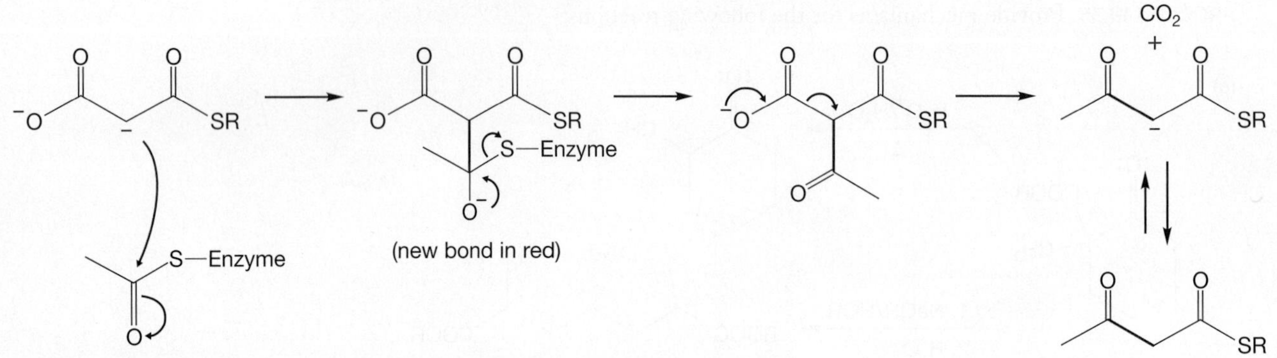

FIGURE 19.110 Loss of CO_2 drives this reaction toward product.

Nature's work is not yet done; first, a reduction of the ketone occurs (Fig. 19.111). The reducing agent is NADPH, a molecule closely related to NADH, which we saw in Chapter 16 (p. 814). Next, there is an enzyme-mediated elimination reaction. Finally, NADPH participates again, reducing the newly formed double bond to give the final product. Repetition of the steps in Figures 19.110 and 19.111 eventually leads to the fatty acid.

FIGURE 19.111 A series of enzyme-mediated reactions leads to an overall reduction of one carbonyl group to a methylene.

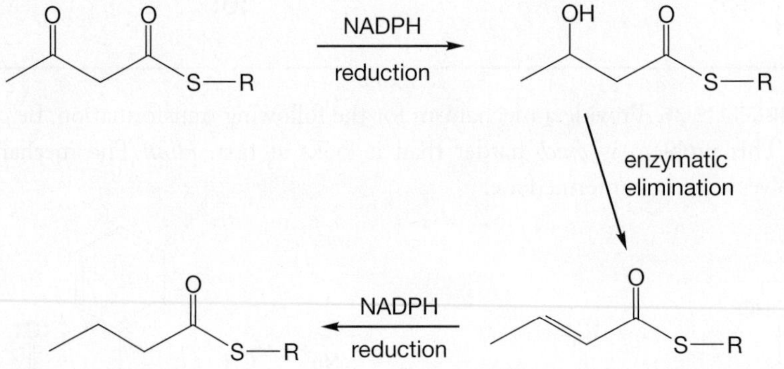

In Chapter 17 (p. 863), we discussed briefly the metabolism of fatty acids, and raised a question: How is a fatty acid cleaved at the α,β bond to give a carbon chain that is two carbons shorter? The problem is that the α,β bond needs to be activated in some way. If the β carbon were oxidized, we would have a β-keto ester all set up for a reverse Claisen condensation. That is exactly the pathway Nature uses. Here, palmitate is first converted into the SCoA derivative. Then three enzymes act in sequence to dehydrogenate, hydrate, and dehydrogenate again to give the β-keto ester (Fig. 19.112). Now the reverse Claisen does the first two-carbon cleavage reaction, and metabolism of this fatty acid has begun.

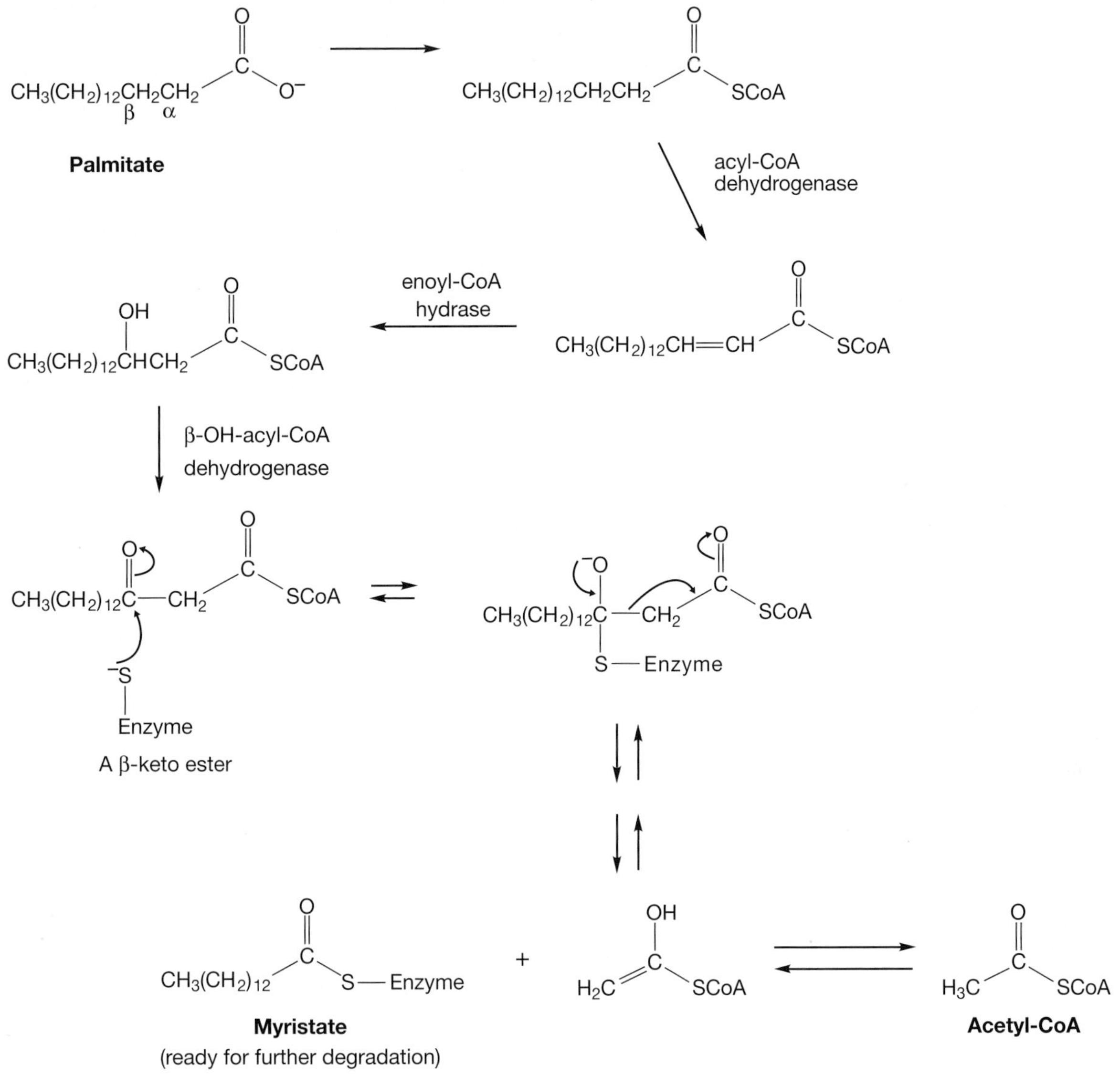

FIGURE 19.112 Metabolism of a fatty acid containing 16 carbons to one containing 14.

Summary

There are several variations of the Claisen condensation. The intramolecular variation is called the Dieckmann condensation. A crossed Claisen condensation is possible between two different esters, but this reaction can lead to multiple products. Claisen condensations can operate in the "forward" direction or in the "reverse" direction, and these two processes lead to the construction and deconstruction of fatty acids.

19.11 Condensation Reactions in Combination

Alas, neither Nature nor a typical chemistry teacher shows much mercy, and one is more likely than not to find condensation reactions in combination. None of these reactions is particularly difficult to understand when encountered by itself. Similarly, the synthetic consequences of the reactions are relatively easy to grasp when they are encountered one at a time, but are by no means simple when combinations of these reactions are involved. Condensation problems, occasionally called "Fun in Base" problems (a term that seems to encompass reactions run in acid as well as base—there are no "Fun in Acid" problems), can be very hard. The best way to become proficient is to work lots of them, but there are also some practical hints that may help.

One common sequence is a combination of the Michael addition and aldol condensations. Indeed, there is a synthetic procedure for constructing six-membered rings, invented by Sir Robert Robinson (1886–1975) and called the **Robinson annulation,** that uses exactly that combination. In this reaction a ketone is treated with methyl vinyl ketone in base (Fig. 19.113). A Michael addition ensues and

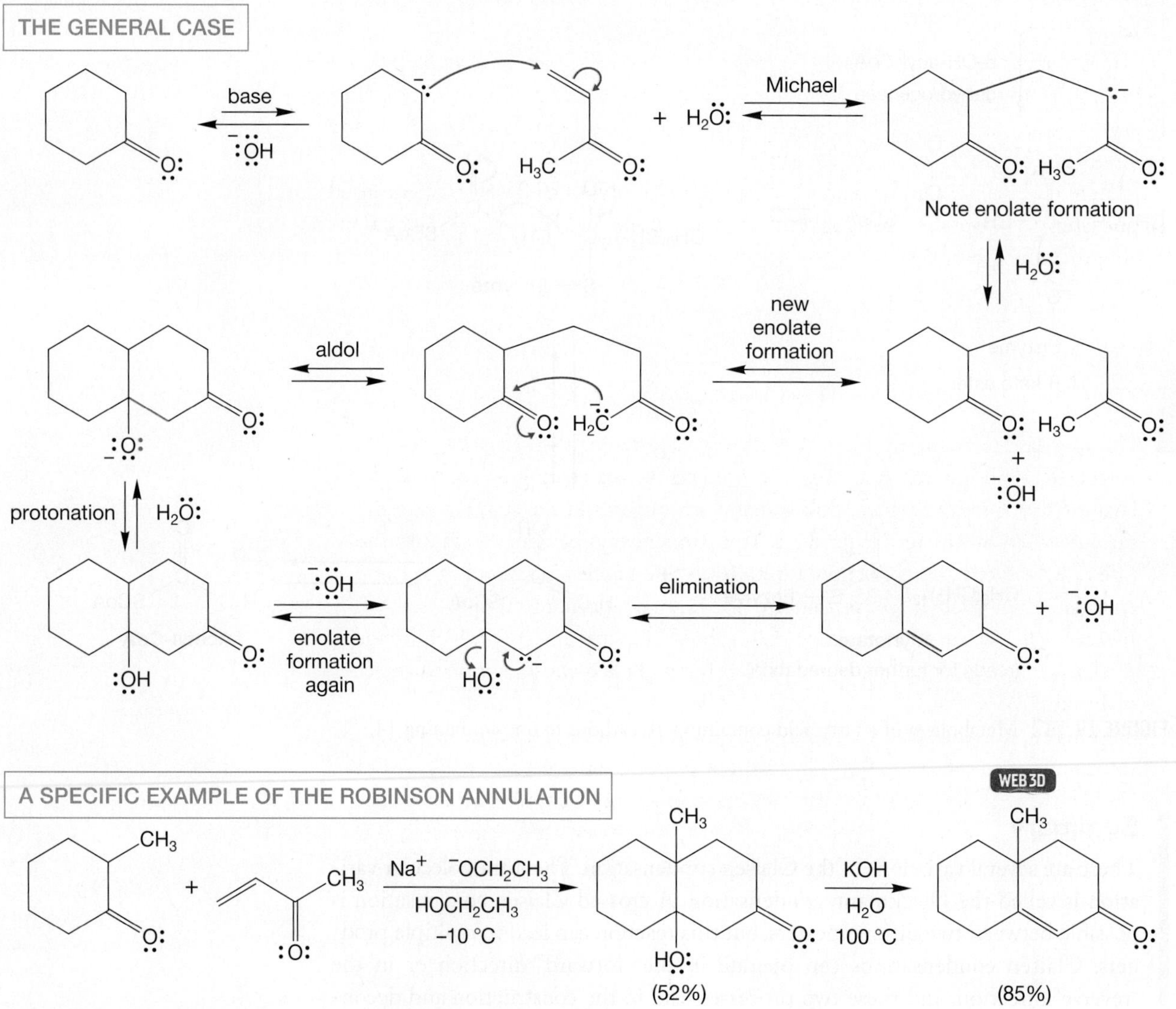

FIGURE 19.113 The Robinson annulation creates a six-membered ring.

is followed by an intramolecular aldol condensation to give the new six-membered ring (Fig. 19.113).

PROBLEM 19.37 Although the mechanism shown in Figure 19.113 is the way the Robinson annulation is always described, there is another (harder) way to write the reaction. Write a mechanism for the general reaction in Figure 19.113 that involves doing the aldol condensation first.

Problem 19.37 brings up a serious practical question. You may well be able to analyze a problem by figuring out the different possible reactions, but choosing among them is more difficult. How do you find the easiest route (there are very often several) to the product you are after? Clearly, experience is invaluable and you are by definition short on that commodity. In such a situation, it is wise to be receptive to good advice, and here is a practical hint from one of the best problem solvers known, Ronald M. Magid (b. 1938), in the form of **Magid's second rule**:[4] When there is a choice, always try the Michael first.

In fact, there are sound reasons for doing the Michael reaction first. Aldol condensation products of ketones are not generally favored at equilibrium, for example, and doing the Michael reaction makes the subsequent aldol condensation intramolecular, and thus more favorable.

There is no substitute for practice, though, and Problem 19.38 offers some good opportunities to practice solving Fun in Base problems. Be sure to look at the Problem Solving box first.

PROBLEM SOLVING

Complex mechanistic problems absolutely require analysis. A good problem solver *always* begins by analyzing the problem before writing any arrows. We suggest first building a map. Ask yourself what atoms in the starting material become what atoms in the product. You won't always be able to identify the fate of every atom, but you should usually be able to see what happens to some of them. Use attached groups that are not changed in the reaction as markers. Next, develop a set of goals. Your map should allow you to spot which bonds in the product must be made and which bonds in the starting material must be broken.

Finally, after building your map and establishing your goals, you can begin to think about how to accomplish those goals. We'll start you off on part (b) of Problem 19.38, but please try the other problems this way. We guarantee that you will get more answers right if you do. And a side benefit is that even if you don't get the problem completely right, the graders will "see" you thinking and partial credit definitely flows to such answers.

---

[4] Magid's first rule is archaic, and we will shortly encounter Magid's third rule.

WORKED PROBLEM 19.38 Provide mechanisms for the reactions shown below.

(a)

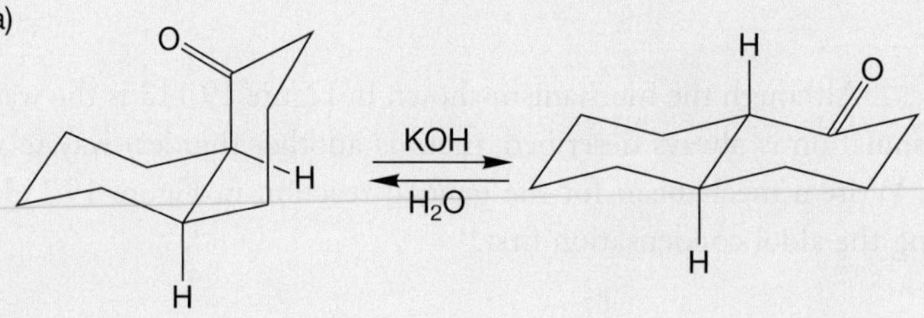

(b)

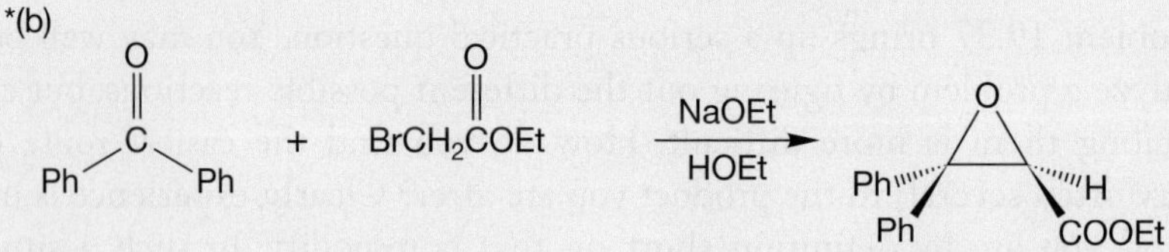

(c)

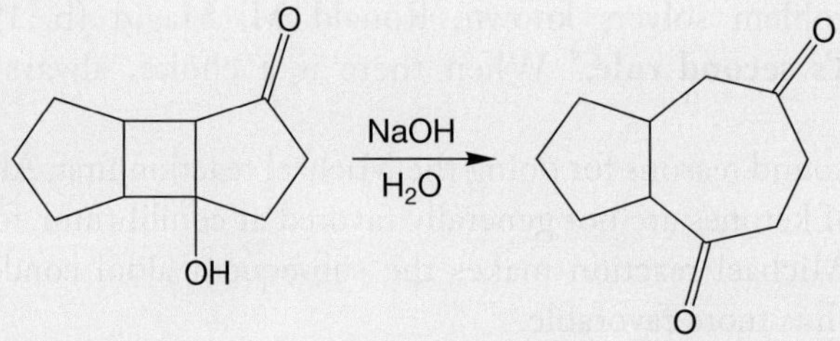

(d)

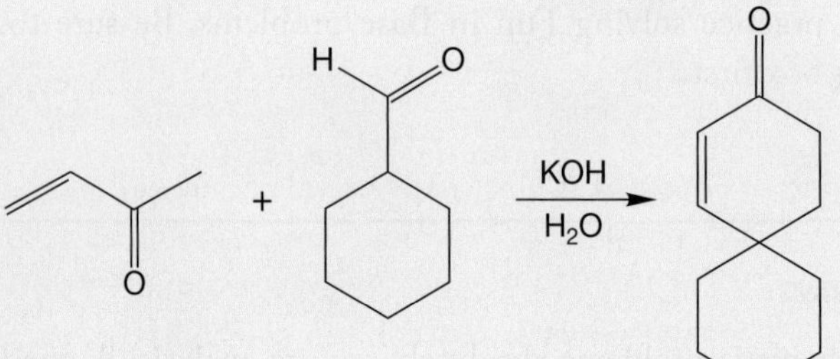

(e)

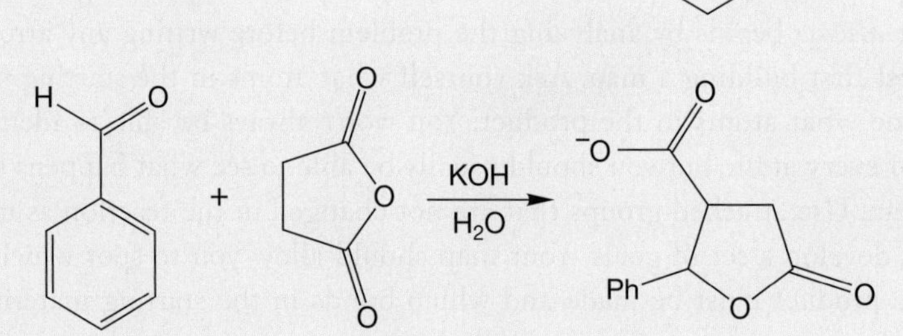

(f) *Hint:* There is a 10-membered ring involved.

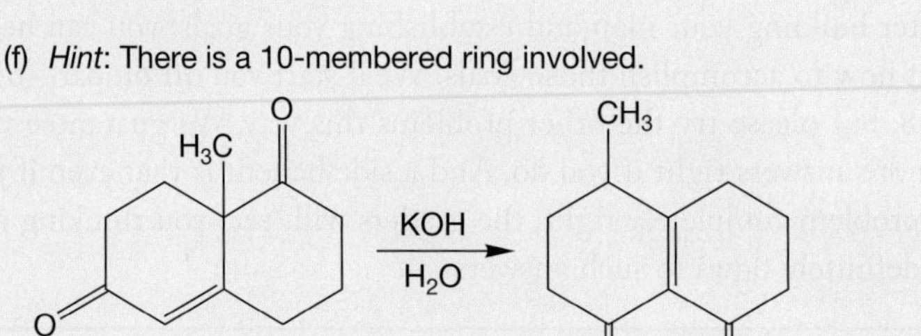

(continued)

ANSWER (b) Those two Ph groups are just along for the ride—nothing happens to them in this reaction. Use them as markers to identify where the carbon attached to them in the starting material winds up in the product (red dot). Similarly, that ester group is preserved. Use it as a marker to find the carbon attached to it in the starting materials (green dot).

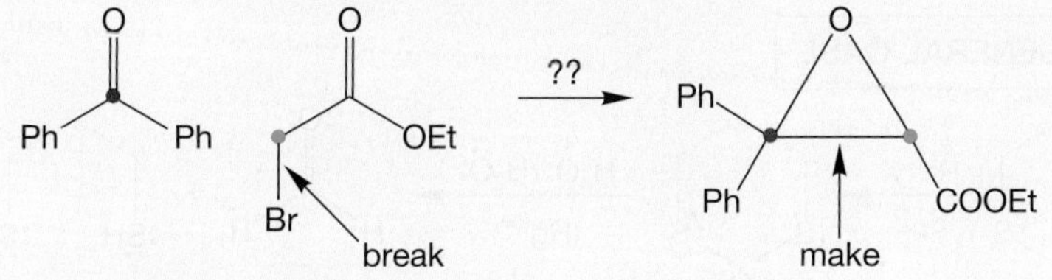

Goals:

1. Our map allows us to see that we must make the red dot–green dot bond (arrow, above), so that's goal number one.

2. We also must make a bond from oxygen to either the red or green dot—but we can't tell which (yet).

3. The bond to bromine must be broken—we know that because there is no Br in the product.

Okay, now you are on your own. Before looking at the *Study Guide*, find a way to make that red–green bond. Once you do that, it will be easy to see how to lose the Br and how to make that oxygen–dot bond.

19.12 Special Topic: Alkylation of Dithianes

Here is a clever way to alkylate carbonyl compounds at the carbonyl carbon atom. Thioacetals, also called **dithianes**, can be made from carbonyl compounds and the sulfur counterparts of diols (Fig. 19.114). Dithianes are more acidic than typical alkanes and can be deprotonated with alkyllithium reagents (Fig. 19.115).

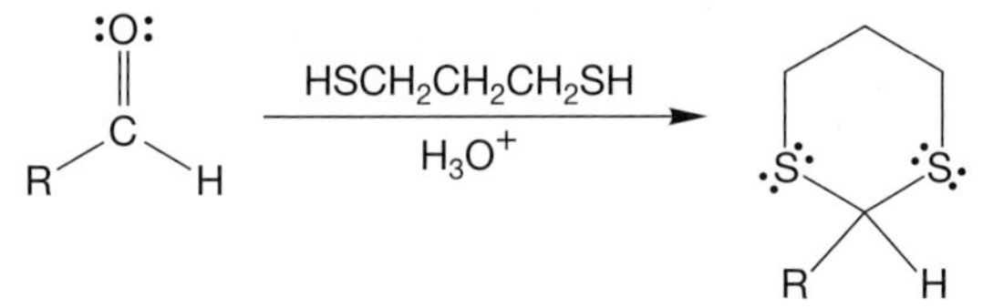

FIGURE 19.114 Formation of a dithiane (a thioacetal).

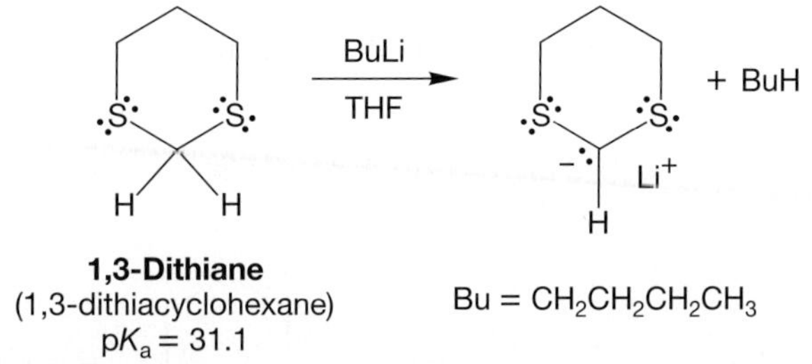

FIGURE 19.115 Deprotonation of 1,3-dithiacyclohexane with butyllithium.

1,3-Dithiane
(1,3-dithiacyclohexane)
pK_a = 31.1

Bu = $CH_2CH_2CH_2CH_3$

PROBLEM 19.39 Explain how the sulfur atoms operate to increase the acidity of the adjacent hydrogens in 1,3-dithiane (1,3-dithiacyclohexane).

The resulting anion is a powerful nucleophile and can undergo both displacement reactions (subject, of course, to the usual restrictions of the S_N2 reaction) and additions to carbonyl compounds (Fig. 19.116). The products of these reactions are dithianes and can be hydrolyzed, usually by mercury salts, to aldehydes or ketones. The dithiane functions as a masked carbonyl group! Here is another synthesis of aldehydes and ketones.

FIGURE 19.116 A synthesis of carbonyl compounds using alkylation of 1,3-dithiane (1,3-dithiacyclohexane).

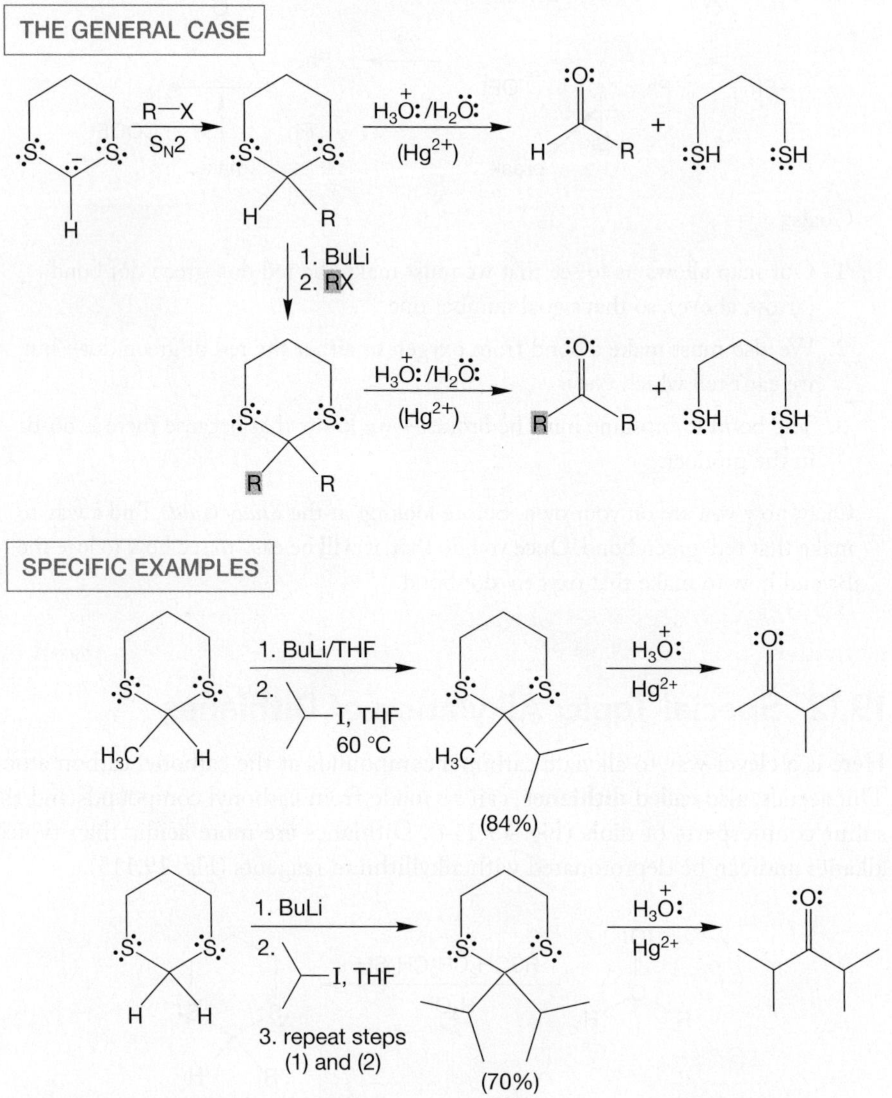

WORKED PROBLEM 19.40 Suggest a synthetic route to alkanes starting from 1,3-dithiane.

ANSWER Easy (and effective in practice). First alkylate the dithiane, then desulfurize with Raney nickel (p. 253).

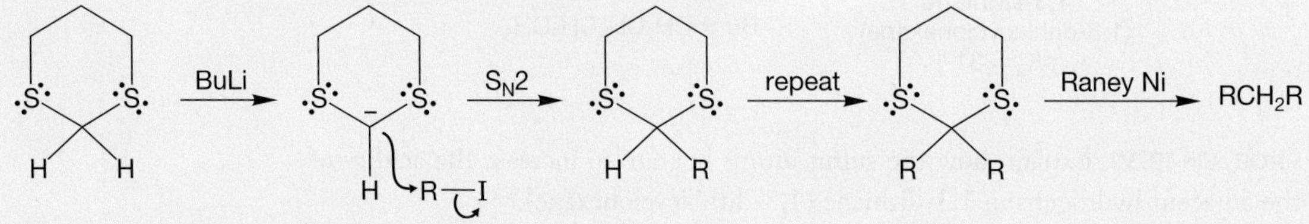

19.13   Special Topic: Amines in Condensation Reactions, the Mannich Reaction

In the **Mannich reaction**, named for Carl Mannich (1877–1947), an aldehyde or ketone is heated with an acid catalyst in the presence of formaldehyde and an amine. The initially formed ammonium ion is then treated with base to liberate the final condensation product, the free β-amino ketone (Fig. 19.117).

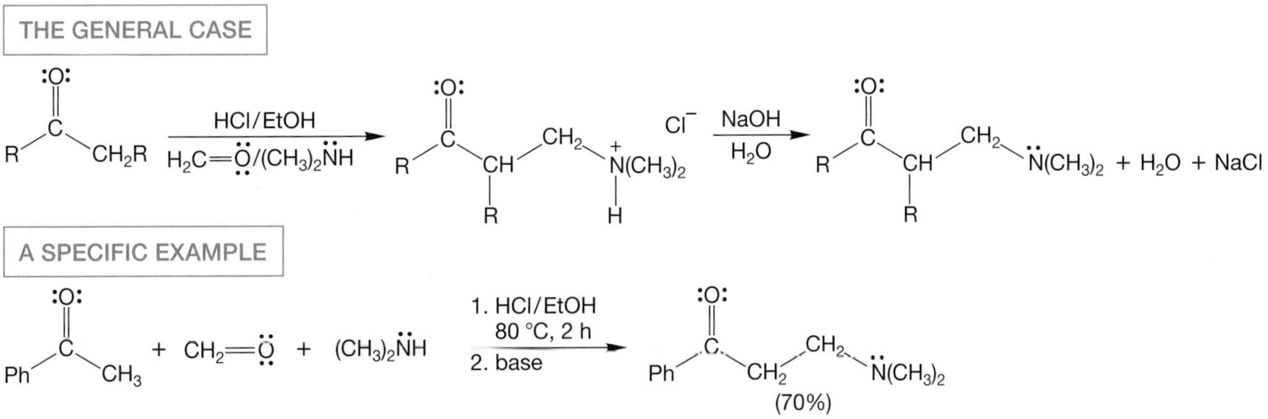

THE GENERAL CASE

A SPECIFIC EXAMPLE

FIGURE 19.117   The Mannich reaction.

The Mannich reaction presents a nice mechanistic problem because two routes to product seem possible, and it is the simpler (and incorrect) process that is the easier to find. The difficulty is that we are now experienced in working out aldol condensations and a reasonable person would probably start that way. An acid-catalyzed crossed aldol condensation (p. 982) leads to a β-hydroxy ketone (Fig. 19.118). Displacement of the hydroxyl group (after protonation, of course) would lead to the initial product, the ammonium ion. Treatment with base would surely liberate the free amine.

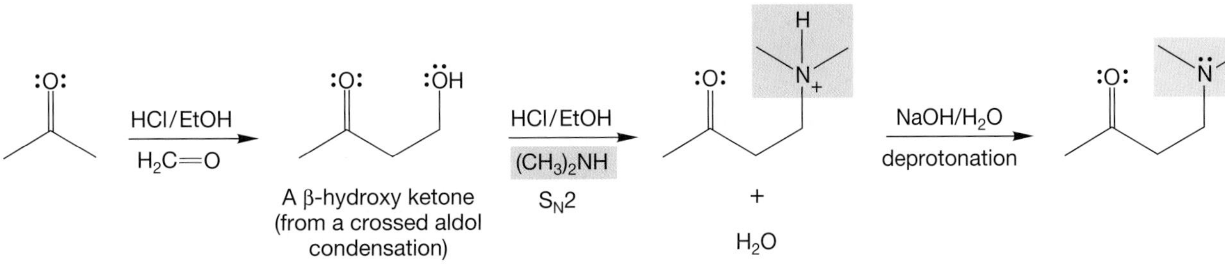

FIGURE 19.118   An attractive, but incorrect, mechanism for the Mannich reaction.

Sadly, this simple mechanism cannot be correct because, when the β-hydroxy ketone is made in other ways, it is not converted into Mannich product under the reaction conditions. What else can happen? Amines are nucleophiles and react rapidly with carbonyl compounds in acid to give iminium ions (p. 795). It is this iminium ion that is the real participant in the Mannich condensation reaction. Formaldehyde is more reactive than the ketone, so the iminium ion is preferentially formed from this carbonyl compound (Fig. 19.119). In an aldol-like process, the enol of the ketone reacts with the iminium ion to give the penultimate product, the ammonium ion. In the second step, the free amine is formed by deprotonation in base.

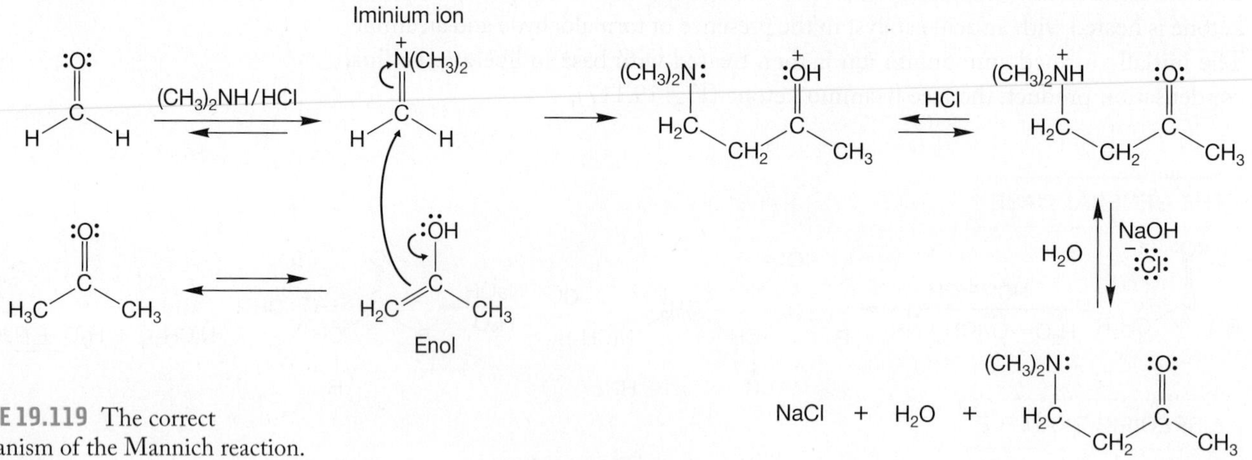

FIGURE 19.119 The correct mechanism of the Mannich reaction.

19.14 Special Topic: Carbonyl Compounds without α Hydrogens

Throughout this chapter we have concentrated on molecules capable of forming either enolates or enols, and the subsequent chemistry has been dependent on those intermediates. It is a legitimate question to ask if there is a base-induced chemistry of carbonyl compounds that do *not* have α hydrogens, and cannot form enolates.

19.14a The Cannizzaro Reaction   Hydroxide ion will give products from nonenolizable carbonyl compounds. If we examine a sample of benzaldehyde that has been allowed to stand in the presence of strong base, we find that a redox reaction has occurred. Some of the aldehyde has been oxidized to benzoate and some reduced to benzyl alcohol (Fig. 19.120).

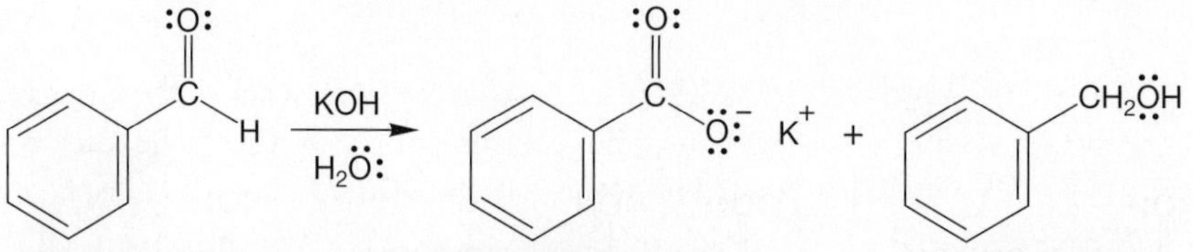

FIGURE 19.120 Benzaldehyde gives benzoate and benzyl alcohol on treatment with hydroxide ion.

This process is known as the **Cannizzaro reaction**, for Stanislao Cannizzaro (1826–1910). If the reaction is carried out in deuterated solvent, no carbon–deuterium bonds appear in the alcohol (Fig. 19.121).

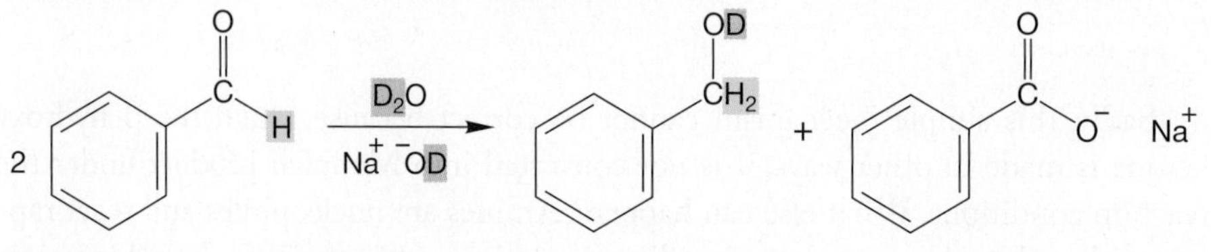

FIGURE 19.121 A labeling experiment shows that no solvent deuterium becomes attached to carbon in the Cannizzaro reaction. The hydrogen that reduces the aldehyde to the alcohol must come from another aldehyde, not the solvent.

Benzaldehyde is especially prone to hydrate formation, and the first step of the Cannizzaro reaction is addition of hydroxide to benzaldehyde. Much of the time this reaction simply reverses to re-form hydroxide and benzaldehyde, but some of the time the intermediate finds a molecule of benzaldehyde in a position to accept a transferred hydride ion (Fig. 19.122).

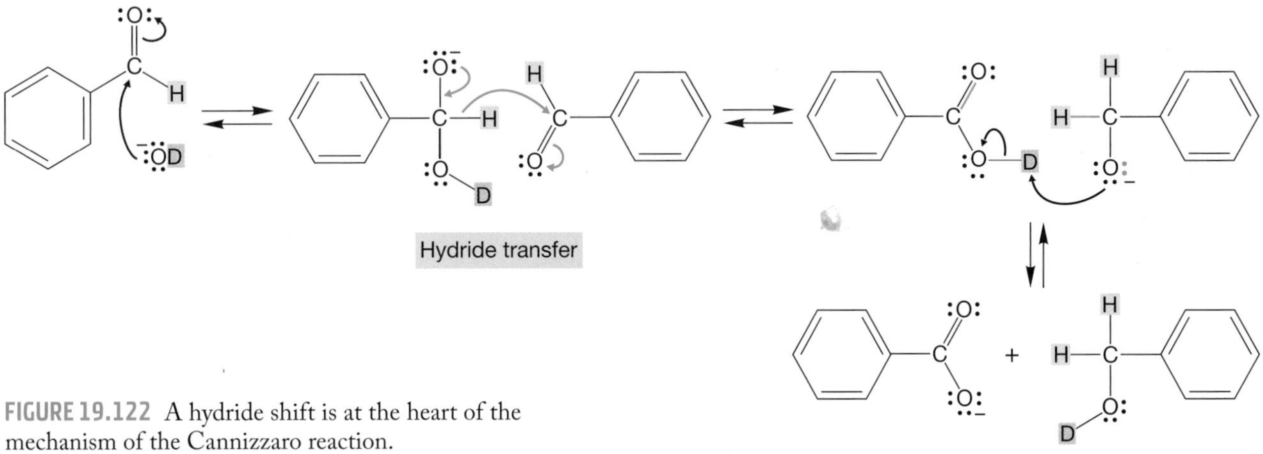

Hydride transfer

FIGURE 19.122 A hydride shift is at the heart of the mechanism of the Cannizzaro reaction.

Do not confuse this hydride transfer step with the expulsion of hydride as a leaving group. Hydride is not a decent leaving group and *cannot* be lost in this way. However, if there is an electrophile positioned in exactly the right place to accept the hydride, it can be transferred in a step that simultaneously reduces the receptor aldehyde and oxidizes the donor aldehyde. It is very important to be aware of the requirement for the Lewis acid acceptor of hydride in the Cannizzaro reaction. The hydride must be *transferred*, and in the transition state for the reaction there is a partial bond between the hydride and the accepting carbonyl group. A subsequent deprotonation of the carboxylic acid and protonation of the alkoxide formed from benzaldehyde completes the reaction. Notice how critical the observation about the nonparticipation of the deuterated solvent is. It identifies the source of the reducing hydrogen atom. Because it cannot come from solvent (all the "hydrogens" in the solvent are deuteriums) it must come from the aldehyde.

PROBLEM 19.41 Draw the transition state for the hydride shift reaction.

If there were a way to hold the hydride acceptor in the right position, the Cannizzaro reaction would be much easier, because the requirements for ordering the reactants in just the correct way (entropy) would be automatically satisfied. The two partners in the reaction would not have to find each other. This notion leads to the next reaction in this section, which is largely devoted to hydride shift reactions with wonderful names.

19.14b The Meerwein–Ponndorf–Verley–Oppenauer Equilibration

In the **Meerwein–Ponndorf–Verley–Oppenauer (MPVO) equilibration** an aluminum atom is used to clamp together the two halves (oxidation and reduction) of the Cannizzaro reaction. An aluminum alkoxide, typically aluminum triisopropoxide, is used as a hydride source to reduce the carbonyl compound (Fig. 19.123).

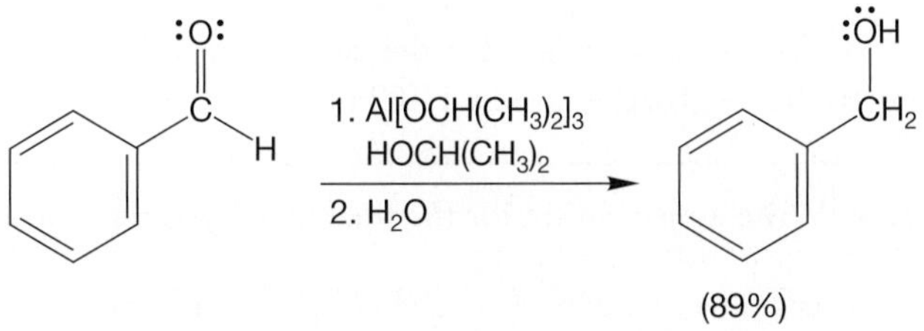

1. Al[OCH(CH₃)₂]₃
   HOCH(CH₃)₂
2. H₂O

(89%)

FIGURE 19.123 The reduction of benzaldehyde using the Meerwein–Ponndorf–Verley–Oppenauer reaction.

The first step in this reaction is the formation of a bond between the carbonyl oxygen and aluminum. Aluminum is a very strong Lewis acid and the bond to the carbonyl oxygen forms easily (Fig. 19.124). This new molecule (**A**) has the two reacting groups clamped together by the aluminum and is nicely set up for an intramolecular transfer of hydride, which simultaneously reduces a molecule of the carbonyl compound and oxidizes an isopropoxide group to acetone.

FIGURE 19.124 The formation of a bond between the nucleophilic carbonyl oxygen and the Lewis acid aluminum alkoxide to give complex **A**, which undergoes an intramolecular hydride transfer facilitated by the aluminum atom chemical clamp.

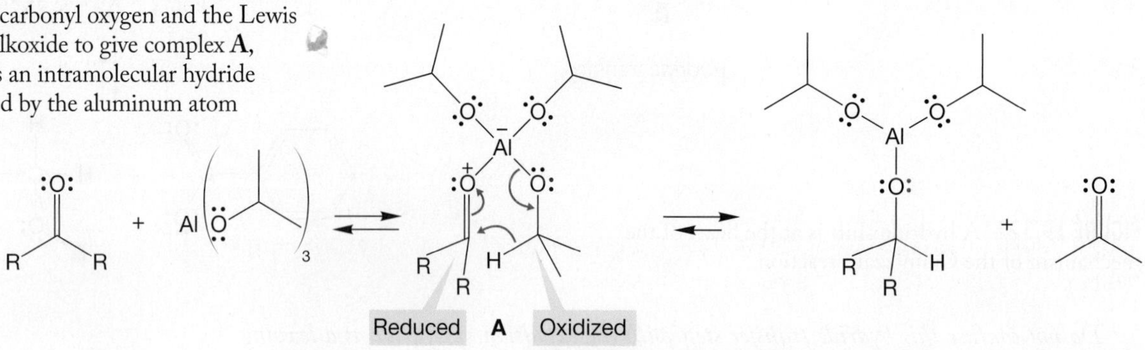

PROBLEM 19.42 Aluminum alkoxides may look very strange to you. Compare the structure of $Al(OR)_3$ to that of a trivalent boron compound such as $B(OH)_3$.

Three repetitions of this process leads to a new aluminum oxide that is hydrolyzed at the end of the reaction to give the corresponding alcohol (Fig. 19.125).

FIGURE 19.125 Repetition and hydrolysis lead to more acetone and the new alcohol.

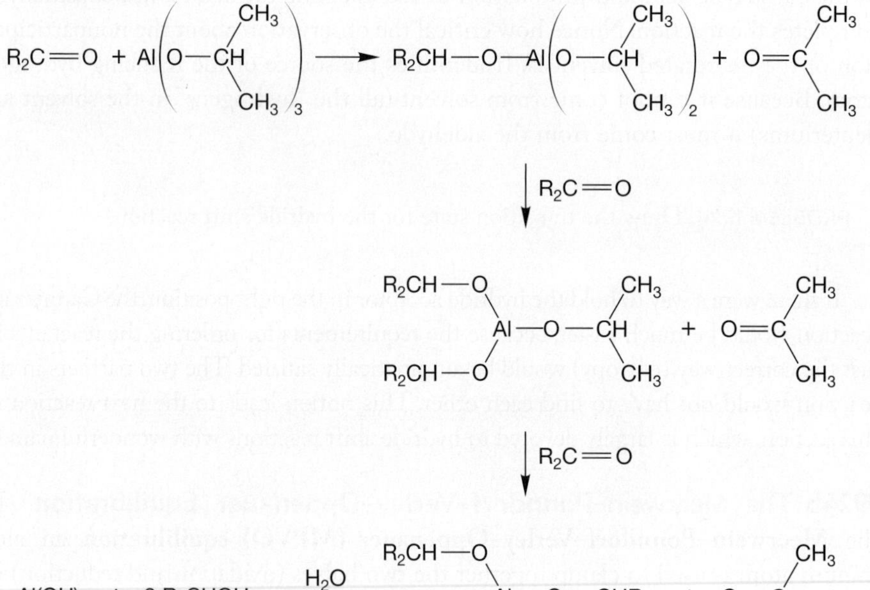

PROBLEM 19.43 Write a mechanism for the second stage in the reaction, the formation of $(R_2CH—O)_2Al—O—CH(CH_3)_2$.

PROBLEM 19.44 Write a mechanism for the final hydrolysis reaction of $Al(OR)_3$ to $Al(OH)_3$.

The effect of the aluminum clamp on the rate of the reaction is profound. This reaction, here shown as a reduction, takes place under very mild conditions and with few side reactions.

The reaction can also be used to oxidize alcohols to carbonyl compounds. In this case, acetone is used as the acceptor, and the alcohol to be oxidized is used to form the aluminium oxide (Fig. 19.126).

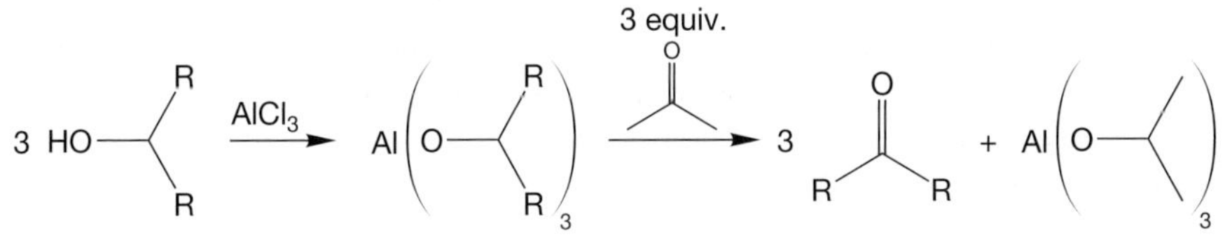

FIGURE 19.126 The Meerwein–Ponndorf–Verley–Oppenauer equilibration can be used to oxidize an alcohol.

Problems involving hydride shifts tend to be hard enough to thwart even excellent problem solvers. That observation leads directly to **Magid's third rule**, which states, When all else fails and desperation is setting in, look for the hydride shift. Problem 19.45 involves some practice hydride shifts.

PROBLEM 19.45 Write mechanisms for the following reactions.

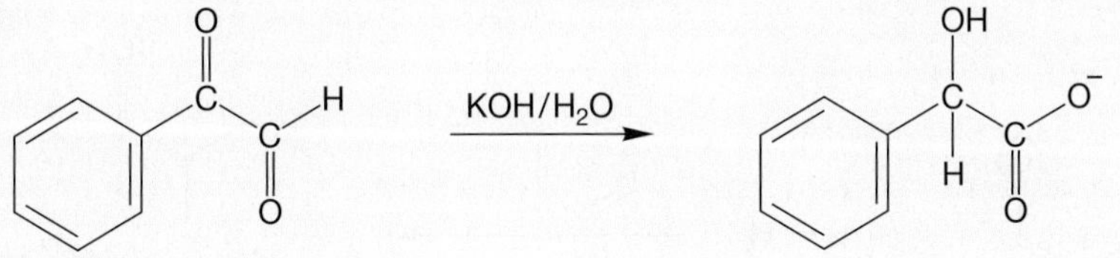

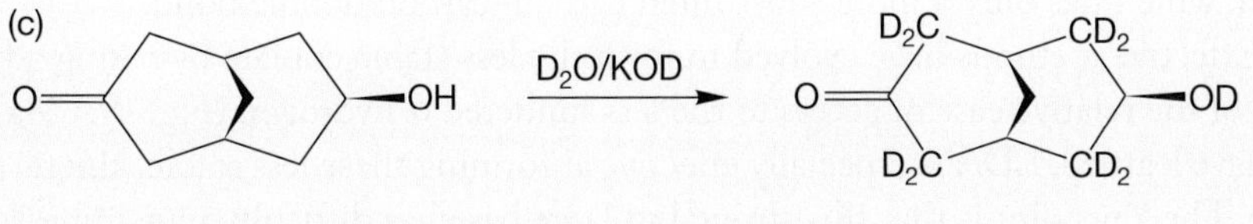

Hints: Draw a good three dimensional picture of this molecule first! The other stereoisomer exchanges only five hydrogens for deuteriums.

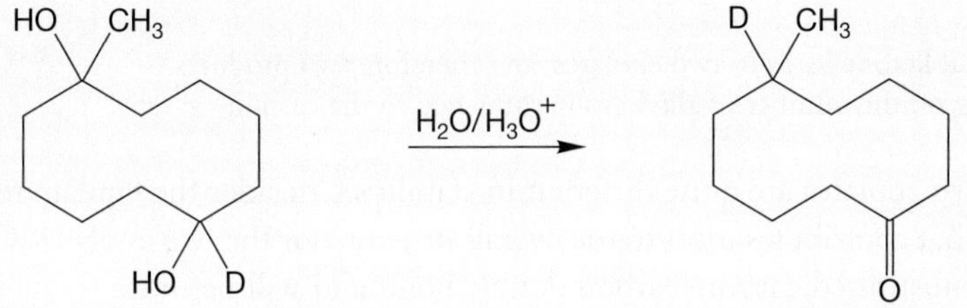

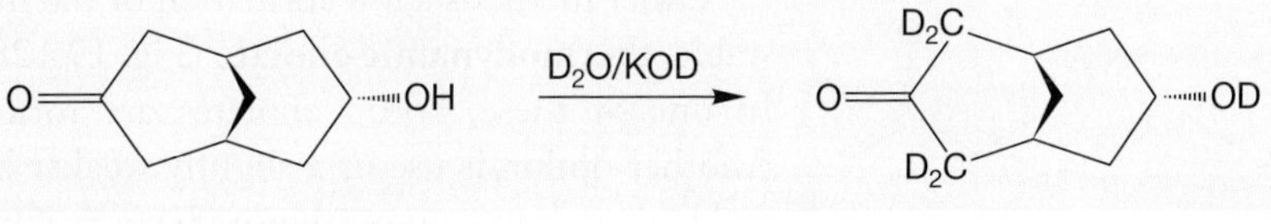

19.15 Special Topic: The Aldol Condensation in the Real World, an Introduction to Modern Synthesis

In this section, we introduce some of the difficulties of real-world chemistry. The detail isn't so important here, although it will be if you go on to become a synthetic chemist! It is important to get an idea of the magnitude of the problems, and of the ways in which chemists try to solve them. The general principles and relatively simple reactions we have studied (e.g., the straightforward aldol condensation) set the stage for a glance at the difficulties encountered when chemists try to do something with these reactions.

In the real world of practical organic synthesis, one rarely needs to do a simple aldol condensation between two identical aldehydes or two identical ketones. Far more common is the necessity to do a crossed aldol between two different aldehydes, two different ketones, or an aldehyde and a ketone. As noted earlier, there are difficulties in doing crossed aldol reactions. Suppose, for example, that we want to condense 2-pentanone with benzaldehyde. Benzaldehyde has no α hydrogen, so no enolate can be formed from it. Some version of the Claisen–Schmidt reaction (p. 984) seems feasible. But 2-pentanone can form two enolates, and the first problem to solve is the specific formation of one or the other enolate (Fig. 19.127).

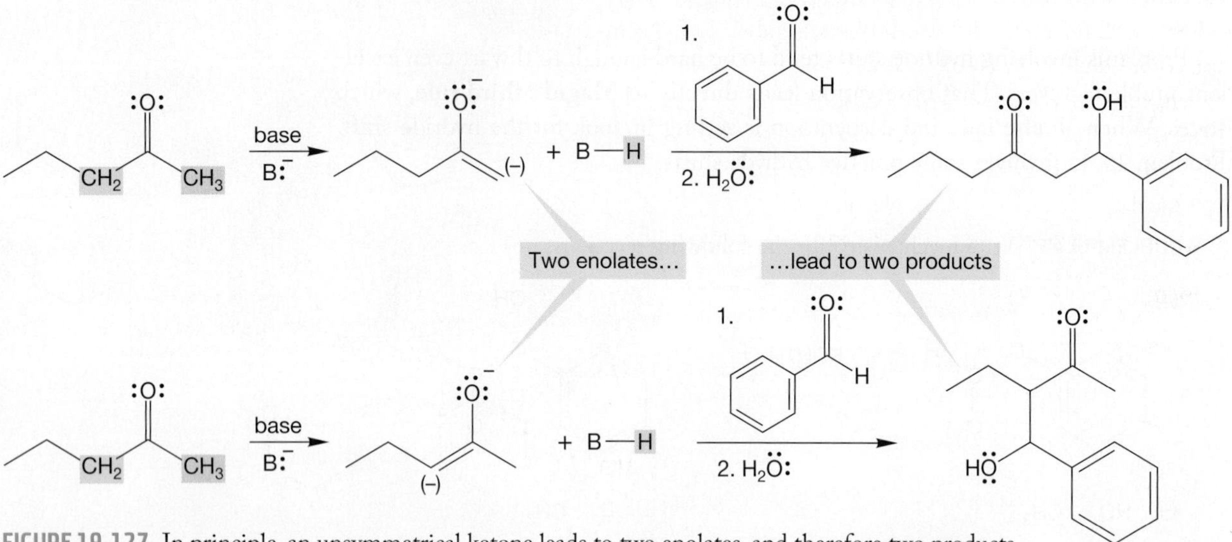

FIGURE 19.127 In principle, an unsymmetrical ketone leads to two enolates, and therefore two products. The two enolates differ in stability, depending on the number of alkyl groups attached to the carbon–carbon double bond.

The two enolates are quite different in stability. Consider the enolate resonance structure that contributes most to the overall structure for the two enolates. One contains a disubstituted carbon–carbon double bond and will be more stable than the other, which has only a monosubstituted carbon–carbon double bond.

Effective methods have evolved to form the less stable enolate by taking advantage of the relative ease of access to the less hindered α hydrogen (Fig. 19.128a). As we have learned, LDA is especially effective at forming these less stable, kinetic enolates. The key point is that this strong but large base has difficulty in gaining access to the more hindered parts of the carbonyl compound. Removal of the less hindered proton leads to the less stable, or kinetic enolate.

FIGURE 19.128 Use of the boron enolate or thermodynamic conditions to make the more stable enolate.

Other methods allow formation of the more stable, **thermodynamic enolate** (Fig. 19.128b). In one of these, boron enolates are formed. Another option is use of a slightly weaker base and higher temperatures (thermodynamic conditions) in making the enolate.

We have described only one of the difficulties encountered in just one important synthetic reaction, the aldol condensation. One goal of all synthetic chemists is selectivity, ideally, specificity. How do we do only one reaction? For example, how do we form one, and only one, stereoisomer of the many often possible?

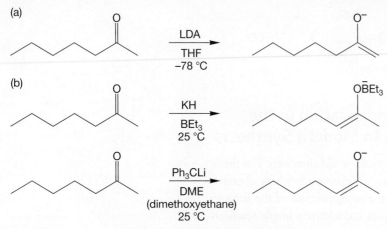

Molecules found in Nature are sometimes quite spectacularly complicated. Yet we are now able to modify the simple reactions we have been describing so as to be able to make breathtakingly complicated molecules, such as the example that follows.

PALYTOXIN

Professor Yoshito Kishi (b. 1937) of Harvard University and his co-workers set out to make a molecule called palytoxin, a compound of the formula $C_{129}H_{223}N_3O_{54}$, found in a coral resident in a small tidal pool on the island of Maui in Hawaii. A derivative of this extraordinarily toxic molecule, palytoxin carboxylic acid, is shown here. Palytoxin is difficult to make, not so much because of its size, but because of its stereochemical complexity and delicacy. It contains no fewer than 61 stereogenic atoms, each of which must be generated specifically in order to produce the real palytoxin.

Several aldol-related reactions were necessary in Kishi's spectacular synthesis. For example, one of the last critical reactions in making palytoxin was a variation of the aldol condensation, a process called the Horner–Emmons reaction, which sewed the molecule together at the red position with an α,β-unsaturated carbonyl group that was later reduced (stereospecifically) to give palytoxin.

19.16 Summary

New Concepts

This chapter deals almost exclusively with the consequences of the acidity of a hydrogen α to a carbonyl group. This concept is really not new, however. You have encountered the idea of resonance stabilization of an enolate before (for example, Problem 9.5), and you could surely have dealt with the question of why an α hydrogen is removable in base whereas other carbon–hydrogen bonds are not acidic.

In base, most carbonyl compounds containing α hydrogens equilibrate with an enolate anion. In acid, it is the enol that is formed in equilibrium with the carbonyl compound (Fig 19.129).

Enol or enolate formation leads to exchange, halogenation, and alkylation at the α position, as well as the more complicated enolate additions to ketones, aldehydes, or esters. Many of the condensation reactions are name reactions. Despite the apparent complexity of these reactions, they all involve fundamentally simple nucleophile plus electrophile chemistry (Fig. 19.130).

This chapter also introduces the notion of forming an enolate indirectly, through Michael addition to an α,β-unsaturated carbonyl group (Fig. 19.82).

The idea of transferring a hydride ion (H⁻) is elaborated in Section 19.14. *Remember*: Hydride is not a good leaving group; it cannot be displaced by a nucleophile, but it can sometimes be transferred, providing only that a suitable acceptor Lewis acid is available.

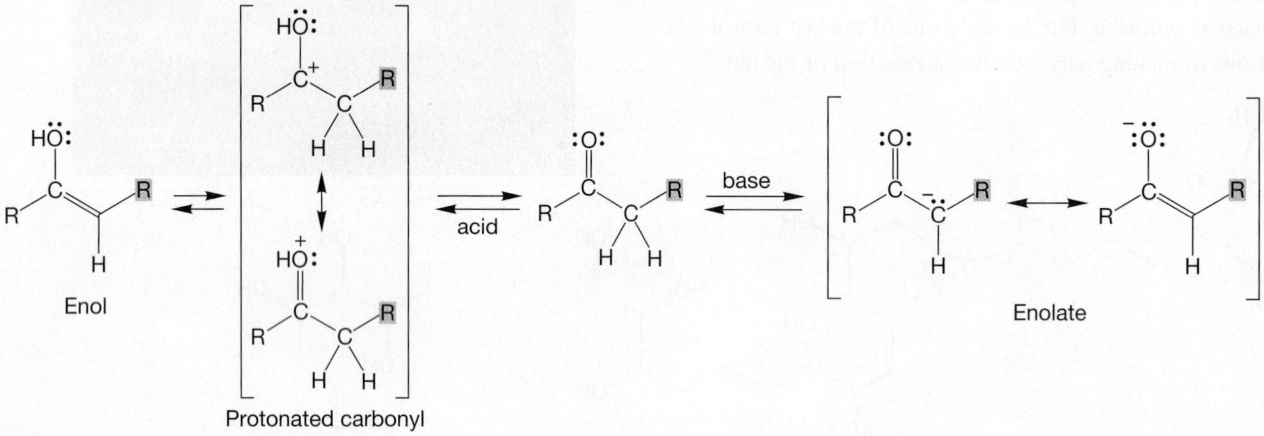

FIGURE 19.129 Enolate formation in base and enol formation in acid are typical reactions of carbonyl compounds bearing α hydrogens.

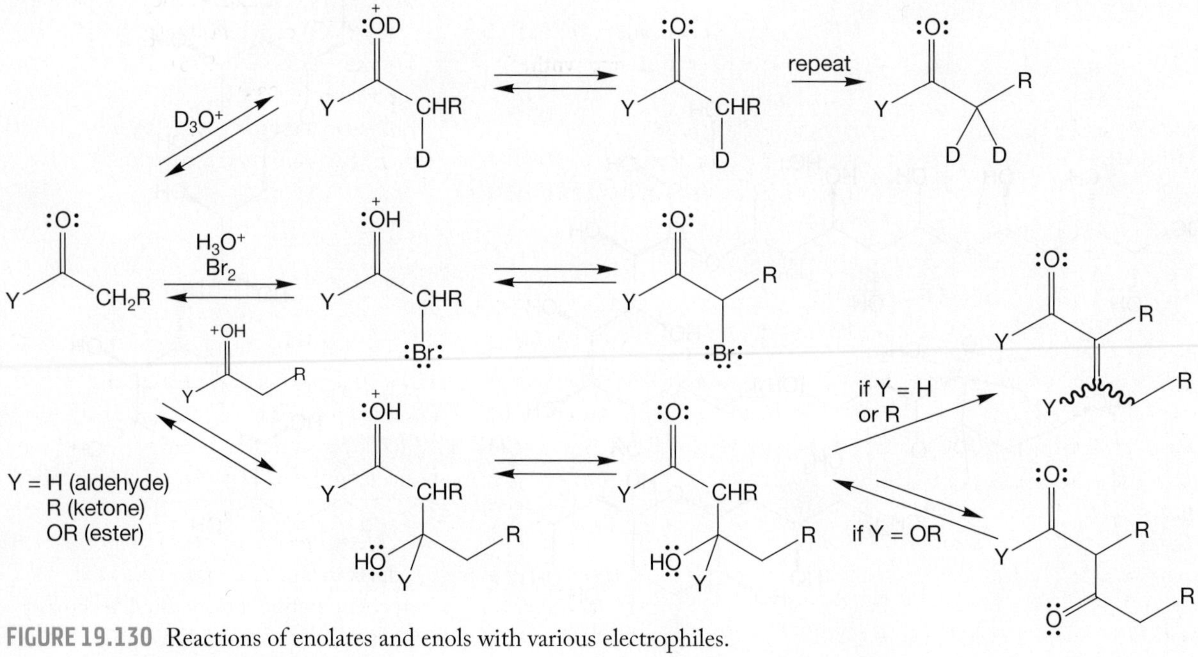

FIGURE 19.130 Reactions of enolates and enols with various electrophiles.

FIGURE 19.130 (*continued*)

Y = H (aldehyde)
    R (ketone)
    OR (ester)

Key Terms

aldol condensation (p. 966)
Cannizzaro reaction (p. 1004)
Claisen condensation (p. 987)
Claisen–Schmidt condensation (p. 984)
crossed (mixed) aldol condensation
    (p. 982)
crossed (mixed) Claisen condensation
    (p. 993)
Dieckmann condensation (p. 992)
dithiane (p. 1001)
enolate (p. 934)

enone (p. 976)
haloform (p. 949)
haloform reaction (p. 948)
Hell–Volhard–Zelinsky (HVZ) reaction
    (p. 950)
keto–enol tautomerization (p. 939)
β-keto ester or acetoacetate synthesis
    (p. 960)
kinetic enolate (p. 985)
Knoevenagel condensation (p. 974)
lithium diisopropylamide (LDA) (p. 944)

Magid's second rule (p. 999)
Magid's third rule (p. 1007)
malonic ester synthesis (p. 961)
Mannich reaction (p. 1003)
Meerwein–Ponndorf–Verley–Oppenauer
    (MPVO) equilibration (p. 1005)
Michael reaction (p. 976)
α position (p. 933)
Robinson annulation (p. 998)
thermodynamic enolate (p. 1008)

Reactions, Mechanisms, and Tools

The beginning point for almost all of the reactions in this chapter, from which almost everything else is derived, is enolate formation in base or enol formation in acid.

The enols and enolates are capable of undergoing many reactions at the α position, among them exchange, racemization, halogenation, alkylation, addition to ketones or aldehydes, and addition to esters. The aldol condensation involves reaction of the enolate, a strong nucleophile, with the electrophilic carbonyl compound, or of the less strongly nucleophilic enol with the powerful electrophile, the protonated carbonyl.

Intramolecular aldol, crossed aldol, and the related Knoevenagel condensations involve similar mechanisms.

Another way of forming an enolate anion is the Michael reaction, in which a nucleophile adds to the β position of a carbon–carbon double bond of an α,β-unsaturated carbonyl group (Fig. 19.82).

Enolates also add to esters. The biologically important Claisen condensation is the standard reaction of this kind. Variations of the Claisen condensation include the Dieckmann condensation and the reverse Claisen condensation.

A variety of reactions involving hydride shifts is described. The Cannizzaro reaction is the most famous of these and involves, as do all hydride shift reactions, a simultaneous oxidation of the hydride donor and reduction of the hydride acceptor (Fig. 19.122).

Enamines, formed through reaction of carbonyl groups with secondary amines, can be used to alkylate α positions. The carbonyl group can be regenerated through hydrolysis of the iminium ion product (Fig. 19.64).

Syntheses

The new synthetic procedures of this chapter are summarized below.

1. Acids

The Cannizzaro reaction; R may not have an α hydrogen

$$+ CO_2$$

2. Alcohols

The MPVO reaction; the mechanism involves complex formation followed by hydride shift

See also the Cannizzaro reaction under "Acids," and "β-Hydroxy Aldehydes and Ketones"

3. Alkylated Acids, Aldehydes, Esters, and Ketones

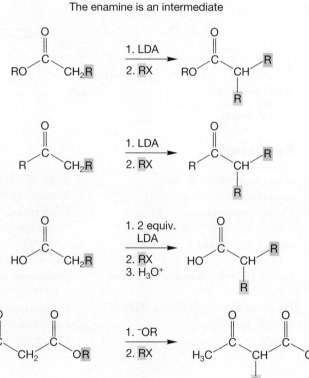

The enamine is an intermediate

R cannot be tertiary

4. β-Aminoketones (and aldehydes)

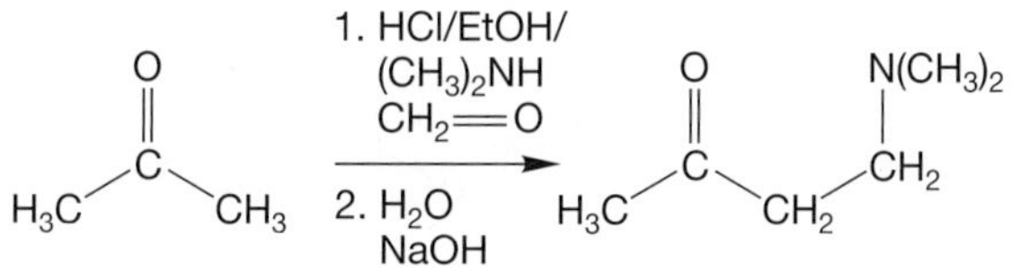

Mannich reaction; there is an intermediate iminium
ion; of course this is also an amine synthesis

5. Deuterated Aldehydes and Ketones

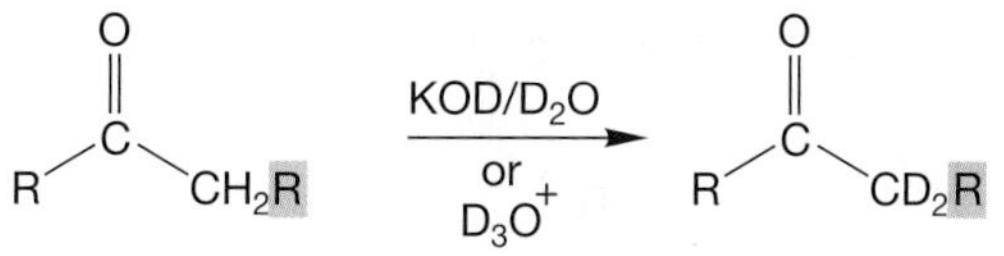

All α hydrogens will be exchanged
in either acid or base

6. α-Halo Acids, Aldehydes, Ketones, and Esters

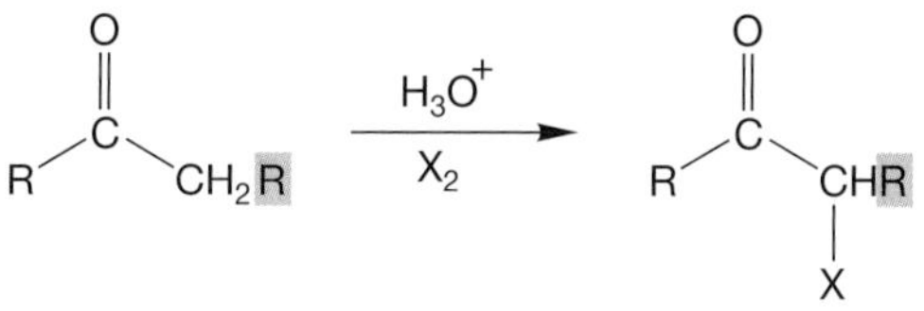

In acid, the reaction stops after one halogenation;
X = Cl, Br, and I

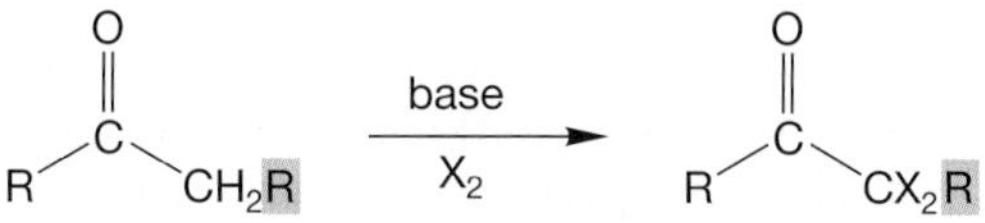

In base, up to three α hydrogens can be
replaced by X; the trihalo carbonyl compounds
react further in the haloform reaction

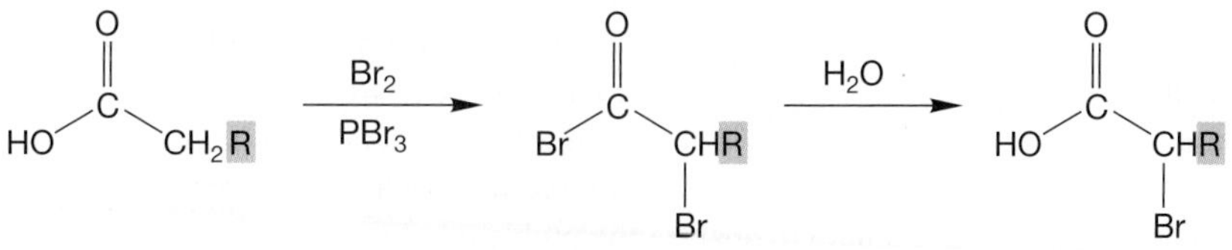

Hell–Volhardt–Zelinsky (HVZ) reaction

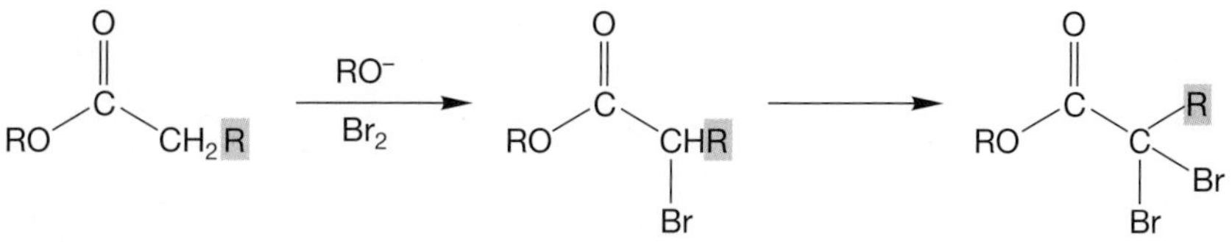

7. Haloforms

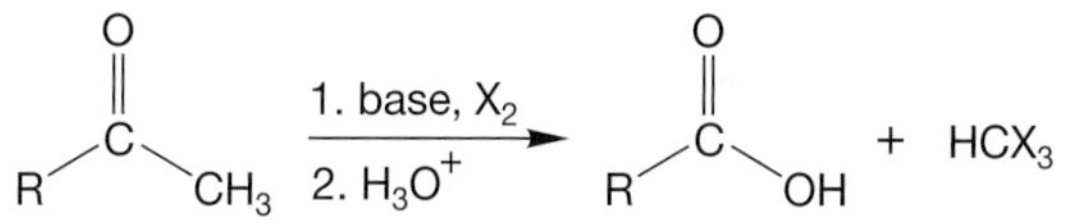

Starting carbonyl compound must have
a methyl group; X = Cl, Br, or I

8. β-Hydroxy Aldehydes and Ketones

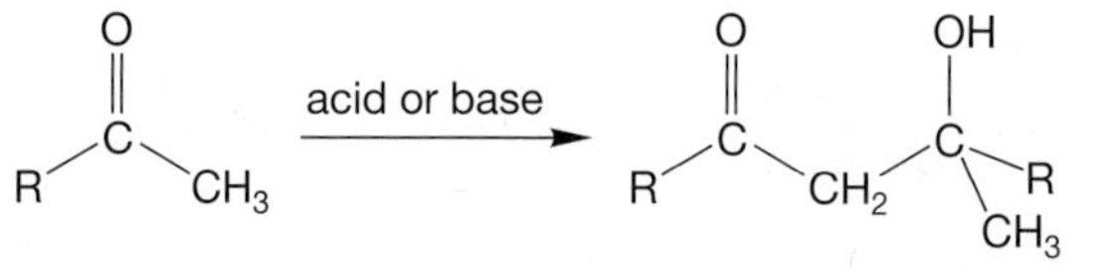

The aldol condensation; dehydration often occurs
in acid, and on occasion in base

9. β-Keto Esters

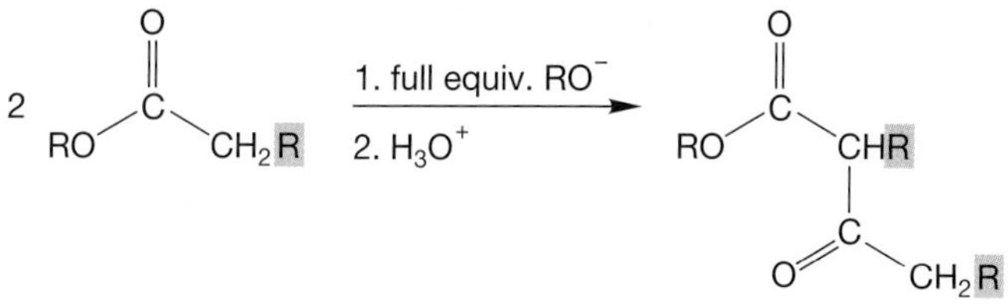

Claisen condensation

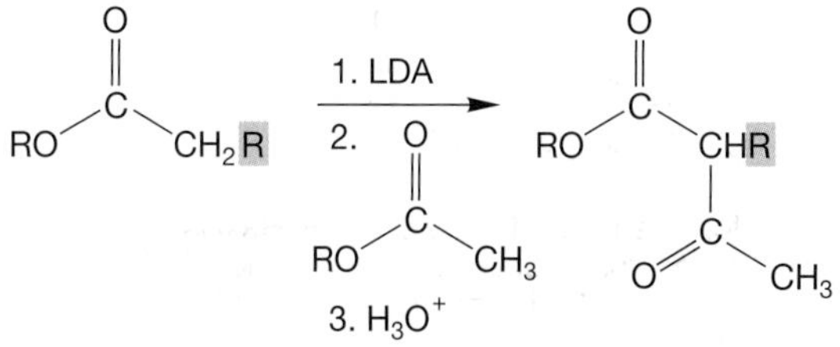

Crossed Claisen condensation

10. Ketones

The MPVO reaction; the first step is formation of
a new aluminum alkoxide; the mechanism involves
complex formation and a hydride shift

Decarboxylation of acetoacetic acid

**11. α,β-Unsaturated Acids, Aldehydes, Esters, and
Ketones**

The dehydration of the products of
aldol condensations is usually acid-catalyzed

Knoevenagel condensation

Common Errors

Condensation reaction problems can be daunting. At the same time, many chemists agree that these mental exercises are fun, and almost everyone has his or her favorites. A common error is to set out without a plan. Don't start on a complicated problem without an analysis of what needs to be accomplished. Make a map and set some goals. Ask yourself if the reaction is done in acid or base. What connections (bonds) must be made? Are rings opened or closed in the reaction? These are the kinds of questions that should be asked before starting on a problem.

Two conventions, used widely in this chapter, have the potential for creating misunderstanding. In the less dangerous of the two, charges or dots in parentheses are used to indicate the atoms sharing the charge, electron, or electrons in a molecule best described by more than one electronic description (resonance form).

The second involves drawings in which arrows of an arrow formalism emanate from only one of several resonance forms.

Although extraordinarily convenient, this practice is dangerous, as it carries the inevitable implication that the resonance form used has some individual reality. It does not. What can be done with one resonance form can always be done from all resonance forms. For simplicity, and bookkeeping purposes, we often draw arrow formalisms using only one of several resonance forms. This practice can be dangerous unless we are very clear about the shortcuts we are taking.

With the appearance of the addition–elimination process, essentially all the major reaction types are in place. Problems and errors become more general as complexity increases. The most common mistake in solving problems is a failure to analyze the problem before starting to "push arrows." The following Problem Solving section makes this point again and gives an example of the kind of analysis we are talking about.

PROBLEM SOLVING

Provide a mechanism for the following transformation. Note the position of the label (dot).

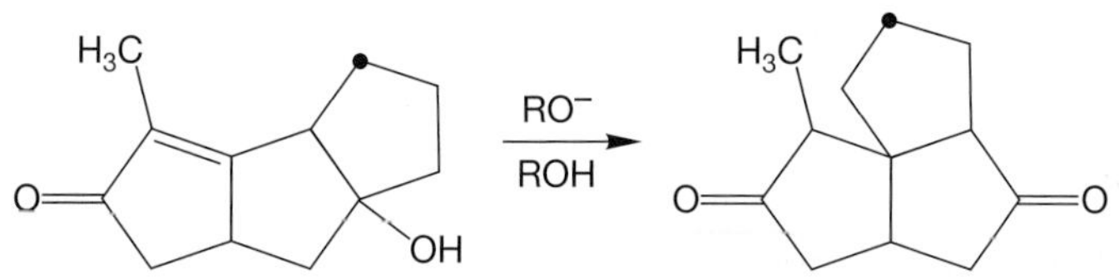

Use that methyl group to help create a map of what atoms in the starting material become what atoms in the product. Methyl groups don't have much chemistry, and are usually unchanged in a reaction (unless they are adjacent to a carbonyl group). If we label that methyl as 1, the adjacent atoms can be called 2 and 3. Note that the carbonyl group "A" on the left remains adjacent to what we have called carbon 2.

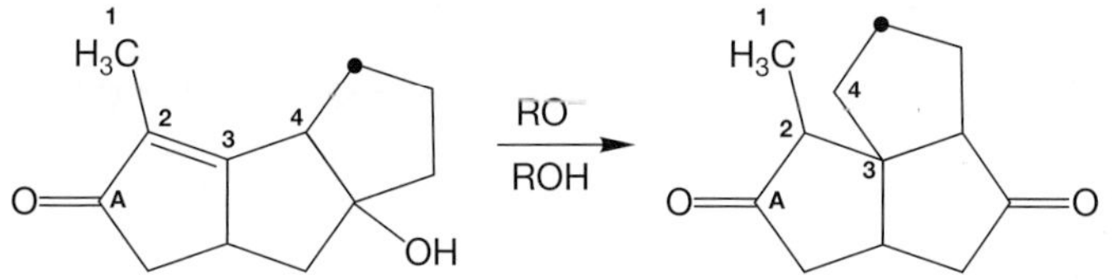

If we count four atoms counterclockwise from A, we come to carbon "E" which is attached to an oxygen. In the product, carbon 4 is still attached to carbon 3 and is adjacent to the label (dot), but it is no longer attached to carbon E as it is in the starting material. We immediately have a goal—we must break the 4—E bond!

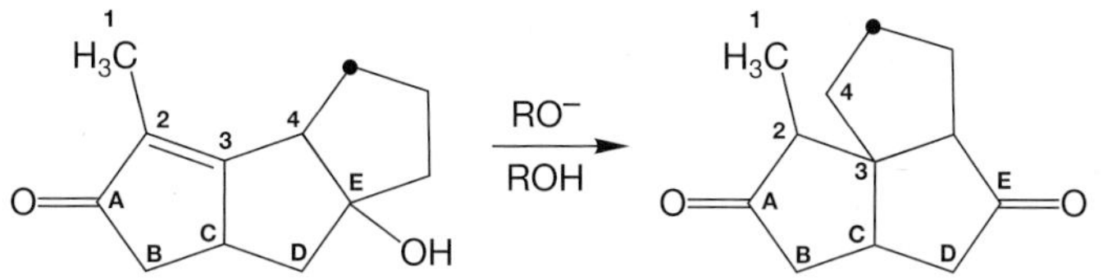

What bond must we make? The map tells you—we must make the F—3 bond. How do we know that? That bond is not present in the starting material, but it is present in the product. So now we have a second goal—make F—3! We also have a third goal—oxidize carbon E from an alcohol to a carbonyl.

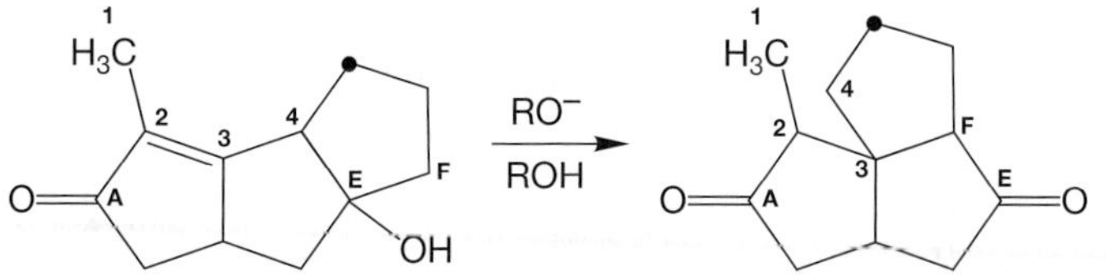

(continued)

So now this problem has evolved from an amorphous "How do we get from starting material to product?" to "How do we break 4—E, make F—3, and do the required oxidation?" That's a much more specific set of questions. This kind of analysis can make very difficult problems doable, and moderate problems almost trivial. As you start to try out reactions, those that accomplish goals feel good and can be followed up, whereas those that do not can be discarded. Here is a set of reactions that solves this problem:

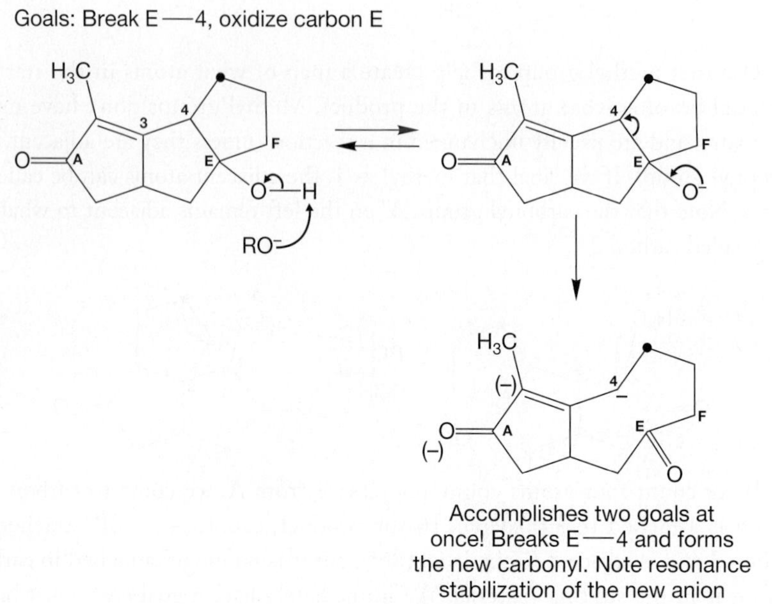

Goals: Break E—4, oxidize carbon E

Accomplishes two goals at once! Breaks E—4 and forms the new carbonyl. Note resonance stabilization of the new anion

Now we are more than halfway there. We only have to make F—3, and a Michael reaction easily does that.

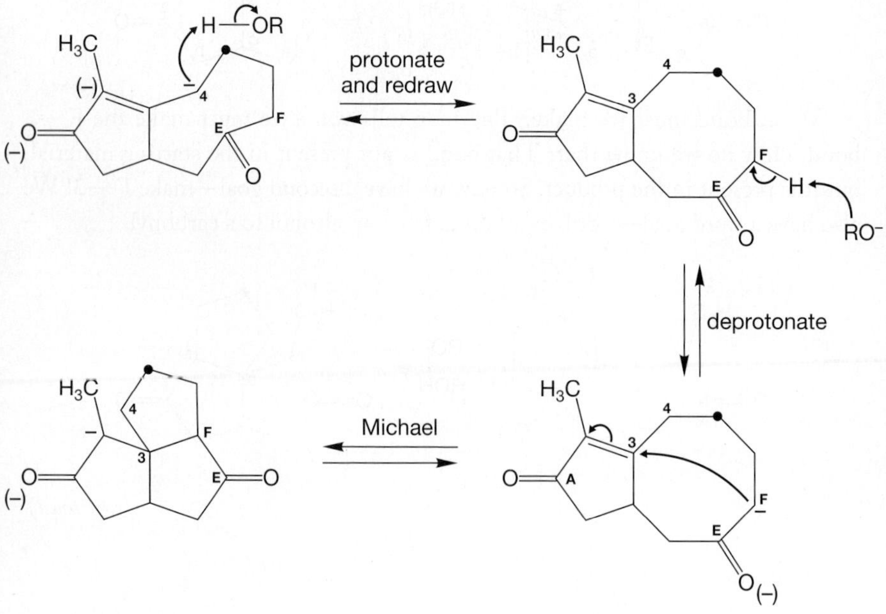

protonate and redraw

deprotonate

Michael

(continued)

A final protonation completes the problem.

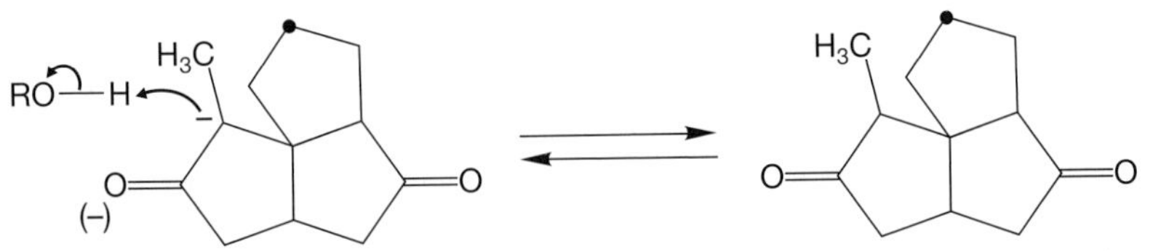

19.17 Additional Problems

PROBLEM 19.46 The "normal" bond length for a carbon–oxygen single bond is 1.43 Å. In hydroxyethene, the enol form of acetaldehyde, it is only 1.38 Å. Why is this carbon–oxygen bond shorter than usual?

PROBLEM 19.47 Write mechanisms for the following acid- and base-catalyzed equilibrations. There is repetition in this drill problem, but being able to do these prototypal reactions quickly and easily is an essential skill. Be certain you can solve this problem easily before going on to more challenging ones.

(a)

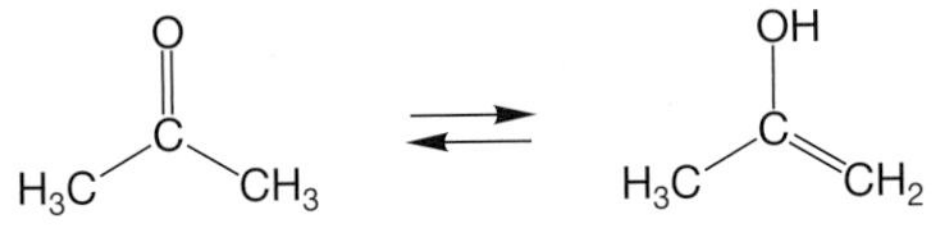

(b)

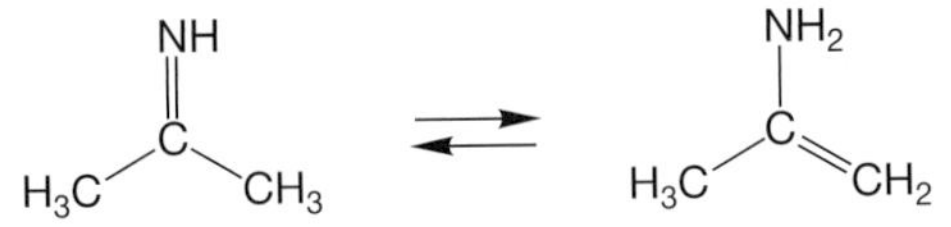

(c)

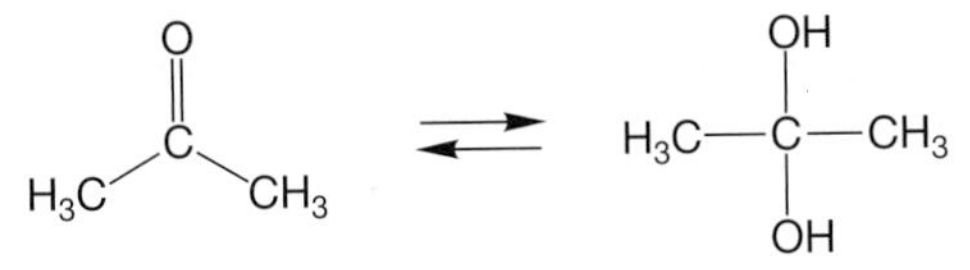

PROBLEM 19.48 The following compound contains two different types of α hydrogens. Removal of which α hydrogen will yield the *more stable* enolate? Which α hydrogen is likely to be removed *faster*? Explain why these questions need not have the same answer. That is, why might the less stable enolate be formed faster than the more stable enolate and vice versa?

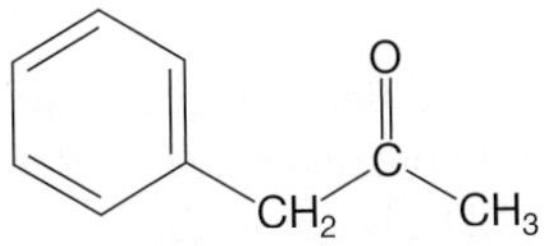

PROBLEM 19.49 Which positions in the following molecules will exchange H for D in D_2O/DO^-?

(a)

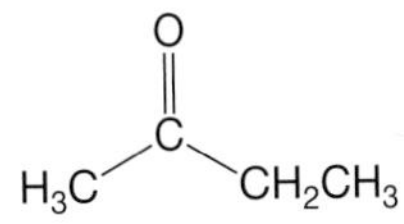

(b)

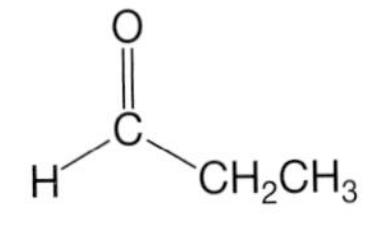

(c)

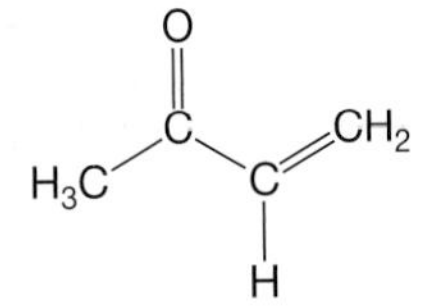

(d)

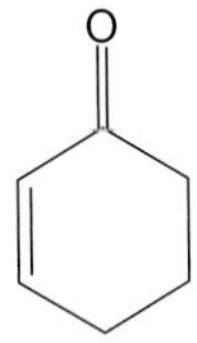

(e)

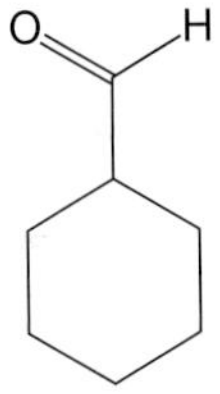

PROBLEM 19.50 As we saw in Problem 19.49c, methyl vinyl ketone exchanges the three methyl hydrogens for deuterium in D_2O/DO^-. Why doesn't the other formally "α" hydrogen exchange as well?

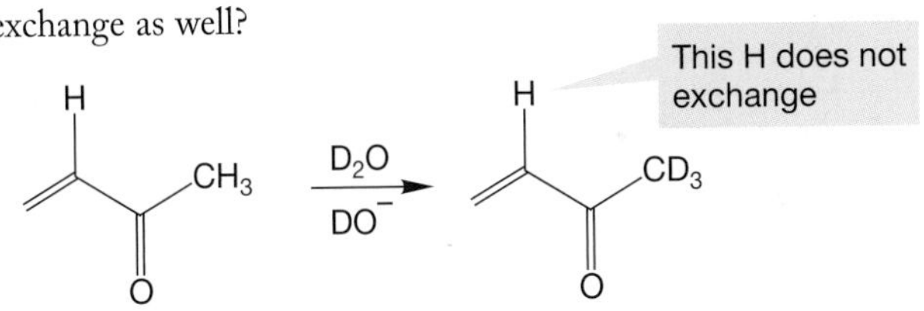

This H does not exchange

PROBLEM 19.51 The proton-decoupled ^{13}C NMR spectrum of 2,4-pentanedione consists of six lines (δ 24.3, 30.2, 58.2, 100.3, 191.4, and 201.9 ppm), not the "expected" three lines. Explain.

PROBLEM 19.52 Provide simple synthetic routes from cyclohexanone to the following compounds. *Hint:* These questions are all review.

(a)

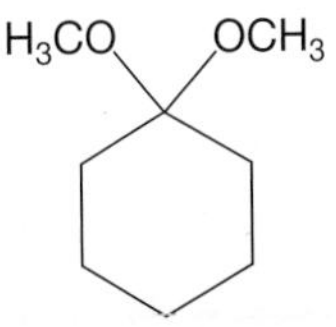

(b)

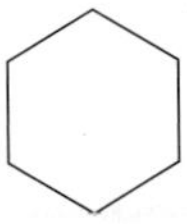

(c)

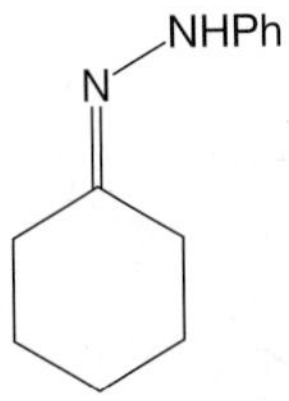

(d)

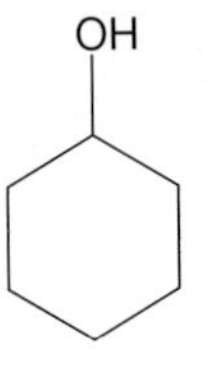

(e)

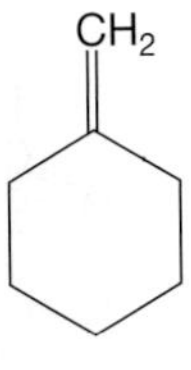

PROBLEM 19.53 Provide simple synthetic routes from cyclohexanone to the following compounds:

(a)

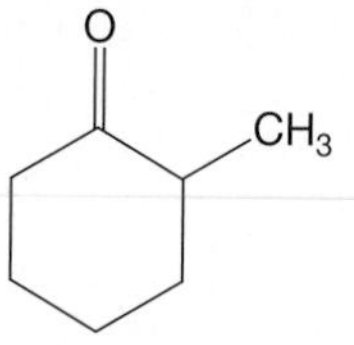

CH₃

(b)

Br

(two ways)

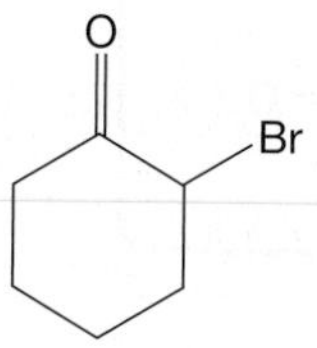

(c)

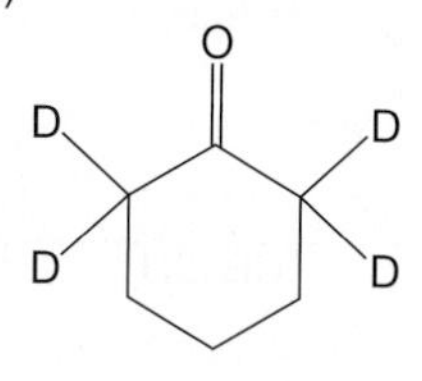

(d)

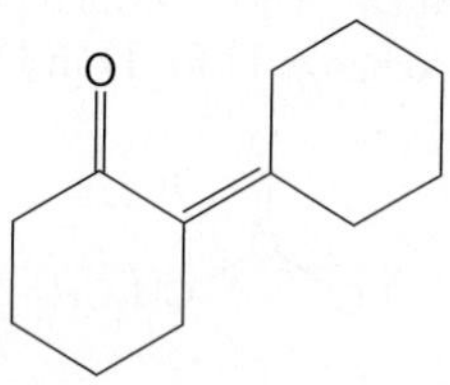

PROBLEM 19.54 Write mechanisms for the following isomerizations:

(a)

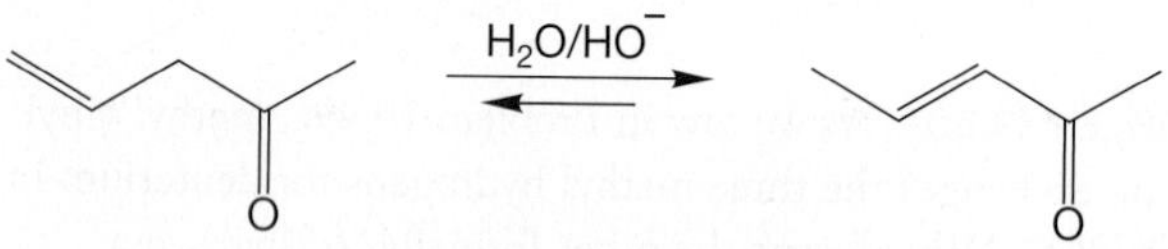

(b)

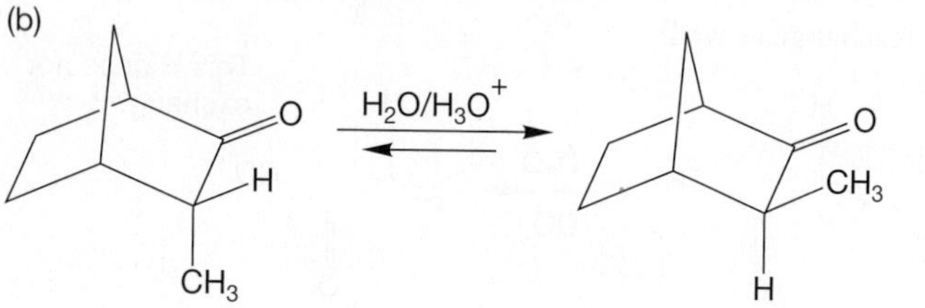

PROBLEM 19.55 Perform retrosynthetic analyses on the following molecules, each of which, in principle, can be synthesized by an aldol condensation–dehydration sequence. Which syntheses will be practical? Explain why some of your suggested routes may not work in practice. What are some of the potential problems in each part?

(a)

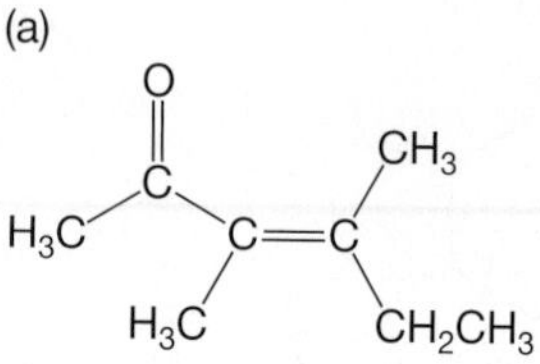

(b)

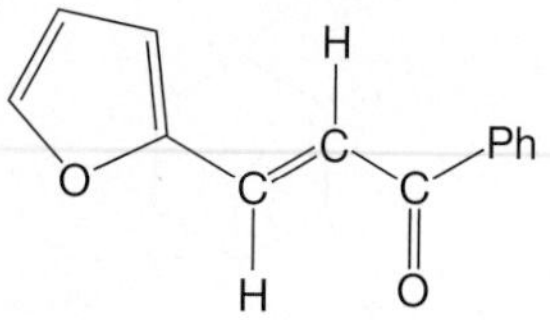

(c)

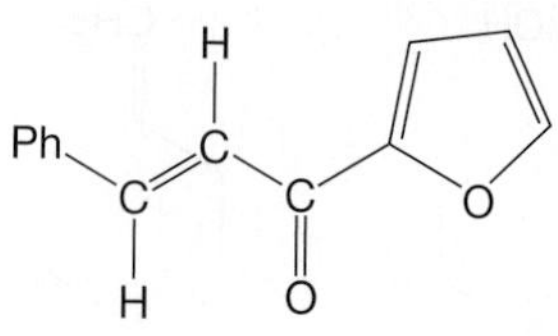

(d)

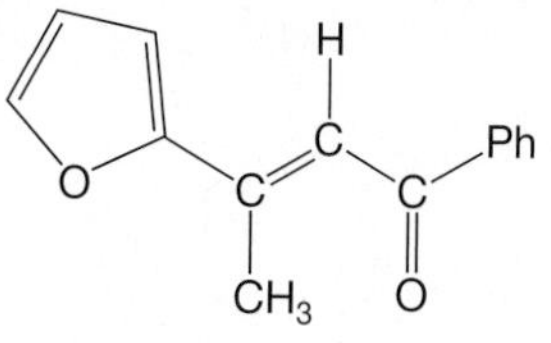

PROBLEM 19.56 Treatment of compound **1** with an aqueous solution of bromine and sodium hydroxide affords, after acidification, pivalic acid (**2**) and bromoform. Deduce the structure of **1** and write an arrow formalism mechanism for its conversion into **2** and bromoform. Use your mechanism to predict the number of equivalents of bromine and sodium hydroxide necessary for this reaction.

$$1 \xrightarrow[\text{2. H}_2\text{O/H}_3\text{O}^+]{\text{1. NaOH/Br}_2\text{/H}_2\text{O}} (CH_3)_3C\text{—COOH} + CHBr_3$$

2

PROBLEM 19.57 Treatment of ketone **1** with LDA in dimethylformamide (DMF) solvent, followed by addition of methyl iodide, leads to two methylated ketones. Write arrow formalisms for their formations and explain why one product is greatly favored over the other.

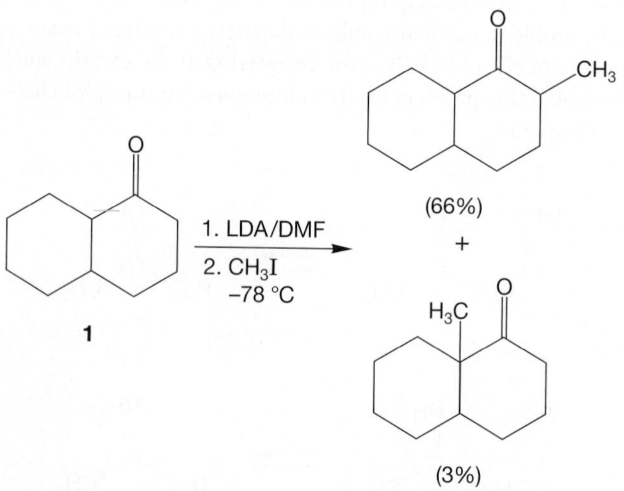

PROBLEM 19.58 As we have seen, alkylation of enolates generally occurs at carbon (see Problem 19.57 for an example). However, this is not always the case. First, use the following estimated bond strengths to explain why treatment of an enolate with trimethylsilyl chloride leads to substitution at oxygen (Si—C ~ 69 kcal/mol; Si—O ~109 kcal/mol).

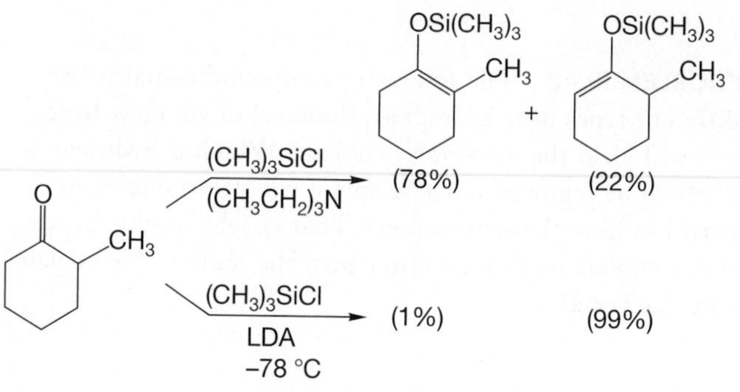

Second, explain the different regiochemical results under the two reaction conditions shown. *Hint*: Remember that enolate formation with LDA is irreversible.

PROBLEM 19.59 When bromo ketone **1** is treated with potassium *tert*-butoxide in *tert*-butyl alcohol at room temperature, it gives exclusively the 5,5-fused bicyclic ketone **2**. In contrast, when **1** is treated with LDA in tetrahydrofuran (THF) at −72 °C, followed by heating, the product is predominately the 5,7-fused ketone **3**. Write arrow formalism mechanisms for these cycloalkylation reactions and explain why the different reaction conditions favor different products.

PROBLEM 19.61 Aldehydes, but not ketones, readily react with 5,5-dimethyl-1,3-cyclohexanedione (dimedone, **1**) in the presence of a base (such as piperidine) to give crystalline dimedone derivatives, **2**. These compounds can be converted into octahydroxanthenes, **3**, upon heating in ethyl alcohol with a trace of acid. Provide arrow formalism mechanisms for the formation of **2** from **1** and an aldehyde (RCHO), as well as for the conversion of **2** into **3**.

PROBLEM 19.60 Condensation of 1,3-diphenylacetone (**1**) and benzil (**2**) in the presence of alcoholic potassium hydroxide affords tetraphenylcyclopentadienone (**3**), a dark purple solid. Write an arrow formalism mechanism for the formation of **3**.

PROBLEM 19.62 Write a mechanism for the following Dieckmann condensation:

1. NaOCH₃ (full equiv.) HOCH₃
2. CH₃COOH

PROBLEM 19.63 Here is a reverse Dieckmann condensation. Write a mechanism for it.

NaOCH₃ | HOCH₃

PROBLEM 19.64 Write an arrow formalism mechanism for the following reaction (the Darzens condensation):

(CH₃)₃COK
(CH₃)₃COH
10 °C, 1.5 h

(89%)

+

PROBLEM 19.65 Account for the formation of both stereoisomers in the Darzens condensation shown below:

K⁺ ⁻OC(CH₃)₃
HOC(CH₃)₃

PROBLEM 19.66 Provide structures for intermediates **A–D**. Mechanisms are not necessary.

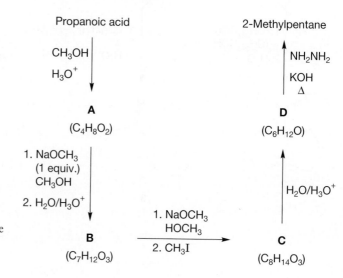

PROBLEM 19.67 Provide structures for compounds **A–D**. Mechanisms are not required.

Propanoic acid

CH₃OH
H₃O⁺

A
(C₄H₈O₂)

1. NaOCH₃ (1 equiv.) CH₃OH
2. H₂O/H₃O⁺

B
(C₇H₁₂O₃)

1. NaOCH₃ HOCH₃
2. CH₃I

C
(C₈H₁₄O₃)

H₂O/H₃O⁺

2-Methylpentane

NH₂NH₂
KOH
Δ

D
(C₆H₁₂O)

PROBLEM 19.68 Here is a pair of relatively simple Knoevenagel condensations. Write arrow formalism mechanisms for the following conversions:

(a)

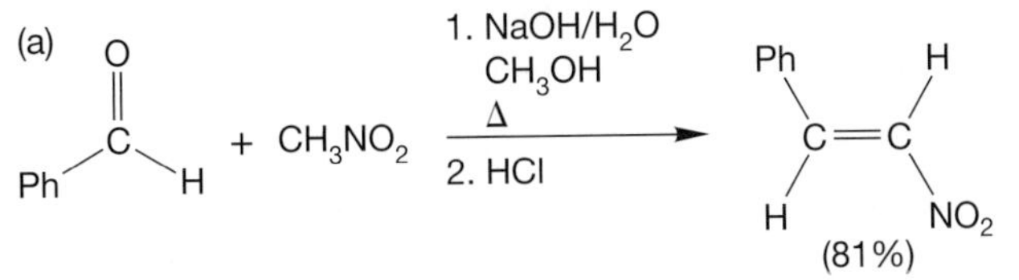

(b)

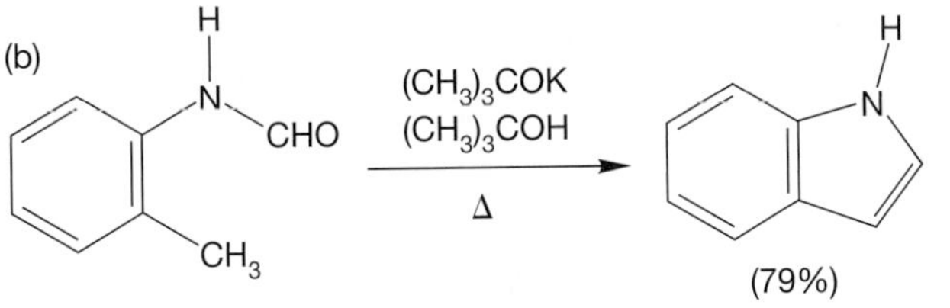

PROBLEM 19.69 Now here is a pair of slightly more complicated Knoevenagel reactions. Provide arrow formalism mechanisms.

(a)

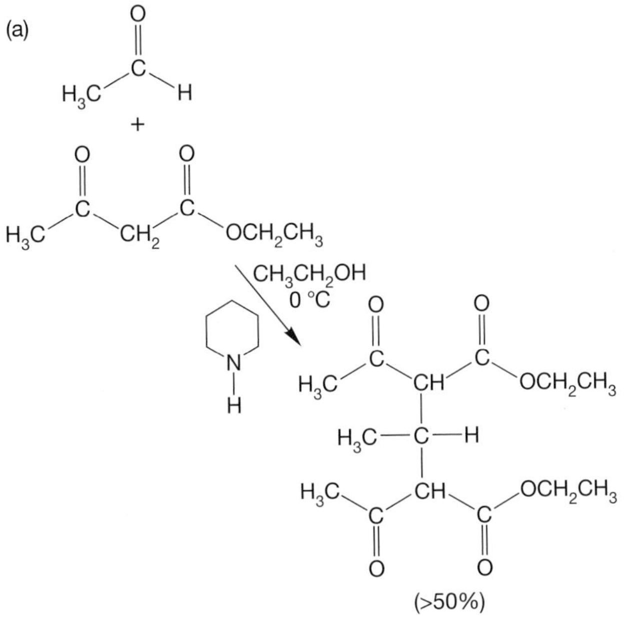

(>50%)

(b)

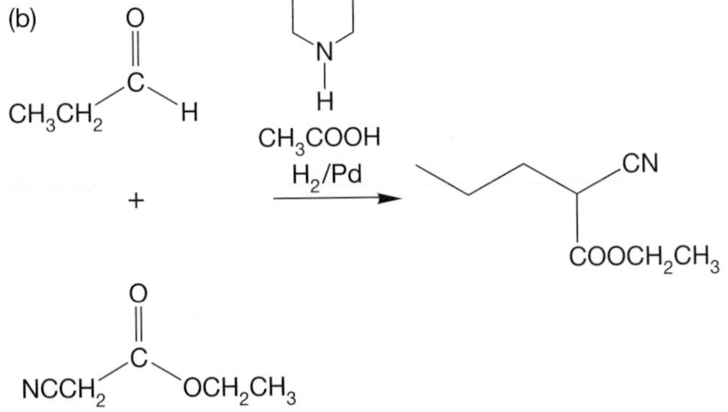

PROBLEM 19.70 Provide structures for compounds A–E. Mechanisms not required.

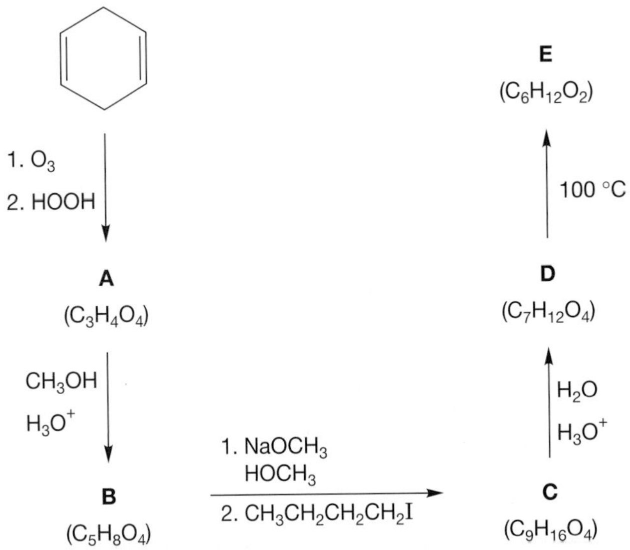

PROBLEM 19.71 Write an arrow formalism mechanism for the conversion of diketone **1** into the important perfumery ingredient *cis*-jasmone (**2**). There is another possible cyclopentenone that could be formed from **1**. What is its structure? Write an arrow formalism mechanism for its formation.

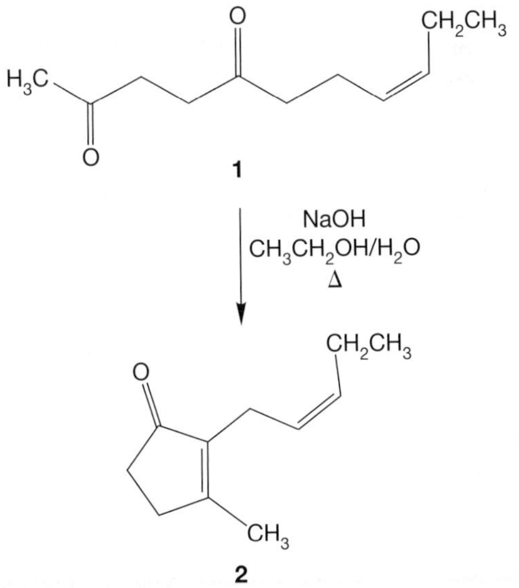

PROBLEM 19.72 Propose syntheses for the following target molecules starting from the indicated material. You may use any other reagents that you need. Mechanisms are not required. Use a retrosynthetic analysis in each case.

(a)

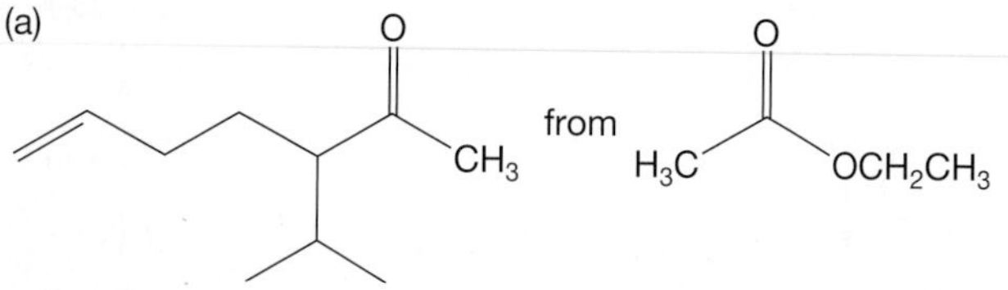
from

(b)

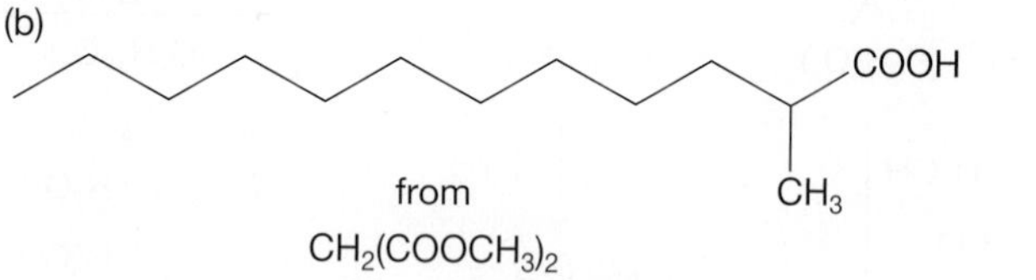
COOH
CH3
from
CH2(COOCH3)2

(c)

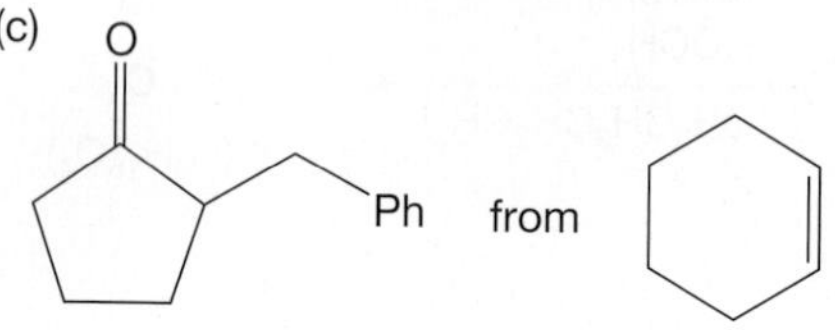
Ph
from

(d)

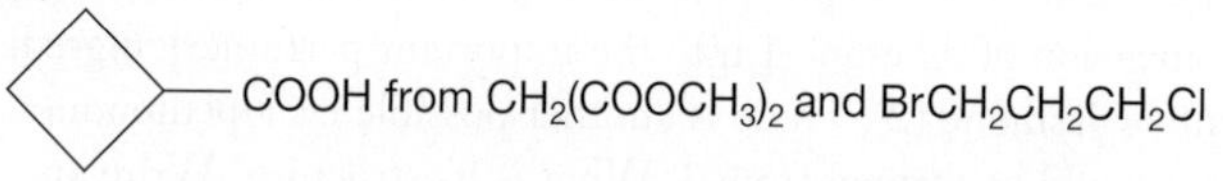
COOH from CH2(COOCH3)2 and BrCH2CH2CH2Cl

PROBLEM 19.73 When benzaldehyde and an excess of acetone react in the presence of aqueous sodium hydroxide, compound **1** is obtained as the major product as well as a small amount of compound **2**. However, when acetone and an excess (at least 2 equiv.) of benzaldehyde react under the same conditions, the predominant product is compound **2**. Spectral data for compounds **1** and **2** are summarized below. Deduce structures for compounds **1** and **2** and rationalize their formation.

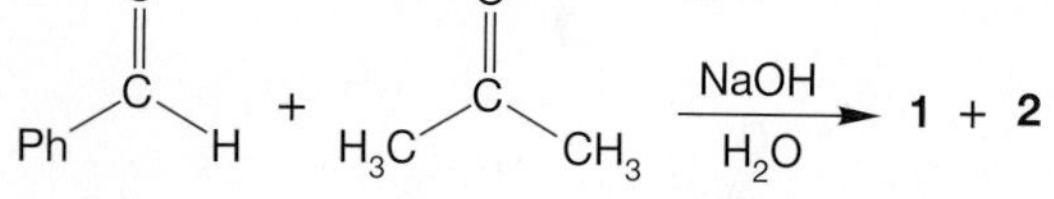
NaOH
H2O
1 + 2

Compound **1**

Mass spectrum: m/z = 146 (p, 75%), 145 (50%), 131 (100%), 103 (80%)

IR (Nujol): 1667 (s), 973 (s), 747 (s), 689 (s) cm^{-1}

^{1}H NMR ($CDCl_3$): δ 2.38 (s 3H), 6.71 (d, J = 16 Hz, 1H), 7.30–7.66 (m, 5H), 7.54 (d, J = 16 Hz, 1H)

Compound **2**

Mass spectrum: m/z = 234 (p, 49%), 233 (44%), 131 (42%), 128 (26%), 103 (90%), 91 (32%), 77 (100%), 51 (49%)

IR (melt): 1651 (s), 984 (s), 762 (s), 670 (s) cm^{-1}

^{1}H NMR ($CDCl_3$): δ 7.10 (d, J = 16 Hz, 1H), 7.30–7.70 (m, 5H), 7.78 (d, J = 16 Hz, 1H)

PROBLEM 19.74 Reaction of 1-morpholinocyclohexene (**1**) and β-nitrostyrene (**2**), followed by hydrolysis, yields the nitro ketone **3**. Write an arrow formalism mechanism for this reaction sequence and be sure to explain the observed regiochemistry.

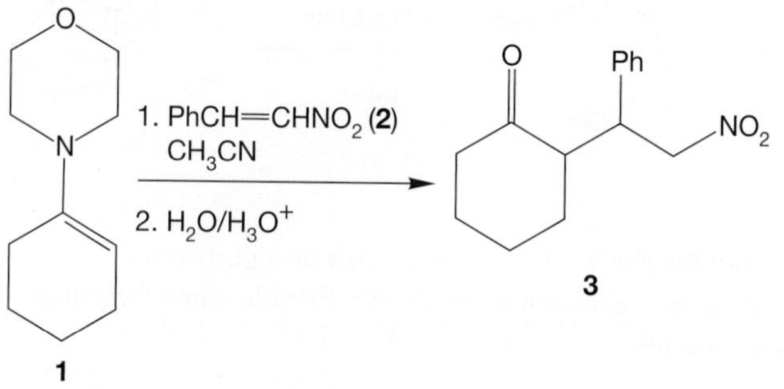
1. PhCH=CHNO2 (2)
CH3CN
2. H2O/H3O+
Ph
NO2
1
3

PROBLEM 19.75 Provide mechanisms for the following changes. *Hint*: Definitely do (a) before (b).

(a)

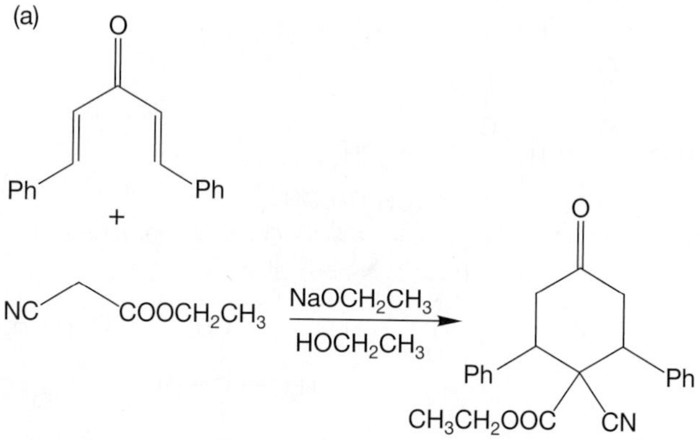
Ph
Ph
+
NC
COOCH2CH3
NaOCH2CH3
HOCH2CH3
Ph
Ph
CH3CH2OOC
CN

(b)

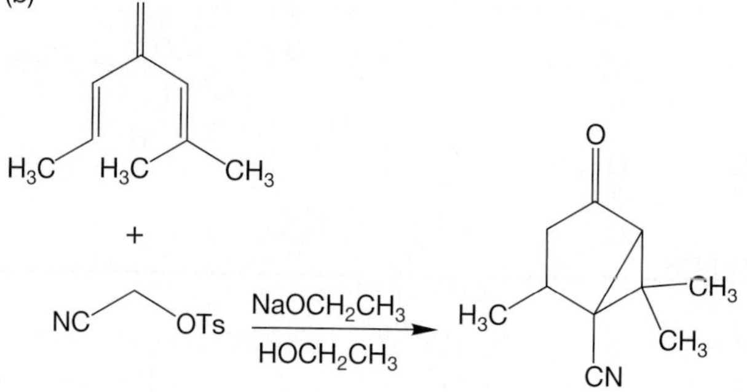
H3C
H3C
CH3
+
NC
OTs
NaOCH2CH3
HOCH2CH3
H3C
CH3
CH3
CN

PROBLEM 19.76 Write a mechanism for the following transformation:

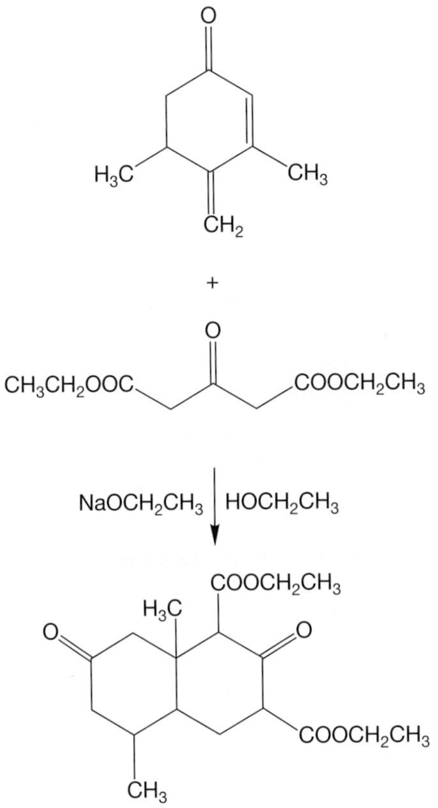

+

NaOCH₂CH₃ | HOCH₂CH₃

PROBLEM 19.77 Explain the striking observation that methylation of the alcohol **1** leads to a methoxy compound in which the stereochemistry of the oxygen has changed. *Hint*: Note that **1** is a β-keto alcohol.

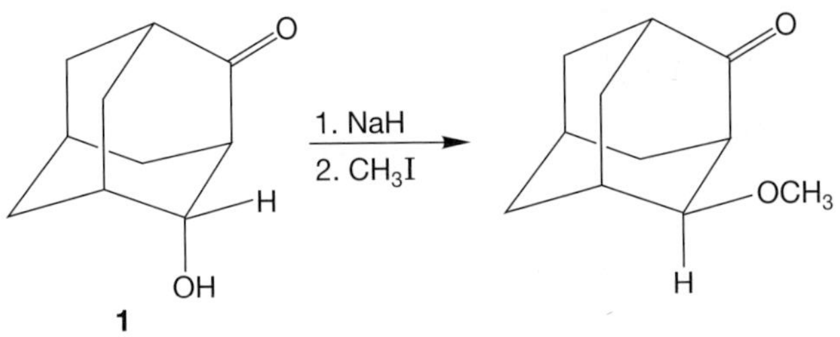

PROBLEM 19.78 In 1917, Robinson reported a synthesis (called "von bewundernswerter Eleganz" by the German chemist Willstätter) of tropinone (**3**) involving the condensation of succindialdehyde (**1**), methylamine, and 1,3-acetone

dicarboxylic acid (**2**). Schöpf subsequently improved the yield of this reaction to about 90%. Propose a plausible mechanism for this reaction. *Hints*: This reaction involves a double condensation of some kind. Acetone can be used in place of **2**, although the yield is reduced.

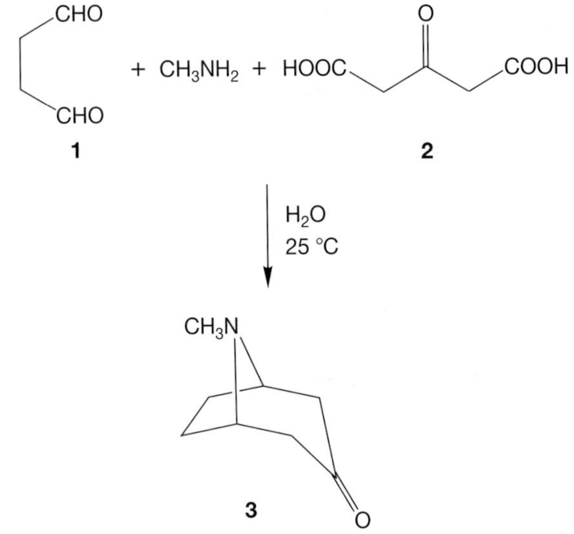

H₂O
25 °C

PROBLEM 19.79 Provide an arrow formalism mechanism for the following reaction:

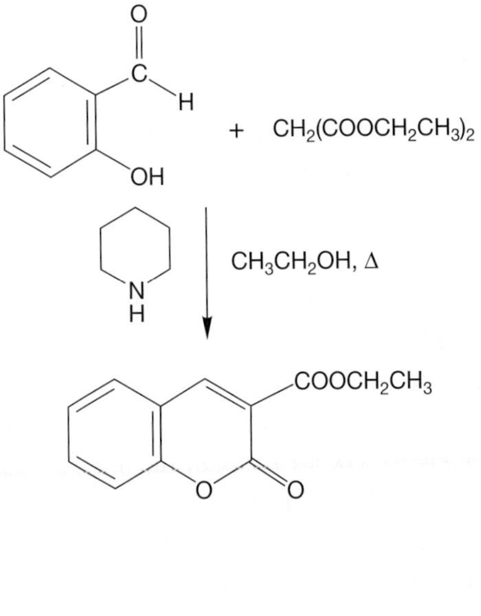

+ CH₂(COOCH₂CH₃)₂

CH₃CH₂OH, Δ

PROBLEM 19.80 Provide structures for compounds **A–F**. Spectral data for compounds **E** and **F** are summarized below.

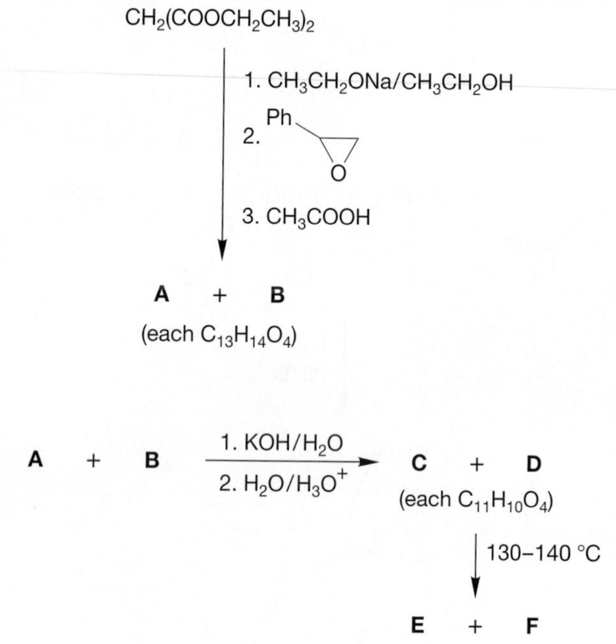

CH2(COOCH2CH3)2
1. CH3CH2ONa/CH3CH2OH
2. Ph, O
3. CH3COOH
A + B
(each C13H14O4)
A + B
1. KOH/H2O
2. H2O/H3O+
C + D
(each C11H10O4)
130–140 °C
E + F

Compound **E**

Mass spectrum: m/z = 162 (M, 100%), 106 (77%), 104 (51%), 43 (61%)

IR (melt): 1778 cm^{-1} (s)

^{1}H NMR (CDCl$_3$): δ 2.1–2.15 (m, 1H), 2.5–2.7 (m, 3H), 5.45–5.55 (m, 1H), 7.2–7.4 (m, 5H)

^{13}C NMR (CDCl$_3$): δ 28.9 (t), 30.9 (t), 81.2 (d), 125.2 (d), 128.4 (d), 128.7 (d), 139.3 (s), 176.9 (s)

Compound **F**

Mass spectrum: m/z = 162 (M, 28%), 104 (100%)

IR (film): 1780 cm^{-1} (s)

^{1}H NMR (CDCl$_3$): δ 2.67 (dd, J = 17.5 Hz, 6.1 Hz, 1H), 2.93 (dd, J = 17.5 Hz, 8.1 Hz, 1H), 3.80 (m, 1H), 4.28 (dd, J = 9.0 Hz, 8.1 Hz), 4.67 (dd, J = 9.0 Hz, 7.9 Hz, 1H), 7.22–7.41 (m, 5H)

^{13}C NMR (CDCl$_3$): δ 35.4 (t), 40.8 (d), 73.8 (t), 126.5 (d), 127.4 (d), 128.9 (d), 139.2 (s), 176.3 (s)

PROBLEM 19.81 Provide an arrow formalism for the following reaction:

COOCH3
COOCH3
HOCH3
NaOCH3
H3COOC
O

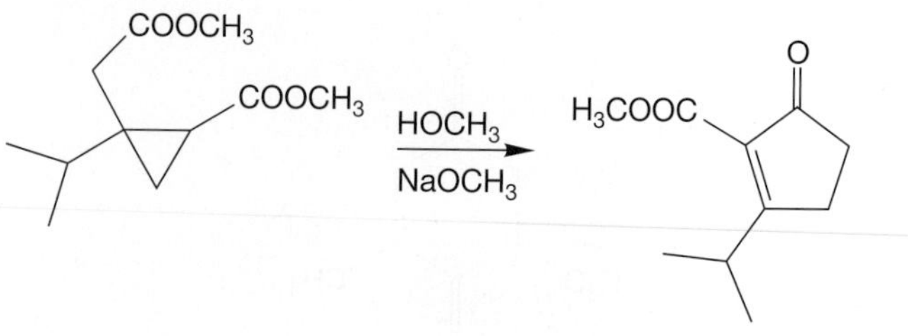

PROBLEM 19.82 Here is an answer to Problem 19.81, which appears in the original paper describing the reaction. Perhaps it is the answer you came up with. First, criticize this proposal—what makes it unlikely? Next, you can either feel good about finding another mechanism already, or, if this was your answer, set about finding a better mechanism. *Hint*: Find a way to remove strain in the starting material.

CH3OOC
COOCH3
CH3OOC
OCH3
C
O
H3COOC
O
H3COOC
O
OCH3
NaOCH3
H3COOC
O
H3COOC
O
O
H3COOC
HOCH3

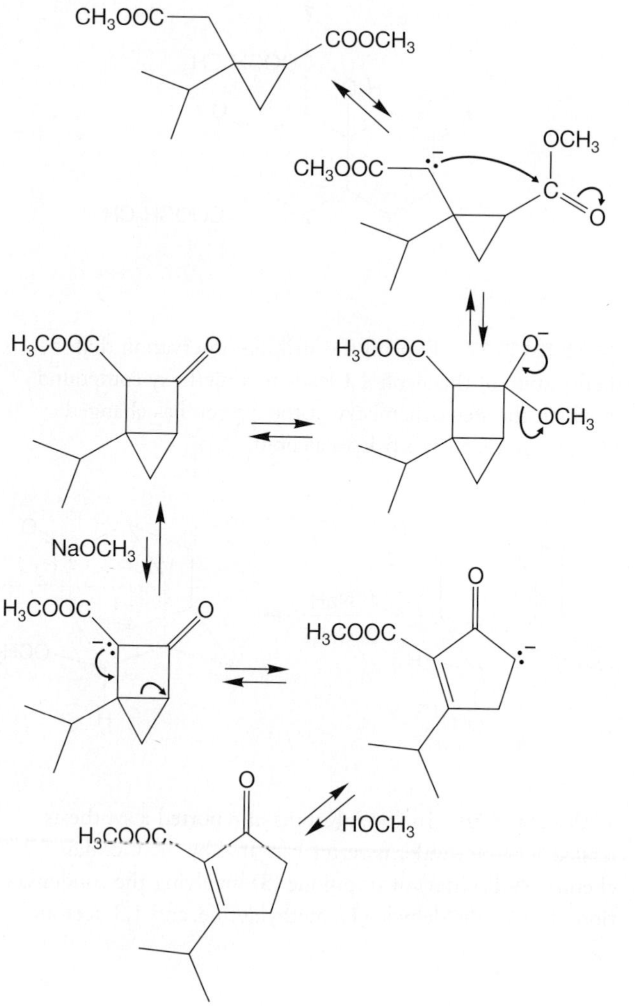

PROBLEM 19.83 Outline mechanisms for the following transformations leading to quinine:[5]

(a)

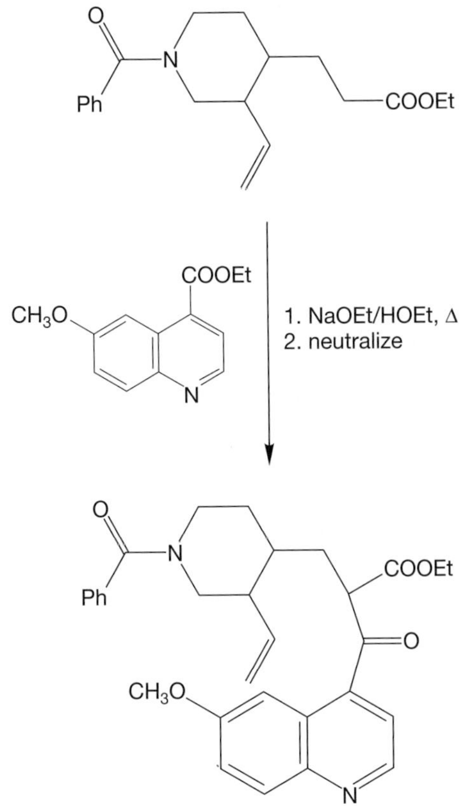
O
N
Ph
COOEt
COOEt
CH_3O
N
1. NaOEt/HOEt, Δ
2. neutralize
O
N
Ph
COOEt
O
CH_3O
N

(b)

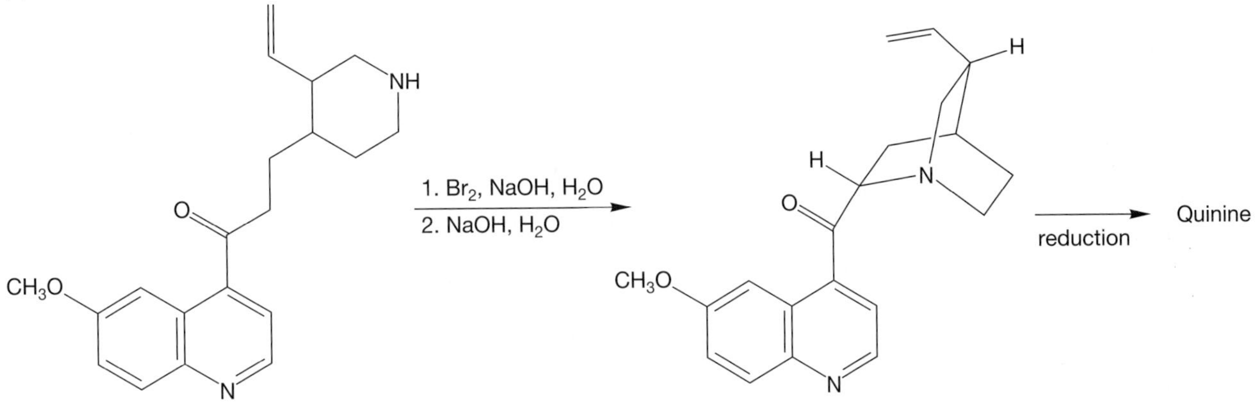
NH
O
CH_3O
N
1. Br_2, NaOH, H_2O
2. NaOH, H_2O
H
H
N
O
CH_3O
N
reduction
Quinine

The structure of this complex molecule presents an extraordinary synthetic challenge and was accomplished in the early 1940s by R. B. Woodward and William von E. Doering. It's worth a look at this synthesis. It was not only an achievement of brilliance, but used many of the reactions whose mechanisms we have studied.

[5]Named for the Countess of Chinchón, wife of a seventeenth century Peruvian viceroy, supposedly cured of the vapors by use of the bark.

PROBLEM 19.84   The reactions shown below were important in the synthesis of LSD. Provide structures for the indicated compounds.

(a)

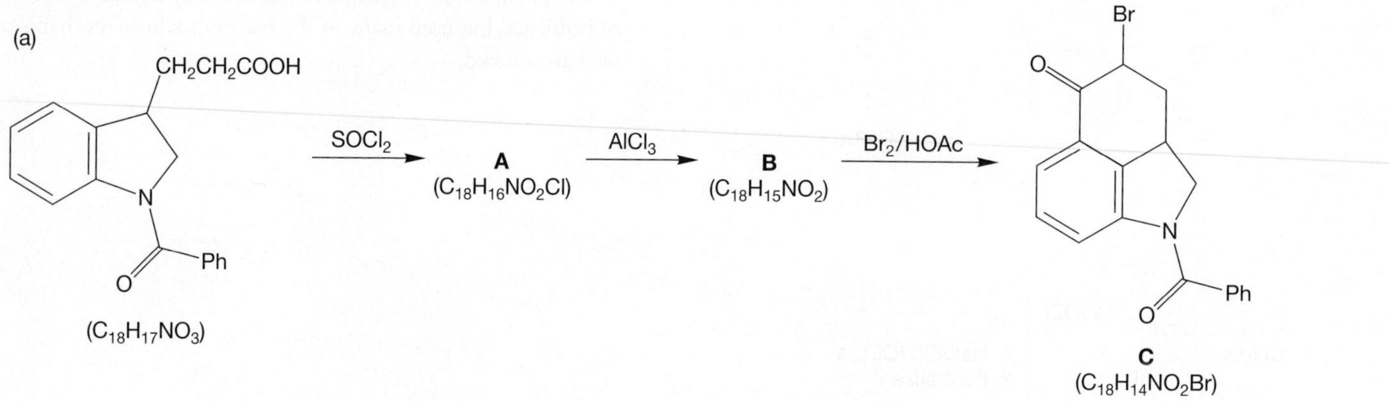

(b)

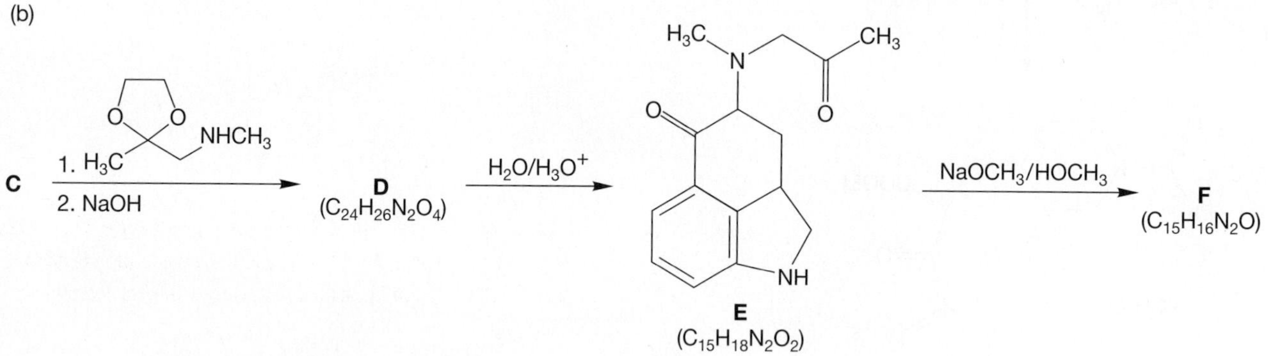

(c)

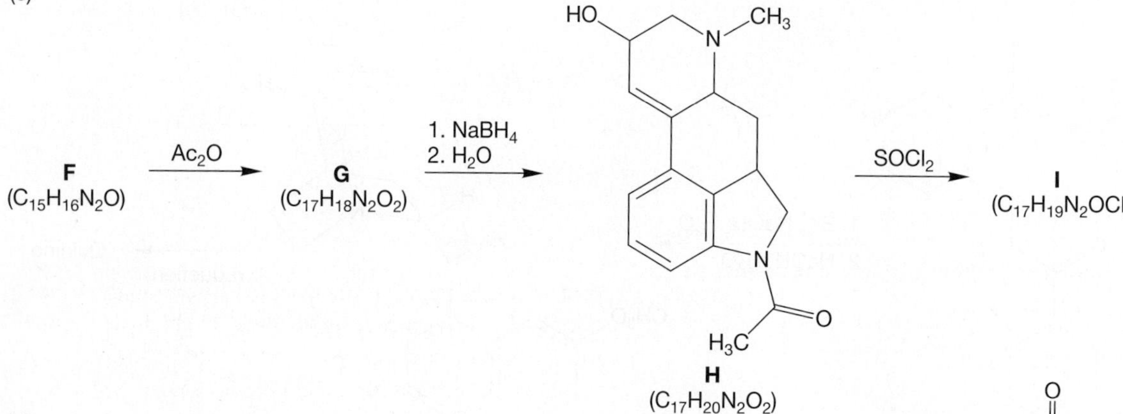

(d)

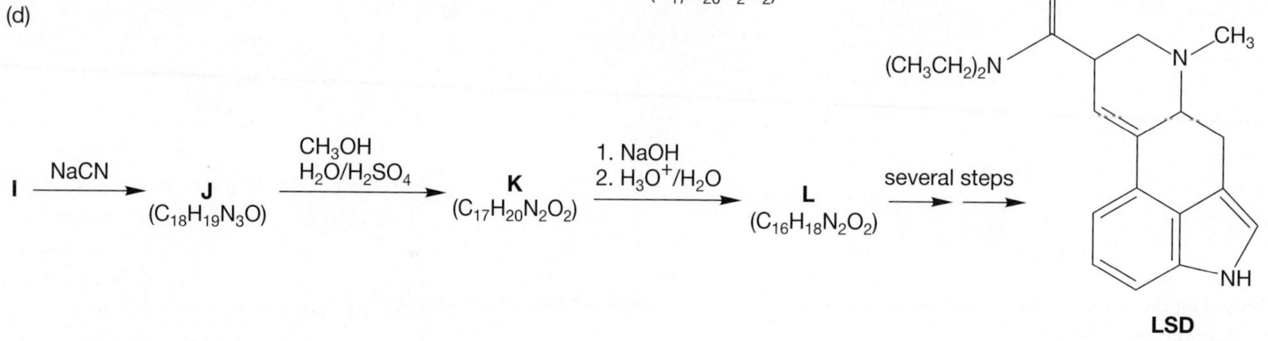

PROBLEM 19.85   Reaction of salicylaldehyde (**1**) and acetic anhydride in the presence of sodium acetate gives compound **2**, a fragrant compound occurring in sweet clover.

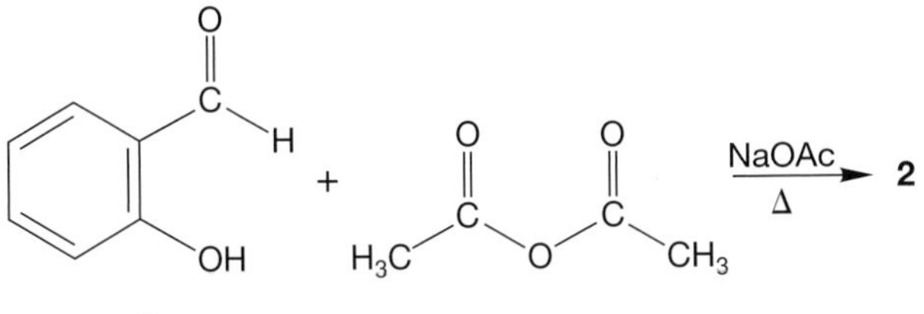

Spectral data for compound **2** are collected below. Deduce the structure of **2**. Naturally, it is an interesting question as to how **2** is formed. This problem turns out to be harder than it looks at first. Nonetheless, you might want to try to formulate a mechanism. However, note that *O*-acetylsalicylaldehyde (**3**) gives only traces of **2** in the absence of acetic anhydride.

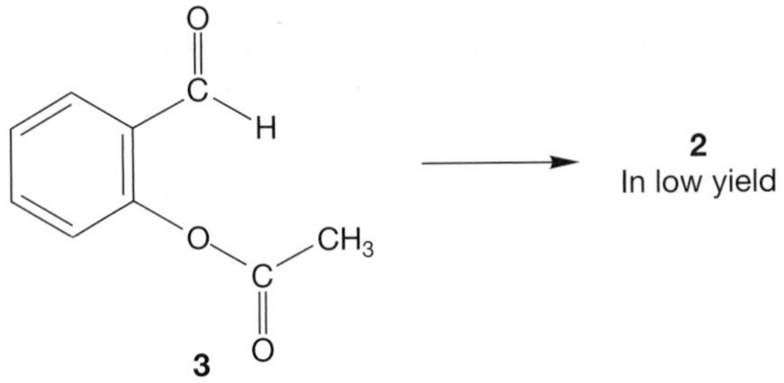

Compound **2**

Mass spectrum: *m/z* = 146 (M, 100%), 118 (80%), 90 (31%), 89 (26%)

IR (Nujol): 1704 cm^{-1} (s)

^{1}H NMR (CDCl$_3$): δ 6.42 (d, *J* = 9 Hz, 1H), 7.23–7.35 (m, 2H), 7.46–7.56 (m, 2H), 7.72 (d, *J* = 9 Hz, 1H)

^{13}C NMR (CDCl$_3$): δ 116.6 (d), 116.8 (d), 118.8 (s), 124.4 (d), 127.9 (d), 131.7 (d), 143.4 (d), 154.0 (s), 160.6 (s)

PROBLEM 19.86   Rationalize the following transformation. *Hint*: See Problem 19.35a.

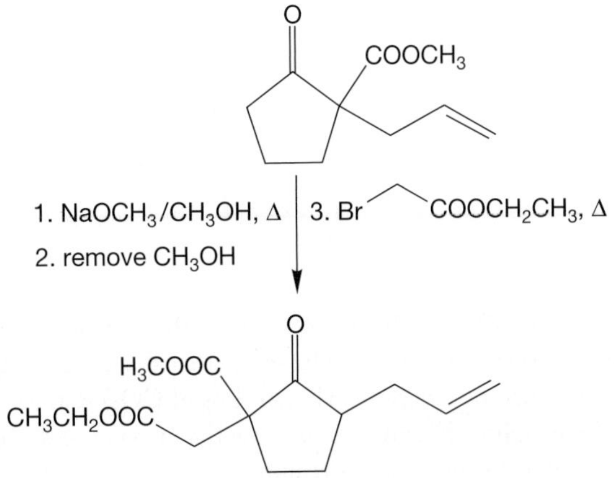

PROBLEM 19.87   Provide an arrow formalism for the following reaction:

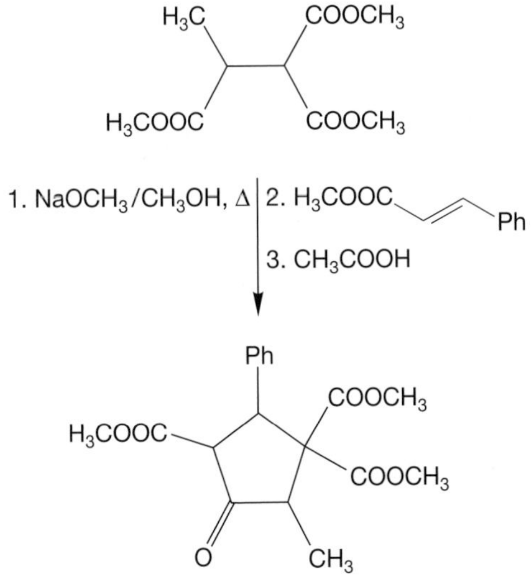

PROBLEM 19.88   Diisobutylaluminum hydride (DIBAL-H), like lithium aluminum hydride (LiAlH$_4$) and sodium borohydride (NaBH$_4$), can reduce aldehydes and ketones to alcohols.

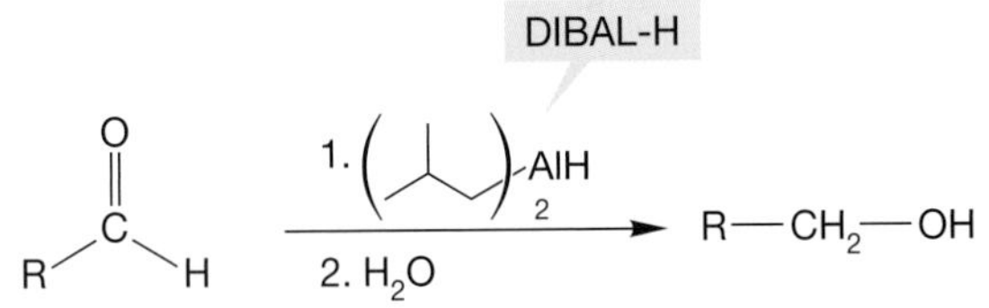

Write a mechanism for this reaction. Interestingly, 1 mol of DIBAL-H can reduce up to 3 mol of aldehyde despite possessing only one apparent hydride equivalent. Propose a mechanism that accounts for the second and third equivalents of hydride in these aldehyde reductions. *Hint*: Remember the Meerwein–Ponndorf–Verley–Oppenauer equilibration.

PROBLEM 19.89   Provide structures for intermediates **A–D** on the next page. A mechanistic analysis is not required for this problem, but will almost certainly be of use in developing an answer. *Hint*: You will discover that there are certain structural ambiguities in this problem. The following experimental observation should prove helpful. Mild hydrolysis of

compound **B** affords compound **E** ($C_{12}H_{18}O_3$), which is characterized by the following spectra:
IR (CHCl$_3$): 1715 cm^{-1}
^{1}H NMR (CDCl$_3$): δ 1.55–1.84 (m, 9H), 2.60 (s, 4H), 3.77 (s, 3H), 5.52 (br d, 1H), 6.08 (d, 1H)

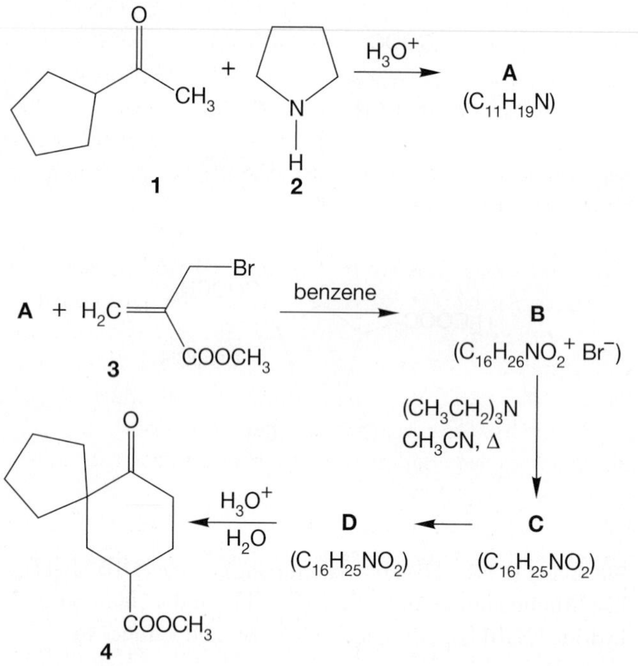

PROBLEM 19.90 When treated with KCN in methyl alcohol, benzaldehyde is converted into a molecule called benzoin. Write a mechanism for this change.

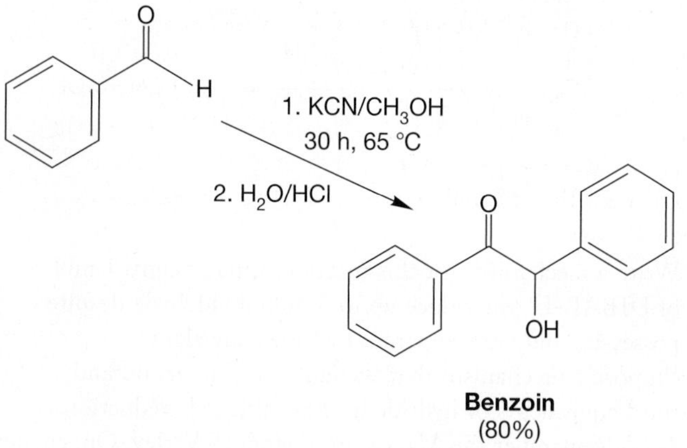

Benzoin
(80%)

Caution! This problem is hard. Do not succumb to the temptation to remove that aldehydic proton—it is *not* acidic. Something else must happen. *Hints:* What will the nucleophile cyanide do to benzaldehyde? How does that reaction alter the nature of the formerly "aldehydic" hydrogen?

PROBLEM 19.91 Show why hydroxide or ethoxide is ineffective in promoting the benzoin condensation. Those anions do add to the carbonyl group of benzaldehyde to form compounds much like a cyanohydrin.

PROBLEM 19.92 A few other catalysts will promote benzoin-like condensations. Deprotonated thiazolium ions are examples. Show how the compound shown can function like cyanide and catalyze the benzoin condensation of pivaldehyde (*tert*-butyl-carboxaldehyde).

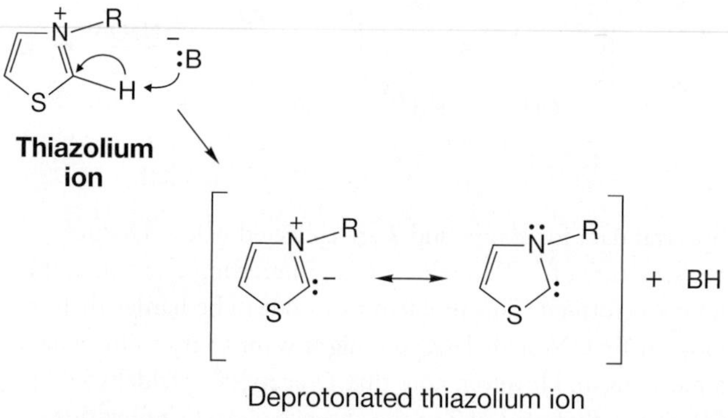

Thiazolium ion

Deprotonated thiazolium ion

PROBLEM 19.93 Acetolactate (**2**) is a precursor of the amino acid valine in bacterial cells. Acetolactate is synthesized by the condensation of two molecules of pyruvate (**1**).

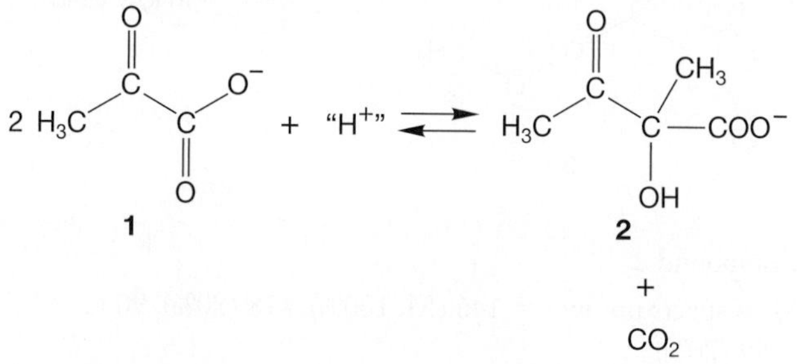

The enzyme that catalyzes this reaction (acetolactate synthetase) requires thiamine pyrophosphate (TPP) as a coenzyme.

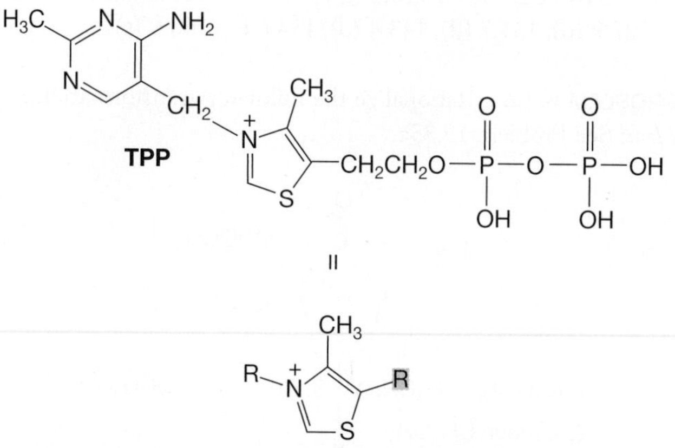

Suggest a plausible mechanism for the synthesis of acetolactate (**2**). *Hints:* Review Problem 19.92. Note that this condensation is accompanied by a decarboxylation (loss of CO$_2$) that actually precedes coupling. Finally, you may use the enzyme to do acid- or base-catalyzed reactions as required.

PROBLEM 19.94   Write a mechanism for the following transformation:

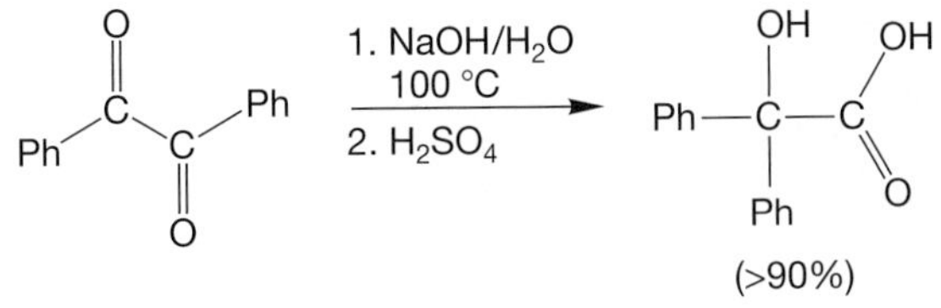

(>90%)

PROBLEM 19.95   A synthesis of pyridines involves the condensation of an α,β-unsaturated ketone (**1**) with a methyl ketone (**2**). This reaction gives a pyrilium salt (**3**), which then reacts with ammonia to produce a 2,4,6-trisubstituted pyridine.

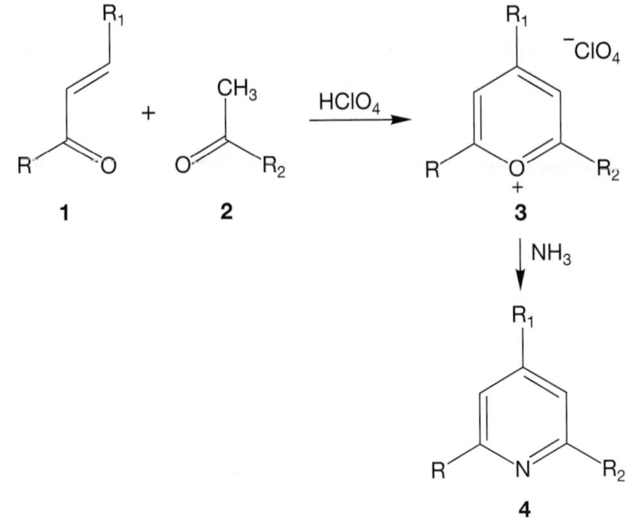

Propose mechanisms for the formation of pyrilium salt **3** from the condensation of **1** and **2**, and for the formation of the pyridine **4** from **3** and ammonia. *Hints*: The formation of **3** occurs best when 1.5–2.0 equivalents of **1** are used. The ketone **5** is a by-product of the condensation; how can it be formed? Think "redox."

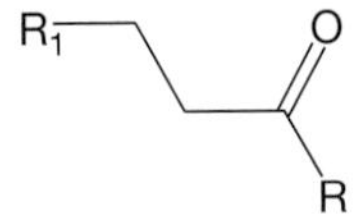

Use Organic Reaction Animations (ORA) to answer the following questions:

PROBLEM 19.96   Select the reaction "Aldol condensation" and observe the reaction that is shown. The reaction stops at the β-hydroxy aldehyde (the aldol). What conditions are necessary to obtain the ultimate condensation product, the α,β-unsaturated aldehyde? Stop the reaction at the first intermediate (the enolate) and select the HOMO track. The enolate has an anion shared between the oxygen and the α carbon. How does the HOMO representation of the enolate help us understand that an electrophile adds at the carbon rather than at the oxygen?

PROBLEM 19.97   Observe the "Mixed aldol condensation" animation. At the first transition state, the lithium coordinates with the enolate and with the new carbonyl species (formaldehyde in this case). What is the size of the ring that is formed in this transition state? What do you suppose is the shape of the ring? How will the shape influence the orientation of more highly substituted enolates and carbonyl electrophiles?

PROBLEM 19.98   The reaction titled "Michael addition" is also known as 1,4-addition or conjugate addition. Select the LUMO track of this reaction and notice the LUMO density of the starting α,β-unsaturated ketone (enone). Is there a large difference between the LUMO density at the β carbon and the carbonyl carbon? Do you think that the nucleophile does 1,2- or 1,4-addition based on the orbital shown? What controls the regioselectivity? Which do you think would be a faster process, 1,2- or 1,4-addition? Why?

PROBLEM 19.99   Observe the "Claisen condensation" and watch the first step closely. This step is the deprotonation of the α hydrogen. What is the orientation of the hydrogen that is deprotonated with respect to the carbonyl? Why? Go back and check the deprotonation step in the "Aldol condensation" and see if the orientation of the α hydrogen matches your prediction.

Special Topic
Reactions Controlled by Orbital Symmetry

20.1 Preview

20.2 Concerted Reactions

20.3 Electrocyclic Reactions

20.4 Cycloaddition Reactions

20.5 Sigmatropic Shift Reactions

20.6 The Cope Rearrangement

20.7 A Molecule with a Fluxional Structure

20.8 How to Work Orbital Symmetry Problems

20.9 Summary

20.10 Additional Problems

CONCERTED ACTIONS In the concerted reaction called a sigmatropic shift, one part of the molecule flies about, eventually being caught at one speficic position within the same molecule.

The fascination of what's difficult
Has dried the sap of my veins, and rent
Spontaneous joy and natural content
Out of my heart.

—WILLIAM BUTLER YEATS,[1] "THE FASCINATION OF WHAT'S DIFFICULT"

20.1 Preview

In Chapters 12–14, we explored the consequences of conjugation—the sideways overlap of $2p$ orbitals. That study led us to the aromatic compounds, the great stability of which can be traced to an especially favorable arrangement of electrons in low-lying bonding molecular orbitals. It can surely be no surprise to find that transition states can also benefit energetically through delocalization, and that most of the effects, including aromaticity, that influence the energies of ground states are important to transition states as well.

In this chapter, we will encounter reactions that were frustratingly difficult for chemists to understand for many, many years. By now, you are used to seeing acid- and base-catalyzed reactions, in which an intermediate is first formed and then produces product with the regeneration of the catalytic agent. Acid-catalyzed additions to alkenes are classic examples and there are already many other such reactions in your notes. There is one class of reactions, however, that is extraordinarily insensitive to catalysis. In these reactions, bases and acids are largely without effect. Even the presence of solvent seems of little relevance, because the reactions proceed as well in the gas phase as in solution. What is one to make of such reactions? How does one describe such a mechanism? We are used to speculating on the structure of an intermediate and then using the postulated intermediate to predict the structures of the transition states surrounding it. In these uncatalyzed reactions the starting material and product are separated by a single transition state (a concerted reaction), and there is very little with which to work in developing a mechanism. Indeed, one may legitimately ask what "mechanism" means in this context. Such processes have been called "no-mechanism" reactions. Some, in which starting material simply rearranges into itself (a **degenerate reaction**), can even be called "no-mechanism, no-reaction, reactions." Figure 20.1 shows an arrow formalism for a typical no-mechanism, no-reaction, reaction.

In 1965, R. B. Woodward (1917–1979) and Roald Hoffmann (b. 1937), both then at Harvard, began to publish a series of papers that ventured a mechanistic description of no-mechanism reactions, and gathered a number of these seemingly different processes under the heading of **pericyclic reactions**, concerted reactions that have a cyclic transition state. Their crucial insight that bonding overlap must be maintained between orbitals during the course of a concerted, pericyclic reaction now seems so simple that you may find it difficult to see why it eluded chemists for so many years. All we can tell you is that simple things are sometimes very hard to see, even for very smart people. **Woodward–Hoffmann theory** had been approached very closely before without the crucial "aha!", without the lightbulb over the head turning on. Perhaps what was lacking was the ability to combine a knowledge of theory with an awareness of the chemical problem, which is precisely what the brilliant experimentalist Woodward and the young theorist Hoffmann brought to the

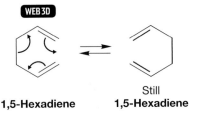

WEB 3D

1,5-Hexadiene   Still
                      1,5-Hexadiene

FIGURE 20.1 A single-barrier, one-transition-state, "no-mechanism, no-reaction" reaction. The starting material and the product are indistinguishable.

[1]William Butler Yeats (1865–1939) was an Irish poet who received the Nobel prize for literature in 1923.

problem. Ultimately, these papers and their progeny resulted in the chemistry Nobel prize in 1981 for Hoffmann [together with Kenichi Fukui (1918–1998) of Kyoto], and would probably have given Woodward his second Nobel prize, had he lived long enough.

The reaction of the chemical community to these papers ranged from enthusiastic admiration to "We knew all this trivial stuff already." We would submit that admiration was by far the more appropriate reaction, and suspect that behind much of the carping was a secret smiting of many foreheads. Woodward–Hoffmann theory is brilliant. It not only answered long-standing and difficult questions, but it had remarkable "legs." Its implications run far into organic and the rest of chemistry, and it makes what are called risky predictions. It is important to separate theory that merely rationalizes—that which explains known phenomena—from theory that demands new experiments. The success of Woodward–Hoffmann theory can be partly judged from the flood of experiments it generated. Of course, not all the experiments were incisive, and not all the interpretations were appropriate; the area entered a rococo phase quite early on. Still, some of the work generated by the early Woodward–Hoffmann papers stands today as an example of the combination of theory and experiment that characterizes the best of our science.

In this chapter, we will see electrocyclic reactions in which rings open and close, cycloaddition reactions in which two partners come together to make a new cyclic compound, and sigmatropic shifts in which one part of a molecule flies about coming to rest at one specific position and no other. Specificity is the hallmark of all these reactions; atoms move as if in lock step in one direction or another, or move from one place to one specific new place, but no other. The chemical world struggled mightily to understand these mysteriously stereospecific reactions, and, thanks to the work by Woodward, Hoffmann, and several others, finally figured it out in a marvelously simple way. This chapter will show you some wonderful chemistry.

ESSENTIAL SKILLS AND DETAILS

1. To be able to understand the course of "no-mechanism" reactions, you need only remember that in any concerted (one barrier, one transition state) reaction, a bonding overlap must be maintained between orbitals. The rest is all detail.

2. The opening and closing reactions of rings—electrocyclic reactions in the jargon of this chapter—are best understood through an analysis of the HOMO (p. 133) of the open-chain partner. Ring closing involves only bonding interactions for allowed reactions.

3. Cycloaddition reactions require an analysis of the interactions of the HOMO and LUMO (p. 133) of the two reacting species. Allowed reactions have two bonding interactions between the reacting partners; "forbidden" reactions do not.

4. In sigmatropic shift reactions, one portion of a molecule detaches from one atom and reattaches at another position of the same molecule. The transition state is best described by an atom (or group of atoms) moving across a polyenyl radical. Once again, an analysis of the symmetry of the HOMO of the polyenyl radical is the key to understanding. A bonding interaction must be achieved at the arrival point for the reaction to be allowed.

20.2 Concerted Reactions

Woodward and Hoffmann provided insights into concerted reactions. As noted on page 389, a reaction is concerted if starting material goes directly to product over a single transition state. Such reactions are "single-barrier" processes. The S_N2 and Diels–Alder reactions are nice examples of concerted reactions you know well (Fig. 20.2).

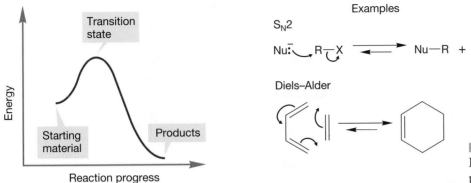

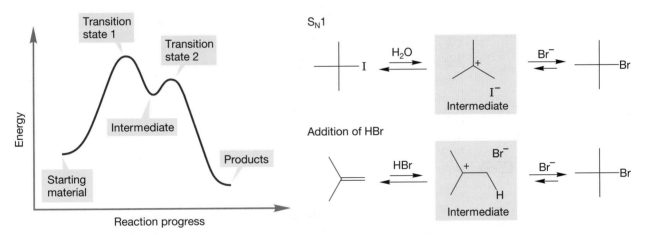

FIGURE 20.2 The S_N2 and Diels–Alder reactions are concerted processes.

Of course, many reactions involve intermediates, and in these reactions more than one transition state is traversed on the way from starting material to product. As noted previously, these are called nonconcerted or stepwise reactions. Typical examples are the S_N1 reaction or the polar addition of HBr to an alkene, two reactions that go through the same intermediate, the *tert*-butyl cation (Fig. 20.3).

FIGURE 20.3 The S_N1 reaction and HBr addition to an alkene are nonconcerted reactions.

It is important to recognize that nonconcerted reactions are made up of series of single-barrier, concerted reactions, each of which could be analyzed independently of whatever other reactions followed or preceded (Fig. 20.4).

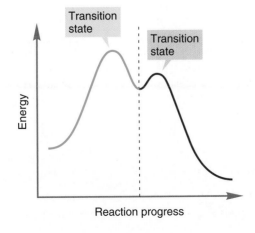

FIGURE 20.4 Any nonconcerted reaction can be separated into a series of concerted processes.

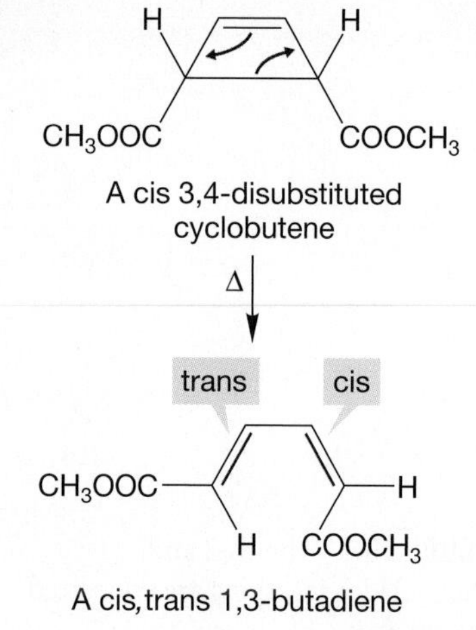

A cis 3,4-disubstituted cyclobutene

A cis,trans 1,3-butadiene

FIGURE 20.5 The thermal opening of a cyclobutene to a 1,3-butadiene. The cis disubstituted cyclobutene rearranges to the cis,trans butadiene.

Woodward and Hoffmann offered explanations of why some reactions are concerted and others not. They found a way to analyze reactions by determining the pathways required to maintain bonding interactions between orbital lobes as the reaction progressed. They used orbital symmetry and the aromaticity of the transition state in their analysis. The following sections take up some examples in which Woodward–Hoffmann theory is particularly useful.

20.3 Electrocyclic Reactions

In 1957, a young German chemist named Emanuel Vogel (b. 1927) was finding his way through the labyrinth of the German academic system, working on the experiments necessary for an entrance ticket for a position at a German university: a super Ph.D. degree called the "Habilitation." His topic included the thermal rearrangements of cyclobutenes. Although cyclobutene had been studied by R. Willstätter (1872–1942, Nobel prize in 1915) as early as 1905, and although the rearrangement to 1,3-butadiene was noted by the American chemist John D. Roberts (b. 1918) in 1949, only Vogel was alert enough to notice the startling stereospecificity of the thermal reaction. For example, the cis 3,4-disubstituted cyclobutene shown in Figure 20.5 rearranges *only* to the cis,trans 1,3-butadiene and no other stereoisomer.

The arrow formalism of Figure 20.5 is easy to write, but the stereochemical outcome of the reaction is anything but obviously predictable. It's worthwhile to see exactly why. There are three possible stereoisomers of the product (Fig. 20.6), and a naive observer would certainly be forgiven for assuming that thermodynamic considerations would dictate the formation of the most energetically favorable isomer, the trans,trans diene, in which the large ester groups are as far apart as possible. Notice that the opening of the cyclobutenes must initially give the dienes in their s-cis conformations rather than in the lower energy s-trans arrangements. Of course, the s-cis molecules will rapidly rotate to the lower energy s-trans conformations.

WEB 3D

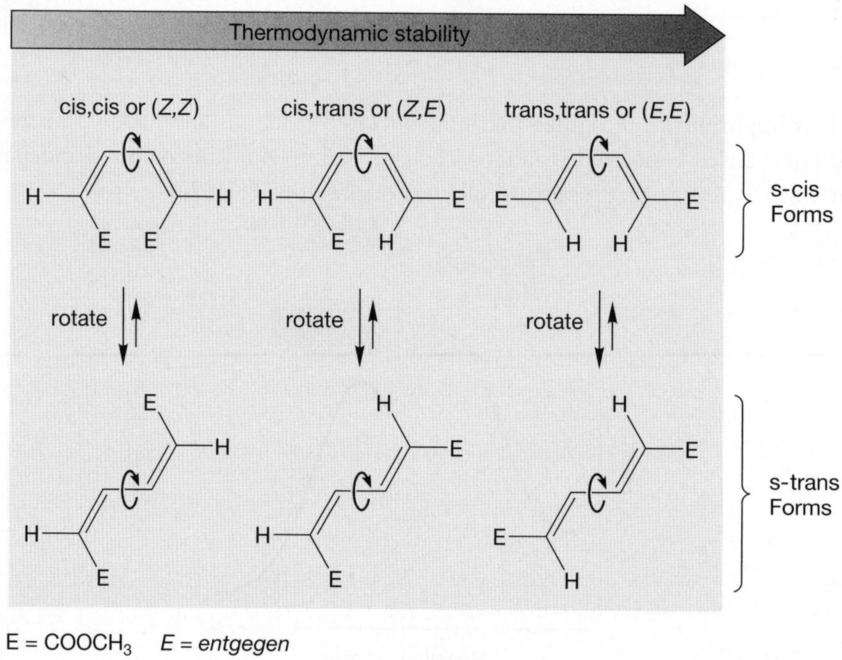

FIGURE 20.6 The three possible stereoisomers of the diene product for the reaction shown in Figure 20.5. Only one, the cis,trans diene of intermediate thermodynamic stability, is formed in the reaction.

E = COOCH₃   *E = entgegen*

As Vogel first found, and as others showed later for many other systems, formation of the trans,trans diene is not the actual result at all; nor is the reaction stereorandom, as only a single stereoisomer is formed. Vogel well understood that this reaction was rather desperately trying to tell us something. In the intervening years, many similar reactions were found and many a seminar hour was spent in a fruitless search for the explanation. What the reaction had to say to us waited for the papers of Woodward and Hoffmann.

Remarkably, it was discovered that the stereochemical outcome of the reaction was dependent on the energy source. Heat gave one stereochemical result and light another. Figure 20.7 sums up the results of the thermal and photochemical interconversions of cyclobutenes and butadienes. In the thermal reaction, a cis 3,4-disubstituted cyclobutene yields the cis,trans diene. By contrast, in the photochemical process the same cis cyclobutene yields a pair of molecules, the cis,cis diene and the trans,trans diene.

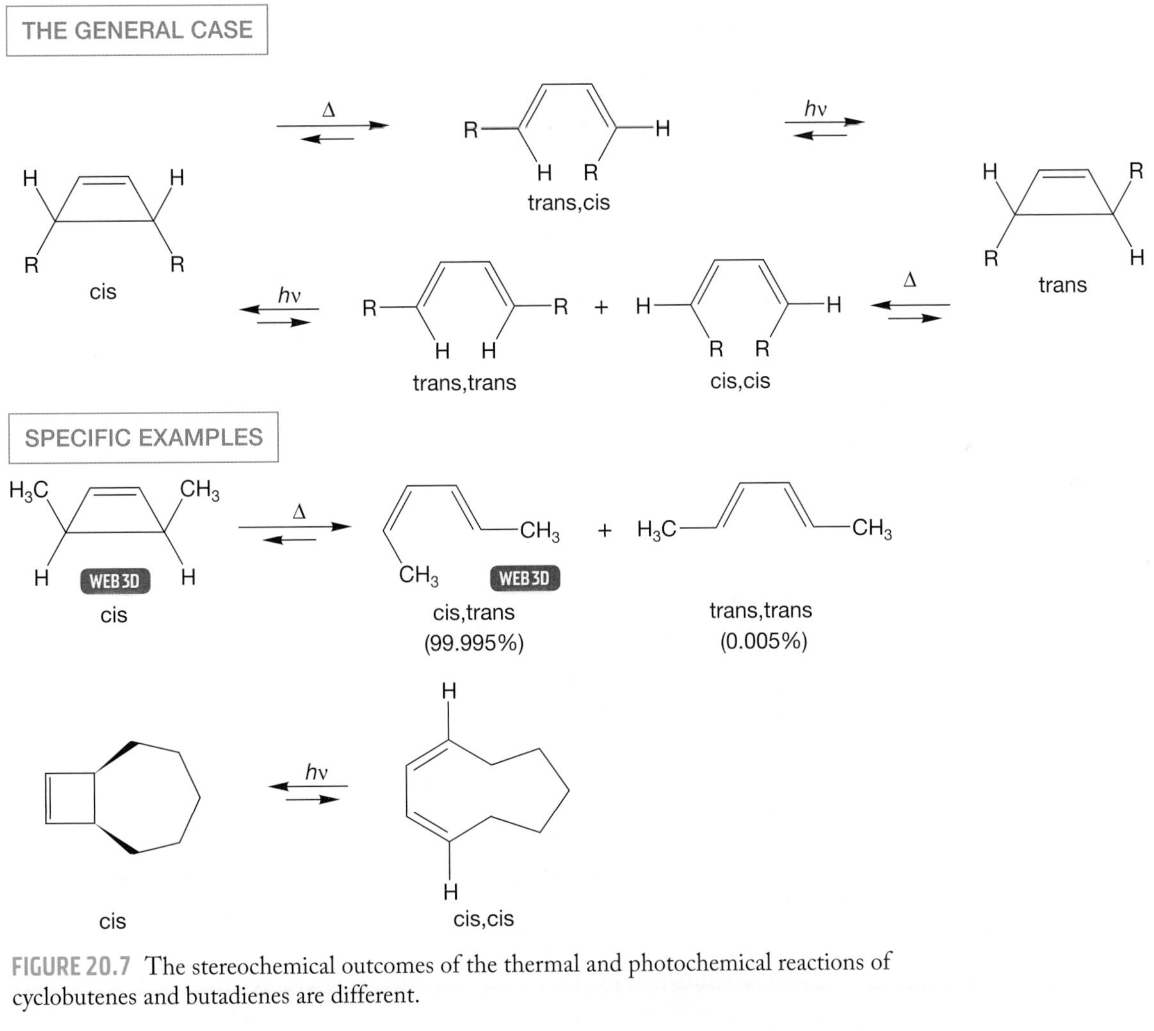

FIGURE 20.7 The stereochemical outcomes of the thermal and photochemical reactions of cyclobutenes and butadienes are different.

Notice that the cis disubstituted cyclobutene can be converted into the trans disubstituted cyclobutene through a thermal reaction and a photochemical reaction (Fig. 20.7). Regardless of the starting material, the thermal reaction favors the more stable 1,3-butadiene. The photochemical process, depending on the wavelength of light used, favors the less stable cyclobutene.

Let's start our analysis by thinking about the thermal opening of cyclobutene to 1,3-butadiene. Here is a general practical hint. It is usually easier to analyze these thermal or photochemical processes, called **electrocyclic reactions**, by examining the *open-chain polyene partner*, regardless of which way the reaction actually runs. Figure 20.8 shows the molecular orbitals involved in the cyclobutene to

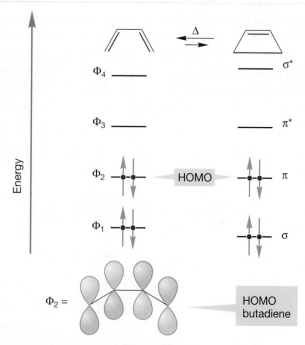

FIGURE 20.8 The molecular orbitals involved in the interconversion of cyclobutene and butadiene.

butadiene interconversion. The four π molecular orbitals of 1,3-butadiene are converted into the σ, σ^*, π, and π^* orbitals of cyclobutene. Now comes a most important point. *The HOMO of the system controls the course of electrocyclic reactions.* This assumption is not ultimately necessary, but it makes predicting the products of these reactions very easy—and it works! It derives from the aromaticity of the transition state, and it can be defended in the following way. It is the electrons of highest energy, the valence electrons located in the highest energy orbital, that control the course of atomic reactions, and it should be no surprise that the same is true for molecular reactions. To do a complete analysis of any reaction, atomic or molecular, we really should follow the changes in energy of all electrons. That process is difficult or impossible in all but a very few simple reactions. Accordingly, theorists have sought out simplifying assumptions, one of which says that the highest energy electrons, those most loosely held, are most important in the reaction. These are the electrons in the HOMO. The HOMO of butadiene is Φ_2, and we assume that it is the controlling molecular orbital. Now ask how Φ_2 must move as 1,3-butadiene becomes cyclobutene. Don't worry that we are analyzing the reaction in the counterthermodynamic sense. If we know about the path in one direction, we know about the path in the other direction as well. In order to create the ring-forming bond between the butadiene end carbons, the *p* orbitals on the end carbons must rotate 90°, and this rotation must be in a fashion that creates a *bond* (σ)

between the atoms, not an *antibond* (σ*) (Fig. 20.9). There are four possible rotations we can consider: both *p* orbitals clockwise, both counterclockwise, left one clockwise and right counterclockwise, and left one counterclockwise and right one clockwise.

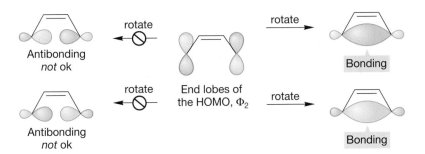

FIGURE 20.9 In the interconversion of cyclobutene and butadiene, the orbitals at the end of the diene must rotate so as to form a bond, not an antibond.

Now look at the symmetry of the end orbitals of Φ_2, the controlling HOMO of butadiene. There are two ways to make a bond: in one the blue lobes overlap, in the other it is the green lobes that overlap (Fig. 20.10). In each case we have created a bonding interaction. In this example, the blue/blue overlap requires that the two ends of the molecule rotate in a clockwise fashion. The green/green overlap requires that the two ends rotate counterclockwise.

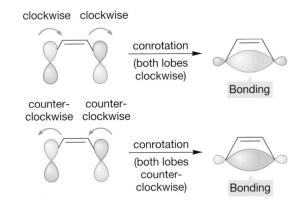

FIGURE 20.10 Formation of a bonding interaction requires what is called conrotation of the ends of the diene. There are two, equivalent conrotatory modes.

These two modes of rotation, in which the two ends rotate in the same direction, are called conrotatory motions, and the whole process is known as **conrotation**.

There is another rotatory mode possible, called **disrotation**, in which the two ends of the molecule rotate in different directions. Figure 20.11 shows the two possible disrotatory motions. In both, an antibond, not a bond, is created between the two end carbons. Disrotation cannot be a favorable process in this case.

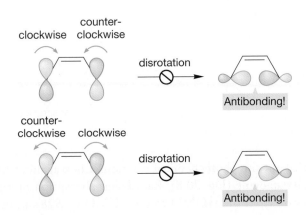

FIGURE 20.11 Disrotation creates an antibonding interaction between the lobes at the ends of the diene. Neither of the two possible disrotatory modes can be a favorable process for butadiene.

An electrocyclic reaction is as easy to analyze as that. Identify the HOMO, and then see whether conrotatory or disrotatory motion is demanded of the end carbons by the lobes of that molecular orbital. All electrocyclic reactions can be understood in this same simple way. The theory tells us that the thermal interconversion of cyclobutene and 1,3-butadiene must take place in a conrotatory way. For the cyclobutene studied by Vogel, conrotation requires the stereochemical relationship that he observed. The cis 3,4-disubstituted cyclobutene can only open in conrotatory fashion, and conrotation *forces* the formation of the cis,trans diene. Note that there are always two possible conrotatory modes (Fig. 20.12), either one giving the same product in this case.

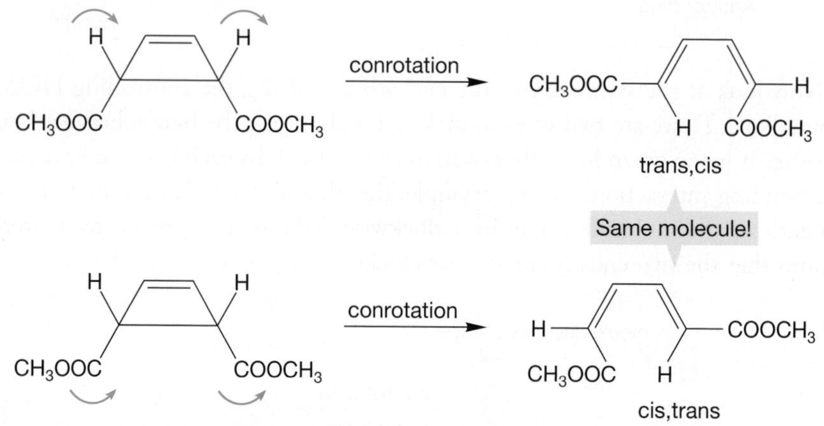

FIGURE 20.12 For the cis disubstituted cyclobutene studied by Vogel, both conrotatory modes demand the formation of the diene with one cis and one trans double bond.

trans,cis

Same molecule!

cis,trans

WORKED PROBLEM 20.1 Show through an equivalent analysis that a mixture of the cis,cis and trans,trans dienes should be formed in the thermal opening of trans 3,4-disubstituted cyclobutenes.

ANSWER Because the opening takes place thermally, it must be conrotatory. There are two conrotatory modes for the trans compound. One leads to the cis,cis (Z,Z) isomer and the other to the trans,trans (E,E) isomer.

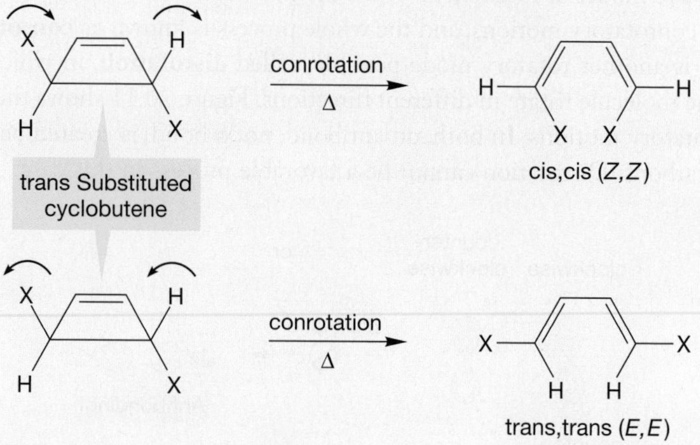

trans Substituted cyclobutene

cis,cis (Z,Z)

trans,trans (E,E)

Now what about the photochemical reaction? Look again at the molecular orbitals of 1,3-butadiene (Fig. 20.8). Recall that absorption of a photon promotes an electron from the HOMO to the LUMO (p. 528), in this case from

Φ_2 to Φ_3, creating a new, "photochemical HOMO" (Fig. 20.13). No longer does conrotation in the HOMO, now Φ_3, create a bond between the end carbons.

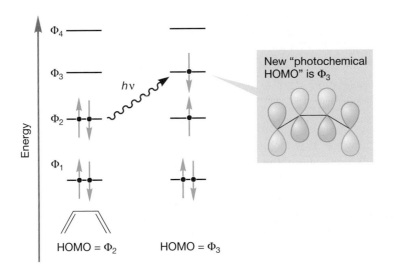

FIGURE 20.13 The HOMO involved in the photochemical reaction of butadiene is Φ_3.

For an orbital of this symmetry, *conrotation creates an antibond.* In the photochemical reaction, it is disrotation that creates the bonding interaction (Fig. 20.14).

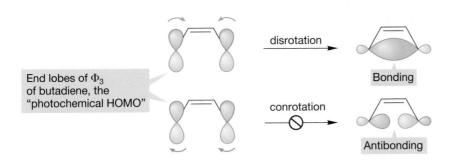

FIGURE 20.14 The photochemical HOMO (Φ_3) demands disrotation. In this molecular orbital, conrotation produces an antibond.

PROBLEM 20.2 Figure 20.14 shows only one disrotatory and one conrotatory mode. Verify that the other possible disrotation produces a bond between the two carbons, and that the other possible conrotation produces an antibond.

Two disrotatory modes interconvert the cis 3,4-disubstituted cyclobutene and the cis,cis and trans,trans isomers of the butadiene (Fig. 20.15).

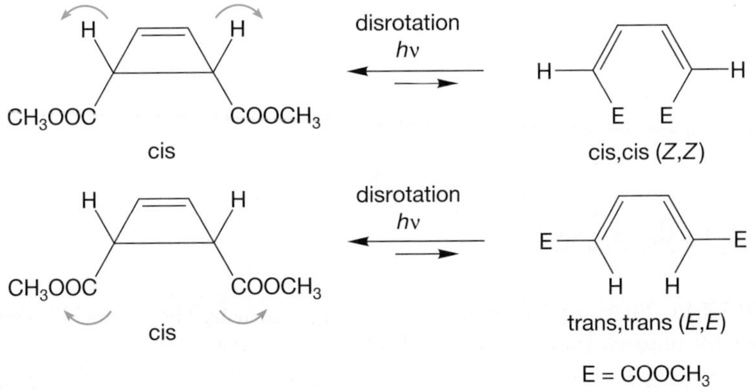

FIGURE 20.15 Disrotation interconverts the cis disubstituted cyclobutene and the cis,cis and trans,trans dienes.

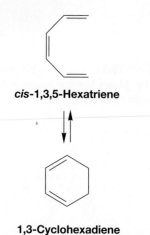

cis-1,3,5-Hexatriene

1,3-Cyclohexadiene

FIGURE 20.16 Another electrocyclic reaction interconverts *cis*-1,3,5-hexatriene and 1,3-cyclohexadiene.

PROBLEM 20.3 Show that disrotation in the photochemical reaction of trans 3,4-disubstituted cyclobutene leads to the cis,trans isomer of the butadiene.

Let's now use this technique for analyzing electrocyclic reactions to make some predictions about the 1,3,5-hexatriene–1,3-cyclohexadiene system (Fig. 20.16).

Because it is always easier to look first at the open polyene system, we will consider the π molecular orbitals for hexatriene (Fig. 20.17). The HOMO for the thermal reaction is Φ_3, and that for the photochemical reaction, Φ_4.

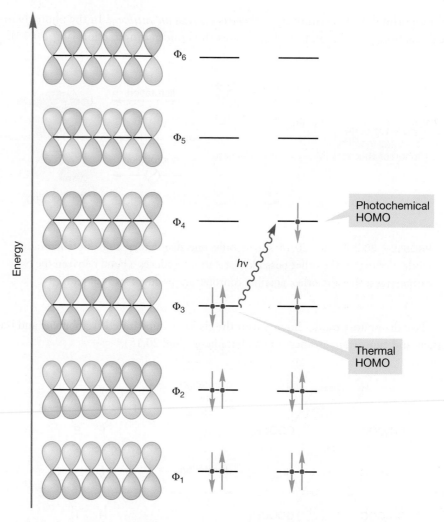

FIGURE 20.17 The π molecular orbitals for 1,3,5-hexatriene. The thermal HOMO is Φ_3 and the photochemical HOMO is Φ_4.

As the open triene closes to the cyclohexadiene (Fig. 20.18), the only way to create a bonding interaction between the end carbons is to close in a disrotatory fashion from Φ_3, and in a conrotatory fashion from Φ_4. So, we predict that the thermal interconversion will occur in a disrotatory way, and that the photochemical reaction must involve conrotation.

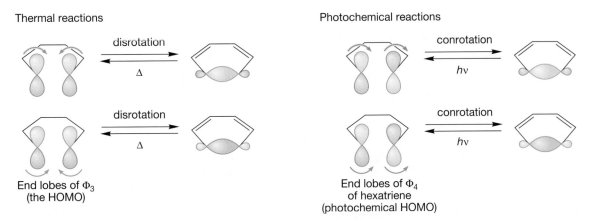

Thermal reactions

disrotation
Δ

Photochemical reactions

conrotation
$h\nu$

disrotation
Δ

conrotation
$h\nu$

End lobes of Φ_3
(the HOMO)

End lobes of Φ_4
of hexatriene
(photochemical HOMO)

FIGURE 20.18 For the hexatriene–cyclohexadiene system, in order to produce a bonding interaction, the thermal reactions must take place in a disrotatory fashion and the photochemical reactions in a conrotatory way.

To test these predictions we need some labeled molecules. The *trans,cis,trans*-2,4,6-octatriene shown in Figure 20.19 will do nicely. A disrotatory thermal closure to *cis*-5,6-dimethylcyclohexa-1,3-diene and a conrotatory photochemical closure to the trans isomer are both known.

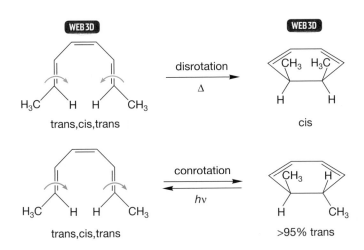

WEB3D

WEB3D

disrotation
Δ

H₃C   H H   CH₃

trans,cis,trans

CH₃ H₃C

H   H

cis

conrotation
$h\nu$

H₃C   H H   CH₃

trans,cis,trans

CH₃   H

H   CH₃

>95% trans

FIGURE 20.19 The experimental results bear out the predictions. The thermal reaction is disrotatory and the photochemical reaction is conrotatory.

Figure 20.19 shows the apparent formation of a single enantiomer of *trans*-5,6-dimethylcyclohexa-1,3-diene from an achiral precursor, which cannot be possible. We have not taken the trouble to draw both enantiomers of the racemic mixture of the *trans*-5,6-dimethylcyclohexa-1,3-diene formed in the reaction. This omission leads directly to Problem 20.4.

CONVENTION ALERT

PROBLEM 20.4 Figure 20.19 shows only one conrotatory and one disrotatory mode. Show that the other disrotation and conrotation give the same relative stereochemical results.

PROBLEM 20.5 Why are cyclohexadienes formed only from the trans,cis,trans isomer shown in Figure 20.19? Why does the trans,trans,trans triene not give a similar reaction?

PROBLEM 20.6 Work out the predicted products from the cis,cis,trans and cis,cis,cis isomers of 2,4,6-octatriene for both photochemical and thermal reactions.

The first two rows of Table 20.1 summarize the results so far. They also point out that the cyclobutene–1,3-butadiene interconversion is a four-electron process, whereas the hexatriene–1,3-cyclohexadiene reaction involves six electrons. In the cyclobutene–butadiene reaction, the four electrons in the σ and π orbitals of the cyclobutene come to occupy the Φ_1 and Φ_2 orbitals of 1,3-butadiene. Similarly, the six electrons in the σ, Φ_1, and Φ_2 of 1,3-cyclohexadiene become the six electrons in Φ_1, Φ_2, and Φ_3 of the 1,3,5-hexatriene molecule.

TABLE 20.1 Rotatory Motions in Electrocyclic Reactions

Reaction	Number of Electrons	Thermal	Photochemical
Cyclobutene–butadiene	4 ($4n$)	Conrotation	Disrotation
Hexatriene–cyclohexadiene	6 ($4n + 2$)	Disrotation	Conrotation
All $4n$ electrocyclic reactions	$4n$	Conrotation	Disrotation
All $4n + 2$ electrocyclic reactions	$4n + 2$	Disrotation	Conrotation

It turns out that all $4n$ systems behave like the four-electron case (thermal reactions conrotatory and photochemical reactions disrotatory) and all $4n + 2$ systems behave like the six-electron case (thermal reactions disrotatory and photochemical reactions conrotatory). This generalization is shown in the second half of Table 20.1, which gives the rules for all electrocyclic reactions. Now, if you can count the number of electrons correctly you can easily predict the stereochemistry of any electrocyclic reaction without even working out the molecular orbitals.

PROBLEM 20.7 Provide mechanisms for the following reactions:

(a)

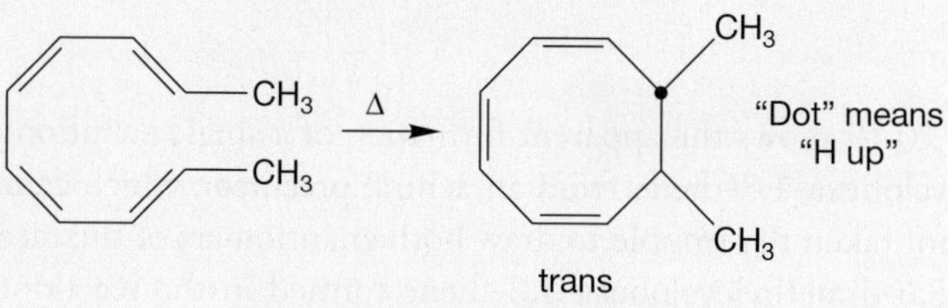

"Dot" means "H up"

(b) Dewar benzene (p. 573) lies perched some 60 kcal/mol higher in energy than its isomer benzene. It would seem that a simple stretching of the central bond in Dewar benzene would inevitably and easily give benzene. Yet Dewar benzene rearranges to benzene only quite slowly. Why?

20.4 Cycloaddition Reactions

Cycloaddition reactions are also orbital symmetry-controlled, pericyclic reactions. We have seen one example already, the Diels–Alder reaction, and we will use it as our prototype. We found the Diels–Alder cycloaddition to be a thermal process that takes place in a concerted (one-step) fashion, passing over a cyclic transition state. Several stereochemical labeling experiments were described in Chapter 12 (p. 549), all of which showed that the reaction involved neither diradical nor polar intermediates. This stereospecificity is important because orbital symmetry considerations apply only to concerted reactions. Of course, all reactions can be subdivided into series of single-step, single-barrier processes, and each of these steps could be analyzed separately by orbital symmetry. However, it makes no sense to speak of the overall orbital symmetry control of a multistep process.

When a diene and dienophile approach each other in the Diels–Alder reaction, the important orbital interactions are between the HOMOs and LUMOs. As Figure 20.20 shows, there are two possible HOMO–LUMO interactions, HOMO$_{(diene)}$–LUMO$_{(dienophile)}$ and HOMO$_{(dienophile)}$–LUMO$_{(diene)}$. The stronger of

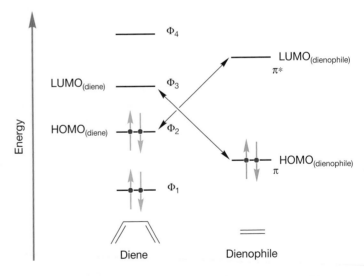

FIGURE 20.20 The two possible HOMO–LUMO interactions in the prototypal Diels–Alder reaction between butadiene and ethylene.

these two will control the reaction. *Remember:* The strength of orbital overlap, and the magnitude of the resulting stabilization, depends on the relative energies of the two orbitals that are mixing. The closer the two are in energy, the stronger the interaction. In the simplest Diels–Alder reaction between ethylene and 1,3-butadiene, the two HOMO–LUMO interactions are of equal magnitude and so we must look at the orbital symmetry of both HOMO$_{(diene)}$–LUMO$_{(dienophile)}$ and HOMO$_{(dienophile)}$–LUMO$_{(diene)}$. Figure 20.21 shows the symmetry relationships for this reaction. Don't forget that in the Diels–Alder and related reactions the two participants approach each other in parallel planes (p. 548).

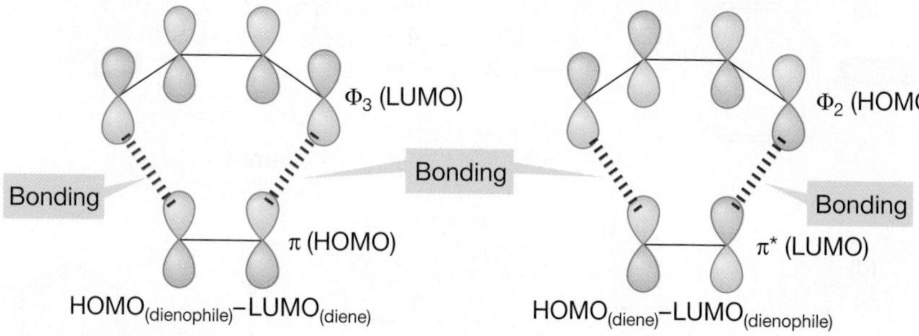

FIGURE 20.21 The orbital overlaps in the two HOMO–LUMO interactions of the Diels–Alder reaction. Either of the HOMO–LUMO interactions can lead to product.

Both interactions involve bonding overlaps at the points of formation of the two new σ bonds. Accordingly, it's no surprise to find that the thermal Diels–Alder reaction occurs. One says that the reaction is "allowed by orbital symmetry."

What about the photochemical Diels–Alder reaction? The observation that this reaction is most uncommon leads us to the immediate suspicion that there is something wrong with it. Usually, the absorption of a photon will promote an electron from the HOMO to the LUMO. In this case, the lower energy HOMO–LUMO gap is that in the diene partner. Absorption of light creates a new photochemical HOMO for the diene, Φ_3, and now the HOMO–LUMO interaction with the dienophile partner involves one antibonding overlap. Both new bonds cannot be formed at the same time (Fig. 20.22). So this photochemical Diels–Alder reaction is said to be "forbidden by orbital symmetry."

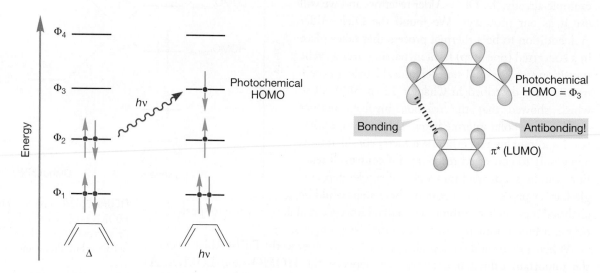

FIGURE 20.22 Absorption of a photon results in the formation of a new HOMO, Φ_3. The HOMO$_{(diene)}$–LUMO$_{(dienophile)}$ interaction involves one antibond. The photo-Diels–Alder reaction is forbidden by orbital symmetry.

PROBLEM 20.8 What would happen if the light were used to promote an electron from the HOMO of the alkene (π) to the LUMO of the alkene (π*)? Would the Diels–Alder reaction be allowed or forbidden? Show your analysis.

Let's now analyze another potential cycloaddition, the joining together of a pair of alkenes to give a cyclobutane (Fig. 20.23). This process can be called a 2 + 2 reaction because it involves a two-electron π bond combining with another two-electron π bond (alkene + alkene). Using this same terminology, a Diels–Alder reaction (diene + dienophile) is a 4 + 2 reaction. But be careful; using this terminology makes most sense in describing the electrons involved in the reaction, whereas another common use of 2 + 2, 4 + 2, etc., is to describe the number of atoms involved in a cyclization.

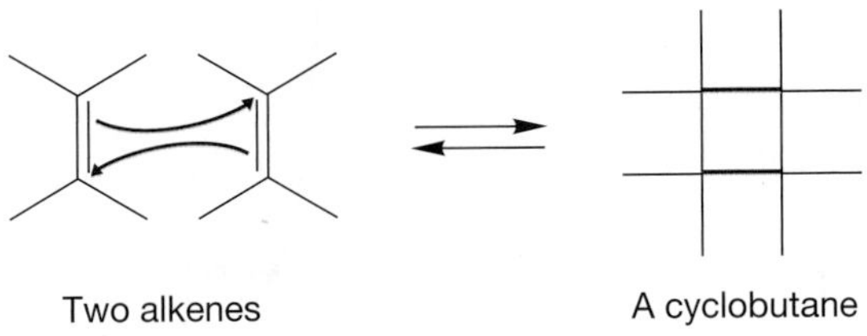

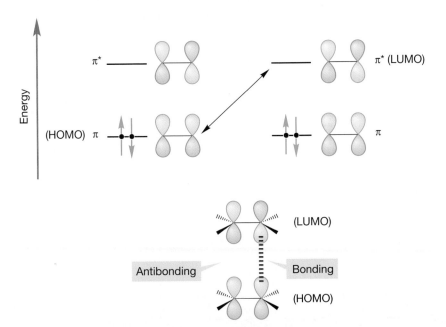

Two alkenes   A cyclobutane

FIGURE 20.23 The hypothetical combination of two alkenes to give a cyclobutane.

PROBLEM 20.9 Will the 2 + 2 reaction between a pair of ethylenes to produce a cyclobutane be exothermic or endothermic? Explain.

The HOMO and LUMO in an alkene are easy to identify, and it is quickly obvious that here, too, the apparently easy bond formation shown by the arrow formalism used in Figure 20.23 is thwarted by an antibonding interaction as the two molecules come together (Fig. 20.24).

FIGURE 20.24 An analysis of the HOMO and LUMO for the reaction of two alkenes to give a cyclobutane shows that it is forbidden by orbital symmetry. There is an antibonding interaction.

Therefore we might guess that the thermal dimerization of ethylenes would be a rare process and we would be exactly correct. Moreover, in most of the reactions of this kind that do occur, it is clear that the two new bonds are formed sequentially and *not* in a concerted manner (Fig. 20.25).

FIGURE 20.25 The known dimerizations of alkenes can be shown to involve diradical intermediates. This reaction is not concerted.

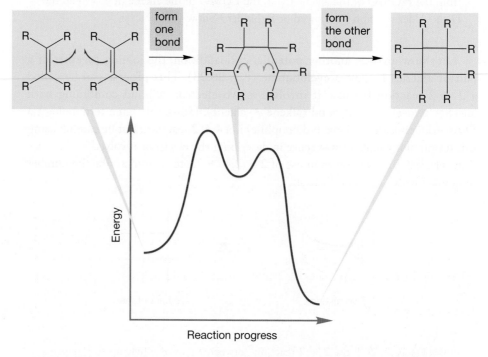

PROBLEM 20.10 Predict the stereochemistry of the product formed in the 2 + 2 cycloaddition of a pair of cis disubstituted ethylenes, X—CH=CH—X (see Fig. 20.25).

What about the photochemical dimerization of alkenes? Promotion of a single electron creates a new photochemical HOMO (Fig. 20.26a), and now the symmetries are perfect for a HOMO–LUMO interaction involving two bonding

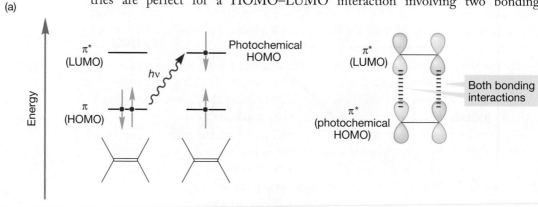

FIGURE 20.26 The photochemical dimerization of two butenes.

overlaps. In contrast to the thermal reaction, the photochemical dimerization of two ethylenes is allowed. For example, the photochemical dimerization of *trans*-2-butene gives the two cyclobutanes shown in Figure 20.26b.

Again, it's time for generalization. The first two rows of Table 20.2 show what we have so far.

TABLE 20.2 Rules for Cycloaddition Reactions

Reaction	Number of π Electrons	Thermal	Photochemical
Alkene + alkene ⇄ cyclobutane	4 (4n)	No	Yes
Alkene + diene ⇄ cyclohexene (Diels–Alder)	6 (4n + 2)	Yes	No
All 4n reactions	4n	No	Yes
All 4n + 2 reactions	4n + 2	Yes	No

Table 20.2 summarizes all 4n and 4n + 2 reactions. Other 4n processes will follow the rules for the 2 + 2 dimerization of a pair of alkenes, and 4n + 2 processes will resemble the 4 + 2 cycloaddition we know as the Diels–Alder reaction. Perhaps you can see the relationship to aromaticity (4n + 2) that plays a role in this analysis. The transition state for these cycloaddition reactions is cyclic and will be allowed only in the cases where the number of electrons makes the transition state aromatic, 4n + 2 electrons for thermal processes and 4n for photochemical reactions.

Now compare Table 20.1 and Table 20.2. This comparison should be a bit disquieting to you. Notice how much more information is available for electrocyclic reactions than for cycloadditions. For the electrocyclic reactions, we have a clear stereochemical prediction of how things must happen. For example, the 4n processes undergo conrotatory thermal ring-closing reactions and disrotatory photochemical ring-closing reactions. On the other hand, for cycloadditions, all we get is a crude "thermal no," "photochemical yes" message; a much less detailed picture.

Electrocyclic reactions are not really different from cycloadditions. Figure 20.27 compares the equilibration of 1,3-butadiene and cyclobutene with the 2 + 2 dimerization of a pair of ethylenes. The only difference is the extra σ bond in butadiene, and this bond is surely not one of the important ones in the reaction—it seems to be just going along for the ride. Why should its presence or absence change the level of detail available to us through an orbital symmetry analysis? It shouldn't, and in fact, it doesn't.

Look again at Figure 20.24, from which we derived the information that the thermal 2 + 2 dimerization of a pair of ethylenes was forbidden. It's only forbidden if we insist on smushing the two ethylenes together head to head exactly as shown in the figure. If we are more flexible, and allow one ethylene to rotate about its carbon–carbon σ bond as the reaction takes place, the offending antibonding overlap vanishes. What orbital symmetry really says is that for the concerted thermal dimerization of two alkenes to take place, there must be this rotation (Fig. 20.28).

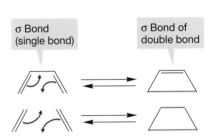

FIGURE 20.27 The equilibration of cyclobutene and 1,3-butadiene and the 2 + 2 dimerization of a pair of ethylenes are very closely related reactions.

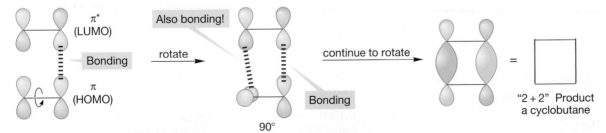

FIGURE 20.28 The thermal 2 + 2 dimerization is allowed if one end of one ethylene can rotate as the reaction occurs.

In practice, this rotation is a very difficult thing to achieve. It is this difficulty that accounts for the absence of the thermal 2 + 2 reaction, not the absolute "No" of Table 20.2. Similarly, the photochemical 4 + 2, Diels–Alder reaction also becomes allowed if we permit one of the partners to undergo a rotation during the reaction or orient the two reagents at an angle so that the bonding overlap is possible (Fig. 20.29).

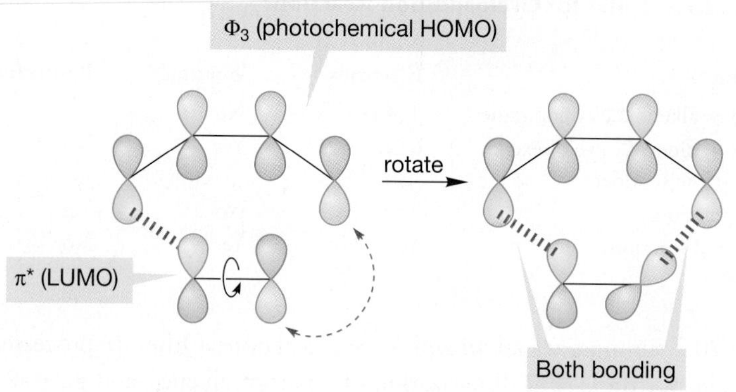

FIGURE 20.29 The photochemical 4 + 2 reaction (the Diels–Alder reaction) is also allowed if a rotation occurs during the cycloaddition or if the orientation is favorable.

Don't forget; these rules do not include any consideration of thermodynamics. The processes allowed with a rotation may well not be very favorable thermodynamically, even though they are permitted electronically.

Summary

Cycloaddition reactions involve the coming together of two (or, rarely, even more) molecules to make a ring. The easiest analysis involves examining the interactions of the HOMOs and LUMOs of the two fragments forming the ring compound.

20.5 Sigmatropic Shift Reactions

When one heats (Z)-1,3-pentadiene nothing obvious happens, as the starting material is recovered, apparently unchanged. Finally, at very high temperatures, radical reactions begin as carbon–hydrogen bonds are broken. However, a deuterium-labeling experiment reveals that much is happening long before the energies necessary for the formation of radicals have been reached (Fig. 20.30).

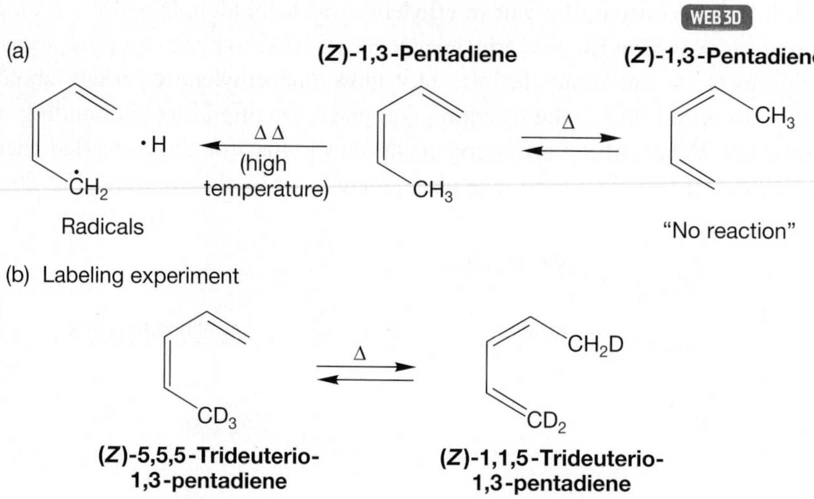

FIGURE 20.30 A labeling experiment reveals the degenerate thermal rearrangement of (Z)-1,3-pentadiene.

This kind of process, in which starting material rearranges into itself, is called a degenerate reaction (p. 1031). Isotopically labeled molecules reveal the degenerate reaction and allow a determination of the kinetic activation parameters. The reaction requires an activation energy of 36–38 kcal/mol. Other labels are possible, of course, and substituted molecules will reveal the reaction as well as the more sophisticated isotopic labels. In most substituted cases the reaction is no longer degenerate and the equilibrium constant *K* for the reaction can no longer equal 1 (Fig. 20.31).

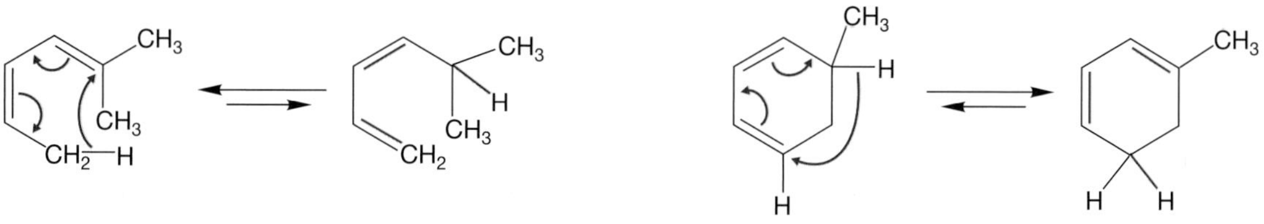

FIGURE 20.31 Substituted molecules also reveal the reaction, but the rearrangement is no longer degenerate, because starting material and product are different.

PROBLEM 20.11 Which of the molecules in each of the two reactions of Figure 20.31 will be favored at equilibrium? Explain your choice. Draw the product of heating (*Z*)-3-methyl-1,3-pentadiene. Is this reaction degenerate or not?

WORKED PROBLEM 20.12 Estimate the bond dissociation energy (BDE) for the migrating carbon–hydrogen bond in 1,3-pentadiene. Why is the observed activation energy of 36–38 kcal/mol for the [1,5] shift of hydrogen shown in Figure 20.30a so much lower than your answer?

ANSWER The carbon–hydrogen bond in ethane has a BDE of 101 kcal/mol (Table 8.2). But the pentadienyl radical is resonance stabilized and that stabilization will begin to be felt as the bond breaks, further decreasing the BDE. If allylic resonance is worth about 13 kcal/mol (p. 475), we might expect the BDE to be about 101 − 13 = 88 kcal/mol.

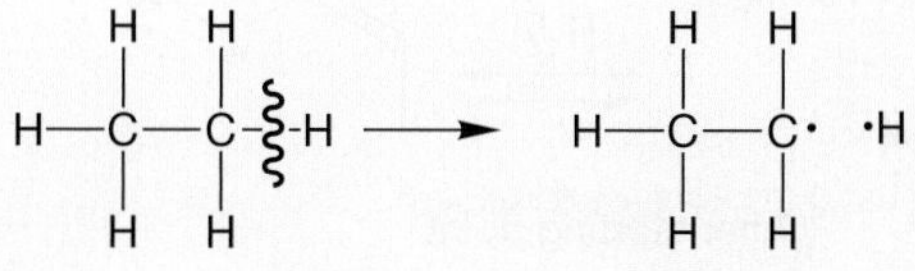

BDE = 101 kcal/mol

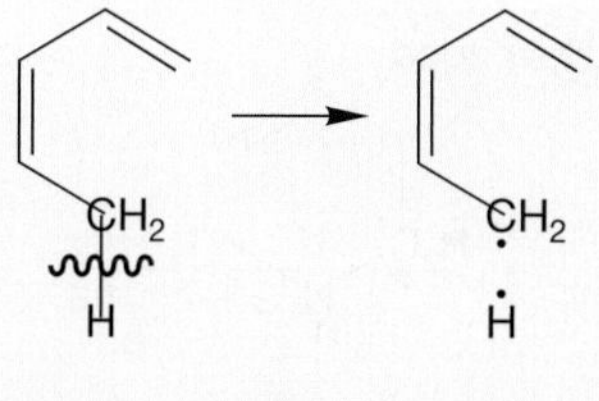

BDE (first guess) = ~101 kcal/mol

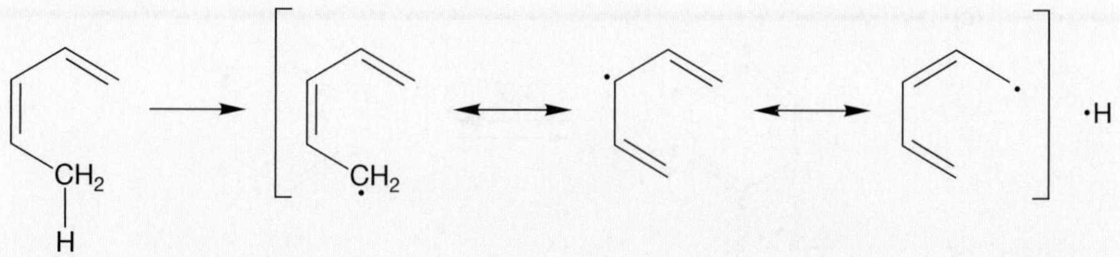

(continued)

Yet the real BDE is much, much lower than this, about 37 kcal/mol. The reason is that as the carbon–hydrogen bond begins to break, the new carbon–hydrogen bond at C(1) is forming. It is *not* necessary to break the carbon–hydrogen bond fully to make the very unstable hydrogen radical.

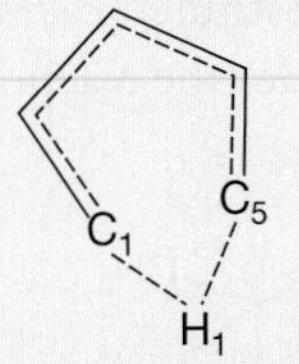

Instead, formation of the new C(5)—H bond is initiated prior to breaking of the original C(1)—H_1 bond. This bond formation decreases the energy required to break the C(1)—H_1 bond.

The deuterium labeling experiment of Figure 20.30b reveals that much is happening before the onset of high-temperature radical reactions. A hydrogen (deuterium, in this case) atom is moving from one position in the molecule to another. This kind of reaction has come to be known as a **sigmatropic shift**.[2] A sigmatropic shift is the movement of a sigma bond from one atom to another. These reactions can involve hydrogens, carbons, or heteroatoms.

CONVENTION ALERT

How are sigmatropic shifts formally described? The first step is to identify the bond that is broken in the reaction and the bond that is made. In other words, write an arrow formalism. The bond that breaks in 1,3-pentadiene could be designated the "1–1" bond. Remember that these numbers have nothing to do with the IUPAC name. Next, note where the new bond is formed. In Figure 20.32a, the new bond is [1,5], in (b) it is [2,3], and in (c) it is [3,3]. The numbers denoting the shift are always enclosed in brackets and separated by a comma.

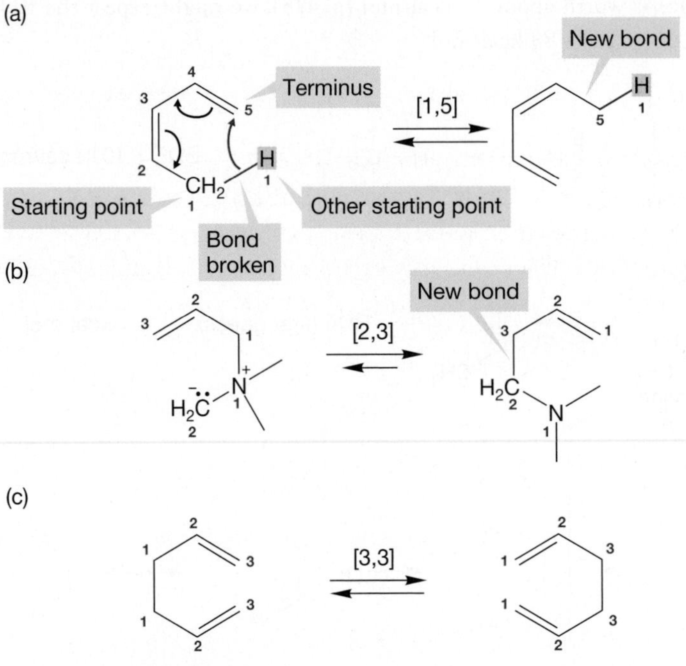

FIGURE 20.32 (a) In a [1,5] sigmatropic shift of hydrogen, the migrating hydrogen atom (atom 1) moves to atom 5. The new sigma bond is between atom 1 and atom 5, hence the shift is [1,5]. (b) In the ammonium ion shown the reaction is a [2,3] sigmatropic shift, and in (c) the shift is [3,3].

[2]It must be admitted that the Woodward–Hoffmann theory is filled with jargon. There is nothing to do but learn it.

PROBLEM 20.13 Classify the following reactions as [x,y] sigmatropic shifts. Which, if any, of the reactions are degenerate?

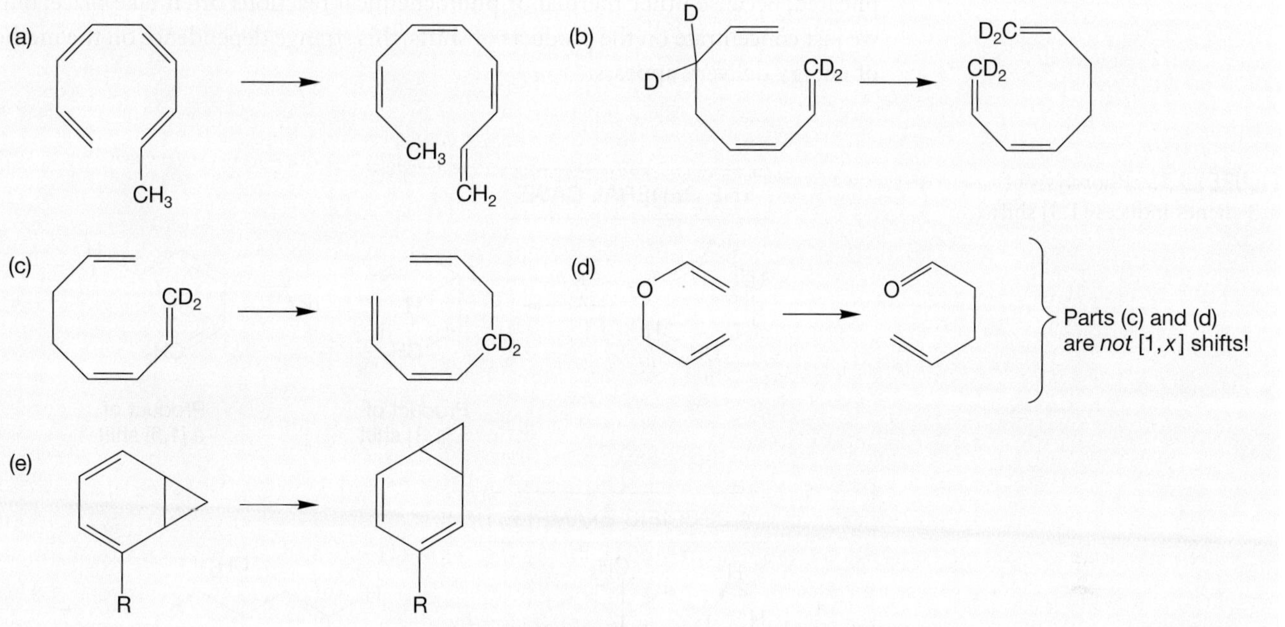

(a)

(b)

(c)

(d)

Parts (c) and (d) are *not* [1,x] shifts!

(e)

Sigmatropic shifts respond to attempted mechanistic experiments in true no-mechanism fashion. Neither acids nor bases strongly affect the reaction, and the solvent polarity, or even the presence of solvent, is usually unimportant. The reaction proceeds quite nicely in the gas phase. It is simple to construct an arrow formalism picture of the reaction, but this device does little more than point out the overall change produced by the migration. In these figures, the arrow formalism is even more of a formalism than usual. For example, the arrows of Figure 20.33 could run in either direction, clockwise or counterclockwise. That is not true for a polar reaction in which the convention is to run the arrows from nucleophile (the Lewis base) toward the electrophile (Lewis acid).

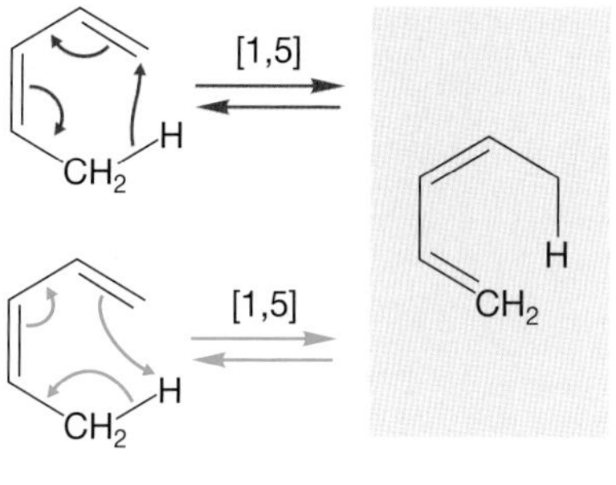

It is already possible to identify one curious facet of this reaction, and a little experimentation reveals others. The product of a [1,3] shift is absent. Why, if the hydrogen is willing to travel to the 5-position, does it never stop off at the 3-position? An arrow formalism can easily be written for a [1,3] shift (Fig. 20.34), and it might reasonably be argued that the [1,3] shift, requiring a shorter path than the [1,5] shift, should be easier. Why are [1,3] shifts not observed?

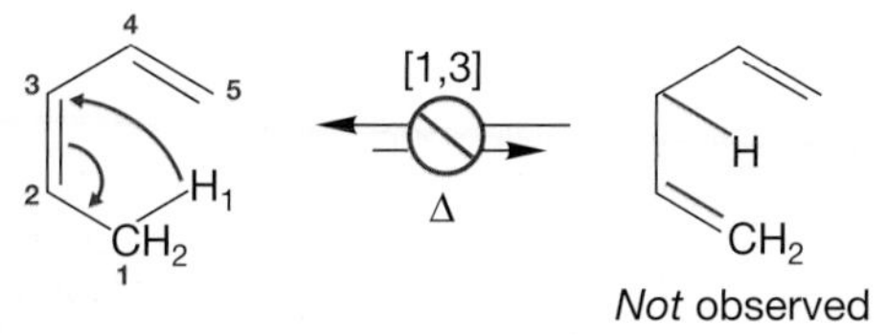

FIGURE 20.33 Two arrow formalism descriptions of the [1,5] shift of hydrogen in (Z)-1,3-pentadiene.

Not observed

FIGURE 20.34 An arrow formalism for the [1,3] shift of hydrogen that does not occur.

A second strange aspect of sigmatropic shifts appears in photochemical experiments. As you already know, it is possible to deliver energy to a molecule in more than one way. One way is to heat up the starting material. However, conjugated molecules absorb light (p. 529), light quanta contain energy, and this gives us another way to transmit energy to the molecule. Curiously, when 1,3-pentadienes are irradiated, the products of the reaction include the molecules formed through [1,3] shifts, but *not* those of [1,5] shifts (Fig. 20.35). The reactions are sometimes complicated, because other thermal or photochemical reactions often take place, but if we just concentrate on the products of shifts, this strange dependence on the method of energy delivery appears.

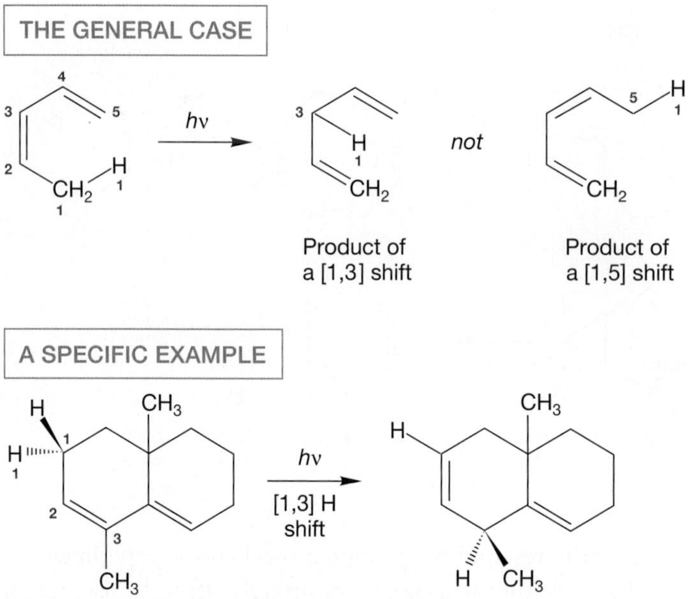

FIGURE 20.35 Photolysis of 1,3-dienes induces [1,3] shifts.

THE GENERAL CASE

Product of a [1,3] shift

not

Product of a [1,5] shift

A SPECIFIC EXAMPLE

[1,3] H shift

WORKED PROBLEM 20.14 Perhaps it could be argued that the loss of conjugation in the product of the [1,3] shift results in an energetic favoring of the [1,5] process. Design a substituted 1,3-pentadiene that would test this (incorrect) surmise.

ANSWER This problem may be hard because you are not used to thinking about designing reactions. The hypothesis is that the [1,5] shift is preferred to the [1,3] shift because it preserves the conjugation present in a 1,3-diene. In order to test this surmise, we need to design a molecule in which both the [1,5] and [1,3] shifts preserve the same amount of conjugation. The deuterium label is needed to reveal the [1,5] or [1,3] nature of the reaction. Of course a [1,7] shift might also occur in this reaction.

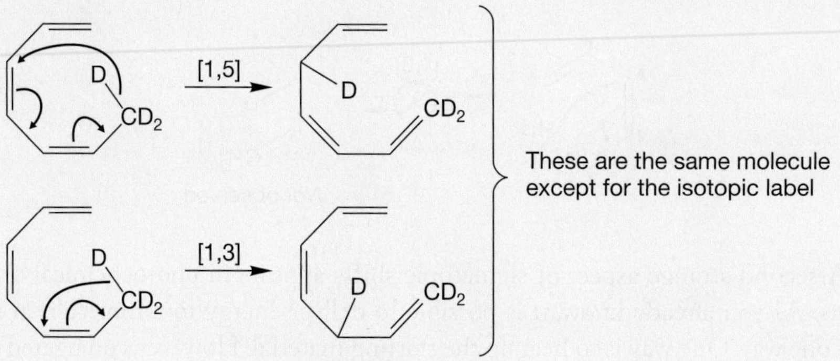

[1,5]

[1,3]

These are the same molecule except for the isotopic label

In summary, thermal energy induces [1,5] shifts and photochemistry favors [1,3] shifts. So, any mechanism must provide an explanation for these observations.

Let's start our analysis by examining the early stages of the [1,5] sigmatropic shift. As the reaction begins, a carbon–hydrogen bond of the methyl group begins to stretch (Fig. 20.36a). If this stretching were continued to its limit, it would produce two radicals, a hydrogen atom and the pentadienyl radical (Fig. 20.36a). We know this cleavage to give radical intermediates does not happen because the products of recombination of these radicals are not found (Fig. 20.36b).

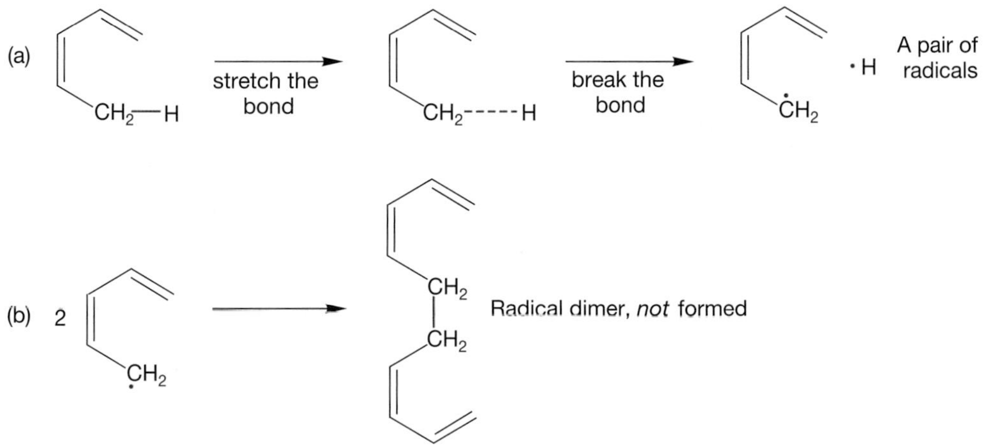

FIGURE 20.36 (a) In the earliest stages of the reaction, a carbon–hydrogen bond begins to stretch. Ultimately a pair of radicals would result. (b) We know this bond breaking does not take place because the radical dimer is not found.

We should stop right here to question the assumption that the carbon–hydrogen bond is breaking in homolytic (radical) fashion. Why not write a heterolytic cleavage to give a pair of ions rather than radicals (Fig. 20.37)?

The stability of ions depends greatly on solvation. It is inconceivable that a heterolytic bond breaking would not be highly influenced by solvent polarity. As we have just seen, a change in solvent polarity affects the rate of this reaction only very slightly. Only homolytic bond breaking is consistent with the lack of a solvent effect.

As the methyl carbon–hydrogen bond begins to stretch (Fig. 20.36a), something must happen that prevents the formation of the pair of radicals. As we have already concluded in Problem 20.12, the developing hydrogen atom is intercepted by reattachment at atom 5. In orbital terms, we would say that as the C(1)—H bond breaks, the developing $1s$ orbital on hydrogen begins to overlap with the p orbital on the carbon at position 5. Now we have a first crude model for the transition state for this sigmatropic shift (Fig. 20.38). Notice that the transition state is cyclic.

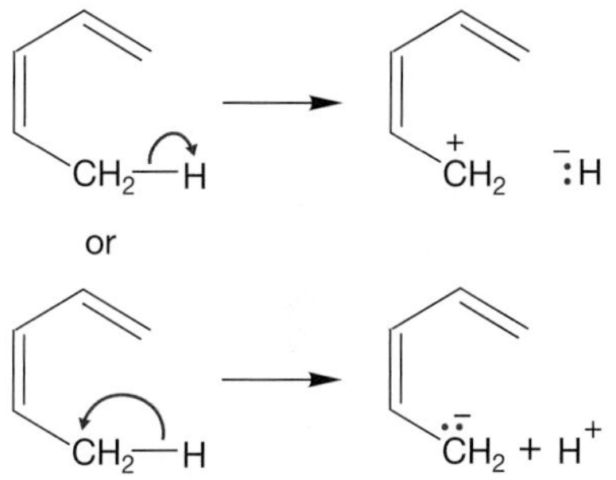

FIGURE 20.37 There are two possible ways to break the carbon–hydrogen bond heterolytically to give a pair of ions. Either way depends strongly on solvent polarity. Because there is essentially no dependence of the rate of this reaction on solvent polarity, this mechanistic possibility is excluded.

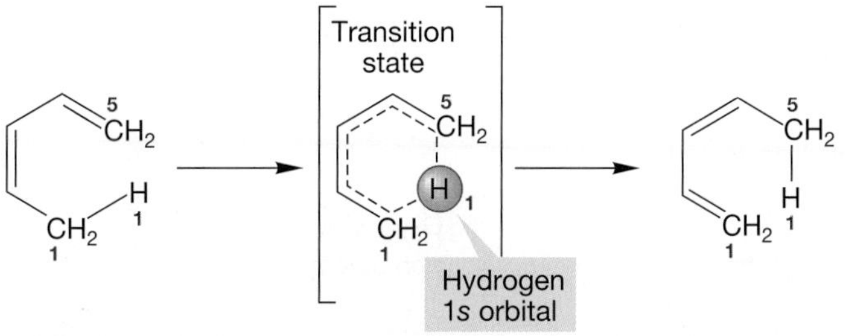

FIGURE 20.38 Radical formation can be thwarted by partial bonding of the migrating hydrogen to the orbitals on carbon atoms 1 and 5 in the transition state. As the C(1)—H bond stretches, the hydrogen $1s$ orbital begins to overlap with an orbital on C(5).

This model leaves unanswered the serious questions of mechanism: Why does the migrating hydrogen reattach at the 5-position in thermal reactions and at the 3-position in photochemical reactions? We have answered neither the questions of regiochemistry ([1,5] or [1,3]) nor of dependence on the mode of energy input (thermal or photochemical).

Let's look more closely at our transition state model. In essence, it describes a hydrogen atom moving across the π system of a pentadienyl radical (Fig. 20.39).

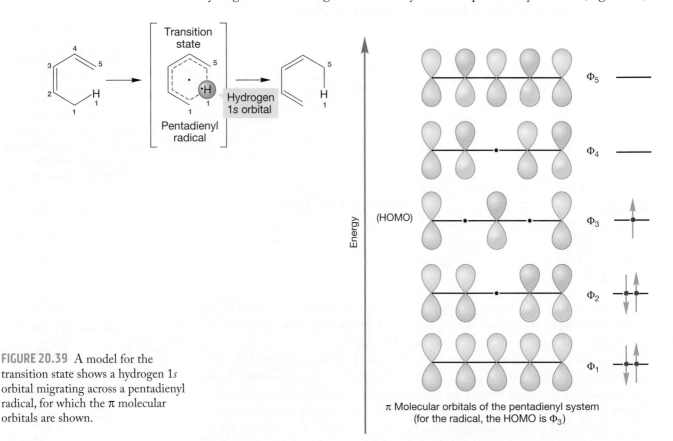

FIGURE 20.39 A model for the transition state shows a hydrogen 1s orbital migrating across a pentadienyl radical, for which the π molecular orbitals are shown.

π Molecular orbitals of the pentadienyl system
(for the radical, the HOMO is Φ_3)

The hydrogen atom is simple to describe—it merely consists of a proton and a single electron in a $1s$ atomic orbital. The pentadienyl molecular orbital system isn't difficult either. There are five π molecular orbitals and these can be constructed by taking combinations of five carbon $2p$ orbitals. We assume again that it is the highest energy electrons, those most loosely held, that are the most important in the reaction. These are the electrons in the HOMO. If we accept this assumption, we can elaborate our picture of the transition state for the reaction to give Figure 20.40. Notice that the transition state for a [1,5] shift involves six electrons ($4n + 2$). They come from the sigma bond that is broken and the two double bonds in the π system.

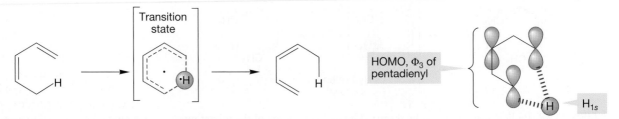

HOMO, Φ_3 of pentadienyl

H_{1s}

FIGURE 20.40 An elaboration of our model for the transition state, showing a crude outline of the interaction of the hydrogen 1s orbital with the lobes of the HOMO for pentadienyl, Φ_3.

Notice a very important thing about the transition state outlined in Figure 20.40. In the starting pentadiene, the bond made between the 1s orbital of hydrogen and the hybrid orbital on carbon atom 1 must of course be made between lobes of the same phase, which is represented by color coding (Fig. 20.41). As this bond stretches, the phases of the lobes do not change: The 1s orbital is always fully or partially bonded to a lobe of the same phase (Fig. 20.41).

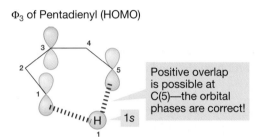

Φ_3 of Pentadienyl (HOMO)

Positive overlap is possible at C(5)—the orbital phases are correct!

1s

FIGURE 20.41 Now we can elaborate our transition state model by filling in the lobes of the HOMO of pentadienyl, Φ_3.

In this way, we know the relative symmetries of some of the lobes in the transition state for the reaction. We also know the symmetry (from Fig. 20.39, or by figuring it out) of Φ_3, the HOMO of pentadienyl. So, we can elaborate the picture of the transition state to include the symmetries of all the orbital lobes. The symmetry of the lobes allows the [1,5] shift to take place as shown.

Notice that the process through which the hydrogen detaches from carbon 1 and reattaches to carbon 5 from the same side allows the maintenance of bonding (same sign) lobal interactions at all times. This reaction is "orbital symmetry allowed." Now look at the [1,3] shift (Fig. 20.42). Notice that it involves only four electrons (4n) during the reaction: two electrons in the sigma bond that breaks and two in the π system. To do the analogous [1,3] shift, keeping the shifting hydrogen on the same side of the molecule requires the overlap of lobes of different symmetry—the formation of an antibond. This reaction is "orbital symmetry forbidden"!

More important jargon enters here: The motions we have been describing in which the migrating atom or group of atoms leaves and reattaches from the *same* side of the π system are called **suprafacial**. If the migrating atom or group of atoms leaves from one side of the π system and arrives at the other, the process is called **antarafacial** (Fig. 20.43). In the [1,5] hydrogen shift, the typical sigmatropic shift involving 4n + 2 electrons, suprafacial motion is allowed by the symmetries of the orbitals; in a [1,3] shift of a hydrogen, a typical 4n electron process, it is not (Figs. 20.41 and 20.42).

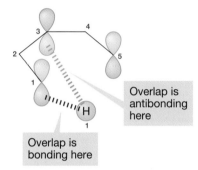

Overlap is antibonding here

Overlap is bonding here

FIGURE 20.42 By contrast, the [1,3] shift, which is a four-electron process, involves the formation of an antibond if the hydrogen reattaches to carbon 3 from the same side.

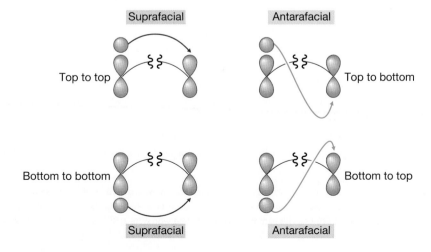

Suprafacial   Antarafacial

Top to top   Top to bottom

Bottom to bottom   Bottom to top

Suprafacial   Antarafacial

FIGURE 20.43 Suprafacial motion involves bond breaking and bond making on the *same* side of the π system. Antarafacial motion involves bond breaking on one side and bond making on the other.

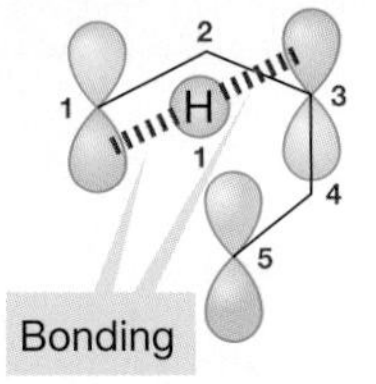

Bonding

Φ_3 of Pentadienyl
(HOMO)

FIGURE 20.44 The antarafacial [1,3] shift is allowed by orbital symmetry, but the small size of the $1s$ orbital makes the stretch impossible.

Why not migrate in a different fashion? Why can't the hydrogen depart from one side of the π system and reattach at the other? This antarafacial migration would result in an orbital symmetry–allowed [1,3] migration (Fig. 20.44). The problem is that the $1s$ orbital is small and cannot effectively span the distance required for an antarafacial migration. There is nothing electronically disallowed or forbidden about the antarafacial [1,3] migration, but there is an insuperable steric problem.

The [1,5] sigmatropic shift is intramolecular and the maintenance of bonding interactions explains why [1,5] hydrogen shifts are possible and [1,3] hydrogen shifts are exceedingly rare. But we have not yet been able to deal with the dependence of the course of this reaction on the mode of energy input—thermal or photochemical.

PROBLEM 20.15 Do you expect the [1,3] shift shown below to occur when propene is heated? Explain carefully, using a molecular orbital argument.

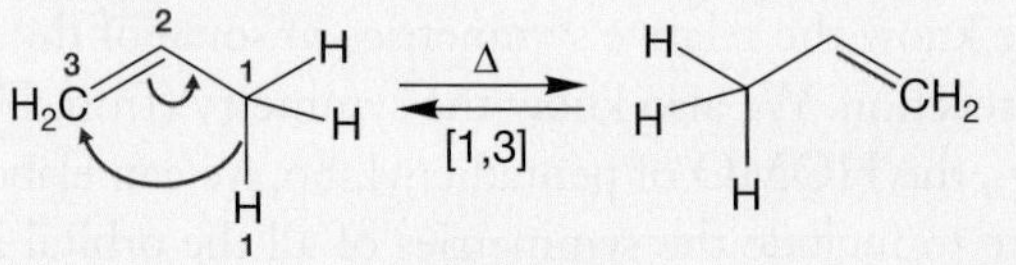

Now comes a most important point. Our model not only rationalizes known experimental data, but includes the roots of a critical prediction. The Woodward–Hoffmann explanation predicts that the [1,5] shift occur in a suprafacial fashion. That is the only way in which bonding interactions can be maintained throughout the migration of a hydrogen atom from carbon 1 to carbon 5. There is *nothing* so far in the available experimental data that allows us to tell whether the observed shifts are suprafacial, antarafacial, or a mixture of both modes.

Let's set about finding an experiment to test the requirement of the theory for suprafacial [1,5] motion. What will we know at the end of the experiment? If we find that the [1,5] shift is indeed strictly suprafacial, the theory will be supported (not proved!), and we will certainly feel better about the mechanistic hypothesis. If we find antarafacial motion, or both suprafacial and antarafacial motions, the theory will be proved (yes, proved) wrong. There is no way our hypothesis can accommodate antarafacial motion; there is an absolute demand for suprafaciality, which nicely illustrates the precarious life of a theory. It can always be disproved by the next experiment, and it can never become free of this state of affairs.

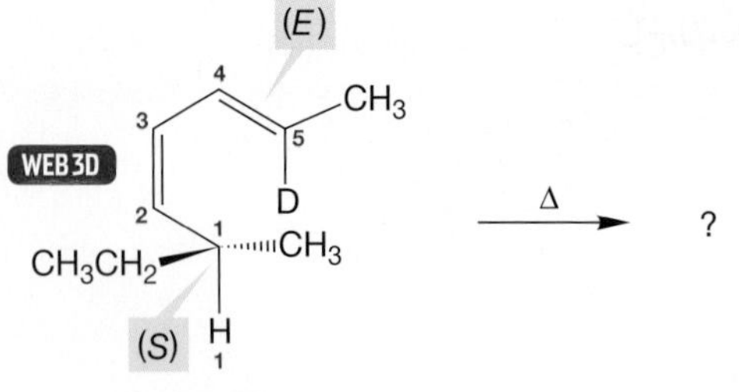

The crucial test was designed by a German chemist, W. R Roth (1930–1997).[3] It is the first of a series of beautiful experiments you will encounter in this chapter. Roth and his co-workers spent some years in developing a synthesis of the labeled molecule shown in Figure 20.45.

Notice that the molecule has the (S) configuration at carbon 1, and the (E) stereochemistry at the C(4)—C(5) double bond.

Now let's work out the possible products from suprafacial and antarafacial [1,5] shifts in this molecule. The situation is a little complicated because there

(E)

Roth's diene

FIGURE 20.45 The 1,3-diene used by Roth and his co-workers to see whether the [1,5] shift proceeded in a strictly suprafacial manner.

---

[3]No typo here. There is no period after the R. Roth claimed that he was advised by his postdoctoral advisor at Yale, William Doering, to take a middle initial in order not to be lost among the myriad "W. Roth's" in German chemical indexes. Others with common last names, and who were at Yale at the time, rejected similar advice.

are two conformational isomers of this molecule that can undergo the reaction, shown in Figure 20.46. In one, the methyl group is aimed at the C(4)—C(5) double bond. In the other, it is the ethyl group that is directed toward the C(4)—C(5) double bond. In suprafacial migration of hydrogen in one conformation, the product has the (R) configuration at C(5) and the (E) stereochemistry at the new double bond located between C(1) and C(2). Another possible suprafacial migration starts from the second conformation. Here it is the opposite stereochemistries that result. The configuration of C(5) is (S) and the new double bond is (Z).

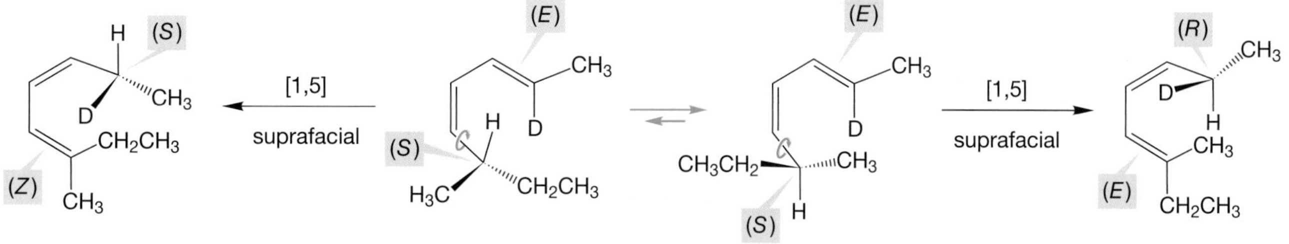

FIGURE 20.46 The predicted products formed by suprafacial migration in the two rotational isomers of Roth's diene.

In antarafacial migrations, the hydrogen leaves from one side and is delivered from the other. Figure 20.47 shows the results of the two possible antarafacial migrations. Note that the products are different from each other and also different from those predicted from suprafacial migration.

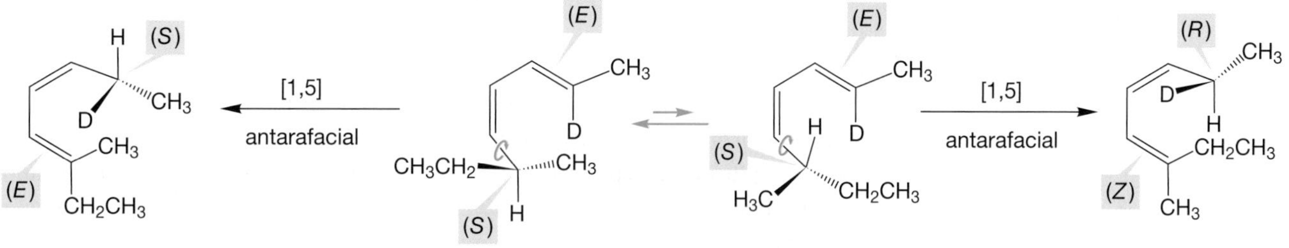

FIGURE 20.47 The possible products of antarafacial migration in Roth's diene.

Roth and his co-workers had the experimental skill necessary to sort out the products and found that only the products of suprafacial migration were formed. The theory is beautifully confirmed by this experiment.

At last we come to the remaining question. Why should the source of energy make a difference in the mode of migration? Remember, the common, thermally induced shifts are [1,5], whereas the photochemical shifts are [1,3]. A molecule absorbs a photon of light and an electron is promoted from the HOMO to the LUMO, producing a new HOMO and changing the symmetry of the lobes at the ends of the π system. What happens when the symmetry of the HOMO changes?

Figure 20.48 shows the situation for the simplest [1,3] shift in propene. Absorption of a photon by propene promotes an electron, shown below in Φ_3 of allyl. Migration can now occur in a suprafacial fashion, and the steric barrier to thermal, necessarily antarafacial [1,3] shifts is gone. The hydrogen can migrate in an easy suprafacial manner.

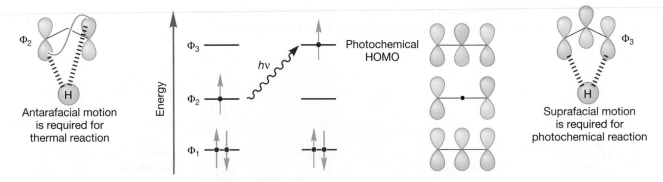

FIGURE 20.48 Suprafacial motion of a hydrogen is possible in Φ_3, the photochemical HOMO for this reaction.

We can now put together Table 20.3 that compares the types of sigmatropic shifts that are possible and links the stereochemical outcome to the number of electrons involved in the process and the type of energy used to drive the reaction.

TABLE 20.3 Rules for Allowed Sigmatropic Reactions

Reaction	Number of Electrons	Thermal	Photochemical
Typical $4n$ reactions	$4n$	Antarafacial	Suprafacial
Typical $4n + 2$ reactions	$4n + 2$	Suprafacial	Antarafacial

Summary

We now understand three seemingly magical concerted reactions. Each is controlled by orbital symmetry. Electrocyclic reactions occur in a conrotatory or disrotatory fashion depending on the number of electrons in the system and whether heat or light is used to drive them. The cycloaddition reactions occur with orbital overlap between the reacting partners and the symmetry of the HOMO and the LUMO determines the outcome of the reaction. Sigmatropic shifts are perhaps the most challenging to understand. As long as you keep in mind the necessity to form bonds and not antibonds, analysis of the HOMO gives you the positions available for reattachment of a migrating hydrogen or group. One can determine whether the reaction is allowed in a suprafacial or antarafacial fashion. Add to this analysis the idea that hydrogen has only a small $1s$ orbital, and therefore any short antarafacial shifts are generally impossible.

PROBLEM 20.16 Provide mechanisms for the reactions shown on the next page. *Hint*: Reactions (a) and (b) each require more than one sigmatropic shift.

(*continued*)

(a)

(b)

(c)

20.6   The Cope Rearrangement

As we saw in Problem 20.13, not all shift reactions are [1,*x*] shifts. One very common sigmatropic shift reaction is called the **Cope rearrangement**, and is named for Arthur C. Cope (1909–1966, the same Cope of the Cope elimination, p. 914) who spent most of his professional life at MIT.[4] The Cope rearrangement is a [3,3] sigmatropic shift that occurs in almost any 1,5-hexadiene. In this sigmatropic shift the new bond is formed between the two number 3 atoms, which is the reason it is called a [3,3] shift (Fig. 20.49). In its simplest form, the Cope rearrangement is degenerate and must be revealed through a labeling experiment.

A [3,3] shift—a Cope rearrangement

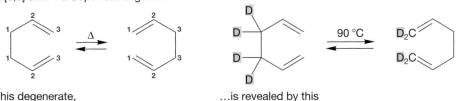

This degenerate, "invisible" reaction...

...is revealed by this labeling experiment

FIGURE 20.49   The degenerate Cope rearrangement, a [3,3] shift, can be revealed by a labeling experiment.

Like the [1,5] shifts we looked at first, the Cope rearrangement is intramolecular, uncatalyzed, and shows no strong dependence on solvent polarity. The activation energy for the Cope rearrangement is about 34 kcal/mol, a value far below what one would estimate for the simple bond breaking into two separated allyl

---

[4]Cope didn't actually discover this reaction; C. Hurd of Northwestern University (1897–1998) apparently did. It was Cope's research group that thoroughly studied the reaction, however.

radicals (Fig. 20.50). If the mechanism of the Cope rearrangement did involve two independent allyl radicals, the labeled 1,5-hexadiene would have to give two products, which it does not.

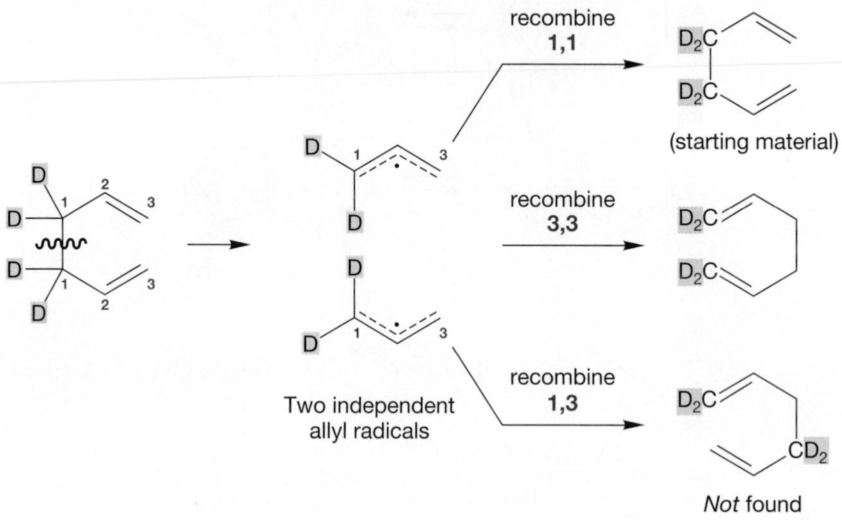

FIGURE 20.50 A labeling experiment shows that the Cope rearrangement does not involve formation of a pair of free allyl radicals because the radical recombination products are not all formed.

recombine 1,1 → (starting material)

recombine 3,3 →

recombine 1,3 → Not found

Two independent allyl radicals

The Cope rearrangement has a nonpolar transition state (Fig. 20.51), but this time it is not a simple hydrogen atom that migrates but a whole allyl radical. An analysis of the symmetry of the orbitals involved shows why this reaction is a relatively facile thermal process but is not commonly observed on photochemical activation. As we break the C(1)—C(1) bond, the phases of the overlapping lobes must be the same. The HOMO of the allyl radical is Φ_2, and that information allows us to fill in the symmetries of the two allyl groups making up the transition state.

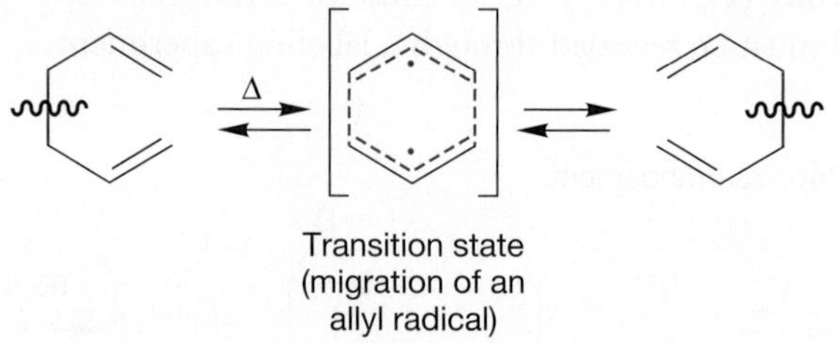

FIGURE 20.51 The transition state for the Cope rearrangement involves the migration of one allyl radical across another.

Transition state (migration of an allyl radical)

Reattachment at the two C(3) positions is allowed because the interaction of the two lobes on the two C(3) carbons is bonding. This picture is consistent with the near ubiquity of the Cope rearrangement when 1,5-dienes of all kinds are heated (Fig. 20.52).

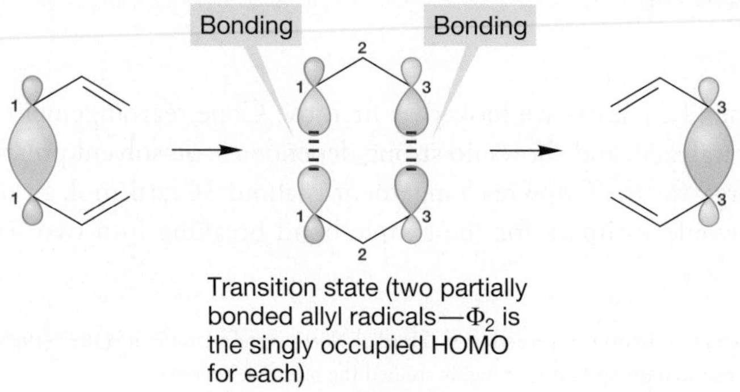

Bonding   Bonding

Transition state (two partially bonded allyl radicals—Φ_2 is the singly occupied HOMO for each)

FIGURE 20.52 An orbital symmetry analysis shows that the thermal Cope rearrangement is allowed.

WORKED PROBLEM 20.17 Use orbital symmetry to analyze the photochemical [3,3] shift in 1,5-hexadiene. Be careful! Photons are absorbed *one at a time*. It requires specialized equipment to form a singly excited molecule and then have it absorb a second photon. So we need only consider the singly excited state.

ANSWER The transition state for the [3,3] shift involves two interacting allyl radicals.

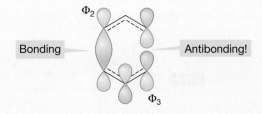

Transition state
for the [3,3] shift

If one electron is promoted on irradiation, the transition state must have one electron in Φ_3, the photochemical HOMO.

Now an orbital analysis shows that there is no easy path for the photochemical [3,3] shift. One antibonding interaction is always present.

Bonding Φ_2 Antibonding!

Φ_3

If a rotation were energetically feasible, if one C(3) carbon could rotate, a bonding interaction could be achieved where the figure notes the antibonding interaction. However, for these sp^2 hybridized carbons, this rotation involves a large energy cost.

So we now have a general picture of the transition state for the Cope rearrangement. But much more detail is possible if we work a bit at it. How might the two migrating allyl radicals of the transition state arrange themselves in space during the reaction? Two things are certain. First, the two C(1) atoms must start out close together, because they are bonded. Second, the two C(3) positions must also be within bonding distance, because in the transition state there is a partial bond between them (Fig. 20.52).

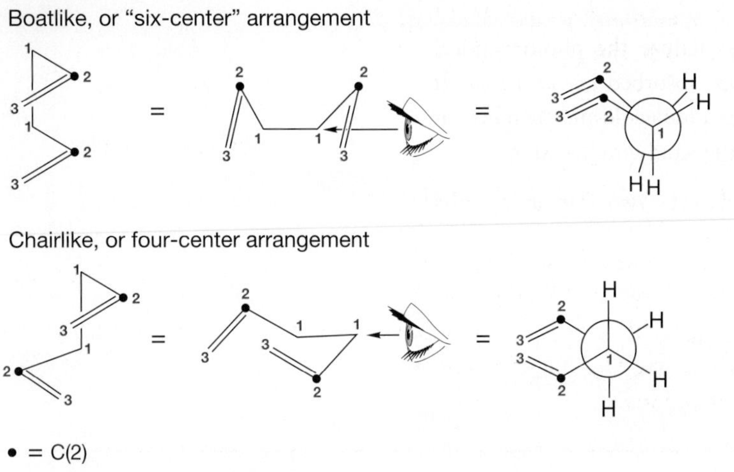

Boatlike, or "six-center" arrangement

Chairlike, or four-center arrangement

• = C(2)

FIGURE 20.53 Several views of the chairlike and boatlike arrangements for the transition state of the Cope rearrangement.

The question now comes down to determining the position of the two C(2) atoms in the transition state. Two possibilities are shown in Figure 20.53. The two C(2) atoms might be close together as in the boatlike transition state originally called the six-center transition state. Or, they can be rather far apart, as in the chairlike or four-center transition state. The figure shows a number of views of these two possibilities, but this is probably a time to get out the models.

In 1962, W. von E. Doering and W. R Roth carried out a brilliant experiment that determined which of these two possible transition states was favored for this [3,3] sigmatropic shift. They used a stereolabeled molecule, 3,4-dimethyl-1,5-hexadiene (Fig. 20.54).

Notice first that there are three forms of this molecule, a meso, achiral form and a racemic pair of enantiomers (p. 156). It all depends on how the methyl groups are attached.

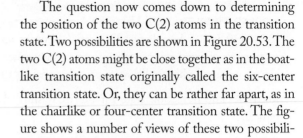

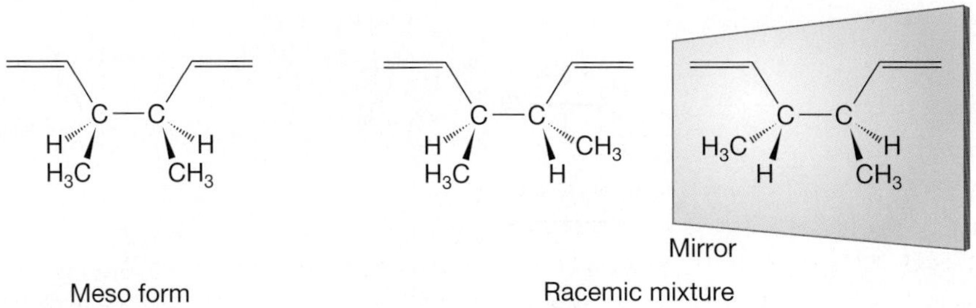

FIGURE 20.54 The meso and chiral forms of 3,4-dimethyl-1,5-hexadiene.

Meso form   Racemic mixture

CONVENTION ALERT   In the following figures, we have drawn only one enantiomer of the racemic pair. Be sure to keep in mind that the actual experiment used a 50:50 racemic mixture of the two enantiomers.

Both the meso and racemic forms of 3,4-dimethyl-1,5-hexadiene can adopt either the boat or chair arrangement. Figure 20.55 shows the results for the

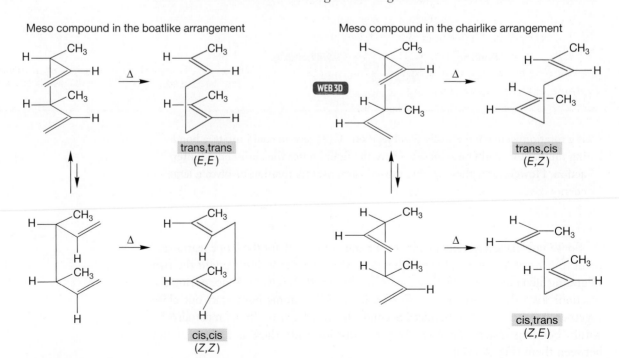

Meso compound in the boatlike arrangement

trans,trans
(E,E)

cis,cis
(Z,Z)

Meso compound in the chairlike arrangement

WEB 3D

trans,cis
(E,Z)

cis,trans
(Z,E)

FIGURE 20.55 The products formed from the boatlike and chairlike arrangements of *meso*-3,4-dimethyl-1,5-hexadiene.

meso compound. The key to Doering and Roth's experiment is that the chair and boat arrangements give different products for both the meso and chiral forms.

The results showed that it is the chairlike form that is favored, and that it is preferred to the boatlike arrangement by at least 6 kcal/mol. Presumably, some of the factors that make chair cyclohexane more stable than the boat form (p. 198) operate to stabilize the related chairlike transition state.

There are many things that are important about this work. First of all, it is a prototypal example of the way labeling experiments can be used to dig out the details of a reaction mechanism. The second reason is more subtle, but perhaps more important. As we will see in Section 20.7, the study of the Cope rearrangement leads directly to an extraordinarily intellectually creative idea.

Cope rearrangement

CHORISMATE TO PREPHENATE: A BIOLOGICAL COPE REARRANGEMENT

The Cope rearrangement is by no means restricted to the laboratory. Nature uses Cope-like processes as well. For example, in bacteria and plants a critical step in the production of the essential amino acids tyrosine and phenylalanine is the rearrangement of chorismate to prephenate, catalyzed by the enzyme chorismate mutase. At the heart of this rearrangement is a Cope-like process called the Claisen[5] rearrangement. The research groups of Jeremy Knowles (1935–2008) at Harvard and Glenn Berchtold (1932–2008) at MIT collaborated on a tritium ([3]H = T, a radioactive isotope of hydrogen) labeling experiment, related to that used by Doering and Roth, to determine that the enzymatic reaction also proceeds through a chairlike transition state. Notice that the chairlike transition state must take the compound with a (Z) double bond to a product in which the stereogenic carbon is (S), not (R). The conversion of chorismate into prephenate is a Cope-like rearrangement. The chair transition state is shown in the top reaction and a boat transition state is shown in the bottom reaction.

(Z) Double bond   This carbon is (S)

Chorismate   **Prephenate**
Chairlike transition
state for chorismate

(Z) Double bond   This carbon would be (R)

If this boatlike transition   ...the tritium-labeled
state were used...   carbon would be (R)

20.7   A Molecule with a Fluxional Structure

We spent some time in Section 20.6 working out the details of the structure of the transition state for the Cope rearrangement. This rearrangement is a most general reaction of 1,5-hexadienes, and appears in almost every molecule incorporating the 1,5-diene substructure. In this section we will examine the consequences of the Cope rearrangement and will work toward a molecule whose structure differs in a fundamental way from anything you have seen so far in this book. At room temperature it has no fixed structure; instead it has a **fluxional structure**, in which nearest neighbors are constantly changing. Along the way to this molecule there is some extraordinary chemistry and some beautiful insights. This section shows not only clever ideas and beautiful chemistry, but also how much fun this subject can be.

---

[5]The same Claisen of the Claisen condensation (p. 987) and the Claisen–Schmidt reaction (p. 984).

The simple 1,5-hexadiene at the heart of the Cope rearrangement can be elaborated in many ways, and in fact we've already seen one. If we add a central π bond we get back to *cis*-1,3,5-hexatriene (Fig. 20.56), a molecule we saw when we considered electrocyclic reactions (p. 1040). The arrow formalism of the Cope rearrangement leads to 1,3-cyclohexadiene.

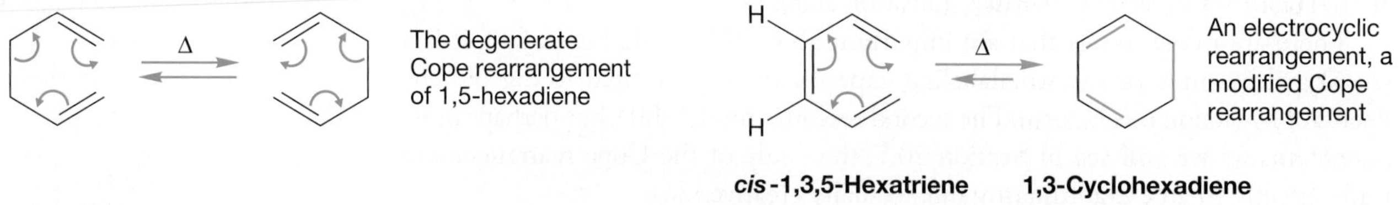

The degenerate
Cope rearrangement
of 1,5-hexadiene

An electrocyclic
rearrangement, a
modified Cope
rearrangement

cis-1,3,5-Hexatriene   1,3-Cyclohexadiene

FIGURE 20.56 The electrocyclic ring closure of *cis*-1,3,5-hexatriene to 1,3-cyclohexadiene is similar to a Cope rearrangement.

But there is a stereochemical problem. This reaction is only possible for the stereoisomer with a cis central double bond. In the trans isomer, the ends of the triene system are too far apart for bond formation (Fig. 20.57a). That's an easy concept, and it appears again in a slightly different modified 1,5-hexadiene: 1,2-divinylcyclopropane (Fig. 20.57b). The Cope rearrangement is possible in this molecule and here, too, it is only the cis molecule that is capable of rearrangement. In the trans compound, the ends cannot reach.

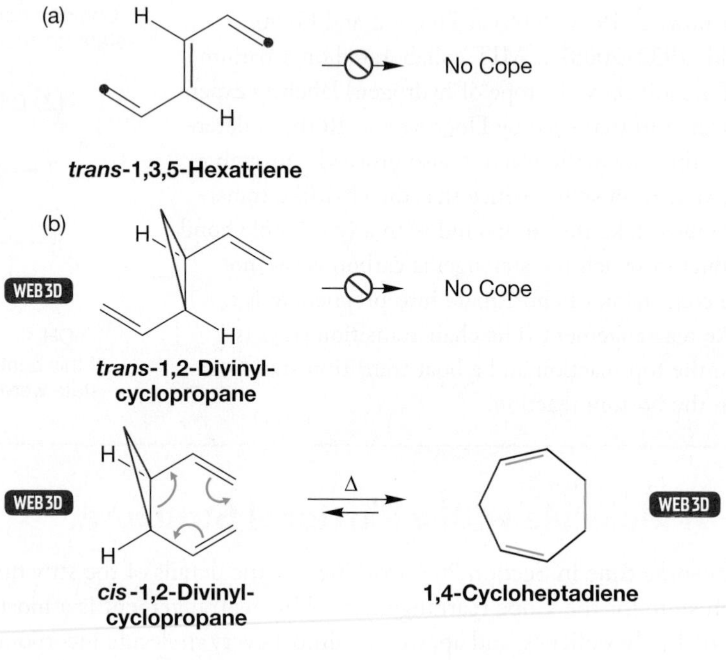

FIGURE 20.57 (a) *trans*-1,3,5-Hexatriene cannot undergo this rearrangement because the ends of the molecule (dots), where bonding must occur, are too far apart. (b) *trans*-1,2-Divinylcyclopropane cannot undergo the Cope rearrangement, but the cis isomer can.

(a)

No Cope

trans-1,3,5-Hexatriene

(b)

No Cope

trans-1,2-Divinyl-
cyclopropane

cis-1,2-Divinyl-
cyclopropane

1,4-Cycloheptadiene

PROBLEM 20.18 The Cope rearrangement of *cis*-1,2-divinylcyclopropane is very easy. It is rapid at room temperature. Why should it be so much easier than the other versions we have seen?

There is another stereochemical problem in the reaction of *cis*-1,2-divinylcyclopropane that is a bit more subtle. The most favorable arrangement for *cis*-1,2-divinylcyclopropane is shown in Figure 20.58. The two vinyl groups are directed away from the ring, and in particular, away from the cis ring hydrogen. Yet this extended form cannot undergo the Cope rearrangement at all! The less stable, coiled form must be adopted where the two vinyl groups point back toward the ring and are closer to the offending cis hydrogen.

The problem is that the extended form must produce a 1,4-cycloheptadiene product with two trans double bonds. Although the paper puts up with these trans double bonds without protest, the molecule will not. Remember, trans double bonds in small rings are most unstable. *trans*-Cycloheptene is a barely detectable reactive intermediate that can be intercepted only at low temperature. The constraints of the ring lead to severe twisting about the π bond and destabilization. A seven-membered ring containing two trans double bonds is unthinkably unstable.

By contrast, the higher energy coiled form yields a seven-membered ring containing two cis double bonds, and that presents no problem. Be certain you see why the two forms give different stereochemistries. Figure 20.59 shows the Energy versus Reaction progress diagram for the Cope rearrangement of *cis*-1,2-divinylcyclopropane.

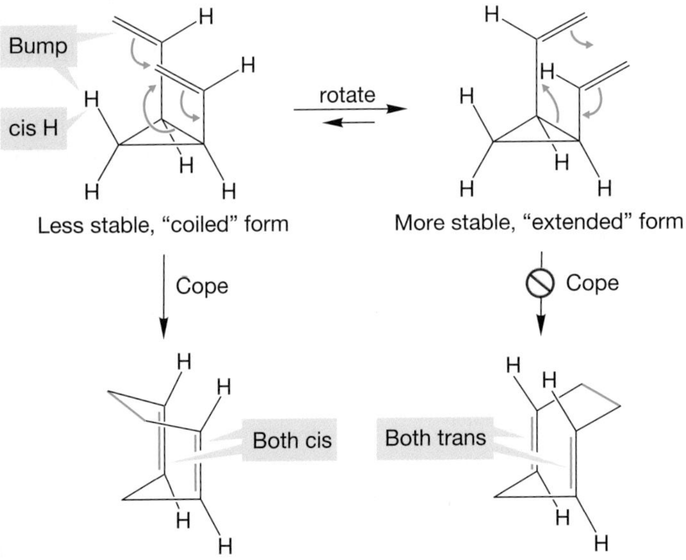

FIGURE 20.58 Only the less stable, coiled arrangement of *cis*-1,2-divinylcyclopropane can undergo the Cope rearrangement. The more stable, extended form must produce a hideously unstable *trans,trans*-cyclohepta-1,4-diene.

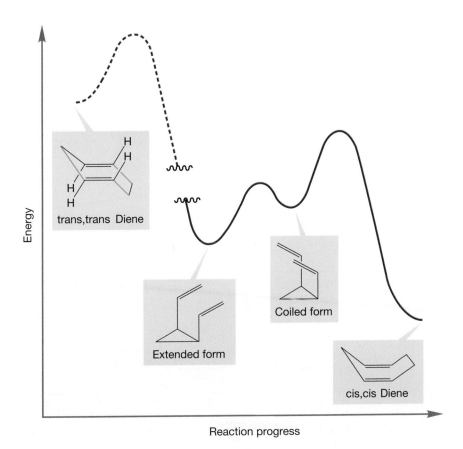

FIGURE 20.59 The energy diagram for the Cope rearrangement of *cis*-1,2-divinylcyclopropane.

Now let's make another addition to the 1,5-diene system. Let's add a methylene group joining the two double bonds to give the molecule named "homotropilidene." We can still draw the arrow formalism, and the Cope rearrangement still proceeds rapidly. As Figure 20.60 shows, we are back to a degenerate Cope rearrangement, because the starting material and product are the same.

FIGURE 20.60 The degenerate Cope rearrangement of homotropilidene.

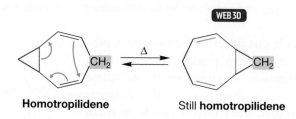

WEB 3D

Homotropilidene   **Still homotropilidene**

PROBLEM 20.19 The rapid Cope rearrangement of homotropilidene led to problems for W. R Roth, who first made the molecule. Can you explain in general terms the three perplexing ^{1}H NMR spectra for this molecule that are shown? Note that both low- and high-temperature spectra have sharp peaks, whereas at an intermediate temperature the spectrum consists of two "blobs." You are not required to analyze the spectra in detail—just give the general picture of what must be happening.

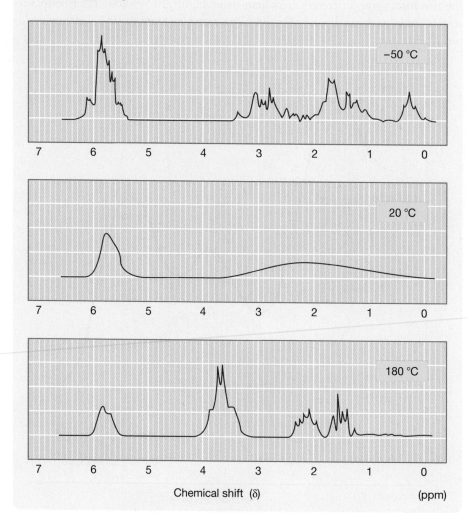

There are also stereochemical problems for homotropilidene (Fig. 20.61). Like 1,2-divinylcyclopropane, the most stable extended form of homotropilidene is unable to undergo the Cope rearrangement, and it must be the higher-energy coiled arrangement that is active in the reaction. Here the destabilization of the coiled arrangement is especially obvious, because the two methylene hydrogens badly oppose each other. As before, the extended form must give an unstable molecule with a pair of trans double bonds incorporated into a seven-membered ring. The less stable, coiled arrangement has no such problem, because it can give a product containing two cis double bonds.

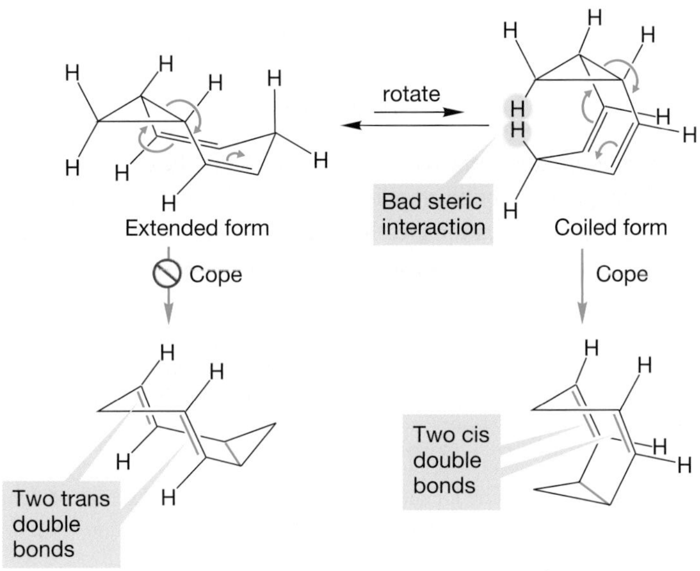

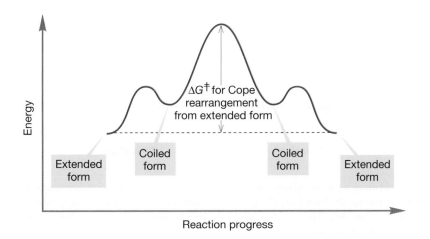

FIGURE 20.61 Only the higher-energy, coiled form of homotropilidene can give a stable product in the Cope rearrangement. The lower-energy, extended form would produce a product containing two trans double bonds in a seven-membered ring.

Although *cis*-1,2-divinylcyclopropane and homotropilidene undergo the Cope rearrangement quite rapidly even at room temperature and below, one might consider how to make the reaction even faster. For the reaction to occur, an unfavorable equilibrium between the more stable but unproductive extended form, and the less stable but productive coiled form must be overcome. The activation energy for the reaction includes this unfavorable equilibrium (Fig. 20.62).

FIGURE 20.62 An Energy versus Reaction progress diagram for the Cope rearrangement of homotropilidene. The activation energy ($\Delta G^{\ddagger}$) for the reaction includes the energy difference between the extended and coiled forms.

If this unfavorable equilibrium could be avoided in some way, and the molecule locked into a coiled arrangement, the activation energy would be lower, and the reaction faster (Fig. 20.63).

FIGURE 20.63 An Energy versus Reaction progress diagram for the Cope rearrangement of a homotropilidene locked into the coiled arrangement. The activation energy is lower than in Figure 20.62 because it is no longer necessary to fight an unfavorable equilibrium from the extended form, shown here as a phantom (dashes) just to remind you of the previous figure.

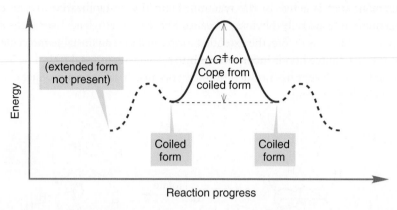

Our next variation of the Cope rearrangement includes the clever notion of bridging the two methylene groups of homotropilidene to lock the molecule in a coiled arrangement (Fig. 20.64). This modification greatly increases the rate of the Cope rearrangement

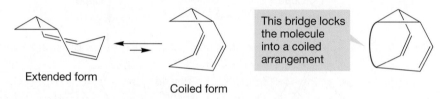

FIGURE 20.64 A homotropilidene locked into the coiled arrangement can be achieved if the two methylene groups are connected with a bridge.

because being locked in the coiled form decreases the activation energy (Fig. 20.63). No longer is it necessary to fight an unfavorable equilibrium in order for the Cope rearrangement to occur; the coiled form is built into the structure of the molecule. Many bridged homotropilidenes are now known, and some are shown in Figure 20.65.

Now bridging is a clever idea, and it has led to some fascinating molecules and some very nice chemistry. But it's only a good idea. What comes next is an extraordinary idea; one of those insights that anyone would envy, and which define through their example what is meant by creativity. Suppose we bridge the two methylene groups with a double bond to give the molecule known as **bullvalene**.[6] At first, this molecule seems like nothing special. As Figure 20.66 shows, the Cope rearrangement is still possible in this highly modified 1,5-hexadiene.

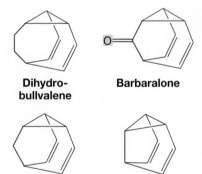

Dihydro-bullvalene   **Barbaralone**

Barbaralane   **Semibullvalene**

FIGURE 20.65 Four bridged homotropilidenes. All undergo very rapid Cope rearrangements.

FIGURE 20.66 If the new bridge is a carbon–carbon double bond, the resulting molecule is known as bullvalene.

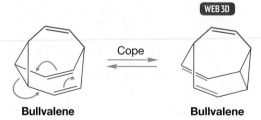

Bullvalene   **Bullvalene**

---

[6]What an odd name! If you are curious about its origin, see Alex Nickon and Ernest F. Silversmith's nice book on chemical naming, *The Name Game* (Pergamon, New York, 1987). You won't find the systematic naming scheme in there, but you will find hundreds of stories about the origins of the names chemists give to their favorite molecules, including bullvalene.

But now perhaps you notice the threefold rotational axis in bullvalene. Remove the color and the three double bonds become indistinguishable. In fact, there is a Cope rearrangement possible in each of the three faces of the molecule, and every Cope rearrangement is degenerate; each re-forms the same molecule (Fig. 20.67). This point is much easier to see if you use a model.

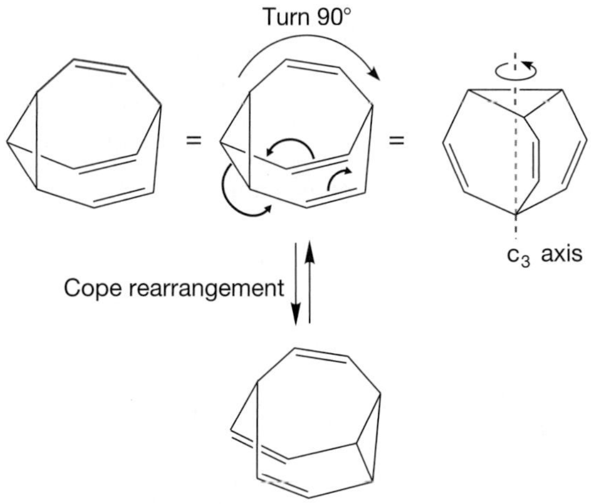

Turn 90°

= =

Cope rearrangement

c_3 axis

FIGURE 20.67 Bullvalene has a threefold rotational axis: All three double bonds are equivalent. The Cope rearrangement can occur in each of the three faces of the molecule and is degenerate in every case.

That's quite interesting, but now begin to look at the structural consequences for bullvalene. As Cope rearrangements occur, *the carbons begin to move apart*! This wandering of atoms is shown in Figure 20.68 using dots to track two arbitrary adjacent carbons. The figure only begins the process, but it can be demonstrated that the Cope rearrangement is eventually capable of completely scrambling the 10 CH units of the molecule.[7] As long as the Cope rearrangement is fast, bullvalene *has no fixed structure*; no carbon has permanent nearest neighbors, but is instead bonded on time average to each of the 9 other methine carbons. It has a fluxional structure.

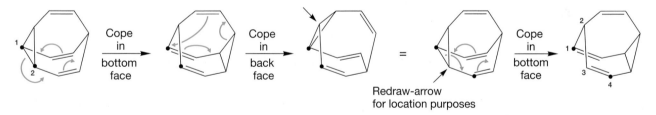

FIGURE 20.68 A sequence of Cope rearrangements serves to move carbon atoms throughout the molecule.

WORKED PROBLEM 20.20 At somewhat above room temperature the ^{1}H NMR spectrum of bullvalene consists of a single broad line at δ 4.2 ppm. Explain.

ANSWER If all the carbon–hydrogen bonds are the same on time average, the NMR spectrum must consist of a single line. We see the averaged spectrum.

---

[7]In an early paper, Doering noted that a sequence of 47 Cope rearrangements sufficed to demonstrate that two adjacent carbons could be exchanged without disturbing the rest of the 10-carbon sequence, and that therefore all 1,209,600 possible isomers could be reached. No claim was made that this represented the minimum path. A later computer-aided effort by two Princeton undergraduates (Allan Fisher and Karl Bennett) located a sequence of 17 Cope rearrangements that could do the same thing.

WORKED PROBLEM 20.21 Explain why the ketone below exchanges *all* its hydrogens for deuteriums when treated with deuterated base.

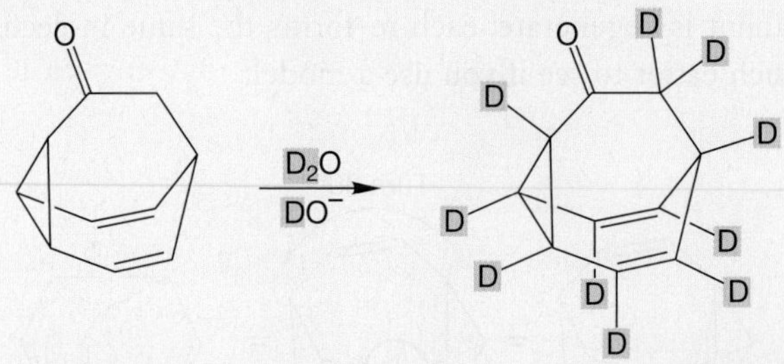

ANSWER Exchange at the α position is routine.

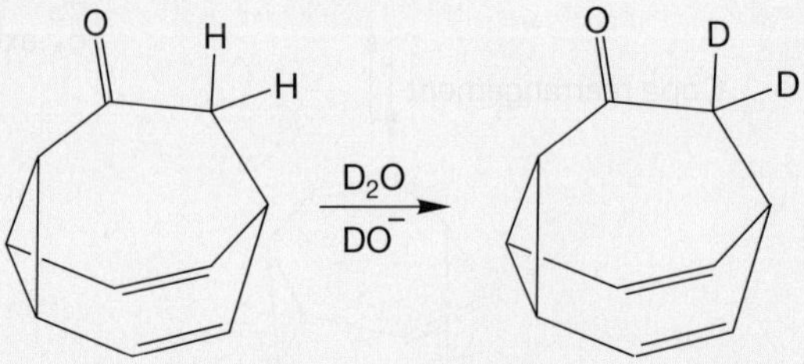

However, simple ketones are in equilibrium with their enol forms, and the enol of this molecule is a substituted bullvalene. The Cope rearrangement makes all the carbons equivalent, and thus, on time average, all carbon–hydrogens will eventually come to occupy the exchangeable α position.

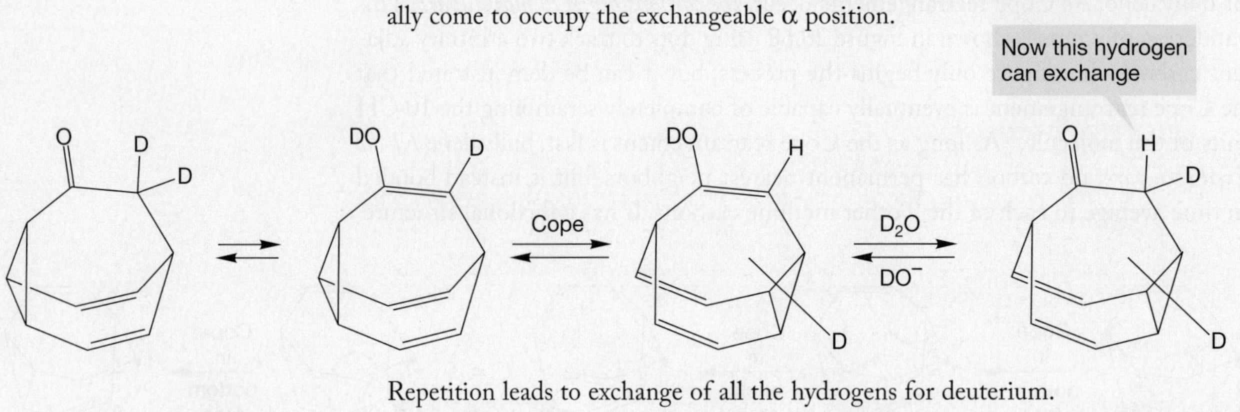

Now this hydrogen can exchange

Repetition leads to exchange of all the hydrogens for deuterium.

Naturally, efforts at synthesis began as soon as the prediction of the properties of bullvalene appeared in the chemical literature. Long before the efforts at a rational synthesis were sucessful, the molecule was discovered by Gerhard Schröder (b. 1929), a German chemist then working in Belgium. Schröder was working out the structures of the several dimers of cyclooctatetraene, and in the course of this effort irradiated one of the dimers. It conveniently fell apart to benzene and bullvalene (Fig. 20.69).

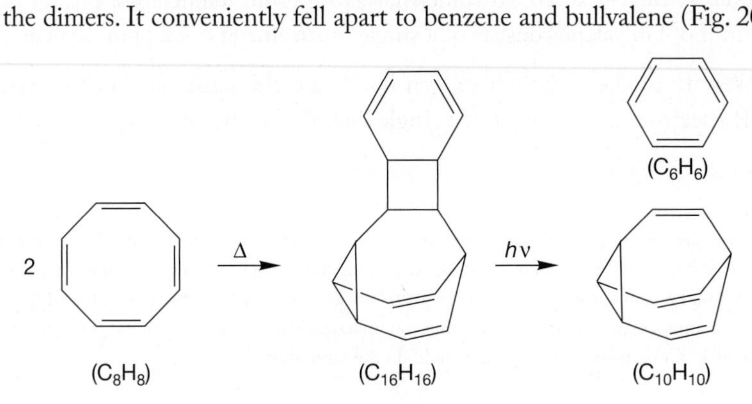

FIGURE 20.69 Schröder's accidental synthesis of bullvalene.

PROBLEM 20.22 Why is the second step of Figure 20.69, the photochemical fragmentation, so easy?

Schröder was alert enough to understand what had happened, and went on to a highly successful career that included an examination of the properties of bullvalene and substituted versions of this molecule. Doering's predictions were completely confirmed by Schröder's synthesis. Other neutral, completely fluxional organic molecules have not appeared, although the phenomenon of fluxionality appears to be rather more common in organic cations and organometallic compounds.

PROBLEM 20.23 Provide mechanisms for the following reactions:

(a)

(b)

Fluxional molecules are the result of electrons and atoms moving much like the blur of a spinning skater.

Summary

The Cope rearrangement, a [3,3] sigmatropic shift, is far more than just a chemical curiosity. It is a general reaction of almost all 1,5-hexadienes, and occurs almost no matter how the skeletal hexadiene form is perturbed and disguised. All sorts of chemical "spinach" can be added, often making it quite difficult to find the bare bones of the hexadiene system, but the reaction will persist. Moreover, as we have just seen, digging deep into the Cope reaction, following the intellectual implications of substitution, leads to the marvelous molecule bullvalene, without precedent in organic chemistry.

20.8 How to Work Orbital Symmetry Problems

PROBLEM SOLVING

All people and all problems are, of course, different. Nonetheless, there are some general principles that apply, and it is worth summarizing ways to approach—and hopefully solve—orbital symmetry problems, which are separable into three varieties.

20.8a Electrocyclic Reactions

1. Look for rings opening or closing within a single molecule. Sound like simple advice? Yes, it is, but in all these problems the major difficulty is one of recognition, and it is well worth your time to be a little "too straightforward," and to "think simple."

2. Once you are sure that the question involves an electrocyclic reaction, draw an arrow formalism.

3. Next, work out the molecular orbitals *of the open-chain partner*.

4. Count the number of electrons involved in the reaction. Be careful to take account of any charges.

5. Knowing the number of electrons allows you to find the HOMO of the open-chain partner.

6. Finally, work out the motion, conrotatory or disrotatory, that allows the ring to close through a bonding—not antibonding—interaction. If there are $4n$ electrons involved in the reaction, then the thermal process will be conrotatory and the photochemical process will be disrotatory. If there are $4n + 2$ electrons, then the thermal process will be disrotatory and the photochemical will be conrotatory.

20.8b Cycloaddition Reactions

1. Look for a ring forming from two molecules or breaking open to give two new molecules. When you find this situation, you know you have a cycloaddition reaction.

2. Draw an arrow formalism.

3. Work out the π molecular orbitals for the two partners.

4. Identify the HOMOs and LUMOs.

5. Examine either possible HOMO–LUMO interaction. It is generally easier to look at ring closing, rather than ring opening. Your final answer will be good for either direction (of course!).

6. Count the number of electrons involved in the reaction.

7. If there are $4n$ electrons involved, then the thermal process will require a rotation in order to occur and the photochemical process will not. If there are $4n + 2$ electrons involved, then the thermal process will not require a rotation and the photochemical will.

20.8c Sigmatropic Shift Reactions   These reactions are by far the hardest for students to "get."

1. The major difficulty is one of recognition. Is some piece of the molecule flying around, departing from one place and reattaching somewhere else? If so, that's probably a sigmatropic shift reaction. Once you are sure it is a shift, identify—mark—the shifting atom or group.

2. Where does it depart from? Where does it arrive? Draw an arrow formalism.

3. Now let's work out a picture of the transition state. Break the bond attaching the migrating group to the position from which it departs in a *homolytic* fashion. Neither ions nor polar transition states are involved. Ask yourself what species is left behind? What species is migrating?

4. Work out the molecular (or atomic) orbitals for both partners.

5. Remember that the migrating group was attached through a bonding interaction. That fact allows you to sketch in the symmetry of the HOMO of the portion of the molecule along which the atom or group is migrating.

6. Can the migrating group reattach reasonably (can it reach?) in a bonding fashion? If so, the reaction is allowed. If it must reattach through an antibonding interaction, the reaction is forbidden.

7. Count the number of electrons involved in the reaction.

8. If there are 4*n* electrons involved, then the thermal hydrogen (or other) shift must be antarafacial in order to occur and the photochemical process will occur in a suprafacial fashion. If there are 4*n* + 2 electrons, then the thermal process will be suprafacial and the photochemical process will be antarafacial.

Now practice! Use the schemes above to work the following three problems.

PROBLEM 20.24 Is the following reaction allowed or disallowed? Give a detailed analysis.

PROBLEM 20.25 Is the following thermal reaction allowed or disallowed? Give a detailed analysis.

PROBLEM 20.26 Is the following thermal reaction allowed suprafacially or antarafacially? Give a detailed analysis.

20.9 Summary

New Concepts

The central insight of Woodward–Hoffmann theory is that one must keep track of the phase relationships of orbital interactions as a concerted reaction proceeds. If bonding overlap can be maintained throughout, the reaction is allowed; if not, the reaction is forbidden by orbital symmetry. This notion leads to a variety of mechanistic explanations for reactions that can otherwise be baffling. Some of these pericyclic reactions are discussed in the section on Reactions, Mechanisms, and Tools.

The notion of a fluxional structure is introduced through the $(CH)_{10}$ molecule bullvalene. In this molecule, a given carbon atom does not have fixed nearest neighbors, but instead is bonded, over time, to each of the other nine carbons in the molecule. In bullvalene, the fluxionality is the result of rapid degenerate Cope rearrangements.

Many stereochemical labeling experiments are described in this chapter. A careful labeling experiment is able to probe deeply into the details of a reaction mechanism. Examples include W. R Roth's determination of the suprafacial nature of the [1,5] shift of hydrogen and Doering and Roth's experiments on the structure of the transition state for the [3,3] sigmatropic shift known as the Cope rearrangement.

Key Terms

antarafacial motion (p. 1055)	degenerate reaction (p. 1031)	sigmatropic shift (p. 1050)
bullvalene (p. 1068)	disrotation (p. 1037)	suprafacial motion (p. 1055)
conrotation (p. 1037)	electrocyclic reaction (p. 1036)	Woodward–Hoffmann theory (p. 1031)
Cope rearrangement (p. 1059)	fluxional structure (p. 1063)	
cycloaddition reaction (p. 1043)	pericyclic reaction (p. 1031)	

Reactions, Mechanisms, and Tools

Certain polyenes and cyclic compounds can be interconverted through a pericyclic process known as an electrocyclic reaction. Examples include the 1,3-butadiene–cyclobutene and 1,3-cyclohexadiene–1,3,5-hexatriene interconversions (Figs. 20.5 and 20.16).

Orbital symmetry considerations dictate that in $4n$-electron reactions the thermal process must use a conrotatory motion, whereas the photochemical reaction must be disrotatory. Just the opposite rules apply for reactions involving $4n + 2$ electrons. The key to analyzing electrocyclic reactions is to look at the way the p orbitals at the end of the open-chain π system must move in order to generate a bonding interaction in the developing σ bond.

Cycloaddition reactions are closely related to electrocyclic reactions. The phases of the lobes in the HOMOs and LUMOs must match so that bonding interactions are preserved in the transition state for the reaction. Sometimes a straightforward head-to-head motion is possible in which the two π systems approach each other in parallel planes to produce two bonding interactions. Such reactions are typically easy and are quite common. An example is the Diels–Alder reaction (Fig. 20.21).

In other reactions, the simple approach of HOMO and LUMO involves an antibonding overlap. In such reactions, a rotation is required in order to bring the lobes of the same sign together (Fig. 20.28). Although in principle this rotation maintains bonding interactions, in practice it requires substantial amounts of energy, and such reactions are rare.

Sigmatropic shifts involve the migration of hydrogen or another atom along a π system. Allowed migrations can be either suprafacial (leave and reattach from the same side of the π system) or antarafacial (leave from one side and reattach from the other) depending on the number of electrons involved in the migration. The important point is that in an allowed shift, bonding overlap must be maintained at all times both to the lobe from which the migrating group departs and to that to which it reattaches. Steric considerations can also be important, especially for hydrogen, which must migrate using a small $1s$ orbital (Figs. 20.43 and 20.44).

Syntheses

The photochemical 2 + 2 dimerization of alkenes.

Common Errors

For many students the big problem in this material is determining the kind of reaction involved. The way to start a problem in this area is to spend some time and thought identifying what kind of reaction is taking place. Cycloadditions are generally easy to find, although even here confusion exists between these reactions and electrocyclic processes. Sigmatropic shifts can be even harder to uncover. Often it is not easy to identify exactly what has happened in a sigmatropic reaction. When one atom or a group of atoms has translocated from one part of the molecule to another, there is often a rather substantial structural change. The product sometimes doesn't look much like the starting material. The temptation is to use other reagents to make the change, but in a sigmatropic shift everything is "in house."

20.10 Additional Problems

PROBLEM 20.27 Analyze the photochemical and thermal [1,7] shifts of hydrogen in 1,3,5-heptatriene. Be sure to look at all possible stereoisomers of the starting material. What kind of [1,7] shift will be allowed thermally? What kind photochemically? How will the starting stereochemistry of the triene affect matters? Top views of the π molecular orbitals of heptatrienyl are given.

$$H_2C{=}CH{-}CH{=}CH{-}CH{=}CH{-}CH_3$$

1,3,5-Heptatriene

Top views of the π molecular orbitals of heptatrienyl (+ means blue, − means green)

PROBLEM 20.28   Vitamin D₃ (**3**) is produced in the skin as a result of UV irradiation. It was once believed that 7-dehydrocholesterol (**1**) was converted directly into **3** upon photolysis. It is now recognized that there is an intermediate, previtamin D₃ (**2**), involved in the reaction. This metabolic process formally incorporates two pericyclic reactions, **1 → 2** and **2 → 3**. Identify and analyze the two reactions.

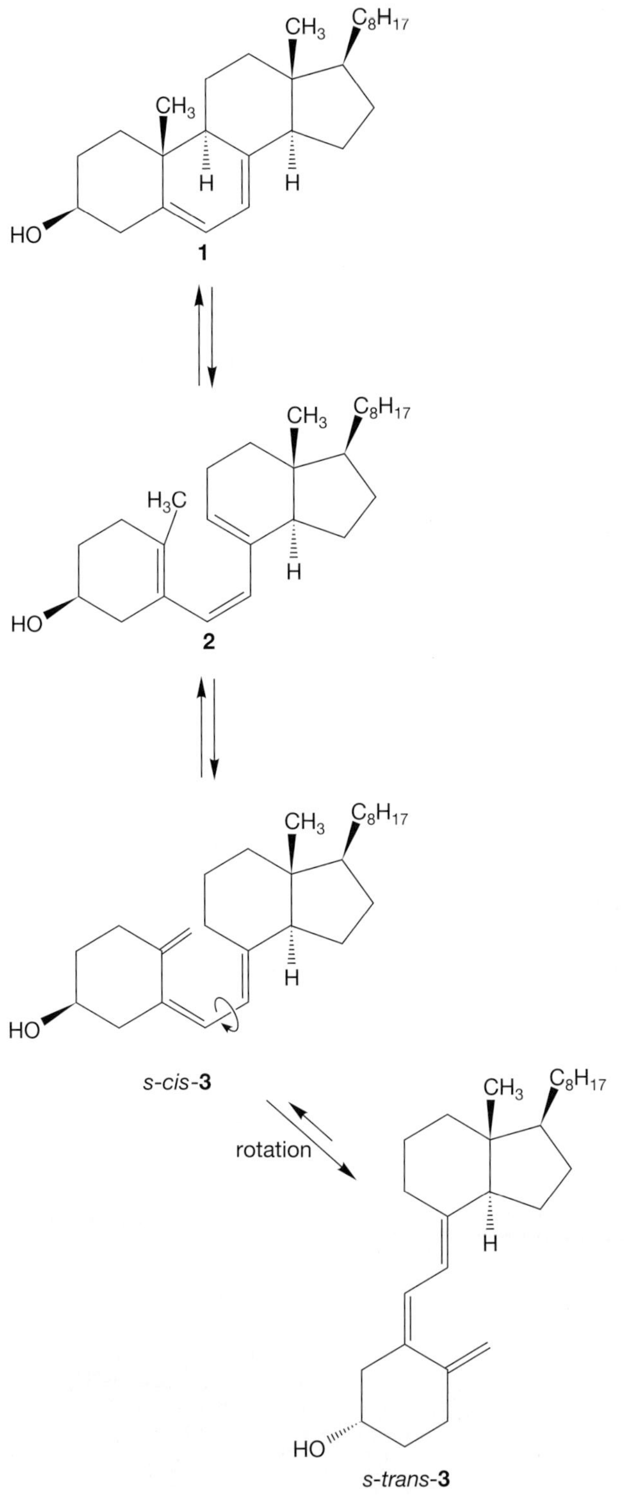

PROBLEM 20.29   The thermally induced interconversion of cyclooctatetraene (**1**) and bicyclo[4.2.0]octa-2,4,7-triene (**2**) can be described with two different sets of arrows. Although these two arrow formalisms may seem to give equivalent results, one of them is, in fact, not a reasonable depiction of the reaction. Explain.

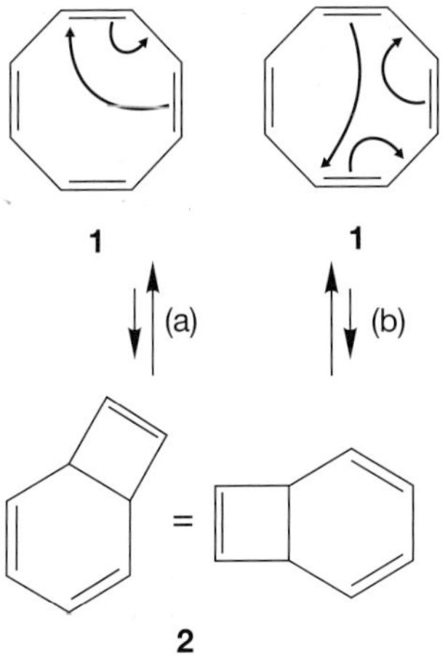

PROBLEM 20.30   Analyze the thermal ring opening to allyl ions of the two cyclopropyl ions shown below. What stereochemistry do you expect in the products? Will the reactions involve conrotation or disrotation? Explain. Do not be put off by the charges! They only serve to allow you to count electrons properly.

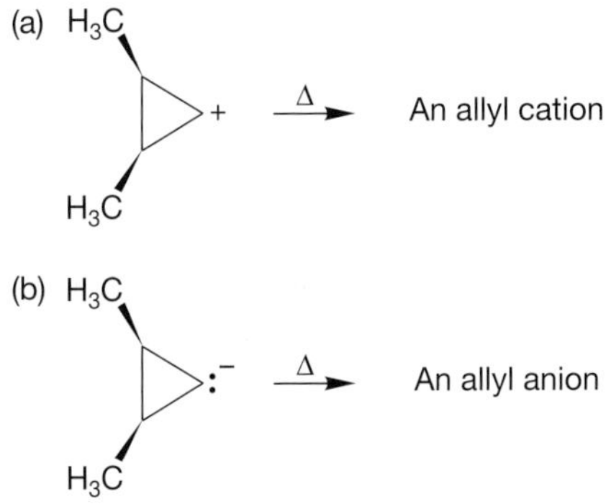

(a)   H₃C ... + → Δ → An allyl cation

(b)   H₃C ... :⁻ → Δ → An allyl anion

PROBLEM 20.31   Orbital symmetry permits one of the two following photochemical cycloaddition reactions to take place in a concerted fashion. What would be the products, which reaction is the concerted one, and why?

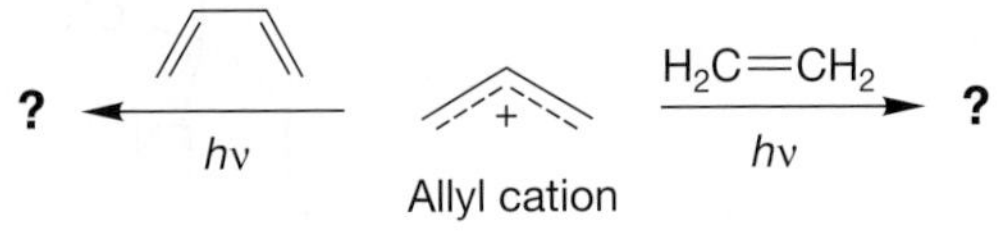

? ← hν ← Allyl cation → hν → H₂C=CH₂ ?

PROBLEM 20.32   The following two cyclobutenes can be opened to butadienes thermally. However, they react at very different rates. Which is the fast one, which is the slow one, and why?

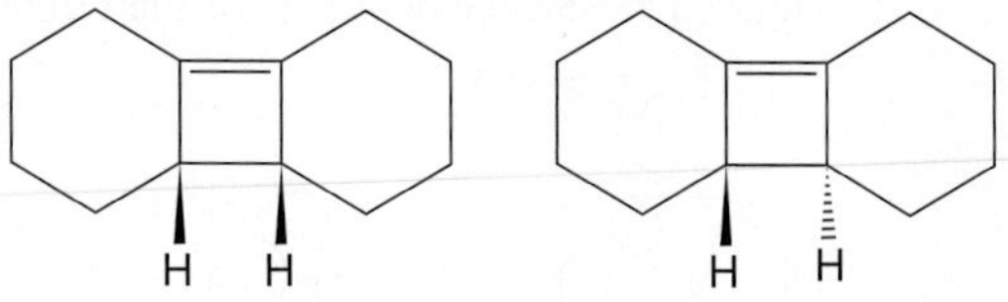

PROBLEM 20.33   These same two cyclobutenes (Problem 20.32) react photochemically to give two monocyclic (one ring) compounds isomeric with starting material. Predict the products, and explain the stereospecificity of the reaction.

PROBLEM 20.34   Benzocyclobutene (**1**) reacts with maleic anhydride when heated at 200 °C to give the single product shown. Write a mechanism for this reaction.

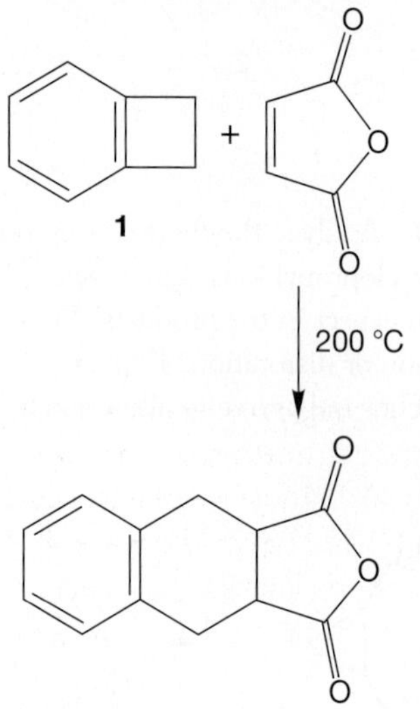

PROBLEM 20.35   Given your answer to Problem 20.34, explain the stereochemistry observed in the following reaction:

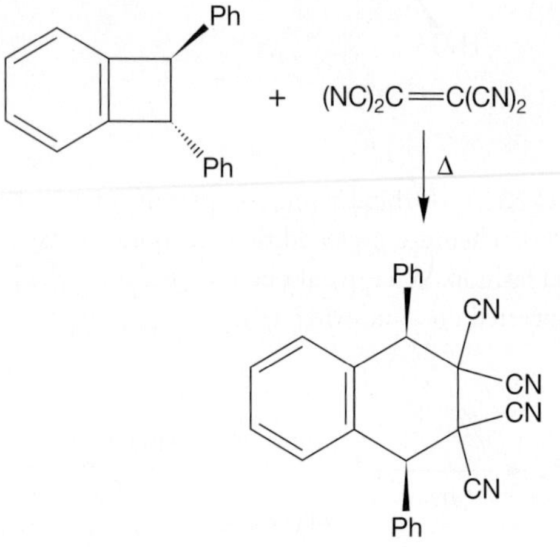

PROBLEM 20.36   Write a mechanism for the following reaction. Note that no similar reaction appears for *cis*-1,3,5-hexatriene. It may be useful to work backward from the product. What molecules must lead to it?

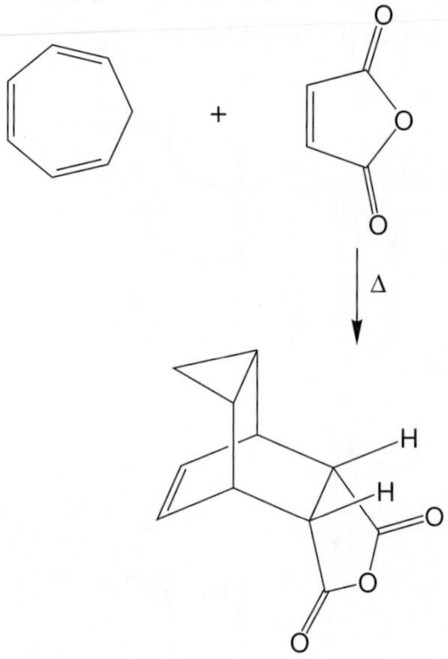

PROBLEM 20.37   Now, here's an interesting variation on the reaction of Problem 20.36. Diels–Alder additions typically take place as shown in Problem 20.36, with the ring-closed partner (**A** in the answer to Problem 20.36) actually doing the addition reaction. However, the molecule shown below is an exception, as the Diels–Alder reaction takes place in the normal way to give the product shown. Draw an arrow formalism for the reaction and then explain why **1** reacts differently from other cycloheptatrienes.

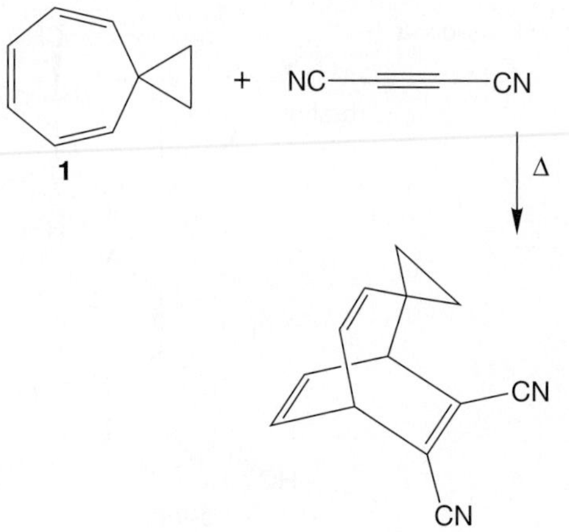

PROBLEM 20.38   Write an arrow formalism mechanism for the following thermal reaction and explain the specific trans stereochemistry. Careful here—note the acid catalyst for this reaction.

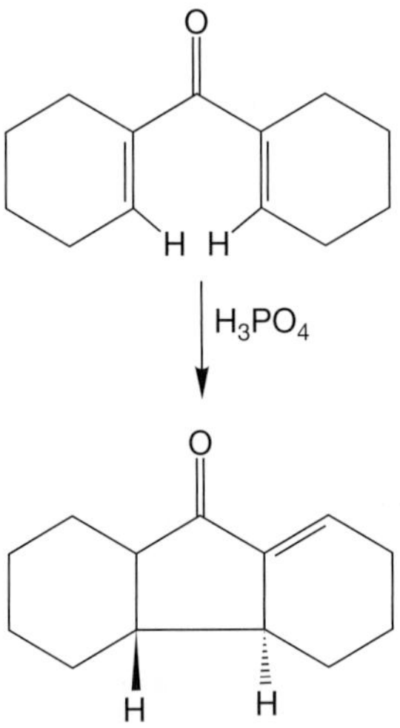

PROBLEM 20.39   Write an arrow formalism mechanism for the following reaction and explain the observed trans stereochemistry:

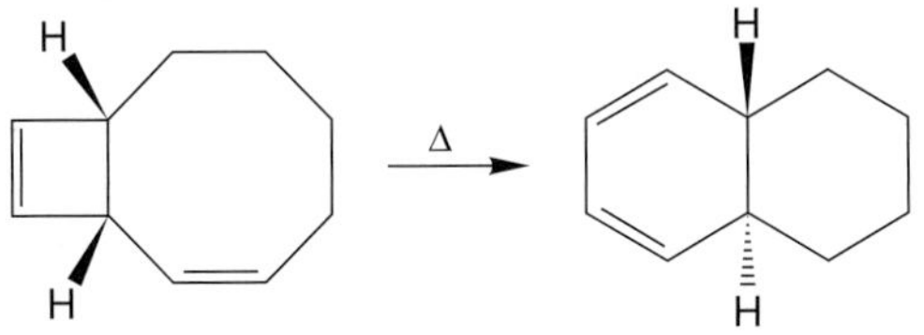

PROBLEM 20.40   Explain the following differences in product stereochemistry:

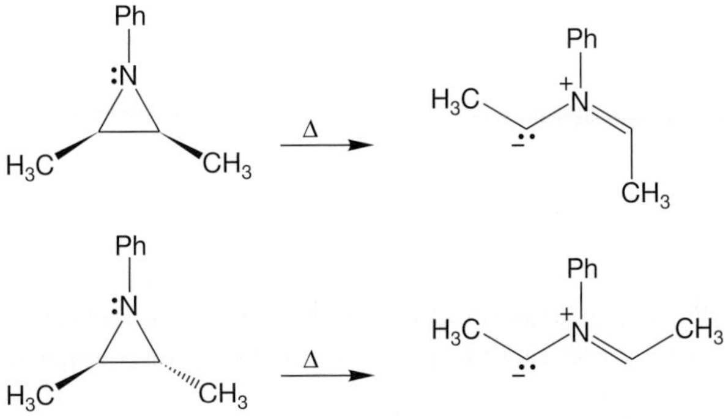

PROBLEM 20.41   Provide mechanisms for the following transformations:

(a)

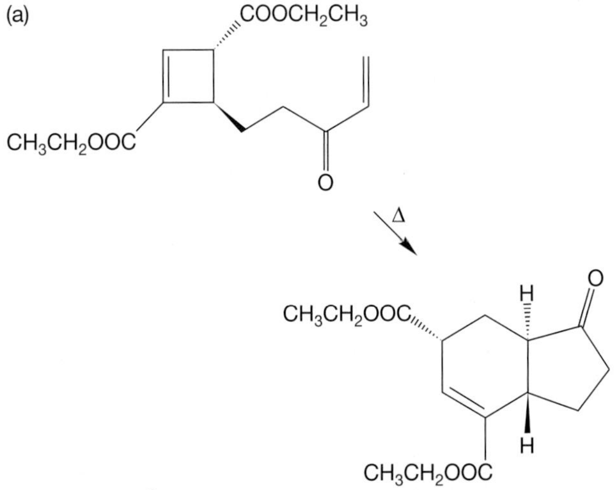

(b)

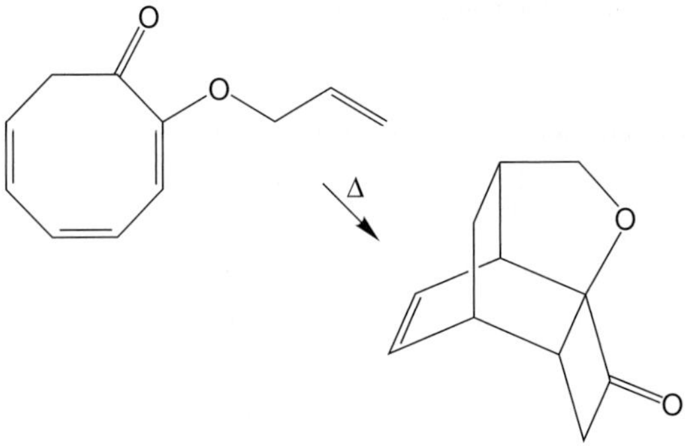

PROBLEM 20.42   Provide mechanisms for the following transformations. Part (a) is harder than it looks (*Hint*: There are two intermediates in which the three-membered ring is gone) and (b) is easier than it looks (*Hint*: What simple photochemical sigmatropic shift do you know?).

(a)

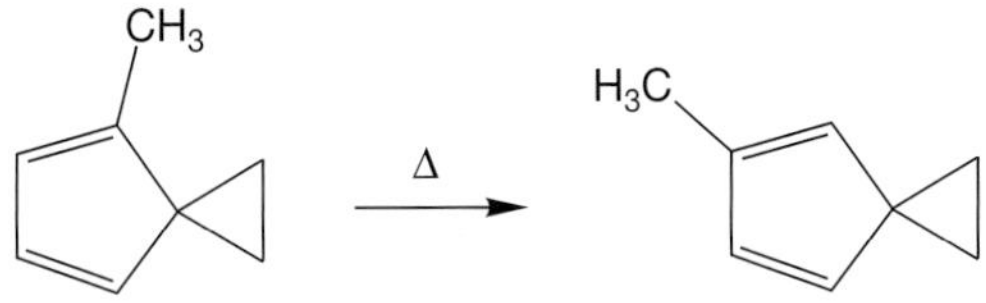

(b)

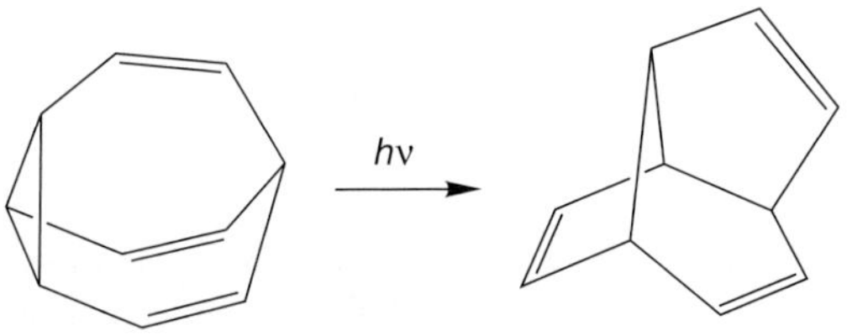

PROBLEM 20.43   In Problem 20.23 (b), we saw an example of a [3,3] sigmatropic shift known as the Claisen rearrangement. This reaction involves the thermally induced intramolecular rearrangement of an allyl phenyl ether to an *o*-allylphenol.

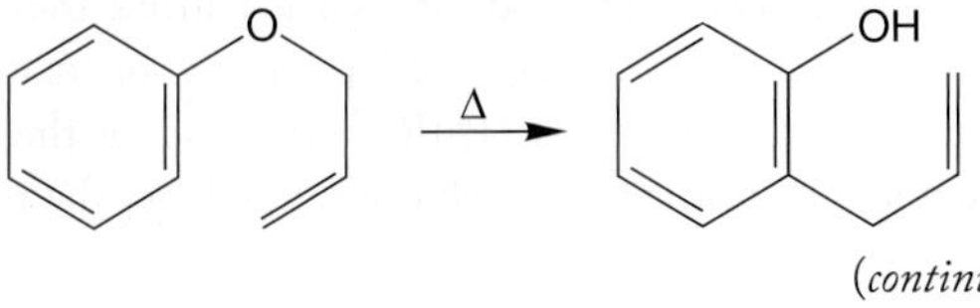

(continued)

(a) It might be argued that this rearrangement is not a concerted process, but instead involves a pair of radical intermediates. Write such a mechanism and design a labeling experiment that would test this hypothesis.

(b) If both the ortho positions of the allyl phenyl ether are blocked with methyl groups, the product of this rearrangement is the *p*-allylphenol. This version of the Claisen reaction is often called the para-Claisen rearrangement.

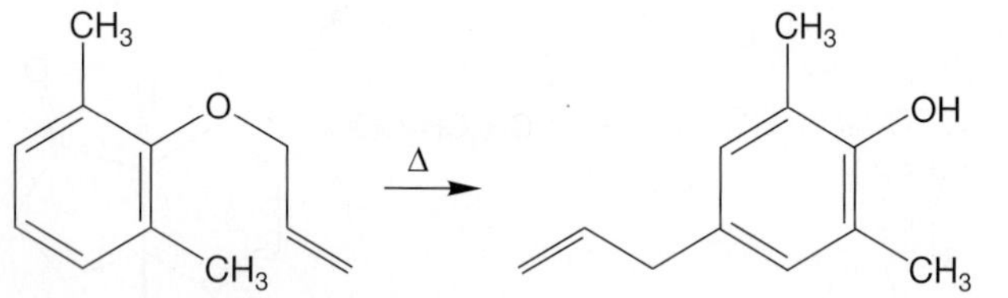

Explain why the rearrangement takes a different course in this case. In addition, would the label you suggested in (a) be useful in this case for distinguishing concerted from non-concerted (radical) mechanisms?

PROBLEM 20.44   Propose an arrow formalism mechanism for the following reaction. Be sure that your mechanism accommodates the specific labeling results shown by the dot ($\bullet = {}^{14}C$). Also, be sure that any sigmatropic shifts you propose do not involve impossibly long distances.

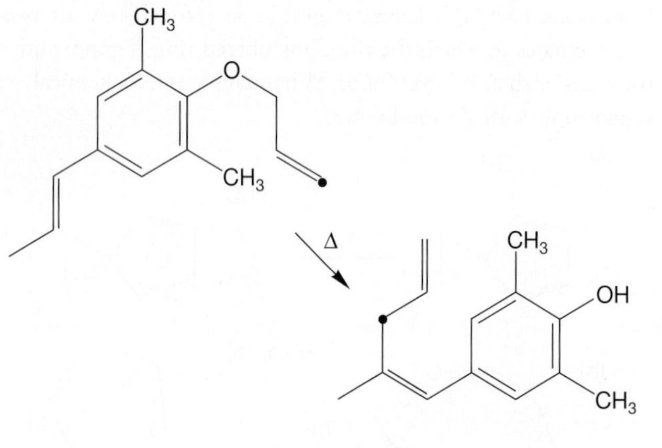

PROBLEM 20.45   We saw on page 1063 that chorismate undergoes a [3,3] sigmatropic shift (Claisen rearrangement) to prephenate catalyzed by the enzyme chorismate mutase. This enzyme-catalyzed reaction occurs more than 1 million times faster than the corresponding nonenzymatic reaction. How does the enzyme effect rate enhancement? Both enzymatic and nonenzymatic reactions proceed through a chairlike transition state. However, the predominant conformer of chorismate in solution, as determined by ^1H NMR spectroscopy, is the pseudodiequatorial conformer **2**, which cannot undergo the reaction.

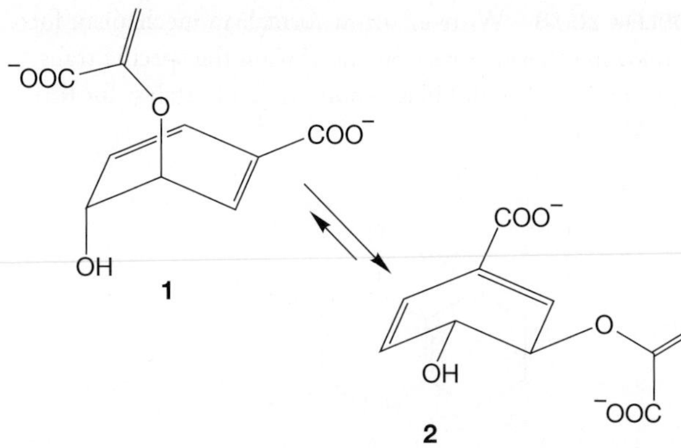

Perhaps the role of the enzyme is to bind the predominant pseudodiequatorial conformer **2**. The enzyme–substrate complex then undergoes a rate-determining conformational change to convert the substrate into the pseudodiaxial form **1**, which rapidly undergoes rearrangement.

(a) Explain why **1** can rearrange to prephenate but **2** cannot.

(b) Suggest a reason why conformer **2** is preferred over conformer **1**. *Hint*: Focus on the OH. What can it do in **2** that it cannot do in **1**?

(c) Draw an appropriate Energy versus Reaction progress diagram that illustrates the situation outlined above, that is, a reaction in which the rate-determining step is the conversion of **2** into **1**. What would the Energy versus Reaction progress diagram look like if the rate-determining step were the conversion of **1** into prephenate?

PROBLEM 20.46   In 1907, Staudinger reported the reaction of diphenylketene (**1**) and benzylideneaniline (**2**) to give β-lactam **3**. Although this reaction is formally a 2 + 2 cycloaddition reaction, there is ample experimental evidence to suggest that the reaction is not concerted, but is instead a two-step process involving a dipolar intermediate.

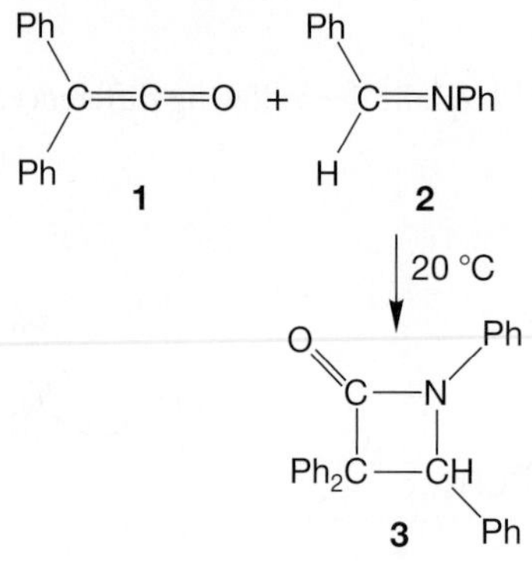

(a) Propose an arrow formalism mechanism for this cycloaddition reaction. Don't forget to include the dipolar intermediate.

(b) Although cycloaddition reactions of this type are not concerted, they are often stereospecific, as in the following example:

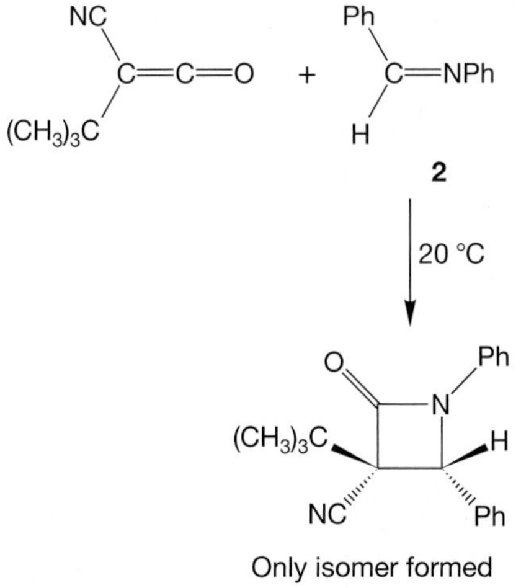

2

20 °C

Only isomer formed

Attempt to explain the stereochemistry observed in the reaction above. Start by minimizing the steric interactions in your proposed dipolar intermediate. Then, see if you can find a concerted process that converts your proposed intermediate into the observed β-lactam. *Hint*: A *tert*-butyl group is much bigger than a cyano group.

PROBLEM 20.47 Heating α-phenylallyl acetate (**1**) affords isomeric compound **2**. Spectral data for compound **2** are summarized below. Deduce the structure of **2** and provide an arrow formalism mechanism for this isomerization.

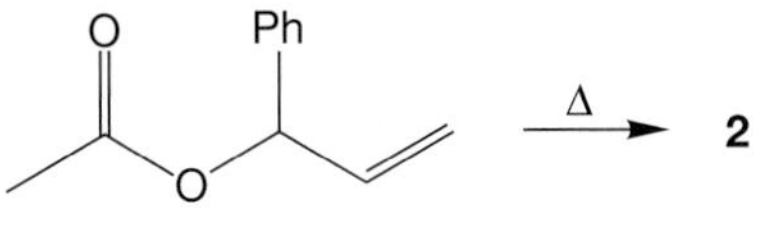

Compound **2**

IR (film): 1740 (s), 1225 (s), 1025 (m), 960 (m), 745 (m), 690 (m) cm⁻1
^{1}H NMR (CDCl$_3$): δ 2.08 (s, 3H)
4.71 (d, J = 5.5 Hz, 2H)
6.27 (dt, J = 16, 5.5 Hz, 1H)
6.63 (d, J = 16 Hz, 1H)
7.2–7.4 (m, 5H)
^{13}C NMR (CDCl$_3$): δ 20.9 (q)
65.0 (t)
123.1 (d)
126.5 (d)
128.0 (d)
128.5 (d)
134.1 (d)
136.1 (s)
170.6 (s)

PROBLEM 20.48 Reaction of bicyclopropenyl (**1**) and *s*-tetrazine (**2**) gives the semibullvalene (**3**). Provide an arrow formalism mechanism for this transformation. *Hint*: There is at least one nonisolable intermediate involved. Also, explain the relatively simple ^{1}H and ^{13}C NMR spectra of compound **3**, summarized below.

Compound **3**

^{1}H NMR (CDCl$_3$): δ 1.13 (s, 3H)
3.73 (s, 3H)
4.79 (s, 2H)
^{13}C NMR (CDCl$_3$): δ 14.9 (q)
51.4 (q)
60.6 (s)
93.7 (d)
127.2 (s)
164.7 (s)

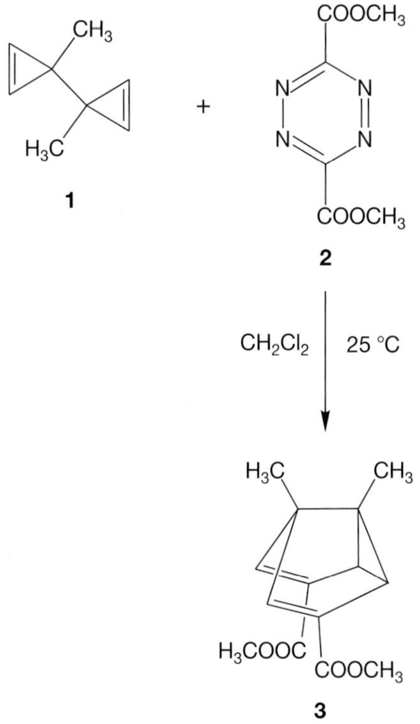

1

2

CH$_2$Cl$_2$ 25 °C

3

PROBLEM 20.49 Figure 20.55 shows the products of the Cope rearrangement of *meso*-3,4-dimethyl-1,5-hexadiene from both the chairlike (four-center) and boatlike (six-center) transition states. Work out the products expected from the other possible form of the diene, racemic 3,4-dimethyl-1,5-hexadiene. Predict what products will actually be formed.

21

Intramolecular Reactions and Neighboring Group Participation

21.1 Preview

21.2 Heteroatoms as Neighboring Groups

21.3 Neighboring π Systems

21.4 Single Bonds as Neighboring Groups

21.5 Coates' Cation

21.6 Summary

21.7 Additional Problems

WATER BARRIER Electrons are like water. There are few barriers that stop movement of available electrons to electrophilic centers in intramolecular reactions.

Our race would not have gotten far,
Had we not learned to bluff it out
And look more certain than we are
Of what our motion is about.

—W. H. AUDEN[1]

21.1 Preview

Here we introduce the chemistry of polyfunctional molecules. This chapter by no means contains an exhaustive treatment. The chemistry of di- and polyfunctional compounds could easily be the subject of an advanced course in organic chemistry, and we will only catch some highlights here.

In principle, two functional groups in a single molecule might not interact at all, and the chemistry of the molecule would be the simple combination of the chemistries of the two separated functional groups. This occurrence is rare, because groups in molecules often do interact, even if they are quite remote from each other. Through inductive and resonance effects, one functional group in a molecule can affect the chemistry of another group at a distant point in the molecule. Sometimes, groups interact even more strongly, and the chemistry is best described not as a sum of the two individual chemistries, but as something quite new. These examples are the most interesting ones.

A **neighboring group effect** can change the rates of formation of products as well as the structures of the molecules produced. Don't be put off by the fancy name. All of intramolecular chemistry involves interactions between neighboring groups. You have already seen lots of intramolecular chemistry. Figure 21.1 gives some familiar, or at least old, examples of such reactions, and we will see more as we go along.

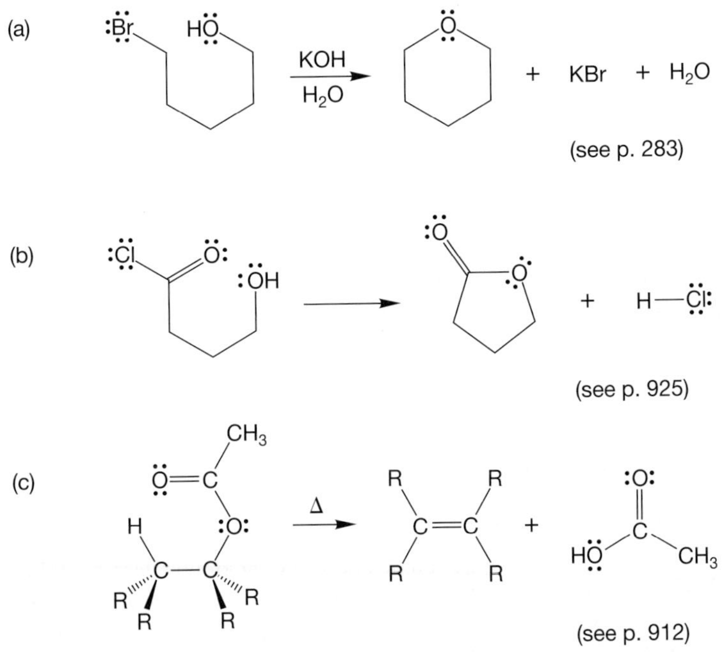

Intramolecular S_N2

(a) (see p. 283)

(b) (see p. 925)

(c) (see p. 912)

FIGURE 21.1 Neighboring group interactions.

---

[1] Wystan Hugh Auden (1907–1973) was one of the most influential British poets of his generation. "Reflections in a Forest," copyright © 1976 by Edward Mendelson, William Meredith, and Monroe K. Spears, Executors of the Estate of W. H. Auden, from *Collected Poems* by W. H. Auden. Used by permission of Curtis Brown, Ltd., Faber and Faber Ltd., and Random House, Inc.

WORKED PROBLEM 21.1 Write mechanisms for the reactions of Figure 21.1.

ANSWER (a) The first step is certainly loss of the hydroxyl proton to give an alkoxide. Intramolecular S_N2 displacement gives the product, a cyclic ether.

(b) Addition–elimination gives the lactone (a cyclic ester).

(c) This reaction is an intramolecular elimination with the carbonyl oxygen playing the part of the nucleophile and the neighboring hydrogen acting as the electrophile for the reaction.

The exploitation of the intuitively obvious notion that groups might interact in a synergistic fashion constituted one of the most fascinating areas of chemical exploration during the middle of the twentieth century, and it continues even today. As we examine this area, we will encounter some sociology as well as chemistry. One of the most heated arguments in the history of organic chemistry centered on this subject, and it is worth our time to see what we can learn about science from a look at this intensely emotional issue. There will be some opinion here, but we'll try to keep it labeled as such.

ESSENTIAL SKILLS AND DETAILS

1. There are three clues to the operation of a neighboring group effect. Learning to recognize them is critical to understanding this chemistry. Fortunately, at least one of these clues is always present, and usually quite easy to find. This material is of the "looks hard—is easy" variety.

   The clues are

   a. An unusual stereochemical result, usually retention where inversion is expected.

   b. A strange rearrangement, often involving a labeled molecule.

   c. An unexpectedly fast rate.

2. Be certain you are completely clear on the difference between an intermediate described by a set of equilibrating structures and an intermediate described by a set of resonance forms. Figure 21.40 gives an example.

21.2 Heteroatoms as Neighboring Groups

In its simplest form, the study of neighboring group effects notes that a heteroatom with a pair of nonbonding electrons can act as a nucleophile in an intramolecular displacement reaction, just as external nucleophiles can participate in intermolecular S_N2 reactions. In an example we have seen (p. 317), oxiranes can be formed from halohydrins in base (Fig. 21.2).

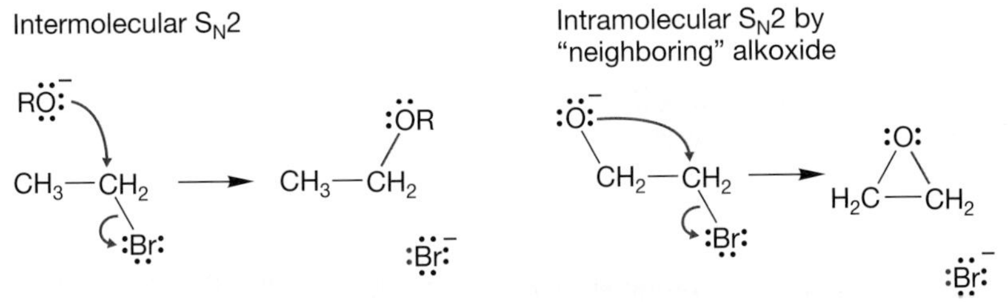

FIGURE 21.2 Formation of an oxirane by treatment of a bromohydrin with hydroxide involves an alkoxide acting as a neighboring group.

First, base removes the hydroxyl hydrogen to make the alkoxide. Intramolecular S_N2 displacement of the bromide generates the oxirane. Here the alkoxide is the neighboring group, and the whole process couldn't be simpler.

Another example is intramolecular hemiacetal formation. If a relatively unstrained ring can be formed, compounds containing both a carbonyl group and a hydroxyl group exist in equilibrium with the cyclic hemiacetal (p. 783; Fig. 21.3).

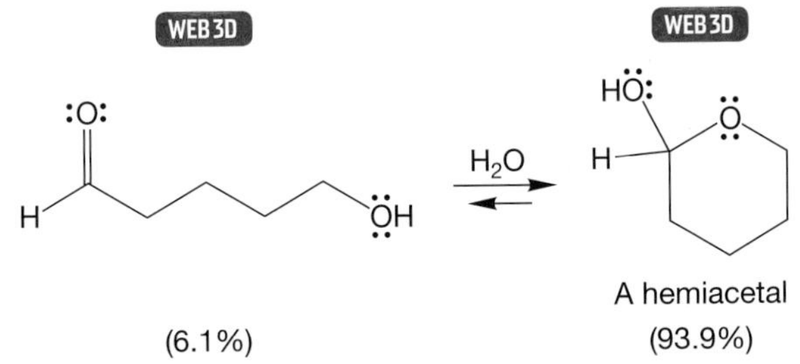

FIGURE 21.3 Cyclic hemiacetal formation involves an intramolecular hydroxyl group acting as a neighboring group on a ketone or aldehyde.

PROBLEM 21.2 Write a mechanism for the reaction of Figure 21.3.

Other reactions are more subtle than these straightforward intramolecular processes, but there are always clues when a neighboring group effect is involved. For example, consider the reaction of the optically active α-bromocarboxylate in methyl alcohol (Fig. 21.4).

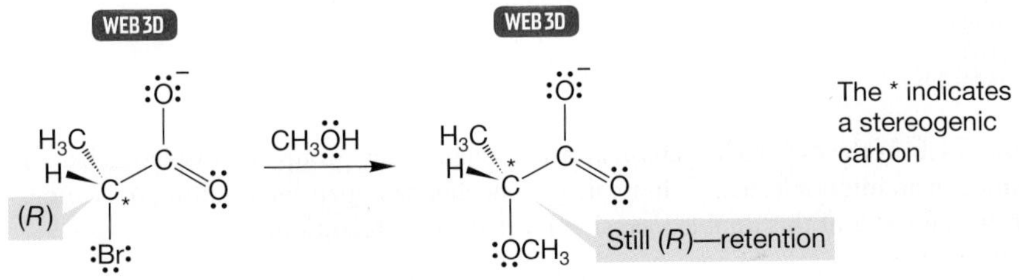

FIGURE 21.4 Treatment of α-bromopropionate in methyl alcohol. Notice that the reaction involves retention of configuration.

The product is formed with the stereochemistry opposite to what we might naively expect. An S_N2 displacement of the bromide from the optically active starting material by methyl alcohol or methoxide must give an optically active product (Fig. 21.5), but notice that the stereochemistry of the product is exactly the reverse of what is observed!

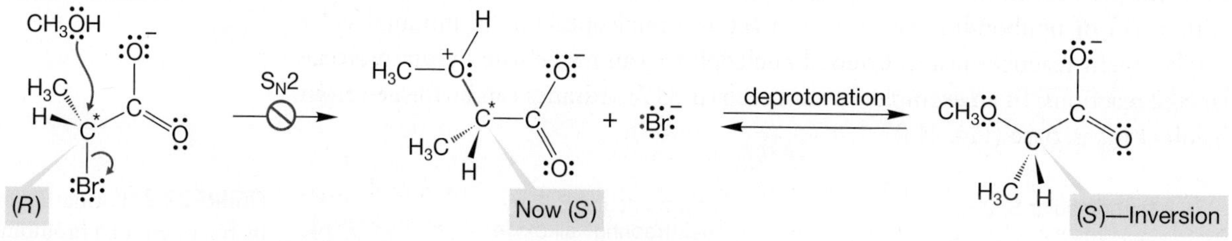

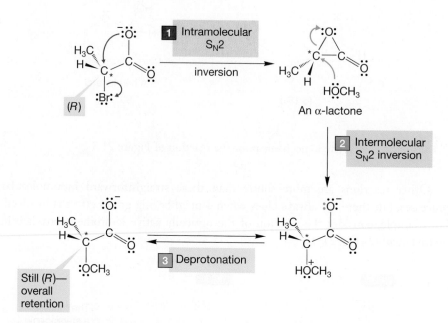

FIGURE 21.5 A straightforward mechanism in which the bromide is displaced by methyl alcohol would give the wrong stereochemical result. This inversion is not observed.

The observation of a strange stereochemical result is the first kind of clue to the operation of a neighboring group effect. The simple mechanism of Figure 21.5 must be wrong, because it cannot accommodate the observed retention of configuration in the product. In this reaction, the carboxylate anion acts as an internal nucleophile, displacing bromide to give an intermediate called an α-lactone (Fig. 21.6). These smallest cyclic esters cannot be isolated, but they are known as reactive intermediates. In this case, methoxide or methyl alcohol adds to the α-lactone in an intermolecular S_N2 displacement to give the actual product of the reaction. Now examine the stereochemistry of the reaction with care. The first

FIGURE 21.6 If the carboxylate anion acts as a neighboring group and displaces the bromide in an intramolecular fashion, methyl alcohol can open the resulting α-lactone. The overall result of this double inversion mechanism is retention.

displacement is an *intra*molecular S_N2 reaction with the carboxylate anion acting as the nucleophile and the bromide as the leaving group. As with all S_N2 reactions, inversion must be the stereochemical result. A second S_N2 displacement, this time *inter*molecular, occurs as the α-lactone is opened by methyl alcohol. Once again inversion is required. The end result of two inversions (or any even number of inversions) is retention of configuration.

Summary

Whenever you see an unexpected stereochemical result (usually retention where inversion is expected), look for an intramolecular displacement reaction—a neighboring group effect. You will almost always find it.

Oxygen need not be negatively charged to participate in an intramolecular displacement reaction. The reaction of the "brosylate" (*p*-bromobenzenesulfonate) of 4-methoxy-1-butanol in acetic acid gives the corresponding acetate (Fig. 21.7).

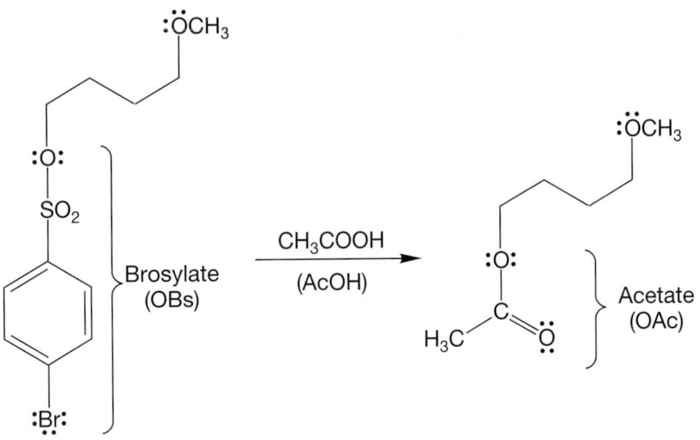

FIGURE 21.7 Acetolysis of the brosylate of 4-methoxy-1-butanol gives the corresponding acetate.

Sulfonates are excellent leaving groups, and a reasonable mechanism for the reaction involves intermolecular displacement by acetic acid or acetate (Fig. 21.8).

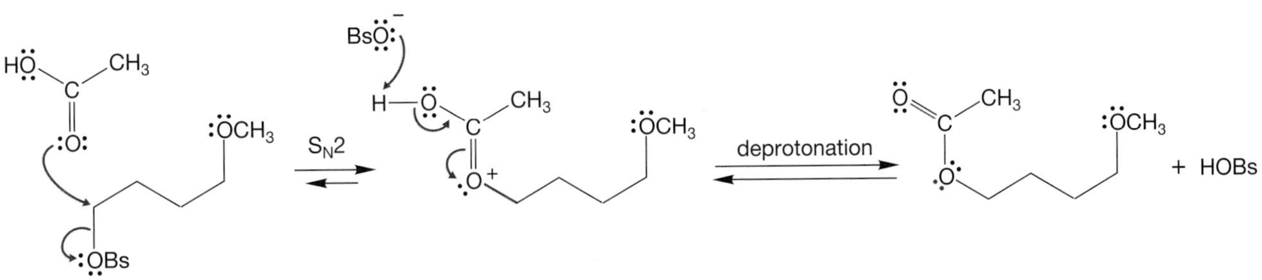

FIGURE 21.8 A simple S_N2 displacement of the brosylate by acetic acid might explain the reaction of Figure 21.7.

Acetic acid, CH_3COOH, is often abbreviated as "HOAc" or "AcOH." The acetates formed as products in many reactions of acetic acid are often written as "ROAc."

CONVENTION ALERT

WORKED PROBLEM 21.3 There are two possible nucleophiles for the S_N2 displacement in Figure 21.8, the acetate ion or acetic acid. In the displacement using acetic acid, why is the carbonyl oxygen, not the hydroxyl oxygen, used as the displacing agent?

ANSWER The carbonyl oxygen is the better nucleophile. Recall that protonation of acetic acid occurs faster on the carbonyl oxygen (p. 837). A resonance-stabilized intermediate is formed in this case, but not from protonation of the hydroxyl oxygen. Reaction of any electrophile with acetic acid is similar. In Figure 21.8, only one resonance form is shown; in each case there is another form in which the carbon atom bears the positive charge—ordinary carbonyl resonance.

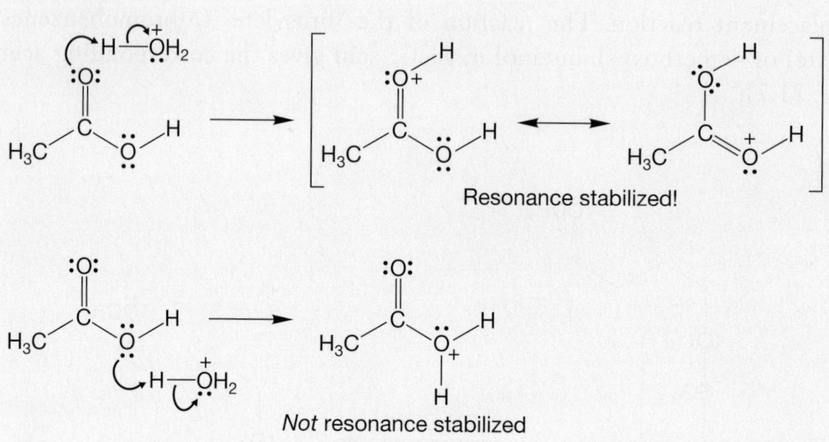

Resonance stabilized!

Not resonance stabilized

When this reaction is run with a similar, but methyl-labeled brosylate (Fig. 21.9), we can see that something more complicated than the simple S_N2 reaction of Figure 21.8 is going on. There are two products formed. Our first mechanism can accommodate one of the products but not the other, which is the second clue to neighboring group participation—the observation of an unusual rearrangement reaction. In this case, it is the ether oxygen that acts as a neighboring group. The sulfonate (OBs) is an excellent leaving group and a five-membered ring is relatively unstrained, so the intramolecular displacement of *p*-bromobenzenesulfonate to give a cyclic oxo-

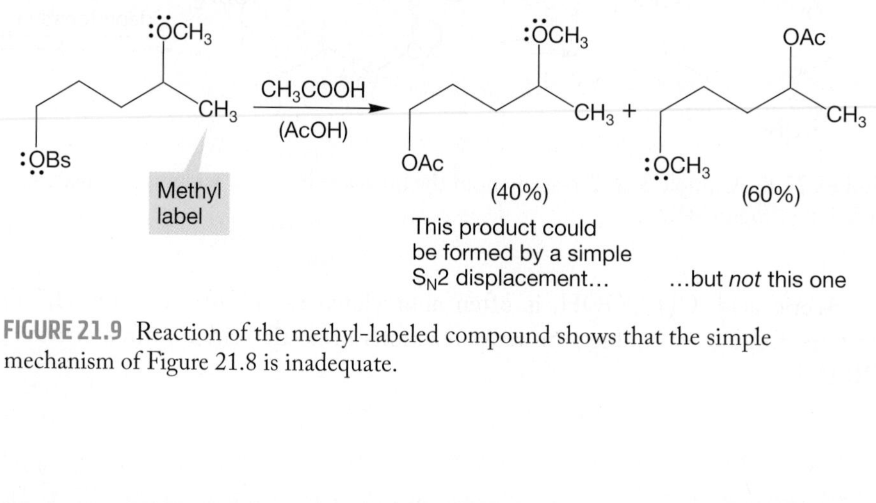

Methyl label

This product could be formed by a simple S_N2 displacement...

(40%)

...but *not* this one

(60%)

FIGURE 21.9 Reaction of the methyl-labeled compound shows that the simple mechanism of Figure 21.8 is inadequate.

nium ion takes place (Fig. 21.10). The cyclic oxonium ion can be opened by acetic acid in two ways to give the two observed products. The result is an apparent rearrangement.

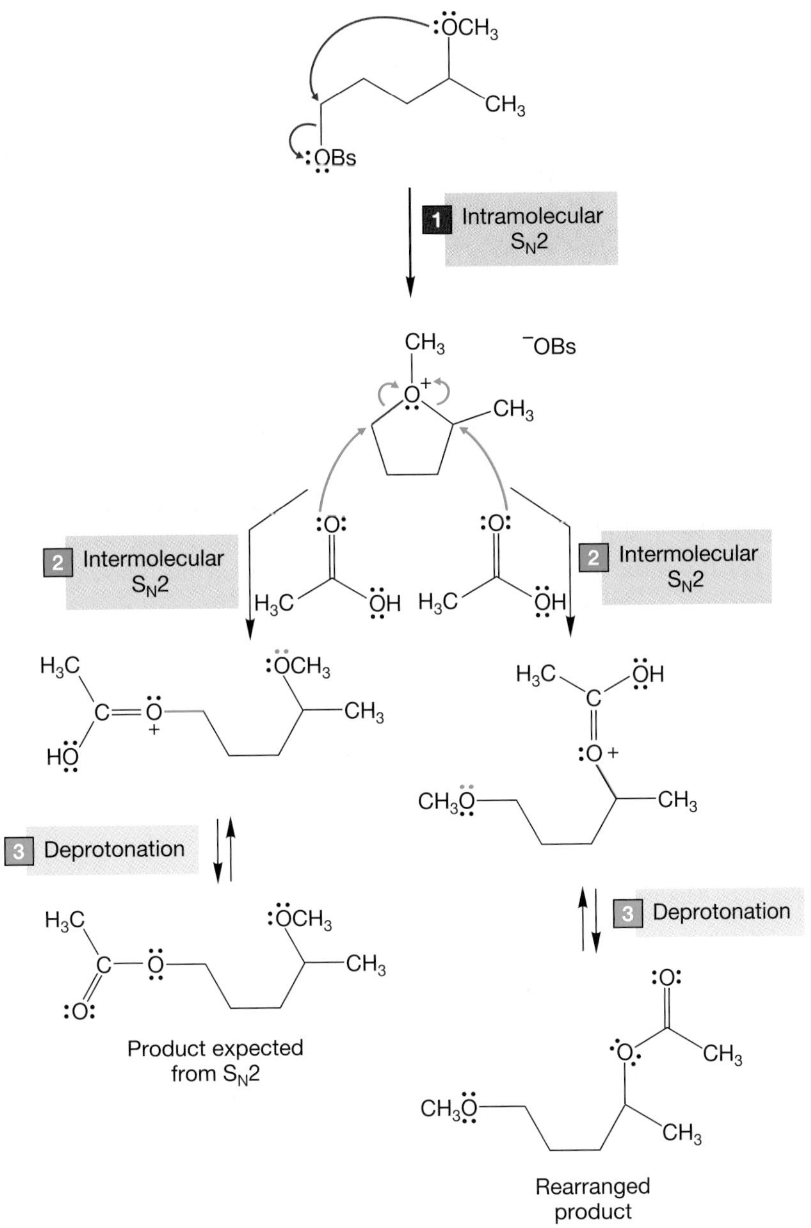

FIGURE 21.10 In this case the methoxyl oxygen acts as neighboring group. A cyclic ion is formed that can open two ways. One mode of opening leads to the product expected from a simple S_N2 reaction. "Rearrangement" is the result of the other possible mode of opening of the intermediate oxonium ion.

PROBLEM 21.4 There is actually a third product in the reaction in Figure 21.9. It is 2-methyltetrahydrofuran (see below). Can you explain the formation of a small amount of this molecule? *Hint*: The intermediate ring is obviously retained in the formation of this product.

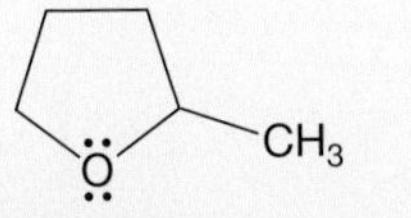

2-Methyltetrahydrofuran

Sulfur atoms are often powerful nucleophiles, and the participation of sulfur as a neighboring group is common. Figure 21.11 shows an example of the third kind of clue, a reaction in which there is an unexpectedly rapid rate. Both hexyl chloride

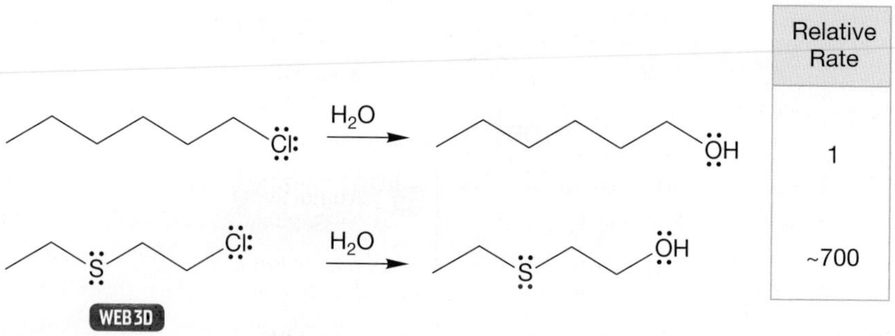

FIGURE 21.11 If a sulfur atom is substituted for the C(3) of hexyl chloride, the formation of the related alcohol is greatly accelerated.

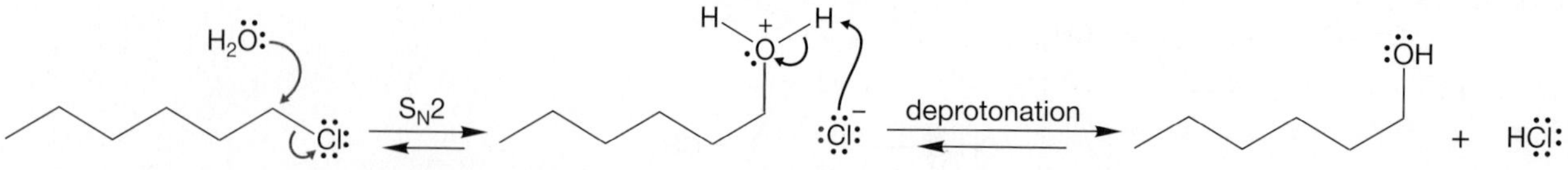

	Relative Rate
	1
	~700

and 2-chloroethyl ethyl sulfide react in water to give the corresponding alcohols. The rate of reaction of the sulfur-containing compound is about 10^3 faster than that of the alkyl chloride, however. The mechanism of alcohol formation from the primary halide must be a simple S_N2 displacement with water acting as nucleophile (Fig. 21.12). It is difficult to see why the sulfide should be so much faster if the reaction mechanism is analogous.

FIGURE 21.12 The reaction of hexyl chloride must be a simple S_N2 displacement of chloride by water, followed by deprotonation.

In the reaction of the sulfide in Figure 21.11, it is the sulfur atom that displaces the leaving chloride (Fig. 21.13). This intramolecular reaction is much faster than the intermolecular displacement of chloride by water either in the alkyl chloride or the chlorosulfide. The intermediate is an **episulfonium ion**. This intermediate is opened by water in a second, intermolecular S_N2 displacement step to give the product.

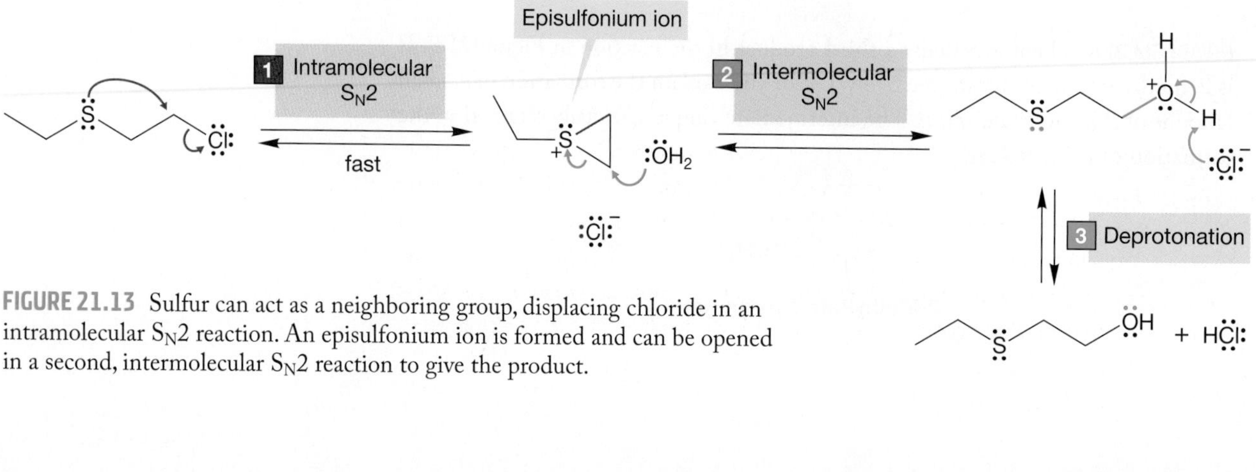

FIGURE 21.13 Sulfur can act as a neighboring group, displacing chloride in an intramolecular S_N2 reaction. An episulfonium ion is formed and can be opened in a second, intermolecular S_N2 reaction to give the product.

PROBLEM 21.5 There are many possible labeling experiments that would support (not prove) the mechanism suggested for the reaction of the chlorosulfide in Figure 21.13. Design two, one using a methyl group as a label, and the other using an isotopic label.

It is important to be certain you understand the source of the rate increase for the sulfide compound. Intramolecular reactions often have a large rate advantage over intermolecular reactions because there is no need for the nucleophile and substrate to find each other in solution. Intramolecular assistance in the rate-determining step of a reaction is called **anchimeric assistance**. Once again, it is important not to be put off by the elaborate name. The concept is quite simple: Intramolecular nucleophiles are often more effective displacing agents than intermolecular nucleophiles. What is somewhat more difficult to anticipate is that an intramolecular displacement by a very weak, but perfectly situated nucleophile can often compete effectively with intermolecular reactions.

MUSTARD GAS

If we make a seemingly trivial change in our sulfur-containing molecule, by adding another chlorine, we come to bis(2-chloroethyl)thioether. This seemingly routine molecule exposes the dark side of our science. Indeed, a review of the history of this compound causes one to reflect on the depths to which human behavior can sink, for this molecule is mustard gas, developed in the early part of the twentieth century and used in World War I as a replacement for other less efficient toxic agents for which defenses had been worked out. Mustard gas penetrated the rubber in use in gas masks at the time and thus was more effective. This mustard is lethal in a few minutes at 0.02–0.05% concentration. Less severe exposure leads to irreversible cell damage and only delays the inevitable. Introduced by the Germans at Ypres in 1917, it was also employed in the China campaign by the Japanese in World War II. Stories of experiments on prisoners of war are easy to find, if not easy to read. Nor is there any reason for national pride in not being the first to use this agent, for we in America were the developers of napalm, essentially jellied hydrocarbons, a chemical agent used by U.S. forces in World War II and the Vietnam War.

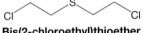
Bis(2-chloroethyl)thioether

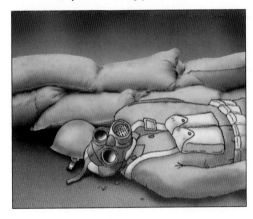

We leave it to you to consider whether it is morally superior to shoot people, or drop an atomic bomb on them, rather than to dose them with mustard gas or burn them with napalm. Even more complicated is the question of the participation by chemists in the development of such agents. Of course, often it is not easy to see the consequences of one's research. But what if one can? Where are the limits?

PROBLEM SOLVING

There are three clues that point to the operation of a neighboring group effect. The first is an unusual, often "backward" stereochemical result. Generally, this involves retention of configuration when inversion might have been anticipated. The second is formation of an unexpected product that appears to be the result of a rearrangement. The unexpected product arises because an intermediate is formed that can react further in more than one way. The third clue is an unexpected increase in the rate of a reaction. The anchimeric assistance provided by an intramolecular nucleophile often reveals itself in the form of a rate increase.

PROBLEM 21.6 Go back and find examples of these three clues in the preceding reactions.

Now let's look at more heteroatoms with nonbonding electrons that might participate in neighboring group reactions. For example, nitrogen bears a lone pair of electrons, and so can act as an internal nucleophile. Let's look at the reactions of the two bicyclic amines in Figure 21.14. Treatment of the endo chloride with ethyl alcohol leads to the endo ether, but the exo chloride gives neither the exo nor endo ether under the reaction conditions. Now we'll see why.

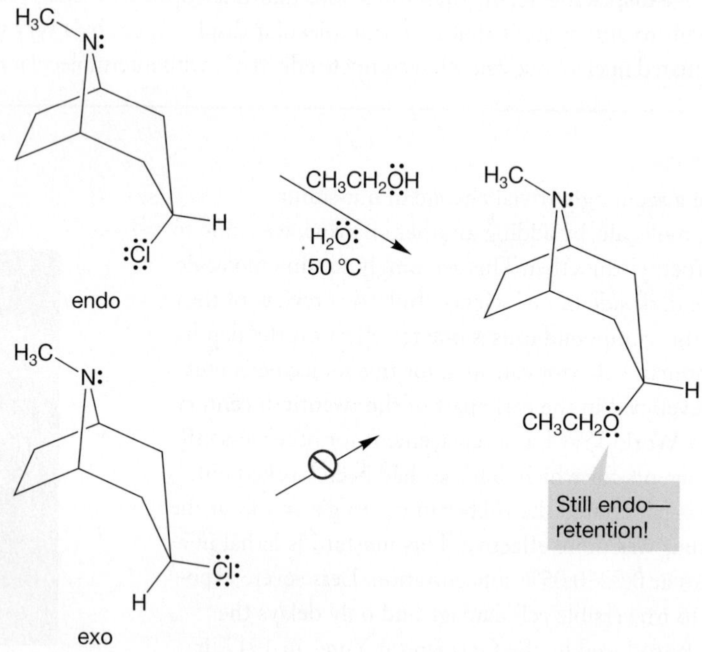

FIGURE 21.14 Treatment of only one isomer of this bicyclic amine gives the related ether. Note the retention of configuration.

In the reaction of the endo isomer, direct displacement of the chloride by ethyl alcohol cannot be happening, as this potential reaction leads to a product that is not observed, the exo ether (Fig. 21.15). The ether *is* formed, but it is formed with *retention* of stereochemistry, not the *inversion* demanded by the simple S_N2 reaction.

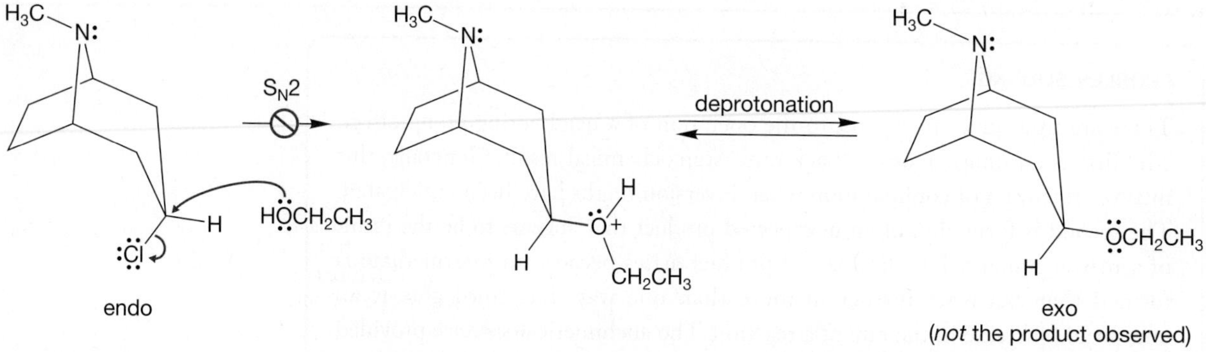

FIGURE 21.15 A simple intermolecular S_N2 displacement demands inversion of configuration, which does not give the observed product.

At least one of the clues to the operation of a neighboring group effect is present in this reaction. The stereochemistry of the product is the opposite to what one would expect from a simple displacement (clue 1).

What's happening? By now you should be alert to the possibility of intramolecular displacement. The nitrogen with its pair of nonbonding electrons is an obvious possible neighboring group, and it is located in the proper position for a backside displacement of the chloride (Fig. 21.16). The result is a cyclic ammonium ion, a ring containing a positively charged nitrogen. If this ring is opened by ethyl alcohol, the product ether must be formed with retention of stereochemistry. The intramolecular S_N2 displacement by nitrogen is the first inversion reaction, and the second, intermolecular displacement, the opening of the ammonium ion by ethyl alcohol, provides the second. Two inversions produce net retention of configuration.

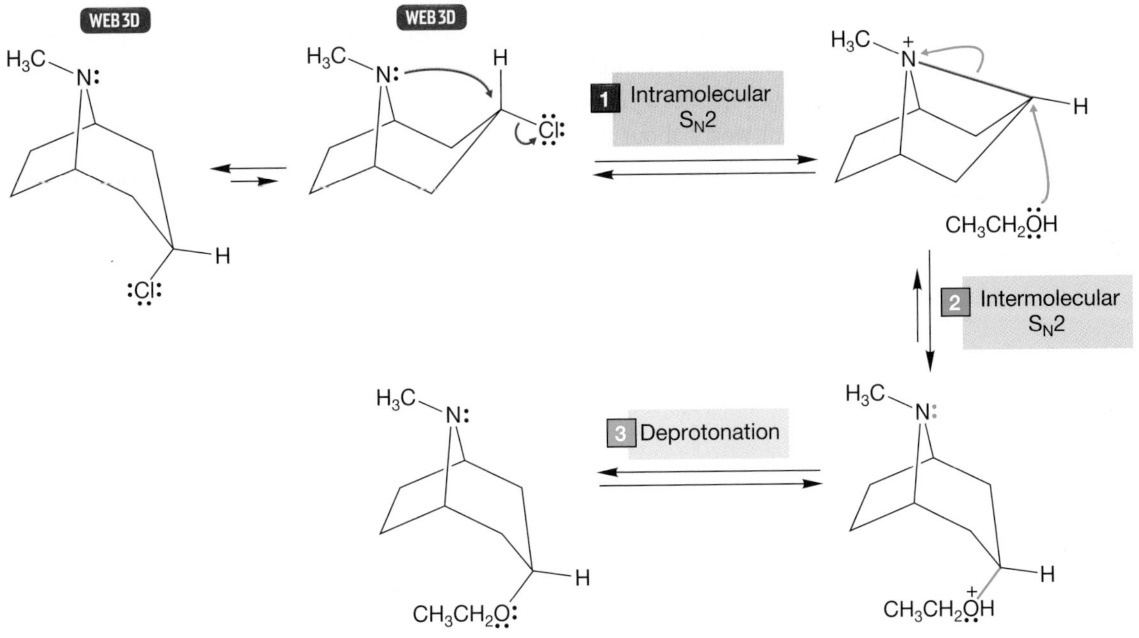

FIGURE 21.16 When nitrogen acts as a neighboring group and displaces the leaving group, a cyclic ammonium ion is formed. The cyclic ion is opened to give the observed product of retention of configuration.

Why does the exo compound not react in the same way? In this molecule, displacement of the leaving chloride would require a frontside S_N2 reaction, and this process is impossible (Fig. 21.17). So, the exo chloride reacts more slowly (and in a different way) from the endo isomer.

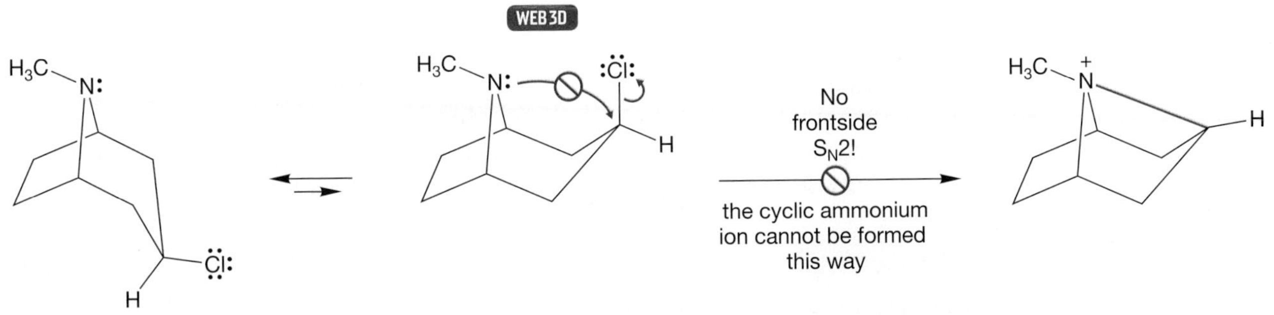

FIGURE 21.17 The exo chloride cannot be displaced by nitrogen, as this reaction would require a frontside S_N2 reaction.

PROBLEM 21.7 The exo isomer of Figure 21.14 gives the product shown below. Suggest a mechanism for this reaction.

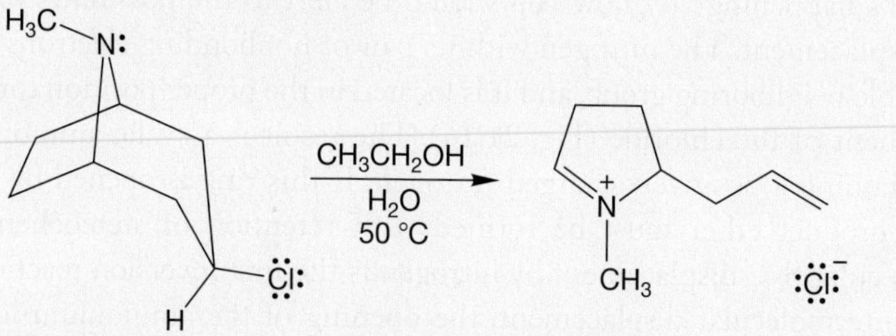

It is often difficult to see these reactions in bicyclic systems. The molecules are not easy to draw, and the extra atoms involved in the cage are distracting. There is nothing *fundamentally* difficult about these reactions, however. In fact, one could argue that once the drawing problems are mastered, the rigidity of the cage is a help. One need not worry about rotations about single bonds, or about finding the proper orientation for the reacting molecules. The stereochemical relationships are defined and preserved by the rigid bicyclic molecules. Here are some problems for practice. It will help to use models.

WORKED PROBLEM 21.8 Write mechanisms for the following reactions:

*(a)

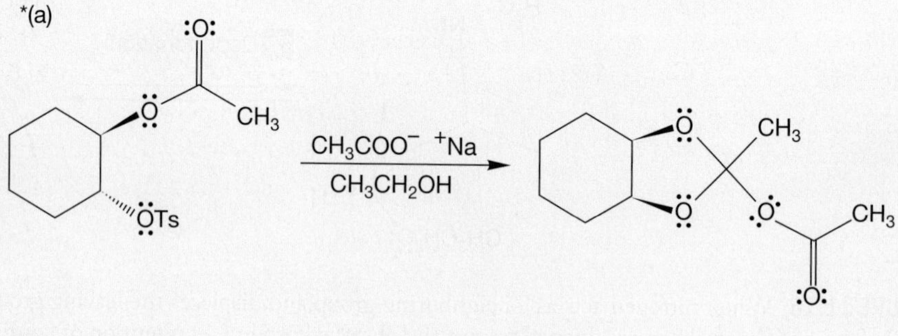

(b)

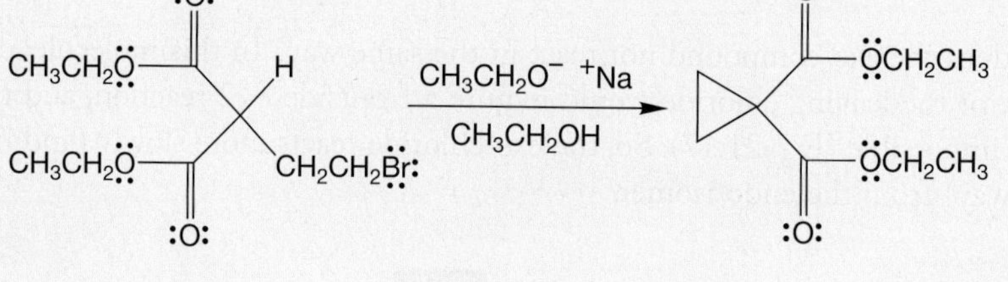

(c)

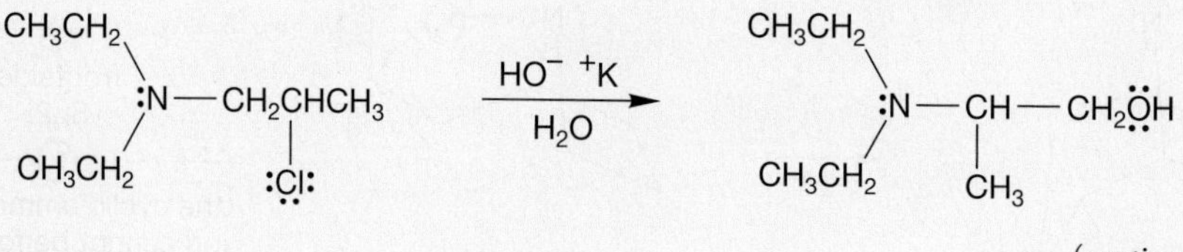

(*continued*)

ANSWER (a) The internal nucleophile is the carbonyl oxygen. Notice that the trans stereochemistry of the starting material allows for a rearside S$_N$2 displacement of the good leaving group, tosylate. Displacement yields a cyclic cation that is nicely stabilized by resonance. Acetate ion adds to the carbon sharing the positive charge, and gives the product.

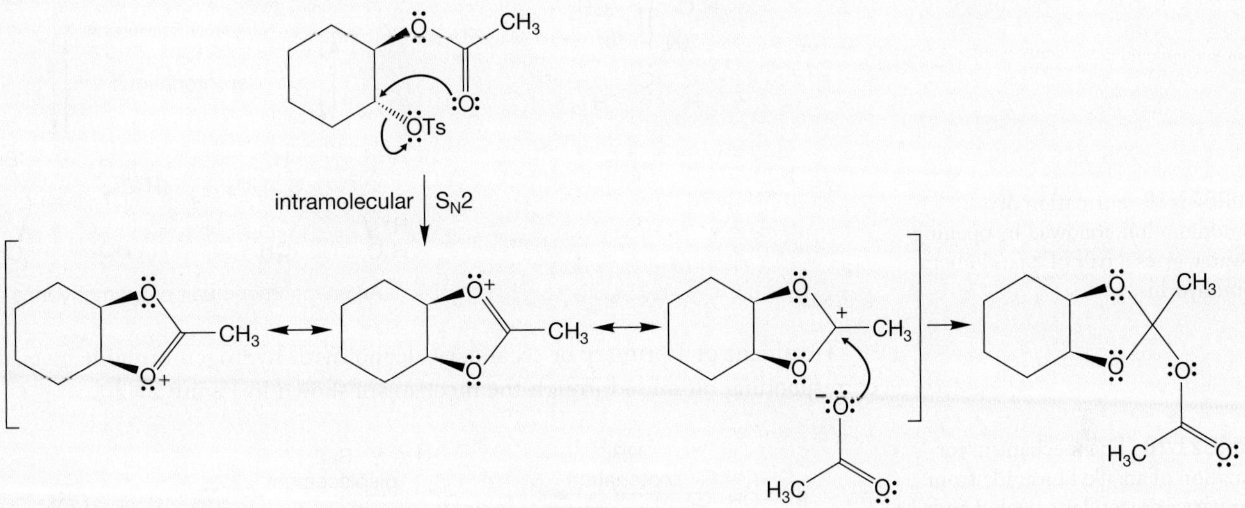

The halogens are another set of heteroatoms with nonbonding electrons. Neighboring group reactions of halogens are almost entirely restricted to iodine, bromine, and chlorine, and very often reveal themselves through the stereochemistry of the resulting products. As an example, we will work through the conversion of *cis*-2-butene into 2,3-dibromobutane through the synthetic scheme shown in Figure 21.18.

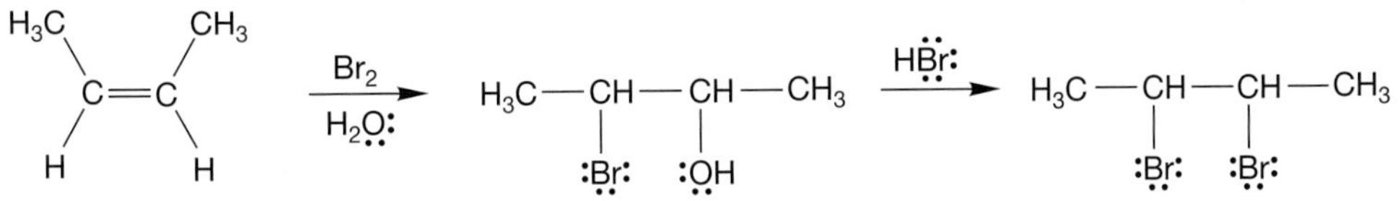

FIGURE 21.18 A possible synthesis of 2,3-dibromobutane.

PROBLEM 21.9 What's wrong with fluorine? Why are neighboring group effects involving fluorine very rare?

PROBLEM 21.10 What is a much easier way of making the 2,3-dibromobutane of Figure 21.18?

In the bromination of *cis*-2-butene, the initial intermediate is the cyclic bromonium ion (p. 415). The ring is opened by water to give the racemic pair of bromohydrin enantiomers shown in Figure 21.19.

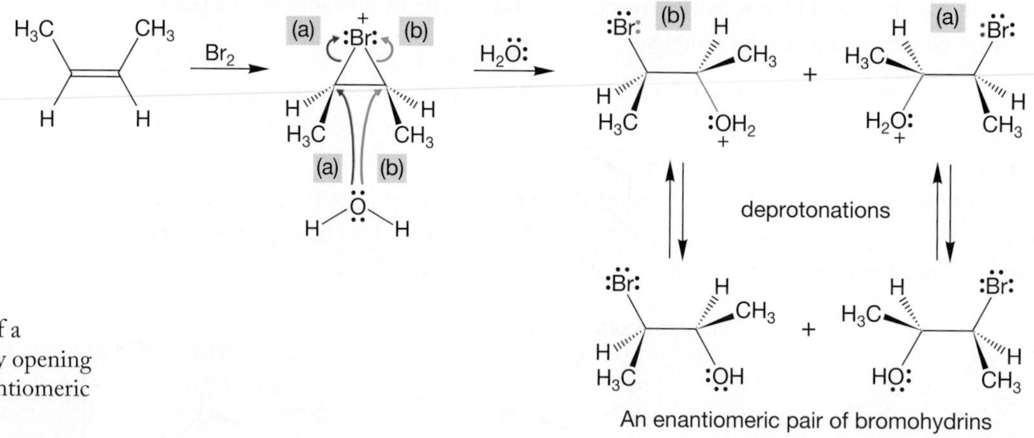

FIGURE 21.19 Formation of a bromonium ion, followed by opening by water, gives a pair of enantiomeric bromohydrins.

An enantiomeric pair of bromohydrins

Treatment of a primary or secondary alcohol with hydrogen bromide gives the corresponding bromide through the mechanism shown in Figure 21.20.

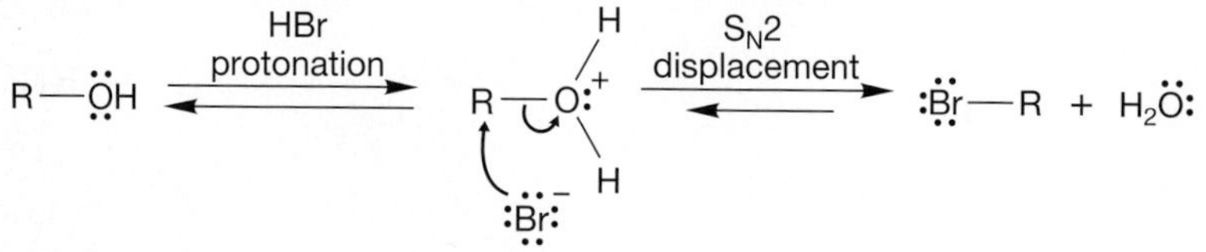

FIGURE 21.20 The mechanism for formation of an alkyl bromide from a primary or secondary alcohol and hydrogen bromide.

The S_N2 reaction of Figure 21.20 would give *meso*-2,3-dibromobutane from the bromohydrins of Figure 21.19, but it is the racemic mixture, not the meso compound that is produced (Fig. 21.21). Be absolutely certain you see this difference. If you are lost, reread Chapter 4, which works through the differences between meso compounds and racemic mixtures.

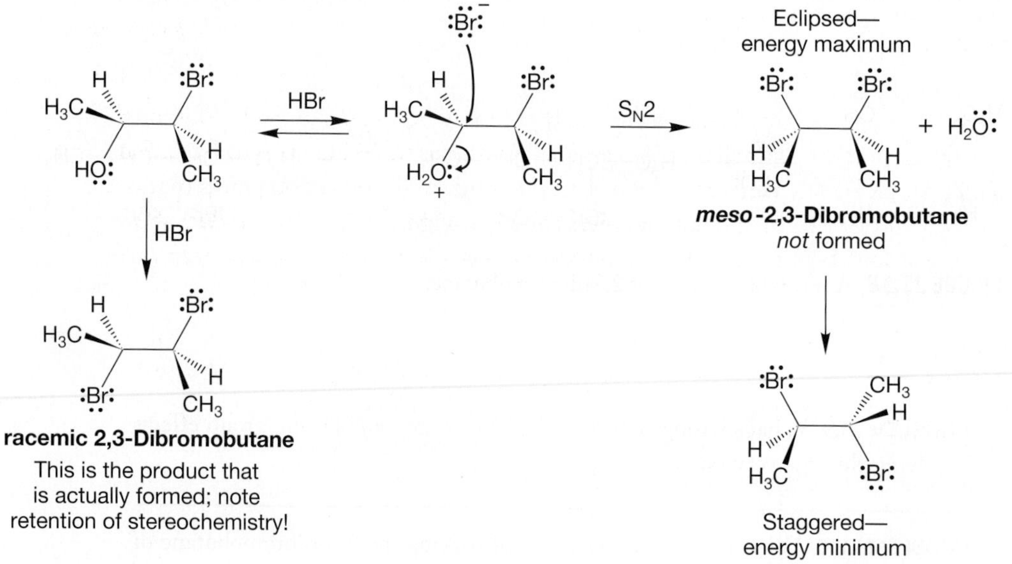

FIGURE 21.21 Reaction of the bromohydrin of Figure 21.19 by S_N2 displacement should give *meso*-2,3-dibromobutane. This compound is *not* the actual product.

racemic 2,3-Dibromobutane
This is the product that is actually formed; note retention of stereochemistry!

meso-2,3-Dibromobutane
not formed

Staggered—
energy minimum

CONVENTION ALERT Two conventions are being used here. First, the figures often carry through with only one enantiomer of a pair. Be sure to do Problem 21.11. Second, for clarity, eclipsed forms are often drawn. Be sure you remember that these are not really energy minima, but energy maxima.

PROBLEM 21.11 Figure 21.21 works the conventional reaction mechanism through with one of the enantiomers of Figure 21.19. You work it through with the other enantiomer.

The formation of the "wrong" stereoisomer is the clue to the operation of a neighboring group effect in the reaction of Figure 21.21. The product is formed not with the inversion of configuration required by the mechanism, but with retention (clue 1). After the hydroxyl group is protonated, it is displaced not by an external bromide ion, but by the bromine atom lurking in the same molecule (Fig. 21.22). It is most important at this point to keep the stereochemical relationships clear. Displacement must be from the rear, as this is an intramolecular S_N2 reaction.

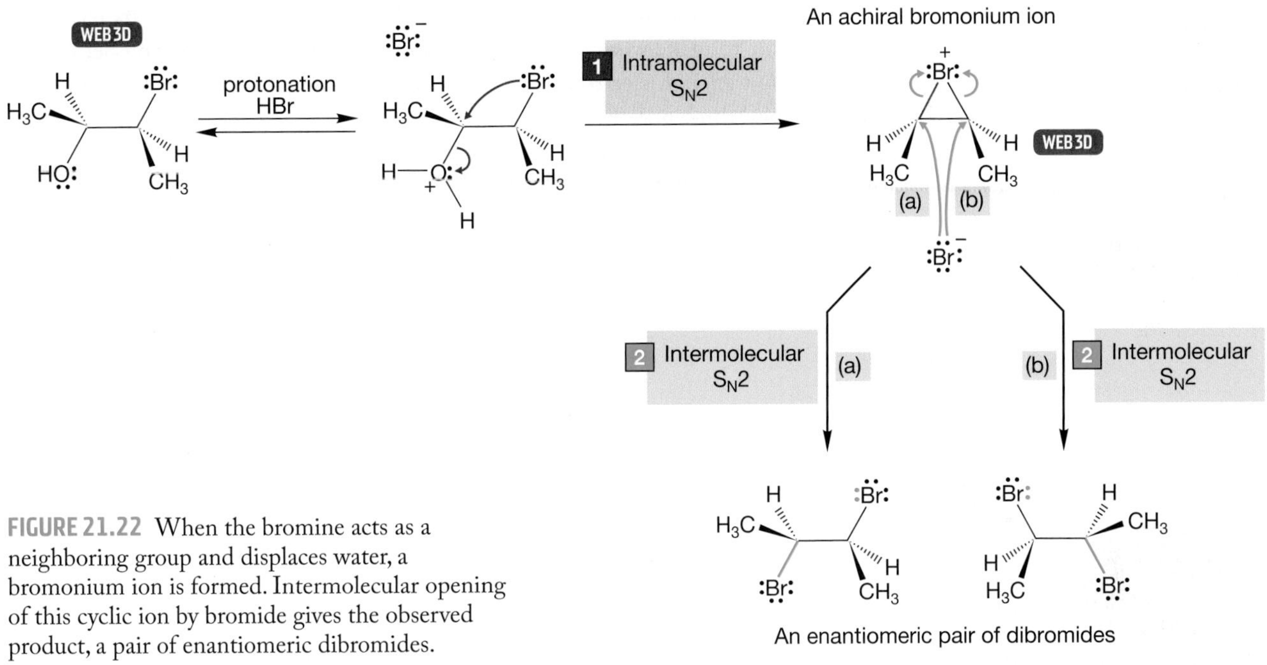

FIGURE 21.22 When the bromine acts as a neighboring group and displaces water, a bromonium ion is formed. Intermolecular opening of this cyclic ion by bromide gives the observed product, a pair of enantiomeric dibromides.

Notice that we have formed the same bromonium ion as in Figure 21.19, though by a different route. Like all bromonium ions, this one is subject to opening by nucleophilic attack. In this case, bromide ion can open the cyclic ion at two points (paths a and b) leading to the pair of enantiomers shown in Figure 21.22. The product shows no optical activity not because it is a meso compound, but because it is a racemic mixture of enantiomers.

PROBLEM 21.12 Work through the stereochemical relationships in the following reactions of *trans*-2-butene in great detail. Follow the reactions of all pairs of enantiomers. Contrast them with what you would expect from a reaction not involving a neighboring group effect.

meso-2,3-Dibromobutane

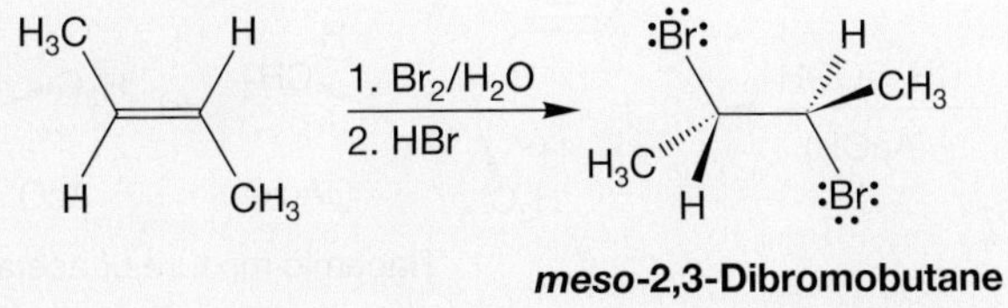

WORKED PROBLEM 21.13 There is a fundamental difference between the formation of a cyclic ion using a pair of nonbonding electrons and forming the cyclic ion using a pair of σ electrons. Carefully draw out the cyclic species for both cases and explain the difference. *Hint*: Count electrons.

ANSWER Problem 21.12 contains many examples of formation of a three-membered ring using a pair of nonbonding electrons. The result is a normal compound in which all bonds are two-electron bonds.

When a ring is formed through participation by a σ bond, everything changes. It isn't possible to make what we could call a normal three-membered ring because there are not enough electrons. The two electrons originally in the σ bond must serve to bind three nuclei.

21.3 Neighboring π Systems

21.3a Aromatic Neighboring Groups
We have progressed from neighboring group reactions involving very strong nucleophiles such as alkoxides and sulfur, to those of less powerful displacing agents such as the halogens. Now we examine even weaker nucleophiles, the π systems of aromatic molecules and alkenes. We will finally move on to very weak nucleophiles indeed, and see σ bonds acting as intramolecular displacing agents.

The π systems of aromatic rings can act as neighboring groups. The reactions of the 3-phenyl-2-butyl tosylates are good examples. D. J. Cram [1919–2001, awarded the Nobel Prize in Chemistry in 1987 for work on crown ethers (p. 254)] and his co-workers showed that in acetic acid, the optically active 3-phenyl-2-butyl tosylate shown in Figure 21.23 gives a racemic mixture of the corresponding 3-phenyl-2-butyl acetates.

But this result is not what we expect from a simple analysis of the possible displacement reactions, either S_N2 or S_N1 substitution. The S_N2 reaction must give

Neighboring pi: Not to be confused with neighborly pies!

FIGURE 21.23 This optically active isomer of 3-phenyl-2-butyl tosylate gives a racemic mixture of the corresponding acetates.

Optically active

$\xrightarrow{\text{CH}_3\text{COOH}}$ (AcOH)

Racemic mixture of acetates

inversion in the substitution step, and would form the optically active acetate shown in Figure 21.24.

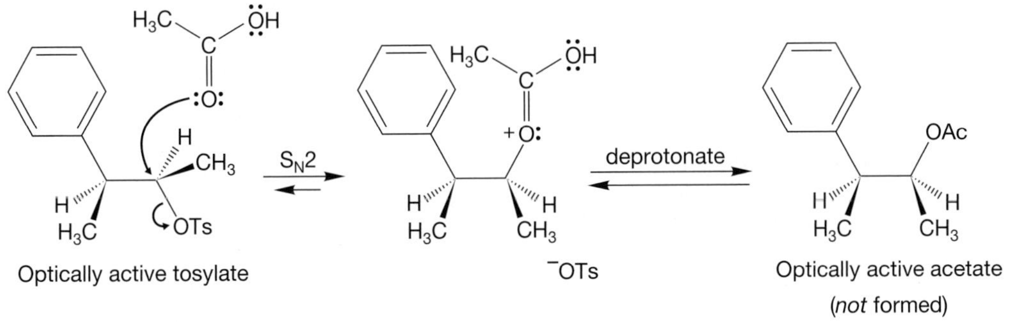

FIGURE 21.24 Simple S_N2 displacement by acetic acid fails to predict the correct product. Note the eclipsed forms drawn in this figure! They are there for clarity only and will be converted into the more stable staggered forms.

Ionization in an S_N1 process gives an intermediate open carbocation that can be attacked at either face by acetic acid to give the two different optically active products of Figure 21.25.

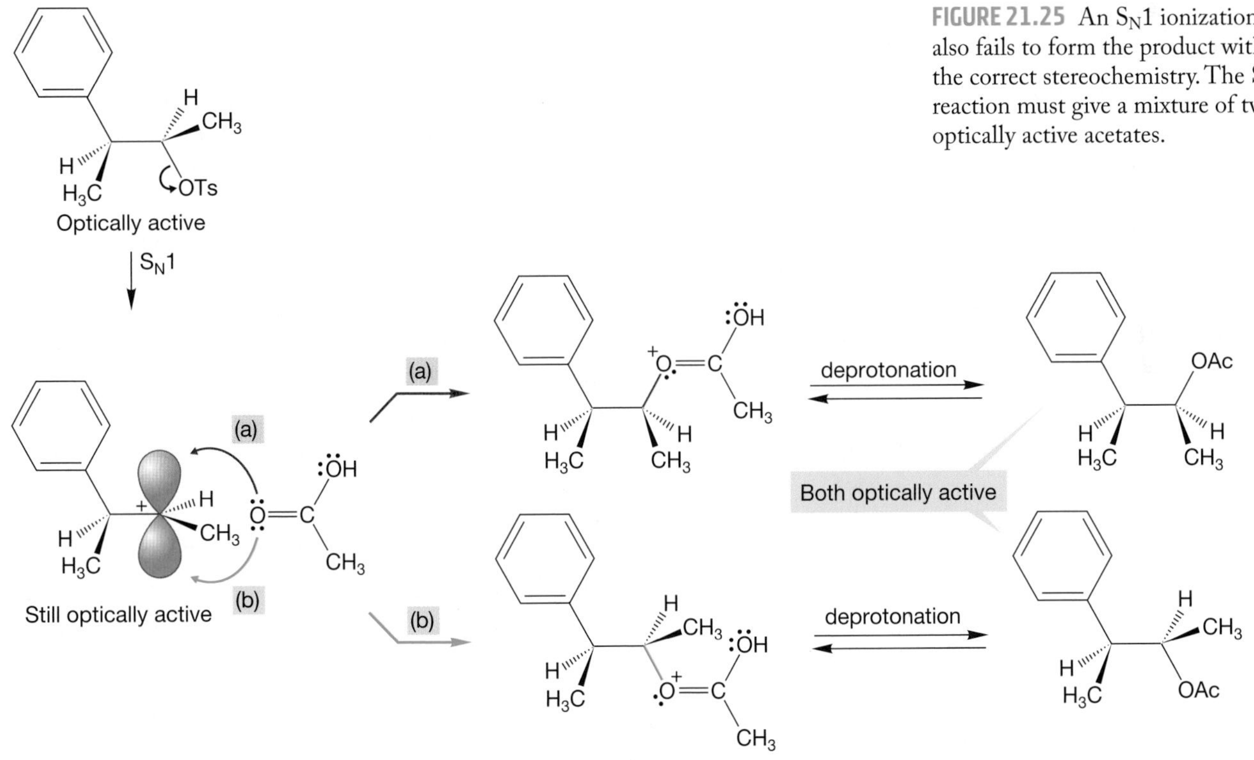

FIGURE 21.25 An S_N1 ionization also fails to form the product with the correct stereochemistry. The S_N1 reaction must give a mixture of two optically active acetates.

PROBLEM 21.14 What kind of stereoisomers are the two products at the right of Figure 21.25?

Because neither the S_N1 nor the S_N2 reaction produces the observed product, some other mechanism must be operating in this reaction. One of the clues for a neighboring group participation is present—an unexpected stereochemical result, clue 1—so let's look around for possible internal nucleophiles.

Aromatic rings act as nucleophiles toward all kinds of electrophiles in the aromatic substitution reaction (Chapter 13), so perhaps we can use the π system of the phenyl ring in 3-phenyl-2-butyl tosylate as a nucleophile. If the π system displaces the tosylate, it must do so from the rear. The result is the **phenonium ion** shown in Figure 21.26.

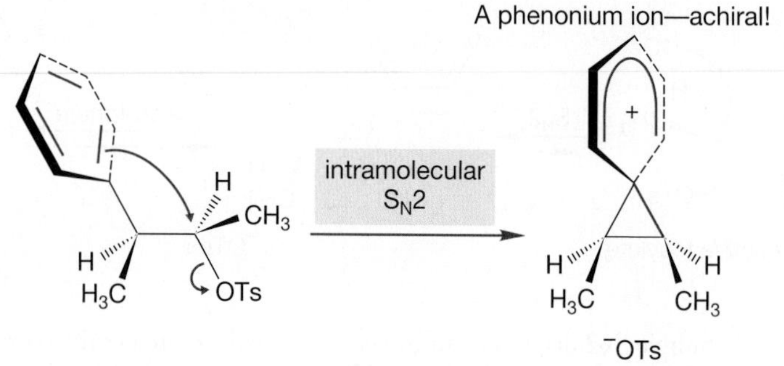

FIGURE 21.26 If the benzene ring acts as a neighboring group, a cyclic phenonium ion intermediate results. Notice that the intermediate has a plane of symmetry and is therefore achiral.

PROBLEM 21.15 Draw resonance forms for the phenonium ion of Figure 21.26.

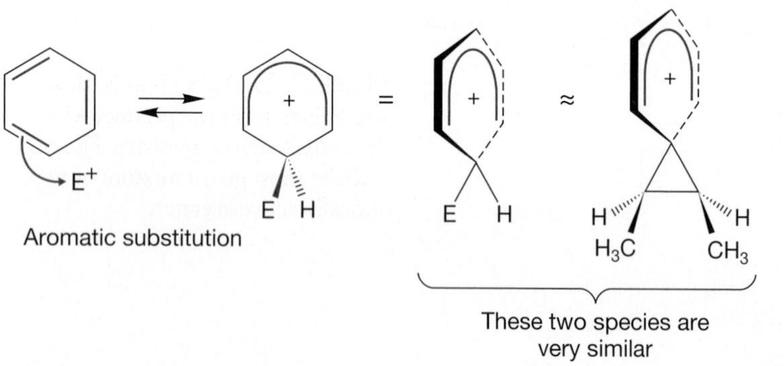

Note carefully the geometry of the phenonium ion. The two rings are perpendicular, not coplanar, and the ion is achiral, a meso compound. All chance of optical activity is lost at this point. Note also the close correspondence between this intermediate and the benzenonium ion formed in an electrophilic aromatic substitution reaction (Fig. 21.27).

There is nothing particularly odd about this spirocyclic species. If it is opened in a second, intermolecular S_N2 reaction, there are two possible sites for attack (Fig. 21.28). The two products are mirror images, and the racemic mixture of enantiomers is exactly what is found experimentally.

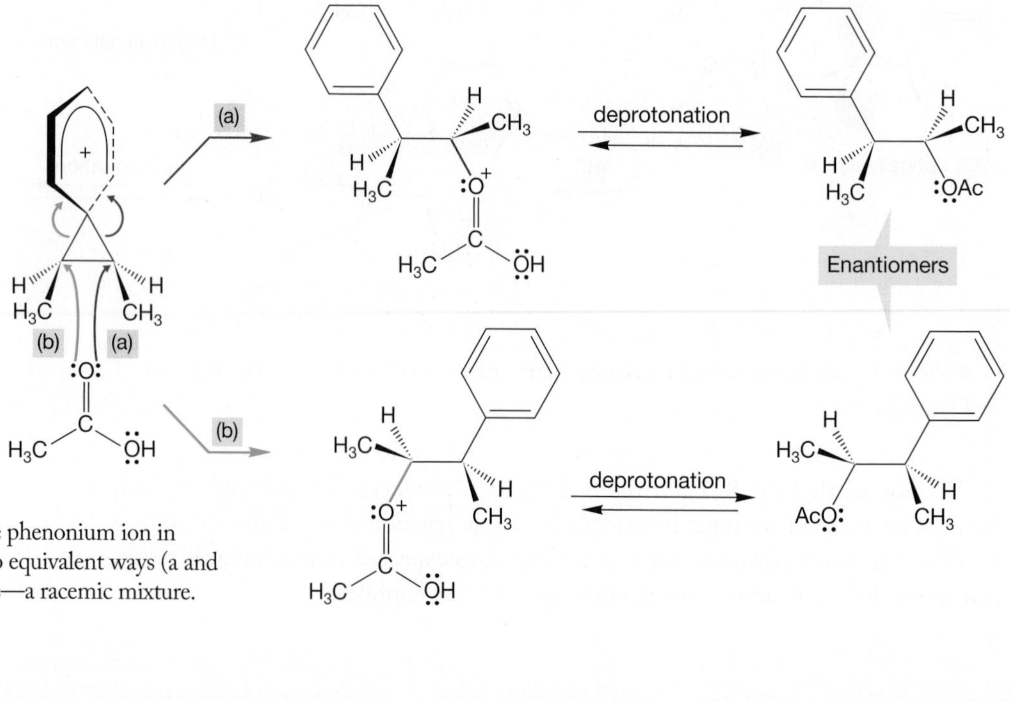

FIGURE 21.27 The phenonium ion is similar to the intermediate ion produced in electrophilic aromatic substitution.

FIGURE 21.28 Opening of the phenonium ion in Figure 21.26 must occur in two equivalent ways (a and b) to give a pair of enantiomers—a racemic mixture.

PROBLEM 21.16 Work through the reaction of the (R,R) optically active isomer of 3-phenyl-2-butyl tosylate in acetic acid (see below). This time the product is the optically active acetate shown.

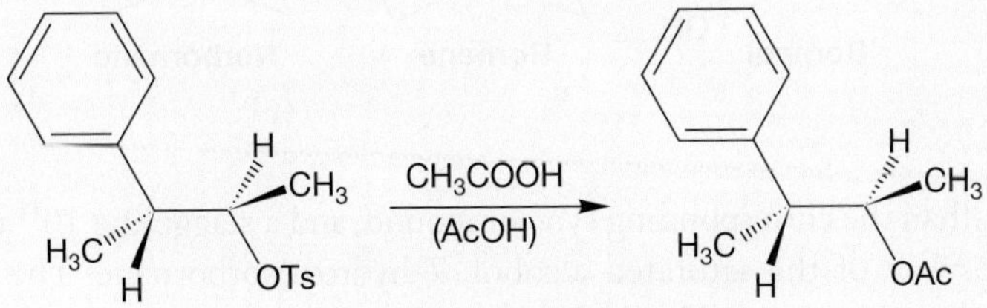

21.3b Carbon–Carbon Double Bonds as Neighboring Groups

Carbon–carbon double bonds are also nucleophiles and can act as neighboring groups in intramolecular displacement reactions. Among the examples you already know are the sequence of intramolecular carbocation additions leading to steroids (p. 560) and intramolecular carbene additions (p. 434). In each of these reactions, a carbon–carbon double bond acts as a nucleophile toward an empty carbon $2p$ orbital as Lewis acid (Fig. 21.29).

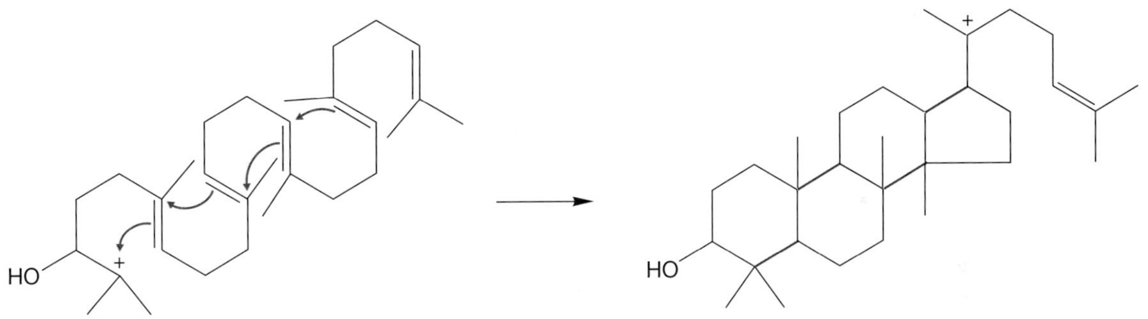

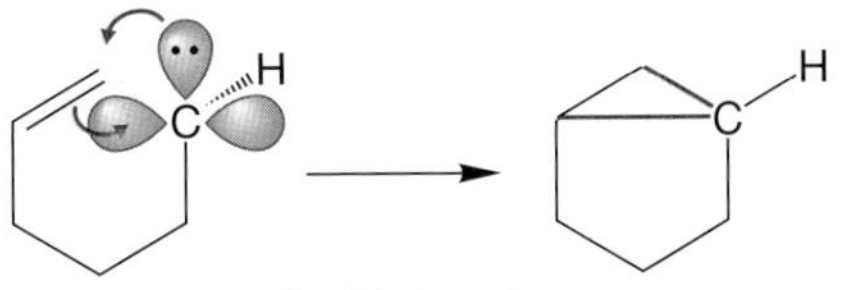

FIGURE 21.29 Two examples of carbon–carbon double bonds acting as intramolecular nucleophiles.

Neighboring carbon–carbon π bonds can influence the rate and control the stereochemistry of a reaction just as a heteroatom can. For example, the tosylate of *anti*-7-hydroxybicyclo[2.2.1]hept-2-ene (7-hydroxynorbornene)[2] reacts in acetic acid 10^7

---

[2]The bicyclo[2.2.1]heptane cage is commonly known as the **norbornyl system**. This obscure name comes from a natural product, borneol, which has the structure shown in Figure 21.30, and comes from the camphor tree, common in Borneo. The prefixes syn and anti locate the direction of the OH or OTs group with respect to the double bond, and nor means "no methyl groups." Thus, 7-hydroxybicyclo[2.2.1]hept-2-ene is 7-hydroxynorborn-2-ene, or, because there is only one possible place for the double bond, 7-hydroxynorbornene.

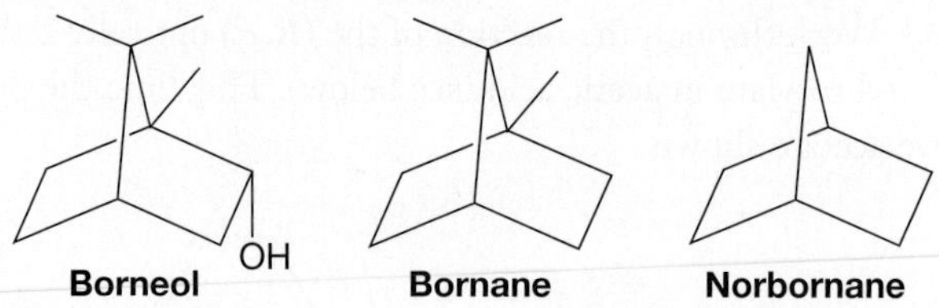

FIGURE 21.30 Norbornane (bicyclo[2.2.1]heptane) is related to the natural product borneol, *endo*-1,7,7-trimethylbicyclo[2.2.1]heptan-2-ol.

Borneol   Bornane   Norbornane

times faster than the corresponding syn compound, and a staggering 10^{11} times faster than the tosylate of the saturated alcohol, 7-hydroxynorbornane. The product is exclusively the anti acetate (Fig. 21.31).

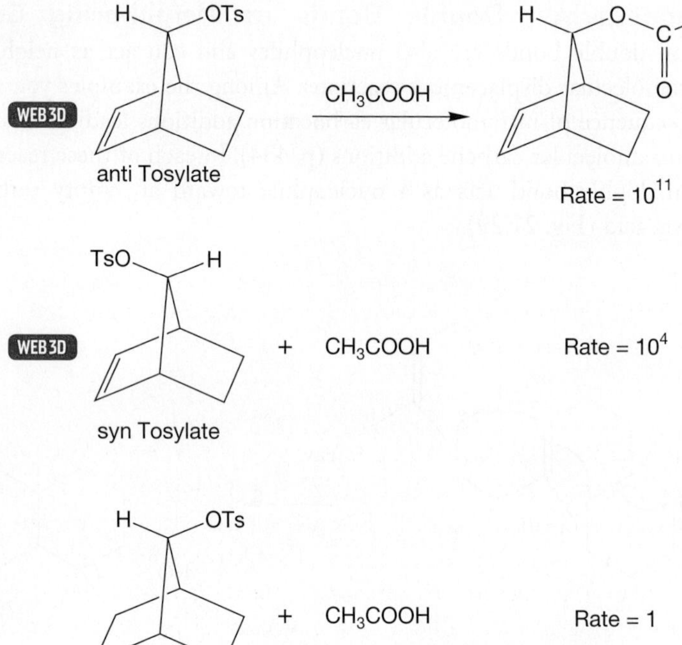

FIGURE 21.31 The anti isomer of 7-norbornenyl tosylates gives the anti acetic acid. The anti tosylate reacts 10^7 times faster than the syn isomer and 10^{11} times faster than the saturated compound!

anti Tosylate

CH₃COOH

Rate = 10^{11}

syn Tosylate   +   CH₃COOH   Rate = 10^4

+   CH₃COOH   Rate = 1

PROBLEM 21.17 There is only one possible place for the double bond in the bicyclo[2.2.1]heptyl ring system. Explain why.

A consistent mechanistic explanation for the huge difference in reaction rates derives from the fact that only in the anti compound is the π system of the double bond in the correct position to assist in ionization of the tosylate. For the double bond in the syn isomer to assist in ionization would require a frontside S_N2 displacement, which, as we know, is an unknown process.

The rate enhancement is attributed to anchimeric assistance by the double bond. Intramolecular displacement of the leaving group is faster than the possible intermolecular reactions and is therefore the dominant process. The stereochemistry of the reaction is explained by noting that addition of acetic acid to the intermediate ion is also an S_N2 reaction and must occur from the rear of the departing leaving group. The net result of the two S_N2 reactions (one intramolecular, one intermolecular) is retention of configuration (Fig. 21.32).

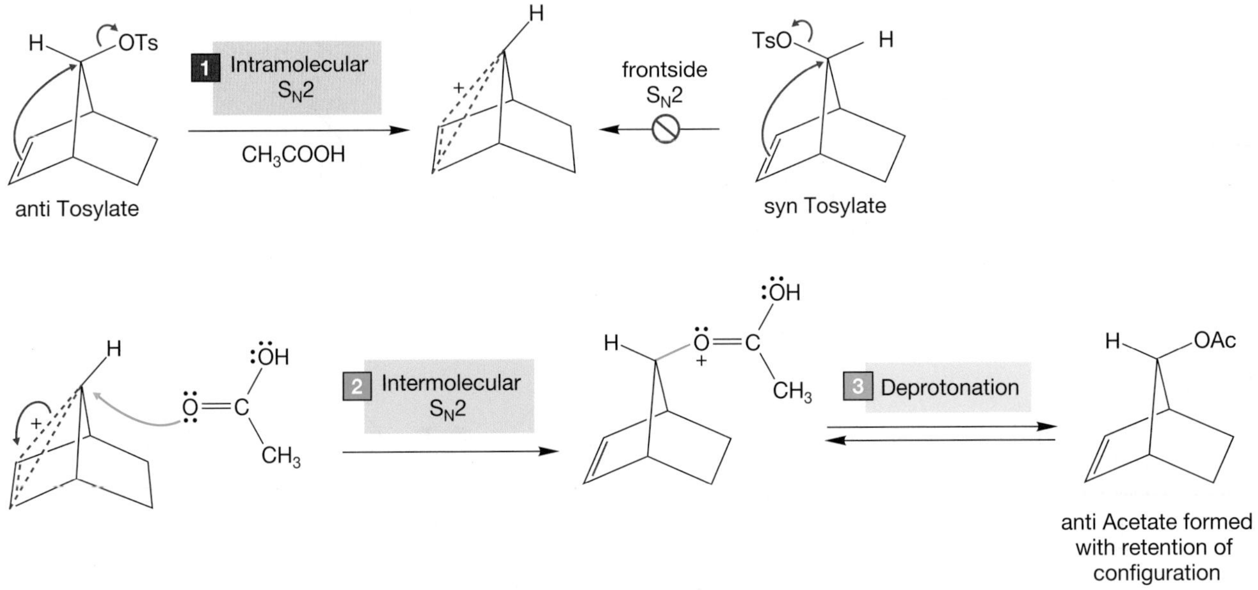

FIGURE 21.32 The carbon–carbon double bond of the anti tosylate, but not of the syn isomer, can react in an intramolecular S_N2 reaction to displace the leaving group. Opening of the cyclic ion must give the anti acetate, the result of net retention.

Think a bit now about the nature of the leaving group in the second, intermolecular S_N2 displacement. Just what is the leaving group, and what do all those dashed bonds in Figure 21.32 mean? To answer these questions well, we must look closely at the structure of the postulated intermediate in this reaction. When we formed three-membered ring intermediates from heteroatoms, there was always a pair of nonbonding electrons available to do the intramolecular displacement. An example is the formation of a bromonium ion by displacement of water by neighboring bromine (Fig. 21.33).

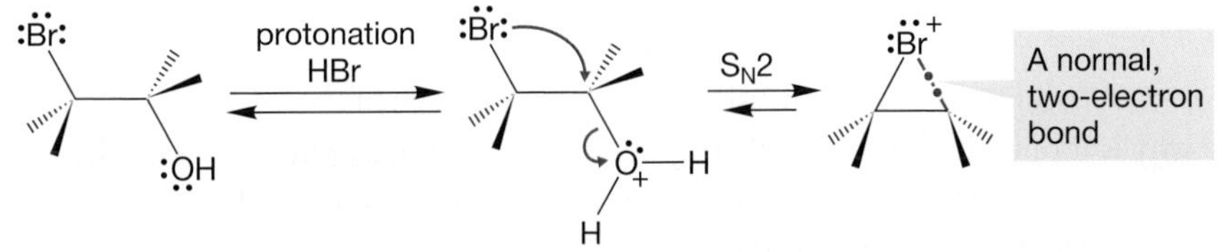

FIGURE 21.33 In a bromonium ion, all bonds are normal two-electron bonds.

Contrast the bromonium ion of Figure 21.33 with the species formed by intramolecular displacement of tosylate using the π bond. In the bromonium ion, there are sufficient electrons to form the new σ bond in the three-membered ring. In the all-carbon species there are not enough electrons. In this cyclic ion, two electrons must suffice to bind three atoms (Fig. 21.34). The ion contains an example of **three-center, two-electron bonding**, and the partial bonds are represented with dashed lines. A resonance formulation shows most clearly how the charge is shared by the three atoms in the ring. The bromonium ion and the cyclic ion of Figure 21.34 are *very* different. When the cyclic bromonium ion is opened, it

FIGURE 21.34 Two views of the cyclic ion from Figure 21.32 and the resonance forms contributing to the ion. Notice the symmetry that is more clearly shown in the lower view.

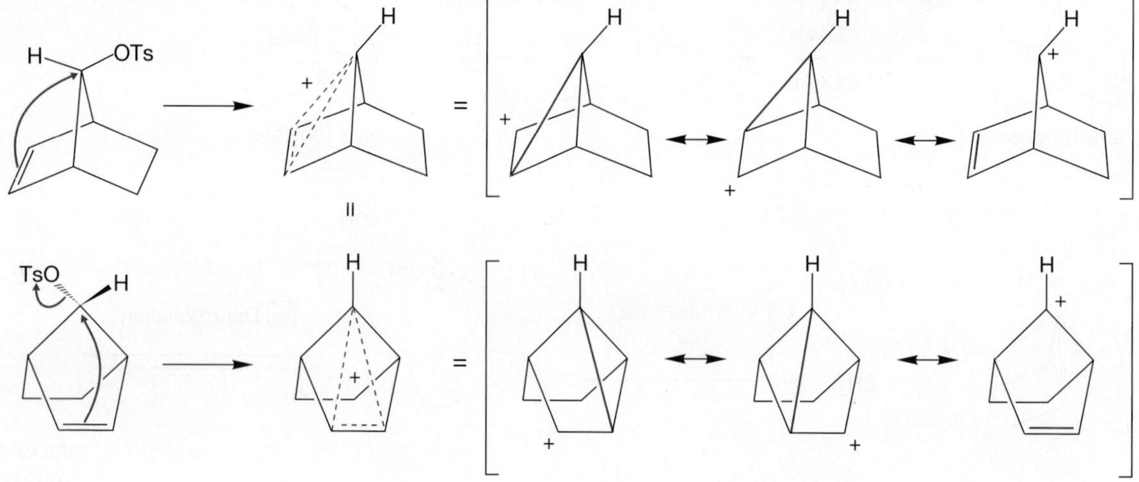

is a normal, two-electron σ bond that is broken. When the all-carbon three-membered ring is attacked, it is the two electrons binding the three carbons that act as the leaving group (Fig. 21.35). The process is still an S_N2 reaction, however, and so there is no reason for softening the requirement for backside displacement.

FIGURE 21.35 When a nucleophile opens a bromonium ion, a pair of electrons is displaced. The same is true for the opening of the two electron–three carbon, cyclic ion.

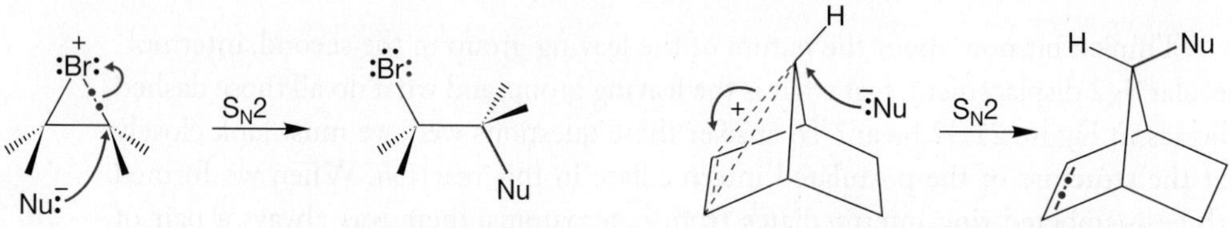

Acceptance of the carbon–carbon double bond as a neighboring group requires you to expand your notions of bonding, and to accept the partial bonds making up the dashed three-center, two-electron delocalized system as a valid bonding arrangement. There are those who resisted this notion mightily, and although they were ultimately shown to be incorrect, their resistance was anything but foolish. The postulated intermediate in this reaction is not a normal species and should be scrutinized with the greatest care.

PROBLEM 21.18 If three atoms share the positive charge of the ion in Figure 21.34, why is it only the 7-position that is attacked by the nucleophile? Explain the lack of another product. *Hint*: Look at the structure of the product formed by addition at the other position in the following reaction:

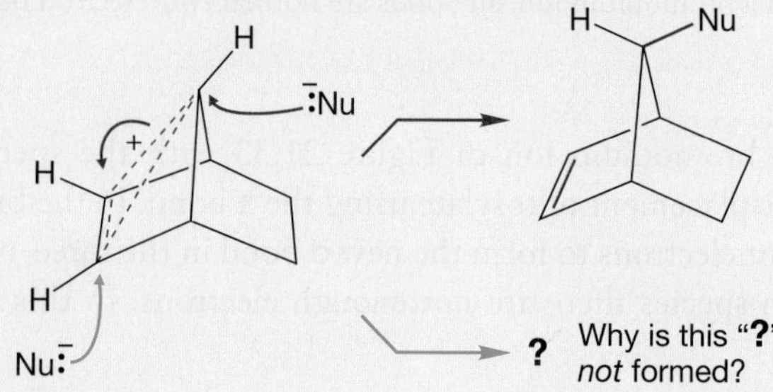

A molecular orbital approach shows how the two electrons can be stabilized through a symmetrical interaction with an empty carbon $2p$ orbital. The intellectual antecedents for this kind of bonding relate to the cyclic H_3^+ molecule in which two electrons are stabilized in a bonding molecular orbital through a cyclic overlap of hydrogen $1s$ orbitals. We can construct this system by allowing a $1s$ orbital to interact with σ and σ* of H_2 as in Figure 21.36 (recall our discussion of the allyl and cyclopropenyl π systems in Chapter 12). The result is a system of three molecular orbitals: one bonding, two antibonding. Two, but only two, electrons can be accommodated in the bonding orbital. The same orbital pattern will emerge for any triangular array of orbitals.

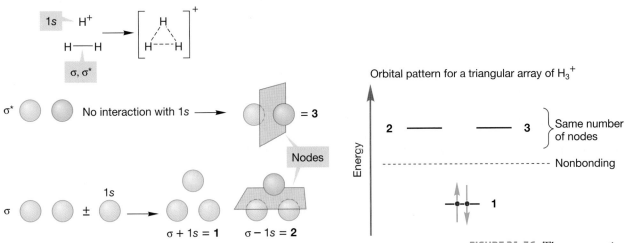

FIGURE 21.36 The construction of the molecular orbitals of cyclic H_3^+.

PROBLEM 21.19 Work out the π molecular orbitals for the cyclopropenium ion shown below.

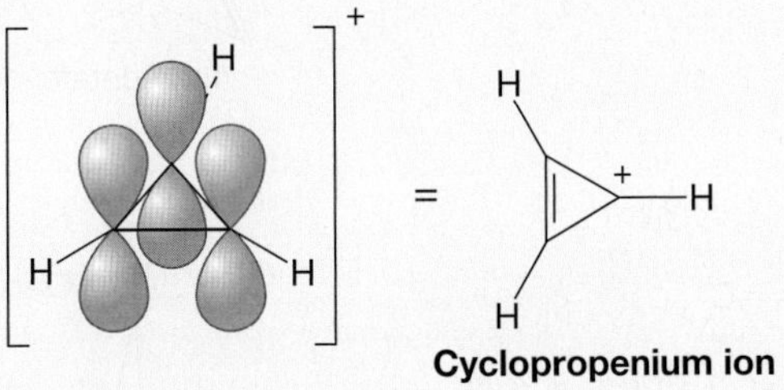

Cyclopropenium ion

Chemists resisting this notion (the "classicists") suggested that there was no need for this new kind of delocalized bonding, and that normal, localized two-electron models would do as well. Let's follow their argument. Remember as we go that they must explain both a rate acceleration for the anti tosylate and the overall retention of stereochemistry.

The two compounds in question here, the syn and anti tosylates, are *different*, and there *must* be some rate difference between them. One can scarcely argue with this point, and one is reduced to relying on the magnitude and direction of the rate acceleration as evidence for the delocalized ion. A classicist might argue that in cases such as this, the rate increase could easily come from the formation of a normal, localized or "classical," carbocationic intermediate. The anti tosylate has the leaving group in an excellent orientation to be attacked by the intramolecular nucleophile, the filled π orbital of the carbon–carbon double bond (Fig. 21.37). The syn tosylate does not.

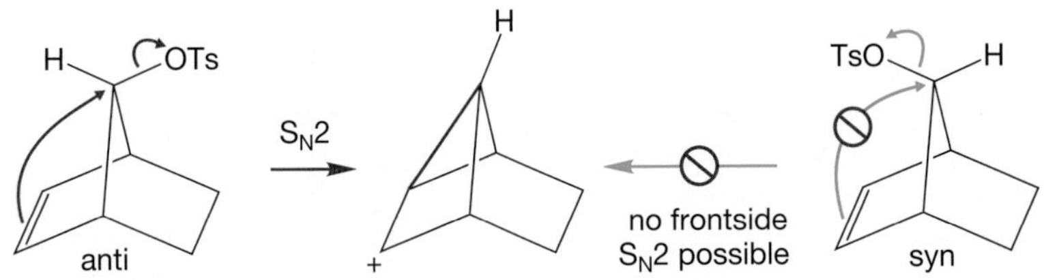

FIGURE 21.37 A "classical" displacement reaction by the π bond of the anti isomer would give a localized ion that could not be formed from the syn isomer.

As in the symmetrical displacement to form the delocalized ion, the required attack on the leaving group from the rear could occur only in the anti compound. Of course, there are two equivalent intermediates possible (**A** and **B**, Fig. 21.38), and it was suggested that they were in equilibrium. The compound with the charge at the 7-position (**C**) might also be a partner in the equilibrium. Be sure you see the difference between the *three* equilibrating structures of Figure 21.38 and the three resonance forms contributing to the *single* structure of Figure 21.34.

FIGURE 21.38 The first-formed ion **A** equilibrates with **B** and, perhaps, **C**.

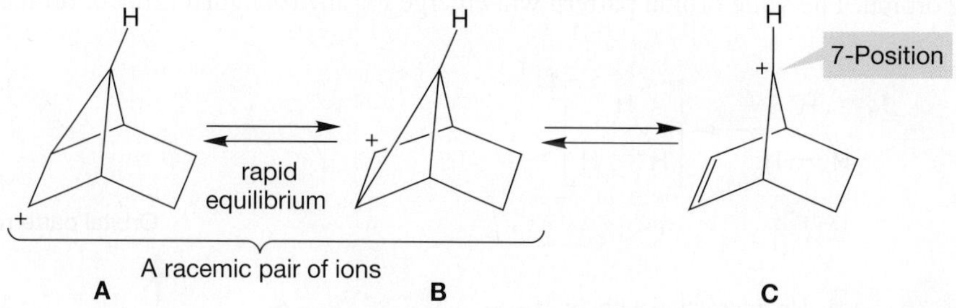

Addition of acetic acid to ions **A** and **B** of Figure 21.38 must give the anti acetate, because the opening is an S$_N$2 reaction. Attack on ion **C** is more problematic, because it is difficult to explain the stereospecificity of the reaction. Still, it must be admitted that the syn and anti faces of the ion are different, and in principle, this difference could lead to a stereospecific reaction (Fig. 21.39).

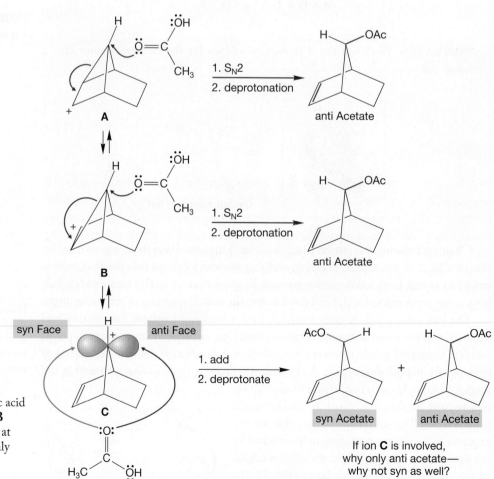

FIGURE 21.39 Addition of acetic acid to the equilibrating ions **A** and **B** must lead to anti acetate. Attack at the 7-position (ion **C**) to give only the anti acetate presents some difficulties.

Stop right now and compare the two mechanistic hypotheses (Fig. 21.40). In one, there is a symmetrical displacement of the leaving group to give a resonance-stabilized delocalized intermediate. In the other, either the leaving group is displaced to give a carbocationic intermediate containing a three-membered ring (**A** and **B**) in equilibrium with at least one other ion, **C**, or a simple S_N1 ionization to **C** takes place.

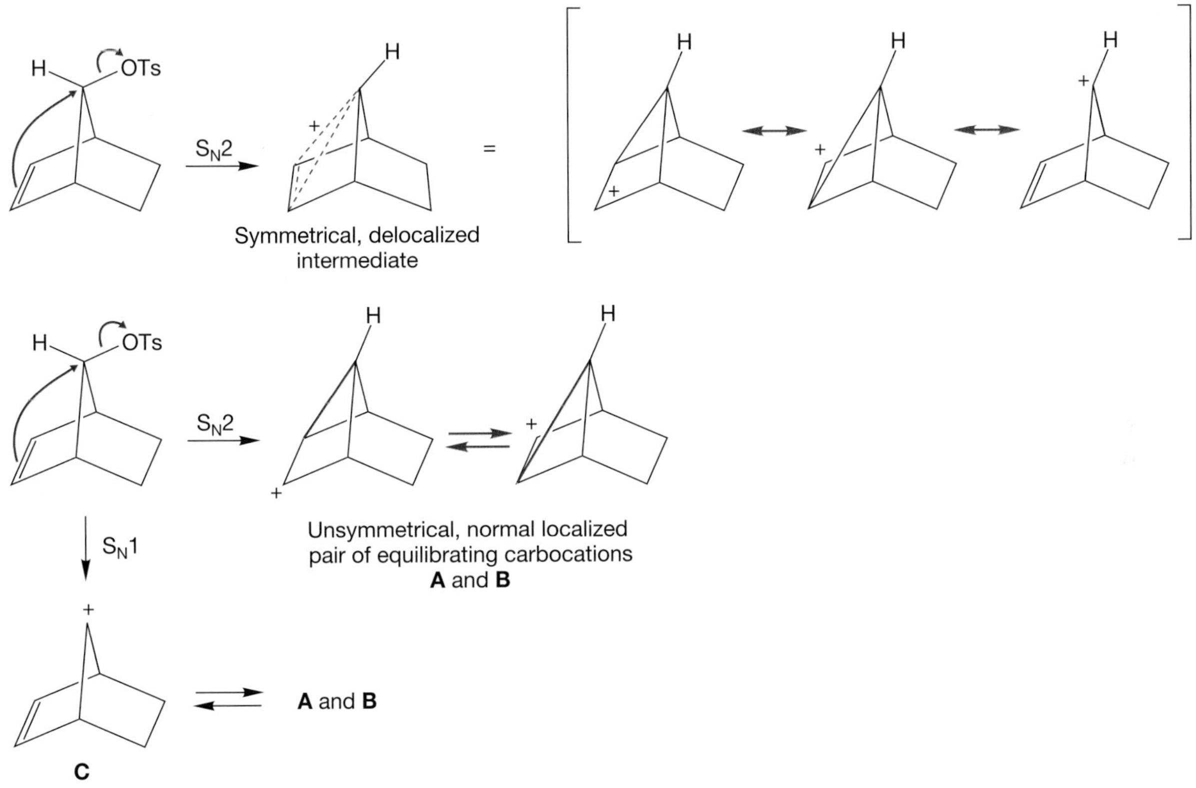

FIGURE 21.40 A comparison of the symmetrical, delocalized intermediate with the localized ions **A**, **B**, and **C**.

The whole difference between the two mechanisms is the formulation of the structure of the intermediate! The structures in the two versions of the mechanism for this reaction are only very subtly different. In one, the constituent structures are contributing resonance forms to a single intermediate delocalized species; in the other they are equilibrating molecules, each with its own separate existence.

Are we now down to questions of chemical trivia? Are we counting angels on pin heads? Many thought so, but we maintain that they were quite misguided in their criticisms of those engaged in the argument. All questions of mechanism come down to the existence and structures of intermediates, and to the number of barriers separating energy minima.

This argument has been recast many times in related reactions. Not all questions have been resolved, but those that have been have generally involved two kinds of evidence. Sometimes a direct observation of the ion in question was made, usually under superacid conditions (p. 679) at low temperature. In other reactions, an experiment like the following one provides evidence for symmetrical participation to produce a delocalized intermediate.

The reactions in acetic acid of the monomethyl and dimethyl derivatives of the *p*-nitrobenzoate of *anti*-bicyclo[2.2.1]hepten-7-ol have been examined. The monomethyl compound reacts faster than the parent (Fig. 21.41), as predicted by either mechanism. A delocalized ion would be stabilized because one of its contributing resonance forms is tertiary. Another way of putting this is to note that the methyl group makes the double bond a stronger nucleophile. In the localized mechanism, a tertiary carbocation is formed and this would surely accelerate the rate of the reaction

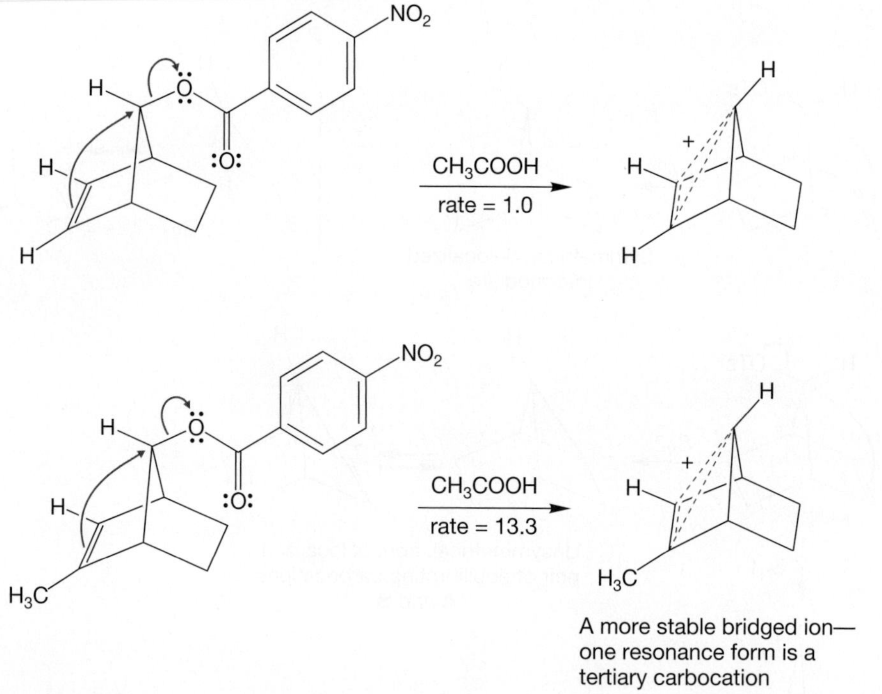

FIGURE 21.41 Formation of a symmetrical, delocalized ion from the methyl-substituted molecule would be faster than ion formation from the parent.

A more stable bridged ion—
one resonance form is a
tertiary carbocation

relative to that of the parent, which can produce only a less stable secondary carbocation (Fig. 21.42). The observed rate increase of 13.3 for the monomethyl analogue is accommodated by both mechanisms.

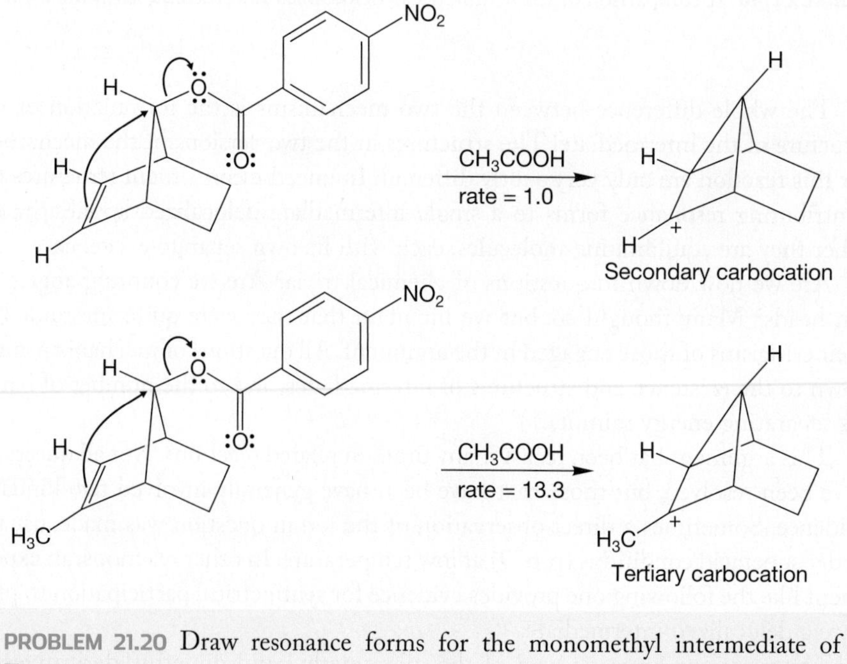

FIGURE 21.42 There would also be a rate increase in the formation of the unsymmetrical, localized ion postulated by the classicists.

Secondary carbocation

Tertiary carbocation

PROBLEM 21.20 Draw resonance forms for the monomethyl intermediate of Figure 21.41.

Now consider the dimethyl compound. If a delocalized intermediate is formed through a symmetrical displacement, both methyl groups exert their influence at the same time. If one methyl group induces a rate effect of 13.3, two should give an effect of 13.3 × 13.3 = 177 (Fig. 21.43).

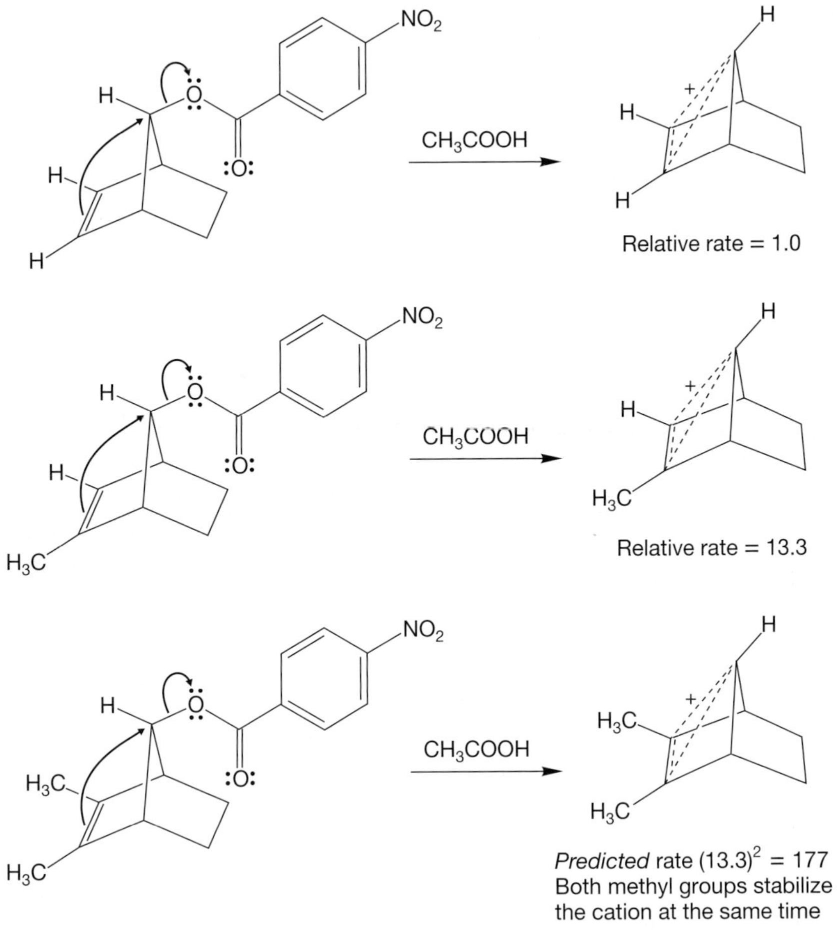

Relative rate = 1.0

Relative rate = 13.3

Predicted rate $(13.3)^2$ = 177
Both methyl groups stabilize
the cation at the same time

FIGURE 21.43 Reaction of the dimethyl compound to give a delocalized ion should be still faster than the reaction of the monomethyl compound. The two methyl groups *simultaneously* stabilize the cation.

On the other hand, if it is the formation of a localized tertiary ion that is important, the second methyl increases the possibilities by a factor of 2, *but the two methyl groups never exert their influence at the same time.* The rate change should approximate a factor of 2 (Fig. 21.44).

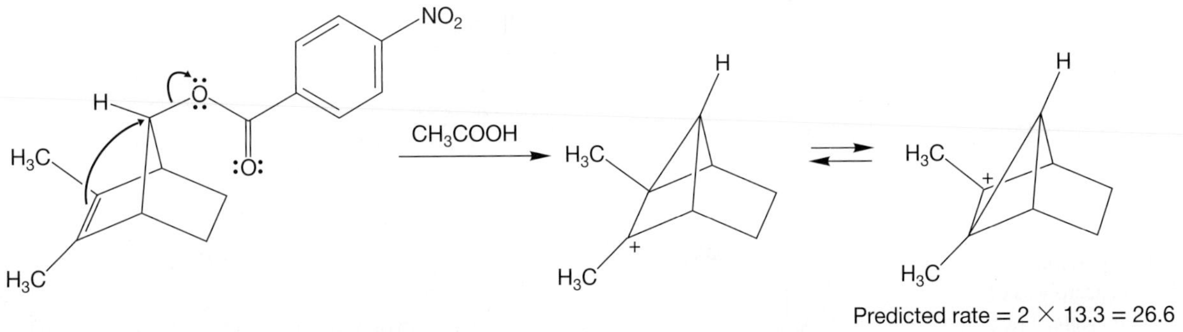

Predicted rate = 2 × 13.3 = 26.6

FIGURE 21.44 In the localized, unbridged ions, the two methyl groups can never stabilize the ion *at the same time.*

The fundamental difference between a mechanism involving a symmetrical delocalized ion and the traditional mechanism, which postulates a localized, less symmetrical ion, is that in the delocalized case both methyl groups exert their stabilizing influence at the same time. In the pair of equilibrating open, localized ions they cannot both act simultaneously.

The observed rate for the dimethyl compound is 148, much closer to that expected for the delocalized intermediate than for the localized ion.

Summary

We have progressed from a discussion of obvious neighboring groups, heteroatoms with nonbonding electrons, to less obvious internal nucleophiles, the filled π orbitals of double bonds. In every case, a pair of internal electrons does an initial intramolecular S_N2 displacement, which is followed by a second, external (intermolecular) S_N2 reaction. The result is overall retention of configuration.

21.4   Single Bonds as Neighboring Groups

Surely, one of the weakest nucleophiles of all would be the electrons occupying the low-energy bonding σ orbitals. Yet there is persuasive evidence that even these tightly held electrons can assist in the ionization of a nearby leaving group. For example, when the tosylate of 3-methyl-2-butanol is treated with acetic acid the result is formation of the rearranged tertiary acetate, not the "expected" secondary acetate (Fig. 21.45).

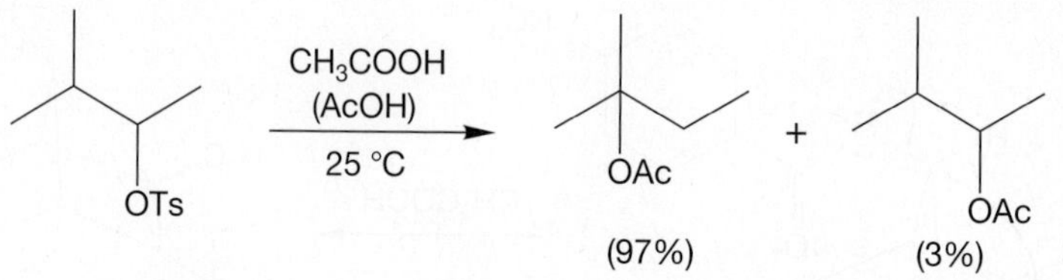

FIGURE 21.45 The reaction of the tosylate of 3-methyl-2-butanol in acetic acid involves a rearrangement.

This result is easily rationalized through a hydride shift to form the relatively stable tertiary carbocation. The interesting mechanistic question is whether the secondary carbocation is an intermediate, or whether the hydride moves *as the leaving group departs* (Fig. 21.46). Thus, the timing of the steps becomes an issue. Is formation of the tertiary carbocation a two- or one-step process? This example recapitulates

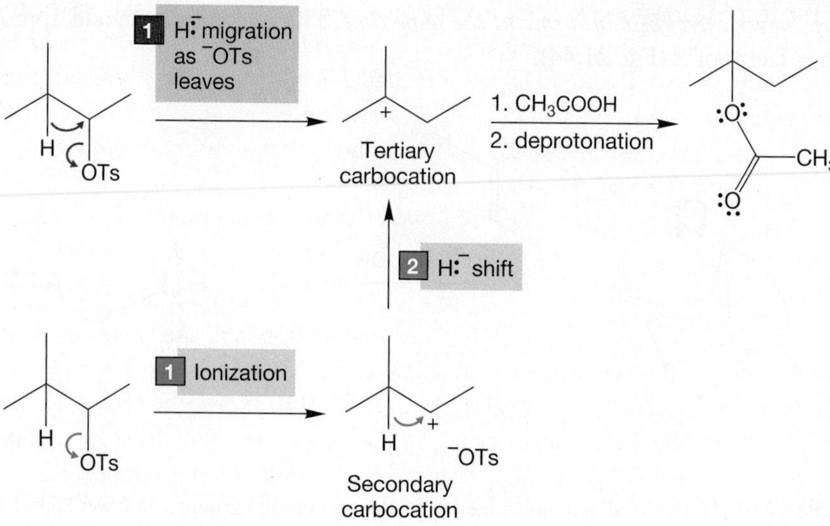

FIGURE 21.46 Two possible mechanisms for the acetolysis of 3-methyl-2-butyl tosylate. The green route is a two-step process. The red path is a one-step process.

the classic mechanistic question: Is there a second intermediate—in this case the secondary carbocation—or is the reaction a concerted formation of the tertiary carbocation (Fig. 21.46)?

> **PROBLEM 21.21** Draw Energy versus Reaction progress diagrams for the two possibilities of Figure 21.46. How many energy barriers separate starting material from the tertiary cation in each mechanism?

The question of the number of steps in the mechanism has been answered in this system by the observation of an **isotope effect**. The rates of reaction of the unlabeled tosylate and the deuterated tosylate of Figure 21.47 are measured. If the mechanism involves slow formation of the secondary carbocation, followed by a faster hydride shift to give the more stable tertiary carbocation, substitution of deuterium for hydrogen would have no great effect on the rate. Even though the carbon–deuterium bond is stronger than the carbon–hydrogen bond, *it is not broken in the rate-determining step of the reaction*. By contrast, if the hydride (or deuteride) shifts as the leaving group departs, the carbon–hydrogen or carbon–deuterium bond *is* broken in the rate-determining step and a substantial effect on the rate would be expected.

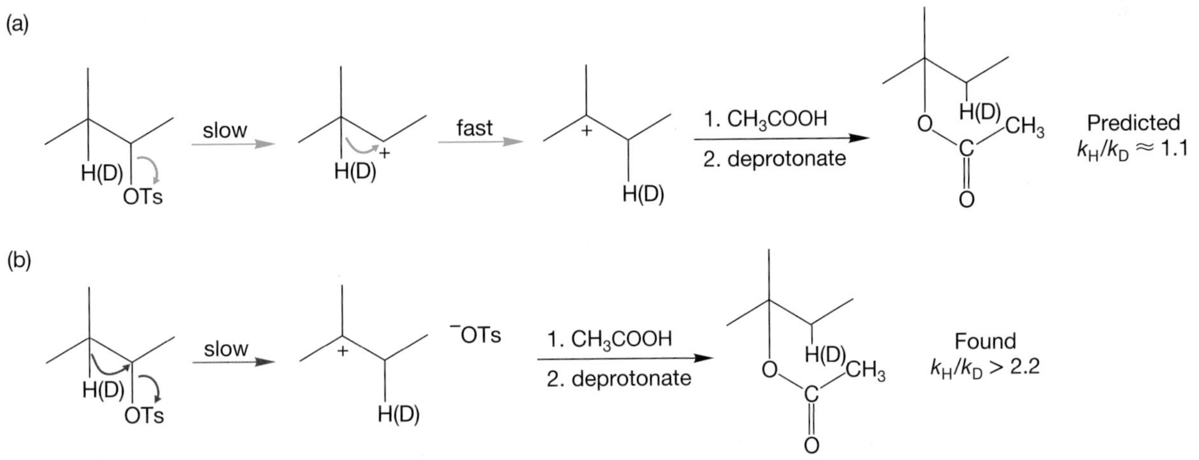

(a)

(b)

FIGURE 21.47 (a) A small isotope effect is predicted for the two-step mechanism. (b) There can only be a substantial isotope effect if breaking the carbon–hydrogen or carbon–deuterium bond occurs in the rate-determining ionization step. A larger isotope effect means that the carbon–hydrogen or carbon–deuterium bond must be breaking in the ionization step of the reaction.

The rate for the unlabeled compound was found to be more than twice as fast as that for the deuterated tosylate. The measured isotope effect ($k_H/k_D = 2.2$) reveals that the hydride must move *as the leaving group departs*. If the slow step in the reaction had involved a simple ionization to a secondary carbocation, *followed* by a hydride shift, only a very small effect of the label would have been observed (precedent suggests $k_H/k_D = 1.1$ to 1.3). The observation of a substantial isotope effect shows that the reaction mechanism must include a neighboring group effect, this time of a neighboring C—H sigma bond!

Now comes an even more subtle question. Given that even single bonds can participate in ionization reactions, are cyclic, delocalized intermediates possible?

> **PROBLEM 21.22** Although it is hard to see why now, phenonium ions (p. 1098) were originally very controversial. Despite their evident similarity to the intermediates of aromatic substitution (Fig. 21.27), there was fierce opposition to the notion of their existence. Explain why a confusion between neighboring π participation and neighboring σ participation may have led to a misunderstanding.

PROBLEM 21.23 There has been something of a controversy over the structure of the ethyl cation, $CH_3CH_2^+$. Contrast orbital pictures of the straightforward open ion with those of a delocalized version, which you might consider making through protonation at the center of the double bond of ethylene. Make analogies! Where have you seen a triangular array of molecular orbitals recently?

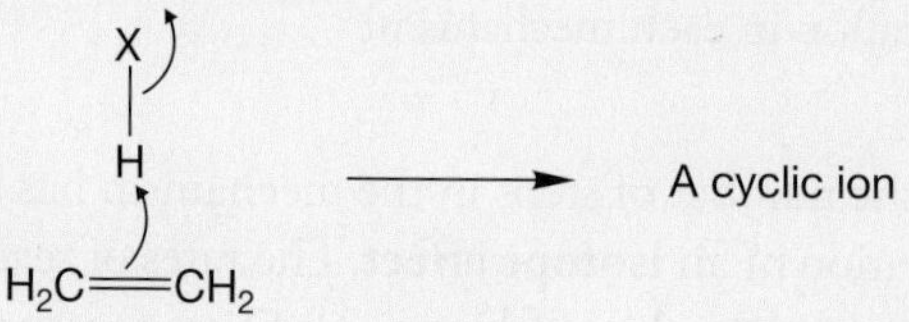

The most famous argument of modern times in organic chemistry concerned the question of delocalized intermediates produced by C—C σ bond participation, and a fierce one it was—and still sometimes is. The reaction in question was the solvolysis (S_N1) reaction of the 2-norbornyl tosylates, the tosylates of *exo-* and *endo-* bicyclo[2.2.1]heptan-2-ol (Fig. 21.48).

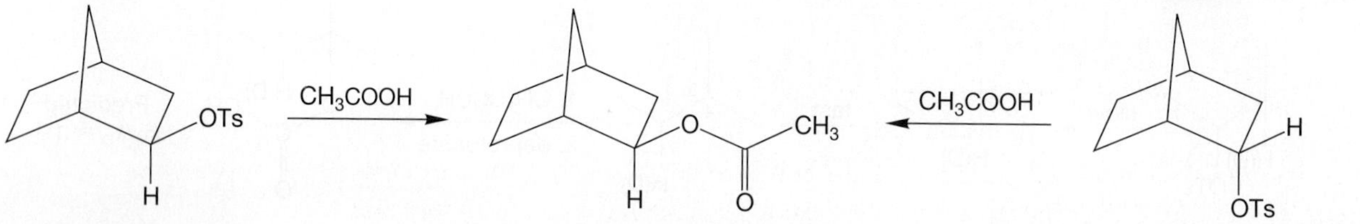

exo-2-Norbornyl tosylate
Relative rate ~ 350

exo-2-Norbornyl acetate

endo-2-Norbornyl tosylate
Relative rate = 1

FIGURE 21.48 Both *endo-* and *exo-*2-norbornyl tosylate react in acetic acid to give *exo-*2-norbornyl acetate. The reaction of the exo isomer is faster.

The exo compound reacted at a faster rate than the endo compound, with the exact difference depending on reaction conditions. At first sight this rate increase would seem to provide evidence for neighboring group participation, but the opponents of participation of C—C σ bonds quite rightly pointed out that the rate acceleration was small (~350) and suggested that this effect could be the result of other differences between the exo and endo tosylates. This argument brings up a most important point. Whenever you are given a rate difference, you must ask yourself, "To what reaction does this rate difference refer?" In this case, the factor of 350 referred to the formation of acetate product. Is that what we are really interested in? Certainly not. When we speak of neighboring group participation, we mean participation in ion formation, and that step is not necessarily equivalent to the product formation step. If ion formation is faster then we need a way to measure the rate acceleration in the ionization step, not the product-forming step (Fig. 21.49).

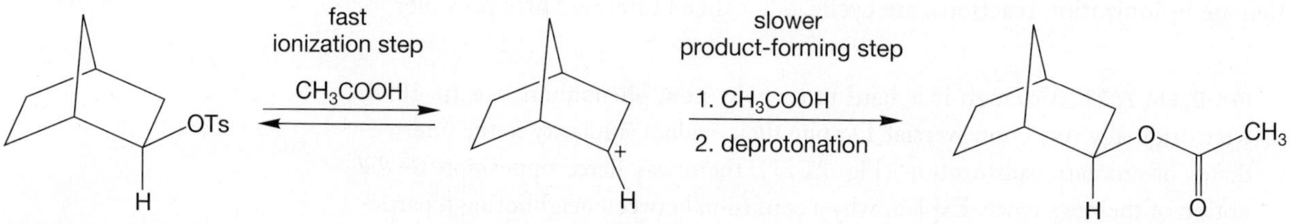

fast
ionization step

slower
product-forming step

1. CH_3COOH
2. deprotonation

FIGURE 21.49 If the rate difference is determined by observing formation of acetate, we may be missing a fast, reversible ionization step.

Perhaps an Energy versus Reaction progress diagram makes the point more clearly. If we measure the rate of appearance of acetate (green), we miss the faster ionization reaction (red) (Fig. 21.50).

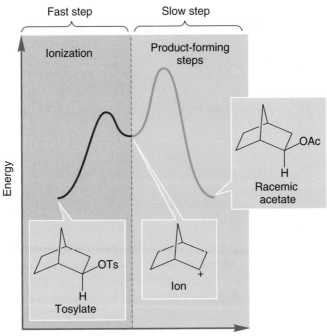

FIGURE 21.50 An Energy versus Reaction progress diagram for the S_N1 solvolysis of *exo*-2-norbornyl tosylate. A rapid ionization precedes a slower formation of acetate.

The stereochemistry of the reaction also points to neighboring group participation. The product from the optically active exo tosylate is not that of a simple S_N2 reaction, which would be the inverted optically active endo acetate (Fig. 21.51). Instead, it is the product of retention, the exo acetate. We have been alerted to look for just this kind of stereochemical result as evidence of a neighboring group effect—it's clue 1. Moreover, optically active tosylate gives racemic acetate. We will see how this is evidence of an apparent rearrangement, which is clue 2.

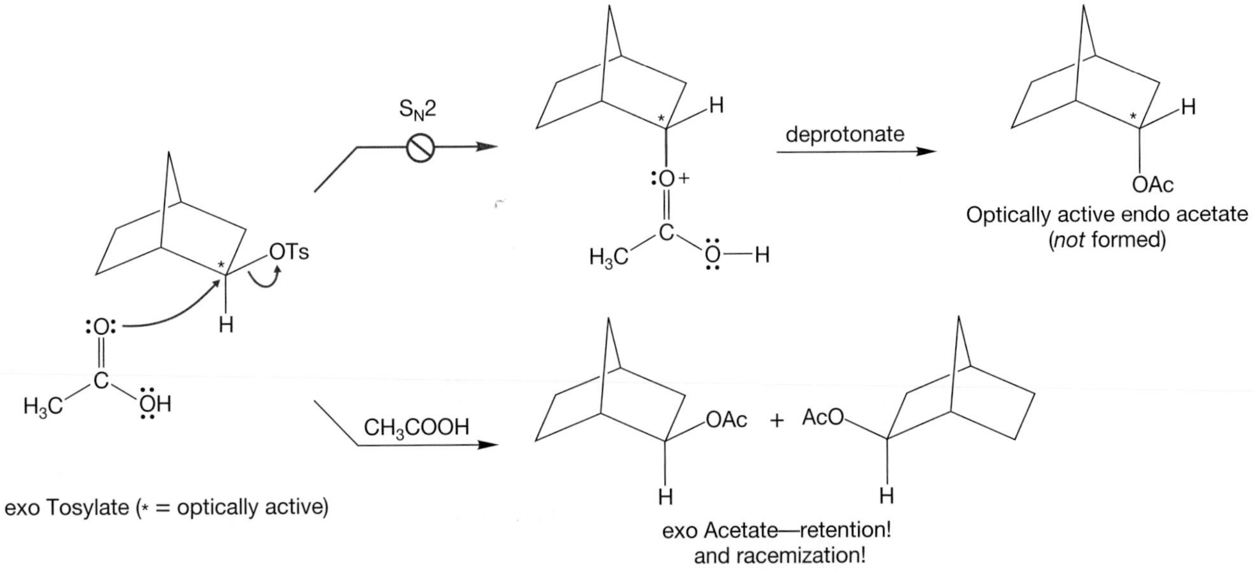

FIGURE 21.51 A simple mechanism involving S_N2 displacement by acetic acid predicts optically active endo acetate, which is not what happens. The product is exo acetate. And it is racemic!

Proponents of neighboring group participation asserted that these pieces of stereo-chemical data constituted powerful evidence for a delocalized ion. First of all, it is only the exo tosylate that can form such an ion directly, and the exo compound did indeed react at a greater rate than the endo isomer (Fig. 21.52).

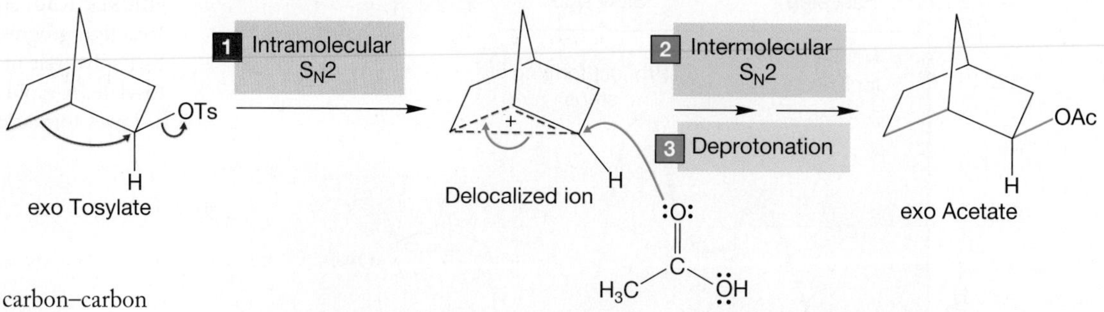

FIGURE 21.52 If the carbon–carbon σ bond assists in ionization (a neighboring group effect), a cyclic, delocalized ion can result. Opening of this ion by acetic acid, followed by deprotonation, would give the observed stereoisomer of the product.

PROBLEM 21.24 Draw resonance forms for the delocalized ion of Figure 21.52.

It was also quite rightly pointed out that the delocalized ion not only explained the observed retention of stereochemistry (exo tosylate gave exo acetate), but demanded a racemic product, because it is achiral (Fig. 21.53). The bridged ion doesn't look symmetrical at first, but if you turn it over, you can see that it clearly is.

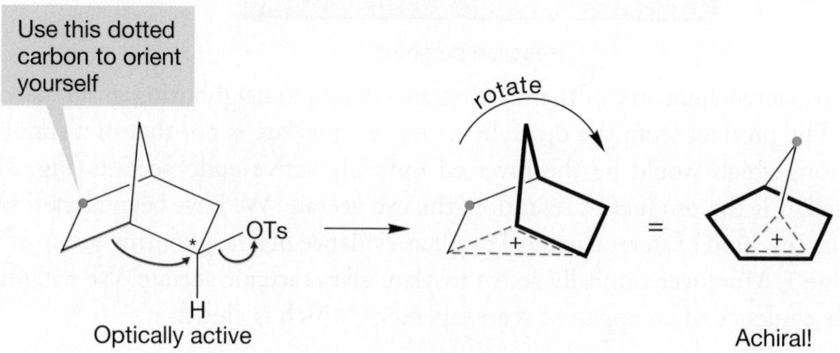

FIGURE 21.53 The observed formation of racemic product (as well as the racemization of recovered starting tosylate) can be explained by the achiral nature of the cyclic ion.

Addition of acetic acid must occur with equal probability at the two equivalent car-bons of the delocalized ion to give both enantiomers of *exo*-2-norbornyl acetate (Fig. 21.54).

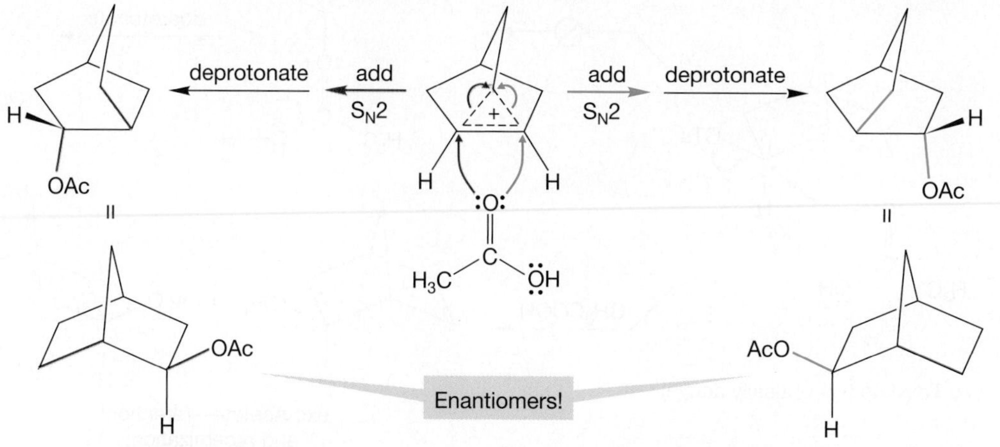

FIGURE 21.54 Addition of acetic acid to the delocalized ion must give a pair of enantiomers in equal amounts—a racemic mixture.

This notion also suggests how to get a better measure of the rate of ionization than simply measuring the rate of acetate formation. *If the initial ionization is reversible*, then starting tosylate will be racemized as soon as the leaving group has sufficiently departed the cation (Fig. 21.55). The tosylate needs to be far enough from the norbornyl framework so that it can't tell which carbon it came from. It needs to become symmetrically disposed with respect to that carbon.

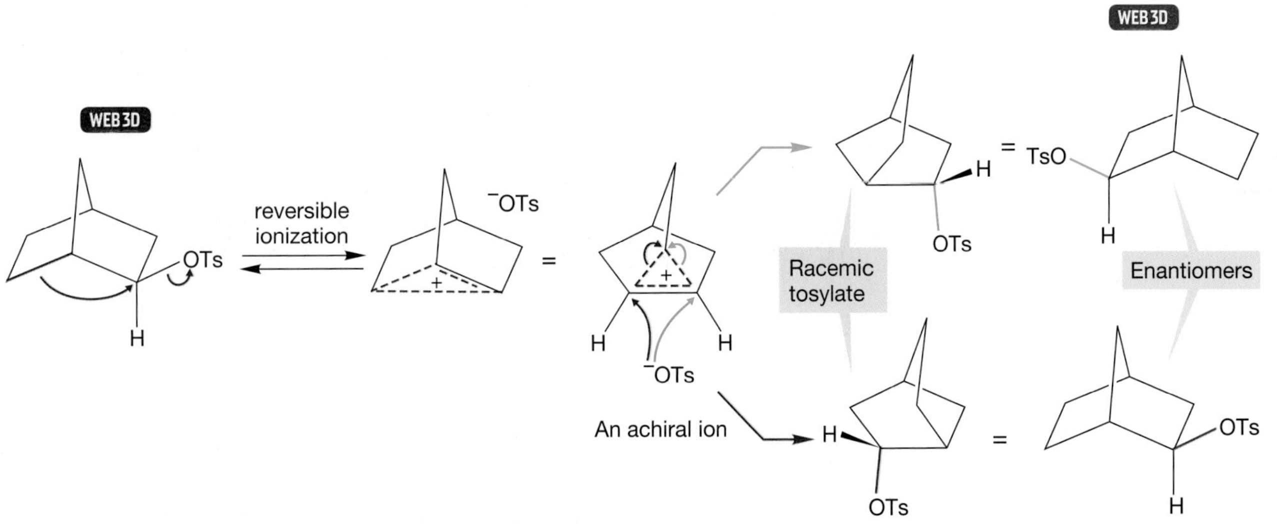

FIGURE 21.55 If the leaving group (tosylate ion) re-adds to the cyclic ion to re-form starting material, racemization results.

So, we should be able to get a better measure of the rate of ionization by determining the rate of racemization of starting tosylate. This clever experiment has now been done and reveals a rate increase of about 1500 for the exo tosylate over the endo, a more impressive rate acceleration than the originally reported factor of 350.

Now we have three traditional kinds of evidence for neighboring group participation: rate acceleration, apparent rearrangement, and retention of stereochemistry. It looks like a strong case can be made for neighboring group participation. What could the opponents of the three-carbon, two-electron system say?

First of all, they pointed out that the rate acceleration was small (they preferred to focus on the original factor of 350, but even 1500 isn't very large), and they were right. Moreover, one could invoke assistance by the σ bond in a simple way, without forming a cyclic ion. That would give some rate acceleration in the exo molecule, as there can be no assistance in the endo isomer (Fig. 21.56).[3]

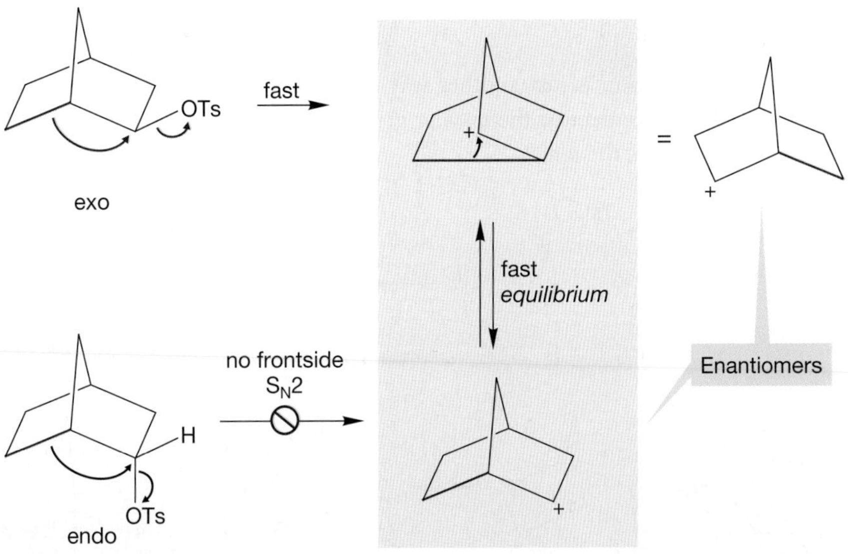

FIGURE 21.56 Formation of two, classical, rapidly equilibrating ions can explain both the fast rate of the exo isomer and the racemization.

---

[3]The classicists actually argued most strongly that the rate of ionization of the endo compound was especially *slow* and that the exo compound was more or less normal.

How would such a species explain the stereochemical results: racemization and retention of the exo configuration? Racemization was easy, one had only to permit the initially formed ion to equilibrate rapidly with the equivalent, mirror-image form as noted in Figure 21.56.

Always keep in mind the difference between the equilibrating open ions of Figure 21.56 and the delocalized ion. The delocalized ion is a single species. But be careful; the resonance forms of the bridged ion do look like the separate, equilibrating forms postulated by the classicists (Fig. 21.57).

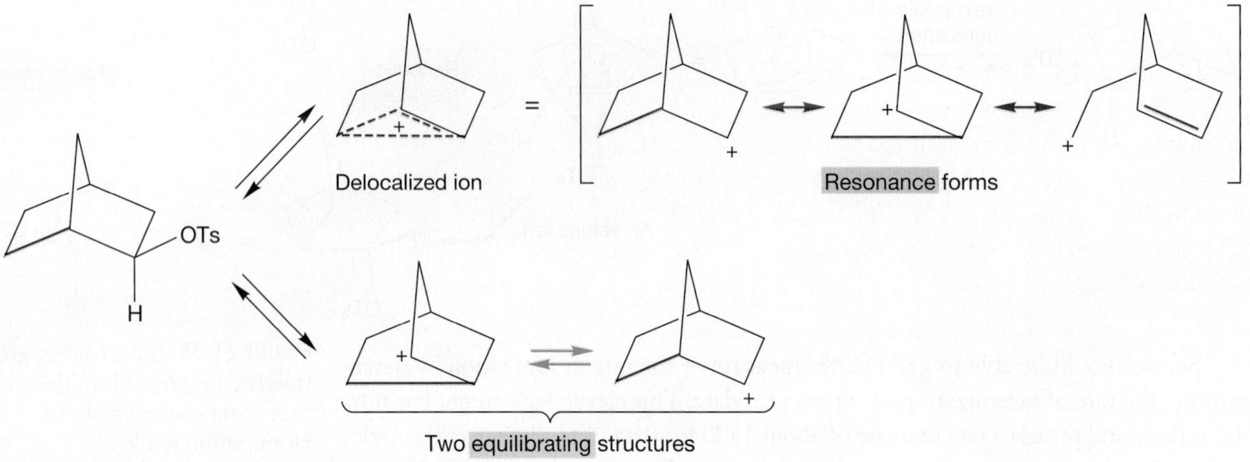

FIGURE 21.57 The two mechanistic proposals compared.

WORKED PROBLEM 21.25 Explain the formation of *exo*-2-norbornyl acetate from the following reaction:

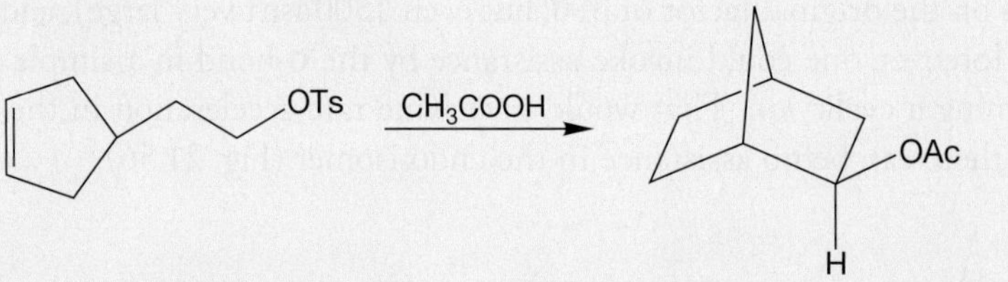

ANSWER This reaction is just another way of forming the 2-norbornyl cation. Symmetrical displacement of the leaving group by the double bond of the five-membered ring gives the ion.

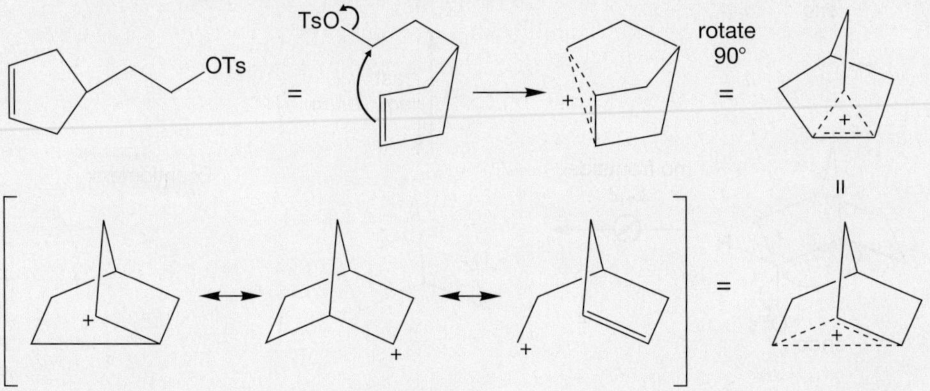

(continued)

Opening by acetic acid in an intermolecular S_N2 reaction must produce the exo acetate.

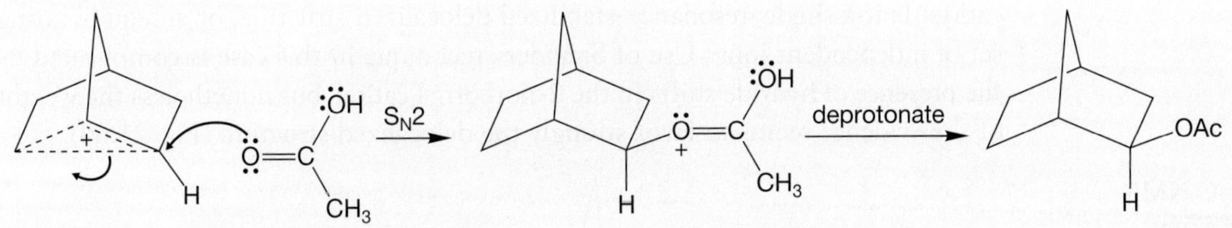

How did the classicists explain retention: the exclusive formation of exo acetate? That task was much harder, and to our taste, never quite achieved. The two faces of the molecule are different, however, and one must admit that in principle, there might be sufficient difference between them to permit the exclusive formation of the exo acetate (Fig. 21.58). Much chemistry of the norbornyl system reveals a real preference for exo reactions, and the question again became one of degree. How much exo addition could be demanded by an open, localized ion?

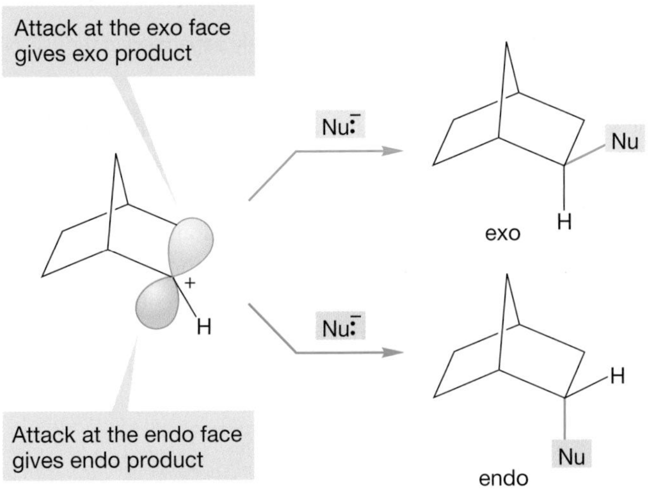

Attack at the exo face gives exo product

Attack at the endo face gives endo product

exo

endo

FIGURE 21.58 The exo and endo faces of the open ion are different. Addition of a nucleophile could, in principle, occur from only the exo side.

In recent years, an increasing number of sophisticated spectroscopic techniques have been applied to this long-standing controversy. The 2-norbornyl cation has actually been detected by NMR spectroscopy at very low temperature in a highly polar but nonnucleophilic solvent. Another spectroscopic technique, pioneered by Martin Saunders of Yale University (b. 1931), focuses on the general differences in the ^{13}C NMR spectra between equilibrating structures and delocalized molecules. For example, the deuteriolabeled cyclohexenyl cation in Figure 21.59, certainly a resonance-stabilized species, shows essentially no difference between the ^{13}C NMR signals for the two carbons that share the positive charge.

FIGURE 21.59 In this resonance-stabilized carbocation, the two "red-dot" carbons sharing the positive charge are not strongly differentiated in the ^{13}C NMR spectrum by deuterium substitution.

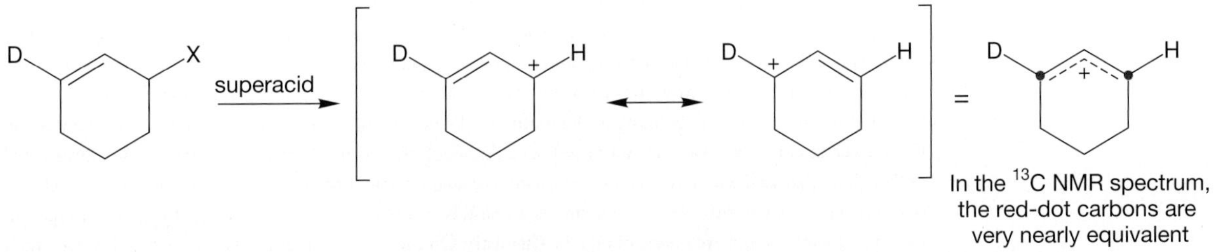

superacid

=

In the ^{13}C NMR spectrum, the red-dot carbons are very nearly equivalent

Saunders demonstrated that in equilibrating structures very great differences, on the order of 100 ppm, appear between such carbons. The Saunders technique gets right to the heart of the matter—to the center of the dispute over the 2-norbornyl cation: Is it a single, resonance-stabilized delocalized structure, or an equilibrating set of independent ions? Use of Saunders' technique in this case is complicated by the presence of hydride shifts in the 2-norbornyl cation, but nonetheless the weight of the evidence seems to favor strongly the delocalized structure (Fig. 21.60).

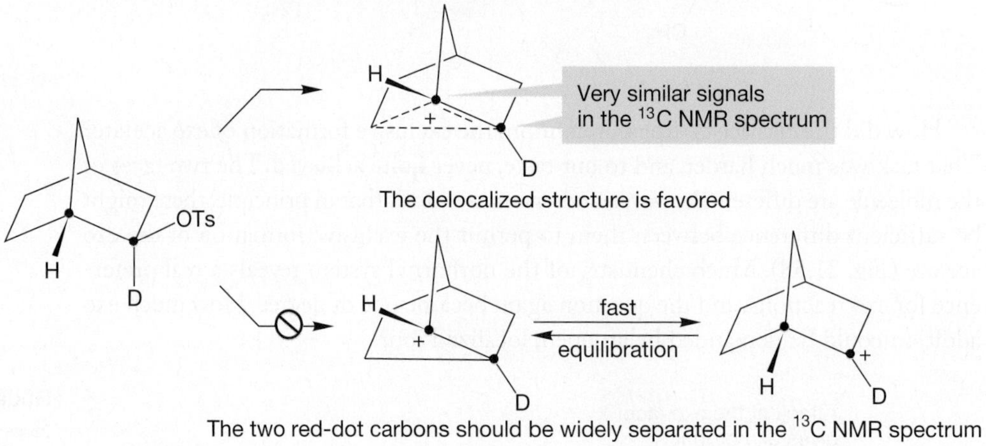

FIGURE 21.60 The ^{13}C NMR spectrum shows only a tiny difference between the deuterated and undeuterated positively charged carbons. The 2-norbornyl cation is delocalized.

Very similar signals in the ^{13}C NMR spectrum

The delocalized structure is favored

fast equilibration

The two red-dot carbons should be widely separated in the ^{13}C NMR spectrum

Are all bicyclic cations delocalized? For a while it certainly seemed so. At least they were drawn that way during what might be called the rococo phase of this theory. But there is no reason why they should *all* be delocalized. It is one form of stabilization, nothing more, and if other forms of stabilization are available, ions will be open, classical species. For example, the classicists were right when they argued that certain tertiary carbocations are localized. Tertiary carbocations are more stable than secondary ones, and so there is no need for the special stabilization of three-center, two-electron bonding. The proponents of delocalization should not have been upset, but they certainly were (Fig. 21.61).[4]

FIGURE 21.61 Not all ions are delocalized. This related but tertiary carbocation is not bridged.

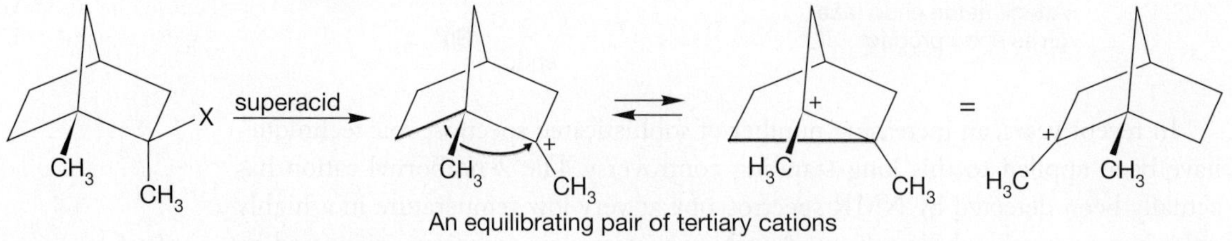

superacid

An equilibrating pair of tertiary cations

[4]The argument over the nature of the 2-norbornyl cation and related molecules was/is remarkable not only for its longevity, but for its intensity as well. The argument became very nasty and both sides definitely lost their cool. Despite the common notion of scientists as Dispassionate Seekers after Truth, we are all too prone to the human vices, in this case overdefending our ideas (our intellectual children). In one corner were the proponents of three-center, two-electron bonding, chief among them Saul Winstein and Donald Cram. In the other was H. C. Brown, who passionately maintained that all could be explained through conventional, localized intermediates. This argument was no battle of lightweights! Brown and Cram have since won the Nobel prize (for other work), and Winstein should have been so honored—and would have been, had he lived long enough. The dispute was raucous and featured many a shouting match at meetings and seminars. It is not easy to be insulting in the body of a scientific paper, because the referees (two or more readers who judge whether the work is publishable) will not stand for it. However, it is much easier to be nasty in the footnotes! There are some really juicy footnotes in the literature on this subject. In our view, both sides went overboard. Although there is much virtue in avoiding new, unwarranted explanations, the classicists were much too resistant to change. There is nothing inherently crazy about a delocalized ion! Three-center, two-electron bonding is a fine way to hold molecules together and is quite common in inorganic chemistry, and even parts of organic chemistry. On the other hand, the delocalizers overdefended their ground. Just because some ions are delocalized in this fashion doesn't mean that all are!

21.5 Coates' Cation

Robert M. Coates (b. 1938) of the University of Illinois has found an ion for which there is unequivocal evidence for a delocalized structure. The ion cannot be represented by a rapidly equilibrating set of localized structures and reacts in acetic acid to give an acetate as shown in Figure 21.62.

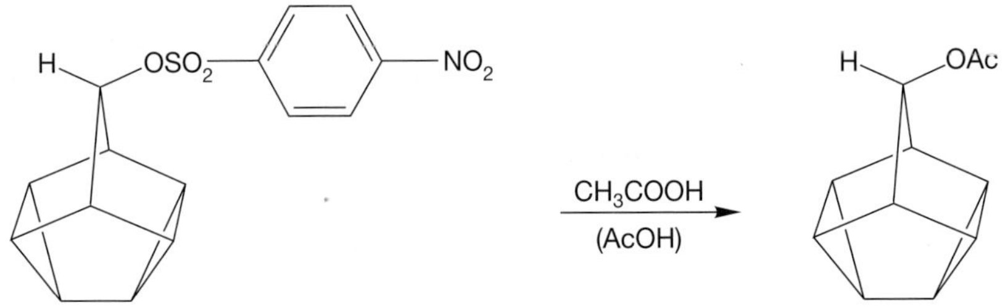

FIGURE 21.62 The solvolysis of this sulfonate in acetic acid gives the related acetate.

At low temperature in superacid, ionization with neighboring group participation of the cyclopropane σ bond gives a delocalized cation that must show three signals in its ^{13}C NMR spectrum (Fig. 21.63).

In this molecule, an equilibrating set of ions must give quite a different ^{13}C NMR spectrum.

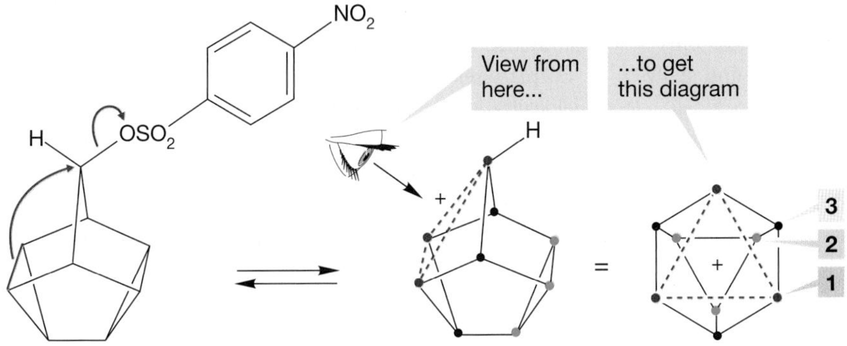

FIGURE 21.63 Participation of the carbon–carbon σ bond to give a symmetrical delocalized ion. Notice that three different kinds of carbons exist. The ^{13}C NMR spectrum will consist of three lines.

PROBLEM 21.26 How many signals would a rapidly equilibrating "Coates' cation" show in its ^{13}C NMR spectrum?

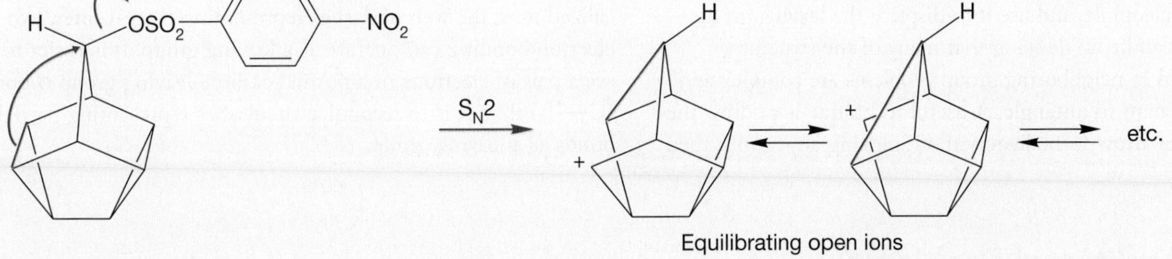

Happily, the sulfonate ionizes to give a stable ion at low temperature in superacid and the case is settled, because the spectrum consists of exactly the three signals required for a delocalized structure: no more and no fewer. Coates' cation is a three-carbon, two-electron system, there is no doubt about it.

21.6 Summary

New Concepts

One concept dominates this chapter. The chemistry of many organic molecules containing leaving groups is strongly influenced by the presence of an internal nucleophile, a neighboring group. Internal nucleophiles can increase the rate of ionization of such compounds by providing anchimeric assistance in the ionization step. Moreover, the structures of the products derived from the ion produced through intramolecular displacement can be subtly different from those expected of simple intermolecular displacement. The three clues to the operation of a neighboring group effect are (1) an unusual stereochemical result, generally retention of configuration where inversion might have been expected; (2) a rearrangement, which is usually the result of formation of a cyclic ion through *intra*molecular displacement of the leaving group; and (3) an unexpectedly fast rate.

The cyclic ions formed by neighboring group displacement are of two kinds. There are classical species in which all bonds are two-electron bonds, and there are more complex, nonclassical or delocalized species in which three-center, two-electron bonding is the rule. Figure 21.40 illustrates the difference.

Key Terms

anchimeric assistance (p. 1089)
episulfonium ion (p. 1088)
isotope effect (p. 1109)

neighboring group effect (p. 1081)
norbornyl system (p. 1099)
phenonium ion (p. 1098)

three-center, two-electron bonding
(p. 1102)

Reactions, Mechanisms, and Tools

This chapter features intramolecular displacement using a variety of internal nucleophiles. Some, such as the heteroatoms bearing lone pairs of electrons, are easy to understand, and really just recapitulate intermolecular chemistry (Fig. 21.2).

Other nucleophiles, such as the electrons in π or σ bonds, are less obvious extensions of old chemistry. Here, the nucleophiles involved are very weak and there often is not a clear analogy to chemistry you already know. One case in which there is a good connection is formation of phenonium ions through the action of a benzene ring as a neighboring group. The analogy is

to the intermolecular reaction of benzenes with electrophiles in the aromatic substitution reaction (Figs. 21.26 and 21.27).

Displacement by carbon–carbon π or σ bonds can lead to delocalized ions about which there has been much controversy. It now appears that such intermediates are involved in some ionizations, and that the three-center, two-electron bonding required is an energetically favorable situation, at least for highly electron-deficient systems such as carbocations (Fig. 21.57).

Common Errors

The problem with this material is mostly one of recognition. Once it is clear that you have a neighboring group problem, essentially every difficulty is resolved by a pair of S_N2 reactions: the first intramolecular, the second intermolecular. Search for the internal nucleophile, and use it to displace the leaving group.

There can be no denying that many of the structures encountered in neighboring group problems are complex and sometimes hard to untangle. A useful technique is to draw the result of the arrow formalism without moving any atoms, then

to relax the structure, probably containing long or even bent bonds, to a more realistic picture. Trying to do both steps at once is dangerous.

A minor problem appears in recognizing that in some delocalized ions, the web of dashes representing three-center, two-electron bonding can operate as a leaving group. It is easier to see a pair of electrons in a normal carbon–leaving group σ bond (C—L) than it is to recognize the dashes representing partial bonds as a leaving group.

21.7 Additional Problems

PROBLEM 21.27 Here comes a problem for rabbits. Provide a mechanism for this simple change.

$$BrCH_2CH_2CH_2CH_2NH_2 \xrightarrow[\substack{2.\ KOH/H_2O \\ (neutralize)}]{1.\ H_2O,\ (CH_3)_2CHOH}$$

PROBLEM 21.28 Similarly, tetrahydrofuran is the product of this next reaction. Provide a mechanism and explain why this process is much faster than the similar reaction of 3-chloro-1-propanol, a molecule that gives the corresponding diol as product.

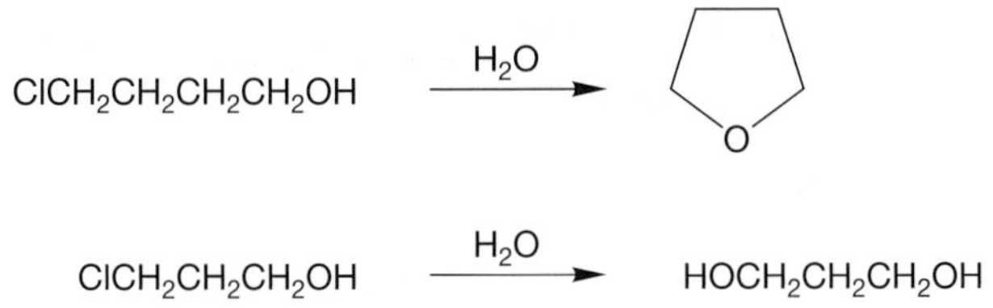
ClCH2CH2CH2CH2OH
H2O
O
ClCH2CH2CH2OH
H2O
HOCH2CH2CH2OH

PROBLEM 21.29 Provide a detailed mechanism showing all intermediates for the following reaction:

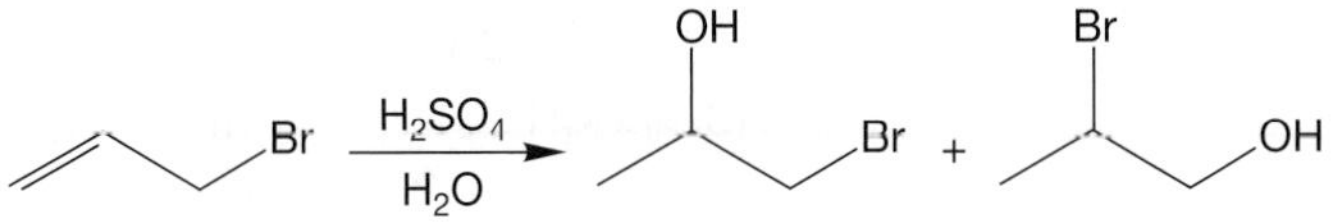
Br
H2SO4
H2O
OH
Br
+
Br
OH

PROBLEM 21.30 Provide a detailed mechansim for the following reaction.

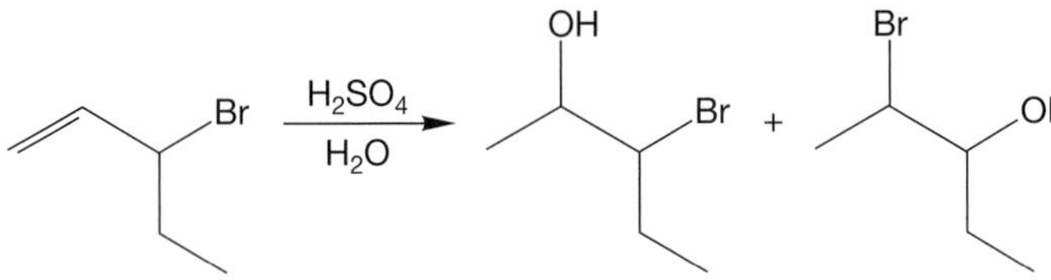
Br
H2SO4
H2O
OH
Br
+
Br
OH

PROBLEM 21.31 Now for something a little more interesting. Explain why the regiochemical results of the following two reactions are so different.

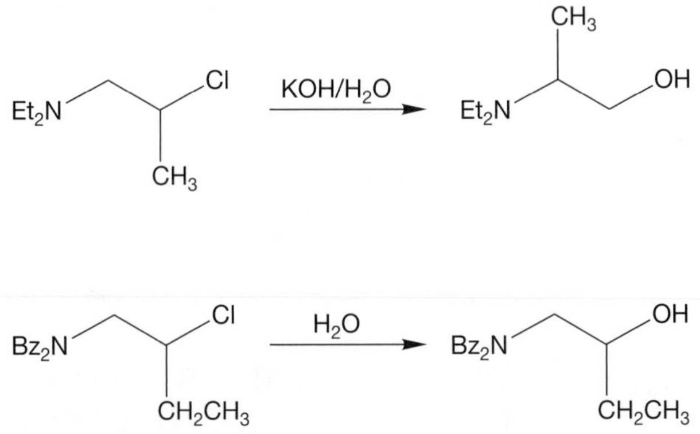
Et2N
Cl
CH3
KOH/H2O
CH3
OH
Et2N
Bz2N
Cl
CH2CH3
H2O
OH
Bz2N
CH2CH3

Et = CH_3CH_2
Bz = $PhCH_2$

PROBLEM 21.32 Addition of HCl to the molecule shown below is 10^4 times slower than addition of HCl to cyclopentene. The product is exclusively (1*R*,2*R*)-1,2-dichloro-1-methylcyclopentane. Provide a mechanism that accounts for the stereospecificity observed. Why is the reaction slow? Remember that chlorine is not a particularly large atom.

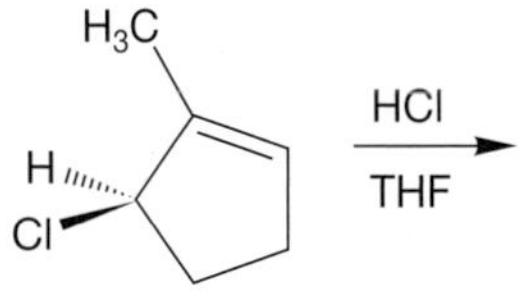
H3C
H
Cl
HCl
THF

PROBLEM 21.33 Explain why compound **1** reacts much faster than **2** (140,000 times) and **3** (190,000 times) in acetic acid. Does this information about rates let you decide whether the reaction proceeds through a delocalized ion?

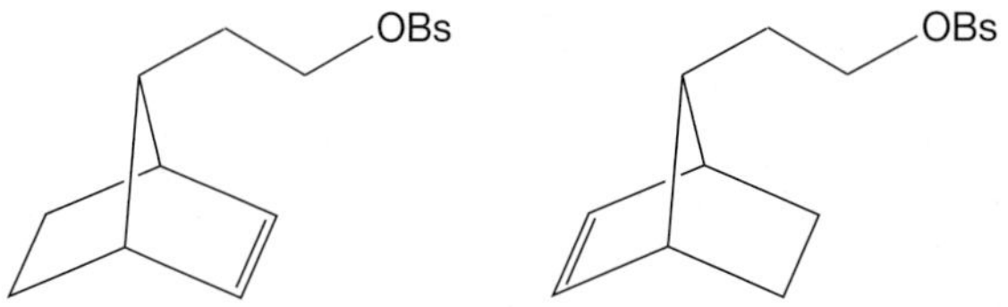
OBs
OBs

1 **2**

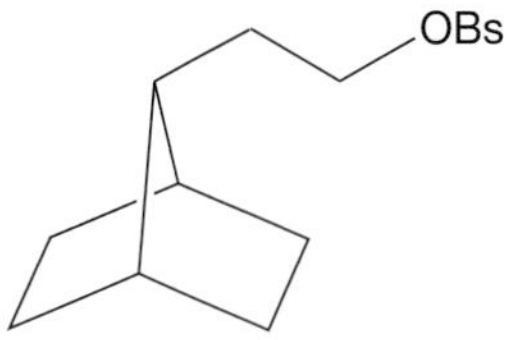
OBs

3

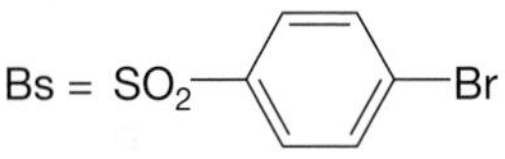
Bs = SO2
Br

PROBLEM 21.34 Compound **1** reacts 10^5–10^6 times faster than its cis isomer **2**. The product is **3**. Explain.

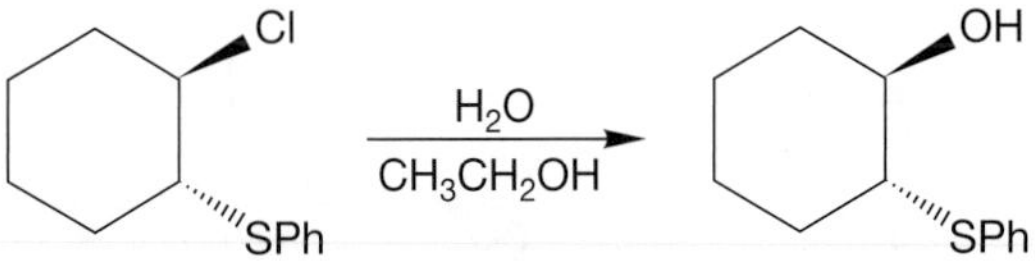
Cl
SPh
H2O
CH3CH2OH
OH
SPh

1 **3**

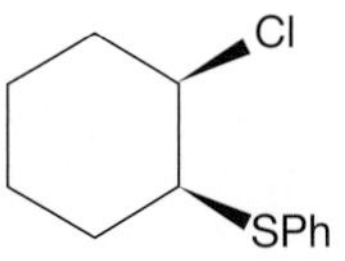
Cl
SPh

2

PROBLEM 21.35 When Z = OCH₃, about 93% of the products of solvolysis of **1** can be attributed to neighboring phenyl group participation. When Z = NO₂, essentially none of the products come from neighboring participation by phenyl. Explain.

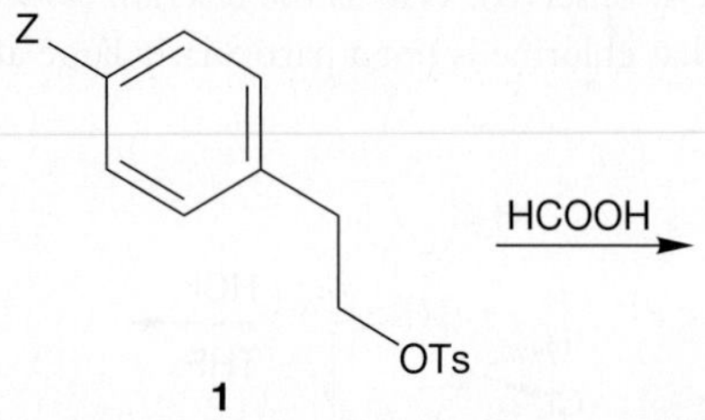

PROBLEM 21.36 In Figure 21.30, we neglected to tell you about the product of reaction of *syn*-7-norbornenyl tosylate (**1**). The product of solvolysis in water is compound **2**. Provide an arrow formalism mechanism and an explanation for the rate acceleration of 10^4 relative to **3**.

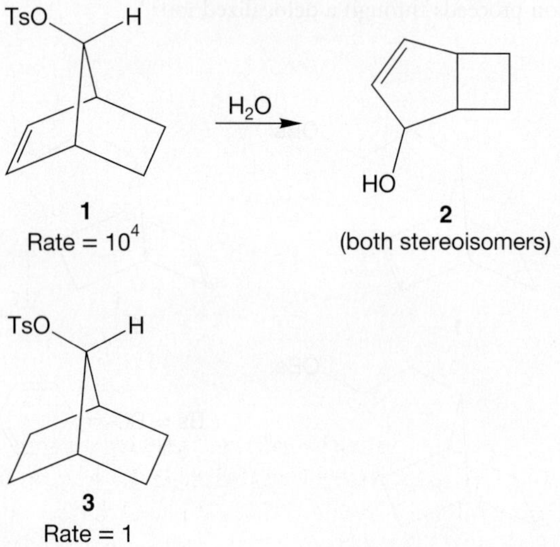

PROBLEM 21.37 Provide a mechanism for the following reaction of **1**. Be sure to explain the observed stereochemistry and the fact that the corresponding trans compound (**2**) does not react in this way.

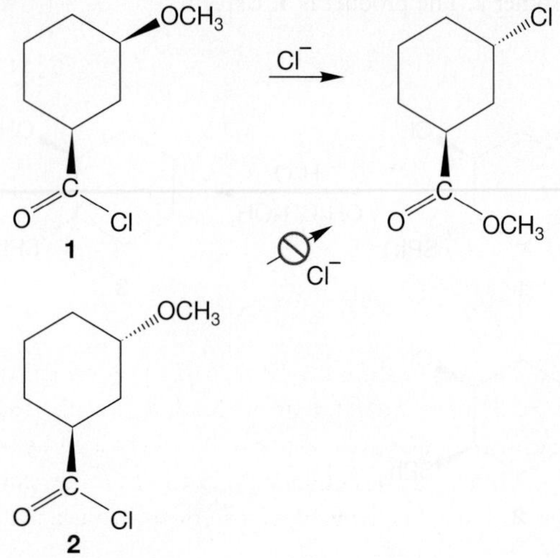

PROBLEM 21.38 Reaction of **1** with bromine leads to a neutral product of the formula $C_{18}H_{17}BrO_2$. Propose a structure for the product and a mechanism for its formation.

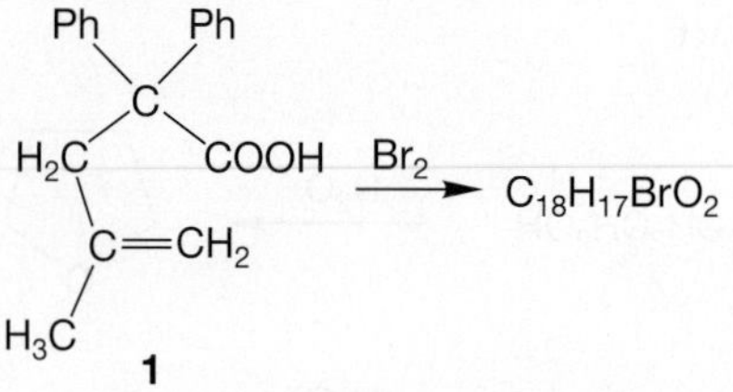

PROBLEM 21.39 Provide an arrow formalism for the following reaction:

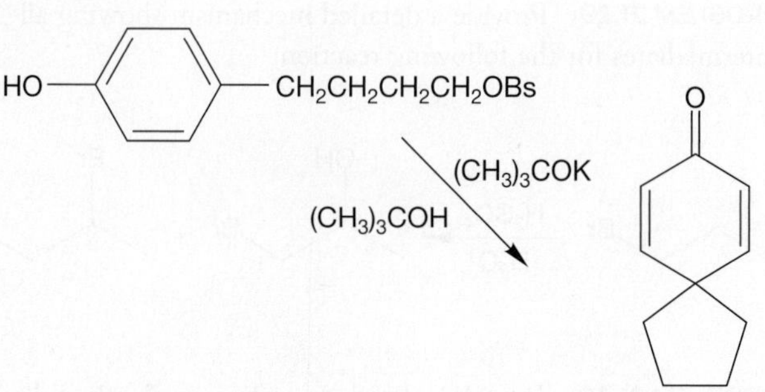

PROBLEM 21.40 Propose a mechanism for the following change. Be sure to explain the observed stereochemistry.

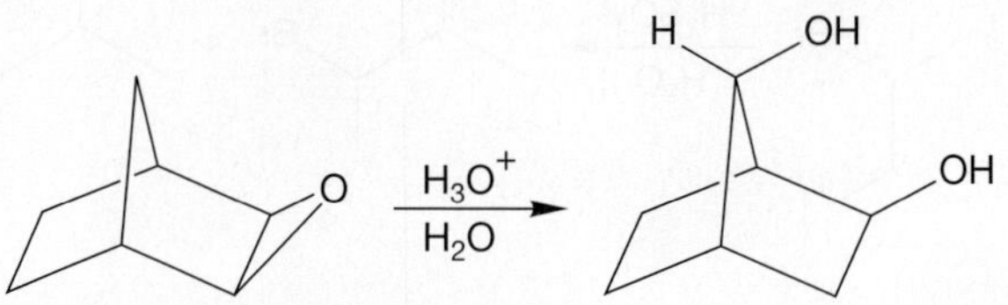

PROBLEM 21.41 What do the relative rate data suggest about the nature of the carbocationic intermediates in the following solvolysis reactions? *Hint*: See Figure 21.43 and Problem 21.23.

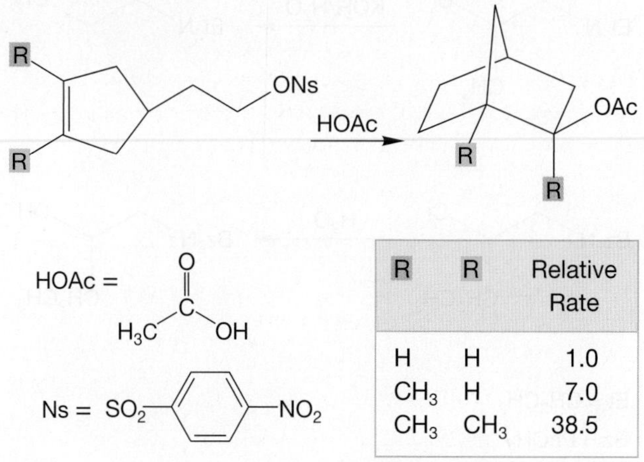

R	R	Relative Rate
H	H	1.0
CH₃	H	7.0
CH₃	CH₃	38.5

PROBLEM 21.42 The concerted rearrangements we saw in Section 18.14 (p. 914) involve neighboring group participation. With that hint, you can do this problem. Provide arrow formalism mechanisms for the following reactions. *Hint*: See pp. 916 and 919.

(a)

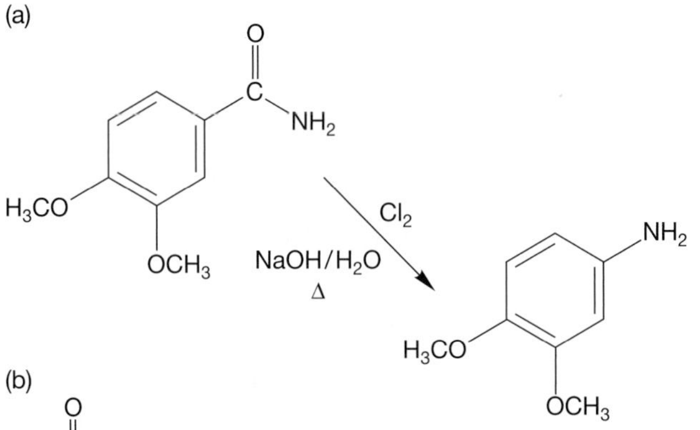

(b)

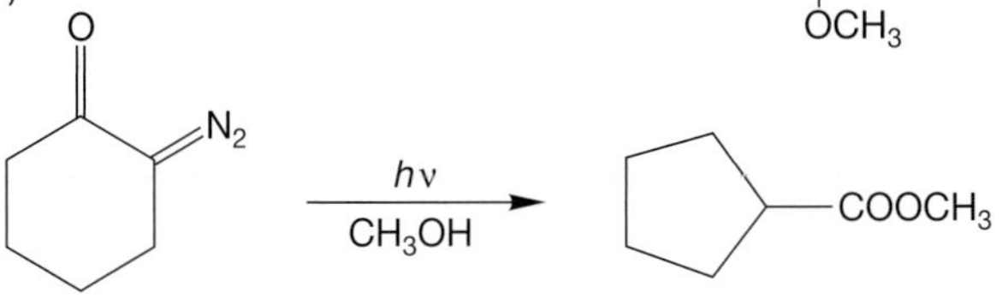

PROBLEM 21.43 Compound **1** reacts about 15 times faster than neopentyl brosylate (**2**). The product is isobutyraldehyde. Provide a mechanism and explain the rate difference.

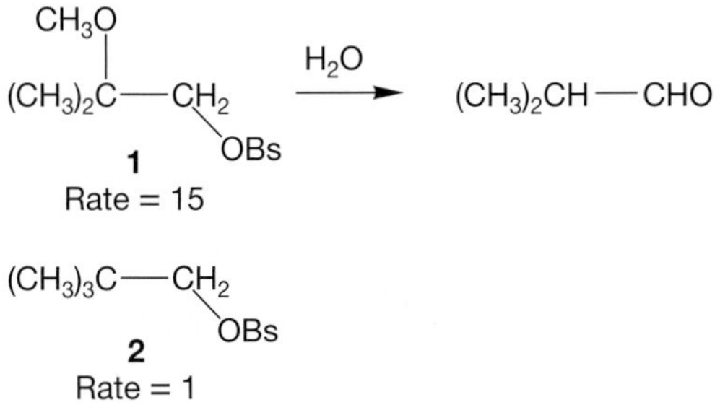

PROBLEM 21.44 Compound **1** is the only product of the reaction of compound **2** under the conditions shown. Provide an arrow formalism mechanism, and deduce the stereochemistry of **2** from the stereochemistry of **1** shown.

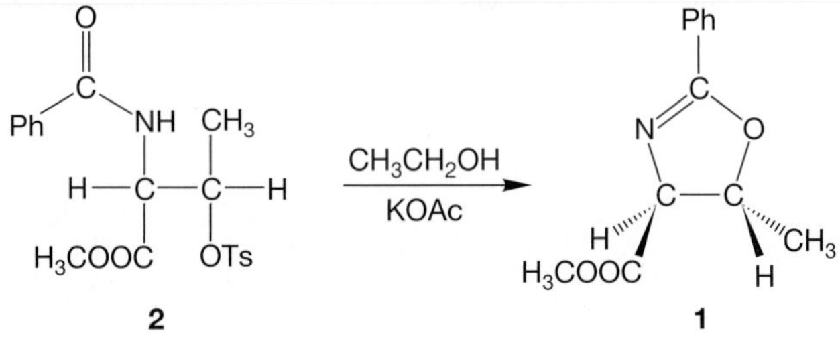

PROBLEM 21.45 Treatment of the following two compounds with HCl leads to the same chloride. Explain, mechanistically. *Hint*: Watch the sulfur atom. What can sulfur do after an initial protonation of the oxygen atom?

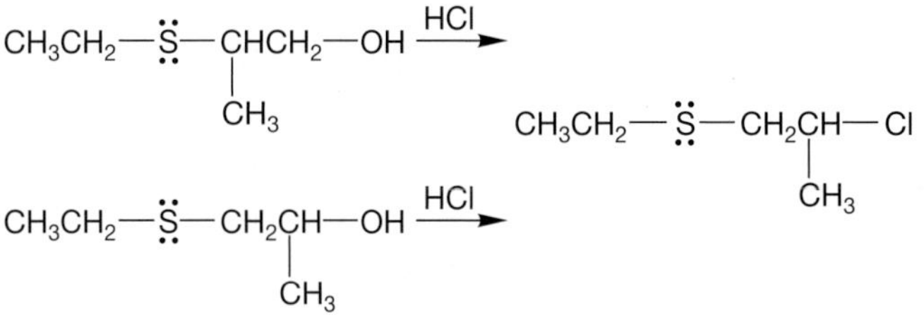

PROBLEM 21.46 Write a mechanism for the formation of the following two products:

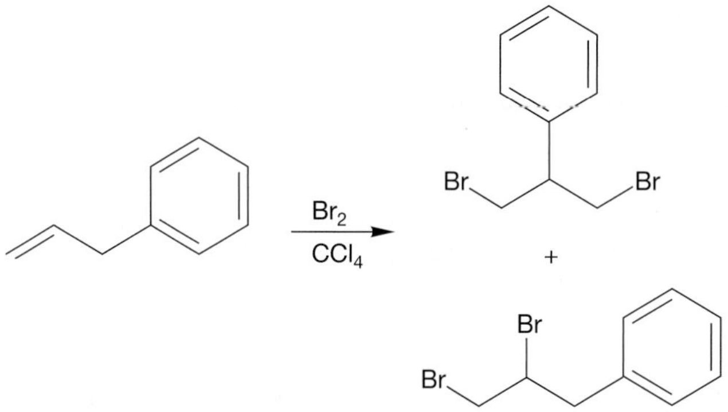

PROBLEM 21.47 Heating β-hydroxyethylamide (**1**) with thionyl chloride affords β-chloroethylamide (**2**). In turn, amide **2** yields oxazoline **3** upon treatment with NaOH.

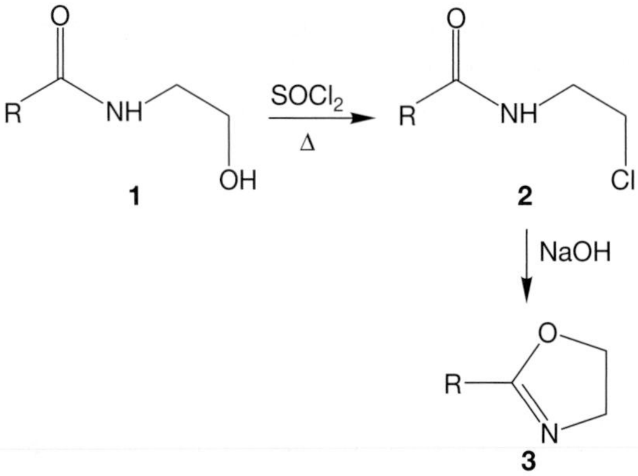

(a) Write arrow formalism mechanisms for the formations of **2** and **3**. Things are not quite as simple as they first appear. For example, if **1** is treated with thionyl chloride below 25 °C, compound **4** is obtained instead of **2**. Compound **4** has the same elemental composition as **2**, but unlike **2**, is soluble in water. Upon heating, **4** gives **2**, and compound **4** yields **3** upon treatment with aqueous sodium carbonate.

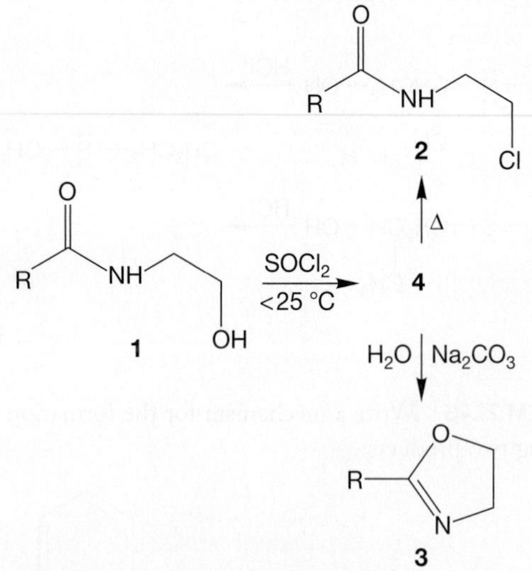

(b) Propose a structure for **4** and show how **4** is formed and converted into **2** and **3**.

PROBLEM 21.48   Addition of iodine (I$_2$) to an alkene does not give a diiodide. But the reaction shown below is a standard procedure in organic synthesis. You have sufficient understanding of alkene chemistry to be able to answer several questions about the reaction shown. What would be the stereochemical outcome? Is the product optically active, a meso compound, or a racemic mixture? Draw the mechanism for this reaction. Your mechanism should explain the stereochemical outcome you have indicated. Why does this reaction make a five-membered ring rather than a six-membered ring?

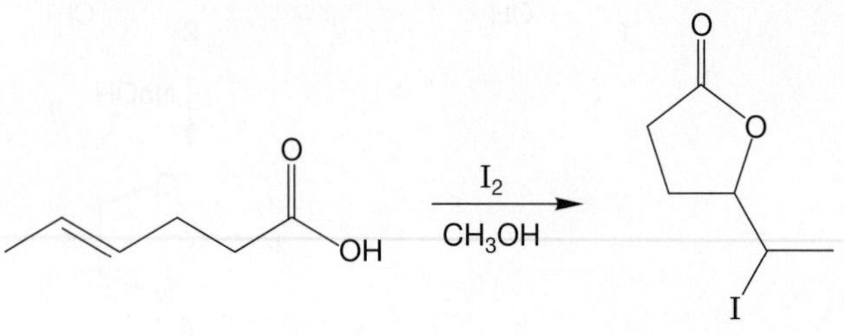

PROBLEM 21.49   Outline the mechanistic steps for the following transformation:

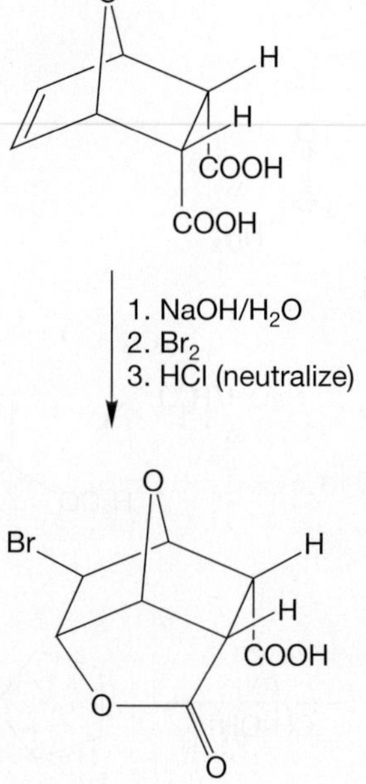

Now explain this more complicated reaction:

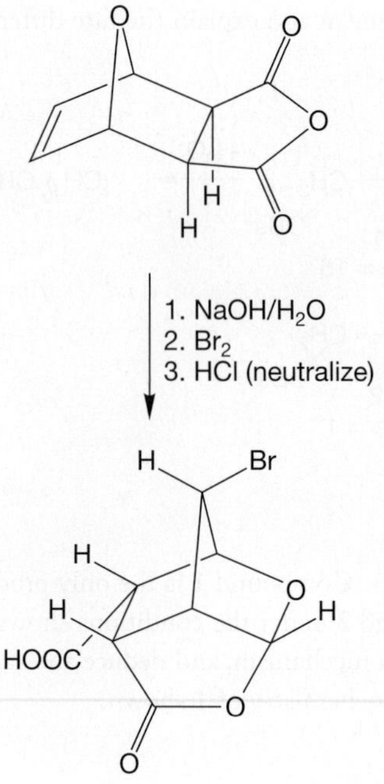

PROBLEM 21.50   The 50-MHz ^{13}C NMR spectrum of the 2-norbornyl cation in $SbF_5/SO_2ClF/SO_2F_2$ solution exhibits an interesting temperature dependence. At $-159\ ^\circ C$ the spectrum consists of five lines at δ 20.4 [C(5)], 21.2 [C(6)], 36.3 [C(3) and C(7)], 37.7 [C(4)], and 124.5 [C(1) and C(2)]. However, at $-80\ ^\circ C$ the spectrum consists of three lines at δ 30.8 [C(3), C(5), and C(7)], 37.7 [C(4)], and 91.7 [C(1), C(2), and C(6)] ppm. Explain. *Hint*: Start with the low-temperature spectrum.

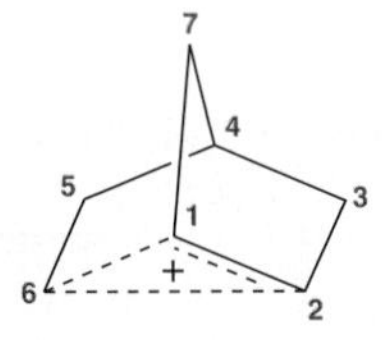

PROBLEM 21.51   2-Amino alcohols can be converted into aziridines if the hydroxyl group is first changed into a better leaving group. A related approach to aziridines might involve treating a 2-amino alcohol with thionyl chloride in the presence of base.

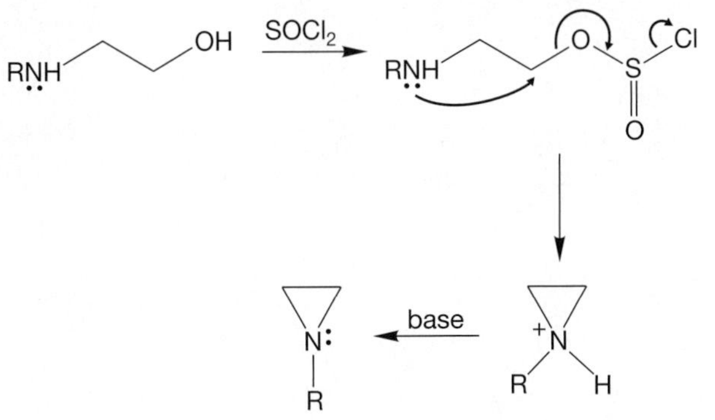

However, when 2-amino alcohol **1** was treated with thionyl chloride in the presence of triethylamine, isomeric compounds **3** and **4** ($C_{12}H_{17}NO_2S$) were obtained rather than the anticipated **2**. Propose structures for isomers **3** and **4** and explain their formation.

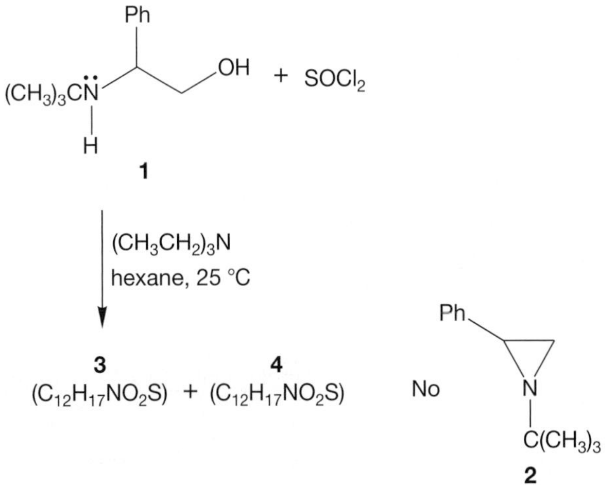

Use Organic Reaction Animations (ORA) to answer the following questions:

PROBLEM 21.52   The animation titled "Intramolecular S_N2" is a nice example of a neighboring group effect. Select that reaction and observe the process. Is the starting material in a boat or a chair conformation? Does the S_N2 process occur from the more stable chair structure? Why or why not?

PROBLEM 21.53   Observe the HOMO and LUMO of the "Intramolecular S_N2" reaction. What atom in the reaction should have the HOMO density prior to epoxide formation? Even more specifically, what orbital on the nucleophilic atom do you expect to be involved in this S_N2 reaction? Verify your answer by viewing the HOMO track of the "Bimolecular nucleophilic substitution: S_N2" animation.

PROBLEM 21.54   Select the reaction titled "Baeyer-Villiger oxidation" and observe the animation. This reaction is also an example of a neighboring group effect. There are several steps in the reaction. Draw out the step that involves a neighboring group. The info page might help. What clues help you recognize that this reaction involves assistance by a neighboring group?

Carbohydrates

22.1 Preview

22.2 Nomenclature and Structure of Carbohydrates

22.3 Formation of Carbohydrates

22.4 Reactions of Carbohydrates

22.5 The Fischer Determination of the Structure of D-Glucose (and the 15 Other Aldohexoses)

22.6 Special Topic: An Introduction to Di- and Polysaccharides

22.7 Summary

22.8 Additional Problems

FIELD OF CARBOHYDRATES Cellulose is a polymer of the monosaccharide glucose found in all plant matter. It is the most common organic compound found in Nature. Converting all that cellulose into biofuel is an active area of chemical research.

There exists no better thing in the world than beer to blur class and social differences and to make men equal.

—EMIL FISCHER[1]

22.1 Preview

In the preceding chapter, we encountered neighboring group effects, the influence of one functional group on the chemistry of another. In this chapter, we discuss molecules that bristle with functionality, and for which intramolecular neighboring group effects play most important roles, influencing both structure and reactivity. These are the **carbohydrates**, which are molecules that have the formula of $C_nH_{2n}O_n$. The name carbohydrate comes from the formulas of these compounds, which can be factored into $C_n(H_2O)_n$. These molecules appear to be hydrates of carbon. Of course they are not literally hydrates, which are compounds having water molecules clustered around an atom, but the classic demonstration experiment in which a carbohydrate sample— usually a sugar cube—is treated with sulfuric acid does yield a spectacular sight as exothermic dehydration reactions produce steam and a cone of carbon.[2] Carbohydrates are also called **sugars** or **saccharides**.

Carbohydrates serve as the storage depots for the energy harvested from the sun through photosynthesis by green plants. Moreover, carbohydrates are present in nucleic acids (Chapter 23) and are thus vital in controlling protein synthesis and in transmission of genetic information. Table sugar is a carbohydrate. More than 100 million tons of this sugar are produced yearly, and each American consumes about 64 lb of table sugar per year (down from over 100 lb just a few years ago). Carbohydrates are vital to human survival, and it is somewhat surprising that their study took so long to emerge as a glamorous area of research. Perhaps the development of carbohydrate chemistry was delayed by the simple technical difficulties of working with goopy syrups rather than nice, well-behaved solids. A full scale renaissance is underway now, however, and carbohydrate chemistry is very much a hot topic.[3]

The carbohydrate chemistry we will see in this chapter involves only a few new reactions, but the polyfunctionality of these compounds can make for some complications. To understand carbohydrate transformations you will have to be able to apply old reactions in new settings, and work with a good deal of stereochemical complexity as well. It's not a bad idea to review the concepts in Chapter 4 before you start on this chapter. That's a hard thing to do—time is probably pressing, and it's always more interesting to move on to new stuff—but at least reread the summary sections of Chapter 4 and be prepared to look back if things get confusing. Dust off your models. There will be moments when they will be essential.

In this chapter, we will also meet Emil Fischer (1852–1919) again. He is surely one of the greatest of organic chemists and a person of heroic theoretical and experimental achievements. Carbohydrates can be hideously frustrating molecules to work with experimentally. They form syrups and are notoriously difficult to crystallize.

---

[1] Emil Fischer (1852–1919) was perhaps the greatest of all organic chemists. He won the 1902 Nobel Prize in Chemistry and made extraordinary contributions to many fields, including the chemistry of carbohydrates and proteins.

[2] **DO NOT** try this experiment unsupervised! The reaction is very exothermic and sometimes violent. Sulfuric acid can be thrown about, and sulfuric acid in the eyes is no picnic!

[3] Although commercial sugars are mostly harvested from plants (mainly sugar beets or sugarcane), there are some marvelously exotic laboratory syntheses. One favorite is that of Philip Shevlin (b. 1939) of Auburn University, who fired bare carbon atoms into water and obtained carbohydrates. Shevlin has suggested that this reaction may well have occurred on the icy surfaces of comets, transforming them into giant orbiting sugar cones.

These properties were especially troublesome in Emil Fischer's time because there were no fancy chromatographs and spectrometers in his basement, and separation was routinely achieved through crystallization. One of Fischer's major contributions was the discovery that phenylhydrazone derivatives of many carbohydrates were rather easily crystallized. That point may seem simple now, but it certainly was not in 1875! One of the great logical triumphs of organic chemistry was the deciphering of the structures of the aldohexoses (we'll see just what this means later, but glucose, a vital source of human energy, is one of the aldohexoses), and Emil Fischer was at the heart of this work for which he won the Nobel prize.

ESSENTIAL SKILLS AND DETAILS

1. Carbohydrate chemistry uses an important structural convention, the Fischer projection. To understand this chapter, you have to be able to move easily from Fischer projections to conventional structures. The directions are right there in Section 22.2.

2. Understanding the logic of structure determination in carbohydrate chemistry is an essential skill.

3. Though neither a skill nor a detail, the Fischer proof of the structure of glucose is something you should know, at least in general, because it is one of the supreme intellectual achievements of human history!

4. Understanding the differences between the anomeric hydroxyl group in an aldose, which is attached to C(1) in the cyclic form, and all the other hydroxyl groups in the molecule is critical to "getting" carbohydrate chemistry.

5. The reactions of carbohydrates can be controlled by use of protecting groups.

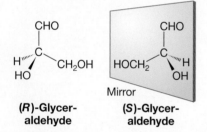

(R)-Glycer-aldehyde   **(S)-Glycer-aldehyde**

FIGURE 22.1 The two enantiomers of the simplest carbohydrate, glyceraldehyde.

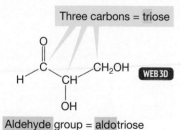

Three carbons = triose

Aldehyde group = aldotriose

Glyceraldehyde

FIGURE 22.2 Glyceraldehyde is an aldotriose because it has a three-carbon backbone and an aldehyde group. Other aldoses have longer carbon backbones.

22.2 Nomenclature and Structure of Carbohydrates

Carbohydrates could all be named systematically, but their history has led to the retention of common names throughout the discipline. There is also a special convention, named after Emil Fischer, that is used to simplify representations of these stereochemically complex molecules.

The simplest carbohydrate is 2,3-dihydroxypropanal, which has the common name *glyceraldehyde* (Fig. 22.1). It contains a single stereogenic carbon and thus can exist in either the (R) or the (S) enantiomeric form.

Glyceraldehyde is an **aldotriose**. The *tri* tells you it is a three-carbon molecule, and the *aldo* tells you that the molecule contains an aldehyde. The *ose* tells us that the molecule is a carbohydrate. The term **aldose** is a more general name that means the molecule has an aldehyde (CHO) at one end, a primary alcohol (CH_2OH) at the other end, and these ends are attached through a series of H—C—OH groups (Fig. 22.2). In glyceraldehyde, of course, there is only one linking H—C—OH group.

WORKED PROBLEM 22.1 Generalize from the definition of an aldotriose to generate the structures of an aldose with four carbons and an aldose with five carbons.

ANSWER The *aldose* means the sugar has an aldehyde group. The aldehyde of an aldose and the primary alcohol at the end of the chain are attached through a series of H—C—OH groups. In the four-carbon sugar there are two

(continued)

H—C—OH groups between the aldehyde and primary alcohol, and the five-carbon sugar has three H—C—OH groups between the aldehyde and the primary alcohol.

If glyceraldehyde is a triose, …   …then this compound is a tetrose…   …and this one is a pentose

In Fischer Projections

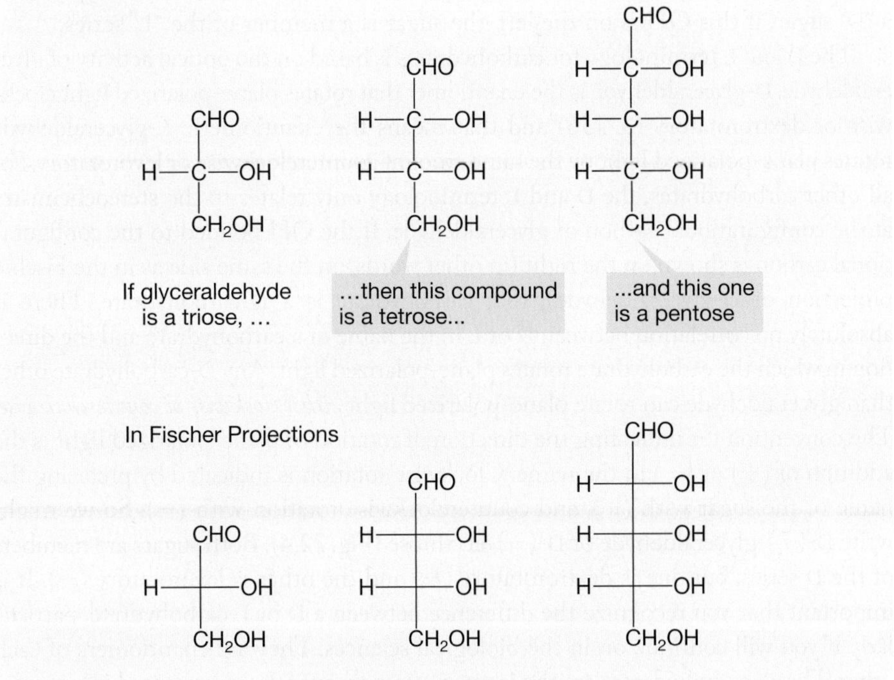

The stereochemical convention called a **Fischer projection** translates the drawings of Figure 22.1 and Problem 22.1 into three-dimensional representations (Fig. 22.3). In a Fischer projection, all horizontal lines are taken as coming out of the plane of the paper toward you, and vertical lines as going away from you. The carbons linking the aldehyde and the primary alcohol ends of the molecule are not explicitly written out. By convention, Fischer projections are written with the most oxidized carbon of the molecule, which in an aldose is the aldehyde carbon, at the top and the primary alcohol at the bottom.

CONVENTION ALERT

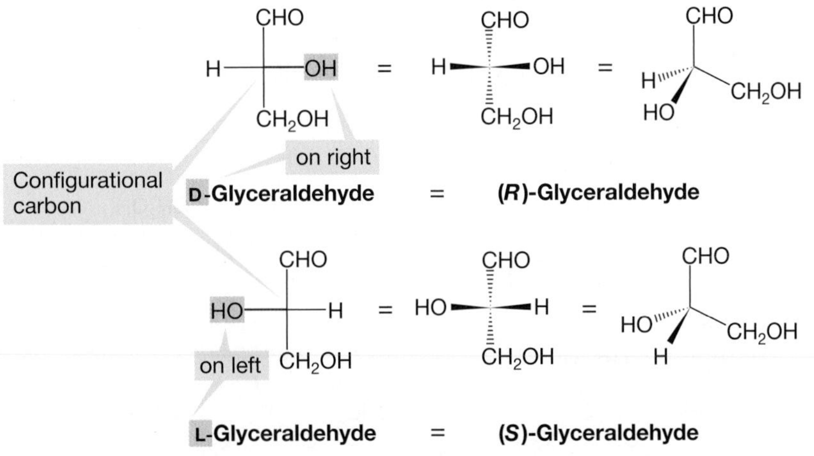

Configurational carbon

on right

D-Glyceraldehyde   =   **(R)-Glyceraldehyde**

on left CH₂OH

L-Glyceraldehyde   =   **(S)-Glyceraldehyde**

FIGURE 22.3 Aldoses are written with the aldehyde group at the top and the primary alcohol at the bottom in the Fischer projection. Horizontal bonds are taken as coming toward the viewer and vertical bonds as going away from the viewer. If the OH group on the configurational carbon is on the right, the carbohydrate is a member of the D series. If it is on the left, the carbohydrate is in the L series. *Notice that to get the designation right, the carbohydrate must be drawn with the aldehyde at the top and the primary alcohol at the bottom.*

Fischer projections do not attempt to represent energy minimum conformations. Instead, they are designed to help distinguish the stereoisomers. An analogy would be the representation of substituted ethanes in an eclipsed conformation, called the *sawhorse* structure (p. 65). Fischer projections are used because they are simple to draw and read. We will see soon how to translate them into more accurate representations of molecules.

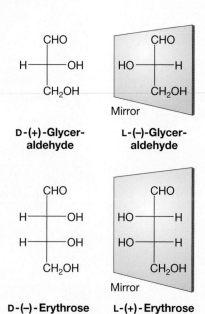

D-(+)-Glycer-aldehyde   **L-(−)-Glycer-aldehyde**

Mirror

D-(−)-Erythrose   **L-(+)-Erythrose**

Mirror

FIGURE 22.4 Other than for glyceraldehyde, there is no relation between D and L in a carbohydrate name and the sign of optical rotation, (+) and (−).

For any sugar shown in Fischer projection, the stereogenic carbon *closest to the primary alcohol* (the stereogenic carbon closest to the bottom of the Fischer projection) is called the **configurational carbon**. If the OH group of the configurational carbon is on the right in the Fischer projection then the molecule is designated as a "D" sugar; if this OH is on the left, the sugar is a member of the "L" series.

The D and L terminology for carbohydrates is based on the optical activity of glyceraldehyde. D-glyceraldehyde is the enantiomer that rotates plane-polarized light clockwise or dextrorotatory (p. 156) and that means the enantiomeric L-glyceraldehyde rotates plane-polarized light by the same amount counterclockwise or levorotatory. For all other carbohydrates, the D and L terminology only relates to the stereochemistry at the configurational carbon of glyceraldehyde. If the OH bonded to the configurational carbon is shown on the right (in other words, on the same side as in the Fischer projection of D-glyceraldehyde), that carbohydrate is a D-carbohydrate. There is absolutely no correlation between D or L in the name of a carbohydrate and the direction in which the carbohydrate rotates plane-polarized light. Any D-carbohydrate other than glyceraldehyde can rotate plane-polarized light *either clockwise or counterclockwise*. The convention for indicating the direction of rotation of plane-polarized light is the addition of (+) or (−) in the name. Clockwise rotation is indicated by prefacing the name of the sugar with (+), and counterclockwise rotation with (−). So we might write D-(+)-glyceraldehyde or D-(−)-erythrose (Fig. 22.4). Both sugars are members of the D series, but one is dextrorotatory (+) and the other is levorotatory (−). It is important that you recognize the difference between a D or L carbohydrate, particularly if you will continue on in the biological sciences. They are enantiomers of each other. The D-carbohydrates are the isomers most frequently encountered in nature.

> **PROBLEM 22.2** Use solid-wedge, dashed-wedge notation to write three-dimensional representations for the following molecules shown in Fischer projection. Label each stereogenic carbon as either (*R*) or (*S*).
>
> CHO   CHO
> H——OH   HO——H
> H——OH   H——OH
> CH₂OH   CH₂OH

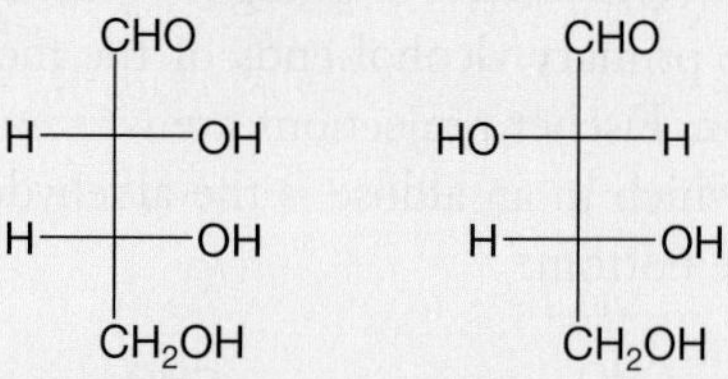

PROBLEM 22.3 Write Fischer projections for the following molecules:

(a)

OHC—C—C—CH₂OH

(b)

OHC—C—C—CH₂OH

(c)

OHC—C—C—CH₂OH

(d)

OHC—C—C—CH₂OH

O
‖
HOCH₂—C—CH₂OH

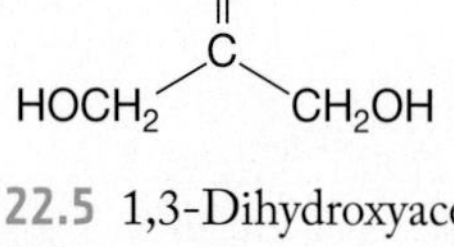

FIGURE 22.5 1,3-Dihydroxyacetone, a ketotriose.

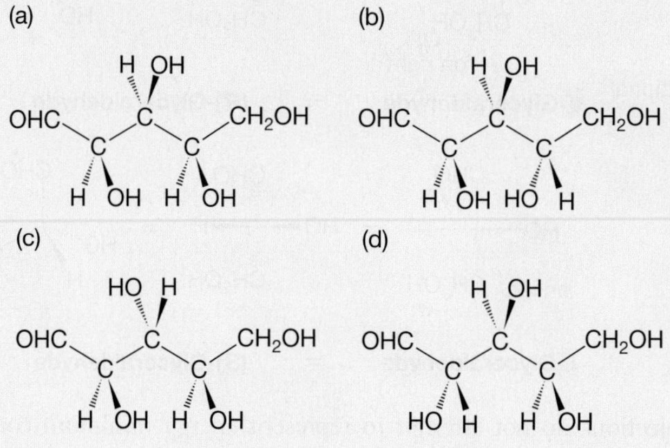

There are also carbohydrates that contain no aldehyde group, but rather a ketone. This kind of sugar is called a **ketose**. Figure 22.5 shows 1,3-dihydroxyacetone, which is a three-carbon ketose, a ketotriose.

Four-carbon sugars are called **tetroses**. There are four possible **aldotetroses,** which are four-carbon sugars with an aldehyde (Fig. 22.6). The aldotetroses are the enantiomeric pair D- and L-erythrose and the enantiomeric pair D- and L-threose. Remember, for a carbohydrate to be D, the OH of the configurational carbon must be on the right-hand side of a Fischer projection.

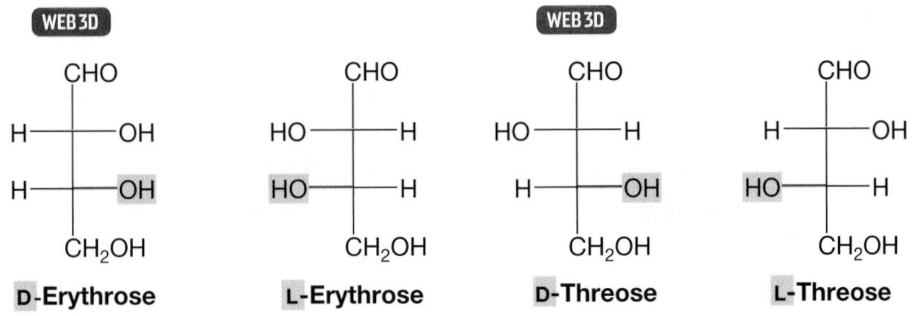

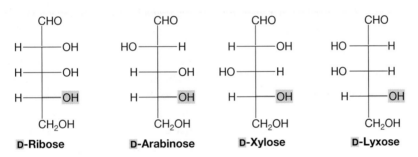

FIGURE 22.6 The four aldotetroses.

Aldopentoses are five-carbon aldoses. These carbohydrates contain three stereogenic carbons, and so there must be a total of eight stereoisomers ($2^3 = 8$). Figure 22.7 shows the four members of the D series of aldopentoses and their common names.

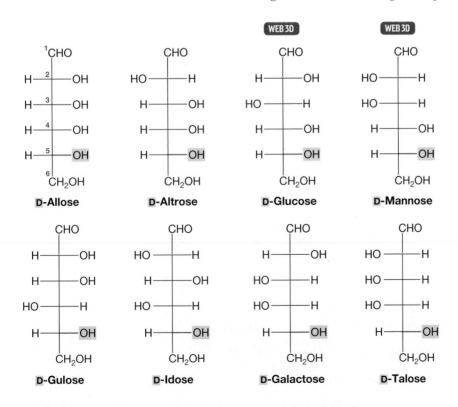

FIGURE 22.7 The four D-aldopentoses and their common names.

Aldohexoses are six-carbon aldoses. They have four stereogenic carbons and so there must be a total of $2^4 = 16$ stereoisomers, eight in the D series (Fig. 22.8) plus

FIGURE 22.8 The eight D-aldohexoses and their common names.

the eight L enantiomers of these molecules. Each carbon of an aldose is numbered starting with the aldehyde carbon as C(1), as shown for allose in Figure 22.8.

Figures 22.7 and 22.8 show only aldopentoses and aldohexoses, but there are also many keto sugars among the **pentoses**, the five-carbon carbohydrates, and the **hexoses**, the six-carbon carbohydrates. Figure 22.9 shows D-fructose, the most common ketohexose.

FIGURE 22.9 D-Fructose, a ketohexose.

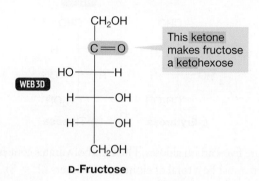

PROBLEM 22.4 Draw Fischer projections for the L-aldopentoses and use Figure 22.7 to provide the common name for each.

PROBLEM 22.5 How many D-aldoheptoses are possible?

PROBLEM 22.6 Treatment of D-glucose (Fig. 22.8) with sodium borohydride (NaBH$_4$) gives D-glucitol (sorbitol), C$_6$H$_{14}$O$_6$. Draw the three-dimensional structure of D-glucitol and write a brief mechanism for this reaction.

One would expect aldoses to combine the chemistries of alcohols and aldehydes, and that expectation is correct. In fact, the proximity of these functional groups leads to important intramolecular chemistry, and here is an example. Although reduction of glucose with sodium borohydride followed by hydrolysis induces the expected change (see Problem 22.6), modern spectroscopic methods such as NMR and IR do *not* reveal the presence of substantial amounts of the aldehyde in the starting carbohydrate (Fig. 22.10). It is important to understand why this is so. The chemistry of the aldose tells us there is an aldehyde and we certainly have been drawing an aldehyde group in our Fischer projections. So why can't we observe the aldehyde spectroscopically?

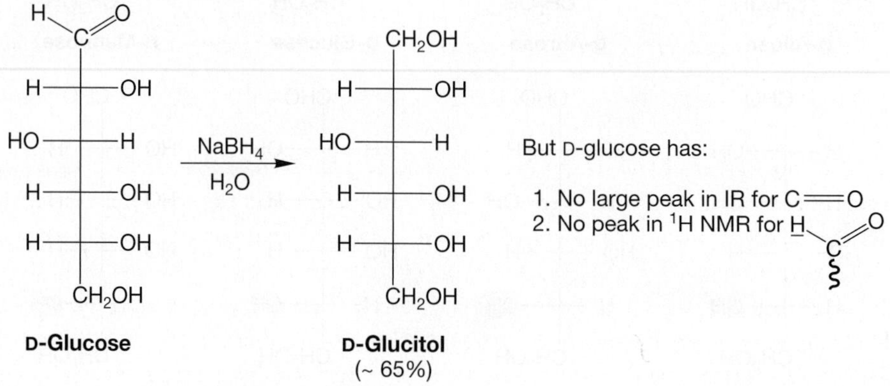

FIGURE 22.10 Although reduction with sodium borohydride, followed by hydrolysis, proceeds normally to give an alcohol, neither NMR nor IR spectroscopy reveals large amounts of an aldehyde in the starting material.

You know already that carbonyl groups react with all manner of nucleophiles (Chapter 16). Hydration and hemiacetal formation are typical examples (Fig. 22.11). A hemiacetal could certainly be formed in an intramolecular reaction in which a hydroxyl group at one location in a molecule reacted with a carbonyl group at another location. Indeed, 4- and 5-hydroxy aldehydes exist mostly in the cyclic hemiacetal forms (p. 783). Figure 22.11 shows a typical reaction of this kind and makes the analogy to both hydration and intermolecular hemiacetal formation.

Hydration

Hemiacetal formation

Intramolecular hemiacetal formation

5-Hydroxyhexanal
(6%)

**2-Hydroxy-6-methyl-
tetrahydropyran**
A cyclic hemiacetal
(94%)

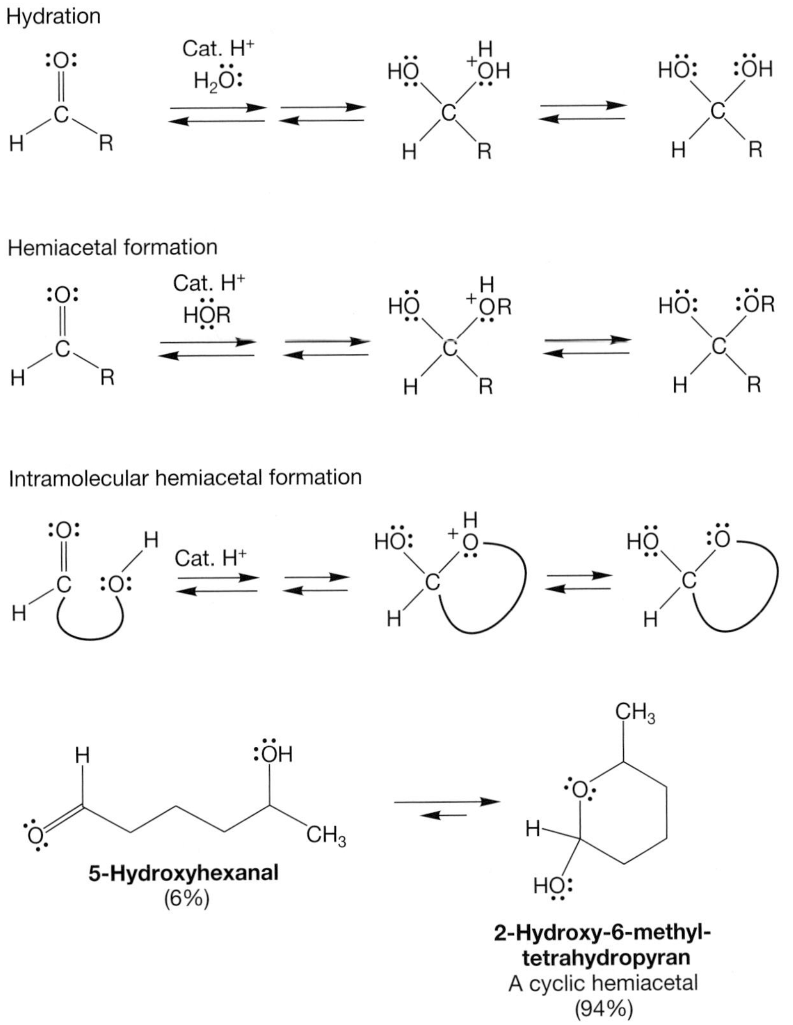

FIGURE 22.11 Intramolecular hemiacetal formation is analogous to hydration and intermolecular hemiacetal formation. Cyclic hemiacetals having five or six atoms in the ring are easily made and are often more stable than their open forms.

The key to the solution of the "missing aldehyde" problem is that for hexoses in an aqueous environment the hemiacetal form is more stable than the open-chain isomer. We can even venture a reasonable guess as to which hydroxyl group of the carbohydrate participates in the ring–forming reaction. For the cyclic hemiacetal to be more stable than the acyclic hydroxy aldehyde, the ring formed surely must not introduce a lot of strain. That eliminates three of the five OH groups and points to either the OH at C(4), which leads to a five-membered ring, or the OH at C(5), which produces a six-membered ring

(Fig. 22.12). These cyclic forms are named as **furanoses** (cyclic carbohydrates having a five-membered ring) or **pyranoses** (cyclic carbohydrates having a six-membered ring) by analogy to the five- and six-membered cyclic ethers: tetrahydrofuran and tetrahydropyran.

FIGURE 22.12 In an aldohexose, intramolecular hemiacetal formation results in either a furanose (five-membered ring) or a pyranose (six-membered ring).

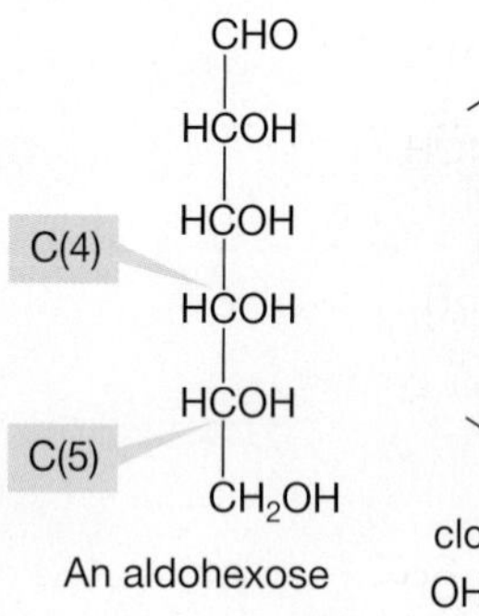

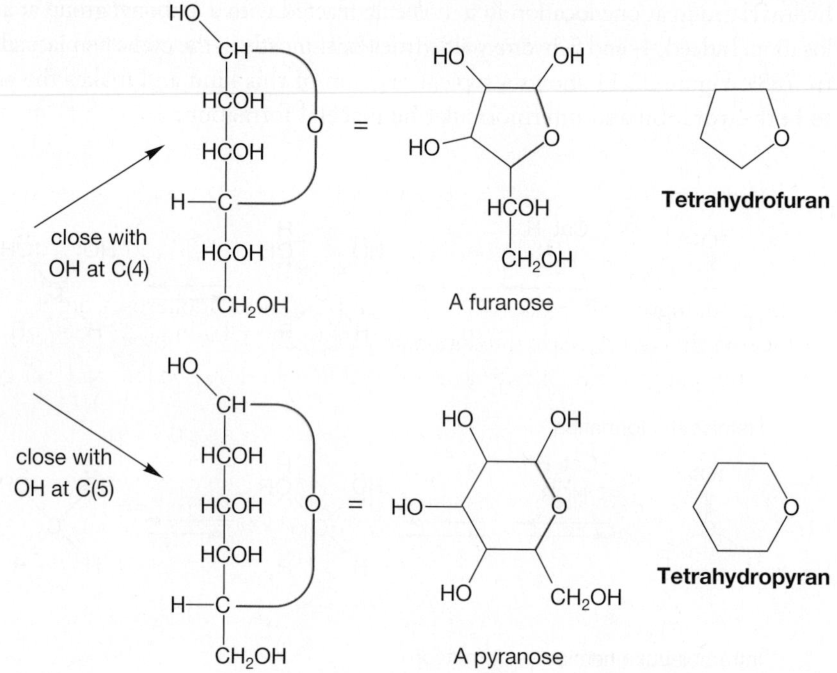

Cyclic forms of carbohydrates are named by replacing the *-se* at the end of the name of the carbohydrate with *-furanose* or *-pyranose* to indicate the size of the ring. D-Glucofuranose and D-glucopyranose can be drawn in Fischer projection and in the more common line representation (Fig. 22.13). Notice that the D-glucopyranose line

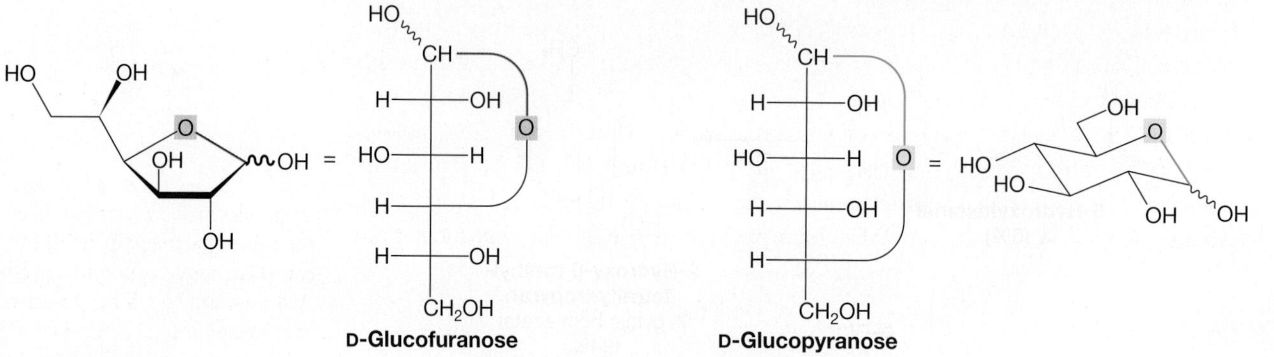

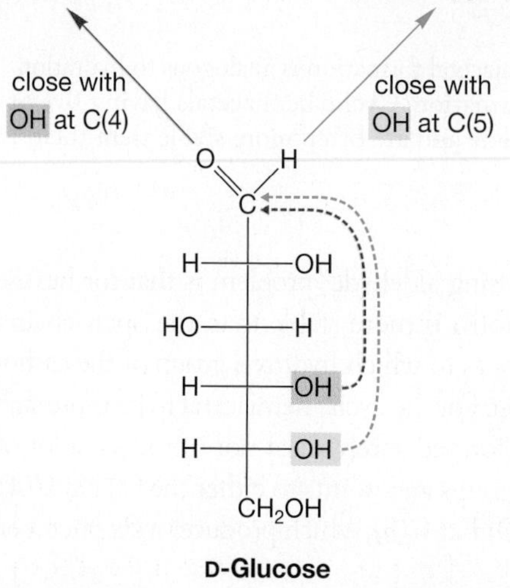

FIGURE 22.13 Fischer projections and corresponding line representations for D-glucofuranose and D-glucopyranose. The squiggled bond means that both stereoisomers are possible.

representation is shown with all groups but the "squiggled" one equatorial. This picture of D-glucopyranose is very important. We can use it as a starting point for drawing all other aldohexopyranoses. For example, D-galactose differs from D-glucose at C(4) (Fig. 22.8), and as a result the OH at the C(4) position in the chair structure for D-galactopyranose is axial rather than equatorial. When two sugars, or any two molecules for that matter, differ only by the stereochemistry at one carbon they are called **epimers**. D-Galactose and D-glucose, for example, are epimers, or C(4) epimers.

> **PROBLEM 22.7** Draw Fischer projections for D-galactofuranose, D-mannopyranose, and L-gulopyranose (see Fig. 22.8).

Both furanoses and pyranoses are formed from most aldohexoses, but it is the six-membered rings that usually predominate (Table 22.1). We'll prove this point a little later in this chapter, but for now accept that this claim is reasonable, and let's go on to

TABLE 22.1   Ratios of Pyranose and Furanose Forms of Aldohexoses at 25 °C in Water

Aldohexose	Pyranose Form (%)	Furanose Form (%)
Allose	92	8
Altrose	70	30
Glucose	~100	<1
Mannose	~100	<1
Gulose	97	3
Idose	75	25
Galactose	93	7
Talose	69	31

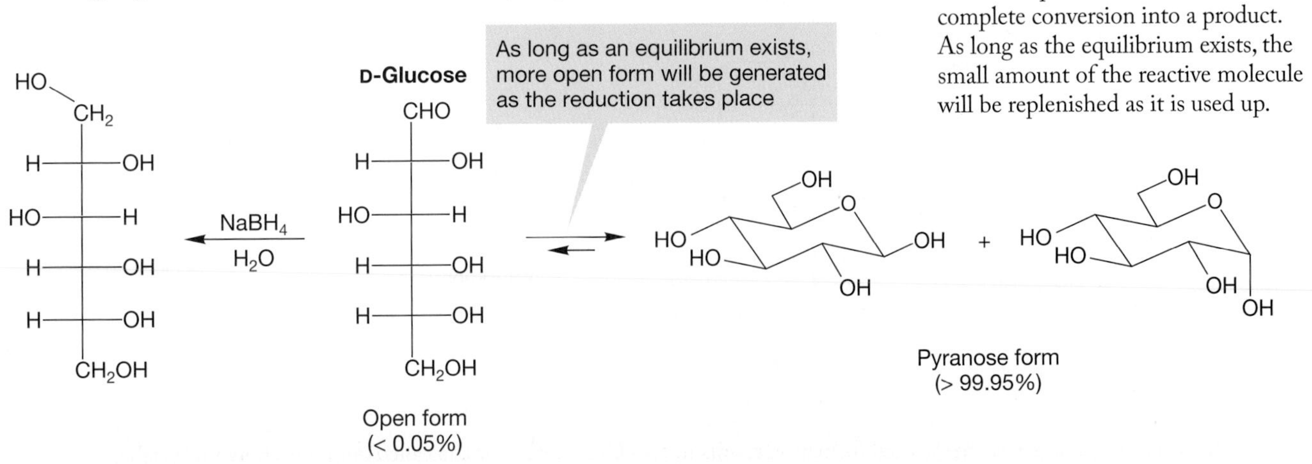

deal with the question of why typical carbonyl reactions (Fig. 22.11) are observed for these cyclic hemiacetals. There is a simple answer to this puzzling question. The molecule isn't locked 100% in the cyclic form, it just exists predominantly in that form. It is in equilibrium with a small amount of the acyclic isomer, which *does* contain a carbonyl group and *does* react with reagents such as sodium borohydride. As the small amount of the reactive carbonyl form is used up, more of it is generated (Fig. 22.14). Sugars that are in the hemiacetal form, and are therefore in equilibrium with an open form, are called **reducing sugars**. We will soon see the chemistry responsible for this descriptive term.

FIGURE 22.14 If two compounds are in equilibrium, irreversible reaction of the minor partner can result in complete conversion into a product. As long as the equilibrium exists, the small amount of the reactive molecule will be replenished as it is used up.

As long as an equilibrium exists, more open form will be generated as the reduction takes place

D-Glucose

Open form
(< 0.05%)

Pyranose form
(> 99.95%)

Table 22.1 lists some common carbohydrates and gives the relative amounts found in pyranose and furanose forms in aqueous solutions. There is very little of the open form present at equilibrium. There is a clear preference for the pyranose rings over the furanose forms, presumably because of the thermodynamic stability of the six-membered ring chair structures (p. 190).

However, because there is always a small amount of the reducible open form present, all of the material can eventually be reduced by sodium borohydride. If we return to our spectrometers, crank up the sensitivity, and look very hard for the aldehyde group present in the open-chain form, there it is, in the tiny amount required by our explanation.

Let's turn our attention to the structure of the D-glucopyranoses. The OH of C(5) bonds with the carbonyl C. The carbonyl O becomes an OH on C(1) and the O from the C(5) OH is the ring O. The first thing to notice is that the intramolecular cyclization reaction creates a new stereogenic carbon at C(1) that can be either (R) or (S). Addition of the nucleophilic OH of C(5) can occur from two possible faces of the aldehyde. The resulting isomers are called **anomers** and the C(1) carbon is called the **anomeric carbon**. Anomers of aldoses differ only in the stereochemistry at the anomeric carbon. If the OH on the anomeric carbon is on the same side in Fischer projection as the OH on the configurational carbon, it is the α anomer. If it is on the opposite side, it is the β anomer. The stereoisomers are shown in Fischer projections and in the more accurate chair structures in Figure 22.15, which also shows the schematic "squiggly bond" used to indicate that the new OH group at the anomeric carbon can be on the right or left in the Fischer projection, which is axial or equatorial in the chair structure.

> **PROBLEM 22.8** Identify the new stereogenic carbon in the α and β anomers of Figure 22.15 as either (R) or (S).

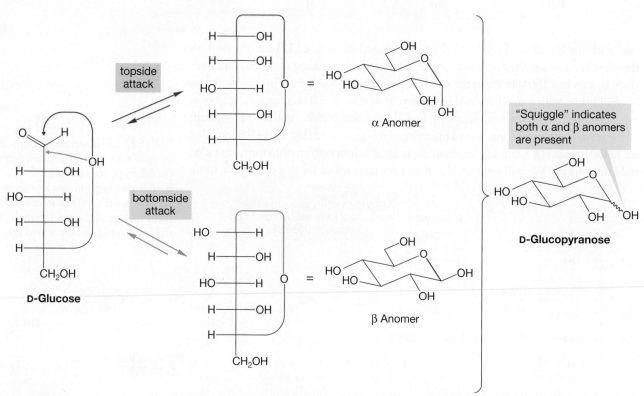

FIGURE 22.15 Intramolecular hemiacetal formation results in two C(1) stereoisomers called anomers shown in Fischer projection and as chair structures. The α anomer has the OH on the anomeric carbon on the same side of the Fischer projection as the OH of the configurational carbon. The β anomer has the OH on the anomeric carbon on the opposite side of the OH of the configurational carbon.

The last few figures have shown Fischer projections with some very peculiar bonds. Notice in particular the "around the corner" bonds sometimes used to show attachment to the ring oxygen (Fig. 22.16a). The squiggly bond is used to show that both anomers can be present (Fig. 22.16b). Because we usually work with D sugars and because we normally draw line representations of the pyranoses and furanoses, it has become conventional to describe the α anomer as having the OH on C(1) down as shown in Figure 22.16. The β anomer in a D sugar has the OH on C(1) up.

CONVENTION ALERT

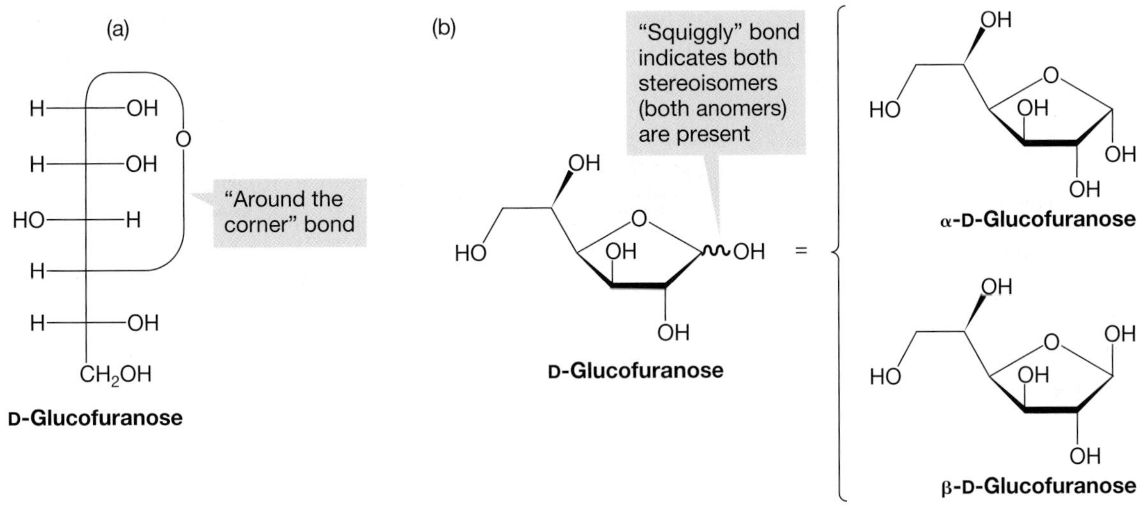

FIGURE 22.16 In Fischer projections, we have drawn "around the corner" bonds. The more common line representations for D-furanose and D-pyranose forms have the OH on the anomeric carbon down for the α anomer and up for the β anomer.

Both α- and β-pyranose and furanose forms exist. Table 22.2 shows the relative amounts for the aldohexoses in aqueous solution.

TABLE 22.2   α- and β-Pyranose and Furanose Forms of the Aldohexoses at 25 °C in Water

Aldohexose	α-Pyranose Form	β-Pyranose Form	α-Furanose Form	β-Furanose Form
Allose	16	76	3	5
Altrose	27	43	17	13
Glucose	36	64	<1	<1
Mannose	66	34	<1	<1
Gulose	16	81	<1	3
Idose	39	36	11	14
Galactose	29	64	3	4
Talose	37	32	17	14

We know that bonds do not go around corners, so let's see how we convert a Fischer projection into a chair structure. There are many ways to do this task, and any way you devise will do as well as the following method. Just be certain that your method works! We will continue to use D-glucose as an example and work through a method of arriving at a three-dimensional structure.

First, redraw the Fischer projection (Fig. 22.17a) so that the horizontal lines are wedges coming out of the plane of the paper (Fig. 22.17b). Second, rotate the molecule 90° clockwise and redraw it with C(1) at the far right of the molecule and going away from the plane of the paper (Fig. 22.17c). Third, rotate the carbon–carbon bond to position the OH that will contribute its O to the ring so that it is pointing toward the C(1) carbon. If a pyranose is being formed, then it is the OH on C(5) that contributes the ring oxygen. Thus, this reaction requires rotation around the C(4)—C(5) bond (Fig. 22.17d). If a furanose is to be formed, then it is the OH on C(4) that contributes the ring oxygen, and the rotation is around the C(3)—C(4) bond. Redraw the molecule with the contributing OH positioned so that it points toward C(1). Fourth, make the pyranose (or furanose) ring by drawing a bond between C(1) and the O bonded to C(5) [or to C(4)], and draw the C(1) OH down for the α anomer and up for the

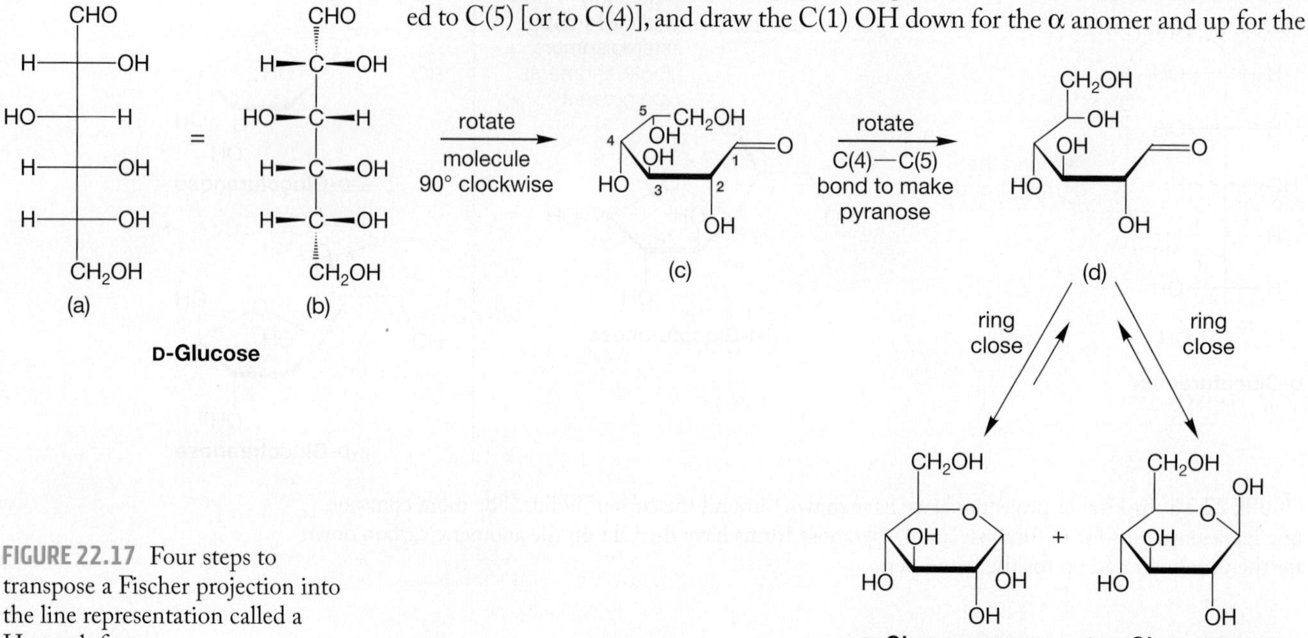

FIGURE 22.17 Four steps to transpose a Fischer projection into the line representation called a Haworth form.

β anomer or a squiggly line to indicate both. This kind of planar representation is called a **Haworth form**, after its inventor, Walter N. Haworth (1883–1950). Figure 22.18 repeats the process, transposing D-ribose into the α and β anomers of D-ribofuranose.

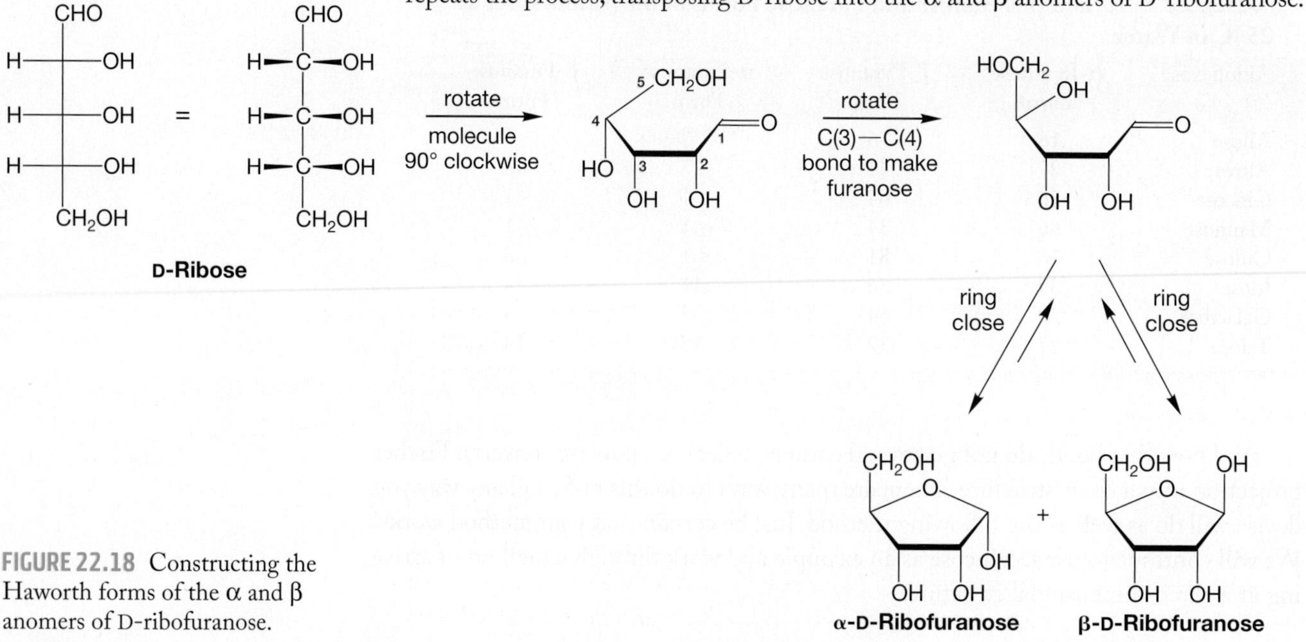

FIGURE 22.18 Constructing the Haworth forms of the α and β anomers of D-ribofuranose.

PROBLEM 22.9 Draw the Haworth form of (a) the α anomer of D-glucofuranose; (b) β-D-ribofuranose, which is the sugar in ribonucleic acid (RNA); and (c) β-D-2-deoxyribofuranose, the sugar in deoxyribonucleic acid (DNA).

Six-membered rings are not flat; they exist in energy-minimum chair forms. The final step in this exercise therefore is to allow the flat Haworth form to relax to a chair (Fig. 22.19). There are two chairs possible for any six-membered ring. They are interconverted by the rotations around carbon–carbon bonds we called a ring flip (p. 192). Be sure to check both forms to see which is lower in energy. Sometimes this evaluation will be difficult, but for β-D-glucopyranose, it is easy. The form with all groups equatorial will be far more stable than the one with all groups axial.

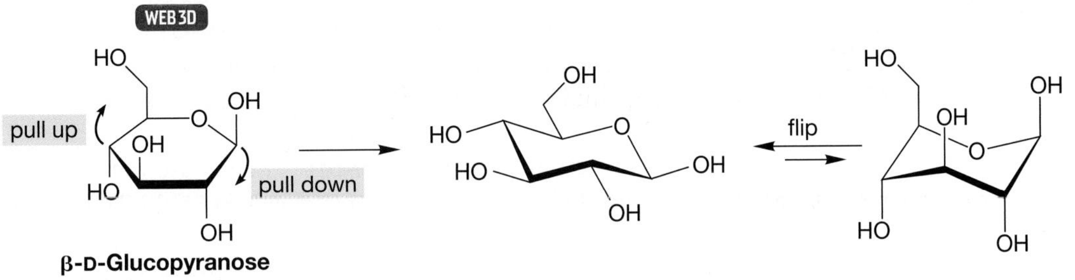

β-D-Glucopyranose

FIGURE 22.19 The Haworth form of a pyranose is much more accurately shown as a chair structure. Don't forget that there are always two possible chair forms.

WORKED PROBLEM 22.10 Transpose the Haworth form of α-D-glucopyranose (Fig. 22.17) into the chair structure.

ANSWER In the α anomer, the hydroxyl group at C(1) is axial, not equatorial as in the β anomer. Otherwise all the substituent groups of the α anomer remain equatorial.

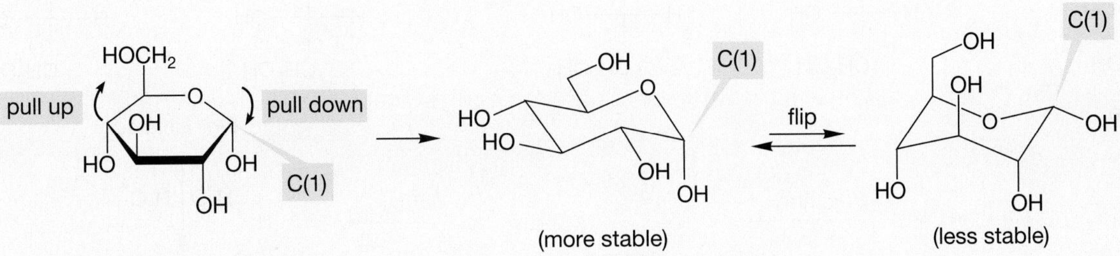

Of course, there are two chair forms, but the one with four equatorial groups is the more stable one.

PROBLEM 22.11 Transform the Fischer projection of D-mannose (Fig. 22.8) into the chair structure of β-D-mannopyranose using the method described in Figures 22.17 and 22.19. Confirm your answer by using the fact that D-mannose is the C(2) epimer of D-glucose.

Summary

Carbohydrate names specify: (1) the length of the carbon chain, (2) whether the molecule is an aldehyde or ketone, (3) whether the configurational carbon is D or L, (4) whether the structure is a cyclic six-membered ring or five-membered ring, and (5) whether the α or β anomer is formed. Some of you will be expected to memorize the common names for all of the sugars, but all students of organic chemistry should be able to draw a reasonable 3-D structure of β-D-glucopyranose!

22.3 Formation of Carbohydrates

Nature has the process for making carbohydrates down to an art. Plants use a number of enzymes to produce sugars. For example, the enzyme ribulose-1,5-bisphosphate carboxylase/oxygenase (Rubisco) is used by plants to assimilate CO_2 from the atmosphere. With the help of this enzyme, the sugar ribose becomes carboxylated by the CO_2 and the resulting six-carbon molecule undergoes a reverse aldol reaction to produce two molecules of glycerate that can be used to synthesize glucose, the carbohydrate used most frequently by living organisms to store chemical energy. Rubisco is perhaps the most prevalent protein on Earth. Its ability to convert CO_2 into the chemical form of energy (sugar) makes it a logical starting point for a Nature-based solution to our burgeoning buildup of atmospheric CO_2, a greenhouse gas.

22.3a Lengthening Chains in Carbohydrates   Plants are not the only sources of carbohydrates, as chemists are able to make sugars in the laboratory. For example, treatment of D-ribose with cyanide followed by catalytic hydrogenation and hydrolysis generates the sugars D-allose and D-altrose (Fig. 22.20).

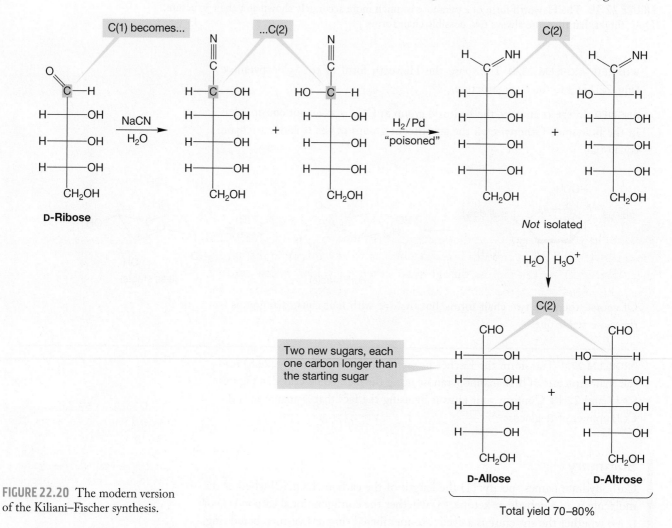

FIGURE 22.20 The modern version of the Kiliani–Fischer synthesis.

This reaction is a modern variation of the **Kiliani–Fischer synthesis** (Emil Fischer, again, and Heinrich Kiliani, 1855–1945). In the first step, sodium cyanide (NaCN) adds to the ribose carbonyl group to make a cyanohydrin (p. 781). The key point is that a new stereogenic carbon is generated in this reaction and both (R) and (S) epimers will be formed. In other words, two products will be formed, one with the new OH group on the right in the Fischer projection and one with it on the left.

The second step of the Kiliani–Fischer synthesis, catalytic reduction with poisoned palladium (p. 452), gives a pair of imines that are hydrolyzed under the reaction conditions to aldoses, each of them one carbon longer than the starting sugar. The new sugars differ from each other only in their stereochemistry at C(2). In other words, they are C(2) epimers.

PROBLEM 22.12   Use the Kiliani–Fischer synthesis starting with D-glyceraldehyde. What new sugars are formed? It is not necessary to write mechanisms for the reactions.

PROBLEM 22.13   The following two sugars are produced by Kiliani–Fischer synthesis from an unknown sugar. What is the structure of that unknown sugar?

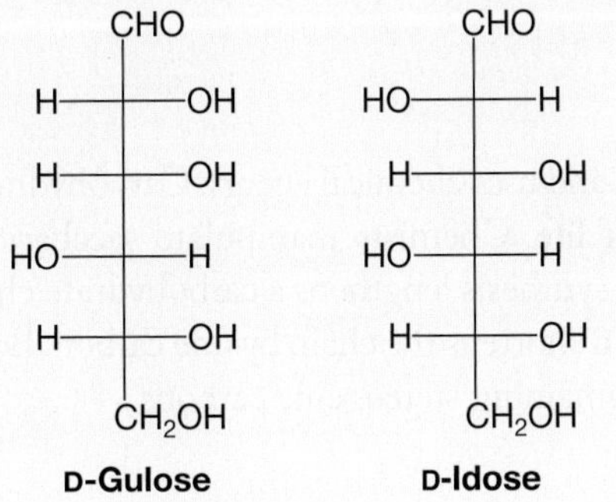

D-Gulose   **D-Idose**

22.3b Shortening Chains in Carbohydrates

The **Ruff degradation** (Otto Ruff, 1871–1939) is a method of shortening the backbone of a sugar by removal of the aldehyde at C(1) and creating a new aldehyde at what was C(2) in the starting sugar (Fig. 22.21). In this process, the aldehyde carbon of an aldose is first oxidized. Then, the calcium salt of the resulting carboxylic acid is treated with ferric ion and hydrogen peroxide to cause decarboxylation and oxidation of what was the C(2) carbon.

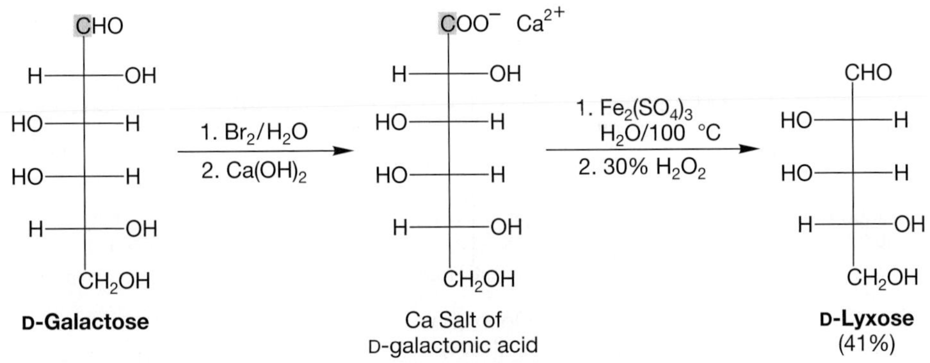

FIGURE 22.21 The Ruff degradation shortens the starting sugar by one carbon. It is the original aldehyde carbon that is lost. In this example, the aldehyde carbon of D-galactose is oxidized to give D-galactonic acid, which is converted into D-lyxose.

The mechanism of this reaction remains obscure (chemistry is a living science—not everything is fully understood) but one imagines that radicals are involved. Metal ions can act as efficient electron-transfer agents, and a guess at the mechanism might involve formation of the carboxyl radical, decarboxylation, capture of the hydroxyl radical formed, and a final transformation of the hydrate into the aldehyde. Whatever the details of the mechanism, the stereochemistry at the original C(2) is lost, but there can be no changes at the old C(3), C(4), or C(5).

Just as Nature can synthesize sugars from smaller carbon chains, there is a natural process called **glycolysis** (Fig. 22.22) that breaks down glucose into two molecules of a three-carbon compound called pyruvate. This chain-shortening pathway is a critical part of the metabolism that occurs in almost all organisms. Glycolysis provides cells with adenosine triphosphate (ATP) and NADH (p. 814).

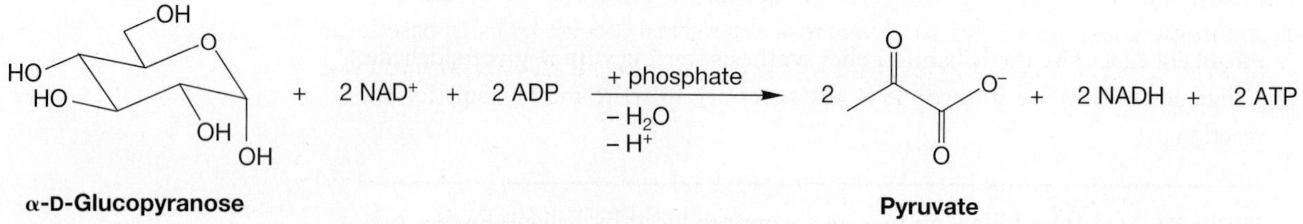

α-D-Glucopyranose   **Pyruvate**

FIGURE 22.22 Glycolysis is a significant metabolic pathway. It involves ten intermediates and ultimately leads to pyruvate. Energy storage is a side reaction in glycolysis that stores the glucose until energy demands bring it back into the metabolic pathway.

Summary

Nature forms, stores, and uses chemical energy. Carbohydrates are the central chemical in this dance of life. Chemists manipulate saccharides many ways, as well. The Kiliani–Fischer synthesis lengthens a carbohydrate chain by one carbon atom. The Ruff degradation shortens the chain by one carbon. Both reactions occur without disturbing the remaining stereogenic carbons.

22.4 Reactions of Carbohydrates

22.4a Mutarotation Pure, crystallized α-D-glucopyranose has a specific rotation (p. 159) of +112°. The pure β anomer has a specific rotation of +18.7°. However, an aqueous solution of either anomer steadily changes specific rotation until the value of +52.7° is reached. This phenomenon is called **mutarotation** and is common for sugars. Our task is to explain it. The answer comes from the realization that both α- and β-D-glucopyranose are hemiacetals and exist in solution in equilibrium with a small amount of the open-chain aldohexose (Fig. 22.23).

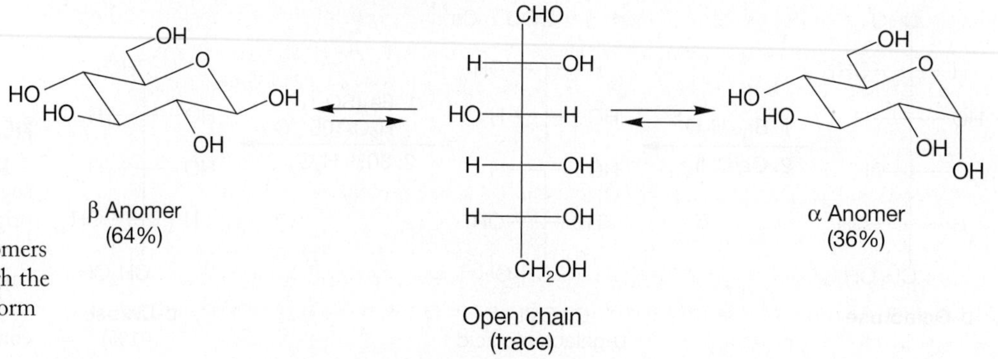

β Anomer (64%)   Open chain (trace)   α Anomer (36%)

FIGURE 22.23 The α and β anomers of a sugar can equilibrate through the very small amount of the open form present at equilibrium.

Once the acyclic form is produced, it can re-form the hemiacetal to make *both* the α- and β-pyranose. Over time, an equilibrium mixture is formed and it is that equilibrated pair of anomers that gives the specific rotation of +52.7°.

PROBLEM 22.14 Write a mechanism for the acid-catalyzed mutarotation of D-glucopyranose.

22.4b Epimerization in Base

In base, sugars rapidly equilibrate with other sugars, which is a more profound change than the equilibration of anomers we saw in the preceding section. Again, let's use D-glucose as an example. As shown in Figure 22.24, D-glucopyranose equilibrates with another D-aldohexose, D-mannopyranose, and a ketose, D-fructose, when treated with aqueous base. This process goes by the absolutely delightful name of the **Lobry de Bruijn–Alberda van Ekenstein reaction**.[4]

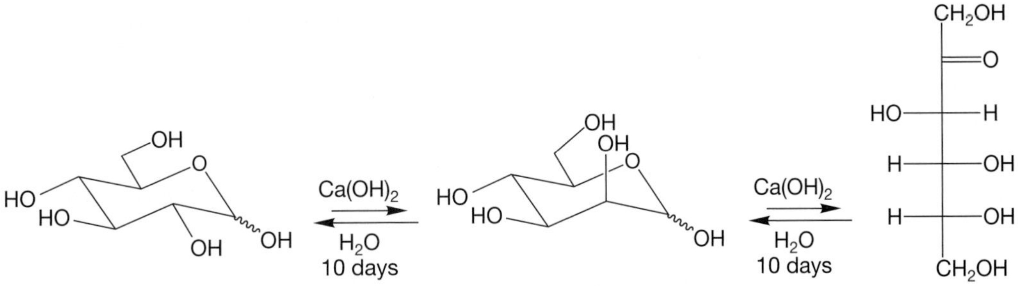

FIGURE 22.24 In base, D-glucopyranose equilibrates with D-mannopyranose and D-fructose.

PROBLEM 22.15 Although D-fructose is shown in Figure 22.24 in the open-chain form, it exists mostly (~67%) in the pyranose form and partly (~31%) in the furanose form in which the hydroxyl group at C(5) is tied up in hemiacetal formation with the ketone. Draw Fischer projections for the pyranose and furanose forms of D-fructose.

The mechanism of the Lobry de Bruijn–Alberda van Ekenstein reaction is simpler than its title. As we have pointed out several times now, the predominant cyclic forms of these sugars are in equilibrium with small amounts of the open-chain isomers. In base, the α hydrogen in the open aldo form can be removed to produce a

---

[4]This "name reaction" is even better than it sounds. One might be forgiven for assuming that someone meeting these two chemists on a Dutch street in 1880 might say, "Goedemorgen, Lobry; hoe gaat het, Alberda?," should he or she be so presumptuously familiar as to address them by their first names. But that assumption would be dead wrong! The reaction title, alas, incorporates only parts of their names. The reaction is named for Cornelius Adriaan van Troostenbery Lobry de Bruijn (1857–1904) and the slightly less spectacularly named Willem Alberda van Ekenstein (1858–1937).

resonance-stabilized enolate (Fig. 22.25). Reprotonation regenerates D-glucose, if reattachment occurs from the same side as proton loss, but D-mannose if reattachment is from the opposite side.

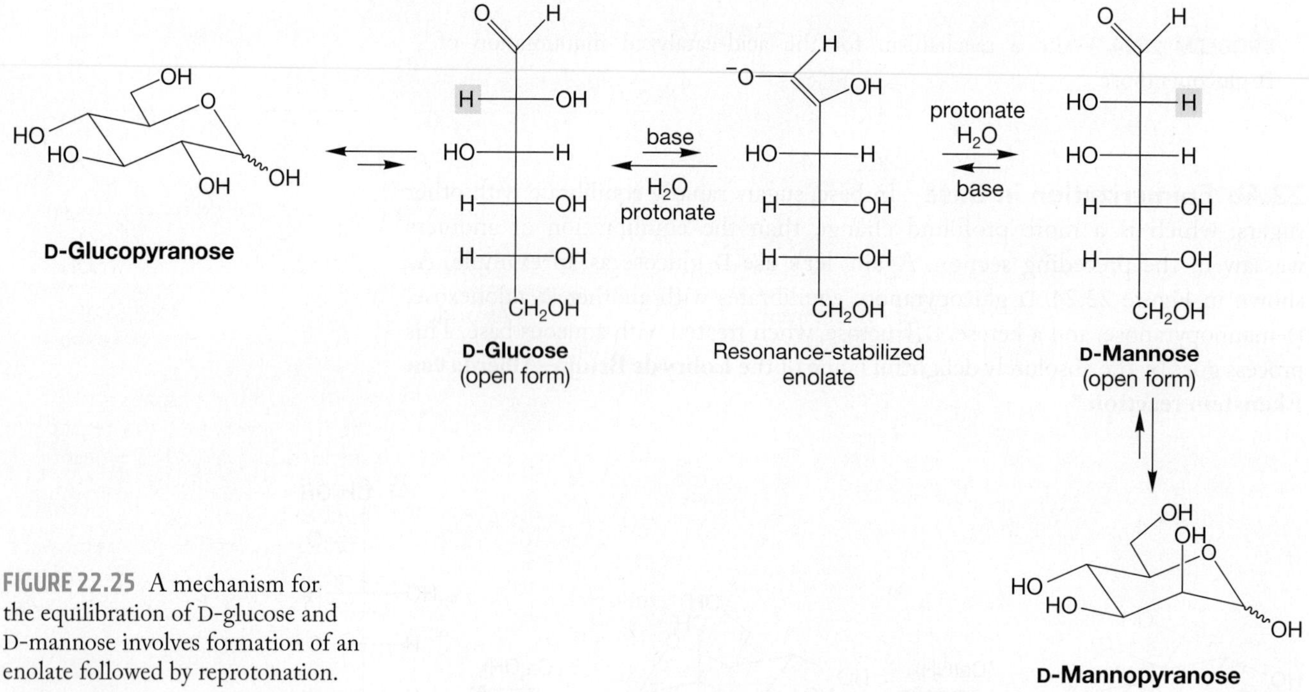

FIGURE 22.25 A mechanism for the equilibration of D-glucose and D-mannose involves formation of an enolate followed by reprotonation.

PROBLEM 22.16 Make a three-dimensional drawing of the enolate in Figure 22.25 to show how both D-glucose and D-mannose can be formed on reprotonation.

If protonation occurs at the enolate oxygen, a "double enol" shown in Figure 22.26 is formed. This compound can re-form a carbon–oxygen double bond in two ways. The double enol can form either an aldehyde or a ketone. If an aldehyde is generated, the epimeric aldohexoses D-glucose and D-mannose shown in Figure 22.25 are produced. If a ketone is formed, it is D-fructose that is the product.

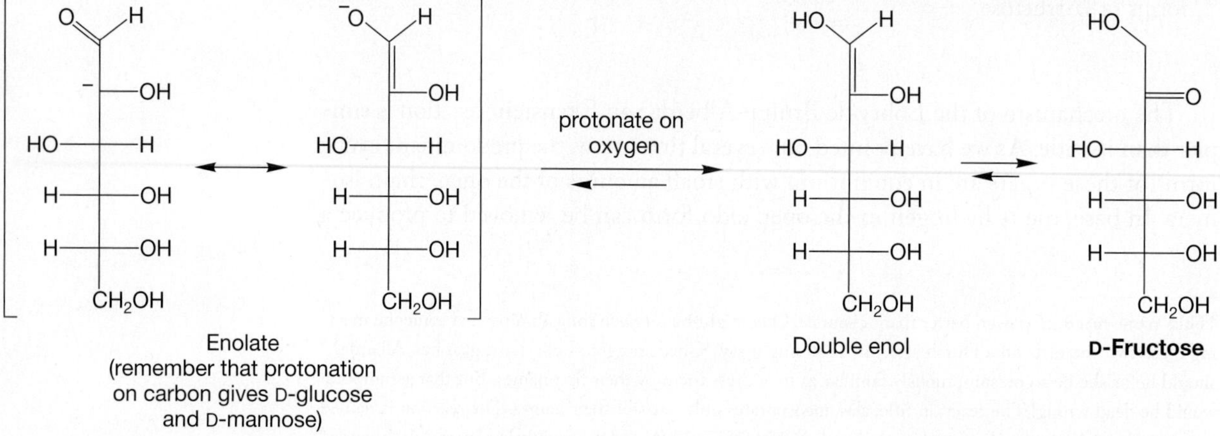

FIGURE 22.26 Protonation on oxygen of the enolate generates a double enol, which leads to D-fructose, D-glucose, or D-mannose.

22.4c Reduction We have already seen one aldose reduction, the reaction of D-glucose with sodium borohydride to give 1,2,3,4,5,6-hexanehexaol (Fig. 22.10). Any sugar that has a hemiacetal can be reduced by these conditions (Fig. 22.27).

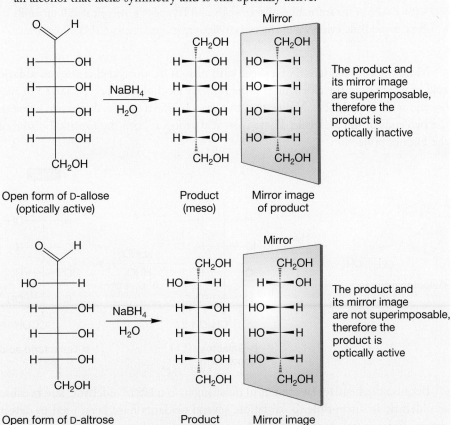

FIGURE 22.27 Reduction of D-galactopyranose to convert the aldehyde into an alcohol proceeds through the small amount of the open, aldo form present at equilibrium. As the open form is used up, it is regenerated through equilibration with the pyranose form.

WORKED PROBLEM 22.17 Reduction of D-altrose with sodium borohydride in water gives an optically active molecule, D-altritol. However, the same procedure applied to D-allose gives an optically inactive product. Explain.

ANSWER This problem is easy, but it is critical for understanding the upcoming material describing the Fischer proof of the stereochemistry of glucose. The important point is to see that reduction of the aldehyde to the alcohol makes the molecule more symmetrical by making *both* ends of the chain CH₂OH groups. D-Allose becomes a meso compound (and therefore optically inactive) because the reduced product now has a plane of symmetry. Reduction of D-altrose gives an alcohol that lacks symmetry and is still optically active.

22.4d Oxidation   Sugars contain numerous oxidizable groups, and methods have been developed in which one or more of them may be oxidized in the presence of the others. Mild oxidation converts only the aldehyde into a carboxylic acid; the primary and secondary hydroxyls are not touched. The reagent of choice is bromine in water, and the product is called an **aldonic acid** (Fig. 22.28).

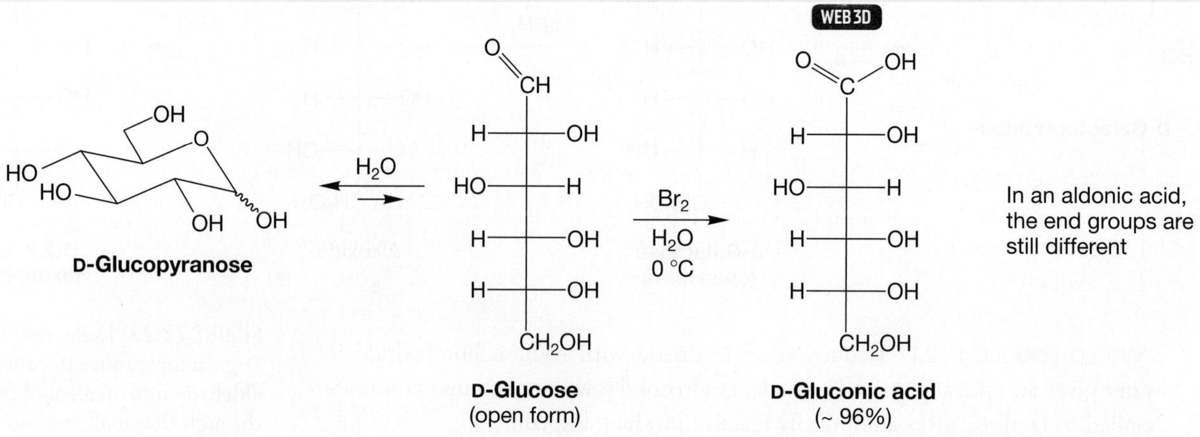

FIGURE 22.28 Oxidation of an aldose with bromine in water gives an aldonic acid. Note that the functional groups at the two ends of the molecule are still different from each other.

PROBLEM 22.18 Write a mechanism for the oxidation shown in Figure 22.28. *Hint for the first step*: What reaction is likely between an aldehyde and water?

PROBLEM 22.19 Examination of the NMR and IR spectra of typical aldonic acids often shows little evidence of the carboxylic acid group. Explain this observation.

More vigorous oxidation of a sugar with nitrous or nitric acid creates an **aldaric acid** in which both the aldehyde and the primary alcohol have been oxidized to carboxylic acids (Fig. 22.29). Notice that this reaction destroys the difference between the two ends of the molecule. There is an acid group at both the top and bottom of the Fischer projection.

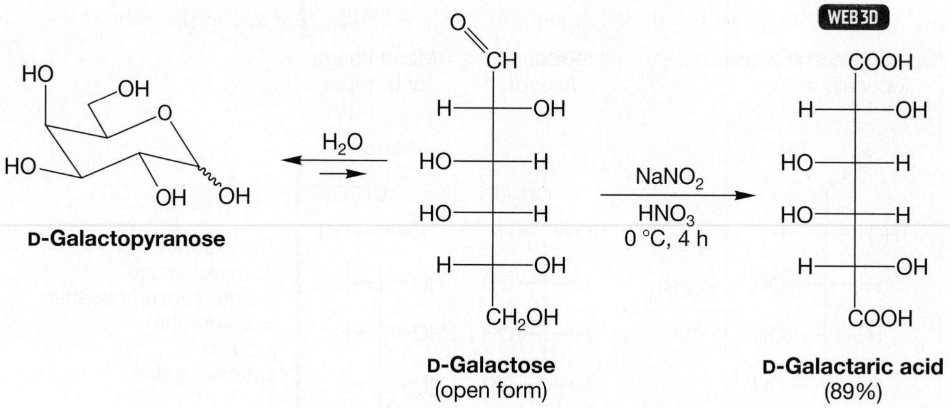

FIGURE 22.29 Oxidation of a sugar with nitric acid generates an aldaric acid, which has a carboxylic acid group at each end of the molecule.

Because the hemiacetal present in most sugars is a latent aldehyde, and because an aldehyde is susceptible to oxidation, several oxidants have been used to detect sugars. These metal reagents (usually copper based) undergo a color change as the aldehyde is oxidized and the metal reduced. It is this kind of redox reaction that has given us the term reducing sugar. If a sugar has a hemiacetal, it is in equilibrium with an aldehyde and so can reduce the metal reagent.

In even more vigorous oxidations, the secondary hydroxyl groups are attacked and the six-carbon backbone can be ruptured. You already know that vicinal diols (1,2-diols) are cleaved on treatment with periodate (p. 807). When sugars are treated with periodate, mixtures of products are obtained, as the various 1,2-diols are cleaved. Treatment with periodate can chop the carbon backbone into small fragments, and the structures of the fragments can often be used in structure determination.

22.4e Osazone Formation   Treatment of aldoses with phenylhydrazine under acidic conditions initiates an odd and very useful reaction. What would we expect? We know that the cyclic forms of aldoses are in equilibrium with the open-chain isomers, which contain a free aldehyde group. Aldehydes react with substituted hydrazines to generate hydrazones (p. 793), and so it is no surprise to see the reaction of Figure 22.30 in which the carbonyl group at C(1) has formed a phenylhydrazone derivative.

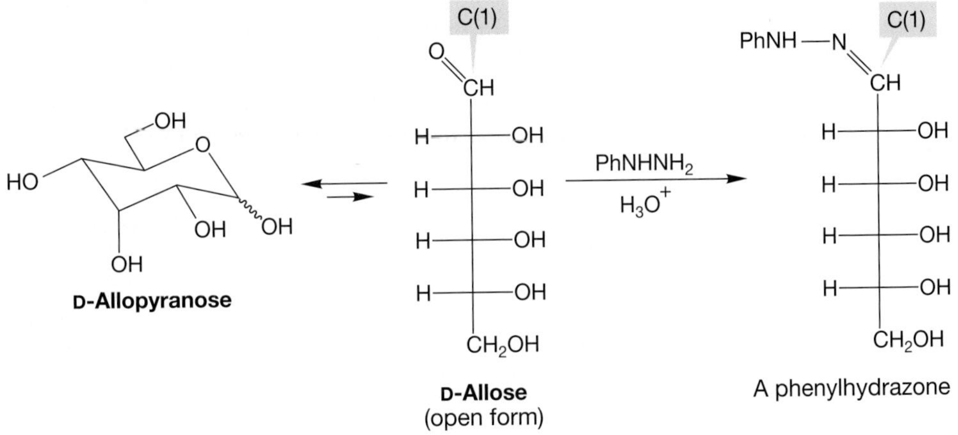

FIGURE 22.30 The small amount of free aldehyde present at equilibrium accounts for phenylhydrazone formation at C(1).

What *is* a surprise is to find that in the presence of excess hydrazine not only C(1) but also C(2) is transformed into a phenylhydrazone. The product is a 1,2-diphenylhydrazone, called an **osazone** (Fig. 22.31). Here are two clues to the mechanism of the reaction. First, it takes three equivalents of phenylhydrazine to complete the reaction and, second, aniline and ammonia are by-products.

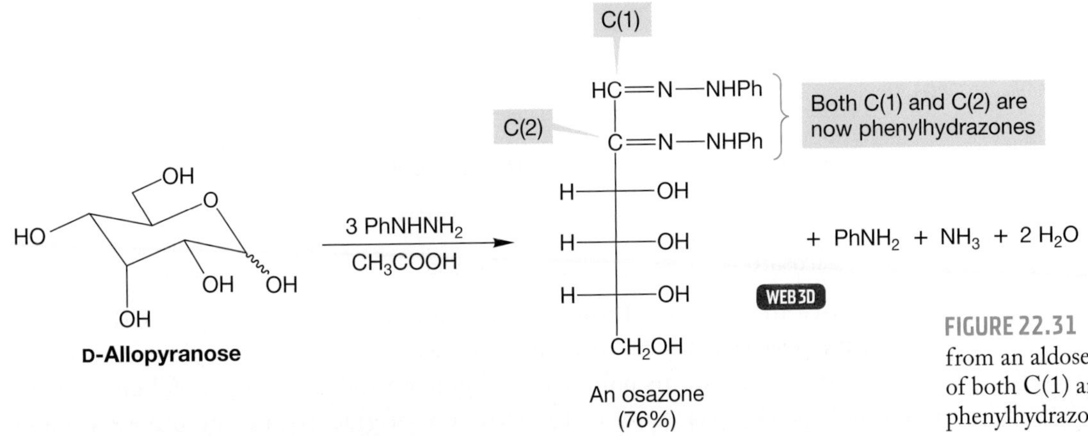

FIGURE 22.31 Osazone formation from an aldose involves conversion of both C(1) and C(2) into phenylhydrazones.

The mechanism begins with the reaction at C(1) to form a phenylhydrazone (p. 793), which consumes the first equivalent of phenylhydrazine. This hydrazone is an imine and imines are in equilibrium with enamines (p. 796), just as ketones are in equilibrium with enols. The imine–enamine equilibrium is shown in Figure 22.32.

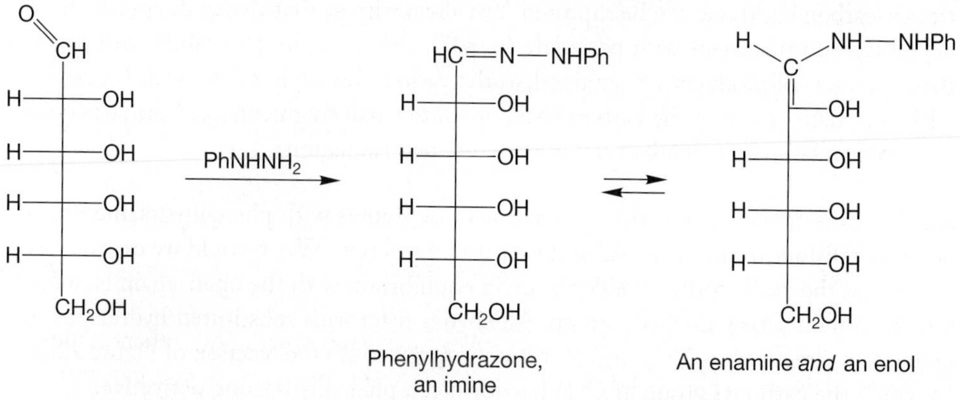

FIGURE 22.32 The phenylhydrazone formed when an aldose reacts with phenylhydrazine is an imine and therefore in equilibrium with an enamine. This enamine is also an enol.

Phenylhydrazone,
an imine

An enamine *and* an enol

In this case, the enamine is also an enol and is therefore in equilibrium with a ketone that can react with a second equivalent of phenylhydrazine (Fig. 22.33). Now aniline (PhNH$_2$) is eliminated through a series of steps, perhaps as shown in Figure 22.33, to give a diimine, which reacts with the third equivalent of phenylhydrazine in an exchange of one imine for another to give the final osazone product.

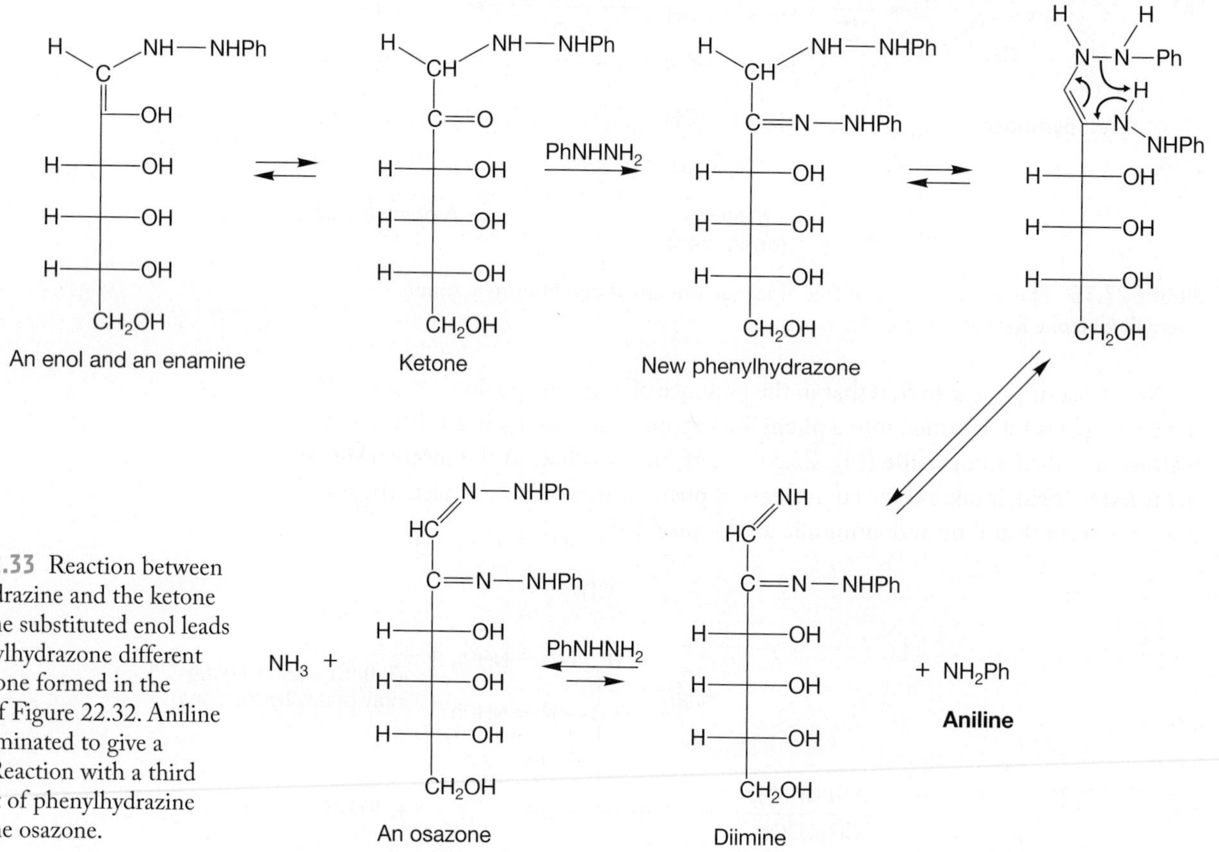

An enol and an enamine

Ketone

New phenylhydrazone

FIGURE 22.33 Reaction between phenylhydrazine and the ketone form of the substituted enol leads to a phenylhydrazone different from the one formed in the reaction of Figure 22.32. Aniline is now eliminated to give a diimine. Reaction with a third equivalent of phenylhydrazine leads to the osazone.

NH$_3$ +

An osazone

Diimine

+ NH$_2$Ph

Aniline

Notice that osazone formation destroys the stereogenicity of C(2). Thus, the same osazone can be formed from two sugars that are epimeric at C(2). Figure 22.34 makes this point using D-glucose and D-mannose. When we work through the Fischer proof of the structure of glucose and the other aldohexoses, we will make use of this point a number of times, so remember it.

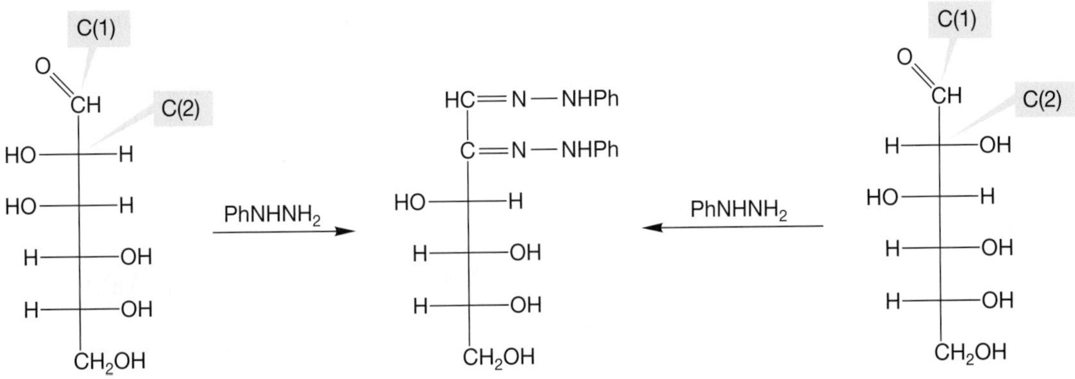

FIGURE 22.34
An osazone can
be formed from
two different
carbohydrates that
are epimeric at C(2).
The stereochemistry
at C(2) is destroyed
in the reaction.

22.4f Ether and Ester Formation

Alcohols can be made into ethers. One classic method in carbohydrate chemistry involves the treatment of a sugar with excess methyl iodide and silver oxide, which leads to methylation at every free hydroxyl group in the molecule to form a polyether (Fig. 22.35).

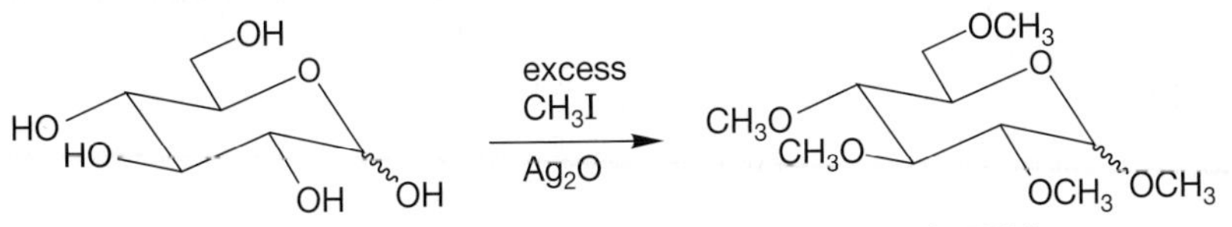

D-Glucopyranose

Methyl 2,3,4,6-tetramethoxy-D-glucopyranoside

(~ 55%)

FIGURE 22.35 Treatment of a sugar with excess methyl iodide and silver oxide leads to formation of a polyether, in this case a hexaether.

One of the classic S$_N$2 reactions is the Williamson ether synthesis (p. 315), the formation of an ether from an alkoxide and an alkyl halide. When an aldohexose is treated with benzyl chloride and potassium hydroxide, a series of Williamson ether syntheses results in formation of a tetraether and an acetal (Fig. 22.36). Be careful to notice the

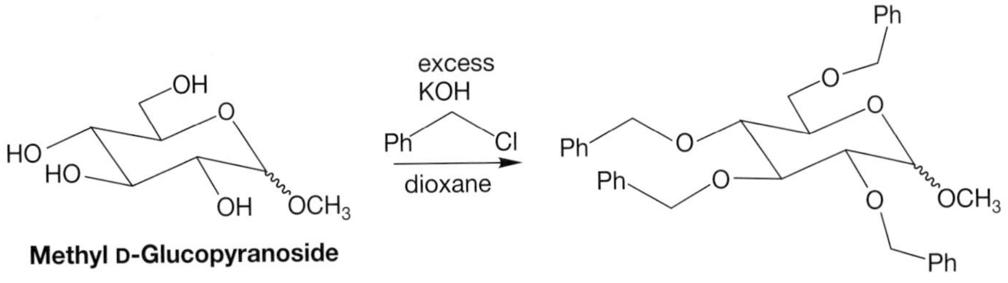

Methyl D-Glucopyranoside

(~ 95%)

FIGURE 22.36 Benzylation through a series of Williamson ether syntheses.

difference between the alkoxy group at C(1) and the four benzyl ethers. The C(1) alkoxy group is part of an acetal. The acetal is stable in basic conditions. If the ether-forming reaction were run on the hemiacetal, then the basic conditions would drive the sugar to the open-chain form and lead to multiple products. But use of the acetal produces the polyether in which neither the alkoxy group at C(1) nor the pyranose ring connection is disturbed in the benzylation reaction.

A carbohydrate reacting with acetic anhydride or acetyl chloride will produce a polyacetylated compound in which all of the free hydroxyl groups are esterified (Fig. 22.37).

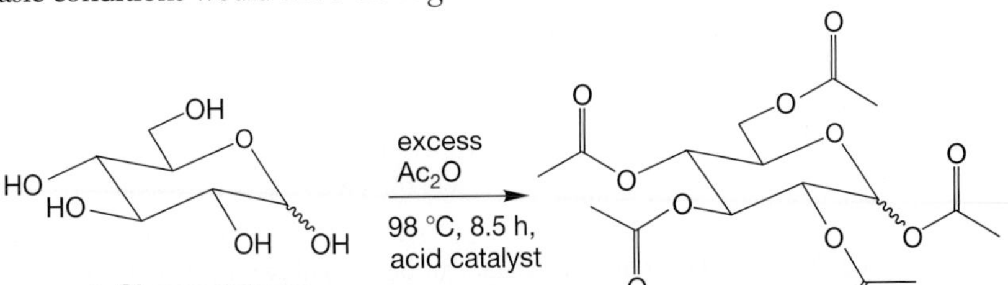

D-Glucopyranose

A pentaester
(58%)

FIGURE 22.37 The free hydroxyl groups of glucopyranose are esterified with acetic anhydride.

Notice that in both the alkylation and acylation reactions the oxygen tied up in the ring remains unaffected by the reaction. It can form neither an ether nor an ester.

The presence of the hemiacetal form allows for selective ether formation when a sugar is treated with dilute acidic alcohol. The product is an acetal (Fig. 22.38).

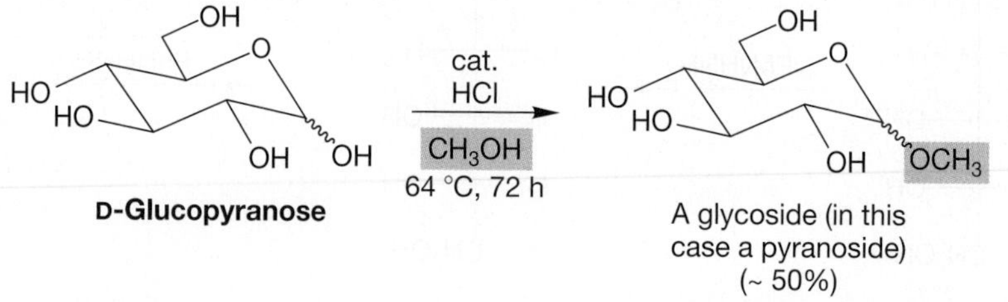

FIGURE 22.38 Treatment of a sugar with dilute HCl and alcohol converts only the anomeric OH into an acetal. Such a molecule is called a glycoside.

The generic name for an acetal of any sugar is **glycoside**. Pyranose glycosides are called **pyranosides**, and furanose glycosides are **furanosides**.

Why is it that only the hydroxyl group at the anomeric carbon is substituted by a methoxyl group in this reaction? The first step is protonation of one of the many OH groups (Fig. 22.39). All the OH groups are reversibly protonated, but only one protonation leads to a resonance-stabilized cation when a molecule of

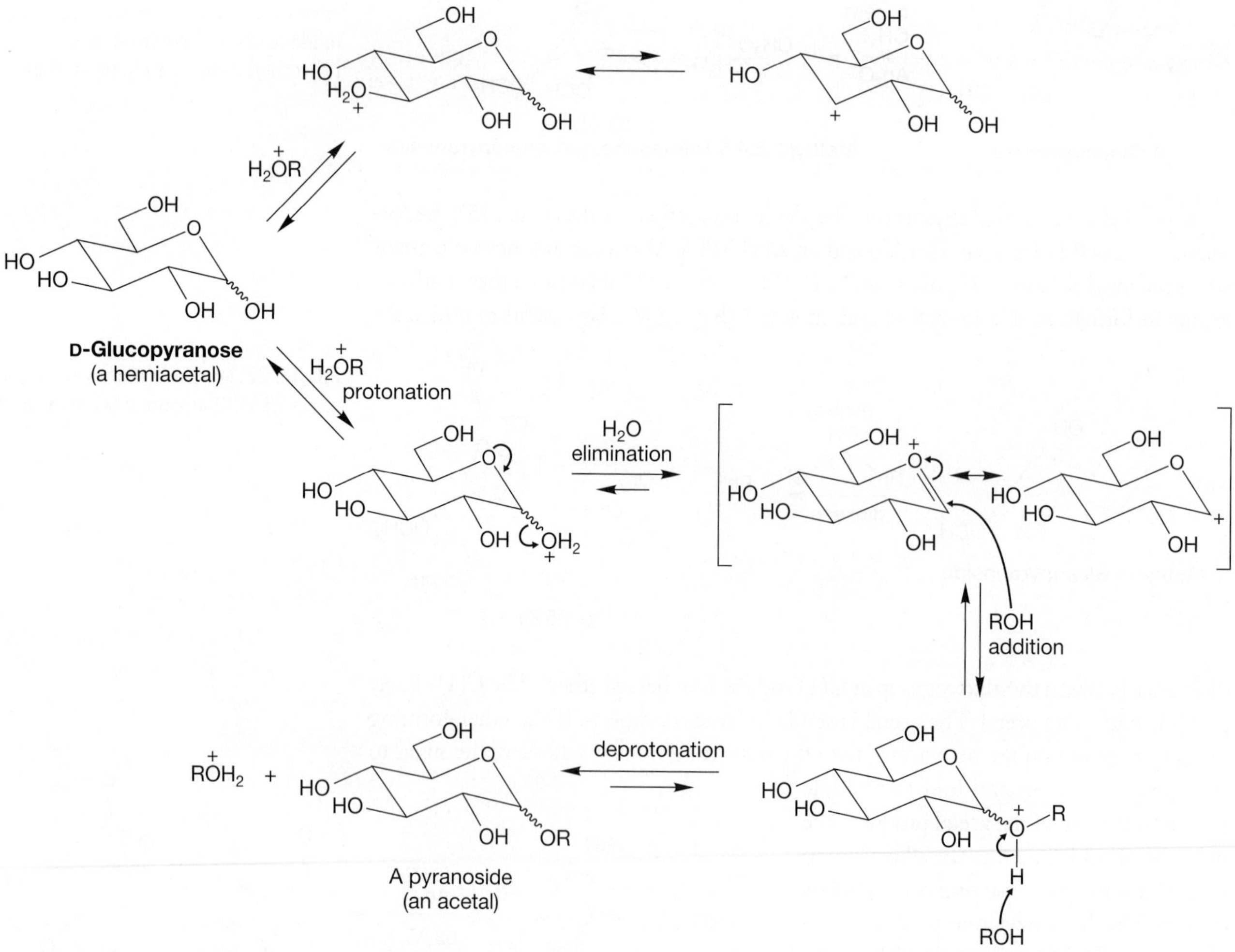

FIGURE 22.39 All OH groups in a sugar can be reversibly protonated. In addition to the protonation at C(3) shown, the top reaction represents protonation at C(2), C(4), and C(6). The bottom reaction shows protonation of the OH at the anomeric C(1). Only in this intermediate can an oxygen "kick out" water and provide resonance stabilization for the resulting cation. Addition of alcohol at C(1) followed by a deprotonation gives the glycoside.

water is eliminated, and so it is this OH that is lost most easily (Fig. 22.39). The ring oxygen helps "kick out" the water molecule. Addition of alcohol at the electrophilic C(1), followed by deprotonation, gives the glycoside. Overall, the hemiacetal OH of the original glucopyranose has been substituted by OR to form an acetal.

Specific glycosides are named by giving the alkyl group name of the alcohol used to make the glycoside followed by the sugar name with the -ose replaced with -oside. For example, if methyl alcohol is used to form an acetal with D-glucopyranose, both methyl α-D-glucopyranoside and methyl β-D-glucopyranoside would be produced (Fig. 22.40a). It is important to note that in water the acetal is more stable than the hemiacetal. Under biological conditions, glycosides do not undergo mutarotation.

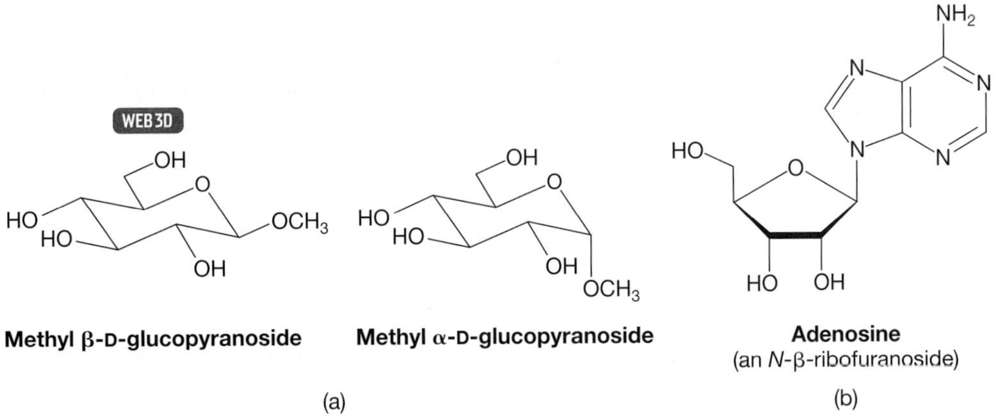

Methyl β-D-glucopyranoside   **Methyl α-D-glucopyranoside**   **Adenosine**
(an N-β-ribofuranoside)

(a)   (b)

FIGURE 22.40 (a) The glycosides methyl α-D-glucopyranoside and methyl β-D-glucopyranoside, acetals made by the reaction of methyl alcohol with D-glucopyranose, and (b) N-β-glycoside adenosine, an amino acetal made by the reaction of D-ribose with the nitrogen-containing compound adenine, a compound we shall see again in Chapter 23.

Nucleophiles other than alcohols can be used to make glycosides. For example, N-glycosides are used in Nature in construction of RNA and DNA. Figure 22.40b shows adenosine, an N-β-glycoside that is a component of RNA.

PROBLEM 22.20 The Koenigs–Knorr reaction is useful for making glycosides. In this reaction, the pentaacetate compound of Figure 22.37 is treated with HBr to give the C(1) bromide. The bromide is then replaced by an alcohol in the presence of Ag₂O. The reaction is very selective for the β anomer, but a series of experiments has established that it is not a simple S_N2 reaction. Show a mechanism that explains the selective formation of the β anomer. *Hint*: Remember Chapter 21.

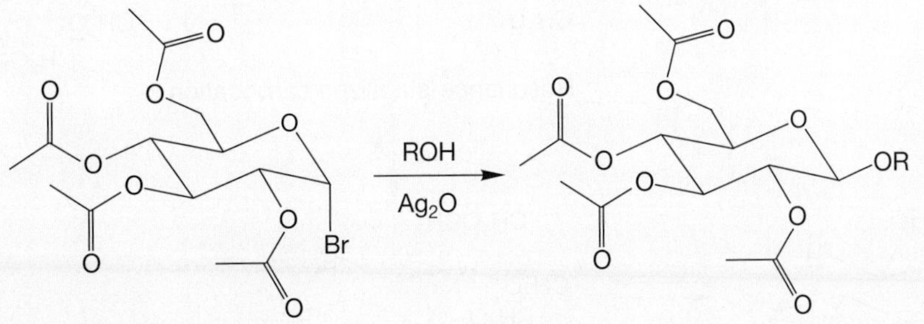

Glycosides are not stable in strongly acidic conditions. The tetramethoxy methyl acetal formed by complete methylation of methyl α-D-glucopyranoside through the

Williamson ether synthesis can be hydrolyzed in acid to a compound in which only the acetal methoxyl group at C(1) has been transformed into a hydroxyl group (Fig. 22.41).

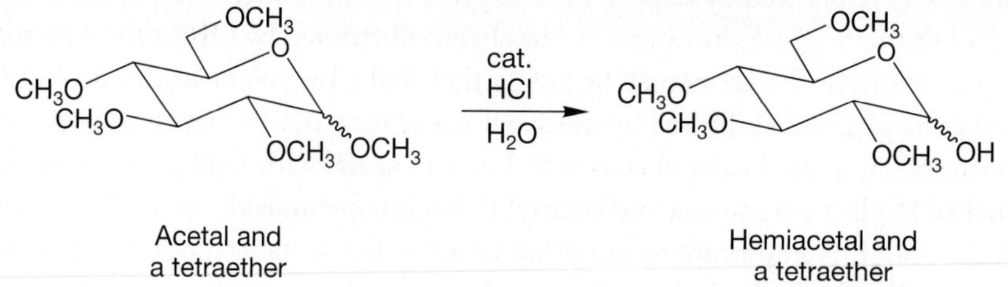

FIGURE 22.41 Acid hydrolysis of the fully methylated acetal leads to a hemiacetal in which only the methoxyl group at the anomeric carbon has been substituted by OH.

Acetal and
a tetraether

Hemiacetal and
a tetraether

WORKED PROBLEM 22.21   Explain why in the reaction shown in Figure 22.41 it is only the methoxyl group at the anomeric carbon that is substituted by a hydroxyl group.

ANSWER   This kind of reaction takes place by protonation of an ether oxygen, loss of alcohol to give a carbocation, and addition of water followed by deprotonation.

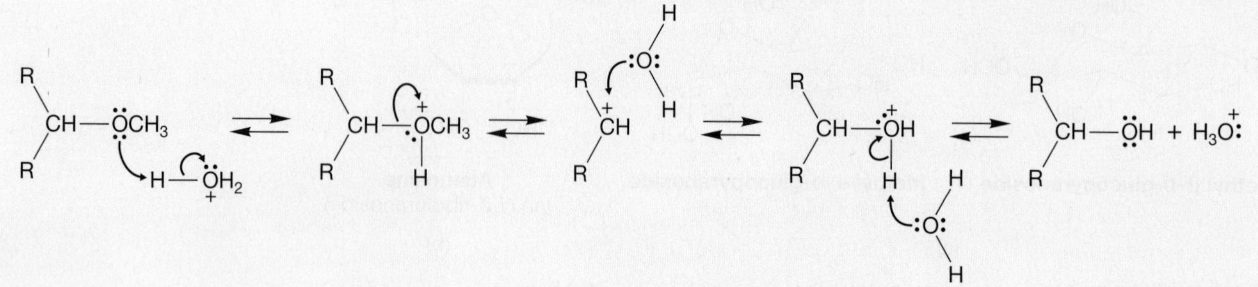

All of the ether oxygens can be protonated, but only in one case can alcohol be lost with assistance from the neighboring group to give a resonance-stabilized carbocation. Addition of water to the cation followed by deprotonation leads to the tetraether hemiacetal shown in Figure 22.41 (both anomers are formed).

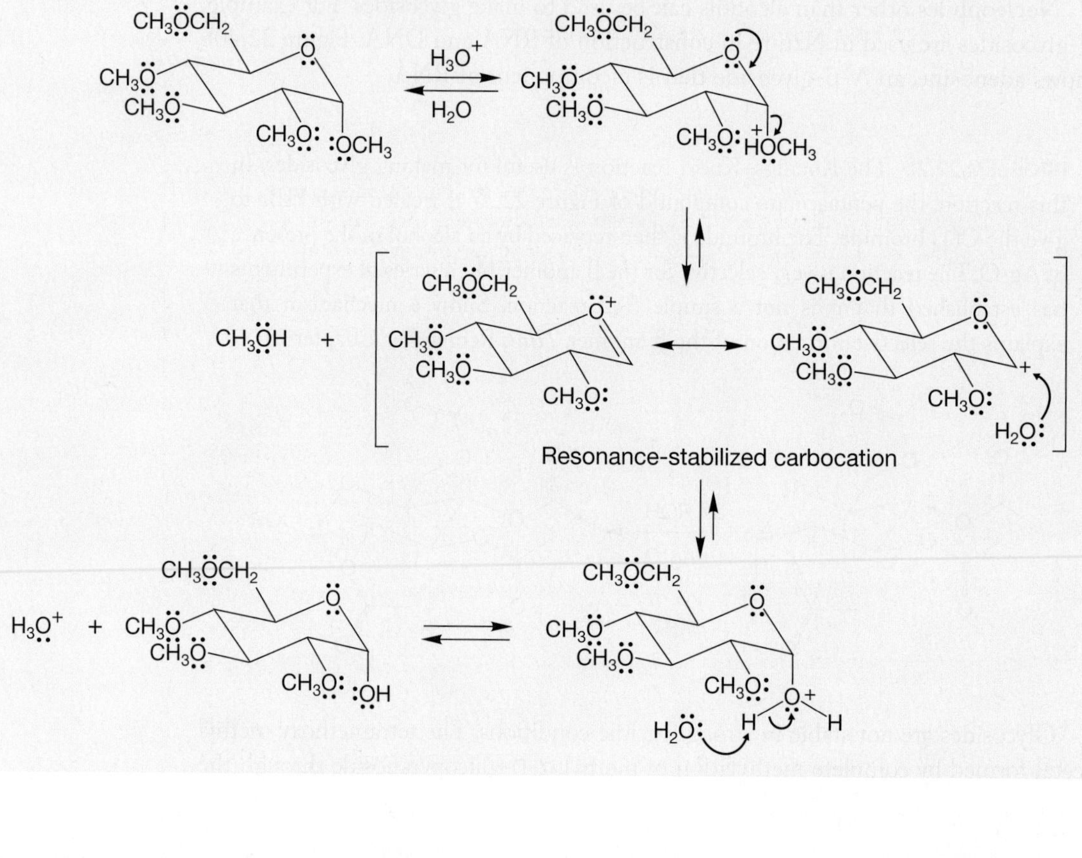

Resonance-stabilized carbocation

We can take advantage of these reactions of sugars, combined with an oxidation, to determine whether the ring in cyclic glucose and other aldohexoses is a furanose or a pyranose. D-Glucose is first converted into its methyl glycoside by treatment with methyl alcohol in acid (Fig. 22.42). Recall that periodic acid cleaves 1,2-diols to dicarbonyl compounds. Therefore, treatment of the methyl D-glucopyranoside with HIO_4 leads to the dialdehyde shown in Figure 22.42. If D-glucose had reacted with methanol to give the furanoside, a quite different product would have been obtained. This procedure is one way in which the predominant ring structure, the six-membered pyranose, was determined.

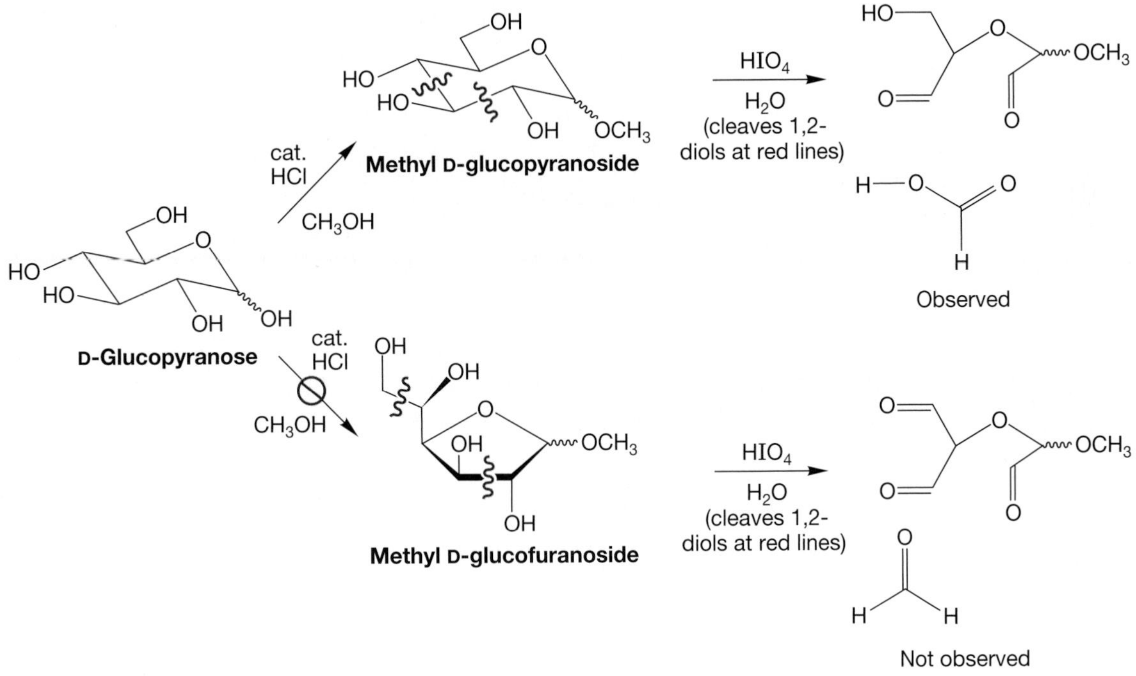

FIGURE 22.42 Formation of methyl D-glucopyranoside and further oxidation with periodic acid gives the observed dialdehyde product, which is evidence that the cyclic form of glucose is (mainly) a pyranose. A furanose would give a different oxidation product.

22.4g Modern Carbohydrate Chemistry   Many natural products and pharmacologically active compounds are covalently attached to carbohydrates. Carbohydrates attached to proteins, for instance, play a major role in cell signaling, gene transcription, and immune response. Attaching a sugar to a substance can be a simple matter of one of the alcohols acting as a nucleophile. Unfortunately there are so many alcohols on any given sugar that simply mixing the reagents will yield multiple products. This procedure is not an acceptable synthetic approach. Nature uses enzymes to control carbohydrate attachments. Chemists control the reactions by employing selective protecting groups (p. 788). For example, if the C(2) OH of D-glucose needs to be attached to a substrate, it stands to reason that we must protect the C(3), C(4), and C(6)

OH groups of the D-glucoside. Figure 22.43 shows the reactions one can use to obtain each of the alcohol groups of α-D-glucopyranoside independently available for further modification. The reactions used for protecting are typically a combination of acetal formation, alkylation, acylation, silylation, and/or reduction. Keep in mind that every carbohydrate is different. So, for example, the method that works for D-glucose may not work for D-mannose.

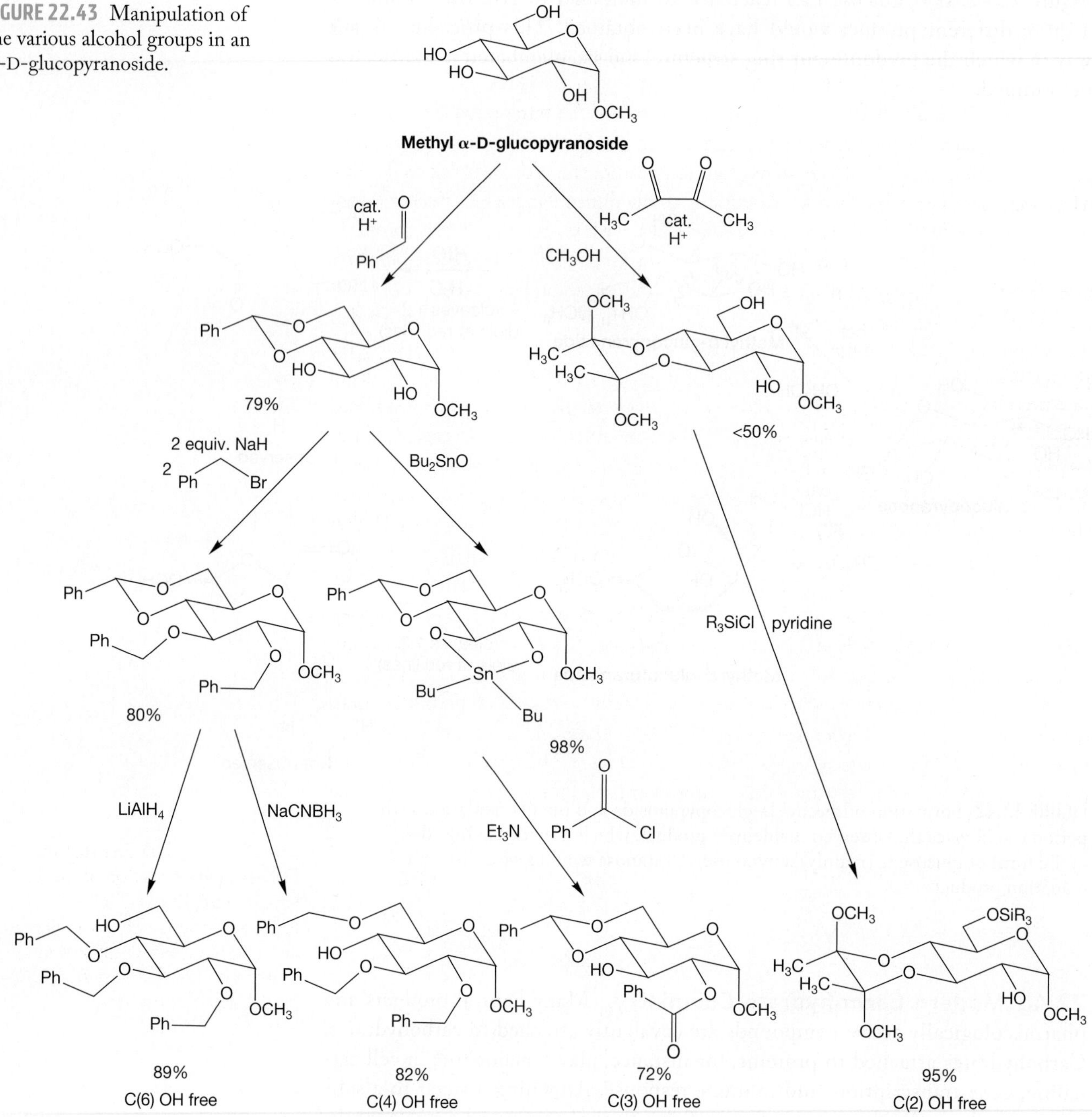

FIGURE 22.43 Manipulation of the various alcohol groups in an α-D-glucopyranoside.

Summary

Reactions in carbohydrates can occur at the carbonyl group or at one of the several alcohols. Aldehydes can be oxidized or reduced. They also can react with amines to form imines. The alcohols can be alkylated or acylated. Selective protection of the glycoside alcohols is possible with careful manipulation of protecting groups.

22.5 The Fischer Determination of the Structure of D-Glucose (and the 15 Other Aldohexoses)

In the following pages, we will approximate the reasoning that led Emil Fischer to the structure of D-glucose. The structures of the other 15 aldohexoses can be determined as we go along. As a bonus, we will also derive the structures of the eight aldopentoses. A risk in this kind of discussion, which straightens out the actual twisted history of the Fischer determination, is that the work will seem too easy, too straightforward. Fischer's papers appeared in 1891, less than two decades after van't Hoff's (Jacobus Henricus van't Hoff, 1852–1911, received the first Nobel Prize in Chemistry in 1901) and Le Bel's (Joseph Achille Le Bel, 1847–1930) independent hypothesis of the tetrahedral carbon atom. Determining stereochemistry was no simple matter, and Fischer's work depended critically on a close analysis of the stereochemical relationships of complex molecules. Moreover, the experimental work was extraordinarily difficult and slow.[5] Fischer's work was monumental and remains an example of the best we humans are able to accomplish.

Fischer recognized that it was impossible for him to solve one of the central problems in this endeavor, the separation of the D series from the enantiomeric L series. In the 1880s there was no way to determine absolute configuration, which is hardly surprising given that the structural hypothesis of the tetrahedral carbon atom was barely 15 years old at the time. However, Fischer recognized that the 16 aldohexoses must be composed of eight pairs of mirror images and that if he knew the structures of one set, he automatically could draw the structures of the mirror-image set. The problem was to tell which was which, and this he knew he could not do. So, he did the next best thing, he was arbitrary; he just guessed. The odds aren't bad, 50:50, and there are times when it is best to accept what is possible and not give up because a perfect solution to the problem at hand isn't available. So Fischer simply *defined* one set of eight isomers as the D series, knowing that it might be necessary for history to correct him. In the event, fortune smiled and he made the right guess. Be sure that you understand that the correctness of Fischer's guess wasn't due to cleverness—his success was not the result of a shrewd guess or the product of a well-developed intuition; there really was no way to know at the time. He was just lucky.

The Fischer proof starts with arabinose, arbitrarily assumed to be the D enantiomer. The first critical observation was that the Kiliani–Fischer synthesis applied to D-arabinose led to a pair of D-aldohexoses (as it must). These were D-glucose and D-mannose (Fig. 22.44). Because this synthesis must produce a pair of sugars that

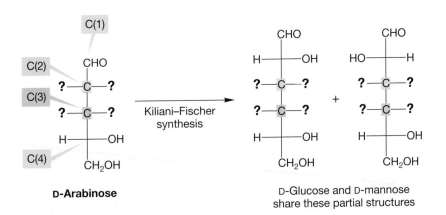

FIGURE 22.44 The first step in Fischer's determination of the structure of glucose. The Kiliani–Fischer synthesis applied to D-arabinose led to D-glucose and D-mannose, which must share the partial structures shown.

---

[5] Here is what Fischer had to say to his mentor, Adolf von Baeyer, about this matter in 1889. "The investigations on sugars are proceeding very gradually. . . . Unfortunately, the experimental difficulties in this group are so great, that a single experiment takes more time in weeks than other classes of compounds take in hours, so only very rarely a student is found who can be used for this work."

have different configurations *only* at the newly generated stereogenic carbon [C(2) in the two hexoses], D-glucose and D-mannose must have the partial structures shown on the right in Figure 22.44.

Let's take a moment to look carefully at Figure 22.44. What do we know, and how do we know it? Fischer has a flask containing one of the enantiomers of the pentose called arabinose, but doesn't know whether this enantiomer is D or L, that is, whether the OH at C(4) is on the right or left. He "solves" this problem by guessing, by assuming the sugar he has in the flask is a member of the D series defined as having the OH at C(4) on the right. He has no way of knowing that almost all natural sugars are of the D series, but he guesses correctly. Fischer also has no idea of the relative arrangement of the OH groups at C(3) and C(2) in the arabinose. Kiliani–Fischer synthesis applied to D-arabinose gives D-glucose and D-mannose, one of which has the newly created OH on the carbon adjacent to the aldehyde on the right, the other of which has it on the left.

The second step in the Fischer proof determines the configuration of C(2) in D-arabinose. Fischer observed that the oxidation of D-arabinose with nitric acid gives an *optically active diacid*. The possibilities are shown in Figure 22.45. There are only three diacids that can be produced from a D-aldopentose, and two of them are meso. In the third, which is the optically active diacid, the OH at C(2) is on the left. Thus, the C(2) in D-arabinose must have OH on the left. Because C(2) in D-arabinose becomes C(3) in the D-glucose and D-mannose of Figure 22.44, we know that C(3) in both hexoses has the OH on the left.

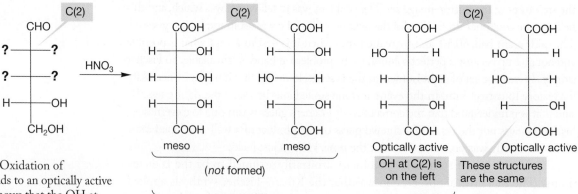

FIGURE 22.45 Oxidation of D-arabinose leads to an optically active diacid, which shows that the OH at C(2) in D-arabinose is on the left.

Now we know a bit more about the partial structures of D-arabinose, D-glucose, and D-mannose. We can specify all but one OH position in D-arabinose and in the two D-aldohexoses. We do not yet know the position of the OH at C(3) of D-arabinose, which becomes C(4) in the D-aldohexoses, and we also do not know which partial structure belongs to D-glucose and which to D-mannose (Fig. 22.46).

FIGURE 22.46 What we now know about the structures of D-arabinose, D-glucose, and D-mannose. For D-arabinose, only the configuration at C(3), which becomes C(4) in D-glucose and D-mannose, is left to be determined.

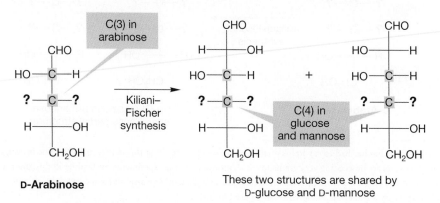

D-Arabinose

C Known from Figure 22.45

In the third step, the position of the last OH in D-arabinose is fixed by the observation that both aldohexoses, D-glucose and D-mannose, give *optically active diacids* when oxidized with nitric acid. There are two possibilities for the OH at C(4) in the aldohexoses: It is either on the left or on the right. If it is on the right, both sugars give an optically active diacid, in agreement with the experimental data (Fig. 22.47).

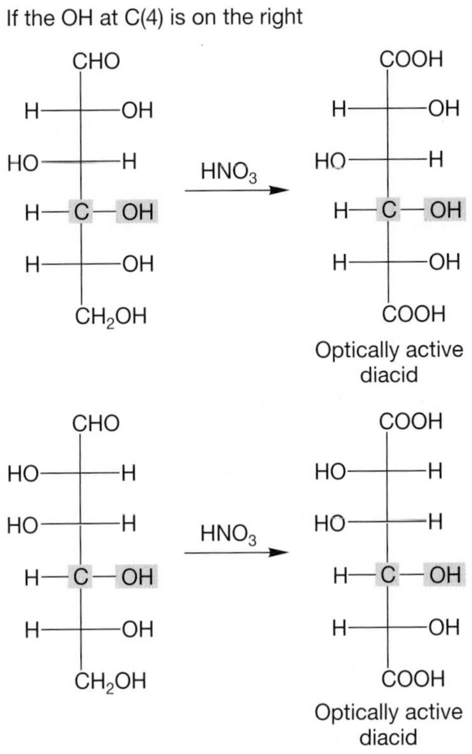

If the OH at C(4) is on the right

Optically active diacid

Optically active diacid

FIGURE 22.47 If the last unknown OH of D-arabinose is on the right, then oxidation of the derived D-glucose and D-mannose with nitric acid produces two optically active diacids. This result matches the experimental results.

On the other hand, if the OH at C(4) is on the left, one of the diacids is meso, not optically active, which does not match the experimental data (Fig. 22.48).

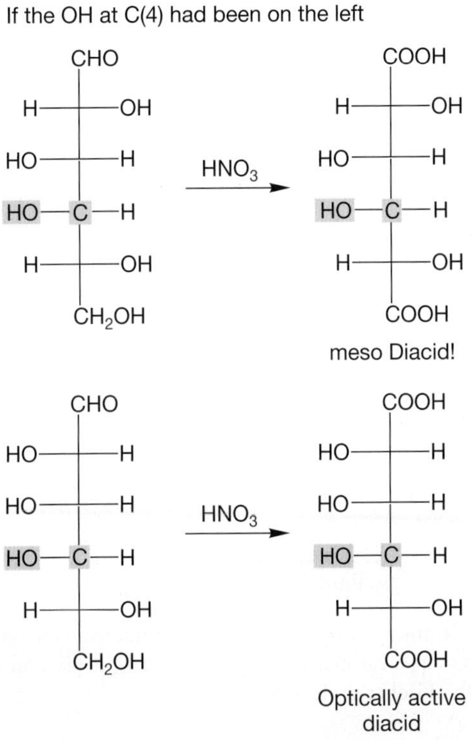

If the OH at C(4) had been on the left

meso Diacid!

Optically active diacid

FIGURE 22.48 However, if the unknown OH had been on the left, one possible diacid is meso, not optically active, which does not match the experimental results.

So, the correct position of the OH at C(3) in D-arabinose and at C(4) in the two D-aldohexoses, D-glucose and D-mannose, must be on the right. D-Glucose and D-mannose share the two structures in Figure 22.49. The next problem is to tell which is which.

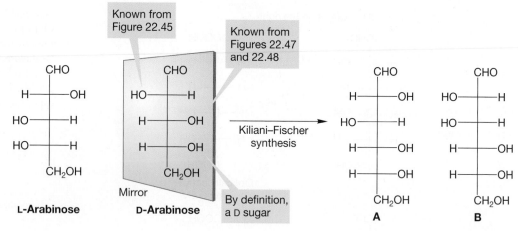

FIGURE 22.49 Now we know the structure of D-arabinose, L-arabinose, and the two structures **A** and **B**, one of which is D-glucose and the other D-mannose.

We now know the full structure of D-arabinose because the positions of all three OH groups have been determined. Of course, because we know the structure of D-arabinose, we also know the structure of its mirror image, L-arabinose (Fig. 22.49).

Now Fischer had to decide which of the two structures **A** and **B** in Figure 22.49 was D-glucose and which was D-mannose. The final observation he needed was that D-glucose *and another hexose*, gulose, gave the same optically active diacid when oxidized with nitric acid (Fig. 22.50). We can find what the structures of D-glucose and the other sugar (gulose) are in the following way. Treatment with nitric acid oxidizes the aldehyde and primary alcohol groups at the two ends of the carbon chain to carboxylic acids. No stereogenic atoms are altered in the process.

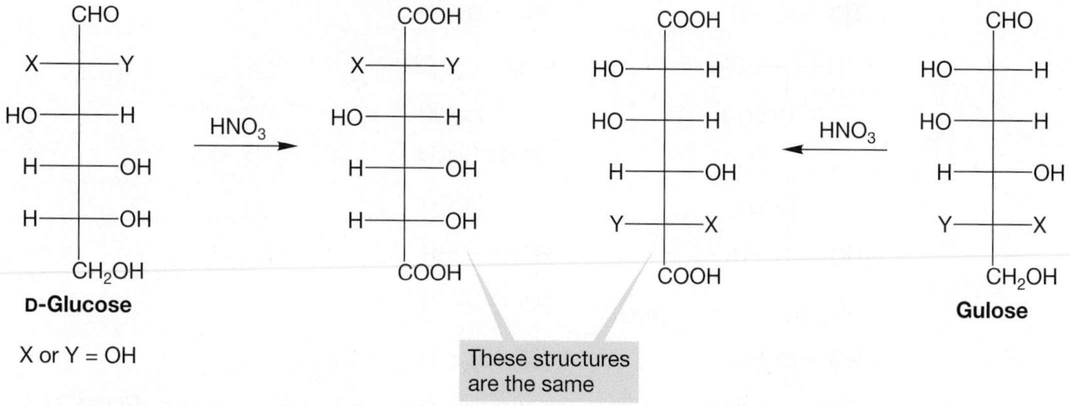

FIGURE 22.50 Oxidation with nitric acid forms an aldaric acid. Both the aldehyde end and the primary alcohol end are converted into the same group, a carboxylic acid. Because D-glucose and gulose give the same aldaric acid, they must differ by turning the Fischer projection end to end.

Therefore the only way two different aldoses can be oxidized to give the same aldaric acid is if those two sugars differ by turning the Fischer projection end to end (Fig. 22.50).

There are two possibilities for the structure of D-glucose, **A** and **B**. If D-glucose has structure **A**, then gulose, the sugar that gives the same aldaric acid on oxidation, must have the structure shown at the right in Figure 22.51.

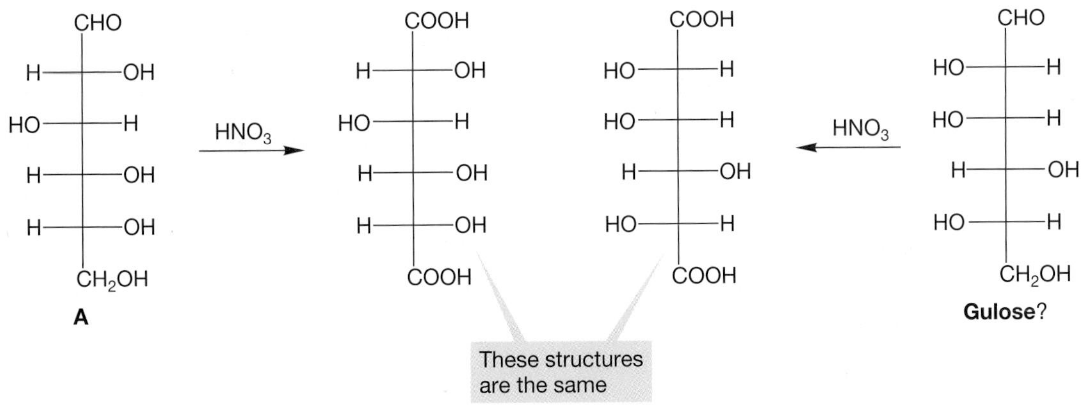

FIGURE 22.51 If D-glucose has structure **A**, then gulose has the structure shown. Oxidation of gulose gives the same aldaric acid as oxidation of D-glucose. Turning the Fischer projection of the oxidized product of gulose 180° shows you the two oxidized products are the same.

Let's see what is required if D-glucose has structure **B**. In this case, gulose must have the structure shown at the right in Figure 22.52 in order to give the same aldaric acid. This time, however, the logic fails, because this gulose would have to be identical with D-glucose, an impossibility. Both would have the structure of **B** (Fig. 22.52)! D-Glucose cannot be **B**, because there can be no second sugar that gives the same diacid. Therefore, D-glucose must have structure **A**, and structure **B** must be D-mannose. We also know the structure of gulose from this logic. Notice that it is an L series sugar!

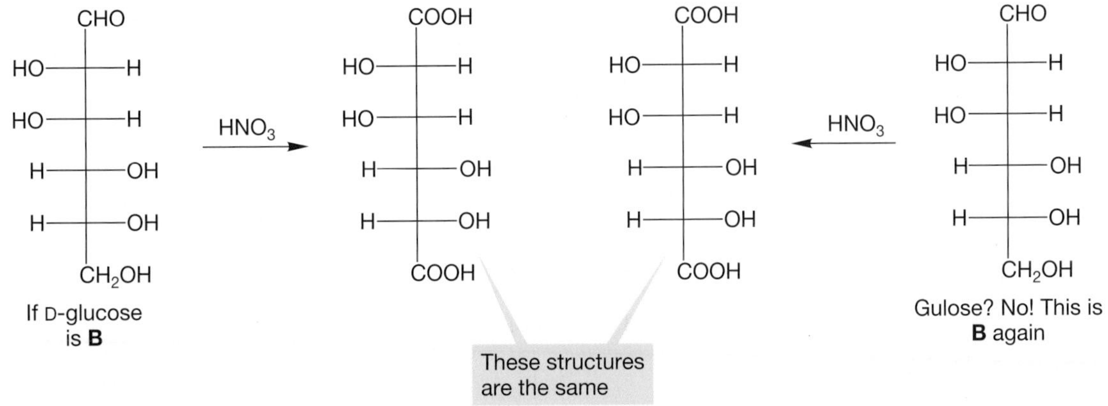

FIGURE 22.52 If D-glucose is structure **B**, then gulose must have the structure on the right so that it can give the same aldaric acid. But this putative structure of gulose is the same as D-glucose. It does not fit Fischer's observation that D-glucose *and another hexose* give the same diacid on oxidation with nitric acid.

Now let's take stock of what we know, which is rather more than we might think at first. We have the structures of 6 of the 16 possible aldohexoses, because we know D-glucose, D-mannose, and L-gulose and their enantiomers L-glucose, L-mannose, and D-gulose. In addition, we know the structures of D- and L-arabinose (Fig. 22.53).

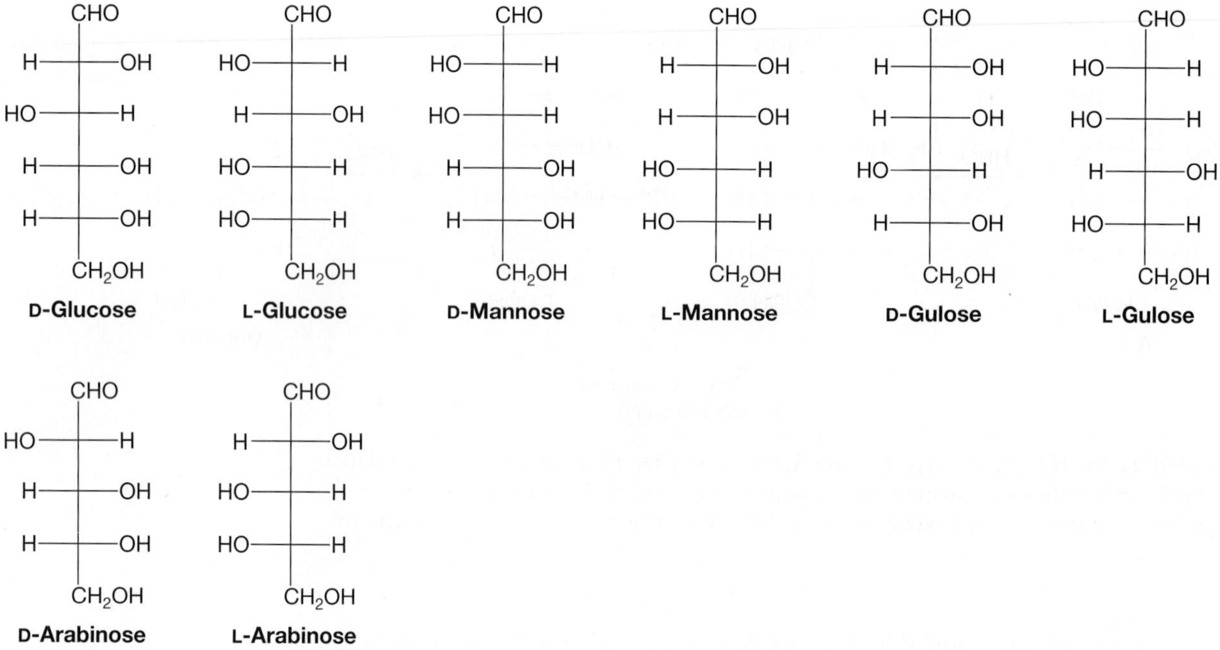

FIGURE 22.53 We now know the structures of these aldohexoses and aldopentoses.

There are many ways to determine the structures of the remaining aldopentoses and aldohexoses. Here is just one possible sequence. Let's first deal with the three remaining D-aldopentoses, given that we know the structure of D-arabinose. The aldopentose D-ribose gives the same osazone as D-arabinose (Fig. 22.54), which means that these two sugars must differ only in the configuration at C(2). Therefore D-ribose must have the structure shown in Figure 22.54.

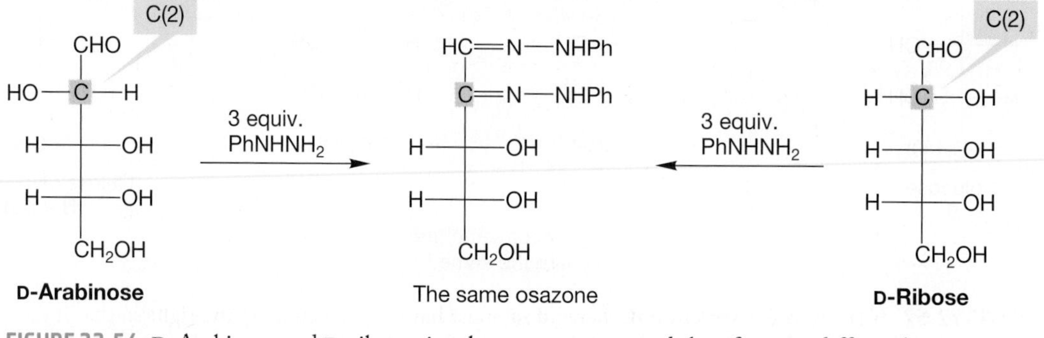

FIGURE 22.54 D-Arabinose and D-ribose give the same osazone and therefore can differ only at C(2). The structure of D-ribose is therefore known. L-Ribose is simply the mirror image of the D isomer.

There are only two remaining D-aldopentoses. One of them, D-xylose, gives a meso diacid on oxidation, whereas the other, D-lyxose, gives an optically active diacid. The structures must be as shown in Figure 22.55.

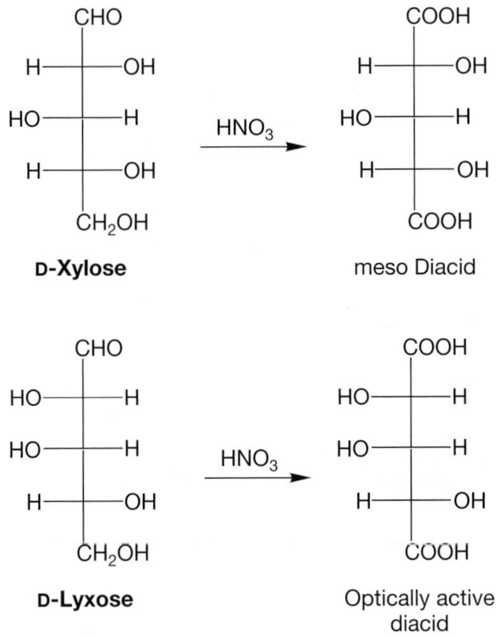

FIGURE 22.55 D-Xylose gives a meso diacid and D-lyxose gives an optically active diacid, which fixes the structures of the remaining D-aldopentoses.

Now we know all four stereoisomers of the D-aldopentoses, and we can construct the L series by drawing the mirror images of the D series. Doing so gives us all eight aldopentoses.

Let's complete the determination of the structures of the D-aldohexoses. D-Gulose and D-idose give the same osazone and must differ only in their configurations at C(2) (Fig. 22.56).

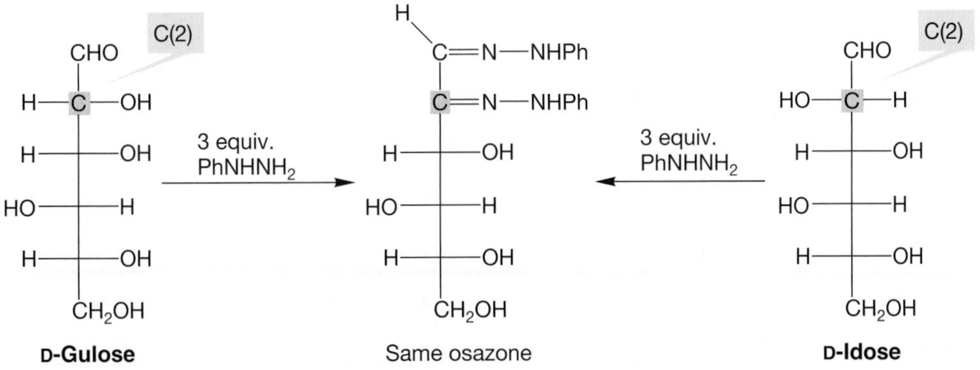

FIGURE 22.56 D-Gulose and D-idose give the same osazone, which determines the structure of D-idose and its mirror image L-idose.

The Kiliani–Fischer synthesis applied to D-ribose gives D-allose and D-altrose, which must have the two structures shown in Figure 22.57. The two structures can be assigned by noting that D-allose gives an optically inactive, meso diacid on treatment with nitric acid, whereas D-altrose gives an optically active diacid.

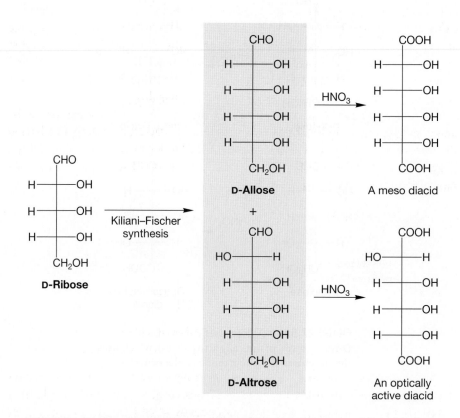

FIGURE 22.57 The Kiliani–Fischer synthesis used to determine the structures of D-allose, D-altrose, and their mirror-image L isomers.

Of the two remaining D-aldohexoses, D-galactose gives a meso diacid on oxidation with nitric acid, but D-talose gives an optically active diacid (Fig. 22.58).

Now we know the structures of all the aldohexoses. The structures of the eight L isomers are simply the mirror images of the D series, and we have at last completed our task.

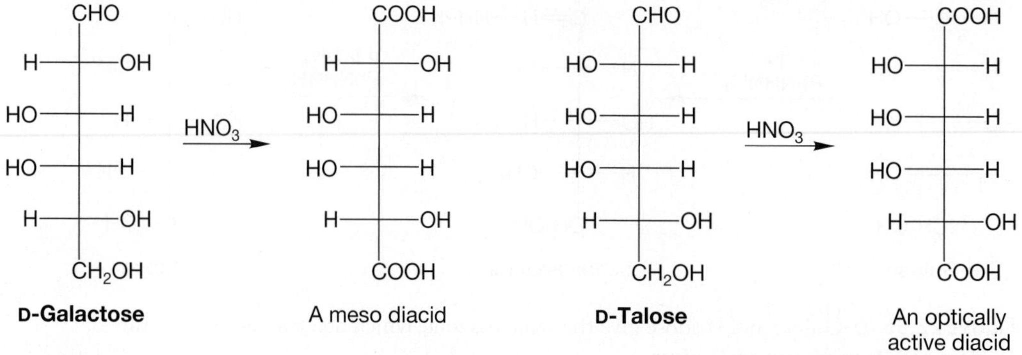

FIGURE 22.58 D-Galactose gives a meso diacid, and D-talose gives an optically active diacid.

22.6 Special Topic: An Introduction to Di- and Polysaccharides

A great many sugars containing 12, 18, and other multiples of 6 carbons are known. Sugars that have the formula $C_nH_{2n}O_n$ are called **monosaccharides** because they are composed of a single sugar molecule. Sugars that come from the combination of two sugar units, the most common being C_{12} carbohydrates, are called **disaccharides** and they have the formula of $C_nH_{2n-2}O_{n-1}$ because the two sugar units are always joined by a dehydration process. Some of the disaccharides have common names. For example, sucrose, lactose, and maltose are all disaccharides. Starch and cellulose are polymers of simple repeating monosaccharide units, less many molecules of water. We call such biopolymers **polysaccharides**.

We will use lactose as an example of how the structure of a disaccharide can be unraveled. The first important observation is that (+)-lactose, a disaccharide having the formula $C_{12}H_{22}O_{11}$, can be hydrolyzed in acid to a pair of simple aldohexoses: D-glucose and D-galactose (Fig. 22.59). This observation identifies the two monosaccharides making up (+)-lactose.

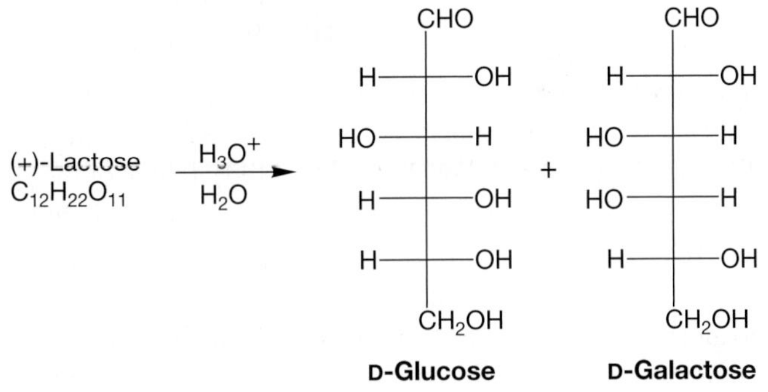

FIGURE 22.59 In acid, (+)-lactose is hydrolyzed to D-glucose and D-galactose. These sugars are C(4) epimers.

But this simple experiment gives us another piece of less obvious information. Remember that only glycosidic linkages, the acetals, are cleaved in acid—simple ether connections are not touched. We saw this distinction first in the hydrolysis of *only* the glycosidic methoxy group of fully methylated D-glucose (p. 1150). D-Glucose and D-galactose must therefore be attached via the C(1) hydroxyl group of one of the sugars (Fig. 22.60). If the attachment were at any other position, dilute acid would not break up the disaccharide.

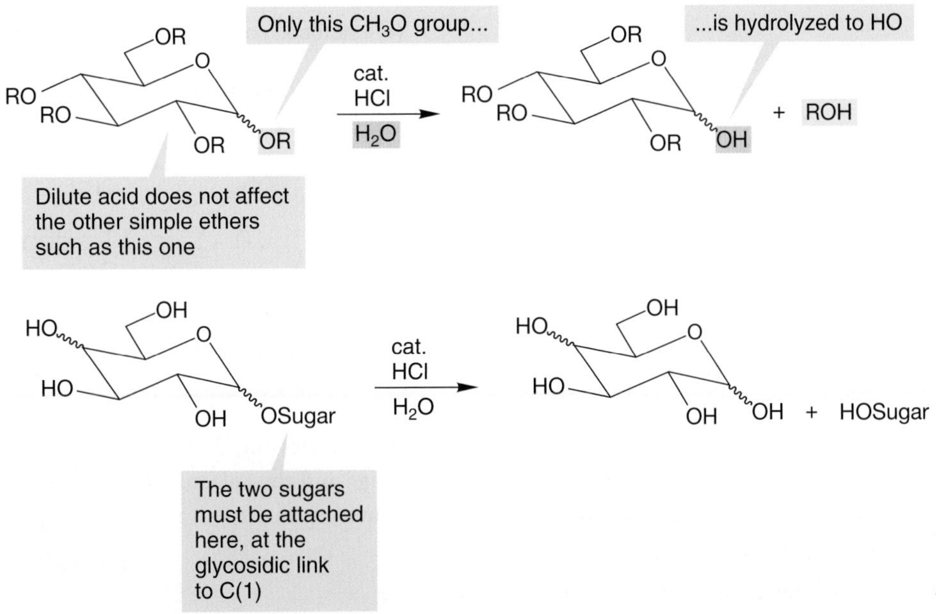

FIGURE 22.60 The fact that (+)-lactose is hydrolyzed in dilute acid to give two monosaccharides means the monosaccharides must be joined via a glycosidic linkage.

We now need to know three details about the dimeric structure of lactose (Fig. 22.61). The issues are:

1. Which sugar is the "attacher" (the ROH) and which is the "attachee" (receives ROH at its anomeric carbon)?
2. Which hydroxyl group of the attacher is used to make the linkage to the anomeric carbon of the attachee?
3. Are the two sugars linked with stereochemistry that is α or β?

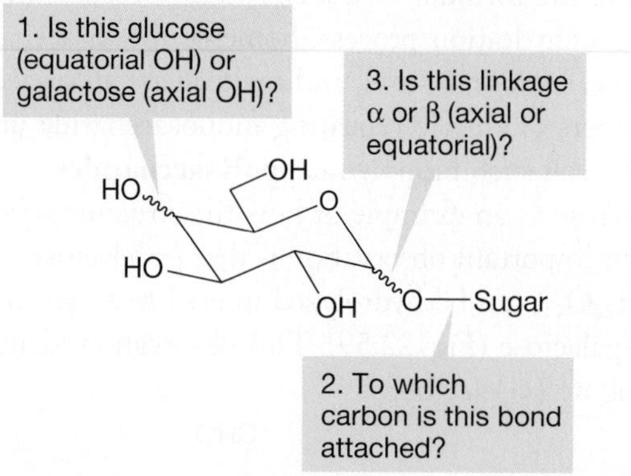

FIGURE 22.61 The issues that must be addressed to determine the structure of (+)-lactose.

1. Is this glucose (equatorial OH) or galactose (axial OH)?

3. Is this linkage α or β (axial or equatorial)?

2. To which carbon is this bond attached?

FIGURE 22.62 The carboxylic acid group in lactobionic acid marks the position of the aldehyde/hemiacetal carbon in (+)-lactose. The fact that hydrolysis of lactobionic acid gives a gluconic acid and not a galactonic acid tells us that it is glucose that has the aldehyde/hemiacetal group in (+)-lactose.

(+)-Lactose can be oxidized by bromine in water to the improbably named lactobionic acid (Fig. 22.62). Bromine in water oxidizes only the aldehydic or hemiacetal carbon of a sugar to give the carboxylic acid (p. 1144). In one of the partners in the disaccharide there is no hemiacetal, and therefore no free aldehyde to be oxidized by bromine/water. Acid hydrolysis of the lactobionic acid gives D-galactopyranose and D-gluconic acid. This result tells us that the aldehyde/hemiacetal group in (+)-lactose must belong to glucose, because it was the monosaccharide that was oxidized. Therefore the galactose must be attached to glucose via an acetal linkage at the galactose C(1).

We don't (yet) know which O is used to make this bond

Hemiacetal link

(+)-**Lactose** in open form

D-Galactopyranose

D-Gluconic acid

Lactobionic acid

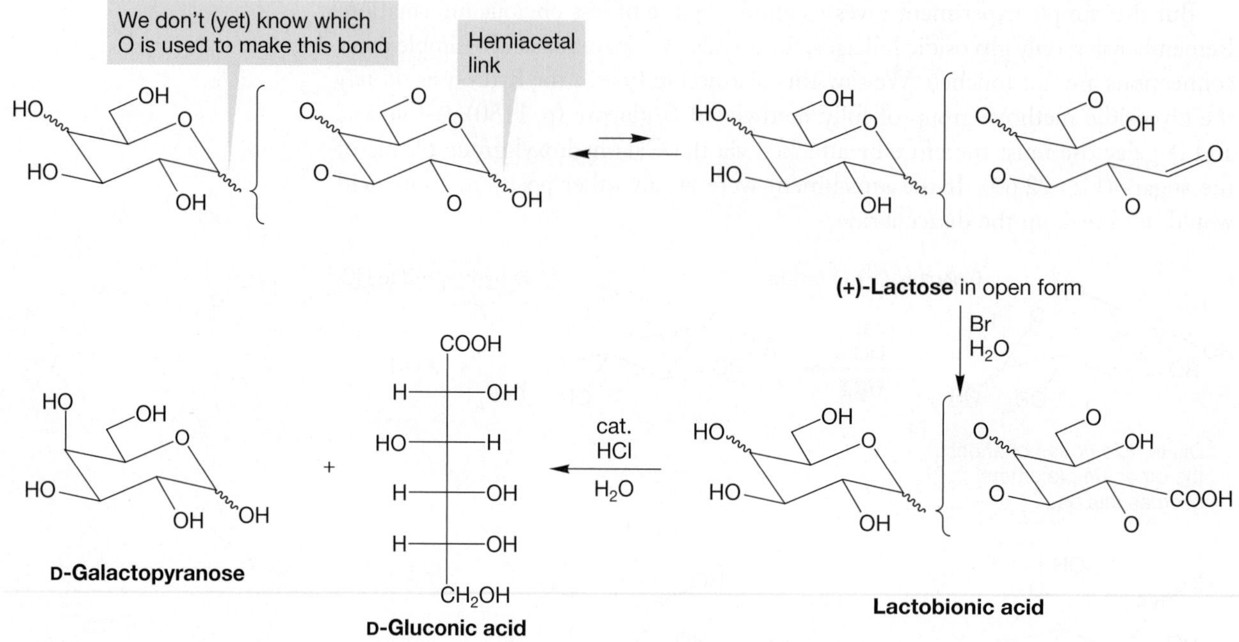

WORKED PROBLEM 22.22 If the two monosaccharides in (+)-lactose were attached with C(1) of glucose as the acetal linkage and the galactose having the free aldehyde/hemiacetal, what would the products of the experiment shown in Figure 22.62 be?

(continued)

ANSWER If the monosaccharides were attached this way, the galactose aldehyde would be oxidized to COOH, and hydrolysis of this *hypothetical* lactobionic acid would give D-galactonic acid and D-glucopyranose.

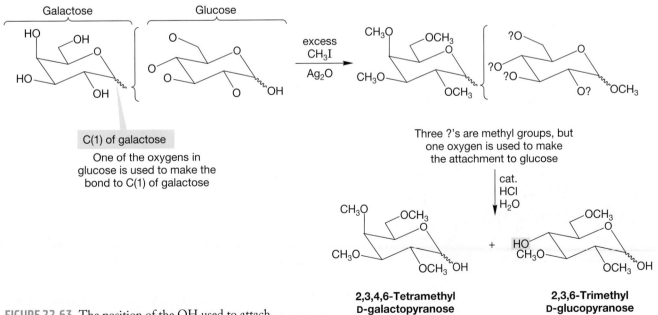

Our next task is to determine which carbon of glucose is attached to C(1) of galactose. To address this issue, we can use a Williamson ether synthesis to methylate all free hydroxyl groups in (+)-lactose (Fig. 22.63). After methylation, the lactose is hydrolyzed in acid to 2,3,4,6-tetramethyl D-galactopyranose and 2,3,6-trimethyl D-glucopyranose.

FIGURE 22.63 The position of the OH used to attach glucose to C(1) of galactose can be determined through a methylation followed by hydrolysis.

This hydrolysis liberates only the OH groups that in (+)-lactose were tied up in acetals; it leaves simple ether groups unchanged. It is the OH at C(4) of glucose that is unmethylated, telling us that this is the glucose position connected to galactose in the lactose molecule.

To determine whether the (+)-lactose glycoside linkage is made as the α or β anomer we turn to enzymes such as lactase[6] that have evolved to cleave only β-glycosidic bonds, and to which α-linked disaccharides are inert. (+)-Lactose is cleaved by lactase and therefore must be β-linked. ^{1}H NMR spectroscopy can also be used to identify whether a disaccharide is linked as the α or β anomer. The hydrogen on the anomeric carbon is more deshielded than any of the other hydrogens, and the C(1) equatorial hydrogen (δ ~5.2 ppm) is further downfield than the C(1) axial hydrogen (δ ~4.6 ppm). Figure 22.64 shows the final three-dimensional representation of (+)-lactose.

FIGURE 22.64 The structure of the disaccharide (+)-lactose. The monosaccharides D-galactose and D-glucose are joined by a β-glycosidic linkage between C(1) of galactose and C(4) of glucose.

SUGAR SUBSTITUTES

The four major sugar substitutes vary widely in structure, but all four are united by a single thread: Saccharin, cyclamate, aspartame, and sucralose were found not by a directed search but by accident. Indeed, each of these four was discovered when a chemist tasted a compound made for other purposes. In fact, cyclamates were discovered because a cigarette smoked in the lab tasted sweet. This observation may seem like courting disaster to you—deliberately ingesting an unknown laboratory chemical, not to mention smoking in a chemistry lab—but it was quite common in the old days. Even as recently as the 1960s it was normal to smell—closely and carefully—just about everything one made. How many chemists destroyed their sense of smell or succumbed to the odor because of this practice, we will never know. Now the laboratory is better ventilated and we gingerly waft the fumes from a flask if we must test for an odor.

After 100 years of research and debate concerning saccharin (discovered in 1879), the consensus is that no ill health effects are associated with moderate consumption. Although cyclamate (discovered in 1937) is banned in the United States, there is no evidence implicating it as a health hazard according to the Food and Drug Administration (FDA). Aspartame (discovered in 1965) may be the single most studied food additive. It has been cleared for use, except for phenylketonurics. People with this disease must avoid phenylalanine, which is formed from the hydrolysis of aspartame. Aspartame also releases methanol when heated, but only extremely small amounts are produced. Sucralose (discovered in 1976) has cleared all toxicology tests. Although halogenated alkanes are known carcinogens, the chlorinated sucralose is not lipophilic enough to be retained in the body.

Cyclamate   **Aspartame**

Saccharin   **Sucralose**

[6]Lactose is found in milk and other dairy products, so it is often called milk sugar. Disaccharides such as lactose need to be cleaved to monosaccharides in order to be absorbed into the bloodstream. The acetal bond in lactose is cleaved by the enzyme lactase. Many adults are lactose intolerant because their bodies don't make lactase. For such individuals, consumed lactose doesn't get absorbed into the bloodstream and the result is abdominal discomfort.

PROBLEM 22.23 Draw a Fischer projection for (+)-lactose. *Hint*: You will need several long, "around the corner" bonds as illustrated in Figure 22.16.

PROBLEM 22.24 Cellobiose is a disaccharide cleaved by hydrolysis into two molecules of D-glucose. Methylation of cellobiose leads to an octamethyl derivative. Hydrolysis with acid converts the octamethyl derivative into 2,3,6-trimethyl-D-glucopyranose and 2,3,4,6-tetramethyl-D-glucopyranose. Cellobiose can be cleaved enzymatically with lactase. There are actually two possible structures for cellobiose. One has ^{1}H NMR signals at δ 4.5 and δ 4.7 ppm. The other isomer has signals at δ 4.5 and δ 5.2 ppm. Draw the two structures of cellobiose.

PROBLEM 22.25 Red blood cells have glycoproteins (carbohydrates covalently attached to a protein, Chapter 23) on the cell surface that play an important role in a person's immune system. There are three versions of the glycoproteins found in humans. These variations lead to what are called *blood types*. You might have blood type A, B, AB, or O. The differences are simply a matter of the number and kinds of carbohydrates in the glycoprotein of your red blood cells. Other animals have different glycoproteins associated with their red blood cells. Horses, for example, have eight blood types. The carbohydrates involved in the human blood types are shown. Type O is a trisaccharide composed of *N*-acetyl-D-glucosamine, D-galactose, and L-fucose. (*Fucose* is the common name for 6-deoxygalactose.) Type B is a tetrasaccharide made up of the three monosaccharides of type O plus a D-galactose. Type A is also a tetrasaccharide that has the basic structure of type O plus an *N*-acetyl-D-galactosamine.

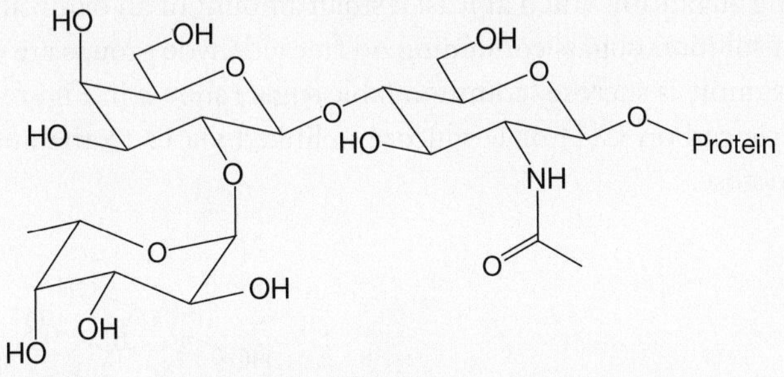

Type O

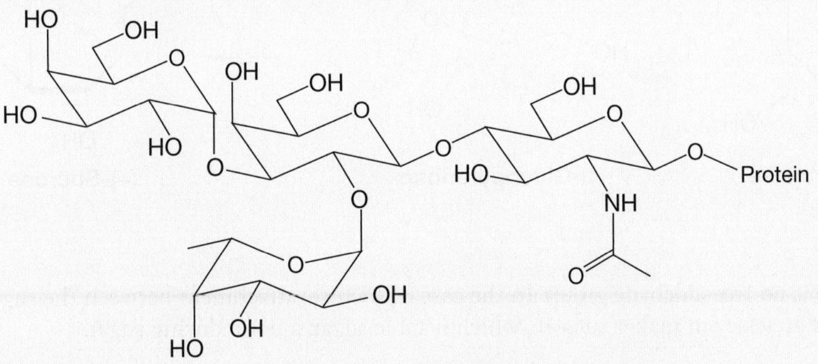

Type B

(continued)

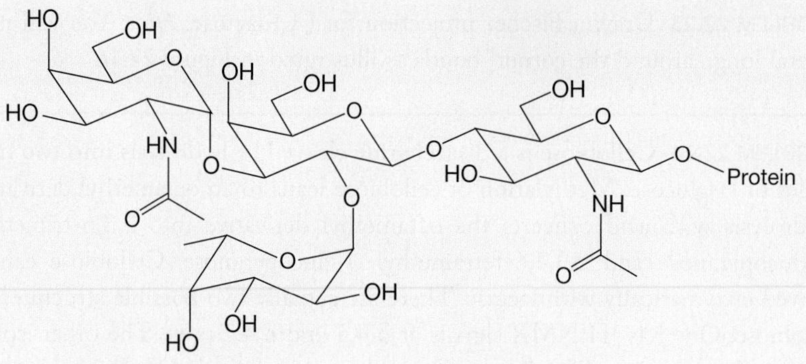

Type A

(a) Draw the Fischer projection of L-fucose. (b) Label the anomeric carbons as α or β. (c) Are these saccharides reducing sugars? Why or why not?

In addition to lactase, the enzyme that cleaves the glycosidic bond in lactose, there are many other glycosidase enzymes in Nature. They play a critical role in a variety of processes related to our diet and our environment. For example, glycosidases are necessary for the degradation of the polysaccharide biomass found in plants. The chemistry a glycosidase facilitates is hydrolysis of the acetal linkage connecting two sugars. The result is a disconnection at the anomeric carbon.

Sometimes two sugars are linked via an oxygen connecting the two anomeric carbons. Because these disaccharides are acetals on *both* sugar units, oxidation to an aldonic acid (Fig. 22.28) is not possible because there is no free aldehyde. *Remember*: In aqueous conditions the cyclic acetals are not in equilibrium with the open form. In contrast to reducing sugars, in which at least a small amount of an oxidizable free aldehyde is present in solution, sugars containing no free aldehyde groups are **nonreducing sugars**. An example is sucrose (common table sugar) shown in Figure 22.65, in which the anomeric carbon C(1) of D-glucose is linked via O to the anomeric carbon C(2) of D-fructose.

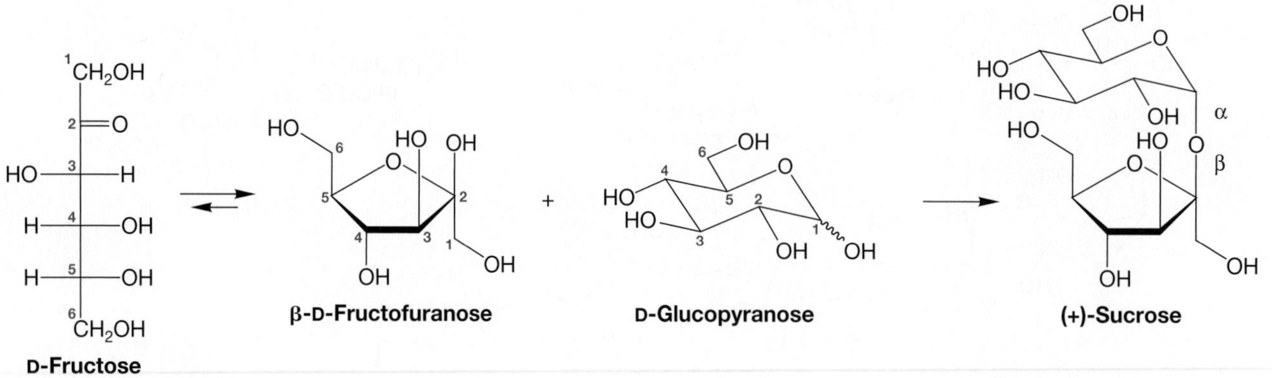

FIGURE 22.65 In nonreducing sugars, there is no free aldehyde group. In the case of sucrose, attachment between the sugars is between the anomeric carbons. This kind of attachment makes sucrose, which is table sugar, a nonreducing sugar.

Synthesis of disaccharides is possible using the protecting group methods discussed earlier (Fig. 22.43). We can use the free alcohol of one monosaccharide to link with the hemiacetal of a protected second monosaccharide. The acetate-protected sugar discussed in Problem 22.20 can be used to make disaccharides. Benzyl protecting groups can also be used. An advantage to employing the benzyl protecting group (PhCH$_2$) is that it can be removed from an oxygen (or nitrogen) under the fairly mild conditions of hydrogenation (H$_2$/Pd). Figure 22.66 uses this protecting group method to make melibiose, a natural isomer of lactose. Melibiose is a galactose–glucose disaccharide that uses the C(6) OH of D-glucose to attach to C(1) of D-galactose.

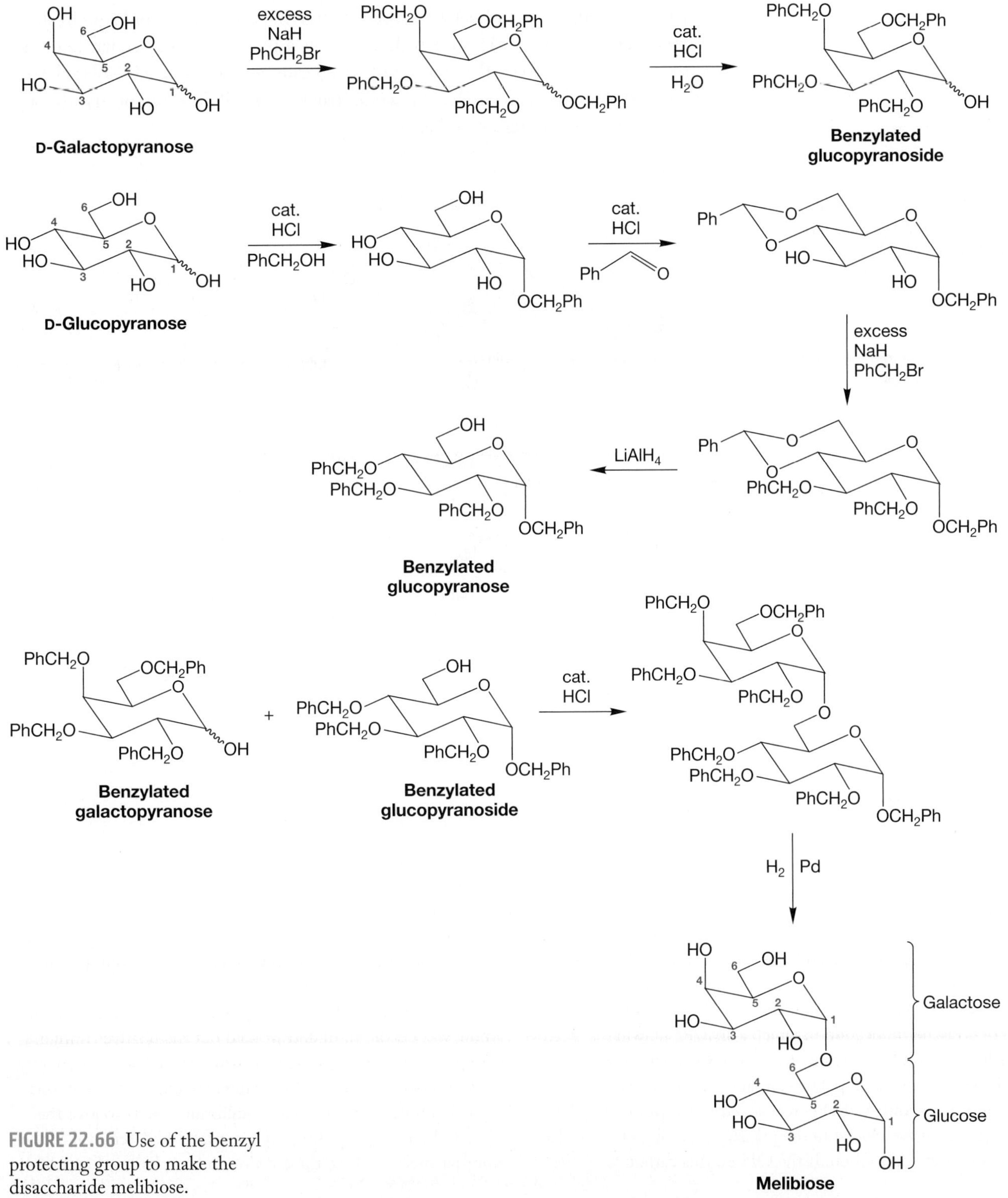

FIGURE 22.66 Use of the benzyl protecting group to make the disaccharide melibiose.

There is no reason that more than two sugars cannot be connected by glycosidic bonds, and larger polysaccharides are common. The carbohydrate **cellulose**, for instance, is a polysaccharide in which thousands of glucose units are connected linearly by β-glycosidic bonds between C(4) of one glucose and the anomeric C(1) of a neighboring glucose (Fig. 22.67a). Cellulose, the stuff plants are made of, contains most of the carbon on Earth! Cotton is 90% cellulose, and wood is about 50%. Most mammals, including humans, are unable to break down cellulose. **Starch** is another polysaccharide made up of glucose units. Starch is the main saccharide in our diet. Corn, rice, potatoes, and wheat are sources of starch. Not all of the linkages in starch are between C(4) and C(1), so it is described as a branched polysaccharide or as a complex sugar. However, the linear polysaccharide amylose can be isolated from starch (Fig. 22.67b). Amylose has the same structure as cellulose except that the sugars are connected through α linkages.

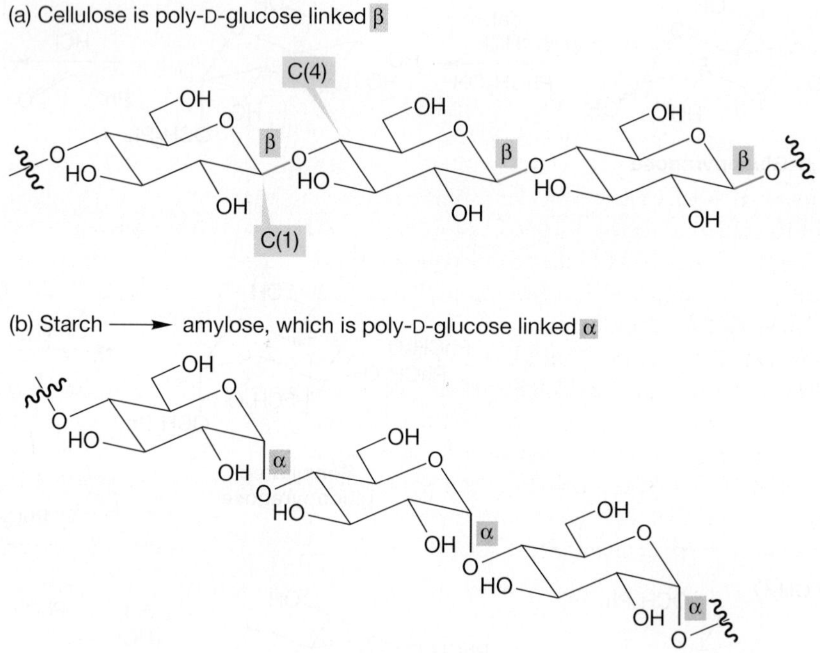

FIGURE 22.67 The two most common natural polysaccharides: (a) cellulose and (b) amylose.

(a) Cellulose is poly-D-glucose linked β

(b) Starch ⟶ amylose, which is poly-D-glucose linked α

22.7 Summary

New Concepts

Carbohydrates, also called sugars or saccharides, are molecules with the general formula of $C_nH_{2n}O_n$. The simplest carbohydrates, the monosaccharides, have an aldehyde at one end in the open form of the molecule and a primary alcohol at the other end. In solution, carbohydrates exist mainly as cyclic hemiacetals.

The stereochemical complications lead to the simplifying tool of a Fischer projection in which a stylized drawing is used to reduce complexity. The molecule is drawn vertically with the aldehyde group at the top. Horizontal bonds are taken as coming toward you and vertical bonds as retreating from you. The stereogenic carbon closest to the primary alcohol group is called the configurational carbon. If the OH on this carbon is on the right in the Fischer projection, the molecule is a D sugar; if it is on the left, the molecule is an L sugar (Fig. 22.3).

When Emil Fischer worked out the structure of glucose in the late 1800s, there was no way to tell absolute configuration. Fischer could not tell the D series from the enantiomeric L series, so he assumed that the OH at C(5), the configurational carbon, was on the right and worked out the relative configurations of the other groups. As it turned out, he guessed correctly.

A number of times in this chapter we encounter reactions in which a minor partner in an equilibrium reacts to give the product. As long as equilibrium is reestablished, removal of the minor partner can drive the reaction to completion (Fig. 22.14).

Key Terms

aldaric acid (p. 1144)	epimers (p. 1133)	mutarotation (p. 1140)
aldohexose (p. 1129)	Fischer projection (p. 1127)	nonreducing sugar (p. 1166)
aldonic acid (p. 1144)	furanose (p. 1132)	osazone (p. 1145)
aldopentose (p. 1129)	furanoside (p. 1148)	pentose (p. 1130)
aldose (p. 1126)	glycolysis (p. 1140)	polysaccharide (p. 1161)
aldotetrose (p. 1129)	glycoside (p. 1148)	pyranose (p. 1132)
aldotriose (p. 1126)	Haworth form (p. 1136)	pyranoside (p. 1148)
anomeric carbon (p. 1134)	hexose (p. 1130)	reducing sugar (p. 1133)
anomers (p. 1134)	ketose (p. 1128)	Ruff degradation (p. 1139)
carbohydrate (p. 1125)	Kiliani–Fischer synthesis (p. 1139)	saccharide (p. 1125)
cellulose (p. 1168)	Lobry de Bruijn–Alberda van Ekenstein	starch (p. 1168)
configurational carbon (p. 1128)	reaction (p. 1141)	sugar (p. 1125)
disaccharide (p. 1161)	monosaccharide (p. 1161)	tetrose (p. 1129)

Reactions, Mechanisms, and Tools

Most of the reactions in this chapter can be found in the chemistry of carbonyl groups (Chapters 16 and 19) and alcohols (Chapters 6 and 7); only a few new reactions appear.

One new twist to a known reaction is *intramolecular* hemiacetal formation. Carbonyl groups react with nucleophiles in the addition reaction (Chapter 16). When the nucleophile is an alcohol, hemiacetals are formed, but generally they are not favored at equilibrium. However, when a relatively strain-free ring can be formed in an intramolecular hemiacetal formation, the cyclic form can be favored. Such is the case for aldohexoses (and many other sugars), which exist mainly in the six-membered pyranose forms (Fig. 22.12).

The equilibration of the cyclic hemiacetal with a small amount of the open aldohexose form allows for either the α or β anomer to produce an equilibrium mixture of the two anomers in a phenomenon called mutarotation (Fig. 22.23).

The reactivity of the hemiacetal OH allows for a differentiation of this OH from the others in the molecule. For example, protonation of the C(1) hemiacetal OH leads to a molecule that can form a resonance-stabilized carbocation on loss of water. Nucleophiles such as alcohols can add to this carbocation to give the acetals. Acetals of carbohydrates are called glycosides. Only the OH at C(1) can react in this way (Fig. 22.38).

Methods have been developed both to lengthen the carbon chain in a carbohydrate (Kiliani–Fischer synthesis) and to shorten it (Ruff degradation).

Reactions at selective positions in a carbohydrate molecule can be accomplished using protecting groups. The unique nature of the anomeric carbon allows for easy conversion of the hemiacetal into an acetal.

Syntheses

Chain lengthening. The Kiliani–Fischer synthesis adds a new C(1) aldehyde and generates a pair of C(2) stereoisomers (Fig. 22.20).

Chain shortening. In the Ruff degradation, the original aldehyde group is lost and a new aldehyde created at the old C(2) (Fig. 22.21).

Osazone formation (Fig. 22.31).

Acetal and tetraether acetal derivatives of carbohydrates (Figs. 22.36 and 22.38).

Aldaric and aldonic acids can be made through oxidation of aldoses (Figs. 22.28 and 22.29).

Common Errors

At the heart of sugar chemistry is the difference between C(1), the anomeric center, and the other carbons. If you lose track of this difference, you lose the ability to follow the labeling studies establishing the ring size of simple sugars and the points of attachment of disaccharides.

It is also sometimes difficult to keep track of open and closed (hemiacetal) forms. We typically write what is convenient, not always good representations of structure. For example, open forms of sugars might be written even though most of the material is really tied up in the cyclic hemiacetal form. Often it is the small amount of the open form that is reactive. This concept can be confusing if you don't have the equilibrium between open and cyclic forms solidly in mind.

22.8 Additional Problems

PROBLEM 22.26 Draw the following molecules:

(a) A Fischer projection of any D-aldoheptose.

(b) A Fischer projection of any L-ketopentose.

(c) A Fischer projection of a D-aldotetrose that gives a meso diacid on oxidation with dilute nitric acid.

(d) Methyl α-D-galactopyranoside in chair form.

(e) The osazone of D-allose in Fischer projection.

(f) Phenyl β-D-ribofuranoside in Haworth form.

PROBLEM 22.27 Make clear three-dimensional chair drawings of all the pyranose forms of D-altrose. Before you start, decide how many will there be.

PROBLEM 22.28

(a) What sugar or sugars would result from the Ruff degradation applied to D-gulose?

(b) What sugar or sugars would result from the Kiliani–Fischer synthesis applied to D-lyxose?

PROBLEM 22.29

(a) What other sugar, if any, would give the same aldaric acid as D-talose when oxidized with nitric acid?

(b) What other sugar, if any, would give the same aldaric acid as D-xylose when oxidized with nitric acid?

(c) What other sugar, if any, would give the same aldaric acid as D-idose when oxidized with nitric acid?

PROBLEM 22.30 What other sugar would give the same osazone as L-talose?

PROBLEM 22.31 What are the principal organic products when D-lyxose is treated under the following conditions? Mechanisms are not necessary. D-Lyxose is drawn in the open form, but it is mostly in the cyclic form in aqueous solution. Keep this fact in mind as you determine the products.

(a) NaBH$_4$, H$_2$O
(b) Br$_2$, H$_2$O
(c) HNO$_3$　　　⟶　?
(d) excess CH$_3$I, Ag$_2$O
(e) cat. HCl, CH$_3$OH

D-Lyxose

PROBLEM 22.32 What are the principal organic products when D-ribose is treated under the following conditions? Mechanisms are not necessary. D-Ribose is drawn in the open form, but it is mostly cyclic.

(a) Ac$_2$O, H$_3$O$^+$
(b) CH$_3$OH, H$_3$O$^+$
(c) PhNHNH$_2$ (3 equiv.)
(d) 1. NaCN
　　2. H$_2$/Pd, poisoned　　⟶　?
　　3. H$_2$O
(e) 1. Br$_2$/H$_2$O
　　2. Ca(OH)$_2$
　　3. Fe$_2$(SO$_4$)$_3$
　　4. H$_2$O$_2$

D-Ribose

PROBLEM 22.33 Glucosamine (2-amino-2-deoxy-D-glucopyranose) is a very important substituted sugar. It is often used for the treatment of osteoarthritis. Draw the β form of glucosamine.

PROBLEM 22.34 Your task is to distinguish between D-talose and D-galactose. The following sequence is suggested. First, oxidize to aldaric acids. Second, make the methyl esters. Then . . . ? Fill in the last step. Oh, we forgot to tell you that at the end of the second step your town's elected officials passed an ordinance forbidding any laboratory work because it would be harmful to their chances for reelection. You are restricted to spectroscopy.

PROBLEM 22.35 Use Table 22.2 (p. 1135) to calculate the energy difference between the α- and β-pyranose forms of D-allose at 25 °C.

PROBLEM 22.36 Calculate the equilibrium ratio of α- and β-D-glucopyranose from the specific rotations of the pure anomers (α = +112°, β = +18.7°; equilibrium value = +52.7°). You might compare your calculation with the values given in Table 22.2.

PROBLEM 22.37 The disaccharide maltose (C$_{12}$H$_{22}$O$_{11}$) can be hydrolyzed in acid to two molecules of D-glucose. It is also hydrolyzed by the enzyme maltase, a molecule known to cleave only α-glycosidic linkages. Maltose can be oxidized by bromine in water to maltobionic acid (MBA, C$_{12}$H$_{22}$O$_{12}$). When MBA is treated with methyl iodide and base, followed by acid hydrolysis, the products are 2,3,4,6-tetramethyl-D-glucopyranose and 2,3,5,6-tetramethyl-D-gluconic acid. Provide a three-dimensional structure for maltose and explain your reasoning.

PROBLEM 22.38 Draw chair structures for the nonreducing disaccharides composed of one molecule of D-allose and one molecule of D-altrose. First determine how many are possible, in principle, and then adopt some scheme for indicating what they are.

PROBLEM 22.39 Write a mechanism for the acid-induced transformation of a D-aldopentose into furfural.

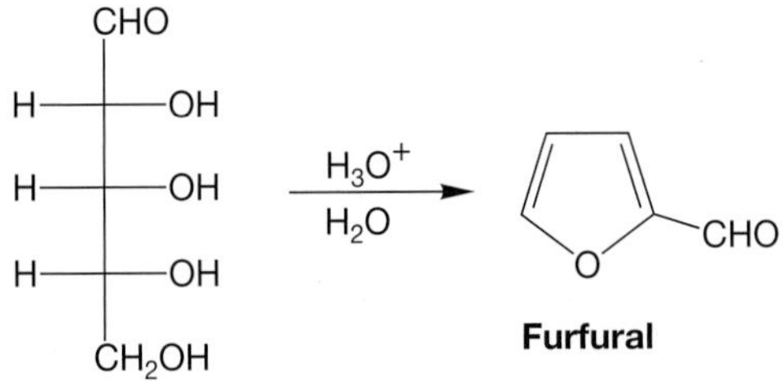

Furfural

PROBLEM 22.40 Figures 22.25 and 22.26 give one mechanism for the Lobry de Bruijn–Alberda van Ekenstein reaction. Use the example given, in which D-glucose equilibrates with D-mannose and D-fructose, and provide another mechanism in which a hydride shift occurs.

PROBLEM 22.41 When α-D-xylopyranose is treated with hydrochloric acid in methyl alcohol, the major products are two isomers of methyl D-xylopyranoside.

(a) Draw the starting material and the two major products and explain how the two major products are formed from the single starting material.

(b) There are two minor products formed as well. Both are methyl furanosides. What are these two products and how are they formed?

PROBLEM 22.42 Outline a mechanism for the following reaction:

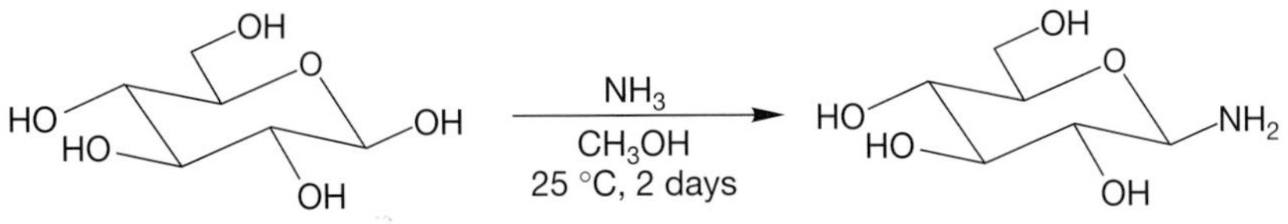

β-D-Glucopyranose

(40%)
Amino β-D-gluco-*N*-pyranoside

PROBLEM 22.43 Outline a mechanism for the following reaction:

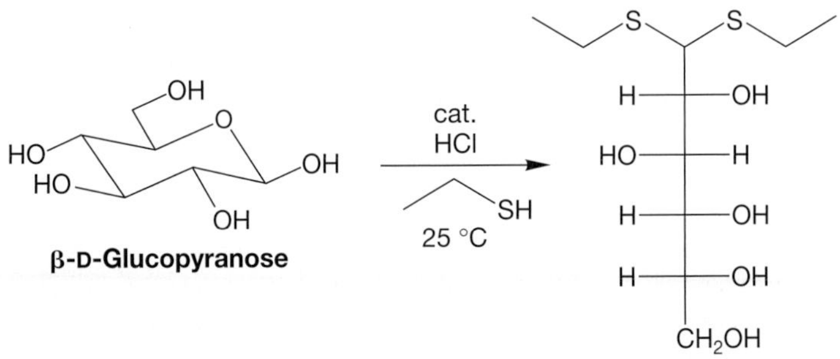

β-D-Glucopyranose

PROBLEM 22.44 When D-glucose, or any polysaccharide containing D-glucose units such as cellulose, is pyrolyzed with an acid catalyst, a new compound, levoglucosan (**1**), is formed. Write a mechanism for the formation of **1** from D-glucose under these conditions.

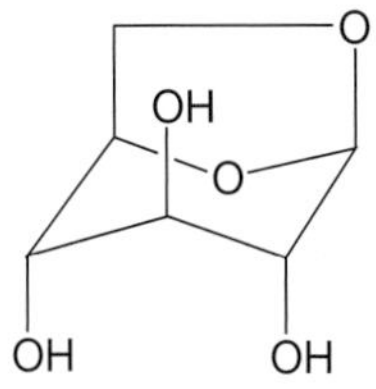

1

PROBLEM 22.45 Treatment of α-D-galactopyranose with acetone and sulfuric acid catalyst leads to a compound that has the formula $C_{12}H_{20}O_6$ and still contains a primary alcohol. Draw a structure for this compound and explain its formation.

PROBLEM 22.46 Draw structures for molecules **A**–**D**, and explain the reactions of the starting material, which is the product formed in Problem 22.45.

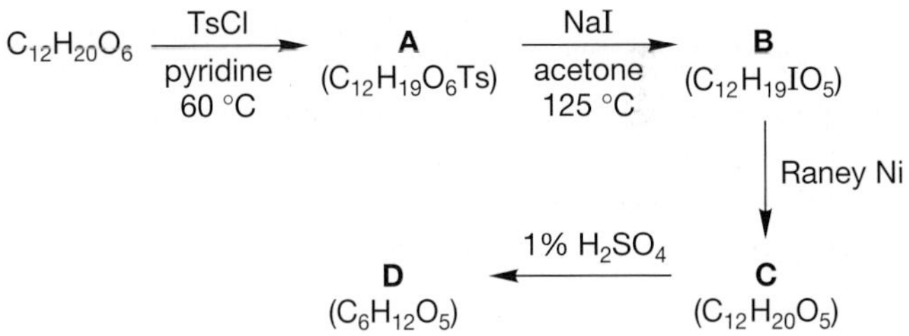

PROBLEM 22.47 The primary alcohol of a sugar will participate in product formation with many carbonyl compounds. For example, many saccharides react with benzaldehyde (PhCHO) in acid to give a six-membered ring in which the primary alcohol is incorporated. However, as shown in Problem 22.45, acetone leads to formation of a five-membered ring rather than a six-membered ring. Draw a detailed three-dimensional structure (a) for the six-membered ring formed when benzaldehyde reacts with the partial sugar shown and (b) for the six-membered ring formed when acetone reacts with the partial sugar. (c) Justify the product you have shown in each case.

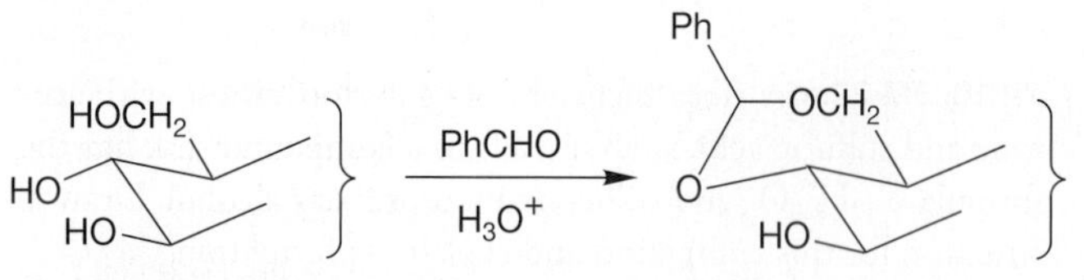

A schematic view
of a sugar

PROBLEM 22.48 The anomeric effect is the observed increase in stabilization of electronegative groups in the α position compared to the β position for D sugars in the pyranose form. One possible explanation for this increase in stabilization is hyperconjugation. Explain how hyperconjugation can be used to justify the anomeric effect.

PROBLEM 22.49 Write a mechanism for the reaction shown in Figure 22.30.

PROBLEM 22.50 The antioxidant vitamin C (shown below) is an essential compound for our diet. The common dietary sources of this vitamin are fruits and vegetables. Interestingly, most other animals are able to biosynthesize the molecule and thus don't rely on a plant-based diet to obtain this necessary substance. The biological pathway for vitamin C formation begins with D-glucose. Based on the structure of vitamin C, do you think glucofuranose is a precursor? Draw glucofuranose and suggest the overall steps (oxidation, reduction, substitution, elimination) that might be used to make this conversion to vitamin C.

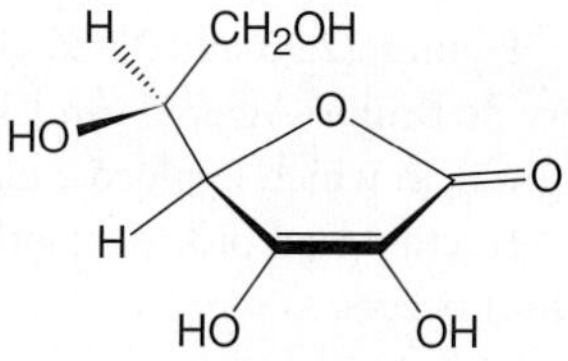

Vitamin C

PROBLEM 22.51 Maltose is a disaccharide that differs very little from cellobiose. It has the same properties as cellobiose described in Problem 22.24, but is not cleaved by lactase (it is cleaved by another enzyme, maltase). Provide a three-dimensional structure for maltose. *Hint*: How else can the two sugars be linked?

Amino Acids and Polyamino Acids (Peptides and Proteins)

23.1 Preview

23.2 Amino Acids

23.3 Reactions of Amino Acids

23.4 Peptide Chemistry

23.5 Nucleosides, Nucleotides, and Nucleic Acids

23.6 Summary

23.7 Additional Problems

AMINO ACIDS ARE US The enzymes that catalyze the bioorganic chemistry in all humans are composed of the same 20 amino acids.

> When I returned from the physical shock of Nagasaki . . . , I tried to persuade my colleagues in governments and the United Nations that Nagasaki should be preserved exactly as it was then. I wanted all future conferences on disarmament . . . to be held in that ashy, clinical sea of rubble. I still think as I did then, that only in this forbidding context could statesmen make realistic judgements of the problems which they handle on our behalf.
>
> Alas, my official colleagues thought nothing of my scheme. On the contrary, they pointed out to me that delegates would be uncomfortable in Nagasaki.
>
> —JACOB BRONOWSKI[1]

23.1 Preview

For many people, fascination with science depends on the relationship between the material they study and biological processes. This connection seems natural and no one can deny the excitement of today's incredible pace of discovery in biological chemistry and molecular biology. It's quite impossible to be a decent molecular biologist without being first a chemist. Organic chemistry is particularly, if not uniquely, important. Molecular biology is organic chemistry (and inorganic and physical chemistry) applied to molecules of a scale not generally encountered in traditional, "small molecule" organic chemistry.[2]

Molecular size and the attendant complexity of structure generate genuinely new chemistry. We have already seen how important molecular shape is in small molecules, and now, in molecules of biological size, stereochemistry becomes more complex and even more vital to understanding mechanism. To a great extent, biological action depends on shape. Much of the immense architecture of amino acid polymers serves "only" to bring a reactive group to an appropriate location on a substrate molecule, which is itself held by the enzyme in just the right position for reaction.

This chapter can only begin to introduce this subject and attempt to show you how important organic chemistry is to it. The choice of subjects is certainly arbitrary—there is a vast number of possible topics. This chapter is held together, somewhat tenuously it must be admitted, by nitrogen atoms. Nitrogen atoms are always components of the alkaloids we saw in Chapter 6, and of amino acids, the subjects of this chapter.

ESSENTIAL SKILLS AND DETAILS

1. In this chapter, we pass from small molecules to big ones. It is necessary to be able to apply old "small molecule" chemistry in these new larger and more complex molecules.

2. Manipulation of blocking and activating groups is necessary for successful peptide synthesis, so you have to know which to use, and when and how to attach and detach them.

3. It is only introduced here, but surely the general mechanism for polynucleotide replication is important. If you go on in biology or biochemistry, you'll learn much more about it.

---

[1]Jacob Bronowski (1908–1974) was a mathematician, poet, and playwright. He was the author and compelling narrator of the brilliant television series, *The Ascent of Man*. Don't miss it if it is replayed.

[2]Scales change depending on your point of view. To a theoretician, a molecule with 10–20 nonhydrogen "heavy atoms" is often too large to be easily approached. Daunting problems are presented, if not of intellectual content, certainly of cost. Even with the decrease in price of computer time, calculations on molecules of "biological" size often simply cost too much. To a molecular biologist, a molecule of only 10 heavy atoms would be laughably tiny—hardly visible on the scale of proteins and enzymes.

23.2 Amino Acids

23.2a Nomenclature There is a systematic nomenclature for amino acids, but as we have so often seen, the real world eschews the system and goes right on using familiar old common names. The generic term *amino acid* itself is shorthand for **α-amino acid**. It is the α that locates the amino group on the carbon adjacent to the carboxylate carbon and distinguishes the α-amino acids from the isomeric but less important β and γ compounds (Fig. 23.1).

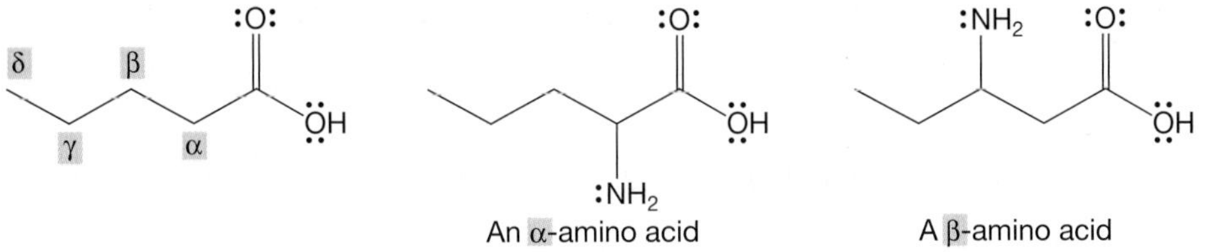

An α-amino acid   A β-amino acid

FIGURE 23.1 The amino group can be located on the hydrocarbon chain with a Greek letter, α, β, γ, or δ.

The IUPAC naming system uses a 2 instead of the α and names the α-amino acids as members of a series of 2-amino carboxylic acids (Fig. 23.2). In practice, the systematic name is rarely used, and common names are retained for most of these molecules. Biochemists tend to view amino acids as substituted α-aminoacetic acids. In this frame of mind, there is an R group on C(2). This R group is called the amino acid **side chain**. Note that as long as R doesn't equal H, the amino acids contain a stereogenic carbon atom.

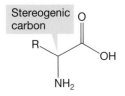

THE GENERAL STRUCTURE

Stereogenic carbon

FIGURE 23.2 An α-amino acid is just an "R-substituted" 2-aminoacetic acid. The IUPAC and the common names are given for several naturally occurring amino acids.

SPECIFIC EXAMPLES

WEB 3D

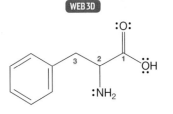

2-Aminoacetic acid (glycine)   **2-Aminopropanoic acid** (alanine)   **2-Amino-3-methylbutanoic acid** (valine)   **2-Amino-3-phenylpropanoic acid** (phenylalanine)

Three-letter codes have been developed for the amino acids and are very commonly used, especially when describing the polymeric amino acids known as peptides or in especially large examples, as proteins. A peptide is a molecule that is composed of two or more amino acids that are connected by an amide bond. The connecting amide bond is referred to as a **peptide bond**. A protein is just a long polypeptide chain. The boundary between the smaller peptides, or polypeptides, and the larger proteins is neither fixed nor always clear, and the terms are sometimes used interchangeably. More recently, an even shorter set of one-letter abbreviations has come into use. Table 23.1 gives the names and abbreviations for the 20 common amino acids and some of their physical properties. Notice that R groups of many kinds appear, from simple hydrocarbon chains to more complicated acidic and basic groups.

TABLE 23.1 Properties of 20 Common Amino Acids

R (Side Chain)	Common Name[a]	Abbreviations	mp (°C) L-Form[b]	pK_a (COOH)	pK_a ($^+NH_3$)	pK_a (Side Chain)	Isoelectric Point (pI)
Simple Alkane Groups							
—H	Glycine	Gly, G	229 (dec)	2.3	9.6		6.0
—CH₃	Alanine	Ala, A	297 (dec)	2.3	9.7		6.0
—CH(CH₃)₂	Valine*	Val, V	315 (dec)	2.3	9.7		6.0
—CH₂CH(CH₃)₂	Leucine*	Leu, L	337 (dec)	2.3	9.6		6.0
—CHCH₃(CH₂CH₃)	Isoleucine*	Ile, I	284 (dec)	2.3	9.6		6.0
Aromatic Groups							
—CH₂— (phenyl)	Phenylalanine*	Phe, F	283 (dec)	1.8	9.2		5.9
—CH₂— (phenol, OH)	Tyrosine	Tyr, Y	316 (dec)	2.2	9.1	10.1	5.7
—CH₂— (imidazole)	Histidine*	His, H	288 (dec)	1.8	9.0	6.0	7.6
—CH₂— (indole)	Tryptophan*	Trp, W	290 (dec)	2.4	9.4		5.9
Groups Containing an OH (see also Tyr)							
—CH₂OH	Serine	Ser, S	228 (dec)	2.2	9.2		5.7
—CH(CH₃)OH	Threonine*	Thr, T	226 (dec)	2.2	9.2		5.6
Groups Containing S							
—CH₂SH	Cysteine	Cys, C		1.7	10.8	8.3	5.0
—CH₂CH₂SCH₃	Methionine*	Met, M	283 (dec)	2.3	9.2		5.8
Groups Containing a Carbonyl Group							
—CH₂COOH	Aspartic Acid	Asp, D	270 (dec)	1.9	9.6	3.7	2.9
—CH₂CH₂COOH	Glutamic Acid	Glu, E	211 (dec)	2.2	9.7	4.3	3.2
—CH₂CONH₂	Asparagine	Asn, N	214 (dec)	2.0	8.8		5.4
—CH₂CH₂CONH₂	Glutamine	Gln, Q	185 (dec)	2.2	9.1		5.7

TABLE 23.1 Properties of 20 Common Amino Acids (*continued*)

R (Side Chain)	Common Name[a]	Abbreviations	mp (°C) L-Form[b]	pK_a (COOH)	pK_a ($^+NH_3$)	pK_a (Side Chain)	Isoelectric Point (pI)
Groups Containing an Amino Group (see also Trp and His)							
—$(CH_2)_4NH_2$	Lysine*	Lys, K	224 (dec)	2.2	8.9	10.3	9.7
—$(CH_2)_3NH$—C=$NH(NH_2)$	Arginine*	Arg, R	207 (dec)	2.2	9.1	12.5	10.8
A Secondary Amine							
(structure: pyrrolidine ring with COOH, N, H)	Proline	Pro, P	221 (dec)	2.0	10.6		6.1

[a] Essential amino acids are marked by an asterisk (*).
[b] Decomposition is abbreviated by dec.

Why pick out this particular set of 20 from the vast number conceivable? Because these are the 20 amino acids encountered in peptides and proteins in Nature. Your body can synthesize 10 of these 20 from the food you eat. One might think that a reasonable way for your insides to proceed would be to break down ingested proteins to their constituent amino acids, and then use these directly. But this is not how it works. Instead, amino acids are metabolized (broken down chemically) into much simpler molecules from which the 10 amino acids are resynthesized. But even your body is not a perfect synthetic chemist—unless you are a superior mutant, there are 10 of these 20 amino acids it is unable to make. These you do use directly. These 10 you must ingest, and if your diet fails to include one or more of these **essential amino acids** you will not last long. The essential amino acids are marked with an asterisk in Table 23.1.

CANAVANINE: AN UNUSUAL AMINO ACID

We tend to think of plants as passive, defenseless creatures, growing quietly at the mercy of herbivorous predators. But this notion seems odd; shouldn't evolutionary pressure lead to defense mechanisms in which the poor plant finds a way to prevent animals from eating it? Indeed it has, and plants are full of all sorts of toxins evolved to discourage predators. Of course, these defenses lead to evolutionary pressure on the predator population, and sometimes one species will work its way around the plant's defenses. The signal that this has happened is the presence of a single predator. All others have been discouraged, but one has found a way to defeat the plant's defense systems. Take, for example, *Dioclea megacarpa*, a vine whose only predator is the beetle *Caryedes brasiliensis* shown in the photo. This plant uses canavanine in place of the closely related arginine (replace the O in canavanine with CH_2 and you get arginine). In most herbivores, replacement of arginine with canavanine leads to incorrect protein folding and arrested development. The single successful beetle has developed an enzyme that can tell the difference between arginine and canavanine, as well as an enzyme specially designed to destroy canavanine. So, *C. brasiliensis* happily munches away on *Dioclea* while all other beetles stay away.

23.2b Structure of Amino Acids All but one of the 20 common amino acids is a primary amine. The only exception is proline, which is an amino acid containing a secondary amine (see Table 23.1). All 20 of the biologically important amino acids but the achiral glycine and cysteine are found in the (*S*) configuration. Many (*R*) amino acids are known in Nature, but most play no important role in human biochemistry. Why? That's a good question and no certain answer exists. Did the first resolution of enantiomers (p. 169) produce an (*S*) amino acid, which then determined the sense of chirality of all the amino acids formed thereafter? Is the preference for (*S*) no more than the result of some primeval accident? No one knows. If it is, we may one day encounter (*R*) civilizations somewhere in the universe, provided we are sufficiently lucky and clever to survive long enough.

> **PROBLEM 23.1** Draw tetrahedral three-dimensional structures for (*R*) and (*S*) valine and aspartic acid.

Amino acids are usually not described as (*S*) or (*R*). Instead, the Fischer projection system is used (p. 1127). The acid group is drawn at the top, and, as with the sugar Fischer projections, vertical lines represent bonds retreating from you and horizontal lines represent bonds coming toward you. The carbon chain is the backbone of the Fischer projection, which means the side chain is placed at the bottom of the drawing. The position of the amino group in the Fischer projection determines whether the amino acid is D or L. If the amino group is on the left, then it is an L-amino acid. If the amino group is on the right, then it is a D-amino acid. As shown in Figure 23.3, typically it is the L-amino acid that is in the (*S*) configuration. It is interesting that Nature depends most on the D-carbohydrates and the L-amino acids.

FIGURE 23.3 An (*S*) amino acid is usually L in the Fischer convention.

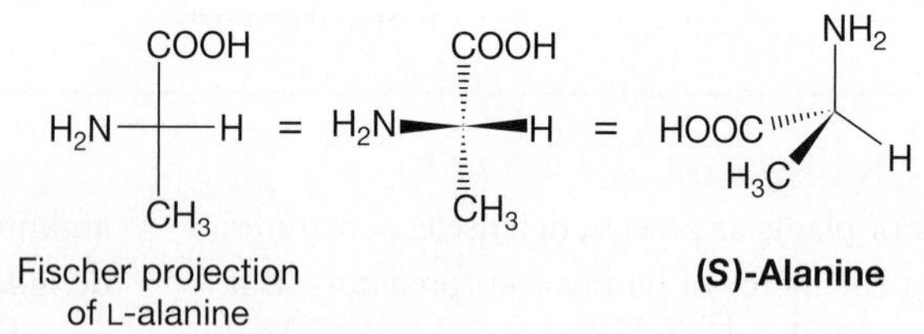

Fischer projection of L-alanine

(*S*)-Alanine

> **WORKED PROBLEM 23.2** Draw the L isomers of valine and cysteine in Fischer projection. Determine the (*R*) or (*S*) configuration of each.
>
> **ANSWER** Remember the convention for Fischer projections. To convert the Fischer projection into a 3-D structure, draw the horizontal bonds as solid wedges coming toward you and vertical bonds as dashed wedges retreating from you. Now the 3-D image can be rotated in space to put the lowest atomic number (the hydrogen) pointing toward the right. Notice that L-cysteine is the only L-amino acid in Table 23.1 that is in the (*R*) configuration.

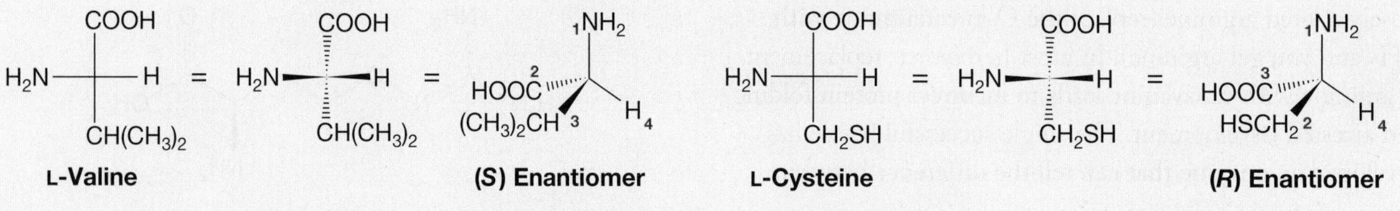

L-Valine   (*S*) Enantiomer   L-Cysteine   (*R*) Enantiomer

As you can see from Table 23.1, amino acids are high-melting solids. Does this not seem curious? Common amino acids have molecular weights between 75 and 204, and this molecular weight does not seem sufficient to produce molecules that are solids. This conundrum leads directly to Section 23.2c.

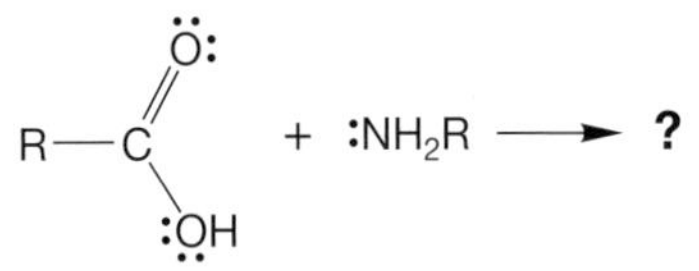

23.2c Acid–Base Properties of Amino Acids

Why are amino acids solids? To answer this question, let's ask another. What reaction do you expect between the generic amine and the generic carboxylic acid of Figure 23.4?

A simple organic acid, such as acetic acid, has a pK_a of about 4.5, and an ammonium ion has a pK_a of about 8–10. An amine and a carboxylic acid must react to give the ammonium salt of Figure 23.5, because the carboxylic acid is a much stronger acid than an ammonium ion. Exactly the same reaction must take place within a single amino acid to form the "inner salt" or **zwitterion**, a molecule containing both positively and negatively charged atoms.

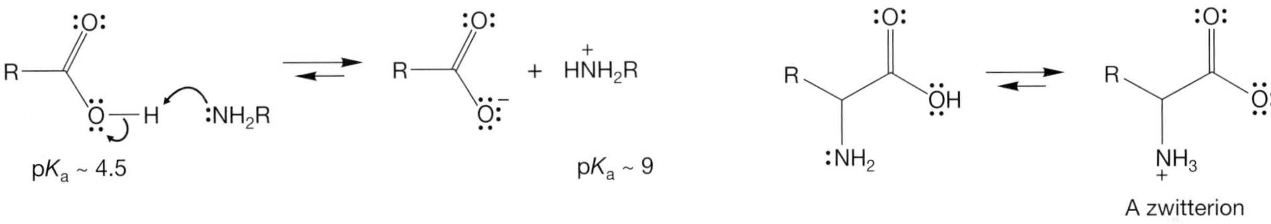

FIGURE 23.4 What reaction must occur between these two molecules?

FIGURE 23.5 An amine and a carboxylic acid must undergo acid–base chemistry to give an ammonium salt of the acid. In amino acids, an intramolecular version of the same reaction gives a zwitterion.

Amino acids are ionic compounds under normal conditions, and therefore have high melting points. Other properties are in accord with this notion. They are miscible with water, but quite insoluble in nonpolar organic solvents. They have high dipole moments. Of course, the charge state of an amino acid depends on the pH of the medium in which it finds itself. Under neutral or nearly neutral conditions, amino acids exist as the zwitterions of Figure 23.5. If we lower the pH, eventually the carboxylate anion will protonate, and the amino acid will exist primarily in the ammonium ion form. Conversely, if we raise the pH, eventually the ammonium ion site will be deprotonated, and the molecule will exist as the carboxylate anion (Fig. 23.6).

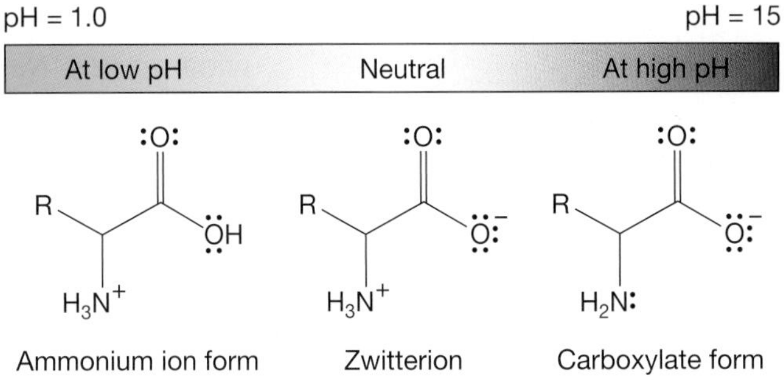

FIGURE 23.6 At low pH (strongly acidic solution), the zwitterion is protonated to give a net positively charged molecule. At high pH (strongly basic solution), the zwitterion is deprotonated to give a net negatively charged molecule.

PROBLEM 23.3 The pK_a for a simple organic acid is 4–5. Note the pK_a values of the amino acids (column 5 in Table 23.1). Why are amino acids much more acidic?

Amino acids can be described by a series of pK_a values. The first refers to the deprotonation of the carboxylic acid group, and it is in the range 1.7–2.4. The second pK_a value is for deprotonation of the ammonium ion, and is in the range 8.8–10.8 (Table 23.1).

The pH at which the concentration of the ammonium ion equals that of the carboxylate ion is called the **isoelectric point (p*I*)**. At this pH the molecule will have a net zero charge. The zwitterion is the predominant form at this pH. This pH is defined as the pI, which is a symbol for isoelectric point. For the amino acids with nonacidic or basic side chains, the pI can be calculated by taking the midpoint between the pK_a of the carboxylic acid and the pK_a of the ammonium group. Thus, p$I = (pK_a + pK_a)/2$.

> PROBLEM 23.4 Use the pK_a data in Table 23.1 to calculate the pI (the isoelectric point) for alanine.

As you can see from Table 23.1, many amino acids bear acidic or basic side chains, which complicates the picture somewhat and can make the determination of the isoelectric point a bit more difficult. These side groups are also involved in acid–base equilibria and there are pK_a values associated with them. For example, arginine will be doubly protonated at very low pH (Fig. 23.7).

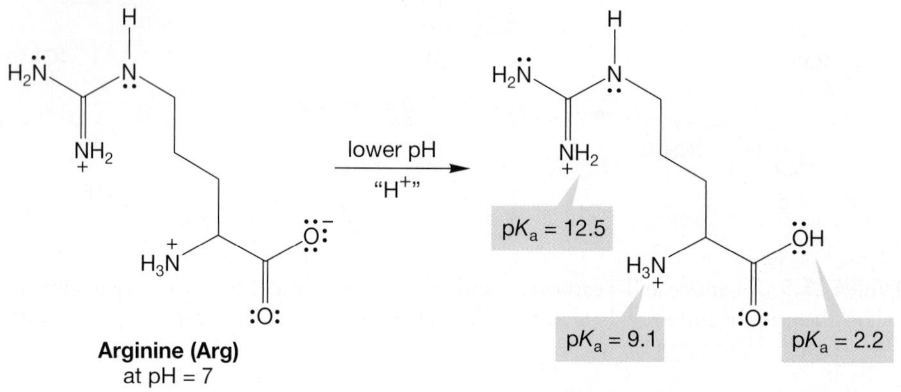

FIGURE 23.7 Arginine is twice protonated at low pH. Because the side chain has a basic group, this amino acid has three pK_a values.

Arginine (Arg) at pH = 7

> PROBLEM 23.5 Notice which nitrogen on the arginine side chain is protonated under acidic conditions. Explain why it is the nitrogen shown that is protonated.

The most acidic group is the carboxylic acid, p$K_a = 2.2$. Second is the ammonium ion, p$K_a = 9.1$. The side chain group has a pK_a of 12.5 and will be the last position deprotonated as the solution becomes less acidic.

> PROBLEM 23.6 Which ring nitrogen in histidine (Table 23.1) will be protonated first as the solution becomes more acidic?

These acid–base properties give rise to an especially powerful separation method widely used in analysis of amino acid mixtures, called **electrophoresis**. It works this way. A mixture of amino acids is placed on moistened paper. At a given pH, each amino acid exists as either a positive ion, a neutral zwitterion, or a negative ion. When a current is applied to the paper, the positively charged ions will migrate to the cathode (negative pole), the neutral zwitterions remain stationary, and the negatively charged ions move to the anode (positive pole) at rates depending on the pH and the strength of the applied current. The net result is a separation, as the amino acids in the mixture migrate at different rates. For example, at pH 5.5, the amino acids Lys, Phe, and Glu exist in the forms shown in Figure 23.8.

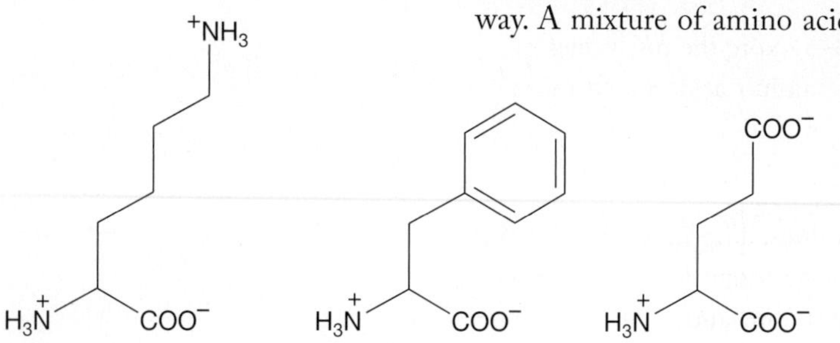

FIGURE 23.8 The charge states of Lys, Phe, and Glu at pH 5.5.

When a current is applied, Phe will not move, Lys will migrate to the cathode, and Glu to the anode (Fig. 23.9).

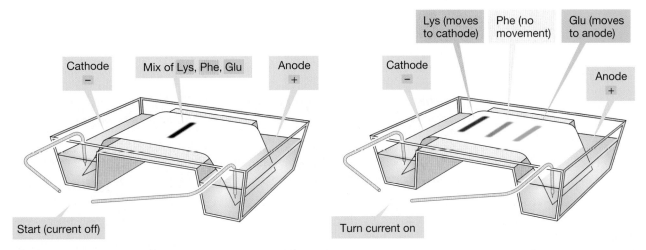

FIGURE 23.9 Separation of three amino acids by electrophoresis.

23.2d Syntheses of Amino Acids   One very simple method for making an amino acid is to alkylate ammonia using the S_N2 reaction of ammonia with an α-halo acid (Fig. 23.10).

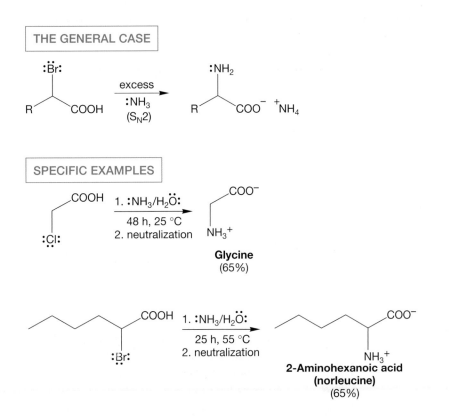

FIGURE 23.10 Simple alkylation of an α-halo acid will give an α-amino acid.

In Chapter 7 (p. 312), a closely related alkylation method appeared as a means of producing alkylamines. A problem for this synthetic procedure is the ease of overalkylation. The simple alkylation reaction is not generally a practical method for producing alkylamines, but it is useful in making α-amino acids. The α-bromo acids are easily available, and a vast excess of ammonia can be used to minimize overalkylation. Moreover, the first formed α-amino acids are not especially effective competitors in the S_N2 reaction, because they are sterically congested and less basic than simple amines. This technique is an effective way of making both "natural" and "unnatural" amino acids.

PROBLEM 23.7 Devise simple syntheses of the following α-bromo acids starting from alcohols containing no more than four carbon atoms and inorganic reagents of your choice (p. 950).

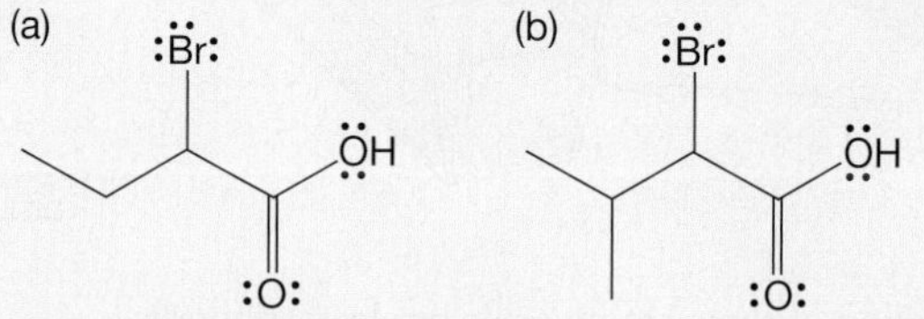

PROBLEM 23.8 Why are α-bromo acids (or their carboxylate salts) more easily alkylated than simple alkyl bromides in the S_N2 reaction (p. 542)?

A successful variation of the simple alkylation synthesis involves a protected, disguised amine, and is called the **Gabriel synthesis** (Siegmund Gabriel, 1851–1924). In this procedure, bromomalonic ester is first treated with potassium phthalimide in an S_N2 displacement of bromide ion (Fig. 23.11).

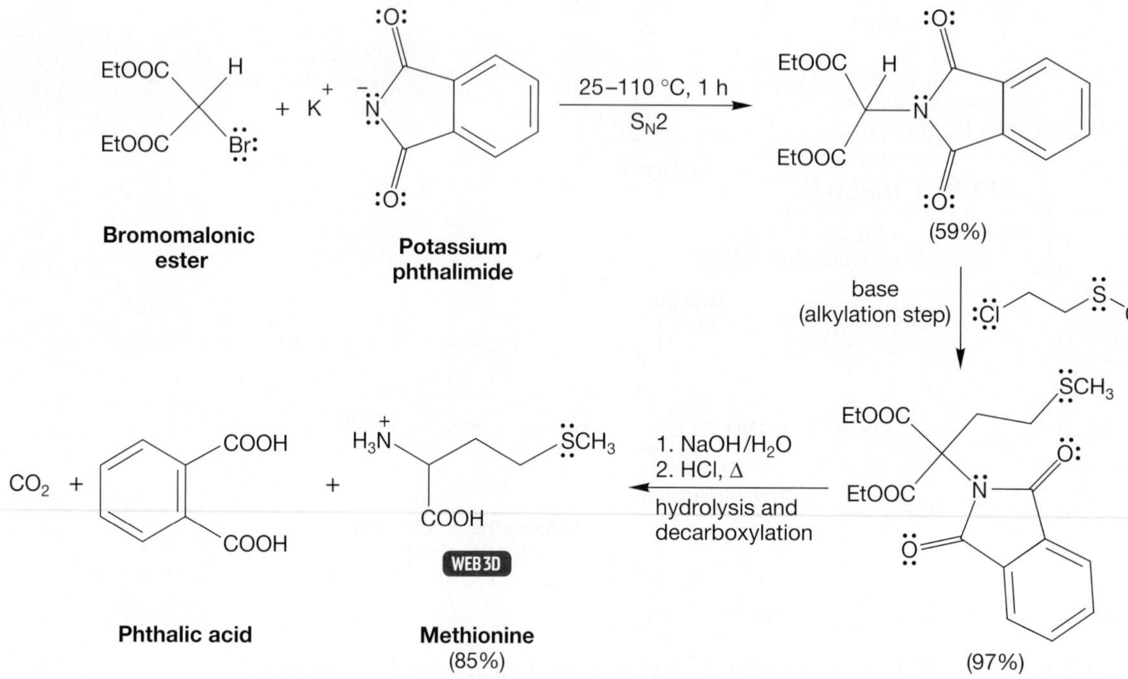

FIGURE 23.11 The Gabriel synthesis uses bromomalonic ester in the first step.

The nitrogen atom destined to become the amine nitrogen of the amino acid is introduced at this point, and rendered nonnucleophilic by the flanking carbonyl groups. In the next step, the malonic ester product is treated with base and an active alkyl halide to introduce the R group of the amino acid. Finally, treatment with acid or base, followed by neutralization, uncovers the amine. The phthalimide group is hydrolyzed to phthalic acid, and the malonic ester is hydrolyzed to a malonic acid. Gentle heating induces decarboxylation to give the amino acid. There are many variations on this general theme.

PROBLEM 23.9 Provide detailed mechanisms for the reactions of Figure 23.11.

WORKED PROBLEM 23.10 In a simpler version of the Gabriel synthesis, acetamidomalonic ester is used rather than bromomalonic ester. Sketch out the steps in this related amino acid synthesis.

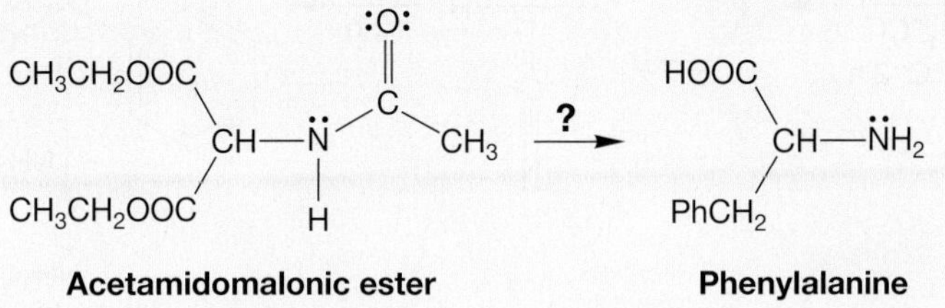

Acetamidomalonic ester   **Phenylalanine**

ANSWER It is essentially the same process as that of Figure 23.11 and Problem 23.9. Advantage is first taken of the acidity of the doubly α hydrogen to introduce an R group.

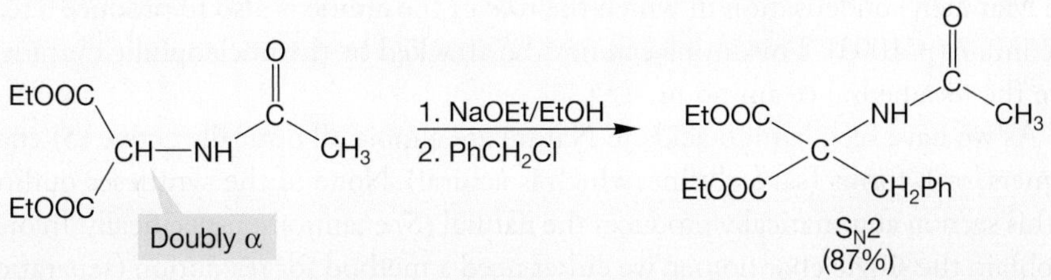

Now acid (or base) is used to convert the esters into carboxylic acids and to hydrolyze the amide. Heating decarboxylates the malonic acid and liberates the amino acid.

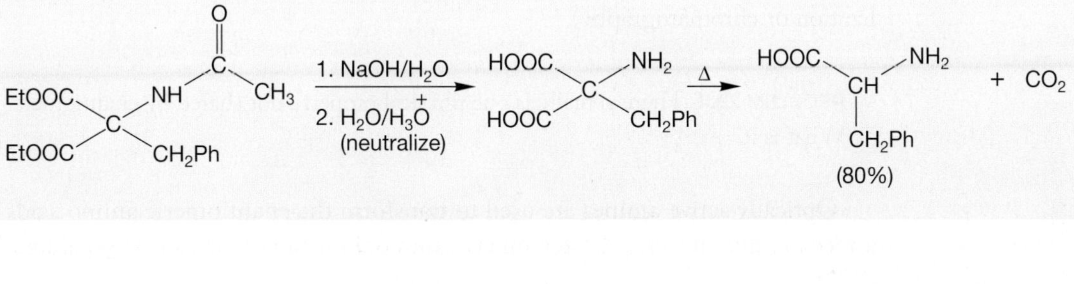

In the **Strecker synthesis** (Adolph Strecker, 1822–1871), a cyanide is used as the source of the carboxylic acid portion of an amino acid (Fig. 23.12). An aldehyde is treated with ammonia, the ultimate source of the amino group, and cyanide ion. In the first step of the reaction, a carbinolamine is formed (p. 790) and water is lost to give an imine. The imine is then attacked by cyanide to give an α-amino cyanide. Acid hydrolysis of the nitrile gives the acid (p. 904).

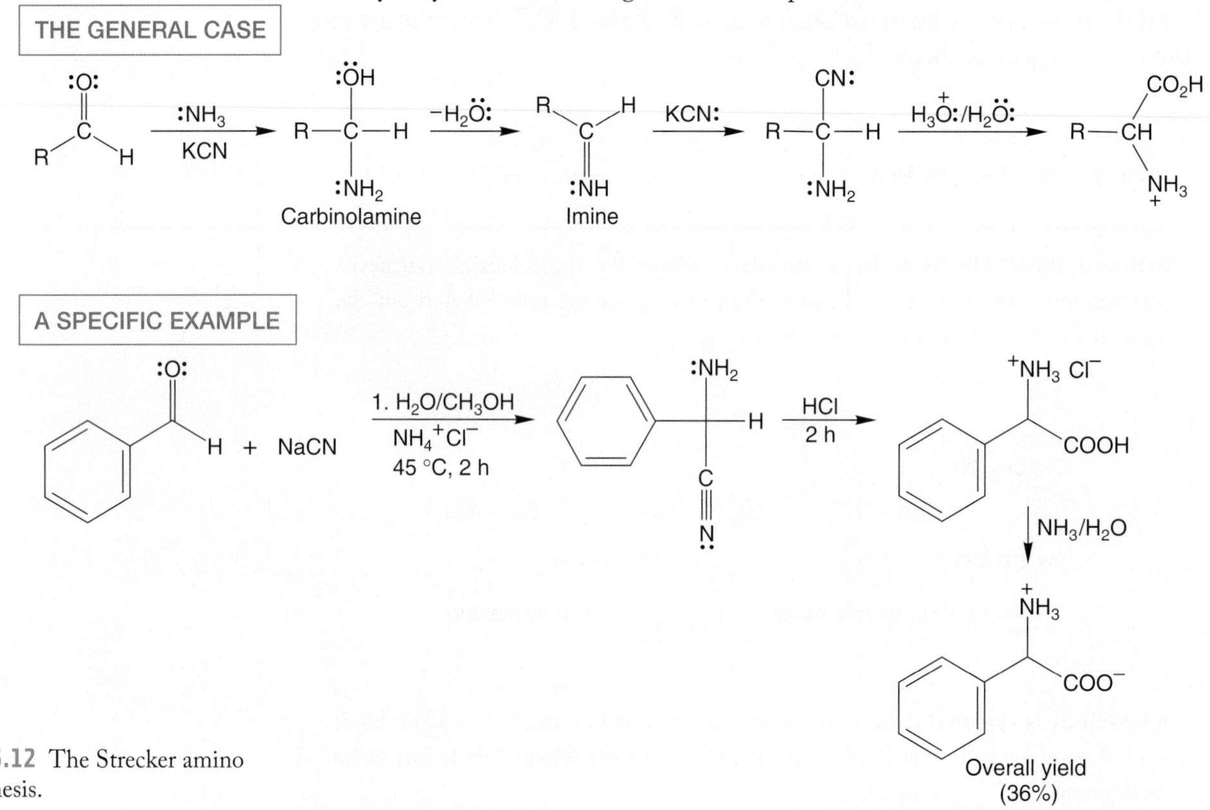

FIGURE 23.12 The Strecker amino acid synthesis.

The first step in this reaction, imine formation, is analogous to the first step in the Mannich condensation in which the role of the amine is also to produce a reactive imine (p. 1003). This imine can next be attacked by the nucleophilic cyanide to give the tetrahedral α-amino nitrile.

As we have seen, amino acids in Nature are almost all optically active (*S*) enantiomers, or L forms (save glycine, which is achiral). None of the syntheses outlined in this section automatically produces the natural (*S*) enantiomer specifically. In order to obtain the single enantiomer, we either need a method for resolution (separation) of the racemic mixture of enantiomers formed in these syntheses or a method of selectively producing the (*S*) enantiomers.

23.2e Resolution of Amino Acids   The standard method is the one we first discussed in our chapter on chirality (p. 169) and involves transformation of the pair of enantiomers, which have identical physical properties, into a pair of diastereomers, which do not, and can be separated by physical means (most often crystallization or chromatography).

> **PROBLEM 23.11** There actually is one physical property not shared by enantiomers. What is it?

Optically active amines are used to transform the enantiomeric amino acids in a racemic mixture into diastereomeric ammonium salts that can be separated by crystallization. The chiral amines of choice are optically pure alkaloids (p. 255), which are readily available in Nature. The process of separation is outlined in Figure 23.13.

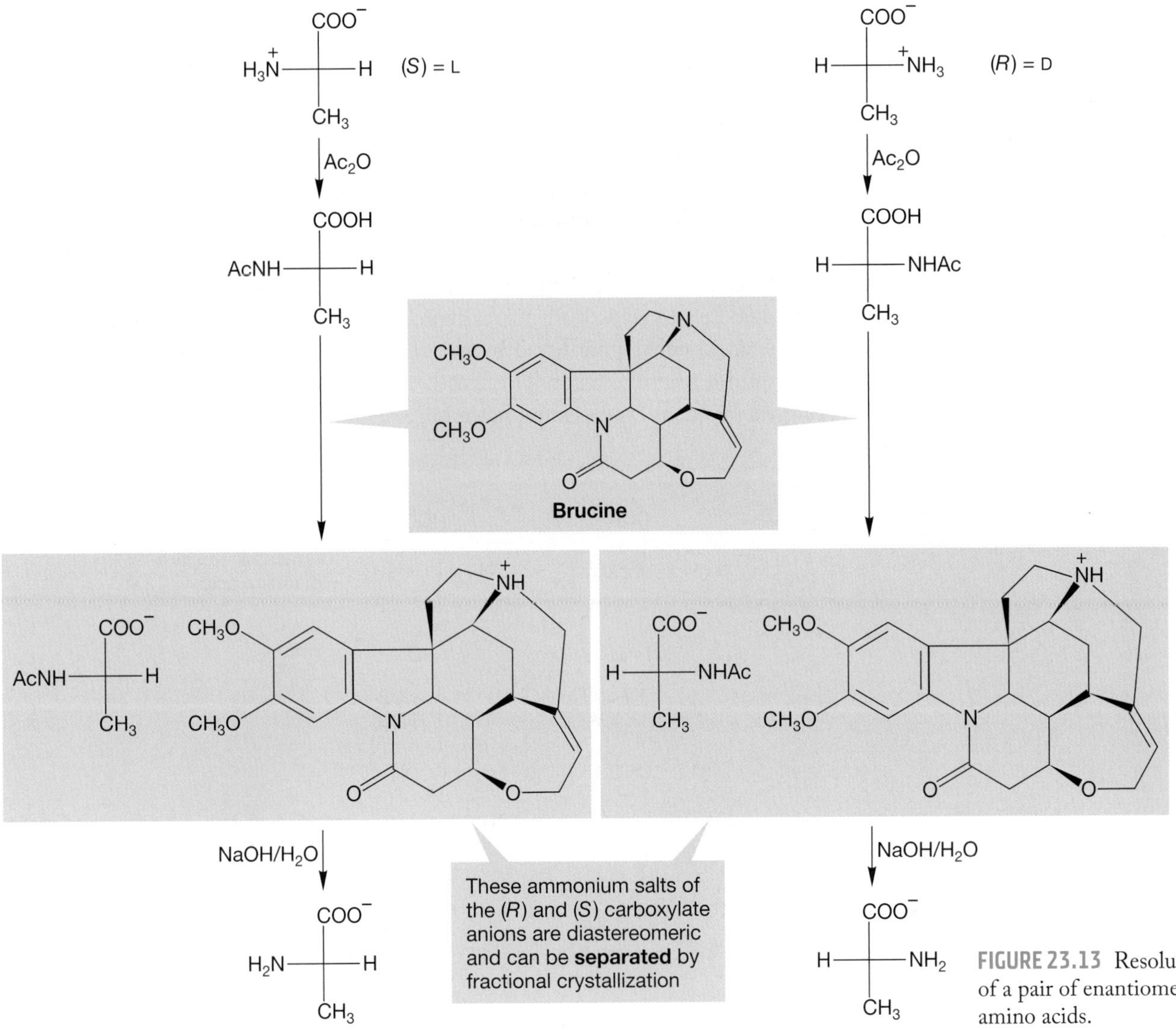

These ammonium salts of the (R) and (S) carboxylate anions are diastereomeric and can be **separated** by fractional crystallization

FIGURE 23.13 Resolution of a pair of enantiomeric amino acids.

However, it seems that there must be a way to avoid the tedium of fractional crystallization and make the optically active compounds directly. After all, Nature does this somehow, and we should be clever enough to find out how this happens and imitate the process. One method Nature uses to make optically active amino acids involves a reductive amination (Fig. 23.14). Ammonia and an α-keto carboxylic

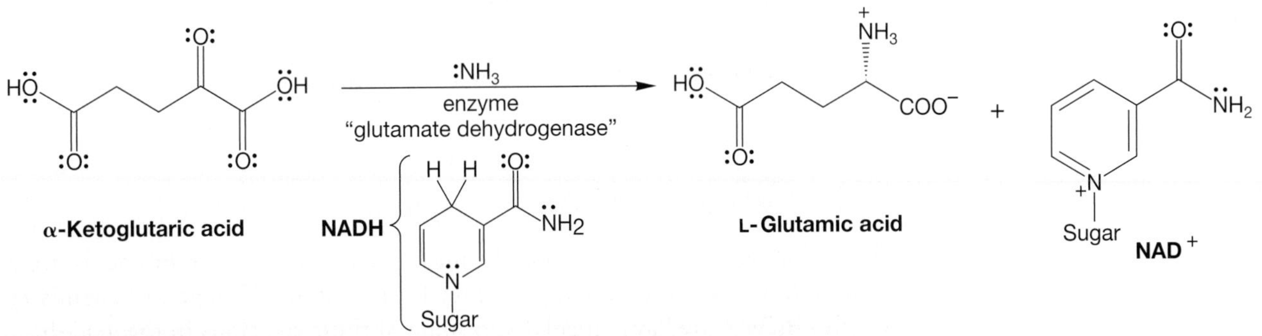

FIGURE 23.14 Synthesis of an optically active L-amino acid through reductive amination.

acid react in the presence of an enzyme and a source of hydride, usually NADH (p. 814), to produce an optically active chiral amino acid.

There are two procedures used by chemists that imitate the natural pathway. In one, actual microorganisms are used in fermentation reactions that can directly produce some L-amino acids. In the other, enzymes are used to react with one enantiomer of a racemic pair. As you might guess, it is usually the L isomer that is "eaten" by the enzyme, because enzymes have evolved in an environment of L isomers. So it is usually the D isomers that are ignored by enzymes. In an extraordinarily clever procedure, this preference for natural, L, enantiomers can be used to achieve a **kinetic resolution** (Fig. 23.15). The mixture of enantiomers is first acetylated to create amide linkages. An enzyme, hog-kidney acylase, then hydrolyzes the amide linkages *of only L-amino acids*. So, treatment of the pair of acetylated amino acids leads to formation of the free L-amino acid. The acylated D enantiomer is unaffected by the enzyme, which has evolved to react only with natural L-amino acids. The free L-amino acid can then easily be separated from the residual D acetylated material.

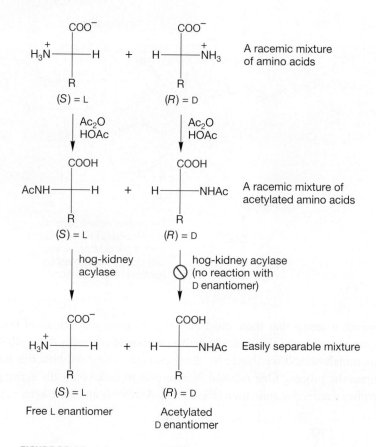

FIGURE 23.15 A kinetic resolution of a pair of enantiomeric amino acids.

23.3 Reactions of Amino Acids

The simple expectation that the chemical reactions of the amino acids are a combination of carboxylic acid (Chapter 17) and amine (Chapter 6) chemistry is right. Indeed, we have already seen some of these reactions in the acid–base chemistry described in Section 23.2c. A few more important and useful examples follow.

23.3a Acylation Reactions

The amino group of amino acids can be acylated using all manner of acylating agents. Typical examples of effective acylating agents are acetic anhydride, benzoyl chloride, and acetyl chloride (Fig. 23.16).

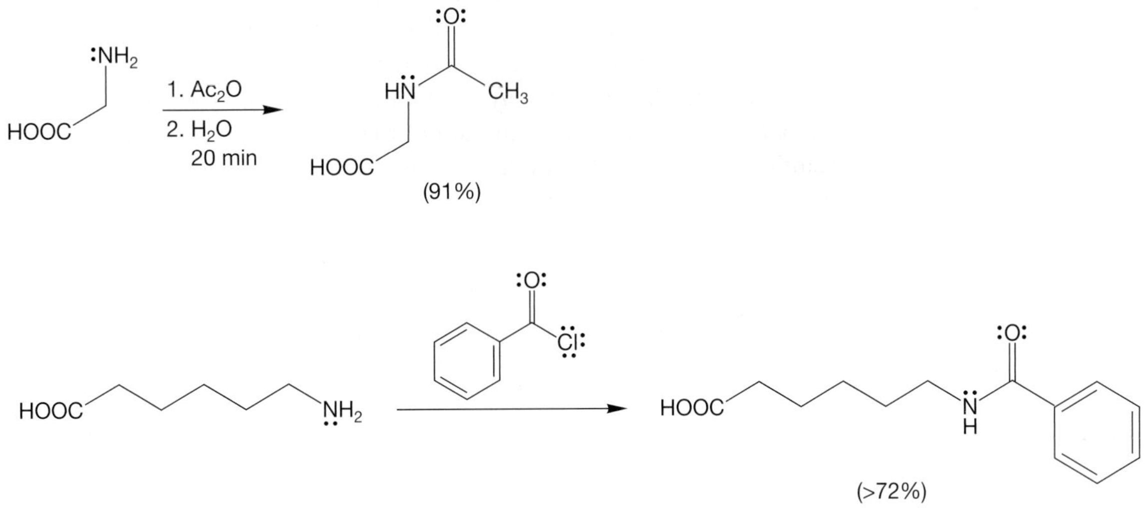

FIGURE 23.16 The amino group of an amino acid can be acylated.

23.3b Esterification Reactions

The carboxylic acid group of amino acids can be esterified using typical esterification reactions. Fischer esterification is the simplest process (Fig. 23.17).

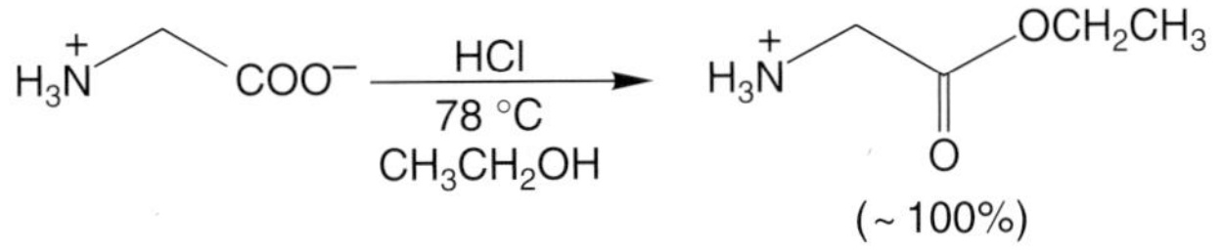

FIGURE 23.17 Fischer esterification of the acid group of an amino acid.

PROBLEM 23.12 Write a mechanism for the acylation reactions of Figure 23.16.

PROBLEM 23.13 If amino acids are typically in their zwitterionic form, how are acylation and esterification reactions, which depend on the presence of free amine or acid groups, possible?

23.3c Reaction with Ninhydrin

The hydrate of indan-1,2,3-trione is called **ninhydrin**. Amino acids react with ninhydrin to form a deep purple molecule. It is this purple color that is used to detect amino acids. Although it may seem curious at first, all amino acids except proline, regardless of the identity of the R group, form the *same* purple molecule on reaction with ninhydrin. How can this be?

Like all hydrates, ninhydrin is in equilibrium with the related carbonyl compound (Fig. 23.18). The amino acid first reacts with the trione to form an imine through the small amount of the free amine form of the amino acid present at equilibrium. We have seen this kind of reaction many times (p. 793). Next, decarboxylation of the imine produces a second imine that is hydrolyzed under the aqueous conditions to give an amine. It is in this hydrolysis step that the R group of the original amino acid is lost. Still another imine, the third in this sequence, is then formed through reaction of the amine with another molecule of the ninhydrin trione. The extensive conjugation in the product leads to a very low energy absorption in the visible region (p. 529), and the observed purple color.

PROBLEM 23.14 Why is the trione of Figure 23.18 hydrated at the middle carbonyl rather than either of the side carbonyl groups?

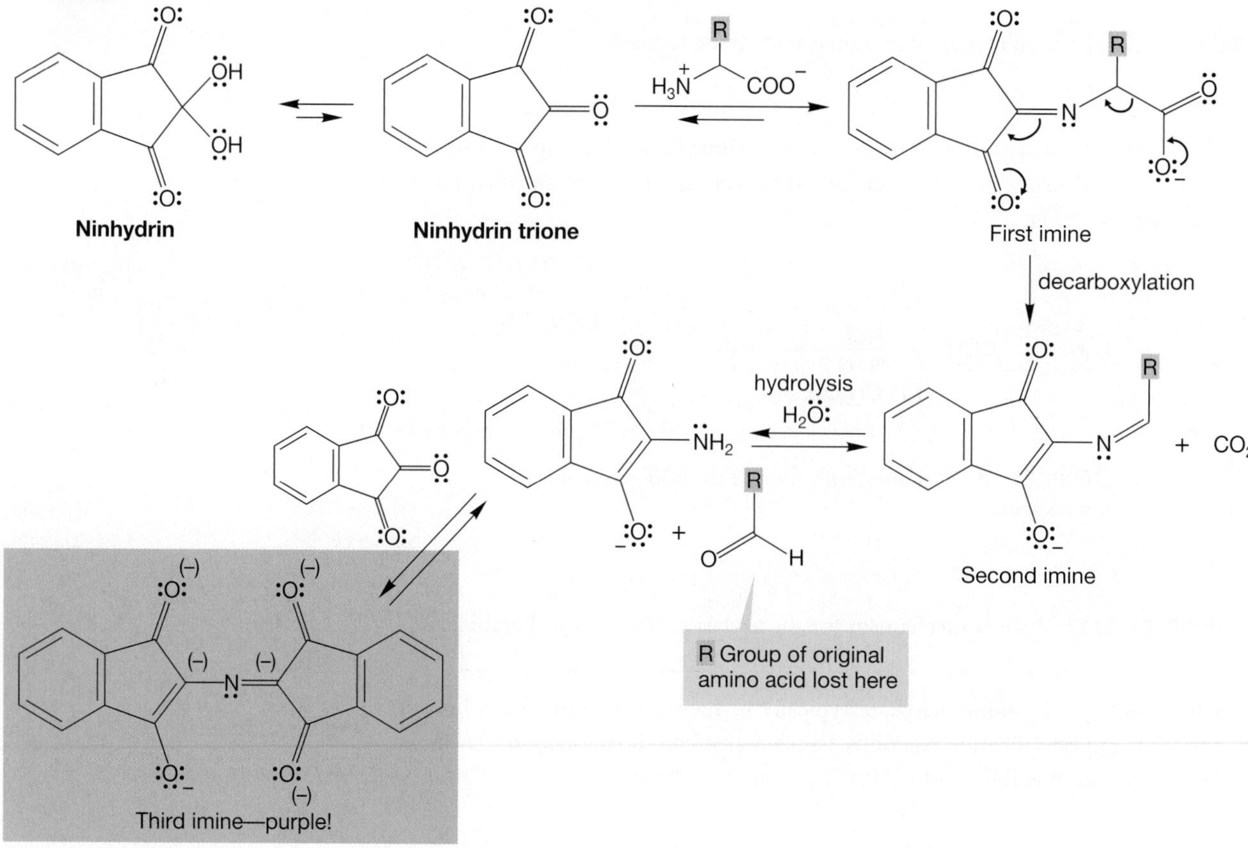

FIGURE 23.18 Formation of a purple molecule through reaction of an amino acid with the ninhydrin trione. Each amino acid in Table 23.1 except proline gives the same purple product.

By far, the most important reaction of amino acids is their polymerization to peptides and proteins through the formation of new amide bonds (Fig. 23.19). The following sections are devoted to this reaction and some of its consequences.

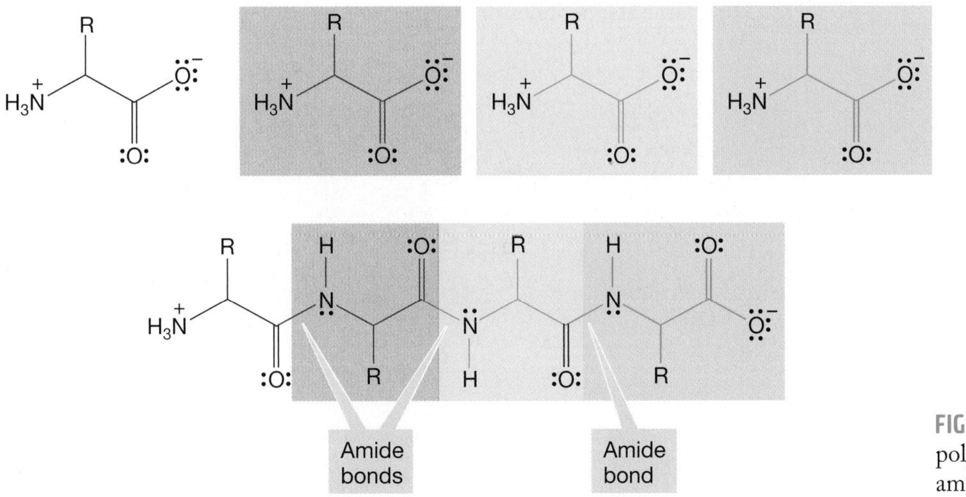

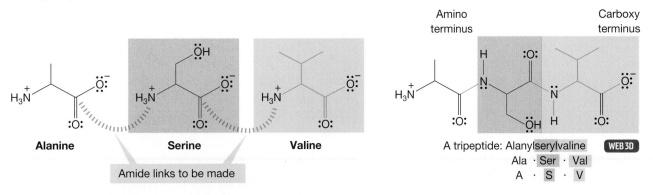

FIGURE 23.19 Peptides are polyamino acids linked through amide bonds.

Summary

Many of the reactions of carboxylic acids and amines appear in the chemistry of amino acids. Acylation and Fischer esterification are two examples. There are some new reactions, however, that involve both the amino and the acid groups. Reaction with ninhydrin is an example.

23.4 Peptide Chemistry

23.4a Nomenclature and Structure   Figure 23.20 shows the schematic construction of a three amino acid peptide (a tripeptide) made from alanine, serine, and valine. Don't confuse this picture with real synthetic procedures—we will get to them soon enough. Peptides are usually written with the **amino terminus** on the left, proceeding toward the **carboxy terminus** on the right. They are always named from the amino terminus to the carboxy terminus. So the tripeptide in Figure 23.20 is alanylserylvaline, or Ala·Ser·Val, or A·S·V.

FIGURE 23.20 The tripeptide alanylserylvaline, Ala·Ser·Val. The amino acid at the amino terminus starts the name, and the amino acid at the carboxy terminus ends it.

PROBLEM 23.15 Draw the structures for Pro·Gly·Tyr, Asp·His·Cys, and Glu·Thr·Phe.

PROBLEM 23.16 Name the peptides drawn below.

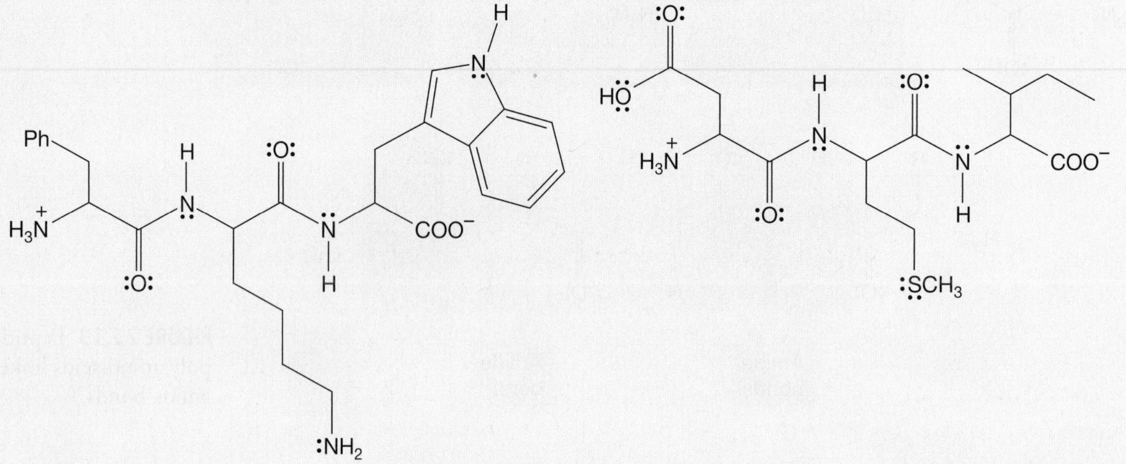

The sequence of amino acids in a peptide or protein constitutes its **primary structure**. These chains of amino acids are linked in another way. Connections between chains or within the same chain are often formed through **disulfide bridges** (Fig. 23.21). Cysteine is usually the amino acid involved in these bridges, which are made through oxidation of the thiol groups (p. 809).

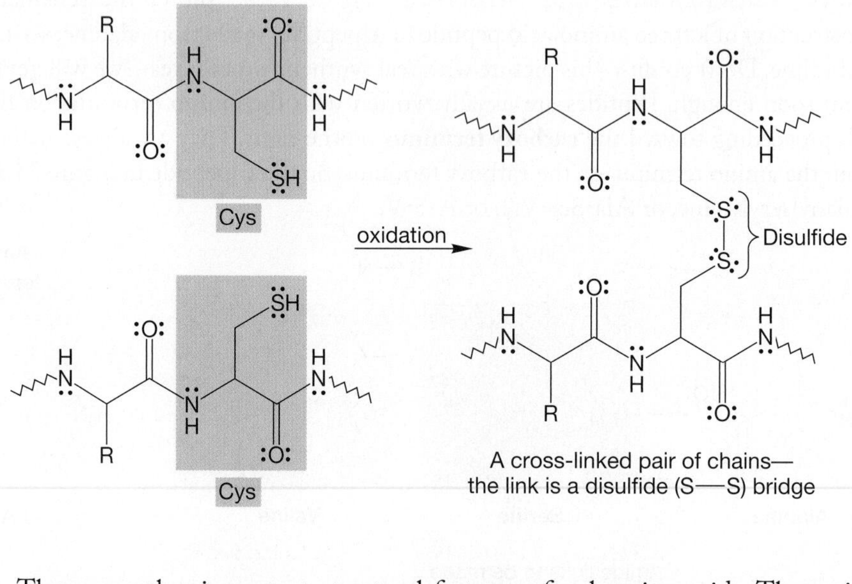

FIGURE 23.21 Formation of disulfide bonds can link one polyamino acid chain to another.

A cross-linked pair of chains— the link is a disulfide (S—S) bridge

There are other important structural features of polyamino acids. The amide groups prefer planarity and the hydrogen bonds between the NH hydrogens and amide carbonyl groups, along with the structural constraints imposed by the disulfide links, lead to regions of order in most large peptides.

WORKED PROBLEM 23.17 Explain why amides prefer planarity.

ANSWER Amides are stabilized by resonance.

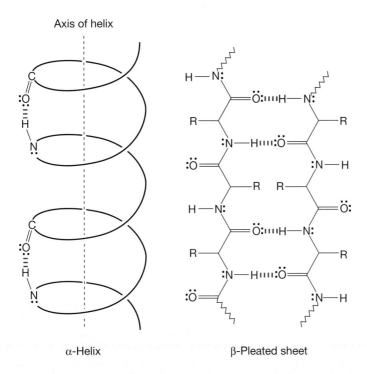

This resonance stabilization is maximized when the $2p$ orbitals on N, C, and O overlap as much as possible, which requires sp^2 hybridization and planarity.

Especially common are regions arranged in an **α-helix** and repeating folded sections known as **β-pleated sheets** (Fig. 23.22). Both of these regions allow regular patterns of hydrogen bonding to develop, thereby imposing order on the sequence of amino acids. In the regions that appear as an α-helix the hydrogen bonds are between the coils of the helix, whereas in the pleated sheets they hold lengths of the chains in roughly parallel lines. Other regions consist of a disordered, **random coil**, series of amino acids. The combination of these structural features determines the folding pattern, or **secondary structure** of the polypeptide.

Axis of helix

α-Helix

β-Pleated sheet

FIGURE 23.22 Two examples of ordered secondary structure in a polypeptide, the α-helix and β-pleated sheet.

There are even higher structural orders for these molecules. It might have been that polypeptides were best described as ordered regions of secondary structure (α-helix or β-pleated sheets) connected by sections of random coil. In such a case,

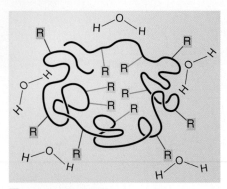

R = Hydrocarbon-like,
     nonpolar side chains
R = Polar side chains

FIGURE 23.23 A globular protein, ordered so as to put the nonpolar side chains in the inside of the "glob" and the polar side chains outside interacting with the polar solvent medium.

a protein would have no fixed overall structure beyond the sections defined by secondary structure. But this idea has been shown not to be correct. X-Ray structure determinations of crystallized proteins have revealed these species to have *specific*, three-dimensional structures.[3] This **tertiary structure** of the protein is determined partly by hydrogen bonding, but also by van der Waals and electrostatic forces. These giant molecules are composed of a carbon backbone to which the amino acid side chains, or R groups, are attached. These R groups are either hydrocarbon-like, nonpolar groups such as methyl, isopropyl, or benzyl, or highly polar amino, hydroxyl, or sulfhydryl groups. In the natural environment of a protein, water, it should be no surprise to find that the protein adjusts its shape so that the polar, hydrophilic groups are aimed outward toward the polar solvent, whereas the nonpolar, "greasy" hydrophobic hydrocarbon portions cluster inside the molecule, maximally protected from the hostile aqueous environment (Fig. 23.23).

Many proteins adopt globular shapes such as that shown in Figure 23.23, which maximizes interior hydrophobic and exterior hydrophilic interactions. Other proteins are fibrous, taking the form of a superhelix composed of ropelike coils of α-helices. Figure 23.24 shows a typical protein, the enzyme lactate dehydrogenase, which incorporates five α-helices and six β-sheets.

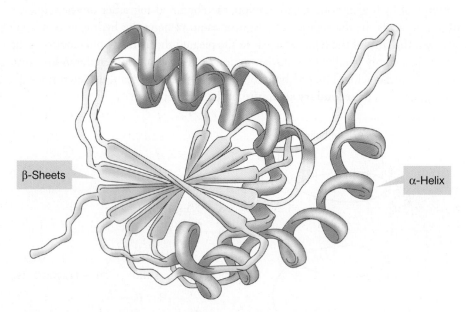

β-Sheets   α-Helix

FIGURE 23.24 Lactate dehydrogenase is a large protein that has well-defined tertiary structure.

---

[3]It is a general problem that until recently the structures of proteins could only be determined by X-ray diffraction studies on crystallized materials. Work was limited by the availability of crystals, and crystallization of these molecules is usually difficult. Successful formation of X-ray quality crystals of a protein is cause for celebration (and publication) even before the work of structure determination begins. But how do we know that the structure of the molecule in solution is the same as that in the solid state crystal? Remember, biological activity is tied intimately to the detailed structure of these molecules—might we not be led astray by a structure that owed its shape to crystal packing forces in the solid state, and that was quite different in solution? Indeed we might. These days, it is becoming possible to use very high-field NMR spectrometers, along with increasingly sophisticated NMR pulse techniques, to determine structures in solution. Remember that high-field spectrometers are anything but cheap; the cost is about 10^3 per megahertz, often of your tax dollars.

This tertiary structure of a protein is extraordinarily important because it determines the shape of these huge molecules and biological activity depends intimately on these shapes. Evolution has produced proteins that have pockets into which specific molecules called substrates fit. This enzyme–substrate binding allows for the proper orientation for reaction of the substrate. The reaction could be complicated, or as simple as the hydrolysis of an ester. In other molecules, substrates are bound only for transportation purposes. These **binding sites** are a direct consequence of the primary structure that determines the secondary and tertiary structures. Figure 23.25 gives a schematic representation of this process.

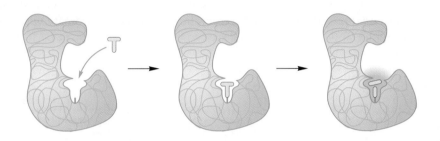

FIGURE 23.25 A binding site in a protein captures (binds) a small molecule for transport, or reaction with appropriately located reactive groups.

How do we know that the *same* tertiary structure is adopted by every protein molecule of a given amino acid sequence? One clue is that synthetic proteins have the same biological activity as the natural proteins. If we reproduce the primary sequence of a protein correctly, the proper biological activity appears. Because this biological activity depends *directly* on the details of shape, we can conclude that the molecule must be folding properly into the appropriate tertiary structure.

Another clue comes from experiments in which the tertiary structure of a protein is deliberately disrupted, which is called **denaturing** the protein. Many ways of denaturing a protein are irreversible. When an egg is cooked, the thermal energy denatures the proteins in the egg white in an irreversible way. Once the egg is fried, there is no unfrying it. Similar chemistry can be induced by pH change. The curdling of milk products is an example. But some denaturing reactions are reversible. For example, disulfide bonds in a protein can be broken through treatment with thiols (Fig. 23.26). This disruption of the secondary structure transforms the ordered sections of the protein into the random coil arrangements.

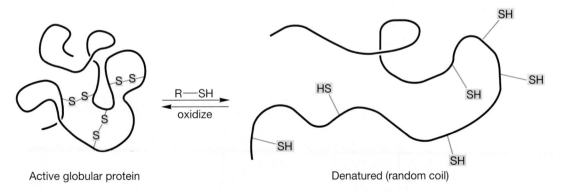

Active globular protein   Denatured (random coil)

FIGURE 23.26 A reversible method of denaturing proteins, of destroying the secondary and tertiary structure, is to break the disulfide bonds. Air oxidization re-forms the disulfides; the higher order structures are regenerated, and biological activity is reestablished.

Similar reversible disruptions of order can be achieved through treatment with urea (in a reaction whose precise mechanism remains unknown), or, sometimes, through heating. If these denatured proteins are allowed to stand, the thiol groups are reoxidized by air to the appropriate disulfide linkages and the original tertiary structure with its biological activity returns!

There is even higher order to some proteins. Particularly large proteins are collections of smaller polypeptides held together by intramolecular attractions—van der Waals and electrostatic forces—into superstructures with **quaternary structure**. The classic example of quaternary structure is hemoglobin, the protein responsible for oxygen absorption and transport in humans. To accomplish these biological functions the protein incorporates four heme units, each consisting of an iron atom surrounded by heterocyclic rings which coordinate with the metal (Fig. 23.27).

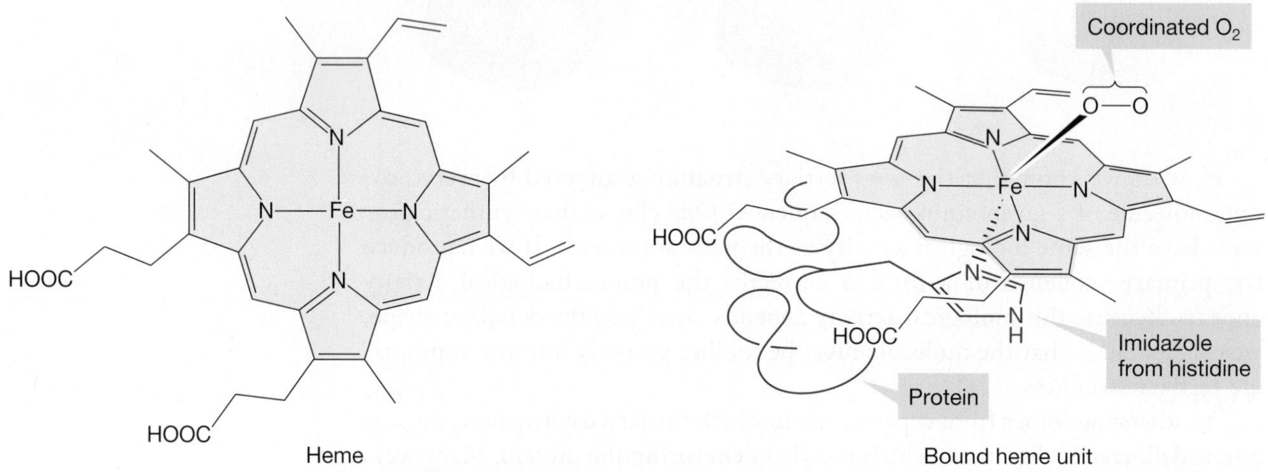

Heme

Coordinated O₂

Imidazole from histidine

Protein

Bound heme unit

FIGURE 23.27 The heme unit and an oxygen-coordinated heme structure of hemoglobin.

FIGURE 23.28 The quaternary structure of hemoglobin. Each heme unit is attached to a globin subunit.

The protein surrounding the heme is called globin, and in each globin subunit a histidine side chain (an imidazole) is poised to hold the heme in position. There are four subunits in hemoglobin, two pairs of slightly different peptide chains (Fig. 23.28). These are held together by electrostatic and van der Waals forces as well as hydrogen bonding, and take the shape of a giant tetrahedron.

Given all this general information about structure, we still face two difficult problems. First, how do we determine the amino acid sequence of an unknown protein, and second, once we know that structure, how can we synthesize it?

23.4b Determination of Protein Structure If suitable crystals can be obtained, and if the X-ray diffraction pattern for the crystal can be solved in a reasonable amount of time, the structure of the protein is apparent. However, this process is not yet a general procedure. Crystallization of these molecules remains difficult, and although the determination of the structure of small molecules by X-ray is now routine and rapid, the determination of the structures of proteins remains difficult and time consuming.

In principle, we can get an overall picture of the composition of a protein by first destroying all disulfide links in the molecule, and then hydrolyzing all the peptide

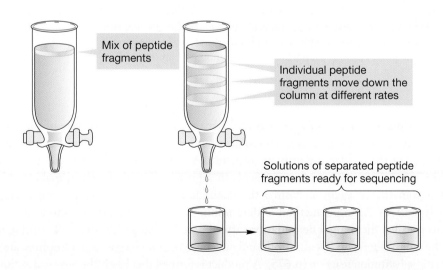

Lys · Cys · Phe

Phe · Ser · Cys

$\xrightarrow[\text{2.H}_3\overset{..}{\text{O}}:/\text{H}_2\overset{..}{\text{O}}:]{\text{1. RSH}}$ 2 Phe + 2 Cys + Ser + Lys

bonds. *Remember*: These peptide bonds are amides which means acid hydrolysis should convert them into acids (p. 902). This procedure regenerates all the amino acids of which the molecule is built. Figure 23.29 shows a hypothetical example. Of course, this technique does nothing to reveal either primary structure or any higher structure.

In practice, the fragments formed when the disulfide bonds are broken are first separated from one another and then hydrolyzed. This separation is no trivial task, but several methods are now available, including electrophoresis (p. 1180) which works for peptides as well as amino acids. Various kinds of chromatography are also effective. In **gel-filtration chromatography** the mixture is passed over a column of polymer beads which, on the molecular level, contain holes into which the smaller peptide fragments fit more easily than the larger pieces. Accordingly, the larger fragments pass more rapidly down the column, and a separation is achieved. **Ion-exchange chromatography** uses electrostatic attractions to hold more highly charged fragments on the column longer than more nearly neutral molecules. As in any chromatographic technique, the more tightly held molecules move more slowly than those less tightly held by the medium (Fig. 23.30).

Mix of peptide fragments

Individual peptide fragments move down the column at different rates

Solutions of separated peptide fragments ready for sequencing

FIGURE 23.30 A schematic description of chromatographic separation of peptide fragments.

Once the peptide fragments are separated, each can be hydrolyzed to its constituent amino acids. Now we need methods for determining how much of each amino acid is present. Again chromatography is used. The amino acids are separated on an ion-exchange column. As each amino acid is eluted from the column, it emerges into a chamber in which it reacts with ninhydrin (p. 1187) to form a purple solution. The intensity of the purple color is proportional to the amount of amino acid present, and can be plotted against time or amount of solvent used for elution from the column. This procedure produces a chromatogram consisting of a series of peaks of various sizes. How do we tell which peak corresponds to which amino acid? This identification is done by running samples of known amino acids through the column, determining their retention times, and matching these with those of the unknown amino acids (Fig. 23.31).

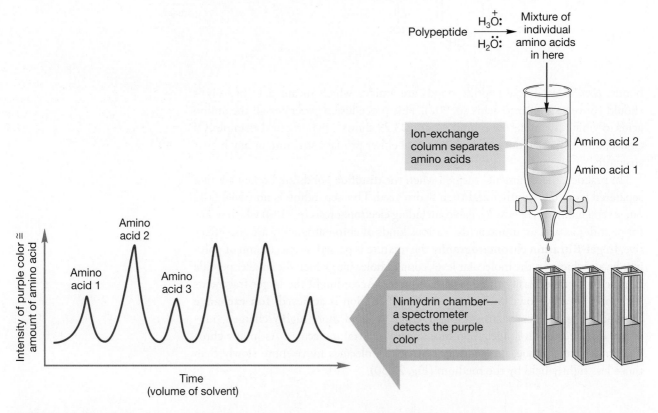

FIGURE 23.31 Peptide fragments can be hydrolyzed to constituent amino acids, which can then be separated by chromatography. Reaction of each amino acid with ninhydrin gives a purple color that can be quantitatively analyzed spectroscopically. The intensity of the purple color is proportional to the amount of amino acid formed.

But this technique only gives us the coarsest picture of the structure of the protein: We know the constituent amino acids and their relative amounts, but we have no idea of their order, the primary structure. An important task is to find a way to determine the sequence of amino acids in a protein. The first step in making this determination is again to destroy the disulfide links and separate the constituent polypeptides. A simple method of determining the amino acid at the amino terminus is called the **Sanger degradation**, after Frederick Sanger (b. 1918). A peptide is allowed to react with 2,4-dinitrofluorobenzene, which undergoes a nucleophilic aromatic substitution reaction (p. 675). A product is formed that labels the terminal amino group with a 2,4-dinitrobenzene group (Fig. 23.32). Hydrolysis now generates a number of amino acids, but only the amino acid terminus is labeled! Unfortunately, we have had to destroy the entire peptide to make this determination!

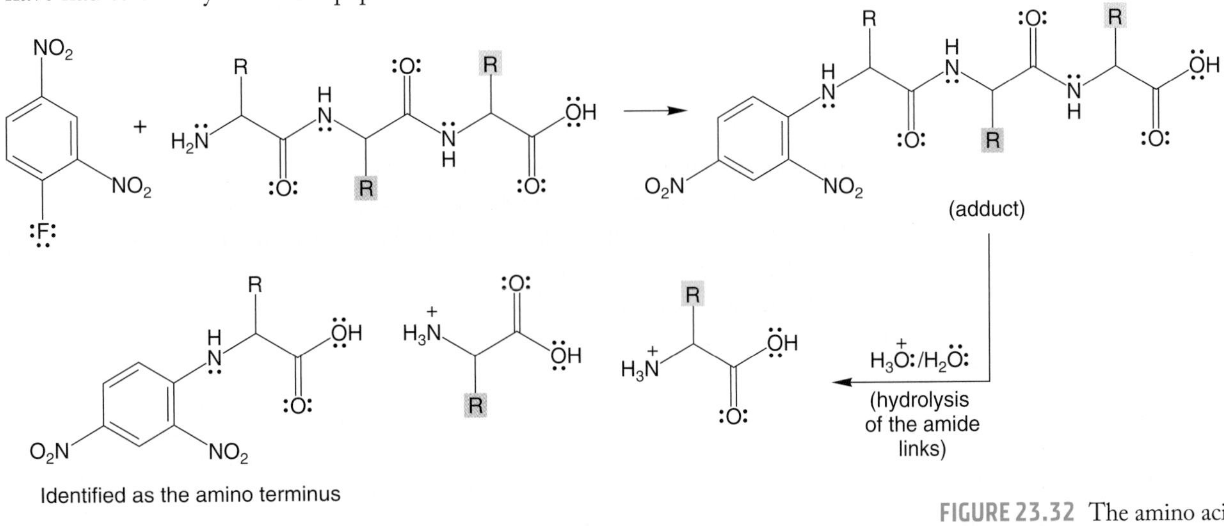

Identified as the amino terminus

It is also possible to determine the amino acid at the other end of a peptide chain, the carboxy terminus. The method of choice uses an enzymatic reaction in which one of a number of carboxypeptidases specifically cleaves the terminal amino acid at the carboxy end of the chain. As each carboxy terminal amino acid reacts with the enzyme, a new amino acid is revealed to be cleaved in turn by more carboxypeptidase. Careful monitoring of the production of amino acids as a function of time can give a good idea of the sequence (Fig. 23.33). Still, it is clear that things will get messy as long peptides are turned into a soup containing an increasingly complex mixture of amino acids.

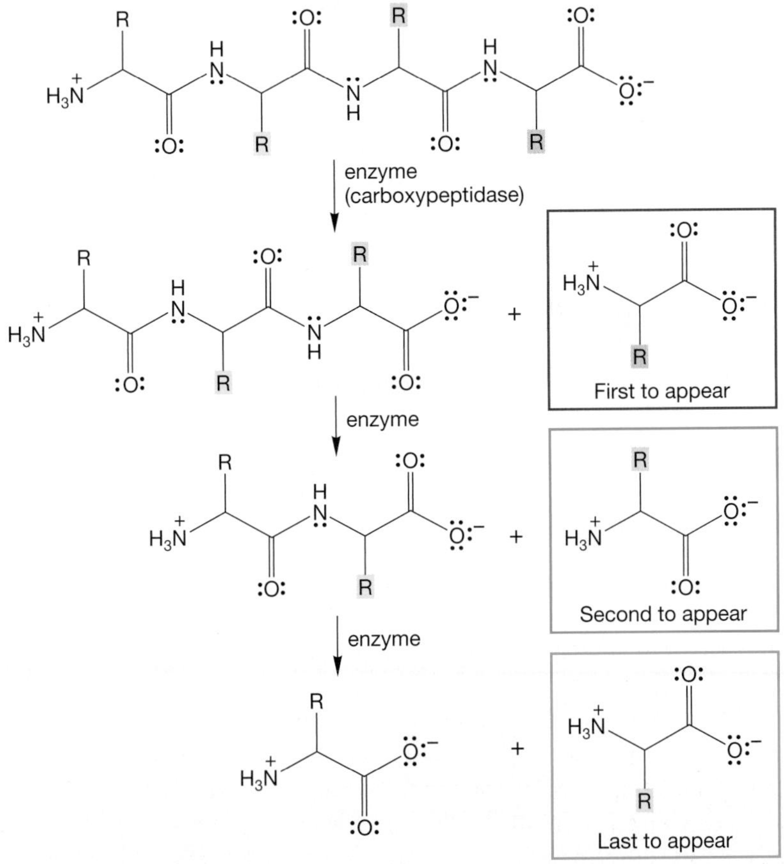

FIGURE 23.32 The amino acid at the amino terminus can be identified by reaction with 2,4-dinitrofluorobenzene, followed by hydrolysis. Only the amino acid at the amino terminus of the chain will be labeled.

FIGURE 23.33 Carboxypeptidase is an enzyme that cleaves only at the carboxy terminus of a polypeptide.

PROBLEM 23.18 Write a mechanism for adduct formation in Figure 23.32. What is the function of the nitro groups?

A much better method would be one that included no overall hydrolysis step and that could be more controlled than the cleavage reactions set in motion by treatment with a carboxypeptidase. We need a method that unzips the protein from one end or the other, revealing at each step the terminal amino acid.

Methods have now been developed that allow just such a step-by-step sequencing of peptides. One effective way is called the **Edman degradation**, after Pehr Edman (1916–1977), and uses the molecule phenylisothiocyanate (Fig. 23.34). Isocyanates have appeared before (p. 918), and an isothiocyanate is just the sulfur analogue of an isocyanate.

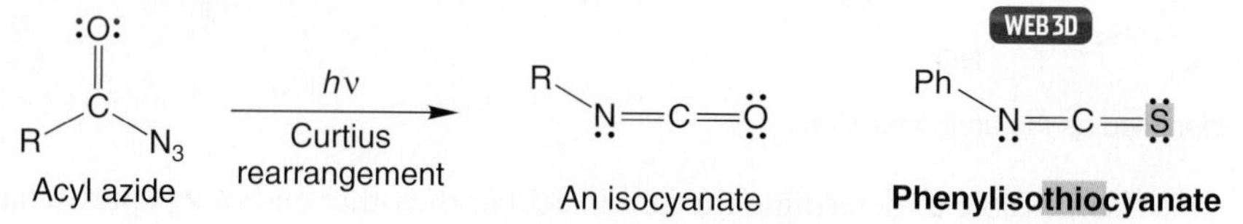

FIGURE 23.34 An isothiocyanate is just the sulfur version of an isocyanate, the product of a Curtius rearrangement.

PROBLEM 23.19 Write a mechanism for the isocyanate formation in Figure 23.34.

Like isocyanates, isothiocyanates react rapidly with nucleophiles. Reactions of isocyanates with ammonia or amines yield ureas. In a similar fashion, isothiocyanates give a **thiourea** as the first formed intermediate (Fig. 23.35).

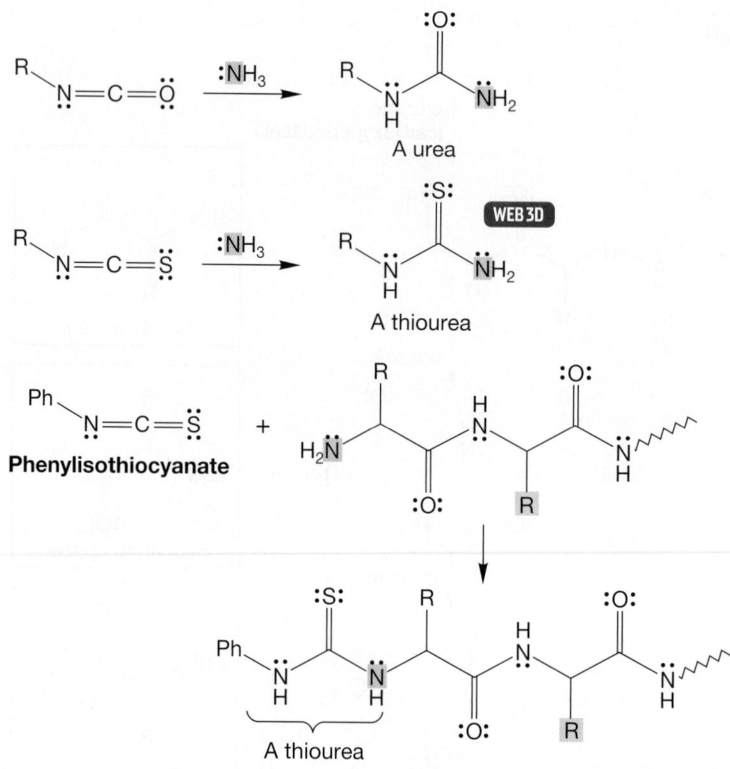

FIGURE 23.35 Isothiocyanates react with amines to give thioureas. Phenylisothiocyanate can be used to label the amino terminus of a polypeptide.

A peptide will react with isothiocyanate at the amino terminus to form a thiourea. The thiourea can be cleaved in acid *without breaking the other amide bonds in the peptide.* As we saw in Chapter 21 (p. 1088), sulfur is an excellent neighboring group, and the introduction of the thiourea places a sulfur atom in just the right position to aid in cleavage of the amino terminus amino acid from the rest of the peptide (Fig. 23.36). The process involves initial protonation of the amino terminus carbonyl. The neighboring sulfur then adds to the carbonyl carbon, triggering the cleavage reaction to give a molecule called a thiazolinone.

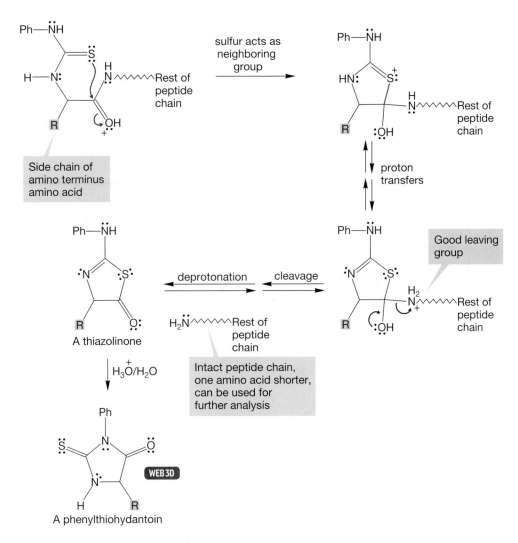

FIGURE 23.36 The Edman degradation procedure for sequencing polyamino acids uses phenylisothiocyanate. The amino terminus is identified by the structure of the phenylthiohydantoin formed. Notice that this method does not destroy the peptide chain as the Sanger procedure does.

Further acid treatment converts the thiazolinone into a phenylthiohydantoin containing the R group of the amino terminus amino acid, which can now be determined based on the identity of the R group. The critical point is that the rest of the peptide remains intact and the Edman procedure can now be repeated, in principle, as many times as are needed to sequence the entire peptide (Fig. 23.37).

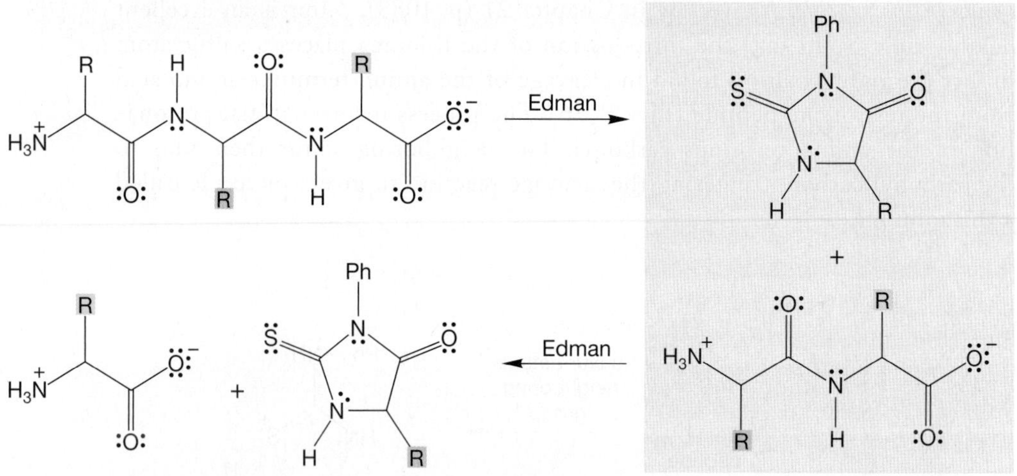

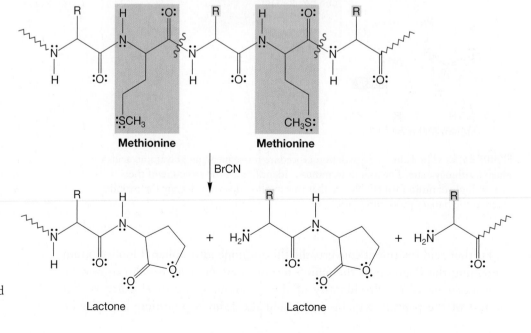

FIGURE 23.37 The primary structures of polyamino acids can be determined through repeated applications of the Edman degradation procedure, which is one method of sequencing.

PROBLEM 23.20 Write a mechanism for the formation of the phenylthiohydantoin from the thiazolinone (Fig. 23.36).

But the practical world intrudes, and even the Edman degradation procedure, good as it is, becomes ineffective for polypeptides longer than about 20–30 amino acids. Impurities build up at each stage in the sequence of determinations and eventually the reaction mixture becomes too complex to yield unequivocal results. Accordingly, methods have been developed to cut proteins into smaller pieces at known positions in the amino acid sequence. These methods take advantage of both biological reactions in which enzymes break peptides only at certain amino acids, and a very few specific cleavages induced by small-molecule laboratory reagents.

Perhaps the best of these "chemical" (enzymes are chemicals, too) cleavages uses **cyanogen bromide (BrCN)** and takes advantage of the nucleophilicity of sulfur and the neighboring group effect to induce cleavage at the carboxy side of each methionine in a polypeptide (Fig. 23.38).

FIGURE 23.38 The carboxy end of each methionine is cleaved on treatment of a polypeptide with cyanogen bromide. A lactone is formed in place of a methionine.

The sulfur atom of methionine first displaces bromide from cyanogen bromide to form a sulfonium ion (Fig. 23.39). Then attack by the proximate carbonyl group of methionine displaces the excellent leaving group, methylthiocyanate, and forms a five-membered ring called a homoserine lactone. Hydrolysis steps follow to produce two smaller peptide fragments, one containing a lactone at its carboxy terminus.

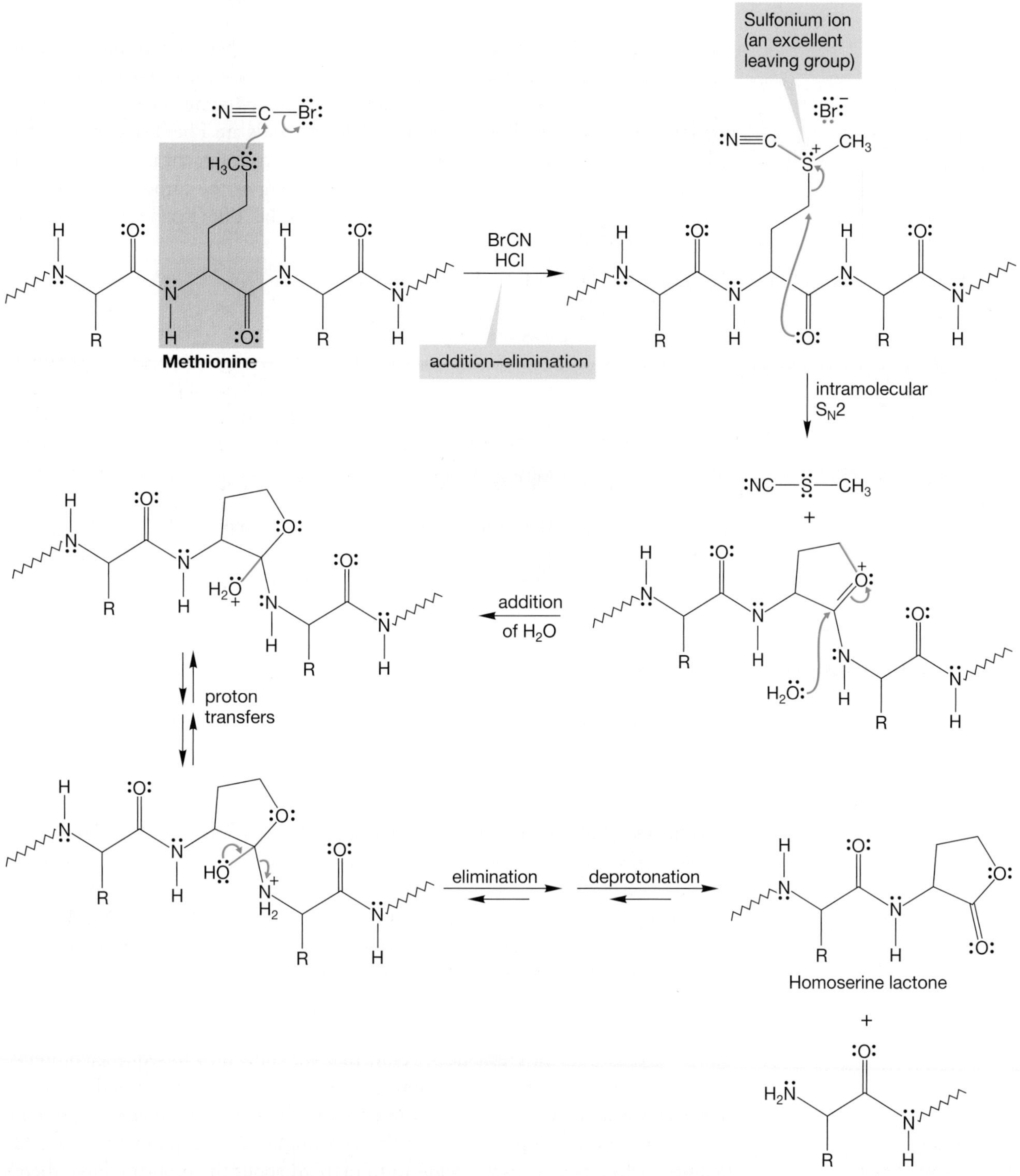

FIGURE 23.39 The mechanism of cleavage using BrCN.

Most methods used to chop large peptides into smaller fragments take advantage of enzymatic reactions. Chymotrypsin, for example, is an enzyme that cleaves peptides at the carboxyl groups of amino acids containing aromatic side chains (phenylalanine, tryptophan, or tyrosine). Trypsin cleaves only at the carboxy end of lysine and arginine. There are several other examples of this kind of specificity. Long peptide chains can be degraded into shorter fragments that can be effectively sequenced by the Edman procedure. A little detective work suffices to put together the entire sequence. Let's look at a simple example and then do a real problem.

Consider the decapeptide in Figure 23.40. We can determine the amino terminus through the Sanger procedure to be Phe. Next, we use the enzyme trypsin in a separate experiment to cleave the decapeptide into three fragments. As we know the peptide must start with Phe, the first four amino acids are Phe·Tyr·Trp·Lys. The only other cleavage point is at Arg, so the Asp·Ile·Arg fragment must be the middle piece. If this tripeptide were at the end, with Arg as the carboxy terminus, no cleavage would be possible. The sequence must be Phe·Tyr·Trp·Lys·Asp·Ile·Arg·Glu·Leu·Val.

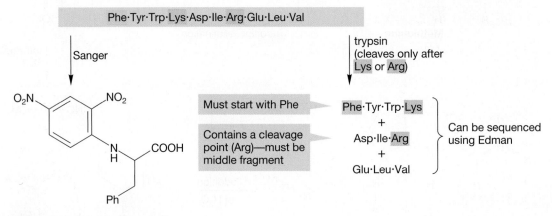

FIGURE 23.40 An example of peptide sequencing using Sanger and Edman procedures as well as enzymatic cleavage reactions.

PROBLEM 23.21 The nonapeptide bradykinin can be completely hydrolyzed in acid to give three molecules of Pro, two molecules each of Arg and Phe, one molecule of Ser, and one molecule of Gly. Treatment with chymotrypsin gives the pentapeptide Arg·Pro·Pro·Gly·Phe, the tripeptide Ser·Pro·Phe, and Arg. End-group analysis shows that the amino acids on both the amino and carboxy termini are the same. Provide the sequence for bradykinin.

23.4c Synthesis of Peptides   Now that we know how to sequence proteins, how about the reverse process—how can we put amino acids together in any sequence we want? Can we synthesize proteins? It is simply a matter of forming amide bonds in the proper order, and this task might seem easy. However, two difficulties surface as soon as we begin to think hard about the problem. First, there's a lot of work to do in order to make even a small protein. Imagine that we only want to create the nine amide linkages of the decapeptide in Figure 23.40. Even if we can make each amide bond in 95% yield, we are in trouble because a series of nine reactions each proceeding in 95% yield, which probably seems quite good to anyone who has spent some time in an organic lab, gives an overall yield of only $(0.95)^9 = 63\%$. Obviously, to get anywhere in the real world of proteins, in which chain lengths of hundreds of amino acids are common, we are going to have to do much better than this.

Second, there is a problem of specificity. Suppose we want to make the trivial dipeptide Ala·Leu. Even if we can avoid simple acid–base chemistry, random reaction between these two amino acids will give us at least the four possible dimeric molecules, and in practice, other larger peptides will be produced as well (Fig. 23.41).

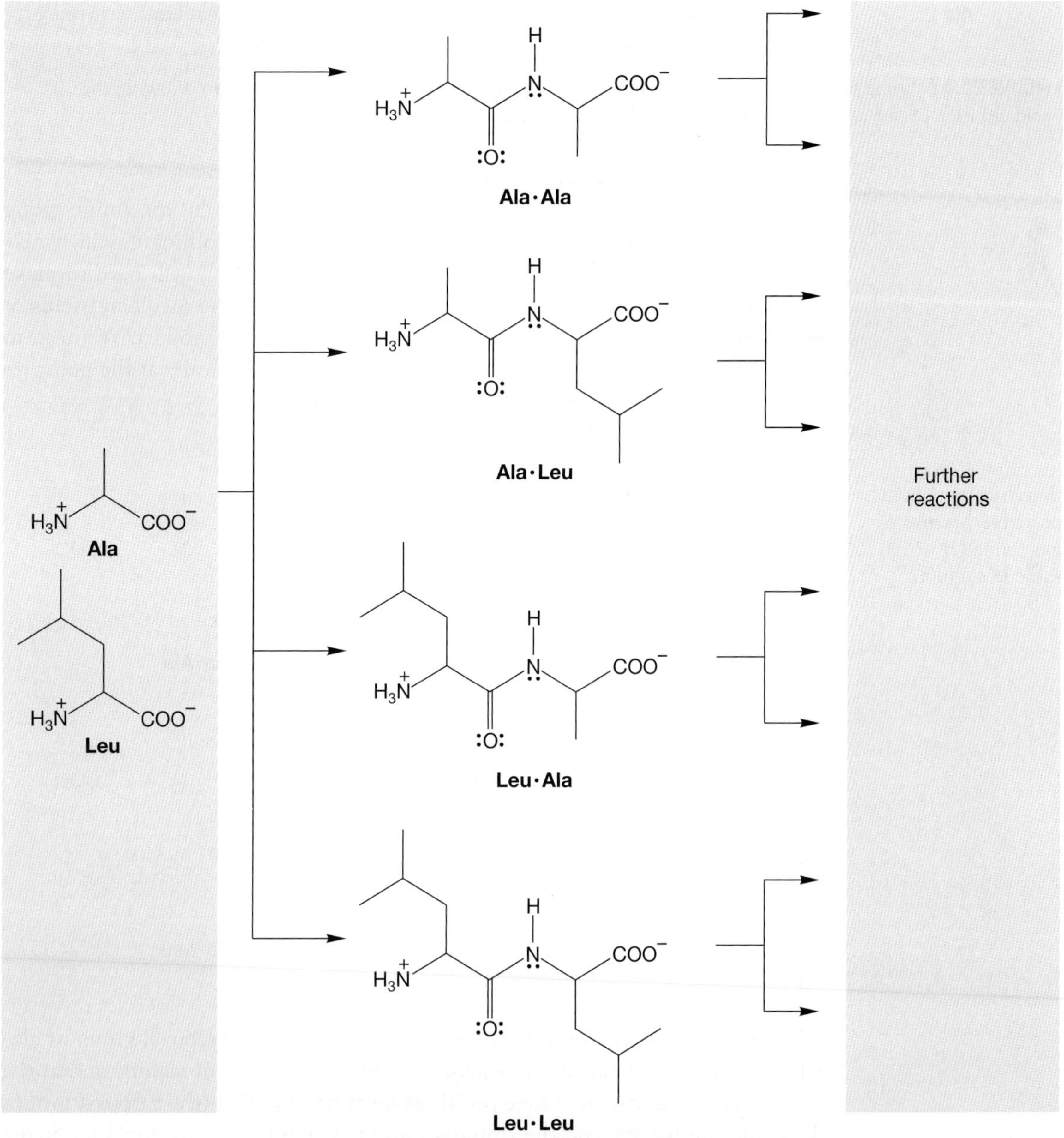

FIGURE 23.41 The possible combinations of two unprotected amino acids.

One simple strategy for avoiding undesired reactions between alanine and leucine would be to activate the carbonyl group of alanine by transforming the acid group of alanine into the acid chloride, and then treating it with leucine (Fig. 23.42).

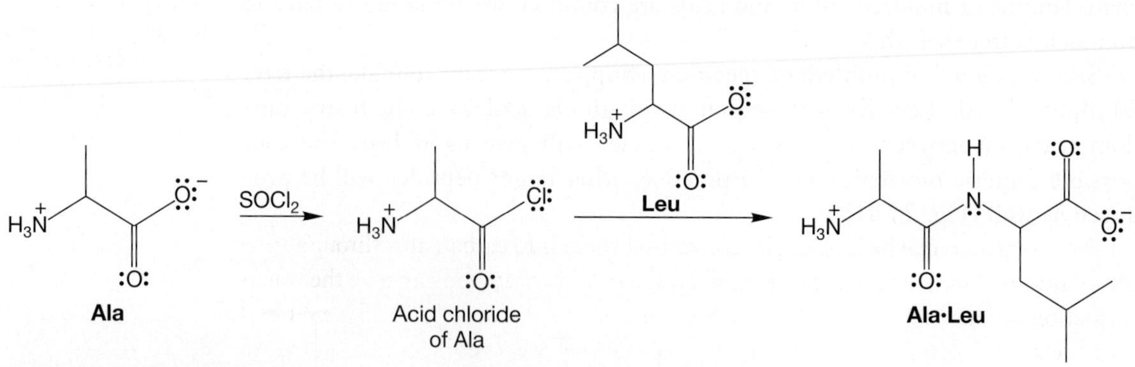

FIGURE 23.42 Some specificity in the reaction between two amino acids might be obtained by allowing the acid chloride of one amino acid to react with the other amino acid.

However, even this procedure is hopeless! What is to prevent the amino group of the original alanine from reacting with the acid chloride? Nothing. Again, a mixture is certain to result (Fig. 23.43). Remember, we ultimately will have to make dozens, perhaps hundreds of amide bonds. We can tolerate no mediocre yields or even worse, mixtures, if we hope to make useful amounts of product. We need to find the most efficient way of making the amide bond specifically at the point we want it, and at the same time avoiding side reactions.

FIGURE 23.43 Here, too, more than one product is inevitable, because as the acid chloride is formed it will react with any amine in solution.

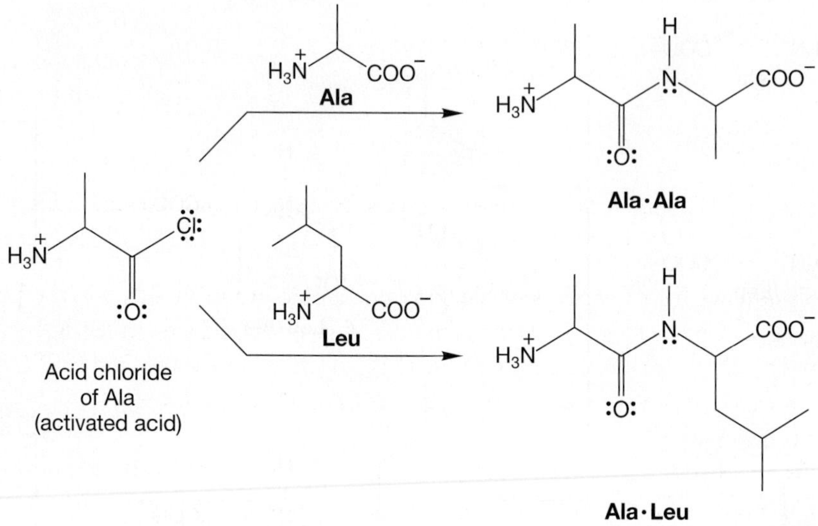

In this trivial case, not only do we need to activate the carboxyl group of alanine, but we need to block the reaction at the amino group of alanine and at the carboxyl group of leucine. Three positions must be modified (the carboxyl groups of alanine and leucine and the amino group of alanine). Left free for reaction are the two positions we hope to join, the activated carboxyl group of alanine and the unmodified amino group of leucine. Remember also that eventually the blocking, or protecting, groups will have to be removed (Fig. 23.44).

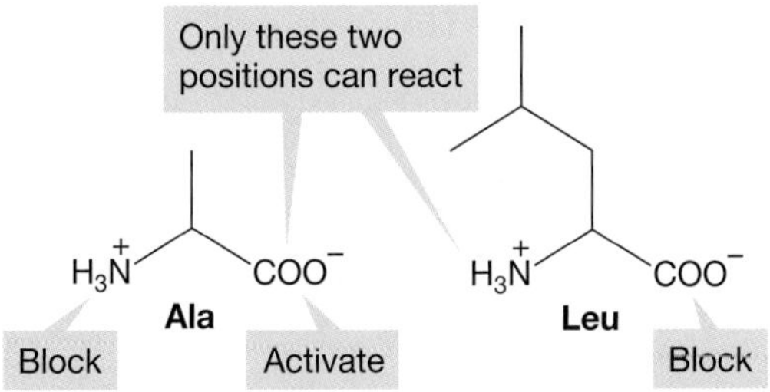

FIGURE 23.44 In any successful strategy, we must activate the carboxy end and block the amino end of one amino acid while blocking reaction at the carboxy end of the other amino acid.

Amino groups can be blocked by converting them into carbamates through simple addition–elimination reactions (p. 901). Two popular methods involve the transformation of the free amine into a carbamate by reaction with either di-*tert*-butyl dicarbonate or benzyl chloroformate. Biochemists are even more addicted to acronyms than are organic chemists, and these protecting groups are called **tBoc** and **Cbz**, respectively (Fig. 23.45).

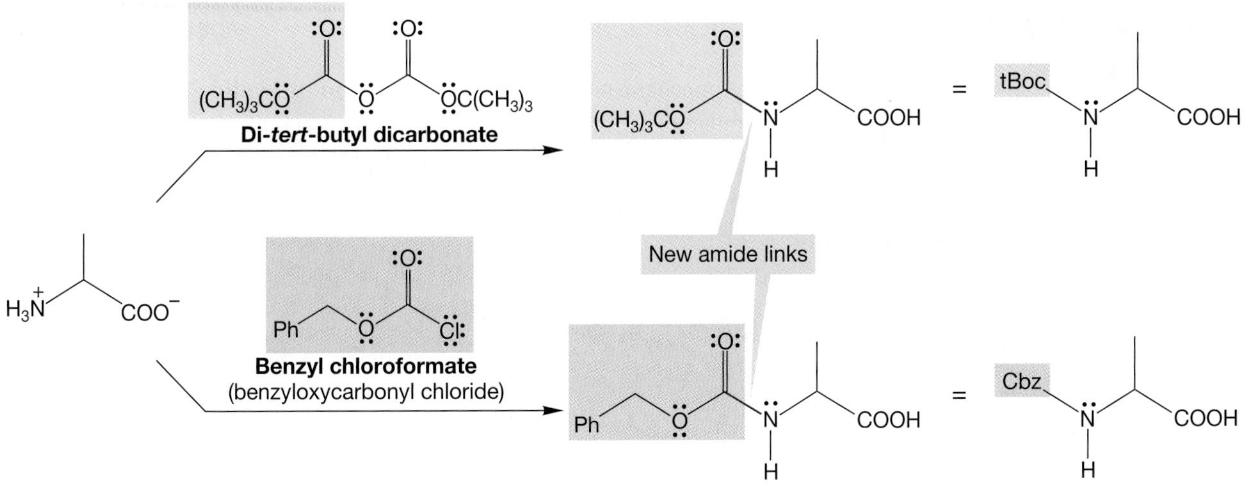

FIGURE 23.45 Two protecting or blocking groups for the amine ends of amino acids.

So now we have two ways of blocking one amino group, in this case the amino group of alanine, so it cannot participate in the formation of the peptide amide bond. Now we need to block the carboxy end of the other amino acid, leucine. This protection can be accomplished by simply transforming the acid into an ester (Fig. 23.46).

We need to do two more things. First, we still have to do the actual joining of the unprotected amino group of leucine to the unprotected carboxyl group of alanine, and we need to do this very efficiently and under the mildest possible conditions. Second, we need to deprotect the blocked amino and carboxyl groups after the amide linkage has been formed.

Dicyclohexylcarbodiimide (DCC) is the reagent of choice for peptide bond making. Formation of the amide bond is a dehydration reaction (water is the other product), and DCC is a powerful dehy-

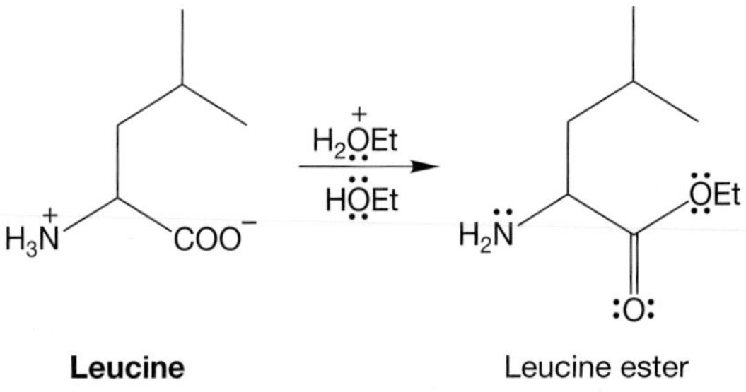

FIGURE 23.46 A carboxy end can be protected by converting it into an ester.

drating agent. The hydrated form of DCC is dicyclohexylurea (DCU), which is a product of the reaction (Fig. 23.47).

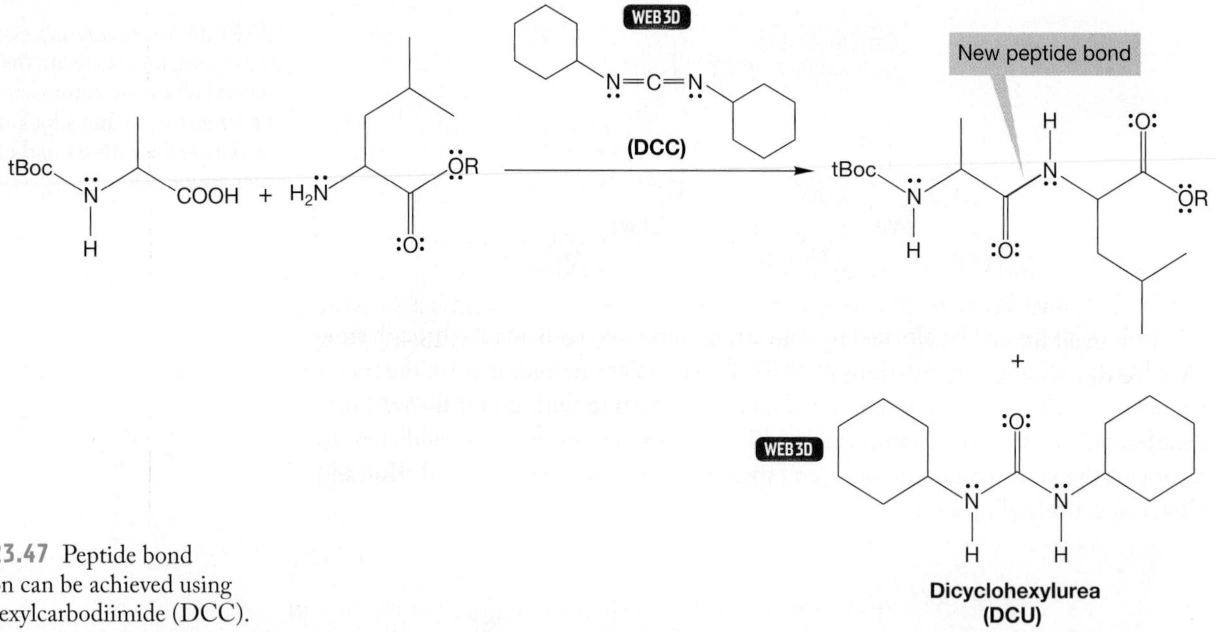

FIGURE 23.47 Peptide bond formation can be achieved using dicyclohexylcarbodiimide (DCC).

What is the mechanism of the coupling reaction of Figure 23.47? Remember that other cumulated double bonds (p. 512) such as ketenes (p. 515), isocyanates (p. 918), and isothiocyanates (p. 1198) all react rapidly with nucleophiles. The related DCC is no exception, and it is attacked by amines to give molecules called guanidines. Carboxylates are nucleophiles too, and will also add to DCC (Fig. 23.48).

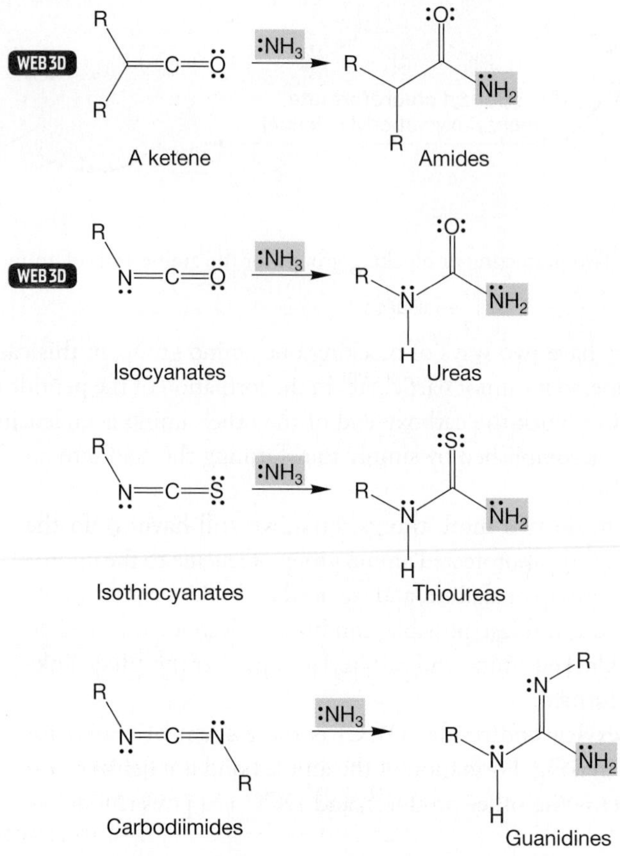

FIGURE 23.48 Cumulated double bonds (DCC in this case) react with all manner of nucleophiles, including amines and carboxylate anions.

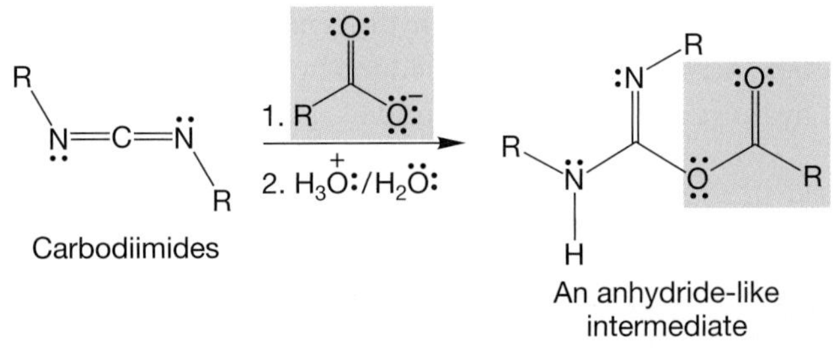

FIGURE 23.48 (*continued*)

Carbodiimides

An anhydride-like intermediate

So in the coupling reaction of Figure 23.47 the unprotected carboxylic acid group of alanine adds to DCC to give an anhydride-like intermediate that can react with a second alanine carboxylic acid to give the anhydride. The amino acid leucine then adds, via an addition–elimination mechanism, to give the dipeptide (Fig. 23.49). Note that the dipeptide is still protected in two places.

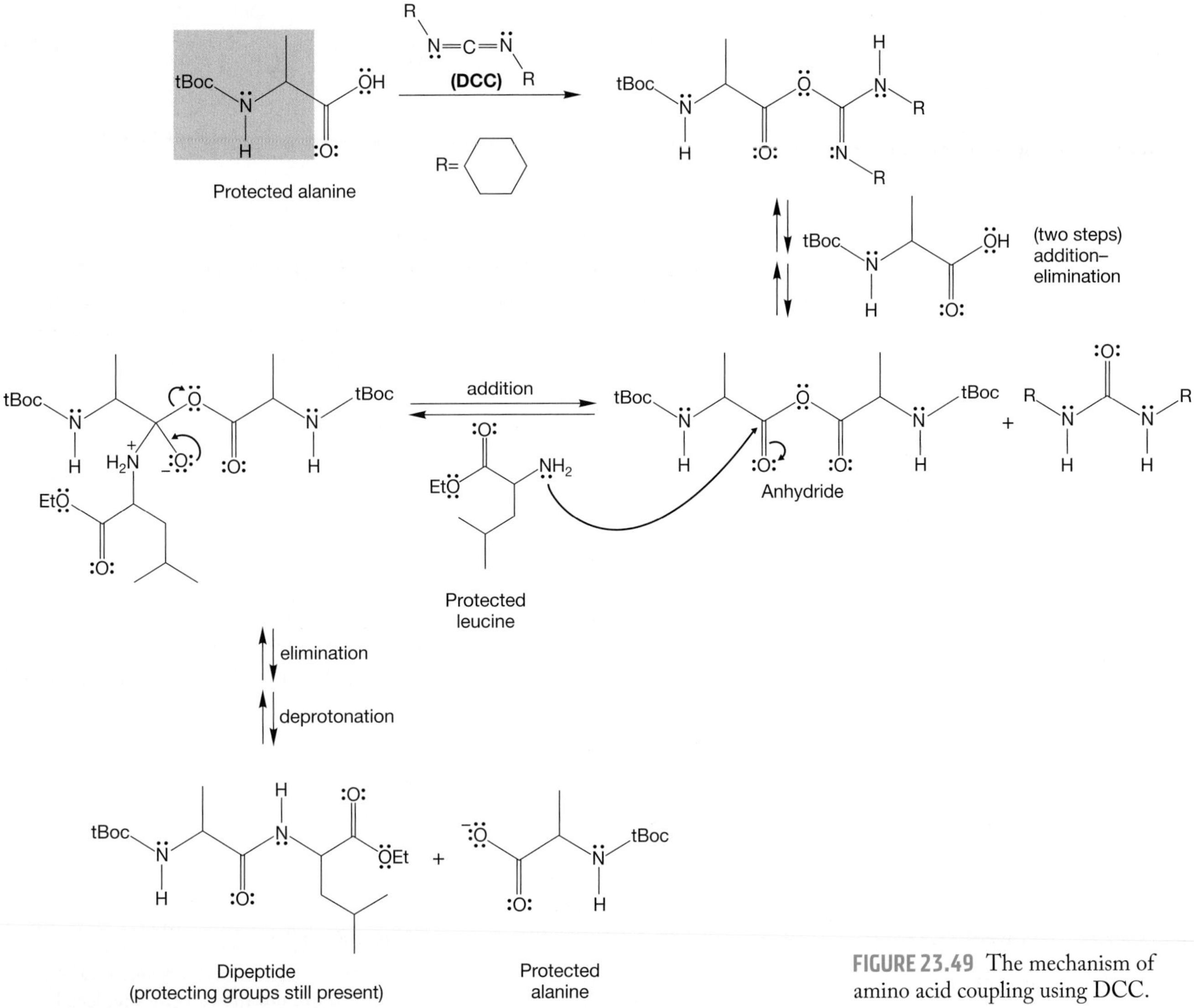

Protected alanine

(DCC)

R =

(two steps) addition–elimination

elimination

deprotonation

Protected leucine

Anhydride

addition

Dipeptide (protecting groups still present)

Protected alanine

FIGURE 23.49 The mechanism of amino acid coupling using DCC.

Now all we have to do is remove these protecting groups and, at last, we will have made our dipeptide, Ala·Leu. The tBoc blocking group is removed by treatment with very mild acid, which does not cleave amide bonds. If the Cbz protecting group is used, it can be easily removed by catalytic hydrogenation (Fig. 23.50). The carboxylic acid on the carboxy terminus can be regenerated by treatment of the ester with base. The ester will hydrolyze faster than the amide.

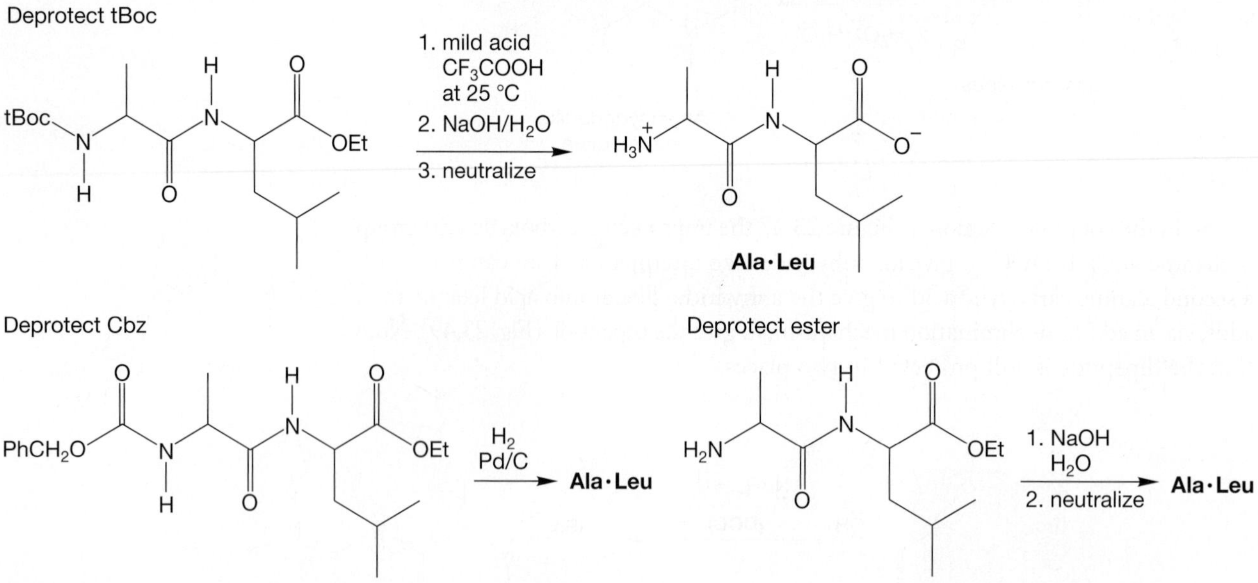

Deprotect tBoc

1. mild acid
   CF$_3$COOH
   at 25 °C
2. NaOH/H$_2$O
3. neutralize

Ala·Leu

Deprotect Cbz

$\xrightarrow{\text{H}_2 \text{ Pd/C}}$ **Ala·Leu**

Deprotect ester

1. NaOH
   H$_2$O
2. neutralize
$\longrightarrow$ **Ala·Leu**

FIGURE 23.50 Methods for removing the protecting groups used in peptide synthesis.

What about the yield for these reactions? Remember, this process, with all its protection and deprotection steps, must be repeated many, many times in the synthesis of a large peptide or a protein. These steps are efficient—they proceed in high yield because there are essentially no side reactions. The problem of tedious repetition has been solved in an old-fashioned way—by automation. Polypeptides can be synthesized by a machine invented by R. Bruce Merrifield (1921–2006). In Merrifield's procedure, the carboxy end of a tBoc-protected amino acid is first anchored to a material constructed of polystyrene in which some phenyl rings have been substituted with chloromethyl groups. The amino acid displaces the chlorine through an S$_N$2 reaction (Fig. 23.51).

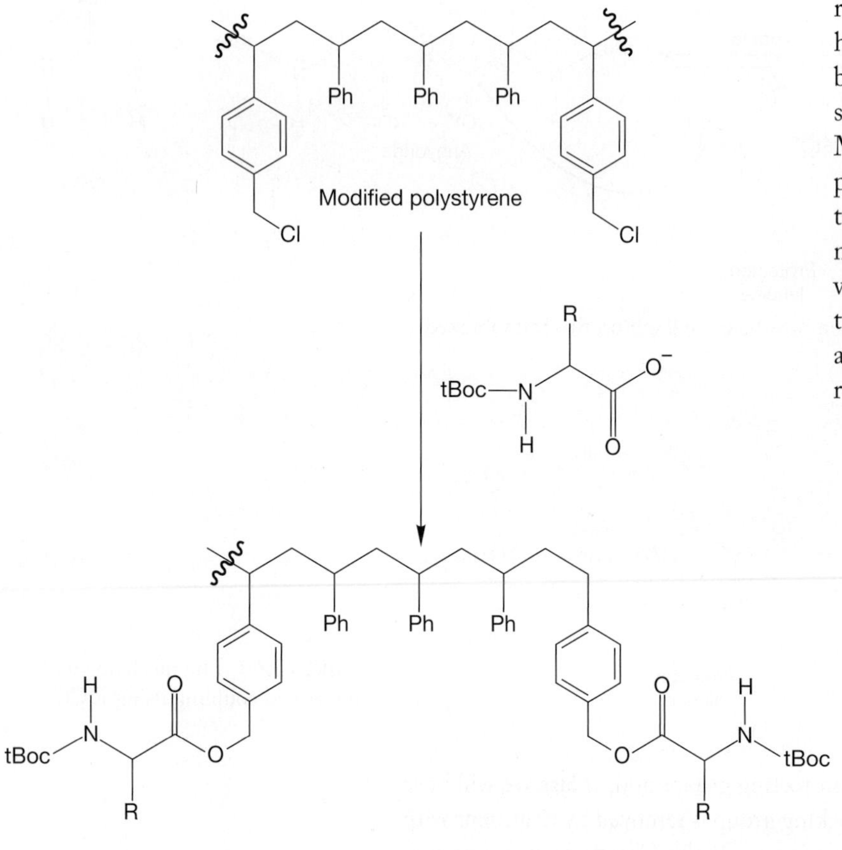

Modified polystyrene

FIGURE 23.51 Polystyrene can be chloromethylated in a Friedel–Crafts procedure, and the chlorines replaced through S$_N$2 reaction with the unprotected carboxy end of an amino acid. This technique anchors an amino acid to the polymer chain.

The tBoc group is then removed with mild acid hydrolysis, and a new, tBoc-protected amino acid is attached to the bound amino acid through DCC-mediated coupling. The multiple steps of peptide synthesis using various amino acids can now be carried out on this immobilized amino acid (Fig. 23.52).

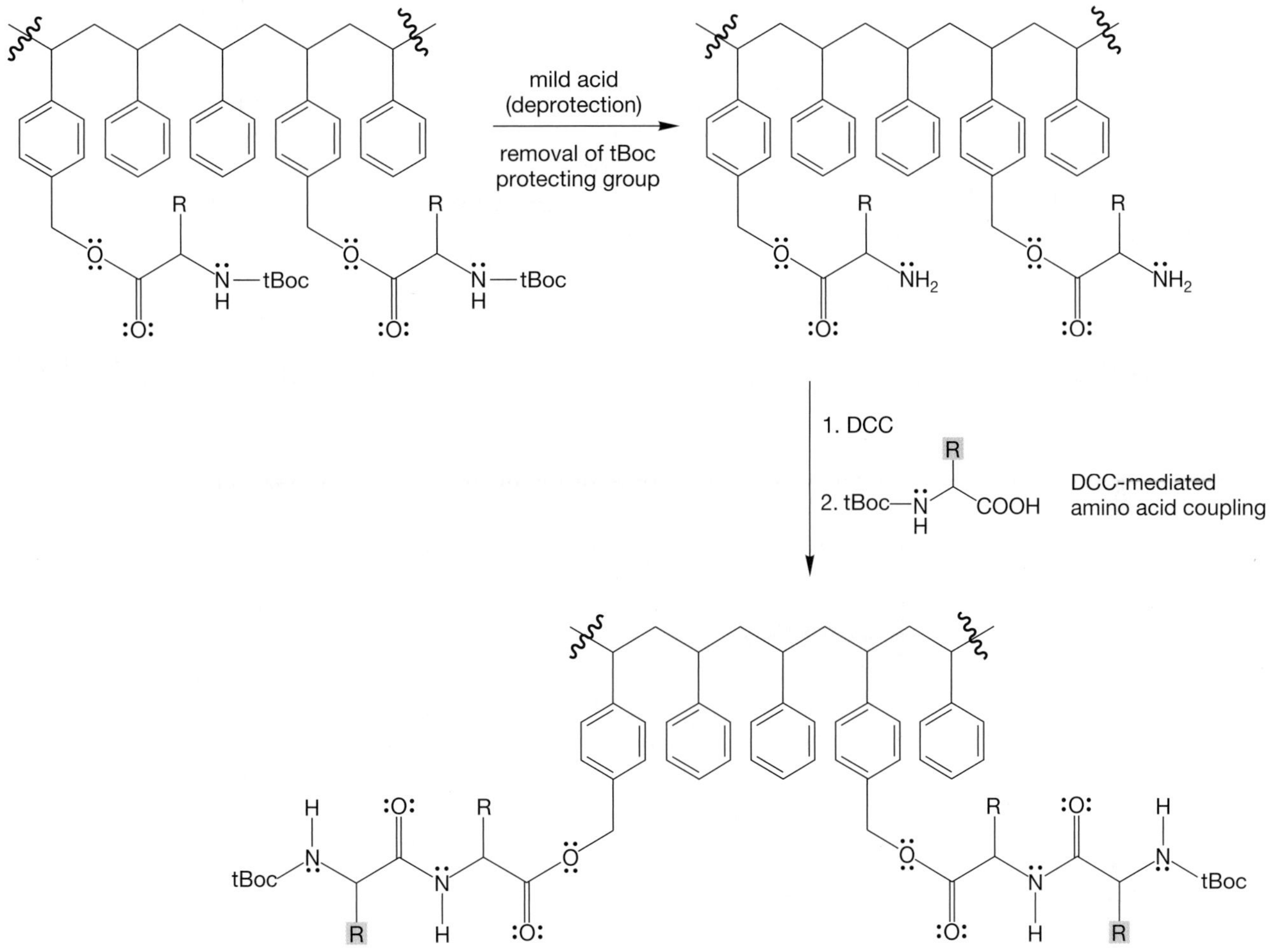

FIGURE 23.52 The steps of peptide synthesis can now be carried out on this immobilized amino acid.

PROBLEM 23.22 What is the function of the benzene ring of polystyrene? Could any ring-containing polymer have been used? Would cyclohexane have done as well, for example?

This procedure can be repeated as many times as necessary. The growing peptide chain cannot escape, because it is securely bound to the polystyrene; reagents can be added and by-products can be washed away after each reaction. The final step is detachment of the peptide from the polystyrene resin, usually through reaction with hydrogen fluoride or another acid (Fig. 23.53).

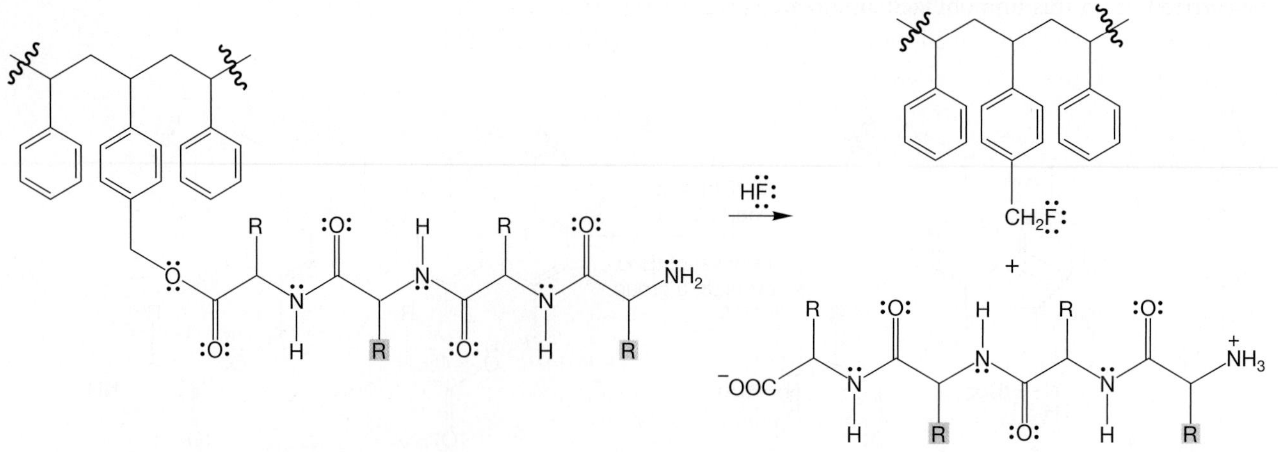

FIGURE 23.53 When synthesis is complete, the polypeptide can be detached from the polymer backbone by reaction with HF.

The first relatively primitive homemade Merrifield machine managed to produce a chain of 125 amino acids in an overall 17% yield, a staggering accomplishment. Merrifield was quite rightly rewarded with the chemistry Nobel prize in 1984.

Summary

Proteins have primary, secondary, tertiary, and quaternary structure. The primary structure of a peptide can be determined by using a combination of enzymes and small molecule reagents. There are several analytical tools that can also be applied to this task. Peptides can be synthesized by employing protecting groups and activating reagents. The common procedures are: amide formation with tBoc or Cbz for amine protection, esterification with ethanol for carboxylic acid protection, and acid activation with DCC.

PROBLEM 23.23 Devise a synthesis of the dipeptide Leu·Ala starting from the amino acids leucine, alanine, and any other reagents described in this chapter. Mechanisms are not necessary.

23.5 Nucleosides, Nucleotides, and Nucleic Acids

In Section 23.4c, the synthesis of proteins carried out by humans was described. In the natural world[4] quite another technique is used, and the uncovering of how it works is one of the great discoveries of human history.

In order to understand the biological synthesis of proteins, we must first learn about the basic building block called a **nucleoside**. A nucleoside is a β-glycoside

---

[4]In this case, "natural" is an artificial distinction, almost as foolish as labels describing natural as opposed to artificial vitamin C. The Merrifield procedure is no less natural than what we are about to describe. The laboratory synthesis is not unnatural, it's just human-mediated.

(p. 1148) formed between a sugar and a heterocyclic molecule (referred to as the base). A ribonucleoside is a nucleoside that has ribose as the attached sugar and a deoxyribonucleoside has deoxyribose as the sugar. Nucleosides that are phosphorylated at the 5′ position are called **nucleotides** (Fig. 23.54). The atoms of the sugar are given the primed numbers (e.g., 5′) because the atoms of the heterocyclic base are numbered first.

Sugars

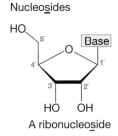

β-Ribofuranose

β-2-Deoxyribofuranose
(no oxygen at the 2-position)

Nucleosides

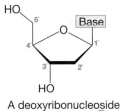

A ribonucleoside

A deoxyribonucleoside

Nucleotides

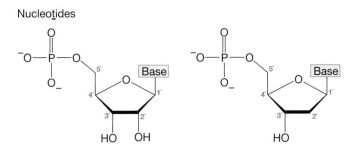

FIGURE 23.54 The sugars ribose and deoxyribose, and the corresponding nucleosides and nucleotides.

The **nucleic acids, deoxyribonucleic acid (DNA)** and **ribonucleic acid (RNA),** are polymers of nucleotides. The sugars in both nucleic acids, ribose in RNA and deoxyribose in DNA, are linked to each other through phosphoric acid groups attached to the 5′ position of one sugar and the 3′ position of the other (Fig. 23.55). Each sugar carries a heterocyclic base attached to the 1′ carbon.

One of the striking features of this replication machine, for that is exactly what we have begun to describe, is the economy with which it operates. So far, we have phosphoric acid and two sugars, ribose and deoxyribose, held together in polymeric fashion. One might be pardoned for imagining that all manner of bases are attached to this backbone, and that the diversity of the world of living things

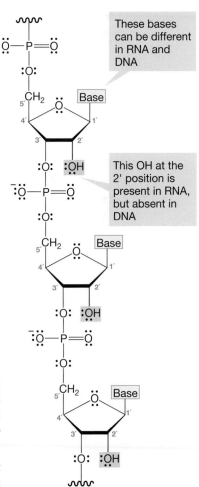

FIGURE 23.55 Nucleic acids (DNA and RNA) are polymeric collections of nucleotides.

results from the introduction of detail in this way. But that idea would be wrong. In fact, there are only five bases used in DNA and RNA, and three of them (cytosine, adenine, and guanine) are common to *both* kinds of nucleic acid. The other two, thymine which is in DNA and uracil which is in RNA, differ only in the presence or absence of a single methyl group (Fig. 23.56). Not much diversity there.

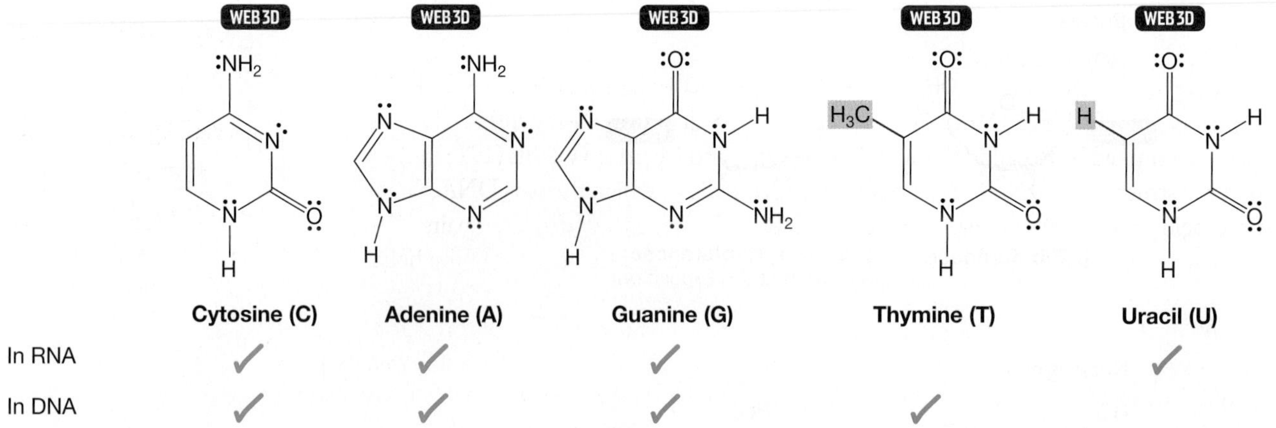

	WEB 3D	WEB 3D	WEB 3D	WEB 3D	WEB 3D
	Cytosine (C)	Adenine (A)	Guanine (G)	Thymine (T)	Uracil (U)
In RNA	✓	✓	✓		✓
In DNA	✓	✓	✓	✓	

FIGURE 23.56 The five bases present in RNA and DNA are shown. Three are common to both kinds of nucleic acid.

The bases are written in shorthand by using a single letter, C, A, G, T, or U. The introduction of detailed information must come in another way.

Notice the similarities between these nucleic acid polymers and the protein polymers. In the polyamino acids, the units are linked by amide bonds. In DNA and RNA the nucleotides are attached by phosphates. The information in proteins is carried by the side chains, the R groups, whose steric and electronic demands determine the higher order structures—secondary, tertiary, and quaternary—of these molecules. The information in DNA and RNA is carried by the bases. There are only 20 common R groups for amino acids, but there are even fewer bases—a mere four for each nucleic acid. Nevertheless, higher ordered structures are also found in the nucleic acids. Our next task is to see how this comes about.

A crucial observation was made by Erwin Chargaff (1905–2002) who noticed in 1950 that in many different DNAs the amounts of the bases adenine (A) and thymine (T) were always equal, and the amount of guanine (G) was always equal to the amount of cytosine (C). The four bases of DNA seemed to be found in pairs. This observation led to the reasonable notion that the members of the pairs must be associated in some way, and to the detailed picture of hydrogen-bonded dimeric structures of **base pairs** in which A is always associated with T and G with C. These combinations are the result of especially facile hydrogen bonding; the molecules fit well together (Fig. 23.57).

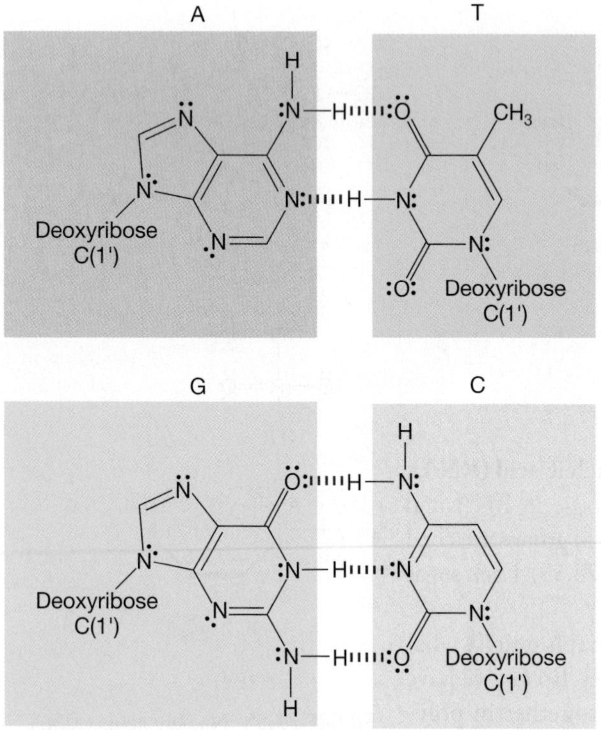

The existence of base pairs shows how the order of bases on one strand of DNA can determine the structure of another strand that is hydrogen bonded to the first. For a sequence

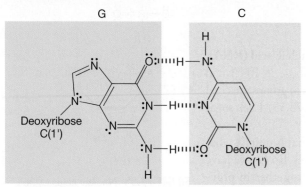

FIGURE 23.57 Adenine–thymine and guanine–cytosine form base pairs through effective hydrogen-bond formation.

of bases C C A T G C T A, the complementary sequence must be G G T A C G A T in order to form the optimal set of hydrogen-bonded base pairs (Fig. 23.58).

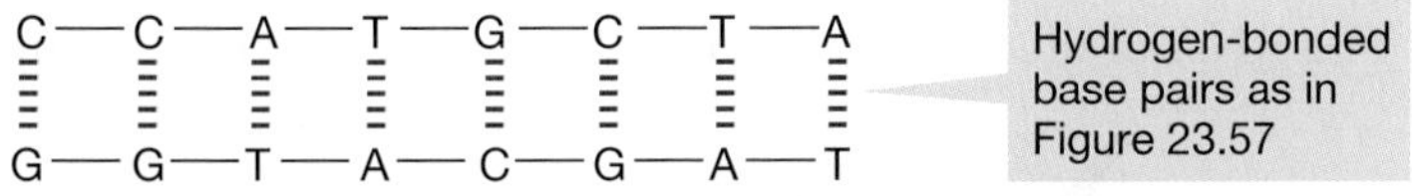

C—C—A—T—G—C—T—A
$\vdots$ $\vdots$ $\vdots$ $\vdots$ $\vdots$ $\vdots$ $\vdots$ $\vdots$
G—G—T—A—C—G—A—T

Hydrogen-bonded base pairs as in Figure 23.57

FIGURE 23.58 The sequence of nucleotides on one polymer determines the sequence in another through formation of hydrogen-bonded base pairs, C-G and A-T.

This notion was also instrumental in the realization by James D. Watson (b. 1928) and Francis H. C. Crick (1916–2004), with immense help from X-ray diffraction patterns determined by Rosalind Franklin (1920–1958) and Maurice Wilkins (1916–2004), that the structure of DNA was a double-stranded helix in which the two DNA chains were held together in the helix by A-T and G-C hydrogen-bonded base pairs. The diameter of the helix is about 20 Å, and there are 10 nucleotides per turn of the helix. Each turn is about 34 Å long (Fig. 23.59). The DNA chain is our genetic code.

We still have two outstanding problems: (1) Where does the density of information come from? How is the seemingly inadequate pool of five bases translated into the wealth of information that must be transferred in the synthesis of proteins containing specific sequences made from the 20 amino acids? (2) What is the mechanism of the information transferal? The second of these two questions is the easier to tackle, so we will start there.

The replication of DNA involves partial unwinding of the double helix to yield two templates for the development of new DNA strands. The order of bases in each strand determines the structure of the new strand. Where there is a T, an A must appear, where there is a G, a C must attach, and so on. *Each strand of the original DNA must induce the formation of an exact duplicate of the strand to which it was originally attached* (Fig. 23.60).

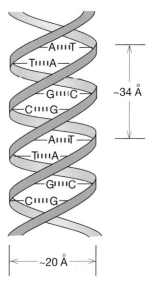

FIGURE 23.59 The famous double helix of DNA is held together by hydrogen bonds between base pairs.

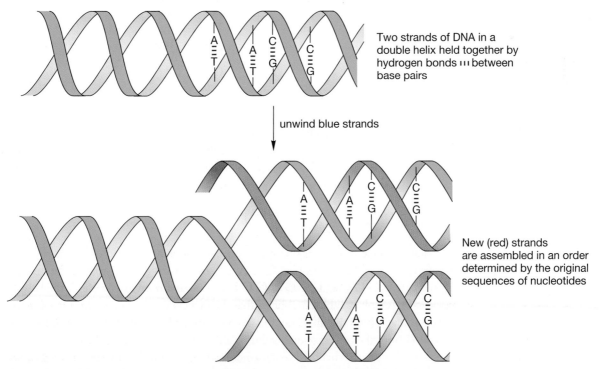

Two strands of DNA in a double helix held together by hydrogen bonds ┉ between base pairs

unwind blue strands

New (red) strands are assembled in an order determined by the original sequences of nucleotides

FIGURE 23.60 The DNA replication machine works by first unraveling to form single-stranded sections. Assembly of new strands is determined by the original base sequence. Where there is an A, a T must be added; a G requires a C.

Protein synthesis starts with a similar, but not quite identical process that involves RNA. A DNA strand can also act as a template for the production of a strand of RNA. In the RNA construction the base uracil is used instead of thymine. The RNA produced is called **messenger RNA (mRNA)** (Fig. 23.61).

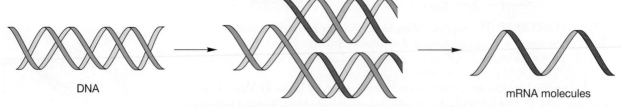

DNA   mRNA molecules

FIGURE 23.61 Synthesis of mRNA.   Now replication uses a ribose, not deoxyribose, based nucleotide. In RNA, the bases are A, C, G, and U *not* T

This new RNA molecule directs protein synthesis in the following spectacularly simple way: Synthesis of each of the 20 common amino acids in proteins is directed by one or more sequences of three bases, called **codons**. How many codons can be constructed from the available four RNA bases? The answer is $4^3 = 64$, more than enough to code for the 20 common amino acids. Indeed, some amino acids are coded by more than one codon. Some extra codons are needed to deal with the problems of when to start amino acid production and when to stop. Table 23.2 gives the three-base codons for the 20 common amino acids and the frequently used commands.

TABLE 23.2 Codons for Amino Acids and Simple Commands

First Base in Sequence	Second Base				Third Base in Sequence
	U	C	A	G	
U	Phe	Ser	Tyr	Cys	U
	Phe	Ser	Tyr	Cys	C
	Leu	Ser	Stop	Stop	A
	Leu	Ser	Stop	Trp	G
C	Leu	Pro	His	Arg	U
	Leu	Pro	His	Arg	C
	Leu	Pro	Gln	Arg	A
	Leu	Pro	Gln	Arg	G
A	Ile	Thr	Asn	Ser	U
	Ile	Thr	Asn	Ser	C
	Ile	Thr	Lys	Arg	A
	Met[a]	Thr	Lys	Arg	G
G	Val	Ala	Asp	Gly	U
	Val	Ala	Asp	Gly	C
	Val	Ala	Glu	Gly	A
	Val	Ala	Glu	Gly	G

[a]The AUG codon translates to Met when in the middle of a message, but it translates as the Start message when it is at the beginning of a sequence.

As an example of how this process works, let's look at a strand of DNA with the sequence T A C C G A A G C A C G A T T. Base pairing must produce the mRNA fragment A U G G C U U C G U G C U A A (remember that in RNA the base U replaces T). A look at Table 23.2 shows that the first codon, A U G, is interpreted as the Start command (find the first letter of the codon in the left column, the middle letter of the codon from the corresponding column, and the final letter from the right column). The second codon of this mRNA fragment is G C U which corresponds to Ala, the next codon is U C G which corresponds to Ser, then codon U G C which corresponds to Cys, and the last codon U A A which instructs the peptide synthesis to Stop. The fragment translates into the following message: "Start-add Ala-add Ser-add Cys-Stop."

WORKED PROBLEM 23.24 Why would 20 amino acids be coded for by 64 three-base sequences? Why is there so much overlap? Consider the alternative method of having two-base sequences coding for amino acid production. How many possible two-base sequences are there for four bases?

ANSWER There are only 2^4 possible combinations of four objects taken two at a time. So there are only 16 things that could be coded by two-base sequences of four different bases, which is not sufficient to code for the 20 common amino acids that make up proteins. Three-base codons turn out to be the most economical way to code for a system of 20 amino acids.

One further problem remains. How does mRNA direct amino acid synthesis? The answer is that another type of RNA is used, called transfer RNA (tRNA), which is relatively small and designed to interact with an enzyme, aminoacyl tRNA synthetase, to acquire one specific amino acid and carry it to the mRNA where it is added to the growing chain at the position of the correct codon.

23.6 Summary

New Concepts

Proteins are α-amino acids assembled into long polymeric chains. The secondary structure of these peptides or proteins involves the regions of α-helical or β-pleated sheet arrangements, which are separated from each other by disordered sections of the chains called random coils. Disulfide bonds, electrostatic forces, van der Waals forces, and hydrogen bonding twist these molecules into shapes characteristic of individual proteins, the tertiary structure. Secondary and tertiary structure can be destroyed, sometimes only temporarily, by any of a number of denaturing processes. Finally, intermolecular forces can hold a number of these protein chains together to form supermolecules, the quaternary structure.

These higher ordered structures are affected by the identities of the R groups on the constituent amino acids. It is the electronic and steric properties of the R groups that generate the particular secondary, tertiary, and quaternary structures of proteins.

These new structures require new analytical and synthetic techniques. Methods for determining the sequence of peptides and proteins involve physical techniques such as X-ray diffraction, as well as chemical techniques for revealing the terminal amino acids of the chains, and even for sequencing entire polymers.

Other biopolymers, the nucleic acids, are constructed not of amino acids, but of nucleotides, which are sugars bearing heterocyclic base groups and joined through phosphoric acid linkages. These nucleic acids also have higher order structures, the most famous being the double-helical arrangement of DNA. Most remarkable of all is the ability of these molecules to carry the genetic code, to uncoil and direct the assembly not only of replicas of themselves, but also, through the mediation of RNA, of the polyamino acids called proteins. In the nucleic acids, the information is not carried in a diverse supply of R groups as it is in peptides and proteins, but in a small number of base groups. Hydrogen bonding in base pairs allows for the replication reactions, and three-base units called codons direct the assembly of amino acids into proteins.

Key Terms

α-amino acids (p. 1175)
amino terminus (p. 1189)
base pair (p. 1212)
binding site (p. 1193)
tBoc (p. 1205)
carboxy terminus (p. 1189)
Cbz (p. 1205)
codon (p. 1214)
cyanogen bromide (BrCN) (p. 1200)
denaturing (p. 1193)
deoxyribonucleic acid (DNA) (p. 1211)
dicyclohexylcarbodiimide (DCC) (p. 1205)
disulfide bridges (p. 1190)

Edman degradation (p. 1198)
electrophoresis (p. 1180)
essential amino acid (p. 1177)
Gabriel synthesis (p. 1182)
gel-filtration chromatography (p. 1195)
α-helix (p. 1191)
ion-exchange chromatography (p. 1195)
isoelectric point (pI) (p. 1179)
kinetic resolution (p. 1186)
messenger RNA (mRNA) (p. 1214)
ninhydrin (p. 1187)
nucleic acid (p. 1211)
nucleoside (p. 1210)
nucleotide (p. 1211)

peptide bond (p. 1175)
β-pleated sheet (p. 1191)
primary structure (p. 1190)
quaternary structure (p. 1194)
random coil (p. 1191)
ribonucleic acid (RNA) (p. 1211)
Sanger degradation (p. 1196)
secondary structure (p. 1191)
side chain (p. 1175)
Strecker synthesis (p. 1184)
tertiary structure (p. 1192)
thiourea (p. 1198)
zwitterion (p. 1179)

Reactions, Mechanisms, and Tools

Reactions important in determining the amino acid sequence of a protein include several techniques for end-group analysis. There are enzymes (proteins themselves) that cleave the amino acid at the carboxy terminus of the polymer. The amino terminus can be found through the Sanger reaction.

In principle, an entire sequence can be unraveled using the Edman degradation in which phenylisothiocyanate is used in a procedure that cleaves the amino acid at the amino terminus and labels it as a phenylthiohydantoin.

Other agents are available for inducing specific cleavages in the amino acid chain that yield smaller fragments for sequencing through the Edman procedure. These agents include various enzymes and small chemical reagents such as cyanogen bromide.

Synthesis of these polymeric molecules requires nothing more than the sequential formation of a large number of amide bonds. The DCC coupling procedure is effective, but it requires that the amino acid monomers be supplied in a form in which the ends to be attached are activated and other reactive points blocked, or protected. Those protecting and activating groups must be added in appropriate places and removed after the reaction is over.

If we need to synthesize the amino acids themselves, the best known routes are the Gabriel and Strecker procedures.

Naturally occurring amino acids are optically active (except for glycine, which is achiral) and occur in the L form. Resolution can be achieved through the classical technique of diastereomer formation followed by crystallization, or through other methods. The most clever of these involves a kinetic resolution in which advantage is taken of the ability of enzymes to attack only the naturally occurring (S) enantiomers, while the (R) isomer is left alone.

Syntheses

1. Amino Acids

Simple alkylation of ammonia using an α-bromo acid; overalkylation is a problem

Gabriel synthesis. The R—X must be S$_N$2 active

Strecker synthesis

2. Disulfide Bridges

Oxidation can convert the SH groups of cysteines into disulfide bonds between or within chains

3. Optically Active Amino Acids

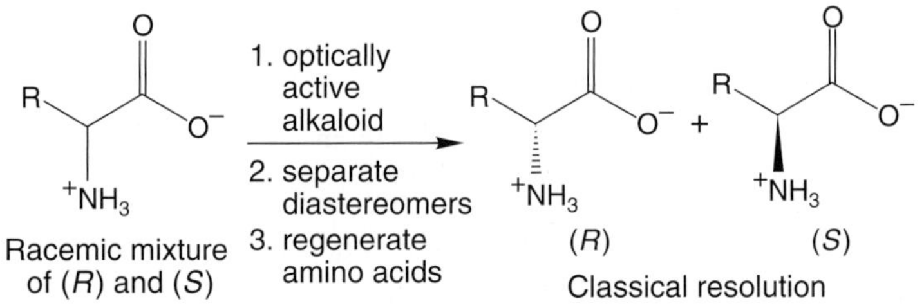

Classical resolution

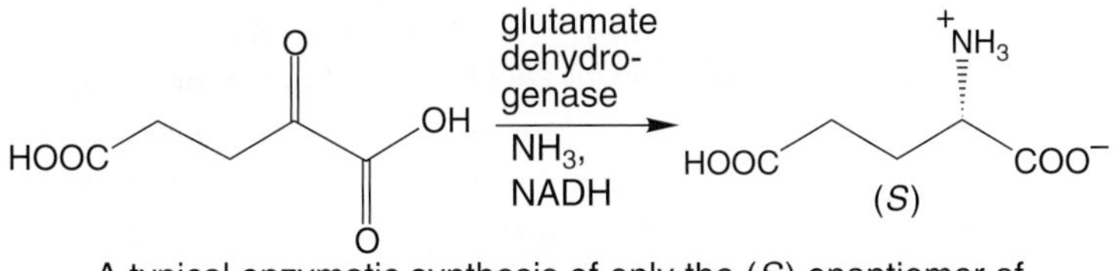

A typical enzymatic synthesis of only the (S) enantiomer of
an amino acid through reductive amination

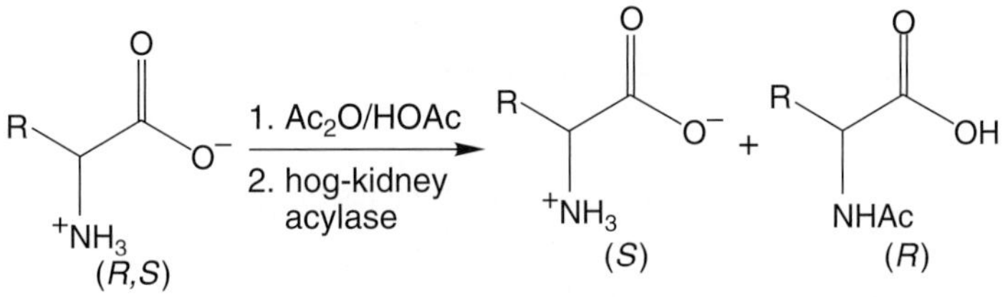

The free (S) amino acid is produced in this kinetic
resolution, along with acylated (R) amino acid

4. Protected Amino Acids

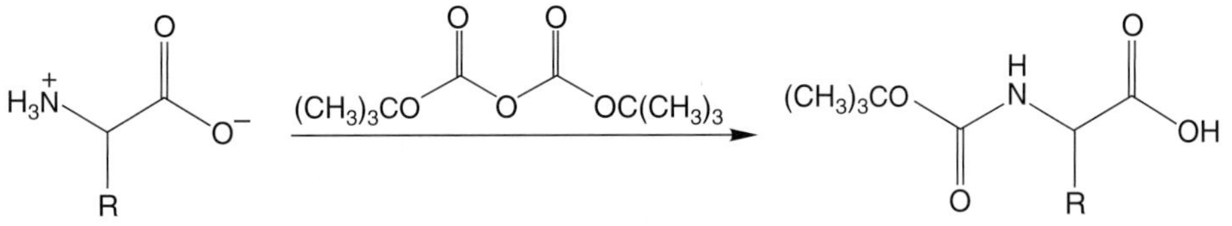

A tBoc-protected
amino acid

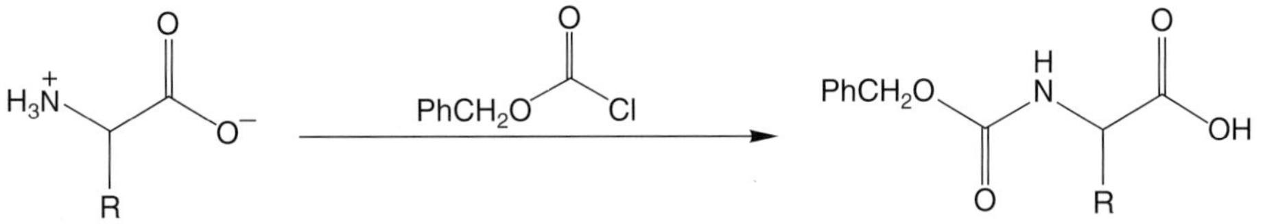

A Cbz-protected amino acid

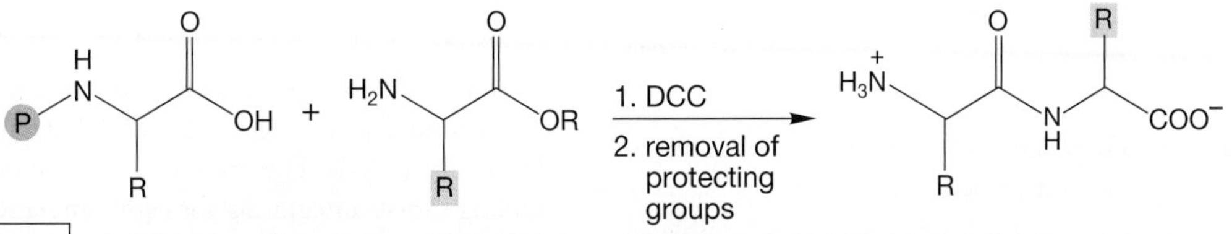

P = tBoc or Cbz   DCC-Mediated coupling; notice that the starting amino acids are prepared
for reaction with protecting groups that are removed in later steps

5. Thiohydantoin

The Edman degradation

6. Thioureas

Addition of an amine to an isothiocyanate

7. Ureas

Addition of an amine to an isocyanate

Hydration of a carbodiimide

23.7 Additional Problems

PROBLEM 23.25 (a) Amides protonate on oxygen rather than nitrogen. Why? (b) Amides (pK_b ~14) are weaker bases than amines (pK_b ~5). Why?

PROBLEM 23.26 Write structures for the following tripeptides and name them:
  (a) Ala·Ser·Cys   (b) Met·Phe·Pro   (c) Val·Asp·His

PROBLEM 23.27 Name the following dipeptides:

(a)

(b)

PROBLEM 23.28 Draw the 3-D perspective of (S)-serine and (S)-proline using solid and dashed wedges.

PROBLEM 23.29 Write Fischer projections for L-serine and L-proline.

PROBLEM 23.30 Give the structures for the amino acids His, Arg, and Phe at pH = 3 and 12.

PROBLEM 23.31 Use the pI data in Table 23.1 to determine which electrode (anode or cathode) the amino acids Asp and Lys will migrate toward at pH 7.

PROBLEM 23.32 All of the amino acids listed in Table 23.1, except the achiral glycine, have the L configuration. Of the 19 amino acids with the L configuration, 18 are (S) amino acids. Which amino acid has an (R) configuration? Explain why the configuration of this L-amino acid is (R) rather than (S).

PROBLEM 23.33 In Section 23.3c, we saw that all the amino acids in Table 23.1, except proline, react with ninhydrin to form the same purple compound. In contrast, proline reacts with ninhydrin to give a yellow compound thought to have structure **1**. Propose a mechanism for the formation of **1**.

1

PROBLEM 23.34 (a) A peptide of unknown structure is treated with the Sanger reagent, 2,4-dinitrofluorobenzene, followed by acid hydrolysis. The result is formation of a compound containing signals integrating for eight aromatic hydrogens in its ^{1}H NMR spectrum. What information can you derive about the structure of this peptide?

(b) A second peptide, similarly treated, gives a compound in which a single hydrogen appears in the high-resolution ^{1}H NMR spectrum as an eight-line signal at about δ 3.5 ppm. What information can you derive about the structure of this peptide?

PROBLEM 23.35 A tripeptide of unknown structure is hydrolyzed in acid to one molecule of Ala, one of Cys, and one of Met. The tripeptide reacts with phenyl isothiocyanate, followed by treatment with acid, to give a phenylthiohydantoin that shows only a three-hydrogen doublet at about δ 2 ppm in the ^{1}H NMR spectrum. The tripeptide is not cleaved by BrCN. Deduce the structure from these data.

PROBLEM 23.36 A dodecapeptide **1** is hydrolyzed in acid to give 2 Arg, Asp, Cys, His, 2 Leu, Lys, 2 Phe, and 2 Val. The Edman procedure yields a phenylthiohydantoin that shows two 3H doublets and a multiplet integrating for 1H at about δ 4.0 ppm in the ^{1}H NMR spectrum. Treatment of **1** with trypsin leads to Val·His·Phe·Leu·Arg, Asp·Cys·Leu·Phe·Lys, and Val·Arg. Treatment of **1** with chymotrypsin leads to Val·His·Phe, Leu·Arg·Asp·Cys·Leu·Phe, and Lys·Val·Arg. Deduce the structure of **1** and explain your reasoning.

PROBLEM 23.37 In Section 23.4, we saw that the tBoc group can be used to protect (or block) the amino group of an amino acid during peptide synthesis. This group can be added by allowing the amino acid to react with di-*tert*-butyl dicarbonate (**1**). Once the amide linkage has been formed, the tBoc protecting group can be removed by mild acid "hydrolysis." Both of these reactions are illustrated below. Provide mechanisms for the addition of the tBoc group and its removal. *Hint*: One of the products of the removal is 2-methylpropene.

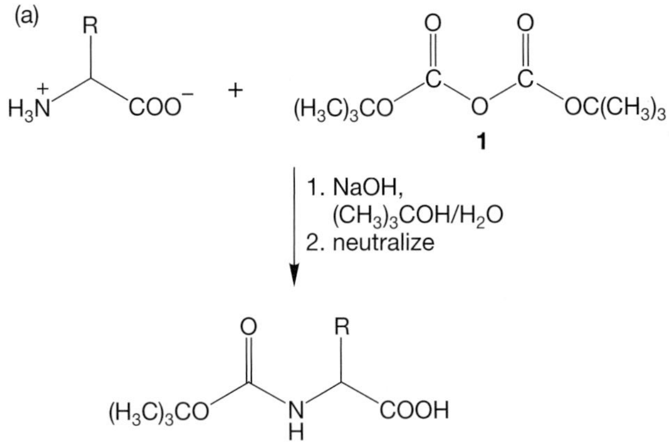

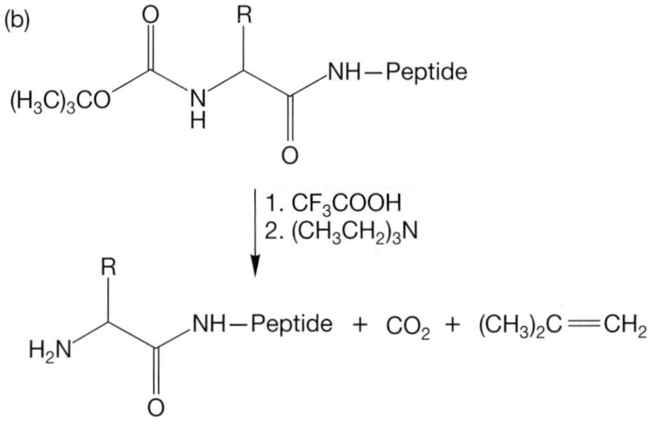

PROBLEM 23.38 In Section 23.4c (Fig. 23.49), we saw that the acylating agent in DCC-mediated peptide bond formation is the anhydride. There is another intermediate that could act as the acylating agent. What is it, and how can it function as the acylating agent? *Hint*: This intermediate appears in Figure 23.49.

PROBLEM 23.39 Reaction of *N*-phenylalanine (**1**) with nitrous acid affords *N*-nitroso amino acid (**2**), which on treatment with acetic anhydride yields "sydnone" (**3**). Sydnones belong to a class of compounds known as mesoionic compounds. These compounds cannot be satisfactorily represented by Lewis structures not involving charge separation. The name sydnone derives from the University of Sydney where the first examples were prepared in 1935. Reaction of **3** and dimethyl acetylenedicarboxylate (DMAD) gives pyrazole (**4**). Propose mechanisms for the formations of **2**, **3**, and **4**. *Hint*: For the formation of **4**, it may be helpful to consider the other possible Lewis structures for sydnone **3**.

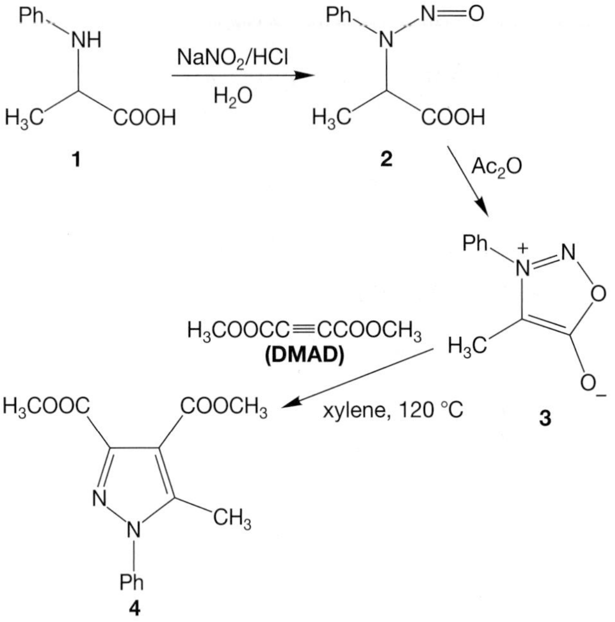

PROBLEM 23.40 In Problem 23.12, you were asked to write a mechanism for the acylation of glycine. The mechanism presented in the answer is actually incomplete, at least when the reaction is run in the presence of excess acetic anhydride. For example, what is the purpose of the water in the second step? There is actually an intermediate involved in this reaction, the 2-oxazolin-5-one (**1**), colloquially referred to as an azlactone.

Propose mechanisms for the formation of azlactone **1** and its hydrolysis to *N*-acetylglycine.

PROBLEM 23.41   In Problem 23.40, we saw that the azlactone (**1**) is an intermediate in the acetylation of glycine in the presence of excess acetic anhydride. The generation of azlactone **1** in this reaction lies at the heart of one of the oldest known amino acid syntheses. For example, if glycine (or *N*-acetylglycine) is treated with acetic anhydride in the presence of benzaldehyde and sodium acetate, the benzylidene azlactone (**2**) is formed. Azlactone **2** can then be converted into racemic phenylalanine (**3**) by the following sequence. Propose a mechanism for the formation of **2**. *Hint*: Azlactone **1** is still an intermediate in this reaction.

PROBLEM 23.42   When optically active *N*-methyl-L-alanine (**1**) is heated with an excess of acetic anhydride, followed by treatment with water, the resulting *N*-acetyl-*N*-methylalanine (**2**) is racemic. If the acetylation reaction is run in the presence of dimethyl acetylenedicarboxylate (DMAD), the pyrrole **3** can be isolated in good yield. Account for the racemization of **2** and the formation of **3**. *Hint*: There is a common intermediate involved in these two reactions. Also, see Problem 23.39.

PROBLEM 23.43   (a) What mRNA sequence does the following DNA sequence produce? T A C G G G T T T A T C (b) What message does that mRNA sequence produce?

Glossary

Absolute configuration (Section 4.3) The arrangement in space of the atoms of an enantiomer, its stereochemical description as (*R*) or (*S*).

Abstraction (Section 11.2) Removal of a hydrogen atom as a result of a reaction with a radical.

Acetal (Section 16.9) The final product in the acid-catalyzed reaction of an aldehyde or ketone with an alcohol.

Acetylenes (Section 3.1) Hydrocarbons of the general formula C_nH_{2n-2}. These molecules, also called alkynes, contain carbon–carbon triple bonds. The parent compound, $HC\equiv CH$, is called acetylene, or ethyne.

Acetylide (Section 3.14) The anion formed by removal by base of a terminal hydrogen from an acetylene.

Achiral (Section 4.2) Not chiral.

Acid anhydride (Section 17.7) A compound formed by formal loss of water from two molecules of a carboxylic acid. The structure is

Acid chloride (Section 14.6) A compound of the structure

Acid derivative (Section 18.1) One of the several functional groups related to carboxylic acids and having the same oxidation level. These are acid halides, acid anhydrides, esters, and amides. We often include nitriles and ketenes in this category.

Acid halide (Section 18.2) A compound of the structure

R = F, Cl, Br, or I

Activating agent (Section 17.7) A reagent that converts a carboxylic acid into a more reactive acid derivative.

Activation energy ($\Delta G^{\ddagger}$) (Section 3.17) The difference in free energy between the starting material and the transition state in a reaction. It is this amount of energy that is required for a molecule of starting material to be transformed into product.

Acyl compound (Section 18.1) A compound of the structure

Acyl group (Section 11.2) A general term for $R-C=O$.

Acylium ion (Section 14.6) The structure $RC^+=O$, which has the resonance form $R-C\equiv O^+$.

Alcohol (Section 3.19) A molecule containing a simple hydroxyl group, $R-OH$.

Aldaric acid (Section 22.4) A diacid derived from an aldohexose by oxidation with nitric acid. It has the overall structure $HOOC-(CHOH)_4-COOH$. In an aldaric acid, the old aldehyde and primary alcohol ends of the sugar have become identical.

Aldehydes (Section 10.5) Compounds containing a monosubstituted carbon–oxygen double bond.

Aldohexose (Section 22.2) A hexose of the structure $O=CH-(CHOH)_4-CH_2OH$.

Aldol condensation (Section 19.6) The acid- or base-catalyzed conversion of an aldehyde or ketone into a β-hydroxy aldehyde or β-hydroxy ketone. In acid, the enol is an intermediate; in base, the intermediate is the enolate. The initial product often loses water to form an α,β-unsaturated ketone or aldehyde.

Aldonic acid (Section 22.4) A monoacid derived from an aldohexose through oxidation with bromine in water. Only the aldehyde group is oxidized to the acid. It has the structure $HOOC-(CHOH)_4-CH_2OH$.

Aldopentose (Section 22.2) A pentose of the structure $O=CH-(CHOH)_3-CH_2OH$.

Aldose (Section 22.2) A sugar molecule that has an aldehyde group.

Aldotetrose (Section 22.2) A tetrose of the structure $O=CH-(CHOH)_2-CH_2OH$.

Aldotriose (Section 22.2) A triose of the structure $O=CH-CHOH-CH_2OH$.

Alkaloid   (Section 4.9) A nitrogen-containing compound, often polycyclic and generally of plant origin. The term is more loosely applied to other naturally occurring amines.

Alkanes   (Section 2.1) The series of saturated hydrocarbons of the general formula C_nH_{2n+2}.

Alkene halogenation   (Section 10.2) Addition of X_2 to a π bond, giving a 1,2-dihalide.

Alkene hydrohalogenation   (Section 9.2) Addition of HX to a π bond, giving an alkyl halide.

Alkenes   (Section 3.1) Hydrocarbons of the general formula, C_nH_{2n}. These molecules, also called olefins, contain carbon–carbon double bonds.

Alkoxide ion   (Section 6.4) The conjugate base of an alcohol, RO^-.

Alkyl compounds   (Section 2.5) Substituted alkanes. One or more hydrogens is replaced by another atom or group of atoms.

Alkyl halide   (Section 6.2) Compound of the formula $C_nH_{2n+1}X$, where X = F, Cl, Br, or I.

Alkynes   (Section 3.1) Hydrocarbons of the general formula C_nH_{2n-2}. These molecules contain carbon–carbon triple bonds.

Allene   (Section 4.13) A 1,2-diene. A compound containing a carbon atom that is part of two double bonds.

Allyl   (Section 3.3) The common name for the $H_2C{=}CH{-}CH_2$ group.

Allylic halogenation   (Section 11.8) Specific formation of a carbon–halogen bond at the position adjacent to a carbon–carbon double bond.

Amide   (Sections 6.7, 17.7) A compound of the structure

$$\underset{\text{R can be H}}{RO{-}\overset{\overset{\displaystyle O}{\|}}{C}{-}NR_2}$$

Remember that this term also refers to the ions H_2N^-, RHN^-, R_2N^-, $RR'N^-$.

Amine   (Section 6.7) A compound of the structure $R_3N:$, where R can be H, alkyl, or aryl but not acyl. Cyclic and aromatic amines are common.

Amine inversion   (Section 6.7) The conversion of one pyramidal form of an amine into the other through a planar, sp^2 hybridized transition state.

α-Amino acids   (Section 23.2) 2-Aminoacetic acids, the monomeric constituents of the polymeric peptides and proteins.

Amino terminus   (Section 23.4) The amino acid at the free amine end of a peptide polymer.

Ammonia   (Section 6.7) $:NH_3$, the simplest of all amines.

Ammonium ion   (Section 6.7) R_4N^+, R = alkyl, aryl, or H.

Anchimeric assistance   (Section 21.2) The increase in rate of a reaction that proceeds through intramolecular displacement over that expected of an intermolecular displacement.

Angle strain   (Section 5.2) The increase in energy caused by the deviation of an angle from the ideal demanded by an atom's hybridization.

Anion   (Section 1.2) A negatively charged atom or molecule.

Annulene   (Section 13.6) A cyclic polyene that is at least formally fully conjugated.

Anomeric carbon   (Section 22.2) The carbon in a cyclic sugar that is the acetal carbon.

Anomers   (Section 22.2) For aldoses, these are sugars differing only in the stereochemistry at C(1). They are C(1) stereoisomers.

Antarafacial motion   (Section 20.5) Migration of a group from one side of a π system to the other in a sigmatropic shift.

Antibonding molecular orbital   (Section 1.5) A molecular orbital that is the result of mixing atomic orbitals in an out-of-phase fashion. Occupation of an antibonding molecular orbital by an electron destabilizes a molecule.

anti Elimination   (Section 7.9) An elimination reaction in which the dihedral angle between the breaking bonds, usually C—H and C—L, is 180°.

anti-Markovnikov addition   (Section 9.11) Addition to a π bond that results in the substituent being on the less substituted carbon of what was the π bond.

Aprotic solvent   (Section 6.5) A solvent that is not a proton donor. An aprotic molecule does not bear a transferable proton.

Arene   (Section 13.8) An aromatic compound containing a benzene ring or rings.

Arndt–Eistert reaction   (Section 18.14) The use of the Wolff rearrangement to elongate the chain of a carboxylic acid by one carbon.

Aromatic character   (Section 13.5) See Aromaticity.

Aromaticity   (Section 13.5) The special stability of planar, cyclic, fully conjugated molecules with $4n + 2$ π electrons. Such molecules will have molecular orbital systems with all bonding molecular orbitals filled and all antibonding molecular orbitals empty. Usually, there will be filled degenerate orbitals.

Arrow formalism, curved arrow formalism, or electron pushing   (Section 1.4) A mapping device for chemical reactions. The electron pairs (lone pairs or bond pairs) are "pushed" using curved arrows that show the bonds that are forming and breaking in the reaction.

Atom   (Section 1.1) A neutral atom consists of a nucleus, or core of protons and neutrons, orbited by a number of electrons equal to the number of protons.

Atomic orbital   (Section 1.1) A three-dimensional representation of the solution of Schrödinger's equation describing the motion of an electron in the vicinity of a nucleus. Atomic orbitals have different shapes, which are determined by quantum numbers. The s orbitals are spherically symmetric, p orbitals roughly dumbbell shaped, and the d and f orbitals are even more complicated.

Aufbau principle   (Section 1.2) When adding electrons to a system of orbitals, first fill the lowest energy orbital available before filling any higher energy orbitals. Electron–electron repulsion is minimized by filling systems of equi-energetic orbitals by singly occupying all orbitals with electrons of the same spin before doubly occupying any of them. See Hund's rule.

Axial hydrogens   (Section 5.2) The set of six straight up and down hydrogens in chair cyclohexane. Ring flip interconverts these hydrogens with the set of equatorial hydrogens.

Aziridine   (Section 6.7) A saturated three-membered ring containing one nitrogen atom. An azacyclopropane.

Azo compounds   (Section 11.2) Compounds of the structure, R—N=N—R'.

Baeyer–Villiger reaction   (Section 18.12) The reaction of a peroxy acid with a carbonyl group, ultimately giving an ester.

Base pair   (Section 23.5) A hydrogen-bonded pair of bases, always adenine–thymine (A-T) in DNA or adenine–uracil (A-U) in RNA, and cytosine–guanine (C-G) in both DNA and RNA.

Base peak   (Section 15.3) The largest peak in a mass spectrum, to which all other peaks are referred.

Beckmann rearrangement   (Section 18.12) The conversion of an imine into an amide via a 1,2 shift followed by hydrolysis of the resulting cation.

Benzaldehyde   (Section 16.3) The common (and always used) name for the simplest aromatic aldehyde, "benzenecarboxaldehyde."

Benzene   (Section 13.1) The archetypal aromatic compound; a planar, regular hexagon of sp^2 hybridized carbons. The six $2p$ orbitals overlap to form a six-electron cycle above and below the plane of the ring. The molecular orbital system has three fully occupied bonding molecular orbitals and three unoccupied antibonding orbitals.

Benzhydryl group   (Section 13.12) The Ph_2CH group.

Benzoic acid   (Section 13.12) Ph—COOH (benzenecarboxylic acid).

Benzyl   (Section 13.7) The $PhCH_2$ group.

Benzyne   (Section 14.14) 1,2-Dehydrobenzene, C_6H_4.

Binding site   (Section 23.4) The location where a substrate resides in an enzyme.

Birch reduction   (Section 13.11) The conversion of aromatic compounds into 1,4-cyclohexadienes through treatment with sodium in liquid ammonia–ethyl alcohol. Radical anions are the first formed intermediates.

tBoc   (Section 23.4) A protecting group for the amino end of an amino acid that works by transforming the amine into a less basic carbamate.

Boltzmann distribution   (Section 8.4) The range of energies of a set of molecules at a given temperature.

Bond dissociation energy (BDE)   (Section 1.6) The amount of energy that must be applied to break a bond into two neutral species. See Homolytic bond cleavage.

Bonding molecular orbital   (Section 1.5) A molecular orbital that is the result of mixing atomic orbitals in an in-phase fashion. Occupation of a bonding molecular orbital by an electron stabilizes a molecule.

Bredt's rule   (Section 3.7) Bredt noticed that there were no examples of bicyclic molecules with double bonds at the bridgehead position.

Bridged   (Section 5.7) In a bridged bicyclic molecule, two rings share more than two atoms.

Bridgehead position   (Section 3.7) The bridgehead positions are shared by the rings in a bicyclic molecule. In a bicyclic molecule, the three bridges emanate from the bridgehead positions.

Bromonium ion   (Section 10.2) A three-membered ring containing bromine that is formed by the reaction of an alkene with Br_2. The bromine atom in the ring is positively charged.

Brønsted acid   (Section 2.15) A proton donor.

Brønsted base   (Section 2.15) A proton acceptor.

Bullvalene   (Section 20.7) Bullvalene is the only known neutral organic molecule with a fluxional structure. Every carbon of this $(CH)_{10}$ compound is bonded on time average to each of the other nine carbons.

Butyl group   (Section 2.8) The group $CH_3CH_2CH_2CH_2$.

sec-Butyl group   (Section 2.8) The group $CH_3CH_2CH(CH_3)$.

tert-Butyl group   (Section 2.8) The group $(CH_3)_3C$.

Cahn–Ingold–Prelog priority system   (Section 3.4) An arbitrary system for naming stereoisomers. It determines a priority system for ordering groups.

Cannizzaro reaction   (Section 19.14) The redox reaction of an aldehyde containing no α hydrogens with hydroxide ion. Addition of hydroxide to the aldehyde is followed by hydride transfer to another aldehyde. Protonation generates a molecule of the carboxylic acid and the alcohol related to the original aldehyde.

Carbamates   (Section 17.7) Esters of carbamic acid. These molecules do not decarboxylate (lose CO_2) easily.

Carbamic acid   (Section 17.7) A compound of the structure

These acids easily decarboxylate to give amines.

Carbanion   (Section 2.4) A compound containing a negatively charged carbon atom. A carbon-based anion.

Carbene   (Section 10.4) A short-lived neutral intermediate containing a divalent carbon atom. See also Singlet carbene and Triplet carbene.

Carbinolamine   (Section 16.11) The initial intermediate in the reaction between a carbonyl-containing molecule, $R_2C=O$, and an amine. It is analogous to a hemiacetal.

Carbocation   (Section 2.4) The compromise and currently widely used name for a molecule containing a trivalent, positively charged carbon atom.

Carbohydrate   (Section 22.1) A molecule whose formula can be factored into $C_x(H_2O)_y$. A sugar or saccharide.

Carbonyl compound   (Section 10.5) A compound containing a carbon–oxygen double bond.

Carboxy terminus   (Section 23.4) The amino acid at the free carboxylic acid end of a peptide polymer.

Carboxylate anion   (Section 17.2) The resonance-stabilized anion formed on deprotonation of a carboxylic acid.

Carboxylic acid   (Section 10.4) A compound of the structure

Catalyst   (Section 3.19) A catalyst functions to increase the rate of a chemical reaction. It is ultimately unchanged by the reaction and functions not by changing the energy of the starting material or product but by providing a lower energy pathway between them. Thus, it operates to lower the energies of the transition states involved in the reaction.

Cation   (Section 1.2) A positively charged atom or molecule.

Cationic polymerization   (Section 9.8) A reaction in which an initially formed carbocation adds to an alkene that in turn adds to another alkene. Repeated additions can lead to polymer formation.

Cbz   (Section 23.4) A protecting group for the amino end of an amino acid that works by transforming the amine into a less basic carbamate.

Cellulose   (Section 22.6) A polymer of glucose in which C(4) of one glucose is linked in β fashion to C(1) of another.

Chain reaction   (Section 11.1) A cycling reaction in which the species necessary for the first step of the reaction is produced in the last step. This intermediate then recycles and starts the process over again.

Chemical shift (δ)   (Section 15.6) The position, on the ppm scale, of a peak in an NMR spectrum. The chemical shift for ^{1}H and ^{13}C is given relative to a standard, TMS, and is determined by the chemical environment surrounding the nucleus.

Chichibabin reaction   (Section 14.12) Nucleophilic addition of an amide ion to pyridine (or a related heteroaromatic compound) leading to an aminopyridine. A key step involves hydride transfer.

Chiral   (Section 4.1) A chiral molecule is not superimposable on its mirror image.

Chirality   (Section 4.1) The ability of a molecule to exist in two nonsuperimposable mirror-image forms; handedness.

cis   (Section 2.12) Hydrogens "on the same side." Applied to specify stereochemical (spatial) relationships in ring compounds and alkenes.

s-cis   (Section 12.6) The less stable, coiled form of a 1,3-diene.

Claisen condensation   (Section 19.8) A condensation reaction of esters in which an ester enolate adds to the carbonyl group of another ester. The result of this addition–elimination process is a β-keto ester.

Claisen–Schmidt condensation   (Section 19.7) A crossed aldol condensation of an aldehyde without α hydrogens with a ketone that does have at least one α hydrogen.

β Cleavage   (Section 11.2) The fragmentation of a radical into a new radical and an alkene through breaking of the α–β bond.

Codon   (Section 23.5) A three-base sequence in a polynucleotide that directs the addition of a particular amino acid to a growing chain of amino acids. Some codons also give directions: "start assembly" or "stop assembly."

Coenzyme   (Section 16.18) A molecule able to carry out a chemical reaction with another molecule only in cooperation with an enzyme. The enzyme's function is often to bring the substrate and the coenzyme together.

Concerted reaction   (Section 9.9) A single-barrier process. In a concerted reaction, starting material is converted into product with no intermediate structures.

Configurational carbon   (Section 22.2) The stereogenic carbon of a carbohydrate that is furthest from C(1) or the carbonyl carbon. This is the carbon whose configuration determines whether the sugar is of the D or L family.

Conformation   (Section 2.5) The three-dimensional structure of a molecule. Conformations are interconverted by rotations about single bonds.

Conformational analysis   (Section 2.8) The study of the relative energies of conformational isomers.

Conformational enantiomers   (Section 4.7) Enantiomers interconvertible through (generally easy) rotations around bonds within the molecule.

Conformational isomers   (Section 2.5) Molecules that can be interconverted by rotation about one or more single bonds.

Conjugate acid   (Section 6.4) Some atom or molecule plus a proton. Conjugate acids and bases are related by the gain and loss of a proton.

Conjugate base   (Section 6.4) Some molecule less a proton. Conjugate acids and bases are related by the gain and loss of a proton.

Conjugated double bonds   (Section 12.1) Double bonds in a 1,3-relationship are conjugated.

Conrotation   (Section 20.3) In a conrotatory process, the end p orbitals of a polyene rotate in the same sense (both clockwise or both counterclockwise).

Constitutional isomers   (Section 4.11) Molecules of the same formula but with different connectivities among the constituent atoms.

Cope rearrangement   (Section 20.6) This [3,3] sigmatropic shift converts one 1,5-diene into another.

Coupling constant (J)   (Section 15.6) The magnitude (in hertz) of J, the measure of the spin–spin interaction between two nuclei.

Covalent bond   (Section 1.2) A bond formed by the sharing of electrons through the overlap of atomic or molecular orbitals.

Crossed (mixed) aldol condensation (Section 19.7) An aldol condensation between two different carbonyl compounds. This reaction is not very useful unless strategies are employed to limit the number of possible products.

Crossed (mixed) Claisen condensation (Section 19.9) A Claisen condensation between two different esters.

Crown ether (Section 6.10) A cyclic polyether often capable of forming complexes with metal ions. The ease of complexation depends on the size of the ring and the number of heteroatoms in the ring.

Cryptand (Section 6.10) A three-dimensional, bicyclic counterpart of a crown ether. Various heteroatoms (O, N, S) act to complex metal ions that fit into the cavity.

Cumulated alkene (cumulene) (Section 12.1) Any molecule containing at least three consecutive double bonds, $R_2C{=}C{=}C{=}CR_2$. The parent cumulene is butatriene.

Curtius rearrangement (Section 18.14) The thermal or photochemical decomposition of an acyl azide to give an isocyanate.

Cyanogen bromide (BrCN) (Section 23.4) A reagent able to cleave peptide chains after a methionine residue.

Cyanohydrin (Section 16.8) The product of addition of hydrogen cyanide to a carbonyl compound.

$$\underset{NC\qquad OH}{\overset{R\qquad R}{C}}$$

Cycloaddition reaction (Section 20.4) A reaction in which two π systems are converted into a ring. The Diels–Alder reaction and the 2 + 2 reaction of a pair of ethylenes to give a cyclobutane are examples.

Cycloalkanes (Section 2.1) Cyclic alkanes.

Cycloalkenes (Section 3.4) Ring compounds containing a double bond within the ring.

Cycloheptatrienylium ion (Section 13.6) See Tropylium ion.

Cyclopentadienyl anion (Section 13.6) A five-carbon aromatic anion containing 6 π electrons ($4n + 2$, $n = 1$).

Daughter ion (Section 15.3) In MS, an ion formed by the fragmentation of the first-formed parent ion.

Decarboxylation (Section 17.7) The loss of carbon dioxide, a common reaction of 1,1-diacids and β-keto acids.

Decoupling (Section 15.6) The removal of coupling between hydrogens or other nuclei (see Coupling constant) through either chemical exchange or electronic means.

Degenerate reaction (Section 20.1) In a degenerate reaction, the starting material and product have the same structure.

Degree of unsaturation (Ω) (Section 3.15) In a hydrocarbon, this is the total number of π bonds and rings.

Delocalization (Section 1.4) The ability of electrons to move between several atoms.

Delocalization energy (Section 13.5) The energy lowering conferred by the delocalization of electrons. In benzene, this is the amount by which benzene is more stable than the hypothetical 1,3,5-cyclohexatriene containing three localized double bonds. See Resonance energy.

Denaturing (Section 23.4) The destruction of the higher order structures of a protein, sometimes reversible, sometimes not.

Deoxyribonucleic acid (DNA) (Section 23.5) A polymer of nucleotides made up of deoxyribose units connected by phosphoric acid links. Each sugar is attached at C(1′) to one of the bases, A, T, G, or C.

DEPT (distortionless enhancement with polarization transfer) (Section 15.7) A spectroscopic method of determining the number of hydrogens attached to a carbon.

Detergent (Section 17.8) A long-chain alkyl sulfonic acid salt.

Dewar benzene (Section 13.2) The unstable C_6H_6 molecule bicyclo[2.2.0]hexa-2,5-diene.

Dewar forms (Section 13.3) Resonance forms for benzene in which overlap between $2p$ orbitals on two para carbons is emphasized. These forms superficially resemble Dewar benzene (bicyclo[2.2.0]hexa-2,5-diene).

Dextrorotatory (Section 4.4) The rotation of the plane of plane-polarized light in the clockwise direction.

Dial (Section 16.3) A molecule containing two aldehyde groups, a dialdehyde.

Diastereomers (Section 4.8) Stereoisomers that are not mirror images.

Diastereotopic (Section 15.6) Diastereotopic hydrogens (or groups) are different both chemically and spectroscopically under all circumstances.

Diazo compounds (Section 10.4) Compounds of the structure $R_2C{=}N_2$.

Diazo ketone (Section 18.14) A compound of the structure

$$\underset{R\qquad\qquad CHN_2}{\overset{O}{\underset{\|}{C}}}$$

Diazonium ion (Section 14.7) The group N_2^+ as in RN_2^+.

Dicyclohexylcarbodiimide (DCC) (Section 23.4) A dehydrating agent effective in the coupling of amino acids through the formation of amide bonds.

Dieckmann condensation (Section 19.9) An intramolecular, or cyclic, Claisen condensation.

Diels–Alder reaction (Section 12.12) The concerted reaction of an alkene or alkyne with a 1,3-diene to form a six-membered ring.

Dienophile   (Section 12.12) A molecule that reacts with a diene in a Diels–Alder fashion.

Dihedral angle   (Section 2.5) The torsional, or twisting, angle between two bonds. In an X—C—C—X system, the dihedral angle is the angle between the X—C—C and C—C—X planes.

Diol   (Section 6.6) A molecule containing two OH groups. Also called a glycol.

Dione   (Section 16.3) A compound containing two ketone groups, a diketone.

1,3-Dipolar reagents, 1,3-dipoles   (Section 10.5) Molecules for which a good neutral structure cannot be written. Ozone is a typical example. These species undergo addition to π systems to give five-membered rings.

Dipole moment   (Section 1.3) A dipole moment in a molecule results when two opposite charges or partial charges are separated.

Diradical   (Section 10.4) A species containing two unpaired electrons, usually on different atoms.

Disaccharide   (Section 22.6) A molecule that is composed of two monosaccharides. The formula for a disaccharide is $C_{12}H_{22}O_{11}$.

Disproportionation   (Section 11.2) The reaction of a pair of radicals to give a saturated and unsaturated molecule by abstraction of a hydrogen by one radical from the position adjacent to the free electron of the other radical.

Disrotation   (Section 20.3) In a disrotatory process, the end p orbitals of a polyene rotate in opposite senses (one clockwise and the other counterclockwise).

Disulfide bridges   (Section 23.4) The attachment of amino acids through sulfur–sulfur bonds formed from the oxidation of cysteine CH_2SH side chains. Disulfide bridges can be formed within a single peptide or between two peptides.

Dithiane   (Section 19.12) Six-membered ring containing a pair of sulfur atoms, usually in the 1,3 positions.

Double bond   (Section 3.1) Two atoms can be attached by a double bond composed of one σ bond and one π bond.

E1cB Reaction   (Section 7.9) An elimination reaction in which the first step is loss of a proton to give an anion. The anion then internally displaces the leaving group in a second step. The product has a new π bond.

E1 Reaction   (Section 7.8) The unimolecular elimination reaction. The ionization of the starting material is followed by the loss of a proton to base. The product has a new π bond.

E2 Reaction   (Section 7.9) The bimolecular elimination reaction. The proton and leaving group are lost in a single, base-induced step. The product has a new π bond.

Eclipsed ethane   (Section 2.5) The conformation of ethane in which all carbon–hydrogen bonds are as close as possible. This conformation is not an energy minimum, but the top of the barrier separating two molecules of the stable, staggered conformation of ethane.

Edman degradation   (Section 23.4) The phenyl isothiocyanate–induced cleavage of the amino acid at the amino terminus of a peptide. Successive applications of the Edman technique can determine the sequence of a peptide.

Electrocyclic reaction   (Section 20.3) The interconversion of a polyene and a ring compound. The end p orbitals of the polyene rotate so as to form the new σ bond of the ring compound.

Electron   (Section 1.1) A particle of tiny mass (1/1845 of a proton) and a single negative charge.

Electron affinity   (Section 1.2) A measure of the tendency for an atom or molecule to accept an electron.

Electronegativity   (Section 1.3) The tendency for an atom to attract electrons.

Electronic spectroscopy   (Section 12.7) The measurement of the absorption of energy when electromagnetic radiation of the proper energy is provided. An electron is promoted from the HOMO to the LUMO.

Electrophile   (Section 1.7) A lover of electrons, a Lewis acid.

Electrophilic aromatic substitution   (Section 14.4) The classic substitution reaction of aromatic compounds with Lewis acids. A hydrogen attached to the benzene ring is replaced by the Lewis acid and the aromatic ring is retained in the overall reaction.

Electrophoresis   (Section 23.2) A technique for separating amino acids or chains of amino acids that takes advantage of the different charge states of different amino acids (or their polymers) at a given pH.

Elimination reaction   (Section 7.8) A reaction that results in a new π bond.

Enamine   (Section 16.11) The nitrogen analogue of an enol, a vinyl amine. These compounds are nucleophilic and useful in alkylation reactions.

Enantiomers   (Section 4.2) Nonsuperimposable mirror images.

Enantiotopic   (Section 15.6) Enantiotopic hydrogens are chemically and spectroscopically equivalent except in the presence of optically active (single enantiomer) reagents.

Endergonic   (Section 8.2) A reaction in which the products are less stable than the starting materials.

endo   (Section 12.12) Aimed "inside" the cage in a bicyclic molecule. In a Diels–Alder reaction, the endo product generally has the substituents aimed toward the newly produced double bond.

Endothermic reaction   (Section 1.6) A reaction in which the bonds in the products are higher energy than those in the starting materials. In an endothermic reaction the product is less stable than the starting material.

Enol   (Section 10.8) A vinyl alcohol. These compounds usually equilibrate with the more stable keto forms.

Enolate   (Section 19.2) The resonance-stabilized anion formed on treatment of an aldehyde or ketone containing an α hydrogen with base.

Enone   (Section 19.6) A molecule with an alkene and a ketone or aldehyde. Usually the π systems are conjugated.

Enthalpy change ($\Delta H°$)   (Section 1.6) The difference in total bond energies between starting material and product in their standard states.

Entropy change ($\Delta S°$) (Section 8.2) The difference in disorder between the starting material and product in their standard states.

Epimers (Section 22.2) Stereoisomers that differ at a single stereogenic atom.

Episulfonium ion (Section 21.2) A three-membered ring containing a trivalent, positively charged sulfur atom.

Epoxidation (Section 10.4) The reaction between a π bond and a peroxy acid to give an epoxide.

Epoxide (Section 7.10) A saturated three-membered ring containing a single oxygen atom. See Oxirane.

Equatorial hydrogens (Section 5.2) The set of six hydrogens more or less in the plane of the ring in chair cyclohexane. These hydrogens are interconverted with the set of axial hydrogens through ring "flipping" of the chair.

Equilibrium constant (K) (Section 8.2) The equilibrium constant is related to the difference in energy between starting material and products ($\Delta G°$) in the following way: $K = e^{-\Delta G°/RT}$.

Essential amino acid (Section 23.2) Any of the 10 amino acids that cannot be synthesized by humans and must be ingested directly.

Ester hydrolysis (Section 18.8) The conversion of an ester into an acid through treatment with an acid catalyst in excess water. The reverse of Fischer esterification. This reaction also occurs in base, and is called saponification in that case.

Ethene (Section 3.2) The simplest alkene is $H_2C{=}CH_2$. Its common name is ethylene.

Ether (Section 6.8) A compound of the general structure ROR or ROR′.

Ethyl compounds (Section 2.5) Substituted ethanes; $CH_3CH_2{-}X$ compounds.

Ethylene (Section 3.2) The simplest alkene, $H_2C{=}CH_2$. It is more properly known as ethene, a name that is rarely used.

Exergonic (Section 8.2) A reaction in which the products are more stable than the starting materials.

exo (Section 12.12) Aimed "outside" the cage in a bicyclic molecule. In a Diels–Alder reaction, the exo product generally has the substituents aimed away from the newly produced double bond.

Exothermic reaction (Section 1.6) A reaction in which the bonds in the products are lower energy than those in the starting materials. In an exothermic reaction, the product is more stable than the starting meerial.

Extinction coefficient (Section 12.7) The proportionality constant ε in Beer's law, $A = \log I_0/I = \varepsilon lc$.

Fatty acids (Section 17.8) Long-chain carboxylic acids generated by the hydrolysis of fats. Fatty acids are derived from acetic acid and always contain an even number of carbons.

First-order reaction (Section 8.4) A reaction for which the rate depends on the product of a rate constant and the concentration of a single reagent.

First-order spectrum (Section 15.6) An NMR spectrum whose coupling and chemical shifts can be interpreted directly. In a first-order spectrum the coupling between hydrogens follows the $n + 1$ rule.

Fischer esterification (Section 17.7) The conversion of a carboxylic acid into an ester by treatment with an acid catalyst in excess alcohol. The reverse of ester hydrolysis.

Fischer projection (Section 22.2) A schematic stereochemical representation. In sugars, the aldehyde group is placed at the top and the primary alcohol at the bottom. Horizontal bonds are taken as coming toward the viewer and vertical bonds as retreating.

Fluxional structure (Section 20.7) In a molecule with a fluxional structure, a given atom does not have fixed nearest neighbors. Instead, the framework atoms move about over time, each being bonded on time average to all the others.

Force constant (Section 15.4) A property of a bond related to the bond strength: to the stiffness of the bond. Bonds with high force constants absorb at high frequency in the IR.

Formal charge (Section 1.3) The charge an atom has when the number of electrons of that atom does not match the number of protons for that atom.

Fragmentation pattern (Section 15.3) The characteristic spectrum of ions formed by decomposition of a parent ion produced in a mass spectrometer when a molecule is bombarded by high-energy electrons.

Free radical (Section 6.3) A neutral molecule containing an odd, unpaired electron. Also simply called radical.

Friedel–Crafts acylation (Section 14.6) The electrophilic substitution of aromatic molecules with acyl chlorides facilitated by strong Lewis acids, usually $AlCl_3$.

Friedel–Crafts alkylation (Section 14.5) The electrophilic substitution of aromatic molecules with alkyl chlorides catalyzed by strong Lewis acids, usually $AlCl_3$.

Frost circle (Section 13.6) A device used to find the relative energies of the molecular orbitals of planar, cyclic, fully conjugated molecules. A polygon corresponding to the ring size of the molecule is inscribed in a circle, vertex down. The intersections of the polygon with the circle give the relative positions of the molecular orbitals.

Functional group (Section 1.4) An atom or group of atoms that generally reacts the same way no matter what molecule it is in.

Furan (Section 13.9) An aromatic five-membered ring compound containing four CH units and one O atom.

Furanose (Section 22.2) A sugar containing a five-membered cyclic ether.

Furanoside (Section 22.4) A furanose in which the anomeric OH has been converted into an acetal.

Fused (Section 5.7) Two rings sharing only two carbons.

Gabriel synthesis (Section 23.2) A synthesis of amino acids that uses phthalimide as a source of the amine nitrogen. Overalkylation is avoided by decreasing the nucleophilicity of the nitrogen in this way.

Gas chromatogaphy (GC) (Section 15.2) A method of separation in which molecules are forced to equilibrate between the moving gas phase and a stationary phase packed in a column. The less easily a molecule is adsorbed in the stationary phase, the faster it moves through the column.

GC/IR (Section 15.2) The combination of gas chromatography and infrared spectroscopy in which the molecules separated by a gas chromatograph are led directly into an infrared spectrometer for analysis.

GC/MS (Section 15.2) The combination of gas chromatography and mass spectrometry in which the molecules separated by a gas chromatograph are led directly into a mass spectrometer for analysis.

Gel-filtration chromatography (Section 23.4) A chromatographic technique that relies on polymeric beads containing molecule-sized holes. Molecules that fit easily into the holes pass more slowly down the column than larger molecules, which fit less well into the holes.

***Gem*-diol** (Section 16.6) A diol with both OH groups on the same carbon.

Geminal (Section 10.6) A disubstituted compound with both substituents on the same carbon.

Gibbs free energy change (ΔG°) (Section 8.2) The difference in free energy during a reaction. The parameter $\Delta G°$ is composed of an enthalpy ($\Delta H°$) term and an entropy ($\Delta S°$) term. $\Delta G° = \Delta H° - T\Delta S°$.

Glycol (Section 6.6) A dialcohol. See Diol.

Glycolysis (Section 22.3) The process of breaking carbohydrates into smaller carbohydrates.

Glycoside (Section 22.4) A sugar in which the anomeric OH at C(1) has been converted into an OR group.

Grignard reagent (Section 6.3) A strongly basic organometallic reagent formed from a halide and magnesium in an ether solvent. An important and characteristic reaction is the addition to carbonyl groups. The simplest formulation is RMgX.

Haloform (Section 19.4) A compound of the structure HCX$_3$, where X = F, Cl, Br, or I.

Haloform reaction (Section 19.4) The conversion of a methyl ketone into a molecule of a carboxylic acid and a molecule of haloform. The trihalo carbonyl compound is formed, base adds to the carbonyl group, and the trihalomethyl anion ($^-$:CX$_3$) is eliminated. The reaction works for X = Cl, Br, or I.

Halohydrin (Section 10.2) A molecule with a halogen and a hydroxy group. Halohydrins are formed when an alkene reacts with X$_2$ in H$_2$O.

Hammond postulate (Section 8.8) The transition state for an endothermic reaction will resemble the product. It can be equivalently stated as, The transition state for an exothermic reaction will resemble the starting material.

Haworth form (Section 22.2) The representation of sugars in which they are shown as planar rings with the ring perpendicular to the plane of the paper.

Heat of formation (ΔH$_f^\circ$) (Section 3.6) The heat evolved or required for the formation of a molecule from its constituent elements in their standard states. The more negative the heat of formation, the more stable the molecule.

Heisenberg uncertainty principle (Section 1.1) For an electron, the uncertainty in position times the uncertainty in momentum (or speed) is a constant. We cannot know the exact position and momentum (speed) of an electron at the same time.

α-Helix (Section 23.4) A right-handed coiled form adopted by many proteins in their secondary structures.

Hell–Volhard–Zelinsky (HVZ) reaction (Section 19.4) The conversion of a carboxylic acid into either the α-bromo acid or the α-bromo acid bromide through reaction with PBr$_3$/Br$_2$.

Hemiacetal (Section 16.9) The initial product when an alcohol adds to an aldehyde or ketone.

Heteroaromatic compound (Section 13.9) See Heterobenzene.

Heterobenzene (Section 13.6) A benzene ring in which one (or more) ring carbons is replaced with another heavy (nonhydrogen) atom.

Heterogeneous catalysis (Section 10.2) A catalytic process in which the catalyst is insoluble.

Heterolytic bond cleavage (Section 1.6) The breaking of a bond to produce a pair of oppositely charged ions.

Hexose (Section 22.2) A six-carbon sugar.

High-performance liquid chromatography (HPLC) (Section 15.2) An especially effective version of column chromatography in which the stationary phase consists of many tiny spheres that provide an immense surface area for absorption.

Hofmann elimination (Section 7.9) The formation of the less substituted alkene in an elimination reaction.

Hofmann rearrangement (Section 18.14) The formation of amines through the treatment of amides with bromine and base. An intermediate isocyanate is hydrolyzed to a carbamic acid that decarboxylates.

HOMO (Section 3.17) Highest occupied molecular orbital.

Homogeneous catalysis (Section 10.2) A catalytic process in which the catalyst is soluble.

Homolytic bond cleavage (Section 1.6) The breaking of a bond to form two neutral species.

Homotopic (Section 15.6) Homotopic hydrogens are identical, both chemically and spectroscopically, under all circumstances.

Hückel's rule   (Section 13.6) All planar, cyclic, fully conjugated molecules with $4n + 2$ π electrons will be aromatic (especially stable). The rule works because such molecules will have molecular orbital systems in which all bonding molecular orbitals are full and all antibonding moleclular orbitals are empty. In such molecules, the bonding degenerate molecular orbitals are filled.

Hund's rule   (Section 1.2) For a given electronic configuration, the state with the greatest number of parallel spins is the lowest in energy. That is,

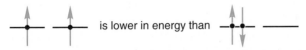

Hunsdiecker reaction   (Section 17.7) The conversion of silver salts of carboxylic acids into alkyl halides, usually bromides.

Hybridization   (Section 2.2) A mathematical model in which atomic orbital wave functions are combined to produce new, combination, or hybrid orbitals. The new orbitals are made up of fractions of the pure atomic orbital wave functions. Thus, an sp^3 hybrid is made of three parts p wave function and one part s wave function.

Hybrid orbitals   (Section 2.2) Orbitals that result from hybridization.

Hydrate   (Section 16.6) The product of the reaction of a carbonyl compound with water.

Hydration   (Section 9.7) The addition of water to a molecule. This reaction is generally acid catalyzed.

Hydride   (Section 2.4) A negatively charged hydrogen ion ($H:^-$) bearing a pair of electrons.

Hydride shift   (Section 9.9) The migration of hydrogen with a pair of electrons ($H:^-$).

Hydroboration   (Section 9.10) The addition of BH_3 (in equilibrium with the dimer B_2H_6) to π systems to form alkylboranes. These alkylboranes can be further converted into alcohols. Addition to an unsymmetrical alkene proceeds so as to give the less substituted alcohol.

Hydrocarbon   (Section 2.1) A molecule containing only carbon and hydrogen.

Hydrocarbon cracking   (Section 11.2) The thermal treatment of high molecular weight hydrocarbons to give lower molecular weight fragments. Bonds are broken to give radicals that abstract hydrogen to give alkanes, undergo β-cleavage, and disproportionate to give alkenes and alkanes.

Hydrogenation   (Section 10.2) Addition of hydrogen (H_2) to the π bond of an alkene to give an alkane. A soluble or insoluble metallic catalyst is necessary. Alkynes also undergo hydrogenation to give alkanes, or under special conditions, alkenes.

Hydrogen bonding   (Section 6.4) A low-energy bond between a pair of electrons, usually on oxygen, and a hydrogen.

Hydrophilic   (Section 17.8) "Water-loving," hence a polar group soluble in water.

Hydrophobic   (Section 17.8) "Water-hating," hence a nonpolar group insoluble in water.

Hyperconjugation   (Section 9.5) The stabilization of carbocations through the overlap of a filled σ orbital with an empty $2p$ orbital on carbon.

Imide   (Section 18.2) A compound containing the structure $O{=}C{-}NH{-}C{=}O$, such as

Imine   (Section 16.11) The nitrogen analogue of a ketone or aldehyde. See Schiff base.

Iminium ion   (Section 16.11) A compound of the structure

Inductive effects   (Section 9.6) Electronic effects transmitted through σ bonds. All bonds between different atoms are polar, and thus many molecules contain dipoles. These dipoles can affect reactions through induction.

Infrared (IR) spectroscopy   (Section 15.4) In IR, absorption of IR energy causes bonds to vibrate and rotate. The characteristic vibrational frequencies can be used to determine the functional groups present in a molecule.

Inhibitors   (Section 11.4) Species that can react with a radical, thus destroying it. Inhibitors interrupt chain reactions.

Initiation   (Section 11.4) The first step in a radical chain reaction in which a free radical is produced that can start the chain-carrying or propagation steps.

Integral   (Section 15.6) The determination of the relative numbers of hydrogens corresponding to the signals in an NMR spectrum.

Inversion of configuration   (Section 7.4) The change in the handedness of a molecule during certain reactions. Generally, inversion of stereochemistry means that an (R) starting material would be transformed into an (S) product.

Ion (Section 1.2) A charged atom or molecule.

Ion-exchange chromatography (Section 23.4) A separation technique that relies on differences in the affinity of molecules for a charged substrate.

Ionic bond (Section 1.2) The electrostatic attraction between a positively charged atom or group of atoms and a negatively charged atom or group of atoms.

Ionization potential (Section 1.2) A measure of the tendency of an atom or molecule to lose an electron.

Ipso attack (Section 14.12) Addition to an aromatic ring at a position already occupied by a nonhydrogen substituent.

Isobutyl group (Section 2.8) The $(CH_3)_2CHCH_2$ group.

Isocyanate (Section 18.14) A compound of the structure

$$R-N=C=O$$

Isoelectric point (p*I*) (Section 23.2) The pH at which the amino acid has no net charge.

Isomers (Section 2.5) Molecules of the same formula but different structures.

Isoprene (Section 12.13) 2-Methyl-1,3-butadiene.

Isoprene rule (Section 12.13) Most terpenes are composed of isoprene units combined in "head-to-tail" fashion.

Isopropyl group (Section 2.8) The $(CH_3)_2CH$ group.

Isotope effect (Section 21.4) The effect on the rate of a reaction of the replacement of an atom with an isotope. The replacement of hydrogen with deuterium is especially common.

Karplus curve (Section 15.6) The dependence between the coupling constant of two hydrogens on adjacent atoms and the dihedral angle between the carbon–hydrogen bonds.

Kekulé forms (Section 13.3) Resonance forms for benzene in which overlap between adjacent carbons is emphasized. These forms superficially resemble 1,3,5-cyclohexatriene.

Ketene (Section 12.3) A compound of the structure

$$R_2C=C=O$$

Keto–enol tautomerization (Section 19.2) The equilibrium between a carbonyl compound and the enol that can be formed if the carbonyl compound has an α hydrogen. The equilibrium almost always favors the carbonyl species.

β-Keto ester or acetoacetate synthesis (Section 19.5) The α hydrogen between two carbonyl groups can be deprotonated with weak base. The anion formed will react with an electrophile in this synthesis.

Ketone (Section 10.5) A compound containing a carbon–oxygen double bond in which the carbon is attached to two other carbons.

Ketose (Section 22.2) A sugar containing not the usual aldehyde group, but a ketone.

Kiliani–Fischer synthesis (Section 22.3) A method of lengthening the chain of an aldose by one carbon atom. A pair of sugars, epimeric at the new C(1), is produced.

Kinetic control (Section 8.8) A reaction in which the product distribution is determined by the heights of the different transition states leading to products.

Kinetic enolate (Section 19.7) The most easily formed enolate. It may or may not be the same as the most stable possible enolate, the thermodynamic enolate.

Kinetic resolution (Section 23.2) A technique for separating a pair of enantiomers based on the selective transformation of one of them, often by an enzyme.

Kinetics (Section 8.8) The study of the rates of reactions.

Knoevenagel condensation (Section 19.6) Any of a number of condensation reactions related to the crossed aldol condensation. A stabilized anion, often an enolate, is first formed and then adds to the carbonyl group of another molecule. Dehydration usually leads to the formation of the final product.

Kolbe electrolysis (Section 17.7) The electrochemical conversion of carboxylate anions into hydrocarbons. The carboxyl radical produced loses CO_2 to give an alkyl radical that dimerizes.

Lactam (Section 17.7) A cyclic amide.

Lactone (Section 17.7) A cyclic ester.

Leaving group (Section 7.3) The departing group in a substitution or elimination reaction.

Le Châtelier's principle (Section 8.2) A system at equilibrium adjusts so as to relieve any stress on it.

Levorotatory (Section 4.4) The rotation of the plane of plane-polarized light in the counterclockwise direction.

Lewis acid (Section 1.7) Any species that reacts with a Lewis base. An electrophile.

Lewis base (Section 1.7) Any species with a reactive pair of electrons. A nucleophile.

Lewis structure (Section 1.3) In a Lewis structure, every electron (except the $1s$ electrons for atoms other than hydrogen or helium) is shown as a dot. In slightly more abstract Lewis structures, electrons in bonds are shown as lines connecting atoms.

Lithium diisopropylamide (LDA) (Section 19.2) An effective base for alkylation reactions of carbonyl compounds. This compound is a strong base but a poor nucleophile, and thus is effective at removing α hydrogens but not at promoting additions to carbonyl compounds.

Lobry de Bruijn–Alberda van Ekenstein reaction (Section 22.4) The base-catalyzed interconversion of aldo and keto sugars. The key intermediate is the double enol formed by protonation of an enolate.

Lone-pair electrons (Section 1.3) Electrons in an orbital that is not involved in binding atoms. See Nonbonding electrons.

Long-range coupling (Section 15.6) Any coupling between nuclei separated by more than three bonds. It is usually observed only in high-field ^1H NMR spectroscopy.

LUMO (Section 3.17) Lowest unoccupied molecular orbital.

M + 1 peak (Section 15.3) The signal in a mass spectrum that comes from a +1 isotope (usually due to ^{13}C) of the molecular ion.

Magid's second rule (Section 19.11) "Always try the Michael reaction first."

Magid's third rule (Section 19.14) "In times of desperation and/or despair, when all attempts at solution seem to have failed, try a hydride shift."

Malonic ester synthesis (Section 19.5) The alkylation of malonic esters to produce substituted diesters, followed by hydrolysis and decarboxylation of the malonic acids to give substituted acetic acids.

Mannich reaction (Section 19.13) The reaction between a ketone or aldehyde with an amine to give an imine, which then reacts with a nucleophile.

Markovnikov's rule (Section 9.5) In additions of Lewis acids to alkenes, the Lewis acid will add to the less substituted end of the alkene. The rule works because additions of Lewis acids in that way produce the more substituted, more stable carbocation.

Mass spectrometry (MS) (Section 15.3) The analysis of the ions formed by bombardment of a molecule with high-energy electrons. High-resolution MS can give the molecular formula of an ion, and an analysis of the fragmentation pattern can give information about the structure.

Meerwein–Ponndorf–Verley–Oppenauer (MPVO) equilibration (Section 19.14) A method of oxidizing alcohols to carbonyl compounds, or, equivalently, of reducing carbonyls to alcohols, that uses an aluminum atom to clamp the partners in the reaction together. A hydride shift is involved in the crucial step.

Meisenheimer complex (Section 14.12) The intermediate in nucleophilic aromatic substitution formed by the addition of a nucleophile to an aromatic compound activated by electron-withdrawing groups, often NO_2.

Mercaptan (Section 6.9) A thiol, RSH. The sulfur counterpart of an alcohol.

Mercaptide (Section 6.9) The sulfur counterpart of an alkoxide, RS^-.

Meso compound (Section 4.8) An achiral compound containing stereogenic atoms.

Messenger RNA (mRNA) (Section 23.5) A polynucleotide produced from DNA whose base sequence codes for amino acid assembly.

Meta (Section 13.7) 1,3-Substitution on a benzene ring.

Methane (Section 2.1) The simplest stable hydrocarbon, CH_4.

Methine group (Section 2.8) The CH group.

Methyl anion (Section 2.4) $^-:CH_3$.

Methyl cation (Section 2.4) $^+CH_3$.

Methyl compounds (Section 2.3) Substituted methanes, CH_3—X compounds.

Methylene group (Section 2.7) The CH_2 group.

Methyl radical (Section 2.4) $\cdot CH_3$.

Micelle (Section 17.8) An aggregated group of fatty acid salts (or other molecules) in which a hydrophobic center consisting of the hydrocarbon chains is protected from an aqueous environment by a spherical skin of polar hydrophilic groups.

Michael reaction (Section 19.6) The addition of a nucleophile, often an enolate, to the β position of a double bond of an α,β-unsaturated carbonyl group.

Microscopic reversibility (Section 8.6) The notion that the lowest-energy path for a reaction in one direction will also be the lowest-energy path in the other direction.

Molecular ion (Section 15.3) The charged species obtained by removing an electron from a molecule.

Molecular orbital (Section 1.1) An orbital not restricted to the region of space surrounding an atom, but extending over several atoms in a molecule. Molecular orbitals are formed through the overlap of atomic or molecular orbitals. Molecular orbitals can be bonding, nonbonding, or antibonding.

Molozonide (Section 10.5) The initial product of the reaction of ozone and an alkene. This molecule contains a five-membered ring with three oxygen atoms in a row. See Primary ozonide.

Monosaccharide (Section 22.6) A simple six-carbon sugar.

Mutarotation (Section 22.4) The interconversion of anomeric sugars in which an equilibrium mixture of α and β forms is reached.

n + 1 rule (Section 15.6) The number of lines for a hydrogen will be $n + 1$, where n is the number of equivalent adjacent hydrogens.

Natural product (Section 2.12) A product formed in Nature, which does not include compounds made in the lab.

Neighboring group effect (Section 21.1) A general term for the influence of an internal nucleophile (very broadly defined) on the rate of a reaction, or on the structures of the products produced.

Newman projection (Section 2.5) A convention used to draw what one would see if one could look down a bond. The groups attached to the front atom are drawn in as three lines. The back atom is represented as a circle to which its attached bonds are drawn. Newman projections are enormously useful in seeing spatial (stereochemical) relationships in molecules.

Ninhydrin (Section 23.3) The hydrate of indan-1,2,3-trione, a molecule that reacts with amino acids to give the purple dye used in quantitative analysis of amino acids.

Nitrene (Section 18.14) A reactive intermediate containing a neutral, monovalent nitrogen atom. The nitrogen counterpart of a carbene.

Nitrile (Section 18.2) A compound of the structure RCN, also commonly called a cyanide.

NMR spectrum (Section 2.14) The result of nuclear magnetic resonance spectroscopy. A spectrum has ppm as the x-axis and signal intensity as the y-axis.

Node (Section 1.2) The region of zero electron density separating regions of opposite sign in an orbital. At a node the sign of the wave function is zero.

Nonbonding electrons (Section 1.3) Electrons in an orbital that is not involved in binding atoms. See Lone-pair electrons.

Nonbonding orbital (Section 1.5) An orbital that is neither bonding nor antibonding. An electron in a nonbonding orbital neither helps to hold the molecule together nor tears it apart.

Nonreducing sugar   (Section 22.6) A carbohydrate containing no amount of an oxidizable aldehyde group.

Norbornyl system   (Section 21.3) The bicyclo[2.2.1]heptyl system.

Nuclear magnetic resonance (NMR) spectroscopy   (Section 2.14) Spectroscopy that detects the absorption of energy as the lower-energy nuclear spin state in which the nuclear spin is aligned with an applied magnetic field flips to the higher-energy spin state in which it is aligned against the field. See also Chemical shift (δ), Coupling constant (J), and Integral.

Nucleic acid   (Section 23.5) The polynucleotides DNA and RNA.

Nucleophile   (Section 1.7) A Lewis base. A strong nucleophile has a high affinity for a carbon $2p$ orbital.

Nucleophilic aromatic substitution   (Section 14.12) The substitution reaction that occurs when a nucleophile replaces a group on an aromatic ring, usually a halogen.

Nucleophilicity   (Section 7.4) The strength of a nucleophile.

Nucleoside   (Section 23.5) A sugar, either ribose or deoxyribose, bonded to a heterocyclic base at its 1′ position.

Nucleotide   (Section 23.5) A phosphorylated nucleoside; one of the monomers of which DNA and RNA are composed.

Nucleus   (Section 1.1) The positively charged core of an atom containing the protons and neutrons.

Octet rule   (Section 1.2) The notion that special stability attends the filling of the $2s$ and $2p$ atomic orbitals to achieve the electronic configuration of neon, a noble gas.

Off-resonance decoupling   (Section 15.7) A technique used in ^{13}C NMR in which coupling between ^{13}C and ^{1}H is restricted to the hydrogens directly attached to the carbon. The number of directly attached hydrogens can be determined from the multiplicity of the observed signal for the carbon.

Olefins   (Section 3.8) Hydrocarbons of the general formula, C_nH_{2n}. These molecules are systematically called alkenes, and they contain carbon–carbon double bonds.

Optical activity   (Section 4.4) The rotation by a molecule of the plane of plane-polarized light.

Orbital   (Section 1.1) A three-dimensional representation of the solution to the Schrödinger equation. See Wave function (ψ).

Orbital interaction diagram   (Section 1.5) A way of analyzing the interaction between atoms that leads to bonding.

Organocuprate   (Section 10.4) An organometallic reagent (R_2CuLi) notable for its ability to react with primary or secondary halides (R′—X) to generate hydrocarbons, R—R′, and to add to the β position of α,β-unsaturated carbonyl compounds.

Organolithium reagent   (Section 6.3) A strongly basic organometallic reagent, R—Li, formed from a halide and lithium. A characteristic reaction is addition to carbonyl groups.

Organometallic reagent   (Section 6.3) Molecule that contains both carbon and a metal. Usually carbon is at least partially covalently bonded to the metal. Examples are Grignard reagents, organolithium reagents, and lithium organocuprates.

Ortho   (Section 13.7) 1,2-Disubstituted on a benzene ring.

Ortho esters   (Section 17.7) Compounds of the structure

Orthogonal orbitals   (Section 1.5) Two noninteracting orbitals. The bonding interactions are exactly balanced by antibonding interactions.

Osazone   (Section 22.4) A 1,2-phenylhydrazone formed by treatment of a sugar with three equivalents of phenylhydrazine.

Oxaphosphetane   (Section 16.17) The four-membered intermediate in the Wittig reaction that contains two carbons, a phosphorus, and an oxygen.

Oxirane   (Section 7.10) A three-membered ring containing oxygen, also called an epoxide or oxacyclopropane.

Oxonium ion   (Section 3.19) A molecule containing a trivalent, positively charged, oxygen atom, R_3O^+.

Oxymercuration   (Section 10.3) The mercury-catalyzed conversion of alkenes into alcohols. Addition is in the Markovnikov sense, and there are no rearrangements. A three-membered ring containing mercury is an intermediate in the reaction. Alkynes also undergo oxymercuration to give enols that are rapidly converted into carbonyl compounds under the reaction conditions.

Ozonide   (Section 10.5) The product of rearrangement of the primary ozonide formed on reaction of ozone with an alkene.

Ozonolysis   (Section 10.5) The reaction of ozone with π systems. The intermediate products are ozonides that can be transformed into carbonyl-containing compounds of various kinds.

Paired spin   (Section 1.2) Two electrons with opposite spins have paired spins.

Para   (Section 13.7) 1,4-Disubstituted on a benzene ring.

[n]Paracyclophane   (Section 14.3) A benzene ring bridged 1,4 (para) with a chain of atoms.

Parallel spin   (Section 1.2) Two electrons with the same spin have parallel spins. These electrons can not occupy the same orbital.

Parent ion (p)   (Section 15.3) An ion formed in the mass spectrometer through loss of a single electron. It can undergo fragmentation reactions before detection.

Pauli principle   (Section 1.2) No two electrons in an atom or molecule may have the same values of the four quantum numbers.

Pentose   (Section 22.2) A five-carbon sugar.

Peptide   (Section 6.1) A polyamino acid in which the constituent amino acids are linked through amide bonds. They are distinguished from proteins only by the length of the polymer.

Peptide bond   (Section 23.2) The amide bond that links amino acids.

Pericyclic reaction   (Section 20.1) A reaction in which the maintenance of bonding overlap between the lobes of the orbitals involved in bond making and breaking is controlling. All single-barrier (concerted) reactions can be regarded as pericyclic reactions.

Phenonium ion   (Section 21.3) The benzenonium ion produced through intramolecular displacement of a leaving group by the π system (not the σ system) of an aromatic ring.

Phenyl (Ph)   (Section 13.7) A common and systematic name for the benzene ring as substituent, C_6H_5—X.

Pi (π) orbitals   (Section 3.2) Molecular orbitals made from the overlap of p orbitals.

pK_a   (Section 6.4) The negative logarithm of the acidity constant. A strong acid has a low pK_a, a weak acid has a high pK_a.

Plane-polarized light   (Section 4.4) Light that has been passed through a polarizing filter.

β-Pleated sheet   (Section 23.4) One of the common forms of secondary peptide structure in which hydrogen bonding holds chains of amino acids roughly in parallel lines.

Polar covalent bond   (Section 1.3) Any shared electron bond between two different atoms must be polar. Two nonidentical atoms must have different electronegativities and will attract the shared electrons to different extents, creating a dipole.

Polarimeter   (Section 4.5) A device for measuring the amount of rotation of the plane of plane-polarized light.

Polyenes   (Section 3.4) Any molecule containing multiple double bonds.

Polynuclear aromatic compound   (Section 13.10) An aromatic molecule composed of two or more fused aromatic rings.

Polysaccharide   (Section 22.6) A carbohydrate that is composed of multiple saccharides. The complex carbohydrates are polysaccharides.

α Position   (Section 19.2) The position next to a functional group. For example, the α position in a ketone is the position adjacent to the carbon–oxygen double bond.

ppm Scale   (Section 15.5) The chemical shift scale commonly used in NMR spectroscopy in which the positions of absorptions are quoted as parts per million (ppm) of applied magnetic field. On the ppm scale, the chemical shift is independent of the magnetic field.

Primary amine   (Section 6.7) An amine bearing only one R group and two hydrogens.

Primary carbon   (Section 2.8) A carbon atom attached to one other carbon.

Primary ozonide   (Section 10.5) The initial product of the reaction of ozone and an alkene. This molecule contains a five-membered ring with three oxygen atoms in a row. See Molozonide.

Primary structure   (Section 23.4) The sequence of amino acids in a protein.

Product-determining step   (Section 7.6) The step in a multistep reaction that determines the structure or structures of the products.

Propagation   (Section 11.4) The product-producing steps of a chain reaction. In the last propagation step, a molecule of product is formed along with a new radical that can carry the chain.

Propargyl   (Section 3.11) The common name for the $HC{\equiv}C$—CH_2 group.

Propyl group   (Section 2.7) The $CH_3CH_2CH_2$ group.

Protecting group   (Section 16.10) Sensitive functional groups in a molecule can be protected by a reaction that converts them into a less reactive group, called a protecting group. A protecting group must be removable to regenerate the original functionality.

Protein   (Section 6.1) A polyamino acid in which the constituent amino acids are linked through amide bonds. They are distinguished from peptides only through the length of the polymer.

Protic solvent   (Section 6.5) A solvent containing a hydrogen easily lost as a proton. Examples are water and most alcohols.

Pseudo-first-order reaction   (Section 8.4) A bimolecular reaction in which the concentration of one reagent (usually the solvent) does not change appreciably.

Pyranose   (Section 22.2) A sugar containing a six-membered cyclic ether.

Pyranoside   (Section 22.4) A pyranose in which the anomeric OH at C(1) has been converted into an OR group.

Pyridine   (Section 13.9) Any compound containing the following "azabenzene" ring system:

Pyrolysis   (Section 11.2) The process of inducing chemical change by supplying heat energy. See Thermolysis.

Pyrrole   (Section 13.6) Any compound containing the following ring structure:

Quantum numbers   (Section 1.3) These numbers evolve from the Schrödinger equation and characterize the various solutions to it. They may have only certain values, and these values determine the distance of an electron from the nucleus (n), the shape (l) and orientation (m_l) of the orbitals, and the electron spin (s).

Quaternary carbon   (Section 2.8) A carbon attached to four other carbons.

Quaternary structure   (Section 23.4) The structure of a self-assembled aggregate of two or more protein units.

Racemic mixture (racemate)   (Section 4.4) A mixture containing equal amounts of two enantiomeric forms of a chiral molecule.

Radical anion   (Section 10.11) A negatively charged molecule containing both a pair of electrons and an odd, unpaired electron.

Radical cation (Section 15.3) A species that is positively charged yet contains a single unpaired electron. These are commonly formed by ejection of an electron when a molecule is bombarded with high-energy electrons in a mass spectrometer.

Random coil (Section 23.4) Disordered portions of a chain of amino acids.

Raney nickel (Section 6.9) A good reducing agent composed of finely divided nickel on which hydrogen has been adsorbed.

Rate constant (k) (Section 8.5) A fundamental property of a reaction that depends on the temperature, pressure, and solvent, but not on the concentrations of the reactants.

Rate-determining step (Section 7.6) The step in a reaction with the highest-energy transition state.

Reaction mechanism (Section 7.2) Loosely speaking, How does the reaction occur? How do the reactants come together? Are there any intermediates? What do the transition states look like? More precisely, a determination in terms of structure and energy of the stable molecules, reaction intermediates, and transition states involved in the reaction, along with a consideration of how the energy changes as the reaction progresses.

Reactive intermediates (Section 2.4) Molecules of great instability, and hence fleeting existence under normal conditions. Most carbon-centered anions, cations, and radicals are examples.

Rearrangement (Section 9.9) The migration of an atom or group of atoms from one place to another in a molecule. Rearrangements are exceptionally common in reactions involving carbocations.

Reducing sugar (Section 22.2) A sugar containing some amount of an oxidizable free aldehyde group.

Regiochemistry (Section 3.18) The orientation of a reaction taking place on an unsymmetrical substrate.

Regioselective reaction (Section 7.9) If a reaction may produce one or more isomers and one predominates, the reaction is called regioselective.

Resolution (Section 4.9) The separation of a racemic mixture into its constituent enantiomeric molecules.

Resonance arrow (Section 1.4) A specific way of indicating that two structures are resonance forms.

Resonance energy (Section 13.5) The energy lowering conferred by the delocalization of electrons. In benzene, this is the amount by which benzene is more stable than the hypothetical 1,3,5-cyclohexatriene containing three localized double bonds. See Delocalization energy.

Resonance forms (Section 1.4) Many molecules cannot be represented adequately by a single Lewis form. Instead, two or more different electronic representations must often be combined to give a good description of the molecule. These different representations are called resonance forms.

Retention of configuration (Section 7.4) The preservation of the handedness of a molecule in a reaction. Generally, retention of configuration means that an (R) starting material would be transformed into an (R) product.

Retrosynthetic analysis (Section 16.15) A way of figuring out how a compound might be synthesized by looking at what compound was its immediate precursor. This process is applied step-by-step until readily available starting materials are reached.

Ribonucleic acid (RNA) (Section 23.5) A polymer of nucleotides made up of ribose units connected by phosphoric acid links. Each sugar is attached at C(1′) to one of the bases, A, U, G, or C.

Robinson annulation (Section 19.11) A classic method for construction of six-membered rings using a ketone and methyl vinyl ketone. It involves a two-step sequence of a Michael reaction between the enolate of the original ketone and methyl vinyl ketone, followed by an intramolecular aldol condensation.

Rosenmund reduction (Section 18.6) The reduction of acid chlorides to aldehydes using a poisoned catalyst and hydrogen.

Ruff degradation (Section 22.3) A method for shortening the carbon backbone of a sugar by one carbon. The aldehyde carbon at C(1) is lost and a new aldehyde created at the old C(2).

Saccharide (Section 22.1) A molecule whose formula can be factored into $C_x(H_2O)_y$. A sugar or carbohydrate.

Sandmeyer reaction (Section 14.7) The reaction of an aromatic diazonium ion with cuprous salts to form substituted aromatic compounds.

Sanger degradation (Section 23.4) A method for determining the amino acid at the amino terminus of a chain using 2,4-dinitrofluorobenzene. Unfortunately, the Sanger degradation hydrolyzes, and thus destroys, the entire peptide.

Saponification (Section 17.8) Base-induced hydrolysis of an ester, usually a fatty acid.

Saturated hydrocarbons (Section 2.12) Alkanes of the molecular formula C_nH_{2n+2}.

Saytzeff elimination (Section 7.8) Formation of the more substituted alkene in an elimination reaction.

Schiff base (Section 16.11) The nitrogen analogue of a ketone or aldehyde. See Imine.

Secondary amine (Section 6.7) An amine bearing one hydrogen and two R groups.

Secondary carbon (Section 2.8) A carbon attached to two other carbons.

Secondary structure (Section 23.4) Ordered regions of a protein chain. The two most common types of secondary structure are the α-helix and the β-pleated sheet.

Second-order reaction (Section 8.4) A reaction for which the rate depends on the product of a rate constant and the concentrations of two reagents.

Side chain (Section 23.2) The group attached to the α carbon of an amino acid.

Sigma bond (Section 2.2) Any bond with cylindrical symmetry.

Sigmatropic shift (Section 20.5) The migration of an atom or group, under orbital symmetry control, along a π system.

Silyl ether (Section 16.10) A molecule of the general form R—O—SiR$_3$.

Singlet carbene (Section 10.4) A singlet carbene contains only paired electrons. In a singlet carbene, the two nonbonding electrons have opposite spins and occupy the same orbital.

S$_N$1 Reaction (Section 7.6) Substitution, nucleophilic, unimolecular. An initial ionization is followed by attack of the nucleophilic solvent.

S$_N$2 Reaction (Section 7.4) Substitution, nucleophilic, bimolecular. In this reaction, the nucleophile dispaces the leaving group by attack from the rear. The stereochemistry of the starting material is inverted.

Soap (Section 17.8) The sodium salt of a fatty acid.

Solvated electron (Section 10.11) The species formed when sodium is dissolved in ammonia. The product is a sodium ion and an electron surrounded by ammonia molecules. This electron can add to alkynes (and some other π systems) in a reduction step.

Solvation (Section 6.4) The stabilizing effects of a solvent, typically the stabilization of ions by a polar solvent.

Solvolysis reaction (Section 7.6) In such S_N1 and S_N2 reactions, the solvent acts as the nucleophile.

Specific rotation (Section 4.5) The rotation of plane-polarized light induced by a solution of a molecule of concentration 1 g/mL in a tube 10 cm long.

Spectroscopy (Section 2.14) The study of the interactions between electromagnetic radiation and atoms and molecules.

sp Hybrid (Section 2.2) A hybrid orbital made by the combination of one s and one p atomic orbital.

sp^2 Hybrid (Section 2.2) A hybrid orbital made by the combination of one s orbital and two p orbitals.

sp^3 Hybrid (Section 2.2) A hybrid orbital made by the combination of one s orbital and three p orbitals.

Spiro (Section 5.7) A structural motif in which two rings share a single carbon.

Staggered ethane (Section 2.5) The energy minimum conformation of ethane in which the carbon–hydrogen bonds (and the electrons in them) are as far apart as possible.

Starch (Section 22.6) A nonlinear polymer of glucose containing amylose, an α-linked linear polymer of glucose.

Stereochemistry (Section 4.1) The physical and chemical consequences of the arrangement in space of the atoms in molecules.

Stereogenic atom (Section 4.3) An atom, usually carbon, of such nature and bearing groups of such nature that it can have two nonequivalent configurations.

Steric requirements (Section 2.9) The space required by an atom or group of atoms.

Steroid (Section 12.14) A class of four-ring compounds, always containing three six-membered rings and one five-membered ring. The ring system can be substituted in many ways, but the rings are always connected in the following fashion:

Strecker synthesis (Section 23.2) A synthesis of amino acids using an aldehyde, cyanide, and ammonia, followed by hydrolysis.

Structural isomers (Section 2.7) Molecules of the same formula that differ structurally.

Substituent (Section 2.3) Any atom or group of atoms other than hydrogen attached to a molecule.

Substitution reaction (Section 7.1) The replacement of a leaving group in R—L by a nucleophile, $Nu:^-$. Typically,

$$Nu:^- + R-L \longrightarrow Nu-R + L:^-$$

Sugar (Section 22.1) A molecule whose formula can be factored into $C_x(H_2O)_y$. A saccharide or carbohydrate.

Sulfide (Section 6.9) The sulfur counterpart of an ether, RSR′. See Thioether.

Sulfone (Section 16.16) $R-SO_2-R$, the final product of oxidation of a sulfide with hydrogen peroxide.

Sulfonium ion (Section 7.10) A positively charged sulfur ion, R_3S^+.

Sulfoxide (Section 16.16) A compound of the structure

Superacid (Section 14.13) A number of highly polar, highly acidic, but weakly or nonnucleophilic solvent systems. Superacids are often useful in stabilizing carbocations, at least at low temperature.

Suprafacial motion (Section 20.5) Migration of a group in a sigmatropic shift in which bond breaking and bond making take place on the same side of the π system.

syn Addition (Section 9.10) The addition of two groups to a π bond in which the two groups end up being on the same side.

syn Elimination (Section 7.9) An elimination reaction in which the dihedral angle between the breaking bonds, usually C—H and C—L, is 0°.

Tautomerization (Section 10.8) The process through which molecules related through the change of position of a single hydrogen interconvert.

Tautomers (Section 1.4) Molecules related through the change of position of a single hydrogen.

Termination (Section 11.4) The mutual annihilation of two radicals. Termination steps destroy chain-carrying radicals and end chains through bond formation.

Terpenes (Section 12.13) Compounds whose carbon skeletons are composed of isoprene units.

Tertiary amine (Section 6.7) An amine bearing no hydrogens and three R groups.

Tertiary carbon (Section 2.8) A carbon attached to three other carbons.

Tertiary structure (Section 23.4) The structure of a protein induced by its folding pattern.

Tetrahedral intermediate (Section 17.7) The first intermediate formed when a nucleophile attacks a carbonyl carbon.

Tetrahydropyranyl (THP) ether (Section 16.10) Protecting groups for alcohols of the structure

Tetramethylsilane (TMS)   (Section 15.5) The standard "zero point" on the ppm scale in NMR spectroscopy. The chemical shift of a nucleus is quoted in parts per million of applied magnetic field relative to the position of TMS.

Tetrose   (Section 22.2) A four-carbon sugar.

Thermodynamic control   (Section 8.8) In a reaction under thermodynamic control, the product distribution depends on the energy differences between the products.

Thermodynamic enolate   (Section 19.15) The most stable enolate. It may or may not be the same as the most rapidly formed enolate, the kinetic enolate.

Thermodynamics   (Section 8.8) The study of energetic relationships.

Thermolysis   (Section 11.2) The process of inducing chemical change by supplying heat energy. See Pyrolysis.

Thioether   (Section 6.9) The sulfur counterpart of an ether, RSR'. A sulfide.

Thiol   (Section 6.9) The sulfur counterpart of an alcohol, RSH.

Thionyl chloride   (Section 7.4) $SOCl_2$, an effective reagent for converting alcohols into chlorides and carboxylic acids into acid chlorides.

Thiourea   (Section 23.4) A compound of the structure

Three-center, two-electron bonding   (Section 21.3) A bonding system in which only two electrons bind three atoms. This arrangement is common in electron-deficient molecules such as boranes and carbocations.

Torsional strain   (Section 5.2) Destabilization caused by the proximity of bonds (usually eclipsing) and the electrons in them.

trans   (Section 2.12) Hydrogens on opposite sides. Used to specify stereochemical (spatial) relationships in ring compounds and alkenes.

s-trans   (Section 12.6) The more stable, extended form of a 1,3-diene:

Transesterification   (Section 18.8) The formation of one ester from another. The reaction can be catalyzed by either acid or base.

Transition state (TS)   (Section 2.5) The high point in energy between starting material and product. The transition state is an energy maximum and not an isolable compound.

Triple bond   (Section 3.5) Two atoms can be attached by a triple bond composed of one σ bond and two π bonds.

Triplet carbene   (Section 10.4) In a triplet carbene, the two nonbonding electrons have the same spin and must occupy different orbitals.

Tropylium ion   (Section 13.6) The 1,3,5-cycloheptatrienylium ion. This ion has $4n + 2$ π electrons ($n = 1$) and is aromatic.

Ultraviolet/visible (UV/vis) spectroscopy   (Section 12.7) Electronic spectroscopy using light of wavelength 200–400 nm.

Unpaired spin   (Section 1.2) Two electrons with the same spin quantum number have unpaired spin.

Unsaturated hydrocarbons   (Section 2.12) Molecules containing π bonds. For example, alkenes or alkynes. This term does not include cycloalkanes.

Urea   (Section 17.7)

Valence electrons   (Section 1.3) The outermost, or most loosely held, electrons.

van der Waals forces   (Section 2.13) Attractive or repulsive intermolecular forces in or between molecules caused by induced dipole–induced dipole interactions.

van der Waals strain   (Section 5.4) Repulsive forces in or between molecules caused by induced dipole–dipole interactions.

Vicinal   (Section 10.2) Groups on adjacent carbons are vicinal.

Vinyl   (Section 3.3) The common name for the $H_2C{=}CH$ group.

Vinylic hydrogen   (Section 15.6) A hydrogen on an alkene.

Wagner–Meerwein rearrangement   (Section 9.9) The 1,2-migration of an alkyl group in a carbocation.

Wave function (ψ)   (Section 1.2) A solution to the Schrödinger equation, another word for orbital.

Wavenumber   (Section 15.4) The wavenumber ($\bar{v}$) equals $1/\lambda$ or v/c.

Weighting factor   (Section 1.4) The relative contribution of a particular resonance structure to the weighted sum of all resonance forms.

Williamson ether synthesis   (Section 7.10) The S_N2 reaction of an alkoxide with R—L to give an ether.

Wittig reaction   (Section 16.17) The reaction of an ylide with a carbonyl group to give, ultimately, an alkene.

Wolff rearrangement   (Section 18.14) The formation of a ketene through the thermal or photochemical decomposition of a diazo ketone.

Woodward–Hoffmann theory   (Section 20.1) The notion that one must take account of the phase relationships between orbitals in order to understand reaction mechanisms. In a concerted reaction, bonding relationships must be maintained at all times.

Xanthate ester   (Section 18.13) A compound of the structure

Ylide   (Section 16.17) A compound containing opposite charges on adjacent atoms.

Zwitterion   (Section 23.2) A dipolar species in which full plus and minus charges coexist within the same molecule.

Credits

Index

Note: Footnotes are indicated by n after the page number. Material in figures or tables is indicated by italic page numbers.

Absolute configuration, 152–55, 172–73, 181
Abstraction, definition, 471
Acenaphthylene, 417
Acetaldehyde (ethanal)
    acid-catalyzed hydrogen exchange reactions,
        938–39, 970
    aldol condensation, 965–66, 970
    base-catalyzed α hydrogen exchange reaction,
        936–38
    from biological oxidation of ethanol, 814–15
    boiling point, 770
    deuterioacetaldehyde, 815
    dipole moment, *766*
    equilibrium between carbonyl and enols forms,
        940–41
    equilibrium in hydration reactions, 777–80
    hydrate formation, 965–66
    infrared stretching frequency, *888*
    NMR properties, *889*
    pK_a, 936
    solubility, 770
    structure, *764, 767*
Acetals
    formation in acid, 786, *791–92, 817*, 846–48
    glycosides, 1148–52, 1210
    hemiacetal intermediate formation in acid,
        784–86, *790, 818*
    hemiketals and ketals, 786n2
    as protecting groups for aldehydes and ketones,
        788–89
    structure, 786, *846*
    tetrahydropyranyl (THP) ethers as protecting
        groups for alcohols, 789–90
    See also Carbonyl compounds, addition
        reactions; Hemiacetals
Acetamide (ethanamide), *882, 943*
Acetamidobenzene (acetanilide), 670, 671
Acetamidomalonic ester, *1183*
Acetate ion resonance stabilization, 421
Acetic acid (ethanoic acid)
    abbreviation, 421, 1085
    acetolysis in intramolecular replacement
        reactions, 1085–87
    from hydrolysis and decarboxylation of
        malonic acid, 961–62
    infrared (IR) spectroscopy, *833, 888*
    physical properties, *830*
    pK_a, 1179
    structure, *849, 879, 880*
    uses, 45
Acetic anhydride (ethanoic anhydride), *881, 888*
Acetic propionic anhydride (ethanoic propanoic
    anhydride)
    structure, *881*

Acetoacetate synthesis, 959–60, 963, *1014*
Acetone (dimethyl ketone, propanone)
    aldol condensation, 971–73
    base-catalyzed carbonyl–enol equilibrium,
        935
    boiling point, 770
    dipole moment, *766*
    enolate, structure, *895*
    equilibrium between carbonyl and enols forms,
        940–41
    equilibrium in hydration reactions, 777–78, 781
    infrared stretching frequency, *888*
    mass spectrometry, 772
    NMR properties, *889*
    solubility, 240, 770
    structure, *768*
Acetonitrile (ethanenitrile, methyl cyanide), 883,
        887, 889, 943
Acetophenone (methyl phenyl ketone), *769*
Acetyl chloride (ethanoyl chloride)
    acylation reactions, 670, 1187
    infrared stretching frequency, *888*
    NMR properties, *889*
    reaction with carbohydrates, 1147
    structure, *880, 893*
Acetylcholine, 217
Acetyl-CoA, 862–63, 997
Acetylene (ethyne)
    bonding, 123–25
    Lewis structure, 18
    molecular formula, 98
    structure, 123–24
    substituted acetylene synthesis from acetylides,
        312–13, 320
    welding, *97*
    See also Alkynes
Acetylide ion, 129–30, 143, 312–13, 320,
        517–18
Acetyl salicylic acid (aspirin), 838
Achiral molecules, definition, 150, 150n2
Acid anhydrides. *See* Anhydrides
Acid–base chemistry
    acid chlorides, 886
    acidity constant (K_a), 234
    acidity of terminal alkynes, 129–30
    acyl compounds, 885–87, *943*
    alcohols, *232*, 233–38
    alkanes, 130, 976
    alkenes, 130
    alkynes, 129–30
    amides, 886–87, 943–44
    amines, 246–49, 834
    amino acids, 1179–81
    ammonium ions, 246–48

    anhydrides, 886–87
    basicity constant (K_b), 248
    carboxylic acids, 834–38
    conjugate acids, 234
    conjugate bases, 234
    esters, 886–87
    ethers, 251, 254
    heteroaromatic compounds, 601–2
    methyl anion, 129, 130
    nitriles (cyanides), 885–86
    overview, 233–35, 320–21
    potassium hydroxide reaction with
        hydrochloric acid, 90, 233–34
    pyridine (azabenzene), 601, 620
    pyrrole, 601, 621
    thiols (mercaptans), 253
    water, 234
    See also Brønsted acids; Brønsted bases; Lewis
        acids; Lewis bases; pK_a
Acid chlorides (acyl chlorides)
    acid–base chemistry, 886
    addition–elimination reaction, 889–91
    AlCl$_3$ complexes, 643
    from carboxylic acids, 643, 853–55, *868*, 917
    conversion to ketenes, 912, *924*
    Friedel–Crafts acylation, 643–46, 893
    general structure, *643, 877*
    infrared spectra, 888
    products of addition–elimination reaction, *890*
    reaction with lithium aluminum hydride,
        891, *922*
    reaction with organometallic reagents,
        891–92
    reduction by Rosenmund reduction, 892–93,
        923
    resonance forms, *885*
    S$_N$Ar reaction of pyridine *N*-oxides, 678–79
    structure, *769*
    See also Acyl compounds
Acid derivatives, 877
    See also Acyl compounds; Carboxylic acids
Acid halides
    α-bromo acid bromides, 951–52
    α-bromo acid halides, 951–52
    α-bromo acids, 950–53, 1182, *1216*
    nomenclature, 880–81
    properties, 880
    See also Acid chlorides
Aconitase, 357
cis-Aconitate, 357
Activating agents, 850, 1204–5

Activation energy (ΔG°)
  addition of hydrogen chloride to alkene, 136
  additions to alkenes, 136
  definition, 136
  effect of solvent polarity, 287
  energy barrier height effect on reaction rate, 334, 339–40
  enthalpy of activation ($\Delta H^\ddagger$), 342, 349, 359
  mountain pass analogy to energy barriers, *331*, 346, 627
  and rate constant, 349
  S_N1 reactions, 342–43
  S_N2 reactions, 273–74, 277, 287, 343–46
  See also Gibbs free energy (ΔG°); Transition states
Acyl azides, 918–19, *922*
Acyl chlorides. *See* Acid chlorides
Acyl compounds
  acid–base chemistry, 885–87, *943*
  acyl benzenes from Friedel–Crafts acylation, 644
  enolates, 942–44
  general formula, 877
  infrared spectra, 887–88
  mechanism of addition–elimination reaction, 877–78
  nomenclature, 879–83
  nuclear magnetic resonance spectra, 888–89
  nuclear magnetic resonance spectroscopy, 888–89
  overview, 877–79, 921–22
  physical properties, 884
  pK_a, *943*
  reaction with lithium aluminum hydride, 891, 900, *922*
  structures, 884–85
  synthesis of other compounds, overview, *890*, 922–25
  tetrahedral intermediate in addition–elimination reaction, 877–78, 889
  See also specific types
Acyl group, structure, 476, *643*
Acylium ion, 643, 645, 646, 704, 772
Adamantanes, 217–19
Adamantylideneadamantane, 416
Adenine, 93, *615*, *1149*, 1212
Adenosine triphosphate (ATP), 1140
S-Adenosylmethionine, 142, 288, 314, 324, 401–2, 455
Adipic acid (hexanedioic acid), *830*, 852
Aflatoxin B$_1$, 86, *212*
Agarospirol, *211*
Ajoene, 252
Alanine, *1175*, *1176*, *1178*, *1189*, *1203*, *1204*
Alcohol dehydrogenase, 400
Alcohols
  acid–base properties, *232*, 233–38
  from acid chlorides, 891, *922*
  from aldehyde and ketone reactions with organometallic reagents, 799–802, *817*
  chemical exchange of OH hydrogens, 733–34, 739
  conversion of hydroxide leaving group to water, 235, 281–82, 322, 555
  conversion to sulfonates, 281–86, 322, 427, 555
  decoupled NMR spectra, 733, 739, 746
  deprotonation, 236, 385
  1-deuterioethyl alcohol, configuration, 152, *154*

  hydrogen bonding, 232–33
  from lithium aluminum hydride reaction with acyl compounds, 891, 900, *922*
  from lithium aluminum hydride reaction with oxiranes, 429–30, *458*
  Meerwein–Ponndorf–Verley–Oppenauer (MPVO) equilibration, 1006–7, *1012*
  nomenclature, 230–31
  oxidation to carbonyl compounds, 802–6, *817*
  oxidation to carboxylic acids, 805, *817*
  oxonium ion formation, 235–36
  physical properties, 232–33
  pK_a values in water, *236*, *249*, 316, 834
  protonation, 235–36, 281–82
  reaction with phosgene, 899
  retrosynthetic alcohol synthesis, 807–9
  structure, 231–32
  synthesis by acid-catalyzed hydration of alkenes, 139–40, *144*, 381–82, *403*
  synthesis by hydroboration, 398–401, *403*
  synthesis by oxymercuration, 421–23, 448, *458*, *459*
  synthesis from acyl compounds, *922*
  synthesis from oxiranes, 429, *458*
  tetrahydropyranyl (THP) ethers as protecting groups for alcohols, 789–90
  thionyl chloride treatment for chloride formation, 284–85
  treatment with phosphorus reagents, 285
  trialkylsilyl ethers as protecting groups, 789–90
  See also specific types
Aldaric acids, 1144, *1156*, 1157
Aldehyde dehydrogenase, 400
Aldehyde hydrogens, 725
Aldehydes
  acetals as protecting groups, 788–89
  from acid chlorides, 891–92, *923*
  aldehyde hydrogens, 725
  aldol condensation, 965–70
  alkylation of enolates, 955–56, *1012*
  Baeyer–Villiger reaction, 907–9, *924*
  bisulfite addition, 781, *818*
  cyanohydrin formation, 781–82, *818*, 1139
  dials (dialdehydes), 767
  enolate formation, 933–42, 965
  from enols, 449, 451, *458*
  formation from ozonolysis, 440–42
  halogenation at α position, in acid, 946–48, *1013*
  halogenation at α position, in base, 948–50, *1013*
  hydrate formation, *818*, 965–66, 1004
  from hydroboration of alkynes, 451, *458*
  infrared (IR) spectroscopy, 770–71
  Knoevenagel condensation, 974–76, *1014*
  nomenclature, 767
  nuclear magnetic resonance (NMR) spectroscopy, 771–72
  from oxidation of alcohols, 802–6, *817*
  oxidation to carboxylic acids, 805, *817*
  pK_a values, *933*
  reaction with enolates, 965–70
  reaction with ester enolates, 974–76
  reaction with metal hydrides to form alcohols, 801–3, *817*
  reaction with organometallic reagents to form alcohols, 800–801, *817*
  resonance forms, *885*

  synthesis by Rosenmund reduction, 892–93, *923*
  ultraviolet spectroscopy, 771, 772
Aldohexoses
  definition, 1129
  epimerization in base, 1141–42
  Fischer determination of structure, 1153–60
  hemiacetal formation, 1132
  α- and β-pyranose and furanose forms, *1134–35*
  ratio of furanose and pyranose forms, *1133*
  structures, *1129*, 1153–60
  Williamson ether syntheses, 1147
  See also Aldoses
Aldol condensation
  acid-catalyzed aldol dehydration, 968, *1013*
  aldehyde reaction with enolates, 965–70
  aldol structure, *966*
  base-catalyzed aldol dehydration, 968–69, *1013*
  comparison of aldol and Claisen condensation, 988–89, *990*
  crossed (mixed) aldol condensation, 982–87
  definition, 966
  β-hydroxy carbonyl products, 966, 968, 973–74
  intramolecular aldol condensations, 980–81, 999
  ketone reaction with enolates, 971–74
  Knoevenagel condensation, 974–76, *1014*
  Michael reaction, 976–80, 998–99, 1016
  problem solving techniques for real-world synthesis, 1007–9
  rate-determining steps in equilibria, 969–70
  reverse aldol reactions, 973
  Robinson annulation, 998–99
  thermodynamic factors, 972, 974, *990*
  α,β-unsaturated carbonyl dehydration products, 968–69, 971–72, 973–75, 981, *1014*
  See also Carbonyl compounds, condensation reactions; Claisen condensation; Claisen–Schmidt condensation
Aldonic acids, 1144, 1166
Aldopentoses, 1129, 1158–59
Aldoses
  anomers, 1134
  definition, 1126
  Fischer projections, 1127, 1129–30
  hemiacetal formation, 1130–34
  osazone formation, 1145–46, *1147*
  oxidation, 1144–45, 1157
  reduction, 1130, 1143
  Ruff degradation, 1139–40
  See also Aldohexoses; Carbohydrates
Aldotetroses, 1129
Aldotriose, 1126
Alkaloids, 170–72, 255–56
Alkanes
  acid–base chemistry, 130, 976
  biochemistry, 92–93
  chemical shift (δ), 722–23
  combustion reactions, 68, 92
  functional group effect on reactivity, 51
  heat of formation (ΔH_f°), 116–18, *126*
  mass spectrometry (MS), 705
  molecular formula, 51, 131
  naming conventions, 78–81
  nomenclature, 78–81
  from organometallic reagent destruction by water, 799–800, *818*

overview, 51–52, 93–94
pK_a, 976
straight-chain alkanes, 79, 86–88
synthesis by Clemmensen reduction, 645, *686*, 802
synthesis by desulfuration, 253, 257
synthesis by disproportionation reactions, 471–72, 473–74, *506*
synthesis by hydrogenation of alkenes and alkynes, 452, *458*
synthesis by Wolff–Kishner reduction, 645, *686*, 802
See also specific types
Alkene halogenation. *See* Halogenation
Alkene hydrohalogenation
 conjugated dienes, 369–70, *404*, 534–35, 537–41
 definition, 365, 414
 HX addition to alkynes, 444–47
 HX hydration of alkenes, 380–84, *404*
 mechanism for hydrogen bromide addition, 414
 radical alkene hydrohalogenation, 485
 radical HX hydration of alkenes, 481–87
 rearrangements during HX addition, 386–89, 501
 See also Alkenes, addition reactions; Alkyl halides; Halogenation
Alkenes
 acid–base chemistry, 130
 biological activity, 142–43
 bonding, 99–107
 cis and trans coupling constant (J_{cis}, J_{trans}), 732
 derivatives, 107–10
 dimerization and polymerization, 384–86, 488–89
 heat of formation (ΔH_f°), 116–18, *126*, *515*
 heat of hydrogenation, *580*
 heats of hydrogenation, *413*
 isomers, 108–10
 molecular formula, 98, 130
 molecular orbitals, 103–4
 nomenclature, 110–12
 overview, 98, 142
 physical properties, *123*
 pi (π) orbitals, 98, 104, 125
 relative stability of isomers, 116–18
 rotational energy barriers, 105–6, 107, 144, 522
 stabilizing effect of 2p/2p overlap, 105–6
 structure, 99, 105–7
 synthesis by alkyne reduction by sodium in ammonia, 452–55, *458*
 synthesis by disproportionation reactions, 471–72, 473–74, *506*
 synthesis by syn hydrogenation of alkynes, 452, *458*
 synthesis by Wittig reaction, 811–13, *817*
 from thermal elimination of esters, 912–14, *923*
 types of carbon–carbon σ bonds in isomers, 116–17
 See also Alkenes, addition reactions; Cycloalkenes
Alkenes, addition reactions
 acid-catalyzed hydration reactions, 139–41, 143, 381–82, *403*, 773, 775–76
 activation energy ($\Delta G^\ddagger$), 136
 bromine addition to alkenes, 396

Brønsted acids, general case, 131
carbocation intermediate, 132–33, 135–36, 140
dimerization and polymerization, 384–86, 488–89, 1045–46, *1074*
HX hydration of alkenes, 380–84, *404*
hydrogen chloride addition to 2,3-dimethyl-2-butene, 131–32, 133–36, 365–67, 374
inductive effects, 378–80
mechanism of hydrogen halide addition, 132–36, 138–39, 365–66
overview, 364–65, 402–4, 410, 456–60
oxidation with permanganate or osmium tetroxide, 443–44
oxymercuration, 421–23, 448, *458*
ozonolysis, 436–40
problem solving techniques, 110, 139, 141
protonation step, 133, 365–66, 385
radical addition to alkenes, 481–87, 484–86
radical HX hydration of alkenes, 481–87
radical-induced tetrahalide addition to alkenes, 487–89
rate-determining step, 365–66
reaction energetics, 135–36
rearrangements during HX addition, 386–89, 501
regiochemistry, 137–39, 141, 366–72
ring formation through addition reactions, *457*
transition states, 136
 See also Alkene hydrohalogenation; Alkenes; Alkynes, addition reactions; Epoxidation; Halogenation; Hydroboration; Hydrogenation
Alkoxide ion
 comparison to carboxylate anion, 834
 from deprotonation of alcohols, 236–37, *249*, 315, 834, *1081–82*
 dimethyl sulfate treatment for methyl ether formation, 285
 formation by sodium hydride, 315
 oxirane formation, 317, 1083
 from reaction of metallic sodium or potassium with alcohols, 315
 from reaction of sodium hydroxide with alcohols, 315–16
 stabilization by alkyl groups in gas phase, 238
 stabilization by solvents, 237
Alkylamines, 241, 247, 318, 1181–82
Alkyl compounds, general, 69, 78–81, 141
Alkyl halides
 alkyl bromides from Hunsdiecker reaction, 861, *869*
 bonding, 226, *227*
 conversion to organometallic reagents, 227–30, 257
 cycloalkyl bromide reactivities in S_N2 reactions, *276*
 dipole moments, *226*
 formation by ether cleavage, 282–83
 formation with phosphorus reagents, 285
 geminal dihalides, 447, *459*
 Grignard reagent formation, 228–29, 257
 HX hydration of alkenes, 380–84, *404*
 hydrocarbon formation, 229–30
 nomenclature, 225–26
 properties, *225*
 radical HX hydration of alkenes, 481–87
 substitution reactions, 267–68
 synthesis by addition of hydrogen halides to alkenes, 132–36, 138–39, 141, 144

synthesis by protonation of alcohol with haloacid, 142, 144, 282
synthesis by S_N1 or S_N2 reactions, 322
tetrahalide formation from alkynes, 448, *459*
from thionyl chloride treatment of alcohols, 284–85
vicinal dihalides, 414, 448, *459*, 490, 499, *507*
 See also Alkene hydrohalogenation; Halogenation
Alkyl shifts, 387–88, 642
Alkynes
 acidity of terminal alkynes, 129–30
 alkyne radical anions, 453–54
 allenes as intermediates in isomerization, 516–18
 derivatives and isomers, 126–28
 heats of formation and relative stability, 126
 molecular formula, 98, 131
 nomenclature, 126, 127
 physical properties, *129*
 pi (π) orbitals, 125
 in rings, 128
 structure and bonding, 123–25, 128
 See also Acetylene; Alkynes, addition reactions; Triple bonds
Alkynes, addition reactions
 HX addition to alkynes, 444–47
 hydration, 448–50
 hydroboration, 450–51
 hydrogenation, 452
 oxymercuration, 448, *459*
 radical-induced anti-Markovnikov addition, 489–90, *507*
 reduction by sodium in ammonia (anti-hydrogenation), 452–55
 tetrahalide formation from alkynes, 448, *459*
 X_2 reagent addition to alkynes, 447–48
 See also Alkenes, addition reactions; Alkynes
Alkynyl hydrogens, 724–25
Allene (propadiene), 178, 513–14, *921*
Allenes, 178, 513–15, 516–18
 See also Cumulated double bonds
D-Allose, *1129*, *1138*, 1143, *1145*, 1160
Allyl alcohol, *231*
Allyl anion (allylic anion), 27–28, 543–44, 589, 934
Allyl cation (allylic cation)
 delocalization of electrons, 27, 369–70
 hydrogen chloride addition to 1,3-butadiene, 369–71
 molecular orbitals, 369–70
 resonance forms, 27, 28, 369–71, 535
 resonance-stabilized formation in S_N1 reactions, 538, 541–42, 564
 S_N1 solvolysis reactions, 541–42
Allyl compounds, *108*, 326
Allyl halides, *404*, 541–43, *564*, 612
Allylic halogenation, 497–501, *507*, 543
Allylic hydrogens, chemical shift (δ), 724
Allyl iodide (3-iodopropene), *541*, 543
Allyl radical, 475, 480, 498–99, *512*, 543
Allyl system, 2p orbital overlap, 369–70, 541
α-helix, 1191, *1192*
α position, definition, 763, 933
 See also Carbonyl compounds, reactions at α position
D-Altrose, *1129*, *1138*, 1143, 1160
Aluminum alkoxides, 1005–6, 1012
Aluminum bromide (AlBr₃), *361*, 640, 646, *687*

Aluminum chloride (AlCl₃)
　electrophilic substitution of furan and
　　thiophene, *654*
　electrophilic substitution of pyridine, *653*
　Friedel–Crafts acylation, 643–44, *687*
　Friedel–Crafts alkylation, 639–43, 679, *686*
Aluminum tri-*tert*-butoxyhydride, 892
Amide ions, 249, 313, 454, 516–17, 680–81
Amides (carbonyl-nitrogen compounds)
　acid–base chemistry, 886–87, 943–44
　acid-induced hydrolysis, 902–3, *922*
　base-induced hydrolysis, 901, 903, *922*
　from Beckmann rearrangement, 909–11, *923*
　biochemistry, 878
　from carboxylic acids, 850–52, *869*
　dicyclohexylcarbodiimide (DCC), 850
　enolates, 943
　formation by dicyclohexylcarbodiimide
　　(DCC), 850, 856, *869–70*, 1205–7, 1209,
　　1217
　general structure, 249n3, *769, 850, 877*
　Hofmann rearrangement, 919–20, *923*
　lactams (cyclic amides), 851, *870, 882, 910, 924*
　nomenclature, 882
　polyamides (nylon), 852, *898*
　reduction by metal hydrides, 903
　resonance stabilization, 884–85, 903, 943
　structure, *769*, 884–85
　See also Acyl compounds
Amines
　acid–base chemistry, 246–49, 834
　N-acylated aromatic amines, 671
　alkylamines, 241, 247, 318, 1181–82
　alkylation reactions, 317–18
　from amide reduction, 903
　amine inversion, 243–44, 257
　from carbanic acid decarboxylation, 860, *869*
　chemical exchange of NH hydrogens, 733–34,
　　739
　condensation reactions, 1003–4
　cyclic amino compounds, 242
　decoupled NMR spectra, 733, 739, 746
　diamines, 246
　from Hofmann rearrangement, 919–20, *923*
　hydrogen bonding, 245–46, 247–48
　intramolecular reactions of endo and exo
　　bicyclic amines, 1090–92
　Mannich reaction, 1003–4, *1013*
　from nitrile reduction, 906, *923*
　nomenclature, 240–42
　odor, 246
　overview, 224
　physical properties, 245–46
　pK_b values, 248, 259, 601
　polyfunctional amines, *241*
　primary amines, 240, 241
　reaction with epoxides, 318
　secondary amines, 240, 241
　structure, 243–44
　synthesis through S$_N$2 reactions, 317–18, 322
　tertiary amines, 240, 241
Amino acids
　abbreviations, 1175, *1176–77*
　acid–base properties, 1179–81
　acylation reactions, 1187
　α-amino acids, definition, 1175
　chromatographic separation and analysis, 1196
　D and L configuration, 1178
　electrophoresis, 1180–81, 1195

essential amino acids, 1063, 1077
esterification reactions, 1187
Fischer esterification, 1187
Fischer projections, 1178
Gabriel synthesis, 1182–83, *1216*
isoelectric point (p *I*), *1176–77*, 1179–80
kinetic resolution, 1185
from methane and ammonia, 93
nomenclature, 1175–77
overview, 1174, 1215–18
properties, *1176–77*
(*R*) and (*S*) configuration, 1178
reaction with 2,4-dinitrofluorobenzene, 1196,
　1197
reaction with ninhydrin, 1187–89, 1196
reductive animation, 1185–86
resolution, 1184–86
RNA codons for amino acids, 1214–15
side chains (R groups), 1175, *1176–77*
simple alkylation synthesis, 1181–82, *1216*
Strecker synthesis, 1184, *1216*
structure, 1178–79
synthesis, 1181–84, 1185–86
zwitterions, 1179–80
Aminoacyl tRNA synthetase, 1214
1-Aminoadamantane, 219
3-Aminobenzoic acid, *670*
γ-Aminobutyric acid (GABA), 400
cis-4-Aminocyclohexanecarboxylic acid, *831*
Amino group, 238, 241, 646, 670–71, 672, *686*
2-Aminohexanoic acid (norleucine), *1181*
2-Aminopropane, *241*
2-Aminopyridine, 676–77
Amino terminus, 1189, 1197
3-Amino-2,4,6-tribromobenzoic acid, *670*
Ammonia, structure, *17*, 240
Ammonium ions
　acid–base chemistry, 246–48, 1179
　alkylammonium ions, 317
　formal charge, 21
　Lewis structure, 21
　nomenclature, 242
　quaternary ammonium ion formation, 318
　quaternary nitrogen compounds, 240
Amyl alcohol (1-pentanol), *230, 231*, 231n2
Analytical chemistry, overview, 695–97, 750–51
Anchimeric assistance, 1089, 1100
Angle strain, 128, 187, 190, 276, 474
Anhydrides
　acid–base chemistry, 886–87
　addition–elimination reactions, 894–95
　definition, 855
　Friedel–Crafts reactions, 895
　general structure, *877*
　infrared spectra, 888
　nomenclature, 881
　structure, *769, 855*
　synthesis, 855–56, *869, 923*
　See also Acyl compounds
Aniline (benzenamine or aminobenzene)
　azo dye production, *647*
　bromination, 670
　from chlorobenzene reaction with strong base,
　　680–81
　conversion to benzenediazonium chloride,
　　647–48
　conversion to pseudomauvine, 648
　nitration, 673
　from nitrobenzene reduction, 646, 651, *686*

reaction with maleic anhydride, *894*
from reduction of nitrobenzene, 646
structure, *243, 595*
Anilinium ion, 673
Anions, 4
　See also Carbanions
Anisaldehyde (4-methoxybenzaldehyde), *768*
Anisole (methyl phenyl ether or methoxybenzene)
　chlorination, 659
　inductive effects, 666–67
　meta substitution, 657
　ortho substitution, 658, 666–67
　para substitution, 657–60, 666–67
　structure, *250, 595, 657, 659*
Annulenes, 594
Anomeric carbon, 1134
Anomers, 1134–35, *1140*
Antarafacial motion, 1055–58
Anthracene, *179*, 603, 621, *692*
Anthranilic acid, *692*
Antibonding molecular orbitals ($\Phi_{antibonding}$, Φ_A),
　32–35, 40, 44, 54
Anti elimination, 305–6, 311, 322
Anti-Markovnikov addition
　alcohol synthesis, 399–400, 403
　definition, 399
　hydration of 2-methylpropene, 382
　hydroboration/oxidation of alkenes, 399–400,
　　403
　hydroboration/oxidation of alkynes, 451
　radical HBr addition to alkenes, 481–82,
　　484–86
　radical HBr addition to alkynes, 489–90, *507*
　See also Markovnikov addition
Antimony hexafluoride anion, 679–80
Antimony pentafluoride, 679–80
Apomorphine, *408*
Aprotic solvents, 238–39, 240, *280*
D-Arabinose, *1129*, 1153–56, 1158
Arenes, 598
Arginine, 1177, *1177*, 1180, 1202
Argon, 4, 6
Arndt–Eistert reaction, 917
Aromatic character, definition, 582
Aromatic compounds
　Birch reduction, 608–9, *618*
　chemical shift (δ), 725
　coupling constant (*J*), 732
　Diels–Alder reaction, 628–31
　hydrogenation, 626–27, *686*
　paracyclophanes, 628–29
　reverse Diels–Alder reactions, 554
　structural features, 582–83
　substitution reactions, 606–8
　See also Hückel's rule; Substitution reactions of
　　aromatic compounds; *specific types*
Aromaticity, definition, 582
　See also Hückel's rule
Arrow formalism, 23, 26
Aryl, 598
Asparagine, *1176*
Aspartame, 1164
Aspartic acid, *1176*
Atomic number, 5
Atomic orbitals
　1*s* orbital, shape, 10–11
　2*p* orbitals, shape, 12–13
　2*s* orbital, shape, 12
　definition, 3

electron density, 10, 11–12, 32–33
electron shells, 4
in first 10 elements, *13*
nodes, 11–12
probability of finding electron in a volume of
    space (ψ^2), 6, 10, 11–12
relationship to quantum numbers, 6–8
See also Electron configurations; Hybrid
    orbitals; Wave functions (ψ)
Atomic weights, *701*
Atoms
    atomic number, 5
    atomic structure, 4–6
    Bohr model, 2
    definition, 2
    history of atomic models, 2
    mass number, 5
    strange atoms, problem solving, 46
Auden, Wystan Hugh, 468, 468n1, 1081, 1081n1
Aufbau principle, 8–9
Axial hydrogens, 191
Azabicyclooctane, 277–78
2-Azacyclobutanone (β-lactam), *882*
2-Azacyclohexanone (δ-lactam), *882*
Azetidine (azacyclobutane), *242*
Azides, *436*, 918–19, *922*
Aziridine (azacyclopropane), *242*, 464–65
Azobisisobutyronitrile (AIBN), 476
Azo compounds, 476, *507, 647, 651*, 689

Babies and quantum mechanics, 42–43
Baeyer–Villiger reaction, 907–9, *924*
Bakkenolide-A, *211*
Barbaralane, *1068*
Base pairs, 288, 615, 1212–13, 1214
Base peak, 702
β (beta) cleavage, 472, 473, 474
Beckmann rearrangement, 909–11, *923, 924*
Beer–Lambert law, 527
Benzaldehyde (benzenecarboxaldehyde)
    Cannizzaro reaction, 1004–5
    Claisen–Schmidt reaction, 983–85, 1008
    hydrate formation, 965–66
    Meerwein–Ponndorf–Verley–Oppenauer
        (MPVO) equilibration, 1005
    mixed aldol condensation with *tert*-butyl
        methyl ketone, 983–84
    structure, *595, 768, 983, 1004*
Benzamide, *882*
Benz[*a*]pyrene, 605, *615*, 616
Benzene
    and bisulfate anion, 633
    catalytic halogenation with $FeCl_3$ or $FeBr_3$,
        638
    catalytic hydrogenation, 580, 627
    chemical properties, 572
    conversion to cyclohexane, 580–81, 626–27,
        686
    and 1,3,5-cyclohexatriene, 572, 573–75,
        580–81, *629*
    delocalization energy, 581–82, 607, 614, 626
    destabilization to lower activation energy,
        628–29
    deuteration, 606–8, *618*, 632–34
    Dewar forms, 576–78, 617, *1043*
    Diels–Alder reaction, 628–29
    energy barriers to substitution reactions, 606,
        627–28
    Frost circle, 584–85

heat of formation, 581–82
heat of hydrogenation, 580–81, 626–27
Kekulé forms, 575–77
methylation reactions, 641, 655–56
molecular orbitals, 578–79
nucleophilic addition to benzenes (S_NAr
    reaction), 674–76
resonance (overlap of $2p$ orbitals), 575–78
resonance stabilization, quantitative evaluation,
    580–82, *629*
structure, 179, 573–75, 582–83
substitution reactions, resistance to, 606–8,
    631–32
See also 1,3,5-Cyclohexatriene;
    Heterobenzenes; Substituted benzenes
Benzenediazonium chloride, 647, *648, 650*
Benzenediazonium ion, *648, 650*
Benzenesulfonic acid, 636–37, *686*
Benzenethiol (phenyl mercaptan, thiophenol), *252*
Benzenol. *See* Phenol
Benzhydryl group, 611
Benzofuran, 605
Benzoic acid (benzenecarboxylic acid)
    anhydride formation, *881*
    oxidation of alkylbenzenes, 613–14, *618*
    physical properties, *830*
    pK_a, 760
    structure, *595, 882*
Benzoic anhydride, *881*
Benzoic butyric anhydride (benzoic butanoic
    anhydride), *881*
Benzoin condensation, 1028
Benzothiophene, 605
3,4-Benzphenanthrene, structure, *179*
Benzvalene (tricyclo[3.1.0.0²·⁶]hex-3-ene), *573,
    574*
    See also Benzene
Benzyl alcohol, *231, 595, 1004*
Benzylamine, *241*
N-Benzylaziridine, *242*
Benzyl cation, 611, 703–4
Benzyl chloride, structure, *595, 611*
Benzylethylmethylamine, *241*
Benzylic compounds
    definition, *595*, 596
    hydrogen abstraction at α position, 613
    oxidation to benzoic acid, 613–14, *618*
    reactivity of benzyl group, 610
    similarity of benzyl and allyl groups, *610*
    S_N1 substitution reactions, 610–11, *618*
    S_N2 substitution reactions, 612, *618*
4-Benzylmorpholine (*N*-benzylmorpholine), *242*
Benzyl radical, 480, 612–13, *618*
Benzyne, 680–82
Beryllium hydride, 54–55
Bicyclic compounds
    bridged bicyclic molecules, 121, 211, 214–16
    from Diels–Alder reaction, 551–54
    fused bicyclic molecules, definition, 121, 211
    fused bicyclic molecules, structure, 212–13
    intramolecular reactions of endo and exo
        bicyclic amines, 1090–92
    neighboring group effect, 1090–93
    nomenclature, 214–15, 551–52
    structure of ring compounds, 211–17
    See also Ring compounds
Bicyclo[4.2.2]deca-2,4,7,9-tetraene, 707
Bicyclo[2.2.1]heptane (norbornyl system), *186*,
    306, 1099n2, *1100*

Bicyclo[3.3.1]non-1-ene, 122–23
Bicyclo[2.2.2]octane, *361*
Bicyclo[3.2.1]octane, 215, *361*
Bicyclo[3.3.0]octane, *361*
Bicyclo[1.1.1]pentane, *216*
3,3′-Bicyclopropenyl, *573, 574*
    See also Benzene
Bierce, Ambrose, 695, 695n1
Bimolecular elimination reactions. *See* E2 reactions
Binding sites, 1193
Biphenyl (phenylbenzene), 179, *595, 688*
Birch reduction, 608–9, *618*
Bisulfite addition to carbonyl compounds, 781,
    782, *818*
Blocking groups, 671
Bohr, Niels, 2, 2n1, 3n4
Bohr model of atom, 2
Boltzman distribution, 340
Bombykol, 534
Bond dissociation energy (BDE)
    average bond dissociation energies, *39*
    bond energy of H_2^+ molecule, 39–40
    butane, 188
    1-butene, 475
    cyclopropane, 188, 474
    definition, 37
    1,5-hexadiene, 475
    homolytic bond cleavage, *337*, 470
    hydrocarbons, *477*
    hydrogen molecule, 37
    methane, 57
    propane, 475
Bonding molecular orbitals ($\Phi_{bonding}$, Φ_B), 32–35,
    44
Borane
    alkylborane formation, 391, 394
    borane–tetrahydrofuran complex, 391
    dialkylboranes, 397, *450*
    diborane, 390–91
    as Lewis acid, 92, 390–91
    sp^2 hybridization, 55–56, 92
    trialkylboranes, 397, 398, *399*
    vinylboranes, 450–51
    See also Hydroboration
Borodin, Alexander, 966n2
Boron, 55–56
Boron enolates, 1008
Boron trifluoride, 251, 264–65, 390
    See also Hydroboration
Bredt, Julius, 121–22
Bredt's rule, 121–22
Bridged bicyclic molecules, 121, 211, 214–16
    See also Bicyclic compounds; Fused
        compounds; Ring compounds
Bridgehead position, 121–22, 212, 213–16
Broad-band decoupling, 740
α-Bromo acids, 950–53, 1182, *1216*
Bromobenzene, *595*
1-Bromobutane, 738
3-Bromobutanoic acid (3-bromobutyric acid), *831*
1-Bromo-2-chlorocyclopropane, 204
2-Bromo-4-chloro-6-fluoroaniline, *597*
1-Bromo-1-chloropropene, *111*
4-Bromo-2-chlorotoluene, *597*
4-Bromo-2,5-dichlorophenol, *597*
Bromoethane, *702*
1-Bromo-3-ethyl-2-nitrobenzene, *597*
Bromoform, 431
Bromohydrins, *463*, 1083, 1094

2-Bromo-3-hydroxypropanal, *154*
2-Bromo-2-methylbutane, 299
Bromonitrobenzene, 672
Bromonium ions, 415–21, 424, 426, 536, 1094–95, 1101–2
Bromophenol, 671
1-Bromopropene, 489
2-Bromopropene, 489
3-Bromopyrrolidine, *242*
N-Bromosuccinimide (NBS), 500–501, *507*, 543, 613, 618
2-Bromotoluene (*o*-bromotoluene), *596*
Bronowski, Jacob, 1174, 1174n1
Brønsted acids, definition, 42, 90, 233
 See also Acid–base chemistry
Brønsted bases
 alkenes in addition reactions, 132, 143
 definition, 42, 90, 233, 298–99
 relationship to nucleophilicity, 278–80
 See also Acid–base chemistry
Brosylate (*p*-bromobenzenesulfonate), 1085–86
Brown, H. C., 390, 398, 1116n4
Brucine, 171
Buckminsterfullerene, 604
Bullvalene, 1068–71
1,2-Butadiene, 516–17
1,3-Butadiene (methylallene)
 addition of Br_2 and Cl_2, 536, 564
 bond lengths, 522
 from electrocyclic opening of cyclobutene, 1034–40, 1042, 1047
 electron transitions and energies, 528–29
 heat of hydrogenation, 524–25
 hydrogen bromide addition to, 406, 534–35
 hydrogen chloride addition to, 369–71, 534–35, 537–39
 hydrogen iodide addition to, 406
 interconversion of s-trans and s-cis conformations, 523–24, *546*
 molecular orbitals, 520–21
 paired electron spins, 28
 physical consequences of conjugation, 521–25
 resonance forms, 26–27, 28, 519, 521, 535
 structure, *110*, *512*
 thermodynamic and kinetic control of addition reactions, 537–41
 See also Conjugated dienes; Dienes
Butanal (butyraldehyde), 767, 779, 933, *984*
Butane
 anti form, 74
 bond dissociation energy, 188
 conformational analysis, 73–75
 dihedral angle, 74
 gauche form, 74, 75, 163–64, 200
 gauche interactions, 74, 74n6, 163–64
 isobutane, 73, 75
 photohalogenation, 494–95
 2,2,3-trimethylbutane, *83*
Butanedial, 767
meso-2,3-Butanediol, 444
1-Butanethiol (butyl mercaptan), *252*, 809
Butanoic acid (butyric acid), *830*, *836*, *849*, *879*, *880*
Butanone (ethyl methyl ketone, 2-butanone), *464*, 768
Butene
 1-butene, bond dissociation energy, 475
 1-butene, heat of hydrogenation, 524
 1-butene, structure, *109*, *110*

cis-2-butene, conversion to 2,3-dibromobutane, 1093–96
cis-2-butene, hydrogenation, 580
cis-2-butene, structure, *109*, *144*, *148*, *177*, *444*
trans-2-butene, structure, *109*, *144*, *148*, *177*
butene radical, mass spectrometry (MS), 705–6
 See also Isobutene
trans-2-Buten-1-ol, *111*
Butlerov, Alexandr M., 574n2
Butyl alcohol (1-butanol), *230*, 705–6
Butylated hydroxy anisole (BHA), 484
Butylated hydroxy toluene (BHT), 484
Butylbenzene (1-phenylbutane), *644*, 645
Butyl compounds, 75–76, 494–95
 See also specific compounds
Butylethylmethylpropylammonium iodide, *242*
sec-Butyl alcohol (2-butanol), 152, *154*, 162, *230*
sec-Butyl radical, *471*, 472
tert-Butyl alcohol (2-methyl-2-propanol)
 from addition of water to 2-methylpropene, 382, 390
 from *tert*-butyl bromide, 289
 from *tert*-butyl iodide solvolysis, 333, 348
 pK_a, 987
 reaction with hydrogen iodide, 343
 solvation, 237, 238
 structure, *141*, *230*
tert-Butylamine, 241
tert-Butylbenzene, *614*, 643
tert-Butyl bromide, 289, 294–95, 301–2
tert-Butyl butyrate (*tert*-butyl butanoate), *879*
tert-Butyl chloride (2-chloro-2-methylpropane), 494, 640, 702, *703*
tert-Butylcyclohexane, 202–4
tert-Butyl cyclopentyl sulfide, *253*
tert-Butyl iodide (2-iodo-2-methylpropane)
 activation energy of ionization reaction, 343–44, 346–47
 energy *versus* reaction progress diagram, in water, *343*
 reaction mechanism in S_N1 reaction, 350–51
 S_N1 solvolysis reaction in water, 333, 343–44, 346–51, 383
 structure, *226*
 transition states in S_N1 reaction, 333, 343, 346–49
1-*tert*-Butyl-4-methylcyclohexane, 749
tert-Butyl methyl ether (2-methoxy-2-methylpropane, MTBE), *250*, 316, 324, 719
tert-Butyl methyl ketone, 983–84
1-Butyne, *126*, *127*, 516–18
2-Butyne, *126*, *128*, 516–18
Butyryl fluoride (butanoyl fluoride), *879*

Cadaverine (1,5-pentanediamine), 246
Cahn–Ingold–Prelog priority system, 111, 112–15, 152–55
Calicheamicin, *214*, 980
Camphor, *558*
Canavanine, 1177
Cannizzaro reaction, 1004–5, 1012
Cantharidin, 568
Caraway [(*S*)-(+)-carvone], 163
Carbamates, 860, 919, *924*, 1205
Carbamic acids, 860, *869*, 920
Carbanions
 allyl anion, 27–28
 2-butyne resonance-stabilized carbanion, 516

 definition, 62
 enolate anion, *27*, 28, 373
 radical anions, 453–54, 609
 stability order, 311
 structure and hybridization, 64
 See also Methyl anion
Carbenes
 concerted reactions of, 432, 434
 definition, 431
 from diazo compounds, 431, 436
 dihalocarbenes, 431
 diradicals, 434, *435*
 formation of cyclopropanes, 431, 432
 ketocarbenes, 915–16
 reaction with alkenes, 431–32
 singlet carbenes, 433–34, 436, 458
 stereospecificity, 432
 triplet carbenes, 433–36, 468
 See also Cyclopropane
Carbenium ion, 62n4
Carbinolamines, 790–91, 794–95, 819–21, 1184
Carbocations
 additions to alkenes, 132–33, 135–36, 140
 definition, 62
 delocalization of electrons, *27*
 deprotonation, 385
 heat of formation (ΔH_f°), *292*, 445
 hyperconjugation, 293–94, 377–78, 478–79
 instability of methyl and primary carbocations, 293, 376
 resonance and carbocation stability, 374–78
 resonance and regiochemistry in addition reactions of alkenes, 366–72
 simple carbocations, 141
 stability in superacid, 679–80
 stability order, 138, 139, 141, 292–93, 376, 478
 structure and hybridization, 64, 477
 transition states, 376–77
 vinyl carbocations, 444–45
Carbodiimides, 516, *1206–7*, *1218*
 See also Dicyclohexylcarbodiimide (DCC)
Carbohydrates
 anomeric carbon, 1134
 anomers, 1134–35, *1140*
 chain lengthening, 1138–39
 configurational carbon, *1127*, 1128, 1129
 D configuration, 1128
 definition, 1125
 disaccharides, 1161–68
 epimerization in base, 1141–42
 epimers, definition, 1133
 ester formation, 1147–51
 exotic laboratory syntheses, 1125n3
 Fischer determination of structures of aldohexoses, 1153–60
 Fischer projections, conventions, 1127–28
 furanoses, 1132–33
 furanosides, 1148
 glycolysis, *1140*
 glycosides, 1148–52, 1210
 Haworth forms, 1136–37
 hemiacetal formation, 784, 1130–34
 ketoses, 1128, 1141
 keto sugars, 1130
 Kiliani–Fischer synthesis, 1138–40, 1153–54, 1160
 L configuration, 1128

Lobry de Bruijn–Alberda van Ekenstein reaction, 1141–42
manipulation of alcohol groups in α-D-glucopyranoside, *1152*
modern carbohydrate chemistry, 1151–52
monosaccharides, definition, 1161
mutarotation, 1140–41
nomenclature and structure, 1126–38
nonreducing sugars, 1166
osazone formation, 1145–46, *1147*, 1158, 1159
overview, 1125–26
oxidation, 1144–45
polysaccharide, 1161, 1166, 1168
protecting groups, 1151–52
α- and β-pyranose and furanose forms of aldohexoses, *1134–35*
pyranoses, 1132–33
pyranosides, 1148
ratio of furanose and pyranose forms of aldohexoses, *1133*
reducing sugars, 1133, 1145, 1166
reduction, 1130, 1134, 1143
rotation of plane-polarized light, (+) or (−) convention, 1128
Ruff degradation, 1139–40
starch, 1161, 1168
synthesis, 1138–40
See also Aldoses; *specific types*
Carbon, 9–10, 700, 740
Carbonate ion, 45
Carbonates from phosgene reaction with alcohols, 899, *924*
Carbon dioxide, 15–16
Carbon disulfide, *914*
Carbonic acid, 859–60
Carbonium ion, 62n4
Carbon monoxide, 47
Carbon nuclear magnetic resonance spectroscopy (¹³C NMR), 88–90, 94, 740–42
Carbon tetrachloride, *16*, 417, 419, 487, 490–93, 500
Carbonyl compounds (overview)
    carbonyl group conversion into methylene group, 645, *686*
    definition, 438
    dipole moments, 766
    equilibrium between carbonyl and enols forms, 940–42
    formation during ozonolysis, 438–39, 440–42
    infrared (IR) spectroscopy, 770–71
    keto–enol tautomerization, 939–40, 944, 961, 966
    mass spectrometry, 704, 772
    meta substitution reactions of aromatic compounds, 664–65
    nomenclature, 767–69
    nuclear magnetic resonance (NMR) spectroscopy, 725, 771–72, 888–89
    oxidation to carbonyl compounds, 802–6
    ozone, role in synthesis, 441
    physical properties, 770
    π (pi) bonding, 764–66
    resonance of carbonyl group, 766
    stability of substituted carbonyl groups, 777–80
    steric destabilization, 778
    steric destabilization of hydrates, 778, 780
    structure of carbon–oxygen double bond, 763, 764–66

ultraviolet spectroscopy, *771*, 772
    See also Carbonyl compounds, addition reactions; Carbonyl compounds, condensation reactions; Carbonyl compounds, reactions at α position; β-Dicarbonyl (1,3-dicarbonyl) compounds; Enones; β-Hydroxy carbonyl compounds
Carbonyl compounds, addition reactions
    acetal formation, 783–89, *817*
    alkane formation by Clemmensen reduction, 645, *686*, 802
    bisulfite addition, 781, 782, *818*
    cyanohydrin formation, 781–82, *818*, 1139
    enamine formation, 795–96, *818*, 821, 963
    equilibrium in addition reactions, 777–81
    hydration, 773–77
    imine formation, 790–94, *818*
    overview, 763–64, 816–19
    reaction with organometallic reagents to form alcohols, 799–802, *817*
    simple reversible additions, 773–77
    substituted imine formation, *793*
    Wittig reaction, 811–13, *817*
    Wolff–Kishner reduction, 645, *686*, 802
    See also Acetals; Carbonyl compounds (overview); Hemiacetals
Carbonyl compounds, condensation reactions
    amines, 1003–4
    benzoin condensation, 1028
    Cannizzaro reaction, 1004–5, 1012
    Claisen–Schmidt condensation, 983–85, 1008
    combination of condensation reactions, 998–1001
    crossed (mixed) Claisen condensation, 993–95
    Dieckmann condensation, 992–93
    Horner–Emmons reaction, 1009
    Knoevenagel condensation, 974–76, *1014*
    Mannich reaction, 1003–4, *1013*
    Meerwein–Ponndorf–Verley–Oppenauer (MPVO) equilibration, 1005–7, *1012, 1014*
    Michael reaction, 976–80, 998–99, 1016
    problem solving techniques, 999–1001, 1007–9, 1014–17
    reverse aldol reactions, 973
    Robinson annulation, 998–99
    α,β-unsaturated aldehyde and ketone from aldol dehydration, 968–69, 971–72, 973–75, 981, *1014*
    α,β-unsaturated aldehyde and ketone from crossed aldol dehydration, 982, *1014*
    α,β-unsaturated aldehyde and ketone in Horner–Emmons reaction, 1009
    α,β-unsaturated aldehyde and ketone in Michael reaction, 976–80
    See also Aldol condensation; Carbonyl compounds; Carbonyl compounds (overview); Carbonyl compounds, reactions at α position; Claisen condensation
Carbonyl compounds, reactions at α position
    acidity of double α hydrogens, 958, 986, 989, 993
    alkylation in the α position, 954–64
    α position, definition, 763, 933
    α-bromo acid bromides, 951–52, 1182, *1216*
    α-bromo acid halides, 951–52
    α-bromo acids, 950–53
    chemical reactions, overview, 932–33, 1010–17

hydrogen abstraction in benzylic compounds, 613
keto–enol tautomerization, 939–40, 944, 961, 966
NMR spectra of α hydrogens or carbons, 714–15, 723, *725*
overview, 932–33, 1010–17
α,β-unsaturated aldehyde and ketone from aldol dehydration, 968–69, 971–72, 973–75, 981, *1014*
α,β-unsaturated aldehyde and ketone from crossed aldol dehydration, 982, *1014*
α,β-unsaturated aldehyde and ketone in Horner–Emmons reaction, 1009
α,β-unsaturated aldehyde and ketone in Michael reaction, 976–80
    See also Carbonyl compounds (overview); Carbonyl compounds, condensation reactions; Carbonyl compounds, overview; Enolates; Enols
N,N'-Carbonyldiimidazole, 873
Carboranes, 217
Carboxylate anion
    ammonium salts, 850, 852, 1179, 1183–84
    comparison to alkoxide ion, 834
    definition, 832
    as displacing agent in S$_N$2 reactions, 848
    electrochemical oxidation (Kolbe electrolysis), 860, 861, 870
    electrostatic stabilization, 835
    metal salts, 832
    methyl ester formation with diazomethane, 848–49
    neighboring group effect on α-carboxylate reaction with methyl alcohol, 1083–85
    nucleophilic addition to carboxylate anion, 836–37
    resonance stabilization, 834–35
Carboxylate salts, 832, 850, 852, 1179, 1183–84
Carboxylic acids
    acid anhydride formation, 855–56
    acid–base chemistry, 834–38
    acid chloride formation, 853–55, *868*
    acidities of substituted carboxylic acids, *836*
    alkylation, 957
    amide formation, 850–52, *869*
    ammonium salts, 850, 852, 1179, 1183–84
    decarboxylation, 858–60, *870*
    definition, 424
    dianion formation, *836, 837*, 857–58, 957
    dimers, 832, 833, 860, *870*
    from esters, *869*, 895–98, 897, 912–14
    and Fischer esterification, 841–49, 852–53, *869*, 896–97, *924*
    formation from ozonolysis, 441–42, *458, 839, 869*
    general formula, 829, *877*
    halogenation, 950–53
    Hell–Volhard–Zelinsky (HVZ) reaction, 950–53
    Hunsdiecker reaction, 861, *869*
    hydrogen bonds, 832, 833
    from hydrolysis and decarboxylation of diesters, 961–62
    from hydrolysis of acid chlorides, *890*
    from hydrolysis of acyl compounds, 877
    from hydrolysis of amides, 901–2, *922*
    from hydrolysis of nitriles, 904–5
    infrared spectra, 833

Carboxylic acids (*continued*)
 β-keto acids, decarboxylation, 858–59, *870*, 959–60
 Kolbe electrolysis, 860, 861, *870*
 lactone formation by intramolecular Fischer esterification, 849, *870*
 NMR spectroscopy, 833–34
 nomenclature, 830–32
 from organometallic reagent reaction with CO_2, 840–41
 overview, 829
 peracids, 424
 pK_a, 834, *836*
 properties, *830*
 protonation of carbonyl group and hydroxyl groups, 837–38
 reaction with metal hydrides, 858
 reaction with organolithium reagents, 856–57, *870*
 s-trans and s-cis conformations, 832–33
 structure, 769, 832–33
 synthesis by oxidative routes, 839–40, *868–69*
 See also Acyl compounds
Carboxypeptidases, 1197–98
Carboxy terminus, 1189
Carcinogenesis and carcinogens, 262, 431, 571, 605, 614–16
Carvone, (*R/S*) enantiomers, 163
Caryedes brasiliensis, 1177
Catalysts
 acid-catalyzed hydration of alkenes, 139–41, 143, 381
 benzene hydrogenation, 627
 chiral catalysts, 426
 definition, 140
 ferric chloride or bromide for halogenation of benzene, 638
 heterogeneous catalysis, 411
 homogeneous catalysis, 411
 Lindlar catalyst, 452, *458*
 palladium on charcoal, 411
 Wilkinson's catalyst (rhodium), 411
Catechol (*o*-dihydroxybenzene), *597*
Cationic polymerization, 385–86
Cations, definition, 4
Cbz (benzyl chloroformate), 1205, 1207–8, *1217*
Cellobiose, 1165
Cellulose, 1168
Cephalosporin C, *532*
Chain reactions, 468, *482*, 483
Chair cyclohexane, overview, 190–92, 220
Chemical shift (δ)
 alkanes, 722–23
 allylic hydrogens, 724
 aromaticity, 725
 definition, 718
 delocalization, 725
 deshielding, 720, 723–24, 725
 effect of magnetic field strength, 716–17, 718, 722, 735, 738
 electronegative groups, 723, 725
 hybridization, 724–25
 hydrogens attached to oxygen or nitrogen, 726
 overview, 720–22
 of various carbons (^{13}C NMR), *741*
 of various hydrogens (^{1}H NMR), *720*
 See also Hydrogen nuclear magnetic resonance spectroscopy (^{1}H NMR); Nuclear magnetic resonance spectroscopy (NMR)

Chichibabin reaction, 676–77, 761
Chiral, definition, 148, 151
Chirality, 148–51, 150, 150n2, 177–80, 181
2-Chloro-2,3-dimethylbutane, 131, 304, 365, 388
trans-1-Chloro-2-butene, 534
M-Chlorobenzaldehyde, 779
Chlorobenzene, 638, 650, 667–68, 680–81
2-Chlorobutane, 80, 165–67
Chlorobutanoic acid, *836*
2-Chloro-1-butanol, *230*
3-Chloro-1-butene, 534
Chlorocyclopropane, 173–74
2-Chloro-2,3-dimethylbutane, *131*, 304, 365, 388
2-Chloro-3,3-dimethylbutane, 388
Chlorodinitrobenzene, 674
2,4-Chlorodinitrobenzene, 674
Chloroethane, *702*
Chloroethyne (chloroacetylene), 126
Chlorofluoromethane, 721
3-Chloro-4-fluoro-2-pentanol, *230*
Chloroform (trichloromethane), *16*, 431, 490–91, 493
cis-2-Chloro-3-hexene, *111*, 394
3-Chloro-3-hexene, *444*
3-Chloro-2-hydroxypropanal, *767*
2-Chloro-2-methylbutane, 138, 386
2-Chloro-3-methylbutane, 386
1-Chloro-2-methyl-1-butene, 113
Chloronitrobenzene, 668, *668*, 674–75, 675
Chloronium ion, 418
trans-4-Chloro-2-pentene, *111*
2-Chloropropanoic acid, 737
2-Chloro-1-propanol, 748–49
1-Chloropropene, *108*
2-Chloropropene, *108*, 446, *447*
3-Chloropropene (allyl chloride), *108*
Chlorosulfite esters, 284–85, *322*, 854, *868*
Cholanthrene, *605*
Cholestanol, 221
Cholesterol
 biosynthesis, 358
 cholesterol benzoate, *497*
 low-density lipoprotein (LDL) cholesterol, 358
 structure, *121*, *216*, *532*, *562*
 synthesis, 358, *532*
Chorismate, 683–84, 1063, 1078
Chorismate mutase, *683*, 1063, 1078
Chromate esters, 804
Chromatography
 column chromatography, 172, 697
 enantiomer separation, 172, 695–96
 gas chromatography (GC), 697–99, 709
 gas chromatography/infrared spectroscopy (GC/IR), 699, 709
 gas chromatography/mass spectrometry (GC/MS), 699
 GC/IR (gas chromatography/infrared spectroscopy), 699, 709
 GC/MS (gas chromatography/mass spectrometry), 699
 gel-filtration chromatography, 1195
 high-performance liquid chromatography (HPLC), 697
 ion-exchange chromatography, 1195
 overview, 695, 697
 separation of enantiomers, 172, 695–96
Chromium trioxide, 803, 804–5
Chymotrypsin, 1202

trans-Cinnamaldehyde [(*E*)-3-phenyl-2-propenal], *768*
cis isomers, 84, *85*, 144
Citrate, 357
Civet cats (*Viverra civetta* and *Viverra zibetha*), 769
Civetone, 769
Claisen, Ludwig, 987–88, 1008, 1063n5
Claisen condensation
 amide base use, 986
 comparison of aldol and Claisen condensations, 988–89, *990*
 crossed (or mixed) Claisen condensations, 993–95, *1013*
 Dieckmann condensation, 992–93
 forward and reverse reactions in fatty acid metabolism, 996–97
 and full equivalent of base, 986, 989, 991, 992
 intermolecular Claisen condensations, 992–93
 ketone enolate reaction with esters, 985–87
 kinetic enolates, *986*, 1008
 mechanism, *985*
 overview, 985–88, *991*, 992
 removal of double α hydrogens, *985*, 986, 989, 991, 994
 reverse Claisen condensations, *924*, 992
 thermodynamic factors, 988–89, *990*, 992
 See also Aldol condensation; Carbonyl compounds, condensation reactions; β-Keto esters
Claisen rearrangement, 1063, 1077–78
Claisen–Schmidt condensation, 983–85, 1008
 See also Carbonyl compounds, condensation reactions
Clemmensen reduction, 645, 686, 802
Coates' cation, 1117
Cocaine, 256
Coenzyme, definition, 814
Combustion, 68, 92, 144
 See also Heat of combustion (ΔH_c°)
Concerted reactions
 comparison to nonconcerted, stepwise reactions, 1033
 definition, 389, 396, 1031
 epoxidation, 424–25
 hydroboration, 391–95
 overview, 1032–34
 ozonolysis, 437
 rearrangements of acyl compounds, 914–20
 See also Diels–Alder reaction; Pericyclic reactions; S_N2 reactions
Configurational carbon, *1127*, 1128, 1129
Conformation, definition, 65
Conformational analysis
 butane, 73–75
 cyclohexane, 197–99
 definition, 73
 ethane, 65–67
 methylcyclohexane, 194–201
 propane, 72
Conformational diastereomers, 177, 208–9
Conformational enantiomers, 164
Conformational isomers, 70, 177
Coniine, 170
Conjugate acid, 234
Conjugate base, 234
Conjugated dienes
 addition reactions, 534–36, *563*, *564*
 bond lengths, 522

chemical properties of acyclic and cyclic dienes, 572
definition, 512
halogenation, 536
heats of hydrogenation, 524–25
hydrohalogenation, 369–70, *404*, 534–35, 537–41
interconversion of s-trans and s-cis conformations, 523–24, 545–46
molecular orbitals, 520–21
physical consequences of conjugation, 521–25
resonance forms, 519, 521
thermodynamic and kinetic control of addition reactions, 537–41
ultraviolet/visible (UV/vis) spectroscopy, 528–32
See also 1,3-Butadiene (methylallene); Diels–Alder reaction; Dienes
Conjugated double bonds, definition, 512
Conrotation, 1037–39, 1041, *1042*
Conrotatory motion, definition, 1037
Cope elimination, 914, *923*
Cope rearrangement
bullvalene, 1068–71
chorismate rearrangement to prephenate, 1063, 1078
definition, 1059
degenerate reaction, 1059
3,4-dimethyl-1,5-hexadiene, 1062–63
1,2-divinylcyclopropane, 1064–65, 1067
cis-1,3,5-hexatriene and 1,3-cyclohexadiene interconversion, 1064
homotropilidene, 1066–68
labeling experiments, 1059–60
photochemical rearrangement, 1061
thermal rearrangement, 1060
transition state, 1060–62
See also Sigmatropic shift reactions
Cortisone, *212*, *562*
Coupling constant (*J*), 718, 728, 731–32
See also Spin–spin coupling
Covalent bonds, overview, 6, 13, 14–16
See also Lewis structures
Cram, Donald J., 254, 1096, 1116n4
Crick, Francis H. C., 1213
Crossed (mixed) aldol condensation, 982–87
See also Aldol condensation
Crossed (mixed) Claisen condensation, 993–95, *1013*
See also Claisen condensation
trans-Crotyl alcohol, *231*
Crown ethers, 254–55
Cryptands, 255
Cubane, *216*
Cumene (isopropylbenzene or 2-phenylpropane), *595*, 639–40, 641–42
Cumulated double bonds
carbodiimides, 516, *1206–7*, *1218*
cumulated alkenes (cumulenes), 512, 515–16
isocyanates, 918–21, *923*, *924*, *1206*
isothiocyanates, 921, 1198–99, *1206*, *1218*
See also Allenes; Ketenes
Curl, R. F., 603–4
Curtius rearrangement, 918–19, 920, *924*
Curved arrow formalism, 23, 24, 44, 263–64
Cyanide addition to carbonyl compounds, 781–82, *818*
Cyanobenzenes, *686*, 719
Cyanogen bromide (BrCN), 1200–1201

Cyanohydrin formation, 781–82, *818*, 1139
Cyclamates, 1164
Cyclic alkanes. *See* Cycloalkanes
Cyclic hydrogen (H_3, H_3^+), 102, 217, 1103
Cycloaddition reactions
definition, 1032, 1043
Diels–Alder reaction, 1033, 1043–45, 1047–48
HOMO–LUMO interactions, 1043–48
photochemical alkene dimerization, 1045–46, 1074
problem solving techniques, 1072
rules for cycloaddition reactions, 1047
See also Pericyclic reactions
Cycloalkanes
1,2-dimethylcyclopropane, 84–85, 148, 205–7
alkylcyclohexanes, axial–equatorial energy differences, 202–4, 277
bicyclopentyl, 85
1,2-dimethylcyclopropane, 208
disubstituted ring compounds, structure, 204–10
formation of ring compounds, *83*
heat of formation (ΔH_f°), *194*

heats of combustion (ΔH_c°), 196–97
molecular formula, 51, 83, 98, 130
monosubstituted cyclohexanes, structure, 194–204
naming conventions, *84*
polycyclic compounds, 85–86
properties, *84*, 86–88
stereochemical analysis, 173–76
strain, 187–93
strain calculation from heat of combustion (ΔH_c°), 194–97
strain calculation from heat of formation (ΔH_f°), 193–94
See also Ring compounds
Cycloalkenes
bicyclic compounds from Diels–Alder reaction, 551–54
bridgehead alkenes, 121–23
heat of hydrogenation, 580
hydrogenation, 412
molecular formula, 131
naming conventions, *110*
natural products, *121*
physical properties, *123*
polycyclic compounds, *121*
stability of cis double bonds, 119–20
structures, 118–23
structures of cyclic polyenes, *118*
twisting of trans double bonds, 119–20
See also Alkenes
Cycloalkynes, 128
Cyclobutadiene, *118*, 572, *573*, 583–84, 585–86
Cyclobutane, 98, 188–89
Cyclobutene, opening to a 1,3-butadiene, 1034–40, 1042, 1047
Cyclodecane, 193
Cyclodecapentaenes, 592–94, *621*
1,3,5-Cycloheptatriene (tropilidine)
bromination, 620
hydride transfer, 587, *618*
orbitals, 583, 584, 588
pK_a, 591
structure, *118*, *573*, 583, 587–88
Cycloheptatrienyl anion, 587–88, 590–91

Cycloheptatrienylium ion (tropylium ion), 587–89, 590, *618*
Cycloheptyne, *128*
1,3-Cyclohexadiene, *118*, *512*, 572, 580–81, 1040–41, 1064
1,4-Cyclohexadiene, *118*, *512*, *564*, 608–9, 617, *618*
1,3-Cyclohexadienone, *942*
See also Phenol
Cyclohexadienyl cation
from benzene, 606–9, 631–33, 635
conversion back to aromaticity, 634, 635, 639
in Friedel–Crafts acylation, 643, 646
in Friedel–Crafts alkylation, 640
in halogenation reactions, 638
in meta substitution of anisole, 657, 658
in nitration reactions, 638
in ortho substitution of anisole, 658
overview of role as intermediate, 626, 685, 688
in sulfonation reactions, 636–37
Cyclohexane
from benzene hydrogenation, 580–81, 626–27, *686*
chair–twist equilibrium, 199
conformational analysis, 197–99
1,1-disubstituted cyclohexanes, 205
1,2-disubstituted cyclohexanes, 205–10
drawing chair form, 190–92, 220
equatorial–axial conformation equilibrium, 201–3, 205, 206
full-boat conformation, 198
half-chair conformation, 197–98
mass spectrometry (MS), 700–701, 702
monosubstituted cyclohexanes, structure, 194–204
nuclear magnetic resonance spectroscopy (NMR), 746–47
ring flip, 192, 198–99, 746–47
rotational energy barrier, 198–99, 747
stereochemistry, 190–92, 197–99
steric interactions in S_N2 reactions, 276–77
strain, 190–92
twist (twist-boat) conformation, *197*, 198, 199
undecadeuteriocyclohexane (cyclohexane-d_{11}), 746
van der Waals strain, 198
Cyclohexanecarboxaldehyde, *767*
1,4-Cyclohexanedione, *768*
Cyclohexanethiol, *252*
Cyclohexanone, *709*, *771*, *812*
1,3,5-Cyclohexatriene, 572, 573–75, 580–81, 629
See also Benzene
Cyclohexene
allylic halogenation, 497–500
from Diels–Alder reaction, 513, 544–48, 550, 554, 563–64
halogenation, 418–19, 497–500
hydrogenation, 580, 626
temperature effect on stability, *338*
See also Hexenes
Cyclohexylamine (cyclohexanamine, aminocyclohexane), *241*
1,2-Cyclononadiene, *514*
Cyclononyne, *128*
1,4-Cyclooctadiene, *512*
1,3,5,7-Cyclooctatetraene, *118*, *573*, 581, 583, 586, 620
cis-Cyclooctene, *120*
trans-Cyclooctene, *120*, 178

Cyclooctyne, *128*
Cyclopentadiene (1,3-cyclopentadiene)
    Diels–Alder reactions, 551, 553
    fulvene formation, 975
    pK_a, 589, 591, 601, 975
    structure, *118, 875*
Cyclopentadienyl anion, 589–90, 591, 599, 618
Cyclopentane, *83, 85, 186,* 190, 414, 416
Cyclopentanol (cyclopentyl alcohol), *230*
Cyclopentene, 119–20, 414
3-Cyclopentylpropyne (propargylcyclopentane),
    127
Cyclopropanated fatty acids, 455
Cyclopropane
    angle strain, 187, 474
    aziridine (azacyclopropane), 242, 464–65
    bent orbitals, *187*
    bond dissociation energy, 188, 474
    cyclopropanated fatty acid formation, 455
    cyclopropyl bromide, 276
    pyrethroids, 456
    structure, *51*
    torsional strain, 187–88, 474
    total strain estimates, 188, 194
    See also Carbenes; Cycloalkanes
Cyclopropenes, 455–56
Cyclopropenium ion, *592*
Cyclopropenylidene, 456
Cysteine, *1176,* 1178, 1190, *1216*
Cytosine, 615, 1212

Dacron, 852
Dalton, John, 2
Daughter ions, 702
Debye, Petrus Josephus Wilhelmus, 15
(*E*)-1,5-Decadiene, *512*
Decalin, 212–14
Decarboxylation
    alcohol formation from bicarbonate esters, *869*
    carbamic acid, 860, *869,* 920
    carboxylic acids, 858–60, *870*
    β-dicarbonyl compounds, 959–62, *1014*
    diesters, 961–62
    Hunsdiecker reaction, 861, *869*
    β-keto acids, 858–59, *870,* 959–60
    β-keto esters, 959–62, 963, *1014*
    malonic acid, 858–59, 961–62
    monoesters of carbonic acid, 859–60
Decoupling of NMR spectra, 733, 739, 740–41,
    746
Degenerate reactions, 1031, 1048–49, 1059
Degree of unsaturation (Ω), 130–31, 751
Dehydroquinate, 683
Dehydroshikimate, 683
Delocalization of electrons
    allyl cation, 27, 369–70
    allyl radical, *475,* 480, 498–99, 543
    benzyl radical, 480, *612*
    carbocations, *27*
    delocalization energy, 581–82, 607, 614, 626
    NMR chemical shift, 725
    radical formation, 475–76, 480, 498
    stabilizing effect in resonance forms, 27, 102
Democritus, 2
Denaturing, 1193–94
Deoxyribonucleic acid (DNA)
    alkylation, 288, 615–16
    alkylation reactions and mutations, 288–89
    base pairs, 288, 615, 1212–13, 1214

bases, 288, 614–15
double helix, 598, 1213
enzymatic epoxidation, 616
hydrogen bonding, *288,* 615, 1212–13
mutation, 288–89, 615–16
π (pi) stacking, 598, 615
and polyaromatic hydrocarbons, 614–16
replication, 1213
S_N2 reactions, 262, 288–89
structure and composition, 614–16, 1149,
    1211–13
See also Nucleic acids
Deoxyribonucleosides, 1211
Deoxyribose, 288, 1211, *1212, 1214*
DEPT (distortionless enhancement with
    polarization transfer), 741
Deshielding, 720, 723–24, 725
Detergents, 866–67
Deuteriobenzene, 607–8, *618,* 632–34, *687*
Deuterium
    abundance, 701
    aldol condensation, 970
    deuterated hydrocarbons, 229, *257*
    deuteride exchange reactions, 606, 936–39,
        944, 949–50, *1013*
    deuteride shift, *402*
    mass, 113
    and NMR spectroscopy, 717, 733–34, 746
    resonance in deuterium-labeled compounds,
        475, 539–40, 606
    sodium borodeuteride, 423
    and syn addition, 394–95, 412
    transfer in S_N2 reactions, 288
Dewar, James, 574
Dewar benzene (bicyclo[2.2.0]hexa-2,5-diene),
    573–74, 578, 617, 1043
    See also Benzene
Dewar forms, 576–78, 617
Dextrorotatory, 156
Diacetone alcohol (4-hydroxy-4-methyl-
    2-pentanone), 971
Dials, 767
Diamantane, *218,* 219
Diamond, 218–19
Diastereomers, 165–67, 169–72, 177, 208–9
Diastereotopic hydrogens, 721–22, 731, 749
Diazo compounds, 431, 436, 620
    See also Carbenes
Diazocyclopentadiene, 431, 620
Diazofluorene, *431*
Diazo ketones, 915–18, *924*
Diazomethane, 431, 848, *875,* 915
Diazonium ions, 329, 647–51, *686, 687*
Diazotic acid, 648
Dibenz[*a,h*]anthracene, 605
Dibenzo[*c,g*]phenanthrene, structure, *179*
Diborane, 390–91
2,3-Dibromobutane, 1093–95
trans-1,2-Dibromocyclopentane, 414, 416
1,2-Dibromo-3,3-dimethylbutane, 415
cis-1,1-Dibromo-2,3-dimethylcyclopropane, *432*
β-Dicarbonyl (1,3-dicarbonyl) compounds
    alkylation, 958, 963
    equilibrium between carbonyl and enols forms,
        941–42, 958–59, 974, 985–86
    hydrogen bonding, 941, 942
    hydrolysis and decarboxylation, 959–62, *1014*

intramolecular hydrogen bonds, *942*
pK_a, *958*
structure, *941*
See also Carbonyl compounds; Carbonyl
    compounds (overview)
1,3-Dichlorobenzene (*m*-dichlorobenzene), *596*
2,3-Dichlorobutane, 176–77
Dichlorocuprate ion, 650
Dichlorocyclopropane, 174–76, 204
1,2-Dichlorofluoroethane, 721
3,5-Dichlorohexanoic acid, *831*
1,3-Dichloro-2-methoxypropane, *732*
1,2-Dichloropropane, *446*
2,2-Dichloropropane, *446, 447*
Dicyanobenzenes, 719
Dicyclohexylcarbodiimide (DCC), 850, 856,
    869–70, 1205–7, 1209, *1217*
Dicyclohexylurea (DCU), 1206
Dicyclopentylamine, *241*
Dicyclopropyl disulfide, *253*
Dieckmann condensation, 992–93
Diels–Alder reaction
    aromatic compounds, 628–31
    benzene, 628–29
    bicyclic compound synthesis, 551–54
    concerted reaction, 547–49, 1033, 1043–45,
        1047–48
    cycloaddition reaction, 1033, 1043–45,
        1047–48
    cyclohexene production, 513, 544–48, 550,
        554, 563–64
    dienophile, definition, 544
    endo and exo products, 551–52, 567
    furans, 629
    orbital overlap and interactions, 544–45, 552,
        1043–45
    overview and background, 544, 563, 564
    photochemical Diels–Alder reaction, 1044–45
    polycyclic products, 552–53
    pyrrole, 630
    reaction mechanism, 547–49
    retention of stereochemistry, 548–49
    reverse Diels–Alder reactions, 554, 564,
        630–31
    s-cis conformation, 545–46, 550
    thermal Diels–Alder reaction, 1043–44
    thermochemistry, 545
Dienes
    allenes, 178, 513–15
    as intermediates in alkyne isomerization,
        516–18
    overview, 512–13
    unconjugated dienes, *512,* 519, 524–25, 528
    See also 1,3-Butadiene (methylallene);
        Conjugated dienes
Dienophile, definition, 544
N,N-Diethylpropanamide, *882*
Diethyl tartrate, 426
Dihedral angle, 65–66, 67, 72, 74, 731
Dihelium (He$_2^+$), 40–41, 479
Dihydrobullvalene, *1068*
Dihydropyran (3,4-dihydro-*2H*-pyran), *700,* 701,
    702, 789
1,3-Dihydroxyacetone, *1128*
Diimide, 411
Diisobutylaluminum hydride (DIBAL-H),
    900–901, 923
2,3-Dimethyl-1-butene, 367
3,3-Dimethyl-1-butene, 388, 392, 415, 501–2

2,3-Dimethyl-2-butene
    catalyzed hydration of, 139–40, 381
    hydroboration, 397
    hydrogen chloride addition, 131–32, 133–36,
        365–67, 374
N,N-Dimethyl acetamide, *889*
Dimethylamine, *241*
3,3-Dimethylbicyclo[3.2.1]octane, 215
2,3-Dimethyl-2-butene, 116, 117, 139–40, 381,
    397
3,3-Dimethyl-1-butyne, *127*
cis-5,6-dimethylcyclohexa-1,3-diene, 1041
1,1-Dimethylcyclohexane, 205
1,2-Dimethylcyclohexane, 205–8
1,2-Dimethylcyclopropane, 84–85, 148, 205–7,
    208
4,4-Dimethyldiazocyclohexa-2,5-diene, 431
1,2-Dimethylenecyclohexane, 546
Dimethyl ether (methoxymethane), *250*, 314
N,N-Dimethylformamide (DMF), *873*, 882, *889*,
    928
6,6-Dimethylfulvene, *975*
3,4-Dimethyl-1,5-hexadiene, 1062–63
Dimethylketene (2-methyl-1-propen-1-one), *883*
1,5-Dimethylnaphthalene, *530*
Dimethylpentane, structures, *82*
4,4-Dimethyl-2-pentanone, 718
2,2-dimethylpropane (neopentane), 705
2,2-Dimethylpropanol, 719
Dimethyl sulfate, 285
Dimethyl sulfide, *253*, 314, 440
Dimethyl sulfoxide (DMSO), 806
Dinitrobenzene, *663*, 668
2,4-Dinitrofluorobenzene, 1196, *1197*
Dinitrogen trioxide, 647
2,4-Dinitrophenol, 674
2,4-Dinitrophenylhydrazone, 793
Dinitrotoluene, *669*
Dioclea megacarpa, 1177
Diols (glycols)
    gem-diols, 776, 781, 783, *790*
    overview, 240
    vicinal diols, oxidative cleavage, 807, *817*, *818*,
        845
    vicinal diols from alkene oxidation, 443, *459*,
        840
Diones, 768
1,4-Dioxane (1,4-dioxacyclohexane, 1,2-dioxan),
    254, *329*
1,3-Dioxolane (1,3-dioxacyclopentane), *254*
Diphenyl ketone (benzophenone), *769*
1,3-Dipolar reagents, 436–37, *459*
    See also Carbenes; Diazo compounds; Ozone;
        Ozonolysis
Dipole moment, definition, 14–16
Diradicals
    calicheamicin antitumor action, 980
    carbenes, 434, *435*
    cyclobutadiene, 585, 586
    cycloheptatrienyl anion, 591
    cyclooctatetraene, 586
    from cyclopropane, 474
    oxygen (O_2), 484
Disaccharides, 1161–68
Disproportionation reaction, 471–72, 473–74, *506*
Disrotation, 1037, 1039–40, 1041, *1042*
Distillation, 695
Disulfide bridges, 1190, 1193, 1194–95, *1216*
Disulfides from thiol oxidation, 809, *818*

Dithianes (thioacetals), alkylation, 1001–2
Dithiolate ions, 324
Ditropyl ether, 620
1,2-Divinylcyclopropane, 1064–65, 1067
DNA. *See* Deoxyribonucleic acid (DNA)
Docosahexaenoic acid (DHA), *865*
Dodecahedrane, *186*, 216, *216*
Doering, William, 274, 1056n3, 1062–63, 1069n7,
    1071
Double bonds, overview
    in alkenes, overview, 18, 98
    Cahn–Ingold–Prelog priority system, 153–54
    carbon–carbon double bonds as intramolecular
        nucleophiles, 1099–1108
    carbon dioxide, 15–16
    conjugated double bonds, definition, 512
    Lewis structures, 18–19
    See also Alkenes; Dienes
Dynamic NMR, 746–49
    See also Nuclear magnetic resonance
        spectroscopy (NMR)

E1cB reactions, 309–10, 968–69
E1 reactions
    comparison to E1cB reactions, 309
    competition between S_N1 and E1 reactions,
        298–99
    overview, 321, 322
    rate law, 298–99
    Saytzeff elimination, 300, 311, 322
    selectivity for more substituted product,
        299–300
E2 reactions
    anti elimination, 305–6, 311, 322
    tert-butyl bromide E2 reactions, 301–2
    competition between S_N2 and E2 reactions,
        301, 321
    effect of substrate branching on E2–S_N2 mix,
        301–2
    Hofmann elimination, 308–9, 310–11
    overview, 321, 322
    rate law, 301
    reaction mechanisms, 301
    Saytzeff elimination, 306, 308, 310, 311
    selectivity in E2 reactions, 306–11
    stereochemistry, 302–6
    syn elimination, 305–6
Eclipsed conformation, 65–67
Eclipsing strain. *See* Torsional strain
Edman degradation, 1198–1200, 1202, *1218*
Eicosapentaenoic acid (EPA), *865*
Electrocyclic reactions
    conrotation, 1037–39, 1041, *1042*
    cyclobutene, opening to a 1,3-butadiene,
        1034–40, 1042, 1047
    definition, 1032
    disrotation, 1037, 1039–40, 1041, *1042*
    cis-1,3,5-hexatriene and 1,3-cyclohexadiene
        interconversion, 1040–41, 1064
    HOMO control of reactions, 1036–41
    photochemical HOMO, 1039–41, 1044, 1046,
        1048, 1058
    photochemical reactions, 1035, 1038–40
    problem solving techniques, 1071–72
    rules for rotary motions, *1042*
    thermal reactions, 1034–38
    See also Pericyclic reactions
Electromagnetic spectrum, *707*
Electron affinity, 4–5, 15

Electron configurations, overview, 3–6, *8*, *10*, 46
    See also Atomic orbitals; Quantum numbers
Electronegativity
    common elements, *15*
    definition, 15, 132
    NRM chemical shift of electronegative groups,
        723, 725
    and polar covalent bonds, 15–16, 132
    and resonance forms, 27–28, 373
Electronic spectroscopy, 526
    See also Ultraviolet/visible (UV/vis)
        spectroscopy
Electron pushing, 23, 24, 25
Electrons, overview, 1, 2, 44
Electron shells, 4
Electron spin, 7–8, 10
Electron volt (eV), 4n5
Electrophiles, definition, 42, 390, 625
Electrophilic aromatic substitution, 635, 685
    See also Substitution reactions of aromatic
        compounds
Electrophoresis, 1180–81, 1195
Elimination, unimolecular, conjugate base (E1cB)
    reactions, 309–10
Elimination reactions
    anti elimination, 305–6, 311, 322
    E1cB reactions, 309–10
    Hofmann elimination, 308–9, 310–11
    overview, 298, 311
    Saytzeff elimination, 300, 306, 308, 310, 322
    syn elimination, 305–6
    See also E1 reactions; E2 reactions
Enamines, 795–96, *818*, 821, 963–64, 1146
Enantiomers
    absolute configuration, 172–73
    chemical differences, 159–63
    definition, 151
    interconversion of conformational enantiomers,
        163–64
    optical activity, 152–55, 1128
    racemic mixture (racemate), 156–57
    reactions with racemic mixtures, 164–69
    resolution, 169–72, 1184–85
    (*R/S*) convention, 152–55, 181
    separation as diastereomeric salts, 170–72
    separation by chromatography, 172, 695–96
    visual resolution, *169*
Enantiotopic hydrogens, 721, 747–48
Endergonic reactions, 335
Endo isomers, definition, 551
Endothermic reactions, 3, 36, 115, 193, 335
Energy barriers. *See* Activation energy (ΔG°);
        Rotational energy barriers; Transition
        states
Enolates
    acyl compounds, 942–44
    alkylation in the α position, 954–64
    base-catalyzed α hydrogen exchange reaction,
        936–38, *944*
    boron enolates, 1008
    comparison of enolate and allyl anions, 934
    definition, 934
    ester enolate reactions with aldehydes and
        ketones (Knoevenagel condensation),
        974–76, *1014*
    ester enolate reaction with esters (Claisen
        condensation), 987–92
    ketone enolate reaction with esters, 985–87
    kinetic enolates, *984*, 985, *986*, 1008

Enolates (*continued*)
　　Michael reaction, 976–80
　　molecular orbitals, 935–36, *955*
　　racemization, 944–46
　　reactions, summary, *1010–14*
　　reaction with aldehydes, 965–71
　　reprotonation, 934–35, 946
　　resonance stabilization, 27–28, 373, 933–34, 955
　　thermodynamic enolates, 1008
　　See also Aldol condensation; Carbonyl compounds, reactions at α position
Enolpyruvylshikimate-3P, *683*
Enols
　　acid-catalyzed α hydrogen exchange reaction, 938–39, *944*
　　conversion to aldehydes, 449, 451
　　conversion to ketones, 448–50, *652*
　　definition, 448
　　from enolate reprotonation, 934–35
　　equilibrium between carbonyl and enols forms, 940–42
　　Hell–Volhard–Zelinsky (HVZ) reaction, 950–53
　　keto–enol tautomerization, 939–40, 944, 961, 966
　　from ketones, *652*
　　from oxymercuration of alkynes, 448–49
　　propanal, 936
　　racemization, 944–46
　　reactions, summary, *1010–14*
　　from vinylvoranes, 451
　　See also Carbonyl compounds, reactions at α position
Enones
　　cuprate addition, 979
　　definition, 976
　　Michael reaction, 976–80
　　thermodynamics of reactions, 978–79
　　α,β-unsaturated aldehyde and ketone from crossed aldol dehydration, 982, *1014*
　　α,β-unsaturated aldehyde and ketone in Horner–Emmons reaction, 1009
　　α,β-unsaturated aldehyde and ketone in Michael reaction, 976–80
　　α,β-unsaturated carbonyl aldol dehydration products, 968–69, 971–72, 973–75, 981, *1014*
　　See also Carbonyl compounds; Carbonyl compounds (overview)
Entgegen (E) isomers, 109, 111, 112–15
　　See also trans isomers
Enthalpy change (ΔH°)
　　definition, 36, 335, 337
　　enthalpy of activation (ΔH‡), 342, 349, 359
　　estimation from bond dissociation energy, 338
　　relationship to equilibrium constant, 335
　　relationship to Gibbs free energy (ΔG°) change, 335, 349
　　sign convention, 3, 37, 40, 193, 195
　　See also Heat of combustion (ΔH°ᶜ); Heat of formation (ΔH°f)
Entropy change (ΔS°), 335, 337–39, 349
Enzymes and reaction rates, 357–58
Epimers, 1133
Episulfonium ion, 1088
Epoxidation
　　of alkenes, 424–25, *459*

asymmetric (Sharpless) epoxidation, 426
DNA, 616
leaving groups, 424, 427, 429
single-step (concerted) mechanism, 424–25, *459*
　　See also Oxiranes
Epoxides
　　definition, 317
　　1,2-diol formation, 426–27
　　mechanisms for opening in acids and bases, 427–29
　　reaction with amines, 318
　　reaction with organometallic reagents, 429–30, 797
　　regiochemistry of ring opening, 427–29
　　synthesis, 317, 424, *459*
　　See also Oxiranes
Equatorial hydrogens, 191
Equilibrium
　　equilibrium arrows, 25
　　overview, 332–33, 358–60
　　resonance *versus* equilibrium, 24–25, 29, 44, 372, 404, 617
　　Sₙ1 reactions, 333, 337, 350
　　Sₙ2 reactions, 267–68, 274, 332
　　in substitution reactions, 268
Equilibrium constant (K), 120, 201–2, 334–37
Erythrose, *1128, 1129*
Essential amino acids, 1063, 1077
Esters
　　acid–base chemistry, 886–87
　　acid-induced ester hydrolysis, 895, 896
　　from Baeyer–Villiger reaction, 907–9, *924*
　　base-induced ester hydrolysis, 862, 869, 895, 896–97
　　α-bromo esters, 954
　　chlorosulfite esters, 284–85, *322*, 854, *868*
　　chromate esters, 804
　　direct alkylation, 963, *1012*
　　enolates, 943
　　and Fischer esterification, 841–49, 852–53, *869*, 896–97, *924*
　　general formula, *877*
　　halogenation, 954
　　ketone enolate reaction with esters, 985–87
　　lactone formation by intramolecular Fischer esterification, 849, *870*
　　lactones (cyclic esters, oxacycloalkanones), 849, *870*, 880, 909, *925*
　　methyl ester formation with diazomethane, 848–49, *869*
　　odors and flavors, 876, 878
　　and organometallic reagents, 898–99
　　ortho esters, 846–48
　　polyesters, 852–53
　　pyrophosphate esters, 555–59
　　resonance forms, *885*
　　structure, *769*
　　thermal elimination reactions, 912–14, *923, 924*
　　transesterification, 896–98, *924*, 987
　　xanthate esters, 914, *923, 925*
　　See also Acyl compounds; β-Keto esters
Estradiol, 182
Estrone, *562*
Ethane
　　¹³C NMR spectroscopy, 88
　　carbon–carbon bond strength, 68
　　combustion, 68
　　conformational analysis, 65–67

1,2-dideuterioethane, 67–68
dihedral angle, 65–66, 67
eclipsed ethane, 65–67
formation from methyl compounds, 64
formation from methyl radicals, 68–69, *83*, 103
Lewis structure, 18
molecular orbitals, 68, 103
properties, 68
rotational energy barrier, 67, 70, 74–75, 93, 105–6, 188, *273*
staggered ethane, 65, 75
structure, 64
structure drawings, *64, 65*, 70–71
Ethanol (ethyl alcohol)
　　alcoholic beverages, 400
　　chemical exchange of OH hydrogens, 733–34, 739
　　decoupled NMR spectra, 733, 739, 746
　　formula, *230*
　　NMR spectroscopy, 733–34, 747–48
　　physiology, 400
Ethene. *See* Ethylene
Ethers
　　acid–base chemistry, 251, 254
　　from alkoxide treatment with dimethyl sulfate, 285
　　bromo ethers and chloro ethers, 419, *458*
　　cleavage by haloacids, 282–83
　　crown ethers, 254–55
　　cryptands, 255
　　cyclic ethers, 251, 254–55, 316, 391
　　and Grignard reagent formation, 228, 251, 254
　　nomenclature, 250
　　and organolithium reagent formation, 228
　　physical properties, 250
　　polyethers, 254
　　structure, 249, 251
　　tetrahydropyranyl (THP) ethers as protecting groups for alcohols, 789–90
　　trialkylsilyl ethers as protecting groups, 789–90
　　Williamson ether synthesis, 254, 315–17, *322*, 1147, 1150, 1163
　　See also Thioethers (sulfides)
Ethoxide ion, *236, 299, 301*, 315–16, *543*
Ethoxyethene (ethyl vinyl ether), *250*
Ethyl acetate, *713, 889, 943*
Ethyl acetoacetate, 959
Ethylamine, *241*
Ethylbenzene (phenylethane), *595, 613*
Ethyl benzoate, 664–65, 741
M-Ethylbenzoic acid (3-ethylbenzoic acid), *831*
Ethyl bromide, 282
Ethyl compounds, structure and properties, 69
Ethyl crotonate, 739
Ethylene (ethene)
　　bonding, 18, 99–107
　　common name for ethene, 99, 99n2
　　electron transitions and energies, 527–29
　　hydration with sulfuric acid, 383
　　Lewis structure, 18
　　molecular orbitals, 103–4
　　pi (π) orbitals, 104
　　plant hormone function, 106
　　protonation, 378–79
　　rotational energy barrier, 105–6, 107, 144
　　stabilizing effect of 2p/2p overlap, 105–6
　　structure, 99, 105–7, *764*
　　torsional strain, 105
　　See also Alkenes

Ethylene glycol (1,2-ethanediol), *231*, 240, 852
Ethylene oxide, *231*
1-Ethyl-4-fluorobenzene (*p*-ethylfluorobenzene), *596*
Ethyl iodide (iodoethane), *226*, *278*, 727–28
Ethyl maleate, 740–41
Ethylmethylamine, 241
Ethylmethylammonium chloride, *242*
Ethylmethylketene (2-methyl-1-buten-1-one), *883*
Ethylmethylketene (2-methyl-1-propen-1-one), *883*
Ethyl 3-methylpentanoate, *879*
Ethyl methyl sulfide, *253*
3-Ethylpentane, structure, *82*
N-Ethylpiperidine, *242*
Ethyl propionate (ethyl propanoate), *879*
Ethyl radical, 471, 473–74
Ethyne. *See* Acetylene
Ethynyl compounds, 126
Ethynylcyclohexane (cyclohexylacetylene), *127*
Eugenol, 582
Exergonic reactions, 335
Exo isomers, definition, 551
Exothermic reactions, 3, 36–37, 40, 115, 193, 335
Extinction coefficient, 527

Farnesol, 557, 558
Farnesyl pyrophosphate (farnesyl–OPP), 556–57, 559
Fats and oils, definition, 862
Fatty acids
    cyclopropanated fatty acids, 455
    definition, 862
    docosahexaenoic acid (DHA), *865*
    eicosapentaenoic acid (EPA), *865*
    α-linolenic acid (ALA; octadecatrienoic acid), *865*
    metabolism, 455, 862–63, 996–97
    nomenclature, *862*
    omega-3 fatty acids, 865
    saturated fats, 863, *864*
    unsaturated fats, 863, *864*
Ferrocene, *531*
Finkelstein reaction, 360
First-order reactions
    E1 reactions, 298, 301
    rate constant, 289, 341
    S_N1 reactions, 268, 289, 294, 340, 341
    substrate and nucleophile in E2 reactions, 301
    substrate and nucleophile in S_N2 reactions, 268, 340
First-order spectrum, 735
Fischer, Emil, 841, 1125–26, 1125n1, 1139, 1153–56
Fischer determination of structures of aldohexoses, 1153–60
Fischer esterification, 841–49, 852–53, *869–70*, *896–97*, *924*, 1187
Fischer projections, 1127–28, 1178
Fluoborate ion, 249–50, *587*, *618*
Fluoride, 4, *266*
Fluorine, 4–5, 14, 17–18, 21, 237
Fluorobenzene, *650*
1-Fluorobutane, 680
Fluxional structure, 1063
    See also Bullvalene
Force constant, 708
Formal charge, 20–22

Formaldehyde (methanal)
    bond angles, *764*
    dipole moment, *766*
    equilibrium in hydration reactions, 777–81
    molecular orbitals, *764*
    reaction with organometallic reagents to form primary alcohols, 800, *817*
    resonance forms, 25
    synthesis, *464*
    in wood products, *765*
Formic acid (methanoic acid), *830*, 832, *879*, *880*
Formyl chloride (methanoyl chloride), *880*
Fractional crystallization, 695
Fragmentation patterns, 702–6
Franklin, Rosalind, 1213
Free radical reactions. *See* Radical reactions
Friedel–Crafts acylation, 643–46, *687*, 893
Friedel–Crafts alkylation, 639–43, 646, *686*
Frost circles
    1,3,5,7-cyclooctatetraene, 586
    benzene, 584–85
    cyclobutadiene, 585
    cyclopentadienyl anion, 589
    overview, 584–85, 618
    tropylium ion, *588*
    See also Hückel's rule; Molecular orbitals
D-Fructose, *1130*, 1141–42, 1166
Fukui, Kenichi, 531, 1032
Fulvenes, 975
Functional groups, 22, 51
Furan, 250, *254*, 599–600, 620, 629, 654–55
Furanoses, 1132–33
Furanosides, 1148
Furfural, *654*
2-Furoic acid, 736–37
Fused compounds, 121, 179–80, 211, 212–13, 592
    See also Polynuclear aromatic compounds

GABA (γ-aminobutyric acid), 400
Gabriel synthesis, 1182–83, *1216*
Galactaric acid, *1144*
D-Galactose
    chain shortening, *1139*
    in lactose, 1161–64
    in melibiose, 1167
    oxidation, *1144*
    reduction, *1143*
    structure, *1129*, 1133, 1160
Garcia, Jerry, 877, 877n1
Garlic (sulfur-containing compounds), *252*
Gas chromatography (GC), 697–99, 709
Gas chromatography/infrared spectroscopy (GC/IR), 699, 709
Gas chromatography/mass spectrometry (GC/MS), 699
Gas constant (*R*), 335, 336n2
Gauche interactions, 74, 74n6, 163–64
GC/IR (gas chromatography/infrared spectroscopy), 699, 709
GC/MS (gas chromatography/mass spectrometry), 699
Gel-filtration chromatography, 1195
Geminal, definition, 447
Geraniol, 557, 558
Geranyl pyrophosphate (geranyl–OPP), 556–57
Gibbs, Josiah Willard, 336n3
Gibbs free energy ($\Delta G°$), 120, 334–39, 349
    See also Activation energy ($\Delta G°$)
Globin, 1194

Glucofuranose structure, *1132*, *1135*
Gluconic acid, *1144*, *1162*
Glucopyranose structure, 1132–33, 1134–38
D-Glucose
    in cellobiose, 1165
    in cellulose, *1124*, 1168
    chemical reactions, 1151–52
    epimerization in base, 1141–42
    glucofuranose structure, *1132*, *1135*
    glucopyranose structure, 1132–33, 1134–38
    hemiacetal formation, *784*, *1132*
    in lactose, 1161–64
    Lobry de Bruijn–Alberda van Ekenstein reaction, 1141–42
    in melibiose, 1167
    oxidation, 1144
    reduction, 1130, 1143
    spectroscopy, 1130
    structure, 222, *784*, *1129*, 1136–38, 1153–58
    in sucrose, 1166
    synthesis, 1138
Glutamic acid, *1176*, *1180*, *1185*
Glutamine, *1176*
Glyceraldehyde (2,3-dihydroxypropanal), 1126, *1127*, 1128
Glycerol (glycerin), *231*, 862
Glycine, *1175*, *1176*, 1178, *1181*, 1184
Glycols (diols), 240
Glycolysis, 1140
Glycoproteins, 1165–66
Glycosides, 1148–52, 1210
Graphite, 115, 219, 603, 604
Grignard, Victor, 228
Grignard reagents
    destruction by water, 799–800, *818*
    deuterolabeled molecule formation, 229–30, 257, 797
    ether or tetrahydrofuran (THF) solvents, 228, 251, 254
    hydrocarbon formation, 229
    as Lewis bases, 228, 429
    mechanism of formation, 228
    protonation, 229
    reaction with aldehydes and ketones to form alcohols, 799–802, *817*
    reaction with CO_2, 840
    reaction with esters, 898–99
    reaction with oxiranes, 429–30, *458*
    reaction with α,β-unsaturated ketones, 979
    structure, *228*
    synthesis from alkyl halides, 228–29, 257, 797, *818*
Guanidines, *1206*
Guanidinium ion, *45*
Guanine, 615, 616, 1212
D-Gulose, *1129*, *1139*, 1156–58, 1159
Gutta percha, 554, 554n3, 557
Gyromagnetic ratio (γ), 715–16, 734, 740

Halobenzenes, 667, 674–75, 680, *681*, *687*
Haloform reaction, 948–50, *1012*, *1013*
Haloforms, 225, 949, *1012*, *1013*
Halogenation
    aldehydes and ketones, in acid, 946–48, *1013*
    aldehydes and ketones, in base, 948–50, *1013*
    alkyne halogenation, 448
    allylic halogenation, 497–501, *507*, 543
    benzene, with $FeCl_3$ or $FeBr_3$, 638
    bromination reaction mechanism, 414–17, 419

Halogenation (*continued*)
　bromonium ions, 415–21, 424, 426, 536
　carboxylic acids, 950–53
　chloronium ion, 418
　conjugated dienes, 536
　cyclohexene, 418–19, 497–500
　cyclopentene, 414
　esters, 954
　geminal dihalides, 447, *459*, 490
　halohydrins, 418–19, *458*
　Hell–Volhard–Zelinsky (HVZ) reaction,
　　　950–53, *1013*
　in nucleophilic solvents, 417–20
　stereochemistry, 396, 414–17
　substituted aromatic compounds, 638–39, 670
　three-membered ring intermediates, 415–16
　vicinal dihalides, 414, 448, *459*, 490, 499, *507*
　See also Alkene hydrohalogenation; Alkenes,
　　　addition reactions; Photohalogenation
Halohydrins, 317, 418–19, *458*, 1083, 1094
Hammond postulate, 355–57, 377, 496
Harp modification of HVZ reaction, 952–53
Haworth forms, 1136–37
Heat of combustion ($\Delta H_c°$), 194–97
　See also Enthalpy change ($\Delta H°$)
Heat of formation ($\Delta H_f°$)
　alkanes, 116–18, *126*
　alkenes, 116–18, *126*
　alkynes, 116–18, *126, 515*
　allenes, *515*
　benzene, 581–82
　carbocations, *292, 445*
　cycloalkanes, *194*
　definition, 115, 193
　hydrogen molecule, 37
　sign convention, 3, 37, 40, 193
　and strain energy, 193–94
　See also Enthalpy change ($\Delta H°$)
Heat of hydrogenation, *413*, 524–25, 580–81, 620,
　　　626–27
Heisenberg, Werner, 2, 2n1, 3n4, 44
Heisenberg uncertainty principle, 2–3, 2n3, 44
Helium, 3, 4, 5, 6, 40–41, 479
α-Helix, 1191, *1192*
Hell–Volhard–Zelinsky (HVZ) reaction, 950–53,
　　　1013
Helprin, Mark, 624, 624n1
Heme units, *1194*
Hemiacetals
　acetal formation in acid, 786, *791–92*
　from carbonyl compounds in acid, 784–86,
　　　790, 818
　from carbonyl compounds in alcohol, 783–84,
　　　818
　from carbonyl compounds in base, 787
　cyclic hemiacetal formation, 783–84, 1083
　hemiketals and ketals, 786n2
　sugars, 784, 1130–34
　tetrahedral intermediate in Fischer
　　　esterification, 842–43
　See also Acetals; Carbonyl compounds, addition
　　　reactions
Hemiketals, 786n2
　See also Hemiacetals
Hemlock, 170
Hemoglobin, 1194
Heptamethylcyclohexadienyl cation
　　　tetrachloroaluminate salt, 655–56

Heptane, *82*
　See also Methylhexanes
Heptanoic acid, *803*
Heroin, 256
Heteroaromatic compounds
　acid–base properties, 601–2
　benzo-fused heterocycles, 605
　Chichibabin reaction, 676–77, 761
　definition, 599
　electrophilic substitution of furan and
　　　thiophene, 654–55
　electrophilic substitution of pyridine and
　　　pyrrole, 652–54
　five-membered aromatic ring structures,
　　　599–600
　fused heterocycles, 605
　resonance energy, *629*
　S_NAr reactions, 676–79
　structures, 598–600
Heterobenzenes, 591, 599
　See also Benzene
Heterogeneous catalysis, 411
Heterolytic bond cleavage, 37, 38, *62, 63*, 470
Hexachlorobenzene, *597*
(E,E)-2,4-Hexadiene, *512*
1,3-Hexadiene, 572
1,5-Hexadiene, 475, 524, 1031, 1040–42, 1064
Hexafluoro-2-butyne, 628
Hexahelicene, 179–80, *604*
Hexamethylbenzene, 655
Hexamethylenediamine, 852
1,2,3,4,5,6-Hexanehexol, *1143*
Hexanes, *149*, 150–51, 158, 473–74
　See also Cyclohexane; Methylhexanes
2-Hexanone, *984*
3-Hexanone, *803*
1,3,5-Hexatriene, 529, 578–79, 583, 584, 1040–41,
　　　1064
2,3,4-Hexatriene (1,4-dimethylbutatriene), 516
Hexenes
　cis-2-chloro-3-hexene, *111, 394*
　3-chloro-3-hexene, *444*
　cis-1,2-dideuterio-1-hexene, 394–95
　heat of formation, 116–18
　cis-3-hexene, *116*
　trans-3-hexene, *116*
　relative stability of isomers, 116–18
　types of carbon–carbon σ bonds in isomers,
　　　116–17
　See also Cyclohexene
E-2-Hexen-4-yne (*trans*-2-hexen-4-yne), *127*
Hexoses, definition, 1130
Highest occupied molecular orbital (HOMO)
　control of electrocyclic reactions, 1036–41
　HOMO–LUMO interactions in cycloaddition
　　　reactions, 1043–48
　hydrogen chloride addition to 2,3-dimethyl-2-
　　　butene, 134
　interactions in Lewis acid–Lewis base
　　　reactions, 266–267
　interactions in S_N2 reactions, 271–72
　overview, 133
　photochemical HOMO, 1039–41, 1044, 1046,
　　　1048, 1058
　ultraviolet/visible (UV/vis) spectroscopy,
　　　528–29
High-performance liquid chromatography
　　　(HPLC), 697
Histidine, *1176*

HMG-CoA reductase, 358
Hoffmann, Roald, 180n6, 531, 1031–32, 1034
　See also Woodward–Hoffmann theory
Hofmann, August Wilhelm, 648
Hofmann elimination, 308–9, 310–11
Hofmann rearrangement, 919–20, *923*
HOMO. *See* Highest occupied molecular orbital
　　　(HOMO)
Homogeneous catalysis, 411
Homolytic bond cleavage
　bond dissociation energy (BDE), *337*, 470
　butane carbon–carbon bonds, 470–71
　definition, 37, 227
　ethane carbon–carbon bonds, *469*, 470
　methane, 57, 63
　single-barbed (fishhook) arrows convention,
　　　37, 38, 63
Homotopic hydrogens, 721, *748*
Homotropilidene, 1066–68
Hooke's law, 708
Horner–Emmons reaction, 1009
Hückel's rule
　and aromaticity, 582–95, 618
　degenerate orbital pairs, 584–85, 586, 588–89,
　　　591, 592, 635
　overview, 583–84, 618
　See also Frost circles
Hund's rule, 9–10
Hunsdiecker reaction, 861, *869*
Hunter, Robert, 877, 877n1
HVZ (Hell–Volhard–Zelinsky) reaction, 950–53,
　　　1013
Hybrid orbitals
　hybridization, overview, 52–53, 57–60
　methyl compounds, 61–62
　and NMR chemical shifts, 724–25
　sp hybrid orbitals, 53–55
　sp^2 hybrid orbitals, 55–56
　sp^3 hybrid orbitals, 56–57
　See also Atomic orbitals
Hydrates
　definition, 773
　gem-diols, 776, 781, 783, *790–91*
　equilibrium constants of carbonyl hydrates,
　　　778, 779
　equilibrium in carbonyl compound hydration
　　　reactions, 777–81
　formation by carbonyl compound hydration,
　　　773, 775–81, *783*, 818
　steric destabilization, 778, 780
Hydration
　acid-catalyzed hydration of alkenes, 139–41,
　　　143, 381–82, *403*, 773, 775–76
　alkynes, 448–50
　carbonyl compounds, equilibrium in hydration
　　　reactions, 777–81
　carbonyl compounds, reaction mechanisms,
　　　773–77
　catalyzed hydration of 2,3-dimethyl-2-butene,
　　　139–40
　comparison of direct, hydroboration, and
　　　oxymercuration methods, 423
　HX addition reactions to alkenes, 380–84, *404*
　oxymercuration, 421–23, 448, *458, 459*
　radical HX hydration of alkenes, 481–87
Hydrazones, *793*, 1145–46
Hydride ions, 62, 264–66
Hydrides
　beryllium hydride, 62

NADH as biological hydride source, 315
poor leaving group, 1005, 1010
reaction with amides, 900
reaction with esters, 900
reaction with organic halides, 798
reaction with water, 798
reduction of carbonyl compounds to alcohols, 801–3, *817*
reduction of nitrile, 906
sodium hydride, 315
See also Lithium aluminum hydride
Hydride shifts
Cannizzaro reaction, 1005, 1012
carbocation rearrangement, 387–88, *392*, 401–2, 642
definition, 387
Magid's third rule, 1007
Meerwein–Ponndorf–Verley–Oppenauer (MPVO) equilibration, 1006–7, 1012
NAD⁺ reduction by hydride transfer from ethanol, 814–15
radical rearrangement, 502
Hydroboration
alcohol synthesis, 398–401
alkylborane formation, 391, 394
alkynes, 450–51
concerted mechanism, 391–95
definition, 390
dialkylboranes, 397
2,3-dimethyl-2-butene, 397
regiochemistry, 392–93, 399–400
steric factors, 392, 397
syn addition, 394–95
trialkylboranes, 397, 398, *399*
See also Alkenes, addition reactions; Borane; Boron trifluoride
Hydrocarbon cracking, 473–74
See also Pyrolysis
Hydrocarbons
bond dissociation energies, *477*
definition, 51
formation from alkyl halides, 229
from Kolbe electrolysis, 860, 861, *870*
solubility, 240
synthesis by reduction with Raney nickel, 253, 257
synthesis from organometallic reagents, 229, 257
Hydrogen
aldehyde hydrogens, 725
bond dissociation energy (BDE), 37
bond energy of H_2^+ molecule, 39–40
bond strength in H_2, 36–41, 57
electron configuration, 6, 10, 14
energy as function of distance between nuclei, *31*
enthalpy ($\Delta H°$) of formation, 37
hydrogen abstraction, definition, 471
Lewis structure, *14*
molecular orbitals, 30–35, *31–35*
orbital interaction diagrams, 33–34
Hydrogenation
alkynes, 452
benzene, heat of hydrogenation, 580–81
benzene catalytic hydrogenation, 580, 627
benzene conversion to cyclohexane, 580–81, 626–27, *686*
cycloalkenes, 412
definition, 410, 411

diimide, 411
energetics, 413–14
heats of hydrogenation, *413*
Lindlar catalyst, 452, *458*
palladium catalysts, 411, 452, *458*
steric factors, 412
Wilkinson's catalyst (rhodium), 411
See also Alkenes, addition reactions; Alkynes, addition reactions
Hydrogen bonding
alcohols, 232–33
amines, 245–46, 247–48
carboxylic acids, 832, 833
definition, 232, 257
β-dicarbonyl (1,3-dicarbonyl) compounds, 941, 942
and NMR chemical shifts, 726
nucleic acids, *288, 615*, 1212–13
peptides and proteins, 1190–91, 1192, 1194
protic solvents, 238–39, 256, 280
solvents, 238–39, 256, 280
Hydrogen bromide
addition to 1,3-butadiene, 406, 534–35
addition to *cis*-2-butene, 406
anti-Markovnikov radical addition to alkenes, 481–82, 484–86, *506*
mechanism for addition to alkenes, 414
Hydrogen chloride
addition to 1,3-butadiene, 369–71, 534–35, 537–39
addition to 2,3-dimethyl-2-butene, 131–32, 133–36, 365–67, 374
addition to 2-methyl-1-butene, 374–75
addition to 3-methyl-1-butene, 386–87
addition to 2-pentene, 375
addition to vinyl chloride, 367, 372
dipole moment, 132
Hydrogen cyanide, 93
Hydrogen fluoride, *14*
Hydrogen iodide, 380–81, 406
Hydrogen nuclear magnetic resonance spectroscopy (¹H NMR)
acyl compounds, 888–89
α hydrogens or carbons, 714–15, 723, *725*
β hydrogens or carbons, 714–15, 723, 725
complicated spectra, 735–39
cyclohexane, 746–47
decoupled spectra, 739
deshielding, 720, 723–24, 725, 771
diastereotopic hydrogens, 721–22, 731, 749
4,4-dimethyl-2-pentanone, 718
enantiotopic hydrogens, 721, 747–48
energy difference between spin states, 714–16, 720
first-order spectra, 735, 737–38
flipping a nuclear spin, 714–15, 720
homotopic hydrogens, 721, *748*
integral, 718–20, 729
magnetic resonance imaging (MRI), *713*
n + 1 rule, 727–29, 735–36
NMR measurements, overview, 717–18
overview, 88, 90, 94
tetramethylsilane (TMS) standard, 716–17, 720–21
See also Chemical shift (δ), Nuclear magnetic resonance spectroscopy (NMR); Spin–spin coupling
Hydrogen peroxide, 398, 424, 441, 602
Hydrophilic groups, 866, 1192

Hydrophobic groups, 866, 1192
Hydroquinone (*p*-dihydroxybenzene), 484, 597
Hydroxybenzene. *See* Phenol
β-Hydroxy carbonyl compounds, 966, 968, 973–74, 982–83, 1003
See also Aldol condensation; Carbonyl compounds; Carbonyl compounds (overview)
7-Hydroxy-5-decanone, *984*
5-Hydroxy-4,5-dimethyl-3-hexane, *982*
5-Hydroxyhexanal, *1131*
Hydroxyl group
as activating group in aromatic compounds, 669–70, 671
hydrogen bonding, 238
leaving group of alcohols, conversion to water, 235, 281–82, 322, 555
naming conventions, 111
protecting hydroxyl groups, 789
protonation in carboxylic acids, 837–38
See also Alcohols; Carbohydrates; Neighboring group effect; Phenols
3-Hydroxy-3-methylcycloheptane, *981*
2-Hydroxy-6-methyltetrahydropyran, *1131*
(*E*)-4-Hydroxy-2-pentenoic acid, *831*
Hyperconjugation, 293–94, 377–78, 478–79

D-Idose, *1129, 1139, 1159*
Imidazole, 621
Imides, 882
Imines (Schiff bases)
equilibrium with enamines, 796, 1146
formation from carbonyl compounds, 790–94, *818*
formation from nitriles, 904–6, *924*
formation of substituted imines, *793*
Iminium ions, 795–96, 903, 964, 1003–4, *1013*
Indene, 406
Indole, 605
Induced dipole, 87
Inductive effects, definition, 379
Infrared (IR) spectroscopy
acyl compounds, 887–88
carbonyl compounds, 770–71
carboxylic acids, 833
characteristic IR absorptions, 709–13
infrared absorptions of functional groups, *710–11*
IR spectra of organic molecules, *709, 712*
IR spectrometers, 708–9
overview, 696, 707–8, 734
See also specific compounds
Inhibitors, 484
Initiation of chain reactions, *482*, 483
Integral, 718–20, 729
Intermediates
allenes in alkyne isomerization, 516–18
benzyne, 681–82
carbocations in alkene addition reactions, 132–33, 135–36, 140
comparison to transition state, 186
cycloalkynes, *128*
definition, 62, 186
dienes in alkyne isomerization, 516–18
hemiacetal intermediate formation in acid, 784–86, 790, *818*
methyl ions and radicals, 62
three-membered ring halogenation, 415–16
See also Tetrahedral intermediate in addition–elimination reactions

Intramolecular reactions. *See* Neighboring group effect
Inversion of configuration
    mechanisms, 270–71, 350
    phosphorus reagents, 285
    S_N1 reactions, 291–92, 297
    S_N2 reactions, 269, 270–75, 284, 288, 297
Iodoform, 949
Iodomethane (methyl iodide), *226*
2-Iodo-2-methylpropane (*tert*-butyl iodide). *See tert*-Butyl iodide
2-Iodopropane (isopropyl iodide), *226*
2-Iodopropene, *446, 447*
3-Iodopropene (allyl iodide), *543*
Ion-exchange chromatography, 1195
Ionic bonds, 5–6, 15
Ionization potential, 4, *5*
Ions, definition, 4
Ipso attack, 677–78
Isobutane, 73, 75
Isobutene (2-methylpropene, isobutylene)
    boiling point, 770
    dimerization, 384–85
    dimerization and polymerization, 385–86
    hydration reactions, 382, 390
    regiochemistry of addition reaction, 137
    solubility, 240
    structure, *109*
    See also Butene
Isobutyl alcohol (2-methyl-1-propanol), *230*
Isobutyl chloride (1-chloro-2-methylpropane), 494
Isobutyl compounds, 75
Isobutyraldehyde (2-methylpropanal), *464, 767*
Isocitrate, 357
Isocyanates, 918–21, *923, 924, 1206*
Isoelectric point (p*I*), *1176–77, 1179–80*
Isoleucine, *1176*
Isomers
    cis isomers, 84, *85*, 144
    conformational isomers, 70, 177
    definition, 69–70
    diastereomers, 165–67, 169–72, 177, 208–9
    drawing isomers, 82–83
    entgegen (*E*) isomers, definition, 109
    overview, 176–77
    problem solving techniques, 82
    relative stability of alkane isomers, 116–18
    stereoisomers, definition, 148, 176
    structural isomers, definition, 73, 174, 176
    trans isomers, 84, *85*, 144
    zusammen (*Z*) isomers, definition, 109
    See also Enantiomers
Isopentane, *76, 77–78*
Isoprene (2-methyl-1,3-butadiene)
    gutta percha, 554, 554n3, 557
    head and tail attachment, 557
    polyisoprene, 554, 557
    steroid biosynthesis, 559
    structure, *512, 554, 557, 565*
    terpene formation, 557–58
Isoprene rule, 558
Isopropyl alcohol (2-propanol), *230*
4-Isopropylbenzaldehyde, *803*
Isopropylbenzene (cumene or 2-phenylpropane), *595*, 639–40, 641–42
Isopropyl bromide, 276, 314
Isopropyl chloride, 138
Isopropyl compounds, structure, 73, *276*
1-Isopropyl-1-methylcyclohexane, 205

1-Isopropyl-2-methylcyclohexane, 208–10
Isopropyl-4-methylphenol, *669*
Isopropyl phenyl ketone (2-methyl-1-phenyl-1-propanone), *769*
p-Isopropyltoluene, 669
Isoquinoline, *602*, 761
Isothiocyanates, 921, 1198–99, *1206, 1218*
1-Isothiocyanato-(4*R*)-(methylsulfinyl)butane, *921*
Isotope effect, 273n3, 1109
Isotopes and their abundances in nature, *700*

Karplus curve, 731–32
Kauzmann, Walter, 42–43, 51, 51n1
Kekulé forms, 575–77
Kekulé von Stradonitz, Friedrich August, 574, 574n2
Ketals. *See* Acetals
Ketene (ethenone), *883, 888, 921*
Ketenes
    from acid chlorides, 912, *924*
    from Arndt–Eistert reaction, 917
    bonding, 887–88
    formation by Wolff rearrangement, 916–18, *924*
    infrared spectra, 888
    nomenclature, 883
    reactions, 907, *922, 1206*
    structure, 515–16, *877, 1206*
    See also Cumulated double bonds
β-Keto acids, decarboxylation, 858–59, *870*, 959–60
Ketocarbenes, 915–16
Keto–enol tautomerization, 939–40, 944, 961, 966
β-Keto esters
    from Claisen condensation, 987–89, 991, 992, 997, *1013*
    cyclic β-keto esters, 992–93
    in fatty acid metabolism, 997
    hydrolysis and decarboxylation, 959–62, 963, *1014*
    in reverse Claisen condensations, 992, 997
    in synthesis of ketones, 959–62, 963, *1014*
    See also Esters
α-Ketoglutaric acid, *1185*
Ketones
    acetals as protecting groups, 788–89
    from acid chlorides, 891–92, 912, *924*
    acyl benzenes, 644
    aldol condensation, 971–74
    alkylation of enolates, 955–56, *1012*
    Baeyer–Villiger reaction, 907–9, *924*
    Beckmann rearrangement, 909–11, *923*
    from carboxylic acids, 856–57, 858–59, *870*
    diazo ketones, 915–18, *924*
    diones (diketones), 768
    enolates, 933–42
    formation from nitriles, 905, *924*
    formation from ozonides, 440–42, *459*
    from Friedel–Crafts acylation, 644, *687*, 893, *924*
    haloform reaction of methyl ketones, 948–50, *1012, 1013*
    halogenation at α position, in acid, 946–48, *1013*
    halogenation at α position, in base, 948–50, *1013*
    from hydration of alkynes, 448–50, 451, *459*
    infrared (IR) spectroscopy, 770–71
    from β-keto acid decarboxylation, 858–59, *870*

Knoevenagel condensation, 974–76, *1014*
    nomenclature, 767–69
    from oxidation of alcohols, 803, 804–6, *818*
    p*K*_a values, *933*
    production from enols, 448–50
    reaction with ester enolates, 974–76
    reaction with metal hydrides to form alcohols, 801–3, *817*
    reaction with organometallic reagents to form alcohols, 801, *817*
    resonance forms, *885*
    synthesis from β-keto esters, 959–62, 963, *1014*
    transformation into enolates, 652
    ultraviolet spectroscopy, *771*, 772
Ketoses, 1128, 1141
Keto sugars, 1130
Ketotriose, 1128
Kharasch, M. S., 482, 482n2
Kiliani–Fischer synthesis, 1138–40, 1153–54, 1160
Kilojoule (kJ), 4n5
Kinetic control, 353, *359*, 537–41
Kinetic enolates, *984, 985, 986*, 1008
Kinetic resolution, 1185
Kinetics
    definition, 351
    energy barrier height effect on reaction rate, 334, 339–40
    Hammond postulate, 355–57, 377
    mountain pass analogy to energy barriers, *331*, 346, 627
    overview, 339–40
    rate constant (*k*), 268, 289, 341, 349
    rate-determining step, 294–95, 350–51, 359
    temperature effect on reaction rate, 340
    thermodynamics *versus* kinetics effect on rates, 351–57, 359–60
Kishi, Yoshito, 1009
Knoevenagel condensation, 974–76, *1014*
    See also Aldol condensation
Koenigs–Knorr reaction, 1149
Kolbe electrolysis, 860, 861, *870*
Kroto, H. W., 603–4

Lactams (cyclic amides), 851, *870, 873, 882*, 910, *924*
Lactase, 1164, 1164n6, 1165, 1166
Lactate dehydrogenase, *1192*
Lactones (cyclic esters, oxacycloalkanones)
    cyanogen bromide cleavage of peptides, 1200–1201
    Fischer esterification, 849, *870*
    α-lactones, 849, 880, 1084–85
    nomenclature, 849, 880
    synthesis, 849, 909, *925, 1081–82*
Lactose, 1161–64, 1164n6
Ladenburg, Albert, 574
Ladenburg benzene
    (tetracyclo[2.2.0.0^{2,6}.0^{3,5}]hexane or prismane), 216, 573–74
Lanosterol, 561
LDA. *See* Lithium diisopropylamide (LDA)
Lead, 650
Lead dioxide, 650n3
Leaving groups
    ammonia, 902
    in benzene halogenation, 638
    effect on S_N1 reactions, *281, 295*, 343, 347

epoxidation reactions, 424, 427, 429
ether cleavage by haloacids, 282–83
in formation of dinitrogen trioxide, 647
good and poor leaving groups and conjugate
   acids, 235, *281*
hydroxide conversion to water, 235, 281–82,
   322, 555
and inversion of configuration, 270–71
nitrite, 647
nitrogen (N₂), 848
pyrophosphates, 555–59
in S_N2 reactions, *262*, 267–68, 281, 343–44, 350
in S_NAr reactions, 677–78
sulfonate reactions, 281–86, 322, 427, 555, 1085
tosylates, 283–84, *305*, *555*, 798, 1093
Le Châtelier, Henri Louis, 336n3
Le Châtelier's principle, 336, 336n3
Lehn, Jean-Marie, 254
Lehrer, Thomas, 224, 224n1
Leucine, *1176*, *1203*, *1204*, *1205*
Levorotatory, 156
Lewis, Gilbert Newton, 14, 42
Lewis acids
   borane, 92
   boron trifluoride, 264–65, 390
   definition, 42, 265
   as electrophiles, 42, 390, 625
   HOMO–LUMO interactions, 266–67
   methyl cation, 91, 264–65
   molecular orbital stabilization, 42, 91–92, 265,
      321
   overview, 264–65
   See also Acid–base chemistry
Lewis bases
   definition, 42, 265
   Grignard reagents, 228, 429
   HOMO–LUMO interactions, 266–67
   leaving group in substitution reactions, 267
   molecular orbital stabilization, 42, 91, 265, 321
   as nucleophiles, 42, 267
   nucleophilicity, 278–81, 463, 605
   organolithium reagents, 229, 429
   overview, 264–65
Lewis structures
   ammonia, *17*
   ammonium ion, 21
   construction, 14, 16–22
   conventions for drawing, 16–17, 44
   double and triple bonds, 18–19
   ethane, 18
   ethene or ethylene, 18
   ethyne or acetylene, 18
   fluorine, *14*
   formal charge, 20–22
   hydrogen, *14*
   hydrogen fluoride, *14*
   methane, *17*
   methyl anion, 21
   nitromethane, *23*
   nonbonding electrons, 14, 17, 19
   See also Covalent bonds
Limonene, *121*, 558
Lindlar catalyst, 452, *458*
A-Linolenic acid (ALA; octadecatrienoic acid), *865*
Lipscomb, William, 217
Lithium aluminum hydride (LiAlH₄, LAH)
   alcohol synthesis from acid chlorides, 891, *922*
   alcohol synthesis from epoxides, 429–30, *458*
   alcohol synthesis from esters, 900

primary alcohol synthesis from carboxylic
   acids, 858, *869*
   reaction with amides, 900
   reaction with organic halides, 798
   reaction with water, 798
   See also Organolithium reagents
Lithium benzoate, *832*
Lithium diisopropylamide (LDA)
   alkylation of carboxylic acids, 957, *1012*
   alkylation of esters, 963
   alkylation of ketones, 956
   Claisen–Schmidt condensation (crossed aldol
      condensation), 984–85, 1008
   crossed aldol condensation of esters and
      ketones, 986
   and deprotonation, 944, *984*
   and enolate formation, 944, 954, 957, 963, 984
Lobry de Bruijn–Alberda van Ekenstein reaction,
   1141–42
Lone-pair electrons, 14, 62
Long-range coupling, 731, 736
Loschmidt, Johann Josef, *574*, 574n2
Lowest unoccupied molecular orbital (LUMO)
   fluoride, *266*
   HOMO–LUMO interactions in cycloaddition
      reactions, 1043–48
   HOMO–LUMO interactions in S_N2 reactions,
      271–72
   hydrogen chloride addition to 2,3-dimethyl-
      2-butene, 134
   interactions in Lewis acid–Lewis base
      reactions, 266–67
   interactions in S_N2 reactions, 271–72
   overview, 133
   ultraviolet/visible (UV/vis) spectroscopy,
      528–29
LSD (lysergic acid diethylamide), *1026*
LUMO. *See* Lowest unoccupied molecular orbital
   (LUMO)
Lycopodine, *214*
Lysine, *1177*, *1180*, 1202
D-Lyxose, *1129*, *1139*, *1159*, *1170*

M + 1 peak, 700–701
M + 2 peak, 706
Magid's second rule, 999
Magid's third rule, 1007
Magnetic fields
   applied magnetic field (B₀) in NMR, 714–15,
      720, 724, *725*, 726, 727
   effect of field strength on chemical shift (δ),
      716–17, 718, 722, 735, 738
   induced magnetic field (Bᵢ) in NMR, 720, 724,
      725, 726
   mass spectrometry, 700, 702
Magnetic resonance imaging (MRI), 88, *713*
   See also Hydrogen nuclear magnetic resonance
      spectroscopy (¹H NMR); Nuclear
      magnetic resonance spectroscopy (NMR)
Maitotoxin, 745
Maleic acid, *830*, *881*
Maleic anhydride, *875*, *881*, *894*
Malic acid, 274
Malonic acid (diethyl malonate or propanedioic
   acid), *830*, 961–62
Malonic ester synthesis, 961
Maltose, 1161, 1170, 1172
Manganese dioxide, 806
Mannich reaction, 1003–4, *1013*

D-Mannopyranose, *1141*, *1142*
D-Mannose, *1129*, 1141–42, 1146, *1147*, 1153–58
Markovnikov addition, 376, 379–84, 394, 399–400
   See also Anti-Markovnikov addition
Markovnikov's rule, 375, 375n2, 380, 399
Mass number, 5
Mass spectrometry (MS)
   acetone, 772
   acylium ion, 704
   alkanes, 705
   base peak, 702
   benzyl cation, 703–4
   bicyclo[4.2.2]deca-2,4,7,9-tetraene, 707
   bromine, 706, 750
   bromoethane, *702*
   1-butanol, 705–6
   butene radical, 705–6
   tert-butyl chloride, 702, *703*
   carbonyl compounds, 704, 772
   chlorine, 702, 706, 750
   chloroethane, *702*
   cyclohexane, 700–701, 702
   daughter ions, 702
   3,4-dihydro-2*H*-pyran, *700*, 701, 702
   fragmentation patterns, 702–6
   history and background, 696
   isotope peaks, 700–701
   isotopes and their abundances in nature, *700*
   M + 1 peak, 700–701
   M + 2 peak, 706
   mass spectrometers, 699–701
   mass-to-charge ratio (*m/z*), 700
   molecular ion, definition, 700
   MS measurements, 701–9
   parent ion (p), 702
   radical cation formation, 699
   toluene, 702–3
Meerwein–Ponndorf–Verley–Oppenauer (MPVO)
   equilibration, 1005–7, *1012*, *1014*
Meisenheimer complex, 675
Menthol, *558*
Menthyl chloride, 328–29
Mercaptans. *See* Thiols
Mercaptides, 253, *267*, 313
Mercuric acetate, 421–22
Mercurinium ion, 423
Merrifield, R. Bruce, 1208, 1210
Meso compounds, 168–69
Messenger RNA (mRNA), 1214–15
Meta (*m*-) substitution, definition, 596, 656
   See also Substitution reactions of aromatic
      compounds
Methanamine (methylamine), *241*
Methane
   ¹³C NMR spectroscopy, 88
   biochemistry, 92–93
   bond angles, 52, 57, 58
   bond dissociation energy (BDE), 57
   bond strength, 57
   clathrates (methane hydrate), 58–59
   combustion, 68, 342
   heat of formation, 115
   homolytic bond cleavage, 57, 63
   Lewis structure, *17*
   molecular orbitals, 57
   photohalogenation, 490–93
   sp³ hybrid orbitals, 56–57
   structure, *51*
   unhybridized model, 58–59

Methanethiol (methyl mercaptan), 252
Methanol (methyl alcohol), 230
Methide ion. *See* Methyl anion
Methine group, 75
Methionine, 106, *1176*, *1182*, *1200–1201*
Methoxide ion, *236*, 238, 307, 315, 316
Methoxyethane (ethyl methyl ether), 250
Methoxymethane (dimethyl ether), 250, 314
2-Methoxypropane, *732*
N-Methylacetamide, *888*
Methyl acetate, *888*, 898
Methyl acetate (methyl ethanoate)
    infrared stretching frequency, *888*
    structure, *879*
Methylamine (methanamine, aminomethane), 241, *243*, 318
Methyl anion
    acid–base chemistry, 129, 130
    formal charge, 21
    formation, 62
    Lewis structure, 21
    sp^3 hybrid orbitals, 63
    structure, 63, 477
N-Methylazacyclooctane
        (heptamethylenemethylamine), *242*
Methyl bromide, *223*, 493, 507
2-Methyl-1,3-butadiene. *See* Isoprene
3-Methyl-1,2-butadiene, *514*
3-Methylbutanamide, *882*
Methyl butanoate, 737–38
3-Methylbutanoic acid, *882*, *918*
2-Methyl-2-butanol, 389
2-Methyl-1-butene, 299–300, 374–75
2-Methyl-2-butene
    addition of hydrogen chloride (HCl), 138
    from 2-bromo-2-methylbutane, 299–300
    hydroboration, 397
    protonation, 354, 356
    protonation and carbocation formation, 354–56
    stereochemistry, 144
    structure, *138*, *144*
3-Methyl-1-butene, 386–87
3-Methyl-3-buten-1-ol, 555
3-Methyl-2-buten-1-ol pyrophosphate, 556
3-Methyl-3-buten-1-ol pyrophosphate
        (isopentenyl pyrophosphate), 555–56
Methyl cation
    formation, 62
    as Lewis acid, 91, 264–65
    sp^2 hybrid orbitals, 63, 133, 377, 477
    structure, 63, 133, 477
    See also Carbocations
Methyl chloride, *91*, 490–93
Methyl compounds, 60–62, 275
Methyl cyanide (acetonitrile), 345–46
(*E*)-1-Methylcycloheptene, 118
(*Z*)-1-Methylcycloheptene, 118
2-Methyl-1,3-cyclohexadiene, *110*
Methylcyclohexane, 194–201
2-Methylcyclohexanone, 955–56
1-Methylcyclohexene, *110*
3-Methylcyclohexene, *110*
1-Methylcyclopentene, 394
Methyl diazoacetate, *431*
Methylene chloride, 490–93, 721
Methylenecyclohexane, *407*, 812–13
3-Methylenecyclohexene, 546
Methylene group, 72, 109

Methylene molecule, 49, 55
Methyl D-glucopyranoside, *1147*, 1149, *1151*, *1152*
3-Methyl-1,3-heptadiene, *110*, 112
Methylhexanes
    2-methylhexane, *82*
    3-methylhexane, absolute configuration, 152–53, *154*, 176
    3-methylhexane, chirality, 151, 160–61
    3-methylhexane, optical activity, 156, 157, *158*
    3-methylhexane, stereochemistry, 149
    structures, *82*
Methyl iodide, 340, 344–46
Methylketene (1-propen-1-one), *883*
2-Methylpentanal, 767
3-Methylpentane, 149, 150–51, 158
3-Methylpentanoic acid, *879*
3-Methyl-2-pentanone (*sec*-butyl methyl ketone), 768
4-Methyl-3-penten-2-one (mesityl oxide), 971
4-Methylphenol, 669
3-Methylpiperidine, *242*
2-Methylpropane, 494–95
Methyl propanoate, *713*
Methyl radical
    bonding models, 99–103
    formation, 63, 469–70
    hybrid orbitals, 99–101
    methyl ions and radicals as reactive intermediates, 62
    molecular orbitals, 101
    structure, 63–64, 100–101
10-Methylstearate, 142–43, 401–2
2-Methyltetrahydrofuran, *1087*
Methyl vinyl ether, 378–80
Mevalonic acid, 358
Micelles, 866, *867*
Michael reaction, 976–80, 998–99, 1016
    See also Aldol condensation
Microscopic reversibility, 346, 359
Molecular ion, 700
Molecular orbitals
    alkenes, 103–4
    allyl cation (allylic cation), 369–70
    antibonding molecular orbitals ($\Phi_{antibonding}$, Φ_A), 32–35, 44
    antibonding sigma orbitals (σ^*), 54
    background, 30–31
    benzene, 578–79
    beryllium hydride, 54–55
    bonding molecular orbitals ($\Phi_{bonding}$, Φ_B), 32–35, 44
    1,3-butadiene (methylallene), 520–21
    conjugated dienes, 520–21
    cyclobutadiene, 585
    definition, 3, 30
    degenerate orbitals, 584–85, 586, 588–89, 591, 592, 635
    empty orbitals, 34, 34n8, 40
    ethane, 68, 103
    ethylene, 103–4
    ethylene (ethene), 103–4
    hydrogen molecule, 31–35
    low-energy orbitals and bond stability, 35, 36–41
    methane, 57
    methyl radical, 101
    nodes, 32
    nonbonding orbitals, 32–33
    orbital interaction diagrams, 33–34

orthogonal orbitals, 34
    rules for constructing, 34–35, 44–45
    See also Frost circles
Molozonides, 437
Monosaccharides, definition, 1161
Morphine, *121*, 255, *256*, *408*
Morpholine, 242
MPVO. *See* Meerwein–Ponndorf–Verley–Oppenauer (MPVO) equilibration
Murexine, *605*
Muscarine, 184
Mustard gas [bis(2-chloroethyl)thioether], 1089
Mycobacterium tuberculosis, 143
Mylar, 852
Myristic acid and myristates, 862, 863, *997*

n + 1 rule, 727–29, 735–36
NADH (reduced nicotinamide adenine dinucleotide), 358, *814*, 1140, *1185*, 1186, *1216*
NAD$^+$ (nicotinamide adenine dinucleotide), 679, 813–15, *1185*
NADPH, 996
Naming conventions. *See* Nomenclature
Napalm, 1089
Naphthalene, *179*, 592, 602, 605, 622
Narbomycin, 182
Natural products, definition, 86
N-Bromosuccinimide (NBS), 500–501, *507*, 543, 613, 618, 953
Neighboring group effect
    acetolysis in intramolecular replacement reactions, 1085–87
    anchimeric assistance, 1089, 1100
    aromatic neighboring groups, 1096–99
    bicyclic systems, 1090–93
    cis-2-butene, conversion to 2,3-dibromobutane, 1093–96
    carbon–carbon double bonds as intramolecular nucleophiles, 1099–1108
    α-carboxylate reaction with methyl alcohol, 1083–85
    Coates' cation, 1117
    cyclic hemiacetal formation, 783–84, 1083
    halogens, 1093–96
    heteroatoms with nonbonding electrons, 1083–96
    intramolecular reactions of endo and exo bicyclic amines, 1090–92
    intramolecular S_N2 reactions, 316–17, 917, 1081–88, 1091, 1095, 1101
    isotope effect, 1109
    3-methyl-2-butyl tosylate reaction with acetic acid, 1108–9
    7-norbornenyl tosylates reaction with acetic acid, 1099–1108
    overview, 1081–83, 1118
    oxirane formation from halohydrins, 1083
    oxygen in intramolecular displacement reactions, 1085–87
    3-phenyl-2-butyl tosylate conversion to 3-phenyl-2-butyl acetate, 1096–99
    problem solving clues, 1082, 1089
    single bonds as neighboring groups, 1108–16
    solvolysis reaction of *exo*- and *endo*-2-norbornyl tosylate, 1110–16
    sulfur compounds, 1085–86, 1088–89
    unexpectedly rapid rate as clue, 1088–89, 1099–1100, 1103–4, 1106–8, 1109–13

unexpected or backward stereochemical results as clue, 1083–85, 1089–91, 1093–94, 1095, 1097, 1111–14
unusual rearrangement reactions as clue, 1086–87, 1089, 1108, 1111, 1113
Neomenthyl chloride, 328–29
Neon, 3, 4, 6
Neopentane, *76, 77*
Neopentyl alcohol, 389
Neopentyl iodide (1-iodo-2,2-dimethylpropane), 389
Neutral methyl. *See* Methyl radical
Newman, Melvin S., 65
Newman projections, 65–66, 94
Nicol prisms, 157–59
Ninhydrin, 1187–89, 1196
Nitrate ion, *45*
Nitration of aromatic compounds, 637–38, *664*, 668–69, 673, *687*
Nitrenes, 918–19
Nitric acid, 806, 809
Nitriles (cyanides)
    acid–base chemistry, 885–86
    acid-induced hydrolysis, 904–6
    base-induced hydrolysis, 905
    cnol like conjugated bases, 942
    infrared spectra, 888
    and metal hydrides, 906
    nomenclature, 883
    and organometallic reagents, 905–6
    reactions, 904–6, *923*
    structure, *877, 887*
    synthesis, 911, *925*
Nitroacetamidobenzene, *670*
Nitrobenzene
    1,3,5-trinitrobenzene, 668
    bromonitrobenzene, 672
    chlorodinitrobenzene, 674, 675
    chloronitrobenzene, 668
    dinitrobenzene, *663*, 668
    inductive effects, 666
    meta substitution, 663–64
    from nitration of benzene, 621, 637, *687*
    nucleophilic substitution reactions of nitrobenzenes, 674–76
    para substitution, 664
    reduction to aniline, 646, 651, *686*
Nitro group
    conversion to amino group, 646, 672, *686*
    effect on electrophilic aromatic substitution, 664, 666, 669, 679
    as a functional group, 22
    in Meisenheimer complex, 675
    in nucleophilic aromatic substitution, 674–75, 679, 680
    positive charge on nitrogen, 663–64
Nitromethane, 22–25, 28, 45, *239*
Nitrones, *436*
Nitronium ion, 621, 637–38
4-Nitrophenol (*p*-nitrophenol), *596*
Nitrosamines, *647*
N-Nitrosoaniline, 647–48
N-Nitroso-*N*-methylurea, *875*
Nitrotoluene, 669
Nitrous oxide, *436*
NMR. *See* Nuclear magnetic resonance spectroscopy (NMR)
Noble gases, 3–6
Nodes, 6, 11–12, 32

Nomenclature
    acid halides, 880–81
    acyl compounds, 879–83
    alcohols, 230–31
    aldehydes, 767
    alkanes, 78–81
    alkenes, 110–12
    alkyl compounds, 78–81
    alkyl halides, 225–26
    alkynes, 126, 127
    amides, 882
    amines, 240–42
    amino acids, 1175–77
    ammonium ions, 242
    anhydrides, 881
    bicyclic compounds, 214–15, 551–52
    Cahn–Ingold–Prelog priority system, 111, 112–15
    carbohydrates, 1126–38
    carbonyl compounds, 767–69
    carboxylic acids, 830–32
    common prefixes, *81*
    cycloalkanes, *84*
    cycloalkenes, *110*
    ethers, 250
    fatty acids, *862*
    hydroxyl group, 111
    International Union of Pure and Applied Chemistry (IUPAC) system, 78–81
    ketenes, 883
    ketones, 767–69
    lactones, 849, 880
    lactones (cyclic esters), 880
    nitriles, 883
    thioethers (sulfides), 252–53
    thiols (mercaptans), 252
Nonbonding electron in Lewis structures, 14, 17, 19
Nonbonding orbitals, 32–33
Nonreducing sugars, 1166
Nonsteroidal antiinflammatory drugs (NSAIDs), 688
Norbornyl system
    norbornane (bicyclo[2.2.1]heptane), 186, 306, 1099n2, *1100*
    7-norbornenyl tosylates reaction with acetic acid, 1099–1108, 1099n2
    solvolysis reaction of *exo*- and *endo*-2-norbornyl tosylate, 1110–16
Nuclear magnetic resonance spectroscopy (NMR)
    applied magnetic field (B_0), 714–15, 720, 724, *725, 726, 727*
    carbon nuclear magnetic resonance spectroscopy (^{13}C NMR), 88–90, 94, 740–42
    convention for reporting NMR signals, 742
    decoupling, 733, 739, 740–41, 746
    DEPT (distortionless enhancement with polarization transfer), 741
    dynamic NMR, 746–49
    energy difference between spin states, 714–16, 720
    flipping a nuclear spin, 714–15, 720
    gyromagnetic ratio (γ), 715–16, 734, 740
    history and background, 526, 696, 696n2, 716
    induced magnetic field (B_i), 720, 724, *725, 726*
    magnetic resonance imaging (MRI), *713*
    NMR spectrum, definition, 88
    nuclear spin (I), 88, 714–15, 727–28

    overview, 88–90, 713–17
    ppm scale, 716–17
    problem solving techniques, 89, 889
    protein structure determination, 1192n3
    tetramethylsilane (TMS) standard, 716–17, 720–21
    See also Chemical shift (δ); Hydrogen nuclear magnetic resonance spectroscopy (^1H NMR); Spin–spin coupling
Nucleic acids
    alkylation of DNA, 288, 615–16
    base pairs, 288, 615, 1212–13, 1214
    bases, 288, 614–15
    DNA double helix, 598, 1213
    DNA replication, 1213
    DNA structure and composition, 614–16, 1149, 1211–13
    enzymatic epoxidation, 616
    hydrogen bonding, *288, 615*, 1212–13
    messenger RNA (mRNA), 1214–15
    mutation, 288–89, 615–16
    nucleosides, 1210–11
    nucleotides, 614–15, 1211–12, 1213, *1214*
    π (pi) stacking in DNA, 598, 615
    and polyaromatic hydrocarbons, 614–16
    RNA codons for amino acids, 1214–15
    RNA structure and composition, 93, 614–15, 1149, 1214
    transfer RNA (tRNA), 1215
Nucleophiles
    effect of solvent polarity, 279–81
    effect on S_N2 reactions, 277–81
    Lewis bases, 42, 267, 278–81, 463, 605
    nitrogen, 317–18
    oxygen, 315–17
    pyridine as nucleophile, 602
    relative nucleophilicity of common species, 279
    sulfur compounds, 313–14
Nucleophilic aromatic substitution. *See* S_NAr reactions
Nucleophilicity, 278–81, 463, 605
Nucleosides, 1210–11
Nucleotides, 614–15, 1211–12, 1213, *1214*
Nucleus, 2
Nylon (polyamide), 852, *898*

Ockham, William of, 590n3
Ockham's razor, 590
Octanes, 195–96
trans,cis,trans-2,4,6-Octatriene, 1041
1-Octene, 380, 753
2-Octene, *753*
Octet rule, 4
Off-resonance decoupling, 740–41
Oils, definition, 862
Olah, George A., 680
Oleate, 142–43
Olefins, 123
    See also Alkenes
Oleum, 636
Opsin, 533
Optical activity, overview, 155–59
Orbital interaction diagrams, 33–34
Orbitals, overview, 3, 4
    See also Atomic orbitals; Molecular orbitals; Wave functions (ϕ)
Organic foods, *312*
Organocuprates, 430, *458*, 797–98, 893, *924*

Organolithium reagents
  aluminum tri-*tert*-butoxyhydride, 892
  amide formation with alkyllithium reagents,
    249
  butyllithium, 1001–2
  destruction by water, 799–800, *818*
  deuterolabeled molecule formation, 229–30,
    257, 797
  dithiane alkylation, 1001–2
  ether or tetrahydrofuran (THF) solvents, 228
  hydrocarbon formation, 229
  as Lewis bases, 229, 429
  lithium triethylborohydride, 798
  protonation, 229
  reaction with acid chlorides, 891
  reaction with aldehydes and ketones to form
    alcohols, 799–802, *817*
  reaction with carboxylic acids, 856–57, *870*
  reaction with esters, 899
  reaction with oxiranes, 429–30, *458*
  S$_N$Ar reaction, 677
  structure, 229
  synthesis from alkyl halides, 228, 229, 257,
    797, *818*
  See also Lithium aluminum hydride
Organometallic reagents
  from alkyl halides, 227–29, 256, 797
  as carbanion equivalents, 429
  definition, 227–28
  destruction by water, 799–800, *818*
  hydrocarbon synthesis, 229, 257
  lithium aluminum hydride, 429–30, *458*
  overview, 227–30, 797–99
  reaction with acid chlorides, 891–92
  reaction with aldehydes and ketones to form
    alcohols, 799–802, *817*
  reaction with CO$_2$, 840–41
  reaction with D$_2$O, 229–30, 797
  reaction with epoxides, 429–30, 797
  reaction with oxiranes, 429–30
  See also Grignard reagents; Organolithium
    reagents
Ortho esters, 846–48
Orthogonal orbitals, 34
Ortho (*o*-) substitution, definition, 596, *656*
  See also Substitution reactions of aromatic
    compounds
Osazones, 1145–46, *1147*, 1158, 1159
Osmium tetroxide, 443–44
Oxabenzene, 599
2-Oxacyclobutanone, *849*
2-Oxacycloheptanone, *880*
2-Oxacyclohexanone, *849*
Oxacyclopentanone, *849*, *880*
Oxacyclopropanone, *849*
Oxalyl chloride, 928
Oxaphosphetanes, 811–12
Oximes, *793*, 909–10, 911
Oxiranes
  definition, 317
  1,2-diol formation, 426–27
  ethylene oxide, *231*
  mechanisms for opening in acids and bases,
    427–29
  reaction with organometallic reagents, 429–30,
    458
  regiochemistry of ring opening, 427–29
  single-step epoxidation, 424–25, *459*
  synthesis by asymmetric (Sharpless)
    epoxidation, 426

  synthesis by Williamson ether synthesis, 317
  synthesis from halohydrins, 317, 1083
  See also Epoxidation; Epoxides
3-Oxohexanal, *768*
Oxonium ions, 140, 235–36, 248–49, 314
2-Oxopropanal (pyruvic aldehyde), *768*
Oxygen, 315–17, 484
Oxymercuration, 421–23, 448, *458*, *459*
Ozone, 46, 437, 441
Ozone depletion, *467*
Ozonides, 437–42, 459
Ozonolysis, 437–42

P-450 monooxygenase, 616
Paige, Leroy Robert "Satchel," 262, 262n1
Paired spin, 8
  See also Electron spin
Palmitic acid and palmitates, 862–63, 997
Palytoxin, 1009
[*n*]Paracyclophane, 628–29
Parallel spins, 10
Para (*p*-) substitution, definition, 596, *656*
  See also Substitution reactions of aromatic
    compounds
Parent ion (p), 702
Pascal's triangle, 729
Patulin, *532*
Pauli, Wolfgang, 8
Pauli principle, 8, 34
Pedersen, Charles J., 254
Penicillin V, *531*
1,3-Pentadiene, 539–40, 1048–52
1,4-Pentadiene, *512*, 519, 524
2,3-Pentadiene (1,3-dimethylallene), *513*, *514*
Pentanal, *767*
Pentane, 76–77, *82*, 87–88, 95
2,4-Pentanedione, *768*
Pentanoic acid (valeric acid), *830*, *849*, *880*
1-Pentanol. *See* Amyl alcohol
2-Pentanone, 771, 1008
Pentanoyl chloride, *880*
Pentoses, definition, 1127, 1130
Pentyl compounds, 77
Peptide and protein synthesis
  automatic procedure using Merrifield machine,
    1208–10
  biological synthesis from RNA, 1214–15
  Cbz (benzyl chloroformate), 1205, 1207–8,
    1217
  conversion of carboxy end into ester, 1205
  dicyclohexylcarbodiimide (DCC), 1205–7,
    1209, *1217*
  protecting groups, 1205, 1207–10
  RNA codons for amino acids, 1214–15
  tBoc (di-*tert*-butyl dicarbonate), 1205–10
  undesirable reactions, overview, 1202–4
Peptides
  α-helix, 1191, *1192*
  amino terminus, 1189, 1197
  carboxy terminus, 1189
  chromatographic separation, 1195–96
  cleavage by carboxypeptidases, 1197–98
  cleavage by chymotrypsin, 1202
  cleavage by cyanogen bromide (BrCN),
    1200–1201
  cleavage by trypsin, 1202
  definition, 1175
  denaturing, 1193–94
  dicyclohexylcarbodiimide (DCC), 1205–7,
    1209, *1217*

  disulfide bridges, 1190, 1193, 1194–95, *1216*
  Edman degradation, 1198–1200, 1202, *1218*
  hydrogen bonds, 1190–91, 1192, 1194
  nitrogen-to-carbon link, *224*
  nomenclature, 1189
  peptide bond, definition, 1175, 1195
  β-pleated sheets, 1191
  primary structure, 1190
  random coil, 1191
  reaction with 2,4-dinitrofluorobenzene, 1196,
    1197
  Sanger degradation, 1196–97, 1202
  secondary structure, 1191–92, 1193
  structure, 1189–94
  tertiary structure, 1192–94
  See also Peptide and protein synthesis
Peracids, 424
Pericyclic reactions
  definition, 1031
  how to work orbital symmetry problems,
    1071–73
  overview, 1031–32, 1073–74
  See also Concerted reactions; Cycloaddition
    reactions; Electrocyclic reactions;
    Sigmatropic shift reactions
Periodic acid, 807, 1151
Perkin, William Henry, 648
Peroxidate ion, 399
Peroxides
  acyl peroxides, 476
  alkyl peroxides, 253, *476*
  anti-Markovnikov HBr addition to alkenes,
    482, 485, *506*
  carboxyl radicals from acyl peroxides, 476
  hydrogen peroxide reaction with boranes, 398,
    451
  instability, 253
  oxidation of ozonides, 441
  radical production from alkyl peroxides, 476
  structure, 253, 424
  weak O–O bond, 476, 482, 483
Phenanthrene, structure, *179*, 603
Phenanthro[3,4-*c*]phenanthrene (hexahelicene),
    179, 180
Phenol (benzenol or hydroxybenzene), *231*, 595,
    670, 942
Phenols, 649, *652*, 670, 687
Phenonium ion, 1098, 1110
Phenoxide, 652
Phenylalanine
  from aspartame hydrolysis, 1164
  biosynthesis, 682–84, 1063
  structure, *1175*, *1176*, *1180*, *1183*
  synthesis, 1183
Phenyl cation, 650, *675*
Phenyl formate (phenyl methanoate), *879*
Phenylhydrazones, *793*, 1126, 1145–46
Phenylisothiocyanate, 1198–99
Phenylketene (2-phenylethen-1-one), *883*
N-Phenylmaleimide, *894*
Phenyl (Ph), 595, 596
Phenyl propyl ketone (butyrophenone), *644*
Phenylpyruvate, 684
Phenyl radical, 650
Phenylthiohydantoins, *1199*, *1200*, *1218*
Pheromones, 534
Phosgene, *880*, 899, *924*
Phosphonium ion, 811
Phosphorus pentachloride, *322*, 853, *868*

Photohalogenation
    allylic halogenation, 497–501, *507*, 543
    butane, 494–95
    chlorination of methane, 490–93
    fluorine reactions, 495
    iodine reactions, 495
    mechanism, 491
    methane, 490–93
    2-methylpropane, 494–95
    photobromination, 493, 494–95, 497–501
    selectivity of allylic bromination, 497–501
    selectivity of reactions with alkanes, 494–97, *506*
    thermochemistry, 491–92
    transition states for hydrogen abstraction, *495*, 496–97
    See also Halogenation
Phthalic acid, 856, *881*, *882*, 894
Phthalic anhydride, 856, *881*, 894
Phthalimide, *882*
Pinacolone, 408
α-Pinene, *558*
β-Pinene, *214*, 413, *558*
Piperidine (azacyclohexane), *242*, 601, *1019*
Pi (π) orbitals, general
    alkenes, 98, 104, 125
    alkynes, 125
    carbonyl compounds, 764–66
    ethylene, 104
    overlap, in alkenes, 98, 104, 125
Pi (π) stacking, 598, 615
pKa
    acetaldehyde, 936
    acetic acid, 1179
    alcohols, *236*, *249*, 316, 834
    aldehydes, *933*
    alkanes, 976
    ammonium ions, 246–48, 1179
    benzoic acid, 760
    tert-butyl alcohol, 987
    carboxylic acids, 834, *836*
    cycloheptatriene, 591
    1,3-cyclopentadiene, 589, 591, 601, 975
    definition, 235, 256, 257
    β-dicarbonyl compounds, *958*
    and hydration of alkenes, 381
    ketones, *933*
    oxonium ions, *236*, *248*
    propane, 589
    propene, 589, 936
    See also Acid–base chemistry
pKb, 248, 259, 601
Planck's constant (*h*), 526, 527
Plane-polarized light, 155–59, 168–69, *762*, 1128
β-pleated sheets, 1191
Poe, Edgar Allan, 148, 148n1
Polar covalent bonds, 14–16, 132
Polarimeters, 159
Polaroid sunglasses, 157–58
Polyamides (nylon), 852, *898*
Polyaromatic hydrocarbons (PAHs), 615
    See also Polynuclear aromatic compounds
Polycyclic compounds
    cycloalkanes, 85–86
    cycloalkenes, *121*
    Diels–Alder reaction, 552–53
    structure, *186*, 216–17
    See also Polynuclear aromatic compounds; Ring compounds

Polyenes, 110, *118*, *511*, 529, 533
Polyesters, 852–53
Polyethylene terephthalate (PET), 853
Polyisoprene, 554, 557
Polynuclear aromatic compounds
    aromaticity, 592
    carcinogens, *571*, 605, 614–16
    definition, 602
    and DNA alkylation, 614–16
    fused benzenes, stereochemistry, 179–80
    structures, 602–5
    See also Aromatic compounds
Polysaccharide, 1161, 1166, 1168
Polystyrene, 1208–10
Potassium benzoate, *1004*
Potassium chloride, 6, *132*
Potassium 2-chlorobutanoate, *832*
Potassium permanganate, 255, 443, 806, 840
Ppm scale, 716–17
Prephenate, 684, 1078, 1083
Primary carbon, 76
Primary ozonides, 437–39
Primary structure, 1190
Prismane (Ladenburg benzene), *216*, 573–74
Product-determining step, 294–95
Progesterone, 86, 562
Proline, *1177*, 1178, 1188, 1218
Propagation of chain reactions, *482*, 483
Propanal (propionaldehyde), *767*, 779, 936
Propane
    ^{13}C NMR spectroscopy, 88
    boiling point, 770
    bond dissociation energy, 475
    conformational analysis, 72
    dihedral angle, 72
    heat of combustion (ΔH_c°), 194
    methylene group, 72
    Newman projection, *71*, 72
    pKa, 589
    rotational energy barrier, 72, 78
    sp^3 hybrid orbitals, 72
    structure, *51*, *71*
Propanedial, *767*
Propane-2-thiol, 314
2-Propanol (isopropyl alcohol). *See* Isopropyl alcohol
1-Propanol (propyl alcohol), *230*, 748
Propargyl compounds, 127
[1.1.1]Propellane (tricyclo[1.1.1.0^{1,3}]pentane), 86, 216
Propene
    addition of HCl, 138
    allyl compounds, 108
    boiling point, 770
    derivatives and isomers, 108–9
    formation, 107
    pKa, 589, 936
    protonation, 378–79
Propionic acid (propanoic acid), *830*, *849*, *879*, *880*
Propionitrile (propanenitrile, ethyl cyanide), 803
Propionyl bromide (propanoyl bromide), *880*
Propylamine (1-aminopropane), *241*
Propylbenzene (1-phenylpropane), 641
Propyl bromide, 314
Propyl compounds, 73
Propylene. *See* Propene
Propylene glycol (1,2-propanediol), 240
Propyl fluoride, 268
3-Propyl-1-nonene, *110*

Propyl radicals, 471–72, 479
Propyne (methylacetylene), 126–27, *446*, 489
1-Propynyl compounds, 127
3-Propynyl compounds, 127
Prostaglandins, 569, 838
Protecting groups
    acetals as protecting groups for aldehydes and ketones, 788–89
    carbohydrate chemistry, 1151–52
    Cbz (benzyl chloroformate), 1205, 1207–8, *1217*
    definition, *788*
    peptide and protein synthesis, 1205, 1207–10, *1217*
    tBoc (di-*tert*-butyl dicarbonate), 1205–10, *1217*
    tetrahydropyranyl (THP) ethers as protecting groups for alcohols, 789–90
    trialkylsilyl ethers as protecting groups for alcohols, 789–90
Proteins
    α-helix, 1191, *1192*
    binding sites, 1193
    chromatographic separation and analysis, 1195–96
    crystallization, 1192, 1192n3, 1194
    definition, 1175
    denaturing, 1193–94
    disulfide bridges, 1190, 1193, 1194–95, *1216*
    Edman degradation, 1198–1200, 1202, *1218*
    fibrous proteins, 1192
    globular proteins, 1192, *1193*
    hemoglobin, 1194
    hydrogen bonds, 1190–91, 1192, 1194
    lactate dehydrogenase, *1192*
    NMR structure determination, 1192n3
    β-pleated sheets, 1191
    primary structure, 1190
    quaternary structure, 1194
    Sanger degradation, 1196–97, 1202
    secondary structure, 1191–92, 1193
    superhelix, 1192
    tertiary structure, 1192–94
    X-ray crystallography structure determination, 1192, 1192n3, 1194
    See also Peptide and protein synthesis
Protic solvents, 238–39, 256, 280, 417–18, 421
Pseudo-first-order reactions, 340
Pseudomauvine, 648
Putrescine (1,4-butanediamine), 246
Pynchon, Thomas, 364, 364n1, 763
Pyranoses, 1132–33
Pyranosides, 1148
Pyrethroids, 456
Pyridine (azabenzene)
    acid–base properties, 601, 620
    Chichibabin reaction, 676–77
    dipole moment, 600
    electron orbitals, 599
    electrophilic substitution reactions, 652–53
    heterobenzene, 598–99
    N-oxides, 678–79
    as nucleophile, 602
    pyridine *N*-oxide, 602
    resonance, 600
    S$_N$Ar reactions, 676–79
    structure, *242*, *592*, *598*, *652*, *804*
Pyridine *N*-oxide, 678

Pyridinium ion, 601–2, 653, 679, 814
Pyrolysis, 470–74, *507*, 852, 856, 913–14, *923*
Pyrophosphate esters, 555–59
Pyrophosphoric acid, 555
Pyrrole
    acid–base properties, 601, 621
    aromaticity, 599–600, 601
    Diels–Alder reaction, 630
    dipole moment, 600
    electrophilic substitution reactions, 652,
        653–54
    resonance, 600
    structure, *242, 592, 599, 652*
Pyrrolidine (azacyclopentane), *242*, 963–64
Pyruvate, *1028*, 1140

Quantum mechanics and babies, 42–43
Quantum model of atoms, *1, 2*
Quantum numbers
    possible combinations of quantum numbers, 7
    principal quantum number (*n*), 6
    relationship between *n, l*, and *m$_l$*, 7
    relationship to atomic orbitals, 6–8
    second quantum number (*l*), 6–7, 10
    spin quantum number (*s*), 7–8
    third quantum number (*m$_l$*), 7
Quaternary carbon, 76
Quaternary structure, 1194
Quinine, 255, *256*, 531, *532*
Quinoline, *602, 892*

Racemic mixture (racemate), 156–57
Radical anions, 453–54, 609
Radical formation
    from azo compounds, 476, *507*
    carboxyl radicals from acyl peroxides, 476
    delocalization of electrons, 475–76, 480, 498
    hydrocarbon cracking, 473–74
    methyl radical, 63, 469–70
    pyrolysis, 470–74, *507*
    radical production from alkyl peroxides, 476
    thermolysis, definition, 470
Radical properties
    delocalization of electrons, 475–76, 480, 498
    resonance forms, 478, *480*, 499
    stability, 475, 477–81
    structure, 63–64, 100–101, 477
Radical reactions
    aging process, 504–5
    anti-Markovnikov HBr addition to alkenes,
        481–82, 484–86, *506*
    anti-Markovnikov HBr addition to alkynes,
        489–90, *507*
    benzyl radical, 612–13, *618*
    β cleavage, 472, 473, 474
    dimerization and polymerization, 472, 473,
        474, 488–89
    disproportionation reaction, 471–72, 473–74,
        506
    formation of H$_2$, methane, and ethane from
        radicals, 31–32, 68–69, *83*, 103, 468
    Hunsdiecker reaction, 861, *869*
    hydrocarbon cracking, 473–74
    hydrogen abstraction, 471, 473
    inhibitors, 484
    mechanism of radical additions to alkenes,
        481–87
    methyl ions and radicals as reactive
        intermediates, 62

    overview, 228, 468, 505–7
    rearrangements, 501–4
    regiospecificity of radical additions to alkenes,
        481–82, 484–86
    steps of chain reactions, *482*, 483
    summary, 473
    tetrahalide addition to alkenes, 487–89
    thermochemistry of radical HX additions to
        alkenes, 486–87
    triphenylmethyl radical dimerization, 480
Random coil, 1191
Raney nickel, 253, 257, 1002
Rates of chemical reactions. *See* Kinetics
Reaction coordinate, 37
Reaction mechanisms
    tert-butyl compounds and reaction mechanism
        studies, 203–4
    definition and overview, 263–64, 349–51
    E2 reactions, 301
    hydration of alkenes, 138–39, 140, 365–66,
        380–84
    impossibility of mechanistic hypotheses, 139,
        382–83
    inversion of configuration, 270–71, 350
    microscopic reversibility, 346, 359
    problem solving techniques, 999–1001,
        1014–17
    rectangular hypothesis, 383
    S$_N$1 reactions, 289–91, 350–51
    S$_N$2 reactions, 270–71, 350
    stereochemistry and, 148–49, 155, 173, 270–71
    use of stereochemistry in determining, 396, 425
Reaction progress graphs, 37
Reactive intermediates, definition, 62
    See also Intermediates
Rearrangements
    alkyl shifts, 387–88, 642
    Beckmann rearrangement, 909–11, *923, 924*
    in biological processes, 401–2
    carbocation rearrangements, 389
    Claisen rearrangement, 1063, 1077–78
    as clue to neighboring group effect, 1086–87,
        1089, 1108, 1111, 1113
    concerted rearrangements of acyl compounds,
        914–20
    Curtius rearrangement, 918–19, 920, *924*
    definition, 387
    Friedel–Crafts alkylation, 642–43, 645, 646,
        686
    Hofmann rearrangement, 919–20, *923*
    during HX addition to alkenes, 386–89, 501
    hydride shifts, 387–88, 401–2, 502, 642
    ozonide formation from primary ozonides,
        437–39
    by phosphorus halides, 285
    radicals, 501–4
    transition states for 1,2-shifts, 502–4
    Wagner–Meerwein rearrangements, 387
    Wolff rearrangement, 916–18, *924*
    See also Cope rearrangement; Sigmatropic shift
        reactions
Rectangular hypothesis, 383
Reducing sugars, 1133, 1145, 1166
Rees, George Owen, 512, 512n1
Regiochemistry
    in addition reactions of alkenes, 366–72
    additions to alkenes, 137–39, 141, 366–72,
        481–82, 484–86
    definition, 137, 307

    hydroboration, 392–93, 399–400
    regiochemistry of epoxide ring opening,
        427–29
    regioselective reactions, 307
Reserpine, 532, *605*
Resolution of enantiomers, 169–72
Resonance and resonance forms
    acetate ion resonance stabilization, 421
    acid chlorides, *885*
    in addition reactions of alkenes, 366–72
    aldehydes, *885*
    allyl anion, 27–28
    allyl cation, 27, 28, 369–71, 535
    amides (carbonyl-nitrogen compounds),
        884–85, 903, 943
    arrow formalism, 23, 26
    benzene, 580–82
    benzene (overlap of 2*p* orbitals), 575–78
    benzyl radical, 480
    1,3-butadiene, 26–27, 28, 519, 521, 535
    2-butyne carbanion, 516
    and carbocation stability, 374–78
    conjugated dienes, 519, 521
    curved arrow formalism, 23, 24, 44
    cycloheptatrienyl anion, *590*
    cyclopentadienyl anion, *590*
    definition, 22, 44
    electronegativity, 27–28, 373
    electron pushing, 23, 24, 25
    enolates, 27–28, 373, 933–34, 955
    equilibrium *versus* resonance, 24–25, 29, 44,
        372, 404, 617
    esters, *885*
    formaldehyde, 25
    ketones, *885*
    nitromethane, 22–25, 28
    overview, 372–74
    paired electron spins, 28
    resonance arrow, 23, 25, 44
    resonance energy, 581, *629*
    resonance hybrids, 22, 25–27, 28, 30, 372
    stabilizing effect of delocalization of electrons,
        27, 102
    tropylium (cycloheptatrienylium) cation, *590*
    weighting factor (*c*), 372–73
    weighting factors, 25, 26–30
Resonance arrow (double-headed arrow), 23, 25, 44
Resorcinol (*m*-dihydroxybenzene), *597*
Retention of configuration, in S$_N$2 reactions, 269,
    270, 271–72
cis-Retinal, 533
trans-Retinal, 533, *762*
Retinal isomerase, 533
Retinol dehydrogenase, 533
trans-Retinol (vitamin A), *121*, 533
Retrosynthetic analysis, 807–9, 973, 982
Reverse Diels–Alder reactions, 554, 564, 630–31
Rhodium (Wilkinson's catalyst), 411
Rhodopsin, 533
Ribonucleic acid (RNA)
    codons for amino acids, 1214–15
    messenger RNA (mRNA), 1214–15
    structure and composition, 93, 614–15, 1149,
        1211–12, 1214
    synthesis, 1214
    transfer RNA (tRNA), 1215
    See also Nucleic acids
Ribonucleosides, 1211

D-Ribose
  chain lengthening, 1138–39, 1160
  D-ribofuranose, *1136*
  in NAD⁺, 814
  in nucleic acids, 1211, 1214
  structure, *1129*, 1158, *1170*, *1211*
Rilke, Rainer Maria, 932, 932n1
Ring compounds
  bicyclic compounds, from Diels–Alder
      reaction, 551–54
  bicyclic compounds, nomenclature, 214–15,
      551–52
  bicyclic compounds, structure, 211–17
  bridged substitution, 121, 211
  bridgehead position, 121–22, 212, 213–16
  chlorocyclopropane, stereochemistry, 173–74
  cyclic polyenes, 110, *118*
  dichlorocyclopropane, stereochemistry, 174–76,
      204
  disubstituted ring compounds, structure,
      204–10
  endo and exo isomers, 551–52
  fused benzenes, stereochemistry, 179–80
  fused bicyclic molecules, definition, 121, 211
  fused bicyclic molecules, structure, 212–13
  fused substitution, 211
  heterocyclic five-membered rings, 436, *459*
  heterocyclic five-membered rings from 1,3
      dipolar reagents, 436
  monosubstituted cyclohexanes, structure,
      194–204
  polycyclic compounds, structure, *186*, 216–17
  polycyclic cycloalkanes, 85–86
  polycyclic cycloalkenes, *121*
  ring formation through additions to alkenes,
      457
  S_N2 reactions, 276
  spiro substitution, 211
  stereochemical analysis, 173–76
  stereochemistry of ring compounds, 214–16
  strain, 187–93
  structural properties, overview, 186, 219–20
  three-membered rings in biochemistry, 455–56
  triple bonds in rings, 128
  See also specific types
RNA. *See* Ribonucleic acid (RNA)
Robbins, Thomas Eugene, 98, 98n1
Robinson, Robert, 172, 998
Robinson annulation, 998–99
Rosenmund reduction, 892, *923*
Rotational energy barriers
  alkenes, 105–6, 107, 144, 522
  1,3-butadiene s-trans and s-cis conformations,
      523–24
  cyclobutane, 189
  cyclohexane, 198–99
  ethane, 67, 70, 74–75, 93, 105–6, 188, *273*
  ethylene, 105–6, 107, 144
  propane, 72, 78
Roth, W. R, 1056, 1057, 1063, 1066
(*R/S*) convention, 152–55, 181, 1178
Rubisco (ribulose-1,5-bisphosphate
      carboxylase/oxygenase), 1138
Ruff degradation, 1139–40
Rutherford, Ernest, 2, 43

Saccharide. *See* Carbohydrates
Saccharin, 1164
Sandmeyer reaction, 649–51, *686*, *687*

Sanger degradation, 1196–97, 1202
Saponification, 862, *869*, 895
Saturated hydrocarbons. *See* Alkanes
Saunders, Martin, 696n2
Saytzeff elimination, 300, 306, 308, 310, 322
Schiemann reaction, 649, *687*
Schiff base. *See* Imines
Schröder, Gerhard, 1070–71
Schrödinger, Erwin, 6
s-cis conformation, 523–24, 545–46, 550
Secondary amines, 240, 241
Secondary carbon, 76
Secondary structure, 1191–92, 1193
Second-order reactions, 268, 273, 340, 341
Semibullvalene, *1068*
Semicarbazones, 793
Serine, *1176*, *1189*
Seth, Vikram, 572, 572n1
Sharpless, Barry, 426
Shikimate-3P, 683
Side chains of amino acids, 1175, *1176–77*
Sigma bonds (σ), definition, 54
Sigmatropic shift reactions
  antarafacial motion, 1055–58
  and bond dissociation energy, 1049–50
  definition, 1032, 1050
  degenerate reactions, 1050–51
  formal description and nomenclature, 1050
  homolytic bond breaking, 1053
  1,3-pentadiene, 1048–52
  photochemical reactions, 1048, 1051–52, 1053,
      1054, 1057–58
  problem solving techniques, 1072–73
  rules for reactions, *1058*
  suprafacial motion, 1055–59
  symmetry effects, 1055
  transition state model, 1053–55
  See also Cope rearrangement; Pericyclic
      reactions
Silyl ethers, 789–90
Singlet carbenes, 433–34, 436, 458
Smalley, R. E., 603–4
S_N1 reactions
  activation energy (Δ*G°*), 342–43
  benzylic compounds, 610–11, *618*
  tert-butyl bromide S_N1 reactions, 289, 294–95
  competition between S_N1 and E1 reactions,
      298–99
  effect of leaving group, *281*, 295, 343, 347
  effect of nucleophile on structure of product,
      290, 294–95
  effect of solvent polarity, 295
  effect of substrate structure, 292–94, 296–97
  equilibrium, 333, 337, 350
  as first-order reactions, 268, 289, 294, 340, 341
  incomplete racemization, 291–92
  inversion of configuration, 291–92, 297
  mechanism, 289–91, 350–51
  problem solving techniques, 293
  product-determining step, 290, 294–95
  rate-determining step, 294–95
  rate law, 289–91, *341*
  relative nucleophilicity of common species, *279*
  relative rates of S_N1 solvolysis of different
      substrates, *542*
  resonance-stabilized allylic cation formation,
      538, 541–42
  solvolysis reactions, 290, 538, 541–42
  stereochemistry, 291–92

steric effects, 329
summary and overview, 296–98, 322
transition states, 333, 343, 346–49
S_N2 reactions
  activation energy, 273–74, 277, 287
  alkylation in the α position of carbonyl
      compounds, 954–64
  allyl halides, 542–43, *564*, 612
  benzylic compounds, 612, *618*
  in biochemistry, 288–89
  Cahn–Ingold–Prelog priority system, 270
  competition between S_N2 and E2 reactions,
      301, 321
  deoxyribonucleic acid (DNA) substitution
      reactions, 262, 288–89
  effect of nucleophile, 277–81
  effect of solvent polarity, 279–81, 286–87, 295
  effect of substrate branching on E2–S_N2 mix,
      301–2
  effect of substrate (R group) structure, 275–77,
      296–97, 301–2
  equilibrium, 267–68, 274, 332
  ether cleavage by haloacids, 282–83
  first-order reactions of substrate and
      nucleophile, 268, 340
  HOMO–LUMO interactions, 271–72
  intramolecular S_N2 reactions, 316–17, 917,
      1081–88, 1091, 1095, 1101
  inversion of configuration, 269, 270–75, 284,
      288, 297
  mechanisms, 270–71, 282–85, 350
  methyl compounds, 249, 275
  nucleophilicity, 278–81
  problem solving techniques, 284
  rate law, 268, *341*, 350
  rates in methyl, primary, and secondary
      substrates, 275, 288
  relative reaction rates various substrates, *543*,
      612
  replacement of ⁻I with ⁻I*, 270–73
  retention of configuration, 269, 270, 271–72
  in ring compounds, 276
  as second-order reactions, 268, 273, 340
  stereochemistry, 268–75
  steric hindrance with tertiary substrates, 275,
      296
  sulfonate reactions, 281–86, 322, 427, 555
  summary and overview, 296–98, 322
  transition states, 272–74, 276–77, 332, 344–46,
      543
  See also Leaving groups
S_NAr reactions
  addition–elimination pathway, 675–76, *677*
  benzyne as intermediate, 681–82
  Chichibabin reaction, 676–77, 761
  elimination–addition pathway, 681
  ipso attack, 677–78
  Meisenheimer complex, 675
  nitro groups, 674–75, 679, 680
  N-oxides, 678–79
  nucleophilic additions to benzenes, 674–76
  nucleophilic aromatic substitution to
      heteroaromatics, 676–79
  organolithium reagents, 677
  pyridine, 676–79
Soaps, 832, *864*, 866–67
Sodamide, 986
Sodium bicarbonate, *859*
Sodium borohydride, 421, 423

Sodium carbonate, *859*
Sodium chloride, solubility, 239
Sodium chromate, 805
Sodium fluoride, 5–6, *132*
Sodium propanoate, *832*
Solvated electrons, 453–54
Solvation, definition, 237
Solvents
  aprotic solvents, 238–39, 240, *280*
  effect on S_N1 reactions, 295
  effect on S_N2 reactions, 279–81, 286–87, 295
  hydrogen bonding, 238–39, 256, 280
  "like dissolves like," 239–40, 256
  polar solvents, 238–39
  protic solvents, 238–39, 256, 280, 417–18, 421
  solvation, definition, 237
Solvolysis reactions
  tert-butyl iodide in water, 333, 343–44, 346–51, 383
  definition, 290
  exo- and *endo*-2-norbornyl tosylate, 1110–16
  methyl iodide in water, 340
  neopentyl iodide (1-iodo-2,2-dimethylpropane) in water, 389
  relative rates of S_N1 solvolysis of different substrates, *542*
  resonance-stabilized allylic cation formation, 538, 541–42
  steric effects, 329
Soxhlet extractors, 972–73, 974
Spaldeens, 159–60, 160n4
Spearmint [(R)-(–)-carvone], 163
Specific rotation, 159
Spectroscopy
  definition, 88, 699
  determining structure by using spectroscopy, 742–45
  overview and history, 525–26, 696, 696n2, 750–51
  See also Mass spectrometry (MS); Nuclear magnetic resonance spectroscopy (NMR); Ultraviolet/visible (UV/vis) spectroscopy
sp hybrid orbitals, 53–55
sp² hybrid orbitals, 55–56
sp³ hybrid orbitals, 56–57
Spin–spin coupling
  chemical exchange of OH and NH hydrogens, 733–34, 739, 746
  cis and trans coupling constant (J_{cis}, J_{trans}) for alkenes, 732
  coupling constant (J), definition, 718, 728
  coupling constant (J) for aromatic compounds, *732*
  dihedral angle, effect on coupling constant, 731
  Karplus curve, 731–32
  long-range coupling, 731, 736
  magnitude of J for two- and three-bond coupling, *731*
  multiple couplings, 730
  n + 1 rule, 727–29, 735–36
  overview, 727
  tree diagrams, *730*
  See also Hydrogen nuclear magnetic resonance spectroscopy (^{1}H NMR)
Spiropentane, 211, 509
Spiro substitution, 211
Squalene, 559
Staggered conformation, 65–67

Starch, 1161, 1168
Stephenson, Neal, 410, 410n1
Stereochemistry
  absolute configuration, 152–55, 172–73, 181
  achiral molecules, definition, 150, 150n2
  bridged bicyclic molecules, 214–16
  1-bromo-2-chlorocyclopropane, 204
  Cahn–Ingold–Prelog priority system, 152–55
  chemical differences between enantiomers, 159–63
  chirality, overview, 148–51
  chirality without four different groups per carbon atom, 177–80
  chlorocyclopropane, 173–74
  cyclohexane, 190–92, 197–99
  dichlorocyclopropane, 174–76, 204
  E2 reactions, 302–6
  fused benzenes, 179–80
  molecules with more than one stereogenic atom, 164–69
  neighboring group effect, 1083–85, 1089–91, 1093–94, 1097, 1111–14
  overview, 148–49, 181
  ring compounds, 173–76, 214–16
  (R/S) convention, 152–55, 181, 1178
  S_N1 reactions, 291–92
  S_N2 reactions, 268–75
  stereochemical analysis of cycloalkanes, 173–76
  stereogenic atoms, 152–55, 167–69, 173, 180
  stereoisomers, definition, 148, 176
  use in determining reaction mechanisms, 396, 425
Stereogenic atoms, 152–55, 167–69, 173, 180
Steric requirements, definition, 78
Steroids, 456, 530, 559–62
Strain
  angle strain, 128, 187, 190, 276, 474
  and heat of combustion (ΔH_c°), 194–97
  and heat of formation (ΔH_f°), 193–94
  quantitative evaluation of strain energy, 193–97
  in ring compounds, 187–93
  S_N2 reactions in ring compounds, 276–77
  torsional strain, 105, 188–89, 190
  van der Waals strain, 198
s-trans conformation, 523–24
Strecker synthesis, 1184, *1216*
Structural isomers, definition, 73, 174, 176
Structure determination, 88, 525–26, 696
  See also Nuclear magnetic resonance spectroscopy (NMR); Spectroscopy; Ultraviolet/visible (UV/vis) spectroscopy
Structure drawings
  alkanes, overview, 70–71
  ball-and-stick models, *78*
  line drawings, 71–72, 77–78
  Newman projection, 65–66, 94
  space-filling models, *78*
Strychnine, 171, 172, 256, *532*
Styrene (vinylbenzene or ethenylbenzene), 489, *595*
Substituents, definition, 60
Substituted benzenes
  aromatic substitution reactions, 606–8
  Birch reduction, 609
  disubstituted benzenes, 596–97
  formation of disubstituted benzenes, 655–66
  meta (*m*-) substitution, definition, 596, *656*
  monosubstituted benzenes, 595–96
  ortho (*o*-) substitution, definition, 596, *656*

  para (*p*-) substitution, definition, 596, *656*
  pattern of further reactions, *656*
  physical properties, 598
  polysubstituted benzenes, 597, 668–73
  See also Benzene; Substitution reactions of aromatic compounds
Substitution, nucleophilic, bimolecular reactions. *See* S_N2 reactions
Substitution, nucleophilic, unimolecular reactions. *See* S_N1 reactions
Substitution reactions of aromatic compounds
  benzene, resistance to substitution reactions, 606–8, 631–32
  benzene deuteration, 606–8, *618*
  blocking groups, 671
  diazonium ions, 647–51, *686*, *687*
  disubstituted benzenes, formation of, 655–66
  electrophilic aromatic substitution, general reaction, 685
  electrophilic aromatic substitution, overview, 635, 685
  electrophilic substitution of furan and thiophene, 654–55
  electrophilic substitution of pyridine and pyrrole, 652–54
  energy barriers to reactions, 624–25, 627–28
  Friedel–Crafts acylation, 643–46, *686*, *687*, 893
  Friedel–Crafts alkylation, 639–43, 646, 655
  halogenation, 638–39, 670
  inductive effects of groups, 666–68
  meta (*m*-) substitution, definition, 596, *656*
  meta substitution of anisole, 657
  meta substitution of ethyl benzoate, 664–65
  meta substitution of halobenzenes, 667–68
  meta substitution of nitrobenzene, 663–64
  meta substitution of toluene, 660–61
  meta substitution of trimethylanilinium ion, 662–63, 666
  nitration, 637–38, *664*, 668–69, 673, *687*
  nucleophilic addition to benzenes (S_NAr reaction), 674–76
  nucleophilic addition to heteroaromatics (S_NAr reaction), 676–79
  ortho (*o*-) substitution, definition, 596, *656*
  ortho substitution of anisole, 658
  overview, 606–8, 624–26, *646*, 685–88
  para (*p*-) substitution, definition, 596, *656*
  para substitution of anisole, 657–60
  para substitution of ethyl benzoate, 665
  para substitution of halobenzenes, 667–68
  para substitution of nitrobenzene, 664
  para substitution of toluene, 660–61
  para substitution of trimethylanilinium ion, 662–63
  pattern of further reactions of substituted benzenes, *656*
  polysubstituted benzenes, formation of, 668–73
  Sandmeyer reaction, 649–51, *686*, *687*
  sulfonation, 636–37, *686*
Substitution reactions of nonaromatic compounds
  alkyl halides, 267–68
  equilibrium, 267–68, 274, 332–33
  overview, 262–63
  synthetic potential of substitution reactions, 312–13
  See also Leaving groups; S_N1 reactions; S_N2 reactions
Succinic acid, *830*, *881*
Succinic anhydride, *881*

Sucralose, 1164
Sucrose, 239, 1125, 1161, 1166
Sugars. *See* Carbohydrates
Sugar substitutes, 1164
Sulfate ion, *45*
Sulfides. *See* Thioethers
Sulfonates
    benzene sulfonation, 636–37, *686*
    brosylate (*p*-bromobenzenesulfonate), 1085–86
    Coates' cation, 1117
    formation from alcohols, 281–86, 322, 427, 555
    sulfonation of phenols, 671
Sulfones, 810, *819*
Sulfonic acids, 636–37, *686*, 809, *819*, 866–67
Sulfonium ions, 313, 314
Sulfoxides, 810, *819*
Sulfur compounds as nucleophiles, 313–14
Sulfuric acid, 45
Superacid, 679–80
Suprafacial motion, 1055–59
Syn addition, 394–95
Syn elimination, 305–6
Synthesis
    building a map as problem-solving technique,
        999–1001, 1014–17
    keeping track of methods, 320
    "real world" difficulties, 318–20, 323
    synthetic potential of substitution reactions,
        312–13
    See also specific substances

Table sugar (sucrose), 239, 1125, 1161, 1166
D-Talose, *1129*, *1160*
Tautomers, 29, 939–40, 944, 961, 966
Taxol, 183
tBoc (di-*tert*-butyl dicarbonate), 1205–10, *1217*
Terephthalic acid, 853
Termination of chain reactions, 483
Terpenes, 554–58
    See also Isoprene
Tertiary carbon, 76
Tertiary structure, 1192–94
Testosterone, *562*
Tetracycline, *531*
Tetraethylammonium bromide, *242*
Tetrahedral intermediate in addition–elimination
        reactions
    acid anhydride formation, 856
    acid chloride formation, 854
    amide formation, 851
    amide hydrolysis, 901
    Claisen condensation, 987
    definition, 842
    ester hydrolysis, 895
    ester reaction with Grignard reagent, *898*
    Fischer esterification reaction, 842–44
    halogenation of carbonyl compounds, *949*
    overview, 868, 877, 878, *890*
    See also Intermediates
Tetrahydrofuran (THF), *250*
    ^{13}C NMR spectroscopy, 89
    dipole moment, 620
    ether–borane complex, 391
    and Grignard reagent formation, 228, 251, 254
    as Lewis base, 391
    and organolithium reagent formation, 228
    solvation of water, 240
    structure, *89*, *250*, *254*

Tetrahydropyran (pentamethylene oxide,
        oxacyclohexane), *254*
Tetrahydropyranyl (THP) ethers, 789–90
2,2,3,3-Tetramethylbutane, 195–96, 195n2
Tetramethylsilane (TMS), 716–17, 720–21
Tetra-*tert*-butyltetrahedrane, *216*
Tetroses, 1127, 1129
Thalidomide, 180, 180n6
Thermodynamic control, 353, *360*, 537–41
Thermodynamic enolates, 1008
Thermodynamics
    definition, 351
    entropy change ($\Delta S°$), 335, 337–39, 349
    Gibbs free energy ($\Delta G°$), 120, 334–39, 349
    and position of equilibrium, 333–34
    thermodynamics *versus* kinetics effect on rates,
        351–57, 359–60
    See also Enthalpy change ($\Delta H°$)
Thermolysis, definition, 470
Thiazolium ion, *1028*
Thioethers (sulfides)
    formation of sulfonium ions, 313
    mustard gas [bis(2-chloroethyl)thioether], 1089
    neighboring group effect in intramolecular
        reactions, 1088–89
    nomenclature, 252–53
    odor, 251–52
    oxidation to sulfoxides, 810
    reduction with Raney nickel, 253, 257
    sulfur-containing compounds in garlic, 252
    synthesis by mercaptide S_N2 reactions, 313
    thioester A and cholesterol formation, 358
    See also Ethers
Thiolate ions, 284, 314, 324
Thiols (mercaptans)
    acidity, 253
    mercaptide formation, 253
    nomenclature, 252
    odor, 251–52
    overview, 251–52
    oxidation reactions, 809–11
    reduction with Raney nickel, 253, 257
    sulfur-containing compounds in garlic, 252
    synthesis by mercaptide S_N2 reactions, 313
    synthesis from alkene, 314
    synthesis from alkyl halide, 314
Thionyl chloride, 284–85, 643, 853–54, 868
Thiophene, *599*, 599–600, 654–55
Thioureas, 1198–99, *1206*, *1218*
Thomson, J. J., 2
Three-center, two-electron bonding, 1102, 1116,
        1116n4
Three-dimensional structures, drawing, 16
Threonine, *1176*
Threose, *1129*
Thujone, *558*
Thurber, James Grove, 332, 332n1
Thymine, *615*, 1212, 1214
Tolkien, John Ronald Reuel, 186, 186n1
p-Tolualdehyde (4-methylbenzaldehyde), *768*
Toluene (methylbenzene or phenylmethane)
    Friedel–Craft alkylation, 641, 655–56
    mass spectrometry (MS), 702–3
    meta substitution, 660–61
    para substitution, 660–61
    radical bromination, 613
    structure, *329*, *592*, *595*
    substitution reactions, 655, 660–62, 663
    synthesis, 641, 655–56

Torsional strain, 105, 188–89, 190
Tosylates
    as leaving group, 283–84, *305*, *555*, 798, 1093
    3-methyl-2-butyl tosylate reaction with acetic
        acid, 1108–9
    7-norbornenyl tosylates reaction with acetic
        acid, 1099–1108
    3-phenyl-2-butyl tosylate conversion to 3-
        phenyl-2-butyl acetate, 1096–99
    solvolysis reaction of *exo*- and *endo*-2-
        norbornyl tosylate, 1110–16
Transesterification, 896–98, *924*, 987
Transfer RNA (tRNA), 1215
trans isomers, 84, *85*, 144
Transition states
    additions to alkenes, 136
    comparison to intermediates, 186
    cyclohexane conformations, 197–98
    definition, 67, 72, 186
    effect of solvent polarity, 286–87
    Hammond postulate, 355–57, 377, 496
    parallelism between transition state and
        product energies, 353–57
    photohalogenation, *495*, 496–97
    rate-determining step, 294–95, 350–51, 359
    relationship to reaction rate, 339
    S_N1 reactions, 333, 343, 346–49
    S_N2 reactions, 272–74, 276–77, 332, 344–46,
        543
    thermodynamics *versus* kinetics, 351–57
    transition state theory, 349
    See also Activation energy ($\Delta G°$)
Trialkylsilyl ethers, 789–90
Trialkylsilyl halides, 789–90
Triamantane, *218*, 219
A [5]Triangulane, *211*
Triangular hydrogen (H_3, H_3^+), 102, 217, 1103
Tribromophenol, 671
2,4,6-Tribromophenol, *670*, 671
1,3,5-Trichlorobenzene, 597
Tricyclo[3.3.1.1^{3,7}]decane (adamantane), 217–19
Tricyclo[4.1.0.0^{4,6}]heptane, *211*
Tricycloillinone, 217
Triethylamine, *241*, 277–78
Trifluoroacetic acid, *424*
Trifluoroperacetic acid, *424*, 672
2,3,3-Trimethyl-1-butene, 419
Trimethylammonium group, 666
Trimethylanilinium ion, 662–64
2,2,3-Trimethylbutane, *83*
Trimethylene oxide (oxacyclobutane, oxetane), *254*
Trimethyloxonium fluoborate, *249*
1,3,5-Trinitrobenzene, 668
1,3,5-Trioxane (1,3,5-trioxacyclohexane,
        paraldehyde), *254*
Triphenylmethyl radical, 480
Triphenylphosphine, 285, 811
Triple bonds
    Cahn–Ingold–Prelog priority system, 114,
        153–54
        Lewis structure, 18–19
        pi (π) orbitals, 125
        in rings, 128
        sigma σ bonding, 123–25, 128
    See also specific compounds
Triplet carbenes, 433–36, 468
Triptycene, 692
Triptycenyl chloride, *611*
Triquinacene, *532*

Trityl cation, 587, 611, *618*
Tropylium fluoborate, 587, *618*
Tropylium ion (cycloheptatrienylium ion), 587–89, 590, *618*
Trypsin, 1202
Tryptophan, *1176*
Twistoflex, *604*
Tyrosine, 684, *1176*

Ultraviolet/visible (UV/vis) spectroscopy
　Beer–Lambert law, 527
　1,3-butadiene, 528–29
　β-carotene, 529
　complex spectra from conjugated molecules, 530
　conjugated dienes, 528–32
　electron transitions and energies, 527–29
　ethylene (ethene), 527–29
　extinction coefficient, 527
　Fiesers' rules, 530
　history and background, 525–26, 696
　HOMO–LUMO gap, 528–29
　spectrometers, 527
　Woodward's rules, 530–31
Unimolecular elimination reactions. *See* E1 reactions
Unsaturated hydrocarbons, 83
　See also Alkenes; Alkynes
Uracil, 1212, 1214
Urea, 860, 1194
　substituted, 1198, *1206*, *1218*

Valence electrons, 3–6, 16
Valine, *955*, *1175*, *1176*, *1178*, *1189*
van der Waals forces, 87
van der Waals strain, 198
Vanillin, 582

van't Hoff, Jacobus Henricus, 1153
Vicinal groups
　definition, 414, 447, 471
　dihalides, 414, 448, *459*, 490, 499, *507*
　vicinal diols, oxidative cleavage, 807, *817*, *818*, 845
　vicinal diols from alkene oxidation, 443, *459*, 840
Vilsmeier reagent, 873, 928
Vinyl alcohol (hydroxyethene), 939–40
Vinyl ammonium ion, *45*
Vinylboranes, 450–51
Vinyl carbocations, 444–45
Vinyl chloride, 45, 367, 372
Vinyl compounds, 107
Vinyl halides, 444, *459*
Vinylic hydrogens, 724–25, 739
Vinyl iodide (iodoethene), *226*
Vinyl radicals, 454, 489
Viquidil, *95*
Vitamin A (*trans*-retinol), *121*, 533
Vitamin B$_{12}$, *532*
Vitamin E, 505
Vogel, Emanuel, 594, 1034–35, 1038
von Baeyer, Adolf, 187
Voodoo lily, 838

Wagner–Meerwein rearrangement, 387
Walden, Paul, 274
Watson, James D., 1213
Wave functions (Ψ)
　antibonding molecular orbitals ($\Phi_{\text{antibonding}}$, Φ_{A}), 32–33
　bonding molecular orbitals (Φ_{bonding}, Φ_{B}), 32–33
　definition, 6

electron density in atomic orbitals, 10, 11–12, 32–33
nodes in atomic orbitals, 11–12
probability of finding electron in a volume of space (Ψ^2), 6, 10, 11–12
Schrödinger's wave equation, 6
See also Orbitals
Wavenumber, 708
Weighting factors, 26–30, 372–73, 419–20
Wiberg, Kenneth B., 216–17
Wilkins, Maurice, 1213
Wilkinson's catalyst (rhodium), 411
Williamson ether synthesis, 254, 315–17, 322, 1147, 1150, 1163
Wittig reaction, 811–13, *817*
Wöhler, Friedrich, 30
Wolff–Kishner reduction, 645, *686*, 802
Wolff rearrangement, 916–18, *924*
Woodward, Robert Burns, 172, 530–31, *532*, 1031–32, 1034
Woodward–Hoffmann theory, 1031–32, 1034, 1035, 1056
Wurtz, Charles Adolphe, 966n2

Xanthate esters, 914, *923*, *925*
Xanturil, *95*
X-ray crystallography, 173, 1192, 1192n3, 1194, 1213
Xylene (dimethylbenzene), isomers, *597*, 598
D-Xylose, *1129*, *1159*

Yeats, William Butler, 1031, 1031n1
Ylides, 811–12

Zusammen (Z) isomers, 109, 111, 112–15
　See also cis isomers
Zwitterions, 1179–80